# FUNDAMENTAL AND PHYSICAL CONSTANT

| quantity | symbol | English | SI |
|---|---|---|---|
| **Charge** | | | |
| electron | $e$ | | $-1.6022 \times 10^{-19}$ C |
| proton | $p$ | | $+1.6021 \times 10^{-19}$ C |
| **Density** | | | |
| air [STP, 32°F, (0°C)] | | 0.0805 $\text{lbm/ft}^3$ | 1.29 $\text{kg/m}^3$ |
| air [70°F, (20°C), 1 atm] | | 0.0749 $\text{lbm/ft}^3$ | 1.20 $\text{kg/m}^3$ |
| earth [mean] | | 345 $\text{lbm/ft}^3$ | 5520 $\text{kg/m}^3$ |
| mercury | | 849 $\text{lbm/ft}^3$ | $1.360 \times 10^4$ $\text{kg/m}^3$ |
| sea water | | 64.0 $\text{lbm/ft}^3$ | 1025 $\text{kg/m}^3$ |
| water [mean] | | 62.4 $\text{lbm/ft}^3$ | 1000 $\text{kg/m}^3$ |
| **Distance** [mean] | | | |
| earth radius | | $2.09 \times 10^7$ ft | $6.370 \times 10^6$ m |
| earth-moon separation | | $1.26 \times 10^9$ ft | $3.84 \times 10^8$ m |
| earth-sun separation | | $4.89 \times 10^{11}$ ft | $1.49 \times 10^{11}$ m |
| moon radius | | $5.71 \times 10^6$ ft | $1.74 \times 10^6$ m |
| sun radius | | $2.28 \times 10^9$ ft | $6.96 \times 10^8$ m |
| first Bohr radius | $a_0$ | $1.736 \times 10^{-10}$ ft | $5.292 \times 10^{-11}$ m |
| **Gravitational Acceleration** | | | |
| earth [mean] | $g$ | 32.174 (32.2) $\text{ft/sec}^2$ | 9.8067 (9.81) $\text{m/s}^2$ |
| moon [mean] | | 5.47 $\text{ft/sec}^2$ | 1.67 $\text{m/s}^2$ |
| **Mass** | | | |
| atomic mass unit | $u$ | $3.66 \times 10^{-27}$ lbm | $1.6606 \times 10^{-27}$ kg |
| earth | | $4.11 \times 10^{23}$ slugs | $6.00 \times 10^{24}$ kg |
| earth [customary U.S.] | | $1.32 \times 10^{25}$ lbm | n.a. |
| electron [rest] | $m_e$ | $2.008 \times 10^{-30}$ lbm | $9.109 \times 10^{-31}$ kg |
| moon | | $1.623 \times 10^{23}$ lbm | $7.36 \times 10^{22}$ kg |
| neutron [rest] | $m_n$ | $3.693 \times 10^{-27}$ lbm | $1.675 \times 10^{-27}$ kg |
| proton [rest] | $m_p$ | $3.688 \times 10^{-27}$ lbm | $1.673 \times 10^{-27}$ kg |
| sun | | $4.387 \times 10^{30}$ lbm | $1.99 \times 10^{30}$ kg |
| **Pressure,** atmospheric | | 14.696 (14.7) $\text{lbf/in}^2$ | $1.0133 \times 10^5$ Pa |
| **Temperature,** standard | | 32°F (492°R) | 0°C (273K) |
| **Velocity** | | | |
| earth escape | | $3.67 \times 10^4$ ft/sec | $1.12 \times 10^4$ m/s |
| light [vacuum] | $c$ | $9.84 \times 10^8$ ft/sec | 2.9979 (3.00) $\times 10^8$ m/s |
| sound [air, STP] | $a$ | 1090 ft/sec | 331 m/s |
| [air, 70°F (20°C)] | | 1130 ft/sec | 344 m/s |
| **Volume,** molal ideal gas [STP] | | 359 $\text{ft}^3$/lbmol | 22.41 $\text{m}^3$/kmol |
| **Fundamental Constants** | | | |
| Avogadro's number | $N_A$ | | $6.022 \times 10^{23}$ $\text{mol}^{-1}$ |
| Bohr magneton | $\mu_B$ | | $9.2732 \times 10^{-24}$ J/T |
| Boltzmann constant | $k$ | $5.65 \times 10^{-24}$ ft-lbf/°R | $1.3807 \times 10^{-23}$ J/K |
| Faraday constant | $F$ | | 96 487 C/mol |
| gravitational constant | $g_c$ | 32.174 lbm-ft/lbf-$\text{sec}^2$ | |
| gravitational constant | $G$ | $3.44 \times 10^{-8}$ $\text{ft}^4/\text{lbf-sec}^4$ | $6.672 \times 10^{-11}$ $\text{N}\cdot\text{m}^2/\text{kg}^2$ |
| nuclear magneton | $\mu_N$ | | $5.050 \times 10^{-27}$ J/T |
| permeability of a vacuum | $\mu_0$ | | $1.2566 \times 10^{-6}$ $\text{N/A}^2$ (H/m) |
| permittivity of a vacuum | $\epsilon_0$ | | $8.854 \times 10^{-12}$ $\text{C}^2/\text{N}\cdot\text{m}^2$ (F/m) |
| Planck's constant | $h$ | | $6.6256 \times 10^{-34}$ J·s |
| Rydberg constant | $R_\infty$ | | $1.097 \times 10^7$ $\text{m}^{-1}$ |
| specific gas constant, air | $R$ | 53.3 ft-lbf/lbm-°R | 287 J/kg·K |
| Stefan-Boltzmann constant | | $1.71 \times 10^{-9}$ $\text{BTU/ft}^2\text{-hr-°R}^4$ | $5.670 \times 10^{-8}$ $\text{W/m}^2\cdot\text{K}^4$ |
| triple point, water | | 32.02°F, 0.0888 psia | 0.01109°C, 0.6123 kPa |
| universal gas constant | $R^*$ | 1545 ft-lbf/lbmol-°R | 8314 J/kmol·K |
| | $R^*$ | 1.986 BTU/lbmol-°R | |

## MATHEMATICAL CONSTANTS

| symbol | name | value (rounded) |
|---|---|---|
| $\pi$ | Archimedes number (pi) | 3.14159 26536 |
| $e$ | base of natural logs | 2.71828 18285 |
| $C$ or $\tau$ | Euler constant | 0.57721 56649 |
| $G$ | Catalan constant | 1.64493 69032 |

# ABOUT THE  SYMBOL



The *EXAM USE* symbol you see on the cover is Professional Publications' way of indicating that this book meets the seven criteria listed below. Professional Publications puts this symbol only on its reference manuals with these characteristics, and does so voluntarily.

1. **This book is primarily concept- and theory-oriented.** Less than 10 percent of the space in this book contains solved problems. The remaining space contains explanations, theory, and data presented at the college level or higher.

2. **This book does not use old examination problems as source material.** Solved example problems in this book are used to illustrate isolated concepts, not to illustrate typical exam problems. The format of examples and practice problems in this book does not imitate that of examination problems.

3. **This book safeguards examination security.** This book contains a highly visible statement that copying examination problems is not permitted. This book discourages problem copying and other subversive activities. All significant blank space is blocked out to discourage you from using that space to copy examination problems.

4. **This book cannot be used to find similar problems.** Solved examples are not indexed, listed, or categorized for fast access. Subjects and explanations are organized logically by theory, not by problem type. Solved examples are spread throughout each chapter, not consolidated for rapid reference.

5. **This book is ethical in authorship.** All authors and contributors are listed, and no pseudonyms are used. Information about examination subjects was obtained from public and published sources. No subversive methods were used.

6. **This book is ethical in presentation.** Primary sources of nonoriginal data and other material are acknowledged. No references are made to actual exam problems. Acceptability to, approval of, or affiliation with the National Council of Examiners for Engineering and Surveying is not claimed. No inflated claims or guarantees about passing the exam related to this book are made.

7. **This is a bona fide book.** It contains standard book elements, such as a title page identifying the author and publisher, a copyright, and an index. This book is permanently bound in softcover, hardcover, or wire format.

Each state's engineering registration board establishes its own rules. The *EXAM USE* symbol does not guarantee you will be able to use this book in an examination. Books from Professional Publications and other publishers without the *EXAM USE* symbol might also be permitted in an examination. Contact your state's engineering licensing board for more information.

# THE NATIONAL SOCIETY OF PROFESSIONAL ENGINEERS

Whether you design water works, consumer goods, or aerospace vehicles; whether you work in private industry, for the U.S. government, or for the public; and whether your efforts are theoretical or practical, you (as an engineer) have a significant responsibility.

Engineers of all types perform exciting and rewarding work, often stretching new technologies to their limits. But those limits are often incomprehensible to nonengineers. As the ambient level of technical sophistication increases, the public has come to depend increasingly and unhesitatingly more on engineers. That is where professional licensing and the National Society of Professional Engineers (NSPE) become important.

NSPE, the leading organization for licensed engineering professionals, is dedicated to serving the engineering profession by supporting activities, such as continuing educational programs for its members, lobbying and legislative efforts on local and national levels, and the promotion of guidelines for ethical service. From local, community-based projects to encourage top-scoring high school students to choose engineering as a career, to hard-hitting lobbying efforts in the nation's capital to satisfy the needs of all engineers, NSPE is committed to you and your profession.

Engineering licensing is a two-way street: it benefits you while it benefits the public and the profession. For you, licensing offers a variety of benefits, ranging from peer recognition to greater advancement and career opportunities. For the profession, licensing establishes a common credential by which all engineers can be compared. For the public, a professional engineering license is an assurance of a recognizable standard of competence.

NSPE has always been a strong advocate of engineering licensing and a supporter of the profession. Professional Publications hopes you will consider membership in NSPE as the next logical step in your career advancement. For more information regarding membership, write to the National Society of Professional Engineers, Information Center, 1420 King Street, Alexandria, VA 22314, or call (703) 684-2800.

# IMPORTANT NAMES, DATES AND ADDRESSES

This book belongs to ______________________

______________________

______________________

phone ______________________

examination date: ____________ hours: ________

examination location: ______________________

______________________

______________________

______________________

tape your cancelled check here

phone number of your registration board: ______________________

address of your registration board: ______________________

______________________

______________________

names of contacts at your registration board: ______________________

______________________

______________________

tape your proof of mailing here

date you sent your application: ______________________

registered/certified mail receipt number: ______________________

date confirmation was received: ______________________

tape quarter here

names of examination proctors: ______________________

______________________

______________________

booklet number: ____________ (A.M.) ____________ (P.M.)

tape quarter here

problems you disagreed with on the examination

problem no. reason

______________________

______________________

______________________

# Civil Engineering Reference Manual

## for the PE Exam

**Sixth Edition**

Michael R. Lindeburg, PE

Professional Publications, Inc. • Belmont, CA

**In the ENGINEERING LICENSING EXAM AND REFERENCE SERIES**

**Engineer-In-Training Reference Manual**
EIT Review Manual
Engineering Fundamentals Quick Reference Cards
Engineer-In-Training Sample Examinations
Mini-Exams for the E-I-T Exam
1001 Solved Engineering Fundamentals Problems
Fundamentals of Engineering Exam Study Guide
Diagnostic FE Exam for the Macintosh
Fundamentals of Engineering Video Series: Thermodynamics
**Civil Engineering Reference Manual**
Quick Reference for the Civil Engineering PE Exam
Civil Engineering Sample Examination
Civil Engineering Review Course on Cassettes
101 Solved Civil Engineering Problems
Seismic Design of Building Structures
Seismic Design Fast
345 Solved Seismic Design Problems
Timber Design for the Civil and Structural Professional Engineering Exams
246 Solved Structural Engineering Problems
**Mechanical Engineering Reference Manual**
Mechanical Engineering Quick Reference Cards
Mechanical Engineering Sample Examination
101 Solved Mechanical Engineering Problems
Mechanical Engineering Review Course on Cassettes
Consolidated Gas Dynamics Tables
Fire and Explosion Protection Systems
**Electrical Engineering Reference Manual**
Quick Reference for the Electrical Engineering PE Exam
Electrical Engineering Sample Examination
**Chemical Engineering Reference Manual for the PE Exam**
Quick Reference for the Chemical Engineering PE Exam
Chemical Engineering Practice Exam Set
**Land Surveyor Reference Manual**
Land Surveyor-In-Training Sample Examination
1001 Solved Surveying Fundamentals Problems

**In the GENERAL ENGINEERING and CAREER ADVANCEMENT SERIES**
How to Become a Professional Engineer
Getting Started as a Consulting Engineer
The Expert Witness Handbook: A Guide for Engineers
Engineering Your Job Search
Engineering Your Start-Up
Engineering Your Writing Success
Intellectual Property Protection: A Guide for Engineers
High-Technology Degree Alternatives
Metric in Minutes
Engineering Economic Analysis
Engineering Law, Design Liability, and Professional Ethics
Engineering Unit Conversions
Calculus Refresher for the Fundamentals of Engineering Exam

**CIVIL ENGINEERING REFERENCE MANUAL**
**Sixth Edition**



Printed in the United States of America

Professional Publications, Inc.
1250 Fifth Avenue, Belmont, CA 94002
(415) 593-9119
www.ppi2pass.com

Current printing of this edition: 5

**Library of Congress Cataloging-in-Publication Data**
Lindeburg, Michael R.
Civil engineering reference manual for the PE exam / Michael R. Lindeburg. -- 6th ed.
p. cm.
Rev. ed. of: Civil engineering reference manual.
Includes index.
ISBN 1-888577-03-7 (hardcover)
1. Civil engineering--United States--Examinations--Study guides.
2. Civil engineering--Handbooks, manuals, etc. I. Lindeburg, Michael R. Civil engineering reference manual. II. Title.
TA159.L54 1996
624'.076--dc20 96-44245
CIP

# TABLE OF CONTENTS

## 4 HYDRAULIC MACHINES



## 5 OPEN CHANNEL FLOW

## 6 HYDROLOGY



## 7 WATER SUPPLY ENGINEERING



## 8 WASTE-WATER ENGINEERING

## 9 SOILS



## 10 FOUNDATIONS AND RETAINING WALLS



## 11 STATICS

## 12 MECHANICS OF MATERIALS



## 13 INDETERMINATE STRUCTURES



## 14 REINFORCED CONCRETE DESIGN

## 15 STEEL DESIGN AND ANALYSIS

## 16 TRAFFIC ANALYSIS, TRANSPORTATION, AND HIGHWAY DESIGN



## 17 SURVEYING

# PREFACE AND ACKNOWLEDGMENTS to the Sixth Edition

When I first wrote the *Civil Engineering Reference Manual*, it seemed complete enough. But, as Stephen King writes in his "Dark Tower" series, "... the world has moved on ..."

Some of the changes in this edition were necessary because of changes to construction codes. The concrete chapter has been revised to make it consistent with *ACI 318-89*. Similarly, the steel chapter has been updated to make it conform with the ninth (green) edition *AISC Manual*. Allowable stress design (ASD) is still the primary treatment for steel design in this book. Load resistance factor design (LRFD) is not yet a significant concern to the primary audience of this book. The number of references to the actual code sections, equations, and tables has been increased significantly in these two chapters. In addition, the practice problems (and their solutions) in the traffic chapter have been updated to reflect the 1985 *Highway Capacity Manual*.

New material has been added to provide fuller coverage of many subjects. New sections on construction staking and earthwork have been added to the surveying chapter. Of course, there are the inevitable changes, reformattings, and improvements resulting from comments, suggestions, and corrections contributed by readers. For example, in response to numerous requests, I have used $\gamma$ for specific weight instead of $\rho$ for mass density in the fluids chapters. I have also tried to make the equations in the fluids chapters dimensionally consistent by distinguishing between $g$ and $g_c$. And, several tables of data have been revised.

Page tabs have been added to this book for the first time. This feature is rapidly becoming a standard among Professional Publications' texts.

A new, typeset version of the solutions manual to this new edition is currently in production. This should eliminate the many complaints that people have made about my handwriting.

There are a lot of other things that I could have changed. For example, I have continued to use the EE format for scientific notation, and I have decided not to add units to the example calculations. The postscripts chapter also remains and continues to grow with each printing. Such changes will have to wait until the next edition.

I want to thank David Fong, P.E., S.E., and Dan Lewin, P.E., S.E., for reviewing and suggesting changes to the concrete and steel chapters. I am also grateful to Jan Van Sickle, P.L.S., for reviewing the new construction staking and earthwork material I wrote for the surveying chapter. Additional thanks go to Peter G. Furth, Ph.D., Department of Civil Engineering at Northeastern University, for his expert analysis and revision of the traffic problems in Chapter 16.

It is probably no surprise to anyone that I see this edition as a maintenance edition—a mere way station on a long journey. With apologies to Stephen King, the seventh edition is already starting to claw its way out of my brain.

Michael R. Lindeburg, P.E.
Belmont, CA
April, 1992

# HOW TO USE THIS BOOK

## QUICK START

If you are in a hurry to use this book and only want to read one paragraph, here it is:

> Start anywhere. The chapters are independent. Solve every practice problem. Use the index extensively. Good luck.

## HOW PRACTICING ENGINEERS AND STUDENTS CAN USE THIS BOOK

How you use this book depends on what you intend to use it for. If you are a practicing engineer or an engineering major and have purchased this book as a general reference handbook for your library, it will probably sit in your bookcase until you have a specific need to remember something you learned a while ago. Then, you can use the index to find material that will help you.

If you are preparing for the NCEES P.E. examination in civil engineering, here are a few suggestions.

- Obtain the companion Solutions Manual. (The solutions manual is separate so that this book can still be used as an examination reference in those states that prohibit collections of solved problems.)
- Become familiar with the format of the P.E. exam. Start your review by reading chapter 21, "Engineering Licensing."
- Become intimately familiar with this book. Know the order of the chapters, the approximate locations of important figures and tables, the contents of the appendices, etc.
- Use the subject title tabs along the side of each page to find the chapter you are looking for.
- Know which subjects in this book are not covered on the P.E. exam. Several chapters (e.g., Mathematics, Systems of Units, and Engineering Economic Analysis) in this book are supportive and do not cover actual exam topics. These chapters provide background for the other chapters and other types of problems.
- Some subjects appear in more than one chapter. You should use the index liberally to learn all there is about a particular subject. Most subjects have secondary or tertiary indexing, which means you should be able to find entries no matter how you look them up.
- Start your review of a subject by skimming through the chapter to familiarize yourself with the subjects before starting the practice problems. That way you will know the location of each subject if you need a quick review.
- It isn't necessary to solve every end-of-chapter practice problem, but I suggest that you try to. The number of practice problems you solve will depend on how much time you have and how skilled you are in each area. To do all the problems requires approximately 15–20 hours of preparation time per chapter.
- Learn to pace yourself in an exam situation by watching the clock when you solve the "Timed" end-of-chapter problems. The timed problems are representative of the subject matter, length, and complexity of the essay-type problems you will find on the actual examination. You don't actually have to start a stopwatch when you do these problems; just be aware of your rate of progression.
- Learn what slows you down. Are you spending time learning to use your calculator? Are you hampered by not having enough reference books? Is the terminology new to you, or is language a barrier?

## HOW INSTRUCTORS CAN USE THIS BOOK

This book started as a series of handouts for my own Professional Engineering exam review course. As with the sets of "notes" I handed out in other courses, the handouts were originally intended as a substitute for all of the long formulas, illustrations, and tables of data that I did not have time to put on the chalkboard. After having taught the course many times, though, I rewrote the chapters to more closely parallel the organization and content of my lectures.

If you are teaching a review course for the P.E. examination without the benefit of recent, first-hand exam experience, you can use the material in this book as a guide to prepare your lectures. You should emphasize the subjects in each chapter and avoid most subjects omitted. You can feel confident that subjects omitted from this book are rarely, if ever, found on the P.E. exam.

I have always tried to overprepare my own students. For that reason, the examples and practice problems in this book are sometimes more difficult and varied than actual examination problems. Also, you will appreciate that it is more efficient in a lecture or while doing practice problems to cover several procedural steps in one practice problem than to ask numerous simple "one-liners." That is the reason there are no multiple-choice problems in this book.

There are many end-of-chapter practice problems for each major examination subject. In my courses, all of the problems are assigned. To do all the problems requires approximately 15–20 hours of preparation per week over the 14-week course.

"Capacity assignment" is the goal of my courses. If you assign 15 hours of practice problems as homework, and a student can only put in 10 hours of preparation that week, that student will have worked to his or her capacity. After the P.E. examination, your students will honestly say that they could not have prepared any more than they did in your course.

Homework assignments in my courses are not individually graded. Instead, the students obtain the accompanying Solutions Manual and have the solutions to all practice problems in advance. However, each student must turn in a completed set of problems for credit each week. I personally address all special needs or questions that are written on the assignments.

I have found that a 14-week format works well for a P.E. review course. Each week there is a 3-hour lecture with a short intermediate break. The table, "Typical P.E. Review Course Format," outlines the basic course format that has worked well for me. You can, of course, add more lecture time or weeks, and that will certainly be appreciated by the students because this is a fast-paced course. However, I don't think you can cover the full breadth of material in much less time or in many fewer weeks.

I have tried to order the course lectures in a logical, progressive manner. The easiest subjects (e.g., fluids) come near the end; fluids has to precede open channel flow and other water subjects; concrete has to come before foundations, and so on.

Lecture coverage of some examination subjects is necessarily brief. For example, lectures on concrete and steel design could go on for months. Other subjects are not covered at all. For example, engineering economic analysis has no lecture. This is consistent with the absence of the subject in the NCEES test plan, even though elements of economic decision-making continue to appear in the examination.

Any skipped chapters and their end-of-chapter practice problems are assigned as floating assignments to be made up in the students' "free time."

I strongly believe in exposing my students to a realistic sample examination, but I no longer administer an in-class mock exam. Since the review course usually ends only a few days before the real P.E. examination, I hesitate to make students sit for several hours in the late evening to take a "final exam." Rather, I distribute and assign a take-home sample examination at the first meeting of the review course. (I use the same *Civil Engineering Sample Examination* publication that is available to your courses from Professional Publications.)

If a practice test is to be used as an indication of preparedness, ask your students not to look at it prior to taking it. Looking at the practice examination or otherwise using it to direct their review may produce unwarranted specialization in subjects contained in the practice examination.

There are many ways to organize a P.E. review course depending on your available time, budget, intended audience, resources, and enthusiasm. However, all good course formats have the same result: the students struggle with the work load during the course, and then they breeze through the examination after the course.

**Typical P.E. Review Course Format**

| meeting | subject covered |
|---|---|
| 1. | Introduction, Exam Format, Math |
| 2. | Statics, Mechanics of Materials |
| 3. | Steel |
| 4. | Concrete |
| 5. | Soils |
| 6. | Foundations |
| 7. | Timber |
| 8. | Masonry (in California substitute Seismic Design) |
| 9. | Surveying |
| 10. | Traffic |
| 11. | Fluids/Hydraulic Machines |
| 12. | Open Channel Flow/Hydrology |
| 13. | Water Supply Engineering |
| 14. | Wastewater Engineering |

# Concentrate Your Studies . . .

Studying for the PE exam requires a significant amount of time and effort on your part. Make every minute count toward your success with these additional study materials by Michael Lindeburg:

## Solutions Manual for the Civil Engineering Reference Manual

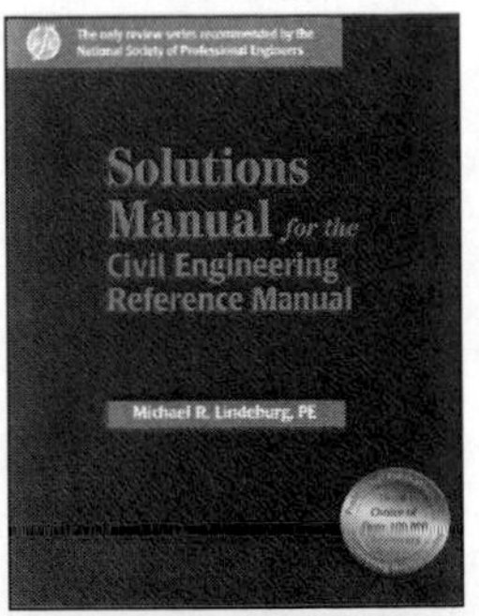


160 pages, 8½ × 11, paperback

Don't forget that there is a companion **Solutions Manual** that provides step-by-step solutions to the practice problems given at the end of each chapter in this reference manual. This important study aid will provide immediate feedback on your progress. Without the **Solutions Manual**, you may never know if your methods are correct.

## Quick Reference for the Civil Engineering PE Exam

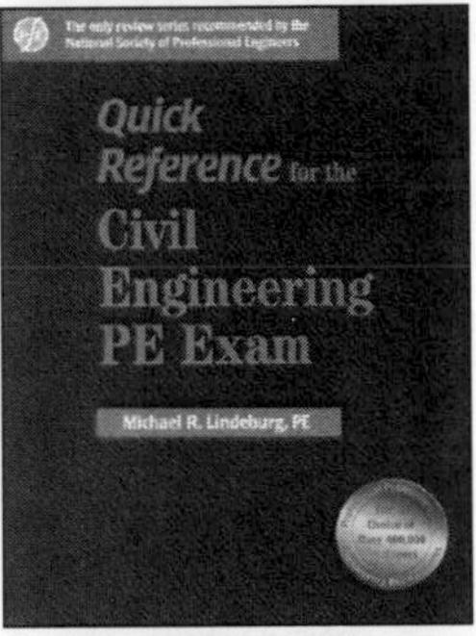


40 pages, 8½ × 11, paperback

Because speed is important during the exam, you will welcome the advantage provided by the **Quick Reference for the Civil Engineering PE Exam**. This handy resource gives you quick access to equations, methods, and data needed during the exam. It is divided into specific exam subjects and follows the **Civil Engineering Reference Manual** in organization and nomenclature.

## Civil Engineering Sample Examination

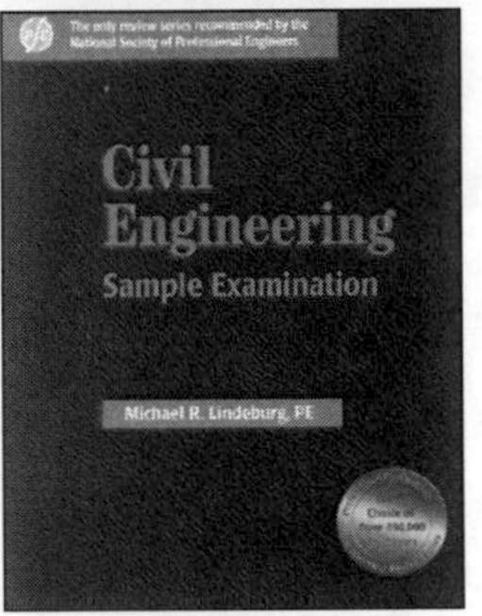


72 pages, 8½ × 11, paperback

Increase your speed and confidence in solving the types of questions that may appear on the exam by taking a facsimile examination. **Civil Engineering Sample Examination** includes 20 typical problems (with complete, worked-out solutions) to acquaint you with the exam format and teach you the correct and most efficient methods for solving PE exam problems.

## Seismic Design of Building Structures

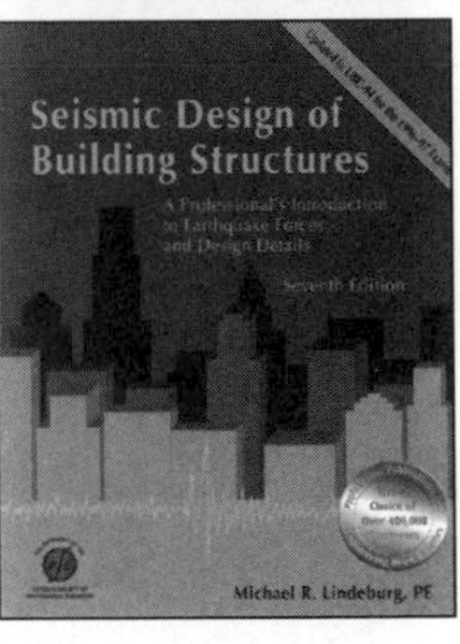


192 pages, 8½ × 11, paperback

Specialized areas of the professional engineering exam require additional review materials for adequate preparation. **Seismic Design of Building Structures** is intended to help you prepare for problems relating to earthquake activity. This compilation of seismic problems (with solutions) is taken from actual previous exams. It presents and illustrates lateral force theory, from simple vibration analysis through the application of modern SEAOC codes. Each concept builds on previous material to gradually introduce you to new ideas.

## Civil Engineering Review Course on Cassettes

The **Civil Engineering Review Course on Cassettes** gives you the discipline of a formal class, taken directly from Michael Lindeburg's outstanding lecture series. By listening to these audio tapes and following along with the accompanying written material, you will benefit from Mr. Lindeburg's insights into exam content and test-taking strategies. This cassette course was taped before a live audience of 100 practicing engineers. Included are case studies, examples, and questions raised during the classes. To save you the time and the inconvenience of taking notes, each lecture is accompanied by detailed notes duplicating material illustrated in class.

The cassette course is divided into two sections so that you can select the topics you need most.

| • Non-Structural Series (14 tapes) | • Structural Series (13 tapes) |
|---|---|
| Approximate lecture time: 19 hours | Approximate lecture time: 19 hours |

**Professional Publications, Inc.**
Department 77 • 1250 Fifth Avenue • Belmont, CA 94002
800-426-1178 • http://www.ppi2pass.com

# Notice to Examinees

Do not copy, memorize, or distribute problems from the Principles and Practice of Engineering (P&P) Examination. These acts are considered to be exam subversion.

The P&P examination is copyrighted by the National Council of Examiners for Engineering and Surveying. Copying and reproducing P&P exam problems for commercial purposes is a violation of federal copyright law. Reporting examination problems to other examinees invalidates the examination process and threatens the health and welfare of the public.

# 1 MATHEMATICS AND RELATED SUBJECTS

## 1 INTRODUCTION

Engineers working in design and analysis encounter mathematical problems on a daily basis. Although algebra and simple trigonometry are often sufficient for routine calculations, there are many instances when a quick review of certain subjects is needed. This chapter, in addition to supporting the calculations used in other chapters, consolidates the mathematical concepts most often needed by engineers in their daily activities.

## 2 SYMBOLS USED IN THIS BOOK

Many symbols, letters, and Greek characters are used to represent variables in the formulas used throughout this book. These symbols and characters are defined in the nomenclature section of each chapter. However, some of the symbols which are used as operators in this book are listed here.

## 3 THE GREEK ALPHABET

**Table 1.2**
The Greek Alphabet

| | | | | | |
|---|---|---|---|---|---|
| $A$ | $\alpha$ | alpha | $N$ | $\nu$ | nu |
| $B$ | $\beta$ | beta | $\Xi$ | $\xi$ | xi |
| $\Gamma$ | $\gamma$ | gamma | $O$ | $o$ | omicron |
| $\Delta$ | $\delta$ | delta | $\Pi$ | $\pi$ | pi |
| $E$ | $\epsilon$ | epsilon | $P$ | $\rho$ | rho |
| $Z$ | $\varsigma$ | zeta | $\Sigma$ | $\sigma$ | sigma |
| $H$ | $\eta$ | eta | $T$ | $\tau$ | tau |
| $\Theta$ | $\theta$ | theta | $\Upsilon$ | $\upsilon$ | upsilon |
| $I$ | $\iota$ | iota | $\Phi$ | $\phi$ | phi |
| $K$ | $\kappa$ | kappa | $X$ | $\chi$ | chi |
| $\Lambda$ | $\lambda$ | lambda | $\Psi$ | $\psi$ | psi |
| $M$ | $\mu$ | mu | $\Omega$ | $\omega$ | omega |

**Table 1.1**
Symbols Used in This Book

| Symbol | Name | Use | Example |
|---|---|---|---|
| $\sum$ | sigma | series addition | $\sum_{i=1}^{3} x_i = x_1 + x_2 + x_3$ |
| $\prod$ | pi | series multiplication | $\prod_{i=1}^{3} x_i = x_1 x_2 x_3$ |
| $\Delta$ | delta | change in quantity | $\Delta h = h_2 - h_1$ |
| $-$ | over bar | average value | $\overline{x}$ |
| $\cdot$ | over dot | per unit time | $\dot{Q}$ = quantity flowing per second |
| ! | factorial | | $x! = x(x-1)(x-2)\cdots(2)(1)$ |
| $\mid\ \mid$ | absolute value | | $\lvert -3 \rvert = +3$ |
| $\approx$ | approximately equal to | | $x \approx 1.5$ |
| $\propto$ | proportional to | | $x \propto y$ |
| $\infty$ | infinity | | $x \rightarrow \infty$ |
| log | base 10 logarithm | | $\log(5.74)$ |
| ln | natural logarithm | | $\ln(5.74)$ |
| EE | scientific notation | | EE–4 |
| exp | exponential power | | $\exp(x) = e^x$ |

## 4 MENSURATION

*Nomenclature*

A total surface area
d distance
h height
p perimeter
r radius
s side (edge) length, arc length
V volume
$\theta$ vertex angle, in radians
$\phi$ central angle, in radians

*Circle*

$$p = 2\pi r \qquad 1.1$$

$$A = \pi r^2 = \frac{p^2}{4\pi} \qquad 1.2$$

*Circular Segment*

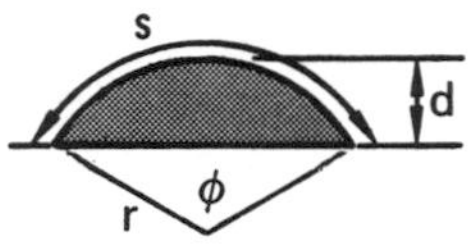

$$A = \frac{1}{2}r^2(\phi - \sin\phi) \qquad 1.3$$

$$\phi = \frac{s}{r} = 2\left(\arccos\frac{r-d}{r}\right) \qquad 1.4$$

*Triangle*

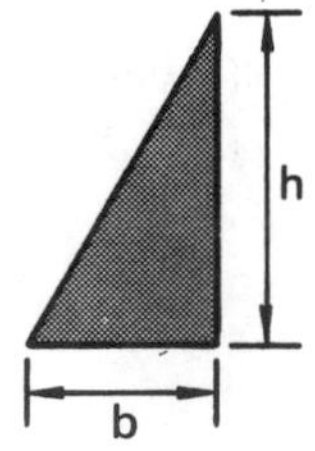


$$A = \frac{1}{2}bh \qquad 1.5$$

*Parabola*

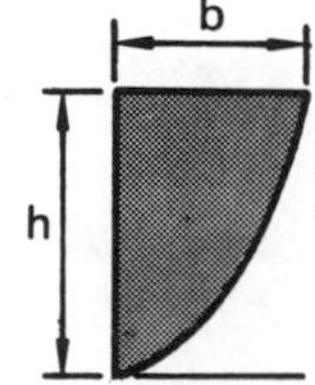


$$A = \frac{2bh}{3} \qquad 1.6$$

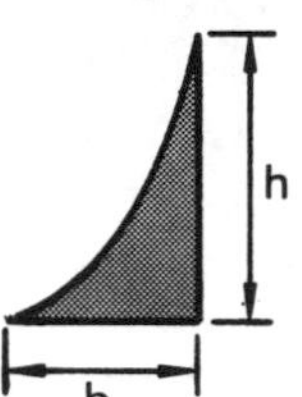


$$A = \frac{1}{3}bh \qquad 1.7$$

*Circular Sector*

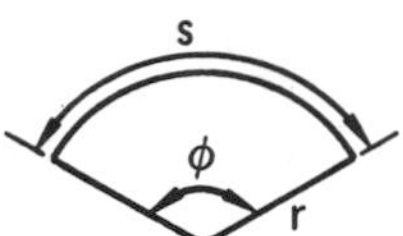


$$A = \frac{1}{2}\phi r^2 = \frac{1}{2}sr \qquad 1.8$$

$$\phi = \frac{s}{r} \qquad 1.9$$

*Ellipse*

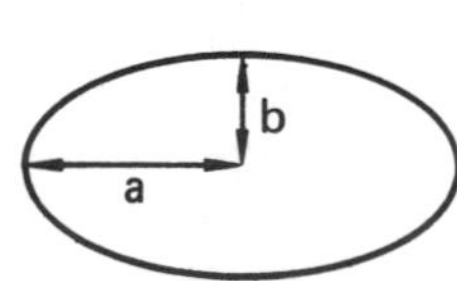


$$A = \pi ab \qquad 1.10$$

$$p = 2\pi\sqrt{\frac{1}{2}(a^2 + b^2)} \qquad 1.11$$

*Trapezoid*

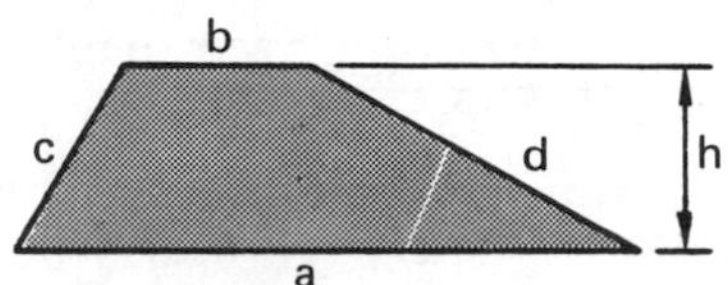


$$p = a + b + c + d \qquad 1.12$$

$$A = \frac{1}{2}h(a + b) \qquad 1.13$$

The trapezoid is isosceles if $c = d$.

*Parallelogram*

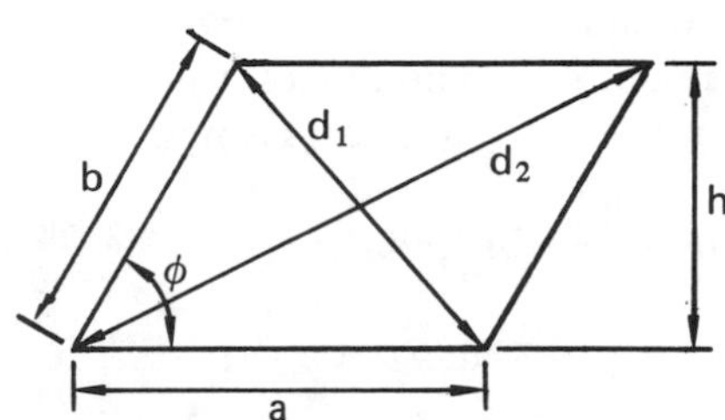


$$p = 2(a + b) \qquad 1.14$$

$$d_1 = \sqrt{a^2 + b^2 - 2ab(\cos\phi)} \qquad 1.15$$

$$d_2 = \sqrt{a^2 + b^2 + 2ab(\cos\phi)} \qquad 1.16$$

$$d_1^2 + d_2^2 = 2(a^2 + b^2) \qquad 1.17$$

$$A = ah = ab(\sin\phi) \qquad 1.18$$

If $a = b$, the parallelogram is a rhombus.

*Regular Polygon*

($n$ equal sides)

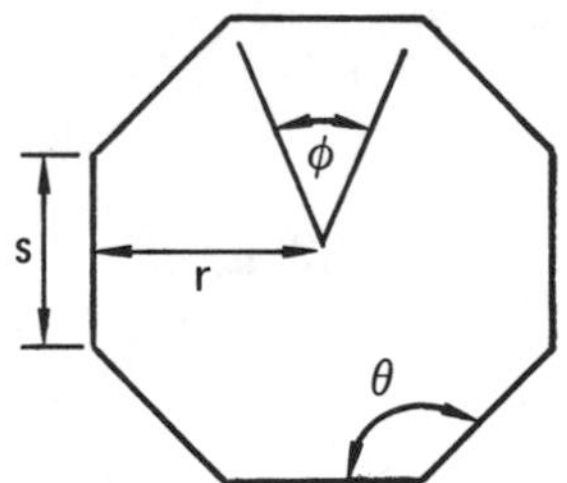


$$\phi = \frac{2\pi}{n} \qquad 1.19$$

$$\theta = \frac{\pi(n-2)}{n} \qquad 1.20$$

$$p = ns \qquad 1.21$$

$$s = 2r\left(\tan\left(\frac{\phi}{2}\right)\right) \qquad 1.22$$

$$A = \frac{1}{2}nsr \qquad 1.23$$

**Table 1.3**
Polygons

| Number of Sides | Name of Polygon |
|---|---|
| 3 | triangle |
| 4 | rectangle |
| 5 | pentagon |
| 6 | hexagon |
| 7 | heptagon |
| 8 | octagon |
| 9 | nonagon |
| 10 | decagon |

*Sphere*

$$V = \frac{4\pi r^3}{3} \qquad 1.24$$

$$A = 4\pi r^2 \qquad 1.25$$

*Right Circular Cone*

$$V = \frac{\pi r^2 h}{3} \qquad 1.26$$

$$A = \pi r\sqrt{r^2 + h^2} \qquad 1.27$$

(does not include base area)

*Right Circular Cylinder*

$$V = \pi h r^2 \qquad 1.28$$

$$A = 2\pi r h \qquad 1.29$$

(does not include end areas)

*Paraboloid of Revolution*

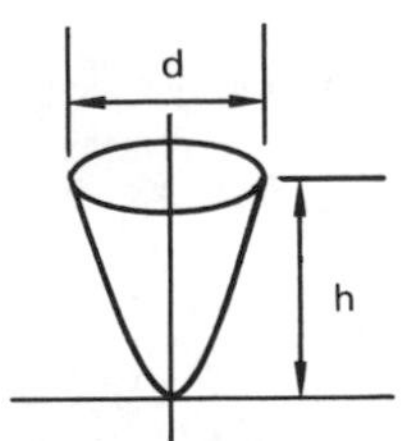


$$V = \frac{\pi h d^2}{8} \qquad 1.30$$

*Regular Polyhedron*

The radius of a sphere inscribed within a regular polyhedron is

$$r = \frac{3V}{A} \qquad 1.31$$

**Table 1.4**
Polyhedrons

| Number of Faces | Form of Faces | Total Surface Area | Volume |
|---|---|---|---|
| 4 | equilateral triangle | $1.7321\ s^2$ | $0.1179\ s^3$ |
| 6 | square | $6.0000\ s^2$ | $1.0000\ s^3$ |
| 8 | equilateral triangle | $3.4641\ s^2$ | $0.4714\ s^3$ |
| 12 | regular pentagon | $20.6457\ s^2$ | $7.6631\ s^3$ |
| 20 | equilateral triangle | $8.6603\ s^2$ | $2.1817\ s^3$ |

*Example 1.1*

What is the hydraulic radius of a 6″ pipe filled to a depth of 2″?

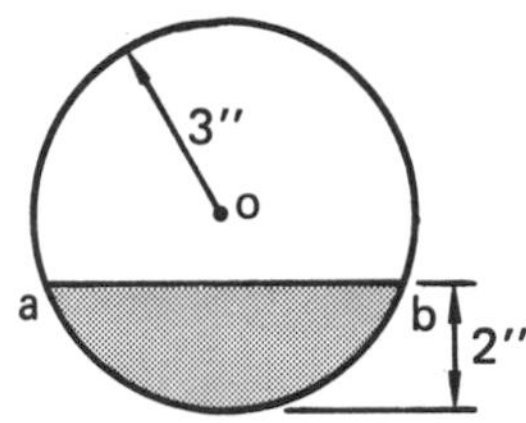


The hydraulic radius is defined as

$$r_h = \frac{\text{area in flow}}{\text{length of wetted perimeter}} = \frac{A}{s}$$

Points $o$, $a$, and $b$ may be used to find the central angle of the circular segment.

$$\frac{1}{2}(\text{angle}\,aob) = \arccos\left(\frac{1}{3}\right) = 70.53°$$
$$\phi = 141.06° = 2.46 \text{ radians}$$

Then,

$$A = \frac{1}{2}(3)^2(2.46 - 0.63) = 8.235 \text{ in}^2$$
$$s = (3)(2.46) = 7.38 \text{ in}$$
$$r_h = \frac{8.235}{7.38} = 1.12 \text{ in}$$

## 5 SIGNIFICANT DIGITS

The significant digits in a number include the left-most, non-zero digits to the right-most digit written. Final answers from computations should be rounded off to the number of decimal places justified by the data. The answer can be no more accurate than the least accurate number in the data. Of course, rounding should be done on final calculation results only. It should not be done on interim results.

| number as written | number of significant digits | implied range |
|---|---|---|
| 341 | 3 | 340.5 to 341.5 |
| 34.1 | 3 | 34.05 to 34.15 |
| 0.00341 | 3 | 0.003405 to 0.003415 |
| 3410. | 4 | 3409.5 to 3410.5 |
| 341 EE7 | 3 | 340.5 EE7 to 341.5 EE7 |
| 3.41 EE−2 | 3 | 3.405 EE−2 to 3.415 EE−2 |

## 6 ALGEBRA

Algebra provides the rules which allow complex mathematical relationships to be expanded or condensed. Algebraic laws may be applied to complex numbers, variables, and numbers. The general rules for changing the form of a mathematical relationship are given here:

*Commutative law for addition:*

$$a + b = b + a \qquad 1.32$$

*Commutative law for multiplication:*

$$ab = ba \qquad 1.33$$

*Associative law for addition:*

$$a + (b + c) = (a + b) + c \qquad 1.34$$

*Associative law for multiplication:*

$$a(bc) = (ab)c \qquad 1.35$$

*Distributive law:*

$$a(b + c) = ab + ac \qquad 1.36$$

### A. POLYNOMIAL EQUATIONS

*1. Standard Forms*

$$(a + b)(a - b) = a^2 - b^2 \qquad 1.37$$
$$(a \pm b)^2 = a^2 \pm 2ab + b^2 \qquad 1.38$$
$$(a \pm b)^3 = a^3 \pm 3a^2b + 3ab^2 \pm b^3 \qquad 1.39$$
$$(a^3 \pm b^3) = (a \pm b)(a^2 \mp ab + b^2) \qquad 1.40$$
$$(a^n + b^n) = (a + b)(a^{n-1} - a^{n-2}b + \cdots + b^{n-1}) \text{(factorable only for } n \text{ odd)} \qquad 1.41$$
$$(a^n - b^n) = (a - b)(a^{n-1} + a^{n-2}b + \cdots + b^{n-1})(n \text{ odd or even}) \qquad 1.42$$

*2. Quadratic Equations*

Given a quadratic equation $ax^2 + bx + c = 0$, the roots $x_1^*$ and $x_2^*$ may be found from

$$x_1^*, x_2^* = \frac{-b \pm \sqrt{b^2 - 4ac}}{2a} \quad 1.43$$

$$x_1^* + x_2^* = -\frac{b}{a} \quad 1.44$$

$$x_1^* x_2^* = \frac{c}{a} \quad 1.45$$

*3. Cubic Equations*

Cubic and higher order equations occur infrequently in most engineering problems. However, they usually are difficult to factor when they do occur. Trial and error solutions are usually unsatisfactory except for finding the general region in which a root occurs. Graphical means can be used to obtain only a fair approximation to the root.

Numerical analysis techniques must be used if extreme accuracy is needed. The more efficient numerical analysis techniques are too complicated to present here. However, the bisection method illustrated in example 1.2 usually can provide the required accuracy with only a few simple iterations.

The bisection method starts out with two values of the independent variable, $L_0$ and $R_0$, which straddle a root. Since the function has a value of zero at a root, $f(L_0)$ and $f(R_0)$ will have opposite signs. The following algorithm describes the remainder of the bisection method:

Let $n$ be the iteration number. Then, for $n = 0, 1, 2, \ldots$ perform the following steps until sufficient accuracy is attained.

Set $m = \frac{1}{2}(L_n + R_n)$

Calculate $f(m)$

If $f(L_n)f(m) \leq 0$, set $L_{n+1} = L_n$ and $R_{n+1} = m$

Otherwise, set $L_{n+1} = m$ and $R_{n+1} = R_n$

$f(x)$ has at least one root in the interval $(L_{n+1}, R_{n+1})$

The estimated value of that root, $x^*$, is

$$x^* \approx \frac{1}{2}(L_{n+1} + R_{n+1}) \quad 1.46$$

The maximum error is $1/2(R_{n+1} - L_{n+1})$. The iterations continue until the maximum error is reasonable for the accuracy of the problem.

*Example 1.2*

Use the bisection method to find the roots of

$$f(x) = x^3 - 2x - 7$$

The first step is to find $L_0$ and $R_0$, which are the values of $x$ which straddle a root and have opposite signs. A table can be made and values of $f(x)$ calculated for random values of $x$.

| $x$ | $-2$ | $-1$ | $0$ | $+1$ | $+2$ | $+3$ |
|---|---|---|---|---|---|---|
| $f(x)$ | $-11$ | $-6$ | $-7$ | $-8$ | $-3$ | $+14$ |

Since $f(x)$ changes sign between $x = 2$ and $x = 3$,

$$L_0 = 2 \text{ and } R_0 = 3$$

*Iteration 0*:

$$m = \frac{1}{2}(2 + 3) = 2.5$$

$$f(2.5) = (2.5)^3 - 2(2.5) - 7 = 3.625$$

Since $f(2.5)$ is positive, a root must exist in the interval (2, 2.5). Therefore,

$$L_1 = 2 \text{ and } R_1 = 2.5$$

At this point, the best estimate of the root is

$$x^* \approx \frac{1}{2}(2 + 2.5) = 2.25$$

The maximum error is $\frac{1}{2}(2.5 - 2) = 0.25$.

*Iteration 1*:

$$m = \frac{1}{2}(2 + 2.5) = 2.25$$

$$f(2.25) = -0.1094$$

Since $f(m)$ is negative, a root must exist in the interval (2.25, 2.5). Therefore,

$$L_2 = 2.25 \text{ and } R_2 = 2.5$$

The best estimate of the root is

$$x^* \approx \frac{1}{2}(2.25 + 2.5) = 2.375$$

The maximum error is $\frac{1}{2}(2.5 - 2.25) = 0.125$.

This procedure continues until the maximum error is acceptable. Of course, this method does not automatically find any other roots that may exist on the real number line.

*4. Finding Roots to General Expressions*

There is no specific technique that will work with all general expressions for which roots are needed. If graphical means are not used, some combination of factoring and algebraic simplification must be used. However, multiplying each side of an equation by a power of a variable may introduce extraneous roots. Such an extraneous root will not satisfy the original equation, even though it was derived correctly according to the rules of algebra.

Although it is always a good idea to check your work, this step is particularly necessary whenever you have squared an expression or multiplied it by a variable.

*Example 1.3*

Find the value of $x$ which will satisfy the following expression:

$$\sqrt{x-2} = \sqrt{x} + 2$$

First, square both sides.

$$x - 2 = x + 4\sqrt{x} + 4$$

Next, subtract $x$ from both sides and combine constants.

$$4\sqrt{x} = -6$$

Solving for $x$ yields $x^* = 9/4$. However $9/4$ does not satisfy the original expression since it is an extraneous root.

B. SIMULTANEOUS LINEAR EQUATIONS

Given $n$ independent equations and $n$ unknowns, the $n$ values which simultaneously solve all $n$ equations can be found by the methods illustrated in example 1.4.

*1. By Substitution (shown by example)*

*Example 1.4*

Solve

$$2x + 3y = 12\,(a)$$
$$3x + 4y = 8\,(b)$$

*step 1*: From equation (a), solve for $x = 6 - 1.5y$

*step 2*: Substitute $(6 - 1.5y)$ into equation (b) wherever $x$ appears. $3(6 - 1.5y) + 4y = 8$ or $y^* = 20$

*step 3*: Solve for $x^*$ from either equation:

$$x^* = 6 - 1.5(20) = -24$$

*step 4*: Check that $(-24,\ 20)$ solves both original equations.

*2. By Reduction (same example)*

*step 1*: Multiply each equation by a number chosen to make the coefficient of one of the variables the same in each equation.

$$3 \times \text{ equation (a): } 6x + 9y = 36\text{(c)}$$
$$2 \times \text{ equation (b): } 6x + 8y = 16\text{(d)}$$

*step 2*: Subtract one equation from the other. Solve for one of the variables.

$$\text{(c)} - \text{(d)}: \ y^* = 20$$

*step 3*: Solve for the remaining variable.

*step 4*: Check that the calculated values of $(x^*, y^*)$ solve both original equations.

*3. By Cramer's Rule*

This method is best for 3 or more simultaneous equations. (The calculation of determinants is covered later in this chapter.)

To find $x^*$ and $y^*$ which satisfy

$$a_1x + b_1y = c_1$$
$$a_2x + b_2y = c_2$$

calculate the determinants

$$\mathbf{D}_1 = \begin{vmatrix} a_1 & b_1 \\ a_2 & b_2 \end{vmatrix} \qquad 1.47$$

$$\mathbf{D}_2 = \begin{vmatrix} c_1 & b_1 \\ c_2 & b_2 \end{vmatrix} \qquad 1.48$$

$$\mathbf{D}_3 = \begin{vmatrix} a_1 & c_1 \\ a_2 & c_2 \end{vmatrix} \qquad 1.49$$

Then, if $\mathbf{D}_1 \neq 0$, the unique numbers satisfying the two simultaneous equations are:

$$x^* = \frac{\mathbf{D}_2}{\mathbf{D}_1} \qquad 1.50$$

$$y^* = \frac{\mathbf{D}_3}{\mathbf{D}_1} \qquad 1.51$$

If $\mathbf{D}_1$ (the determinant of the coefficients matrix) is zero, the system of simultaneous equations may still have a solution. However, Cramer's rule cannot be used to find that solution. If the system is homogeneous (i.e., has the general form $\mathbf{A}x = \mathbf{0}$), then a non-zero solution exists if and only if $\mathbf{D}_1$ is zero.

*Example 1.5*

Solve the following system of simultaneous equations:

$$2x + 3y - 4z = 1$$
$$3x - y - 2z = 4$$
$$4x - 7y - 6z = -7$$

Calculate the determinants:

$$\mathbf{D}_1 = \begin{vmatrix} 2 & 3 & -4 \\ 3 & -1 & -2 \\ 4 & -7 & -6 \end{vmatrix} = 82$$

$$\mathbf{D}_2 = \begin{vmatrix} 1 & 3 & -4 \\ 4 & -1 & -2 \\ -7 & -7 & -6 \end{vmatrix} = 246$$

$$\mathbf{D}_3 = \begin{vmatrix} 2 & 1 & -4 \\ 3 & 4 & -2 \\ 4 & -7 & -6 \end{vmatrix} = 82$$

$$\mathbf{D}_4 = \begin{vmatrix} 2 & 3 & 1 \\ 3 & -1 & 4 \\ 4 & -7 & -7 \end{vmatrix} = 164$$

Then,

$$x^* = \frac{\mathbf{D}_2}{\mathbf{D}_1} = 3$$
$$y^* = \frac{\mathbf{D}_3}{\mathbf{D}_1} = 1$$
$$z^* = \frac{\mathbf{D}_4}{\mathbf{D}_1} = 2$$

## C. SIMULTANEOUS QUADRATIC EQUATIONS

Although simultaneous non-linear equations are best solved graphically, a specialized method exists for simultaneous quadratic equations. This method is known as *Eliminating the Constant Term.*

*step 1*: Isolate the constant terms of both equations on the right-hand side of the equalities.

*step 2*: Multiply both sides of one equation by a number chosen to make the constant terms of both equations the same.

*step 3*: Subtract one equation from the other to obtain a difference equation.

*step 4*: Factor the difference equation into terms.

*step 5*: Solve for one of the variables from one of the factor terms.

*step 6*: Substitute the formula for the variable into one of the original equations and complete the solution.

*step 7*: Check the solution.

*Example 1.6*

Solve for the simultaneous values of $x$ and $y$:

*step 1*:

$$2x^2 - 3xy + y^2 = 15$$
$$x^2 - 2xy + y^2 = 9$$

*steps 2 & 3*:

$$\begin{array}{r} 6x^2 - 9xy + 3y^2 = 45 \\ -(5x^2 - 10xy + 5y^2) = 45 \\ \hline x^2 + xy - 2y^2 = 0 \end{array}$$

*steps 4 & 5*: $x^2+xy-2y^2$ factors into $(x+2y)(x-y)$ from which we obtain $x = -2y$.

*step 6*: Substituting $x = -2y$ into $(2x^2 - 3xy + y^2 = 15)$ gives $y^* = \pm 1$, from which $x^* = \pm 2$ can be derived by further substitution.

## D. EXPONENTIATION

($x$ is any variable or constant)

$$x^m x^n = x^{(n+m)} \quad 1.52$$
$$\frac{x^m}{x^n} = x^{(m-n)} \quad 1.53$$
$$(x^n)^m = x^{(mn)} \quad 1.54$$
$$a^{m/n} = \sqrt[n]{a^m} \quad 1.55$$
$$\left(\frac{a}{b}\right)^n = \frac{a^n}{b^n} \quad 1.56$$
$$\sqrt[n]{x} = (x)^{1/n} \quad 1.57$$
$$x^{-n} = \frac{1}{x^n} \quad 1.58$$
$$x^0 = 1 \quad 1.59$$

## E. LOGARITHMS

Logarithms are exponents. That is, the exponent $x$ in the expression $b^x = n$ is the logarithm of $n$ to the base $b$. Therefore, $(\log_b n) = x$ is equivalent to $(b^x = n)$.

The base for common logs is 10. Usually, *log* will be written when common logs are desired, although $\log_{10}$ appears occasionally. The base for *natural (Napierian) logs* is 2.718..., a number which is given the symbol $e$. When natural logs are desired, usually *ln* will be written, although $\log_e$ is also used.

Most logarithms will contain an integer part (the *characteristic*) and a fractional part (the *mantissa*). The logarithm of any number less than one is negative. If the number is greater than one, its logarithm is positive.

Although the logarithm may be negative, the mantissa is always positive.

For common logarithms of numbers greater than one, the characteristics will be positive and equal to one less than the number of digits in front of the decimal. If the number is less than one, the characteristic will be negative and equal to one more than the number of zeros immediately following the decimal point.

*Example 1.7*

**What is $\log_{10}$ (0.05)?**

Since the number is less than one and there is one leading zero, the characteristic is −2. From the logarithm tables, the mantissa of 5.0 is 0.699. Two ways of combining the mantissa and characteristic are possible:

*Method 1*: $\bar{2}.699$

*Method 2*: $8.699 - 10$

If the logarithm is to be used in a calculation, it must be converted to operational form: $-2+0.699 = -1.301$. Notice that $-1.301$ is not the same as $\bar{1}.301$.

## F. LOGARITHM IDENTITIES

$$x^a = \text{antilog}[a \log(x)] \qquad 1.60$$

$$\log(x^a) = a \log(x) \qquad 1.61$$

$$\log(xy) = \log(x) + \log(y) \qquad 1.62$$

$$\log\left(\frac{x}{y}\right) = \log(x) - \log(y) \qquad 1.63$$

$$ln(x) = \frac{\log_{10} x}{\log_{10} e} \approx 2.3(\log_{10} x) \qquad 1.64$$

$$\log_b(b) = 1 \qquad 1.65$$

$$\log(1) = 0 \qquad 1.66$$

$$\log_b(b^n) = n \qquad 1.67$$

*Example 1.8*

The surviving fraction, $x$, of a radioactive isotope is given by

$$x = e^{-0.005t}$$

For what value of $t$ will the surviving fraction be 7%?

$$0.07 = e^{-0.005t}$$

Taking the natural log of both sides,

$$ln(0.07) = ln(e^{-0.005t})$$

$$-2.66 = -0.005t$$

$$t = 532$$

## G. PARTIAL FRACTIONS

Given some rational fraction $H(x) = P(x)/Q(x)$ where $P(x)$ and $Q(x)$ are polynomials, the polynomials and constants $A_i$ and $Y_i(x)$ are needed such that

$$H(x) = \sum_i \frac{A_i}{Y_i(x)} \qquad 1.68$$

*Case 1*: $Q(x)$ factors into $n$ different linear terms. That is,

$$Q(x) = (x - a_1)(x - a_2) \cdots (x - a_n) \qquad 1.69$$

Then,

$$H(x) = \sum_{i=1}^{n} \frac{A_i}{x - a_i} \qquad 1.70$$

*Case 2*: $Q(x)$ factors into $n$ identical linear terms. That is,

$$Q(x) = (x - a)(x - a) \cdots (x - a) \qquad 1.71$$

Then,

$$H(x) = \sum_{i=1}^{n} \frac{A_i}{(x - a)^i} \qquad 1.72$$

*Case 3*: $Q(x)$ factors into $n$ different quadratic terms, $(x^2 + p_i x + q_i)$. Then,

$$H(x) = \sum_{i=1}^{n} \frac{A_i x + B_i}{x^2 + p_i x + q_i} \qquad 1.73$$

*Case 4*: $Q(x)$ factors into $n$ identical quadratic terms, $(x^2 + px + q)$. Then,

$$H(x) = \sum_{i=1}^{n} \frac{A_i x + B_i}{(x^2 + px + q)^i} \qquad 1.74$$

*Case 5*: $Q(x)$ factors into any combination of the above. The solution is illustrated by example 1.9.

*Example 1.9*

Resolve

$$H(x) = \frac{x^2 + 2x + 3}{x^4 + x^3 + 2x^2}$$

into partial fractions.

Here, $Q(x) = x^4 + x^3 + 2x^2$ which factors into $x^2(x^2 + x + 2)$. This is a combination of cases 2 and 3. We set

$$H(x) = \frac{A_1}{x} + \frac{A_2}{x^2} + \frac{A_3 + A_4 x}{x^2 + x + 2}$$

Cross multiplying to obtain a common denominator yields

$$\frac{(A_1+A_4)x^3 + (A_1+A_2+A_3)x^2 + (2A_1+A_2)x + 2A_2}{x^4 + x^3 + 2x^2}$$

Since the original numerator is known, the following simultaneous equations result:

$$\begin{aligned} A_1 + A_4 &= 0 \\ A_1 + A_2 + A_3 &= 1 \\ 2A_1 + A_2 &= 2 \\ 2A_2 &= 3 \end{aligned}$$

The solutions are: $A_1^* = 0.25$; $A_2^* = 1.5$; $A_3^* = -0.75$; $A_4^* = -0.25$. So,

$$H(x) = \frac{1}{4x} + \frac{3}{2x^2} - \frac{x+3}{4(x^2+x+2)}$$

## H. LINEAR AND MATRIX ALGEBRA

A matrix is a rectangular collection of variables or scalars contained within a set of square or round brackets. In the discussion that follows, matrix $\mathbf{A}$ will be assumed to have $m$ rows and $n$ columns. There are several classifications of matrices:

If $n = m$, the matrix is *square.*

A *diagonal* matrix is a square matrix with all zero values except for the $a_{ij}$ values, for all $i = j$.

An *identity* matrix is a diagonal matrix with all non-zero entries equal to '1'. (This usually is designated as '$\mathbf{I}$'.)

A *scalar* matrix is a square diagonal matrix with all non-zero entries equal to some constant.

A *triangular* matrix has zeros in all positions above or below the diagonal. This is not the same as an *echelon* matrix since the diagonal entries are non-zero.

Matrices are used to simplify the presentation and solution of sets of linear equations (hence the name 'linear algebra'). For example, the system of equations in example 1.4 can be written in matrix form as:

$$\begin{pmatrix} 2 & 3 \\ 3 & 4 \end{pmatrix} \begin{pmatrix} x \\ y \end{pmatrix} = \begin{pmatrix} 12 \\ 8 \end{pmatrix}$$

The above expression implies that there is a set of algebraic operations that can be performed with matrices. The important algebraic operations are listed here, along with their extensions to linear algebra.

(a) *Equality of Matrices*: For two matrices to be equal, they must have the same number of rows and columns. Corresponding entries must all be the same.

(b) *Inequality of Matrices*: There are no 'less-than' or 'greater than' relationships in linear algebra.

(c) *Addition and Subtraction of Matrices*: Addition (or subtraction) of two matrices can be accomplished by adding (or subtracting) the corresponding entries of two matrices which have the same shape.

(d) *Multiplication of Matrices*: Multiplication can be done only if the left-hand matrix has the same number of columns as the right-hand matrix has rows. Multiplication is accomplished by multiplying the elements in each left-hand matrix row by the elements in each right-hand matrix column, adding the products, and then placing the sum at the intersection point of the involved row and column. This is illustrated by example 1.10.

(e) *Division of Matrices*: Division can be accomplished only by multiplying by the inverse of the denominator matrix.

*Example 1.10*

$$\begin{pmatrix} 1 & 4 & 3 \\ 5 & 2 & 6 \end{pmatrix} \begin{pmatrix} 7 & 12 \\ 11 & 8 \\ 9 & 10 \end{pmatrix} = \mathbf{C}$$

$$\begin{aligned} [(1)(7) + (4)(11) + (3)(9)] &= 78 \\ [(1)(12) + (4)(8) + (3)(10)] &= 74 \\ [(5)(7) + (2)(11) + (6)(9)] &= 111 \\ [(5)(12) + (2)(8) + (6)(10)] &= 136 \end{aligned}$$

$$\mathbf{C} = \begin{pmatrix} 78 & 74 \\ 111 & 136 \end{pmatrix}$$

Other operations which can be performed on a matrix are described and illustrated below.

1. The *transpose* is an $(n \times m)$ matrix formed from the original $(m \times n)$ matrix by taking the ith row and making it the ith column. The diagonal is unchanged in this operation. The transpose of a matrix $\mathbf{A}$ is indicated as $\mathbf{A}^t$.

*Example 1.11*

What is the transpose of

$$\mathbf{A} = \begin{pmatrix} 1 & 6 & 9 \\ 2 & 3 & 4 \\ 7 & 1 & 5 \end{pmatrix}?$$

$$\mathbf{A}^t = \begin{pmatrix} 1 & 2 & 7 \\ 6 & 3 & 1 \\ 9 & 4 & 5 \end{pmatrix}$$

2. The *determinant*, **D**, is a scalar calculated from a square matrix. The determinant of a matrix is indicated by enclosing the matrix in vertical lines.

For a (2 × 2) matrix,

$$\mathbf{A} = \begin{pmatrix} a & b \\ c & d \end{pmatrix}$$

$$\mathbf{D} = \begin{vmatrix} a & b \\ c & d \end{vmatrix} = ad - bc \quad 1.75$$

For a (3 × 3) matrix,

$$\mathbf{A} = \begin{pmatrix} a & b & c \\ d & e & f \\ g & h & i \end{pmatrix}$$

$$D = a\begin{vmatrix} e & f \\ h & i \end{vmatrix} - d\begin{vmatrix} b & c \\ h & i \end{vmatrix} + g\begin{vmatrix} b & c \\ e & f \end{vmatrix} \quad 1.76$$

There are several rules governing the calculation of determinants:

- If **A** has a row or column of zeros, the determinant is zero.
- If **A** has two identical rows or columns, the determinant is zero.
- If **A** is triangular, the determinant is equal to the product of the diagonal entries.
- If **B** is obtained from **A** by multiplying a row or column by a scalar $k$, then $\mathbf{D}_B = k(\mathbf{D}_A)$.
- If **B** is obtained from **A** by switching two rows or columns, then $\mathbf{D}_B = -\mathbf{D}_A$.
- If **B** is obtained from **A** by adding a multiple of a row or column to another, then $\mathbf{D}_B = \mathbf{D}_A$.

*Example 1.12*

What is the determinant of

$$\begin{pmatrix} 2 & 3 & -4 \\ 3 & -1 & -2 \\ 4 & -7 & -6 \end{pmatrix}?$$

$$\begin{aligned} \mathbf{D} &= 2\begin{vmatrix} -1 & -2 \\ -7 & -6 \end{vmatrix} - 3\begin{vmatrix} 3 & -4 \\ -7 & -6 \end{vmatrix} + 4\begin{vmatrix} 3 & -4 \\ -1 & -2 \end{vmatrix} \\ &= 2(6-14) - 3(-18-28) + 4(-6-4) \\ &= 82 \end{aligned}$$

3. The *cofactor* of an entry in a matrix is the determinant of the matrix formed by omitting the entry's row and column in the original matrix. The sign of the cofactor is determined from the following positional matrices:

For a (2 × 2) matrix,

$$\begin{pmatrix} + & - \\ - & + \end{pmatrix}$$

For a (3 × 3) matrix,

$$\begin{pmatrix} + & - & + \\ - & + & - \\ + & - & + \end{pmatrix}$$

*Example 1.13*

What is the cofactor of the (−3) in the following matrix?

$$\begin{pmatrix} 2 & 9 & 1 \\ -3 & 4 & 0 \\ 7 & 5 & 9 \end{pmatrix}$$

The resulting matrix is

$$\begin{pmatrix} 9 & 1 \\ 5 & 9 \end{pmatrix}$$

with determinant 76. The cofactor is −76.

4. The *classical adjoint* is a matrix formed from the transposed cofactor matrix with the conventional sign arrangement. The resulting matrix is represented as $\mathbf{A}_{adj}$.

*Example 1.14*

What is the classical adjoint of

$$\begin{pmatrix} 2 & 3 & -4 \\ 0 & -4 & 2 \\ 1 & -1 & 5 \end{pmatrix}?$$

The matrix of cofactors (considering the sign convention) is

$$\begin{pmatrix} -18 & 2 & 4 \\ -11 & 14 & 5 \\ -10 & -4 & -8 \end{pmatrix}$$

The transposed cofactor matrix is

$$\mathbf{A}_{adj} = \begin{pmatrix} -18 & -11 & -10 \\ 2 & 14 & -4 \\ 4 & 5 & -8 \end{pmatrix}$$

5. The *inverse*, $\mathbf{A}^{-1}$, of **A** is a matrix such that $(\mathbf{A})(\mathbf{A}^{-1}) = \mathbf{I}$. (**I** is a square matrix with ones along the left-to-right diagonal and zeros elsewhere.)

For a (2 × 2) matrix

$$\begin{pmatrix} a & b \\ c & d \end{pmatrix}$$

the inverse is

$$\frac{1}{\mathbf{D}}\begin{pmatrix} d & -b \\ -c & a \end{pmatrix} \qquad 1.77$$

For larger matrices, the inverse is best calculated by dividing every entry in the classical adjoint by the determinant of the original matrix.

*Example 1.15*

What is the inverse of

$$\begin{pmatrix} 4 & 5 \\ 2 & 3 \end{pmatrix}?$$

The determinant is 2. The inverse is

$$\frac{1}{2}\begin{pmatrix} 3 & -5 \\ -2 & 4 \end{pmatrix} = \begin{pmatrix} \frac{3}{2} & -\frac{5}{2} \\ -1 & 2 \end{pmatrix}$$

## 7 TRIGONOMETRY

### A. DEGREES AND RADIANS

360 degrees $=$ one complete circle $= 2\pi$ radians

90 degrees $=$ right angle $= \frac{1}{2}\pi$ radians

one radian $=$ 57.3 degrees

one degree $=$ 0.0175 radians

multiply degrees by $\left(\frac{\pi}{180}\right)$ to obtain radians

multiply radians by $\left(\frac{180}{\pi}\right)$ to obtain degrees

### B. RIGHT TRIANGLES

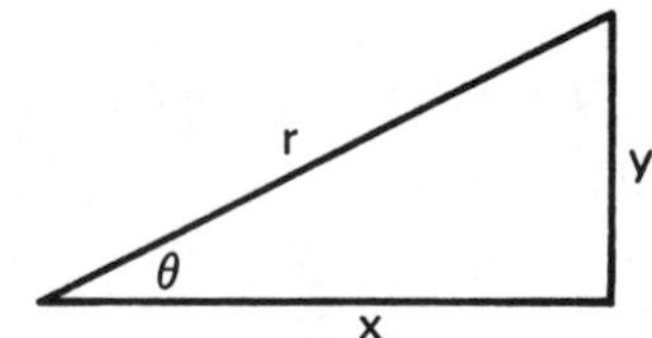


**Figure 1.1** A Right Triangle

*1. Pythagorean Theorem*

$$x^2 + y^2 = r^2 \qquad 1.78$$

*2. Trigonometric Functions*

$$\sin\theta = \frac{y}{r} \qquad 1.79$$

$$\cos\theta = \frac{x}{r} \qquad 1.80$$

$$\tan\theta = \frac{y}{x} \qquad 1.81$$

$$\cot\theta = \frac{x}{y} \qquad 1.82$$

$$\csc\theta = \frac{r}{y} \qquad 1.83$$

$$\sec\theta = \frac{r}{x} \qquad 1.84$$

*3. Relationship of the Trigonometric Functions to the Unit Circle*

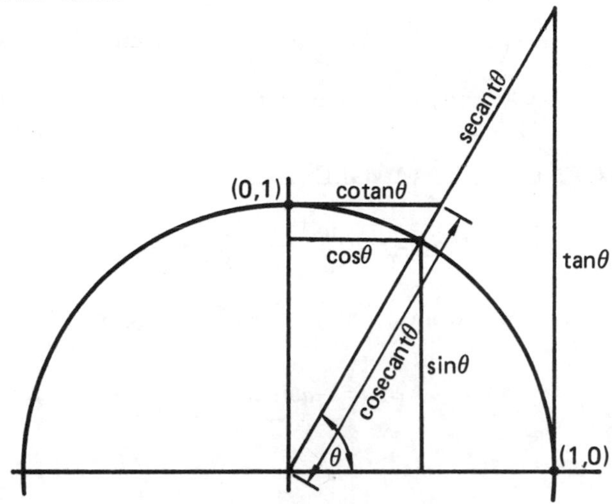


**Figure 1.2** The Unit Circle

*4. Signs of the Trigonometric Functions*

| quadrants | | quadrant | I | II | III | IV |
|---|---|---|---|---|---|---|
| II | I | sin | + | + | − | − |
| | | cos | + | − | − | + |
| III | IV | tan | + | − | + | − |

*5. Functions of the Related Angles*

| $f(\theta)$ | $-\theta$ | $90-\theta$ | $90+\theta$ | $180-\theta$ | $180+\theta$ |
|---|---|---|---|---|---|
| sin | $-\sin\theta$ | $\cos\theta$ | $\cos\theta$ | $\sin\theta$ | $-\sin\theta$ |
| cos | $\cos\theta$ | $\sin\theta$ | $-\sin\theta$ | $-\cos\theta$ | $-\cos\theta$ |
| tan | $-\tan\theta$ | $\cot\theta$ | $-\cot\theta$ | $-\tan\theta$ | $\tan\theta$ |

*6. Trigonometric Identities*

$$\sin^2\theta + \cos^2\theta = 1 \qquad 1.85$$

$$1 + \tan^2\theta = \sec^2\theta \qquad 1.86$$

$$1 + \cot^2\theta = \csc^2\theta \qquad 1.87$$

$$\sin 2\theta = 2(\sin\theta)(\cos\theta) \qquad 1.88$$

$$\cos 2\theta = \cos^2\theta - \sin^2\theta = 1 - 2\sin^2\theta \qquad 1.89$$

$$\sin\theta = 2\left[\sin\left(\frac{\theta}{2}\right)\cos\left(\frac{\theta}{2}\right)\right] \qquad 1.90$$

$$\sin\left(\frac{\theta}{2}\right) = \pm\sqrt{\frac{1}{2}(1-\cos\theta)} \quad 1.91$$

*7. Two-Angle Formulas*

$$\sin(\theta+\phi) = [\sin\theta][\cos\phi] + [\cos\theta][\sin\phi] \quad 1.92$$

$$\sin(\theta-\phi) = [\sin\theta][\cos\phi] - [\cos\theta][\sin\phi] \quad 1.93$$

$$\cos(\theta+\phi) = [\cos\theta][\cos\phi] - [\sin\theta][\sin\phi] \quad 1.94$$

$$\cos(\theta-\phi) = [\cos\theta][\cos\phi] + [\sin\theta][\sin\phi] \quad 1.95$$

C. GENERAL TRIANGLES

Law of Sines: $\frac{\sin A}{a} = \frac{\sin B}{b} = \frac{\sin C}{c}$ 1.96

Law of Cosines: $a^2 = b^2 + c^2 - 2bc(\cos A)$ 1.97

$$\text{Area} = \frac{1}{2}ab(\sin C) \quad 1.98$$

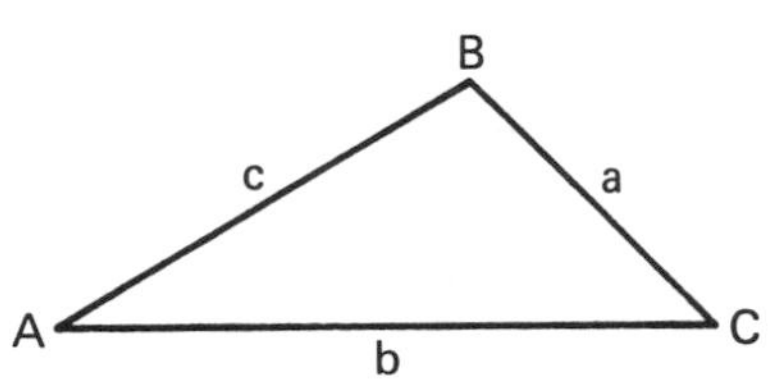


**Figure 1.3** A General Triangle

D. HYPERBOLIC FUNCTIONS

Hyperbolic functions are specific equations containing the terms $e^x$ and $e^{-x}$. These combinations of $e^x$ and $e^{-x}$ appear regularly in certain types of problems. In order to simplify the mathematical equations in which they appear, these hyperbolic functions are given special names and symbols.

$$\sinh x = \frac{e^x - e^{-x}}{2} \quad 1.99$$

$$\cosh x = \frac{e^x + e^{-x}}{2} \quad 1.100$$

$$\tanh x = \frac{e^x - e^{-x}}{e^x + e^{-x}} = \frac{\sinh x}{\cosh x} \quad 1.101$$

$$\coth x = \frac{e^x + e^{-x}}{e^x - e^{-x}} = \frac{\cosh x}{\sinh x} \quad 1.102$$

$$\operatorname{sech} x = \frac{2}{e^x + e^{-x}} = \frac{1}{\cosh x} \quad 1.103$$

$$\operatorname{csch} x = \frac{2}{e^x - e^{-x}} = \frac{1}{\sinh x} \quad 1.104$$

The hyperbolic identities are somewhat different from the standard trigonometric identities. Several of the most common identities are presented below.

$$\cosh^2 x - \sinh^2 x = 1 \quad 1.105$$

$$1 - \tanh^2 x = \operatorname{sech}^2 x \quad 1.106$$

$$1 - \coth^2 x = \operatorname{csch}^2 x \quad 1.107$$

$$\cosh x + \sinh x = e^x \quad 1.108$$

$$\cosh x - \sinh x = e^{-x} \quad 1.109$$

$$\sinh(x+y) = [\sinh x][\cosh y] + [\cosh x][\sinh y] \quad 1.110$$

$$\cosh(x+y) = [\cosh x][\cosh y] + [\sinh x][\sinh y] \quad 1.111$$

## 8 STRAIGHT LINE ANALYTIC GEOMETRY

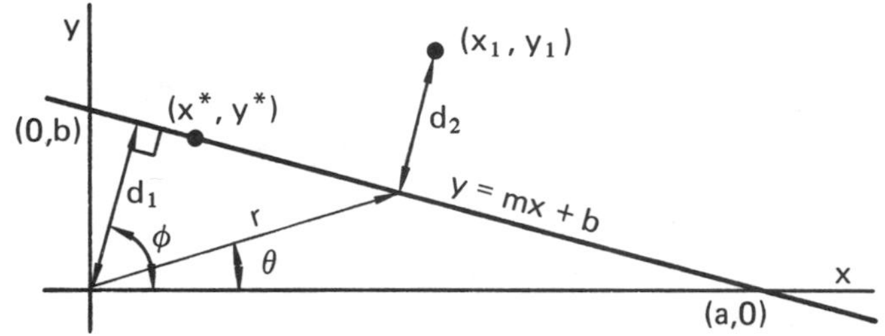


**Figure 1.4** A Straight Line

A. EQUATIONS OF A STRAIGHT LINE

General Form: $Ax + By + C = 0$ 1.112

Slope Form: $y = mx + b$ 1.113

Point-Slope Form: $(y - y^*) = m(x - x^*)$ 1.114

$(x^*, y^*)$ is any point on the line.

Intercept Form: $\frac{x}{a} + \frac{y}{b} = 1$ 1.115

Two-Point Form: $\frac{y - y_1^*}{x - x_1^*} = \frac{y_2^* - y_1^*}{x_2^* - x_1^*}$ 1.116

Normal Form: $x(\cos\phi) + y(\sin\phi) - d_1 = 0$ 1.117

Polar Form: $r = \frac{d_1}{\cos(\phi - \theta)}$ 1.118

B. POINTS, LINES, AND DISTANCES

The distance $d_2$ between a point and a line is:

$$d_2 = \frac{|Ax_1 + By_1 + C|}{\sqrt{A^2 + B^2}} \quad 1.119$$

The distance between two points is:

$$d = \sqrt{(x_2 - x_1)^2 + (y_2 - y_1)^2} \qquad 1.120$$

Parallel lines:

$$\frac{A_1}{A_2} = \frac{B_1}{B_2} \qquad 1.121$$

$$m_1 = m_2 \qquad 1.122$$

Perpendicular lines:

$$A_1A_2 = -B_1B_2 \qquad 1.123$$

$$m_1 = \frac{-1}{m_2} \qquad 1.124$$

Point of intersection of two lines:

$$x_1 = \frac{B_2C_1 - B_1C_2}{A_2B_1 - A_1B_2} \qquad 1.125$$

$$y_1 = \frac{A_1C_2 - A_2C_1}{A_2B_1 - A_1B_2} \qquad 1.126$$

Smaller angle between two intersecting lines:

$$\tan\phi = \frac{A_1B_2 - A_2B_1}{A_1A_2 + B_1B_2} = \frac{m_2 - m_1}{1 + m_1m_2} \qquad 1.127$$

$$\phi = \left|\arctan(m_1) - \arctan(m_2)\right|. \qquad 1.128$$

*Example 1.16*

What is the angle between the lines?

$$y_1 = -0.577x + 2$$
$$y_2 = +0.577x - 5$$

*method 1*:

$$\arctan\left[\frac{m_2 - m_1}{1 + m_1m_2}\right] = \arctan\left[\frac{0.577 - (-0.577)}{1 + (0.577)(-0.577)}\right] = 60^\circ$$

*method 2*: Write both equations in general form:

$$-0.577x - y_1 + 2 = 0$$

$$0.577x - y_2 - 5 = 0$$

$$\arctan\left[\frac{A_1B_2 - A_2B_1}{A_1A_2 + B_1B_2}\right] = \arctan\left[\frac{(-0.577)(-1) - (0.577)(-1)}{(-0.577)(0.577) + (-1)(-1)}\right] = 60^\circ$$

*method 3*:

$$\phi = \left|\arctan(-0.577) - \arctan(0.577)\right| = \left|-30^\circ - 30^\circ\right| = 60^\circ$$

## C. LINEAR AND CURVILINEAR REGRESSION

If it is necessary to draw a straight line through $n$ data points $(x_1, y_1), (x_2, y_2), \ldots, (x_n, y_n)$, the following method based on the theory of least squares can be used:

*step 1*: Calculate the following quantities.

$$\sum x_i \quad \sum x_i^2 \quad \left(\sum x_i\right)^2 \quad \overline{x} = \left(\frac{\sum x_i}{n}\right) \quad \sum x_iy_i$$

$$\sum y_i \quad \sum y_i^2 \quad \left(\sum y_i\right)^2 \quad \overline{y} = \left(\frac{\sum y_i}{n}\right)$$

*step 2*: Calculate the slope of the line $y = mx + b$.

$$m = \frac{n\sum(x_iy_i) - \left(\sum x_i\right)\left(\sum y_i\right)}{n\sum x_i^2 - \left(\sum x_i\right)^2} \qquad 1.129$$

*step 3*: Calculate the $y$ intercept.

$$b = \overline{y} - m\overline{x} \qquad 1.130$$

*step 4*: To determine the goodness of fit, calculate the correlation coefficient.

$$r = \frac{n\sum(x_iy_i) - \left(\sum x_i\right)\left(\sum y_i\right)}{\sqrt{\left[n\sum x_i^2 - \left(\sum x_i\right)^2\right]\left[n\sum y_i^2 - \left(\sum y_i\right)^2\right]}} \qquad 1.131$$

If $m$ is positive, $r$ will be positive. If $m$ is negative, $r$ will be negative. As a general rule, if the absolute value of $r$ exceeds .85, the fit is good. Otherwise, the fit is poor. $r$ equals 1.0 if the fit is a perfect straight line.

*Example 1.17*

An experiment is performed in which the dependent variable ($y$) is measured against the independent variable ($x$). The results are as follows:

| $x$ | $y$ |
|---|---|
| 1.2 | 0.602 |
| 4.7 | 5.107 |
| 8.3 | 6.984 |
| 20.9 | 10.031 |

What is the least squares straight line equation which represents this data?

*step 1*:

$$\sum x_i = 35.1$$
$$\sum y_i = 22.72$$
$$\sum x_i^2 = 529.23$$
$$\sum y_i^2 = 175.84$$
$$\left(\sum x_i\right)^2 = 1232.01$$
$$\left(\sum y_i\right)^2 = 516.38$$
$$\overline{x} = 8.775$$
$$\overline{y} = 5.681$$
$$\sum x_i y_i = 292.34$$
$$n = 4$$

*step 2*:

$$m = \frac{(4)(292.34) - (35.1)(22.72)}{(4)(529.23) - (35.1)^2} = 0.42$$

*step 3*:

$$b = 5.681 - (0.42)(8.775) = 2.0$$

*step 4*: From equation 1.131, $r = 0.91$.

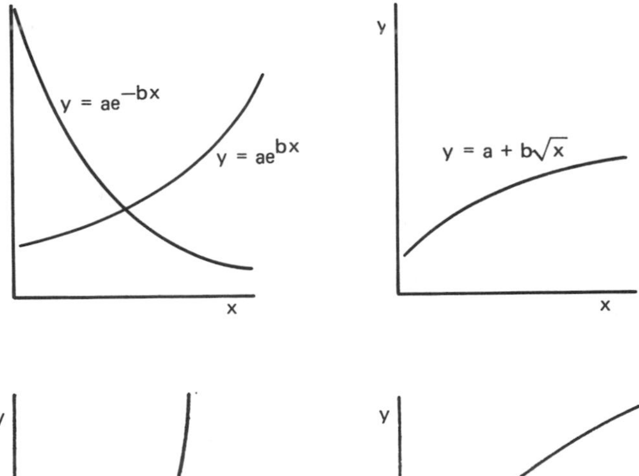


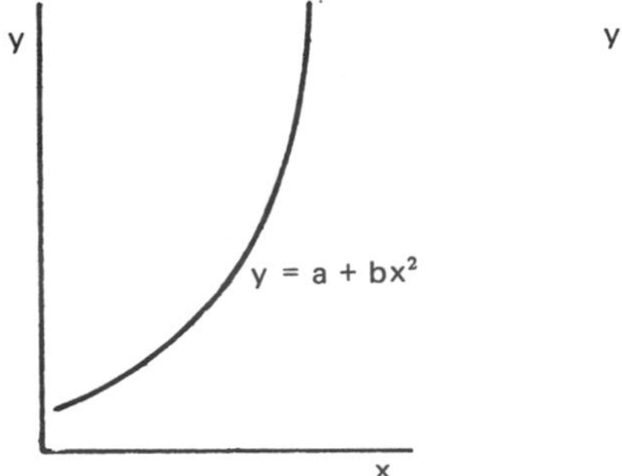


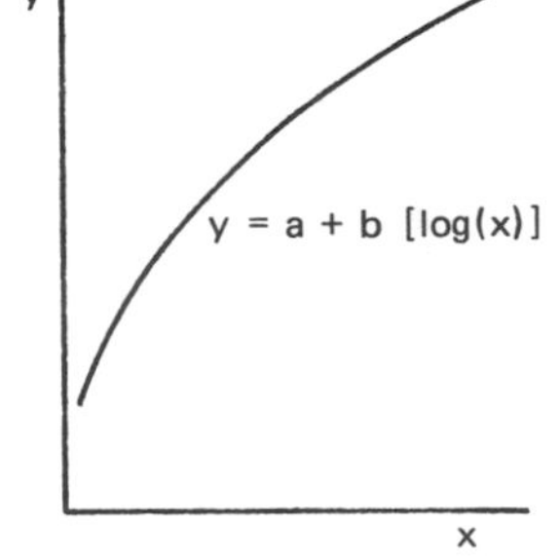


**Figure 1.5** Non-Linear Data Plots

A low value of $r$ does not eliminate the possibility of a non-linear relationship existing between $x$ and $y$. It is possible that the data describes a parabolic, logarithmic, or other non-linear relationship. (Usually this will be apparent if the data are graphed.) It may be necessary to convert one or both variables to new variables by taking squares, square roots, cubes, or logs, to name a few of the possibilities.

The apparent shape of the line through the data will give a clue to the type of variable transformation that is required. The following curves may be used as guides to some of the simpler variable transformations.

*Example 1.18*

Repeat example 1.17 assuming that the relationship between the variables is non-linear.

The first step is to graph the data. Since the graph has the appearance of the fourth case, it can be assumed that the relationship between the variables has the form of $y = a + b[\log(x)]$. Therefore, the variable change $z = \log(x)$ is made, resulting in the following set of data:

| z | $y$ |
|---|---|
| 0.0792 | 0.602 |
| 0.672 | 5.107 |
| 0.919 | 6.984 |
| 1.32 | 10.031 |

If the regression analysis is performed on this set of data, the resulting equation and correlation coefficient are:

$$y = -0.036 + 7.65z$$
$$r = 0.999$$

This is a very good fit. The relationship between the variable $x$ and $y$ is approximately

$$y = -0.036 + 7.65[\log(x)]$$

Figure 1.6 illustrates several common problems encountered in trying to fit and evaluate curves from experimental data. Figure 1.6(a) shows a graph of clustered data with several extreme points. There will be moderate correlation due to the weighting of the extreme points, although there is little actual correlation at low values of the variables. The extreme data should be excluded or the range should be extended by obtaining more data.

Figure 1.6(b) shows that good correlation exists in general, but extreme points are missed, and the overall correlation is moderate. If the results within the small linear range can be used, the extreme points should be

excluded. Otherwise, additional data points are needed, and curvilinear relationships should be investigated.

Figure 1.6(c) illustrates the problem of drawing conclusions of cause and effect. There may be a predictable relationship between variables, but that does not imply a cause and effect relationship. In the case shown, both variables are functions of a third variable, the city population. But, there is no direct relationship between the plotted variables.

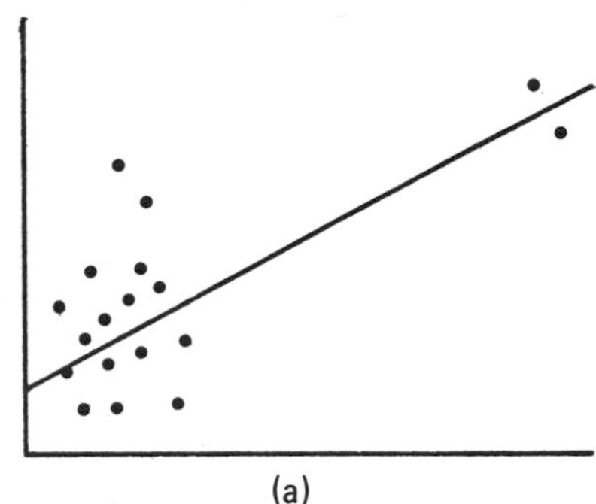


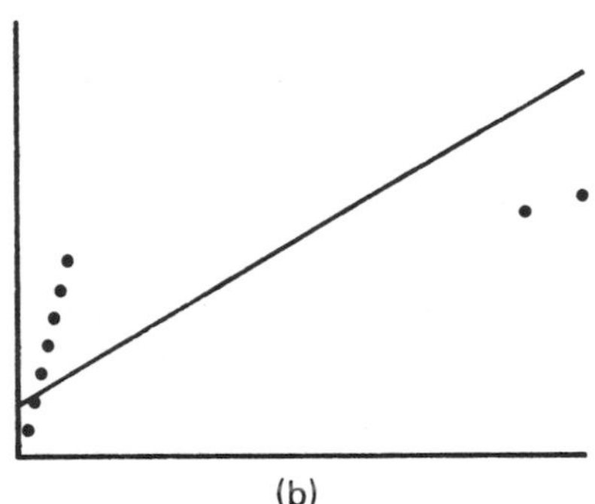


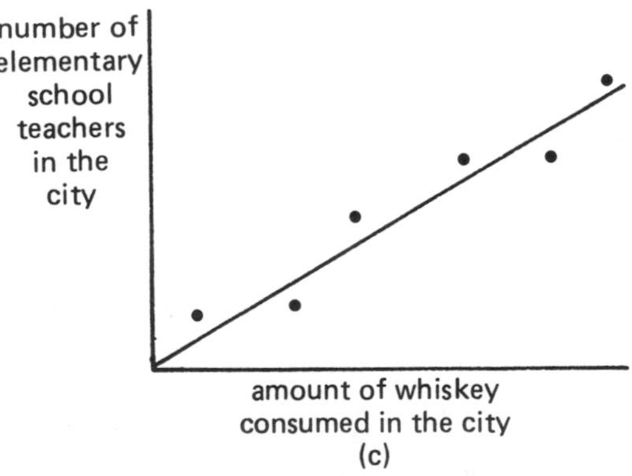


**Figure 1.6** Common Regression Difficulties

## D. VECTOR OPERATIONS

A vector is a directed straight line of a given magnitude. Two directed straight lines with the same magnitudes and directions are said to be *equivalent*. Thus, the actual end points of a vector often are irrelevant as long as the direction and magnitude are known.

A vector defined by its end points and direction is designated as

$$\mathbf{V} = \vec{p_1}\vec{p_2}$$

Usually, $p_1$ will be the origin, in which case $\mathbf{V}$ will be designated by its end-point, $p_2 = (x, y)$. Such a zero-based vector is equivalent to all other vectors of the same magnitude and direction. Any vector $p_1p_2$ can be transformed into a zero-based vector by subtracting $(x_1, y_1)$ from all points along the vector line.

A vector also can be specified in the terms of the unit vectors $(\mathbf{i}, \mathbf{j}, \mathbf{k})$. Thus,

$$\mathbf{V} = (x, y) = (x\mathbf{i} + y\mathbf{j})$$

Or,

$$\mathbf{V} = (x, y, z) = (x\mathbf{i} + y\mathbf{j} + z\mathbf{k})$$

Important operations on vectors based at the origin are:

$$c\mathbf{V} = (cx, cy) \quad \text{(vector multiplication by a scalar)}$$

$$\mathbf{V}_1 + \mathbf{V}_2 = (x_1 + x_2, y_1 + y_2) \qquad 1.132$$

$$|\mathbf{V}| = \sqrt{x^2 + y^2} \quad \text{(vector magnitude)} \qquad 1.133$$

$$\alpha = \text{angle between vector } \mathbf{V} \text{ and x axis}$$

$$= \arccos\left(\frac{x}{|\mathbf{V}|}\right) = \arcsin\left(\frac{y}{|\mathbf{V}|}\right) \qquad 1.134$$

$$m = \text{slope of vector} = \frac{y}{x} \qquad 1.135$$

$$\theta = \left\{\begin{matrix}\text{angle between}\\ \text{two vectors}\end{matrix}\right\}$$

$$= \arccos\frac{(x_1x_2 + y_1y_2)}{|\mathbf{V}_1||\mathbf{V}_2|} \qquad 1.136$$

$$\mathbf{V}_1 \cdot \mathbf{V}_2 = \text{dot product}$$

$$= |\mathbf{V}_1||\mathbf{V}_2|\cos\theta = x_1x_2 + y_1y_2 \qquad 1.137$$

When equation 1.137 is solved for $\cos\theta$, it is known as the *Cauchy-Schwartz theorem.*

$$\cos\theta = \frac{x_1x_2 + y_1y_2}{|\mathbf{V}_1||\mathbf{V}_2|} \qquad 1.138$$

$$\mathbf{V}_1 \times \mathbf{V}_2 = \text{cross product} = \begin{vmatrix} \mathbf{i} & x_1 & x_2 \\ \mathbf{j} & y_1 & y_2 \\ \mathbf{k} & z_1 & z_2 \end{vmatrix}$$

$$= (y_1 z_2 - y_2 z_1)\mathbf{i} - (x_1 z_2 - x_2 z_1)\mathbf{j} + (x_1 y_2 - x_2 y_1)\mathbf{k} \quad 1.139$$

$$\mathbf{V}_1 \times \mathbf{V}_2 = -\mathbf{V}_2 \times \mathbf{V}_1 \quad 1.140$$

$$|\mathbf{V}_1 \times \mathbf{V}_2| = |\mathbf{V}_1||\mathbf{V}_2| \sin\theta \quad 1.141$$

*Example 1.19*

What is the angle between the vectors $\mathbf{V}_1 = (-\sqrt{3}, 1)$ and $\mathbf{V}_2 = (2\sqrt{3}, 2)$?

$$\cos\theta = \frac{\mathbf{V}_1 \cdot \mathbf{V}_2}{|\mathbf{V}_1||\mathbf{V}_2|} = \frac{(-\sqrt{3})(2\sqrt{3}) + (1)(2)}{\sqrt{3+1}\sqrt{12+4}} = -\frac{1}{2}$$

$$\theta = 120°$$

(Graph and compare this result to example 1.16, in which the lines were not directed.)

*Example 1.20*

Find a unit vector orthogonal to $\mathbf{V}_1 = \mathbf{i} - \mathbf{j} + 2\mathbf{k}$ and $\mathbf{V}_2 = 3\mathbf{j} - \mathbf{k}$.

The cross product is orthogonal to $\mathbf{V}_1$ and $\mathbf{V}_2$, although its length may not be equal to one.

$$\mathbf{V}_1 \times \mathbf{V}_2 = \begin{vmatrix} \mathbf{i} & 1 & 0 \\ \mathbf{j} & -1 & 3 \\ \mathbf{k} & 2 & -1 \end{vmatrix} = -5\mathbf{i} + \mathbf{j} + 3\mathbf{k}$$

Since the length of $|\mathbf{V}_1 \times \mathbf{V}_2|$ is $\sqrt{35}$, it is necessary to divide $\mathbf{V}_1 \times \mathbf{V}_2$ by this amount to obtain a unit vector. Thus,

$$\mathbf{V}_3 = \frac{-5\mathbf{i} + \mathbf{j} + 3\mathbf{k}}{\sqrt{35}}$$

The orthogonality can be proved from

$$\mathbf{V}_1 \cdot \mathbf{V}_3 = 0 \text{ and } \mathbf{V}_2 \cdot \mathbf{V}_3 = 0$$

That $\mathbf{V}_3$ is a unit vector can be proved from

$$\mathbf{V}_3 \cdot \mathbf{V}_3 = +1$$

## E. DIRECTION NUMBERS, DIRECTION ANGLES, AND DIRECTION COSINES

Given a directed line from $(x_1, y_1, z_1)$ to $(x_2, y_2, z_2)$, the direction numbers are:

$$L = x_2 - x_1 \quad 1.142$$

$$M = y_2 - y_1 \quad 1.143$$

$$N = z_2 - z_1 \quad 1.144$$

The distance between two points is:

$$d = \sqrt{L^2 + M^2 + N^2} \quad 1.145$$

The direction cosines are:

$$\cos\alpha = \frac{L}{d} \quad 1.146$$

$$\cos\beta = \frac{M}{d} \quad 1.147$$

$$\cos\gamma = \frac{N}{d} \quad 1.148$$

Note that

$$\cos^2\alpha + \cos^2\beta + \cos^2\gamma = 1 \quad 1.149$$

The direction angles are the angles between the axes and the lines. They are found from the inverse functions of the direction cosines. That is,

$$\alpha = \arccos\left(\frac{L}{d}\right) \quad 1.150$$

$$\beta = \arccos\left(\frac{M}{d}\right) \quad 1.151$$

$$\gamma = \arccos\left(\frac{N}{d}\right) \quad 1.152$$

Once the direction cosines have been found, they can be used to write the equation of the straight line in terms of the unit vectors. The line $\mathbf{R}$ would be defined as

$$\mathbf{R} = \mathbf{i}\cos\alpha + \mathbf{j}\cos\beta + \mathbf{k}\cos\gamma \quad 1.153$$

Similarly, the line may be written in terms of its direction numbers,

$$\mathbf{R} = L\mathbf{i} + M\mathbf{j} + N\mathbf{k} \quad 1.154$$

Given two directed lines, $\mathbf{R}_1$ and $\mathbf{R}_2$, the angle between $\mathbf{R}_1$ and $\mathbf{R}_2$ is defined as the angle between the two arrow heads.

$$\cos\phi = \cos\alpha_1\cos\alpha_2 + \cos\beta_1\cos\beta_2 + \cos\gamma_1\cos\gamma_2 = \frac{L_1L_2 + M_1M_2 + N_1N_2}{d_1d_2} \quad 1.155$$

If $\mathbf{R}_1$ and $\mathbf{R}_2$ are parallel and in the same direction, then

$$\alpha_1 = \alpha_2$$

$$\beta_1 = \beta_2$$

$$\gamma_1 = \gamma_2$$

If $\mathbf{R}_1$ and $\mathbf{R}_2$ are parallel but in opposite directions, then

$$\alpha_1 + \alpha_2 = 180 \text{ (etc.)}$$

If $\mathbf{R}_1$ and $\mathbf{R}_2$ are normal to each other, then

$$\phi = 90^\circ \text{ and } \cos\phi = 0$$

*Example 1.21*

A line passes through the points (4,7,9) and (0,1,6). Write the equation of the line in terms of its direction cosines and direction numbers.

$$L = 4 - 0 = 4$$
$$M = 7 - 1 = 6$$
$$N = 9 - 6 = 3$$

Now, the line may be written in terms of its direction numbers.

$$\mathbf{R} = 4\mathbf{i} + 6\mathbf{j} + 3\mathbf{k}$$

The distance between the two points is

$$d = \sqrt{(4)^2 + (6)^2 + (3)^2} = 7.81$$

The line now may be written in terms of its direction cosines.

$$\begin{aligned}\mathbf{R} &= \frac{4\mathbf{i} + 6\mathbf{j} + 3\mathbf{k}}{7.81} \\ &= 0.512\mathbf{i} + 0.768\mathbf{j} + 0.384\mathbf{k}\end{aligned}$$

### F. CURVILINEAR INTERPOLATION

A situation which occurs frequently is one in which a function value must be interpolated from other data along the curve. Straight-line interpolation typically is used because of its simplicity and speed. However, straight-line interpolation ignores all but two of the points on the curve and is, therefore, unable to include any effects of curvature.

A more powerful technique is the *Lagrangian Interpolating Polynomial.* It is assumed that $(n+1)$ values of $f(x)$ are known (for $x_0, x_1, x_2, \ldots, x_n$) and that $f(x)$ is a continuous, real-valued function on the interval $(x_0, x_n)$. The value of $f(x)$ at $x^*$ can be estimated from the following equations:

$$f(x^*) = \sum_{k=0}^{n} f(x_k) L_k(x^*) \qquad 1.156$$

where the Lagrangian Interpolating Polynomial is

$$L_k(x^*) = \prod_{\substack{i=0 \\ i \neq k}}^{n} \frac{x^* - x_i}{x_k - x_i} \qquad 1.157$$

Example 1.22 illustrates use of the Lagrangian Interpolating Polynomial.

## 9 TENSORS

A scalar has magnitude only. A vector has magnitude and a definite direction. A *tensor* has magnitude in a specific direction, but the direction is not unique. An example of a tensor is stress. From the combined stress equation, stress at a point in a solid depends on the direction of the plane passing through that point. Tensors frequently are associated with ***anisotropic materials*** which have different properties in different directions. Other examples are dielectric constant and magnetic susceptibility.

A vector in a three-dimensional space is defined completely by three quantities, $F_x$, $F_y$, and $F_z$. A tensor in three-dimensional space requires nine quantities for complete definition. These nine values are given in ma-

*Example 1.22*

A real-valued function has the following values:

$$f(1) = 1.5709 \quad f(4) = 1.5727 \quad f(6) = 1.5751$$

What is $f(3.5)$?

$$\begin{array}{llllll} & & \underline{i=0} & \underline{i=1} & \underline{i=2} & \\ \underline{k=0}: & L_0(3.5) = & \left(\cancel{\frac{3.5-1}{1-1}}\right) & \left(\frac{3.5-4}{1-4}\right) & \left(\frac{3.5-6}{1-6}\right) & = 0.08333 \\ \underline{k=1}: & L_1(3.5) = & \left(\frac{3.5-1}{4-1}\right) & \left(\cancel{\frac{3.5-4}{4-4}}\right) & \left(\frac{3.5-6}{4-6}\right) & = 1.04167 \\ \underline{k=2}: & L_2(3.5) = & \left(\frac{3.5-1}{6-1}\right) & \left(\frac{3.5-4}{6-4}\right) & \left(\cancel{\frac{3.5-6}{6-6}}\right) & = -0.12500 \end{array}$$

$$\begin{aligned}f(x^*) &= (1.5709)(0.08333) + (1.5727)(1.04167) \\ &\quad + (1.5751)(-0.12500) \\ &= 1.57225\end{aligned}$$

trix form. The tensor definition for stress at a point is

$$\begin{pmatrix} \sigma_{xx} & \sigma_{xy} & \sigma_{xz} \\ \sigma_{yx} & \sigma_{yy} & \sigma_{yz} \\ \sigma_{zx} & \sigma_{zy} & \sigma_{zz} \end{pmatrix} \qquad 1.158$$

## 10 PLANES

A plane **P** is uniquely determined by one of three combinations of parameters:

1. three non-collinear points in space
2. two non-parallel vectors ($\mathbf{V}_1$and$\mathbf{V}_2$) and their intersection point $p_0$
3. a point $p_0$ and a normal vector **N**

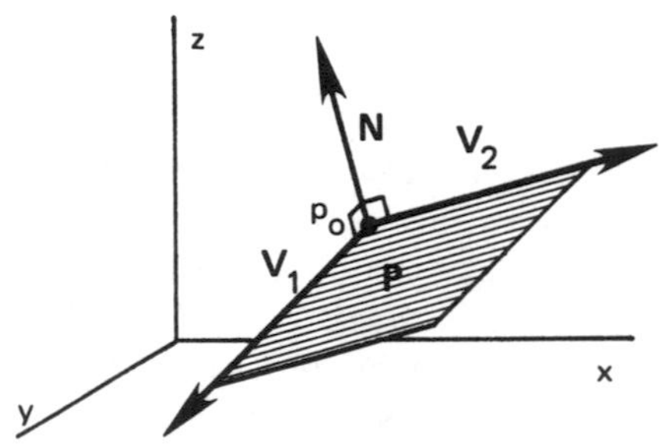


**Figure 1.7** A Plane in 3-Space

The plane consists of all points such that the coordinates can be written as a linear combination of $\mathbf{V}_1$ and $\mathbf{V}_2$. That is, points in the plane can be written as

$$(x, y, z) = s\mathbf{V}_1 + t\mathbf{V}_2 \qquad 1.159$$

where $s$ and $t$ are constants and

$$\mathbf{V}_1 = a_1\mathbf{i} + b_1\mathbf{j} + c_1\mathbf{k} \qquad 1.160$$

$$\mathbf{V}_2 = a_2\mathbf{i} + b_2\mathbf{j} + c_2\mathbf{k} \qquad 1.161$$

If the intersection point $p_0 = (x_0, y_0, z_0)$ is known, then points in the plane can be represented by the parametric equations given below. Notice the similarity to the slope form of an equation for a straight line.

$$x = sa_1 + ta_2 + x_0 \qquad 1.162$$

$$y = sb_1 + tb_2 + y_0 \qquad 1.163$$

$$z = sc_1 + tc_2 + z_0 \qquad 1.164$$

The plane also is defined by its rectangular equations:

$$A(x - x_0) + B(y - y_0) + C(z - z_0) = 0 \qquad 1.165$$

or

$$Ax + By + Cz + D = 0 \qquad 1.166$$

where

$$D = -(Ax_0 + By_0 + Cz_0) \qquad 1.167$$

Constants A, B, and C are found from the cross product giving the normal vector **N**.

$$\mathbf{N} = \mathbf{V}_1 \times \mathbf{V}_2 = A\mathbf{i} + B\mathbf{j} + C\mathbf{k} \qquad 1.168$$

*Example 1.23*

A plane is defined by a point $(2, 1, -4)$ and two vectors:

$$\mathbf{V}_1 = (2\mathbf{i} - 3\mathbf{j} + \mathbf{k}) \qquad \mathbf{V}_2 = (2\mathbf{j} - 4\mathbf{k})$$

Find the parametric and rectangular plane equations.

The parametric equations (for any values of $s$ and $t$) are:

$$x = 2 + 2s$$
$$y = 1 - 3s + 2t$$
$$z = -4 + s - 4t$$

The normal vector is found by evaluating the determinant

$$\mathbf{N} = \begin{vmatrix} \mathbf{i} & 2 & 0 \\ \mathbf{j} & -3 & 2 \\ \mathbf{k} & 1 & -4 \end{vmatrix}$$

$$= \mathbf{i}\,(12 - 2) - 2\,(-4\mathbf{j} - 2\mathbf{k}) = 10\mathbf{i} + 8\mathbf{j} + 4\mathbf{k}$$

One form of the rectangular equation is

$$10(x - 2) + 8(y - 1) + 4(z + 4) = 0$$

Another form can be derived from equations 1.166 and 1.167

$$D = -[(10)(2) + (8)(1) + (4)(-4)] = -12$$
$$\mathbf{P} = 10x + 8y + 4z - 12 = 0$$

Three noncollinear points can be used to describe a plane with the following procedure:

*step 1*: Form vectors $\mathbf{V}_1$ and $\mathbf{V}_2$ from two pairs of the points.

*step 2*: Find the normal vector $\mathbf{N} = \mathbf{V}_1 \times \mathbf{V}_2$.

*step 3*: Write the rectangular form of the plane using A, B, and C from the normal vector and any one of the three points.

If the rectangular form of the plane is known, it can be used to write parametric equations. In this case, two of the three variables $(x, y, z)$ replace the parameters $s$ and $t$.

*Example 1.24*

Find the rectangular and parametric equations of a plane containing the following points: $(2,1,-4)$; $(4,-2,-3)$; $(2,3,-8)$.

Use the first two points to find $\mathbf{V}_1$:

$$\begin{aligned}\mathbf{V}_1 &= (4-2)\mathbf{i} + (-2-1)\mathbf{j} + (-3-(-4))\mathbf{k}\\ &= 2\mathbf{i} - 3\mathbf{j} + \mathbf{k}\end{aligned}$$

Similarly,

$$\begin{aligned}\mathbf{V}_2 &= (2-2)\mathbf{i} + (3-1)\mathbf{j} + (-8-(-4))\mathbf{k}\\ &= 2\mathbf{j} - 4\mathbf{k}\end{aligned}$$

From the previous example,

$$\begin{aligned}\mathbf{N} &= 10\mathbf{i} + 8\mathbf{j} + 4\mathbf{k}\\ \mathbf{P} &= 10x + 8y + 4z - 12 = 0\end{aligned}$$

Dividing the rectangular form by 4 gives

$$2.5x + 2y + z - 3 = 0$$

or

$$z = 3 - 2y - 2.5x$$

Using $x$ and $y$ as the parameters, the parametric equations are

$$\begin{aligned}x &= x\\ y &= y\\ z &= 3 - 2y - 2.5x\end{aligned}$$

The angle between two planes is the same as the angle between their normal vectors, as calculated from the following equation:

$$\begin{aligned}\cos\phi &= \frac{|\mathbf{N}_1\cdot\mathbf{N}_2|}{|\mathbf{N}_1||\mathbf{N}_2|}\\ &= \frac{|A_1A_2 + B_1B_2 + C_1C_2|}{\sqrt{A_1^2+B_1^2+C_1^2}\sqrt{A_2^2+B_2^2+C_2^2}}\end{aligned} \quad 1.169$$

A vector equation of the line formed by the intersection of two planes is given by the cross product $(\mathbf{N}_1 \times \mathbf{N}_2)$. The distance from a point $(x', y', z')$ to a plane is given by

$$d = \frac{Ax' + By' + Cz' + D}{\sqrt{A^2+B^2+C^2}} \quad 1.170$$

## 11 CONIC SECTIONS

### A. CIRCLE

The center-radius form of a circle with radius $r$ and center at $(h,k)$ is

$$(x-h)^2 + (y-k)^2 = r^2 \quad 1.171$$

The $x$-intercept is found by letting $y = 0$ and solving for $x$. The $y$-intercept is found similarly.

The general form is

$$x^2 + y^2 + Dx + Ey + F = 0 \quad 1.172$$

This can be converted to the center-radius form.

$$\left(x + \frac{D}{2}\right)^2 + \left(y + \frac{E}{2}\right)^2 = \frac{1}{4}(D^2 + E^2 - 4F) \quad 1.173$$

If the right-hand side is greater than zero, the equation is that of a circle with center at $(-\frac{1}{2}D, -\frac{1}{2}E)$ and radius given by the square root of the right-hand side. If the right-hand side is zero, the equation is that of a point. If the right-hand side is negative, the plot is imaginary.

### B. PARABOLA

A parabola is formed by a locus of points equidistant from point F and the *directrix.*

$$(y-k)^2 = 4p(x-h) \quad 1.174$$

Equation 1.174 represents a parabola with *vertex* at $(h,k)$, focus at $(p+h,k)$, and directrix equation $x = h-p$. The parabola points to the left if $p > 0$ and points to the right if $p < 0$.

$$(x-h)^2 = 4p(y-k) \quad 1.175$$

Equation 1.175 represents a parabola with vertex at $(h,k)$, focus at $(h,p+k)$, and directrix equation $y = k-p$. The parabola points down if $p > 0$ and points up if $p < 0$.

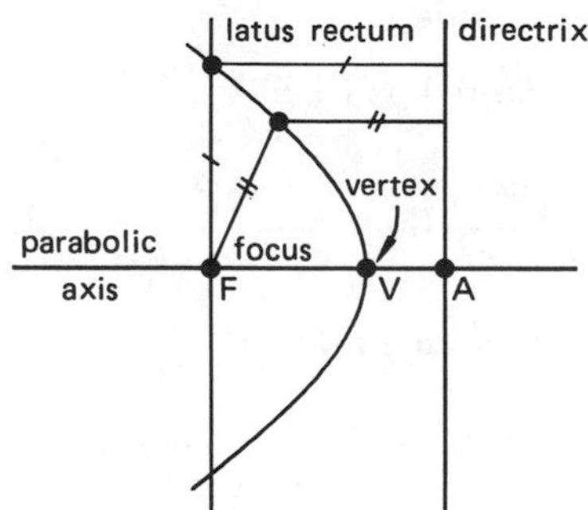


**Figure 1.8** A Parabola

An alternate form of the vertically-oriented parabola is

$$Ax^2 + Bx + C = 0 \quad 1.176$$

This parabola has a vertex at

$$\left(\frac{-B}{2A}, C - \frac{B^2}{4A}\right)$$

and points down if $A > 0$ and points up if $A < 0$.

## C. ELLIPSE

An ellipse is formed from a locus of points such that the sum of distances from the two foci is constant. The distance between the two foci is $2c$. The sum of those distances is

$$F_1P + PF_2 = 2a \quad 1.177$$

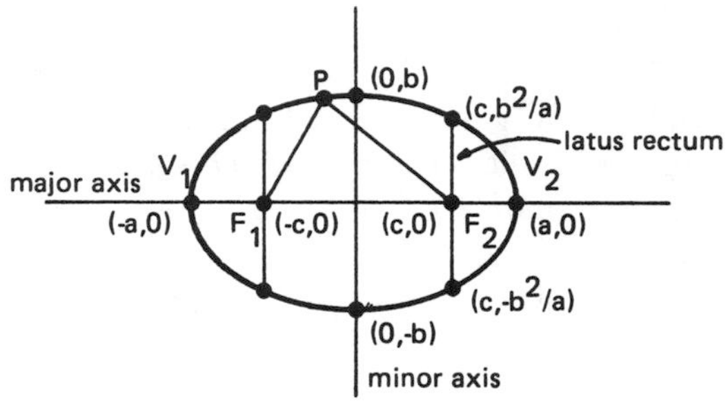


**Figure 1.9** An Ellipse

The eccentricity of an ellipse is less than 1, and is equal to

$$e = \frac{\sqrt{a^2 - b^2}}{a} \quad 1.178$$

For an ellipse centered at the origin,

$$\left(\frac{x}{a}\right)^2 + \left(\frac{y}{b}\right)^2 = 1 \quad 1.179$$

$$b^2 = a^2 - c^2 \quad 1.180$$

If $a > b$, the ellipse is wider than it is tall. If $a < b$, it is taller than it is wide.

For an ellipse centered at $(h, k)$,

$$\frac{(x-h)^2}{a^2} + \frac{(y-k)^2}{b^2} = 1 \quad 1.181$$

The general form of an ellipse is

$$Ax^2 + Cy^2 + Dx + Ey + F = 0 \quad 1.182$$

If $A \neq C$ and both have the same sign, the general form can be written as

$$A\left(x + \frac{D}{2A}\right)^2 + C\left(y + \frac{E}{2C}\right)^2 = M \quad 1.183$$

$$M = \frac{D^2}{4A} + \frac{E^2}{4C} - F \quad 1.184$$

If $M = 0$, the graph is a single point at

$$\left(\frac{-D}{2A}, \frac{-E}{2C}\right).$$

If $M < 0$, the graph is the null set.

If $M > 0$, then the ellipse is centered at

$$\left(-\frac{D}{2A}, -\frac{E}{2C}\right)$$

and the equation can be rewritten

$$\frac{\left(x + \frac{D}{2A}\right)^2}{\frac{M}{A}} + \frac{\left(y + \frac{E}{2C}\right)^2}{\frac{M}{C}} = 1 \quad 1.185$$

## D. HYPERBOLA

A hyperbola is a locus of points such that $F_1P - PF_2 = 2a$. The distance between the foci is $2c$, and $a < c$.

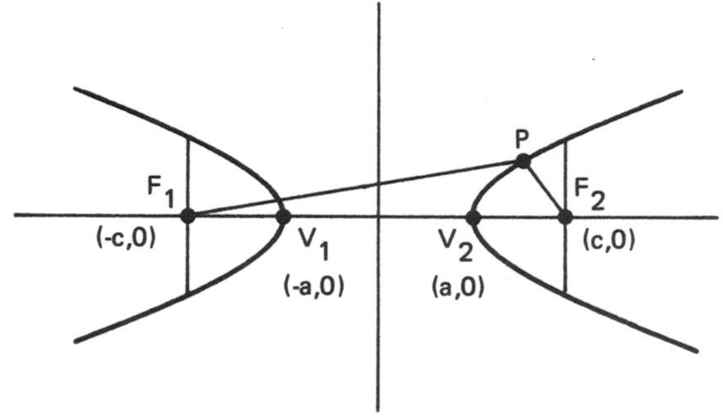


**Figure 1.10** A Hyperbola

For a hyperbola centered at the origin with foci on the x-axis,

$$\left(\frac{x}{a}\right)^2 - \left(\frac{y}{b}\right)^2 = 1 \quad \text{with} \quad b^2 = c^2 - a^2 \quad 1.186$$

If the foci are on the y-axis,

$$\left(\frac{y}{a}\right)^2 - \left(\frac{x}{b}\right)^2 = 1 \quad 1.187$$

The coordinates and length of the *latus recta* are the same as for the ellipse. The hyperbola is asymptotic to the lines

$$y = \pm\left(\frac{b}{a}\right)x \quad 1.188$$

The asymptotes need not be perpendicular, but if they are, the hyperbola is known as a *rectangular hyperbola.* If the asymptotes are the $x$ and $y$ axes, the equation of the hyperbola is

$$xy = \pm a^2 \quad 1.189$$

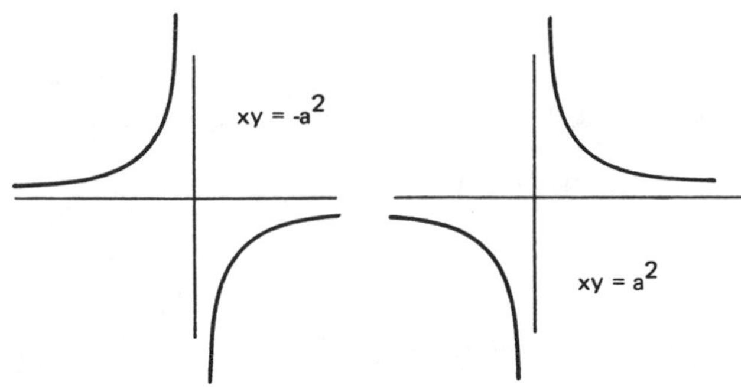


**Figure 1.11** Rectangular Hyperbolas

In general, for a hyperbola with transverse axis parallel to the x-axis and center at $(h,k)$,

$$\frac{(x-h)^2}{a^2} - \frac{(y-k)^2}{b^2} = 1 \qquad 1.190$$

The general form of the hyperbolic equation is

$$Ax^2 + Cy^2 + Dx + Ey + F = 0 \qquad 1.191$$

If $AC < 0$, the equation can be rewritten as

$$A\left(x + \frac{D}{2A}\right)^2 + C\left(y + \frac{E}{2C}\right)^2 = M \qquad 1.192$$

where

$$M = \frac{D^2}{4A} + \frac{E^2}{4C} - F \qquad 1.193$$

If $M = 0$, the graph is two intersecting lines.

If $M \neq 0$, the graph is a hyperbola with center at

$$\left(-\frac{D}{2A}, -\frac{E}{2C}\right)$$

The transverse axis is horizontal if $(M/A)$ is positive. It is vertical if $(M/C)$ is positive.

## 12 SPHERES

The equation of a sphere whose center is at the point $(h,k,l)$ and whose radius is $r$ is

$$(x-h)^2 + (y-k)^2 + (z-l)^2 = r^2 \qquad 1.194$$

If the sphere is centered at the origin, its equation is

$$x^2 + y^2 + z^2 = r^2 \qquad 1.195$$

## 13 PERMUTATIONS AND COMBINATIONS

Suppose you have $n$ objects, and you wish to work with a subset of $r$ of them. An order-conscious arrangement of $n$ objects taken $r$ at a time is known as *permutation*. The permutation is said to be order-conscious because the arrangement of two objects (say A and B) as AB is different from the arrangement BA. There are a number of ways of taking $n$ objects $r$ at a time. The total number of possible permutations is

$$P(n,r) = \frac{n!}{(n-r)!} \qquad 1.196$$

*Example 1.25*

A shelf has room for only three vases. If four different vases are available, how many ways can the shelf be arranged?

$$P(4,3) = \frac{4!}{(4-3)!} = \frac{(4)(3)(2)(1)}{(1)} = 24$$

The special cases of $n$ objects taken $n$ at a time are illustrated by the following examples.

*Example 1.26*

How many ways can seven resistors be connected end-to-end into a single unit?

$$P(7,7) = \frac{7!}{(7-7)!} = \frac{7!}{0!} = 7! = 5040$$

*Example 1.27*

Five people are to sit at a round table with five chairs. How many ways can these five people be arranged so that they all have different companions?

This is known as a *ring permutation.* Since the starting point of the arrangement around the circle does not affect the number of permutations, the answer is

$$(5-1)! = 4! = 24$$

An arrangement of $n$ objects taken $r$ at a time is known as a *combination* if the arrangement is not order-conscious. The total number of possible combinations is

$$C(n,r) = \frac{n!}{(n-r)!r!} \qquad 1.197$$

*Example 1.28*

How many possible ways can four out of six people fit into a four-seat boat?

$$C(6,4) = \frac{6!}{(6-4)!4!} = \frac{(6)(5)(4)(3)(2)(1)}{(2)(1)(4)(3)(2)(1)} = 15$$

## 14 PROBABILITY AND STATISTICS

### A. PROBABILITY RULES

The following rules are applied to sample spaces **A** and **B**:

$\mathbf{A} = [A_1, A_2, A_3, \ldots, A_n]$ and $\mathbf{B} = [B_1, B_2, B_3, \ldots, B_n]$

where the $A_i$ and $B_i$ are independent.

Rule 1:

$$p\{\emptyset\} = \text{probability of an impossible event} = 0 \quad 1.198$$

*Example 1.29*

An urn contains five white balls, two red balls, and three green balls. What is the probability of drawing a blue ball from the urn?

$$p\{\text{blue ball}\} = p\{\emptyset\} = 0$$

Rule 2:

$$p\{A_1 \text{ or } A_2 \text{ or } \ldots \text{ or } A_n\} = p\{A_1\} + p\{A_2\} + \cdots + p\{A_n\} \quad 1.199$$

*Example 1.30*

Returning to the urn described in example 1.29, what is the probability of getting either a white ball or a red ball in one draw from the urn?

$$p\{\text{red or white}\} = p\{\text{red}\} + p\{\text{white}\} = 0.2 + 0.5 = 0.7$$

Rule 3:

$$p\{A_i \text{ and } B_i \text{ and } \ldots Z_i\} = p\{A_i\}p\{B_i\}\ldots p\{Z_i\} \quad 1.200$$

*Example 1.31*

Given two identical urns (as described in example 1.29), what is the probability of getting a red ball from the first urn and a green ball from the second urn, given one draw from each urn?

$$p\{\text{red and green}\} = p\{\text{red}\}\,p\{\text{green}\} = (0.2)(0.3) = 0.06$$

Rule 4:

$$p\{\text{not A}\} = \text{probability of event A not occurring} = 1 - p\{A\} \quad 1.201$$

*Example 1.32*

Given the urn of example 1.29, what is the probability of not getting a red ball from the urn in one draw?

$$p\{\text{not red}\} = 1 - p\{\text{red}\} = 1 - 0.2 = 0.8$$

Rule 5:

$$p\{A_i \text{ or } B_i\} = p\{A_i\} + p\{B_i\} - p\{A_i\}p\{B_i\} \quad 1.202$$

*Example 1.33*

Given one urn as described in example 1.29 and a second urn containing eight red balls and two black balls, what is the probability of drawing either a white ball from the first urn or a red ball from the second urn, given one draw from each?

$$\begin{aligned} p\{\text{white or red}\} &= p\{\text{white}\} + p\{\text{red}\} - p\{\text{white}\}p\{\text{red}\} \\ &= 0.5 + 0.8 - (0.5)(0.8) = 0.9 \end{aligned}$$

Rule 6:

$p\{A|B\}$ = probability that A will occur given that B has already occurred, where the two events are dependent.

$$= \frac{p\{A \text{ and } B\}}{p\{B\}} \quad 1.203$$

The above equation is known as *Bayes Theorem.*

### B. PROBABILITY DENSITY FUNCTIONS

Probability density functions are mathematical functions giving the probabilities of numerical events. A *numerical event* is any occurrence that can be described by an integer or real number. For example, obtaining heads in a coin toss is not a numerical event. However, a concrete sample having a compressive strength less than 5000 psi is a numerical event.

Discrete density functions give the probability that the event $x$ will occur. That is,

$$f\{x\} = \text{probability of a process having a value of } x$$

Important discrete functions are the binomial and Poisson distributions.

*1. Binomial*

$n$ is the number of trials

$x$ is the number of successes

$p$ is the probability of a success in a single trial

$q$ is the probability of failure, $1 - p$

$\binom{n}{x}$ is the binomial coefficient $= \frac{n!}{(n-x)!x!}$

$x! = x(x-1)(x-2)\cdots(2)(1)$

Then, the probability of obtaining $x$ successes in $n$ trials is

$$f\{x\} = \binom{n}{x} p^x q^{(n-x)} \tag{1.204}$$

The mean of the binomial distribution is $np$. The variance of the distribution is $npq$.

*Example 1.34*

In a large quantity of items, 5% are defective. If seven items are sampled, what is the probability that exactly three will be defective?

$$f\{3\} = \binom{7}{3} (0.05)^3 (0.95)^4 = 0.0036$$

*2. Poisson*

Suppose an event occurs, on the average, $\lambda$ times per period. The probability that the event will occur $x$ times per period is

$$f\{x\} = \frac{e^{-\lambda}\lambda^x}{x!} \tag{1.205}$$

$\lambda$ is both the distribution mean and the variance. $\lambda$ must be a number greater than zero.

*Example 1.35*

The number of customers arriving in some period is distributed as Poisson with a mean of eight. What is the probability that six customers will arrive in any given period?

$$f\{6\} = \frac{e^{-8}8^6}{6!} = 0.122$$

Continuous probability density functions are used to find the cumulative distribution functions, $F\{x\}$. Cumulative distribution functions give the probability of event $x$ or less occurring.

$x$ = any value, not necessarily an integer

$$f\{x\} = \frac{dF\{x\}}{dx} \tag{1.206}$$

$F\{x\}$ = probability of $x$ or less occurring

*3. Exponential*

$$f\{x\} = u(e^{-ux}) \tag{1.207}$$

$$F\{x\} = 1 - e^{-ux} \tag{1.208}$$

The mean of the exponential distribution is $\frac{1}{u}$. The variance is $\left(\frac{1}{u}\right)^2$.

*Example 1.36*

The reliability of a unit is exponentially distributed with mean time to failure (MTBF) of 1000 hours. What is the probability that the unit will be operational at $t = 1200$ hours?

The reliability of an item is (1− probability of failing before time $t$). Therefore,

$$R\{t\} = 1 - F\{t\} = 1 - (1 - e^{-ut}) = e^{-ut}$$

$$u = \frac{1}{\textbf{MTBF}} = \frac{1}{1000} = 0.001$$

$$R\{1200\} = e^{-(0.001)(1200)} = 0.3$$

*4. Normal*

Although $f\{x\}$ can be expressed mathematically for the normal distribution, tables are used to evaluate $F\{x\}$ since $f\{x\}$ cannot be easily integrated. Since the $x$ axis of the normal distribution will seldom correspond to actual sample variables, the sample values are converted into standard values. Given the mean, $u$, and the standard deviation, $\sigma'$, the standard normal variable is

$$z = \frac{\text{sample value} - u}{\sigma'} \tag{1.209}$$

Then, the probability of a sample exceeding the given sample value is equal to the area in the tail past point $z$.

*Example 1.37*

Given a population that is normally distributed with mean of 66 and standard deviation of five, what percent of the population exceeds 72?

$$z = \frac{72 - 66}{5} = 1.2$$

Then, from table 1.5,

$$p\{\text{exceeding } 72\} = 0.5 - 0.3849 = 0.1151 \text{ or } 11.5\%$$

## C. STATISTICAL ANALYSIS OF EXPERIMENTAL DATA

Experiments can take on many forms. An experiment might consist of measuring the weight of one cubic foot of concrete. Or, an experiment might consist of measuring the speed of a car on a roadway. Generally, such experiments are performed more than once to increase the precision and accuracy of the results.

Of course, the intrinsic variability of the process being measured will cause the observations to vary, and we would not expect the experiment to yield the same result each time it was performed. Eventually, a collection of experimental outcomes (observations) will be available for analysis.

One fundamental technique for organizing random observations is the *frequency distribution.* The frequency distribution is a systematic method for ordering the observations from small to large, according to some convenient numerical characteristic.

*Example 1.38*

The number of cars that travel through an intersection between 12 noon and 1 p.m. is measured for 30 consecutive working days. The results of the 30 observations are:

79, 66, 72, 70, 68, 66, 68, 76, 73, 71, 74, 70, 71, 69, 67, 74, 70, 68, 69, 64, 75, 70, 68, 69, 64, 69, 62, 63, 63, 61

What is the frequency distribution using an interval of 2 cars per hour?

| cars per hour | frequency of occurrence |
|---|---|
| 60–61 | 1 |
| 62–63 | 3 |
| 64–65 | 2 |
| 66–67 | 3 |
| 68–69 | 8 |
| 70–71 | 6 |
| 72–73 | 2 |
| 74–75 | 3 |
| 76–77 | 1 |
| 78–79 | 1 |

In example 1.38, two cars per hour is known as the *step interval.* The step interval should be chosen so that the data is presented in a meaningful manner. If there are too many intervals, many of them will have zero frequencies. If there are too few intervals, the frequency distribution will have little value. Generally, 10 to 15 intervals are used.

Once the frequency distribution is complete, it can be represented graphically as a histogram. The procedure in drawing a histogram is to mark off the interval limits on a number line and then draw bars with lengths that are proportional to the frequencies in the intervals. If it is necessary to show the continuous nature of the data, a frequency polygon can be drawn.

*Example 1.39*

Draw the frequency histogram and frequency polygon for the data given in example 1.38.

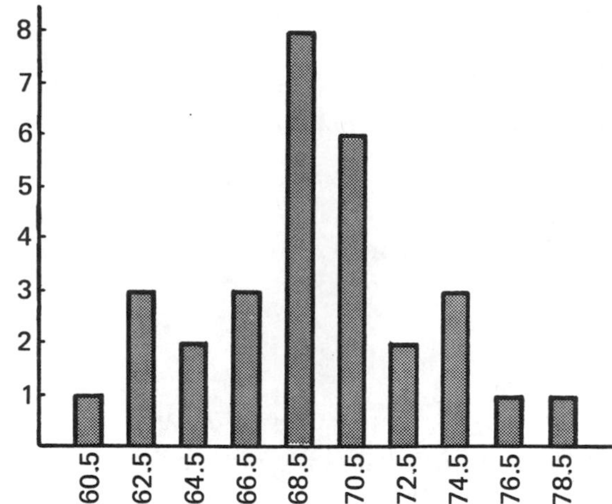


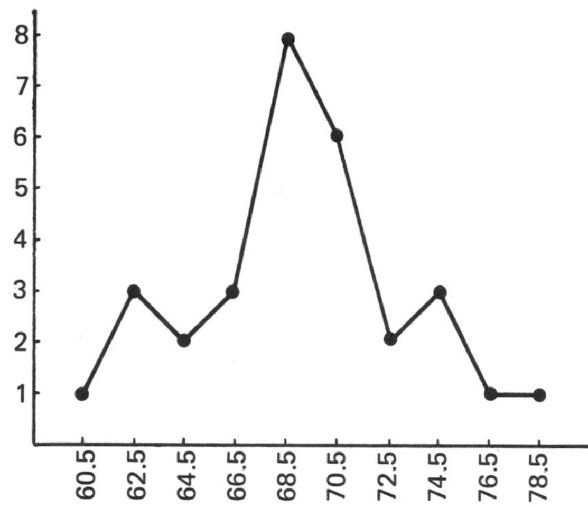


If it is necessary to know the number or percentage of observations that occur up to and including some value, the cumulative frequency table can be formed. This procedure is illustrated in the following example.

*Example 1.40*

Form the cumulative frequency distribution and graph for the data given in example 1.38.

| cars per hour | frequency | cumulative frequency | cumulative percent |
|---|---|---|---|
| 60–61 | 1 | 1 | 3 |
| 62–63 | 3 | 4 | 13 |
| 64–65 | 2 | 6 | 20 |
| 66–67 | 3 | 9 | 30 |
| 68–69 | 8 | 17 | 57 |
| 70–71 | 6 | 23 | 77 |
| 72–73 | 2 | 25 | 83 |
| 74–75 | 3 | 28 | 93 |
| 76–77 | 1 | 29 | 97 |
| 78–79 | 1 | 30 | 100 |

**Table 1.5**

Areas Under The Standard Normal Curve

(0 to z)

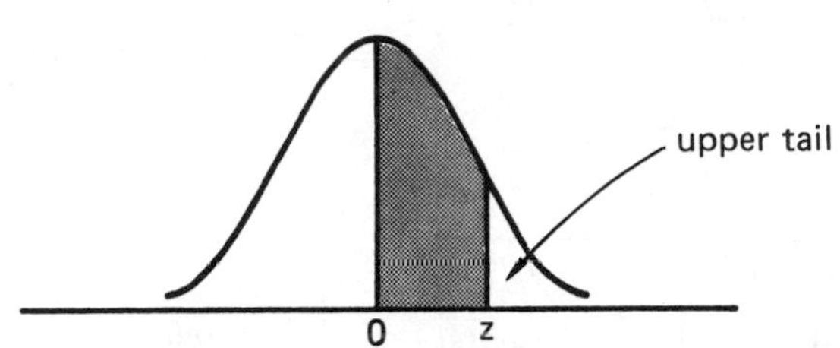


| z | 0 | 1 | 2 | 3 | 4 | 5 | 6 | 7 | 8 | 9 |
|---|---|---|---|---|---|---|---|---|---|---|
| 0.0 | .0000 | .0040 | .0080 | .0120 | .0160 | .0199 | .0239 | .0279 | .0319 | .0359 |
| 0.1 | .0398 | .0438 | .0478 | .0517 | .0557 | .0596 | .0636 | .0675 | .0714 | .0754 |
| 0.2 | .0793 | .0832 | .0871 | .0910 | .0948 | .0987 | .1026 | .1064 | .1103 | .1141 |
| 0.3 | .1179 | .1217 | .1255 | .1293 | .1331 | .1368 | .1406 | .1443 | .1480 | .1517 |
| 0.4 | .1554 | .1591 | .1628 | .1664 | .1700 | .1736 | .1772 | .1808 | .1844 | .1879 |
| 0.5 | .1915 | .1950 | .1985 | .2019 | .2054 | .2088 | .2123 | .2157 | .2190 | .2224 |
| 0.6 | .2258 | .2291 | .2324 | .2357 | .2389 | .2422 | .2454 | .2486 | .2518 | .2549 |
| 0.7 | .2580 | .2612 | .2642 | .2673 | .2704 | .2734 | .2764 | .2794 | .2823 | .2852 |
| 0.8 | .2881 | .2910 | .2939 | .2967 | .2996 | .3023 | .3051 | .3078 | .3106 | .3133 |
| 0.9 | .3159 | .3186 | .3212 | .3238 | .3264 | .3289 | .3315 | .3340 | .3365 | .3389 |
| 1.0 | .3413 | .3438 | .3461 | .3485 | .3508 | .3531 | .3554 | .3577 | .3599 | .3621 |
| 1.1 | .3643 | .3665 | .3686 | .3708 | .3729 | .3749 | .3770 | .3790 | .3810 | .3830 |
| 1.2 | .3849 | .3869 | .3888 | .3907 | .3925 | .3944 | .3962 | .3980 | .3997 | .4015 |
| 1.3 | .4032 | .4049 | .4066 | .4082 | .4099 | .4115 | .4131 | .4147 | .4162 | .4177 |
| 1.4 | .4192 | .4207 | .4222 | .4236 | .4251 | .4265 | .4279 | .4292 | .4306 | .4319 |
| 1.5 | .4332 | .4345 | .4357 | .4370 | .4382 | .4394 | .4406 | .4418 | .4429 | .4441 |
| 1.6 | .4452 | .4463 | .4474 | .4484 | .4495 | .4505 | .4515 | .4525 | .4535 | .4545 |
| 1.7 | .4554 | .4564 | .4573 | .4582 | .4591 | .4599 | .4608 | .4616 | .4625 | .4633 |
| 1.8 | .4641 | .4649 | .4656 | .4664 | .4671 | .4678 | .4686 | .4693 | .4699 | .4706 |
| 1.9 | .4713 | .4719 | .4726 | .4732 | .4738 | .4744 | .4750 | .4756 | .4761 | .4767 |
| 2.0 | .4772 | .4778 | .4783 | .4788 | .4793 | .4798 | .4803 | .4808 | .4812 | .4817 |
| 2.1 | .4821 | .4826 | .4830 | .4834 | .4838 | .4842 | .4846 | .4850 | .4854 | .4857 |
| 2.2 | .4861 | .4864 | .4868 | .4871 | .4875 | .4878 | .4881 | .4884 | .4887 | .4890 |
| 2.3 | .4893 | .4896 | .4898 | .4901 | .4904 | .4906 | .4909 | .4911 | .4913 | .4916 |
| 2.4 | .4918 | .4920 | .4922 | .4925 | .4927 | .4929 | .4931 | .4932 | .4934 | .4936 |
| 2.5 | .4938 | .4940 | .4941 | .4943 | .4945 | .4946 | .4948 | .4949 | .4951 | .4952 |
| 2.6 | .4953 | .4955 | .4956 | .4957 | .4959 | .4960 | .4961 | .4962 | .4963 | .4964 |
| 2.7 | .4965 | .4966 | .4967 | .4968 | .4969 | .4970 | .4971 | .4972 | .4973 | .4974 |
| 2.8 | .4974 | .4975 | .4976 | .4977 | .4977 | .4978 | .4979 | .4979 | .4980 | .4981 |
| 2.9 | .4981 | .4982 | .4982 | .4983 | .4984 | .4984 | .4985 | .4985 | .4986 | .4986 |
| 3.0 | .4987 | .4987 | .4987 | .4988 | .4988 | .4989 | .4989 | .4989 | .4990 | .4990 |
| 3.1 | .4990 | .4991 | .4991 | .4991 | .4992 | .4992 | .4992 | .4992 | .4993 | .4993 |
| 3.2 | .4993 | .4993 | .4994 | .4994 | .4994 | .4994 | .4994 | .4995 | .4995 | .4995 |
| 3.3 | .4995 | .4995 | .4995 | .4996 | .4996 | .4996 | .4996 | .4996 | .4996 | .4997 |
| 3.4 | .4997 | .4997 | .4997 | .4997 | .4997 | .4997 | .4997 | .4997 | .4997 | .4998 |
| 3.5 | .4998 | .4998 | .4998 | .4998 | .4998 | .4998 | .4998 | .4998 | .4998 | .4998 |
| 3.6 | .4998 | .4998 | .4999 | .4999 | .4999 | .4999 | .4999 | .4999 | .4999 | .4999 |
| 3.7 | .4999 | .4999 | .4999 | .4999 | .4999 | .4999 | .4999 | .4999 | .4999 | .4999 |
| 3.8 | .4999 | .4999 | .4999 | .4999 | .4999 | .4999 | .4999 | .4999 | .4999 | .4999 |
| 3.9 | .5000 | .5000 | .5000 | .5000 | .5000 | .5000 | .5000 | .5000 | .5000 | .5000 |

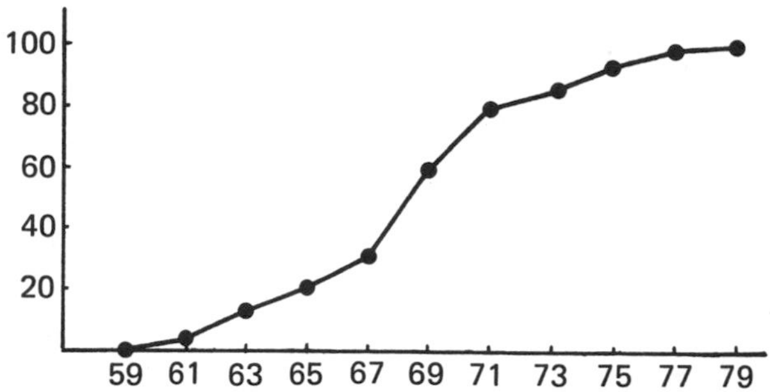


It is often unnecessary to present the experimental data in its entirety, either in tabular or graphical form. In such cases, the data and distribution can be represented by various parameters. One type of parameter is a measure of *central tendency.* Mode, median, and mean are measures of central tendency. The other type of parameter is a measure of dispersion. Standard deviation and variance are measures of dispersion.

The *mode* is the observed value which occurs most frequently. The mode may vary greatly between series of observations. Therefore, its main use is as a quick measure of the central value since no computation is required to find it. Beyond this, the usefulness of the mode is limited.

The *median* is the point in the distribution which divides the total observations into two parts containing equal numbers of observations. It is not influenced by the extremity of scores on either side of the distribution. The median is found by counting up (from either end of the frequency distribution) until half of the observations have been accounted for. The procedure is more difficult if the median falls within an interval, as illustrated in example 1.41.

Similar in concept to the median are *percentile ranks, quartiles,* and *deciles.* The median could also have been called the *50th percentile* observation. Similarly, the 80th percentile would be the number of cars per hour for which the cumulative frequency was 80%. The quartile and decile points on the distribution divide the observations or distribution into segments of 25% and 10%, respectively.

The *arithmetic mean* is the arithmetic average of the observations. The *mean* may be found without ordering the data (which was necessary to find the mode and median). The mean can be found from the following formula:

$$\overline{x} = \left(\frac{1}{n}\right)(x_1 + x_2 + \cdots + x_n) = \frac{\sum x_i}{n} \qquad 1.210$$

The *geometric mean* is used occasionally when it is necessary to average ratios. The geometric mean is calculated as

$$\text{geometric mean} = \sqrt[n]{x_1 x_2 x_3 \ldots x_n} \qquad 1.211$$

The *harmonic mean* is defined as

$$\text{harmonic mean} = \frac{n}{\frac{1}{x_1} + \frac{1}{x_2} + \cdots + \frac{1}{x_n}} \qquad 1.212$$

The *root-mean-squared* (*rms*) *value* of a series of observations is defined as

$$x_{rms} = \sqrt{\frac{\sum x_i^2}{n}} \qquad 1.213$$

*Example 1.41*

Find the mode, median, and arithmetic mean of the distribution represented by the data given in example 1.38.

The mode is the interval 68–69, since this interval has the highest frequency. If 68.5 is taken as the interval center, then 68.5 would be the mode.

Since there are 30 observations, the median is the value which separates the observations into two groups of 15. From example 1.40, the median occurs someplace within the 68–69 interval. Up through interval 66–67, there are nine observations, so six more are needed to make 15. Interval 68–69 has eight observations, so the median is found to be $\left(\frac{6}{8}\right)$ or $\left(\frac{3}{4}\right)$ of the way through the interval. Since the real limits of the interval are 67.5 and 69.5, the median is located at

$$67.5 + \frac{3}{4}(69.5 - 67.5) = 69$$

The mean can be found from the raw data or from the grouped data using the interval center as the assumed observation value. Using the raw data,

$$\overline{x} = \frac{\sum x}{n} = \frac{2069}{30} = 68.97$$

The simplest statistical parameter which describes the variation in observed data is the *range.* The range is found by subtracting the smallest value from the largest. Since the range is influenced by extreme (low probability) observations, its use as a measure of variability is limited.

The *standard deviation* is a better estimate of variability because it considers every observation. The standard deviation can be found from:

$$\sigma = \sqrt{\frac{\sum (x_i - \overline{x})^2}{n}} = \sqrt{\frac{\sum x_i^2}{n} - (\overline{x})^2} \qquad 1.214$$

The above formula assumes that $n$ is a large number, such as above 50. Theoretically, $n$ is the size of the entire population. If a small sample (less than 50) is used to calculate the standard deviation of the distribution, the formulas are changed. The *sample standard deviation* is

$$s = \sqrt{\frac{\sum (x_i - \overline{x})^2}{n-1}} = \sqrt{\frac{\sum x_i^2 - \frac{\left(\sum x_i\right)^2}{n}}{n-1}} \qquad 1.215$$

The difference is small when $n$ is large, but care must be taken in reading the problem. If the *standard deviation of the sample* is requested, calculate $\sigma$. If an estimate of the *population standard deviation* or *sample standard deviation* is requested, calculate $s$. (Note that the standard deviation of the sample is not the same as the sample standard deviation.)

The *relative dispersion* is defined as a measure of dispersion divided by a measure of central tendency. The *coefficient of variation* is a relative dispersion calculated from the standard deviation and the mean. That is,

$$\text{coefficient of variation} = \frac{s}{\overline{x}} \qquad 1.216$$

*Skewness* is a measure of a frequency distribution's lack of symmetry. It is calculated as

$$\text{skewness} = \frac{\overline{x} - \text{mode}}{s} \qquad 1.217$$

$$\approx \frac{3(\overline{x} - \text{median})}{s} \qquad 1.218$$

*Example 1.42*

Calculate the range, standard deviation of the sample, and population variance from the data given in example 1.38.

$$\sum x = 2069 \quad \left(\sum x\right)^2 = 4{,}280{,}761 \quad \sum x^2 = 143{,}225$$

$$n = 30 \quad \overline{x} = 68.967$$

$$\sigma = \sqrt{\frac{143{,}225}{30} - \left(\frac{2069}{30}\right)^2} = 4.21$$

$$s = \sqrt{\frac{143{,}225 - \frac{4{,}280{,}761}{30}}{29}} = 4.29$$

$$s^2 = 18.4 \quad \text{(sample variance)}$$

$$\sigma^2 = 17.8 \quad \text{(population variance)}$$

$$R = 79 - 61 = 18$$

Referring again to example 1.38, suppose that the hourly through-put for 15 similar intersections is measured over a 30 day period. At the end of the 30 day period, there will be 15 ranges, 15 medians, 15 means, 15 standard deviations, and so on. These parameters themselves constitute distributions.

The mean of the sample means is an excellent estimator of the average hourly through-put of an intersection, $\mu$.

$$\mu = \left(\frac{1}{15}\right) \sum \overline{x}$$

The standard deviation of the sample means is known as the *standard error of the mean* to distinguish it from the standard deviation of the raw data. The standard error is written as $\sigma_{\overline{x}}$.

The standard error is not a good estimator of the population standard deviation, $\sigma'$.

In general, if $k$ sets of $n$ observations each are used to estimate the population mean ($\mu$) and the population standard deviation ($\sigma'$), then

$$\mu \approx \left(\frac{1}{k}\right) \sum \overline{x} \qquad 1.219$$

$$\sigma' \approx \sqrt{k}\,\sigma_{\overline{x}} \qquad 1.220$$

## 15 BASIC HYPOTHESIS TESTING

Suppose a distribution is $\sim N(\mu, \sigma'^2)$.[1] If samples of size $n$ are taken $k$ times, the values of the sample means $\overline{x}$ will form a distribution themselves. These means also will be distributed normally with the form

$$\sim N\left(\mu, \frac{\sigma'^2}{n}\right)$$

That is, the mean of the sample means will be identical to the original population, but the variance and standard deviation will be much smaller.

Thus, the probability that $\overline{x}$ exceeds some value, say $x^*$, is

$$P\left\{z > \frac{x^* - \mu}{\frac{\sigma'}{\sqrt{n}}}\right\} \qquad 1.221$$

This can be solved as an *exceedance problem* (see example 1.37), or a hypothesis test can be performed. A *hypothesis test* has the following characteristics:

- a sample is taken in an experiment
- a parameter (usually $\overline{x}$) is measured
- it is desired to know if the sample could have come from a population $\sim N(\mu, \sigma'^2)$

There are many types of hypothesis tests, depending on the type of population (i.e., whether or not normal), the parameter being tested (i.e., central tendency or dispersion), and the size of the sample.

[1] This is the standard method of saying the distribution is normally distributed with mean $\mu$ and variance $\sigma'^2$.

If the sample size is not much greater than 30, if the native population is assumed to be normal, and if $\mu$ and $\sigma'$ are known, the following procedure can be used.

*step 1*: Assume random sampling from a normal population.

*step 2*: Choose the desired confidence level, C. Usually, a 95% confidence level result is said to be ***significant.*** 99% test results are said to be ***highly significant.***

*step 3*: Decide on a 1-tail or 2-tail test. If the question is worded as "Has the population mean changed?" or "Are the populations the same?", then a ***2-tail test*** is needed. If the question is "Has the mean increased?" or "... decreased?", then a ***1-tail test*** is needed.

*step 4*: From the normal table, find the value $z'$ for a table entry equal to

$$\frac{1-C}{\#\text{tails in the test}} \qquad 1.222$$

*step 5*: Calculate

$$z = \left|\frac{\overline{x}-\mu}{\frac{\sigma'}{\sqrt{n}}}\right| \qquad 1.223$$

If $z \geq z'$, then the distributions are not the same.

*Example 1.43*

When operating properly, a chemical plant has a product output which is normally distributed with mean 880 tons/day and standard deviation of 21 tons. The output is measured on 50 consecutive days, and the mean output is 871 tons/day. Is the plant operating correctly?

*step 1*: Assume random sampling from the normal distribution

*step 2*: Choose $C = 0.95$ for significant results.

*step 3*: Wanting to know if the plant is operating correctly is the same as asking, "Has anything changed?" There is no mention of ***direction*** (i.e., the question was not, "Has the output decreased?"). Therefore, choose a 2-tail test.

*step 4*: $\frac{1}{2}(1-C) = 0.025$. The 0.025 outside lower limit in table 1.5 is $z' = 1.96$. (This corresponds to an area under the curve of $0.5 - 0.025 = 0.475$.)

*step 5*:

$$z = \left|\frac{871-880}{\frac{21}{\sqrt{50}}}\right| = 3.03$$

Since $3.03 > 1.96$, the distributions are not the same. There is a 95% chance that the plant is not operating correctly.

## 16 DIFFERENTIAL CALCULUS

### A. TERMINOLOGY

Given $y$, a function of $x$, the first derivative with respect to $x$ may be written as

$$\mathbf{D}y,\ y',\ \text{or}\ \left(\frac{dy}{dx}\right)$$

The first derivative corresponds to the slope of the line described by the function $y$. The second derivative may be written as

$$\mathbf{D}^2y,\ \left(\frac{d^2y}{dx^2}\right),\ \text{or}\ y''$$

### B. BASIC OPERATIONS

In the formulas that follow, $f$ and $g$ are functions of $x$. $\mathbf{D}$ is the derivative operator. $a$ is a constant.

$$\mathbf{D}(a) = 0 \qquad 1.224$$

$$\mathbf{D}(af) = a\mathbf{D}(f) \qquad 1.225$$

$$\mathbf{D}(f+g) = \mathbf{D}(f) + \mathbf{D}(g) \qquad 1.226$$

$$\mathbf{D}(f-g) = \mathbf{D}(f) - \mathbf{D}(g) \qquad 1.227$$

$$\mathbf{D}(f\cdot g) = f\mathbf{D}(g) + g\mathbf{D}(f) \qquad 1.228$$

$$\mathbf{D}\left(\frac{f}{g}\right) = \frac{g\mathbf{D}(f) - f\mathbf{D}(g)}{g^2} \qquad 1.229$$

$$\mathbf{D}(x^n) = nx^{n-1} \qquad 1.230$$

$$\mathbf{D}(f^n) = nf^{n-1}\mathbf{D}(f) \qquad 1.231$$

$$\mathbf{D}(f(g)) = \frac{df(g)}{dg}\mathbf{D}(g) \qquad 1.232$$

$$\mathbf{D}(lnx) = \frac{1}{x} \qquad 1.233$$

$$\mathbf{D}(e^{ax}) = ae^{ax} \qquad 1.234$$

*Example 1.44*

A function is given as $f(x) = x^3 - 2x$. What is the slope of the line at $x = 3$?

$$y' = 3x^2 - 2$$
$$y'(3) = 27 - 2 = 25$$

### C. TRANSCENDENTAL FUNCTIONS

$$\mathbf{D}(\sin x) = \cos x \qquad 1.235$$

$$\mathbf{D}(\cos x) = -\sin x \qquad 1.236$$

$$\mathbf{D}(\tan x) = \sec^2 x \qquad 1.237$$

$$\mathbf{D}(\cot x) = -\csc^2 x \qquad 1.238$$

$$\mathbf{D}(\sec x) = (\sec x)(\tan x) \qquad 1.239$$

$$\mathbf{D}(\csc x) = (-\csc x)(\cot x) \qquad 1.240$$

$$\mathbf{D}(\arcsin x) = \frac{1}{\sqrt{1-x^2}} \qquad 1.241$$

$$\mathbf{D}(\arctan x) = \frac{1}{1+x^2} \qquad 1.242$$

$$\mathbf{D}(\text{arcsec}\, x) = \frac{1}{x\sqrt{x^2-1}} \qquad 1.243$$

$$\mathbf{D}(\arccos x) = -\mathbf{D}(\arcsin x) \qquad 1.244$$

$$\mathbf{D}(\text{arccot}\, x) = -\mathbf{D}(\arctan x) \qquad 1.245$$

$$\mathbf{D}(\text{arccsc}\, x) = -\mathbf{D}(\text{arcsec}\, x) \qquad 1.246$$

## D. VARIATIONS ON DIFFERENTIATION

### *1. Partial Differentiation*

If the function has two or more independent variables, a partial derivative is found by considering all extraneous variables as constants. The geometric interpretation of the partial derivative $(\partial z/\partial x)$ is the slope of a line tangent to the 3-dimensional surface in a plane of constant $y$ and parallel to the $x$ axis. Similarly, the interpretation of $(\partial z/\partial y)$ is the slope of a line tangent to the surface in a plane of constant $x$ and parallel to the $y$ axis.

*Example 1.45*

A surface has the equation $x^2 + y^2 + z^2 = 9$. What is the slope of a line tangent to (1,2,2) and parallel to the $x$ axis?

$$z = \sqrt{9 - x^2 - y^2}$$

$$\frac{\partial z}{\partial x} = \frac{-x}{\sqrt{9 - x^2 - y^2}}$$

At the point (1,2,2),

$$\frac{\partial z}{\partial x} = -\frac{1}{2}$$

### *2. Implicit Differentiation*

If a relationship between $n$ variables cannot be manipulated to yield an explicit function of $(n-1)$ independent variables, the relationship implicitly defines the $n$th remaining variable. The derivative of the implicit variable taken with respect to any other variable is found by a process known as implicit differentiation.

If $f(x,y) = 0$ is a function, the implicit derivative is

$$\frac{dy}{dx} = -\frac{\partial f}{\partial x} \Big/ \frac{\partial f}{\partial y} \qquad 1.247$$

If $f(x,y,z) = 0$ is a function, the implicit derivatives are

$$\frac{\partial z}{\partial x} = -\frac{\partial f}{\partial x} \Big/ \frac{\partial f}{\partial z} \qquad 1.248$$

$$\frac{\partial z}{\partial y} = -\frac{\partial f}{\partial y} \Big/ \frac{\partial f}{\partial z} \qquad 1.249$$

*Example 1.46*

If $f = x^2 + xy + y^3$, what is $\frac{dy}{dx}$?

Since this function cannot be written as an explicit function of $x$, implicit differentiation is required.

$$\frac{\partial f}{\partial x} = 2x + y$$

$$\frac{\partial f}{\partial y} = x + 3y^2$$

$$\frac{dy}{dx} = \frac{-(2x+y)}{x + 3y^2}$$

*Example 1.47*

Solve example 1.45 using implicit differentiation.

$$f = x^2 + y^2 + z^2 - 9$$

$$\frac{\partial f}{\partial x} = 2x$$

$$\frac{\partial f}{\partial z} = 2z$$

$$\frac{\partial z}{\partial x} = \frac{-2x}{2z} = -\frac{x}{z}$$

and at (1,2,2),

$$\frac{\partial z}{\partial x} = -\frac{1}{2}$$

### *3. The Gradient Vector*

The slope of a function is defined as the change in one variable with respect to a distance in another direction. Usually, this direction is parallel to an axis. However, the maximum slope at a point on a 3-dimensional object may not be in a direction parallel to one of the coordinate axes.

The gradient vector function $\nabla f(x,y,z)$ (pronounced "del f") gives the maximum rate of change of the function f(x,y,z). The gradient vector function is defined as

$$\nabla f(x,y,z) = \frac{\partial f(x,y,z)}{\partial x}\mathbf{i} + \frac{\partial f(x,y,z)}{\partial y}\mathbf{j} + \frac{\partial f(x,y,z)}{\partial z}\mathbf{k} \qquad 1.250$$

*Example 1.48*

Find the maximum slope of $f(x,y) = 2x^2 - y^2 + 3x - y$ at the point $(1,-2)$. What is the equation of the maximum-slope tangent?

This is a 2-dimensional problem.

$$\frac{\partial f(x,y)}{\partial x} = 4x + 3$$
$$\frac{\partial f(x,y)}{\partial y} = -2y - 1$$
$$\nabla f(x,y) = (4x+3)\mathbf{i} + (-2y-1)\mathbf{j}$$

The equation of the maximum-slope tangent is

$$\nabla f(1,-2) = 7\mathbf{i} + 3\mathbf{j}$$

The magnitude of the slope is

$$\sqrt{(7)^2 + (3)^2} = \sqrt{58}$$

*4. The Directional Derivative*

The rate of change of a function in the direction of some given vector $\mathbf{U}$ can be found from the directional derivative function, $\nabla_u f(x,y,z)$. This directional derivative function depends on the gradient vector and the direction cosines of the vector $\mathbf{U}$.

$$\nabla_u f(x,y,z) = \frac{\partial f(x,y,z)}{\partial x}\cos\alpha + \frac{\partial f(x,y,z)}{\partial y}\cos\beta + \frac{\partial f(x,y,z)}{\partial z}\cos\gamma \quad 1.251$$

*Example 1.49*

What is the rate of change of $f(x,y) = 3x^2 + xy - 2y^2$ at the point $(1,-2)$ in the direction $4\mathbf{i} + 3\mathbf{j}$?

$$\cos\alpha = \frac{4}{\sqrt{(4)^2 + (3)^2}} = \frac{4}{5}$$
$$\cos\beta = \frac{3}{5}$$
$$\frac{\partial f(x,y)}{\partial x} = 6x + y$$
$$\frac{\partial f(x,y)}{\partial y} = x - 4y$$
$$\nabla_u f(x,y) = \left(\frac{4}{5}\right)(6x+y) + \left(\frac{3}{5}\right)(x-4y)$$
$$\nabla_u f(1,-2) = \left(\frac{4}{5}\right)[(6)(1) - 2] + \left(\frac{3}{5}\right)[1 - (4)(-2)] = 8.6$$

*5. Tangent Plane Function*

Partial derivatives can be used to find the tangent plane to a 3-dimensional surface at some point $p_o$. If the surface is defined by the function $f(x,y,z) = 0$, the equation of the tangent plane is

$$T(x_o,y_o,z_o) = (x - x_o)\frac{\partial f(x,y,z)}{\partial x}\Big|_{p_o} + (y - y_o)\frac{\partial f(x,y,z)}{\partial y}\Big|_{p_o} + (z - z_o)\frac{\partial f(x,y,z)}{\partial z}\Big|_{p_o} \quad 1.252$$

*Example 1.50*

What is the equation of the plane tangent to $f(x,y,z) = 4x^2 + y^2 - 16z$ at the point (2,4,2)?

$$\frac{\partial f(x,y,z)}{\partial x}\Big|_{p_o} = 8x\,\Big|_{(2,4,2)} = (8)(2) = 16$$

$$\frac{\partial f(x,y,z)}{\partial y}\Big|_{p_o} = 2y\,\Big|_{(2,4,2)} = (2)(4) = 8$$

$$\frac{\partial f(x,y,z)}{\partial z}\Big|_{p_o} = -16\,\Big|_{(2,4,2)} = -16$$

Therefore,

$$T(2,4,2) = 16(x-2) + 8(y-4) - 16(z-2) = 2x + y - 2z - 4$$

*6. Normal Line Function*

Partial derivatives can be used to find the equation of a straight line normal to a 3-dimensional surface at some point $p_o$. If the surface is defined by the function $f(x,y,z) = 0$, the equation of the normal line is

$$\mathbf{N} = A\mathbf{i} + B\mathbf{j} + C\mathbf{k} \quad 1.253$$

where

$$A = \frac{\partial f(x,y,z)}{\partial x}\Big|_{p_o} \quad 1.254$$
$$B = \frac{\partial f(x,y,z)}{\partial y}\Big|_{p_o} \quad 1.255$$
$$C = \frac{\partial f(x,y,z)}{\partial z}\Big|_{p_o} \quad 1.256$$

*7. Extrema and Optimization*

Derivatives can be used to locate local *maxima*, *minima*, and *points of inflection*. No distinction is made between local and global extrema. The end points of the interval always should be checked against the local extrema located by the method below. The following rules define the extreme points.

$$\begin{aligned} f'(x) &= 0 \text{ at any extrema} \\ f''(x) &= 0 \text{ at an inflection point} \\ f''(x) &\quad \text{is negative at a maximum} \\ f''(x) &\quad \text{is positive at a minimum} \end{aligned}$$

There is always an inflection point between a maximum and a minimum.

*Example 1.51*

Find the global extreme points of the function $f(x) = x^3 + x^2 - x + 1$ on the interval $[-2, +2]$.

$$\begin{aligned} f'(x) &= 3x^2 + 2x - 1 \\ f'(x) &= 0 \text{ at } x = \frac{1}{3} \text{ and } x = -1 \\ f''(x) &= 6x + 2 \\ f(-1) &= 2 \\ f''(-1) &= -4 \end{aligned}$$

So, $x = -1$ is a maximum.

$$\begin{aligned} f\left(\frac{1}{3}\right) &= \frac{22}{27} \\ f''\left(\frac{1}{3}\right) &= +4 \end{aligned}$$

So, $x = -\frac{1}{3}$ is a minimum.

Checking the end points,

$$\begin{aligned} f(-2) &= -1 \\ f(+2) &= +11 \end{aligned}$$

Therefore, the absolute extreme points are the end points.

## 17 INTEGRAL CALCULUS

### A. FUNDAMENTAL THEOREM

The *Fundamental Theorem of Calculus* is

$$\int_{x_1}^{x_2} f'(x)dx = f(x_2) - f(x_1) \qquad 1.257$$

### B. INTEGRATION BY PARTS

If $f$ and $g$ are functions, then

$$\int f\,dg = fg - \int g\,df \qquad 1.258$$

*Example 1.52*

Evaluate the following integral: $\int xe^x dx$

Use integration by parts.

Let $f = x$. Then, $df = dx$.

Let $dg = e^x dx$. Then, $g = \int e^x dx = e^x$.

Therefore,

$$\begin{aligned} \int xe^x dx &= xe^x - \int e^x dx + C \\ &= xe^x - e^x + C \end{aligned}$$

### C. INDEFINITE INTEGRALS ("$\cdots + C$" omitted)

$$\int dx = x \qquad 1.259$$

$$\int au\;dx = a\int u\;dx \qquad 1.260$$

$$\int (u+v)dx = \int u\;dx + \int v\;dx \qquad 1.261$$

$$\int x^m dx = \frac{x^{(m+1)}}{m+1} \qquad m \neq -1 \qquad 1.262$$

$$\int \frac{dx}{x} = ln|x| \qquad 1.263$$

$$\int e^{ax} dx = \frac{1}{a}e^{ax} \qquad 1.264$$

$$\int xe^{ax} dx = \frac{1}{a^2}e^{ax}(ax-1) \qquad 1.265$$

$$\int \cosh x\;dx = \sinh x \qquad 1.266$$

$$\int \sinh x\;dx = \cosh x \qquad 1.267$$

$$\int \sin x\;dx = -\cos x \qquad 1.268$$

$$\int \cos x\;dx = \sin x \qquad 1.269$$

$$\int \tan x\;dx = ln(\sec x) \qquad 1.270$$

$$\int \cot x\;dx = ln(\sin x) \qquad 1.271$$

$$\int \sec x\;dx = ln(\sec x + \tan x) \qquad 1.272$$

$$\int \csc x\;dx = ln(\csc x - \cot x) \qquad 1.273$$

$$\int \frac{dx}{1+x^2} = \arctan x \qquad 1.274$$

$$\int \frac{dx}{\sqrt{1-x^2}} = \arcsin x \qquad 1.275$$

$$\int \frac{dx}{x\sqrt{x^2-1}} = \text{arcsec}\;x \qquad 1.276$$

### D. USES OF INTEGRALS

*1. Finding Areas*

The area bounded by $x = a$, $x = b$, $f_1(x)$ above, and $f_2(x)$ below is given by

$$A = \int_a^b [f_1(x) - f_2(x)]dx \qquad 1.277$$

*2. Surfaces of Revolution*

The surface area obtained by rotating $f(x)$ about the $x$ axis is

$$A_s = 2\pi \int_a^b f(x)\sqrt{1 + [f'(x)]^2}dx \qquad 1.278$$

*3. Rotation of a Function*

The volume of a function rotated about the $x$ axis is

$$V = \pi \int_a^b (f(x))^2 dx \qquad 1.279$$

The volume of a function rotated about the $y$ axis is

$$V = 2\pi \int_a^b x f(x)dx \qquad 1.280$$

*4. Length of a Curve*

The length of a curve given by f(x) is

$$L = \int_a^b \sqrt{1 + (f'(x))^2}dx \qquad 1.281$$

*Example 1.53*

For the shaded area shown, find (a) the area, and (b) the volume enclosed by the curve rotated about the $x$ axis.

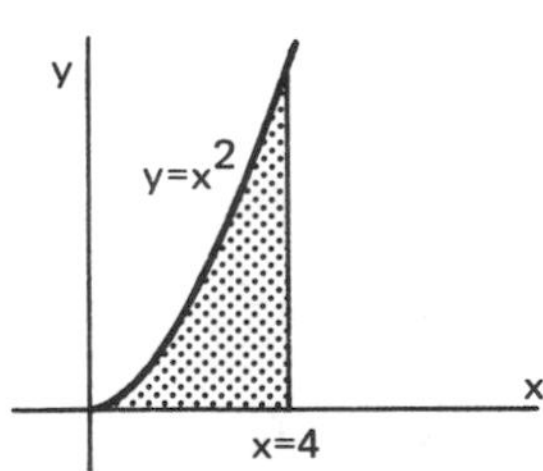


(a)

$$f_2(x) = 0 \qquad f_1(x) = x^2$$

$$A = \int_0^4 x^2 dx = \left[\frac{x^3}{3}\right]_0^4 = 21.33$$

(b)

$$V = \pi \int_0^4 (x^2)^2 dx = \pi \left[\frac{x^5}{5}\right]_0^4 = 204.8\pi$$

## 18 DIFFERENTIAL EQUATIONS

A differential equation is a mathematical expression containing a dependent variable and one or more of that variable's derivatives. First order differential equations contain only the first derivative of the dependent variable. Second order equations contain the second derivative.

The differential equation is said to be *linear* if all terms containing the dependent variable are multiplied only by real scalars. The equation is said to be *homogeneous* if there are no terms which do not contain the dependent variable or one of its derivatives.

Most differential equations are difficult to solve. However, there are several forms which are fairly simple. These are presented here.

### A. FIRST ORDER LINEAR

The first order linear differential equation has the general form given by equation 1.282. $p(t)$ and $g(t)$ may be constants or any function of $t$.

$$x' + p(t)x = g(t) \qquad 1.282$$

The solution depends on an *integrating factor* defined as

$$u = \exp\left[\int p(t)\ dt\right] \qquad 1.283$$

The solution to the first order linear differential equation is

$$x = \frac{1}{u}\left[\int ug(t)\ dt + c\right] \qquad 1.284$$

*Example 1.54*

Find a solution to the differential equation

$$x' - x = 2te^{2t} \qquad x(0) = 1$$

This meets the definition of a first order linear equation with

$$p(t) = -1 \text{ and } g(t) = 2te^{2t}$$

The integrating constant is

$$u = \exp\left[\int -1\ dt\right] = e^{-t}$$

Then, $x$ is

$$\begin{aligned} x &= \left(\frac{1}{e^{-t}}\right)\left[\int e^{-t} 2te^{2t} dt + c\right] \\ &= e^t\left[\int 2te^t dt + c\right] \\ &= e^t[2te^t - 2e^t + c] \end{aligned}$$

But, $x(0) = 1$, so

$$c = +3, \text{ and}$$
$$x = e^t[2e^t(t-1)+3]$$

## B. SECOND ORDER HOMOGENEOUS WITH CONSTANT COEFFICIENTS

This type of differential equation has the following general form:

$$c_1x'' + c_2x' + c_3x = 0 \quad 1.285$$

The solution can be found by first solving the characteristic quadratic equation for its roots $k_1^*$ and $k_2^*$. This characteristic equation is derived directly from the differential equation:

$$c_1k^2 + c_2k + c_3 \quad 1.286$$

The form of the solution depends on the values of $k_1^*$ and $k_2^*$. If $k_1^* \neq k_2^*$ and both are real, then

$$x = a_1\left(e^{k_1^*t}\right) + a_2\left(e^{k_2^*t}\right) \quad 1.287$$

If $k_1^* = k_2^*$, then

$$x = a_1\left(e^{k_1^*t}\right) + a_2t\left(e^{k_2^*t}\right) \quad 1.288$$

If $k^* = (r \pm iu)$, then

$$x = a_1(e^{rt})\cos(ut) + a_2(e^{rt})\sin(ut) \quad 1.289$$

In all three cases, $a_1$ and $a_2$ must be found from the given initial conditions.

*Example 1.55*

Solve the following differential equation for $x$.

$$x'' + 6x' + 9x = 0 \quad x(0) = 0, \; x'(0) = 1$$

The characteristic equation is

$$k^2 + 6k + 9 = 0$$

This has roots of $k_1^* = k_2^* = -3$; therefore, the solution has the form

$$x(t) = a_1e^{-3t} + a_2te^{-3t}$$

But, $x(0) = 0$,

$$0 = a_1(e^0) + a_2(0)(e^0)$$
$$0 = a_1(1) + 0$$
$$0 = a_1$$

Also, $x'(0) = 1$. The derivative of x(t) is

$$x'(t) = -3a_2te^{-3t} + a_2e^{-3t}$$
$$1 = -3a_2(0)(e^0) + a_2e^0$$
$$1 = 0 + a_2$$
$$1 = a_2$$

The final solution is

$$x = te^{-3t}$$

## 19 LAPLACE TRANSFORMS

Traditional methods of solving non-homogeneous differential equations are very difficult. The Laplace transformation can be used to reduce the solution of many complex differential equations to simple algebra.

Every mathematical function can be converted into a Laplace function by use of the following transformation definition.

$$\mathcal{L}[f(t)] = \int_0^\infty e^{-st}f(t)dt \quad 1.290$$

The variable $s$ is equivalent to the derivative operator. However, it may be thought of as a simple variable.

*Example 1.56*

Let f(t) be the unit step. That is, $f(t) = 0$ for $t < 0$ and $f(t) = 1$ for $t \geq 0$.

Then, the Laplace transform of $f(t) = 1$ is

$$\mathcal{L}[f(t)] = \int_0^\infty e^{-st}(1)dt = -\frac{e^{-st}}{s}\Bigg]_0^\infty$$
$$= 0 - \left(-\frac{1}{s}\right) = \frac{1}{s}$$

*Example 1.57*

What is the Laplace transformation of $f(t) = e^{at}$?

$$\mathcal{L}[e^{at}] = \int_0^\infty e^{-st}e^{at}dt$$
$$= \int_0^\infty e^{-(s-a)t}dt$$
$$= -\frac{e^{-(s-a)t}}{s-a}\Bigg]_0^\infty$$
$$= \frac{1}{s-a}$$

Generally it is unnecessary to actually obtain a function's Laplace transform by use of equation 1.290. Tables of these transforms are readily available. A small collection of the most frequently required transforms is given at the end of this chapter.

The Laplace transform method can be used with any linear differential equation with constant coefficients. Assuming the dependent variable is $x$, the basic procedure is as follows:

*step 1*: Put the differential equation in standard form.

*step 2*: Use superposition and take the Laplace transform of each term.

*step 3*: Use the following relationships to expand terms.

$$\mathcal{L}(x'') = s^2\mathcal{L}(x) - sx_0 - x_0' \quad 1.291$$

$$\mathcal{L}(x') = s\mathcal{L}(x) - x_0 \quad 1.292$$

*step 4*: Solve for $\mathcal{L}(x)$. Simplify the resulting expression using partial fractions.

*step 5*: Find $x$ by applying the inverse transform.

This method reduces the solutions of differential equations to simple algebra. However, a complete set of transforms is required.

Working with Laplace transforms is simplified by the following two theorems:

*Linearity Theorem*: If $c$ is constant, then

$$\mathcal{L}[cf(t)] = c\mathcal{L}[f(t)] \quad 1.293$$

*Superposition Theorem*: If $f(t)$ and $g(t)$ are different functions, then

$$\mathcal{L}[f(t) \pm g(t)] = \mathcal{L}[f(t)] \pm \mathcal{L}[g(t)] \quad 1.294$$

*Example 1.58*

Suppose the following differential equation results from the analysis of a mechanical system:

$$x'' + 2x' + 2x = \cos(t)$$

$$x_0 = 1, \; x_0' = 0$$

$x$ is the dependent variable. Start by taking the Laplace transform of both sides:

$$\mathcal{L}(x'') + 2\mathcal{L}(x') + 2\mathcal{L}(x) = \mathcal{L}(\cos(t))$$

$$s^2\mathcal{L}(x) - sx_0 - x_0' + 2s\mathcal{L}(x) - 2x_0 + 2\mathcal{L}(x) = \mathcal{L}\cos(t)$$

But, $x_0 = 1$ and $x_0' = 0$. Also, the Laplace transform of $\cos(t)$ can be found from the appendix of this chapter.

$$s^2\mathcal{L}(x) - s + 2s\mathcal{L}(x) - 2 + 2\mathcal{L}(x) = \frac{s}{s^2+1}$$

$$\mathcal{L}(x)[s^2 + 2s + 2] - s - 2 = \frac{s}{s^2+1}$$

$$\mathcal{L}(x) = \frac{s^3 + 2s^2 + 2s + 2}{(s^2+1)(s^2+2s+2)}$$

This is now expanded by partial fractions:

$$\begin{aligned} &\frac{s^3 + 2s^2 + 2s + 2}{(s^2+1)(s^2+2s+2)} \\ &= \frac{A_1 s + B_1}{s^2+1} + \frac{A_2 s + B_2}{s^2+2s+2} \\ &= [s^3(A_1 + A_2) + s^2(2A_1 + B_1 + B_2) \\ &\quad + s(2A_1 + 2B_1 + A_2) + 2B_1 + B_2] \\ &\quad \div [(s^2+1)(s^2+2s+2)] \end{aligned}$$

The following simultaneous equations result:

$$\begin{array}{lllllllll} A_1 & + & A_2 & & & & & = & 1 \\ 2A_1 & & & + & B_1 & + & B_2 & = & 2 \\ 2A_1 & + & A_2 & + & 2B_1 & & & = & 2 \\ & & & & 2B_1 & + & B_2 & = & 2 \end{array}$$

These equations have the solutions

$$A_1^* = \frac{1}{5} \quad A_2^* = \frac{4}{5}$$

$$B_1^* = \frac{2}{5} \quad B_2^* = \frac{6}{5}$$

Therefore, $x$ can be found by taking the following inverse transform:

$$x = \mathcal{L}^{-1}\left[\frac{\frac{s}{5} + \frac{2}{5}}{s^2+1} + \frac{\frac{4s}{5} + \frac{6}{5}}{s^2+2s+2}\right]$$

The solution is

$$x = \frac{1}{5}\cos(t) + \frac{2}{5}\sin(t) + \frac{4}{5}e^{-t}\cos(t) + \frac{2}{5}e^{-t}\sin(t)$$

## 20 APPLICATIONS OF DIFFERENTIAL EQUATIONS

### A. FLUID MIXTURE PROBLEMS

The typical fluid mixing problem involves a tank containing some liquid. There may be an initial solute in the liquid, or the liquid may be pure. Liquid and solute are added at known rates. A drain usually removes some of the liquid which is assumed to be thoroughly mixed. The problem is to find the weight or concentration of solute in the tank at some time $t$. The following symbols are used.

$C(t)$ concentration of solute in tank at time $t$
$I(t)$ liquid inflow rate from all sources at time $t$
$k$ a constant
$\Phi(t)$ liquid outflow rate due to all drains at time $t$

$S_1(t)$ solute inflow rate at time $t$ (this may have to be calculated from the incoming concentration and $I(t)$)
$S_2(t)$ solute outflow rate at time $t$
$V_o$ original volume of tank at time = 0
$V(t)$ volume of tank at time = $t$ (equal to $V_o + \int I(t)\,dt - \int \Phi(t)\,dt$)
$W_o$ initial weight of solute in tank at $t = 0$
$W(t)$ weight of solute in tank at time = $t$

*Case 1*: Constant Volume

$$\begin{aligned} W'(t) &= S_1(t) - S_2(t) \\ &= S_1(t) - \frac{\Phi(t)W(t)}{V_o} \end{aligned} \qquad 1.295$$

The differential equation is

$$W'(t) + \frac{\Phi W(t)}{V_o} = S_1(t) \qquad 1.296$$

This is a first order linear equation because $\Phi$, $V_o$, and $S_1$ are constants.

*Case 2*: Changing Volume

$$W'(t) = S_1(t) - S_2(t) \qquad 1.297$$

$$= S_1(t) - \frac{\Phi(t)W(t)}{V(t)} \qquad 1.298$$

The differential equation is

$$W'(t) + \frac{\Phi(t)W(t)}{V(t)} = S_1(t) \qquad 1.299$$

*Example 1.59*

A tank contains 100 gallons of pure water at the beginning of an experiment. 1 gpm of pure water flows into the tank, as does 1 gpm of water containing $\frac{1}{4}$ pound of salt per gallon. A perfectly mixed solution drains from the tank at the rate of 2 gpm. How much salt is in the tank 8 minutes after the experiment has begun?

Choose $W(t)$ as the variable giving the weight of salt in the tank at time $t$. $\frac{1}{4}$ pound of salt enters the tank per minute. What goes out depends on the concentration in the tank. Specifically, the leaving salt is

$$\begin{aligned} \text{salt leaving} &= (2\ \text{gpm})(\#\ \text{lbs salt per gallon}) \\ &= (2\ \text{gpm})\left(\frac{\#\ \text{lbs salt total}}{100}\right) = 0.02W(t) \end{aligned}$$

The difference between the inflow and the outflow is given by equation 1.295.

$$W'(t) = \frac{1}{4} - 0.02W(t)$$

This is a first order linear differential equation. It can be solved using the integrating factor (equation 1.283), simple constant coefficient methods, or Laplace transforms. The solution is

$$W(t) = 12.5 - 12.5e^{-0.02t}$$

At $t = 8$, $W(t) = 1.85$ lbs.

B. DECAY PROBLEMS

A given quantity is known to decrease at a rate proportional to the amount present. The original amount is known, and the amount at some time $t$ is desired.

$k$ a negative proportionality constant
$Q_o$ original amount present
$Q(t)$ amount present at time t
$t$ time
$t_{1/2}$ half-life

The differential equation is

$$Q'(t) = kQ(t) \qquad 1.300$$

The solution is

$$Q(t) = Q_o e^{kt} \qquad 1.301$$

If $Q^*$ is known for some time $t^*$, $k$ can be found from

$$k = \left(\frac{1}{t^*}\right) ln \left(\frac{Q^*}{Q_o}\right) \qquad 1.302$$

k also can be found from the half-life:

$$k = \frac{-0.693}{t_{1/2}} \qquad 1.303$$

C. SURFACE TEMPERATURE

$k$ a constant
$t$ time
$T$ absolute temperature of the surface
$T_o$ ambient temperature

Assuming that the surface temperature changes at a rate proportional to the difference in surface and ambient temperatures, the differential equation is

$$\frac{dT}{dt} = k(T - T_o) \qquad 1.304$$

Equation 1.304 is known as *Newton's Law of Cooling.*

D. SURFACE EVAPORATION

| | |
|---|---|
| $A$ | exposed surface area |
| $k$ | proportionality constant |
| $r$ | radius |
| $s$ | side length |
| $t$ | time |
| $V$ | object volume |

The equation is

$$\frac{dV}{dt} = -kA \tag{1.305}$$

For a spherical drop, this reduces to

$$\frac{dr}{dt} = -k \tag{1.306}$$

For a cube, this reduces to

$$\frac{ds}{dt} = -2k \tag{1.307}$$

## 21 FOURIER ANALYSIS

Any periodic waveform can be written as the sum of an infinite number of sinusoidal terms. In practice, it is possible to obtain a close approximation to the original waveform with a limited number of sinusoidal terms since most series converge rapidly.

Fourier's theorem is given in equation 1.308. The object of Fourier analysis is to determine the coefficients $a_n$ and $b_n$.

$$\begin{aligned} f(t) = a_o &+ a_1 \cos\omega_o t + a_2 \cos 2\omega_o t + \cdots \\ &+ b_1 \sin\omega_o t + b_2 \sin 2\omega_o t + \cdots \end{aligned} \tag{1.308}$$

$\omega_o$ is known as the *fundamental frequency* of the waveform. It depends on the actual waveform period.

$$\omega_o = \frac{2\pi}{T} \tag{1.309}$$

To simplify the analysis, the time domain can be normalized to the radian scale. The normalized scale is obtained by dividing all frequencies by $\omega_o$. Then, the Fourier series becomes

$$\begin{aligned} f(t) = a_o &+ a_1 \cos t + a_2 \cos 2t + \cdots \\ &+ b_1 \sin t + b_2 \sin 2t + \cdots \end{aligned} \tag{1.310}$$

The coefficients $a_n$ and $b_n$ can be found from the following relationships:

$$a_o = \frac{1}{2\pi}\int_o^{2\pi} f(t)\,dt \tag{1.311}$$

$$a_n = \frac{1}{\pi}\int_o^{2\pi} f(t)\cos\ nt\ dt \tag{1.312}$$

$$b_n = \frac{1}{\pi}\int_o^{2\pi} f(t)\sin\ nt\ dt \tag{1.313}$$

Notice that $a_o$ is the average value of the function. Usually, this average value can be determined by observation without having to go through the integration process. The equation for $a_n$ cannot be used to find $a_o$.

*Example 1.60*

Find the Fourier series for

$$f(t) = \left\{ \begin{matrix} 1 & 0 < t < \pi \\ 0 & \pi < t < 2\pi \end{matrix} \right\}$$

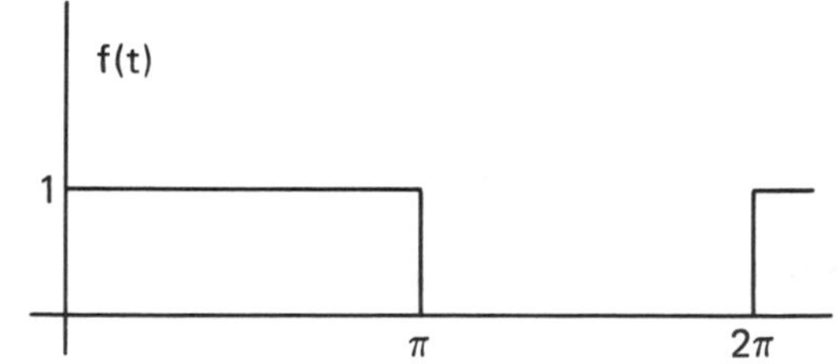


$$a_o = \frac{1}{2\pi}\int_o^{\pi}(1)dt + \frac{1}{2\pi}\int_\pi^{2\pi}(0)dt = \frac{1}{2}$$

This value of $\frac{1}{2}$ corresponds to the average value of $f(t)$. It could have been found by observation.

$$\begin{aligned} a_1 &= \frac{1}{\pi}\int_o^{\pi}(1)\cos t\ dt + \frac{1}{\pi}\int_\pi^{2\pi}(0)\cos t\ dt \\ &= \frac{1}{\pi}\left[\sin t\right]_o^\pi + 0 = 0 \end{aligned}$$

In general,

$$a_n = \frac{1}{\pi}\left[\frac{\sin\ nt}{n}\right]_o^\pi = 0$$

$$\begin{aligned} b_1 &= \frac{1}{\pi}\int_o^{\pi}(1)\sin t\ dt + \frac{1}{\pi}\int_\pi^{2\pi}(0)\sin t\ dt \\ &= \frac{1}{\pi}\left[-\cos t\right]_o^\pi = \frac{2}{\pi} \end{aligned}$$

In general,

$$b_n = \frac{1}{\pi}\left[\frac{-\cos\ nt}{n}\right]_0^\pi = \left\{ \begin{matrix} 0 \text{ for n even} \\ \frac{2}{\pi n} \text{ for n odd} \end{matrix} \right\}$$

The series is

$$f(t) = \frac{1}{2} + \frac{2}{\pi}\left[\sin t + \frac{1}{3}\sin\ 3t + \frac{1}{5}\sin\ 5t + \cdots\right]$$

The sum of the first few terms is illustrated.

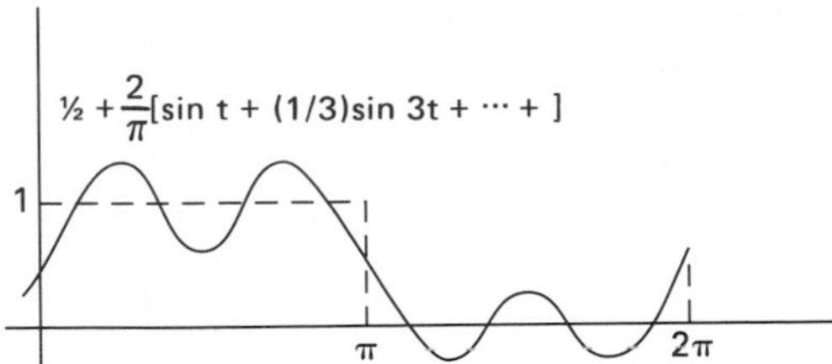


It may be possible to eliminate some of the $a_n$ or $b_n$ coefficients if the function $f(t)$ is symmetrical. There are four types of *symmetry.*

A function is said to have *even symmetry* if $f(t) = f(-t)$. The cosine is an example of this type of waveform. Even symmetry can be detected from the graph of the function. The function to the left of $t = 0$ is a reflection of the function to the right of $t = 0$. With even symmetry, all $b_n$ terms are zero.

A function is said to have *odd symmetry* if $f(t) = -f(-t)$. The sine is an example of this type of waveform. With odd symmetry, all $a_n$ terms are zero (but not necessarily $a_o$).

A function is said to have *rotational symmetry* or *half-wave symmetry* if $f(t) = -f(t + \pi)$. Functions of this type are identical on alternate $\frac{1}{2}$-cycles, except for a sign reversal. All $a_n$ and $b_n$ are zero for even values of $n$.

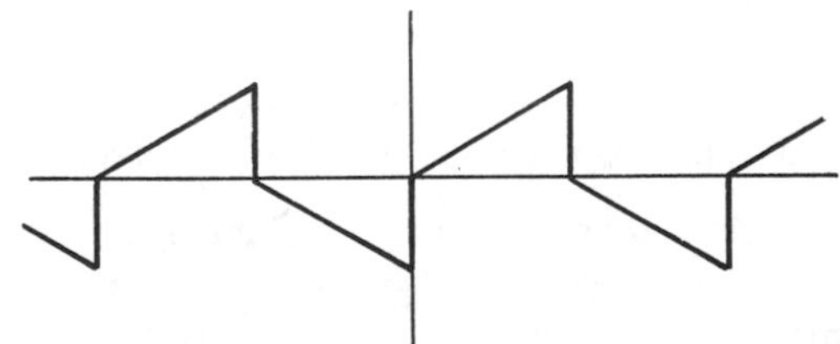

**Figure 1.12** Rotational Symmetry

These types of symmetry are not mutually exclusive. For example, it is possible for a function with rotational symmetry to have either odd or even symmetry also. Such a case is known as *quarter-wave symmetry.*

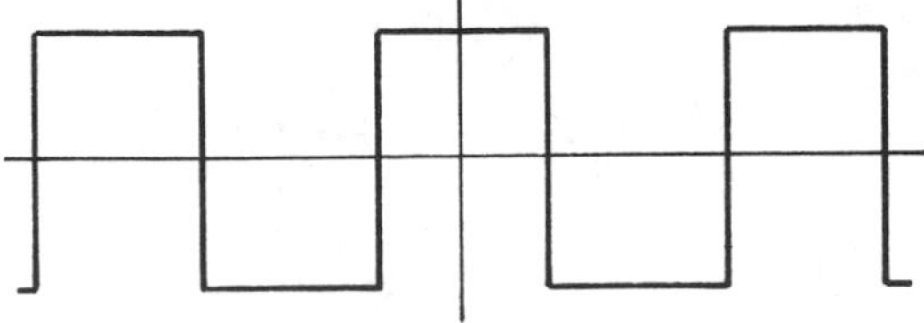

**Figure 1.13** Quarter-Wave Symmetry

## 22 CRITICAL PATH TECHNIQUES

### A. INTRODUCTION

Critical path techniques are used to represent graphically the multiple relationships between stages in complicated projects. The graphical networks show the dependencies or *precedence relationships* between the various activities and can be used to control and monitor the progress, cost, and resources of projects. Critical path techniques will also identify the most critical activities in projects.

### B. DEFINITIONS

Activity: Any subdivision of a project whose execution requires time and other resources.

Critical path: A path connecting all activities which have minimum or zero slack times. The critical path is the longest path through the network.

Duration: The time required to perform an activity. All durations are *normal durations* unless otherwise referred to as *crash durations.*

Event: The beginning or completion of an activity.

Event time: Actual time at which an event occurs.

Float: Same as slack time.

Slack time: The maximum time that an activity can be delayed without causing the project to fall behind schedule. Slack time is always minimum or zero along the critical path.

Critical path techniques use directed graphs to represent a project. These graphs are made up of arcs (arrows) and nodes (junctions). The placement of the arcs and nodes completely specifies the precedences of the project. Durations and precedences usually are given in a *precedence table* or matrix.

One specific technique is known as the *Critical Path Method, CPM.* This deterministic method is applicable when all activity durations are known in advance. CPM usually is represented as an *activity-on-node model.* Arcs are used to specify precedence, and the nodes actually represent the activities. Events are not present on the graph, other than as the heads and tails of the arcs. Two dummy nodes taking zero time can be used to specify the start and finish of the project.

*Example 1.61*

Given the project listed in the precedence table, construct the precedence matrix and draw an activity-on-node-network.

| Activity | Time (days) | Predecessors |
|---|---|---|
| A, start | 0 | – |
| B | 7 | A |
| C | 6 | A |
| D | 3 | B |
| E | 9 | B,C |
| F | 1 | D,E |
| G | 4 | C |
| H, finish | 0 | F,G |

The precedence matrix is given below.

| predecessor \ successor | A | B | C | D | E | F | G | H |
|---|---|---|---|---|---|---|---|---|
| A | | X | X | | | | | |
| B | | | | X | X | | | |
| C | | | | | X | | X | |
| D | | | | | | X | | |
| E | | | | | | X | | |
| F | | | | | | | | X |
| G | | | | | | | | X |
| H | | | | | | | | |

The activity-on-node-network is shown.

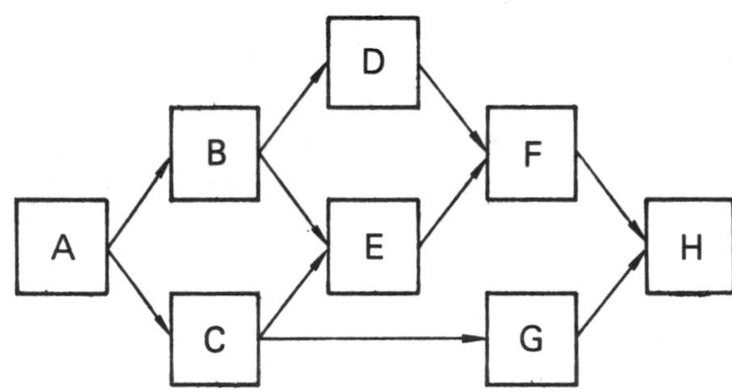


## C. SOLVING A CPM PROBLEM

The solution of a critical path problem results in a knowledge of the earliest and latest times that an activity can be started and finished. It also identifies the critical path and generates the slack time for each activity.

To facilitate the solution method, each node should be replaced by a square which has been quartered. The compartments have the meanings indicated by the following key.

| ES | EF |
|---|---|
| LS | LF |

Key
ES: Earliest Start
EF: Earliest Finish
LS: Latest Start
LF: Latest Finish

The following procedure will find the earliest and latest starts and finishes of each node.

1. Place the project start time or date in the ES and EF positions of the start activity. The start time is zero for relative calculations.
2. Consider any unmarked activity whose predecessors have all been marked in the EF and ES positions. (Go to step 4 if there is none.) Mark in its ES position the largest number marked in the EF position of those predecessors.
3. Add the activity time to the ES time and write this in the EF box. Return to step 2.
4. Place the value of the latest finish date in the LS and LF boxes of the finish node.
5. Consider any unmarked activity whose successors have all been marked in the LS and LF positions. The LF is the smallest LS of the successors. Go to step 7 if there are no unmarked activities.
6. The LS for the new node is LF minus its activity time. Return to step 5.
7. The slack for each node is (LS-ES) or (LF-EF).
8. The critical path encompasses nodes for which the slack equals (LS-ES) from the start node. There may be more than one critical path.

*Example 1.62*

Complete the network for the previous example and find the critical path. Assume the desired completion date is in 19 days.

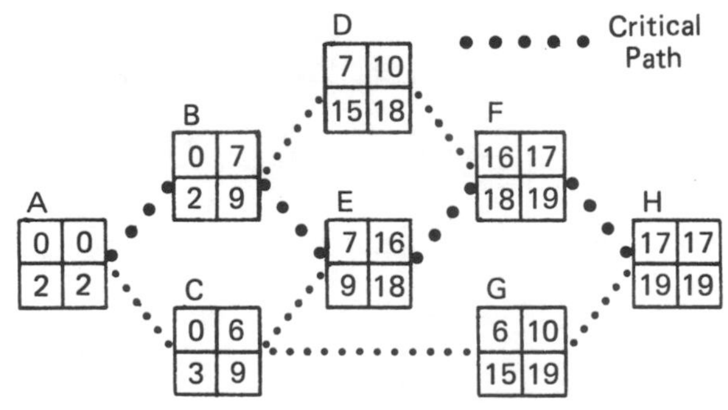

## D. PROBABILISTIC CRITICAL PATH MODELS

Probabilistic networks differ from deterministic networks only in the way in which the activity durations are found. Whereas durations are known explicitly for a deterministic network, the time for a probabilistic activity is distributed as a random variable.

This variable nature complicates the problem greatly since the actual distribution of times often is unknown. For this reason, such a problem usually is solved as a deterministic model using the mean of the duration distribution as the activity duration.

The most common probabilistic critical path model is *PERT*, which stands for *Program Evaluation and Review Technique*. In PERT, all duration variables are assumed to come from a beta distribution, with mean and standard deviation given as:

$$t_{\text{mean}} = \left(\frac{1}{6}\right)(t_{\text{minimum}} + 4t_{\text{most likely}} + t_{\text{maximum}}) \quad 1.314$$

$$\sigma = \frac{1}{6}(t_{\text{maximum}} - t_{\text{minimum}}) \quad 1.315$$

The project completion time for large projects is assumed to be distributed normally with mean ($\mu$) equal to the critical path length and overall variance ($\sigma^2$) equal to the sum of the variances along the critical path.

If necessary, the probability that a project duration will exceed some length ($D$) can be found from the normal table and the following relationship:

$$P\{\text{duration} > D\} = p\{x > z\} \quad 1.316$$

where $z$ is the standard normal variable equal to

$$z = \frac{D - \mu}{\sigma} \quad 1.317$$

*Example 1.63*

Determine the probability that the project illustrated will be completed on or before 19 days.

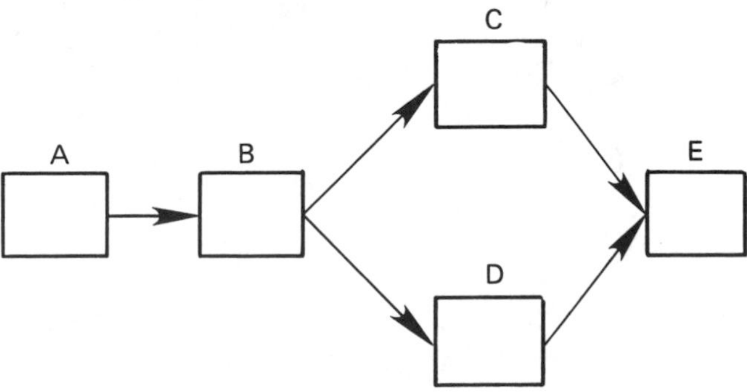


| event | $t_{\text{min}}$ | $t_{\text{likely}}$ | $t_{\text{max}}$ |
|---|---|---|---|
| A | 2 days | 3 days | 5 days |
| B | 4 | 6 | 7 |
| C | 1 | 3 | 5 |
| D | 2 | 3 | 7 |
| E | 3 | 5 | 10 |

Start by calculating the mean time of completion and the variance for each event. Using equations 1.314 and 1.315 for event A,

$$t_{A,\text{mean}} = \frac{2 + 4 \times 3 + 5}{6} = 3.2$$

$$\sigma_A^2 = \left[\frac{5-2}{6}\right]^2 = 0.25$$

The remaining times and variances are similarly calculated.

| event | $t_{\text{mean}}$ | $\sigma^2$ |
|---|---|---|
| A | 3.2 | 0.25 |
| B | 5.8 | 0.25 |
| C | 3.0 | 0.44 |
| D | 3.5 | 0.69 |
| E | 5.5 | 1.36 |

Since the mean time for event D is greater than the mean time for event C, the critical path is the sequence of events A-B-D-E. The sums of mean times and variances along the critical path are

$$\mu = 3.2 + 5.8 + 3.5 + 5.5 = 18$$

$$\sigma^2 = 0.25 + 0.25 + 0.69 + 1.36 = 2.55$$

The project has a 50% probability of being completed within 18 days, since 18 is the distribution mean. To calculate the probability of being completed in 19 days, the standard normal variable is calculated from equation 1.317.

$$z = \frac{19 - 18}{\sqrt{2.55}} = 0.63$$

The area under the standard normal curve corresponding to $z = 0.63$ is 0.2357. Therefore, the probability of finishing the project on or before 19 days is $0.5 + 0.2357 = 0.7357$ (73.6%).

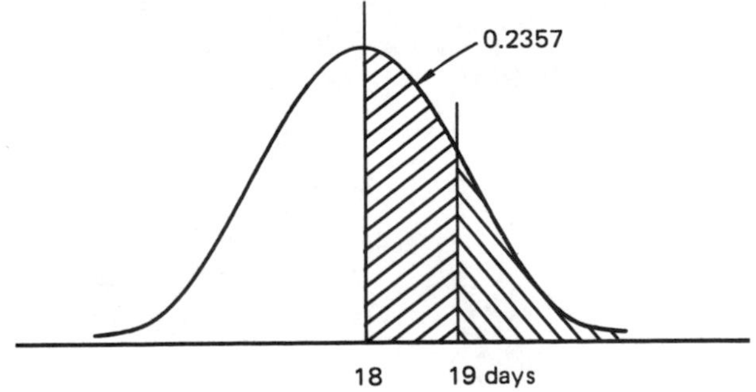

# Appendix A: Laplace Transforms

| f(t) | $\mathscr{L}[f(t)]$ |
|---|---|
| Unit impulse at t=0 | 1 |
| Unit impulse at t=c | $e^{-cs}$ |
| Unit step at t=0 | (1/s) |
| Unit step at t=c | $\frac{e^{-cs}}{s}$ |
| t | $\frac{1}{s^2}$ |
| $\frac{t^{n-1}}{(n-1)!}$ | $\frac{1}{s^n}$ |
| sinAt | $\frac{A}{s^2+A^2}$ |
| At − sinAt | $\frac{A^3}{s^2(s^2+A^2)}$ |
| sinh(At) | $\frac{A}{s^2-A^2}$ |
| t sinAt | $\frac{2As}{(s^2+A^2)^2}$ |
| cosAt | $\frac{s}{s^2+A^2}$ |
| 1 − cosAt | $\frac{A^2}{s(s^2+A^2)}$ |
| cosh(At) | $\frac{s}{s^2-A^2}$ |
| t cosAt | $\frac{s^2-A^2}{(s^2+A^2)^2}$ |
| $t^n$ (n is a positive integer) | $\frac{n!}{s^{(n+1)}}$ |
| $e^{At}$ | $\frac{1}{s-A}$ |
| $e^{At}\sin Bt$ | $\frac{B}{(s-A)^2+B^2}$ |
| $e^{At}\cos Bt$ | $\frac{s-A}{(s-A)^2+B^2}$ |
| $e^{At}t^n$ (n is positive integer) | $\frac{n!}{(s-A)^{n+1}}$ |
| $1 - e^{-At}$ | $\frac{A}{s(s+A)}$ |
| $e^{-At} + At - 1$ | $\frac{A^2}{s^2(s+A)}$ |
| $\frac{e^{-At} - e^{-Bt}}{B-A}$ | $\frac{1}{(s+A)(s+B)}$ |
| $\frac{(C-A)e^{-At} - (C-B)e^{-Bt}}{B-A}$ | $\frac{s + C}{(s+A)(s+B)}$ |
| $\frac{1}{AB} + \frac{Be^{-At} - Ae^{-Bt}}{AB(A-B)}$ | $\frac{1}{s(s+A)(s+B)}$ |

# Appendix B: Conversion Factors

| To Convert | Into | Multiply by |
|---|---|---|
| Acres | hectares | 0.4047 |
| Acres | square feet | 43,560.0 |
| Acres | square miles | 1.562 EE-3 |
| Ampere hours | coulombs | 3,600.0 |
| Angstrom units | inches | 3.937 EE-9 |
| Angstrom units | microns | 1 EE-4 |
| Astronomical units | kilometers | 1.495 EE8 |
| Atmospheres | cms of mercury | 76.0 |
| BTU's | horsepower-hrs | 3.931 EE-4 |
| BTU's | kilowatt-hrs | 2.928 EE-4 |
| BTU/hr | watts | 0.2931 |
| Bushels | cubic inches | 2,150.4 |
| Calories, gram (mean) | BTU (mean) | 3.9685 EE-3 |
| Centares | square meters | 1.0 |
| Centimeters | kilometers | 1 EE-5 |
| Centimeters | meters | 1 EE-2 |
| Centimeters | millimeters | 10.0 |
| Centimeters | feet | 3.281 EE-2 |
| Centimeters | inches | 0.3937 |
| Chains | inches | 792.0 |
| Coulombs | faradays | 1.036 EE-5 |
| Cubic centimeters | cubic inches | 0.06102 |
| Cubic centimeters | pints (U.S. liq.) | 2.113 EE-3 |
| Cubic feet | cubic meters | 0.02832 |
| Cubic feet/min. | pounds water/min. | 62.43 |
| Cubic feet/sec. | gallons/min. | 448.831 |
| Cubits | inches | 18.0 |
| Days | seconds | 86,400.0 |
| Degrees (angle) | radians | 1.745 EE-2 |
| Degrees/sec. | revolutions/min. | 0.1667 |
| Dynes | grams | 1.020 EE-3 |
| Dynes | joules/meter (newtons) | 1 EE-5 |
| Ells | inches | 45.0 |
| Ergs | BTU's | 9.480 EE-11 |
| Ergs | foot-pounds | 7.3670 EE-8 |
| Ergs | kilowatt-hours | 2.778 EE-14 |
| Faradays/sec. | amperes (absolute) | 96,500 |
| Fathoms | feet | 6.0 |
| Feet | centimeters | 30.48 |
| Feet | meters | 0.3048 |
| Feet | miles (nautical) | 1.645 EE-4 |
| Feet | miles (statute) | 1.894 EE-4 |
| Feet/min. | centimeters/sec. | 0.5080 |
| Feet/sec. | knots | 0.5921 |
| Feet/sec. | miles/hour | 0.6818 |
| Foot-pounds | BTU's | 1.286 EE-3 |
| Foot-pounds | kilowatt-hours | 3.766 EE-7 |
| Furlongs | miles (U.S.) | 0.125 |
| Furlongs | feet | 660.0 |
| Gallons | liters | 3.785 |
| Gallons of water | pounds of water | 8.3453 |
| Gallons/min. | cubic feet/hour | 8.0208 |
| Grams | ounces (avoirdupois) | 3.527 EE-2 |
| Grams | ounces (troy) | 3.215 EE-2 |
| Grams | pounds | 2.205 EE-3 |
| Hectares | acres | 2.471 |
| Hectares | square feet | 1.076 EE5 |
| Horsepower | BTU/min. | 42.42 |
| Horsepower | kilowatts | 0.7457 |
| Horsepower | watts | 745.7 |
| Hours | days | 4.167 EE-2 |
| Hours | weeks | 5.952 EE-3 |
| Inches | centimeters | 2.540 |
| Inches | miles | 1.578 EE-5 |
| Joules | BTU's | 9.480 EE-4 |
| Joules | ergs | 1 EE7 |
| Kilograms | pounds | 2.205 |
| Kilometers | feet | 3,281.0 |
| Kilometers | meters | 1,000.0 |
| Kilometers | miles | 0.6214 |
| Kilometers/hr. | knots | 0.5396 |
| Kilowatts | horsepower | 1.341 |
| Kilowatt-hours | BTU'S | 3,413.0 |
| Knots | feet/hour | 6,080.0 |
| Knots | nautical miles/hr. | 1.0 |
| Knots | statute miles/hr. | 1.151 |
| Light years | miles | 5.9 EE12 |
| Links (surveyor's) | inches | 7.92 |
| Liters | cubic centimeters | 1,000.0 |
| Liters | cubic inches | 61.02 |
| Liters | gallons (U.S. liq.) | 0.2642 |
| Liters | milliliters | 1,000.0 |
| Liters | pints (U.S. liq.) | 2.113 |
| Meters | centimeters | 100.0 |
| Meters | feet | 3.281 |
| Meters | kilometers | 1 EE-3 |
| Meters | miles (nautical) | 5.396 EE-4 |
| Meters | miles (statute) | 6.214 EE-4 |
| Meters | millimeters | 1,000.0 |
| Microns | meters | 1 EE-6 |
| Miles (nautical) | feet | 6,080.27 |
| Miles (statute) | feet | 5,280.0 |
| Miles (nautical) | kilometers | 1.853 |
| Miles (statute) | kilometers | 1.609 |
| Miles (nautical) | miles (statute) | 1.1516 |
| Miles (statute) | miles (nautical) | 0.8684 |
| Miles/hour | feet/min. | 88.0 |
| Milligrams/liter | parts/million | 1.0 |
| Milliliters | liters | 1 EE-3 |
| Millimeters | inches | 3.937 EE-2 |
| Newtons | dynes | 1 EE5 |
| Ohms (international) | ohms (absolute) | 1.0005 |
| Ounces | grams | 28.349527 |
| Ounces | pounds | 6.25 EE-2 |
| Ounces (troy) | ounces (avoirdupois) | 1.09714 |
| Parsecs | miles | 19 EE12 |
| Parsecs | kilometers | 3.084 EE13 |
| Pints (liq.) | cubic centimeters | 473.2 |
| Pints (liq.) | cubic inches | 28.87 |
| Pints (liq.) | gallons | 0.125 |
| Pints (liq.) | quarts (liq.) | 0.5 |
| Pounds | kilograms | 0.4536 |
| Pounds | ounces | 16.0 |
| Pounds | ounces (troy) | 14.5833 |
| Pounds | pounds (troy) | 1.21528 |
| Quarts (dry) | cubic inches | 67.20 |

| To Convert | Into | Multiply by |
|---|---|---|
| Quarts (liq.) | cubic inches | 57.75 |
| Quarts (liq.) | gallons | 0.25 |
| Quarts (liq.) | liters | 0.9463 |
| Radians | degrees | 57.30 |
| Radians | minutes | 3,438.0 |
| Revolutions | degrees | 360.0 |
| Revolutions/min. | degrees/sec. | 6.0 |
| Rods | meters | 5.029 |
| Rods | feet | 16.5 |
| Rods (surveyor's measure) | yards | 5.5 |
| Seconds | minutes | 1.667 EE − 2 |
| Slugs | pounds | 32.17 |
| Tons (long) | kilograms | 1,016.0 |
| Tons (short) | kilograms | 907.1848 |
| Tons (long) | pounds | 2,240.0 |
| Tons (short) | pounds | 2,000.0 |
| Tons (long) | tons (short) | 1.120 |
| Tons (short) | tons (long) | 0.89287 |
| Volt (absolute) | statvolts | 3.336 EE − 3 |
| Watts | BTU/hour | 3.4129 |
| Watts | horsepower | 1.341 EE − 3 |
| Yards | meters | 0.9144 |
| Yards | miles (nautical) | 4.934 EE − 4 |
| Yards | miles (statute) | 5.682 EE − 4 |

# Appendix C: Computational Values of Fundamental Constants

| Constant | SI | English |
|---|---|---|
| charge on electron | −1.602 EE−19 C | |
| charge on proton | +1.602 EE−19 C | |
| atomic mass unit | 1.66 EE−27 kg | |
| electron rest mass | 9.11 EE−31 kg | |
| proton rest mass | 1.673 EE−27 kg | |
| neutron rest mass | 1.675 EE−27 kg | |
| earth weight | | 1.32 EE25 lb |
| earth mass | 6.00 EE24 kg | 4.11 EE23 slug |
| mean earth radius | 6.37 EE3 km | 2.09 EE7 ft |
| mean earth density | 5.52 EE3 kg/m$^3$ | 345 lbm/ft$^3$ |
| earth escape velocity | 1.12 EE4 m/s | 3.67 EE4 ft/sec |
| distance from sun | 1.49 EE11 m | 4.89 EE11 ft |
| Boltzmann constant | 1.381 EE−23 J/°K | 5.65 EE−24 $\frac{\text{ft}-\text{lbf}}{°\text{R}}$ |
| permeability of a vacuum | 1.257 EE−6 H/m | |
| permittivity of a vacuum | 8.854 EE−12 F/m | |
| Planck constant | 6.626 EE−34 J·s | |
| Avogadro's number | 6.022 EE23 molecules/gmole | 2.73 EE26 $\frac{\text{molecules}}{\text{pmole}}$ |
| Faraday's constant | 9.648 EE4 C/gmole | |
| Stefan-Boltzmann constant | 5.670 EE−8 W/m$^2$−K$^4$ | 1.71 EE−9 $\frac{\text{BTU}}{\text{ft}^2-\text{hr}-°\text{R}^4}$ |
| gravitational constant (G) | 6.672 EE−11 m$^3$/s$^2$−kg | 3.44 EE−8 $\frac{\text{ft}^4}{\text{lbf}-\text{sec}^4}$ |
| universal gas constant | 8.314 J/°K−gmole | 1545 $\frac{\text{ft}-\text{lbf}}{°\text{R}-\text{pmole}}$ |
| speed of light | 3.00 EE8 m/s | 9.84 EE8 ft/sec |
| speed of sound, air, STP | 3.31 EE2 m/s | 1.09 EE3 ft/sec |
| speed of sound, air, 70°F, one atmosphere | 3.44 EE2 m/s | 1.13 EE3 ft/sec |
| standard atmosphere | 1.013 EE5 N/m$^2$ | 14.7 psia |
| standard temperature | 0°C | 32°F |
| molar ideal gas volume (STP) | 22.4138 EE−3 m$^3$/gmole | 359 ft$^3$/pmole |
| standard water density | 1 EE3 kg/m$^3$ | 62.4 lbm/ft$^3$ |
| air density, STP | 1.29 kg/m$^3$ | 8.05 EE−2 lbm/ft$^3$ |
| air density, 70°F, 1 atm | 1.20 kg/m$^3$ | 7.49 EE−2 lbm/ft$^3$ |
| mercury density | 1.360 EE4 kg/m$^3$ | 8.49 EE2 lbm/ft$^3$ |
| gravity on moon | 1.67 m/s$^2$ | 5.47 ft/sec$^2$ |
| gravity on earth | 9.81 m/s$^2$ | 32.17 ft/sec$^2$ |

## Practice Problems: MATHEMATICS

Untimed

1. A state law requires a statistical analysis of the average speed driven by motorists on a road prior to the use of radar speed control. The following speeds were observed in a random sample of 40 cars: 44, 48, 26, 25, 20, 43, 40, 42, 29, 39, 23, 26, 24, 47, 45, 28, 29, 41, 38, 36, 27, 44, 42, 43, 29, 37, 34, 31, 33, 30, 42, 43, 28, 41, 29, 36, 35, 30, 32, 31 (all in mph).

(a) Tabulate the frequency distribution of the above data. (b) Draw the frequency histogram. (c) Draw the frequency polygon. (d) Tabulate the cumulative frequency distribution. (e) Draw the cumulative frequency graph. (f) What is the upper quartile speed? (g) What are the mode, median, and mean speeds? (h) What is the standard deviation of the sample data? (i) What is the sample standard deviation? (j) What is the sample variance?

2. Activities constituting a project are given. The project starts at time zero.

| activity | predecessors | successors | duration |
|---|---|---|---|
| start | – | A | 0 |
| A | start | B,C,D | 7 |
| B | A | G | 6 |
| C | A | E,F | 5 |
| D | A | G | 2 |
| E | C | H | 13 |
| F | C | H,I | 4 |
| G | D,B | I | 18 |
| H | E,F | finish | 7 |
| I | F,G | finish | 5 |
| finish | H,I | – | 0 |

(a) Draw the CPM network. (b) Indicate the critical path. (c) What is the earliest finish? (d) What is the latest finish? (e) What is the slack along the critical path? (f) What is the float along the critical path?

3. Activities constituting a short project are given.

| activity | predecessors | successors | $t_{min}$ | $t_{likely}$ | $t_{max}$ |
|---|---|---|---|---|---|
| start | – | A | 0 | 0 | 0 |
| A | start | B,D | 1 | 2 | 5 |
| B | A | C | 7 | 9 | 20 |
| C | B | D | 5 | 12 | 18 |
| D | A,C | finish | 2 | 4 | 7 |
| finish | D | – | 0 | 0 | 0 |

If the project starts at $t=15$, what is the probability that the project will be completed by $t=42$ or sooner?

4. A pipe with an inside diameter of 18.812″ contains fluid to a depth of 15.7″. What is the hydraulic radius?

5. What is the determinant of the following matrix?

$$\begin{bmatrix} 8 & 2 & 0 & 0 \\ 2 & 8 & 2 & 0 \\ 0 & 2 & 8 & 2 \\ 0 & 0 & 2 & 4 \end{bmatrix}$$

6. The number of cars entering a toll plaza on a bridge during the hour following midnight is distributed as Poisson with a mean of 20. What is the probability that 17 cars will pass through the toll plaza during that hour on any given night? What is the probability that 3 or fewer cars will pass through the toll plaza at that hour on any given night?

7. The time taken by a toll taker to collect the toll from vehicles crossing a bridge is an exponential distribution with mean of 23 seconds when a line of vehicles exists waiting to enter the toll booth. What is the probability that a random vehicle will be processed in 25 seconds or more (i.e., will take longer than 25 seconds)?

8. The average number of vehicles lining up behind a flashing railroad crossing has been observed for five trains of different lengths. What is the mathematical formula which relates the two variables?

| # cars in train | # vehicles |
|---|---|
| 2 | 14.8 |
| 5 | 18.0 |
| 8 | 20.4 |
| 12 | 23.0 |
| 27 | 29.9 |

9. Holes drilled in structural steel parts for bolts are normally distributed with a mean of 0.502″ and standard deviation of .005″. Holes are defective if their diameters are less than 0.497″ or more than 0.507″. (a) What is the probability that a hole chosen at random will be defective? (b) What is the probability that 2 holes out of a sample of 15 will be defective?

10. The oscillation exhibited by a certain 1-story building in free motion is given by the following differential equation:

$$x'' + 2x' + 2x = 0 \qquad x(0) = 0; \quad x'(0) = 1$$

(a) What is $x$ as a function of time? (b) What is the building's natural frequency of vibration? (c) What is the amplitude of oscillation? (d) What is $x$ as a function of time if a lateral wind load is applied with form of sin(t)?

11. Using the bisection method, find the root of the equation

$$x^3 + 2x^2 + 8x - 2 = 0$$

12. Consider the yield data from five treatment plants. Develop a mathematical equation to correlate the data.

| treatment plant | average yield | average temperature |
|---|---|---|
| 1 | 92.30 | 207.1 |
| 2 | 92.58 | 210.3 |
| 3 | 91.56 | 200.4 |
| 4 | 91.63 | 201.1 |
| 5 | 91.83 | 203.4 |

13. The following data is given from a waterflow experiment. What is the mathematical formula which relates the two variables?

| x | y |
|---|---|
| −1 | 0 |
| 0 | 1 |
| 1 | 1.4 |
| 2 | 1.7 |
| 3 | 2 |
| 4 | 2.2 |
| 5 | 2.4 |
| 6 | 2.6 |
| 7 | 2.8 |
| 8 | 3 |

14. Four recruits whose shoe sizes are 7, 8, 9, and 10 report to the supply clerk to be issued shoes. The supply clerk selects one pair of shoes in each of the four required sizes and hands them at random to the men. What is the probability that (a) no man will receive the correct size? (b) exactly 3 men will receive the correct size?

15. In an ammeter, two resistances are connected in parallel. Most of the current passing through the meter goes through the shunt. In order to determine the accuracy of the resistance of shunts being made for ammeters, a manufacturer tested a sample of 100 shunts. The resistance of each, to the nearest hundredth of an ohm, is indicated in the following data (number of shunts followed by resistance in ohms).

1–.200, 2–.210, 2–.290, 3–.280, 1–.210, 2–.220, 3–.230, 2–.270, 4–.260, 28–.250, 3–.220, 3–.230, 7–.240, 5–.260, 12–.250, 4–.230, 5–.240, 4–.260, 4–.270, 5–.240.

(a) Draw the frequency distribution. (b) Find the arithmetic mean. (c) Find the sample standard deviation. (d) Draw a frequency polygon. (e) Find the median. (f) Find the sample variance.

16. Listed is a set of activities and sequence requirements to start a warehouse construction project. Prepare a CPM project diagram.

| activity | letter code | code of immediate predecessor |
|---|---|---|
| move-in | A | |
| job layout | B | A |
| excavations | C | B |
| make-up forms | D | A |
| shop drawing, order rebar | E | A |
| erect forms | F | C,D |
| rough in plumbing | G | F |
| install rebars | H | E,F |
| pour, finish concrete | I | G,H |

17. Activities constituting a bridge construction project are listed. (a) Draw the CPM network showing the critical path. (b) Compute ES, EF, LS, and LF for each of the activities. Assume a target time for completing the project which, for the bridge is 3 days after the EF time.

| job number | | immediate predecessors | time (days) |
|---|---|---|---|
| a | Start | | 0 |
| b | | a | 4 |
| c | | b | 2 |
| d | | c | 4 |
| e | | d | 6 |
| f | | c | 1 |
| g | | f | 2 |
| h | | f | 3 |
| i | | d | 2 |
| j | | d,g | 4 |
| k | | i,j,h | 10 |
| l | | k | 3 |
| m | | l | 1 |
| n | | l | 2 |
| o | | l | 3 |
| p | | e | 2 |
| q | | p | 1 |
| r | | c | 1 |
| s | | o,t | 2 |
| t | | m,n | 3 |
| u | | t | 1 |
| v | | q,r | 2 |
| w | | v | 5 |
| x | Finish | s,u,w | 0 |

18. Prepare a PERT diagram for the activities and sequence requirements given in problem 17.

19. Listed is a set of activities, sequence requirements, and estimated activity times required for the renewal of a pipeline. (a) Prepare a PERT project diagram.

| activity | letter code | code of immediate predecessor | activity time requirement (days) |
|---|---|---|---|
| assemble crew for job | A | | 10 |
| use old line to build inventory | B | D | 28 |
| measure and sketch old line | C | A | 2 |
| develop materials list | D | C | 1 |
| erect scaffold | E | D | 2 |
| procure pipe | F | D | 30 |
| procure valves | G | D | 45 |
| deactivate old line | H | B | 1 |
| remove old line | I | E,H | 6 |
| prefabricate new pipe | J | F | 5 |
| place valves | K | I,G | 1 |
| place new pipe | L | I,J | 6 |
| weld pipe | M | L | 2 |
| connect valves | N | K,M | 1 |
| insulate | O | N | 4 |
| pressure test | P | N | 1 |
| remove scaffold | Q | O,P | 1 |
| clean up and turn over to operating crew | R | O,Q | 1 |

(b) There is additional information in the form of optimistic, most likely, and pessimistic time estimates for the project. Compute the expected mean time, $t_m$, and the variance, $\sigma^2$, for the activities. Which activities have the greatest uncertainty in their completion schedules?

| activity code | optimistic time ($t_{min}$) | most likely time ($t_{ml}$) | pessimistic time ($t_{max}$) |
|---|---|---|---|
| A | 8 | 10 | 12 |
| B | 26 | 26.5 | 36 |
| C | 1 | 2 | 3 |
| D | 0.5 | 1 | 1.5 |
| E | 1.5 | 1.63 | 4 |
| F | 28 | 28 | 40 |
| G | 40 | 42.5 | 60 |
| H | 1 | 1 | 1 |
| I | 4 | 6 | 8 |
| J | 4 | 4.5 | 8 |
| K | 0.5 | 0.9 | 2 |
| L | 5 | 5.25 | 10 |
| M | 1 | 2 | 3 |
| N | 0.5 | 1 | 1.5 |
| O | 3 | 3.75 | 6 |
| P | 1 | 1 | 1 |
| Q | 1 | 1 | 1 |
| R | 1 | 1 | 1 |

(c) Suppose that due to penalties in the contract each day the pipeline renewal project can be shortened is worth \$100. Which of the following possibilities would you follow and why?

- Shorten $t_m$ of activity B by 4 days at a cost of \$100.
- Shorten $t_{ml}$ of activity G by 5 days at a cost of \$50.
- Shorten $t_m$ of activity O by 2 days at a cost of \$150.
- Shorten $t_m$ of activity O by 2 days by drawing resources from activity N, thereby lengthening its $t_e$ by 2 days.

20. An 8% solution, a 10% solution, and a 20% solution of nitric acid are to be mixed in order to get 100 ml of a 12% solution. If the volume of acid from the 8% solution equals half the volume of acid from the other two solutions, how much of each is needed?

21. A tank contains 100 gallons of brine made by dissolving 60 pounds of salt in water. Salt water containing 1 pound of salt per gallon runs in at the rate of 2 gallons per minute. A well-stirred mixture runs out at the rate of 3 gallons per minute. Find the amount of salt in the tank at the end of 1 hour.

## Timed

1. A survey field crew measures one leg of a traverse four times. The following results are obtained:

| repetition | measurement | direction |
|---|---|---|
| 1 | 1249.529 | forward |
| 2 | 1249.494 | backward |
| 3 | 1249.384 | forward |
| 4 | 1249.348 | backward |

The crew chief is under orders to obtain readings with confidence limits of 90%. (a) Which readings are acceptable? (b) Which readings are not acceptable? (c) Explain how to determine which readings are not acceptable. (d) What is the most probable value of the distance? (e) What is the error in the most probable value (at 90% confidence)? (f) If the distance is one side of a square traverse whose sides are all equal, what is the most probable closure error? (g) What is the probable error of part (f) expressed as a fraction? (h) Is this error of closure within the second order of accuracy? (i) Define accuracy and distinguish it from precision. (j) Give an example of a systematic error.

2. A 90-pound bag of a chemical is accidentally dropped in an aerating lagoon. The chemical is water soluble and non-reacting. The lagoon is 120 feet in diameter and filled to a depth of 10 feet. The aerators circulate and distribute the chemical evenly throughout the lagoon.

Water enters the lagoon at the rate of 30 gallons per minute. Fully mixed water is pumped into a reservoir at the rate of 30 gpm.

The established safe concentration of this chemical is 1 ppb (part per billion). How long will it take (in days) for the concentration of the discharge water to reach this level?

# 2 ENGINEERING ECONOMIC ANALYSIS



*Nomenclature*

| | | |
|---|---|---|
| A | annual amount or annuity | \$ |
| B | present worth of all benefits | \$ |
| $BV_j$ | book value at the end of the *j*th year | \$ |
| C | cost, or present worth of all costs | \$ |
| d | declining balance depreciation rate | decimal |
| $D_j$ | depreciation in year *j* | \$ |
| D.R. | present worth of after-tax depreciation recovery | \$ |
| e | natural logarithm base (2.718) | – |
| EAA | equivalent annual amount | \$ |
| EUAC | equivalent uniform annual cost | \$ |
| f | federal income tax rate | decimal |
| F | future amount or future worth | \$ |
| G | uniform gradient amount | \$ |
| i | effective rate per period (usually per year) | decimal |
| k | number of compounding periods per year | – |
| n | number of compounding periods, or life of asset | – |
| P | present worth or present value | \$ |
| $P_t$ | present worth after taxes | \$ |
| ROR | rate of return | decimal |
| ROI | return on investment | \$ |
| r | nominal rate per year (rate per annum) | decimal |
| s | state income tax rate | decimal |
| $S_n$ | expected salvage value in year *n* | \$ |
| t | composite tax rate, or time | decimal,- |
| z | a factor equal to $\frac{1+i}{1-d}$ | decimal |
| $\phi$ | effective rate per period | decimal |

## 1 EQUIVALENCE

Industrial decision makers using engineering economics are concerned with the timing of a project's cash flows as well as with the total profitability of that project. In this situation, a method is required to compare projects involving receipts and disbursements occurring at different times.

By way of illustration, consider \$100 placed in a bank account which pays 5% effective annual interest at the end of each year. After the first year, the account will have grown to \$105. After the second year, the account will have grown to \$110.25.

Assume that you will have no need for money during the next two years and that any money received would immediately go into your 5% bank account. Then, which of the following options would be more desirable?

**option a:** \$100 now

**option b:** \$105 to be delivered in one year

**option c:** \$110.25 to be delivered in two years

In light of the previous illustration, none of the options is superior under the assumptions given. If the first option is chosen, you will immediately place \$100 into a 5% account, and in two years the account will have grown to \$110.25. In fact, the account will contain \$110.25 at the end of two years regardless of the option chosen. Therefore, these alternatives are said to be *equivalent*.

## 2 CASH FLOW DIAGRAMS

Although they are not always necessary in simple problems (and they are often unwieldy in very complex problems), *cash flow diagrams* may be drawn to help visualize and simplify problems having diverse receipts and disbursements.

The conventions below are used to standardize cash flow diagrams.

- The horizontal (time) axis is marked off in equal increments, one per period, up to the duration or horizon of the project.

- All disbursements and receipts (cash flows) are assumed to take place at the end of the year in which they occur. This is known as the *year-end convention.* The exception to the year-end convention is any initial cost (purchase cost) which occurs at $t = 0$.
- Two or more transfers in the same year are placed end-to-end, and these may be combined.
- Expenses incurred before $t = 0$ are called *sunk costs.* Sunk costs are not relevant to the problem.
- Receipts are represented by arrows directed upward. Disbursements are represented by arrows directed downward. The arrow length is proportional to the magnitude of the cash flow.

*Example 2.1*

A mechanical device will cost \$20,000 when purchased. Maintenance will cost \$1000 each year. The device will generate revenues of \$5000 each year for 5 years after which the salvage value is expected to be \$7000. Draw and simplify the cash flow diagram.

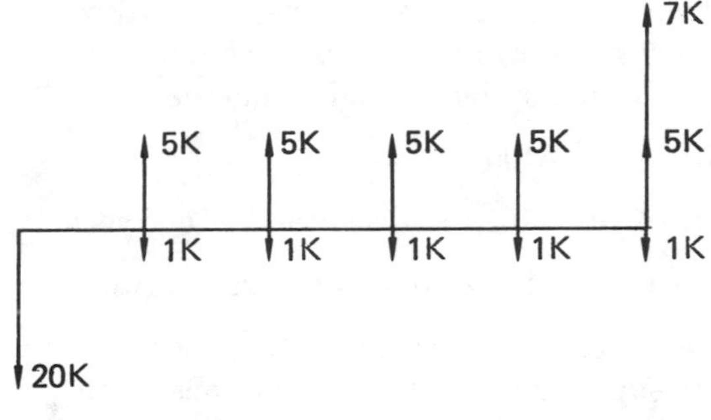


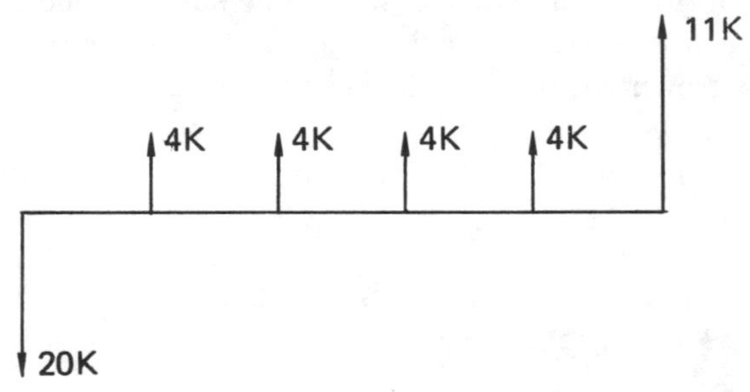


## 3 TYPICAL PROBLEM FORMAT

With the exception of some investment and rate of return problems, the typical problem involving engineering economics will have the following characteristics:

- An interest rate will be given.
- Two or more alternatives will be competing for funding.
- Each alternative will have its own cash flows.
- It is necessary to select the best alternative.

*Example 2.2*

Investment **A** costs \$10,000 today and pays back \$11,500 two years from now. Investment **B** costs \$8000 today and pays back \$4500 each year for two years. If an interest rate of 5% is used, which alternative is superior?

The solution to this example is not difficult, but it will be postponed until methods of calculating equivalence have been covered.

## 4 CALCULATING EQUIVALENCE

It was previously illustrated that \$100 now is equivalent at 5% to \$105 in one year. The equivalence of any present amount, $P$, at $t = 0$ to any future amount, $F$, at $t = n$ is called the *future worth* and can be calculated from equation 2.1.

$$F = P(1+i)^n \qquad 2.1$$

The factor $(1+i)^n$ is known as the *compound amount factor* and has been tabulated at the end of this chapter for various combinations of $i$ and $n$. Rather than actually writing the formula for the compound amount factor, the convention is to use the standard functional notation $(F/P, i\%, n)$. Thus,

$$F = P(F/P, i\%, n) \qquad 2.2$$

Similarly, the equivalence of any future amount to any present amount is called the *present worth* and can be calculated from

$$P = F(1+i)^{-n} = F(P/F, i\%, n) \qquad 2.3$$

The factor $(1+i)^{-n}$ is known as the *present worth factor*, with functional notation $(P/F, i\%, n)$. Tabulated values are also given for this factor at the end of this chapter.

*Example 2.3*

How much should you put into a 10% savings account in order to have \$10,000 in 5 years?

This problem could also be stated: What is the equivalent present worth of \$10,000 5 years from now if money is worth 10%?

$$P = F(1+i)^{-n} = 10{,}000(1+0.10)^{-5} = 6209$$

The factor 0.6209 would usually be obtained from the tables.

A cash flow which repeats regularly each year is known as an *annual amount.* When annual costs are incurred due to the functioning of a piece of equipment, they are often known as *operating and maintenance (O&M) costs.* The annual costs associated with operating a business in general are known as *general, selling, and administrative (GS&A) expenses.* Although the equivalent value for each of the $n$ annual amounts could be calculated and then summed, it is much easier to use one of the *uniform series factors*, as illustrated in example 2.4.

*Example 2.4*

Maintenance costs for a machine are $250 each year. What is the present worth of these maintenance costs over a 12 year period if the interest rate is 8% ?

Notice that

$$(P/A, 8\%, 12) = (P/F, 8\%, 1) + (P/F, 8\%, 2) + \cdots + (P/F, 8\%, 12)$$

Then,

$$P = A(P/A, i\%, n) = -250(7.5361) = -1884$$

A common complication involves a uniformly increasing cash flow. Such an increasing cash flow should be handled with the *uniform gradient factor*, $(P/G, i\%, n)$. The uniform gradient factor finds the present worth of a uniformly increasing cash flow which starts in year 2 (not year 1) as shown in example 2.5.

*Example 2.5*

Maintenance on an old machine is $100 this year but is expected to increase by $25 each year thereafter. What is the present worth of 5 years of maintenance? Use an interest rate of 10%.

In this problem, the cash flow must be broken down into parts. Notice that the 5-year gradient factor is used even though there are only 4 non-zero gradient cash flows.

$$P = A(P/A, 10\%, 5) + G(P/G, 10\%, 5) = -100(3.7908) - 25(6.8618) = -551$$

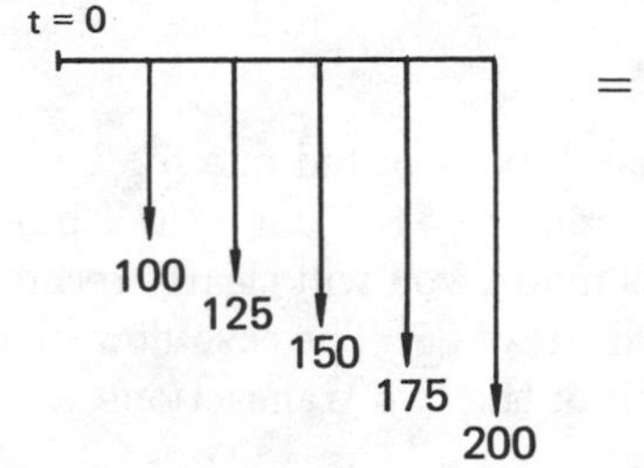


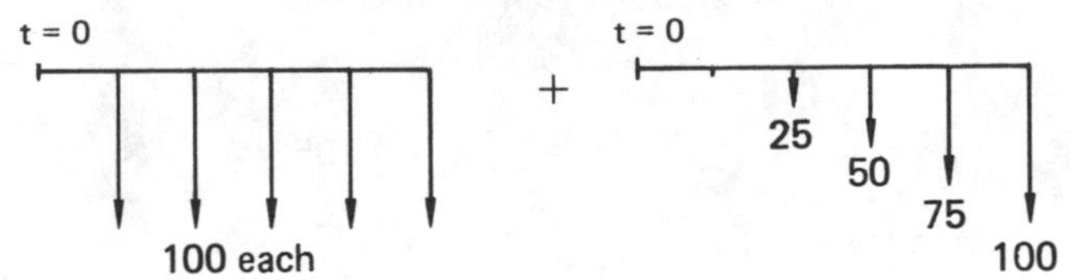


**Table 2.1**
Discount Factors for Discrete Compounding

| factor name | converts | symbol | formula |
|---|---|---|---|
| single payment compound amount | $P$ to $F$ | $(F/P, i\%, n)$ | $(1+i)^n$ |
| present worth | $F$ to $P$ | $(P/F, i\%, n)$ | $(1+i)^{-n}$ |
| uniform series Sinking Fund | $F$ to $A$ | $(A/F, i\%, n)$ | $\frac{i}{(1+i)^n-1}$ |
| capital recovery | $P$ to $A$ | $(A/P, i\%, n)$ | $\frac{i(1+i)^n}{(1+i)^n-1}$ |
| compound amount | $A$ to $F$ | $(F/A, i\%, n)$ | $\frac{(1+i)^n-1}{i}$ |
| equal series present worth | $A$ to $P$ | $(P/A, i\%, n)$ | $\frac{(1+i)^n-1}{i(1+i)^n}$ |
| uniform gradient | $G$ to $P$ | $(P/G, i\%, n)$ | $\frac{(1+i)^n-1}{i^2(1+i)^n} - \frac{n}{i(1+i)^n}$ |

Various combinations of the compounding and discounting factors are possible. For instance, the annual cash flow that would be equivalent to a uniform gradient may be found from

$$A = G(P/G, i\%, n)(A/P, i\%, n) \tag{2.4}$$

Formulas for all of the compounding and discounting factors are contained in table 2.1. Normally, it will not be necessary to calculate factors from the formulas. The tables at the end of this chapter are adequate for solving most problems.

## 5 THE MEANING OF "PRESENT WORTH" AND "$i$"

It is clear that \$100 invested in a 5% bank account will allow you to remove \$105 one year from now. If this investment is made, you will clearly receive a *return on investment* (ROI) of \$5. The cash flow diagram and the present worth of the two transactions are

$$\begin{aligned} P &= -100 + 105(P/F, 5\%, 1) \\ &= -100 + 105(0.9524) = 0 \end{aligned}$$

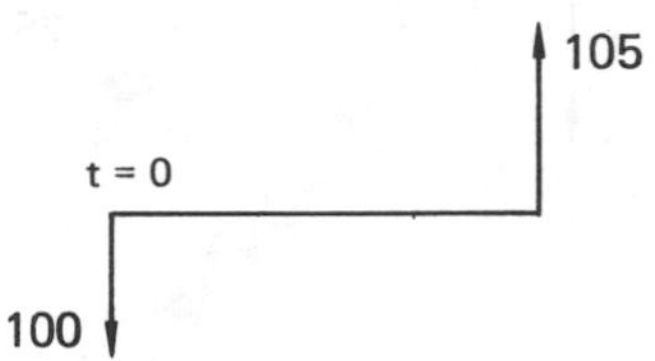


**Figure 2.1**

Notice that the present worth is zero even though you did receive a 5% return on your investment.

However, if you are offered \$120 for the use of \$100 over a one-year period, the cash flow diagram and present worth (at 5%) would be

$$\begin{aligned} P &= -100 + 120(P/F, 5\%, 1) \\ &= -100 + 120(0.9524) = 14.29 \end{aligned}$$

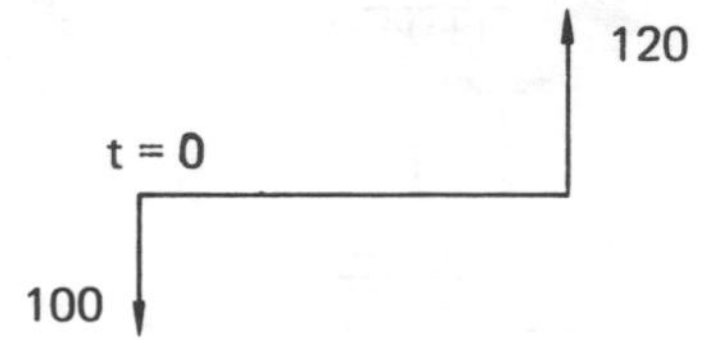


**Figure 2.2**

Therefore, it appears that the present worth of an alternative is equal to the equivalent value at $t = 0$ of the increase in return above that which you would be able to earn in an investment offering $i$% per period. In the above case, \$14.29 is the present worth of (\$20–\$5), the difference in the two *ROI's*.

Alternatively, the actual earned interest rate, called *rate of return* (*ROR*), can be defined as the rate which makes the present worth of the alternative zero.

The *present worth* is also the amount that you would have to be given to dissuade you from making an investment, since placing the initial investment amount along with the present worth into a bank account earning $i$% will yield the same eventual *ROI*. Relating this to the previous paragraphs, you could be dissuaded against investing \$100 in an alternative which would return \$120 in one year by a $t = 0$ payment of \$14.29. Clearly, (\$100 + \$14.29) invested at $t = 0$ will also yield \$120 in one year at 5%.

The selection of the interest rate is difficult in engineering economics problems. Usually it is taken as the average rate of return that an individual or business organization has realized in past investments. Fortunately, an interest rate is usually given. A company may not know what effective interest rate to use in an economic analysis. In such a case, the company can establish a minimum acceptable return on its investment. This *minimum attractive rate of return* (*MARR*) should be used as the effective interest rate $i$ in economic analyses.

It should be obvious that alternatives with negative present worths are undesirable, and that alternatives with positive present worths are desirable because they increase the average earning power of invested capital.

## 6 CHOICE BETWEEN ALTERNATIVES

A variety of methods exists for selecting a superior alternative from among a group of proposals. Each method has its own merits and applications.

### Present Worth Method

The *Present Worth Method* has already been implied. When two or more alternatives are capable of performing the same functions, the superior alternative will have the largest present worth. This method is suitable for ranking the desirability of alternatives. The present worth method is restricted to evaluating alternatives that are mutually exclusive and which have the same lives.

Returning to example 2.2, the present worth of each alternative should be found in order to determine which alternative is superior.

*Example 2.2, continued*

$$P(\mathbf{A}) = -10{,}000 + 11{,}500(P/F, 5\%, 2) = 431$$
$$P(\mathbf{B}) = -8000 + 4500(P/A, 5\%, 2) = 367$$

Alternative **A** is superior and should be chosen.

### Capitalized Cost Method

The present worth of a project with an infinite life is known as the *capitalized cost* or life cycle cost. Capitalized cost is the amount of money at $t = 0$ needed to perpetually support the project on the earned interest only. Capitalized cost is a positive number when expenses exceed income.

$$\text{Capitalized Cost} = \text{Initial Cost} + \frac{\text{Annual Costs}}{i} \quad 2.5$$

Capitalized cost is the present worth of an infinitely-lived project. Normally, it would be difficult to work with an infinite stream of cash flows since most economics tables don't list factors for periods in excess of 100 years. However, the $(A/P)$ discounting factor approaches the interest rate as $n$ becomes large. Since the $(P/A)$ and $(A/P)$ factors are reciprocals of each other, we would expect to divide an infinite series of equal cash flows by the interest rate in order to calculate the present worth of the infinite series. This is the basis of equation 2.5.

Equation 2.5 can be used when the annual costs are equal in every year. The "Annual Cost" in that equation is assumed to be the same each year. If the operating and maintenance costs occur irregularly instead of annually, or if the costs vary from year to year, it will be necessary to somehow determine a cash flow of equal annual amounts (EAA) which is equivalent to the stream of original costs.

The equal annual amount may be calculated in the usual manner by first finding the present worth of all the actual costs, and then multiplying the present worth by the interest rate (the $(A/P)$ factor for an infinite series). However, it is not even necessary to convert the present worth to an equal annual amount, since equation 2.6 will convert the equal annual amount back to the present worth.

$$\text{Capitalized Cost} = \text{Initial Cost} + \frac{\text{EAA}}{i} \quad 2.6$$

In comparing two alternatives, each of which is infinitely lived, the superior alternative will have the lowest capitalized cost.

### Annual Cost Method

Alternatives which accomplish the same purpose but which have unequal lives must be compared by the *Annual Cost Method.* The annual cost method assumes that each alternative will be replaced by an identical twin at the end of its useful life (infinite renewal). This method, which may also be used to rank alternatives according to their desirability, is also called the *Annual Return Method* and *Capital Recovery Method.*

Restrictions are that the alternatives must be mutually exclusive and infinitely renewed up to the duration of the longest-lived alternative. The calculated annual cost is known as the *Equivalent Uniform Annual Cost, EUAC.* Cost is a positive number when expenses exceed income.

ECONOMICS

*Example 2.6*

Which of the following alternatives is superior over a 30 year period if the interest rate is 7%?

| | **A** | **B** |
|---|---|---|
| type | brick | wood |
| life | 30 years | 10 years |
| cost | $1800 | $450 |
| maintenance | $5/year | $20/year |

$$\text{EUAC}(A) = 1800(A/P, 7\%, 30) + 5 = 150$$
$$\text{EUAC}(B) = 450(A/P, 7\%, 10) + 20 = 84$$

Alternative **B** is superior since its annual cost of operation is the lower. It is assumed that three wood facilities, each with a life of 10 years and a cost of $450, will be built to span the 30 year period.

### Benefit-Cost Ratio Method

The *Benefit-Cost Ratio Method* is often used in municipal project evaluations where benefits and costs accrue to different segments of the community. With this method, the present worth of all benefits (regardless of the beneficiary) is divided by the present worth of all costs. The project is considered acceptable if the ratio exceeds *one.*

When the benefit-cost ratio method is used, disbursements by the initiators or sponsors are *costs.* Disbursements by the users of the project are known as *disbenefits.* It is often difficult to determine whether a cash flow is a cost or a disbenefit (whether to place it in the numerator or denominator of the benefit-cost ratio calculation).

Regardless of where the cash flow is placed, an acceptable project will always have a benefit-cost ratio greater than one, although the actual numerical result will depend on the placement. For this reason, the benefit-cost ratio method should not be used to rank competing projects.

The benefit-cost ratio method may be used to rank alternative proposals only if an *incremental analysis* is used. First, determine that the ratio is greater than one for each alternative. Then, calculate the ratio of benefits to costs

$$\frac{B_2 - B_1}{C_2 - C_1}$$

for each possible pair of alternatives. If the ratio exceeds one, alternative 2 is superior to alternative 1. Otherwise, alternative 1 is superior.

### Rate of Return Method

Perhaps no method of analysis is less understood than the *Rate of Return* (ROR) *Method.* As was stated previously, the ROR is the interest rate that would yield identical profits if all money were invested at that rate. The present worth of any such investment is zero.

The ROR is defined as the interest rate that will discount all cash flows to a total present worth equal to the initial required investment. This definition is used to determine the ROR of an alternative. The advantage of the ROR method is that no knowledge of an interest rate is required.

To find the ROR of an alternative, proceed as follows:

*step 1*: Set up the problem as if to calculate the present worth.

*step 2*: Arbitrarily select a reasonable value for $i$. Calculate the present worth.

*step 3*: Choose another value of $i$ (not too close to the original value) and again solve for the present worth.

*step 4*: Interpolate or extrapolate the value of $i$ which gives a zero present worth.

*step 5*: For increased accuracy, repeat steps (2) and (3) with two more values that straddle the value found in step (4).

A common, although incorrect, method of calculating the ROR involves dividing the annual receipts or returns by the initial investment. However, this technique ignores such items as salvage, depreciation, taxes, and the time value of money. This technique also fails when the annual returns vary.

Once a rate of return is known for an investment alternative, it is typically compared to the *minimum attractive rate of return* (*MARR*) specified by a company. However, ROR should not be used to rank alternatives. When two alternatives have ROR's exceeding the MARR, it is not sufficient to select the alternative with the higher ROR.

An *incremental analysis*, also known as a *rate of return on added investment study*, should be performed if ROR is to be used to select between investments. In an incremental analysis, the cash flows for the investment with the lower initial cost are subtracted from the cash flows for the higher-priced alternative on a year-by-year basis. This produces, in effect, a third alternative representing the cost and benefits of the added investment. The added expense of the higher-priced investment is not warranted unless the ROR of this third alternative exceeds the MARR as well.

*Example 2.7*

What is the return on invested capital if $1000 is invested now with $500 being returned in year 4 and $1000 being returned in year 8?

First, set up the problem as a present worth calculation.

$$P = -1000 + 500(P/F, i\%, 4) + 1000(P/F, i\%, 8)$$

Arbitrarily select $i = 5\%$. The present worth is then found to be $88.15. Next take a higher value of $i$ to reduce the present worth. If $i = 10\%$, the present worth is $-\$192$. The ROR is found from simple interpolation to be approximately 6.6%. (A more exact solution is 6.37%.)

## 7 TREATMENT OF SALVAGE VALUE IN REPLACEMENT STUDIES

An investigation into the retirement of an existing process or piece of equipment is known as a *replacement study.* Replacement studies are similar in most respects to other alternative comparison problems: an interest rate is given, two alternatives exist, and one of the previously mentioned methods of comparing alternatives is used to choose the superior alternative.

In replacement studies, the existing process or piece of equipment is known as the *defender.* The new process or piece of equipment being considered for purchase is known as the *challenger.*

Because most defenders still have some market value when they are retired, the problem of what to do with the salvage arises. It seems logical to use the salvage value of the defender to reduce the initial purchase cost of the challenger. This is consistent with what would actually happen if the defender were to be retired.

By convention, however, the salvage value is subtracted from the defender's present value. This does not seem logical, but it is done to keep all costs and benefits related to the defender with the defender. In this case, the salvage value is treated as an opportunity cost which would be incurred if the defender is not retired.

If the defender and the challenger have the same lives and a present worth study is used to choose the superior alternative, the placement of the salvage value will have no effect on the net difference between present worths for the challenger and defender. Although the values of the two present worths will be different depending on the placement, the difference in present worths will be the same.

If the defender and the challenger have different lives, an annual cost comparison must be made. Since the salvage value would be 'spread over' a different number of years depending on its placement, it is important to abide by the conventions listed in this section.

There are a number of ways to handle salvage value. The best way is to think of the EUAC of the defender as the cost of keeping the defender from now until next year. In addition to the usual operating and maintenance costs, that cost would include an opportunity interest cost incurred by not selling the defender and also a drop in the salvage value if the defender is kept for one additional year. Specifically,

$$\begin{aligned}\text{EUAC(defender)} &= \text{maintenance costs}\\ &\quad + i(\text{current salvage value})\\ &\quad + (\text{current salvage–next}\\ &\quad \text{year's salvage})\end{aligned} \qquad 2.7$$

It is important in retirement studies not to double count the salvage value. That is, it would be incorrect to add the salvage value to the defender and at the same time subtract it from the challenger.

## 8 BASIC INCOME TAX CONSIDERATIONS

Assume that an organization pays $f\%$ of its profits to the federal government as income taxes. If the organization also pays a state income tax of $s\%$, and if state taxes paid are recognized by the federal government as expenses, then the composite tax rate is

$$t = s + f - sf \qquad 2.8$$

The basic principles used to incorporate taxation into economic analyses are listed below.

a. Initial purchase cost is unaffected by income taxes.

b. Salvage value is unaffected by income taxes.

c. Deductible expenses, such as operating costs, maintenance costs, and interest payments, are reduced by $t\%$ (e.g., multiplied by the quantity $(1-t)$).

d. Revenues are reduced by $t\%$ (e.g., multiplied by the quantity $(1-t)$).

e. Depreciation is multiplied by $t$ and added to the appropriate year's cash flow, increasing that year's present worth.

Income taxes and depreciation have no bearing on municipal or governmental projects since municipalities, states, and the U.S. Government pay no taxes.

*Example 2.8*

A corporation which pays 53% of its revenue in income taxes invests \$10,000 in a project which will result in \$3000 annual revenue for 8 years. If the annual expenses are \$700, salvage after 8 years is \$500, and 9% interest is used, what is the after-tax present worth? Disregard depreciation.

$$\begin{aligned}P_t &= -10{,}000 + 3000(P/A, 9\%, 8)(1-0.53)\\ &\quad - 700(P/A, 9\%, 8)(1-0.53)\\ &\quad + 500(P/F, 9\%, 8)\\ &= -3766\end{aligned}$$

## 9 DEPRECIATION

Although depreciation calculations may be considered independently in examination questions, it is important to recognize that depreciation has no effect on engineering economic calculations unless income taxes are also considered.

Generally, tax regulations do not allow the cost of equipment[1] to be treated as a deductible expense in the year of purchase. Rather, portions of the cost may be allocated to each of the years of the item's economic life (which may be different from the actual useful life). Each year, the book value (which is initially equal to the purchase price) is reduced by the depreciation in that year. Theoretically, the book value of an item will equal the market value at any time within the economic life of that item.

Since tax regulations allow the depreciation in any year to be handled as if it were an actual operating expense, and since operating expenses are deductible from the income base prior to taxation, the after-tax profits will be increased. If $D$ is the depreciation, the net result to the after-tax cash flow will be the addition of $tD$.

[1] The IRS tax regulations allow depreciation on almost all forms of ***property*** except land. The following types of property are distinguished: ***real*** (e.g., buildings used for business), ***residential*** (e.g., buildings used as rental property), and ***personal*** (e.g., equipment used for business). Personal property does *not* include items for personal use, despite its name. ***Tangible*** personal property is distinguished from ***intangible property*** (e.g., goodwill, copyrights, patents, trademarks, franchises, and agreements not to compete).

The present worth of all depreciation over the economic life of the item is called the *depreciation recovery.* Although originally established to do so, depreciation recovery can never fully replace an item at the end of its life.

Depreciation is often confused with amortization and depletion. While depreciation spreads the cost of a fixed asset over a number of years, *amortization* spreads the cost of an intangible asset (e.g., a patent) over some basis such as time or expected units of production.

*Depletion* is another artificial deductible operating expense designed to compensate mining organizations for decreasing mineral reserves. Since original and remaining quantities of minerals are seldom known accurately, the *depletion allowance* is calculated as a fixed percentage of the organization's gross income. These percentages are usually in the 10%–20% range and apply to such mineral deposits as oil, natural gas, coal, uranium, and most metal ores.

There are four common methods of calculating depreciation. The book value of an asset depreciated with the *Straight Line* (SL) *Method* (also known as the *Fixed Percentage Method*) decreases linearly from the initial purchase at $t = 0$ to the estimated salvage at $t = n$. The depreciated amount is the same each year. The quantity $(C - S_n)$ in equation 2.9 is known as the *depreciation base.*

$$D_j = \frac{C - S_n}{n} \qquad 2.9$$

*Double Declining Balance*[2] (DDB) depreciation is independent of salvage value. Furthermore, the book value never stops decreasing, although the depreciation decreases in magnitude. Usually, any remaining book value is written off in the last year of the asset's estimated life. Unlike any of the other depreciation methods, DDB depends on accumulated depreciation.

$$D_j = \frac{2(C - \sum_{i=1}^{j-1} D_i)}{n} \qquad 2.10$$

In *Sum-of-the-Years'-Digits* (SOYD) depreciation, the digits from 1 to $n$ inclusive, are summed. The total, $T$, can also be calculated from

$$T = \frac{1}{2}n(n+1) \qquad 2.11$$

[2] Double declining balance depreciation is a particular form of *declining balance depreciation,* as defined by the IRS tax regulations. Declining balance depreciation also includes 125% declining balance and 150% declining balance depreciations which can be calculated by substituting 1.25 and 1.50 respectively for the 2 in equation 2.10.

The depreciation can be found from

$$D_j = \frac{(C - S_n)(n - j + 1)}{T} \qquad 2.12$$

The *Sinking Fund Method* is seldom used in industry because the initial depreciation is low. The formula for sinking fund depreciation (which increases each year) is

$$D_j = (C - S_n)(A/F, i\%, n)(F/P, i\%, j-1) \qquad 2.13$$

The above discussion gives the impression that any form of depreciation may be chosen regardless of the nature and circumstances of the purchase. In reality, the IRS tax regulations place restrictions on the higher-rate ("accelerated") methods such as DDB and SOYD. Furthermore, the *Economic Recovery Act of 1981* substantially changed the laws relating to personal and corporate income taxes.

Property placed into service in 1981 or after must use the *Accelerated Cost Recovery System* (*ACRS*). Other methods (straight-line, declining balance, etc.) cannot be used except in special cases.

Property placed into service in 1980 or before must continue to be depreciated according to the method originally chosen (e.g., straight-line, declining balance, or sum-of-years-digits). *ACRS* cannot be used.

Under *ACRS*, the cost recovery amount in the $j$th year of an asset's cost recovery period is calculated by multiplying the initial cost by a factor.

$$D_j = (\text{initial cost})(\text{factor}) \qquad 2.14$$

The initial cost used is not reduced by the asset's salvage value for either the regular or alternate *ACRS* calculations. The factor used depends on the asset's cost recovery period. Such factors are subject to continuing legislation changing them. Current tax publications should be consulted before using the *ACRS* method.

Three other depreciation methods should be mentioned, not because they are currently accepted or in widespread use, but because they are occasionally called for by name.

The ***sinking-fund plus interest on first cost*** depreciation method, like the following two methods, is an attempt to include the *opportunity interest cost* on the purchase price with the depreciation. That is, the purchasing company not only incurs an annual loss due to the drop in book value, but it also loses the interest on the purchase price. The formula for this method is

$$D_j = (C - S_n)(A/F, i\%, n) + (C)(i) \qquad 2.15$$

The *straight-line plus interest on first cost* method is similar. Its formula is

$$D_j = \left(\frac{1}{n}\right)(C - S_n) + (C)(i) \qquad 2.16$$

The *straight-line plus average interest method* assumes that the opportunity interest cost should be based on the book value only, not on the full purchase price. Since the book value changes each year, an average value is used. The depreciation formula is

$$D_j = \left(\frac{1}{n}\right)(C - S_n) + \frac{1}{2}(i)(C - S_n)\left(\frac{n+1}{n}\right) + iS_n \qquad 2.17$$

These three depreciation methods are not to be used in the usual manner (e.g., in conjunction with the income tax rate). These methods are attempts to calculate a more accurate annual cost of an alternative. Sometimes they work, and sometimes they give misleading answers. Their use cannot be recommended. They are included in this chapter only for the sake of completeness.

*Example 2.9*

An asset is purchased for \$9000. Its estimated economic life is 10 years, after which it will be sold for \$1000. Find the depreciation in the first three years using SL, DDB, and SOYD.

**SL:** $D = \frac{(9000 - 1000)}{10} = 800$ each year

**DDB:** $D_1 = \frac{2(9000)}{10} = 1800$ in year 1

$D_2 = \frac{2(9000 - 1800)}{10} = 1440$ in year 2

$D_3 = \frac{2(9000 - 3240)}{10} = 1152$ in year 3

**SOYD:** $T = \frac{1}{2}(10)(11) = 55$

$D_1 = \left(\frac{10}{55}\right)(9000 - 1000) = 1455$ in year 1

$D_2 = \left(\frac{9}{55}\right)(8000) = 1309$ in year 2

$D_3 = \left(\frac{8}{55}\right)(8000) = 1164$ in year 3

*Example 2.10*

For the asset described in example 2.9, calculate the book value during the first three years if SOYD depreciation is used.

The book value at the beginning of year 1 is \$9000. Then,

$$BV_1 = 9000 - 1455 = 7545$$
$$BV_2 = 7545 - 1309 = 6236$$
$$BV_3 = 6236 - 1164 = 5072$$

*Example 2.11*

For the asset described in example 2.9, calculate the after-tax depreciation recovery with SL and SOYD depreciation methods. Use 6% interest with 48% income taxes.

**SL:** $D.R. = 0.48(800)(P/A, 6\%, 10) = 2826$

**SOYD:** The depreciation series can be thought of as a constant 1,454 term with a negative 145 gradient.

$$D.R. = 0.48(1454)(P/A, 6\%, 10) - 0.48(145)(P/G, 6\%, 10) = 3076$$

Finding book values, depreciation, and depreciation recovery is particularly difficult with DDB depreciation, since all previous years' quantities seem to be required. It appears that the depreciation in the 6th year cannot be calculated unless the values of depreciation for the first five years are calculated first. Questions asking for depreciation or book value in the middle or at the end of an asset's economic life may be solved from the following equations:

$$d = \frac{2}{n} \qquad 2.18$$

$$z = \frac{1+i}{1-d} \qquad 2.19$$

$$(P/EG) = \frac{z^n - 1}{z^n(z-1)} \qquad 2.20$$

Then, the present worth of the depreciation recovery is

$$D.R. = t\left[\frac{(d)(C)}{(1-d)}(P/EG)\right] \qquad 2.21$$

$$D_j = (d)(C)(1-d)^{j-1} \qquad 2.22$$

$$BV_j = C(1-d)^j \qquad 2.23$$

ECONOMICS

*Example 2.12*

What is the after-tax present worth of the asset described in example 2.8 if SL, SOYD, and DDB depreciation methods are used?

The after-tax present worth, neglecting depreciation, was previously found to be −3766.

Using SL, the depreciation recovery is

$$\begin{aligned} D.R. &= (0.53)\frac{(10{,}000 - 500)}{8}(P/A, 9\%, 8) \\ &= 3483 \end{aligned}$$

Using SOYD, the depreciation recovery is calculated as follows:

$$T = \frac{1}{2}(8)(9) = 36$$

$$\text{Depreciation base} = (10{,}000 - 500) = 9500$$

$$\begin{aligned} D_1 &= \frac{8}{36}(9500) = 2111 \\ G &= \text{gradient} = \frac{1}{36}(9500) \\ &= 264 \end{aligned}$$

$$\begin{aligned} D.R. &= (0.53)\,[2111(P/A, 9\%, 8) \\ &\quad - 264\,(P/G, 9\%, 8)] \\ &= 3829 \end{aligned}$$

Using DDB, the depreciation recovery is calculated as follows:

$$d = \frac{2}{8} = 0.25$$

$$z = \frac{1.09}{0.75} = 1.4533$$

$$(P/EG) = \frac{(1.4533)^8 - 1}{(1.4533)^8(0.453)} = 2.095$$

$$\begin{aligned} D.R. &= 0.53\left[\frac{(0.25)(10{,}000)}{0.75}(2.095)\right] \\ &= 3701 \end{aligned}$$

The after-tax present worths including depreciation recovery are:

**SL:** $P_t = -3766 + 3483 = -283$
**SOYD:** $P_t = -3766 + 3829 = 63$
**DDB:** $P_t = -3766 + 3701 = -65$

## 10 ADVANCED INCOME TAX CONSIDERATIONS

There are a number of specialized techniques that are needed infrequently. These techniques are related more to the accounting profession than to the engineering profession. Nevertheless, it is occasionally necessary to use these techniques.

### A. INVESTMENT TAX CREDIT

An investment tax credit (also known as a tax credit or an investment credit) is a one-time credit against income taxes. The investment tax credit is calculated as a fraction of the initial purchase price of certain types of equipment purchased for industrial, commercial, and manufacturing use.

$$\text{credit} = (\text{initial cost})(\text{fraction}) \qquad 2.24$$

The fraction is subject to continuing legislation changing its value and applicability. The fraction, which typically is taken as 10% for initial estimates, actually depends on the asset life, year of acquisition, and number of years the asset is held before being disposed of. The current tax laws should be studied before using the investment tax credit.

Since the investment tax credit reduces the buyer's tax liability, the credit should only be used in after-tax analyses.

### B. GAIN ON THE SALE OF A DEPRECIATED ASSET

If an asset is sold for more than its current book value, the difference between selling price and book value is taxable income. The gain is taxed at capital gains rates. Excluded from this preferential treatment is non-residential real property depreciated under regular *ACRS* provisions. However, non-residential real property depreciated under the straight-line alternate method qualifies for the capital gains rate.

### C. CAPITAL GAINS AND LOSSES

A *gain* is defined as the difference between selling and purchase prices of a capital asset. The gain is called a *regular gain* if the item sold has been kept less than one year. The gain is called a *capital gain* if the item sold has been kept for longer than one year. Capital gains are taxed at the taxpayer's usual rate, but 60% of the gain is excluded from taxation.

*Regular* (as defined above) *losses* are fully deductible in the year of their occurrence. The IRS tax regulations should be consulted to determine the treatment of *capital losses*.

## 11 RATE AND PERIOD CHANGES

All of the foregoing calculations were based on compounding once a year at an *effective interest rate*, $i$. However, some problems specify compounding more frequently than annually. In such cases, a *nominal interest rate*, $r$, will be given. The nominal rate does not include the effect of compounding and is not the same as the effective rate, $i$. A nominal rate may be used to calculate the effective rate by using equation 2.25 or 2.26.

$$i = \left(1 + \frac{r}{k}\right)^k - 1 \qquad 2.25$$

$$= (1 + \phi)^k - 1 \qquad 2.26$$

A problem may also specify an effective rate per period, $\phi$, (e.g., per month). However, that will be a simple problem since compounding for $n$ periods at an effective rate per period is not affected by the definition or length of the period.

The following rules may be used to determine which interest rate is given in a problem:

- Unless specifically qualified in the problem, the interest rate given is an annual rate.
- If the compounding is annually, the rate given is the effective rate. If compounding is other than annually, the rate given is the nominal rate.
- If the type of compounding is not specified, assume annual compounding.

In the case of continuous compounding, the appropriate discount factors may be calculated from the formulas in table 2.2.

**Table 2.2**
Discount Factors for
Continuous Compounding

| | |
|---|---|
| (F/P) | $e^{rn}$ |
| (P/F) | $e^{-rn}$ |
| (A/F) | $(e^r - 1)/(e^{rn} - 1)$ |
| (F/A) | $(e^{rn} - 1)/(e^r - 1)$ |
| (A/P) | $(e^r - 1)/(1 - e^{-rn})$ |
| (P/A) | $(1 - e^{-rn})/(e^r - 1)$ |

*Example 2.13*

A savings and loan offers $5\frac{1}{4}\%$ compounded daily. What is the annual effective rate?

method 1: $r = 0.0525$, $k = 365$

$$i = \left(1 + \frac{0.0525}{365}\right)^{365} - 1 = 0.0539$$

method 2: Assume daily compounding is the same as continuous compounding.

$$i = (F/P) - 1 = e^{0.0525} - 1 = 0.0539$$

## 12 PROBABILISTIC PROBLEMS

Thus far, all of the cash flows included in the examples have been known exactly. If the cash flows are not known exactly but are given by some implicit or explicit probability distribution, the problem is ***probabilistic.***

Probabilistic problems typically possess the following characteristics:

- There is a chance of extreme loss that must be minimized.
- There are multiple alternatives that must be chosen from. Each alternative gives a different degree of protection against the loss or failure.
- The outcome is independent of the alternative chosen. Thus, as illustrated in example 2.15, the size of the dam that is chosen for construction will not alter the rainfall in successive years. However, it will alter the effects on the down-stream watershed areas.

Probabilistic problems are typically solved using annual costs and expected values. An *expected value* is similar to an 'average value' since it is calculated as the mean of the given probability distribution. If cost 1 has a probability of occurrence of $p_1$, cost 2 has a probability of occurrence of $p_2$, and so on, the expected value is

$$E(\text{cost}) = p_1(\text{cost } 1) + p_2(\text{cost } 2) + \cdots \qquad 2.27$$

*Example 2.14*

Flood damage in any year is given according to the table below. What is the present worth of flood damage for a 10-year period? Use 6%.

| Damage | Probability |
|---|---|
| 0 | 0.75 |
| $10,000 | 0.20 |
| $20,000 | 0.04 |
| $30,000 | 0.01 |

The expected value of flood damage is

$$\begin{aligned} E(\text{damage}) &= (0)(0.75) + (10{,}000)(0.20) \\ &\quad + (20{,}000)(0.04) + (30{,}000)(0.01) \\ &= 3100 \\ \text{present worth} &= 3100(P/A, 6\%, 10) \\ &= 22{,}816 \end{aligned}$$

Probabilities in probabilistic problems may be given to you in the problem (as in the example above) or you may have to obtain them from some named probability distribution. In either case, the probabilities are known explicitly and such problems are known as *explicit probability problems.*

*Example 2.15*

A dam is being considered on a river which periodically overflows and causes $600,000 damage. The damage is essentially the same each time the river causes flooding. The project horizon is 40 years. A 10% interest rate is being used.

Three different designs are available, each with different costs and storage capacities.

| design alternative | cost | maximum capacity |
|---|---|---|
| **A** | 500,000 | 1 unit |
| **B** | 625,000 | 1.5 units |
| **C** | 900,000 | 2.0 units |

The U.S. Weather Service has provided a statistical analysis of annual rainfall in the area draining into the river.

| units annual rainfall | probability |
|---|---|
| 0 | 0.10 |
| 0.1–0.5 | 0.60 |
| 0.6–1.0 | 0.15 |
| 1.1–1.5 | 0.10 |
| 1.6–2.0 | 0.04 |
| 2.1 or more | 0.01 |

Which design alternative would you choose assuming the dam is essentially empty at the start of each rainfall season?

The sum of the construction cost and the expected damage needs to be minimized. If alternative **A** is chosen, it will have a capacity of 1 unit. Its capacity will be exceeded (causing $600,000 damage) when the annual rainfall exceeds 1 unit. Therefore, the annual cost of **A** is

$$\begin{aligned}\text{EUAC}(\mathbf{A}) &= 500{,}000(A/P, 10\%, 40)\\ &\quad + 600{,}000(0.10 + 0.04 + 0.01)\\ &= 141{,}150\end{aligned}$$

Similarly,

$$\begin{aligned}\text{EUAC}(\mathbf{B}) &= 625{,}000(A/P, 10\%, 40)\\ &\quad + 600{,}000(0.04 + 0.01)\\ &= 93{,}940\end{aligned}$$

$$\begin{aligned}\text{EUAC}(\mathbf{C}) &= 900{,}000(A/P, 10\%, 40)\\ &\quad + 600{,}000(0.01)\\ &= 98{,}070\end{aligned}$$

Alternative **B** should be chosen.

In other problems, a probability distribution will not be given even though some parameter (such as the life of an alternative) is not known with certainty. Such problems are known as *implicit probability problems* since they require a reasonable assumption about the probability distribution.

Implicit probability problems typically involve items whose expected times to failure are known. The key to such problems is in recognizing that an expected time to failure is not the same as a fixed life.

Reasonable assumptions can be made about the form of probability distributions in implicit probability problems.

One such reasonable assumption is that of a *rectangular distribution.* A rectangular distribution is one which is assumed to give an equal probability of failure in each year. Such an assumption is illustrated in example 2.16.

*Example 2.16*

A bridge is needed for 20 years. Failure of the bridge at any time will require a 50% reinvestment. Assume that each alternative has an annual probability of failure that is inversely proportional to its expected time to failure. Evaluate the two design alternatives below using 6% interest.

| design alternative | initial cost | expected time to failure | annual costs | salvage at $t = 20$ |
|---|---|---|---|---|
| **A** | 15,000 | 9 years | 1200 | 0 |
| **B** | 22,000 | 43 years | 1000 | 0 |

For alternative **A**, the probability of failure in any year is $\left(\frac{1}{9}\right)$. Similarly, the annual failure probability for alternative **B** is $\left(\frac{1}{43}\right)$.

$$\begin{aligned}\text{EUAC}(\mathbf{A}) &= 15{,}000(A/P, 6\%, 20)\\ &\quad + 15{,}000(0.5)\left(\frac{1}{9}\right) + 1200\\ &= 3341\end{aligned}$$

$$\begin{aligned}\text{EUAC}(\mathbf{B}) &= 22{,}000(A/P, 6\%, 20)\\ &\quad + 22{,}000(0.5)\left(\frac{1}{43}\right) + 1000\\ &= 3174\end{aligned}$$

Alternative **B** should be chosen.

## 13 ESTIMATING ECONOMIC LIFE

As assets grow older, their operating and maintenance costs typcially increase each year. Eventually, the cost to keep an asset in operation becomes prohibitive, and the asset is retired or replaced. However, it is not always obvious when an asset should be retired or replaced.

As the asset's maintenance is increasing each year, the amortized cost of its initial purchase is decreasing. It is the sum of these two costs that should be evaluated to determine the point at which the asset should be retired or replaced. Since an asset's initial purchase price is likely to be high, the amortized cost will be the controlling factor in those years when the maintenance costs are low. Therefore, the EUAC of the asset will decrease in the initial part of its life.

However, as the asset grows older, the change in its amortized cost decreases while maintenance increases. Eventually the sum of the two costs reaches a minimum and then starts to increase. The age of the asset at the minimum cost point is known as the *economic life* of the asset. The economic life is, generally, less than the mission and technological lifetimes of the asset.

The determination of an asset's economic life is illustrated by example 2.17.

*Example 2.17*

A bus in a municipal transit system has the characteristics listed below. When should the city replace its buses if money can be borrowed at 8% ?

Initial cost: \$120,000

| year | maintenance cost | salvage value |
|---|---|---|
| 1 | 35,000 | 60,000 |
| 2 | 38,000 | 55,000 |
| 3 | 43,000 | 45,000 |
| 4 | 50,000 | 25,000 |
| 5 | 65,000 | 15,000 |

If the bus is kept for 1 year and then sold, the annual cost will be

$$\begin{aligned}\text{EUAC}(1) &= 120{,}000(A/P, 8\%, 1) + 35{,}000(A/F, 8\%, 1)\\ &\quad - 60{,}000(A/F, 8\%, 1)\\ &= 104{,}600\end{aligned}$$

If the bus is kept for 2 years and then sold, the annual cost will be

$$\begin{aligned}\text{EUAC}(2) &= [120{,}000 + 35{,}000(P/F, 8\%, 1)](A/P, 8\%, 2)\\ &\quad + (38{,}000 - 55{,}000)(A/F, 8\%, 2)\\ &= 77{,}300\end{aligned}$$

If the bus is kept for 3 years and then sold, the annual cost will be

$$\begin{aligned}\text{EUAC}(3) &= [120{,}000 + 35{,}000(P/F, 8\%, 1)\\ &\quad + 38{,}000(P/F, 8\%, 2)](A/P, 8\%, 3)\\ &\quad + (43{,}000 - 45{,}000)(A/F, 8\%, 3)\\ &= 71{,}200\end{aligned}$$

This process is continued until EUAC begins to increase. In this example, EUAC(4) is 71,700. Therefore, the bus should be retired after 3 years.

ECONOMICS

## 14 BASIC COST ACCOUNTING

*Cost accounting* is the system which determines the cost of manufactured products. Cost accounting is called *job cost accounting* if costs are accumulated by part number or contract. It is called *process cost accounting* if costs are accumulated by departments or manufacturing processes.

Three types of costs (direct material, direct labor, and all indirect costs) make up the total manufacturing cost of a product.

*Direct material costs* are the costs of all materials that go into the product, priced at the original purchase cost.

*Indirect material and labor costs* are generally limited to costs incurred in the factory, excluding costs incurred in the office area. Examples of indirect materials are cleaning fluids, assembly lubricants, and temporary routing tags. Examples of indirect labor are stock-picking, inspection, expediting, and supervision labor.

Here are some important points concerning basic cost accounting:

- The sum of direct material and direct labor costs is known as the *prime cost.*
- Indirect costs may be called *indirect manufacturing expenses* (IME).
- Indirect costs may also include the overhead sector of the company (e.g., secretaries, engineers, and corporate administration). In this case, the indirect cost is usually called *burden* or *overhead.* Burden may also include the EUAC of non-regular costs which must be spread evenly over several years.
- The cost of a product is usually known in advance from previous manufacturing runs or by estimation. Any deviation from this known cost is called a *variance.* Variance may be broken down into *labor variance* and *material variance.*

- Indirect cost per item is not easily measured. The method of allocating indirect costs to a product is as follows:

  **step 1**: Estimate the total expected indirect (and overhead) costs for the upcoming year.

  **step 2**: Decide on some convenient vehicle for allocating the overhead to production. Usually, this vehicle is either the number of units expected to be produced or the number of direct hours expected to be worked in the upcoming year.

  **step 3**: Estimate the quantity or size of the overhead vehicle.

  **step 4**: Divide expected overhead costs by the expected overhead vehicle to obtain the unit overhead.

  **step 5**: Regardless of the true size of the overhead vehicle during the upcoming year, one unit of overhead cost is allocated per product.

- Although estimates of production for the next year are always somewhat inaccurate, the cost of the product is assumed to be independent of forecasting errors. Any difference between true cost and calculated cost goes into a variance account.
- *Burden (overhead) variance* will be caused by errors in forecasting both the actual overhead for the upcoming year and the vehicle size. In the former case, the variance is called *burden budget variance*; in the latter, it is called *burden capacity variance.*

*Example 2.18*

A small company expects to produce 8000 items in the coming year. The current material cost is \$4.54 each. Sixteen minutes of direct labor are required per unit. Workers are paid \$7.50 per hour. 2133 direct labor hours are forecast for the product. Miscellaneous overhead costs are estimated at \$45,000.

Find the expected direct material cost, the direct labor cost, the prime cost, the burden as a function of production and direct labor, and the total cost.

- The direct material cost was given as \$4.54.
- The direct labor cost is $\left(\frac{16}{60}\right)(\$7.50) = \$2.00$.
- The prime cost is $\$4.54 + \$2.00 = \$6.54$.
- If the burden vehicle is production, the burden rate is $\$\frac{45{,}000}{8000} = \$5.63$ per item, making the total cost $\$4.54 + \$2.00 + \$5.63 = \$12.17$.
- If the burden vehicle is direct labor hours, the burden rate is $\left(\frac{45{,}000}{2133}\right) = \$21.10$ per hour, making the total cost $\$4.54 + \$2.00 + \frac{16}{60}(\$21.10) = \$12.17$.

*Example 2.19*

The actual performance of the company in example 2.18 is given by the following figures:

actual production: 7560
actual overhead costs: \$47,000

What are the burden budget variance and the burden capacity variance?

The burden capacity variance is

$$\$45{,}000 - 7560(\$5.63) = \$2437$$

The burden budget variance is

$$\$47{,}000 - \$45{,}000 = \$2000$$

The overall burden variance is

$$\$47{,}000 - 7560(\$5.63) = \$4437$$

## 15 BREAK-EVEN ANALYSIS

*Break-even analysis* is a method of determining when costs exactly equal revenue. If the manufactured quantity is less than the break-even quantity, a loss is incurred. If the manufactured quantity is greater than the break-even quantity, a profit is incurred.

Consider the following special variables:

- $f$ a fixed cost which does not vary with production
- $a$ an incremental cost which is the cost to produce one additional item. It may also be called the *marginal cost* or *differential cost.*
- $Q$ the quantity sold
- $p$ the incremental revenue
- $R$ the total revenue
- $C$ the total cost

Assuming no change in the inventory, the *break-even point* can be found from $C = R$, where

$$C = f + aQ \quad 2.28$$

$$R = pQ \quad 2.29$$

An alternate form of the break-even problem is to find the number of units per period for which two alternatives have the same total costs. Fixed costs are to be spread over a period longer than one year. One of the alternatives will have a lower cost if production is less than the break-even point. The other will have a lower cost for production greater than the break-even point.

The *cost per unit* problem is a variation of the breakeven problem. In the typical cost per unit problem, data will be available to determine the direct labor and material costs per unit, but some method is needed to additionally allocate part of the annual overhead (burden) and initial facility purchase/construction costs.

Annual overhead is allocated to the unit cost simply by dividing the overhead by the number of units produced each year. The initial purchase/construction cost is multiplied by the appropriate $(A/P)$ factor before similarly dividing by the production rate. The total unit cost is the sum of the direct labor, direct material, prorata share of overhead, and prorata share of the equivalent annual facility investment costs.

*Example 2.20*

Two plans are available for a company to obtain automobiles for its salesmen. How many miles must the cars be driven each year for the two plans to have the same costs? Use an interest rate of 10%.

*Plan A* Lease the cars and pay $0.15 per mile

*Plan B* Purchase the cars for $5000. Each car has an economic life of three years, after which it can be sold for $1,200. Gas and oil cost $0.04 per mile. Insurance is $500 per year. (Assume the year-end convention applies to the insurance.)

Let $x$ be the number of miles driven per year. Then, the EUAC for both alternatives is:

$$\begin{aligned}\text{EUAC}(A) &= 0.15x \\ \text{EUAC}(B) &= 0.04x + 500 + 5000(A/P, 10\%, 3) \\ &\quad - 1200(A/F, 10\%, 3) \\ &= 0.04x + 2148\end{aligned}$$

Setting EUAC(A) and EUAC(B) equal and solving for $x$ yields 19,527 miles per year as the break-even point.

## 16 HANDLING INFLATION

It is important to perform economic studies in terms of *constant value dollars.* One method of converting all cash flows to constant value dollars is to divide the flows by some annual *economic indicator* or price index. Such indicators would normally be given to you as part of a problem.

If indicators are not available, this method can still be used by assuming that inflation is relatively constant at a decimal rate $e$ per year. Then, all cash flows can be converted to '$t = 0$' dollars by dividing by $(1 + e)^n$ where $n$ is the year of the cash flow.

*Example 2.21*

**What is the uninflated present worth of a $2000 future value in two years if the average inflation rate is 6% and $i$ is 10%?**

$$P = \frac{\$2000}{(1.10)^2(1.06)^2} = \$1471.07$$

An alternative is to replace $i$ with a value corrected for inflation. This corrected value, $i'$, is

$$i' = i + e + ie \qquad 2.30$$

This method has the advantage of simplifying the calculations. However, pre-calculated factors may not be available for the non-integer values of $i'$. Therefore, table 2.1 will have to be used to calculate the factors.

*Example 2.22*

Repeat example 2.21 using $i'$.

$$i' = 0.10 + 0.06 + (0.10)(0.06) = 0.166$$

$$P = \frac{\$2000}{(1.166)^2} = \$1471.07$$

## 17 LEARNING CURVES

The more products that are made, the more efficient the operation becomes due to experience gained. Therefore, direct labor costs decrease. Usually, a ***learning curve*** is specified by the decrease in cost each time the cumulative quantity produced doubles. If there is a 20% decrease per doubling, the curve is said to be an 80% learning curve.

Consider the following special variables:

$T_1$ time or cost for the first item
$T_n$ time or cost for the $n$th item
$n$ total number of items produced
$b$ learning curve constant

Then, the time to produce the $n$th item is given by

$$T_n = \frac{T_1}{n^b} \qquad 2.31$$

**Table 2.3**
Learning Curve Constants

| learning curve | b |
|---|---|
| 80% | 0.322 |
| 85% | 0.234 |
| 90% | 0.152 |
| 95% | 0.074 |

The total time to produce units from quantity $n_1$ to $n_2$ inclusive is given by equation 2.32. Notice that $T_1$ is the first item's time and does not correspond to $n_1$.

$$\int_{n_1}^{n_2} T_n dn \approx \frac{T_1}{(1-b)}\left[\left(n_2+\frac{1}{2}\right)^{1-b}-\left(n_1-\frac{1}{2}\right)^{1-b}\right] \qquad 2.32$$

The average time per unit over the production from $n_1$ to $n_2$ is the above total time from equation 2.32 divided by the quantity produced, $(n_2 - n_1 + 1)$.

It is important to remember that learning curve reductions apply only to direct labor costs. They are not applied to indirect labor or direct material costs.

*Example 2.23*

A 70% learning curve is used with an item whose first production time was 1.47 hours. How long will it take to produce the 11th item? How long will it take to produce the 11th through 27th items?

First, use log identities to find $b$.

$$\frac{T_2}{T_1} = 0.7 = (2)^{-b} \quad \text{or } b = 0.515$$

Then, $T_{11} = 1.47(11)^{-0.515} = 0.428$ hours.

The time to produce the 11th item through 27th item is approximately

$$T = \frac{1.47}{1-0.515}\left[(27.5)^{1-0.515} - (10.5)^{1-0.515}\right]$$

$$= 5.643 \text{ hours}$$

## 18 ECONOMIC ORDER QUANTITY

The *economic order quantity* (EOQ) is the order quantity which minimizes the inventory costs per unit time. Although there are many different EOQ models, the simplest is based on the following assumptions:

- Reordering is instantaneous. The time between order placement and receipt is zero.
- Shortages are not allowed.
- Demand for the inventory item is deterministic (i.e., is not a random variable).
- Demand is constant with respect to time.
- An order is placed when the on-hand quantity is zero.

The following special variables are used:

| | |
|---|---|
| $a$ | the constant depletion rate $\left(\frac{\text{items}}{\text{unit time}}\right)$ |
| $h$ | the inventory storage cost $\left(\frac{\$}{\text{item-unit time}}\right)$ |
| $H$ | the total inventory storage cost between orders (\$) |
| $K$ | the fixed cost of placing an order (\$) |
| $Q_0$ | the order quantity |

If the original quantity on hand is $Q_0$, the stock will be depleted at

$$t^* = \frac{Q_0}{a}$$

The total inventory storage cost between $t_0$ and $t^*$ is

$$H = \frac{1}{2}h\frac{Q_o^2}{a} \qquad 2.33$$

The total inventory and ordering cost per unit time is

$$C_t = \frac{aK}{Q_0} + \frac{1}{2}hQ_0 \qquad 2.34$$

$C_t$ can be minimized with respect to $Q_0$. The EOQ and time between orders are:

$$Q_0^* = \sqrt{2\frac{aK}{h}} \qquad 2.35$$

$$t^* = \frac{Q_0^*}{a} \qquad 2.36$$

## 19 CONSUMER LOANS

Many consumer loans cannot be handled by the equivalence formulas presented up to this point. Many different arrangements can be made between lender and borrower. Four of the most common consumer loan arrangements are presented below. Refer to a real estate or investment analysis book for more complex loans.

### A. SIMPLE INTEREST

Interest due does not compound with a *simple interest* loan. The interest due is merely proportional to the length of time the principal is outstanding. Because of this, simple interest loans are seldom made for long periods (e.g., longer than one year).

*Example 2.24*

A \$12,000 simple interest loan is taken out at 16% per annum. The loan matures in one year with no intermediate payments. How much will be due at the end of the year?

$$\text{Amount due} = (1 + 0.16)(\$12{,}000) = \$13{,}920$$

For loans less than one year, it is commonly assumed that a year consists of 12 months of 30 days each.

*Example 2.25*

\$4000 is borrowed for 75 days at 16% per annum simple interest. How much will be due at the end of 75 days?

$$\text{Amount due} = \$4000 + (0.16)\left(\frac{75}{360}\right)(4000) = \$4133$$

B. LOANS WITH CONSTANT AMOUNT PAID TOWARDS PRINCIPAL

With this loan type, the payment is not the same each period. The amount paid towards the principal is constant, but the interest varies from period to period. The following special symbols are used.

| | |
|---|---|
| $BAL_j$ | principal balance after the $j$th payment |
| $LV$ | total value loaned (cost minus down payment) |
| $j$ | payment or period number |
| $N$ | total number of payments to pay off the loan |
| $PI_j$ | $j$th interest payment |
| $PP_j$ | $j$th principal payment |
| $PT_j$ | $j$th total payment |
| $\phi$ | effective rate per period ($r/k$) |

The equations which govern this type of loan are

$$BAL_j = LV - (j)(PP) \qquad 2.37$$

$$PI_j = \phi(BAL_{j-1}) \qquad 2.38$$

$$PT_j = PP + PI_j \qquad 2.39$$

C. DIRECT REDUCTION LOANS

This is the typical 'interest paid on unpaid balance' loan. The amount of the periodic payment is constant, but the amounts paid towards the principal and interest both vary.

The same symbols are used with this type of loan as are listed above.

$$N = -\frac{ln\left[\frac{-\phi(LV)}{PT} + 1\right]}{ln(1+\phi)} \qquad 2.40$$

$$BAL_{j-1} = PT\left[\frac{1-(1+\phi)^{j-1-N}}{\phi}\right] \qquad 2.41$$

$$PI_j = \phi(BAL_{j-1}) \qquad 2.42$$

$$PP_j = PT - PI_j \qquad 2.43$$

$$BAL_j = BAL_{j-1} - PP_j \qquad 2.44$$

*Example 2.26*

A \$45,000 loan is financed at 9.25% per annum. The monthly payment is \$385. What are the amounts paid toward interest and principal in the 14th period? What is the remaining principal balance after the 14th payment has been made?

The effective rate per month is

$$\phi = \frac{r}{k} = \frac{0.0925}{12} = 0.007708$$

$$N = -\frac{ln\left[\frac{-(0.007708)(45{,}000)}{385} + 1\right]}{ln(1+0.007708)} = 301$$

$$BAL_{13} = 385\left[\frac{1-(1+0.007708)^{14-1-301}}{0.007708}\right] = \$44{,}476.39$$

$$PI_{14} = (0.007708)(\$44{,}476.39) = \$342.82$$

$$PP_{14} = \$385 - \$342.82 = \$42.18$$

$$BAL_{14} = \$44,476.39 - \$42.18 = \$44,434.21$$

Equation 2.40 calculates the number of payments necessary to pay off a loan. This equation can be solved with effort for the total periodic payment (PT) or the initial value of the loan (LV). It is easier, however, to use the $(A/P, i\%, n)$ factor to find the payment and loan value.

$$PT = (LV)(A/P, \phi\%, n)$$

If the loan is repaid in yearly installments, then $i$ is the effective annual rate. If the loan is paid off monthly, then $i$ should be replaced by the effective rate per month ($\phi$ from equation 2.26). For monthly payments, $n$ is the number of months in the payback period.

D. DIRECT REDUCTION LOAN WITH BALLOON PAYMENT

This type of loan has a constant periodic payment, but the duration of the loan is insufficient to completely pay back the principal. Therefore, all remaining unpaid principal must be paid back in a lump sum when the loan matures. This large payment is known as a *balloon payment.*

Equations 2.40 through 2.44 can also be used with this type of loan. The remaining balance after the last payment is the balloon payment. This balloon payment must be repaid along with the last regular payment calculated.

## 20 SENSITIVITY ANALYSIS

Data analysis and forecasts in economic studies represent judgment on costs which will occur in the future. There are always uncertainties about these costs. However, these uncertainties are insufficient reason not to make the best possible estimates of the costs. Nevertheless, a decision between alternatives often can be made more confidently if it is known whether or not the conclusion is sensitive to moderate changes in data forecasts. Sensitivity analysis provides this extra dimension to an economic analysis.

The sensitivity of a decision is determined by inserting a range of estimates for critical cash flows. If radical changes can be made to a cash flow without changing the decision, the decision is said to be insensitive to uncertainties regarding that cash flow. However, if a small change in the estimate of a cash flow will alter the decision, that decision is said to be very sensitive to changes in the estimate.

An established semantic tradition distinguishes between risk analysis and uncertainty analysis. Risk analysis addresses variables which have a known or estimated probability distribution. In this regard, statistics and probability theory can be used to determine the probability of a cash flow varying between given limits. On the other hand, uncertainty analysis is concerned with situations in which there is not enough information to determine the probability or frequency distribution for the variables involved.

As a first step, sensitivity analysis should be applied one at a time to the dominant cost factors. Dominant cost factors are those which have the most significant impact on the present value of the alternative. If warranted, additional investigation can be used to determine the sensitivity to several cash flows varying simultaneously. Significant judgment is needed, however, to successfully determine the proper combinations of cash flows to vary.

It is common to plot the dependency of the present value on the cash flow being varied in a two-dimensional graph. Simple linear interpolation is used (within reason) to determine the critical value of the cash flow being varied.

## STANDARD CASH FLOW FACTORS

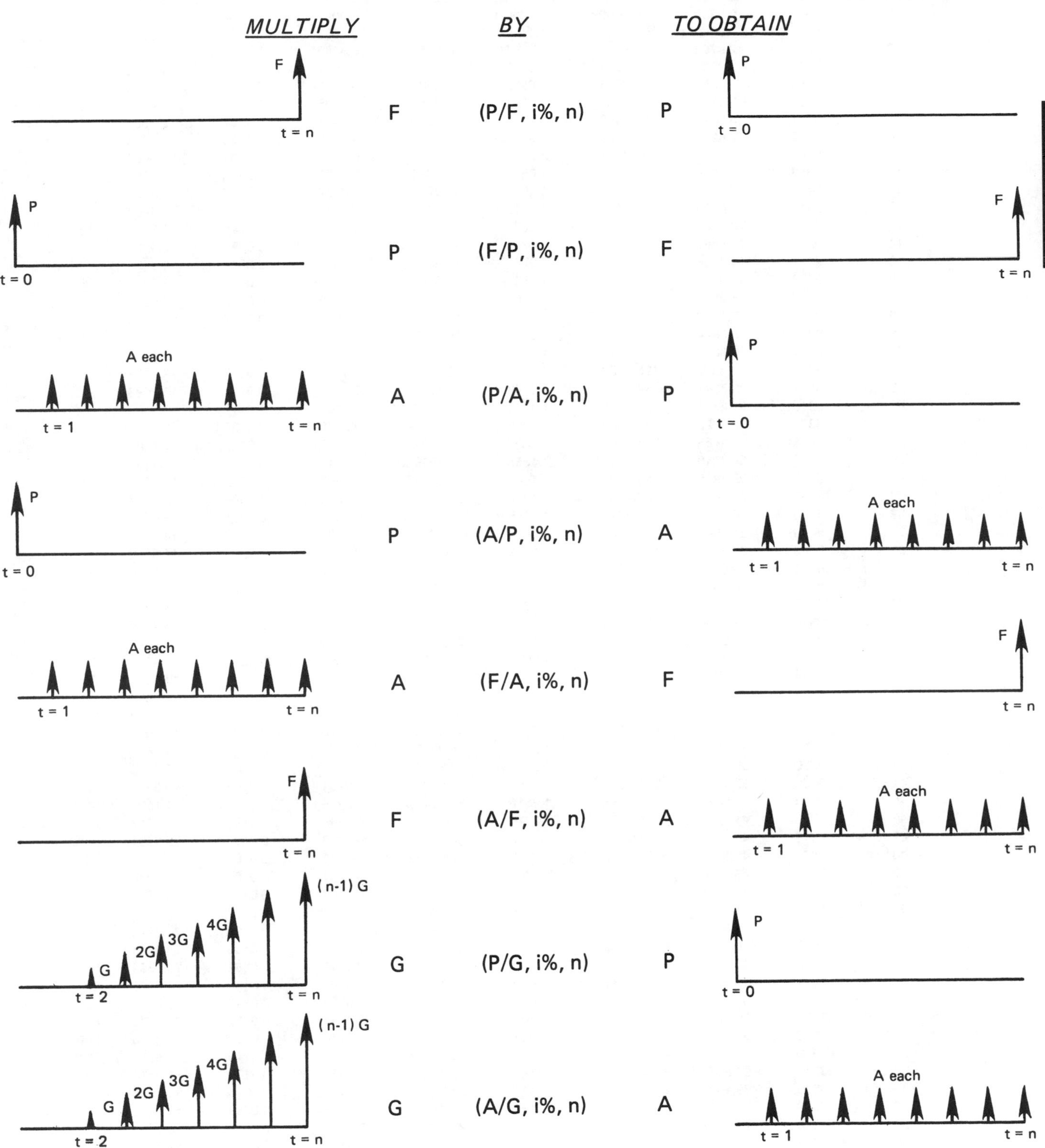

| MULTIPLY | BY | TO OBTAIN |
|---|---|---|
| F | (P/F, i%, n) | P |
| P | (F/P, i%, n) | F |
| A | (P/A, i%, n) | P |
| P | (A/P, i%, n) | A |
| A | (F/A, i%, n) | F |
| F | (A/F, i%, n) | A |
| G | (P/G, i%, n) | P |
| G | (A/G, i%, n) | A |

**I = 0.50 %**

| N | (P/F) | (P/A) | (P/G) | (F/P) | (F/A) | (A/P) | (A/F) | (A/G) | N |
|---|---|---|---|---|---|---|---|---|---|
| 1 | .9950 | 0.9950 | −0.0000 | 1.0050 | 1.0000 | 1.0050 | 1.0000 | −0.0000 | 1 |
| 2 | .9901 | 1.9851 | 0.9901 | 1.0100 | 2.0050 | 0.5038 | 0.4988 | 0.4988 | 2 |
| 3 | .9851 | 2.9702 | 2.9604 | 1.0151 | 3.0150 | 0.3367 | 0.3317 | 0.9967 | 3 |
| 4 | .9802 | 3.9505 | 5.9011 | 1.0202 | 4.0301 | 0.2531 | 0.2481 | 1.4938 | 4 |
| 5 | .9754 | 4.9259 | 9.8026 | 1.0253 | 5.0503 | 0.2030 | 0.1980 | 1.9900 | 5 |
| 6 | .9705 | 5.8964 | 14.6552 | 1.0304 | 6.0755 | 0.1696 | 0.1646 | 2.4855 | 6 |
| 7 | .9657 | 6.8621 | 20.4493 | 1.0355 | 7.1059 | 0.1457 | 0.1407 | 2.9801 | 7 |
| 8 | .9609 | 7.8230 | 27.1755 | 1.0407 | 8.1414 | 0.1278 | 0.1228 | 3.4738 | 8 |
| 9 | .9561 | 8.7791 | 34.8244 | 1.0459 | 9.1821 | 0.1139 | 0.1089 | 3.9668 | 9 |
| 10 | .9513 | 9.7304 | 43.3865 | 1.0511 | 10.2280 | 0.1028 | 0.0978 | 4.4589 | 10 |
| 11 | .9466 | 10.6770 | 52.8526 | 1.0564 | 11.2792 | 0.0937 | 0.0887 | 4.9501 | 11 |
| 12 | .9419 | 11.6189 | 63.2136 | 1.0617 | 12.3356 | 0.0861 | 0.0811 | 5.4406 | 12 |
| 13 | .9372 | 12.5562 | 74.4602 | 1.0670 | 13.3972 | 0.0796 | 0.0746 | 5.9302 | 13 |
| 14 | .9326 | 13.4887 | 86.5835 | 1.0723 | 14.4642 | 0.0741 | 0.0691 | 6.4190 | 14 |
| 15 | .9279 | 14.4166 | 99.5743 | 1.0777 | 15.5365 | 0.0694 | 0.0644 | 6.9069 | 15 |
| 16 | .9233 | 15.3399 | 113.4238 | 1.0831 | 16.6142 | 0.0652 | 0.0602 | 7.3940 | 16 |
| 17 | .9187 | 16.2586 | 128.1231 | 1.0885 | 17.6973 | 0.0615 | 0.0565 | 7.8803 | 17 |
| 18 | .9141 | 17.1728 | 143.6634 | 1.0939 | 18.7858 | 0.0582 | 0.0532 | 8.3658 | 18 |
| 19 | .9096 | 18.0824 | 160.0360 | 1.0994 | 19.8797 | 0.0553 | 0.0503 | 8.8504 | 19 |
| 20 | .9051 | 18.9874 | 177.2322 | 1.1049 | 20.9791 | 0.0527 | 0.0477 | 9.3342 | 20 |
| 21 | .9006 | 19.8880 | 195.2434 | 1.1104 | 22.0840 | 0.0503 | 0.0453 | 9.8172 | 21 |
| 22 | .8961 | 20.7841 | 214.0611 | 1.1160 | 23.1944 | 0.0481 | 0.0431 | 10.2993 | 22 |
| 23 | .8916 | 21.6757 | 233.6768 | 1.1216 | 24.3104 | 0.0461 | 0.0411 | 10.7806 | 23 |
| 24 | .8872 | 22.5629 | 254.0820 | 1.1272 | 25.4320 | 0.0443 | 0.0393 | 11.2611 | 24 |
| 25 | .8828 | 23.4456 | 275.2686 | 1.1328 | 26.5591 | 0.0427 | 0.0377 | 11.7407 | 25 |
| 26 | .8784 | 24.3240 | 297.2281 | 1.1385 | 27.6919 | 0.0411 | 0.0361 | 12.2195 | 26 |
| 27 | .8740 | 25.1980 | 319.9523 | 1.1442 | 28.8304 | 0.0397 | 0.0347 | 12.6975 | 27 |
| 28 | .8697 | 26.0677 | 343.4332 | 1.1499 | 29.9745 | 0.0384 | 0.0334 | 13.1747 | 28 |
| 29 | .8653 | 26.9330 | 367.6625 | 1.1556 | 31.1244 | 0.0371 | 0.0321 | 13.6510 | 29 |
| 30 | .8610 | 27.7941 | 392.6324 | 1.1614 | 32.2800 | 0.0360 | 0.0310 | 14.1265 | 30 |
| 31 | .8567 | 28.6508 | 418.3348 | 1.1672 | 33.4414 | 0.0349 | 0.0299 | 14.6012 | 31 |
| 32 | .8525 | 29.5033 | 444.7618 | 1.1730 | 34.6086 | 0.0339 | 0.0289 | 15.0750 | 32 |
| 33 | .8482 | 30.3515 | 471.9055 | 1.1789 | 35.7817 | 0.0329 | 0.0279 | 15.5480 | 33 |
| 34 | .8440 | 31.1955 | 499.7583 | 1.1848 | 36.9606 | 0.0321 | 0.0271 | 16.0202 | 34 |
| 35 | .8398 | 32.0354 | 528.3123 | 1.1907 | 38.1454 | 0.0312 | 0.0262 | 16.4915 | 35 |
| 36 | .8356 | 32.8710 | 557.5598 | 1.1967 | 39.3361 | 0.0304 | 0.0254 | 16.9621 | 36 |
| 37 | .8315 | 33.7025 | 587.4934 | 1.2027 | 40.5328 | 0.0297 | 0.0247 | 17.4317 | 37 |
| 38 | .8274 | 34.5299 | 618.1054 | 1.2087 | 41.7354 | 0.0290 | 0.0240 | 17.9006 | 38 |
| 39 | .8232 | 35.3531 | 649.3883 | 1.2147 | 42.9441 | 0.0283 | 0.0233 | 18.3686 | 39 |
| 40 | .8191 | 36.1722 | 681.3347 | 1.2208 | 44.1588 | 0.0276 | 0.0226 | 18.8359 | 40 |
| 41 | .8151 | 36.9873 | 713.9372 | 1.2269 | 45.3796 | 0.0270 | 0.0220 | 19.3022 | 41 |
| 42 | .8110 | 37.7983 | 747.1886 | 1.2330 | 46.6065 | 0.0265 | 0.0215 | 19.7678 | 42 |
| 43 | .8070 | 38.6053 | 781.0815 | 1.2392 | 47.8396 | 0.0259 | 0.0209 | 20.2325 | 43 |
| 44 | .8030 | 39.4082 | 815.6087 | 1.2454 | 49.0788 | 0.0254 | 0.0204 | 20.6964 | 44 |
| 45 | .7990 | 40.2072 | 850.7631 | 1.2516 | 50.3242 | 0.0249 | 0.0199 | 21.1595 | 45 |
| 46 | .7950 | 41.0022 | 886.5376 | 1.2579 | 51.5758 | 0.0244 | 0.0194 | 21.6217 | 46 |
| 47 | .7910 | 41.7932 | 922.9252 | 1.2642 | 52.8337 | 0.0239 | 0.0189 | 22.0831 | 47 |
| 48 | .7871 | 42.5803 | 959.9188 | 1.2705 | 54.0978 | 0.0235 | 0.0185 | 22.5437 | 48 |
| 49 | .7832 | 43.3635 | 997.5116 | 1.2768 | 55.3683 | 0.0231 | 0.0181 | 23.0035 | 49 |
| 50 | .7793 | 44.1428 | 1035.6966 | 1.2832 | 56.6452 | 0.0227 | 0.0177 | 23.4624 | 50 |
| 51 | .7754 | 44.9182 | 1074.4670 | 1.2896 | 57.9284 | 0.0223 | 0.0173 | 23.9205 | 51 |
| 52 | .7716 | 45.6897 | 1113.8162 | 1.2961 | 59.2180 | 0.0219 | 0.0169 | 24.3778 | 52 |
| 53 | .7677 | 46.4575 | 1153.7372 | 1.3026 | 60.5141 | 0.0215 | 0.0165 | 24.8343 | 53 |
| 54 | .7639 | 47.2214 | 1194.2236 | 1.3091 | 61.8167 | 0.0212 | 0.0162 | 25.2899 | 54 |
| 55 | .7601 | 47.9814 | 1235.2686 | 1.3156 | 63.1258 | 0.0208 | 0.0158 | 25.7447 | 55 |
| 60 | .7414 | 51.7256 | 1448.6458 | 1.3489 | 69.7700 | 0.0193 | 0.0143 | 28.0064 | 60 |
| 65 | .7231 | 55.3775 | 1675.0272 | 1.3829 | 76.5821 | 0.0181 | 0.0131 | 30.2475 | 65 |
| 70 | .7053 | 58.9394 | 1913.6427 | 1.4178 | 83.5661 | 0.0170 | 0.0120 | 32.4680 | 70 |
| 75 | .6879 | 62.4136 | 2163.7525 | 1.4536 | 90.7265 | 0.0160 | 0.0110 | 34.6679 | 75 |
| 80 | .6710 | 65.8023 | 2424.6455 | 1.4903 | 98.0677 | 0.0152 | 0.0102 | 36.8474 | 80 |
| 85 | .6545 | 69.1075 | 2695.6389 | 1.5280 | 105.5943 | 0.0145 | 0.0095 | 39.0065 | 85 |
| 90 | .6383 | 72.3313 | 2976.0769 | 1.5666 | 113.3109 | 0.0138 | 0.0088 | 41.1451 | 90 |
| 95 | .6226 | 75.4757 | 3265.3298 | 1.6061 | 121.2224 | 0.0132 | 0.0082 | 43.2633 | 95 |
| 100 | .6073 | 78.5426 | 3562.7934 | 1.6467 | 129.3337 | 0.0127 | 0.0077 | 45.3613 | 100 |

**I = 0.75 %**

| N | (P/F) | (P/A) | (P/G) | (F/P) | (F/A) | (A/P) | (A/F) | (A/G) | N |
|---|---|---|---|---|---|---|---|---|---|
| 1 | .9926 | 0.9926 | -0.0000 | 1.0075 | 1.0000 | 1.0075 | 1.0000 | -0.0000 | 1 |
| 2 | .9852 | 1.9777 | 0.9852 | 1.0151 | 2.0075 | 0.5056 | 0.4981 | 0.4981 | 2 |
| 3 | .9778 | 2.9556 | 2.9408 | 1.0227 | 3.0226 | 0.3383 | 0.3308 | 0.9950 | 3 |
| 4 | .9706 | 3.9261 | 5.8525 | 1.0303 | 4.0452 | 0.2547 | 0.2472 | 1.4907 | 4 |
| 5 | .9633 | 4.8894 | 9.7058 | 1.0381 | 5.0756 | 0.2045 | 0.1970 | 1.9851 | 5 |
| 6 | .9562 | 5.8456 | 14.4866 | 1.0459 | 6.1136 | 0.1711 | 0.1636 | 2.4782 | 6 |
| 7 | .9490 | 6.7946 | 20.1808 | 1.0537 | 7.1595 | 0.1472 | 0.1397 | 2.9701 | 7 |
| 8 | .9420 | 7.7366 | 26.7747 | 1.0616 | 8.2132 | 0.1293 | 0.1218 | 3.4608 | 8 |
| 9 | .9350 | 8.6716 | 34.2544 | 1.0696 | 9.2748 | 0.1153 | 0.1078 | 3.9502 | 9 |
| 10 | .9280 | 9.5996 | 42.6064 | 1.0776 | 10.3443 | 0.1042 | 0.0967 | 4.4384 | 10 |
| 11 | .9211 | 10.5207 | 51.8174 | 1.0857 | 11.4219 | 0.0951 | 0.0876 | 4.9253 | 11 |
| 12 | .9142 | 11.4349 | 61.8740 | 1.0938 | 12.5076 | 0.0875 | 0.0800 | 5.4110 | 12 |
| 13 | .9074 | 12.3423 | 72.7632 | 1.1020 | 13.6014 | 0.0810 | 0.0735 | 5.8954 | 13 |
| 14 | .9007 | 13.2430 | 84.4720 | 1.1103 | 14.7034 | 0.0755 | 0.0680 | 6.3786 | 14 |
| 15 | .8940 | 14.1370 | 96.9876 | 1.1186 | 15.8137 | 0.0707 | 0.0632 | 6.8606 | 15 |
| 16 | .8873 | 15.0243 | 110.2973 | 1.1270 | 16.9323 | 0.0666 | 0.0591 | 7.3413 | 16 |
| 17 | .8807 | 15.9050 | 124.3887 | 1.1354 | 18.0593 | 0.0629 | 0.0554 | 7.8207 | 17 |
| 18 | .8742 | 16.7792 | 139.2494 | 1.1440 | 19.1947 | 0.0596 | 0.0521 | 8.2989 | 18 |
| 19 | .8676 | 17.6468 | 154.8671 | 1.1525 | 20.3387 | 0.0567 | 0.0492 | 8.7759 | 19 |
| 20 | .8612 | 18.5080 | 171.2297 | 1.1612 | 21.4912 | 0.0540 | 0.0465 | 9.2516 | 20 |
| 21 | .8548 | 19.3628 | 188.3253 | 1.1699 | 22.6524 | 0.0516 | 0.0441 | 9.7261 | 21 |
| 22 | .8484 | 20.2112 | 206.1420 | 1.1787 | 23.8223 | 0.0495 | 0.0420 | 10.1994 | 22 |
| 23 | .8421 | 21.0533 | 224.6682 | 1.1875 | 25.0010 | 0.0475 | 0.0400 | 10.6714 | 23 |
| 24 | .8358 | 21.8891 | 243.8923 | 1.1964 | 26.1885 | 0.0457 | 0.0382 | 11.1422 | 24 |
| 25 | .8296 | 22.7188 | 263.8029 | 1.2054 | 27.3849 | 0.0440 | 0.0365 | 11.6117 | 25 |
| 26 | .8234 | 23.5422 | 284.3888 | 1.2144 | 28.5903 | 0.0425 | 0.0350 | 12.0800 | 26 |
| 27 | .8173 | 24.3595 | 305.6387 | 1.2235 | 29.8047 | 0.0411 | 0.0336 | 12.5470 | 27 |
| 28 | .8112 | 25.1707 | 327.5416 | 1.2327 | 31.0282 | 0.0397 | 0.0322 | 13.0128 | 28 |
| 29 | .8052 | 25.9759 | 350.0867 | 1.2420 | 32.2609 | 0.0385 | 0.0310 | 13.4774 | 29 |
| 30 | .7992 | 26.7751 | 373.2631 | 1.2513 | 33.5029 | 0.0373 | 0.0298 | 13.9407 | 30 |
| 31 | .7932 | 27.5683 | 397.0602 | 1.2607 | 34.7542 | 0.0363 | 0.0288 | 14.4028 | 31 |
| 32 | .7873 | 28.3557 | 421.4675 | 1.2701 | 36.0148 | 0.0353 | 0.0278 | 14.8636 | 32 |
| 33 | .7815 | 29.1371 | 446.4746 | 1.2796 | 37.2849 | 0.0343 | 0.0268 | 15.3232 | 33 |
| 34 | .7757 | 29.9128 | 472.0712 | 1.2892 | 38.5646 | 0.0334 | 0.0259 | 15.7816 | 34 |
| 35 | .7699 | 30.6827 | 498.2471 | 1.2989 | 39.8538 | 0.0326 | 0.0251 | 16.2387 | 35 |
| 36 | .7641 | 31.4468 | 524.9924 | 1.3086 | 41.1527 | 0.0318 | 0.0243 | 16.6946 | 36 |
| 37 | .7585 | 32.2053 | 552.2969 | 1.3185 | 42.4614 | 0.0311 | 0.0236 | 17.1493 | 37 |
| 38 | .7528 | 32.9581 | 580.1511 | 1.3283 | 43.7798 | 0.0303 | 0.0228 | 17.6027 | 38 |
| 39 | .7472 | 33.7053 | 608.5451 | 1.3383 | 45.1082 | 0.0297 | 0.0222 | 18.0549 | 39 |
| 40 | .7416 | 34.4469 | 637.4693 | 1.3483 | 46.4465 | 0.0290 | 0.0215 | 18.5058 | 40 |
| 41 | .7361 | 35.1831 | 666.9144 | 1.3585 | 47.7948 | 0.0284 | 0.0209 | 18.9556 | 41 |
| 42 | .7306 | 35.9137 | 696.8709 | 1.3686 | 49.1533 | 0.0278 | 0.0203 | 19.4040 | 42 |
| 43 | .7252 | 36.6389 | 727.3297 | 1.3789 | 50.5219 | 0.0273 | 0.0198 | 19.8513 | 43 |
| 44 | .7198 | 37.3587 | 758.2815 | 1.3893 | 51.9009 | 0.0268 | 0.0193 | 20.2973 | 44 |
| 45 | .7145 | 38.0732 | 789.7173 | 1.3997 | 53.2901 | 0.0263 | 0.0188 | 20.7421 | 45 |
| 46 | .7091 | 38.7823 | 821.6283 | 1.4102 | 54.6898 | 0.0258 | 0.0183 | 21.1856 | 46 |
| 47 | .7039 | 39.4862 | 854.0056 | 1.4207 | 56.1000 | 0.0253 | 0.0178 | 21.6280 | 47 |
| 48 | .6986 | 40.1848 | 886.8404 | 1.4314 | 57.5207 | 0.0249 | 0.0174 | 22.0691 | 48 |
| 49 | .6934 | 40.8782 | 920.1243 | 1.4421 | 58.9521 | 0.0245 | 0.0170 | 22.5089 | 49 |
| 50 | .6883 | 41.5664 | 953.8486 | 1.4530 | 60.3943 | 0.0241 | 0.0166 | 22.9476 | 50 |
| 51 | .6831 | 42.2496 | 988.0050 | 1.4639 | 61.8472 | 0.0237 | 0.0162 | 23.3850 | 51 |
| 52 | .6780 | 42.9276 | 1022.5852 | 1.4748 | 63.3111 | 0.0233 | 0.0158 | 23.8211 | 52 |
| 53 | .6730 | 43.6006 | 1057.5810 | 1.4859 | 64.7859 | 0.0229 | 0.0154 | 24.2561 | 53 |
| 54 | .6680 | 44.2686 | 1092.9842 | 1.4970 | 66.2718 | 0.0226 | 0.0151 | 24.6898 | 54 |
| 55 | .6630 | 44.9316 | 1128.7869 | 1.5083 | 67.7688 | 0.0223 | 0.0148 | 25.1223 | 55 |
| 60 | .6387 | 48.1734 | 1313.5189 | 1.5657 | 75.4241 | 0.0208 | 0.0133 | 27.2665 | 60 |
| 65 | .6153 | 51.2963 | 1507.0910 | 1.6253 | 83.3709 | 0.0195 | 0.0120 | 29.3801 | 65 |
| 70 | .5927 | 54.3046 | 1708.6065 | 1.6872 | 91.6201 | 0.0184 | 0.0109 | 31.4634 | 70 |
| 75 | .5710 | 57.2027 | 1917.2225 | 1.7514 | 100.1833 | 0.0175 | 0.0100 | 33.5163 | 75 |
| 80 | .5500 | 59.9944 | 2132.1472 | 1.8180 | 109.0725 | 0.0167 | 0.0092 | 35.5391 | 80 |
| 85 | .5299 | 62.6838 | 2352.6375 | 1.8873 | 118.3001 | 0.0160 | 0.0085 | 37.5318 | 85 |
| 90 | .5104 | 65.2746 | 2577.9961 | 1.9591 | 127.8790 | 0.0153 | 0.0078 | 39.4946 | 90 |
| 95 | .4917 | 67.7704 | 2807.5694 | 2.0337 | 137.8225 | 0.0148 | 0.0073 | 41.4277 | 95 |
| 100 | .4737 | 70.1746 | 3040.7453 | 2.1111 | 148.1445 | 0.0143 | 0.0068 | 43.3311 | 100 |

**I = 1.00 %**

| N | (P/F) | (P/A) | (P/G) | (F/P) | (F/A) | (A/P) | (A/F) | (A/G) | N |
|---|---|---|---|---|---|---|---|---|---|
| 1 | .9901 | 0.9901 | -0.0000 | 1.0100 | 1.0000 | 1.0100 | 1.0000 | -0.0000 | 1 |
| 2 | .9803 | 1.9704 | 0.9803 | 1.0201 | 2.0100 | 0.5075 | 0.4975 | 0.4975 | 2 |
| 3 | .9706 | 2.9410 | 2.9215 | 1.0303 | 3.0301 | 0.3400 | 0.3300 | 0.9934 | 3 |
| 4 | .9610 | 3.9020 | 5.8044 | 1.0406 | 4.0604 | 0.2563 | 0.2463 | 1.4876 | 4 |
| 5 | .9515 | 4.8534 | 9.6103 | 1.0510 | 5.1010 | 0.2060 | 0.1960 | 1.9801 | 5 |
| 6 | .9420 | 5.7955 | 14.3205 | 1.0615 | 6.1520 | 0.1725 | 0.1625 | 2.4710 | 6 |
| 7 | .9327 | 6.7282 | 19.9168 | 1.0721 | 7.2135 | 0.1486 | 0.1386 | 2.9602 | 7 |
| 8 | .9235 | 7.6517 | 26.3812 | 1.0829 | 8.2857 | 0.1307 | 0.1207 | 3.4478 | 8 |
| 9 | .9143 | 8.5660 | 33.6959 | 1.0937 | 9.3685 | 0.1167 | 0.1067 | 3.9337 | 9 |
| 10 | .9053 | 9.4713 | 41.8435 | 1.1046 | 10.4622 | 0.1056 | 0.0956 | 4.4179 | 10 |
| 11 | .8963 | 10.3676 | 50.8067 | 1.1157 | 11.5668 | 0.0965 | 0.0865 | 4.9005 | 11 |
| 12 | .8874 | 11.2551 | 60.5687 | 1.1268 | 12.6825 | 0.0888 | 0.0788 | 5.3815 | 12 |
| 13 | .8787 | 12.1337 | 71.1126 | 1.1381 | 13.8093 | 0.0824 | 0.0724 | 5.8607 | 13 |
| 14 | .8700 | 13.0037 | 82.4221 | 1.1495 | 14.9474 | 0.0769 | 0.0669 | 6.3384 | 14 |
| 15 | .8613 | 13.8651 | 94.4810 | 1.1610 | 16.0969 | 0.0721 | 0.0621 | 6.8143 | 15 |
| 16 | .8528 | 14.7179 | 107.2734 | 1.1726 | 17.2579 | 0.0679 | 0.0579 | 7.2886 | 16 |
| 17 | .8444 | 15.5623 | 120.7834 | 1.1843 | 18.4304 | 0.0643 | 0.0543 | 7.7613 | 17 |
| 18 | .8360 | 16.3983 | 134.9957 | 1.1961 | 19.6147 | 0.0610 | 0.0510 | 8.2323 | 18 |
| 19 | .8277 | 17.2260 | 149.8950 | 1.2081 | 20.8109 | 0.0581 | 0.0481 | 8.7017 | 19 |
| 20 | .8195 | 18.0456 | 165.4664 | 1.2202 | 22.0190 | 0.0554 | 0.0454 | 9.1694 | 20 |
| 21 | .8114 | 18.8570 | 181.6950 | 1.2324 | 23.2392 | 0.0530 | 0.0430 | 9.6354 | 21 |
| 22 | .8034 | 19.6604 | 198.5663 | 1.2447 | 24.4716 | 0.0509 | 0.0409 | 10.0998 | 22 |
| 23 | .7954 | 20.4558 | 216.0660 | 1.2572 | 25.7163 | 0.0489 | 0.0389 | 10.5626 | 23 |
| 24 | .7876 | 21.2434 | 234.1800 | 1.2697 | 26.9735 | 0.0471 | 0.0371 | 11.0237 | 24 |
| 25 | .7798 | 22.0232 | 252.8945 | 1.2824 | 28.2432 | 0.0454 | 0.0354 | 11.4831 | 25 |
| 26 | .7720 | 22.7952 | 272.1957 | 1.2953 | 29.5256 | 0.0439 | 0.0339 | 11.9409 | 26 |
| 27 | .7644 | 23.5596 | 292.0702 | 1.3082 | 30.8209 | 0.0424 | 0.0324 | 12.3971 | 27 |
| 28 | .7568 | 24.3164 | 312.5047 | 1.3213 | 32.1291 | 0.0411 | 0.0311 | 12.8516 | 28 |
| 29 | .7493 | 25.0658 | 333.4863 | 1.3345 | 33.4504 | 0.0399 | 0.0299 | 13.3044 | 29 |
| 30 | .7419 | 25.8077 | 355.0021 | 1.3478 | 34.7849 | 0.0387 | 0.0287 | 13.7557 | 30 |
| 31 | .7346 | 26.5423 | 377.0394 | 1.3613 | 36.1327 | 0.0377 | 0.0277 | 14.2052 | 31 |
| 32 | .7273 | 27.2696 | 399.5858 | 1.3749 | 37.4941 | 0.0367 | 0.0267 | 14.6532 | 32 |
| 33 | .7201 | 27.9897 | 422.6291 | 1.3887 | 38.8690 | 0.0357 | 0.0257 | 15.0995 | 33 |
| 34 | .7130 | 28.7027 | 446.1572 | 1.4026 | 40.2577 | 0.0348 | 0.0248 | 15.5441 | 34 |
| 35 | .7059 | 29.4086 | 470.1583 | 1.4166 | 41.6603 | 0.0340 | 0.0240 | 15.9871 | 35 |
| 36 | .6989 | 30.1075 | 494.6207 | 1.4308 | 43.0769 | 0.0332 | 0.0232 | 16.4285 | 36 |
| 37 | .6920 | 30.7995 | 519.5329 | 1.4451 | 44.5076 | 0.0325 | 0.0225 | 16.8682 | 37 |
| 38 | .6852 | 31.4847 | 544.8835 | 1.4595 | 45.9527 | 0.0318 | 0.0218 | 17.3063 | 38 |
| 39 | .6784 | 32.1630 | 570.6616 | 1.4741 | 47.4123 | 0.0311 | 0.0211 | 17.7428 | 39 |
| 40 | .6717 | 32.8347 | 596.8561 | 1.4889 | 48.8864 | 0.0305 | 0.0205 | 18.1776 | 40 |
| 41 | .6650 | 33.4997 | 623.4562 | 1.5038 | 50.3752 | 0.0299 | 0.0199 | 18.6108 | 41 |
| 42 | .6584 | 34.1581 | 650.4514 | 1.5188 | 51.8790 | 0.0293 | 0.0193 | 19.0424 | 42 |
| 43 | .6519 | 34.8100 | 677.8312 | 1.5340 | 53.3978 | 0.0287 | 0.0187 | 19.4723 | 43 |
| 44 | .6454 | 35.4555 | 705.5853 | 1.5493 | 54.9318 | 0.0282 | 0.0182 | 19.9006 | 44 |
| 45 | .6391 | 36.0945 | 733.7037 | 1.5648 | 56.4811 | 0.0277 | 0.0177 | 20.3273 | 45 |
| 46 | .6327 | 36.7272 | 762.1765 | 1.5805 | 58.0459 | 0.0272 | 0.0172 | 20.7524 | 46 |
| 47 | .6265 | 37.3537 | 790.9938 | 1.5963 | 59.6263 | 0.0268 | 0.0168 | 21.1758 | 47 |
| 48 | .6203 | 37.9740 | 820.1460 | 1.6122 | 61.2226 | 0.0263 | 0.0163 | 21.5976 | 48 |
| 49 | .6141 | 38.5881 | 849.6237 | 1.6283 | 62.8348 | 0.0259 | 0.0159 | 22.0178 | 49 |
| 50 | .6080 | 39.1961 | 879.4176 | 1.6446 | 64.4632 | 0.0255 | 0.0155 | 22.4363 | 50 |
| 51 | .6020 | 39.7981 | 909.5186 | 1.6611 | 66.1078 | 0.0251 | 0.0151 | 22.8533 | 51 |
| 52 | .5961 | 40.3942 | 939.9175 | 1.6777 | 67.7689 | 0.0248 | 0.0148 | 23.2686 | 52 |
| 53 | .5902 | 40.9844 | 970.6057 | 1.6945 | 69.4466 | 0.0244 | 0.0144 | 23.6823 | 53 |
| 54 | .5843 | 41.5687 | 1001.5743 | 1.7114 | 71.1410 | 0.0241 | 0.0141 | 24.0945 | 54 |
| 55 | .5785 | 42.1472 | 1032.8148 | 1.7285 | 72.8525 | 0.0237 | 0.0137 | 24.5049 | 55 |
| 60 | .5504 | 44.9550 | 1192.8061 | 1.8167 | 81.6697 | 0.0222 | 0.0122 | 26.5333 | 60 |
| 65 | .5237 | 47.6266 | 1358.3903 | 1.9094 | 90.9366 | 0.0210 | 0.0110 | 28.5217 | 65 |
| 70 | .4983 | 50.1685 | 1528.6474 | 2.0068 | 100.6763 | 0.0199 | 0.0099 | 30.4703 | 70 |
| 75 | .4741 | 52.5871 | 1702.7340 | 2.1091 | 110.9128 | 0.0190 | 0.0090 | 32.3793 | 75 |
| 80 | .4511 | 54.8882 | 1879.8771 | 2.2167 | 121.6715 | 0.0182 | 0.0082 | 34.2492 | 80 |
| 85 | .4292 | 57.0777 | 2059.3701 | 2.3298 | 132.9790 | 0.0175 | 0.0075 | 36.0801 | 85 |
| 90 | .4084 | 59.1609 | 2240.5675 | 2.4486 | 144.8633 | 0.0169 | 0.0069 | 37.8724 | 90 |
| 95 | .3886 | 61.1430 | 2422.8811 | 2.5735 | 157.3538 | 0.0164 | 0.0064 | 39.6265 | 95 |
| 100 | .3697 | 63.0289 | 2605.7758 | 2.7048 | 170.4814 | 0.0159 | 0.0059 | 41.3426 | 100 |

**I = 1.50 %**

| N | (P/F) | (P/A) | (P/G) | (F/P) | (F/A) | (A/P) | (A/F) | (A/G) | N |
|---|---|---|---|---|---|---|---|---|---|
| 1 | .9852 | 0.9852 | -0.0000 | 1.0150 | 1.0000 | 1.0150 | 1.0000 | -0.0000 | 1 |
| 2 | .9707 | 1.9559 | 0.9707 | 1.0302 | 2.0150 | 0.5113 | 0.4963 | 0.4963 | 2 |
| 3 | .9563 | 2.9122 | 2.8833 | 1.0457 | 3.0452 | 0.3434 | 0.3284 | 0.9901 | 3 |
| 4 | .9422 | 3.8544 | 5.7098 | 1.0614 | 4.0909 | 0.2594 | 0.2444 | 1.4814 | 4 |
| 5 | .9283 | 4.7826 | 9.4229 | 1.0773 | 5.1523 | 0.2091 | 0.1941 | 1.9702 | 5 |
| 6 | .9145 | 5.6972 | 13.9956 | 1.0934 | 6.2296 | 0.1755 | 0.1605 | 2.4566 | 6 |
| 7 | .9010 | 6.5982 | 19.4018 | 1.1098 | 7.3230 | 0.1516 | 0.1366 | 2.9405 | 7 |
| 8 | .8877 | 7.4859 | 25.6157 | 1.1265 | 8.4328 | 0.1336 | 0.1186 | 3.4219 | 8 |
| 9 | .8746 | 8.3605 | 32.6125 | 1.1434 | 9.5593 | 0.1196 | 0.1046 | 3.9008 | 9 |
| 10 | .8617 | 9.2222 | 40.3675 | 1.1605 | 10.7027 | 0.1084 | 0.0934 | 4.3772 | 10 |
| 11 | .8489 | 10.0711 | 48.8568 | 1.1779 | 11.8633 | 0.0993 | 0.0843 | 4.8512 | 11 |
| 12 | .8364 | 10.9075 | 58.0571 | 1.1956 | 13.0412 | 0.0917 | 0.0767 | 5.3227 | 12 |
| 13 | .8240 | 11.7315 | 67.9454 | 1.2136 | 14.2368 | 0.0852 | 0.0702 | 5.7917 | 13 |
| 14 | .8118 | 12.5434 | 78.4994 | 1.2318 | 15.4504 | 0.0797 | 0.0647 | 6.2582 | 14 |
| 15 | .7999 | 13.3432 | 89.6974 | 1.2502 | 16.6821 | 0.0749 | 0.0599 | 6.7223 | 15 |
| 16 | .7880 | 14.1313 | 101.5178 | 1.2690 | 17.9324 | 0.0708 | 0.0558 | 7.1839 | 16 |
| 17 | .7764 | 14.9076 | 113.9400 | 1.2880 | 19.2014 | 0.0671 | 0.0521 | 7.6431 | 17 |
| 18 | .7649 | 15.6726 | 126.9435 | 1.3073 | 20.4894 | 0.0638 | 0.0488 | 8.0997 | 18 |
| 19 | .7536 | 16.4262 | 140.5084 | 1.3270 | 21.7967 | 0.0609 | 0.0459 | 8.5539 | 19 |
| 20 | .7425 | 17.1686 | 154.6154 | 1.3469 | 23.1237 | 0.0582 | 0.0432 | 9.0057 | 20 |
| 21 | .7315 | 17.9001 | 169.2453 | 1.3671 | 24.4705 | 0.0559 | 0.0409 | 9.4550 | 21 |
| 22 | .7207 | 18.6208 | 184.3798 | 1.3876 | 25.8376 | 0.0537 | 0.0387 | 9.9018 | 22 |
| 23 | .7100 | 19.3309 | 200.0006 | 1.4084 | 27.2251 | 0.0517 | 0.0367 | 10.3462 | 23 |
| 24 | .6995 | 20.0304 | 216.0901 | 1.4295 | 28.6335 | 0.0499 | 0.0349 | 10.7881 | 24 |
| 25 | .6892 | 20.7196 | 232.6310 | 1.4509 | 30.0630 | 0.0483 | 0.0333 | 11.2276 | 25 |
| 26 | .6790 | 21.3986 | 249.6065 | 1.4727 | 31.5140 | 0.0467 | 0.0317 | 11.6646 | 26 |
| 27 | .6690 | 22.0676 | 267.0002 | 1.4948 | 32.9867 | 0.0453 | 0.0303 | 12.0992 | 27 |
| 28 | .6591 | 22.7267 | 284.7958 | 1.5172 | 34.4815 | 0.0440 | 0.0290 | 12.5313 | 28 |
| 29 | .6494 | 23.3761 | 302.9779 | 1.5400 | 35.9987 | 0.0428 | 0.0278 | 12.9610 | 29 |
| 30 | .6398 | 24.0158 | 321.5310 | 1.5631 | 37.5387 | 0.0416 | 0.0266 | 13.3883 | 30 |
| 31 | .6303 | 24.6461 | 340.4402 | 1.5865 | 39.1018 | 0.0406 | 0.0256 | 13.8131 | 31 |
| 32 | .6210 | 25.2671 | 359.6910 | 1.6103 | 40.6883 | 0.0396 | 0.0246 | 14.2355 | 32 |
| 33 | .6118 | 25.8790 | 379.2691 | 1.6345 | 42.2986 | 0.0386 | 0.0236 | 14.6555 | 33 |
| 34 | .6028 | 26.4817 | 399.1607 | 1.6590 | 43.9331 | 0.0378 | 0.0228 | 15.0731 | 34 |
| 35 | .5939 | 27.0756 | 419.3521 | 1.6839 | 45.5921 | 0.0369 | 0.0219 | 15.4882 | 35 |
| 36 | .5851 | 27.6607 | 439.8303 | 1.7091 | 47.2760 | 0.0362 | 0.0212 | 15.9009 | 36 |
| 37 | .5764 | 28.2371 | 460.5822 | 1.7348 | 48.9851 | 0.0354 | 0.0204 | 16.3112 | 37 |
| 38 | .5679 | 28.8051 | 481.5954 | 1.7608 | 50.7199 | 0.0347 | 0.0197 | 16.7191 | 38 |
| 39 | .5595 | 29.3646 | 502.8576 | 1.7872 | 52.4807 | 0.0341 | 0.0191 | 17.1246 | 39 |
| 40 | .5513 | 29.9158 | 524.3568 | 1.8140 | 54.2679 | 0.0334 | 0.0184 | 17.5277 | 40 |
| 41 | .5431 | 30.4590 | 546.0814 | 1.8412 | 56.0819 | 0.0328 | 0.0178 | 17.9284 | 41 |
| 42 | .5351 | 30.9941 | 568.0201 | 1.8688 | 57.9231 | 0.0323 | 0.0173 | 18.3267 | 42 |
| 43 | .5272 | 31.5212 | 590.1617 | 1.8969 | 59.7920 | 0.0317 | 0.0167 | 18.7227 | 43 |
| 44 | .5194 | 32.0406 | 612.4955 | 1.9253 | 61.6889 | 0.0312 | 0.0162 | 19.1162 | 44 |
| 45 | .5117 | 32.5523 | 635.0110 | 1.9542 | 63.6142 | 0.0307 | 0.0157 | 19.5074 | 45 |
| 46 | .5042 | 33.0565 | 657.6979 | 1.9835 | 65.5684 | 0.0303 | 0.0153 | 19.8962 | 46 |
| 47 | .4967 | 33.5532 | 680.5462 | 2.0133 | 67.5519 | 0.0298 | 0.0148 | 20.2826 | 47 |
| 48 | .4894 | 34.0426 | 703.5462 | 2.0435 | 69.5652 | 0.0294 | 0.0144 | 20.6667 | 48 |
| 49 | .4821 | 34.5247 | 726.6884 | 2.0741 | 71.6087 | 0.0290 | 0.0140 | 21.0484 | 49 |
| 50 | .4750 | 34.9997 | 749.9636 | 2.1052 | 73.6828 | 0.0286 | 0.0136 | 21.4277 | 50 |
| 51 | .4680 | 35.4677 | 773.3629 | 2.1368 | 75.7881 | 0.0282 | 0.0132 | 21.8047 | 51 |
| 52 | .4611 | 35.9287 | 796.8774 | 2.1689 | 77.9249 | 0.0278 | 0.0128 | 22.1794 | 52 |
| 53 | .4543 | 36.3830 | 820.4986 | 2.2014 | 80.0938 | 0.0275 | 0.0125 | 22.5517 | 53 |
| 54 | .4475 | 36.8305 | 844.2184 | 2.2344 | 82.2952 | 0.0272 | 0.0122 | 22.9217 | 54 |
| 55 | .4409 | 37.2715 | 868.0285 | 2.2679 | 84.5296 | 0.0268 | 0.0118 | 23.2894 | 55 |
| 60 | .4093 | 39.3803 | 988.1674 | 2.4432 | 96.2147 | 0.0254 | 0.0104 | 25.0930 | 60 |
| 65 | .3799 | 41.3378 | 1109.4752 | 2.6320 | 108.8028 | 0.0242 | 0.0092 | 26.8393 | 65 |
| 70 | .3527 | 43.1549 | 1231.1658 | 2.8355 | 122.3638 | 0.0232 | 0.0082 | 28.5290 | 70 |
| 75 | .3274 | 44.8416 | 1352.5600 | 3.0546 | 136.9728 | 0.0223 | 0.0073 | 30.1631 | 75 |
| 80 | .3039 | 46.4073 | 1473.0741 | 3.2907 | 152.7109 | 0.0215 | 0.0065 | 31.7423 | 80 |
| 85 | .2821 | 47.8607 | 1592.2095 | 3.5450 | 169.6652 | 0.0209 | 0.0059 | 33.2676 | 85 |
| 90 | .2619 | 49.2099 | 1709.5439 | 3.8189 | 187.9299 | 0.0203 | 0.0053 | 34.7399 | 90 |
| 95 | .2431 | 50.4622 | 1824.7224 | 4.1141 | 207.6061 | 0.0198 | 0.0048 | 36.1602 | 95 |
| 100 | .2256 | 51.6247 | 1937.4506 | 4.4320 | 228.8030 | 0.0194 | 0.0044 | 37.5295 | 100 |

**I = 2.00 %**

| N | (P/F) | (P/A) | (P/G) | (F/P) | (F/A) | (A/P) | (A/F) | (A/G) | N |
|---|---|---|---|---|---|---|---|---|---|
| 1 | .9804 | 0.9804 | -0.0000 | 1.0200 | 1.0000 | 1.0200 | 1.0000 | -0.0000 | 1 |
| 2 | .9612 | 1.9416 | 0.9612 | 1.0404 | 2.0200 | 0.5150 | 0.4950 | 0.4950 | 2 |
| 3 | .9423 | 2.8839 | 2.8458 | 1.0612 | 3.0604 | 0.3468 | 0.3268 | 0.9868 | 3 |
| 4 | .9238 | 3.8077 | 5.6173 | 1.0824 | 4.1216 | 0.2626 | 0.2426 | 1.4752 | 4 |
| 5 | .9057 | 4.7135 | 9.2403 | 1.1041 | 5.2040 | 0.2122 | 0.1922 | 1.9604 | 5 |
| 6 | .8880 | 5.6014 | 13.6801 | 1.1262 | 6.3081 | 0.1785 | 0.1585 | 2.4423 | 6 |
| 7 | .8706 | 6.4720 | 18.9035 | 1.1487 | 7.4343 | 0.1545 | 0.1345 | 2.9208 | 7 |
| 8 | .8535 | 7.3255 | 24.8779 | 1.1717 | 8.5830 | 0.1365 | 0.1165 | 3.3961 | 8 |
| 9 | .8368 | 8.1622 | 31.5720 | 1.1951 | 9.7546 | 0.1225 | 0.1025 | 3.8681 | 9 |
| 10 | .8203 | 8.9826 | 38.9551 | 1.2190 | 10.9497 | 0.1113 | 0.0913 | 4.3367 | 10 |
| 11 | .8043 | 9.7868 | 46.9977 | 1.2434 | 12.1687 | 0.1022 | 0.0822 | 4.8021 | 11 |
| 12 | .7885 | 10.5753 | 55.6712 | 1.2682 | 13.4121 | 0.0946 | 0.0746 | 5.2642 | 12 |
| 13 | .7730 | 11.3484 | 64.9475 | 1.2936 | 14.6803 | 0.0881 | 0.0681 | 5.7231 | 13 |
| 14 | .7579 | 12.1062 | 74.7999 | 1.3195 | 15.9739 | 0.0826 | 0.0626 | 6.1786 | 14 |
| 15 | .7430 | 12.8493 | 85.2021 | 1.3459 | 17.2934 | 0.0778 | 0.0578 | 6.6309 | 15 |
| 16 | .7284 | 13.5777 | 96.1288 | 1.3728 | 18.6393 | 0.0737 | 0.0537 | 7.0799 | 16 |
| 17 | .7142 | 14.2919 | 107.5554 | 1.4002 | 20.0121 | 0.0700 | 0.0500 | 7.5256 | 17 |
| 18 | .7002 | 14.9920 | 119.4581 | 1.4282 | 21.4123 | 0.0667 | 0.0467 | 7.9681 | 18 |
| 19 | .6864 | 15.6785 | 131.8139 | 1.4568 | 22.8406 | 0.0638 | 0.0438 | 8.4073 | 19 |
| 20 | .6730 | 16.3514 | 144.6003 | 1.4859 | 24.2974 | 0.0612 | 0.0412 | 8.8433 | 20 |
| 21 | .6598 | 17.0112 | 157.7959 | 1.5157 | 25.7833 | 0.0588 | 0.0388 | 9.2760 | 21 |
| 22 | .6468 | 17.6580 | 171.3795 | 1.5460 | 27.2990 | 0.0566 | 0.0366 | 9.7055 | 22 |
| 23 | .6342 | 18.2922 | 185.3309 | 1.5769 | 28.8450 | 0.0547 | 0.0347 | 10.1317 | 23 |
| 24 | .6217 | 18.9139 | 199.6305 | 1.6084 | 30.4219 | 0.0529 | 0.0329 | 10.5547 | 24 |
| 25 | .6095 | 19.5235 | 214.2592 | 1.6406 | 32.0303 | 0.0512 | 0.0312 | 10.9745 | 25 |
| 26 | .5976 | 20.1210 | 229.1987 | 1.6734 | 33.6709 | 0.0497 | 0.0297 | 11.3910 | 26 |
| 27 | .5859 | 20.7069 | 244.4311 | 1.7069 | 35.3443 | 0.0483 | 0.0283 | 11.8043 | 27 |
| 28 | .5744 | 21.2813 | 259.9392 | 1.7410 | 37.0512 | 0.0470 | 0.0270 | 12.2145 | 28 |
| 29 | .5631 | 21.8444 | 275.7064 | 1.7758 | 38.7922 | 0.0458 | 0.0258 | 12.6214 | 29 |
| 30 | .5521 | 22.3965 | 291.7164 | 1.8114 | 40.5681 | 0.0446 | 0.0246 | 13.0251 | 30 |
| 31 | .5412 | 22.9377 | 307.9538 | 1.8476 | 42.3794 | 0.0436 | 0.0236 | 13.4257 | 31 |
| 32 | .5306 | 23.4683 | 324.4035 | 1.8845 | 44.2270 | 0.0426 | 0.0226 | 13.8230 | 32 |
| 33 | .5202 | 23.9886 | 341.0508 | 1.9222 | 46.1116 | 0.0417 | 0.0217 | 14.2172 | 33 |
| 34 | .5100 | 24.4986 | 357.8817 | 1.9607 | 48.0338 | 0.0408 | 0.0208 | 14.6083 | 34 |
| 35 | .5000 | 24.9986 | 374.8826 | 1.9999 | 49.9945 | 0.0400 | 0.0200 | 14.9961 | 35 |
| 36 | .4902 | 25.4888 | 392.0405 | 2.0399 | 51.9944 | 0.0392 | 0.0192 | 15.3809 | 36 |
| 37 | .4806 | 25.9695 | 409.3424 | 2.0807 | 54.0343 | 0.0385 | 0.0185 | 15.7625 | 37 |
| 38 | .4712 | 26.4406 | 426.7764 | 2.1223 | 56.1149 | 0.0378 | 0.0178 | 16.1409 | 38 |
| 39 | .4619 | 26.9026 | 444.3304 | 2.1647 | 58.2372 | 0.0372 | 0.0172 | 16.5163 | 39 |
| 40 | .4529 | 27.3555 | 461.9931 | 2.2080 | 60.4020 | 0.0366 | 0.0166 | 16.8885 | 40 |
| 41 | .4440 | 27.7995 | 479.7535 | 2.2522 | 62.6100 | 0.0360 | 0.0160 | 17.2576 | 41 |
| 42 | .4353 | 28.2348 | 497.6010 | 2.2972 | 64.8622 | 0.0354 | 0.0154 | 17.6237 | 42 |
| 43 | .4268 | 28.6616 | 515.5253 | 2.3432 | 67.1595 | 0.0349 | 0.0149 | 17.9866 | 43 |
| 44 | .4184 | 29.0800 | 533.5165 | 2.3901 | 69.5027 | 0.0344 | 0.0144 | 18.3465 | 44 |
| 45 | .4102 | 29.4902 | 551.5652 | 2.4379 | 71.8927 | 0.0339 | 0.0139 | 18.7034 | 45 |
| 46 | .4022 | 29.8923 | 569.6621 | 2.4866 | 74.3306 | 0.0335 | 0.0135 | 19.0571 | 46 |
| 47 | .3943 | 30.2866 | 587.7985 | 2.5363 | 76.8172 | 0.0330 | 0.0130 | 19.4079 | 47 |
| 48 | .3865 | 30.6731 | 605.9657 | 2.5871 | 79.3535 | 0.0326 | 0.0126 | 19.7556 | 48 |
| 49 | .3790 | 31.0521 | 624.1557 | 2.6388 | 81.9406 | 0.0322 | 0.0122 | 20.1003 | 49 |
| 50 | .3715 | 31.4236 | 642.3606 | 2.6916 | 84.5794 | 0.0318 | 0.0118 | 20.4420 | 50 |
| 51 | .3642 | 31.7878 | 660.5727 | 2.7454 | 87.2710 | 0.0315 | 0.0115 | 20.7807 | 51 |
| 52 | .3571 | 32.1449 | 678.7849 | 2.8003 | 90.0164 | 0.0311 | 0.0111 | 21.1164 | 52 |
| 53 | .3501 | 32.4950 | 696.9900 | 2.8563 | 92.8167 | 0.0308 | 0.0108 | 21.4491 | 53 |
| 54 | .3432 | 32.8383 | 715.1815 | 2.9135 | 95.6731 | 0.0305 | 0.0105 | 21.7789 | 54 |
| 55 | .3365 | 33.1748 | 733.3527 | 2.9717 | 98.5865 | 0.0301 | 0.0101 | 22.1057 | 55 |
| 60 | .3048 | 34.7609 | 823.6975 | 3.2810 | 114.0515 | 0.0288 | 0.0088 | 23.6961 | 60 |
| 65 | .2761 | 36.1975 | 912.7085 | 3.6225 | 131.1262 | 0.0276 | 0.0076 | 25.2147 | 65 |
| 70 | .2500 | 37.4986 | 999.8343 | 3.9996 | 149.9779 | 0.0267 | 0.0067 | 26.6632 | 70 |
| 75 | .2265 | 38.6771 | 1084.6393 | 4.4158 | 170.7918 | 0.0259 | 0.0059 | 28.0434 | 75 |
| 80 | .2051 | 39.7445 | 1166.7868 | 4.8754 | 193.7720 | 0.0252 | 0.0052 | 29.3572 | 80 |
| 85 | .1858 | 40.7113 | 1246.0241 | 5.3829 | 219.1439 | 0.0246 | 0.0046 | 30.6064 | 85 |
| 90 | .1683 | 41.5869 | 1322.1701 | 5.9431 | 247.1567 | 0.0240 | 0.0040 | 31.7929 | 90 |
| 95 | .1524 | 42.3800 | 1395.1033 | 6.5617 | 278.0850 | 0.0236 | 0.0036 | 32.9189 | 95 |
| 100 | .1380 | 43.0984 | 1464.7527 | 7.2446 | 312.2323 | 0.0232 | 0.0032 | 33.9863 | 100 |

## I = 3.00 %

| N | (P/F) | (P/A) | (P/G) | (F/P) | (F/A) | (A/P) | (A/F) | (A/G) | N |
|---|---|---|---|---|---|---|---|---|---|
| 1 | .9709 | 0.9709 | -0.0000 | 1.0300 | 1.0000 | 1.0300 | 1.0000 | -0.0000 | 1 |
| 2 | .9426 | 1.9135 | 0.9426 | 1.0609 | 2.0300 | 0.5226 | 0.4926 | 0.4926 | 2 |
| 3 | .9151 | 2.8286 | 2.7729 | 1.0927 | 3.0909 | 0.3535 | 0.3235 | 0.9803 | 3 |
| 4 | .8885 | 3.7171 | 5.4383 | 1.1255 | 4.1836 | 0.2690 | 0.2390 | 1.4631 | 4 |
| 5 | .8626 | 4.5797 | 8.8888 | 1.1593 | 5.3091 | 0.2184 | 0.1884 | 1.9409 | 5 |
| 6 | .8375 | 5.4172 | 13.0762 | 1.1941 | 6.4684 | 0.1846 | 0.1546 | 2.4138 | 6 |
| 7 | .8131 | 6.2303 | 17.9547 | 1.2299 | 7.6625 | 0.1605 | 0.1305 | 2.8819 | 7 |
| 8 | .7894 | 7.0197 | 23.4806 | 1.2668 | 8.8923 | 0.1425 | 0.1125 | 3.3450 | 8 |
| 9 | .7664 | 7.7861 | 29.6119 | 1.3048 | 10.1591 | 0.1284 | 0.0984 | 3.8032 | 9 |
| 10 | .7441 | 8.5302 | 36.3088 | 1.3439 | 11.4639 | 0.1172 | 0.0872 | 4.2565 | 10 |
| 11 | .7224 | 9.2526 | 43.5330 | 1.3842 | 12.8078 | 0.1081 | 0.0781 | 4.7049 | 11 |
| 12 | .7014 | 9.9540 | 51.2482 | 1.4258 | 14.1920 | 0.1005 | 0.0705 | 5.1485 | 12 |
| 13 | .6810 | 10.6350 | 59.4196 | 1.4685 | 15.6178 | 0.0940 | 0.0640 | 5.5872 | 13 |
| 14 | .6611 | 11.2961 | 68.0141 | 1.5126 | 17.0863 | 0.0885 | 0.0585 | 6.0210 | 14 |
| 15 | .6419 | 11.9379 | 77.0002 | 1.5580 | 18.5989 | 0.0838 | 0.0538 | 6.4500 | 15 |
| 16 | .6232 | 12.5611 | 86.3477 | 1.6047 | 20.1569 | 0.0796 | 0.0496 | 6.8742 | 16 |
| 17 | .6050 | 13.1661 | 96.0280 | 1.6528 | 21.7616 | 0.0760 | 0.0460 | 7.2936 | 17 |
| 18 | .5874 | 13.7535 | 106.0137 | 1.7024 | 23.4144 | 0.0727 | 0.0427 | 7.7081 | 18 |
| 19 | .5703 | 14.3238 | 116.2788 | 1.7535 | 25.1169 | 0.0698 | 0.0398 | 8.1179 | 19 |
| 20 | .5537 | 14.8775 | 126.7987 | 1.8061 | 26.8704 | 0.0672 | 0.0372 | 8.5229 | 20 |
| 21 | .5375 | 15.4150 | 137.5496 | 1.8603 | 28.6765 | 0.0649 | 0.0349 | 8.9231 | 21 |
| 22 | .5219 | 15.9369 | 148.5094 | 1.9161 | 30.5368 | 0.0627 | 0.0327 | 9.3186 | 22 |
| 23 | .5067 | 16.4436 | 159.6566 | 1.9736 | 32.4529 | 0.0608 | 0.0308 | 9.7093 | 23 |
| 24 | .4919 | 16.9355 | 170.9711 | 2.0328 | 34.4265 | 0.0590 | 0.0290 | 10.0954 | 24 |
| 25 | .4776 | 17.4131 | 182.4336 | 2.0938 | 36.4593 | 0.0574 | 0.0274 | 10.4768 | 25 |
| 26 | .4637 | 17.8768 | 194.0260 | 2.1566 | 38.5530 | 0.0559 | 0.0259 | 10.8535 | 26 |
| 27 | .4502 | 18.3270 | 205.7309 | 2.2213 | 40.7096 | 0.0546 | 0.0246 | 11.2255 | 27 |
| 28 | .4371 | 18.7641 | 217.5320 | 2.2879 | 42.9309 | 0.0533 | 0.0233 | 11.5930 | 28 |
| 29 | .4243 | 19.1885 | 229.4137 | 2.3566 | 45.2189 | 0.0521 | 0.0221 | 11.9558 | 29 |
| 30 | .4120 | 19.6004 | 241.3613 | 2.4273 | 47.5754 | 0.0510 | 0.0210 | 12.3141 | 30 |
| 31 | .4000 | 20.0004 | 253.3609 | 2.5001 | 50.0027 | 0.0500 | 0.0200 | 12.6678 | 31 |
| 32 | .3883 | 20.3888 | 265.3993 | 2.5751 | 52.5028 | 0.0490 | 0.0190 | 13.0169 | 32 |
| 33 | .3770 | 20.7658 | 277.4642 | 2.6523 | 55.0778 | 0.0482 | 0.0182 | 13.3616 | 33 |
| 34 | .3660 | 21.1318 | 289.5437 | 2.7319 | 57.7302 | 0.0473 | 0.0173 | 13.7018 | 34 |
| 35 | .3554 | 21.4872 | 301.6267 | 2.8139 | 60.4621 | 0.0465 | 0.0165 | 14.0375 | 35 |
| 36 | .3450 | 21.8323 | 313.7028 | 2.8983 | 63.2759 | 0.0458 | 0.0158 | 14.3688 | 36 |
| 37 | .3350 | 22.1672 | 325.7622 | 2.9852 | 66.1742 | 0.0451 | 0.0151 | 14.6957 | 37 |
| 38 | .3252 | 22.4925 | 337.7956 | 3.0748 | 69.1594 | 0.0445 | 0.0145 | 15.0182 | 38 |
| 39 | .3158 | 22.8082 | 349.7942 | 3.1670 | 72.2342 | 0.0438 | 0.0138 | 15.3363 | 39 |
| 40 | .3066 | 23.1148 | 361.7499 | 3.2620 | 75.4013 | 0.0433 | 0.0133 | 15.6502 | 40 |
| 41 | .2976 | 23.4124 | 373.6551 | 3.3599 | 78.6633 | 0.0427 | 0.0127 | 15.9597 | 41 |
| 42 | .2890 | 23.7014 | 385.5024 | 3.4607 | 82.0232 | 0.0422 | 0.0122 | 16.2650 | 42 |
| 43 | .2805 | 23.9819 | 397.2852 | 3.5645 | 85.4839 | 0.0417 | 0.0117 | 16.5660 | 43 |
| 44 | .2724 | 24.2543 | 408.9972 | 3.6715 | 89.0484 | 0.0412 | 0.0112 | 16.8629 | 44 |
| 45 | .2644 | 24.5187 | 420.6325 | 3.7816 | 92.7199 | 0.0408 | 0.0108 | 17.1556 | 45 |
| 46 | .2567 | 24.7754 | 432.1856 | 3.8950 | 96.5015 | 0.0404 | 0.0104 | 17.4441 | 46 |
| 47 | .2493 | 25.0247 | 443.6515 | 4.0119 | 100.3965 | 0.0400 | 0.0100 | 17.7285 | 47 |
| 48 | .2420 | 25.2667 | 455.0255 | 4.1323 | 104.4084 | 0.0396 | 0.0096 | 18.0089 | 48 |
| 49 | .2350 | 25.5017 | 466.3031 | 4.2562 | 108.5406 | 0.0392 | 0.0092 | 18.2852 | 49 |
| 50 | .2281 | 25.7298 | 477.4803 | 4.3839 | 112.7969 | 0.0389 | 0.0089 | 18.5575 | 50 |
| 51 | .2215 | 25.9512 | 488.5535 | 4.5154 | 117.1808 | 0.0385 | 0.0085 | 18.8258 | 51 |
| 52 | .2150 | 26.1662 | 499.5191 | 4.6509 | 121.6962 | 0.0382 | 0.0082 | 19.0902 | 52 |
| 53 | .2088 | 26.3750 | 510.3742 | 4.7904 | 126.3471 | 0.0379 | 0.0079 | 19.3507 | 53 |
| 54 | .2027 | 26.5777 | 521.1157 | 4.9341 | 131.1375 | 0.0376 | 0.0076 | 19.6073 | 54 |
| 55 | .1968 | 26.7744 | 531.7411 | 5.0821 | 136.0716 | 0.0373 | 0.0073 | 19.8600 | 55 |
| 60 | .1697 | 27.6756 | 583.0526 | 5.8916 | 163.0534 | 0.0361 | 0.0061 | 21.0674 | 60 |
| 65 | .1464 | 28.4529 | 631.2010 | 6.8300 | 194.3328 | 0.0351 | 0.0051 | 22.1841 | 65 |
| 70 | .1263 | 29.1234 | 676.0869 | 7.9178 | 230.5941 | 0.0343 | 0.0043 | 23.2145 | 70 |
| 75 | .1089 | 29.7018 | 717.6978 | 9.1789 | 272.6309 | 0.0337 | 0.0037 | 24.1634 | 75 |
| 80 | .0940 | 30.2008 | 756.0865 | 10.6409 | 321.3630 | 0.0331 | 0.0031 | 25.0353 | 80 |
| 85 | .0811 | 30.6312 | 791.3529 | 12.3357 | 377.8570 | 0.0326 | 0.0026 | 25.8349 | 85 |
| 90 | .0699 | 31.0024 | 823.6302 | 14.3005 | 443.3489 | 0.0323 | 0.0023 | 26.5667 | 90 |
| 95 | .0603 | 31.3227 | 853.0742 | 16.5782 | 519.2720 | 0.0319 | 0.0019 | 27.2351 | 95 |
| 100 | .0520 | 31.5989 | 879.8540 | 19.2186 | 607.2877 | 0.0316 | 0.0016 | 27.8444 | 100 |

**I = 4.00 %**

| N | (P/F) | (P/A) | (P/G) | (F/P) | (F/A) | (A/P) | (A/F) | (A/G) | N |
|---|---|---|---|---|---|---|---|---|---|
| 1 | .9615 | 0.9615 | -0.0000 | 1.0400 | 1.0000 | 1.0400 | 1.0000 | -0.0000 | 1 |
| 2 | .9246 | 1.8861 | 0.9246 | 1.0816 | 2.0400 | 0.5302 | 0.4902 | 0.4902 | 2 |
| 3 | .8890 | 2.7751 | 2.7025 | 1.1249 | 3.1216 | 0.3603 | 0.3203 | 0.9739 | 3 |
| 4 | .8548 | 3.6299 | 5.2670 | 1.1699 | 4.2465 | 0.2755 | 0.2355 | 1.4510 | 4 |
| 5 | .8219 | 4.4518 | 8.5547 | 1.2167 | 5.4163 | 0.2246 | 0.1846 | 1.9216 | 5 |
| 6 | .7903 | 5.2421 | 12.5062 | 1.2653 | 6.6330 | 0.1908 | 0.1508 | 2.3857 | 6 |
| 7 | .7599 | 6.0021 | 17.0657 | 1.3159 | 7.8983 | 0.1666 | 0.1266 | 2.8433 | 7 |
| 8 | .7307 | 6.7327 | 22.1806 | 1.3686 | 9.2142 | 0.1485 | 0.1085 | 3.2944 | 8 |
| 9 | .7026 | 7.4353 | 27.8013 | 1.4233 | 10.5828 | 0.1345 | 0.0945 | 3.7391 | 9 |
| 10 | .6756 | 8.1109 | 33.8814 | 1.4802 | 12.0061 | 0.1233 | 0.0833 | 4.1773 | 10 |
| 11 | .6496 | 8.7605 | 40.3772 | 1.5395 | 13.4864 | 0.1141 | 0.0741 | 4.6090 | 11 |
| 12 | .6246 | 9.3851 | 47.2477 | 1.6010 | 15.0258 | 0.1066 | 0.0666 | 5.0343 | 12 |
| 13 | .6006 | 9.9856 | 54.4546 | 1.6651 | 16.6268 | 0.1001 | 0.0601 | 5.4533 | 13 |
| 14 | .5775 | 10.5631 | 61.9618 | 1.7317 | 18.2919 | 0.0947 | 0.0547 | 5.8659 | 14 |
| 15 | .5553 | 11.1184 | 69.7355 | 1.8009 | 20.0236 | 0.0899 | 0.0499 | 6.2721 | 15 |
| 16 | .5339 | 11.6523 | 77.7441 | 1.8730 | 21.8245 | 0.0858 | 0.0458 | 6.6720 | 16 |
| 17 | .5134 | 12.1657 | 85.9581 | 1.9479 | 23.6975 | 0.0822 | 0.0422 | 7.0656 | 17 |
| 18 | .4936 | 12.6593 | 94.3498 | 2.0258 | 25.6454 | 0.0790 | 0.0390 | 7.4530 | 18 |
| 19 | .4746 | 13.1339 | 102.8933 | 2.1068 | 27.6712 | 0.0761 | 0.0361 | 7.8342 | 19 |
| 20 | .4564 | 13.5903 | 111.5647 | 2.1911 | 29.7781 | 0.0736 | 0.0336 | 8.2091 | 20 |
| 21 | .4388 | 14.0292 | 120.3414 | 2.2788 | 31.9692 | 0.0713 | 0.0313 | 8.5779 | 21 |
| 22 | .4220 | 14.4511 | 129.2024 | 2.3699 | 34.2480 | 0.0692 | 0.0292 | 8.9407 | 22 |
| 23 | .4057 | 14.8568 | 138.1284 | 2.4647 | 36.6179 | 0.0673 | 0.0273 | 9.2973 | 23 |
| 24 | .3901 | 15.2470 | 147.1012 | 2.5633 | 39.0826 | 0.0656 | 0.0256 | 9.6479 | 24 |
| 25 | .3751 | 15.6221 | 156.1040 | 2.6658 | 41.6459 | 0.0640 | 0.0240 | 9.9925 | 25 |
| 26 | .3607 | 15.9828 | 165.1212 | 2.7725 | 44.3117 | 0.0626 | 0.0226 | 10.3312 | 26 |
| 27 | .3468 | 16.3296 | 174.1385 | 2.8834 | 47.0842 | 0.0612 | 0.0212 | 10.6640 | 27 |
| 28 | .3335 | 16.6631 | 183.1424 | 2.9987 | 49.9676 | 0.0600 | 0.0200 | 10.9909 | 28 |
| 29 | .3207 | 16.9837 | 192.1206 | 3.1187 | 52.9663 | 0.0589 | 0.0189 | 11.3120 | 29 |
| 30 | .3083 | 17.2920 | 201.0618 | 3.2434 | 56.0849 | 0.0578 | 0.0178 | 11.6274 | 30 |
| 31 | .2965 | 17.5885 | 209.9556 | 3.3731 | 59.3283 | 0.0569 | 0.0169 | 11.9371 | 31 |
| 32 | .2851 | 17.8736 | 218.7924 | 3.5081 | 62.7015 | 0.0559 | 0.0159 | 12.2411 | 32 |
| 33 | .2741 | 18.1476 | 227.5634 | 3.6484 | 66.2095 | 0.0551 | 0.0151 | 12.5396 | 33 |
| 34 | .2636 | 18.4112 | 236.2607 | 3.7943 | 69.8579 | 0.0543 | 0.0143 | 12.8324 | 34 |
| 35 | .2534 | 18.6646 | 244.8768 | 3.9461 | 73.6522 | 0.0536 | 0.0136 | 13.1198 | 35 |
| 36 | .2437 | 18.9083 | 253.4052 | 4.1039 | 77.5983 | 0.0529 | 0.0129 | 13.4018 | 36 |
| 37 | .2343 | 19.1426 | 261.8399 | 4.2681 | 81.7022 | 0.0522 | 0.0122 | 13.6784 | 37 |
| 38 | .2253 | 19.3679 | 270.1754 | 4.4388 | 85.9703 | 0.0516 | 0.0116 | 13.9497 | 38 |
| 39 | .2166 | 19.5845 | 278.4070 | 4.6164 | 90.4091 | 0.0511 | 0.0111 | 14.2157 | 39 |
| 40 | .2083 | 19.7928 | 286.5303 | 4.8010 | 95.0255 | 0.0505 | 0.0105 | 14.4765 | 40 |
| 41 | .2003 | 19.9931 | 294.5414 | 4.9931 | 99.8265 | 0.0500 | 0.0100 | 14.7322 | 41 |
| 42 | .1926 | 20.1856 | 302.4370 | 5.1928 | 104.8196 | 0.0495 | 0.0095 | 14.9828 | 42 |
| 43 | .1852 | 20.3708 | 310.2141 | 5.4005 | 110.0124 | 0.0491 | 0.0091 | 15.2284 | 43 |
| 44 | .1780 | 20.5488 | 317.8700 | 5.6165 | 115.4129 | 0.0487 | 0.0087 | 15.4690 | 44 |
| 45 | .1712 | 20.7200 | 325.4028 | 5.8412 | 121.0294 | 0.0483 | 0.0083 | 15.7047 | 45 |
| 46 | .1646 | 20.8847 | 332.8104 | 6.0748 | 126.8706 | 0.0479 | 0.0079 | 15.9356 | 46 |
| 47 | .1583 | 21.0429 | 340.0914 | 6.3178 | 132.9454 | 0.0475 | 0.0075 | 16.1618 | 47 |
| 48 | .1522 | 21.1951 | 347.2446 | 6.5705 | 139.2632 | 0.0472 | 0.0072 | 16.3832 | 48 |
| 49 | .1463 | 21.3415 | 354.2689 | 6.8333 | 145.8337 | 0.0469 | 0.0069 | 16.6000 | 49 |
| 50 | .1407 | 21.4822 | 361.1638 | 7.1067 | 152.6671 | 0.0466 | 0.0066 | 16.8122 | 50 |
| 51 | .1353 | 21.6175 | 367.9289 | 7.3910 | 159.7738 | 0.0463 | 0.0063 | 17.0200 | 51 |
| 52 | .1301 | 21.7476 | 374.5638 | 7.6866 | 167.1647 | 0.0460 | 0.0060 | 17.2232 | 52 |
| 53 | .1251 | 21.8727 | 381.0686 | 7.9941 | 174.8513 | 0.0457 | 0.0057 | 17.4221 | 53 |
| 54 | .1203 | 21.9930 | 387.4436 | 8.3138 | 182.8454 | 0.0455 | 0.0055 | 17.6167 | 54 |
| 55 | .1157 | 22.1086 | 393.6890 | 8.6464 | 191.1592 | 0.0452 | 0.0052 | 17.8070 | 55 |
| 60 | .0951 | 22.6235 | 422.9966 | 10.5196 | 237.9907 | 0.0442 | 0.0042 | 18.6972 | 60 |
| 65 | .0781 | 23.0467 | 449.2014 | 12.7987 | 294.9684 | 0.0434 | 0.0034 | 19.4909 | 65 |
| 70 | .0642 | 23.3945 | 472.4789 | 15.5716 | 364.2905 | 0.0427 | 0.0027 | 20.1961 | 70 |
| 75 | .0528 | 23.6804 | 493.0408 | 18.9453 | 448.6314 | 0.0422 | 0.0022 | 20.8206 | 75 |
| 80 | .0434 | 23.9154 | 511.1161 | 23.0498 | 551.2450 | 0.0418 | 0.0018 | 21.3718 | 80 |
| 85 | .0357 | 24.1085 | 526.9384 | 28.0436 | 676.0901 | 0.0415 | 0.0015 | 21.8569 | 85 |
| 90 | .0293 | 24.2673 | 540.7369 | 34.1193 | 827.9833 | 0.0412 | 0.0012 | 22.2826 | 90 |
| 95 | .0241 | 24.3978 | 552.7307 | 41.5114 | 1012.7846 | 0.0410 | 0.0010 | 22.6550 | 95 |
| 100 | .0198 | 24.5050 | 563.1249 | 50.5049 | 1237.6237 | 0.0408 | 0.0008 | 22.9800 | 100 |

**I = 5.00 %**

| N | (P/F) | (P/A) | (P/G) | (F/P) | (F/A) | (A/P) | (A/F) | (A/G) | N |
|---|---|---|---|---|---|---|---|---|---|
| 1 | .9524 | 0.9524 | -0.0000 | 1.0500 | 1.0000 | 1.0500 | 1.0000 | -0.0000 | 1 |
| 2 | .9070 | 1.8594 | 0.9070 | 1.1025 | 2.0500 | 0.5378 | 0.4878 | 0.4878 | 2 |
| 3 | .8638 | 2.7232 | 2.6347 | 1.1576 | 3.1525 | 0.3672 | 0.3172 | 0.9675 | 3 |
| 4 | .8227 | 3.5460 | 5.1028 | 1.2155 | 4.3101 | 0.2820 | 0.2320 | 1.4391 | 4 |
| 5 | .7835 | 4.3295 | 8.2369 | 1.2763 | 5.5256 | 0.2310 | 0.1810 | 1.9025 | 5 |
| 6 | .7462 | 5.0757 | 11.9680 | 1.3401 | 6.8019 | 0.1970 | 0.1470 | 2.3579 | 6 |
| 7 | .7107 | 5.7864 | 16.2321 | 1.4071 | 8.1420 | 0.1728 | 0.1228 | 2.8052 | 7 |
| 8 | .6768 | 6.4632 | 20.9700 | 1.4775 | 9.5491 | 0.1547 | 0.1047 | 3.2445 | 8 |
| 9 | .6446 | 7.1078 | 26.1268 | 1.5513 | 11.0266 | 0.1407 | 0.0907 | 3.6758 | 9 |
| 10 | .6139 | 7.7217 | 31.6520 | 1.6289 | 12.5779 | 0.1295 | 0.0795 | 4.0991 | 10 |
| 11 | .5847 | 8.3064 | 37.4988 | 1.7103 | 14.2068 | 0.1204 | 0.0704 | 4.5144 | 11 |
| 12 | .5568 | 8.8633 | 43.6241 | 1.7959 | 15.9171 | 0.1128 | 0.0628 | 4.9219 | 12 |
| 13 | .5303 | 9.3936 | 49.9879 | 1.8856 | 17.7130 | 0.1065 | 0.0565 | 5.3215 | 13 |
| 14 | .5051 | 9.8986 | 56.5538 | 1.9799 | 19.5986 | 0.1010 | 0.0510 | 5.7133 | 14 |
| 15 | .4810 | 10.3797 | 63.2880 | 2.0789 | 21.5786 | 0.0963 | 0.0463 | 6.0973 | 15 |
| 16 | .4581 | 10.8378 | 70.1597 | 2.1829 | 23.6575 | 0.0923 | 0.0423 | 6.4736 | 16 |
| 17 | .4363 | 11.2741 | 77.1405 | 2.2920 | 25.8404 | 0.0887 | 0.0387 | 6.8423 | 17 |
| 18 | .4155 | 11.6896 | 84.2043 | 2.4066 | 28.1324 | 0.0855 | 0.0355 | 7.2034 | 18 |
| 19 | .3957 | 12.0853 | 91.3275 | 2.5270 | 30.5390 | 0.0827 | 0.0327 | 7.5569 | 19 |
| 20 | .3769 | 12.4622 | 98.4884 | 2.6533 | 33.0660 | 0.0802 | 0.0302 | 7.9030 | 20 |
| 21 | .3589 | 12.8212 | 105.6673 | 2.7860 | 35.7193 | 0.0780 | 0.0280 | 8.2416 | 21 |
| 22 | .3418 | 13.1630 | 112.8461 | 2.9253 | 38.5052 | 0.0760 | 0.0260 | 8.5730 | 22 |
| 23 | .3256 | 13.4886 | 120.0087 | 3.0715 | 41.4305 | 0.0741 | 0.0241 | 8.8971 | 23 |
| 24 | .3101 | 13.7986 | 127.1402 | 3.2251 | 44.5020 | 0.0725 | 0.0225 | 9.2140 | 24 |
| 25 | .2953 | 14.0939 | 134.2275 | 3.3864 | 47.7271 | 0.0710 | 0.0210 | 9.5238 | 25 |
| 26 | .2812 | 14.3752 | 141.2585 | 3.5557 | 51.1135 | 0.0696 | 0.0196 | 9.8266 | 26 |
| 27 | .2678 | 14.6430 | 148.2226 | 3.7335 | 54.6691 | 0.0683 | 0.0183 | 10.1224 | 27 |
| 28 | .2551 | 14.8981 | 155.1101 | 3.9201 | 58.4026 | 0.0671 | 0.0171 | 10.4114 | 28 |
| 29 | .2429 | 15.1411 | 161.9126 | 4.1161 | 62.3227 | 0.0660 | 0.0160 | 10.6936 | 29 |
| 30 | .2314 | 15.3725 | 168.6226 | 4.3219 | 66.4388 | 0.0651 | 0.0151 | 10.9691 | 30 |
| 31 | .2204 | 15.5928 | 175.2333 | 4.5380 | 70.7608 | 0.0641 | 0.0141 | 11.2381 | 31 |
| 32 | .2099 | 15.8027 | 181.7392 | 4.7649 | 75.2988 | 0.0633 | 0.0133 | 11.5005 | 32 |
| 33 | .1999 | 16.0025 | 188.1351 | 5.0032 | 80.0638 | 0.0625 | 0.0125 | 11.7566 | 33 |
| 34 | .1904 | 16.1929 | 194.4168 | 5.2533 | 85.0670 | 0.0618 | 0.0118 | 12.0063 | 34 |
| 35 | .1813 | 16.3742 | 200.5807 | 5.5160 | 90.3203 | 0.0611 | 0.0111 | 12.2498 | 35 |
| 36 | .1727 | 16.5469 | 206.6237 | 5.7918 | 95.8363 | 0.0604 | 0.0104 | 12.4872 | 36 |
| 37 | .1644 | 16.7113 | 212.5434 | 6.0814 | 101.6281 | 0.0598 | 0.0098 | 12.7186 | 37 |
| 38 | .1566 | 16.8679 | 218.3378 | 6.3855 | 107.7095 | 0.0593 | 0.0093 | 12.9440 | 38 |
| 39 | .1491 | 17.0170 | 224.0054 | 6.7048 | 114.0950 | 0.0588 | 0.0088 | 13.1636 | 39 |
| 40 | .1420 | 17.1591 | 229.5452 | 7.0400 | 120.7998 | 0.0583 | 0.0083 | 13.3775 | 40 |
| 41 | .1353 | 17.2944 | 234.9564 | 7.3920 | 127.8398 | 0.0578 | 0.0078 | 13.5857 | 41 |
| 42 | .1288 | 17.4232 | 240.2389 | 7.7616 | 135.2318 | 0.0574 | 0.0074 | 13.7884 | 42 |
| 43 | .1227 | 17.5459 | 245.3925 | 8.1497 | 142.9933 | 0.0570 | 0.0070 | 13.9857 | 43 |
| 44 | .1169 | 17.6628 | 250.4175 | 8.5572 | 151.1430 | 0.0566 | 0.0066 | 14.1777 | 44 |
| 45 | .1113 | 17.7741 | 255.3145 | 8.9850 | 159.7002 | 0.0563 | 0.0063 | 14.3644 | 45 |
| 46 | .1060 | 17.8801 | 260.0844 | 9.4343 | 168.6852 | 0.0559 | 0.0059 | 14.5461 | 46 |
| 47 | .1009 | 17.9810 | 264.7281 | 9.9060 | 178.1194 | 0.0556 | 0.0056 | 14.7226 | 47 |
| 48 | .0961 | 18.0772 | 269.2467 | 10.4013 | 188.0254 | 0.0553 | 0.0053 | 14.8943 | 48 |
| 49 | .0916 | 18.1687 | 273.6418 | 10.9213 | 198.4267 | 0.0550 | 0.0050 | 15.0611 | 49 |
| 50 | .0872 | 18.2559 | 277.9148 | 11.4674 | 209.3480 | 0.0548 | 0.0048 | 15.2233 | 50 |
| 51 | .0831 | 18.3390 | 282.0673 | 12.0408 | 220.8154 | 0.0545 | 0.0045 | 15.3808 | 51 |
| 52 | .0791 | 18.4181 | 286.1013 | 12.6428 | 232.8562 | 0.0543 | 0.0043 | 15.5337 | 52 |
| 53 | .0753 | 18.4934 | 290.0184 | 13.2749 | 245.4990 | 0.0541 | 0.0041 | 15.6823 | 53 |
| 54 | .0717 | 18.5651 | 293.8208 | 13.9387 | 258.7739 | 0.0539 | 0.0039 | 15.8265 | 54 |
| 55 | .0683 | 18.6335 | 297.5104 | 14.6356 | 272.7126 | 0.0537 | 0.0037 | 15.9664 | 55 |
| 60 | .0535 | 18.9293 | 314.3432 | 18.6792 | 353.5837 | 0.0528 | 0.0028 | 16.6062 | 60 |
| 65 | .0419 | 19.1611 | 328.6910 | 23.8399 | 456.7980 | 0.0522 | 0.0022 | 17.1541 | 65 |
| 70 | .0329 | 19.3427 | 340.8409 | 30.4264 | 588.5285 | 0.0517 | 0.0017 | 17.6212 | 70 |
| 75 | .0258 | 19.4850 | 351.0721 | 38.8327 | 756.6537 | 0.0513 | 0.0013 | 18.0176 | 75 |
| 80 | .0202 | 19.5965 | 359.6460 | 49.5614 | 971.2288 | 0.0510 | 0.0010 | 18.3526 | 80 |
| 85 | .0158 | 19.6838 | 366.8007 | 63.2544 | 1245.0871 | 0.0508 | 0.0008 | 18.6346 | 85 |
| 90 | .0124 | 19.7523 | 372.7488 | 80.7304 | 1594.6073 | 0.0506 | 0.0006 | 18.8712 | 90 |
| 95 | .0097 | 19.8059 | 377.6774 | 103.0347 | 2040.6935 | 0.0505 | 0.0005 | 19.0689 | 95 |
| 100 | .0076 | 19.8479 | 381.7492 | 131.5013 | 2610.0252 | 0.0504 | 0.0004 | 19.2337 | 100 |

**I = 6.00 %**

| N | (P/F) | (P/A) | (P/G) | (F/P) | (F/A) | (A/P) | (A/F) | (A/G) | N |
|---|---|---|---|---|---|---|---|---|---|
| 1 | .9434 | 0.9434 | -0.0000 | 1.0600 | 1.0000 | 1.0600 | 1.0000 | -0.0000 | 1 |
| 2 | .8900 | 1.8334 | 0.8900 | 1.1236 | 2.0600 | 0.5454 | 0.4854 | 0.4854 | 2 |
| 3 | .8396 | 2.6730 | 2.5692 | 1.1910 | 3.1836 | 0.3741 | 0.3141 | 0.9612 | 3 |
| 4 | .7921 | 3.4651 | 4.9455 | 1.2625 | 4.3746 | 0.2886 | 0.2286 | 1.4272 | 4 |
| 5 | .7473 | 4.2124 | 7.9345 | 1.3382 | 5.6371 | 0.2374 | 0.1774 | 1.8836 | 5 |
| 6 | .7050 | 4.9173 | 11.4594 | 1.4185 | 6.9753 | 0.2034 | 0.1434 | 2.3304 | 6 |
| 7 | .6651 | 5.5824 | 15.4497 | 1.5036 | 8.3938 | 0.1791 | 0.1191 | 2.7676 | 7 |
| 8 | .6274 | 6.2098 | 19.8416 | 1.5938 | 9.8975 | 0.1610 | 0.1010 | 3.1952 | 8 |
| 9 | .5919 | 6.8017 | 24.5768 | 1.6895 | 11.4913 | 0.1470 | 0.0870 | 3.6133 | 9 |
| 10 | .5584 | 7.3601 | 29.6023 | 1.7908 | 13.1808 | 0.1359 | 0.0759 | 4.0220 | 10 |
| 11 | .5268 | 7.8869 | 34.8702 | 1.8983 | 14.9716 | 0.1268 | 0.0668 | 4.4213 | 11 |
| 12 | .4970 | 8.3838 | 40.3369 | 2.0122 | 16.8699 | 0.1193 | 0.0593 | 4.8113 | 12 |
| 13 | .4688 | 8.8527 | 45.9629 | 2.1329 | 18.8821 | 0.1130 | 0.0530 | 5.1920 | 13 |
| 14 | .4423 | 9.2950 | 51.7128 | 2.2609 | 21.0151 | 0.1076 | 0.0476 | 5.5635 | 14 |
| 15 | .4173 | 9.7122 | 57.5546 | 2.3966 | 23.2760 | 0.1030 | 0.0430 | 5.9260 | 15 |
| 16 | .3936 | 10.1059 | 63.4592 | 2.5404 | 25.6725 | 0.0990 | 0.0390 | 6.2794 | 16 |
| 17 | .3714 | 10.4773 | 69.4011 | 2.6928 | 28.2129 | 0.0954 | 0.0354 | 6.6240 | 17 |
| 18 | .3503 | 10.8276 | 75.3569 | 2.8543 | 30.9057 | 0.0924 | 0.0324 | 6.9597 | 18 |
| 19 | .3305 | 11.1581 | 81.3062 | 3.0256 | 33.7600 | 0.0896 | 0.0296 | 7.2867 | 19 |
| 20 | .3118 | 11.4699 | 87.2304 | 3.2071 | 36.7856 | 0.0872 | 0.0272 | 7.6051 | 20 |
| 21 | .2942 | 11.7641 | 93.1136 | 3.3996 | 39.9927 | 0.0850 | 0.0250 | 7.9151 | 21 |
| 22 | .2775 | 12.0416 | 98.9412 | 3.6035 | 43.3923 | 0.0830 | 0.0230 | 8.2166 | 22 |
| 23 | .2618 | 12.3034 | 104.7007 | 3.8197 | 46.9958 | 0.0813 | 0.0213 | 8.5099 | 23 |
| 24 | .2470 | 12.5504 | 110.3812 | 4.0489 | 50.8156 | 0.0797 | 0.0197 | 8.7951 | 24 |
| 25 | .2330 | 12.7834 | 115.9732 | 4.2919 | 54.8645 | 0.0782 | 0.0182 | 9.0722 | 25 |
| 26 | .2198 | 13.0032 | 121.4684 | 4.5494 | 59.1564 | 0.0769 | 0.0169 | 9.3414 | 26 |
| 27 | .2074 | 13.2105 | 126.8600 | 4.8223 | 63.7058 | 0.0757 | 0.0157 | 9.6029 | 27 |
| 28 | .1956 | 13.4062 | 132.1420 | 5.1117 | 68.5281 | 0.0746 | 0.0146 | 9.8568 | 28 |
| 29 | .1846 | 13.5907 | 137.3096 | 5.4184 | 73.6398 | 0.0736 | 0.0136 | 10.1032 | 29 |
| 30 | .1741 | 13.7648 | 142.3588 | 5.7435 | 79.0582 | 0.0726 | 0.0126 | 10.3422 | 30 |
| 31 | .1643 | 13.9291 | 147.2864 | 6.0881 | 84.8017 | 0.0718 | 0.0118 | 10.5740 | 31 |
| 32 | .1550 | 14.0840 | 152.0901 | 6.4534 | 90.8898 | 0.0710 | 0.0110 | 10.7988 | 32 |
| 33 | .1462 | 14.2302 | 156.7681 | 6.8406 | 97.3432 | 0.0703 | 0.0103 | 11.0166 | 33 |
| 34 | .1379 | 14.3681 | 161.3192 | 7.2510 | 104.1838 | 0.0696 | 0.0096 | 11.2276 | 34 |
| 35 | .1301 | 14.4982 | 165.7427 | 7.6861 | 111.4348 | 0.0690 | 0.0090 | 11.4319 | 35 |
| 36 | .1227 | 14.6210 | 170.0387 | 8.1473 | 119.1209 | 0.0684 | 0.0084 | 11.6298 | 36 |
| 37 | .1158 | 14.7368 | 174.2072 | 8.6361 | 127.2681 | 0.0679 | 0.0079 | 11.8213 | 37 |
| 38 | .1092 | 14.8460 | 178.2490 | 9.1543 | 135.9042 | 0.0674 | 0.0074 | 12.0065 | 38 |
| 39 | .1031 | 14.9491 | 182.1652 | 9.7035 | 145.0585 | 0.0669 | 0.0069 | 12.1857 | 39 |
| 40 | .0972 | 15.0463 | 185.9568 | 10.2857 | 154.7620 | 0.0665 | 0.0065 | 12.3590 | 40 |
| 41 | .0917 | 15.1380 | 189.6256 | 10.9029 | 165.0477 | 0.0661 | 0.0061 | 12.5264 | 41 |
| 42 | .0865 | 15.2245 | 193.1732 | 11.5570 | 175.9505 | 0.0657 | 0.0057 | 12.6883 | 42 |
| 43 | .0816 | 15.3062 | 196.6017 | 12.2505 | 187.5076 | 0.0653 | 0.0053 | 12.8446 | 43 |
| 44 | .0770 | 15.3832 | 199.9130 | 12.9855 | 199.7580 | 0.0650 | 0.0050 | 12.9956 | 44 |
| 45 | .0727 | 15.4558 | 203.1096 | 13.7646 | 212.7435 | 0.0647 | 0.0047 | 13.1413 | 45 |
| 46 | .0685 | 15.5244 | 206.1938 | 14.5905 | 226.5081 | 0.0644 | 0.0044 | 13.2819 | 46 |
| 47 | .0647 | 15.5890 | 209.1681 | 15.4659 | 241.0986 | 0.0641 | 0.0041 | 13.4177 | 47 |
| 48 | .0610 | 15.6500 | 212.0351 | 16.3939 | 256.5645 | 0.0639 | 0.0039 | 13.5485 | 48 |
| 49 | .0575 | 15.7076 | 214.7972 | 17.3775 | 272.9584 | 0.0637 | 0.0037 | 13.6748 | 49 |
| 50 | .0543 | 15.7619 | 217.4574 | 18.4202 | 290.3359 | 0.0634 | 0.0034 | 13.7964 | 50 |
| 51 | .0512 | 15.8131 | 220.0181 | 19.5254 | 308.7561 | 0.0632 | 0.0032 | 13.9137 | 51 |
| 52 | .0483 | 15.8614 | 222.4823 | 20.6969 | 328.2814 | 0.0630 | 0.0030 | 14.0267 | 52 |
| 53 | .0456 | 15.9070 | 224.8525 | 21.9387 | 348.9783 | 0.0629 | 0.0029 | 14.1355 | 53 |
| 54 | .0430 | 15.9500 | 227.1316 | 23.2550 | 370.9170 | 0.0627 | 0.0027 | 14.2402 | 54 |
| 55 | .0406 | 15.9905 | 229.3222 | 24.6503 | 394.1720 | 0.0625 | 0.0025 | 14.3411 | 55 |
| 60 | .0303 | 16.1614 | 239.0428 | 32.9877 | 533.1282 | 0.0619 | 0.0019 | 14.7909 | 60 |
| 65 | .0227 | 16.2891 | 246.9450 | 44.1450 | 719.0829 | 0.0614 | 0.0014 | 15.1601 | 65 |
| 70 | .0169 | 16.3845 | 253.3271 | 59.0759 | 967.9322 | 0.0610 | 0.0010 | 15.4613 | 70 |
| 75 | .0126 | 16.4558 | 258.4527 | 79.0569 | 1300.9487 | 0.0608 | 0.0008 | 15.7058 | 75 |
| 80 | .0095 | 16.5091 | 262.5493 | 105.7960 | 1746.5999 | 0.0606 | 0.0006 | 15.9033 | 80 |
| 85 | .0071 | 16.5489 | 265.8096 | 141.5789 | 2342.9817 | 0.0604 | 0.0004 | 16.0620 | 85 |
| 90 | .0053 | 16.5787 | 268.3946 | 189.4645 | 3141.0752 | 0.0603 | 0.0003 | 16.1891 | 90 |
| 95 | .0039 | 16.6009 | 270.4375 | 253.5463 | 4209.1042 | 0.0602 | 0.0002 | 16.2905 | 95 |
| 100 | .0029 | 16.6175 | 272.0471 | 339.3021 | 5638.3681 | 0.0602 | 0.0002 | 16.3711 | 100 |

**I = 7.00 %**

| N | (P/F) | (P/A) | (P/G) | (F/P) | (F/A) | (A/P) | (A/F) | (A/G) | N |
|---|---|---|---|---|---|---|---|---|---|
| 1 | .9346 | 0.9346 | -0.0000 | 1.0700 | 1.0000 | 1.0700 | 1.0000 | -0.0000 | 1 |
| 2 | .8734 | 1.8080 | 0.8734 | 1.1449 | 2.0700 | 0.5531 | 0.4831 | 0.4831 | 2 |
| 3 | .8163 | 2.6243 | 2.5060 | 1.2250 | 3.2149 | 0.3811 | 0.3111 | 0.9549 | 3 |
| 4 | .7629 | 3.3872 | 4.7947 | 1.3108 | 4.4399 | 0.2952 | 0.2252 | 1.4155 | 4 |
| 5 | .7130 | 4.1002 | 7.6467 | 1.4026 | 5.7507 | 0.2439 | 0.1739 | 1.8650 | 5 |
| 6 | .6663 | 4.7665 | 10.9784 | 1.5007 | 7.1533 | 0.2098 | 0.1398 | 2.3032 | 6 |
| 7 | .6227 | 5.3893 | 14.7149 | 1.6058 | 8.6540 | 0.1856 | 0.1156 | 2.7304 | 7 |
| 8 | .5820 | 5.9713 | 18.7889 | 1.7182 | 10.2598 | 0.1675 | 0.0975 | 3.1465 | 8 |
| 9 | .5439 | 6.5152 | 23.1404 | 1.8385 | 11.9780 | 0.1535 | 0.0835 | 3.5517 | 9 |
| 10 | .5083 | 7.0236 | 27.7156 | 1.9672 | 13.8164 | 0.1424 | 0.0724 | 3.9461 | 10 |
| 11 | .4751 | 7.4987 | 32.4665 | 2.1049 | 15.7836 | 0.1334 | 0.0634 | 4.3296 | 11 |
| 12 | .4440 | 7.9427 | 37.3506 | 2.2522 | 17.8885 | 0.1259 | 0.0559 | 4.7025 | 12 |
| 13 | .4150 | 8.3577 | 42.3302 | 2.4098 | 20.1406 | 0.1197 | 0.0497 | 5.0648 | 13 |
| 14 | .3878 | 8.7455 | 47.3718 | 2.5785 | 22.5505 | 0.1143 | 0.0443 | 5.4167 | 14 |
| 15 | .3624 | 9.1079 | 52.4461 | 2.7590 | 25.1290 | 0.1098 | 0.0398 | 5.7583 | 15 |
| 16 | .3387 | 9.4466 | 57.5271 | 2.9522 | 27.8881 | 0.1059 | 0.0359 | 6.0897 | 16 |
| 17 | .3166 | 9.7632 | 62.5923 | 3.1588 | 30.8402 | 0.1024 | 0.0324 | 6.4110 | 17 |
| 18 | .2959 | 10.0591 | 67.6219 | 3.3799 | 33.9990 | 0.0994 | 0.0294 | 6.7225 | 18 |
| 19 | .2765 | 10.3356 | 72.5991 | 3.6165 | 37.3790 | 0.0968 | 0.0268 | 7.0242 | 19 |
| 20 | .2584 | 10.5940 | 77.5091 | 3.8697 | 40.9955 | 0.0944 | 0.0244 | 7.3163 | 20 |
| 21 | .2415 | 10.8355 | 82.3393 | 4.1406 | 44.8652 | 0.0923 | 0.0223 | 7.5990 | 21 |
| 22 | .2257 | 11.0612 | 87.0793 | 4.4304 | 49.0057 | 0.0904 | 0.0204 | 7.8725 | 22 |
| 23 | .2109 | 11.2722 | 91.7201 | 4.7405 | 53.4361 | 0.0887 | 0.0187 | 8.1369 | 23 |
| 24 | .1971 | 11.4693 | 96.2545 | 5.0724 | 58.1767 | 0.0872 | 0.0172 | 8.3923 | 24 |
| 25 | .1842 | 11.6536 | 100.6765 | 5.4274 | 63.2490 | 0.0858 | 0.0158 | 8.6391 | 25 |
| 26 | .1722 | 11.8258 | 104.9814 | 5.8074 | 68.6765 | 0.0846 | 0.0146 | 8.8773 | 26 |
| 27 | .1609 | 11.9867 | 109.1656 | 6.2139 | 74.4838 | 0.0834 | 0.0134 | 9.1072 | 27 |
| 28 | .1504 | 12.1371 | 113.2264 | 6.6488 | 80.6977 | 0.0824 | 0.0124 | 9.3289 | 28 |
| 29 | .1406 | 12.2777 | 117.1622 | 7.1143 | 87.3465 | 0.0814 | 0.0114 | 9.5427 | 29 |
| 30 | .1314 | 12.4090 | 120.9718 | 7.6123 | 94.4608 | 0.0806 | 0.0106 | 9.7487 | 30 |
| 31 | .1228 | 12.5318 | 124.6550 | 8.1451 | 102.0730 | 0.0798 | 0.0098 | 9.9471 | 31 |
| 32 | .1147 | 12.6466 | 128.2120 | 8.7153 | 110.2182 | 0.0791 | 0.0091 | 10.1381 | 32 |
| 33 | .1072 | 12.7538 | 131.6435 | 9.3253 | 118.9334 | 0.0784 | 0.0084 | 10.3219 | 33 |
| 34 | .1002 | 12.8540 | 134.9507 | 9.9781 | 128.2588 | 0.0778 | 0.0078 | 10.4987 | 34 |
| 35 | .0937 | 12.9477 | 138.1353 | 10.6766 | 138.2369 | 0.0772 | 0.0072 | 10.6687 | 35 |
| 36 | .0875 | 13.0352 | 141.1990 | 11.4239 | 148.9135 | 0.0767 | 0.0067 | 10.8321 | 36 |
| 37 | .0818 | 13.1170 | 144.1441 | 12.2236 | 160.3374 | 0.0762 | 0.0062 | 10.9891 | 37 |
| 38 | .0765 | 13.1935 | 146.9730 | 13.0793 | 172.5610 | 0.0758 | 0.0058 | 11.1398 | 38 |
| 39 | .0715 | 13.2649 | 149.6883 | 13.9948 | 185.6403 | 0.0754 | 0.0054 | 11.2845 | 39 |
| 40 | .0668 | 13.3317 | 152.2928 | 14.9745 | 199.6351 | 0.0750 | 0.0050 | 11.4233 | 40 |
| 41 | .0624 | 13.3941 | 154.7892 | 16.0227 | 214.6096 | 0.0747 | 0.0047 | 11.5565 | 41 |
| 42 | .0583 | 13.4524 | 157.1807 | 17.1443 | 230.6322 | 0.0743 | 0.0043 | 11.6842 | 42 |
| 43 | .0545 | 13.5070 | 159.4702 | 18.3444 | 247.7765 | 0.0740 | 0.0040 | 11.8065 | 43 |
| 44 | .0509 | 13.5579 | 161.6609 | 19.6285 | 266.1209 | 0.0738 | 0.0038 | 11.9237 | 44 |
| 45 | .0476 | 13.6055 | 163.7559 | 21.0025 | 285.7493 | 0.0735 | 0.0035 | 12.0360 | 45 |
| 46 | .0445 | 13.6500 | 165.7584 | 22.4726 | 306.7518 | 0.0733 | 0.0033 | 12.1435 | 46 |
| 47 | .0416 | 13.6916 | 167.6714 | 24.0457 | 329.2244 | 0.0730 | 0.0030 | 12.2463 | 47 |
| 48 | .0389 | 13.7305 | 169.4981 | 25.7289 | 353.2701 | 0.0728 | 0.0028 | 12.3447 | 48 |
| 49 | .0363 | 13.7668 | 171.2417 | 27.5299 | 378.9990 | 0.0726 | 0.0026 | 12.4387 | 49 |
| 50 | .0339 | 13.8007 | 172.9051 | 29.4570 | 406.5289 | 0.0725 | 0.0025 | 12.5287 | 50 |
| 51 | .0317 | 13.8325 | 174.4915 | 31.5190 | 435.9860 | 0.0723 | 0.0023 | 12.6146 | 51 |
| 52 | .0297 | 13.8621 | 176.0037 | 33.7253 | 467.5050 | 0.0721 | 0.0021 | 12.6967 | 52 |
| 53 | .0277 | 13.8898 | 177.4447 | 36.0861 | 501.2303 | 0.0720 | 0.0020 | 12.7751 | 53 |
| 54 | .0259 | 13.9157 | 178.8173 | 38.6122 | 537.3164 | 0.0719 | 0.0019 | 12.8500 | 54 |
| 55 | .0242 | 13.9399 | 180.1243 | 41.3150 | 575.9286 | 0.0717 | 0.0017 | 12.9215 | 55 |
| 60 | .0173 | 14.0392 | 185.7677 | 57.9464 | 813.5204 | 0.0712 | 0.0012 | 13.2321 | 60 |
| 65 | .0123 | 14.1099 | 190.1452 | 81.2729 | 1146.7552 | 0.0709 | 0.0009 | 13.4760 | 65 |
| 70 | .0088 | 14.1604 | 193.5185 | 113.9894 | 1614.1342 | 0.0706 | 0.0006 | 13.6662 | 70 |
| 75 | .0063 | 14.1964 | 196.1035 | 159.8760 | 2269.6574 | 0.0704 | 0.0004 | 13.8136 | 75 |
| 80 | .0045 | 14.2220 | 198.0748 | 224.2344 | 3189.0627 | 0.0703 | 0.0003 | 13.9273 | 80 |
| 85 | .0032 | 14.2403 | 199.5717 | 314.5003 | 4478.5761 | 0.0702 | 0.0002 | 14.0146 | 85 |
| 90 | .0023 | 14.2533 | 200.7042 | 441.1030 | 6287.1854 | 0.0702 | 0.0002 | 14.0812 | 90 |
| 95 | .0016 | 14.2626 | 201.5581 | 618.6697 | 8823.8535 | 0.0701 | 0.0001 | 14.1319 | 95 |
| 100 | .0012 | 14.2693 | 202.2001 | 867.7163 | 12381.6618 | 0.0701 | 0.0001 | 14.1703 | 100 |

**I = 8.00 %**

| N | (P/F) | (P/A) | (P/G) | (F/P) | (F/A) | (A/P) | (A/F) | (A/G) | N |
|---|---|---|---|---|---|---|---|---|---|
| 1 | .9259 | 0.9259 | -0.0000 | 1.0800 | 1.0000 | 1.0800 | 1.0000 | -0.0000 | 1 |
| 2 | .8573 | 1.7833 | 0.8573 | 1.1664 | 2.0800 | 0.5608 | 0.4808 | 0.4808 | 2 |
| 3 | .7938 | 2.5771 | 2.4450 | 1.2597 | 3.2464 | 0.3880 | 0.3080 | 0.9487 | 3 |
| 4 | .7350 | 3.3121 | 4.6501 | 1.3605 | 4.5061 | 0.3019 | 0.2219 | 1.4040 | 4 |
| 5 | .6806 | 3.9927 | 7.3724 | 1.4693 | 5.8666 | 0.2505 | 0.1705 | 1.8465 | 5 |
| 6 | .6302 | 4.6229 | 10.5233 | 1.5869 | 7.3359 | 0.2163 | 0.1363 | 2.2763 | 6 |
| 7 | .5835 | 5.2064 | 14.0242 | 1.7138 | 8.9228 | 0.1921 | 0.1121 | 2.6937 | 7 |
| 8 | .5403 | 5.7466 | 17.8061 | 1.8509 | 10.6366 | 0.1740 | 0.0940 | 3.0985 | 8 |
| 9 | .5002 | 6.2469 | 21.8081 | 1.9990 | 12.4876 | 0.1601 | 0.0801 | 3.4910 | 9 |
| 10 | .4632 | 6.7101 | 25.9768 | 2.1589 | 14.4866 | 0.1490 | 0.0690 | 3.8713 | 10 |
| 11 | .4289 | 7.1390 | 30.2657 | 2.3316 | 16.6455 | 0.1401 | 0.0601 | 4.2395 | 11 |
| 12 | .3971 | 7.5361 | 34.6339 | 2.5182 | 18.9771 | 0.1327 | 0.0527 | 4.5957 | 12 |
| 13 | .3677 | 7.9038 | 39.0463 | 2.7196 | 21.4953 | 0.1265 | 0.0465 | 4.9402 | 13 |
| 14 | .3405 | 8.2442 | 43.4723 | 2.9372 | 24.2149 | 0.1213 | 0.0413 | 5.2731 | 14 |
| 15 | .3152 | 8.5595 | 47.8857 | 3.1722 | 27.1521 | 0.1168 | 0.0368 | 5.5945 | 15 |
| 16 | .2919 | 8.8514 | 52.2640 | 3.4259 | 30.3243 | 0.1130 | 0.0330 | 5.9046 | 16 |
| 17 | .2703 | 9.1216 | 56.5883 | 3.7000 | 33.7502 | 0.1096 | 0.0296 | 6.2037 | 17 |
| 18 | .2502 | 9.3719 | 60.8426 | 3.9960 | 37.4502 | 0.1067 | 0.0267 | 6.4920 | 18 |
| 19 | .2317 | 9.6036 | 65.0134 | 4.3157 | 41.4463 | 0.1041 | 0.0241 | 6.7697 | 19 |
| 20 | .2145 | 9.8181 | 69.0898 | 4.6610 | 45.7620 | 0.1019 | 0.0219 | 7.0369 | 20 |
| 21 | .1987 | 10.0168 | 73.0629 | 5.0338 | 50.4229 | 0.0998 | 0.0198 | 7.2940 | 21 |
| 22 | .1839 | 10.2007 | 76.9257 | 5.4365 | 55.4568 | 0.0980 | 0.0180 | 7.5412 | 22 |
| 23 | .1703 | 10.3711 | 80.6726 | 5.8715 | 60.8933 | 0.0964 | 0.0164 | 7.7786 | 23 |
| 24 | .1577 | 10.5288 | 84.2997 | 6.3412 | 66.7648 | 0.0950 | 0.0150 | 8.0066 | 24 |
| 25 | .1460 | 10.6748 | 87.8041 | 6.8485 | 73.1059 | 0.0937 | 0.0137 | 8.2254 | 25 |
| 26 | .1352 | 10.8100 | 91.1842 | 7.3964 | 79.9544 | 0.0925 | 0.0125 | 8.4352 | 26 |
| 27 | .1252 | 10.9352 | 94.4390 | 7.9881 | 87.3508 | 0.0914 | 0.0114 | 8.6363 | 27 |
| 28 | .1159 | 11.0511 | 97.5687 | 8.6271 | 95.3388 | 0.0905 | 0.0105 | 8.8289 | 28 |
| 29 | .1073 | 11.1584 | 100.5738 | 9.3173 | 103.9659 | 0.0896 | 0.0096 | 9.0133 | 29 |
| 30 | .0994 | 11.2578 | 103.4558 | 10.0627 | 113.2832 | 0.0888 | 0.0088 | 9.1897 | 30 |
| 31 | .0920 | 11.3498 | 106.2163 | 10.8677 | 123.3459 | 0.0881 | 0.0081 | 9.3584 | 31 |
| 32 | .0852 | 11.4350 | 108.8575 | 11.7371 | 134.2135 | 0.0875 | 0.0075 | 9.5197 | 32 |
| 33 | .0789 | 11.5139 | 111.3819 | 12.6760 | 145.9506 | 0.0869 | 0.0069 | 9.6737 | 33 |
| 34 | .0730 | 11.5869 | 113.7924 | 13.6901 | 158.6267 | 0.0863 | 0.0063 | 9.8208 | 34 |
| 35 | .0676 | 11.6546 | 116.0920 | 14.7853 | 172.3168 | 0.0858 | 0.0058 | 9.9611 | 35 |
| 36 | .0626 | 11.7172 | 118.2839 | 15.9682 | 187.1021 | 0.0853 | 0.0053 | 10.0949 | 36 |
| 37 | .0580 | 11.7752 | 120.3713 | 17.2456 | 203.0703 | 0.0849 | 0.0049 | 10.2225 | 37 |
| 38 | .0537 | 11.8289 | 122.3579 | 18.6253 | 220.3159 | 0.0845 | 0.0045 | 10.3440 | 38 |
| 39 | .0497 | 11.8786 | 124.2470 | 20.1153 | 238.9412 | 0.0842 | 0.0042 | 10.4597 | 39 |
| 40 | .0460 | 11.9246 | 126.0422 | 21.7245 | 259.0565 | 0.0839 | 0.0039 | 10.5699 | 40 |
| 41 | .0426 | 11.9672 | 127.7470 | 23.4625 | 280.7810 | 0.0836 | 0.0036 | 10.6747 | 41 |
| 42 | .0395 | 12.0067 | 129.3651 | 25.3395 | 304.2435 | 0.0833 | 0.0033 | 10.7744 | 42 |
| 43 | .0365 | 12.0432 | 130.8998 | 27.3666 | 329.5830 | 0.0830 | 0.0030 | 10.8692 | 43 |
| 44 | .0338 | 12.0771 | 132.3547 | 29.5560 | 356.9496 | 0.0828 | 0.0028 | 10.9592 | 44 |
| 45 | .0313 | 12.1084 | 133.7331 | 31.9204 | 386.5056 | 0.0826 | 0.0026 | 11.0447 | 45 |
| 46 | .0290 | 12.1374 | 135.0384 | 34.4741 | 418.4261 | 0.0824 | 0.0024 | 11.1258 | 46 |
| 47 | .0269 | 12.1643 | 136.2739 | 37.2320 | 452.9002 | 0.0822 | 0.0022 | 11.2028 | 47 |
| 48 | .0249 | 12.1891 | 137.4428 | 40.2106 | 490.1322 | 0.0820 | 0.0020 | 11.2758 | 48 |
| 49 | .0230 | 12.2122 | 138.5480 | 43.4274 | 530.3427 | 0.0819 | 0.0019 | 11.3451 | 49 |
| 50 | .0213 | 12.2335 | 139.5928 | 46.9016 | 573.7702 | 0.0817 | 0.0017 | 11.4107 | 50 |
| 51 | .0197 | 12.2532 | 140.5799 | 50.6537 | 620.6718 | 0.0816 | 0.0016 | 11.4729 | 51 |
| 52 | .0183 | 12.2715 | 141.5121 | 54.7060 | 671.3255 | 0.0815 | 0.0015 | 11.5318 | 52 |
| 53 | .0169 | 12.2884 | 142.3923 | 59.0825 | 726.0316 | 0.0814 | 0.0014 | 11.5875 | 53 |
| 54 | .0157 | 12.3041 | 143.2229 | 63.8091 | 785.1141 | 0.0813 | 0.0013 | 11.6403 | 54 |
| 55 | .0145 | 12.3186 | 144.0065 | 68.9139 | 848.9232 | 0.0812 | 0.0012 | 11.6902 | 55 |
| 60 | .0099 | 12.3766 | 147.3000 | 101.2571 | 1253.2133 | 0.0808 | 0.0008 | 11.9015 | 60 |
| 65 | .0067 | 12.4160 | 149.7387 | 148.7798 | 1847.2481 | 0.0805 | 0.0005 | 12.0602 | 65 |
| 70 | .0046 | 12.4428 | 151.5326 | 218.6064 | 2720.0801 | 0.0804 | 0.0004 | 12.1783 | 70 |
| 75 | .0031 | 12.4611 | 152.8448 | 321.2045 | 4002.5566 | 0.0802 | 0.0002 | 12.2658 | 75 |
| 80 | .0021 | 12.4735 | 153.8001 | 471.9548 | 5886.9354 | 0.0802 | 0.0002 | 12.3301 | 80 |
| 85 | .0014 | 12.4820 | 154.4925 | 693.4565 | 8655.7061 | 0.0801 | 0.0001 | 12.3772 | 85 |
| 90 | .0010 | 12.4877 | 154.9925 | 1018.9151 | 12723.9386 | 0.0801 | 0.0001 | 12.4116 | 90 |
| 95 | .0007 | 12.4917 | 155.3524 | 1497.1205 | 18701.5069 | 0.0801 | 0.0001 | 12.4365 | 95 |
| 100 | .0005 | 12.4943 | 155.6107 | 2199.7613 | 27484.5157 | 0.0800 | 0.0000 | 12.4545 | 100 |

**I = 9.00 %**

| N | (P/F) | (P/A) | (P/G) | (F/P) | (F/A) | (A/P) | (A/F) | (A/G) | N |
|---|---|---|---|---|---|---|---|---|---|
| 1 | .9174 | 0.9174 | -0.0000 | 1.0900 | 1.0000 | 1.0900 | 1.0000 | -0.0000 | 1 |
| 2 | .8417 | 1.7591 | 0.8417 | 1.1881 | 2.0900 | 0.5685 | 0.4785 | 0.4785 | 2 |
| 3 | .7722 | 2.5313 | 2.3860 | 1.2950 | 3.2781 | 0.3951 | 0.3051 | 0.9426 | 3 |
| 4 | .7084 | 3.2397 | 4.5113 | 1.4116 | 4.5731 | 0.3087 | 0.2187 | 1.3925 | 4 |
| 5 | .6499 | 3.8897 | 7.1110 | 1.5386 | 5.9847 | 0.2571 | 0.1671 | 1.8282 | 5 |
| 6 | .5963 | 4.4859 | 10.0924 | 1.6771 | 7.5233 | 0.2229 | 0.1329 | 2.2498 | 6 |
| 7 | .5470 | 5.0330 | 13.3746 | 1.8280 | 9.2004 | 0.1987 | 0.1087 | 2.6574 | 7 |
| 8 | .5019 | 5.5348 | 16.8877 | 1.9926 | 11.0285 | 0.1807 | 0.0907 | 3.0512 | 8 |
| 9 | .4604 | 5.9952 | 20.5711 | 2.1719 | 13.0210 | 0.1668 | 0.0768 | 3.4312 | 9 |
| 10 | .4224 | 6.4177 | 24.3728 | 2.3674 | 15.1929 | 0.1558 | 0.0658 | 3.7978 | 10 |
| 11 | .3875 | 6.8052 | 28.2481 | 2.5804 | 17.5603 | 0.1469 | 0.0569 | 4.1510 | 11 |
| 12 | .3555 | 7.1607 | 32.1590 | 2.8127 | 20.1407 | 0.1397 | 0.0497 | 4.4910 | 12 |
| 13 | .3262 | 7.4869 | 36.0731 | 3.0658 | 22.9534 | 0.1336 | 0.0436 | 4.8182 | 13 |
| 14 | .2992 | 7.7862 | 39.9633 | 3.3417 | 26.0192 | 0.1284 | 0.0384 | 5.1326 | 14 |
| 15 | .2745 | 8.0607 | 43.8069 | 3.6425 | 29.3609 | 0.1241 | 0.0341 | 5.4346 | 15 |
| 16 | .2519 | 8.3126 | 47.5849 | 3.9703 | 33.0034 | 0.1203 | 0.0303 | 5.7245 | 16 |
| 17 | .2311 | 8.5436 | 51.2821 | 4.3276 | 36.9737 | 0.1170 | 0.0270 | 6.0024 | 17 |
| 18 | .2120 | 8.7556 | 54.8860 | 4.7171 | 41.3013 | 0.1142 | 0.0242 | 6.2687 | 18 |
| 19 | .1945 | 8.9501 | 58.3868 | 5.1417 | 46.0185 | 0.1117 | 0.0217 | 6.5236 | 19 |
| 20 | .1784 | 9.1285 | 61.7770 | 5.6044 | 51.1601 | 0.1095 | 0.0195 | 6.7674 | 20 |
| 21 | .1637 | 9.2922 | 65.0509 | 6.1088 | 56.7645 | 0.1076 | 0.0176 | 7.0006 | 21 |
| 22 | .1502 | 9.4424 | 68.2048 | 6.6586 | 62.8733 | 0.1059 | 0.0159 | 7.2232 | 22 |
| 23 | .1378 | 9.5802 | 71.2359 | 7.2579 | 69.5319 | 0.1044 | 0.0144 | 7.4357 | 23 |
| 24 | .1264 | 9.7066 | 74.1433 | 7.9111 | 76.7898 | 0.1030 | 0.0130 | 7.6384 | 24 |
| 25 | .1160 | 9.8226 | 76.9265 | 8.6231 | 84.7009 | 0.1018 | 0.0118 | 7.8316 | 25 |
| 26 | .1064 | 9.9290 | 79.5863 | 9.3992 | 93.3240 | 0.1007 | 0.0107 | 8.0156 | 26 |
| 27 | .0976 | 10.0266 | 82.1241 | 10.2451 | 102.7231 | 0.0997 | 0.0097 | 8.1906 | 27 |
| 28 | .0895 | 10.1161 | 84.5419 | 11.1671 | 112.9682 | 0.0989 | 0.0089 | 8.3571 | 28 |
| 29 | .0822 | 10.1983 | 86.8422 | 12.1722 | 124.1354 | 0.0981 | 0.0081 | 8.5154 | 29 |
| 30 | .0754 | 10.2737 | 89.0280 | 13.2677 | 136.3075 | 0.0973 | 0.0073 | 8.6657 | 30 |
| 31 | .0691 | 10.3428 | 91.1024 | 14.4618 | 149.5752 | 0.0967 | 0.0067 | 8.8083 | 31 |
| 32 | .0634 | 10.4062 | 93.0690 | 15.7633 | 164.0370 | 0.0961 | 0.0061 | 8.9436 | 32 |
| 33 | .0582 | 10.4644 | 94.9314 | 17.1820 | 179.8003 | 0.0956 | 0.0056 | 9.0718 | 33 |
| 34 | .0534 | 10.5178 | 96.6935 | 18.7284 | 196.9823 | 0.0951 | 0.0051 | 9.1933 | 34 |
| 35 | .0490 | 10.5668 | 98.3590 | 20.4140 | 215.7108 | 0.0946 | 0.0046 | 9.3083 | 35 |
| 36 | .0449 | 10.6118 | 99.9319 | 22.2512 | 236.1247 | 0.0942 | 0.0042 | 9.4171 | 36 |
| 37 | .0412 | 10.6530 | 101.4162 | 24.2538 | 258.3759 | 0.0939 | 0.0039 | 9.5200 | 37 |
| 38 | .0378 | 10.6908 | 102.8158 | 26.4367 | 282.6298 | 0.0935 | 0.0035 | 9.6172 | 38 |
| 39 | .0347 | 10.7255 | 104.1345 | 28.8160 | 309.0665 | 0.0932 | 0.0032 | 9.7090 | 39 |
| 40 | .0318 | 10.7574 | 105.3762 | 31.4094 | 337.8824 | 0.0930 | 0.0030 | 9.7957 | 40 |
| 41 | .0292 | 10.7866 | 106.5445 | 34.2363 | 369.2919 | 0.0927 | 0.0027 | 9.8775 | 41 |
| 42 | .0268 | 10.8134 | 107.6432 | 37.3175 | 403.5281 | 0.0925 | 0.0025 | 9.9546 | 42 |
| 43 | .0246 | 10.8380 | 108.6758 | 40.6761 | 440.8457 | 0.0923 | 0.0023 | 10.0273 | 43 |
| 44 | .0226 | 10.8605 | 109.6456 | 44.3370 | 481.5218 | 0.0921 | 0.0021 | 10.0958 | 44 |
| 45 | .0207 | 10.8812 | 110.5561 | 48.3273 | 525.8587 | 0.0919 | 0.0019 | 10.1603 | 45 |
| 46 | .0190 | 10.9002 | 111.4103 | 52.6767 | 574.1860 | 0.0917 | 0.0017 | 10.2210 | 46 |
| 47 | .0174 | 10.9176 | 112.2115 | 57.4176 | 626.8628 | 0.0916 | 0.0016 | 10.2780 | 47 |
| 48 | .0160 | 10.9336 | 112.9625 | 62.5852 | 684.2804 | 0.0915 | 0.0015 | 10.3317 | 48 |
| 49 | .0147 | 10.9482 | 113.6661 | 68.2179 | 746.8656 | 0.0913 | 0.0013 | 10.3821 | 49 |
| 50 | .0134 | 10.9617 | 114.3251 | 74.3575 | 815.0836 | 0.0912 | 0.0012 | 10.4295 | 50 |
| 51 | .0123 | 10.9740 | 114.9420 | 81.0497 | 889.4411 | 0.0911 | 0.0011 | 10.4740 | 51 |
| 52 | .0113 | 10.9853 | 115.5193 | 88.3442 | 970.4908 | 0.0910 | 0.0010 | 10.5158 | 52 |
| 53 | .0104 | 10.9957 | 116.0593 | 96.2951 | 1058.8349 | 0.0909 | 0.0009 | 10.5549 | 53 |
| 54 | .0095 | 11.0053 | 116.5642 | 104.9617 | 1155.1301 | 0.0909 | 0.0009 | 10.5917 | 54 |
| 55 | .0087 | 11.0140 | 117.0362 | 114.4083 | 1260.0918 | 0.0908 | 0.0008 | 10.6261 | 55 |
| 60 | .0057 | 11.0480 | 118.9683 | 176.0313 | 1944.7921 | 0.0905 | 0.0005 | 10.7683 | 60 |
| 65 | .0037 | 11.0701 | 120.3344 | 270.8460 | 2998.2885 | 0.0903 | 0.0003 | 10.8702 | 65 |
| 70 | .0024 | 11.0844 | 121.2942 | 416.7301 | 4619.2232 | 0.0902 | 0.0002 | 10.9427 | 70 |
| 75 | .0016 | 11.0938 | 121.9646 | 641.1909 | 7113.2321 | 0.0901 | 0.0001 | 10.9940 | 75 |
| 80 | .0010 | 11.0998 | 122.4306 | 986.5517 | 10950.5741 | 0.0901 | 0.0001 | 11.0299 | 80 |
| 85 | .0007 | 11.1038 | 122.7533 | 1517.9320 | 16854.8003 | 0.0901 | 0.0001 | 11.0551 | 85 |
| 90 | .0004 | 11.1064 | 122.9758 | 2335.5266 | 25939.1842 | 0.0900 | 0.0000 | 11.0726 | 90 |
| 95 | .0003 | 11.1080 | 123.1287 | 3593.4971 | 39916.6350 | 0.0900 | 0.0000 | 11.0847 | 95 |
| 100 | .0002 | 11.1091 | 123.2335 | 5529.0408 | 61422.6755 | 0.0900 | 0.0000 | 11.0930 | 100 |

**I = 10.00 %**

| N | (P/F) | (P/A) | (P/G) | (F/P) | (F/A) | (A/P) | (A/F) | (A/G) | N |
|---|---|---|---|---|---|---|---|---|---|
| 1 | .9091 | 0.9091 | −0.0000 | 1.1000 | 1.0000 | 1.1000 | 1.0000 | −0.0000 | 1 |
| 2 | .8264 | 1.7355 | 0.8264 | 1.2100 | 2.1000 | 0.5762 | 0.4762 | 0.4762 | 2 |
| 3 | .7513 | 2.4869 | 2.3291 | 1.3310 | 3.3100 | 0.4021 | 0.3021 | 0.9366 | 3 |
| 4 | .6830 | 3.1699 | 4.3781 | 1.4641 | 4.6410 | 0.3155 | 0.2155 | 1.3812 | 4 |
| 5 | .6209 | 3.7908 | 6.8618 | 1.6105 | 6.1051 | 0.2638 | 0.1638 | 1.8101 | 5 |
| 6 | .5645 | 4.3553 | 9.6842 | 1.7716 | 7.7156 | 0.2296 | 0.1296 | 2.2236 | 6 |
| 7 | .5132 | 4.8684 | 12.7631 | 1.9487 | 9.4872 | 0.2054 | 0.1054 | 2.6216 | 7 |
| 8 | .4665 | 5.3349 | 16.0287 | 2.1436 | 11.4359 | 0.1874 | 0.0874 | 3.0045 | 8 |
| 9 | .4241 | 5.7590 | 19.4215 | 2.3579 | 13.5795 | 0.1736 | 0.0736 | 3.3724 | 9 |
| 10 | .3855 | 6.1446 | 22.8913 | 2.5937 | 15.9374 | 0.1627 | 0.0627 | 3.7255 | 10 |
| 11 | .3505 | 6.4951 | 26.3963 | 2.8531 | 18.5312 | 0.1540 | 0.0540 | 4.0641 | 11 |
| 12 | .3186 | 6.8137 | 29.9012 | 3.1384 | 21.3843 | 0.1468 | 0.0468 | 4.3884 | 12 |
| 13 | .2897 | 7.1034 | 33.3772 | 3.4523 | 24.5227 | 0.1408 | 0.0408 | 4.6988 | 13 |
| 14 | .2633 | 7.3667 | 36.8005 | 3.7975 | 27.9750 | 0.1357 | 0.0357 | 4.9955 | 14 |
| 15 | .2394 | 7.6061 | 40.1520 | 4.1772 | 31.7725 | 0.1315 | 0.0315 | 5.2789 | 15 |
| 16 | .2176 | 7.8237 | 43.4164 | 4.5950 | 35.9497 | 0.1278 | 0.0278 | 5.5493 | 16 |
| 17 | .1978 | 8.0216 | 46.5819 | 5.0545 | 40.5447 | 0.1247 | 0.0247 | 5.8071 | 17 |
| 18 | .1799 | 8.2014 | 49.6395 | 5.5599 | 45.5992 | 0.1219 | 0.0219 | 6.0526 | 18 |
| 19 | .1635 | 8.3649 | 52.5827 | 6.1159 | 51.1591 | 0.1195 | 0.0195 | 6.2861 | 19 |
| 20 | .1486 | 8.5136 | 55.4069 | 6.7275 | 57.2750 | 0.1175 | 0.0175 | 6.5081 | 20 |
| 21 | .1351 | 8.6487 | 58.1095 | 7.4002 | 64.0025 | 0.1156 | 0.0156 | 6.7189 | 21 |
| 22 | .1228 | 8.7715 | 60.6893 | 8.1403 | 71.4027 | 0.1140 | 0.0140 | 6.9189 | 22 |
| 23 | .1117 | 8.8832 | 63.1462 | 8.9543 | 79.5430 | 0.1126 | 0.0126 | 7.1085 | 23 |
| 24 | .1015 | 8.9847 | 65.4813 | 9.8497 | 88.4973 | 0.1113 | 0.0113 | 7.2881 | 24 |
| 25 | .0923 | 9.0770 | 67.6964 | 10.8347 | 98.3471 | 0.1102 | 0.0102 | 7.4580 | 25 |
| 26 | .0839 | 9.1609 | 69.7940 | 11.9182 | 109.1818 | 0.1092 | 0.0092 | 7.6186 | 26 |
| 27 | .0763 | 9.2372 | 71.7773 | 13.1100 | 121.0999 | 0.1083 | 0.0083 | 7.7704 | 27 |
| 28 | .0693 | 9.3066 | 73.6495 | 14.4210 | 134.2099 | 0.1075 | 0.0075 | 7.9137 | 28 |
| 29 | .0630 | 9.3696 | 75.4146 | 15.8631 | 148.6309 | 0.1067 | 0.0067 | 8.0489 | 29 |
| 30 | .0573 | 9.4269 | 77.0766 | 17.4494 | 164.4940 | 0.1061 | 0.0061 | 8.1762 | 30 |
| 31 | .0521 | 9.4790 | 78.6395 | 19.1943 | 181.9434 | 0.1055 | 0.0055 | 8.2962 | 31 |
| 32 | .0474 | 9.5264 | 80.1078 | 21.1138 | 201.1378 | 0.1050 | 0.0050 | 8.4091 | 32 |
| 33 | .0431 | 9.5694 | 81.4856 | 23.2252 | 222.2515 | 0.1045 | 0.0045 | 8.5152 | 33 |
| 34 | .0391 | 9.6086 | 82.7773 | 25.5477 | 245.4767 | 0.1041 | 0.0041 | 8.6149 | 34 |
| 35 | .0356 | 9.6442 | 83.9872 | 28.1024 | 271.0244 | 0.1037 | 0.0037 | 8.7086 | 35 |
| 36 | .0323 | 9.6765 | 85.1194 | 30.9127 | 299.1268 | 0.1033 | 0.0033 | 8.7965 | 36 |
| 37 | .0294 | 9.7059 | 86.1781 | 34.0039 | 330.0395 | 0.1030 | 0.0030 | 8.8789 | 37 |
| 38 | .0267 | 9.7327 | 87.1673 | 37.4043 | 364.0434 | 0.1027 | 0.0027 | 8.9562 | 38 |
| 39 | .0243 | 9.7570 | 88.0908 | 41.1448 | 401.4478 | 0.1025 | 0.0025 | 9.0285 | 39 |
| 40 | .0221 | 9.7791 | 88.9525 | 45.2593 | 442.5926 | 0.1023 | 0.0023 | 9.0962 | 40 |
| 41 | .0201 | 9.7991 | 89.7560 | 49.7852 | 487.8518 | 0.1020 | 0.0020 | 9.1596 | 41 |
| 42 | .0183 | 9.8174 | 90.5047 | 54.7637 | 537.6370 | 0.1019 | 0.0019 | 9.2188 | 42 |
| 43 | .0166 | 9.8340 | 91.2019 | 60.2401 | 592.4007 | 0.1017 | 0.0017 | 9.2741 | 43 |
| 44 | .0151 | 9.8491 | 91.8508 | 66.2641 | 652.6408 | 0.1015 | 0.0015 | 9.3258 | 44 |
| 45 | .0137 | 9.8628 | 92.4544 | 72.8905 | 718.9048 | 0.1014 | 0.0014 | 9.3740 | 45 |
| 46 | .0125 | 9.8753 | 93.0157 | 80.1795 | 791.7953 | 0.1013 | 0.0013 | 9.4190 | 46 |
| 47 | .0113 | 9.8866 | 93.5372 | 88.1975 | 871.9749 | 0.1011 | 0.0011 | 9.4610 | 47 |
| 48 | .0103 | 9.8969 | 94.0217 | 97.0172 | 960.1723 | 0.1010 | 0.0010 | 9.5001 | 48 |
| 49 | .0094 | 9.9063 | 94.4715 | 106.7190 | 1057.1896 | 0.1009 | 0.0009 | 9.5365 | 49 |
| 50 | .0085 | 9.9148 | 94.8889 | 117.3909 | 1163.9085 | 0.1009 | 0.0009 | 9.5704 | 50 |
| 51 | .0077 | 9.9226 | 95.2761 | 129.1299 | 1281.2994 | 0.1008 | 0.0008 | 9.6020 | 51 |
| 52 | .0070 | 9.9296 | 95.6351 | 142.0429 | 1410.4293 | 0.1007 | 0.0007 | 9.6313 | 52 |
| 53 | .0064 | 9.9360 | 95.9679 | 156.2472 | 1552.4723 | 0.1006 | 0.0006 | 9.6586 | 53 |
| 54 | .0058 | 9.9418 | 96.2763 | 171.8719 | 1708.7195 | 0.1006 | 0.0006 | 9.6840 | 54 |
| 55 | .0053 | 9.9471 | 96.5619 | 189.0591 | 1880.5914 | 0.1005 | 0.0005 | 9.7075 | 55 |
| 60 | .0033 | 9.9672 | 97.7010 | 304.4816 | 3034.8164 | 0.1003 | 0.0003 | 9.8023 | 60 |
| 65 | .0020 | 9.9796 | 98.4705 | 490.3707 | 4893.7073 | 0.1002 | 0.0002 | 9.8672 | 65 |
| 70 | .0013 | 9.9873 | 98.9870 | 789.7470 | 7887.4696 | 0.1001 | 0.0001 | 9.9113 | 70 |
| 75 | .0008 | 9.9921 | 99.3317 | 1271.8954 | 12708.9537 | 0.1001 | 0.0001 | 9.9410 | 75 |
| 80 | .0005 | 9.9951 | 99.5606 | 2048.4002 | 20474.0021 | 0.1000 | 0.0000 | 9.9609 | 80 |
| 85 | .0003 | 9.9970 | 99.7120 | 3298.9690 | 32979.6903 | 0.1000 | 0.0000 | 9.9742 | 85 |
| 90 | .0002 | 9.9981 | 99.8118 | 5313.0226 | 53120.2261 | 0.1000 | 0.0000 | 9.9831 | 90 |
| 95 | .0001 | 9.9988 | 99.8773 | 8556.6760 | 85556.7605 | 0.1000 | 0.0000 | 9.9889 | 95 |
| 100 | .0001 | 9.9993 | 99.9202 | 13780.6123 | 137796.1234 | 0.1000 | 0.0000 | 9.9927 | 100 |

**I = 12.00 %**

| N | (P/F) | (P/A) | (P/G) | (F/P) | (F/A) | (A/P) | (A/F) | (A/G) | N |
|---|---|---|---|---|---|---|---|---|---|
| 1 | .8929 | 0.8929 | -0.0000 | 1.1200 | 1.0000 | 1.1200 | 1.0000 | -0.0000 | 1 |
| 2 | .7972 | 1.6901 | 0.7972 | 1.2544 | 2.1200 | 0.5917 | 0.4717 | 0.4717 | 2 |
| 3 | .7118 | 2.4018 | 2.2208 | 1.4049 | 3.3744 | 0.4163 | 0.2963 | 0.9246 | 3 |
| 4 | .6355 | 3.0373 | 4.1273 | 1.5735 | 4.7793 | 0.3292 | 0.2092 | 1.3589 | 4 |
| 5 | .5674 | 3.6048 | 6.3970 | 1.7623 | 6.3528 | 0.2774 | 0.1574 | 1.7746 | 5 |
| 6 | .5066 | 4.1114 | 8.9302 | 1.9738 | 8.1152 | 0.2432 | 0.1232 | 2.1720 | 6 |
| 7 | .4523 | 4.5638 | 11.6443 | 2.2107 | 10.0890 | 0.2191 | 0.0991 | 2.5515 | 7 |
| 8 | .4039 | 4.9676 | 14.4714 | 2.4760 | 12.2997 | 0.2013 | 0.0813 | 2.9131 | 8 |
| 9 | .3606 | 5.3282 | 17.3563 | 2.7731 | 14.7757 | 0.1877 | 0.0677 | 3.2574 | 9 |
| 10 | .3220 | 5.6502 | 20.2541 | 3.1058 | 17.5487 | 0.1770 | 0.0570 | 3.5847 | 10 |
| 11 | .2875 | 5.9377 | 23.1288 | 3.4785 | 20.6546 | 0.1684 | 0.0484 | 3.8953 | 11 |
| 12 | .2567 | 6.1944 | 25.9523 | 3.8960 | 24.1331 | 0.1614 | 0.0414 | 4.1897 | 12 |
| 13 | .2292 | 6.4235 | 28.7024 | 4.3635 | 28.0291 | 0.1557 | 0.0357 | 4.4683 | 13 |
| 14 | .2046 | 6.6282 | 31.3624 | 4.8871 | 32.3926 | 0.1509 | 0.0309 | 4.7317 | 14 |
| 15 | .1827 | 6.8109 | 33.9202 | 5.4736 | 37.2797 | 0.1468 | 0.0268 | 4.9803 | 15 |
| 16 | .1631 | 6.9740 | 36.3670 | 6.1304 | 42.7533 | 0.1434 | 0.0234 | 5.2147 | 16 |
| 17 | .1456 | 7.1196 | 38.6973 | 6.8660 | 48.8837 | 0.1405 | 0.0205 | 5.4353 | 17 |
| 18 | .1300 | 7.2497 | 40.9080 | 7.6900 | 55.7497 | 0.1379 | 0.0179 | 5.6427 | 18 |
| 19 | .1161 | 7.3658 | 42.9979 | 8.6128 | 63.4397 | 0.1358 | 0.0158 | 5.8375 | 19 |
| 20 | .1037 | 7.4694 | 44.9676 | 9.6463 | 72.0524 | 0.1339 | 0.0139 | 6.0202 | 20 |
| 21 | .0926 | 7.5620 | 46.8188 | 10.8038 | 81.6987 | 0.1322 | 0.0122 | 6.1913 | 21 |
| 22 | .0826 | 7.6446 | 48.5543 | 12.1003 | 92.5026 | 0.1308 | 0.0108 | 6.3514 | 22 |
| 23 | .0738 | 7.7184 | 50.1776 | 13.5523 | 104.6029 | 0.1296 | 0.0096 | 6.5010 | 23 |
| 24 | .0659 | 7.7843 | 51.6929 | 15.1786 | 118.1552 | 0.1285 | 0.0085 | 6.6406 | 24 |
| 25 | .0588 | 7.8431 | 53.1046 | 17.0001 | 133.3339 | 0.1275 | 0.0075 | 6.7708 | 25 |
| 26 | .0525 | 7.8957 | 54.4177 | 19.0401 | 150.3339 | 0.1267 | 0.0067 | 6.8921 | 26 |
| 27 | .0469 | 7.9426 | 55.6369 | 21.3249 | 169.3740 | 0.1259 | 0.0059 | 7.0049 | 27 |
| 28 | .0419 | 7.9844 | 56.7674 | 23.8839 | 190.6989 | 0.1252 | 0.0052 | 7.1098 | 28 |
| 29 | .0374 | 8.0218 | 57.8141 | 26.7499 | 214.5828 | 0.1247 | 0.0047 | 7.2071 | 29 |
| 30 | .0334 | 8.0552 | 58.7821 | 29.9599 | 241.3327 | 0.1241 | 0.0041 | 7.2974 | 30 |
| 31 | .0298 | 8.0850 | 59.6761 | 33.5551 | 271.2926 | 0.1237 | 0.0037 | 7.3811 | 31 |
| 32 | .0266 | 8.1116 | 60.5010 | 37.5817 | 304.8477 | 0.1233 | 0.0033 | 7.4586 | 32 |
| 33 | .0238 | 8.1354 | 61.2612 | 42.0915 | 342.4294 | 0.1229 | 0.0029 | 7.5302 | 33 |
| 34 | .0212 | 8.1566 | 61.9612 | 47.1425 | 384.5210 | 0.1226 | 0.0026 | 7.5965 | 34 |
| 35 | .0189 | 8.1755 | 62.6052 | 52.7996 | 431.6635 | 0.1223 | 0.0023 | 7.6577 | 35 |
| 36 | .0169 | 8.1924 | 63.1970 | 59.1356 | 484.4631 | 0.1221 | 0.0021 | 7.7141 | 36 |
| 37 | .0151 | 8.2075 | 63.7406 | 66.2318 | 543.5987 | 0.1218 | 0.0018 | 7.7661 | 37 |
| 38 | .0135 | 8.2210 | 64.2394 | 74.1797 | 609.8305 | 0.1216 | 0.0016 | 7.8141 | 38 |
| 39 | .0120 | 8.2330 | 64.6967 | 83.0812 | 684.0102 | 0.1215 | 0.0015 | 7.8582 | 39 |
| 40 | .0107 | 8.2438 | 65.1159 | 93.0510 | 767.0914 | 0.1213 | 0.0013 | 7.8988 | 40 |
| 41 | .0096 | 8.2534 | 65.4997 | 104.2171 | 860.1424 | 0.1212 | 0.0012 | 7.9361 | 41 |
| 42 | .0086 | 8.2619 | 65.8509 | 116.7231 | 964.3595 | 0.1210 | 0.0010 | 7.9704 | 42 |
| 43 | .0076 | 8.2696 | 66.1722 | 130.7299 | 1081.0826 | 0.1209 | 0.0009 | 8.0019 | 43 |
| 44 | .0068 | 8.2764 | 66.4659 | 146.4175 | 1211.8125 | 0.1208 | 0.0008 | 8.0308 | 44 |
| 45 | .0061 | 8.2825 | 66.7342 | 163.9876 | 1358.2300 | 0.1207 | 0.0007 | 8.0572 | 45 |
| 46 | .0054 | 8.2880 | 66.9792 | 183.6661 | 1522.2176 | 0.1207 | 0.0007 | 8.0815 | 46 |
| 47 | .0049 | 8.2928 | 67.2028 | 205.7061 | 1705.8838 | 0.1206 | 0.0006 | 8.1037 | 47 |
| 48 | .0043 | 8.2972 | 67.4068 | 230.3908 | 1911.5898 | 0.1205 | 0.0005 | 8.1241 | 48 |
| 49 | .0039 | 8.3010 | 67.5929 | 258.0377 | 2141.9806 | 0.1205 | 0.0005 | 8.1427 | 49 |
| 50 | .0035 | 8.3045 | 67.7624 | 289.0022 | 2400.0182 | 0.1204 | 0.0004 | 8.1597 | 50 |
| 51 | .0031 | 8.3076 | 67.9169 | 323.6825 | 2689.0204 | 0.1204 | 0.0004 | 8.1753 | 51 |
| 52 | .0028 | 8.3103 | 68.0576 | 362.5243 | 3012.7029 | 0.1203 | 0.0003 | 8.1895 | 52 |
| 53 | .0025 | 8.3128 | 68.1856 | 406.0273 | 3375.2272 | 0.1203 | 0.0003 | 8.2025 | 53 |
| 54 | .0022 | 8.3150 | 68.3022 | 454.7505 | 3781.2545 | 0.1203 | 0.0003 | 8.2143 | 54 |
| 55 | .0020 | 8.3170 | 68.4082 | 509.3206 | 4236.0050 | 0.1202 | 0.0002 | 8.2251 | 55 |
| 60 | .0011 | 8.3240 | 68.8100 | 897.5969 | 7471.6411 | 0.1201 | 0.0001 | 8.2664 | 60 |
| 65 | .0006 | 8.3281 | 69.0581 | 1581.8725 | 13173.9374 | 0.1201 | 0.0001 | 8.2922 | 65 |
| 70 | .0004 | 8.3303 | 69.2103 | 2787.7998 | 23223.3319 | 0.1200 | 0.0000 | 8.3082 | 70 |
| 75 | .0002 | 8.3316 | 69.3031 | 4913.0558 | 40933.7987 | 0.1200 | 0.0000 | 8.3181 | 75 |
| 80 | .0001 | 8.3324 | 69.3594 | 8658.4831 | 72145.6925 | 0.1200 | 0.0000 | 8.3241 | 80 |
| 85 | .0001 | 8.3328 | 69.3935 | 15259.2057 | 127151.7140 | 0.1200 | 0.0000 | 8.3278 | 85 |
| 90 | .0000 | 8.3330 | 69.4140 | 26891.9342 | 224091.1185 | 0.1200 | 0.0000 | 8.3300 | 90 |
| 95 | .0000 | 8.3332 | 69.4263 | 47392.7766 | 394931.4719 | 0.1200 | 0.0000 | 8.3313 | 95 |
| 100 | .0000 | 8.3332 | 69.4336 | 83522.2657 | 696010.5477 | 0.1200 | 0.0000 | 8.3321 | 100 |

**I = 15.00 %**

| N | (P/F) | (P/A) | (P/G) | (F/P) | (F/A) | (A/P) | (A/F) | (A/G) | N |
|---|---|---|---|---|---|---|---|---|---|
| 1 | .8696 | 0.8696 | -0.0000 | 1.1500 | 1.0000 | 1.1500 | 1.0000 | -0.0000 | 1 |
| 2 | .7561 | 1.6257 | 0.7561 | 1.3225 | 2.1500 | 0.6151 | 0.4651 | 0.4651 | 2 |
| 3 | .6575 | 2.2832 | 2.0712 | 1.5209 | 3.4725 | 0.4380 | 0.2880 | 0.9071 | 3 |
| 4 | .5718 | 2.8550 | 3.7864 | 1.7490 | 4.9934 | 0.3503 | 0.2003 | 1.3263 | 4 |
| 5 | .4972 | 3.3522 | 5.7751 | 2.0114 | 6.7424 | 0.2983 | 0.1483 | 1.7228 | 5 |
| 6 | .4323 | 3.7845 | 7.9368 | 2.3131 | 8.7537 | 0.2642 | 0.1142 | 2.0972 | 6 |
| 7 | .3759 | 4.1604 | 10.1924 | 2.6600 | 11.0668 | 0.2404 | 0.0904 | 2.4498 | 7 |
| 8 | .3269 | 4.4873 | 12.4807 | 3.0590 | 13.7268 | 0.2229 | 0.0729 | 2.7813 | 8 |
| 9 | .2843 | 4.7716 | 14.7548 | 3.5179 | 16.7858 | 0.2096 | 0.0596 | 3.0922 | 9 |
| 10 | .2472 | 5.0188 | 16.9795 | 4.0456 | 20.3037 | 0.1993 | 0.0493 | 3.3832 | 10 |
| 11 | .2149 | 5.2337 | 19.1289 | 4.6524 | 24.3493 | 0.1911 | 0.0411 | 3.6549 | 11 |
| 12 | .1869 | 5.4206 | 21.1849 | 5.3503 | 29.0017 | 0.1845 | 0.0345 | 3.9082 | 12 |
| 13 | .1625 | 5.5831 | 23.1352 | 6.1528 | 34.3519 | 0.1791 | 0.0291 | 4.1438 | 13 |
| 14 | .1413 | 5.7245 | 24.9725 | 7.0757 | 40.5047 | 0.1747 | 0.0247 | 4.3624 | 14 |
| 15 | .1229 | 5.8474 | 26.6930 | 8.1371 | 47.5804 | 0.1710 | 0.0210 | 4.5650 | 15 |
| 16 | .1069 | 5.9542 | 28.2960 | 9.3576 | 55.7175 | 0.1679 | 0.0179 | 4.7522 | 16 |
| 17 | .0929 | 6.0472 | 29.7828 | 10.7613 | 65.0751 | 0.1654 | 0.0154 | 4.9251 | 17 |
| 18 | .0808 | 6.1280 | 31.1565 | 12.3755 | 75.8364 | 0.1632 | 0.0132 | 5.0843 | 18 |
| 19 | .0703 | 6.1982 | 32.4213 | 14.2318 | 88.2118 | 0.1613 | 0.0113 | 5.2307 | 19 |
| 20 | .0611 | 6.2593 | 33.5822 | 16.3665 | 102.4436 | 0.1598 | 0.0098 | 5.3651 | 20 |
| 21 | .0531 | 6.3125 | 34.6448 | 18.8215 | 118.8101 | 0.1584 | 0.0084 | 5.4883 | 21 |
| 22 | .0462 | 6.3587 | 35.6150 | 21.6447 | 137.6316 | 0.1573 | 0.0073 | 5.6010 | 22 |
| 23 | .0402 | 6.3988 | 36.4988 | 24.8915 | 159.2764 | 0.1563 | 0.0063 | 5.7040 | 23 |
| 24 | .0349 | 6.4338 | 37.3023 | 28.6252 | 184.1678 | 0.1554 | 0.0054 | 5.7979 | 24 |
| 25 | .0304 | 6.4641 | 38.0314 | 32.9190 | 212.7930 | 0.1547 | 0.0047 | 5.8834 | 25 |
| 26 | .0264 | 6.4906 | 38.6918 | 37.8568 | 245.7120 | 0.1541 | 0.0041 | 5.9612 | 26 |
| 27 | .0230 | 6.5135 | 39.2890 | 43.5353 | 283.5688 | 0.1535 | 0.0035 | 6.0319 | 27 |
| 28 | .0200 | 6.5335 | 39.8283 | 50.0656 | 327.1041 | 0.1531 | 0.0031 | 6.0960 | 28 |
| 29 | .0174 | 6.5509 | 40.3146 | 57.5755 | 377.1697 | 0.1527 | 0.0027 | 6.1541 | 29 |
| 30 | .0151 | 6.5660 | 40.7526 | 66.2118 | 434.7451 | 0.1523 | 0.0023 | 6.2066 | 30 |
| 31 | .0131 | 6.5791 | 41.1466 | 76.1435 | 500.9569 | 0.1520 | 0.0020 | 6.2541 | 31 |
| 32 | .0114 | 6.5905 | 41.5006 | 87.5651 | 577.1005 | 0.1517 | 0.0017 | 6.2970 | 32 |
| 33 | .0099 | 6.6005 | 41.8184 | 100.6998 | 664.6655 | 0.1515 | 0.0015 | 6.3357 | 33 |
| 34 | .0086 | 6.6091 | 42.1033 | 115.8048 | 765.3654 | 0.1513 | 0.0013 | 6.3705 | 34 |
| 35 | .0075 | 6.6166 | 42.3586 | 133.1755 | 881.1702 | 0.1511 | 0.0011 | 6.4019 | 35 |
| 36 | .0065 | 6.6231 | 42.5872 | 153.1519 | 1014.3457 | 0.1510 | 0.0010 | 6.4301 | 36 |
| 37 | .0057 | 6.6288 | 42.7916 | 176.1246 | 1167.4975 | 0.1509 | 0.0009 | 6.4554 | 37 |
| 38 | .0049 | 6.6338 | 42.9743 | 202.5433 | 1343.6222 | 0.1507 | 0.0007 | 6.4781 | 38 |
| 39 | .0043 | 6.6380 | 43.1374 | 232.9248 | 1546.1655 | 0.1506 | 0.0006 | 6.4985 | 39 |
| 40 | .0037 | 6.6418 | 43.2830 | 267.8635 | 1779.0903 | 0.1506 | 0.0006 | 6.5168 | 40 |
| 41 | .0032 | 6.6450 | 43.4128 | 308.0431 | 2046.9539 | 0.1505 | 0.0005 | 6.5331 | 41 |
| 42 | .0028 | 6.6478 | 43.5286 | 354.2495 | 2354.9969 | 0.1504 | 0.0004 | 6.5478 | 42 |
| 43 | .0025 | 6.6503 | 43.6317 | 407.3870 | 2709.2465 | 0.1504 | 0.0004 | 6.5609 | 43 |
| 44 | .0021 | 6.6524 | 43.7235 | 468.4950 | 3116.6334 | 0.1503 | 0.0003 | 6.5725 | 44 |
| 45 | .0019 | 6.6543 | 43.8051 | 538.7693 | 3585.1285 | 0.1503 | 0.0003 | 6.5830 | 45 |
| 46 | .0016 | 6.6559 | 43.8778 | 619.5847 | 4123.8977 | 0.1502 | 0.0002 | 6.5923 | 46 |
| 47 | .0014 | 6.6573 | 43.9423 | 712.5224 | 4743.4824 | 0.1502 | 0.0002 | 6.6006 | 47 |
| 48 | .0012 | 6.6585 | 43.9997 | 819.4007 | 5456.6047 | 0.1502 | 0.0002 | 6.6080 | 48 |
| 49 | .0011 | 6.6596 | 44.0506 | 942.3108 | 6275.4055 | 0.1502 | 0.0002 | 6.6146 | 49 |
| 50 | .0009 | 6.6605 | 44.0958 | 1083.6574 | 7217.7163 | 0.1501 | 0.0001 | 6.6205 | 50 |
| 51 | .0008 | 6.6613 | 44.1360 | 1246.2061 | 8301.3737 | 0.1501 | 0.0001 | 6.6257 | 51 |
| 52 | .0007 | 6.6620 | 44.1715 | 1433.1370 | 9547.5798 | 0.1501 | 0.0001 | 6.6304 | 52 |
| 53 | .0006 | 6.6626 | 44.2031 | 1648.1075 | 10980.7167 | 0.1501 | 0.0001 | 6.6345 | 53 |
| 54 | .0005 | 6.6631 | 44.2311 | 1895.3236 | 12628.8243 | 0.1501 | 0.0001 | 6.6382 | 54 |
| 55 | .0005 | 6.6636 | 44.2558 | 2179.6222 | 14524.1479 | 0.1501 | 0.0001 | 6.6414 | 55 |
| 60 | .0002 | 6.6651 | 44.3431 | 4383.9987 | 29219.9916 | 0.1500 | 0.0000 | 6.6530 | 60 |
| 65 | .0001 | 6.6659 | 44.3903 | 8817.7874 | 58778.5826 | 0.1500 | 0.0000 | 6.6593 | 65 |
| 70 | .0001 | 6.6663 | 44.4156 | 17735.7200 | 118231.4669 | 0.1500 | 0.0000 | 6.6627 | 70 |
| 75 | .0000 | 6.6665 | 44.4292 | 35672.8680 | 237812.4532 | 0.1500 | 0.0000 | 6.6646 | 75 |
| 80 | .0000 | 6.6666 | 44.4364 | 71750.8794 | 478332.5293 | 0.1500 | 0.0000 | 6.6656 | 80 |
| 85 | .0000 | 6.6666 | 44.4402 | 144316.6470 | 962104.3133 | 0.1500 | 0.0000 | 6.6661 | 85 |
| 90 | .0000 | 6.6666 | 44.4422 | 290272.3252 | 1935142.1680 | 0.1500 | 0.0000 | 6.6664 | 90 |
| 95 | .0000 | 6.6667 | 44.4433 | 583841.3276 | 3892268.8509 | 0.1500 | 0.0000 | 6.6665 | 95 |
| 100 | .0000 | 6.6667 | 44.4438 | 1174313.4507 | 7828749.6713 | 0.1500 | 0.0000 | 6.6666 | 100 |

**I = 20.00 %**

| N | (P/F) | (P/A) | (P/G) | (F/P) | (F/A) | (A/P) | (A/F) | (A/G) | N |
|---|---|---|---|---|---|---|---|---|---|
| 1 | .8333 | 0.8333 | -0.0000 | 1.2000 | 1.0000 | 1.2000 | 1.0000 | -0.0000 | 1 |
| 2 | .6944 | 1.5278 | 0.6944 | 1.4400 | 2.2000 | 0.6545 | 0.4545 | 0.4545 | 2 |
| 3 | .5787 | 2.1065 | 1.8519 | 1.7280 | 3.6400 | 0.4747 | 0.2747 | 0.8791 | 3 |
| 4 | .4823 | 2.5887 | 3.2986 | 2.0736 | 5.3680 | 0.3863 | 0.1863 | 1.2742 | 4 |
| 5 | .4019 | 2.9906 | 4.9061 | 2.4883 | 7.4416 | 0.3344 | 0.1344 | 1.6405 | 5 |
| 6 | .3349 | 3.3255 | 6.5806 | 2.9860 | 9.9299 | 0.3007 | 0.1007 | 1.9788 | 6 |
| 7 | .2791 | 3.6046 | 8.2551 | 3.5832 | 12.9159 | 0.2774 | 0.0774 | 2.2902 | 7 |
| 8 | .2326 | 3.8372 | 9.8831 | 4.2998 | 16.4991 | 0.2606 | 0.0606 | 2.5756 | 8 |
| 9 | .1938 | 4.0310 | 11.4335 | 5.1598 | 20.7989 | 0.2481 | 0.0481 | 2.8364 | 9 |
| 10 | .1615 | 4.1925 | 12.8871 | 6.1917 | 25.9587 | 0.2385 | 0.0385 | 3.0739 | 10 |
| 11 | .1346 | 4.3271 | 14.2330 | 7.4301 | 32.1504 | 0.2311 | 0.0311 | 3.2893 | 11 |
| 12 | .1122 | 4.4392 | 15.4667 | 8.9161 | 39.5805 | 0.2253 | 0.0253 | 3.4841 | 12 |
| 13 | .0935 | 4.5327 | 16.5883 | 10.6993 | 48.4966 | 0.2206 | 0.0206 | 3.6597 | 13 |
| 14 | .0779 | 4.6106 | 17.6008 | 12.8392 | 59.1959 | 0.2169 | 0.0169 | 3.8175 | 14 |
| 15 | .0649 | 4.6755 | 18.5095 | 15.4070 | 72.0351 | 0.2139 | 0.0139 | 3.9588 | 15 |
| 16 | .0541 | 4.7296 | 19.3208 | 18.4884 | 87.4421 | 0.2114 | 0.0114 | 4.0851 | 16 |
| 17 | .0451 | 4.7746 | 20.0419 | 22.1861 | 105.9306 | 0.2094 | 0.0094 | 4.1976 | 17 |
| 18 | .0376 | 4.8122 | 20.6805 | 26.6233 | 128.1167 | 0.2078 | 0.0078 | 4.2975 | 18 |
| 19 | .0313 | 4.8435 | 21.2439 | 31.9480 | 154.7400 | 0.2065 | 0.0065 | 4.3861 | 19 |
| 20 | .0261 | 4.8696 | 21.7395 | 38.3376 | 186.6880 | 0.2054 | 0.0054 | 4.4643 | 20 |
| 21 | .0217 | 4.8913 | 22.1742 | 46.0051 | 225.0256 | 0.2044 | 0.0044 | 4.5334 | 21 |
| 22 | .0181 | 4.9094 | 22.5546 | 55.2061 | 271.0307 | 0.2037 | 0.0037 | 4.5941 | 22 |
| 23 | .0151 | 4.9245 | 22.8867 | 66.2474 | 326.2369 | 0.2031 | 0.0031 | 4.6475 | 23 |
| 24 | .0126 | 4.9371 | 23.1760 | 79.4968 | 392.4842 | 0.2025 | 0.0025 | 4.6943 | 24 |
| 25 | .0105 | 4.9476 | 23.4276 | 95.3962 | 471.9811 | 0.2021 | 0.0021 | 4.7352 | 25 |
| 26 | .0087 | 4.9563 | 23.6460 | 114.4755 | 567.3773 | 0.2018 | 0.0018 | 4.7709 | 26 |
| 27 | .0073 | 4.9636 | 23.8353 | 137.3706 | 681.8528 | 0.2015 | 0.0015 | 4.8020 | 27 |
| 28 | .0061 | 4.9697 | 23.9991 | 164.8447 | 819.2233 | 0.2012 | 0.0012 | 4.8291 | 28 |
| 29 | .0051 | 4.9747 | 24.1406 | 197.8136 | 984.0680 | 0.2010 | 0.0010 | 4.8527 | 29 |
| 30 | .0042 | 4.9789 | 24.2628 | 237.3763 | 1181.8816 | 0.2008 | 0.0008 | 4.8731 | 30 |
| 31 | .0035 | 4.9824 | 24.3681 | 284.8516 | 1419.2579 | 0.2007 | 0.0007 | 4.8908 | 31 |
| 32 | .0029 | 4.9854 | 24.4588 | 341.8219 | 1704.1095 | 0.2006 | 0.0006 | 4.9061 | 32 |
| 33 | .0024 | 4.9878 | 24.5368 | 410.1863 | 2045.9314 | 0.2005 | 0.0005 | 4.9194 | 33 |
| 34 | .0020 | 4.9898 | 24.6038 | 492.2235 | 2456.1176 | 0.2004 | 0.0004 | 4.9308 | 34 |
| 35 | .0017 | 4.9915 | 24.6614 | 590.6682 | 2948.3411 | 0.2003 | 0.0003 | 4.9406 | 35 |
| 36 | .0014 | 4.9929 | 24.7108 | 708.8019 | 3539.0094 | 0.2003 | 0.0003 | 4.9491 | 36 |
| 37 | .0012 | 4.9941 | 24.7531 | 850.5622 | 4247.8112 | 0.2002 | 0.0002 | 4.9564 | 37 |
| 38 | .0010 | 4.9951 | 24.7894 | 1020.6747 | 5098.3735 | 0.2002 | 0.0002 | 4.9627 | 38 |
| 39 | .0008 | 4.9959 | 24.8204 | 1224.8096 | 6119.0482 | 0.2002 | 0.0002 | 4.9681 | 39 |
| 40 | .0007 | 4.9966 | 24.8469 | 1469.7716 | 7343.8578 | 0.2001 | 0.0001 | 4.9728 | 40 |
| 41 | .0006 | 4.9972 | 24.8696 | 1763.7259 | 8813.6294 | 0.2001 | 0.0001 | 4.9767 | 41 |
| 42 | .0005 | 4.9976 | 24.8890 | 2116.4711 | 10577.3553 | 0.2001 | 0.0001 | 4.9801 | 42 |
| 43 | .0004 | 4.9980 | 24.9055 | 2539.7653 | 12693.8263 | 0.2001 | 0.0001 | 4.9831 | 43 |
| 44 | .0003 | 4.9984 | 24.9196 | 3047.7183 | 15233.5916 | 0.2001 | 0.0001 | 4.9856 | 44 |
| 45 | .0003 | 4.9986 | 24.9316 | 3657.2620 | 18281.3099 | 0.2001 | 0.0001 | 4.9877 | 45 |
| 46 | .0002 | 4.9989 | 24.9419 | 4388.7144 | 21938.5719 | 0.2000 | 0.0000 | 4.9895 | 46 |
| 47 | .0002 | 4.9991 | 24.9506 | 5266.4573 | 26327.2863 | 0.2000 | 0.0000 | 4.9911 | 47 |
| 48 | .0002 | 4.9992 | 24.9581 | 6319.7487 | 31593.7436 | 0.2000 | 0.0000 | 4.9924 | 48 |
| 49 | .0001 | 4.9993 | 24.9644 | 7583.6985 | 37913.4923 | 0.2000 | 0.0000 | 4.9935 | 49 |
| 50 | .0001 | 4.9995 | 24.9698 | 9100.4382 | 45497.1908 | 0.2000 | 0.0000 | 4.9945 | 50 |
| 51 | .0001 | 4.9995 | 24.9744 | 10920.5258 | 54597.6289 | 0.2000 | 0.0000 | 4.9953 | 51 |
| 52 | .0001 | 4.9996 | 24.9783 | 13104.6309 | 65518.1547 | 0.2000 | 0.0000 | 4.9960 | 52 |
| 53 | .0001 | 4.9997 | 24.9816 | 15725.5571 | 78622.7856 | 0.2000 | 0.0000 | 4.9966 | 53 |
| 54 | .0001 | 4.9997 | 24.9844 | 18870.6685 | 94348.3427 | 0.2000 | 0.0000 | 4.9971 | 54 |
| 55 | .0000 | 4.9998 | 24.9868 | 22644.8023 | 113219.0113 | 0.2000 | 0.0000 | 4.9976 | 55 |
| 60 | .0000 | 4.9999 | 24.9942 | 56347.5144 | 281732.5718 | 0.2000 | 0.0000 | 4.9989 | 60 |
| 65 | .0000 | 5.0000 | 24.9975 | 140210.6469 | 701048.2346 | 0.2000 | 0.0000 | 4.9995 | 65 |
| 70 | .0000 | 5.0000 | 24.9989 | 348888.9569 | 1744439.7847 | 0.2000 | 0.0000 | 4.9998 | 70 |
| 75 | .0000 | 5.0000 | 24.9995 | 868147.3693 | 4340731.8466 | 0.2000 | 0.0000 | 4.9999 | 75 |

**I = 25.00 %**

| N | (P/F) | (P/A) | (P/G) | (F/P) | (F/A) | (A/P) | (A/F) | (A/G) | N |
|---|---|---|---|---|---|---|---|---|---|
| 1 | .8000 | 0.8000 | 0.0 | 1.2500 | 1.0000 | 1.2500 | 1.0000 | 0.0 | 1 |
| 2 | .6400 | 1.4400 | 0.6400 | 1.5625 | 2.2500 | 0.6944 | 0.4444 | 0.4444 | 2 |
| 3 | .5120 | 1.9520 | 1.6640 | 1.9531 | 3.8125 | 0.5123 | 0.2623 | 0.8525 | 3 |
| 4 | .4096 | 2.3616 | 2.8928 | 2.4414 | 5.7656 | 0.4234 | 0.1734 | 1.2249 | 4 |
| 5 | .3277 | 2.6893 | 4.2035 | 3.0518 | 8.2070 | 0.3718 | 0.1218 | 1.5631 | 5 |
| 6 | .2621 | 2.9514 | 5.5142 | 3.8147 | 11.2588 | 0.3388 | 0.0888 | 1.8683 | 6 |
| 7 | .2097 | 3.1611 | 6.7725 | 4.7684 | 15.0735 | 0.3163 | 0.0663 | 2.1424 | 7 |
| 8 | .1678 | 3.3289 | 7.9469 | 5.9605 | 19.8419 | 0.3004 | 0.0504 | 2.3872 | 8 |
| 9 | .1342 | 3.4631 | 9.0207 | 7.4506 | 25.8023 | 0.2888 | 0.0388 | 2.6048 | 9 |
| 10 | .1074 | 3.5705 | 9.9870 | 9.3132 | 33.2529 | 0.2801 | 0.0301 | 2.7971 | 10 |
| 11 | .0859 | 3.6564 | 10.8460 | 11.6415 | 42.5661 | 0.2735 | 0.0235 | 2.9663 | 11 |
| 12 | .0687 | 3.7251 | 11.6020 | 14.5519 | 54.2077 | 0.2684 | 0.0184 | 3.1145 | 12 |
| 13 | .0550 | 3.7801 | 12.2617 | 18.1899 | 68.7596 | 0.2645 | 0.0145 | 3.2437 | 13 |
| 14 | .0440 | 3.8241 | 12.8334 | 22.7374 | 86.9495 | 0.2615 | 0.0115 | 3.3559 | 14 |
| 15 | .0352 | 3.8593 | 13.3260 | 28.4217 | 109.6868 | 0.2591 | 0.0091 | 3.4530 | 15 |
| 16 | .0281 | 3.8874 | 13.7482 | 35.5271 | 138.1085 | 0.2572 | 0.0072 | 3.5366 | 16 |
| 17 | .0225 | 3.9099 | 14.1085 | 44.4089 | 173.6357 | 0.2558 | 0.0058 | 3.6084 | 17 |
| 18 | .0180 | 3.9279 | 14.4147 | 55.5112 | 218.0446 | 0.2546 | 0.0046 | 3.6698 | 18 |
| 19 | .0144 | 3.9424 | 14.6741 | 69.3889 | 273.5558 | 0.2537 | 0.0037 | 3.7222 | 19 |
| 20 | .0115 | 3.9539 | 14.8932 | 86.7362 | 342.9447 | 0.2529 | 0.0029 | 3.7667 | 20 |
| 21 | .0092 | 3.9631 | 15.0777 | 108.4202 | 429.6809 | 0.2523 | 0.0023 | 3.8045 | 21 |
| 22 | .0074 | 3.9705 | 15.2326 | 135.5253 | 538.1011 | 0.2519 | 0.0019 | 3.8365 | 22 |
| 23 | .0059 | 3.9764 | 15.3625 | 169.4066 | 673.6264 | 0.2515 | 0.0015 | 3.8634 | 23 |
| 24 | .0047 | 3.9811 | 15.4711 | 211.7582 | 843.0329 | 0.2512 | 0.0012 | 3.8861 | 24 |
| 25 | .0038 | 3.9849 | 15.5618 | 264.6978 | 1054.7912 | 0.2509 | 0.0009 | 3.9052 | 25 |
| 26 | .0030 | 3.9879 | 15.6373 | 330.8722 | 1319.4890 | 0.2508 | 0.0008 | 3.9212 | 26 |
| 27 | .0024 | 3.9903 | 15.7002 | 413.5903 | 1650.3612 | 0.2506 | 0.0006 | 3.9346 | 27 |
| 28 | .0019 | 3.9923 | 15.7524 | 516.9879 | 2063.9515 | 0.2505 | 0.0005 | 3.9457 | 28 |
| 29 | .0015 | 3.9938 | 15.7957 | 646.2349 | 2580.9394 | 0.2504 | 0.0004 | 3.9551 | 29 |
| 30 | .0012 | 3.9950 | 15.8316 | 807.7936 | 3227.1743 | 0.2503 | 0.0003 | 3.9628 | 30 |
| 31 | .0010 | 3.9960 | 15.8614 | 1009.7420 | 4034.9678 | 0.2502 | 0.0002 | 3.9693 | 31 |
| 32 | .0008 | 3.9968 | 15.8859 | 1262.1774 | 5044.7098 | 0.2502 | 0.0002 | 3.9746 | 32 |
| 33 | .0006 | 3.9975 | 15.9062 | 1577.7218 | 6306.8872 | 0.2502 | 0.0002 | 3.9791 | 33 |
| 34 | .0005 | 3.9980 | 15.9229 | 1972.1523 | 7884.6091 | 0.2501 | 0.0001 | 3.9828 | 34 |
| 35 | .0004 | 3.9984 | 15.9367 | 2465.1903 | 9856.7613 | 0.2501 | 0.0001 | 3.9858 | 35 |
| 36 | .0003 | 3.9987 | 15.9481 | 3081.4879 | 12321.9516 | 0.2501 | 0.0001 | 3.9883 | 36 |
| 37 | .0003 | 3.9990 | 15.9574 | 3851.8599 | 15403.4396 | 0.2501 | 0.0001 | 3.9904 | 37 |
| 38 | .0002 | 3.9992 | 15.9651 | 4814.8249 | 19255.2994 | 0.2501 | 0.0001 | 3.9921 | 38 |
| 39 | .0002 | 3.9993 | 15.9714 | 6018.5311 | 24070.1243 | 0.2500 | 0.0000 | 3.9935 | 39 |
| 40 | .0001 | 3.9995 | 15.9766 | 7523.1638 | 30088.6554 | 0.2500 | 0.0000 | 3.9947 | 40 |
| 41 | .0001 | 3.9996 | 15.9809 | 9403.9548 | 37611.8192 | 0.2500 | 0.0000 | 3.9956 | 41 |
| 42 | .0001 | 3.9997 | 15.9843 | 11754.9435 | 47015.7740 | 0.2500 | 0.0000 | 3.9964 | 42 |
| 43 | .0001 | 3.9997 | 15.9872 | 14693.6794 | 58770.7175 | 0.2500 | 0.0000 | 3.9971 | 43 |
| 44 | .0001 | 3.9998 | 15.9895 | 18367.0992 | 73464.3969 | 0.2500 | 0.0000 | 3.9976 | 44 |
| 45 | .0000 | 3.9998 | 15.9915 | 22958.8740 | 91831.4962 | 0.2500 | 0.0000 | 3.9980 | 45 |
| 46 | .0000 | 3.9999 | 15.9930 | 28698.5925 | 114790.3702 | 0.2500 | 0.0000 | 3.9984 | 46 |
| 47 | .0000 | 3.9999 | 15.9943 | 35873.2407 | 143488.9627 | 0.2500 | 0.0000 | 3.9987 | 47 |
| 48 | .0000 | 3.9999 | 15.9954 | 44841.5509 | 179362.2034 | 0.2500 | 0.0000 | 3.9989 | 48 |
| 49 | .0000 | 3.9999 | 15.9962 | 56051.9386 | 224203.7543 | 0.2500 | 0.0000 | 3.9991 | 49 |
| 50 | .0000 | 3.9999 | 15.9969 | 70064.9232 | 280255.6929 | 0.2500 | 0.0000 | 3.9993 | 50 |
| 51 | .0000 | 4.0000 | 15.9975 | 87581.1540 | 350320.6161 | 0.2500 | 0.0000 | 3.9994 | 51 |
| 52 | .0000 | 4.0000 | 15.9980 | 109476.4425 | 437901.7701 | 0.2500 | 0.0000 | 3.9995 | 52 |
| 53 | .0000 | 4.0000 | 15.9983 | 136845.5532 | 547378.2126 | 0.2500 | 0.0000 | 3.9996 | 53 |
| 54 | .0000 | 4.0000 | 15.9986 | 171056.9414 | 684223.7658 | 0.2500 | 0.0000 | 3.9997 | 54 |
| 55 | .0000 | 4.0000 | 15.9989 | 213821.1768 | 855280.7072 | 0.2500 | 0.0000 | 3.9997 | 55 |
| 60 | .0000 | 4.0000 | 15.9996 | 652530.4468 | 2610117.7872 | 0.2500 | 0.0000 | 3.9999 | 60 |

$I = 40.00\ \%$

| n | (P/F) | (P/A) | (P/G) | (F/P) | (F/A) | (A/P) | (A/F) | (A/G) | n |
|---|---|---|---|---|---|---|---|---|---|
| 1 | 0.7143 | 0.7143 | 0.0000 | 1.4000 | 1.0000 | 1.4000 | 1.0000 | 0.000 | 1 |
| 2 | 0.5102 | 1.2245 | 0.5102 | 1.9600 | 2.4000 | 0.8167 | 0.4167 | 0.416 | 2 |
| 3 | 0.3644 | 1.5889 | 1.2391 | 2.7440 | 4.3600 | 0.6294 | 0.2294 | 0.779 | 3 |
| 4 | 0.2603 | 1.8492 | 2.0200 | 3.8416 | 7.1040 | 0.5408 | 0.1408 | 1.092 | 4 |
| 5 | 0.1859 | 2.0352 | 2.7637 | 5.3782 | 10.9456 | 0.4914 | 0.0914 | 1.358 | 5 |
| 6 | 0.1328 | 2.1680 | 3.4278 | 7.5295 | 16.3238 | 0.4613 | 0.0613 | 1.581 | 6 |
| 7 | 0.0949 | 2.2628 | 3.9970 | 10.5414 | 23.8534 | 0.4419 | 0.0419 | 1.766 | 7 |
| 8 | 0.0678 | 2.3306 | 4.4713 | 14.7579 | 34.3947 | 0.4291 | 0.0291 | 1.918 | 8 |
| 9 | 0.0484 | 2.3790 | 4.8585 | 20.6610 | 49.1526 | 0.4203 | 0.0203 | 2.042 | 9 |
| 10 | 0.0346 | 2.4136 | 5.1696 | 28.9255 | 69.8137 | 0.4143 | 0.0143 | 2.141 | 10 |
| 11 | 0.0247 | 2.4383 | 5.4166 | 40.4957 | 98.7391 | 0.4101 | 0.0101 | 2.221 | 11 |
| 12 | 0.0176 | 2.4559 | 5.6106 | 56.6939 | 139.2348 | 0.4072 | 0.0072 | 2.284 | 12 |
| 13 | 0.0126 | 2.4685 | 5.7618 | 79.3715 | 195.9287 | 0.4051 | 0.0051 | 2.334 | 13 |
| 14 | 0.0090 | 2.4775 | 5.8788 | 111.1201 | 275.3002 | 0.4036 | 0.0036 | 2.372 | 14 |
| 15 | 0.0064 | 2.4839 | 5.9688 | 155.5681 | 386.4202 | 0.4026 | 0.0026 | 2.403 | 15 |
| 16 | 0.0046 | 2.4885 | 6.0376 | 217.7953 | 541.9883 | 0.4018 | 0.0018 | 2.426 | 16 |
| 17 | 0.0033 | 2.4918 | 6.0901 | 304.9135 | 759.7837 | 0.4013 | 0.0013 | 2.444 | 17 |
| 18 | 0.0023 | 2.4941 | 6.1299 | 426.8789 | 1064.6971 | 0.4009 | 0.0009 | 2.457 | 18 |
| 19 | 0.0017 | 2.4958 | 6.1601 | 597.6304 | 1491.5760 | 0.4007 | 0.0007 | 2.468 | 19 |
| 20 | 0.0012 | 2.4970 | 6.1828 | 836.6826 | 2089.2064 | 0.4005 | 0.0005 | 2.476 | 20 |
| 21 | 0.0009 | 2.4979 | 6.1998 | 1171.3556 | 2925.8889 | 0.4003 | 0.0003 | 2.482 | 21 |
| 22 | 0.0006 | 2.4985 | 6.2127 | 1639.8978 | 4097.2445 | 0.4002 | 0.0002 | 2.486 | 22 |
| 23 | 0.0004 | 2.4989 | 6.2222 | 2295.8569 | 5737.1423 | 0.4002 | 0.0002 | 2.490 | 23 |
| 24 | 0.0003 | 2.4992 | 6.2294 | 3214.1997 | 8032.9993 | 0.4001 | 0.0001 | 2.492 | 24 |
| 25 | 0.0002 | 2.4994 | 6.2347 | 4499.8796 | 11247.1990 | 0.4001 | 0.0001 | 2.494 | 25 |
| 26 | 0.0002 | 2.4996 | 6.2387 | 6299.8314 | 15747.0785 | 0.4001 | 0.0001 | 2.495 | 26 |
| 27 | 0.0001 | 2.4997 | 6.2416 | 8819.7640 | 22046.9099 | 0.4000 | 0.0000 | 2.496 | 27 |
| 28 | 0.0001 | 2.4998 | 6.2438 | 12347.6696 | 30866.6739 | 0.4000 | 0.0000 | 2.497 | 28 |
| 29 | 0.0001 | 2.4999 | 6.2454 | 17286.7374 | 43214.3435 | 0.4000 | 0.0000 | 2.498 | 29 |
| 30 | 0.0000 | 2.4999 | 6.2466 | 24201.4324 | 60501.0809 | 0.4000 | 0.0000 | 2.498 | 30 |
| 31 | 0.0000 | 2.4999 | 6.2475 | 33882.0053 | 84702.5132 | 0.4000 | 0.0000 | 2.499 | 31 |
| 32 | 0.0000 | 2.4999 | 6.2482 | 47434.8074 | 118584.5185 | 0.4000 | 0.0000 | 2.499 | 32 |
| 33 | 0.0000 | 2.5000 | 6.2487 | 66408.7304 | 166019.3260 | 0.4000 | 0.0000 | 2.499 | 33 |
| 34 | 0.0000 | 2.5000 | 6.2490 | 92972.2225 | 232428.0563 | 0.4000 | 0.0000 | 2.499 | 34 |
| 35 | 0.0000 | 2.5000 | 6.2493 | 130161.1116 | 325400.2789 | 0.4000 | 0.0000 | 2.499 | 35 |
| 36 | 0.0000 | 2.5000 | 6.2495 | 182225.5562 | 455561.3904 | 0.4000 | 0.0000 | 2.499 | 36 |
| 37 | 0.0000 | 2.5000 | 6.2496 | 255115.7786 | 637786.9466 | 0.4000 | 0.0000 | 2.499 | 37 |
| 38 | 0.0000 | 2.5000 | 6.2497 | 357162.0901 | 892902.7252 | 0.4000 | 0.0000 | 2.499 | 38 |
| 39 | 0.0000 | 2.5000 | 6.2498 | 500026.9261 | 1250064.8153 | 0.4000 | 0.0000 | 2.499 | 39 |
| 40 | 0.0000 | 2.5000 | 6.2498 | 700037.6966 | 1750091.7415 | 0.4000 | 0.0000 | 2.499 | 40 |
| 41 | 0.0000 | 2.5000 | 6.2499 | 980052.7752 | 2450129.4381 | 0.4000 | 0.0000 | 2.500 | 41 |
| 42 | 0.0000 | 2.5000 | 6.2499 | 1372073.8853 | 3430182.2133 | 0.4000 | 0.0000 | 2.500 | 42 |
| 43 | 0.0000 | 2.5000 | 6.2499 | 1920903.4394 | 4802256.0986 | 0.4000 | 0.0000 | 2.500 | 43 |
| 44 | 0.0000 | 2.5000 | 6.2500 | 2689264.8152 | 6723159.5381 | 0.4000 | 0.0000 | 2.500 | 44 |
| 45 | 0.0000 | 2.5000 | 6.2500 | 3764970.7413 | 9412424.3533 | 0.4000 | 0.0000 | 2.500 | 45 |

**Practice Problems: ENGINEERING ECONOMY**

Untimed

1. A structure costing $10,000 has the operating costs and salvage values given. (a) What is the economic life of the structure? (b) Assuming that the structure has been owned and operated for 4 years, what is the cost of owning the structure for exactly one more year? Use 20% as the interest rate.

Year 1: maintenance $2000, salvage $8000
Year 2: maintenance $3000, salvage $7000
Year 3: maintenance $4000, salvage $6000
Year 4: maintenance $5000, salvage $5000
Year 5: maintenance $6000, salvage $4000

2. A man purchases a car for $5000 for personal use, intending to drive 15,000 miles per year. It costs him $200 per year for insurance and $150 per year for maintenance. He gets 15 mpg and gasoline costs $.60 per gallon. The resale value after 5 years is $1000. Because of unexpected business driving (5000 miles per year extra), his insurance is increased to $300 per year and maintenance to $200. Salvage is reduced to $500. Use 10% to answer the following questions. (a) The man's company offers $.10 per mile reimbursement. Is that adequate? (b) How many miles must be driven per year at $.10 per mile to justify the company buying a car for its use? The cost would be $5000, but insurance, maintenance, and salvage would be $250, $200, and $800 respectively.

3. A shredder installed at the entrance to a sewage treatment plant can remove 7 pounds per hour of debris from the incoming flow. The economic life of this shredder is 20 years. Any debris left in the flow will cause $25,000 damage to the wet-well pumps. Several investments are available to increase the capacity of the shredder. At 10% interest, what should be done?

| debris rate (pounds per hour) | probability of exceeding debris rate | required investment to meet debris rate |
|---|---|---|
| 7 | .15 per year | no cost |
| 8 | .10 per year | $15,000 |
| 9 | .07 per year | $20,000 |
| 10 | .03 per year | $30,000 |

4. A new machine will cost $17,000 and will have a value of $14,000 in 5 years. Special tooling will cost $5000 and it will have a resale value of $2500 after 5 years. Maintenance will be $200 per year. What will be the average cost of ownership during the next 5 years if interest is at 6%?

5. An old highway bridge can be strengthened at a cost of $9000, or it can be replaced for $40,000. The present salvage value of the old bridge is $13,000. It is estimated that the reinforced bridge will last for 20 years with an annual cost of $500 and will have a salvage value of $10,000 at the end of 20 years. The estimated salvage of the new bridge after 25 years is $15,000. The maintenance for the new bridge will be $100 annually. Which is the best alternative at 8% interest?

6. A firm expects to receive $32,000 each year for 15 years from the sale of a product. It will require an initial investment of $150,000. Expenses will run $7530 per year. Salvage is zero and straight-line depreciation is used. The tax rate is 48%. What is the after-tax rate of return?

7. A public works project has initial costs of $1,000,000, benefits of $1,500,000, and disbenefits of $300,000. (a) What is the benefit/cost ratio? (b) What is the excess of benefits over costs?

8. An apartment complex is purchased for $500,000. What is the depreciation in each of the first 3 years if the salvage value is $100,000 in 25 years? Use (a) straight-line, (b) sum-of-the-year's digits, and (c) double declining balance depreciations.

9. Equipment is purchased for $12,000 which is expected to be sold after 10 years for $2000. The estimated maintenance is $1000 the first year, but is expected to increase $200 each year thereafter. Using 10%, find the present worth and the annual cost.

10. One of 5 grades of pipe with average lives (in years) and costs (in dollars) of (9,1500), (14,1600), (30,1750), (52,1900), and (86,2100) is to be chosen for a 20-year project. A failure of the pipe at any time during the project will result in a cost equal to 35% of the original cost. Annual costs are 4% of the initial cost, and the pipes are not recoverable. At 6%, which pipe is superior? HINT: The lives are average lives, not absolute replacement times. Assume a rectangular failure distribution, with probability of failure in any year inversely proportional to the average life.

11. Make a recommendation to your client to accept one of the following alternatives. Use the present worth comparison method. (Initial costs are the same.)

- A 25 year annuity paying $4800 at the end of each year, where the interest rate is a nominal 12% per annum.
- A 25 year annuity paying $1200 every quarter at 12% nominal annual interest.

12. A firm has two alternatives for improvement of its existing production line. The data are as follows:

| | A | B |
|---|---|---|
| initial installment cost | 1500 | 2500 |
| annual operating cost | 800 | 650 |
| service life | 5 years | 8 years |
| salvage value | 0 | 0 |

Determine the best alternative using an interest rate of 15%.

13. Two mutually exclusive alternatives requiring different investments are being considered. The life of both alternatives is estimated at 20 years with no salvage values. The minimum rate of return that is considered acceptable is 4%. Which alternative is best?

| | A | B |
|---|---|---|
| investment required | 70,000 | 40,000 |
| net income per year | 5620 | 4075 |
| rate of return on total investment | 5% | 8% |

14. Compare the costs of two plant renovation schemes A and B. Assume equal lives of 25 years, no salvage values, and interest at 25%. Make the comparison on the basis of (a) present worth, (b) capitalized cost, and (c) annual cost.

| | A | B |
|---|---|---|
| first cost | $20,000 | $25,000 |
| annual expenditure | $ 3000 | $ 2500 |

15. With interest at 8%, obtain the solutions to the following to the nearest dollar. (a) A machine costs $18,000 and has a salvage value of $2000. It has a useful life of 8 years. What is its book value at the end of 5 years using straight-line depreciation? (b) Using data from part (a), find the depreciation in the first three years using the sinking fund method. (c) Repeat part (a) using double declining-balance depreciation to find the first five years' depreciation.

16. A chemical pump motor unit is purchased for $14,000. The estimated life is 8 years, after which it will be sold for $1800. Find the depreciation in the first two years by the sum-of-the-year's digits method. Calculate the after-tax depreciation recovery using 15% interest with 52% income tax.

17. A soda ash plant has the water effluent from the processing equipment treated in a large settling basin. The settling basin eventually discharges into a river that runs alongside the basin. Recently enacted environmental regulations require all rainfall on the plant to be diverted and treated in the settling basin. A heavy rainfall will cause the entire basin to overflow. An uncontrolled overflow will cause environmental damage and heavy fines. The construction of additional height on the existing basin walls is under consideration.

Data on the costs of construction and expected costs for environmental clean up and fines are shown. Data on 50 typical winters has been collected. The soda ash plant management considers 12% to be their minimum rate of return and it is felt that after 15 years the plant will be closed. The company wants to select the alternative that minimizes its total expected costs.

| additional basin height | number of winters with basin overflow | expense for environmental clean up per year | construction cost |
|---|---|---|---|
| 0 feet | 24 | | 0 |
| 5 | 14 | $600,000 | $ 600,000 |
| 10 | 8 | $650,000 | $ 710,000 |
| 15 | 3 | $700,000 | $ 900,000 |
| 20 | 1 | $800,000 | $1,000,000 |
| | 50 | | |

18. A wood processing plant installed a waste gas scrubber at a cost of $30,000 to remove pollutants from the exhaust discharged into the atmosphere. The scrubber has no salvage value and will cost $18,700 to operate next year, with operating costs expected to increase at the rate of $1200 per year thereafter. When should the company consider replacing the scrubber? Money can be borrowed at 12%.

19. Two alternative piping schemes are being considered by a water treatment facility. On the basis of a 10 year life and an interest rate of 12%, determine the number of hours of operation for which the two installations will break even.

| | A | B |
|---|---|---|
| pipe diameter | 4 in | 6 in |
| head loss for required flow | 48 ft | 26 ft |
| size motor required | 20 hp | 7 hp |
| energy cost per hour operation | $ .30 | $ .10 |
| cost of motor installed | $3600 | $2800 |
| cost of pipes and fittings | $3050 | $5010 |
| salvage value at end of 10 years | $ 200 | $ 280 |

20. An 88% learning curve is used with an item whose first production time was 6 weeks. How long will it take to produce the 4th item? How long will it take to produce the 6th through 14th items?

## Timed

1. A company is considering two alternatives, only one of which can be selected.

| alternative | initial investment | salvage value | annual net profit | life |
|---|---|---|---|---|
| A | 120,000 | 15,000 | 57,000 | 5 yrs |
| B | 170,000 | 20,000 | 67,000 | 5 yrs |

The net profit is after operating and maintenance costs, but before taxes. The company pays 45% of its year-end profit as income taxes. Use straight-line depreciation. Do not use investment tax credit. Find the best alternative if the company's minimum attractive rate of return is 15%.

2. A company is considering the purchase of equipment to expand its capacity. The equipment cost is $300,000. The equipment is needed for 5 years, after which it will be sold for $50,000. The company's before-tax cash flow will be improved $90,000 annually by the purchase of the asset.

The corporate tax rate is 48% and straight-line depreciation will be used. The company will take the investment tax credit of 6.67%. What is the after-tax rate of return associated with this equipment purchase?

3. A 120-room hotel is purchased for $2,500,000. A 25-year loan is available for 12%. A study was conducted to determine the various occupancy rates.

| occupancy | probability |
|---|---|
| 65% full | .40 |
| 70% | .30 |
| 75% | .20 |
| 80% | .10 |

The operating costs of the hotel are:

| | |
|---|---|
| taxes & insurance | $ 20,000 annually |
| maintenance | 50,000 annually |
| operating | 200,000 annually |

The life of the hotel is figured to be 25 years when operating 365 days per year. The salvage after 25 years is $500,000.

Neglect tax credit and income taxes. Determine the average rate which should be charged per room per night to return 15% of the initial cost each year.

4. A company is insured for $3,500,000 against fire. The insurance rate is $0.69/1000. The insurance company will decrease the rate to $0.47/1000 if fire sprinklers are installed. The initial cost of the sprinklers is $7500. Annual costs are $200; additional taxes are $100 annually. The system life is 25 years. What is the rate of return on this investment?

5. Heat losses through the walls in an existing building cost a company $1,300,000 per year. This amount is considered excessive, and two alternatives are being evaluated. None of the alternatives will increase the life of the existing building beyond the current expected life of 6 years. None of the alternatives will produce a salvage value.

Alternative A: Do nothing. Continue with current losses.

Alternative B: Spend $2,000,000 immediately to upgrade the building and reduce the loss by 80%. This alternative will require annual maintenance of $150,000.

Alternative C: Spend $1,200,000 immediately. Repeat the $1,200,000 expenditure 3 years from now. Heat loss the first year will be reduced 80%. Due to deterioration, the reduction will be 55% and 20% in the second and third years. (The pattern is repeated starting after the second expenditure.) There are no maintenance costs.

All energy and maintenance costs are regarded as expenses for tax purposes. The company's tax rate is 48%, and straight-line depreciation is used. 15% is regarded as the effective annual interest rate. Evaluate each alternative on an after-tax basis, and recommend the best alternative.

6. You have been asked to determine if a 7-year-old machine should be replaced. Give a full explanation for your recommendation. Base your decision on a before-tax interest rate of 15%.

The existing machine is presumed to have a 10-year life. It has been depreciated on a straight-line basis from its original value of $1,250,000 to a current book value of $620,000. Its ultimate salvage value was assumed to be $350,000 for purposes of depreciation. Its present salvage value is estimated at $400,000, and this is not expected to change over the next 3 years. The current operating costs are not expected to change from $200,000 per year.

A new machine costs $800,000, with operating costs of $40,000 the first year, and increasing by $30,000 each year thereafter. The new machine has an expected life of 10 years. The salvage value depends on the year the new machine is retired:

| year retired | salvage |
|---|---|
| 1 | $600,000 |
| 2 | $500,000 |
| 3 | $450,000 |
| 4 | $400,000 |
| 5 | $350,000 |
| 6 | $300,000 |
| 7 | $250,000 |
| 8 | $200,000 |
| 9 | $150,000 |
| 10 | $100,000 |

# 3 FLUID STATICS AND DYNAMICS



*Nomenclature*

| | | |
|---|---|---|
| a | acceleration | ft/sec$^2$ |
| bhp | brake horsepower | hp |
| A | area | ft$^2$ |
| c | speed of sound in fluid | ft/sec |
| C | compressibility, Hazen-Williams constant, or coefficient | ft$^2$/lbf, –, – |
| d | depth, diameter | ft |
| D | diameter, drag | ft, lbf |
| ehp | electrical horsepower | hp |
| E | bulk modulus, energy | lbf/ft$^2$, ft-lbf |
| f | Darcy friction factor | – |
| fhp | friction horsepower | hp |
| F | force | lbf |
| $F_{va}$ | velocity of approach factor | – |
| g | local gravitational acceleration | ft/sec$^2$ |
| $g_c$ | gravitational constant (32.2) | lbm-ft/lbf-sec$^2$ |
| G | mass flow rate per unit area | lbm/sec-ft$^2$ |
| h | fluid height, head, depth | ft |
| H | total head | ft |
| I | moment of inertia | ft$^4$ |
| k | ratio of specific heats | – |
| K | minor loss coefficient | – |
| L | length of pipe, lift | ft, lbf |
| m | mass | lbm |
| $\dot{m}$ | mass flow rate | lbm/sec |
| n | rotational speed | rpm |
| $n_s$ | specific speed | – |
| $N_{Fr}$ | Froude number | |
| $N_{Re}$ | Reynolds number | – |
| $N_W$ | Weber number | – |
| p | pressure | lbf/ft$^2$ |
| P | power | ft-lbf/sec |
| Q | flow rate | gpm |
| r | radius | ft |
| $r_h$ | hydraulic radius | ft |
| R | specific gas constant | ft-lbf/lbm-°R |
| s | length | ft |
| S.G. | specific gravity | – |
| t | time | sec |
| T | absolute temperature | °R |
| v | velocity | ft/sec |
| V | volume | ft$^3$ |
| $\dot{V}$ | volumetric flow rate | ft$^3$/sec |
| w | weight | lbf |
| whp | water horsepower | hp |
| x | x-coordinate | ft |
| y | distance, y-coordinate | ft |
| z | height above datum | ft |

*Symbols*

| | | |
|---|---|---|
| $\beta$ | contact angle, beta ratio | ° |
| $\gamma$ | specific weight | lbf/ft$^3$ |
| $\epsilon$ | specific roughness | ft |
| $\eta$ | efficiency | – |
| $\theta$ | angle | ° |
| $\mu$ | absolute viscosity | lbf-sec/ft$^2$ |
| $\nu$ | kinematic viscosity | ft$^2$/sec |
| $\rho$ | density | lbm/ft$^3$ |
| $\tau$ | shear stress | lbf/ft$^2$ |
| $T$ | surface tension | lbf/ft |
| $v$ | specific volume | ft$^3$/lbm |
| $\phi$ | angle, deflection angle | ° |
| $\omega$ | rotational speed | rad/sec |

*Subscripts*

| | |
|---|---|
| a | atmospheric |
| A | added |
| b | blade |
| c | centroid, contraction |
| d | discharge |
| D | drag |
| e | equivalent, entrance |
| E | English, extracted |
| f | friction, flow |
| i | inside, inlet |

| | |
|---|---|
| j | jet |
| k | kinetic |
| L | lift |
| m | manometer fluid, metacentric, model, motor |
| M | metric |
| n | nozzle |
| o | outside, outlet |
| p | static pressure, pump |
| r | ratio |
| R | resultant |
| s | stagnation |
| STP | standard temperature and pressure |
| t | total, tank, true |
| v | velocity |
| vp | vapor pressure |

# PART 1: Fluid Properties

Fluids are generally divided into two categories: ideal and real. *Ideal fluids* are those which have zero viscosity and shearing forces, are incompressible, and have uniform velocity distributions when flowing.

*Real fluids* are divided into Newtonian and non-Newtonian fluids. *Newtonian fluids* are typified by gases, thin liquids, and most fluids having simple chemical formulas. *Non-Newtonian fluids* are typified by gels, emulsions, and suspensions. Both Newtonian and non-Newtonian fluids exhibit finite viscosities and nonuniform velocity distributions. However, Newtonian fluids exhibit viscosities which are independent of the rate of change of shear stress, while non-Newtonian fluids exhibit viscosities dependent on the rate of change of shear stress.

Most fluid problems assume Newtonian fluid characteristics.

## 1 FLUID DENSITY

Most fluid flow calculations are based on an inconsistent system of units which measures density in pounds-mass per cubic foot. That convention is followed in this chapter when the symbol $\rho$ is employed.

$$\rho = \gamma\left(\frac{g_c}{g}\right) \equiv \text{fluid density in lbm/ft}^3 \qquad 3.1$$

Hydrostatic pressure and energy conservation equations given in this chapter require a standard local gravity of 32.2 ft/sec$^2$. However, a short discussion of situations with non-standard gravity is given at the ends of parts 2 and 4 in this chapter.

The density of a fluid in liquid form is usually given, known in advance, or easily obtained from a table (see end of this chapter). The density of a gas can be found from the following formula, which has been derived from the ideal gas law:

$$\rho = \frac{p}{RT} \qquad 3.2$$

## 2 SPECIFIC VOLUME

Specific volume is the volume occupied by a pound of fluid. It is the reciprocal of the density.

$$v = \frac{1}{\rho} \qquad 3.3$$

## 3 SPECIFIC GRAVITY

Specific gravity is the ratio of a fluid's density to some specified reference density. For liquids, the reference density is the density of pure water. There is some confusion about this reference since the density of water varies with temperature, and various reference temperatures have been used (e.g., 39°F, 60°F, 70°F, etc.).

Strictly speaking, specific gravity of a liquid cannot be given without specifying the reference temperature at which the water's density was evaluated. However, the reference temperature is often omitted since water's density is fairly constant over the normal ambient temperature range. Using three significant digits, this reference density is 62.4 lbm/ft$^3$.

$$S.G._{\text{liquid}} = \frac{\rho}{62.4} \qquad 3.4$$

Specific gravities of petroleum products and aqueous acid solutions can be found from hydrometer readings. There are two basic hydrometer scales. The Baumé scale has been used widely in the past. Now, however, the API (American Petroleum Institute) scale is recommended for use with all liquids.

For liquids lighter than water, specific gravity may be found from the Baumé hydrometer reading:

$$S.G. = \frac{140.0}{130.0 + \ {}^\circ\text{Baumé}} \qquad 3.5$$

For liquids heavier than water, specific gravity may be found from the Baumé hydrometer reading:

$$S.G. = \frac{145.0}{145.0 - \ {}^\circ\text{Baumé}} \qquad 3.6$$

The modern API scale may be used with all liquids:

$$S.G. = \frac{141.5}{131.5 + \,^\circ\text{API}} \qquad 3.7$$

Specific gravities also can be given for gases. The reference density is the density of air at specified conditions of pressure and temperature. The density of air evaluated at STP is approximately 0.075 lbm/ft$^3$. Therefore,

$$S.G._{\text{gas}} = \frac{\rho_{\text{STP}}}{0.075} \qquad 3.8$$

If the gas and air densities both are evaluated at the same temperature and pressure, the specific gravity is the inverse ratio of specific gas constants.

$$S.G._{\text{gas}} = \frac{R_{\text{air}}}{R_{\text{gas}}} = \frac{53.3}{R_{\text{gas}}} \qquad 3.9$$

*Example 3.1*

Determine the specific gravity of carbon dioxide (150°F, 20 psia) using STP air as a reference.

The specific gas constant for carbon dioxide is approximately 35.1 ft-lbf/lbm-°R. Using equation 3.2 and converting temperature to °R, the density is

$$\rho = \frac{(20)(144)}{(35.1)(150+460)} = 0.135 \text{ lbm/ft}^3$$

From equation 3.8,

$$S.G. = \frac{0.135}{0.075} = 1.8$$

## 4 VISCOSITY

Viscosity of a fluid is a measure of its resistance to flow. Consider two plates separated by a viscous fluid with thickness equal to $y$. The bottom plate is fixed. The top plate is kept in motion at a constant velocity $v$ by a constant force $F$.

Experiments with Newtonian fluids have shown that the force required to maintain the velocity is proportional to the velocity and inversely proportional to the separation of the plates. That is,

$$\frac{F}{A} \propto \frac{dv}{dy} \qquad 3.10$$

The constant of proportionality is known as the *absolute*[1] *viscosity*. Recognizing that the quantity $(F/A)$ is the *fluid shear stress* allows the following equation to be written.

$$\tau = \frac{F}{A} = \mu \frac{dv}{dy} \qquad 3.11$$

[1] Another name for *absolute* viscosity is *dynamic* viscosity. The term *absolute* is preferred.

Another quantity using the name *viscosity* is the combination of units given by equation 3.12. This combination of units, known as the *kinematic viscosity*, appears sufficiently often in fluids problems to warrant its own symbol and name.

$$\nu = \frac{\mu g_c}{\rho} \qquad 3.12$$

There are a number of different units used to measure viscosity. Table 3.1 lists the most commonly used units in the English and SI systems.

**Table 3.1**
Typical Viscosity Units

| | Absolute | Kinematic |
|---|---|---|
| English | lbf-sec/ft$^2$ (slug/ft-sec) | ft$^2$/sec |
| Conventional Metric | dyne-sec/cm$^2$ (poise) | cm$^2$/sec (stoke) |
| SI | Pascal-second (N-s/m$^2$) | m$^2$/s |

Conversions between the two types of viscosities and between the English and various metric systems can be accomplished with table 3.2.

*Example 3.2*

Water at 60°F has a specific gravity of 0.999 and a kinematic viscosity of 1.12 centistokes. What is the absolute viscosity in lbf-sec/ft$^2$?

$$\nu_M = \frac{1.12}{100} = 0.0112 \text{ stokes}$$

$$\mu_M = (0.0112)(0.999) = 0.01119 \text{ poise}$$

$$\mu_E = \frac{0.01119}{478.8} = 2.34 \text{ EE} - 5 \text{ lbf-sec/ft}^2$$

Viscosity also can be measured by a viscometer. A viscometer essentially is a container which allows the fluid to leak out through a small hole. The more viscous the fluid, the more time will be required to leak out a given quantity. Viscosity measured in this indirect manner has the units of seconds. Seconds Saybolt Universal (SSU) and Seconds Saybolt Furol (SSF) are two systems of indirect viscosity measurement.

In liquids, molecular cohesion is the dominating cause of viscosity. As the temperature of a liquid increases, these cohesive forces decrease, resulting in an absolute viscosity decrease.

In gases, the dominating cause of viscosity is random collisions between gas molecules. This molecular agita-

**Table 3.2**
Viscosity Conversions

| to obtain | multiply | by | and divide by |
|---|---|---|---|
| $ft^2$/sec | lbf-sec/$ft^2$ | 32.2 | density |
| $ft^2$/sec | stokes | 1.076 EE–3 | 1 |
| lbf-sec/$ft^2$ | $ft^2$/sec | density | 32.2 |
| lbf-sec/$ft^2$ | poise | 1 | 478.8 |
| $m^2$/s | centistokes | 1 EE–6 | 1 |
| $m^2$/s | stokes | 1 EE–4 | 1 |
| $m^2$/s | $ft^2$/sec | 9.29 EE–2 | 1 |
| pascal-sec | centipoise | 1 EE–3 | 1 |
| pascal-sec | lbm/ft-sec | 1.488 | 1 |
| pascal-sec | lbf-sec/$ft^2$ | 47.88 | 1 |
| pascal-sec | poise | 0.1 | 1 |
| pascal-sec | slug/ft-sec | 47.88 | 1 |
| poise | lbf-sec/$ft^2$ | 478.8 | 1 |
| poise | stokes | specific gravity | 1 |
| stokes | $ft^2$/sec | 929 | 1 |
| stokes | poise | 1 | specific gravity |



tion increases with increases in temperature. Therefore, viscosity in gases increases with temperature.

The absolute viscosity of both gases and liquids is independent of changes in pressure. Of course, kinematic viscosity greatly depends on both temperature and pressure since these variables affect density.

## 5 VAPOR PRESSURE

Molecular activity in a liquid tends to free some surface molecules. This tendency toward vaporization is dependent on temperature. The partial pressure exerted at the surface by these free molecules is known as the *vapor pressure.* Boiling occurs when the vapor pressure is increased (by increasing the fluid temperature) to the local ambient pressure. Thus, a liquid's boiling point depends on both the temperature and the external pressure. Liquids with low vapor pressures are used in accurate barometers.

Vapor pressure is a function of temperature only. Typical values are given in table 3.3. Appendix A lists values for water.

**Table 3.3**
Typical Vapor Pressures at 68°F

| Fluid | Vapor Pressure |
|---|---|
| Ethyl alcohol | 122.4 psf |
| Turpentine | 1.115 |
| Water | 48.9 |
| Ether | 1231. |
| Mercury | 0.00362 |

## 6 SURFACE TENSION

The skin which seems to form on the free surface of a fluid is due to the intermolecular cohesive and adhesive forces known as *surface tension.* Surface tension is the amount of work required to form a new unit of surface area. The units, therefore, are ft-lbf/$ft^2$ or just lbf/ft.

Surface tension can be measured as the tension between two points on the surface separted by a foot. It decreases with temperature increases and depends on the gas contacting the free surface. Surface tension values usually are quoted for air contact.

**Table 3.4**
Typical Surface Tensions
(68°F, air contact)

| Fluid | T |
|---|---|
| ethyl alcohol | 0.001527 lbf/ft |
| turpentine | 0.001857 |
| water | 0.004985 |
| mercury | 0.03562 |
| n-octane | 0.00144 |
| acetone | 0.00192 |
| benzene | 0.00192 |
| carbon tetrachloride | 0.00180 |

The relationship between surface tension and the pressure in a bubble surrounded by gas is given by equation 3.13. $r$ is the radius of the bubble.

$$T = \frac{1}{4} r \left( p_{\text{inside}} - p_{\text{outside}} \right) \qquad 3.13$$

The surface tension in a full spherical droplet or in a bubble in a liquid is given by equation 3.14.

$$T = \frac{1}{2} r \left( p_{\text{inside}} - p_{\text{outside}} \right) \qquad 3.14$$

## 7 CAPILLARITY

Surface tension is the cause of capillarity which occurs whenever a liquid comes into contact with a vertical solid surface. In water, adhesive forces dominate. They cause water to attach itself readily to a vertical surface, to climb the wall. In a thin-bore tube, water will rise above the general level as it tries to wet the interior surface.

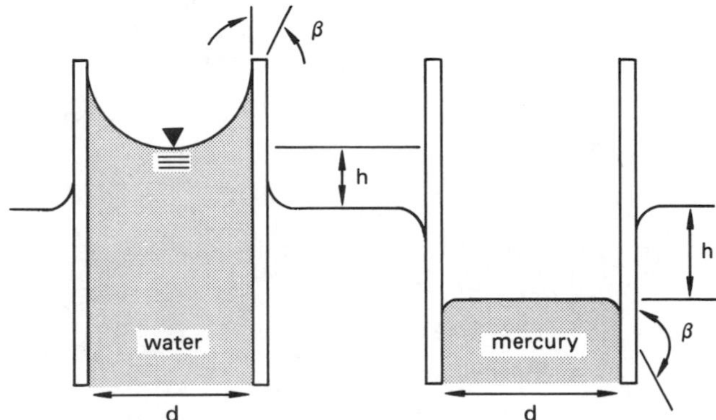


**Figure 3.1** Capillarity in Thin-Wall Tubes

On the other hand, cohesive forces dominate in mercury, since mercury molecules have a great affinity for each other. The curved surface called the *meniscus* formed inside a thin-bore tube inserted into a container of mercury will be below the general level.

Whether adhesive or cohesive forces dominate can be determined by the *angle of contact*, $\beta$, as shown in figure 3.1. For a contact angle less than 90°, adhesive forces dominate. Typical values of $\beta$ are given in table 3.5.

If the tube diameter is less than 0.1″, the surface tension inside a circular capillary tube can be approximated by equation 3.15. The meniscus is assumed spherical with radius $r$. Equation 3.15 also can be used to estimate the capillary rise in a capillary tube, as illustrated in figure 3.1.

$$T = \frac{h\rho d}{4\cos\beta} \left( \frac{g}{g_c} \right) \qquad 3.15$$

$$r = \frac{d}{2\cos\beta} \qquad 3.16$$

Table 3.5 can be used to determine T and $\beta$ for various combinations of contacting liquids.

## 8 COMPRESSIBILITY

Usually, fluids are considered to be incompressible. Actually, fluids are somewhat compressible. Compressibility is the percentage change in a unit volume per unit change in pressure.

$$C = \frac{\frac{\Delta V}{V}}{\Delta p} \qquad 3.17$$

## 9 BULK MODULUS

The bulk modulus of a liquid is the reciprocal of the compressibility.

$$E = \frac{1}{C} \qquad 3.18$$

The bulk modulus of an ideal gas is given by equation 3.19, where $k$ is the ratio of specific heats. $k$ is equal to 1.4 for air.

$$E = kp \qquad 3.19$$

## 10 SPEED OF SOUND

The speed of sound in a pure liquid or gas is given by equations 3.20 and 3.21.

$$c_{\text{liquid}} = \sqrt{\frac{Eg_c}{\rho}} = \sqrt{\frac{g_c}{C\rho}} \qquad 3.20$$

$$c_{\text{gas}} = \sqrt{kg_cRT} = \sqrt{\frac{kpg_c}{\rho}} \qquad 3.21$$

**Table 3.5**
Capillary Constants

| combination | surface tension, $T$ | contact angle, $\beta$ |
|---|---|---|
| Mercury-vacuum-glass | 3.29 EE–2 lbf/ft | 140° |
| Mercury-air-glass | 2.02 EE–2 | 140° |
| Mercury-water-glass | 2.60 EE–2 | 140° |
| Water-air-glass | 5.00 EE–3 | 0° |

The temperature term in equation 3.21 must be in degrees absolute.

*Example 3.3*

What is the velocity of sound in 150°F water?

From the tables at the end of this chapter, the density of water at 150°F is 61.2 $\text{lbm/ft}^3$. Similarly, the bulk modulus is 328 EE3 psi. From equation 3.20,

$$c = \sqrt{\frac{(328\ \text{EE3})(144)(32.2)}{61.2}} = 4985\ \text{ft/sec}$$

*Example 3.4*

What is the velocity of sound in 150°F air at atmospheric pressure?

The specific gas constant for air is 53.3 ft-lbf/lbm-°R. Using equation 3.21,

$$\begin{aligned} c &= \sqrt{(1.4)(32.2)(53.3)(150+460)} \\ &= 1210.7\ \text{ft/sec} \end{aligned}$$

# PART 2: Fluid Statics

## 1 MEASURING PRESSURES

The value of pressure, regardless of the device used to measure it, is dependent on the reference point chosen. Two such reference points exist: zero absolute pressure and standard atmospheric pressure.

If standard atmospheric pressure (approximately 14.7 psia) is chosen as the reference, pressures are known as *gage* pressures. Positive gage pressures always are pressures above atmospheric pressure. Vacuum (negative gage pressure) is the pressure below atmospheric. Maximum vacuum, according to this convention, is −14.7 psig. The term *gage* is somewhat misleading, as a mechanical gauge may not be used to measure gage pressures.

If zero absolute pressure is chosen as the reference, the pressures are known as *absolute* pressures. The barometer is a common device for measuring the absolute pressure of the atmosphere. It is constructed by filling a long, hollow tube, open at one end, with mercury, and inverting it such that the open end is below the level of a mercury-filled container. If the vapor pressure is neglected, the mercury will be supported only by the atmospheric pressure transmitted through the container fluid at the lower, open end. The equation balancing the weight of the fluid against the atmospheric force is:

$$p_a = 0.491(h)(144) \qquad 3.22$$

$h$ is the height of the mercury column in inches, and 0.491 is the density of mercury in pounds per cubic inch.

Any fluid can be used to measure atmospheric pressure, although vapor pressure may be significant. For any fluid used in a barometer,

$$p_a = [(0.0361)(S.G.)(h) + p_v](144) \qquad 3.23$$

0.0361 is the density of water in pounds per cubic inch. $p_v$ should be given in psi in equation 3.23.

*Example 3.5*

A vacuum pump is used to drain a flooded mine shaft of 68°F water. The pump is incapable of lifting the water beyond 400 inches. What is the atmospheric pressure?

From table 3.3, the vapor pressure of 68°F water is

$$\frac{48.9}{144} = 0.34 \text{ psi}$$

Then, the atmospheric pressure is

$$p_a = [(0.0361)(1)(400) + 0.34](144) = 2128.3 \text{ psf}$$

This is 14.78 psia.

## 2 MANOMETERS

Manometers are used frequently to measure pressure differentials. Figure 3.2 shows a simple U-tube manometer whose ends are connected to two pressure vessels. Often, one end will be open to the atmosphere, which then determines that end's pressure.

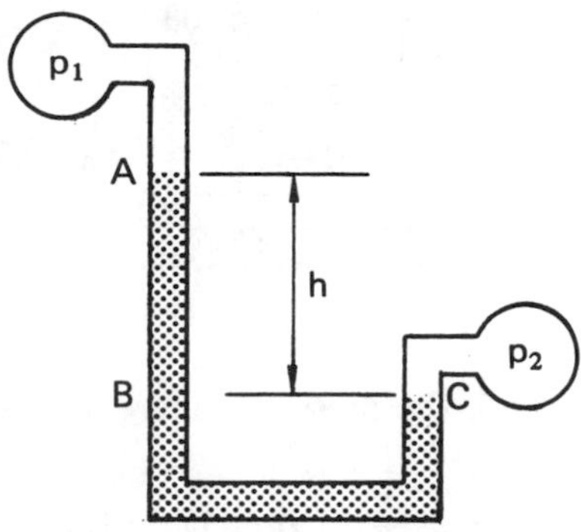


**Figure 3.2** A Simple Manometer

Since the pressure at point B is the same as at point C, the pressure differential produces the fluid column of height $h$.

$$\Delta p = p_2 - p_1 = \gamma_m h \qquad 3.24$$

Equation 3.24 assumes that the manometer is small and that only low density gases fill the tubes above the measuring fluid. If a high density fluid (such as water) is present above the measuring fluid, or if the gas columns $h_1$ or $h_2$ are very long, corrections must be made:

$$\Delta p = \gamma_m h + \gamma_1 h_1 - \gamma_2 h_2 \qquad 3.25$$

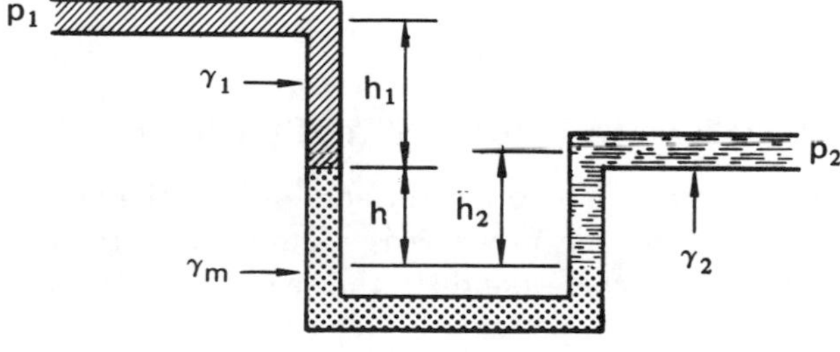


**Figure 3.3** A Manometer Requiring Corrections

Corrections for capillarity are seldom needed since manometer tubes generally are large in diameter.

FLUIDS

*Example 3.6*

What is the pressure at the bottom of the water tank?

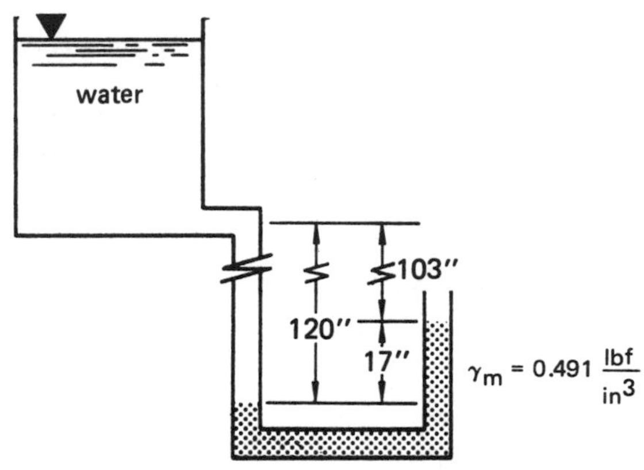


Using equation 3.25, the pressure differential is

$$\begin{aligned}\Delta p = p_{\text{tank bottom}} - p_a &= (0.491)(17) - (0.0361)(120) \\ &= 8.347 - 4.332 \\ &= 4.015 \text{ psig}\end{aligned}$$

The third term in equation 3.25 was omitted because the density of air is much smaller than that of water or mercury.

## 3 HYDROSTATIC PRESSURE DUE TO INCOMPRESSIBLE FLUIDS

Hydrostatic pressure is the pressure which a fluid exerts on an object or container walls. It always acts through the center of pressure and is normal to the exposed surface, regardless of the object's orientation or shape. It varies linearly with depth and is a function of depth and density only.

### A. FORCE ON A HORIZONTAL PLANE SURFACE

In the case of a horizontal surface, such as the bottom of a container, the pressure is uniform, and the center of pressure corresponds to the centroid of the plane surface. The gage pressure is

$$p = \gamma h \tag{3.26}$$

The total vertical force on the horizontal plane is

$$R = pA \tag{3.27}$$

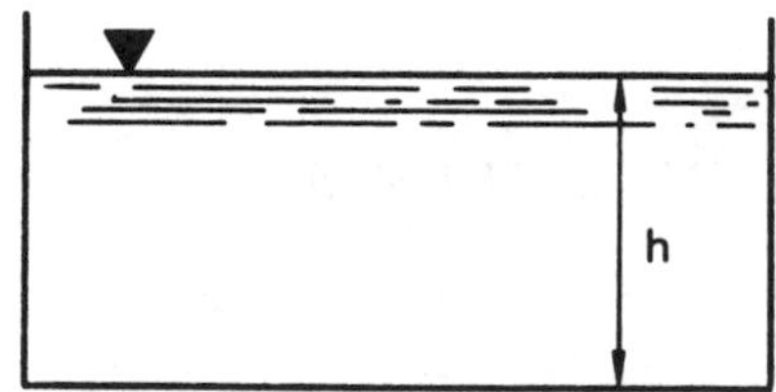


**Figure 3.4** Horizontal Plane Surface

### B. VERTICAL AND INCLINED RECTANGULAR PLANE SURFACES

If a rectangular plate is vertical or inclined within a fluid body, the linear variation in pressure with depth is maintained. The pressures at the top and bottom of the plate are

$$p_1 = \gamma h_1 \sin\theta \tag{3.28}$$

$$p_2 = \gamma h_2 \sin\theta \tag{3.29}$$

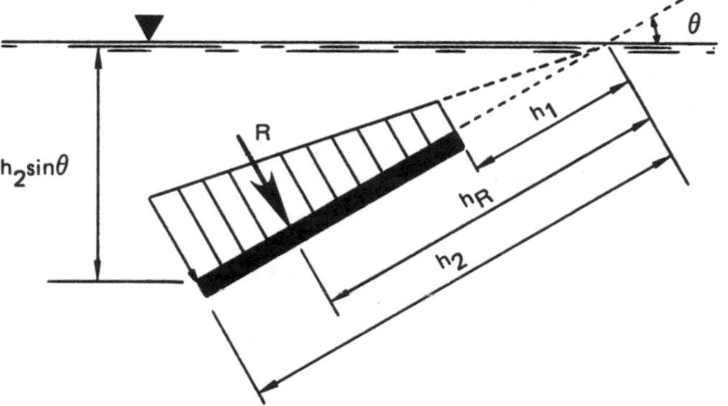


**Figure 3.5** Inclined Rectangular Plate

The average pressure occurs at the average depth $(\frac{1}{2})(h_1 + h_2)\sin\theta$. The average pressure over the entire vertical or inclined surface is

$$\bar{p} = \frac{1}{2}\gamma(h_1 + h_2)\sin\theta \tag{3.30}$$

The total resultant force on the inclined plane is

$$R = \bar{p}A \tag{3.31}$$

The center of pressure is not located at the average depth but is located at the centroid of the triangular or trapezoidal pressure distribution. That depth is

$$h_R = \frac{2}{3}\left[h_1 + h_2 - \frac{h_1 h_2}{(h_1 + h_2)}\right] \tag{3.32}$$

If the object is inclined, $h_R$ must be measured parallel to the object's surface (e.g., an inclined length).

*Example 3.7*

The tank shown is filled with water. What is the force on a one-foot width of the inclined portion of the wall? Where is the resultant located on the inclined section?

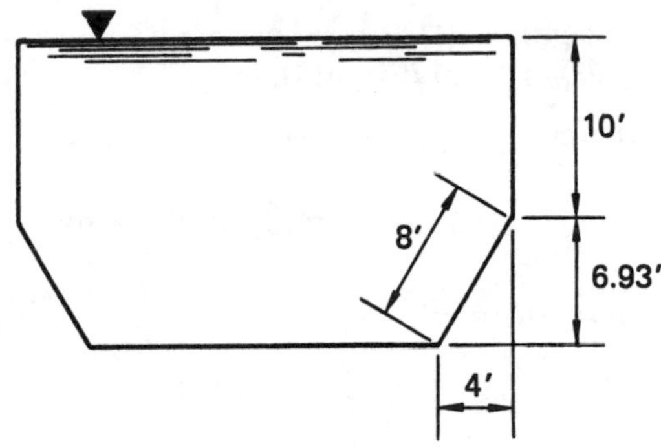


The $\sin\theta$ terms in equations 3.28, 3.29, and 3.30 convert the inclined distances to vertical distances. Therefore, the $\sin\theta$ terms may be omitted if the vertical distances are known. The average pressure in the inclined section is

$$\overline{p} = \frac{1}{2}(62.4)(10 + 16.93) = 840.2 \text{ psf}$$

The total force is

$$R = (840.2)(8)(1) = 6721.6 \text{ lbf}$$

To determine $h_R$, $\theta$ must be known in order to calculate $h_1$ and $h_2$.

$$\theta = \arctan\left(\frac{6.93}{4}\right) = 60°$$

$$h_1 = \frac{10}{\sin\ 60°} = 11.55 \text{ ft}$$

$$h_2 = \frac{16.93}{\sin\ 60°} = 19.55 \text{ ft}$$

$h_R$ can be calculated from equation 3.32 by substituting the inclined distances.

$$h_R = \frac{2}{3}\left[11.55 + 19.55 - \frac{(11.55)(19.55)}{11.55 + 19.55}\right]$$
$$= 15.89 \text{ ft}$$

## C. GENERAL PLANE SURFACE

For any non-rectangular plane surface, the average pressure depends on the location of the surface's centroid, $h_c$.

$$\overline{p} = \gamma h_c \sin\theta \qquad 3.33$$

$$R = \overline{p}A \qquad 3.34$$

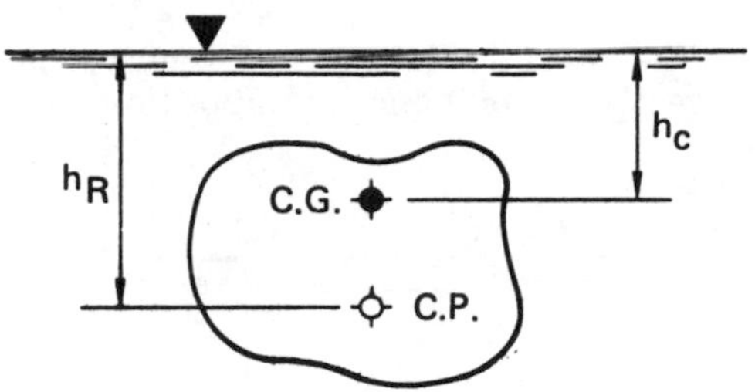


**Figure 3.6** General Plane Surface

The resultant is normal to the surface, acting at depth $h_R$.

$$h_R = h_c + \frac{I_c}{Ah_c} \qquad 3.35$$

$I_c$ is the moment of inertia about an axis parallel to the surface through the area's centroid. As with the previous case, $h_c$ and $h_r$ must be measured parallel to the area's surface. That is, if the plane is inclined, $h_c$ and $h_R$ also must be the inclined distances.

*Example 3.8*

What is the force on a one-foot diameter circular sight-hole whose top edge is located 4′ below the water surface? Where does the resultant act?

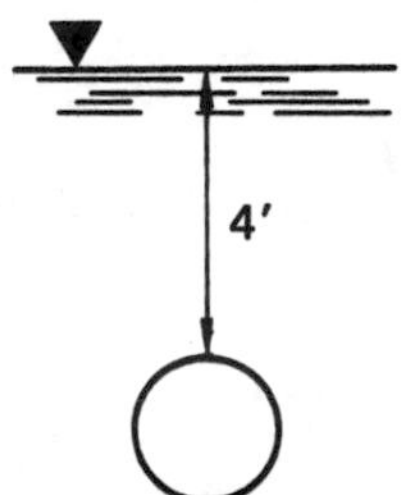


$$h_c = 4.5 \text{ ft}$$
$$A = \frac{1}{4}\pi(1)^2 = 0.7854 \text{ ft}^2$$
$$I_c = \frac{1}{4}\pi r^4 = 0.049 \text{ ft}^4$$
$$\overline{p} = (62.4)(4.5) = 280.8 \text{ psf}$$
$$R = (280.8)(0.7854) = 220.5 \text{ lb}$$
$$h_R = 4.5 + \frac{0.049}{(0.7854)(4.5)} = 4.514 \text{ ft}$$

## D. CURVED SURFACES

The horizontal component of the resultant force acting on a curved surface can be found by the same method used for a vertical plane surface. The vertical component of force on an area usually will equal the weight of the liquid above it. In figure 3.7, the vertical force on length AB is the weight of area ABCD, with a line of action passing through the centroid of the area ABCD.

The resultant magnitude and direction may be found from conventional component composition.

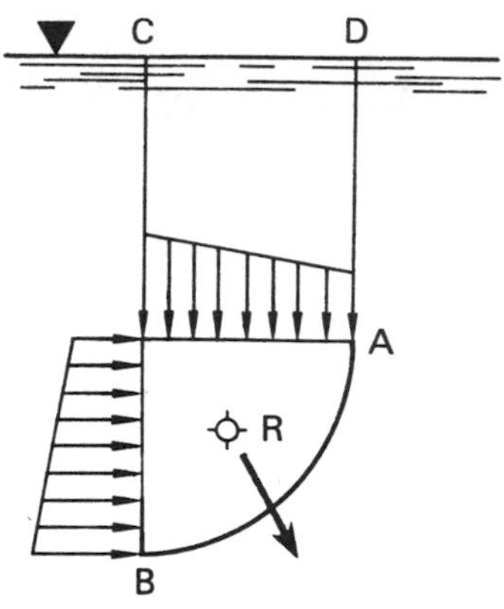


**Figure 3.7** Forces On a Curved Surface



*Example 3.9*

What is the total force on a one-foot section of the wall DAB described in example 3.7?

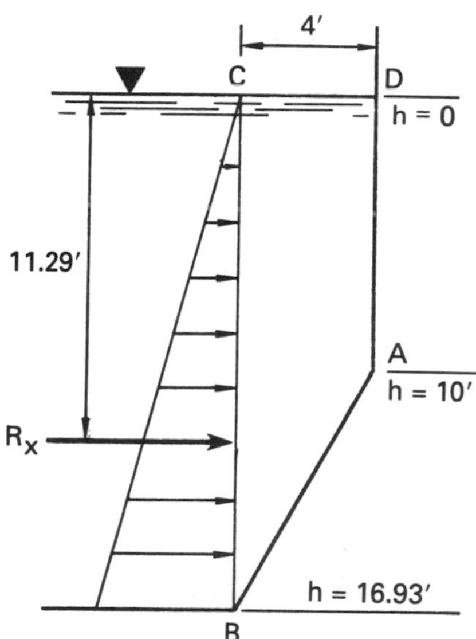


The average depth is $\frac{1}{2}(0 + 16.93) = 8.465$

Using equations 3.30 and 3.31, the average pressure and horizontal component of the resultant are

$$\bar{p} = 62.4(8.465) = 528.2 \text{ psf}$$
$$R_x = (16.93)(1)(528.2) = 8942.4 \text{ lbf}$$

From equation 3.32, the horizontal component acts $(\frac{2}{3})(16.93) = 11.29$ ft from the top.

The volume of a one foot section of area ABCD is

$$(1)\left[(4)(10) + \frac{1}{2}(4)(6.93)\right] = 53.86 \text{ ft}^3$$

Therefore, the vertical component is

$$R_y = (62.4)(53.86) = 3360.9 \text{ lbf}$$

The resultant of $R_x$ and $R_y$ is

$$R = \sqrt{(8942.4)^2 + (3360.9)^2} = 9553.1 \text{ lbf}$$
$$\phi = \arctan\left(\frac{3360.9}{8942.4}\right) = 20.6°$$

In general, it is not correct to calculate the vertical component of force on a submerged surface as being the weight of the fluid above it, as was done in example 3.9. This procedure is valid only when there is no change in the cross section of the tank area.

The *hydrostatic paradox* is illustrated by figure 3.8. The pressure anywhere on the bottom of either container is the same. This pressure is dependent on only the maximum height of the fluid, not the volume.

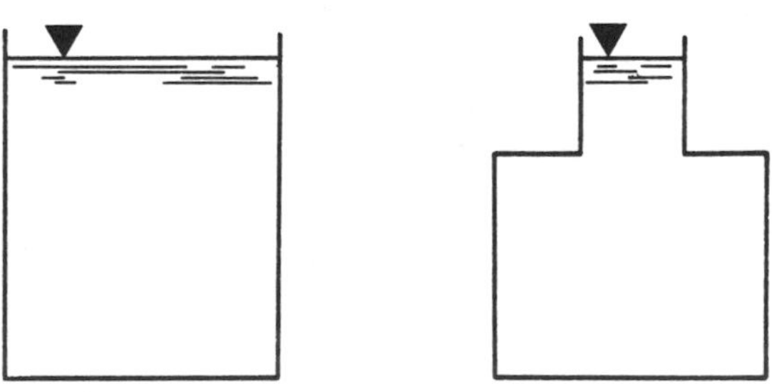

**Figure 3.8** Hydrostatic Paradox

## 4 HYDROSTATIC PRESSURE DUE TO COMPRESSIBLE FLUIDS

Equation 3.26 is a special case of a more general equation known as the *Fundamental Equation of Fluid Statics*, presented as equation 3.36. As defined in the nomenclature, $h$ is a variable representing height, and it is assumed that $h_2$ is greater than $h_1$. The minus sign in equation 3.36 indicates that pressure decreases when height increases.

$$\int_1^2 \frac{g_c dp}{\rho g} = -(h_2 - h_1) \qquad 3.36$$

If the fluid is a compressible layer of perfect gas, and if compression is assumed to be isothermal, equation 3.36 becomes

$$h_2 - h_1 = \frac{g_c RT}{g} ln\left(\frac{p_1}{p_2}\right) \qquad 3.37$$

The pressure at height $h_2$ in a layer of gas which has been isothermally compressed is

$$p_2 = p_1 \left[e^{\frac{g(h_1-h_2)}{g_c RT}}\right] \qquad 3.38$$

The following relationships assume a polytropic compression of the gas layer. These three relationships can be used for adiabatic compression by substituting $k$ for $n$.

$$h_2 - h_1 = \left(\frac{n}{n-1}\right)\left(\frac{g_c}{g}\right) RT_1 \left[1 - \left(\frac{p_2}{p_1}\right)^{\frac{n-1}{n}}\right] \qquad 3.39$$

$$p_2 = p_1 \left[1 - \left(\frac{n-1}{n}\right)\left(\frac{g}{g_c}\right)\left(\frac{h_2 - h_1}{RT_1}\right)\right]^{\frac{n}{n-1}} \qquad 3.40$$

$$T_2 = T_1 \left[1 - \left(\frac{n-1}{n}\right)\left(\frac{g}{g_c}\right)\left(\frac{h_2 - h_1}{RT_1}\right)\right] \qquad 3.41$$

*Example 3.10*

The pressure at sea level is 14.7 psia. Assume 70°F isothermal compression, and calculate the pressure at 5000 feet altitude.

$R$ = 53.3 ft-lbf/lbm-°R for air. $T$ = (70 + 460) = 530 °R.

From equation 3.38,

$$p_{5000} = 14.7 \left[e^{\frac{(32.2)(0-5000)}{(32.2)(53.3)(530)}}\right]$$
$$= 12.32 \text{ psia}$$

## 5 BUOYANCY

The buoyancy theorem, also known as *Archimedes' principle*, states that the upward force on an immersed object is equal to the weight of the displaced fluid. A buoyant force due to displaced air also is relevant in the case of partially-submerged objects. For lighter-than-air crafts, the buoyant force results entirely from displaced air.

$$F_{\text{buoyant}} = \left(\begin{array}{c}\text{displaced}\\\text{volume}\end{array}\right)\left(\begin{array}{c}\text{density of}\\\text{displaced fluid}\end{array}\right) \qquad 3.42$$

In the case of floating or submerged objects not moving vertically, the buoyant force and weight are equal. If the forces are not in equilibrium, the object will rise or fall until some equilibrium is reached. The object will sink until it is supported by the bottom or until the density of the supporting fluid increases sufficiently. It will rise until the weight of the displaced fluid is reduced, either by a decrease in the fluid density or by breaking the surface.

FLUIDS

*Example 3.11*

An empty polyethylene telemetry balloon with payload has a mass of 500 pounds. It is charged with helium when the atmospheric conditions are 60°F and 14.8 psia. What volume of helium is required for liftoff from a sea level platform? The specific gas constant of helium is 386.3 ft-lbf/lbm-°R.

Using equation 3.2 and converting temperature to °R, the gas densities are

$$\rho_{\text{air}} = \frac{p}{RT} = \frac{(14.8)(144)}{(53.3)(60+460)} = 0.07689 \text{ lbm/ft}^3$$

$$\rho_{\text{helium}} = \frac{(14.8)(144)}{(386.3)(60+460)} = 0.01061 \text{ lbm/ft}^3$$

The total mass of the balloon, payload, and helium is

$$m = 500 + (0.01061)\left(\begin{array}{c}\text{helium}\\\text{volume}\end{array}\right)$$

The buoyant force is the weight of the displaced air.

$$F = (0.07689)\left(\begin{array}{c}\text{helium}\\\text{volume}\end{array}\right)$$

Equating $F$ and $m$ results in a helium volume of 7544 ft$^3$.

*Example 3.12*

A 6-foot diameter sphere floats half-submerged in sea water. How much concrete (in lbm) is required as an external anchor to just submerge the sphere completely?

Assume the densities are 64.0 lbm/ft$^3$ for sea water, 150 lbm/ft$^3$ for concrete, and 0.075 lbm/ft$^3$ for air.

The weight of the sphere can be calculated from the buoyant force required to support it when half-submerged. Both the displaced sea water and the displaced air contribute to the buoyant force.

$$V_{\text{sphere}} = \frac{4\pi(3)^3}{3} = 113.1\ \text{ft}^3$$

$$w_{\text{sphere}} = \left(\frac{1}{2}\right)(113.1)(64) + \left(\frac{1}{2}\right)(113.1)(0.075) = 3623.4\ \text{lbf}$$

The bouyant force equation for a fully-submerged sphere and anchor can be solved for the concrete volume.

$$w_{\text{sphere}} + w_{\text{concrete}} = (V_{\text{sphere}} + V_{\text{concrete}})\,(64.0)$$

$$3623.4 + 150V_{\text{concrete}} = (113.1 + V_{\text{concrete}})\,(64.0)$$

$$V_{\text{concrete}} = 42.0\ \text{ft}^3$$

$$m_{\text{concrete}} = (42)\,(150) = 6305\ \text{lbm}$$



## 6 STABILITY OF FLOATING OBJECTS

The bouyant force on a floating object acts upward through the center of gravity of the displaced *volume*, known as the *center of buoyancy*. The weight acts downward through the center of gravity of the *object*. For totally submerged objects such as balloons and submarines, the center of buoyancy must be above the center of gravity for stability. For partially submerged vessels, the metacenter must be above the center of gravity. Stability exists because a righting moment is created if the vessel heels over, since the center of buoyancy moves outboard of the center of gravity.

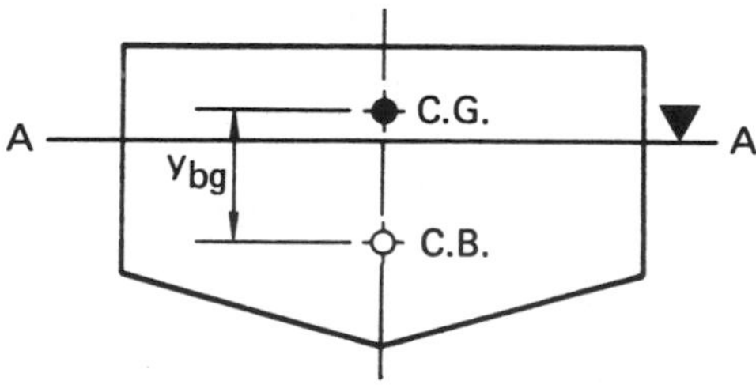


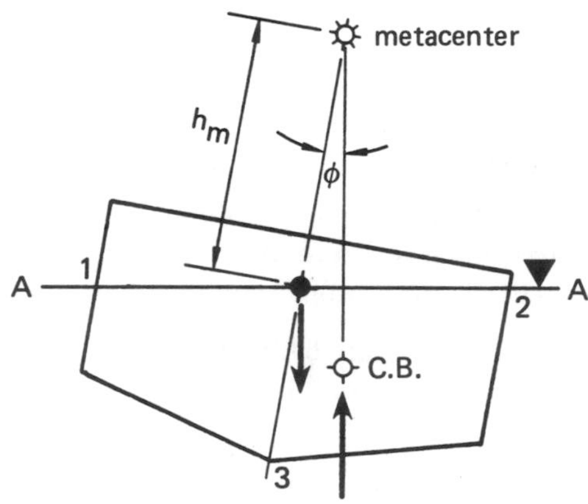


**Figure 3.9** Locating the Metacenter

Refer to figure 3.9. If the floating body heels through an angle $\phi$, the location of the center of gravity does not change. However, the center of buoyancy will shift to the center of gravity of the new submerged section 123. The center of buoyancy and gravity are no longer in line. The couple thus formed tends to resist further overturning.

This righting couple exists when the extension of the buoyant force F intersects line O-O above the center of gravity at M, the *metacenter*. If M lies below the center of gravity, an overturning couple will exist. The distance between the center of gravity and the metacenter is called the *metacentric height*, and it is reasonably constant for heel angles less than 10 degrees.

The metacentric height, $h_m$, can be found from equation 3.43 where $I$ is the area moment of inertia of the above- and below-surface cross section about a longitudinal (fore-and-aft) waterline axis, and $V$ is the displaced volume.

$$h_m = \frac{I}{V} \pm y_{bg} \qquad 3.43$$

## 7 FLUID MASSES UNDER ACCELERATION

The pressures obtained thus far have assumed that the fluid is subjected only to gravitational acceleration. As soon as the fluid is subjected to any other acceleration, additional forces which change hydrostatic pressures are imposed.

If the fluid is subjected to constant accelerations in the vertical and/or horizontal directions, the fluid behavior will be given by equations 3.44 and 3.45.

$$\theta = \arctan\left(\frac{a_x}{a_y + g}\right) \qquad 3.44$$

$$p_h = \gamma h\left(1 + \frac{a_y}{g}\right) \qquad 3.45$$

$a_y$ is negative if the acceleration is downward. Notice that a plane of equal pressure also is inclined if the fluid mass experiences a horizontal acceleration.

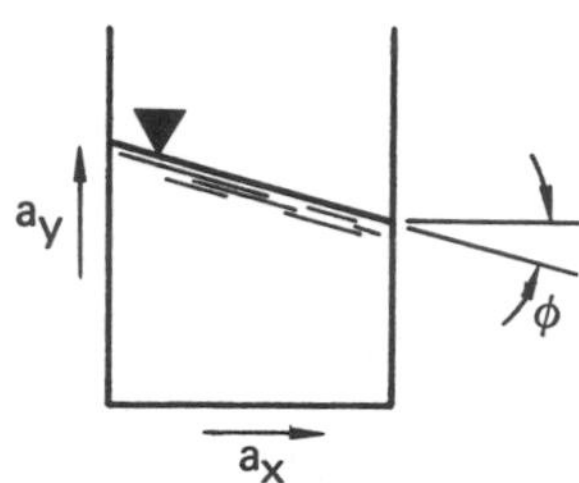


**Figure 3.10** Constant Linear Acceleration

If the fluid mass is rotated about a vertical axis, a parabolic fluid surface will result. The distance $h$ is measured from the lowermost part of the fluid during rotation. $h$ is not measured from the original level of the stationary fluid. $h$ is the height of the fluid at a distance $r$ from the center of rotation.

$$\theta = \arctan\left(\frac{\omega^2 r}{g}\right) \quad 3.46$$

$$h = \frac{(\omega r)^2}{2g} = \frac{v^2}{2g} \quad 3.47$$

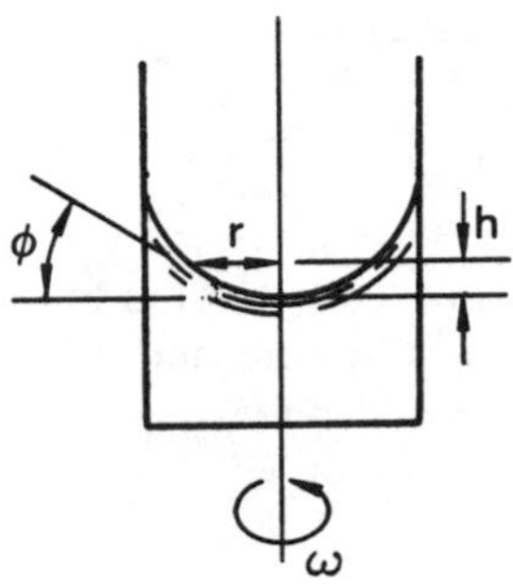


**Figure 3.11** Constant Rotational Acceleration

## 8 MODIFICATION FOR OTHER GRAVITIES

All of the equations in Part 2 of this chapter can be used in a standard gravitational field of 32.2 ft/sec$^2$. In such a standard field, the terms *lbm* and *lbf* can be cancelled freely. However, a modification to the formulas is needed if the local gravity deviates from standard.

The modification that must be made is that of converting the mass density in lbm/ft$^3$ to a *specific weight* in lbf/ft$^3$. This can be done through the use of equation 3.48.

$$\gamma = \frac{\rho g_{\text{local}}}{g_c} \quad 3.48$$

$g_{\text{local}}$ is the actual gravitational acceleration in ft/sec$^2$. $g_c$ is the dimensional conversion factor presented in chapter 22. $g_c$ is approximately equal to 32.2 lbm-ft/lbf-sec$^2$.

The various hydrostatic pressure formulas can be used in non-standard gravitational fields by substituting $\gamma$ for $\rho$.

*Example 3.13*

What is the maximum height that a vacuum pump can lift 60°F water if the atmospheric pressure is 14.6 psia and the local gravity is 28 ft/sec$^2$?

From appendix A, the vapor pressure head at 60°F is approximately 0.59 feet, corresponding to 0.26 psi. The effective pressure which can be used to lift water is (14.6 − 0.26) = 14.34 psi.

$$p = \left(\frac{g}{g_c}\right)\rho h$$

$$h = \frac{(14.34)(144)\left(\frac{32.2}{28.0}\right)}{62.4} = 38.06 \text{ ft}$$

FLUIDS

# PART 3: Fluid Flow Parameters



## 1 INTRODUCTION

Pressure commonly is measured in pounds per square inch (psi) or pounds per square foot (psf). However, pressure may be changed into a new variable called *head* by dividing by the specific weight of the fluid. This operation does more than just scale down the pressure by a factor equal to the reciprocal of the specific weight. Since specific weight itself possesses dimensional units, the units of head are not the same as the units of pressure.

$$h \text{ in ft} = \frac{p \text{ in lbf/ft}^2}{\gamma \text{ in lbf/ft}^3} \qquad 3.49$$

Equation 3.49 is, of course, the same as equation 3.26. As long as the fluid density and local gravitational acceleration remain constant, there is complete interchangeability between the variables of pressure and head.

When Bernoulli's equation is introduced in this chapter, head also will be used as a measure of energy. Actually, head is used as a measure of *specific* energy. This is commonly justified by equation 3.50.

$$h \text{ in ft} = \frac{(E \text{ in ft-lbf})}{(m \text{ in lbm})}\left(\frac{g_c}{g}\right) \qquad 3.50$$

A certain amount of care in the use of equations 3.49 and 3.50 is required since lb$f$ is being cancelled completely by lb$m$. The actual operation being performed is given by equation 3.51.

$$h \text{ in ft} = \frac{\left(g_c \text{ in } \frac{\text{lbm-ft}}{\text{lbf-sec}^2}\right)(p \text{ in lbf/ft}^2)}{\left(g \text{ in } \frac{\text{ft}}{\text{sec}^2}\right)\left(\rho \text{ in } \frac{\text{lbm}}{\text{ft}^3}\right)} \qquad 3.51$$

As $g_c$ always equals 32.2, it can be seen from equation 3.51 that equations 3.49 and 3.50 will give the correct numerical value for head as long as the local gravitational acceleration is 32.2 ft/sec$^2$.

## 2 FLUID ENERGY

A fluid can possess energy in three forms[2]—as pressure, kinetic, or potential energies. The energy (work) that must be put into a fluid to raise its pressure is known as the *pressure energy* or *static energy*. (This form of energy also is known as *flow work* or *flow energy*). In keeping with the common convention to put fluid energy terms into units of feet, the *pressure head* or *static head* is defined by equation 3.52.

$$h_p = \frac{p}{\gamma} \qquad 3.52$$

Energy also is required to accelerate fluid to velocity $v$. The specific kinetic energy with units of feet is known as the *velocity head* or *dynamic head*. Although the units in equation 3.53 actually do yield feet, you should remember that specific energy is in foot-pounds per pound mass of fluid.

$$h_v = \frac{v^2}{2g} \qquad 3.53$$

Potential energy also is given with units of feet. This results in a very simple expression for *potential head* or *gravitational head*. $z$ in equation 3.54 is the height of the fluid above some arbitrary reference point.

$$h_z = z \qquad 3.54$$

## 3 BERNOULLI'S EQUATION

The Bernoulli equation is an energy conservation equation. It states that the total head of a fluid flowing without losses in a pipe cannot change. The total head possessed by a fluid is the sum of its pressure, kinetic, and potential heads.

$$\frac{p_1}{\gamma} + \frac{v_1^2}{2g} + z_1 = \frac{p_2}{\gamma} + \frac{v_2^2}{2g} + z_2 \qquad 3.55$$

Equation 3.55 is valid for laminar and turbulent flow. It can be used for gases as well as liquids if the gases are incompressible.[3] It is assumed that the flow between points 1 and 2 is frictionless and adiabatic.

The sum of the three head terms is known as the *total head*, $H$. The *total pressure* can be calculated from the total head.

$$H = \frac{p}{\gamma} + \frac{v^2}{2g} + z \qquad 3.56$$

$$p_t = \gamma H \qquad 3.57$$

The total energy of the fluid stream has two definitions. The *total specific energy* is the same as total head, $H$. The total energy of all fluid flowing can be calculated from equation 3.58.

$$E_t = mH\left(\frac{g}{g_c}\right) \qquad 3.58$$

A graph of the total specific energy versus distance along a pipe is known as a *total energy line*. In a friction-

[2] Another important energy form is *thermal* energy. Thermal energy terms (internal energy and enthalpy) are not included in this analysis since it is assumed that the temperature of the fluid remains constant.

[3] A gas can be considered to be incompressible as long as its pressure does not change more than 10% between points 1 and 2, and its velocity is less than Mach 0.3 everywhere.

less pipe without pumps or turbines, the total specific energy will remain constant. Total specific energy will decrease if fluid friction is present.

*Example 3.14*

A pipe takes water from the reservoir as shown and discharges it freely 100 feet below. The flow is frictionless. (a) What is the total specific energy at point B? (b) What is the velocity at point C?

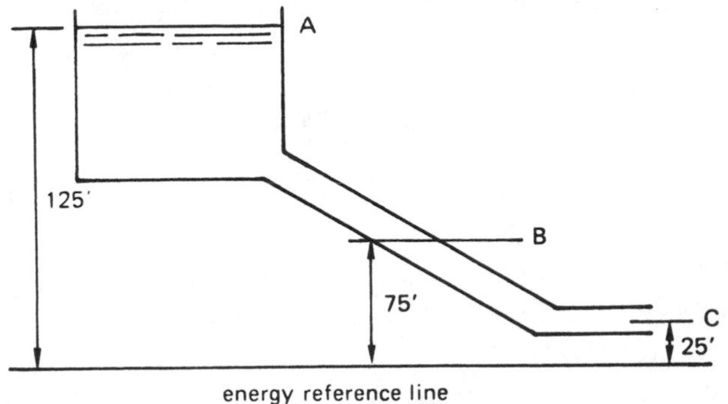


(a) At point A, the velocity and the gage pressure both are zero, so the total specific energy with respect to the reference line is

$$H_A = 0 + 0 + 125 = 125 \text{ ft}$$

At point B, the fluid is moving and possesses kinetic energy. The fluid is also under hydrostatic pressure. However, the flow is frictionless, and the total specific energy is constant (see equation 3.56). The velocity and static heads have increased at the expense of the potential head. Therefore,

$$H_B = H_A = 125 \text{ ft}$$

(b) At point C, the pressure head is again zero since the discharge is at atmospheric pressure. The potential head with respect to the energy reference line is 25 ft. From equation 3.56,

$$125 = 0 + \frac{v^2}{2g} + 25$$
$$v^2 = (2)(32.2)(100)$$
$$v = 80.2 \text{ ft/sec}$$

*Example 3.15*

Water is pumped at a rate of 3 cfs through the piping system illustrated. If the pump has a discharge pressure of 150 psig, to what elevation can the tank be raised? Assume the head loss due to friction is 10 feet.

$$h_{p,1} = \frac{(150)(144)}{62.4} = 346.15 \text{ ft}$$

From appendix F, the internal area of 4″ pipe is 0.0884 ft$^2$.

$$v_1 = \frac{3}{0.0884} = 33.94 \text{ fps}$$
$$h_{v,1} = \frac{(33.94)^2}{(2)(32.2)} = 17.89 \text{ ft}$$
$$z_1 = 0$$
$$h_{p,2} = 0 \text{ (at free surface)}$$
$$v_2 = 0 \text{ (at free surface)}$$

From equation 3.55, taking point 2 as the free water surface,

$$346.15 + 17.89 = z_2 + 10$$
$$z_2 = 354.0 \text{ ft}$$

The tank bottom can be raised to $(354.0 - 10 + 1) = 345$ feet above the ground.

## 4 IMPACT ENERGY

Impact energy (also known as stagnation or total[4] energy) is the sum of the kinetic and pressure energies. The impact head is

$$h_s = \frac{p}{\gamma} + \frac{v^2}{2g} = h_p + h_v \qquad 3.59$$

Impact head represents the effective head in a fluid which has been brought to rest (stagnated) in an adiabatic and reversible manner. Equation 3.59 can be used with a gas as long as the velocity is low—less than 400 ft/sec.

Impact head can be measured directly by using a pitot tube. This is illustrated in figure 3.12. Equation 3.59 can be used with a pitot tube.

A mercury manometer must be used if the stagnation properties of a gas or high-pressure liquid are being measured. Measurement of stagnation properties is covered in greater detail in part 4 of this chapter.

[4] There is confusion about ***total head*** as defined by equations 3.56 and 3.59. The effective pressure in a fluid which has been brought to rest adiabatically does not depend on the potential energy term, $z$. The application will determine which definition of total head is intended.

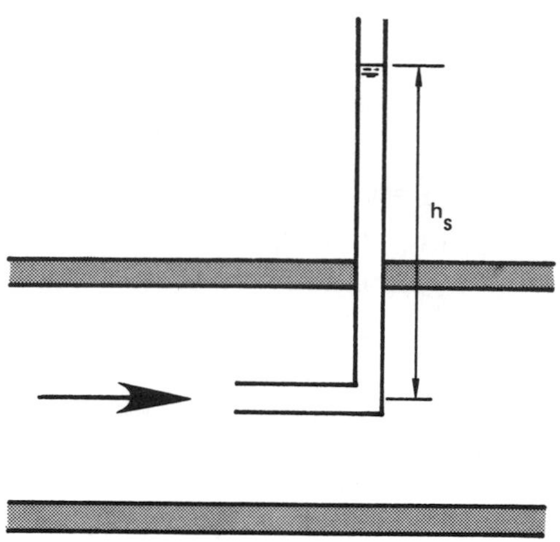


**Figure 3.12** A Pitot Tube

*Example 3.16*

The static pressure of air ($\gamma = 0.075$ lbf/ft$^3$) flowing in a pipe is measured by a precision gage to be 10.0 psig. A pitot tube manometer indicates 20.6 inches of mercury. What is the velocity of the air in the pipe?

The pitot tube measures stagnation pressure. From equation 3.24, using 0.491 lbm/in$^3$ as the density of mercury,

$$p_s = (20.6)(0.491) = 10.11 \text{ psig}$$

Since stagnation pressure is the sum of static and velocity pressures, the velocity pressure is

$$p_v = p_s - p_p = 10.11 - 10.0 = 0.11 \text{ psig}$$

The velocity head is

$$h_v = \frac{p_v}{\gamma} = \frac{(0.11)(144)}{0.075} = 211.1 \text{ ft}$$

From equation 3.53,

$$v = \sqrt{(2)(32.2)(211.2)} = 116.6 \text{ ft/sec}$$

## 5 HYDRAULIC GRADE LINE

The hydraulic grade line is a graphical representation of the sum of the static and potential heads versus position along the pipeline

$$\text{hydraulic grade} = z + h_p \qquad 3.60$$

Since the pressure head can increase at the expense of the velocity head, the hydraulic grade line can increase if an increase in flow area is encountered.

## 6 REYNOLDS' NUMBER

The Reynolds number is a dimensionless ratio of the inertial flow forces to the viscous forces within the fluid. Two expressions for Reynolds' number are used, one requiring absolute viscosity, the other kinematic viscosity:

$$N_{Re} = \frac{D_e v \rho}{\mu g_c} \qquad 3.61$$

$$= \frac{D_e v}{\nu} \qquad 3.62$$

The Reynolds number also can be calculated from the mass flow rate per unit area, $G$. $G$ must have the units of lbm/sec-ft$^2$.

$$N_{Re} = \frac{D_e G}{\mu g_c} \qquad 3.63$$

The Reynolds number is an important indicator in many types of problems. In addition to being used quantitatively in many equations, the Reynolds number also is used to determine whether fluid flow is laminar or turbulent.

A Reynolds number of 2000 or less indicates *laminar flow*. Fluid particles in laminar flow move in straight paths parallel to the flow direction. Viscous effects are dominant, resulting in a parabolic velocity distribution with a maximum velocity along the fluid flow centerline.

The fluid is said to be *turbulent* if the Reynolds number is greater than 2000.[5] Turbulent flow is characterized by random movement of fluid particles. The velocity distribution is essentially uniform with turbulent flow.

## 7 EQUIVALENT DIAMETER

The equivalent diameter, $D_e$, used in equations 3.61 and 3.62, is equal to the inside diameter of a circular pipe. The equivalent diameters of other cross sections in flow are given by table 3.6.

*Example 3.17*

Determine the equivalent diameter of the open trapezoidal channel shown.

[5] The beginning of the turbulent region is difficult to predict. There actually is a transition region between Reynolds numbers 2000 to 4000. In most fluid problems, however, flow is well within the turbulent region.

**Table 3.6**
Equivalent Diameters

| conduit cross section | $D_e$ |
|---|---|
| flowing full | |
| annulus | $D_o - D_i$ |
| square | $L$ |
| rectangle | $\frac{2L_1L_2}{L_1 + L_2}$ |
| flowing partially full | |
| half-filled circle | $D$ |
| rectangle ($h$ deep, $L$ wide) | $\frac{4hL}{L + 2h}$ |
| wide, shallow stream ($h$ deep) | $4h$ |
| triangle ($h$ deep, $L$ broad, $s$ side) | $\frac{hL}{s}$ |
| trapezoid ($h$ deep, $a$ wide at top, $b$ wide at bottom, $s$ side) | $\frac{2h(a + b)}{b + 2s}$ |



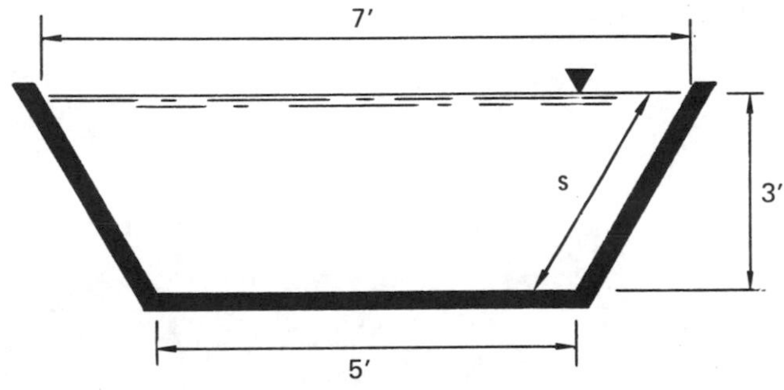


$$s = \sqrt{3^2 + 1^2} = 3.16 \text{ feet}$$

$$D_e = \frac{2(3)(7 + 5)}{5 + 2(3.16)} = 6.36 \text{ feet}$$

## 8 HYDRAULIC RADIUS

The equivalent diameter also can be found from the hydraulic radius, which is defined as the area in flow divided by the wetted perimeter. The wetted perimeter does not include free fluid surface.

$$D_e = 4r_h \qquad 3.64$$

$$r_h = \frac{\text{area in flow}}{\text{wetted perimeter}} \qquad 3.65$$

Consider a circular pipe flowing full. The area in flow is $\pi r^2$. The wetted perimeter is the entire circumference, $2\pi r$. The hydraulic radius is

$$\left(\frac{\pi r^2}{2\pi r}\right) = \frac{1}{2}r$$

Therefore, the hydraulic radius and the pipe radius are not the same. (The hydraulic radius of a pipe flowing half full is also $\frac{1}{2}r$, as the flow area and the wetted perimeter both are halved.)

The hydraulic radius of a pipe flowing less than full can be found from table 3.7 or appendix E.

*Example 3.18*

What is the hydraulic radius of the trapezoidal channel described in example 3.17?

From equation 3.65,

$$r_h = \frac{(5)(3) + (3)(1)}{3.16 + 5 + 3.16} = 1.59 \text{ feet}$$

Using the results of the previous example and equation 3.64,

$$r_h = \frac{6.36}{4} = 1.59 \text{ feet}$$

**Table 3.7**
Hydraulic Radius
of Partially Filled Circular Pipes
(Also see appendix E)

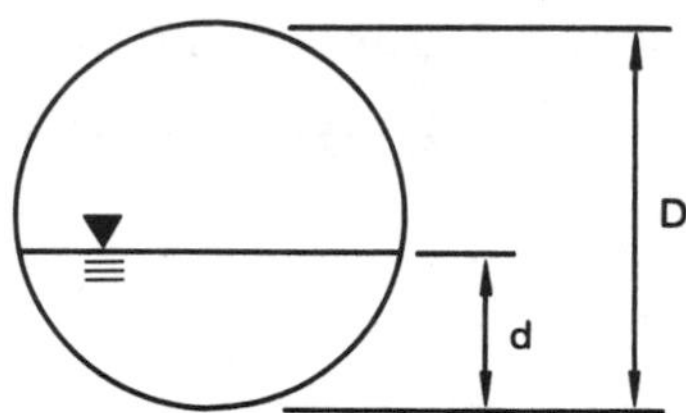


| $\frac{d}{D}$ | $\frac{\text{hyd. rad.}}{D}$ | $\frac{d}{D}$ | $\frac{\text{hyd. rad.}}{D}$ |
|---|---|---|---|
| 0.05 | 0.0326 | 0.55 | 0.2649 |
| 0.10 | 0.0635 | 0.60 | 0.2776 |
| 0.15 | 0.0929 | 0.65 | 0.2881 |
| 0.20 | 0.1206 | 0.70 | 0.2962 |
| 0.25 | 0.1466 | 0.75 | 0.3017 |
| 0.30 | 0.1709 | 0.80 | 0.3042 |
| 0.35 | 0.1935 | 0.85 | 0.3033 |
| 0.40 | 0.2142 | 0.90 | 0.2980 |
| 0.45 | 0.2331 | 0.95 | 0.2864 |
| 0.50 | 0.2500 | 1.00 | 0.2500 |

# PART 4: Fluid Dynamics

## 1 FLUID CONSERVATION LAWS

Many fluid flow problems can be solved by using the principles of conservation of mass and energy.

When applied to fluid flow, the principle of mass conservation is known as the *continuity equation*:

$$\rho_1 A_1 v_1 = \rho_2 A_2 v_2 \qquad 3.66$$

$$\dot{m}_1 = \dot{m}_2 \qquad 3.67$$

If the fluid is incompressible, $\rho_1 = \rho_2$, so

$$A_1 v_1 = A_2 v_2 \qquad 3.68$$

$$\dot{V}_1 = \dot{V}_2 \qquad 3.69$$

The energy conservation principle is based on the Bernoulli equation. However, terms for friction loss and hydraulic machines must be included.

$$\left(\frac{p_1}{\gamma} + \frac{v_1^2}{2g} + z_1\right) + h_A = \left(\frac{p_2}{\gamma} + \frac{v_2^2}{2g} + z_2\right) + h_E + h_f \qquad 3.70$$

## 2 HEAD LOSS DUE TO FRICTION

The most common expression for calculating head loss due to friction ($h_f$) is the *Darcy formula*:

$$h_f = \frac{fLv^2}{2Dg} \qquad 3.71$$

The *Moody friction factor chart* (figure 3.13) probably is the most convenient method of determining the friction factor, $f$.

The basic parameter required to use the Moody friction factor chart is the Reynolds number. If the Reynolds number is less than 2000, the friction factor is given by equation 3.72.

$$f = \frac{64}{N_{Re}} \qquad 3.72$$

For turbulent flow ($N_{Re} > 2000$), the friction factor depends on the relative roughness of the pipe. This roughness is expressed by the ratio $\frac{\epsilon}{D}$, where $\epsilon$ is the specific surface roughness and $D$ is the inside diameter. Values of $\epsilon$ for various types of pipe are found in table 3.8.

Another method for finding the friction head loss is the *Hazen-Williams formula*. The Hazen-Williams formula gives good results for liquids that have kinematic viscosities around 1.2 EE−5 ft$^2$/sec (corresponding to 60°F water). At extremely high and low temperatures, the Hazen-Williams formula can be as much as 20% in error for water. The Hazen-Williams formula should be used only for turbulent flow.

The Hazen-Williams head loss is

$$h_f = \frac{(3.022)(v)^{1.85}L}{(C)^{1.85}(D)^{1.165}} \qquad 3.73$$

Or, in terms of other units,

$$h_f = (10.44)(L)\,\frac{(\text{gpm})^{1.85}}{(C)^{1.85}(d_{\text{inches}})^{4.8655}} \qquad 3.74$$

Use of these formulas requires a knowledge of the Hazen-Williams coefficient, $C$, which is assumed to be independent of the Reynolds number. Table 3.8 gives values of $C$ for various types of pipe.

Values of $f$ and $h_f$ are appropriate for clean, new pipe. As some pipes age, it is not uncommon for scale build-up to decrease the equivalent flow diameter. This diameter decrease produces a dramatic increase in the friction loss.

$$\frac{h_{f,\text{scaled}}}{h_{f,\text{new}}} = \left(\frac{D_{\text{new}}}{D_{\text{scaled}}}\right)^5 \qquad 3.75$$

Because of this scale effect, an uprating factor of 10-30% is commonly applied to $f$ or $h_f$ in anticipation of future service conditions.

*Example 3.19*

50°F water is pumped through 4″ schedule 40 welded steel pipe ($\epsilon = 0.0002$) at the rate of 300 gpm. What is the friction head loss calculated by the Darcy formula for 1000 feet of pipe?

First, it is necessary to collect data on the pipe and water. The fluid viscosity and pipe dimensions can be found from tables at the end of the chapter.

kinematic viscosity = 1.41 EE − 5 ft$^2$/sec

inside diameter = 0.3355 ft

flow area = 0.0884 ft$^2$

The flow quantity is

$$(300)(0.002228) = 0.6684 \text{ cfs}$$

The velocity is

$$v = \frac{\dot{V}}{A} = \frac{0.6684}{0.0884} = 7.56 \text{ fps}$$

The Reynolds number is

$$N_{Re} = \frac{(0.3355)(7.56)}{1.41 \text{ EE} - 5} = 1.8 \text{ EE5}$$

FLUIDS

**Table 3.8**
Specific Roughness and Hazen-Williams Constants for Various Pipe Materials

| type of pipe or surface | $\epsilon$(ft) range | $\epsilon$(ft) design | $C$ range | $C$ clean/ new | $C$ design/ 10–20 years old |
|---|---|---|---|---|---|
| STEEL | | | | | |
| welded and seamless | 0.0001–0.0003 | 0.0002 | 150–80 | 140 | 100 |
| interior riveted, no projecting rivets | | | | 139 | 100 |
| projecting girth rivets | | | | 130 | 100 |
| projecting girth and horizontal rivets | | | | 115 | 100 |
| vitrified, spiral-riveted, flow with lap | | | | 110 | 100 |
| vitrified, spiral-riveted, flow against lap | | | | 100 | 90 |
| corrugated | | | | 60 | 60 |
| MINERAL | | | | | |
| concrete | 0.001–0.01 | 0.004 | 150–85 | 120 | 100 |
| cement-asbestos | | | 160–140 | 150 | 140 |
| vitrified clays | | | | | 110 |
| brick sewer | | | | | 100 |
| IRON | | | | | |
| cast, plain | 0.0004–0.002 | 0.0008 | 150–80 | 130 | 100 |
| cast, tar (asphalt) coated | 0.0002–0.0006 | 0.0004 | 145–50 | 130 | 100 |
| cast, cement lined | 0.000008 | 0.000008 | | 150 | 140 |
| cast, bituminous lined | 0.000008 | 0.000008 | 160–130 | 150 | 140 |
| cast, centrifugally spun | 0.00001 | 0.00001 | | | |
| galvanized, plain | 0.0002–0.0008 | 0.0005 | | | |
| wrought, plain | 0.0001–0.0003 | 0.0002 | 150–80 | 130 | 100 |
| MISCELLANEOUS | | | | | |
| fiber | | | | 150 | 140 |
| copper and brass | 0.000005 | 0.000005 | 150–120 | 140 | 130 |
| wood stave | 0.0006–0.003 | 0.002 | 145–110 | 120 | 110 |
| transite | 0.000008 | 0.000008 | | | |
| lead, tin, glass | | 0.000005 | 150–120 | 140 | 130 |
| plastic (PVC and ABS) | | 0.000005 | 150–120 | 140 | 130 |

The relative roughness is

$$\frac{\epsilon}{D} = \frac{0.0002}{0.3355} = 0.0006$$

From the Moody friction factor chart, $f = 0.0195$.

From equation 3.71,

$$h_f = \frac{(0.0195)(1000)(7.56)^2}{(2)(0.3355)(32.2)} = 51.6 \text{ ft}$$

*Example 3.20*

Repeat example 3.19 using the Hazen-Williams formula. Assume $C = 100$.

Using equation 3.73,

$$h_f = \frac{(3.022)(7.56)^{1.85}(1000)}{(100)^{1.85}(0.3355)^{1.165}} = 90.8 \text{ ft}$$

Using equation 3.74,

$$h_f = (10.44)(1000)\frac{(300)^{1.85}}{(100)^{1.85}(4.026)^{4.8655}} = 90.9 \text{ ft}$$

## 3 MINOR LOSSES

In addition to the head loss caused by friction between the fluid and the pipe wall, losses also are caused by obstructions in the line, changes in direction, and changes in flow area. These losses are named *minor losses* because they are much smaller in magnitude than the $h_f$ term. Two methods are used to determine these losses: the method of equivalent lengths and the method of loss coefficients.

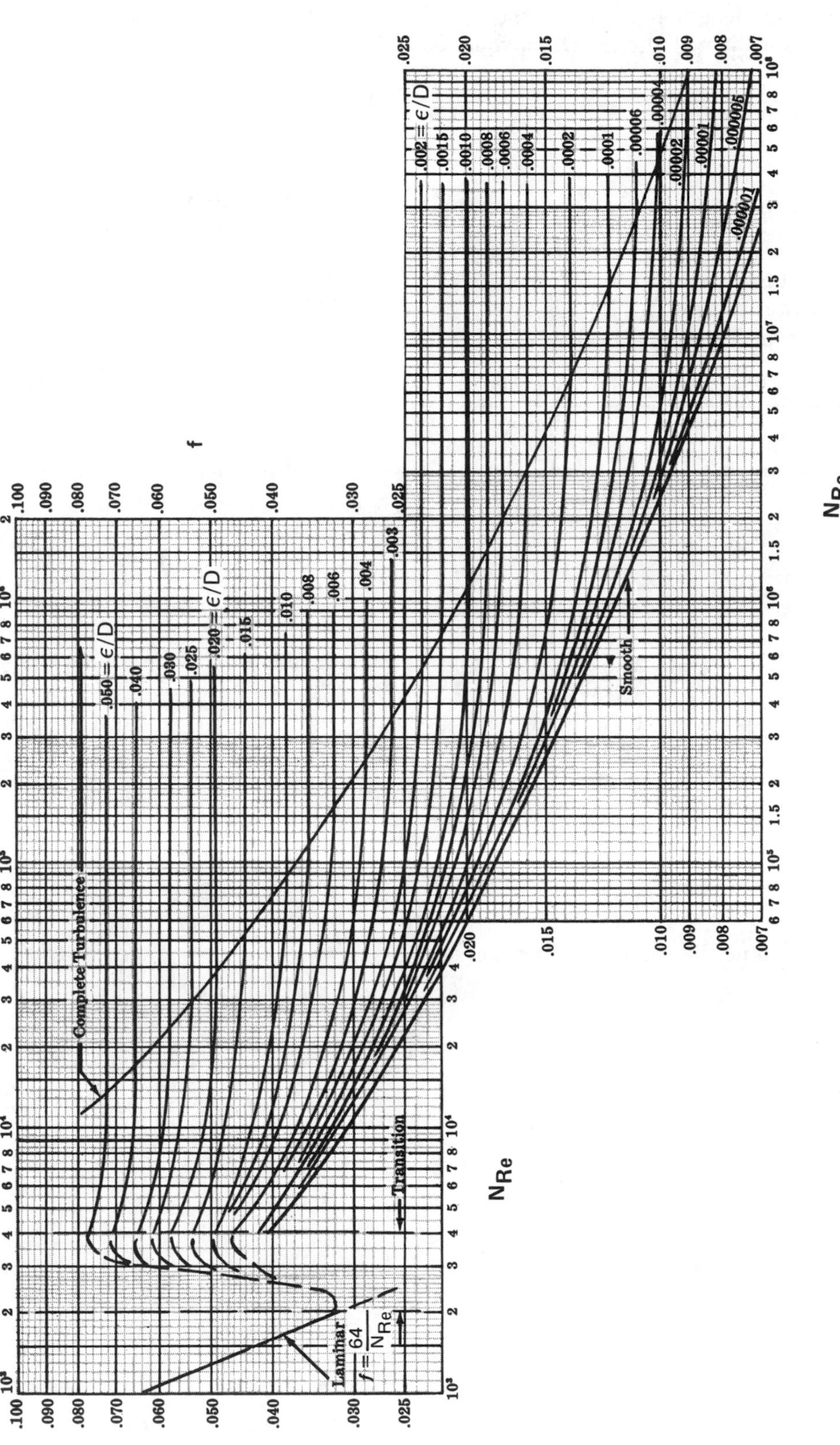


**Figure 3.13** Moody Friction Factor Chart

FLUIDS

The method of *equivalent lengths* uses a table to convert each valve and fitting into an equivalent length of straight pipe. This length is added to the actual pipeline length and substituted into the Darcy equation for $L_e$.

$$h_f = \frac{fL_e v^2}{2Dg} \qquad 3.76$$

**Table 3.9**
Typical Equivalent Lengths of Schedule 40 Straight Pipe For Screwed Steel Fittings and Valves
(Also, see appendix L)
(For any fluid in turbulent flow)

| | equivalent length, ft | | |
|---|---|---|---|
| | pipe size | | |
| fitting type | 1″ | 2″ | 4″ |
| short radius, regular 90° elbow | 5.2 | 8.5 | 13.0 |
| long radius 90° elbow | 2.7 | 3.6 | 4.6 |
| regular 45° elbow | 1.3 | 2.7 | 5.5 |
| tee, flow through line (run) | 3.2 | 7.7 | 17.0 |
| tee, flow through stem | 6.6 | 12.0 | 21.0 |
| 180° return bend | 5.2 | 8.5 | 13.0 |
| globe valve | 29.0 | 54.0 | 110.0 |
| gate valve | 0.84 | 1.5 | 2.5 |
| angle valve | 17.0 | 18.0 | 18.0 |
| swing check valve | 11.0 | 19.0 | 38.0 |
| coupling or union | 0.29 | 0.45 | 0.65 |

*Example 3.21*

Using table 3.9, determine the equivalent length of the piping network shown.

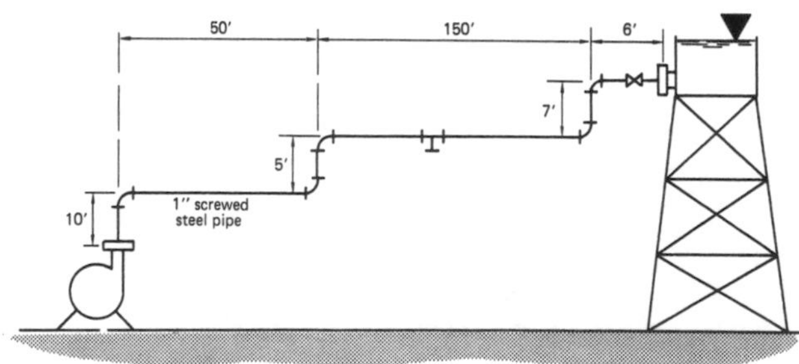


The line consists of:

| | |
|---|---|
| 1 gate valve | 0.84 |
| 5 90° standard elbows | 5.2×5 |
| 1 tee run | 3.2 |
| straight pipe | 228 |
| $L_e$ = | 258 feet |

The alternative is to use a loss coefficient, $K$. This loss coefficient, when multiplied by the velocity head, will give the head loss in feet. This method must be used to find exit and entrance losses.

$$h_f = K\frac{v^2}{2g} \qquad 3.77$$

Values of $K$ are widely tabulated, but they also can be calculated from the following formulas.

*Valves and Fittings*: Refer to the manufacturer's data, or calculate from the equivalent length.

$$K = \frac{fL_e}{D} \qquad 3.78$$

*Sudden Enlargements*: $(D_1 < D_2)$

$$K = \left[1 - \left(\frac{D_1}{D_2}\right)^2\right]^2 \qquad 3.79$$

*Sudden Contractions*: $(D_1 < D_2)$

$$K = \frac{1}{2}\left[1 - \left(\frac{D_1}{D_2}\right)^2\right] \qquad 3.80$$

*Pipe Exit*: (projecting exit, sharp-edged, and rounded)

$$K = 1.0$$

*Pipe Entrance*:

*Reentrant*: $K = 0.78$

*Sharp edged*: $K = 0.5$

*Rounded*:

| $\frac{r}{D}$ | $K$ |
|---|---|
| 0.02 | 0.28 |
| 0.04 | 0.24 |
| 0.06 | 0.15 |
| 0.10 | 0.09 |
| 0.15 | 0.04 |

*Tapered Diameter Changes*:

$$\beta = \frac{\text{small diameter}}{\text{large diameter}}$$

$\phi$ = wall-to-horizontal angle

| | Gradual, $\phi < 22°$ | Sudden, $\phi > 22°$ | |
|---|---|---|---|
| Enlargement | $2.6(\sin\phi)(1-\beta^2)^2$ | $(1-\beta^2)^2$ | 3.81 |
| Contraction | $0.8(\sin\phi)(1-\beta^2)$ | $\frac{1}{2}(1-\beta^2)\sqrt{\sin\phi}$ | 3.82 |

## 4 HEAD ADDITIONS/EXTRACTIONS

A pump adds head (energy) to the fluid stream. A turbine extracts head from the fluid stream. The amount of head added or extracted can be found by evaluating Bernoulli's equation (equation 3.55) on both sides of the device.

$$h_A = H_2 - H_1 \quad \text{(pumps)} \tag{3.83}$$

$$h_E = H_1 - H_2 \quad \text{(turbines)} \tag{3.84}$$

The head increase from a pump is given by equation 3.85.

$$h_A = \frac{(550)(\text{pump input horsepower})\eta_{\text{pump}}}{\dot{m}} \left(\frac{g_c}{g}\right) \tag{3.85}$$

Bernoulli's equation also can be used to calculate the power available to a turbine in a fluid stream by multiplying the total energy by the mass flow rate. This is called the *water horsepower.*

$$P = \dot{m}H\left(\frac{g}{g_c}\right) = \dot{m}\left(\frac{p}{\gamma} + \frac{v^2}{2g} + z\right)\left(\frac{g}{g_c}\right) \tag{3.86}$$

$$\dot{m} = \rho A v \tag{3.87}$$

$$whp = \frac{P}{550} \tag{3.88}$$

Pumps and turbines are covered in greater detail in chapter 4.

## 5 DISCHARGE FROM TANKS

Flow from a tank discharging liquid to the atmosphere through an opening in the tank wall (figure 3.14) is affected by both the area and the shape of the opening. At the orifice, the total head of the fluid is converted into kinetic energy according to equation 3.89.[6]

$$v_o = C_v\sqrt{2gh} \tag{3.89}$$

[6] Although the term $g_c$ appears in the equation for velocity head (equation 3.53), here it is the local gravity, $g$, which appears in equation 3.89. An analysis of equation 3.197 will show you why this is so.

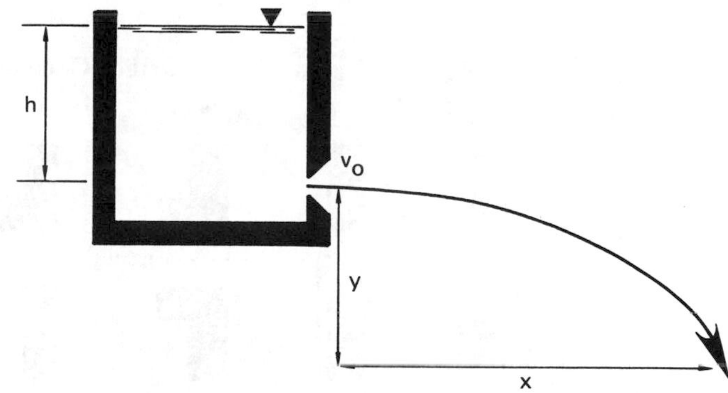


**Figure 3.14** Discharge From a Tank

$C_v$ is the *coefficient of velocity* which can be calculated from the *coefficients of discharge* and *contraction.* Typical values of $C_v$, $C_d$, and $C_c$ are given in table 3.10.

$$C_v = \frac{C_d}{C_c} \tag{3.90}$$

The discharge from the orifice is

$$\dot{V} = (C_c A_o)v_o = C_c A_o C_v\sqrt{2gh} = C_d A_o\sqrt{2gh} \tag{3.91}$$

The head loss due to turbulence at the orifice is

$$h_f = \left(\frac{1}{C_v^2} - 1\right)\frac{v_o^2}{2g} \tag{3.92}$$

The discharge stream coordinates (see figure 3.14) are

$$x = v_o t = v_o\sqrt{\frac{2y}{g}} = 2C_v\sqrt{hy} \tag{3.93}$$

$$y = \frac{gt^2}{2} = \frac{g}{2}\left(\frac{x}{v_o}\right)^2 \tag{3.94}$$

The fluid velocity at a point downstream of the orifice is

$$v_x = v_o \tag{3.95}$$

$$v_y = gt \tag{3.96}$$

If the liquid in a tank is not being replenished constantly, the static head forcing discharge through the orifice will decrease. For a tank with a constant cross-sectional area, the time required to lower the fluid level from level $h_1$ to $h_2$ is calculated from equation 3.97.

$$t = \frac{2A_t(\sqrt{h_1} - \sqrt{h_2})}{C_d A_o\sqrt{2g}} \tag{3.97}$$

If the tank has a varying cross section, the following basic relationship holds.

$$Q\,dt = -A_t\,dh \tag{3.98}$$

FLUIDS

**Table 3.10**
**Orifice Coefficients for Water**
(fully turbulent)

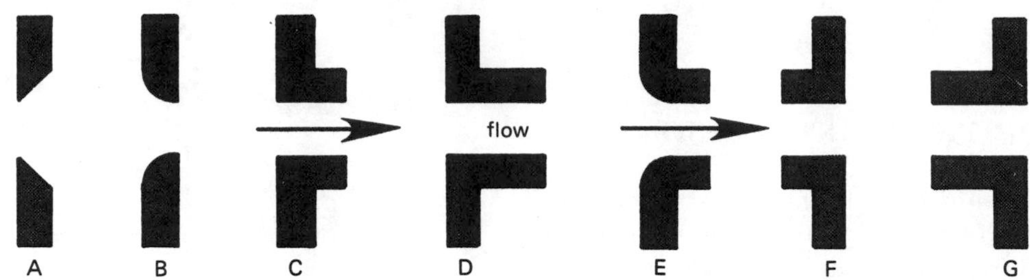


| illustration | description | $C_d$ | $C_c$ | $C_v$ |
|---|---|---|---|---|
| A | sharp-edged | 0.62 | 0.63 | 0.98 |
| B | round-edged | 0.98 | 1.00 | 0.98 |
| C | short tube (fluid separates from walls) | 0.61 | 1.00 | 0.61 |
| D | short tube (no separation) | 0.82 | 1.00 | 0.82 |
| E | short tube with rounded entrance | 0.97 | 0.99 | 0.98 |
| F | reentrant tube, length less than one-half of pipe diameter | 0.54 | 0.55 | 0.99 |
| G | reentrant tube, length 2 to 3 pipe diameters | 0.72 | 1.00 | 0.72 |
| not shown | smooth, well-tapered nozzle | 0.98 | 0.99 | 0.99 |



An expression for the tank area, $A_t$, as a function of $h$, must be determined. Then, the time to empty the tank from height $h_1$ to lower height $h_2$ is

$$t = \int_{h_1}^{h_2} \frac{A_t\, dh}{C_d A_o \sqrt{2gh}} \qquad 3.99$$

For a tank being fed at a rate, $\dot{V}_{\text{in}}$, which is less than the discharge through the orifice, the time to empty expression is

$$t = \int_{h_1}^{h_2} \frac{A_t dh}{(C_d A_o \sqrt{2gh}) - \dot{V}_{\text{in}}} \qquad 3.100$$

When a tank is being fed at a rate greater than the discharge, equation 3.100 will become positive, indicating a rising head. $t$ then will be the time it takes to raise the fluid level from $h_1$ to $h_2$.

The preceding discussion has assumed that the tank has been open or vented to the atmosphere. If the fluid is discharging from a pressurized tank, the total head will be increased by the gage pressure converted to head of fluid by means of equation 3.52.

*Example 3.22*

A 15′ diameter tank discharges 150°F water through a sharp edged 1″ diameter orifice. If the original water depth is 12′ and the tank is continually pressurized to 50 psig, find the time to empty the tank.

At 150°F, $\rho = 61.20$ lbm/ft$^3$

For the orifice,

$$A_o = 0.00545 \text{ ft}^2, \; C_d = 0.62$$

$$h_1 = 12 + \frac{(50)(144)}{61.2} = 129.65 \text{ ft}$$

$$h_2 = \frac{(50)(144)}{61.20} = 117.65 \text{ ft}$$

From equation 3.97,

$$t = \frac{2\left[\pi(7.5)^2\right](\sqrt{129.65} - \sqrt{117.65})}{(0.62)(0.00545)\sqrt{(2)(32.2)}} = 7035 \text{ seconds}$$

## 6 CULVERTS AND SIPHONS

A culvert is a water path used to drain runoff from an obstructing geographical feature. Most culvert designs are empirical. However, if the entrance and exit of the culvert both are submerged, the discharge will be independent of the barrel slope. (Chapter 5 treats this subject in greater detail.) In that case, equation 3.101 can be used to evaluate the discharge. $h$ is the difference in surface levels of the headwater and tailwater.

$$\dot{V} = C_d A \sqrt{2gh} \qquad 3.101$$

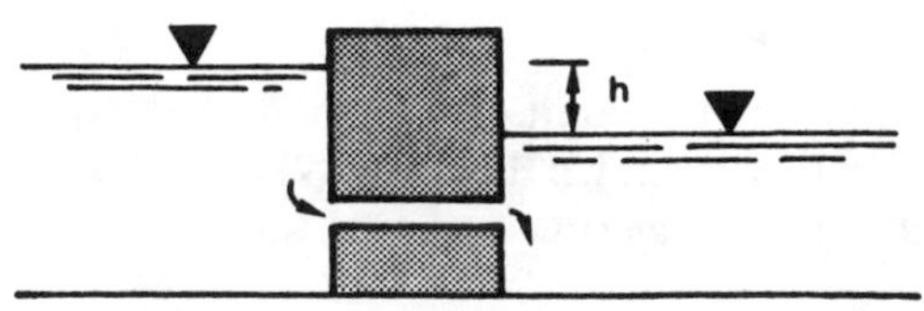


**Figure 3.15** A Simple Pipe Culvert

If the culvert length is greater than 50 feet or if the entrance is not smooth, the available energy will be divided between friction and velocity heads. The effective head to be used in equation 3.101 is:

$$h' = h - h_{\text{entrance}} - h_f \qquad 3.102$$

The entrance head loss is calculated using loss coefficients:

$$h_{\text{entrance}} = K_e\left(\frac{v^2}{2g}\right) \qquad 3.103$$

Typical values of $K_e$ are:

0.08 for a smooth and tapered entrance

0.10 for a flush concrete groove or bell design

0.15 for a projecting concrete groove or bell design

0.50 for a flush square-edged entrance

0.90 for a projecting square-edged entrance

The friction loss, $h_f$, can be found in the usual manner, either from the Darcy equation and Moody friction factor chart or from the Hazen-Williams equation. A trial and error solution may be necessary since $v$ is not known, but is needed to find the friction factor.

## 7 MULTIPLE PIPE SYSTEMS

### A. SERIES PIPE SYSTEMS

A series pipe system has one or more diameters along its run. If $Q$ or $v$ is known in any part of the system, the friction loss can be found easily as the sum of the friction losses in the sections. (The pipe discharges to air at point 2. The solution procedure is easily changed by adding $p_b/\gamma$ to equation 3.108 if the discharge is to a reservoir at pressure $p_b$.)

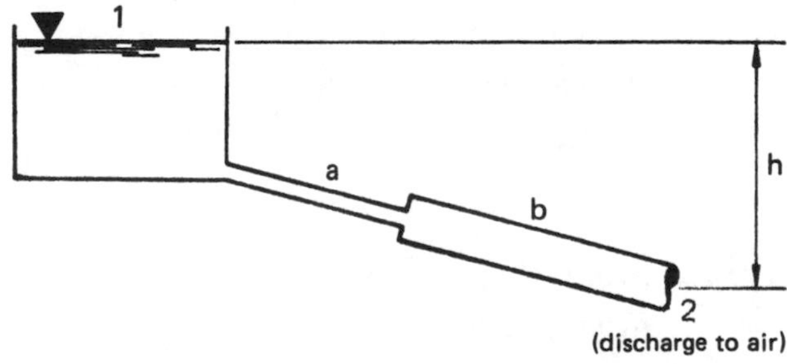


**Figure 3.16** A Series Pipe System

If both $v$ and $Q$ are unknown, a trial and error solution method is required. The following procedure can be used with the Darcy friction factor.

*step 1*: Using the Moody diagram with $\epsilon_a$, $\epsilon_b$, $D_a$, and $D_b$, find $f_a$ and $f_b$ for fully turbulent flow.

*step 2*: Write all of the velocities in terms of one unknown velocity.

$$Q_a = Q_b \qquad 3.104$$

$$v_b = \left(\frac{A_a}{A_b}\right) v_a \qquad 3.105$$

*step 3*: Write the friction loss in terms of the unknown velocity.

$$h_{f,\text{total}} = \frac{f_a L_a v_a^2}{2D_a g} + \frac{f_b L_b}{2D_b g}\left(\frac{A_a}{A_b}\right)^2 v_a^2 \qquad 3.106$$

$$= \frac{v_a^2}{2g}\left[\frac{f_a L_a}{D_a} + \frac{f_b L_b}{D_b}\left(\frac{A_a}{A_b}\right)^2\right] \qquad 3.107$$

*step 4*: Solve for the unknown velocity using Bernoulli's equation between points 1 and 2. Include the pipe friction but, for convenience, ignore minor losses.

$$h = \frac{v_b^2}{2g} + h_f \qquad 3.108$$

$$= \frac{v_a^2}{2g}\left[\left(\frac{A_a}{A_b}\right)^2\left(1 + \frac{f_b L_b}{D_b}\right) + \frac{f_a L_a}{D_a}\right] \qquad 3.109$$

*step 5*: Using the values of $v_a$ and $v_b$ found from step 4, check the values of $f_a$ and $f_b$. Repeat steps 3 and 4 using the new values of $f_a$ and $f_b$ if necessary.

If the Hazen-Williams coefficients are given for the pipe sections, the procedure for finding unknown velocities and flow quantities is similar although considerably more difficult since the $v^2$ and $v^{1.85}$ terms cannot be combined. A first approximation, however, can be obtained by replacing $v^{1.85}$ with $v^2$ in the Hazen-Williams formula for friction loss. A trial and error method can then be used to find $v$.

### B. PARALLEL PIPE SYSTEMS

A common method of increasing the capacity of an existing line is to install a second line parallel to the first.

If that is done, the flow will divide in such a manner as to make the friction loss the same in both branches.

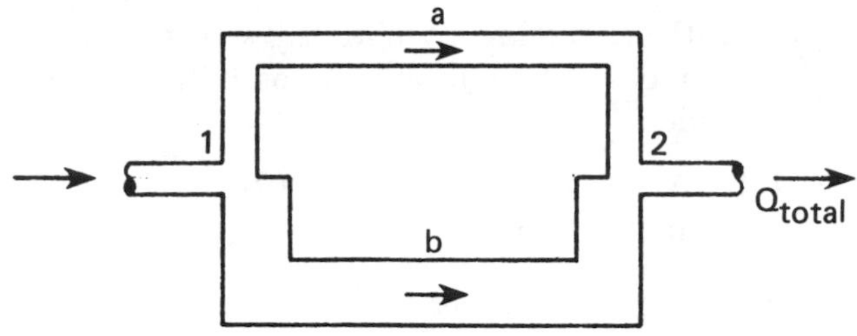


**Figure 3.17** A Parallel Pipe System



If the parallel system has only two branches, a simultaneous solution approach can be taken.

$$h_{f,a} = h_{f,b} = \frac{f_a L_a v_a^2}{2D_a g} = \frac{f_b L_b v_b^2}{2D_b g} \qquad 3.110$$

$$Q_a + Q_b = Q_c \qquad 3.111$$

$$\frac{1}{4}\pi\left(D_a^2 v_a + D_b^2 v_b\right) = Q_c \qquad 3.112$$

However, if the parallel system has three or more branches, it is easier to use the following procedure, which can also be modified for use with the Hazen-Williams loss formula, as demonstrated in example 3.23.

*step 1*: Write $h_f = \dfrac{fLv^2}{2Dg}$ for each branch. Both $h_f$ and $v$ will be unknown.

*step 2*: Solve for $v$ for each branch.

$$v = \sqrt{\frac{2Dg}{fL} h_f} \qquad 3.113$$

*step 3*: Solve for $Q$ for each branch. (There will be a different value of $k'$ for each branch.)

$$Q = Av = A\sqrt{\frac{2Dg}{fL} h_f} = k'\sqrt{h_f} \qquad 3.114$$

*step 4*:

$$\begin{aligned} Q_{\text{total}} &= Q_1 + Q_2 + Q_3 \\ &= (k_1' + k_2' + k_3')\sqrt{h_f} \end{aligned} \qquad 3.115$$

Since $Q_{\text{total}}$, $k_1'$, $k_2'$, and $k_3'$ are known, it is possible to solve for the friction loss.

*step 5*: Check the values of $f$ and repeat as necessary.

*Example 3.23*

3 cubic feet per second of water enter the schedule 40 piping network shown below. What is the head loss between the connecting points $A$ and $B$?

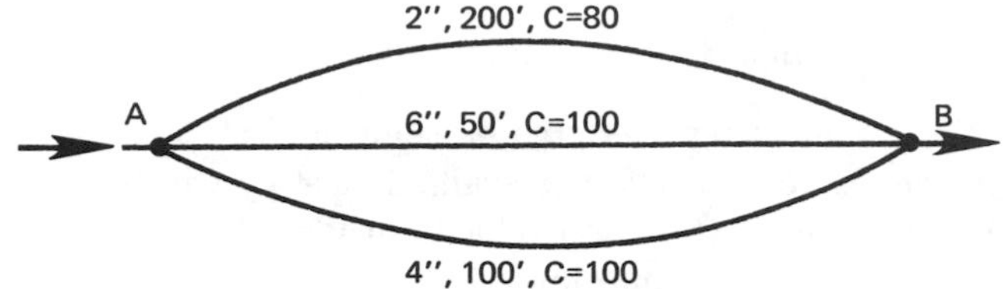


The pipe dimensions are determined from a table of schedule 40 pipe. (See appendix F.)

| | 2″ | 4″ | 6″ |
|---|---|---|---|
| flow area | 0.0233 | 0.0884 | 0.2006 |
| diameter | 0.1723 | 0.3355 | 0.5054 |

*step 1*: Since the Hazen-Williams loss coefficients are given for each branch, the Hazen-Williams friction loss equation will be used.

$$h_f = \frac{(3.022)(v)^{1.85} L}{(C)^{1.85}(D)^{1.165}}$$

*step 2*: The Hazen-Williams friction loss equation can be solved for the velocity term.

$$v = \frac{(0.550)(C)(D)^{0.63}}{L^{0.54}}(h_f)^{0.54}$$

The velocity in the 2″ pipe section can be written in terms of the constant friction loss term.

$$v_{2''} = \frac{(0.550)(80)(0.1723)^{0.63}}{(200)^{0.54}}(h_f)^{0.54} = 0.831(h_f)^{0.54}$$

Similarly,

$$v_{6''} = 4.327(h_f)^{0.54}$$
$$v_{4''} = 2.299(h_f)^{0.54}$$

*step 3*: Since the pipe areas and flow velocities (in terms of the friction loss) are known, the flow quantities can be calculated.

$$\begin{aligned} Q &= Av \\ Q_{2''} &= (0.0233)(0.831)(h_f)^{0.54} = 0.0194(h_f)^{0.54} \\ Q_{6''} &= (0.2006)(4.327)(h_f)^{0.54} = 0.8680(h_f)^{0.54} \\ Q_{4''} &= (0.0884)(2.299)(h_f)^{0.54} = 0.2032(h_f)^{0.54} \end{aligned}$$

*step 4*: The three flow quantities must total the quantity flowing through the system, 3 cfs. Since the friction loss is the same across all of the branches, it can be calculated directly.

$$Q_{\text{total}} = Q_{2''} + Q_{6''} + Q_{4''}$$
$$3 = (0.0194 + 0.8680 + 0.2032)(h_f)^{0.54}$$
$$h_f = \left(\frac{3}{1.0906}\right)^{\frac{1}{0.54}} = (2.751)^{1.85} = 6.50 \text{ ft}$$

## C. RESERVOIR BRANCHING SYSTEMS

The three-reservoir problem requires a trial and error solution method which is easily proceduralized. Although there are many possible choices for the unknown variable (e.g., pipe length, diameter, head, flow rate, etc.), there are three common types of problems.

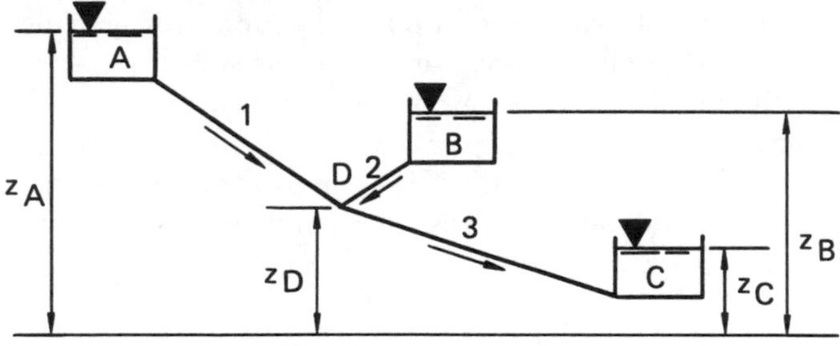


**Figure 3.18** 3-Reservoir System

*Case 1*: Given: all lengths, diameters, and elevations

Find: $Q_1$, $Q_2$, $Q_3$

Although an analytical solution method is possible, this problem is best solved by trial and error.

*step 1*: Assume $Q_1$ and use the Bernoulli equation to find the assumed pressure at point $D$. Ignore minor losses and velocity head. The friction term depends on the assumed value of $Q_1$.

$$v_1 = \frac{Q_1}{A_1} \qquad 3.116$$

$$z_A = z_D + (p_D/\gamma) + h_{f,1} \qquad 3.117$$

*step 2*: Once the assumed $p_D$ is known, use it to find $Q_2$.

$$z_B = z_D + (p_D/\gamma) + h_{f,2} \qquad 3.118$$

$$Q_2 = v_2 A_2 \qquad 3.119$$

If $z_D + (p_D/\gamma)$ is greater than $z_B$, flow will be into reservoir $B$. In that case, the friction term should be subtracted, not added.

*step 3*: Find $Q_3$.

$$z_C = z_D + (p_D/\gamma) - h_{f,3} \qquad 3.120$$

$$Q_3 = v_3 A_3 \qquad 3.121$$

*step 4*: Check that $Q_1 + Q_2 = Q_3$. If it does not, return to step 1. After two iterations, plot $Q_1 + Q_2 - Q_3$ versus $Q_1$. Estimate $Q_1$ by interpolation or extrapolation.

*Alternative Solution Method*: Assume a value of $p_D$ and solve for the flow quantities. Repeat with different values of $p_D$.

*Case 2*: Given: $Q_1$, all lengths, diameters, $z_A$, $z_B$, and $z_D$.

Find: $z_C$

*step 1*: $v_1 = Q_1/A_1$ 3.122

*step 2*: Solve for $p_D$ from

$$z_A = z_D + (p_D/\gamma) + h_{f,1} \qquad 3.123$$

*step 3*: Solve for $v_2$ from

$$z_B = z_D + (p_D/\gamma) + h_{f,2} \qquad 3.124$$

If $z_D + (p_D/\gamma)$ is greater than $z_B$, the flow will be into reservoir $B$. In that case, the friction $h_{f,2}$ should be subtracted, not added.

*step 4*: $Q_2 = A_2 v_2$ 3.125

*step 5*: $Q_3 = Q_1 \pm Q_2$ 3.126

*step 6*: $v_3 = Q_3/A_3$ 3.127

*step 7*: Calculate $h_{f,3}$ from $v_3$, $L_3$, and $D_3$.

*step 8*: Find $z_C$ from

$$z_C = z_D + (p_D/\gamma) - h_{f,3} \qquad 3.128$$

*Case 3*: Given: All lengths, elevations, $Q_1$, and diameters $D_1$ and $D_2$

Find: $D_3$

*step 1–5*: Repeat steps 1–5 from case 2.

*step 6*: Find $h_{f,3}$ from

$$z_C = z_D + (p_D/\gamma) - h_{f,3} \qquad 3.129$$

*step 7*: Find $D_3$ from $h_{f,3}$.

D. PIPE NETWORKS

Network flows in multi-loop systems can be determined with the Hardy Cross method when a manual solution is necessary. This is a systematic trial-and-error method which first assumes flows and then adds consecutive adjustments to the assumed flows. The Hardy Cross method is easy to apply. It is based on the following principles:

- The flows entering a junction must equal the flows leaving the junction.
- The algebraic sum of friction losses around any closed loop is zero.

If $Q_a$ is the assumed flow in a pipe, and $Q_t$ is the true flow, the difference is $\delta$, where

$$\delta = Q_t - Q_a \qquad 3.130$$

The true flow can be written in terms of the assumed flow and the correction:

$$Q_t = Q_a + \delta \qquad 3.131$$

The friction loss in the pipe has the form of $h_f = K'Q_t^n$, where $n = 2$ if the Darcy equation is used, and $n = 1.85$ if the Hazen-Williams equation is used. (Note that $k'$ used in equation 3.114 is equal to $1/\sqrt{K'}$.)

- For $Q$ in cfs, $L$ in feet, and $D$ in feet, the Hazen-Williams friction coefficient is

$$K' = \frac{(4.727)L}{D^{4.87}C^{1.85}} \qquad 3.132$$

- For $Q$ in gpm, $L$ in feet, and $d$ in inches, the Hazen-Williams friction coefficient is

$$K' = \frac{(10.44)L}{C^{1.85}d^{4.87}} \qquad 3.133$$

- For $Q$ in cfs, $L$ in feet, and $D$ in feet, the Darcy friction coefficient is

$$K' = \frac{(0.0252)fL}{D^5} \qquad 3.134$$

$f$ is usually assumed to be the same (such as 0.02) in all parts of the network.

- For $Q$ in gpm, $L$ in feet, and $D$ in feet, the Darcy friction coefficient is

$$K' = (1.251\,\text{EE–}7)\frac{fL}{D^5} \qquad 3.135$$

Combining and expanding the friction loss as a series,

$$h_f = K'(Q_a + \delta)^n \simeq K'Q_a^n + nK'\delta Q_a^{n-1} + \cdots \qquad 3.136$$

Subsequent higher order terms can be omitted because it is assumed that the correction is small.

Around a complete loop in a network, the sum of the friction drops is zero. Therefore,

$$\sum h_f = \sum K'Q_a^n + \delta \sum nK'Q_a^{n-1} = 0 \qquad 3.137$$

$\delta$ has been taken outside of the summation because all branches in the loop have the same correction. If $n$ is the same for all pipes, it can be taken out of the summation also.

This equation can be solved for $\delta$.

$$\delta = \frac{-\sum K'Q_a^n}{n\sum \left|K'Q_a^{n-1}\right|} = \frac{-\sum h_f}{n(\sum h_f/Q_a)} \qquad 3.138$$

To use this equation, it is necessary to first assume the flow directions as well as the flow rates. The numerator is the sum of head losses around the loop, taking signs into consideration. Because the denominator is a sum of the absolute values, $\delta$ must be applied in the same sense to each branch in the loop. If clockwise is assumed as the positive direction (an arbitrary decision), then $\delta$ is added to clockwise flows and subtracted from counterclockwise flows.

The application of the Hardy Cross method is as follows:

*step 1*: Choose between the Darcy and Hazen-Williams friction loss equations. The Darcy equation results in an easier expression to evaluate because the exponent $(n-1)$ is 1.

*step 2*: Choose a positive direction (e.g., clockwise).

*step 3*: Number all pipes in the network or identify all nodes.

*step 4*: Divide the network into independent loops such that each branch is included in at least one loop.

*step 5*: Calculate $K'$ for each pipe in the network.

*step 6*: Assume flow rates and directions. This may seem like a difficult step, but it is not. Most inaccurate first assumptions yield good results after several iterations.

*step 7*: Calculate $\delta$ for each independent loop.

*step 8*: Apply $\delta$ to each pipe in its loop using the previously mentioned sign convention.

*step 9*: Return to step 7.

*Example 3.24*

Use a Moody friction factor of $f = 0.02$ to calculate the flow in each pipe in the network shown. (Pipes are shown as straight lines for convenience only.)

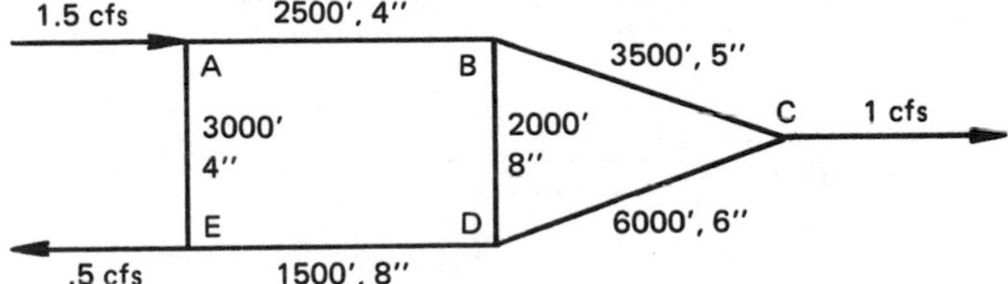


*step 1*: The Moody friction factor is given, so the Darcy friction loss equation will be used.

*step 2*: Choose clockwise as the positive direction.

*step 3*: Use the identification system shown on the network.

*step 4*: Work with loops $ABDE$ and $BCD$. Notice that loop $ABCDE$ is not independent if the other two loops are used.

*step 5*:

$$\text{pipe } AB : D = (4/12) = 0.3333$$

$$K' = \frac{(0.0252)(0.02)(L)}{D^5} = \frac{(0.0252)(0.02)(2500)}{(0.3333)^5} = 306.2$$

$$\text{pipe } BC : K' = 140.5$$
$$\text{pipe } DC : K' = 96.8$$
$$\text{pipe } BD : K' = 7.7$$
$$\text{pipe } ED : K' = 5.7$$
$$\text{pipe } AE : K' = 367.4$$

*step 6*: Assume the directions and flows shown in the figure.

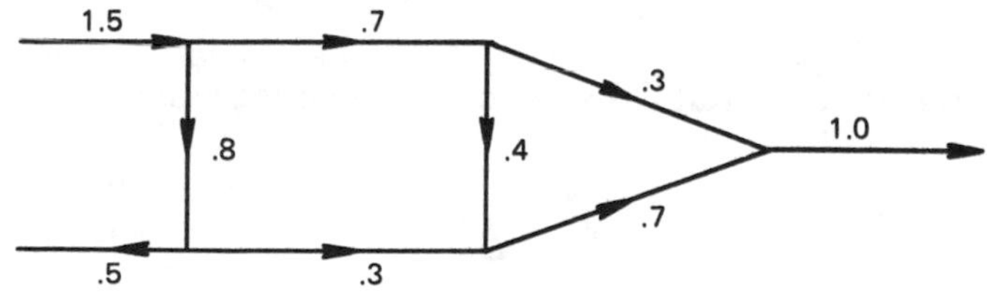


*step 7*:

$\delta_{ABDE}$

$$= \frac{-[(306.2)(0.7)^2+(7.7)(0.4)^2-(5.7)(0.3)^2-(367.4)(0.8)^2]}{2[(306.2)(0.7)+(7.7)(0.4)+(5.7)(0.3)+(367.4)(0.8)]} = +0.08$$

$\delta_{BCD}$

$$= \frac{-[(140.5)(0.3)^2-(96.8)(0.7)^2-(7.7)(0.4)^2]}{2[(140.5)(0.3)+(96.8)(0.7)+(7.7)(0.4)]} = +0.16$$

*step 8*: The corrected flows are:

$$\text{pipe } AB : \quad 0.7 + (0.08) = 0.78$$
$$\text{pipe } BC : \quad 0.3 + (0.16) = 0.46$$
$$\text{pipe } DC : \quad 0.7 - (0.16) = 0.54$$
$$\text{pipe } BD : \quad 0.4 + (0.08) - (0.16) = 0.32$$
$$\text{pipe } ED : \quad 0.3 - (0.08) = 0.22$$
$$\text{pipe } AE : \quad 0.8 - (0.08) = 0.72$$

*step 9*: The procedure is repeated using the corrected flows.

FLUIDS

## 8 FLOW MEASURING DEVICES

The total energy in a fluid flow is the sum of pressure head, velocity head, and gravitational head.

$$H = \frac{p}{\gamma} + \frac{v^2}{2g} + z \qquad 3.139$$

Change in gravitational head within a flow-measuring instrument is negligible. Therefore, if two of the three remaining variables ($H$, $p$, or $v$) are known, the third can be found from subtraction. The flow measuring devices discussed in this section are capable of measuring total head ($H$) or pressure head ($p$).

### A. VELOCITY MEASUREMENT

Velocity of a fluid stream is determined by measuring the difference between the static and the stagnation pressures, then solving for the velocity head.

A ***piezometer tap*** can be used to measure the pressure head directly in feet of fluid.

$$h_p = \frac{p}{\gamma} \qquad 3.140$$

For liquids with pressures higher than the capability of the direct reading tap, a manometer can be used with a piezometer tap or with a ***static probe*** as shown in figure 3.20.

For either configuration of figure 3.20, the static pressure is

$$p = \gamma_m \Delta h_m - \gamma y \qquad 3.141$$

$$h_p = \frac{\gamma_m \Delta h_m}{\gamma} - y \qquad 3.142$$

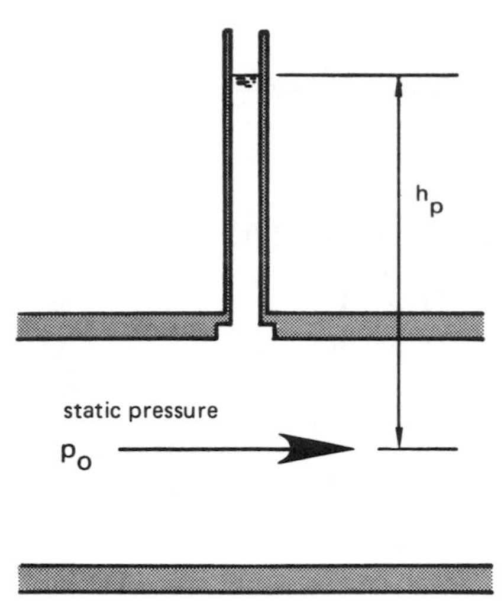


**Figure 3.19** Piezometer Tap

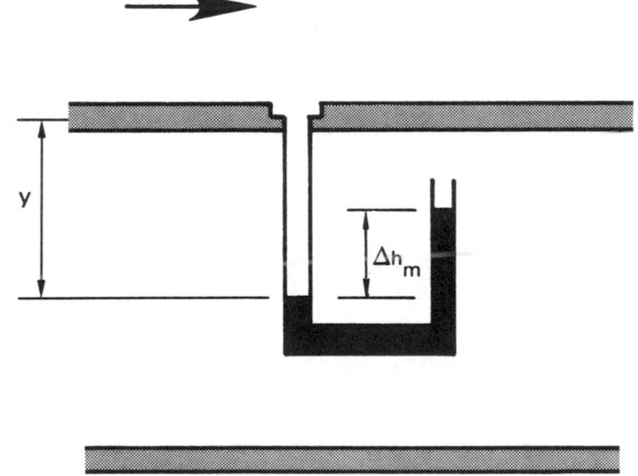


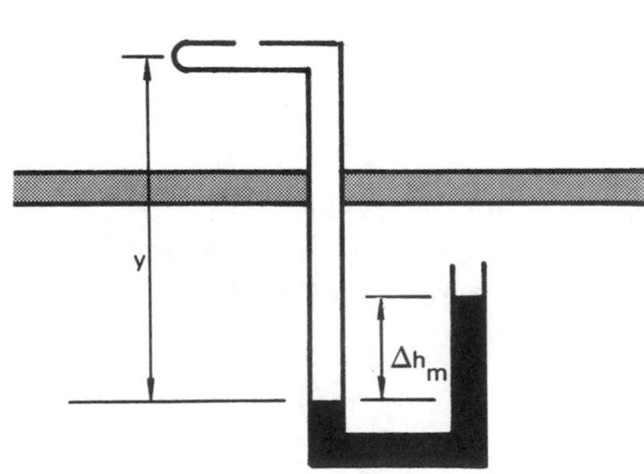


**Figure 3.20** Use of Manometers to Measure Static Pressure

*Stagnation pressure*, also known as *total pressure* or *impact pressure*, can be measured directly in feet of fluid by using a pitot tube as shown in figure 3.12.

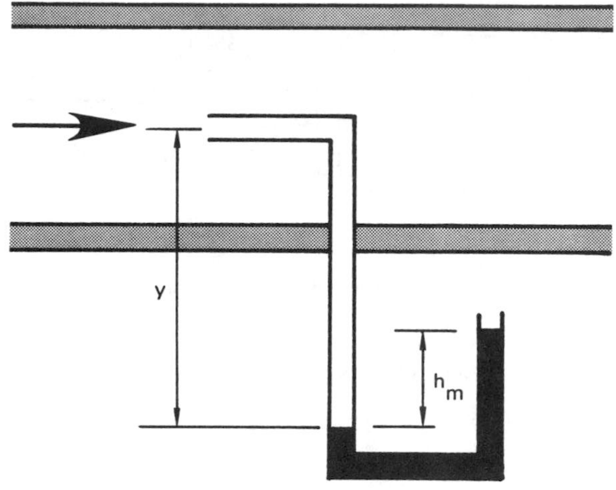


**Figure 3.21** Use of Manometer to Measure Total Pressure

Using the results of the measurements above, the velocity head can be calculated from equation 3.145.

For high-pressure fluids, a manometer must be used to measure stagnation pressure:

$$p_s = \gamma h_s = \gamma_m \Delta h_m - \gamma y \qquad 3.143$$

$$h_s = \frac{p}{\rho} + \frac{v^2}{2g} = \frac{\gamma_m \Delta h_m}{\gamma} - y \qquad 3.144$$

$$v = \sqrt{2g(h_s - h_p)} = \sqrt{\frac{2g(p_s - p)}{\gamma}} \qquad 3.145$$

If the piezometer tap of figure 3.19 and the pitot tube of figure 3.12 are placed at the same point as in figure 3.22, the velocity head in feet of fluid can be read directly.

$$\frac{v^2}{2g} = \Delta h \qquad 3.146$$

$$v = \sqrt{2g\Delta h} \qquad 3.147$$

The instrumentation arrangement of figures 3.20 and 3.21 can be combined into a single instrument to provide a measurement of velocity head as shown in figure 3.23.

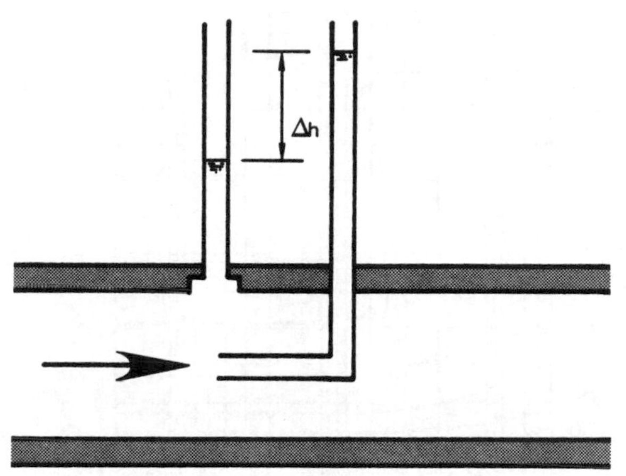


**Figure 3.22** Comparative Velocity Head Measurement

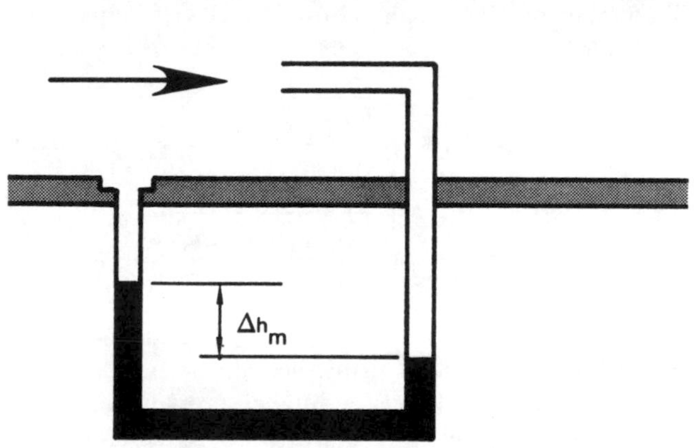


**Figure 3.23** Velocity Head Measurement

$$\frac{v^2}{2g} = \frac{\Delta h_m(\gamma_m - \gamma)}{\gamma} \qquad 3.148$$

$$v = \sqrt{\frac{2g(\gamma_m - \gamma)\Delta h_m}{\gamma}} \qquad 3.149$$

*Example 3.25*

50°F water is flowing through a pipe. A pitot-static gage registers a 3″ deflection of mercury. What is the velocity within the pipe? (The density of mercury is 848.6 pcf.)

Using equation 3.149,

$$v = \sqrt{\frac{2\,(32.2)(848.6 - 62.4)\,\left(\frac{3}{12}\right)}{62.4}} = 14.24 \text{ fps}$$

B. FLOW MEASUREMENT

Using the techniques described in the preceding section, the flow rate in a line can be determined by measuring the pressure drop across a restriction. Once the geometry of the restriction is known, the Bernoulli equation, along with empirically determined correction coefficients, can be applied to obtain an expression directly relating flow rate with pressure drop.

If potential head is neglected, Bernoulli's equation becomes

$$\frac{p_1}{\gamma} + \frac{v_1^2}{2g} = \frac{p_2}{\gamma} + \frac{v_2^2}{2g} \qquad 3.150$$

But, $v_1$ and $v_2$ are related. From equation 3.68,

$$v_1 = v_2\left(\frac{A_2}{A_1}\right) \qquad 3.151$$

Combining equations 3.150 and 3.151 yields the standard flow measurement equation.

$$v_2 = \frac{\sqrt{2g\left(\frac{p_1 - p_2}{\gamma}\right)}}{\sqrt{1 - \left(\frac{A_2}{A_1}\right)^2}} \qquad 3.152$$

The flow quantity can be found from

$$\dot{V} = v_2 A_2 \qquad 3.153$$

The reciprocal of the denominator of equation 3.152 is known as the *velocity of approach factor*, $F_{va}$. The *beta ratio* can be incorporated into the formula for $F_{va}$.

$$F_{va} = \frac{1}{\sqrt{1 - \beta^4}} \qquad 3.154$$

$$\beta = \frac{D_2}{D_1} \qquad 3.155$$

The simplest fluid flow measuring device is the *orifice plate.* This consists of a thin plate or diaphragm with a central hole through which the fluid flows.

Equations 3.156 and 3.157 are the governing orifice plate equations for liquid flow.

$$\dot{V} = F_{va}C_dA_o\sqrt{\frac{2g(p_1 - p_2)}{\gamma}} \qquad 3.156$$

$$= F_{va}C_dA_o\sqrt{\frac{2g(\gamma_m - \gamma)\Delta h_m}{\gamma}} \qquad 3.157$$

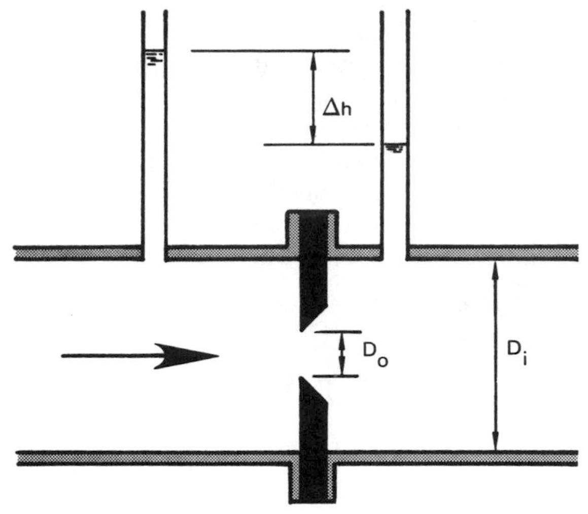


**Figure 3.24** Comparative Reading Orifice Plate

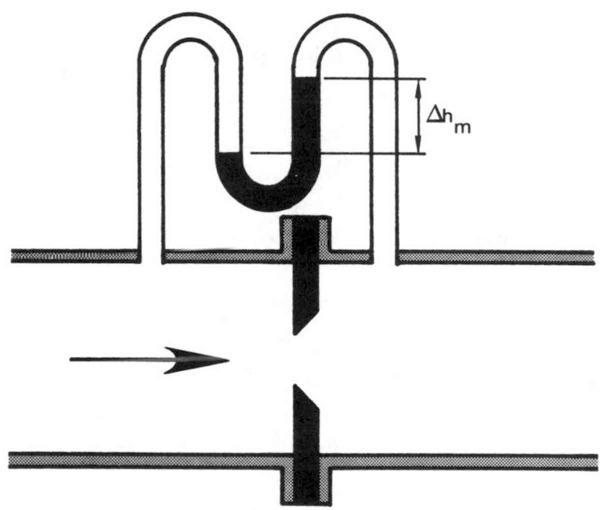


**Figure 3.25** Direct Reading Orifice Plate

The definition of the velocity of approach factor is modified slightly for the orifice plate.

$$F_{va} = \frac{1}{\sqrt{1 - \left(\frac{C_c A_o}{A_i}\right)^2}} \quad 3.158$$

The *flow coefficient* depends on the velocity of approach factor and the discharge coefficient. It also can be obtained from figure 3.26.

$$C_f = F_{va} C_d \quad 3.159$$

$$C_d = C_v C_c \quad 3.160$$

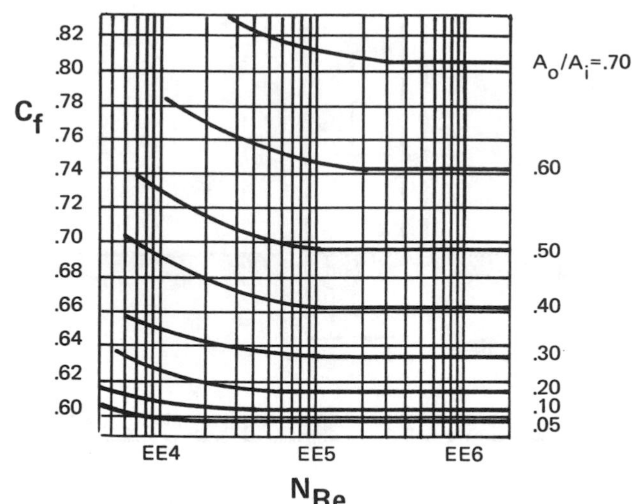


**Figure 3.26** Flow Coefficients for I.S.A. Orifice Plates

The flow coefficients can be used to rewrite equations 3.156 and 3.157.

$$\dot{V} = C_f A_o \sqrt{\frac{2g(p_1 - p_2)}{\gamma}} \quad 3.161$$

$$= C_f A_o \sqrt{\frac{2g(\gamma_m - \gamma)\Delta h_m}{\gamma}} \quad 3.162$$

Operating on the same principles as the orifice plate, the *venturi meter* induces a smaller pressure drop. It is, however, mechanically more complex, as shown by figure 3.27.

The governing equations are similar to those for orifice plates. $C_c$ usually is 1.0 for venturi meters.

$$v_2 = F_{va} \sqrt{\frac{2g(p_1 - p_2)}{\gamma}} \quad 3.163$$

$$\dot{V} = F_{va} C_d A_2 \sqrt{\frac{2g(p_1 - p_2)}{\gamma}}$$

$$= C_f A_2 \sqrt{\frac{2g(p_1 - p_2)}{\gamma}} \quad 3.164$$

$$F_{va} = \frac{1}{\sqrt{1 - \left(\frac{A_2}{A_1}\right)^2}} \quad 3.165$$

$$C_d = C_v C_c \quad 3.166$$

$$C_f = F_{va} C_d \quad 3.167$$

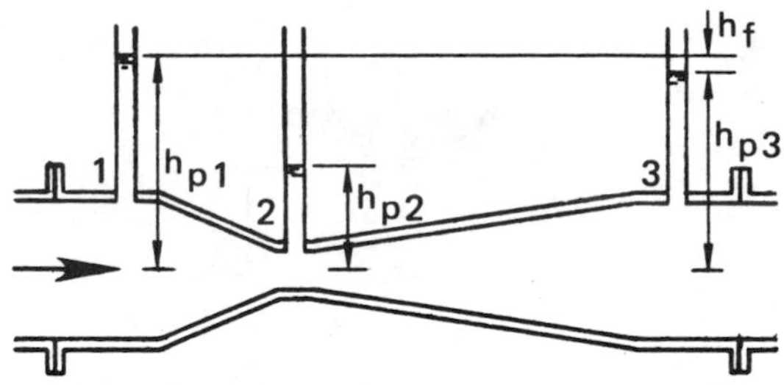


**Figure 3.27** Venturi Meter with Wall Taps

**Table 3.11**
$C_d$ for Venturi Meters

| $2 < (A_1/A_2) < 3$ | |
|---|---|
| $C_d$ | $N_{Re}$ |
| 0.94 | 6000 |
| 0.95 | 10,000 |
| 0.96 | 20,000 |
| 0.97 | 50,000 |
| 0.98 | 200,000 |
| 0.99 | 2000,000 |

*Example 3.26*

150°F water is flowing in an 8″ schedule 40 steel pipe at 2.23 cfs. If a 7 inch sharp edged orifice plate is bolted across the line, what manometer deflection in inches of mercury would be expected? (Mercury has a density of 848.6 pcf.) Neglect the correction for water in the manometer tubes.

| | 7″ orifice | 8″ schedule 40 |
|---|---|---|
| flow area | 0.267 ft$^2$ | 0.3474 ft$^2$ |
| diameter | 0.583 ft | 0.6651 ft |

From table 3.10 for the orifice: $C_c = 0.63$, $C_d = 0.62$.

Using equation 3.158,

$$F_{va} = \left[1 - \frac{C_c A_o}{A_i}\right]^{-\frac{1}{2}}$$

$$= \left[1 - \left(\frac{(0.63)(0.267)}{0.3474}\right)^2\right]^{-\frac{1}{2}} = 1.14$$

From equation 3.157,

$$\Delta h_m = \left(\frac{\dot{V}}{F_{va} C_d A_o}\right)^2 \frac{\gamma}{2g(\gamma_m - \gamma)}$$

$$= \left(\frac{2.23}{(1.14)(0.62)(0.267)}\right)^2 \times \frac{61.2}{(2)(32.2)(848.6 - 61.2)} \times 12\frac{\text{in}}{\text{ft}}$$

$$= 2.02''$$

## 9 THE IMPULSE/MOMENTUM PRINCIPLE

A force is required to cause a direction or velocity change in a flowing fluid. Conventions necessary to determine such a force are:

1. $\Delta v = v_2 - v_1$
2. A positive $\Delta v$ indicates an increase in velocity. A negative $\Delta v$ indicates a decrease in velocity.
3. $F$ and $x$ are positive to the right. $F$ and $y$ are positive upward.
4. $F$ is the force on the fluid. The force on the walls or support has the same magnitude but opposite direction.
5. The fluid is assumed to flow horizontally from left to right and is assumed to possess no $y$-component of velocity.

The *momentum* possessed by a moving fluid is defined as the product of mass (in slugs) and velocity (in ft/sec).

The $g_c$ term in equation 3.168 is needed to convert pounds-mass into slugs.

$$\text{momentum} = \frac{mv}{g_c} \tag{3.168}$$

*Impulse* is defined as the product of a force and the length of time the force is applied.

$$\text{impulse} = F\Delta t \tag{3.169}$$

The *impulse-momentum principle* states that the impulse applied to a moving body is equal to the change in momentum. This is expressed by equation 3.170.

$$F\Delta t = \frac{m\Delta v}{g_c} \tag{3.170}$$

Solving for $F$ and combining $m$ and $\Delta t$ yields equation 3.171.

$$F = \frac{m\Delta v}{g_c \Delta t} = \frac{\dot{m}\Delta v}{g_c} \tag{3.171}$$

Since $F$ is a vector, it can be broken into its components

$$F_x = \frac{\dot{m}\Delta v_x}{g_c} \tag{3.172}$$

$$F_y = \frac{\dot{m}\Delta v_y}{g_c} \tag{3.173}$$

If the fluid flow is directed through an angle $\phi$,

$$\Delta v_x = v(\cos\phi - 1) \tag{3.174}$$

$$\Delta v_y = v \sin\phi \tag{3.175}$$

There are several fluid applications of the impulse-momentum principle.

FLUIDS

A. JET PROPULSION

$$\dot{m}_2 = \dot{m}_1 + \dot{m}_{\text{fuel}} = \dot{V}_1\rho_1 + \dot{V}_{\text{fuel}}\rho_{\text{fuel}} \qquad 3.176$$

$$F_x = \frac{\dot{V}_2\rho_2 v_{2x} - \dot{V}_1\rho_1 v_{1x}}{g_c} \qquad 3.177$$

$$F_y = \frac{\dot{V}_2\rho_2 v_{2y} - \dot{V}_1\rho_1 v_{1y}}{g_c} \qquad 3.178$$

**Figure 3.28** Jet Propulsion

B. OPEN JET ON VERTICAL FLAT PLATE

$$\Delta v_y = 0 \qquad 3.179$$

$$\Delta v_x = -v \qquad 3.180$$

$$F_x = \frac{-\dot{m}v}{g_c} = \frac{-\dot{V}\rho v}{g_c} \qquad 3.181$$

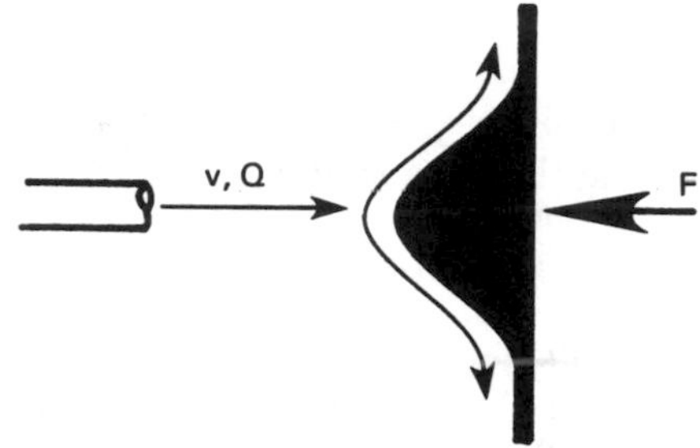


**Figure 3.29** Open Jet on Vertical Plate

C. OPEN JET ON HORIZONTAL FLAT PLATE

As the jet travels upwards, its velocity decreases since gravity is working against it. By the time the liquid has reached the plate, the velocity has become

$$v_y = \sqrt{v_0^2 - 2gh} \qquad 3.182$$

$$\Delta v_x = 0 \qquad 3.183$$

$$\Delta v_y = -\sqrt{v_0^2 - 2gh} \qquad 3.184$$

$$F = \left(\frac{-\dot{m}}{g_c}\right)\sqrt{v_o^2 - 2gh}$$

$$= \left(\frac{\dot{V}\rho}{g_c}\right)\sqrt{v_o^2 - 2gh} \qquad 3.185$$

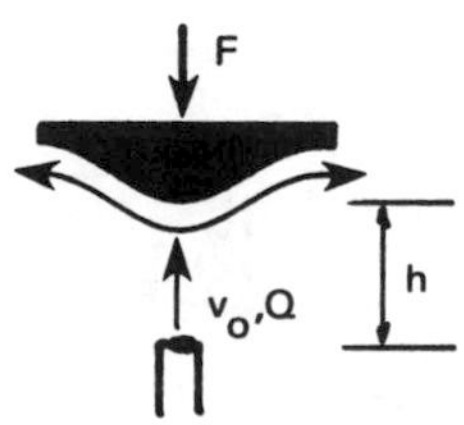


**Figure 3.30** Open Jet on Horizontal Plate

D. OPEN JET ON SINGLE STATIONARY BLADE

$v_2$ may not be the same as $v_1$ if friction is present. If no information is given, assume that $v_2 = v_1$.

$$\Delta v_x = v_2\cos\phi - v_1 \qquad 3.186$$

$$\Delta v_y = v_2\sin\phi \qquad 3.187$$

$$F_x = \left(\frac{\dot{m}}{g_c}\right)(v_2\cos\phi - v_1)$$

$$= \left(\frac{\dot{V}\rho}{g_c}\right)(v_2\cos\phi - v_1) \qquad 3.188$$

$$F_y = \left(\frac{\dot{m}}{g_c}\right)(v_2\sin\phi) = \left(\frac{\dot{V}\rho}{g_c}\right)(v_2\sin\phi) \qquad 3.189$$

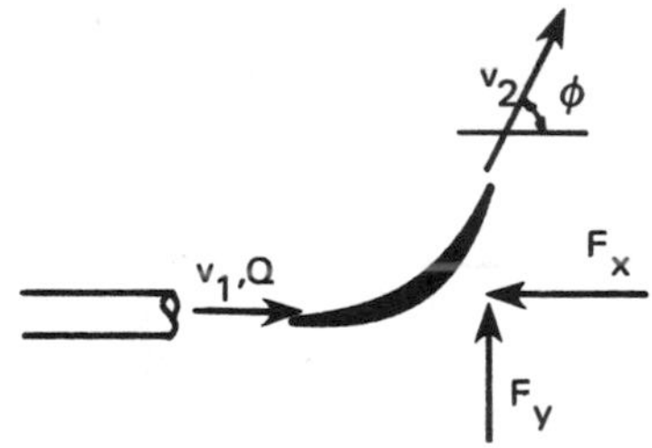


**Figure 3.31** Open Jet on Stationary Blade

E. OPEN JET ON SINGLE MOVING BLADE

$v_b$ is the blade velocity. For simplicity, friction is ignored. The discharge overtaking the moving blade is $\dot{V}'$.

$$\dot{V}' = \left(\frac{v - v_b}{v}\right)\dot{V} \qquad 3.190$$

$$\Delta v_x = (v - v_b)(\cos\phi - 1) \qquad 3.191$$

$$\Delta v_y = (v - v_b)(\sin\phi) \qquad 3.192$$

$$F_x = \frac{\dot{m}'\Delta v_x}{g_c} = \left(\frac{\dot{V}'\rho}{g_c}\right)\Delta v_x \qquad 3.193$$

$$F_y = \frac{\dot{m}'\Delta v_y}{g_c} = \left(\frac{\dot{V}'\rho}{g_c}\right)\Delta v_y \qquad 3.194$$

The power transferred to the blade is given by equation 3.195. Power is maximized when $\phi = 180°$ and $v_b = \frac{1}{2}v$.

$$P = F_x v_b \qquad 3.195$$

Equations 3.193 and 3.194 can be used with a *multiple-bladed* wheel by using the full $\dot{V}$ instead of $\dot{V}'$.

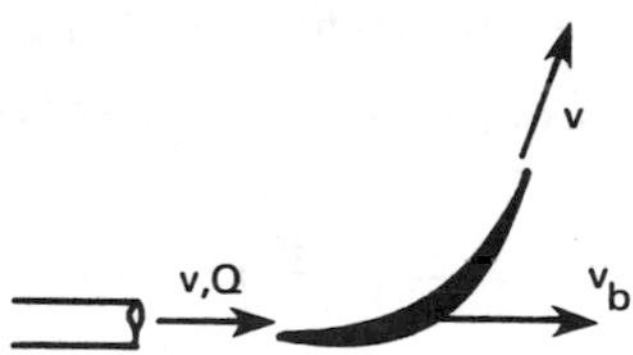


**Figure 3.32** Open Jet on Moving Blade

## F. CONFINED STREAMS IN PIPE BENDS

Since the fluid is confined, the forces caused by static pressure must be included along with the force from momentum changes. Using gage pressures and neglecting the fluid weight,

$$F_x = p_2 A_2 \cos\phi - p_1 A_1 + \left(\frac{\dot{V}\rho}{g_c}\right)(v_2 \cos\phi - v_1) \qquad 3.196$$

$$F_y = \left[p_2 A_2 + \frac{\dot{V}\rho v_2}{g}\right] \sin\phi \qquad 3.197$$

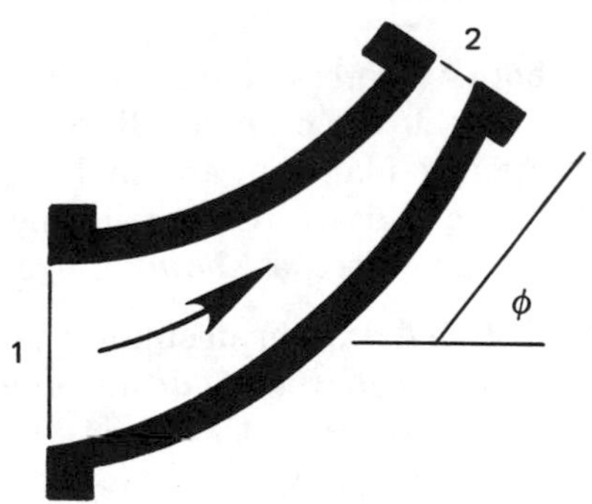


**Figure 3.33** A Pipe Bend

## G. WATER HAMMER

Water hammer is an increase in pressure in a pipe caused by a sudden velocity decrease. The sudden velocity decrease usually will be caused by a valve's closing.

Assuming the pipe material is inelastic, the time required for the water hammer shock wave to travel from a valve to the end of a pipe and back is given by

$$t = \frac{2L}{c} \qquad 3.198$$

The fluid pressure increase resulting from this shock wave is

$$\Delta p = \frac{\rho c \Delta v}{g_c} = \frac{\gamma c \Delta v}{g} \qquad 3.199$$

Water hammer or *surge* can be handled by one or more of the following methods:

- Moderating the valve closure time by either a manual or automatic valve controller
- A surge tank with a free water surface
- An air chamber on the discharge line
- A surge surpressor
- A surge relief valve

*Example 3.27*

60°F water at 40 psig flowing at 8 ft/sec enters a 12″×8″ reducing elbow as shown and is turned 30°. (a) What is the resultant force on the water? (b) What other forces should be considered in the design of supports for the fitting?

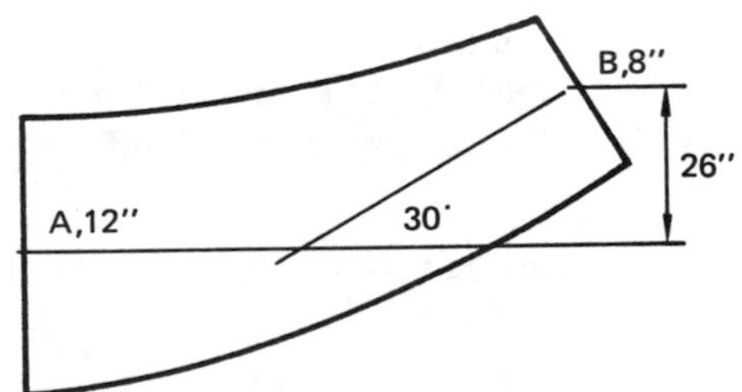


(a) The total head at point A is

$$\frac{(40)(144)}{(62.4)} + \frac{(8)^2}{(2)(32.2)} + 0 = 93.3 \text{ ft}$$

At point B, the velocity is

$$(8)\left(\frac{12}{8}\right)^2 = 18 \text{ ft/sec}$$

The pressure at B can be found from Bernoulli's equation.

$$93.3 = \frac{p_B(144)}{62.4} + \frac{(18)^2}{(2)(32.2)} + \frac{26}{12}$$

So, $p_B = 37.3$ psig

$$\dot{V} = vA = (8)\left(\frac{1}{4}\right)\pi\left(\frac{12}{12}\right)^2 = 6.28 \text{ cfs}$$

From equation 3.196,

$$\begin{aligned} F_x &= +(37.3)(144)\left(\frac{1}{4}\right)\pi\left(\frac{8}{12}\right)^2 \cos 30° \\ &\quad - (40)(144)\left(\frac{1}{4}\right)\pi\left(\frac{12}{12}\right)^2 \\ &\quad + \left(\frac{(6.28)(62.4)}{32.2}\right)[(18)(\cos 30°) - 8] \\ &= -2808 \text{ lbf} \end{aligned}$$

From equation 3.197,

$$\begin{aligned} F_y &= \left[(37.3)(144)\left(\frac{1}{4}\right)\pi\left(\frac{8}{12}\right)^2 \right. \\ &\quad \left. + \left(\frac{(6.28)(62.4)(18)}{32.2}\right)\right] \sin 30° \\ &= 1047 \text{ lbf} \end{aligned}$$

The resultant force on the water is

$$R = \sqrt{(-2808)^2 + (1047)^2} = 2997 \text{ lbf}$$

(b) The support also should be designed to carry the weight of the water in the pipe and the bend, and the weight of the pipe and the bend itself.

*Example 3.28*

40°F water is flowing at 10 ft/sec through a 4″ schedule 40 welded steel pipe. A valve suddenly is closed. Assuming rigid pipe, what increase in fluid pressure will occur?

Assume that the closing valve completely stops the flow. Therefore, $\Delta v$ is 10 ft/sec.

At 40°F, $E$ = 294 EE3 psi, and $\rho$ = 62.43 lbm/ft$^3$. From equation 3.20, the speed of sound in the water is

$$c = \sqrt{\frac{(294 \text{ EE3})(144)(32.2)}{62.43}} = 4673 \text{ fps}$$

From equation 3.199,

$$\Delta p = \frac{(62.43)(4673)(10)}{32.2} = 90{,}600 \text{ psf}$$

## H. OPEN JET ON INCLINED PLATE

An open jet will be diverted both up and down a stationary, inclined flat plate. The velocity in each diverted flow will be $v$, the same as in the approaching jet. The fractions $f_1$ and $f_2$ of the jet which are diverted up and down can be found from equations 3.200 and 3.201.

$$f_1 = \frac{1 + \cos\phi}{2} \quad 3.200$$

$$f_2 = \frac{1 - \cos\phi}{2} \quad 3.201$$

$$f_1 - f_2 = \cos\phi \quad 3.202$$

$$f_1 + f_2 = 1 \quad 3.203$$

As the flow along the plate is assumed to be frictionless, there will be no force component parallel to the plate.

The force perpendicular to the plate is

$$F = \left(\frac{\dot{V}\rho}{g_c}\right) v \sin\phi \quad 3.204$$

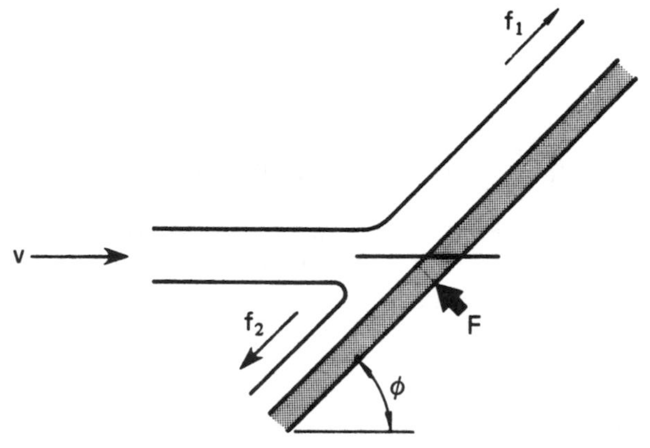


**Figure 3.34** Open Jet on Inclined Plate

## 10 LIFT AND DRAG

Lift and drag both are forces exerted on an object as it passes through a fluid. For example, lift on the wing of an airplane forces the plane upward, and drag tries to slow it down. Lift and drag are the components of the resultant force on an object, as shown in figure 3.35.

The amounts of lift and drag on an object depend on the shape of the object. *Coefficients of lift* and *drag* are used to measure the effectiveness of the object in producing lift and drag. Lift and drag may be calculated from equations 3.205 and 3.206.

$$L = \frac{C_L A \gamma v^2}{2g} \quad 3.205$$

$$D = \frac{C_D A \gamma v^2}{2g} \quad 3.206$$

$A$, in equation 3.205, is the object's area projected onto a plane parallel to the direction of motion. In equation 3.206, $A$ is the area projected onto a plane normal to the direction of motion.

FLUIDS

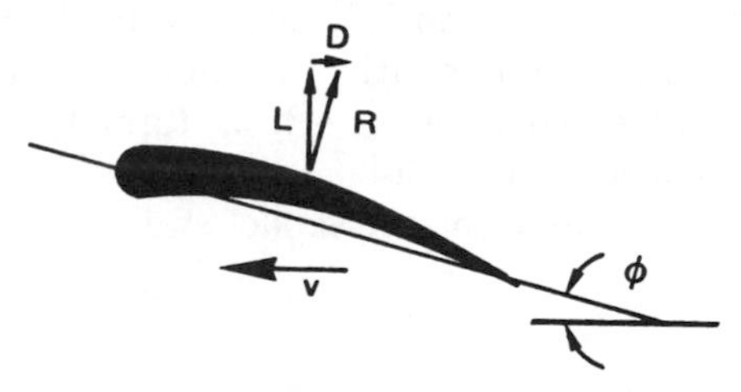


**Figure 3.35** Lift and Drag on an Airfoil

Values of $C_L$ for various airfoil sections have been correlated with $N_{Re}$. No simple relationship can be given for airfoils in general. However, the theoretical relationship for a thin flat plate inclined at an angle $\phi$ is

$$C_L = 2\pi \sin\phi \qquad 3.207$$

The drag coefficient for a sphere moving with $N_{Re}$ less than 0.4 is predicted by *Stokes' law*, equation 3.208. The same equation may be used for circular disks. Values of $C_D$ for other shapes are given in table 3.12.

$$C_D = \frac{24}{N_{Re}} \qquad 3.208$$



**Table 3.12**
Approximate Drag Coefficients

(Do not interpolate between $N_{Re}$ = EE5 and $N_{Re}$ = EE6)

Reynolds Number, $N_{Re}$

| body shape, and (characteristic dimension) | EE0 | EE1 | EE2 | EE3 | EE4 | EE5 | EE6 | fully turbulent EE6-EE7 |
|---|---|---|---|---|---|---|---|---|
| sphere (diameter) | (a) | 4 | 1.0 | 0.45 | 0.40 | 0.55 | 0.25 | 0.2 |
| flat disk (diameter) | (a) | 4 | 1.5 | 1.9(b) | 1.1 | 1.1 | 1.1 | 1.1 |
| flat plate, normal to flow, (short side) | | | | | | | | |
| length/breadth = 1 | | | | (b) | 1.16 | 1.16 | 1.16 | 1.16 |
| 4 | | | | (b) | 1.17 | 1.17 | 1.17 | 1.17 |
| 8 | | | | (b) | 1.23 | 1.23 | 1.23 | 1.23 |
| 12.5 | | | | (b) | 1.34 | 1.34 | 1.34 | 1.34 |
| 20 | | | | (b) | 1.50 | 1.50 | 1.50 | 1.50 |
| 25 | | | | (b) | 1.57 | 1.57 | 1.57 | 1.57 |
| 50 | | | | (b) | 1.76 | 1.76 | 1.76 | 1.76 |
| ∞ | | | | (b) | 2.0 | 2.0 | 2.0 | 2.0 |
| circular cylinder, axis normal to flow (diameter) | | | | | | | | |
| length/diameter = 1 | | | | 0.6 | 0.6 | 0.6 | | 0.35 |
| 5 | | | | 0.7 | 0.9 | 0.9 | | |
| 20 | | | | 0.9 | 0.9 | 0.9 | | |
| ∞ | 10 | 2.5 | 1.3 | 0.9 | 1.1 | 1.4 | 0.37 | 0.33 |
| circular cylinder, axis parallel to flow (diameter) | | | | | | | | |
| length/diameter = 1 | | | | 0.91 | 0.91 | 0.91 | | |
| 2 | | | | 0.85 | 0.85 | 0.85 | | |
| 4 | | | | 0.87 | 0.87 | 0.87 | | |
| 7 | | | | 0.99 | 0.99 | 0.99 | | |

Note a: Use Stokes' law, equation 3.208 Note b: Becomes fully turbulent at $N_{Re}$ = 3 EE3

## 11 SIMILARITY

*Similarity* between a model (subscript $m$) and a full-sized object (subscript $t$) implies that the model can be used to predict the performance of the full-sized object. Such a model is said to be *mechanically similar* to the full-sized object.

Complete mechanical similarity requires geometric and dynamic similarity.[7] *Geometric similarity* means that the model is true to scale in length, area, and volume. *Dynamic similarity* means that the ratios of all types of forces are equal. These forces result from inertia, gravity, viscosity, elasticity (fluid compressibility), surface tension, and pressure.

The *model scale* or *length ratio* is

$$L_r = \frac{\text{size of model}}{\text{full size}} \qquad 3.209$$

The area and volume ratios are based on the length ratio.

$$\frac{A_m}{A_t} = (L_r)^2 \qquad 3.210$$

$$\frac{V_m}{V_t} = (L_r)^3 \qquad 3.211$$

The number of possible ratios of forces is large. Fortunately, some force ratios may be ignored because the forces are negligible or self-canceling. Three important cases where the analysis can be simplified are dominant viscous and inertial forces, dominant inertial and gravitational forces, and dominant surface tension.

### A. VISCOUS AND INERTIAL FORCES DOMINATE

Consider the testing of a completely submerged object such as a submarine. Surface tension effects are negligible. The fluid is assumed incompressible for low velocities. Gravity does not change the path of the fluid particles significantly during the time the submarine is near.

Only inertial, viscous, and pressure forces are significant. Being the only significant ones, these three forces are in equilibrium. Since they are in equilibrium, knowing any two will define the third completely. Since it is not an independent force, pressure is omitted from the similarity analysis.

The ratio of the inertial forces to the viscous forces is the Reynolds number. Setting the model and full-size Reynolds numbers equal will ensure similarity. That is,

$$(N_{Re})_m = (N_{Re})_t \qquad 3.212$$

This approach works for problems involving fans, pumps, turbines, drainage through holes in tanks, closed-pipe flow with no free surfaces (in the turbulent region with the same relative roughness), and for completely submerged objects such as torpedoes, airfoils, and submarines. It is assumed that the drag coefficients are the same for the model and for the full-size object.

*Example 3.29*

A 1/30th size model is tested in a wind tunnel at 120 mph. The wind tunnel conditions are 50 psia and 100°F. What would be the equivalent speed of a prototype traveling at 14.0 psia, 40°F still air?

Start by setting the Reynolds number of the model and its prototype.

$$\frac{v_m L_m}{\nu_m} = \frac{v_p L_p}{\nu_p}$$

$$v_p = v_m\left(\frac{L_m}{L_p}\right)\left(\frac{\nu_p}{\nu_m}\right) = (120)\left(\frac{1}{30}\right)\left(\frac{\nu_p}{\nu_m}\right)$$

Air viscosity terms must be evaluated at the respective temperatures and pressures. As tables of viscosities are not readily available, the viscosities must be calculated.

Absolute viscosity essentially is independent of pressure. In Appendix B, the absolute viscosity of air is

$$\mu_p \,@\, 40^\circ = 3.62 \text{ EE}-7$$

$$\mu_m \,@\, 100^\circ F = 3.96 \text{ EE}-7$$

The density of air at the two conditions is

$$\rho_p = \frac{(14.0)(144)}{(53.3)(460+40)} = 0.0756 \text{ lbm/ft}^3$$

$$\rho_m = \frac{(50)(144)}{(53.3)(460+100)} = 0.2412$$

The kinematic viscosity can be calculated from the absolute viscosity. (The $g_c$ terms are omitted as they ultimately cancel out.)

$$\nu = \frac{\mu g_c}{\rho}$$

$$\nu_p = \frac{3.62 \text{ EE}-7}{0.0756} = 4.79 \text{ EE}-6$$

$$\nu_m = \frac{3.96 \text{ EE}-7}{0.2412} = 1.64 \text{ EE}-6$$

Then, the prototype velocity is

$$v_p = (120)\left(\frac{1}{30}\right)\left(\frac{4.79}{1.64}\right) = 11.7 \text{ mph}$$

[7] Complete mechanical similarity also requires kinematic and thermal similarity, which are not discussed in this book.

FLUIDS

B. INERTIAL AND GRAVITATIONAL FORCES DOMINATE

Elasticity and surface tension can be neglected in the analysis of large surface vessels. This leaves pressure, inertia, viscosity, and gravity. Pressure, again, is omitted as being dependent.

There are only two possible combinations of the remaining three forces. The ratio of inertial and viscous forces is recognized again as the Reynolds number. The ratio of the inertial forces to the gravitational forces is known as the *Froude number.*

$$N_{Fr} = \frac{v^2}{Lg} \qquad 3.213$$

Similarity is ensured when equations 3.214 and 3.215 are satisfied.

$$(N_{Re})_m = (N_{Re})_t \qquad 3.214$$

$$(N_{Fr})_m = (N_{Fr})_t \qquad 3.215$$

As an alternative, equations 3.213 and 3.62 can be solved simultaneously. This results in the following requirement for complete similarity.

$$\frac{\nu_m}{\nu_t} = \left(\frac{L_m}{L_t}\right)^{3/2} = (L_r)^{3/2} \qquad 3.216$$

Sometimes it is not possible to satisfy equation 3.215 or 3.216. This occurs when a model viscosity that is not available is called for. If only equation 3.214 is satisfied, the model is said to be *partially similar.*

This analysis is valid for surface ships, seaplane hulls, and open channels with varying surface levels such as weirs and spillways.

C. SURFACE TENSION DOMINATES

Problems involving waves, droplets, bubbles, and air entrainment can be solved by setting the *Weber numbers* equal.

$$N_W = \frac{v^2 L\rho}{T} \qquad 3.217$$

$$(N_W)_m = (N_W)_t \qquad 3.218$$

## 12 EFFECTS OF NON-STANDARD GRAVITY

Most of the equations in part 4 are based on Bernoulli's equation. This equation can be modified to allow for non-standard gravities. Assuming an incompressible fluid, Bernoulli's equation becomes

$$\frac{p_1}{\rho} + \frac{v_1^2}{2g_c} + \frac{gz_1}{g_c} = \frac{p_2}{\rho} + \frac{v_2^2}{2g_c} + \frac{gz_2}{g_c} \qquad 3.219$$

## 13 CHOICE OF PIPING MATERIALS

Steel and copper are commonly used in pressure piping systems. Each material is available in several configurations. For example, steel can be uncoated or galvanized. Copper tubing can be hard or soft. Table 3.13 can be used to select an appropriate pipe material.

Steel pipe is specified by its nominal size and schedule. In the past, steel pipe was designated as *standard* (S), *extra-strong* (X), and *double extra-strong* (XX). However, these designations have been replaced by a numerical rating. For example, schedule 40 now corresponds to a standard wall steel pipe in most cases.

The approximate schedule required can be found from equation 3.220.

$$\text{schedule} \approx \frac{(1000)(p)}{SE} \qquad 3.220$$

$p$ is the operating pressure in psig; $S$ is the allowable material stress in psi; $E$ is the joint efficiency. A value of 6500 psi can be used for the product $SE$ with low carbon steel in butt-welded lines and temperatures less than 650°F.

When copper is used as a pipe material, there is a potential for confusion, as there are two different sets of dimensions for copper pipe. Copper pipe in the K, L, and M categories is available in both annealed rolls ("tubing") and hardened straight lengths. Dimensions for such copper pipe, commonly referred to as *copper water tubing*, are given in Appendix G.

Type DWV copper drainage tube also is available. It is recommended for sanitary drainage installations above ground. The tube walls are thinner than type M, making it lighter and less expensive. It is strictly for non-pressure applications.

Copper and brass can also be formed into pipe with the dimensions given in Appendix H. Since the term "copper pipe" is ambiguous, the application must be used to determine the correct dimensions. Type L tubing in straight lengths is used principally in domestic and commercial plumbing because of its cost and the availability of soldered fittings. However, brass piping may be used with high-temperature water and corrosive fluids.

## 14 FLOW OF COMPRESSIBLE FLUIDS AND STEAM

Under certain conditions, Compressible fluids can be handled as incompressible flow. Specifically, a compressible gas, such as air or steam, can be treated as

FLUIDS

incompressible if the pressure drop along the pipe run is not excessive.

If the pressure drop, based on the entrance pressure, is less than 10%, the fluid properties can be evaluated at any known point long the pipe run. If the pressure drop is greater than 10% but less than 40%, use of the midpoint properties will yield reasonably close friction loss calculations. If the pressure drop is greater than 40%, exact compressible gas dynamics equations should be used.

Since the pressure drop is being used to determine if the pressure drop is excessive, several iterations may be necessary to determine the pressures by trial and error.

**Table 3.13**
Recommended Pipe Materials for Various Services

| SERVICE | | PIPE |
|---|---|---|
| REFRIGERANTS 12, 22, 500 and 502 | Suction Line | Hard copper tubing, Type L*<br>Steel pipe, standard wall Lap welded or seamless |
| | Liquid Line | Hard copper tubing, Type L*<br>Steel pipe, standard wall Lap welded or seamless |
| | Hot Gas Line | Hard copper tubing, Type L*<br>Steel pipe, standard wall Welded or seamless |
| CHILLED WATER | | Plain or Galvanized steel pipe†<br>Hard copper tubing† |
| CONDENSER OR MAKE-UP WATER | | Galvanized steel pipe†<br>Hard copper tubing† |
| DRAIN OR CONDENSATE LINES | | Galvanized steel pipe†<br>Hard copper tubing† |
| STEAM OR CONDENSATE | | Steel pipe†<br>Hard copper tubing† |
| HOT WATER | | Steel pipe<br>Hard copper tubing† |

* Except for sizes 1/4″ and 3/8″ OD where wall thicknesses of 0.30 and 0.32 in. are required. Soft copper refrigeration tubing may be used for sizes 1 3/8″ OD and smaller. Mechanical joints must not be used with soft copper tubing in sizes larger than 7/8″ OD.

† Normally, standard wall steel pipe or Type M hard copper tubing is satisfactory for air conditioning applications. However, the piping material selected should be checked for the design temperature-pressure ratings.

# Appendix A: Properties of Water at Atmospheric Pressure

| temp. °F | density lbm/ft³ | absolute viscosity lbf-sec/ft² | kinematic viscosity ft²/sec | surface tension lbf/ft | vapor pressure head[a] ft | bulk modulus lbf/in² |
|---|---|---|---|---|---|---|
| 32 | 62.42 | 3.746 EE−5 | 1.931 EE−5 | 0.518 EE−2 | 0.20 | 293 EE3 |
| 40 | 62.43 | 3.229 EE−5 | 1.664 EE−5 | 0.514 EE−2 | 0.28 | 294 EE3 |
| 50 | 62.41 | 2.735 EE−5 | 1.410 EE−5 | 0.509 EE−2 | 0.41 | 305 EE3 |
| 60 | 62.37 | 2.359 EE−5 | 1.217 EE−5 | 0.504 EE−2 | 0.59 | 311 EE3 |
| 70 | 62.30 | 2.050 EE−5 | 1.059 EE−5 | 0.500 EE−2 | 0.84 | 320 EE3 |
| 80 | 62.22 | 1.799 EE−5 | 0.930 EE−5 | 0.492 EE−2 | 1.17 | 322 EE3 |
| 90 | 62.11 | 1.595 EE−5 | 0.826 EE−5 | 0.486 EE−2 | 1.62 | 323 EE3 |
| 100 | 62.00 | 1.424 EE−5 | 0.739 EE−5 | 0.480 EE−2 | 2.21 | 327 EE3 |
| 110 | 61.86 | 1.284 EE−5 | 0.667 EE−5 | 0.473 EE−2 | 2.97 | 331 EE3 |
| 120 | 61.71 | 1.168 EE−5 | 0.609 EE−5 | 0.465 EE−2 | 3.96 | 333 EE3 |
| 130 | 61.55 | 1.069 EE−5 | 0.558 EE−5 | 0.460 EE−2 | 5.21 | 334 EE3 |
| 140 | 61.38 | 0.981 EE−5 | 0.514 EE−5 | 0.454 EE−2 | 6.78 | 330 EE3 |
| 150 | 61.20 | 0.905 EE−5 | 0.476 EE−5 | 0.447 EE−2 | 8.76 | 328 EE3 |
| 160 | 61.00 | 0.838 EE−5 | 0.442 EE−5 | 0.441 EE−2 | 11.21 | 326 EE3 |
| 170 | 60.80 | 0.780 EE−5 | 0.413 EE−5 | 0.433 EE−2 | 14.20 | 322 EE3 |
| 180 | 60.58 | 0.726 EE−5 | 0.385 EE−5 | 0.426 EE−2 | 17.87 | 313 EE3 |
| 190 | 60.36 | 0.678 EE−5 | 0.362 EE−5 | 0.419 EE−2 | 22.29 | 313 EE3 |
| 200 | 60.12 | 0.637 EE−5 | 0.341 EE−5 | 0.412 EE−2 | 27.61 | 308 EE3 |
| 212 | 59.83 | 0.593 EE−5 | 0.319 EE−5 | 0.404 EE−2 | 35.38 | 300 EE3 |

[a] based on actual densities, not standard "cold, clear water"

FLUIDS

# Appendix B: Properties of Air at Atmospheric Pressure

| temp. °F | density lbm/ft³ | kinematic viscosity ft²/sec | absolute viscosity lbf-sec/ft² |
|---|---|---|---|
| 0 | 0.0862 | 12.6 EE−5 | 3.28 EE−7 |
| 20 | 0.0827 | 13.6 EE−5 | 3.50 EE−7 |
| 40 | 0.0794 | 14.6 EE−5 | 3.62 EE−7 |
| 60 | 0.0763 | 15.8 EE−5 | 3.74 EE−7 |
| 68 | 0.0752 | 16.0 EE−5 | 3.75 EE−7 |
| 80 | 0.0735 | 16.9 EE−5 | 3.85 EE−7 |
| 100 | 0.0709 | 18.0 EE−5 | 3.96 EE−7 |
| 120 | 0.0684 | 18.9 EE−5 | 4.07 EE−7 |
| 250 | 0.0559 | 27.3 EE−5 | 4.74 EE−7 |

# Appendix C: Viscosity of Water

| | absolute viscosity | kinematic viscosity | | |
|---|---|---|---|---|
| temperature (°F) | centipoise | centistokes | SSU | $ft^2/sec$ |
| 32 | 1.79 | 1.79 | 33.0 | 0.00001931 |
| 50 | 1.31 | 1.31 | 31.6 | 0.00001410 |
| 60 | 1.12 | 1.12 | 31.2 | 0.00001217 |
| 70 | 0.98 | 0.98 | 30.9 | 0.00001059 |
| 80 | 0.86 | 0.86 | 30.6 | 0.00000930 |
| 85 | 0.81 | 0.81 | 30.4 | 0.00000869 |
| 100 | 0.68 | 0.69 | 30.2 | 0.00000739 |
| 120 | 0.56 | 0.57 | 30.0 | 0.00000609 |
| 140 | 0.47 | 0.48 | 29.7 | 0.00000514 |
| 160 | 0.40 | 0.41 | 29.6 | 0.00000442 |
| 180 | 0.35 | 0.36 | 29.5 | 0.00000385 |
| 212 | 0.28 | 0.29 | 29.3 | 0.00000319 |



# Appendix D: Important Fluid Conversions

| multiply | by | to obtain |
|---|---|---|
| cubic feet | 7.4805 | gallons |
| cfs | 448.83 | gpm |
| cfs | 0.64632 | MGD |
| gallons | 0.1337 | cubic feet |
| gpm | 0.002228 | cfs |
| inches of mercury | 0.491 | psi |
| inches of mercury | 70.7 | psf |
| inches of mercury | 13.60 | inches of water |
| inches of water | 5.199 | psf |
| inches of water | 0.0361 | psi |
| inches of water | 0.0735 | inches of mercury |
| psi | 144 | psf |
| psi | 2.308 | feet of water |
| psi | 27.7 | inches of water |
| psi | 2.037 | inches of mercury |
| psf | 0.006944 | psi |

# Appendix E: Area, Wetted Perimeter and Hydraulic Radius of Partially Filled Circular Pipes

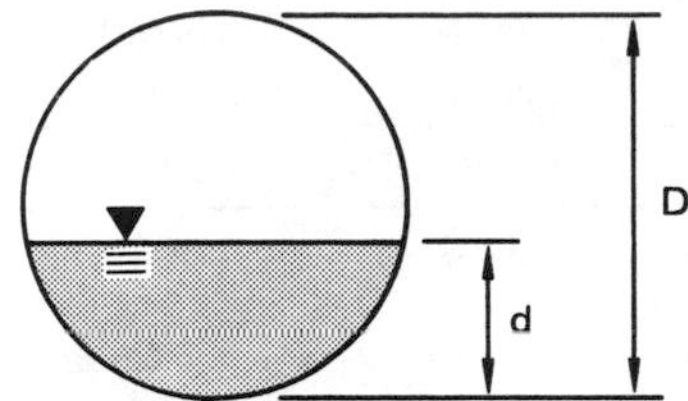


| $\frac{d}{D}$ | $\frac{\text{area}}{D^2}$ | $\frac{\text{wet. per.}}{D}$ | $\frac{\text{hyd. rad.}}{D}$ | $\frac{d}{D}$ | $\frac{\text{area}}{D^2}$ | $\frac{\text{wet. per.}}{D}$ | $\frac{\text{hyd. rad.}}{D}$ |
|---|---|---|---|---|---|---|---|
| 0.01 | 0.0013 | 0.2003 | 0.0066 | 0.51 | 0.4027 | 1.5908 | 0.2531 |
| 0.02 | 0.0037 | 0.2838 | 0.0132 | 0.52 | 0.4127 | 1.6108 | 0.2561 |
| 0.03 | 0.0069 | 0.3482 | 0.0197 | 0.53 | 0.4227 | 1.6308 | 0.2591 |
| 0.04 | 0.0105 | 0.4027 | 0.0262 | 0.54 | 0.4327 | 1.6509 | 0.2620 |
| 0.05 | 0.0147 | 0.4510 | 0.0326 | 0.55 | 0.4426 | 1.6710 | 0.2649 |
| 0.06 | 0.0192 | 0.4949 | 0.0389 | 0.56 | 0.4526 | 1.6911 | 0.2676 |
| 0.07 | 0.0242 | 0.5355 | 0.0451 | 0.57 | 0.4625 | 1.7113 | 0.2703 |
| 0.08 | 0.0294 | 0.5735 | 0.0513 | 0.58 | 0.4723 | 1.7315 | 0.2728 |
| 0.09 | 0.0350 | 0.6094 | 0.0574 | 0.59 | 0.4822 | 1.7518 | 0.2753 |
| 0.10 | 0.0409 | 0.6435 | 0.0635 | 0.60 | 0.4920 | 1.7722 | 0.2776 |
| 0.11 | 0.0470 | 0.6761 | 0.0695 | 0.61 | 0.5018 | 1.7926 | 0.2797 |
| 0.12 | 0.0534 | 0.7075 | 0.0754 | 0.62 | 0.5115 | 1.8132 | 0.2818 |
| 0.13 | 0.0600 | 0.7377 | 0.0813 | 0.63 | 0.5212 | 1.8338 | 0.2839 |
| 0.14 | 0.0688 | 0.7670 | 0.0871 | 0.64 | 0.5308 | 1.8546 | 0.2860 |
| 0.15 | 0.0739 | 0.7954 | 0.0929 | 0.65 | 0.5404 | 1.8755 | 0.2881 |
| 0.16 | 0.0811 | 0.8230 | 0.0986 | 0.66 | 0.5499 | 1.8965 | 0.2899 |
| 0.17 | 0.0885 | 0.8500 | 0.1042 | 0.67 | 0.5594 | 1.9177 | 0.2917 |
| 0.18 | 0.0961 | 0.8763 | 0.1097 | 0.68 | 0.5687 | 1.9391 | 0.2935 |
| 0.19 | 0.1039 | 0.9020 | 0.1152 | 0.69 | 0.5780 | 1.9606 | 0.2950 |
| 0.20 | 0.1118 | 0.9273 | 0.1206 | 0.70 | 0.5872 | 1.9823 | 0.2962 |
| 0.21 | 0.1199 | 0.9521 | 0.1259 | 0.71 | 0.5964 | 2.0042 | 0.2973 |
| 0.22 | 0.1281 | 0.9764 | 0.1312 | 0.72 | 0.6054 | 2.0264 | 0.2984 |
| 0.23 | 0.1365 | 1.0003 | 0.1364 | 0.73 | 0.6143 | 2.0488 | 0.2995 |
| 0.24 | 0.1449 | 1.0239 | 0.1416 | 0.74 | 0.6231 | 2.0714 | 0.3006 |
| 0.25 | 0.1535 | 1.0472 | 0.1466 | 0.75 | 0.6318 | 2.0944 | 0.3017 |
| 0.26 | 0.1623 | 1.0701 | 0.1516 | 0.76 | 0.6404 | 2.1176 | 0.3025 |
| 0.27 | 0.1711 | 1.0928 | 0.1566 | 0.77 | 0.6489 | 2.1412 | 0.3032 |
| 0.28 | 0.1800 | 1.1152 | 0.1614 | 0.78 | 0.6573 | 2.1652 | 0.3037 |
| 0.29 | 0.1890 | 1.1373 | 0.1662 | 0.79 | 0.6655 | 2.1895 | 0.3040 |
| 0.30 | 0.1982 | 1.1593 | 0.1709 | 0.80 | 0.6736 | 2.2143 | 0.3042 |
| 0.31 | 0.2074 | 1.1810 | 0.1755 | 0.81 | 0.6815 | 2.2395 | 0.3044 |
| 0.32 | 0.2167 | 1.2025 | 0.1801 | 0.82 | 0.6893 | 2.2653 | 0.3043 |
| 0.33 | 0.2260 | 1.2239 | 0.1848 | 0.83 | 0.6969 | 2.2916 | 0.3041 |
| 0.34 | 0.2355 | 1.2451 | 0.1891 | 0.84 | 0.7043 | 2.3186 | 0.3038 |
| 0.35 | 0.2450 | 1.2661 | 0.1935 | 0.85 | 0.7115 | 2.3462 | 0.3033 |
| 0.36 | 0.2546 | 1.2870 | 0.1978 | 0.86 | 0.7186 | 2.3746 | 0.3026 |
| 0.37 | 0.2642 | 1.3078 | 0.2020 | 0.87 | 0.7254 | 2.4038 | 0.3017 |
| 0.38 | 0.2739 | 1.3284 | 0.2061 | 0.88 | 0.7320 | 2.4341 | 0.3008 |
| 0.39 | 0.2836 | 1.3490 | 0.2102 | 0.89 | 0.7384 | 2.4655 | 0.2996 |
| 0.40 | 0.2934 | 1.3694 | 0.2142 | 0.90 | 0.7445 | 2.4981 | 0.2980 |
| 0.41 | 0.3032 | 1.3898 | 0.2181 | 0.91 | 0.7504 | 2.5322 | 0.2963 |
| 0.42 | 0.3130 | 1.4101 | 0.2220 | 0.92 | 0.7560 | 2.5681 | 0.2944 |
| 0.43 | 0.3229 | 1.4303 | 0.2257 | 0.93 | 0.7612 | 2.6061 | 0.2922 |
| 0.44 | 0.3328 | 1.4505 | 0.2294 | 0.94 | 0.7662 | 2.6467 | 0.2896 |
| 0.45 | 0.3428 | 1.4706 | 0.2331 | 0.95 | 0.7707 | 2.6906 | 0.2864 |
| 0.46 | 0.3527 | 1.4907 | 0.2366 | 0.96 | 0.7749 | 2.7389 | 0.2830 |
| 0.47 | 0.3627 | 1.5108 | 0.2400 | 0.97 | 0.7785 | 2.7934 | 0.2787 |
| 0.48 | 0.3727 | 1.5308 | 0.2434 | 0.98 | 0.7816 | 2.8578 | 0.2735 |
| 0.49 | 0.3827 | 1.5508 | 0.2467 | 0.99 | 0.7841 | 2.9412 | 0.2665 |
| 0.50 | 0.3927 | 1.5708 | 0.2500 | 1.00 | 0.7854 | 3.1416 | 0.2500 |

# Appendix F: Dimensions of Welded and Seamless Steel Pipe

| nominal diameter inches | schedule | outside diameter inches | wall thickness inches | internal diameter inches | internal area sq inches | internal diameter feet | internal area sq feet |
|---|---|---|---|---|---|---|---|
| $\frac{1}{8}$ | 40 (S) | 0.405 | 0.068 | 0.269 | 0.0568 | 0.0224 | 0.00039 |
| | 80 (X) | | 0.095 | 0.215 | 0.0363 | 0.0179 | 0.00025 |
| $\frac{1}{4}$ | 40 (S) | 0.540 | 0.088 | 0.364 | 0.1041 | 0.0303 | 0.00072 |
| | 80 (X) | | 0.119 | 0.302 | 0.0716 | 0.0252 | 0.00050 |
| $\frac{3}{8}$ | 40 (S) | 0.675 | 0.091 | 0.493 | 0.1909 | 0.0411 | 0.00133 |
| | 80 (X) | | 0.126 | 0.423 | 0.1405 | 0.0353 | 0.00098 |
| $\frac{1}{2}$ | 40 (S) | 0.840 | 0.109 | 0.622 | 0.3039 | 0.0518 | 0.00211 |
| | 80 (X) | | 0.147 | 0.546 | 0.2341 | 0.0455 | 0.00163 |
| | 160 | | 0.187 | 0.466 | 0.1706 | 0.0388 | 0.00118 |
| | (XX) | | 0.294 | 0.252 | 0.0499 | 0.0210 | 0.00035 |
| $\frac{3}{4}$ | 40 (S) | 1.050 | 0.113 | 0.824 | 0.5333 | 0.0687 | 0.00370 |
| | 80 (X) | | 0.154 | 0.742 | 0.4324 | 0.0618 | 0.00300 |
| | 160 | | 0.219 | 0.612 | 0.2942 | 0.0510 | 0.00204 |
| | (XX) | | 0.308 | 0.434 | 0.1479 | 0.0362 | 0.00103 |
| 1 | 40 (S) | 1.315 | 0.133 | 1.049 | 0.8643 | 0.0874 | 0.00600 |
| | 80 (X) | | 0.179 | 0.957 | 0.7193 | 0.0798 | 0.00500 |
| | 160 | | 0.250 | 0.815 | 0.5217 | 0.0679 | 0.00362 |
| | (XX) | | 0.358 | 0.599 | 0.2818 | 0.0499 | 0.00196 |
| $1\frac{1}{4}$ | 40 (S) | 1.660 | 0.140 | 1.380 | 1.496 | 0.1150 | 0.01039 |
| | 80 (X) | | 0.191 | 1.278 | 1.283 | 0.1065 | 0.00890 |
| | 160 | | 0.250 | 1.160 | 1.057 | 0.0967 | 0.00734 |
| | (XX) | | 0.382 | 0.896 | 0.6305 | 0.0747 | 0.00438 |
| $1\frac{1}{2}$ | 40 (S) | 1.900 | 0.145 | 1.610 | 2.036 | 0.1342 | 0.01414 |
| | 80 (X) | | 0.200 | 1.500 | 1.767 | 0.1250 | 0.01227 |
| | 160 | | 0.281 | 1.338 | 1.406 | 0.1115 | 0.00976 |
| | (XX) | | 0.400 | 1.100 | 0.9503 | 0.0917 | 0.00660 |
| 2 | 40 (S) | 2.375 | 0.154 | 2.067 | 3.356 | 0.1723 | 0.02330 |
| | 80 (X) | | 0.218 | 1.939 | 2.953 | 0.1616 | 0.02051 |
| | 160 | | 0.344 | 1.687 | 2.235 | 0.1406 | 0.01552 |
| | (XX) | | 0.436 | 1.503 | 1.774 | 0.1253 | 0.01232 |
| $2\frac{1}{2}$ | 40 (S) | 2.875 | 0.203 | 2.469 | 4.788 | 0.2058 | 0.03325 |
| | 80 (X) | | 0.276 | 2.323 | 4.238 | 0.1936 | 0.02943 |
| | 160 | | 0.375 | 2.125 | 3.547 | 0.1771 | 0.02463 |
| | (XX) | | 0.552 | 1.771 | 2.464 | 0.1476 | 0.01711 |
| 3 | 40 (S) | 3.500 | 0.216 | 3.068 | 7.393 | 0.2557 | 0.05134 |
| | 80 (X) | | 0.300 | 2.900 | 6.605 | 0.2417 | 0.04587 |
| | 160 | | 0.438 | 2.624 | 5.408 | 0.2187 | 0.03755 |
| | (XX) | | 0.600 | 2.300 | 4.155 | 0.1917 | 0.02885 |
| $3\frac{1}{2}$ | 40 (S) | 4.000 | 0.226 | 3.548 | 9.887 | 0.2957 | 0.06866 |
| | 80 (X) | | 0.318 | 3.364 | 8.888 | 0.2803 | 0.06172 |

| nominal diameter inches | schedule | outside diameter inches | wall thickness inches | internal diameter inches | internal area sq inches | internal diameter feet | internal area sq feet |
|---|---|---|---|---|---|---|---|
| | 40 (S) | 4.500 | 0.237 | 4.026 | 12.73 | 0.3355 | 0.08841 |
| 4 | 80 (X) | | 0.337 | 3.826 | 11.50 | 0.3188 | 0.07984 |
| | 120 | | 0.438 | 3.624 | 10.32 | 0.3020 | 0.07163 |
| | 160 | | 0.531 | 3.438 | 9.283 | 0.2865 | 0.06447 |
| | (XX) | | 0.674 | 3.152 | 7.803 | 0.2627 | 0.05419 |
| 5 | 40 (S) | 5.563 | 0.258 | 5.047 | 20.01 | 0.4206 | 0.1389 |
| | 80 (X) | | 0.375 | 4.813 | 18.19 | 0.4011 | 0.1263 |
| | 120 | | 0.500 | 4.563 | 16.35 | 0.3803 | 0.1136 |
| | 160 | | 0.625 | 4.313 | 14.61 | 0.3594 | 0.1015 |
| | (XX) | | 0.750 | 4.063 | 12.97 | 0.3386 | 0.09004 |
| 6 | 40 (S) | 6.625 | 0.280 | 6.065 | 28.89 | 0.5054 | 0.2006 |
| | 80 (X) | | 0.432 | 5.761 | 26.07 | 0.4801 | 0.1810 |
| | 120 | | 0.562 | 5.501 | 23.77 | 0.4584 | 0.1650 |
| | 160 | | 0.719 | 5.187 | 21.13 | 0.4323 | 0.1467 |
| | (XX) | | 0.864 | 4.897 | 18.83 | 0.4081 | 0.1308 |
| 8 | 20 | 8.625 | 0.250 | 8.125 | 51.85 | 0.6771 | 0.3601 |
| | 30 | | 0.277 | 8.071 | 51.16 | 0.6726 | 0.3553 |
| | 40 (S) | | 0.322 | 7.981 | 50.03 | 0.6651 | 0.3474 |
| | 60 | | 0.406 | 7.813 | 47.94 | 0.6511 | 0.3329 |
| | 80 (X) | | 0.500 | 7.625 | 45.66 | 0.6354 | 0.3171 |
| | 100 | | 0.594 | 7.437 | 43.44 | 0.6198 | 0.3017 |
| | 120 | | 0.719 | 7.187 | 40.57 | 0.5989 | 0.2817 |
| | 140 | | 0.812 | 7.001 | 38.50 | 0.5834 | 0.2673 |
| | (XX) | | 0.875 | 6.875 | 37.12 | 0.5729 | 0.2578 |
| | 160 | | 0.906 | 6.813 | 36.46 | 0.5678 | 0.2532 |
| 10 | 20 | 10.75 | 0.250 | 10.250 | 82.52 | 0.85417 | 0.5730 |
| | 30 | | 0.307 | 10.136 | 80.69 | 0.84467 | 0.5604 |
| | 40 (S) | | 0.365 | 10.020 | 78.85 | 0.83500 | 0.5476 |
| | 60 (X) | | 0.500 | 9.750 | 74.66 | 0.8125 | 0.5185 |
| | 80 | | 0.594 | 9.562 | 71.81 | 0.7968 | 0.4987 |
| | 100 | | 0.719 | 9.312 | 68.11 | 0.7760 | 0.4730 |
| | 120 | | 0.844 | 9.062 | 64.50 | 0.7552 | 0.4479 |
| | 140 (XX) | | 1.000 | 8.750 | 60.13 | 0.7292 | 0.4176 |
| | 160 | | 1.125 | 8.500 | 56.75 | 0.7083 | 0.3941 |
| 12 | 20 | 12.75 | 0.250 | 12.250 | 117.86 | 1.0208 | 0.8185 |
| | 30 | | 0.330 | 12.090 | 114.80 | 1.0075 | 0.7972 |
| | (S) | | 0.375 | 12.000 | 113.10 | 1.0000 | 0.7854 |
| | 40 | | 0.406 | 11.938 | 111.93 | 0.99483 | 0.7773 |
| | (X) | | 0.500 | 11.750 | 108.43 | 0.97917 | 0.7530 |
| | 60 | | 0.562 | 11.626 | 106.16 | 0.96883 | 0.7372 |
| | 80 | | 0.688 | 11.374 | 101.61 | 0.94783 | 0.7056 |
| | 100 | | 0.844 | 11.062 | 96.11 | 0.92183 | 0.6674 |
| | 120 (XX) | | 1.000 | 10.750 | 90.76 | 0.89583 | 0.6303 |
| | 140 | | 1.125 | 10.500 | 86.59 | 0.87500 | 0.6013 |
| | 160 | | 1.312 | 10.126 | 80.53 | 0.84383 | 0.5592 |

| nominal diameter inches | schedule | outside diameter inches | wall thickness inches | internal diameter inches | internal area sq inches | internal diameter feet | internal area sq feet |
|---|---|---|---|---|---|---|---|
| 14 OD | 10 | 14.00 | 0.250 | 13.500 | 143.14 | 1.1250 | 0.9940 |
| | 20 | | 0.312 | 13.376 | 140.52 | 1.1147 | 0.9758 |
| | 30 (S) | | 0.375 | 13.250 | 137.89 | 1.1042 | 0.9575 |
| | 40 | | 0.438 | 13.124 | 135.28 | 1.0937 | 0.9394 |
| | (X) | | 0.500 | 13.000 | 132.67 | 1.0833 | 0.9213 |
| | 60 | | 0.594 | 12.812 | 128.92 | 1.0677 | 0.8953 |
| | 80 | | 0.750 | 12.500 | 122.72 | 1.0417 | 0.8522 |
| | 100 | | 0.938 | 12.124 | 115.45 | 1.0104 | 0.8017 |
| | 120 | | 1.094 | 11.812 | 109.58 | 0.98433 | 0.7610 |
| | 140 | | 1.250 | 11.500 | 103.87 | 0.95833 | 0.7213 |
| | 160 | | 1.406 | 11.188 | 98.31 | 0.93233 | 0.6827 |
| 16 OD | 10 | 16.00 | 0.250 | 15.500 | 188.69 | 1.2917 | 1.3104 |
| | 20 | | 0.312 | 15.376 | 185.69 | 1.2813 | 1.2895 |
| | 30 (S) | | 0.375 | 15.250 | 182.65 | 1.2708 | 1.2684 |
| | 40 (X) | | 0.500 | 15.000 | 176.72 | 1.2500 | 1.2272 |
| | 60 | | 0.656 | 14.688 | 169.44 | 1.2240 | 1.1767 |
| | 80 | | 0.844 | 14.312 | 160.88 | 1.1927 | 1.1172 |
| | 100 | | 1.031 | 13.938 | 152.58 | 1.1615 | 1.0596 |
| | 120 | | 1.219 | 13.562 | 144.46 | 1.1302 | 1.0032 |
| | 140 | | 1.438 | 13.124 | 135.28 | 1.0937 | 0.9394 |
| | 160 | | 1.594 | 12.812 | 128.92 | 1.0677 | 0.8953 |
| 18 OD | 10 | 18.00 | 0.250 | 17.500 | 240.53 | 1.4583 | 1.6703 |
| | 20 | | 0.312 | 17.376 | 237.13 | 1.4480 | 1.6467 |
| | (S) | | 0.375 | 17.250 | 233.71 | 1.4375 | 1.6230 |
| | 30 | | 0.438 | 17.124 | 230.00 | 1.4270 | 1.5993 |
| | (X) | | 0.500 | 17.000 | 226.98 | 1.4167 | 1.5762 |
| | 40 | | 0.562 | 16.876 | 223.68 | 1.4063 | 1.5533 |
| | 60 | | 0.750 | 16.500 | 213.83 | 1.3750 | 1.4849 |
| | 80 | | 0.938 | 16.124 | 204.19 | 1.3437 | 1.4180 |
| | 100 | | 1.156 | 15.688 | 193.30 | 1.3073 | 1.3423 |
| | 120 | | 1.375 | 15.250 | 182.65 | 1.2708 | 1.2684 |
| | 140 | | 1.562 | 14.876 | 173.81 | 1.2397 | 1.2070 |
| | 160 | | 1.781 | 14.438 | 163.72 | 1.2032 | 1.1370 |
| 20 OD | 10 | 20.00 | 0.250 | 19.500 | 298.65 | 1.6250 | 2.0739 |
| | 20 (S) | | 0.375 | 19.250 | 291.04 | 1.6042 | 2.0211 |
| | 30 (X) | | 0.500 | 19.000 | 283.53 | 1.5833 | 1.9689 |
| | 40 | | 0.594 | 18.812 | 277.95 | 1.5677 | 1.9302 |
| | 60 | | 0.812 | 18.376 | 265.21 | 1.5313 | 1.8417 |
| | 80 | | 1.031 | 17.938 | 252.72 | 1.4948 | 1.7550 |
| | 100 | | 1.281 | 17.438 | 238.83 | 1.4532 | 1.6585 |
| | 120 | | 1.500 | 17.000 | 226.98 | 1.4167 | 1.5762 |
| | 140 | | 1.750 | 16.500 | 213.83 | 1.3750 | 1.4849 |
| | 160 | | 1.969 | 16.062 | 202.62 | 1.3385 | 1.4071 |

| nominal diameter inches | schedule | outside diameter inches | wall thickness inches | internal diameter inches | internal area sq inches | internal diameter feet | internal area sq feet |
|---|---|---|---|---|---|---|---|
| 24 OD | 10 | 24.00 | 0.250 | 23.500 | 433.74 | 1.9583 | 3.0121 |
| | 20 (S) | | 0.375 | 23.250 | 424.56 | 1.9375 | 2.9483 |
| | (X) | | 0.500 | 23.000 | 415.48 | 1.9167 | 2.8852 |
| | 30 | | 0.562 | 22.876 | 411.01 | 1.9063 | 2.8542 |
| | 40 | | 0.688 | 22.624 | 402.00 | 1.8853 | 2.7917 |
| | 60 | | 0.969 | 22.062 | 382.28 | 1.8385 | 2.6547 |
| | 80 | | 1.219 | 21.562 | 365.15 | 1.7802 | 2.5358 |
| | 100 | | 1.531 | 20.938 | 344.32 | 1.7448 | 2.3911 |
| | 120 | | 1.812 | 20.376 | 326.92 | 1.6980 | 2.2645 |
| | 140 | | 2.062 | 19.876 | 310.28 | 1.6563 | 2.1547 |
| | 160 | | 2.344 | 19.312 | 292.92 | 1.6093 | 2.0342 |
| 30 OD | 10 | 30.00 | 0.312 | 29.376 | 677.76 | 2.4480 | 4.7067 |
| | (S) | | 0.375 | 29.250 | 671.62 | 2.4375 | 4.6640 |
| | 20 (X) | | 0.500 | 29.000 | 660.52 | 2.4167 | 4.5869 |
| | 30 | | 0.625 | 28.750 | 649.18 | 2.3958 | 4.5082 |

S = Wall thickness, formerly designated "standard weight"

X = Wall thickness, formerly designated "extra strong"

XX = Wall thickness, formerly designated "double extra strong"

Actual wall thickness may vary slightly.

Extracted from American Standard Wrought Steel and Wrought Iron Pipe (ASA B36, 10—1959), The American Society of Mechanical Engineers.

# Appendix G: Dimensions of Copper Water Tubing

| CLASSIFICATION | NOM. TUBE SIZE (in.) | OUTSIDE DIAM (in.) | WALL THICK-NESS (in.) | INSIDE DIAM (in.) | TRANS-VERSE AREA (sq. in.) | SAFE WORKING PRESSURE (psi) |
|---|---|---|---|---|---|---|
| HARD | 1/4 | 3/8 | .025 | .325 | .083 | 1000 |
| | 3/8 | 1/2 | .025 | .450 | .159 | 1000 |
| | 1/2 | 5/8 | .028 | .569 | .254 | 890 |
| | 3/4 | 7/8 | .032 | .811 | .516 | 710 |
| | 1 | 1 1/8 | .035 | 1.055 | .874 | 600 |
| | 1 1/4 | 1 3/8 | .042 | 1.291 | 1.309 | 590 |
| Type "M" 250 psi Working Pressure | 1 1/2 | 1 5/8 | .049 | 1.527 | 1.831 | 580 |
| | 2 | 2 1/8 | .058 | 2.009 | 3.17 | 520 |
| | 2 1/2 | 2 5/8 | .065 | 2.495 | 4.89 | 470 |
| | 3 | 3 1/8 | .072 | 2.981 | 6.98 | 440 |
| | 3 1/2 | 3 5/8 | .083 | 3.459 | 9.40 | 430 |
| | 4 | 4 1/8 | .095 | 3.935 | 12.16 | 430 |
| | 5 | 5 1/8 | .109 | 4.907 | 18.91 | 400 |
| | 6 | 6 1/8 | .122 | 5.881 | 27.16 | 375 |
| | 8 | 8 1/8 | .170 | 7.785 | 47.6 | 375 |
| HARD | 3/8 | 1/2 | .035 | .430 | .146 | 1000 |
| | 1/2 | 5/8 | .040 | .545 | .233 | 1000 |
| | 3/4 | 7/8 | .045 | .785 | .484 | 1000 |
| | 1 | 1 1/8 | .050 | 1.025 | .825 | 880 |
| | 1 1/4 | 1 3/8 | .055 | 1.265 | 1.256 | 780 |
| Type "L" 250 psi Working Pressure | 1 1/2 | 1 5/8 | .060 | 1.505 | 1.78 | 720 |
| | 2 | 2 1/8 | .070 | 1.985 | 3.094 | 640 |
| | 2 1/2 | 2 5/8 | .080 | 2.465 | 4.77 | 580 |
| | 3 | 3 1/8 | .090 | 2.945 | 6.812 | 550 |
| | 3 1/2 | 3 5/8 | .100 | 3.425 | 9.213 | 530 |
| | 4 | 4 1/8 | .110 | 3.905 | 11.97 | 510 |
| | 5 | 5 1/8 | .125 | 4.875 | 18.67 | 460 |
| | 6 | 6 1/8 | .140 | 5.845 | 26.83 | 430 |
| HARD | 1/4 | 3/8 | .032 | .311 | .076 | 1000 |
| | 3/8 | 1/2 | .049 | .402 | .127 | 1000 |
| | 1/2 | 5/8 | .049 | .527 | .218 | 1000 |
| | 3/4 | 7/8 | .065 | .745 | .436 | 1000 |
| | 1 | 1 1/8 | .065 | .995 | .778 | 780 |
| | 1 1/4 | 1 3/8 | .065 | 1.245 | 1.217 | 630 |
| Type "K" 400 psi Working Pressure | 1 1/2 | 1 5/8 | .072 | 1.481 | 1.722 | 580 |
| | 2 | 2 1/8 | .083 | 1.959 | 3.014 | 510 |
| | 2 1/2 | 2 5/8 | .095 | 2.435 | 4.656 | 470 |
| | 3 | 3 1/8 | .109 | 2.907 | 6.637 | 450 |
| | 3 1/2 | 3 5/8 | .120 | 3.385 | 8.999 | 430 |
| | 4 | 4 1/8 | .134 | 3.857 | 11.68 | 420 |
| | 5 | 5 1/8 | .160 | 4.805 | 18.13 | 400 |
| | 6 | 6 1/8 | .192 | 5.741 | 25.88 | 400 |
| SOFT | 1/4 | 3/8 | .032 | .311 | .076 | 1000 |
| | 3/8 | 1/2 | .049 | .402 | .127 | 1000 |
| | 1/2 | 5/8 | .049 | .527 | .218 | 1000 |
| | 3/4 | 7/8 | .065 | .745 | .436 | 1000 |
| | 1 | 1 1/8 | .065 | .995 | .778 | 780 |
| | 1 1/4 | 1 3/8 | .065 | 1.245 | 1.217 | 630 |
| Type "K" 250 psi Working Pressure | 1 1/2 | 1 5/8 | .072 | 1.481 | 1.722 | 580 |
| | 2 | 2 1/8 | .083 | 1.959 | 3.014 | 510 |
| | 2 1/2 | 2 5/8 | .095 | 2.435 | 4.656 | 470 |
| | 3 | 3 1/8 | .109 | 2.907 | 6.637 | 450 |
| | 3 1/2 | 2 5/8 | .120 | 3.385 | 8.999 | 430 |
| | 4 | 4 1/8 | .134 | 3.857 | 11.68 | 420 |
| | 5 | 5 1/8 | .160 | 4.805 | 18.13 | 400 |
| | 6 | 6 1/8 | .192 | 5.741 | 25.88 | 400 |

# Appendix H: Dimensions of Brass and Copper Tubing

**regular**

| pipe size in. | nominal dimensions in. | | | cross sectional area of bore sq. in. | lb per ft | |
|---|---|---|---|---|---|---|
| | O.D. | I.D. | wall | | red brass | copper |
| 1/8 | .405 | .281 | .062 | .062 | .253 | .259 |
| 1/4 | .540 | .376 | .082 | .110 | .447 | .457 |
| 3/8 | .675 | .495 | .090 | .192 | .627 | .641 |
| 1/2 | .840 | .626 | .107 | .307 | .934 | .955 |
| 3/4 | 1.050 | .822 | .114 | .531 | 1.270 | 1.300 |
| 1 | 1.315 | 1.063 | .126 | .887 | 1.780 | 1.820 |
| 1 1/4 | 1.660 | 1.368 | .146 | 1.470 | 2.630 | 2.690 |
| 1 1/2 | 1.900 | 1.600 | .150 | 2.010 | 3.130 | 3.200 |
| 2 | 2.375 | 2.063 | .156 | 3.340 | 4.120 | 4.220 |
| 2 1/2 | 2.875 | 2.501 | .187 | 4.910 | 5.990 | 6.120 |
| 3 | 3.500 | 3.062 | .219 | 7.370 | 8.560 | 8.750 |
| 3 1/2 | 4.000 | 3.500 | .250 | 9.620 | 11.200 | 11.400 |
| 4 | 4.500 | 4.000 | .250 | 12.600 | 12.700 | 12.900 |
| 5 | 5.562 | 5.062 | .250 | 20.100 | 15.800 | 16.200 |
| 6 | 6.625 | 6.125 | .250 | 29.500 | 19.000 | 19.400 |
| 8 | 8.625 | 8.001 | .312 | 50.300 | 30.900 | 31.600 |
| 10 | 10.750 | 10.020 | .365 | 78.800 | 45.200 | 46.200 |
| 12 | 12.750 | 12.000 | .375 | 113.000 | 55.300 | 56.500 |

**extra strong**

| pipe size in. | nominal dimensions in. | | | cross sectional area of bore sq. in. | lb per ft | |
|---|---|---|---|---|---|---|
| | O.D. | I.D. | wall | | red brass | copper |
| 1/8 | .405 | .205 | .100 | .033 | .363 | .371 |
| 1/4 | .540 | .294 | .123 | .068 | .611 | .625 |
| 3/8 | .675 | .421 | .127 | .139 | .829 | .847 |
| 1/2 | .840 | .542 | .149 | .231 | 1.230 | 1.250 |
| 3/4 | 1.050 | .736 | .157 | .425 | 1.670 | 1.710 |
| 1 | 1.315 | .951 | .182 | .710 | 2.460 | 2.510 |
| 1 1/4 | 1.660 | 1.272 | .194 | 1.270 | 3.390 | 3.460 |
| 1 1/2 | 1.900 | 1.494 | .203 | 1.750 | 4.100 | 4.190 |
| 2 | 2.375 | 1.933 | .221 | 2.94 | 5.670 | 5.800 |
| 2 1/2 | 2.875 | 2.315 | .280 | 4.21 | 8.660 | 8.850 |
| 3 | 3.500 | 2.892 | .304 | 6.57 | 11.600 | 11.800 |
| 3 1/2 | 4.000 | 3.358 | .321 | 8.86 | 14.100 | 14.400 |
| 4 | 4.500 | 3.818 | .341 | 11.50 | 16.900 | 17.300 |
| 5 | 5.562 | 4.812 | .375 | 18.20 | 23.200 | 23.700 |
| 6 | 6.625 | 5.751 | .437 | 26.00 | 32.200 | 32.900 |
| 8 | 8.625 | 7.625 | .500 | 45.70 | 48.400 | 49.500 |
| 10 | 10.750 | 9.750 | .500 | 74.70 | 61.100 | 62.400 |

FLUIDS

# Appendix I: Typical Dimensions and Weights of Concrete Sewer Pipe

(All dimensions in inches)

(Weights given are for reinforced tongue and groove pipe. Reinforced bell and spigot pipe is heavier. Weights are based on 150 pcf concrete.)

| 3000 psi | | | 3500 psi | | | 4000 psi | | |
|---|---|---|---|---|---|---|---|---|
| internal diameter, inches | minimum shell thickness, inches | weight per foot, in pounds | internal diameter, inches | minimum shell thickness, inches | weight per foot, in pounds | internal diameter, inches | minimum shell thickness, inches | weight per foot, in pounds |
| 12 | 2 | 93 | 12 | 13/4 | 79 | 12 | | |
| 15 | 2 1/4 | 127 | 15 | 2 | 111 | 15 | | |
| 18 | 2 1/2 | 168 | 18 | | | 18 | 2 | 131 |
| 21 | 2 3/4 | 214 | 21 | | | 21 | 2 1/4 | 171 |
| 24 | 3 | 264 | 24 | 2 5/8 | 229 | 24 | 2 1/2 | 217 |
| 27 | 3 | 295 | 27 | 2 3/4 | 268 | 27 | 2 5/8 | 255 |
| 30 | 3 1/2 | 384 | 30 | 3 | 324 | 30 | 2 3/4 | 295 |
| 33 | 3 3/4 | 451 | 33 | 3 1/4 | 396 | 33 | 2 3/4 | 322 |
| 36 | 4 | 524 | 36 | 3 3/8 | 435 | 36 | 3 | 383 |
| 42 | 4 1/2 | 686 | 42 | 3 3/4 | 561 | 42 | 3 3/8 | 500 |
| 48 | 5 | 867 | 48 | 4 1/4 | 727 | 48 | 3 3/4 | 635 |
| 54 | 5 1/2 | 1068 | 54 | 4 5/8 | 887 | 54 | 4 1/4 | 810 |
| 60 | 6 | 1295 | 60 | 5 | 1064 | 60 | 4 1/2 | 950 |
| 66 | 6 1/2 | 1542 | 66 | 5 3/8 | 1256 | 66 | 4 3/4 | 1100 |
| 72 | 7 | 1811 | 72 | 5 3/4 | 1463 | 72 | 5 | 1260 |
| 78 | 7 1/2 | 2100 | | | | | | |
| 84 | 8 | 2409 | | | | | | |
| 90 | 8 | 2565 | | | | | | |
| 96 | 8 1/2 | 2906 | | | | | | |
| 108 | 9 | 3446 | | | | | | |

# Appendix J: Cast Iron Pipe Dimensions

(all dimensions in inches)

| nominal diameter | class A 100 foot head 43 psig | | class B 200 foot head 86 psig | | class C 300 foot head 130 psig | | class D 400 foot head 173 psig | |
|---|---|---|---|---|---|---|---|---|
| | outside diameter | inside diameter | outside diameter | inside diameter | outside diameter | inside diameter | outside diameter | inside diameter |
| 3 | 3.80 | 3.02 | 3.96 | 3.12 | 3.96 | 3.06 | 3.96 | 3.00 |
| 4 | 4.80 | 3.96 | 5.00 | 4.10 | 5.00 | 4.04 | 5.00 | 3.96 |
| 6 | 6.90 | 6.02 | 7.10 | 6.14 | 7.10 | 6.08 | 7.10 | 6.00 |
| 8 | 9.05 | 8.13 | 9.05 | 8.03 | 9.30 | 8.18 | 9.30 | 8.10 |
| 10 | 11.10 | 10.10 | 11.10 | 9.96 | 11.40 | 10.16 | 11.40 | 10.04 |
| 12 | 13.20 | 12.12 | 13.20 | 11.96 | 13.50 | 12.14 | 13.50 | 12.00 |
| 14 | 15.30 | 14.16 | 15.30 | 13.98 | 15.65 | 14.17 | 15.65 | 14.01 |
| 16 | 17.40 | 16.20 | 17.40 | 16.00 | 17.80 | 16.20 | 17.80 | 16.02 |
| 18 | 19.50 | 18.22 | 19.50 | 18.00 | 19.92 | 18.18 | 19.92 | 18.00 |
| 20 | 21.60 | 20.26 | 21.60 | 20.00 | 22.06 | 20.22 | 22.06 | 20.00 |
| 24 | 25.80 | 24.28 | 25.80 | 24.02 | 26.32 | 24.22 | 26.32 | 24.00 |
| 30 | 31.74 | 29.98 | 32.00 | 29.94 | 32.40 | 30.00 | 32.74 | 30.00 |
| 36 | 37.96 | 35.98 | 38.30 | 36.00 | 38.70 | 39.98 | 39.16 | 36.00 |
| 42 | 44.20 | 42.00 | 44.50 | 41.94 | 45.10 | 42.02 | 45.58 | 42.02 |
| 48 | 50.50 | 47.98 | 50.80 | 47.96 | 51.40 | 47.98 | 51.98 | 48.06 |
| 54 | 56.66 | 53.96 | 57.10 | 54.00 | 57.80 | 54.00 | 58.40 | 53.94 |
| 60 | 62.80 | 60.02 | 63.40 | 60.06 | 64.20 | 60.20 | 64.82 | 60.06 |
| 72 | 75.34 | 72.10 | 76.00 | 72.10 | 76.88 | 72.10 | | |
| 84 | 87.54 | 84.10 | 88.54 | 84.10 | | | | |

| nominal diameter | class E 500 foot head 217 psig | | class F 600 foot head 260 psig | | class G 700 foot head 304 psig | | class H 800 foot head 347 psig | |
|---|---|---|---|---|---|---|---|---|
| | outside diameter | inside diameter | outside diameter | inside diameter | outside diameter | inside diameter | outside diameter | inside diameter |
| 6 | 7.22 | 6.06 | 7.22 | 6.00 | 7.38 | 6.08 | 7.38 | 6.00 |
| 8 | 9.42 | 8.10 | 9.42 | 8.00 | 9.60 | 8.10 | 9.60 | 8.00 |
| 10 | 11.60 | 10.12 | 11.60 | 10.00 | 11.84 | 10.12 | 11.84 | 10.00 |
| 12 | 13.78 | 12.14 | 13.78 | 12.00 | 14.08 | 12.14 | 14.08 | 12.00 |
| 14 | 15.98 | 14.18 | 15.98 | 14.00 | 16.32 | 14.18 | 16.32 | 14.00 |
| 16 | 18.16 | 16.20 | 18.16 | 16.00 | 18.54 | 16.18 | 18.54 | 16.00 |
| 18 | 20.34 | 18.20 | 20.34 | 18.00 | 20.78 | 18.22 | 20.78 | 18.00 |
| 20 | 22.54 | 20.24 | 22.54 | 20.00 | 23.02 | 20.24 | 23.02 | 20.00 |
| 24 | 26.90 | 24.28 | 26.90 | 24.00 | 27.76 | 24.26 | 27.76 | 24.00 |
| 30 | 33.10 | 30.00 | 33.46 | 30.00 | | | | |
| 36 | 39.60 | 36.00 | 40.04 | 36.00 | | | | |

# Appendix K: American Standard Piping Symbols

| | Flanged | Screwed | Bell & Spigot | Welded | Soldered |
|---|---|---|---|---|---|
| Joint | | | | | |
| Elbow—90° | | | | | |
| Elbow—45° | | | | | |
| Elbow—Turned Up | | | | | |
| Elbow—Turned Down | | | | | |
| Elbow—Long Radius | | | | | |
| Reducing Elbow | | | | | |
| Tee | | | | | |
| Tee—Outlet Up | | | | | |
| Tee—Outlet Down | | | | | |
| Side Outlet Tee—Outlet Up | | | | | |
| Cross | | | | | |
| Reducer—Concentric | | | | | |
| Reducer—Eccentric | | | | | |
| Lateral | | | | | |
| Gate Valve | | | | | |
| Globe Valve | | | | | |
| Check Valve | | | | | |
| Stop Cock | | | | | |
| Safety Valve | | | | | |
| Expansion Joint | | | | | |
| Union | | | | | |
| Sleeve | | | | | |
| Bushing | | | | | |

# Appendix L: Equivalent Length of Straight Pipe for Various Fittings (feet)

(turbulent flow only, for any fluid)
c.i. = cast iron

| fittings | | | pipe size $\frac{1}{4}$ | $\frac{3}{8}$ | $\frac{1}{2}$ | $\frac{3}{4}$ | 1 | $1\frac{1}{4}$ | $1\frac{1}{2}$ | 2 | $2\frac{1}{2}$ | 3 | 4 | 5 | 6 | 8 | 10 | 12 | 14 | 16 | 18 | 20 | 24 |
|---|---|---|---|---|---|---|---|---|---|---|---|---|---|---|---|---|---|---|---|---|---|---|---|
| regular 90° ell | screwed | steel | 2.3 | 3.1 | 3.6 | 4.4 | 5.2 | 6.6 | 7.4 | 8.5 | 9.3 | 11.0 | 13.0 | | | | | | | | | | |
| | | c.i. | | | | | | | | | | 9.0 | 11.0 | | | | | | | | | | |
| | flanged | steel | | | 0.92 | 1.2 | 1.6 | 2.1 | 2.4 | 3.1 | 3.6 | 4.4 | 5.9 | 7.3 | 8.9 | 12.0 | 14.0 | 17.0 | 18.0 | 21.0 | 23.0 | 25.0 | 30.0 |
| | | c.i. | | | | | | | | | | 3.6 | 4.8 | | 7.2 | 9.8 | 12.0 | 15.0 | 17.0 | 19.0 | 22.0 | 24.0 | 28.0 |
| long radius 90° ell | screwed | steel | 1.5 | 2.0 | 2.2 | 2.3 | 2.7 | 3.2 | 3.4 | 3.6 | 3.6 | 4.0 | 4.6 | | | | | | | | | | |
| | | c.i. | | | | | | | | | | 3.3 | 3.7 | | | | | | | | | | |
| | flanged | steel | | | 1.1 | 1.3 | 1.6 | 2.0 | 2.3 | 2.7 | 2.9 | 3.4 | 4.2 | 5.0 | 5.7 | 7.0 | 8.0 | 9.0 | 9.4 | 10.0 | 11.0 | 12.0 | 14.0 |
| | | c.i. | | | | | | | | | | 2.8 | 3.4 | | 4.7 | 5.7 | 6.8 | 7.8 | 8.6 | 9.6 | 11.0 | 11.0 | 13.0 |
| regular 45° ell | screwed | steel | 0.34 | 0.52 | 0.71 | 0.92 | 1.3 | 1.7 | 2.1 | 2.7 | 3.2 | 4.0 | 5.5 | | | | | | | | | | |
| | | c.i. | | | | | | | | | | 3.3 | 4.5 | | | | | | | | | | |
| | flanged | steel | | | 0.45 | 0.59 | 0.81 | 1.1 | 1.3 | 1.7 | 2.0 | 2.6 | 3.5 | 4.5 | 5.6 | 7.7 | 9.0 | 11.0 | 13.0 | 15.0 | 16.0 | 18.0 | 22.0 |
| | | c.i. | | | | | | | | | | 2.1 | 2.9 | | 4.5 | 6.3 | 8.1 | 9.7 | 12.0 | 13.0 | 15.0 | 17.0 | 20.0 |
| tee-line flow | screwed | steel | 0.79 | 1.2 | 1.7 | 2.4 | 3.2 | 4.6 | 5.6 | 7.7 | 9.3 | 12.0 | 17.0 | | | | | | | | | | |
| | | c.i. | | | | | | | | | | 9.9 | 14.0 | | | | | | | | | | |
| | flanged | steel | | | 0.69 | 0.82 | 1.0 | 1.3 | 1.5 | 1.8 | 1.9 | 2.2 | 2.8 | 3.3 | 3.8 | 4.7 | 5.2 | 6.0 | 6.4 | 7.2 | 7.6 | 8.2 | 9.6 |
| | | c.i. | | | | | | | | | | 1.9 | 2.2 | | 3.1 | 3.9 | 4.6 | 5.2 | 5.9 | 6.5 | 7.2 | 7.7 | 8.8 |
| tee-branch flow | screwed | steel | 2.4 | 3.5 | 4.2 | 5.3 | 6.6 | 8.7 | 9.9 | 12.0 | 13.0 | 17.0 | 21.0 | | | | | | | | | | |
| | | c.i. | | | | | | | | | | 14.0 | 17.0 | | | | | | | | | | |
| | flanged | steel | | | 2.0 | 2.6 | 3.3 | 4.4 | 5.2 | 6.6 | 7.5 | 9.4 | 12.0 | 15.0 | 18.0 | 24.0 | 30.0 | 34.0 | 37.0 | 43.0 | 47.0 | 52.0 | 62.0 |
| | | c.i. | | | | | | | | | | 7.7 | 10.0 | | 15.0 | 20.0 | 25.0 | 30.0 | 35.0 | 39.0 | 44.0 | 49.0 | 57.0 |
| 180° return bend | screwed | steel | 2.3 | 3.1 | 3.6 | 4.4 | 5.2 | 6.6 | 7.4 | 8.5 | 9.3 | 11.0 | 13.0 | | | | | | | | | | |
| | | c.i. | | | | | | | | | | 9.0 | 11.0 | | | | | | | | | | |
| | reg. flanged | steel | | | 0.92 | 1.2 | 1.6 | 2.1 | 2.4 | 3.1 | 3.6 | 4.4 | 5.9 | 7.3 | 8.9 | 12.0 | 14.0 | 17.0 | 18.0 | 21.0 | 23.0 | 25.0 | 30.0 |
| | | c.i. | | | | | | | | | | 3.6 | 4.8 | | 7.2 | 9.8 | 12.0 | 15.0 | 17.0 | 19.0 | 22.0 | 24.0 | 28.0 |
| | long rad. flanged | steel | | | 1.1 | 1.3 | 1.6 | 2.0 | 2.3 | 2.7 | 2.9 | 3.4 | 4.2 | 5.0 | 5.7 | 7.0 | 8.0 | 9.0 | 9.4 | 10.0 | 11.0 | 12.0 | 14.0 |
| | | c.i. | | | | | | | | | | 2.8 | 3.4 | | 4.7 | 5.7 | 6.8 | 7.8 | 8.6 | 9.6 | 11.0 | 11.0 | 13.0 |
| globe valve | screwed | steel | 21.0 | 22.0 | 22.0 | 24.0 | 29.0 | 37.0 | 42.0 | 54.0 | 62.0 | 79.0 | 110.0 | | | | | | | | | | |
| | | c.i. | | | | | | | | | | 65.0 | 86.0 | | | | | | | | | | |
| | flanged | steel | | | 38.0 | 40.0 | 45.0 | 54.0 | 59.0 | 70.0 | 77.0 | 94.0 | 120.0 | 150.0 | 190.0 | 260.0 | 310.0 | 390.0 | | | | | |
| | | c.i. | | | | | | | | | | 77.0 | 99.0 | | 150.0 | 210.0 | 270.0 | 330.0 | | | | | |
| gate valve | screwed | steel | 0.32 | 0.45 | 0.56 | 0.67 | 0.84 | 1.1 | 1.2 | 1.5 | 1.7 | 1.9 | 2.5 | | | | | | | | | | |
| | | c.i. | | | | | | | | | | 1.6 | 2.0 | | | | | | | | | | |
| | flanged | steel | | | | | | | | 2.6 | 2.7 | 2.8 | 2.9 | 3.1 | 3.2 | 3.2 | 3.2 | 3.2 | 3.2 | 3.2 | 3.2 | 3.2 | 3.2 |
| | | c.i. | | | | | | | | | | 2.3 | 2.4 | | 2.6 | 2.7 | 2.8 | 2.9 | 2.9 | 3.0 | 3.0 | 3.0 | 3.0 |
| angle valve | screwed | steel | 12.8 | 15.0 | 15.0 | 15.0 | 17.0 | 18.0 | 18.0 | 18.0 | 18.0 | 18.0 | 18.0 | | | | | | | | | | |
| | | c.i. | | | | | | | | | | 15.0 | 15.0 | | | | | | | | | | |
| | flanged | steel | | | 15.0 | 15.0 | 17.0 | 18.0 | 18.0 | 21.0 | 22.0 | 28.0 | 38.0 | 50.0 | 63.0 | 90.0 | 120.0 | 140.0 | 160.0 | 190.0 | 210.0 | 240.0 | 300.0 |
| | | c.i. | | | | | | | | | | 23.0 | 31.0 | | 52.0 | 74.0 | 98.0 | 120.0 | 150.0 | 170.0 | 200.0 | 230.0 | 280.0 |
| swing check valve | screwed | steel | 7.2 | 7.3 | 8.0 | 8.8 | 11.0 | 13.0 | 15.0 | 19.0 | 22.0 | 27.0 | 38.0 | | | | | | | | | | |
| | | c.i. | | | | | | | | | | 22.0 | 31.0 | | | | | | | | | | |
| | flanged | steel | | | 3.8 | 5.3 | 7.2 | 10.0 | 12.0 | 17.0 | 21.0 | 27.0 | 38.0 | 50.0 | 63.0 | 90.0 | 120.0 | 140.0 | | | | | |
| | | c.i. | | | | | | | | | | 22.0 | 31.0 | | 52.0 | 74.0 | 98.0 | 120.0 | | | | | |
| coupling or union | screwed | steel | 0.14 | 0.18 | 0.21 | 0.24 | 0.29 | 0.36 | 0.39 | 0.45 | 0.47 | 0.53 | 0.65 | | | | | | | | | | |
| | | c.i. | | | | | | | | | | 0.44 | 0.52 | | | | | | | | | | |
| | bell mouth inlet | steel | 0.04 | 0.07 | 0.10 | 0.13 | 0.18 | 0.26 | 0.31 | 0.43 | 0.52 | 0.67 | 0.95 | 1.3 | 1.6 | 2.3 | 2.9 | 3.5 | 4.0 | 4.7 | 5.3 | 6.1 | 7.6 |
| | | c.i. | | | | | | | | | | 0.55 | 0.77 | | 1.3 | 1.9 | 2.4 | 3.0 | 3.6 | 4.3 | 5.0 | 5.7 | 7.0 |
| | square mouth inlet | steel | 0.44 | 0.68 | 0.96 | 1.3 | 1.8 | 2.6 | 3.1 | 4.3 | 5.2 | 6.7 | 9.5 | 13.0 | 16.0 | 23.0 | 29.0 | 35.0 | 40.0 | 47.0 | 53.0 | 61.0 | 76.0 |
| | | c.i. | | | | | | | | | | 5.5 | 7.7 | | 13.0 | 19.0 | 24.0 | 30.0 | 36.0 | 43.0 | 50.0 | 57.0 | 70.0 |
| | re-entrant pipe | steel | 0.88 | 1.4 | 1.9 | 2.6 | 3.6 | 5.1 | 6.2 | 8.5 | 10.0 | 13.0 | 19.0 | 25.0 | 32.0 | 45.0 | 58.0 | 70.0 | 80.0 | 95.0 | 110.0 | 120.0 | 150.0 |
| | | c.i. | | | | | | | | | | 11.0 | 15.0 | | 26.0 | 37.0 | 49.0 | 61.0 | 73.0 | 86.0 | 100.0 | 110.0 | 140.0 |

# Appendix M: Hazen-Williams Nomograph

($C = 100$)

For values of $C$ other than 100, multiply the nomograph values for head loss by $\left(\frac{100}{C}\right)^{1.85}$

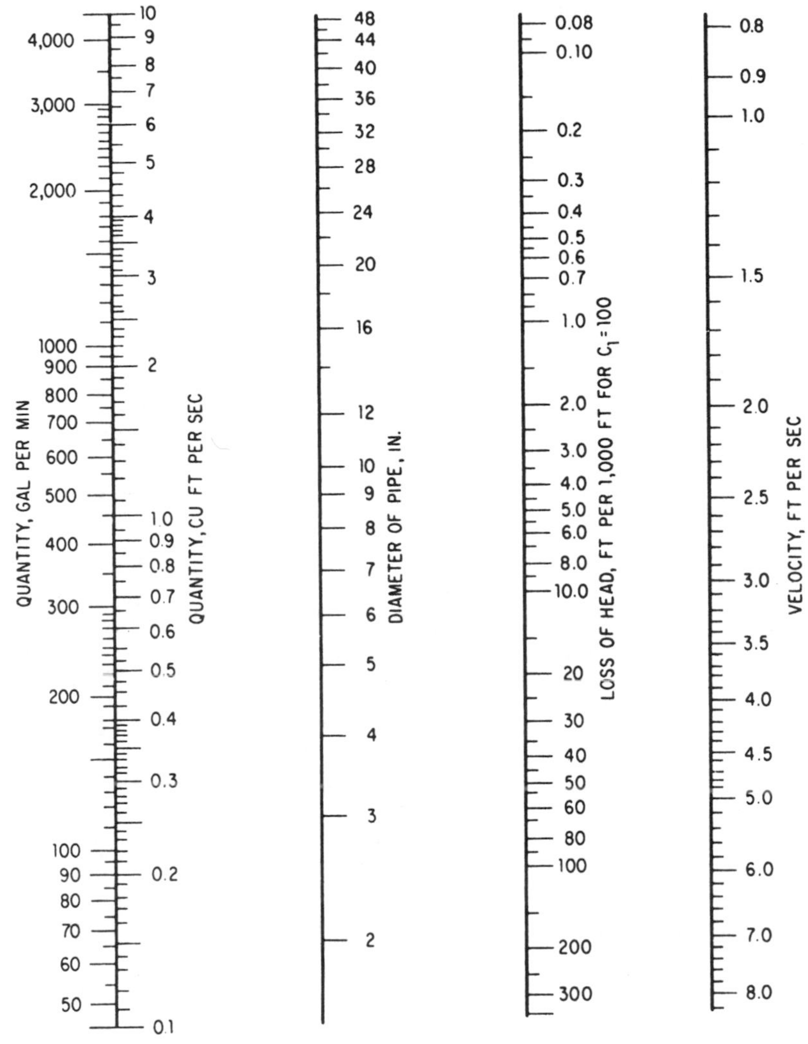

## Appendix N: Manning Nomograph

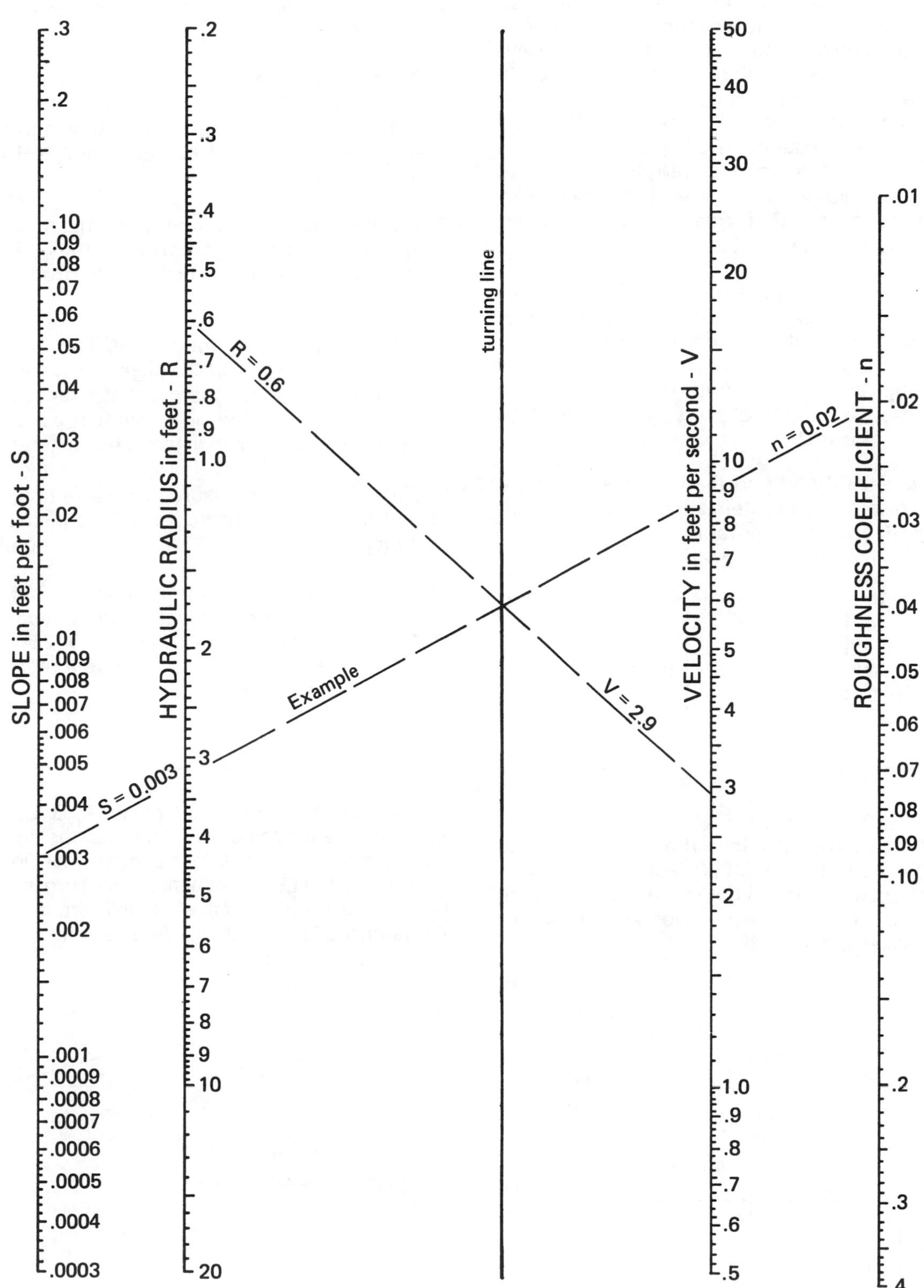

## Practice Problems: HYDRAULICS

### Untimed

1. Three reservoirs (A, B, and C) are interconnected with a common junction at elevation 25 feet above some arbitrary reference point. The water surface levels for reservoirs A, B, and C are at elevations of 50, 40, and 22 feet respectively. The pipe from reservoir A to the junction is 800 feet of 3″ steel pipe. The pipe from reservoir B to the junction is 500 feet of 10″ steel pipe. The pipe from reservoir C to the junction is 1000 feet of 4″ schedule-40 steel pipe. Is the flow into or out of the reservoir B? Neglect minor losses and velocity heads. Assume f = .02.

2. A class 20 cast iron pipe has an inside diameter of 24″ and a wall thickness of 1.16″. Water is flowing at 6 fps. (a) If a valve is closed instantaneously, what will be the pressure created? (b) If the pipe is 500 feet long, over what length of time must the valve be closed to create a pressure equivalent to instantaneous closure?

3. Assume C = 100 and find the flow in each of the pipes in the distribution system shown below. An accuracy of ± 10 gpm is acceptable.

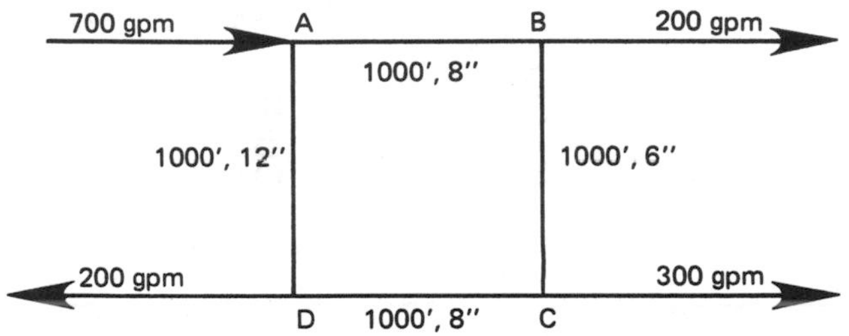


4. The distance between each labeled point on the distribution graph below is 1000 feet. Assume C = 100 for each pipe section. What is the pressure at all labeled points? If the pump receives 20 psig water, what horsepower is required?

| point | pressure | elevation |
|---|---|---|
| A | | 200 ft |
| B | | 150 |
| C | 40 psig | 300 |
| D | | 150 |
| E | | 200 |
| F | | 150 |

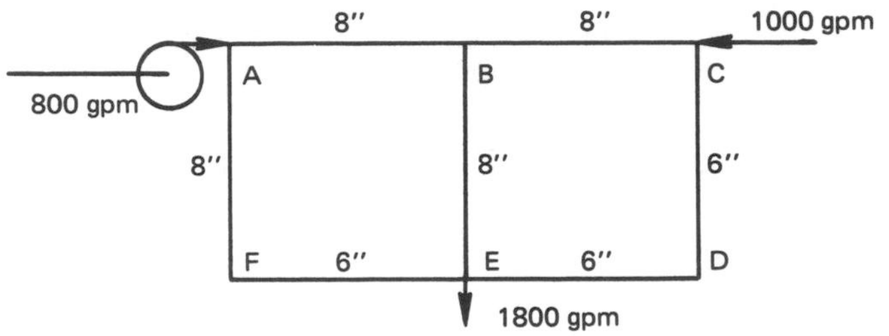


5. 70°F water is carried in a schedule-40 steel pipe which changes gradually from 6″ at point A to 18″ at point B. B is 15 feet higher than A. 5 cfs are flowing and the respective pressures at A and B are 10 psia and 7 psia. What is the direction of the flow?

6. Points A and B are 3000 feet apart along a new 6″ steel pipe. B is 60 feet above A. 750 gpm of 60°F water flow in the pipe. The flow direction is from A to B. What pressure must be maintained at A if the pressure at B is to be 50 psig?

7. A cylindrical tank 20 feet in diameter and 40 feet high has a 4″ hole in the bottom with $C_d$ = .98. How long will it take for the water level to drop from 40′ to 20′?

8. A venturi water meter with an 8″ diameter throat is installed in a 12″ diameter water line. Assume the venturi discharge coefficient is equal to one. What is the flow in cubic feet per second if a mercury manometer registers a 4″ differential?

9. What will be the measured pressure drop across a .2 foot diameter sharp-edged orifice installed in a 1 foot diameter pipe if 70°F water is flowing at 2 fps?

10. A pipe necks down from 24″ at point A to 12″ at point B. The discharge is 8 cfs in the direction of A to B. The pressure head at A is 20 feet. Assume no friction. Find the resultant force and its direction on the fluid if water is flowing.

### Timed

1. A pipe network connects points A, B, C, and D as shown in the figure below. Water can be added or removed at any of these four points. The flow is from point A to point D. The minimum pressure is 20 psi. (a) What is the elevation of the hydraulic grade line at point A referenced to point D? (b) What are the pressures (in psi) for the four points?

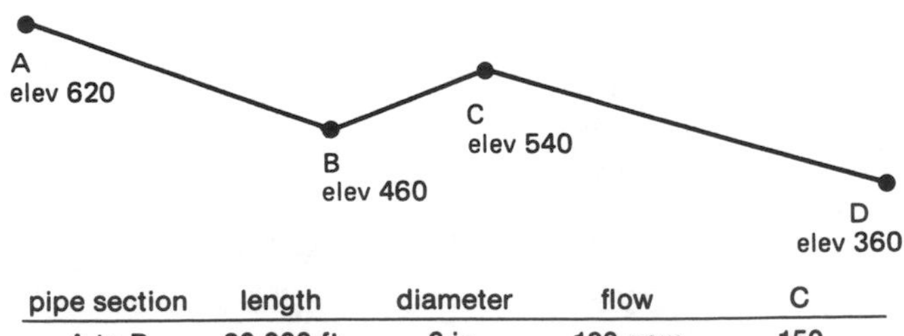


| pipe section | length | diameter | flow | C |
|---|---|---|---|---|
| A to B | 20,000 ft | 6 in. | 120 gpm | 150 |
| B to C | 10,000 ft | 6 in. | 160 gpm | 150 |
| C to D | 30,000 ft | 4 in. | 120 gpm | 150 |

# 4 HYDRAULIC MACHINES

*Nomenclature*

| | | |
|---|---|---|
| bhp | brake horsepower | hp |
| $C_v$ | coefficient of velocity | – |
| d | diameter | inches |
| D | diameter | ft |
| ehp | electrical horsepower | hp |
| E | energy | ft-lbf/lbm |
| f | Darcy friction factor, or frequency | –, hertz |
| fhp | friction horsepower | hp |
| g | acceleration due to gravity (32.2) | ft/sec$^2$ |
| $g_c$ | gravitational constant (32.2) | $\frac{\text{lbm-ft}}{\text{lbf-sec}^2}$ |
| h | head | ft |
| H | dynamic head | ft |
| L | pipe length | ft |
| m | mass | lbm |
| n | rotational speed | rpm |
| $n_s$ | specific speed | rpm |
| $n_{ss}$ | suction specific speed | rpm |
| NPSHA | net positive suction head available | ft |
| NPSHR | net positive suction head required | ft |
| p | pressure | psf |
| Q | flow quantity | gpm |
| s | slip | – |
| S.G. | specific gravity | – |
| T | temperature | °F |
| v | velocity | ft/sec |
| V | volume | ft$^3$ |
| w | weight | lbf |
| whp | hydraulic ("water") horsepower | hp |
| z | height above datum | ft |

*Symbols*

| | | |
|---|---|---|
| $\gamma$ | specific weight | lbf/ft$^3$ |
| $\eta$ | efficiency | – |
| $\epsilon$ | specific pipe roughness | ft |
| $\rho$ | density | lbm/ft$^3$ |
| $\beta$ | blade angle | ° |
| $\sigma$ | cavitation coefficient | – |

*Subscripts*

| | |
|---|---|
| a | atmospheric |
| A | added by pump |
| d | discharge |
| f | friction |
| i | intake |
| j | jet |
| m | motor |
| n | nozzle |
| p | pump or pressure |
| s | suction |
| sd | static discharge |
| th | theoretical |
| ts | total static |
| T | turbine |
| v | velocity |
| vp | vapor pressure |

## 1 INTRODUCTION

Pumps and turbines are the two types of hydraulic machines discussed in this chapter. Pumps convert mechanical energy into fluid energy. Turbines convert fluid energy into mechanical energy.

## 2 TYPES OF PUMPS

Pumps can be classified in several ways. The clearest categorization is based on the method by which pumping energy is transmitted to the fluid. This approach separates pumps into *kinetic pumps* and *positive displacement pumps.*

The two most common forms of positive displacement pumps are *reciprocating action pumps* (which use pistons, plungers, diaphragms, or bellows) and *rotary action pumps* (using vanes, screws, lobes, or progressing cavities). Such pumps discharge a given volume for each stroke or revolution. Energy is added intermittently to the fluid flow.

Kinetic pumps rely on a transformation of kinetic energy to static pressure. Jet and ejector pumps fall into this category, but centrifugal pumps are the primary examples of kinetic pumps. Pumps covered in this chapter are assumed to be centrifugal pumps.

**Table 4.1**
Characteristics of Kinetic and Displacement Pumps

| characteristic | displacement | kinetic |
|---|---|---|
| flow rate | low | high |
| pressure rise per stage | high | low |
| constant variable over operating range | flow rate | pressure rise |
| self-priming | yes | no |
| outlet stream | pulsing | steady |
| works with high-viscosity fluids | yes | no |



Liquid flowing into the *suction side* (the *inlet*) of a centrifugal pump is captured by the *impeller* and thrown to the outside of the pump casing. Within the casing, the velocity imparted to the fluid by the impeller is converted into pressure energy.

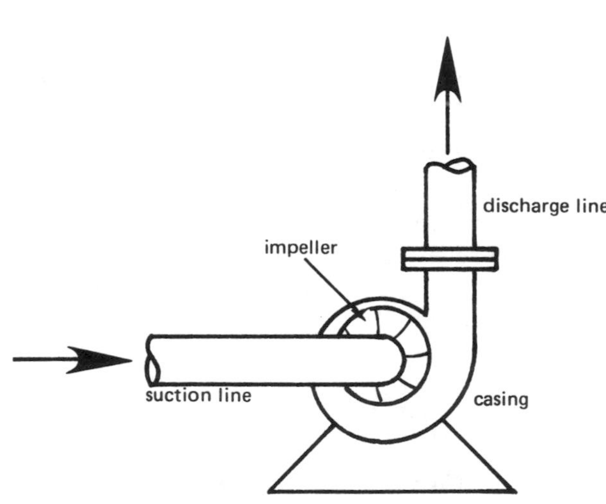


**Figure 4.1** A Centrifugal Pump

## 3 TYPES OF CENTRIFUGAL PUMPS

Centrifugal pumps can be classified into three general categories according to the way the impeller imparts energy to the fluid. Each of these categories has its own range of specific speeds and appropriate applications.

*Radial flow impellers* impart energy primarily by centrifugal force. Liquid enters the impeller at the hub and flows radially to the outside of the casing. Single suction impellers have a specific speed less than 5000. Double suction impellers have a specific speed less than 6000.

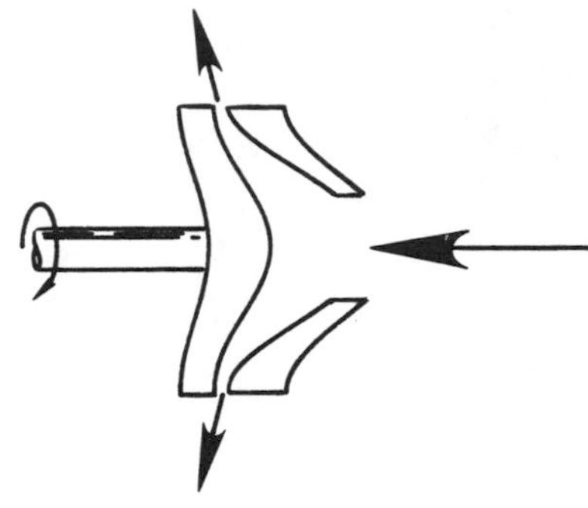

**Figure 4.2** Centrifugal (Radial) Flow Pump

*Mixed flow impellers* impart energy partially by centrifugal force and partially by axial force, since the vanes act partially as an axial compressor. Liquid enters the impeller at the hub and flows both radially and axially to discharge. Specific speeds of mixed flow pumps range from 4200 to 9000.

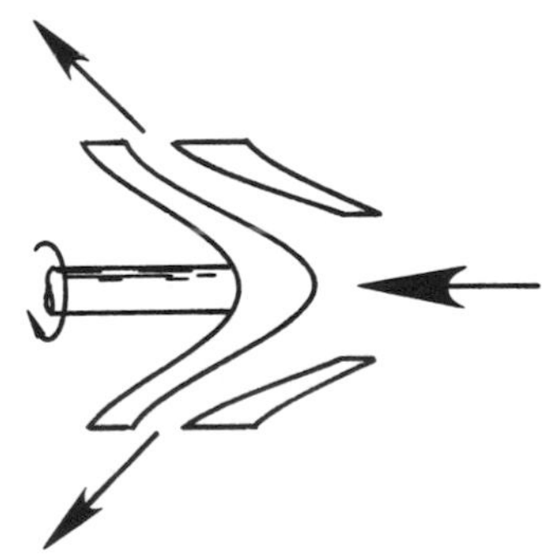

**Figure 4.3** Mixed Flow Pump

*Axial flow impellers* impart energy to the fluid by acting as axial flow compressors. Fluid enters and exits along the axis of rotation. Specific speed is greater than 9000.

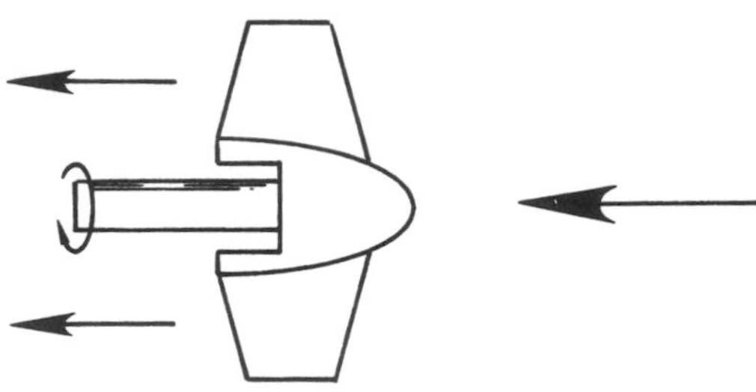

**Figure 4.4** Axial Flow Pump

Radial flow and mixed flow centrifugal pumps can be designed for either single or double suction operation. In a ***single suction pump***, fluid enters only one side of the impeller. In a ***double suction pump***, fluid enters both sides of the impeller. Thus, for an impeller with a given specific speed, a greater flow rate can be expected from a double suction pump. In addition, a double suction pump has a lower NPSHR for a given flow than a single suction pump.

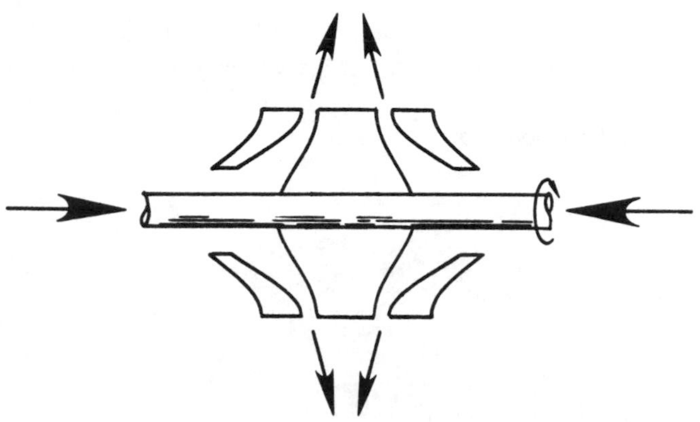

**Figure 4.5** Radial Flow Pump (Double Suction)

A *multiple stage pump* consists of two or more impellers within a single casing. The discharge of one stage is the input of the next stage. In this manner, higher heads are achieved than would be possible with a single impeller.

## 4 PUMP AND HEAD TERMINOLOGY

Like most specialized subjects, the centrifugal pump field has developed its own terminology. It is essential that this terminology be understood, since its interpretation will often affect an installation's physical configuration.

All of the terms which follow are *head terms*, and as such, have units of feet. Of course, any head term can be converted to pressure by using equation 4.1.

$$p = \gamma h \qquad 4.1$$

*Friction head* ($h_f$): The head required to overcome resistance to flow in the pipe, fittings, valves, entrances, and exits.

$$h_f = \frac{fL_e v^2}{2Dg} \qquad 4.2$$

*Velocity head* ($h_v$): The head of a fluid as a result of its kinetic energy.

$$h_v = \frac{v^2}{2g} \qquad 4.3$$

*Atmospheric head* ($h_a$): Atmospheric pressure converted to feet of fluid being pumped.

$$h_a = \frac{p_a}{\gamma} \qquad 4.4$$

*Pressure head* ($h_p$): Pressure converted to feet of fluid being pumped.

$$h_p = \frac{p}{\gamma} \qquad 4.5$$

*Vapor pressure head* ($h_{vp}$): Fluid vapor pressure converted to feet of fluid being pumped. Steam tables can be used to evaluate the vapor pressure of water. Figure 4.9 can be used with hydrocarbons.

$$h_{vp} = \frac{p_{vp}}{\gamma} \qquad 4.6$$

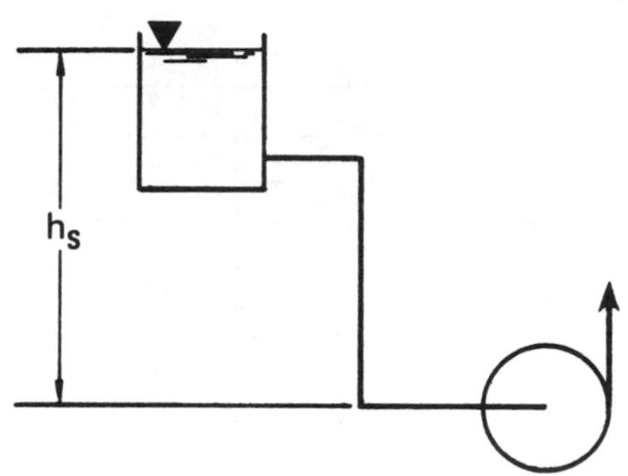


**Figure 4.6** Static Suction Head

*Static suction head* ($h_s$): The vertical distance in feet above the centerline of the inlet to the free level of the fluid source. If the free level of the fluid source is below the inlet, $h_s$ will be negative. In this case, $h_s$ is known as *static suction lift*.

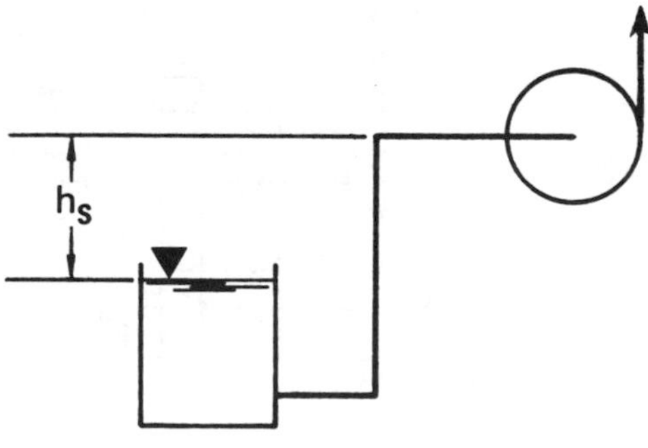


**Figure 4.7** Static Suction Lift

HYD MACHINE

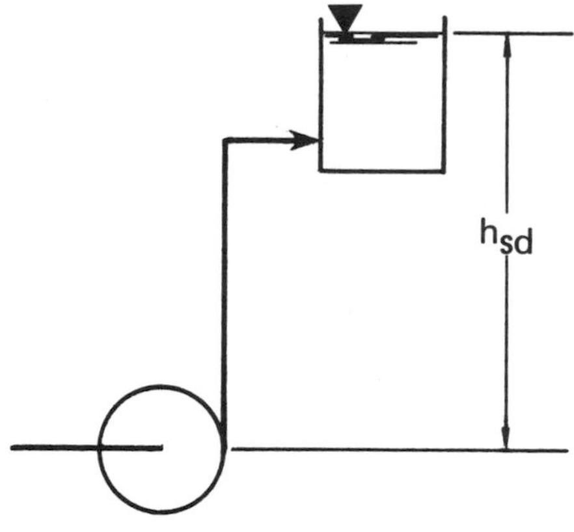


**Figure 4.8** Static Discharge Head

*Total (dynamic) suction head* ($H_s$): The static suction head minus the friction head in the suction line (i.e., the total energy of the fluid entering the impeller). If $h_s$ is negative (i.e., the free level of the fluid source is below the inlet), then $H_s$ will be negative. In this case, $H_s$ is known as *total (dynamic) suction lift.*

$$H_s = h_s - h_{f\,(s)} \tag{4.7}$$

*Static discharge head* ($h_{sd}$): The vertical distance in feet above the pump inlet centerline to the free level of the discharge tank or point of free discharge.

*Total (dynamic) discharge head* ($H_d$): The static discharge head plus the discharge velocity head plus the friction head in the discharge line (i.e., the total energy of the fluid leaving the pump).

$$H_d = h_{sd} + h_{vd} + h_{f\,(d)} \tag{4.8}$$

$$= h_{sd} + \frac{v_d^2}{2g} + h_{f(d)} \tag{4.9}$$



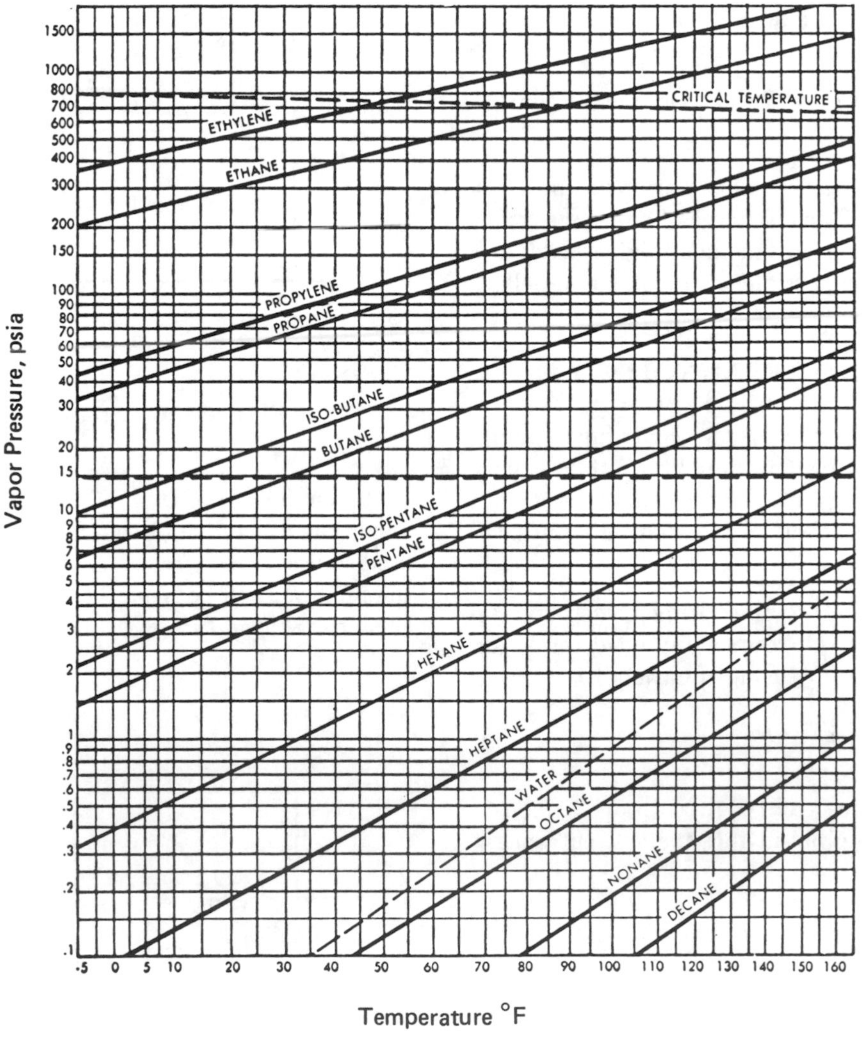


**Figure 4.9** Vapor Pressure of Hydrocarbons

*Total static head* ($h_{ts}$): The vertical distance in feet between the free level of the supply and either the point of free discharge or the free level of the discharge tank.

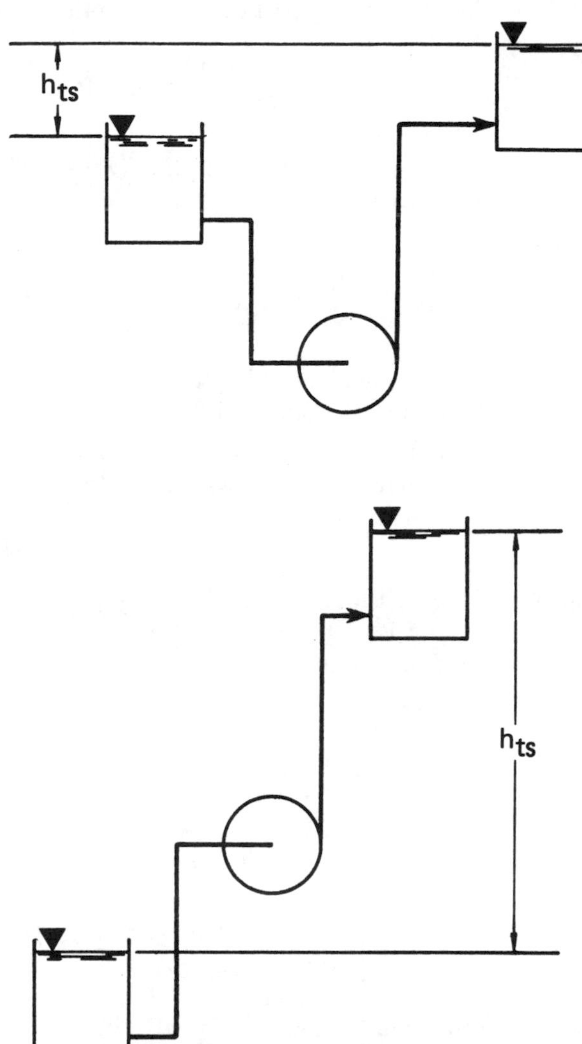


**Figure 4.10** Total Static Head

*Total (dynamic) head* ($H$): The total discharge head less the total suction head.

$$H = H_d - H_s \quad 4.10$$

$$= h_{sd} - h_s + \frac{v_d^2}{2g} + h_{f(s)} + h_{f(d)} \quad 4.11$$

## 5 NET POSITIVE SUCTION HEAD AND CAVITATION

Liquid is not sucked into a pump. A positive head (normally atmospheric pressure) must push the liquid into the impeller (i.e., "flood" the impeller). *Net Positive Suction Head Required* (*NPSHR*) is the minimum fluid energy required at the inlet by the pump for satisfactory operation. NPSHR is usually specified by the pump manufacturer.[1] *Net Positive Suction Head Available* (*NPSHA*) is the actual fluid energy at the inlet.[2]

$$\text{NPSHA} = h_a + h_s - h_{f\,(s)} - h_{vp} \quad 4.12$$

$$= h_{p\,(i)} + h_{v\,(i)} - h_{vp} \quad 4.13$$

If NPSHA is less than NPSHR, the fluid will cavitate. *Cavitation* is the vaporization of fluid within the casing or suction line. If the fluid pressure is less than the vapor pressure, pockets of vapor will form. As vapor pockets reach the surface of the impeller, the local high fluid pressure will collapse them, causing noise, vibration, and possible structural damage to the pump.

Cavitation may be caused by any of the following conditions:

1. Discharge heads far below the pump's calibrated head at peak efficiency.
2. Suction lift higher or suction head lower than the manufacturer's recommendation.
3. Speeds higher than the manufacturer's recommendation.
4. Liquid temperatures (thus, vapor pressures) higher than that for which the system was designed.

The following steps can be used to check for cavitation:

*step 1*: Determine the minimum NPSHR for the given pump. This should be given as part of the pump performance data. NPSHR follows the $Q^2$ law. If NPSHR is known for one flow rate, it can be determined for another flow rate from equation 4.14.

$$\frac{\text{NPSHR}_2}{\text{NPSHR}_1} = \left(\frac{Q_2}{Q_1}\right)^2 \quad 4.14$$

*step 2*: Calculate NPSHA from either equation 4.12 or 4.13.

*step 3*: If NPSHA is greater than NPSHR, cavitation will not occur. A good safety margin is

[1] If NPSHR is multiplied by the fluid density, it is known as *NIPR*, the *Net Inlet Pressure Required.* Similarly, NPSHA can be converted to *NIPA*.

[2] Equations 4.12 and 4.13 represent two totally different methods, both of which are correct, for calculating NPSHA. Equation 4.12 is based on the conditions at the fluid surface at the top of a tank. There is potential energy (the $h_s$ term) but no kinetic energy. Equation 4.13 is based on the conditions at the immediate entrance to the pump. At that point, some of the potential head has been converted to velocity head.

2–3 feet of fluid. If NPSHA is insufficient, it should be increased or the NPSHR should be decreased. NPSHA can be increased by:

a. increasing the height of the free fluid level of the supply tank

b. reducing the distance and minor losses between the supply tank and the pump, or by using a larger pipe size

c. reducing the temperature of the fluid

d. pressurizing the supply tank

e. reducing the flow rate or velocity

NPSHR can be reduced by:

a. placing a throttling valve in the discharge line (i.e., this will increase the total head, thereby reducing the capacity of the pump and driving its operating point into a region of lower NPSHR)

b. using a double suction pump

Applications which require very high NPSHR, such as boiler feed pumps needing 150 to 250 feet, should use booster pumps in front of the high NPSHR pumps. Such booster pumps are typically single stage, double suction pumps running at low speed. Their NPSHR can be 25 feet or less.

It is important to note that throttling the input line to a pump and venting or evacuating the receiving tank both work to increase cavitation. Throttling the input line increases the friction head term and decreases NPSHA. Evacuating the receiving tank increases the flow rate, which also increases the friction head term.

*Example 4.1*

2 cfs of water are pumped from a feed tank mounted on a platform to an open reservoir through 6″ schedule 40 ($\epsilon/D = 0.000293$) steel pipe. Determine the NPSHA.

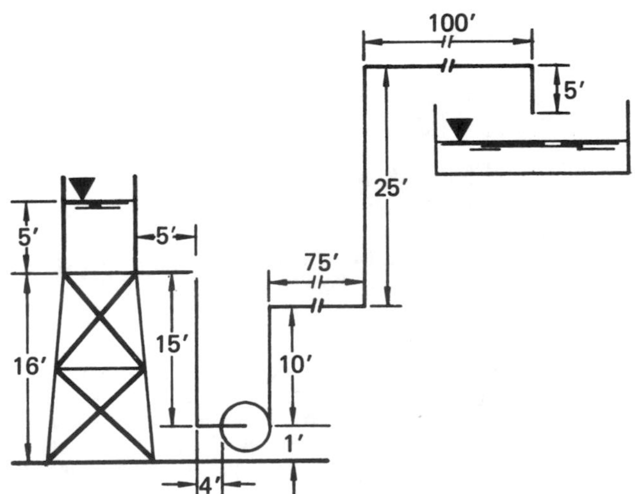


*step 1*: Assume 60°F and 14.7 psia. From equation 4.4,

$$h_a = \frac{(14.7)(144)}{62.4} = 33.9 \text{ ft}$$

*step 2*: For 6″ schedule 40 steel pipe, $D = 0.505$ ft, $A = 0.201 \text{ ft}^2$.

*step 3*:

$$v = \frac{Q}{A} = \frac{2}{0.201} = 9.95 \text{ ft/sec}$$

*step 4*: From appendix L, chapter 3, the equivalent lengths of the pipe and flanged fittings are:

| | |
|---|---|
| square entrance loss | 1 × 16 = 16 |
| 90° long radius elbows | 2 × 5.7 = 11.4 |
| pipe run (5 + 15 + 4) | 24 |
| | 51.4 ft |

*step 5*: At 60°, the kinematic viscosity of water is 1.217 EE−5 ft²/sec. The vapor pressure is 0.6 feet of water.

*step 6*: The Reynolds number is

$$N_{Re} = \frac{(0.505)(9.95)}{1.217 \text{ EE} - 5} = 4.13 \text{ EE5}$$

*step 7*: From the Moody friction factor chart, $f = 0.0165$, so

$$h_f = \frac{(0.0165)(51.4)(9.95)^2}{(2)(0.505)(32.2)} = 2.6 \text{ ft}$$

*step 8*: From equation 4.12,

$$\text{NPSHA} = 33.9 + 20 - 2.6 - 0.6 = 50.7 \text{ ft}$$

## 6 PUMPING HYDROCARBONS AND OTHER LIQUIDS

NPSHR is specified by pump manufacturers for use with cold (85°F) water.[3] Minor variations in the water temperature do not change the NPSHR appreciably. Experiments have shown, however, that the NPSHR can be reduced from the cold water values when hydrocarbons are pumped. This reduction apparently is due to the slow vapor release of complex organic liquids. Figure 4.11 gives the percentage correction to be applied to NPSHR values for cold water. Notice that the vapor pressure at the pumping temperature must be known. This can be obtained from figure 4.9.

[3] The term "cold *clear* water" is frequently used. This chapter omits the *clear* qualification.

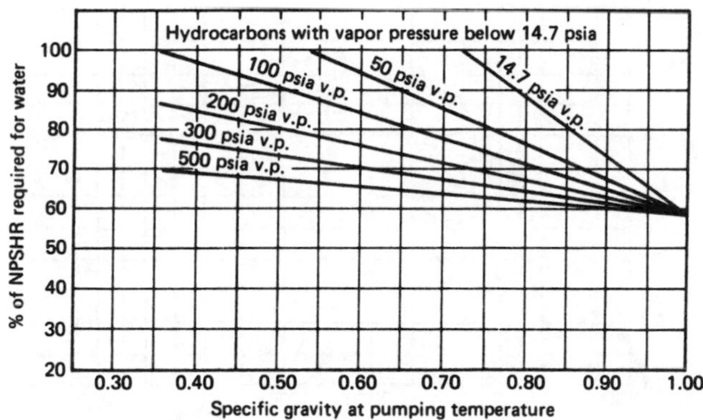


**Figure 4.11** Hydrocarbon NPSHR Correction Factor

Head developed by a pump is independent of the liquid being pumped, although the required horsepower is dependent on the fluid specific gravity. Because of this independence, pump performance curves from water tests can be used with other Newtonian fluids (e.g., gasoline, alcohol, and saline solutions).

High viscosity fluids, however, result in decreased head and capacity, as well as in an increase in input horsepower when compared to the pumping of water. Therefore, corrections are required when using water-based pump curves with viscous fluids. The parameters that would be used with water curves are calculated from equations 4.15 through 4.17.

$$H_{\text{water}} = \frac{H_{\text{viscous}}}{C_H} \tag{4.15}$$

$$Q_{\text{water}} = \frac{Q_{\text{viscous}}}{C_Q} \tag{4.16}$$

$$\eta_{\text{water}} = \frac{\eta_{\text{viscous}}}{C_E} \tag{4.17}$$

No exact method for determining the correction factors, other than actual tests of an installation with both fluids, exists. Nevertheless, several sources have produced correction factor charts based on experiments in limited viscosity and size ranges. One such chart is reproduced as Appendix E of this chapter.

*Example 4.2*

10°F liquid iso-butane (vapor pressure = 15 psia, specific gravity = 0.58) is to be used with a centrifugal pump whose NPSHR is 12 psia with cold water. What NPSHR should be used with the butane?

From figure 4.11, the intersection of 0.58 specific gravity and 15 psia is above the horizontal line. Therefore, the NPSHR is a full 12 psia.

## 7 RECIRCULATION

Cavitation is a high-volume problem. If a pump is forced to operate a flow rate higher than originally intended, one of the results could be cavitation.

Operating a pump at a flow rate much less than it was designed for, on the other hand, can result in *recirculation*. Recirculation of the fluid at both the impeller inlet and the pump outlet is possible. Such recirculation produces characteristics similar to cavitation. Specifically, vibration and noise are produced when the fluid energy is reduced through internal friction and fluid shear.

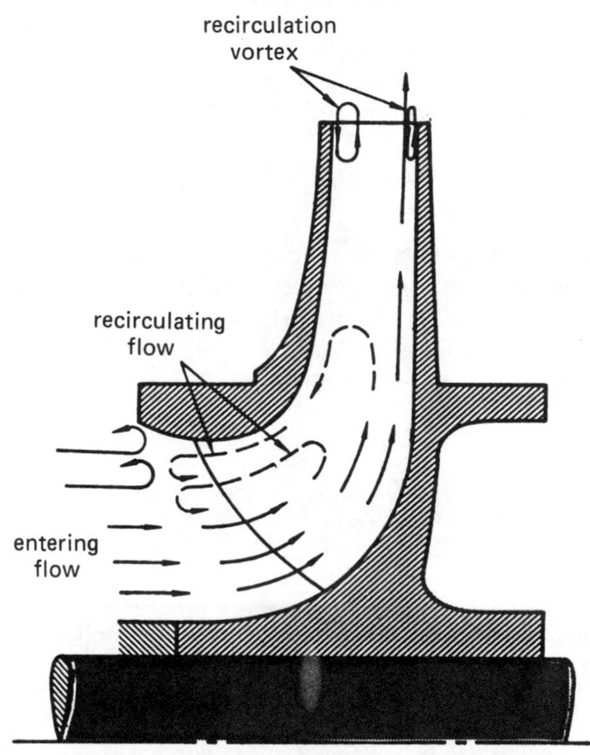


**Figure 4.12** Low-Volume Recirculation

## 8 PUMPING POWER AND EFFICIENCY

The energy (head) added by a pump can be determined by evaluating Bernoulli's equation on either side of the pump. Writing Bernoulli's equation for the discharge and inlet conditions produces equation 4.18.

$$h_A = \frac{p_d}{\gamma} - \frac{p_i}{\gamma} + \frac{v_d^2}{2g} - \frac{v_i^2}{2g} + z_d - z_i \tag{4.18}$$

The work performed by a pump is a function of the total head and the mass of the liquid pumped in a given time. Pump output is measured in *hydraulic horsepower, whp.*[4] Relationships for finding the hydraulic horsepower are given in table 4.2.

**Table 4.2**
Hydraulic Horsepower Equations

| | $Q$ in gpm | $\dot{m}$ in lbm/sec | $\dot{V}$ in cfs |
|---|---|---|---|
| $h_A$ is added head in feet | $\frac{h_A Q\,(S.G.)}{3956}$ | $\frac{h_A \dot{m}}{550}$ | $\frac{h_A \dot{V}\,(S.G.)}{8.814}$ |
| p is added head in psf | $\frac{pQ}{2.468\text{EE}5}$ | $\frac{p\dot{m}}{(34320)(S.G.)}$ | $\frac{p\dot{V}}{550}$ |

The input horsepower delivered to the pump shaft is the *brake horsepower.*

$$bhp = \frac{whp}{\eta_p} \qquad 4.19$$

The difference between hydraulic horsepower and brake horsepower is the power lost within the pump due to mechanical and hydraulic friction. This is referred to as *heat horsepower* or *friction horsepower* and is determined from equation 4.20.

$$fhp = bhp - whp \qquad 4.20$$

*Electrical horsepower* to the motor is

$$ehp = \frac{bhp}{\eta_m} \qquad 4.21$$

*Overall efficiency* is the pump efficiency multiplied by the motor efficiency:

$$\eta = (\eta_p)(\eta_m) = \frac{whp}{ehp} \qquad 4.22$$

Ideal pump efficiency is a function of the flow rate and specific speed. Figure 4.13 can be used if both quantities are known.

Larger horsepower pumps usually are driven by three-phase *induction motors.* The *synchronous speed* of such a motor is the speed of the rotating field.

$$n = \frac{(120)(f)}{\text{no. of poles}} \qquad 4.23$$

The frequency, $f$, is either 60 Hz (cycles per second) or 50 Hz, depending on the location of the installation.[5] The number of poles can be two or usually some multiple of four, but it must be an even number.

[4] This term also is known as *water horsepower.*

[5] 50 Hz is used in Europe.

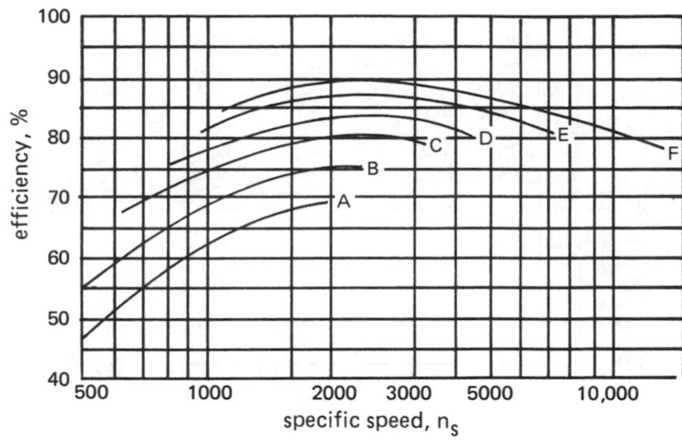


curve A: 100 gpm
curve B: 200 gpm
curve C: 500 gpm
curve D: 1000 gpm
curve E: 3000 gpm
curve F 10,000 gpm

**Figure 4.13** Efficiency versus Specific Speed

**Table 4.3**
Synchronous Speeds

| Number of Poles | Synchronous Speed |
|---|---|
| 2 | 3600 |
| 4 | 1800 |
| 6 | 1200 |
| 8 | 900 |
| 12 | 600 |
| 18 | 400 |
| 24 | 300 |
| 48 | 150 |

Induction motors do not run at their synchronous speeds. Rather, they run at slightly less than synchronous speed. The percentage deviation is known as the *slip.* Slip is seldom greater than 10%, and it is usually much less than that. 4% is a good estimate for evaluation studies.

$$s = \frac{n_{\text{synchronous}} - n_{\text{actual}}}{n_{\text{synchronous}}} \qquad 4.24$$

Of course, special gear or belt drives can be used with various reduction ratios to obtain any required operating speed. For belt drive applications, most motors are of the 1800 and 1200 rpm varieties. Running speeds of 1750 and 1150 rpm are typically assumed for these motors when fully loaded.

Table 4.4 lists common motor sizes in horsepower. It is always best to specify standard motor sizes when possible.

**Table 4.4**
Standard Motor Sizes (BHP)

| |
|---|
| $\frac{1}{8}, \frac{1}{6}, \frac{1}{4}, \frac{1}{3}$ |
| 0.5, 0.75, 1, 1.5, 2, 3, 5, 7.5 |
| 10, 15, 20, 25, 30, 40, 50, 60 |
| 75, 100, 125, 150, 200, 250 |

Induction motors are usually specified in terms of their KVA (kilo-volt-amp) ratings. KVA is not the same as the motor power in kilowatts, although one can be derived from the other if the motor's power factor is known. Such power factors can range from 0.8 to 0.9, depending on the installation and motor size.

$$\text{KVA rating} = \frac{\text{kilowatt power}}{\text{power factor}} \qquad 4.25$$

$$\text{kilowatt power} = (0.7457)\,ehp \qquad 4.26$$

If exact flow is not critical when changes in system and pump conditions occur, a constant speed drive can be used. Integral gear motors and motor-reducer drives are rugged, self-contained drives generally using 1800 rpm (typically taken as 1750 rpm to account for slip), three-phase induction motors and helical gear reducers. Horsepowers up to 50 are commonly available. With the gear reductions, the following approximate stock speeds are obtained: 37, 45, 56, 68, 84, 100, 125, 155, 180, 230, 280, 350, 420, 520, and 640 rpm.

V-belts are usually the lowest initial cost constant speed drive. The V-belt drive provides some flexibility for changing pump speeds through changes in sheave size. Using readily available standard motors of 1200 and 1800 (taken as 1150 and 1750) rpm, a range of pump speeds is possible. Due to sheave size and space limitations, the useful range of pump speeds is generally 200 to 600 rpm.

One disadvantage of a V-belt drive is the side load or overhung load it puts on both the pump and motor shafts and bearings. This problem is particularly significant with low speeds and high horsepowers. If it is determined that the overhung load is excessive, it will be necessary to use a jack drive or outboard bearing installation.

The integral gear motor is generally more compact, lower in cost, and easier to install. However, the motor and separate reducer is sometimes preferred because of its flexibility, especially in changing standard motors for maintenance.

If variable flow will require variations in motor speed, there are many types of packaged variable speed drives available. Such drives offer the ability to adjust pump speed to control flow, as well as adjusting for changes in the system and eventual pump wear. It is important to match the torque capacity of the drive with the required pump torque with all types of variable speed drives.

Belt type variable speed drives are available in a wide choice of horsepower and speed ranges, and they provide a compact drive at reasonable cost. Traction type drives are infinitely variable from zero speed, and they are reversible. Electronic variable speed drives, using both DC and AC motors with variable voltage or frequency to vary speed, generally require a speed reducer to obtain the required torque at lower pump speeds. Hydraulic drives, either packaged or custom designed, have excellent high torque capacity over a broad speed range.

When the calculated speed and required horsepower are known, a conservative approach is to select the next lower stock speed and a stock horsepower equal to or greater than the requirement. If a minimum flow must be maintained, even with system changes and pump wear, the next higher speed may be needed. In this case, the system should be recalculated, as the higher speed and resulting higher flow and pressure drop will require more horsepower. The motor/drive selected must be able to supply this horsepower.

*Example 4.3*

Recommend a 6-pole induction motor size for the pump in example 4.1. The friction loss in the discharge line is 12.0 feet. Neglect the electrical motor efficiency. Assume the pump is primed.

The Bernoulli equation can be used to find the head added by the pump.

$$0 + 0 + 20 + h_A = 0 + \frac{(9.95)^2}{2\,(32.2)} + 30 + 2.6 + 12$$

$$h_A = 26.14 \text{ ft}$$

From table 4.2,

$$whp = \frac{(26.14)(2)(1)}{8.814} = 5.93 \text{ hp}$$

The flow rate is

$$\frac{(2)}{(0.002228)} = 897.7 \text{ gpm}$$

Assuming a motor speed of 1150 rpm, equation 4.27 gives the specific speed

$$n_s = \frac{1150\sqrt{897.7}}{(26.14)^{0.75}} = 2980 \text{ rpm}$$

From figure 4.13, a pump efficiency of about 82% can be expected. So, the minimum motor horsepower would be

$$\frac{5.93}{0.82} = 7.23$$

From table 4.4, choose a 7.5 hp or larger motor.

## 9 SPECIFIC SPEED

The capacity or flow rate of a centrifugal pump is governed by the impeller thickness. For a given impeller diameter, the deeper the vanes, the greater the capacity of the pump.

For a desired flow rate or a desired discharge head, there will be one optimum impeller design. The impeller that is best for developing a high discharge pressure will have different proportions from an impeller designed to produce a high flow rate. The quantitative index of this optimization is called *specific speed* ($n_s$).



Specific speed is a function of the a pump's capacity, head, and rotational speed at peak efficiency. For a given pump and impeller configuration, the specific speed remains essentially constant over a range of flow rates and heads. Theoretically, specific speed is the speed in rpm at which a homologous pump would have to turn in order to put out 1 gpm at 1 foot total head. (For double suction pumps, $Q$ in equation 4.27 is not divided by 2.)

$$n_s = \frac{n\sqrt{Q}}{(h_A)^{0.75}} \quad 4.27$$

Specific speed is used as a guide to selecting the most efficient pump type. Given a desired flow rate, pipeline geometry, and motor speed, $n_s$ is calculated from equation 4.27. The type of impeller is chosen from table 4.5

**Table 4.5**
Specific Speed versus Impeller Types

| approximate range of specific speed (rpm) | impeller type |
|---|---|
| 500–1000 | radial vane |
| 2000–3000 | Francis (mixed) vane |
| 4000–7000 | mixed flow |
| 9000 and above | axial flow |

Highest heads per stage are developed at low specific speeds. However, for best efficiency, a centrifugal pump's specific speed should be greater than 650 at its operating point. At low specific speeds, the impeller diameter is large with high mechanical friction and hydraulic losses. If the specific speed for a given set of conditions drops below 650, a multiple stage pump should be selected.[6]

As the specific speed increases, the ratio of the impeller diameter to the inlet diameter decreases. As this ratio decreases, the pump is capable of developing less head. Best efficiencies are usually obtained from pumps with specific speeds between 1500 and 3000. At specific speeds of 10,000 or higher, the pump is suitable for high flow rates but low discharge heads.

Other uses for specific speed are:

- If the pump and impeller are known, a maximum specific speed can be determined from table 4.5. This maximum specific speed can be translated into maximum values of rpm and flow rate, as well as a minimum value of total head added.
- If specific speed is known, an approximate pump efficiency can be found from figure 4.13.
- Specific speed limits have been established by the Hydraulic Institute.[7] These limits are presented in graphical form at the end of this chapter.

If NPSHR is substituted for total head in the expression for specific speed, a formula for *suction specific speed* results.

Suction specific speed is an index of the suction characteristics of the impeller.

$$n_{ss} = \frac{n\sqrt{Q}}{\text{NPSHR}^{0.75}} \quad 4.28$$

Ideally, $n_{ss}$ should be approximately 7900 for single suction pumps and 11,200 for double suction pumps. (The value of $Q$ in equation 4.28 should be halved for double suction pumps.) If these ideal values are assumed, equation 4.28 can be solved for approximate values of NPSHR.

$$\text{NPSHR}_{\text{single suction}} \approx (6.36\ \text{EE}-6)\, n^{1.33} Q^{0.67} \quad 4.29$$

$$\text{NPSHR}_{\text{double suction}} \approx (3.99\ \text{EE}-6)\, n^{1.33} Q^{0.67} \quad 4.30$$

*Example 4.4*

A 3600 rpm pump ($Q$ = 150 gpm) is used to increase fluid pressure from 35 psi to 220 psi. The pump adds negligible velocity or potential energy. If efficiency is to

[6] Relatively recent advances in *partial emission, forced vortex centrifugal pumps* allow operation down to specific speeds of 150. Such partial emission pumps use radial vanes and a single tangential discharge. Discharge is through a conical diffuser in which the energy conversion occurs. Partial emission pumps have been used for low flow, high head applications, such as petrochemical high-pressure cracking processes.

[7] Hydraulic Institute, 30200 Detroit Road, Westlake, OH 44145.

be maximized, how many stages should be used? Assume that each stage adds its proportionate share of head.

The head added is

$$h_A = \frac{(220 - 35)\,(144)}{62.4} = 426.9 \text{ ft}$$

From equation 4.27, the specific speed is

$$n_s = \frac{(3600)\sqrt{150}}{(426.9)^{0.75}} = 470$$

From figure 4.13, the approximate efficiency is 45%.

Try a two stage pump. The head is split between the stages.

$$n_s = \frac{(3600)\sqrt{150}}{\left(\frac{426.9}{2}\right)^{0.75}} = 790$$

From figure 4.13, the efficiency is approximately 60%.

This process continues until the specific speed reaches 2000, at which time the number of stages will be approximately seven. It is questionable, however, whether the cost of multi-staging is worthwhile in this low-volume application.

*Example 4.5*

A direct driven pump is to discharge 150 gpm against a 300 foot total head when turning at the fully-loaded speed of 3500 rpm. What type of pump should be selected?

Calculate the specific speed from equation 4.27.

$$n_s = \frac{3500\sqrt{150}}{(300)^{0.75}} = 594.7 \text{ rpm}$$

From table 4.5, the pump should be a radial vane type. However, pumps with best efficiencies have $n_s$ greater than 650. To increase the specific speed, the rotational speed can be increased or the total head can be decreased. Since 3600 rpm is the maximum practical speed for induction motors, the better choice would be to divide the total head between two stages (or to use two pumps in a series).

In a two stage system, the specific speed would be:

$$n_s = \frac{3500\sqrt{150}}{(150)^{0.75}} = 1000$$

This is satisfactory for a radial vane pump.

## 10 AFFINITY LAWS—CENTRIFUGAL PUMPS

Most parameters (impeller diameter, speed, and flow rate) determining a pump's performance can vary. If the impeller diameter is held constant and the speed varied, the following ratios are maintained with no change of efficiency:

$$\frac{Q_2}{Q_1} = \frac{n_2}{n_1} \qquad 4.31$$

$$\frac{h_2}{h_1} = \left(\frac{n_2}{n_1}\right)^2 = \left(\frac{Q_2}{Q_1}\right)^2 \qquad 4.32$$

$$\frac{bhp_2}{bhp_1} = \left(\frac{n_2}{n_1}\right)^3 = \left(\frac{Q_2}{Q_1}\right)^3 \qquad 4.33$$

These relationships assume that the efficiencies of the larger and smaller pumps are the same. In reality, larger pumps will be more efficient than smaller pumps. Therefore, extrapolations to much larger or much smaller sizes should be avoided.

Equation 4.34 can be used to predict the efficiency of a larger or smaller pump based on homologous data.

$$\frac{1 - \eta_{\text{smaller}}}{1 - \eta_{\text{larger}}} = \left(\frac{d_{\text{larger}}}{d_{\text{smaller}}}\right)^{0.2} \qquad 4.34$$

If the speed is held constant and the impeller size varied,

$$\frac{Q_2}{Q_1} = \frac{d_2}{d_1} \qquad 4.35$$

$$\frac{h_2}{h_1} = \left(\frac{d_2}{d_1}\right)^2 \qquad 4.36$$

$$\frac{bhp_2}{bhp_1} = \left(\frac{d_2}{d_1}\right)^3 \qquad 4.37$$

*Example 4.6*

A pump delivers 500 gpm against a total head of 200 feet operating at 1770 rpm. Changes have increased the total head to 375 feet. At what rpm should this pump be operated to achieve this new head at the same efficiency?

From equation 4.32,

$$n_2 = 1770\sqrt{\frac{375}{200}} = 2424 \text{ rpm}$$

## 11 PUMP SIMILARITY

The performance of one pump can be used to predict the performance of a *dynamically similar* (*homologous*) *pump*. This can be done by using equations 4.38 through 4.40.

$$\frac{n_1 d_1}{\sqrt{h_1}} = \frac{n_2 d_2}{\sqrt{h_2}} \qquad 4.38$$

$$\frac{Q_1}{d_1^2 \sqrt{h_1}} = \frac{Q_2}{d_2^2 \sqrt{h_2}} \qquad 4.39$$

$$\frac{bhp_1}{\rho_1 d_1^2 h_1^{1.5}} = \frac{bhp_2}{\rho_2 d_2^2 h_2^{1.5}} \qquad 4.40$$

$$\frac{Q_1}{n_1 d_1^3} = \frac{Q_2}{n_2 d_2^3} \qquad 4.41$$

$$\frac{bhp_1}{\rho_1 n_1^3 d_1^5} = \frac{bhp_2}{\rho_2 n_2^3 d_2^5} \qquad 4.42$$

$$\frac{n_1 \sqrt{Q_1}}{(h_1)^{0.75}} = \frac{n_2 \sqrt{Q_2}}{(h_2)^{0.75}} \qquad 4.43$$

HYD MACHINE

These so-called *similarity laws* assume that both pumps

- operate in the turbulent region.
- have the same operating efficiency.
- operate with the same percentage of wide-open flow.

Similar pumps also will have the same specific speed and cavitation number.

*Example 4.7*

A 6″ pump operating at 1770 rpm discharges 1500 gpm of cold water (S.G. = 1.0) against an 80 foot head at 80% efficiency. A homologous 8″ pump operating at 1170 rpm is being considered as a replacement. What capacity and total head can be expected from the new pump? What would be the new power requirement?

From equation 4.38,

$$H_2 = \left[\frac{(8)(1170)}{(6)(1770)}\right]^2 (80) = 62.14 \text{ ft}$$

From equation 4.41,

$$Q_2 = \left[\frac{(1170)(8)^3}{(1770)(6)^3}\right] (1500) = 2350.3 \text{ gpm}$$

$$whp_2 = \frac{(2350.3)(62.14)(1.0)}{3956} = 36.92 \text{ hp}$$

$$bhp_2 = \frac{36.92}{0.8} = 46.2 \text{ hp}$$

## 12 THE CAVITATION NUMBER

Although it is difficult to predict when cavitation will occur, the *cavitation number* (*cavitation coefficient*) can be used in modeling and in comparing experimental results. The actual cavitation number, given by equation 4.44, is compared with the critical cavitation number from experimental results. If the critical cavitation number is larger, it is concluded that cavitation will result.

$$\sigma = \frac{p - p_{vp}}{\frac{\rho v^2}{2g}} = \frac{\text{NPSHA}}{h_A \text{ per stage}} \qquad 4.44$$

The two forms of equation 4.44 yield slightly different results. The first form is essentially the ratio of the net pressure available for collapsing a vapor bubble to the velocity pressure creating the vapor. The first form is useful in model experiments, whereas the second form is applicable to tests of production model pumps.

## 13 PUMP PERFORMANCE CURVES

Evaluating the performance of a pump is often simplified by examining a graphical representation of its operating characteristics. For a given impeller diameter and constant speed, the head added by a centrifugal pump will decrease as the flow rate increases. This is illustrated by figure 4.14. Other operating characteristics also vary with the flow rate. These can be presented on individual graphs. However, since the independent variable (flow rate) is the same for all, common practice is to plot all charcteristics together on a single graph. A pump *performance curve* is illustrated in figure 4.15.

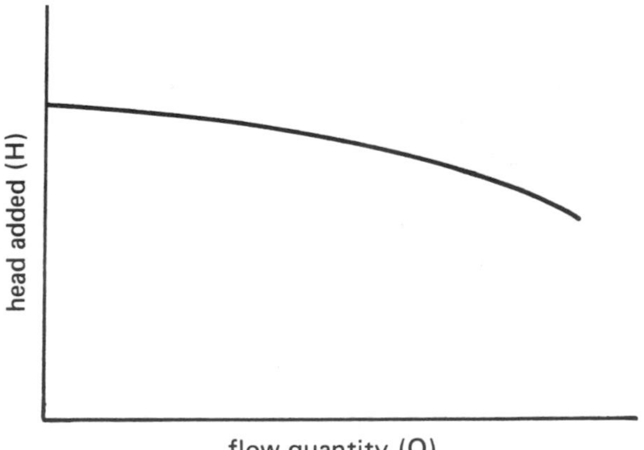


**Figure 4.14** Head versus Flow Rate

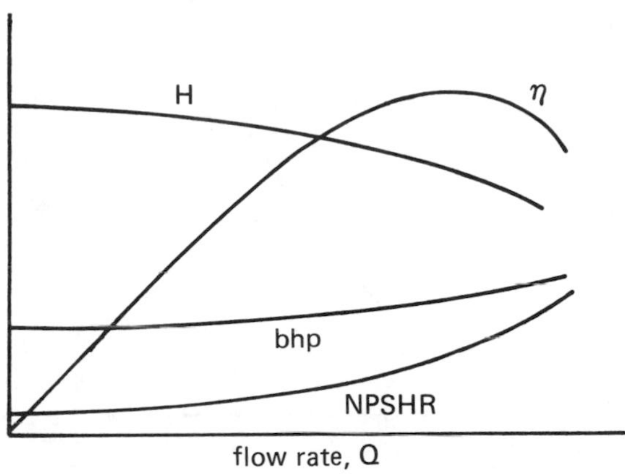


**Figure 4.15** Pump Performance Curves

Figures 4.14 and 4.15 are for a pump with a fixed impeller diameter and rotational speed. The characteristics of a pump operated over a range of speeds are illustrated in figure 4.16. For maximum efficiency, the operating point should fall along the dotted line.

Manufacturers' performance curves show pump performance at a limited number of calibration speeds. The desired operating point can be outside the range of the published curves. It is then necessary to estimate a speed at which the pump would give the required performance. This is done by using the affinity laws, as illustrated in example 4.8.

*Example 4.8*

A pump with the 1750 rpm performance curve shown is required to pump 500 gpm at 425 feet total head. At what speed must this pump be driven to achieve the desired performance with no change in efficiency or impeller size?

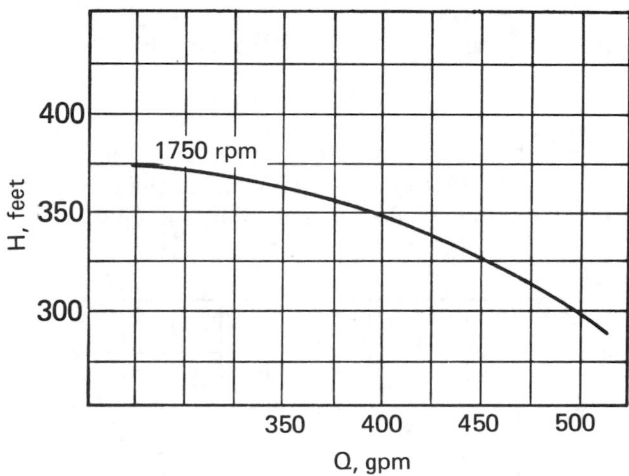




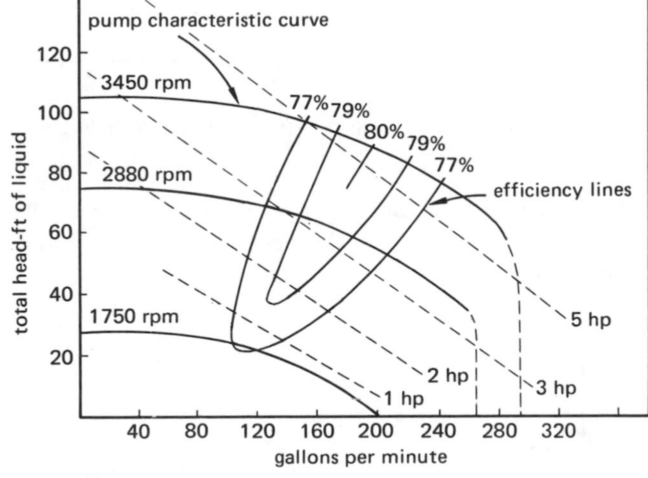


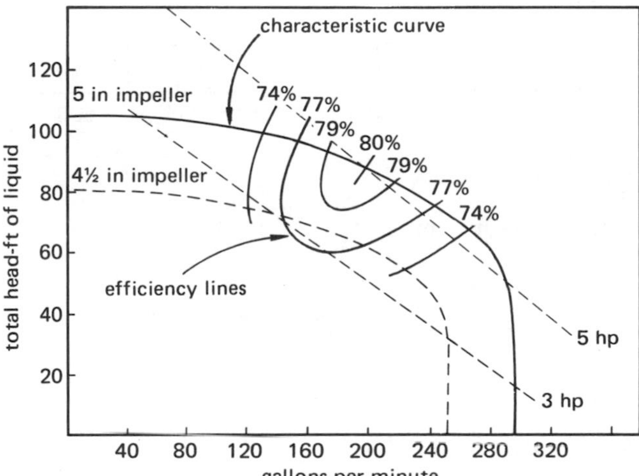


**Figure 4.16** Two Types of Characteristic Curves

From equation 4.32, the quantity $(H/Q^2)$ is constant for a pump with a given impeller size. In this case,

$$\frac{425}{(500)^2} = 1.7 \text{ EE} - 3$$

In order to apply an affinity equation, it is necessary to know the operating point on the 1750 rpm curve. To find the operating point, choose random values of Q and solve for H such that $(H/Q^2) = 1.7 \text{ EE} - 3$.

| Q | H |
|---|---|
| 475 | 383 |
| 450 | 344 |
| 425 | 307 |
| 400 | 272 |

These four points are plotted on the performance curve graph. The intersection of the constant efficiency line and the original 1750 rpm curve is at 440 gpm.

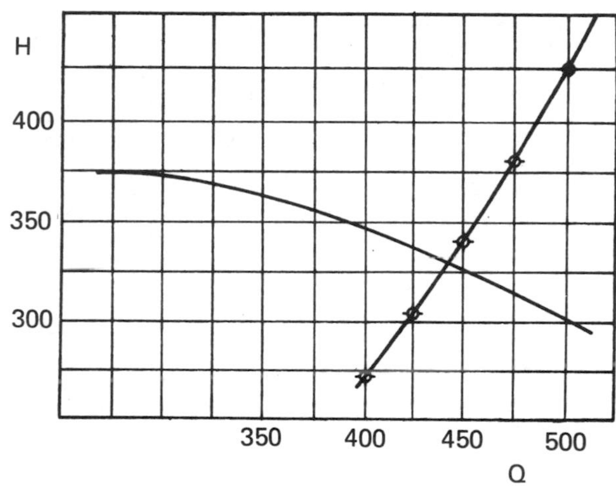


Then, from equation 4.31,

$$n_2 = 1750\left(\frac{500}{440}\right) = 1989 \text{ rpm}$$

## 14 SYSTEM CURVES

A *system curve* graph can also be made from the resistance to flow of the piping system. This resistance varies with the square of the flow rate since $h_f$ varies with $v^2$ in the Darcy friction formula.

$$\frac{H_1}{Q_1^2} = \frac{H_2}{Q_2^2} \qquad 4.45$$

Equation 4.45 is illustrated by figure 4.17, in which there is no static head ($h_{ts}$) to overcome.

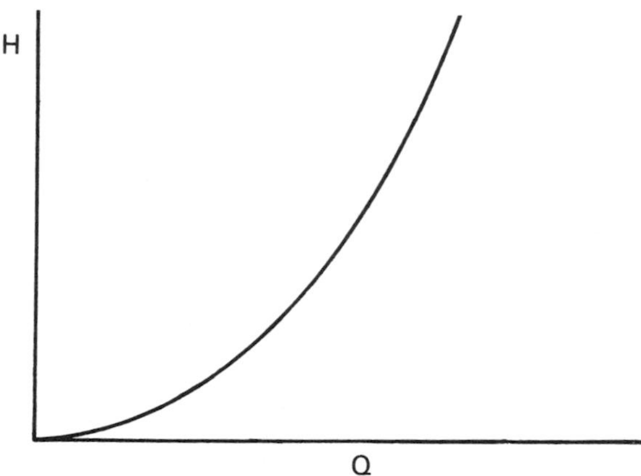


**Figure 4.17** System Performance Curve (Dynamic Losses Only)

When a static head ($h_{ts}$) exists in a system, the loss curve is displaced upward an amount equal to the static head. This is illustrated in figure 4.18.

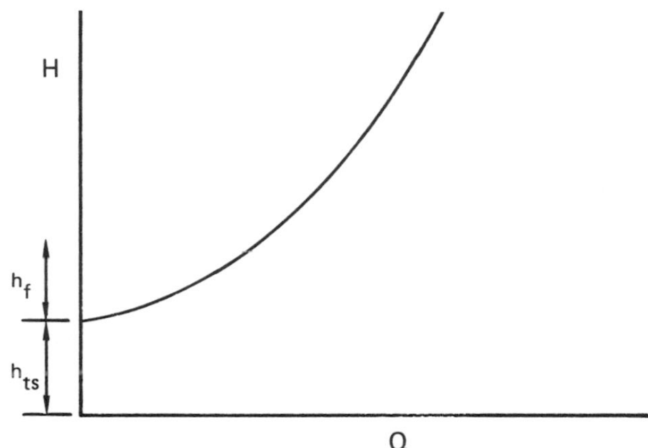


**Figure 4.18** System Performance Curve

The intersection of the pump characteristic curve with the system curve defines the *operating point* as shown in figure 4.19.

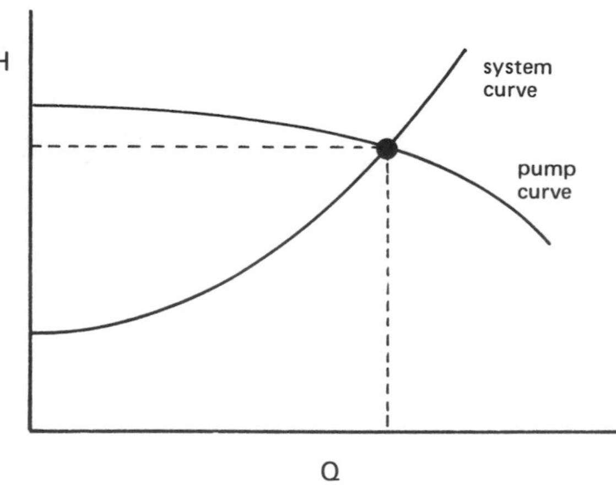


**Figure 4.19** Operating Point

After a pump is installed, it may be desired to vary the pump's performance. If a valve is placed in the discharge line, the operating point may be moved along the performance curve by opening or closing the valve. This is illustrated in figure 4.20. (A throttling valve should never be placed in the intake line since that would reduce NPSHA.)

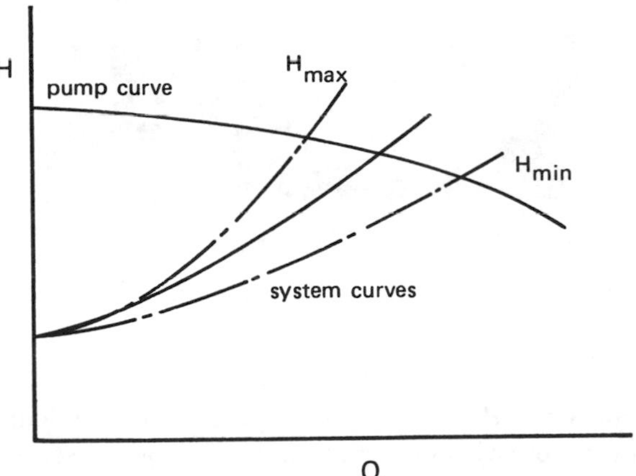


**Figure 4.20** Effect of Throttling the Discharge

In most systems, the static head will vary as the feed tank is drained or as the discharge tank fills. The system head is then defined by a pair of parallel curves intersecting the performance curve. The two intercept points are the maximum and the minimum capacity requirements.

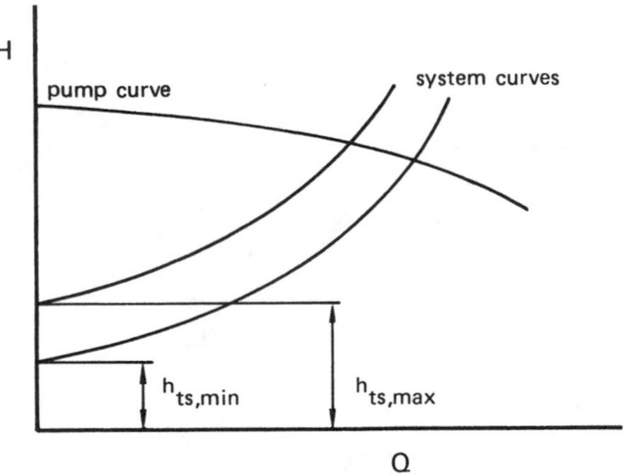


**Figure 4.21** Extreme Operating Points

## 15 PUMPS IN SERIES AND PARALLEL

Parallel operation is obtained by having two pumps discharging into a common header. This type of connection is advantageous when the system demand varies greatly. A single pump providing total flow would have to operate far from its optimum efficiency at one point or another. With two pumps in parallel, one can be shut down during low demand. This allows the remaining pump to operate close to its optimum efficiency point.

Figure 4.22 illustrates that parallel operation increases the capacity of the system while maintaining the same total head.

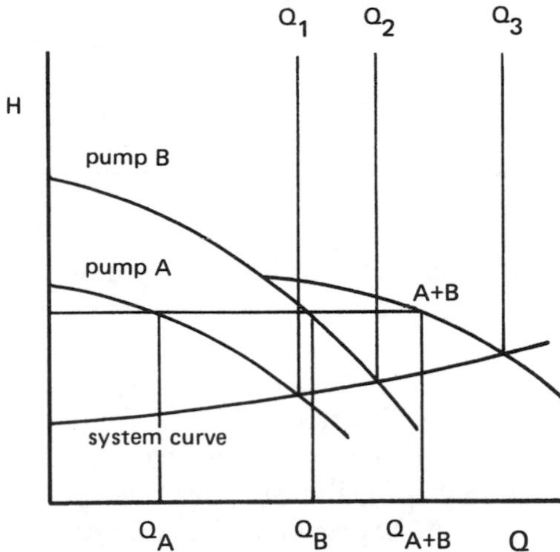


**Figure 4.22** Pumps Operating in Parallel

The performance curve for a set of pumps in parallel can be plotted by adding the capacities of the two pumps at various heads. Capacity does not increase at heads above the maximum head of the smaller pump. Furthermore, a second pump will operate only when its discharge head is greater than the discharge head of the pump already running.

When the parallel performance curve is plotted with the system head curve, the operating point is the intersection of the system curve with the $A + B$ curve. With pump A operating alone, the capacity is given by $Q_1$. When pump B is added, the capacity increases to $Q_3$ with a slight increase in total head.

Series operation is achieved by having one pump discharge into the suction of the next. This arrangement is used primarily to increase the discharge head, although a small increase in capacity also results.

The performance curve for a set of pumps in series can be plotted by adding the heads of the two pumps at various capacities.

## 16 CONSIDERATION FOR WASTEWATER PLANTS

The primary consideration in choosing a pump to lift sewage is the pump's tendency to clog. Centrifugal pumps for sewage and liquids with large solids should always be of the single-suction type with non-clog, open

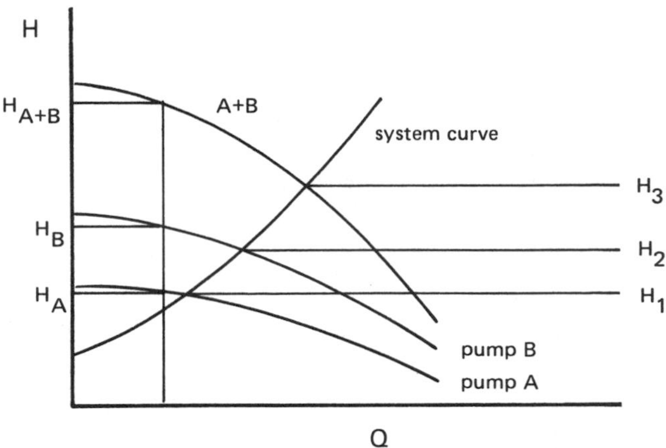


**Figure 4.23** Pumps Operating in Series

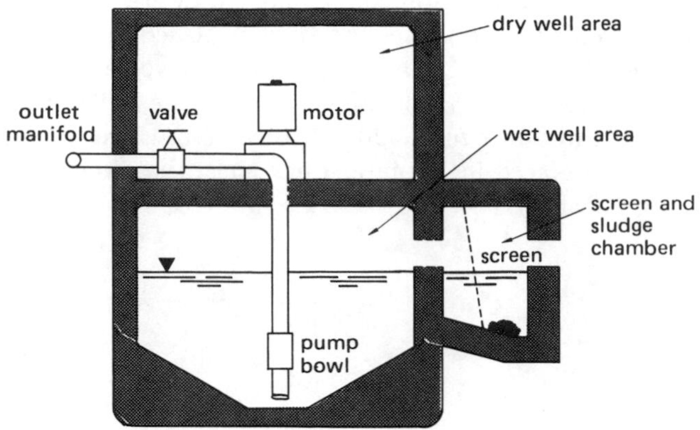


**Figure 4.24** Typical Wastewater Pump Installation (greatly simplified)

impellers. (Double suction pumps are prone to clogging because rags will catch and wrap around the shaft which extends through the impeller eye.) Clogging can be further minimized by limiting the number of impeller vanes to two or three, providing for large passageways, and using a bar screen ahead of the pump.



Non-clog pumps are of heavy construction but are constructed for ease of cleaning and repair. Horizontal pumps usually have a split casing, one-half of which can be removed for maintenance. A hand-sized cleanout opening may also be built into the casing. Although designed for long life, a sewage pump should normally be used with a grit chamber for prolonged bearing life.

The solid-handling capacity of a pump may be given in terms of the largest sphere which can pass through it without clogging. For example, a wastewater pump with a 6″ inlet should be able to pass a 4″ sphere. Of course, the pump should also be capable of handling spheres with diameters slightly larger than the bar screen spacing.

Figure 4.24 shows a simplified wastewater pump installation. Not shown are instrumentation and water level measurement devices, baffles, lighting, drains for the dry well, electrical power, pump lubrication equipment, and access holes. In addition, the multiplicity of pumping equipment is not apparent from the figure. (Totally submerged pumps do not require dry wells. However, such pumps may be more difficult to access, maintain, and repair.)

The number of pumps used in a wastewater installation is largely dependent on expected demand, pump capacity, and design criteria for backup operation. Although there may be state and federal regulations affecting the design, it is considered good practice to install pumps in sets, with a backup pump being available for each set of pumps that performs the same function. The number of pumps and their capacities should be able to handle the peak flow with one pump in the set out of service.

## 17 IMPULSE TURBINES

As shown in figure 4.25, an *impulse turbine* converts the energy of a fluid stream into kinetic energy by use of a nozzle which directs the stream jet against the turbine blades. Impulse turbines are generally employed where the available head exceeds 800 feet. Their efficiencies are typically in the 80% to 90% range.

The *total head available* to an impulse turbine is given by equation 4.46. ($p$ is the pressure of the fluid at the nozzle entrance.)

$$H' = H - h_n = \frac{p}{\gamma} + \frac{v^2}{2g} - h_n \qquad 4.46$$

$$= h_s - \frac{fL_e v^2}{2Dg} - h_n \qquad 4.47$$

The *nozzle loss* is

$$h_n = \left(\frac{p}{\gamma} + \frac{v^2}{2g}\right)(1 - C_v^2) \qquad 4.48$$

The velocity of the fluid jet is

$$v_j = \sqrt{2gH'} = C_v\sqrt{2gH} \qquad 4.49$$

The energy transmitted by each pound of fluid to the turbine runner is given by equation 4.50.

$$E = \frac{v_T(v_j - v_T)}{g}(1 - \cos\beta) \qquad 4.50$$

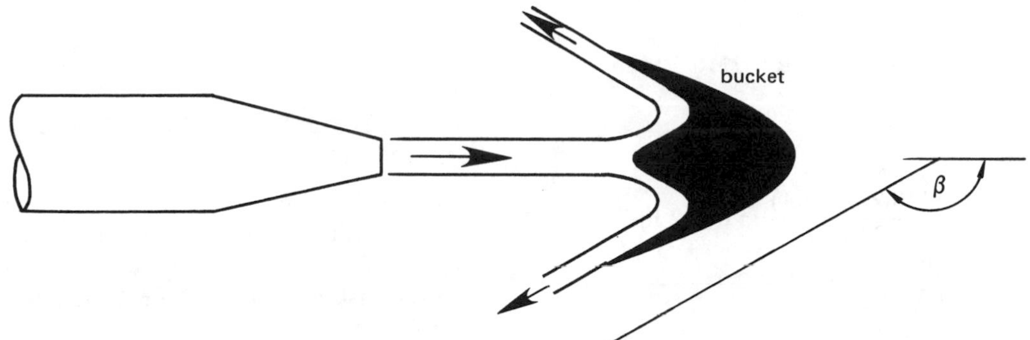


**Figure 4.25** A Simple Impulse Turbine

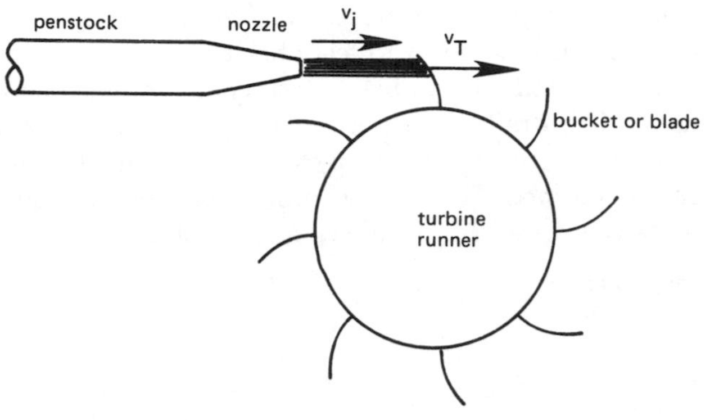


**Figure 4.26** Turbine Blade Geometry

Multiplying the energy by the fluid flow rate gives an expression for the theoretical horsepower output of the turbine.

$$bhp_{th} = \frac{Q\gamma(v_j - v_T)v_T(1 - \cos\beta)}{(2.47\ \text{EE5})g} \qquad 4.51$$

The actual output will be less than the theoretical output. Typical efficiencies range from 80% to 90%.

*Example 4.9*

A Pelton wheel impulse turbine developing 100 bhp (net) is driven by a water stream from an 8″ schedule 40 penstock. Total head (before nozzle loss) is 200 feet. If the turbine runner is rotating at 500 rpm and its efficiency is 80%, determine the area of the jet, the flow rate, and the pressure head at the nozzle entrance. ($C_v = 0.95$)

From equation 4.49, the jet velocity is

$$v_j = 0.95\sqrt{2\,(32.2)\,(200)} = 107.8 \text{ ft/sec}$$

From equation 4.48, the nozzle loss is

$$h_n = 200[1 - (0.95)^2] = 19.5 \text{ ft}$$

Using table 4.2, the flow rate is

$$\dot{V} = \frac{(8.814)\,(100)}{(200 - 19.5)\,(1)\,(0.8)} = 6.10 \text{ cfs}$$

The jet area is

$$A_j = \frac{6.10}{107.8} = 0.0566 \text{ ft}^2$$

The flow area of 8″ schedule 40 pipe is 0.3474 ft$^2$.

The velocity at the nozzle entrance is

$$\frac{6.10}{0.3474} = 17.56 \text{ ft/sec}$$

The pressure head at the nozzle entrance is

$$200 - \frac{(17.56)^2}{2(32.2)} = 195.2 \text{ ft}$$

## 18 REACTION TURBINES

*Reaction turbines* are essentially centrifugal pumps in reverse. They are used when the total head is small, typically below 800 feet. However, their energy conversion efficiency is higher than for impulse turbines, typically 90% to 95%.

Since reaction turbines are centrifugal pumps in reverse, all of the affinity and similarity relationships (equations 4.31 through 4.42) can be used when comparing homologous turbines.

*Example 4.10*

A reaction turbine develops 500 bhp. Flow through the turbine is 50 cfs. Water enters at 20 fps with a 100′ pressure head. Elevation of the turbine above tailwater level is 10′. Find the effective head and turbine efficiency.

The effective (total) fluid head is

$$H = 100 + \frac{(20)^2}{2\,(32.2)} + 10 = 116.2 \text{ ft}$$

From table 4.2,

$$(whp)_{\text{in}} = \frac{(116.2)\,(50)\,(1)}{8.814} = 659.2 \text{ hp}$$

$$\eta_T = \frac{500}{659.2} = 0.758$$

## 19 TURBINE SPECIFIC SPEED

Like centrifugal pumps, turbines are classified according to the manner in which the impeller extracts energy from the fluid flow. This is measured by the turbine specific speed equation, which is different from the equation used to calculate specific speed for pumps.

$$n_s = \frac{n\sqrt{bhp}}{H^{1.25}} \qquad 4.52$$

Each of the three types of turbines is associated with a range of specific speeds.

- Axial-flow turbines are used for low heads, high rotational speeds, and large flow rates. The propeller turbines operate with specific speeds in the 80 to 200 range. Their best efficiencies, however, are produced between 120 and 160.
- For reaction turbines, the specific speed varies from 10 to 100. Best efficiencies are found in the 40 to 60 range with heads between 80 and 600 feet.
- Radial flow turbines have the lowest specific speeds. These impulse wheels have specific speeds below 5.

## 20 TURBINE EFFICIENCY

A turbine extracts energy from fluid. Its efficiency is the ratio of actual energy extracted to ideal energy extracted—the opposite of the definition of pump efficiency.

$$\eta_{\text{turbine}} = \frac{bhp}{whp} \qquad 4.53$$

## 21 HYDROELECTRIC GENERATING PLANTS

The turbine is generally housed in a *powerhouse*, with water conducted to the turbine through the *penstock* piping. Water originates in a reservoir, dam, or *forebay* (in the instance where the reservoir is a large distance from the turbine).

After the water passes through the turbine, it is discharged through the draft tube to the receiving reservoir, known as the *tail water*. The *draft tube* is used to keep the turbine up to 15 feet above the tail water surface, while still being able to extract the total available head. If a draft tube is not employed, water may be returned to the tail water by way of a channel known as the *tail race*. The turbine, draft tube, and all related parts comprise what is known as the *setting*.

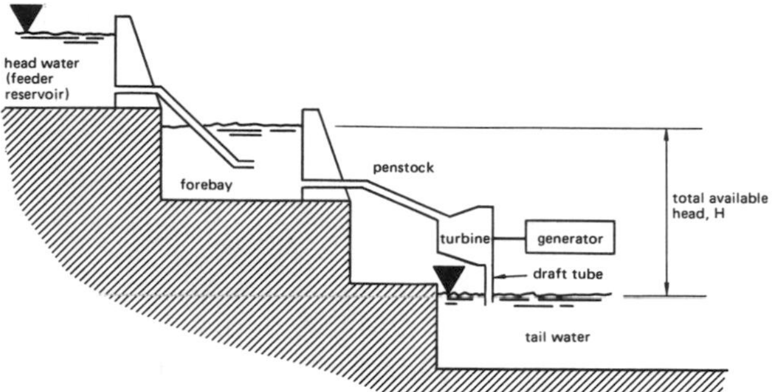


**Figure 4.27** A Typical Hydroelectric Plant

When a forebay is not part of the generating plant's design, it will be desirable to provide a *surge chamber* in order to relieve the effects of rapid changes in flow rate. In the case of a sudden power demand, the surge chamber would provide an immediate source of water, without waiting for a contribution from the feeder reservoir. Similarly, in the case of a sudden decrease in discharge through the turbine, the excess water would surge back into the surge chamber.

HYD MACHINE

# Appendix A: Atmospheric Pressure versus Altitude

| altitude ft | pressure psia | altitude ft | pressure psia |
|---|---|---|---|
| 0 | 14.696 | 33000 | 3.797 |
| 1000 | 14.175 | 34000 | 3.625 |
| 2000 | 13.664 | 35000 | 3.458 |
| 3000 | 13.168 | 36000 | 3.296 |
| 4000 | 12.692 | 37000 | 3.143 |
| 5000 | 12.225 | 38000 | 2.996 |
| 6000 | 11.778 | 39000 | 2.854 |
| 7000 | 11.341 | 40000 | 2.721 |
| 8000 | 10.914 | 41000 | 2.593 |
| 9000 | 10.501 | 42000 | 2.475 |
| 10000 | 10.108 | 43000 | 2.358 |
| 11000 | 9.720 | 44000 | 2.250 |
| 12000 | 9.347 | 45000 | 2.141 |
| 13000 | 8.983 | 46000 | 2.043 |
| 14000 | 8.630 | 47000 | 1.950 |
| 15000 | 8.291 | 48000 | 1.857 |
| 16000 | 7.962 | 49000 | 1.768 |
| 17000 | 7.642 | 50000 | 1.690 |
| 18000 | 7.338 | 51000 | 1.611 |
| 19000 | 7.038 | 52000 | 1.532 |
| 20000 | 6.753 | 53000 | 1.464 |
| 21000 | 6.473 | 54000 | 1.395 |
| 22000 | 6.2 | 55000 | 1.331 |
| 23000 | 5.943 | 56000 | 1.267 |
| 24000 | 5.693 | 57000 | 1.208 |
| 25000 | 5.452 | 58000 | 1.154 |
| 26000 | 5.216 | 59000 | 1.100 |
| 27000 | 4.990 | 60000 | 1.046 |
| 28000 | 4.774 | 61000 | 0.997 |
| 29000 | 4.563 | 62000 | 0.953 |
| 30000 | 4.362 | 63000 | 0.909 |
| 31000 | 4.165 | 64000 | 0.864 |
| 32000 | 3.978 | 65000 | 0.825 |

# Appendix B: Upper Limits of Specific Speeds

Single Stage, Single and Double Suction Pumps
Handling Clear Water at 85°F at Sea Level

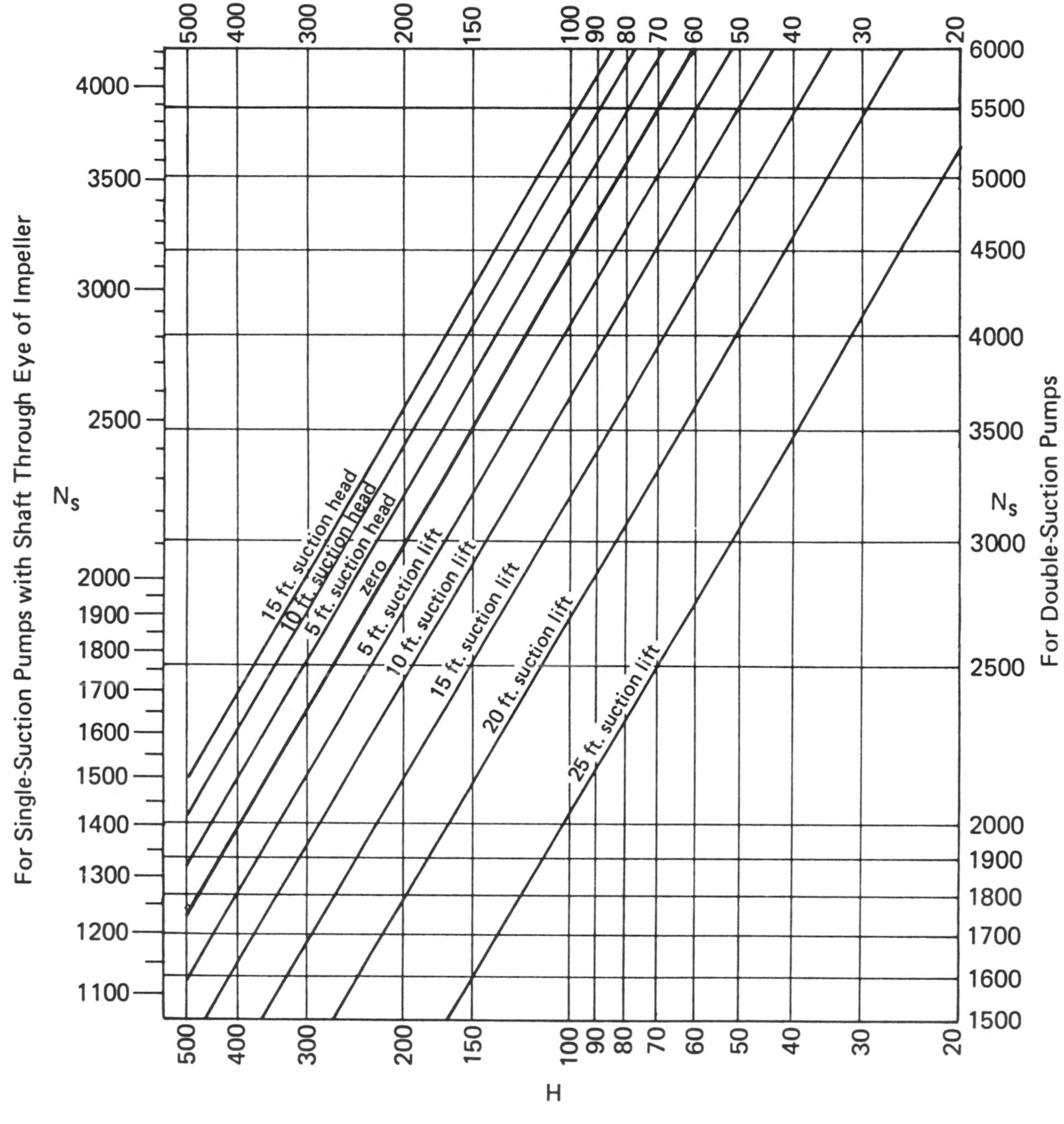

# Appendix C: Upper Limits of Specific Speeds

Single Stage, Single Suction, Mixed and Axial Flow Pumps
Handling Clear Water at 85°F at Sea Level

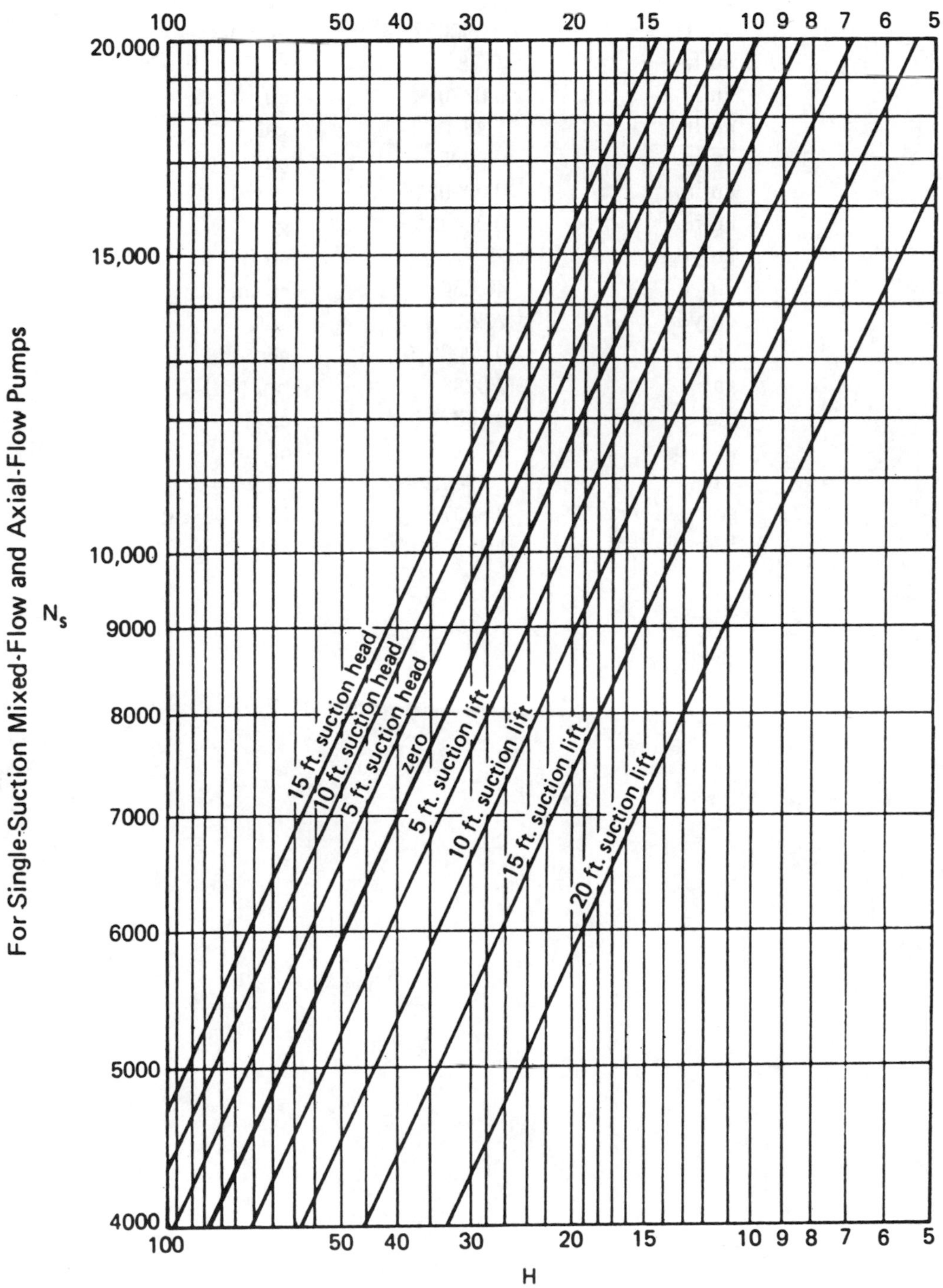

# Appendix D: Volumetric Conversion Factors

| multiply | by | to obtain |
|---|---|---|
| acre-ft | 43,560 | cu ft |
| acre-ft | 325,851 | gal |
| bbl (oil) | 42 | gal |
| cu ft | 0.0000229 | acre-ft |
| cu ft | 1728 | cu in. |
| cu ft | 0.0370 | cu yd |
| cu ft | 7.48 | gal |
| cu in. | 0.000579 | cu ft |
| cu in. | 0.0000214 | cu yd |
| cu in. | 0.00433 | gal |
| cu yd | 27 | cu ft |
| cu yd | 46,656 | cu in. |
| cu yd | 202.2 | gal |
| gal | 0.00000307 | acre-ft |
| gal | 0.0238 | bbl (oil) |
| gal | 0.1337 | cu ft |
| gal | 231 | cu in. |
| gal | 0.00495 | cu yd |
| gal | 0.8327 | Imperial gal |
| Imperial gal | 1.2 | gal |

# Appendix E: Pump Performance Correction Factor Chart

(Prepared from tests on 1″ and smaller pumps)

*Instructions for use*: Start with the actual fluid flow rate on the bottom horizontal scale. Move vertically until the required head is reached on the diagonal head scale. Move horizontally to the left until the fluid viscosity is reached on the diagonal viscosity scale. Move upward and read the correction factors.

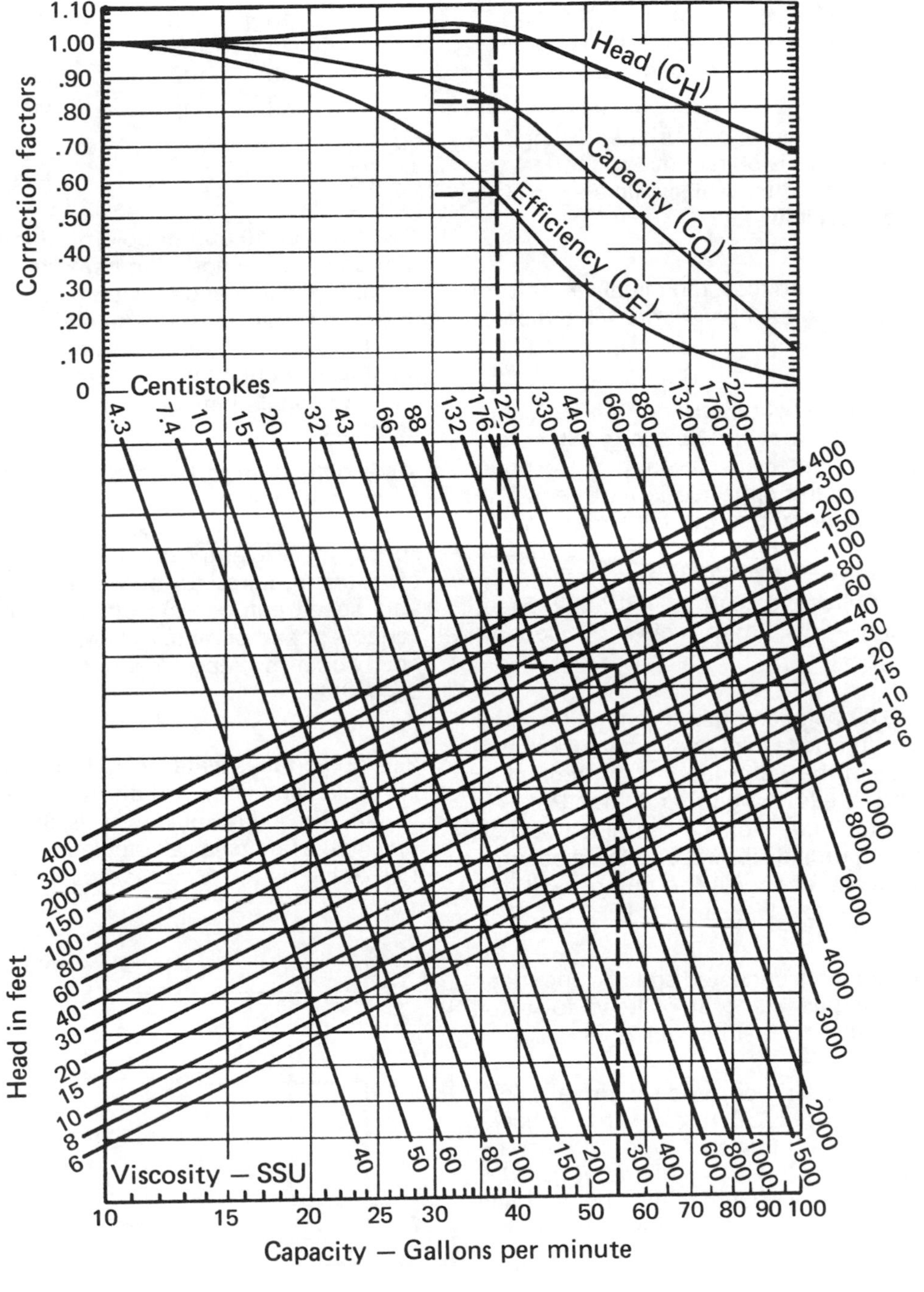

## Practice Problems: HYDRAULIC MACHINES

Untimed

1. A sludge slurry with a specific gravity of 1.2 is pumped at the rate of 2000 gpm through an inlet of 12″ and out an 8″ discharge at the same level. The inlet gage reads 8″ of mercury below atmospheric. The discharge gage reads 20 psig and is located 4 feet above the centerline of the pump outlet. If the pump efficiency is 85%, what is the input power?

2. A pump discharges 60°F water at 12 fps through a 6″ line. The inlet is a section of 8″ line. Suction is 5 psig below atmospheric. If the pump is 20 horsepower and 70% efficient, what is the maximum height at which water at atmospheric pressure is available? Assume all friction losses add up to 10 feet of fluid.

3. Water flows from a source to a turbine, exiting 625 feet lower. The head loss is 58 feet due to friction, the flow rate is 1000 cfs, and the turbine efficiency is 89%. What is the output in kilowatts?

4. A horizontal turbine reduces 100 cfs of water from 30 psia to 5 psia. Neglecting friction, what horsepower is generated?

5. A Francis hydraulic reaction turbine with 22″ diameter blades runs at 610 rpm and develops 250 horsepower when 25 cfs of water flow through it. The pressure head at the turbine entrance is 92.5 feet. The elevation of the turbine above the tail water level is 5.26 feet. The inlet and outlet velocities are 12 fps. Find the (a) effective head, (b) turbine efficiency, (c) rpm at 225 feet effective head, (d) BHP at 225 feet effective head, and (e) discharge in cfs at 225 feet effective head.

6. Water (180°F, 80 psia) empties through 30 feet of 1½″ pipe by a pump whose inlet and outlet are 20 feet below the surface of the water level when the tank is full. The pumping rate is 100 gpm and the NPSHR is 10 feet for that rate. If the inlet line contains two gate valves and two long radius elbows, and the discharge is into a 2 psig tank, when will the pump cavitate? Neglect entrance and exit losses.

7. What is the maximum suggested specific speed for a 2-stage pump adding 300 feet of head to water pulled through an inlet 10 feet below it?

8. Water at 500 psig will be used to drive a 250 horsepower turbine at 1750 rpm against a backpressure of 30 psig. (a) What type of turbine would you suggest? (b) If a 4″ diameter jet discharging at 35 fps is deflected 80° by a single moving vane with velocity of 10 fps, what is the total force acting on the blade?

9. A 1750 rpm pump is normally splined to a ½ horsepower motor. What horsepower is required if the pump is to run at 2000 rpm?

10. The inlet of a centrifugal pump is 7 feet above a water surface level. The inlet is 12 feet of 2″ pipe and contains one long-radius elbow and one check valve. The 2″ outlet contains two long-radius elbows in its 80 feet of length. The discharge is 20 feet above the surface. The following pump curve data is available. What is the flow if water is at 70°F?

| gpm | head | gpm | head |
|---|---|---|---|
| 0 | 110 | 50 | 93 |
| 10 | 108 | 60 | 87 |
| 20 | 105 | 70 | 79 |
| 30 | 102 | 80 | 66 |
| 40 | 98 | 90 | 50 |

Timed

1. A town of 10,000 people produces an average of 100 gpcd of sewage. The peak flow, however, is 250 gpcd. The pipeline to the pumping station is 5000 feet in length with C = 130. The elevation drop along the length is 48 feet. Minor losses are insignificant. The maximum suction lift is 10 feet inside the pump. Neglecting population growth, design the pumping station.

(a) Size the pipe to the station. (b) If constant speed pumps are to be used, specify the size and capacity of each. (c) If variable speed pumps are to be used, specify the size and capacity of each. (d) Find the size of motors required for both constant speed and variable speed pumps. (e) State six ways of controlling variable speed pumps. (f) How many air changes per hour would you recommend for the wet well? For the dry well?

2. A pump takes water from the clear well of a 10′ × 20′ rapid sand filter and discharges it at a higher elevation. The pump efficiency is 85%, and the motor driving the pump has an efficiency of 90%. It is desired to size the motor. Minor losses are to be ignored.

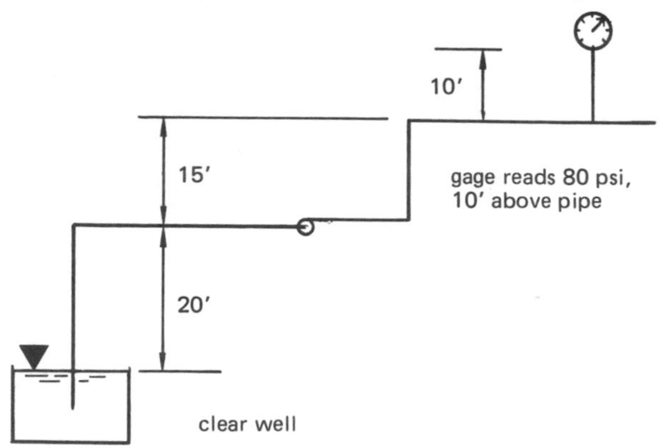

(a) What is the static suction lift? (b) What is the static discharge head? (c) What is the total dynamic head? (d) Determine the motor horsepower required to maintain the clear well at its current level if the filter output is 3.5 MGD. (e) If the filter output is increased 25%, how much surplus will there be in 8 hours?

3. A water distribution network is 10–15 years old, and is constructed of cast iron pipe. There is a requirement for 40 psi minimum pressure at point D. The inlet is at point A and the outlet is at point B.

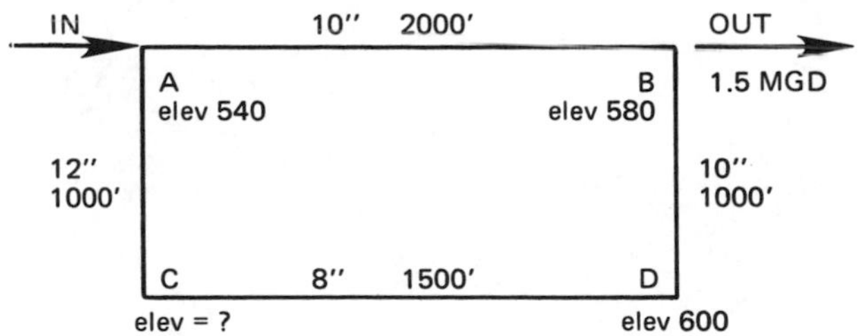


(a) Determine the flow rates into the two branches.
(b) Determine the pressures at points A and B.

# 5 OPEN CHANNEL FLOW

*Nomenclature*

| | | |
|---|---|---|
| A | area | $ft^2$ |
| b | weir width | ft |
| C | coefficient | $ft^{\frac{1}{2}}/sec$ |
| d | depth of flow | ft |
| $d_H$ | hydraulic depth | ft |
| D | pipe diameter | ft |
| E | specific energy | ft |
| f | Darcy friction factor | – |
| g | acceleration due to gravity (32.2) | $ft/sec^2$ |
| $g_c$ | gravitational constant | $\frac{lbm\text{-}ft}{lbf\text{-}sec^2}$ |
| h | head | ft |
| H | total hydraulic head | ft |
| k | minor loss coefficient | – |
| K | conveyance | $ft^3/sec$ |
| L | channel length | ft |
| n | Manning roughness coefficient | – |
| N | number of end contractions | – |
| p | pressure | $lbf/ft^2$ |
| P | wetted perimeter, or weir height | ft, ft |
| Q | flow quantity | $ft^3/sec$ |
| $r_H$ | hydraulic radius | ft |
| S | slope of energy line (energy gradient) | (decimal) |
| $S_o$ | channel slope | (decimal) |
| v | velocity | ft/sec |
| w | channel width | ft |
| Y | weir height | ft |
| z | height above datum | ft |

*Symbols*

| | | |
|---|---|---|
| $\gamma$ | specific weight | $lbf/ft^3$ |
| $\rho$ | density | $lbm/ft^3$ |

*Subscripts*

| | |
|---|---|
| b | brink |
| c | critical, or composite |
| e | equivalent, or entrance |
| f | friction |
| H | hydraulic |
| n | normal |
| o | uniform, or culvert barrel |
| s | spillway |
| t | total |
| w | weir |

## 1 INTRODUCTION

An open channel is a fluid passageway which allows part of the fluid to be exposed to the atmosphere. This type of channel includes natural waterways, canals, culverts, flumes, and pipes flowing under the influence of gravity (as opposed to pressure conduits which always flow full).

There are difficulties in evaluating open channel flow. The unlimited geometric cross sections and variations in roughness have contributed to a small number of scientific observations upon which to estimate the required coefficients and exponents. Therefore, the analysis of open channel flow is more empirical and less exact than that of pressure conduit flow. This lack of precision, however, is more than offset by the percentage error in runoff calculations that generally precede the channel calculations.

Flow in open channels is almost always turbulent. However, within that are many somewhat confusing categories of flow. Flow can also be categorized on the basis of the channel material. Except for a short discussion of erodible canals, this chapter assumes the channel is non-erodible.

Flow can be a function of time and location. If the flow quantity is invariant, it is said to be *steady.* If the flow cross section does not depend on the location along the channel, the flow is said to be *uniform.* Steady flow can be *non-uniform,* as in the case of a river with a varying cross section or on a steep slope. Furthermore, uniform channel construction does not ensure uniform flow, as will be seen in the case of hydraulic jumps.

Due to the adhesion between the wetted surface of the channel and the water, the velocity will not be uniform across the area in flow. The velocity term used in this chapter is the *mean velocity*. The mean velocity, when multiplied by the flow area, gives the flow quantity.

$$Q = Av \tag{5.1}$$

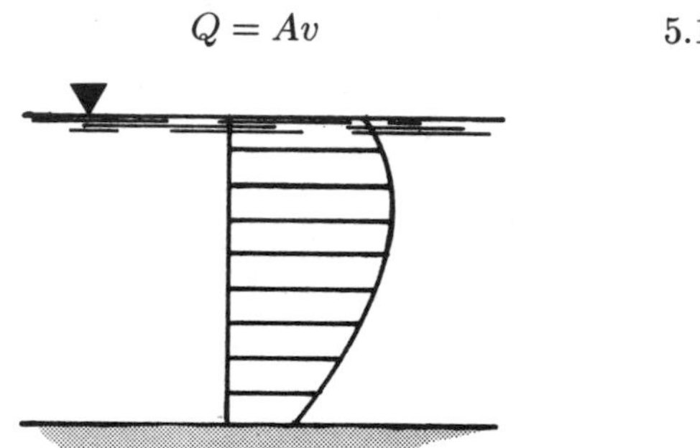

**Figure 5.1** Distribution of Velocities in a Channel

The location of the mean velocity depends on the distribution of velocities in the waterway, which is generally quite complex. The procedure for measuring the velocity of a channel (called *stream gaging*) involves measuring the average channel velocity at multiple locations across the channel width. These sub-average velocities are then themselves averaged to give a grand average (mean) flow velocity.



## 2 DEFINITIONS

Accelerated Flow: A form of varied flow in which the velocity is increasing and the depth is decreasing.

Apron: An underwater 'floor' constructed along the channel bottom to prevent scour. Aprons are almost always extensions of spillways and culverts.

Backwater: Water upstream from a dam or other obstruction which is deeper than it would normally be without the obstruction.

Backwater Curve: A plot of depth versus location along the channel containing backwater.

Check: A short section of built-up channel placed in a canal or irrigation ditch and provided with gates or flashboards to control flow or raise upstream level for diversion.

Colloid: An extremely fine sediment which will not easily settle out.

Conjugate Depths: The depths on either side of a hydraulic jump.

Contraction: A decrease in the width or depth of flow caused by the geometry of a weir, orifice, or obstruction.

Critical Flow: Flow at the critical depth and velocity. Critical flow minimizes the specific energy and maximizes discharge.

Critical Depth: The depth which minimizes the specific energy of flow.

Critical Slope: The slope which produces critical flow.

Critical Velocity: The velocity which minimizes specific energy. When water is moving at its critical velocity, a disturbance wave cannot move upstream since it moves at the critical velocity.

Downpull: A force on a gate, typically less at lower depths than at upper depths due to increased velocity, when the gate is partially open.

Energy Gradient: The slope of the specific energy line (i.e., the sum of the potential and velocity heads).

Flume: An open channel constructed above the earth's surface, usually supported on a trestle or on piers.

Forebay: A reservoir holding water for use after it has been discharged from a dam.

Freeboard: The height of the channel side above the water level.

Gradient: See 'Slope.'

Headwall: Entrance to a culvert or sluiceway.

Hydraulic Jump: A spontaneous increase in flow depth from a velocity higher than critical to a velocity lower than critical.

Hydraulic Mean Depth: Same as hydraulic radius.

Normal Depth: The depth of uniform flow. This is a unique depth of flow for any combination of channel conditions. Normal depth is found from the Chezy-Manning equation.

Overchute: A flume passing over a canal to carry floodwaters away without contaminating the canal water. An elevated culvert.

*q*-curve: A plot of depth of flow versus quantity flowing for a channel with a constant specific energy.

Rapid Flow: Flow at less than the critical depth, as typically occurs on steep slopes.

Rating Curve: A plot of quantity flowing versus depth for a natural watercourse.

Reach: A section of channel.

Retarded Flow: A form of varied flow in which the velocity is decreasing and the depth increasing.

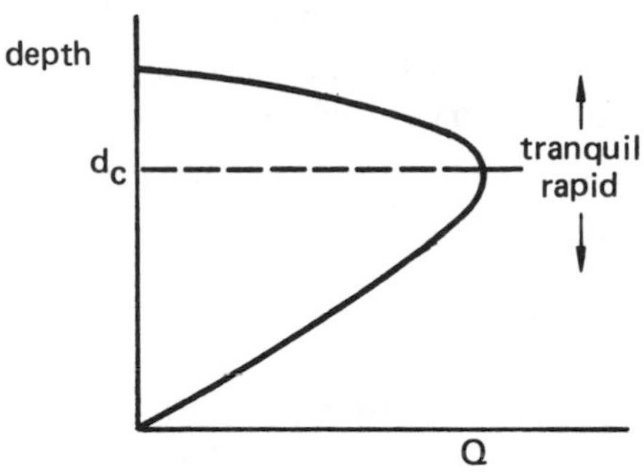


**Figure 5.2** A $q$-Curve

Sand Trap: A section constructed deeper than the rest of the channel to allow sediment to settle out.

Scour: Erosion at the exit of an open channel or toe of a spillway.

Settling Basin: A large, shallow basin through which water passes at low velocity, causing most of the suspended sediment to settle out.

Shooting Flow: See 'Rapid Flow.'

Sill: A submerged wall or weir.

Slope: The head loss per foot. For almost-level channels in uniform flow, the slope is equal to the tangent of the angle made by the channel bottom.

Standing Wave: A stationary wave caused by an obstruction in a water-course. The wave cannot move (propagate) because the water is flowing at its critical speed.

Steady Flow: Flow which does not vary with time.

Stilling Basin: An excavated pool downstream from a spillway used to decrease tailwater depth and to ensure an energy-dissipating hydraulic jump.

Stream Gaging: A method of determining the velocity in an open channel.

Subcritical Flow: Flow at greater than the critical depth (less than the critical velocity).

Supercritical Flow: Flow at less than the critical depth (greater than the critical velocity).

Tail Race: An open waterway leading water out of a dam spillway back to a natural channel.

Tail Water: The water into which a spillway or outfall discharges.

Tranquil Flow: Flow at greater than the critical depth.

Turnout: A pipe placed through a canal embankment to carry water from the canal for other uses.

Uniform Flow: Flow which has a constant depth, volume, and shape along its course.

Varied Flow: Flow that has a changing depth along the water course. The variation is with respect to location, not time.

Wasteway: A canal or pipe which returns excess irrigation water back to the main channel.

Wetted Perimeter: The length of the channel which has water contact. The air-water interface is not included in the wetted perimeter.

## 3 PARAMETERS USED IN OPEN CHANNEL FLOW

The *hydraulic radius* is the ratio of area in flow to wetted perimeter.

$$r_H = \frac{A}{P} \qquad 5.2$$

For a circular channel flowing either full or half-full, the hydraulic radius is $(D/4)$. Hydraulic radii of other channel shapes are easily calculated from the basic definition.

The *hydraulic mean depth* is the ratio of area in flow to the width of the channel at the fluid surface.[1]

$$d_H = \frac{A}{w} \qquad 5.3$$

The *slope*, $S$, in open channel equations is the slope of the energy line. If the flow is uniform, the slope of the energy line will parallel the water surface and channel bottom, and the energy gradient will equal the geometric slope, $S_o = \Delta z/L$. In general, the slope can be calculated from the Bernoulli equation as the energy loss per unit length of channel.

$$S = \frac{dE}{dL} \qquad 5.4$$

## 4 GOVERNING EQUATIONS FOR UNIFORM FLOW

Since water is incompressible, the continuity equation is

$$A_1 v_1 = A_2 v_2 \qquad 5.5$$

The more common equation used to calculate the flow velocity in open channels is the *Chezy equation.*[2]

$$v = C\sqrt{r_H S} \qquad 5.6$$

Various equations for evaluating the coefficient $C$ have been proposed. If it is assumed that the channel is

[1] For a rectangular channel, $d_H = d$.

[2] Pronounced "Shay'-zee".

large, then the friction loss will not depend so much on the Reynolds number as on the channel roughness. The *Manning formula* is frequently used to evaluate the constant $C$. Notice that the value of $C$ depends only on the channel roughness and geometry.[3]

$$C = \frac{1.49}{n}(r_H)^{1/6} \qquad 5.7$$

$n$ is the Manning roughness constant. Combining equations 5.6 and 5.7 produces the *Chezy-Manning equation.*

$$v = \frac{1.49}{n}(r_H)^{2/3}\sqrt{S} \qquad 5.8$$

All of the coefficients and constants in the Chezy-Manning equation may be combined into the *conveyance*, $K$.

$$\begin{aligned} Q = vA &= \frac{1.49}{n}(A)(r_H)^{2/3}\sqrt{S} \\ &= K\sqrt{S} \end{aligned} \qquad 5.9$$

*Example 5.1*

A rectangular channel on a 0.002 slope is constructed of finished concrete and is 8 feet wide. What is the uniform flow if water is at a depth of 5 feet?

The hydraulic radius is: $r_H = \frac{(8)(5)}{5+8+5} = 2.22$ ft

From Appendix A, the roughness coefficient for finished concrete is 0.012. The Manning coefficient is

$$C = \frac{1.49}{0.012}(2.22)^{1/6} = 141.8$$

The discharge from equation 5.6 is

$$Q = vA = (141.8)(8)(5)\sqrt{(2.22)(0.002)} = 377.9 \text{ cfs}$$

## 5 VARIATIONS IN THE MANNING ROUGHNESS CONSTANT, $n$

For most calculations, $n$ is assumed to be constant. The accuracy of other parameters used in open flow calculations often does not warrant considering the variation of $n$ with depth, and the choice to use varying $n$ values is left to the individual designer.

If it is desired to acknowledge variations in $n$ with respect to depth, it is expedient to make use of tables or graphs of hydraulic elements. Table 5.1 lists such hydraulic elements under the assumption that $n$ varies. (Appendix C can be used for both varying and non-varying $n$.)

[3] Originally proposed in 1868 by an investigator with the name of Gaukler. In Europe, the Manning equation may be known as *Strickler's equation.*

**Table 5.1**
Circular Channel Ratios
($n$ varying with depth)

| $d/D$ | $Q/Q_{\text{full}}$ | $v/v_{\text{full}}$ |
|---|---|---|
| 0.1 | 0.02 | 0.31 |
| 0.2 | 0.07 | 0.48 |
| 0.3 | 0.14 | 0.61 |
| 0.4 | 0.26 | 0.71 |
| 0.5 | 0.41 | 0.80 |
| 0.6 | 0.56 | 0.88 |
| 0.7 | 0.72 | 0.95 |
| 0.8 | 0.87 | 1.01 |
| 0.9 | 0.99 | 1.04 |
| 0.95 | 1.02 | 1.03 |
| 1.00 | 1.00 | 1.00 |

*Example 5.2*

2.5 cfs of water are in uniform flow in a 20″ sewer line ($n = 0.015$, $S = 0.001$). What are the depth and velocity? (Assume $n$ varies with depth.)

If the pipe were to flow full, it would carry $Q_{\text{full}}$.

$$\begin{aligned} D &= 20/12 = 1.667 \text{ ft} \\ r_H &= \frac{1}{4}D = \frac{1}{4}(1.667) = 0.417 \text{ ft} \\ Q_{\text{full}} &= \frac{1}{4}\pi(1.667)^2\left(\frac{1.49}{0.015}\right)(0.417)^{2/3}\sqrt{0.001} \\ &= 3.83 \text{ cfs} \\ v_{\text{full}} &= \frac{3.83}{\frac{1}{4}\pi(1.667)^2} = 1.75 \text{ ft/sec} \end{aligned}$$

$Q/Q_{\text{full}} = 2.5/3.83 = 0.65$. From Appendix C, $(d/D) = 0.66$ and $(v/v_{\text{full}}) = 0.92$. So,

$$\begin{aligned} v &= (0.92)(1.75) = 1.61 \text{ ft/sec} \\ d &= (0.66)(20) = 13.2 \text{ inches} \end{aligned}$$

## 6 NORMAL DEPTH

When the depth of flow is constant along the length of the channel (i.e., depth is neither increasing nor decreasing), the flow is said to be *uniform.* The depth of flow in that case is known as the *normal depth*, $d_o$. If the normal depth is known or can be calculated, it can be compared with the actual depth of flow to determine if the flow is uniform.[4]

[4] In reality, there are two normal depths for any given discharge. This is apparent from the specific energy curves presented elsewhere in this chapter.

The difficulty or ease with which the normal depth is calculated depends on the cross section of the channel. If the width of a rectangular channel is very large compared to the depth, the Chezy-Manning equation can be used. (Equation 5.10 assumes that the hydraulic radius equals the normal depth.)

$$d_o = 0.79\left(\frac{nQ}{\sqrt{S}w}\right)^{3/5} \quad [w \gg d_o] \qquad 5.10$$

Normal depth in circular channels can be calculated directly only under limited conditions. If the circular channel is flowing full, the normal depth is the inside pipe diameter.

$$D = d_o = 1.33\left(\frac{nQ}{\sqrt{S}}\right)^{3/8} \qquad 5.11$$

If a circular channel is flowing half-full, the normal depth is half the inside pipe diameter.

$$D = 2d_o = 1.73\left(\frac{nQ}{\sqrt{S}}\right)^{3/8} \qquad 5.12$$

For other cases of uniform flow, it is more difficult to determine normal depth. Various researchers have prepared tables and figures to assist in the calculations. For example, table 5.1 is derived from Appendix C, and can be used for circular channels flowing other than full or half-full.

In the absence of tables or figures, trial and error solutions are required. The appropriate expression for the hydraulic radius is used in the Chezy-Manning equation. Trial values are used in conjunction with graphical techniques or linear interpolation to determine the normal depth. For example, for a rectangular channel whose width is not large compared to the depth, the hydraulic radius depends on the normal depth.

$$r_H = \frac{wd_o}{w + 2d_o} \qquad 5.13$$

The Chezy-Manning equation in terms of the normal depth can then be compared to the actual known flow quantity.

$$Q = \frac{1.49}{n}(wd_o)\left(\frac{wd_o}{w + 2d_o}\right)^{2/3}\sqrt{S} \qquad 5.14$$

## 7 ENERGY AND FRICTION RELATIONSHIPS

The Bernoulli equation can be written for two points along the bottom of an open channel experiencing uniform flow.

$$\frac{p_1}{\gamma} + \frac{v_1^2}{2g} + z_1 = \frac{p_2}{\gamma} + \frac{v_2^2}{2g} + z_2 + h_f \qquad 5.15$$

However, $(p/\gamma) = d$. And since $d_1 = d_2$ and $v_1 = v_2$ for uniform flow,

$$h_f = z_1 - z_2 \qquad 5.16$$

For small slopes typical of almost all natural waterways, the channel length and horizontal run are essentially identical. Then, the hydraulic slope is

$$S = S_o = \frac{z_1 - z_2}{L} \qquad 5.17$$

Therefore, the friction loss in a length of channel is

$$h_f = LS \qquad 5.18$$

The friction loss can also be calculated from equation 5.19.

$$h_f = \frac{Ln^2v^2}{2.21(r_H)^{4/3}} \qquad 5.19$$

*Example 5.3*

Both $v_2$ and $v_1$ are unknown in the sluice gate shown. What is $v_2$?

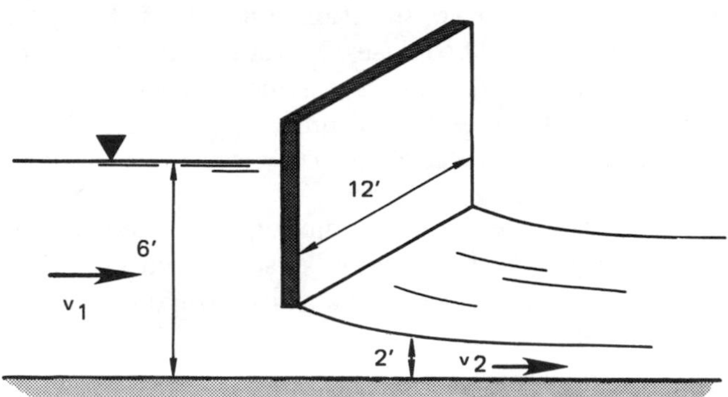


This problem can be solved using Bernoulli's equation and the continuity of flow equation. Since the channel bottom is essentially level, $z_1 = z_2$. Bernoulli's equation reduces to

$$6 + \frac{v_1^2}{2g} = 2 + \frac{v_2^2}{2g}$$

$v_1$ and $v_2$ are related by continuity.

$$\begin{aligned} Q_1 &= Q_2 \\ A_1v_1 &= A_2v_2 \\ (6)(12)v_1 &= (2)(12)v_2 \\ v_1 &= \frac{1}{3}v_2 \end{aligned}$$

Substituting the expression for $v_1$ into the Bernoulli equation,

$$\begin{aligned} 6 + \frac{v_2^2}{(3)^2(2)(32.2)} &= 2 + \frac{v_2^2}{(2)(32.2)} \\ 6 + 0.00173\,v_2^2 &= 2 + 0.0155\,v_2^2 \\ v_2 &= 17.0 \text{ fps} \end{aligned}$$

*Example 5.4*

In example 5.1, an open channel in normal flow had the following characteristics: $S = 0.002$, $n = 0.012$, $v = 9.447$ ft/sec, $r_H = 2.22$ ft. What is the energy loss per 1000 feet?

From equation 5.18, $h_f = (1000)(0.002) = 2$ feet

From equation 5.19, $h_f = \dfrac{1000(0.012)^2(9.447)^2}{2.21(2.22)^{4/3}} = 2$ feet

## 8 MOST EFFICIENT CROSS SECTION

The most efficient open channel should minimize the cross-sectional area, $A$, for any given Manning coefficient, $n$, slope, $S_0$, and discharge, $Q$. For area to be minimum, the velocity, $v$, must be maximum. Accordingly, the Chezy-Manning equation, Eq. 5.7, requires the hydraulic radius to be maximum. For a given flow area, $A$, the wetted perimeter, $P$, will be minimum.

The most efficient cross section is also generally assumed to minimize construction cost. This is true only in the most simplified cases, however, since the labor and material costs of excavation and formwork must be considered. (Trapezoidal channels are much easier to form than semicircular channels.)

Semicircular cross sections have the smallest wetted perimeter, and therefore the cross section with the highest efficiency is the semicircle. Although such a shape can be constructed with concrete, it cannot be used with earth channels.

Rectangular channels are frequently used with wooden flumes. The most efficient rectangle is one which has a depth equal to one-half of the width.

For trapezoidal channels, the most efficient cross section will be one in which the depth is twice the hydraulic radius. The sides of such a trapezoid will be inclined at 60° from the horizontal, and the flow area is half a hexagon.

A semicircle with its center at the middle of the water surface can always be inscribed in a cross section with maximum efficiency.

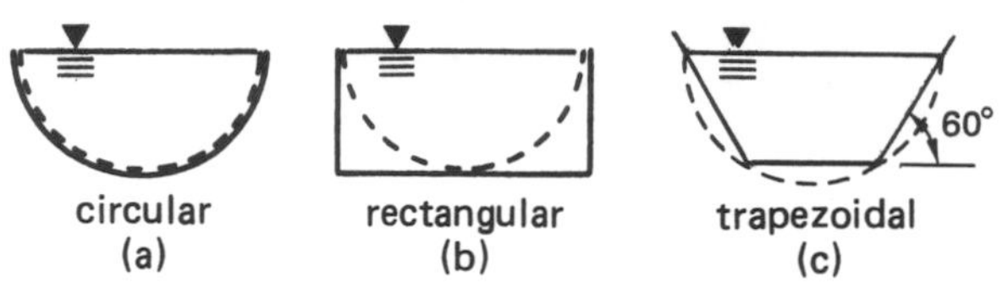


**Figure 5.3** Circles Inscribed in Efficient Channels

*Example 5.5*

A rubble masonry open channel is being designed to carry 500 cfs of water on a 0.0001 slope. Using $n = 0.017$, find the most efficient dimensions for a rectangular channel.

Let the depth and width be $d$ and $w$ respectively. For an efficient rectangle, $d = \frac{1}{2}w$.

$$A = dw = \frac{1}{2}w^2$$

$$P = d + w + d = 2w$$

$$r_H = \frac{\frac{1}{2}w^2}{2w} = \frac{1}{4}w$$

Using equation 5.9,

$$500 = \left(\frac{1}{2}w^2\right)\left(\frac{1.49}{0.017}\right)\left(\frac{1}{4}w\right)^{2/3}(0.0001)^{1/2}$$

$$500 = (0.1739)w^{8/3}$$

$$w = 19.82$$

$$d = \frac{1}{2}w = 9.91 \text{ ft}$$

## 9 ANALYSIS OF NATURAL WATERCOURSES

Natural watercourses do not have uniform paths or cross sections. This complicates their analysis considerably. Frequently, analyzing the flow from a river is a case of doing the best you can. Some types of problems can be solved with a reasonable amount of error.

As was seen in equation 5.19, the friction loss (and hence the hydraulic gradient) depends on the square of the roughness coefficient. Therefore, an attempt must be made to evaluate $n$ as accurately as possible. If the channel consists of a river with overbank *flood plains*, it should be treated as parallel channels. The flow from each subdivision can be calculated independently, and the separate values added to obtain the total flow. Alternatively, a composite value of the roughness coefficient, $n_c$, can be approximated from the individual values of $n$ and the corresponding wetted perimeters.

$$n_c = \left[\frac{\sum P_i(n_i)^{3/2}}{\sum P_i}\right]^{2/3} \qquad 5.20$$

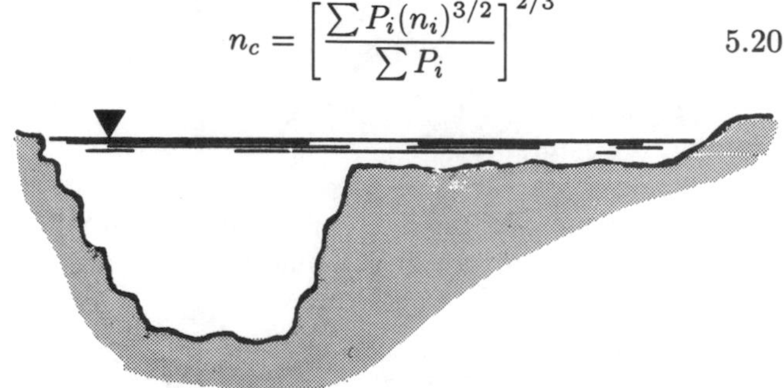

**Figure 5.4** River with Flood Plain

OPEN CHANNEL

If the channel is divided by an island into two channels (figure 5.5), $Q$ will usually be known. It may be necessary to calculate $Q_1$ and $Q_2$ in that case, or, if $Q_1$ and $Q_2$ are known, it may be necessary to find the slope.

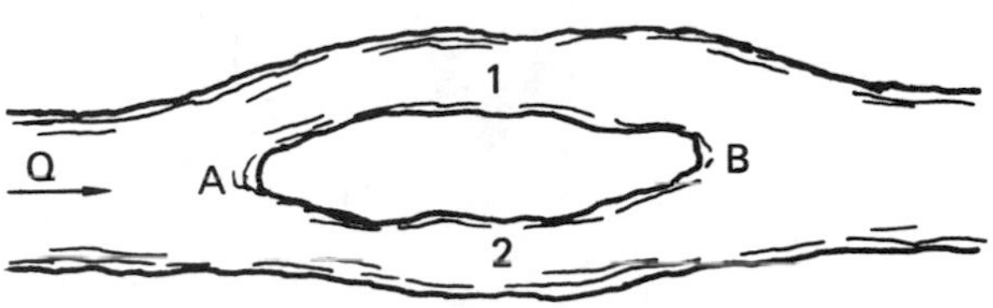


**Figure 5.5** Divided Channel

Since the drop $(z_B - z_A)$ between points $A$ and $B$ is the same regardless of flow path,

$$S_1 = \frac{z_B - z_A}{L_1} \quad 5.21$$

$$S_2 = \frac{z_B - z_A}{L_2} \quad 5.22$$

Once the slopes are known, $Q_1$ and $Q_2$ can be found from equation 5.9. The sum of $Q_1$ and $Q_2$ will probably not be the same as the given flow quantity, $Q$. In that case, $Q$ should be prorated according to the ratios of $Q_1$ and $Q_2$ to $(Q_1 + Q_2)$.

If the lengths $L_1$ and $L_2$ are the same or almost so, the Chezy-Manning equation may be solved for the slope by writing equation 5.23.

$$Q = Q_1 + Q_2 = 1.49\left[\frac{A_1}{n_1}(r_{H,1})^{2/3} + \frac{A_2}{n_2}(r_{H,2})^{2/3}\right]\sqrt{S} \quad 5.23$$

## 10 FLOW MEASUREMENT WITH WEIRS

A *weir* is an obstruction in an open channel over which flow occurs. Although a dam spillway is a specific type of weir, most weirs are designed for flow measurement. These weirs consist of a vertical flat plate with sharpened edges. Because of their construction, they are called *sharp-crested weirs.*

Sharp-crested weirs are most frequently rectangular, consisting of a straight, horizontal crest. However, weirs may also have trapezoidal and triangular openings.

If a rectangular weir is constructed with an opening width less than the channel width, the overfalling liquid sheet (called the *nappe*) decreases in width as it falls. This *contraction* of the nappe causes these weirs to be called *contracted weirs,* although it is the nappe that is actually contracted. If the opening of the weir extends the full channel width, the weir is called a *suppressed weir,* since the contractions are suppressed.

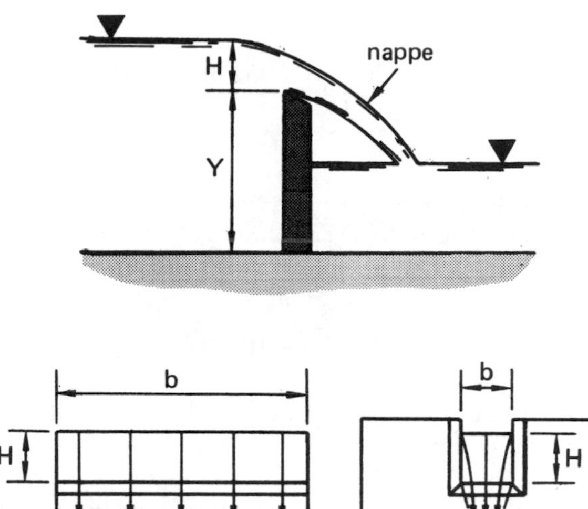


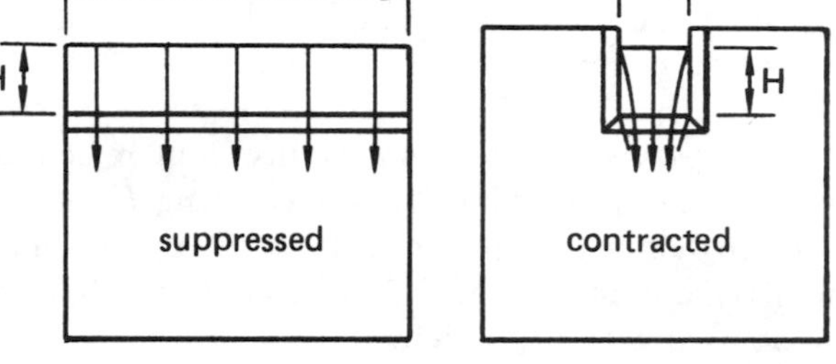


**Figure 5.6** Contracted and Suppressed Weirs

The derivation of the basic weir equation is not particularly difficult, but it is dependent on many simplifying assumptions. The basic weir equation (equation 5.24 or 5.25) is, therefore, an approximate result requiring correction by the inclusion of experimental coefficients.

If it is assumed that the contractions are suppressed, upstream velocity is uniform, flow is laminar over the crest, nappe pressure is zero, the nappe is fully ventilated, and viscosity, turbulence, and surface tension effects are negligible, then the following equation may be derived from the Bernoulli equation:

$$Q = \frac{2}{3}b\sqrt{2g}\left[\left(H + \frac{v_1^2}{2g}\right)^{3/2} - \left(\frac{v_1^2}{2g}\right)^{3/2}\right] \quad 5.24$$

If $v_1$ is negligible, then

$$Q = \frac{2}{3}b\sqrt{2g}(H)^{3/2} \quad 5.25$$

Equation 5.25 must be corrected for all of the assumptions made. This is done by introducing a coefficient, $C_1$, to account primarily for a non-uniform velocity distribution.

$$Q = \frac{2}{3}(C_1)b\sqrt{2g}(H)^{3/2} \quad 5.26$$

A number of investigations have been done to evaluate $C_1$. Perhaps the most widely known is the coefficient formula developed by *Rehbock:*[5]

$$C_1 = \left[0.6035+0.0813\left(\frac{H}{Y}\right)+\frac{0.000295}{Y}\right]\left[1+\frac{0.00361}{H}\right]^{3/2} \quad 5.27$$

If the contractions are not suppressed (i.e., one or both sides do not extend to the channel sides), then the actual width, $b$, should be replaced with the *effective width* when calculating the flow area.

$$b_{\text{effective}} = b_{\text{actual}} - (0.1)(N)(H) \quad 5.28$$

$N$ is one if one side is contracted, and $N$ is two if there are two end contractions.

A submerged rectangular weir requires a more complex analysis, due to the difficulty in measuring $H$, and because the discharge depends on both the upstream and downstream depths. The following equation, however, may be used with little difficulty.

$$Q_{\text{submerged}} = Q_{\text{free flow}}\left[1-\left(\frac{H_{\text{downstream}}}{H_{\text{upstream}}}\right)^{3/2}\right]^{0.385} \quad 5.29$$

Equation 5.29 is used by first finding $Q$ from equation 5.26 and then correcting it with the bracketed quantity.

OPEN CHANNEL

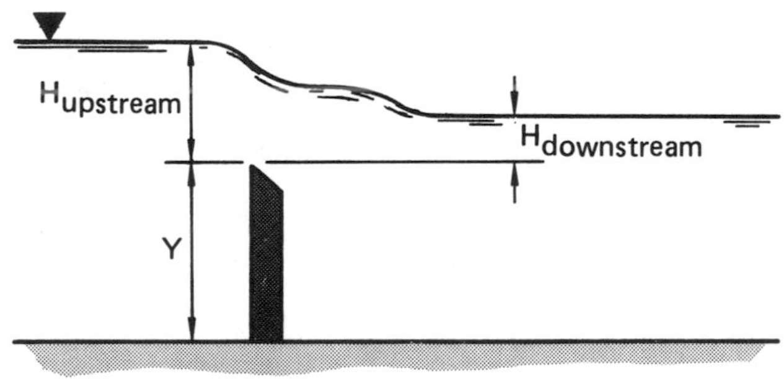


**Figure 5.7** Submerged Weir

*Triangular (V-notch) weirs* should be used when small flow rates are to be measured. The flow over a triangular weir depends on the notch angle, $\theta$. For a 90° weir, $C_2 \approx 0.593$.

$$Q = C_2\left(\frac{8}{15}\right)\tan\left(\frac{1}{2}\theta\right)\sqrt{2g}(H)^{5/2} \quad 5.30$$

$$Q \approx 2.5H^{2.5}(90° \text{ weir}) \quad 5.31$$

[5] There is much variation in how different investigators calculate the discharge coefficient, $C_1$. For ratios of $H/b$ less than 5, $C_1 = 0.622$ gives a reasonable value. With the questionable accuracy of some of the other variables used in open channel flow problems, the pursuit of greater accuracy is of dubious value.

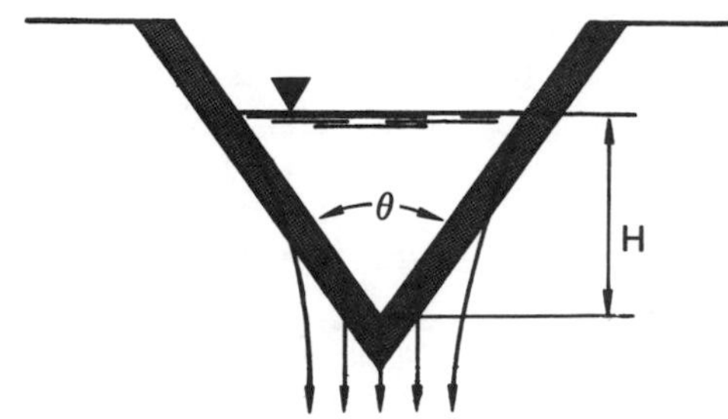


**Figure 5.8** Triangular Weir

A *trapezoidal weir* is essentially a rectangular weir with a triangular weir on either side. If the angle of the sides from the vertical is approximately 14° (i.e., 4 vertical and 1 horizontal) the weir is known as a *Cipoletti weir.* The discharge from the triangular ends of a Cipoletti weir approximately make up for the contractions that reduce rectangular flow. Therefore, no correction is theoretically necessary. The discharge from a Cipoletti weir is given by equation 5.32.

$$Q = 3.367(b)(H)^{3/2} \quad 5.32$$

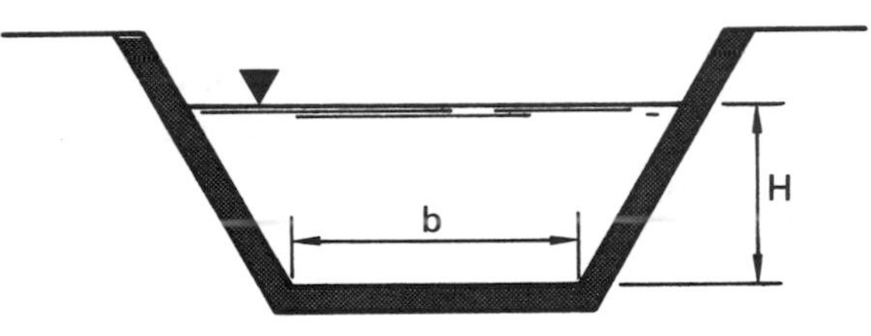


**Figure 5.9** Trapezoidal Weir

Equation 5.26 can also be used for *broad-crested weirs* ($C = 0.5$ to $0.57$) and *ogee spillways* ($C = 0.60$ to $0.75$).

*Example 5.6*

A sharp-crested, rectangular weir with two contractions is $2\frac{1}{2}$ feet high and 4 feet long. A 4″ head exists upstream from the weir. What is the velocity of approach?

$$H = 4/12 = 0.333 \text{ ft}$$

From equation 5.28, $N = 2$ and the effective width is

$$b_{\text{effective}} = 4 - (0.1)(2)(0.333) = 3.93$$

The Rehbock coefficient (from equation 5.27) is

$$C_1 = \left(0.6035 + 0.0813\left(\frac{0.333}{2.5}\right) + \frac{0.000295}{2.5}\right)\left(1 + \frac{0.00361}{0.333}\right)^{3/2}$$
$$= 0.624$$

From equation 5.26, the flow is

$$Q = \frac{2}{3}(0.624)(3.93)\sqrt{(2)(32.2)}(0.333)^{3/2}$$
$$= 2.52 \text{ cfs}$$
$$v = \frac{Q}{A} = \frac{2.52}{(4)(2.5 + 0.333)} = 0.222 \text{ ft/sec}$$

## 11 FLOW MEASUREMENT WITH PARSHALL FLUMES

The Parshall flume is one of the most widely used devices for measuring open channel wastewater flows. It performs well even in instances where head loss must be kept to a minimum or when there is a high concentration of suspended solids.

The Parshall flume is constructed with a converging upstream section, a throat, and a diverging downstream section. The walls of the flume are vertical, and the floor of the throat section drops. The length, width, and height of the flume are essentially predefined by the flow rate anticipated.[6]

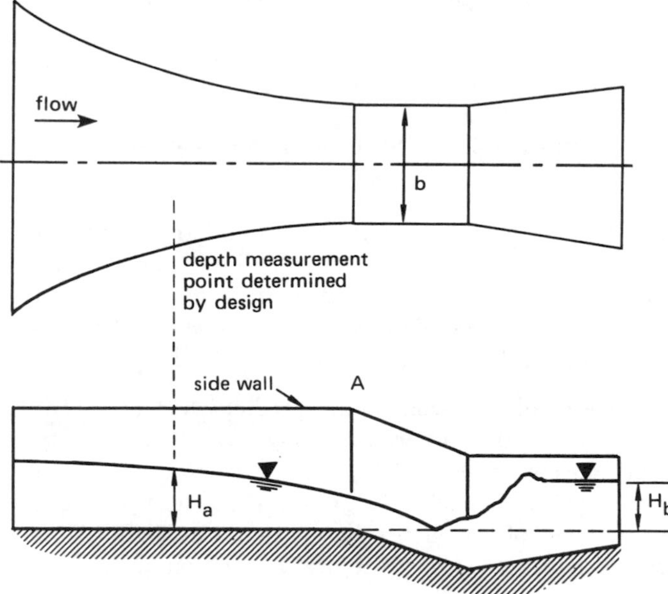


**Figure 5.10** The Parshall Flume

[6] This chapter does not attempt to design the Parshall flume, only to predict flow rates through its use.

The throat geometry in a Parshall flume has been chosen so as to force the occurrence of a critical flow there. Following the critical section is a short length of supercritical flow followed by a hydraulic jump. This construction does not produce a dead water region where debris and silt can accumulate (as is the case with broad crested and other flat-top weirs).

The discharge relationship for a Parshall flume is given by equation 5.33 for submergence ratios of $H_b/H_a$ up to 0.7. Above 0.7, the true discharge is less than predicted by the equation. Values of $K$ are given in the table 5.2, although using a value of 4.0 is accurate for most purposes.

$$Q = Kb\,(H_a)^n \qquad 5.33$$

$$n = 1.522(b)^{0.026} \qquad 5.34$$

**Table 5.2**
$K$ values for the Parshall Flume

| $b$ | $K$ |
|---|---|
| 0.25 ft | 3.97 |
| 0.50 | 4.12 |
| 0.75 | 4.09 |
| 1.0 | 4.00 |
| 1.5 | 4.00 |
| 2.0 | 4.00 |
| 3.0 | 4.00 |
| 4.0 | 4.00 |

## 12 STEADY FLOW

Steady open flow is one of constant-volume flow. However, the flow may be uniform or non-uniform. Figure 5.11 illustrates that three definitions of "slope" exist for open channel flow. These three slopes are the slope of the channel bottom, the slope of the water surface, and the slope of the energy gradient line.

Under conditions of uniform flow, all of these three slopes are equal, since the flow quantity and flow depth are constant along the length of flow.[7] With non-uniform flow, however, the flow velocity and depth vary along the length of channel, and the three slopes are not necessarily equal.

[7] As a simplification, this chapter deals only with channels of constant width. If the width is varied, changes in flow depth may not coincide with changes in flow quantity.

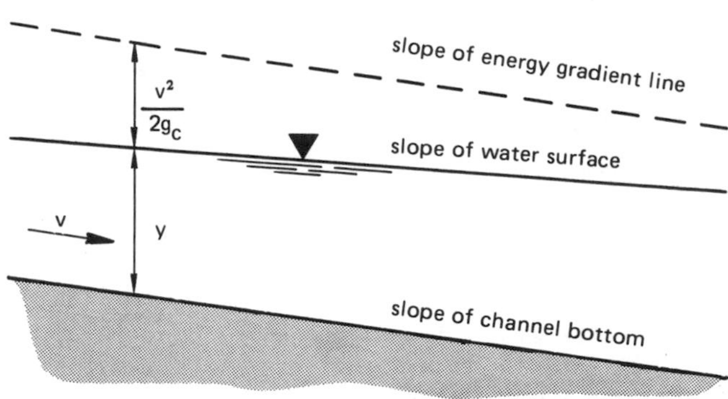


**Figure 5.11** Slopes Used in Open Channel Flow

If water is introduced down a path with a steep slope (as down a spillway), the effect of gravity will be to cause an increasing velocity. This velocity will be opposed by friction. Since the gravitational force is constant but friction varies as the square of velocity, these two forces eventually become equal. When they become equal, the velocity stops increasing, the depth stops decreasing, and the flow becomes uniform. Until they become equal, however, the flow is non-uniform (varied).

The total head is given by the Bernoulli equation.

$$H_t = z + \frac{p}{\gamma} + \frac{v^2}{2g} \qquad 5.35$$

*Specific energy* is the total head with respect to the channel bottom. In this case, $z = 0$ and $(p/\gamma) = d$.

$$E = d + \frac{v^2}{2g} \qquad 5.36$$



Equation 5.36 is not meant to imply that the potential energy is not an important factor in open channel flow problems. The concept of specific energy is used for convenience only, and it should be clear that the Bernoulli equation is still valid. Indeed, for problems in which there is a step in the channel bottom, the Bernoulli equation written in terms of the specific energy is invaluable in perceiving the behavior of the flow.

$$E_1 + z_1 = E_2 + z_2 \qquad 5.37$$

$$E_1 - E_2 = z_2 - z_1 \qquad 5.38$$

In uniform flow, total head decreases due to the frictional effects, but specific energy is constant. In non-uniform flow, total head decreases, but specific energy may increase or decrease.

Since $v = Q/A$, equation 5.36 can be written

$$E = d + \frac{Q^2}{2gA^2} \qquad 5.39$$

For a rectangular channel, the velocity can be written in terms of the width and flow depth.

$$v = \frac{Q}{A} = \frac{Q}{wd} \qquad 5.40$$

The specific energy equation for a rectangular channel is, then,

$$E = d + \frac{Q^2}{2g(wd)^2} \qquad 5.41$$

Solving for $d$ from equation 5.41 requires working with a cubic equation. There are three values of $d$ which will satisfy equation 5.41. One of them is negative, as figure 5.12 shows. Since depth cannot be negative, that value can be discarded. The two remaining values are known as *alternate depths.*

Since the area depends on the depth, fixing the channel shape and slope and assuming a depth will determine $Q$. This also will determine the specific energy, as illustrated in the *specific energy diagram.*

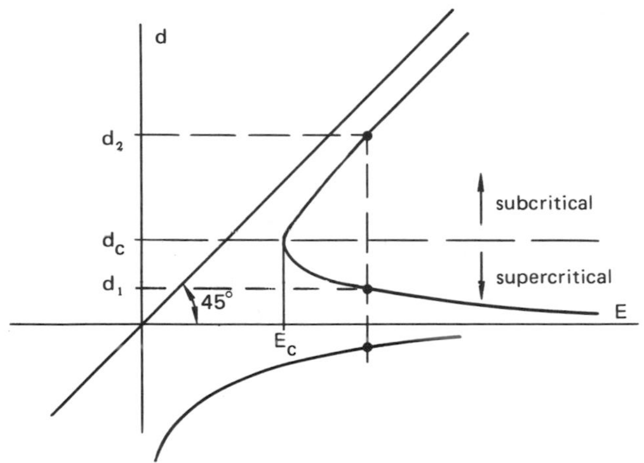


**Figure 5.12** Specific Energy Diagram

For a given flow rate, there are two different depths of flow that have the same energy—a high velocity with low depth and a low velocity with high depth. The former is called *rapid* (*supercritical*) *flow*; the latter is called *tranquil* (*subcritical*) *flow.*

The Bernoulli equation doesn't predict which of the two alternate depths will occur for any given flow quantity. The concept of *accessibility* is required to evaluate the two depths. Specifically, we say that the upper and lower limbs of the energy curve are not accessible from each other unless there is a local restriction in the flow.

Energy curves can be drawn for different flow quantities, as shown in figure 5.13 for flow quantities $Q_A$ and $Q_B$. Suppose that flow is initially at point 1. (Since the flow is on the upper limb, the flow is initially subcritical.) If there is a step up in the channel bottom, equation 5.38 predicts that the specific energy will decrease.

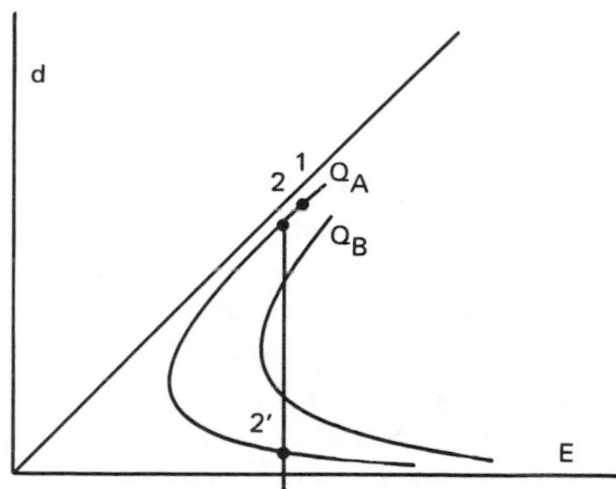


**Figure 5.13** Specific Energy Curve Families

However, the flow cannot arrive at point 2′ without changing flow quantity (i.e., going through a specific energy curve for a different flow quantity). Therefore, we say that point 2′ is not accessible from point 1 without going through point 2 first.[8]

*Example 5.7*

An open channel flow is initially flowing at 4 fps in a 7 foot wide channel 6 feet deep. The flow encounters a 1.0 foot step in the channel bottom. What is the depth of flow above the step?

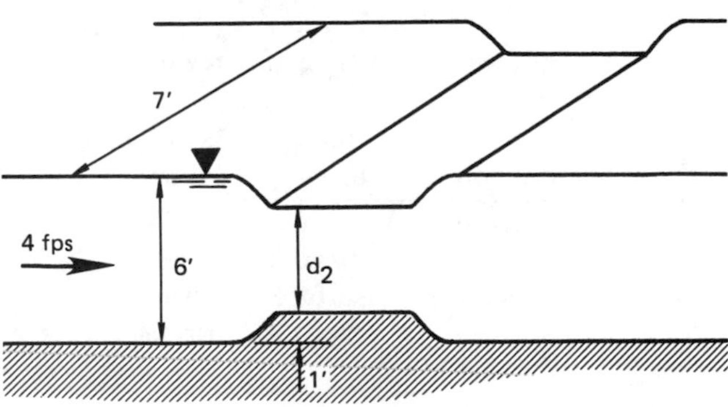


The initial specific energy is found from equation 5.36.

$$E_1 = 6 + \frac{(4)^2}{(2)(32.2)} = 6.25$$

From equation 5.38, the specific energy over the step is

$$\begin{aligned} E_2 &= E_1 + z_1 - z_2 \\ &= 6.25 + 0 - 1 = 5.25 \end{aligned}$$

The flow quantity is

$$Q = Av = (4)(6)(7) = 168 \text{ cfs}$$

Substituting $Q$ into equation 5.41,

$$5.25 = d_2 + \frac{(168)^2}{(2)(32.2)(7)^2(d_2)^2}$$

By trial and error,

$$d_2 = 1.55, 4.9 \quad \text{(alternate depths)}$$

Since the 1.55 foot depth is not accessible from the initial depth of 6 feet, the depth over the step is 4.9 feet. Notice that the drop in the water level is $6-(4.9+1) = 0.1$ feet.

Since the water level dropped only slightly, it is apparent that the flow is well up on the top limb of the specific energy curve. Since the upper limb is asymptotic to a 45° line, any change in specific energy will result in almost the same change in depth.[9] That is,

$$\Delta d \approx \Delta E \qquad 5.42$$

Therefore, the surface level will remain almost the same.

However, if the initial point is close to the critical point, then a small change in the specific energy (such as might be caused by a small variation in the channel floor) will cause a large change in depth. That is why there commonly is severe turbulence near critical flow.

## 13 CRITICAL FLOW IN RECTANGULAR CHANNELS

There is one depth, known as the *critical depth*, which minimizes the energy of flow. (The depth is not minimized, however.) The critical depth for a given flow depends on the shape of the channel.

For a rectangular channel, if equation 5.41 is differentiated with respect to depth in order to minimize the specific energy, equation 5.43 results.

$$d_c^3 = \frac{Q^2}{gw^2} \qquad 5.43$$

[8] Actually, specific energy curves are typically plotted for $q = Q/w$. If that is the case, a jump from one limb to the other could take place if the width was allowed to change as well as depth. However, the width is constant in most open channel flow problems.

[9] A rise in the channel bottom does not always produce a drop in the water surface. Only if the flow is initially subcritical will the water surface drop upon encountering a step. The water surface will rise if the flow is initially supercritical.

Geometrical and analytical methods can be used to correlate the critical depth and the minimum specific energy.

$$d_c = \frac{2}{3}E_c \qquad 5.44$$

For a rectangular channel, $Q = d_c w v_c$. Substituting this into equation 5.43 produces an equation for the *critical velocity*.

$$v_c = \sqrt{g d_c} \qquad 5.45$$

The expression for critical velocity also coincides with the expression for the velocity of a *surge (surface) wave* moving in liquid of depth $d_c$. Since surface disturbances are transmitted as ripples upstream (and downstream) at velocity $v_c$, it is apparent that a surge wave will be stationary in a channel moving at the critical velocity. Such a motionless wave is known as a *standing wave*.

If the flow velocity is less than the surge wave velocity (for the actual depth), then a ripple can make its way upstream. If the flow velocity exceeds the surge wave velocity, the ripple will be swept downstream.

## 14 CRITICAL FLOW IN NON-RECTANGULAR CHANNELS

For non-rectangular shapes, the critical depth can be found by trial and error from the following equation in which $b$ is the surface width. To use equation 5.46, assume trial values of the critical depth, use it to calculate dependent quantities in the equation, and then verify the equality.

$$\frac{Q^2}{g} = \frac{A^3}{b} \qquad 5.46$$

Equation 5.46 is particularly difficult to use with circular channels. Appendix D is a convenient method of determining critical depth in that case.

## 15 THE FROUDE NUMBER

The Froude number, when sufficient information is available to calculate it, can be used to determine if the flow is subcritical or supercritical. This dimensionless number can be calculated from equation 5.47. $d$ is the depth corresponding to velocity $v$. For a rectangular channel, $d_H = d$.

$$N_{Fr} = \frac{v}{\sqrt{g d_H}} \qquad 5.47$$

When the Froude number is less than one, the flow is subcritical. That is, the depth of flow is greater than the critical depth, and the velocity is less than critical velocity.

When the Froude number is greater than one, the flow is supercritical. The depth is less than critical depth, and the flow velocity is greater than critical velocity.

When the Froude number is equal to one, the flow is critical.[10]

## 16 PREDICTING OPEN CHANNEL FLOW BEHAVIOR

Upon encountering a variation in the channel bottom, the behavior of an open channel flow is dependent on whether the flow is initially subcritical or supercritical. The Froude number is usually used as the quantitative determination of flow regime. Open channel flow is governed by the following equation, in which the Froude number is the primary independent variable.

$$\frac{dd}{dx}\left[1 - N_{Fr}^2\right] + \frac{dz}{dx} = 0 \qquad 5.48$$

The quantity $\frac{dd}{dx}$ is the slope of the surface. (That is, it is the derivative of the depth with respect to the channel length.) The quantity $\frac{dz}{dx}$ is the slope of the channel bottom.

For an upward step, $\frac{dz}{dx} > 0$. If the flow is initially subcritical (i.e., $N_{FR} < 1$), then equation 5.48 requires that $\frac{dd}{dx} < 0$, a drop in depth. This logic can be repeated for other combinations of the terms.

Table 5.3 lists the various behaviors of open channel flow surface levels based on equation 5.48.

**Table 5.3**
Surface Level Change Behavior

| initial flow | step up | step down |
|---|---|---|
| subcritical | surface drops | surface rises |
| supercritical | surface rises | surface drops |

If $\frac{dz}{dx} = 0$ (i.e., a horizontal slope), then either the depth must be constant, or the Froude number must be unity. The former case is obvious. The latter case predicts critical flow. Such critical flow actually occurs where the slope is horizontal over broad crested weirs (see figure 5.16) and at the top of a rounded spillway. Since broad crested weirs and spillways produce critical flow, they represent a class of *controls on flow*.

[10] The similarity of the Froude number to the Mach number used to classify gas flows is more than coincidental. Both bodies of knowledge employ similar concepts.

## 17 OCCURRENCES OF CRITICAL FLOW

For any given discharge and cross section, there is a unique slope that will produce and maintain flow at critical depth. Once $d_c$ is known, this critical slope can be found from the Chezy-Manning equation.

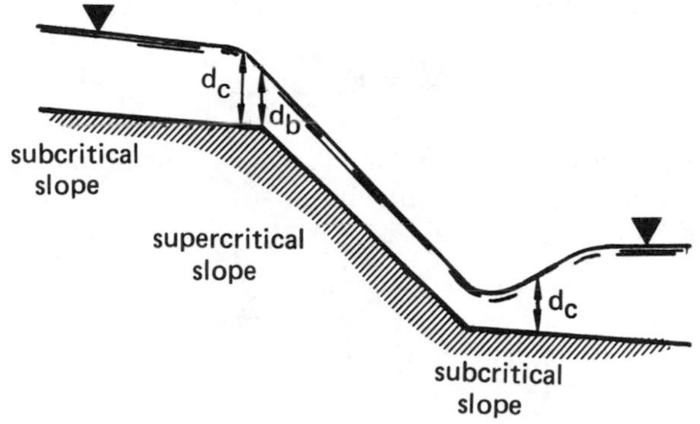


**Figure 5.14** Occurrence of Critical Depth

Critical depth occurs at free outfall from a channel of mild slope. The occurrence is at the point of curvature inversion, just upstream from the brink. For mild slopes, the brink depth is approximately

$$d_b = (0.715)d_c \qquad 5.49$$

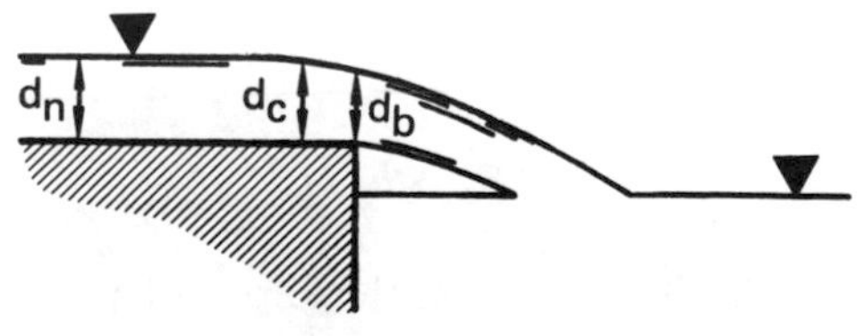


**Figure 5.15** Free Outfall

Critical flow can occur across a broad-crested weir.[11] With no obstruction to hold the water, it falls from the normal depth to the critical depth, but it can fall no more than that because there is no source to increase the specific energy (to increase the velocity). This is not a contradiction of the previous free outfall case where the brink depth is less than the critical depth. The flow curvatures in free outfall are a result of the constant gravitational acceleration.

[11] Figure 5.16 is an example of a ***hydraulic drop***, the opposite of a hydraulic jump. A hydraulic drop can be recognized by the sudden decrease in depth over a short length of channel.

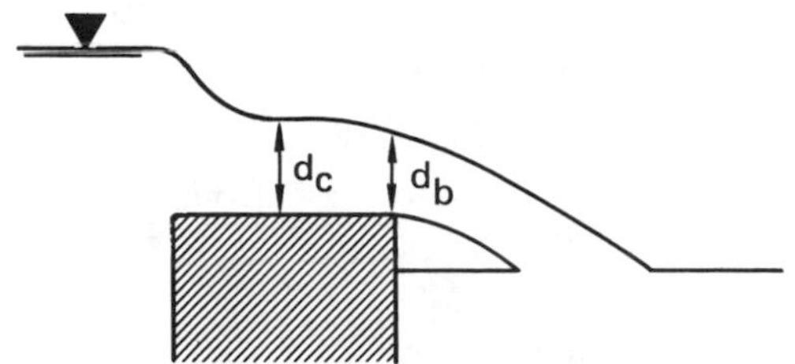


**Figure 5.16** Broad-Crested Weir

Critical depth can also occur when a channel bottom has been raised to choke the flow. A raised channel bottom is essentially a broad-crested weir.

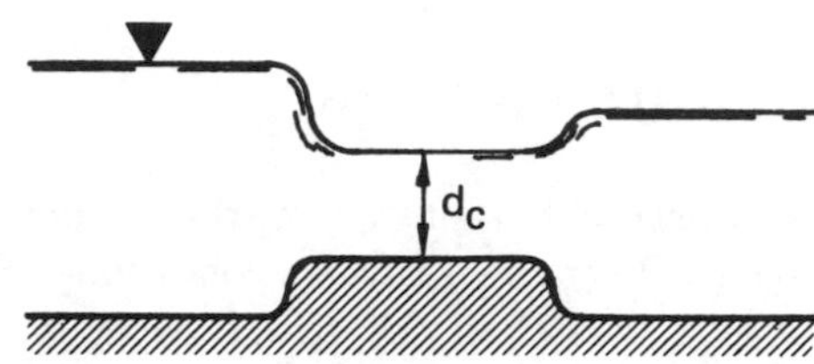


**Figure 5.17** Raised Channel Bottom with Choked Flow

The critical depth is important, not because it minimizes the energy of flow, but because it maximizes the quantity flowing for a given cross section and slope. Critical flow is generally quite turbulent because of the large changes in energy that occur with small elevation and depth changes. Critical depth flow is characterized by water surface undulations.

In all of the previous instances of critical depth, equation 5.41 can be used to calculate the actual velocity.

*Example 5.8*

At a particular point in an open rectangular channel ($n = 0.013$, $S = 0.002$, $w = 10$ feet) the flow is 250 cfs and the depth is 4.2 feet.

(a) Is the flow tranquil, critical, or rapid?

(b) What is the normal depth?

(c) If the flow ends in a free outfall, what is the brink depth?

(a) From equation 5.43, the critical depth is

$$d_c = \sqrt[3]{(250)^2/(32.2)(10)^2} = 2.69$$

Since the actual depth exceeds the critical depth, the flow is tranquil.

(b)

$$A = (d_n)(10)$$
$$P = 2d_n + 10$$
$$r_H = \frac{10d_n}{2d_n + 10} = \frac{5d_n}{d_n + 5}$$

From equation 5.9,

$$250 = (10)(d_n)\left(\frac{1.49}{0.013}\right)\left(\frac{5d_n}{d_n+5}\right)^{2/3}\sqrt{0.002}$$

By trial and error, $d_n = 3.1$ feet. Since the actual and normal depths are different, the flow is non-uniform.

(c) $d_b = 0.715(2.69) = 1.92$ ft

## 18 CONTROLS ON FLOW

If flow is subcritical, then a disturbance downstream will be able to affect the upstream conditions. Since the flow velocity is less than the critical velocity, a ripple will be able to propagate upstream to signal a change in the downstream conditions. Any object downstream which affects the flow rate, velocity, or depth upstream is known as a *downstream control.*

If a flow is supercritical, then a downstream obstruction will have no effect, since disturbances cannot propagate upstream faster than the flow velocity. The only effect on supercritical flow is from an upstream obstruction. Such an obstruction is said to be an *upstream control.*



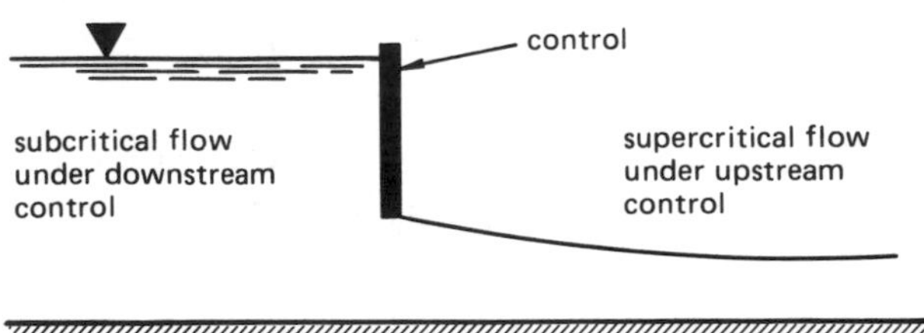


**Figure 5.18** A Control on Flow

A downstream control may also be an upstream control, as figure 5.18 shows.

In general, any feature which determines depth and discharge rate is known as a *control on flow.* Controls may consist of control structures (weirs, gates, sluices, etc.), forced flow through critical depth (as in a free outfall), sudden changes of slope (which forces a hydraulic jump or hydraulic drop to the new normal depth), or free flow between reservoirs of different surface elevations.

## 19 FLOW CHOKING

A control which is severe enough to influence the upstream flow is known as a *choke*, and the corresponding flow through the control (and any flow thereafter) is known as *choked flow.* Choked flow will occur, in the case of upward or downward steps in the channel bottom, when the step size becomes equal to the difference between the upstream specific energy and the critical flow energy.

$$\Delta z = E_1 - E_c \qquad 5.50$$

In the case of a rectangular channel, the maximum variation in channel bottom will be

$$\begin{aligned}\Delta z &= E_1 - \left(d_c + \frac{v_c^2}{2g}\right) \\ &= E_1 - \frac{3}{2}d_c \qquad 5.51\end{aligned}$$

If critical depth is achieved on the step, the flow further downstream can be subcritical or supercritical depending on the downstream conditions. If there is a downstream control, such as a sluice gate, the flow will be subcritical. If there is no downstream control, then the step will serve as an upstream control, and the flow will be supercritical.

## 20 VARIED FLOW CALCULATIONS

*Accelerated flow* occurs in any channel where the actual slope exceeds the friction loss per foot.

$$S_o > h_f/L \qquad 5.52$$

*Retarded flow* occurs when the actual slope is less than the unit friction loss.

$$S_o < h_f/L \qquad 5.53$$

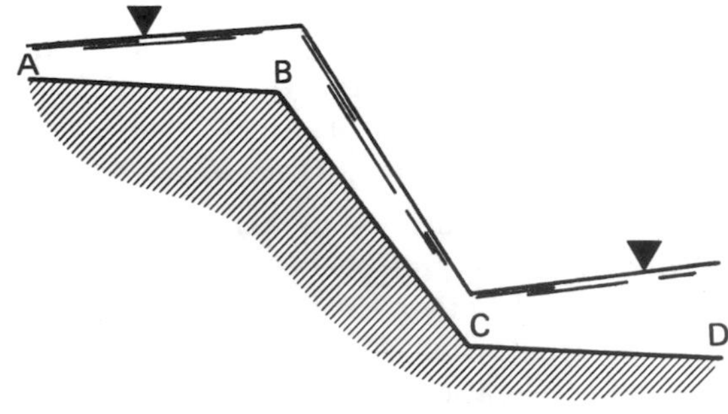


**Figure 5.19** Varied Flow

In sections $AB$ and $CD$ of figure 5.19, the slopes are less than the energy gradient, so the flows are retarded. In section $BC$, the slope is greater than the energy gradient, and the velocity increases (i.e., the flow is accelerated). If section $BC$ were long enough, the friction loss would eventually become equal to the accelerating energy, and the flow would become uniform.

Cases of accelerated and retarded flow (except for hydraulic jump) can be evaluated from the following procedure which will give the distance between points of two known or assumed depths. Assuming that the friction losses are the same for varied flow as for uniform flow, the following equations are needed:

$$S = \left(\frac{n v_{\text{ave}}}{1.486(r_{H,\text{ave}})^{2/3}}\right)^2 \qquad 5.54$$

$$v_{\text{ave}} = \frac{1.486}{n}(r_{H,\text{ave}})^{2/3}(S)^{1/2} \qquad 5.55$$

$S$ is the slope of the energy gradient from equation 5.54, not the channel slope $S_o$. The usual method of finding the depth profile is to start at a point in the channel where $d_2$ and $v_2$ are known. Then, assume a depth $d_1$, find $v_1$ and $S$, and solve equations 5.56 for $L$.

$$L = \frac{\left(d_1 + \frac{v_1^2}{2g}\right) - \left(d_2 + \frac{v_2^2}{2g}\right)}{S - S_o} = \frac{E_1 - E_2}{S - S_o} \qquad 5.56$$

*Example 5.9*

How far from the point in example 5.8 will the depth be 4 feet?

The difference between 4 feet and 4.2 feet is small, so a one-step calculation will probably be sufficient.

$$d_1 = 4 \text{ feet}$$

$$v_1 = Q/A = \frac{250}{4(10)} = 6.25 \text{ ft/sec}$$

$$E_1 = 4 + \frac{(6.25)^2}{(2)(32.2)} = 4.607 \text{ ft}$$

$$r_H = \frac{(4)(10)}{4 + 10 + 4} = 2.22$$

$$d_2 = 4.2$$

$$v_2 = \frac{250}{(4.2)(10)} = 5.95$$

$$E_2 = 4.2 + \frac{(5.95)^2}{(2)(32.2)} = 4.75$$

$$r_H = \frac{(4.2)(10)}{4.2 + 10 + 4.2} = 2.28$$

$$v_{\text{ave}} = \frac{1}{2}(6.25 + 5.95) = 6.1$$

$$r_{H,\text{ave}} = \frac{1}{2}(2.22 + 2.28) = 2.25$$

From equation 5.54,

$$S = \left(\frac{(0.013)(6.1)}{1.486(2.25)^{0.667}}\right)^2 = 0.000965$$

From equation 5.56,

$$L = \frac{4.607 - 4.75}{0.000965 - 0.002} = 138 \text{ ft}$$

## 21 HYDRAULIC JUMP

If water is introduced at high (supercritical) velocity to a section of slow-moving (subcritical) flow (as in section C in figure 5.19), the velocity will be reduced through a hydraulic jump. A hydraulic jump is an abrupt rise in the water surface. The increase in depth is always from below the critical depth to above the critical depth.[12]

The hydraulic jump has practical applications in the design of stilling basins where supercritical velocities are reduced to slower velocities by having the flow cross a series of baffles on the channel bottom. Although energy is lost across the baffles, momentum is conserved. Momentum considerations can be used to predict the downstream depth.



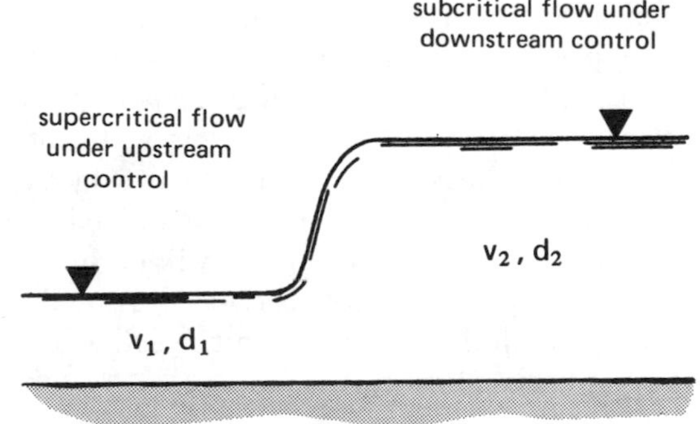


**Figure 5.20** Conjugate Depths

[12] This gives us a way to determine if a hydraulic jump can occur in a channel. If the original depth is above the critical depth, the flow is already subcritical. Therefore, a hydraulic jump cannot form.

If the depths $d_1$ and $d_2$ are known, then the velocity $v_1$ can be found from equation 5.57.

$$v_1^2 = \frac{g\,d_2}{2d_1}(d_1 + d_2) \qquad 5.57$$

$d_1$ and $d_2$ are known as *conjugate depths*, because they occur on either side of the jump. These depths are:

$$d_1 = -\frac{1}{2}d_2 + \sqrt{\frac{2v_2^2 d_2}{g} + \frac{d_2^2}{4}} \qquad 5.58$$

$$d_2 = -\frac{1}{2}d_1 + \sqrt{\frac{2v_1^2 d_1}{g} + \frac{d_1^2}{4}} \qquad 5.59$$

The specific energy lost in the jump is the energy lost per pound of water flowing.

$$\Delta E = \left[\left(d_1 + \frac{v_1^2}{2g}\right) - \left(d_2 + \frac{v_2^2}{2g}\right)\right] \qquad 5.60$$

In the case of an apron at the bottom of a dam spillway, the apron is usually insufficient to overcome friction, and the water depth will gradually increase.

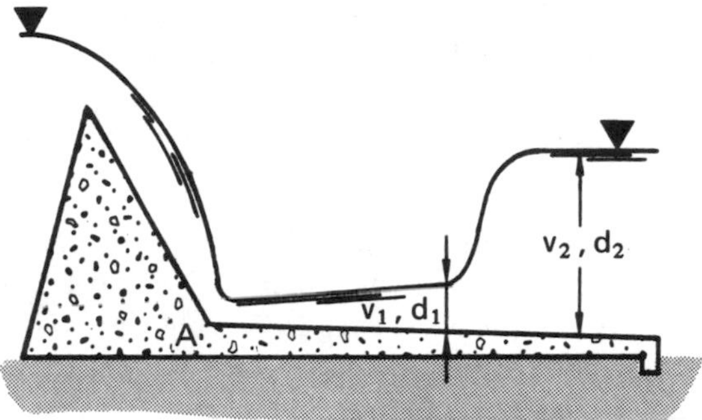


**Figure 5.21** Hydraulic Jump to Reach Tailwater Level

If the tailwater depth $d_2$ is less than critical, no hydraulic jump will occur. If the tailwater depth at the toe is less than the conjugate depth corresponding to $d_2$ but greater than the critical depth, flow will continue until the depth increases to $d_1$, and then a hydraulic jump will form. If the tailwater depth is equal to the conjugate depth at the toe, the jump will occur at the toe. If the tailwater depth exceeds the conjugate depth, the hydraulic jump may occur up on the spillway, or it may be completely submerged.

*Example 5.10*

A hydraulic jump is produced at a point in a 10 foot wide channel where the depth is 1 foot and the flow is 200 cfs. (a) What is the depth after the jump? (b) What is the total power dissipated?

(a) $v_1 = Q/A = 200/(10)(1) = 20 \text{ ft/sec}$

From equation 5.59,

$$d_2 = -\frac{1}{2}(1) + \sqrt{\frac{(2)(20)^2(1)}{32.2} + \frac{(1)^2}{4}} = 4.51$$

(b) The flow rate is

$$(200)(62.4) = 12{,}480 \text{ lbm/sec}$$

The velocity after the jump is

$$v_2 = 200/(10)(4.51) = 4.43$$

From equation 5.60, the change in specific energy is

$$\left(1 + \frac{(20)^2}{2(32.2)}\right) - \left(4.51 + \frac{(4.43)^2}{2(32.2)}\right) = 2.4 \text{ feet}$$

The total power dissipated is

$$(12{,}480)(2.4) = 29{,}952 \ \frac{\text{ft-lbf}}{\text{sec}}$$

## 22 LENGTH OF HYDRAULIC JUMP

The length of a hydraulic jump is somewhat difficult to measure (and investigate) due to the difficulty in defining the downstream termination of the jump. The majority of the evidence collected indicates that the length of the jump varies within the limits of $5 < L/d_2 < 6.5$, in which $L$ is the length of the jump. Where greater accuracy is warranted, table 5.4 can be used. This table correlates the length of the jump to the upstream Froude number.

**Table 5.4**
Approximate Lengths of Hydraulic Jumps

| $N_{Fr,1}$ | $L/d_2$ |
|---|---|
| 3 | 5.25 |
| 4 | 5.8 |
| 5 | 6.0 |
| 6 | 6.1 |
| 7 | 6.15 |
| 8 | 6.15 |

## 23 SPILLWAYS

A spillway is designed for a capacity based on the dam's inflow hydrograph, turbine capacity, and storage capacity. *Overflow spillways* frequently have a cross section known as an *ogee*, which closely approximates the underside of a nappe from a sharp-crested weir. This cross section reduces cavitation which is likely to occur when the water surface breaks contact with the spillway at heads higher than were designed for.[13]

Discharge from an overflow spillway is the same as for a weir:

$$Q = C_s b (H)^{3/2} \tag{5.61}$$

$C_s$ is the spillway coefficient, which varies from about 3 to 4 for an ogee spillway. (Use 3.97 in the absence of other information.) $C_s$ is dependent on $H$, the upstream design head above the spillway top. $C_s$ increases as $H$ increases.

If the velocity of approach is significant, the flow quantity is

$$Q = C_s b \left( H + \frac{v^2}{2g} \right)^{3/2} \tag{5.62}$$

*Scour protection* is usually needed at the toe of a spillway to protect the area exposed to a hydraulic jump. This protection usually takes the form of an extended horizontal or sloping apron. Other measures, however, are needed if the tailwater exhibits large variations in depth.

## 24 SLUICEWAYS

A sluiceway carries water away from a dam or reservoir, and is usually constructed in such a manner as to allow withdrawal at various reservoir levels. Its intake is submerged, but the inlet level is seldom below the minimum reservoir level. Sluiceways are generally round or square in shape.

Analysis and design of sluiceways is the same as for culverts.

## 25 ERODIBLE CANALS

Design of erodible channels is similar to that of concrete or pipe channels, except for the added considerations of maximum velocities and permissible side slopes.

The sides of the channel should not be slopes exceeding the natural *angle of repose* for the material used. Although there are other factors which determine the maximum permissible side slope, table 5.5 lists some guidelines.

[13] No cavitation or separation is expected as long as the actual head, $H$, is less than twice the design value. The shape of the ogee spillway will be a function of the design head.

**Table 5.5**
Recommended Side Slopes

| Type of Channel | (horizontal : vertical) |
|---|---|
| firm rock | vertical to $\frac{1}{4}:1$ |
| concrete lined stiff clay | $\frac{1}{2}:1$ |
| fissured rock | $\frac{1}{2}:1$ |
| firm earth with stone lining | $1:1$ |
| firm earth, large channels | $1:1$ |
| firm earth, small channels | $1\frac{1}{2}:1$ |
| loose, sandy earth | $2:1$ |
| sandy, porous loam | $3:1$ |

Maximum velocities that should be used with erodible channels are given in table 5.6.

## 26 APPLICATIONS OF OPEN CHANNEL FLOW TO CULVERT DESIGN[14]

Culverts are classified as having either *inlet* or *outlet control*. Specifically, either the inlet or outlet will control the discharge capacity. The most important feature of a culvert, particularly a culvert under inlet control, is whether or not the culvert flows full. Since length of culvert is one of the most important factors in determining the degree of filling, a culvert may be known as "hydraulically long" if it runs full and "hydraulically short" if it does not.[15]

Due to the numerous variables involved, no single formula can be given to design a culvert. Culvert design is often an empirical, trial and error process. Figure 5.22 illustrates some of the important variables which affect culvert performance.

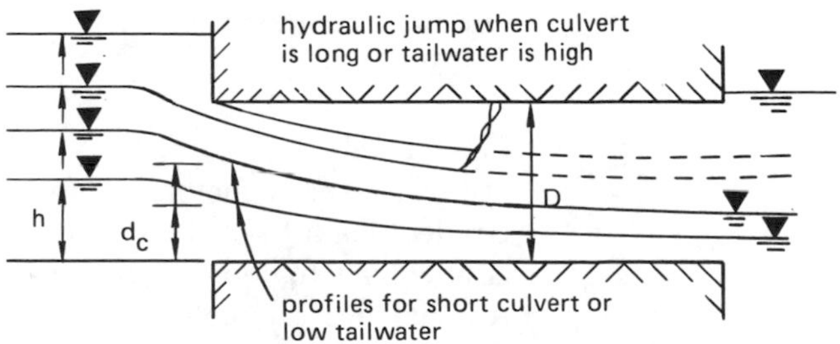


**Figure 5.22** Flow Profiles in Culvert Design

[14] The methods of culvert flow analysis in this chapter are based on "Measurement of Peak Discharge at Culverts by Indirect Methods," U.S. Department of the Interior, 1968.

[15] Proper design of culvert entrances can reduce the importance of length on culvert filling.

The culvert can operate with its entrance totally submerged or partially submerged. Similarly, the exit can be totally submerged, partially submerged, or free outfall. The upstream head, $h$, is the water surface level above the lowest part of the culvert barrel, known as the *invert*.[16]

In figure 5.22, the three lowermost profiles are of the type which would be produced with conditions of *inlet control.* Such a situation could occur when the culvert is short and steep. Flow at the entrance would be critical (hence, the inlet control) as the water falls over the brink.

If the tailwater covers the culvert completely (i.e., a submerged exit), the culvert will be full at that point, even though the inlet control forces the culvert to be only partially full at the inlet. The transition from partially full to full occurs in a hydraulic jump, the location of which depends on the flow resistance and water levels. If the flow resistance is very high, or if the headwater and tailwater levels are high enough, the jump will occur close to or at the entrance.

[16] The highest part of the culvert barrel is known as the *soffit* or *crown.*

If the flow in a culvert is full for its entire length, then the flow is said to be under *outlet control.* The discharge will be a function of the differences in tailwater and headwater levels, as well as the flow resistance along the barrel length.

For convenience, culvert flow is classified into six different types on the basis of the type of control and the relative tailwater and headwater heights.[18] These six types are illustrated in figure 5.23.

## 27 DETERMINING TYPE OF CULVERT FLOW

The types of flow are categorized according to the steepness of the barrel, the elevations of the tailwater and headwater, and in some cases, the relationship between critical depth and culvert size. These parameters are quantified through the use of the ratios in table 5.7.

[18] It should be cautioned that the six cases which are presented here do not exhaust the various possibilities for entrance and exit control. Culvert design is complicated by this multiplicity of possible flows. Each situation needs to be carefully evaluated, since only the easiest problems can be immediately categorized as one of the six cases which follow.



**Table 5.6**
Suggested Maximum Velocities[17]

| Soil type or lining (earth; no vegetation) | Maximum permissible velocities (fps) clear water | water carrying fine silts | water carrying sand and gravel |
|---|---|---|---|
| Fine sand (noncolloidal) | 1.5 | 2.5 | 1.5 |
| Sandy loam (noncolloidal) | 1.7 | 2.5 | 2.0 |
| Silt loam (noncolloidal) | 2.0 | 3.0 | 2.0 |
| Ordinary firm loam | 2.5 | 3.5 | 2.2 |
| Volcanic ash | 2.5 | 3.5 | 2.0 |
| Fine gravel | 2.5 | 5.0 | 3.7 |
| Stiff clay (very colloidal) | 3.7 | 5.0 | 3.0 |
| Graded, loam to cobbles (noncolloidal) | 3.7 | 5.0 | 5.0 |
| Graded, silt to cobbles (colloidal) | 4.0 | 5.5 | 5.0 |
| Alluvial silts (noncolloidal) | 2.0 | 3.5 | 2.0 |
| Alluvial silts (colloidal) | 3.7 | 5.0 | 3.0 |
| Coarse gravel (noncolloidal) | 4.0 | 6.0 | 6.5 |
| Cobbles and shingles | 5.0 | 5.5 | 6.5 |
| Shales and hard pans | 6.0 | 6.0 | 5.0 |

[17] As recommended by Special Committee on Irrigation Research, ASCE, 1926.

**Table 5.7**
Flow Type Classification Parameters

| Flow type | $\frac{h_1-z}{D}$ | $\frac{h_4}{h_c}$ | $\frac{h_4}{D}$ | culvert slope | barrel flow | location of control | kind of control |
|---|---|---|---|---|---|---|---|
| 1 | $< 1.5$ | $< 1.0$ | $\leq 1.0$ | steep | partial | inlet | critical depth |
| 2 | $< 1.5$ | $< 1.0$ | $\leq 1.0$ | mild | partial | outlet | critical depth |
| 3 | $< 1.5$ | $> 1.0$ | $\leq 1.0$ | mild | partial | outlet | backwater |
| 4 | $> 1.0$ | | $> 1.0$ | any | full | outlet | backwater |
| 5 | $\geq 1.5$ | | $\leq 1.0$ | any | partial | inlet | entrance geometry |
| 6 | $\geq 1.5$ | | $\leq 1.0$ | any | full | outlet | entrance and barrel geometry |

Identification of the type of flow beyond the guidelines in table 5.7 requires a trial and error procedure.

In the following cases, several variables appear repeatedly. $C_d$ is the *discharge coefficient*, a function of the barrel inlet geometry. Use orifice data if specific information is unavailable. $v_1$ is the average velocity of the water approaching the culvert entrance, often an insignificant quantity. Not all of the kinetic energy of the approaching water survives the inlet transition, and the *velocity-head coefficient*, $\alpha$, for the approach section accounts for the loss. $d_c$ is the critical depth, which may not correspond to the actual depth of flow. (It will have to be calculated from the flow conditions.) Finally, $h_f$ is the friction loss in the identified section. The friction head loss between sections 1 and 2, for example, can be calculated from equation 5.63.

$$h_{f,1-2} = \frac{LQ^2}{K_1 K_2} \qquad 5.63$$

$$K = \frac{1.486}{n}(r_H)^{2/3}A \qquad 5.64$$

The friction loss can be found in the usual manner, from the Darcy equation and the Moody friction factor chart. The *Manning equation* can also be used, and it is particularly useful since it eliminates the need for trial and error solutions.

$$h_f = \frac{v^2 n^2 L}{(2.21)(r_H)^{4/3}} \qquad 5.65$$



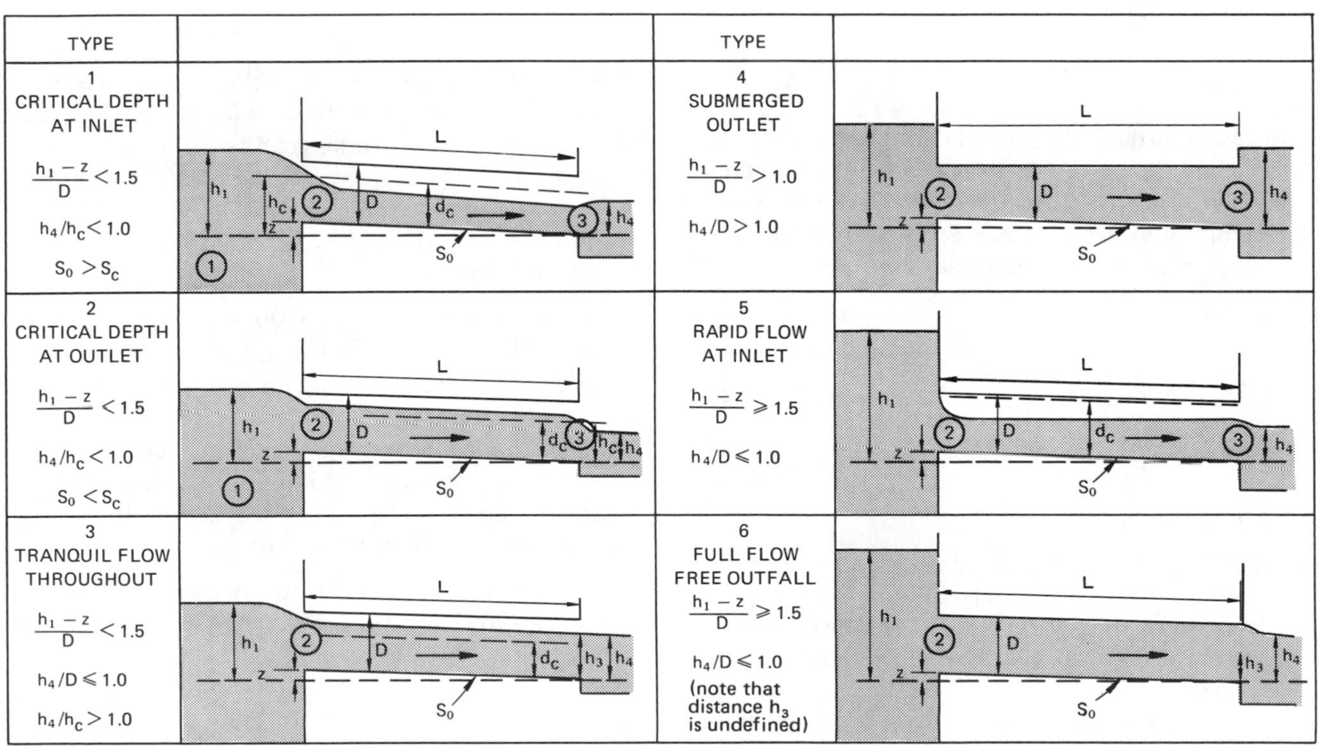


**Figure 5.23** Culvert Flow Classifications

For example, suppose the total hydraulic head available was $H$. This head would be divided between the velocity head in the culvert, the entrance loss (if considered), and the friction.

$$H = \frac{v^2}{2g} + k_e\left(\frac{v^2}{2g}\right) + \frac{v^2 n^2 L}{(2.21)(r_H)^{4/3}} \qquad 5.66$$

This can be solved directly for the velocity.

$$v = \sqrt{\frac{H}{\frac{(1+k_e)}{2g} + \frac{n^2 L}{(2.21)(r_H)^{4/3}}}} \qquad 5.67$$

**Table 5.8**
Minor Entrance Loss Coefficients

| $k_e$ | condition of entrance |
|---|---|
| 0.08 | smooth, tapered |
| 0.10 | flush concrete groove |
| 0.10 | flush concrete bell |
| 0.15 | projecting concrete groove |
| 0.15 | projecting concrete bell |
| 0.50 | flush, square-edged |
| 0.90 | projecting, squared-edged |

The area, $A$, used in the discharge equations which follow, is not always the culvert area, particularly when the culvert does not flow full. $A_c$ is the area in flow at the critical section, while $A_i$ is the area in flow at numbered section i.

OPEN CHANNEL

### A. TYPE 1 FLOW

Water passes through the critical depth near the culvert entrance, and the culvert flows partially full. The slope of the culvert barrel is greater than the critical slope, and the tailwater elevation is less than the elevation of the water surface at the control section.

The discharge is

$$Q = C_d A_c \sqrt{2g\left(h_1 - z + \frac{\alpha_1 v_1^2}{2g} - d_c - h_{f1-2} - h_{f2-3}\right)} \qquad 5.68$$

### B. TYPE 2 FLOW

Flow passes through the critical depth at the culvert outlet, and the barrel flows partially full. The slope of the culvert is less than critical, and the tailwater elevation does not exceed the elevation of the water surface at the control section.

$$Q = C_d A_c \sqrt{2g\left(h_1 + \frac{\alpha_1 v_1^2}{2g} - d_c - h_{f1-2} - h_{f2-3}\right)} \qquad 5.69$$

### C. TYPE 3 FLOW

When backwater is the controlling factor in culvert flow, the critical depth cannot occur. The upstream water-surface elevation for a given discharge is a function of the height of the tailwater.

For type 3 flow, flow is subcritical for the entire length of the culvert, with the flow being partial. The outlet is not submerged, but the tailwater elevation does exceed the elevation of critical depth at the terminal section.

$$Q = C_d A_3 \sqrt{2g\left(h_1 + \frac{\alpha_1 v_1^2}{2g} - h_3 - h_{f1-2} - h_{f2-3}\right)} \qquad 5.70$$

### D. TYPE 4 FLOW

As in type 3 flow, the backwater elevation is the controlling factor in this case. Critical depth cannot occur, and the upstream water surface elevation for a given discharge is a function of the tailwater elevation. Discharge is independent of barrel slope.

The culvert is submerged at both the headwater and tailwater. No differentiation between low-head and high-head is made for this case. If the velocity head at section 1, the entrance friction loss and the exit friction loss are neglected, the discharge can be calculated.

$$Q = C_d A_o \sqrt{\frac{2g(h_1 - h_4)}{1 + \frac{29 C_d^2 n^2 L}{(r_H)^{4/3}}}} \qquad 5.71$$

The complicated term in the denominator corrects for friction. For rough estimates and for culverts less than 50 feet long, the friction loss can be ignored.

$$Q = C_d A_o \sqrt{2g(h_1 - h_4)} \qquad 5.72$$

### E. TYPE 5 FLOW

Partially-full flow under a high head is classified as type 5 flow. The flow pattern is similar to the flow downstream from a sluice gate with rapid flow near the entrance. Usually, type 5 flow requires a relatively square entrance that causes contraction of the flow area to less than the culvert area. In addition, the barrel length, roughness, and bed slope must be sufficient to maintain the flow area less than the culvert area.

It is difficult to distinguish in advance between type 5 and type 6 flow. Within a range of the important parameters, either flow can occur.[19]

$$Q = C_d A_o \sqrt{2g(h_1 - z)} \qquad 5.73$$

[19] If the water surface ever touches the top of the culvert, the passage of air to the culvert will be prevented, and the culvert will flow full everywhere. This is type 6 flow.

F. TYPE 6 FLOW

Type 6 flow, like type 5, is considered a high-head flow. The culvert is full under pressure with free outfall. The discharge is

$$Q = C_d A_o \sqrt{2g(h_1 - h_3 - h_{f2-3})} \quad 5.74$$

Equation 5.74 is inconvenient because $h_3$ (the true piezometric head at the outfall) is difficult to evaluate without special graphical aids. The actual hydraulic head driving the culvert flow is a function of the Froude number. For conservative first approximations, $h_3$ can be taken as the barrel diameter. In reality, it varies from somewhat less than half the barrel diameter to the full diameter. This will give the minimum hydraulic head.

If $h_3$ is taken as the barrel diameter, the total hydraulic head ($H = h_1 - h_3$) will be split between the velocity head and friction. In that case, equation 5.67 can be used to calculate the velocity. The discharge is easily calculated from equation 5.75.[20]

$$Q = A_o v \quad 5.75$$

*Example 5.11*

Size a square culvert which has the following characteristics:

slope = 0.01
length = 250 feet
capacity = 45 cfs
$n = 0.013$
entrance fluid level 5 feet above barrel top
free exit

Since the $h_1$ dimension in all six cases is measured from the culvert invert, it is difficult to classify the type of flow at this point. However, either type 5 or type 6 is likely.

*step 1*: Assume a trial culvert size. Select a square opening with 1.0 foot sides.

*step 2*: Calculate the flow assuming case 5. With entrance control, the culvert will not flow full. The entrance will act like an orifice.

$$A_o = (1)(1) = 1\ \text{ft}^2$$
$$H = h_1 - z = 5 + 1 = 6\ \text{ft}$$
$$C_d = 0.62 \text{ for square-edged openings}$$
$$Q = (0.62)(1)\sqrt{(2)(32.2)(6)} = 12.2\ \text{cfs}$$

Since this size has insufficient capacity, try a larger culvert. Choose 2.0 foot sides.

$$A_o = (2)(2) = 4$$
$$H = 5 + 2 = 7$$
$$Q = (0.62)(4)\sqrt{(2)(32.2)(7)} = 52.7\ \text{cfs}$$

*step 3*: Begin checking the entrance control assumption by calculating the maximum hydraulic radius. Because the flow is entrance controlled, the upper surface of the culvert is not wetted. The hydraulic radius is maximum at the entrance.

$$r_H = \frac{A_o}{p_w} = \frac{4}{2+2+2} = 0.667\ \text{ft}$$

*step 4*: Calculate the velocity using the Manning equation for open channel flow. Since the hydraulic radius is maximum, the velocity will also be maximum.

$$v = \frac{1.486}{0.013}(0.667)^{2/3}\sqrt{0.01}$$
$$= 8.72\ \text{ft/sec}$$

*step 5*: Calculate the normal depth, $d_n$.

$$d_n = \frac{Q}{(v)(w)} = \frac{45}{(8.72)(2)} = 2.58$$

Since the normal depth is greater than the culvert size, the discharge will be full pipe flow under pressure. (It was not necessary to calculate the critical depth since the flow is implicitly subcritical.) The entrance control assumption was, therefore, not valid for this size culvert.[21] At this point, two things can be done. A larger culvert can be chosen if entrance control is desired. Or, the solution can continue by checking to see if the culvert has the required capacity as a pressure conduit.

*step 6*: Check the capacity as a pressure conduit. $H$ is the total available head.

$$H = h_1 - h_3$$
$$= [5 + 2 + (0.01)(250)] - 2$$
$$= 7.5\ \text{ft}$$

[20] Equation 5.75 does not include the discharge coefficient. $v$, when calculated from equation 5.67, is implicitly the velocity in the barrel.

[21] If the normal depth had been less than the barrel diameter, it would still be necessary to determine the critical depth of flow. If the normal depth was less than the critical depth, the entrance control assumption would have been valid.

*step 7*: Since the pipe is flowing full, the hydraulic radius is

$$r_H = \frac{4}{8} = 0.5$$

*step 8*: Equation 5.67 can be used to calculate the flow velocity. Since the culvert has a square-edged entrance, a loss coefficient of $k_e = 0.5$ is used. However, this doesn't affect the velocity greatly.

$$v = \sqrt{\frac{7.5}{\frac{(1+0.5)}{2(32.2)} + \frac{(0.013)^2(250)}{(2.21)(0.5)^{4/3}}}}$$
$$= 10.24\ \text{ft/sec}$$

*step 9*: Check the capacity.

$$Q = vA_o = (10.24)(4) = 40.96\ \text{cfs}$$

The culvert size is not acceptable since its discharge under the maximum head does not have a capacity of 45 cfs.

*step 10*: Repeat from step 2, trying a larger size culvert. With a 2.5 foot side, the following values are obtained:

$$A_o = (2.5)(2.5) = 6.25\ \text{ft}^2$$
$$H = 5 + 2.5 = 7.5\ \text{ft}$$
$$Q = (0.62)(6.25)\sqrt{(2)(32.2)(7.5)}$$
$$= 85.2\ \text{cfs}$$
$$r_H = \frac{6.25}{7.5} = 0.833\ \text{ft}$$
$$v = \frac{1.486}{0.013}(0.833)^{2/3}\sqrt{0.01} = 10.12\ \text{ft/sec}$$
$$d_n = \frac{45}{(10.12)(2.5)} = 1.78$$

*step 11*: Calculate the critical depth. For rectangular channels, equation 5.43 can be used.

$$d_c = \sqrt[3]{\frac{Q^2}{gw^2}} = \sqrt[3]{\frac{(45)^2}{32.2(2.5)^2}} = 2.16\ \text{ft}$$

Since the normal depth is less than the critical depth, the flow is supercritical. The entrance control assumption was correct for the culvert. The culvert has sufficient capacity to carry 45 cfs.

# Appendix A: Design Use Values of Manning's $n$

| channel material | $n$ |
|---|---|
| clean, uncoated cast iron | 0.013–0.015 |
| clean, coated cast iron | 0.012–0.014 |
| dirty, tuberculated cast iron | 0.015–0.035 |
| riveted steel | 0.015–0.017 |
| lock-bar and welded | 0.012–0.013 |
| galvanized iron | 0.015–0.017 |
| brass and glass | 0.009–0.013 |
| wood stave | |
| small diameter | 0.011–0.012 |
| large diameter | 0.012–0.013 |
| concrete | |
| with rough joints | 0.016–0.017 |
| dry mix, rough forms | 0.015–0.016 |
| wet mix, steel forms | 0.012–0.014 |
| very smooth, finished | 0.011–0.012 |
| vitrified sewer | 0.013–0.015 |
| common-clay drainage tile | 0.012–0.014 |
| asbestos | 0.011 |
| planed timber | 0.011 |
| canvas | 0.012 |
| unplaned timber | 0.014 |
| brick | 0.016 |
| rubble masonry | 0.017 |
| smooth earth | 0.018 |
| firm gravel | 0.023 |
| corrugated metal pipe | 0.022 |
| natural channels, good condition | 0.025 |
| natural channels with stones and weeds | 0.035 |
| very poor natural channels | 0.060 |

# Appendix B: Manning Nomograph

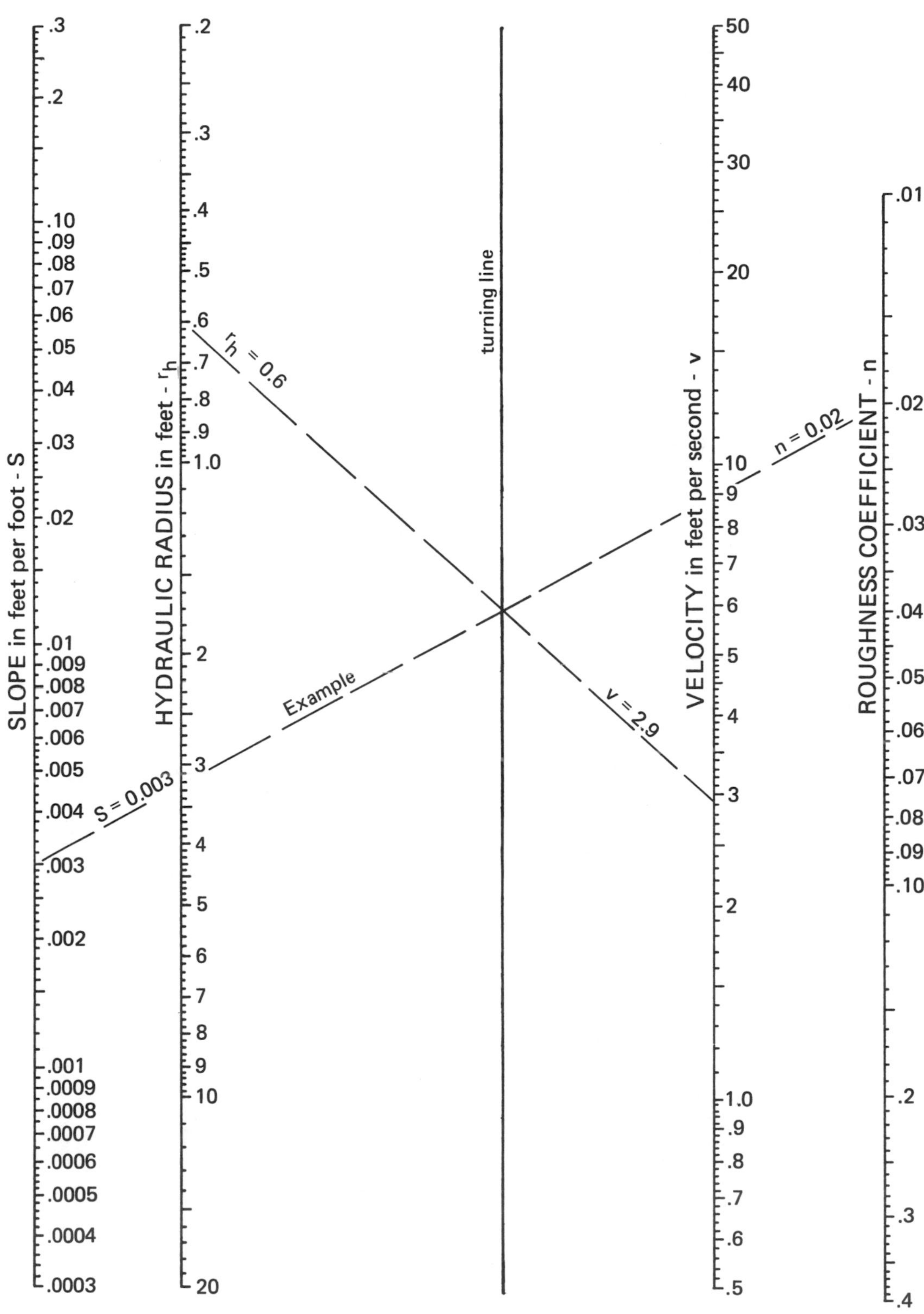

# Appendix C: Circular Channel Ratios

Experiments have shown that $n$ varies slightly with depth. This figure gives velocity and flow rate ratios for varying $n$ (solid line) and constant $n$ (broken line) assumptions.

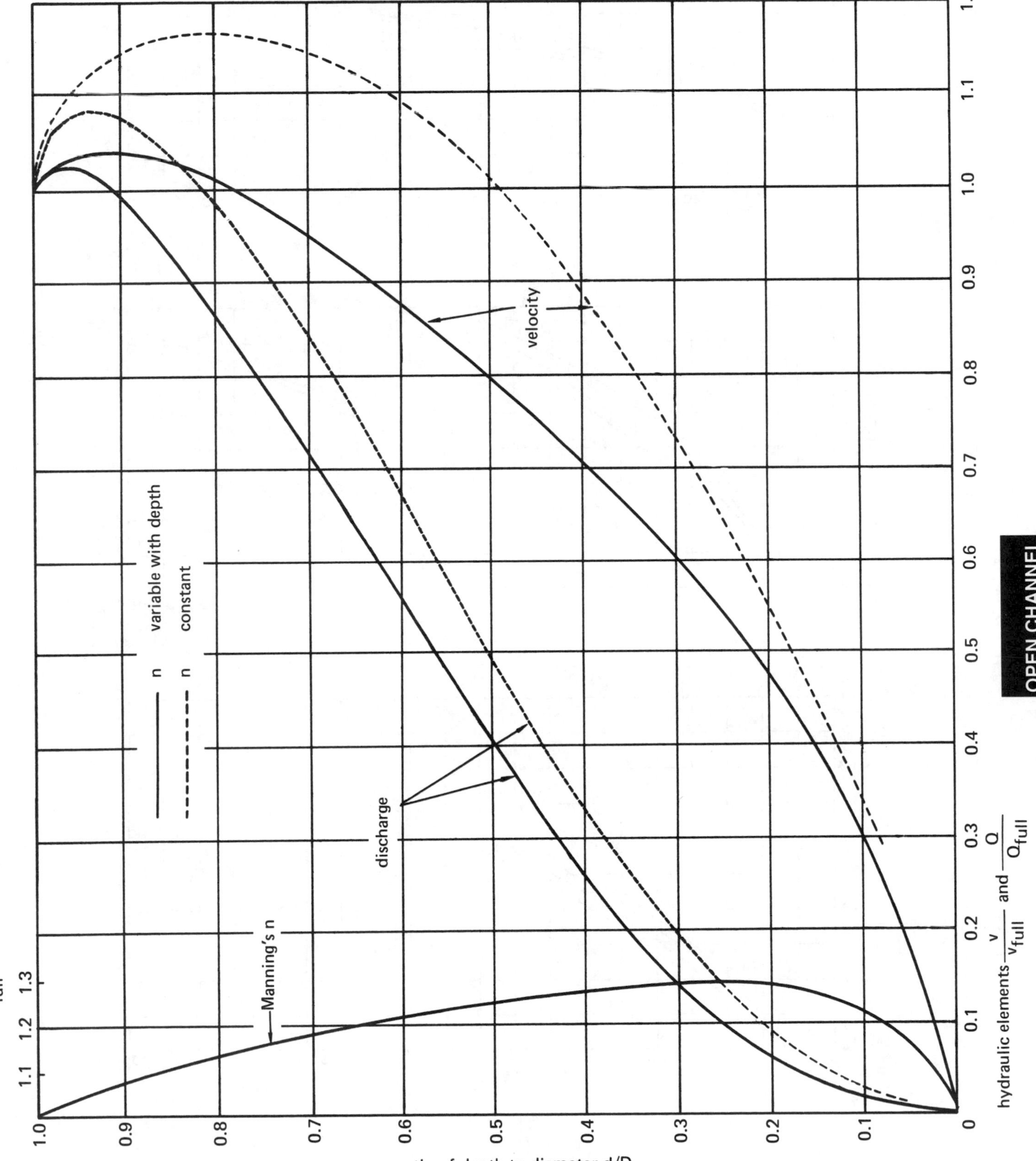

# Appendix D: Critical Depths in Circular Channels

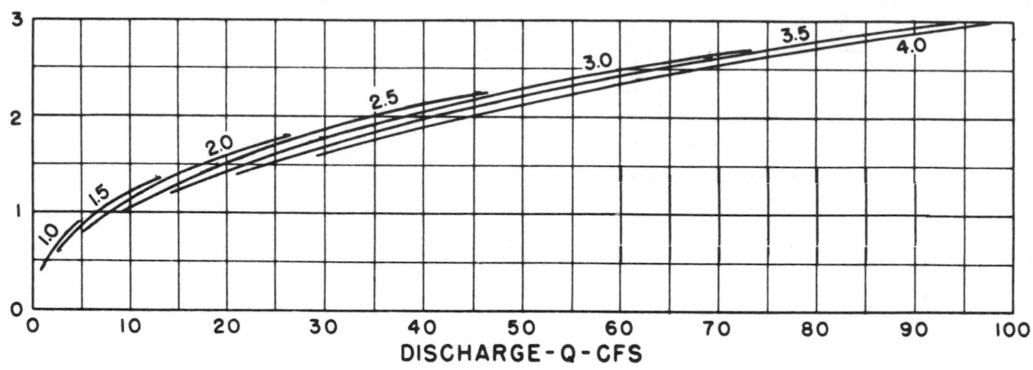


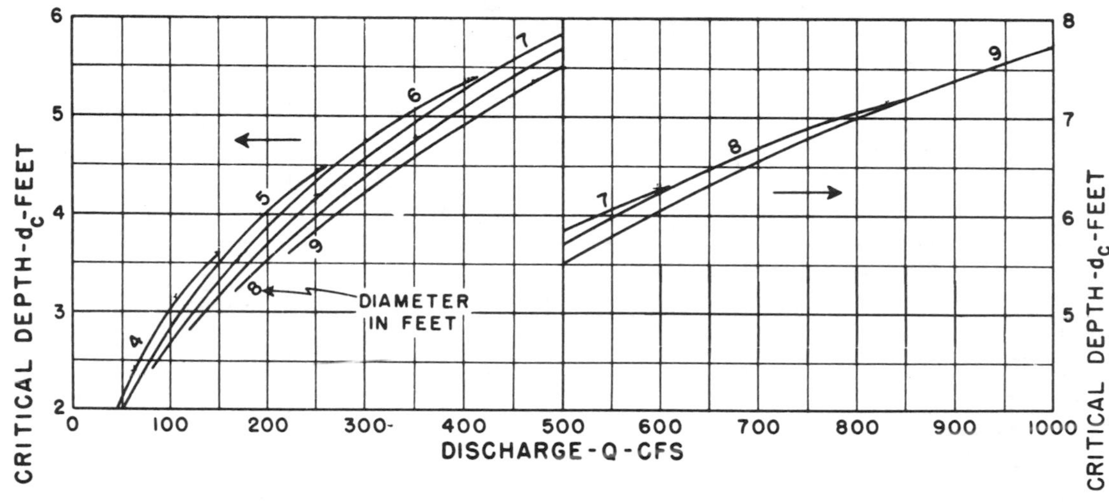


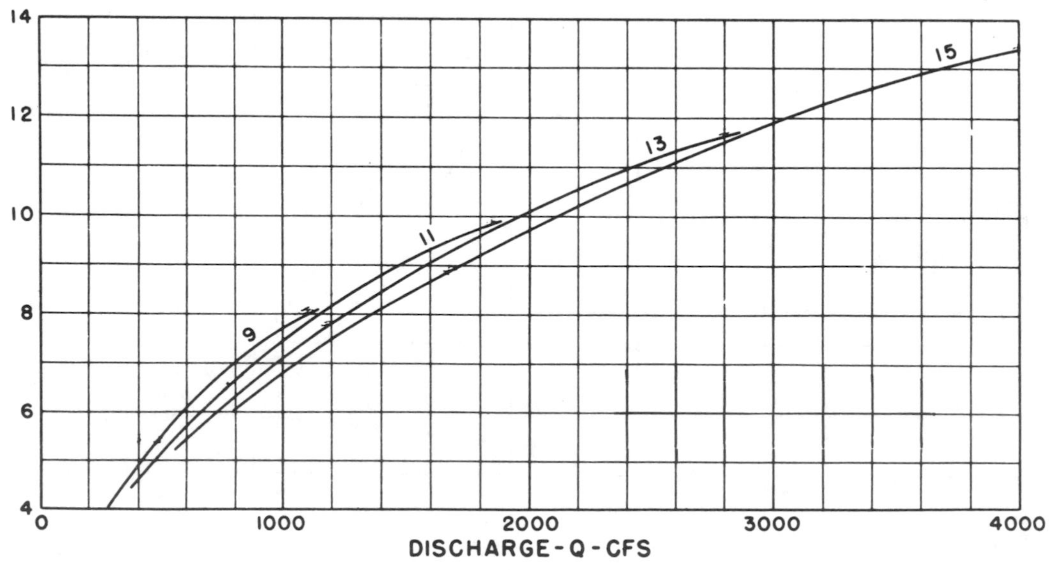

**Practice Problems: OPEN CHANNEL FLOW**

Untimed

1. 30 years ago, a 24 inch diameter pipe ($n$ = .013) was installed on a .001 slope. Recent tests indicate that the full-flow capacity of the pipe is 6.0 cfs. Find (a) the original velocity when full, (b) the present velocity when full, (c) the present value of $n$, and (d) the original capacity.

2. A sewer is to be installed on a 1% grade. Its roughness coefficient is $n$ = .013. Maximum full capacity is to be 3.5 cfs. (a) What size pipe would you recommend? (b) What is the capacity of the pipe size you have chosen when flowing full? (c) What is the velocity when flowing full? (d) What is the depth of flow when the flow is .7 cfs? (e) What minimum velocity will prevent solids settling out in the sewer?

3. The depth of flow upstream from a hydraulic jump is 1 foot. The depth of flow after the jump is 2.4 feet. The channel is rectangular with a 5 foot width. What is the discharge rate of the channel?

4. A wooden flume ($n$ = .012) of rectangular cross section is 2 feet wide. The flume carries 3 cfs of water on a 1% slope. What is the depth?

5. A 4-foot diameter concrete storm drain ($n$ = .013, slope = .02) carries water at a depth of 1.5 feet. (a) What is the velocity of the water in the pipe? (b) What is the maximum velocity of water flowing in the pipe? (c) What is the maximum capacity of the pipe?

6. A hydraulic jump forms at the toe of a spillway. The depths of flow are 0.2 feet and 6 feet on either side of the jump. The velocity before the jump is 54.7 fps. What is the energy loss in the jump?

7. A spillway operates with 2 feet of head. The toe of the spillway is 40 feet below the top of the spillway. Use a spillway coefficient of 3.5. (a) What is the discharge per foot of crest? (b) What is the depth of flow at the toe?

8. An ogee spillway operates with $C_s$ = 3.5 and a head of 5 feet. The spillway crest is 10 feet above the toe. What is the discharge per foot of crest?

9. 10,000 cfs of water flow down a 100-foot wide spillway placed on a 5% grade. The spillway surface has a roughness coefficient of $n$ = .012. (a) What is the depth of flow (normal depth) down the spillway? (b) What is the critical depth? (c) Is the flow tranquil or shooting? (d) A hydraulic jump forms at the junction of the 5% slope and a horizontal toe. What is the depth after the jump?

10. A sharp-crested rectangular weir with two end contractions is 5 feet wide. The weir height is 6 feet. What is the flow rate if the head is measured as .43 feet?

11. The trapezoidal channel shown is laid at a slope of .002. The depth of flow is 2.0 ft. Determine the flow rate. Use $n$ = .013.

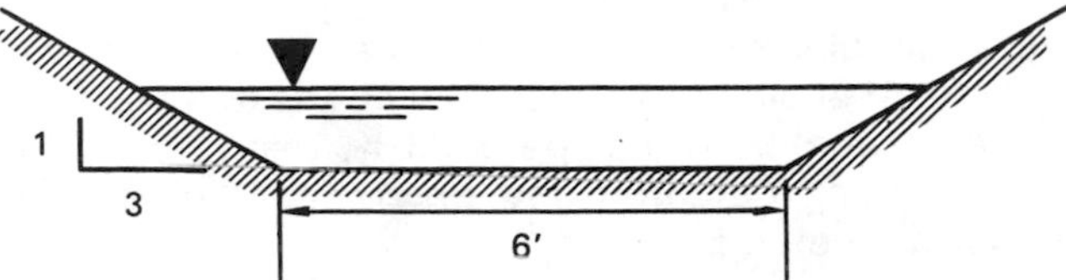


12. The spillway shown has a crest length of 60.0 ft. The stilling basin is 60 ft wide with a level bottom. The head loss in the chute is 20% of the difference in level between the reservoir surface and the stilling basin bottom. Use a spillway coefficient of 3.7. (a) Determine the flow rate. (b) Determine the depth of flow at the toe of the spillway. (c) What tailwater depth is required to cause a hydraulic jump at the toe? (d) What is the energy loss in the hydraulic jump?

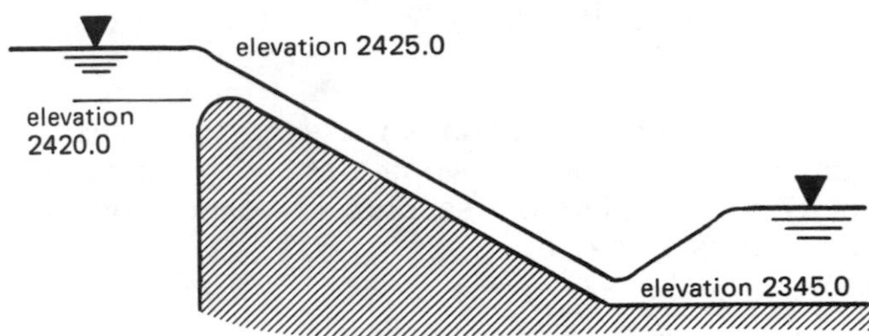


13. A Parshall flume has a throat width of 6.0 ft. The upstream head measured from the throat floor is 18.0 in. (a) What is the flow rate? (b) What are the advantages of the Parshall flume?

14. The weir shown has an 18.0 in. base and a 4-to-1 side slope. The depth of flow over the weir is 9.0 in. What is the rate of discharge?

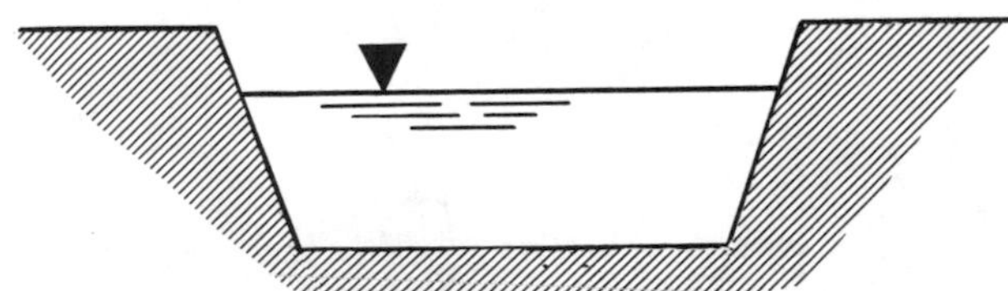

15. Water is flowing at the rate of 50 cfs and a depth of 3.0 ft in a 6.0 ft wide rectangular channel. How high a hump must be placed in the channel to produce critical flow over the hump?

16. A 42 in. diameter concrete culvert is 250 ft long and laid at a slope of .006. The culvert entrance is flush and square-edged. Determine the capacity of the culvert when the outlet is submerged to the crown of the barrel and the headwater is 5.0 ft above the crown of the culvert's inlet.

17. A rectangular channel 8.0 ft wide carries a flow of 150 cfs. The channel slope is .0015 and the Manning coefficient is $n$ = .015. A weir installed across the channel raises the depth at the weir to 6.0 ft. (a) Determine the normal depth of flow. (b) Draw the backwater curve.

18. A circular storm sewer is to carry a peak flow of 5.0 cfs. At peak flow the depth of flow is to be 0.75 the sewer diameter. The slope is 2.0%. Use $n$ = .012 in the Manning equation and determine the required sewer diameter.

19. Water is flowing in a 6.0 ft wide rectangular channel. A hydraulic jump takes place as shown. (a) Determine the flow rate. (b) What is the energy loss in the jump?

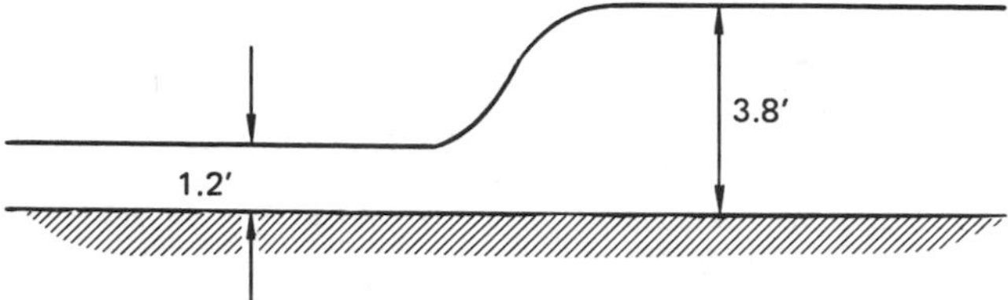


Timed

1. Dimension an optimum rectangular channel (smooth lined concrete, slope = .08) to handle a flow of 17 cubic meters per second. The dimensions can be in units of feet.

2. An overflow structure has three inlets sized 1′ high by 2′ wide as shown. The coefficient of discharge for each orifice is 0.7. The water standing in the intake structure is 4 feet above the top of the box culvert. The coefficient of discharge of the box culvert entrance is 2/3. Assume $n$ = 0.013 (Manning's coefficient) and slope = 0.05 for the culvert which is 100 feet long. Size the square box culvert to handle the flow. Assume the water level behind the dam is constant.

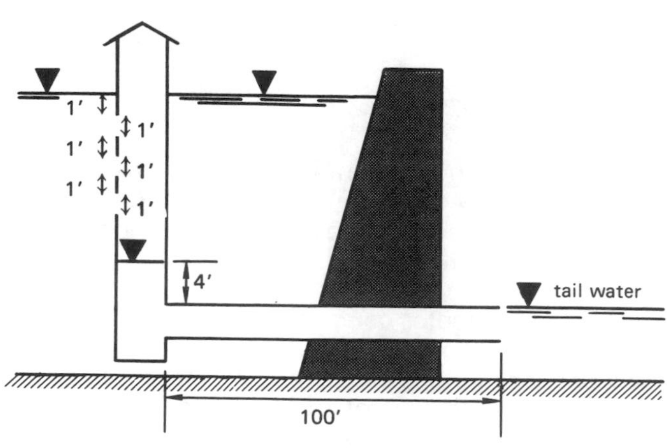


3. A dam spills 70,000 cfs of water over its crest as shown. The width of the dam is 500 feet. The depth at point B is 2 feet. (a) What is the friction loss between points A and B? (b) What is the depth of water at point C? (c) Sketch the energy gradient line from point A to point B. (d) What else needs to be calculated to determine if the hydraulic jump occurs at the toe (point B)?

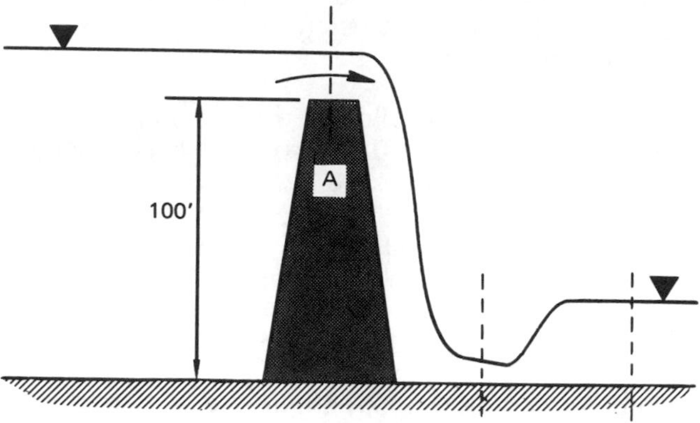


4. Flow in a 6′ diameter, newly-formed concrete culvert is 150 cfs at a point where the depth of flow is 3′. (a) What is the hydraulic radius of the flow? (b) What is the slope of the pipe? (c) Is the flow subcritical, critical, or supercritical? (d) What would be the flow if the culvert were full? (e) Assuming that you know the entrance type and discharge rate, list 3 factors to be considered in selecting a culvert size.

5. A 17′ × 20′ rectangular tank is fed from the bottom and overflows through a double-constricted rectangular weir (to be designed) into a trough.

The following specifications are known:

- discharge rate is 2 MGD minimum
  4 MGD average
  8 MGD maximum
- minimum freeboard in tank = 4″
- minimum head over weir = 10″
- maximum weir elevation = 590′
- elevation of tank bottom = 580′

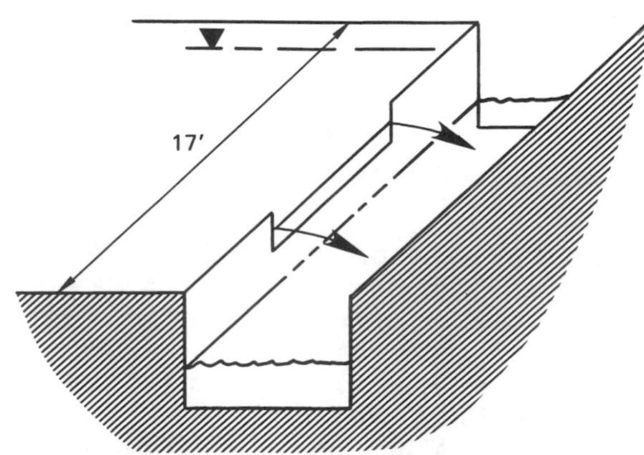


(a) Specify the length of the weir opening. (b) What is the surface elevation at maximum flow? (c) What is the water surface elevation at minimum flow?

OPEN CHANNEL

# 6 HYDROLOGY

*Nomenclature*

| | | |
|---|---|---|
| A | area | $ft^2$ |
| $A_d$ | drainage area | acres |
| B | aquifer width | ft |
| C | rational runoff coefficient, or other coefficient | – |
| $C_p$ | pan coefficient | – |
| $C_r$ | retardance coefficient | – |
| CN | curve number | – |
| d | diameter | ft |
| d | drawdown, or distance between stations | ft, or miles |
| E | evaporation | in/day |
| F | frequency of occurrence, or infiltration | –/yrs, inches |
| $G_H$ | hydraulic gradient | ft/ft |
| H | total hydraulic head | ft |
| I | rainfall intensity | in/hr |
| $I_a$ | initial abstraction | inches |
| $K_p$ | constant of permeability | $ft^3$/day-$ft^2$ |
| $L_c$ | centroidal stream length | miles |
| $L_o$ | overland flow path length | ft |
| $L_s$ | main stream length | miles |
| N | average precipitation per year, or time from peak to end of runoff | inches, or hours |
| P | precipitation over a short period | inches |
| Q | flow quantity | cfs |
| r | radial distance from well | ft |
| R | regional factor | – |
| S | slope of the surface, or storage | –, inches |
| $t_c$ | storm duration (time of concentration) | minutes |
| $t_p$ | time from start of storm to peak runoff | hrs |
| $t_r$ | rain storm duration | hrs |
| T | transmissivity | $ft^3$/day-ft |
| v | flow velocity | ft/sec |
| V | volume | $ft^3$ |
| y | aquifer thickness after drawdown | ft |
| Y | original aquifer thickness | ft |

*Subscripts*

| | |
|---|---|
| f | flow |
| g | gross |
| N | net |
| o | at well |
| p | peak, pan, or equipotential |
| r | radius r, rainfall |
| R | reservoir |
| t | total |
| u | unit |
| x | unknown |
| y | years |

## 1 IMPORTANT CONVERSIONS

| Multiply | By | To Get |
|---|---|---|
| acre | 43,560 | $ft^2$ |
| acre-ft | 43,560 | $ft^3$ |
| acre-ft | 325,851 | gallons (U.S.) |
| acre-inches/hr | 1.008 | cfs |
| cubic feet | 7.4805 | gallons |
| cubic feet/sec | 1.9834 | acre-ft/day |
| cubic feet/sec | 448.83 | gpm |
| cubic feet/sec | 0.64632 | MGD |
| darcy | 1.062 EE–11 | $ft^2$ |
| gallons (U.S.) | 0.1337 | $ft^3$ |
| gallons | 3.07 EE–6 | acre-ft |
| gallons/day | 1.547 EE–6 | $ft^3$/sec |
| gpm | 1440 | gallon/day |
| gpm | 0.002228 | cfs |
| gpm | 192.5 | $ft^3$/day |
| hectare | 2.471 | acres |
| horsepower | 0.7457 | kw |
| horsepower | 550 | ft-lbf/sec |
| inch of runoff | 53.3 | acre-ft/sq. mile |
| inch of runoff | 2.323 EE6 | $ft^3$/sq. mile |
| Meinzer unit | 1.00 | gal/day-$ft^2$ |
| MGD | 1 EE6 | gallon/day |

| | | |
|---|---|---|
| sq. mile | 640 | acres |
| sq. mile | 2.788 EE7 | $ft^2$ |
| sq. mile-inch | 53.3 | acre-ft |
| sq. mile-inch/day | 26.88 | cfs |

## 2 DEFINITIONS

Anabranch: The intertwining channels of a braided stream.

Anticlinal spring: A portion of an exposed aquifer (usually on a slope) between two impervious layers.

Aquiclude: An underground source of water with insufficient porosity to support any sufficient removal rate.

Aquifer: An underground source of water capable of supplying a well or other use.

Aquifuge: An underground geological formation which has no porosity or openings at all through which water can enter or be removed.

Artesian well: A spring in which water flows naturally out of the earth's surface due to pressure placed on the water by an impervious overburden and hydrostatic head.

Artesian formation: An aquifer in which the piezometric height is greater than the aquifer thickness.

Base flow: Runoff which percolates down to the water table and then discharges into a stream. Up to 2 years may elapse between precipitation and discharge.

Bifurcation ratio: The average number of streams feeding into the next side (order) waterway. The range is usually 2 to 4.

Blind drainage: Geographically large (with respect to the drainage basin) depressions which store water during a storm and, therefore, stop it from contributing to surface runoff.

Braided stream: A wide, shallow stream with many anabranches.

Capillary water: Water just above the water table which is drawn up out of an aquifer due to capillary action of the soil.

Cone of depression: The shape of the water table around a well during and immediately after use. The cone's apex differs from the original water table by the well's drawdown.

Confined water: Artesian water overlaid with an impervious layer, usually under pressure.

Connate water: Water, frequently saline, present in rock at its formation.

Depression storage: Initial storage of rain in small surface puddles.

Depth-Area-Duration analysis: A study made to determine the maximum amounts of rain within a given time period over a given area.

Dimple spring: A depression in the earth below the water table.

Drainage density: The total length of streams in a watershed divided by the drainage area.

Drawdown: The difference in water table level at a well head and far from it.

Dry weather flow: See 'Base flow.'

Effluent stream: A stream which intersects the water table and receives groundwater. Effluent streams seldom go completely dry during the rainless periods.

Ephemeral stream: A stream which goes dry during rainless periods.

Evapotranspiration: Evaporation of water from a study area due to all sources including water, soil, snow, ice, vegetation, and transpiration.

Flowing well: A well which flows on its own accord to the surface. See also 'Artesian well.'

Forebay: An area which recharges an aquifer.

Gravitational water: Water in transit downward through the earth.

Groundwater: Subsurface water flowing in an aquifer towards a stream. Groundwater is not water flowing on the ground. It is water flowing underground.

Hydrological cycle: The cycle experienced by water in its travel from the ocean, through evaporation and precipitation, percolation, runoff, and return to the ocean.

Hydrometeor: Any form of water falling from the sky.

Hygroscopic water: Moisture adhering in a thin film to soil grains.

Impervious layer: A geologic layer through which no water can pass.

Infiltration: The movement of water through the upper soil.

Influent stream: A stream above the water table. Influent streams may go dry during the rainless season.

Initial loss: The sum of interception and depression loss, but excluding blind drainage.

Interception: Rain which falls on vegetation and other impervious objects and which evaporates without contributing to runoff.

Interflow: Infiltrated subsurface water which travels to a stream without percolating down to the water level.

Juvenile water: Water formed chemically within the earth.

Lysimeter: A soil container used to observe and measure evaptotranspiration.

Meandering stream: A stream which flows in large loops, not in a straight line.

Meteoric water: See 'Hydrometeor.'

Negative boundaries: A fault or similar geologic structure.

Net rain: Rain which contributes to surface runoff.

Overland flow: Water which travels over the ground surface to a stream.

Pan: A container used to measure surface evaporation rates.

Perched spring: A localized saturated area which occurs above an impervious layer on a slope.

Percolation: The travel of water down through the soil to the water table.

Phreatic zone: The layer below the water table down to an impervious layer.

Phreatophytes: Trees with root systems which extend into the water table.

Piezometric level: The level to which water will rise in a pipe due to its own pressure.

Plat: A small plot of land.

Porosity: The ratio of pore volume to total formation volume.

Probable maximum rainfall: The rainfall corresponding to some given probability (e.g., 1 in 100 years).

Safe yield: The maximum rate of water withdrawal which is economically and ecologically feasible.

Seep: See 'Spring.'

Sinuosity: The stream length divided by the valley length.

Specific yield: The ratio of water volume which will drain freely from a substance sample to the total volume. Specific yield is always less than porosity. Specific yield is also defined as the volume of water obtained by lowering one square foot of the water table by one foot.

Spring: A place where the earth surface and aquifer coincide.

Stream order: An artificial categorization of stream geneology. Small streams are 1st order. 2nd order streams are fed by 1st order streams, 3rd order streams are fed by 2nd order streams, etc.

Subsurface runoff: See 'Interflow.'

Surface detention: Rain water which collects as a film and runs off of the saturated surface during a storm.

Surface retention: That part of a storm which does not immediately appear as infiltration or surface runoff. Retention is made up of depression storage, interception, and evaporation.

Surface runoff: Water flow over the surface which reaches a stream after a storm.

Time of concentration: The time required for water to flow from the most distant point on a runoff area to the measurement or collection point.

Transpiration: The process in which plants give off internal moisture to the atmosphere.

Unit stream power: The product of velocity and slope, representing the rate of energy expenditure per pound of water.

Vadose water: All water above the water table, including soil water, gravitational water, and capillary water.

Vadose zone: A zone above the water table containing both saturated and empty soil pores.

Water table: The top level of an aquifer, defined as the locus of points where the water pressure is equal to the atmospheric pressure.

Xerophytes: Drought-resistant plants, typically existing with root systems well above the water table.

Zone of aeration: See 'Vadose zone.'

Zone of saturation: See 'Phreatic zone.'

## 3 PRECIPITATION

Although the word *precipitation* encompasses all hydrometeoric forms, it is often applied only to rainfall in the liquid form. Precipitation data may be collected in a number of ways, but the open 8-inch precipitation rain gage is quite common.

If a rain measurement is lost or is not available, it can be estimated by one of the following procedures:

*Method 1*: Choose three stations evenly spaced around and close to the location which has missing data. If the normal annual precipitations at the three sites do not vary more than 10% from the missing station's normal annual precipitation, the rainfall is estimated as the arithmetic mean of the three neighboring stations' precipitations for the period in question.

*Method 2*: If the difference is more than 10%, the Normal-Ratio Method can be used:

$$P_x = \left(\frac{1}{3}\right)\left[\left(\frac{N_x}{N_A}\right)P_A + \left(\frac{N_x}{N_B}\right)P_B + \left(\frac{N_x}{N_C}\right)P_C\right] \quad 6.1$$

*Method 3*: A method used by the U.S. National Weather Service for river forecasting is to use data from stations in the four nearest quadrants (North, South, East, and West of the unknown station) and to weight the data with the square of the distance between stations.

$$P_x = \frac{d^2_{A-x}P_A + d^2_{B-x}P_B + d^2_{C-x}P_C + d^2_{D-x}P_D}{d^2_{A-x} + d^2_{B-x} + d^2_{C-d} + d^2_{D-x}} \quad 6.2$$

The average precipitation over a specific area or time basis can be found from station data in several ways.

*Method 1*: If the stations are uniformly distributed over a flat site, their precipitations can be averaged. This also requires that the individual precipitation records not vary too much from the mean.

*Method 2*: The *Thiessen method* calculates the average by weighting station measurements by the area of the assumed basin for each station. These assumed basin areas are found by drawing dotted lines between all stations and bisecting these dotted lines with solid lines (which are extended outward until they connect with other solid lines). The solid lines will form a polygon whose area is the assumed basin area.

*Method 3*: The most accurate method is the *Isohyetal method.* This method requires plotting lines of constant precipitation (isohyets) and weighting the isohyet values by the areas enclosed by the lines. Station data is used to draw isohyets, but not in the calculation of average rainfall.

HYDROLOGY

Effective design of a structure will depend on the geographical location and degree of protection required. Once the location of a structure has been chosen, it is up to the engineer to design for the area's most probable maximum rainfall or probable maximum flood. Both of these require some judgment since the maximum is not a deterministic number.

***Rainfall intensity*** is the rate of precipitation per hour. Intensity will be low for most storms, but it will be high for some storms. These high intensity storms can be expected very infrequently—say every 20, 50, or 100 years. The average number of years between storms of a given magnitude is known as the ***design-storm frequency of occurrence*** or ***recurrence interval.***

The instantaneous intensity of a storm can be calculated from equation 6.3.

$$I = \frac{K'(F)^a}{(t_c + b)^c} \quad 6.3$$

$K'$, $a$, $b$, and $c$ are constants which depend on the conditions and location.

The *Steel formula* is a simplification of equation 6.3.

$$I = \frac{K}{t_c + b} \quad 6.4$$

$K$ and $b$ are dependent on the storm frequency and geographical location. $t_c$ can be either the time of concentration (in runoff studies) or the rainfall duration.

Values of the constants $K$ and $b$ in equation 6.4 are not difficult to obtain once the intensity-duration-frequency curve is established. Although a logarithmic transformation could be used to obtain the data in straight line form, an easier method exists. This method starts by taking the reciprocal of equation 6.4, converting the equation to a straight line.

$$\frac{1}{I} = \frac{t_c + b}{K} = \frac{t_c}{K} + \frac{b}{K} = C_1 t_c + C_2 \quad 6.5$$

As equation 6.5 shows, it is possible to use linear regression to determine the relationship between $1/I$ and $t_c$. Once $C_1$ and $C_2$ have been found, $K$ and $b$ can be calculated.

$$K = \frac{1}{C_1} \quad 6.6$$

$$b = \frac{C_2}{C_1} \quad 6.7$$

For very rough estimates, published values of $K$ and $b$ can be obtained from the geographical location. Table 6.1 is typical of some of this general data.

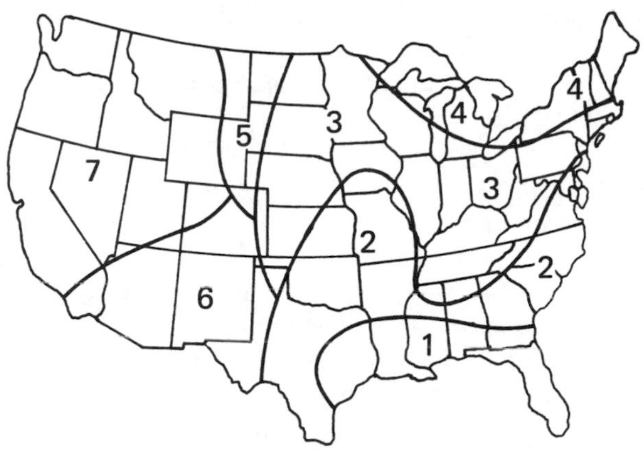


**Figure 6.1** Steel Formula Rainfall Regions

**Table 6.1**
Steel Formula Coefficients

| frequency in years | coefficients | region 1 | 2 | 3 | 4 | 5 | 6 | 7 |
|---|---|---|---|---|---|---|---|---|
| 2 | K | 206 | 140 | 106 | 70 | 70 | 68 | 32 |
| | b | 30 | 21 | 17 | 13 | 16 | 14 | 11 |
| 4 | K | 247 | 190 | 131 | 97 | 81 | 75 | 48 |
| | b | 29 | 25 | 19 | 16 | 13 | 12 | 12 |
| 10 | K | 300 | 230 | 170 | 111 | 111 | 122 | 60 |
| | b | 36 | 29 | 23 | 16 | 17 | 23 | 13 |
| 25 | K | 327 | 260 | 230 | 170 | 130 | 155 | 67 |
| | b | 33 | 32 | 30 | 27 | 17 | 26 | 10 |
| 50 | K | 315 | 350 | 250 | 187 | 187 | 160 | 65 |
| | b | 28 | 38 | 27 | 24 | 25 | 21 | 8 |
| 100 | K | 367 | 375 | 290 | 220 | 240 | 210 | 77 |
| | b | 33 | 36 | 31 | 28 | 29 | 26 | 10 |

Since the intensity, $I$, is the average intensity, the total rainfall (in inches) can be calculated from the storm duration.

$$P = It_c \tag{6.8}$$

*Example 6.1*

A storm has an instantaneous intensity given by the equation:

$$I = \frac{100}{t_c + 10}$$

15 minutes are required for runoff from the farthest corner of a 5-acre plat to reach a discharge culvert. What is the design intensity?

$$I = \frac{100}{15 + 10} = 4\,\text{inches/hour}$$

Note that at $t > 15$ minutes, the intensity will be less. Thus, the intensity at $t = 15$ minutes is the maximum (i.e., is the design intensity).

## 4 FLOOD CONSIDERATIONS

A flood occurs when more water than a river can carry is introduced. When a water course is too small to contain the flow, the water overflows the banks. The results of this overflow may produce nuisance flooding, damaging flooding, or devastating flooding.

*Nuisance flooding* may result in inconveniences such as wet feet, tire spray, and soggy lawns. *Damaging floods* go on to soak flooring, carpeting, and first-floor furniture. *Devastating floods* can wash buildings and vehicles downstream, as well as take lives.

Although rain causes flooding, large rain storms do not always create flooding. The size of a flood depends not only on the amount of rainfall, but also on the conditions within the watershed before and during the storm. When rain falls on a very wet watershed that is unable to absorb more water, or when a very large amount of rain falls on a dry watershed faster than it can be absorbed, the water runs off.

It is not economically justifiable to attempt to provide protection against the largest flood that could occur. Such protection would be too costly for such a rare event. Therefore, governmental and private institutions have agreed that a *one percent flood* is the most infrequent event to be protected against. A one percent flood, also called a *hundred year flood*, is a flood that would be exceeded in severity only once every hundred years on the average.

Planning for the one percent flood has proved to be a good compromise between not doing enough and spending too much. The U.S. Army Corps of Engineers, Soil Conservation Service, Department of Transportation, and others use the one percent flood to design channels and bridges and to protect facilities in which they have an interest.

Specific terms are sometimes used to designate the degree of protection required. For example, the *probable maximum flood* (*PMF*) is a hypothetical flood which can be expected to occur as a result of the most severe combination of critical meteorlogic and hydrologic conditions possible within a region. Since many structures are built with a limited life expectancy, it may be extremely uneconomical to design against the PMF.

The *standard project flood* (*SPF*) is a flood which can be selected from the most severe combinations of meteorological and hydrological conditions reasonably characteristic of the region, excluding extremely rare combinations of events. SPF volumes are commonly 40% to 60% of the PMF volumes.

The *design basis flood* (*DBF*) depends on the site. It is the flood that is adopted as the basis for design of a particular project. The DBF is usually determined from economic considerations, or is specified as part of the contract document.

Although the one percent flood is a common choice for the design basis flood, shorter recurrence intervals are often used, particularly in low value areas such as cropland. For example, 5-year storm curves are used in residential areas, 10-year curves for business sections, and 15-year frequencies for high-value districts where flooding will result in more extensive damage. The ultimate choice of a recurrence interval, however, must be made on the basis of economic considerations and tradeoffs.

The probability that a flooding event in any year will equal or exceed the design basis flood based on a recurrence interval or frequency, $F$, is

$$p\{F \text{ event in one year}\} = \frac{1}{F} \qquad 6.9$$

The probability of an $F$ event occurring in $n$ years is

$$p\{F \text{ event in } n \text{ years}\} = 1 - \left(1 - \frac{1}{F}\right)^n \qquad 6.10$$

*Example 6.2*

A wastewater treatment plant has been designed to be in use for 40 years. What is the probability that a one percent flood will occur within the useful lifetime of the plant?

$$F = 100$$

$$p\{100 \text{ year flood in 40 years}\} = 1 - \left(1 - \frac{1}{100}\right)^{40} = 0.33 \quad (33\%)$$

## 5 SUBSURFACE WATER

Subsurface water is a major source of all water used in the United States. In dry areas, it may be the only source of water used for domestic and irrigation uses. Subsurface zones are divided into two parts by the water table. The *vadose zone* exists above the water table, and pores in the vadose zone may be either empty or full. Below the water table is the *phreatic zone*, whose pores are always full.

Soil moisture content is measured in pounds per cubic foot. Soil moisture is usually determined by oven drying a sample of soil and measuring the weight loss. The moisture content can also be determined with a tensiometer, which measures the vapor pressure of the moisture in the soil.

Movement of water through an aquifer is given by equation 6.11 in which $K_p$ is known as the ***coefficient of permeability*** (or ***hydraulic conductivity***).[1] (Also, see Sec. 9.6(H).)

$$Q = (1.157\,\text{EE–5}) K_p A G_H \qquad 6.11$$

In the past, $K_p$ in the United States has been given in (now obsolete) *Meinzer units* (gallons per day per square foot). Permeability is now widely given in units of cm/sec. (See table 9.9.) The aquifer flow area, $A$, is given by equation 6.13. The hydraulic gradient, $G_H = \Delta H/L$, is the change in hydraulic head, $\Delta H$, over some length, $L$, measured in the direction of seepage.

**Table 6.2**

Approximate Coefficients of Permeability, $K_p$

| material | ft / day or ft$^3$/day-ft$^2$ | gal/day-ft$^2$ | darcys |
|---|---|---|---|
| Clay | 1.3 EE–3 | EE–2 | EE–3 |
| Sand | 1.3 EE2 | EE3 | EE2 |
| Gravel | 1.3 EE4 | EE5 | EE4 |
| Gravel/Sand | 1.3 EE3 | EE4 | EE3 |
| Sandstone | 1.3 EE1 | EE2 | EE1 |
| Dense shale & limestone | 1.3 EE–1 | 1.0 | EE–1 |
| Quartzite & granite | 1.3 EE–3 | EE–2 | EE–3 |

The *transmissivity* of flow from a saturated aquifer of thickness $Y$ and width $B$ is given by equation 6.12.

$$T = K_p Y \qquad 6.12$$

The cross sectional area in flow depends on the aquifer dimensions.[2]

$$A = BY = \frac{BT}{K_p} \qquad 6.13$$

Combining equation 6.11 with equation 6.13,

$$Q = (1.157\,\text{EE–5}) B T G_H \qquad 6.14$$

## 6 WELL DRAWDOWN IN AQUIFERS

A virgin aquifer with a well is shown in figure 6.2 (a). Once pumping of the well starts, the water table will be lowered in the vicinity of the well, and the resulting water table surface is known as the *cone of depression*. The decrease in water level at the well is known as the *drawdown*, $d_o$. The drawdown at some distance $r$ from the well is $d_r$.

If the drawdown is small with respect to the aquifer thickness, $Y$, and the well completely penetrates the aquifer, then the equilibrium (steady state) well dis-

[1] This is one form of *Darcy's law*. The 1.157 EE–5 is a conversion factor from cubic feet per day to cfs. Also, see Eq. 9.24.

[2] Since water can only flow between rock, not all of the aquifer area contributes to flow. It is important to know that the area, $A$, and the permeability, $K_p$, are consistent. If only the clear flow area is known, it will be necessary to multiply by the porosity to reduce the area in flow.

charge is given by the *Dupuit equation*, equation 6.15. The 86,400 converts cubic feet per day to cubic feet per second.

$$Q = \frac{\pi K_p(y_1^2 - y_2^2)}{86{,}400\, ln(r_1/r_2)} \qquad 6.15$$

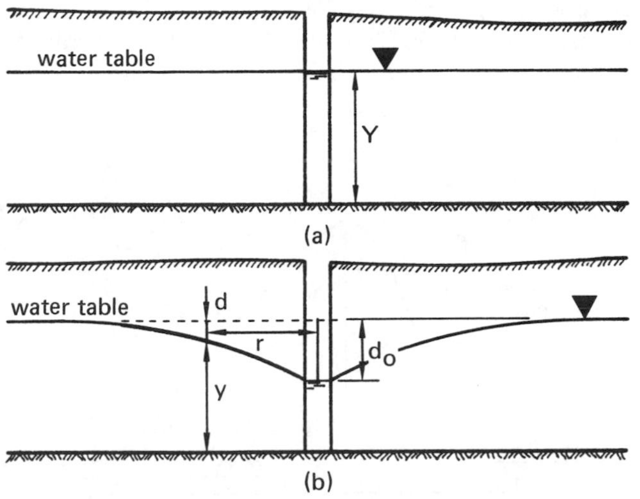


**Figure 6.2** Well Drawdown

In equation 6.15, $y_1$ and $y_2$ are the aquifer depths at any two radii, $r_1$ and $r_2$, respectively. $y_1$ can also be taken as the original aquifer depth, $Y$, if $r_1$ is the well's *radius of influence*.

For an artesian well fed by a confined aquifer of thickness $Y$, the discharge is

$$Q = \frac{2\pi K_p(y_1 - y_2)Y}{86{,}400\, ln(r_1/r_2)} \qquad 6.16$$

When pumping first begins, the removed water also comes from the aquifer above the equilibrium cone of depression. Therefore, equation 6.15 cannot be used, and a non-equilibrium analysis is required.

*Example 6.3*

A 9″ diameter well is pumped at the rate of 50 gpm. The aquifer is 100 feet thick. The well sides cave in and are replaced with an 8″ diameter tube. Assuming a 6 foot drawdown, what will be the steady flow from the new well? Assume the water table recovers its original thickness 2500 feet from the well.

$$(50 \text{ gpm})\left(0.002228\,\frac{\text{cfs}}{\text{gpm}}\right) = 0.1114 \text{ cfs}$$

From equation 6.15,

$$y_2 = 100 - 6 = 94\,\text{ft}$$

$$r_2 = 9/(2)(12) = 0.375\,\text{ft}$$

$$0.1114 = \frac{\pi K_p((100)^2 - (94)^2)}{(86{,}400) ln(2500/0.375)}$$

$$K_p = 23.17\,\text{ft}^3/\text{day-ft}^2$$

For an 8″ (r = 0.333 ft) pipe,

$$Q = \frac{\pi(23.17)((100)^2 - (94)^2)}{(86{,}400)\, ln(2500/0.333)} = 0.110\,\text{cfs}$$

## 7 SEEPAGE AND FLOW NETS

Groundwater flows from locations of high hydraulic head to locations of lower hydraulic head. Problems requiring the actual calculation of the flow quantity are best solved with the flow net concept, particularly when a manual solution is required.

*Flow nets* are constructed from streamlines and equipotential lines. *Streamlines* (*flow lines*) show the path taken by the seepage. *Equipotential lines* connect points of constant pressure. The flow net concept is limited to cases where the flow is steady, two-dimensional, incompressible, through a homogeneous medium, and where the liquid has a constant viscosity.

The flow net is constructed according to the following rules.

- Streamlines enter and leave pervious surfaces perpendicular to those surfaces.
- Streamlines approach the *line of seepage* (above which there is no hydrostatic pressure) asymptotically to that surface.
- Streamlines are parallel to but cannot touch impervious surfaces which are streamlines.
- Streamlines are parallel to the flow direction.
- Equipotential lines are drawn perpendicular to streamlines, such that the resulting cells are square and the intersections are 90° angles.
- Equipotential lines enter and leave impervious surfaces perpendicular to those surfaces.

The size of the cells formed by the intersection of the streamlines and equipotential lines is not important, although the more cells formed, the greater will be the accuracy.

The object of a graphical solution is to construct a network of flow paths (outlined by the streamlines) and equal pressure drops (bordered by equipotential lines). No fluid flows across streamlines, and a constant amount of fluid flows between any two streamlines.

Figure 6.3 shows flow nets for several common situations. A careful study of the flow nets will help to clarify the rules and conventions previously listed. In drawing flow nets, remember that the number of streamlines and equipotential lines drawn is up to the investigator. There are many flow nets that can be drawn, all more or less correct, differing only in their accuracy.

Once the flow net is drawn, it can be used to calculate the seepage. First, the number of flow channels, $N_f$, between the streamlines is counted. Then, the number of equipotential drops, $N_p$, between equipotential lines is counted. The total hydraulic head is determined as a function of the water surface levels.

$$Q = K_p H \left(\frac{N_f}{N_p}\right) \qquad 6.17$$

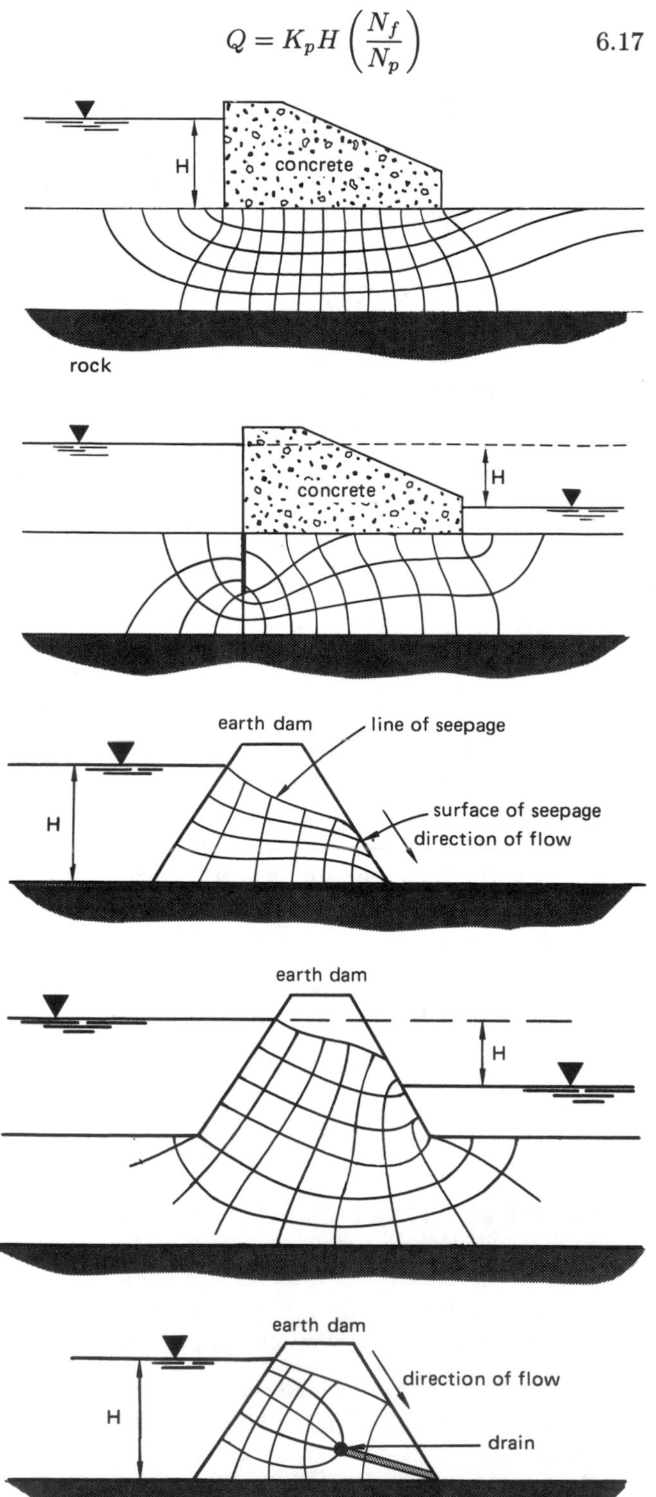


**Figure 6.3** Typical Flow Nets

## 8 TOTAL SURFACE RUNOFF FROM STREAM HYDROGRAPHS

After a rain, water makes its way to a stream. A plot of stream discharge versus time is known as a *hydrograph.* Hydrograph periods may be very long (such as a year) down to very short (hours). A typical hydrograph is shown in figure 6.4, which illustrates the *time base* and the time from peak runoff to cessation.

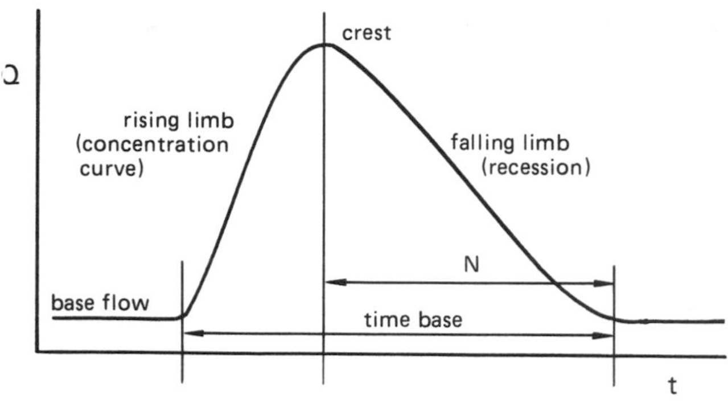


**Figure 6.4** A Stream Hydrograph

In figure 6.4, the portion of the hydrograph to the left of the crest is known as the *rising limb.* To the right of the crest, the curve is known as the *recession* or *falling limb.*

The stream discharge is assumed to consist of *overland flow* and *groundwater flow.* Since culverts do not have to be designed to carry groundwater, a procedure called *hydrograph separation* or *hydrograph analysis* is necessary to separate surface and groundwater.[3]

There are several methods of separating baseflow from overland flow. All of the methods are somewhat arbitrary. Three of the methods easily carried out manually are presented here.

*Method 1*: (Straight Line Method): A horizontal line is drawn from the start of the rising limb to the falling limb. All of the flow under the horizontal line is considered base flow. This is illustrated in figure 6.5.

[3] The total rain dropped by the storm is the *gross rain.* The rain which actually appears as immediate runoff is known as *surface runoff, overland flow, surface flow,* and *net rain.* The water which is absorbed by the soil and which does not contribute to the surface runoff is known as *base flow, groundwater, infiltration,* and *dry weather flow.*

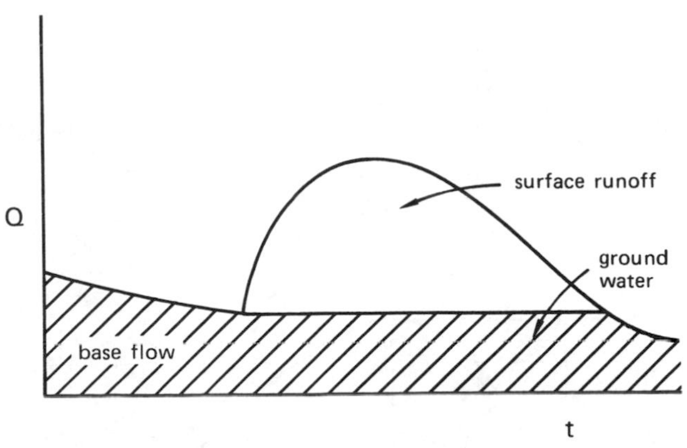


**Figure 6.5** Hydrograph Separation Method 1

*Method 2*: (Fixed Base Method): The base flow existing before the storm is projected or continued down to a point directly under the crest of the hydrograph. Then, a straight line is used to connect the projection to the falling limb $N$ days later. $N$ is determined by inspection or it can be calculated from the rule-of-thumb equation 6.18.

$$N = (\text{area in sq. mi})^{0.2} \text{ in days}$$
$$= 6.59\,(\text{area in acres})^{0.2} \text{ in hours} \qquad 6.18$$

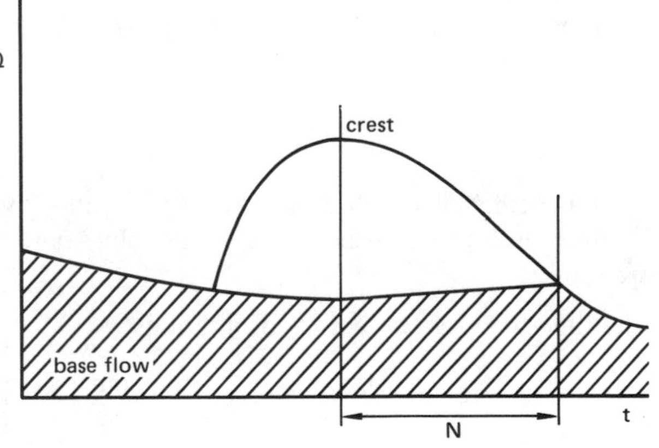


**Figure 6.6** Hydrograph Separation Method 2

*Method 3*: (Variable Slope Method): This method recognizes that the shape of the base flow curve before the storm will probably match the shape of the base flow curve after the storm. The groundwater curve after the storm is projected back under the hydrograph to a point under the inflection point of the falling limb. The separation line under the rising limb is drawn arbitrarily.

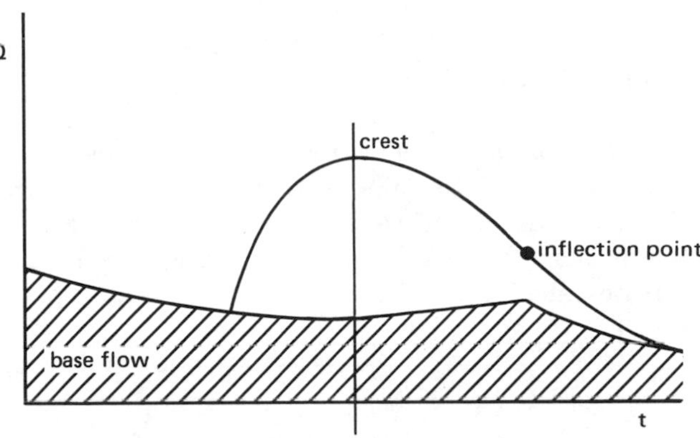


**Figure 6.7** Hydrograph Separation Method 3

Once the base flow is separated out, the hydrograph of overland flow will take on the appearance of figure 6.8.

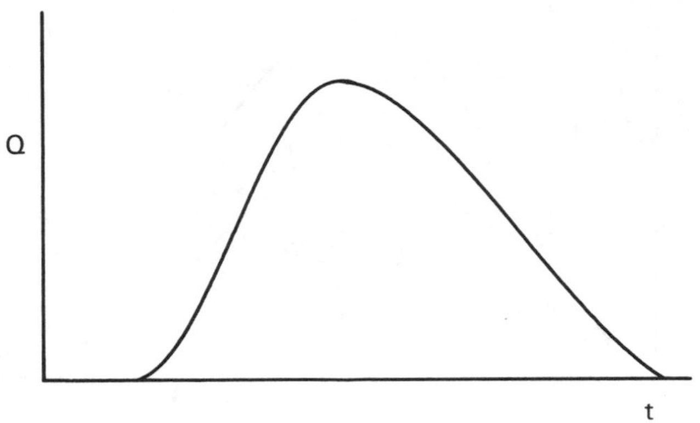


**Figure 6.8** Overland Flow Hydrograph

The total volume of the storm can be found from the area under the hydrograph. Although this can be accomplished by integration or planimetry, it is often sufficiently accurate to approximate the hydrograph with a histogram.

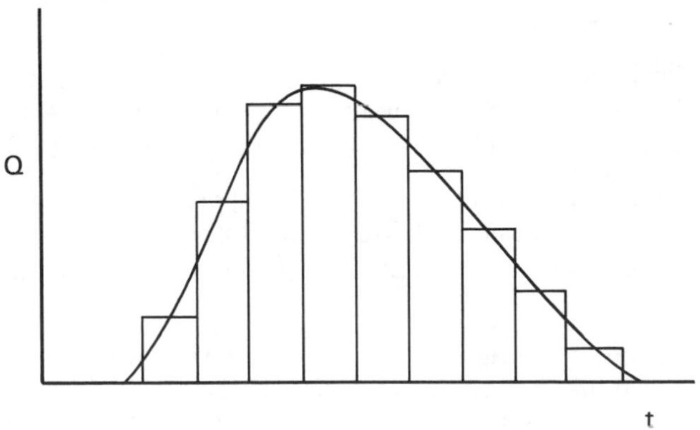


**Figure 6.9** Hydrograph Histogram

## 9 PEAK RUNOFFS FROM THE UNIT HYDROGRAPH

Once the overland flow hydrograph for a storm has been developed, the total rainfall volume can be found from the area under the curve. Furthermore, since the area of the drainage basin is known, the average precipitation can be calculated. (Use consistent units in equation 6.19.)

$$P = \frac{V}{A_d} \tag{6.19}$$

This precipitation will be some number of inches (e.g., 1.7″ or 0.92″, etc.). If every point on the hydrograph is divided by this average precipitation, a unit hydrograph will be derived. This is a hydrograph of a storm dropping 1″ of rain on the entire basin. Figure 6.10 shows how a unit hydrograph compares to a similar storm dropping an average of 1.7″ of rain.

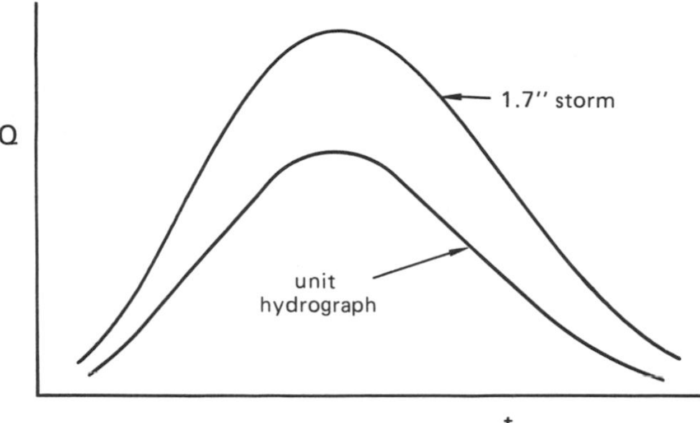


**Figure 6.10** A Storm and Unit Hydrograph

Direct use of a unit hydrograph assumes that:

- All storms in that basin have the same duration.
- The time base, $N$, is constant for all storms.
- The shape of the distribution is the same for all storms.
- Only the total amount of rainfall varies from storm to storm.

The hydrograph of a storm producing more or less than 1″ of rain is found by multiplying all ordinates on the unit hydrograph by the total precipitation of the storm.

A unit hydrograph can be used to predict the runoff for storms which have durations differing as much as ±25% from the storm duration used to derive the unit hydrograph.

If a basin is ungaged so that no records are available to produce a unit hydrograph, important hydrograph parameters can be derived analytically (with some success). Knowing these parameters and recognizing that the total precipitation from a unit hydrograph must be one inch permits sketching a rough approximation to a unit hydrograph.

The *Snyder synthetic hydrograph* is shown in figure 6.11.

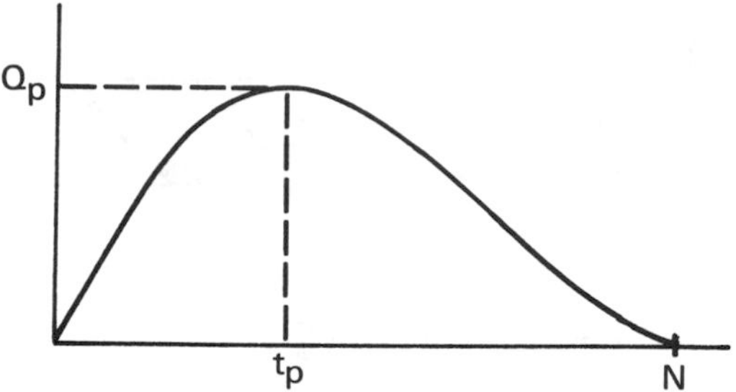


**Figure 6.11** Snyder Synthetic Hydrograph

An estimate for the time to peak runoff for the Snyder hydrograph is

$$t_p = C_t(L_sL_c)^{0.3} \tag{6.20}$$

$C_t$ is a coefficient which varies from 1.8 to 2.2, depending on slope, averaging 2 for steep slopes. $L_s$ is the main stream length from outlet to divide in miles. $L_c$ is the distance in miles from the outlet to a point on the stream nearest the basin centroid. Equation 6.20 is valid for basin areas of 10 to 10,000 square miles.

The peak flow for the unit hydrograph is

$$Q_p = C_pA_d/t_p \tag{6.21}$$

$C_p$ is a coefficient which varies between 0.4 and 0.8 depending on slope and other factors, but values near 0.60 are typical.

The time base for the unit hydrograph is

$$N = 72 + 3t_p \tag{6.22}$$

Since the value of $N$ can never be less than 72 hours, the Snyder hydrograph is not suitable for small basins which reach their peaks in a matter of hours.

Equations 6.21 and 6.22 are valid if the storm rain duration is

$$t_r = t_p/5.5 \tag{6.23}$$

If the duration is known to be otherwise, say $t'_r$, the actual time to peak runoff that should be used in equations 6.21 and 6.22, is $t'_p$.

$$t'_p = t_p + \frac{t'_r - \frac{t_p}{5.5}}{4} \tag{6.24}$$

**Table 6.3**
Representative Snyder Hydrograph Constants

| location | $C_t$ | $C_p$ |
|---|---|---|
| Southern California | 0.4 | 0.94 |
| Appalachian highlands | 2.0 | 0.63 |
| Gulf of Mexico | | |
| eastern portions | 8.0 | 0.61 |

*Example 6.4*

After a 2-hour storm, a station downstream from a 45 square mile drainage basin measures 9400 cfs as a peak discharge and 3300 acre-feet as total runoff. Find the 2-hour unit hydrograph peak discharge. What would be the peak runoff and design flood volume if a 2-hour storm dropped $2\frac{1}{2}$ inches net precipitation?

1 inch of runoff from 45 square miles is

$$\begin{aligned} V &= (45)\,\text{mile}^2\,(1)\,\text{inch}\,(53.3)\frac{\text{acre-ft}}{\text{sq. mile-in}} \\ &= 2399\,\text{acre-feet} \end{aligned}$$

The runoff ratio is $3300/2399 = 1.38$.

The unit hydrograph peak discharge is $9400/1.38 = 6812$ cfs.

For a $2\frac{1}{2}''$ storm, peak runoff would be $(2.5)(6812) = 17{,}030$ cfs.

The design flood would be $(2.5)(2399) = 5998$ acre-feet.

*Example 6.5*

A storm drops its significant rainfall in 6 hours on a 25 square mile basin. The resulting surface runoff is listed. (a) Construct the unit hydrograph of this 6-hour storm. (b) Find the runoff at $t = 15$ hours from a two storm system (both at 6 hours duration) if the first storm drops $2''$ net starting at $t = 0$ and the second storm drops $5''$ net starting at $t = 12$ hours.

| hours after rainfall starts | runoff (cfs) | hours after rainfall starts | runoff (cfs) |
|---|---|---|---|
| 0 | 0 | 21 | 600 |
| 3 | 400 | 24 | 400 |
| 6 | 1300 | 27 | 300 |
| 9 | 2500 | 30 | 200 |
| 12 | 1700 | 33 | 100 |
| 15 | 1200 | 36 | 0 |
| 18 | 800 | | |
| | | TOTAL | 9500 |

(a) The actual runoff is plotted below:

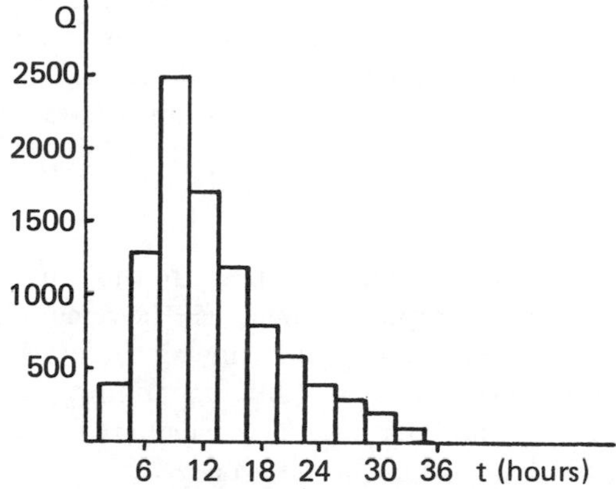


Finding the total area under the curve is equivalent to the following calculation:

$$\begin{aligned} V &= (9500)\,\text{cfs}(3)\,\text{hours}(3600)\frac{\text{sec}}{\text{hr}} \\ &= 1.026\,\text{EE8 ft}^3\,\text{runoff} \end{aligned}$$

The basin area is

$$(25)\ \text{miles}^2\left(5280\,\frac{\text{feet}}{\text{mile}}\right)^2 = 6.97\ \text{EE8 ft}^2$$

The net precipitation is

$$\frac{(1.026\,\text{EE8})\text{ft}^3\,(12)\text{in/ft}}{(6.97\,\text{EE8})\,\text{ft}^2} = 1.767\,\text{inches}$$

The unit hydrograph has the same shape as the actual hydrograph, with all ordinates reduced by 1.767.

(b) The flow at $t = 15 - 12 = 3$ is 400 cfs. The runoff at $t = 15$ is

$$2\left(\frac{1200}{1.767}\right) + 5\left(\frac{400}{1.767}\right) = 2490\,\text{cfs}$$

## 10 THE SCS UNIT HYDROGRAPH (1957)

The U.S. Soil Conservation Service has developed a synthetic unit hydrograph based on dimensionless variables.[4] In order to use this method, it is necessary to know the time to peak flow ($t_p$) and the peak discharge ($Q_p$). Provisions for calculating both of these important parameters are included in the method.

$$t_r = 0.133 t_c \tag{6.25}$$

$$t_p = 0.5 t_r + t_1 \tag{6.26}$$

$$Q_p = \frac{0.756 A_d}{t_p} \tag{6.27}$$

[4] A hydrograph can also be constructed from the SCS curve number (CN) method used to calculate peak flow. This method is described elsewhere in this chapter.

$t_1$ in equation 6.26 is the lag time from the centroid of the rainfall distribution to the peak discharge, in hours. If unknown, the lag time can be determined from correlations with geographical region and drainage area. As a last resort, $t_1$ can be estimated from equation 6.28.

$$t_1 = 0.6t_c \qquad 6.28$$

$Q_p$ and $t_p$ only contribute one point to the construction of the unit hydrograph. To construct the remainder, figure 6.12 must be used. Using time as the independent variable, selections of time (different from $t_p$) are arbitrarily made, and the ratio $t/t_p$ is calculated. The curve then is used to obtain the ratio of $Q_t/Q_p$.

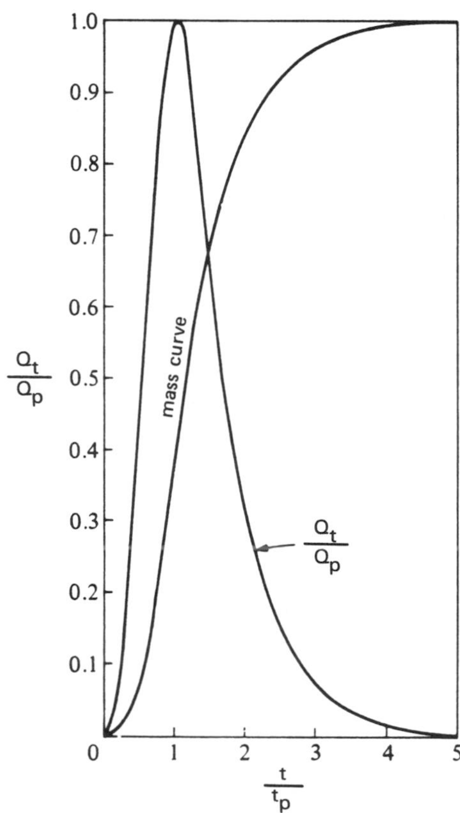


**Figure 6.12** Dimensionless Unit Hydrograph

## 11 HYDROGRAPH SYNTHESIS

If a storm's duration is not the same as the hydrograph's base length, the unit hydrograph cannot be used to predict runoff. For example, the runoff from a 6-hour storm cannot be predicted from a unit hydrograph derived from a 2-hour storm. However, the technique of hydrograph synthesis can be used to construct the hydrograph of the longer storm from the unit hydrograph of a shorter storm.

### A. LAGGING STORM METHOD

If a unit hydrograph for a storm of duration $t_r$ is available, it can be used to construct the hydrograph of a storm whose duration is a whole multiple of $t_r$.[5] For example, a 6-hour storm hydrograph can be constructed from a 2-hour unit hydrograph since $6/2 = 3$ is a whole number.

Let the whole multiple number be $N$. To construct the longer hydrograph, draw $N$ unit hydrographs, each separated by time $t_r$. Then add the ordinates to obtain a hydrograph for an $Nt_r$ duration storm. Since the total rainfall from this new hydrograph is $N$ inches (having been constructed from $N$ unit hydrographs), the curve will have to be reduced (i.e., divided) by $N$ everywhere to produce a unit hydrograph.

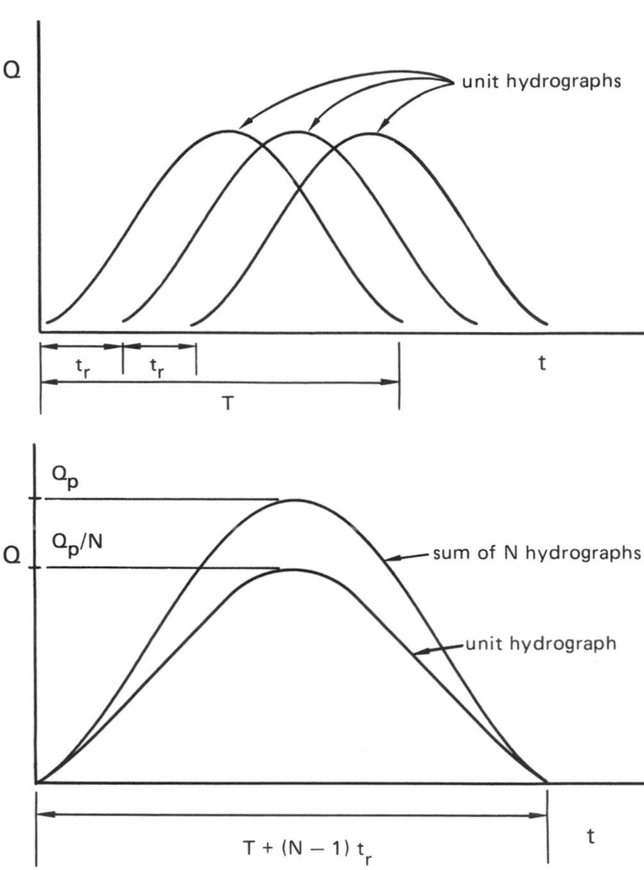


**Figure 6.13** Hydrograph Synthesis by the Lagging Storm Method

### B. S-CURVE METHOD

The S-curve method can be used to construct hydrographs from unit hydrographs with longer or shorter durations, and when the storm durations are not multiples. This method begins by adding the ordinates of

[5] Distinction is made between the *storm duration*, $t_r$ and the *time base* of the hydrograph. A 2-hour storm may produce a hydrograph with runoff continuing over a time base of many days.

many unit hydrographs, each lagging the other by time $t$, the duration of the storm which produced the unit hydrograph. After a sufficient number of lagging unit hydrographs have been added together, the accumulation will level off and remain constant. At that point, the lagging can be stopped. The resulting accumulation is known as an *S-curve.*

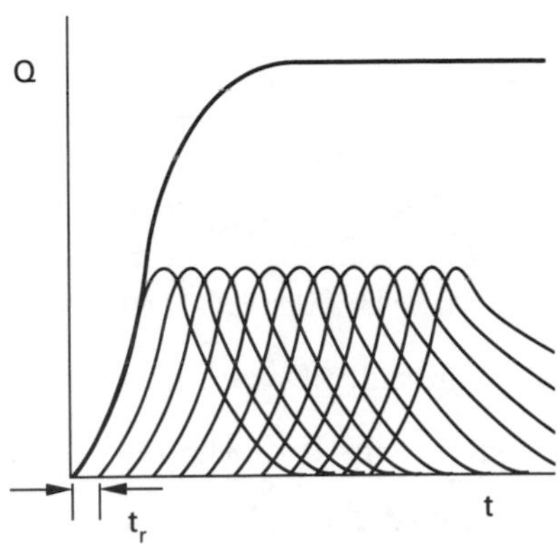


**Figure 6.14** Constructing the S-Curve

If two S-curves are drawn, one lagging the other by time $t'$, the area between the two curves represents a hydrograph area for a storm of duration $t'_r$. The differences between the two curves can be plotted and scaled to a unit hydrograph by multiplying by the ratio of $(t_r/t'_r)$.

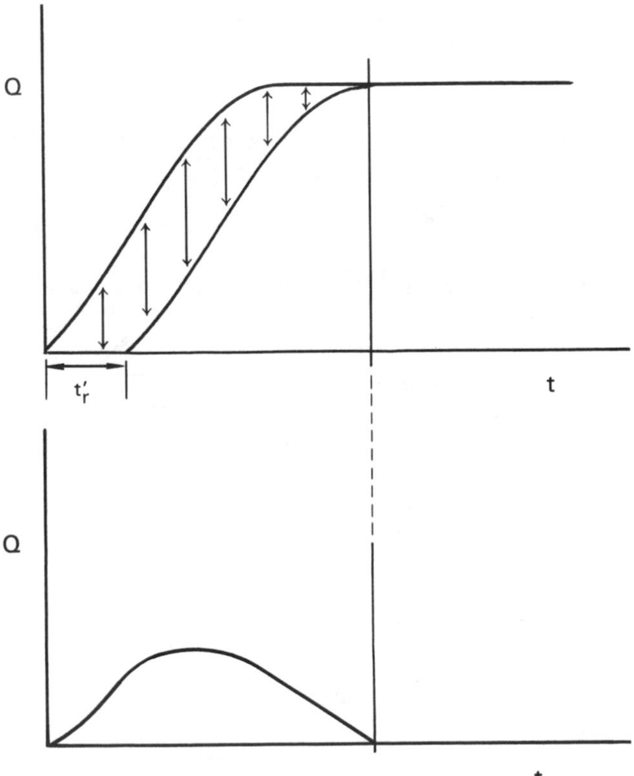


**Figure 6.15** Using S-Curves to Construct a $t'$ hydrograph

## 12 PEAK RUNOFF FROM THE RATIONAL METHOD

Although total runoff data are required for reservoir and dam design, the instantaneous peak runoff is needed to size culverts and storm drains.

In the closed-form equations used to find peak runoff, it is usually assumed that the rainfall is applied to a surface at a constant rate. If this assumption is true and the surface is largely impervious, the runoff will eventually equal the rate of rainfall. The time between the start of rainfall and the time of peak flow is known as the *time of concentration,* $t_c$. Typical values of $t_c$ for areas less than 50 acres range from 5 to 30 minutes.

The *rational formula* is based on the assumptions given and has been in widespread use for small areas (i.e., less than several hundred acres or so) but is seldom used for areas greater than five square miles. Values of $C$ are found in appendix A.

$$Q_p = CIA_d \tag{6.29}$$

Strictly speaking, $Q_p$ is in acre-inches/hour, but it is typically taken as cubic feet per second since the conversion between these two units is 1.008. For a small drainage area, $t_c$ is taken as the largest combination of overland flow time and channel time. *Channel time* is found by dividing the channel length by the (usually assumed) channel velocity.[6] Assuming laminar flow, the overland flow time for small areas without defined channels can be found from the *Izzard formula.* Equation 6.30 should be used only if the product $(IL_o)$ is less than 500.

$$t = \frac{(41)(b)(L_o)^{1/3}}{(CI)^{2/3}} \tag{6.30}$$

$$b = \frac{0.0007(I) + C_r}{S^{1/3}} \tag{6.31}$$

The retardance coefficient, $C_r$, is given in table 6.4.

**Table 6.4**
Retardance Coefficient

| Type of Surface | $C_r$ |
|---|---|
| smooth asphalt | 0.007 |
| concrete pavement | 0.012 |
| tar & gravel pavement | 0.017 |
| closely clipped lawn | 0.046 |
| dense bluegrass turf | 0.060 |

[6] If the pipe or channel size is known, the velocity can be found from $Q = Av$. If the pipe size is not known, then $A$ will have to be estimated. In that case, one might as well estimate $v$ instead. 5 fps is a reasonable velocity. The minimum velocity for a self-cleansing pipe is 2 fps.

The overland flow time has been predicted by the Federal Aviation Administration after analyses of airport drainage areas.

$$t = \frac{1.8(1.1 - C)\sqrt{L_o}}{S^{1/3}} \qquad 6.32$$

The distance $L_o$ in equations 6.30 and 6.32 is the longest distance to the collection point, as shown in figure 6.16.

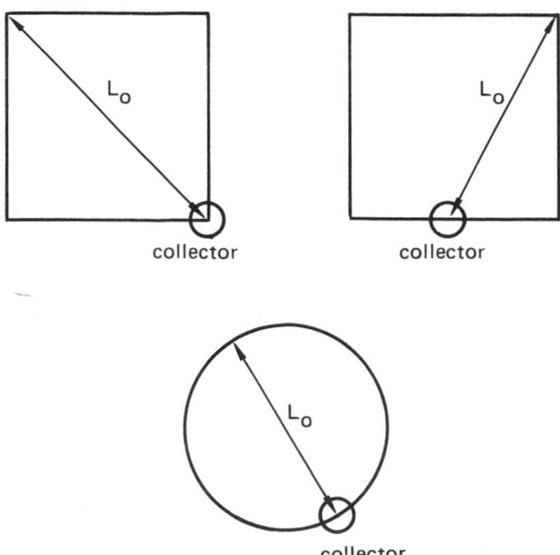


**Figure 6.16** Overland Flow Distances

For irregularly-shaped drainage areas, it may be necessary to evaluate several alternative overland flow distances. For example, figure 6.17 shows a drainage area with a long tongue. Although the tongue area contributes to the drainage area, it does lengthen the overland flow time. Depending on the intensity-duration-frequency curve, the longer overland flow time (resulting in a lower rainfall intensity) may offset the increase in area due to the tongue. Therefore, two runoffs need to be compared, one ignoring and the other including the tongue, and taking the larger.

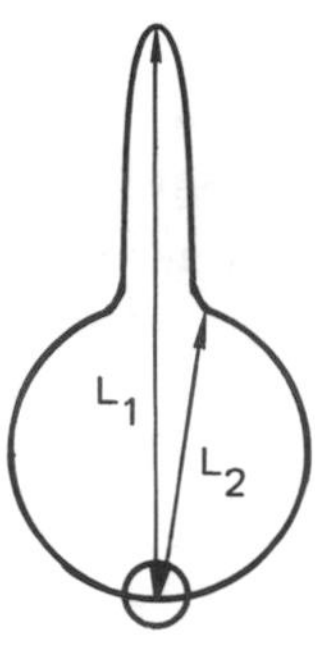


**Figure 6.17** An Irregular Drainage Area

HYDROLOGY

The most important part of equation 6.29 is the rainfall intensity. Rainfall data can be compiled into intensity-duration-frequency curves similar to those in figure 6.18. The intensity used in equation 6.29 will depend on the time of concentration and the degree of protection desired.[7]

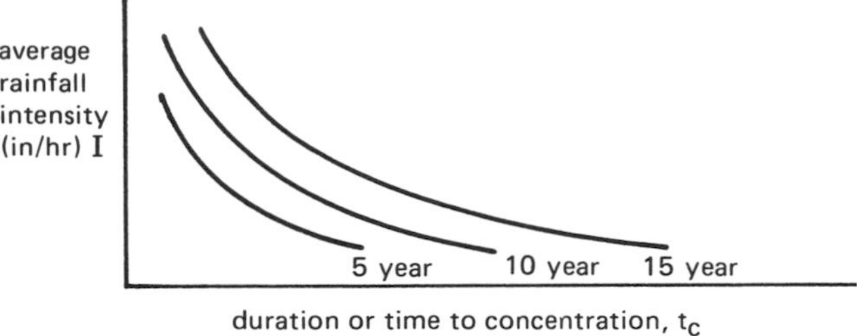


**Figure 6.18** Intensity-Duration-Frequency Curves

The following steps constitute the rational method.

*step 1*: Estimate $t_c$. This is the sum of the overland flow and conduit flow times. $t_c$ will increase the farther you get from the drainage area.

*step 2*: Choose a value of $C$. If more than one area contributes to the runoff, $C$ is weighted by the areas.

*step 3*: Select a frequency or return period for the storm.

*step 4*: Calculate or determine the average storm intensity from intensity-duration-frequency curves.

*step 5*: Use equation 6.29 to calculate the peak flow.

*step 6*: Use open channel flow design techniques to size the channel carrying surface water away.

*Example 6.6*

Two adjacent fields contribute runoff to a collector whose capacity is to be determined. The intensity for a 25 minute duration is 3.9 in/hr. Use the rational method to calculate the peak flow.

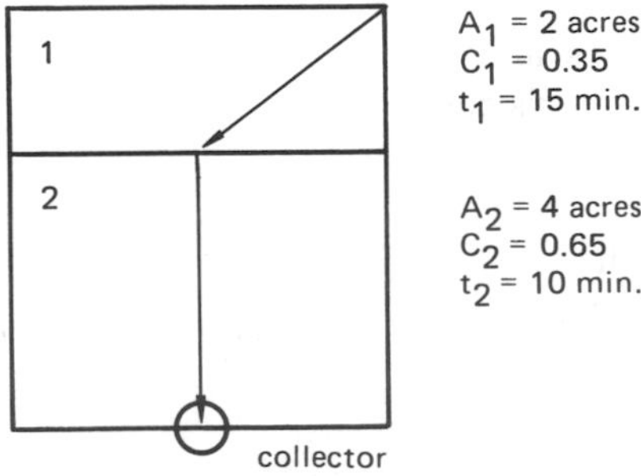


[7] By using intensity-duration-frequency curves to size storm sewers, culverts, and other channels, it is assumed that the frequencies and probabilities of flood damage and storms are identical. This is not generally true, but the assumption is made when other supporting data are not available.

*step 1*: The overland flow time is given for both areas. The time for water from the farthest corner to reach the collector is

$$t_c = 15 + 10 = 25\,\text{minutes}$$

*step 2*: The runoff coefficients are given for each area. Since we want to size the pipe carrying the total runoff, the coefficients are weighted by their respective contributing areas.

$$C = \frac{(2)(0.35) + (4)(0.65)}{2 + 4} = 0.55$$

*step 3 & 4*: The intensity for a 25 minute duration was given as 3.9 in/hr.

*step 5*: The total area is $4 + 2 = 6$ acres. The peak flow is found from equation 6.29.

$$Q_p = (0.55)(3.9)(6) = 12.9\,\text{cfs}$$

*Example 6.7*

A drainage area has the following characteristics:

| area | size (acres) | overland flow time (min) | $C$ |
|---|---|---|---|
| A | 10 | 20 | 0.3 |
| B | 2 | 5 | 0.7 |
| C | 15 | 25 | 0.4 |

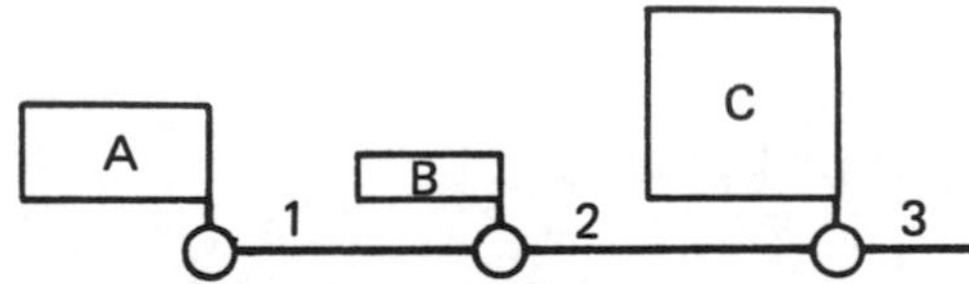


The instantaneous rainfall intensity for the area (and a specified storm frequency) is given by

$$I = \frac{115}{t_c + 15}\ \text{(in/hr)}$$

The manholes are 300 feet apart. The pipe slope is 0.009. Manning's roughness is 0.015. Assume the flow velocity is 5 ft/sec in all sections. Assume the storm duration is long enough to permit all three areas to contribute to the combined flow through section 3. What should be the pipe size in section 2? What is the maximum flow through section 3? Use the rational method.

For area $A$:

$$t_c = 20$$
$$I = \frac{115}{20 + 15} = 3.29\,\text{in/hr}$$
$$Q = (0.3)(10)(3.29) = 9.87\,\text{cfs}$$

This value of $Q$ (9.87 cfs) should be used to size line #1.

To find the flow time between manholes, a flow velocity is needed. This would normally be calculated from $Q$ and equation 5.9 or be obtained graphically (appendix 5.C). However, v is given in this example.

Using a flow velocity of 5 ft/sec, the flow time between drainage inlets is

$$t = \frac{300}{5} = 60\,\text{seconds} = 1\,\text{minute}$$

For area $B$:

At $t = 5$, $I = 5.75$ in/hr. The runoff from 2 acres is (0.7)(2)(5.75) or 8.05 cfs. However, at $t = (20 + 1)$, the peak flow from area $A$ will reach the second manhole. At $t = 21$, $I = 3.19$ in/hr.

The sum of $CA$ values is $(0.3)(10) + (0.7)(2) = 4.4$

$$Q = (4.4)(3.19) = 14.0\,\text{cfs}$$

This value of $Q$ should be used to design section 2 of the pipe. If the pipe is assumed to flow full, the required diameter is found from open channel flow considerations.

$$d = (1.33)\left[\frac{(14.0)(0.015)}{\sqrt{0.009}}\right]^{3/8} = 1.79\,\text{ft}\ \ \text{(round to 2.0 ft)}$$

For area $C$:

Theoretically the larger diameter would require finding the actual flow velocity, since the pipe does not flow full. But, the velocity is given in this example.

Assuming 5 fps as the flow velocity in the pipe, the time from the start of the storm for water from plat $A$ to reach the 3rd manhole is

$$20 + 1 + 1 = 22$$

Since 25 is larger than 22, the maximum runoff will occur 25 minutes after the start of the storm. The 22 minute datum is not used.

$$I = \frac{115}{25 + 15} = 2.875$$

The sum of $CA$ values is $(0.3)(10) + (0.7)(2) + (0.4)(15) = 10.4$.

$$Q = (10.4)(2.875) = 29.9\,\text{cfs}$$

## 13 PEAK RUNOFF BY SCS METHODS

Several methods of calculating peak runoff have been suggested by the U.S. Soil Conservation Service. These methods have generally been well correlated with actual experience, but the classification of terrain and soil conditions into SCS categories is difficult. Correct application of these SCS methods is dependent on being able to determine the various coefficients and parameters used. SCS literature should be consulted in most cases.

## A. THE COOK EQUATION (1940; OBSOLETE)

The Cook equation is used for watersheds of less than 2500 acres.

*step 1*: Determine the frequency factor, $F$. For 50-year protection, use $F = 1.00$; for 25-year storms, use $F = 0.83$; for 10-year storms, use $F = 0.71$.

*step 2*: Determine the rainfall factor, $R$, from the geology of the region. This factor varies from 0.6 to 1.1.

*step 3*: Determine the watershed characteristics. Based on the relief, soil infiltration, vegetation, and surface storage conditions, calculate the total watershed coefficient, $W = \sum w_i$. Various values of the weights, $w_i$, are listed in table 6.5. $W$ varies between 25 and 100.

*step 4*: Determine $Q_{p,50}$, the standard 50-year peak flow, from figure 6.19.

*step 5*: Calculate the peak discharge associated with the given return period and geological region.

$$Q_p = Q_{p,50}RF \qquad 6.33$$

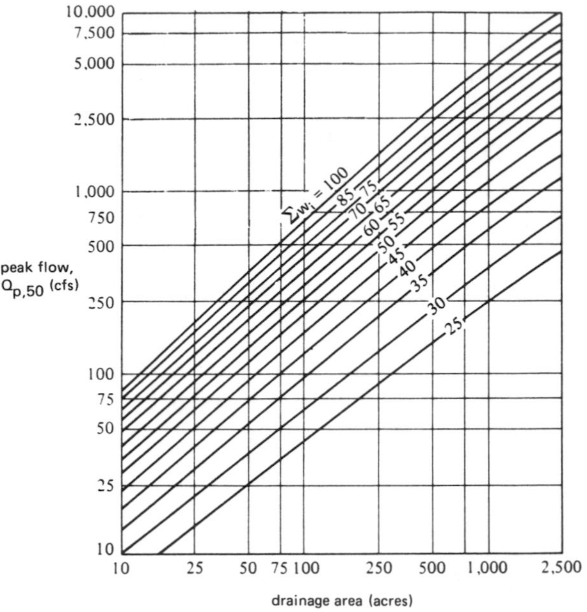


**Figure 6.19** Standard 50-year Peak Flows

**Table 6.5**
Cook Equation Watershed Weights ($w_i$)

| | | | | |
|---|---|---|---|---|
| Relief | (40) Steep, rugged, terrain with average slopes generally above 30% | (30) Hilly, with average slopes of 10–30% | (20) Rolling, with average slopes of 5–10% | (10) Relatively flat land, with average slopes of 0–5% |
| Soil infiltration | (20) No effective soil cover; either rock or thin soil mantle of negligible infiltration capacity | (15) Slow to take up water; clay or other soil of low infiltration capacity, such as heavy gumbo. | (10) Normal, deep loam with infiltration about equal to that of typical prairie soil | (5) High; deep sand or other soil that takes up water readily and rapidly |
| Vegetal cover | (20) No effective plant cover; bare except for very sparse cover | (15) Poor to fair; clean-cultivated crops or poor natural cover; less than 10% of drainage area under good cover | (10) Fair to good; about 50% of drainage area in good grassland, woodland, or equivalent cover; not more than 50% of area in clean-cultivated crops | (5) Good to excellent, about 90% of drainage area in good grassland, woodland, or equivalent cover |
| Surface storage | (20) Negligible; surface depressions are few and shallow; drainage-ways steep and small; no ponds or marshes | (15) Low; well-defined system of small drainage-ways; no ponds or marshes | (10) Normal; considerable surface depression storage; drainage system similar to that of typical prairie lands; lakes, ponds and marshes less than 2% of drainage area | (5) High; surface-depression storage high; drainage system not sharply defined; large flood plain storage or a large number of lakes, ponds or marshes |

### B. THE SCS DRAINAGE COEFFICIENT METHOD (1939)

A simple method used by the Soil Conservation Service for calculating surface drainage (i.e., removal rates) from flatlands depends on the characteristics of the watershed to be protected. This method applies to areas where the natural land slopes are about 1 percent or less. The formula can be used for minor portions of steeper land in a watershed which is predominantly flatland. Flow from uplands in the watershed should be computed by other procedures and added to the flatlands flow.

$$Q_p = C\left(\frac{A_d}{640}\right)^{5/6} \qquad 6.34$$

The 640 term in equation 6.34 converts the drainage area from acres to square miles. The coefficient $C$ depends on use of the flatland. Table 6.6 lists typical values of the coefficient $C$.

**Table 6.6**
SCS Drainage Coefficients

| use and location | $C$ |
|---|---|
| SE coastal plain, cultivated | 45 |
| SE delta, cultivated | 40 |
| NE and cornbelt, cultivated | 37 |
| SE coastal plain, pasture | 30 |
| NE and cornbelt, pasture | 25 |
| SE delta and SE coastal ricelands | 22.5 |
| Red River Valley, cultivated | 20 |
| SW range | 15 |
| SE coastal plain, woodland | 10 |

### C. CURVE NUMBER METHOD (1975)

This SCS method is dependent on being able to classify the land use and soil type by a single parameter called the *curve number*, $CN$. This method can be used for any size homogeneous watershed with a known percentage of imperviousness. If the watershed varies in soil type or in cover, it should be divided into regions to be analyzed separately.[8] This method uses precipitation records and an assumed distribution of rainfall to construct a *synthetic storm*.

*step 1*: Classify the soil according to *runoff potential*. Soil is classified into types $A$ (low runoff potential) through $D$ (high runoff potential).

- Type A: High infiltration rates (0.30–0.45 in/hr) even if thoroughly saturated; chiefly sands and gravels with good drainage and high moisture transmission.
- Type B: Moderate infiltration rates if thoroughly wetted (0.15–0.30 in/hr), moderate rates of moisture transmission, and consisting chiefly of coarse to moderately fine textures.
- Type C: Slow infiltration rates (0.05–0.15 in/hr) if thoroughly wetted, and slow moisture transmission; soils which have moderately fine to fine textures, or which impede the downward movement of water.
- Type D: Very slow infiltration rates (less than 0.05 in/hr) if thoroughly wetted, very slow water transmission, and consisting primarily of clay soils with high potential for swelling, soils with permanent high water tables, or soils with an impervious layer near the surface.

*step 2*: Determine the pre-existing soil conditions. The soil condition is classified into *antecedent moisture conditions* (*AMC*) I through III.

- AMC I: Dry soils, as prior to or after plowing or cultivation.
- AMC II: Typical condition existing before maximum annual flood.
- AMC III: Saturated soil due to heavy rainfall (or light rainfall with low temperatures) during 5 days prior to storm.

*step 3*: Classify the hydrologic condition of the soil-cover complex. The condition is *good* if it is lightly grazed or has plant cover over 75% or more of its area. The condition is *fair* if plant coverage is 50%–75% or not heavily grazed. The condition is *poor* if the area is heavily grazed, has no mulch, or has plant cover over less than 50% of the area. These definitions are applicable to pasture or range, and may be difficult to apply to row crops and grains.

*step 4*: Use table 6.7 to determine the curve number, $CN$, corresponding to the soil classification for AMC II.

*step 5*: If the soil is AMC I or AMC III, convert the curve number from step 4 using table 6.8.

*step 6*: If any significant fraction of the watershed is impervious (i.e., $CN = 100$), or if the watershed consists of areas with different curve numbers, calculate the composite curve number by weighting by the runoff areas.

[8] If the net rain and time to concentration are known, it may be possible to jump immediately to step 11 in this procedure.

**Table 6.7**
Runoff Curve Numbers for Soil Use and Condition (AMC II)

| use of land | treatment | hydrologic condition | runoff potential *A* | *B* | *C* | *D* |
|---|---|---|---|---|---|---|
| fallow | straight row | – | 77 | 86 | 91 | 94 |
| row crops | straight row | poor | 72 | 81 | 88 | 91 |
| | straight row | good | 67 | 78 | 85 | 89 |
| | contoured | poor | 70 | 79 | 84 | 88 |
| | contoured | good | 65 | 75 | 82 | 86 |
| | contoured and terraced | poor | 66 | 74 | 80 | 82 |
| | contoured and terraced | good | 62 | 71 | 78 | 81 |
| small grain | straight row | poor | 65 | 76 | 84 | 88 |
| | | good | 63 | 75 | 83 | 87 |
| | contoured | poor | 63 | 74 | 82 | 85 |
| | | good | 61 | 73 | 81 | 84 |
| | contoured and terraced | poor | 61 | 72 | 79 | 82 |
| | | good | 59 | 70 | 78 | 81 |
| close-seeded legumes | straight row | poor | 66 | 77 | 85 | 89 |
| or rotation meadow | straight row | good | 58 | 72 | 81 | 85 |
| | contoured | poor | 64 | 75 | 83 | 85 |
| | contoured | good | 55 | 69 | 78 | 83 |
| | contoured and terraced | poor | 63 | 73 | 80 | 83 |
| | contoured and terraced | good | 51 | 67 | 76 | 80 |
| pasture or range | | poor | 68 | 79 | 86 | 89 |
| | | fair | 49 | 69 | 79 | 84 |
| | | good | 39 | 61 | 74 | 80 |
| | contoured | poor | 47 | 67 | 81 | 88 |
| | contoured | fair | 25 | 59 | 75 | 83 |
| | contoured | good | 6 | 35 | 70 | 79 |
| meadow | | good | 30 | 58 | 71 | 78 |
| woods | | poor | 45 | 66 | 77 | 83 |
| | | fair | 36 | 60 | 73 | 79 |
| | | good | 25 | 55 | 70 | 77 |
| farmsteads | | – | 59 | 74 | 82 | 86 |
| roads(dirt) | | – | 72 | 82 | 87 | 89 |
| (hard surface) | | – | 74 | 84 | 90 | 92 |

**Table 6.8**
Curve Numbers for AMC I and AMC III

| *CN* for AMC II | Corresponding *CN*'s AMC I | AMC III |
|---|---|---|
| 100 | 100 | 100 |
| 95 | 87 | 98 |
| 90 | 78 | 96 |
| 85 | 70 | 94 |
| 80 | 63 | 91 |
| 75 | 57 | 88 |
| 70 | 51 | 85 |
| 65 | 45 | 82 |
| 60 | 40 | 78 |
| 55 | 35 | 74 |
| 50 | 31 | 70 |
| 45 | 26 | 65 |
| 40 | 22 | 60 |
| 35 | 18 | 55 |
| 30 | 15 | 50 |
| 25 | 12 | 43 |
| 20 | 9 | 37 |
| 15 | 6 | 30 |
| 10 | 4 | 22 |
| 5 | 2 | 13 |

*step 7*: Estimate the time to concentration of the watershed.

*step 8*: Determine the gross rain from the storm. To do this, it is necessary to assume the storm length and frequency. A 24-hour storm is typically chosen, probably because of the 24-hour thunderstorms experienced in all but the Pacific Coast states. Maps from the U.S. Weather Bureau can be used to read gross point rainfalls for storms with durations from 30 minutes to 24 hours, and with frequencies from 1 to 100 years.[9]

*step 9*: Multiply the gross rain point value from step 8 by a factor from figure 6.20 to make the gross rain representative of larger areas. This is the ***areal gross rain.***

*step 10*: Calculate the ***net rain*** (***precipitation excess***) from the ***gross rain.*** Subtract losses from interception, storm period evaporation, depression storage, and infiltration from the gross rain to obtain the net rain.

This step is difficult. The SCS procedure is to first determine the distribution of the storm. Type I storms are applicable to Hawaii, Alaska, and the coastal side of the Sierra Nevada and Cascade Mountains in California, Oregon, and Washington. Type II distributions are typical of the rest of the United States, Puerto Rico, and the Virgin Islands. Table 6.9 shows the cumulative rainfall over 24 hours for type I and II storms.

[9] "Rainfall Frequency Atlas of the United States," U.S. Weather Bureau, Technical Paper 40.

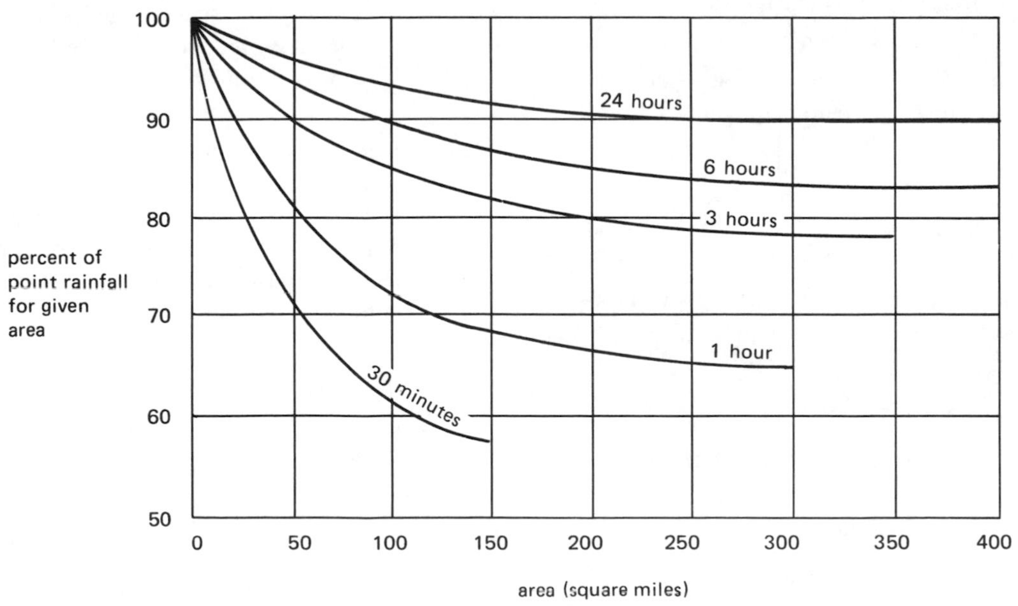


**Figure 6.20** Point to Areal Depth Conversion Factors

**Table 6.9**
Cumulative Distribution of Type I and II 24-hour Storms

| time(hours) | type I | type II |
|---|---|---|
| 0 | 0 | 0 |
| 2.0 | 0.035 | 0.022 |
| 4.0 | 0.076 | 0.048 |
| 6.0 | 0.125 | 0.080 |
| 7.0 | 0.156 | – |
| 8.0 | 0.194 | 0.120 |
| 8.5 | 0.219 | – |
| 9.0 | 0.254 | 0.147 |
| 9.5 | 0.303 | 0.163 |
| 9.75 | 0.362 | – |
| 10.0 | 0.515 | 0.181 |
| 10.5 | 0.583 | 0.204 |
| 11.0 | 0.624 | 0.235 |
| 11.5 | 0.654 | 0.283 |
| 11.75 | – | 0.387 |
| 12.0 | 0.682 | 0.663 |
| 12.5 | – | 0.735 |
| 13.0 | 0.727 | 0.772 |
| 13.5 | – | 0.799 |
| 14.0 | 0.767 | 0.820 |
| 16.0 | 0.830 | 0.880 |
| 20.0 | 0.926 | 0.952 |
| 24.0 | 1.000 | 1.000 |

Next, the soil complex curve number ($CN$) is used to calculate the excess of rain over infiltration. The following assumptions are common:

- Infiltration follows an exponential decay curve with time.
- *Storage capacity* of the soil, $S$, can be calculated from the curve number by using equation 6.35.

$$S = \frac{1000}{CN} - 10 \qquad 6.35$$

$$CN = \frac{1000}{10 + S} \qquad 6.36$$

- *Initial abstraction* (depression storage, evaporation, and interception losses) is equal to 20% of the storage capacity.

$$I_a = 0.2S \qquad 6.37$$

- The gross rain equals or exceeds the initial abstraction.

$$P \geq I_a \qquad 6.38$$

- The storage capacity equals or exceeds the initial abstraction plus the infiltration.

$$S \geq I_a + F \qquad 6.39$$

The net rain is then calculated from the areal gross rain, $P_g$, by using equation 6.40.

$$P_N = \frac{(P_g - 0.2S)^2}{P_g + 0.8S} \qquad 6.40$$

If it is necessary to draw a stream flow hydrograph, calculate the increments in accumulated rainfall excess.[10]

*step 11*: Use figure 6.21 to determine the peak discharge. Notice that the peak discharge is given per inch of net rain with units of cubic feet per second per square mile of drainage area. Figure 6.21 is valid only for type II, 24-hour storms. Refer to SCS literature for other storm types, or determine the peak flow directly from the hydrograph drawn from the incremental data in step 10.

## 14. RESERVOIR YIELD AND RESERVOIR SIZING

The reservoir yield problem is an accounting problem (i.e., keeping track of what comes in and what goes out). The purpose of the reservoir yield analysis is to determine the proper size of a dam or reservoir, or to evaluate the ability of an existing dam to meet water demands. There are three basic methods of solving the reservoir yield problem.

*Method #1*: Tabular simulation is the easiest method to apply, although its validity is dependent on choosing time increments as small as possible. It is also necessary to make some assumption about the distribution of annual water inflows.

*step 1*: Determine the starting storage volume, $V_n$. If $V_n$ is zero or considerably different than the average steady-state storage, a large number of iterations will be required before the simulation reaches its steady-state results.

*step 2*: For the next iteration, determine the inflow, discharge, evaporation, and seepage. Determine the starting storage volume for the next iteration by solving equation 6.41.

[10] In effect, the runoff hydrograph is simulated. For example, the areal gross rain is multiplied consecutively by the factors in table 6.9 to give the accumulated areal gross rainfall at the various times. The accumulated excess (net) is calculated from equation 6.40. The consecutive excesses are subtracted to get the incremental excesses. These incremental excesses can be used to draw the hydrograph.

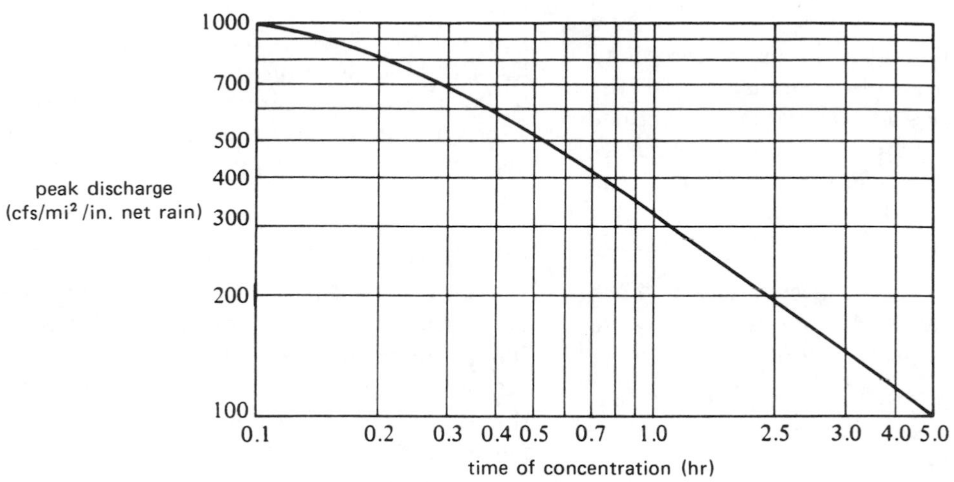


**Figure 6.21** Peak Runoff versus Time of Concentration

$$V_{n+1} = V_n + (\text{inflow})_n - (\text{discharge})_n - (\text{seepage})_n - (\text{evaporation})_n \qquad 6.41$$

Repeat step 2 as many times as necessary.

Reservoir seepage is generally very small compared to inflow and discharge. It is, therefore often neglected. Reservoir evaporation can be estimated from analytical relationships or by evaluating data from *evaporation pans*. Pan data is extended to reservoir evaporation by the pan coefficient formula. In equation 6.42, the summation is taken over the number of days in the simulation period.

$$E_R = \sum C_p E_p \qquad 6.42$$

Units of $E$ are typically inches per day. A typical value of $C_p$ is 0.7.

Inflow can be taken as actual past history. However, since it is not very likely that history will repeat, the next method is preferred.

*Method #2*: This method is the same as method #1 except for its method of determining the inflow. Method #2 uses a *Monte Carlo simulation* which is dependent on enough historical data to establish a cumulative inflow distribution. A Monte Carlo simulation is suitable if long periods are to be simulated. If short periods are to be simulated, the simulation should be performed several times and the results averaged.

*step 1*: Tabulate or otherwise determine a frequency distribution of inflow quantities.

*step 2*: Form a cumulative distribution of inflow quantities.

*step 3*: Multiply the cumulative x-axis (which runs from 0 to 1) by 100 or 1000, depending on the accuracy needed.

*step 4*: Generate random numbers between 0 and 100 (or 0 and 1000). Use of a random number table (appendix B) is adequate for hand simulation.

*step 5*: Locate the inflow quantity corresponding to the random number from the cumulative distribution x-axis.

*Method #3*: The *non-sequential drought method* is more complex, but it has the advantage of giving an estimate of the required reservoir size, rather than evaluating a trial size as the first two methods do.

In the absence of synthetic drought information, it is first necessary to develop intensity-duration-frequency curves from stream flow records.

*step 1*: Choose a duration. Usually, the first duration used will be 7 days, although choosing 15 days will not introduce too much error.

*step 2*: Search the streamflow records to find the smallest flow during the duration chosen. (The first time through, for example, find the smallest discharge totaled over any 7 days.) The days do not have to be sequential.

*step 3*: Continue searching the discharge records to find the next smallest discharge over the number of days in the period. Continue searching and finding the smallest discharges (which gradually increase) until all of the days in the record have been used up. Do not use the same day more than once.

*step 4*: Give the values of smallest discharge order numbers. That is, $M = 1$ is given to the smallest discharge, $M = 2$ to the next smallest, etc.

*step 5*: For each observation, calculate the recurrence interval as

$$F = n_y/M \qquad 6.43$$

$n_y$ is the number of years of streamflow data that was searched to find the smallest discharges.

*step 6*: Plot the points as discharge on the $y$-axis versus $F$ in years on the $x$-axis. Draw a reasonably continuous curve through the points.

*step 7*: Return to step 1 for the next duration. Repeat for all of the following durations: 7, 15, 30, 60, 120, 183, and 365 days.

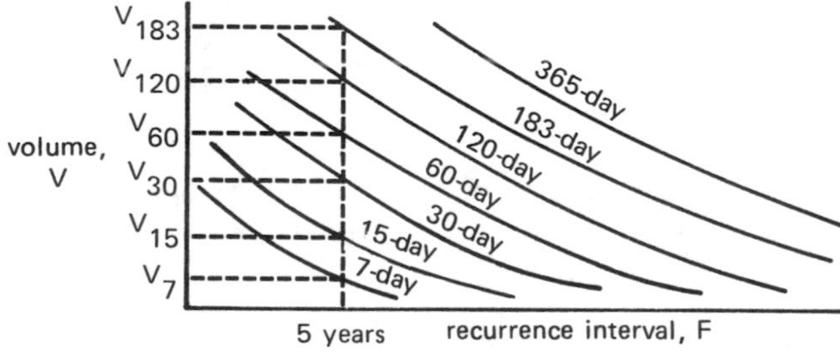


**Figure 6.22** Sample Family of Synthetic Inflow Curves

A synthetic drought can be constructed for any recurrence interval. For example, if a 5-year drought is to be planned for, the discharges $V_7, V_{15}, V_{30}, \ldots, V_{365}$ are read from the appropriate curves for $F = 5$ years.

The next step is to plot the *mass diagram* (also known as a *Rippl diagram*) for the reservoir. This is a simultaneous plot of the cumulative demand (known as ***draft***) and cumulative inflow. The mass diagram is used to graphically determine the reservoir storage requirements (i.e., size).

As long as the slopes of the cumulative demand and inflow lines are equal, the water reserve in the reservoir will not change. When the slope of the inflow is less than the slope of the demand, the inflow cannot by itself satisfy the community's water needs, and the reservoir is drawn down to make up the difference. A peak followed by a trough is, therefore, a drought condition.

If the reservoir is to be sized so that the community will not run dry during a drought, the required capacity is the maximum separation between two parallel lines (pseudo-demand lines with slopes of the demand rate) drawn tangent to a peak and a subsequent trough. If the mass diagram covers enough time so that multiple droughts are present, the largest separation between peaks and subsequent troughs is the capacity.

In order for the reservoir to supply enough water during a drought condition, the reservoir must be full prior to the start of the drought. This fact is not represented when the mass diagram is drawn, hence the need to draw a pseudo-demand line parallel to the peak.

After a drought equal to the capacity of the reservoir, the reservoir will be empty. At the trough, however, the reservoir will begin to fill up again. When the cumulative excess exceeds the reservoir capacity, the reservoir will have to spill (i.e., release) water. This occurs when the cumulative inflow line crosses the prior peak's pseudo-demand line as shown in figure 6.23.

A *flood-control dam* is built to keep water in and must be sized so that water is not spilled. The mass diagram can still be used, but the maximum separation between troughs and subsequent peaks (not peaks followed by troughs) becomes the capacity.

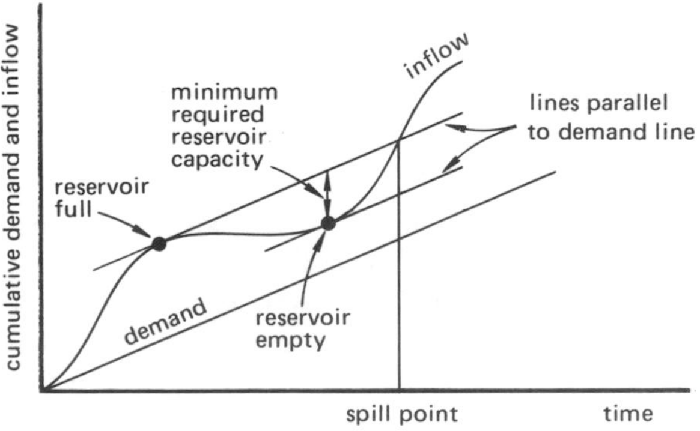


**Figure 6.23** Reservoir Mass Diagram (Rippl Diagram)

HYDROLOGY

*Example 6.8*

A well-monitored stream has been observed for 50 years and has the following frequency distribution of total annual discharges.

| discharge (units) | frequency (years) | % of time |
|---|---|---|
| 0 to 0.5 | 5 | 0.10 |
| 0.5 to 1.0 | 21 | 0.42 |
| 1.0 to 1.5 | 17 | 0.34 |
| 1.5 to 2.0 | 7 | 0.14 |
| | 50 | 1.00 total |

It is proposed to dam the stream and create a reservoir with a capacity of 1.8 units. The reservoir is to support a town which will draw 1.2 units per year. Simulate 10 years of reservoir operation assuming it starts with 1.5 units.

*step 1*: The frequency distribution was given.

*steps 2 and 3*: The cumulative distribution is

| discharge | cumulative frequency | cumulative frequency × 100 |
|---|---|---|
| 0 to 0.5 | 0.10 | 10 |
| 0.5 to 1.0 | 0.52 | 52 |
| 1.0 to 1.5 | 0.86 | 86 |
| 1.5 to 2.0 | 1.00 | 100 |

*step 4*: From appendix B, choose ten 2-digit numbers. Their choice is arbitrary, but they must come sequentially from a row or column. Use the first row for this simulation:

$$78, 46, 68, 33, 26, 96, 58, 98, 87, 27$$

*step 5*: For the first year, the random number is 78. Since 78 is greater than 52 but less than 86, the inflow is in the 1.0 to 1.5 unit range. The mid-point of this range is taken as the inflow which would be 1.25. The reservoir volume after the first year would be $1.5 + 1.25 - 1.20 = 1.55$. The remaining years can be similarly simulated.

No shortages are experienced; one spill is required.

| year | starting volume | + inflow | − usage | = ending volume | + spill |
|---|---|---|---|---|---|
| 1 | 1.5 | 1.25 | 1.2 | 1.55 | |
| 2 | 1.55 | 0.75 | 1.2 | 1.1 | |
| 3 | 1.1 | 1.25 | 1.2 | 1.15 | |
| 4 | 1.15 | 0.75 | 1.2 | 0.7 | |
| 5 | 0.7 | 0.75 | 1.2 | 0.25 | |
| 6 | 0.25 | 1.75 | 1.2 | 0.8 | |
| 7 | 0.8 | 1.25 | 1.2 | 0.85 | |
| 8 | 0.85 | 1.75 | 1.2 | 1.4 | |
| 9 | 1.4 | 1.75 | 1.2 | 1.95 | 0.15 |
| 10 | 1.8 | 0.75 | 1.2 | 1.35 | |

# Appendix A: Rational Method Runoff Coefficients

| categorized by surface | |
|---|---|
| forested | 0.059–0.2 |
| asphalt | 0.7–0.95 |
| brick | 0.7–0.85 |
| concrete | 0.8–0.95 |
| shingle roof | 0.75–0.95 |
| lawns, well drained (sandy soil) | |
| up to 2% slope | 0.05–0.1 |
| 2% to 7% slope | 0.10–0.15 |
| over 7% slope | 0.15–0.2 |
| lawns, poor drainage (clay soil) | |
| up to 2% slope | 0.13–0.17 |
| 2% to 7% slope | 0.18–0.22 |
| over 7% slope | 0.25–0.35 |
| driveways, walkways | 0.75–0.85 |

| categorized by use | |
|---|---|
| farmland | 0.05–0.3 |
| pasture | 0.05–0.3 |
| unimproved | 0.1–0.3 |
| parks | 0.1–0.25 |
| cemeteries | 0.1–0.25 |
| railroad yard | 0.2–0.40 |
| playgrounds (except asphalt or concrete) | 0.2–0.35 |
| business districts | |
| neighborhood | 0.5–0.7 |
| city (downtown) | 0.7–0.95 |
| residential | |
| single family | 0.3–0.5 |
| multi-plexes, detached | 0.4–0.6 |
| multi-plexes, attached | 0.6–0.75 |
| suburban | 0.25–0.4 |
| apartments, condominiums | 0.5–0.7 |
| industrial | |
| light | 0.5–0.8 |
| heavy | 0.6–0.9 |

# Appendix B: Random Numbers

| | | | | |
|---|---|---|---|---|
| 78466 83326 | 96589 88727 | 72655 49682 | 82338 28583 | 01522 11248 |
| 78722 47603 | 03477 29528 | 63956 01255 | 29840 32370 | 18032 82051 |
| 06401 87397 | 72898 32441 | 88861 71803 | 55626 77847 | 29925 76106 |
| 04754 14489 | 39420 94211 | 58042 43184 | 60977 74801 | 05931 73822 |
| 97118 06774 | 87743 60156 | 38037 16201 | 35137 54513 | 68023 34380 |
| 71923 49313 | 59713 95710 | 05975 64982 | 79253 93876 | 33707 84956 |
| 78870 77328 | 09637 67080 | 49168 75290 | 50175 34312 | 82593 76606 |
| 61208 17172 | 33187 92523 | 69895 28284 | 77956 45877 | 08044 58292 |
| 05033 24214 | 74232 33769 | 06304 54676 | 70026 41957 | 40112 66451 |
| 95983 13391 | 30369 51035 | 17042 11729 | 88647 70541 | 36026 23113 |
| 19946 55448 | 75049 24541 | 43007 11975 | 31797 05373 | 45893 25665 |
| 03580 67206 | 09635 84610 | 62611 86724 | 77411 99415 | 58901 86160 |
| 56823 49819 | 20283 22272 | 00114 92007 | 24369 00543 | 05417 92251 |
| 87633 31761 | 99865 31488 | 49947 06060 | 32083 47944 | 00449 06550 |
| 95152 10133 | 52693 22480 | 50336 49502 | 06296 76414 | 18358 05313 |
| 05639 24175 | 79438 92151 | 57602 03590 | 25465 54780 | 79098 73594 |
| 65927 55525 | 67270 22907 | 55097 63177 | 34119 94216 | 84861 10457 |
| 59005 29000 | 38395 80367 | 34112 41866 | 30170 84658 | 84441 03926 |
| 06626 42682 | 91522 45955 | 23263 09764 | 26824 82936 | 16813 13878 |
| 11306 02732 | 34189 04228 | 58541 72573 | 89071 58066 | 67159 29633 |
| 45143 56545 | 94617 42752 | 31209 14380 | 81477 36952 | 44934 97435 |
| 97612 87175 | 22613 84175 | 96413 83336 | 12408 89318 | 41713 90669 |
| 97035 62442 | 06940 45719 | 39918 60274 | 54353 54497 | 29789 82928 |
| 62498 00257 | 19179 06313 | 07900 46733 | 21413 63627 | 48734 92174 |
| 80306 19257 | 18690 54653 | 07263 19894 | 89909 76415 | 57246 02621 |
| 84114 84884 | 50129 68942 | 93264 72344 | 98794 16791 | 83861 32007 |
| 58437 88807 | 92141 88677 | 02864 02052 | 62843 21692 | 21373 29408 |
| 15702 53457 | 54258 47485 | 23399 71692 | 56806 70801 | 41548 94809 |
| 59966 41287 | 87001 26462 | 94000 28457 | 09469 80416 | 05897 87970 |
| 43641 05920 | 81346 02507 | 25349 93370 | 02064 62719 | 45740 62080 |
| 25501 50113 | 44600 87433 | 00683 79107 | 22315 42162 | 25516 98434 |
| 98294 08491 | 25251 26737 | 00071 45090 | 68628 64390 | 42684 94956 |
| 52582 89985 | 37863 60788 | 27412 47502 | 71577 13542 | 31077 13353 |
| 26510 83622 | 12546 00489 | 89304 15550 | 09482 07504 | 64588 92562 |
| 24755 71543 | 31667 83624 | 27085 65905 | 32386 30775 | 19689 41437 |
| 38399 88796 | 58856 18220 | 51056 04976 | 54062 49109 | 95563 48244 |
| 18889 87814 | 52232 58244 | 95206 05947 | 26622 01381 | 28744 38374 |
| 51774 89694 | 02654 63161 | 54622 31113 | 51160 29015 | 64730 07750 |
| 88375 37710 | 61619 69820 | 13131 90406 | 45206 06386 | 06398 68652 |
| 10416 70345 | 93307 87360 | 53452 61179 | 46845 91521 | 32430 74795 |

# Appendix C: Map of the U.S. Showing Average Annual Precipitation in Inches for the Period 1889–1938

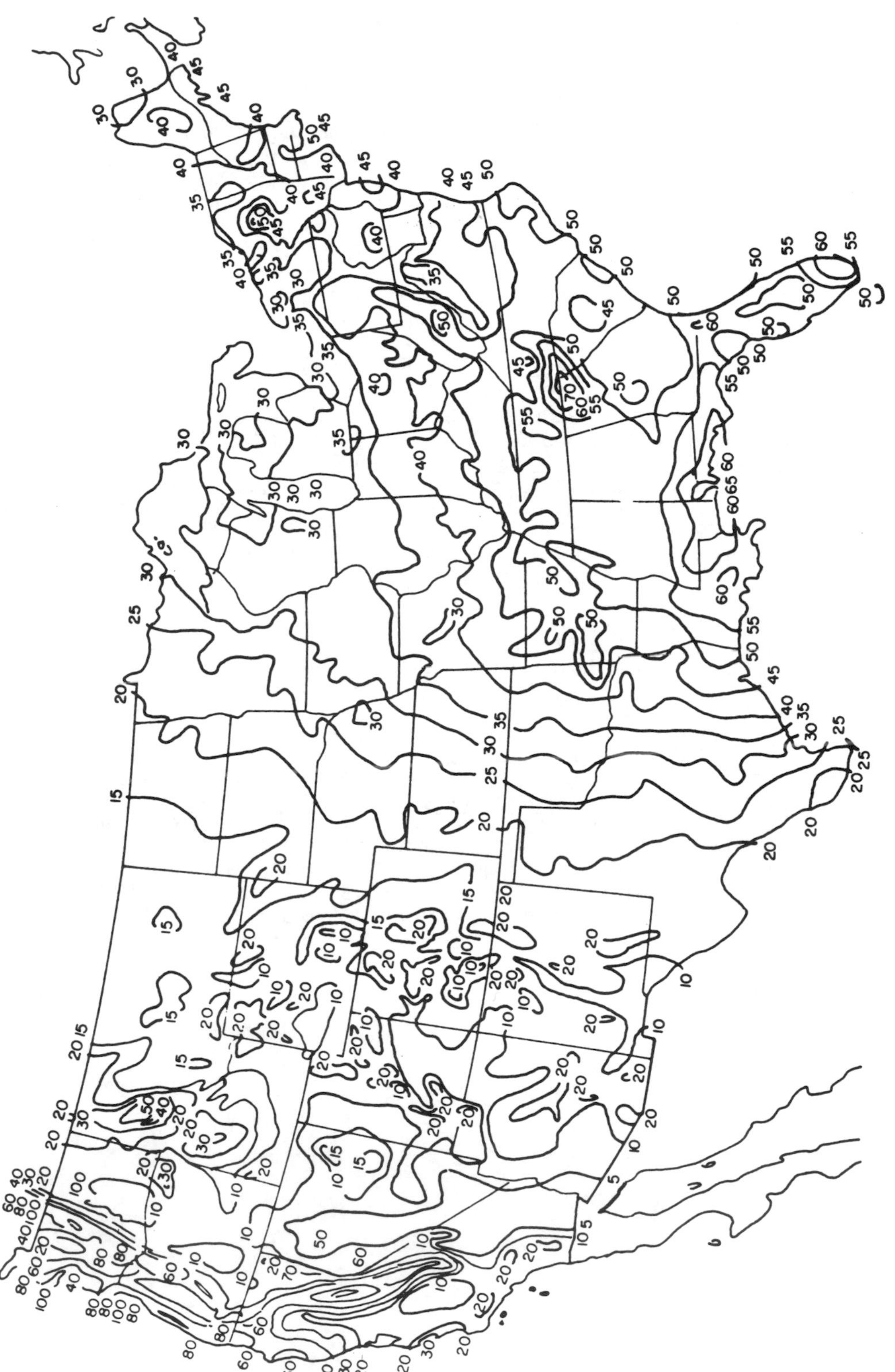

## Practice Problems: HYDROLOGY

Untimed

1. Four 5-acre areas are served by a 1200 foot storm drain ($n = .013$ and slope $= .005$). Inlets to the storm drain are placed every 300 feet along the storm drain. The inlet time for each area served by an inlet is 15 minutes, and the run-off coefficient is 0.55. A storm to be used for design purposes has the following characteristics:

$$I = \frac{100}{t_c + 10}$$

I is in inches/hr
t is in minutes

What is the size of the last section of storm drain assuming that all flows are maximum, and all pipe sizes are available?

2. An aquifer has a water table level of 100 feet below the ground surface. An 18-inch diameter well extends 200 feet into the aquifer, for a total depth of 300 feet. The aquifer transmissivity is 10,000 gallons/day-foot. The well's radius of influence is 900 feet with a 20-foot drawdown at the well. (a) What steady discharge is possible? (b) What horsepower motor is required to achieve the steady discharge?

3. A 2-hour storm over a 43 square mile area produced a flood volume of 3300 acre-feet with a peak discharge of 9300 cfs. (a) What is the unit hydrograph peak discharge? (b) If a 2-hour storm producing 2.5 inches of runoff is to be used to design a culvert, what is the design flood hydrograph volume? (c) What is the design discharge?

4. A 0.5 square mile drainage area has a suggested run-off coefficient of 0.6 and a time of concentration of 60 minutes. The drainage area is in Steel region #3, and a 10-year storm is to be used for design purposes. What is the run-off?

5. A screened well extends from the ground surface (elevation 383 feet) through a gravel bed aquifer to a layer of bedrock at elevation 289 feet. 1500 feet from the well, the undisturbed (zero drawdown) aquifer elevation is 363 feet. The well is pumped by a 10″ schedule-40 steel pipe that draws 120,000 gallons per day. The permeability of the acquifer is 1600 gallons/day-ft$^2$. The pump discharges into a piping network whose friction head is 100 feet. What net horsepower is required for steady flow?

6. A measurement station on a stream recorded the following discharges from a 1.2 square mile drainage area.

| hour | cfs | hour | cfs | hour | cfs |
|---|---|---|---|---|---|
| 0 | 102 | 6 | 455 | 12 | 55 |
| 1 | 99 | 7 | 325 | 13 | 49 |
| 2 | 101 | 8 | 205 | 14 | 43 |
| 3 | 215 | 9 | 145 | 15 | 38 |
| 4 | 507 | 10 | 100 | | |
| 5 | 625 | 11 | 70 | | |

(a) Draw the actual hydrograph. (b) Draw the unit hydrograph. (c) Determine the time from peak runoff to cessation of runoff. (d) Separate the groundwater and surface water.

7. A watershed has an area of 100 square miles. The length of the main stream channel is 20 miles and the distance to a point opposite the centroid is 11 miles. Assume that $C_t = 1.8$ and find (a) the time lag, (b) the duration of the synthetic hydrograph, and (c) the peak discharge.

8. A reservoir has a total capacity of 7 units. At the beginning of a study, the reservoir contains 5.5 units. The monthly demand on the reservoir from a nearby city is 0.7 units. The monthly inflow to the reservoir is normally distributed with a mean of 0.9 units and standard deviation of 0.2 units. Simulate one year of reservoir operation.

9. Repeat problem 8 assuming that the monthly demand on the reservoir is normally distributed with a mean of 0.7 units and a standard deviation of 0.2 units.

10. A class A pan located near a reservoir shows an evaporation loss of 0.8 inches in one day. If the pan coefficient is 0.7, what is the approximate evaporation loss in the reservoir?

Timed

1. A reservoir is needed to provide 20 acre-ft of water each month. The inflow for each of 13 representative months is given below. Size the reservoir by whatever means you choose. Assume the reservoir starts full.

| month | F | M | A | M | J | J | A | S | O | N | D | J | F |
|---|---|---|---|---|---|---|---|---|---|---|---|---|---|
| inflow (acre-ft) | 30 | 60 | 20 | 10 | 5 | 10 | 5 | 10 | 20 | 90 | 85 | 75 | 50 |

2. An impervious concrete dam is shown. The dam reduces seepage by use of two impervious sheets extending 15 feet below the dam bottom. (a) Sketch the flow net. (b) Determine the seepage in cubic feet per minute. (c) What is the uplift pressure on the dam at point A, midway between the left and right edges?

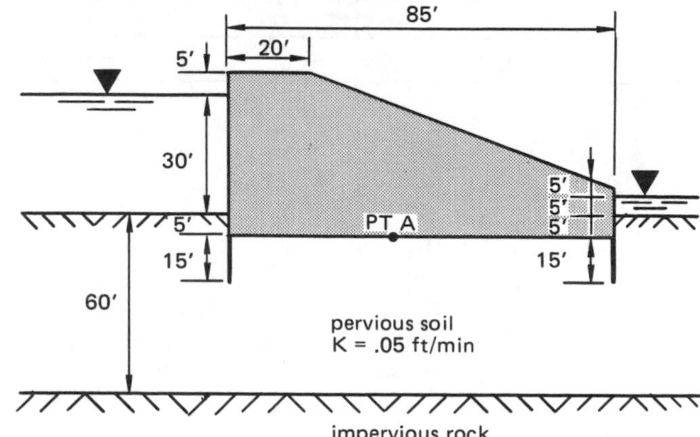

3. A standard 4-hour storm produces 2 inches net of runoff. A stream gauging report is produced from successive sampling every few hours. Two weeks later, the first four hours of an 8-hour storm over the same basin produces 1 inch of runoff. The second four hours produce 2 inches of runoff. Neglecting ground water, draw a hydrograph of the 8-hour storm.

| hour | flow (cfs) | hour | flow (cfs) |
|---|---|---|---|
| 0 | 0 | 14 | 150 |
| 2 | 100 | 16 | 100 |
| 4 | 350 | 18 | — |
| 6 | 600 | 20 | 50 |
| 8 | 420 | 22 | — |
| 10 | 300 | 24 | 0 |
| 12 | 250 | | |

4. A reservoir with constant draft of 240 MG/-mi$^2$-yr is being designed. Given the rainfall distribution, what should be the minimum reservoir size? When would the reservoir start to spill?

| month | inflow (MG/mi$^2$) | month | inflow (MG/mi$^2$) |
|---|---|---|---|
| J | 20 | J | 15 |
| F | 30 | A | 5 |
| M | 45 | S | 15 |
| A | 30 | O | 60 |
| M | 40 | N | 90 |
| J | 30 | D | 40 |

5. A 75-acre urbanized section of land drains into a rectangular 5′ x 7′ channel which directs runoff through a round culvert under a roadway. The culvert is concrete, and 60″ in diameter. It is 36′ long and placed on a 1% slope.

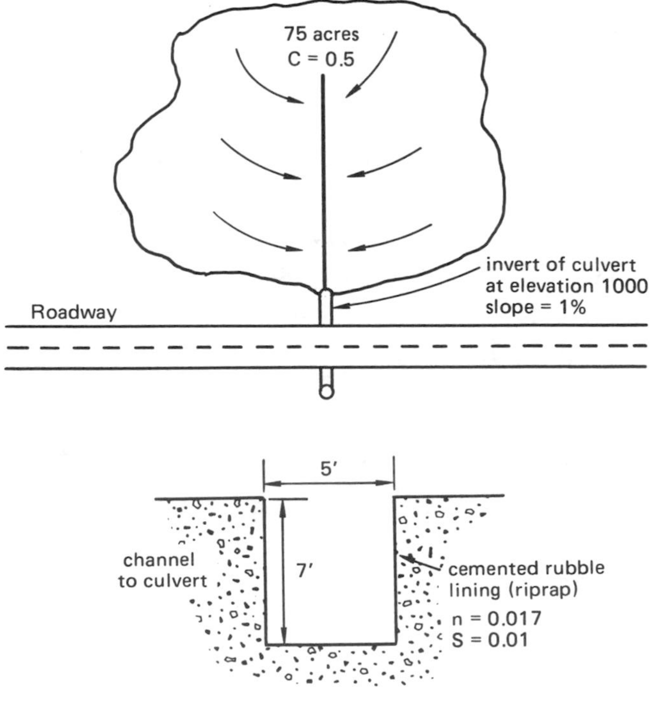


It is desired to evaluate the culvert design based on a 50-year storm. It is known that the time of concentration to the head of the culvert is 30 minutes.

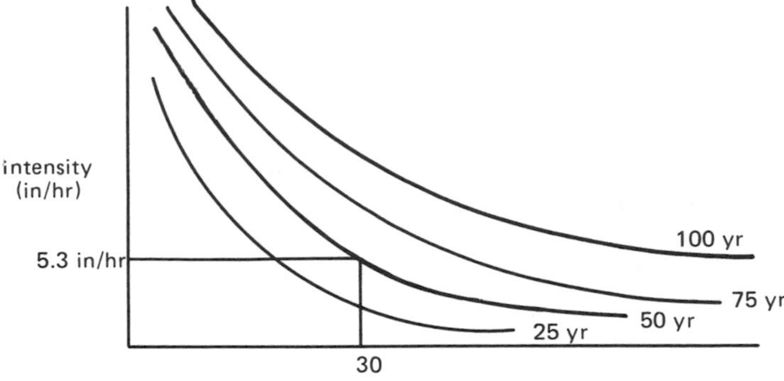


(a) If the minimum road surface elevation is 1010′, will the road surface be flooded? (b) What is the water elevation upstream from the culvert entrance after 30 minutes? (c) What is the water elevation downstream of the culvert exit after 30 minutes?

6. Draw the flow net for the coffer-dam shown. Sheet piles extend below the mud line, and the coffer-dam floor is unlined. The figure is drawn to scale.

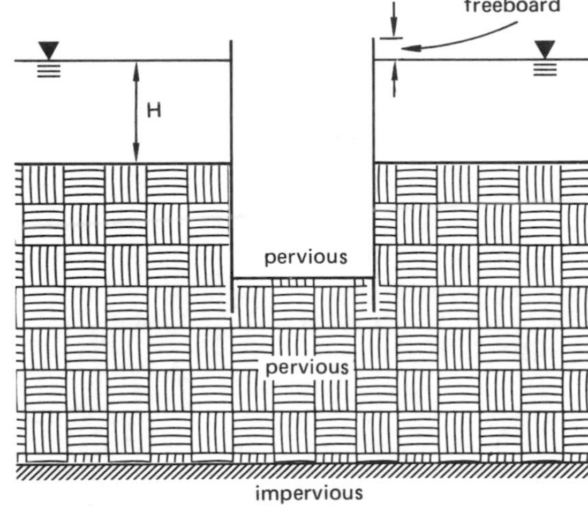


7. The masonry dam shown is drawn to scale. Sketch the flow net for the dam and estimate the seepage rate. The silt has a permeability coefficient of $k = 0.15$ ft/hr.

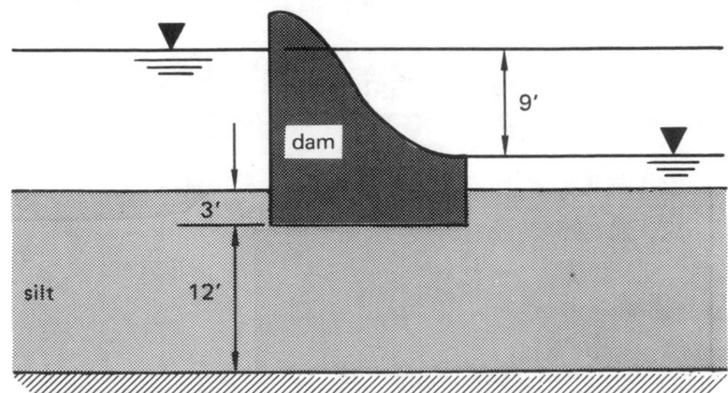

# 7 WATER SUPPLY ENGINEERING

*Nomenclature*

| | | |
|---|---|---|
| A | area | $ft^2$ |
| B | width | ft |
| C | constant, or concentration | –, or mg/l |
| $C_D$ | drag coefficient | – |
| d | particle diameter | ft |
| D | outside pipe diameter, or drag force | ft, lbf |
| f | change in charge | – |
| $F_I$ | impact factor | – |
| g | acceleration due to gravity (32.2) | $ft/sec^2$ |
| $g_c$ | gravitational constant | $ft\text{-}lbm/sec^2\text{-}lbf$ |
| G | mean velocity gradient | 1/sec |
| h | height or depth | ft |
| H | height | ft |
| K | rate constant | 1/sec |
| $K_{eq}$ | equilibrium constant | – |
| $K_{sp}$ | solubility product | – |
| L | length | ft |
| LF | load factor | – |
| $N_{Re}$ | Reynolds number | – |
| p | pressure | psf |
| P | population, force, or power | 1000's of people, pounds, or ft-lbf/sec |
| Q | flow | cfs |
| SG | specific gravity | – |
| t | time | seconds |
| v | velocity, or rate of reaction | ft/sec,– |
| v* | surface loading | ft/sec |
| V | volume | $ft^3$ |
| w | pipe load | lbf/ft |
| x | mole fraction | – |
| z | diagonal distance | ft |

*Symbols*

| | | |
|---|---|---|
| $\gamma$ | specific weight | $lbf/ft^3$ |
| $\rho$ | density | $lbm/ft^3$ |
| $\nu$ | kinematic viscosity | $ft^2/sec$ |
| $\mu$ | absolute viscosity | $lbf\text{-}sec/ft^2$ |
| $\eta$ | efficiency | – |

*Subscripts*

| | |
|---|---|
| d | detention |
| DL | dead load |
| f | flow-through |
| i | incoming |
| o | outgoing |
| s | settling |

## 1 CONVERSIONS

| multiply | by | to obtain |
|---|---|---|
| Clark degrees | 1 | grains/Imperial gallon |
| cubic feet | 7.481 | gallons |
| cubic feet | 28.32 | liters |
| cubic feet/sec | 0.6463 | million gallons/day |
| cubic feet/sec | 448.8 | gpm |
| feet | 0.3048 | meters |
| gallons | 3785 | cubic centimeters |
| gallons | 0.1337 | cubic feet |
| gallons | 3.785 | liter |
| gpm | 0.002228 | cfs |
| grains | 1.429 EE–4 | pounds |
| grains/gallon | 142.9 | pounds/million gallons |
| grains/gallon | 17.1 | ppm |
| grains/gallon | 17.1 | mg/l |
| grains/gallon | 1.2 | Clark degrees |
| grams | 0.03527 | ounces |
| grams | 0.002205 | pounds |
| Imperial gallons | 1.2 | U.S. gallons |
| inches | 2.540 | centimeters |
| kilograms | 2.205 | pounds |
| $kg/m^3$ | 0.06243 | $pounds/ft^3$ |
| liters | 1000 | cubic centimeters |
| liters | 0.03532 | cubic feet |
| liters | 0.2642 | gallons |
| liter/sec | 15.85 | gpm |
| meters | 3.281 | feet |
| meters | 39.37 | inches |
| microns | 0.001 | millimeters |
| MGD | 1.547 | cubic feet/sec |
| mg/l | 1.0 | ppm |

| | | |
|---|---|---|
| mg/l | 0.0583 | grains/gallon |
| mg/l | 8.345 | pounds/million gallons |
| pounds | 7000 | grains |
| pounds | 453.6 | grams |
| ppm | 0.0583 | grains/gallon |
| ppm | 0.07 | Clark degrees |

## 2 DEFINITIONS

Acid: A compound containing hydrogen ions ($H^+$ or $H_3O^+$) in an aqueous solution. Acids have a sour taste, conduct electricity, turn blue litmus paper red, and have a pH between 0 and 7.

Aeration: Mixing water with air, either by spraying water or diffusing air through it.

Aerobic: Requiring oxygen.

Air break: A means by which drinking water can be used for fire fighting without contaminating the drinking supply. Drinking water is freely discharged into the top of a tank which feeds the fire main.

Algae: One-celled plant life.

Altitude valve: A valve that automatically opens to prevent overflow in storage tanks.

AMU: Atomic mass unit. One AMU is 1/12th the atomic weight of carbon.

Anaerobic: Not requiring oxygen.

Anion: Negative ions that migrate to the positive electrode in an electrolytic solution.

Atomic number: The number of protons in the nucleus of an atom.

Atomic weight: Approximately the number of protons and neutrons in the nucleus of an element.

Avogadro's law: A gram-mole of any substance contains 6.023 EE23 molecules.

B. coli: Bacteria coli. See 'E. coli.'

Backfill: The soil that is used to cover a pipe put into a trench.

Base: A compound containing hydroxide ions ($OH^-$) combined with alkali metals and alkali earths. Bases in aqueous solutions have a bitter taste, conduct electricity, turn red litmus paper blue, and have a pH between 7 and 14.

Belt-line layout: See 'Grid iron layout.'

Breakpoint chlorination: Application of chlorine which results in a minimum of chloramine residuals. No free chlorine residuals are produced unless the breakpoint is exceeded.

Capita: Person.

Carbonate hardness: Hardness caused by bicarbonates.

Cation: A positive ion that migrates to the negative electrode in an electrolytic solution.

Chloramine: Compounds of chlorine and ammonia (e.g., $NH_2Cl$, $NHCl_2$, or $NCl_3$).

Chlorine demand: The difference between applied chlorine and the chlorine residual. Chlorine demand is chlorine that has been reduced in chemical reactions and is no longer available for purification.

Clear well: Storage in a water treatment plant which normally takes water from the filters.

Coagulation: Floc formation as the result of adding coagulating chemicals. Coagulants destabilize (by reducing repulsive forces) suspended particles, allowing them to agglomerate.

Coliform: See 'Colon bacilli.'

Colloid: A fine particle ranging in size from 1 to 500 millimicrons. Colloids cause turbidity.

Colon bacilli: Bacteria residing normally in the intestinal tract. These are not necessarily dangerous if present in drinking water, but they are indicators of other possible pathogens.

Combined residuals: Compounds of an additive (such as chlorine) that have combined with something else. Chloramines are combined residuals.

Compound: A homogeneous substance composed of two or more elements that can be decomposed by chemical changes only.

Confirmed test: A second test used if the presumptive test for coliforms is positive.

Cross connection: Connecting fire and drinking water supplies together.

Detention time: The average time spent by water in a settling basin.

Detritus: See 'Grit.'

Distillation: Salt removal in water by boiling and condensation.

Domestic use: Water use by the public (home use).

Double main system: Separate water mains for domestic and fire fighting use.

E. coli: See 'Escherichia coli.'

Element: A pure substance that cannot be decomposed by chemical means.

Electrodialysis: A method of using induced currents

and direction-selective membranes to remove dissolved salts from water.

Electrolysis: The production of an oxidation-reduction reaction by a D.C. current.

Enteric: Intestinal.

Equivalent weight: The molecular weight divided by the change in charge or oxidation number.

Escherichia coli: A common bacterium found in the digestive tract.

Eutrophication: Aging of a lake due to plant growth and sedimentation.

Facultative: Able to live under different or changing conditions.

Floc: Agglomerated colloidal particles.

Flush hydrant: Hydrants located in pits below street level.

Free residuals: Ions or compounds not combined or reduced. The presence of free residuals signifies excess dosage.

Fungus: Multi-cellular plant growth common to humid areas.

Gravity distribution: A water supply that uses natural flow from an elevated tank or mountain reservoir to supply pressure.

Grid-iron layout: A system of distribution pipes in which there are alternative paths through which water can flow in case one path is disturbed.

Grit: Sand-like particles mixed with mud and other debris.

Hard water: Water containing bicarbonates of calcium and magnesium, as well as chlorides and sulfates.

Hydrogen ion: Positively charged combination of a proton and water molecule ($H_3O^+$).

Hydrophilic: Seeking or liking water.

Hydrophobic: Disliking water.

Hydronium ion: See 'hydrogen ion.'

Ions: Atoms that have lost or gained one or more electrons, giving them a charge.

Intrinsic water: Ultra-pure water with essentially no mineral or ion content. Typically used in electronic industries.

Isotopes: Atoms of the same atomic number but with different atomic weights due to a variable number of neutrons.

Mixture: A heterogeneous physical combination of two or more substances, each of which retains its identity and specific properties.

Mole: A quantity of substance equal to its molecular weight in grams (*gmole* or *gram-mole*) or in pounds (*pmole* or *pound-mole*).

Molecule: The smallest division of an element or compound.

Molecular weight: The sum of the atomic weights of all atoms in the molecule.

Non-pathogenic: Not biologically harmful.

Osmosis: The flow of a solvent through a permeable membrane separating two solutions of different concentrations.

Oxidation: The loss of electrons.

Oxidation number: An arbitrarily assigned number used in redox calculations to balance molecular charges.

Pathogenic: Biologically harmful.

Permanent hardness: Hardness that cannot be removed by heating.

pH: A measure of a solution's hydrogen ion concentration (acidity).

pOH: A measure of a solution's hydroxyl ion concentration (alkalinity).

Polished water: See 'Intrinsic water.'

Post hydrant: A fire hydrant that rises from the sidewalk surface.

Potable: Drinkable.

Presumptive test: A first stage test which is inconclusive if positive, but is conclusive if negative.

Protozoa: Single-celled aquatic animals.

Radical: A charged group of atoms that combines as a single element.

Recarbonation: Addition of carbon dioxide to water to neutralize excess lime.

Redox reaction: A reaction in which oxidation and reduction occur.

Reduction: A gain in electrons.

Residual: Additive that is left over after some of the additive has been combined or inactivated.

Retention period: See 'Detention time.'

Reverse osmosis: A process that uses pressure to force a solvent to flow against the ionization gradient.

Salt: An ionic compound formed by direct union of elements, reactions between acids and bases, reactions of acids and salts, and reactions between different salts.

Single main system: One main supplies both potable and fire fighting water.

Solution: A homogeneous mixture of solute and solvent.

Stoichiometry: The study of how elements combine in predetermined quantities to form compounds.

Superchlorination: Chlorination past the breakpoint.

Syndets: Synthetic detergents containing phosphates.

Temporary hardness: Hardness that can be removed by heating water.

Thermocline: A layer of water in a lake or reservoir where temperature declines rapidly with depth.

Turbidity: Cloudiness of water.

Valence: The relative combining capacity of an atom or group of atoms compared to that of the standard hydrogen atom. Valence of an ion is the same as the ion's charge.

Zeolite: A natural or synthetic resin that has an affinity for ions.

## 3 CHEMISTRY REVIEW

### A. VALENCE

The charge of any element in its free state is zero. A charged condition will occur when the element has lost or gained one or more electrons. The *valence* of an ion is equal to its charge.

Common elements with one valence number are:

| cations | | | anions | | |
|---|---|---|---|---|---|
| +1 | +2 | +3 | −1 | −2 | ±4 |
| H | Mg | Al | F | O | C |
| Li | Ca | B | Cl | S | Si |
| Na | Sr | | Br | | |
| K | Ba | | I | | |
| Ag | Zn | | | | |
| | Cd | | | | |

Some elements have two valence numbers. The lower valence is associated with the generic ending 'ous'. The higher valence is associated with the ending 'ic'.

| +1/+2 | +2/+3 | +2/+4 | +3/+5 |
|---|---|---|---|
| Cu | Fe | Pb | As |
| Hg | Ni | Sn | Sb |
| | Co | | Bi |

The following elements have more than two valence numbers:

| | | | | | | |
|---|---|---|---|---|---|---|
| Cr: | +2, | +3, | +6 | | | |
| Mn: | +2, | +3, | +4, | +6, | +7 | |
| S: | −2 | +4, | +6 | | | |
| N: | −3, | +1, | +2, | +3, | +4, | +5 |
| P: | −3, | +3, | +5 | | | |

Common radicals are:

| | | |
|---|---|---|
| $NH_4$ | Ammonium | +1 |
| $ClO_3$ | Chlorate | −1 |
| $ClO_2$ | Chlorite | −1 |
| ClO | Hypochlorite | −1 |
| $NO_3$ | Nitrate | −1 |
| $NO_2$ | Nitrite | −1 |
| $C_2H_3O_2$ | Acetate | −1 |
| $MnO_4$ | Permanganate | −1 |
| OH | Hydroxide | −1 |
| $HSO_3$ | Bisulfite | −1 |
| $HSO_4$ | Bisulfate | −1 |
| $HCO_3$ | Bicarbonate | −1 |
| $CO_3$ | Carbonate | −2 |
| $SO_4$ | Sulfate | −2 |
| $SO_3$ | Sulfite | −2 |
| $CrO_4$ | Chromate | −2 |
| $Cr_2O_7$ | Dichromate | −2 |
| $BO_3$ | Borate | −3 |
| $PO_4$ | Phosphate | −3 |
| $Fe(CN)_6$ | Ferricyanide | −3 |
| $Fe(CN)_6$ | Ferrocyanide | −4 |

Compounds always form in such a manner as to obtain a neutral charge. For example, $MgNO_3$ would not be allowed because the magnesium has a charge of +2 and the nitrate ion has a charge of −1. However, $Mg(NO_3)_2$ is a valid compound.

### B. CHEMICAL REACTIONS

During chemical changes, bonds between atoms are broken and new bonds are formed. Reactants are either converted to simpler products or are synthesized into more complex compounds. There are five common types of chemical reactions.

- **Direct combination or synthesis.** This is the simplest type of reaction where two elements or compounds combine directly to form a compound.

$$2H_2 + O_2 \rightarrow 2H_2O$$

$$SO_2 + H_2O \rightarrow H_2SO_3$$

- **Decomposition.** Bonds uniting a compound are disrupted by heat or other energy source to yield simpler compounds or elements.

$$2HgO \rightarrow 2Hg + O_2$$
$$H_2CO_3 \rightarrow H_2O + CO_2$$

- **Single displacements.** This type of reaction is identified by one element and one compound as reactants.

$$2Na + 2H_2O \rightarrow 2NaOH + H_2$$
$$2KI + Cl_2 \rightarrow 2KCl + I_2$$

- **Double decomposition.** These are reactions characterized by having two compounds as reactants and forming two new compounds.

$$AgNO_3 + NaCl \rightarrow AgCl + NaNO_3$$
$$H_2SO_4 + ZnS \rightarrow H_2S + ZnSO_4$$

- **Oxidation-Reduction (Redox).** These reactions involve oxidation of one substance and reduction of another. In the example, calcium loses electrons and is oxidized; oxygen gains electrons and is reduced.

$$2Ca + O_2 \rightarrow 2CaO$$

Balancing chemical equations is largely a matter of deductive trial and error. The coefficients in front of each compound can be thought of as the number of molecules or moles taking part in the reaction. The number of atoms for each element must be equal on both sides of the equation. Total atomic weights must also be equal on both sides.

*Example 7.1*

Balance the following reaction equation:

$$Al + H_2SO_4 \rightarrow Al_2(SO_4)_3 + H_2$$

As written, the reaction is not balanced. For example, there is one aluminum on the left, but there are two on the right. The starting element in the balancing procedure is chosen somewhat arbitrarily.

*step 1*: Since there are two aluminums on the right, multiply (Al) by 2.

$$2Al + H_2SO_4 \rightarrow Al_2(SO_4)_3 + H_2$$

*step 2*: Since there are three sulfate radicals ($SO_4$) on the right, multiply ($H_2SO_4$) by 3.

$$2Al + 3H_2SO_4 \rightarrow Al_2(SO_4)_3 + H_2$$

*step 3*: Now there are six hydrogens on the left, so multiply $H_2$ by 3 to balance the equation.

$$2Al + 3H_2SO_4 \rightarrow Al_2(SO_4)_3 + 3H_2$$

C. STOICHIOMETRY

Stoichiometric problems are known as 'weight and proportion' problems. Their solutions use simple ratios to determine the weight of reactants required to produce some given amount of products. The procedure for solving these problems is essentially the same regardless of the complexity of the reaction.

*step 1*: Write and balance the chemical equation.

*step 2*: Calculate the molecular weight of each compound or element in the equation.

*step 3*: Multiply the molecular weights by their respective coefficients and write the products under the formulas.

*step 4*: Write the given weight data under the molecular weights calculated from step 3.

*step 5*: Fill in missing information by using simple ratios.

*Example 7.2*

Caustic soda (NaOH) is made from sodium carbonate ($Na_2CO_3$) and slaked lime ($Ca(OH)_2$). How many pounds of caustic soda can be made from one ton of sodium carbonate?

The balanced chemical equation is

| | $Na_2CO_3$ + | $Ca(OH)_2$ → | $2NaOH$ + | $CaCO_3$ |
|---|---|---|---|---|
| MW's: | 106 | 74 | 80 | 100 |
| given: | 2000 | | X | |

The ratio used is

$$\frac{NaOH}{Na_2CO_3} = \frac{80}{106} = \frac{X}{2000}$$

$$X = 1509\,\text{pounds}$$

D. EQUIVALENT WEIGHTS

The *equivalent weight* of a molecule that takes place in a chemical reaction is the molecular weight divided by the change in charge (or oxidation number) which is experienced by the molecule or ion.

*Example 7.3*

What are the equivalent weights of the following compounds?

(a) Al in the reaction: $Al^{+++} + 3e^{-} \rightarrow Al$

(b) $H_2SO_4$ in the reaction: $H_2SO_4 + H_2O \rightarrow 2H^{+} + SO_4^{--} + H_2O$

(c) NaOH in the reaction: $NaOH + H_2O \rightarrow Na^{+} + OH^{-} + H_2O$

(a) The atomic weight of aluminum is 27. Since the change in charge is three, the equivalent weight is $27/3 = 9$.

(b) The molecular weight of sulfuric acid is 98.1. Since it went from neutral to ions with 2 charges each, the equivalent weight is $98.1/2 = 49.05$.

(c) Sodium hydroxide has a molecular weight of 40. The molecule went from neutral to a singly-charged state. So, the equivalent weight is $40/1 = 40$.

## E. SOLUTIONS OF SOLIDS IN LIQUIDS

Various methods exist for measuring the strengths of solutions.

N – **Normality**: The number of gram equivalent weights of solute per liter of solution. A solution is 'normal' if there is exactly one gram equivalent weight per liter.

M – **Molarity**: The number of gram moles of solute per liter of solution. A 'molar' solution contains one gram mole per liter of solution. Hydrated water adds to molecular weight.

F – **Formality:** The number of gram formula weights (i.e., molecular weights in grams) per liter of solution. Hydrated water does not add to formula weight.

m – **Molality**: The number of gram moles of solute per 1000 grams of solvent. A 'molal' solution contains 1 gmole per 1000 grams.

x – **Mole fraction**: The number of moles of solute divided by the number of moles of solvent and all solutes.

mg/l – **milligrams per liter**: The number of milligrams per liter. Same as ppm for solutions of water.

ppm – **parts per million**: The number of pounds (or grams) of solute per million pounds (or grams) of solution. Same as mg/l for solutions of water.

meq/l – **milligram equivalent weights** of solute per liter of solution: Calculated by multiplying normality by 1000.

*Example 7.4*

A solution is made by dissolving 0.353 grams of $Al_2(SO_4)_3$ in 730 grams of water. Assuming 100% ionization, what is the concentration expressed as normality, molarity, and mg/l?

The molecular weight of $Al_2(SO_4)_3$ is

$$\text{MW} = (2)(26.98) + 3[32.06 + (4)(16)] = 342.14$$

The equivalent weight is $342.14/6 = 57.02$

The number of gram equivalent weights used is $0.353/57.02 = 6.19$ EE–3.

The number of liters of solution (same as the solvent volume if the small amount of solute is neglected) is 0.73.

The normality is

$$\text{N} = \frac{6.19\,\text{EE–3}}{0.73} = 8.48\,\text{EE–3}$$

The number of moles of solute used is $0.353/342.14 = 1.03$ EE–3.

The molarity is

$$M = \frac{1.03\,\text{EE–3}}{0.73} = 1.41\,\text{EE–3}$$

The number of milligrams is $0.353/0.001 = 353$.

$$\text{mg/l} = \frac{353}{0.73} = 483.6$$

For the purpose of water supply calculations, however, the most useful measure of solution strength is the *$CaCO_3$ equivalent* measurement. In this method, substance quantities are reported in mg/l *as $CaCO_3$*, even when $CaCO_3$ is unrelated to the substance or reaction that produced the substance.

Actual gravimetric amounts of a substance can be converted to amounts as $CaCO_3$ by use of the conversion factors in appendix A. These factors are easily derived from stoichiometric principles.

The value of converting all substance quantities to amounts as $CaCO_3$ is that equal $CaCO_3$ amounts constitute stoichiometric reaction quantities. For example, 100 mg/l *as $CaCO_3$* of sodium ion ($Na^{+}$) reacts with 100 mg/l *as $CaCO_3$* of chloride ion ($Cl^{-}$) to produce 100 mg/l *as $CaCO_3$* of salt (NaCl), even though the gravimetric quantities differ and $CaCO_3$ is not part of the reaction.

*Example 7.5*

Lime is added to water to remove carbon dioxide gas.

$$CO_2 + Ca(OH)_2 \rightarrow CaCO_3 \downarrow + H_2O$$

If water contains 5 mg/l of $CO_2$, how much lime is required for its removal?

From appendix A, the factor which converts $CO_2$ *as substance* to $CO_2$ *as* $CaCO_3$ is 2.27.

$$\begin{aligned} CO_2 \text{ as } CaCO_3 \text{ equivalent } &= (2.27)(5) \\ &= 11.35 \text{ mg/l as } CaCO_3 \end{aligned}$$

Therefore, the $CaCO_3$ equivalent of lime required will also be 11.35 mg/l.

From appendix A, the factor which converts lime as $CaCO_3$ to lime as substance is (1/1.35).

$$Ca(OH)_2 \text{ substance} = \frac{11.35}{1.35} = 8.41 \text{ mg/l as substance}$$

## F. SOLUTIONS OF GASES IN LIQUIDS

*Henry's law* states that the amount of gas dissolved in a liquid is proportional to the partial pressure of the gas. This applies separately to each gas to which the liquid is exposed.

*Example 7.6*

At 20 °C and 760 mm Hg, one liter of water can absorb 0.043 grams of oxygen or 0.019 grams of nitrogen. Find the mass of oxygen and nitrogen in one liter of water exposed to air at 20 °C and 760 mm Hg total pressure.

Partial pressure is volumetrically weighted. Air is 20.9% oxygen by volume. The remainder is taken as nitrogen.

$$\begin{aligned} \text{oxygen dissolved } &= (0.209)(0.043) = 0.009 \text{ g/l} \\ \text{nitrogen dissolved } &= (0.791)(0.019) = 0.015 \text{ g/l} \end{aligned}$$

## G. ACIDS AND BASES

An *acid* is any substance that dissociates in water into $H^+$ (or $H_3O^+$). A *base* dissociates in water and gives up $OH^-$. A measure of the strength of the acid or base is the number of these hydrogen or hydroxide ions in a liter of solution. Since these are large (small) numbers, a logarithmic scale is used.

$$pH = -\log_{10}[H^+] \qquad 7.1$$

$$pOH = -\log_{10}[OH^-] \qquad 7.2$$

The following relationship always exists between pH and pOH:

$$pH + pOH = 14 \qquad 7.3$$

[X] is defined as the *ionic concentration* in moles per liter. The number of moles can be calculated from Avogadro's law by dividing the actual number of ions by 6.023 EE23. An easier method is to use equation 7.4.

$$[X] = (\text{fraction ionized})(\text{molarity}) \qquad 7.4$$

*Example 7.7*

Calculate the concentrations of $H^+$, $OH^-$, pH, and pOH in 4.2% ionized 0.010M ammonia solution prepared from ammonium hydroxide ($NH_4OH$).

$$\begin{aligned} [OH^-] &= (0.042)(0.010) = 4.2 \text{ EE–4 moles/liter} \\ pOH &= -\log(4.2 \text{ EE–4}) = 3.38 \\ pH &= 14 - 3.38 = 10.62 \\ [H^+] &= 10^{-10.62} = 2.4 \text{ EE–11 moles/liter} \end{aligned}$$

Since $H^+ + OH^- \rightarrow H_2O$, acids and bases neutralize each other by forming water. The volumes required for complete neutralization are:

$$\begin{aligned} &(\text{vol. base})(\text{normality of base}) \\ &\quad = (\text{vol. acid})(\text{normality of acid}) \qquad 7.5 \end{aligned}$$

If the concentrations are expressed as molarities,

$$\begin{aligned} &(\text{vol. base})(\text{base molarity})(\text{change in base charge}) = \\ &(\text{vol. acid})(\text{acid molarity})(\text{change in acid charge}) \qquad 7.6 \end{aligned}$$

Both equation 7.5 and 7.6 assume 100% ionization of the solute.

## H. REVERSIBLE REACTIONS

Some reactions are capable of going in either direction. Such reactions are called *reversible reactions* and are characterized by the presence of all reactants and all products simultaneously.

$$N_2 + 3H_2 \rightleftharpoons 2NH_3 + 24{,}500 \text{ calories}$$

*Le Chatelier's principle* predicts the direction the reaction will go when some property or condition is changed. That principle says that when a reversible reaction has resulted in an equilibrium condition, and the reactants and products are further stressed by changing the pressure, temperature, or concentration, a new equilibrium will be formed in the direction that reduces the stress.

Consider the reaction between nitrogen and hydrogen. When the reaction proceeds in the forward direction, heat is given off (an *exothermic reaction*). If the reaction proceeds in the reverse direction, heat is absorbed (*endothermic reaction*). If the system is stressed by increasing the temperature, the reaction will proceed in the reverse direction because that direction will absorb heat and reduce the temperature.

For reactions that involve gases, the coefficients in front of the molecules can be interpreted as volumes. In the nitrogen-hydrogen reaction, 4 volumes combine to form

2 volumes. If the equilibrium system is stressed by increasing the pressure, then the forward reaction will occur because this direction reduces the volume and thereby reduces the pressure.

If the concentration of any substance is increased, the reaction proceeds in a direction away from the substance with the increased concentration.

If a *catalyst* is introduced, the equilibrium will not be changed. However, the reaction speed is increased, and equilibrium is reached more quickly. This reaction speed is known as the *rate of reaction* or *reaction velocity*.

The *law of mass action* says that the reaction speed is proportional to the concentrations of reactants. Consider the following reaction.

$$A + B \rightleftharpoons C + D$$

The rates of reaction can be calculated as the products of the reactants' concentrations.

$$v_{\text{forward}} = C_1[A][B] \qquad 7.7$$

$$v_{\text{reverse}} = C_2[C][D] \qquad 7.8$$

The constants, $C_1$ and $C_2$, are needed to obtain the proper units. For reversible reactions, the *equilibrium constant* is defined by equation 7.9. This equilibrium constant is essentially independent of pressure, but will depend on temperature.

$$K_{\text{eq}} = \frac{[C][D]}{[A][B]} \qquad 7.9$$

Consider the following complex reaction,

$$aA + bB \rightleftharpoons cC + dD$$

The equilibrium constant is calculated from the *mass action equation.*

$$K_{\text{eq}} = \frac{[C]^c[D]^d}{[A]^a[B]^b} \qquad 7.10$$

The only special rule to be observed is that if any of $A, B, C,$ or $D$ are pure solids or liquids, then their concentrations are omitted in the calculation of the equilibrium constant.

*Example 7.8*

Acetic acid dissociates according to the following equation. What is the equilibrium constant if the ion concentrations are as given?

$$HC_2H_3O_2 + H_2O \rightleftharpoons H_3O^+ + C_2H_3O_2^-$$

$$[HC_2H_3O_2] = 0.09866 \text{ moles/liter}$$

$$[H_2O] = 55.5555 \text{ moles/liter}$$

$$[H_3O^+] = 0.00134 \text{ moles/liter}$$

$$[C_2H_3O_2^-] = 0.00134 \text{ moles/liter}$$

From equation 7.9, omitting the water's concentration,

$$K_{eq} = \frac{(0.00134)(0.00134)}{0.09866} = 1.82 \text{ EE}{-5}$$

For weak aqueous solutions, the concentration of water is very large and essentially constant. Therefore, it is omitted. The new constant is known as the *ionization constant* to distinquish it from the equilibrium constant.

$$K_{\text{ionization}} = K_{\text{equilibrium}} \qquad 7.11$$

Pure water is a very weak electrolyte and it ionizes only slightly by itself.

$$2H_2O \rightleftharpoons H_3O^+ + OH^- \qquad 7.12$$

At equilibrium, the ion concentrations can be measured to be

$$[H_3O^+] = \text{EE–7}$$

$$[OH^-] = \text{EE–7}$$

The ionization constant for pure water is

$$\begin{aligned} K_{\text{ionization}} &= [H_3O^+][OH^-] \\ &= (\text{EE–7})(\text{EE–7}) = \text{EE–14} \qquad 7.13 \end{aligned}$$

Notice that the concentration of water was omitted because the ionization constant, not the equilibrium constant, was calculated.

Taking logs of both sides of equation 7.13 will derive equation 7.3.

*Example 7.9*

A 0.1M acetic acid solution is 1.34% ionized. Find the

(a) hydrogen ion concentration

(b) acetate ion concentration

(c) un-ionized acid concentration

(d) ionization constant

(a) From equation 7.4,

$[H_3O^+] = (0.0134)(0.1) = 0.00134$ moles/liter.

(b) Since every hydrogen ion has a corresponding acetate ion (see example 7.8), the acetate ion concentration is also 0.00134 moles per liter.

(c) The concentration of un-ionized acid can be derived from equation 7.4:

$$[HC_2H_3O_2] = (\text{fraction not ionized})(\text{molarity}) = (1 - 0.0134)(0.1) = 0.09866\ \text{moles/liter}$$

(d) The ionization constant is

$$K_{\text{ionization}} = \frac{(0.00134)(0.00134)}{0.09866} = 1.82\ \text{EE}-5$$

A faster way to find the ionization constant is to use equation 7.14.

$$K_{\text{ionization}} = \frac{(\text{molarity})(\text{fraction ionized})^2}{1 - \text{fraction ionized}} \qquad 7.14$$

*Example 7.10*

Find the concentration of the hydrogen ion for a 0.2M acetic acid solution if $K_{\text{ionization}} = 1.8$ EE–5.

From equation 7.14, letting $X$ be the fraction ionized,

$$1.8\,\text{EE–5} = \frac{(0.2)(X)^2}{1 - X}$$

If $X$ is small, then $(1 - X) \approx 1$. So, $X = 9.49$ EE–3.

From equation 7.4, the concentration is

$$[H_3O^+] = (9.49\,\text{EE–3})(0.2) = 1.9\,\text{EE–3}$$

The *common ion effect law* is a form of Le Chatelier's law. If a salt containing a common ion is added to a weak acid or base solution, ionization will be suppressed. This is a consequence of the need to have an unchanged ionization constant.

*Example 7.11*

What is the hydrogen ion concentration of a solution with 0.1 gmole of 80% ionized ammonium acetate ($NH_4C_2H_3O_2$) in one liter of 0.1M acetic acid ($HC_2H_3O_2$)?

As before, $1.8\ \text{EE–5} = \frac{[H_3O^+][C_2H_3O_2^-]}{[HC_2H_3O_2]}$.

Let $X = [H_3O^+]$.

Then, $[HC_2H_3O_2] = 0.1 - X \approx 0.1$

$$[C_2H_3O_2^-] = X + (0.8)(0.1) \approx (0.8)(0.1) = 0.08$$

So, from the ionization constant equation,

$$1.8\,\text{EE–5} = \frac{(X)(0.08)}{0.1}$$

$$X = 2.2\,\text{EE–5}$$

## I. SOLUBILITY PRODUCT

When an ionic solid is dissolved in a solvent, it dissociates and ionizes.

$$AgCl \rightleftharpoons Ag^+ + Cl^- \quad \text{(in water)}$$

When the equation for the equilibrium constant is written, the terms for solid components are omitted. When the equation for the ionization constant is written, the term for water concentration is also omitted. Thus, when an ionic solid is placed in water, the ionization constant will consist only of the ion concentrations. This ionization constant is known as the ***solubility product***.

$$K_{sp} = [Ag^+][Cl^-]$$

As with the general case of ionization constants, the solubility product is essentially constant for slightly soluble solutes. Any time that the product of terms exceeds the standard value of the solubility product, solute will

**Table 7.1**
Approximate Ionization Constants (moles/liter)

| substance | 0°C | 5°C | 10°C | 15°C | 20°C | 25°C |
|---|---|---|---|---|---|---|
| $HClO$ | 2.0 EE–8 | 2.3 EE–8 | 2.6 EE–8 | 3.0 EE–8 | 3.3 EE–8 | 3.7 EE–8 |
| $HC_2H_3O_2$ | 1.67 EE–5 | 1.70 EE–5 | 1.73 EE–5 | 1.75 EE–5 | 1.75 EE–5 | 1.75 EE–5 |
| $HBrO$ | | | | | ≈ 2 EE–9 | |
| $H_2CO_3$ ($K_1$) | 2.6 EE–7 | 3.04 EE–7 | 3.44 EE–7 | 3.81 EE–7 | 4.16 EE–7 | 4.45 EE–7 |
| $HClO_2$ | | | | | ≈ 1.1 EE–2 | |
| $NH_3$ | 1.37 EE–5 | 1.48 EE–5 | 1.57 EE–5 | 1.65 EE–5 | 1.71 EE–5 | 1.77 EE–5 |
| $NH_4OH$ | | | | | | 1.79 EE–5 |
| $Ca(OH)_2$ | | | | | | 3.74 EE–3 |
| water* | 14.9435 | 14.7338 | 14.5346 | 14.3463 | 14.1669 | 13.9965 |

*$-\log_{10}(K)$ given.

precipitate out until the product of the remaining ion concentrations attains the standard value. If the product is less than the standard value, the solution is not saturated.

**Table 7.2**
Approximate Solubility Products

(moles/liter)

| substance | 12°F | 15°C | 18°C | 25°C |
|---|---|---|---|---|
| $Al(OH)_3$ | | 4 EE–13 | 1.1 EE–15 | 3.7 EE–15 |
| $CaCO_3$ | | 0.99 EE–8 | | 0.87 EE–8 |
| $CaF_2$ | | | 3.4 EE–11 | 4.0 EE–11 |
| $Fe(OH)_3$ | | | 1.1 EE–36 | |
| $Mg(OH)_2$ | | | 1.2 EE–11 | 1 EE–11 |
| $MgCO_3$ | 2.6 EE–5 | | | 1 EE–5 |
| $BaCO_3$ | | 7 EE–9 | | 8.1 EE–9 |
| $BaSO_4$ | | | 0.87 EE–10 | 1.1 EE–10 |
| $CaSO_4$ | 2 EE–4 | | | |
| $Fe(OH)_2$ | | | 1.6 EE–14 | 8 EE–6 |
| $MnCO_3$ | | | | 1.8 EE–11 |
| $Mn(OH)_2$ | | | 4 EE–14 | |
| $MgF_2$ | | | 7.1 EE–9 | 6.6 EE–9 |

*Example 7.12*

What is the solubility product of lead sulfate ($PbSO_4$) if its solubility is 38 mg/l?

$$PbSO_4 \rightleftharpoons Pb^{++} + SO_4^{--} \quad \text{(in water)}$$

The molecular weight of $PbSO_4$ is 207.19 + 32.06 + (4)(16) = 303.25

The number of moles of $PbSO_4$ in a liter of saturated solution is

$$0.038/303.25 = 1.25\,\text{EE–4}$$

This is also the number of moles of $Pb^{++}$ and $SO_4^{--}$ that will form in the solution. Therefore,

$$K_{sp} = [Pb^{++}][SO_4^{--}] = (1.25\,\text{EE–4})^2 = 1.56\,\text{EE–8}$$

The method used in example 7.12 to find the solubility product works well with chromates ($CrO_4^{--}$), halides ($F^-$, $Cl^-$, $Br^-$, $I^-$), sulfates ($SO_4^{--}$), and iodates ($IO_3^-$). However, sulfides ($S^{--}$), carbonates ($CO_3^{--}$), phosphates ($PO_4^{---}$), and the salts of transition elements (such as iron) hydrolyze and must be treated differently.

WATER SUPPLY

## 4 QUALITIES OF SUPPLY WATER

### A. ACIDITY AND ALKALINITY

*Acidity* is a measure of acids in solution. Acidity in surface water is caused by formation of *carbonic acid* ($H_2CO_3$) from carbon dioxide in the air.[1]

$$CO_2 + H_2O \rightarrow H_2CO_3 \qquad 7.15$$

$$H_2CO_3 + H_2O \rightarrow HCO_3^- + H_3O^+ \;(\text{pH} > 4.5) \qquad 7.16$$

$$HCO_3^- + H_2O \rightarrow CO_3^{--} + H_3O^+ \;(\text{pH} > 8.3) \qquad 7.17$$

Measurement of acidity is done by titration with a standard basic measuring solution. Acidity in water is typically given in terms of the $CaCO_3$ equivalent that would neutralize the acid.

$$\text{acidity (mg/l of } CaCO_3) = \frac{(\text{vol. titrant})(\text{titrant normality})(50{,}000)}{(\text{vol. sample})} \qquad 7.18$$

*Alkalinity* is a measure of the amount of negative ions in the water. Specifically, $OH^-$, $CO_3^{--}$, and $HCO_3^-$ all contribute to alkalinity.[2] The measure of alkalinity is the sum of concentrations of each of the substances measured as $CaCO_3$.

*Example 7.13*

Water from a city well is analyzed and is found to carry 20 mg/l as substance of $HCO_3^-$ and 40 mg/l as substance of $CO_3^{--}$. What is the alkalinity of this water?

From appendix A, the factors converting $HCO_3^-$ and $CO_3^{--}$ ions to $CaCO_3$ equivalents are 0.82 and 1.67 respectively.

$$\text{alkalinity} = (0.82)(20) + (1.67)(40) = 83.2\,\text{mg/l as } CaCO_3$$

Alkalinity can also be found by using an acidic titrant.

$$\text{alkalinity (mg/l of } CaCO_3) = \frac{(\text{vol. titrant})(\text{titrant normality})(50{,}000)}{(\text{vol. sample})} \qquad 7.19$$

Alkalinity and acidity of a titrated sample is determined from color changes in indicators added to the titrant. Table 7.3 lists indicators that are commonly used.

[1] Carbonic acid is very aggressive and must be neutralized to eliminate the cause of water pipe corrosion. If the pH of water is greater than 4.5, carbonic acid ionizes to form bicarbonate (equation 7.16). If the pH is greater than 8.3, carbonate ions form which cause water hardness by combining with calcium. (See equation 7.17.)

[2] Other radicals, such as $NO_3^-$, also contribute to alkalinity, but their presence is rare. If detected, they should be included in the calculation of alkalinity.

**Table 7.3**
Commonly Used Indicator Solutions

| common name | pH visual transition interval | color acidic | color basic |
|---|---|---|---|
| cresol red | 0.2–1.8 | red | yellow |
| thymol blue | 1.2–2.8 | red | yellow |
| methyl yellow | 2.4–4.0 | red | yellow |
| bromophenol blue | 3.0–4.6 | yellow | blue |
| methyl orange | 3.2–4.4 | red | yellow-orange |
| methyl orange + xylene cyanole FF, 40 : 56 | (3.8–4.1)[a] | violet | green |
| bromocresol green | 3.9–5.4 | yellow | blue |
| methyl red | 4.2–6.2 | pink | yellow |
| methyl red + methylene blue, 1 : 1 | (~ 5.3)[a] | red-violet | green |
| bromocresol purple | 5.2–6.8 | yellow | purple |
| bromothymol blue | 6.0–7.6 | yellow | blue |
| cresol red | 7.2–8.8 | yellow | red |
| phenol red | 6.8–8.2 | yellow | red |
| thymol blue | 8.0–9.2 | yellow | blue |
| phenol-phthalein | (8.0–9.8)[b] | colorless | red-violet |
| phenol-phthalein + methylene green, 1 : 2 | (8.8)[c] | green | violet |
| thymol-phthalein | (9.0–10.5)[b] | colorless | blue |
| eriochrome black T | 7–10 | blue | wine-red |
| alizarin yellow | 10.1–12 | yellow | red |

[a] Screened indicator, neutral gray at stated pH.

[b] Based on addition of 1 or 2 drops of a 0.1% indicator solution to 10 ml of aqueous solution.

[c] Screened indicator, pale blue at stated pH.

For strongly basic samples (pH > 8.3), consecutive titration with both phenolphthalein and methyl orange is often done. The alkalinity is the sum total of the *phenolphthalein alkalinity* and the *methyl orange alkalinity*. (Note: Phenolphthalein alkalinity is not the same as carbonate alkalinity since $CO_3^{--}$ has been converted to $HCO_3^-$ but not neutralized. See equation 7.17. The *carbonate alkalinity* is twice the phenolphthalein alkalinity.)

*Example 7.14*

0.02N sulfuric acid is used to titrate 110 ml of water. 3.3 ml of titrant is needed to reach the phenolphthalein point, and 13.2 ml is needed to reach the methyl orange point. What are the total and phenolphthalein alkalinities?

From equation 7.19, the phenolphthalein alkalinity is

$$(3.3)(0.02)(50{,}000)/(110) = 30\,\text{mg/l as } CaCO_3$$

The total alkalinity is

$$(13.2 + 3.3)(0.02)(50{,}000)/(110) = 150\,\text{mg/l as } CaCO_3$$

The alkalinity of 150 mg/l is caused by carbonates ($2 \times 30 = 60$ mg/l) and bicarbonates ($150 - 60 = 90$ mg/l).

B. HARDNESS

Water hardness is caused by multi-valent (doubly-charged, triply-charged, etc., but not singly-charged) positive metallic ions such as calcium, magnesium, iron, and manganese. (Iron and manganese are not as common, however.) Hardness reacts with soap to reduce its cleansing effectiveness, and to form scum on the water surface and ring around the bathtub.

Water containing bicarbonate ($HCO_3^-$) ions can be heated to precipitate a carbonate molecule.[3] This hardness is known as *temporary hardness* or *carbonate hardness*.[4]

$$Ca^{++} + 2HCO_3^- + \text{heat} \rightarrow CaCO_3 \downarrow + CO_2 + H_2O \qquad 7.20$$

$$Mg^{++} + 2HCO_3^- + \text{heat} \rightarrow MgCO_3 \downarrow + CO_2 + H_2O \qquad 7.21$$

Remaining hardness due to sulfates, chlorides, and nitrates is known as *permanent hardness* or *non-carbonate hardness* because it cannot be removed by heating. Permanent hardness can be calculated numerically by causing precipitation, drying, and then weighing the precipitate.

$$Ca^{++} + SO_4^{--} + Na_2CO_3 \rightarrow 2Na^+ + SO_4^{--} + CaCO_3 \downarrow \qquad 7.22$$

$$Mg^{++} + 2Cl^- + 2NaOH \rightarrow 2Na^+ + 2Cl^- + Mg(OH)_2 \downarrow \qquad 7.23$$

***Total hardness*** is the sum of temporary and permanent hardnesses, both expressed in mg/l as $CaCO_3$.

Hardness can also be measured by the titration method using a titrant (complexione, versene, EDTA, or BDH)

[3] Hard water forms scale when heated. This scale, if it forms in pipes, eventually restricts water flow. Even in small quantities, the scale insulates boiler tubes. Therefore, water used in steam-producing equipment must be essentially hardness-free.

[4] The hardness is known as *carbonate* hardness even though it is caused by *bicarbonate* radicals, not carbonate radicals.

and an indicator (such as eriochrome black T). The standard hardness reagent used for titration has an equivalent hardness of 1 mg/l per ml used.

The hardness of water can be classified according to table 7.4. Although high values of hardness are not organically dangerous, public acceptance of the water supply requires a hardness of well below 150 mg/l. Except for special industrial uses, potable water should have the carbonate hardness reduced to at least 40 mg/l, and the total hardness should be below 75 mg/l. Where it is economically feasible, the carbonate hardness should be reduced to 25 mg/l.

**Table 7.4**
Hardness Classifications

| class | type | hardness |
|---|---|---|
| A | soft | below 60 mg/l |
| B | medium hard | 60–120 |
| C | hard | 120–180 |
| D | very hard | 180–350 |
| E | saline, brackish | above 350 |

*Example 7.15*

A 75 ml water sample required 8.1 ml of EDTA. What is the hardness?

$$\text{hardness} = \frac{(8.1)\,\text{ml}(1)\,\text{mg/l}}{(75/1000)} = 108\,\text{mg/l}$$

*Example 7.16*

Water is found to contain sodium ($Na^+$, 15 mg/l), magnesium ($Mg^{++}$, 70 mg/l), and calcium ($Ca^{++}$, 40 mg/l). What is the hardness?

Sodium is singly-charged, so it does not contribute to hardness. The approximate equivalent weights of the relevant compounds and elements are:

$$\text{Mg}: 12 \quad \text{Ca}: 20 \quad \text{CaCO}_3: 50$$

The equivalent hardness is

$$(70)\left(\frac{50}{12}\right) + (40)\left(\frac{50}{20}\right) = 392\,\text{mg/l as CaCO}_3$$

Alternatively, appendix A could have been used to convert the ionic concentrations to $CaCO_3$ equivalents.

$$(70)(4.10) + (40)(2.50) = 387\,\text{mg/l as CaCO}_3$$



## C. IRON CONTENT

Even in low concentrations, iron is objectionable because it stains bathroom fixtures, causes a brown color in laundered clothing, and affects taste. Water originally pumped from anaerobic sources may contain ($Fe^{++}$) ferrous ions which are invisible and soluble. When exposed to oxygen, insoluble ($Fe^{+++}$) ferric oxides form which give water the rust coloration.

Iron is measured optically by comparing the color of a sample with standard colors. The comparison can be made by eye or with a photoelectric *colorimeter*. Iron concentrations greater than 0.3 mg/l are undesirable.

## D. MANGANESE CONTENT

Manganese ions are similar in effect, detection, and measurement to iron ions. Manganous manganese ($Mn^{++}$) oxidizes to manganic manganese ($Mn^{++++}$) to give water a rust color. An undesirable concentration is 0.05 mg/l.

## E. FLUORIDE CONTENT

An optimum concentration of fluoride in the form of a fluoride ion, $F^-$, is between 0.8 mg/l for hot climates (80°F–90°F average) to 1.2 mg/l for cold climates (50°F average). These amounts reduce the population cavity rate to a minimum without producing significant fluorosis (staining) of the teeth. The actual amount of fluoridation depends on the average outside temperature since the temperature affects the amount of water that is ingested by the population.

**Table 7.5**
Maximum Fluoride Concentrations

(Note that the 1974 Safe Drinking Water Act and its 1986 amendments set the maximum fluoride concentration at 4 mg/l for all temperatures.)

| 5-year average of maximum daily air temperatures | fluoride concentrations |
|---|---|
| (deg F) | (mg/l) |
| 50.0–53.7 | 2.4 |
| 53.8–58.3 | 2.2 |
| 58.4–63.8 | 2.0 |
| 63.9–70.6 | 1.8 |
| 70.7–79.2 | 1.6 |
| 79.3–90.5 | 1.4 |

Fluoridation can be obtained by the readily dissociating compounds in table 7.6.

**Table 7.6**
Fluoridation Chemicals

$(NH_4)_2SiF_6$ (ammonium silicofluoride)
$CaF_2$ (calcium fluoride)
$H_2SiF_6$ (fluosilic acid)
NaF (sodium fluoride)
$Na_2SiF_6$ (sodium silicofluoride)

Fluoride content is measured by colorimetric and electrical methods.

## F. CHLORIDE CONTENT

Chlorine is used as a disinfectant for water, but its strong oxidation potential allows it also to be used to remove iron and manganese ions. Chlorine gas in water forms hypochlorous and hydrochloric acids.

$$Cl_2 + H_2O \underset{pH<4}{\overset{pH\geq4}{\rightleftharpoons}} HCl + HOCl \underset{pH<9}{\overset{pH\geq9}{\rightleftharpoons}} H^+ + OCl^- \qquad 7.24$$

Free chlorine, hypochlorous acid, and hypochlorite ions are known as *free chlorine residuals.* Hypochlorous acid reacts with ammonia (if it is present) to form mono-, di-, and trichloramines. Chloramines are known as *combined residuals.* Chloramines are more stable than free residuals, but their disinfecting ability is weaker. Their action may extend for a considerable distance into the distribution system.

The amount of chlorine to be added depends on the organic and inorganic matter present in the water. However, most waters are effectively treated within 10 minutes if a free residual of 0.2 mg/l is maintained. Larger residual concentrations may cause objectional odor and taste.

If the water contains phenol, it and the chlorine will form chlorophenol compounds which produce an objectionable taste. This may be stopped by adding ammonia to the water before chlorination.

Both free and combined residual chlorine can be detected by color comparison. However, organic matter in waste water makes it necessary to use a test based on water conductivity. The color comparison test with supply water, however, is adequate.

## G. PHOSPHORUS CONTENT

*Orthophosphates* ($H_2PO_4^-$, $HPO_4^{--}$, and $PO_4^{---}$) and *polyphosphates* (such as $Na_3(PO_3)_6$) result from the use of *synthetic detergents* (*syndets*). Phosphate content is more of a concern in waste water than in supply water. Excessive phosphate discharge contributes to aquatic plant growth and subsequent *eutrophication.*

Phosphates are measured by a variety of means, including colorimetry and filtered precipitation analysis. Care should be taken not to confuse phosphates with phosphorus. Multiply mg/l of phosphate by 0.326 to obtain mg/l of phosphorus.

## H. NITROGEN CONTENT

Nitrogen is present in water in many forms, including organic (protein), ammonia, nitrate, and gaseous ammonia. As with phosphates, nitrogen contamination is more of a problem with waste water than with supply water. Nitrogen pollution promotes algae growth. Ammonia is toxic to fish.

Drinking water is typically tested only for nitrates. The following tests are used:

**Table 7.7**
Tests for Nitrogen

| to test for | procedure |
|---|---|
| ammonia | distillation |
| organic nitrogen | digestion with distillation |
| nitrate, nitrite | colorimetry |

Gaseous nitrogen is of little concern since it is not normally metabolized by plants and it is of no danger to animal or human life.

## I. COLOR

Color in domestic water is undesirable aesthetically, and it may dull the color of clothes and stain bathroom fixtures. Some industries (such as beverage production, dairy, food processing, paper manufacturing, and textile production) also have strict water color standards.

Water color is measured with a colorimeter or comparitively with tubes containing standard platinum/cobalt solutions. Color is graded on a scale of 0 (clear) to 70 color units.

## J. TURBIDITY

Turbidity is a measure of the insoluble solids (soil, organics, and microorganisms) in water which impede light passage. Completely clean water measures 0 *turbidity units* (NTU). 5 NTU is noticeable to an average consumer, and this is a practical upper limit for drinking water. Muddy water exceeds 100 NTU. A TU is equivalent to 1 mg/l of silica in suspension.

Turbidity is usually measured with a *nephelometer*, *Jackson candle apparatus*, or *Baylis turbidimeter*. Units can be NTU (nephelometer turbidity units) or JTU (Jackson turbidity units).

K. SUSPENDED AND DISSOLVED SOLIDS

Solids present in a sample of drinking water can be divided into several categories, not all of which are mutually exclusive.

- **Suspended solids**: Suspended solids, the same as *filterable solids*, are measured by filtering a sample of water and weighing the residue.
- **Dissolved solids**: Dissolved solids, same as *nonfilterable solids*, are measured as the difference between total solids and suspended solids.
- **Total solids**: Total solids are made up of suspended and dissolved solids. They are measured by drying a sample of water and weighing the residue.
- **Volatile solids**: Volatile solids are measured as the decrease in weight of total solids which have been ignited in an electric furnace.
- **Fixed solids**: The fixed solids can be found as the difference between total solids and volatile solids.
- **Settleable solids**: The volume (ml/l) of settleable solids is measured by allowing a sample to stand for one hour in a graduated conical container (*Imhoff cone*).

An upper limit of 500 mg/l of total solids is recommended.

**Table 7.8**
Water-Borne Organisms

| this organism | causes this disease |
|---|---|
| BACTERIA | |
| Salmonella | salmonellosis |
| Bacillus typhosa | typhoid fever |
| Vibrio comma | cholera |
| Shigella dysenteriae | dysentery |
| Escherichia coli | enteric problems |
| fecal streptococci | enteric problems |
| VIRUSES | |
| Poliomyelitis | polio |
| Infectious hepatitis | hepatitis |
| PROTOZOA | |
| Entamoeba histolytica | amoebic dysentery |
| PARASITES | |
| flatworms | Schistosomiasis |
| flatworms | Bilharziasis |
| Giardia lamblia | giardiasis |
| Cryptosporidium | — |



L. WATER-BORNE DISEASES

Organisms that are present in water consist of bacteria, fungi, viruses, algae, protozoa, and multicellular animals. Not all of these are dangerous, but some are. Important organisms are listed in table 7.8.

## 5 TRIHALOMETHANES

Trihalomethanes (THM's) are organic chemicals produced during the disinfection of water. The chemically active elements of chlorine, iodine, and bromine react with various organic precursors to produce THM's. However, iodine is seldom used in disinfection. Therefore, only four THM's are found in significant quantities:[5]

- $CHCl_3$ – trichloromethane (chloroform)
- $CHBr_3$ – tribromomethane (bromoform)
- $CHBrCl_2$ – bromodichloromethane
- $CHBr_2Cl$ – dibromochloromethane

The *organic precursors* which react with chlorine to produce THM's tend to be naturally occurring. For example, decaying vegetation produces humic and fulvic acids which are natural precursors. The precursors in themselves are not harmful, but the THM's produced from them have been shown to be carcinogenic.[6]

Table 7.9 lists the maximum contaminant level (MCL) for THM's in drinking water. Communities with populations of 10,000 and above are covered by the MCL if they add a disinfectant to their drinking water supply. The MCL reported in table 7.9 is for total THM (TTHM), not just chloroform. Precursors are not limited or monitored.

When THM levels need to be reduced, several options are available. These options fall into two categories, depending on whether the precursors are removed prior to chlorination, or a disinfectant is chosen that does not produce THM's. The first category includes the following options:

- Using *granular activated carbon* (GAC), other adsorbents, or filters, including weak-base resins, to remove precursors
- Selecting a water source with fewer precursors

[5] Bromine can be present in gaseous chlorine as an impurity. Bromine also results from reacting chlorine with the bromide present in high-salinity water.

[6] Actually, tests have shown that only chloroform in high doses is carcinogenic to rats and mice. The other THM's are considered carcinogenic by association. Some haloacetic acids (HAA) are also carcinogenic.

- Moving the chlorination point to the end of the treatment process so that most precursors are removed prior to disinfection (70–75% TTHM reduction)
- Optimizing coagulation and settling processes to improve precursor removal

Within the second category are the following options:

- Using ozone, chlorine dioxide, or potassium permanganate to disinfect without THM formation (60–90% TTHM reduction)
- Dechlorination using *sodium metabisulfate* or other methods after chlorination to prevent the reaction of chlorine with precursors
- Adding ammonia to water prior to discharge to induce chloramine formation, since chloramines suppress the formation of THM's

25–60% TTHM's can also be removed after formation by contacting with granular activated carbon.

Caution is required when changing to alternate disinfectants. The disinfectant and its byproducts should be evaluated to determine disinfecting power, residual power, toxicity, and other health effects.[7]

Costs of operation will increase when alternate disinfectants are used. Moving the point of application may not result in any significant operating costs after a modest capital expenditure is made. Cost of using ozone as an alternative disinfectant is often less than using chlorine dioxide, but it is more than using chloramines or changing the points of chlorine application.

[7] Ozone does not form any potentially dangerous byproducts.

## 6 COMPARISON OF ALKALINITY AND HARDNESS

Hardness measures the presence of $Mg^{++}$, $Ca^{++}$, $Fe^{++}$, and other multi-valent ions. Alkalinity measures the presence of $HCO_3^-$, $SO_4^{--}$, $Cl^-$, $NO_3^-$, and $OH^-$ ions. Both positive and negative ions can exist side by side, so an alkaline water can also be hard.

If certain assumptions are made, then it is possible to draw conclusions about the water composition. For example, $Fe^{++}$ is an unlikely ion in most water supplies, and it is often neglected in comparing alkalinity and hardness.

Figure 7.1 gives an easy method of comparing hardness and alkalinity, and using the comparison to deduce other compounds in the water.

If hardness (as $CaCO_3$) and alkalinity (also as $CaCO_3$) are the same, then there are no $SO_4^{--}$, $Cl^-$, or $NO_3^-$ ions present. (That is, there is no non-carbonate, permanent hardness.) If hardness is greater than alkalinity, however, then non-carbonate, permanent hardness is present, and the temporary carbonate hardness is equal to the alkalinity. If hardness is less than alkalinity, then all hardness is carbonate, temporary hardness, and the extra $HCO_3^-$ comes from other sources (such as $NaHCO_3$).

M = alkalinity
H = total hardness
Ca = calcium
O = hydroxides
S = sulfate hardness
L = free lime

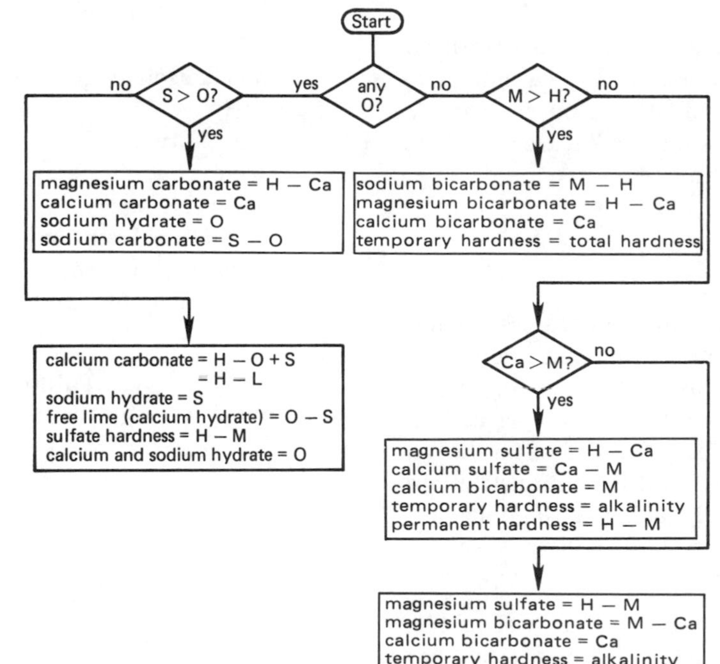


**Figure 7.1** Hardness and Alkalinity
(All results expressed as $CaCO_3$)

*Example 7.17*

A sample of water has been found to contain the following:

alkalinity: 220 mg/l as $CaCO_3$

hardness: 180 mg/l as $CaCO_3$

calcium ($Ca^{++}$): 140 mg/l as $CaCO_3$

(a) What is the non-carbonate hardness?

(b) What is the $Mg^{++}$ content in mg/l as substance?

To use figure 7.1, the absence of any significant hydroxides must be assumed. Since the alkalinity is greater than the hardness, the figure indicates the following compounds in the water:

$$NaHCO_3 = 220 - 180 = 40 \text{ mg/l as } CaCO_3$$
$$Mg(HCO_3)_2 = 180 - 140 = 40 \text{ mg/l as } CaCO_3$$
$$Ca(HCO_3)_2 = 140 \text{ mg/l}$$

There is no non-carbonate hardness in the water.

The $Mg^{++}$ ion content in $CaCO_3$ is equal to the $Mg(HCO_3)_2$ content as $CaCO_3$. Appendix A can be used to convert $CaCO_3$ equivalents to amounts as substance.

$$Mg^{++} = \frac{40}{4.1} = 9.6 \text{ mg/l as substance}$$

## 7 WATER QUALITY STANDARDS

Minimum drinking water quality standards have been set by the Safe Drinking Water Act. Typical minimum standards are given in table 7.9 as *Maximum Contaminant Levels (MCL's).* Values in table 7.9 are subject to change as new legislation is enacted.

**Table 7.9**
Typical Water Quality Standards
(subject to change)

| contaminant/quality | MCL |
|---|---|
| inorganic compounds | |
| arsenic | 0.05 mg/l |
| barium | 1.0 mg/l |
| cadmium | 0.01 mg/l |
| chloride | 250 mg/l |
| chromium | 0.05 mg/l |
| copper | 1.0 mg/l |
| cyanide | 0.005 mg/l |
| iron | 0.3 mg/l |
| lead | 0.05 mg/l |
| manganese | 0.05 mg/l |
| mercury | 0.002 mg/l |
| nitrate | 10 mg/l |
| selenium | 0.01 mg/l |
| silver | 0.05 mg/l |
| sulfate | 250 mg/l |
| zinc | 5.0 mg/l |
| organic compounds | |
| trihalomethanes (total) | 0.1 mg/l |
| organic pesticides | |
| endrin | 0.0002 mg/l |
| lindane | 0.004 mg/l |
| methoxychlor | 0.1 mg/l |
| toxaphene | 0.005 mg/l |
| 2,4-D | 0.1 mg/l |
| 2,4,5-TP(silvex) | 0.01 mg/l |
| miscellaneous regulations | |
| pH | 6.5–8.5 |
| turbidity | 1 NTU |
| color | 15 units |
| microbiological | 1 coliform/100 ml |
| total dissolved solids | 500 mg/l |
| odor (in threshold odor numbers) | 3 T.O.N. |

## 8 WATER DEMAND

Water demand comes from a number of sources, including residential, commercial, industrial, and public consumers, as well unavoidable loss and waste. In project planning, a minimum of about 165 gallons per capita-day should be considered. This 165 gpcd is a total of all demands, as given in table 7.10. If large industries are present (such as canning, steel making, automobile production, electronics, etc.), then those industries' special needs must also be considered.

For ordinary domestic use, the water pressure should be 25 to 40 psi. A minimum of 60 psi at the fire hydrant is usually adequate, since that allows for up to 20 psi pressure drop in fire hoses. 75 psi and higher is common in commercial and industrial districts.

**Table 7.10**
Annual Average Water Requirements (gpcd)

(Excluding fire fighting)

| | |
|---|---|
| residential | 75–130 |
| commercial & industrial | 70–100 |
| public | 10–20 |
| loss & waste | 10–20 |
| | 165–270 total |

Variations can be expected with the time of day and season. If the average daily demand is to be used to estimate peak demands or fluctuation, then table 7.11 lists some multipliers. These multipliers are to be used against the 165 gpcd (or whatever other average is available).

**Table 7.11**
Demand Multipliers For Peak Periods

| consumption time/period | multiplier |
|---|---|
| winter | 0.80 |
| summer | 1.30 |
| maximum daily | 1.50–1.80 |
| maximum hourly | 2.00–3.00 |
| early morning | 0.25–0.40 |
| noon | 1.50–2.0 |

Water demand (in gpcd) must be multiplied by the population to obtain the total demand. Since a population changes, a supply system must be designed to handle demands through a reasonable time into the future. Several methods exist for estimating future demand.

- Uniform growth rate
- Constant percentage growth rate (e.g., 40% per decade)
- Decreasing growth rate (e.g., 6% each decade)
- Straight-line graphical extension
- Comparison with neighboring cities

Typical growth factors are 1.25 for large systems and 1.50 for small systems. Economic aspects of the project dictate the number of years into the future which should be designed for.

Since the maximum demand can be up to 3 times the average daily demand (table 7.11), the design rate should be 3 times the average daily rate plus an allowance for fire fighting. If the water treatment plant's capacity is fixed, the distribution system should be able to handle the plant's capacity plus an allowance for fire fighting (which can be passed around the treatment plant).

The requirements for fire fighting at any point will vary between 500 gpm (a minimum) to 12,000 gpm for a single fire.[8] Multiple fires will place a greater demand on the distribution system. A municipality must continue to serve its domestic, commercial, and industrial customers during a fire, however. The Insurance Services Office recommends that the fire system be able to operate with the remainder of the potable water system operating at the maximum daily rate, as taken over all 24 hour periods within the last three years.

Recommended fire flow in a neighborhood will depend on construction type, occupancy, and floor area. An estimate for a neighborhood can be found from equation 7.25, as proposed by the Insurance Services Office.

$$\begin{aligned} Q &= 0.04C\sqrt{A} \text{ cfs} \\ &= 18C\sqrt{A} \text{ gpm} \end{aligned} \qquad 7.25$$

$C$ is a constant which depends on construction: 1.5 for wood frame, 1.0 for ordinary construction, 0.8 for noncombustible construction, and 0.6 for fire resistant construction. $A$ is the area (in square feet) of all stories in the building, except for basements. Special rules are used to find $A$ for multi-story fire-resistant structures, buildings with various fire loadings, or buildings with sprinkler systems. $Q$ is rounded to the nearest 250 gpm, but it should not be less than 500 gpm or more than 8000 gpm for a single building.

In estimating the water requirements for fire fighting on a population basis, the American Insurance Association has recommended the following formula:

$$\begin{aligned} Q &= 2.27\sqrt{P}(1 - 0.01\sqrt{P}) \text{ cfs} \\ &= 1020\sqrt{P}(1 - 0.01\sqrt{P}) \text{ gpm} \end{aligned} \qquad 7.26$$

**(P in thousands of people)**

Most insurance requirements will be met if the flow rate can be maintained for $T$ hours, where $T$ is the flow rate in 1000's of gallons per minute, with a maximum of 10 hours.

Fire hydrants are spaced at a maximum distance of 500 feet. They are ordinarily located at street corners where use from four directions is possible. The actual separation of hydrants can be calculated from standards presented by the Insurance Services Offices. These standards are presented in table 7.12.

The Insurance Services Office also suggests durations that various fire flow rates must be able to be maintained. Critical equipment such as power sources, pumps, supply mains, and treatment processes should be able to provide several days of peak load in addition to the flow for fire fighting for the duration specified.[9]

## 9 METHODS OF WATER DISTRIBUTION

Several methods are used to distribute water, depending on terrain, economics, and other local conditions.

*Gravity distribution* is available when a lake or reservoir is located significantly higher in elevation than the

[8] A standard 1.125″ smooth nozzle discharges approximately 250 gpm of water.

[9] It should be possible to maintain peak flow for 2–5 days, with the actual duration depending on the component, redundancy, and expected repair time.

**Table 7.12**
Standard Fire Hydrant Distribution

| fire flow required (gpm) | minimum average area per hydrant (sq ft) |
|---|---|
| 1000 or less | 160,000 |
| 1500 | 150,000 |
| 2000 | 140,000 |
| 2500 | 130,000 |
| 3000 | 120,000 |
| 3500 | 110,000 |
| 4000 | 100,000 |
| 4500 | 95,000 |
| 5000 | 90,000 |
| 5500 | 85,000 |
| 6000 | 80,000 |
| 6500 | 75,000 |
| 7000 | 70,000 |
| 7500 | 65,000 |
| 8000 | 60,000 |
| 8500 | 57,500 |
| 9000 | 55,000 |
| 10,000 | 50,000 |
| 11,000 | 45,000 |
| 12,000 | 40,000 |

**Table 7.13**
Required Durations for Fire Fighting

| required fire flow (gpm) | required duration (hr) |
|---|---|
| 10,000 and greater | 10 |
| 9500 | 9 |
| 9000 | 9 |
| 8500 | 8 |
| 8000 | 8 |
| 7500 | 7 |
| 7000 | 7 |
| 6500 | 6 |
| 6000 | 6 |
| 5500 | 5 |
| 5000 | 5 |
| 4500 | 4 |
| 4000 | 4 |
| 3500 | 3 |
| 3000 | 3 |
| 2500 or less | 2 |



city. The high hydraulic head that results can be easily maintained for domestic and fire fighting mains.

Distribution by pumping with storage is the most desirable option if gravity distribution is not used. Excess water is pumped into elevated storage during periods of low consumption. During periods of high consumption, water is drawn from the storage to augment the pumped water. More uniform pumping rates result, and the pumps are able to run near their rated capacity most of the time.

Using pumps without storage to force water directly into mains is the least desirable option. Pumps must be able to maintain the flow during peak consumption, and pumps will not always run in their economical capacity ranges. In the event of a power outage, all water supply will be lost unless backup power is available.

Water mains in a distribution system should be at least 6 inches in diameter.

## 10 STORAGE OF WATER

Water is stored to equalize pumping rates, to equalize supply and demand over periods of high consumption, and to furnish extraordinary volumes during emergencies such as fires.

To equalize the pumping rate during the day will ordinarily require storage of 15 to 30 percent of the maximum daily use. Storage for emergencies is more difficult to determine, and is dictated by economic benefits to the public. Fire insurance rates are generally lower the greater the emergency storage capacity is.

## 11 PIPE MATERIALS

Many types of pipe are available. Successful pipe materials used for distribution must have adequate strength to withstand external loads from backfill, traffic, and earth movement; high burst strength to withstand water pressure; smooth interior surfaces; corrosion resistant exteriors; and tight joints.

The types of pipes listed in table 7.14 are suitable for use in the water distribution system.

## 12 LOADS ON BURIED PIPES

If a pipe is buried (placed in an excavated trench and backfilled), it must support an external vertical load in addition to its internal pressure load. The magnitude of the load depends on the amount of backfill, type of soil, and type of pipe. For rigid pipes (concrete and cast iron) which cannot deform and which are placed in narrow trenches (2 or 3 diameters), the load in pounds per foot pipe is given by equation 7.27.[10]

$$w = C\gamma B^2 \qquad 7.27$$

[10] Equation 7.27 is known as *Marston's formula*.

**Table 7.14**
Water Pipe Materials

| type | comments |
|---|---|
| ductile and gray cast iron | Long life, strong, impervious. High cost and heavy. May be coated to resist exterior and interior corrosion. Available in 4″ to 54″ standard sizes. 350 psi working pressure. |
| asbestos-cement | Immune to electrolysis and corrosion. Low flexural strength. Smooth interior surface. Available 4″ to 42″ diameter. Up to 200 psi working pressure. |
| concrete | Durable, watertight, low maintenance, smooth interior surface. Diameters 16″ to 144″, 50 psi (plain) and 250 psi (reinforced) working pressures. |
| steel | High strength, good yielding and shock resistance, but susceptible to corrosion. Exterior may be tarred, painted, or wrapped. Interior may have enamel or cement mortar lining. Smooth interior surface. 16″ to 120″ diameters, 250 psi pressures. |
| plastic | Chemically inert, corrosion resistant, smooth interior. PVC most popular. PVC available in rating to 315 psi, diameters 1/2″ to 16″. |

Typical values of $C$ and $\gamma$ are given in table 7.15.[11] $B$ is the trench width in feet at the top of the pipe. (A minimum trench width to allow working room is commonly estimated at 4/3 times the pipe diameter plus 8″.)

[11] There is considerable literature on the coefficients used in equations 7.27, 7.29, and 7.30. The values listed in this chapter are representative, but are not intended to cover every case. In most instances, other factors may be necessary to correctly select the coefficients.

The dead load pressure is simply the experienced load divided by the trench width.

$$p_{DL} = \frac{w}{B} \qquad 7.28$$

*Flexible pipes* (steel, plastic, copper) are sufficiently flexible to develop horizontal restraining pressures equal to the vertical pressures if the backfill is well compacted.

$$w = C\gamma BD \qquad 7.29$$

**Table 7.15**
Approximate Pipe Load Correction Coefficients

| backfill material | cohesionless granular material | sand & gravel | saturated topsoil | clay | saturated clay |
|---|---|---|---|---|---|
| density (pcf) | 100 | 100 | 100 | 120 | 130 |
| h/B | values of C | | | | |
| 1 | 0.82 | 0.84 | 0.86 | 0.88 | 0.90 |
| 2 | 1.40 | 1.45 | 1.50 | 1.55 | 1.62 |
| 3 | 1.80 | 1.90 | 2.00 | 2.10 | 2.20 |
| 4 | 2.05 | 2.22 | 2.33 | 2.49 | 2.65 |
| 5 | 2.20 | 2.45 | 2.60 | 2.80 | 3.03 |
| 6 | 2.35 | 2.60 | 2.78 | 3.04 | 3.33 |
| 7 | 2.45 | 2.75 | 2.95 | 3.23 | 3.57 |
| 8 | 2.50 | 2.80 | 3.03 | 3.37 | 3.76 |
| 10 | 2.55 | 2.92 | 3.17 | 3.56 | 4.04 |
| 12 | 2.60 | 2.97 | 3.24 | 3.68 | 4.22 |
| ∞ | 2.60 | 3.00 | 3.25 | 3.80 | 4.60 |

If a pipe is placed on undisturbed ground and covered with fill (*broad fill* or *embankment fill*) the load is

$$w = C_p \gamma D^2 \tag{7.30}$$

**Table 7.16**
Representative Values of $C_p$

| h/D | rigid pipe, rigid surface, noncohesive backfill | flexible pipe, average conditions |
|---|---|---|
| 1 | 1.2 | 1.1 |
| 2 | 2.8 | 2.6 |
| 3 | 4.7 | 4.0 |
| 4 | 6.7 | 5.4 |
| 6 | 11.0 | 8.2 |
| 8 | 16.0 | 11.0 |

Equation 7.27 shows that the trench width is an important parameter in determining whether or not the pipe is overloaded. There is a depth for each conduit size beyond which no additional load is transmitted to the conduit, regardless of trench width. This limiting value is known as the *transition width.* Specifically, the load on a rigid pipe (e.g., cast iron, concrete, ductile iron) can never exceed the value calculated from equation 7.30. The transition width can be calculated by equating equations 7.27 and 7.30.

$$B_{\text{transition}} = D\sqrt{\frac{C_p}{C}} \tag{7.31}$$

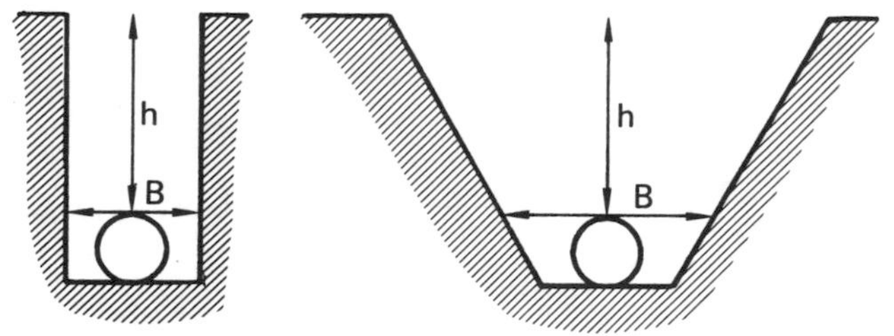


**Figure 7.2** Backfilled Trenches

*Boussinesq's equation* should be used to calculate the load on a pipe due to a superimposed line load, $P$, at the surface (figure 7.3). This load should be added to the loadings calculated from equations 7.27, 7.29, and 7.30.

$$p = \frac{3h^3P}{2\pi z^5} = \frac{3P}{2\pi h^2}\left[\frac{1}{1+\left(\frac{r}{h}\right)^2}\right]^{5/2} \tag{7.32}$$

$$w = Dp \tag{7.33}$$

If the pipe has less than 3 feet of cover, a multiplicative impact factor should also be used.

$$w = F_I D p \tag{7.34}$$

**Table 7.17**
Impact Factors

| | |
|---|---|
| BY DEPTH (general use) | |
| less than 1 foot | 1.3 |
| 1 to 2 feet | 1.2 |
| 2 to 3 feet | 1.1 |
| more than 3 feet | 1.00 |
| BY USE | |
| highway | 1.50 |
| railway | 1.75 |
| airfield runway | 1.00 |
| airfield taxiway, apron | 1.50 |

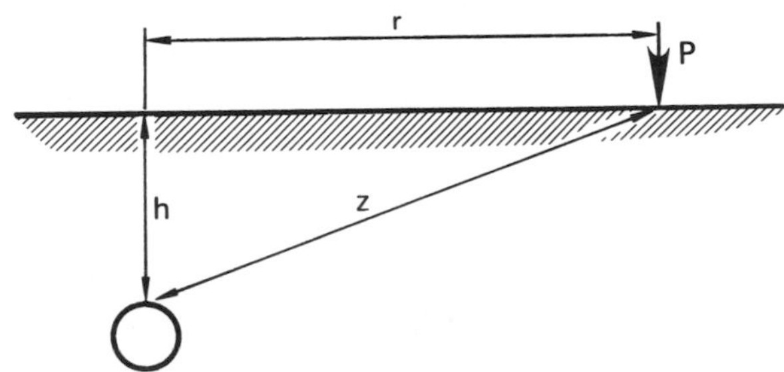


**Figure 7.3** External Loads on Buried Pipes

Concrete pipes are tested in a 3-edge bearing mechanism, as shown in figure 7.4, to determine the *crushing strength.* Crushing load (taken as the load which produces a 0.01″ crack) is given in pounds per foot of pipe per foot of inside diameter. Therefore, crushing strength (in pounds per linear foot of pipe) is

$$\text{crushing strength} = (\text{load per unit diameter})(\text{diameter}) \tag{7.35}$$

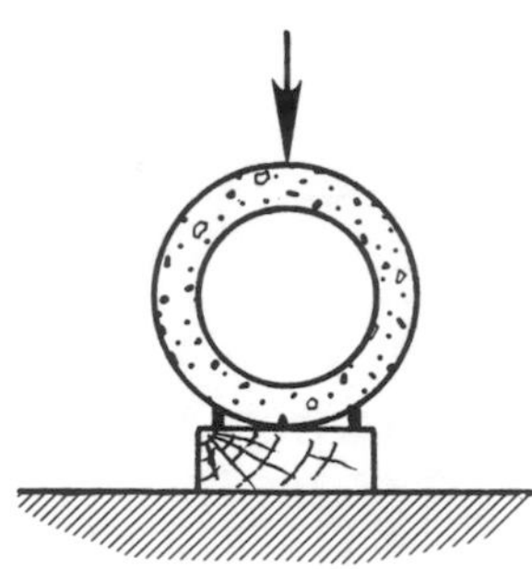

**Figure 7.4** Three-Edge Bearing Test

The load on the pipe can be decreased below the crushing strength by proper bedding. Different bedding methods and their load factors are given in figure 7.5. The allowable (safe) load per foot of pipe is given by equation 7.36. The factor of safety varies from 1.25 for flexible pipe to 1.50 for rigid pipe. For reinforced concrete, the factor of safety is 1.5.

$$\text{allowable load per foot of pipe} = \frac{(\text{crushing strength})(\text{load factor, LF})}{\text{factor of safety}} \tag{7.36}$$

## 13 WATER SUPPLY TREATMENT METHODS

### A. AERATION

Aeration can be used where there is a high concentration of carbon dioxide, where tastes and odors are objectionable, and where iron and manganese are present in amounts above 0.3 ppm.

Typical aerating devices are listed in table 7.18. Greatest efficiencies can be achieved by designing for increased water surface exposed to air, rapid change of air in contact with the water, and increased aeration periods.

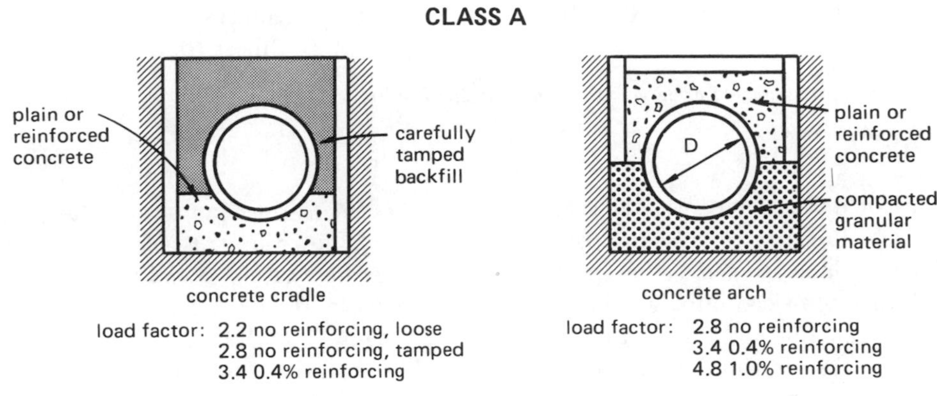


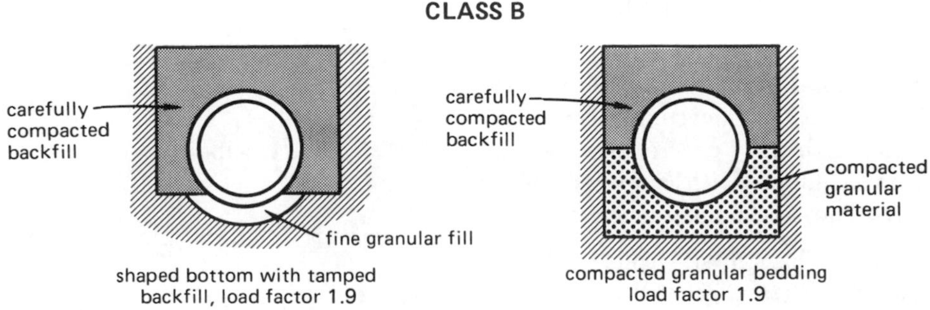


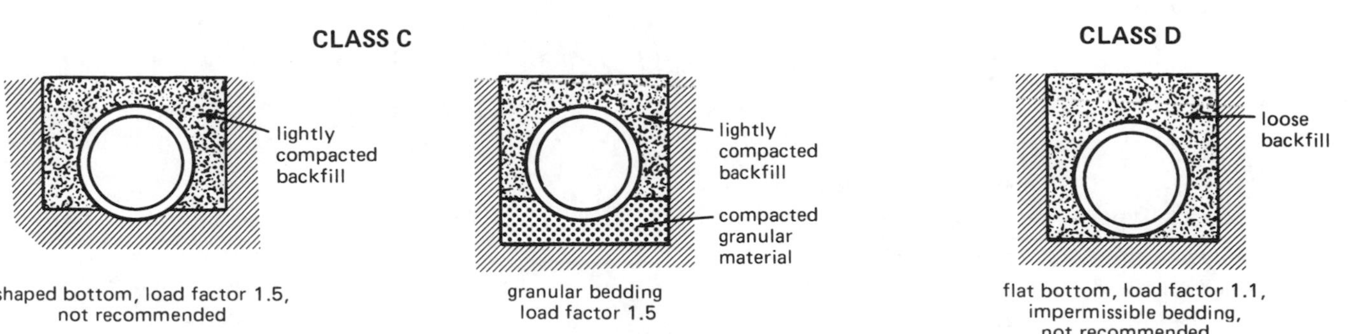


**Figure 7.5** Bedding Classes and Load Factors

**Table 7.18**
Characteristics of Aerators

| type | operating head (ft) | loading capacity | efficiency in $CO_2$ removal | remarks |
|---|---|---|---|---|
| spray | 8–28 | 4–180 gpm nozzle | high | requires protection from loss of water by wind; ice hazard in cold climates |
| cascade | 3–10 | 20–50 gpm/sq ft | low to | requires large space |
| perforated tray | | | fair | requires larger space and higher head than coke tray |
| coke tray | 6–10 | <35 gpm sq ft | high | used also for iron and manganese removal |
| forced draft | 10–25 | 16–18 gpm sq ft | high | compact but more complex than above types |
| diffused air | 5–10 psi[1] | 0.02–0.2 cfm/gpm[2] | high | requires compressed air; most complex |

[1]Air pressure depends upon water depth and pipe friction losses.

[2]Air requirement.

Transfer efficiencies of diffused air systems vary with depth and bubble size. If coarse bubbles are produced, only 4% to 8% of the available oxygen will be transferred to the water. With medium-sized bubbles, the *transfer efficiency* varies between 6% and 15%. Fine bubble systems are capable of transferring 10% to 30% of the supplied oxygen to the water.

## B. PLAIN SEDIMENTATION (CLARIFICATION)

Water contaminated with sand, dirt, mud, etc., can be treated in a sedimentation basin or tank. Up to 80% of the incoming sediment can be removed in this manner. Sedimentation basins are usually concrete, rectangular or circular in plan, and equipped with scrapers or raking arms to periodically remove accumulated sludge.

Settlement of water-borne particles depends on the water temperature (which affects viscosity), particle size, and particle specific gravity. Typical specific gravities are given in table 7.19.

**Table 7.19**
Properties of Particles

| particle type | specific gravity | particle size, mm |
|---|---|---|
| coarse sand | 2.65 | 0.50–2.00 |
| medium sand | 2.65 | 0.25–0.50 |
| fine sand | 2.65 | 0.10–0.25 |
| very fine sand | 2.65 | 0.05–0.10 |
| silt | 2.65 | 0.005–0.05 |
| clay | – | 0.001–0.005 |
| flocculated mud | 1.03 | – |

Settlement time can be calculated from the settling velocity, flow-through velocity, and depth of the tank. Tank depths are typically 6 to 15 feet. The settling velocities for spherical particles in 68°F water are given in figure 7.6. Of course, settling velocities will be much less than those shown in figure 7.6 because actual sediment particles are not spherical.

If it is necessary to calculate the settling velocity of a particle of diameter $d$ (in feet), the following procedure can be used.

*step 1*: Assume $v_s$.

*step 2*: Calculate the Reynolds number.

$$N_{Re} = \frac{v_s d}{\nu} \tag{7.37}$$

*step 3*: If $N_{Re} < 1$, use Stoke's law.

$$v_s = \frac{(\rho_{\text{particle}} - \rho_{\text{water}})d^2 g}{18\mu g_c} = \frac{(\text{SG}_{\text{particle}} - 1)d^2 g}{18\nu} \tag{7.38}$$

If $1 < N_{Re} < 2000$, use figure 7.6.

If $N_{Re} > 2000$, use Newton's law.

$$v_s = \sqrt{\frac{4g(\rho_{\text{particle}} - \rho_{\text{water}})d}{3(\rho_{\text{water}})C_D}} \tag{7.39}$$

Values of $C_D$ are given in table 7.20.

WATER SUPPLY

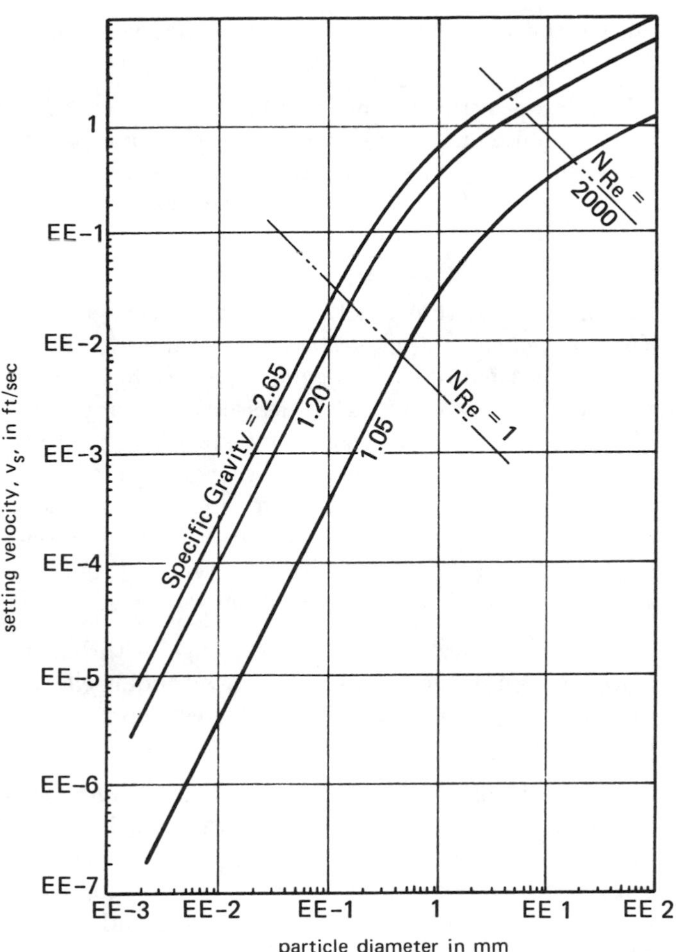


**Figure 7.6** Settling Velocities (spherical particles, 68°F water)

**Table 7.20**
Approximate Drag Coefficients for Spheres

| $N_{Re}$ | $C_D$ |
|---|---|
| 2000 | 0.4 |
| 10,000 | 0.4 |
| 50,000 | 0.5 |
| 100,000 | 0.5 |
| 200,000 | 0.4 |

If it is assumed that the water velocity is a uniform $v_f$, then all particles with $v_s > v^*$ will be removed. $v^*$ is known as the *overflow rate* (*surface loading* or *critical velocity*) and typically has a value of 600 to 1000 gpd/ft$^2$ for rectangular basins. For square and circular basins, the surface loading should be within 500–750 gpd/ft$^2$. $B$ is the tank width, typically 30 to 40 feet, and $L$ is the length, typically 100 to 200 feet. (For radial flow circular basins, a typical diameter is 100 feet.)

$$v^* = \frac{Q}{A_{\text{surface}}} = \frac{Q}{BL} \qquad 7.40$$

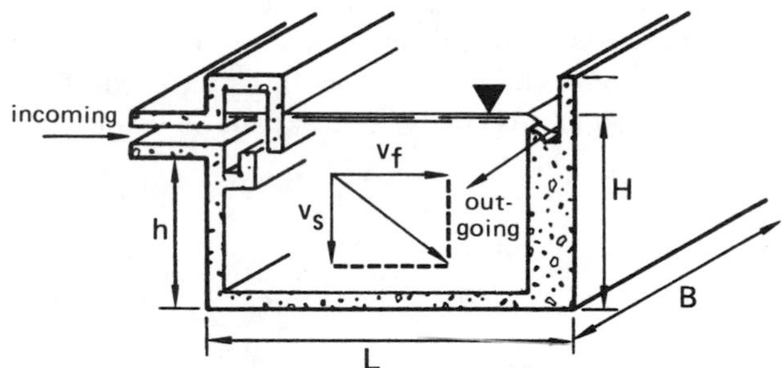


**Figure 7.7** Rectangular Settling Basin

If water enters at some level other than the surface, such as at level $h$ in figure 7.7, all particles will be removed that have

$$v_s > \frac{hQ}{HBL} \qquad 7.41$$

Rectangular basins are preferred, even if sludge removal means emptying and taking them out of service. Rectangular basins should be constructed with aspect ratios of greater than 3:1, and preferably in excess of 4:1. Slope the bottom toward the drain at no less than 1 percent. Use multiple inlets along the entire inlet wall, if possible. If fewer than four ports are used, an inlet baffle should be provided.

In the case of square or circular basins, the slope toward the drain should be greater, typically on the order of 1 : 12. A baffled center inlet should be provided. Square and circular basins are appropriate when space is limited. Otherwise, rectangular basins or solid contact units are preferred.

In using equation 7.40, divide the total flow to be treated into at least two basins. That way, one basin can be out of service for cleaning without interruption of operation.

Basins should be constructed from concrete for all permanent installations. Steel should be used only for small or temporary installations. Where steel parts are unavoidable, as in the case of internal parts of rotors, adequate corrosion resistance is necessary.

The time spent by water in the basin is known as the *detention* (or *retention*) *time* (or *period*). The detention time is given by equation 7.42. A minimum time recommended is 3–4 hours, although periods from 1 to 10 hours are used.[12]

$$t = \frac{\text{tank volume}}{Q} \qquad 7.42$$

[12] Long detention times, up to 12 hours, are required to remove fine particles.

The *weir loading* is the daily flow rate divided by the total effluent weir length, usually expressed in gallons/ft-day. A recommended weir loading is 15,000–20,000 gpd/ft, but should certainly not exceed 50,000 gpd/ft.

The *basin efficiency* is

$$\eta = \frac{\text{flow-through period}}{\text{detention time}} \qquad 7.43$$

The *flow-through period* is found by using color dye in the basin. The flow velocity should not exceed 1 fpm.

*Sludge* needs to be removed according to the demand placed on the basin. Removal is called for when the sludge has reached approximately 25 mg/l or is organic. Various methods of removing sludge are available, including scrapers and pumps. A linear velocity less than 15 fpm should be used with sludge scrapers.

## C. MIXING AND FLOCCULATION

Chemicals can be added to obtain a desired water quality. These chemicals are added to the water in mixing basins. There are two types of mixing basins: complete mixing and plug flow mixing.

*Complete mixing basins* dispense the chemical immediately throughout the volume by using mixing paddles. If the volume of water being mixed is small, the tank is known as a *flash* or *quick mixer*. Quick mixing detention time is often less than 60 seconds.

The retention time required for complete mixing in a tank of volume $V$ depends on the rate constant, $K$, and is given by equation 7.44.

$$t = \frac{V}{Q} = \frac{1}{K}\left(\frac{C_i}{C_o} - 1\right) \qquad 7.44$$

*Plug flow mixing* is accomplished by allowing the water and additive to flow through a long chamber at a uniform rate without mechanical agitation. The retention time in a plug flow mixer of length $L$ is given by equation 7.45.

$$t = \frac{V}{Q} = \frac{L}{v} = \frac{1}{K} ln(C_i/C_o) \qquad 7.45$$

If the mixer adds coagulant for the removal of colloidal sediment, it is known as a *flocculator*. *Floc* is a precipitate that forms when the coagulant allows the colloidal particles to agglomerate. Flocculation is enhanced by gentle agitation, but the floc disintegrates with violent agitation.

Common coagulants among the *hydrolyzing metal ions* are aluminium sulfate ('alum,' $Al_2(SO_4)_3 \cdot 14H_2O$), ferrous sulfate ('copperas,' $FeSO_4 \cdot 7H_20$), ferric chloride ($FeCl_3$), or chlorinated copperas (a mixture of ferrous sulfate and ferric chloride). The most common coagulant is alum. The usual dosage is 10 to 40 mg/l.

Alum reacts with alkalinity in the incoming water to form an aluminum hydroxide floc. If the water is not sufficiently alkaline, an auxiliary chemical such as CaO (lime) or $Na_2CO_3$ (soda ash) is used along with the alum.[13] Lime is also used as an auxiliary chemical with $FeSO_4$ coagulant.

*Polymers* constitute another group of coagulants. *Organic polymers* (*polyelectrolytes*) such as starches and polysaccharides and *synthetic polymers* such as polysacylimides are used. Polymers are chains or groups of repeating identical molecules (*mers*) with many available active adsorption sites. Their molecular weights range from a few hundred to a few million. Polymers can be positively charged (*cationic polymers*), negatively charged (*anionic polymers*), or neutral (*nonionic polymers*). The charge can vary with water pH.

Polymers are effective in a narrow range of turbidity. Turbidity can be increased artificially by adding clay or alum (to produce $Al(OH)_3$ floc). Polymers also require an incoming water with alkalinity.

Another category of chemicals are *flocculation additives.* These additives improve the coagulation efficiency by increasing or decreasing the floc size. *Weighting agents* (e.g., bentonite clays), *adsorbents* (e.g., activated carbon), and *oxidants* (e.g., chlorine, ozone, and potassium permanganate) are used. Polymers are also used in conjunction with metallic ion coagulants.

After flash mixing, a 20- to 60-minute period of gentle mixing is used to permit flocculation. During this period, the flow-through velocity should be between 0.5 and 1.5 ft/min. The peripheral speed of mixing paddles should range from 0.5 ft/sec for fragile, cold water floc to 3.0 ft/sec for tough, warm water floc. The flocculation is followed by sedimentation for 2 to 8 hours in a low-velocity basin.

The *drag force* on a paddle is given by the standard fluid drag force equation. For flat plates, $C_D \approx 1.8$.

$$D = \frac{C_D A \gamma v^2}{2g} \qquad 7.46$$

The power requirement is easily calculated from the drag force and mixing velocity. (The *mixing velocity*, also known as the *relative paddle velocity*, is approximately 0.7 to 0.8 times the tip speed.)

$$P = Dv \qquad 7.47$$

$$\text{horsepower} = \frac{P}{550} \qquad 7.48$$

[13] One mg/l of alum requires $\frac{1}{2}$ mg/l of alkalinity.

WATER SUPPLY

The *mean velocity gradient*, $G$, varies from 20 to 75 $\text{sec}^{-1}$ for a 15 to 30 minute mixing period. However, $G$ is often used in conjunction with the power requirement and detention time.[14]

$$G = \sqrt{\frac{P}{\mu V_{\text{tank}}}} \qquad 7.49$$

$$P = \mu G^2 V_{\text{tank}} \qquad 7.50$$

$$Gt_d = \frac{V_{\text{tank}}}{Q}\sqrt{\frac{P}{\mu V_{\text{tank}}}} = \frac{1}{Q}\sqrt{\frac{PV_{\text{tank}}}{\mu}} \qquad 7.51$$

Typical values of $Gt_d$ range from $10^4$ to $10^5$. $Gt_d$ is the *mixing opportunity parameter.*

## D. CLARIFICATION WITH FLOCCULATION

The *flocculation clarifier* combines mixing, flocculation, and sedimentation into a single tank. These units are called *solid contact units* and *upflow tanks.* They are generally round in construction, with mixing and flocculation taking place near the central hub, and sedimentation occurring at the periphery. Flocculation clarifiers are most suitable when combined with softening since the precipitated solids help seed the floc.

The following general guidelines can be used to design a flocculation clarifier.

- minimum flocculation and mixing time: 30 minutes
- minimum retention time: 1.5 hours (2.0 hours typical)
- maximum weir loading: 10 gpm/ft
- upflow rate: 0.8–1.7 gpm/$\text{ft}^2$ (1.0 typical)
- maximum sludge formation rate: 5% of water flow

[14] For mixing units, such as rapid-mix flash units, which do not have to preserve the integrity of floc, a mean velocity gradient in the range of 1000–3000 1/sec can be achieved. For gentle mixing in a flocculator, however, the mean velocity gradient must be much less.

## E. FILTRATION

Nonsettling floc can be removed by filtering. The *rapid sand (gravity) filter* is the most common filter for this use. Rapid sand filters are essentially beds of granulated gravel and sand. Although the box depth can be 10 feet, the sand bed will be only about 3 feet deep. Since the water has been coagulant treated, it may be passed through the filter quickly—hence the name 'rapid.' Rapid filters are usually square or nearly square in design, and operate with a one to eight foot hydraulic head.

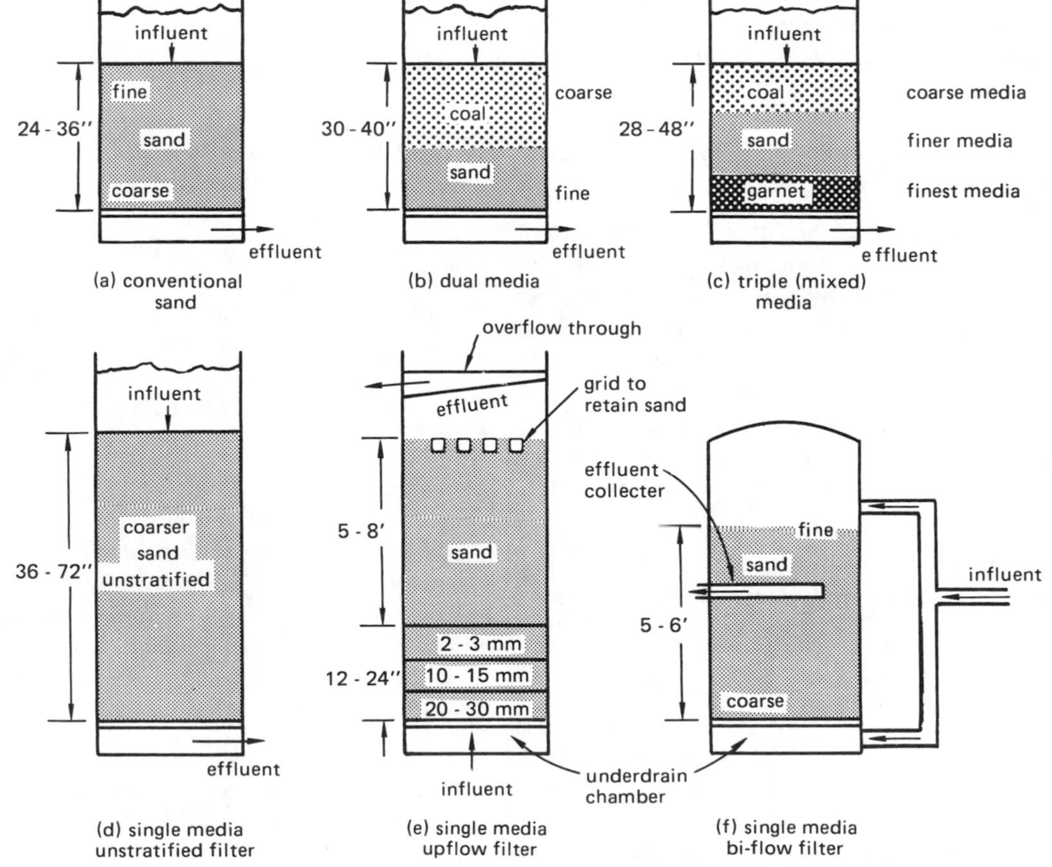


**Figure 7.8** Granular Media Filters

Optimum filter operation will occur when the top layer of sand is slightly more coarse than the rest of the sand. During backwashing, however, the finest sand rises to the top. Various designs using coal and garnet layers in conjunction with sand layers (i.e., *dual-* and *triple-media* filters) overcome this difficulty due to the differences in specific gravity.

Historically, the flow rate has been 2 gpm/ft$^2$ in rapid sand filter design, although some current filters operate at 8 gpm/ft$^2$. 4 gpm/ft$^2$ is a reasonable rate for modern designs.

A water treatment plant should have at least three filters so that two can be in operation when one is being cleaned. Typical total through-puts per filter range from 350 gpm for small plants to 3500 gpm for large plants.

Optimum design of a filtering system includes discharge into a *clearwell.* Clearwells are storage reservoirs with capacities of 30% to 60% of the daily output with a minimum of 12 hours of maximum daily consumption. Demand can be satisfied by the clearwell if one or more of the filters is serviced.

The most common type of service needed by filters is *backwashing.* Filters require backwashing when the pores between sand particles clog up. Typically, this occurs after 1 to 3 days of operation. Backwashing is done when the head loss through the filter bed reaches approximately 8 feet. Backwashing with filtered water expands the sand layer up to 50%, which dislodges the trapped material. Backwashing for 3 to 5 minutes at 8–15 gpm/ft$^2$ is a reasonable design standard. The head loss is reduced to 1 foot after washing.

Water is pumped through the filter from the bottom during backwashing. The rate at which the water rises in the filter housing varies between 12 and 36 inches per minute. This rise rate should not exceed the settling velocity of the smallest particle which is to be retained in the filter. Backwashing usually takes between 3 and 5 minutes. Water which is collected in troughs for disposal and used for backwashing, constitutes between 1% and 5% of the total processed water (approximately 75–100 gal/ft$^2$ total).

The actual amount of backwash water can be found from equation 7.52. Be sure to use consistent units.

$$\text{water volume} = \begin{pmatrix}\text{backwash}\\ \text{time}\end{pmatrix}\begin{pmatrix}\text{filter}\\ \text{area}\end{pmatrix}\begin{pmatrix}\text{rise}\\ \text{rate}\end{pmatrix} \qquad 7.52$$

*Slow sand filters* are similar in design to rapid sand filters, except that the sand layer is thicker (24″ to 48″), the gravel layer is thinner (6″ to 12″), and the flow rate is much lower (0.05 to 0.1 gpm/ft$^2$). Slow sand filters are limited to low-turbidity applications not requiring chemical treatment. Cleaning is usually accomplished by removing a few inches of sand. Slow sand filters operate with a 0.2 to 4.0 foot head loss.

Other types of filters are seeing limitea use.

- **Pressure filters**: Similar to rapid sand filters except incoming water is pressurized up to 25 feet (hydraulic). Filter rates of 2 to 4 gpm/ft$^2$. Not used in large installations.
- **Diatomaceous earth filters**: 1 to 3 gpm/ft$^2$; short (20-hour) cycle life.
- **Microstrainers**: Woven stainless steel fabric, usually mounted on a rotating drum.

## F. DISINFECTION

Chlorination is used for disinfection and oxidation. As a disinfectant, chlorine destroys bacteria and microorganisms. As an oxidant, it removes iron, manganese, and ammonia nitrogen.

Chlorine can be added as a gas or a solid. (If it is added to the water as a gas, it is stored as a liquid which vaporizes around −35°C.) Liquid chlorine is the predominant form since it is cheaper than hypochlorite solid ($Ca(OCl)_2$). If chlorine liquid or gas is added to water, the following reaction occurs to form hypochlorous acid, which itself ionizes to hypochlorite and hydrogen ions.

$$Cl_2 + H_2O \rightarrow HCl + HOCl \underset{pH<7}{\overset{pH>8}{\rightleftharpoons}} H^+ + OCL^- + HCL \qquad 7.53$$

If calcium hypochlorite solid is added to water, the ionization follows immediately.

$$Ca(OCl)_2 + H_2O \longrightarrow Ca^{++} + 2OCl^- + H_2O \qquad 7.54$$

Chlorine existing in water as hypochlorous acid and hypochlorite ions is known as *free available chlorine* (*free residuals*). Chlorine in combination with ammonia is known as *combined available chlorine* (*combined residuals*).

The average chlorine dose is in the 1 to 2 mg/l range. Minimum chlorine residuals for 70°F water are given in table 7.21. Reliable chlorination requires a pH of water below 9.0. However, inactivated viruses (such as might be present in surface water) require a heavier chlorine concentration. Since treatment of water is by both free and combined residuals, ammonia can be added to the water to produce chloramines.

Excess chlorine can be removed with a reducing agent, usually called a *dechlor.* Sulfur dioxide and sodium bisulfate (sodium metabisulfate) are used in this manner. Aeration also reduces chlorine content, as does passing the water through an activated charcoal filter.

**Table 7.21**
Minimum Chlorine Residuals (mg/l)

| pH Value | Free residuals after 10–20 minutes | Combined residuals after 1–2 hours |
|---|---|---|
| 6.0 | 0.2 | 1.0 |
| 7.0 | 0.2 | 1.5 |
| 8.0 | 0.4 | 1.8 |
| 9.0 | 0.8 | 3.0* |
| 10.0 | 0.8 | 3.0* |

*not recommended

Alternatives to chlorination have become popular since THM's were traced to the chlorination process. Chlorine dioxide can be used in place of chlorine, but it is expensive. In high dosages, chlorine dioxide is also thought to produce its own toxic byproducts. Ozone is a more powerful disinfectant than chlorine, but it is expensive to generate and requires costly contact chambers. Ozone, which is used extensively in Western Europe, Canada, the USSR, and Japan, is generated onsite by running high voltage electrical currents through dry air or pure oxygen.

## G. FLUORIDATION

Fluoridation can occur any time after filtering. Smaller utilities almost always choose liquid solution and a volumetric feeding mechanism, with solutions being manually prepared. Larger utilities use gravimetric dry feeders with sodium silicofluoride or solution feeders with fluorsilic acid. The characteristics and dose rates of common fluorine compounds are given in table 7.22.

**Table 7.22**
Dose Rates for Fluorine Compounds

| Formula | $H_2SiF_6$ | NaF | $Na_2SiF_6$ |
|---|---|---|---|
| Form | liquid | solid | solid |
| Typical purity | 22–30% | 90–98% | 98–99% |
| Dose to obtain 1.0 mg/l (in pounds per million gallons) | 35.2 (with 30% purity) | 18.8 (with 98% purity) | 14.0 (with 98.5% purity) |

*Defluoridation* (required if the fluoride exceeds 1.5 mg/l) can be achieved with calcined alumina or bone char (tricalcium phosphate). Softening using lime can also be used when waters contain smaller amounts of fluoride. Each 45 to 65 mg/l reduction in magnesium will result in a 1.0 mg/l reduction in fluoride.

## H. IRON AND MANGANESE REMOVAL

Several methods of removing iron exist. (Manganese is not easily removed by aeration alone. However, the remaining methods work.)

- Aeration, followed by sedimentation and filtration

$$Fe^{++} + O_x \rightarrow FeO_x \downarrow \qquad 7.55$$

- Aeration, followed by chemical oxidation, sedimentation, and filtration. Chlorine or potassium permanganate may be used as an oxidizer.
- Manganese zeolite process: Manganese dioxide removes soluble iron ions.
- Lime water softening

Table 7.23 lists the characteristics of iron and manganese removal processes.

## I. WATER SOFTENING

- Lime and Soda Ash Softening

Water softening can be accomplished with lime and soda ash to precipitate calcium and magnesium ions from the solution. Lime treatment has added benefits of disinfection, iron removal, and clarification.

*Lime* (CaO) is available as *granular quicklime* (minimum purity: 90% CaO) or *hydrated lime*, $Ca(OH)_2$. *Quicklime* is slaked prior to use, which means that water is added to form a lime slurry in an exothermic reaction.

$$CaO + H_2O \rightarrow Ca(OH)_2 + \text{heat} \qquad 7.56$$

The hydrated lime is delivered to the water supply as a suspension (i.e., *milk of lime*).

*Soda ash* is usually available as 98% pure sodium carbonate ($Na_2CO_3$).

**FIRST STAGE TREATMENT**: In the first stage treatment, lime added to water reacts with free carbon dioxide to form calcium carbonate precipitate.

$$CO_2 + Ca(OH)_2 \rightarrow CaCO_3 \downarrow + H_2O \qquad 7.57$$

Next, the lime reacts with calcium bicarbonate.

$$Ca(HCO_3)_2 + Ca(OH)_2 \rightarrow 2CaCO_3 \downarrow + 2H_2O \qquad 7.58$$

Any magnesium hardness is also removed at this time.

$$Mg(HCO_3)_2 + Ca(OH)_2 \rightarrow CaCO_3 \downarrow + 2H_2O + MgCO_3 \qquad 7.59$$

To remove the soluble $MgCO_3$, the pH must be above 10.8. This is accomplished by adding an excess of approximately 35 mg/l of CaO or 50 mg/l of $Ca(OH)_2$ plus lime to satisfy equation 7.60.

$$MgCO_3 + Ca(OH)_2 \rightarrow CaCO_3 \downarrow + Mg(OH)_2 \downarrow \qquad 7.60$$

**Table 7.23**
Iron/Manganese Removal Processes

| processes | iron and/or manganese removed | pH required | remarks |
|---|---|---|---|
| aeration, settling, and filtration | ferrous bicarbonate | 7.5 | provide aeration unless incoming water contains adequate dissolved oxygen |
| | ferrous sulfate | 8.0 | |
| | manganous bicarbonate | 10.3 | |
| | manganous sulfate | 10.0 | |
| aeration, free residual chlorination, settling, and filtration | ferrous bicarbonate | 5.0 | provide aeration unless incoming water contains adequate dissolved oxygen |
| | manganous bicarbonate | 9.0 | |
| aeration, lime softening, settling, and filtration | ferrous bicarbonate | 8.5–9.6 | |
| | manganous bicarbonate | | |
| aeration, coagulation, lime softening, settling, and filtration | colloidal or organic iron | 8.5–9.6 | require lime, and alum or iron coagulant |
| | colloidal or organic manganese | 10.0 | |
| ion exchange | ferrous bicarbonate | 6.5± | water must be devoid of oxygen |
| | manganous bicarbonate | | iron and manganese in raw water not to exceed 2.0 mg/l |
| | | | consult manufacturers for type of ion exchange resin to be used |

**FIRST STAGE RECARBONATION**: Lime added to precipitate hardness removes itself. This is desirable because any calcium that remains in the water has the potential for forming scale. Further stabilization can be achieved by *recarbonation* (treatment with carbon dioxide).

$$Ca(OH)_2 + CO_2 \rightarrow CaCO_3 \downarrow + H_2O \qquad 7.61$$

Excess recarbonation should be avoided. If the pH is allowed to drop below 9.5, then carbonate hardness reappears.

$$CaCO_3 + CO_2 + H_2O \rightarrow Ca(HCO_3)_2 \qquad 7.62$$

At this time, any unsettled $Mg(OH)_2$ can be returned to a soluble state.

$$2Mg(OH)_2 + 2CO_2 \rightarrow 2MgCO_3 + 2H_2O \qquad 7.63$$



**SECOND STAGE TREATMENT**: The second stage treatment removes calcium noncarbonate hardness (sulfates and chlorides) which needs soda ash for precipitation.

$$CaSO_4 + Na_2CO_3 \rightarrow CaCO_3 \downarrow + Na_2SO_4 \qquad 7.64$$

Magnesium noncarbonate hardness needs both lime and soda ash.

$$MgSO_4 + Ca(OH)_2 \rightarrow Mg(OH)_2 \downarrow + CaSO_4 \qquad 7.65$$

$$CaSO_4 + Na_2CO_3 \rightarrow CaCO_3 \downarrow + Na_2SO_4 \qquad 7.66$$

Excess soda ash leaves sodium ions in the water. However, noncarbonate hardness is a small part of total hardness. Soda ash is also costly, so the actual dose might be slightly reduced from what is needed.

**SECOND STAGE RECARBONATION**: Second stage recarbonation is needed to remove $CaCO_3$. $CO_2$ is added until the pH is about 8.6, at which time no further precipitation will occur because $[Ca^{++}][CO_3^{--}] <$ $K_{sp}$ of $CaCO_3$.

Sodium polyphosphate can be added at this time to inhibit crusting on filter sand and scale formation in pipes.

A *split process* can be used to reduce the amount of lime that is neutralized by recarbonation (and is wasted). Excess lime is added in the first stage. This forces precipitation of magnesium in the first stage instead of in the second stage. Excess lime reacts with calcium hardness in the second stage. The amount of bypass depends on the allowable hardness of water leaving the plant. A typical split process is shown in figure 7.9.

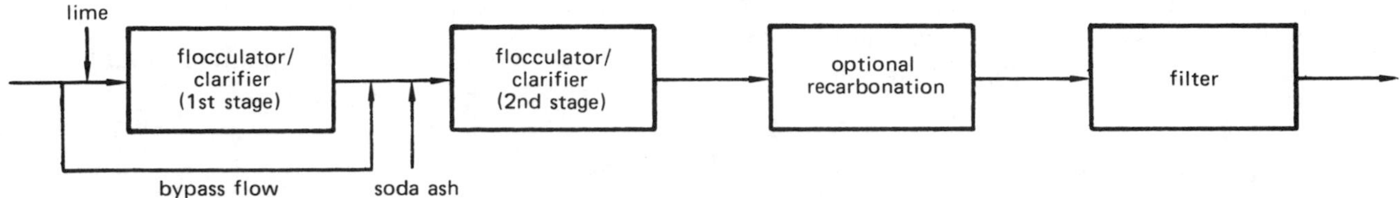


**Figure 7.9** A Split (Bypass) Process

Practical limits of *precipitation softening* by the lime process are 30 to 35 mg/l of $CaCO_3$ and 8 to 10 mg/l of $Mg(OH)_2$ as $CaCO_3$ due to intrinsic solubilities. Water softened with this process usually leaves the apparatus with a hardness of between 50 and 80.

*Example 7.18*

Water contains 130 mg/l as $CaCO_3$ of $Ca(HCO_3)_2$. How much slaked lime ($Ca(OH)_2$) is required to remove the hardness?

Since the $Ca(HCO_3)_2$ is given in $CaCO_3$ equivalents, 130 mg/l of lime (as $CaCO_3$) is implicitly required. It only remains to convert the $CaCO_3$ equivalent to a substance measurement using appendix A.

$$Ca(OH)_2 = \frac{130}{1.35} = 96.3 \text{ mg/l as substance}$$

*Example 7.19*

How much slaked lime (90% pure), soda ash, and carbon dioxide are required to reduce the hardness of the water evaluated below to zero using the lime-soda ash process? Neglect the fact that this process cannot really produce zero hardness, and base your answer on stoichiometric considerations.

| | |
|---|---|
| total hardness: | 250 mg/l as $CaCO_3$ |
| alkalinity: | 150 mg/l as $CaCO_3$ |
| carbon dioxide: | 5 mg/l |

Using appendix A, the $CO_2$ is first converted to its $CaCO_3$ equivalent.

$$(2.27)(5) = 11.35 \text{ mg/l as } CaCO_3$$

The alkalinity of 150 mg/l is already in $CaCO_3$ equivalent form. Therefore, the total $CaCO_3$ equivalent from substances requiring lime for neutralization is $11.35 + 150 = 161.35$ mg/l as $CaCO_3$.

From appendix A, the amount of 90% pure slaked lime ($Ca(OH)_2$) is

$$\frac{161.35}{(1.35)(0.90)} = 132.8 \text{ mg/l as substance}$$

50 mg/l of lime is arbitrarily added to raise the pH above 10.8. The total lime requirement is then

$$132.8 + \frac{50}{0.90} = 188.4 \text{ mg/l as substance}$$

The noncarbonate hardness is $250 - 150 = 100$ mg/l as $CaCO_3$. The soda ash ($Na_2CO_3$, 98% pure) requirement is

$$\frac{100}{(0.94)(0.98)} = 108.6 \text{ mg/l as substance}$$

The first stage recarbonation $CO_2$ requirement depends on the excess lime added.

$$\frac{(50)(1.35)}{(2.27)} = 29.7 \text{ mg/l as substance}$$

• Ion Exchange Method

In the ion exchange process (also known as *zeolite process*, *resin exchange process*, or *ion exchange method*), water is passed through a filter bed of exchange material. This exchange material is known as *zeolite*. Ions in the insoluble exchange material are displaced by ions in the water. When the exchange material is spent, it is regenerated with a rejuvenating solution such as sodium chloride (salt), or, in the case of common cationic resins, sulfuric and hydrochloric acids are used as *regenerants*. Soda ash is used as a regenerant in weakly-basic exchangers.

The processed water will have a zero hardness. However, since there is no need for water with zero hardness, some water is usually bypassed around the unit.

There are three types of ion exchange materials. *Greensand* (*glauconite*) is a natural substance that is mined and treated with manganese dioxide. *Siliceous-gel zeolite* is an artificial solid used in small volume deionizer columns. *Polystyrene resins* are also synthetic. Polystyrene resins currently dominate the softening field.[15]

[15] Differences in the polymerization step can result in polymers with gel or macroporous structures. *Gel polymers* have low cross linking, high capacity, and fast reaction kinetics. *Macroporous polymers* have high cross linking, reduced capacity, and lower kinetics. Gel resins have historically been used in water softening. However, the chemical resistance of macroporous forms is advantageous in special applications.

During operation, the calcium and magnesium ions are removed in reactions similar to the following reaction. Z is the zeolite anion. The resulting sodium compounds are soluble.

$$\left\{\begin{matrix} Ca \\ Mg \end{matrix}\right\}\left\{\begin{matrix} (HCO_3)_2 \\ SO_4 \\ Cl_2 \end{matrix}\right\} + Na_2Z$$

$$\rightarrow \left\{\begin{matrix} 2NaHCO_3 \\ Na_2SO_4 \\ 2NaCl \end{matrix}\right\} + \left\{\begin{matrix} Ca \\ Mg \end{matrix}\right\} Z \qquad 7.67$$

Typical characteristics of an ion exchange unit are expressed per 1000 grains of hardness removed.[16]

- exchange capacity: 3000 grains hardness/$ft^3$ zeolite for natural; 5000–30,000 (20,000 typical) for synthetic.
- flow rate: 2 to 6 gpm/$ft^3$ (2 gpm/$ft^3$ standard)

  6 gpm/$ft^2$ of filter bed
- backwash flow: 5 to 6 gpm/$ft^2$
- salt dosage: 5 to 20 pounds/$ft^3$. Alternatively, 0.3 to 0.7 pound of salt per 1000 grains of hardness removed
- brine contact time: 25 to 45 minutes
- depth of ion exchange bed: 2 ft (minimum) to 9 ft (maximum)

[16] 1000 grains of hardness is also known as a *kilograin.*

*Example 7.20*

A municipal plant receives water with a total hardness of 200 mg/l. The designed discharge hardness is 50 mg/l. If an ion exchange unit is used, what is the bypass factor?

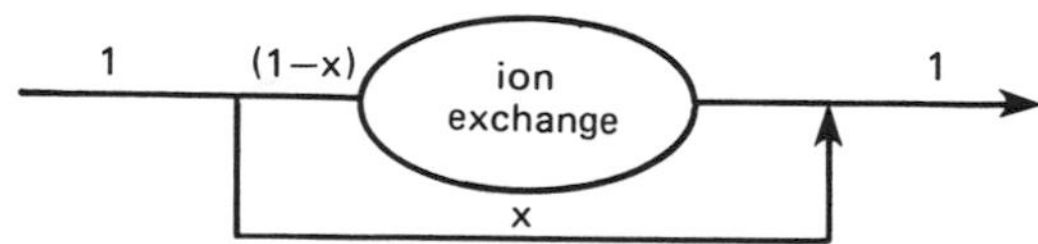


Let $x$ be the bypass factor. Since the water passing through the ion exchange unit is reduced to zero hardness,

$$(1-x)0 + x(200) = 50$$
$$x = 0.25$$

## J. TURBIDITY REMOVAL

*Coagulants* can be based on aluminum (e.g., aluminum sulfate, sodium aluminate, potash alum, or ammonia alum) or iron (e.g., ferric sulfate, ferrous sulfate, chlorinated copperas, or ferric chloride). If significant hydrolysis of iron and aluminum salts is ignored, the re-

**Table 7.24**
Types of Synthetic Exchange Materials

| type of resin | drained density lbm/$ft^3$ | operating pH range | regeneration | characteristics |
|---|---|---|---|---|
| strong acid | 49–53 | 0–14 | excess strong acid | high exchange rates; are stable; low swelling; long life, up to 20 years or more; can split strong and weak salts |
| weak acid | 45 | 7–14 | weak or strong acid | capacities double of strong acid; resistant to chlorine and other oxidants; high (90%) swell; not effective for electrolytic salt cations |
| strong base | 45 | 0–14 | excess strong base | irreversibly fouled by humic acids from decaying vegetation; can split strong or weak salts; less stable than cation resins (life less than 3 years); can remove silica; often used with food processing |
| weak base | 32 | 0–6 | weak or strong base | resistant to organic fouling; does not remove $CO_2$ or silica; capacity double of strong base; can remove color |
| intermediate base | 43 | 0–14 | strong base | can absorb $CO_2$ silica and phenol. Useful as substitutes for weak base resins in multiple-bed processes |

lationships in this section can be used to calculate the approximate stoichiometric quantities.

The most-used coagulant is aluminum sulfate ($Al_2(SO_4)_3 \cdot 14H_2O$). Filter alum is about 17% soluble material. The hydrolysis of the aluminum ion is complex. Assuming that the aluminum floc is $Al(OH)_3$ and the water pH is near neutral, then 1 mg/l of alum with a molecular weight of 600 removes the following quantities:

0.5 mg/l ($CaCO_3$) of natural alkalinity

0.39 mg/l of 95% hydrated lime ($Ca(OH)_2$)

0.33 mg/l of 85% quicklime ($CaO$)

0.53 mg/l of soda ash ($Na_2CO_3$)

If the alum has a molecular weight that is different than 600 (due to the variation in the number of waters of hydration), multiply the above quantities by (600/actual molecular weight).

Typical doses of alum are 5 to 50 mg/l, depending on turbidity. Alum flocculation is effective within pH limits of 5.5 to 8.0.

Ferrous sulfate ($FeSO_4 \cdot 7H_2O$), also known as *copperas*, reacts with lime ($Ca(OH)_2$) to flocculate ferric hydroxide ($Fe(OH)_3$). This is an effective method of clarifying turbid waters at higher pH, as in lime softening. 1 mg/l of ferrous sulfate with a molecular weight of 278 will react with 0.27 mg/l of lime.

Ferric sulfate ($Fe_2(SO_4)_3$) reacts with natural alkalinity or lime to create floc. 1 mg/l of ferric sulfate will react with

1.22 mg/l of $Ca(HCO_3)_2$

0.56 mg/l of $Ca(OH)_2$

0.62 mg/l of natural alkalinity (as $CaCO_3$)

Ferric sulfate can be used for color removal at low pH; at high pH, it is useful for iron and manganese removal, as well as a coagulant with precipitation softening.

## K. TASTE AND ODOR CONTROL

- **Copper Sulfate Treatment**. This treatment is used in impounding reservoirs, lakes, storage reservoirs, and occasionally in settling basins or treated water, to prevent biological growths. Dosages may vary from 0.5 to 2.0 milligrams per liter; the lower dosage ordinarily suffices for soft water. For very hard water, a dosage above 2.0 milligrams per liter may be used after laboratory tests to determine the necessary algicidal dose. Effects on fish life should be monitored.

- **Aeration**. This process can be used to improve tastes and odors in water where the cause is hydrogen sulfide or the absence of dissolved oxygen. This method has little effect on most tastes and odors.

- **Activated Carbon**. This material removes most tastes and odors. Dosages may vary from 0.5 to 200 milligrams per liter, ordinarily ranging from 2 to 10 milligrams per liter.

- **Superchlorination and Dechlorination**. This treatment will improve tastes and odors caused by organic matter and industrial wastes, especially phenolic wastes. Normally, the dosage required will be several times greater than those for ordinary disinfection (as determined by testing). Provide chlorinating equipment capable of dosing at these high values; allow a minimum of 20 minutes contact time; furnish equipment for dechlorinating with sulfur dioxide or other reducing agent.

- **Chlorine-Ammonia Treatment**. Where chlorosubstitution products cause tastes and odors, the chlorine-ammonia treatment can be used to prevent them. It can also be used for maintaining the combined residual chlorine for an extended period as, for example, in reservoirs or distribution systems.

  - Chloramines are less active disinfectants than free chlorine and, therefore, may not be substituted where adequate disinfection requires free residual chlorine.

  - The ratio of chlorine to ammonia required for disinfection varies from 3:1 to 7:1.

  - Periodic laboratory tests should be conducted to determine the proper dosage. Apply chlorine after ammonia has been properly dispersed in the water.

- **Free Residual Chlorination**. Use this method before filtration to reduce tastes and odors caused by organic matter at locations where experience shows it to be effective and acceptable. Increase the chlorine dosage until the residual consists solely of free available chlorine.

- **Chlorine Dioxide**. In some cases, this chemical can be used to destroy phenolic and other organic tastes and odors in raw water. The dosage varies from 0.2 to 0.3 milligram per liter, as determined by testing.

- **Microstraining**. This method is used as a means of reducing the number of algae and other organisms in the water, and thus reduces the subsequent production of tastes and odors. The microstrainer removes no dissolved or colloidal organic matter. It utilizes monel metal cloth with 35 micron (0.0014 inch) openings. Finer mesh can be obtained.

L. DEMINERALIZATION/DESALINATION

If dissolved salts are to be removed, one of the following methods must be used.

- **Distillation**: The water is vaporized, leaving the salt behind. The vapor is reclaimed by condensation.
- **Electrodialysis**: Positive and negative ions flow through selective membranes under the influence of an induced electrical current.
- **Ion exchange**: This is the same process as described for water softening.
- **Reverse osmosis**: This is the least expensive method of demineralization. In operation, a thin membrane of cellulose acetate plastic separates two salt solutions of different concentrations. Although ions would normally flow through the membrane into the solution with the lower concentration, the migration direction can be reversed by applying pressure to the low concentration fluid. Typical reverse osmosis units operate at 400 psi and produce about 2 gallons per day of fresh water for each square foot of surface.

## 14 TYPICAL MUNICIPAL SYSTEMS

The processes employed in treating incoming water will depend on the characteristics of the water. However, some sequences work better than others due to the physical and chemical nature of the processes. Listed in this section are some typical sequences. Not present in the lists are the usual system hardware items such as intake screens, pumps, pipes, hydrants, reservoirs, and holding basins.

Table 7.25 provides guidelines for choosing processes required to achieve satisfactory water quality. This table bases the required processes on the incoming water quality.

Additives and chemicals can be applied to the water supply at various points along the treatment path. Figure 7.10 indicates typical application points.

- **For Well Ground Water** (typically cleaner than surface water)

sequence #1: intake
chlorination
fluoridation

sequence #2: intake
aeration
oxidation(chlorine or potassium permanganate)
settling
filtering
chlorination
fluoridation

sequence #3: intake
aeration
lime addition
soda ash addition
rapid mix
flocculation
settling
recarbonation
filtering
chlorination
fluoridation

- **For Lake or Surface Water** (typically turbid, and carrying odor and color)

sequence #1: intake
chlorination
coagulation
rapid mixing
flocculation
optional chlorination
addition of activated carbon
settling
addition of activated carbon
filtering
chlorination
fluoridation

- **For River Surface Water** (very turbid)

sequence #1: intake
presedimentation (holding basin)
chlorination
coagulation
rapid mix
flocculation
settling
coagulation
rapid mix
flocculation
addition of activated carbon
settling
addition of activated carbon
filtering
chlorination
fluoridation

**Table 7.25**
Applicability of Treatment Methods

| water quality | | | pretreatment | | | treatment | | | | | special treatments | | | |
|---|---|---|---|---|---|---|---|---|---|---|---|---|---|---|
| constituents | concentration, mg/l | screening | prechlorination | plain settling | aeration | lime softening | coagulation and sedimentation | rapid sand filtration | slow sand filtration | postchlorination | superchlorination[1] or chlor-ammoniation | active carbon | special chemical treatment | salt water conversion[2] |
| coliform monthly avg mpn/100 ml[5] | 0–20 | | | | | | | | | E | | | | |
| | 20–100 | | | O | | | O | O | O | E | | | | |
| | 100–5000 | | E | | | | E | E | O | E | | | | |
| | >5000 | | E | O[3] | | | E | E | | E | O | | | |
| suspended solids | 0–100 | O | | | | | | | O | | | | | |
| | 100–200 | O | | | | | E | E | | | | | | |
| | > 200 | O | | O[4] | | | E | E | | | | | | |
| color, mg/l | 20–70 | | | | | | O | O | | | O | | | |
| | > 70 | | | | | | E | E | | | O | | | |
| tastes and odors | noticeable | | O | | O | | | | O | | O | E | | |
| $CaCO_3$, mg/l | > 200 | | | | | E | E | E | | | | | E | |
| pH | < 5.0–9.0< | | | | | | | | | | | | | |
| iron and manganese mg/l | ≤ 0.3 | | O | O | | | | | | | | | | |
| | 0.3–1.0 | | | | O | | E | E | O | | | | | |
| | > 1.0 | | E | | E | | E | E | O | | | | O | |
| chloride, mg/l | 0–250 | | | | | | | | | | | | | |
| | 250–500 | | | | | | | | | | | | | O |
| | 500 + | | | | | | | | | | | | | E |
| phenolic compounds, mg/l | 0–0.005 | | | | | | O | O | | | O | O | | |
| | > 0.005 | | | | | | E | E | | | O | E | O | |
| toxic chemicals | | | | | | | E | E | | | | E | O | |
| less critical chemicals | | | | | | | O | O | | | | O | O | |

Note: E = essential, O = optional

[1]Superchlorination shall be followed by dechlorination.

[2]As alternate, dilute with low chloride water.

[3]Double settling shall be provided for coliform exceeding 20,000 mpn/100 ml[5].

[4]For extremely muddy water, presedimentation by plain settling may be provided.

[5]mpn = most probable number

- **For Hard Water**

sequence #1:
intake (bypass to second flocculator)
lime addition
alum addition
rapid mixing
flocculation
sedimentation
oxidation (chlorine or potassium permanganate)
second flocculation
sedimentation
filtering
fluoridation
chlorination

sequence #2:
intake
presedimentation (in a basin)
chlorination
mixing
addition of activated carbon
lime addition
alum addition
flocculation
sedimentation
addition of activated carbon
mixing
filtering
fluoridation
chlorination
soda ash addition

sequence #3: intake
lime addition
alum addition
addition of activated carbon
mixing
flocculation
chlorination
sedimentation
recarbonation
filtering (bypass to discharge)
zeolite treatment

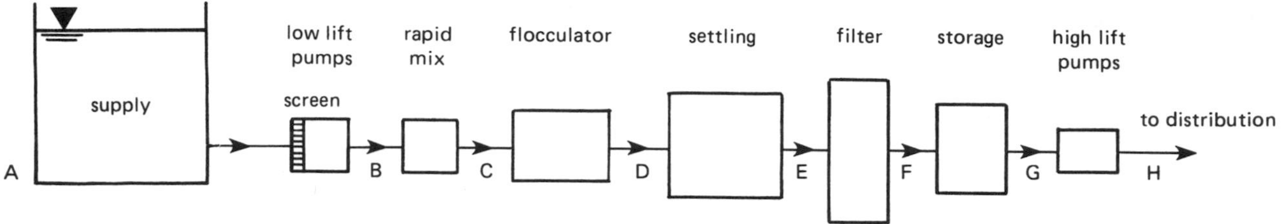


Typical flow diagram of water treatment plant

| category of chemicals | possible points of application | | | | | | | |
|---|---|---|---|---|---|---|---|---|
| | A | B | C | D | E | F | G | H |
| algicide | X | | | | X | | | |
| disinfectant | | X | X | | X | X | X | X |
| activated carbon | | X | X | X | X | | | |
| coagulants | | X | X | | | | | |
| coagulation aids | | X | X | | X | | | |
| alkali: | | | | | | | | |
| for flocculation | | | X | | | | | |
| for corrosion control | | | | | | X | | |
| for softening | | | X | | | | | |
| acidifier | | | X | | | X | | |
| fluoride | | | | | | X | | |
| cupric-chloramine | | | | | | X | | |
| dechlorinating agent | | | | | | X | | X |

Note: With solids contact reactors, point C is same as point D.

**Figure 7.10** Application Points for Chemicals

# Appendix A: Conversions from mg/l as a Substance to mg/l as $CaCO_3$

Multiply the mg/l of the substances listed below by the corresponding factors to obtain mg/l as $CaCO_3$. For example, 70 mg/l of $Mg^{++}$ would be (70)(4.10) = 287 mg/l as $CaCO_3$.

| Substance | Factor | Substance | Factor |
|---|---|---|---|
| $Al^{+++}$ | 5.56 | $HCO_3^-$ | 0.82 |
| $Al_2(SO_4)_3$* | 0.88 | $K^+$ | 1.28 |
| $AlCl_3$ | 1.13 | KCl | 0.67 |
| $Al(OH)_3$ | 1.92 | $K_2CO_3$ | 0.72 |
| $Ba^{++}$ | 0.73 | $Mg^{++}$ | 4.10 |
| $Ba(OH)_2$ | 0.59 | $MgCl_2$ | 1.05 |
| $BaSO_4$ | 0.43 | $MgCO_3$ | 1.19 |
| $Ca^{++}$ | 2.50 | $Mg(HCO_3)_2$ | 0.68 |
| $CaCl_2$ | 0.90 | MgO | 2.48 |
| $CaCO_3$ | 1.00 | $Mg(OH)_2$ | 1.71 |
| $Ca(HCO_3)_2$ | 0.62 | $Mg(NO_3)_2$ | 0.67 |
| CaO | 1.79 | $MgSO_4$ | 0.83 |
| $Ca(OH)_2$ | 1.35 | $Mn^{++}$ | 1.82 |
| $CaSO_4$* | 0.74 | $Na^+$ | 2.18 |
| $Cl^-$ | 1.41 | NaCl | 0.85 |
| $CO_2$ | 2.27 | $Na_2CO_3$ | 0.94 |
| $CO_3^{--}$ | 1.67 | $NaHCO_3$ | 0.60 |
| $Cu^{++}$ | 1.57 | $NaNO_3$ | 0.59 |
| $Cu^{+++}$ | 2.36 | NaOH | 1.25 |
| $CuSO_4$ | 0.63 | $Na_2SO_4$* | 0.70 |
| $F^-$ | 2.66 | $NH_3$ | 2.94 |
| $Fe^{++}$ | 1.79 | $NH_4^+$ | 2.78 |
| $Fe^{+++}$ | 2.69 | $NH_4OH$ | 1.43 |
| $Fe(OH)_3$ | 1.41 | $(NH_4)_2SO_4$ | 0.76 |
| $FeSO_4$* | 0.66 | $NO_3^-$ | 0.81 |
| $Fe_2(SO_4)_3$ | 0.75 | $OH^-$ | 2.94 |
| $FeCl_3$ | 0.93 | $PO_4^{---}$ | 1.58 |
| $H^+$ | 50.0 | $SO_4^{--}$ | 1.04 |
| | | $Zn^{++}$ | 1.54 |

* anhydrous

WATER SUPPLY

# Appendix B: Atomic Weights of Elements Referred to Carbon (12)

| Element | Symbol | Atomic Weight | Element | Symbol | Atomic Weight |
|---|---|---|---|---|---|
| Actinium | Ac | (227) | Mercury | Hg | 200.59 |
| Aluminum | Al | 26.9815 | Molybdenum | Mo | 95.94 |
| Americium | Am | (243) | Neodymium | Nd | 144.24 |
| Antimony | Sb | 121.75 | Neon | Ne | 20.183 |
| Argon | Ar | 39.948 | Neptunium | Np | (237) |
| Arsenic | As | 74.9216 | Nickel | Ni | 58.71 |
| Astatine | At | (210) | Niobium | Nb | 92.906 |
| Barium | Ba | 137.34 | Nitrogen | N | 14.0067 |
| Berkelium | Bk | (249) | Osmium | Os | 190.2 |
| Beryllium | Be | 9.0122 | Oxygen | O | 15.9994 |
| Bismuth | Bi | 208.980 | Palladium | Pd | 106.4 |
| Boron | B | 10.811 | Phosphorus | P | 30.9738 |
| Bromine | Br | 79.909 | Platinum | Pt | 195.09 |
| Cadmium | Cd | 112.40 | Plutonium | Pu | (242) |
| Calcium | Ca | 40.08 | Polonium | Po | (210) |
| Californium | Cf | (251) | Potassium | K | 39.102 |
| Carbon | C | 12.01115 | Praseodymium | Pr | 140.907 |
| Cerium | Ce | 140.12 | Promethium | Pm | (145) |
| Cesium | Cs | 132.905 | Protactinium | Pa | (231) |
| Chlorine | Cl | 35.453 | Radium | Ra | (226) |
| Chromium | Cr | 51.996 | Radon | Rn | (222) |
| Cobalt | Co | 58.9332 | Rhenium | Re | 186.2 |
| Copper | Cu | 63.54 | Rhodium | Rh | 102.905 |
| Curium | Cm | (247) | Rubidium | Rb | 85.47 |
| Dysprosium | Dy | 162.50 | Ruthenium | Ru | 101.07 |
| Einsteinium | Es | (254) | Samarium | Sm | 150.35 |
| Erbium | Er | 167.26 | Scandium | Sc | 44.956 |
| Europium | Eu | 151.96 | Selenium | Se | 78.96 |
| Fermium | Fm | (253) | Silicon | Si | 28.086 |
| Fluorine | F | 18.9984 | Silver | Ag | 107.870 |
| Francium | Fr | (223) | Sodium | Na | 22.9898 |
| Gadolinium | Gd | 157.25 | Strontium | Sr | 87.62 |
| Gallium | Ga | 69.72 | Sulfur | S | 32.064 |
| Germanium | Ge | 72.59 | Tantalum | Ta | 180.948 |
| Gold | Au | 196.967 | Technetium | Tc | (99) |
| Hafnium | Hf | 178.49 | Tellurium | Te | 127.60 |
| Helium | He | 4.0026 | Terbium | Tb | 158.924 |
| Holmium | Ho | 164.930 | Thallium | Tl | 204.37 |
| Hydrogen | H | 1.00797 | Thorium | Th | 232.038 |
| Indium | In | 114.82 | Thulium | Tm | 168.934 |
| Iodine | I | 126.9044 | Tin | Sn | 118.69 |
| Iridium | Ir | 192.2 | Titanium | Ti | 47.90 |
| Iron | Fe | 55.847 | Tungsten | W | 183.85 |
| Krypton | Kr | 83.80 | Uranium | U | 238.03 |
| Lanthanum | La | 138.91 | Vanadium | V | 50.942 |
| Lead | Pb | 207.19 | Xenon | Xe | 131.30 |
| Lithium | Li | 6.939 | Ytterbium | Yb | 173.04 |
| Lutetium | Lu | 174.97 | Yttrium | Y | 88.905 |
| Magnesium | Mg | 24.312 | Zinc | Zn | 65.37 |
| Manganese | Mn | 54.9380 | Zirconium | Zr | 91.22 |
| Mendelevium | Md | (256) | | | |

# Appendix C: Inorganic Chemicals Used in Water Treatment

| Chemical Name | Formula | Use | Molecular Weight | Equivalent Weight |
|---|---|---|---|---|
| Activated carbon | $C$ | Taste and odor control | 12.0 | ---- |
| Aluminum sulfate (filter alum) | $Al_2(SO_4)_3 \cdot 14.3H_2O$ | Coagulation | 600 | 100 |
| Aluminum hydroxide | $Al(OH)_3$ | (Hypothetical combination) | 78.0 | 26.0 |
| Ammonia | $NH_3$ | Chloramine disinfection | 17.0 | ---- |
| Ammonium fluosilicate | $(NH_4)_2SiF_6$ | Fluoridation | 178 | ---- |
| Ammonium sulfate | $(NH_4)_2SO_4$ | Coagulation | 132 | 66.1 |
| Calcium bicarbonate | $Ca(HCO_3)_2$ | (Hypothetical combination) | 162 | 81.0 |
| Calcium carbonate | $CaCO_3$ | Corrosion control | 100 | 50.0 |
| Calcium fluoride | $CaF_2$ | Fluoridation | 78.1 | ---- |
| Calcium hydroxide | $Ca(OH)_2$ | Softening | 74.1 | 37.0 |
| Calcium hypochlorite | $Ca(ClO)_2 \cdot 2H_2O$ | Disinfection | 179 | ---- |
| Calcium oxide (lime) | $CaO$ | Softening | 56.1 | 28.0 |
| Carbon dioxide | $CO_2$ | Recarbonation | 44.0 | 22.0 |
| Chlorine | $Cl_2$ | Disinfection | 71.0 | ---- |
| Chlorine dioxide | $ClO_2$ | Taste and odor control | 67.0 | ---- |
| Copper sulfate | $CuSO_4$ | Algae control | 160 | 79.8 |
| Ferric chloride | $FeCl_3$ | Coagulation | 162 | 54.1 |
| Ferric hydroxide | $Fe(OH)_3$ | (Hypothetical combination) | 107 | 35.6 |
| Ferric sulfate | $Fe_2(SO_4)_3$ | Coagulation | 400 | 66.7 |
| Ferrous sulfate (copperas) | $FeSO_4 \cdot 7H_2O$ | Coagulation | 278 | 139 |
| Fluosilicic acid | $H_2SiF_6$ | Fluoridation | 144 | ---- |
| Hydrochloric acid | $HCl$ | pH adjustment | 36.5 | 36.5 |
| Magnesium hydroxide | $Mg(OH)_2$ | Defluoridation | 58.3 | 29.2 |
| Oxygen | $O_2$ | Aeration | 32.0 | 16.0 |
| Potassium permanganate | $KMnO_4$ | Oxidation | 158 | ---- |
| Sodium aluminate | $NaAlO_2$ | Coagulation | 82.0 | ---- |
| Sodium bicarbonate (baking soda) | $NaHCO_3$ | pH adjustment | 84.0 | 84.0 |
| Sodium carbonate (soda ash) | $Na_2CO_3$ | Softening | 106 | 53.0 |
| Sodium chloride (common salt) | $NaCl$ | Ion-exchange regeneration | 58.4 | 58.4 |
| Sodium fluoride | $NaF$ | Fluoridation | 42.0 | ---- |
| Sodium hexametaphosphate | $(NaPO_3)_n$ | Corrosion control | ---- | ---- |
| Sodium hydroxide | $NaOH$ | pH adjustment | 40.0 | 40.0 |
| Sodium hypochlorite | $NaClO$ | Disinfection | 74.4 | ---- |
| Sodium silicate | $Na_4SiO_4$ | Coagulation aid | 184 | ---- |
| Sodium fluosilicate | $Na_2SiF_6$ | Fluoridation | 188 | ---- |
| Sodium thiosulfate | $Na_2S_2O_3$ | Dechlorination | 158 | ---- |
| Sulfur dioxide | $SO_2$ | Dechlorination | 64.1 | ---- |
| Sulfuric acid | $H_2SO_4$ | pH adjustment | 98.1 | 49.0 |
| Water | $H_2O$ | ---- | 18.0 | ---- |

**Practice Problems: WATER SUPPLY ENGINEERING**

Untimed

1. A water treatment plant has four rapid sand filters, each of which has a capacity of 600,000 gallons per day. Each filter is backwashed once a day for eight minutes. (a) Determine the inside dimensions of the sand filter. (b) What percentage of the filtered water is used for backwashing?

2. (a) Design a circular, mechanically-cleaned clarifier using the following specifications:

| | |
|---|---|
| flow rate: | 2.8 million gallons/day |
| detention period: | 2 hours |
| surface loading: | 700 gallons/ft²-day |

(b) If the initial flow rate is only 1.1 million gallons per day, what are the surface loading and average detention periods?

3. A town's water supply is to be taken from a river with the following quality characteristics:

| | |
|---|---|
| turbidity: | varies between 20 and 100 units |
| total hardness: | less than 60 mg/l (as $CaCO_3$) |
| coliform count: | varies between 200 and 1000 per 100 ml |

The town has a design population of 15,000 people and an average consumption of 110 gpcd. (a) What rate (gpm) should the distribution be designed to carry? (b) What total filter area would you recommend? (c) Is softening required? (d) If 2 mg/l of chlorine are required to obtain the necessary chlorine residual, how many pounds per 24 hours of chlorine are required?

4. A town's water supply has the following hypothetical ion concentrations. (a) What is the total hardness in mg/l (as $CaCO_3$)? (b) How much lime ($Ca(OH)_2$) and soda ash are required to react with the carbonate hardness?

| | | | |
|---|---|---|---|
| $Ca^{++}$ | 80.2 mg/l | $CO_3^{--}$ | 0 |
| $Na^{+}$ | 46.0 mg/l | $Mg^{++}$ | 24.3 mg/l |
| $NO_3^{-}$ | 0 | $Fl^{-}$ | 0 |
| $Cl^{-}$ | 85.9 mg/l | $SO_4^{--}$ | 125 mg/l |
| $CO_2$ | 19 mg/l | $Fe^{++}$ | 1.0 mg/l |
| $Al^{+++}$ | 0.5 mg/l | $HCO_3^{-}$ | 185 mg/l |

5. The following concentrations of inorganic compounds are found during a routine analysis of a city's water supply.

| | |
|---|---|
| $Ca(HCO_3)_2$ | 137 mg/l (as $CaCO_3$) |
| $CO_2$ | 0 mg/l |
| $MgSO_4$ | 72 mg/l (as $CaCO_3$) |

(a) How many pounds of lime ($Ca(OH)_2$) and soda ash ($Na_2CO_3$) are required to soften one million gallons of this water to 100 mg/l if 30 mg/l excess lime is required for a complete reaction? (b) How many pounds of salt would be required if a zeolite process is used with the following characteristics:

| | |
|---|---|
| exchange capacity: | 10,000 grains hardness/ft³ |
| salt requirement: | .5 pound/1000 grains hardness removed |

6. A water treatment plant has five square rapid sand filters, each of which has a capacity of one million gallons per day. (a) What are the recommended dimensions for the filters? (b) If each filter is backwashed each day for 5 minutes, what percentage of the plant's filtered water is used for backwashing?

7. Water from an underground aquifer is to be reduced from 245 mg/l hardness to 80 mg/l hardness by the zeolite process. (a) Draw a line schematic of the process used to accomplish this reduction. (b) What is the time between regenerations of the softener if the exchanger has the following characteristics:

| | |
|---|---|
| flow volume: | 20,000 gallons per day |
| exchanger resin volume: | 2 cubic feet |
| resin exchange capacity: | 20,000 grains per cubic foot |

8. A 12″ standard strength clay sewer pipe is to be installed under a backfill of 11 feet of saturated topsoil which has a density of 120 pounds per cubic foot. The pipe strength is 1,500 pounds per foot. Design a bedding using a safety factor of 1.5.

9. A settling tank has an overflow rate of 100,000 gal/ft²-day. Water carrying grit of various sizes is introduced. The grit has the following distribution of settling velocities:

| settling velocity (fpm) | weight fraction remaining |
|---|---|
| 10.0 | .54 |
| 5.0 | .45 |
| 2.0 | .35 |
| 1.0 | .20 |
| .75 | .10 |
| .50 | .03 |

What is the percentage by weight of the grit removed?

10. What is the settling velocity of a spherical sand particle which has a specific gravity of 2.6 and a diameter of 1 millimeter?

Timed

1. A flocculator tank with a volume of 200,000 ft³ uses a paddle wheel to disperse chemicals throughout the mixture. Use the values listed to determine (a) the required paddle area, (b) the drag force on the paddle, and (c) the theoretical power requirement to drive the paddle.

| | |
|---|---|
| mean velocity gradient: | 45 1/sec |
| water temperature: | 60°F |
| drag coefficient: | 1.75 |
| paddle-tip velocity: | 2 ft/sec |
| relative water/paddle velocity: | 1.5 ft/sec |

2. The figure shows the cumulative per capita water demand for a peak day in an area expected to attain a population of 40,000 after 20 years.

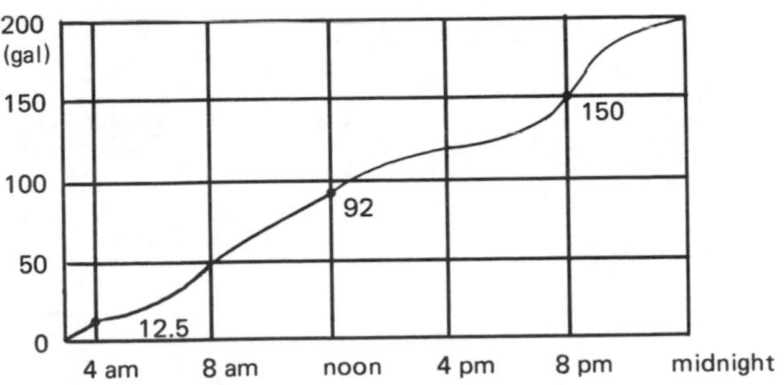


(a) Determine the daily per capita demand for a peak day. (b) Assuming uniform operation, what storage volume is required in the treatment plant for all uses, including fire fighting demand? Assume 24 hour per day operation. (c) Assume that the pumping station only runs from 4 AM to 8 AM. If pumping is uniform during this period, what storage is required to meet all uses including fire fighting?

3. You are to determine whether two sedimentation basins have been correctly designed. The current design features are:

| | |
|---|---|
| design average daily flow | 1.5 MGD (2 basins) |
| configuration | 2 basins, 90′ × 16′ × 12′ deep |
| total weir length per basin | 48 ft |
| 3-month sustained average low | .7 (design average daily flow) |
| 3-month sustained average high | 2.0 (design average daily flow) |

The basins must meet the following government standards.

| | |
|---|---|
| minimum retention time | 4.0 hours |
| maximum weir load | 20,000 gpd/ft |
| maximum velocity | .5 ft/min |

4. Two boreholes are located at points *A* and *B*. The soil between them has a permeability of 3 EE-5 ft/sec and a porosity of .4. Water flows from *B* to *A* in the water table. A 500′ × 100′ hazardous waste containment cell will be built between points *A* and *B* as indicated. The bottom of the containment cell must be at least 5 feet above the water table at all points. (a) What is the hydraulic (ground water) gradient between points *A* and *B*? (b) What is the minimum elevation of the containment cell? (c) Assume that the protective lining of the cell fails and contaminant reaches the groundwater. How long will it take for the contaminated water to reach point *A?*

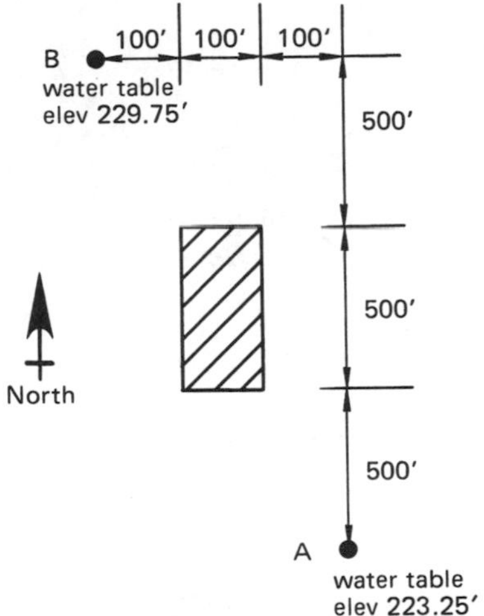

# 8 WASTE-WATER ENGINEERING

*Nomenclature*

| | | |
|---|---|---|
| A | area | $ft^2$ |
| BOD | biochemical oxygen demand | mg/l |
| C | concentration | mg/l |
| COD | chemical oxygen demand | mg/l |
| d | stream flow depth, or particle diameter | ft, or mm |
| D | oxygen deficit | mg/l |
| DO | dissolved oxygen | mg/l |
| f | Darcy friction factor | – |
| F | effective number of passes | – |
| g | acceleration due to gravity (32.2) | $ft/sec^2$ |
| k | rate constant | –/days |
| $K_D$ | deoxygenation coefficient | –/days |
| $K_R$ | reoxygenation coefficient | –/days |
| $K_t$ | oxygen transfer coefficient | –/hrs |
| L | loading | varies |
| MLSS | mixed liquor suspended solids | mg/l |
| P | population | 1000's of people |
| Q | flow quantity | gallons/day |
| Q′ | flow quantity | cfs |
| R | ratio | – |
| ROT | rate of oxygen transfer | mg/l-hr |
| s | sludge suspended solids | decimal |
| SA | sludge age | days |
| SG | specific gravity | – |
| SS | suspended solids | mg/l |
| SVI | sludge volume index | |
| t | time | days |
| T | temperature | °C |
| v | velocity | ft/sec |
| V | volume | ml |
| w | weighting factor | – |
| W | sludge removal rate | lbm/day (dry) |
| x | a fraction | decimal |

*Symbols*

| | | |
|---|---|---|
| $\eta$ | efficiency | – |
| $\beta$ | oxygen saturation coefficient | – |

*Subscripts*

| | |
|---|---|
| a | aeration |
| as | activated sludge |
| c | critical |
| d | discharge or detention |
| D | deoxygenation |
| e | equivalent |
| f | final |
| F–M | food to microorganism |
| H | hydraulic |
| i | initial |
| ML | mixed liquor |
| o | immediately after mixing, or original |
| p | particle, or primary |
| R | recirculation, return, or reoxygenation |
| RS | return sludge |
| req | required at discharge |
| s | standard 5-day, or secondary |
| sat | saturated |
| ss | suspended solids |
| t | at time t |
| T | at temperature T |
| u | ultimate carbonaceous |
| w | raw wastewater |

## 1 CONVERSIONS

| multiply | by | to obtain |
|---|---|---|
| acre-feet | 43.56 | 1000's of cubic feet |
| cubic feet | 7.48 | gallons |
| cubic feet/sec (cfs) | 0.6463 | MGD |
| cubic feet/sec (cfs) | 448.8 | gpm |
| gallons | 0.1337 | cubic feet |

(continued)

(continued from previous page)

| multiply | by | to obtain |
|---|---|---|
| gallons/day (gpd) | 1.547 EE–6 | cfs |
| gallons/min (gpm) | 0.002228 | cfs |
| gallons/acre-day (gad) | 2.296 EE–5 | gallons/day-ft$^2$ |
| gallons/ft$^2$-day (gpd/ft$^2$) | 0.04356 | million gallons/ acre-day |
| million gallons/ acre-day (mgad) | 22.96 | gpd/ft$^2$ |
| million gallons/ day (MGD) | 1.547 | cfs |
| milligrams (mg) | 2.205 EE–6 | pounds |
| milligrams/liter (mg/l) | 8.345 | pounds/million gallons |
| meters | 3.281 | feet |
| millimeters/meter | 0.012 | in/foot |
| m$^3$/m$^2$-day | 24.54 | gallons/ft$^2$-day |
| m$^3$/m-day | 80.52 | gallons/ft-day |
| miles per hour (mph) | 1.4667 | ft/sec |
| pounds | 4.536 EE5 | milligrams |
| pounds/acre-ft-day | 0.02296 | lbm/1000 ft$^3$-day |
| pounds/1000 ft$^3$-day | 43.56 | lbm/acre-ft-day |
| pounds/1000 ft$^3$-day | 133.7 | lbm/million gal-day |
| pounds/million gallons | 0.1198 | mg/l |
| pounds/million gallons-day | 0.00748 | lbm/1000 ft$^3$-day |

## 2 DEFINITIONS

Activated sludge: Solids from aerated settling tanks which are rich in bacteria.

Aerated lagoon: A holding basin into which air is mechanically introduced to speed up aerobic decomposition.

Appurtenance: A thing which belongs with (or is designed to complement) something else. For example, a manhole is a sewer appurtenance.

Bioactivation process: A process using sedimentation, trickling filter, and secondary sedimentation before adding activated sludge. Aeration and final sedimentation are the follow-up processes.

Biosorption process: A process which mixes raw sewage and sludge which have been pre-aerated in a separate tank.

Biota: The flora and fauna of a region, process, or tank.

Branch sewer: A sewer off the main sewer.

Bulking: See 'Sludge bulking.'

Carbonaceous demand: Oxygen demand due to biological activity in a water sample.

Chemical precipitation: Causing suspended solids to settle out by adding coagulating chemicals.

Clean-out: A pipe through which snakes can be pushed to unplug a sewer.

Combined system: A system using a single sewer for domestic waste and storm water.

Comminutor: A device which cuts solid waste into small pieces.

Complete mixing: Mixing accomplished by mechanical means (stirring).

Cunette: A small channel in the invert of a large combined sewer for dry weather flow.

Deoxygenation: The act of removing dissolved oxygen from water.

Dewatering: Removal of excess moisture from sludge waste.

Digestion: Conversion of sludge solids to gas.

Dilution disposal: Relying on a large water volume (lake or stream) to dilute waste to an acceptable concentration.

Domestic waste: Waste which originates from households.

Effluent: That which flows out of a process.

Elutriation: A counter-current sludge washing process used to remove dissolved salts.

First-stage demand: See 'Carbonaceous demand.'

Floatation: Adding chemicals or bubbling air through waste to get solids to float to the top as scum.

Force main: A sewer line which is pressurized.

Humus: A greyish brown sludge consisting of relatively large particle biological debris, as is the material sloughed off from a trickling filter.

Infiltration: Ground water which enters sewer pipes through cracks and joints.

Influent: Flow entering a process.

Inverted siphon: A sewer line which drops below the hydraulic gradient.

Kraus process: Mixing raw sewage, activated sludge, and material from sludge digesters.

Lamp holes: Sewer inspection holes large enough to lower a lamp into but too small for a man.

Lateral: A sewer line which goes off at right angles to another.

Main: A large sewer at which all other branches terminate.

Malodorous: Offensive smelling.

Mesophilic bacteria: Bacteria growing between 10 and 40°C , with an optimum temperature of 37°C. 40°C is, therefore, the upper limit for most wastewater processes.

Mohlman index: Same as the 'Sludge volume index.'

Nitrogenous demand: Oxygen demand from nitrogen-consuming bacteria.

Outfall: The pipe which discharges completely treated wastewater into a lake, stream, or ocean.

Partial treatment: Primary treatment only.

Post-chlorination: Addition of chlorine after all other processes have been completed.

Pre-chlorination: Addition of chlorine prior to sedimentation to help control odors and to aid in grease removal.

Putrefaction: Anaerobic decomposition of organic matter with accompanying foul odors.

Refractory: Dissolved organic materials which are biologically resistant and difficult to remove.

Regulator: A device or weir which deflects large volume flows into a special high-capacity sewer.

Sag pipe: See 'Inverted siphon.'

Second stage demand: See 'Nitrogenous demand.'

Seed: The activated sludge initially taken from the secondary settling tank and returned to the aeration tank to start the activated sludge process.

Separate system: Separate sewers for domestic and storm waste water.

Septic: Produced by putrefaction.

Sludge bulking: Failure of suspended solids to completely settle out.

Split chlorination: Addition of chlorine prior to sedimentation and after final processing.

Submain: See 'Branch.'

Supernatant: The clarified liquid floating on top of a digesting sludge layer.

Thermophilic bacteria: Bacteria which thrive in the 45°C to 75°C range (optimum near 55°C).

Volatile solid: Solid material in a water sample or in sludge which can be burned or vaporized at high temperature.

Wet well: A short-term storage tank containing a pump or pump entrance, and into which the raw influent is brought.

Zooglea: The gelatinous film of aerobic organisms which cover the rocks in a trickling filter.

## 3 WASTEWATER QUALITY CHARACTERISTICS

### A. DISSOLVED OXYGEN

Fish and most aquatic life require oxygen.[1] The biological decomposition of organic solids is also dependent on oxygen. If the dissolved oxygen content of water is less than the saturated values given in appendix B, there is good reason to believe that the water is organically polluted. Other reasons for measuring the dissolved oxygen concentration are for aerobic treatment monitoring, aeration process monitoring, BOD testing, and pipe corrosion studies.

The difference between the saturated and actual dissolved oxygen concentrations is known as the *oxygen deficit.*

$$D = DO_{\text{sat}} - DO \qquad 8.1$$

The oxygen deficit is reduced by aerating the water (i.e., the dissolved oxygen concentration is increased). An exponential decay is traditionally used to predict the oxygen deficit as a function of time. Equation 8.2 assumes that oxygen is not being depleted during the reoxygenation process.

$$D_t = D_o 10^{-K_R t} \qquad 8.2$$

$K_R$ is the *reoxygenation (reaeration) coefficient*, which depends on the type of flow and temperature.[2] Reoxygenation coefficients are also given for use with a different logarithmic base.

$$D_t = D_o e^{-K'_R t} \qquad 8.3$$

[1] 4–6 mg/l is the generally accepted range of dissolved oxygen required to support fish populations. 5 mg/l is adequate, as is verifiable from high-altitude trout lakes. However, 6 mg/l is preferable, particularly for large fish populations.

[2] $K_R$ may be written as $K_2$ in the literature.

The constants $K_R$ and $K'_R$ are not the same, but they are related.[3]

$$K'_R = 2.3K_R \qquad 8.4$$

Table 8.12 lists representative values of $K_R$.

## B. BIOCHEMICAL OXYGEN DEMAND

When oxidizing organic waste material in water, biological organisms remove oxygen from the water. Therefore, oxygen use is an indication of the organic waste content. The biochemical oxygen demand (BOD) of a biologically active sample is given by equation 8.5:

$$\text{BOD}_s = \frac{DO_i - DO_f}{\dfrac{V_{\text{sample}}}{V_{\text{sample}} + V_{\text{dilution}}}} \qquad 8.5$$

BOD is determined by adding a measured amount of wastewater (which supplies the organic material) to a measured amount of dilution water (which reduces toxicity and supplies dissolved oxygen). An oxygen use curve similar to that in figure 8.1 will result. (More than one identical sample must be prepared in order to determine initial and final concentrations of dissolved oxygen.)

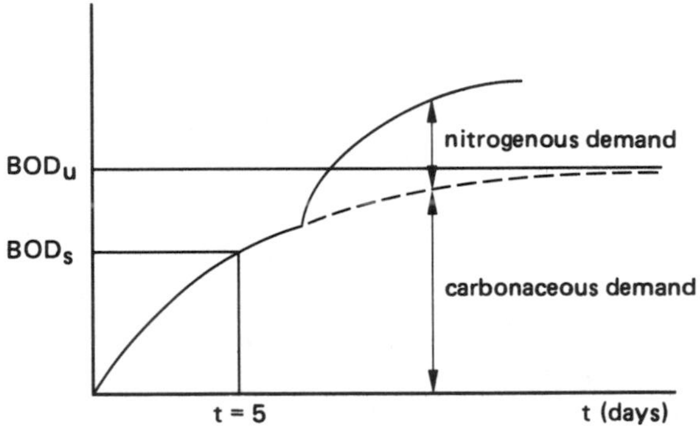


**Figure 8.1** BOD Time Curve

The deviation from the expected exponential growth curve in figure 8.1 is due to *nitrification* or *nitrogenous demand.* Nitrification is the use of oxygen by *autotrophic bacteria.*[4] Such bacteria use fixed carbon as food. (For example, the carbon in carbon dioxide is used by autotrophic bacteria.) Autotrophic bacteria oxidize ammonia to nitrites and nitrates. However, the number of autotrophic bacteria is small. Generally, six to ten days are required for the autotrophic population to become sufficiently large enough to affect a BOD test. Therefore, the standard BOD test is terminated before the autotrophic contribution to BOD becomes significant.

The standard BOD test typically calls for a 5-day incubation period at 20°C. The BOD at any time can be found from equation 8.6.

$$\text{BOD}_t = \text{BOD}_u(1 - 10^{-K_D t}) \qquad 8.6$$

$K_D$ is the *deoxygenation rate constant,* typically taken as 0.1. The ultimate BOD cannot be found from long term studies due to the effect of nitrogen-consuming bacteria in the sample. However, if $K_D$ is 0.1, the ultimate BOD can be found from equation 8.7.

$$\text{BOD}_u \approx 1.47\,\text{BOD}_s \qquad 8.7$$

$K_D$ for other temperatures can be found from equation 8.8. (The 1.047 constant is often quoted in literature. Recent research suggests 1.135 for 4°C to 20°C, and 1.056 for 20°C to 30°C.)

$$K_{D,T} = (1.047)^{T-20} K_{D,20°\text{C}} \qquad 8.8$$

The variation in BOD with temperature is given by equation 8.9.

$$\text{BOD}_T = \text{BOD}_{20°\text{C}}(0.02T + 0.6) \qquad 8.9$$

**Table 8.1**
Typical Values of $K_D$

| | |
|---|---|
| treatment plant effluents | 0.05–0.10 |
| highly polluted shallow streams | 0.25 |

*Example 8.1*

Ten 5-ml samples of wastewater are placed in 300 ml BOD bottles. Half of the bottles are titrated immediately with an average initial concentration of dissolved oxygen of 7.9 mg/l. The remaining bottles are incubated for 5 days, after which the average dissolved oxygen is determined to be 4.5 mg/l. What is the standard BOD and ultimate carbonaceous BOD assuming $K_D = 0.13$?

From equation 8.5:

$$\text{BOD}_s = \frac{7.9 - 4.5}{\dfrac{5}{300}} = 204\,\text{mg/l}$$

[3] $K'_R$ may be written as $K_R$ (*base e*) in the literature.

[4] Most bacteria in wastewater are heterotrophic. *Heterotrophic bacteria* use organic carbon as food.

From equation 8.6, the ultimate BOD is

$$\mathrm{BOD}_u = \frac{204}{1 - 10^{(-0.13)(5)}} = 263\,\mathrm{mg/l}$$

If a sample of industrial wastewater is taken, it will probably lack sufficient microorganisms to metabolize the organic matter. In such a case, seed organisms must be added. The BOD for seeded experiments is found by measuring dissolved oxygen in the seeded sample after 15 minutes ($DO_i$) and after 5 days ($DO_f$), as well as the dissolved oxygen of the seed material itself after 15 minutes ($DO_i^*$) and after 5 days ($DO_f^*$).

$$\mathrm{BOD}_s = \frac{DO_i - DO_f - x[DO_i^* - DO_f^*]}{\dfrac{V_{\mathrm{sample}}}{V_{\mathrm{sample}} + V_{\mathrm{dilution}}}} \qquad 8.10$$

$$x = \frac{\text{volume of seed added to sample}}{\text{volume of seed used to find } DO^*} \qquad 8.11$$

The BOD of domestic waste is typically taken as 0.17 to 0.20 pounds per capita-day, excluding industrial wastes. This makes it possible to calculate the *population equivalent* of any BOD loading.

$$P_e = \frac{\left(\mathrm{BOD}\,\frac{\mathrm{mg}}{\mathrm{l}}\right)\left(Q\,\frac{\mathrm{gal}}{\mathrm{day}}\right)\left(8.345\,\frac{\mathrm{lbm\text{-}l}}{\mathrm{MG\text{-}mg}}\right)}{\left(1{,}000{,}000\,\frac{\mathrm{gal}}{\mathrm{MG}}\right)(1000)\left(0.17\,\frac{\mathrm{lbm}}{\mathrm{person\text{-}day}}\right)}$$

(in 1000's of people) 8.12

Values of BOD for various industrial wastewaters are given in table 8.2.

BOD of 100 mg/l is considered a *weak wastewater*; BOD of 200 to 250 mg/l is considered a ***medium strength wastewater***; above 300 mg/l, it is considered to be a ***strong wastewater.***

C. RELATIVE STABILITY

The relative stability test is much easier to perform than the BOD test, although it is much less accurate. The relative stability of an effluent is defined as the percent of initial BOD that has been satisfied. The test consists of taking a sample of effluent and adding a small amount of methylene blue dye. When all oxygen has been removed from the water, anaerobic bacteria start to remove the dye. The time for the color to start degrading is known as the *stabilization time* or *decoloration time.*

The relative stability can be found from the stabilization time by using table 8.3.

**Table 8.3**
Relative Stability (at 20°C)

| stabilization time (days) | relative stability % | stabilization time (days) | relative stability % |
|---|---|---|---|
| 1/2 | 11 | 8 | 84 |
| 1 | 21 | 9 | 87 |
| 1 1/2 | 30 | 10 | 90 |
| 2 | 37 | 11 | 92 |
| 2 1/2 | 44 | 12 | 94 |
| 3 | 50 | 13 | 95 |
| 4 | 60 | 14 | 96 |
| 5 | 68 | 16 | 97 |
| 6 | 75 | 18 | 98 |
| 7 | 80 | 20 | 99 |

**Table 8.2**
Typical BOD and COD of Industrial Wastewaters

| industry/type of waste | BOD | COD |
|---|---|---|
| canning | | |
| corn | 19.5 lbm/ton corn | |
| tomatoes | 8.4 lbm/ton tomatoes | |
| dairy milk processing | 1150 lbm/ton raw milk | 1900 mg/l |
| | 1000 mg/l | |
| beer brewing | 1.2 lbm/barrel beer | |
| commercial laundry | 1250 lbm/1000 pounds dry | 2400 mg/l |
| | 700 mg/l | |
| slaughterhouse | 7.7 lbm/animal | 2100 mg/l |
| (meat packing) | 1400 mg/l | |
| papermill | 121 lbm/ton pulp | |
| synthetic textile | 1500 mg/l | 3300 mg/l |
| chlorophenolic | | |
| manufacturing | 4300 mg/l | 5400 mg/l |
| milk bottling | 230 mg/l | 420 mg/l |
| cheese production | 3200 mg/l | 5600 mg/l |
| candy production | 1600 mg/l | 3000 mg/l |

*Example 8.2*

A sample treatment plant effluent begins to clarify after 13 days. What percent of the original BOD remains unsatisfied?

From table 8.3, the relative stability is 95%. Therefore, only 5% of the initial BOD remains unsatisfied.

## D. CHEMICAL OXYGEN DEMAND

Unlike BOD, which is a measure of oxygen removed by biological organisms, chemical oxygen demand (COD) is a measure of maximum oxidizable substances. Therefore, COD is an excellent measure of *effluent strength.*

COD testing is required in environments of chemical pollution. In such environments, the organisms necessary to metabolize organic compounds may not exist. Furthermore, the toxicity of the water may make the standard BOD test impossible to carry out. The COD test also produces results faster than the BOD test. COD test results are usually available in a matter of hours.

If the toxicity is low, BOD and COD test results can be correlated. The $BOD_s/COD$ ratio typically varies from 0.4 to 0.8. This is a wide range, but for any given treatment plant and waste type, the correlation is essentially constant. The correlation can, however, vary along the treatment path.

## E. CHLORINE DEMAND

Chlorination destroys bacteria, hydrogen sulfide, and other noxious substances. For example, hydrogen sulfide is oxidized according to equation 8.13.

$$H_2S + 4H_2O + 4Cl_2 \rightarrow H_2SO_4 + 8HCl \qquad 8.13$$

*Chlorine demand* is the amount of chlorine (or its chloramine or hypochlorite equivalent) required to give a 0.5 mg/l residual after 15 minutes of contact time. 15 minutes is the recommended contact and mixing time prior to discharge since this period will kill nearly all pathogenic bacteria in the water. Typical doses for wastewater effluent are given in table 8.4.

**Table 8.4**
Typical Chlorine Doses

| final process | dose (mg/l) |
|---|---|
| no treatment (straight discharge) | 10–30 |
| secondary sand filter | 2–6 |
| secondary activated sludge | 2–8 |
| secondary trickling filter | 3–15 |
| primary sedimentation | 5–25 |

In actuality, the chlorine dose needs to be determined by careful monitoring of coliform counts and free residuals, since there are several ways that chlorine can be used up without producing significant disinfection. Only after uncombined (free) chlorine starts showing up is it assumed that all chemical reactions and disinfection are complete.[5]

Because of their reactivity, chlorine is initially used up in the neutralization of hydrogen sulfide and the rare ferrous and manganous ($Fe^{++}$ and $Mn^{++}$) ions. The resulting HCl, $FeCl_2$, and $MnCl_2$ ions do not contribute to disinfection. They are known as *unavailable combined residuals.*

Plants and animals use nitrogen. Bacterial decomposition and the hydrolysis of urea produces ammonia, $NH_3$. This ammonia, once it enters the wastewater stream, forms ammonium ion, $NH_4^+$, also known as *ammonia nitrogen.*

Ammonia nitrogen combines with chlorine to form the family of *chloramines.* Depending on the water pH, *monochloramines* ($NH_2Cl$), *dichloramines* ($NHCl_2$), or *trichloramines* (nitrogen trichloride, $NCl_3$) may form.[6] Chloramines have long-term disinfection capabilities, and chloramines are therefore known as *available combined residuals.* Equation 8.14 is a typical chloramine formation reaction.

$$NH_4^+ + HOCl \rightleftharpoons NH_2Cl + H_2O + H^+ \qquad 8.14$$

The continued addition of chlorine after chloramine formation changes the pH, and *chloramine destruction* begins. Chloramines are converted to nitrogen gas ($N_2$) and nitrous oxide ($N_2O$). The destruction of chloramines continues with the continued application of chlorine, until no ammonia remains in the water. The point at which all ammonia has been removed is known as the *breakpoint.* Equation 8.15 is a typical chloramine destruction reaction.

$$2NH_2Cl + HOCl \rightleftharpoons N_2 + 3HCl + H_2O \qquad 8.15$$

In the *breakpoint chlorination* method, additional chlorine is added after the breakpoint in order to obtain free chlorine residuals. The free residuals have a high disinfection capacity. Typical free residuals are free chlorine ($Cl_2$), hypochlorous acid (HOCl), and hypochlorite ions. Equations 8.16 and 8.17 illustrate the formation of these free residuals.

$$Cl_2 + H_2O \rightarrow HCl + HOCl \qquad 8.16$$

$$HOCl \rightarrow H^+ + ClO^- \qquad 8.17$$

[5] Chlorine kills most bacteria, but many viruses are resistant.

[6] Lower pH favors the formation of di- and trichloramines.

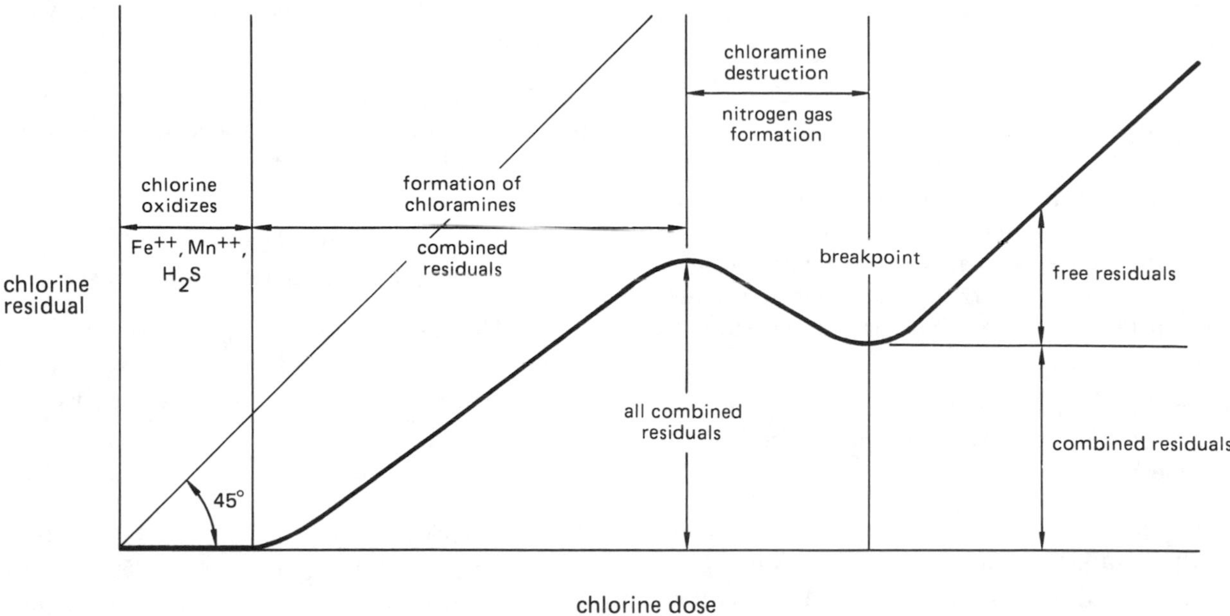


**Figure 8.2** Breakpoint Chlorination

There are several problems associated with breakpoint chlorination.

- It may not be economical to use breakpoint chlorination unless the ammonia nitrogen has been reduced.
- Free chlorine residuals favor the formation of trihalomethanes. Where free residuals are not permitted, the water may need to be dechlorinated using sulfur dioxide gas or sodium bisulfate. (Where small concentrations of free residuals are permitted, dechlorination may be needed only during the dry months. During winter storm months, the chlorine residuals may be adequately diluted with rain water.)

## F. GREASE

Greases are organic substances including fats, vegetable and mineral oils, waxes, fatty acids from soaps, and other hydrocarbons. Grease's low solubility causes adhesion problems in pipes and tanks, reduces contact area during various filtering processes, and produces sludge which is difficult to dispose of.

## G. VOLATILE ACIDS

Volatile acids (acetic, propionic, and butyric) occur in anaeorobically digested sludge. These acids can be used to indicate the completion of a sludge digestion process. Acid content is given in mg/l as acetic acid.

## H. SUSPENDED SOLIDS

Suspended solids, as in water supply engineering, can be categorized in several ways. Generally, suspended solids constitute only a small amount of the incoming flow, less than 1/10%. Together with *dissolved solids*, suspended solids constitute *total solids*. Figure 8.3 illustrates the relationships between the various solids categories. A further division of each category into organic and inorganic solids is possible.

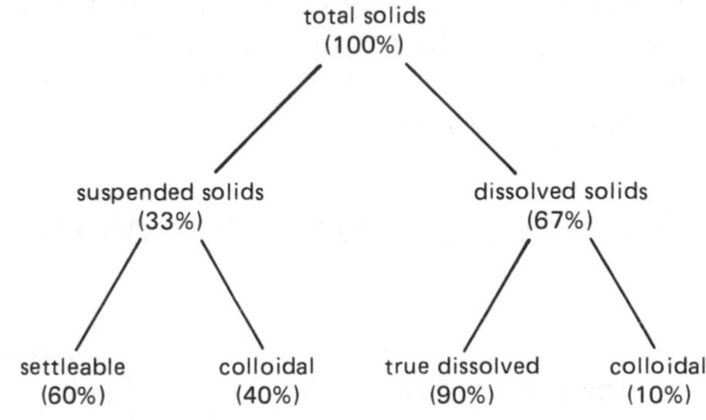


**Figure 8.3** Family of Solids
(Typical percentages given.)

The term *volatile solids* can be used as a measure of organic pollutants capable of affecting the oxygen content of the flow. Volatile solids, as in water supply testing, are measured by igniting filtered solids, and measuring the decrease in mass.

*Refractory pollutants* are solids which are difficult to remove by common processes. In this case, the term "refractory" is used to mean "stubborn."

## 4 DISINFECTION

Chlorine gas is the least expensive, and therefore the most common, method of disinfecting wastewater. However, chlorine gas is toxic, corrosive, and displaces oxygen since it is heavier than air. Chlorine gas also lowers the pH of the water, favoring the formation of combined residuals.

Because of these disadvantages, alternatives to chlorine gas need to be considered when recommending the disinfection method.

- **Hypochlorites**: Both sodium and calcium hypochlorite are solids that dissolve in water. They have a limited shelf life, and are susceptible to photodecomposition. Hypochlorites are less effective and slightly more expensive than chlorine gas.
- **Chlorine Dioxide Gas**: $Cl_2O$ is explosive and must be generated on-site. It reacts with many compounds, requiring larger doses than chlorine gas. However, it combines with organics without combining with ammonia.
- **Ozone Gas**: Ozone is one of the most effective oxidizing agents. In addition to its disinfection capabilities, ozone also increases the dissolved oxygen content of water. Ozone is toxic and corrosive. It must be generated at the point of application. Because of its very short half-life, step feeding is required to obtain the necessary contact period.
- **Exotics**: Other methods exist, but are typically high in cost. These exotic alternatives include bromine (bromine chloride), iodine, silver oxide, gamma radiation, and ultraviolet radiation.

## 5 TYPICAL COMPOSITION OF DOMESTIC SEWAGE

Not all sewage flows are the same. Some sewages are stronger than others. Table 8.5 lists typical values for strong and weak domestic sewages. A medium classification would be approximately midway between the values for strong and weak.

## 6 WASTEWATER QUALITY STANDARDS

Applicable wastewater quality standards have been set by both the Water Pollution Control Act and the Environmental Protection Agency. General standards set by the Water Pollution Control Act are given in table 8.6. These standards must be met by facilities that receive federal funding.

The EPA's standards for secondary treatment are given in terms of 5-day BOD, suspended solids, coliform count, and pH. Table 8.7 presents typical values.

**Table 8.5**
Strong and Weak Domestic Sewages

(All concentrations in mg/l unless noted.)

| constituent | strong | weak |
|---|---|---|
| solids, total | 1200 | 350 |
| dissolved, total | 850 | 250 |
| fixed | 525 | 145 |
| volatile | 325 | 105 |
| suspended, total | 350 | 100 |
| fixed | 75 | 30 |
| volatile | 275 | 70 |
| settleable solids, (ml/liter) | 20 | 5 |
| biochemical oxygen demand, 5-day, 20°C | 300 | 100 |
| total organic carbon | 300 | 100 |
| chemical oxygen demand | 1000 | 250 |
| nitrogen, (total as N) | 85 | 20 |
| organic | 35 | 8 |
| free ammonia | 50 | 12 |
| nitrites | 0 | 0 |
| nitrates | 0 | 0 |
| phosphorus (total as P) | 20 | 6 |
| organic | 5 | 2 |
| inorganic | 15 | 4 |
| chlorides | 100 | 30 |
| alkalinity (as $CaCO_3$) | 200 | 50 |
| grease | 150 | 50 |

## 7 DESIGN FLOW QUANTITY

Approximately 70 to 80% of a community's domestic and industrial water use will return as wastewater. This water is discharged into the sewer systems, which may be different or the same as storm drains. Therefore, the nature of the return system must be known before sizing can occur.

*Sanitary sewer* sizing can often be based on an average of 100–125 gpcd. There will be variations with time in the flow, although the variations are not as pronounced as they are for water supply. Table 8.8 lists peak multipliers for treatment plant influent volume. Due to storage in ponds, clarifiers, and sedimentation basins, these multipliers may not be applicable throughout all processes in the treatment plant.

**Table 8.6**
Typical Surface Water Standards

| water use | minimum dissolved oxygen (mg/l) | maximum dissolved solids | maximum coliforms per 100 ml |
|---|---|---|---|
| domestic use (food preparation) | 6 | * | none |
| water contact recreation | 4 to 5 | * | 1000 total ave. 200 fecal ave. not more than 10% exceeding 400 fecal (2000 total) |
| fisheries | 4 to 6 | * | 5000 ave. |
| industrial supply | 3 to 5 | 750 to 1500 mg/l | – |
| agricultural irrigation | 3 to 5 | 750 to 1500 mg/l | – |
| shellfish harvesting | 4 to 6 | * | 70 total ave. not more than 10% exceeding 230 |

* No floating solids or settling solids that form deposits.

Design codes established by state boards of health frequently specify a design loading of 400 gpcd (laterals and submains) and 250 gpcd (mains, trunks, and outfall). Both of these include the effect of *infiltration*. Infiltration due to cracks and poor joints is limited by many municipal codes to 500 gallons per day per mile of pipe per inch of diameter. Modern piping materials and joints should be able to reduce this quantity to 200 gpd/inch-mile. Infiltration may also be roughly estimated at 3%–5% of the peak hourly domestic rate, or as 10% of the average rate.

## 8 COLLECTION SYSTEMS

### A. STORM DRAINS AND INLETS

Curb inlets to storm drains should be placed no more than 600 feet apart, and a limit of 300 feet is advisable. Inlets are required at all low points where pondage could occur. A common practice is to install 3 inlets in a sag vertical curve—one at the lowest point and one on each side with an elevation of 0.2 feet above the center inlet. Openings may be of the covered grate type or the curb inlet type.

**Table 8.7**
Typical Secondary Effluent Standards

| quality | average over | discharge maximum |
|---|---|---|
| BOD (5-day) | 30 days | 30 mg/l |
| | 7 days | 45 mg/l |
| | 30 days | 15% of incoming BOD |
| suspended solids | 30 days | 30 mg/l |
| | 7 days | 45 mg/l |
| | 30 days | 15% of incoming SS |
| fecal coliforms** | 30 days | 200 per 100 ml |
| | 7 days | 400 per 100 ml |
| pH | at all times | within 6 to 9 |

**A geometric mean is used, not arithmetic

**Table 8.8**
Variations in Wastewater Flow
(based on the average daily flow)

| description | when/where | variation |
|---|---|---|
| daily peak | 10–12 a.m. (residential) | 225% |
| | constant during day (commercial) | 150% |
| | 12 noon at the outfall | 150% |
| daily minimum | 4–5 a.m. | 40% |
| seasonal peak | late summer | 125% |
| seasonal minimum | winter's end | 90% |
| seasonal average | May, June | 100% |
| maximum peaks | in laterals | 300% |
| | treatment plant influent | 200% |

The capacity of a curb-type opening which diverts 100% of gutter flow is given by equation 8.18. A typical curb depression is 5 inches.

$$Q' = (0.7)\begin{pmatrix}\text{curb}\\\text{opening}\\\text{length, ft}\end{pmatrix}\begin{pmatrix}\text{inlet flow}\\\text{depth, ft}\end{pmatrix} + \begin{pmatrix}\text{curb inlet}\\\text{depression, ft}\end{pmatrix}^{3/2} \quad 8.18$$

Grate inlets accepting flows less than 0.4 feet deep have a capacity given by equation 8.19. The bars should be parallel to the flow and at least 18 inches long. Equation 8.19 should also be used for combined curb-grate inlets.

$$Q' = 3\begin{pmatrix}\text{grate}\\\text{perimeter, ft}\end{pmatrix}\begin{pmatrix}\text{inlet flow}\\\text{depth, ft}\end{pmatrix}^{3/2} \quad 8.19$$

B. MANHOLES

Manholes should be provided at junctions and at changes in elevation, direction, size, diameter, and slope of sewers. If the sewer is too small for a man to enter, manholes should be placed every 400 feet to allow for cleaning. A maximum recommended spacing is every 700 feet.

**Table 8.9**
Recommended Manhole Spacing

| pipe diameter | spacing |
|---|---|
| less than 18″ | 400 ft |
| 18″–48″ | 500 |
| more than 48″ | 600 |

C. PIPES

Concrete pipe is commonly used for storm sewers. Circular pipe is used in most applications, although special shapes (arch, egg, elliptical, etc.) are available at extra cost. Concrete pipe in diameters up to 24″ are usually not reinforced, and are available in standard 3 and 4 foot lengths. Reinforced pipe in diameters ranging from 12″ to 144″ is available in lengths from 4 to 12 feet.

Sewer pipes are constructed from clay, concrete, asbestos-cement, steel, cast iron, and plastic derivatives. *Vitrified clay pipe* is especially resistant to acids, alkalines, hydrogen sulfide (septic sewage), erosion, and scour. Clay is typically used for diameters less than 36″. Clay pipe is available in standard diameters of 4″ and 6″ in 2 foot lengths; 8″, 10″, 12″, 15″, 18″, 21″, and 24″ diameters in $2\frac{1}{2}$ foot lengths; and 27″, 30″, 33″, and 36″ diameters in 3 foot lengths.

Two strengths of clay pipe are available. The standard strength is suitable for pipes less than 12″ in diameter for any depth of cover if the '4/3D + 8″' trench width rule is observed. Double strength pipe is recommended for large pipe deeply trenched.

*Asbestos-cement pipe* can be used for both gravity and pressure sewers carrying non-septic and non-corrosive wastes through non-corrosive soils. Light weight and longer laying lengths are the inherent advantages of asbestos-cement pipe.

*Concrete pipe* is primarily used for large diameter (16 inches or larger) trunk and interceptor sewers. In some geographical regions, concrete pipe is used for domestic sewers in smaller pipe sizes. However, concrete domestic lines should be selected only where stale or septic sewage is not anticipated. Concrete pipe can be used for gravity as well as pressure mains. However, it should not be used with corrosive wastes or soils.

*Cast iron pipe* is particularly suited to installations where scour, high velocity waste, and high external loads are anticipated. It can be used for domestic connections, although it is more expensive than clay pipe. Special linings, coatings, wrappings, or encasements are required for corrosive wastes and soils.

*Polyvinyl chloride (PVC)* and *acrylonitrile-butadiene-styrene (ABS)* are two plastic compositions that can be used for normal domestic sewage and industrial wastewaters. They have excellent resistance to corrosive soils. However, special care must be given to trench loadings and pipe beddings.

ABS plastic may also be combined with concrete reinforcement for collector lines for corrosive domestic sewage and industrial wastes. Such pipe is known as *truss pipe* due to its construction.[7]

For pressure lines, welded steel pipe with an epoxy liner, and cement-lined and coated steel pipe are also used occasionally.

In general, sewers in the collection system (including laterals, interceptors, trunks, and mains) should be at least 8″ in diameter. Building service connections can be as small as 4″ in diameter.

## 9 PIPE FLOW VELOCITIES

Sewer flow velocities greater than 15 ft/sec require special provisions to protect against erosion and momentum effects. 2 ft/sec is often quoted as the minimum *self-cleansing velocity*.[8] However, the minimum design velocity depends on the particulate matter size. Ta-

[7] The plastic is extruded with inner and outer web-connected pipe walls. The voids between the inner and outer walls are filled with lightweight concrete.

[8] Even 1.5 ft/sec is acceptable if the main is occasionally flushed out by peak flow.

ble 8.10 can be used to select velocity for the collection system as well as for the plant treatment system.

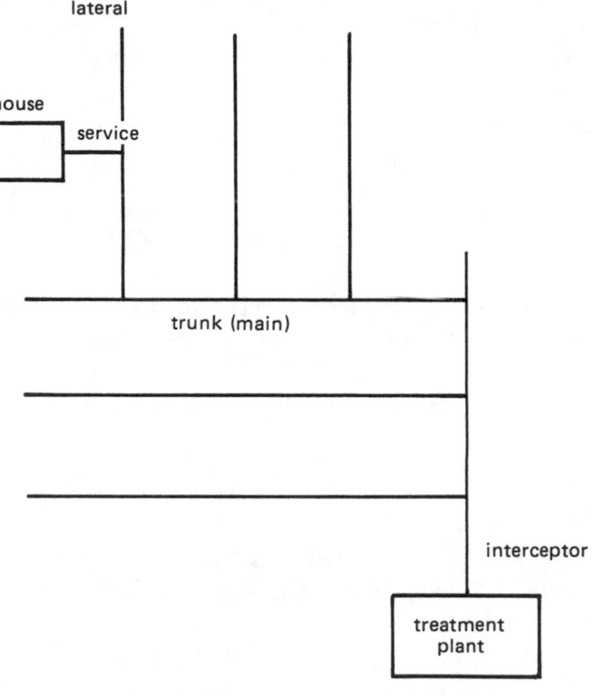


**Figure 8.4** Types of Sewer Lines

**Table 8.10**
Minimum Flow Velocities

| pipe carrying | minimum velocity to keep particles in suspension | minimum resuspension velocity |
|---|---|---|
| raw sewage | 2.5 | 3.5 |
| grit tank effluent | 2 | 2.5 |
| primary settling tank effluent | 1.5 | 2 |
| mixed liquor | 1.5 | 2 |
| trickling filter effluent | 1.5 | 2 |
| secondary settling tank effluent | 0.5 | 1 |

## 10 PUMPS USED IN WASTEWATER PLANTS

Wastewater plant flows should be gravity fed wherever possible. However, there are still many instances where pumping assistance is required. Table 8.11 lists pump types as a function of type of material to be pumped.

**Table 8.11**
Pumps Used in Wastewater Plants

| type of material being pumped | flow rate (gpm) | pump type |
|---|---|---|
| raw sewage | 0 to 50 | pneumatic ejector |
| | 50 to 200 | submersible or end-suction non-clog centrifugal |
| | 200 up | end-suction non-clog centrifugal |
| settled sewage | 0 to 500 | end-suction non-clog centrifugal |
| | 500 up | vertical axial or mixed flow centrifugal |
| sludge (primary, thickened, or digested) | | plunger pump |
| secondary sludge | | end-suction non-clog centrifugal |
| scum | | plunger pump or recessed impeller |
| grit | | recessed impeller centrifugal, pneumatic ejector, or conveyor rake |

## 11 DILUTION PURIFICATION

Dilution purification (also known as *self purification*) refers to discharge of partially treated sewage into a body of water such as a stream or river. If the body is large and is adequately oxygenated, the sewage's BOD may be satisfied without putrefaction. Other conditions which must be monitored besides BOD are oxygen content and suspended solids.

Equation 8.20 can be used to calculate the final concentration of BOD, oxygen, and sediment when the two flows are mixed. Dilution requirements may be expressed in terms of ratios (e.g., 23 stream volumes per discharge volume) or absolute flow quantities (e.g., 4 to 7 cfs per 1000 population).

$$C_1Q_1 + C_2Q_2 = C_f(Q_1 + Q_2) \qquad 8.20$$

*Example 8.3*

Wastewater ($DO$ = 0.9 mg/l, 6 MGD) is discharged into a 50°F stream flowing at 40 cfs. Assuming the stream is saturated with oxygen, what is the oxygen content of the stream immediately after mixing?

From appendix B, the saturated oxygen content at 50°F (10°C) is 11.3 mg/l.

$$(6 \text{ MGD})\left(1.547\ \frac{\text{cfs}}{\text{MGD}}\right) = 9.28 \text{ cfs}$$

$$C = \frac{(0.9)(9.28) + (11.3)(40)}{9.28 + 40} = 9.34 \text{ mg/l}$$

The *oxygen deficit* is the difference between actual and saturated oxygen concentrations. Since reoxygenation and deoxygenation of a polluted river occur simultaneously, an oxygen deficit will occur only if the reoxygenation rate is less than the deoxygenation rate. If the oxygen content goes to zero, anaerobic decomposition and putrefaction will occur.

The oxygen deficit at any time $t$ is given by the *Streeter-Phelps equation*:

$$D_t = DO_{\text{sat}} - DO_t$$

$$= \frac{K_D \text{BOD}_u}{K_R - K_D}\left(10^{-K_D t} - 10^{-K_R t}\right)$$

$$+ D_o\left(10^{-K_R t}\right) \qquad 8.21$$

$D_t$ is the dissolved oxygen deficit, $t$ is in days, and $\text{BOD}_u$ is the ultimate carbonaceous BOD of the stream immediately after mixing. $K_D$ and $K_R$ are the deoxygenation and reoxygenation rate constants respectively, and $D_o$ is the dissolved oxygen deficit immediately after mixing.[9]

$K_R$ can be approximated by equation 8.22 if field test data is not available.[10]

$$K_{R,20°\text{C}} \approx \frac{3.3v}{d^{1.33}} \qquad 8.22$$

$K_R$ for different temperatures is given by equation 8.23. Typical values of $K_R$ are given in table 8.12.

$$K_{R,T} = (1.016)^{T-20} K_{R,20°\text{C}} \qquad 8.23$$

**Table 8.12**
Typical Reoxygenation Constants (base 10, 1/days)

| | |
|---|---|
| white water | 0.5 and above |
| swiftly flowing | 0.3 to 0.5 |
| large streams | 0.15 to 0.3 |
| large lakes | 0.10 to 0.15 |
| sluggish streams | 0.10 to 0.15 |
| small ponds | 0.05 to 0.10 |

Equations 8.6 and 8.21 can be plotted simultaneously as shown in figure 8.5. The plot of equation 8.21 is known as the *oxygen sag curve.* The difference between the two curves is the effect of reoxygenation.

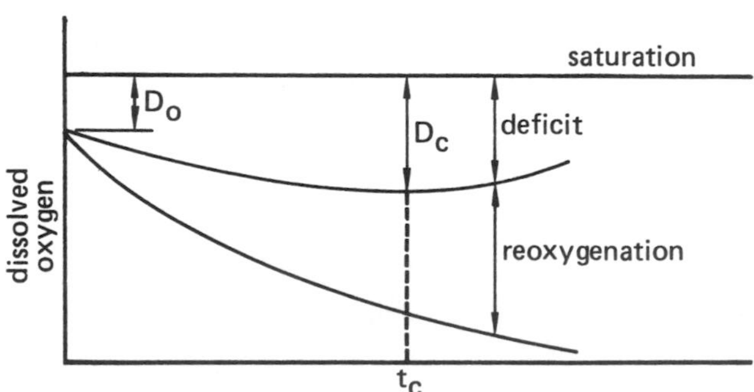


**Figure 8.5** The Oxygen Sag Curve

[9] $K_R$ and $K_D$ may be written as $K_1$ and $K_2$ by other authors.

[10] Equation 8.22 is the O'Connor and Dobbins formula for natural streams.

The time to the minimum or *critical point* on the sag curve is given by equation 8.24.

$$t_c = \frac{1}{K_R - K_D} \times \log_{10}\left[\left(\frac{K_D \text{BOD}_u - K_R D_o + K_D D_o}{K_D \text{BOD}_u}\right)\left(\frac{K_R}{K_D}\right)\right] \quad 8.24$$

The ratio of $K_R/K_D$ is known as the *self-purification coefficient.* The *critical oxygen deficit* is given by equation 8.25.

$$D_c = \left(\frac{K_D \text{BOD}_u}{K_R}\right) 10^{-K_D t_c} \quad 8.25$$

Knowing $t_c$ and the stream flow velocity will locate the point where the oxygen level is the lowest.

*Example 8.4*

A treatment plant discharge has the following characteristics:

15 cfs
45 mg/l BOD (5 day, 20°C)
2.9 mg/l DO
24°C
$K_{D,20°C} = 0.1$ per day (when mixed with river water)

The outfall is located in a river with the following characteristics:

0.55 ft/sec velocity
4.0 feet average depth
120 cfs
4 mg/l BOD (5 day, 20°C)
8.3 mg/l DO
16°C

Determine the distance downstream where the oxygen level is minimum, and predict if the river can support fish life at that point.

*step 1*: Find the river conditions immediately after mixing. Use equation 8.20 three times.

$$(\text{BOD})_{20°C} = \frac{(15)(45) + (120)(4)}{15 + 120} = 8.56 \text{ mg/l}$$

$$DO = \frac{(15)(2.9) + (120)(8.3)}{135} = 7.7 \text{ mg/l}$$

$$T = \frac{(15)(24) + (120)(16)}{135} = 16.89°\text{C}$$

*step 2:* The river may begin to return to its original equilibrium temperature. However, no information about the temperature recovery is given. Assume that the mixture temperature is constant at least to the critical point. Calculate the rate constants. From equation 8.8,

$$K_{D,16.89°C} = (0.1)(1.047)^{16.89-20} = 0.0867$$

From equation 8.22,

$$K_{R,20°C} \approx 3.3(0.55/(4)^{1.33}) = 0.287$$

From equation 8.23,

$$K_{R,16.89°C} = (0.287)(1.016)^{16.89-20} = 0.273$$

*step 3:* Estimate $\text{BOD}_u$. Using equation 8.6,

$$(\text{BOD}_u)_{20°C} = \frac{8.56}{1 - (10)^{-(0.0867)(5)}} = 13.56$$

From equation 8.9,

$$(\text{BOD}_u)_{16.89°C} = (13.56)[(0.02)(16.89) + 0.6] = 12.72$$

*step 4:* Calculate $D_o$. From appendix B, the saturated oxygen cencentration at 16.89°C is approximately 9.7 mg/l. So, $D_o = 9.7 - 7.7 = 2.0$.

*step 5:* Calculate $t_c$ from equation 8.24.

$$t_c = \frac{1}{0.273 - 0.0867} \times \log\left[\frac{(0.0867)(12.72) - (0.273)(2) + (0.0867)(2)}{(0.0867)(12.72)} \times \left(\frac{0.273}{0.0867}\right)\right] = 1.70 \text{ days}$$

*step 6:* The distance downstream is

$$\frac{(1.70 \text{ days})(0.55 \text{ ft/sec})(86{,}400 \text{ sec/day})}{5280 \text{ ft/mile}} = 15.3 \text{ miles}$$

*step 7:* The critical oxygen deficit is found from equation 8.25.

$$D_c = \frac{(0.0867)(12.72)}{0.273}(10)^{-(0.0867)(1.77)} = 3.0$$

*step 8:* If the temperature 15.3 miles downstream is 16.89°C, the saturated oxygen content is approximately 9.7 mg/l. Since the critical deficit is 3, the minimum oxygen content is 6.7 mg/l. This is adequate for fish life.

## 12 SMALL VOLUME DISPOSAL

### A. CESSPOOLS

A cesspool is a form of sub-surface disposal consisting of a lined and covered underground cavern into which sewage is discharged. Cesspools may be watertight if they are for temporary storage only. *Leaching cesspools* allow seepage of sewage into the ground. Cesspools are acceptable only for small volumes (1 or 2 families) of sewage.

### B. SEPTIC TANK

A septic tank is a simple tank which allows both sedimentation and digestion to occur. Typical detention times are 8 to 24 hours. Effluent is discharged into underground tile fields which allow the water to percolate into the soil. Only 30–50% of the suspended solids are removed by septic tanks. The remaining solids eventually clog the tank and must be mechanically removed.

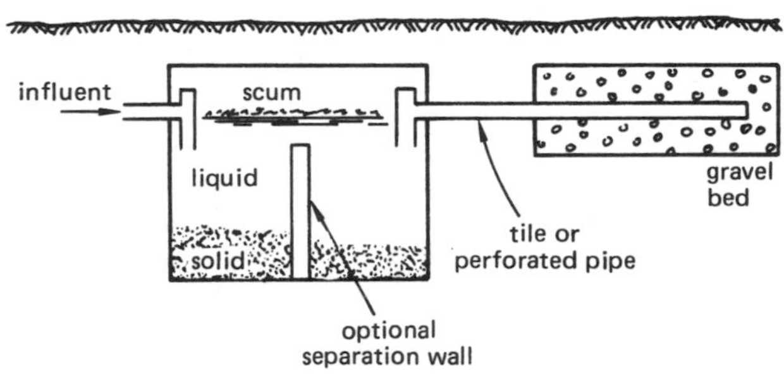


**Figure 8.6** Typical Septic Tank

Typical design parameters of a domestic septic tank are:

- minimum capacity below flow line — 300 gal. for 5 persons or less; 500 gal. preferred. No garbage disposals. Add 30 gal. per additional person.
- plan aspect ratio — 1:2
- minimum depth below flow line — 3 to 4 feet
- minimum freeboard — 1 foot
- tank burial depth — 1 to 2 feet
- tile length — 30 feet per person
- maximum tile run length — 60 feet
- minimum tile depth — 15 inches (30″ preferred)
- lateral line spacing — 6 feet
- gravel bed size — 4 inch radius around tile, 12″ below
- minimum soil layer below tile bed — 10 feet

Municipal septic tanks should be designed to hold 12–24 hours of flow plus stored sludge. A general rule is to allow at least 25 gallons per person served by the tank.

### C. IMHOFF TANK

An Imhoff tank is similar to a septic tank in that sedimentation and sludge digestion both occur. However, these two processes occur in different parts of the tank, and Imhoff vessels are larger in capacity than simple domestic septic tanks. Wastewater enters the tank at the top where sedimentation occurs. The sediment slides down the sloped inner baffles.

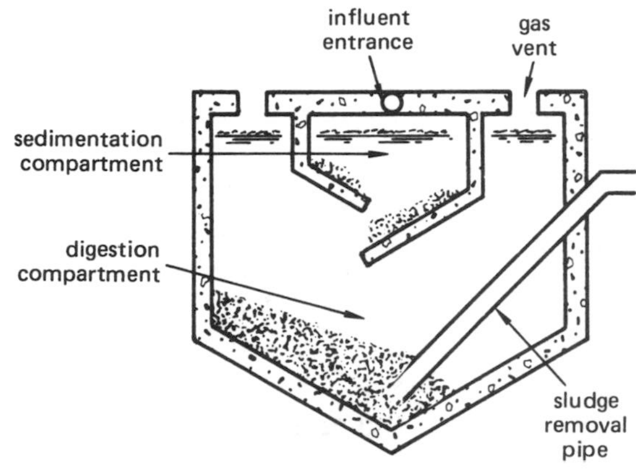


**Figure 8.7** Simple Imhoff Tank

One of the inner baffles extends past the other so that gas produced in the digester chamber will not enter the sedimentation chamber.

Imhoff tanks are usually very large (i.e., 2 stories high) and have been used in the past where the loading is between 250,000 gpd and 1,000,000 gpd. Imhoff tanks can remove up to 60% of the suspended solids during the 1 to 2 hour retention time. They are more efficient than septic tanks, but require very frequent (up to hourly) attention.

Typical characteristics of an Imhoff tank are:

- sludge chamber capacity — 2.5 $ft^3$/person
- slope of inner baffles — 2 vertical: 1 horizontal
- total depth — 15 feet minimum
- gas vent area — 15% to 25% of top area
- fall-through slot width — 8″ minimum
- baffle overhang — 8″
- sludge pipe diameter — 8″ minimum
- distance from slot to sludge — 18″minimum

## 13 WASTEWATER PLANT SITING CONSIDERATIONS

Wastewater plants should be located as far as possible from inhabited areas. A minimum distance of 1000 feet for uncovered plants is desired. Uncovered plants should be located downwind when a definite wind direction prevails. Foundation conditions need to be evaluated, as does the elevation of the water table. Elevation in relationship to the need for sewage pumping (and for dikes around the site) is relevant. Furthermore, the plant must be protected against flooding. 100-year storms are typically chosen as the design flood when designing dikes and similar facilities. Distance to the outfall and possible effluent pumping need to be considered.

Table 8.13 lists the approximate acreage for preliminary engineering estimates. Of course, an estimate of expansion is proper when evaluating acreage requirements.

**Table 8.13**
Treatment Plant Acreage Requirements

| type of treatment | acres per MGD |
|---|---|
| activated sludge plants | 2 |
| trickling filter plants | 3 |
| aerated lagoons | 16 |
| stabilization basins | 20 |
| physical-chemical plants | 1.5 |

## 14 PRETREATMENT OF INDUSTRIAL WASTES

Industrial wastes that would harm collection or treatment facilities or upset subsequent biological processes need to be pretreated. The guidelines which follow should be evaluated for applicability and conformance to local, state, and federal codes. If possible, eliminate the contaminants at their sources.

- **chromium removal**: If hexavalent chromium is greater than 2 mg/l in the influent, use chemical reduction followed by chemical precipitation.
- **heavy metal removal**: Use chemical precipitation if total heavy metals exceed 1 mg/l. If recovery of the metals is desired, use ion exchange methods.
- **cyanide removal**: Use chemical oxidation if the concentration exceeds 2 mg/l. Use electrolysis for high-strength, low-flow waste streams.
- **phenol removal**: Biochemical oxidation and chemical oxidation can both be used, although separate biological treatment for phenol may be uneconomical. Maximum concentration needs to be determined empirically.
- **pH adjustment**: The pH of water entering biological treatment should be between 6.0 and 9.0. Neutralize with acid or alkalai additives.
- **emulsified oil removal**: Use coagulation and flotation adsorption on activated carbon.
- **hydrogen sulfide removal**: Preaerate if sulfides exceed 50 mg/l.
- **oil separation**: Use gravity separation.

## 15 WASTEWATER PROCESSES

### A. PRELIMINARY TREATMENT

Preliminary preparation of the wastewater stream is essentially a mechanical process. It removes large objects, rags, and wood from the flow. Heavy solids and excessive oils and grease are also eliminated. Damage to pumps and other equipment would be expected without preliminary treatment.

**Screens**: Trash racks or coarse screens with openings 2 inches or larger should precede pumps to prevent clogging. Screenings usually consist of paper, wood, and rags. Medium screens ($\frac{1}{2}''$ to $1\frac{1}{2}''$ openings) and fine screens ($\frac{1}{16}''$ to $\frac{1}{8}''$) are also used to relieve the load on grit chambers and sedimentation basins.[11] Screens are cleaned by automatic scraping arms. Screen capacities and head losses are specified by the manufacturer. In general, however, flow through screens should be limited to 3 fps or less.

**Grit Chambers**: Grit is an abrasive that wears pumps, clogs pipes, and accumulates in excessive volumes. A grit chamber (also known as *grit clarifier* or *detritus tank*) slows the wastewater down to approximately 1 ft/sec. This velocity allows the grit to settle out but moves the organic matter through. The grit can be manually or mechanically removed with buckets or screw conveyors.

[11] Fine screens are rare except when used with some industrial waste processing plants.

Typical design standards for grit chambers are:

- grit removal rate 1 to 5 $ft^3$/MG
- grit size 0.2 mm or larger
- grit specific gravity 2.65
- depth 3 to 4 feet
- length (width not critical) 40 to 100 feet
- detention time 45 to 90 seconds
- horizontal velocity 0.75 to 1.25 ft/sec

In actual practice, grit chambers are designed to keep the flow velocity as close to 1.0 ft/sec as possible. If an analytical design based on settling velocity is required, the *scouring velocity* should not be exceeded. Scouring of the minimum-sized particles which have already settled will be prevented if the velocity is kept below that in equation 8.26. ($SG_p$ is the specific gravity of the particle. $d_p$ is the particle diameter in millimeters.) The friction factor is approximately 0.03 for grit chambers. The scouring velocity has been converted to ft/sec even though d is in mm.[12]

$$v = 2.2\sqrt{\frac{gd_p}{f}(SG_p - 1)} \approx 1.3\sqrt{d_p(SG_p - 1)} \qquad 8.26$$

**Aerated Grit Chambers**: In smaller plants, the grit chamber may be a hopper-bottomed tank with a small detention time. Diffused aeration from one side of the tank rolls the water and is employed to keep the organics in suspension while the grit settles out. Influent enters on one side of the tank, and an effluent weir on the opposite side removes the degritted wastewater. The water spirals or rolls through the tank.

Solids are removed by airlift pump, screw conveyer, bucket elevator, or gravity flow. However, since the scouring velocity is not maintained, the grit will have a significant organic content. A grit washer or cyclone separator may be used to clean the grit.

Typical characteristics of aerated grit chambers are:

- detention time 2 to 5 minutes at peak flow (3 typical)
- air supply:
  - shallow tanks 1.5 to 5.0 cfm/ft length (3.0 typical)
  - deep tanks 3.0 to 8.0 cfm/ft length (5.0 typical)
- grit and scum quantities 0.5 to 25.0 $ft^3$/MG (2.0 typical)
- length-width ratio 2.5:1 to 5:1 (3:1 typical)
- depth 6 to 15 feet
- length 20 to 60 feet
- width 7 to 20 feet

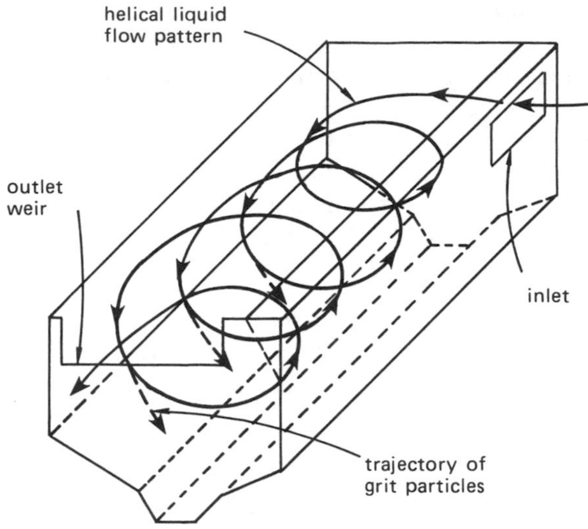


**Figure 8.8** Basic Aerated Grit Chamber

**Skimming Tanks**: If the sewage has excessive grease or oil, a basin 8 to 10 feet deep providing 5 to 15 minutes of detention time will allow the grease to rise to the surface.[13] An aerating device below will help coagulate and float grease to the surface. Approximately 0.01 to 0.1 cubic feet of 40 to 80 psig air per gallon of influent should be used. (The actual air pressure will depend on the tank depth.) Surface grease can be mechanically removed by skimming troughs. The tank outlet is submerged, and it is lower than the inlet.

If the skimming tank is enclosed and the air evacuated to approximately 9″ of mercury, rising bubbles in the sewage will expand and help float grease upwards without the need for mechanical aeration.

A small fraction (e.g., 30%) of the influent may be recycled in some cases.

**Shredders**: Shredders (also called *comminutors*) cut waste solids to approximately 1/4″ in size. They reduce the amount of screenings which must be disposed of. Shreddings generally stay with the flow for later settling.

**Other Pretreatment Processes**: Odor control through chlorination, aeration, or ozonation, freshening of septic waste by aeration, and flow equalization

[12] Equation 8.26 is known as the *Camp formula.*

[13] 50 mg/l or more of total floatables should be considered as requiring skimming.

in holding basins can also be loosely categorized as pre-treatment processes.

## B. PRIMARY TREATMENT

Primary treatment is a mechanical (settling) process used to remove most of the settleable solids. A 25 to 35% reduction in BOD is also achieved, but BOD reduction is not the goal of primary treatment.

**Plain Sedimentation**: Plain sedimentation basins are described in chapter 7. Design characteristics for wastewater treatment are:

- BOD reduction — 20% to 40%
- total suspended solids reduction — 35% to 65%
- bacteria reduction — 50% to 60%
- organic content of settled solids — 50% to 75%
- specific gravity of settled solids — 1.2 or less
- typical settling velocity — above 4 feet/hr
- plan shape — rectangular or circular
- basin depth — 6 to 15 feet (12 typical)
- basin width — 10 to 50 feet
- minimum freeboard — 18 inches
- minimum hopper wall angle — 60°
- aspect ratio (rectangular) — 3:1 to 5:1
- detention time — 1.5 to 2.5 hours
- circular diameter — 30 to 150 feet (100 common)
- flow-through velocity — 0.005 ft/sec
- flow-through time — at least 30% of detention time
- overflow rate — 400 to 2000 gpd/ft$^2$ (800 to 1200 typical)
- bottom — slight slope (8%) towards hopper
- inlet — baffled for uniform velocity
- scum removal — mechanical or manual
- weir loading — 10,000 to 20,000 gpd/ft

**Chemical Sedimentation**: Chemical flocculation (*clarification* or *coagulation*) is similar to that described in chapter 7 except that the coagulant doses are greater. Typically, the most economical coagulant used is ferric chloride. Lime and sulfuric acid may be used to adjust the pH for proper coagulation. Chemical precipitation is used when the stream into which the outfall discharges is running low, when there is a large increase in sewage flow, and generally when plain sedimentation is insufficient.

## C. SECONDARY TREATMENT

Secondary treatment is a biological treatment. It became mandatory for all publicly owned water treatment plants as of July 1977 under the Federal Water Pollution Control Act ammendments of 1972.

**Trickling Filters**: Trickling filters (also known as *biological beds*) consist of beds of 2″ to 5″ rocks up to 9 feet thick (6 feet typical) over which influent is sprayed. The biological and microbial slime growth attached to the rocks purify the wastewater as it trickles through the rocks. The water is introduced into the filter by rotating arms which move by virtue of the spray reaction. The clarified water is collected by an underdrain system. Some water may be returned to the filter for a longer contact time.

On the average, one acre of standard filter area is needed for each 20,000 people served. Trickling filters can remove 70% to 90% of the suspended solids, 65% to 85% of the BOD, and 70% to 95% of the bacteria. Most of the reduction occurs in the first few feet of bed, and organisms in the lower part of the bed may be in a near-starvation condition. The bed will periodically slough off (unload) parts of its slime coating, and sedimentation after filtering is necessary.

Since there are limits to the heights of trickling filters, longer contact times can be achieved by returning some of the collected filter water to the filter. This is known as *recirculation*. Recirculation is also used to keep the filter medium from drying out and to smooth out fluctuations in the hydraulic loading.

*High rate filters* are now in use by most modern facilities. The higher hydraulic loading flushes the rockpile and inhibits excess biological growth. High rate filters may be only 3 to 4 feet deep. The high rate is possible because much of the filter discharge is recirculated.

The *hydraulic loading* of a trickling filter is the water flow divided by the plan area. Typical values of hydraulic loading are 25 to 100 gpd/ft$^2$ for standard filters, and up to 1000 gpd/ft$^2$ for high-rate filters.

$$L_H = \frac{Q_w + Q_R}{A_{\text{filter}}} = \frac{Q_w + R_R Q_w}{A_{\text{filter}}} = \frac{Q_w(1+R_R)}{A_{\text{filter}}} \quad (Q \text{ in gpd}) \qquad 8.27$$

The *recirculation ratio* is given by equation 8.28. It can be as high as 3 for high rate filters, although it is zero for standard low-rate filters.

$$R_R = Q_R/Q_w = \frac{L_H A_{\text{filter}}}{Q_w} - 1 \qquad 8.28$$

The *BOD loading* (same as *organic loading*) is calculated without considering any recirculated flow. BOD

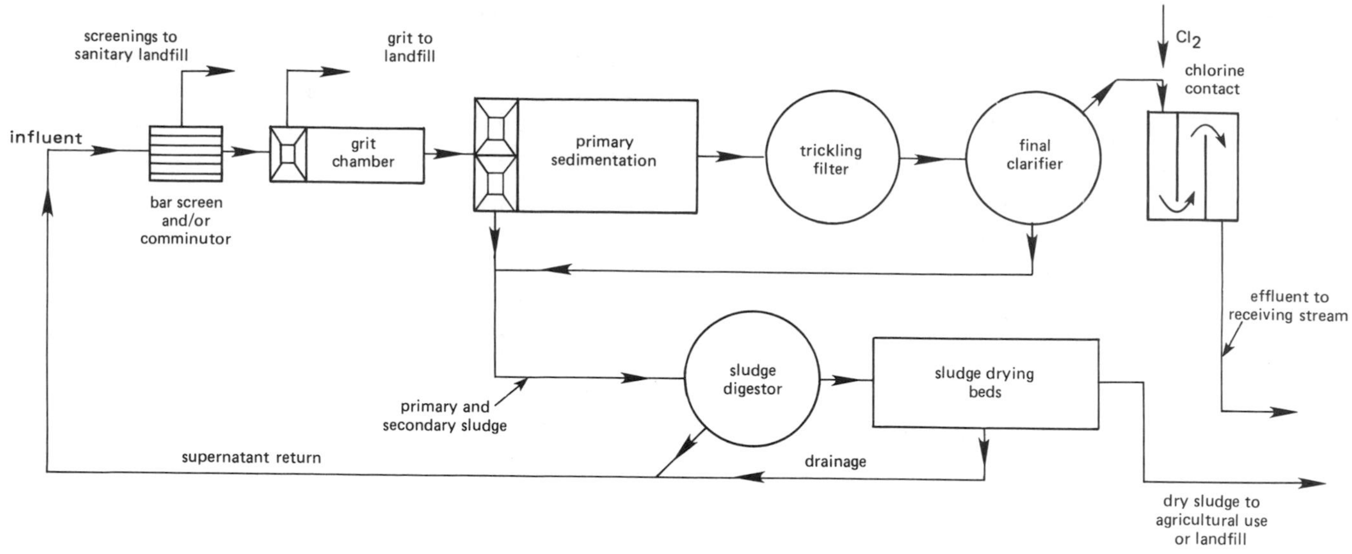


**Figure 8.9** A Typical Trickling Filter Plant

loading for the filter/clarifier combination is essentially the BOD of the applied wastewater divided by the filter volume.

$$L_{\text{BOD}} = \frac{(Q_{w,\text{MGD}})(\text{BOD}_{s,i,\text{mg/l}})(8.345)(1000)}{\text{filter volume in ft}^3} \qquad 8.29$$

BOD loading is given in pounds per 1000 cubic feet per day. Typical values are 5 to 25 lbm/1000 cubic feet-day (low rate) and 30 to 90 lbm/1000 cubic feet-day (high rate.)

Significant reduction in BOD occurs in a trickling filter. Standard rate filters produce an 80%–85% reduction. Because they offer less contact area and time, high rate filters only remove 65%–80% of the BOD.

If it is assumed that the biological layer and hydraulic loading are uniform, the water is at 20°C, and the filter is single-stage rock followed by a settling tank, then the following equation developed by the National Research Council can be used to calculate the BOD removal efficiency of the filter/clarifier combination.[14] Equation 8.30 is easily solved from figure 8.10.[15]

$$\eta = \frac{\text{BOD}_{\text{removed}}}{\text{BOD}_{\text{entering}}} = \frac{1}{1 + 0.0561\sqrt{L_{\text{BOD}}/F}} \qquad 8.30$$

[14] The National Research Council did studies in 1946 on sewage treatment plants at military installations. It concluded that the organic loading had a greater effect on removal efficiency than did volumetric loading.

[15] The constant 0.0561 in equation 8.30 is also reported as 0.0085 in the literature. However, that value is for use with media volumes expressed in acre-feet, not 1000's of ft³.

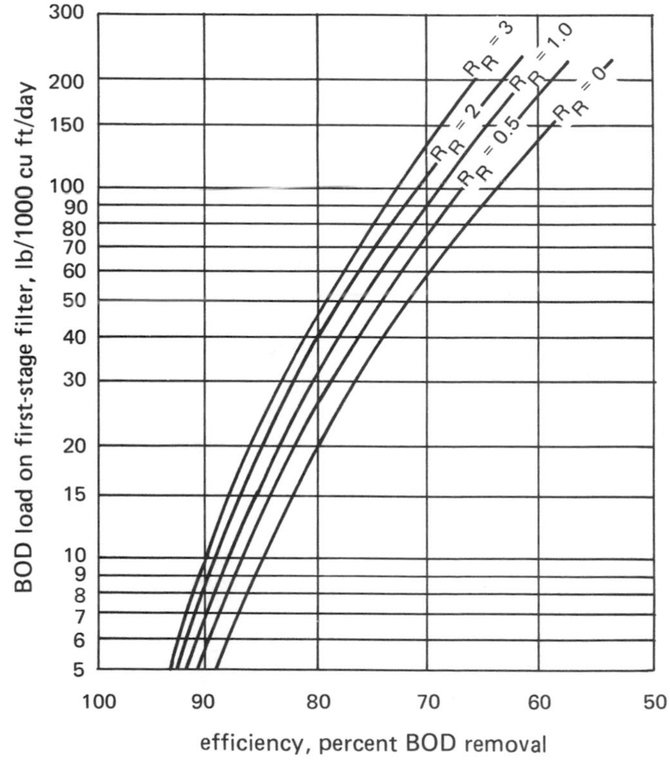


**Figure 8.10** Solution of Equation 8.30

For single-stage trickling filters operating at other temperatures, calculate the efficiency assuming 20°C, and then correct the efficiency using equation 8.31.

$$\eta_T = \eta_{20^\circ\text{C}}(1.01)^{T-20} \qquad 8.31$$

WASTE-WATER

If the incoming and required BOD are known, the approximate BOD loading required is given by equation 8.32.

$$L_{\text{BOD}} = 317.7\left(\frac{\text{BOD}_{\text{req}}}{\text{BOD}_i - \text{BOD}_{\text{req}}}\right)^2 \qquad 8.32$$

The BOD loading for a high rate, single stage filter with recirculation is given by equation 8.34, where $F$ is the effective number of passes through the filter. The *weighting factor*, $w$, is typically assigned a value of 0.1.

$$F = \frac{1+R_R}{(1+wR_R)^2} \qquad 8.33$$

$$L_{\text{BOD}} = (317.7)F\left(\frac{\text{BOD}_{\text{req}}}{\text{BOD}_i - \text{BOD}_{\text{req}}}\right)^2 \qquad 8.34$$

There are a number of ways to recirculate water back to the filter. Some methods are shown in figure 8.11. The variations in performance are not significant. Equation 8.34 should be used only with the first four recirculation schemes shown in figure 8.11.

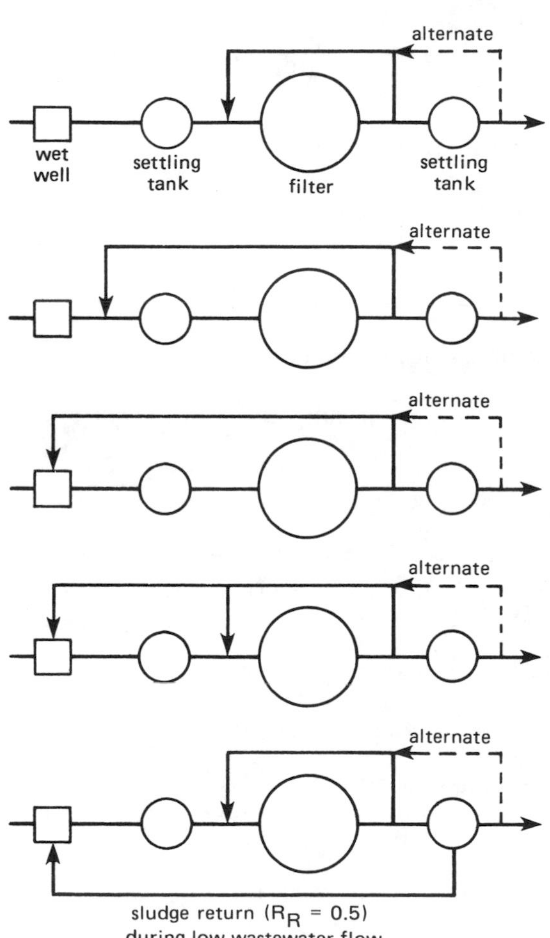


**Figure 8.11** Typical One-Stage Recirculation Methods

If even more BOD and solids removal is needed, two filters can be connected in series to form a two-stage filter system. A two-stage filter system is shown in figure 8.12(a) with an optional intermediate settling tank.

Typical two-stage performance is:

- BOD loading 45 to 70 lbm/1000 $\text{ft}^3$-day
- hydraulic loading 0.16 to 0.48 gpm/$\text{ft}^2$
- recirculation ratio 0.5 to 4.0
- final effluent BOD 30 mg/l

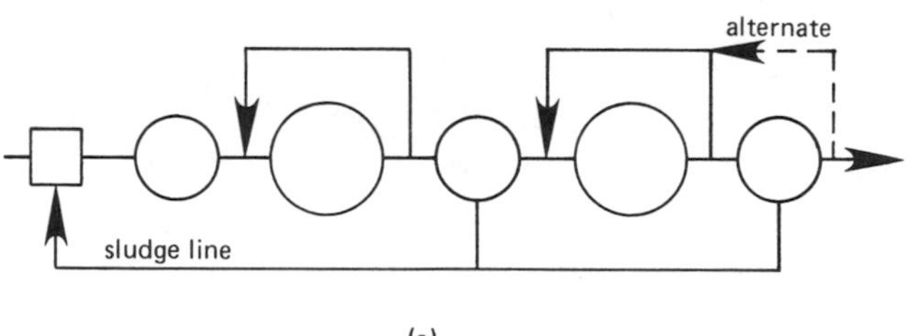


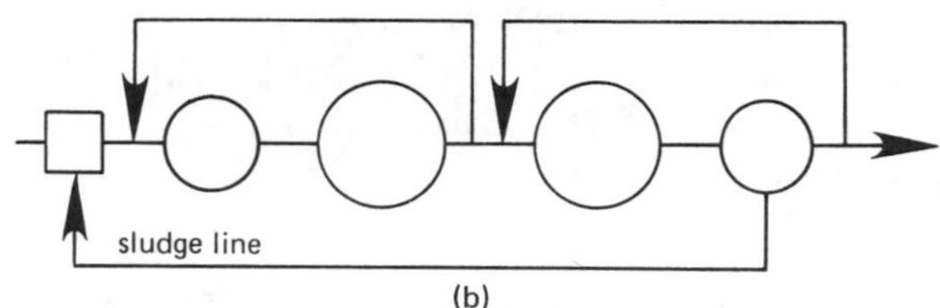


**Figure 8.12** Typical Two-Stage Filter Systems

The approximate BOD loading that is required to accomplish a given reduction in BOD is given by equation 8.35.

$$L_{\text{BOD}} = (317.7)F_{\text{2nd stage}} \times \left(\frac{\text{BOD}_{\text{out,1st}}}{\text{BOD}_{\text{in,1st}}}\right)^2 \left(\frac{\text{BOD}_{\text{out,2nd}}}{\text{BOD}_{\text{in,2nd}} - \text{BOD}_{\text{out,2nd}}}\right)^2 \qquad 8.35$$

The efficiency of a second-stage filter is considerably less than that of a first-stage filter because much of the biological food has been removed. This lowered efficiency can be considered as an increase in BOD loading, as given by equation 8.36.

$$L_{\text{BOD, adjusted 2nd stage}} = \frac{L_{\text{BOD, actual second stage}}}{[1 - \text{first stage efficiency}]^2} \qquad 8.36$$

The actual second stage load is calculated from equation 8.29 using the incoming BOD from the intermediate clarifier. Equation 8.30 is then used to find the 2nd stage efficiency.

The overall BOD efficiency of the two-stage system is:

$$\eta_{\text{overall}} = \frac{\text{BOD}_{\text{in}} - \text{BOD}_{\text{out}}}{\text{BOD}_{\text{in}}} = 1 - (1 - \eta_{\text{settling basin}})(1 - \eta_{\text{1st stage}})(1 - \eta_{\text{2nd stage}}) \quad 8.37$$

For two-stage trickling filters operating at temperatures other than 20°C, equation 8.38 can be used.

$$\eta_T = \eta_{20°\text{C}}(1.035)^{T-20} \quad 8.38$$

*Example 8.5*

A single-stage trickling filter plant is to process 1.4 MGD of domestic waste with 170 mg/l BOD.

| | |
|---|---|
| primary clarifier | 50 feet diameter |
| | 7 feet wet depth |
| | peripheral weir |
| trickling filter | 90 feet diameter |
| | 7 feet wet depth |
| | rock media |
| | 50% recirculation |
| final clarifier | same dimensions as primary |

Determine if the units have been sized correctly, and estimate the final BOD if the plant operates at 16°C.

Primary Clarifier

circumference: $\pi(50) = 157\,\text{ft}$

surface area: $\frac{1}{4}\pi(50)^2 = 1963\,\text{ft}^2$

volume: $(1963)(7) = 13{,}740\,\text{ft}^3$

The surface loading is

$$\frac{1.4\,\text{EE6}}{1963} = 713\,\text{gpd/ft}^2 \text{ (ok)}$$

The retention time is

$$t = \frac{V}{Q} = \frac{(13{,}740)\,\text{ft}^3\,(24)\,\text{hr/day}}{(1.4\,\text{EE6})\,\text{gal/day}(0.1337)\,\text{ft}^3/\text{gal}} = 1.76\,\text{hrs (ok)}$$

The weir loading is $\frac{1.4\,\text{EE6}}{157} = 8917$ gpd/ft (ok)

Assume a 30% reduction in BOD at 20°C. The effluent BOD is

$$\text{BOD} = (0.7)(170) = 119\,\text{mg/l}$$

Trickling Filter

area: $\frac{1}{4}\pi(90)^2 = 6362\,\text{ft}^2$

volume: $(6362)(7) = 44{,}534\,\text{ft}^3$

From equation 8.27, the hydraulic load on the filter is

$$\frac{(1.5)(1.4\,\text{EE6})}{6362} = 330\,\text{gpd/ft}^2 \quad \text{(ok for high-rate filter)}$$

From equation 8.29, the BOD loading is

$$\frac{(1.4\text{ MGD})\left(119\ \frac{\text{mg}}{\text{l}}\right)\left(8.345\ \frac{\text{lbm-l}}{\text{MG-mg}}\right)(1000)}{44{,}534\ \text{ft}^3} = 31.2\,\text{lbm/day-1000 ft}^3 \quad \text{(ok for high-rate filter)}$$

From equation 8.33,

$$F = \frac{1 + 0.5}{(1 + (0.1)(0.5))^2} = 1.36$$

From equation 8.30, the approximate BOD removal efficiency is

$$\eta = \frac{1}{1 + 0.0561\sqrt{31.2/1.36}} = 0.788 \text{ at } 20°\text{C}$$

The final clarifier effluent BOD is

$$(1 - 0.788)(119) = 25.2\,\text{mg/l}$$

Final Settling Tank

Same area, volume, surface loading, weir loading, and retention time as the primary clarifier. The BOD removal effect is included in equation 8.30.

16°C Performance

The 20°C overall efficiency is $\frac{170 - 25.2}{170} = 0.852$

From equation 8.31, the efficiency at 16°C is

$$\eta_{16°\text{C}} = (0.852)(1.01)^{16-20} = 0.819$$
$$\text{BOD}_{16°\text{C}} = (1 - 0.819)(170) = 30.8\text{ mg/l}$$

**Rotating Biological Contactors**: *Rotating biological contactors* (also known as *rotating biological reactors*) consist of large-diameter plastic disks mounted on a shaft which turns slowly. The rotation progressively wets the disks, on which a microorganism population grows. This biomass population, since it is well oxygenated, is efficient at removing organic solids from the wastewater.

Several stages of RBC's typically constitute the process. The process can be placed in series or parallel with existing trickling filter or activated sludge processes. Recirculation is not common with RBC processes.

The primary design criterion is hydraulic loading, not organic (BOD) loading. For a specific hydraulic loading, the BOD removal efficiency will be essentially constant, regardless of variations in BOD. Other design criteria are listed here:

- plant design flows — all
- hydraulic loading:
  - (secondary treatment) — 2 to 4 gpd/ft$^2$ of disk area
  - (denitrification) — 0.75 to 2 gpd/ft$^2$
- optimum peripheral rotational velocity — 1 ft/sec
- tank volume — 0.12 gal/ft$^2$ of biomass area
- temperature — 55°–90°F
- BOD removed — 70–80%
- retention time — fixed by tank volume and hydraulic loading
- number of stages — 2–4
- diameter of disks — 10–12 feet
- immersion — 40% at any instant

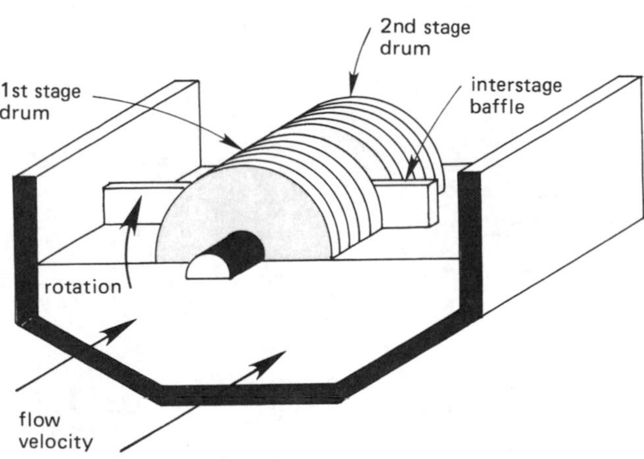


**Figure 8.13** Rotating Biological Contactor

**Intermittent Sand Filters**: For small populations, a *slow sand filter* or *intermittent sand filter* can be used. Because of the lower flow rate, the filter area per person is higher than for a trickling filter. Roughly one acre is needed for a population of 1000. The filter is constructed as a sand bed 2 to 3 feet deep over a 6″ to 12″ gravel bed. Application rates are usually 2 to $2\frac{1}{2}$ gpd/ft$^2$. The filter is cleaned by removing the top layer of clogged sand.

The filter is alternately exposed to water from a settling tank and to air (hence the term *intermittent*). Straining and aerobic decomposition clean the water. If the water is applied continuously as a final process from a secondary treatment plant, the filter is known as a *polishing filter*. The water rate of a polishing filter may be as high as 10 gpd/ft$^2$. Up to 95% of the BOD can be satisfied in an intermittent sand filter.

**Stabilization Ponds**: A stabilization pond (also known as an *oxidation pond*) holds partially treated water for 3 to 6 weeks (up to 6 months in cold weather) at a depth of 2 to 6 feet (4 typical). Aquatic plants, weeds, algae, and microorganisms are used to stabilize the organic matter. The algae gives off oxygen from growth in sunlight. The oxygen is used by microorganisms to digest organic matter. The microorganisms give off $CO_2$, ammonia, and phosphates which the algae use.

Areas required are large—up to one acre per 50 pounds BOD per day (about one acre per 300 people) for warm southern states. In cold climates, twice this area may be required. The BOD loading is 15 to 35 lbm/acre-day.

**Aerated Lagoons**: If a stabilization pond has air added mechanically, it is known as an *aerated lagoon*. Such a lagoon typically is deeper and has a shorter detention time than a stabilization pond. With floating aerators, one acre can support 500 to 1000 pounds of BOD per day. In cold climates, twice this area is required.

The lagoon volume must be sufficient to hold the design flow during the detention period. (Since $V/Q = t_d$, equation 8.39 can be used to determine $t_d$ if the BOD removal efficiency is known.)

$$V = Qt_d \tag{8.39}$$

The BOD reduction can be calculated if the overall, first-order BOD *removal rate constant* is known. Typical values of $k$ are 0.25 to 1.0 $\frac{1}{\text{day}}$ (base-10).

$$\frac{\text{BOD}_f}{\text{BOD}_i} = \frac{1}{1 + \frac{kV}{Q}} \tag{8.40}$$

If the removal rate constant is quoted for a different temperature than that at which the lagoon is to operate, it must be corrected using equation 8.41.

$$k_T = k_{20^\circ\text{C}}(1.06)^{\text{T}-20} \tag{8.41}$$

Other common design characteristics of a mechanically aerated lagoon are:

- aspect ratio — less than 3:1
- depth — 10 to 12 feet
- detention time — 4 to 10 days
- BOD loading — 20–400 lbm/day-acre (220 typical)
- temperature range — 32° to 100°F (70°F optimum)
- typical effluent BOD — 20 to 70 mg/l
- oxygen requirements — 0.7 to 1.4 times BOD removed

**Activated Sludge Processes**: Sludge produced during the oxidation process has an extremely high concentration of active aerobic bacteria. For this reason, partially oxidized sludge is called *activated sludge*. Purification of raw sewage can be speeded up considerably if the raw sewage is mixed (seeded) with activated sludge. The mixture of raw sewage and activated sludge is known as *mixed liquor* (*ML*). The biological systems in the mixed liquor are known as *mixed liquor suspended solids* (*MLSS*).

In operation, an activated sludge process takes raw water and allows it to settle. The settled effluent is mixed with activated sludge in the approximate ratio of 1 part sludge per 3 or 4 parts effluent. Mechanical aeration is used. The effluent is then settled in a second sedimentation tank, chlorinated, and discharged. Settled sludge from this last tank supplies the continuous seed for the activation.

Activated sludge processes are highly efficient, with the following typical characteristics for a conventional system. (Also, see table 8.14).

- BOD reduction — 90 to 95%
- BOD loading — 0.25 to 1 lbm/lbm MLSS
- maximum aeration chamber volume — 5000 $ft^3$
- aeration chamber depth — 10 to 15 feet
- aeration chamber width — 20 ft
- air rate — 1/2 to 2 $ft^3$/gal raw sewage
- minimum dissolved oxygen — 2 mg/l
- biological mass density — 1000 to 4000 mg/l
- sedimentation basin depth — 15 ft
- sedimentation basin detention time — 2 hours
- basin overflow rate — 400 to 2000 gpd/$ft^2$ (1000 typical)
- % sludge returned — 20 to 30%
- frequency of sludge transfer — once each hour
- activated sludge volume index — 50 to 150
- weir loading — 10,000 gpd/ft
- maximum tank volume — 2500 $ft^2$

The *rate of oxygen transfer* (ROT) from the air to the mixed liquor during aeration is given by equation 8.42.

$$\text{ROT} = K_t D \qquad 8.42$$

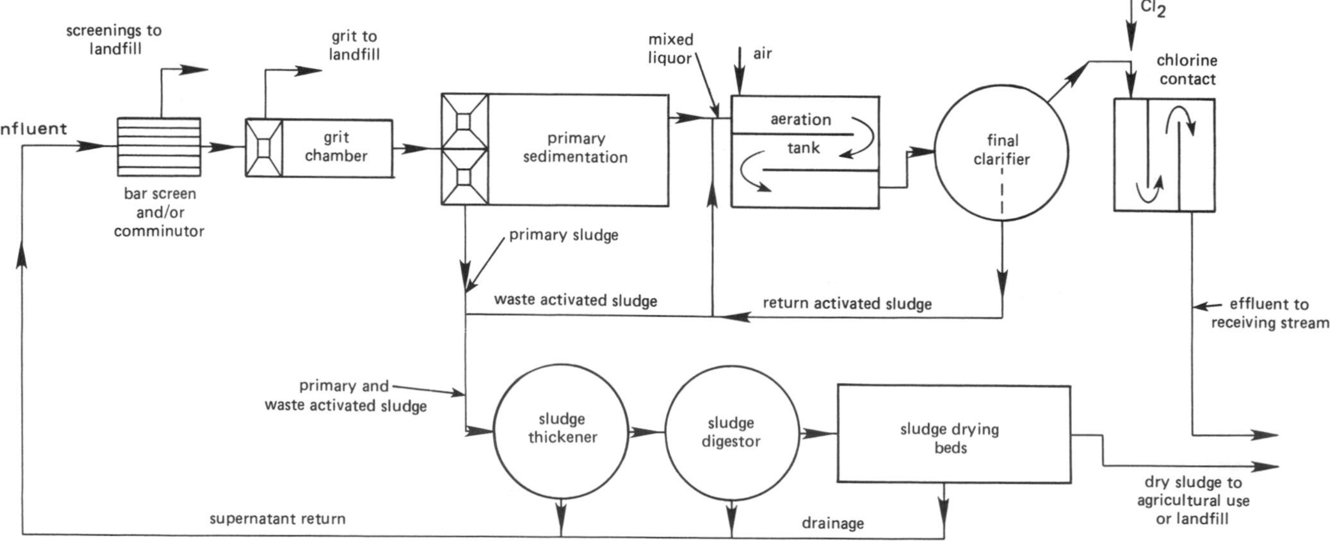


**Figure 8.14** A Typical Activated Sludge Plant

$K_t$ is a transfer coefficient which depends on the equipment and waste characteristics. The oxygen deficit, $D$, is given by equation 8.43.

$$D = \beta DO_{\text{saturated}} - DO_{\text{actual ML}} \quad 8.43$$

$\beta$ is the water's *oxygen saturation coefficient*, usually 0.8 or 0.9.

*Example 8.6*

20°C wastewater ($\beta = 0.9$) is processed in an aerating system with an oxygen transfer coefficient of 2.7 per hour. If the wastewater dissolved oxygen is 3 mg/l, what is the rate of oxygen transfer?

At 20°C, the saturated oxygen content is 9.2 mg/l. So, the oxygen deficit is

$$(0.9)(9.2) - 3 = 5.28\,\text{mg/l}$$

From equation 8.42,

$$\text{ROT} = (2.7)(5.28) = 14.3\,\text{mg/l-hr}$$

The *aerating period* is calculated from equation 8.44. Recirculation is not a factor in calculating the aerating period.

$$t_A = \frac{\text{aeration tank volume, gallons}}{Q_w} \quad 8.44$$

The BOD loading $L_{\text{BOD}}$ in lbm/1000 ft$^3$-day, on the aeration tank is

$$L_{\text{BOD}} = \frac{(Q_{w,\text{gal/day}})(\text{BOD}_{w,\text{mg/l}})\left(0.0624\,\frac{\text{lbm-l}}{\text{mg-1000 ft}^3}\right)}{\text{aeration tank volume, gallons}} \quad 8.45$$

The *food-to-microorganism ratio* in pounds of BOD per day per pound of MLSS is defined by equation 8.46.

$$R_{F-M} = \frac{(Q_w)(\text{BOD}_w)}{(MLSS)(\text{aeration tank volume, gallons})} \quad 8.46$$

The *rate of return sludge* is

$$R_{RS} = Q_{RS}/Q_w \quad 8.47$$

The BOD efficiency of an activated sludge process is

$$\eta_{\text{BOD}} = \frac{\text{BOD}_w - \text{BOD}_{\text{after settling}}}{\text{BOD}_w} \quad 8.48$$

*Sludge bulking* refers to a condition in which the sludge will not settle out. Since the solids do not settle, they leave the sedimentation tank and cause problems in subsequent processes. The *sludge volume index* (also known as the *Mohlman index*) can be calculated by taking one liter of mixed liquor and measuring the volume of settled solids after 30 minutes. Then, the sludge volume index (SVI) is

$$\text{SVI} = \frac{(\text{settled volume in ml})(1000)}{\text{MLSS}} \quad 8.49$$

If SVI is above 150, the sludge is bulking. Remedies include the addition of lime, chlorine, more aeration, and the reduction in MLSS.

SVI is related to the concentration of suspended solids in the activated sludge by equation 8.50.

$$SS_{as} = \frac{1{,}000{,}000}{\text{SVI}} \quad 8.50$$

The theoretical quantity of return sludge required can be calculated from the SVI test results:

$$Q_R/Q_w = \frac{(\text{settled volume in ml/l})}{(1000 - \text{settled volume in ml/l})} \quad 8.51$$

Another important parameter is *sludge age*. Although the water passes through the system in a matter of hours, the sludge is recycled continuously and has an average stay much longer in duration. Two measures of sludge age are used: age of the suspended solids and age of BOD. Sludge age ($SA_{ss}$) is typically 3 to 5 days.

$$SA_{ss} = \frac{\text{pounds MLSS in aerating basin}}{\text{pounds SS in effluent and waste sludge per day}} \quad 8.52$$

$$SA_{\text{BOD}} = 1/R_{\text{F-M}} = \frac{\text{pounds MLSS in aerating basin}}{\text{pounds BOD applied to basin per day}} \quad 8.53$$

*Example 8.7*

One liter of liquid is taken from an aerating lagoon near its discharge point. After settling for 30 minutes in a one liter graduated cylinder, 250 ml of solids have settled out. A second water sample is taken and the suspended solids concentration determined to be 2300 mg/l. Find the SVI and percentage of required return sludge.

From equation 8.49, $\text{SVI} = \frac{(250)(1000)}{2300} = 109$

From equation 8.51, $Q_R/Q_w = \frac{250}{1000 - 250} = 0.33$

**Activated Sludge Aeration Methods:** Various methods of aeration are used, each having its own characteristic ranges of operating parameters. Table 8.14 lists typical parameters for selected aeration methods.

*Extended Aeration*: Small flows can be treated with the extended aeration method. This method uses mechanical floating or fixed sub-surface aerators to oxygenate the mixed liquor for 24 to 36 hours. There is no primary clarification and there are generally no sludge wasting facilities. Sludge is allowed to accumulate at the bottom of the lagoon for several months. Then, the system is shut down and the lagoon is pumped out.

Sedimentation basins are sized very small with low overflow rates (200 to 600 gpd/ft$^2$), and they have long retention times. All sludge is returned to the aerating basin.

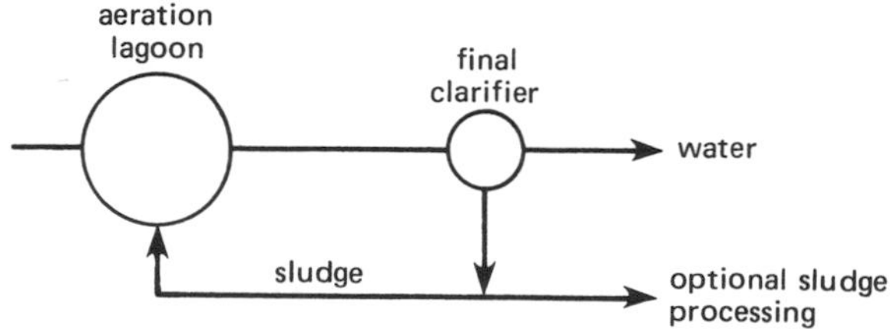


**Figure 8.15** Extended Aeration

*Conventional Aeration*: In this method, (also known as *plug flow*), the influent is taken from a primary clarifier and then aerated. The amount of aeration may be decreased (i.e., *tapered aeration*) as the wastewater travels through the circuitous route since the BOD also decreases along the route.

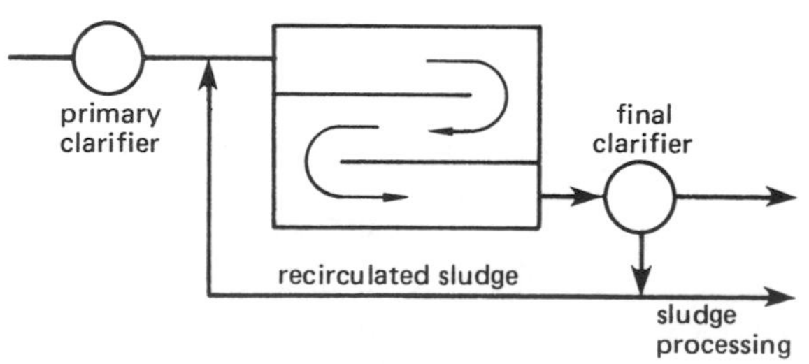


**Figure 8.16** Conventional Aeration

*Step Flow*: In step flow, aeration is constant along the length of the aeration path, but influent is introduced at various points along the path.

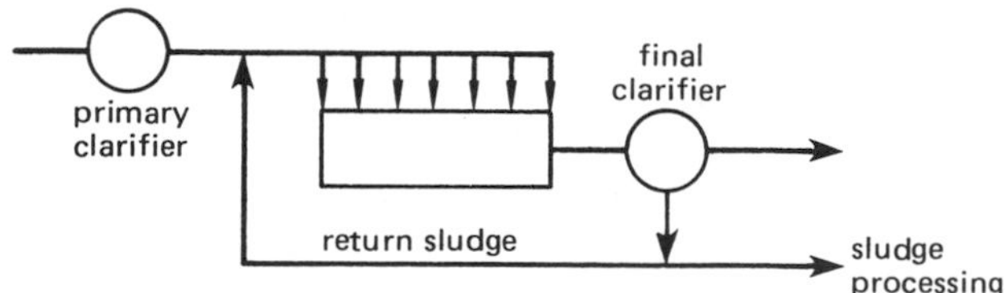


**Figure 8.17** Step Aeration

*Complete Mix*: Waste is added uniformly and the mixed liquor is removed uniformly over the length of the tank. This method is often used for industrial waste processing.

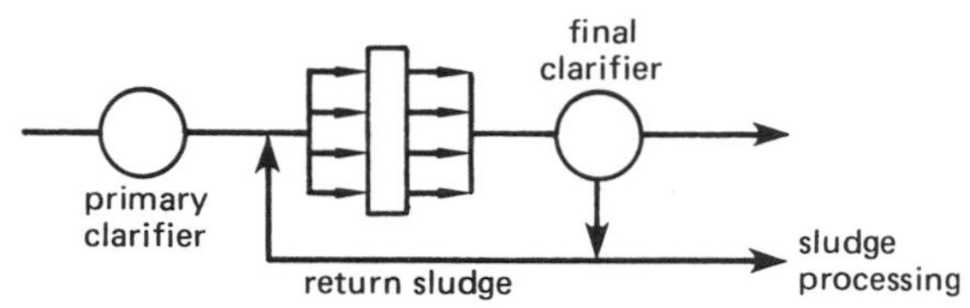


**Figure 8.18** Complete Mixing

*Contact Stabilization*: Units for the contact stabilization (*biosorption*) method are typically factory-built and erected on a site. They are compact, but not as economical or efficient as a regular plant. The units may be constructed as concentric compartmentalized cylinders. This method is designed primarily to handle colloidal wastes.

In this method, the aeration tank is called a *contact tank*. The stabilization tank takes the sludge from the clarifier and aerates it. Less time and space is required for this process because the sludge stabilization is done when the sludge is still concentrated.

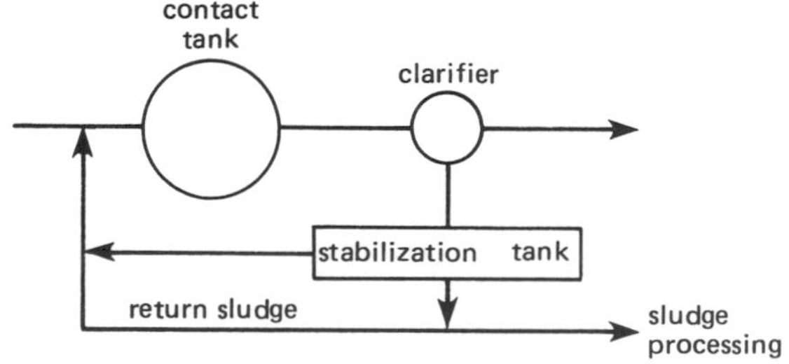


**Figure 8.19** Contact Stabilization

*High Rate Aeration*: This method uses mechanical mixing along with aeration to decrease the aeration period and to increase the BOD load per unit volume.

*High Purity Oxygen Aeration*: This method requires the use of bottled or manufactured oxygen which is introduced to closed aerating tanks in place of atmospheric air. Mechanical mixing is needed to take full advantage of the oxygen.

**Intermediate Clarifiers**: Sedimentation tanks located between trickling filter stages (see figure 8.12) or between a filter and subsequent aeration are known as *intermediate clarifiers*. Recommended standards are:

- maximum overflow rate 1000 gpd/ft$^2$
- minimum water depth 7 feet
- maximum weir loading 10,000 gpd/ft (plants 1 MGD or less); 15,000 gpd/ft (plants over 1 MGD)

**Final Clarifiers**: Final sedimentation in secondary treatment is done in final clarifiers. The purpose of final clarifiers is to collect sloughed off filter material (trickling filter processes) or to collect sludge and return it for aeration (activated sludge processes).

General characteristics for clarifiers following trickling filters are:

- BOD removal See equation 8.30
- minimum depth 7 feet
- maximum overflow rate 800 gpd/ft$^2$
- maximum weir loading Same as for intermediate clarifiers, but lower preferred.

If the final clarifier follows an activated sludge process, the sludge should be removed rapidly from the entire bottom of the clarifier. Characteristics of clarifiers following an activated sludge process are given in table 8.15.

## D. ADVANCED TERTIARY TREATMENT

*Suspended Solids*: Suspended solids are removed by microstrainers or polishing filter beds.

*Phosphorus Removal*: Phosphorus can be removed by chemical precipitation. Aluminum and iron coagulants, as well as lime, are effective in removing phosphates.

*Nitrogen Conversion and Removal*: In the *ammonia stripping* (*air stripping*) method, lime is added to water to increase its pH to above 10. The water is then passed through a packed tower into which air is blown. The air (at the rate of approximately 400 ft$^3$/gallon) strips the ammonia out of the water. Recarbonation follows to remove the excess lime.

$$NH_4 + OH^- \underset{pH<10}{\overset{pH\geq 11}{\rightleftharpoons}} NH_4OH \xrightarrow{\text{air}} H_2O + NH_3 \qquad 8.54$$

In the *nitrification and denitrification process*, bacteria oxidize ammonium ions to nitrate and nitrite in an aeration tank kept at low BOD. Nitrate and nitrite ions do not absorb further oxygen and may be discharged.

$$NH_3^+ \xrightarrow[\text{oxygen}]{\text{bacteria}} NO_2^- + NO_3^- \qquad 8.55$$

**Table 8.14**
Representative Operating Conditions for Aeration

| type of aeration | plant flow rate (MGD) | $t_A$ (hrs) | oxygen required (lbm/lbm BOD removed) | waste sludge (lbm/lbm BOD removed) | total plant BOD load (lbm/day) | aerator BOD load, $L_{BOD}$ (lbm/1000 ft$^3$-day) | $R_{F/M}$ (lbm/lbm) | MLSS (mg/l) | $R_R$ (%) | $\eta_{BOD}$ (%) |
|---|---|---|---|---|---|---|---|---|---|---|
| conventional | 0–0.5 | 7.5 | 0.8–1.1 | 0.4–0.6 | 0–1000 | 30 | 0.2–0.5 | 1500–3000 | 30 | 90–95 |
| | 0.5–1.5 | 7.5–6.0 | | | 1000–3000 | 30–40 | | | | |
| | 1.5 up | 6.0 | | | 3000 up | 40 | | | | |
| contact stabilization | 0–0.5 | 3.0* | 0.8–1.1 | 0.4–0.6 | 0–1000 | 30 | 0.2–0.5 | 1000–3000* | 100 | 85–90 |
| | 0.5–1.5 | 3.0–2.0* | 0.4–0.6 | | 1000–3000 | 30–50 | | | | |
| | 1.5 up | 1.5–2.0* | 0.4–0.6 | | 3000 up | 50 | | | | |
| extended | 0–0.05 | 24 | 1.4–1.6 | 0.15–0.3 | all | 10.0 | 0.05–0.1 | 3000–6000 | 100 | 85-95 |
| | 0.05–0.15 | 20 | | | | 12.5 | | | | |
| | 0.15 up | 16 | | | | 15.0 | | | | |
| high rate | 0–0.5 | 4.0 | 0.7–0.9 | 0.5–0.7 | 2000 up | 100 | 1.0 or less | 4000–10,000 | 100 | 80–85 |
| | 0.5–1.5 | 3.0 | | | | | | | | |
| | 1.5 up | 2.0 | | | | | | | | |
| step aeration | 0–0.5 | 7.5 | | | 0–1000 | 30 | 0.2–0.5 | 2000–3500 | 50 | 85–95 |
| | 0.5–1.5 | 7.5–5.0 | | | 1000–3000 | 30-50 | | | | |
| | 1.5 up | 5.0 | | | 3000 up | 50 | | | | |
| high purity oxygen | | 1.0–3.0 | | | | above 120 | 0.6–1.5 | 6000–8000 | 50 | 90–95 |

* in contact unit only

**Table 8.15**
Final Clarifiers for Activated Sludge Processes

| type of aeration | design flow (MGD) | minimum detention time (hr) | maximum overflow rate (gpd/$ft^2$) |
|---|---|---|---|
| conventional, high rate, and step | $< 0.5$ | 3.0 | 600 |
| | 0.5 to 1.5 | 2.5 | 700 |
| | $> 1.5$ | 2.0 | 800 |
| contact stabilization | $< 0.5$ | 3.6 | 500 |
| | 0.5 to 1.5 | 3.0 | 600 |
| | $> 1.5$ | 2.5 | 700 |
| extended aeration | $< 0.05$ | 4.0 | 300 |
| | 0.05 to 0.15 | 3.6 | 300 |
| | $> 0.15$ | 3.0 | 600 |

Following sedimentation and sludge recirculation, the effluent may be treated to convert nitrates and nitrites to nitrogen gas. Methanol supplies the energy required by the denitrification bacteria.

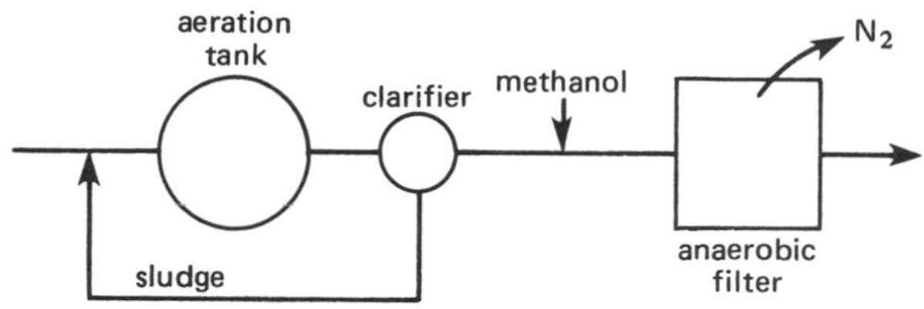


**Figure 8.20** Nitrification and Denitrification Process

Ammonia can also be removed by breakpoint chlorination. Other methods of nitrogen/ammonia removal include anion ion exchange and algae ponds.

*Inorganic Salt Removal*: Ions from inorganic salts can be economically removed by electrodialysis and reverse osmosis.

*Dissolved Solids Removal*: The so-called *trace organics* or *refractory substances* are dissolved organic solids that are biologically resistant. They can be removed by filtering through activated carbon or ozonation.

## 16 SLUDGE DISPOSAL

### A. SLUDGE QUANTITIES

Sludge removed from sedimentation basins is 95% to 99% moisture. With primary treatment only, about 0.1 pounds of dried sludge can be expected per capita day. With secondary treatment, the total sludge load will be about 0.2 pounds per capita day (when dried). This amounts to approximately 2 quarts of sludge per 100 gallons of wastewater processed.

**Table 8.16**
Typical Characteristics of Domestic Sewage Sludge
(also, see table 8.18)

| origin of sludge | solids content of wet sludge (s in percent) | dry solids (lbm/day/capita) |
|---|---|---|
| primary settling tank | 6 | 0.12 |
| trickling filter secondary | 4 | 0.04 |
| mixed primary and trickling filter secondary | 5 | 0.16 |
| high rate activated sludge secondary | 2.5–5 | 0.06 |
| mixed primary and high rate activated sludge secondary | 5 | 0.18 |
| conventional activated sludge secondary | 0.5–1 | 0.07 |
| mixed primary and conventional activated sludge secondary | 2–3 | 0.19 |
| extended aeration secondary | 2 | 0.02 |

The dry weight of solids from primary settling basins is:

$$W_p = (\text{decrease in SS})(Q_{\text{gpd}})(8.345\,\text{EE}{-6}) \quad 8.56$$

The dry weight of solids from secondary aeration lagoons and biological filters is:

$$W_s = K(\text{BOD})_{\text{removed}}(Q_{\text{gpd}})(8.345\,\text{EE}{-6}) \quad 8.57$$

$K$ in equation 8.57 depends on the food-to-microorganism ratio, as given in table 8.17. $K$ is the fraction of BOD that appears as excess biological solids. For trickling filters and extended aeration, $K$ ranges from 0.2 to 0.33. For conventional and step aeration, $K$ ranges from 0.33 to 0.42. $K$ is known as the *cell yield* or *yield coefficient.*

**Table 8.17**
Cell Yield (Yield Coefficient)

| $R_{F\text{-}M}$ | $K$ |
|---|---|
| 0.05 | 0.2 |
| 0.07 | 0.21 |
| 0.1 | 0.24 |
| 0.15 | 0.28 |
| 0.2 | 0.33 |
| 0.3 | 0.37 |
| 0.4 | 0.4 |
| 0.5 | 0.43 |

Equations 8.56 and 8.57 give dry weight of sludge. Assuming the sludge specific gravity is near 1, the volume of wet sludge with a solids concentration, $s$, can be found from the dry weight by using equation 8.58.

$$\begin{array}{c}\text{gallons of}\\ \text{sludge/day}\end{array} = \frac{\text{dried weight per day, lbm}}{(s)(8.345)} \qquad 8.58$$

Typical values of $s$ are given in table 8.18.

**Table 8.18**
Total Sludge Solids, s
(also, see table 8.16)

| source or type of sludge | s, as fraction |
|---|---|
| primary settling tank sludge | 0.06 to 0.08 |
| primary settling tank sludge mixed with filter sludge | 0.04 to 0.06 |
| primary settling tank sludge mixed with activated sludge from aeration lagoons | 0.03 to 0.04 |
| excess activated sludge | 0.005 to 0.02 |
| filter backwashing water | 0.01 to 0.1 |
| softening sludge | 0.02 to 0.15 |

The specific gravity of the wet sludge is:

$$\frac{1}{(SG)_{\text{total}}} = \frac{\text{fraction moisture}}{1.0} + \frac{\text{fraction solids}}{(SG)_{\text{solids}}} \qquad 8.59$$

The volume of sludge is:

$$V = \frac{W}{(SG)_{\text{total}}(62.4)} \text{ in ft}^3\text{/day} \qquad 8.60$$

*Example 8.8*

A trickling filter plant processes domestic waste with the following characteristics: 190 mg/l BOD, 230 mg/l SS, and 4,000,000 gpd.

(a) What is the wet sludge volume from the primary sedimentation tank and trickling filter? Assume the combined sludge solids content is 5%.

(b) What is the approximate weight of dry solids produced per person-day?

*step 1*: Find the weight of the dry solids obtained from the primary settling basin. From equation 8.56, assuming 50% of solids can be removed,

$$\begin{aligned} W_p &= (0.5)(230)(4\,\text{EE6})(8.345\,\text{EE–6}) \\ &= 3839\,\text{pounds/day} \end{aligned}$$

Assume a 30% BOD reduction, so the BOD leaving the basin is $(0.7)(190) = 133$ mg/l.

*step 2*: Assume a cell yield value of 0.25. Then, the weight of dry solids from the filter is given by equation 8.57.

$$\begin{aligned} W_s &= (0.25)(133)(4\,\text{EE6})(8.345\,\text{EE–6}) \\ &= 1110\,\text{pounds/day} \end{aligned}$$

*step 3*: The wet sludge volume can be found from equation 8.58.

$$\frac{3839 + 1110}{(0.05)(8.345)} = 11{,}860\,\text{gallons/day}$$

*step 4*: The equivalent population is given by equation 8.12.

$$P_e = \frac{(190)(8.345)(4)}{0.17} = 37{,}310\,\text{people}$$

*step 5*: The per capita dried solids rate is

$$\frac{3839 + 1110}{37{,}310} = 0.13\frac{\text{lbm dry solids}}{\text{person-day}}$$

## B. SLUDGE THICKENING

Since the volume of wet sludge is inversely proportional to its solids content (equation 8.58), thickening of sludge is desirable. Thickening is required to at least 4% solids if dewatering is to be feasible. *Gravity thickening* uses a stirred sedimentation tank into which sludge is fed. A doubling of solids content is usually possible with a gravity thickener.

Thickening can also be accomplished through the *dissolved air flotation* method. Air is bubbled through a tank containing sludge. The solid particles adhere to the air bubbles, float to the surface, and are skimmed away. The skimmed scum has a solids content of approximately 4%. Up to 85% of the total solids may be recovered, although chemical flocculants may be used to increase this to 95%. Two to four pounds of solids are obtained each hour for each square foot of surface area.

Floatation thickening, as well as centrifuge thickening, of activated sludge may be desirable where unusually large quantities are produced in proportion to primary sludge. Gravity thickening is usually best for primary or mixtures of primary and secondary sludges.

The following criteria can be used to design gravity thickening tanks.

| | |
|---|---|
| • required overflow rate, gpd per sq ft | 600 to 800 |
| • maximum solids loading, lbm of dry solids/sq ft per day: | |
| primary sludge | 22 |
| primary plus trickling filter sludge | 15 |
| primary plus modified aeration activated sludge | 12 |
| primary plus conventional activated sludge | 8 |
| waste activated sludge | 4 |
| • shape | circular |
| • minimum side water depth, feet | 10 |
| • minimum detention time, hours | 6 |
| • minimum floor slope, vertical to horizontal | 2.75 to 12 |
| • minimum number of tanks (unless alternate means is available for thickening or storing sludge) | 2 |
| • minimum angle between hopper wall and horizontal plane | 60° |
| • minimum freeboard | 18 in |
| • sludge collection and stirring mechanism | rotary scraper collector |

## C. SLUDGE DEWATERING

Once the sludge has been thickened, it can be digested or dewatered prior to disposal. Several methods of dewatering are available, including vacuum filtration, pressure filtration, centrifugation, drying beds of sand or gravel, and lagooning.

A common method of dewatering is vacuum filtration in a *rotary drum filter.* Suction is applied from within the drum to attract solids to the filter and to extract moisture. The dried cake is scraped off of the drum in the discharge section of the device. Chemical flocculants are used to collect fine particles on the filter drum. A final solids content of 20% to 25% is attained. This is sufficient for sanitary landfill. A solids content of 30% is needed, however, for direct incineration.

Sand drying beds are preferable when digested sludge is to be disposed of in a landfill. Sand beds produce cake with a high dryness. Generally, vacuum filtration and centrifugation are used with undigested sludge. (Gritty sludge should not be centrifuged.) Vacuum filtering presents more odor and cleanliness problems than centrifugation. Lagooning of sludge prior to landfilling should be considered only in isolated areas where the average temperature is above 60°F.

Sludge drying beds can be designed according to the following criteria:

| | |
|---|---|
| • depth of application | 8″ |
| • number of applications | 8/year |
| • area required (open beds): | |
| primary | 1 $ft^2$/cap |
| trickling filter | 1.5 |
| activated sludge | 2 |
| chemical coagulation | 2 |
| • typical dimensions | 20 to 30 ft wide<br>50 to 100 ft long |
| • enclosing wall height | 18 inches |

## D. DIGESTION AND STABILIZATION

Much of the organic material in sludge is easily digested (i.e., "stabilized") by anaerobic microbes. Solids which are capable of being digested are known as *volatile solids.* Digestion of volatile solids results in methane, carbon dioxide, and hydrogen sulfide gases.

Anaerobic digestion is more complex and more easily upset than aerobic digestion. However, it has a lower operating cost. Aerobic digestion is preferable with stabilized primary and mixed primary-secondary sludge. Chloride concentrations must be monitored with both types of digestion.

A third alternative, that of *wet oxidation* with low pressure and low temperature, conditions sludge as well as stabilizes it. However, below plant flows of 2 MGD, this process may be too complex. Wet oxidation is not subject to toxic upsets as are anaerobic digesters. This method, as well as anaerobic digestion, yields a byproduct liquor with a high BOD and ammonia content. Separate treatment or recirculation may be required. Where there are nitrogenous oxygen demand or nutrient restrictions on the plant effluent, wet oxidation (and thermal conditioning and anaerobic digestion) may be impractical.

**Anaerobic Digestion**: If the digestion takes place in the absense of oxygen, it is known as *anaerobic digestion.* Two types of bacteria are involved: acid forming and acid splitting. The pH must be kept above 6.5 for the methane producing bacteria to function.

In a single-stage, floating-cover digester (figure 8.21) raw sludge is brought into the tank at the cover and top of the dome. The contents of the digester stratify into four layers: scum on top, *supernatant*, a layer of actively digesting sludge, and a bottom layer of concentrated sludge. Some sludge may be withdrawn, heated, and returned to keep the temperature up.

Supernatant is removed along the periphery of the digester and returned to the inlet of the processing plant. Digested sludge is removed from the bottom and is then dewatered. Gas is removed from the gas dome and is burned. The heat from the burning methane can be used to warm the sludge that is withdrawn, or to warm raw sludge initially introduced.

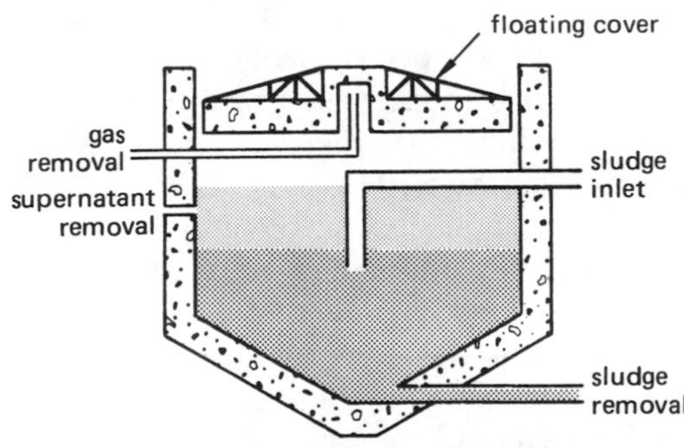


**Figure 8.21** Floating Cover Digester

The following characteristics are relevant to a single-stage digester:

| | |
|---|---|
| • sludge loading | 0.13–0.2 lbm of volatile solids/ft$^3$-day |
| • optimum temperature | 95 to 98°F |
| • optimum pH | 6.7 to 7.8 (7.0 to 7.1 preferred) |
| • gas production | 7 to 10 ft$^3$/lbm volatile solids added<br>0.5 to 1.0 ft$^3$/capita day |
| • gas composition | 65% methane, 35% $CO_2$ |
| • retention time | 30 to 90 days (conventional)<br>15 to 25 days (high rate) |
| • final moisture content | 90 to 95% |
| • gas heat content | 600 BTU/ft$^3$ |
| • depth | 20 to 45 feet |
| • depth to diameter ratio | use manufacturer's recommendations |
| • minimum freeboard | 2 feet |
| • slope of tank bottom | 1:12 to 1:4 |
| • minimum number of tanks | 2 |
| • typical diameter | 20 to 115 feet |

A single stage digester performs the functions of digestion, gravity thickening, and storage in one tank. In a two-stage process, two digesters in series are used. Heating and mechanical mixing occur in the first digester. Since the sludge is continually mixed, it will not settle. Settling and further digestion occur in the unheated second tank.

Historically, rules of thumb were used to size digesters. For single-stage, heated digesters taking primary waste from a trickling filter, 3 to 4 cubic feet of digester volume per equivalent capita is required. For primary and secondary waste, 6 cubic feet are required. Equation 8.61 can also be used to size the digester.

$$\begin{aligned}\text{digester volume} &= \tfrac{1}{2}(Q_{\text{raw sludge}} + Q_{\text{digested sludge}})t_{\text{digestion}} \\ &\quad + (Q_{\text{digested sludge}})t_{\text{storage}} \qquad 8.61\end{aligned}$$

**Aerobic Digestion**: Aerobic digestion in a holding tank or digester uses mechanical aerators in a manner similar to aerated lagoons. Construction of an aerobic digester is similar to that of an aerated lagoon.

Aerobic digesters have the ability to significantly reduce volatile solids. Typical operating and physical characteristics are:

| | |
|---|---|
| • sizing | 2 to 3 ft$^3$/capita |
| • aeration period: | |
| only activated sludge | 10 to 15 days |
| activated sludge without primary settling | 12 to 18 |
| mixture of primary sludge and activated or trickle filter sludge | 15 to 20 |
| • maximum volatile solids loading | 0.1 to 0.3 |
| • sludge age: | |
| primary sludge | 25 to 30 days |
| activated sludge | 15 to 20 |
| • maximum tank depth | 15 feet |
| • oxygen required | 1.5 to 1.9 lbm $O_2$/lbm BOD removed |
| • volatile solids reduction | 35 to 45% typical<br>45 to 70% maximum |
| • dissolved oxygen level | 1 to 2 mg/l |
| • mixing energy required | 0.75 to 1.5 hp/1000 ft$^3$ |

### E. ADDITIONAL SLUDGE PROCESSING METHODS

Sludge may be heated when an economical source of combustion heat is available. Even then, other methods may be more economical. Incineration produces the smallest quantity of inert residue. Sludge storage is not really a process at all, but it can be considered for short-term solutions prior to mechanical dewatering.

### F. RESIDUAL DISPOSAL

Stabilized sludge and clean grit may be eligible for ocean dumping. However, this disposal method is discouraged. Where isolated land is available, land surface spreading is an option. Water pollution from sludge must be prevented. (Harrowing sludge into the soil and other runoff-erosion methods must be used.) Sludge must also be free of heavy metals and other toxic materials in excess of concentrations permitted.

Stabilized sludge and clean grit can be disposed of by lagooning, and restrictions to those for land surface spreading apply. In addition, lagooning should be considered only where eventual restoration of the lagoon area to other uses is possible.

Where a satisfactory landfill site is available, sludge cake, grit, screenings, and scum can be disposed of. However the possible effects on leachate quality will affect the treatment methods and sludge quality.

Incineration is the most expensive treatment, requiring the near total destruction of volatile solids in sludge, scum, and grit. Land disposal of ash should be considered only when lower-cost options are unavailable.

## 17 EFFLUENT DISPOSAL[16]

Wherever possible, effluent should be discharged to receiving waters. Flowing surface waters (such as streams, estuaries, and oceans) are desired. Disposal to lakes and reservoirs should be avoided.

Wastes from very small installations (up to about 30 people) in rural or remote locations can be discharged to groundwater through soil absorption systems. *Deep well injection* can be used for difficult-to-treat industrial wastes if permitted by regulatory agencies and if soil strata are suitable for the discharge type.

If economical, treated effluent can be used for industrial purposes, irrigation, flushing, or (with sufficient treatment) for potable water.

[16] Regulatory agencies should be consulted to determine the applicability of any of these disposal methods.

## 18 SANITARY LANDFILLS

### A. INTRODUCTION

Sanitary landfilling is the disposing of waste solids by thin layering, compacting to minimum volume, and daily soil covering. There is no burning of waste solids.

Sanitary landfill sites need to be selected on the basis of many relevant characteristics. Some of these characteristics are:

- economics and availability of cheap land
- location, including ease of access, distance, acceptance by the population, and aesthetics
- availability of cover soil, if required
- wind direction and wind speed, including odor, dust, and erosion considerations
- flat topography
- dry climate, or low water table and low infiltration rates
- low risk of acquifer contamination
- type and permeability of underlying strata
- absence of winter freezing

Each day's solids are covered by a soil layer, producing a *cell*. Cells can be constructed in a number of ways, but the area, trench, and progressive methods are most common. In the *area method*, waste is spread and compacted on the natural slope of the ground before being covered. With the *trench method*, waste is placed in a trench and covered with its own trench soil. A new trench extension is then dug for the subsequent day's waste. In the *progressive method*, cover soil is taken from the front (toe) of the working face.

Soils and cover soil of clay are preferred, since clay is both hard and strong when dry.

The cell height including the soil cover is known as the *lift*. Lift and cell thickness are illustrated in figure 8.22.

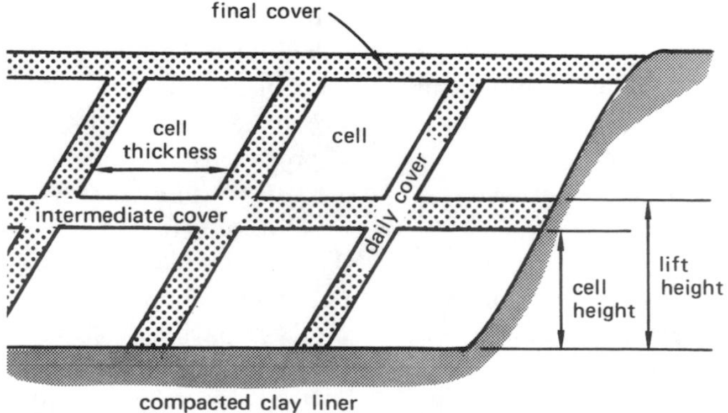


**Figure 8.22** Cells and Landfilling Terms

The daily, intermediate, and final soil covers are essential to proper landfill operation. The cover prevents fly emergence, discourages rats and rodents, and controls entering moisture. The soil cover also serves a purpose when gases of decomposition are to be controlled or collected. In the event of ignition, the soil covers form fire stops within the landfill.

Once covered, the generous food supply within a cell hosts considerable biological decomposition activity. The waste exhausts the oxygen from the cell quickly. Digestion continues anaerobically, producing methane ($CH_4$) gas.[17] The temperature may increase to up to 150°F.

Once closed, the landfill can be used for grassed green areas, shallow-rooted agriculture, and recreational areas (golf, ball fields, etc.). Some light construction uses are also possible, although there may be problems with gas accumulation, corrosion of pipe and foundation piles, and settlement.

### B. DESIGN OF SANITARY LANDFILLS

Sanitary landfills should be designed for a 5-year minimum life, during which time the design volume will be 5 to 8 lbm/capita-day. (5 lbm/capita-day is a typical conservative estimate of incoming volume which is commonly used.)

The waste is placed in layers typically 2 to 3 feet thick. If the cell is sloped, the slope will be less than 40°, typically 20 to 30°. The daily soil cover can be assumed to be 6 inches; the intermediate soil cover will be approximately 1 foot; and the final cover of 2 feet should be applied and compacted in 4 to 6 inch layers. (However, the cover may be up to 4 feet thick, with compaction in 1 foot layers.) When the landfill layer is complete, the ratio of solid waste volume to cover volume will be between 4:1 and 3:1.

Although the *cell height* can be taken as 8 feet for design studies, it can actually be up to 30 feet. The height should be chosen to minimize the amount of cover material required.

Once placed, the solid waste can be compacted to a density of 400 to 1500 lbm/cu. yd., although 1000 lbm/cu. yd. is typical. Generally, a density of 800 lbm/cu. yd. is the minimum acceptable density. Compaction is achieved by 2 to 5 passes by a tracked bulldozer. A 50% reduction in volume can be expected.

The daily increase in landfill volume can be predicted by equation 8.62. The loading factor is 1.25 for a 4:1 ratio of solid waste to cover volumes. It is 1.33 for a 3:1 ratio. (The compacted density can be calculated as a weighted average of soil and solid waste densities. Fine grained soils have densities in the 70 to 120 lbm/ft$^3$ range; coarse grained soils have densities in the 100 to 135 lbm/ft$^3$ range.)[18]

$$\frac{\text{cubic yards}}{\text{day}} = \frac{\begin{pmatrix}\text{population}\\ \text{size}\end{pmatrix}\begin{pmatrix}\text{waste in}\\ \text{lbm/capita-day}\end{pmatrix}\begin{pmatrix}\text{loading}\\ \text{factor}\end{pmatrix}}{\text{compacted density, lbm/cu. yd.}} \quad 8.62$$

Once a landfill site has been closed, settlement can be expected. In areas with high rainfall, there will be greater waste decomposition, with up to 20% settlement based on the overall landfill height. In dry areas, the settlement may be limited to 2% or 3%. Little information is available on bearing capacity of the landfill. Some studies have suggested capacities in the 500 to 800 lbf/ft$^2$ range.

### C. GAS PRODUCTION

Little or no gas will be produced during the first year. However, the peak gas production rate will be reached quickly within 12 to 18 months after closing. Steady gas production can be expected for up to 5 years. Over the gas-producing life of the landfill, 40 cubic feet of gas per cubic yard of landfill can be expected. The gas will be approximately 50% methane.

Gas can be allowed to escape through a semi-permeable cover material such as sand or gravel. It can also be vented through a gravel trench or inclined gravel vent. Vent pipes with collection laterals are expensive, as is vacuum collection.

Compacted clay liners and clay trenches outside the landfill should be placed to prevent the widespread lateral distribution of gas. At least 5 feet should separate the water table and the bottom clay liner. This amount of separation will be sufficient to remove decomposed organisms and living bacteria from the leachate before it reaches the groundwater.

## 19 HAZARDOUS WASTE DISPOSAL

### A. INTRODUCTION

Hazardous wastes can be handled by disposal, dispersal, encapsulation, treatment, or recovery and recycling.[19]

[17] The products of *aerobic decomposition* are $CO_2$, $H_2O$, and nitrates. The products of *anaerobic decomposition* include $CH_4$, CO and $CO_2$, $H_2O$, $N_2$, ammonia, organic acids, and various metallic sulfides.

[18] Multiply cubic yards by $\frac{27}{43,560}$ to get acre-feet.

[19] The 1984 amendments to the *Resource Conservation Recovery Act* (*RCRA*) of 1976 eliminate land disposal for many wastes and keep such disposal at a minimum for many other wastes. The types of disposal methods prohibited unless approved by the Environmental Protection Agency include landfills, surface impoundments, waste piles, injection wells, land treatment facilities, salt dome formations, salt bed formations, and underground mines.

Disposal can be at landfills, impoundment of liquids at the surface, underground injection wells, aboveground piles, and storage in containers and tanks. Dispersal (dilution) is suitable only for degradable substances. Land spreading and ocean dumping are examples of dispersal.

Encapsulation is occasionally used with disposal. Hazardous substances can be encapsulated with cement or potted in plastic (polymerization).

Treatment is a method of modifying the hazard. Organic substances can be incinerated or treated by high temperature decomposition. Chemical reduction in tanks or biological processes (digestion) in biocontactors can also be used.

Recovery/recycling methods include filtration, electrolysis, distillation/evaporation, sedimentation/precipitation, ion exchange and reverse osmosis, and adsorption.

## B. HAZARDOUS WASTE LANDFILLS

Hazardous waste disposal sites must be designed to hold wastes so that they do not adversely affect the air and water quality. Primarily, this means that free liquids within the landfill must be controlled against migration.

Hazardous waste disposal sites can be designed for almost any type of waste, but they are best at controlling soluble and volatile wastes. The most difficult wastes are those that are soluble, toxic, and persistent.[20]

A disposal site's performance will be a function of the reliability and longevity of the leachate collection systems, the hydrogeological characteristics of the site, the type of waste, and the site use, management, and monitoring.

## C. HYDROGEOLOGICAL SITE CHARACTERISTICS

An ideal disposal site would have unfractured bedrock topped with many feet of native, low-permeability clay. The rate of evaporation for the area would exceed precipitation, and the water table would be low or nonexistent. The disposal site would not extend over acquifers which serve as a sole source water supply. The site would not be within the 100-year flood plain.

## D. PRETREATMENT REQUIRED

Ignitable, reactive, and corrosive wastes must be pretreated to reduce those characteristics. When placed in a disposal site, the wastes must be compatible. Containerized liquids should be rendered non-free flowing.

[20] Arsenic, chromium, spent solvents, dioxins, and other highly toxic wastes may not be disposed of in landfills.

Serious thought should be given to disposing of toxic, persistent, and highly mobile wastes when allowed.

## E. LINER AND DISPOSAL SITE DESIGNS

Two alternative liner schemes can be used, both of which require a double bottom liner. The top liner for both alternatives is a synthetic membrane with minimum thickness of 30 mils. The two designs have different bottom liners, however. For the first, the bottom is of compacted clay. The other design is preferred by the EPA. It has a synthetic sheet above the clay liner.

Both designs require two leachate collection systems. One of the collection systems is within the landfill itself to limit the hydraulic head of liquid that has reached the bottom of the landfill. The other collection system catches and recycles leachate which has passed through the top liner.

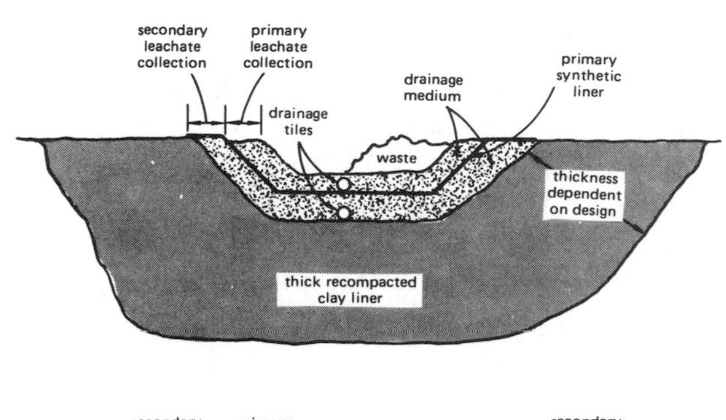


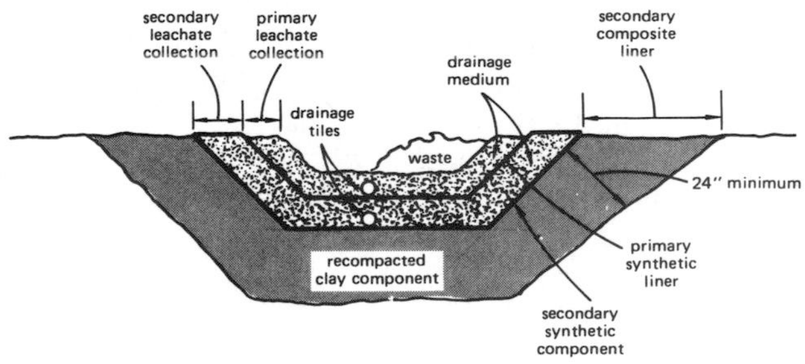


**Figure 8.23** Alternative Liner Designs

## F. LINERS FOR DISPOSAL SITES

Bottom liners are designed to reduce the rate of leachate migration into the subsoil. Clay liners 2 to 20 feet thick have been used frequently in the past. Synthetic *flexible membrane liners* (*FML's*) 30 to 80 mils thick are now required. Other liner materials that have been used include soil cement, asphalt concrete, sprayed-on asphalt and urethane membranes, and soil sealants. Liners fail by faulty installation, structural failure due to hydrostatic pressure, settling of the soil, and chemical decomposition of the liner from wastes. Clay liners are particularly susceptible to dessication from concentrated organic chemical such as solvents.

Migration through undamaged liners is proportional to the conductivity and pressure. The pressure is a function of the leachate depth. Conductivities of clay liners vary between EE–9 and EE–11 $m^3/m^2$-sec. For synthetic liners, the conductivity ranges between EE–11 and EE–14 $m^3/m^2$-sec.

Double liner systems require ground water monitoring with monitoring wells extending past the bottom of the disposal site into the acquifer.

## G. COVERS FOR DISPOSAL SITES

Covers are required to prevent infiltration of rain into disposal cells. Clay or synthetic materials, or combinations of the two, can be used. (Liners are effective at diverting water off of mounded clay covers.) Covers fail by dessication (drying out), penetration by roots, animal activity, freezing and thawing cycles, erosion, settling, and breakdown of synthetic liners from sunlight exposure. Failure will also occur when the cover collapses into the disposal site due to open voids below or excess loading (rainwater ponding) above.

## H. LEACHATE COLLECTION/RECOVERY SYSTEMS

Leachate will be prevented from leaving the landfill by a leachate collection system. By removing the leachate above the first liner, the hydrostatic pressure on the liner will be reduced. A pump is used to raise collected leachate to the surface once a predetermined level has been reached. Tracer compounds (e.g., lithium compounds or radioactive hydrogen) can be buried with the wastes to signal migration and leakage.[21]

Leachate collection and recovery systems fail by clogging drainage layers and pipes, crushing of the collection pipes due to waste load, and pump failures.

## 20 INJECTION WELLS

Liquid or low-viscous wastes can be injected under high pressure into appropriate strata 2000 to 6000 feet below the surface. Wastes will displace native fluids under pressure. The injection well is capped to maintain the pressure.

[21] Tritium has a half-life of 12.3 years. It is essentially completely decayed after 120 years. It is easily detected, and is not affected chemically by other wastes.

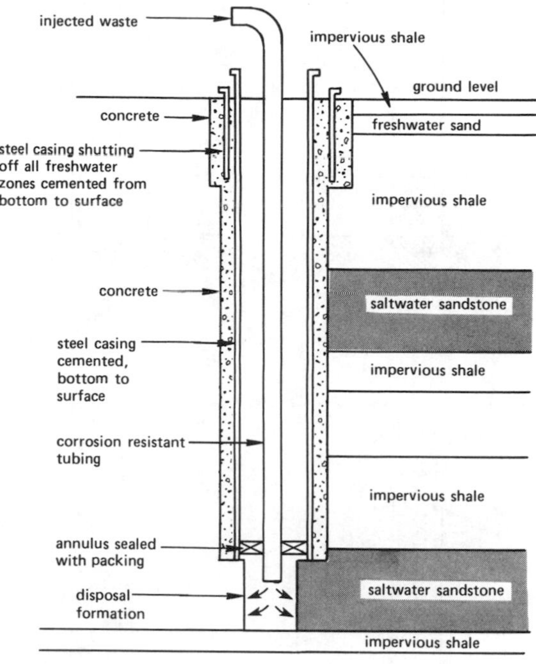


**Figure 8.24** A Typical Injection Well

Injection wells fail primarily by waste plumes (capillary action) through fractures, shrinkage cracks, pressure fractures, dissolution channels, fault slips in strata, and seepage around the borehole.

## 21 SLURRY TRENCH CONTAINMENT

Slurry trenches 4 to 12 inches wide are opened hydraulically using a bentonite slurry trenching fluid. (The slurry contains *bentonite* 1 to 6% by weight.) When the slurry hardens on the walls of the trench, the trench is backfilled. If backfilled with soil, the trench is known as a *soil bentonite* trench. If the trench is filled with bentonite slurry as the trench is opened, it is known as a *pure bentonite* or *cement bentonite trench.*

Bentonite is not very chemical resistant, particularly when it is fully hydrated. Furthermore, it's permeability is on the order of only EE–10 $m^3/m^2$-sec. So, a synthetic liner is required with exposure to hazardous wastes. Asphaltic cutoffs can also be used.

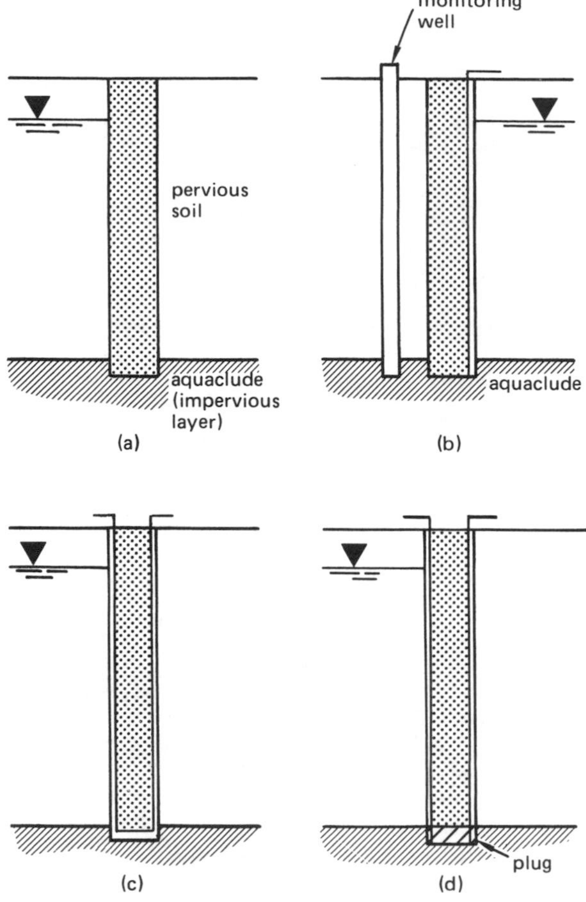


(a) standard trench

(b) trench with single sheet and monitoring well

(c) envelope design double sheet

(d) double sheet with bottom plug

**Figure 8.25** Slurry Trench Containment Walls

## 22 STEPS IN CLEANING UP HAZARDOUS WASTE SPILLS

The steps required to clean up a hazardous waste spill depend on the type of waste spilled, the weather, environment, geology, proximity to population, and other factors. Also, some steps will need to be performed as quickly as possible, while others are long term. Not all of the steps listed here will be appropriate for every case.

- immediate confinement: sandbag dams, bentonite clay linings, slurry walls
- physical removal: removal of solids, liquids, bulk containers, and contaminated soil and vegetation to hazardous waste disposal sites
- pumping through purge wells: removing contaminated water from the acquifer
- holding contaminated water in basins for later treatment
- dilution: mixing clean or processed water with contaminated water through spraying, injection into the soil or acquifer, or deep well disposal injection.
- adsorption: mixing surface soil with granular carbon, soda ash, and other chemicals to raise pH, elevate temperature, and initiate decay and decomposition of remaining waste
- coverage: covering the contaminated area by a surface layer to decrease rain water forming leachate and driving contaminant further down

# Appendix A: Selected "10-States' Standards"

The following selected standards are derived from "Recommended Standards for Sewage Works", originally developed by the Great Lakes-Upper Mississippi River Board of State Sanitary Engineers. Since there are ten states in this board, the publication is commonly referred to as "10-States' Standards."

The standards are subject to change, and this summary is not complete. The items listed here are merely meant to be representative of good wastewater plant design in colder areas of the U.S.

- **Hydraulic Load:** 100 gallons per capita-day for new systems in undeveloped areas unless other information is available.
- **Pumps**: At least 2 pumps are required, and 3 are required when the design flow exceeds 1 MGD. Both pumps should have the same capacity if only 2 pumps are used. This capacity must exceed the design flow for each pump. If 3 or more pumps are used, the capacities may vary, but capacity pumping must be possible with one pump out of service.
- **Racks and Bar Screens**: All racks and screens shall have openings less than 1.75″ wide. The smallest opening for manually-cleaned screens is 1.0″. The smallest opening for automatically-cleaned screens is 0.625″. Flow velocity should be 1.25 to 3.0 fps.
- **Grinders and Shredders**: Grinder/shredders are required if there is no primary sedimentation or fine screens. Gravel traps or grit-removal equipment should precede comminutors.
- **Grit Chambers**: Grit chambers are required when combined storm and sanitary sewers are used. A minimum of two grit chambers in parallel should be used, with a provision for bypassing. The optimum velocity is 1.0 fps throughout. The detention time is dependent on particle sizes to be removed.
- **Plain Sedimentation Tanks**: Multiple units are desirable and must be provided if the flow exceeds 100,000 gpd. For primary settling, the depth should be 7.0′ or greater. The maximum peak overflow rate is 1500 gpd/ft$^2$.[22] 15,000 gpd/ft is the maximum weir loading. If the flow rate is less than 1.0 MGD, the weir loading should be reduced to 10,000 gpd/ft.
- **Trickle Filters**: Rock media should have a depth of 5′ to 10′. Manufactured media should have a depth of 10′ to 30′. The rock media should be 1″ to $4\frac{1}{2}''$ in size, with no fines. Freeboard of 4′ or more is required. The drain should slope at 1% or more.
- **Activated Sludge Processes**: For sedimentation basins, the following hydraulic loading maximums are specified: 1200 gpd/ft$^2$ for conventional, step, and contact units; 1000 gpd/ft$^2$ for extended aeration units; 800 gpd/ft$^2$ for separate nitrification units. The maximum BOD loading shall be: 40 lbm/day-1000 ft$^3$ for conventional, step, and complete mix units: 50 lbm/day-1000 ft$^3$ for contact stabilization units; 15 lbm/day-1000 ft$^3$ for extended aeration units.

  Aeration tank depths should be between 10′ and 30′. At least two aeration tanks should be used. The dissolved oxygen content should not be allowed to drop below 2.0 mg/l at any time. The aeration rate should be 1500 ft$^3$ oxygen per pound of $BOD_s$. For extended aeration, the rate should be 2000 ft$^3$ oxygen per pound of $BOD_s$.
- **Final Clarifiers**: Maximum surface settling rate is 800 gpd/ft$^2$ for separate nitrification stages, 1000 gpd/ft$^2$ for extended aeration, and 1200 gpd/ft$^2$ for all other cases, including fixed film biological processes.
- **Lagoons**: Maximum BOD application is 15 to 35 lbm $BOD_s$ per acre-day for both controlled-discharge and flow-through stabilization ponds.
- **Chlorination**: Requires a 15 minute contact period at peak flow.
- **Anaerobic Digesters**: For completely mixed digesters, up to 80 lbm of volatile solids per day per 1000 ft$^3$ of digester. For moderately mixed digesters, the limit is 40 lbm/day-1000 ft$^3$. Multiple units. Minimum 20 feet sidewater depth.
- **Sludge Drying Beds**: Requires 2 ft$^2$/capita-day, if drying beds are the primary dewatering method, and 1 ft$^2$/capita-day if beds are a back-up dewatering method.

[22] The basin size shall also be calculated based on the average design flow rate and a maximum overflow rate of 1000 gpd/ft$^2$. The larger of the two sizes shall be used.

# Appendix B: Saturated Oxygen Concentrations

To convert °C to °F:

$$°F = \frac{9}{5}°C + 32°$$

To convert °F to °C:

$$°C = \frac{5}{9}(°F - 32°)$$

(1 atmosphere)

| temperature °C | dissolved oxygen, mg/l | subtract for each 100 mg/l chloride |
|---|---|---|
| 0 | 14.6 | 0.017 |
| 1 | 14.2 | 0.016 |
| 2 | 13.8 | 0.015 |
| 3 | 13.5 | 0.015 |
| 4 | 13.1 | 0.014 |
| 5 | 12.8 | 0.014 |
| 6 | 12.5 | 0.014 |
| 7 | 12.2 | 0.013 |
| 8 | 11.9 | 0.013 |
| 9 | 11.6 | 0.012 |
| 10 | 11.3 | 0.012 |
| 11 | 11.1 | 0.011 |
| 12 | 10.8 | 0.011 |
| 13 | 10.6 | 0.011 |
| 14 | 10.4 | 0.010 |
| 15 | 10.2 | 0.010 |
| 16 | 10.0 | 0.010 |
| 17 | 9.7 | 0.010 |
| 18 | 9.5 | 0.009 |
| 19 | 9.4 | 0.009 |
| 20 | 9.2 | 0.009 |
| 21 | 9.0 | 0.009 |
| 22 | 8.8 | 0.008 |
| 23 | 8.7 | 0.008 |
| 24 | 8.5 | 0.008 |
| 25 | 8.4 | 0.008 |
| 26 | 8.2 | 0.008 |
| 27 | 8.1 | 0.008 |
| 28 | 7.9 | 0.008 |
| 29 | 7.8 | 0.008 |
| 30 | 7.6 | 0.008 |

# Appendix C: Typical Sequences Used in Wastewater Plants

The following partial sequences are used to construct a complete treatment plant:

| | |
|---|---|
| I | intake and preconditioning |
| P | primary treatment |
| S | secondary treatment |
| T | tertiary treatment |
| D | discharge |
| SP | sludge processing |
| SD | sludge disposal |

## I: Intake and Preconditioning

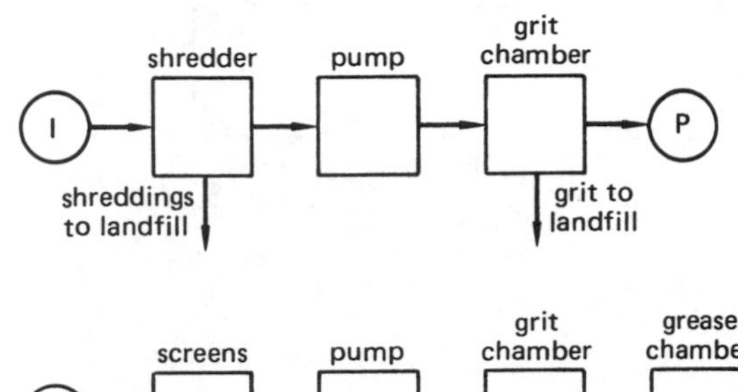


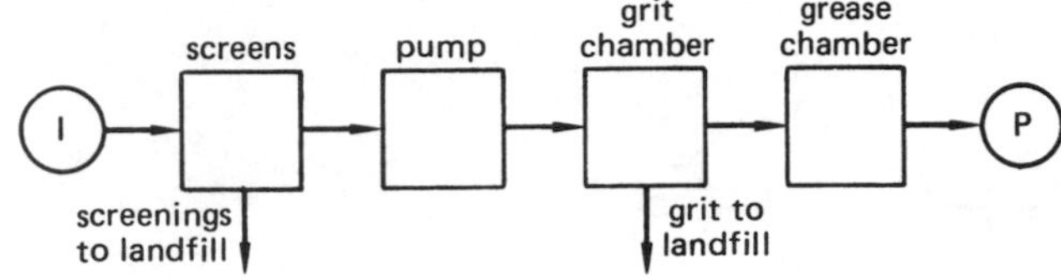


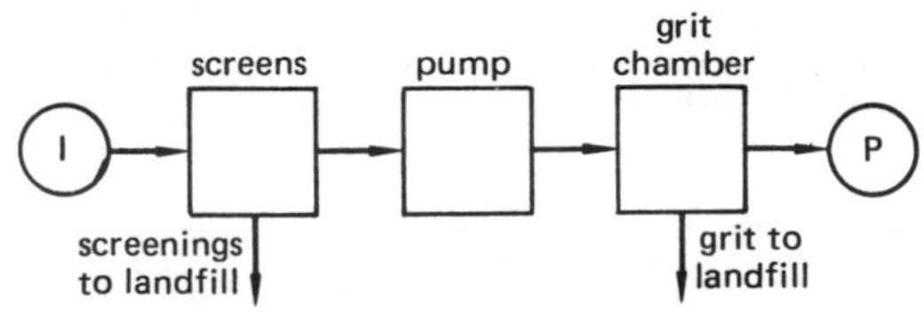


## P: Primary Treatment

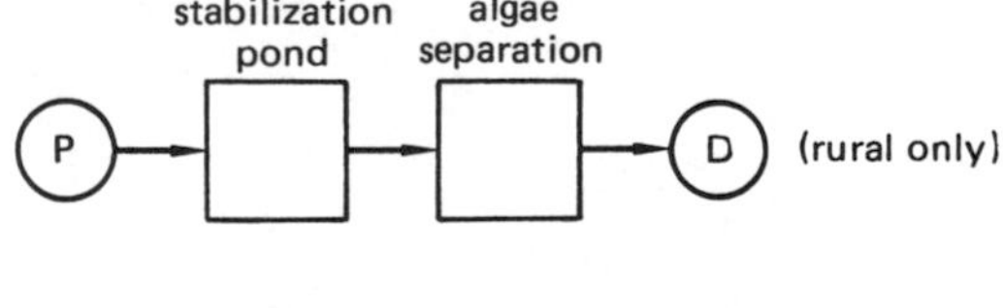


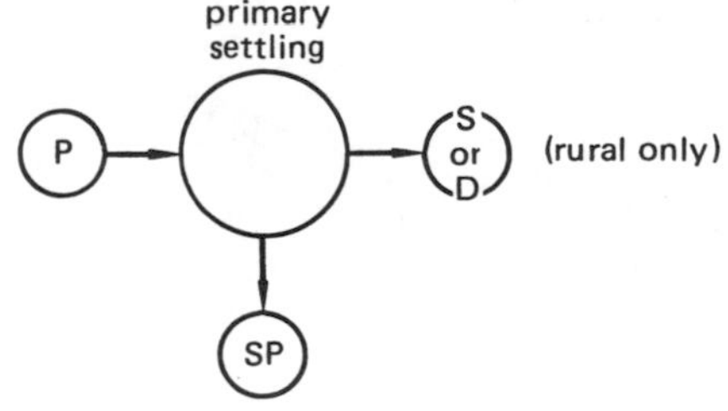


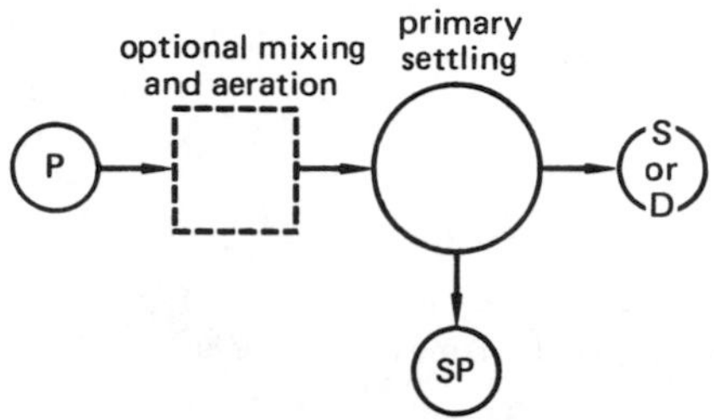


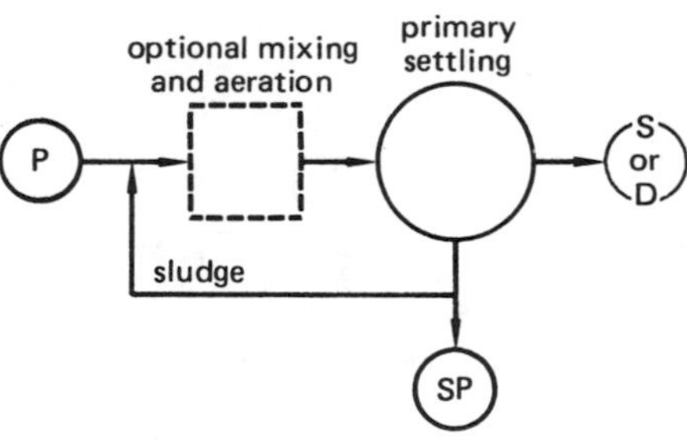


## S: Secondary Treatment

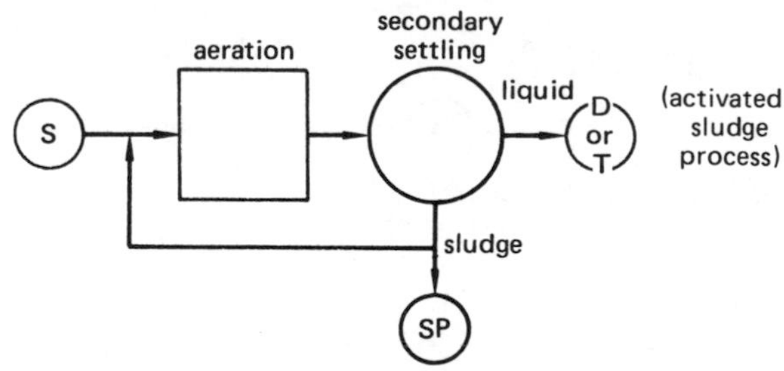


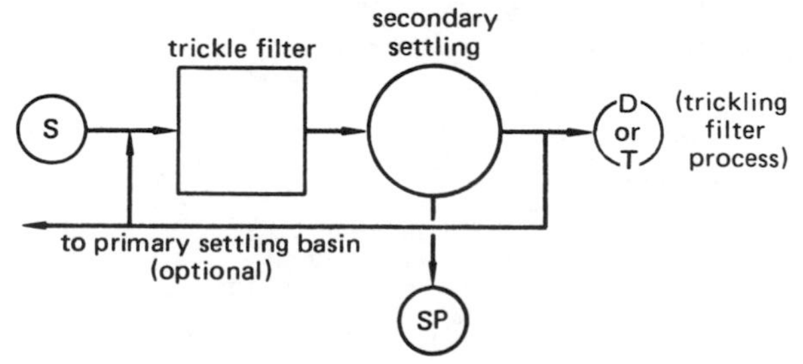

### T: Tertiary Treatment

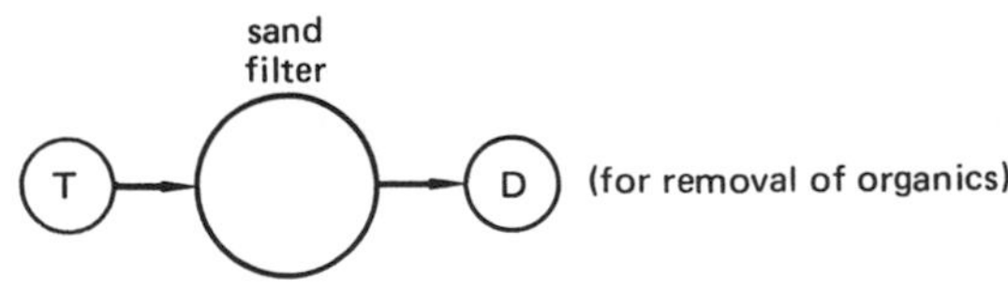


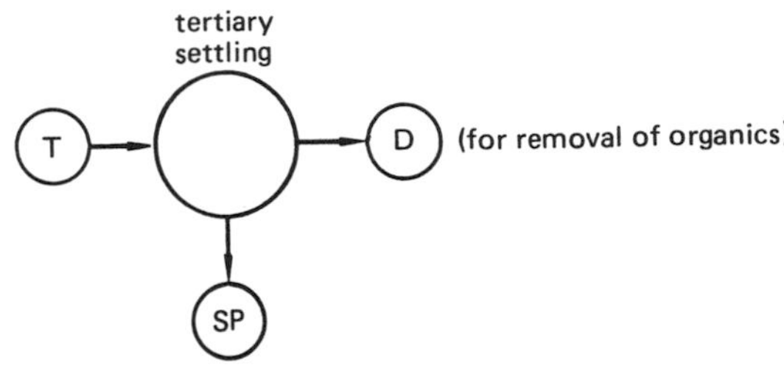


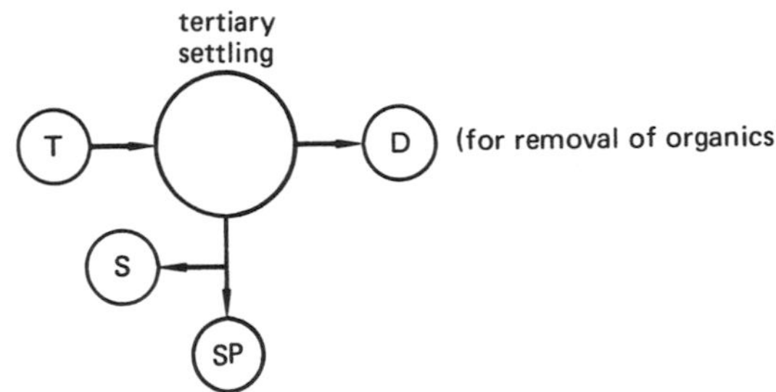


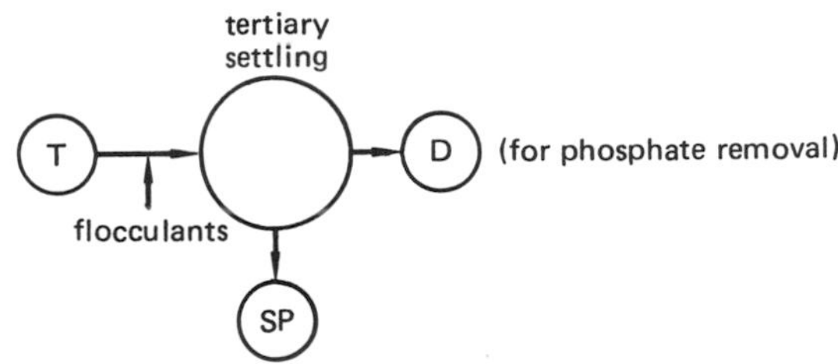


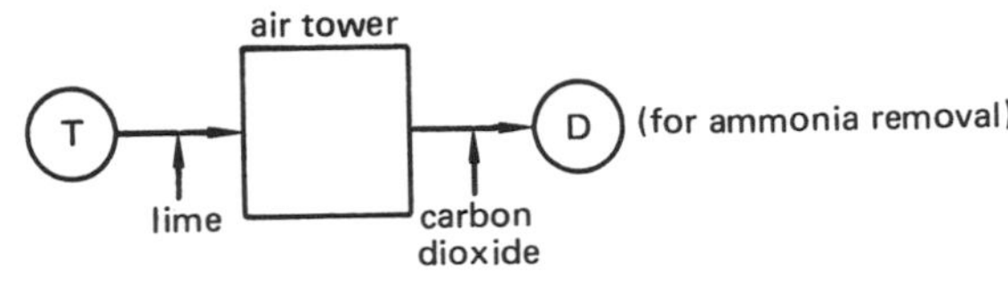


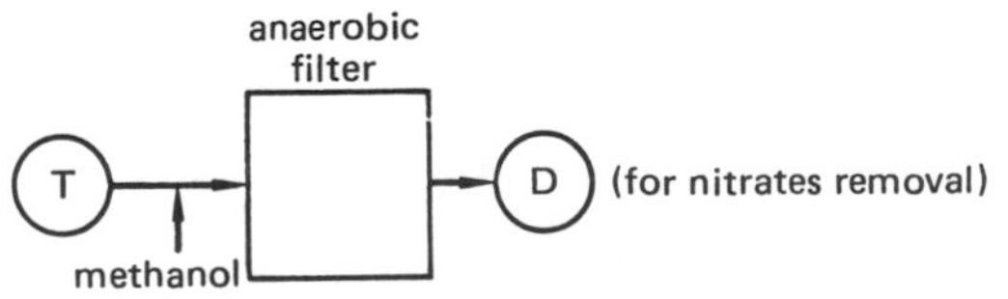


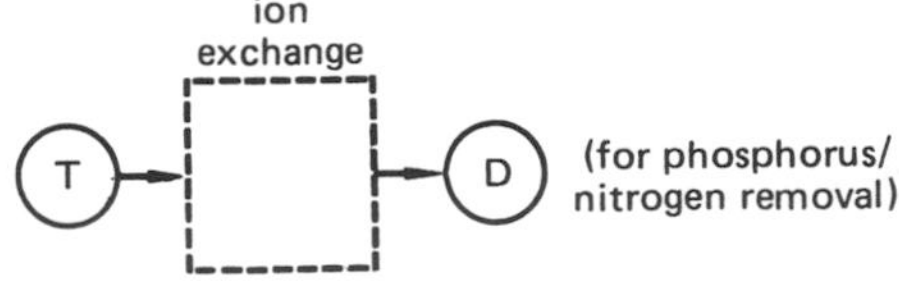


### SP: Sludge Processing

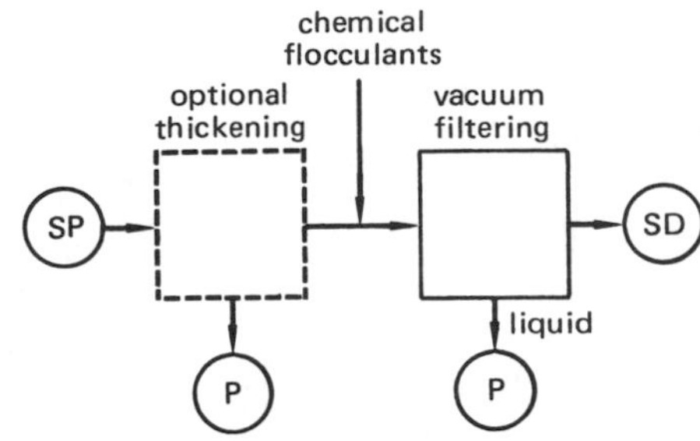


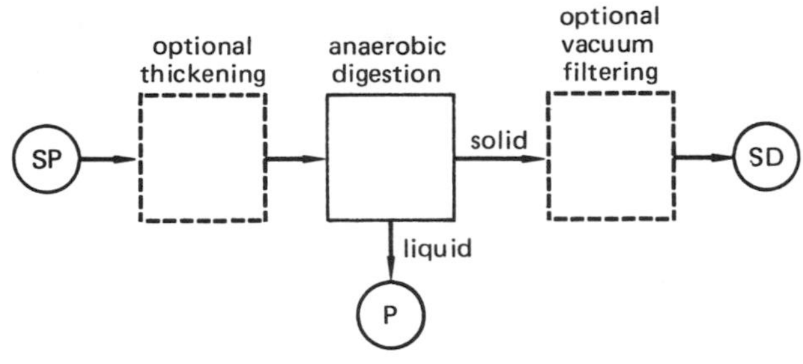


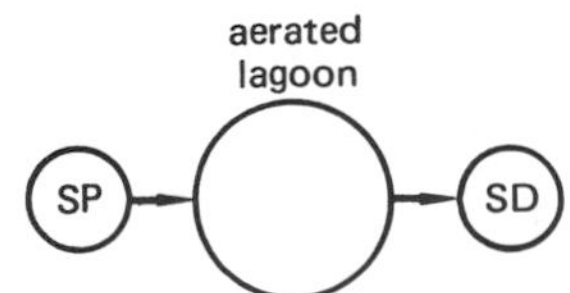


### SD: Solids Disposal

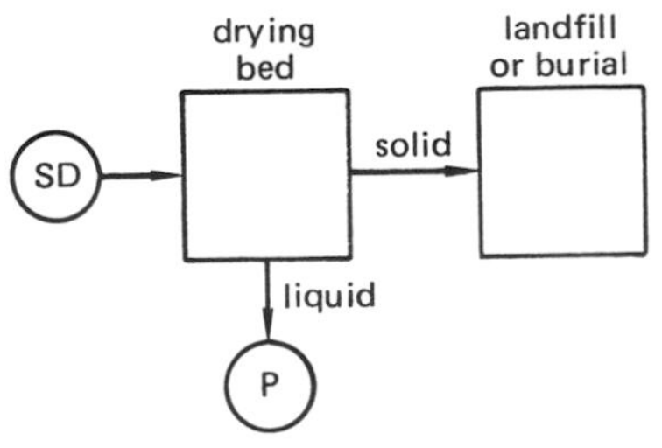


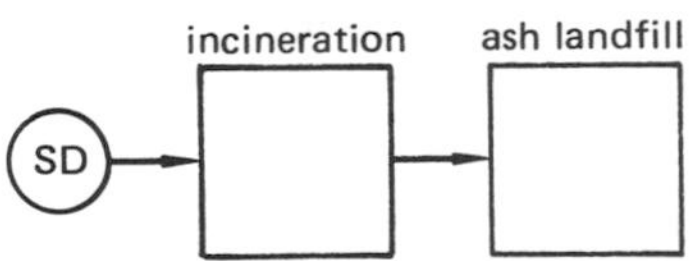


### D: Discharge

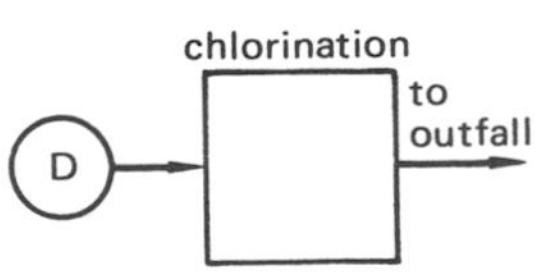

**Practice Problems: WASTE WATER ENGINEERING**

Untimed

1. It is estimated that the BOD of raw sewage received at a treatment plant you are designing will be 300 mg/l from a population of 20,000. A single-stage, high-rate filter is to be used to reduce the plant effluent to 50 mg/l. 30% of the raw sewage BOD is removed by settling. Recirculation is from the filter effluent to the primary settling influent. Use '10-State Standards'. (a) What is the design flow rate? (b) If a round filter is used, what should be its depth and diameter? (c) What volume should be recirculated? (d) Draw a flow diagram of the process. (e) What is the overall plant efficiency?

2. The average waste-water flow from a community of 20,000 is 125 gpcd. The 5-day, 20°C BOD is 250 mg/l. The suspended solids content is 300 mg/l. The final plant effluent is to be 50 mg/l BOD through use of settling tanks and trickling filters. The settling tanks are to have a surface settling rate of 1000 gpd/ft². The trickling filters are to be 6 feet deep and used without recirculation. (a) Design circular settling tanks for this plant. (b) Estimate the BOD removal in the settling tanks. (c) Determine the size of the trickling filters.

3. A small town of 10,000 people discharges its wastes directly into a stream which has the following characteristics:

| | |
|---|---|
| minimum flow rate: | 120 cfs |
| minimum dissolved oxygen: (at 15°C) | 7.5 mg/l |
| velocity: | 3 mph |
| temperature: | 15°C |
| reaeration coefficient of stream and wastes: | 0.2 @ 20°C |
| BOD reaction coefficient of stream and wastes: | 0.1 @ 20°C |

The town's waste consists of the following:

| | volume | BOD @ 20°C | temperature |
|---|---|---|---|
| domestic | 122 gpcd | 0.191 lb/cd | 64°F |
| infiltration | 116,000 gpd | | 51°F |
| industrial #1 | 180,000 gpd | 800 mg/l | 95°F |
| industrial #2 | 76,000 gpd | 1700 mg/l | 84°F |

(a) What is the domestic waste BOD in mg/l? (b) What is the overall waste BOD in mg/l just before discharge into the stream? (c) What is the temperature of the waste just before discharge into the stream? (d) What is the theoretical minimum dissolved oxygen concentration in the stream? (e) How far downstream would you expect the minimum dissolved oxygen concentration to occur?

4. A sewage treatment plant is being designed to handle both domestic and industrial waste with the following characteristics:

| | |
|---|---|
| industrial #1: | 1.3 MGD; 1100 mg/l BOD |
| industrial #2: | 1.0 MGD; 500 mg/l BOD |
| domestic: | 100 gpcd; 0.18 lb/cd BOD |

The city has a population of 20,000, and an excess capacity factor of 15% is to be used. (a) What is the design population equivalent for the plant? (b) What is the plant's organic loading? (c) What is the plant's hydraulic loading?

5. Sewage from a city of 40,000 has an average daily flow of 4.4 MGD. An analysis of the raw sewage shows the following characteristics:

| | |
|---|---|
| pH | 7.8 |
| suspended solids | 180 mg/l |
| 5-day BOD | 160 mg/l |
| COD | 800 mg/l |
| total solids | 900 mg/l |
| volatile solids | 320 mg/l |
| settleable solids | 8 mg/l |

(a) What are the dimensions of a circular primary sedimentation basin that would remove about 30% of the BOD? (b) How many basins are required? (c) What are the dimensions of square sedimentation basins that would remove about 30% of the BOD? (d) How many square basins would be required? (e) Design a single-stage trickling filter and final sedimentation basin (circular) which would produce, in combination with the primary unit, a final effluent having a 5-day BOD of not more than 20 mg/l.

6. A cheese factory discharges 35,000 gpd with the following effluent characteristics: 10,000 gpd with 1000 mg/l BOD; 25,000 gpd with 250 mg/l BOD. (a) Assuming an average depth of 4 feet, what lagoon size is required for stabilization? (b) What is the detention time?

7. An activated sludge plant processes 10 MGD of influent with 240 mg/l BOD and 225 mg/l of suspended solids. The discharge from the final clarifier contains 15 mg/l BOD and 20 mg/l suspended solids. The following assumptions may be used:

- Primary clarification removes 60% of the suspended solids and 35% of the BOD.
- Sludge has a specific gravity of 1.02 and consists of 6% solids by weight.
- The cell yield (BOD conversion to biological solids) in the aeration basin is 60%.
- The final clarifier does not reduce BOD.

(a) What is the daily weight of sludge produced? (b) Assuming the sludge is completely dried prior to disposal, determine the daily sludge volume.

Timed

1. Your client has just completed a subdivision and his sewage hooks up to a collector which goes into a trunk. The problem is that the first manhole up from the trunk on the collector pipe overflows periodically. An industrial plant is hooked directly into the trunk and it is the plant's volume which is making your cli-

ent's line back up. The subdivision is in a flood plain and the sewer lines are very flat and cannot be steepened. The rim on the manhole cannot be raised. (a) Give two solutions to this problem. (b) Sketch a plan view showing what your solutions would look like.

2. A high-rate, two-stage trickle filter system processes 5.2 MGD of sewage with a 5-day BOD of 320 mg/l. The effluent is to have a 5-day BOD of 28 mg/l, within a ± 20 mg/l range. (a) What is the final BOD of the effluent? (b) Determine the required filter size (diameters). Both filters have the same size.

hydraulic load to filter: 32 mgad including recycle
30% BOD removal in primary clarifier
filter depth: 5 feet for both filters
Use NRC standards.

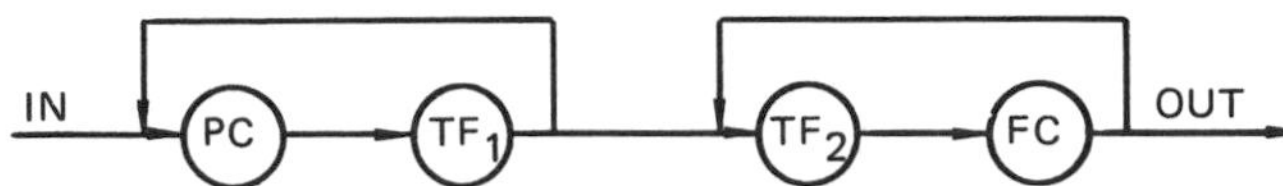


3. A town has a current population of 10,000 and is expected to double in size in 15 years. The town has plans to deposit its solid waste in a 30-acre landfill that will be used as a park in 20 years. (a) Describe the general requirements for a sanitary landfill. (b) What factors should be considered in selecting a site? (c) If the average person generates 5 pounds of solid waste per day, how long will it take to produce a 6-foot lift? Hint: The town continues to grow after 15 years.

4. A town of 10,000 people has selected a 50 acre site to deposit its solid waste. The minimum side borders are 50 feet. The maximum trench depth is ± 20 feet. There is a minimum requirement of 10 feet of earth cover for the lift. (a) What is the service life of the disposal site? (b) What is the daily annual solid waste volume? (c) What is the volumetric capacity of the disposal site? (d) Discuss the environmental considerations relating to traffic, pollution, aesthetics, and other factors.

5. A town of 10,000 people has its own primary treatment plant. (a) What mass of total solids (in pounds per day) should the treatment plant expect? (b) If the town is 4 miles from the treatment plant and 400 feet above it in elevation, what size pipe should be used between the town and the plant?

6. A small community has a projected average flow of 1 MGD. Incoming wastewater has the following properties:

BOD of 250 mg/l
grit specific gravity of 2.65
total suspended solids of 400 mg/l

The community wishes to have a wastewater treatment system consisting of a single aerated grit chamber, a single primary clarifier, two trickling filters, and a single secondary clarifier. The final effluent is to have a final BOD of 30 mg/l. The recirculation ratio of the system is 100%. (a) Size the aerated grit chamber with one foot of freeboard. (b) Determine the air requirements for the grit chamber in order to capture 95% of the grit. (c) Size the primary clarifier. (d) Size the trickling filters in acre-feet. (e) Size the secondary clarifier.

7. Wastewater enters the recirculating biological contactor (RBC) treatment process shown with 250 mg/l BOD at the rate of 1.5 MGD. The RBC removes a fraction of the incoming BOD.

$$\text{Fraction BOD removal in RBC (expressed as a decimal)} = \frac{1}{\left[1+\frac{kA}{Q}\right]^3}$$

$$k = 2.45\ \frac{\text{gpd}}{\text{ft}^2}$$

Q does not include recirculation.

(a) Find the area of the RBC such that $BOD_{out}$ is 30 mg/l. (b) Find the recirculation ratio such that the $BOD_{out}$ is 30 mg/l. Keep the RBC area the same as in (a). (c) Find the sludge volume produced from the clarifiers if the yield is 0.4 lbs/lb BOD removed. S.G. of raw sludge – 1.0. (d) What is the organic loading to the RBC under the flow scheme outlined in part (a)?

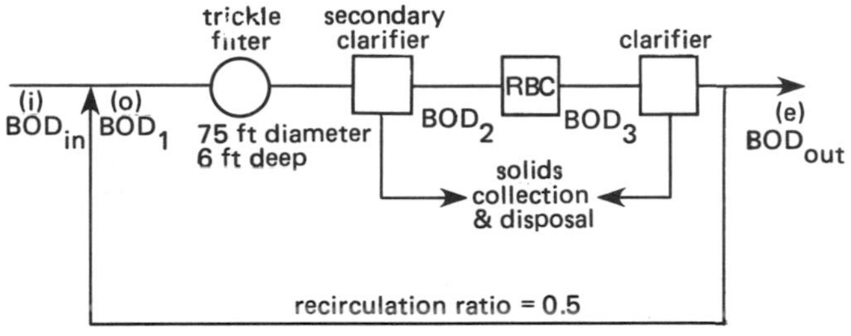


WASTE-WATER

# 9 SOILS

*Nomenclature*

| | | |
|---|---|---|
| A | area | various |
| c | cohesion | psf |
| $C_c$ | compression index | – |
| $C_u$ | uniformity coefficient | – |
| $C_z$ | coefficient of curvature | – |
| CBR | California bearing ratio | – |
| D | diameter | mm |
| e | void ratio | – |
| F | percent passing through the sieve, or shape factor | –, various |
| $G_H$ | hydraulic gradient | – |
| h | head | cm |
| $I_d$ | density index | – |
| $I_g$ | group index | – |
| $I_l$ | liquidity index | – |
| $I_p$ | plasticity index | – |
| k | coefficient of permeability | cm/sec |
| L | flow path length | cm |
| n | porosity | – |
| N | number of blows | – |
| p | pressure | psi |
| P | load | lbf |
| PPS | percent pore space | – |
| Q | flow quantity | $cm^3$/sec |
| r | radius | various |
| R | overconsolidation ratio, or Hveem's resistance | –, – |
| s | degree of saturation | – |
| S | strength | psi |
| SG | specific gravity[1] | – |
| t | time | seconds |
| v | velocity | cm/sec |
| V | volume | $cm^3$ |
| w | water content | – |
| W | weight | grams |
| $w_l$ | liquid limit | – |
| $w_p$ | plastic limit | – |

*Symbols*

| | | |
|---|---|---|
| $\epsilon$ | strain | – |
| $\rho$ | mass density | $g/cm^3$ or $lbm/ft^3$ |
| $\gamma$ | specific weight | $lbf/ft^3$ |
| $\sigma$ | normal stress | psi |
| $\phi$ | angle of internal friction | degrees |
| $\tau$ | shear stress | psf |
| $\theta$ | angle of principal stress plane | |

*Subscripts*

| | |
|---|---|
| A | axial |
| B | borrow |
| c | compressive |
| d | dry |
| eq | equilibrium |
| f | final |
| F | fill |
| g | air |
| i | ith component, or initial |
| n | unconfined |
| o | consolidated |
| R | radial |
| s | soil |
| sat | saturated |
| t | total |
| u | ultimate |
| uc | ultimate compressive |
| us | ultimate shear |
| v | void or volumetric |
| w | water |
| z | zero air voids |

## 1 CONVERSIONS

| multiply | by | to obtain |
|---|---|---|
| centimeters | 0.3937 | inches |
| centimeters squared | 0.155 | square inches |
| cubic yards | 27 | cubic feet |
| cubic yards | 202.2 | gallons |
| dynes | EE–5 | newtons |
| cubic feet | 7.48 | gallons |
| cubic feet | 0.03704 | cubic yards |

[1] As a peculiarity of soils engineering, the specific gravity is usually given the symbol *G*, as opposed to this book which uses *SG* throughout.

| | | |
|---|---|---|
| feet/min | 0.508 | cm/sec |
| gallons | 0.1337 | cubic feet |
| gallons | 4.95 EE–3 | cubic yards |
| gallons of water | 8.345 | pounds of water |
| grams/cubic centimeter | 62.428 | pounds/cubic foot |
| inches | 2.54 | centimeters |
| square inches | 6.4516 | square centimeters |
| kilograms | 2.20462 | pounds |
| newtons | 0.22481 | pounds |
| newtons | EE5 | dynes |
| pascals | 0.145 EE–3 | psi |
| pounds | 0.4536 | kilograms |
| pounds | 453.59 | grams |
| pounds | 4.448 | newtons |
| pounds/square inch | 144 | psf |
| pounds/square inch | 6.894 EE3 | pascals |
| pounds/cubic foot | 0.01602 | grams/cubic centimeter |

## 2 DEFINITIONS

Admixture: Material added to soil to increase its workability, strength, or imperviousness, or to lower its freezing point.

Adsorption: Absorption characterized by a higher concentration of water at the surface of the solid than throughout.

Adsorbed water: Water held near the surface of a material by electrochemical forces.

Aggregate: A mixture of various soil components (e.g., sand, gravel, and silt).

Bentonite: A volcanic clay which exhibits extremely large volume changes with moisture content changes.

Caisson: An air- and water-tight chamber used to work or excavate below the water level.

Catena: A group of different soils which frequently occur together.

Cation: Positively charged ion.

Dilatancy: Property of increasing in volume when changing shape.

Fine: Combined silt and clay.

Friable: Easily crumbled.

Frost susceptibility: Susceptible to having water continually drawn up from the water table by capillary action, forming ice crystals below the surface (but above the frost line).

Gap graded: Having large particles and small particles, but no medium-sized particles.

Glacial till: Soil resulting from a receding glacier, consisting of mixed clay, sand, gravel, and boulders.

Gumbo: Silty soil that becomes soapy, sticky, or waxy when wet.

Horizon: Dividing line between layers of soil with different colors or compositions.

Loess: A deposit of wind-blown silt.

Normally loaded soil: A soil which has never been loaded to a greater extent than at present.

Pycnometer: A stoppered flask with graduations.

Pedology: The study of the soil constituting the upper 4 or 5 feet of the earth's crust.

Rock flour: Fine grained, rounded quartz grains with little plasticity, characteristic of glacial activity.

Stratum: Layer.

Thixotropic: Gradually increasing in strength as absorbed water distributes itself through the soil.

Till: See 'Glacial till'.

## 3 SOIL TYPES

Soil is an aggregate of loose mineral and organic particles. This definition distinguishes soil from rock, which exhibits strong and permanent cohesive forces between the mineral particles.

Proper calculations for foundations and retaining walls require that the nature of the soil be known. This can be done either quantitatively or qualitatively. The primary components of any soil are gravel, sand, silt, and clay. Organic material can also be present in surface samples. If the soil is a mixture of two or more of these components, the soil is given the name of the constituent that has the greatest influence on its mechanical behavior (e.g., silty clay or sandy loam).

Particle size limits for defining gravel and sand have been suggested, and these are given in table 9.1. Whereas sand and gravel are classified as coarse grained soils, inorganic silt and clay are classified as fine grained soils. The clay-silt distinction cannot be made on the basis of size alone. Silt possesses little plasticity and cohesion. Clay, on the other hand, is very plastic and cohesive.

*Clays* may be distinguished from *silts* by using the following simple tests:

**Table 9.1**
Soil Classification by Particle Size

| system | date | sizes(mm) gravel | sand | silt | clay |
|---|---|---|---|---|---|
| Bureau of Soils | 1890 | 1–100 | 0.05–1 | 0.005–0.05 | $< 0.005$ |
| Atterberg | 1905 | 2–100 | 0.2–2 | 0.002–0.2 | $< 0.002$ |
| MIT | 1931 | 2–100 | 0.06–2 | 0.002–0.06 | $< 0.002$ |
| USDA | 1938 | 2–100 | 0.05–2 | 0.002–0.05 | $< 0.002$ |
| Unified (or AC) | 1953 | 4.75–75 | 0.075–4.75 | $< 0.075$ | |
| ASTM | 1967 | 4.75–75 | 0.075–4.75 | $< 0.075$ | |
| AASHTO | 1970 | 2–75 | 0.075–2 | 0.002–0.075 | 0.001–0.002 |

Dry Strength Test: Mold a small brick of soil and allow it to air dry. Break the brick and place a small (1/8″) fragment between thumb and finger. A silt fragment will break easily, whereas clay will not.

Dilatancy Test: Mix a small sample with water to form a thick slurry. When the sample is squeezed, water will flow back into a silty sample quickly. The return rate will be much lower for clay.

Plasticity Test: Roll a moist soil sample into a thin (1/8″) thread. As the thread dries, silt will be weak and friable, but clay will be tough.

Dispersion Test: Disperse a sample of soil in water. Measure the time for the particles to settle. Sand will settle in 30 to 60 seconds. Silt will settle in 15 to 60 minutes, and clay will remain in suspension for a long time.

*Organic matter* can also be present in soil, and this presence can have a significant effect on the mechanical properties of the soil. Organic material is classified into *organic silt* and *organic clay.* Generally, the greater the organic content, the darker will be the soil color.

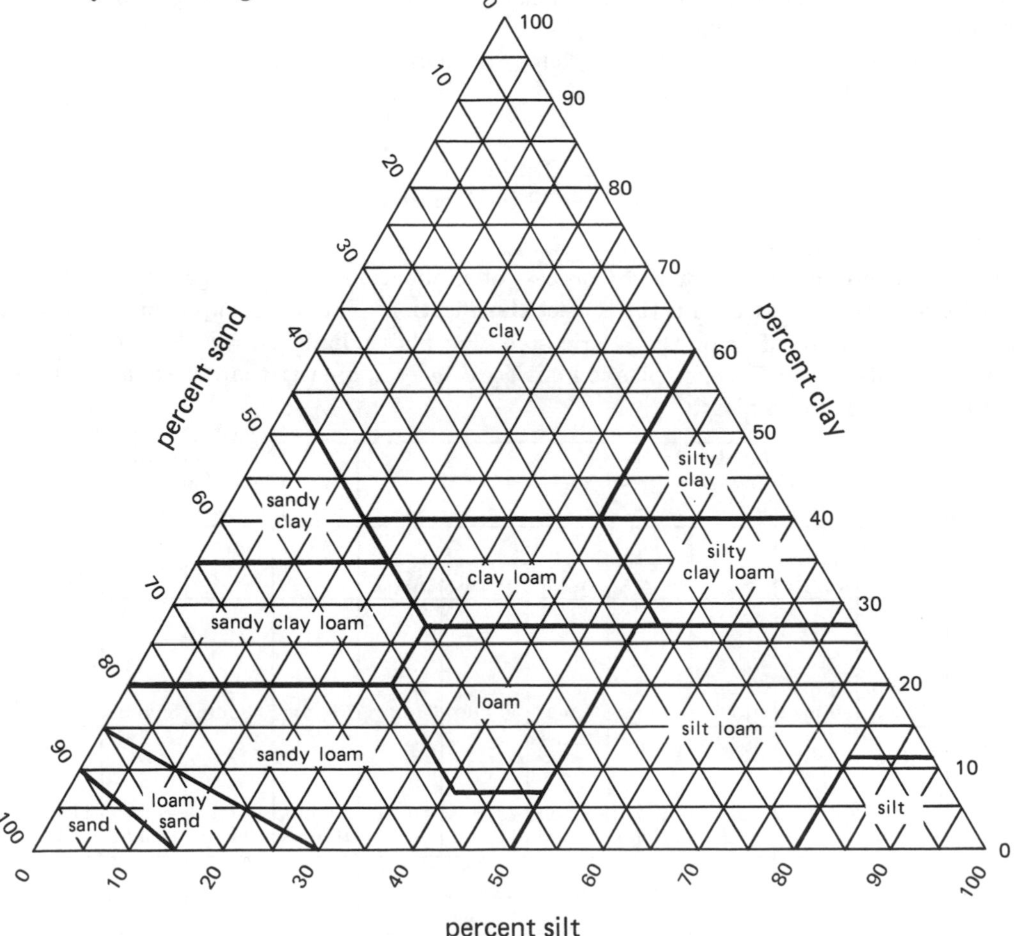


**Figure 9.1** USDA Triangle Chart

General soil classifications have been established by a number of organizations. The most common single index used in classification is particle size. The various classification schemes are presented in table 9.1. The actual classification of a soil will depend on the percentages of each constituent. For example, a soil might be classified as "4% gravel, 45% sand, 15% silt, and 36% clay (MIT)."

One method of giving qualitative descriptions to the soil is by using the *USDA triangular chart*, figure 9.1. This classification ignores the presence of any gravel, although the adjective 'stony' can be used in conjunction with the chart classifications.

Since the qualitative description obtained from the USDA classification chart does not necessarily reflect the mechanical properties of the soil, other systems have been developed. The American Association of State Highway Transportation Officials (AASHTO) has developed a system based on the sieve analysis, liquid limit, and plasticity index.

Soils excellent for roadway subgrade construction are classified as A-1. Highly organic soils not suitable for roadway subgrade construction are classified as A-8. Subgroup classifications are also used, as well as group indexes for fine grained soil. The *group index* of a fine grained soil is given by equation 9.1. $w_l - 40$ and $I_p - 10$ may be negative. However, the group index, $I_g$, is reported as zero if it is calculated to be negative. For A-2-6 and A-2-7 subgroups, only the second term in Eq. 9.1 is used in calculating the group index. $I_g$ is reported to the nearest whole number.

$$I_g = (F_{200} - 35)[0.2 + 0.005(w_l - 40)] + 0.01(F_{200} - 15)(I_p - 10) \quad 9.1$$

$$(I_p = w_l - w_p)$$

$F_{200}$ is the percentage of soil that passes through a #200 sieve. The AASHTO classification system is given in table 9.2. A group index of zero is a good subgrade material. Group indexes of 20 or higher represent poor subgrade materials.

*Example 9.1*

Determine the AASHTO classification of an inorganic soil with the following characteristics:

| soil size | fraction retained on sieve | |
|---|---|---|
| $< 0.002$ | 0.19 | $w_l = 53\%$ |
| 0.002–0.005 | 0.12 | |
| 0.005–0.05 | 0.36 | $w_p = 22\%$ |
| 0.05–0.075 | 0.04 | $F_{200} = 0.04 + 0.36 + 0.12 + 0.19 = 71\%$ |
| 0.075–2.0 | 0.29 | |
| $> 2.0$ | 0 | |

**Table 9.2**
AASHTO Soil Classification System

**Classification procedure: Using the test data, proceed from left to right in the chart. The correct group will be found by process of elimination. The first group from the left consistent with the test data is the correct classification. The A-7 group is subdivided into** A-7-5 or A-7-6 depending on the plastic limit. For plastic limit $w_p = w_l - I_p$ less than 30, the classification is A-7-6. For plastic limit $w_p = w_l - I_p$ greater than or equal to 30, it is A-7-5. NP means non-plastic.

| | granular materials (35% or less passing no. 200 sieve) | | | | | | | silt-clay materials (more than 35% passing no. 200 sieve) | | | | |
|---|---|---|---|---|---|---|---|---|---|---|---|---|
| | A-1 | | A-3 | A-2 | | | | A-4 | A-5 | A-6 | A-7 | A-8 |
| | A-1-a | A-1-b | | A-2-4 | A-2-5 | A-2-6 | A-2-7 | | | | A-7-5 or A-7-6 | |
| sieve analysis: % passing | | | | | | | | | | | | |
| no. 10 | 50 max | | | | | | | | | | | |
| no. 40 | 30 max | 50 max | 51 min | | | | | | | | | |
| no. 200 | 15 max | 25 max | 10 max | 35 max | 35 max | 35 max | 35 max | 36 min | 36 min | 36 min | 36 min | |
| characteristics of fraction passing no. 40: | | | | | | | | | | | | |
| $w_l$: liquid limit | | | | 40 max | 41 min | 40 max | 41 min | 40 max | 41 min | 40 max | 41 min | |
| $I_p$: plasticity index | 6 max | | NP | 10 max | 10 max | 11 min | 11 min | 10 max | 10 max | 11 min | 11 min | |
| usual types of significant constituents | stone fragments gravel and sand | | fine sand | silty or clayey gravel and sand | | | | silty soils | | clayey soils | | peat, highly organic soils |
| general subgrade rating | excellent to good | | | | | fair to poor | | | | | | unsatisfactory |

From table 9.2, the classification is A-7-5 or A-7-6. Since the plastic limit is 22 (less than 30), the classification is A-7-6. The group index is

$$\begin{aligned} I_g &= (71-35)[0.2+0.005(53-40)] \\ &\quad +0.01(71-15)(31-10) \\ &= 21.3 \end{aligned}$$

The soil would be classified as A-7-6 (21).

ASTM standards also provide a method of classifying soils based on qualitative descriptions. Coarse-grained soils are divided into two categories: gravel soils (symbol $G$) and sand soils (symbol $S$). Sands and gravels are further subdivided into 4 subcategories:

symbol $W$: well-graded, fairly clean

symbol $C$: well-graded, with excellent clay binder

symbol $P$: poorly graded, fairly clean

symbol $M$: coarse materials with fines, not in preceding 3 groups

Fine-grained soils are divided into three categories: inorganic silty and very fine sandy soils (symbol $M$), inorganic clays (symbol $C$), and organic silts and clays (symbol $O$). These three are subdivided into two subcategories:

symbol $L$: low compressibilities ($w_l$ 50 or less)

symbol $H$: high compressibilities ($w_l$ greater than 50)

Table 9.7 illustrates the use of some of these classification symbols.

## 4 SOIL INDEXING

Indexing of the soil is necessary in order to apply some of the quantitative property relationships contained in this chapter. Indexing is accomplished by performing various classification tests on the soil. Index properties are of two types: grain properties and aggregate properties.

Soil grain properties include particle size distribution, density, and mineral composition. Density is found from a hydrometer test. Particle size distribution is found from a sieve test for coarse soils, and from a dispersion test for fine soils.

**Table 9.3**
Sieve Sizes

| this size sieve | has this size openings |
|---|---|
| 4″ | 100 mm |
| 3″ | 75 |
| 2″ | 50 |
| $1\frac{1}{2}$″ | 37.5 |
| 1″ | 25 |
| $\frac{3}{4}$″ | 19 |
| $\frac{1}{2}$″ | 12.5 |
| $\frac{3}{8}$″ | 9.5 |
| no.4 | 4.75 |
| 8 | 2.36 |
| 10 | 2.00 |
| 16 | 1.18 |
| 20 | 0.850 |
| 30 | 0.600 |
| 40 | 0.425 (425 μm) |
| 50 | 0.300 |
| 60 | 0.250 |
| 70 | 0.212 |
| 100 | 0.150 |
| 140 | 0.106 |
| 200 | 0.075 (75 μm) |

The results of particle size tests are graphed as a *particle size distribution.*

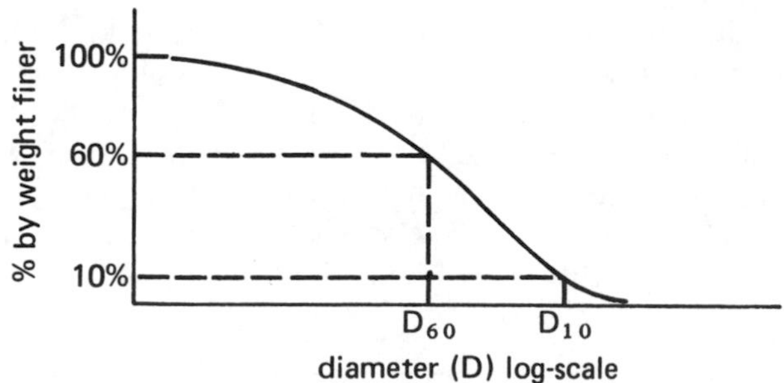


**Figure 9.2** Particle Size Distribution

The *effective grain size* is defined as the diameter for which only 10% of the particles are finer ($D_{10}$). The *Hazen uniformity coefficient* is given by equation 9.2.

$$C_u = \frac{D_{60}}{D_{10}} \qquad 9.2$$

If the uniformity coefficient is less than 4 or 5, the soil is considered uniform in particle size. Well graded soils have uniformity coefficients greater than 10.

The *coefficient of curvature* is

$$C_z = \frac{D_{30}^2}{D_{10}\,D_{60}} \qquad 9.3$$

**Table 9.4**
Typical Soil Coefficients

| soil | $C_u$ | $C_z$ |
|---|---|---|
| gravel | $>4$ | 1–3 |
| fine sand | 5–10 | 1–3 |
| coarse sand | 4–6 | |
| mixture of silty sand and gravel | 15–300 | |
| mixture of clay, sand, silt, and gravel | 25–1000 | |

## 5 AGGREGATE SOIL PROPERTIES

The aggregate index properties are essentially weight-volume relationships. In any sample of soil, there will be some air-filled voids, water-filled voids, and solid material. The percentages of these constituents (by both volume and weight) are used to calculate the aggregate properties.

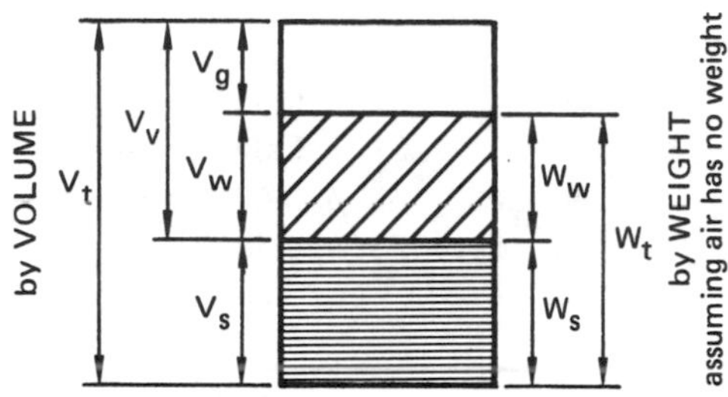


**Figure 9.3** Soil Sample Constituents

The *porosity* is

$$n = \frac{V_v}{V_t} = \frac{V_g + V_w}{V_g + V_w + V_s} \quad 9.4$$

The *void ratio* is

$$e = \frac{V_v}{V_s} = \frac{V_g + V_w}{V_s} \quad 9.5$$

The *water content* is

$$w = \frac{W_w}{W_s} \quad 9.6$$

The volume of the sample will decrease as the water content is reduced down to the *shrinkage limit*, $w = SL$. Below the shrinkage limit, air enters the voids and the water content decreases are not accompanied by decreases in volume.

The *degree of saturation* is

$$s = \frac{V_w}{V_v} = \frac{V_w}{V_g + V_w} \quad 9.7$$

The *soil density* is[2]

$$\rho = \frac{W_t}{V_t} = \frac{W_w + W_s}{V_g + V_w + V_s} \quad 9.8$$

The *dry density* is

$$\rho_d = \frac{W_s}{V_t} = \frac{W_s}{V_g + V_w + V_s} \quad 9.9$$

If the water content is known, the dry density (also known as the *bulk density*) of a moist sample can be found from equation 9.10.

$$\rho_d = \frac{W_t}{(1+w)V_t} = \frac{\rho}{1+w} \quad 9.10$$

The *compacted density* (*zero air voids density*) is

$$\rho_z = \frac{W_s}{V_w + V_s} \quad 9.11$$

The density of the solid constituents is

$$\rho_s = \frac{W_s}{V_s} \quad 9.12$$

The *percent pore space* is

$$PPS = V_v/V_t = 1 - (V_s/V_t) = 1 - (\rho_d/\rho_s) \quad 9.13$$

The specific gravity of the solid constituents is given by equation 9.14. The specific gravity of sand is approximately 2.65, and for clay it ranges from 2.5 to 2.9 with an average around 2.7.

$$SG_s = \frac{\rho_s}{\rho_w} = \frac{\rho_s}{62.4} \quad 9.14$$

Typical values of these soil parameters are given in table 9.6

**Table 9.6**
Typical Soil Indexes

| description | n | e | $w_{sat}$ | $\rho_d$ | $\rho_{sat}$ |
|---|---|---|---|---|---|
| sand, loose and uniform | 0.46 | 0.85 | 0.32 | 90 | 118 |
| sand, dense and uniform | 0.34 | 0.51 | 0.19 | 109 | 130 |
| sand, loose and mixed | 0.40 | 0.67 | 0.25 | 99 | 124 |
| sand, dense and mixed | 0.30 | 0.43 | 0.16 | 116 | 135 |
| glacial clay, soft | 0.55 | 1.20 | 0.45 | 76 | 110 |
| glacial clay, stiff | 0.37 | 0.60 | 0.22 | 106 | 129 |

[2] Soils engineering uses the symbol $\gamma$ and the term *unit weight*. This chapter is consistent with the rest of the book in specifying mass density with symbol $\rho$.

**Table 9.5**
Consolidated Soil Indexing Formulas
(Specific Gravity = G)

| | property | | saturated sample ($W_s, W_w, G$, are known) | unsaturated sample ($W_s, W_w, G, V$ are known) | supplementary formulas relating measured and computed factors | | | |
|---|---|---|---|---|---|---|---|---|
| volume components | $V_s$ | volume of solids | | $\frac{W_s}{G\rho_w}$ | $V-(V_a+V_w)$ | $V(1-n)$ | $\frac{V}{1+e}$ | $\frac{V_v}{e}$ |
| | $V_w$ | volume of water | | $\frac{W_w}{\rho_w{}^*}$ | $V_v-V_a$ | $SV_v$ | $\frac{SVe}{1+e}$ | $Sv_se$ |
| | $V_a$ | volume of air or gas | zero | $V-(V_s+V_w)$ | $V_v-V_w$ | $(1-S)V_v$ | $\frac{(1-S)Ve}{1+e}$ | $(1-S)V_se$ |
| | $V_v$ | volume of voids | $\frac{W_w}{\rho_w{}^*}$ | $V-\frac{W_s}{G\rho_w}$ | $V-V_s$ | $\frac{V_sn}{1-n}$ | $\frac{Ve}{1+e}$ | $V_se$ |
| | $V$ | total volume of sample | $V_s+V_w$ | measured | $V_s+V_a+V_w$ | $\frac{V_s}{1-n}$ | $V_s(1+e)$ | $\frac{V_v(1+e)}{e}$ |
| | $n$ | porosity | | $\frac{V_v}{V}$ | $1-\frac{V_s}{V}$ | $1-\frac{W_s}{GV\rho_w}$ | $\frac{e}{1+e}$ | |
| | $e$ | void ratio | | $\frac{V_v}{V_s}$ | $\frac{V}{V_s}-1$ | $\frac{GV\rho_w}{W_s}-1$ | $\frac{W_wG}{W_sS}$ | $\frac{n}{1-n}$ \| $\frac{wG}{S}$ |
| weights for specific sample | $W_s$ | weight of solids | | measured | $\frac{W_t}{1+w}$ | $GV\rho_w(1-n)$ | $\frac{W_wG}{eS}$ | |
| | $W_w$ | weight of water | | measured | $wW_s$ | $S\rho_wV_v$ | $\frac{eW_sS}{G}$ | $V\cdot\rho_D\cdot w$ |
| | $W_t$ | total weight of sample | | $W_s+W_w$ | $W_s(1+w)$ | | | |
| weights for sample of unit volume | $\rho_D$ | dry unit weight | $\frac{W_s}{V_s+V_w}$ | $\frac{W_s}{V}$ | $\frac{W_t}{V(1+w)}$ | $\frac{G\rho_w}{1+e}$ | $\frac{G\rho_w}{1+wG/S}$ | |
| | $\rho_T$ | wet unit weight | $\frac{W_s+W_w}{V_s+V_w}$ | $\frac{W_s+W_w}{V}$ | $\frac{W_t}{V}$ | $\frac{(G+Se)\rho_w}{1+e}$ | $\frac{(1+w)\rho_w}{w/S+1/G}$ | $\rho_D(1+w)$ |
| | $\rho_{SAT}$ | saturated unit weight | $\frac{W_s+W_w}{V_s+V_w}$ | $\frac{W_s+V_v\rho_w}{V}$ | $\frac{W_s}{V}+\left(\frac{e}{1+e}\right)\rho_w$ | $\frac{(G+e)\rho_w}{1+e}$ | $\frac{(1+w)\rho_w}{w+1/G}$ | |
| | $\rho_{SUB}$ | submerged (buoyant) unit weight | | $\rho_{SAT}-\rho_w{}^*$ | $\frac{W_s}{V}-\left(\frac{1}{1+e}\right)\rho_w{}^*$ | $\left(\frac{G+e}{1+e}-1\right)\rho_w{}^*$ | $\left(\frac{1-1/G}{w+1/G}\right)\rho_w{}^*$ | |
| combined relations | $w$ | moisture content | | $\frac{W_w}{W_s}$ | $\frac{W_t}{W_s}-1$ | $\frac{Se}{G}$ | $S\left[\frac{\rho_w{}^*}{\rho_D}-\frac{1}{G}\right]$ | |
| | $S$ | degree of saturation | 1.00 | $\frac{V_w}{V_v}$ | $\frac{W_w}{V_v\rho_w{}^*}$ | $\frac{wG}{e}$ | $\frac{w}{\frac{\rho_w{}^*}{\rho_D}-\frac{1}{G}}$ | |
| | $G$ | specific gravity | | $\frac{W_s}{V_s\rho_w}$ | $\frac{Se}{w}$ | | | |

$\rho_w$ is the density of water. Where noted with an asterisk (*) use the actual density of water. In other cases, use 62.4 lbm/ft$^3$.

The *density index* (also known as *relative density*) is a measure of the tendency or ability of granular soils (not clays) to compact during loading. The density index is equal to 1 for a very dense soil; it is equal to 0 for a very loose soil.

$$I_d = \frac{e_{max} - e}{e_{max} - e_{min}} \tag{9.15}$$

The many relationships between the above soil indexes and parameters are listed in table 9.5.

*Example 9.2*

What is the degree of saturation for a sand sample with $SG = 2.65$, $\rho = 115$ lbm/ft$^3$, and $w = 17\%$?

In these problems it is always a good idea to keep track of the various weight and volume phases on a phase diagram.

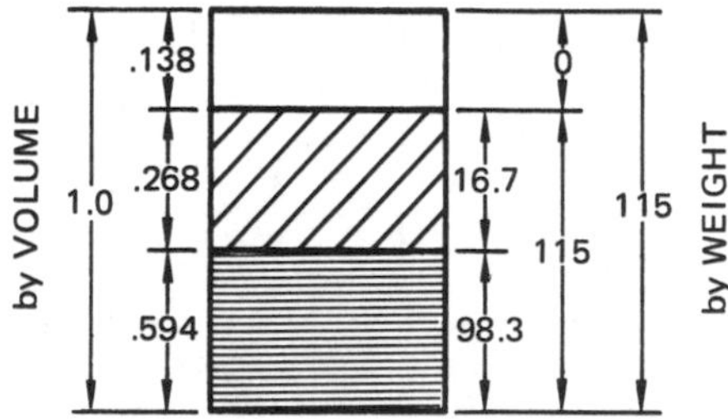


From equation 9.6,

$$W_w = wW_s = (0.17)(W_s)$$

But since $W_w + W_s = 115$,

$$\begin{aligned}(0.17)W_s + W_s &= 115\\ W_s &= 98.3\\ W_w &= 16.7\end{aligned}$$

The solids volume is given by equation 9.12:

$$V_s = W_s/\rho_s = W_s/(SG)_s\rho_w = \frac{98.3}{(2.65)(62.4)} = 0.594\,\text{ft}^3$$

Similarly, the water volume is

$$V_w = \frac{16.7}{62.4} = 0.268$$

The air volume is $1 - 0.594 - 0.268 = 0.138$.

The degree of saturation is

$$s = \frac{0.268}{0.268 + 0.138} = 0.66$$

*Example 9.3*

Borrow soil is used to fill a 100,000 cubic yard depression. The borrow soil has the following characteristics: density = 96.0 lbm/ft$^3$; water content = 8%; specific gravity of the solids = 2.66. The final in-place dry density should be 112.0 lbm/ft$^3$, and the final water content should be 13% (dry basis).

(a) How many cubic yards of borrow soil are needed? (b) Assuming no evaporation loss, how many pounds of water are needed to achieve 13% moisture? (c) What will be the density of the in-place fill after a long rain?

The first step in borrow problems is to draw the phase diagrams for both the borrow and compacted fill soils. Use subscript $B$ for borrow soil and $F$ for fill soil, and work with 1 cubic foot of fill material.

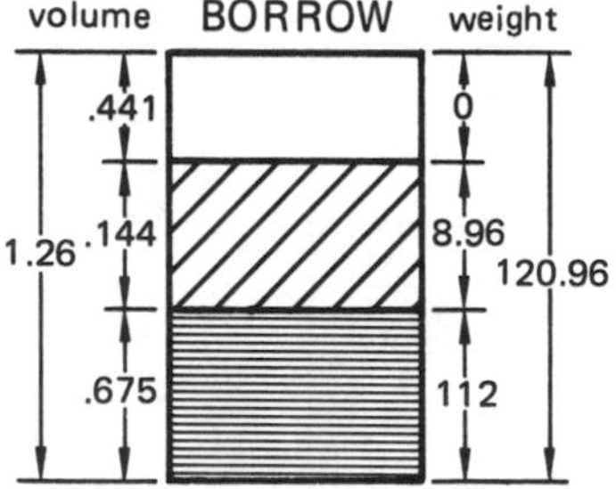


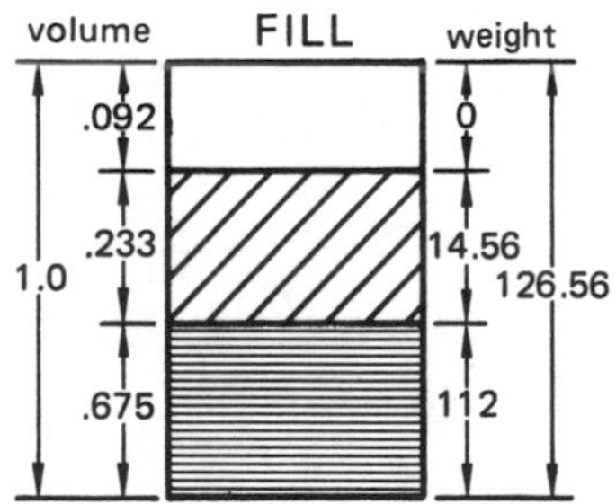


*step 1*: The air has no weight. Dry density precludes water. The soil content (weight) is the same at both locations. (That is, getting 112 pounds of soil in the fill requires taking 112 pounds of borrow soil.) Per cubic foot of fill,

$$W_{sB} = W_{sF} = 112$$

*step 2*: The weight of the water in one cubic foot of fill is

$$W_{wF} = wW_s = (0.13)(112) = 14.56$$

*step 3*: Total weight and density of the fill are

$$W_{tF} = 112 + 14.56 = 126.56$$

$$\rho_F = 126.56/1 = 126.56 \text{ lbm/ft}^3 \text{ fill}$$

*step 4*: The solid volume of the solids in the fill and borrow is

$$V_{sF} = \frac{112}{(2.66)(62.4)} = 0.675 \text{ ft}^3$$

*step 5*: The volume of the water in the fill is

$$V_{wF} = \frac{14.56}{62.4} = 0.233 \text{ ft}^3$$

*step 6*: The air volume in the fill is

$$1 - 0.233 - 0.675 = 0.092 \text{ ft}^3$$

*step 7*: The weight of the water (per cubic foot of fill) in the borrow soil is

$$W_{wB} = wW_{sB} = (0.08)(112) = 8.96$$

(Note that the weight of water per cubic foot of borrow soil is 8.96/1.26.)

*step 8*: The total weight of the borrow soil per cubic foot of fill is

$$W_{tB} = 112 + 8.96 = 120.96$$

*step 9*: The total volume of the borrow soil per cubic foot of fill is

$$V_{tB} = \frac{120.96}{96} = 1.26$$

*step 10*: From step 4, the volume of solids in the borrow soil is

$$V_{sB} = V_{sF} = 0.675$$

*step 11*: The volume of water in the borrow soil per cubic foot of fill is

$$V_{wB} = \frac{8.96}{62.4} = 0.144$$

*step 12*: The air volume in the borrow soil per cubic foot of fill is

$$1.26 - 0.144 - 0.675 = 0.441$$

*step 13* (*a*):

$$V_{\text{required},B} = \left(\frac{1.26}{1}\right)(100{,}000)$$
$$= 126{,}000 \text{ cubic yards}$$

(b): The actual moisture in the compacted borrow soil is

$$\frac{(126{,}000)(27)(8.96)}{1.26} = 2.42 \text{ EE7 pounds}$$

The required moisture in the fill soil is

$$(100{,}000)(27)(14.56) = 3.93 \text{ EE7 pounds}$$

The required additional moisture is

$$3.93 \text{ EE7} - 2.42 \text{ EE7} = 1.51 \text{ EE7 pounds}$$

(c): The void fraction is the sum of water and air volumes per cubic foot of fill (in ft$^3$/ft$^3$). The saturated density is the sum of the water and solid weights.

$$(0.233 + 0.092)(62.4) + 112 = 132.3 \text{ lbm/ft}^3$$

## 6 SOIL TESTING AND MECHANICAL PROPERTIES

### A. PENETRATION RESISTANCE TEST

The most common test is the *standard penetration test* (SPT), which measures resistance to the penetration of a standard split spoon sampler.[3] The number of blows required to drive the sampler a distance of 12 inches (after an initial penetration of 6 inches) is referred to as the *N-value*, in blows per foot.

The N-value has been correlated with many other mechanical properties, including shear modulus, unconfined compressive strength, and effective vertical stress. Figure 9.4 relates $N$ and $S_{nc}$.

### B. MOISTURE-DENSITY RELATIONSHIPS

Soils are compacted to increase their stability, decrease permeability, enhance resistance to erosion, and decrease compressibility. The laboratory test to determine the optimum moisture content and dry density in clay soils is known as the (modified) *Proctor test.* (Nuclear gauges are used in the field to measure density and moisture content in situ.)

A soil sample is compacted in 3 layers by a specific number of hammer blows. The actual density is then given by equation 9.8. The dry density of the sample can be found from equation 9.10. This procedure is repeated for various water contents, and a graph similar to figure 9.5 is obtained. $\rho_d^*$ is known as the *maximum dry density*, or *density at 100% compaction.* $w^*$ is known as the *optimum water content.*

[3] *Cone penetrometer* tests are also performed.

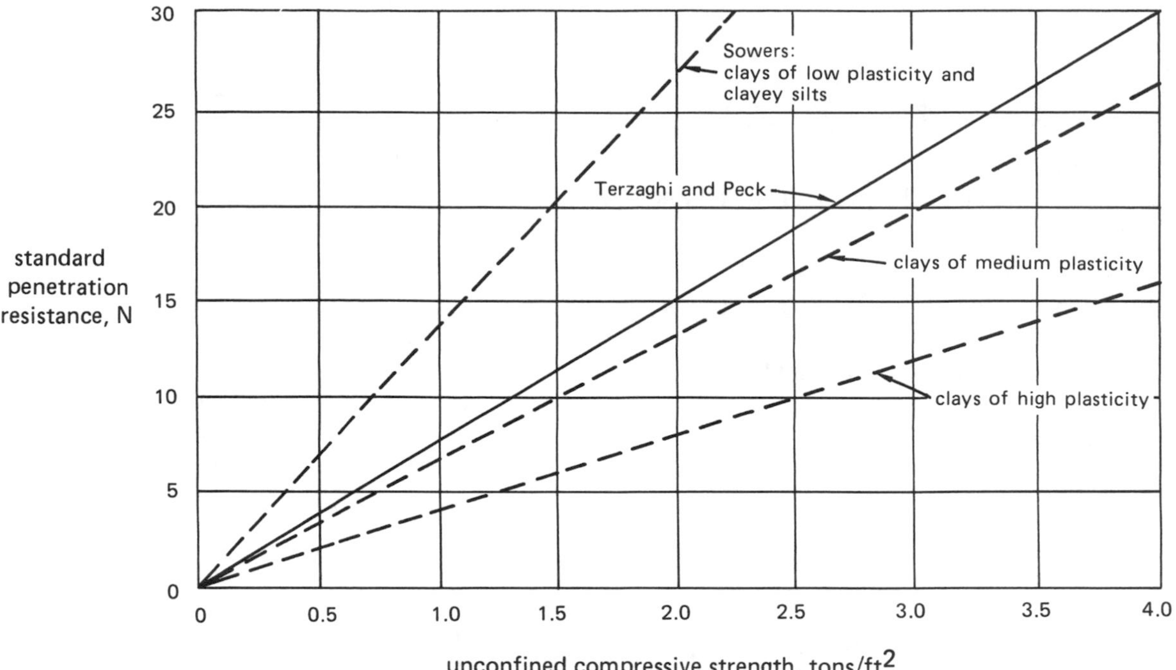


**Figure 9.4** Approximate Relationships Between N and the Unconfined Compressive Strength for Clay

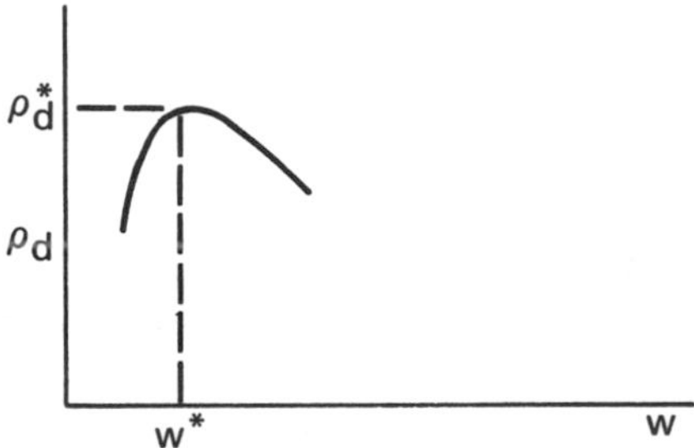


**Figure 9.5** Proctor Test Results

Since the actual compacted density will usually be below (or above) the maximum dry density, the *percentage of compaction* is defined as $\rho_d/\rho_d^*$.

It is not usually feasible to compact soil to the optimum value derived from the Proctor test. Construction compaction methods do not parallel the compaction method used in the test. Usually, some percentage of the maximum Proctor dry density is specified. Table 9.7 lists typical values of optimum moisture content for various soil types, as well as the suggested degree of compaction.

The degree of compaction suggested in table 9.7 depends on the category of soil use. Class 1 uses include the upper 9 feet of fills supporting 1- or 2-story buildings, the upper 3 feet of subgrade under pavements, and the upper 1 foot of subgrade under floors. Class 2 uses include deeper parts of fills under buildings and pavements, as well as earth dams. All other fills requiring some degree of strength or incompressibility are classified as class 3.

For a given water content, saturation will result from perfect compaction, since all air will be removed. The densities resulting from saturation at each water content can be plotted versus water content, and the result is known as a *zero air voids curve*.[4] The theoretical maximum density of the zero air voids curve is calculated from equation 9.16.

$$\rho_z = \frac{\rho_w}{w + \left(\frac{1}{SG}\right)} = \frac{62.4}{w + \left(\frac{1}{SG}\right)} \qquad 9.16$$

The maximum value of the zero air voids density occurs at $w = 0$. At that point, the maximum dry zero air voids density is equal to the density of the solid itself (as calculated from the solid specific gravity).

$$\rho_{zd,\max} = (62.4)(SG) \qquad 9.17$$

$\rho_{zd,\max}$ and $\rho_d^*$ are not the same, however, since air voids exist in the $\rho_d^*$ case.

[4] The zero air voids curve always lies above the Proctor test curve, since that test cannot expel all air.

**Table 9.7**
Typical Values of Optimum Moisture Content and Suggested Compactions (Based on Standard Proctor Test)

| class group symbol | description | range of maximum dry densities lbm/ft$^3$ | range of optimum moisture content % | recommended percentage of Proctor maximum class 1 | 2 | 3 |
|---|---|---|---|---|---|---|
| GW | well-graded, clean gravels, gravel-sand mixtures | 125–135 | 11–8 | 97 | 94 | 90 |
| GP | poorly-graded clean gravels, gravel-sand mixtures | 115–125 | 14–11 | 97 | 94 | 90 |
| GM | silty gravels, poorly graded gravel-sand silt | 120–135 | 12–8 | 98 | 94 | 90 |
| GC | clayey gravels, poorly-graded gravel-sand-clay | 115–130 | 14–9 | 98 | 94 | 90 |
| SW | well-graded clean sands, gravely sands | 110–130 | 16–9 | 97 | 95 | 91 |
| SP | poorly-graded clean sands, sand-gravel mix | 100–120 | 21–12 | 98 | 95 | 91 |
| SM | silty sands, poorly-graded sand-silt mix | 110–125 | 16–11 | 98 | 95 | 91 |
| SM-SC | sand-silt-clay mix with slightly plastic fines | 110–130 | 15–11 | 99 | 96 | 92 |
| SC | clayey sands, poorly-graded sand-clay mix | 105–125 | 19–11 | 99 | 96 | 92 |
| ML | inorganic silts and clayey silts | 95–120 | 24–12 | 100 | 96 | 92 |
| ML-CL | mixture of organic silt and clay | 100–120 | 22–12 | 100 | 96 | 92 |
| CL | inorganic clays of low-to-medium plasticity | 95–120 | 24–12 | 100 | 96 | 92 |
| OL | organic silts and silt-clays, low plasticity | 80–100 | 33–21 | – | 96 | 93 |
| MH | inorganic clayey silts, elastic silts | 70–95 | 40–24 | – | 97 | 93 |
| CH | inorganic clays of high plasticity | 75–105 | 36–19 | – | – | 93 |
| OH | organic and silty clays | 65–100 | 45–21 | – | 97 | 93 |

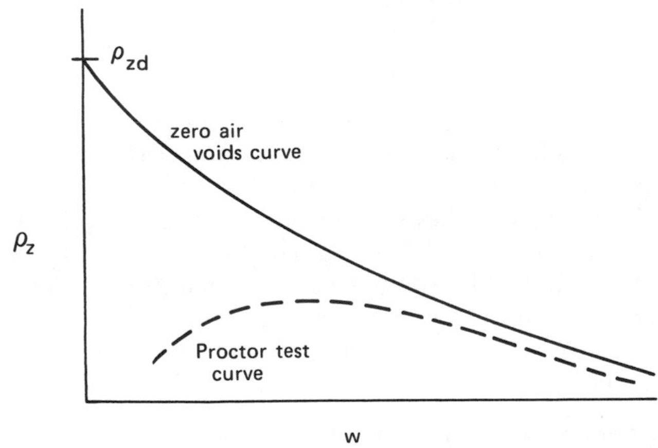


**Figure 9.6** A Typical Zero Air Voids Curve

*Example 9.4*

A Proctor test using a 1/30 ft$^3$ mold is performed on a sample of soil.

| test no. | sample net mass (lbm) | water content (%) |
|---|---|---|
| 1 | 4.28 | 7.3 |
| 2 | 4.52 | 9.7 |
| 3 | 4.60 | 11.0 |
| 4 | 4.55 | 12.8 |
| 5 | 4.50 | 14.4 |

If 0.032 cubic feet of compacted soil tested at a construction site weighed 3.87 pounds wet and 3.74 pounds dry, what is the percent of compaction?

The actual density of sample 1 is (4.28)(30) = 128.4 lbm/ft$^3$. From equation 9.10, the dry density is

$$\rho_d = \frac{\rho}{1+w} = \frac{128.4}{1+0.073} = 119.7 \text{ lbm/ft}^3$$

The following table is constructed from the results of all 5 tests.

| test no. | dry density |
|---|---|
| 1 | 119.7 |
| 2 | 123.6 |
| 3 | 124.3 |
| 4 | 121.0 |
| 5 | 118.0 |

If these data are graphed, the following figure results. The peak is near point 3, so take the maximum density to be 124.3 lbm/ft$^3$. The sample density is 3.74/0.032 = 116.9, so the percentage of compaction is 116.9/124.3 = 0.94.

### C. MODIFIED PROCTOR TEST

This test is similar to the standard Proctor test except that the soil is compacted in 5 layers with a heavier hammer falling a greater distance. The result is a denser soil which is more representative of compaction densities available from modern equipment. Table 9.7 can be used by adding 10 to 20 lbm/ft$^3$ to the densities and taking 3 to 10% from the moisture contents.

### D. IN-PLACE DENSITY TEST

This test, also known as the *field density test*, starts by compacting soil in the field and digging a 3″ to 5″ deep hole with smooth sides. All soil taken from the hole is saved and weighed before the water content can change. The hole volume is determined by filling the hole with sand or a water-filled rubber balloon. The required densities are given by equations 9.8 and 9.10.

### E. UNCONFINED COMPRESSIVE STRENGTH TEST

A cylinder of cohesive soil (usually clay) is loaded to compressive failure. (Failure of elastic soils is taken as a 20% strain.) The unconfined compressive strength is given by equation 9.18. The ultimate shear strength is taken as one half of the unconfined compressive strength.

$$S_{nc} = P/A \tag{9.18}$$

$$S_{us} = \frac{S_{nc}}{2} \tag{9.19}$$

### F. SENSITIVITY TESTS

Clay will become softer as it is worked, and clay soils may turn into viscous liquids during construction. This tendency is determined by measuring the ultimate strength of two unconfined samples, one of which has been packed and extruded.

$$\text{sensitivity} = \frac{S_{nc,\text{undisturbed}}}{S_{nc,\text{remolded}}} \tag{9.20}$$

**Table 9.8**
Sensitivity Classifications

| sensitivity | class |
|---|---|
| 1–8 | natural clays |
| 4–8 | sensitive |
| 8–15 | extra sensitive |
| > 15 | quick |

### G. ATTERBERG LIMIT TESTS (CONSISTENCY TESTS)

Clay soils can be either solid, plastic, or liquid depending on the water content. The water contents corresponding to the transitions from solid to plastic or plastic to liquid are known as the *Atterberg limits.* These transitions are called the *plastic limit* ($w_p$) and *liquid limit* ($w_l$) respectively.

When a soil has a liquid limit of 100, the weight of moisture equals the weight of the dry soil (i.e., $w = 1$). Alternatively, at the liquid limit, the soil is half water and half solids. A liquid limit of 50 means that the soil at the liquid limit is two-thirds soil and one-third water.

Sandy soils have low liquid limits—on the order of 20. In such soils, the test is of little significance in judging load carrying capacities. Silts and clays can have significant liquid limits—as high as 100. Most clays, however, have liquid limits between 40 and 60. High liquid limits indicate high clay content and low load carrying capacity.

The plastic limit depends on the clay content. Some silt and sand soils have no plastic limit at all. They are known as *non-plastic soils.* The test is of no value in judging the relative load carrying capacities of such soils.

The difference between the liquid and plastic limits is known as the *plasticity index.*

$$I_p = w_l - w_p \tag{9.21}$$

The plasticity index gives the range in moisture content over which the soil is in a plastic condition. A small plasticity index shows that a small change in moisture content will change the soil from a semisolid to a liquid condition. Such a soil is sensitive to moisture. A large plasticity index (i.e., greater than 20) shows that considerable water can be added before the soil becomes liquid.

*Atterberg limits* vary with the clay content, type of clay, and the ions (cations) contained in the clay.

The *liquidity index* of a clay soil is

$$I_l = \frac{w - w_p}{I_p} \tag{9.22}$$

The Atterberg *liquid limit* is found by taking a soil sample and placing it in a shallow container. The sample is parted in half with a special grooving tool. The container is dropped 25 times. At the liquid limit, the sample will have rejoined for a length of $\frac{1}{2}$″.

The *plastic limit* test consists of rolling a soil sample

into a 1/8″ thread. The sample will crumble when it is at the plastic limit when rolled to that diameter.

*Example 9.5*

A clay has the following Atterberg limits: liquid limit = 60%; plastic limit = 40%; shrinkage limit = 25%. The clay shrinks from 15 cubic centimeters to 9.57 cubic centimeters when the moisture content is decreased from the liquid limit to the shrinkage limit in the Atterberg tests. What is the clay's specific gravity (dry)?

The water reduction is 15 − 9.57 = 5.43 cubic centimeters. Since 1 cubic centimeter of water weighs 1 gram, the weight loss is 5.43 grams. The percentage weight loss (dry basis) is 60% − 25% = 35%. Therefore, from equation 9.6, the solid weight is

$$W_s = \frac{\Delta W_w}{\Delta w} = \frac{5.43}{0.35} = 15.5 \text{ g}$$

The water volume at the shrinkage limit is

$$V_w = (0.25)(15.5) = 3.875$$

Since at and above the shrinkage limit there are no air voids, the volume of solid at the shrinkage limit is

$$9.57 - 3.875 = 5.695$$

The density of the solid is

$$\rho = \frac{15.5}{5.695} = 2.72 \text{ g/cm}^3$$

$$SG = 2.72$$

## H. PERMEABILITY TESTS

Permeability of a soil is a measure of continuous voids. A permeable material permits a significant flow of water. The flow of water through a permeable acquifer or soil is given by equation 9.23, known as *Darcy's law.*

$$v = kG_H/n \qquad 9.23$$

$$Q = nAv \qquad 9.24$$

The area $A$ in equation 9.24 is the cross sectional area of the aquifer, not the actual area in flow. Water can only flow through the area between the solids. This open area is $nA$.

Typical values of the coefficient of permeability, $k$, are given in table 9.9. Soils with permeabilities of less than EE−6 cm/sec are essentially impervious. The soil is considered pervious if $k$ is greater than EE−4 cm/sec.

**Table 9.9**
Typical Permeabilities

| group symbol | typical coefficient of permeability (cm/sec) |
|---|---|
| GW | 2.5 EE–2 |
| GP | 5 EE–2 |
| GM | > 5 EE–7 |
| GC | > 5 EE–8 |
| SW | > 5 EE–4 |
| SP | > 5 EE–4 |
| SM | > 2.5 EE–5 |
| SM-SC | > EE–6 |
| SC | > 2.5 EE–7 |
| ML | > 5 EE–6 |
| ML-CL | > 2.5 EE–7 |
| CL | > 5 EE–8 |
| OL | – |
| MH | > 2.5 EE–7 |
| CH | > 5 EE–8 |
| OH | – |

For loose filter sands, $k$ is given approximately by equation 9.25. $D_{10}$ is the *effective grain size*—the diameter for which only 10% of the particles are finer.

$$k \approx 100(D_{10})^2 \qquad 9.25$$

Actual numerical values can be calculated from controlled permeability tests using constant- or falling-head *permeators* (figure 9.7). For *constant-head tests,* $k$ can be found from equation 9.26. ($V$ is the water volume.)

$$k = \frac{VL}{hAt} \qquad 9.26$$

For *falling-head tests,* $k$ can be found from equation 9.27.

$$k = \frac{A'L}{At} ln(h_i/h_f) \qquad 9.27$$

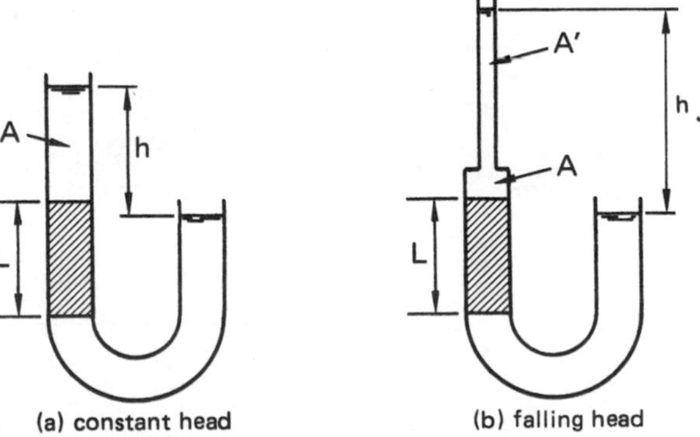


**Figure 9.7** Permeators

For the *auger-hole method* (i.e., in-field, falling-head tests), the combination of area and length variables may be known as the *shape factor* or *conductivity coefficient*, $F$. For example, for a cased hole below the water table of length $L$ and radius $r$ whose impervious casing extends all the way to the hole bottom and whose liquid level rises from $h_i$ to $h_f$ in time $t$ (approaching the water table level), the shape factor and permeability are:

$$F = \frac{11r}{2} \qquad 9.28$$

$$k = \frac{\pi r^2}{Ft} ln\left(\frac{h_i}{h_f}\right) = \frac{2\pi r}{11t} ln\left(\frac{h_i}{h_f}\right) \qquad 9.29$$

Other in-field tests can use cased holes with constant head, or uncased holes with constant and variable head. The shape factors for these tests are not the same as that in equation 9.28.

*Example 9.6*

The permeability of a semi-impervious soil was evaluated in a falling-head permeator whose head decreased from 100 to 40 cm in 5 minutes. The body diameter was 13 cm; the standpipe diameter was 0.3 cm; and the sample length was 8 cm. What was the permeability of the soil?

From equation 9.27:

$$k = \frac{\frac{1}{4}\pi(0.3)^2(8)}{\frac{1}{4}\pi(13)^2(5)(60)} ln(100/40) = 1.3\,\text{EE–5 cm/sec}$$

I. CONSOLIDATION TESTS

*Consolidation tests* (also known as *confined compression* and *oedometer tests*) start with a disc of cohesive soil (usually clay) confined by a metal ring. The faces of the disc are covered with porous plates. The disc sandwich is loaded in a water tank. The testing time is very long, since the water out-seepage is very slow. The load versus the void ratio is plotted as an *e-log p curve.*

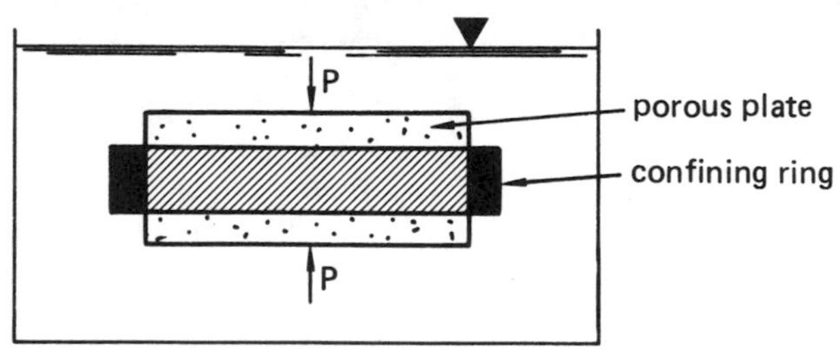


**Figure 9.8** Consolidation Test

Figure 9.9 shows an e-log p curve for a soil sample from which the load has been removed at $m$ allowing the clay to recover.

The line segment $m$–$r$ is known as the *virgin branch* or *virgin consolidation line.* This type of behavior is typical of *normally consolidated clay.* Normally consolidated clay can either be virgin, previously unloaded clay, or it can be clay carrying a load which has never been removed or exceeded.

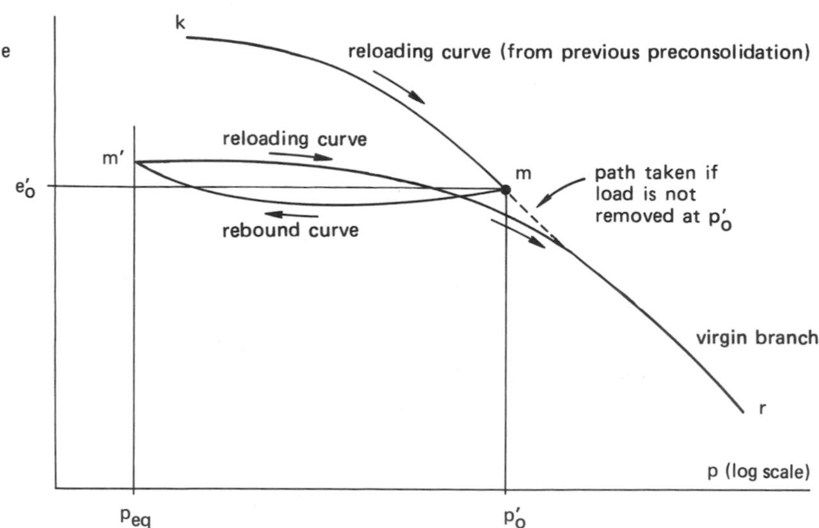


**Figure 9.9** e-log p Curve

Line $m$–$m'$ is a *rebound curve*. Line $m'$–$r$ is known as a *reloading curve*. Such curves result when a normally loaded clay is relaxed and restressed.

Notice that point $m'$ can only be reached by loading the soil to a pressure of $p'_o$ and then removing the pressure. This clay is said to have been *preloaded* or *overconsolidated*.[5] Although the pressure of the clay is essentially the same as when it started, its void ratio has been reduced. The *overconsolidation ratio* is defined by equation 9.30.

$$R_o = p'_o/p_{eq} \qquad 9.30$$

The *overconsolidation pressure*, $p'_o$, can be estimated by eye as a point slightly above the point of maximum curvature. Graphical means are also used.

The shape of the e-log p curve will depend on the degree of previous overconsolidation, as shown in figure 9.10.

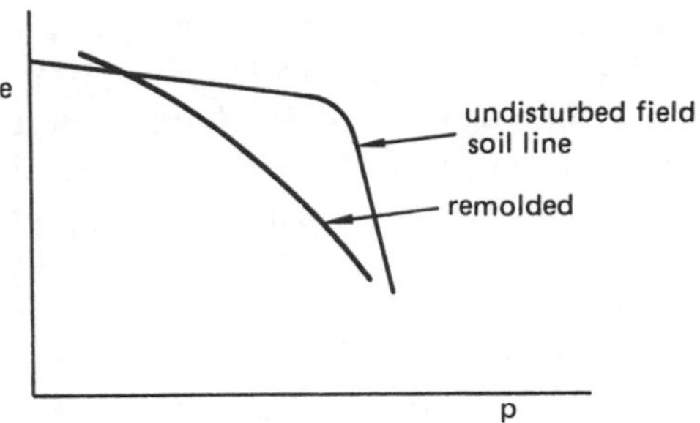


**Figure 9.10** Consolidation of Various Soils

Laboratory results can be used to find the *preconsolidation pressure* (point $m$ in figure 9.9). This is illustrated in the following procedure.

> Draw 2 lines—a tangent line and a horizontal line—through the point of maximum curvature, which is determined by eye. Bisect the resulting angle. Draw a tangent (line $k$) to the tail of the field soil line. The intersection of this tangent and the bisection line defines $e_o$ and $p_o$.[6]

Line $k$ can be used to predict consolidation of the soil under various loadings. The *compression index* is the (negative of the) logarithmic slope of line $k$ and is given by equation 9.31, where points 1 and 2 correspond to any two points on line $k$.

$$C_c = \frac{e_2 - e_1}{\log_{10}(p_1/p_2)} = \frac{e_o - e_1}{\log_{10}(p_1/p_o)} \qquad 9.31$$

If the clay is soft and near its liquid limit, the compression index can be approximated by equation 9.32. (In equation 9.32, $w_l$ is a whole number, not a decimal percentage.)

$$C_c \approx 0.009(w_l - 10) \qquad 9.32$$

The *reconsolidation index* and *swelling index* can be found from the (negative of the) logarithmic slopes of the rebound and reloading curves, respectively.

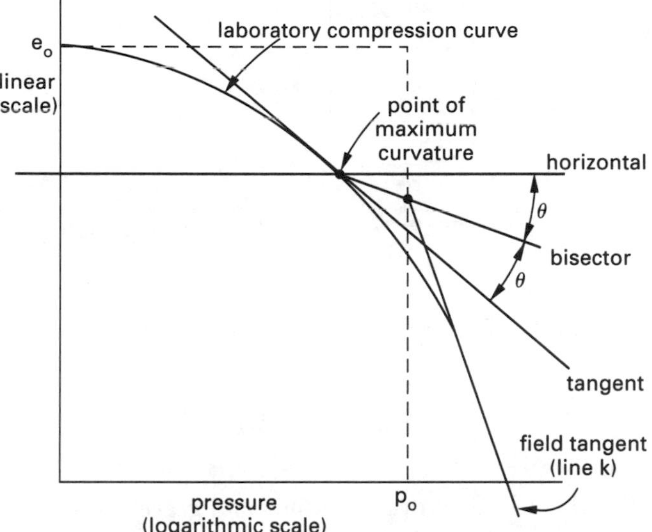


**Figure 9.11** Casagrande Method

## J. TRIAXIAL STRESS TESTS

In a triaxial stress test, a cylindrical sample is loaded on both ends and all around its surface. Usually, the radial stress ($\sigma_R$) is kept constant and the axial stress ($\sigma_A$) is varied. The normal and shear stresses on a plane of any angle can be found from the combined stress equations. (Consider compression positive.)

$$\sigma_\theta = \tfrac{1}{2}(\sigma_A + \sigma_R) + \tfrac{1}{2}(\sigma_A - \sigma_R)\cos 2\theta \qquad 9.33$$

$$\tau_\theta = +\tfrac{1}{2}(\sigma_A - \sigma_R)\sin 2\theta \qquad 9.34$$

These equations represent points on Mohr's circle, which can easily be constructed once $\sigma_A$ and $\sigma_R$ are known. (Care must be taken when plotting this graph. The sample is usually exposed to a pressure $p_R$ over all of its surface, including the ends. Thus, $p_R$ is equal to $\sigma_R$. The pressure applied to the ends, $p_A$, is in addition to radial pressure. Therefore, $\sigma_A = p_R + p_A$.) Test results are shown in figure 9.12 for two different samples which were both tested to failure. The ultimate shear strength, $S_{us}$, can be read directly from the y-axis.[7]

The equation for the *rupture line* (also known as the *envelope of rupture*) is given by Coulomb's equation.[8]

$$\tau = S_{us} = c + \sigma(\tan\phi) \qquad 9.35$$

[5] *Preloaded clay* is also known as *preconsolidated clay*, as well as *overconsolidated clay*.

[6] This method of finding the preconsolidation pressure is known as the *Casagrande method*.

[7] The ultimate shear strength is given the symbol $s$ in most soils books.

[8] Equation 9.35 is also known as the *Mohr-Coulomb equation*.

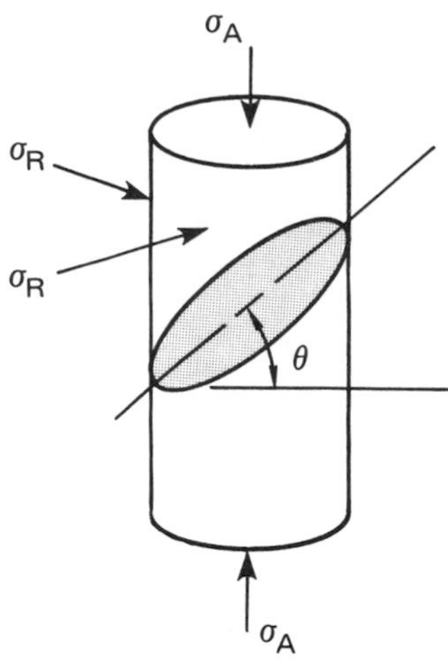


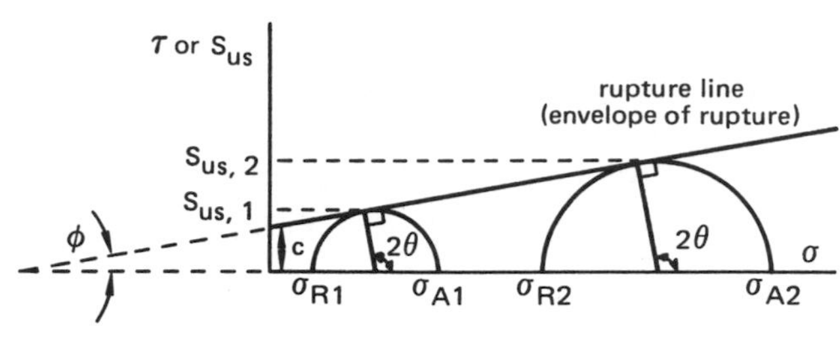


**Figure 9.12** Mohr's Circle of Stress

The *plane of failure* is inclined at the angle

$$\theta = 45^\circ + \tfrac{1}{2}\phi \qquad 9.36$$

For slow shear of drained sands and gravels, the *cohesion*, $c$, is zero. Therefore, it is possible to draw the rupture line with only one test. Typically, $c$ varies from 200 to 2000 psf for very soft and very stiff clays, respectively.

For saturated clays in quick shear, it is commonly assumed that $\phi = 0$. This would be represented as a horizontal rupture line.

Representative values of $\phi$ are given in table 9.10. $\phi$ is known as the *angle of internal friction.*[9]

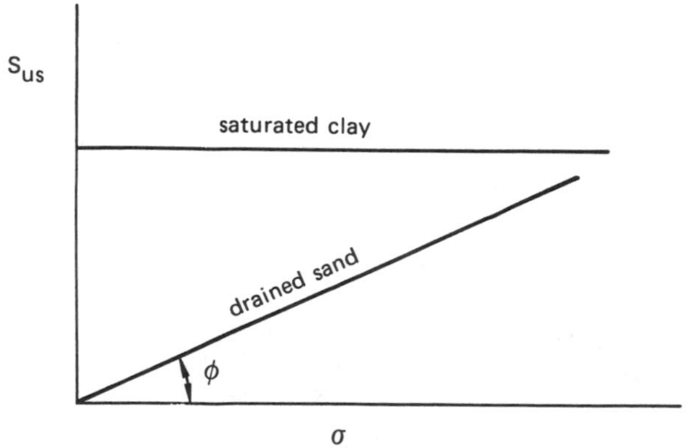


**Figure 9.13** Rupture Lines for Ideal Sand and Saturated Clay

[9] For cohesionless soils, the angle of internal friction is the angle from the horizontal naturally formed by a pile. For example, sand makes a pile with a slope of approximately 30°. Saturated clay, on the other hand, ideally behaves like a liquid, with $\phi = 0$. For most soils, the natural angle of repose will not be the same as the angle of internal friction, due to the effects of cohesion.

Presence of water in the pores of a sample will not affect these results much if the triaxial test is conducted in such a manner as to allow pore water pressure to dissipate (i.e., pore water to flow freely). Such triaxial tests are known as *S-tests* or *consolidated-drained tests.* However, testing of slow-draining soils may require several weeks time. If the test is peformed quickly so that the pore pressure does not have a chance to dissipate, the test is known as an *R-test* or *consolidated-undrained test.* In such a case, much of the axial load will be carried by the pore moisture. In the *Q-test* (*quick test*), the water content of the specimen is not allowed to change. Such a test is justified only with low permeability (e.g., EE-3 cm/sec) soils.

The *effective soil pressure* is the pressure that soil grains exert on each other. This pressure provides the shear strength of granular materials. It can be calculated from equation 9.37, where $c'$ and $\phi'$ are the *effective stress parameters.*

$$S_{us} = c' + \sigma' \tan\phi' \qquad 9.37$$

The *total pressure* also includes the *pore pressure*, $\mu$. The pore pressure can be found from the rise in a capillary tube, or it can be measured directly in a triaxial shear test.

$$\sigma' = \sigma - \mu \qquad 9.38$$

R-tests are used to determine the effective stress parameters, $c'$ and $\phi'$. In the absence of pore pressure measurements, R-tests can only record the total stress parameters $c$ and $\phi$.[10]

[10] If a soil is always going to be saturated, the total stress parameters can be used for foundation design. In cases where the soil is not always saturated, only the effective stress parameters should be used.

**Table 9.10**
Typical Strength Characteristics

| group symbol | cohesion (as compacted) psf $c$ | cohesion (saturated) psf $c_{sat}$ | effective stress envelope degrees $\phi$ |
|---|---|---|---|
| GW | 0 | 0 | > 38 |
| GP | 0 | 0 | > 37 |
| GM | – | – | > 34 |
| GC | – | – | > 31 |
| SW | 0 | 0 | 38 |
| SP | 0 | 0 | 37 |
| SM | 1050 | 420 | 34 |
| SM-SC | 1050 | 300 | 33 |
| SC | 1550 | 230 | 31 |
| ML | 1400 | 190 | 32 |
| ML-CL | 1350 | 460 | 32 |
| CL | 1800 | 270 | 28 |
| OL | – | – | – |
| MH | 1500 | 420 | 25 |
| CH | 2150 | 230 | 19 |
| OH | – | – | – |

*Example 9.7*

A sample of dry sand is taken and a triaxial test performed. The added axial stress causing failure was 5.43 tons/ft$^2$ when the radial stress was 1.5 tons/ft$^2$. What is the angle of internal friction? What is the angle of the failure plane?

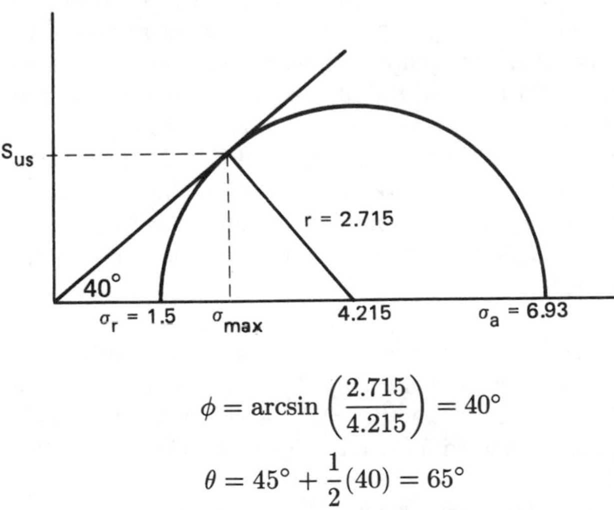


$$\phi = \arcsin\left(\frac{2.715}{4.215}\right) = 40°$$

$$\theta = 45° + \frac{1}{2}(40) = 65°$$

For any given radial pressure, a stress-strain curve can be plotted. This is illustrated in figure 9.14. The strain is volumetric strain due to the axial load only. The stress is the difference between the axial and radial stresses. The ultimate compressive stress ($S_{uc}$) can be read directly from the chart. $S_{uc}$ is usually taken as the stress difference for which the strain is 20%. The initial slope of the line is the *elastic modulus.*

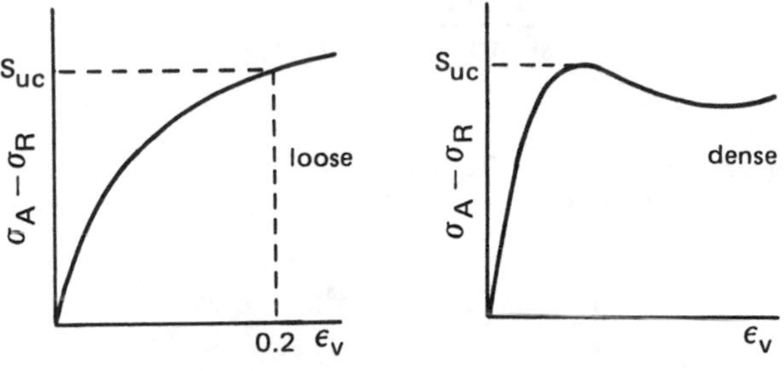


**Figure 9.14** Stress Strain Curves

## K. CALIFORNIA BEARING RATIO TEST: SHEARING RESISTANCE[11]

The *California Bearing Ratio* (*CBR*) test consists of measuring the relative load required to cause a standard (3 square inches) plunger to penetrate a water-saturated soil specimen at a specific rate to a specific depth. The word 'relative' is used because the actual load is compared to a standard load derived from a sample of crushed stone. The ratio is multiplied by 100 and the percent omitted.

The resulting data will be in the form of inches of penetration versus load. This data can be plotted as shown in figure 9.15. If the plot is concave upward (curve B), the steepest slope is extended downward to the x-axis. This point is taken as the zero penetration point and all penetration values adjusted accordingly.

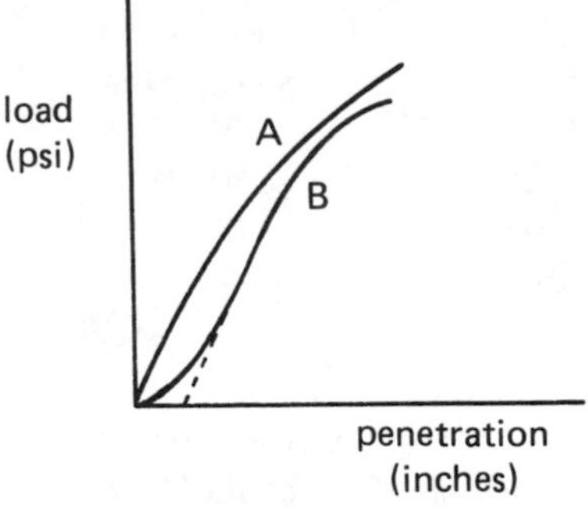


**Figure 9.15** Plotting CBR Test Data

Standard loads for crushed stone are given in table 9.11. For a plunger of 3 square inches, the CBR is the ratio of the load for a 0.1 inch penetration divided by 1000 psi. The CBR for 0.2 inches should also be calculated.

[11] California's Department of Transportation was the first to make use of the CBR test. However, other states and the Corps of Engineers have adopted CBR testing techniques. These states have, generally, retained the California Bearing Ratio name.

The test should be repeated if $CBR_{0.2} > CBR_{0.1}$. If the results are similar, use $CBR_{0.2}$.

$$CBR = \frac{\text{actual load (psi)}}{\text{standard load (psi)}} \times (100) \qquad 9.39$$

**Table 9.11**
Standard CBR Loads

| inches of penetration | standard load (psi) |
|---|---|
| 0.1 | 1000 |
| 0.2 | 1500 |
| 0.3 | 1900 |
| 0.4 | 2300 |
| 0.5 | 2600 |

**Table 9.12**
Typical CBR Values

| group symbol | range of CBR values |
|---|---|
| GW | 40 to 80 |
| GP | 30 to 60 |
| GM | 20 to 60 |
| GC | 20 to 40 |
| SW | 20 to 40 |
| SP | 10 to 40 |
| SM | 10 to 40 |
| SM-SC | 5 to 30 |
| SC | 5 to 20 |
| ML | $\leq$ 15 |
| ML-CL | – |
| CL | $\leq$ 15 |
| OL | $\leq$ 5 |
| MH | $\leq$ 10 |
| CH | $\leq$ 15 |
| OH | $\leq$ 5 |

*Example 9.8*

The following load data is collected for a 3 square inch plunger test. What is the California Bearing Ratio?



| penetration (inches) | load (psi) |
|---|---|
| 0.025 | 20 |
| 0.050 | 130 |
| 0.075 | 230 |
| 0.100 | 320 |
| 0.125 | 380 |
| 0.150 | 470 |
| 0.175 | 530 |
| 0.200 | 600 |
| 0.250 | 700 |
| 0.300 | 830 |

Upon graphing the data, it is apparent that a 0.02 inch correction is required. Therefore, the 0.1″ load is read from the graph as a 0.12 inch load.

$$CBR_{0.1} = \frac{(368)(100)}{1000} = 36.8 \text{ (percent omitted)}$$

$$CBR_{0.2} = \frac{(645)(100)}{1500} = 43$$

Since $CBR_{0.2}$ is greater than $CBR_{0.1}$, the test should be repeated.

## L. PLATE BEARING VALUE TEST: THE SUBGRADE MODULUS

A standard diameter round steel plate is set over soil on a bed consisting of fine sand and/or plaster of paris. Smaller diameter plates are placed on top of the bottom plate to ensure rigidity. After the plate is seated by a quick but temporary load, it is loaded to a deflection of about 0.04 inches. This load is maintained until the deflection rate decreases to 0.01 inch/minute. Then the load is released. The deflection prior to loading, the final deflection, and the deflection each minute are recorded.

The test is repeated 10 times. For each repetition of each load, the *end-point deflection* is found for which the deflection rate is exactly 0.001 inch/minute. The loads are then corrected for dead weights of jacks, plates, etc.

The corrected load versus the corrected deflection is graphed for the 10th repetition. The *bearing value* is the interpolated load which would produce a deflection of 0.5 inches. Figure 9.16 can be used to find the *subgrade modulus*, or *modulus of subgrade reaction*, $k$, which is the slope of the line (in psi/inch) in the loading range encountered by the soil.

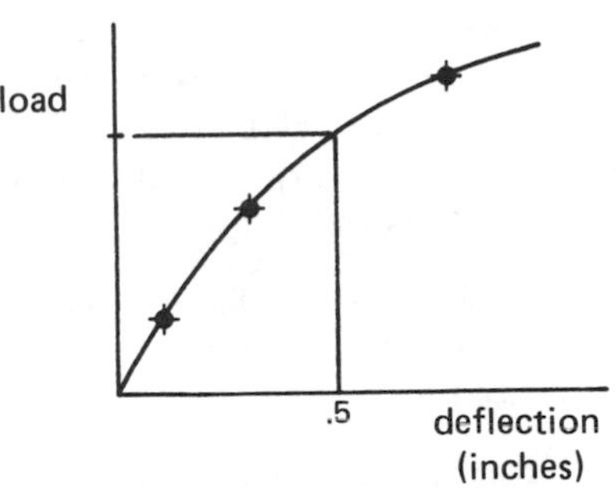


**Figure 9.16** 10th Repetition Bearing Load

## M. HVEEM'S RESISTANCE VALUE TEST: THE R-VALUE

The term 'resistance' refers to the ability of a soil to resist lateral deformation when a vertical load acts upon it. When displacement does occur, the soil moves out and away from the applied load.

Measuring the *R-value* of a soil is done with a *stabilometer test.* The R-value will range from zero (the resistance of water) to 100 (the approximate resistance of steel). R-values of soil and aggregate usually range from 5 to 85.

**Table 9.13**
Typical Values of Subgrade Modulus

| group symbol | range of subgrade modulus $k$ 1 psi/in |
|---|---|
| GW | 300–500 |
| GP | 250–400 |
| GM | 100–400 |
| GC | 100–300 |
| SW | 200–300 |
| SP | 200–300 |
| SM | 100–300 |
| SM-SC | 100–300 |
| SC | 100–300 |
| ML | 100–200 |
| ML-CL | – |
| CL | 50–200 |
| OL | 50–100 |
| MH | 50–100 |
| CH | 50–150 |
| OH | 25–100 |

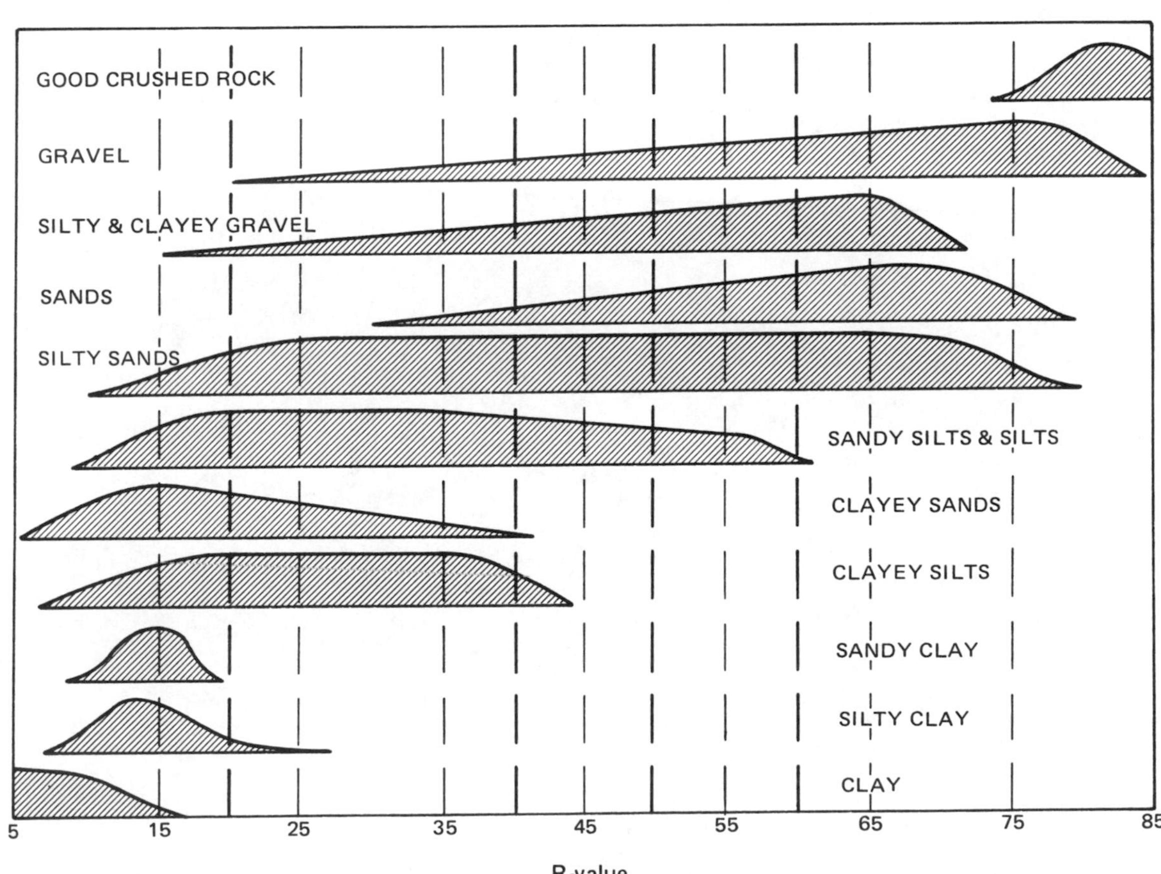


**Figure 9.17** R-Values of Various Soils

The R-value is determined using soil samples which are compacted as they would be during normal construction. They are tested as near to saturation as possible to give the lowest expected R-value. Thus, the R-value represents the worst possible state the soil might attain during use.

The procedure also takes into account the fact that some soils are expansive. When a compacted soil expands due to the absorption of water, the R-value also decreases. The test procedure accounts for this by lowering the R-value.

In the absence of stabilometer testing, rough estimates of the R-value can be made using simple soil classification tests. The soil type can be found from sieve analyses, hydrometer tests, or figure 9.1.

After the soil has been classified, figure 9.17 can be used to determine an approximate R-value. It will be seen that each soil type covers an R-value range. The curves respresenting the various soils are approximate frequency distributions.

For fine-grained soils, the upper tail (high R-value) represents a lower plasticity; the lower tail represents soils with higher plasticity.

The curves for coarse-grained materials are affected in the same manner. Lower tails represent materials with either more clay or clay with a higher activity. In coarse-grained materials with little or no clay, the lower tail represents hard, smooth-surfaced and uniformly sized material. The upper tail represents rough-surfaced material with a distribution of sizes.

## Practice Problems: SOILS

### Untimed

1. A sample of moist soil was found to have the following characteristics:

| | |
|---|---|
| volume: | .01456 m³ (as sampled) |
| mass: | 25.74 kg (as sampled) |
| | 22.10 kg (after oven drying) |
| specific gravity of solids: | 2.69 |

Find the density, unit weight, void ratio, porosity, and degree of saturation for the soil.

2. A Proctor test was performed on a soil which has a specific gravity for its solids of 2.71. For the data below, (a) plot the moisture-dry density curve. (b) Find the maximum density and optimum moisture. (c) What range of moisture is permitted if a contractor must achieve 90% compaction?

| water content % | actual density pcf | water content % | actual density pcf |
|---|---|---|---|
| 10 | 98 | 20 | 129 |
| 13 | 106 | 22 | 128 |
| 16 | 119 | 25 | 123 |
| 18 | 125 | | |

3. For the soil in problem #2, how many gallons of water need to be added to obtain 1 yd³ of soil at the maximum density if the soil is originally at 10% water content (dry basis)?

4. A triaxial shear test is performed on a well-drained sand sample. Failure occurred when the normal stress was 6260 psf and the shear stress was 4175 psf. What is the angle of internal friction? What are the principal stresses?

5. The results of a sieve test below give the percentage passing through the sieve. (a) Plot the particle size distribution. (b) Calculate the uniformity coefficient. (c) Calculate the coefficient of curvature.

| sieve | % finer by weight |
|---|---|
| ½″ | 52 |
| #4 | 37 |
| #10 | 32 |
| #20 | 23 |
| #40 | 11 |
| #60 | 7 |
| #100 | 4 |

6. A permeability test is conducted with a sample of soil which is 60 mm in diameter and 120 mm long. The head is kept constant at 225 mm. The flow is 1.5 ml in 6.5 minutes. What is the coefficient of permeability in units of meter/year?

7. A consolidation test is performed on a soil with the following results:

| pressure (psf) | e | pressure (psf) | e |
|---|---|---|---|
| 250 | .755 | 8350 | .724 |
| 520 | .754 | 16,700 | .704 |
| 1040 | .753 | 33,400 | .684 |
| 2090 | .750 | 8350 | .691 |
| 4180 | .740 | 250 | .710 |

(a) Graph the curve of stress versus void ratio. (Hint: use log or semi-log paper.) (b) What is the compression index? (c) If the initial pressure on the soil layer is 1400 psf, how much stress can a 10 foot thickness of this soil carry without settling more than 1.0 inch?

8. A sample of soil has the following characteristics:

| | |
|---|---|
| % passing #40 screen: | 95 |
| % passing #200 screen: | 57 |
| liquid limit: | 37 |
| plastic limit: | 18 |

Use the AASHTO system to classify this soil. Include the group index number.

9. A sample of sand has a relative density of 40% with a solids specific gravity of 2.65. The minimum void ratio is .45; the maximum void ratio is .97. (a) What is the density of this sand in a saturated condition? (b) If the sand is compacted to a relative density of 65%, what will be the decrease in thickness of a 4 foot thick layer?

10. Specifications on a job require a fill using borrow soil to be compacted to 95% of its standard Proctor maximum dry density. Tests indicate that this maximum is 124.0 pcf when dry. The soil now has 12% moisture. The borrow material has a void ratio of .60 and a solid specific gravity of 2.65. What is the minimum volume of borrow soil required to fill 1.0 cubic foot?

### Timed

1. Two choices for borrow soil are available.

| borrow A | | borrow B |
|---|---|---|
| 115 lb/ft³ | density in place | 120 |
| ? | density in transport | 95 |
| .92 | void ratio in transport | ? |
| 25% | water content in place | 20% |
| $.20/yd³ | cost to excavate | $.10/yd³ |
| $.30/yd³ | cost to haul | $.40/yd³ |
| 2.7 | S.G. of solids | 2.7 |
| 112 lb/ft³ | max Proctor dry density | 110 |

It will be necessary to fill a 200,000 yd³ depression, and the fill material must be compacted to 95% of the standard Proctor (maximum) density. A final 10% moisture content (dry basis) is desired in either case. (a) What soil would be cheaper to use? (b) What is the volume of borrow from each site? (c) What is the minimum quantity (volume) of material to haul?

2. Two series of triaxial shear tests on a soil were performed with the following results:

| undrained series | | drained series | |
|---|---|---|---|
| confining pressure | total axial pressure | confining pressure | total axial pressure |
| 0 psi | 60 psi | 50 psi | 250 psi |
| 50 | 110 | 100 | 400 |
| 100 | 160 | 150 | 550 |

(a) Determine the angle of internal friction for both series. (b) What is the cohesion for both series? (c) What is the angle of the failure plane (with respect to the horizontal axis) for both series? (d) Given a fourth test of the drained sample with radial confining pressure = 300 psi, what is the expected axial load at failure?

# 10 FOUNDATIONS AND RETAINING WALLS

FOUNDATIONS

*Nomenclature*

| | | |
|---|---|---|
| $a_v$ | coefficient of compressibility | $ft^2/lbf$ |
| A | area | $ft^2$ |
| B | width or diameter | ft |
| c | cohesion (undrained shear strength) | $lbf/ft^2$ |
| $c_a$ | adhesion | $lbf/ft^2$ |
| C | multiplicative correction factor | – |
| $C_c$ | compression index | – |
| $C_v$ | coefficient of consolidation | $ft^2/sec$ |
| d | depth factor | – |
| D | depth | ft |
| e | void ratio | – |
| $f_o$ | skin friction coefficient | – |
| F | factor of safety | – |
| h | depth | ft |
| H | soil layer thickness or depth | ft |
| k | permeability coefficient, or a constant | ft/sec,– |
| $k_o$ | coefficient of earth pressure at rest | – |
| L | length | ft |
| M | moment | ft-lbf |
| N | capacity factor, or number of blows | –, – |
| $N_o$ | stability number | – |
| p | pressure | $lbf/ft^2$ |
| P | load or force | lbf |
| q | uniform surcharge | lbf/ft, or $lbf/ft^2$ |
| r | distance (moment arm) | ft |
| R | force (resistance) | lbf/ft of wall |
| S | strength or settlement | $lbf/ft^2$, or ft |
| SG | specific gravity | – |
| t | time | various |
| $T_v$ | time factor | – |
| $U_z$ | percent of total consolidation | – |
| w | water content | – |
| $w_l$ | liquid limit | – |
| W | weight (mass) | lbm |
| z | depth | ft |

*Symbols*

| | | |
|---|---|---|
| $\phi$ | angle of internal friction | degrees |
| $\delta$ | angle of wall friction | degrees |
| $\alpha$ | secondary compression index | – |
| $\rho$ | density | $lbm/ft^3$ |
| $\mu$ | pore pressure | $lbf/ft^2$ |
| $\epsilon$ | eccentricity | ft |
| $\eta$ | efficiency | – |
| $\beta$ | cut angle | degrees |
| $\gamma$ | specific weight | $lbf/ft^3$ |

*Subscripts*

| | |
|---|---|
| a | allowable |
| A | active |
| b | below mudline |
| c | compressive |
| f | footing |
| g | gross |
| h | horizontal |
| i | the ith component |
| n | unconfined |
| o | at rest |
| P | passive |
| q | surcharge |
| s | shear, or sliding |
| v | vertical |
| w | water or water table |
| $\gamma$ | density (as a subscript) |

## 1 CONVERSIONS

| multiply | by | to obtain |
|---|---|---|
| kips | 1000 | pounds |
| pounds | 5 EE−4 | tons |
| pounds | 0.001 | kips |
| pounds/square foot | 5 EE−4 | $tons/ft^2$ |
| pounds/square inch | 0.072 | $tons/ft^2$ |
| tons | 2000 | pounds |
| tons/square foot | 2000 | $pounds/ft^2$ |
| tons/square foot | 13.889 | $pounds/inch^2$ |
| Newtons/square meter | 0.021 | $pounds/ft^2$ |

## 2 DEFINITIONS

Abutment: A retaining wall which also supports a vertical load.

Active pressure: Pressure causing a wall to move away from the soil.

Batter pile: A pile inclined from the vertical.

Bell: An enlarged section at the base of a pile or pier used as an anchor.

Berm: A shelf, ledge, or pile.

Cased hole: An excavation whose sides are lined or sheeted.

Dead load: An inert, inactive load, primarily due to the structure's own weight.

Dredge level: See 'Mud line.'

Freeze (of piles): A large increase in the ultimate capacity (and required driving energy) of a pile after it has been driven some distance.

Grillage: A footing or part of a footing consisting of horizontally laid timbers or steel beams.

Lagging: Heavy planking used to construct walls in excavations and braced cuts.

Live load: The weight of all non-permanent objects in a structure, including people and furniture. Live load does not include seismic or wind loading.

Mud line: The lower surface of an excavation or braced cut.

Passive pressure: A pressure acting to counteract active pressure.

Pier shaft: The part of a pier structure which is supported by the pier foundation.

Ranger: See 'Wale.'

Rip rap: Pieces of broken stone used to protect the sides of waterways from erosion.

Sheeted pit: See 'Cased hole.'

Slickenside: A surface (plane) in stiff clay which is a potential slip plane.

Soldier pile: An upright pile used to hold lagging.

Stringer: See 'Wale.'

Surcharge: A surface loading in addition to the soil load behind a retaining wall.

Wale: A horizontal brace used to hold timbers in place against the sides of an excavation, or to transmit the braced loads to the lagging.

## 3 COMPARISON OF SAND AND CLAY AS FOUNDATION MATERIALS

Ordinarily, sand makes a good foundation material. It doesn't settle after its initial loading. It drains quickly. However, it behaves poorly in excavations. When sand is fine and saturated, it can become quick, and a major loss in supporting strength occurs.

Care must be taken when distinguishing between "moist" and "saturated" sands. Sand which has been allowed to drain may be "moist" in the normal sense of the word. However, if the water is not captive, pore pressure will not develop, and the sand can be considered dry. However, special considerations are required if the sand is below the water table. Such sand is saturated, not moist.

Clay, on the other hand, is good in excavations, but is poor for foundations. It continues to settle indefinitely. It retains water for a long time, and large volume changes can result when large changes in moisture content occur.

## 4 GENERAL CONSIDERATIONS FOR FOOTINGS

A *footing* is an enlargement at the base of a load-supporting column designed to transmit forces to the soil. The area of the footing will depend on the load and the soil characteristics. The following types of footings are used.

- spread footing: A footing used to support a single column. This is also known as an *individual column footing* and *isolated footing.*
- continuous footing: A long footing supporting a continuous wall. Also known as *wall footing.*
- combined footing: A footing carrying more than one column.
- cantilever footing: A combined footing that supports a column and an exterior wall or column.

If possible, footings should be designed according to the following general considerations:

- The footing should be located below the frost line and below the level which is affected by moisture content changes.
- Footings need not be any lower than the highest-adequate stratum.
- The centroid of the footing should coincide with the centroid of the applied load.
- Allowable soil pressures should not be exceeded.

- Below-grade footings should be equipped with a drainage system.
- Footings on fill over loose sand should be densified with piles.
- If possible, footings should be placed in excavations made in compacted fill. They should not be put in place prior to compaction.
- Size footings to the nearest 3″ above or equal to the theoretical size.

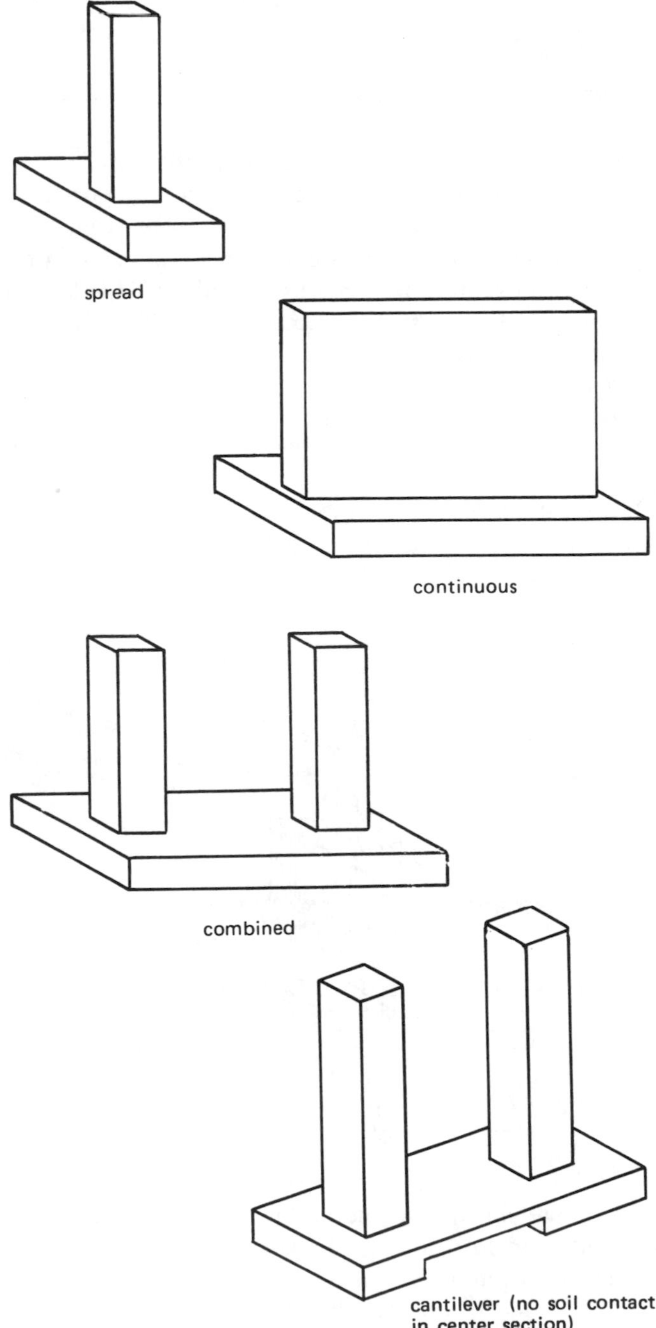


**Figure 10.1** Types of Footings

## 5 ALLOWABLE SOIL PRESSURES

When data from soil tests are unavailable, table 10.1 can be used for preliminary calculations.

**Table 10.1**
Typical Allowable Soil Bearing Pressures[1]

| type of soil | allowable pressure |
|---|---|
| massive crystalline bedrock | 4000 lbf/ft$^2$ |
| sedimentary and foliated rock | 2000 |
| sandy gravel and/or gravel (GW and GP) | 2000 |
| sand, silty sand, clayey sand, silty gravel, and clayey gravel (SW, SP, SM, SC, GM, GC) | 1500 |
| clay, sandy clay, silty clay, and clayey silt (CL, ML, MH, and CH) | 1000 |



## 6 GENERAL FOOTING DESIGN EQUATION

The *gross* (or *ultimate*) *bearing capacity* or *gross pressure* for a soil is given by equation 10.1, which is known as the *Terzaghi-Meyerhoff equation.* The equation is good for both sandy and clayey soils. It is specifically valid for continuous wall footings. ($p_q$ is a surface surcharge.)

$$p_g = \frac{1}{2}\gamma B N_\gamma + cN_c + (p_q + \gamma D_f)N_q \qquad 10.1$$

Various researchers have made improvements on this theory, leading to somewhat different terms and sophistication in evaluating $N_\gamma$, $N_c$, and $N_q$.[2] However, the general form remains valid for design, with corrections for various footing geometries.

Figure 10.2 and table 10.3 can be used to evaluate the capacity factors $N_\gamma$, $N_c$, and $N_q$ in equation 10.1.

**Table 10.2**
$N_c$ Bearing Capacity Factor Multipliers for Various Values of B/L
(See figure 10.3)

| B/L | multiplier |
|---|---|
| 1 (square) | 1.25 |
| 0.5 | 1.12 |
| 0.2 | 1.05 |
| 0.0 | 1.00 |
| 1 (circular) | 1.20 |

[1] As in the definition of $p_a$, the term 'allowable' implies that a factor of safety has already been applied.

[2] Differences in reported values of $N_\gamma$, $N_c$, and $N_q$ may also be due to the different units used by researchers.

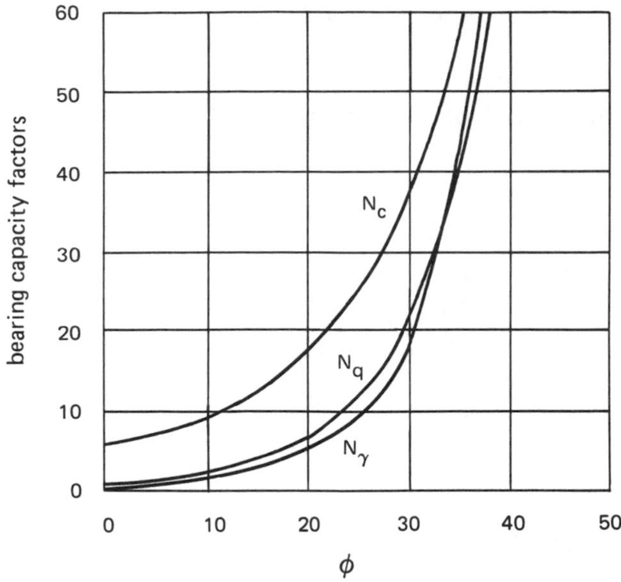


**Figure 10.2** Bearing Capacity Factors

**Table 10.3**
Terzaghi Bearing Capacity Factors for General Shear[3]

| $\phi$ | $N_c$ | $N_q$ | $N_\gamma$ |
|---|---|---|---|
| 0 | 5.7 | 1.0 | 0.0 |
| 5 | 7.3 | 1.6 | 0.5 |
| 10 | 9.6 | 2.7 | 1.2 |
| 15 | 12.9 | 4.4 | 2.5 |
| 20 | 17.7 | 7.4 | 5.0 |
| 25 | 25.1 | 12.7 | 9.7 |
| 30 | 37.2 | 22.5 | 19.7 |
| 34 | 52.6 | 36.5 | 35.0 |
| 35 | 57.8 | 41.4 | 42.4 |
| 40 | 95.7 | 81.3 | 100.4 |
| 45 | 172.3 | 173.3 | 297.5 |
| 48 | 258.3 | 287.9 | 780.1 |
| 50 | 347.5 | 415.1 | 1153.2 |

**Table 10.4**
$N_\gamma$ Multipliers for Various Values of B/L (See figure 10.3)

| B/L | multiplier |
|---|---|
| 1.0 (square) | 0.85 |
| 1.0 (circular) | 0.70 |
| 0.5 | 0.90 |
| 0.2 | 0.95 |
| 0.0 | 1.0 |

[3] In *general shear*, the soil resists an increased load until failure is sudden. There is another case, that of *local shear*, which results with looser soil. However, it is unlikely that foundations would be designed for loose soil without compaction. With compaction, the general shear case holds.

Once a gross pressure is determined, it is corrected by the ***overburden***, giving the ***net soil pressure***.

$$p_{\text{net}} = p_g - \gamma D_f \tag{10.2}$$

The ***allowable soil pressure*** is determined by dividing the net pressure by a factor of safety.[4] A safety factor of 3 (based on $p_{\text{net}}$) should be used for average conditions. Exceptional loadings and improbable combinations of snow, wind, and seismic forces may be allowed to reduce the safety factor to 2.

$$p_a = \frac{p_{\text{net}}}{F} \tag{10.3}$$

## 7 FOOTINGS ON CLAY AND PLASTIC SILT

Clay is normally soft, fairly impermeable, and highly preloaded. When loads are first applied to saturated clay, the pore pressure increases. For a short time, this pore pressure does not dissipate and the angle of internal friction should be taken as $\phi = 0°$. This is known as the $\phi = 0°$ or ***undrained case***. The undrained clay shear strength is one-half of the unconfined compressive strength.

$$c = \frac{1}{2} S_{nc} \tag{10.4}$$

If $\phi = 0°$, then $N_\gamma = 0$ and $N_q = 1$. If there is no surface surcharge, the gross bearing capacity is given by equation 10.5

$$p_g = cN_c + \gamma D_f \tag{10.5}$$

$$p_{\text{net}} = p_g - \gamma D_f = cN_c \tag{10.6}$$

$$p_a = \frac{p_{\text{net}}}{F} \tag{10.7}$$

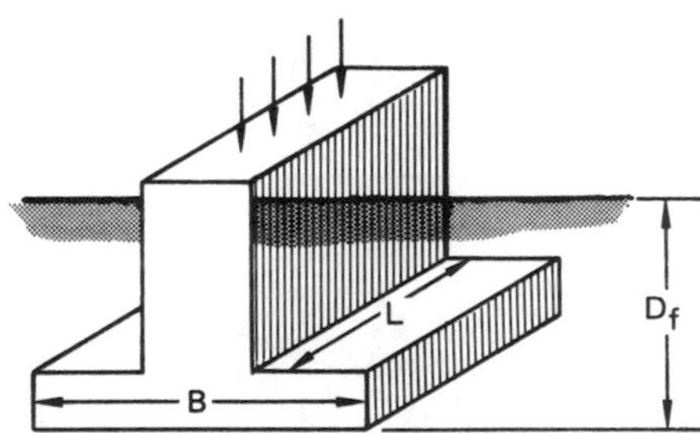


**Figure 10.3** A Spread Footing

*Example 10.1*

An individual square column footing carries an 83,800 pound dead load and a 75,400 pound live load. The unconfined compressive strength of the supporting clay is 0.84 tons/ft$^2$ and its density is 115 lbm/ft$^3$. The

[4] The term 'allowable' implies that the safety factor has been included. This pressure is also known as "safe" and "net allowable" pressure.

footing is covered by a 6″ basement slab. The footing thickness is initially unknown. Neglect depth correction factors. Do not design the structural steel. Specify the footing size and thickness.

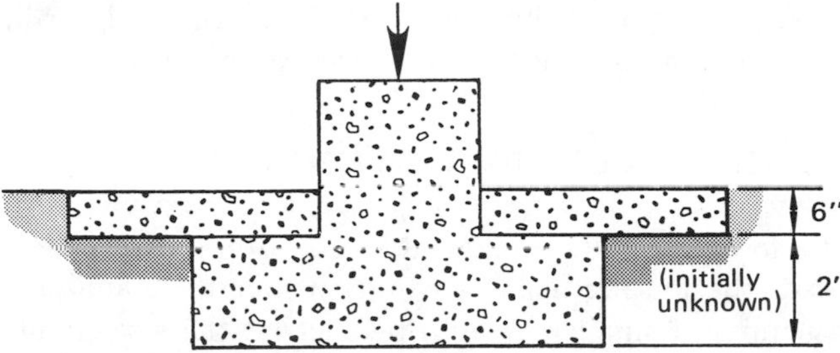


The total load on the column is

$$\frac{83{,}800 + 75{,}400}{2000} = 79.6 \text{ tons}$$

From tables 10.2 and 10.3, for square footings,

$$N_c \approx (1.25)(5.7) = 7.1$$

The cohesion is estimated from the unconfined compressive strength and equation 10.4.

$$c = \frac{0.84}{2} = 0.42 \text{ tons/ft}^2$$

Using a factor of safety of 3, the allowable pressure is

$$p_a = \frac{(0.42)(7.1)}{3} = 0.99 \text{ tons/ft}^2$$

The approximate area required is

$$A = \frac{79.6}{0.99} = 80.4 \text{ ft}^2$$

So, try a 9′3″ square footing (area = 85.6 ft$^2$). (At this point, a footing thickness would be determined based on concrete design considerations.) Assume a 2.0 foot footing thickness.

The actual pressure under the footing due to applied load is

$$p = \frac{79.6}{85.6} = 0.93 \text{ tons/ft}^2$$

This first iteration did not consider the concrete weight. The concrete density is approximately 150 lbm/ft$^3$. Therefore, the pressure surcharge due to one square foot of concrete floor is

$$p_q = \frac{1 \times 1 \times \frac{6}{12} \times 150}{2000} = 0.04 \text{ tons/ft}^2$$

Similarly, the footing itself has weight. The footing extends 2 feet down.

$$p_f = \frac{1 \times 1 \times 2 \times 150}{2000} = 0.15 \text{ tons/ft}^2$$

Therefore, the total pressure under the footing is

$$p_{\text{total}} = 0.93 + 0.04 + 0.15 = 1.12 \text{ tons/ft}^2$$

Equation 10.2 gives the allowable pressure in excess of the soil surcharge. The footing bottom is 2.5 feet below the original grade, so the soil surcharge is[5]

$$\frac{(2.5)(115)}{2000} = 0.14 \text{ tons/ft}^2$$

The net actual pressure to be compared to the allowable pressure is

$$p = 1.12 - 0.14 = 0.98$$

This is essentially the same as $p_a$.

## 8 FOOTINGS ON SAND

The cohesion, $c$, is zero in sand. The gross ultimate bearing capacity can be derived from equation 10.1 by setting $c = 0$.

$$p_g = \frac{1}{2}B\gamma N_\gamma + (p_q + \gamma D_f)N_q \qquad 10.8$$

The net ultimate bearing capacity when there is no surface surcharge (i.e., $p_q = 0$) is

$$p_{\text{net}} = p_g - \gamma D_f = \frac{1}{2}B\gamma N_\gamma + \gamma D_f(N_q - 1) \qquad 10.9$$

If the water table level is above the footing face (*submerged condition*), $p_{\text{net}}$ should be reduced by 50%.

The allowable sand loading is based on a factor of safety, which is typically taken as 2 for sand.

$$\begin{aligned} p_a &= \frac{p_{\text{net}}}{F} \\ &= \frac{B}{F}\left[\frac{1}{2}\gamma N_\gamma + \gamma(N_q - 1)\frac{D_f}{B}\right] \end{aligned} \qquad 10.10$$

Since sand is permeable and rapidly adjusts to changes in loading, design the footing based on the maximum instantaneous load. Determine the allowable soil pressure based on the footing with the maximum load, smallest $N$, deepest (highest) surface water, etc. Use this soil pressure for all footings in the building foundation.

[5] A depth of 2 feet could also be used if the basement slab was constructed on the original grade. This calculation assumes the slab is poured 6″ below the original grade.

Since the quantity in brackets in equation 10.10 is constant for specific $D_f/B$ ratios, and since the bearing capacity factors depend on $\phi$ (which can be correlated to $N$), equation 10.11 can be derived.[6] This equation assumes $F = 2, \gamma = 100\ \text{lbf/ft}^3$, and $D_f < B$.

$$p_a = (0.11)C_n N \qquad 10.11$$

No correction is usually made if the density is different from 100 lbm/ft$^3$. However, the equation assumes that the overburden load ($D_f\gamma$) is approximately 1 ton/ft$^2$. This means that the $N$ values were derived from data corresponding to depths of 10 to 15 feet below the original surface (not the basement surface). If the footing is to be installed close to the original surface, then a correction factor is required.[7]

**Table 10.5**
Overburden Corrections

| overburden | $C_n$ |
|---|---|
| 0 tons/ft$^2$ | 2 |
| 0.25 | 1.45 |
| 0.5 | 1.21 |
| 1.0 | 1.00 |
| 1.5 | 0.87 |
| 2.0 | 0.77 |
| 2.5 | 0.70 |
| 3.0 | 0.63 |
| 3.5 | 0.58 |
| 4.0 | 0.54 |
| 4.5 | 0.50 |
| 5.0 | 0.46 |

For a given sand settlement, the soil pressure will be greatest in intermediate width ($B$ = 2 to 4 feet) footings. This is illustrated in figure 10.4. Equation 10.11 should not be used for small-width footings, since bearing pressure governs. For wide footings, (i.e., B > 2 to 4 feet), settlement governs.

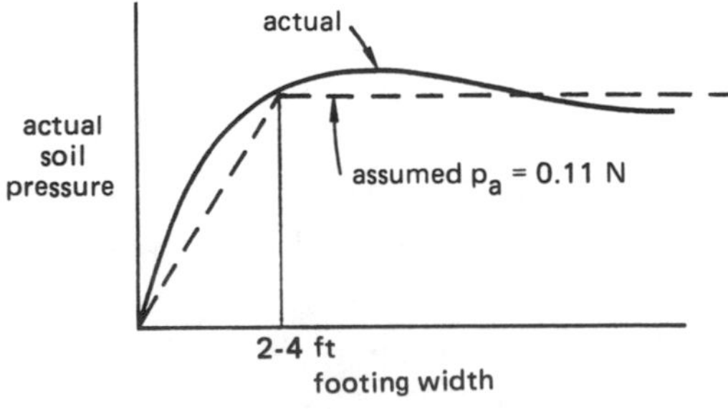


**Figure 10.4** Soil Pressure on Sand with Constant Settlement

[6] $N$ is the number of blows per foot from a standard penetration test.

[7] The correction is actually a correction for $N$. If corrected $N$ values are known, $C_n$ may be neglected.

## 9 FOOTINGS ON ROCK

If bedrock can be reached by excavation, the allowable pressure is likely to be determined by local codes. A safety factor of 5 based on the unconfined compressive strength is typical. For most rock beds, the design will be based on settlement characteristics, not strength.

## 10 MOMENTS ON FOOTINGS

If a footing carries a moment in addition to its vertical load, the footing bearing capacity should be analyzed assuming a smaller area.[8] Specifically, the size should be reduced by twice the eccentricity.[9]

$$\epsilon_B = \frac{M_B}{P}; \quad \epsilon_L = \frac{M_L}{P} \qquad 10.12$$

$$L' = L - 2\epsilon_L; \quad B' = B - 2\epsilon_B \qquad 10.13$$

$$A' = L'B' \qquad 10.14$$

This area reduction places the equivalent force at the centroid of the reduced area, $A'$. The actual value of $B$ should be used in calculating $D_f/B$ ratios used in finding capacity and depth factors. However, $B'$ should be used in equation 10.1.

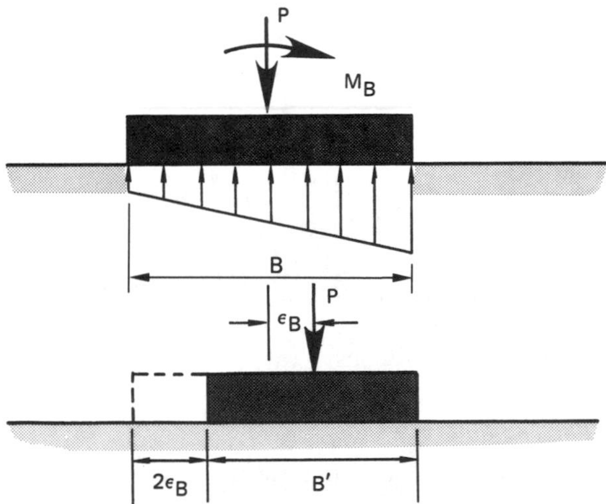


**Figure 10.5** A Footing with an Overturning Moment

Although the eccentricity is independent of the footing dimensions, a trial and error solution may be necessary when designing footings. Trial and error is not required when analyzing a footing of known dimensions.

[8] Usually, there will be no $M_L$ moment.

[9] This discussion is for rectangular footings. It is much more difficult to construct an equivalent footing for circular shapes.

Assuming $M_L = 0$, $\epsilon_B = \epsilon$, and disregarding the concrete and overburden weights, the actual soil pressure distribution is given by equation 10.15. $B'$ and $L'$ should not be used in equation 10.15 because these variables place the load at the centroid of the reduced area, producing a uniform pressure distribution.

$$p_{\max}, p_{\min} = \frac{P}{BL}\left(1 \pm \frac{6\epsilon}{B}\right) \qquad 10.15$$

If the eccentricity, $\epsilon$, is sufficiently large, a negative soil pressure will result. Since soil cannot carry a tensile stress, such stresses are neglected. This results in a reduced area to carry the load.

If the resultant force is within the middle third of the footing, all of the footing will contribute. That is, the maximum eccentricity without incurring a reduction footing area will be $B/6$.

## 11 GENERAL CONSIDERATIONS FOR RAFTS

A *raft* or *mat* is a combined footing-slab that covers the entire area beneath a building and supports all walls and columns. A raft foundation should be used (at least for economic reasons) any time the individual footings would constitute half or more of the area beneath a building.

## 12 RAFTS ON CLAY

The net ultimate bearing pressure for rafts on clay can be found in the same manner as for footings. Since the size of the raft is essentially fixed by the building size (plus or minus a few feet), the only method available to increase the loading is to lower the elevation (increase $D_f$) of the raft.

The factor of safety produced by a raft construction is given by equation 10.16, which can also be solved to give the required $D_f$ if the factor of safety is known. The factor of safety should be at least 3 for normal loadings, but may be reduced to 2 during temporary extreme loading.

$$F = \frac{cN_c}{\frac{\text{total load}}{\text{raft area}} - \gamma D_f} \qquad 10.16$$

If the denominator in equation 10.16 is small, the factor of safety is very large. If the denominator is zero, the raft is said to be a *fully compensated foundation.* For $D_f$ less than the fully-compensated depth, the raft is said to be *partially compensated.*

*Example 10.2*

A raft foundation is to be designed for a $120' \times 200'$ building with a total loading of 5.66 EE7 pounds. The clay density is 115 lbm/ft$^3$, and the clay has an average unconfined compressive strength of 0.3 tons/ft$^2$. (a) What should be the raft depth, $D_f$, for full compensation? (b) What should be the raft depth for a factor of safety of 3? Neglect depth correction factors.

The loading pressure is

$$p_{\text{load}} = \frac{5.66\ \text{EE7}}{(120)(200)} = 2.36\ \text{EE3 lbf/ft}^2$$

(a) For full compensation, $p_{\text{load}} = \gamma D_f$.

$$D_f = \frac{2.36\ \text{EE3}}{115} = 20.5\ \text{ft}$$

(b) From table 10.3, $N_c = 5.7$. Since $B/L = 120/200 = 0.6$, use an $N_c$ multiplier from table 10.2 of 1.15.

$$3 = \frac{\left(\frac{1}{2}\right)(0.3)(2000)(1.15)(5.7)}{2.36\ \text{EE3} - (115)D_f}$$

$$D_f = 14.8\ \text{ft}$$

## 13 RAFTS ON SAND

Rafts on sand are always well protected against bearing capacity failure. Therefore, settlement will govern the design. Since differential settlement will be much smaller for various locations on the raft (due to the raft's rigidity), the allowable soil pressure may be doubled.

$$p_a = 0.22C_nN\ (\text{tons/ft}^2) \qquad 10.17$$

$N$ should always be at least 5 after correcting for overburden. Otherwise, the sand should be compacted or a pier/pile foundation used.

The net soil pressure should be compared with the allowable pressure. The net soil pressure is

$$p_{\text{net}} = \frac{\text{total load}}{\text{raft area}} - \gamma D_f \qquad 10.18$$

## 14 GENERAL CONSIDERATIONS FOR PIERS

A pier is a large underground structure with a length (depth) greater than its width (diameter). It differs from a pile in its diameter, load carrying capacity, and installation method. A pier is usually constructed within an excavation.

## 15 PIERS IN CLAY

If $D_f/B > 4$, then $N_c$ is constant at approximately $N_c = 9.0$. The design of a pier foundation is similar in other respects to a footing design. A factor of safety of 3 is usually used.

$$p_{\text{net}} = p_g - \gamma_{\text{clay}} D_f - \gamma_w h \qquad 10.19$$

$$p_a = \frac{cN_c}{F} \qquad 10.20$$

$c$ is the undrained shear strength defined by equation 10.4.

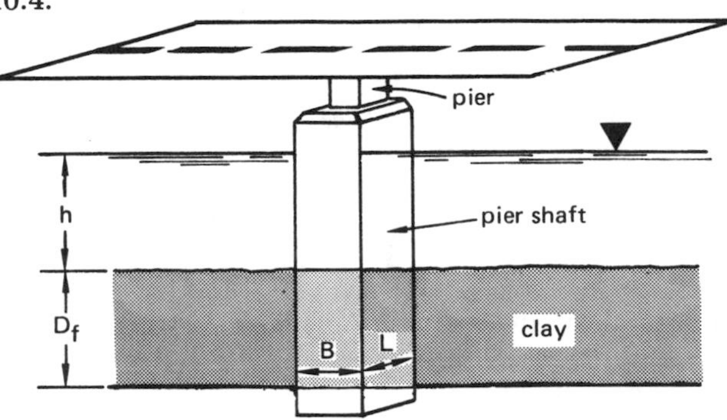


**Figure 10.6** A Pier in Saturated Clay

Piers derive additional supporting strength from skin friction. Skin friction strength is 0.3 to 0.5 times the average undrained shear strength.[10] The additional load that the pier can support is the skin friction times the surface area of the pier shaft. If the pier is belled, only the straight part of the pier is used to calculate the skin friction capacity.

If the skin friction is used to support any of the applied load, $N_c$ should be conservatively taken as approximately 6.2.

## 16 PIERS IN SAND

Piers in sand are designed similarly to footings, since skin friction is insignificant. A conservative estimate of bearing capacity can be found from equation 10.21.

$$p_{\text{net}} = \frac{1}{2} B\gamma N_\gamma + \gamma D_f (N_q - 1) \qquad 10.21$$

$$p_a = (0.11) C_w (C_n N) \qquad 10.22$$

$C_w$ is a correction for water table height, which is neglected when $D_w \geq D_f + B$.

$$C_w = 0.5 + 0.5 \left( \frac{D_w}{D_f + B} \right) \qquad 10.23$$

[10] The skin friction strength should not be greater than 1 ton/ft$^2$. If it is, use 1 ton/ft$^2$.

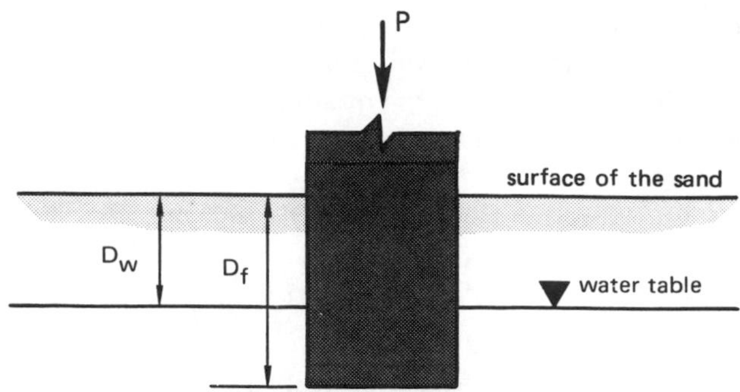


**Figure 10.7** A Pier in Partially Drained Sand

## 17 GENERAL CONSIDERATIONS FOR PILES

Piles are small-area members that are usually hammered or vibrated into place. They provide strength to soils that are too weak or compressible to otherwise support a foundation. Piles are often grouped together to provide the required strength to support a column or wall.

Two major pile classifications exist: friction and point-bearing piles. *Friction piles* derive their load-bearing ability from the friction between the soil and pile. *Point-bearing piles* derive their strength from the support of the soil near the point.

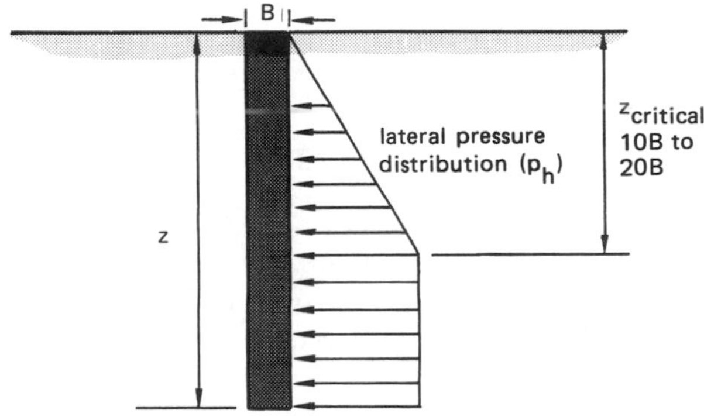


**Figure 10.8** Pressure Distribution on a Typical Pile

### A. CAPACITY OF INDIVIDUAL PILES

In reality, friction bearing capacity and end bearing capacity are both present in friction piles and point-bearing piles. However, one mode of bearing capacity may be predominant. The weight, $W$, which a pile can support is

$$\begin{aligned} W &= [\text{end capacity}] + [\text{friction capacity}] \\ &= \left[ \left( \frac{\pi B^2}{4} \right) \left( \frac{1}{2} \gamma B N_\gamma + cN_c + \gamma z (N_q - 1) \right) \right] \\ &\quad + [\pi B z f_o] \end{aligned} \qquad 10.24$$

If the pile diameter, $B$, is small, the $\frac{1}{2}\gamma BN_\gamma$ term can be omitted. There is also some evidence that the $\gamma z(N_q-1)$ term does not increase without bound, but rather, has as upper limit of $N_q \tan\phi$.

The *skin friction coefficient*, $f_o$, includes both cohesive and adhesive terms. In evaluating $f_o$ and the bearing capacity factors, the friction angle $\phi$ should be increased by 2° to 5° for piles driven into sand. For drilled or jetted piles, no increase is necessary.

$$f_o = \text{smaller of} \begin{Bmatrix} c + p_h \tan\phi \\ c_a + p_h \tan\delta \end{Bmatrix} \qquad 10.25$$

The friction angle, $\delta$, is generally taken as $\frac{2}{3}\phi$, but selected values can be obtained from table 10.6. ($\delta$ is 0° when $\phi$ is 0°.) The lateral earth pressure depends on the depth, down to a *critical depth*, $z_{\text{critical}}$, after which it is essentially constant.[11]

$$p_h = k(\gamma z - \mu) \qquad 10.26$$

[11] Between relative densities of 30% and 70%, the critical depth can be interpolated between 10 and 20 diameters.

$$z_{\text{critical}} = \begin{Bmatrix} 10B \text{ for relative density} < 30\% \\ 20B \text{ for relative density} > 70\% \end{Bmatrix} \qquad 10.27$$

The *adhesion*, $c_a$, should be obtained from testing. In the absence of such tests, it can be approximated as a fraction of the cohesion. For rough concrete, rusty steel, and corrugated metal, $c_a = c$. For wood, $0.9c \le c_a \le c$. For smooth concrete, $0.8c \le c_a \le c$. For clean steel, $0.5c \le c_a \le 0.9c$.

For driven piles, the *coefficient of lateral earth pressure at failure*, $k$, also depends on the relative density. For loose sands (relative density < 30%), $2 \le k \le 3$. For driven piles in dense sand (relative density > 70%), $3 \le k \le 4$. For drilled piles, the coefficients of lateral earth pressure are approximately 50% of the values for driven piles. For jetted piles, the coefficients are approximately 25% of the driven values.

Of course, the pore pressure will not develop in drained sandy soils. For sand below the water table, the pore pressure will be

$$\mu = 62.4 \times \text{depth of water} \qquad 10.28$$

**Table 10.6**
Friction Angles

| interface materials* | friction angle, $\delta$, degrees** |
|---|---|
| concrete or masonry on the following foundation materials: | |
| clean, sound rock | 35 |
| clean gravel, gravel-sand mixtures, and coarse sand | 29–31 |
| clean fine to medium sand, silty medium to coarse sand, and silty or clayey gravel | 24–29 |
| clean fine sand, and silty or clayey fine to medium sand | 19–24 |
| fine sandy silt, and non-plastic silt | 17–19 |
| very stiff clay, and hard residual or preconsolidated clay | 22–26 |
| medium stiff clay, stiff clay, and silty clay | 17–19 |
| steel sheet piles against the following soils: | |
| clean gravel, gravel-sand mixtures, and well-graded rock fill with spalls | 22 |
| clean sand, silty sand-gravel mixtures, and single-size hard rock fill | 17 |
| silty sand, gravel or sand mixed with silt or clay | 14 |
| fine sandy silt, and non-plastic silt | 11 |
| formed concrete or concrete sheet piling against the following soils: | |
| clean gravel, gravel-sand mixtures, and well-graded rock fill with spalls | 22–26 |
| clean sand, silty sand-gravel mixtures, and single-size hard rock fill | 17–22 |
| silty sand, and gravel or sand mixed with silt or clay | 17 |
| fine sandy silt, and non-plastic silt | 14 |
| miscellaneous combinations of structural materials: | |
| masonry on masonry, igneous and metamorphic rocks: | |
| dressed soft rock on dressed soft rock | 35 |
| dressed hard rock on dressed soft rock | 33 |
| dressed hard rock on dressed hard rock | 29 |
| masonry on wood (cross grain) | 26 |
| steel on steel at sheet-steel interlocks | 17 |

* Angles given are ultimate values. Sufficient movement is required before failure will occur.

** For materials not listed, use $\delta = \frac{2}{3}\phi$.

FOUNDATIONS

*Example 10.3*

An 11″, smooth concrete pile with a blunt end is driven 60 feet into clay. The clay's cohesion and density are 1400 psf and 120 lbm/ft$^3$ respectively. The water table extends to the ground surface. What is the ultimate bearing capacity of the pile?

*step 1*: The pile diameter and areas are:

$$B = \frac{11}{12} = 0.917\,\text{ft}$$
$$A_{\text{end}} = \frac{\pi}{4}(0.917)^2 = 0.66\,\text{ft}^2$$
$$A_{\text{surface}} = \pi(0.917)(60) = 172.9\,\text{ft}^2$$

*step 2*: Assume $\phi = 0°$ for saturated clay. From table 10.3, $N_q = 1$. From table 10.2 and table 10.3, $N_c = (1.2)(5.7) = 6.8$. (The contribution of $N_\gamma$ is ignored since $B$ is small.)

*step 3*: The point bearing capacity is

$$(0.66)(1400 \times 6.8) = 6283 \text{ lbf}$$

*step 4*: The lateral earth pressure is disregarded in evaluating the friction capacity, since $\phi = 0°$ and $\tan 0° = 0$ in equation 10.25. Estimate $c_a = 0.9c$ for smooth concrete. Then, the friction capacity is

$$(172.9)(0.9)(1400) = 217{,}850 \text{ lbf}$$

*step 5*: The total capacity is 6283 + 217,850 = 224,133 lbf.

## B. CAPACITY OF PILE GROUPS

The capacity of a pile group will generally be more or less than the sum of the individual piles.[12] The *pile group efficiency* is the ratio of actual capacity to the sum of individual capacities.

$$\eta = \frac{\text{group capacity}}{\sum \text{individual capacities}} \qquad 10.29$$

The group capacity of a large number of piles can be approximated by assuming that the piles form a large footing. This large footing extends from the surface to the depth of the pile points. The length and width of the large footing are the length and width of the pile group. The group bearing capacity is computed from the general bearing capacity equation, equation 10.1. The shearing-related capacity includes the effects of cohesion (or adhesion), but the increase in friction capacity due to lateral soil pressure is disregarded.

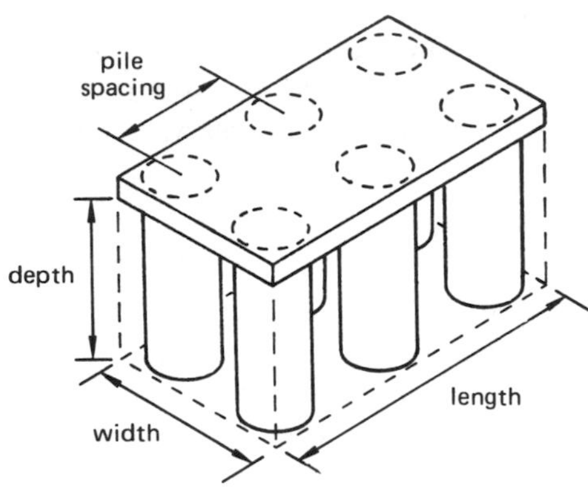


**Figure 10.9** A Pile Group

## 18 SOIL PRESSURES DUE TO APPLIED LOADS

### A. BOUSSINESQ'S EQUATION

Figure 10.10 illustrates a load applied through a footing to a soil below. The increase in pressure, $p_v$, due to the application of the building load can be found from *Boussinesq's equation.* This equation requires the footing width to be small compared to the depth, $h$, at which the increase in pressure is desired (i.e., $h > 2B$).

$$p_v = \frac{3h^3P}{2\pi z^5} = \frac{3P}{2\pi(h^2)}\left[\frac{1}{1+(r/h)^2}\right]^{5/2} \qquad 10.30$$

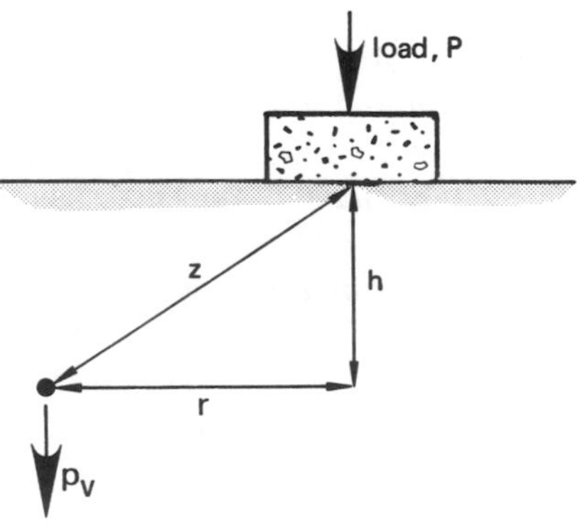


**Figure 10.10** Application of Boussinesq's Equation

[12] For sand, the efficiency is maximized ($\eta = 200\%$) with pile spacings of approximately 2B center-to-center. For clay, the efficiency is less than 100% up to a spacing of 2B, after which the efficiency of 100% is reached and maintained for all reasonable pile spacings above 2B.

B. INFLUENCE CHART METHOD

If a footing or mat foundation is large compared to the depth where the pressure is wanted, the vertical pressure can be found by use of an *influence chart*, similar to figure 10.11. This chart is used in the following manner.

Let the distance A–B on the chart correspond to the depth at which the pressure is wanted. Using this scale, draw a plan view of the footing on a piece of tracing paper.

Place the tracing paper over the influence chart. Locate the footing tracing so that the center of the chart coincides with the location under the footing where the pressure is wanted.

Count the number of squares seen under the footing drawing. Count partial squares as fractions. Count the pie-shaped areas in the center circle as squares.

Calculate the pressure from equation 10.31[13]

$$p = (\#\,\text{squares})(0.005)(\text{applied pressure}) \qquad 10.31$$

[13] Equation 10.31 assumes the influence chart's *influence value* is 0.005, as it is in figure 10.11. Other charts may have other influence values.



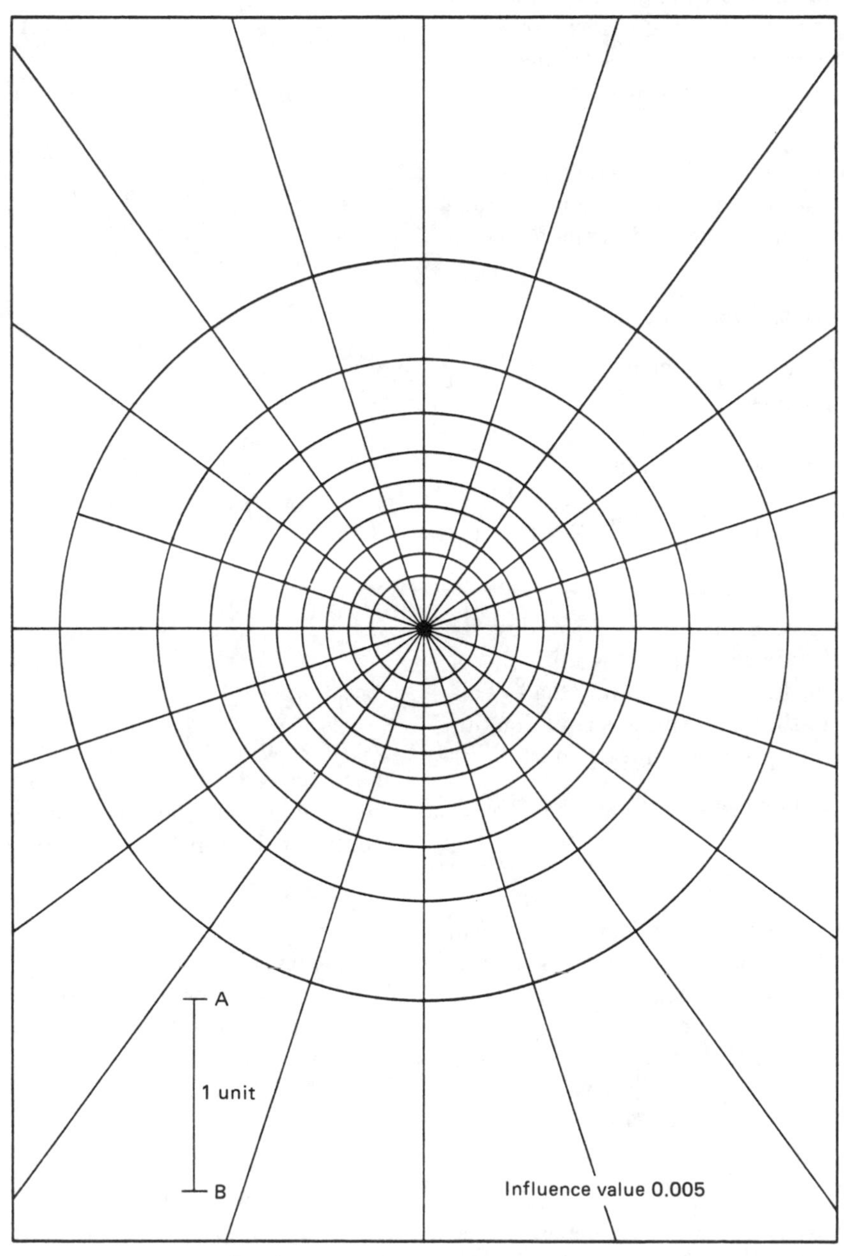


**Figure 10.11** Influence Chart

C. STRESS CONTOUR CHARTS

If a soil is assumed to be semi-infinite, elastic, isotropic, and homogeneous, stress contour charts based on the *Boussinesq case* can be used to obtain subsurface pressures.[14] Such charts exist for a variety of surface loadings, including point loads, square or rectangular footings, circular footings, corners of mats, etc. Appendixes C and D are typical of such contour charts.

## 19 PRIMARY SETTLEMENT IN CLAY

The purpose of pre-construction settlement calculations is to determine the magnitude of expected settlement due to an increase in surface loading. However, these calculations can also be used to find the settlement due to change in any variable, such as a drop in the water table.

*step 1*: Find the original *effective pressure*, $p_o$, at the mid-height of the clay layer. The average effective pressure is the sum of the following items:

- For layers above the water table:

$$p = \begin{pmatrix}\text{layer}\\\text{thickness}\end{pmatrix}\begin{pmatrix}\text{layer}\\\text{density}\end{pmatrix} \qquad 10.32$$

- For layers below the water table:

$$p = \begin{pmatrix}\text{layer}\\\text{thickness}\end{pmatrix}\left(\begin{matrix}\text{saturated layer}\\\text{density}\end{matrix} - 62.4\right) \qquad 10.33$$

*step 2*: Find the increase in pressure, $p_v$, directly below the foundation and at the midpoint of the clay layer due to the building load. This can be done using Boussinesq's equation, influence charts, or stress contour charts.

*step 3*: Estimate the unconsolidated voids ratio, $e_o$.

$$e_o \approx w(SG) \qquad 10.34$$

*step 4*: Estimate the *compression index* (*coefficient of consolidation*) of the soil. This is the logarithmic slope of the primary consolidation curve.

$$C_c = \frac{e_2 - e_1}{\log_{10}\left(\frac{p_1}{p_2}\right)} \qquad 10.35$$

If necessary, the compression index can be estimated from other soil parameters. For inorganic soils with sensitivities less than 4, equation 10.36 applies.

$$C_c = 0.009(w_l - 10) \qquad 10.36$$

For organic soils, such as peat, the compression index can be estimated from the natural moisture content.

$$C_c = 0.0155w_n \qquad 10.37$$

For clays, a general expression is

$$C_c = 1.15(e_o - 0.35) \qquad 10.38$$

*step 5*: Calculate the settlement. For any clay (normally loaded or pre-loaded) for which the original void ratio and change in void ratio are known, equation 10.39 should be used. $H$ is the thickness of the clay layer, regardless of any overlying layers or surcharges.

$$S = H\left(\frac{\Delta e}{1 + e_o}\right) \qquad 10.39$$

For normally loaded clays only, equation 10.40 can be used. $p_v$ is the change in vertical pressure due to applied loads. Therefore, $p_o + p_v$ is the total pressure after load application or removal.[15]

$$S = \left(\frac{C_c}{1 + e_o}\right) H \log_{10}\left(\frac{p_o + p_v}{p_o}\right) \qquad 10.40$$

*Example 10.4*

A 40′ × 60′ raft is constructed as shown. The building rests on sand which has already settled. What long-term settlement can be expected in the clay (a) at the center of the raft? (b) at a corner of the raft?

*step 1*:

| | | |
|---|---|---|
| silt layer: | (5)(90) | = 450 lbf/ft² |
| dry sand layer: | (14)(120) | = 1680 |
| wet sand layer: | (22)(130 − 62.4) | = 1490 |
| clay layer: | $\frac{1}{2}$(14)(110 − 62.4) | = 330 |
| | | $p_o$ = 3950 lbf/ft² |

*step 2*: Use the influence chart (figure 10.11) to calculate the pressure increase due to the building. The distance from the bottom of the

[14] The Boussinesq case is not the only case possible. The *Westergaard case* assumes layered or anisotropic foundation soil, consisting of alternating layers of soft and stiff materials. The effect of such layering is to reduce the stresses substantially below those obtained from the Boussinesq case. The Westergaard case is typically used in the analysis of wheel loads over multi-layered highway pavement sections.

[15] $p_o$ in equation 10.40 is the *effective pressure*, excluding the pore pressure. Experience has indicated that effective stress alone can cause consolidation.

raft to the mid-point of the clay layer is $36 - 3 + 7 = 40$ feet. Using a scale of 1 inch = 40 feet, the raft is $1'' \times 1\frac{1}{2}''$. 86 squares are covered.

The net pressure at the base of the raft is the applied pressure minus the overburden pressure due to the excavated sand and silt.

$$p_{\text{net}} = 2400 - 5(90) - 3(120) = 1590\,\text{lbf/ft}^2$$

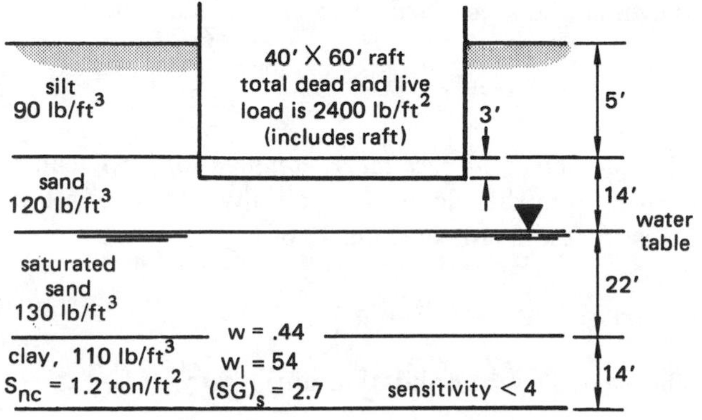


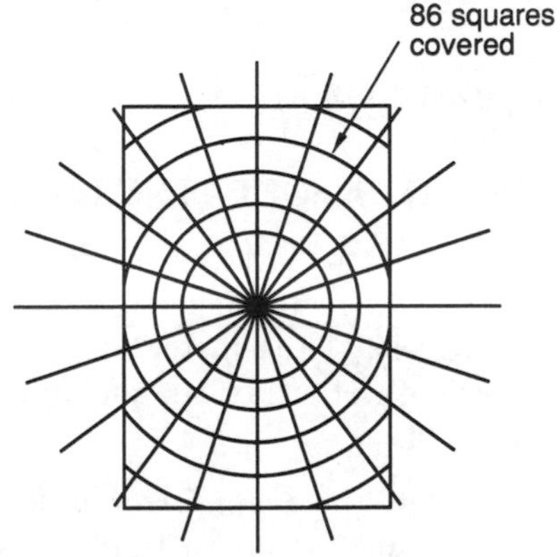


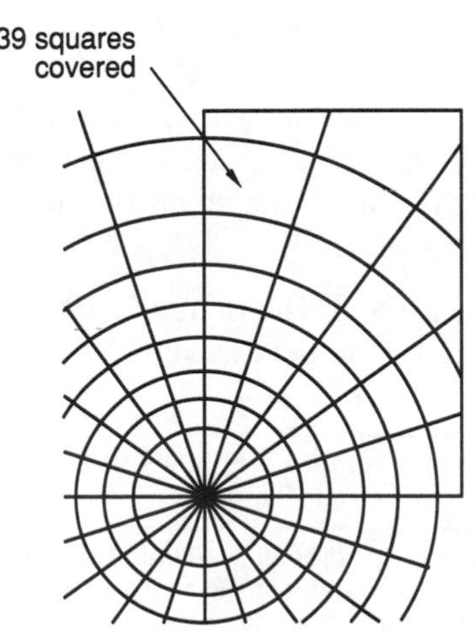


The mid-layer increase in pressure below the centroid of the raft is

$$p_v = (1590)(86)(0.005) = 680\ \text{lbf/ft}^2$$

Similarly, the mid-layer increase in pressure at the corner is

$$p_v = (1590)(39)(0.005) = 310\ \text{lbf/ft}^2$$

*step 3*: The original void content of the clay is found from equation 10.34.

$$e_o = (0.44)(2.7) = 1.188$$

*step 4*: The compression index is estimated from equation 10.36.

$$C_c = 0.009(54 - 10) = 0.396$$

*step 5*: The settlements are

$$S_{\text{center}} = \left(\frac{0.396}{1 + 1.188}\right)(14)\log\left(\frac{3950 + 680}{3950}\right)$$
$$= 0.175\ \text{ft}$$

$$S_{\text{corner}} = \left(\frac{0.396}{1 + 1.188}\right)(14)\log\left(\frac{3950 + 310}{3950}\right)$$
$$= 0.083\ \text{ft}$$

## 20 TIME RATE OF PRIMARY CONSOLIDATION

Settling in clay is a continuous process. The time to reach a specific settlement is given by equation 10.41. $z$ is the layer's *half-thickness* if drainage is through the top and bottom surfaces (i.e., *two-way drainage*). If drainage is from one surface only (i.e., *one-way drainage*), $z$ is the layer's full thickness. Units of $t$ will depend on units of $C_v$.

$$t = \frac{T_v z^2}{C_v} \qquad 10.41$$

The *coefficient of consolidation* is assumed to remain constant over small variations in the void ratio, $e$.

$$C_v = \frac{k(1 + e)}{\gamma_w (a_v)} \qquad 10.42$$

The *coefficient of compressibility*, $a_v$, can be found from the void ratio and effective stress for any two different loadings.

$$a_v = \frac{e_1 - e_2}{p_2 - p_1} \qquad 10.43$$

$T_v$ is a dimensionless number known as the *time factor*. $T_v$ depends on the *degree of consolidation*, $U_z$. $U_z$ is the percent of the total consolidation (settlement) expected. For $U_z$ less than 0.60 (60%), $T_v$ is given by equation 10.44.

$$T_v = \frac{1}{4}\pi U_z^2 \qquad 10.44$$

Table 10.7 should be used to find $T_v$ for larger values of $U_z$.

**Table 10.7**
Approximate Time Factors

| $U_z$ | $T_v$ |
|---|---|
| 0.60 | 0.28 |
| 0.70 | 0.40 |
| 0.75 | 0.48 |
| 0.80 | 0.55 |
| 0.85 | 0.70 |
| 0.90 | 0.85 |
| 0.95 | 1.3 |
| 1.0 | $\infty$ |

Following the end of *primary consolidation* (approximately $T_v = 1$ in equation 10.44), the rate of consolidation will decrease considerably. The continued consolidation is known as *secondary consolidation.*

## 21 SECONDARY CONSOLIDATION

*Secondary consolidation* is a gradual consolidation which continues long after the majority of the initial consolidation has occurred. Secondary consolidation may not occur at all, as in the case of granular soils. However, secondary consolidation may be a major factor for inorganic clays and silts, as well as for highly-organic soils.

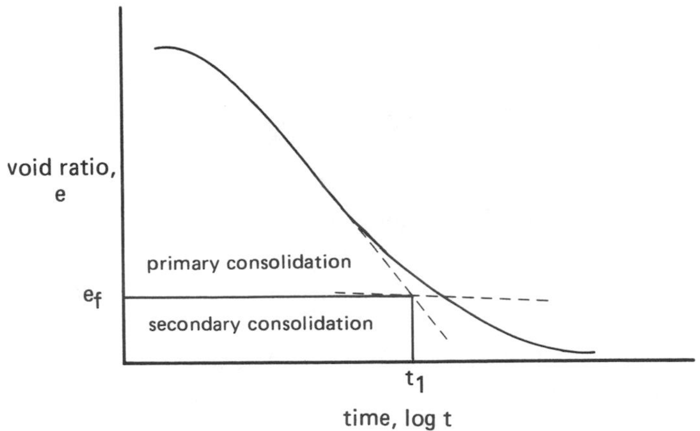


**Figure 10.12** Primary and Secondary Consolidation

Secondary consolidation can be identified by a plot of void ratio (or settlement) versus time on a logarithmic scale. The region of secondary consolidation is characterized by a slope reduction on the plot. The plot can be used to obtain important parameters necessary to calculate the magnitude and progression of secondary consolidation.

The final void ratio, $e_f$, at the end of the primary consolidation period is read from the intersection of the projections of the primary and secondary curves. The logarithmic slope of the secondary compression line is known as the *secondary compression index*, $\alpha$.[16]

$$\alpha = \frac{-(\log_{10} t_2 - \log_{10} t_1)}{e_2 - e_1} \qquad 10.45$$

The *coefficient of secondary compression*, $C_\alpha$, can be derived from this slope. The initial voids ratio, $e_o$, can be estimated from equation 10.34.

$$C_\alpha = \frac{\alpha}{1 + e_o} \qquad 10.46$$

The secondary consolidation during any period $t_2 - t_1$ is

$$S_{\text{secondary}} = C_\alpha H \log\left(\frac{t_2}{t_1}\right) \qquad 10.47$$

## 22 SLOPE STABILITY

### A. HOMOGENEOUS, SOFT CLAY ($\phi = 0°$)

For homogeneous, soft clay, the *Taylor chart* can be used to determine the factor of safety against slope failure.[17] Alternatively, if the factor of safety is known, the maximum depth of cut or the maximum cut angle can be determined.

*step 1*: Calculate the *depth factor*, $d$, from the slope height and the depth from the slope toe to the lowest point on the slip circle.

$$d = \frac{D}{H} \qquad 10.48$$

The *slope height*, $H$, is essentially the depth of the cut. $D$ is the vertical distance from the toe of the slope to the firm base below the clay.

*step 2*: Based on the depth factor, $d$, and the angle of the slope, determine the *stability number*, $N_o$.

[16] The secondary compression index generally ranges from 0 to 0.03, and seldom exceeds 0.04.

[17] The *circular arc method* and *method of slices* can also be used to analyze a particular failure surface. However, if the failure surface is unknown, these methods may require trial and error solutions to locate the critical failure plane.

*step 3*: Calculate the factor of safety in cohesion, $F$, from equation 10.49. For submerged clays, the buoyant force should be subtracted from the clay density.

$$F = \frac{N_o c}{\gamma_{\text{eff}} H} \tag{10.49}$$

$$\gamma_{\text{eff}} = \gamma_{\text{sat}} - 62.4 \text{ (if submerged)} \tag{10.50}$$

Minimum factors of safety are difficult to specify, but limited research seems to imply a range of 1.3 to 1.5 as minimum acceptable values.

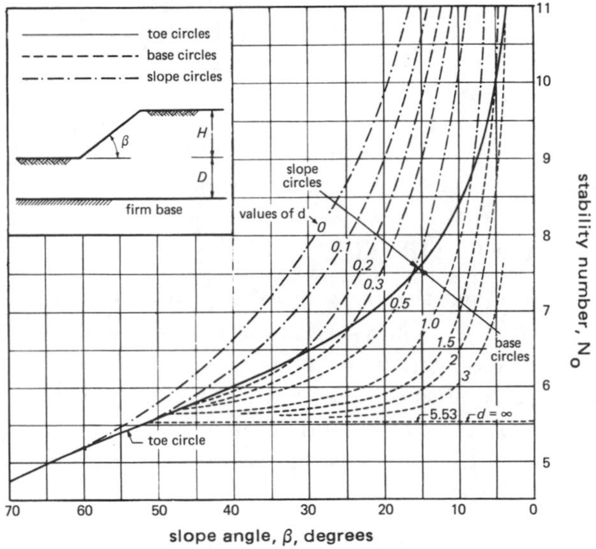


**Figure 10.13** The Taylor Chart for Slope Stability ($\phi = 0$)

Figure 10.13 shows that *toe circle failures* occur for all slopes steeper than 53°. For slopes less than 53°, *slope circle failure*, toe circle failure, or *base circle failure* may occur. These possibilities are illustrated in figure 10.14.

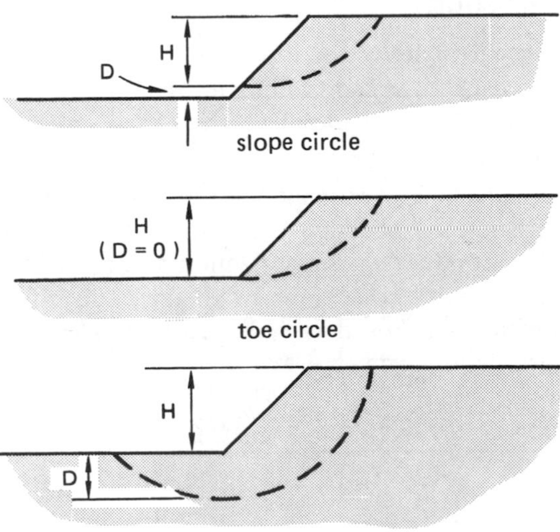


**Figure 10.14** Failure Modes in Clay

*Example 10.5*

An underwater trench is excavated in soft bay mud. The walls of the trench are sloped at 4.5 vertical:3 horizontal. The mud has a saturated density of 100 lbm/ft$^3$ and a cohesion of 400 lbf/ft$^2$. There is a layer of dense sand and gravel 59 feet below the surface of the mud. What is the depth of cut that can be used while maintaining a 1.5 safety factor?

The cut angle is

$$\beta = \arctan\left(\frac{4.5}{3}\right) = 56°$$

Since $\beta > 53°$, the failure will be toe circle. From figure 10.13, the stability number for $\beta = 56°$ is approximately 5.4.

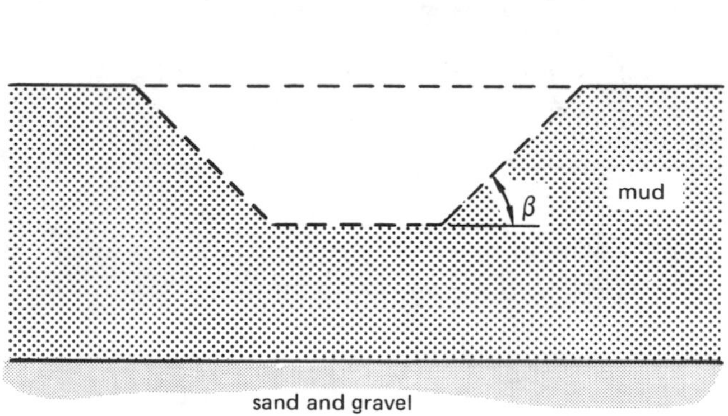


Since the clay is submerged, the effective unit weight is $(100 - 62.4) = 37.6$.

From equation 10.49, the maximum cut depth is

$$H = \frac{N_o c}{F\gamma} = \frac{(5.4)(400)}{(1.5)(37.6)} = 38.3 \text{ ft}$$

B. HOMOGENEOUS, COHESIVE SOIL ($c > 0$, $\phi > 0°$)

For cohesive, non-granular soils, failure will be through the slope toe for $\phi > 5°$.

FOUNDATIONS

C. COHESIONLESS SAND ($c = 0$)

The maximum slope angle for cohesionless sand is the angle of internal friction, $\phi$.

## 23 EARTH PRESSURE THEORIES

The general equation for horizontal *active earth pressure* is:

$$p_{\text{horizontal}} = p_{\text{vertical}} \tan^2\left(45^\circ - \frac{\phi}{2}\right) - 2c\tan\left(45^\circ - \frac{\phi}{2}\right) \quad 10.51$$

$c$ in equation 10.51 is the soil's cohesion.

$p_{\text{vertical}}$ can be due to surcharge, externally applied loads, or the soil's own mass.

If $\phi = 0^\circ$, as in the limiting case for saturated clay, then

$$p_{\text{horizontal}} = p_{\text{vertical}} - 2c \quad 10.52$$

If $c = 0$, as in the limiting case for drained sand,

$$p_{\text{horizontal}} = p_{\text{vertical}}\left[\tan^2\left(45^\circ - \frac{\phi}{2}\right)\right] \quad 10.53$$

The quantity in brackets in equation 10.53 is known as the *coefficient of active earth pressure.*

$$k_A = \tan^2\left(45^\circ - \frac{\phi}{2}\right) = \frac{1 - \sin\phi}{1 + \sin\phi} \quad 10.54$$

The general equation for horizontal *passive earth pressure* is:

$$p_{\text{horizontal}} = p_{\text{vertical}} \tan^2\left(45^\circ + \frac{\phi}{2}\right) + 2c\tan\left(45^\circ + \frac{\phi}{2}\right) \quad 10.55$$

The *coefficient of passive earth pressure* for sand is

$$k_P = \frac{1}{k_A} = \tan^2\left(45^\circ + \frac{\phi}{2}\right) = \frac{1 + \sin\phi}{1 - \sin\phi} \quad 10.56$$

A. THE RANKINE THEORY

If it is assumed that the backfill soil is dry, cohesionless sand, then the *Rankine theory* can be used. At any depth, $H$, the vertical pressure is

$$p_{\text{vertical}} = \gamma H \quad 10.57$$

The horizontal pressure depends on the *coefficient of earth pressure at rest*, $k_o$, which varies from 0.4 to 0.5 for untamped sand.[18]

$$p_{\text{horizontal}} = k_o\gamma H \quad 10.58$$

$$k_o \approx 1 - \sin\phi \quad 10.59$$

$$R_o = \tfrac{1}{2}k_o\gamma H^2$$

Equations 10.57 and 10.58 apply only to a sand deposit of infinite depth and extent. For sand that is compressed or tensioned (as in around a retaining wall) the reactions are given by equations 10.60 and 10.61.

$$R_A = (p_{\text{horizontal}})\left(\frac{H}{2}\right) = \frac{1}{2}k_A\gamma H^2 \quad 10.60$$

$$R_P = \frac{1}{2}k_P\gamma H^2 \quad 10.61$$

$R_A$ and $R_P$ are horizontal if the soil above the heel and toe is horizontal. (See figure 10.19.)

B. WEDGE THEORIES

The Rankine theory is based on infinite, cohesionless soil. It also requires that the soil above the heel be level. Modifications can be made to lift these restrictions, as well as to allow a water table above the foundation base. Several modifications are known as *wedge theories. Coulomb's earth-pressure theory* is one such wedge theory.

The wedge methods are based on the observation that retaining walls fail when the active soil shears. Although the shear plane is actually a slightly curved surface, it is assumed to be linear (line $*$–$*$ in figure 10.15). However, since the actual shear plane is not known in advance, several trial planes need to be taken. This is known as the *trial wedge method.*

## 24 SLOPED AND BROKEN SLOPE BACKFILL

It is possible to derive equations for the active force with a sloped backfill, as shown in figure 10.16. However, the complexity of these equations usually makes a graphical solution a better choice.

With sloped or broken slope backfill, the active force is not horizontal. Appendix A and appendix B provide a method of evaluating the horizontal and vertical earth pressure. Notice that $k_h$ and $k_v$ have units of lbf/ft$^2$ per foot of wall. Soil density is not used. $H$ is defined as shown in appendices A and B.

$$R_{A,h} = \frac{1}{2}k_h H^2 \quad 10.62$$

$$R_{A,v} = \frac{1}{2}k_v H^2 \quad 10.63$$

[18] It is appropriate to use the at-rest soil case whenever the wall does not move. Bridge abutments and basement walls are examples where movement is essentially nonexistent.

$$R_A = \sqrt{R_h^2 + R_v^2} \quad 10.64$$

$$\theta = \arctan\left(\frac{R_v}{R_h}\right) \quad 10.65$$

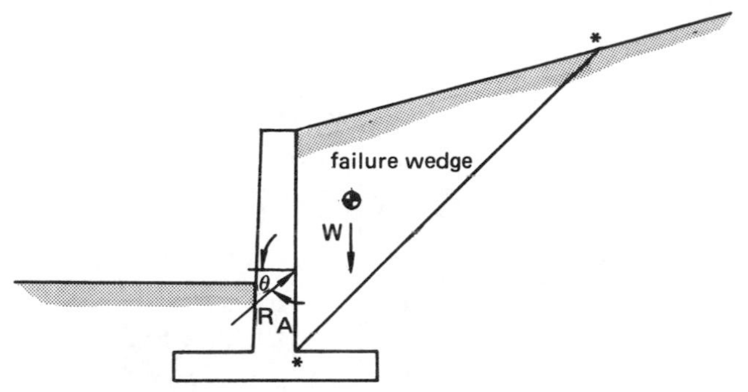


**Figure 10.15** Failure Wedge

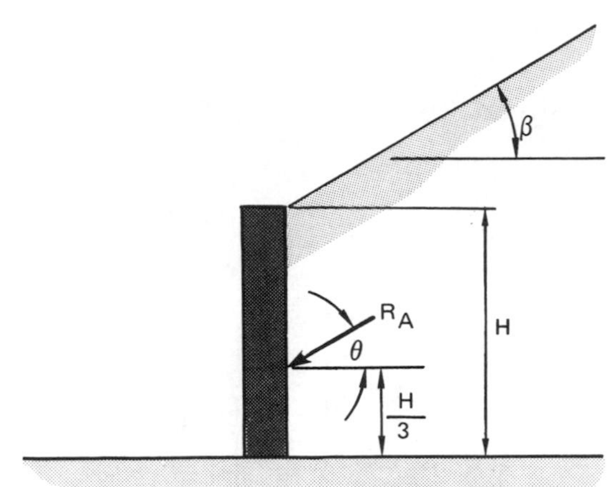


**Figure 10.16** Sloped Backfill

## 25 SURCHARGE LOADING

### A. UNIFORM SURCHARGE

If there is a uniform surcharge load of $q$ lbf/ft$^2$ above the backfill, there will be an additional active force, $R_q$. $R_q$ acts at $H/2$ above the base. This force is in addition to the regular active force which acts at $H/3$.

$$R_q = k_A q H \times \text{wall width} \quad 10.66$$

### B. POINT LOAD

If a point load is applied a distance $x$ back from the wall face, as in figure 10.18, the distribution of pressure behind the wall can be found from equations 10.67 through 10.70.[19]

[19] These equations are based on elastic theories with a Poisson's ratio of $\mu = 0.5$. The coefficients have been adjusted to bring the theory into agreement with observed values.

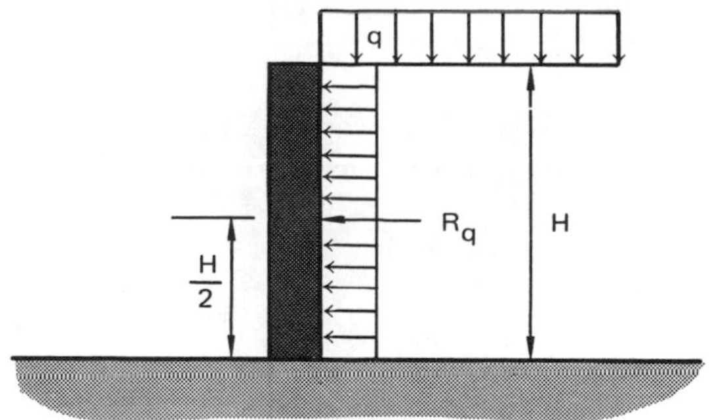


**Figure 10.17** A Uniform Surcharge

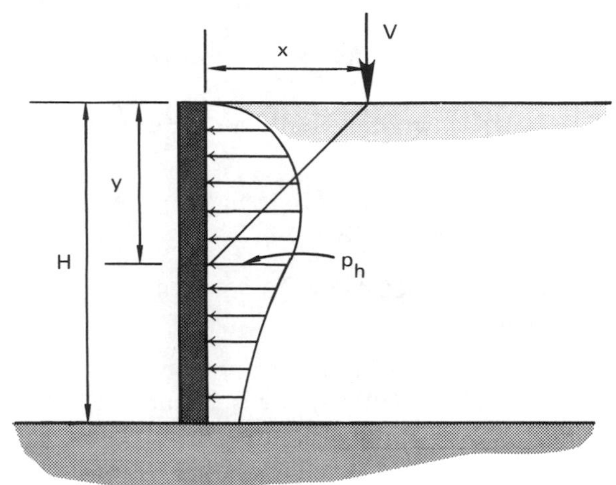


**Figure 10.18** A Point Load Distribution

$$m = \frac{x}{H} \quad 10.67$$

$$n = \frac{y}{H} \quad 10.68$$

$$p_h = \frac{1.77V}{H^2}\frac{m^2n^2}{(m^2+n^2)^3} \quad (m > 0.4) \quad 10.69$$

$$p_h = \frac{0.28V}{H^2}\frac{n^2}{(0.16+n^2)^3} \quad (m \le 0.4) \quad 10.70$$

### C. LINE LOAD

For a line load of $q$ lbf/ft, the distribution of pressure behind the wall is given by equations 10.71 and 10.72. Figure 10.18 applies if $V$ is replaced with $q$.

$$p_h = \frac{4}{\pi}\frac{q}{H}\left(\frac{m^2n}{(m^2+n^2)^2}\right) \quad (m > 0.4) \quad 10.71$$

$$p_h = \frac{q}{H}\frac{0.203n}{(0.16+n^2)^2} \quad (m \le 0.4) \quad 10.72$$

## 26 HORIZONTAL PRESSURES FROM SATURATED SAND

Sand, being porous, allows trapped water to exert a horizontal pressure against vertical retaining walls. Therefore, the sand and water both exert a horizontal force. The hydrostatic pressure is

$$p_{\text{hydrostatic}} = (62.4)H \qquad 10.73$$

However, there is also a buoyant force, since each sand particle is submerged below the water table. The soil pressure depends on its saturated density and the buoyant force.

$$p_{\text{soil, vertical}} = (\gamma_{\text{sat}} - 62.4)H \qquad 10.74$$

$$p_{\text{soil, horizontal}} = k_A(\gamma_{\text{sat}} - 62.4)H \qquad 10.75$$

The total horizontal pressure from saturated soil is the sum of equations 10.73 and 10.75.

$$p_{\text{horizontal}} = [62.4 + k_A(\gamma_{\text{sat}} - 62.4)]H \qquad 10.76$$

The increase in the horizontal pressure over the dry condition is $(1 - k_A)(62.4)$. This product is known as the *effective hydrostatic loading* of the fluid. If it is convenient to do so, equation 10.76 can be interpreted as a pressure from a fluid with *effective hydrostatic density* $\gamma_{\text{eff}}$.[20]

$$\gamma_{\text{eff}} = 62.4(1 - k_A) \qquad 10.77$$

$$p_{\text{horizontal}} = [k_A\gamma_{\text{sat}} + \gamma_{\text{eff}}]H \qquad 10.78$$

The saturated soil density used in the above equations can be calculated if the dry soil density and either the porosity or void ratio is known.

$$\gamma_{\text{sat}} = \gamma_{\text{dry}} + 62.4n = \gamma_{\text{dry}} + 62.4\left(\frac{e}{1+e}\right) \qquad 10.79$$

## 27 RETAINING WALLS

Retaining walls must be safe against settlement. In this regard, their design is similar to footings. They must have sufficient resistance against overturning and sliding. Retaining walls must also possess adequate structural strength. The method of meeting these requirements is one of trial and error.

The analysis of a retaining wall's stability requires knowledge of at least six different distributed force systems. These forces result from the distributions shown in figure 10.20. They are the forward (active or tensioned) earth reaction, $R_A$, the backward (passive or compressed) earth reaction, $R_P$, the soil force, $R_v$, the shear resistance, $R_s$, the weights of the earth masses, $R_T$ and $R_H$, and the weight of the wall itself.

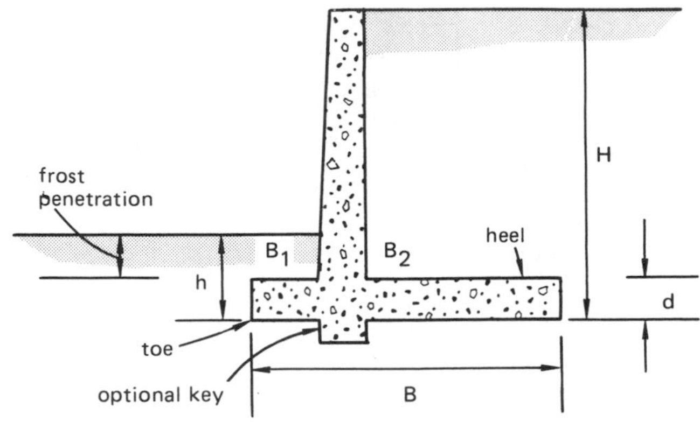


**Figure 10.19** A Cantilever Retaining Wall

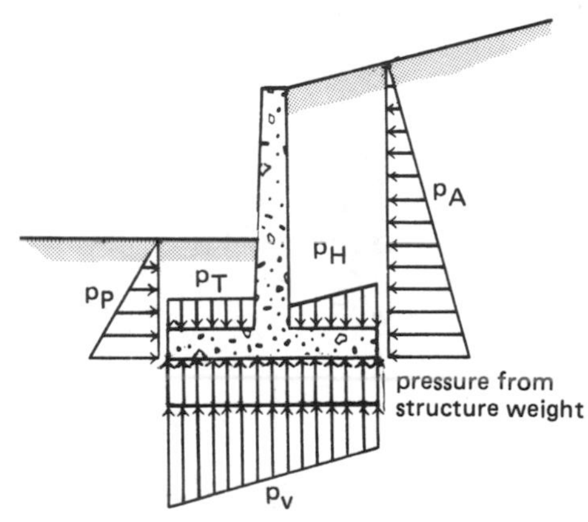


**Figure 10.20** Pressure Distributions on a Retaining Wall

The important factors to be considered when evaluating the design of a retaining wall are factor of safety against overturning, the maximum soil pressure under the base, and the factor of safety against sliding.

The following procedure can be used to analyze a retaining wall. (Sizing a retaining wall is covered in chapter 14.)

*step 1*: Determine the active reaction and its point of application. Include the reactions from all point, line, and distributed surcharges.[21]

[20] 45 lbm/ft³ is typically taken as the effective hydrostatic density of the fluid behind a retaining wall.

[21] Retaining walls should be analyzed and designed for a minimum density of 30 lbm/ft³ of fill, regardless of the actual load.

*step 2*: Determine the passive reaction and its point of application.[22]

*step 3*: Find the vertical forces against the base. These forces are the weights, $W_i$, of the areas shown in figure 10.21. Find the centroid of each area and the moment arm, $r_i$, from the centroid of the $i$th area to point $G$.[23] Calculate the moments of each area about point $G$.

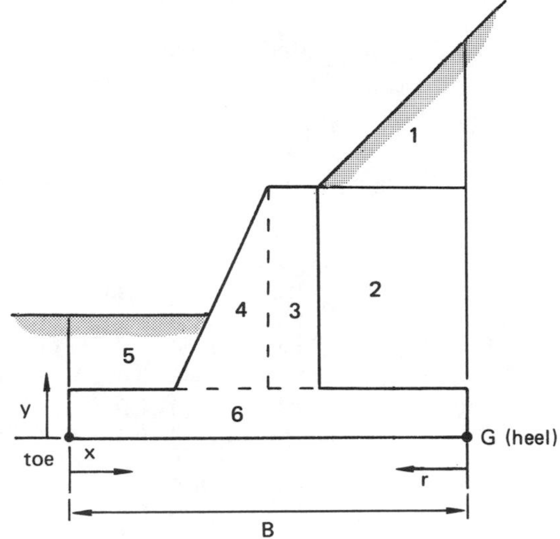


**Figure 10.21** Elements Contributing to Vertical Force

*step 4*: Find the moment about point $G$ of all the vertical forces and the active pressure.

$$M_G = \sum W_i r_i + R_{A,h} y_A \qquad 10.80$$

*step 5*: Find the location and eccentricity of the vertical force resultant. The eccentricity is the distance from the center of the base to the vertical force resultant. Eccentricity should be less than $B/6$ for the entire base to be in compression.

$$r_R = \frac{M_G}{\sum W_i + R_{A,v}} \qquad 10.81$$

$$\epsilon = r_R - \frac{1}{2}B \qquad 10.82$$

[22] The passive pressure is usually disregarded on the assumption that the backfill will be in place prior to the front fill, or that the front fill will be removed at some future date for repairs to the wall.

[23] Moments can also be taken about the toe.

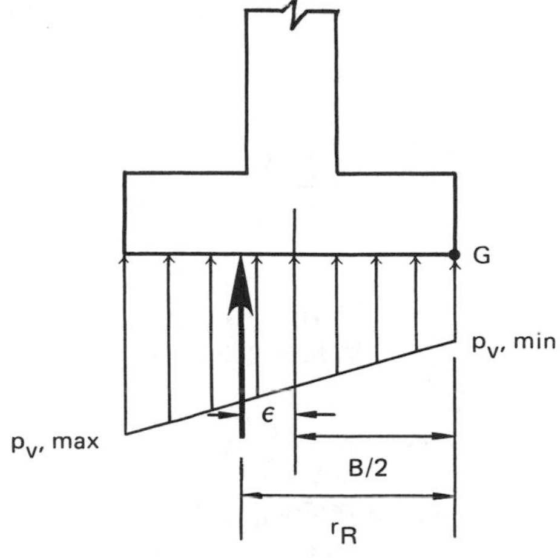


**Figure 10.22** Resultant Forces on Base

*step 6*: Check the *factor of safety against overturning* by summing moments about the toe. The moment arms $x$ and $y$ must be measured from the toe.[24] The factor of safety should generally exceed 1.5 for cohesionless soils, and 2.0 for cohesive soils.

$$F_{\text{overturning}} = \frac{\sum W_i x_i + R_{A,v} x_{A,v}}{R_{A,h} y_{A,h}} \qquad 10.83$$

*step 7*: Find the maximum (at the toe) and minimum (at the heel) foundation pressure on the base.

$$p_{v,\text{max}}, p_{v,\text{min}} = \frac{\sum W_i + R_{A,v}}{B}\left(1 \pm \frac{6\epsilon}{B}\right) \qquad 10.84$$

The maximum pressure should not exceed the allowable soil pressure.

*step 8*: Calculate the *resistance against sliding*. Disregarding the passive pressure (which may not be present throughout the life of the wall), the active pressure must be resisted by the shearing strength of the soil or the friction between the base and the soil, plus the adhesion between the two materials. Equation 10.85 is for use when the wall has a key, and then only for the soil to the left of the key. Equation 10.86 is for use with the soil to the right of a key and for flat-bottomed walls.

$$R_s = \left(\sum W_i + R_{A,v}\right) \tan\phi + c_a B \qquad 10.85$$

$$R_s = \left(\sum W_i + R_{A,v}\right) \tan\delta + c_a B \qquad 10.86$$

[24] Other interpretations may include $R_{A,v}$ as a negative term in the denominator.

In the absense of friction coefficient data, the resistance can be found from equation 10.87 and table 10.8.

$$R_s = k_s\left(\sum W_i + R_{A,v}\right) + c_a B \qquad 10.87$$

**Table 10.8**
Values of $k_s$

| | |
|---|---|
| coarse grained soil without silt | 0.55 |
| coarse grained soil with silt | 0.45 |
| silt | 0.35 |

*step 9*: Calculate the *factor of safety against sliding*. In cases where the passive pressure is disregarded, a factor of safety in excess of 1.5 is desired. If passive pressure is included in the resistance against sliding, the factor of safety should exceed 2.0. (Neglecting the passive force will give a minimum factor of safety.)

$$F_{\text{sliding}} = \frac{R_s}{R_{A,h}} \qquad 10.88$$

If the factor of safety is too low, increase the base length, $B$, or include a vertical key in the design. (See figure 10.19.)

*Example 10.6*

A retaining wall is being designed for a soil with $\phi = 30°$, $\delta = 17°$, and a maximum allowable pressure of 3000 psf. The backfill is coarse-grained sand with silt, with a density of 125 lbm/ft$^3$. Using an adhesion of 950 psf, check the tentative design for stability against sliding. Do not check for factor of safety against overturning.

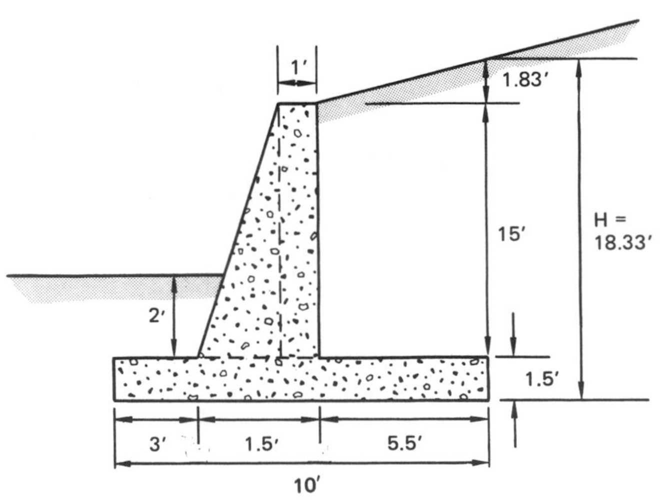


*step 1*: The backfill is sloped. Assume a type-2 fill and use appendix A.

$$\beta = \arctan\left(\frac{1.83}{5.5}\right) = 18.4° \quad (3:1)$$

From appendix A, $k_v \approx 10$ and $k_h \approx 40$.

$$R_{A,v} = \frac{1}{2}(10)(18.33)^2 = 1680\,\text{lbf}$$

$$R_{A,h} = \frac{1}{2}(40)(18.33)^2 = 6720\,\text{lbf}$$

$R_{A,h}$ is located 18.33/3 = 6.11 ft above the bottom of the base. (See footnote 22.)

*step 2*: Assume $R_P = 0$.

*step 3*: (Disregard the sloped face of area 5.)

| $i$ | area | $\gamma$ | $W_i$(lbf) | $r_i$ | moment | |
|---|---|---|---|---|---|---|
| 1 | $\frac{1}{2}$(5.5)(1.83)=5.03 | 125 | 629 | 1.83 | 1151 | |
| 2 | 5.5(15)=82.5 | 125 | 10,313 | 2.75 | 28,361 | |
| 3 | 1(15)=15 | 150 | 2250 | 6.0 | 13,500 | |
| 4 | $\frac{1}{2}$(0.5)(15)=3.75 | 150 | 563 | 6.67 | 3755 | |
| 6 | 1.5(10)=15 | 150 | 2250 | 5.0 | 11,250 | |
| 5 | 2(3)=6 | 125 | 750 | 8.5 | 6375 | |
| | | | 16,755 | | 64,392 | totals |
| $R_{A,v}$ | | | 1680 | 0 | 0 | |
| $R_{A,h}$ | | | 6720 | 6.11 | 41,060 | |

(Notice $R_{A,v}$ goes through point G.)

*step 4*: $M_G = 64{,}392 + 41{,}060 = 105{,}450$ ft-lbf

*step 5*:

$$r = \frac{105{,}450}{16{,}755 + 1680} = 5.72\,\text{ft}$$

$$\epsilon = 5.72 - \frac{1}{2}(10) = 0.72\,\text{ft}$$

Since 0.72 is less than 10/6, the base is in compression everywhere.

*step 6*: Skipped.

*step 7*: The maximum pressure at the base is

$$p_{\max} = \left(\frac{16{,}755 + 1680}{10}\right)\left(1 + \frac{6(0.72)}{10}\right) = 2640\,\text{lbf/ft}^2$$

This is less than 3000 lbf/ft$^2$.

*step 8*: The resisting force against sliding is

$$R_s = (16{,}755 + 1680)\tan 17° + (950)(10) = 15{,}136\,\text{lbf}$$

*step 9*: From equation 10.88,

$$F_{s,\min} = \frac{15{,}136}{6720} = 2.25 \text{ (O.K.)}$$

## 28 FLEXIBLE ANCHORED BULKHEADS

*Anchored bulkheads* are supported at their bases by having been sunk into the ground. They are anchored further up with rods projecting back into the soil. These rods can terminate at deadmen, piles, walls, or beams.

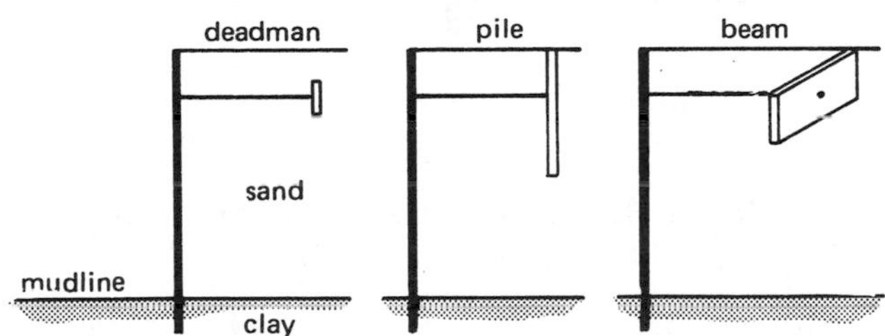


**Figure 10.23** Anchored Bulkheads

Anchored bulkheads can fail in the following ways:

- The base clay can fail due to inadequate bearing capacity. The clay will shear along a circular arc passing under the bulkhead.
- The anchorage can fail. There are several ways for this to occur, including rod tension failure and deadman movement.
- The toe embedment at the mud line can fail.
- The sheeting can fail, although this is rare.

The depth of embedment that is required is found by taking moments about the anchor attachment point on the bulkhead. This anchor pull is found by summation of all horizontal loads on the bulkhead.

The maximum bending moment in the bulkhead itself should be found by taking moments about a point of counterflexure listed in table 10.9.

**Table 10.9**
Points of Counterflexure

| embedment material | point |
|---|---|
| firm and dense | mud line |
| loose and weak | 1 or 2 feet below mud line |
| soft over hard layer | at hard layer depth |

## 29 BRACED CUTS

### A. INTRODUCTION

A braced cut is an excavation in which the load from one bulkhead is used to support the opposite bulkhead's load. Failure in dry soils above the water line generally occurs by wale crippling followed by strut buckling. Planned excavations below the water line should be dewatered prior to cutting. The analysis of braced cuts is approximate due to extensive bending of the sheeting.

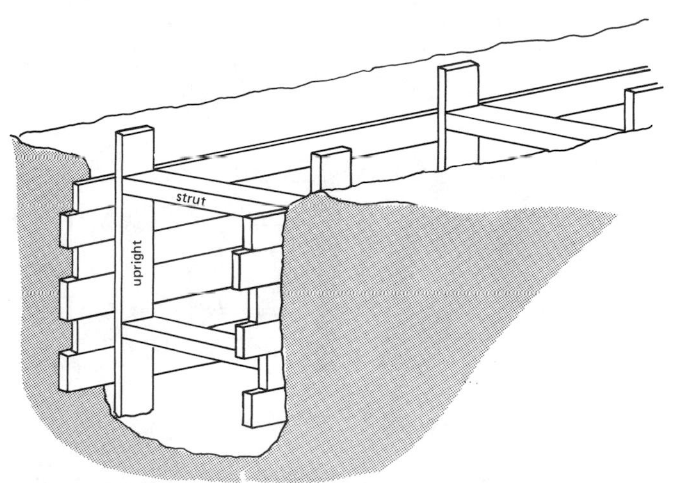


**Figure 10.24** A Braced Cut with Box Shoring

Since the struts are installed as the excavation goes down, the upper part of the wall deflects very little due to the strut restraints. Therefore, the final pressure on the upper part of the wall will be considerably higher than would be normally predicted by the active pressure equation. These larger than expected loads near the top have resulted in failure of the upper strut in braced cuts.

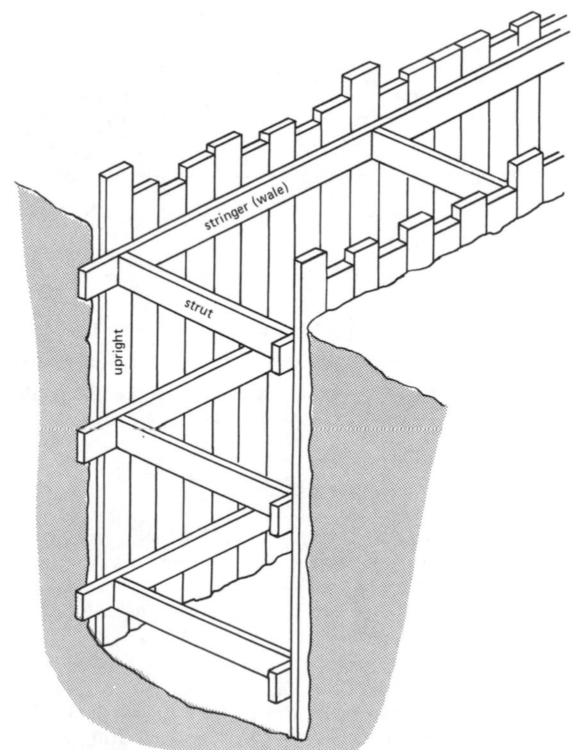


**Figure 10.25** A Braced Cut with Close Sheeting

## B. PRESSURE DISTRIBUTIONS IN BRACED CUTS

The pressure distribution depends on the cut material. For dry or drained sand, the pressure distribution is similar to that in figure 10.26.[25] The maximum pressure is given by equation 10.89.

$$p_{max} = (0.65)k_A\gamma H \qquad 10.89$$

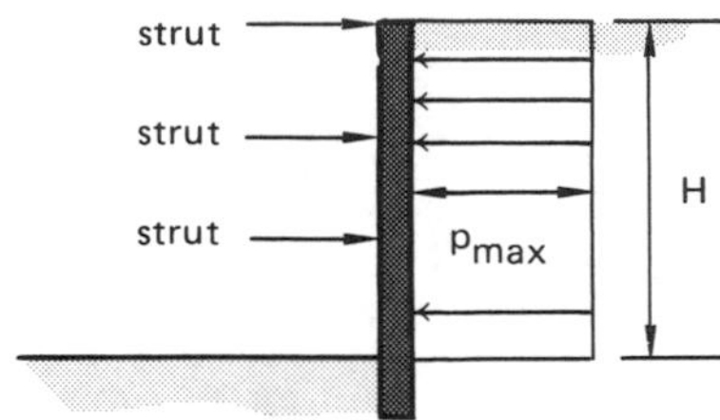


**Figure 10.26** Cuts in Sand

For undrained clay (typical of cuts made rapidly in comparison to drainage times) where $\phi = 0°$, the pressure envelope depends on the average undrained shear strength of the clay in the cut zone. If $\gamma H/c \leq 4$, (e.g., stiff clay), the pressure envelope is shown by figure 10.27.

$$0.2\gamma H \leq p_{max} \leq 0.4\gamma H \qquad 10.90$$

Use the lower values of $p_{max}$ when movement is minimal or when the construction period is short.

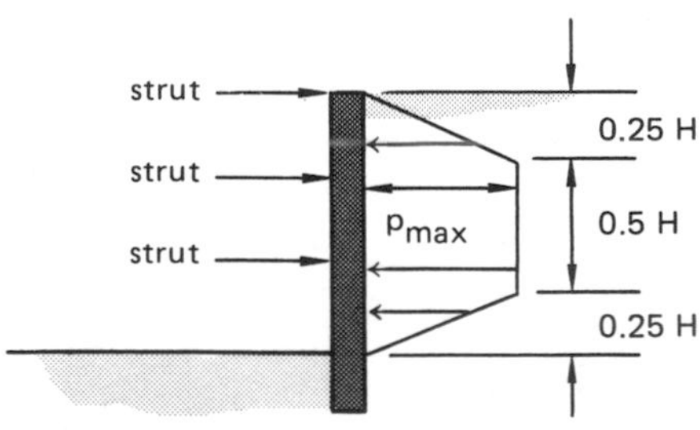


**Figure 10.27** Cuts in Stiff Clay

If $\gamma H/c \geq 6$ (e.g., soft to medium clay), the pressure distribution is as appears in figure 10.28. Except for cuts underlain by deep, soft, normally consolidated clays, equation 10.91 can be used.[26]

$$p_{max} = \gamma H - 4c \qquad 10.91$$

[25] This is not the only distribution that has been proposed for sand. The Tschebotarioff trapezoidal pressure distribution increases (starting at the surface and working down) from 0 to 0.8 $k_A\gamma H$ in the first 0.1$H$ of depth. It remains constant for 0.7$H$, and decreases linearly to zero in the lower 0.2$H$.

[26] If the shearing strength of the clay below the cut, $S_b$, is known, the quantity $\gamma H/S_b$ should be checked. If it is below 6, the bearing capacity of the soil is sufficient to prevent shearing and upward heave. Simple braced cuts should not be attempted if this quantity exceeds 8.

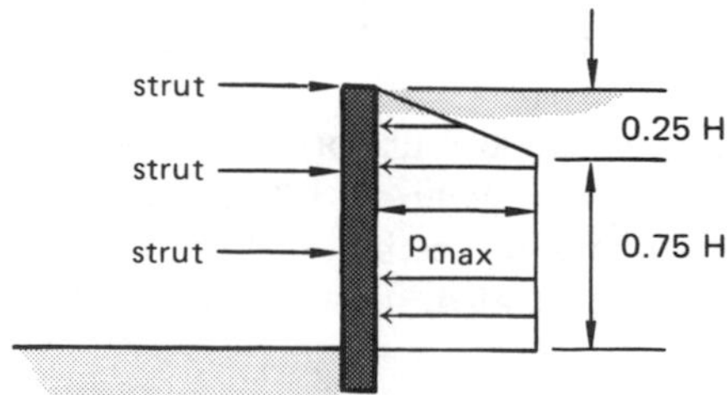


**Figure 10.28** Cuts in Soft Clay

If $4 < \gamma H/c < 6$, use either the soft or stiff clay case, whichever results in the stronger design.

## C. DESIGN OF BRACED CUTS

The first step in the design of braced cuts is to determine the pressure distribution based on equations 10.60 and 10.76. The passive pressure reduces the active pressure below the mudline. The point at which the active pressure distribution is zero is taken as the hinge point. The passive pressure below the hinge point is disregarded.[27]

If there are no struts, the wall is designed as a cantilever beam.

If there is one strut, moments are taken about the hinge point to determine the strut load.

With multiple struts, the moment distribution method can be used to determine strut loads. An easier method is to assign portions of the active distribution to the struts based on areas between the mid-points of the strut separation distances.[28]

Once a tentative design has been reached, the design should be checked using the distributions presented in the previous section. Strut loads are computed by tributary areas. Vertical members are designed as beams on unyielding supports.

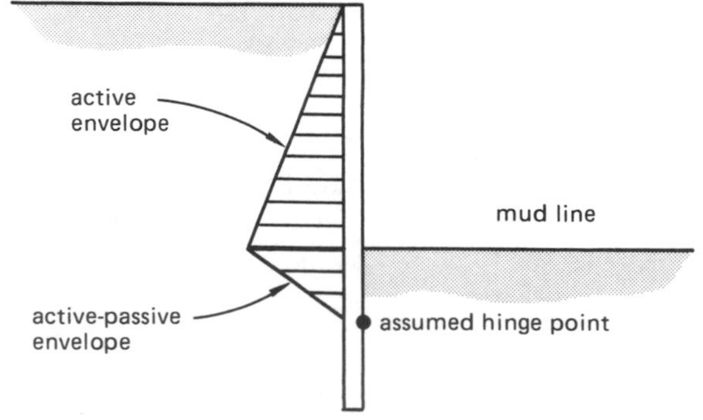


**Figure 10.29** Preliminary Design of Braced Cuts

[27] Passive pressures greater than the active pressure are disregarded.

[28] That is, the sheeting is assumed to be hinged at the strut points.

# Appendix A: Active Components for Retaining Walls with Straight Slope Backfill

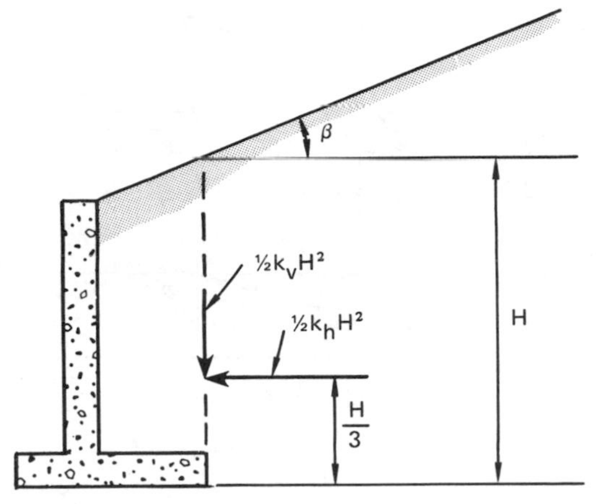


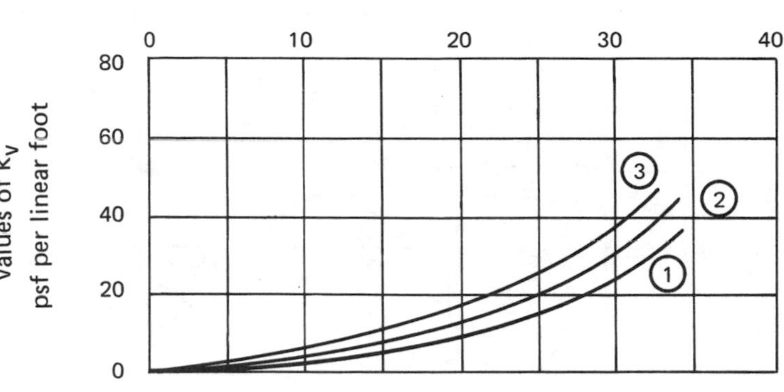


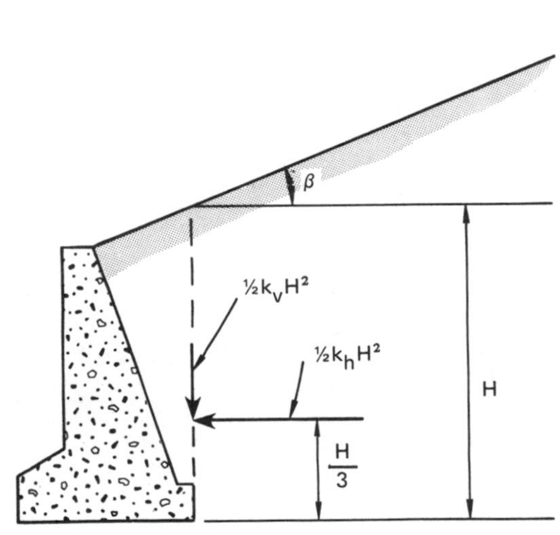


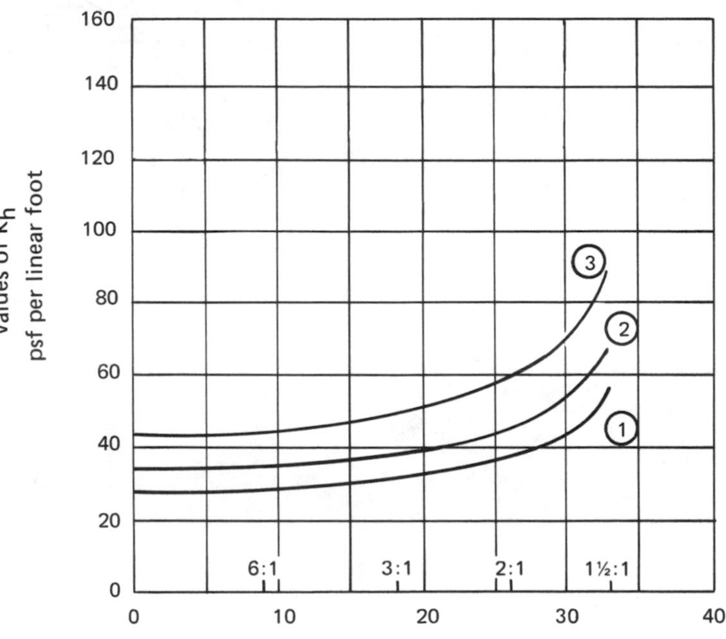


Circled numbers indicate the following soil types.

1. Clean sand and gravel: GW, GP, SW, SP.
2. Dirty sand and gravel of restricted permeability: GM, GM–GP, SM, SM–SP.
3. Stiff residual silts and clays, silty fine sands, and clayey sands and gravels: CL, ML, CH, MH, SM, SC, GC.

# Appendix B: Active Components for Retaining Walls with Broken Slope Backfill

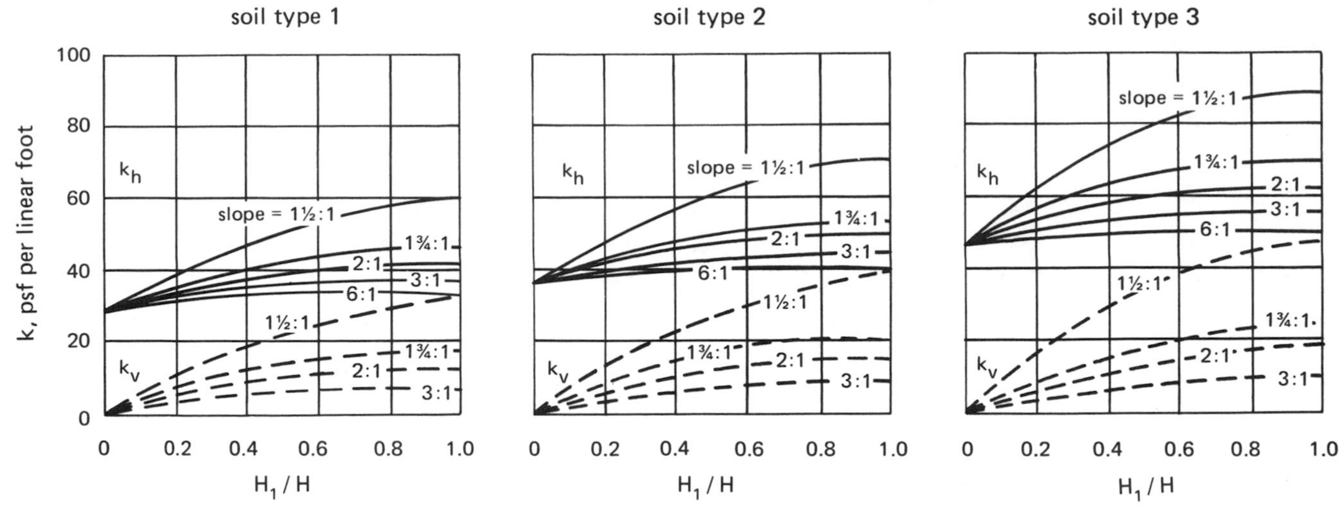


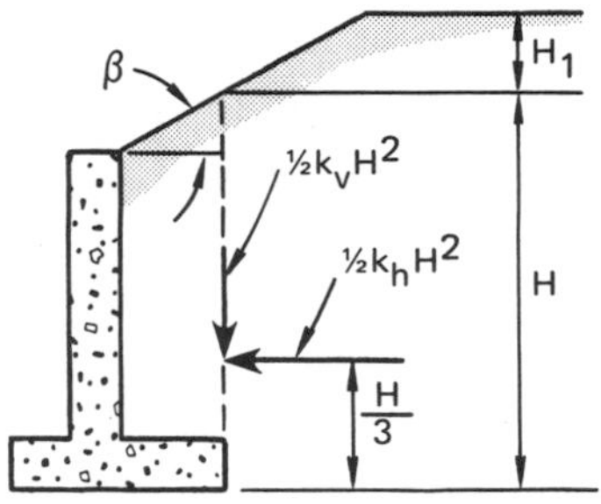

# Appendix C: Boussinesq Stress Contour Charts Infinitely Long and Square Footings

$p$ = uniform foundation pressure

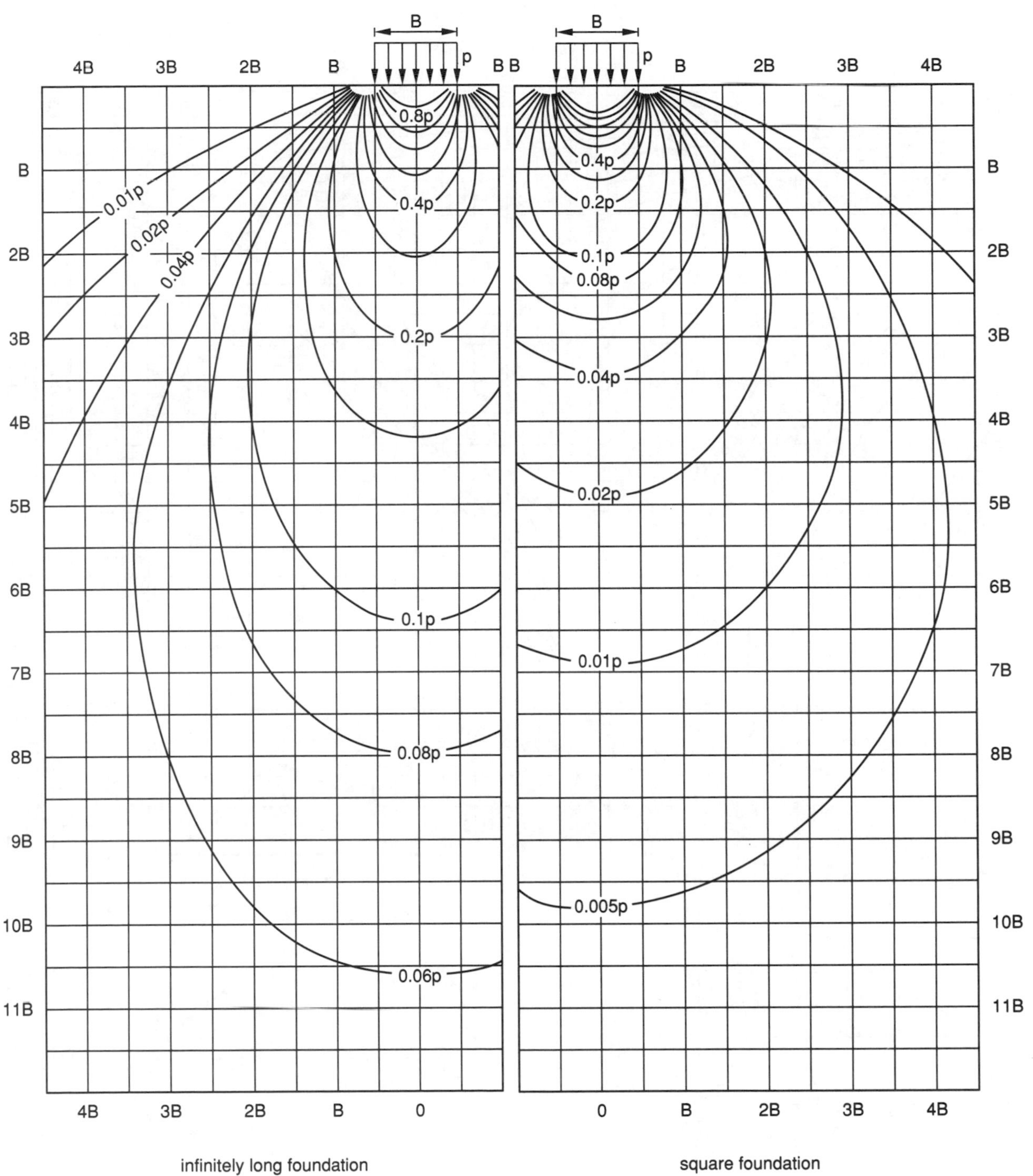

# Appendix D: Boussinesq Stress Contour Chart Uniformly Loaded Circular Footings

$p$ = uniform foundation pressure

$\sigma_{z,\text{vertical}} = p \times$ chart influence value

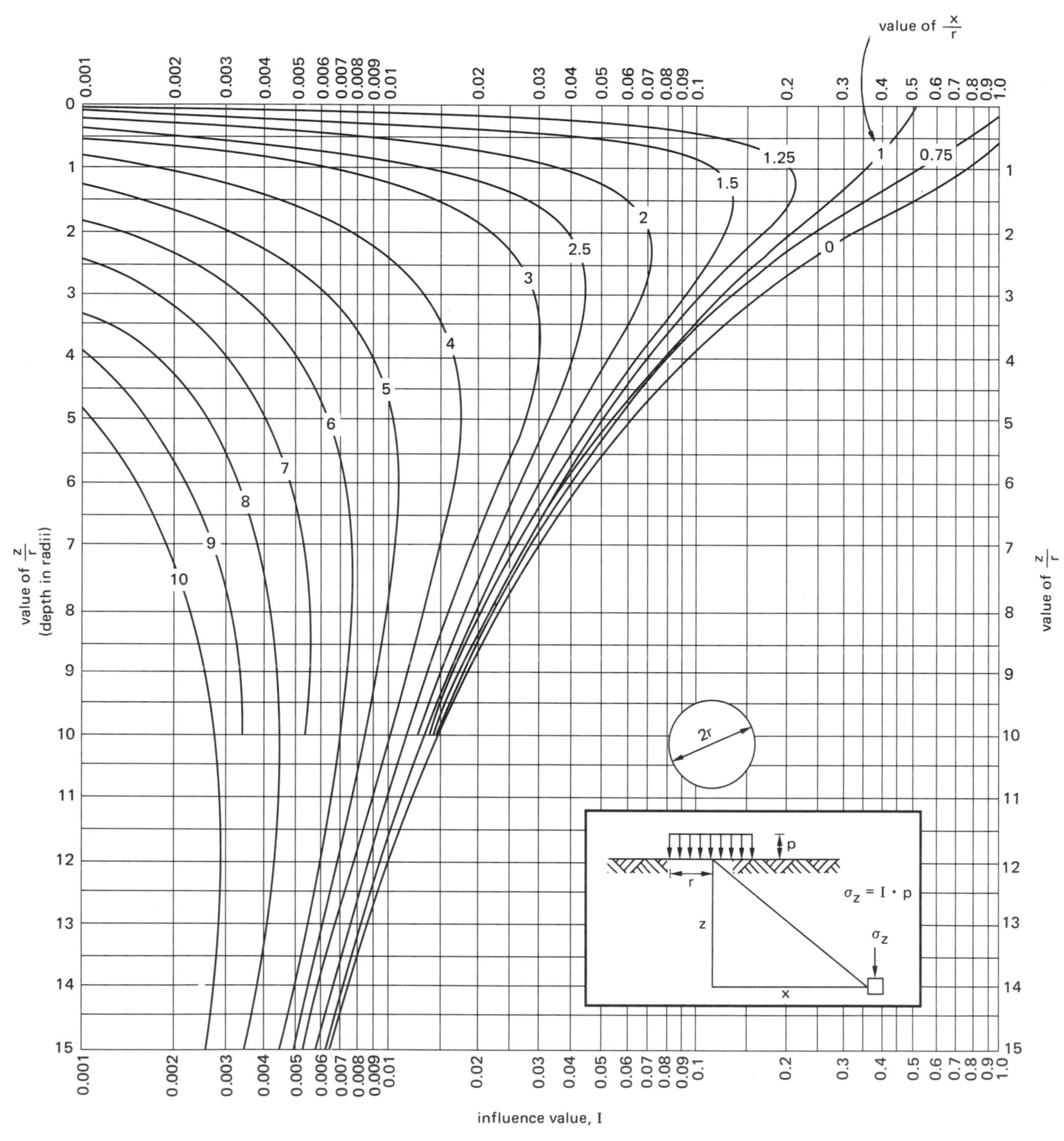

## Practice Problems: FOUNDATIONS AND RETAINING WALLS

Untimed

1. A large manufacturing plant with a floor load of 500 psf is supported on a large mat foundation. A 38 foot thick layer of silty soil is underlain by sand and gravel. The plant was constructed on the silty soil 10 feet above the water table. After a number of years and after all settlement of the building had stopped, a series of wells were drilled which dropped the water table from 10 feet below the surface to 18 feet below the surface. The unit weight of the soil above the water table is 100 pcf. It is 120 pcf below the water table. The silt compression index is .02. The void ratio before the water table was lowered was 0.6. How much can the building be expected to settle?

2. 360 kips are to be supported by a square footing. The footing is to rest directly on sand which has a density of 121 pcf and an angle of internal friction of 38°. (a) Size the footing. (b) Size the footing if it is to be placed 4 feet below the surface.

3. A reinforced concrete retaining wall is used to support a 14 foot cut in sandy soil. The top surface is horizontal but supports a surcharge of 500 psf. The soil has the following characteristics:

density: 130 pcf (sandy soil)
angle of internal friction: 35°
coefficient of friction with concrete: .5
allowable soil pressure: 4500 psf
frost line: 4 feet below grade

Use a factor of safety of 1.5 against sliding and overturning to design the dimensions of the retaining wall. Neglect structural details.

4. A 26 foot high wall holds back sand with a 96 pcf dry density. The water table is permanently 10 feet below the top of the wall. The saturated density of the sand is 121 pcf. The angle of internal friction is estimated to be 36°. (a) Disregarding capillary rise, calculate the active earth resultant. (b) Where is the active earth resultant located? (c) Assuming the water table level could be dropped 16 feet to the bottom of the wall, what would be the reduction in overturning moment?

5. The compression index of a normally-loaded clay soil is 0.31. When the effective stress on the soil is 2600 psf, the void ratio is 1.04 and the permeability is 4 EE-7 mm/sec. The stress is increased gradually to 3900 psf. (a) What is the change in the void ratio? (b) Compute the settlement of a 16-foot layer. (c) How much time is required for the settlement to be 75% complete assuming two-way drainage?

6. A 30 foot deep excavation is being planned for sand ($\phi = 40°$, density of 121 pcf) with a drained water table during construction. The excavation will be 40 feet square. The bracing is to consist of horizontal lagging supported by 8-inch H-beam soldier piles. (a) What is the pressure diagram? (b) If the soldier piles are 8 feet apart, what bending moment is placed on the lagging?

7. A 2:1 (horizontal:vertical) slope is cut in homogeneous, saturated clay which has a density of 112 pcf and a shear strength of 1.1 EE3 psf. The cut is 43 feet deep, and the clay extends 15 feet below the cut bottom to a rock layer. Compute the safety of this slope.

8. What is the increase in stress at a depth of 10 feet below and 8 feet from the center of a 10 foot square footing that exerts a pressure of 3000 psf on the soil?

9. A concentrated vertical load of 6000 pounds is applied at the ground surface. What is the vertical pressure caused by this load at a point 3.5 feet below the surface and 4 feet from the action line of the force?

10. A 10.75″ O.D. steel pile is driven 65 feet into a stiff, insensitive clay with a shear strength of 1.3 EE3 psf. The pile has a flat-plate, closed end. The soil density is 115 pcf, and the water table is at the ground surface. What is the ultimate bearing capacity?

Timed

1. Use Terzaghi's equations to size the three foundations illustrated below. The soil is sandy clay with the following characteristics: density – 108 lbm/ft³; angle of internal friction – 25°; cohesion – 400 psf. The water table is 35 feet below the ground surface. Use a safety factor of 2.5. Neglect footing weight.

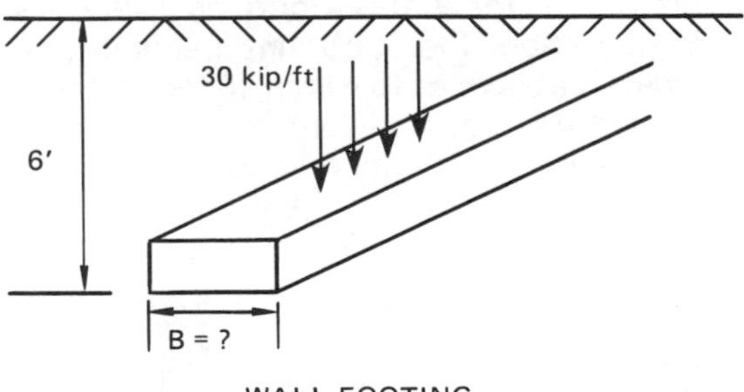


WALL FOOTING

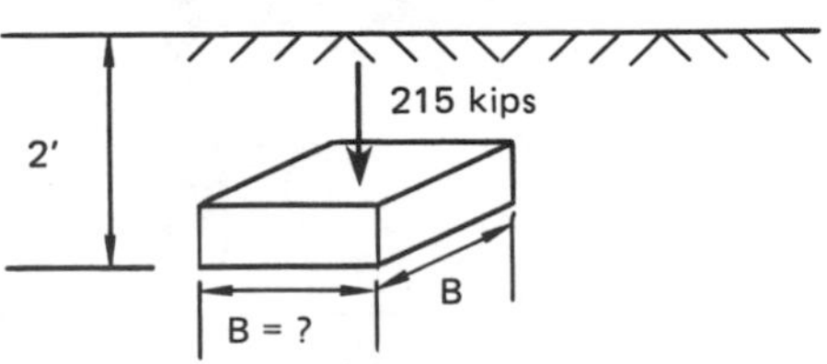


SQUARE FOOTING

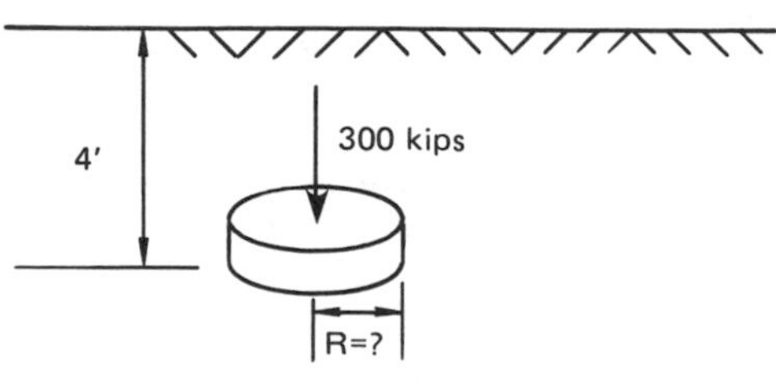

ROUND FOOTING

2. A construction firm wants to consolidate a clay layer at a building site. The soil consists of 8 feet of soft clay on top of a rock layer. The clay is covered with 15 feet of silty sand. The water table is 18 feet above the rock layer. It is proposed to consolidate the clay layer by surcharging the site with 10 more feet of sandy fill (density = 110 pcf) and by dewatering the sand 5 feet (i.e., lowering the water table 5 feet). What is the settlement caused by the surcharging and dewatering?

SITE AS FOUND

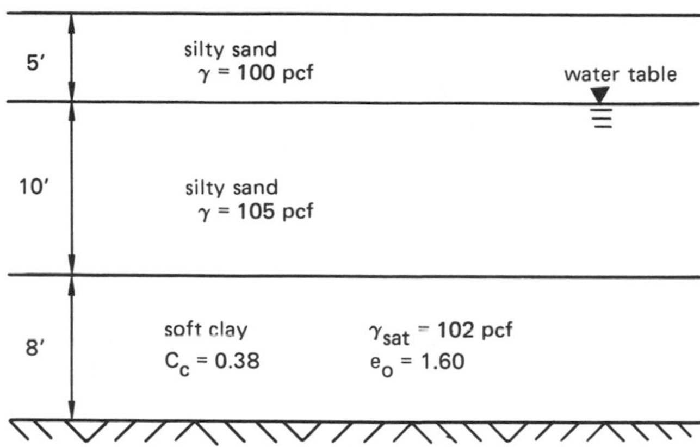


3. A retaining wall is designed for free-draining granular backfill with adequate subdrains. After several years of operation, the subdrains become plugged and the water table rises to within 10 feet of the top of the wall. Find the resultant force and its location for both the drained and plugged conditions.

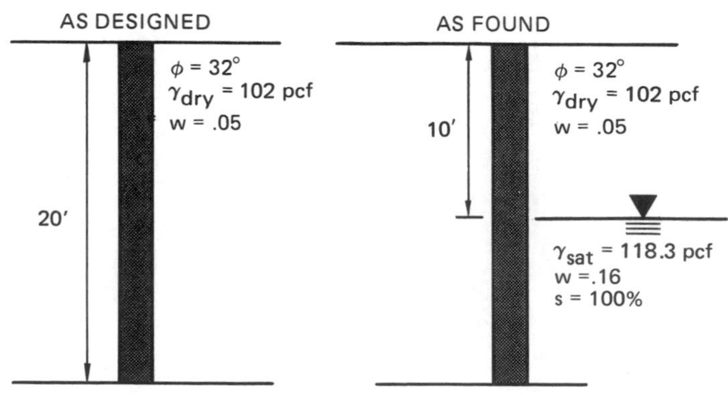


4. A sheet pile is driven through 10 feet of clay to support a 25 foot vertical cut through sand. The total height of the sheet pile is 35 feet, and bedrock is located below that depth. No penetration of the bedrock is made. A tieback is located 8 feet below the surface, and it terminates at a deadman. There is no significant water table. What is the tensile force in the tie rod?

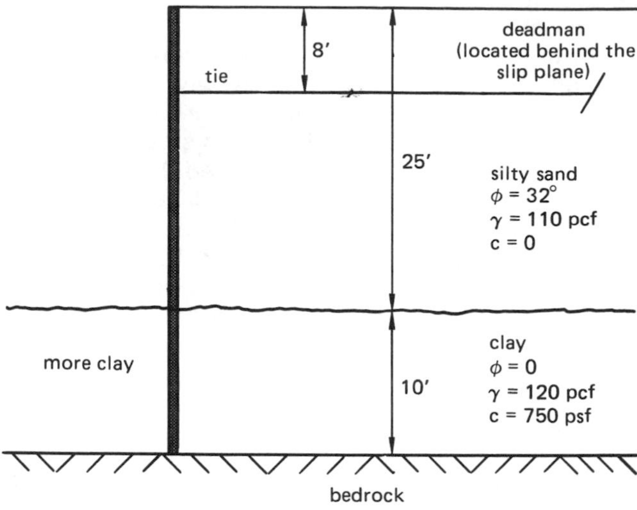


5. A test has shown that a typical pile in the group of 36 piles shown has an uplift capacity of 150 tons and a compressive capacity of 500 tons. The pile grouping is capped by a thick concrete slab producing an axial load of 600 tons. (a) Find the maximum moment that the pile group can take in the x- and y-directions. (b) If the concrete slab is separated along the y-axis into two separate pieces, and then reattached by a steel plate, what is the maximum moment that the pile group can take about the y-axis?

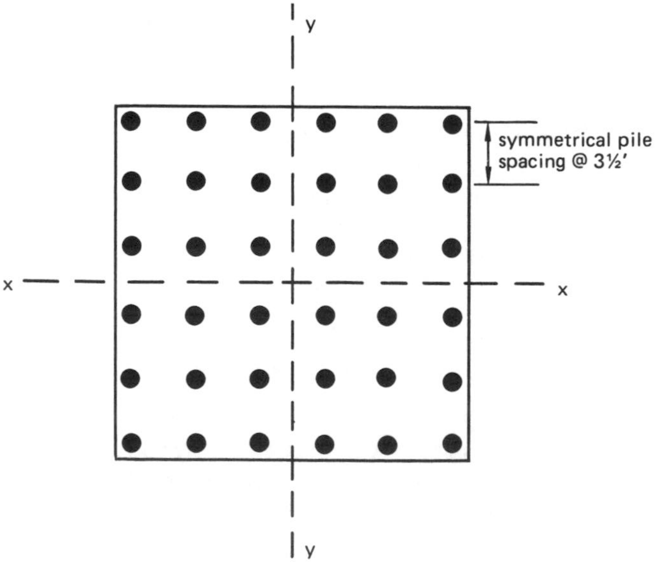


6. A small PVC pipe (2″ diameter) is to be buried in a trench 18 inches wide and 3½ feet deep. The angle of internal friction of the cohesive soil is 18°, cohesion is 200 psf. The moisture content is 20%, dry density is 115 pcf. (a) Will the soil stand or

slide during trenching if the spoils are placed directly alongside the trench sides? (b) How far away from the trench must the spoils be placed to satisfy OSHA? (c) When the spoils are spread evenly for a distance of 2 feet from the trench, will the soil remain stable? (d) Recommend initial bedding and backfill materials. (e) What additional precautions would you recommend to protect the workers?

7. A building weighing 20 tons is placed on a basement slab as shown. The soil below the basement is dense sand of 100 feet thickness. Find the pressure produced by the building weight (a) at a depth 30 feet below point A, and (b) at a depth of 45 feet below point B.

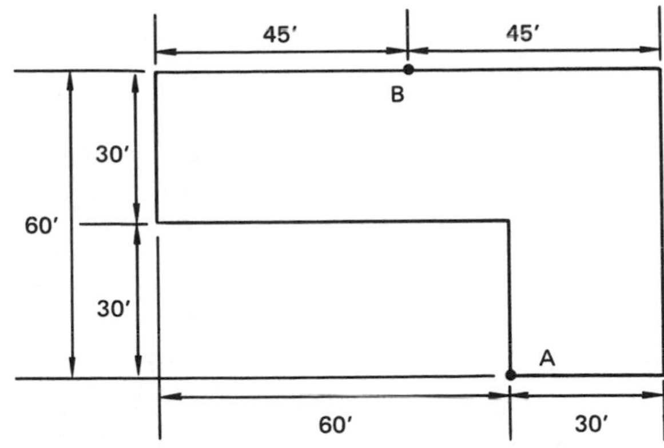

# 11 STATICS

*Nomenclature*

| | | |
|---|---|---|
| a | horizontal distance from point of maximum sag | ft |
| A | area | $ft^2$ |
| c | parameter in catenary equations | ft |
| d | distance | ft |
| E | modulus of elasticity | $lbf/ft^2$ |
| f | coefficient of friction | – |
| F | vertical force | lbf |
| H | horizontal component of tension | lbf |
| I | moment of inertia | $ft^4$ |
| J | polar moment of inertia | $ft^4$ |
| L | length, cable length from point of maximum sag | ft |
| m | mass | slugs |
| M | moment | ft-lbf |
| N | normal force | lbf |
| p | pressure | $lbf/ft^2$ |
| P | load, or product of inertia | lbf, $ft^4$ |
| r | radius of gyration | ft |
| S | maximum cable sag | ft |
| T | tension, or temperature | lbf, °F |
| w | load per unit weight, or weight | lbf/ft, lbf |

*Symbols*

| | | |
|---|---|---|
| $\alpha$ | coefficient of linear thermal expansion | 1/°F |
| $\delta$ | deflection | ft |
| $\rho$ | density | $lbm/ft^3$ |

*Subscripts*

| | |
|---|---|
| c | centroidal, or concrete |
| i | the ith component |
| R | resultant |
| s | steel |

## 1 CONCENTRATED FORCES AND MOMENTS

Forces are vector quantities having magnitude, direction, and location in 3-dimensional space. The direction of a force **F** is given by its *direction cosines*, which are cosines of the true angles made by the force vector with the $x$, $y$, and $z$ axes. The components of the force are given by equations 11.1, 11.2, and 11.3.

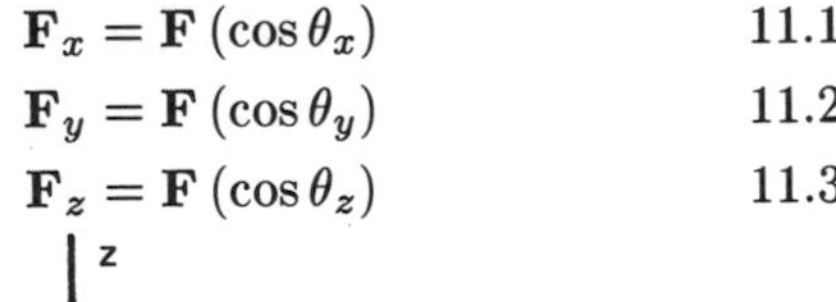

$$\mathbf{F}_x = \mathbf{F}\,(\cos\theta_x) \qquad 11.1$$

$$\mathbf{F}_y = \mathbf{F}\,(\cos\theta_y) \qquad 11.2$$

$$\mathbf{F}_z = \mathbf{F}\,(\cos\theta_z) \qquad 11.3$$

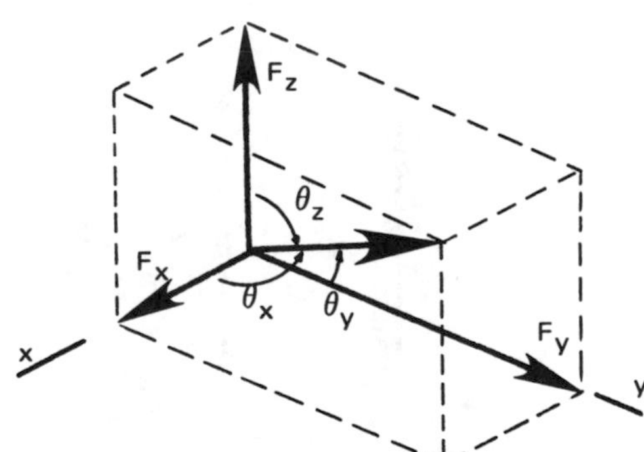


**Figure 11.1** Components of a Force **F**

A force which would cause an object to rotate is said to contribute a *moment* to the object. The magnitude of a moment can be found by multiplying the magnitude of the force times the appropriate moment arm. That is, **M**=**F**·d.

The *moment arm* is a perpendicular distance from the force's line of application to some arbitrary reference point. This reference point should be chosen to eliminate one or more unknowns. This can be done by choosing the reference as a point at which unknown reactions are applied.

Moments also can be treated as vector quantities, and they are shown as double-headed arrows. Using the *right-hand rule* as shown, the direction cosines again are used to give the $x$, $y$, and $z$ components of a moment vector.[1]

$$\mathbf{M}_x = \mathbf{M}(\cos\theta_x) \qquad 11.4$$

$$\mathbf{M}_y = \mathbf{M}(\cos\theta_y) \qquad 11.5$$

$$\mathbf{M}_z = \mathbf{M}(\cos\theta_z) \qquad 11.6$$

$$|\mathbf{M}| = \sqrt{\mathbf{M}_x^2 + \mathbf{M}_y^2 + \mathbf{M}_z^2} \qquad 11.7$$

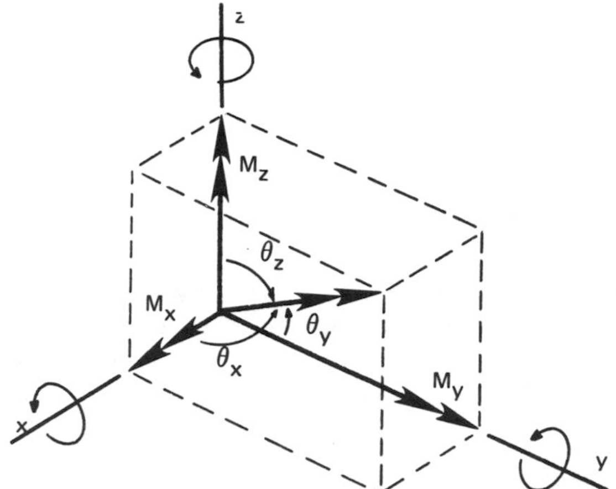


**Figure 11.2** Components of a Moment **M**

Moment vectors have the properties of magnitude and direction, but not of location (point of application). Moment vectors can be moved from one location to another without affecting the equilibrium of solid bodies.

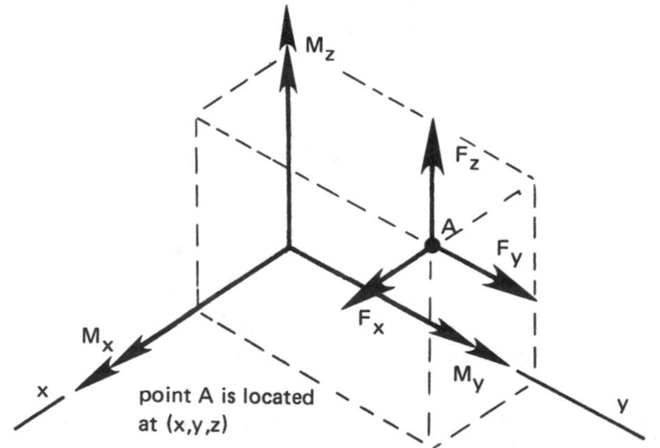


**Figure 11.3** Coordinates of a Point A

If a force is not parallel to an axis, it produces a moment around that axis. The moment is evaluated by finding the components of the force and their respective distances to the axis. In figure 11.3, a force acts through point A located at ($x$, $y$, $z$) and produces moments given by equations 11.8, 11.9, and 11.10.

$$\mathbf{M}_x = y\mathbf{F}_z - z\mathbf{F}_y \qquad 11.8$$

$$\mathbf{M}_y = z\mathbf{F}_x - x\mathbf{F}_z \qquad 11.9$$

$$\mathbf{M}_z = x\mathbf{F}_y - y\mathbf{F}_x \qquad 11.10$$

Any two equal, opposite, and parallel forces constitute a *couple.* A couple is statically equivalent to a single moment vector. In figure 11.4, the two forces, $\mathbf{F}_1$ and $\mathbf{F}_2$, of equal magnitude produce a moment vector $\mathbf{M}_z$ of magnitude **Fy**. The two forces can be replaced by this moment vector which then can be moved to any location on the object.

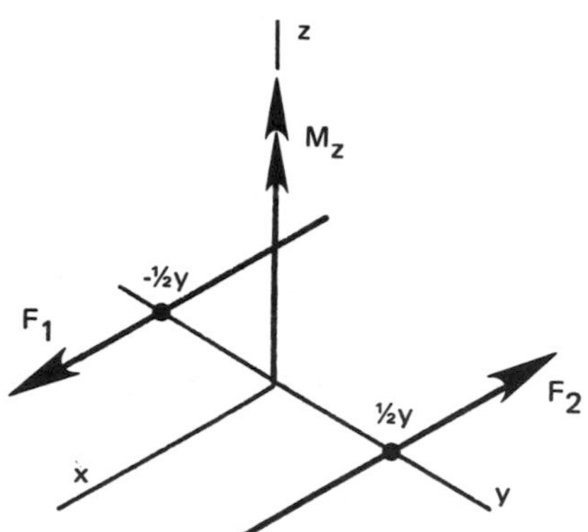


**Figure 11.4** A Couple

## 2 DISTRIBUTED LOADS

If an object is loaded by its own weight or by another type of continuous loading, it is said to be subjected to a *distributed load.* Provided that the load per unit length, $w$, is acting in the same direction everywhere, the statically equivalent concentrated load can be found from equation 11.11 by integrating over the line of application.

$$\mathbf{F}_R = \int w\,dx \qquad 11.11$$

The location of the resultant is given by equation 11.12.

$$\overline{x} = \frac{\int (wx)\,dx}{\mathbf{F}_R} \qquad 11.12$$

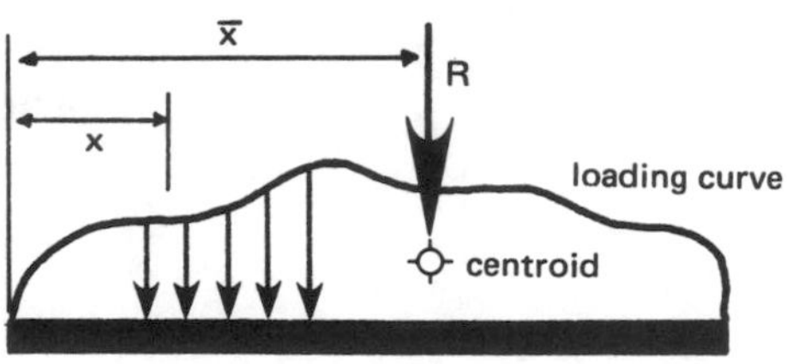


**Figure 11.5** A Distributed Load and Resultant

[1] The right-hand rule: Close your right hand in such a way that your fingers curl in the direction of the force (i.e., of rotation). Your thumb will point along the axis of the moment.

In the case of a straight beam under *transverse loading*, the magnitude of **F** equals the area under the loading curve.[2] The location of the resultant force coincides with the centroid of that area. If the distributed load is uniform so that $w$ is constant along the beam,

$$\mathbf{F} = wL \quad 11.13$$

$$\bar{x} = \tfrac{1}{2}L \quad 11.14$$

If the distribution is triangular and increases to a maximum of $w$ pounds per unit length as $x$ increases,

$$\mathbf{F} = \tfrac{1}{2}wL \quad 11.15$$

$$\bar{x} = \tfrac{2}{3}L \quad 11.16$$

*Example 11.1*

Find the magnitude and the location of the resultant of the distributed loads on each span of the beam shown.

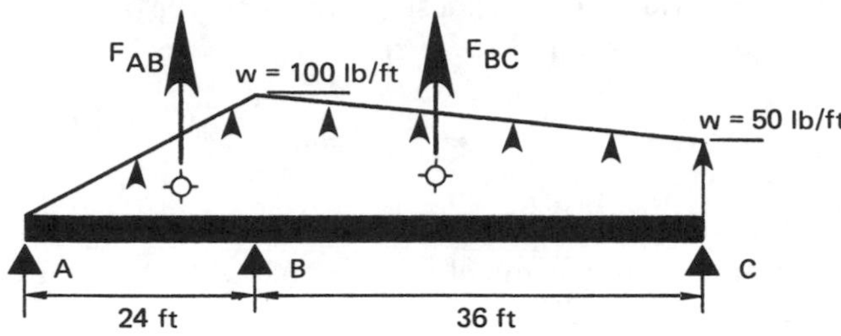


*For span A-B*: The area under the loading curve is $\frac{1}{2}(100)(24) = 1200$ pounds. The centroid of the loading triangle is $\left(\frac{2}{3}\right)(24) = 16$ feet from point $A$. Therefore, the triangular load on the span A-B can be replaced (for the purposes of statics) with a concentrated load of 1200 pounds located 16 feet from the left end.

*For span B-C*: The area under the loading curve is

$$(50)(36) + \tfrac{1}{2}(50)(36) = 2700 \text{ lbs}$$

The centroid of the trapezoid is

$$\frac{36\,[(2)(100) + 50]}{3\,(100 + 50)} = 20 \text{ ft from pt } C$$

Therefore, the distributed load on span B-C can be replaced (for the purposes of statics) with a concentrated load of 2700 pounds located 20 feet to the left of point $C$.

[2] Loading is said to be *transverse* if its line of action is perpendicular to the length of the beam.

## 3 PRESSURE LOADS

Hydrostatic pressure is an example of a pressure load that is distributed over an area. The pressure is denoted as $p$ pounds per unit area of surface. It is normal to the surface at every point. If the surface is plane, the statically equivalent concentrated load can be found by integrating over the area. The resultant is numerically equal to the average pressure times the area. The point of application will be the centroid of the area over which the integration was performed.

## 4 RESOLUTION OF FORCES AND MOMENTS

Any system (collection) of forces and moments is statically equivalent to a single resultant force vector plus a single resultant moment vector in 3-dimensional space. Either or both of these resultants may be zero.

The $x$-component of the resultant force is the sum of all the $x$-components of the individual forces, and similarly for the $y$- and the $z$-components of the resultant force.

$$\mathbf{F}_{Rx} = \sum \mathbf{F}_i(\cos\theta_{x,i}) \quad 11.17$$

$$\mathbf{F}_{Ry} = \sum \mathbf{F}_i(\cos\theta_{y,i}) \quad 11.18$$

$$\mathbf{F}_{Rz} = \sum \mathbf{F}_i(\cos\theta_{z,i}) \quad 11.19$$

The determination of the resultant moment vector is more complex. The resultant moment vector includes the moments of all system forces around the reference axes plus the components of all system moments.

$$\mathbf{M}_{Rx} = \sum(y_i\mathbf{F}_{z,i} - z_i\mathbf{F}_{y,i}) + \sum \mathbf{M}_i(\cos\theta_{x,i}) \quad 11.20$$

$$\mathbf{M}_{Ry} = \sum(z_i\mathbf{F}_{x,i} - x_i\mathbf{F}_{z,i}) + \sum \mathbf{M}_i(\cos\theta_{y,i}) \quad 11.21$$

$$\mathbf{M}_{Rz} = \sum(x_i\mathbf{F}_{y,i} - y_i\mathbf{F}_{x,i}) + \sum \mathbf{M}_i(\cos\theta_{z,i}) \quad 11.22$$

## 5 CONDITIONS OF EQUILIBRIUM

An object which is not moving is said to be static. All forces on a static object are in equilibrium. For an object to be in equilibrium, it is necessary that the resultant force vector and the resultant moment vectors be equal to zero.

$$\mathbf{F}_R = \sqrt{\mathbf{F}_{Rx}^2 + \mathbf{F}_{Ry}^2 + \mathbf{F}_{Rz}^2} = 0 \quad 11.23$$

$$\mathbf{M}_R = \sqrt{\mathbf{M}_{Rx}^2 + \mathbf{M}_{Ry}^2 + \mathbf{M}_{Rz}^2} = 0 \quad 11.24$$

Since the square of any quantity cannot be negative, equations 11.25 through 11.30 follow directly from equations 11.23 and 11.24.

$$\mathbf{F}_{Rx} = \sum \mathbf{F}_x = 0 \qquad 11.25$$

$$\mathbf{F}_{Ry} = \sum \mathbf{F}_y = 0 \qquad 11.26$$

$$\mathbf{F}_{Rz} = \sum \mathbf{F}_z = 0 \qquad 11.27$$

$$\mathbf{M}_{Rx} = \sum \mathbf{M}_x = 0 \qquad 11.28$$

$$\mathbf{M}_{Ry} = \sum \mathbf{M}_y = 0 \qquad 11.29$$

$$\mathbf{M}_{Rz} = \sum \mathbf{M}_z = 0 \qquad 11.30$$



## 6 FREE-BODY DIAGRAMS

A *free-body diagram* is a representation of an object in equilibrium, showing all external forces, moments, and support reactions. Since the object is in equilibrium, the resultant of all forces and moments on the free-body is zero.

If any part of the object is removed and replaced by the forces and moments which are exerted on the cut surface, a free-body of the remaining structure is obtained, and the conditions of equilibrium will be satisfied by the new free-body.

By dividing the object into a sufficient number of free-bodies, the internal forces and moments can be found at all points of interest, providing that the conditions of equilibrium are sufficient to give a static solution.

## 7 REACTIONS

A typical first step in solving statics problems is to determine the supporting reaction forces. The manner in which the structure is supported will determine the type, the location, and the direction of the reactions. Conventional symbols can be used to define the type of reactions which occur at each point of support. Some examples are shown in table 11.1.

*Example 11.2*

Find the reactions $\mathbf{R}_1$ and $\mathbf{R}_2$.

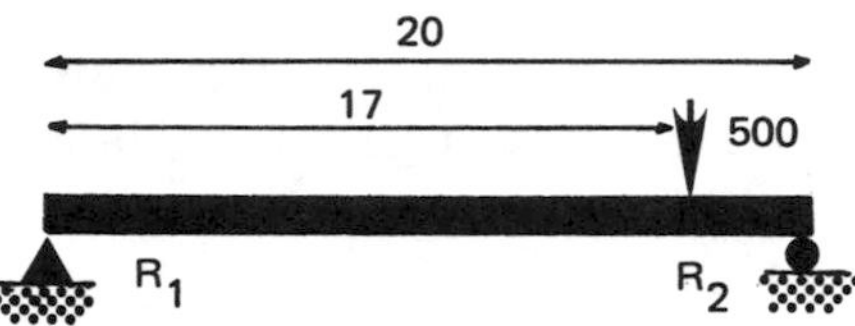


Since the left support is a simple support, its reaction can have any direction. $\mathbf{R}_1$ can, therefore, be written in terms of its $x$ and $y$ components, $\mathbf{R}_{1,x}$ and $\mathbf{R}_{1,y}$, respectively. $\mathbf{R}_2$ is a roller support which cannot sustain an $x$ component.

From equation 11.25, choosing forces to the right as positive, $\mathbf{R}_{1,x}$ is found to be zero.

$$\sum \mathbf{F}_x = \mathbf{R}_{1,x} = 0$$

Equation 11.26 can be used to obtain a relationship between the $y$ components of force. Forces acting upward are considered positive.

$$\sum \mathbf{F}_y = \mathbf{R}_{1,y} + \mathbf{R}_2 - 500 = 0$$

Since both $\mathbf{R}_{1,y}$ and $\mathbf{R}_2$ are unknown, a second equation is needed. Equation 11.30 is used. The reference point is chosen as the left end to make the moment arm for $\mathbf{R}_{1,y}$ equal to zero. This eliminates $\mathbf{R}_{1,y}$ as an unknown, allowing $\mathbf{R}_2$ to be found directly. Clockwise moments are considered positive.

$$\sum \mathbf{M}_{\text{left end}} = (500)(17) - (\mathbf{R}_2)(20) = 0$$

$$\mathbf{R}_2 = 425$$

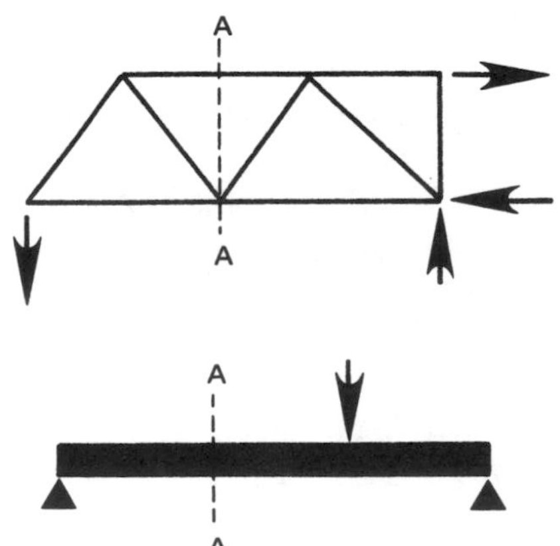


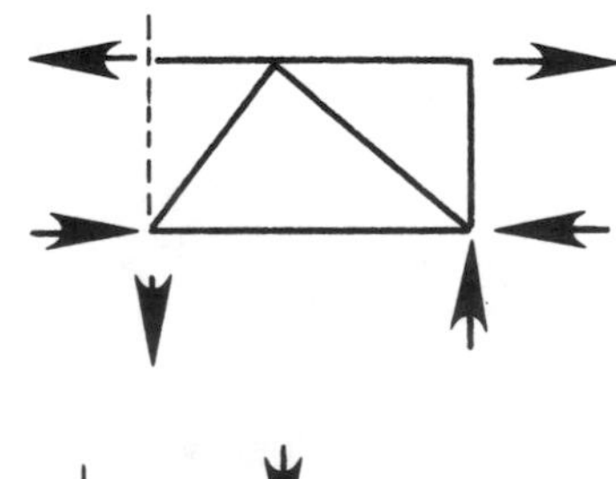

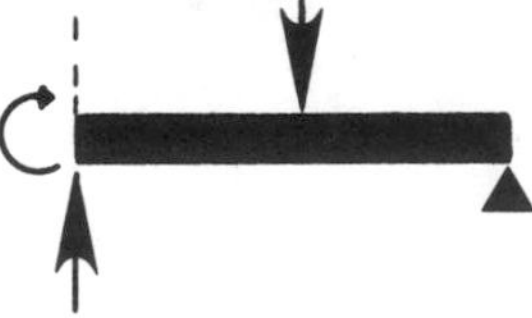

**Figure 11.6** Original and Cut Free-bodies

**Table 11.1**
Common Support Symbols

| type of support | symbol | characteristics |
|---|---|---|
| Built-in | | Moments and forces in any direction |
| Simple | | Load in any direction; no moment |
| Roller | | Load normal to surface only; no moment |
| Cable | | Load in cable direction; no moment |
| Guide | | No load or moment in guide direction |
| Hinge | | Load in any direction; no moment |



Once $\mathbf{R}_2$ is known, $\mathbf{R}_{1,y}$ is found easily from equation 11.26.

$$\sum \mathbf{F}_y = \mathbf{R}_{1,y} + 425 - 500 = 0$$

$$\mathbf{R}_{1,y} = 75$$

## 8 INFLUENCE LINES FOR REACTIONS

An *influence line* (*influence graph*) is an $x$-$y$ plot of the magnitude of a reaction (any reaction on the object) as it would vary as the load is placed at different points on the object. The x-axis corresponds to the location along the object (as along the length of a beam); the y-axis corresponds to the magnitude of the reaction. For uniformity, the load is taken as 1 unit. Therefore, for an actual load of $\mathbf{P}$ units, the actual reaction would be given by equation 11.31.

$$\text{actual reaction} = \mathbf{P}\left[\begin{array}{c}\text{influence graph}\\ \text{ordinate}\end{array}\right] \qquad 11.31$$

*Example 11.3*

Draw the influence graphs of the left and right reactions for the beam shown.

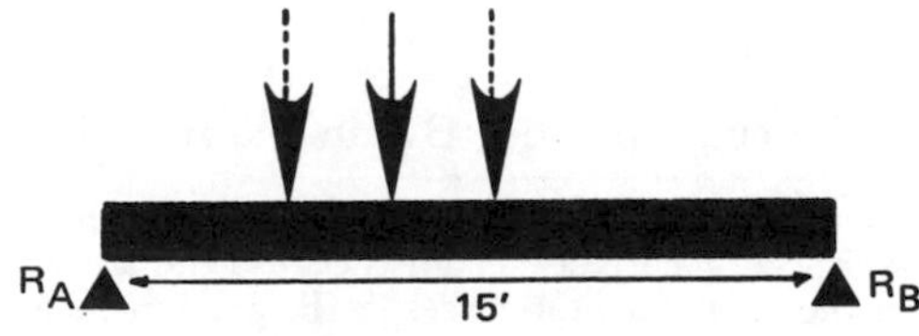


If a unit load is at the left end, reaction $\mathbf{R}_A$ will be equal to 1. If the unit load is at the right end, it will be supported entirely by $\mathbf{R}_B$, so $\mathbf{R}_A$ will be zero. The influence line for $\mathbf{R}_A$ is

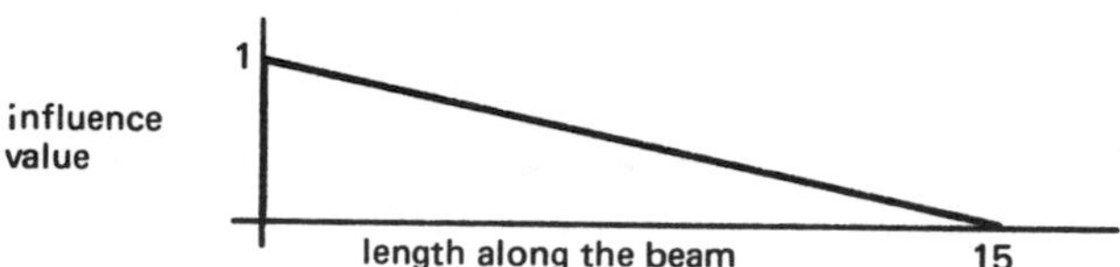


The influence line for $\mathbf{R}_B$ is found similarly.

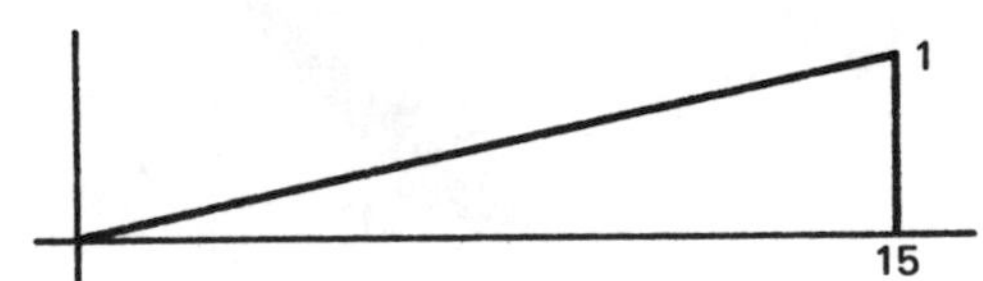


## 9 AXIAL MEMBERS

A member which is in equilibrium when acted upon by forces at each end and by no other forces or moments is an *axial member*. For equilibrium to exist, the resultant forces at the ends must be equal, opposite, and collinear. In an actual truss, this type of loading can be approached through the use of frictionless bearings or pins at the ends of the axial members. In simple truss analysis, the members are assumed to be axial members, regardless of the end conditions.

A typical inclined axial member is illustrated in figure 11.7. For that member to be in equilibrium, the following equations must hold.

$$\mathbf{F}_{Rx} = \mathbf{F}_{Bx} - \mathbf{F}_{Ax} = 0 \quad 11.32$$

$$\mathbf{F}_{Ry} = \mathbf{F}_{By} - \mathbf{F}_{Ay} = 0 \quad 11.33$$

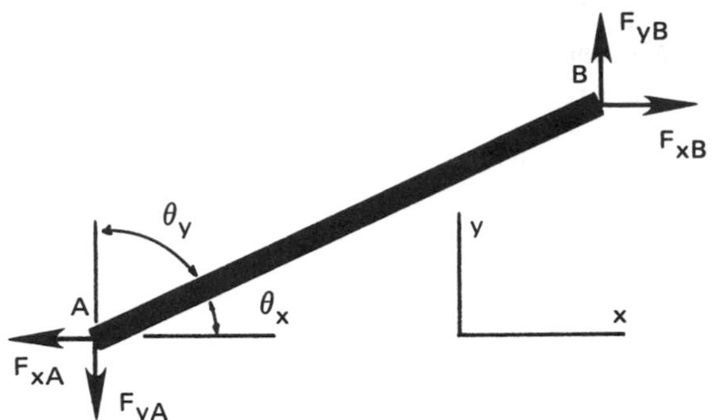


**Figure 11.7** An Axial Member

The resultant force, $\mathbf{F}_R$, can be derived from the components by trigonometry and direction cosines.

$$\mathbf{F}_{Rx} = \mathbf{F}_R \cos\theta_x \quad 11.34$$

$$\mathbf{F}_{Ry} = \mathbf{F}_R \cos\theta_y = \mathbf{F}_R \sin\theta_x \quad 11.35$$

If, however, the geometry of the axial member is known, similar triangles can be used to find the resultant and/or the components. This is illustrated in example 11.4.

*Example 11.4*

A 12′ long axial member carrying an internal load of 180 pounds is inclined as shown. What are the $x$- and $y$-components of the load?

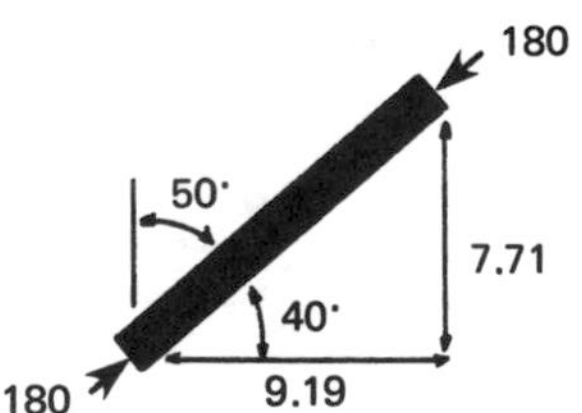


*method 1: Direction Cosines*

$$\mathbf{F}_x = 180\,(\cos\ 40°) = 137.9$$

$$\mathbf{F}_y = 180\,(\cos\ 50°) = 115.7$$

*method 2: Similar Triangles*

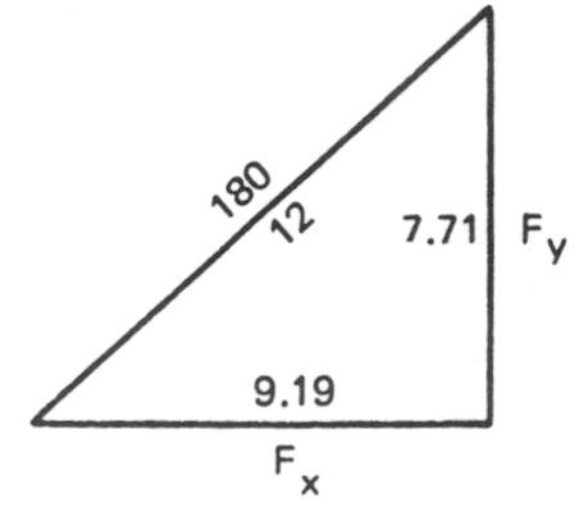


$$\mathbf{F}_x = \left(\frac{9.19}{12}\right)(180) = 137.9$$

$$\mathbf{F}_y = \left(\frac{7.71}{12}\right)(180) = 115.7$$

## 10 TRUSSES

This discussion is directed toward 2-dimensional trusses. The loads in truss members are represented by arrows pulling away from the joints for tension, and by arrows pushing toward the joints for compression.[3]

The equations of equilibrium can be used to find the external reactions on a truss. To find the internal resultants in each axial member, three methods can be used. These methods are *method of joints*, *cut-and-sum*, and *method of sections.*

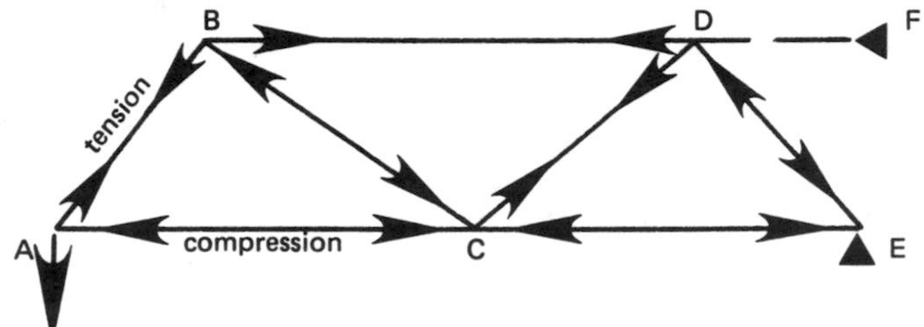


**Figure 11.8** Truss Notation

All joints in a truss will be *determinate*, and all member loads can be found if equation 11.36 holds.

$$\#\ \text{truss members} = 2\,(\#\ \text{joints}) - 3 \quad 11.36$$

If the left-hand side is greater than the right-hand side, indeterminate methods must be used to solve the truss. If the left-hand side is less than the right-hand side, the truss is not rigid and will collapse under certain types of loading.

### A. METHOD OF JOINTS

The *method of joints* is a direct application of equations 11.25 and 11.26. The sums of forces in the $x$- and $y$-directions are taken at consecutive joints in the truss. At each joint, there may be up to two unknown axial forces, each of which may have two components. Since there are two equations of equilibrium, a joint with two unknown forces will be determinate.

*Example 11.5*

Find the force in member **BD** in the truss shown.

[3] The method of showing tension and compression on a truss drawing appears incorrect. This is because the arrows show the forces on the joints, not the forces in the axial members.

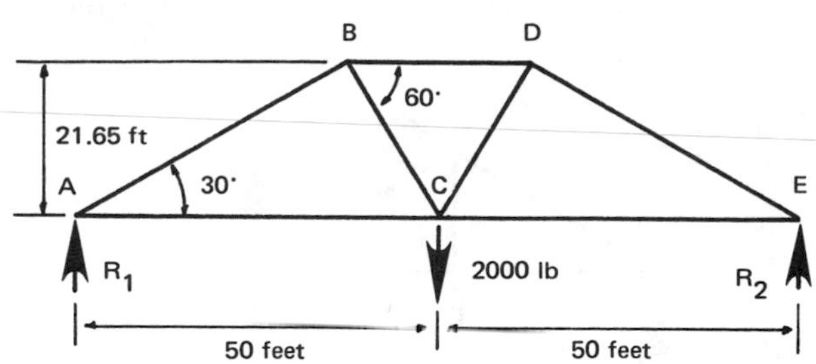


*step 1*: Find the reactions $\mathbf{R}_1$ and $\mathbf{R}_2$. From equation 11.29, the sum of moments must be zero. Taking moments (counterclockwise as positive) about point A gives $\mathbf{R}_2$.

$$\sum \mathbf{M}_A = 100\,(\mathbf{R}_2) - 2000\,(50) = 0$$
$$\mathbf{R}_2 = 1000$$

From equation 11.26, the sum of the forces (vertical positive) in the $y$ direction must be zero.

$$\sum \mathbf{F}_y = \mathbf{R}_1 - 2000 + 1000 = 0$$
$$\mathbf{R}_1 = 1000$$

*step 2*: Although we want the force in member **BD**, there are three unknowns at joints B and D. Therefore, start with joint A where there are only two unknowns (**AB** and **AC**). The free-body of joint A is shown below. The direction of $\mathbf{R}_1$ is known. However, the directions of the member forces usually are not known and need to be found by inspection or assumption. If an incorrect direction is assumed, the force will show up with a negative sign in later calculations.

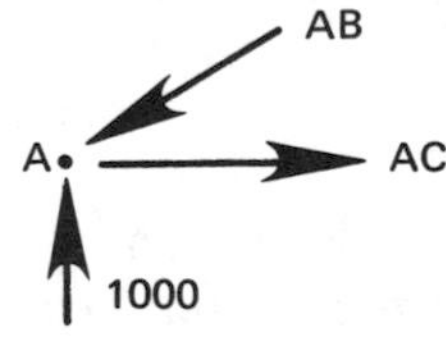


*step 3*: Resolve all inclined forces on joint A into horizontal and vertical components using trigonometry or similar triangles. $\mathbf{R}_1$ and **AC** already are parallel to the $y$ and $x$ axes, respectively. Only **AB** needs to be resolved into components. By observation, it is clear that $\mathbf{AB}_y = 1000$. If this were not true, equation 11.26 would not hold.

$$\mathbf{AB}_y = \mathbf{AB}\,(\sin\ 30°)$$
$$1000 = \mathbf{AB}\,(0.5) \text{ or } \mathbf{AB} = 2000$$
$$\mathbf{AB}_x = \mathbf{AB}\,(\cos\ 30°) = 1732$$

*step 4*: Draw the free-body diagram of joint B. Notice that the direction of force **AB** is toward the joint, just as it was for joint A. The direction of load **BC** is chosen to counteract the vertical component of load **AB**. The direction of load **BD** is chosen to counteract the horizontal components of loads **AB** and **BC**.

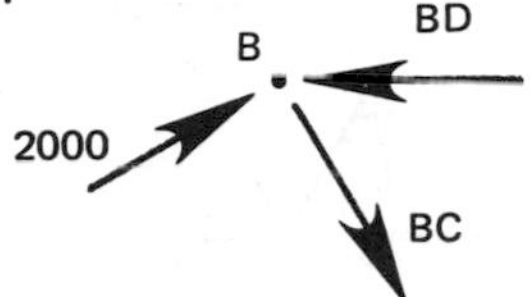


*step 5*: Resolve all inclined forces into horizontal and vertical components.

$$\mathbf{AB}_x = 1732$$
$$\mathbf{AB}_y = 1000$$
$$\mathbf{BC}_x = \mathbf{BC}\,(\sin\ 30°) = 0.5\mathbf{BC}$$
$$\mathbf{BC}_y = \mathbf{BC}\,(\cos\ 30°) = 0.866\mathbf{BC}$$

*step 6*: Write the equations of equilibrium for joint B.

$$\sum \mathbf{F}_x = 1732 + 0.5\mathbf{BC} - \mathbf{BD} = 0$$
$$\sum \mathbf{F}_y = 1000 - 0.866\mathbf{BC} = 0$$

**BC** from the second equation is found to be 1155. Substituting 1155 into the first equilibrium condition equation gives

$$1732 + 0.5\,(1155) - \mathbf{BD} = 0$$
$$\mathbf{BD} = 2310$$

Since **BD** turned out to be positive, its direction was chosen correctly. The direction of the arrow indicates that the member is compressing the pin joint. Consequently, the pin is compressing the member, and member **BD** is in compression.

B. CUT-AND-SUM METHOD

The *cut-and-sum method* can be used if a load in an inclined member in the middle of a truss is wanted. The method is strictly an application of the equilibrium condition requiring the sum of forces in the vertical direction to be zero. The method is illustrated in example 11.6.

*Example 11.6*

Find the force in member **BC** for the truss shown in example 11.5.

*step 1*: Find the external reactions. This is the same step as in example 11.5. $\mathbf{R}_1 = \mathbf{R}_2 = 1000$.

*step 2*: Cut the truss through, making sure that the cut goes through only one member with a vertical component. In this case, that member is **BC**.

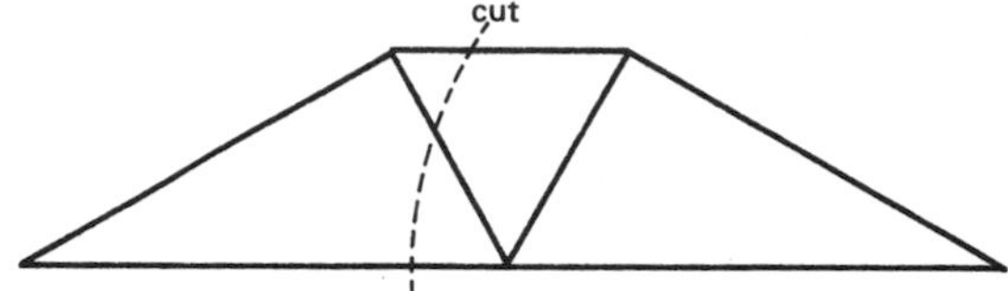


*step 3*: Draw the free-body of either part of the remaining truss.

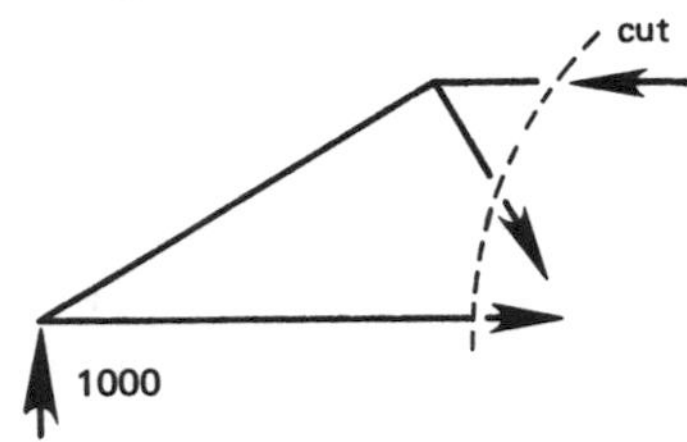


*step 4*: Resolve the unknown inclined force into vertical and horizontal components.

$$\mathbf{BC}_x = 0.5\,(\mathbf{BC})$$
$$\mathbf{BC}_y = 0.866\,(\mathbf{BC})$$

*step 5*: Sum forces in the $y$ direction for the entire free-body.

$$\sum \mathbf{F}_y = 1000 - 0.866\ (\mathbf{BC}) = 0$$
$$\mathbf{BC} = 1155$$

C. METHOD OF SECTIONS

The cut-and-sum method will work only if it is possible to cut the truss without going through two members with vertical components.

The *method of sections* is a direct approach for finding member loads at any point in a truss. In this method, the truss is cut at an appropriate section, and the conditions of equilibrium are applied to the resulting free-body. This is illustrated in example 11.7.

*Example 11.7*

For the truss shown, find the load in members **CE** and **CD**.

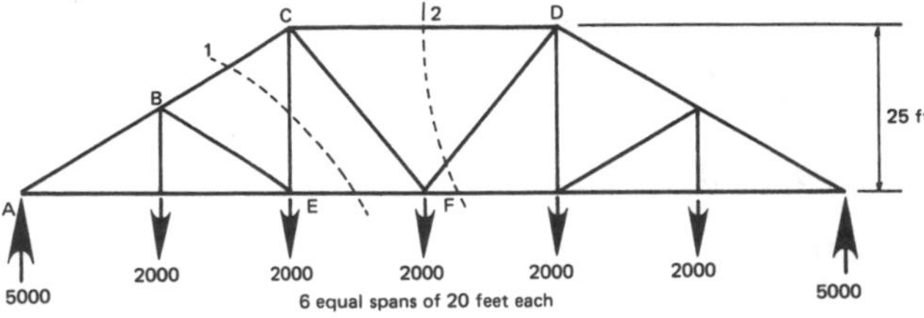


For member force **CE**, the truss is cut at section 1.

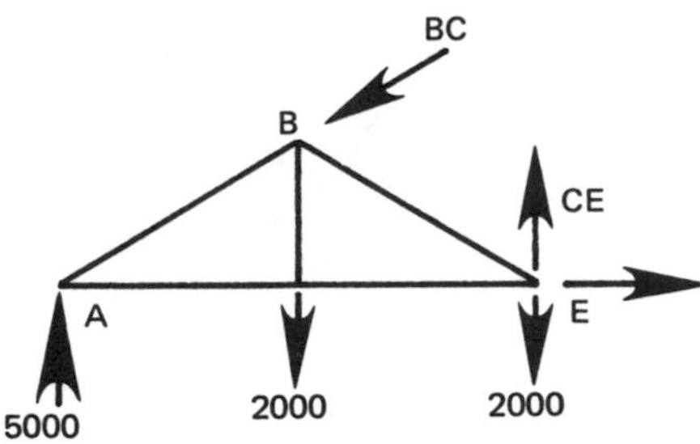


Taking moments about A will eliminate all unknowns except force **CE**.

$$\sum \mathbf{M}_A = \mathbf{CE}\,(40) - 2000\,(20) - 2000\,(40) = 0$$
$$\mathbf{CE} = 3000$$

For member **CD**, the truss is cut at section 2. Taking moments about point F will eliminate all unknowns except **CD**.

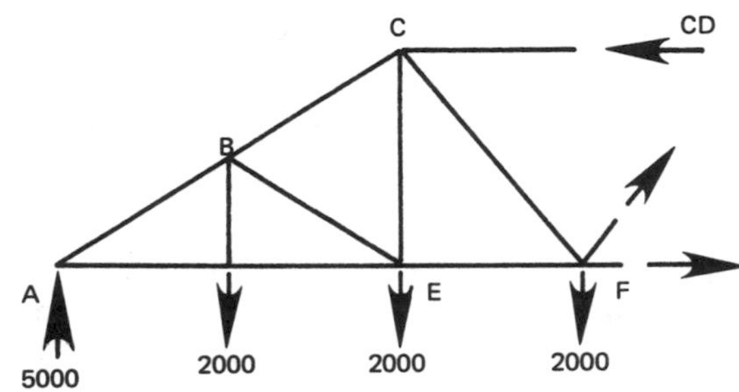


$$\sum \mathbf{M}_F = \mathbf{CD}(25) + 2000(20) + 2000(40) - 5000(60) = 0$$
$$\mathbf{CD} = 7200$$

## 11 SUPERPOSITION OF LOADINGS

For any group of forces and moments which satisfies the conditions of equilibrium, the resultant force and moment vectors are zero. The resultant of two zero vectors is another zero vector. Therefore, any number of such equilibrium systems can be combined without disturbing the equilibrium.

Superposition methods must be used with discretion in working with actual structures since some structures change shape significantly under load. If the actual structure were to deflect so that the points of application of loads were quite different from those in the undeflected structure, superposition would not be applicable.

In simple truss analysis, the change of shape under load is neglected when finding the member loads. Superposition, therefore, can be assumed to apply.

## 12 CABLES

### A. CABLES UNDER CONCENTRATED LOADS

An ideal cable is assumed to be completely flexible. It acts as an axial member in tension between any two points of concentrated load application.

The method of joints and sections used in truss analysis applies equally well to cables under concentrated loads. However, no compression members will be found. As in truss analysis, if the cable loads are unknown, some information concerning the geometry of the cable must be known in order to solve for the axial tension in the segments.

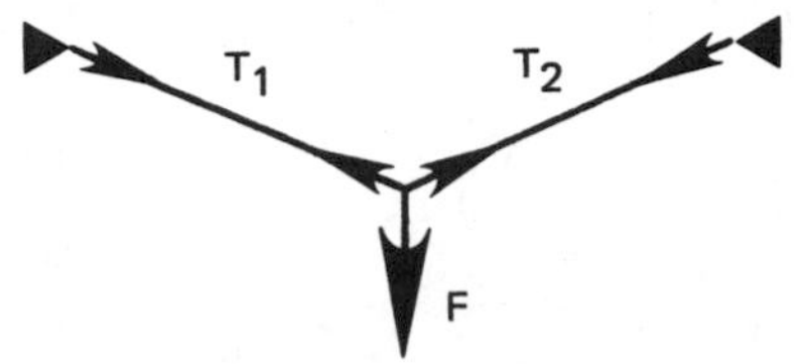


**Figure 11.9** Cable Under Transverse Loading

*Example 11.8*

Find the tension $\mathbf{T}_2$ between points B and C.

*step 1*: Take moments about point A to find $\mathbf{T}_3$.

$$\sum \mathbf{M}_A = a\mathbf{F}_1 + b\mathbf{F}_2 + d\mathbf{T}_3 \cos\theta_3 - c\mathbf{T}_3 \sin\theta_3 = 0$$

*step 2*: Sum forces in the $x$ direction to find $\mathbf{T}_1$.

$$\sum \mathbf{F}_x = \mathbf{T}_1 \cos\theta_1 - \mathbf{T}_3 \cos\theta_3 = 0$$

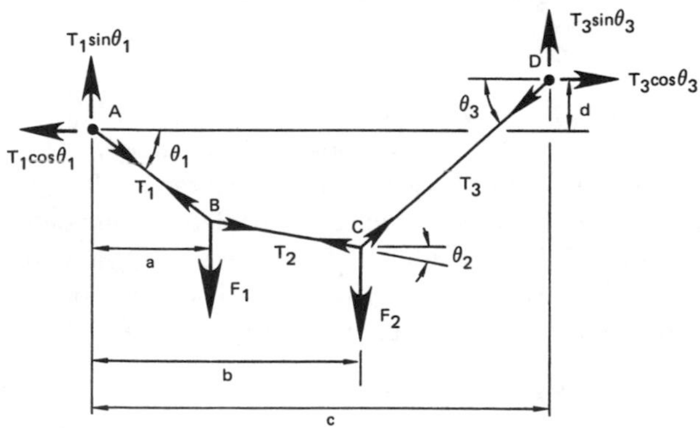


*step 3*: Sum forces in the $x$ direction at point B to find $\mathbf{T}_2$.

$$\sum \mathbf{F}_x = \mathbf{T}_2 \cos\theta_2 - \mathbf{T}_1 \cos\theta_1 = 0$$

### B. CABLES UNDER DISTRIBUTED LOADS

An idealized tension cable under a distributed load is similar to a linkage made up of a very large number of axial members. The cable is an axial member in the sense that the internal tension acts in a direction which is along the centerline of the cable everywhere.

Figure 11.10 illustrates a cable under a unidirectionally distributed load. A free-body diagram of segment B-C of the cable also is known. $\mathbf{F}$ is the vertical resultant of the distributed load on the segment.

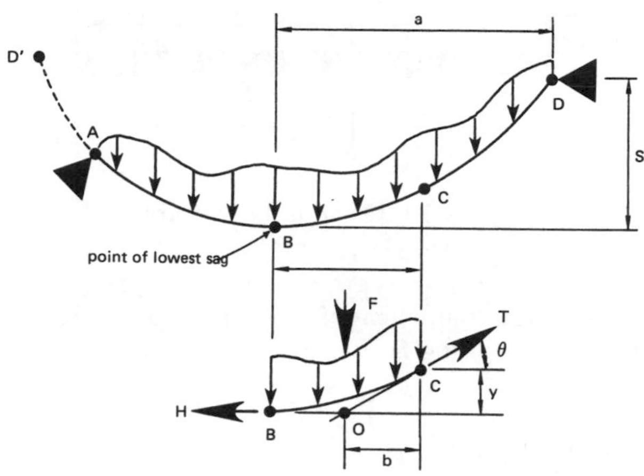


**Figure 11.10** Cable Under Distributed Load

$\mathbf{T}$ is the cable tension at point C, and $\mathbf{H}$ is the cable tension at the point of lowest sag. From the conditions of equilibrium for free-body B-C, it is apparent that the three forces, $\mathbf{H}$, $\mathbf{F}$, and $\mathbf{T}$, must be concurrent at point O. Taking moments about point C, the following equations are obtained.

$$\sum \mathbf{M}_C = \mathbf{F}b - \mathbf{H}y = 0 \qquad 11.37$$

$$\mathbf{H} = \frac{\mathbf{F}b}{y} \qquad 11.38$$

But $\tan\theta = \left(\frac{y}{b}\right)$. So,

$$\mathbf{H} = \frac{\mathbf{F}}{\tan\theta} \qquad 11.39$$

From the summation of forces in the vertical and horizontal directions,

$$\mathbf{T}\cos\theta = \mathbf{H} \qquad 11.40$$

$$\mathbf{T}\sin\theta = \mathbf{F} \qquad 11.41$$

$$\mathbf{T} = \sqrt{\mathbf{H}^2 + \mathbf{F}^2} \qquad 11.42$$

The shape of the cable is a function of the relative amount of sag at point B and the relative distribution (not the absolute magnitude) of the applied running load.

## C. PARABOLIC CABLES

If the distribution load per unit length, $w$, is constant with respect to a horizontal line (as is the load from a bridge floor), the cable will be parabolic in shape. This is illustrated in figure 11.11.

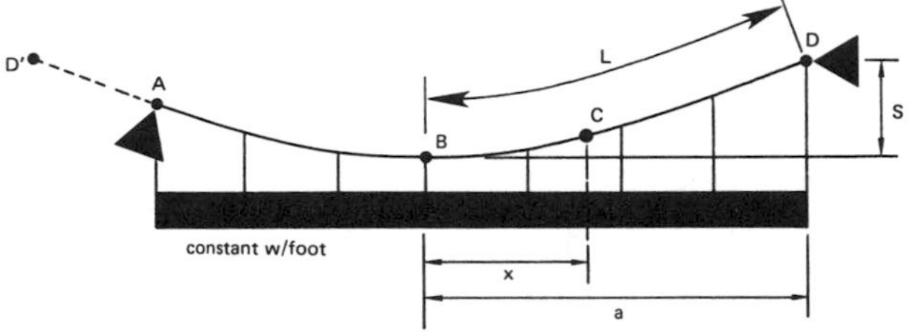


**Figure 11.11** Parabolic Cable



The horizontal component of tension can be found from equation 11.38 using **F**=wa, b= $\frac{1}{2}$a, and y=S.

$$\mathbf{H} = \frac{\mathbf{F}b}{y} = \frac{wa^2}{2S} \qquad 11.43$$

$$\mathbf{T} = \sqrt{\mathbf{H}^2 + \mathbf{F}^2} = \sqrt{\left(\frac{wa^2}{2S}\right)^2 + (wx)^2} \qquad 11.44$$

$$= w\sqrt{x^2 + \left(\frac{a^2}{2S}\right)^2} \qquad 11.45$$

The shape of the cable is given by equation 11.46.

$$y = \frac{wx^2}{2\mathbf{H}} \qquad 11.46$$

The approximate length of the cable from the lowest point to the support is given by equation 11.47.

$$L \approx (a)\left[1 + \tfrac{2}{3}\left(\frac{S}{a}\right)^2 - \tfrac{2}{5}\left(\frac{S}{a}\right)^4\right] \qquad 11.47$$

*Example 11.9*

A pedestrian bridge has two suspension cables and a flexible floor. The floor weighs 28 pounds per foot. The span of the bridge is 100 feet between the two end supports. When the bridge is empty, the tension at point A is 1500 pounds. What is the cable sag, S, at the center? What is the approximate cable length?

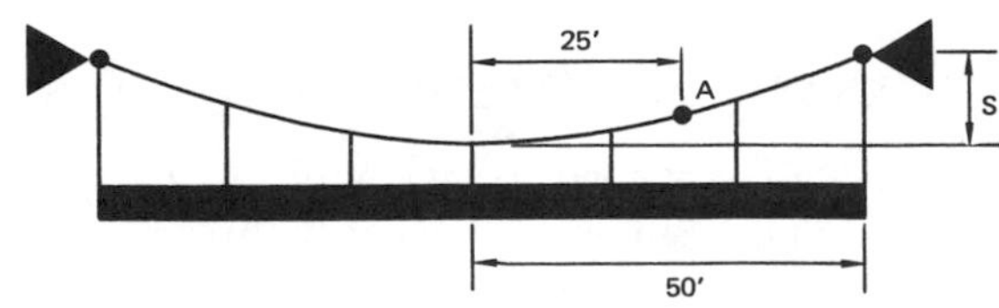


The floor weight per cable is 28/2= 14 lb/ft. From equation 11.45,

$$1500 = 14\sqrt{[25]^2 + \left[\frac{(50)^2}{2S}\right]^2}$$

$$S = 12 \text{ feet}$$

From equation 11.47,

$$L = 50\left[1 + \left(\frac{2}{3}\right)\left(\frac{12}{50}\right)^2 - \left(\frac{2}{5}\right)\left(\frac{12}{50}\right)^4\right]$$

$$= 51.9 \text{ feet}$$

The cable length is $2 \times 51.9 = 103.8$ ft.

## D. THE CATENARY

If the distributed load, $w$, is constant along the length of the cable (as in the case of a cable loaded by its own weight), the cable will have the shape of a *catenary.* This is illustrated in figure 11.12.

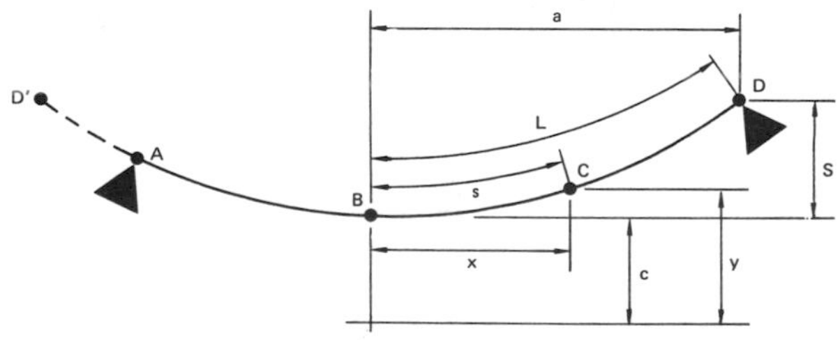


**Figure 11.12** A Catenary

As shown in figure 11.12, $y$ is measured from a reference plane located a distance $c$ below the lowest point of the cable, point B. The location of this reference plane is a parameter of the cable which must be determined before equations 11.48 through 11.53 are used. The value of $c$ does not correspond to any physical distance, nor does the reference plane correspond to the ground.

The equations of the catenary are presented below. Some judgment usually is necessary to determine which equations should be used and in which order. To define cable shape, it is necessary to have some initial information which can be entered into the equations. For example, if $a$ and the sag $S$ are given, equation 11.51 can be solved by trial and error to obtain $c$. Once $c$

is known, the cable geometry and forces are defined by the remaining equations.

$$y = c\left[\cosh\left(\frac{x}{c}\right)\right] \quad 11.48$$

$$s = c\left[\sinh\left(\frac{x}{c}\right)\right] \quad 11.49$$

$$y = \sqrt{s^2 + c^2} \quad 11.50$$

$$S = c\left[\cosh\left(\frac{a}{c}\right) - 1\right] \quad 11.51$$

$$\tan\theta = \frac{s}{c} \quad 11.52$$

$$\mathbf{H} = wc \quad 11.53$$

$$\mathbf{F} = ws \quad 11.54$$

$$\mathbf{T} = wy \quad 11.55$$

$$\tan\theta = \frac{ws}{\mathbf{H}} \quad 11.56$$

$$\cos\theta = \frac{\mathbf{H}}{\mathbf{T}} \quad 11.57$$

*Example 11.10*

A cable 100′ long is loaded by its own weight. The sag is 25′, and the supports are on the same level. What is the distance between the supports?

From equation 11.50 at point $D$, with $S = 25$,

$$c + S = \sqrt{s^2 + c^2}$$

$$c + 25 = \sqrt{(50)^2 + c^2}$$

$$c = 37.5$$

From equation 11.49,

$$50 = 37.5\left[\sinh\left(\frac{a}{37.5}\right)\right]$$

$$a = 41.2 \text{ feet}$$

The distance between supports is

$$(2a) = (2)(41.2) = 82.4 \text{ feet}$$

Providing that the lowest point, B, is known or can be found, the location of the cable supports at different levels does not significantly affect the analysis of cables. The same procedure is used in proceeding from point B to either support. In fact, once the theoretical shape of the cable has been determined, the supports can be relocated anywhere along the cable line without affecting the equilibrium of the supporting segment.

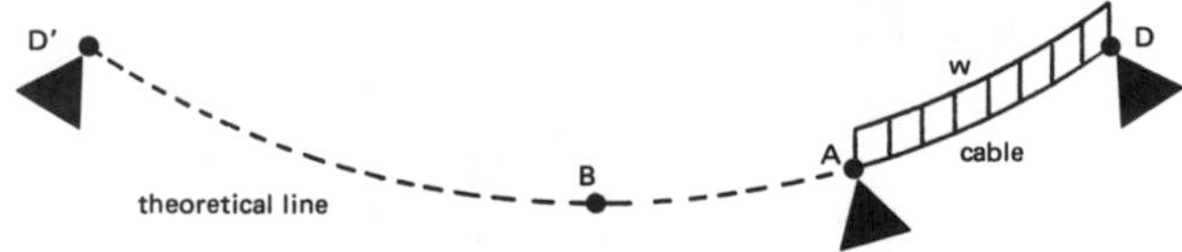


**Figure 11.13** Non-Symmetrical Segment of Symmetrical Cable

## 13 3-DIMENSIONAL STRUCTURES

The static analysis of 3-dimensional structures usually requires the following steps.

*step 1*: Determine the components of all loads and reactions. This usually is accomplished by finding the $x$, $y$, and $z$ coordinates of all points and then using direction cosines.

*step 2*: Draw three free-bodies of the structure—one each for the $x$, $y$, and $z$ components of loads and reactions.

*step 3*: Solve for unknowns using $\sum\mathbf{F} = 0$ and $\sum\mathbf{M} = 0$.

*Example 11.11*

Beam ABC is supported by the two cables as shown. The connection at A is pinned (hinged). Find the cable tensions $\mathbf{T}_1$ and $\mathbf{T}_2$.

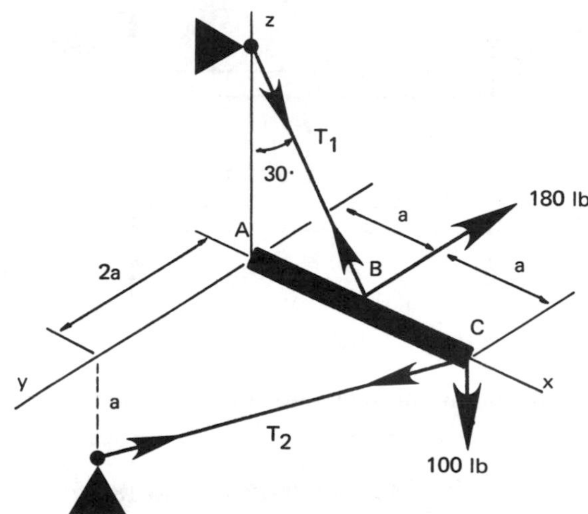


*step 1*: For the 180 pound load,

$$\mathbf{F}_x = 0$$

$$\mathbf{F}_y = -180$$

$$\mathbf{F}_z = 0$$

For the 100 pound load,

$$\mathbf{F}_x = 0$$

$$\mathbf{F}_y = 0$$

$$\mathbf{F}_z = -100$$

For cable 1:

$$\cos\theta_x = \cos\ 120° = -0.5$$

$$\cos\theta_y = 0$$

$$\cos\theta_z = \cos\ 30° = 0.866$$

$$\mathbf{T}_{1x} = -0.5\mathbf{T}_1$$

$$\mathbf{T}_{1y} = 0$$

$$\mathbf{T}_{1z} = 0.866\mathbf{T}_1$$

For cable 2: The length of the cable is

$$L = \sqrt{(2a)^2 + (-2a)^2 + (-a)^2} = 3a$$

$$\cos\theta_x = \frac{-2a}{3a} = -0.667$$

$$\cos\theta_y = \frac{2a}{3a} = 0.667$$

$$\cos\theta_z = \frac{-a}{3a} = -0.333$$

$$\mathbf{T}_{2x} = -0.667\mathbf{T}_2$$

$$\mathbf{T}_{2y} = 0.667\mathbf{T}_2$$

$$\mathbf{T}_{2z} = -0.333\mathbf{T}_2$$

*step 2*:

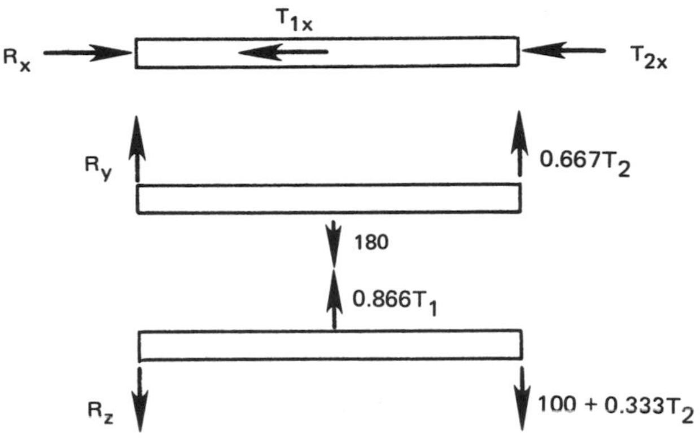


*step 3*: Summing moments about point A for the $y$ case gives $\mathbf{T}_2$.

$$\sum \mathbf{M}_{Az} = 0.667\mathbf{T}_2(2a) - 180(a) = 0$$

$$\mathbf{T}_2 = 135$$

Summing moments about point A for the z case gives $\mathbf{T}_1$.

$$\sum \mathbf{M}_{Ay} = 0.866\mathbf{T}_1(a) - 0.333(135)(2a) - 100(2a) = 0$$

$$\mathbf{T}_1 = 335$$

## 14 GENERAL TRIPOD SOLUTION

The procedure given in the preceding section will work with a tripod consisting of three axial pin-ended members with a load in any direction applied at the apex. However, the tripod problem occurs frequently enough to develop a specialized procedure for solution.

*step 1*: Use the direction cosines of the force, $\mathbf{F}$, to find its components.

$$\mathbf{F}_x = \mathbf{F}(\cos\theta_x) \quad 11.58$$

$$\mathbf{F}_y = \mathbf{F}(\cos\theta_y) \quad 11.59$$

$$\mathbf{F}_z = \mathbf{F}(\cos\theta_z) \quad 11.60$$

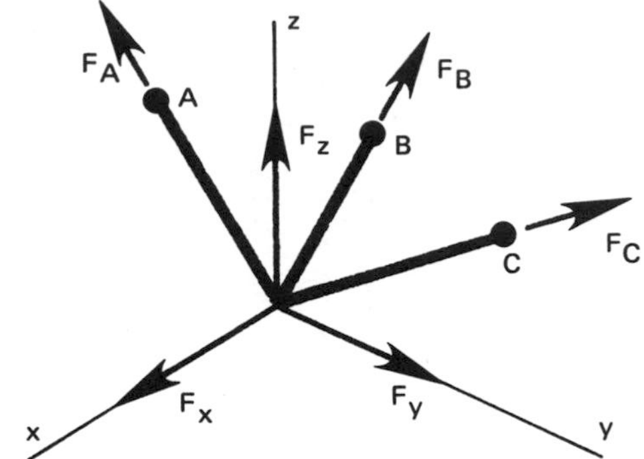


**Figure 11.14** A General Tripod

*step 2*: Using the $x$, $y$, and $z$ coordinates of points A, B, and C (taking the origin at the apex), find the direction cosines for the legs. Repeat the following four equations for each member, observing algebraic signs of $x$, $y$, and $z$.

$$L^2 = x^2 + y^2 + z^2 \quad 11.61$$

$$\cos\theta_x = \frac{x}{L} \quad 11.62$$

$$\cos\theta_y = \frac{y}{L} \quad 11.63$$

$$\cos\theta_z = \frac{z}{L} \quad 11.64$$

*step 3*: Write the equations of equilibrium for joint O. The following simultaneous equations assume tension in all three members. A minus sign in the solution for any member indicates compression instead of tension.

$$\mathbf{F}_A\cos\theta_{xA} + \mathbf{F}_B\cos\theta_{xB} + \mathbf{F}_C\cos\theta_{xC} + \mathbf{F}_x = 0 \quad 11.65$$

$$\mathbf{F}_A\cos\theta_{yA} + \mathbf{F}_B\cos\theta_{yB} + \mathbf{F}_C\cos\theta_{yC} + \mathbf{F}_y = 0 \quad 11.66$$

$$\mathbf{F}_A\cos\theta_{zA} + \mathbf{F}_B\cos\theta_{zB} + \mathbf{F}_C\cos\theta_{zC} + \mathbf{F}_z = 0 \quad 11.67$$

*Example 11.12*

Find the load on each leg of the tripod shown.

| Point | x | y | z |
|---|---|---|---|
| O | 0 | 0 | 6 |
| A | 2 | −2 | 0 |
| B | 0 | 3 | −4 |
| C | −3 | 3 | 2 |

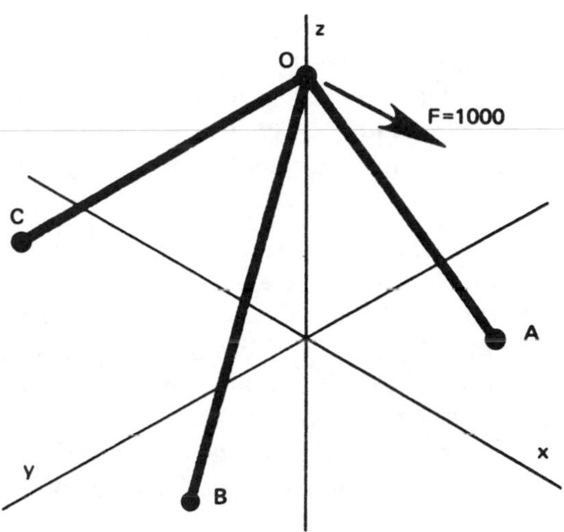


Since the apex is not at point (0, 0, 0), it is necessary to transfer the origin to the apex. This is done by the following equations. Only the $z$ values are affected.

$$x' = x - x_0$$
$$y' = y - y_0$$
$$z' = z - z_0$$

The new coordinates with the origin at the apex are

| Point | x | y | z |
|---|---|---|---|
| O | 0 | 0 | 0 |
| A | 2 | −2 | −6 |
| B | 0 | 3 | −10 |
| C | −3 | 3 | −4 |

The components of the applied force are

$$\mathbf{F}_x = \mathbf{F}\,(\cos\theta_x) = \mathbf{F}\,[\cos(0°)] = 1000$$
$$\mathbf{F}_y = 0$$
$$\mathbf{F}_z = 0$$

The direction cosines of the legs are found from the following table.

| Member | $x^2$ | $y^2$ | $z^2$ | $L^2$ | L | $\cos\theta_x$ | $\cos\theta_y$ | $\cos\theta_z$ |
|---|---|---|---|---|---|---|---|---|
| O–A | 4 | 4 | 36 | 44 | 6.63 | 0.3015 | −0.3015 | −0.9046 |
| O–B | 0 | 9 | 100 | 109 | 10.44 | 0 | 0.2874 | −0.9579 |
| O–C | 9 | 9 | 16 | 34 | 5.83 | −0.5146 | 0.5146 | −0.6861 |

From equations 11.65, 11.66, and 11.67, the equilibrium equations are

$$0.3015\mathbf{F}_A + \qquad -0.5146\mathbf{F}_C + 1000 = 0$$
$$-0.3015\mathbf{F}_A + 0.2874\mathbf{F}_B + 0.5146\mathbf{F}_C + \quad 0 = 0$$
$$-0.9046\mathbf{F}_A - 0.9579\mathbf{F}_B - 0.6861\mathbf{F}_C + \quad 0 = 0$$

The solution to this set of simultaneous equations is

$$\mathbf{F}_A = +1531 \text{ (tension)}$$
$$\mathbf{F}_B = -3480 \text{ (compression)}$$
$$\mathbf{F}_C = +2841 \text{ (tension)}$$

## 15 PROPERTIES OF AREAS

### A. CENTROIDS

The location of the *centroid* of a 2-dimensional area which is defined mathematically as $y = f(x)$ can be found from equations 11.68 and 11.69.[4] This is illustrated in example 11.13.

$$\overline{x} = \frac{\int x\,dA}{A} \qquad 11.68$$

$$\overline{y} = \frac{\int y\,dA}{A} \qquad 11.69$$

$$A = \int f(x)\,dx \qquad 11.70$$

$$dA = f(x)\,dx = f(y)\,dy \qquad 11.71$$

*Example 11.13*

Find the $x$ component of the centroid of the area bounded by the $x$ and $y$ axes, $x = 2$, and $y = e^{2x}$.

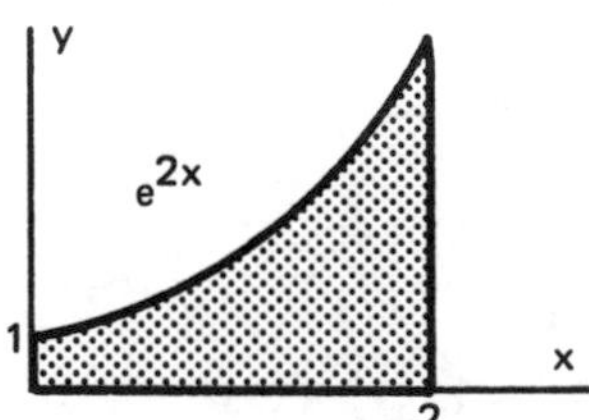


*step 1*: Find the area.

$$A = \int_0^2 e^{2x}dx = \left[\tfrac{1}{2}e^{2x}\right]_0^2$$
$$= 27.3 - 0.5 = 26.8$$

*step 2*: Put $dA$ in terms of $dx$.

$$dA = f(x)\,dx = e^{2x}dx$$

*step 3*: Use equation 11.68 to find $\overline{x}$.

$$\overline{x} = \frac{1}{26.8}\int_0^2 xe^{2x}dx = \frac{1}{26.8}\left[\tfrac{1}{2}xe^{2x} - \tfrac{1}{4}e^{2x}\right]_0^2$$
$$= 1.54$$

[4] The centroid also is known as the *first moment of the area.*

STATICS

With few exceptions, most areas for which the centroidal location is needed will be either rectangular or triangular. The locations of the centroids for these and other common shapes are given as an appendix of this chapter.

The centroid of a complex 2-dimensional area which can be divided into the simple shapes in appendix A can be found from equations 11.72 and 11.73.

$$\overline{x}_{\text{composite}} = \frac{\sum(A_i\overline{x}_i)}{\sum A_i} \quad 11.72$$

$$\overline{y}_{\text{composite}} = \frac{\sum(A_i\overline{y}_i)}{\sum A_i} \quad 11.73$$

*Example 11.14*

Find the y-coordinate of the centroid for the object shown.

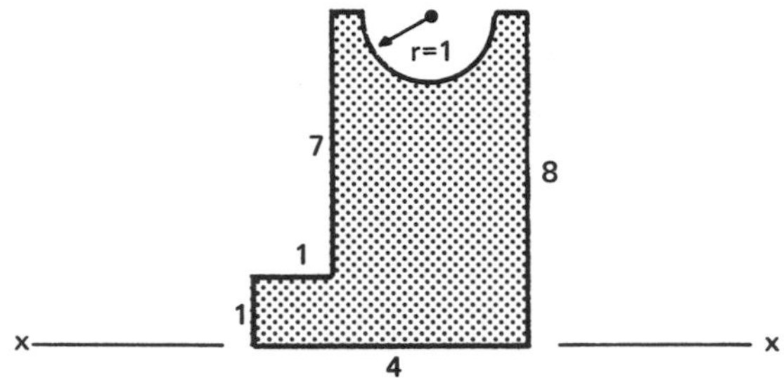


The object is divided into three parts: a $1 \times 4$ rectangle, a $3 \times 7$ rectangle, and a half-circle of radius 1. Then, the areas and distances from the $x$-$x$ axis to the individual centroids are found.

$$A_1 = (1)(4) = 4$$
$$A_2 = (3)(7) = 21$$
$$A_3 = \left(-\tfrac{1}{2}\right)\pi(1)^2 = -1.57$$
$$\overline{y}_1 = \tfrac{1}{2}$$
$$\overline{y}_2 = 4\tfrac{1}{2}$$
$$\overline{y}_3 = 8 - 0.424 = 7.576$$

Using equation 11.73,

$$\overline{y} = \frac{(4)\left(\frac{1}{2}\right) + (21)\left(4\frac{1}{2}\right) - (1.57)(7.576)}{4 + 21 - 1.57} = 3.61$$

## B. MOMENT OF INERTIA

The moment of inertia, $I$, of a 2-dimensional area is a parameter which often is needed in mechanics of materials problems. It has no simple geometric interpretation, and its units (length to the fourth power) add to the mystery of this quantity. However, it is convenient to think of the moment of inertia as a resistance to bending.

If the moment of inertia is a resistance to bending, it is apparent that this quantity always must be positive. Since bending of an object (e.g., a beam) can be in any direction, the resistance to bending must depend on the direction of bending. Therefore, a reference axis or direction must be included when specifying the moment of inertia.

In this chapter, $I_x$ is used to represent a moment of inertia with respect to the $x$ axis. Similarly, $I_y$ is with respect to the $y$ axis. $I_x$ and $I_y$ are not components of the "resultant" moment of inertia. The moment of inertia taken with respect to a line passing through the area's centroid is known as the *centroidal moment of inertia*, $I_c$. The centroidal moment of inertia is the smallest possible moment of inertia for the shape.

The moments of inertia of a function which can be expressed mathematically as $y = f(x)$ are given by equations 11.74 and 11.75.

$$I_x = \int y^2\, dA \quad 11.74$$

$$I_y = \int x^2\, dA \quad 11.75$$

In general, however, moments of inertia will be found from appendix A.

*Example 11.15*

Find $I_y$ for the area bounded by the y axis, $y = 8$, and $y^2 = 8x$.

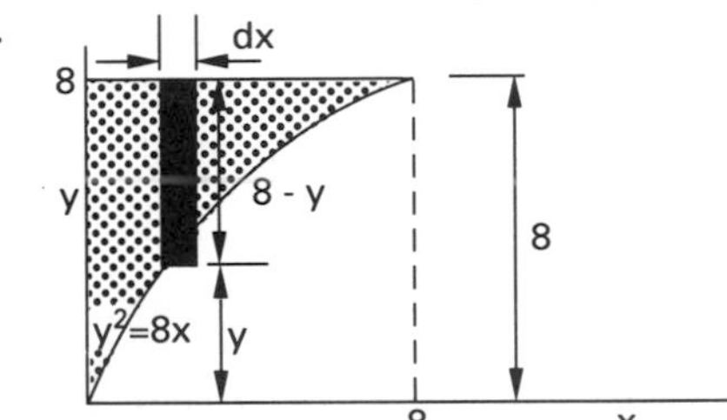


$$dA = (dx)(\text{height of block})$$

$$I_y = \int_0^8 x^2\, dA = \int_0^8 x^2(8-y)dx$$

But $y = \sqrt{8x}$.

$$I_y = \int_0^8 \left(8x^2 - \sqrt{8}x^{\frac{5}{2}}\right) dx = 195.04 \text{ inches}^4$$

*Example 11.16*

What is the centroidal moment of inertia of the area shown?

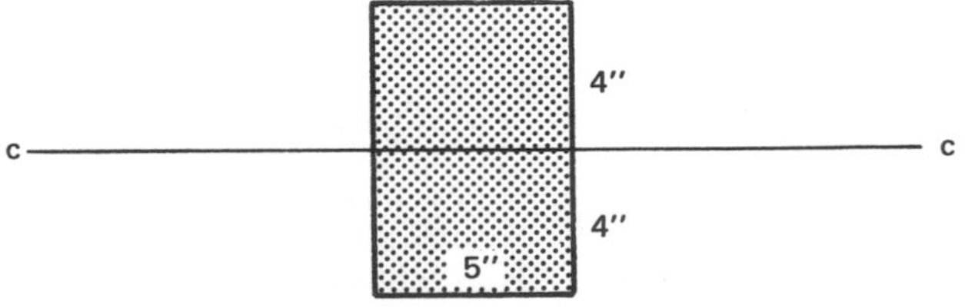

From appendix A,

$$I_c = \frac{(5)(8)^3}{12} = 213.3 \text{ inches}^4$$

The *polar moment of inertia* of a 2-dimensional area can be thought of as a measure of the area's resistance to torsion (twisting). Although the polar moment of inertia can be evaluated mathematically by equation 11.76, it is more expedient to use equation 11.77 if $I_x$ and $I_y$ are known.

$$J_z = \int (x^2 + y^2)\, dA \qquad 11.76$$

$$J_z = I_x + I_y \qquad 11.77$$

The *radius of gyration*, $r$, is a distance at which the entire area can be assumed to exist. The distance is measured from the axis about which the moment of inertia was taken.

$$I = r^2 A \qquad 11.78$$

$$r = \sqrt{\frac{I}{A}} \qquad 11.79$$

*Example 11.17*

What is the radius of gyration for the section shown in example 11.16? What is the significance of this value?

$$A = (5)(4+4) = 40$$

From equation 11.79,

$$r = \sqrt{\frac{213.3}{40}} = 2.31 \text{ inches}$$

2.31″ is the distance from the axis c-c that an infinitely long strip (with area of 40 square inches) would have to be located to have a moment of inertia of 213.3 inches$^4$.

The *parallel axis theorem* usually is needed to evaluate the moment of inertia of a composite object made up of several simple 2-dimensional shapes.[5] The parallel axis theorem relates the moment of inertia of an area taken with respect to any axis to the centroidal moment of inertia. In equation 11.80, $A$ is the 2-dimensional object's area, and $d$ is the distance between the centroidal and new axes.

$$I_{\text{any parallel axis}} = I_c + Ad^2 \qquad 11.80$$

*Example 11.18*

Find the moment of inertia about the $x$ axis for the 2-dimensional object shown.

[5] This theorem also is known as the *transfer axis theorem*.

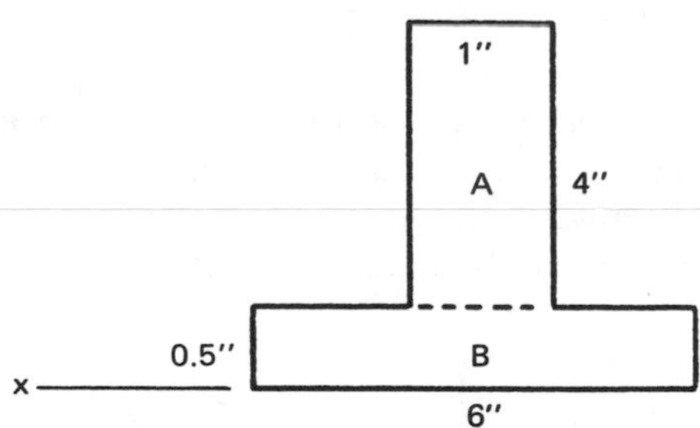


The T-section is divided into two parts: A and B. The moment of inertia of section B can be evaluated readily by using appendix A.

$$I_{x\text{-}x} = \frac{(6)(0.5)^3}{3} = 0.25$$

The moment of inertia of the stem about its own centroidal axis is

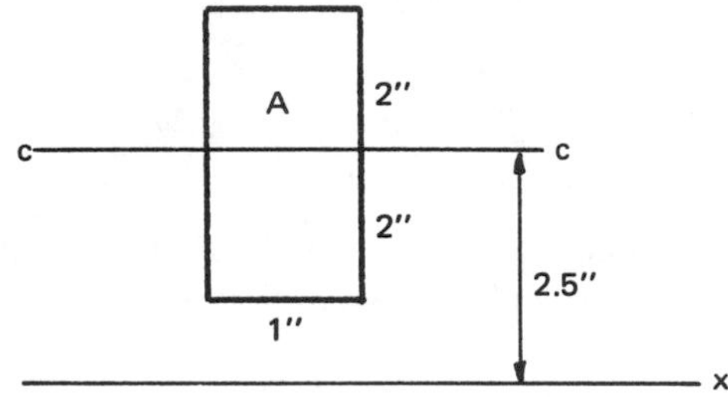


$$I_{c\text{-}c} = \frac{(1)(4)^3}{12} = 5.33$$

Using equation 11.80, the moment of inertia of the stem about the x-x axis is

$$I_{x\text{-}x} = 5.33 + (4)(2.5)^2 = 30.33$$

The total moment of inertia of the T-section is

$$0.25 + 30.33 = 30.58 \text{ inches}^4$$

## C. PRODUCT OF INERTIA

The *product of inertia* of a 2-dimensional object is found by multiplying each differential element of area times its $x$ and $y$ coordinates, and then integrating over the entire area.

$$P_{xy} = \int xy\, dA \qquad 11.81$$

The product of inertia is zero when either axis is an axis of symmetry. The product of inertia may be negative.

STATICS

## 16 ROTATION OF AXES

Suppose the various properties of an area are known for one set of axes, $x$ and $y$. If the axes are rotated through an angle without rotating the area itself, the new properties can be found from the old properties.

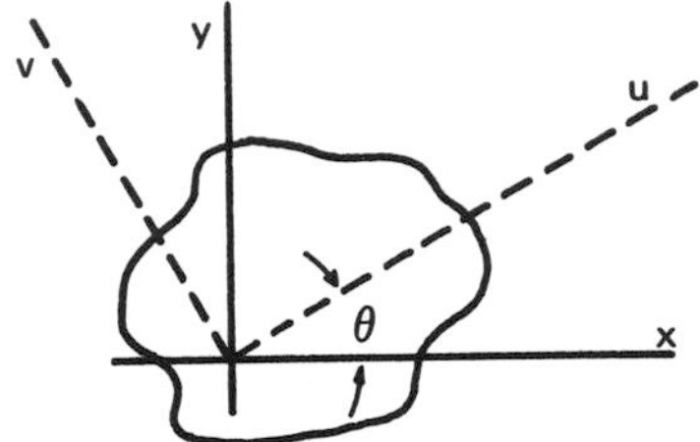


**Figure 11.15** Rotation of Axes

$$I_u = I_x \cos^2\theta - 2P_{xy}\sin\theta\,\cos\theta + I_y \sin^2\theta \quad 11.82$$

$$= \tfrac{1}{2}(I_x + I_y) + \tfrac{1}{2}(I_x - I_y)\cos 2\theta - P_{xy}\sin 2\theta \quad 11.83$$

$$I_v = I_x \sin^2\theta + 2P_{xy}\sin\theta\,\cos\theta + I_y\cos^2\theta \quad 11.84$$

$$= \tfrac{1}{2}(I_x + I_y) - \tfrac{1}{2}(I_x - I_y)\cos 2\theta + P_{xy}\sin 2\theta \quad 11.85$$

$$P_{uv} = I_x \sin\theta\,\cos\theta + P_{xy}(\cos^2\theta - \sin^2\theta) - I_y\sin\theta\,\cos\theta \quad 11.86$$

$$= \tfrac{1}{2}(I_x - I_y)\sin 2\theta + P_{xy}\cos 2\theta \quad 11.87$$

Since the polar moment of inertia about a fixed axis is constant, the sum of the two area moments of inertia is also constant.

$$I_x + I_y = I_u + I_v \quad 11.88$$

There is one angle that will maximize the moment of inertia, $I_u$. This angle can be found from calculus by setting $dI_u/d\theta = 0$. The resulting equation defines two angles, one of which maximizes $I_u$, the other of which minimizes $I_u$.

$$\tan 2\theta = -\frac{2P_{xy}}{I_x - I_y} \quad 11.89$$

The two angles which satisfy equation 11.89 are 90° apart. These are known as the *principal axes*. The moments of inertia about these two axes are known as the *principal moments of inertia*. These principal moments are given by equation 11.90.

$$I_{\text{max, min}} = \tfrac{1}{2}(I_x + I_y) \pm \sqrt{\tfrac{1}{4}(I_x - I_y)^2 + P_{xy}^2} \quad 11.90$$

## 17 PROPERTIES OF MASSES

### A. CENTER OF GRAVITY

The *center of gravity* in 3-dimensional objects is analogous to centroids in 2-dimensional areas. The center of gravity can be located mathematically if the object can be described by a mathematical function.

$$\overline{x} = \frac{\int x\,dm}{m} \quad 11.91$$

$$\overline{y} = \frac{\int y\,dm}{m} \quad 11.92$$

$$\overline{z} = \frac{\int z\,dm}{m} \quad 11.93$$

The location of the center of gravity often is obvious for simple objects. It always is located on an axis of symmetry. If the object is complex or composite, the overall center of gravity can be found from the individual centers of gravity of the constituent objects.

$$\overline{x}_{\text{composite}} = \frac{\sum(m_i\overline{x}_i)}{\sum m_i} \quad 11.94$$

$$\overline{y}_{\text{composite}} = \frac{\sum(m_i\overline{y}_i)}{\sum m_i} \quad 11.95$$

$$\overline{z}_{\text{composite}} = \frac{\sum(m_i\overline{z}_i)}{\sum m_i} \quad 11.96$$

### B. MASS MOMENT OF INERTIA

The mass moment of inertia can be thought of as a measure of resistance to rotational motion. Although it can be found mathematically from equations 11.97, 11.98, and 11.99, it is more expedient to use appendix B to evaluate simple objects.

$$I_x = \int (y^2 + z^2)\,dm \quad 11.97$$

$$I_y = \int (x^2 + z^2)\,dm \quad 11.98$$

$$I_z = \int (x^2 + y^2)\,dm \quad 11.99$$

The *centroidal mass moment of inertia* is found by evaluating the moment of inertia about an axis passing through the object's center of gravity. Once this centroidal mass moment of inertia is known, the parallel axis theorem can be used to find the moment of inertia about any parallel axis.

$$I_{\text{any parallel axis}} = I_c + md^2 \quad 11.100$$

The radius of gyration of a 3-dimensional object is defined by equation 11.101.

$$r = \sqrt{\frac{I}{m}} \quad 11.101$$

$$I = r^2 m \quad 11.102$$

## 18 FRICTION

Friction is a force which resists motion or attempted motion. It always acts parallel to the contacting surfaces. The frictional force exerted on a stationary object is known as *static friction* or *coulomb friction.* If the object is moving, the friction is known as *dynamic friction.* Dynamic friction is less than static friction in most situations.

The actual magnitude of the frictional force depends on the *normal force* and the *coefficient of friction*, $f$, between the object and the surface. For an object resting on a horizontal surface, the normal force is the weight, $w$.

$$\mathbf{F}_f = f\mathbf{N} = fw \qquad 11.103$$

If the object is resting on an inclined surface, the normal force will be

$$\mathbf{N} = w \cos\theta \qquad 11.104$$

The frictional force again is equal to the product of the normal force and the coefficient of friction.

$$\mathbf{F}_f = f\mathbf{N} = fw \cos\theta \qquad 11.105$$

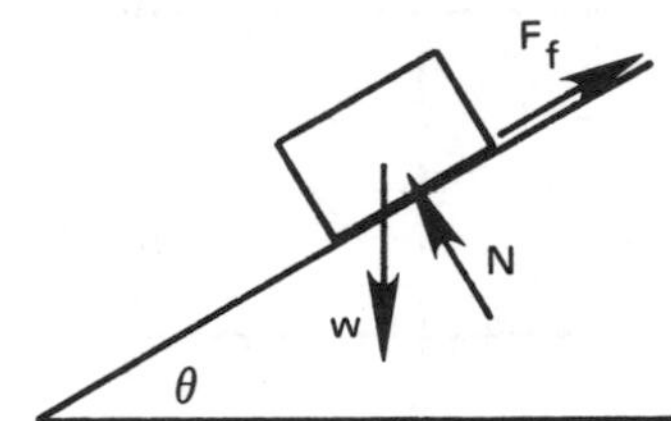


**Figure 11.16** Frictional Force

The object shown in figure 11.16 will not slip down the plane until the angle reaches a critical angle known as the *angle of repose.* This angle is given by equation 11.106.

$$\tan\theta = f \qquad 11.106$$

Typical values of the coefficient of friction are given in table 11.2.

Friction also exists between a belt, rope, or band wrapped around a drum, pulley, or sheave. If $\mathbf{T}_1$ is the tight side tension, if $\mathbf{T}_2$ is the slack side tension, and if $\phi$ is the contact angle in *radians*, the relationship governing *belt friction* is given by equation 11.107.

$$\frac{\mathbf{T}_1}{\mathbf{T}_2} = e^{f\phi} \qquad 11.107$$

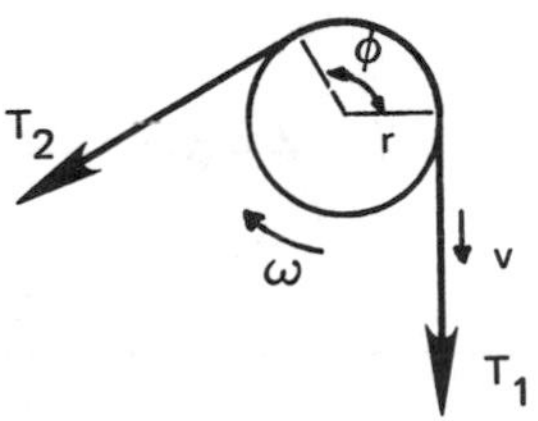


**Figure 11.17** Belt Friction

The transmitted torque in ft-lbf is

$$\text{torque} = (\mathbf{T}_1 - \mathbf{T}_2)\,r \qquad 11.108$$

The power in ft-lbf/sec transmitted by the belt running at speed $v$ in ft/sec is

$$\text{power} = (\mathbf{T}_1 - \mathbf{T}_2)\,v \qquad 11.109$$

**Table 11.2**
Approximate Coefficients of Friction

| material | condition | dynamic | static |
|---|---|---|---|
| cast iron on cast iron | dry | 0.15 | 1.10 |
| plastic on steel | dry | 0.35 | |
| grooved rubber on pavement | dry | 0.40 | 0.55 |
| bronze on steel | oiled | | 0.09 |
| steel on graphite | dry | | 0.21 |
| steel on steel | dry | 0.42 | 0.78 |
| steel on steel | oiled | 0.08 | 0.10 |
| steel on asbestos-faced steel | dry | | 0.15 |
| steel on asbestos-faced steel | oiled | | 0.12 |
| press fits (shaft in hole) | oiled | | 0.10–0.15 |

# Appendix A: Centroids and Area Moments of Inertia

| SHAPE | DIMENSIONS | CENTROID $(x_c, y_c)$ | AREA MOMENT OF INERTIA |
|---|---|---|---|
| Rectangle | 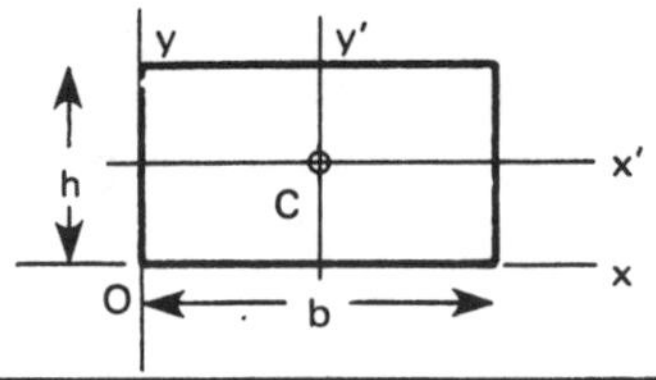 | $(\frac{1}{2}b, \frac{1}{2}h)$ | $I_{x'} = (1/12)bh^3$<br>$I_{y'} = (1/12)hb^3$<br>$I_x = (1/3)bh^3$<br>$I_y = (1/3)hb^3$<br>$J_C = (1/12)bh(b^2 + h^2)$ |
| Triangle | 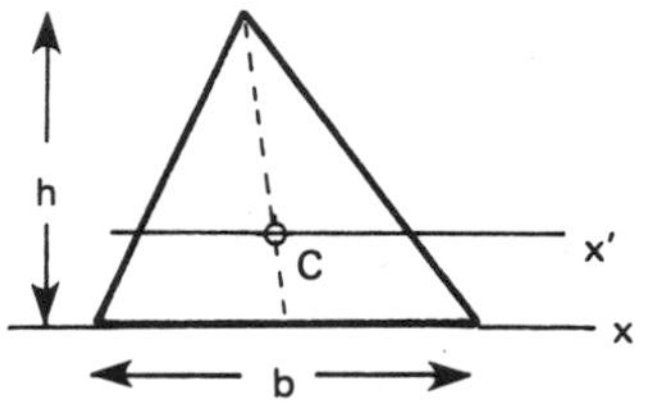 | $y_c = (h/3)$ | $I_{x'} = (1/36)bh^3$<br>$I_x = (1/12)bh^3$ |
| Trapezoid | 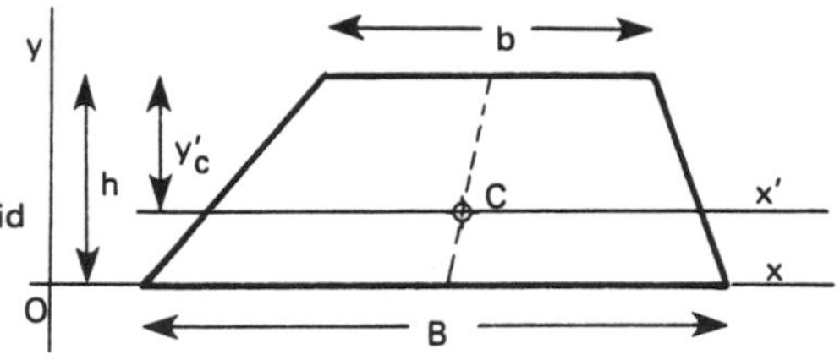 | $y'_c = \frac{h(2B + b)}{3(B + b)}$<br>Note that this is measured from the top surface. | $I_{x'} = \frac{h^3(B^2+4Bb+b^2)}{36(B+b)}$<br>$I_x = \frac{h^3(B+3b)}{12}$ |
| Quarter-Circle, of radius r | 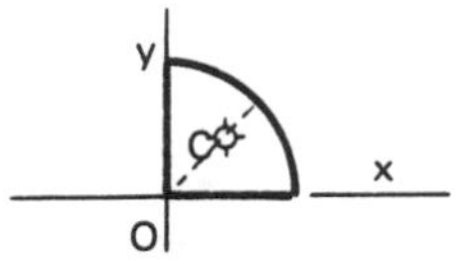 | $((4r/3\pi), (4r/3\pi))$ | $I_x = I_y = (1/16)\pi r^4$<br>$J_O = (1/8)\pi r^4$ |
| Half Circle, of radius r | 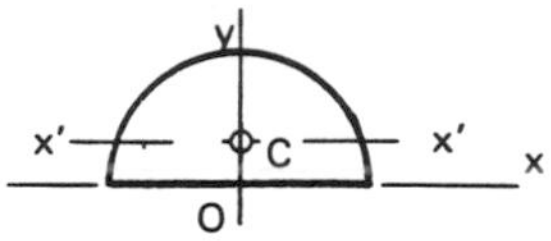 | $(0, (4r/3\pi))$ | $I_x = I_y = (1/8)\pi r^4$<br>$J_O = \frac{1}{4}\pi r^4$ $I_{x'} = .11r^4$ |
| Circle, of radius r | 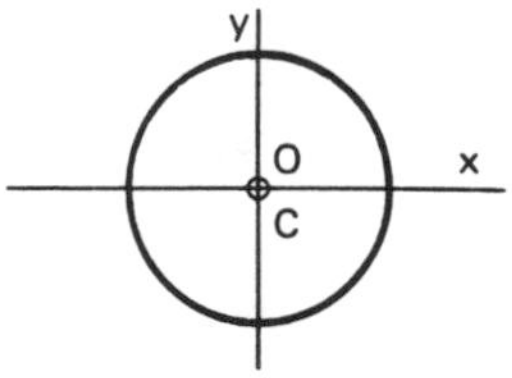 | $(0,0)$ | $I_x = I_y = \frac{1}{4}\pi r^4$<br>$J_O = \frac{1}{2}\pi r^4$ |
| Parabolic Area | 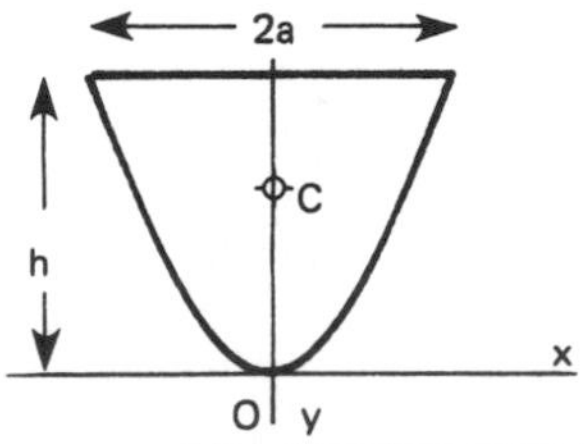 | $(0, (3h/5))$ | $I_x = 4h^3a/7$<br>$I_y = 4ha^3/15$ |
| Parabolic Spandrel | 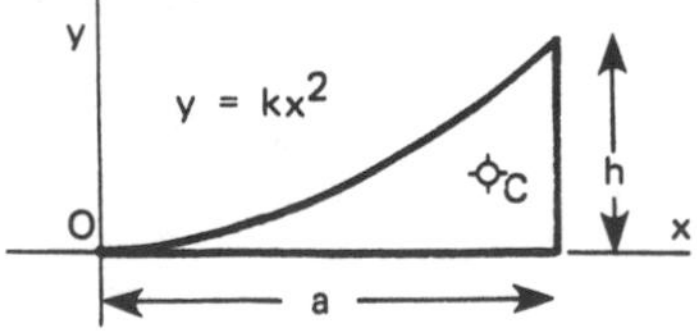 $y = kx^2$ | $((3a/4),(3h/10))$ | $I_x = ah^3/21$<br>$I_y = 3ha^3/15$ |

## Appendix B: Mass Moments of Inertia

($m$ is in slugs; lengths are in feet)

| | | |
|---|---|---|
| Slender rod | 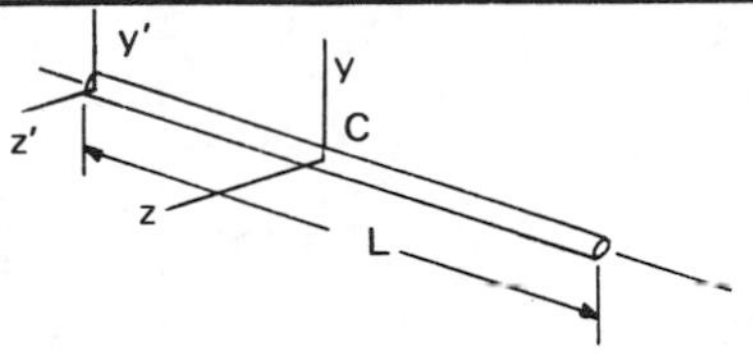 | $I_y = I_z = (1/12)mL^2$<br>$I_{y'} = I_{z'} = (1/3)mL^2$ |
| Thin rectangular plate | 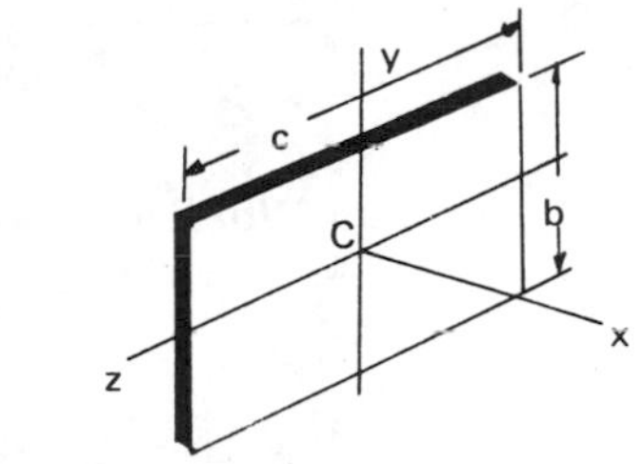 | $I_x = (1/12)(b^2+c^2)\,m$<br>$I_y = (1/12)mc^2$<br>$I_z = (1/12)mb^2$ |
| Rectangular Parallelepiped | 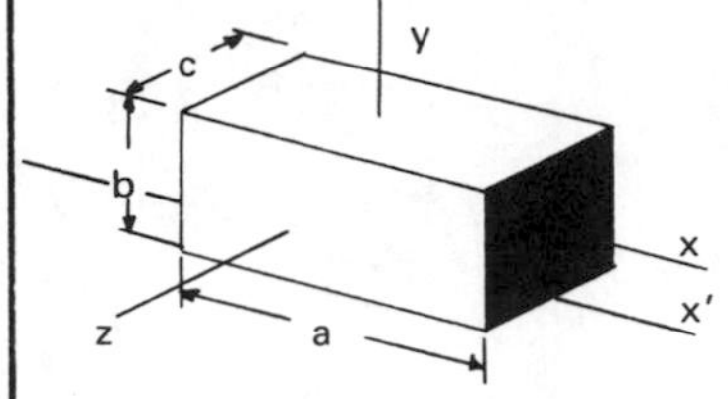 | $I_x = (1/12)m(b^2+c^2)$<br>$I_y = (1/12)m(c^2+a^2)$<br>$I_z = (1/12)m(a^2+b^2)$<br>$I_{x'} = (1/12)m(4b^2+c^2)$ |
| Thin disk, radius r | 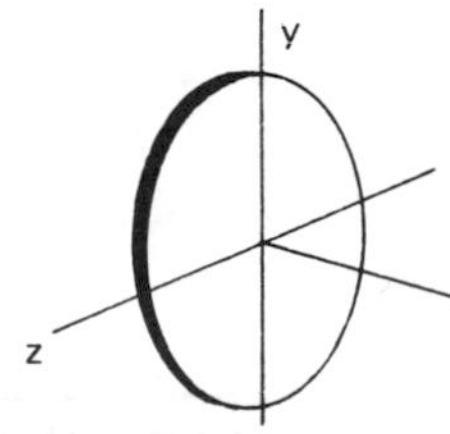 | $I_x = \frac{1}{2}mr^2$<br>$I_y = I_z = \frac{1}{4}mr^2$ |
| Circular cylinder, radius r | 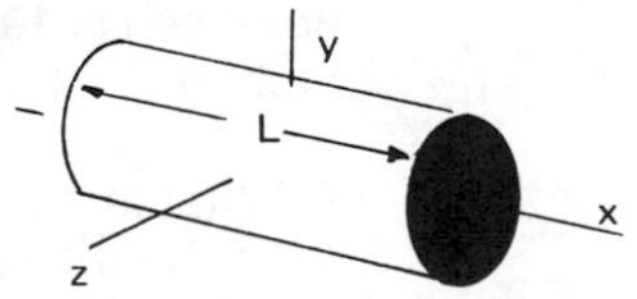 | $I_x = \frac{1}{2}mr^2$<br>$I_y = I_z = (1/12)m(3r^2+L^2)$ |
| Circular cone, base radius r | 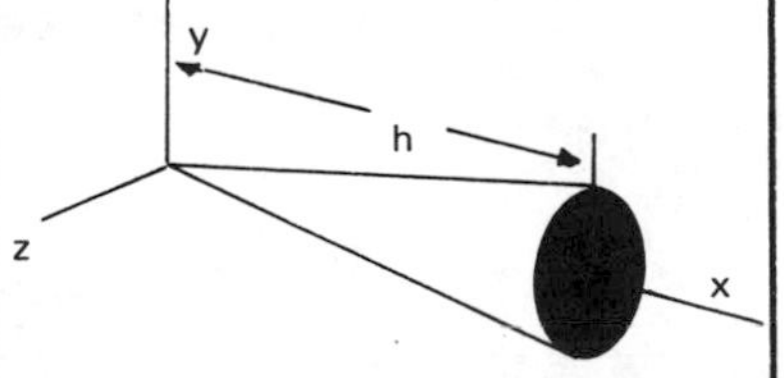 | $I_x = (3/10)mr^2$<br>$I_y = I_z = (3/5)m(\frac{1}{4}r^2+h^2)$ |
| Sphere, radius r | 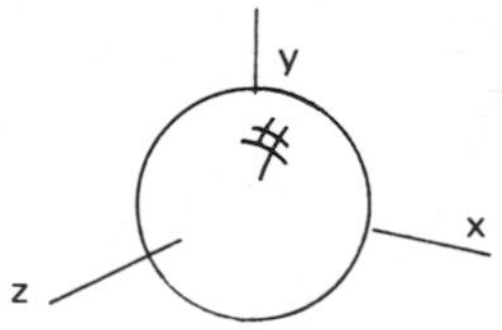 | $I_x = I_y = I_z = (2/5)mr^2$ |
| Hollow circular cylinder | 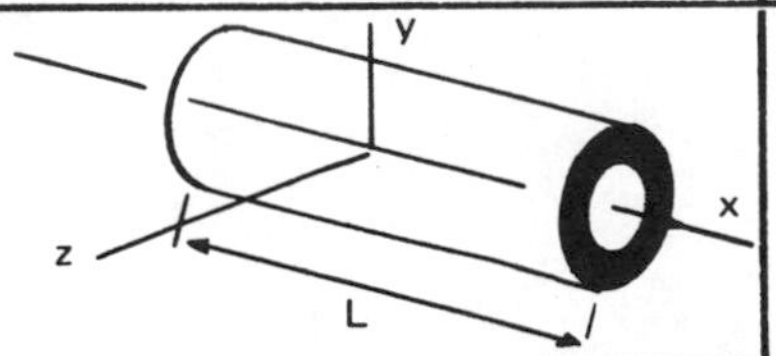 | $I_x = \frac{1}{2}m(r_o^2 + r_i^2) = \frac{\pi\rho L}{2g_c}(r_o^4 - r_i^4)$ |

**Practice Problems: STATICS**

Untimed

1. Find the forces in all members of the truss shown.

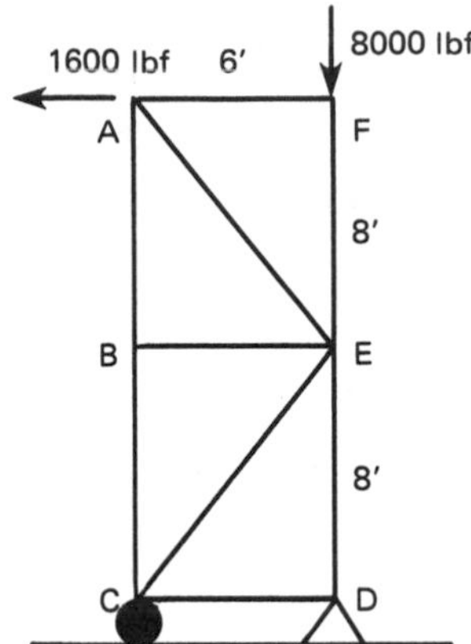


2. Find the forces in each of the legs.

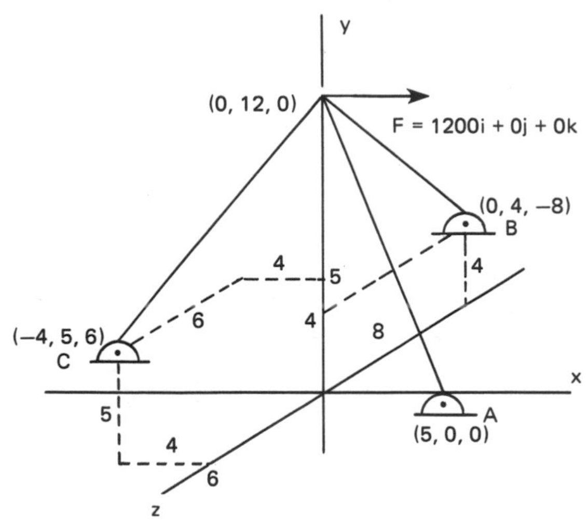


3. Find the centroidal moment of inertia about an axis parallel to the x-axis.

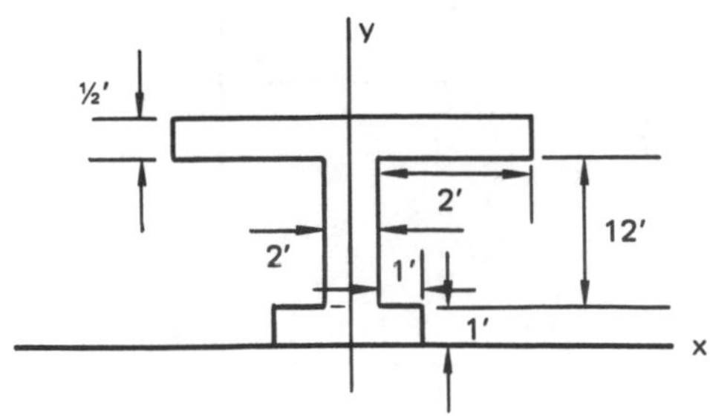


4. Find the forces in members DE and HJ.

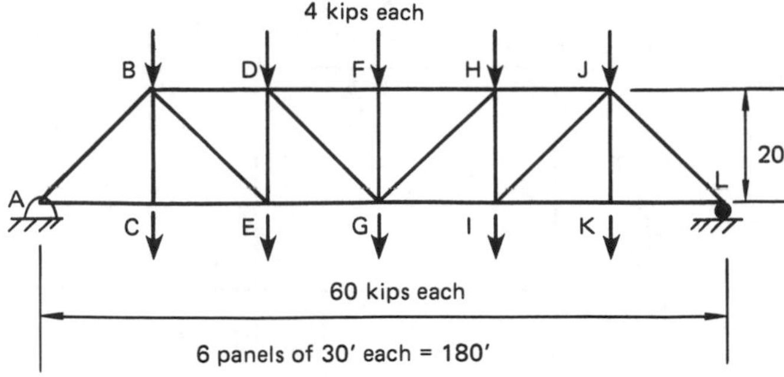


5. What are the x, y, and z components of the forces at A, B, and C?

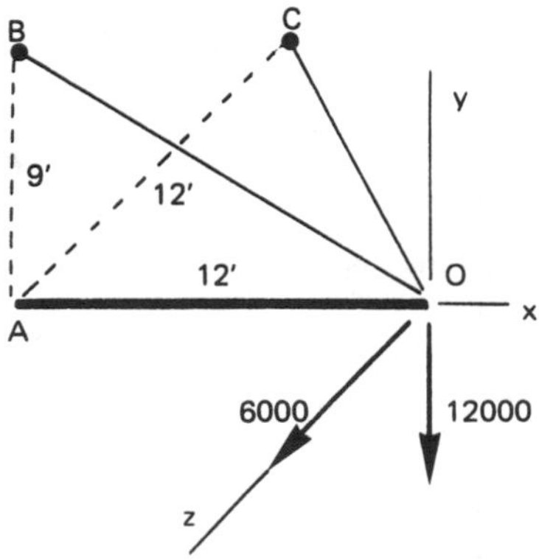


6. A power line weighs 2 pounds per foot of length. It is supported by two equal height towers over a level forest. The tower spacing is 100 feet and the mid-point sag is 10 feet. What are the maximum and minimum tensions?

7. What is the sag for the cable described in problem #6 if the maximum tension is 500 pounds?

8. Locate the centroid of the object shown.

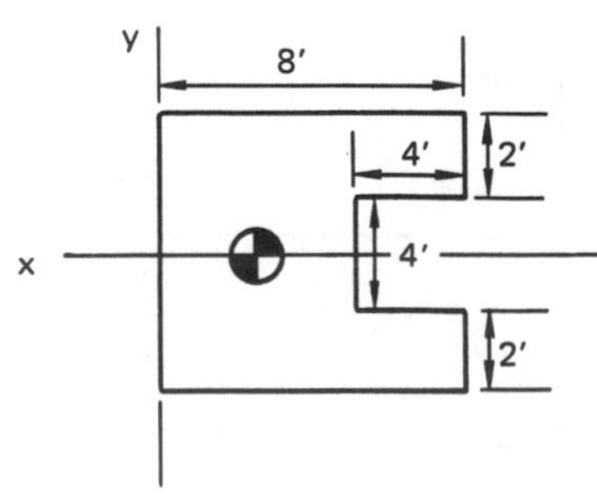

9. Replace the distributed load with three concentrated loads. Indicate the points of application.

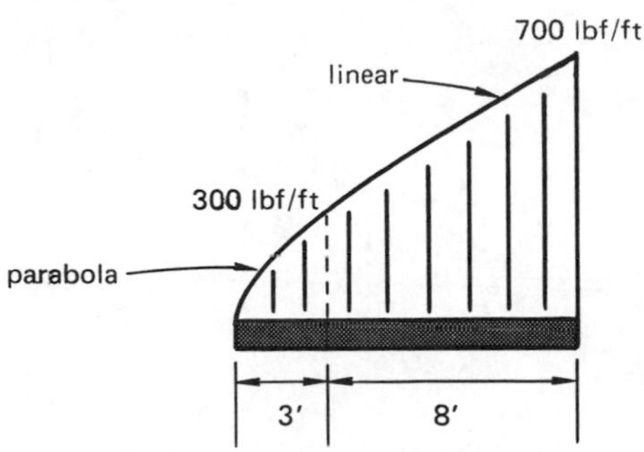


10. Find the forces in all members of the truss shown.

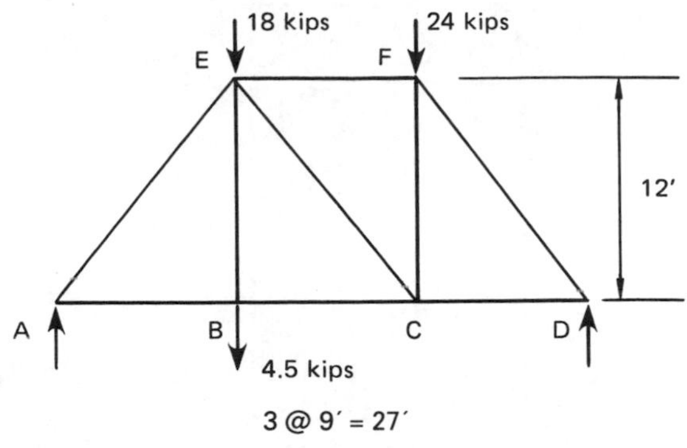

# 12 MECHANICS OF MATERIALS

## PART 1: Strength of Materials

*Nomenclature*

| | | |
|---|---|---|
| A | area | $in^2$ |
| b | width | in |
| c | distance from neutral axis to extreme fiber | in |
| C | correction | – |
| D | diameter | in |
| e | eccentricity | in |
| E | modulus of elasticity | psi |
| F | force, or load | lbf |
| F.S. | factor of safety | |
| g | local gravitational acceleration | $ft/sec^2$ |
| $g_c$ | gravitational constant (32.2) | $\frac{lbm\text{-}ft}{lbf\text{-}sec^2}$ |
| G | shear modulus | psi |
| I | moment of inertia | $in^4$ |
| J | polar moment of inertia | $in^4$ |
| k | spring constant | lbf/in |
| K | stress concentration factor, or end restraint coefficient | |
| L | length | in |
| m | mass | lbm |
| M | moment | in-lbf |
| n | ratio, rotational speed, or number | –, rpm, – |
| N | number of cycles | – |
| p | pressure | psi |
| Q | statical moment | $in^3$ |
| r | radius, or radius of gyration | in |
| S | strength, or axial load | psi, lbf |
| t | thickness | in |
| T | temperature, or torque | °F, in-lbf |
| u | virtual truss load | lbf |
| U | energy | in-lbf |
| V | shear, or volume | lbf, $in^3$ |
| w | load per unit length, or width | lbf/in, in |
| W | work | in-lbf |
| x | distance, or displacement | in |
| y | deflection, or distance | in |
| Z | section modulus | $in^3$ |

*Symbols*

| | | |
|---|---|---|
| $\delta$ | elongation, or displacement | in |
| $\theta$ | angle | degrees |
| $\phi$ | angle | radians |
| $\sigma$ | normal stress | psi |
| $\alpha$ | coefficient of linear thermal expansion | 1/°F |
| $\beta$ | coefficient of volumetric thermal expansion | 1/°F |
| $\gamma$ | coefficient of area thermal expansion | 1/°F |
| $\tau$ | shear stress | psi |
| $\epsilon$ | strain | – |
| $\mu$ | Poisson's ratio | – |

*Subscripts*

| | |
|---|---|
| a | allowable |
| b | bending |
| br | bearing |
| c | centroidal, or compressive |
| e | endurance, Euler, or equivalent |
| ext | external |
| h | hoop |
| i | inside |
| L | long |
| o | original, or outside |
| p | pull |
| s | shear |
| t | transformed, tension, or temperature |
| th | thermal |
| T | torsion |
| u | ultimate |
| y | yield |

## 1 PROPERTIES OF STRUCTURAL MATERIALS

### A. THE TENSILE TEST

Many material properties can be derived from the standard tensile test. In a tensile test, a material sample is loaded axially in tension, and the elongation is measured as the load is increased. A graphical representation of typical test data for steel is shown in figure 12.1, in which the elongation, $\delta$, is plotted against the applied load, $F$.

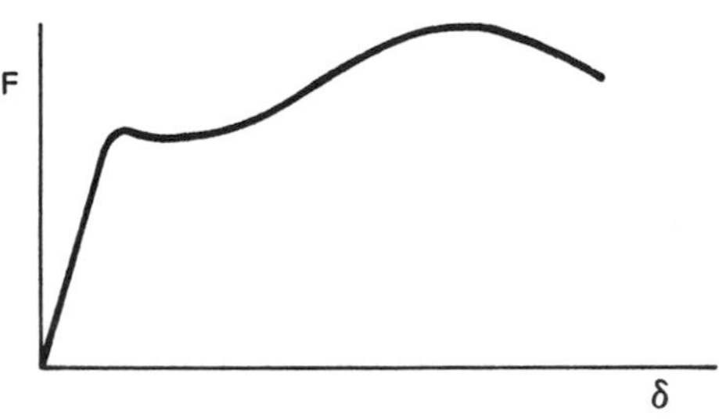


**Figure 12.1** Typical Tensile Test Results for Steel

Since this graph is applicable only to an object with the same length and area as the test sample, the data are converted to *stresses* and *strains* by use of equations 12.1 and 12.2. $\sigma$ is known as the *normal stress*, and $\epsilon$ is known as the *strain*. Strain is the percentage elongation of the sample.

$$\sigma = \frac{F}{A} \qquad 12.1$$

$$\epsilon = \frac{\delta}{L} \qquad 12.2$$

The stress-strain data also can be graphed, and the shape of the resulting curve will be the same as figure 12.1 with the scales changed.

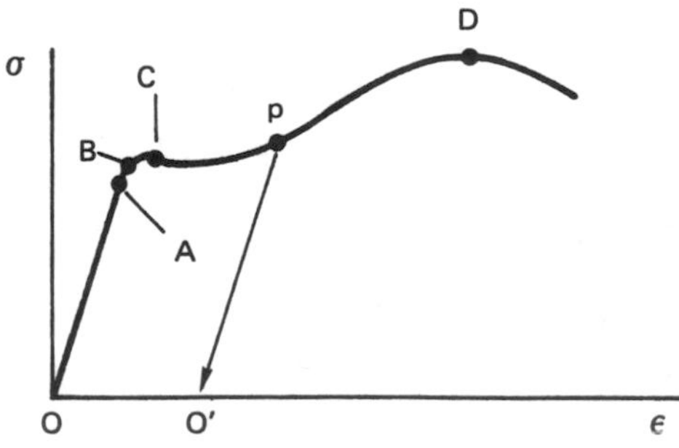


**Figure 12.2** A Typical Stress-Strain Curve for Steel

The line O-A in figure 12.2 is a straight line. The relationship between the stress and the strain is given by *Hooke's law*, equation 12.3. $E$ is the *modulus of elasticity* (*Young's modulus*) and is the slope of the line segment O-A. The stress at point $A$ is known as the *proportionality limit.* The modulus of elasticity for steel is approximately 3 EE7 psi.

$$\sigma = E\epsilon \qquad 12.3$$

Slightly above the proportionality limit is the *elastic limit* (point $B$). As long as the stress is kept below the elastic limit, there will be no permanent strain when the applied stress is removed. The strain is said to be *elastic*, and the stress is said to be in the *elastic region.*

If the elastic limit stress is exceeded before the load is removed, recovery will be along a line parallel to the straight line portion of the curve, as shown in the line segment p-O′. The strain that results (line O-O′) is permanent and is known as *plastic strain* or *permanent set.*

The *yield point* (point $C$) is very close to the elastic limit. For all practical purposes, the *yield stress*, $S_y$, can be taken as the stress which accompanies the beginning of plastic strain. Since permanent deformation is to be avoided, the yield stress is used in calculating safe stresses in ductile materials such as steel. A36 structural steel has a minimum yield strength of 36,000 psi.

$$\sigma_a = \frac{S_y}{F.S.} \qquad 12.4$$

Some materials, such as aluminum, do not have a well-defined yield point. This is illustrated in figure 12.3. In such cases, the yield point is taken as the stress which will cause a 0.2% *parallel offset.*

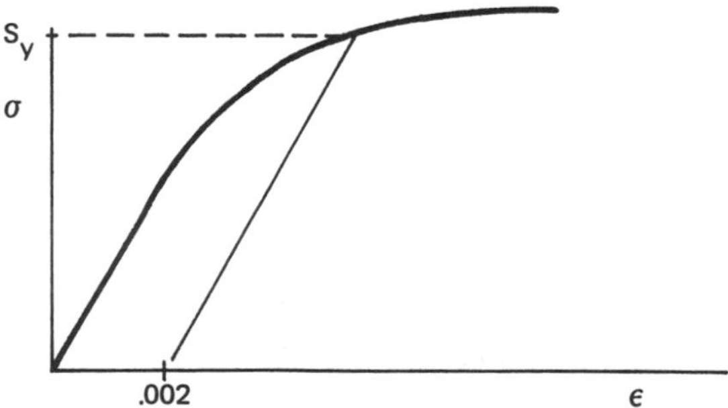


**Figure 12.3** A Typical Stress-Strain Curve for Aluminum

The *ultimate tensile strength*, point $D$ in figure 12.2, is the maximum load-carrying ability of the material. However, since stresses near the ultimate strength are accompanied by large plastic strains, this parameter should not be used for the design of ductile materials such as steel and aluminum.

As the sample is elongated during a tensile test, it also will decrease in thickness (width or diameter). The ratio of the lateral strain to the axial strain is known as *Poisson's ratio*, $\mu$. $\mu$ typically is taken as 0.3 for steel and as 0.33 for aluminum.

$$\mu = \frac{\epsilon_{\text{lateral}}}{\epsilon_{\text{axial}}} = \frac{\frac{\Delta D}{D_o}}{\frac{\Delta L}{L_o}} \qquad 12.5$$

B. FATIGUE TESTS

A part may fail after repeated stress loading even if the stress never exceeds the ultimate fracture strength of the material. This type of failure is known as *fatigue failure*.

The behavior of a material under repeated loadings can be evaluated in a fatigue test. A sample is loaded repeatedly to a known stress, and the number of applications of that stress is counted until the sample fails. This procedure is repeated for different stress levels. The results of many of these tests can be graphed, as is done in figure 12.4.

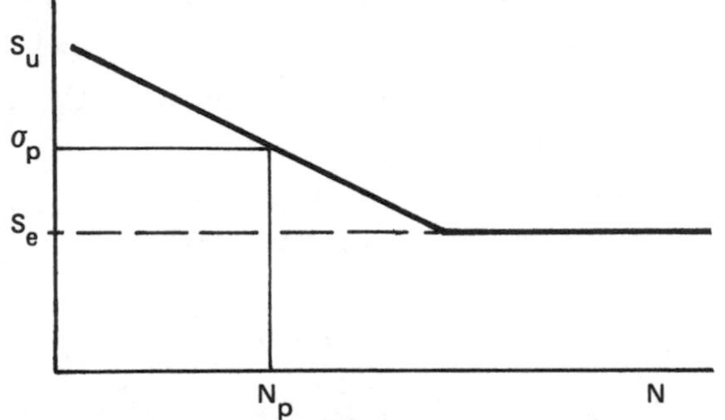


**Figure 12.4** Results of Many Fatigue Tests for Steel

For any given stress level, say $\sigma_p$ in figure 12.4, the corresponding number of applications of the stress which will cause failure is known as the *fatigue life*. That is, the fatigue life is just the number of cycles of stress required to cause failure. If the material is to fail after only one application of stress, the required stress must equal or exceed the ultimate strength of the material.

Below a certain stress level, called the *endurance limit* or the *endurance strength*, the part will be able to withstand an infinite number of stress applications without experiencing failure. Therefore, if a dynamically loaded part is to have an infinite life, the applied stress must be kept below the endurance limit.

Some materials, such as aluminum, do not have a well-defined endurance limit. In such cases, the endurance limit is taken as the stress that will cause failure at EE8 or 5 EE8 applications of the stress.

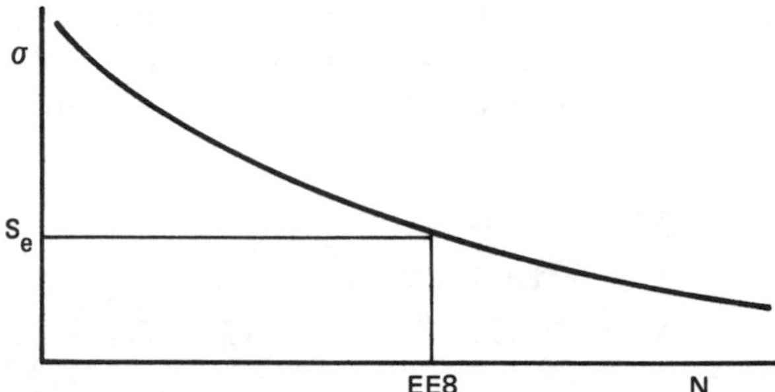


**Figure 12.5** Fatigue Test Results for Aluminum

C. ESTIMATES OF MATERIAL PROPERTIES

Although the properties of a material will depend on its classification (ASTM, AISC, etc.), average values are given in table 12.1.

## 2 DEFORMATION UNDER LOADING

Equation 12.2 can be rearranged to give the elongation of an axially loaded member in compression or tension.

$$\delta = L\epsilon = \frac{L\sigma}{E} = \frac{LF}{AE} \qquad 12.6$$

A tension load is taken as positive, and a compressive load is taken as negative. The actual length of a member under loading is given by equation 12.7 where the

**Table 12.1**
Typical Material Properties

| material | E (psi) | G (psi) | $\mu$ | $\rho$ (pcf) | $\alpha$ (1/°F) |
|---|---|---|---|---|---|
| steel (hard) | 30 EE6 | 11.5 EE6 | 0.30 | 489 | 6.5 EE−6 |
| steel (soft) | 29 EE6 | 11.5 EE6 | 0.30 | 489 | 6.5 EE−6 |
| aluminum alloy | 10 EE6 | 3.9 EE6 | 0.33 | 173 | 12.8 EE−6 |
| magnesium alloy | 6.5 EE6 | 2.4 EE6 | 0.35 | 112 | 14.5 EE−6 |
| titanium alloy | 15.4 EE6 | 6.0 EE6 | 0.34 | 282 | 4.9 EE−6 |
| cast iron (class 20) | 20 EE6 | 8 EE6 | 0.27 | 442 | 5.6 EE−6 |

algebraic sign of the deformation must be observed.

$$L_{\text{actual}} = L_o + \delta \qquad 12.7$$

The energy stored in a loaded member is equal to the work required to deform it. Below the proportionality limit, this energy is given by equation 12.8.

$$U = \frac{1}{2}F\delta = \frac{1}{2}\left(\frac{F^2 L}{AE}\right) \qquad 12.8$$

## 3 THERMAL DEFORMATION

If the temperature of an object is changed, the object will experience length, area, and volume changes. These changes can be predicted by equations 12.9, 12.10, and 12.11.

$$\Delta L = \alpha L_o (T_2 - T_1) \qquad 12.9$$

$$\Delta A = \gamma A_o (T_2 - T_1) \approx 2\alpha A_o (T_2 - T_1) \qquad 12.10$$

$$\Delta V = \beta V_o (T_2 - T_1) \approx 3\alpha V_o (T_2 - T_1) \qquad 12.11$$

If equation 12.9 is rearranged, an expression for the *thermal strain* is obtained. Thermal strain is handled in the same manner as strain due to an applied load.

$$\epsilon_{th} = \frac{\Delta L}{L_o} = \alpha\,(T_2 - T_1) \qquad 12.12$$

For example, if a bar is heated but is not allowed to expand, the stress will be given by equation 12.13.

$$\sigma_{th} = E\,\epsilon_{th} \qquad 12.13$$

## 4 SHEAR AND MOMENT DIAGRAMS

It was illustrated in chapter 11 that, for an object in equilibrium, the sums of forces and moments are equal to zero everywhere. For example, the sum of moments about point $A$ for the beam shown in figure 12.6 is zero.

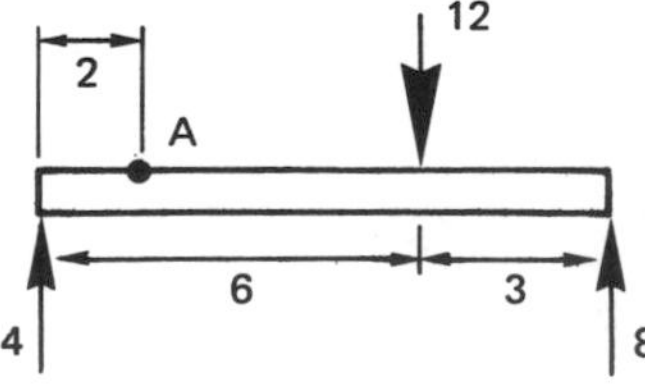


**Figure 12.6** A Beam in Equilibrium

Nevertheless, the beam shown in figure 12.6 will bend under the influence of the forces. This bending is evidence of the stress experienced by the beam. Since the sum of moments about any point is zero, the moment used to find stresses and deflection is taken from the point in question to one end of the beam only. This is called the *one-way moment.* The absolute value of the moment will not depend on the end used. This can be illustrated by the beam shown in figure 12.6.

$$\sum M_{A\,(\text{to right end})} = -8(7) + 4(12) = -8$$

$$\sum M_{A\,(\text{to left end})} = 4(2) = +8$$

The moment obtained will depend on the location chosen. A graphical representation of the one-way moment at every point along a beam is known as a *moment diagram.* The following guidelines should be observed in constructing moment diagrams.

- Moments should be taken from the left end to the point in question. If the beam is cantilever, place the built-in end at the right.
- Clockwise moments are positive.
- Concentrated loads produce linearly increasing lines on the moment diagram.
- Uniformly distributed loads produce parabolic lines on the moment diagram.
- The maximum moment will occur when the shear ($V$) is zero.
- The moment at any point is equal to the area under the shear diagram up to that point. That is,

$$M = \int V\,dx \qquad 12.14$$

- The moment is zero at a free end or hinge.

Similarly, the sum of forces in the $y$ direction on a beam in equilibrium is zero. However, the shearing stress at a point along the beam will depend on the sum of forces and reactions from the point in question to one end only.

A *shear diagram* is drawn to represent graphically the shear at any point along a beam. The following guidelines should be observed in constructing a shear diagram.

- Loads and reactions acting up are positive.
- The shear at any point is equal to the sum of the loads and reactions from the left end to the point in question.
- Concentrated loads produce straight vertical lines on the shear diagram.

- Uniformly distributed loads produce straight sloping lines on the shear diagram.
- The magnitude of the shear at any point is equal to the slope of the moment diagram at that point.

$$V = \frac{dM}{dx} \qquad 12.15$$

*Example 12.1*

Draw the shear and moment diagrams for the following beam.

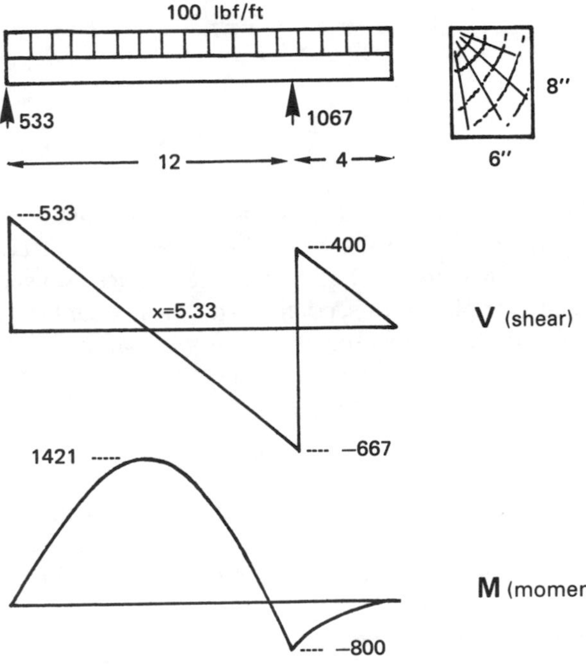


## 5 STRESSES IN BEAMS

### A. NORMAL STRESS

*Normal stress* is the type of stress experienced by a member which is axially loaded. The normal stress is the load divided by the area.

$$\sigma = \frac{F}{A} \qquad 12.16$$

Normal stress also occurs when a beam bends, as shown in figure 12.7. The lower part of the beam experiences normal tensile stress (which causes lengthening). The upper part of the beam experiences a normal compressive stress (which causes shortening). There is no stress along a horizontal plane passing through the centroid of the cross section. This plane is known as the *neutral plane* or the *neutral axis.*

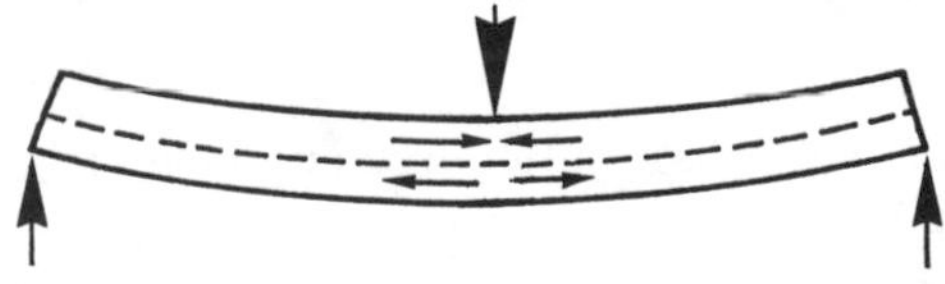

**Figure 12.7** Normal Stress Due to Bending

Although it is a normal stress, the stress produced by the bending usually is called *bending stress* or *flexure stress.* Bending stress varies with position within the beam. It is zero at the neutral axis, but it increases linearly with distance from the neutral axis.

$$\sigma_b = \frac{-My}{I_c} \qquad 12.17$$

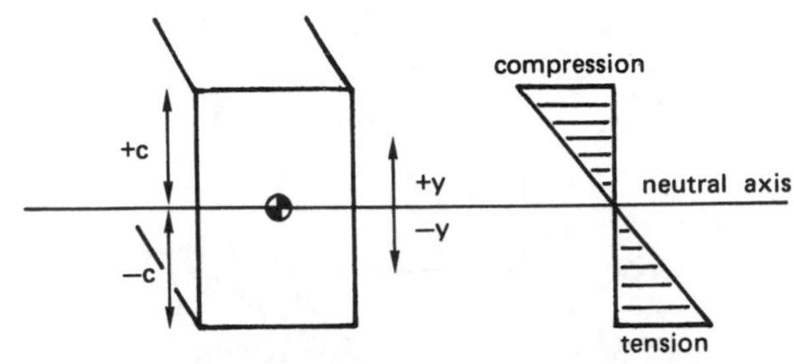


**Figure 12.8** Bending Stress Distribution in a Beam

The moment, $M$, used in equation 12.17 is the *one-way moment* previously discussed. $I_c$ is the centroidal moment of inertia of the beam's cross sectional area. The negative sign in equation 12.17 typically is omitted. However, it is required to be consistent with the convention that compression is negative.

Since the maximum stress will govern the design, $y$ can be set equal to $c$ to obtain the maximum stress. $c$ is the distance from the neutral axis to the *extreme fiber.*

$$\sigma_{b,\text{max}} = \frac{Mc}{I_c} \qquad 12.18$$

For any given structural shape, $c$ and $I_c$ are fixed. Therefore, these two terms can be combined into the *section modulus,* $Z$.

$$\sigma_{b,\text{max}} = \frac{M}{Z} \qquad 12.19$$

$$Z = \frac{I_c}{c} \qquad 12.20$$

For most beams, the section modulus, $Z$, is constant along the length of the beam. Equation 12.19 shows that the maximum stress along the length of a beam is proportional to the moment at that point. The location of the maximum bending moment is called the *dangerous section.* The dangerous section can be found directly from a moment or shear diagram of the beam.

If an axial member is loaded eccentrically, it will experience axial stress (equation 12.16) as well as bending stress (equation 12.17). This is illustrated by figure 12.9, in which a load is not applied to the centroid of a column's cross sectional area.

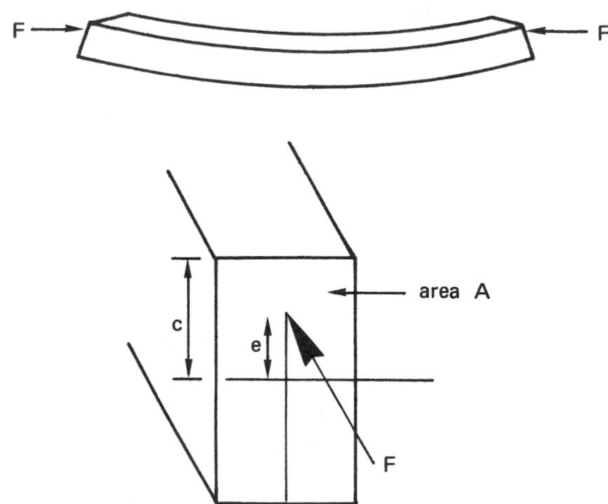


**Figure 12.9** Eccentric Loading of an Axial Member

Because the beam bends and supports a compressive load, the stress produced is a sum of bending and normal stress.

$$\sigma_{\text{max, min}} = \frac{F}{A} \pm \frac{Mc}{I_c} = \frac{F}{A} \pm \frac{Fec}{I_c} \qquad 12.21$$

If a cross section is loaded with an eccentric compressive load, part of the section can be in tension. This is illustrated in example 12.3. There will be no stress sign reversal, however, as long as the load is applied within a diamond-shaped area formed from the middle-thirds of the centroidal axes. This area is known as the *kern* or the *kernel*. It is particularly important to keep eccentric compressive loads within the kern on concrete and masonry piers since these materials do not tolerate tension. (The kern of a circular shaft or a round beam is outlined by a circle whose radius is one-quarter of the shaft radius.)

The *elastic strain energy* stored in a beam experiencing a moment (bending) is

$$U = \frac{1}{2} \int \frac{M^2}{EI} dx \qquad 12.22$$

B. SHEAR STRESS

Normal stress is produced when a load is absorbed by an area normal to it. *Shear stress* is produced by a load being carried by an area parallel to the load. This is illustrated in figure 12.11.

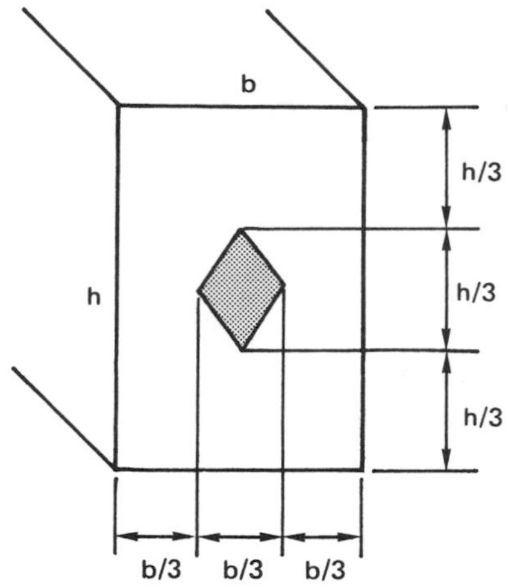


**Figure 12.10** The Kern

The average shear stress experienced by a pin, a bolt, or a rivet in single shear (as illustrated in figure 12.11) is given by equation 12.23. Because it gives an average value over the cross section of the shear member, this equation should be used only when the loading is low or when there is multiple redundancy in the shear group.

$$\tau = \frac{F}{A} \qquad 12.23$$

The actual shear stress in a beam is dependent on the location within the beam, just as was the bending stress. Shear stress is zero at the top and bottom surfaces of a beam and maximum at the neutral axis. This is illustrated in figure 12.12.

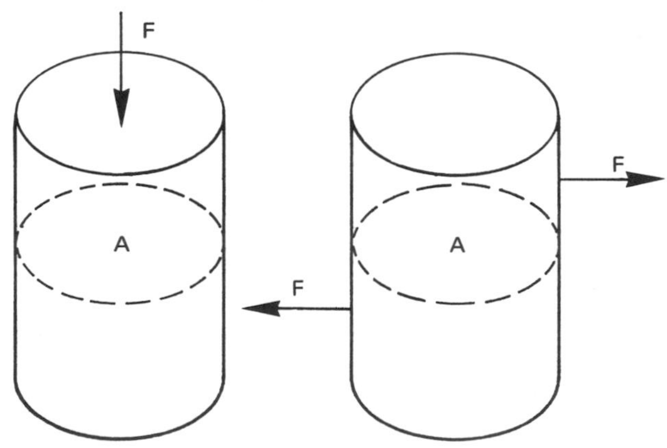


**Figure 12.11** Normal and Shear Stresses

The shear stress distribution within a beam is given by equation 12.24.

$$\tau = \frac{QV}{Ib} \qquad 12.24$$

MECH OF MATL

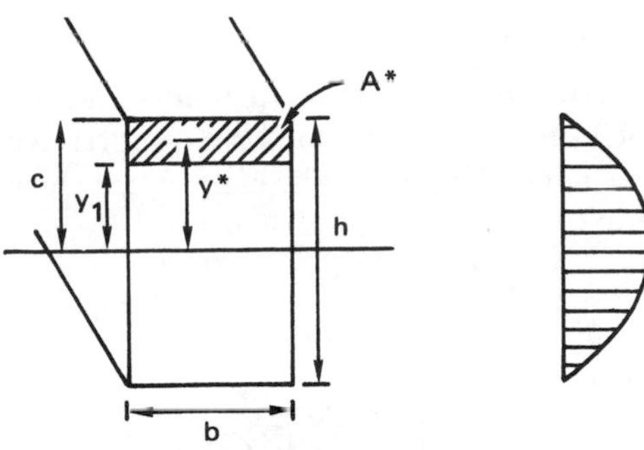


**Figure 12.12** Shear Stress Distribution in a Rectangular Beam

$V$ is the shear (in pounds) at the section where the shear stress is wanted. $V$ can be found from a shear diagram. $I$ is the beam's centroidal moment of inertia. $b$ is the width of the beam at the depth $y_1$ within the beam where the shear stress is wanted. $Q$ is the *statical moment*[1], as defined by equation 12.25.

$$Q = \int_{y_1}^{c} y\, dA \qquad 12.25$$

For rectangular beams, $dA = bdy$. Equation 12.25 can be simplified to equation 12.26 for rectangular beams.

$$Q = y^* A^* \qquad 12.26$$

Equation 12.26 says that the statical moment at a location $y_1$ within a rectangular beam is equal to the product of the area above $y_1$ and the distance from the centroidal axis to the centroid of $A^*$.

The maximum shear stress in a rectangular beam is

$$\tau_{\text{max}} = \frac{3V}{2A} = \frac{3V}{2bh} \qquad 12.27$$

For a round beam of radius $r$ and area $A$, the maximum shear stress is

$$\tau_{\text{max}} = \frac{4V}{3A} = \frac{4V}{3\pi r^2} \qquad 12.28$$

The shear, $V$, used in equations 12.27 and 12.28 is the *one-way shear*.

*Example 12.2*

What are the maximum shear and bending stresses for the beam shown in example 12.1?

From the shear diagram, the maximum shear is −667 pounds. From equation 12.27, the maximum shear stress is

$$\tau_{\text{max}} = \frac{(3)(-667)}{(2)(6)(8)} = -20.8 \text{ psi}$$

[1] The statical moment also is known as the *first moment of the area*.

From the moment diagram, the maximum moment is +1421 ft-lbf. The centroidal moment of inertia is

$$I_c = \frac{(6)(8)^3}{12} = 256 \text{ in}^4$$

From equation 12.18, the maximum bending stress is

$$\sigma_{b,\text{max}} = \frac{(1421)(12)(4)}{256} = 266.4 \text{ psi}$$

*Example 12.3*

The chain hook shown carries a load of 500 pounds. What are the minimum and maximum stresses in the vertical portion of the hook?

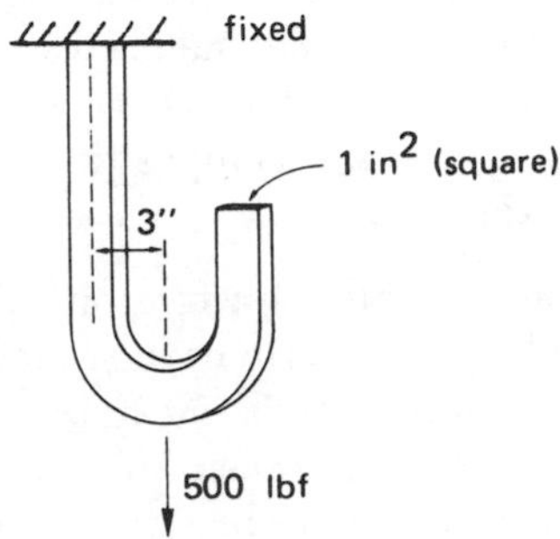


The hook is loaded eccentrically because the load and the supporting force are not in line. The centroidal moment of inertia of the 1″ × 1″ section is

$$I_c = \frac{bh^3}{12} = \frac{(1)(1)^3}{12} = 0.0833 \text{ in}^4$$

From equation 12.21,

$$\begin{aligned}\sigma_{\text{max, min}} &= \frac{500}{1} \pm \frac{(500)(3)(0.5)}{0.0833} \\ &= 500 \pm 9000 \\ &= +9500 \text{ and } -8500\end{aligned}$$

The 500 psi direct stress is tensile. However, the flexural compressive stress of 9000 psi counteracts this tensile stress, resulting in 8500 psi compressive stress at the outer face of the hook. The stress is 9500 psi tension at the inner face.

## 6 STRESSES IN COMPOSITE STRUCTURES

A *composite structure* is one in which two or more different materials are used, each carrying a part of the load. Unless all the various materials used have the same modulus of elasticity, the stress analysis will be dependent on the assumptions made.

Some simple composite structures can be analyzed using the assumption of *consistent deformations.* This is illustrated in examples 12.4 and 12.5. The technique used to analyze structures for which the strains are consistent is known as the *transformation method.*

*step 1*: Determine the modulus of elasticity for each of the materials used in the structure.

*step 2*: For each of the materials used, calculate the ratio

$$n = \frac{E}{E_{\text{weakest}}} \qquad 12.29$$

$E_{\text{weakest}}$ is the smallest modulus of elasticity of any of the materials used in the composite structure.

*step 3*: For all of the materials except the weakest, multiply the actual material stress area by $n$. Consider this expanded (*transformed*) area to have the same composition as the weakest material.

*step 4*: If the structure is a tension or compression member, the distribution or placement of the transformed area is not important. Just assume that the transformed areas carry the axial load. For beams in bending, the transformed area can add to the width of the beam, but it cannot change the depth of the beam or the thickness of the reinforcing.

*step 5*: For compression or tension numbers, calculate the stresses in the weakest and stronger materials.

$$\sigma_{\text{weakest}} = \frac{F}{A_t} \qquad 12.30$$

$$\sigma_{\text{stronger}} = \frac{nF}{A_t} \qquad 12.31$$

*step 6*: For beams in bending, proceed through step 9. Find the centroid of the transformed beam.

*step 7*: Find the centroidal moment of inertia of the transformed beam, $I_{ct}$.

*step 8*: Find $V_{\text{max}}$ and $M_{\text{max}}$ by inspection or from the shear and moment diagrams.

*step 9*: Calculate the stresses in the weakest and stronger materials.

$$\sigma_{\text{weakest}} = \frac{Mc_{\text{weakest}}}{I_{ct}} \qquad 12.32$$

$$\sigma_{\text{stronger}} = \frac{nMc_{\text{stronger}}}{I_{ct}} \qquad 12.33$$

*Example 12.4*

Find the stress in the steel inner cylinder and the copper tube which surrounds it if a uniform compressive load of 100 kips is applied axially. The copper and the steel are well bonded. Use $E_{\text{steel}} = 3$ EE7 psi and $E_{\text{copper}} = 1.75$ EE7 psi.

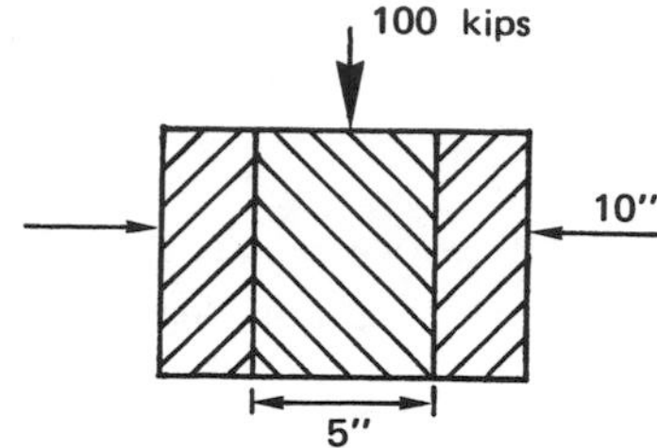


$$n = \frac{3 \text{ EE7}}{1.75 \text{ EE7}} = 1.714$$

The actual steel area is $\frac{1}{4}\pi(5)^2 = 19.63$ in$^2$.

The actual copper area is $\frac{1}{4}\pi[(10)^2 - (5)^2] = 58.9$ in$^2$.

The transformed area is $A_t = 58.9 + 1.714\,(19.63) = 92.55$ in$^2$.

$$\sigma_{\text{copper}} = \frac{100{,}000}{92.55} = 1080.5 \text{ psi}$$

$$\sigma_{\text{steel}} = (1.714)(1080.5) = 1852.0 \text{ psi}$$

*Example 12.5*

Find the maximum bending stress in the steel-reinforced wood beam shown at a point where the moment is 40,000 ft-lbf. Use $E_{\text{steel}} = 3$ EE7 psi and $E_{\text{wood}} = 1.5$ EE6 psi.

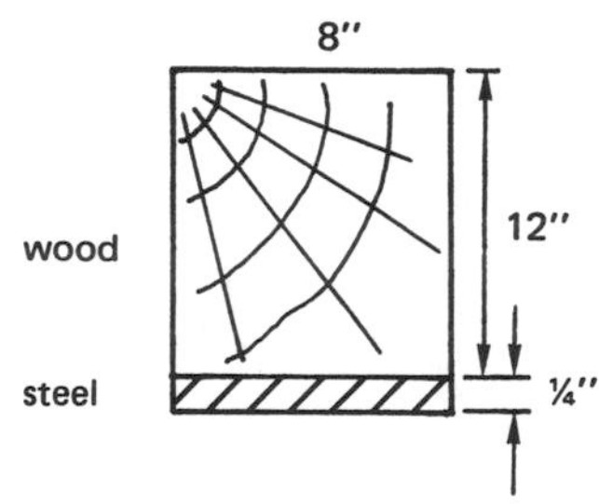


$$n = \frac{3 \text{ EE7}}{1.5 \text{ EE6}} = 20$$

The actual steel area is $(0.25)(8) = 2$.

The area of the steel is expanded to $20(2) = 40$. Since the depth of beam and reinforcement cannot be increased, the width must increase. The 160″ dimension is arrived at by dividing the area of 40 square inches by the thickness of $\frac{1}{4}''$.

MECH OF MATL

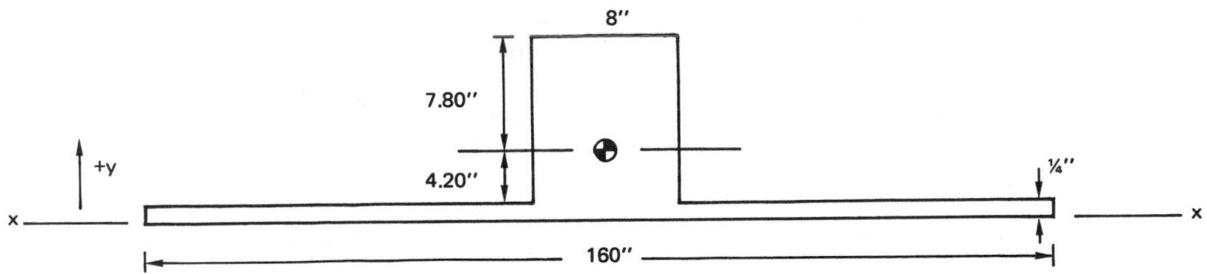


The centroid is located at $\overline{y} = 4.45$ inches from the x-x axis. The centroidal moment of inertia of the transformed section is

$$I_c = \frac{(8)(7.8)^3}{3} + \frac{(8)(4.2)^3}{3} + \frac{(160)(0.25)^3}{12} + (160)(0.25)\left(4.2 + \frac{0.25}{2}\right)^2 = 2211.5 \text{ in}^4$$

Then, from equations 12.32 and 12.33,

$$\sigma_{\text{max, wood}} = \frac{(40,000)(12)(7.8)}{(2211.5)} = 1692 \text{ psi}$$

$$\sigma_{\text{max, steel}} = \frac{(20)(40,000)(12)(4.45)}{2211.5} = 19,320 \text{ psi}$$

## 7 ALLOWABLE STRESSES

Once the actual stresses are known, they must be compared to allowable stresses. If the allowable stress is calculated, it should be based on the yield stress and a reasonable factor of safety. This is known as the *allowable stress design method* or the *working stress design method.*

$$\sigma_a = \frac{S_y}{F.S.} \qquad 12.34$$

For steel, the factor of safety ranges from 1.5 to 2.5, depending on the type of steel and application.

The allowable stress method is being replaced in structural work by the *load factor design method*, also known as the *ultimate strength method* and the *plastic design method.* In this method, the applied loads are multiplied by a load factor. The product must be less than the structural member's ultimate strength, usually determined from a table.

## 8 BEAM DEFLECTIONS

### A. DOUBLE INTEGRATION METHOD

The deflection and the slope of a loaded beam are related to the applied moment and shear by equations 12.35 through 12.38.

$$y = \text{deflection} \qquad 12.35$$

$$y' = \frac{dy}{dx} = \text{slope} \qquad 12.36$$

$$y'' = \frac{d^2y}{dx^2} = \frac{M}{EI} \qquad 12.37$$

$$y''' = \frac{d^3y}{dx^3} = \frac{V}{EI} \qquad 12.38$$

If the moment function, $M(x)$, is known for a section of the beam, the deflection at any point can be found from equation 12.39.

$$y = \frac{1}{EI}\int\left[\int M(x)\,dx\right]dx \qquad 12.39$$

In order to find the deflection, constants must be introduced during the integration process. These constants can be found from table 12.2.



**Table 12.2**
Beam Boundary Conditions

| end condition | y | y′ | y″ | V | M |
|---|---|---|---|---|---|
| simple support | 0 | | | | 0 |
| built-in support | 0 | 0 | | | |
| free end | | | 0 | 0 | 0 |
| hinge | | | | | 0 |

*Example 12.6*

Find the tip deflection of the beam shown. $EI$ is 5 EE10 lbf-in$^2$ everywhere on the beam.

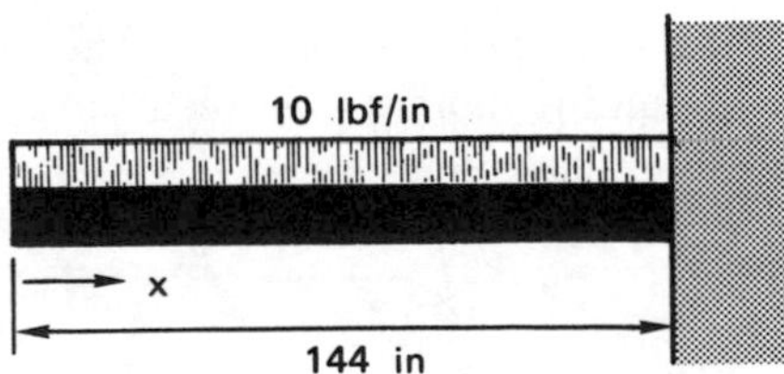


The moment at any point $x$ from the left end of the beam is

$$M(x) = (-10)(x)\left(\frac{1}{2}x\right) = -5x^2$$

This is negative by the left-hand rule convention. From equation 12.37,

$$y'' = \frac{M}{EI}$$

So,

$$EIy'' = -5x^2$$

$$EIy' = \int -5x^2\,dx = -\frac{5}{3}x^3 + C_1$$

Since $y' = 0$ at a built-in support (table 12.2) and $x = 144$ inches at the built-in support,

$$0 = -\frac{5}{3}(144)^3 + C_1$$

$$C_1 = 4.98 \text{ EE6}$$

$$EIy = \int \left(-\frac{5}{3}x^3 + 4.98 \text{ EE6}\right) dx$$

$$= -\frac{5}{12}x^4 + (4.98 \text{ EE6})x + C_2$$

Again, $y = 0$ at $x = 144$, so $C_2 = -5.38$ EE8.

Therefore, the deflection as a function of $x$ is

$$y = \left(\frac{1}{EI}\right)\left[\left(-\frac{5}{12}\right)x^4 + (4.98 \text{ EE6})x - 5.38 \text{ EE8}\right]$$

At the tip $x = 0$, so the deflection is

$$y_{\text{tip}} = \frac{-5.38 \text{ EE8}}{5 \text{ EE10}} = -0.0108 \text{ inches}$$

B. MOMENT AREA METHOD

The moment area method is a semi-graphical technique which is applicable whenever slopes of deflection beams are not too great. This method is based on the following two theorems.

**Theorem I**: The angle between tangents at any two points on the *elastic line* of a beam is equal to the area of the moment diagram between the two points divided by $EI$. That is,

$$\theta = \int \frac{M(x)\,dx}{EI} \qquad 12.40$$

**Theorem II**: One point's deflection away from the tangent of another point is equal to the *statical moment* of the bending moment between those two points divided by $EI$. That is,

$$y = \int \frac{xM(x)\,dx}{EI} \qquad 12.41$$

The application of these two theorems is aided by the following two comments.

- If $EI$ is constant, the statical moment $\int xM(x)\,dx$ can be calculated as the product of the total moment diagram area times the distance from the point whose deflection is wanted to the centroid of the moment diagram.
- If the moment diagram has positive and negative parts (areas above and below the zero line), the statical moment should be taken as the sum of two products, one for each part of the moment diagram.

*Example 12.7*

Find the deflection, $y$, and the angle, $\phi$ (in radians), at the free end of the cantilever beam shown.

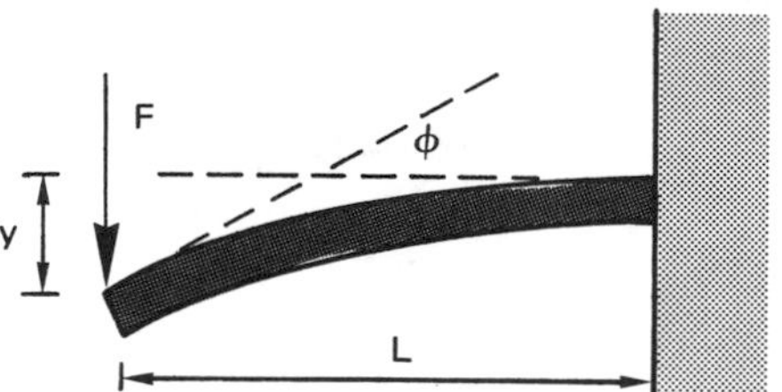


The deflection angle, $\phi$, is the angle between the tangents at the free and built-in ends (Theorem I). The moment diagram is

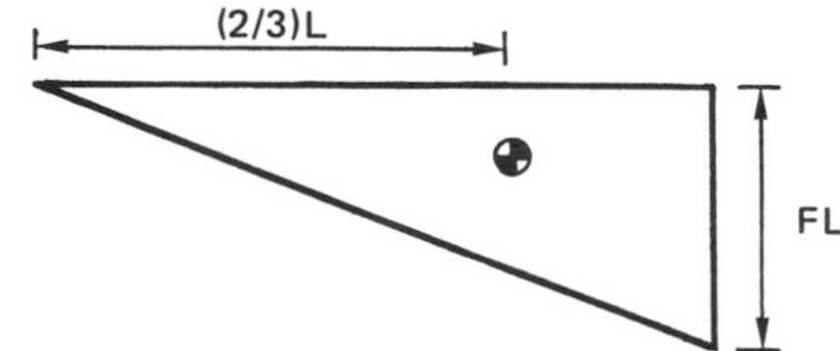


The area of the moment diagram is

$$\frac{1}{2}(FL)(L) = \frac{1}{2}FL^2$$

From Theorem I,

$$\phi = \frac{FL^2}{2EI}$$

From Theorem II,

$$y = \frac{FL^2}{2EI}\left(\frac{2}{3}L\right) = \frac{FL^3}{3EI}$$

*Example 12.8*

Find the deflection of the free end of the cantilever beam shown.

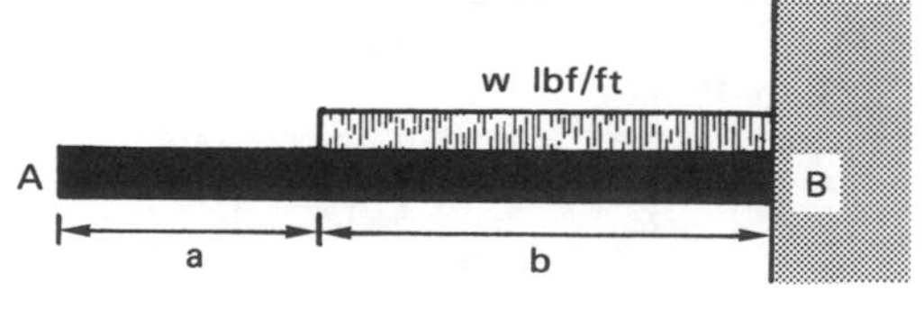

The distance from point $A$ (where the deflection is wanted) to the centroid is $(a + 0.75b)$. The area of the moment diagram is $(wb^3/6)$. From Theorem II,

$$y = \frac{wb^3}{6EI}(a + 0.75b)$$

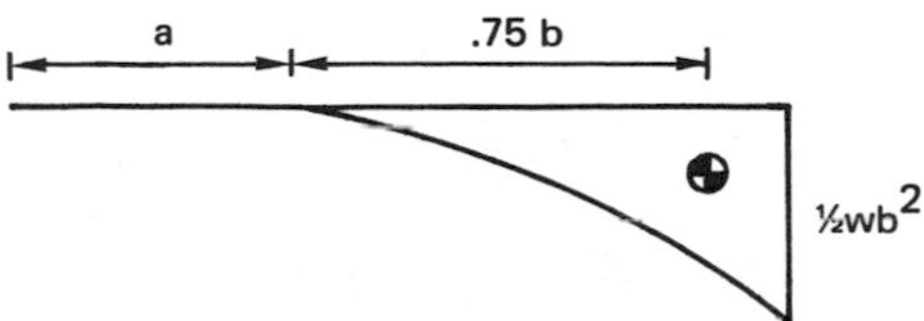


## C. STRAIN ENERGY METHOD

The deflection at a point of load application can be found by the strain energy method. This method equates the external work to the total internal strain energy as given by equations 12.8, 12.22, and 12.81. Since work is a force moving through a distance (which in this case is the deflection) we can write equation 12.42.

$$\frac{1}{2}Fy = \sum U \qquad 12.42$$

*Example 12.9*

Find the deflection at the tip of the stepped beam shown.

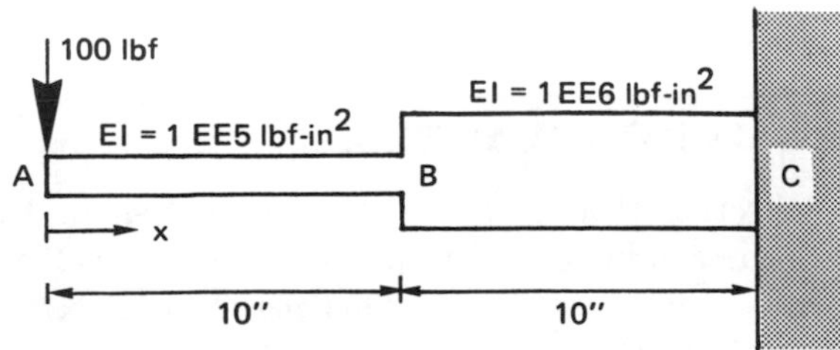


In section A–B: $M = 100x$ in-lbf

From equation 12.22,

$$U = \frac{1}{2}\int_0^{10} \frac{(100x)^2}{1 \text{ EE5}} dx = 16.67 \text{ in-lbf}$$

In section B-C: $M = 100x$

$$U = \frac{1}{2}\int_{10}^{20} \frac{(100x)^2}{1 \text{ EE6}} dx = 11.67 \text{ in-lbf}$$

Equating the internal work ($U$) and the external work,

$$16.67 + 11.67 = \frac{1}{2}(100)y$$
$$y = 0.567 \text{ in}$$

## D. CONJUGATE BEAM METHOD

The *conjugate beam method* changes a deflection problem into one of drawing moment diagrams. The method has the advantage of being able to handle beams of varying cross sections and materials. It has the disadvantage of not easily being able to handle beams with two built-in ends. The following steps constitute the conjugate beam method.

*step 1*: Draw the moment diagram for the beam as it is actually loaded.

*step 2*: Construct the $M/EI$ diagram by dividing the value of $M$ at every point along the beam by the product of $EI$ at that point. If the beam is of constant cross section, $EI$ will be constant, and the $M/EI$ diagram will have the same shape as the moment diagram. However, if the beam cross section varies with $x$, $I$ will change. In that case, the $M/EI$ diagram will not look the same as the moment diagram.

*step 3*: Draw a conjugate beam of the same length as the original beam. The material and the cross sectional area of this conjugate beam are not relevant.

(a) If the actual beam is simply supported at its ends, the conjugate beam will be simply supported at its ends.

(b) If the actual beam is simply supported away from its ends, the conjugate beam has hinges at the support points.

(c) If the actual beam has free ends, the conjugate beam has built-in ends.

(d) If the actual beam has built-in ends, the conjugate beam has free ends.

*step 4*: Load the conjugate beam with the $M/EI$ diagram. Find the conjugate reactions by methods of statics. Use the superscript, *, to indicate conjugate parameters.

*step 5*: Find the conjugate moment at the point where the deflection is wanted. The deflection is numerically equal to the moment as calculated from the conjugate beam forces.

*Example 12.10*

Find the deflections at the two load points. $EI$ has a constant value of 2.356 EE7 lbf-in$^2$.

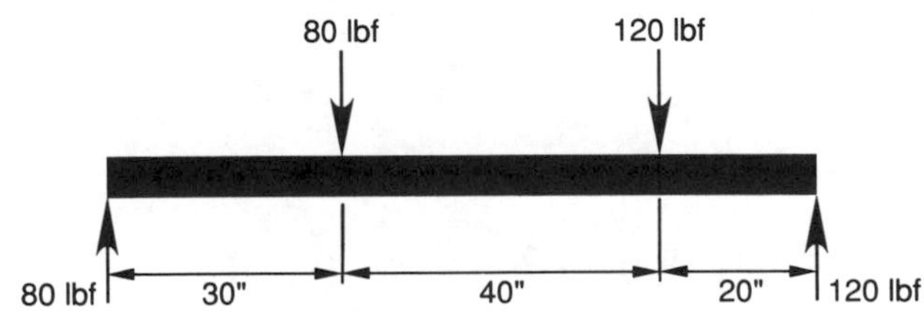

*step 1*: The moment diagram for the actual beam is

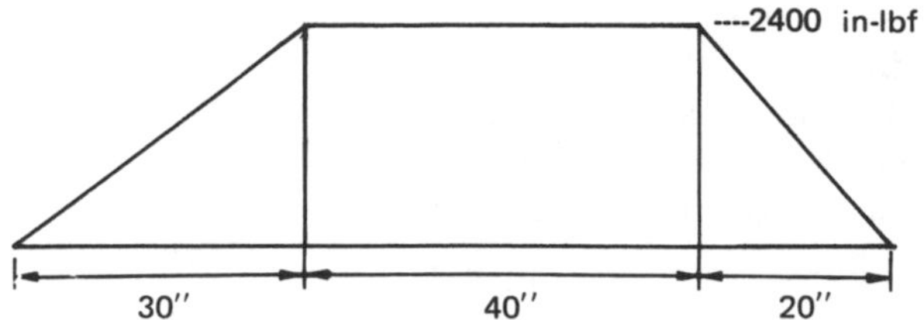


*steps 2, 3, and 4*: Since the cross section is constant, the conjugate load has the same shape as the original moment diagram. The peak load on the conjugate beam is

$$\frac{2400 \text{ in-lbf}}{2.356 \text{ EE7 lbf-in}^2} = 1.019 \text{ EE-4 (1/in)}$$

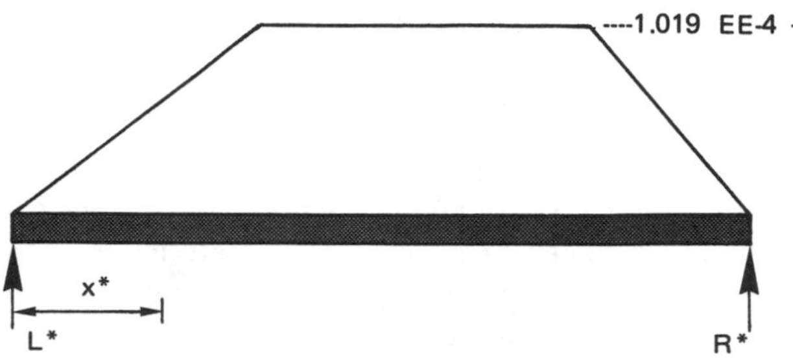


The conjugate reaction, $L^*$, is found by the following method. The loading diagram is assumed to be made up of a rectangular load and two "negative" triangular loads. The area of the rectangular load (which has a centroid at $x^* = 45$) is (90)(1.019 EE–4) = 9.171 EE–3.

Similarly, the area of the left dashed triangle (which has a centroid at $x^* = 10$) is $\frac{1}{2}$(30)(1.019 EE–4) = 1.529 EE–3. The area of the right dashed triangle (which has a centroid at $x^* = 83.33$) is $\frac{1}{2}$(20)(1.019 EE–4) = 1.019 EE–3.

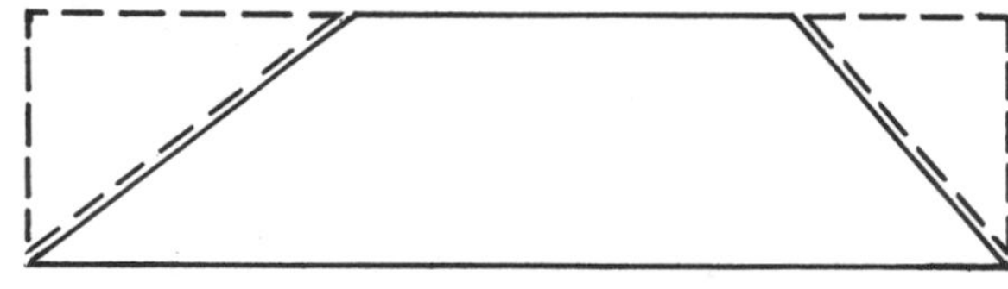

$$\begin{aligned}\sum M^*_{L^*} &= 90R^* + (1.019 \text{ EE-3})(83.3) \\ &\quad + (1.529 \text{ EE-3})(10) - (9.171 \text{ EE-3})(45) \\ &= 0\end{aligned}$$

$$R^* = 3.472 \text{ EE-3 } \frac{1}{\text{in}}$$

Then,

$$\begin{aligned}L^* &= (9.171 - 1.019 - 1.529 - 3.472) \text{ EE-3} \\ &= 3.151 \text{ EE-3 } \frac{1}{\text{in}}\end{aligned}$$

*step 5*: The conjugate moment at $x^* = 30$ is

$$\begin{aligned}M^* &= (3.151 \text{ EE-3})(30) + (1.529 \text{ EE-3})(30 - 10) \\ &\quad - (9.171 \text{ EE-3})\left(\frac{30}{90}\right)(15) \\ &= 7.926 \text{ EE-2 in}\end{aligned}$$

The conjugate moment at the right-most load is

$$\begin{aligned}M^* &= (3.472 \text{ EE-3})(20) + (1.019 \text{ EE-3})(13.3) \\ &\quad - (9.171 \text{ EE-3})\left(\frac{20}{90}\right)(10) \\ &= 6.266 \text{ EE-2 in}\end{aligned}$$

### E. TABLE LOOK-UP METHOD

Appendix A is an extensive listing of the most commonly needed beam formulas. The use of these formulas is recommended whenever they can be applied singly or as part of a superposition solution.

### F. METHOD OF SUPERPOSITION

If the deflection at a point is due to the combined action of two or more loads, the deflections at that point due to the individual loads can be added to find the total deflection.

## 9 TRUSS DEFLECTIONS

### A. STRAIN-ENERGY METHOD

The deflection of a truss at the point of a single load application can be found by the *strain-energy method* if all member forces are known. This method is illustrated by example 12.11.

*Example 12.11*

Find the vertical deflection of point $A$ under the external load of 707 pounds. $AE$ = 10 EE5 pounds for all members. The internal forces have been determined.

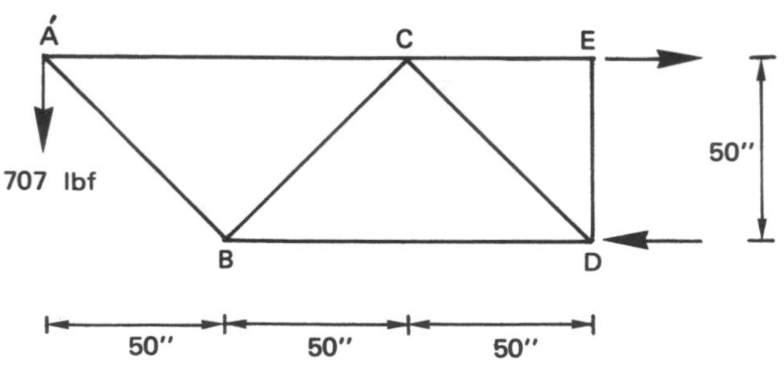

The length of member $AB$ is $\sqrt{(50)^2+(50)^2} = 70.7$ inches. From equation 12.8, the internal strain energy in member $AB$ is

$$U = \frac{(-1000)^2(70.7)}{2\,(10\text{ EE5})} = 35.4\text{ in-lbf}$$

Similarly, the energy in all members can be determined.

| Member | L | F | U |
|---|---|---|---|
| AB | 70.7 | −1000 | +35.4 |
| BC | 70.7 | +1000 | +35.4 |
| AC | 100 | +707 | +25.0 |
| BD | 100 | −1414 | +100.0 |
| CD | 70.7 | −1000 | +35.4 |
| CE | 50 | +2121 | +112.5 |
| DE | 50 | +707 | +12.5 |
| | | | 356.2 |

The work done by a constant force $F$ moving through a distance $y$ is $Fy$. In this case, the force increases with $y$. The average force is $\frac{1}{2}F$. The external work is $W_{\text{ext}} = \frac{1}{2}(707)y$, so

$$\left(\frac{1}{2}\right)(707)y = 356.2$$

$$y = 1\text{ inch}$$

## B. VIRTUAL WORK METHOD (HARDY CROSS METHOD)

An extension of the strain-energy method results in an easy procedure for computing the deflection of *any* point on a truss.

*step 1*: Draw the truss twice.

*step 2*: On the first truss, place all the actual loads.

*step 3*: Find the forces, $S$, due to the actual applied loads in all the members.

*step 4*: On the second truss, place a dummy one pound load in the direction of the desired displacement.

*step 5*: Find the forces, $u$, due to the one pound dummy load in all members.

*step 6*: Find the desired displacement from equation 12.43.

$$\delta = \sum \frac{SuL}{AE} \qquad 12.43$$

In equation 12.43, the summation is over all truss members which have non-zero forces in *both* trusses.

*Example 12.12*

What is the horizontal deflection of joint $F$ on the truss shown? Use $E = 3$ EE7 psi. Joint $A$ is restrained horizontally. (Member areas are given in the table below and have been chosen for convenience.)

*steps 1 and 2*: Use the truss as drawn.

*step 3*: The forces in all the truss members are summarized in step 5.

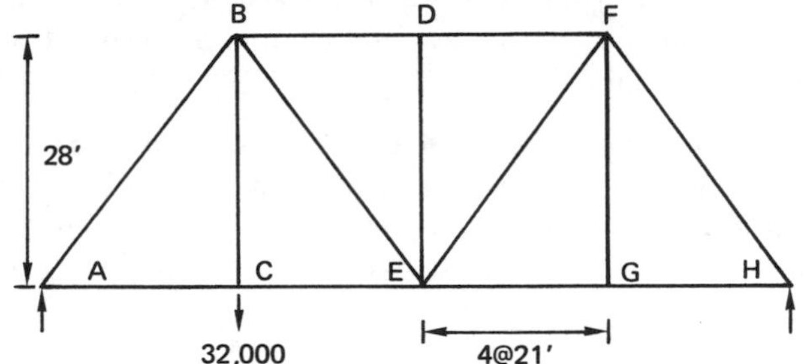


*step 4*: Draw the truss and load it with a unit horizontal force at point $F$.

*step 5*: Find the forces, $u$, in all members of the second truss. These are summarized in the following table. Notice the sign convention: + for tension and − for compression.

Table from example 12.12

| member | S(lbf) | u | L(ft) | A(in$^2$) | $\frac{SuL}{AE}$(ft) |
|---|---|---|---|---|---|
| AB | −30,000 | 5/12 | 35 | 17.5 | −8.33 EE−4 |
| CB | 32,000 | 0 | 28 | 14 | 0 |
| EB | −10,000 | −5/12 | 35 | 17.5 | 2.78 EE−4 |
| ED | 0 | 0 | 28 | 14 | 0 |
| EF | 10,000 | 5/12 | 35 | 17.5 | 2.78 EE−4 |
| GF | 0 | 0 | 28 | 14 | 0 |
| HF | −10,000 | −5/12 | 35 | 17.5 | 2.78 EE−4 |
| BD | −12,000 | 1/2 | 21 | 10.5 | −4.00 EE−4 |
| DF | −12,000 | 1/2 | 21 | 10.5 | −4.00 EE−4 |
| AC | 18,000 | 3/4 | 21 | 10.5 | 9.00 EE−4 |
| CE | 18,000 | 3/4 | 21 | 10.5 | 9.00 EE−4 |
| EG | 6000 | 1/4 | 21 | 10.5 | 1.00 EE−4 |
| GH | 6000 | 1/4 | 21 | 10.5 | 1.00 EE−4 |
| | | | | | 12.01 EE−4 (ft) |

MECH OF MATL

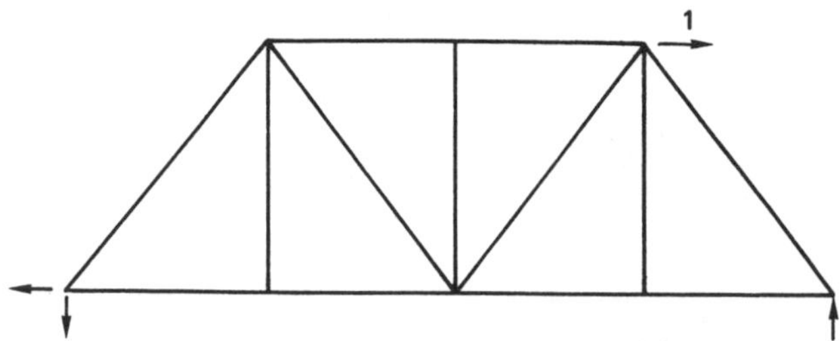


Since 12.01 EE−4 is positive, the deflection is in the direction of the dummy unit load. In this case, the deflection is to the right.

## 10 COMBINED STRESSES

Most practical cases of combined stresses have normal stresses on two perpendicular planes and a known shear stress acting parallel to these two planes. Based on knowledge of these stresses, the shear and the normal stresses on all other planes can be found from conditions of equilibrium.

Under any condition of stress at a point, a plane can be found where the shear stress is zero. The normal stresses on this plane are known as the *principal stresses*. The principal stresses are the maximum and the minimum stresses at the point in question.

The normal and shear stresses on a plane whose normal line is inclined an angle $\theta$ from the horizontal are given by equations 12.44 and 12.45.

$$\sigma_\theta = \frac{1}{2}(\sigma_x + \sigma_y) + \frac{1}{2}(\sigma_x - \sigma_y)\cos 2\theta + \tau \sin 2\theta \qquad 12.44$$

$$\tau_\theta = -\frac{1}{2}(\sigma_x - \sigma_y)\sin 2\theta + \tau\cos 2\theta \qquad 12.45$$

The maximum and minimum values of $\sigma_\theta$ and $\tau_\theta$ (as $\theta$ is varied) are the principal stresses. These are given by equations 12.46 and 12.47.

$$\sigma(\text{max, min}) = \tfrac{1}{2}(\sigma_x + \sigma_y) \pm \tau(\text{max}) \qquad 12.46$$

$$\tau(\text{max, min}) = \pm\tfrac{1}{2}\sqrt{(\sigma_x - \sigma_y)^2 + (2\tau)^2} \qquad 12.47$$

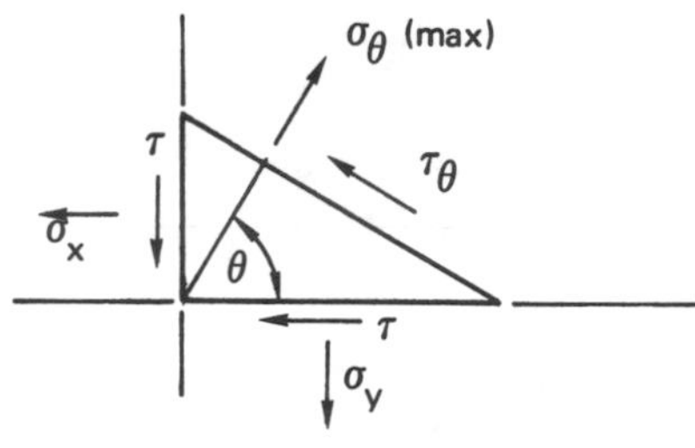


**Figure 12.13** Plane of Principal Stresses

The angles of the planes on which the normal stresses are minimum and maximum are given by equation 12.48. $\theta$ is measured from the $x$ axis, clockwise if negative and counterclockwise if positive. Equation 12.48 will yield two angles. These angles must be used in equation 12.44 to determine which angle corresponds to the minimum normal stress and which angle corresponds to the maximum normal stress.

$$\theta = \frac{1}{2}\arctan\left(\frac{2\tau}{\sigma_x - \sigma_y}\right) \qquad 12.48$$

The angles of the planes on which the shear stress is minimum and maximum are given by equation 12.49. The same angle sign convention used for equation 12.48 applies to equation 12.49.

$$\theta = \frac{1}{2}\arctan\left(\frac{\sigma_x - \sigma_y}{-2\tau}\right) \qquad 12.49$$

Proper sign convention must be adhered to when using equations 12.44 through 12.49. Normal tensile stresses are positive; normal compressive stresses are negative. In two dimensions, shear stresses are designated as clockwise (positive) or counterclockwise (negative).

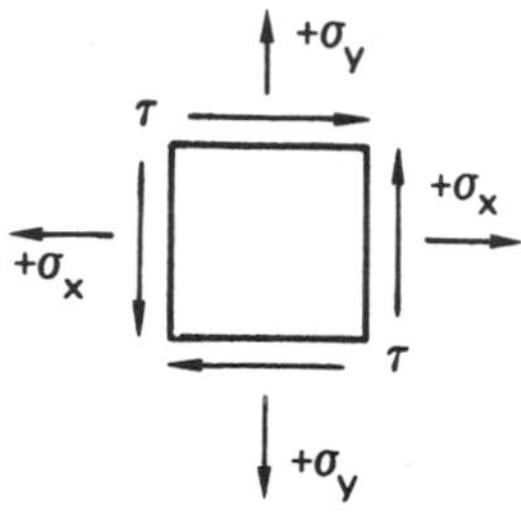


**Figure 12.14** Sign Convention

*Example 12.13*

Find the maximum shear stress and the maximum normal stress on the object shown.

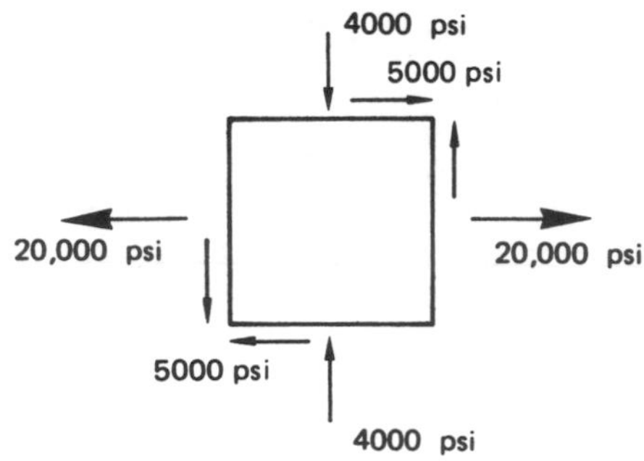

By the sign convention of figure 12.14, the 4000 psi is negative. From equation 12.47, the maximum shear stress is

$$\tau_{max} = \frac{1}{2}\sqrt{[20{,}000 - (-4000)]^2 + [(2)(5000)]^2}$$
$$= 13{,}000 \text{ psi}$$

From equation 12.46, the maximum normal stress is

$$\sigma_{max} = \frac{1}{2}[20{,}000 + (-4000)] + 13{,}000$$
$$= 21{,}000 \text{ psi (tension)}$$

## 11 DYNAMIC LOADING

If a load is applied suddenly to a structure, the transient response may create stresses greater than would normally be calculated from the concepts of statics and mechanics of materials alone. Although a *dynamic analysis* of the structure is appropriate, the procedure is extremely lengthy and complicated. Therefore, arbitrary dynamic factors are applied to the static stress. For example, if the load is applied quickly compared to the natural period of the structure, a dynamic factor of 2 can be used. This assumes that the load is applied as a ramp function.

## 12 INFLUENCE DIAGRAMS

Shear, moment, and reaction influence diagrams (influence lines) can be drawn for any point on a beam or truss. This is a necessary first step in the evaluation of stresses induced by moving loads. It is important to realize, however, that the influence diagram applies only to one point on the beam or truss.

### A. INFLUENCE DIAGRAMS FOR BEAM REACTIONS

In a typical problem, the load is fixed in position, and the reactions do not change. If a load is allowed to move across a beam, the reactions will vary. An influence diagram can be used to investigate the value of a chosen reaction as the load position varies.

To make the influence diagram as general in application as possible, the load is taken as one pound. As an example, consider a 20 foot, simply supported beam, and determine the effect on the left reaction of moving a one-pound load across the beam.

If the load is directly over the right reaction ($x = 0$), the left reaction will not carry any load. Therefore, the ordinate of the influence diagram is zero at that point. (Even though the right reaction supports one pound, this influence diagram is being drawn for one point only—the left reaction.) Similarly, if the load is directly over the left reaction ($x = L$), the ordinate of the influence diagram will be 1. Basic statics can be used to complete the rest of the diagram, as shown in figure 12.15.

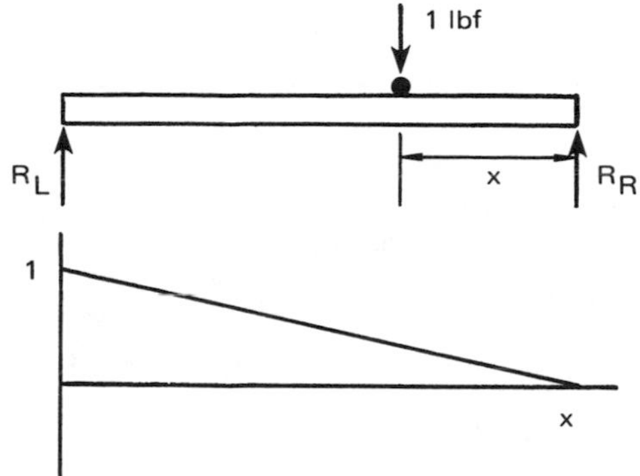


**Figure 12.15** Influence Diagram for Reaction of Simple Beam

We can use this rudimentary example of an influence diagram to calculate the left reaction for any placement of any load (not just one-pound loads) by multiplying the actual load by the ordinate of the influence diagram.

$$R_L = P \times \text{ordinate} \qquad 12.50$$

Even though the influence diagram was drawn for a point load, it can still be used when the beam carries a uniformly distributed load. In the case of a uniform load of $w$ lbf/ft distributed over the beam from $x_1$ to $x_2$, the left reaction can be calculated from equation 12.51.

$$R_L = \int_{x_1}^{x_2} (w \times \text{ordinate})dx$$
$$= w \times \text{area under curve} \qquad 12.51$$

*Example 12.14*

A 500 lbf load is placed 15 feet from the right end of a 20 foot, simply supported beam. Use the influence diagram to determine the left reaction.

Since the influence line increases linearly from 0 to 1, the ordinate is the ratio of position to length. That is, the ordinate is $15/20 = 0.75$. The left reaction is

$$R_L = (0.75)(500) = 375 \text{ lbf}$$

*Example 12.15*

A uniform load of 15 lbf/ft is distributed between $x = 4$ and $x = 10$ along a 20 foot, simply supported beam. What is the left reaction?

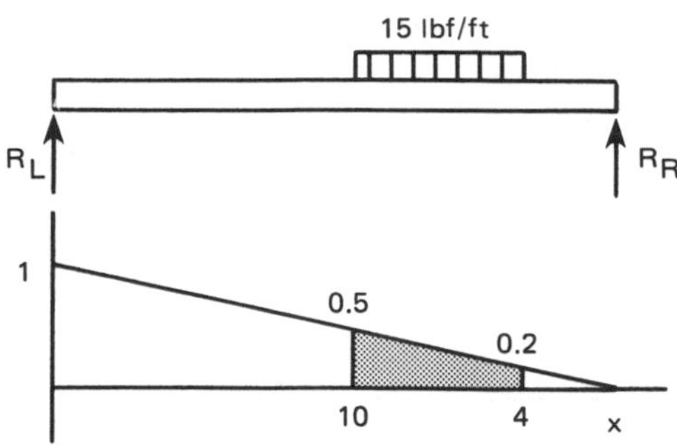


From equation 12.51, the left reaction can be calculated from the area under the influence diagram between the limits of loading.

$$\text{area} = \frac{1}{2}(10)(0.5) - \frac{1}{2}(4)(0.2) = 2.1$$

The left reaction is

$$R_L = (15)(2.1) = 31.5\,\text{lbf}$$

## B. FINDING REACTION INFLUENCE DIAGRAMS GRAPHICALLY

Since the reaction will always have a value of one when the unit load is directly over the reaction, and since the reaction is always directly proportional to the distance $x$, the reaction influence diagram can be easily determined from the following steps:

*step 1*: Remove the support being investigated

*step 2*: Displace (lift) the beam upward a distance of one unit at the support point. The resulting beam shape will be the shape of the reaction influence diagram.

*Example 12.16*

What is the approximate shape of the reaction influence diagram for reaction 2?

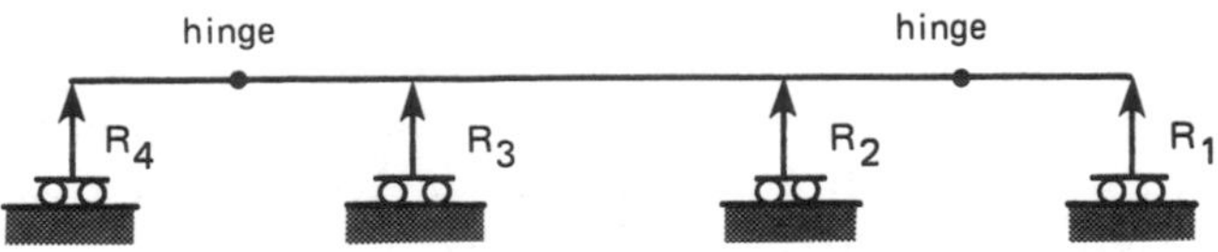


Pushing up at reaction 2 such that the deflection is one unit results in the shown shape.

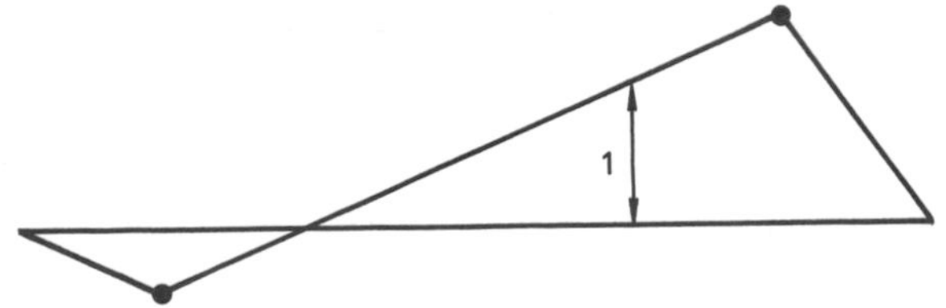


## C. INFLUENCE DIAGRAMS FOR BEAM SHEARS

A shear influence diagram (not the same as a shear diagram) illustrates the effect on the shear at a particular point in the beam of moving a load along the beam's length. As an illustration, consider point $A$ along the length of a 20 foot, simply supported beam.

In all cases, principles of statics can be used to calculate the shear at point $A$ as the sum of loads and reactions on the beam from point $A$ to the left end. (With the appropriate sign convention, summation to the right end could be used as well.) If the unit load is placed between the right end ($x = 0$) and point $A$, the shear at point $A$ will consist only of the left reaction, since there are no other loads between point $A$ and the left end. From the reaction influence diagram, we know that the left reaction varies linearly. At $x = 12$, the location of point $A$, the shear is $V = R_L = 12/20 = 0.6$.

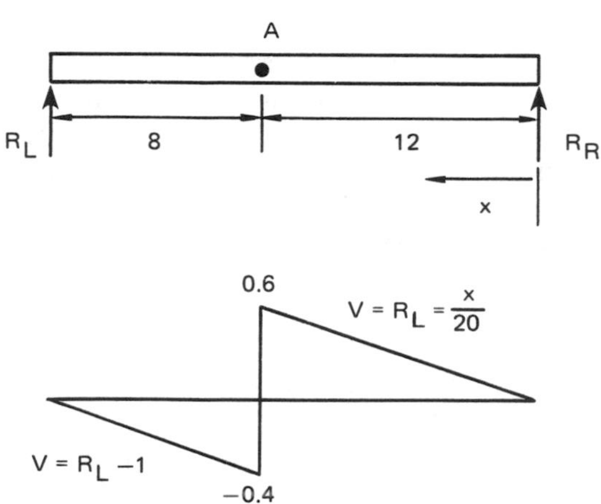


**Figure 12.16** Shear Influence Diagram for Simple Beam

When the unit load is between point $A$ and the left end, the shear at point $A$ is the sum of the left reaction (upward and positive) and the unit load itself (downward and negative). Therefore, $V = R_L - 1$. At $x = 12$, the shear is $V = 0.6 - 1 = -0.4$

Figure 12.16 is the shear influence diagram. In the diagram, notice that the shear goes through a reversal of 1. It is also helpful to note that the slopes of the two inclined sections are the same.

Shear influence diagrams are used in the same manner as reaction influence diagrams. The shear at point $A$ for any position of the load can be calculated by multiplying the ordinate of the diagram by the actual load. Distributed loads are found by multiplying the uniform load by the area under the diagram between the limits of loading. If the loading extends over positive and negative parts of the curve, the sign of the area is considered when performing the final summation.

If it is necessary to determine the distribution of loading which will produce the *maximum shear* at a point whose influence diagram is available, the load should be positioned in order to maximize the area under the diagram.[2] This can be done by "covering" either all of the positive area or all of the negative area.[3]

## D. SHEAR INFLUENCE DIAGRAMS BY VIRTUAL DISPLACEMENT

A difficulty in drawing shear influence diagrams for continuous beams on more than two supports is finding the reactions. The method of *virtual displacement* or *virtual work* can be used to find the influence diagram without going through that step.

*step 1*: Replace the point being investigated (i.e., point $A$) with an imaginary link with unit length. (It may be necessary to think of the link as having a length of 1 foot, but the link does not add to or subtract from any length of the beam.) If the point being investigated is a reaction, place a hinge at that point and lift the hinge upward a unit distance.

*step 2*: Push the two ends of the beam (with the link somewhere in between) a very small amount until the linkage is vertical. The distance between supports does not change, but the linkage allows the beam sections to assume a slope. The sections to the left and right of the linkage displace $\delta_1$ and $\delta_2$, respectively, from their equilibrium positions. The slope of both sections is the same. Points of support remain in contact with the beam.

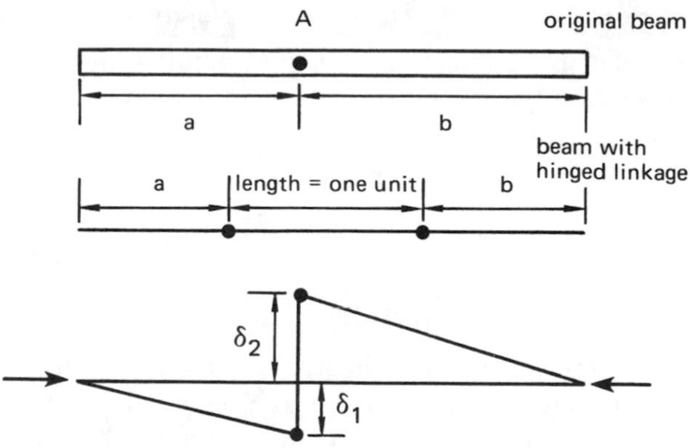


**Figure 12.17** Virtual Beam Displacements

[2] If the *minimum shear* is requested, the maximum negative shear is implied. The minimum shear is not zero in most cases.

[3] Usually, the dead load is assumed to extend over the entire length of the beam. The uniform live loads are distributed in any way which will case the maximum shear.

*step 3*: Determine the ratio of $\delta_1$ and $\delta_2$. Since the slope on the two sections is the same, the longer section will have the larger deflection. If $L = a + b$ is the length of the beam, the relationships between the deflections can be determined from equations 12.52 through 12.54.

$$\delta_1 + \delta_2 = 1$$

$$\frac{\delta_1}{\delta_2} = \frac{a}{b} \qquad 12.52$$

$$\delta_1 = \left(\frac{a}{L}\right)\delta \qquad 12.53$$

$$\delta_2 = \left(\frac{b}{L}\right)\delta \qquad 12.54$$

Since $\delta = \delta_1 + \delta_2$ was chosen as one, equations 12.53 and 12.54 really give the relative proportions of the unit link which extend below and above the reference line in figure 12.17.

Knowing that the total shear reversal through point $A$ is one unit, and that the slopes are the same, the relative proportions of the reversal below and above the line will determine the shape of the displaced beam. The shape of the influence diagram is the shape taken on by the beam.

*step 4*: As required, use equations of straight lines to obtain the shear influence ordinate as a function of position along the beam.

*Example 12.17*

For the simply supported beam shown, draw the shear influence diagram for a point 10 feet from the right end.

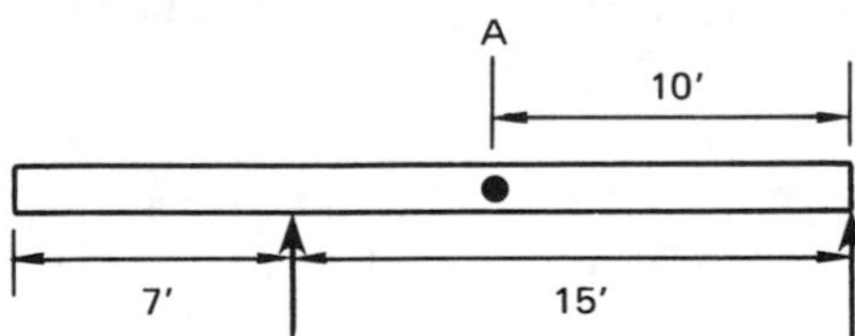


If a unit link is placed at point $A$ and the beam ends are pushed together, the following shape will result. Notice that the beam must remain in contact with the points of support, and that the two slopes are the same.

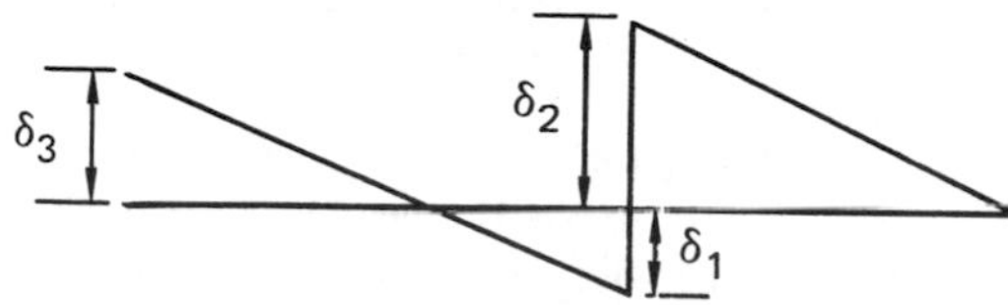


The overhanging seven feet of beam don't change the shape of the shear influence diagram between the sup-

MECH OF MATL

ports. The deflections can be evaluated assuming a 15′ long beam.

$$\delta_1 = \frac{5}{15} = 0.33$$

$$\delta_2 = \frac{10}{15} = 0.67$$

The slope in both sections of the beam is the same. This slope can be used to calculate $\delta_3$.

$$m = \frac{\delta_1}{a} = \frac{0.33}{5} = 0.066$$

$$\delta_3 = (7)(0.066) = 0.46$$

*Example 12.18*

Where should a uniformly distributed load be placed on the beam shown below to maximize the shear at section $A$?

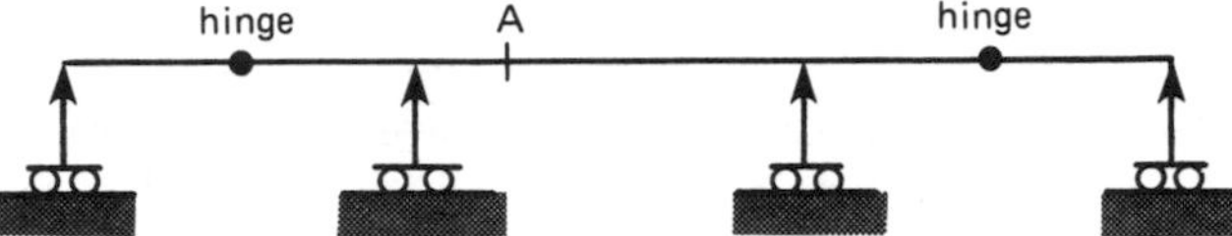


Using the principle of virtual displacement, the following shear influence diagram results by inspection. (It is not necessary to calculate the relative displacements to answer this question. It is only necessary to identify the positive and negative parts of the influence diagram.)

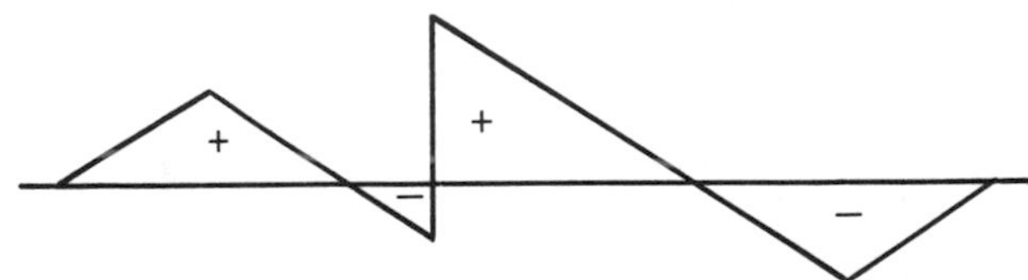


To maximize the shear, the uniform load should be distributed either over all positive or all negative sections of the influence diagram.

## E. MOMENT INFLUENCE DIAGRAMS BY VIRTUAL DISPLACEMENT

A moment influence diagram (not the same as a moment diagram) gives the moment at a particular point for any location of a unit load. The method of virtual displacement can be used in this situation to simplify finding the moment influence diagram.

*step 1*: Replace the point being investigated (i.e., point $A$) with an imaginary hinge.

*step 2*: Rotate the beam a unit rotation by applying equal but opposite moments to each of the two beam sections. Except where the point being investigated is at a support, this unit rotation can be achieved simply by "pushing up" on the beam at the hinge point.

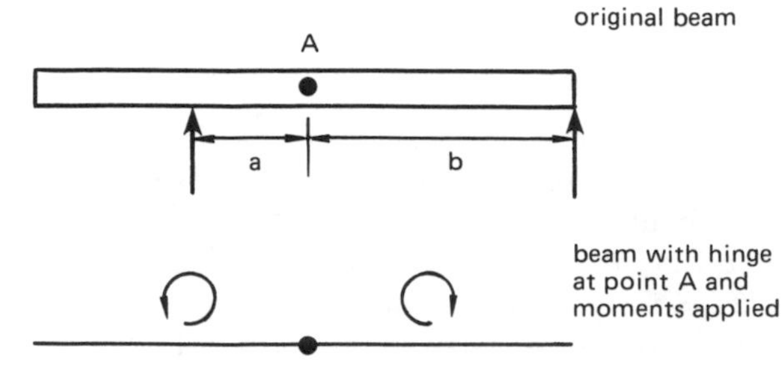


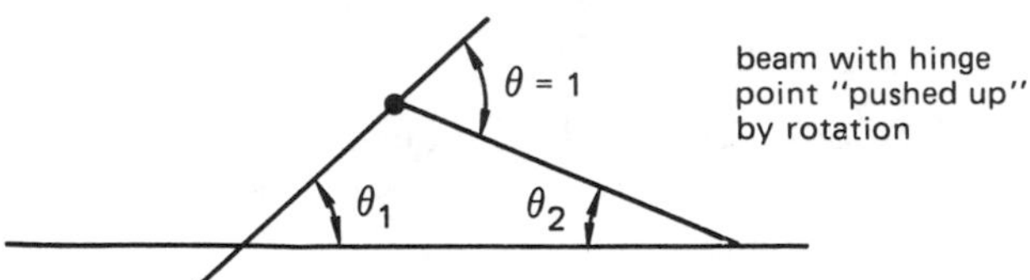


**Figure 12.18** Moment Influence Diagram by Virtual Displacement

*step 3*: The angles made by the sections on either side of the hinge will be proportional to the lengths of the opposite sections. (Since the angle is small for a virtual displacement, the angle and its tangent, or slope, are the same.)

$$\theta_1 = \frac{b}{L} \qquad 12.55$$

$$\theta_2 = \frac{a}{L} \qquad 12.56$$

$$L = a + b \qquad 12.57$$

*Example 12.19*

What are the approximate shapes of the moment influence diagrams for points $A$ and $B$ on the beam shown?

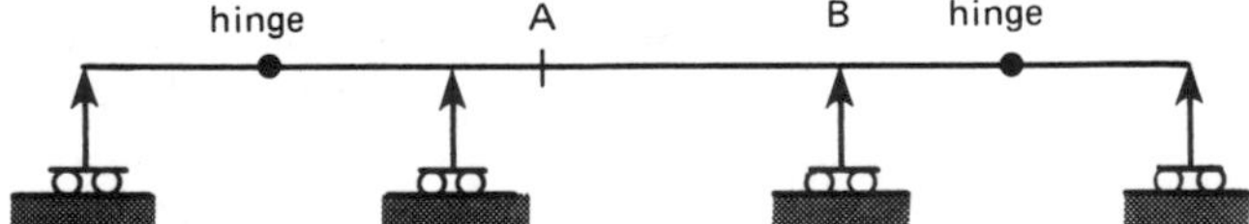


By placing an imaginary hinge at point $A$ and rotating the two adjacent sections of the beam, the following shape results.

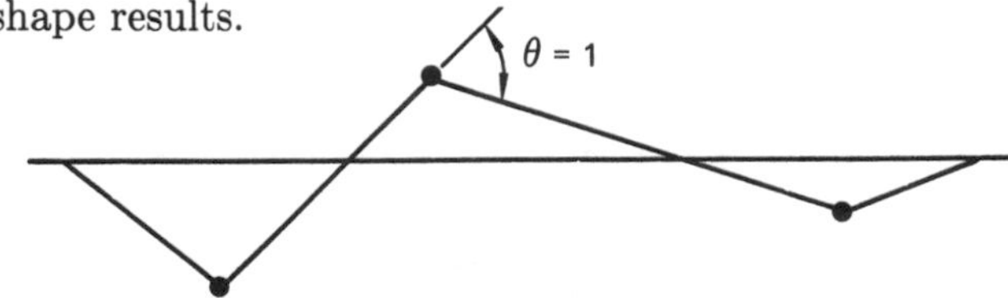


The moment influence diagram for point $B$ is found by placing an imaginary hinge at point $B$ and applying a rotating moment. Since the beam must remain in contact with all supports, and since there is no hinge between the two middle supports, the moment influence diagram must be horizontal in that region.

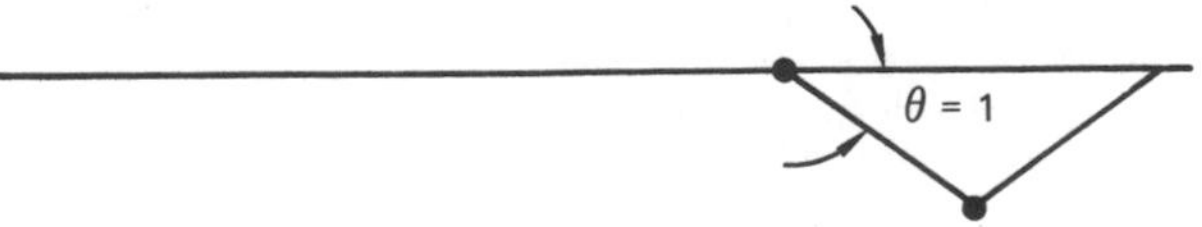

## F. SHEAR INFLUENCE DIAGRAMS ON CROSS-BEAM DECKS

When girder type construction is used to construct a road or bridge deck, the loads probably will not be applied directly to the girder. Rather, the loads will be transmitted to the girder at panel points from cross beams. Figure 12.19 shows a typical construction detail involving girders and cross beams.

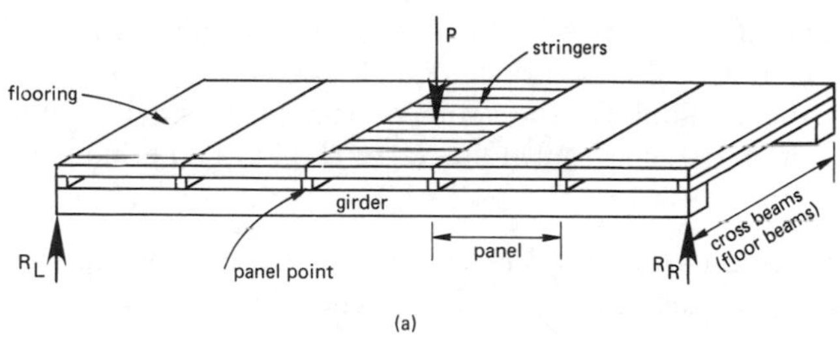


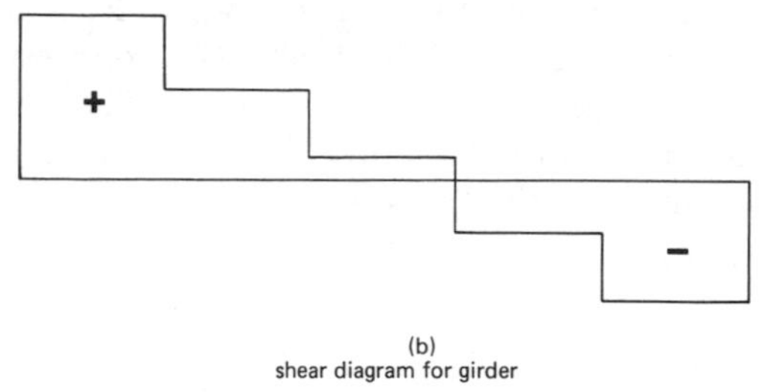


**Figure 12.19** Cross Beam Decking

A load applied to the deck stringers will be transmitted to the girder only at the panel points. Because the girder experiences a series of concentrated loads, the shear between panel points is horizontal. Since the shear is always constant between panel points, we speak of *panel shear* rather than shear at a point. Accordingly, shear influence diagrams are drawn for a panel, not for a point. Moment influence diagrams are similarly drawn for a panel.

## G. INFLUENCE DIAGRAMS ON CROSS-BEAM DECKS

Shear and moment influence diagrams for girders with cross beams are identical to simple beams, except for the panel being investigated. Once the influence diagram has been drawn for the simple beam, the influence diagram ordinates at the ends of the panel being investigated are connected to obtain the influence diagram for the girder. This is illustrated in figure 12.20.

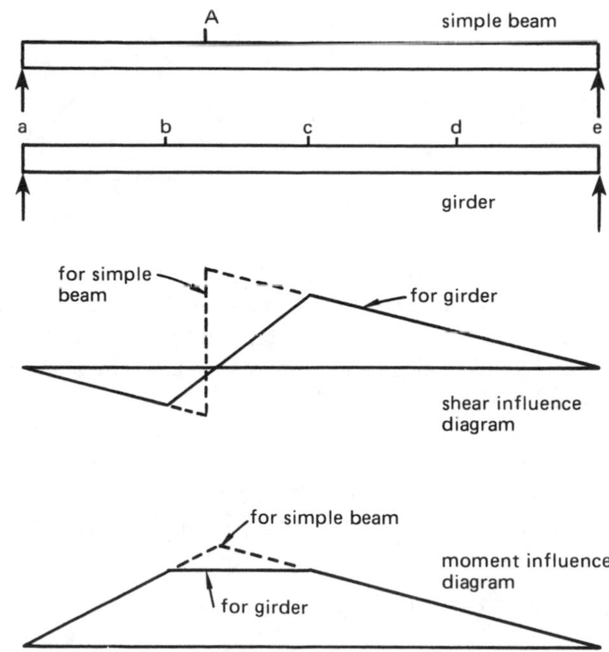


**Figure 12.20** Comparison of Influence Diagrams for Simple Beams and Girders

## H. INFLUENCE DIAGRAMS FOR TRUSS MEMBERS

Since members in trusses are initially assumed to be axial members, they cannot carry shears or moments. Therefore, shear and moment influence diagrams don't exist for truss members. However, it is possible to obtain an influence diagram showing the variation in axial force in a given truss member as the load varies in position.

There are two general cases for finding forces in truss members. The force in a horizontal truss member is proportional to the moment across the member's panel. The force in an inclined truss member is proportional to the shear across that member's panel.

So, even though we may only want the axial load in a truss member, it is still necessary to construct the shear and moment influence diagrams for the entire truss in order to determine the applications of loading on the truss which produce the maximum shear and moment across the member's panel.

*Example 12.20*

Draw the influence diagram for vertical shear in panel $DF$ of the through truss shown. What is the maximum force in member $DG$ if a 1000 pound load moves across the truss?

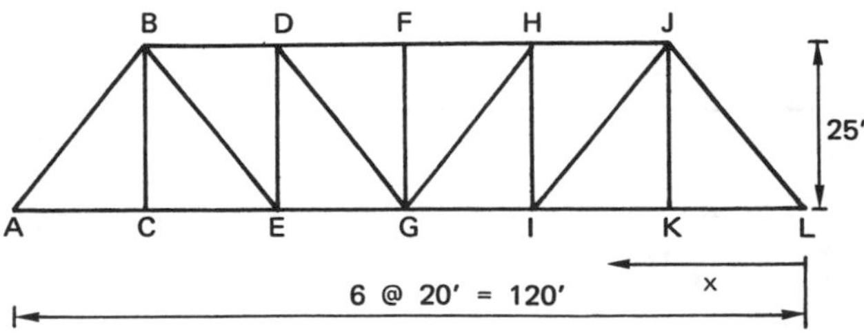


Allow a unit load to move from joint $L$ to joint $G$ along the lower chords. If the unit vertical load is at a distance $x$ from point $L$, the right reaction will be $+[1-(x/120)]$. The unit load itself has a value of $(-1)$, so the shear at distance $x$ is just $(-x/120)$.

Allow a unit load to move from joint $A$ to joint $E$ along the lower chords. If the unit load is a distance $x$ from point $L$, the left reaction will be $(x/120)$, and the shear at distance $x$ will be $[(x/120)-1]$.

These two lines can be graphed.

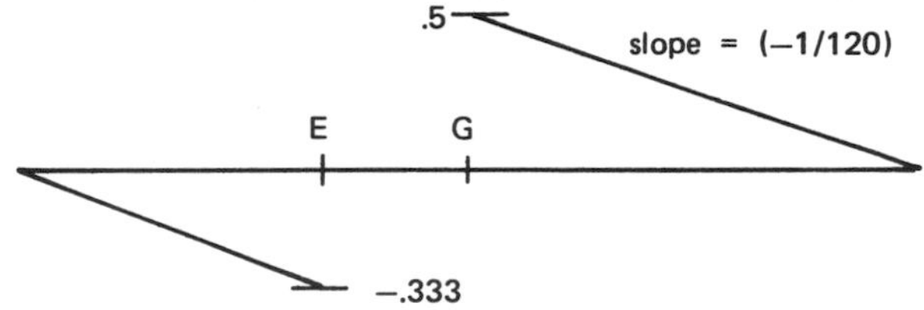


The influence line is completed by connecting the two lines as shown. Therefore, the maximum shear in panel $DF$ will occur when a load is at point $G$ on the truss.

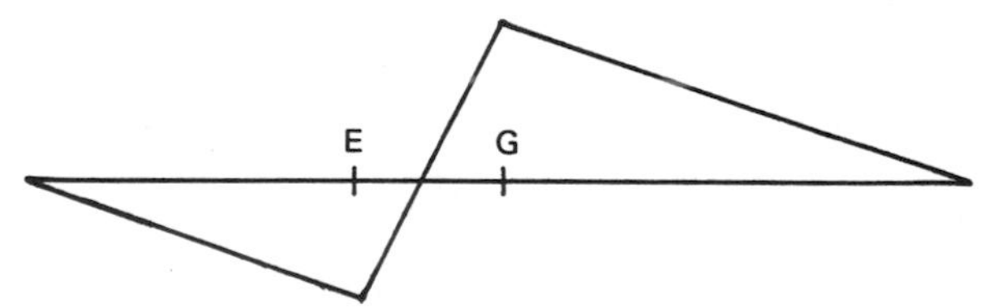


If the 1000 pound load is at point $G$, the two reactions at points $A$ and $L$ will each be 500 pounds. The cut-and-sum method can be used to calculate the force in member $DG$ simply by evaluating the vertical forces on the freebody to the left of point $G$.

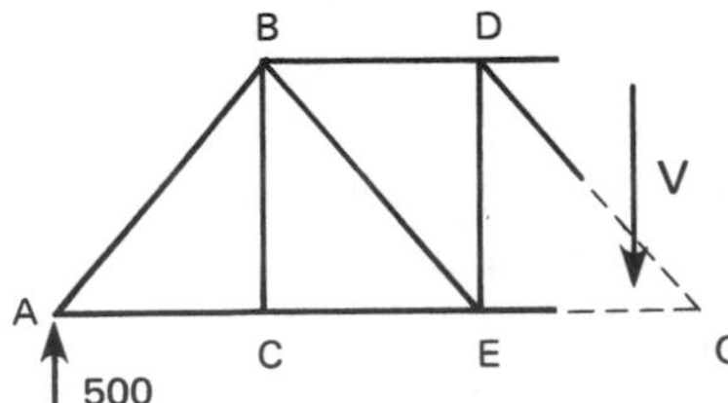


For equilibrium to occur, $V$ must be 500. This vertical shear is entirely carried by member $DG$. The length of member $DG$ is

$$\sqrt{(20)^2+(25)^2}-32$$

The force in member $DG$ is

$$DG=\left(\frac{32}{25}\right)500=640$$

*Example 12.21*

Draw the moment influence diagram for panel $DF$ on the truss shown in example 12.20. What is the maximum force in member $DF$ if a 1000 pound load moves across the truss?

The left reaction is $(x/120)$ where $x$ is the distance from the unit load to the right end. If the unit load is to the right of point $G$, the moment can be found by summing moments from point $G$ to the left. The moment is $(x/120)(60)=0.5x$.

If the unit load is to the left of point $E$, the moment will again be found by summing moments about point $G$.

$$\left(\frac{x}{120}\right)(60)-(1)(x-60)=60-0.5x$$

These two lines can be graphed. The moment for a unit load between points $E$ and $G$ is obtained by connecting the two end points of the lines derived above. Therefore, the maximum moment in panel $DF$ will occur when the load is at point $G$ on the truss.

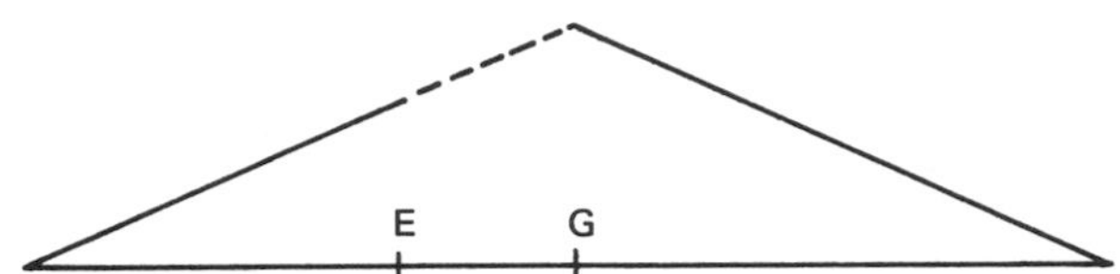


If the 1000 pound load is at point $G$, the two reactions at points $A$ and $L$ will each be 500 pounds. The method of sections can be used to calculate the force in member $DF$ by taking moments about joint $G$.

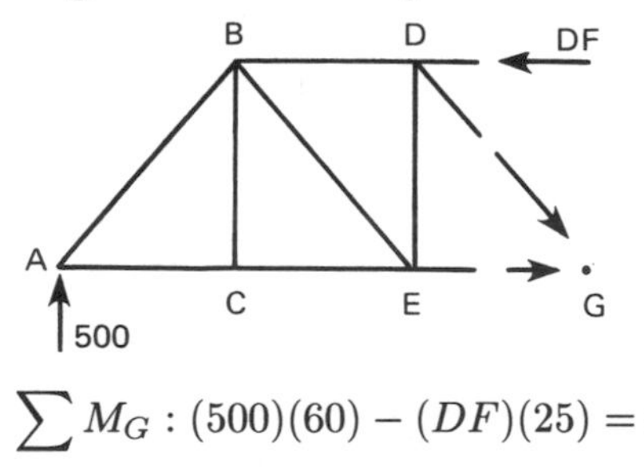


$$\sum M_G:(500)(60)-(DF)(25)=0$$

$$DF=1200$$

## 13 MOVING LOADS ON BEAMS

### A. GLOBAL MAXIMUM MOMENT ANYWHERE ON BEAM

If a beam supports a single moving load, the maximum bending and shearing stresses at any point can be found

MECH OF MATL

by drawing the moment and shear influence diagrams for that point. Once the positions of maximum moment and maximum shear are known, the stresses at the point in question can be found from equations 12.18 and 12.24.

If a simply-supported beam carries a set of moving loads (which remain equidistant as they travel across the beam), the following procedure can be used to find the *dominant load*. The dominant load is the one which occurs directly over the point of maximum moment.

*step 1*: Calculate and locate the resultant of the load group.

*step 2*: Assume that one of the loads is dominant. Place the group on the beam such that the distance from one support to the assumed dominant load is equal to the distance from the other support to the resultant of the load group.

*step 3*: Check to see that all loads are on the span and that the shear changes sign under the assumed dominant load. If the shear does not change sign under the assumed dominant load, the maximum moment may occur when only some of the load group is on the beam. If it does change sign, calculate the bending moment under the assumed dominant load.

*step 4*: Repeat steps 2 and 3, assuming that the other loads are dominant.

*step 5*: Find the maximum shear by placing the load group such that the resultant is a minimum distance from a support.

### B. PLACEMENT OF LOAD GROUP TO MAXIMIZE LOCAL MOMENT

In the design of specific members or connections, it is necessary to place the load group in a position which will maximize the load on those members or connections. The procedure for finding these positions of local maximum loadings is different from the global maximum procedures.

The solution to the problem of local maximization is somewhat trial and error oriented. It is aided by use of the influence diagram. In general, the variable being evaluated (reaction, shear, or moment) is maximum when one of the wheels is at the location or section of interest.

When there are only two or three wheels in the load group, the various alternatives can be simply evaluated by using the influence diagram for the variable being evaluated. When there are many loads in the load group (e.g., a train loading), it may be advantageous to use heuristic rules for predicting the dominant wheel.

## 14 COLUMNS

The *Euler load* is the theoretical maximum load that an initially straight column can support without buckling. For columns with pinned ends, this load is given by equation 12.58.

$$F_e = \frac{\pi^2 EI}{L^2} = \frac{(r\pi)^2 EA}{L^2} \qquad 12.58$$

The corresponding column stress is

$$\sigma_e = \frac{F_e}{A} = \frac{\pi^2 E}{\left(\frac{L}{r}\right)^2} \qquad 12.59$$

Equations 12.58 and 12.59 assume that the column is long so that the Euler stress is reached before the yield stress is reached. If the column is short, the yield stress of the material may be less than the Euler stress. In that case, short-column curves based on test data are used to predict the allowable column stress.

The value of $L/r$ at the point of intersection of the short column and the Euler curves is known as the critical *slenderness ratio*. The critical slenderness ratio becomes smaller as the compressive yield stress increases. The region in which the short column formulas apply is determined by tests for each particular type of column and material. Typical critical slenderness ratios range from 80 to 120.

In general, the Euler allowable stress formulas can be used if the stress obtained from equation 12.59 does not exceed the compressive yield stress.

*Example 12.22*

An S-type, 4 × 9.5 A36 steel I-beam 8.5′ long is used as a column. What is the working stress for a safety factor of 3? Use $E = 2.9$ EE7 psi. The yield stress for A36 steel is 36,000 psi. The required properties of the I beam are $A = 2.79$ in$^2$, $I = 0.903$ in$^4$, and $r = 0.569$ in.

From equation 12.59, the Euler stress is

$$\sigma_e = \frac{\pi^2(2.9 \text{ EE7})}{\left[\frac{(8.5)(12)}{0.569}\right]^2} = 8907 \text{ psi}$$

Since 8907 is less than 36,000, the Euler formula is valid. The allowable working stress is

$$\sigma_a = \frac{8907}{3} = 2969 \text{ psi}$$

MECH OF MATL

An ultimate load for any column can be found by using the *secant formula*. The secant formula is particularly suited for use when the column is intermediate in length. The maximum stress, $\sigma_{max}$, at the extreme compressive fiber should be kept below the compressive yield strength.

$$\begin{aligned}\sigma_{max} &= \sigma_{ave}(1 + \text{amplification factor})\\ &= \frac{F}{A}\left(1 + \frac{ec}{k^2}\sec\left[\frac{\pi}{2}\sqrt{\frac{F}{F_e}}\right]\right)\\ &= \frac{F}{A}\left(1 + \frac{ec}{k^2}\sec\left[\frac{L}{2k}\sqrt{\frac{F}{AE}}\right]\right)\\ &= \frac{F}{A}\left(1 + \frac{ec}{k^2}\sec\phi\right) \qquad 12.60\end{aligned}$$

$$\phi = \frac{1}{2}\left(\frac{L}{k}\right)\sqrt{\frac{F}{AE}} \qquad 12.61$$

The formula is solved by trial and error for $F$ with the given eccentricity, $e$. If the value of $e$ is not known, the eccentricity ratio $(ec/r^2)$ is taken as 0.25. Substituting this value and $E$ = 2.9 EE7 for steel and 1.00 EE7 for aluminum, respectively, the following formulas result which converge quickly to the known $L/r$ ratio when assumed values of $F$ are substituted.

$$\phi = \arccos\left(\frac{0.25F}{S_yA - F}\right) \qquad 12.62$$

$$\frac{L}{r} = 2\phi\sqrt{\frac{EA}{F}} \qquad 12.63$$

$$\left(\frac{L}{r}\right)_{steel} = \frac{10{,}770(\phi)}{\sqrt{\frac{F}{A}}} \qquad 12.64$$

$$\left(\frac{L}{r}\right)_{aluminum} = \frac{6325(\phi)}{\sqrt{\frac{F}{A}}} \qquad 12.65$$

*Example 12.23*

A steel member ($S_y$ = 36,000 psi, $A$ = 17.9 in$^2$, least $r$ = 2.45 in) is used as a 20-foot column. Use the secant formula and a factor of safety of 2.5 to determine the maximum concentric load.

Even though the loading is intended to be concentric, use $ec/r^2$ = 0.25 to account for uncertainties in construction and loading.

The slenderness ratio is

$$\frac{L}{r} = \frac{(20)(12)}{2.45} = 98$$

Assume a critical load of $F$ = 300,000 lbf. From equation 12.62,

$$\phi = \arccos\left[\frac{(0.25)(300)}{(36)(17.9) - 300}\right] = 1.35 \text{ radians}$$

From equation 12.64,

$$\frac{L}{r} = \frac{(10{,}770)(1.35)}{\sqrt{\frac{300{,}000}{17.9}}} = 112.3$$

Since $L/r$ will be smaller when $F$ is larger, try $F$ = 350,000 lbf.

$$\phi = \arccos\left[\frac{(0.25)(350)}{(36)(17.9) - 350}\right] = 1.27$$

$$\frac{L}{r} = \frac{(10{,}770)(1.27)}{\sqrt{\frac{350{,}000}{17.9}}} = 97.8 \quad \text{(close enough)}$$

$$F_{allowable} = \frac{350{,}000}{2.5} = 140{,}000 \text{ lbf}$$

All the preceding column formulas are for columns with frictionless round or pinned ends. For other end conditions, the *effective length* $L'$ should be used in place of $L$.

$$L' = KL \qquad 12.66$$

$K$ is the *end restraint coefficient* which varies from 0.5 to 2. For practical columns, $K$ smaller than 0.7 should not be used because infinite stiffness of the support structure is not normally achievable.

**Table 12.3**
Theoretical End-Restraint Coefficients
(Also see table 15.5)

| illus. | end conditions | ideal $K$ | design $K$ |
|---|---|---|---|
| (a) | both ends pinned | 1 | 1.0* |
| (b) | both ends built in | 0.5 | 0.65*–0.90 |
| (c) | one end pinned, one end built in | 0.707 | 0.80*–0.90 |
| (d) | one end built in, one end free | 2 | 2.0–2.1* |
| (e) | one end built in, one end fixed against rotation but free | 1 | 1.2* |
| (f) | one end pinned, one end fixed against rotation but free | 2 | 2.0* |

* AISC values

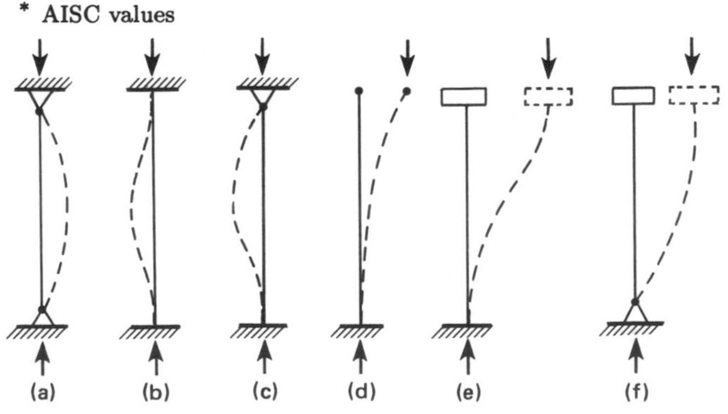

# PART 2: Application to Design

## 1 SPRINGS

Springs are assumed to be perfectly elastic within their working range. *Hooke's law* can be used to predict the amount of compression experienced when a load is placed on a spring.

$$\mathbf{F} = kx \qquad 12.67$$

$k$ is the *spring constant.* It has units of pounds per unit length.

When a spring is compressed, it stores energy. This energy can be recovered by restoring the spring's original length. It is assumed that no energy is lost through friction or hysteresis when a spring returns to its original length. The energy storage in a spring is given by equation 12.68. This energy is the same as the work required to compress the spring.

$$W = \Delta U = \tfrac{1}{2}kx^2 \qquad 12.68$$

If a weight is dropped from height $h$ onto a spring, the compression can be found by equating the change in potential energy to the energy storage.

$$m\left(\frac{g}{g_c}\right)(h+x) = \tfrac{1}{2}kx^2 \qquad 12.69$$

## 2 THIN-WALLED CYLINDERS

A cylinder can be considered *thin-walled* if its wall thickness-to-diameter ratio is less than 0.1. The circumferential *hoop stress* for internal pressure can be derived easily from the free-body diagram of a cylinder half.[4] This hoop stress is

$$\sigma_h = \frac{pr}{t} \qquad 12.70$$

Since the cylinder is assumed to be thin-walled, the radius used in equation 12.70 is taken as the inside radius.

If the cylinder is part of a tank, the axial force on the end plates produces an axial stress. The axial force is equal to the tank pressure times the end plate area. The stress produced is at right angles to the hoop stress. Accordingly, it is called *longitudinal stress* or *long stress.*

$$\sigma_L = \frac{pr}{2t} \qquad 12.71$$

Equation 12.71 also gives the stress in a spherical tank. In a spherical tank, the hoop and long stresses are the same.

The hoop and long stresses are principal stresses. They do not combine into a larger stress.

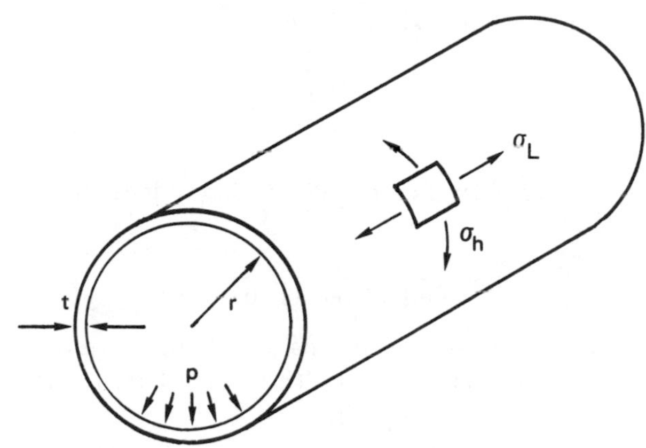


**Figure 12.21** Hoop and Long Stresses



## 3 RIVET AND BOLT CONNECTIONS

A *tension splice* using rivets or bolts can fail in one of three ways: bearing failure, shear failure, or tension failure. All three failure mechanisms must be checked to determine the maximum load the splice can carry.

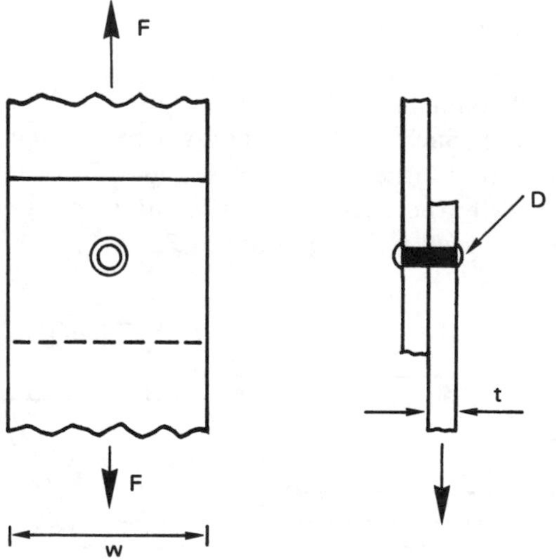


**Figure 12.22** A Tension Splice

The plate can fail in bearing. For one connector, the *bearing stress* in the plate is

$$\sigma_{br} = \frac{F}{Dt} \qquad 12.72$$

[4] There is no easy method of evaluating stresses in thin-walled cylinders under external pressure, since failure is by collapse, not elongation. However, empirical equations exist for predicting the collapsing pressure.

The number of rivets required to keep the actual bearing stress below the allowable bearing stress is

$$n_{br} = \frac{\sigma_{br}}{\text{allowable bearing stress}} \qquad 12.73$$

The rivet can fail in shear. The shear stress in the rivet is

$$\tau = \frac{F}{\frac{1}{4}\pi D^2} \qquad 12.74$$

The number of rivets required, as determined by shear, is

$$n_s = \frac{\tau}{\text{allowable shear stress}} \qquad 12.75$$

The plate also can fail in tension. If there are $n$ rivet holes in a line across the width of the plate, the minimum area in tension will be

$$A_t = t(w - nD) \qquad 12.76$$

The tensile stress in the plate at the minimum section is

$$\sigma_t = \frac{F}{A_t} \qquad 12.77$$

The maximum number of rivets across the plate width must be chosen to keep the tensile stress less than the allowable stress.



## 4 FILLET WELDS

The most common weld type is the *fillet weld*, shown in figure 12.23. Such welds commonly are used to connect one plate to another. The applied load is assumed to be carried by the *effective weld throat* which is related to the weld size, $y$, by equation 12.78.

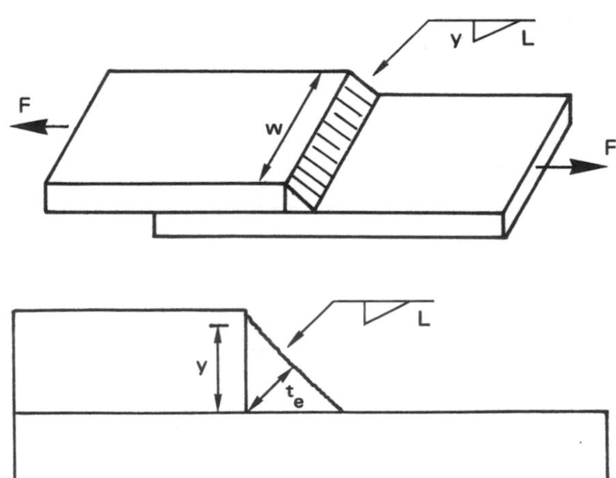


**Figure 12.23** Fillet Lap Weld and Symbol

The effective weld throat size is

$$t_e = (0.707)y \qquad 12.78$$

Weld sizes ($y$) of $\frac{3}{16}''$, $\frac{1}{4}''$, and $\frac{5}{16}''$ are desirable because they can be made in a single pass. However, fillet welds from $\frac{3}{16}''$ to $\frac{1}{2}''$ in $\frac{1}{16}''$ increments are available. The increment is $\frac{1}{8}''$ for larger welds.

Neglecting any effects due to eccentricity, the shear stress in the fillet lap weld shown in figure 12.23 is

$$\tau = \frac{F}{wt_e} \qquad 12.79$$

## 5 SHAFT DESIGN

Shear stress occurs when a shaft is placed in torsion. The shear stress at the outer surface of a bar of radius, $r$, which is torsionally loaded by a torque, $T$, is[5]

$$\tau = G\phi = \frac{Tr}{J} \qquad 12.80$$

The total strain energy due to torsion is

$$U = \frac{T^2 L}{2GJ} \qquad 12.81$$

$J$ is the shaft's polar moment of inertia, as defined in chapter 11. For a solid round shaft, $J$ is

$$J = \frac{\pi r^4}{2} = \frac{\pi D^4}{32} \qquad 12.82$$

For a hollow round shaft, the polar moment of inertia is

$$J = \frac{\pi}{2}[r_o^4 - r_i^4] \qquad 12.83$$

If a shaft of length $L$ carries a torque $T$, the angle of twist (in radians) will be

$$\phi = \frac{\tau}{G} = \frac{TL}{GJ} \qquad 12.84$$

$G$ is the *shear modulus*, approximately equal to 11.5 EE6 psi for steel. The shear modulus can be calculated from the modulus of elasticity by using equation 12.85.

$$G = \frac{E}{2(1+\mu)} \qquad 12.85$$

[5] Shear stress in steel shafts commonly is limited to approximately 6000 psi. This represents a factor of safety of approximately 3 based on the torsional yield strength.

The torque, $T$, carried by a shaft spinning at $n$ revolutions per minute is related to the transmitted horsepower.[6]

$$T = \frac{(63{,}025)(\text{horsepower})}{n} \qquad 12.86$$

## 6 ECCENTRIC CONNECTOR ANALYSIS

An eccentric torsion connection is illustrated in figure 12.24. This type of connection gets its name from the load's tendency to rotate the bracket. This rotation must be resisted by the shear stress in the connectors.

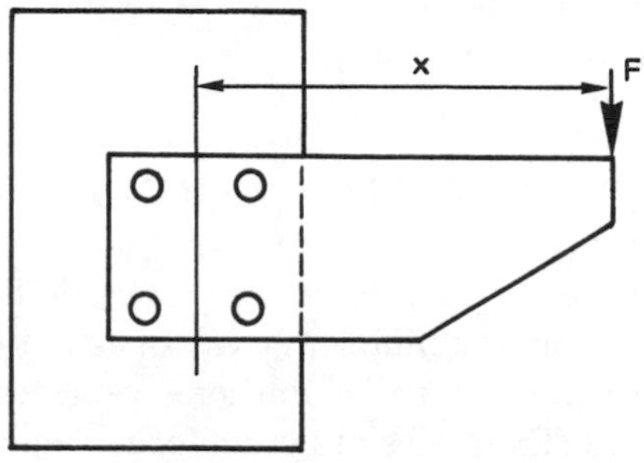


**Figure 12.24** Torsion Resistance

An extension of equation 12.80 can be used to evaluate the maximum stresses in the connector group. To use equation 12.80, the following changes in definition must be made.

- The torque, $T$, is replaced by the moment on the bracket. This moment is the product of the eccentric load, $F$, and the distance from the load to the centroid of the fastener group, $x$.
- $r$ is taken as the distance from the centroid of the fastener group to the critical fastener. The critical fastener is the one for which the vector sum of the vertical and torsional shear stresses is the greatest.
- $J$ is based on the parallel-axis theorem. As bolts and rivets have little resistance to twisting in their holes, their polar moments of inertia, $J_i$, are omitted.

$$J = \sum r_i^2 A_i \qquad 12.87$$

$r_i$ is the distance from the fastener group centroid to the $i$th fastener, which has an area of $A_i$.

- The vertical shear stress in the critical fastener must be added in a vector sum to the torsional shear stress. This vertical shear stress is

$$\text{vertical shear stress} = \frac{F/\#\text{ of fasteners}}{A_{\text{critical}}} \qquad 12.88$$

*Example 12.24*

For the bracket shown, find the load on the most critical fastener. All fasteners have a nominal $\frac{1}{2}''$ diameter.

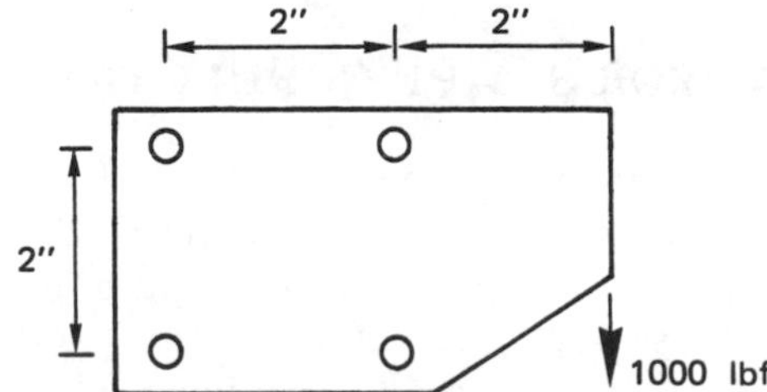


Since the fastener group is symmetrical, the group centroid is centered within the 4 fasteners. This makes the eccentricity of the load equal to 3 inches. Each fastener is located $r$ from the centroid, where

$$r = \sqrt{(1)^2 + (1)^2} = 1.414$$

The area of each fastener is

$$A_i = \tfrac{1}{4}\pi(0.5)^2 = 0.1963$$

Using the parallel axis theorem for polar moments of inertia,

$$J = 4[0.1963(1.414)^2] = 1.570 \text{ in}^4$$

The torsional stress on each fastener is

$$\tau_T = \frac{(1000)(3)(1.414)}{(1.570)} = 2702 \text{ psi}$$

This torsional shear stress is directed perpendicularly to a line connecting each fastener with the centroid.

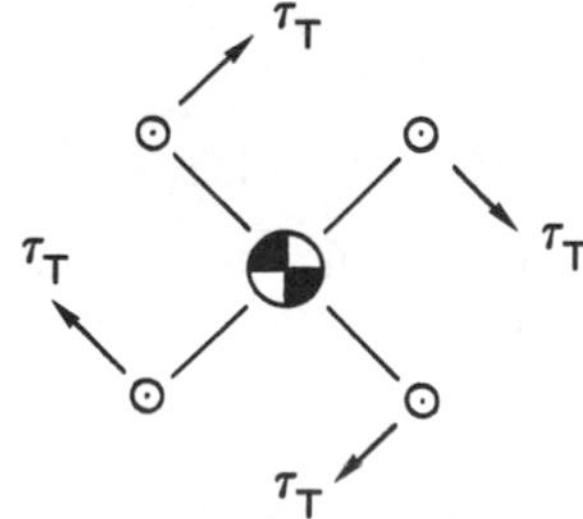


$\tau_T$ can be divided into horizontal stresses of $\tau_{T,x}$ and vertical stresses of $\tau_{T,y}$. Both of these components

[6] The torque is assumed to be steady, as would be supplied by a belt or a pulley. If the load varies, or if the shaft also carries a bending moment, a more complex method is required.

are equal to 1911 psi. In addition, each fastener carries a vertical shear load equal to $(1000/4) = 250$ pounds. The vertical shear stress due to this load is $(250/.1963) = 1274$.

The two right fasteners have vertical downward components of $\tau_T$ which add to the vertical downward stress of 1274. Thus, both of the two right fasteners are critical. The total stress in each of these fasteners is

$$\tau = \sqrt{(1911)^2 + (1911 + 1274)^2} = 3714 \text{ psi}$$

## 7 SURVEYOR'S TAPE CORRECTIONS

The standard surveyor's tape consists of a flat steel ribbon with a length very close to 100 feet. Such tapes are standardized at a particular temperature and with a specific tension and type of support. Since the tape cannot be used in the conditions under which it was standardized, corrections are needed.



### A. TEMPERATURE CORRECTION

If the tape is not used at the standardized temperature, the change in length will be

$$C_t = \alpha (T - T_{std})L \qquad 12.89$$

$\alpha$ for steel has an approximate value of 6.5 EE−6 1/°F, although low-coefficient tapes containing nickel can reduce this expansion 75%. The correction given by equation 12.89 can be positive or negative, depending on the values of $T$ and $T_{std}$. The correction is applied to the distance according to the algebraic operations listed in table 12.4.

**Table 12.4**
Corrections for Surveyors' Tapes

| measuring a distance | setting out points |
|---|---|
| add $C_t$ | subtract $C_t$ |
| add $C_p$ | subtract $C_p$ |

### B. TENSION CORRECTION

The correction due to non-standard pull (tension) can be found from equation 12.90. It can be either positive or negative.

$$C_p = \frac{(F - F_{std})L}{AE} \qquad 12.90$$

The correction is applied to the distance according to the algebraic conventions listed in table 12.4.

*Example 12.25*

A steel surveyor's tape is standardized at 68°F. It is used at 50°F to place two monuments exactly 79 feet apart. What should be the tape reading used to place the monuments?

From equation 12.89,

$$C_t = (6.5 \text{ EE–6})(50 - 68)(79) = -9.2 \text{ EE–3}$$

From table 12.4,

$$\begin{aligned}\text{tape reading} &= 79.0000 - (-9.2 \text{ EE–3})\\ &= 79.0092\end{aligned}$$

(The tape cannot be read to the degree of precision indicated by this answer.)

## 8 STRESS CONCENTRATION FACTORS

Stress concentration factors are correction factors used to account for nonuniform stress distributions within objects.[7] Nonuniform distributions result from nonuniform shapes. Examples of nonuniform shapes requiring stress concentration factors are stepped shafts, plates with holes, shafts with keyways, etc.

The actual stress experienced is the product of the stress concentration factor and the ideal stress. Values of $K$ always are greater than 1.0, and they typically range from 1.2 to 2.5 for most designs. The exact values must be determined graphically from published results of extensive experimentation.

$$\sigma' = K\sigma \qquad 12.91$$

## 9 CABLES

Cables (*wire ropes*) can be obtained in a wide variety of materials and cross sections to suit the application. Strength and weight properties of steel *standard hoisting rope* (6 strands of 19 wires each) are given in table 12.5.

In addition to the primary tension load, the design of cables should include the significant effects of bending, friction, and the weight of the cable. Appropriate dynamic factors should be applied to allow for acceleration, deceleration, stops, and starts. In general, the working stress should not exceed 20% of the breaking strength (i.e., a factor of safety of 5).

The stress due to bending a cable, such as bending around a drum, is included as an equivalent added tension load. (For good design, the diameter of the drum on which a cable is wound should be 45 to 90 times the

[7] Stress concentration factors also are known as *stress risers.*

cable diameter.) If $d$ is the cable diameter in inches, $R$ is the bending radius in inches, and $N$ is the number of wires in the cable (114 for a 6×19 cable), the equivalent tensile load from bending is approximately

$$F = \frac{2.8\ \text{EE9}d^3}{N^2R} \qquad 12.92$$

**Table 12.5**
6 × 19 (Standard Hoisting) Wire Ropes

| diam. inches | approx. weight per ft., pounds | breaking strength tons of 2000 pounds impr. plow steel | plow steel | mild plow steel |
|---|---|---|---|---|
| 1/4 | 0.10 | 2.74 | 2.39 | 2.07 |
| 5/16 | 0.16 | 4.26 | 3.71 | 3.22 |
| 3/8 | 0.23 | 6.10 | 5.31 | 4.62 |
| 7/16 | 0.31 | 8.27 | 7.19 | 6.25 |
| 1/2 | 0.40 | 10.7 | 9.35 | 8.13 |
| 9/16 | 0.51 | 13.5 | 11.8 | 10.2 |
| 5/8 | 0.63 | 16.7 | 14.5 | 12.6 |
| 3/4 | 0.90 | 23.8 | 20.7 | 18.0 |
| 7/8 | 1.23 | 32.2 | 28.0 | 24.3 |
| 1 | 1.60 | 41.8 | 36.4 | 31.6 |
| 1 1/8 | 2.03 | 52.6 | 45.7 | 39.8 |
| 1 1/4 | 2.50 | 64.6 | 56.2 | 48.8 |
| 1 3/8 | 3.03 | 77.7 | 67.5 | 58.8 |
| 1 1/2 | 3.60 | 92.0 | 80.0 | 69.6 |
| 1 5/8 | 4.23 | 107 | 93.4 | 81.2 |
| 1 3/4 | 4.90 | 124 | 108 | 93.6 |
| 1 7/8 | 5.63 | 141 | 123 | 107 |
| 2 | 6.40 | 160 | 139 | 121 |
| 2 1/8 | 7.23 | 179 | 156 | |
| 2 1/4 | 8.10 | 200 | 174 | |
| 2 1/2 | 10.0 | 244 | 212 | |
| 2 3/4 | 12.1 | 292 | 254 | |

For ropes with steel cores, add 7 1/2% to the above strengths.

For galvanized ropes, deduct 10% from the above strengths.

*Example 12.26*

What is the factor of safety when a $\frac{1}{2}$ inch, mild plow steel, 6 × 19 standard hoisting cable carrying 10,000 pounds is bent around a 24 inch sheave? Is the factor of safety adequate?

$$F_{\text{bending}} = \frac{(2.8\ \text{EE9})(0.5)^3}{(114)^2(12)} = 2240\ \text{lbf}$$

$$F_{\text{total}} = 10{,}000 + 2240 = 12{,}240\ \text{lbf}$$

$$\text{breaking strength} = 8.13\,(2000) = 16{,}260\ \text{lbf}$$

$$\text{factor of safety} = \frac{16{,}260}{12{,}240} = 1.33$$

Even without including the cable weight, this is not adequate, since a factor of safety of at least 5 is recommended.

## 10 THICK-WALLED CYLINDERS UNDER EXTERNAL AND INTERNAL PRESSURE

### A. STRESSES

The theory of thick-walled cylinders, *Lamé's solution*, is a continuation of the theory of thin-walled cylinders. The thick-walled cylinder is assumed to be made up of thin laminar rings. The strain variation through the wall is determined such that all the rings are in equilibrium, and the stresses and the deformations are consistent at the boundaries between the rings.

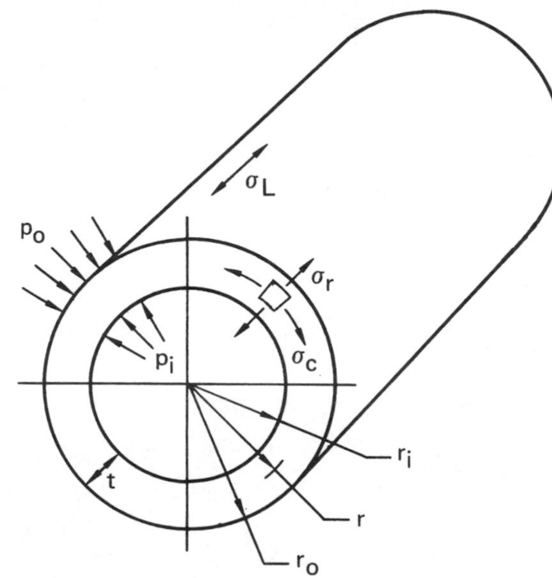


**Figure 12.25** Thick-Walled Cylinder

The general equations for stress are given here.[8]

$$\sigma_c = \frac{r_i^2p_i - r_o^2p_o + \frac{(p_i-p_o)r_i^2r_o^2}{r^2}}{r_o^2 - r_i^2} \qquad 12.93$$

$$\sigma_r = \frac{r_i^2p_i - r_o^2p_o - \frac{(p_i-p_o)r_i^2r_o^2}{r^2}}{r_o^2 - r_i^2} \qquad 12.94$$

The cases of main interest are those of internal or external pressure only. The stress formulas for these cases are summarized in table 12.6.

The maximum shear and normal stresses occur at the inner surface for both external and internal pressure.

[8] It is essential that compressive stresses be given a negative sign in all thick-walled cylinder equations, including those for deflection.

MECH OF MATL

When a longitudinal stress due to internal pressure acting against end plates exists, the longitudinal stress is

$$\sigma_L = \frac{pr_i^2}{(r_o + r_i)t} \qquad 12.95$$

The longitudinal stress is assumed to be uniform across the wall. The magnitude and the location of the maximum shear and normal stress is not changed due to the addition of the longitudinal stress.

At every point in the cylinder, $\sigma_c$, $\sigma_r$, and $\sigma_L$ are the principal stresses.

**Table 12.6**
Stresses in Thick-Walled Cylinders

| stress | external pressure, $p$ | internal pressure, $p$ |
|---|---|---|
| $\sigma_{co}$ | $\frac{-(r_o^2 + r_i^2)p}{(r_o + r_i)t}$ | $\frac{2r_i^2 p}{(r_o + r_i)t}$ |
| $\sigma_{ro}$ | $-p$ | 0 |
| $\sigma_{ci}$ | $\frac{-2r_o^2 p}{(r_o + r_i)t}$ | $\frac{(r_o^2 + r_i^2)p}{(r_o + r_i)t}$ |
| $\sigma_{ri}$ | 0 | $-p$ |
| $\tau_{\max}$ | $\left(\frac{1}{2}\right)\sigma_{ci}$ | $\left(\frac{1}{2}\right)(\sigma_{ci} + p)$ |

MECH OF MATL

B. STRAINS

The *diametral strain*, $\Delta D/D$, and the *circumferential strain*, $\Delta C/C$, are equal in a circular cylinder under pressure loading.

$$\frac{\Delta D}{D} = \frac{\Delta C}{C} = \frac{\Delta r}{r} = \frac{\sigma_c - \mu(\sigma_r + \sigma_L)}{E} \qquad 12.96$$

*Example 12.27*

A steel cylinder of 1″ I.D. and 2″ O.D. is pressurized internally to 10,000 psi. (a) There are no end caps. What is the change in the inside diameter? (b) What would be the effect on the inside diameter of adding end caps? Use $E = 2.9$ EE7 and $\mu = 0.3$.

From table 12.6,

$$\sigma_{ci} = \frac{(1^2 + 0.5^2)(10{,}000)}{(1 + 0.5)0.5} = 16{,}667 \text{ psi}$$

$$\sigma_{ri} = -10{,}000 \text{ psi}$$

$$\sigma_L = 0$$

From equation 12.96,

$$\frac{\Delta D}{D} = \frac{[16{,}667 - 0.3(-10{,}000 + 0)]}{2.9 \text{ EE7}} = 0.000678$$

$$\Delta D = 0.000678\,(1) = 0.000678 \text{ inch}$$

(b)

$$\sigma_L = \frac{(10{,}000)(0.5)^2}{(1 + 0.5)(0.5)} = 3333 \text{ psi}$$

$$\frac{\Delta D}{D} = \frac{[16{,}667 - 0.3(-10{,}000 + 3333)]}{2.9 \text{ EE7}} = 0.000506$$

$$\Delta D = 0.000506\,(1) = 0.000506 \text{ in}$$

C. PRESS FITS

If two cylinders are pressed together with an initial interference, $I$, the pressure, $p$, acting between them expands the outer cylinder and compresses the inner one. *Interference* usually means diametral interference. Although the total interference can be allocated to inner and outer cylinders in almost any combination, the total interference usually is given to the outer disk in the case of shafts with disks.

$$I = |\Delta D_i|_{\text{outer cylinder}} + |\Delta D_o|_{\text{inner cylinder}} \qquad 12.97$$

If the cylinders are the same length, this method can be used to find the stress conditions after assembly. A stress concentration factor as high as 4 may be needed if the lengths are different.

In the special case where the shaft and the hub have the same modulus of elasticity and the shaft is solid, the total interference can be found from equation 12.98. The interference pressure, $p$, can be found if $I$ is known.

$$I = \frac{4r_{\text{shaft}}p}{E}\left[\frac{1}{1 - \left(\frac{r_{\text{shaft}}}{r_{\text{hub}}}\right)^2}\right] \qquad 12.98$$

When pieces are pressed together, the assembly force can be calculated as a sliding frictional force based on the normal force.

$$F_{\max,\text{ assembly}} = fN = 2\pi r_{\text{shaft}} Lpf \qquad 12.99$$

The coefficient of friction for press fits is highly variable, having been reported in the range of 0.03 to 0.33.

If the hub is acted upon by a torque or a torque-causing force, the maximum resisting torque can be calculated.

$$T_{\max} = 2\pi r_{\text{shaft}}^2 Lpf \qquad 12.100$$

*Example 12.28*

A brass inner cylinder and an aluminum alloy outer cylinder (both 2.0″ long) have been pressed together with an interference of 0.004 in. What is the maximum shear stress in the brass? Assuming a coefficient of friction of 0.25, what is the force required to press them apart?

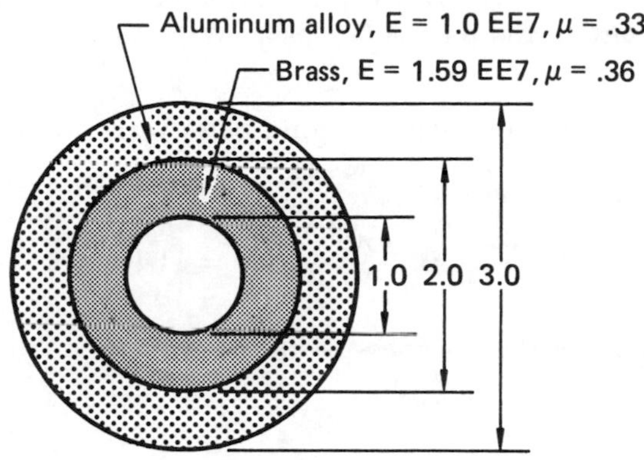


Aluminum alloy, internal pressure:

$$\sigma_{ci} = \frac{(1.5^2 + 1^2)p}{(1.5+1)(0.5)} = 2.6p$$

$$\sigma_{ri} = -p$$

$$\left(\frac{\Delta D}{D}\right)_i = \frac{[2.6p - 0.33(-p)]}{1.0\ \text{EE7}} = (2.93\ \text{EE–7})p$$

$$\Delta D_i = (2.93\ \text{EE–7})p(2) = (5.86\ \text{EE–7})p$$

Brass, external pressure:

$$\sigma_{co} = \frac{-(1^2 + 0.5^2)p}{(1+0.5)(0.5)} = -1.667p$$

$$\sigma_{ro} = -p$$

$$\left(\frac{\Delta D}{D}\right)_o = \frac{[-1.667p - 0.36(-p)]}{1.59\ \text{EE7}} = (-0.822\ \text{EE–7})p$$

$$\Delta D_o = (-0.822\ \text{EE–7})p(2) = (-1.644\ \text{EE–7})p$$

From equation 12.97,

$$|(5.86\ \text{EE–7})|p + |(-1.644\ \text{EE–7})|p = 0.004$$
$$p = 5330\ \text{psi}$$

From table 12.6,

$$\sigma_{\text{ci, brass}} = \frac{-2(1^2)5330}{(1+0.5)(0.5)} = 14{,}213\ \text{psi}$$

$$\tau_{\max} = \frac{1}{2}(14{,}213) = 7106.5\ \text{psi}$$

The force to separate the cylinders is

$$F = (5330)[2\pi(1)(2)](0.25) = 16{,}745\ \text{lbf}$$

**Table 12.7**
Flat Plates Under Uniform Pressure of p psi

| shape | edge condition | maximum stress | deflection at center |
|---|---|---|---|
| circular | simply supported | $(3/8)pr^2(3+\mu)/t^2$ at center | $(3/16)pr^4(1-\mu)(5+\mu)/(Et^3)$ |
| | built-in | $(3/4)pr^2/t^2$ at edge | $(3/16)pr^4(1-\mu^2)/(Et^3)$ |
| rectangular | simply supported | $C_1pb^2/t^2$ at center | $C_2pb^4/(Et^3)$ |
| | built-in | $C_3pb^2/t^2$ at centers of long edges | $C_4pb^4/(Et^3)$ |

| a/b | 1.0 | 1.2 | 1.4 | 1.6 | 1.8 | 2 | 3 | 4 | 5 | ∞ |
|---|---|---|---|---|---|---|---|---|---|---|
| $C_1$ | 0.287 | 0.376 | 0.453 | 0.517 | 0.569 | 0.610 | 0.713 | 0.741 | 0.748 | 0.750 |
| $C_2$ | 0.044 | 0.062 | 0.077 | 0.091 | 0.102 | 0.111 | 0.134 | 0.140 | 0.142 | 0.142 |
| $C_3$ | 0.308 | 0.383 | 0.436 | 0.487 | 0.497 | 0.500 | 0.500 | 0.500 | 0.500 | 0.500 |
| $C_4$ | 0.0138 | 0.0188 | 0.023 | 0.025 | 0.027 | 0.028 | 0.028 | 0.028 | 0.028 | 0.028 |

## 11 FLAT PLATES UNDER UNIFORM PRESSURE LOADING

Formulas for stresses and deflections of flat plates are given in table 12.7. Their application is subject to the following constraints.

- The plates are of medium thickness, meaning that the thickness is equal to or less than 1/4 the minimum width dimension, and the maximum deflection is equal to or less than 1/2 the thickness.
- The plates are constructed of isentropic, elastic material, and the elastic limit is not exceeded under the applied loading.

*Example 12.29*

The end of a 10″ inside diameter pipe is capped by welding on an end plate made from mild steel. The safe stress in the end cap is 11,100 psi. The internal pressure in the pipe is 500 psia. What plate thickness is required?

The welding produces a round plate with fixed edges. From table 12.7, the stress is

$$\sigma = \frac{3pr^2}{4t^2} = \frac{(3)(500)(5)^2}{4t^2} = 11{,}100 \text{ psi}$$

Solving for the thickness results in $t = 0.92''$.

# Appendix A: Elastic Beam Formulas

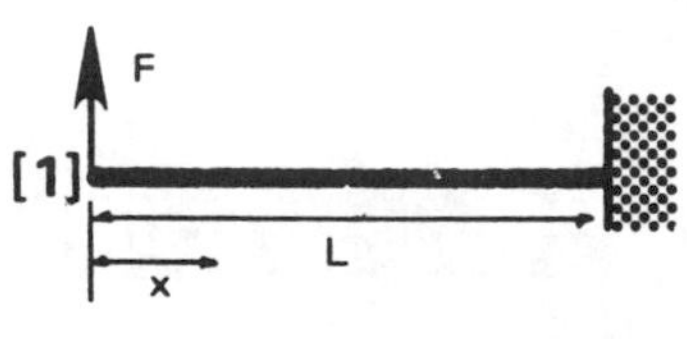


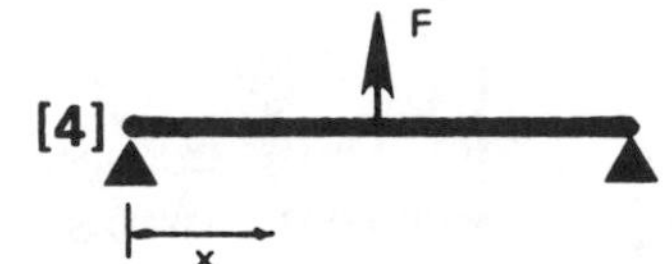


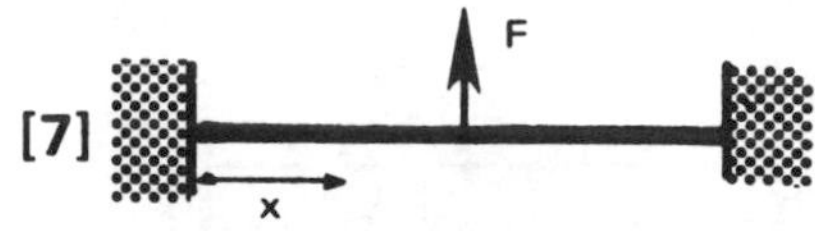

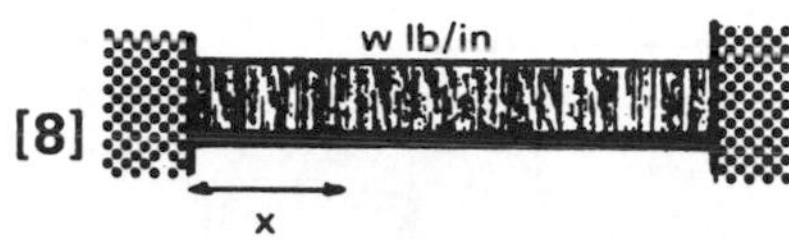


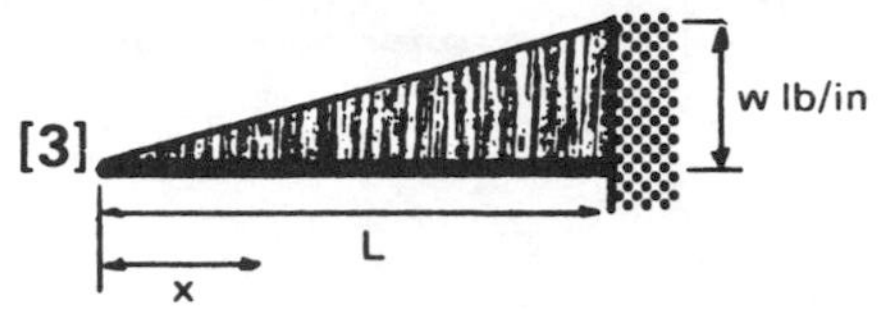

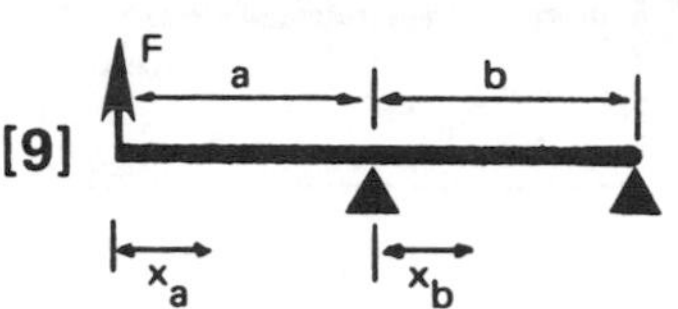


| case | moment | deflection |
|---|---|---|
| 1 | $M = Fx$<br>$M_{\text{max}} = FL$ | $y = (F/6EI)(2L^3 - 3L^2x + x^3)$<br>$y_{\text{max}} = FL^3/3EI$ |
| 2 | $M = \frac{1}{2}wx^2$<br>$M_{\text{max}} = \frac{1}{2}wL^2$ | $y = (w/24EI)(3L^4 - 4L^3x + x^4)$<br>$y_{\text{max}} = wL^4/8EI$ |
| 3 | $M = wx^3/6L$<br>$M_{\text{max}} = wL^2/6$ | $y = (w/120EIL)(4L^5 - 5L^4x + x^5)$<br>$y_{\text{max}} = wL^4/30EI$ |
| 4 | $M = -\frac{1}{2}Fx$<br>$M_{\text{max}} = -\frac{1}{4}FL$ | $y = (Fx/48EI)(3L^2 - 4x^2)$<br>$y_{\text{max}} = FL^3/48EI$ |
| 5 | $M = \left(\frac{1}{2}wx\right)(x - L)$<br>$M_{\text{max}} = -wL^2/8$ | $y = (wx/24EI)(L^3 - 2Lx^2 + x^3)$<br>$y_{\text{max}} = 5wL^4/384EI$ |
| 6 | $M = (-wx/6L)(L^2 - x^2)$<br>$M_{\text{max}} = -0.064wL^2$ at $x = 0.5774L$ | $y = (wx/360EIL)(7L^4 - 10L^2x^2 + 3x^4)$<br>$y_{\text{max}} = 0.00652wL^4/EI$ at $x = 0.5193L$ |
| 7 | $M = \frac{1}{2}F\left[\left(\frac{1}{4}L\right) - x\right]$<br>$M_{\text{max}} = FL/8$ at $x = 0$<br>$M_{\text{max}} = -FL/8$ at $x = \frac{1}{2}L$ | $y = (Fx^2/48EI)(3L - 4x)$<br>$y_{\text{max}} = FL^3/192EI$ |
| 8 | $M = \left(\frac{1}{2}wL^2\right)\left[(1/6) - (x/L) + (x/L)^2\right]$<br>$M_{\text{max}} = wL^2/12$ at $x = 0$ and $x = L$<br>$M = -wL^2/24$ at $x = \frac{1}{2}L$ | $y = (wx^2/24EI)(L - x)^2$<br>$y_{\text{max}} = wL^4/384EI$ |
| 9 | $M_a = Fx_a$<br>$M_b = (Fa/b)(b - x_b)$<br>$M_{\text{max}} = Fa$ at $x_a = a$ | $y_a = (F/3EI)[(a^2 + ab)(a - x_a) + (x_a/2)(x_a^2 - a^2)]$<br>$y_b = (Fax_b/6EI)[3x_b - (x_b^2/b) - 2b]$<br>$y_{\text{tip}} = (Fa^2/3EI)(a + b)$ (max up)<br>$y_{\text{max}} = (0.06415)Fab^2/EI$ at $x_b = 0.4226b$ (max down) |

## Appendix A, continued

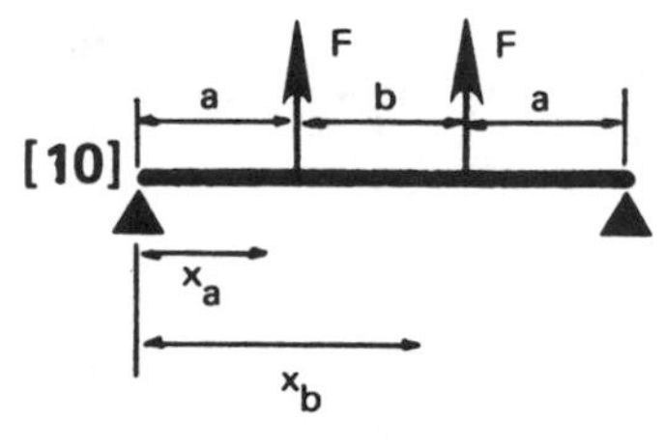


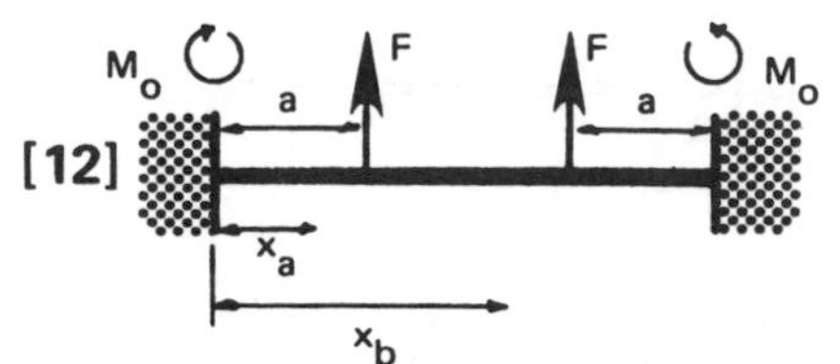


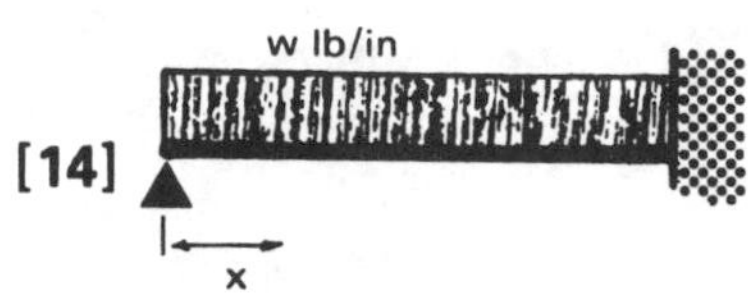


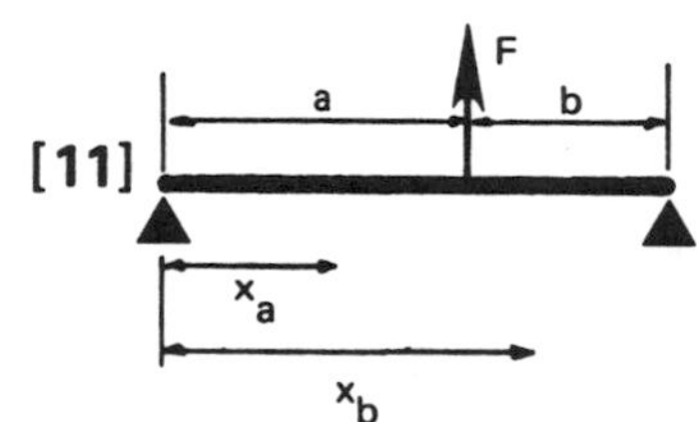


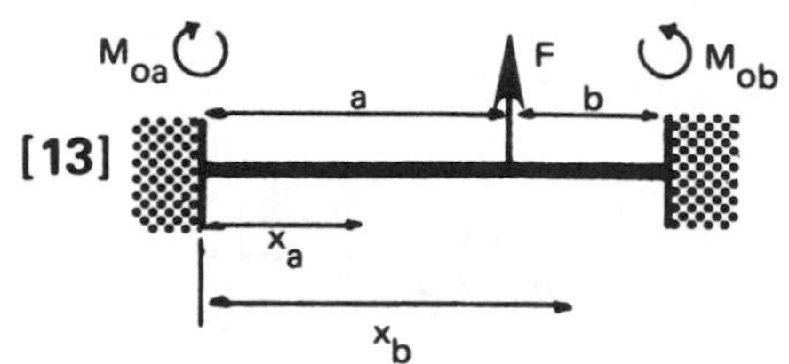


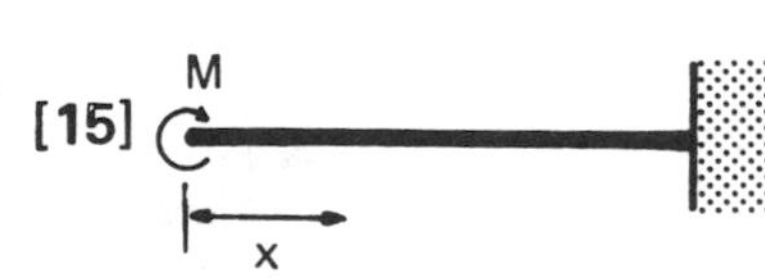




| case | moment | deflection |
|---|---|---|
| 10 | $M_a = -Fx_a$<br>$M_b = -Fa$<br>$M_{\max} = -Fa$ (everywhere between loads) | $y_a = (Fx_a/6EI)[(3a)(L-a) - x_a^2]$<br>$y_b = (Fa/6EI)[3x_b(L-x_b) - a^2]$<br>$y_{\max} = (Fa/24EI)[3L^2 - 4a^2]$ |
| 11 | $M_a = -Fbx_a/L$<br>$M_b = -Fa(L-x_b)/L$<br>$M_{\max} = -Fab/L$ at $x_a = a$ | $y_a = (Fbx_a/6EIL)(L^2 - b^2 - x_a^2)$<br>$y_b = (Fb/6EIL)[(L/b)(x_b-a)^3 + (L^2-b^2)x_b - x_b^3]$<br>$y = Fa^2b^2/3EIL$ at $x_a = a$<br>$y_{\max} = (0.06415Fb/EIL)(L^2-b^2)^{3/2}$ at $x = \sqrt{a(L+b)/3}$ |
| 12 | $M_a = (Fa/L)(L-a) - Fx_a$<br>$M_b = Fa^2/L$<br>$M_o = (Fa/L)(L-a)$ | $y_a = (Fx_a^2/2EI)[a(1-(a/L)) - (x_a/3)]$<br>$y_b = (Fa^2/2EI)[x_b - (x_b^2/L) - (a/3)]$<br>$y_{\max} = (Fa^2/24EI)(3L-4a)$ at $x = \frac{1}{2}L$ |
| 13 | $M_a = (Fb^2/L^3)[aL - x_a(L+2a)]$<br>$M_b = (Fa^2/L^3)[bL - (L-x_b)(L+2b)]$<br>$M_{oa} = Fab^2/L^2$ (max when $a < b$)<br>$M_{ob} = Fa^2b/L^2$ (max when $a > b$)<br>$M = -2Fa^2b^2/L^3$ at $x_a = a$ | $y_a = (Fx_a^2b^2/6EIL^3)[3aL - x_a(3a+b)]$<br>$y_b = (F(L-x_b)^2a^2/6EIL^3)[3bL - (L-x_b)(3b+a)]$<br>$y = Fa^3b^3/3EIL^3$ at $x_a = a$<br>$y_{\max} = 2Fa^3b^2/[3EI(L+2a)^2]$ at $x = 2aL/(L+2a)$ |
| 14 | $M = (3wLx/8) - \frac{1}{2}wx^2$<br>$M_{\max} = wL^2/8$ at $x = L$ | $y = (wx/48EI)[L^3 - 3Lx^2 + 2x^3]$<br>$y_{\max} = wL^4/185EI$ at $x = 0.4215L$ |
| 15 | $M = M$ everywhere | $y = (M/2EI)(L^2 - 2xL + x^2)$<br>$y_{\max} = ML^2/2EI$ at free end |

# Appendix B: Centroids and Area Moments of Inertia

| SHAPE | DIMENSIONS | CENTROID $(x_c, y_c)$ | AREA MOMENT OF INERTIA |
|---|---|---|---|
| Rectangle | y, y′, h, C, x′, x, O, b | $(\frac{1}{2}b, \frac{1}{2}h)$ | $I_{x'} = (1/12)bh^3$<br>$I_{y'} = (1/12)hb^3$<br>$I_x = (1/3)bh^3$<br>$I_y = (1/3)hb^3$<br>$J_C = (1/12)bh(b^2 + h^2)$ |
| Triangle | h, C, x′, x, b | $y_c = (h/3)$ | $I_{x'} = (1/36)bh^3$<br>$I_x = (1/12)bh^3$ |
| Trapezoid | y, b, $y'_c$, h, C, x′, x, O, B | $y'_c = \frac{h(2B + b)}{3(B + b)}$<br>Note that this is measured from the top surface. | $I_{x'} = \frac{h^3(B^2+4Bb+b^2)}{36(B+b)}$<br>$I_x = \frac{h^3(B+3b)}{12}$ |
| Quarter-Circle, of radius r | y, x, O | $((4r/3\pi), (4r/3\pi))$ | $I_x = I_y = (1/16)\pi r^4$<br>$J_O = (1/8)\pi r^4$ |
| Half Circle, of radius r | y, x′, C, x′, x, O | $(0, (4r/3\pi))$ | $I_x = I_y = (1/8)\pi r^4$<br>$J_O = \frac{1}{4}\pi r^4$ $I_{x'} = .11r^4$ |
| Circle, of radius r | y, O, C, x | $(0,0)$ | $I_x = I_y = \frac{1}{4}\pi r^4$<br>$J_O = \frac{1}{2}\pi r^4$ |
| Parabolic Area | 2a, C, h, x, O, y | $(0, (3h/5))$ | $I_x = 4h^3a/7$<br>$I_y = 4ha^3/15$ |
| Parabolic Spandrel | y, $y = kx^2$, C, h, O, x, a | $((3a/4),(3h/10))$ | $I_x = ah^3/21$<br>$I_y = 3ha^3/15$ |

# Appendix C: Typical Properties of Structural Steel, Aluminum, and Magnesium (ksi)

| Designation | Application | $S_u$ | $S_y$ | Approximate $S_e$ |
|---|---|---|---|---|
| A36-70a | shapes | 58–80 | 36 | 29–40 |
| | plates | 58–80 | 36 | 29–40 |
| A53-72a | pipe | 60 | 35 | 30 |
| A242-70a | shapes | 70 | 50 | 35 |
| | plates to 3/4″ | 70 | 50 | 35 |
| A440-70a | shapes | 70 | 50 | 35 |
| | plates to 3/4″ | 70 | 50 | 35 |
| A441-70a | shapes | 70 | 50 | 35 |
| | plates to 3/4″ | 70 | 50 | 35 |
| A500-72 | tubes | 45 | 33 | 22 |
| A501-71a | tubes | 58 | 36 | 29 |
| A514-70 | plates to 3/4″ | 115–135 | 100 | 55 |
| A529-72 | shapes | 60–85 | 42 | 30–42 |
| | plates to 1/2″ | 60–85 | 42 | 30–42 |
| A570-72 | sheet/strip | 55 | 40 | 27 |
| A572-72 | shapes | 60 | 42 | 30 |
| | plates | 60 | 42 | 30 |
| A588-71 | shapes | 70 | 50 | 35 |
| | plates to 4″ | 70 | 50 | 35 |
| A606-71 | hot rolled sheet | 70 | 50 | 35 |
| | cold rolled sheet | 65 | 45 | 32 |
| A607-70 | sheet | 60 | 45 | 30 |
| A618-71 | shapes | 70 | 50 | 35 |
| | tubes | 70 | 50 | 35 |

Typical Properties of Structural Aluminum (ksi)

| Designation | Application | $S_u$ | $S_y$ | Approximate $S_e$ (EE8 cyc.) |
|---|---|---|---|---|
| 2014-T6 | shapes/bars | 63 | 55 | 19 |
| 6061-T6 | all | 42 | 35 | 14.5 |

Typical Properties of Structural Magnesium (ksi)

| Designation | Application | $S_u$ | $S_y$ | Approximate $S_e$ (EE7 cyc.) |
|---|---|---|---|---|
| AZ31 | shapes | 38 | 29 | 19 |
| AZ61 | shapes | 45 | 33 | 19 |
| AZ80 | shapes | 55 | 40 | |

**Practice Problems: MECHANICS OF MATERIALS**

Untimed

1. A 14 foot simple beam is uniformly loaded with 200 pounds per foot over its entire length. If the beam is 3.625″ wide and 7.625″ deep, what is the maximum bending stress? What is the maximum shear stress?

2. A reinforced concrete beam is illustrated. It is subjected to a maximum moment of 8125 ft-lbf. The total cross-sectional area of steel is 1 square inch. Take the modulus of elasticity for concrete to be 2 EE6 psi. Disregard the area of concrete in tension. Use the transformation method to calculate the maximum stress in the steel and concrete.

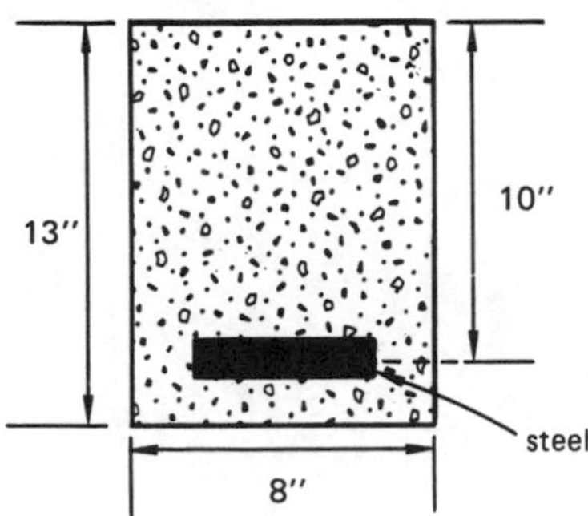


3. A steel truss with pinned joints is constructed as shown. The length/area ratio for each member is 50 in$^{-1}$. Support $R_1$ is a pin. Support $R_2$ is a roller. What is the vertical deflection at the point where the 10 kip load is applied? Use E = 2.9 EE7 psi.

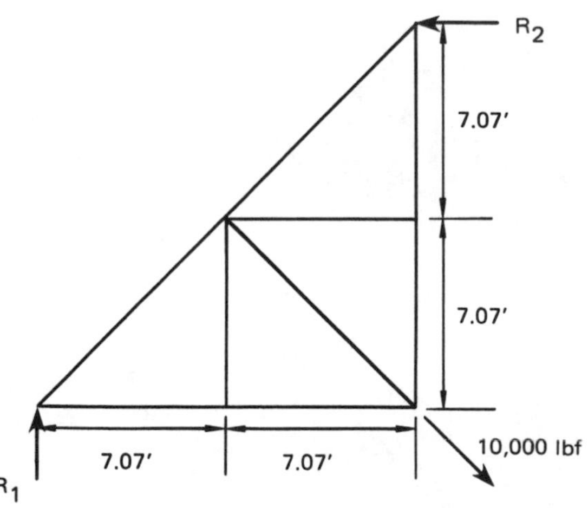


4. A beam 25 feet long is simply supported at the left end and 5 feet from the right end. A uniform load of 2 kips/ft extends over a 10 foot length starting from the left end. There is also a concentrated 10 kip load at the right end. Draw the shear and moment diagrams.

5. A 40 foot long simply supported steel beam with moment of inertia I is reinforced along the middle 20 feet, leaving the two 10 foot ends unreinforced. The moment of inertia of the reinforced section is 2I. A 20,000 pound load is applied mid-span along the beam, 20 feet from each end. What is the deflection at the midpoint of the beam in terms of EI?

6. The truss shown carries a moving uniform live load of 2 kips per foot and a moving concentrated live load of 15 kips. What are the maximum forces in members Bb and BC?

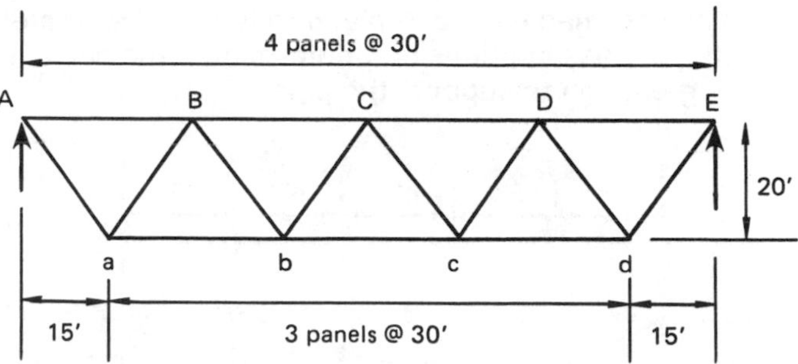


7. A 24 foot, 4 inch × 4 inch white oak timber (E = 1.5 EE6 psi) is used to support a sign. One end is fixed in a deep concrete base, the other end supports a sign 9 feet above the ground. Neglect wind effects and find the critical buckling sign weight.

8. What is the mid-span deflection for the steel beam shown? The cross sectional moment of inertia is 200 inches$^4$.

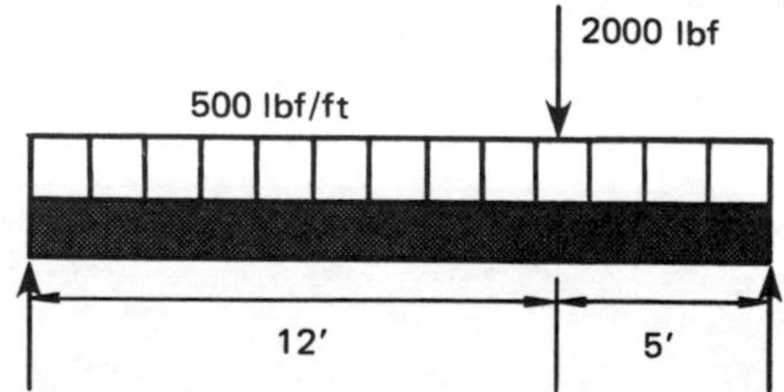


9. The truss shown carries a group of live loads along its bottom chord. What is the maximum force in member CD?

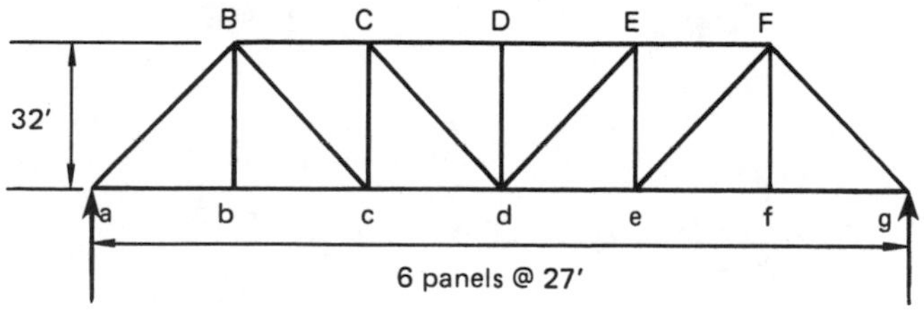


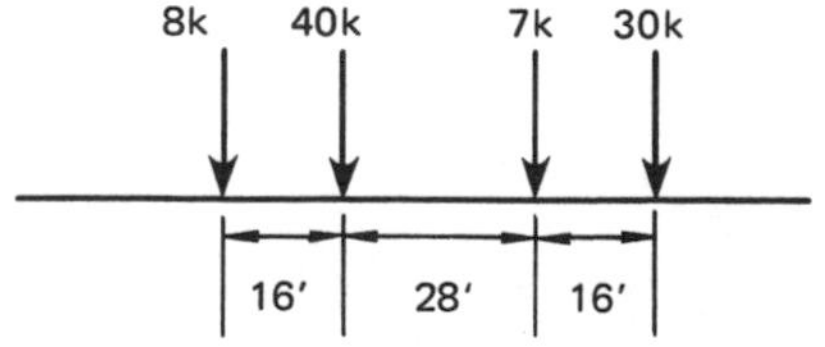


MECH OF MATL

10. The allowable stress in a steel beam is 20,000 psi. If the maximum applied moment is 1.5 EE5 ft-lb, what is the required section modulus?

## Timed

1. Wet concrete is needed 200 feet from the nearest pump location, which is separated from the pour location by a deep ravine. It is decided to use a series of ½″ steel cables to support a 6″ steel pipe. The pipe can be assumed to be completely filled with concrete. State your assumptions and determine if the cable is strong enough to support the pipe.

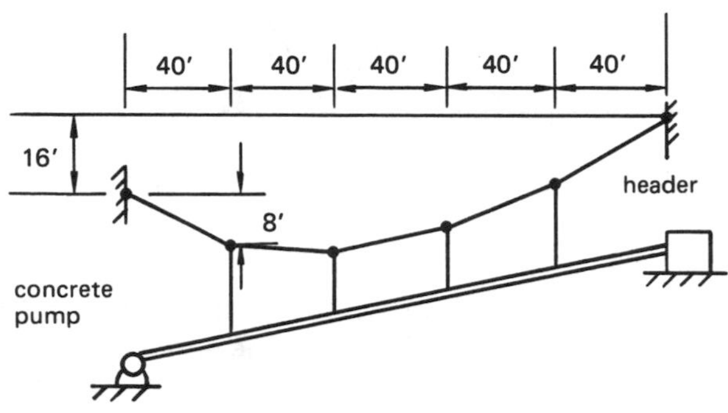


2. A moving load, consisting of two 30 kip loads separated by a constant 6 feet, travels over a two-span bridge. The bridge has an interior expansion joint that can be considered to be a hinge. Find the maximum moment and shear at point B.

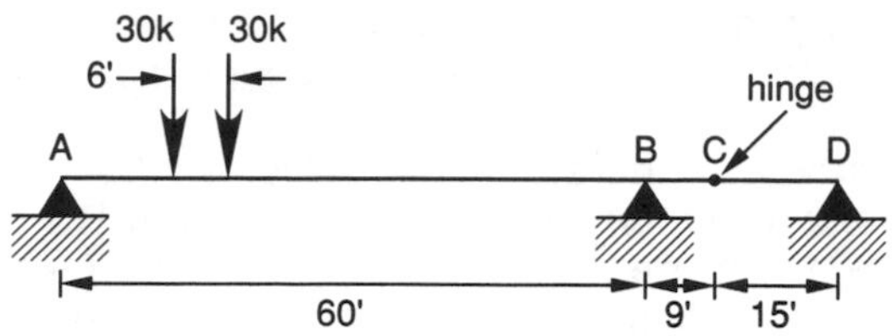

# 13 INDETERMINATE STRUCTURES

*Nomenclature*

| | | |
|---|---|---|
| A | area | $ft^2$ |
| C | carry-over factor | – |
| COM | carry-over moment | ft–lbf |
| D | distribution factor | – |
| E | modulus of elasticity | $lbf/ft^2$ |
| F | fixity factor | – |
| I | moment of inertia | $in^4$ |
| k | spring constant | ft–lbf/radian |
| K | stiffness factor | ft–lbf |
| L | length | ft |
| M | moment | ft-lbf |
| P | load | lbf |
| R | relative stiffness, or reaction | ft–lbf, or lbf |
| S | force in truss member | lbf |
| T | temperature | °F |
| u | dummy force in truss member | lbf |
| w | continuous load | lbf/ft |
| $x^*$ | distance to centroid of moment diagram | ft |

*Symbols*

| | | |
|---|---|---|
| $\delta$ | deflection | ft |
| $\sigma$ | stress | $lbf/ft^2$ |
| $\alpha$ | coefficient of thermal expansion | 1/°F |
| $\theta$ | angle | radians |
| $\Delta$ | deflection | ft |

*Subscripts*

| | |
|---|---|
| BM | bending moment |
| c | concrete |
| M | due to moment M |
| P | due to force P |
| s | steel |

## 1 INTRODUCTION

A structure that is statically indeterminate is one for which the equations of statics are not sufficient to determine all reactions, moments, and internal stress distributions. Additional formulas involving deflection are required to completely determine these unknowns. Although there are a large number of problem types which are statically indeterminate, this chapter is primarily concerned with the following cases:

- Beams on more than two supports
- Trusses with more than 2(# joints)–3 members
- Rigid frames

The *degree of redundancy* is equal to the number of reactions or members that would have to be removed in order to make the structure statically determinate. For example, a two-span beam on three simple supports is redundant to the first degree.



## 2 CONSISTENT DEFORMATION

The method of *consistent deformation* can be used to evaluate simple structures consisting of two or three members in tension or compression. This method is simple to learn and apply. The method makes use of geometry to develop relationships between the deflections (deformations) between different members or locations on the structure.

*Example 13.1*

A pile is constructed of concrete with a steel jacket. What are the stresses in the steel and concrete if a load $P$ is applied? Assume the end caps are rigid and the steel-concrete bond is perfect.

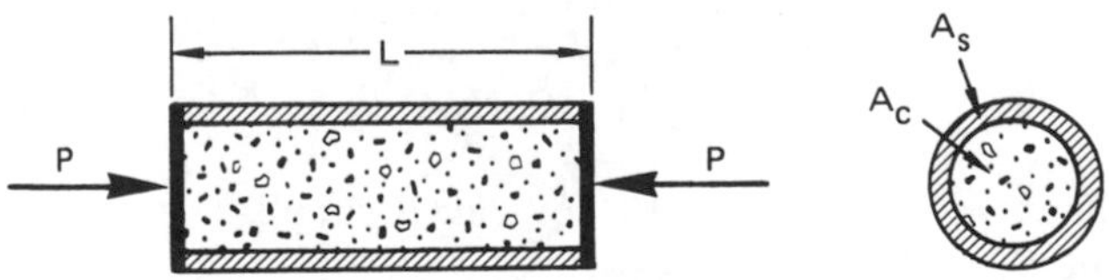

Let $P_c$ and $P_s$ be the loads carried by the concrete and steel respectively. Then,

$$P_c + P_s = P \qquad 13.1$$

The deformation of the steel is

$$\delta_s = \frac{P_s L}{A_s E_s} \qquad 13.2$$

Similarly, the deflection of the concrete is

$$\delta_c = \frac{P_c L}{A_c E_c} \qquad 13.3$$

But, $\delta_c = \delta_s$ since the bonding is perfect. Therefore,

$$\frac{P_c L}{A_c E_c} - \frac{P_s L}{A_s E_s} = 0 \qquad 13.4$$

Equations 13.1 and 13.4 are solved simultaneously to determine $P_c$ and $P_s$. The respective stresses are:

$$\sigma_s = \frac{P_s}{A_s} \qquad 13.5$$

$$\sigma_c = \frac{P_c}{A_c} \qquad 13.6$$

*Example 13.2*

A uniform bar is clamped at both ends and the axial load applied near one of the supports. What are the reactions?

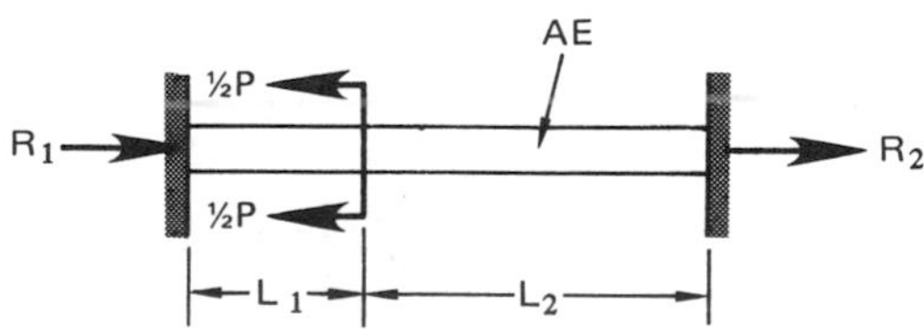


The first required equation is

$$R_1 + R_2 = P \qquad 13.7$$

The shortening of section 1 due to the reaction $R_1$ is

$$\delta_1 = \frac{-R_1 L_1}{AE} \qquad 13.8$$

The elongation of section 2 due to the reaction $R_2$ is

$$\delta_2 = \frac{R_2 L_2}{AE} \qquad 13.9$$

However, the bar is continuous, so $\delta_1 = -\delta_2$. Therefore,

$$R_1 L_1 = R_2 L_2 \qquad 13.10$$

Equations 13.7 and 13.10 are solved simultaneously to find $R_1$ and $R_2$.

*Example 13.3*

The non-uniform bar shown is clamped at both ends. What are the reactions if a temperature change of $\Delta T$ is experienced?

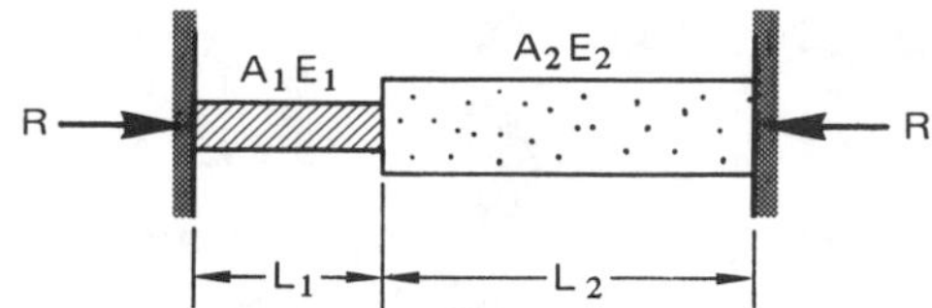


The thermal deformations of sections 1 and 2 can be calculated directly.

$$\delta_1 = \alpha_1 L_1 \Delta T \qquad 13.11$$

$$\delta_2 = \alpha_2 L_2 \Delta T \qquad 13.12$$

The total deformation is $\delta = \delta_1 + \delta_2$. However, the deformation can also be calculated from the principles of mechanics.

$$\delta = \frac{RL_1}{A_1 E_1} + \frac{RL_2}{A_2 E_2} \qquad 13.13$$

Combining equations 13.11 through 13.13 produces an equation that can be solved directly for $R$.

$$(\alpha_1 L_1 + \alpha_2 L_2)\Delta T = \left(\frac{L_1}{A_1 E_1} + \frac{L_2}{A_2 E_2}\right) R \qquad 13.14$$

*Example 13.4*

The beam shown is supported by dissimilar members. What are the forces in the members? Assume the bar is rigid and remains horizontal.[1]

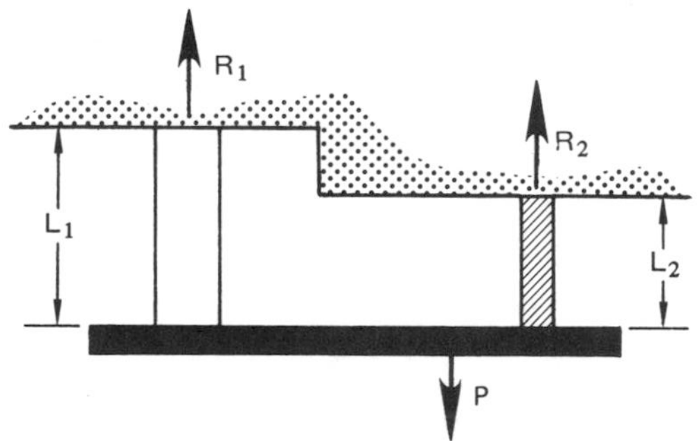


The required equilibrium condition is

$$R_1 + R_2 = P \qquad 13.15$$

The elongations of the two tension members are

$$\delta_1 = \frac{R_1 L_1}{A_1 E_1} \qquad 13.16$$

$$\delta_2 = \frac{R_2 L_2}{A_2 E_2} \qquad 13.17$$

[1] Notice that the location of force $P$ is not specified.

IND STRUCT

If the horizontal bar remains horizontal, then $\delta_1 = \delta_2$.

$$\frac{R_1L_1}{A_1E_1} = \frac{R_2L_2}{A_2E_2} \qquad 13.18$$

Equations 13.15 and 13.18 are solved simultaneously to find $R_1$ and $R_2$.

*Example 13.5*

The beam shown is supported by dissimilar members. The bar is rigid, but is not constrained to remain horizontal. What are the reactions in the vertical members?

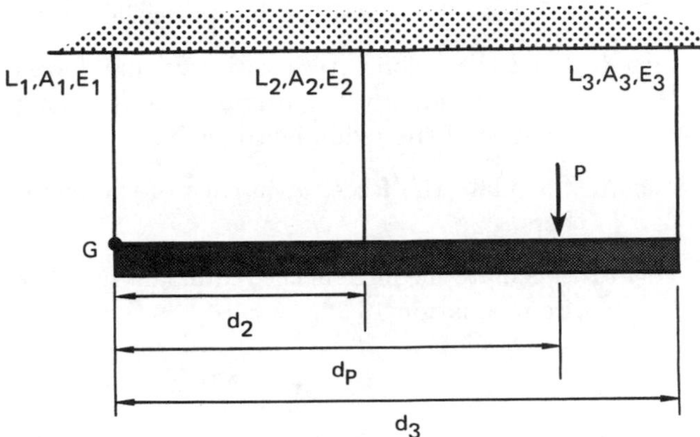


The forces in the supports are $R_1$, $R_2$, and $R_3$. Any of these may be tensile (positive) or compressive (negative).

$$R_1 + R_2 + R_3 = P \qquad 13.19$$

The changes in lengths are

$$\delta_1 = \frac{R_1L_1}{A_1E_1} \qquad 13.20$$

$$\delta_2 = \frac{R_2L_2}{A_2E_2} \qquad 13.21$$

$$\delta_3 = \frac{R_3L_3}{A_3E_3} \qquad 13.22$$

Since the bar is rigid, the deflections will be proportional to the distance from point $G$.

$$\delta_2 = \delta_1 + \frac{d_2}{d_3}(\delta_3 - \delta_1) \qquad 13.23$$

Moments can be summed about point $G$ to give a third equation.

$$M_G = R_3d_3 + R_2d_2 - Pd_P = 0 \qquad 13.24$$

*Example 13.6*

Write two equilibrium conditions for the three tension members.

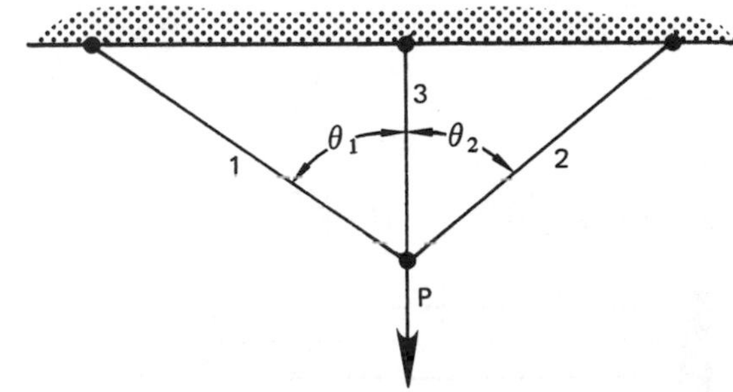


The equilibrium requirement is

$$P_{1y} + P_3 + P_{2y} = P \qquad 13.25$$

$$P_1 \cos\theta_1 + P_3 + P_2 \cos\theta_2 = P \qquad 13.26$$

Assuming the elongations are small compared to the member lengths, the angles $\theta_1$ and $\theta_2$ are unchanged. Then,

$$\frac{P_1L_1}{A_1E_1\,\cos\theta_1} = \frac{P_3L_3}{A_3E_3} = \frac{P_2L_2}{A_2E_2\,\cos\theta_2} \qquad 13.27$$

Equations 13.26 and 13.27 can be solved simultaneously to find $P_1$, $P_2$, and $P_3$. (It may be necessary to work with the $x$-components of the deflections in order to find a third equation.)

## 3 USING SUPERPOSITION WITH STATICALLY INDETERMINATE BEAMS

Continuous beams and propped cantilevers that are indeterminate to the first degree can often be solved by superposition.[2] This method requires finding the deflections with one or more of the supports removed, and then satisfying the given boundary conditions.

*step 1*: Remove the redundant supports to reduce the structure to a statically determinate condition.

*step 2*: Calculate the deflections at the previous locations of redundant supports. Use consistent sign conventions.

*step 3*: Apply each redundant support as a load, and find the deflections at the redundant support points as functions of the redundant support forces.

[2] Actually, this method can also be used with higher order indeterminate problems. However, the simultaneous equations that must be solved may make this method unattractive for manual solutions.

*step 4*: Use superposition to combine (i.e., add) the deflections due to the actual loads and the redundant support "loads." The total deflections must agree with the known deflections (usually zero) at the redundant support points.

*Example 13.7*

Determine the reaction, $S$, at the prop.

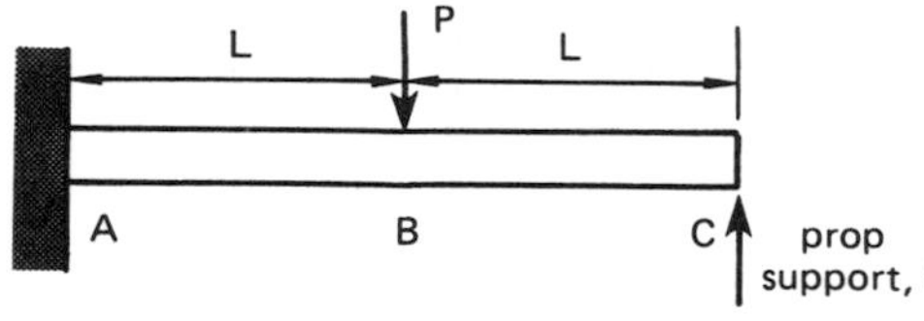


Start by removing the unknown prop reaction at point $C$. The cantilever that results is statically determinate. The deflection and slope at point $B$ can be found or derived from the beam equations.

$$\delta_B = \frac{-PL^3}{3EI} \qquad 13.28$$

$$y'_B = \frac{-PL^2}{2EI} \qquad 13.29$$

The slope remains constant to the right of point $B$. Therefore, the deflection at point $C$ due to the load at point $B$ is

$$\delta_{C,P} = \delta_B + y'_B L$$

$$= \frac{-5PL^3}{6EI} \qquad 13.30$$

The upward deflection at the cantilever tip due to the prop support, $S$, alone is

$$\delta_{C,S} = \frac{S(2L)^3}{3EI} = \frac{8SL^3}{3EI} \qquad 13.31$$

Now, it is known that the actual deflection at point $C$ is zero (the boundary condition). Therefore, the prop support, $S$, can be determined as a function of the applied load.

$$\delta_{C,S} + \delta_{C,P} = 0 \qquad 13.32$$

$$\frac{8SL^3}{3EI} - \frac{5PL^3}{6EI} = 0 \qquad 13.33$$

$$S = \frac{5P}{16} \qquad 13.34$$

## 4 INDETERMINATE TRUSSES: DUMMY UNIT LOAD METHOD

Due to the time required, it is unlikely that you would manually work with an indeterminate truss with more than one redundant member. Therefore, the following method is written specifically for trusses that are indeterminate to the first degree.

*step 1*: Draw the truss twice. Omit the redundant member on both trusses. (There may be a choice of redundant members.)

*step 2*: Load the first truss (which is now determinate) with the actual loads.

*step 3*: Calculate the forces, $S$, in all of the members. Assign a positive sign to tensile forces.

*step 4*: Load the second truss with two unit forces acting colinearly towards each other along the line of the redundant member.

*step 5*: Calculate the force, $u$, in each of the members.

*step 6*: Calculate the force in the redundant member from equation 13.35.

$$S_{\text{redundant}} = \frac{-\sum\left(\frac{SuL}{AE}\right)}{\sum\left(\frac{u^2L}{AE}\right)} \qquad 13.35$$

If $AE$ is the same for all members,

$$S_{\text{redundant}} = \frac{-\sum SuL}{\sum u^2L} \qquad 13.36$$

The true force in member $j$ of the truss is

$$P_{j,\text{true}} = S_j + (S_{\text{redundant}})u_j \qquad 13.37$$

*Example 13.8*

Find the force in members $BC$ and $BD$. $AE = 1$ for all members except for $CB$, which is 2, and $AD$, which is 1.5.

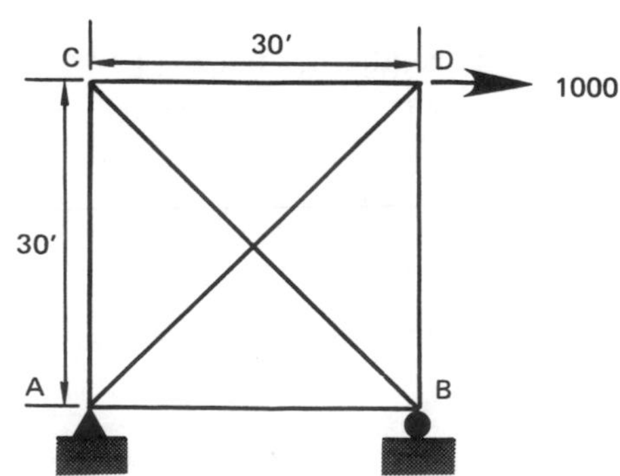

The two trusses are shown appropriately loaded.

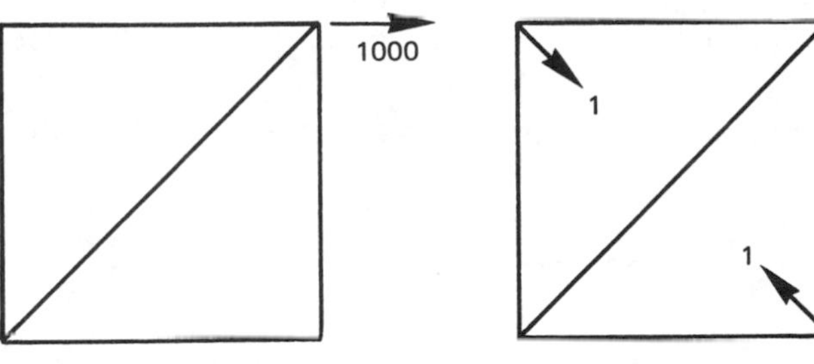


| member | L | AE | S | u | $\frac{SuL}{AE}$ | $\frac{u^2L}{AE}$ |
|---|---|---|---|---|---|---|
| $AB$ | 30 | 1 | 0 | −0.707(C) | 0 | 15 |
| $BD$ | 30 | 1 | −1000(C) | −0.707 | 21,210 | 15 |
| $DC$ | 30 | 1 | 0 | −0.707 | 0 | 15 |
| $CA$ | 30 | 1 | 0 | −0.707 | 0 | 15 |
| $CB$ | 42.43 | 2 | 0 | 1.0 | 0 | 21.22 |
| $AD$ | 42.43 | 1.5 | 1414(T) | 1.0 | 39,997 | 28.29 |
| | | | | | 61,207 | 109.51 |

From equation 13.35,

$$S_{BC} = \frac{-61{,}207}{109.51} = -558.9\,(\text{C})$$

From equation 13.37,

$$P_{BD,\text{true}} = -1000 + (-558.9)(-0.707) = -604.9\,(\text{C})$$

## 5 INTRODUCTION TO THE MOMENT DISTRIBUTION METHOD[3]

The moment distribution method is extremely powerful. It can be used on beams of almost any complexity. It converges rapidly to a solution, despite the fact that the method is essentially an iterative process. Furthermore, the moment distribution method is easily learned.

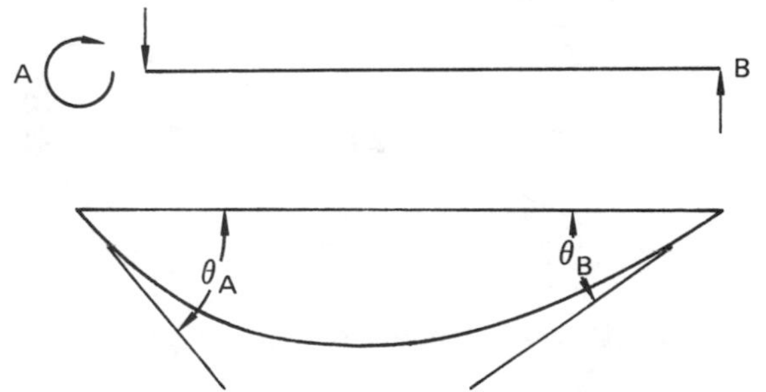


**Figure 13.1** Simply-Supported Beam with a Couple

Consider the simply-supported beam shown in figure 13.1. (The type of loading is not important, as only general relationships are being developed.) The deflection angles $\theta_A$ and $\theta_B$ can be found from the moment-area method.

If a simply-supported beam is acted upon by a clockwise couple (a moment) at one end as shown in figure 13.1, the angles of rotation will be given by equations 13.38 and 13.39.

$$\theta_A = \frac{M_A L}{3EI} \qquad 13.38$$

$$\theta_B = \frac{-M_A L}{6EI} \qquad 13.39$$

Figure 13.1 is particularly important because the equilibrium of the beam is not dependent on whether the moment is applied and the forces react, or whether the forces are applied and the moment reacts. If the forces are applied to the beam in figure 13.1, the moment develops to keep the beam end horizontal. A moment that is required to keep a beam end horizontal is known as a *fixed-end moment*.[4]

Any moment that would rotate an end of a beam clockwise is positive. Similarly, any moment that would rotate an interior joint clockwise is positive.

Being able to find the fixed-end moments at both ends of a loaded beam is absolutely necessary to the success of the moment distribution method. Extensive tables are available that contain the fixed-end moments for almost every conceivable loading. Appendix B is one such collection.

*Example 13.9*

The beam shown is acted upon by a moment at the right end. What is the relationship between the applied moment, $M_B$, and the fixed-end moment, $M_A$?

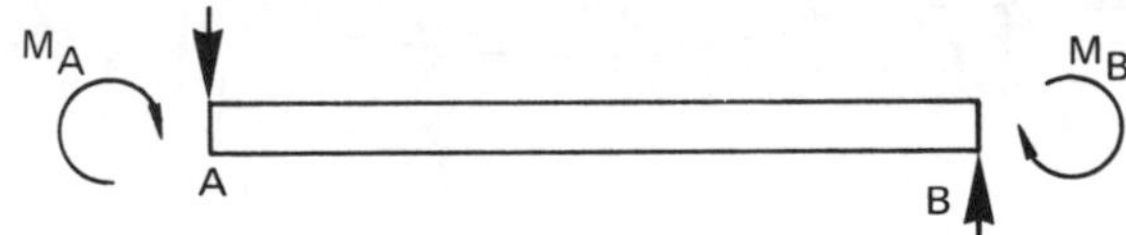


From equation 13.39, the angle of rotation at the left end due to a moment at the right end is

$$\theta_{A,M_B} = \frac{-M_B L}{6EI} \qquad 13.40$$

The angle at the left end due to the fixed-end moment at the left end is given by equation 13.38.

$$\theta_{A,M_A} = \frac{M_A L}{3EI} \qquad 13.41$$

[3] The moment distribution method was, at one time, also known as the *Cross method*, having been named after Professor Hardy Cross.

[4] Fixed-end moments are usually given double subscripts to indicate the location of the moment as well as the loaded span. For example, $M_{AB}$ would be a fixed-end moment at end $A$ caused by forces along the span $AB$.

However, $\theta_A = 0$, so $\theta_{A,M_B} + \theta_{A,M_A} = 0$. Or,

$$\frac{M_B L}{6EI} = \frac{M_A L}{3EI} \qquad 13.42$$

$$M_A = \tfrac{1}{2} M_B \qquad 13.43$$

## 6 SIGNS OF FIXED-END MOMENTS

Determining the proper sign for fixed-end moments can sometimes be confusing. The phrase "positive moments make the beam smile" simply does not work with end moments.

End moments are the moments that the support must apply to the beam to keep the beam (or joint) from rotating. The common convention is that clockwise moments acting on the beam at supports are positive. What is confusing is that a beam acted upon by a single force can have both positive and negative end moments.

Consider the beam in figure 13.2(a). If the ends were simply supported, the deflection curve would appear as figure 13.2(b). Since the actual deflection curve is as shown in figure 13.2(c), there must be moments acting on the ends (i.e., end moments). The moment at the left end is counterclockwise, and therefore, negative. The moment at the right end is clockwise, and therefore, positive.



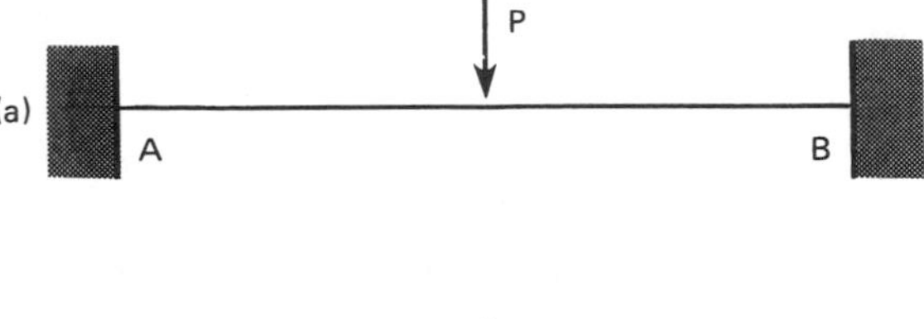


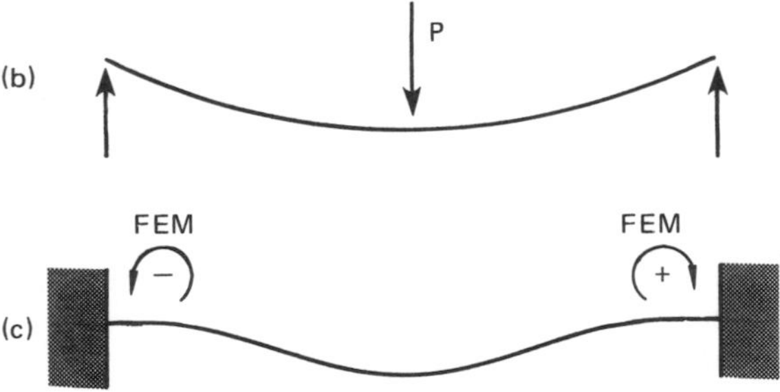


**Figure 13.2** A Simple Case of End Moments

Even though the beam in figure 13.2 "smiles," and we usually say it is acted upon by a positive moment, the moments exerted by the fixed supports are both positive and negative.

Another case is that of a moment applied directly to a beam, as in figure 13.3(a). The same rules and sign conventions can be used to determine the signs of the end moments.

If the beam had simply supported ends, the deflection curve would be as in figure 13.3(b). Since the ends are actually horizontal due to the fixed support, there must be end moments preventing the rotation. Both of these end moments are positive, as shown in figure 13.3(c).

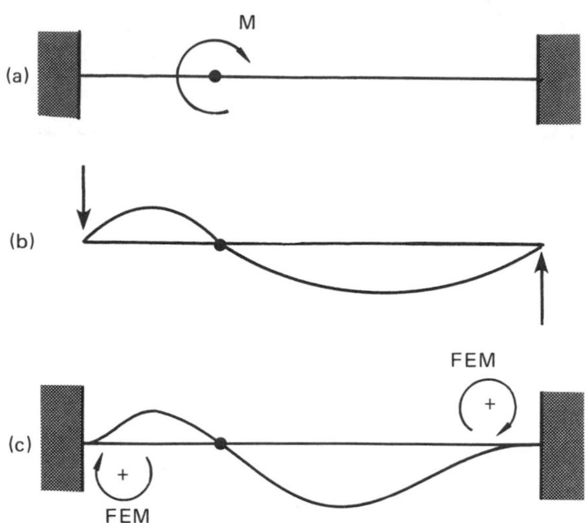


**Figure 13.3** End Moments Due to Direct Moment

Finally, consider the case of a force on a cantilever span. There is no moment at the free end, of course. However, the built-in end is fixed against rotation. Therefore, there is a fixed-end moment acting. As is the convention, clockwise moments are positive.

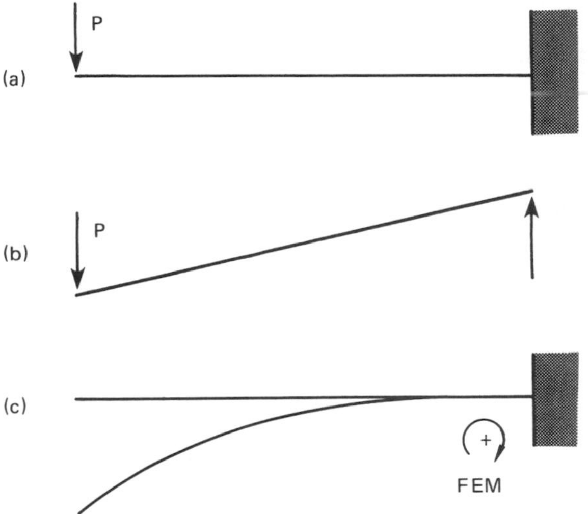


**Figure 13.4** End Moment on a Cantilever Support

## 7 BEAM STIFFNESS

From example 13.9, it is seen that a moment at one end of a simply-supported beam is partially transmitted to a fixed end. This transmitted moment is known as a *carry-over moment*, and the ratio of the carry-over moment to the original moment is known as the *carry-over factor*, $C$. In example 13.9, $C$ was $\frac{1}{2}$.

The carry-over factor can be determined for various degrees of *fixity*. The fixity is specified by a *fixity factor*, $F$, which varies from 0 (for a pin end) to 1 (for a built-in end).[5]

$$C = \frac{2F}{3+F} \qquad 13.44$$

The product $EI$ is known as the *stiffness* of the beam. The *relative stiffness* is

$$R = \frac{EI}{L} \qquad 13.45$$

The *angular spring constant* is known as the *stiffness factor*, $K$.

$$K = \frac{M}{\theta} = (3+F)R \qquad 13.46$$

It is possible to generalize about the moment required to produce a given angle of rotation. Equation 13.47 gives the moment for a simply supported beam.

$$M_{AB} = \frac{3EI\theta}{L} \qquad 13.47$$

Equation 13.47 could have been derived by using equation 13.46. Since $F = 0$ for a pinned end or simply-supported end,

$$M_{AB} = K\theta = 3R\theta = \frac{3EI\theta}{L} \qquad 13.48$$

If the beam is built-in at an opposite end, the moment required to produce an angle $\theta$ can be derived using equation 13.46 with $F = 1$.

$$M_{AB} = \frac{4EI\theta}{L} \qquad 13.49$$

Therefore, beams with built-in ends are 4/3 as stiff as beams with pinned ends.

## 8 DISTRIBUTION FACTORS

Consider the joint $B$ illustrated in figure 13.5(a). The joint rigidly connects a complex beam that is simply supported at ends $A$ and $D$, but is fixed at end $C$. The relative stiffnesses are $R_1, R_2$, and $R_3$ respectively for beams $BA, BC$, and $BD$. A moment $U$ is applied to the joint $B$. The joint will rotate as shown in figure 13.5(b).

[5] This is fixity against rotation, not fixity against translation. A simply-supported beam is completely fixed against translation, but it is still free to rotate.

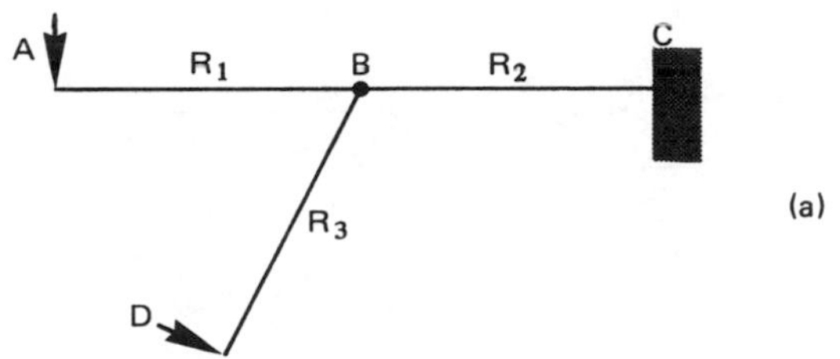


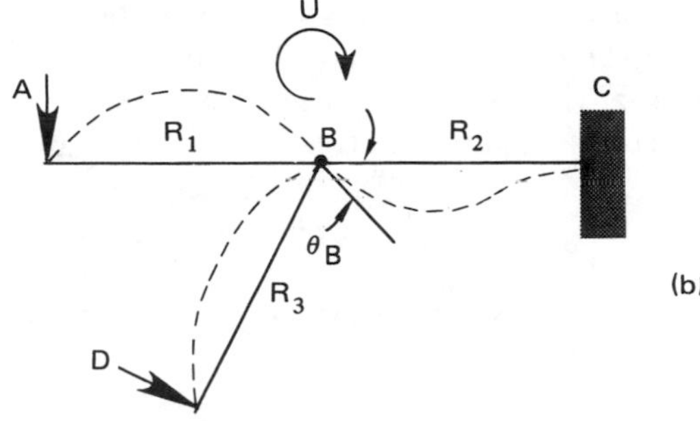


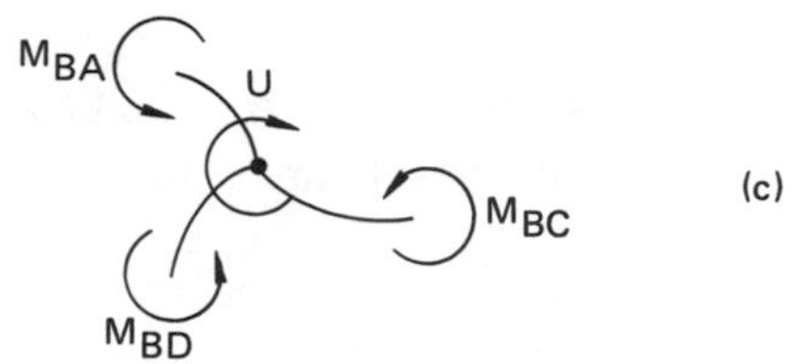


**Figure 13.5** A General Joint

A free body diagram of joint $B$ is shown in figure 13.5(c). It is clear that the rotation will continue until the moment $U$ is resisted by the stiffness of the beams.

$$U = M_{BA} + M_{BC} + M_{BD} \qquad 13.50$$

However, knowing $\theta_B$ gives the values of $M_{BA}, M_{BC}$, and $M_{BD}$ from equations 13.48 and 13.49.

$$U = 3R_1\theta_B + 4R_2\theta_B + 3R_3\theta_B \qquad 13.51$$

The individual resisting moments, $M_{BA}, M_{BC}$, and $M_{BD}$, can be found. $M_{BA}$, for example, is

$$\frac{M_{BA}}{U} = \frac{3R_1\theta_B}{3R_1\theta_B + 4R_2\theta_B + 3R_3\theta_B} \qquad 13.52$$

Since $\theta_B$ cancels,

$$M_{BA} = U\left(\frac{K_1}{K_1 + K_2 + K_3}\right) \qquad 13.53$$

IND STRUCT

Similarly,

$$M_{BC} = U\left(\frac{K_2}{K_1 + K_2 + K_3}\right) \qquad 13.54$$

$$M_{BD} = U\left(\frac{K_3}{K_1 + K_2 + K_3}\right) \qquad 13.55$$

From equations 13.53, 13.54, and 13.55, the moment $U$ is distributed to the three spans in proportion to the ratios of the stiffness factors. The quantities in parentheses are known as *distribution factors* since they distribute the applied moment, $U$, over all resisting beams.

## 9 MOMENT DISTRIBUTION METHOD PROCEDURE

The following steps constitute the moment distribution method.[6]

*step 1*: Divide the beam into independent spans, with all ends assumed to be built-in.

*step 2*: Calculate the relative stiffness of each member. Overhanging spans have $R = 0$ by definition. Both $E$ and $I$ can be given a value of 1.0 if they are constant along the entire span.

$$R = \frac{EI}{L} \qquad 13.56$$

*step 3*: Determine the fixity factors, $F$, for the beam joints. $F = 0$ when the joint is simply supported or is otherwise free to rotate (i.e., the joint cannot impart a resisting moment). $F = 1$ when the joint is built-in or is a continuous support.

*step 4*: Calculate the stiffness factors, $K$. $F$ in equation 13.57 is the value associated with the opposite or facing joint, not the joint's own value of $F$. Therefore, $F = 0$ is used when the joint faces a simply-supported free end.

$$K = (3 + F_{\text{opposite}})R \qquad 13.57$$

*step 5*: Calculate the distribution factor based on the stiffness factors for all spans tributary to the joint. By definition, $D = 0$ at built-in ends, since the support absorbs all the moment. Also, $D = 1$ at any simply-supported end.

$$D = \frac{K}{\sum K} \qquad 13.58$$

*step 6*: Determine the carry-over factors for ends. As in step 4, the values of $F$ used are from the beam's opposite ends.

$$C = \frac{2F_{\text{opposite}}}{3 + F_{\text{opposite}}} \qquad 13.59$$

*step 7*: Calculate the fixed-end moments using appendix B and the actual transverse loading on each span.

*step 8*: At each joint, balance any unbalanced moments by distributing a counter moment among the connecting members according to their distribution factors. Add the following counter moments for balancing:

| | |
|---|---|
| at a built-in end: | zero |
| at an interior joint: | –(unbalance) |
| at a pinned end: | –(unbalance) |
| at a partially restrained end: | –(1–$F$)(unbalance) |

*step 9*: These distributed balancing moments produce carry-over moments at the opposite ends of members equal to the distributed moment times the carry-over factors, with the same signs as the distributed balancing moment. The carry-over moment appears at the opposite end from the location where the balancing moment was applied.

*step 10*: Repeat steps 6 and 7 until sufficient accuracy is obtained. Unless there is just one distribution (as in a 2-span beam), end the distribution process with a distribution, not a carry-over. The final fixed-end moment at the end of a member is equal to the algebraic sum of the original fixed-end moments and all distributed balancing and carry-over moments. The final moment is the moment on the beam end, not the moment on the support due to the loads.

[6] There are several different versions of the moment distribution method. The standard method is to set $C = \frac{1}{2}$ for all joints, and to always calculate $K = 4R$. The procedure presented here uses modified stiffness factors (i.e., modified by $F$) and converges more quickly than the standard method. However, the distribution factors calculated by the two methods will not be the same, even though the final fixed-end moments will be.

*Example 13.10*

Find the moments at points $A, B$, and $C$.

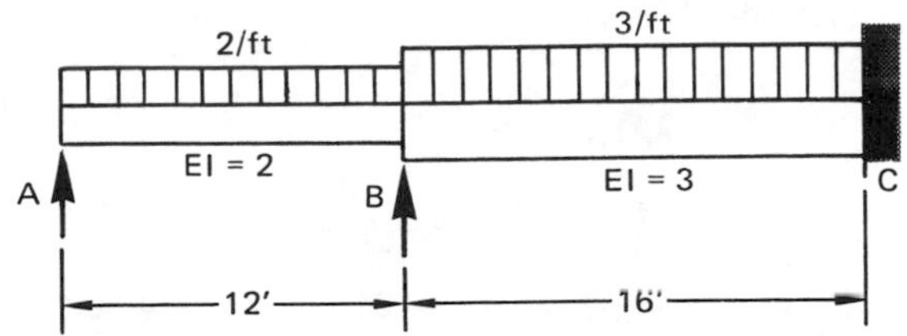


*step 1*: Divide the beam into spans $A$–$B$ and $B$–$C$.

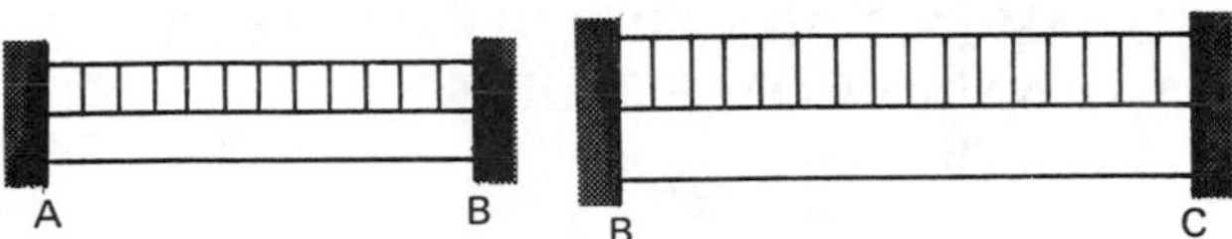


*step 2*:

$$R_{BA} = 2/12 = 1/6$$
$$R_{BC} = 3/16$$

*step 3*:

$F_{AB} = 0$ since the end is free to rotate.

$$F_{BA} = 1$$
$$F_{BC} = 1$$
$$F_{CB} = 1$$

*step 4*: Calculate $K_{AB}$ based on $F_{BA}$ (the fixity factor for the opposite end of span $AB$).

$$K_{AB} = (3+1)\left(\tfrac{1}{6}\right) = \tfrac{2}{3}$$

Calculate $K_{BA}$ based on $F_{AB}$.

$$K_{BA} = (3+0)\left(\tfrac{1}{6}\right) = \tfrac{1}{2}$$

$K_{BC}$ and $K_{CB}$ are both the same.

$$K_{BC} = K_{CB} = (3+1)\left(\tfrac{3}{16}\right) = \tfrac{12}{16} = \tfrac{3}{4}$$

*step 5*: $D = 1$ for joint $AB$, since any moment there must be transmitted to some other joint. $D = 0$ for joint $CB$, since the built-in support absorbs all moments. The only other joint in the structure is $B$.

$$D_{BC} = \frac{\frac{3}{4}}{\frac{3}{4}+\frac{1}{2}} = \tfrac{3}{5} = 0.6$$

$$D_{BA} = \frac{\frac{1}{2}}{\frac{3}{4}+\frac{1}{2}} = \tfrac{2}{5} = 0.4$$

*step 6*:

$$C_{AB} = \tfrac{1}{2}$$
$$C_{BA} = 0$$
$$C_{BC} = \tfrac{1}{2}$$
$$C_{CB} = \tfrac{1}{2}$$

*step 7*: Work with span $BC$, assuming both ends are fixed. The moment at $B$ is caused by the applied load. The applied moment is clockwise, and therefore it is positive. However, fixed-end moments are the resisting moments, and the resisting moment at $B$ is counterclockwise. From appendix B, the fixed-end moment at joint B is

$$M_{BC} = \frac{-wL^2}{12} = \frac{-(3)(16)^2}{12} = -64$$
$$M_{CB} = +64$$

Now, work with span $BA$, still assuming both ends are fixed.

$$M_{AB} = \frac{-(2)(12)^2}{12} = -24$$
$$M_{BA} = +24$$

At this time, it is necessary to start a table to keep track of the results of the calculations.

| | $M_{AB}$ | $M_{BA}$ | $M_{BC}$ | $M_{CB}$ |
|---|---|---|---|---|
| *FEM* | −24 | 24 | −64 | 64 |

*step 8*: At joint $A$, there is a −24 moment that is not balanced. Since joint $A$ is a pinned end that cannot carry any moment, the moment there must be removed. This is done by adding +24 to that joint. Since joint $A$ has only one 'side,' the entire +24 is added to $M_{AB}$.

| | $M_{AB}$ | $M_{BA}$ | $M_{BC}$ | $M_{CB}$ |
|---|---|---|---|---|
| *FEM* | −24 | 24 | −64 | 64 |
| Balance | +24 | | | |

At joint $B$, there is an unbalance of $(24 - 64) = -40$. So, this must be balanced by adding +40 to joint $B$. However, joint $B$ goes in two directions (towards $A$ and $C$), so the +40 must be distributed to spans $BA$ and $BC$ in proportion to their stiffnesses. The distribution factors were previously found to be 0.4 and 0.6 for $M_{BA}$ and $M_{BC}$ respectively.

$$\text{Balance } M_{BA} = (0.4)(40) = 16$$

$$\text{Balance } M_{BC} = (0.6)(40) = 24$$

IND STRUCT

| | $M_{AB}$ | $M_{BA}$ | $M_{BC}$ | $M_{CB}$ |
|---|---|---|---|---|
| *FEM* | −24 | 24 | −64 | 64 |
| Balance | 24 | 16 | 24 | |

The support at joint $C$ was assumed fixed in step 1. Since it actually is fixed, no correction is required.

| | $M_{AB}$ | $M_{BA}$ | $M_{BC}$ | $M_{CB}$ |
|---|---|---|---|---|
| *FEM* | −24 | 24 | −64 | 64 |
| Balance | 24 | 16 | 24 | 0 |

*step 9*: The balancing moments produce carry-over moments. If we add +24 to $M_{AB}$, we must also add $\frac{1}{2}(24) = 12$ to $M_{BA}$ since the carry-over factor $C_{AB} = \frac{1}{2}$. Similarly, the rest of the following table is prepared.

| | $M_{AB}$ | $M_{BA}$ | $M_{BC}$ | $M_{CB}$ |
|---|---|---|---|---|
| *FEM* | −24 | 24 | −64 | 64 |
| Balance | 24 | 16 | 24 | 0 |
| COM | 0 | 12 | 0 | 12 |

*step 10*: However, now the addition of the carry-over moments has unbalanced the beam again. So, the process must be repeated from step 8.

*step 8, repeated*: *At joint A*: No carry-over moment was applied, so no balancing is needed.

*At joint B*: The total carry-over moment applied was $(12 + 0) = 12$. So, a balancing moment of −12 is needed. The −12 is distributed to $M_{BA}$ and $M_{BC}$ according to

$$M_{BA} = (0.4)(-12) = -4.8$$
$$M_{BC} = (0.6)(-12) = -7.2$$

*At joint C*: Joint $C$ is actually fixed, so the carry-over moment of 12 is balanced by the beam's support. No balancing moment is needed.

*step 9, repeated*: Using the carry-over factors, we add 0 to $M_A$ and −3.6 to $M_{CB}$.

| | $M_{AB}$ | $M_{BA}$ | $M_{BC}$ | $M_{CB}$ |
|---|---|---|---|---|
| *FEM* | −24 | 24 | −64 | 64 |
| 1st Balance | 24 | 16 | 24 | 0 |
| 1st COM | 0 | 12 | 0 | 12 |
| 2nd Balance | 0 | −4.8 | −7.2 | 0 |
| 2nd COM | 0 | 0 | 0 | −3.6 |

*step 10, repeated*: Since the −3.6 is balanced by the external support at joint $C$, and since all other terms are zero, no further work is required.

$$M_{AB} = -24 + 24 = 0$$

$$M_{BA} = 24 + 16 + 12 - 4.8 = 47.2$$

$$M_{BC} = -64 + 24 - 7.2 = -47.2$$

$$M_{CB} = 64 + 12 - 3.6 = 72.4$$

## 10 DRAWING SHEAR AND MOMENT DIAGRAMS FROM FIXED-END MOMENTS

It is generally easy to obtain the shear and moment diagrams from the distributed fixed-end moments. Calculate the reactions at the ends of each span by taking moments about each end in turn. Be sure to consider the signs of the span end-moments as determined by the distribution.

*Example 13.11*

Determine the reactions, moment diagram, and shear-diagram for the beam evaluated in example 13.10.

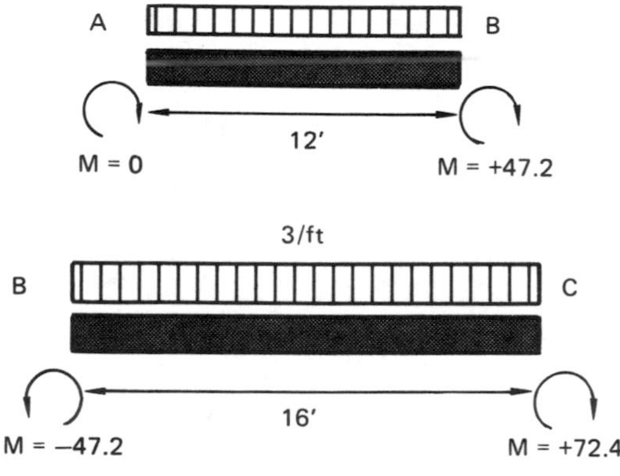


First, work with the left span. Sum moments about point $A$.

$$\sum M_A : 0 + 47.2 + \left(\tfrac{1}{2}\right)(2)(12)^2 - 12R_B = 0$$
$$R_B = 15.93$$

To find the reaction at point $A$, sum moments about point $B$.

$$\sum M_B : 0 + 47.2 - \left(\tfrac{1}{2}\right)(2)(12)^2 + 12R_A = 0$$
$$R_A = 8.07$$

Now, work with the right span. Sum moments about point $B$ to find the reaction at point $C$.

$$\sum M_B : -47.2 + 72.4 + \left(\tfrac{1}{2}\right)(3)(16)^2 - 16R_C = 0$$
$$R_C = 25.58$$

Sum moments about point $C$ to find the reaction at point $B$.

$$\sum M_C : -47.2 + 72.4 - \left(\tfrac{1}{2}\right)(3)(16)^2 + 16R_B = 0$$
$$R_B = 22.43$$

*Example 13.12*

Draw the moment diagram for the beam shown. All sections of the beam have the same value of $EI$.

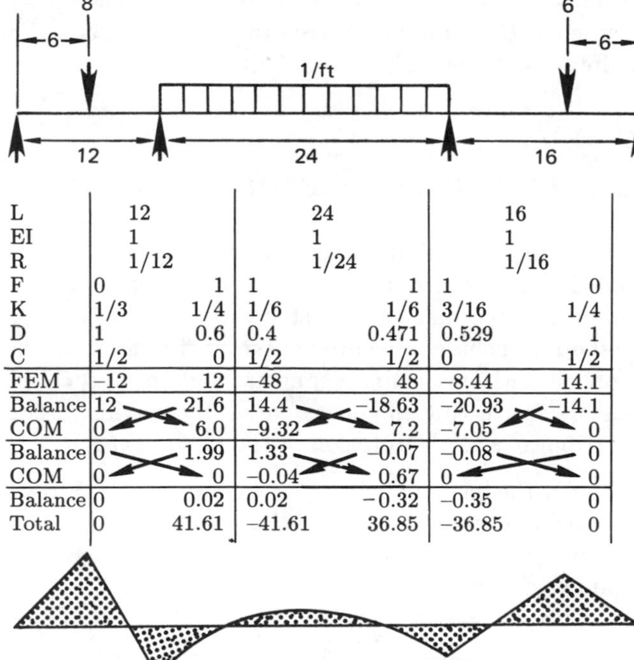


| | | | | | | |
|---|---|---|---|---|---|---|
| L | 12 | | 24 | | 16 | |
| EI | 1 | | 1 | | 1 | |
| R | 1/12 | | 1/24 | | 1/16 | |
| F | 0 | 1 | 1 | 1 | 1 | 0 |
| K | 1/3 | 1/4 | 1/6 | 1/6 | 3/16 | 1/4 |
| D | 1 | 0.6 | 0.4 | 0.471 | 0.529 | 1 |
| C | 1/2 | 0 | 1/2 | 1/2 | 0 | 1/2 |
| FEM | −12 | 12 | −48 | 48 | −8.44 | 14.1 |
| Balance | 12 | 21.6 | 14.4 | −18.63 | −20.93 | −14.1 |
| COM | 0 | 6.0 | −9.32 | 7.2 | −7.05 | 0 |
| Balance | 0 | 1.99 | 1.33 | −0.07 | −0.08 | 0 |
| COM | 0 | 0 | −0.04 | 0.67 | 0 | 0 |
| Balance | 0 | 0.02 | 0.02 | −0.32 | −0.35 | 0 |
| Total | 0 | 41.61 | −41.61 | 36.85 | −36.85 | 0 |

## 11 DIRECT MOMENTS AND FREE ENDS

When a moment is applied directly to a beam or joint, or when the beam has a cantilever end-span, the moment distribution procedure is changed slightly. If a concentrated moment is applied mid-span, the fixed-end moments at the span ends needed to keep the beam from rotating are added to fixed-end moments from other loads. The distances to the span ends on either side of the point of application determine the fraction of the direct moment distributed to each span end.

If a moment is applied directly to a joint, it may be unclear which direction the fixed-end moment acts to counter the rotation. In this case, it may be easier to consider the applied moment as an initial unbalance (i.e., place it in the BALANCE row) to be distributed immediately to both sides of the joint without change in sign.[7]

Spans with loaded cantilever sections can be replaced with an equivalent load and moment. All of the vertical load is carried by the adjacent support. The moment can be considered to be applied directly to the BALANCE row, without change of sign. Since the cantilever section is not included in the moment distribution process, the adjacent support becomes the end of the beam. The entire applied moment, therefore, is given to the inward-facing side of the first support.

After distribution, the moment at the first support will be zero, since it was assumed to be a free end. The moment at this support will have to be corrected by the amount of the original applied moment, since that moment is known to exist at that point.

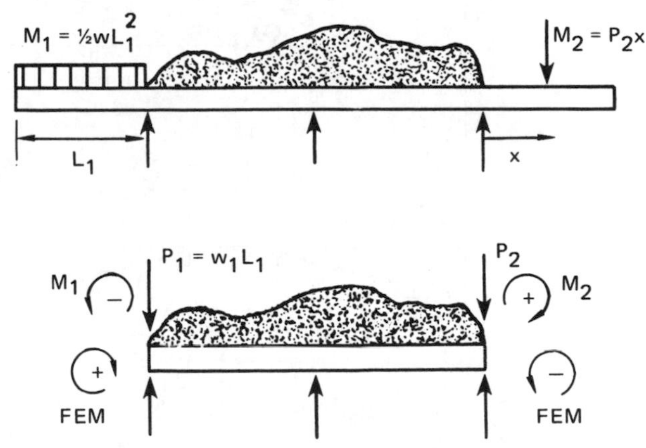


**Figure 13.6** Beam with Cantilever Ends and Equivalent Beam

[7] Obviously, putting the fixed-end moment in the FEM row or reversing the sign and putting it in the BAL column have the same effect. However, the rule to "... put the applied moment in the BAL row ..." is easy to remember.

*Example 13.13*

Find the bending moments at the ends of members $OA$ and $OB$. Joint $O$ is rigid. The vertical load of 6000 lbf is applied on a lever arm of length 3 feet.

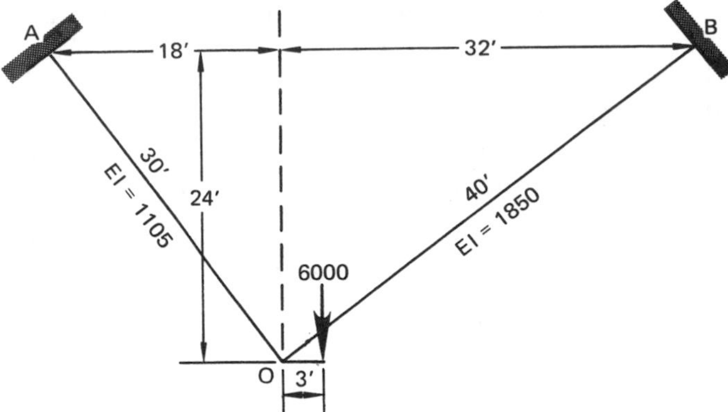


The applied moment is positive. This will require positive fixed-end moments about connections $A$ and $B$, as shown, to counteract the beam's rotation.

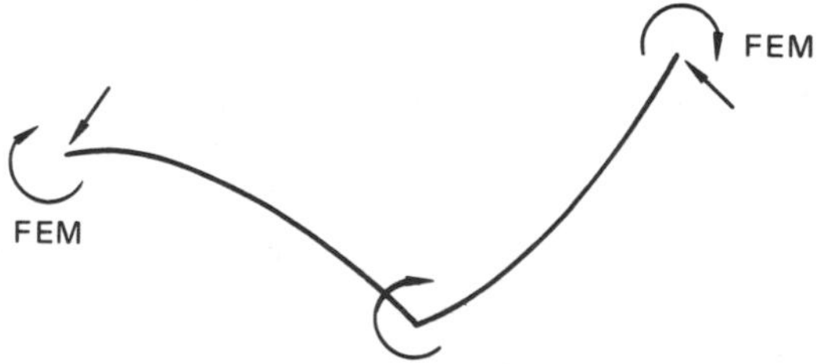


The amount of the applied moment distributed to legs $OA$ and $OB$ depends on the distribution factors, $D$. Since the fixed-end moments are known to be positive, a positive moment is distributed to each leg.

| | AO | OA | OB | BO |
|---|---|---|---|---|
| L | 30 | | 40 | |
| EI | 1105 | | 1850 | |
| R | 36.83 | | 46.25 | |
| F | 1 | 1 | 1 | 1 |
| K | 147.3 | 147.3 | 185.0 | 185.0 |
| D | 0 | 0.443 | 0.557 | 0 |
| C | 1/2 | 1/2 | 1/2 | 1/2 |
| Applied Moment | +18,000 | | | |
| Distributed BAL | | 7974 | 10,026 | |
| COM | 3987 | | | 5013 |
| Total | 3987 | 7974 | 10,026 | 5013 |

*Example 13.14*

Distribute the moments on the continuous beam shown. $EI$ is a constant 1 EE6 lbf-ft$^2$ along all spans.

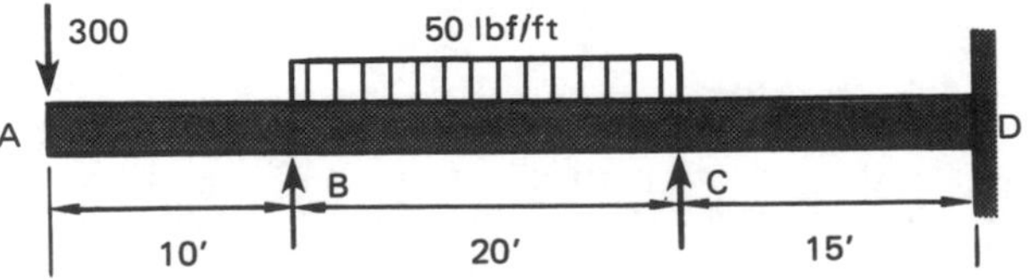


The magnitude of the moment at point $B$ due to the cantilever load is $(300)(10) = 3000$. The fixed-end moment at $B$ required to keep the continuous joint $B$ horizontal is clockwise, and therefore positive.

The fixed-end moments for the central span are

$$M_{BC} = -\frac{(50)(20)^2}{12} = -1667$$
$$M_{CB} = +1667$$

Span $AB$ can be omitted from the analysis. Its fixed-end moment is added to $M_{BC}$.

$$M'_{BC} = -1667 + 3000 = 1333$$

| | BC | CB | CD | DC |
|---|---|---|---|---|
| L | 20 | | 15 | |
| EI | 1 EE6 | | 1 EE6 | |
| R | 5 | | 6.67 | |
| F | 0 | 1 | 1 | 1 |
| K | 20 | 15 | 26.67 | 26.67 |
| D | 1 | 0.36 | 0.64 | 0 |
| C | 1/2 | 0 | 1/2 | 1/2 |
| FEM | 1333 | 1667 | 0 | 0 |
| Balance | −1333 | −600 | −1067 | 0 |
| COM | 0 | −667 | 0 | −533.5 |
| Balance | 0 | 240 | 427 | 0 |
| COM | 0 | 0 | 0 | 213.5 |
| Total | 0 | 640 | −640 | −320 |

$M_{BC}$ must be corrected by the amount of the original moment, since joint $B$ is not actually the end of the beam. On span $AB$, a clockwise moment is required to keep end $B$ horizontal. Therefore, $M_{BA} = +3000$. To maintain balance, $M_{BC} = -3000$.

## 12 FLEXIBLE AND YIELDING SUPPORTS

The moment distribution procedure presented assumes that the supports are unyielding. If any support moves, the distributed moments must be adjusted. The usual method of including yielding effects is first to solve the problem with unyielding supports, and then to calculate the correcting moments, since there will be two contributions to the rotation of the beam ends.

Consider the beam in figure 13.7 whose far end, $B$, rests on a yielding support. End $A$ is built-in (i.e., fixed), and end $B$ is simply supported. From equation 13.46, the stiffness is

$$K = \frac{M_A}{\theta} = 3R = \frac{3EI}{L} \tag{13.60}$$

IND STRUCT

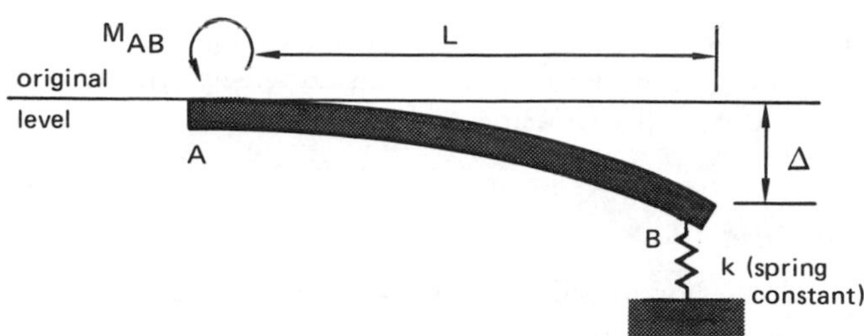


**Figure 13.7** Yielding Support Free to Rotate

However, $\theta \approx \tan\theta = \Delta/L$, so the moment causing deflection is

$$M_{AB} = \frac{3EI\Delta}{L^2} \qquad 13.61$$

Other relationships can be derived using the basic spring equations. Notice that the spring constant, $k$, is not the same as the beam stiffness, $K$.

$$M = PL \qquad 13.62$$

$$P = k\Delta \qquad 13.63$$

If ends $A$ and $B$ are both fixed, but end $B$ is free to translate (i.e., deflect due to the yielding support), the moment causing deflection will be

$$M_{AB} = M_{BA} = \frac{6EI\Delta}{L^2} \qquad 13.64$$

This moment appears at ends $A$ and $B$ with the same sign at both.

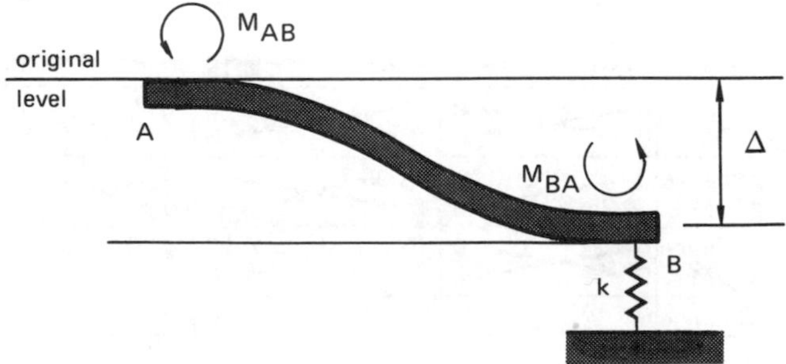


**Figure 13.8** Yielding Support Constrained to Horizontal

The fixed-end moments required to keep settled beam ends horizontal are no different than any other fixed-end moments, and the same sign conventions apply.

*Example 13.15*

Repeat example 13.14 assuming that support $C$ settles 0.5 inch.

Example 13.14 has already distributed the moments due to the applied loads. The settlement will not change these distributed moments, and it is not necessary to repeat those steps.

The settlement does, however, allow the beam to rotate. Therefore, additional moments are produced in the joints. Section $CD$ is fixed at both ends, and equation 13.64 is used to calculate the fixed-end moments. The deflection curve shows that positive moments are required to keep the beam ends horizontal.

$$M_{DC} = M_{CD} = \frac{(6)(1\,\text{EE}6)(0.5/12)}{(15)^2} = 1111$$

The support at joint $B$ is essentially a simple support, and span $AB$ is ignored. Therefore, equation 13.61 is used. $M_{CB}$ is negative because a counterclockwise moment is required to keep joint $C$ horizontal.

$$M_{CB} = \frac{-(3)(1\,\text{EE}6)(0.5/12)}{(20)^2} = -313$$

These moments are distributed in the usual manner.

| | BC | CB | CD | DC |
|---|---|---|---|---|
| L | 20 | | 15 | |
| EI | 1 EE6 | | 1 EE6 | |
| R | 5 | | 6.67 | |
| F | 0 | 1 | 1 | 1 |
| K | 20 | 15 | 26.67 | 26.67 |
| D | 1 | 0.36 | 0.64 | 0 |
| C | 1/2 | 0 | 1/2 | 1/2 |
| FEM | 0 | –313 | 1111 | 1111 |
| Balance | 0 | –287 | –511 | 0 |
| COM | 0 | 0 | 0 | –255.5 |
| Total | 0 | –600 | 600 | 855.5 |

The moments from this distribution are added to the moments from the distribution of example 13.14.

$$M_{AB} = 0$$
$$M_{BA} = +3000$$
$$M_{BC} = -3000 + 0 = -3000$$
$$M_{CB} = 640 - 600 = 40$$
$$M_{CD} = -640 + 600 = -40$$
$$M_{DC} = -320 + 855.5 = 535.5$$

## 13 FRAMES WITHOUT SIDESWAY

The rigid *portal frame* or *bent* is a fundamental structural unit. Therefore, an ability to analyze portal frames is essential.

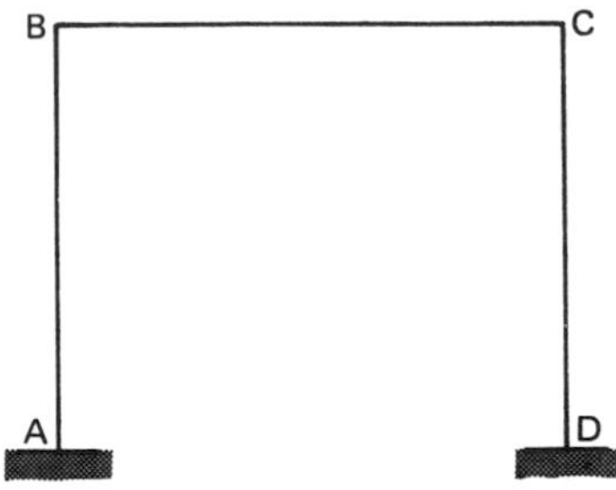


**Figure 13.9** A Rigid Frame

Because of the rigid joints at $B$ and $C$ in figure 13.9, it follows that all members meeting at a joint rotate through the same angle at that joint. Furthermore, any moment generated on a member will be transmitted through the joint to a connecting member. The moment distribution method can be applied directly in most cases, considering the frame to be a "bent" beam.

If a frame has an inclined member, the fixed-end moments depend only on the horizontal moment arm and the vertical component of force. The actual lengths, $L$, should be used in calculating rigidities, however.



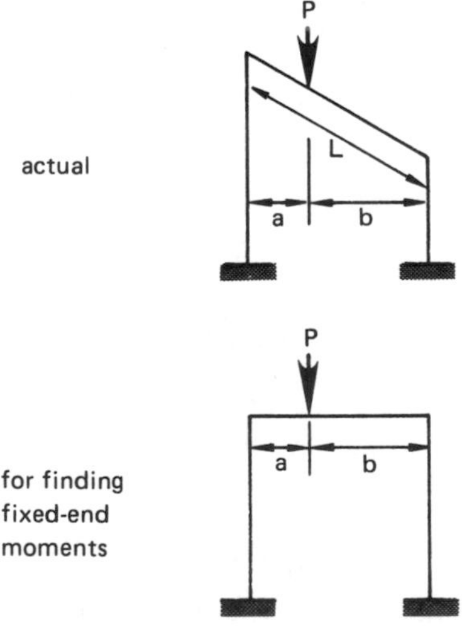


**Figure 13.10** Frame with Inclined Member

If a portal frame is symmetrical with respect to its leg lengths and loading, it will not sway or tend to move horizontally. If the frame or loading is unsymmetrical, it will sway unless prevented from doing so. Frames that are prevented from swaying are easily solved with the moment distribution method.

*Example 13.16*

Find the horizontal reactions at points $A$ and $D$. The frame is kept from moving laterally by a passive force at $C$. That is, the reaction at $C$ resists lateral movement, but it does not add any load or moment to the structure. Determine the moments and reactions at all joints, including the wall reaction at $C$.

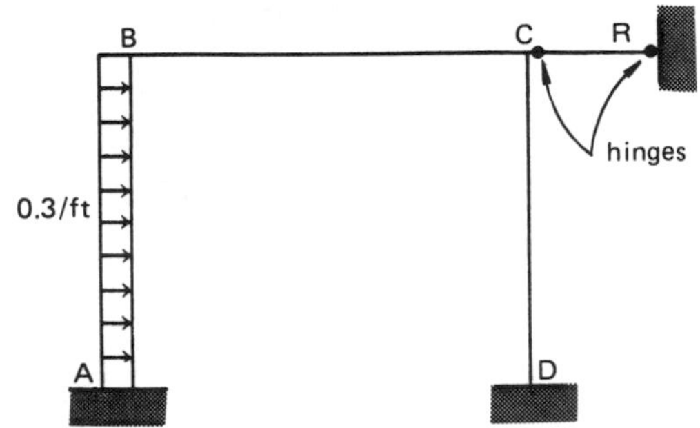


| | AB | BA | BC | CB | CD | DC |
|---|---|---|---|---|---|---|
| L | 25 | | 70 | | 25 | |
| EI | 1000 | | 10,500 | | 1000 | |
| R | 40 | | 150 | | 40 | |
| F | 1 | 1 | 1 | 1 | 1 | 1 |
| K | 160 | 160 | 600 | 600 | 160 | 160 |
| D | 0 | 0.21 | 0.79 | 0.79 | 0.21 | 0 |
| C | 1/2 | 1/2 | 1/2 | 1/2 | 1/2 | 1/2 |
| FEM | −15.63 | 15.63 | 0 | 0 | 0 | 0 |
| Balance | 0 | −3.28 | −12.35 | 0 | 0 | 0 |
| COM | −1.64 | 0 | 0 | −6.18 | 0 | 0 |
| Balance | 0 | 0 | 0 | 4.88 | 1.30 | 0 |
| COM | 0 | 0 | 2.44 | 0 | 0 | 0.65 |
| Balance | 0 | −0.51 | −1.93 | 0 | 0 | 0 |
| COM | −0.25 | 0 | 0 | −0.97 | 0 | 0 |
| Balance | 0 | 0 | 0 | 0.77 | 0.20 | 0 |
| COM | 0 | 0 | 0.38 | 0 | 0 | 0.10 |
| Balance | 0 | −0.08 | −0.30 | 0 | 0 | 0 |
| Total | −17.52 | 11.76 | −11.76 | −1.50 | 1.50 | 0.75 |

The free-bodies of the sections are used to determine the reactions.

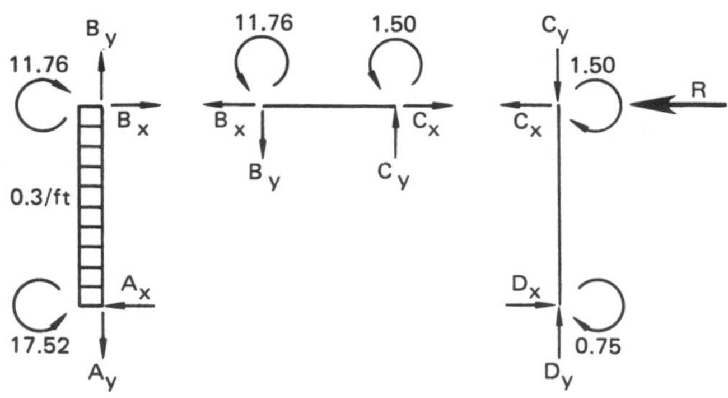

*Member AB.* To find $B_x$, sum moments about point $A$.

$$\sum M_A = \tfrac{1}{2}(25)^2(0.3) + 11.76 - 17.52 + 25(B_x) = 0$$
$$B_x = -3.52 \text{ (to the left)}$$

To find $A_x$, sum moments about point $B$.

$$\sum M_B = -\tfrac{1}{2}(25)^2(0.3) + 11.76 - 17.52 + 25(A_x) = 0$$
$$A_x = 3.98$$

*Member DC.* $C_x = -B_x = 3.52$ (to the right). To find $D_x$, sum moments about point $C$.

$$\sum M_C = 1.50 + 0.75 - 25(D_x) = 0$$
$$D_x = 0.09 \text{ (to the right)}$$

The wall reaction, $R$, is found from the horizontal equilibrium of freebody $CD$. (Alternatively, moments could be taken about point $D$.)

$$\sum F_x = 3.52 + 0.09 - R = 0$$
$$R = -3.61 \text{ (to the left)}$$

## 14 FRAMES WITH SIDESWAY

If an unsymmetrically-loaded frame is not braced against sidesway, lateral movement will occur until the generated shear force in the vertical columns just equals the horizontal force needed to prevent sidesway.

Superposition is used to evaluate frames with sidesway. First, the frame is analyzed based on the assumption that sidesway is prevented. This results in moments at all the joints. These moments and the artificial restraining force can be evaluated using moment distribution.

Then, all loads are removed from the frame and the restraining force is assumed to act alone opposite to its original direction. The moments from this distribution are added to the moments from the restrained distribution to obtain the true joint moments.

The following alternate procedure can also be used to calculate the fixed-end moments for a frame with sidesway.

*step 1*: Analyze the frame assuming sidesway is prohibited.

*step 2*: Calculate the reaction, $R$, needed to prevent sidesway.

*step 3*: Remove the actual loads, and deflect the frame an arbitrary amount, $\Delta$. There will be fixed-end moments $M_1$ and $M_2$ set up at tops and fixed bottoms of the vertical members to resist this deflection.[8]

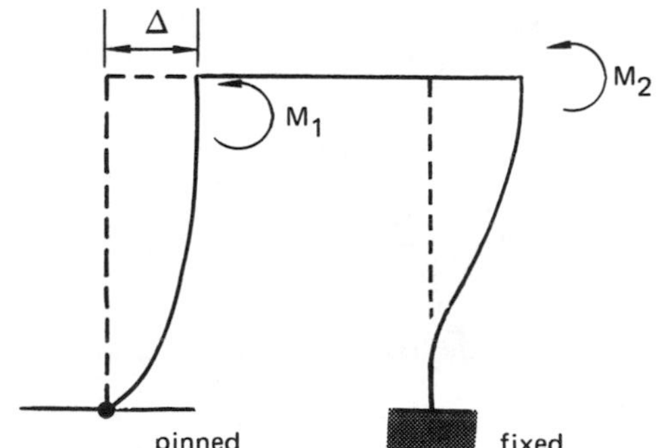


**Figure 13.11** Frame with Arbitrary Deflection

Equations 13.61 and 13.64 relate the fixed-end moments to the deflection. For the pinned member,

$$M_1 = \frac{3EI\Delta}{L^2} \tag{13.65}$$

For the built-in member,

$$M_2 = \frac{6EI\Delta}{L^2} \tag{13.66}$$

The total resisting moment, $M_1 + M_2$ is unknown, but it can be assumed to be some value (e.g., 100 or 1000). This total resisting moment is divided between the vertical members in proportion to the values of $M_1$ and $M_2$. The signs of the divided moments at the ends of both vertical members will be the same. (No moment is given to pinned bases.)

*step 4*: Analyze the frame using the moment distribution method.

*step 5*: Find the equivalent force, $R'$, causing the sidesway. If the directions of the reaction $R$ and the force $R'$ are the same, then the sign of the moments $M_1$ and $M_2$ used in step 3 was chosen incorrectly. In that case, merely reverse the sign of $M_{\text{from step 4}}$ in equation 13.67.

*step 6*: Calculate the true moment at each joint.

$$M_{\text{true}} = M_{\text{restrained}} + \left(\frac{R}{R'}\right) M_{\text{from step 4}} \tag{13.67}$$

[8] In effect, we have applied an equal but opposite force, $-R$ (to the right), which causes the deflection, $\Delta$. This equal but opposite force cancels the imaginary force used to prohibit sidesway initially.

*Example 13.17*

Find the moment at end $D$ for the frame in example 13.16. Sidesway is permitted.

*step 1*: See example 13.16.

*step 2*: See example 13.16.

*step 3*: Load the frame with a total moment of 100.

For member $AB$, the fixed-end moments are proportional to

$$M_1 = \frac{(6)(1000)}{(25)^2} = 9.6$$

For member $DC$, the fixed-end moments are the same.

$$M_2 = 9.6$$

If the total resisting moment is assumed to be $-100$, the fraction taken by member $AB$ is

$$M_{AB} = (-100)\left(\frac{9.6}{9.6+9.6}\right)$$

Member $DC$ takes the other 50 percent of the total resisting moment.

The sign of the FEM's applied to each span end are determined by observing the unconstrained deflection shape. In order to keep the ends of members $AB$ and $CD$ vertical, counterclockwise (negative) moments need to be applied. Therefore, $-50$ is the moment distributed to each member.

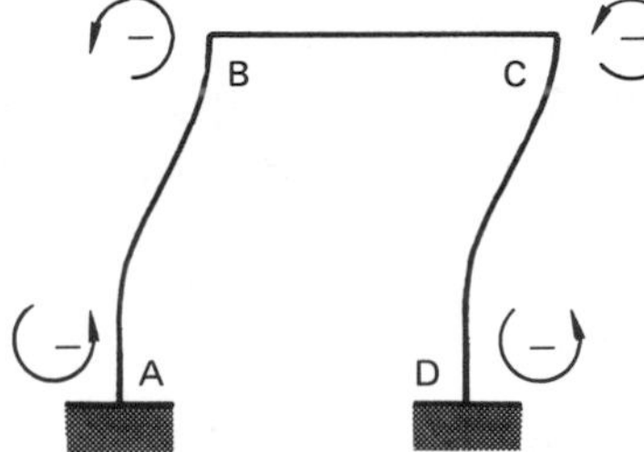


*step 4*:

| | AB | BA | BC | CB | CD | DC |
|---|---|---|---|---|---|---|
| L | 25 | | 70 | | 25 | |
| EI | 1000 | | 10,500 | | 1000 | |
| R | 40 | | 150 | | 40 | |
| F | 1 | 1 | 1 | 1 | 1 | 1 |
| K | 160 | 160 | 600 | 600 | 160 | 160 |
| D | 0 | 0.21 | 0.79 | 0.79 | 0.21 | 0 |
| C | 1/2 | 1/2 | 1/2 | 1/2 | 1/2 | 1/2 |
| $EI/L^2$ | 1.6 | | not loaded | | 1.6 | |
| FEM | −50 | −50 | 0 | 0 | −50 | −50 |
| Balance | 0 | 10.5 | 39.5 | 39.5 | 10.5 | 0 |
| COM | 5.25 | 0 | 19.75 | 19.75 | 0 | 5.25 |
| Balance | 0 | −4.15 | −15.60 | −15.60 | −4.15 | 0 |
| COM | −2.07 | 0 | −7.8 | −7.8 | 0 | −2.07 |
| Balance | 0 | 1.64 | 6.16 | 6.16 | 1.64 | 0 |
| COM | 0.82 | 0 | 3.08 | 3.08 | 0 | 0.82 |
| Balance | 0 | −0.65 | −2.43 | −2.43 | −0.65 | 0 |
| Total | −46.00 | −42.66 | 42.66 | 42.66 | −42.66 | −46.00 |

*step 5*: Take member $AB$ as a freebody.

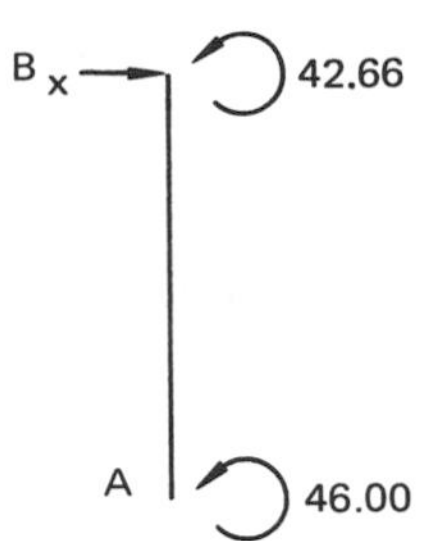


$$\sum M_A = -42.66 - 46.00 + B_x(25) = 0$$
$$B_x = 3.55 \text{ (to the right)}$$

Take member $DC$ as a freebody. $C_x = -B_x = -3.55$ (to the left).

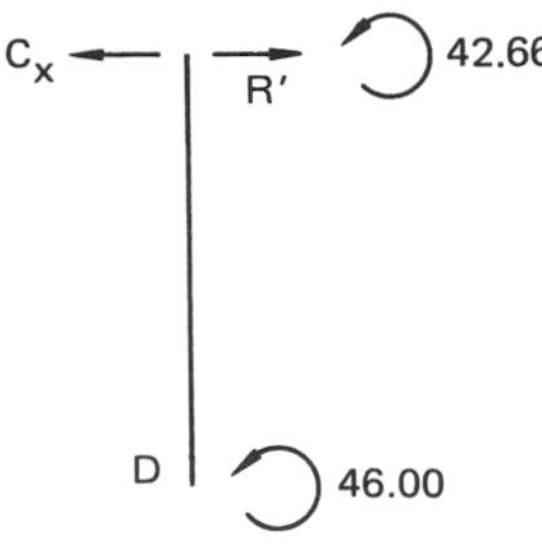


$$\sum M_D = -42.66 - 46.00 - 3.55(25) + R'(25)$$
$$R' = 7.10 \text{ (to the right)}$$

*step 6*:

$$M_{DC} = 0.75 + \left(\frac{3.61}{7.10}\right)(-46.0) = -22.6$$

## 15 FRAMES WITH CONCENTRATED JOINT LOADS

A frame that carries a concentrated load at a joint, as in figure 13.12, does not develop any fixed-end moments due to transverse loading since there are no transverse loads. However, the load does cause the frame to sway, and hence, moments are generated at the joints.

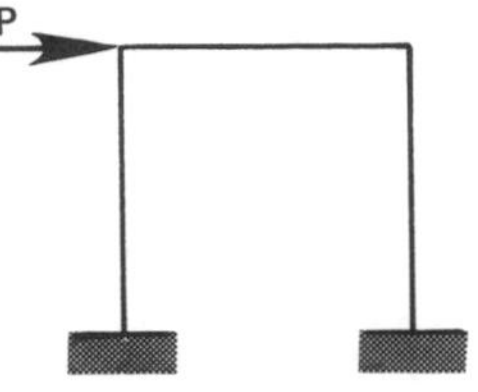


**Figure 13.12** Joint Loaded Frame

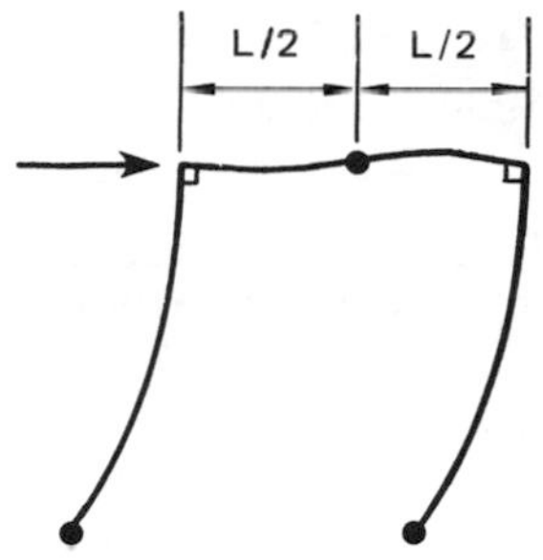


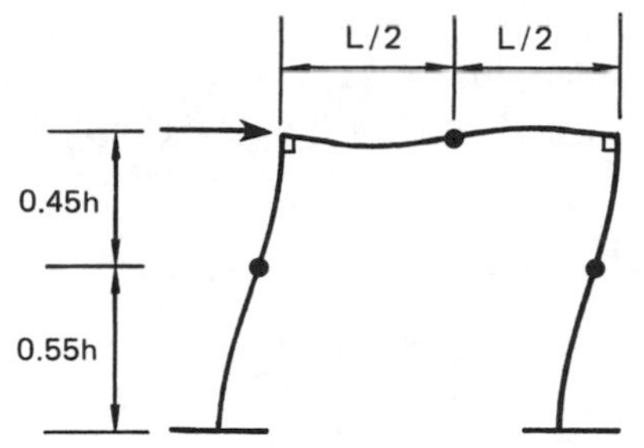


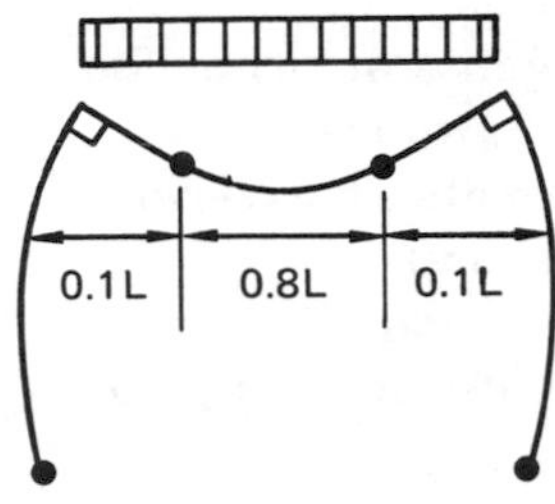


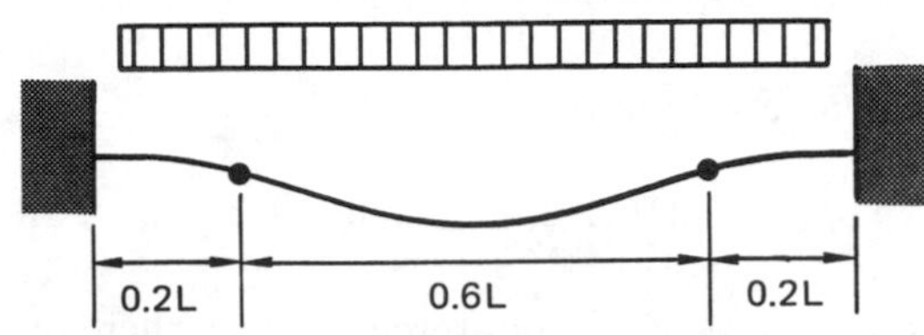


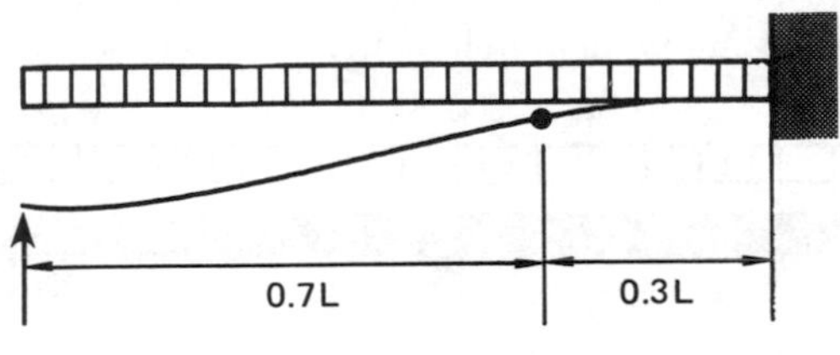


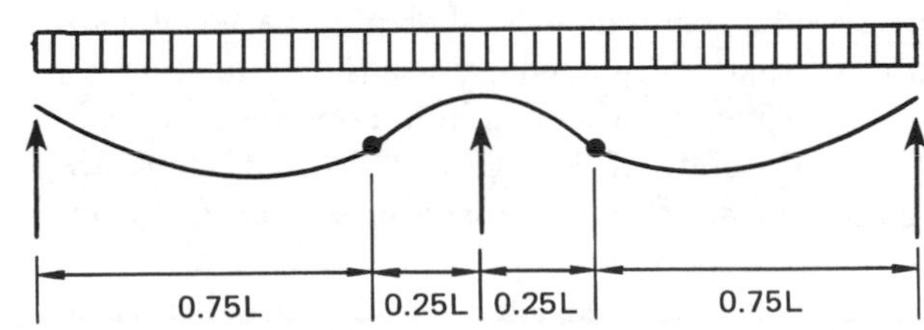


**Figure 13.13** Approximate Inflection Points

The solution procedure is similar to that used for frames with sidesway, except that the initial step is skipped. It is unnecessary to distribute moments from transverse loading, because there is no transverse loading. In effect, the applied load is equal to the force needed to prevent sidesway.

## 16 APPROXIMATE METHODS

When an exact solution is unnecessary, when time is short, or when it is desired to quickly check an exact solution, approximate methods may be used.

### A. ASSUMED INFLECTION POINTS

At points of inflection, the curvature will be changing from positive to negative. Accordingly, the moment at such a point will be zero. It is possible to assume a hinge exists at an inflection point, and that no moment is transferred across the hinge.

The accuracy of this method depends on being able to accurately predict the location of the inflection point. Figure 13.13 provides reasonable predictions of these locations.

*Example 13.18*

Determine the fixed-end moment for joint $A$.

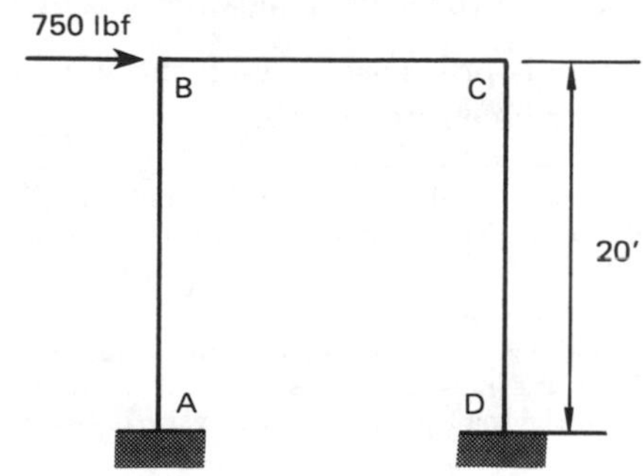

It is assumed that a point of inflection occurs on each vertical member $(0.55)(20) = 11$ feet above the supports. By symmetry and equilibrium, the shear at the inflection point is

$$V = \frac{750}{2} = 375\text{ lbf}$$

The moment at the base is

$$M = (11)(375) = 4125\text{ ft-lbf}$$

The fixed-end moment is counterclockwise; therefore, $M_{AB} = -4125$.

## B. USE OF DESIGN MOMENTS (MOMENT COEFFICIENTS)[9]

For continuous beams and other similar structures meeting specific requirements, shortcuts based on theory can be taken when constructing moment envelopes.[10] Specifically, the maximum moments at the ends and mid-points of continuously-loaded spans are taken as some fraction of the distributed load function, $wL^2$, where $L$ is the span length between supports, and $w$ is the ultimate factored load. For example, the moment at some point along the beam would be taken as

$$M = C_1 wL^2 \qquad 13.68$$

The coefficient, $C_1$, in equation 13.68 is known as the *moment coefficient*. The method of moment coefficients can be used in concrete design when the following conditions are met:

- The load is continuously distributed.
- Construction is not prestressed.
- There are two or more spans.
- All spans are the same length, ±20% of $L$.
- The beam is prismatic, having the same cross section along its entire length.

[9] This method is allowed in ACI-318 for continuous beams and one-way slabs provided the conditions are met. This method is sometimes referred to as the *direct design method*.

[10] Strictly speaking, it is improper to draw the moment envelope using both the maximum positive and maximum negative moments, since the load placement will vary between these two extremes.

**Table 13.1**
ACI Moment Coefficients

| condition | $C_1$ |
|---|---|
| **positive moments near mid-span** | |
| end spans | |
| simple support | 1/11 |
| built-in | 1/14 |
| interior spans | 1/16 |
| **negative moments at exterior face of first interior support** | |
| 2 spans | 1/9 |
| 3 or more spans | 1/10 |
| **negative moments at other faces or interior supports** | 1/11 |
| **negative moments at exterior built-in supports** | |
| support is a column | 1/16 |
| support is a cross beam or girder | 1/24 |

*Example 13.19*

Draw the critical moment diagram for the uniformly-loaded, continuous beam shown.

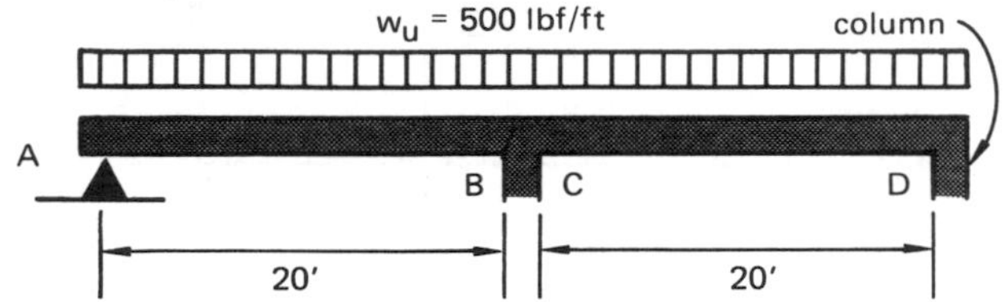


Point $A$ is a simple support. The moment there must be zero.

Point $B$ is an exterior face of the first interior support over a 2-span beam. The moment there is

$$M_B = -\frac{1}{9}(500)(20)^2 = -22,222\text{ ft-lbf}$$

Point $C$ is also an exterior face of the first interior support (counting from the opposite end).

$$M_C = M_B = -22,222\text{ ft-lbf}$$

Point $D$ is an exterior column support.

$$M_D = -\frac{1}{16}(500)(20)^2 = -12,500\text{ ft-lbf}$$

The left span is a simply-supported end span. The maximum positive moment is

$$M_L = \frac{1}{11}(500)(20)^2 = 18,182\text{ ft-lbf}$$

The right span is an end span with a built-in support. The maximum positive moment for it is

$$M_R = \frac{1}{14}(500)(20)^2 = 14,286\,\text{ft-lbf}$$

The critical moment diagram is

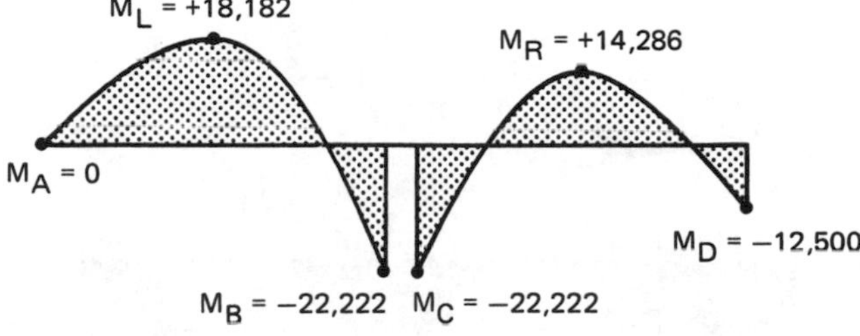


C. USE OF DESIGN SHEARS (SHEAR COEFFICIENTS)

When the conditions of the preceding section are met, it is possible to predict the critical shear in continuous beams and slabs on the basis of design coefficients. For shear, the design coefficient is used with the average span loading, $\frac{1}{2}wL$. That is,

$$V = C_2\left(\frac{wL}{2}\right) \qquad 13.69$$

$C_2$ for shear in end members at the first interior support is 1.15. For shear at the face of all other supports, $C_2 = 1.0$.

## Appendix A: Moment Distribution Worksheet

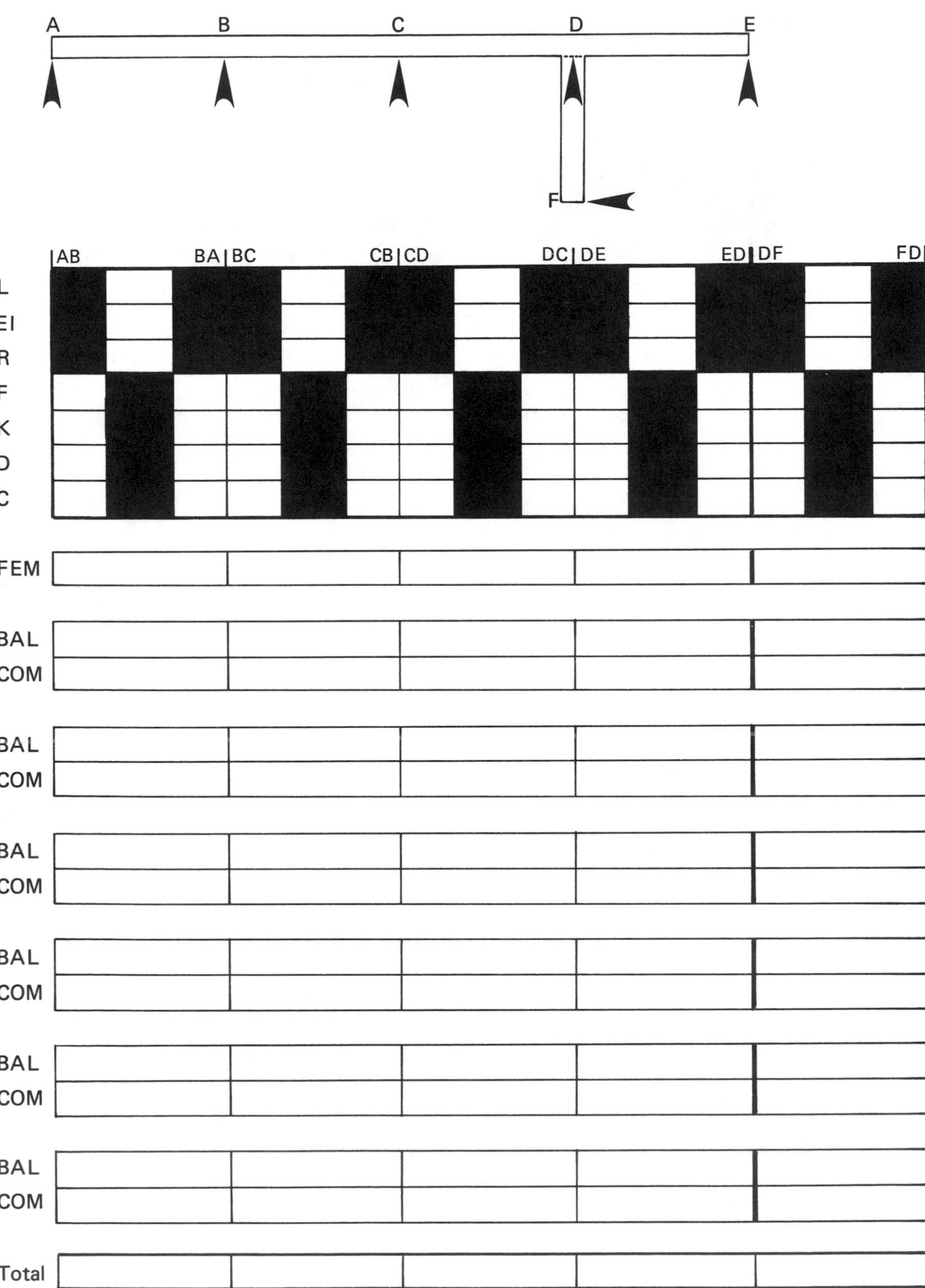

# Appendix B: Elastic Fixed-End Moments

(clockwise is positive)

1.
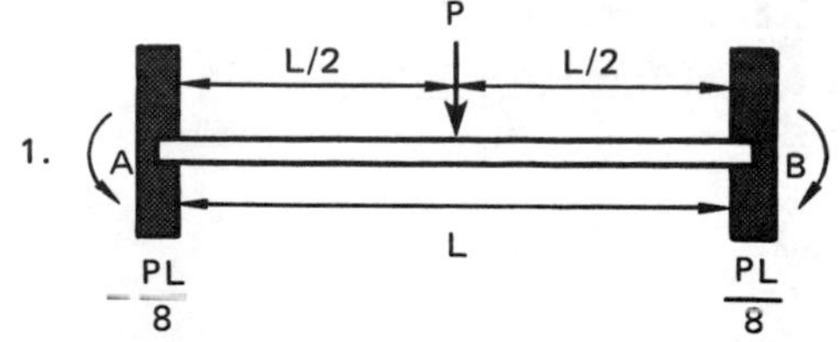


2.
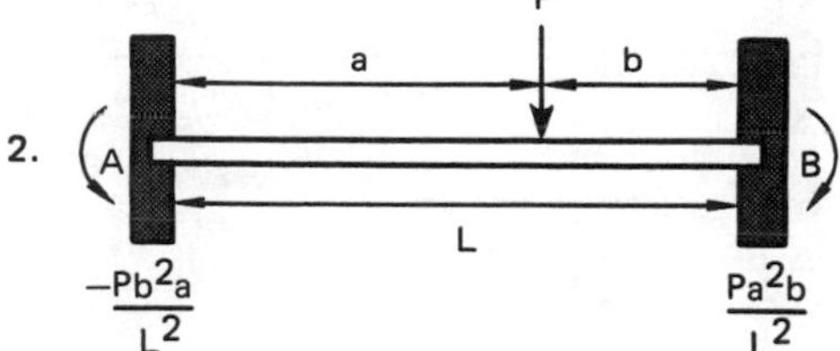


3.
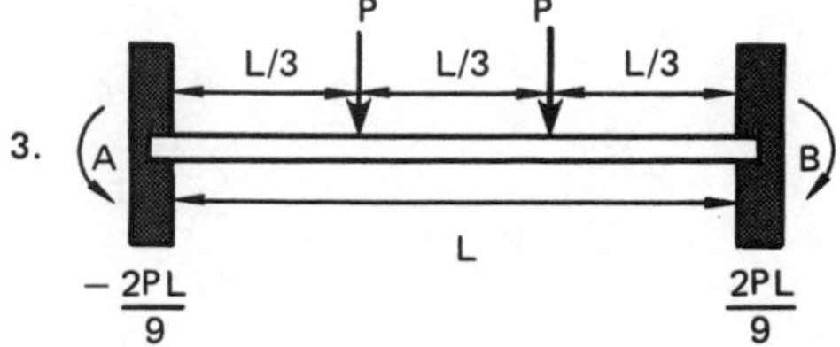


4.
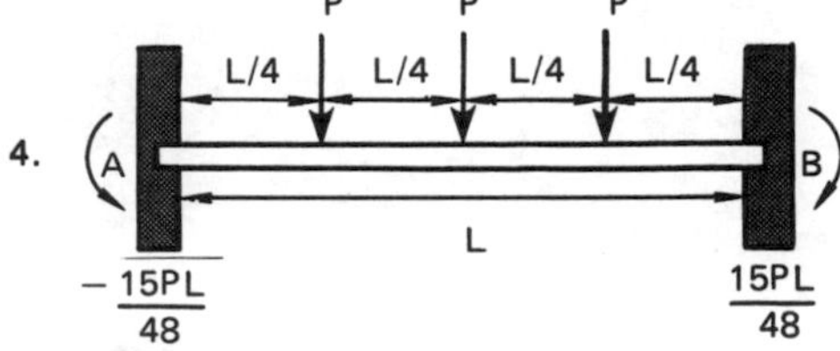


5.
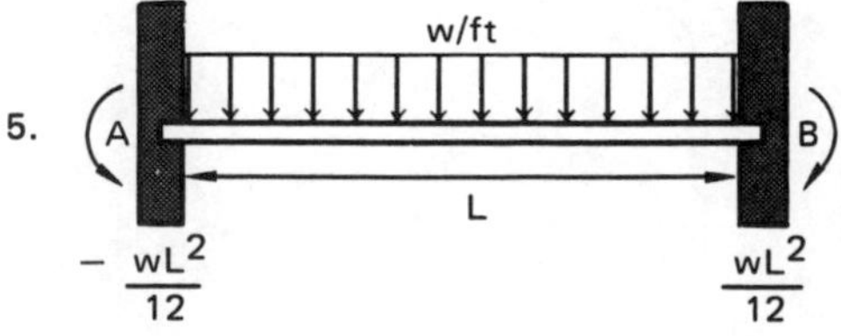


6.
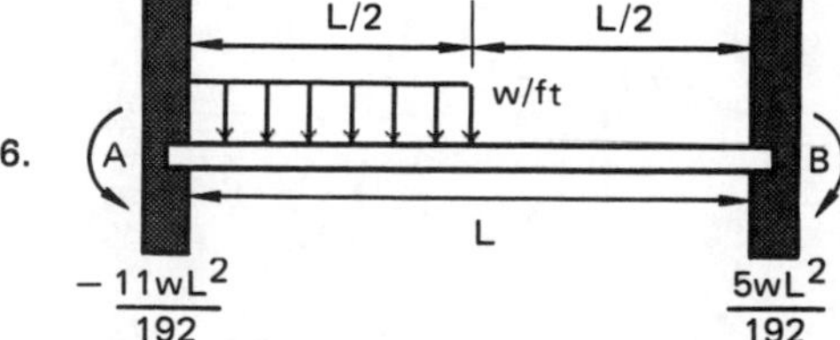


7.
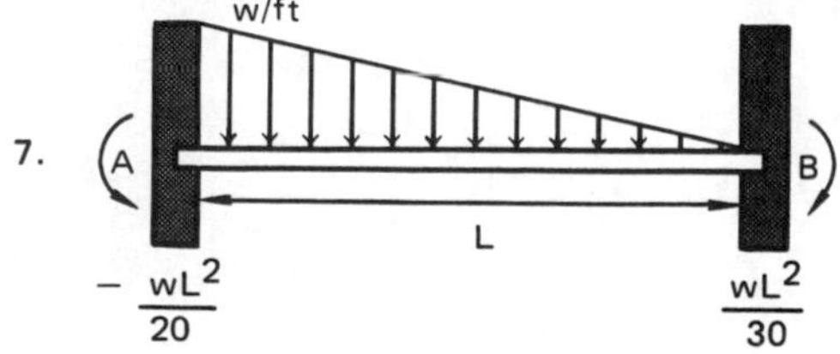


8.
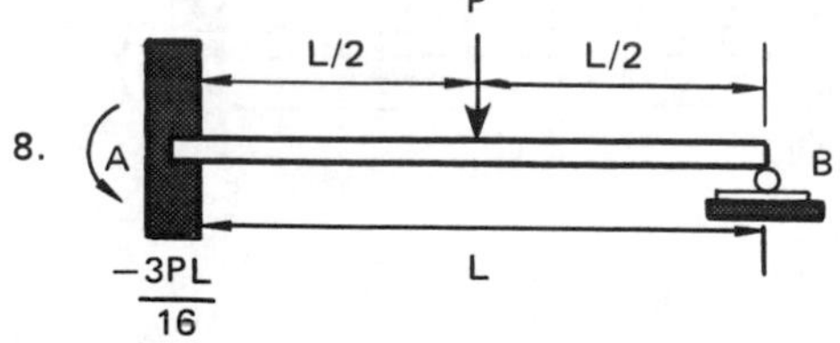


9.
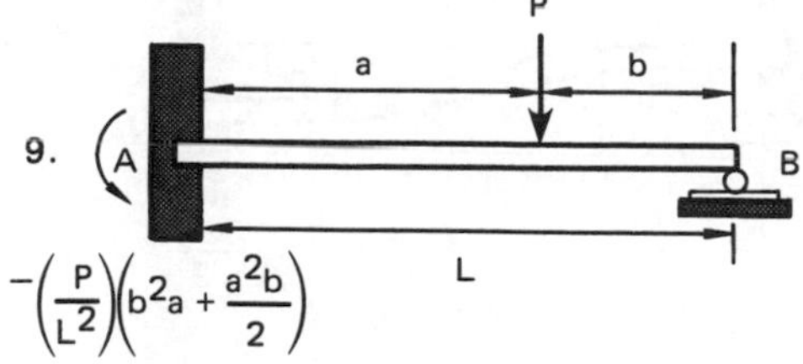


10.
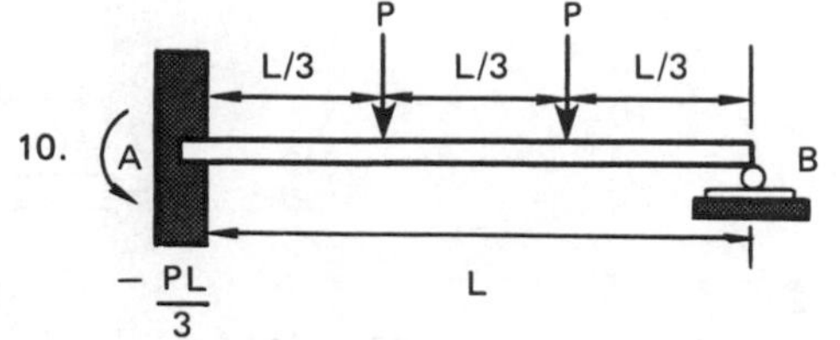


11.
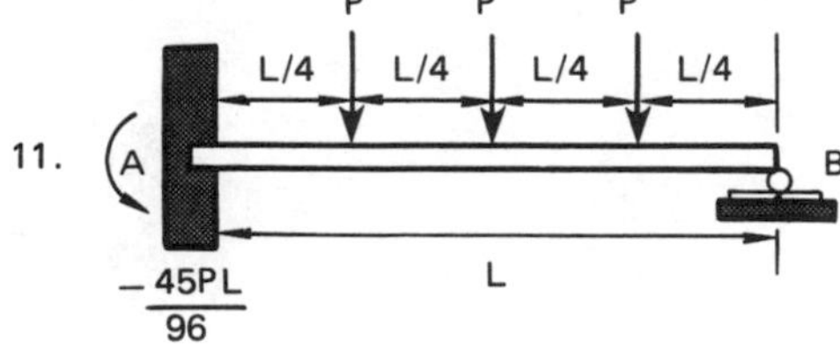


12.
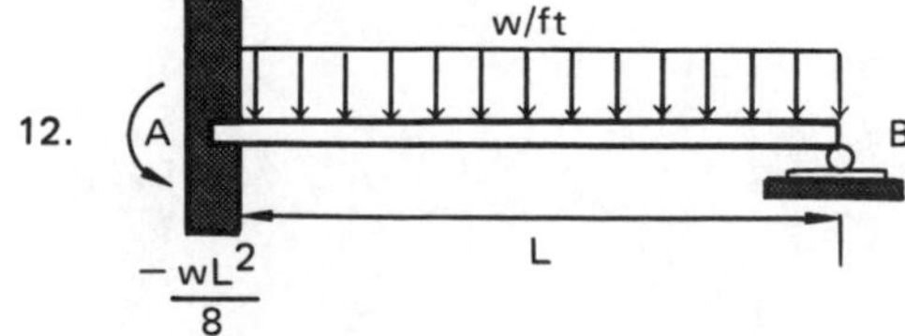


13.
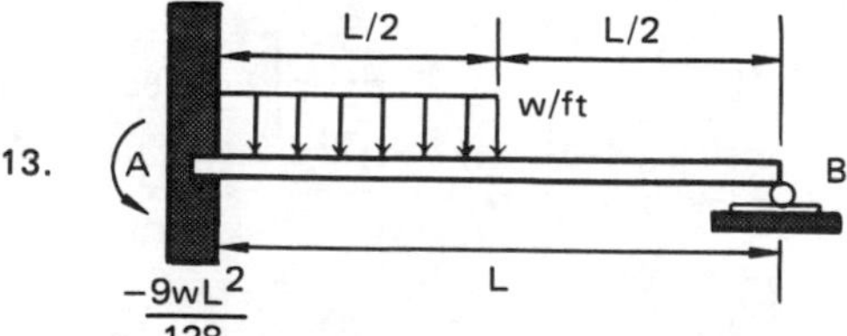


14.
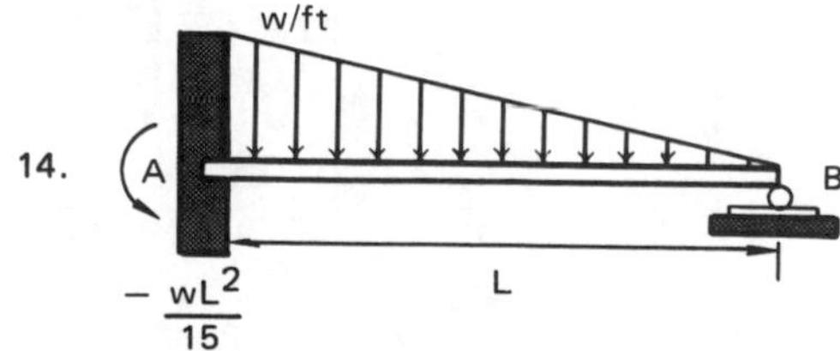

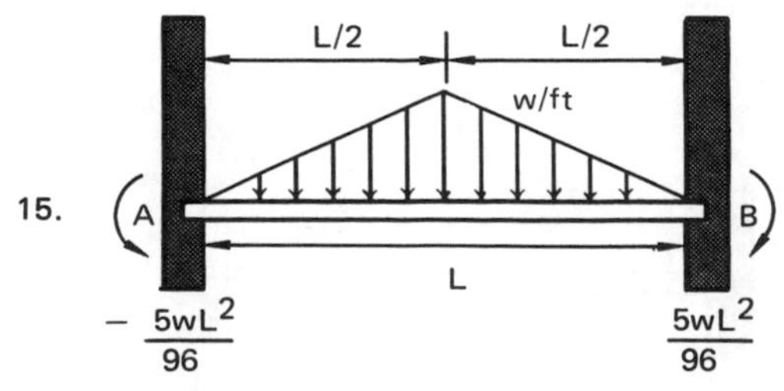
15.
L/2
L/2
w/ft
A
B
L
$-\frac{5wL^2}{96}$
$\frac{5wL^2}{96}$

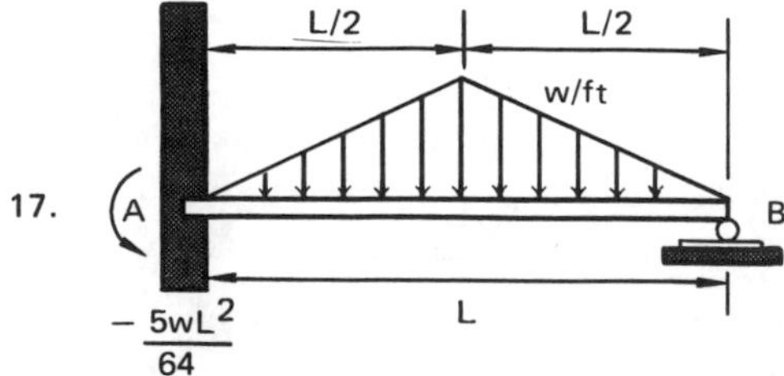
17.
L/2
L/2
w/ft
A
B
L
$-\frac{5wL^2}{64}$

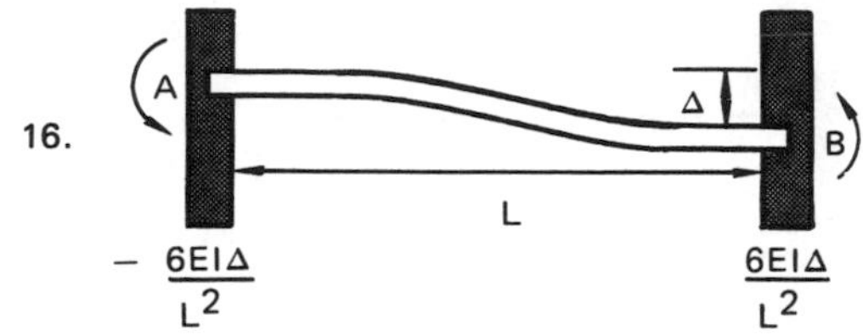
16.
A
Δ
B
L
$-\frac{6EI\Delta}{L^2}$
$\frac{6EI\Delta}{L^2}$

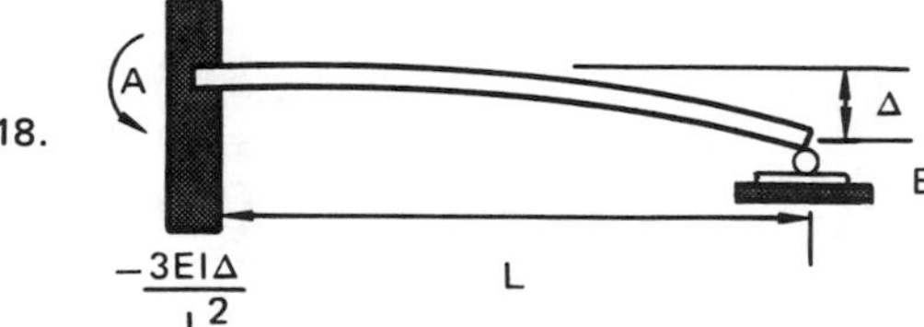
18.
A
Δ
B
L
$\frac{-3EI\Delta}{L^2}$

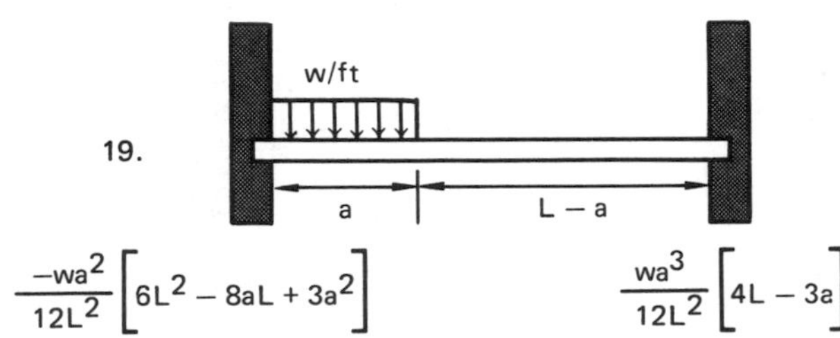
19.
w/ft
a
L − a
$\frac{-wa^2}{12L^2}\left[6L^2 - 8aL + 3a^2\right]$
$\frac{wa^3}{12L^2}\left[4L - 3a\right]$

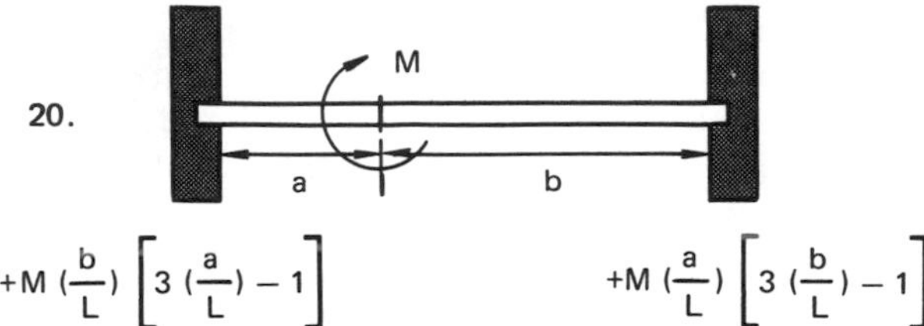
20.
M
a
b
$+M\left(\frac{b}{L}\right)\left[3\left(\frac{a}{L}\right) - 1\right]$
$+M\left(\frac{a}{L}\right)\left[3\left(\frac{b}{L}\right) - 1\right]$

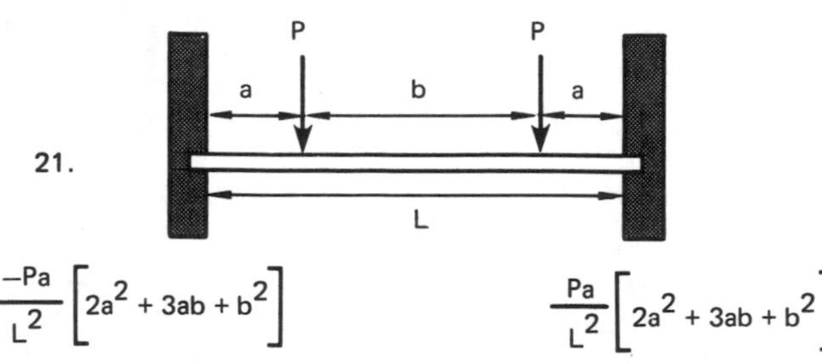
21.
P
P
a
b
a
L
$\frac{-Pa}{L^2}\left[2a^2 + 3ab + b^2\right]$
$\frac{Pa}{L^2}\left[2a^2 + 3ab + b^2\right]$

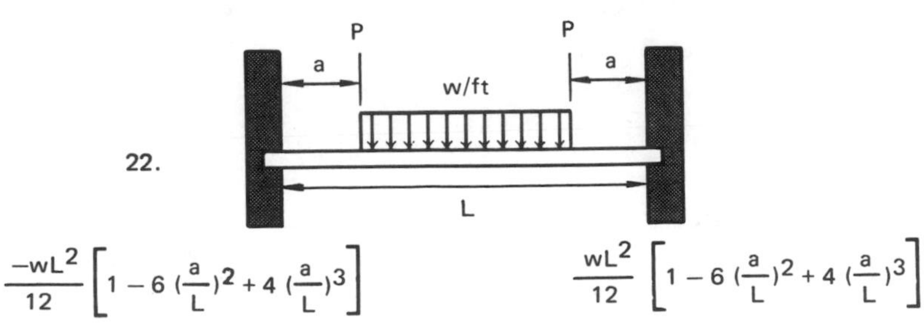
22.
P
P
a
a
w/ft
L
$\frac{-wL^2}{12}\left[1 - 6\left(\frac{a}{L}\right)^2 + 4\left(\frac{a}{L}\right)^3\right]$
$\frac{wL^2}{12}\left[1 - 6\left(\frac{a}{L}\right)^2 + 4\left(\frac{a}{L}\right)^3\right]$

# Appendix C: Beam Formulas

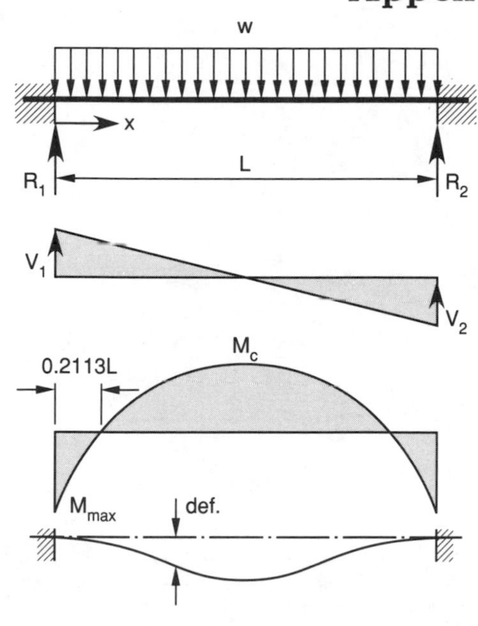


Uniformly distributed load, $w$ lb/ft
Total load $W = wL$

Reactions: $R_1 = R_2 = \frac{W}{2}$

Shear forces: $V_1 = +\frac{W}{2}$

$V_2 = -\frac{W}{2}$

Maximum (negative) bending moment

$$M_{\text{max}} = -\frac{wL^2}{12} = -\frac{WL}{12}, \text{ at end}$$

Maximum (positive) bending moment

$$M_c = \frac{wL^2}{24} = \frac{WL}{24}, \text{ at center}$$

$$\text{Maximum deflection} = \frac{wL^4}{384\,EI} = \frac{WL^3}{384\,EI}, \text{ at center}$$

$$\text{def.} = \frac{wx^2}{24\,EI}(L-x)^2,\ 0 \le x \le L$$

---

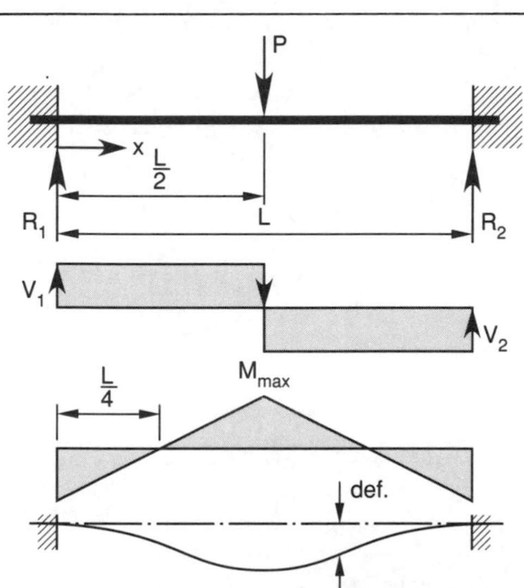


Concentrated load, $P$, at center

Reactions: $R_1 = R_2 = \frac{P}{2}$

Shear forces: $V_1 = +\frac{P}{2}$; $V_2 = -\frac{P}{2}$

Maximum bending moment

$$M_{\text{max}} = \frac{PL}{8}, \text{ at center}$$

$$M_{\text{max}} = -\frac{PL}{8}, \text{ at ends}$$

$$\text{Maximum deflection} = \frac{PL^3}{192\,EI}, \text{ at center}$$

$$\text{def.} = \frac{Px^2}{48\,EI}(3L-4x),\ 0 \le x \le \frac{L}{2}$$

---

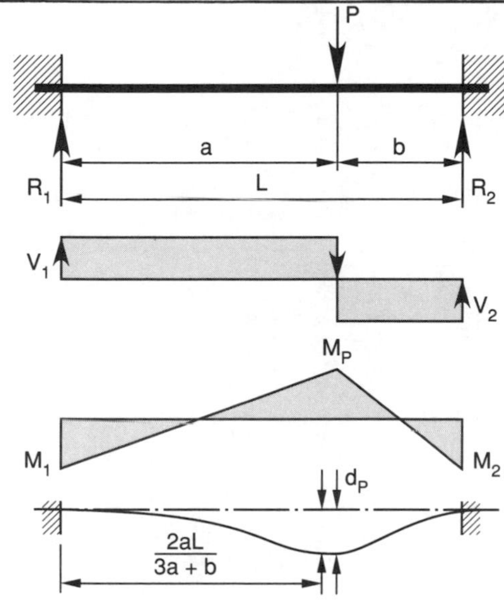


Concentrated load, $P$, at any point

Reactions: $R_1 = \frac{Pb^2}{L^3}(3a+b)$

$R_2 = \frac{Pa^2}{L^3}(3b+a)$

Shear forces: $V_1 = R_1$; $V_2 = -R_2$

Bending moments:

$$M_1 = -\frac{Pab^2}{L^2}, \text{ max. when } a < b$$

$$M_2 = -\frac{Pa^2b}{L^2}, \text{ max. when } a > b$$

$$M_p = +\frac{2Pa^2b^2}{L^3}, \text{ at point of load}$$

$$\text{Deflection} = \frac{Pa^3b^3}{3\,EIL^3}, \text{ at point of load}$$

$$\text{Max. def.} = \frac{2Pa^3b^2}{3\,EI(3a+b)^2}, \text{ at } x = \frac{2aL}{3a+b}, \text{ for } a > b$$

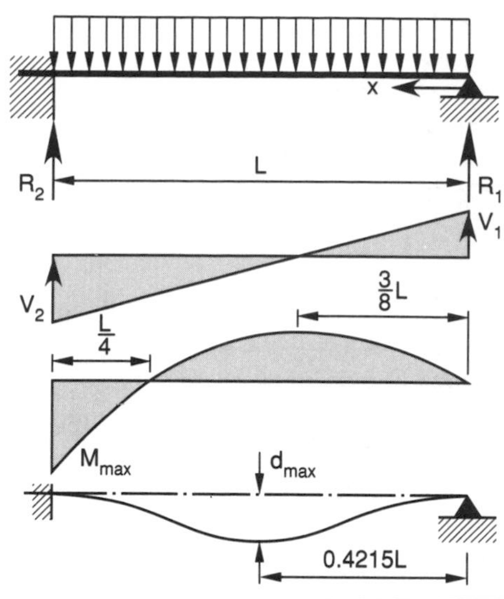


Uniformly distributed load, $w$ lb/ft

Total load $W = wL$

Reactions: $R_1 = \frac{3wL}{8}$, $R_2 = \frac{5wL}{8}$

Shear forces: $V_1 = +R_1$; $V_2 = -R_2$

Bending moments:

Max. negative moment $= -\frac{wL^2}{8}$, at left end

Max. positive moment $= \frac{9}{128}wL^2, x = \frac{3}{8}L$

$$M = \frac{3wLx}{8} - \frac{wx^2}{2},\ 0 \le x \le L$$

Maximum deflection $= \frac{wL^4}{185\,EI}$, $x = 0.4215L$

$$\text{def.} = \frac{wx}{48\,EI}(L^3 - 3Lx^2 + 2x^3),\ 0 \le x \le L$$

---

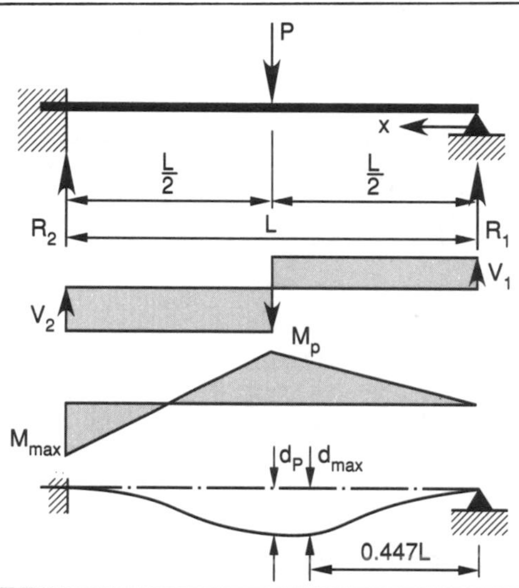


Concentrated load, $P$, at center

Reactions: $R_1 = \frac{5}{16}P$; $R_2 = \frac{11}{16}P$

Shear forces: $V_1 = R_1$; $V_2 = -R_2$

Bending moments:

Max. negative moment $= -\frac{3PL}{16}$, at fixed end

Max. positive moment $= \frac{5PL}{32}$, at center

Maximum deflection $= 0.009317\frac{PL^3}{EI}$, at $x = 0.447L$

Deflection at center under load $= \frac{7PL^3}{768\,EI}$

---

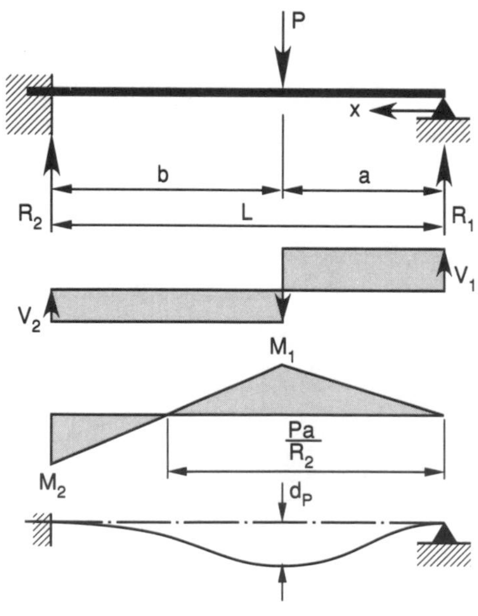


Concentrated load, $P$, at any point

Reactions: $R_1 = \frac{Pb^2}{2L^3}(a + 2L)$, $R_2 = \frac{Pa}{2L^3}(3L^2 - a^2)$

Shear forces: $V_1 = R_1$; $V_2 = -R_2$

Bending moments:

Max. negative moment, $M_2 = -\frac{Pab}{2L^2}(a + L)$, at fixed end

Max. positive moment, $M_1 = \frac{Pab^2}{2L^3}(a + 2L)$, at load

Deflections: $d_p = \frac{Pa^2b^3}{12\,EIL^3}(3L + a)$, at load

$$d_{\text{max}} = \frac{Pa(L^2 - a^2)^3}{3\,EI(3L^2 - a^2)^2},\ \text{at } x = \frac{L^2 + a^2}{3L^2 - a^2}L,\ \text{when } a < 0.414L$$

$$d_{\text{max}} = \frac{Pab^2}{6\,EI}\sqrt{\frac{a}{2L + a}},\ \text{at } x = L\sqrt{\frac{a}{2L + a}},\ \text{when } a > 0.414L$$

IND STRUCT

Continuous beam of two equal spans—equal concentrated loads, $P$, at center of each span

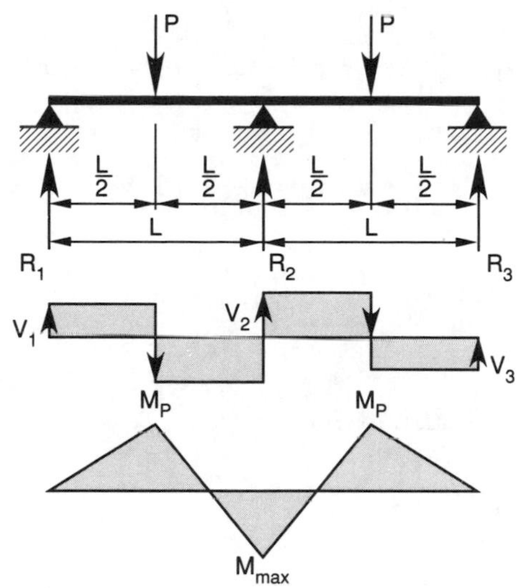


Reactions: $R_1 = R_3 = \frac{5}{16}P$

$R_2 = 1.375P$

Shear forces: $V_1 = -V_3 = \frac{5}{16}P$

$V_2 = \pm\frac{11}{16}P$

Bending moments:

$M_{\text{max}} = -\frac{6}{32}PL$, at $R_2$

$M_P = \frac{5}{32}PL$, at point of load

---

Continuous beam of two equal spans—concentrated loads, $P$, at third points of each span

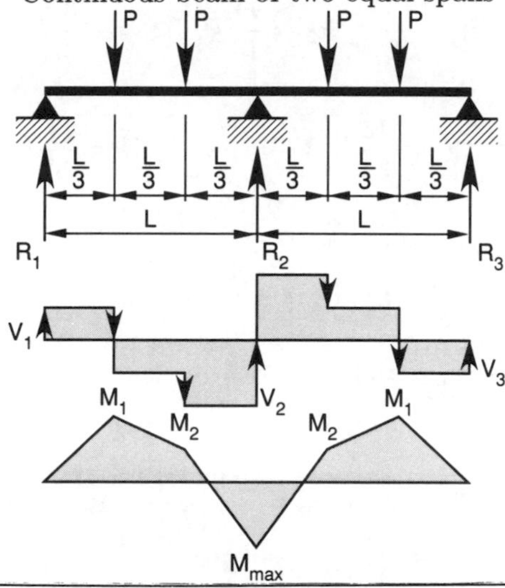


Reactions: $R_1 = R_3 = \frac{2}{3}P$

$R_2 = \frac{8}{3}P$

Shear forces: $V_1 = -V_3 = \frac{2}{3}P$

$V_2 = \pm\frac{4}{3}P$

Bending moments:

$M_{\text{max}} = -\frac{1}{3}PL$, at $R_2$

$M_1 = \frac{2}{9}PL$

$M_2 = \frac{1}{9}PL$

---

Continuous beam of two equal spans—uniformly distributed load of $w$ lb/ft

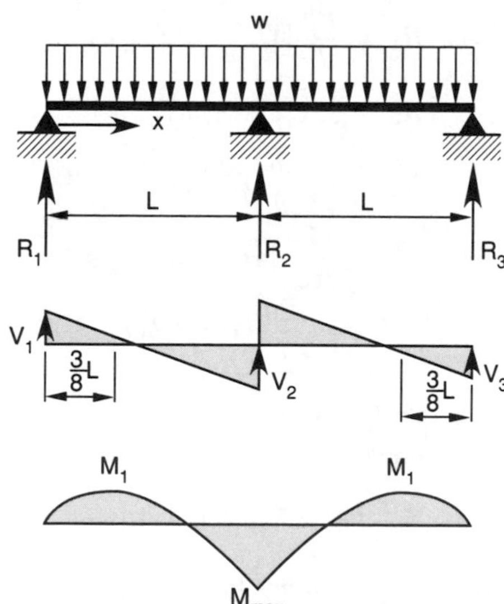


Reactions: $R_1 = R_3 = \frac{3}{8}wL$

$R_2 = 1.25wL$

Shear forces: $V_1 = -V_3 = \frac{3}{8}wL$

$V_2 = \pm\frac{5}{8}wL$

Bending moments:

$M_{\text{max}} = -\frac{1}{8}wL^2$

$M_1 = \frac{9}{128}wL^2$

Maximum deflection $= 0.00541\frac{wL^4}{EI}$

at $x = 0.4215L$

Def. $= \frac{w}{48\,EI}(L^3x - 3Lx^3 + 2x^4)$, $0 \le x \le L$

Continuous beam of two equal spans—uniformly distributed load of $w$ lb/ft on one span

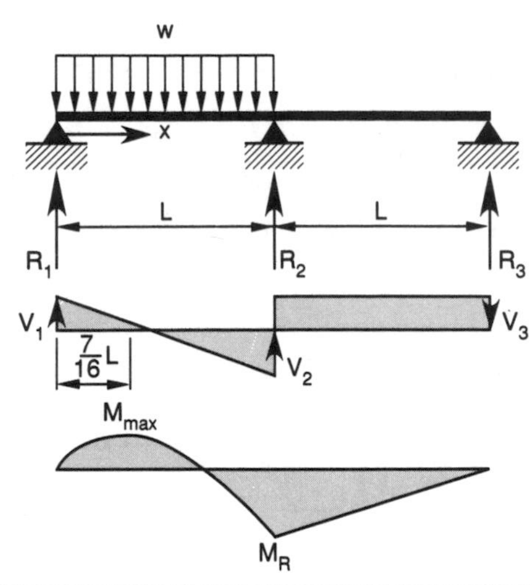


Reactions: $R_1 = \frac{7}{16}wL$, $R_2 = \frac{5}{8}wL$, $R_3 = -\frac{1}{16}wL$

Shear forces: $V_1 = \frac{7}{16}wL$, $V_2 = -\frac{9}{16}wL$, $V_3 = \frac{1}{16}wL$

Bending moments:

$$M_{\text{max}} = \frac{49}{512}wL^2, \text{ at } x = \frac{7}{16}L$$

$$M_R = -\frac{1}{16}wL^2, \text{ at } R_2$$

$$M = \frac{wx}{16}(7L - 8x),\ 0 \le x \le L$$

---

Continuous beam of two equal spans—concentrated load, $P$, at center of one span.

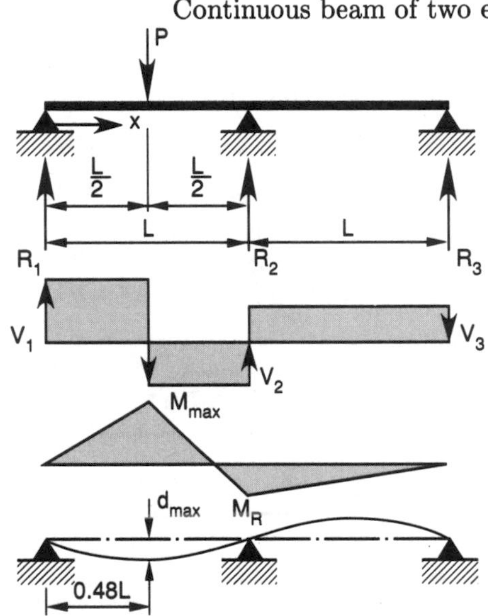


Reactions: $R_1 = \frac{13}{32}P$, $R_2 = \frac{11}{16}P$, $R_3 = -\frac{3}{32}P$

Shear forces: $V_1 = \frac{13}{32}P$, $V_2 = -\frac{19}{32}P$, $V_3 = \frac{3}{32}P$

Bending moments:

$$M_{\text{max}} = \frac{13}{64}PL, \text{ at point of load}$$

$$M_R = -\frac{3}{32}PL, \text{ at support } R_2$$

Maximum deflection: $d_{\text{max}} = \frac{0.96\,PL^3}{64\,EI}$, at $x = 0.48L$

---

Continuous beam of two equal spans—concentrated load, $P$, at any point on one span.

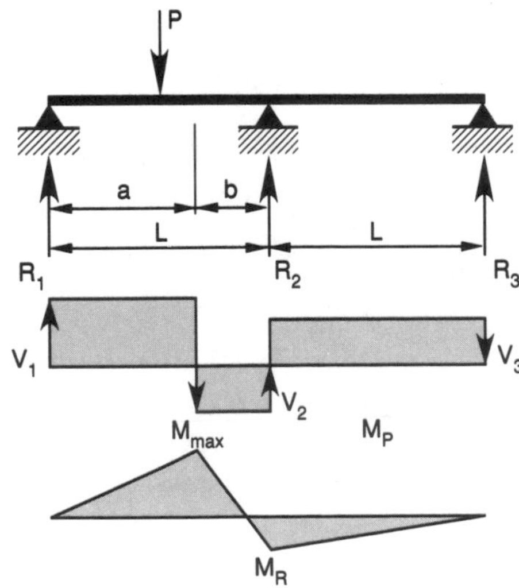


Reactions: $R_1 = \frac{Pb}{4L^3}[4L^2 - a(L+a)]$

$$R_2 = \frac{Pa}{2L^3}[2L^2 + b(L+a)]$$

$$R_3 = -\frac{Pab}{4L^3}(L+a)$$

Shear forces: $V_1 = \frac{Pb}{4L^3}[4L^2 - a(L+a)]$

$$V_2 = -\frac{Pa}{4L^3}[4L^2 + b(L+a)]$$

$$V_3 = \frac{Pab}{4L^3}(L+a)$$

Bending moments:

$$M_{\text{max}} = \frac{Pab}{4L^3}[4L^2 - a(L+a)]$$

$$M_R = -\frac{Pab}{4L^2}(L+a)$$

Continuous beam of three equal spans—concentrated load, $P$, at center of each span

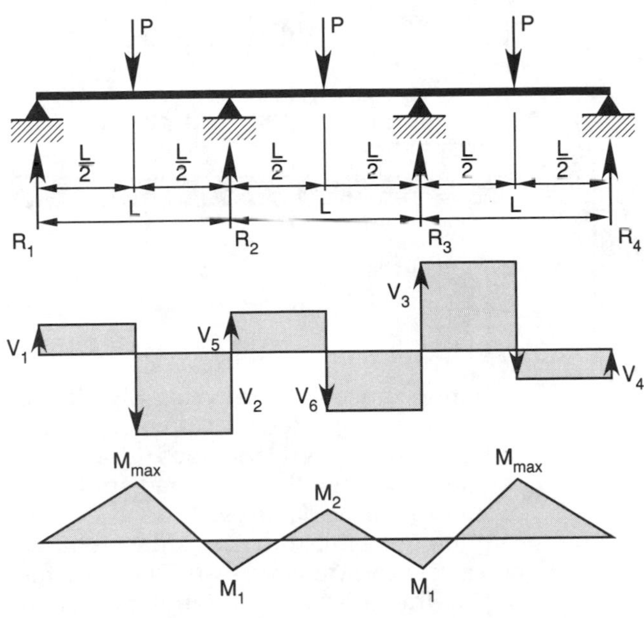


Reactions:

$$R_1 = R_4 = \frac{7}{20}P$$

$$R_2 = R_3 = \frac{23}{20}P$$

Shear forces:

$$V_1 = -V_4 = \frac{7}{20}P$$

$$V_3 = -V_2 = \frac{13}{20}P$$

$$V_5 = -V_6 = \frac{P}{2}$$

Bending moments:

$$M_{\text{max}} = \frac{7}{40}PL$$

$$M_1 = -\frac{3}{20}PL$$

$$M_2 = \frac{1}{10}PL$$

---

Continuous beam of three equal spans—concentrated loads, $P$, at third points of each span

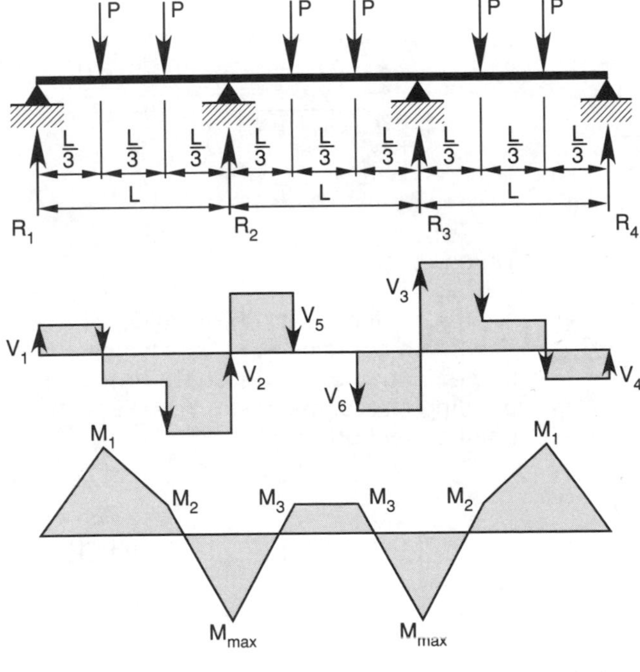


Reactions:

$$R_1 = R_4 = \frac{11}{15}P$$

$$R_2 = R_3 = \frac{34}{15}P$$

Shear forces:

$$V_1 = -V_4 = \frac{11}{15}P$$

$$V_3 = -V_2 = \frac{19}{15}P$$

$$V_5 = -V_6 = P$$

Bending moments:

$$M_{\text{max}} = -\frac{12}{45}PL$$

$$M_1 = \frac{11}{45}PL$$

$$M_2 = \frac{7}{45}PL$$

$$M_3 = \frac{3}{45}PL$$

## Practice Problems: INDETERMINATE STRUCTURES

Untimed

1. What are the reactions supporting the continuous beam shown? The reactions are all simple.

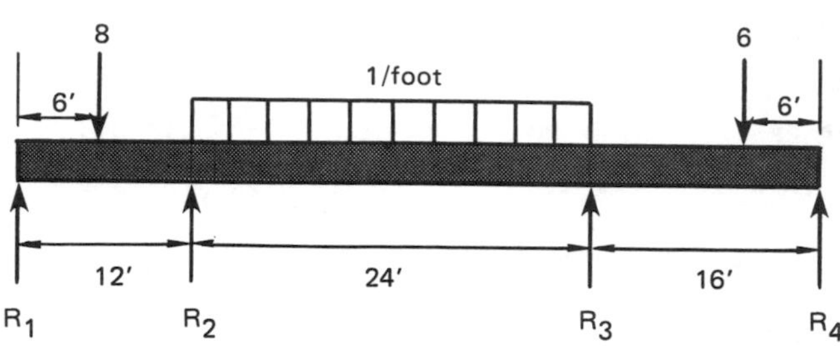


2. What are the joint moments and the reactions for the rigid frame shown? The supports may be assumed to be pinned.

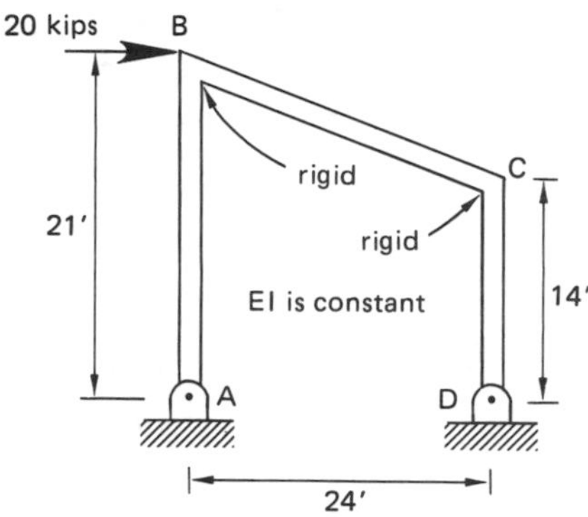


3. A frame with rigid joints is shown. Draw the moment diagrams for members AB, BC, CE, and CD.

$$I_{AB} = 240 \text{ in}^4$$
$$I_{BE} = 300 \text{ in}^4$$
$$I_{CD} = 300 \text{ in}^4$$

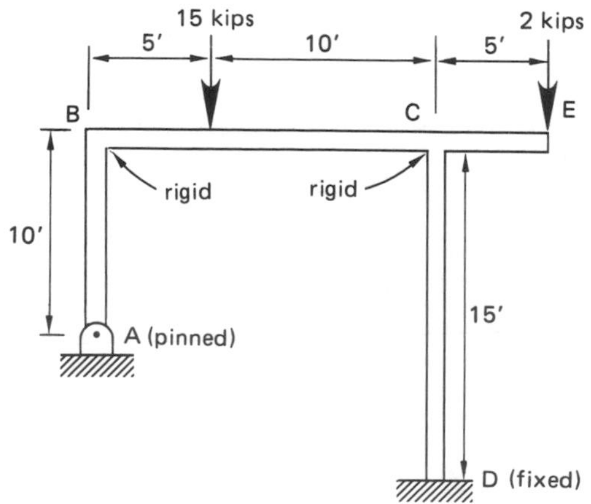


4. What are the maximum moments on the spans AB, BC, and CD?

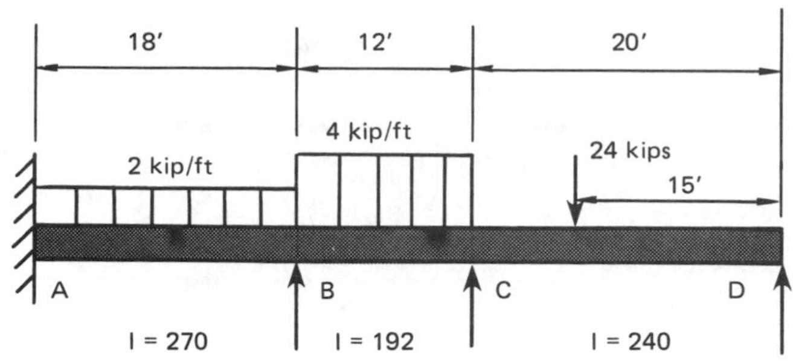


5. During the construction of Shasta dam, access ramps were supported by a steel frame which projected from the concrete dam face. The horizontal members were sufficiently embedded so that they may be considered as having fixed bases.

At the free ends, the horizontal members were tied with a vertical strut as shown. Calculate the deflection of the frame assuming the tied connections are (a) pinned and (b) rigid.

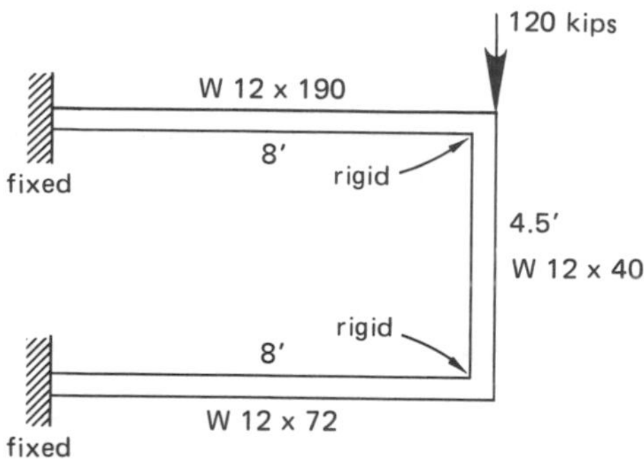


Timed

1. An A36 steel beam (W24 × 76) is loaded as shown. The beam contains two joints which can be assumed to be frictionless hinges. Determine the maximum bending stress, maximum shear stress, and the midpoint deflection.

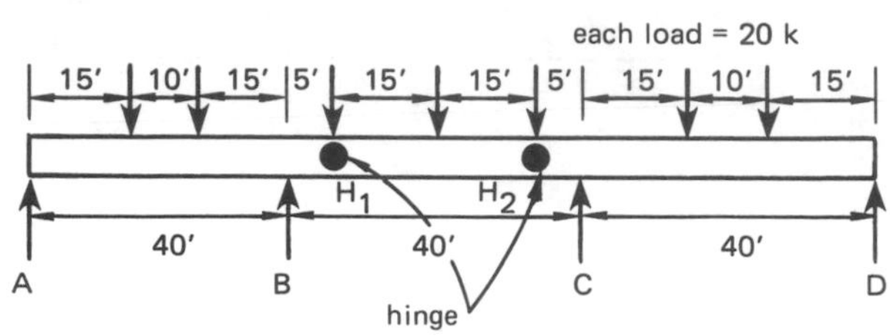

2. A steel beam is loaded as shown and is laterally supported only at the three reaction points. The beam carries 1400 lbf/ft over its total span. (a) Determine the reactions. (b) Draw the shear and moment diagrams. (c) Choose the lightest A36 × W14 beam that is capable of developing a minimum of $0.60F_y$ bending stress for this application. (d) Specify the maximum unbraced length for the beam you choose.

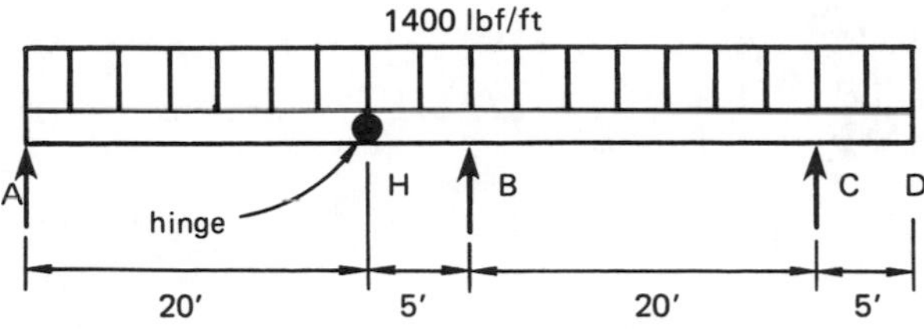


3. For the rigid-joint structure shown, (a) determine the reactions at A, C, and D, and (b) draw the moment diagram on the tension side of the frame. (Neglect the beam weights. Reaction at C is passive and does not load the frame.)

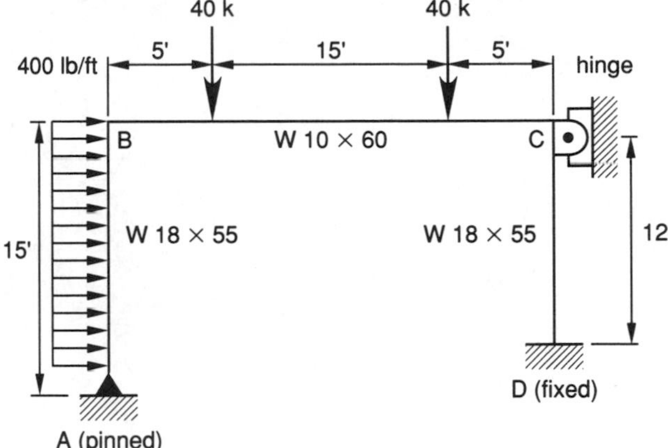


IND STRUCT

# 14 REINFORCED CONCRETE DESIGN[1]

*Nomenclature*

| | | |
|---|---|---|
| a | short side length, or height of stress block | in |
| A | area | $in^2$ |
| b | long side length, or width | in |
| B | width | ft |
| c | distance from neutral axis to extreme compression fiber | in |
| C | compressive force | lbf |
| d | depth or diameter | in |
| D | depth of footing | ft |
| e | eccentricity | in |
| E | modulus of elasticity | psi |
| f | stress or strength | psi |
| $f'_c$ | compressive strength of concrete | psi |
| $f_r$ | modulus of rupture | psi |
| $f_y$ | yield strength of steel | psi |
| h | height | in |
| H | height | ft |
| I | moment of inertia | $in^4$ |
| j | a fraction | – |
| k | a fraction, or end condition factor | – |
| l | length | in |
| $l_d$ | development length | in |
| m | a factor | – |
| M | moment | in-lbf |
| n | modular ratio ($E_{st}/E_c$) | – |
| p | factored soil pressure | psi |
| P | load or force | lbf |
| r | radius of gyration | in |
| $R_u$ | coefficient of resistance | psi |
| s | spacing | in |
| SR | slenderness ratio | – |
| t | thickness | in |
| T | tensile force | lbf |
| v | shear stress | psi |
| V | shear, or shear strength | lbf |
| w | density, or load per unit length | pcf or lbf/in |
| z | crack parameter | kips/in |

*Symbols*

| | | |
|---|---|---|
| $\alpha$ | ratio of column to footing area | – |
| $\beta$ | ratio of long side to short side | – |
| $\beta_1$ | a factor | – |
| $\gamma$ | unit weight | pcf |
| $\delta$ | magnification factor | – |
| $\epsilon$ | fraction eccentricity | – |
| $\xi$ | timespan factor | – |
| $\rho$ | reinforcement ratio | – |
| $\lambda$ | long term deflection factor | – |
| $\theta$ | angle from the horizontal | – |
| $\phi$ | capacity reduction factor | – |

*Subscripts*

| | |
|---|---|
| A | active |
| b | base, bar, bending, or bearing |
| c | concrete, column, cover, or core |
| cr | cracked |
| D | dead |
| e | effective |
| f | footing or flexure |
| g | gross |
| h | horizontal |
| i | initial |
| L | live |
| n | nominal or clear |
| o | zero eccentricity |
| p | prestress |
| r | rupture |
| s | stem or spiral |
| st | steel |
| t | tension |
| u | ultimate |
| v | shear |
| w | service load |
| y | yield |

[1] This chapter is no substitute for the ACI code, and covers only a fraction of the relevant material.

## 1 CONVERSIONS

| multiply | by | to obtain |
|---|---|---|
| ft-kips | 1.356 | kN-m |
| ft-lbf | 1.356 | N-m |
| in-lbf | 0.113 | N-m |
| kips | 4.448 | kN |
| kips | 1000 | lbf |
| kips/ft | 14.59 | kN/m |
| kips/ft$^2$ | 47.88 | kN/m$^2$ |
| kN | 0.2248 | kips |
| kN/m | 0.06852 | kips/ft |
| kPa | 1000 | Pa |
| ksi | 6.895 EE6 | Pa |
| lbf | 0.001 | kips |
| lbf | 4.448 | N |
| N-m | 0.7376 | ft-lbf |
| N-m | 8.851 | in-lbf |
| N-m$^2$ | 1.0 | Pa |
| Pa | 1.0 | N/m$^2$ |
| Pa | EE–6 | MPa |
| Pa | 0.001 | kPa |
| Pa | 1.45 EE–7 | ksi |
| Pa | 1.45 EE–4 | psi |
| Pa | 0.02089 | psf |
| psf | 47.88 | Pa |
| psi | 6.895 EE3 | Pa |

## 2 DEFINITIONS

Absorption: The process by which a liquid is drawn into and tends to fill permeable pores in a porous body. Also, the increase in weight of a porous solid body resulting from the penetration of liquid into its permeable pores.

Admixture: A material other than water, aggregates, and portland cement that is used as an ingredient in concrete and is added to the batch immediately before or during its mixing.

Aggregate: Inert material which is mixed with portland cement and water to produce concrete.

Aggregate, coarse: Aggregate which is retained on a #4 sieve.

Aggregate, fine: Aggregate passing the #4 sieve and retained on the #200 sieve.

Aggregate, lightweight: Aggregate having a dry density of 70 pounds per cubic foot or less.

Bleeding: The autogenous flow of mixing water within, or its emergence from, recently placed concrete.

Cement factor: The number of bags or cubic feet of cement per cubic yard of concrete.

Column: An upright compression member, the length of which exceeds three times its least lateral dimension.

Combination column: A column in which a structural steel member, designed to carry the principal part of the load, is encased in concrete which carries the remainder of the load.

Composite column: A column in which a steel structural member is completely encased in concrete containing spiral and longitudinal reinforcement.

Composite concrete flexural construction: A precast concrete member and cast-in-place reinforced concrete so interconnected that the component elements act together as a flexural member.

Concrete: A mixture of portland cement, fine aggregate, coarse aggregate, and water.

Concrete, normal weight: Concrete having a hardened density of approximately 150 pcf which is made from aggregate of approximately the same density.

Concrete, plain: Concrete that is not reinforced with steel.

Concrete, precast: A plain or reinforced concrete element cast in other than its final position in the structure.

Concrete, prestressed: Reinforced concrete in which there have been introduced internal stresses of such magnitude and distribution that the stresses resulting from service loads are counteracted to a desired degree.

Concrete, reinforced: Concrete containing steel reinforcement.

Concrete, structural lightweight: A concrete containing lightweight aggregate.

Crushed gravel: The product resulting from artificial crushing of gravel with substantially all fragments having at least one fracture face.

Crushed stone: The product resulting from the artificial crushing of rocks where substantially all faces have resulted from the crushing operation.

Deformed bar: A reinforcing bar with ridges to increase bonding with the concrete.

Double reinforcement: A concrete beam with steel on both sides of the neutral axis to resist tension and compression.

Fineness modulus: An empirical factor obtained by adding the total percentages of a sample of the aggregate retained on each of a specified series of sieves, and dividing the sum by 100.

Formwork: The wood molds used to hold concrete during the curing and pouring processes.

Gravel: Granular material retained on a #4 sieve which is the result of crushing or natural disintegration of rock.

Heat of hydration: The heat evolved as a result of an exothermic reaction between cement and water.

Hydration: The exothermic reaction between cement constituents and water, resulting in hydrated compounds (i.e., compounds bound to water molecules.)

Monolithic construction: Constructed as one piece.

Pedestal: An upright compression member whose height does not exceed three times its average least lateral dimension.

Plain bar: Reinforcement that does not conform to the definition of a deformed bar. That is, reinforcing bars without raised ridges.

Rebar: Steel reinforcing bar.

Sand: Granular material passing through a #4 sieve but predominantly retained on a #200 sieve.

Saturated Surface Dry: A condition of an aggregate which holds as much water as it can without having any free surface water between the aggregate particles.

Slab: A cast concrete member of uniform thickness.

Slump: The decrease in height of wet concrete when a supporting mold is removed.

Spiral column: A column with a continuous spiral of wire around the longitudinal steel.

SSD: See 'Saturated Surface Dry.'

Surface water: Water carried by an aggregate in addition to that held by absorption within the aggregate particles themselves. Water in addition to SSD water.

Tied column: A column which has individual loops of wire around the longitudinal steel.

Two-way: Construction with steel reinforcing running in two perpendicular directions.

Water-cement ratio: The ratio of water weight to cement weight. Alternatively, the number of gallons of water per 94 pound sack of cement.

## 3 CONCRETE MIXING

### A. TYPES OF CONCRETE

Concrete is a mixture of mineral aggregates locked into a solid structure by a binding material. The concrete is produced by adding water to the aggregate and binder, and then casting the mixture in place. The semi-fluid mixture hardens by chemical action.

The usual binding material is portland cement, which is manufactured from lime, silica, and alumina (with a small amount of plaster of paris) in the appropriate amounts. There are five types of portland cement, and the choice is dependent on the application.

- **Type I: Normal portland cement:** This is a general-purpose cement used whenever sulfate hazards are absent and when the heat of hydration will not produce objectionable rises in temperature. Typical uses are sidewalks, pavement, beams, columns, and culverts.
- **Type II: Modified portland cement:** This cement has a moderate sulfate resistance, but is generally used in hot weather in the construction of large concrete structures. Its heat rate and total heat generation are lower than for normal portland cement.
- **Type III: High-early strength portland cement:** This type develops its strength quickly. It is suitable for use when the structure must be put into early use or when long-term protection against cold temperatures is not feasible. Its shrinkage rate, however, is higher than for types I and II; and extensive cracking may result.
- **Type IV: Low-heat portland cement:** For extensive concrete structures, such as gravity dams, low-heat cement is required to minimize the curing heat. The ultimate strength also develops more slowly than for the other types.
- **Type V: Sulfate-resistant portland cement:** This type of cement is applicable when exposure to severe sulfate concentration is expected. This typically occurs in states having highly alkaline soils.

**Table 14.1**
Relative Strengths of Concrete Types

| | compressive strength, % of normal strength concrete | | |
|---|---|---|---|
| | 3 days | 28 days | 3 months |
| type I | 100 | 100 | 100 |
| type II | 80 | 85 | 100 |
| type III | 190 | 130 | 115 |
| type IV | 50 | 65 | 90 |
| type V | 65 | 65 | 85 |

CONCRETE

## B. AGGREGATES

The bulk of concrete consists of sand and rock particles that have been added to the cement-water mixture to increase the weight and volume. These sand and rock particles are known as *aggregate*. Sand and other particles that will pass through a #4 sieve (less than 0.25″) are known as *fine aggregate*. Any particles that are larger than this are known as *coarse aggregate*.

Aggregate having a density of 70 pcf or less is known as *lightweight aggregate*. This type is used in the production of *lightweight concrete*. Lightweight concrete in which only the coarse aggregate is lightweight is known as *sand-lightweight concrete*. If both the coarse and fine aggregates are lightweight, it is known as *all-lightweight concrete*. Unless noted otherwise, concrete in this chapter is assumed to be *normal weight concrete*.

## C. ADMIXTURES

Anything added to the concrete to improve its workability, hardening, or strength characteristics is known as an *admixture*. Hydrated lime, diatomaceous silica, fly ash, and bentonite are added to concrete which has too little fine aggregate. These admixtures separate the coarse aggregate and reduce the friction of the mixture. Calcium chloride can be added as a curing accelerator. It is also an anti-freeze, but its use as such is not recommended.[2]

Sulfonated soaps and oils, as well as natural resins, increase *air entrainment*. This increases the durability of the concrete while decreasing the strength and weight only slightly.

Styrofoam can be added to increase the insulating values of the concrete.



## D. PROPORTIONING CONCRETE

The proportions of a concrete mixture are usually designated as a ratio of cement, fine aggregate, and coarse aggregate, in that order. For example, 1 : 2 : 3 means that one part of cement, two parts of fine aggregate, and three parts of coarse aggregate are to be combined. The ratio can be either in terms of weight or volume. Weight ratios are more common.

[2] The ACI code does not allow chloride to be added to concrete used in prestressed construction or in concrete with aluminum embedments. Natural chloride ions are also limited.

The amount of water used is called out in terms of gallons of water per 94 pound sack of cement. This is known as the *water-cement ratio*.[3]

## E. IN-PLACE VOLUME

The amount of concrete that can be made from mixing known quantities of ingredients can be found from the *absolute* or *solid volume method*. This assumes that there will be no voids in the placed concrete. Therefore, the amount of concrete is the sum of the solid volumes of the cement, sand, coarse aggregate, and water.

To use the absolute volume method, it is necessary to know the solid densities of the constituents. In the absence of other information, the following data can be used for solid densities: cement 195 pcf; fine aggregate, 165 pcf; coarse aggregate, 165 pcf; water, 62.4 pcf. A sack of cement weighs 94 pounds, and 7.48 gallons of water make a cubic foot. Alternatively, there are 239.7 gallons per ton of water.

If the mix proportions are volumetric, it will be necessary to multiply the ratio values by the bulk densities to get the weights of the constituents. Then, weight ratios may be calculated and the absolute volume method applied directly.

The problem may be complicated by *air entrainment* and/or the water content of the aggregate. The following guidelines should be observed,

- The yield is increased by the addition of air. This can be accounted for by dividing the solid yield by (1 − air percentage).
- Any water content in the aggregate above the *saturated surface dry* (SSD) water content must be subtracted from the water requirements.
- Any porosity (water affinity) below the SSD water content must be added to the water requirement.
- The densities used in the calculation of yield should be the SSD densities.

*Example 14.1*

A mix is designed as 1:1.9:2.8 by weight. The water-cement ratio is 7 gallons of water per sack of cement. (a) What is the concrete yield in cubic feet per sack of cement? (b) How much of each constituent is needed to make 45 cubic yards of concrete?

[3] Alternatively, the water-cement ratio may be given in pounds of water to pounds of cement.

(a) The solution can be tabulated as follows:

| material | ratio | weight per sack cement | solid density | absolute volume ($ft^3$/sack) |
|---|---|---|---|---|
| cement | 1.0 | $1 \times 94 = 94$ | 195 | $94/195 = 0.48$ |
| sand | 1.9 | $1.9 \times 94 = 179$ | 165 | $179/165 = 1.08$ |
| coarse | 2.8 | $2.8 \times 94 = 263$ | 165 | $263/165 = 1.60$ |
| water | | | | $7/7.48 = \underline{0.94}$ |
| | | | | 4.10 |

The yield is 4.1 cubic feet of concrete per sack cement.

(b) The number of one-sack batches is

$$\frac{(45)\,\text{yd}^3(27)\,\text{ft}^3/\text{yd}^3}{(4.1)\,\text{ft}^3/\text{sack}} = 296.3\,\text{sacks} \quad (\text{say } 297)^4$$

Order 297 sacks of cement.

$$\frac{(297)(1.9)(94)}{2000} = 26.5 \text{ tons of sand}$$

$$\frac{(297)(2.8)(94)}{2000} = 39.1 \text{ tons of coarse aggregate}$$

$$(297)(7) = 2079 \text{ gallons of water}$$

*Example 14.2*

50 cubic feet of 1:2½:4 (by weight) concrete are to be produced. The constituents have the following properties:

| constituent | SSD density (pcf) | moisture (dry basis from SSD) |
|---|---|---|
| cement | 197 | – |
| sand | 164 | 5% excess |
| coarse aggregate | 168 | 2% deficit |

**What are the required order quantities if the design calls for 5.5 gallons of water per sack and 6% entrained air?**

| constituent | ratio | weight per sack cement | SSD density | absolute volume |
|---|---|---|---|---|
| cement | 1.0 | 94 | 197 | 0.477 |
| sand | 2.5 | 235 | 164 | 1.433 |
| coarse | 4.0 | 376 | 168 | 2.238 |
| water | | 5.5/7.48 | | $= \underline{0.735}$ |
| | | | | 4.883 $ft^3$/sack |

The yield with 6% air is

$$\frac{4.883}{1-0.06} = 5.19\,\text{ft}^3/\text{sack}$$

The number of one sack batches is

$$\frac{50}{5.19} = 9.63$$

(In practice, this would be rounded up.)

The required sand weight (ordered as is, not SSD) is

$$\frac{(9.63)(1.05)(94)(2.5)}{2000} = 1.19\,\text{tons}$$

The required coarse aggregate weight (ordered as is, not SSD) is

$$\frac{(9.63)(0.98)(94)(4)}{2000} = 1.77\,\text{tons}$$

The excess water contained in the sand is

$$(1.19)\left(\frac{0.05}{1.05}\right)(239.7)\,\text{gal/ton} = 13.58\,\text{gallons}$$

The water needed to bring the coarse aggregate to SSD conditions is

$$(1.77)\left(\frac{0.02}{0.98}\right)(239.7)\,\text{gal/ton} = 8.66\,\text{gallons}$$

The total water needed is

$$(5.5)(9.63) + 8.66 - 13.58 = 48.0\,\text{gallons}$$

## 4 PROPERTIES OF CONCRETE

### A. SLUMP

The *slump test* is a measure of consistency of the plastic concrete mass prior to hardening. This test consists of filling a slump cone in three layers of about one-third of the mold volume. Each layer is rodded with a round rod. The mold is then removed by raising it carefully in the vertical direction. The slump is the difference in the mold height and the resulting concrete pile height.

Concrete mixtures that do not slump appreciably are known as *stiff mixtures.* Stiff mixtures are inexpensive because of the large amount of coarse aggregate. However, placing time and workability are impaired. Mixtures with large slumps are known as *wet* or *watery mixtures.* Such mixtures are needed for thin castings and structures with extensive reinforcing.

Recommended slumps for concrete that is hand vibrated are given in table 14.2. If high-frequency vibrators are used, the values given should be reduced by about one-third.

[4] Don't round down, or the volume will be short.

**Table 14.2**
Recommended Slumps

| application | slumps, inch maximum | minimum |
|---|---|---|
| reinforced foundations and footings | 3 | 1 |
| plain footings and substructure walls | 3 | 1 |
| slabs, beams, and reinforced walls | 4 | 1 |
| columns, reinforced | 4 | 1 |
| pavement and slabs | 3 | 1 |
| heavy mass construction | 2 | 1 |

### B. COMPRESSIVE STRENGTH

Cylinders of the concrete are cast for each mixture ratio to be tested. After curing, the cylinders are placed in a compressive tester. A load is applied at a constant rate. The compressive strength is found by dividing the maximum load carried (at failure) by the area. Typical compressive strengths, $f'_c$, vary from 2000 psi to 8000 psi at 28 days, although 6000 psi is a common upper limit.

Since the strength of concrete increases with time, all values of $f'_c$ should be referenced to the age. If no age is given, a standard 28 day age is assumed.

The ultimate compressive strength is highly dependent on the water/cement ratio, and it is fairly independent of the proportion of the mixes. The compressive strength of concrete varies directly with the cement/water ratio, provided that the mix is of a workable consistency.[5]

### C. TENSILE STRENGTH

Because concrete is not used to resist tension, tensile tests are seldom performed on normal weight concrete. However, the ACI code does specify a splitting tensile test for structural lightweight concrete. The ultimate tensile strength usually varies between 7% and 10% of the ultimate compressive strength.

### D. SHEAR STRENGTH

The shear strength of concrete can be determined by torsion tests. Such tests vary widely in method and results. The shear strength will be between one-sixth and one-quarter of the ultimate compressive strength.

[5] This is *Abram's strength law*, named after Dr. Duff Abrams who formulated it in 1918.

### E. DENSITY

The weight density (also known as *unit weight*) of concrete can vary from 100 pcf to 160 pcf, depending on the mixture ratios and the specific gravities of the constituents. Generally, the range will be 140 pcf to 160 pcf. Steel reinforced concrete will be between 3% and 5% higher than similar plain concrete. An average of 150 pcf can be used in most calculations.

### F. MODULUS OF ELASTICITY

Typical results of a standard compressive test are shown in figure 14.1. The slope of the stress-strain line varies with the applied stress, and there are several ways of evaluating the modulus of elasticity. These methods give the *initial tangent modulus*, *actual tangent modulus*, and *secant modulus*. The secant modulus is most frequently used in design work.

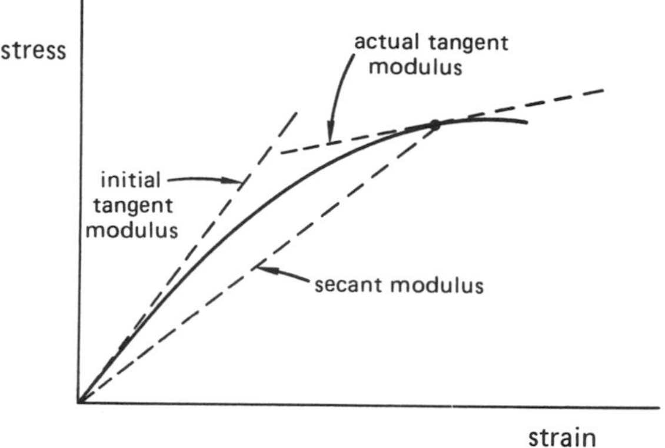


**Figure 14.1** Compressive Test Results

The ACI code calculates the secant modulus of elasticity based on the compressive strength. Equation 14.1 is for normal weight concrete only, with a density of $w = 90$ to 155 pcf. [ACI 318 8.5.1]

$$E_c = w^{1.5}(33)\sqrt{f'_c} \approx 57{,}000\sqrt{f'_c} \qquad 14.1$$

The modulus of elasticity is greatly dependent on age, quality, and proportions. It can vary from 1 EE6 to 5 EE6 psi at 28 days.

## 5 STEEL REINFORCING

Since concrete is essentially incapable of resisting tension, steel reinforcing is used. Figure 14.2 shows typical reinforcing in concrete beams.

Straight bars resist flexural tension in the central part of the beam. Since the bending moment is smaller near the ends of the beam, less reinforcing is necessary. Some of the bar may be bent up to resist diagonal shear near

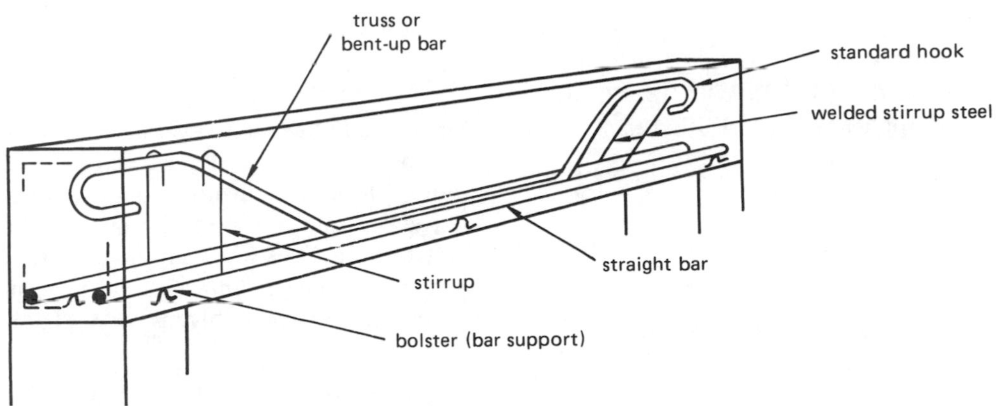


**Figure 14.2** Beam Reinforcement

the beam ends. (In continuous beams, the horizontal upper parts of the bent-up bars are continued on to the next span.)

*Stirrups* are used to resist the diagonal tension resulting from shear stress. These pass underneath the bottom steel for anchoring or, much less frequently, are welded to the bottom steel.

The horizontal steel is supported on *bolsters* (*chairs*), of which there are a variety of designs.

Reinforcing steel, known as *rebar*, is available in a number of sizes, as well as in the form of wire for spiral wrapping, and wire mesh for shrinkage and thermal expansion control.

Steel grade is specified by its yield strength. Grade 60 is most common, although grade 40 was widely used in the past, and grades 50 and 80 are also available. However, not all sizes may be available in every grade.

**Table 14.3**
ASTM Standard Reinforcing Bars[6]

| size | weight (lb/ft) | diameter (in) | area ($in^2$) | perimeter (in) |
|---|---|---|---|---|
| #2 | 0.167 | 0.250 | 0.05 | 0.786 |
| #3 | 0.376 | 0.375 | 0.11 | 1.178 |
| #4 | 0.668 | 0.500 | 0.20 | 1.571 |
| #5 | 1.043 | 0.625 | 0.31 | 1.963 |
| #6 | 1.502 | 0.750 | 0.44 | 2.356 |
| #7 | 2.044 | 0.875 | 0.60 | 2.749 |
| #8 | 2.670 | 1.000 | 0.79 | 3.142 |
| #9 | 3.400 | 1.128 | 1.00 | 3.544 |
| #10 | 4.303 | 1.270 | 1.27 | 3.990 |
| #11 | 5.313 | 1.410 | 1.56 | 4.430 |
| #14 | 7.65 | 1.693 | 2.25 | 5.32 |
| #18 | 13.60 | 2.257 | 4.00 | 7.09 |

[6] There is no difference between ACI and ASTM sizes and grades.

## 6 DEVELOPMENT LENGTH

The bond between the concrete and the reinforcement must be sufficient to keep the reinforcement from being pulled or pushed through the concrete. This is accomplished, in the ACI code, by specifying the minimum length of bar required to develop the full strength. This minimum length is known as the *development length.*

The *basic development length* for unhooked bars in tension with sizes up to and including #11 is[7] [ACI 318 12.2.1]

$$l_d = \max \begin{cases} \dfrac{0.04A_b f_y}{\sqrt{f'_c}} \\ 12'' \end{cases} \qquad 14.2$$

If at least $2.5d_b$ of clear cover is provided and the spacing is $5d_b$ or more, the basic development length may be multiplied by 0.8. [ACI 12.2.3.4] If more steel reinforcing is used than is required, the development length may also be reduced. [ACI 318 12.2.5]

$$l'_d = l_d \left( \frac{A_{st,\text{required}}}{A_{st,\text{actual}}} \right) \qquad 14.3$$

If standard hooks are used, the basic development length is [ACI 318 12.5]

$$l_d = \max \begin{cases} \dfrac{1200 d_b f_y}{60{,}000\sqrt{f'_c}} \\ 8d_b \\ 6'' \end{cases} \qquad 14.4$$

[7] Different rules apply for larger bars. The basic development length given in equation 14.2 must be modified for use with top steel, lightweight concrete, epoxy-coated bars, and steel yield strengths in excess of 60,000 psi.

If the cover on the hook is at least $2\frac{1}{2}''$ (normal to hook plane) and 2″ (behind the hook), the basic development length may be multiplied by 0.7. [ACI 318 12.5.3.2] Equation 14-3 is also applicable.[8] [ACI 318 12.5.3.4]

The basic development length for bars in compression is [ACI 318 12.3.2]

$$l_d = \max \begin{cases} \dfrac{0.02 d_b f_y}{\sqrt{f'_c}} \\ 0.0003 d_b f_y \\ 8'' \end{cases} \qquad 14.5$$

Equation 14.3 applies. [ACI 318 12.3.3.1] However, hooks are not considered effective in reducing development length for bars in compression.

## 7 ULTIMATE STRENGTH DESIGN

The ultimate strength design method is known as the *strength design method* in the ACI code. The actual (i.e., service) loads are increased by multiplicative safety factors. These *factored loads* are compared to the loads that would cause failure.[9]

The ultimate strength is also multiplied by a *capacity reduction factor*, $\phi$, to account for workmanship and understrength in the materials. Values of $\phi$ are given in table 14.4.

**Table 14.4**
Strength Reduction Factors

| type of stress | $\phi$ |
|---|---|
| flexure | 0.90 |
| axial tension | 0.90 |
| shear | 0.85 |
| torsion | 0.85 |
| axial compression with spiral reinforcement | 0.75 |
| axial compression with tied reinforcement | 0.70 |
| bearing on concrete | 0.70 |

[8] Other rules apply for hooks enclosed by ties or stirrups and lightweight concrete.

[9] Actually, the ACI code determines beam size and dimensions on the basis of ultimate strength, but calculates the moments on beams based on elastic theory. Redistribution of moments, which is taken into account in plastic steel design, is not considered in concrete design.

## 8 BEAMS

### A. INTRODUCTION

Figure 14.3 illustrates several types of beams. The normal beam shown in figure 14.3(a) defines the beam width and depth. Notice that the depth is measured to the centroid of the steel group. If the ratio of beam length to depth is less than 5 (i.e., $l/d < 5$), the beam is defined as a *deep beam*. Special shear provisions apply to the design of deep beams.

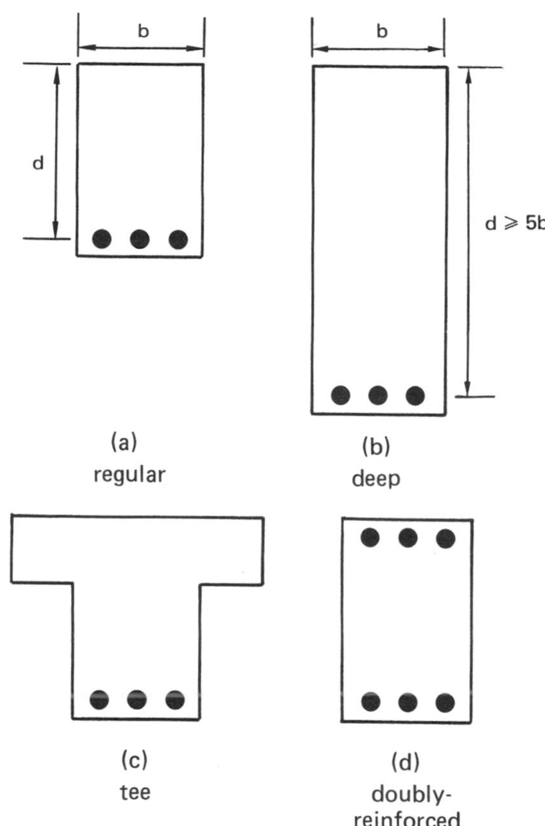


**Figure 14.3** Types of Beams

The *t-beam* shown in figure 14.3(c) is usually a portion of a larger monolithic slab-and-girder system. The *doubly-reinforced beam* shown in figure 14.3(d) is needed in continuous construction where the moment near supports is negative. Not shown are beams of prestressed concrete construction.

### B. FACTORED LOADS

The ultimate strength multipliers (i.e., *overload factors*) of 1.4 and 1.7 are used for dead and live loads, respectively. Thus, the factored moment and factored shear at a particular point are

$$M_u = 1.4M_D + 1.7M_L \qquad 14.6$$

$$V_u = 1.4V_D + 1.7V_L \qquad 14.7$$

CONCRETE

C. GENERAL PROVISIONS OF BEAM DESIGN

- The *reinforcement ratio*, $\rho$, is defined by equation 14.8.

$$\rho = \frac{A_{st}}{A_c} = \frac{A_{st}}{bd} \qquad 14.8$$

The ACI code places limits on the reinforcement ratio, as defined by equation 14.9 (ACI sections [10.3.3] and [10.5.1]).[10]

$$\frac{200}{f_y} \le \rho \le 0.75\rho_{\text{balanced}} \qquad 14.9$$

The balanced steel ratio represents a design such that the steel and concrete yield simultaneously. By limiting the steel ratio to 75% of the balanced value, a ductile failure of the steel is assured. The steel will, by design, gradually yield before the onset of a sudden brittle concrete failure.

The *balanced reinforcement ratio* is given by equation 14.10.

$$\rho_{\text{balanced}} = \frac{(0.85)\beta_1 f'_c}{f_y}\left[\frac{87{,}000}{87{,}000 + f_y}\right] \qquad 14.10$$

The factor $\beta_1$ is defined by equations 14.11 and 14.12. $\beta_1$ may not be less than 0.65 [ACI 318 10.2.7.3]

$$\beta_1 = 0.85 - 0.05\left(\frac{f'_c - 4000}{1000}\right) \qquad 4000 < f'_c \le 8000 \qquad 14.11$$

$$\beta_1 = 0.85 \qquad f'_c \le 4000 \qquad 14.12$$

- $b$ and $d$ can take on any reasonable values. However, small values of $d$ will produce small moments of inertia, leading to large deflections. Since large deflections produce cracks, most beams are designed to satisfy equation 14.13.[11]

$$1.75 < \frac{d}{b} < 2 \qquad 14.13$$

- All steel, including shear reinforcement, must be adequately covered by concrete. Appendix A lists the cover on reinforcing steel in detail. However, a minimum of $1\frac{1}{2}''$ of cover is generally required.

- The clear distance between bars is the maximum of one bar diameter or 1″.

- The minimum clear distance between layers is 1″.

D. SPACING OF REINFORCEMENT IN BEAMS

Table 14.5 is easily derived from the clear spacing and depth of cover requirements of the ACI code.[12] The table assumes that #3 stirrups will be used, and cover is provided for them. (The bar diameter, $d_b$, for bars #4 through #10 is less than the minimum bend diameter of $1\frac{1}{2}''$ for #3 stirrups. [ACI 318 7.2] A small allowance has been included in the beam widths to achieve the full $1\frac{1}{2}''$ bend radius.) If no stirrups are used, deduct 3/4″ from the table values. For additional bars horizontally, increase the beam width by adding the value in the last column.

E. LOCATING THE NEUTRAL AXIS

Figure 14.4 shows a simple concrete beam. The neutral axis is located a distance $c$ down from the top of the beam.[13] To determine $c$, it is necessary to balance the areas above and below the neutral axis.

[10] Actually, it is possible to have flexural reinforcement ratios less than $200/f_y$ as long as the reinforcement provided is one-third more than is required for strength.

[11] However, this is not an ACI code requirement.

[12] See appendix A for spacing and cover requirements.

[13] With the older working stress method, it was common to refer to the distance $c$ as $kd$.

**Table 14.5**
Minimum Beam Widths
(not exposed to weather, with #3 stirrups)

| size of bars | number of bars in single layer of reinforcement | | | | | | | add for each added bar |
|---|---|---|---|---|---|---|---|---|
| | 2 | 3 | 4 | 5 | 6 | 7 | 8 | |
| #4 | 6.1 | 7.6 | 9.1 | 10.6 | 12.1 | 13.6 | 15.1 | 1.50 |
| #5 | 6.3 | 7.9 | 9.6 | 11.2 | 12.8 | 14.4 | 16.1 | 1.63 |
| #6 | 6.5 | 8.3 | 10.0 | 11.8 | 13.5 | 15.3 | 17.0 | 1.75 |
| #7 | 6.7 | 8.6 | 10.5 | 12.4 | 14.2 | 16.1 | 18.0 | 1.88 |
| #8 | 6.9 | 8.9 | 10.9 | 12.9 | 14.9 | 16.9 | 18.9 | 2.00 |
| #9 | 7.3 | 9.5 | 11.8 | 14.0 | 16.3 | 18.6 | 20.8 | 2.26 |
| #10 | 7.7 | 10.2 | 12.8 | 15.3 | 17.8 | 20.4 | 22.9 | 2.54 |
| #11 | 8.0 | 10.8 | 13.7 | 16.5 | 19.3 | 22.1 | 24.9 | 2.82 |
| #14 | 8.9 | 12.3 | 15.6 | 19.0 | 22.4 | 25.8 | 29.2 | 3.39 |
| #18 | 10.5 | 15.0 | 19.5 | 24.0 | 28.6 | 33.1 | 37.6 | 4.51 |

The "strength" of the concrete compressive block is $E_c cb$. The centroid of the compressive block is $c/2$ from the neutral axis.

Similarly, the "strength" of the steel tensile area is $E_{st}A_{st}$, and the steel's centroid is located $d - c$ from the neutral axis.

It is necessary to equate these two strengths and solve for $c$ to find the neutral axis.

$$E_c cb\left(\frac{c}{2}\right) = E_{st}A_{st}(d - c) \qquad 14.14$$

If the *modular ratio* is known, it can be incorporated into equation 14.14. $n$ is typically 8 for non-prestressed construction.

$$\frac{1}{2}bc^2 = nA_{st}(d - c) \qquad 14.15$$

$$n = \frac{E_{st}}{E_c} \qquad 14.16$$

As an alternative, equation 14.17 can be used to solve for $c$ directly.

$$c = \frac{nA_{st}}{b}\left[\sqrt{1 + \frac{2bd}{nA_{st}}} - 1\right] \qquad 14.17$$

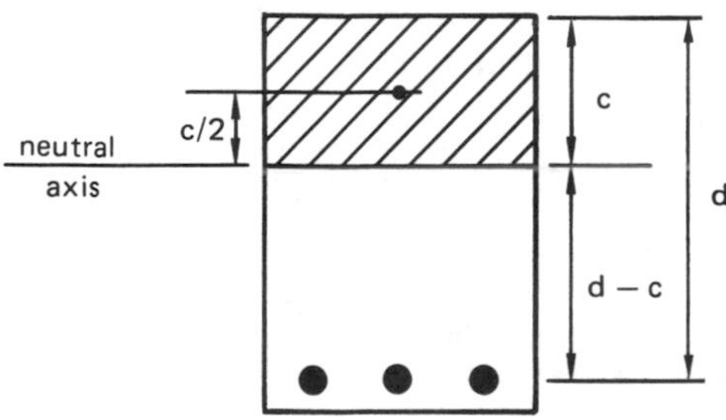


**Figure 14.4** The Neutral Axis

## F. INTRODUCTION TO STRENGTH THEORY[14]

Failure of concrete beams does not occur when steel reaches its yield point. Therefore, inelastic effects must be considered to predict the ultimate strength of beams. Since the location of the neutral axis shifts as the beam is stressed inelastically, much of the strength design theory is empirically derived.

The ACI code permits the assumption that the compressive stress distribution in the concrete is any shape that produces results in agreement with tests. It is commonly assumed that the distribution is rectangular with a value of $0.85f'_c$ at failure.[15]

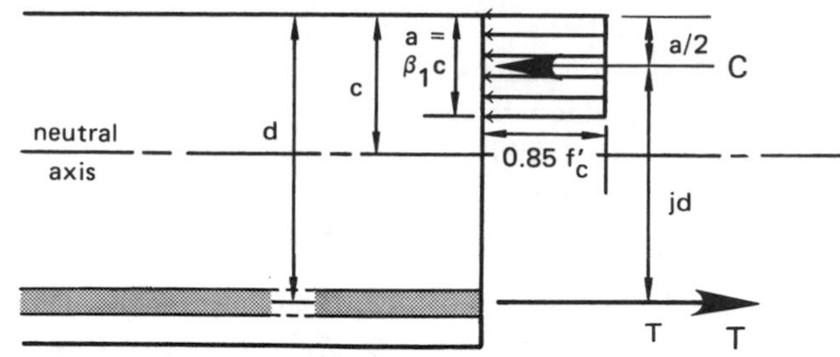


**Figure 14.5** Ultimate Strength Stress Distribution

In figure 14.5, the tensile and compressive forces are equal. It follows from $C = T$ that the height of the compressive block is

$$a = \frac{A_{st}f_y}{0.85f'_c b} = \frac{\rho f_y d}{0.85f'_c} \qquad 14.18$$

$$\rho = \frac{A_{st}}{A_c} = \frac{A_{st}}{bd} \qquad 14.19$$

The *nominal moment carrying ability* of the beam is

$$M_n = Tjd = Cjd \qquad 14.20$$

In equation 14.20, $jd$ is easily calculated from $d$ and $a$ (see figure 14.5). The steel is assumed to be stressed to yielding at failure. That is,

$$T = A_{st}f_y \qquad 14.21$$

$$C = 0.85f'_c ab \qquad 14.22$$

$$jd = d - \tfrac{1}{2}a \qquad 14.23$$

The *nominal strength* of the beam can be rewritten as

$$M_n = A_{st}f_y(d - \tfrac{1}{2}a) = 0.85f'_c ab(d - \tfrac{1}{2}a) \qquad 14.24$$

Substituting equation 14.18 into 14.24,

$$M_n = A_{st}f_y\left[d - 0.59\frac{A_{st}f_y}{f'_c b}\right]$$
$$= \rho bd^2 f_y\left[1 - \frac{0.59\rho f_y}{f'_c}\right] \qquad 14.25$$

14 T-beams and doubly-reinforced beams are not covered in this section.

15 This is known as the *Whitney assumption*. The rectangular block is an approximation to a parabolic block that extends over the entire distance, $c$. Equations 14.11 and 14.12 give the factor $\beta_1$ used in figure 14.5.

The ACI code builds on this theory. It adds the *capacity reduction factor*, $\phi$, and restricts $\rho$.[16] (See equation 14.9.) The ultimate moment is calculated according to equation 14.26.

$$M_u = 1.4M_D + 1.7M_L \tag{14.26}$$

The strength requirement for flexure is

$$M_n \geq \frac{M_u}{\phi} = \frac{1.4M_D + 1.7M_L}{\phi} \tag{14.27}$$

*Example 14.3*

Based solely on flexural requirements, what is the maximum uniform live load that the beam shown can carry? The span is 120 inches long, and the ends are built-in. Use $f'_c$ = 3500 psi and $f_y$ = 40,000 psi. (The cross section of the beam shown is near the ends. Near the mid-span, steel will be required near the bottom.)

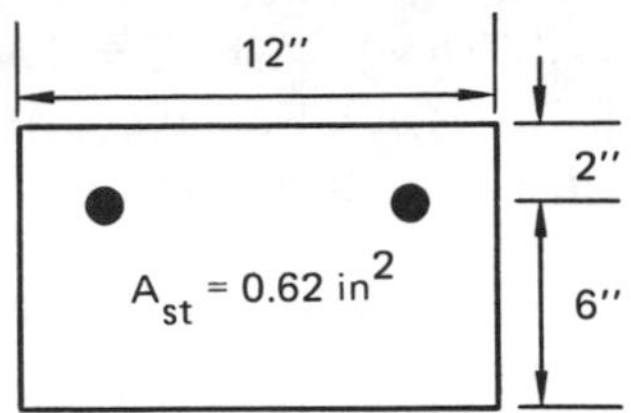


Near the ends, the beam is bent in reverse curvature, and $d = 6''$. From equation 14.19, the reinforcement ratio is

$$\rho = \frac{0.62}{(12)(6)} = 0.00861$$

From equation 14.18,

$$a = \frac{(0.00861)(40{,}000)(6)}{(0.85)(3500)} = 0.695 \text{ inch}$$

From equation 14.24, the nominal strength of the beam is

$$\begin{aligned} M_n &= (0.62)(40{,}000)\left(6 - \frac{0.695}{2}\right) \\ &= 140{,}200 \text{ in-lbf} \end{aligned}$$

For a beam fixed at both ends, the maximum moment occurs at the ends.

$$M_{\text{max}} = \frac{wL^2}{12}$$

Solving for the total nominal unit loading,

$$\begin{aligned} w_{\text{total}} &= \frac{12M_{\text{max}}}{L^2} = \frac{(12)(140{,}200)}{(120)^2} = 116.8 \text{ lbf/in} \\ &= 1402 \text{ lbf/ft} \end{aligned}$$

Assuming 150 pcf concrete and ignoring the steel weight, the dead load per foot of beam length is

$$w_D = \frac{(12)(8)(150)}{(144)(1)} = 100 \text{ lbf/ft}$$

From equation 14.6,

$$\begin{aligned} w_L &= \frac{w_{\text{total}}\phi - 1.4w_D}{1.7} \\ &= \frac{(1402)(0.90) - (1.4)(100)}{1.7} = 660 \text{ lbf/ft} \end{aligned}$$

## G. STRENGTH DESIGN OF BEAMS

The following procedure can be used to design a rectangular beam with tension reinforcement (only).

*step 1*: Select a tension reinforcement ratio, $\rho$, according to equation 14.9. Cracking will be minimized by keeping $\rho$ to less than half the maximum allowable (i.e., 0.375 of $\rho_{\text{balanced}}$), particularly when grade 40 or higher steel is used. A general rule used to design economical beams is to try to satisfy equation 14.28.

$$\frac{\rho f_y}{f'_c} \approx 0.18 \text{ (beams only)} \tag{14.28}$$

*step 2*: Calculate the ***coefficient of resistance***, $R_u$, defined as $M_n/bd^2$.

$$R_u = \rho f_y(1 - \tfrac{1}{2}\rho m) \tag{14.29}$$

$$m = \frac{f_y}{0.85f'_c} \tag{14.30}$$

*step 3*: Calculate the required value of $bd^2$ from equation 14.31. $M_u$ is the factored moment from equation 14.6. $\phi = 0.90$ for flexure. Be sure to use consistent units.

$$bd^2 = \frac{M_u}{\phi R_u} = \frac{M_n}{R_u} \tag{14.31}$$

*step 4*: Size the member so that $bd^2$ is approximately equal to the value calculated from

[16] $\phi$ =0.90 for flexure.

CONCRETE

equation 14.31. A good choice is to keep $d/b$ between 1.75 and 2.0. Size the beam to the nearest $\frac{1}{4}''$.[17]

*step 5*: If the actual $bd^2$ quantity is greatly different from the $bd^2$ value calculated in step 3, recalculate $\rho$.[18]

$$R_{u,\text{revised}} = \frac{M_n}{bd^2} = \frac{M_u}{\phi bd^2} \qquad 14.32$$

$$\rho_{\text{revised}} \approx \rho_{\text{old}}\left(\frac{R_{u,\text{revised}}}{R_{u,\text{old}}}\right) \qquad 14.33$$

*step 6*: Calculate the required steel area.

$$A_{st} = \rho_{\text{revised}} bd \qquad 14.34$$

*step 7*: Select the reinforcement steel bars to satisfy the distribution and placement requirements of the ACI code. Refer to tables 14.5 and 14.6.

[17] Actually, final beam dimensions ($b$ and $h$) are usually rounded to the nearest whole inch (slabs to the nearest half-inch). Regardless of the choice of $b$ and $d$, the section must ultimately be checked for excessive cracking and deflection.

[18] The relationship between $R_u$ and $\rho$ is linear only over small variations in either variable. However, it is not difficult to extract $\rho$ from equation 14.29 and obtain an exact value of the revised reinforcement ratio. See equation 14.142.

*step 8*: Specify the concrete cover thickness for steel protection to satisfy the ACI code. (See Appendix A, "Minimum Cover," in this chapter.) Cover thickness should be measured from the lowest stirrup surface. Note that concrete protection and cover depth, $d_c$, are different.

*step 9*: Design shear reinforcement (stirrups).

*step 10*: Verify the capacity of the beam. (See example 14.3.)

*step 11*: Check for cracking.

*step 12*: Check deflection.

*Example 14.4*

Design a rectangular beam with tension reinforcement (only) to carry service moments of 34,300 ft-lbf (dead) and 30,000 ft-lbf (live). Use $f'_c = 3500$ psi and $f_y = 40{,}000$ psi. #3 bars are available for shear reinforcing. (Do not design stirrup placement.)

*step 1*: From equation 14.10, the balanced reinforcement ratio is

$$\rho_{\text{balanced}} = \frac{(0.85)(0.85)(3500)}{40{,}000}\left(\frac{87{,}000}{87{,}000 + 40{,}000}\right)$$

$$= 0.0433$$

**Table 14.6**
Total Steel Areas for Various Numbers of Bars[19]

| bar size | nominal diameter (in.) | weight (lbm/ft) | number of bars: 1 | 2 | 3 | 4 | 5 | 6 | 7 | 8 | 9 | 10 |
|---|---|---|---|---|---|---|---|---|---|---|---|---|
| #3 | 0.375 | 0.376 | 0.11 | 0.22 | 0.33 | 0.44 | 0.55 | 0.66 | 0.77 | 0.88 | 0.99 | 1.10 |
| #4 | 0.500 | 0.668 | 0.20 | 0.40 | 0.60 | 0.80 | 1.00 | 1.20 | 1.40 | 1.60 | 1.80 | 2.00 |
| #5 | 0.625 | 1.043 | 0.31 | 0.62 | 0.93 | 1.24 | 1.55 | 1.86 | 2.17 | 2.48 | 2.79 | 3.10 |
| #6 | 0.750 | 1.502 | 0.44 | 0.88 | 1.32 | 1.76 | 2.20 | 2.64 | 3.08 | 3.52 | 3.96 | 4.40 |
| #7 | 0.875 | 2.044 | 0.60 | 1.20 | 1.80 | 2.40 | 3.00 | 3.60 | 4.20 | 4.80 | 5.40 | 6.00 |
| #8 | 1.000 | 2.670 | 0.79 | 1.58 | 2.37 | 3.16 | 3.95 | 4.74 | 5.53 | 6.32 | 7.11 | 7.90 |
| #9 | 1.128 | 3.400 | 1.00 | 2.00 | 3.00 | 4.00 | 5.00 | 6.00 | 7.00 | 8.00 | 9.00 | 10.00 |
| #10 | 1.270 | 4.303 | 1.27 | 2.54 | 3.81 | 5.08 | 6.35 | 7.62 | 8.89 | 10.16 | 11.43 | 12.70 |
| #11 | 1.410 | 5.313 | 1.56 | 3.12 | 4.68 | 6.24 | 7.80 | 9.36 | 10.92 | 12.48 | 14.04 | 15.60 |
| #14 | 1.693 | 7.65 | 2.25 | 4.50 | 6.75 | 9.00 | 11.25 | 13.50 | 15.75 | 18.00 | 20.25 | 22.50 |
| #18 | 2.257 | 13.60 | 4.00 | 8.00 | 12.00 | 16.00 | 20.00 | 24.00 | 28.00 | 32.00 | 36.00 | 40.00 |

[19] #14 and #18 bars are typically used in columns only.

The maximum and minimum reinforcement ratios (from equation 14.9) are

$$\rho_{\text{max}} = 0.75 \times 0.0433 = 0.0325$$

$$\rho_{\text{min}} = \frac{200}{40{,}000} = 0.005$$

From equation 14.28, choose an approximate reinforcement ratio.

$$\rho = \frac{(0.18)(3500)}{40{,}000} = 0.0158 \quad \text{(ok)}$$

*step 2*: The coefficient of resistance is

$$m = \frac{40{,}000}{(0.85)(3500)} = 13.45$$

$$R_u = (0.0158)(40{,}000)\left(1 - \frac{(0.0158)(13.45)}{2}\right) = 564.8$$

*step 3*: The factored moment is

$$M_u = [(1.4)(34{,}300) + (1.7)(30{,}000)] \times 12 = 1.19\,\text{EE6 in lbf}$$

$$bd^2 = \frac{(1.19\,\text{EE6})}{(0.90)(564.8)} = 2341\,\text{in}^3$$

*step 4*: Choose a ratio of $d/b = 1.8$.

$$d = \sqrt[3]{\frac{d}{b}bd^2} = \sqrt[3]{(1.8)(2341)} = 16.15 \quad \text{(say 16.25'')}$$

$$b = \frac{16.15}{1.8} = 8.97 \quad \text{(say 9'')}$$

*step 5*: The revised reinforcement ratio will not be significantly different.

$$bd^2 = (9)(16.25)^2 = 2377$$

$$R_{u,\text{revised}} = \frac{1.19\,\text{EE6}}{(0.90)(2377)} = 556.3$$

$$\rho_{\text{revised}} = \frac{(0.0158)(556.3)}{564.8} = 0.0156$$

*step 6*: The required steel area is

$$A_{st} = (0.0156)(9)(16.25) = 2.28\,\text{in}^2$$

*step 7*: Select 3 #8 bars (2.37 in$^2$).

*step 8*: Use 1.5'' of concrete protection. The cover depth (using #3 stirrups and measuring from the center of the steel layer) is

$$d_c = \frac{1.00}{2} + 0.375 + 1.5 = 2.375$$

The total beam depth will be

$$h = d + d_c = 16.25 + 2.375 = 18.625 \quad \text{(say 19'')}$$

## H. CRACK CHECKING

When the ultimate strength design method is used with steels having a yield strength in excess of 40,000 psi, cracking can be a problem. The ACI code provides a simplified method of determining if cracking will be a significant factor in a beam's performance.[20]

The check is performed by calculating the limiting parameter, $z$, from equation 14.35. If $z$ exceeds 145 kips/in (exterior service) or 175 kips/in (interior service), cracking will be significant and the beam should be redesigned. [ACI 318 10.6.4]

$$z = f_{st}\sqrt[3]{d_c A_e} = f_{st}\sqrt[3]{\frac{d_c(2d_{st}b)}{\#\text{ of bars}}} \qquad 14.35$$

$d_c$ is the *cover depth*—the distance from the center of the bottom steel layer to the bottom of the beam. For a beam with $1\frac{1}{2}''$ of cover over a stirrup, the cover depth would be

$$d_c = 1.5'' + d_{\text{stirrup}} + \tfrac{1}{2}d_b \qquad 14.36$$

$A_e$ in equation 14.35 is the effective concrete area in the tension block, divided by the total number of steel bars.

$f_{st}$ in equation 14.35 is the steel stress at the service level. It may be found in one of two ways. The first method is an obvious simplification.

$$f_{st} = 0.60f_y \qquad 14.37$$

If $z$ is acceptable using $f_{st}$ as defined in equation 14.37, no further checking will be required. However, if $z$ exceeds the limits specified, it may still be possible that the beam is adequate, and $f_{st}$ should be properly calculated.

[20] Actually, this method checks to see if the steel is sufficiently distributed throughout the tension region of the beam, rather than being clustered in one spot.

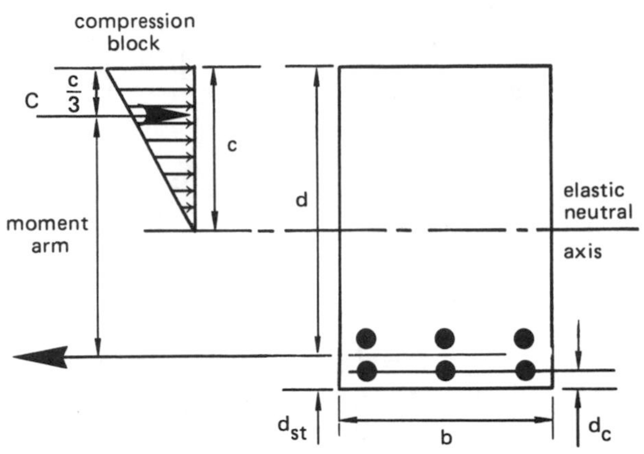


**Figure 14.6** Parameters in Crack Checking

In order to calculate the steel service stress accurately, it is necessary to know the location of the neutral axis. $M_w$ in equation 14.38 is the *service moment* (i.e., the actual moment on the beam at the location being investigated, not the factored moment).

$$f_{st} = \frac{M_w}{A_{st} \times \text{moment arm}} \quad 14.38$$

The moment arm in equation 14.38 is the distance between the tension and compression resultants. The moment arm can be derived from the location of the neutral axis.

$$\text{moment arm} = d - \frac{c}{3} \quad 14.39$$

If it is concluded that unacceptable cracking will occur, then smaller reinforcement bars must be used to redistribute the steel around a greater area.

CONCRETE

## I. BEAM DEFLECTIONS

The ACI code controls beam deflections only in general terms. If deflections are not calculated at all, the limits on beam thickness in table 14.7 apply. If the deflections are calculated by appropriate methods, the limits in appendix A apply. These limits are presented as a fraction of the beam length between supports, and are designed to avoid the appearance of excessive deflection.

Beam deflections are calculated for concrete beams in the same way as for any beam—with the standard beam equations. Beam deflection equations contain two important parameters: $E$ and $I$. The modulus of elasticity, $E$, is the *secant modulus* specified by equation 14.1. The moment of inertia is calculated by considering only the portion of the beam which is uncracked and contributes to bending strength.

**Table 14.7**
Minimum Beam Thicknesses[21]
[ACI 318 Table 9.5(a)]
(deflections not computed)
($f_y$ = 60,000 psi)

| construction | minimum $h$ (fraction of span length) |
|---|---|
| simply supported | 1/16 |
| one end continuous | 1/18.5 |
| both ends continuous | 1/21 |
| cantilever | 1/8 |

(For $f_y$ other than 60,000 psi, multiply the minimum thicknesses by $0.4 + f_y/100{,}000$.)

The uncracked portion of the beam varies along the beam's length. Near supports in continuous beams, the cracks are above the neutral axis. Near beam centers, the cracks are below the neutral axis. At inflection points where there is no moment, there are no cracks. To circumvent the problem of deciding what the uncracked moment of inertia should be, the ACI codes specifies an *effective moment of inertia*, as defined by equation 14.40, to be used with the maximum moment and the standard beam deflection equations.[22] [ACI 318 9.5.2.3]

$$I_e = \left(\frac{M_{cr}}{M_{\max}}\right)^3 I_g + \left[1 - \left(\frac{M_{cr}}{M_{\max}}\right)^3\right] I_{cr} \le I_g \quad 14.40$$

$$M_{cr} = \frac{f_r I_g}{y_t} = 1.25bh^2\sqrt{f'_c} \quad 14.41$$

$$f_r = 7.5\sqrt{f'_c} \quad 14.42$$

The *cracked transformed moment of inertia*, $I_{cr}$, is the moment of inertia neglecting the concrete below the neutral axis of the service section, but including the transformed steel. As equation 14.40 indicates, $I_e$ must be less than the *gross moment of inertia*, $I_g$, ignoring reinforcing.

$$I_g = \frac{bh^3}{12} \quad 14.43$$

$M_{cr}$ is the *cracking moment*. $M_{\max}$ is the maximum unfactored service moment on the beam. $f_r$ is the *modulus of rupture*. $y_t$ is the distance from the neutral axis of the gross section to the extreme concrete fiber in tension. That is,

$$y_t = \frac{1}{2}h \quad 14.44$$

[21] The same limits apply to ribbed one-way slabs. However, different limits apply to solid one-way slabs. See Appendix A.

[22] Refer to the ACI code if lightweight concrete is used.

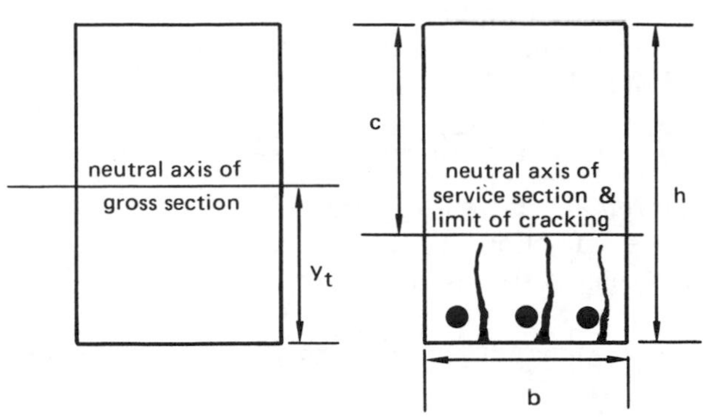


**Figure 14.7** Parameters for Calculating the Effective Moment of Inertia

The deflection calculated from $I_e$ and the secant modulus gives the *instantaneous deflection* (due to sustained loads). The additional *long-term deflection* resulting from creep and shrinkage effects can be calculated from equation 14.45. [ACI 318 9.5.2.5]

$$\delta_{\text{long term}} = \lambda \delta_{\text{instantaneous}} \qquad 14.45$$

$$\lambda = \frac{\xi}{1 + 50\rho'} \qquad 14.46$$

$$\rho' = \frac{A_{st,\text{compression}}}{bd} \qquad 14.47$$

In equation 14.46, $\xi$ depends on the timespan being considered, and has values of 1.0 (3 months), 1.2 (6 months), 1.4 (12 months), and 2.0 (5 years or more). $\rho'$ is the reinforcement ratio for compression steel only. [ACI 318 9.5.2.5]

The total deflection is the sum of the instantaneous and long term deflections.

*Example 14.5*

Determine the gross, cracked, and effective moments of inertia for the beam shown. The service moment is 100 ft-kips. Use $f'_c = 4000$ psi and $n = 8$.

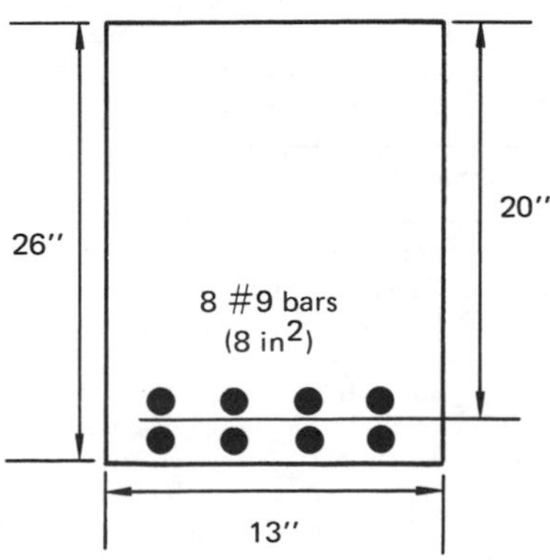


The gross moment of inertia is

$$I_g = \frac{(13)(26)^3}{12} = 19{,}041 \text{ in}^4$$

To find the cracked moment of inertia, it is necessary to locate the neutral axis. Tension cracks are expected below the neutral axis, and the concrete in the tension region should be disregarded when calculating the cracked moment of inertia. Equation 14.17 can be used to quickly locate the neutral axis.

$$c = \frac{(8)(8)}{13}\left[\sqrt{1 + \frac{(2)(13)(20)}{(8)(8)}} - 1\right] = 9.95 \text{ in}$$

From the parallel axis theorem with the 8 in$^2$ steel expanded by $n$ (8), the moment of inertia considering the compression concrete above the neutral axis and the transformed steel below the neutral axis is

$$I_{cr} = \frac{(13)(9.95)^3}{3} + (8)(8)(20 - 9.95)^2 = 10{,}733 \text{ in}^4$$

The modulus of rupture is

$$f_r = 7.5\sqrt{4000} = 474 \text{ psi}$$

$y_t$ is half of the beam thickness, 13″. The cracking moment is

$$M_{cr} = \frac{(474)(19{,}041)}{(13)(12)(1000)} = 57.9 \text{ ft-kips}$$

$$\left(\frac{M_{cr}}{M_{\text{max}}}\right)^3 = \left(\frac{57.9}{100}\right)^3 = 0.194$$

From equation 14.40, the effective moment of inertia to be used in calculating deflections is

$$I_e = (0.194)(19{,}041) + [1 - 0.194](10{,}733) = 12{,}345 \text{ in}^4$$

## J. SHEAR REINFORCEMENT IN BEAMS

Consider a simply-supported, single-span beam. The moment will be close to zero near the supports, while the shear will be maximum. Where the moment is zero, the flexural stress is also zero. Near the beam ends, a diagonal tension stress equal to the shear stress and inclined at 45° exists.[23] The stress on the beam at those points is in a state of nearly pure shear (tension).

[23] This is easily shown with the combined stress equation using $\sigma_x = 0$ and $\sigma_y = 0$.

Figure 14.8 shows a concrete beam with cracks that have formed below the neutral axis due to the diagonal shear. To reduce crack propagation, *shear reinforcement* in the form of stirrups attached to the longitudinal reinforcement and extending into the compression region are used.[24] This reinforcement is also known as *web reinforcement* and *diagonal reinforcement.*

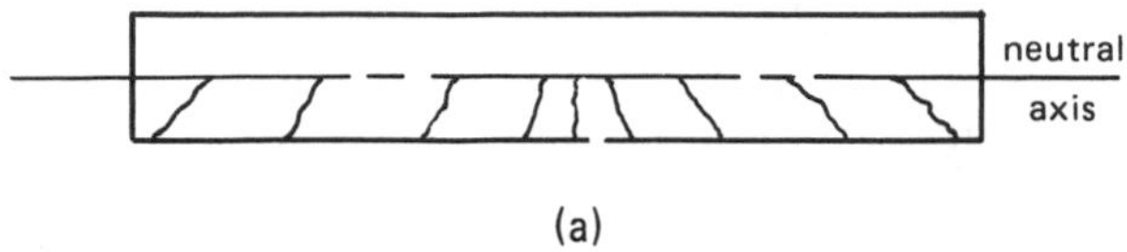


(a)

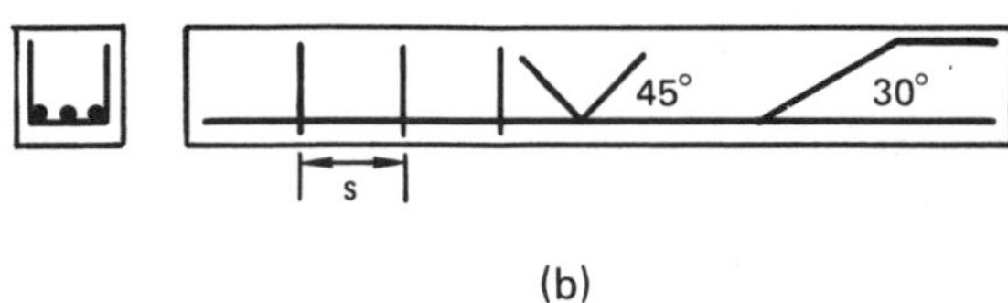


(b)

**Figure 14.8** Shear Cracking and Reinforcement

Shear reinforcement can take on several forms. Vertical stirrups are used most often, but stirrups inclined as little as 45° from the horizontal are occasionally seen, as is the use of welded wire fabric. Longitudinal steel that is bent up at 30° or more and enters the compression zone also contributes to shear reinforcement.[25]

If stirrups in the shape of a U are used, each bent stirrup contributes twice the stirrup bar area.

For the purpose of designing shear reinforcement, the live load should be considered variable in location, and should be moved to positions which will maximize the shear on the beam. An approximate *shear envelope* can be drawn. This envelope is a straight line drawn between the maximum possible shear at the midpoint of the support (determined by placing the live load to maximize the support reaction) and the maximum possible shear at midspan (determined by placing the live load to maximize the midspan shear).[26] Influence diagrams are valuable in determining the proper placement. Of course, the dead load shear must be added to the live load shear envelope.

Concrete has a nominal amount of shear strength of its own, even when no shear reinforcement is present.[27] [ACI 318 11.3.1.1]

$$V_c = 2\sqrt{f'_c}\,bd \qquad 14.48$$

Shear reinforcement is required when the factored shear exceeds a fraction of this nominal strength. That is, reinforcement should be provided when[28]

$$\phi V_c \le V_u \qquad 14.49$$

Reinforcement is used to make up the difference between $V_u$ and $\phi V_c$. [ACI 318 11.1.1] The nominal shear strength to be provided by the steel is

$$V_{st} = V_n - V_c = \frac{V_u}{\phi} - V_c \qquad 14.50$$

If the required shear reinforcement is too great, the beam should be redesigned. [ACI 318 11.5.6.8] The maximum shear reinforcement is

$$V_{st,\max} = 8\sqrt{f'_c}\,bd \qquad 14.51$$

A minimum amount of shear reinforcement is also required when[29] [ACI 318 11.5.5.1]

$$\phi V_c > V_u > \tfrac{1}{2}\phi V_c \qquad 14.52$$

This minimum shear reinforcement is [ACI 318 formula 11-14]

$$A_v = \frac{50bs}{f_y} \qquad 14.53$$

Equation 14.53 implies that reinforcement must supply a minimum of 50 psi of shear strength. The corresponding shear contribution is

$$V_{st} = 50bd \qquad 14.54$$

To satisfy equation 14.49, checking for critical shear can begin a distance $d$ from the support face.[30] [ACI 318

[24] Shear reinforcement does not prevent crack formation. It is used to limit the length of cracks.

[25] When inclined stirrups and bent longitudinal bars are used, every 45° line extending downward from depth $\frac{1}{2}$d to the tension reinforcement must be crossed by at least one line of that reinforcement. However, only the center three-fourths of the inclined sections of longitudinal steel bars are considered effective in resisting shear.

[26] With a uniform live load, the maximum midspan shear typically occurs when only one-half of the beam is loaded.

[27] The ACI code contains an alternate method of calculating the nominal concrete shear strength. However, this method is usually only used when justifying larger spacings at critical sections than would normally be permitted otherwise.

[28] $\phi = 0.85$ for shear.

[29] **Thin slab-like members, such as footings, slabs, and beams with a total depth of less than 10" are exempt from this minimum requirement. [ACI 318 section 11.5.5]**

[30] Care must be taken in applying this concession. The implication is that there will be no cracks in the compression region closer to the support than $d$. This requires the reaction on the beam to induce compression in the end regions of the beams. Not all beam supports do so. Beams supported from above are particularly suspect.

section 11.1.3.1] Whatever shear exists at that location can be used to design the shear reinforcement and spacing between the support and that point.[31]

For nonprestressed members, the maximum spacing between stirrups is[32] [ACI 11.5.4.1]

$$s = \min\begin{Brematrix} 24'' \\ d/2 \end{Bmatrix} \qquad 14.55$$

The maximum spacing given by equation 14.55 is to be reduced by one-half when the shear reinforcement requirement is greater than twice the nominal concrete shear strength. [ACI 318 11.5.4.3] Specifically, when

$$V_{st} > 4\sqrt{f'_c}bd \qquad 14.56$$

The actual spacing required increases toward midspan. However, spacing is not varied continuously. Groups of stirrups should be evenly spaced based on the maximum shear within the region covered by that spacing. Judgment is used to determine when the spacing should be changed, but minimum-cost designs use three spacings: a starting value, an intermediate spacing, and a maximum spacing, usually $d/2$.

If $A_v$ is known, equation 14.53 can be used to calculate the spacing.

For shear reinforcement perpendicular to the beam's axis and with area $A_v$ within the spacing distance $s$, the shear strength contribution is given by equation 14.57.[33]

$$V_{st} = \frac{A_v f_y d}{s} \qquad 14.57$$

For straight stirrups inclined at an angle $\theta$ from the horizontal, the shear strength contribution is [ACI 318 11.5.6.3]

$$V_{st} = \frac{A_v f_y(\sin\theta + \cos\theta)d}{s} \qquad 14.58$$

If shear reinforcement is provided by a single bar or a single group of parallel bars, all bent up at the same distance from the support, the shear strength contribution is [ACI 318 11.5.6.4]

$$V_{st} = A_v f_y \sin\theta \leq 3\sqrt{f'_c}bd \qquad 14.59$$

## K. ANCHORAGE OF SHEAR REINFORCEMENT

In order to develop its full capacity, shear reinforcement must be anchored at both ends by one of two means.[34]

In a U-shaped stirrup, one end of the shear reinforcement is bent around the longitudinal steel, anchoring that end. For #5 or smaller stirrups, and for stirrups constructed of #6, #7, and #8 bars with $f_y \leq 40{,}000$ psi, the other end may be considered to be developed with a standard hook around longitudinal reinforcement. [ACI 318 12.13.2.1]

For #6, #7, and #8 stirrups with $f_y > 40{,}000$ psi, the other end may be considered to be developed with a standard hook around longitudinal reinforcement plus an embedment distance between midheight of the member and the outside of the hook given by equation 14.60. [ACI 318 12.13.2.2]

$$\text{embedment length} = \frac{0.14 d_b f_y}{\sqrt{f'_c}} \qquad 14.60$$

Figure 14.9 illustrates these two cases.

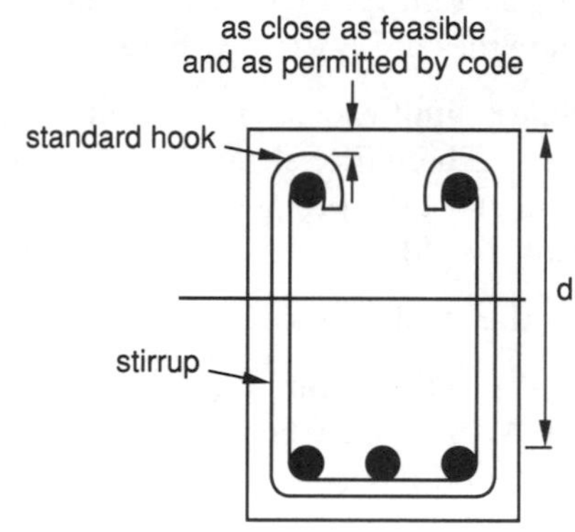


(a)
#5, or #6, #7, #8 stirrups with $f_y \leq 40{,}000$ psi

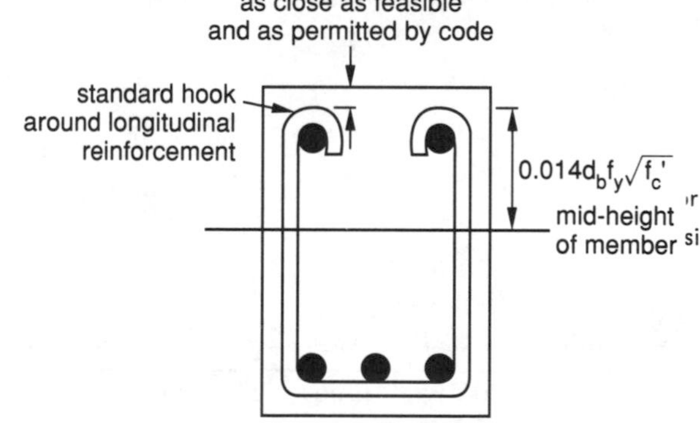


(b)
#5, or #6, #7, #8 stirrups with $f_y > 40{,}000$ psi

**Figure 14.9** Anchorage Methods for Shear Reinforcement

[31] It is common to place the first stirrup a distance $s/2$ from the face of the support, but not closer than 2″. However, the first stirrup can be placed a full space from the face.

[32] Though not specified by the code, the minimum practical spacing is about 3″. If stirrups need to be closer than 3″, a larger bar size should be used.

[33] The yield strength of shear reinforcement should not exceed 60,000 psi.

[34] The ACI code requires that the shear reinforcement extend into the compression and tension regions as far as cover requirements and proximity to other bars permit.

[35] (This footnote omitted in this edition.)

## 9 ONE-WAY FLOOR SLABS

A floor slab is typically supported on all four sides. If the slab length is more than twice the slab width, it can be assumed that the primary support will be from the beams along the length of the slab, and the slab is known as a *one-way slab*.[36]

One-way slabs can be designed and evaluated as beams. The beam width, $b$, is taken as 12″, which fixes one of the dimensions. The ultimate moment is computed per foot of slab width. Table 14.8 is used to determine the steel per foot of slab width.

If table 14.8 doesn't cover a specific application, equation 14.61 can be used to calculate the average steel per foot of width.

$$A_{st} = \frac{12A_b}{s} \qquad 14.61$$

Although slabs are designed as beams (the procedure given for beams can be used), there are some important differences. Important ACI code requirements and general design guidelines are presented here.

- Due to thickness limitations, shear reinforcement is seldom used. Therefore, shear stress is limited to the nominal strength provided by the concrete itself (equation 14.48).
- The cover on the steel may be as low as 3/4″.
- The ACI code specifies minimum slab thicknesses. See appendix A. Slab thicknesses are rounded to the next higher half-inch.
- In slabs with several continuous spans, the maximum negative moment will occur at the exterior face of the first interior support. This moment should be used to calculate the slab thickness.
- Span length is measured from the centers of supports for the purpose of determining fixed-end moments and stiffness factors. However, moment and shear coefficients may be used to design slabs, in which case the clear span (distance between faces of integral supporting beams) is used to calculate the moment on the slab.
- To limit deflection, the reinforcement ratio should be approximately one-half of the maximum allowed. That is,

$$\rho \approx 0.375\rho_{\text{balanced}} \qquad 14.62$$

The effective moment of inertia (equation 14.40) is used to calculate deflection.

- Development length may be a factor in parts of the reinforcement design.
- Where the slab is supported by a beam, some of the reinforcement is bent up to carry tension in the upper half of the slab. (For thin slabs, under 5″, straight bars are used.)
- The minimum bar size used is commonly #4.
- Shrinkage and temperature reinforcement perpendicular to the main reinforcement is required. For slabs reinforced in one direction only, the minimum reinforcement ratio for shrinkage and temperature control is 0.0014 (up to but excluding grade 40 steel), 0.0020 (grade 40), and 0.0018 (grade 60).[37] The maximum steel spacing is [ACI 318 7.12]

$$s = \min\begin{Bmatrix} 18'' \\ 5 \times t_{\text{slab}} \end{Bmatrix} \qquad 14.63$$

[36] The primary reinforcement also runs perpendicular to the long direction, in one direction. However, this is not why the slabs are called *one-way*.

[37] The code has other provisions for higher grade steel as well.

**Table 14.8**
Average Steel Area per Foot of Width

| bar size number | nominal diameter (in.) | spacing of bars in inches | | | | | | | | | | | | | |
|---|---|---|---|---|---|---|---|---|---|---|---|---|---|---|---|
| | | 2 | $2\frac{1}{2}$ | 3 | $3\frac{1}{2}$ | 4 | $4\frac{1}{2}$ | 5 | $5\frac{1}{2}$ | 6 | 7 | 8 | 9 | 10 | 12 |
| 3 | 0.375 | 0.66 | 0.53 | 0.44 | 0.38 | 0.33 | 0.29 | 0.26 | 0.24 | 0.22 | 0.19 | 0.17 | 0.15 | 0.13 | 0.11 |
| 4 | 0.500 | 1.18 | 0.94 | 0.78 | 0.67 | 0.59 | 0.52 | 0.47 | 0.43 | 0.39 | 0.34 | 0.29 | 0.26 | 0.24 | 0.20 |
| 5 | 0.625 | 1.84 | 1.47 | 1.23 | 1.05 | 0.92 | 0.82 | 0.74 | 0.67 | 0.61 | 0.53 | 0.46 | 0.41 | 0.37 | 0.31 |
| 6 | 0.750 | 2.65 | 2.12 | 1.77 | 1.51 | 1.32 | 1.18 | 1.06 | 0.96 | 0.88 | 0.76 | 0.66 | 0.59 | 0.53 | 0.44 |
| 7 | 0.875 | 3.61 | 2.88 | 2.40 | 2.06 | 1.80 | 1.60 | 1.44 | 1.31 | 1.20 | 1.03 | 0.90 | 0.80 | 0.72 | 0.60 |
| 8 | 1.000 | | 3.77 | 3.14 | 2.69 | 2.36 | 2.09 | 1.88 | 1.71 | 1.57 | 1.35 | 1.18 | 1.05 | 0.94 | 0.78 |
| 9 | 1.128 | | 4.80 | 4.00 | 3.43 | 3.00 | 2.67 | 2.40 | 2.18 | 2.00 | 1.71 | 1.50 | 1.33 | 1.20 | 1.00 |
| 10 | 1.270 | | | 5.06 | 4.34 | 3.80 | 3.37 | 3.04 | 2.76 | 2.53 | 2.17 | 1.89 | 1.69 | 1.52 | 1.27 |
| 11 | 1.410 | | | 6.25 | 5.36 | 4.69 | 4.17 | 3.75 | 3.41 | 3.12 | 2.68 | 2.34 | 2.08 | 1.87 | 1.56 |

## 10 T-BEAMS

### A. INTRODUCTION

Although a t-shaped beam can be constructed individually, it usually occurs as part of a monolithic beam-slab system. Since the flexural stresses in the beam "flange" are due to slab action, the slab flexural stresses are at right angles to the beam flexural stresses. Any interaction between the two stress distributions is ignored, and the slab (thickness) design and t-beam design are independent.

Equation 14.64 can be used to determine if a beam is rectangular. This equation is based on several approximations (i.e., $\beta_1 = 0.85$ and $d - \frac{1}{2}a = 0.90d$), but is sufficiently accurate for most purposes.[38] The beam is rectangular if

$$t \geq \frac{1.71 M_u}{f'_c b_e d} \qquad 14.64$$

$M_u$ must be expressed in consistent (inch) units in equation 14.64.

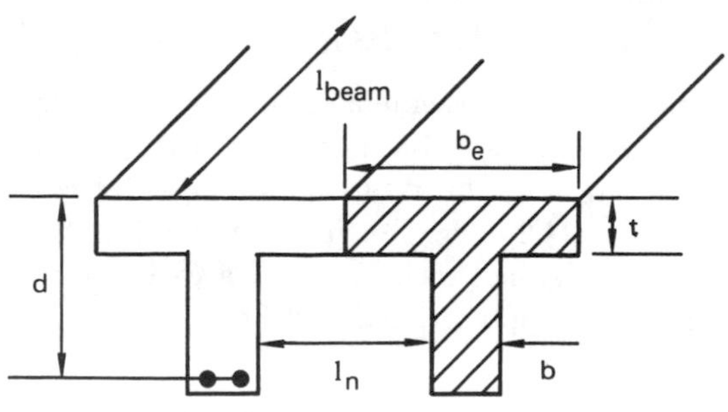


**Figure 14.10** T-Beam Dimensions

What appears to be a t-beam may not be a t-beam. To be a t-beam, the neutral axis must fall within the web, placing part of the web in compression. If the neutral axis falls within the flange (i.e., the slab), none of the web is in compression. Since concrete in tension is ignored in the calculation of moment of inertia, and since only the steel-to-neutral axis distance is relevant, the beam is not t-shaped.[39]

On the other hand, beams that appear to be odd-shaped may, indeed, be t-shaped. Figure 14.11 shows an I-beam shape that is actually a t-beam. Since all of the concrete below the neutral axis is in tension, only the steel-to-neutral axis distance is relevant in the design.

[38] The worst case occurs when the neutral axis is near the bottom of the flange. However, in that case, rectangular beam and t-team equations produce the same design.

[39] The beam should be designed as a rectangular beam in this instance.

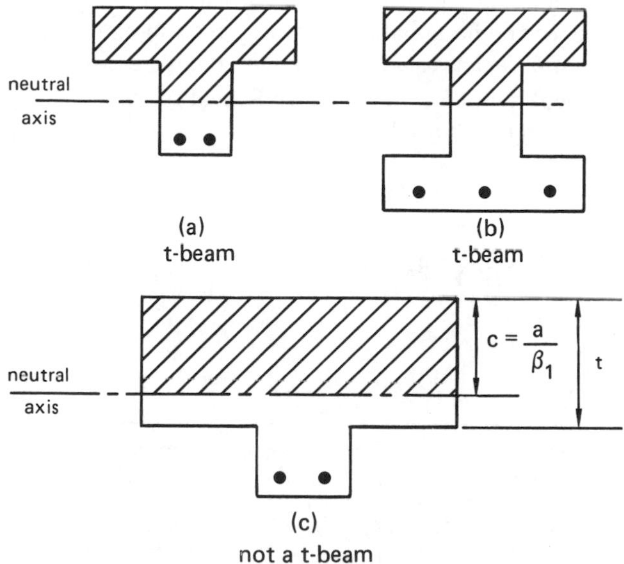


**Figure 14.11** T-Beams and Non-T-Beams

### B. FLANGE WIDTH LIMITATIONS

Although it is assumed that the stress is uniform across the entire flange width, the stress actually decreases at great distances from the web. Therefore, the ACI code limits the width of flanges for the purpose of calculating properties of the shape. [ACI 318 8.10.2] Referring to figure 14.10, the effective width is[40]

$$b_e \leq \min \begin{Bmatrix} \frac{1}{4} l_{\text{beam}} \\ b + 16t \\ b + l_n \end{Bmatrix} \qquad 14.65$$

### C. T-BEAM DESIGN AND ANALYSIS

Design of T-beams is similar, in many respects, to the design of rectangular beams. The moment at ultimate strength (see figure 14.10) is

$$M_u = \phi f_y A_{st} \left(d - \tfrac{1}{2}a\right) \qquad 14.66$$

If $c = \dfrac{a}{\beta_1}$ is less than $t$, the neutral axis falls within the flange, and the beam is rectangular. If $c$ is greater than $t$, the neutral axis falls within the web, and the beam is T-shaped. Ignoring the small compressive strength contributed by the web, the strength of the beam is

$$M_u = \phi f_y A_{st} (d - \tfrac{1}{2}t) \qquad 14.67$$

[40] L-shaped beams, such as would occur at the edges of a beam-slab system, and isolated t-beams are governed by different limitations. ACI 318 8.10.2 states the limitations in equation 14.65 in a different, but equivalent, manner.

In designing a t-beam, it may be desirable to separate the moment-carrying abilities of the overhanging portions of the flange and remaining rectangular beam. If this approach is taken, the flange's compressive strength is calculated, and this strength is used to determine part of the steel required.

$$C_f = 0.85f'_c(b_e - b)t \qquad 14.68$$

$$A_{st,f} = \frac{C_f}{f_y} \qquad 14.69$$

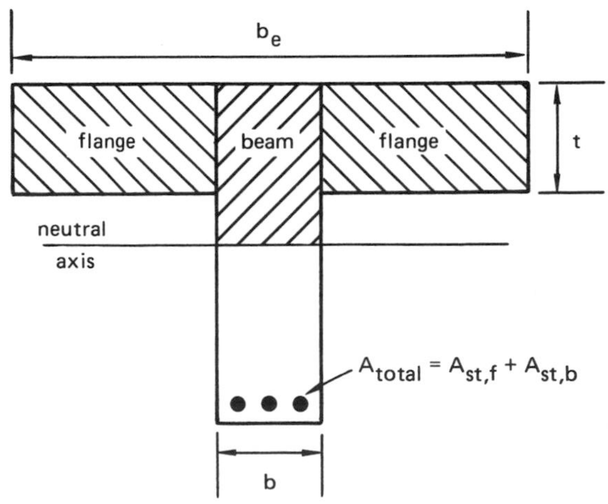


**Figure 14.12** Dual Beam Approach to T-Beam Design

Next, the design moment which the flange is capable of carrying alone is determined.

$$\phi M_{u,f} = \phi[A_{st,f} f_y(d - \tfrac{1}{2}t)] \qquad 14.70$$

The "beam" must carry the remaining design load.

$$M_{u,b} = M_u - \phi M_{u,f} \qquad 14.71$$

The remainder of the design procedure is the same as for any other rectangular beam, including that of determining the location of the neutral axis, height of stress block, and steel requirement, $A_{st,b}$.

The total steel required is

$$A_{\text{total}} = A_{st,f} + A_{st,b} \qquad 14.72$$

Additional steps required are: checking the reinforcement quantity against the maximum permitted, checking for excessive cracking, and checking required web width to meet cover requirements.

## 11 DEEP BEAMS

Deep beams are defined by the ACI code [section 10.7.1] to have depth-to-clear span $(d/l_n)$ ratios greater than 0.4 for continuous spans and 0.8 for simple spans.[41] Because of the expected orientation of cracks in deep beams, shear reinforcement is required horizontally as well as vertically.

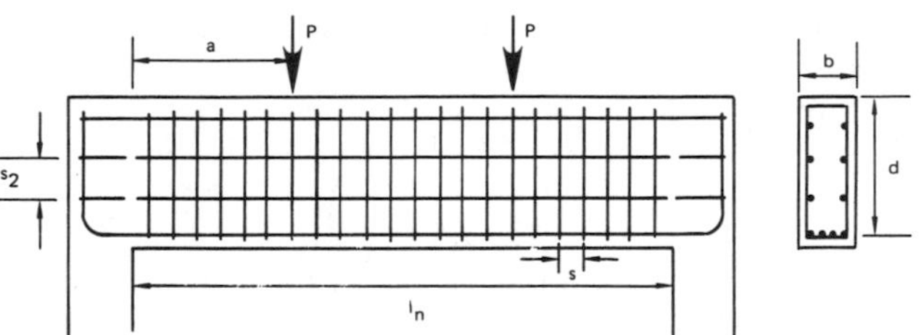


**Figure 14.13** A Deep Beam

The concrete's shear strength is the same as for normal-depth beams.[42]

$$V_c = 2\sqrt{f'_c}bd \qquad 14.73$$

The same shear reinforcement spacing is used throughout the beam. The spacing is determined from the *design shear force*, $V_u$, located at the *critical section*, as defined by the ACI code. For uniform loading, the critical section is defined to be located a distance $l_c$ from the face of the support. [ACI 318 11.8.5]

$$l_c = \min\begin{Bmatrix} 0.15l_n \\ d \end{Bmatrix} \qquad 14.74$$

For concentrated loading, the location of the critical section is given by equation 14.75. In equation 14.75, $a$ is the *shear span*, the distance from the face of the support to the closest concentrated load. [ACI 318 11.8.5]

$$l_c = \min\begin{Bmatrix} 0.50a \\ d \end{Bmatrix} \qquad 14.75$$

If $V_n$ is greater than $\frac{1}{2}V_c$, the ACI code requires a minimum amount of shear reinforcement. (Below $\frac{1}{2}V_c$, no shear reinforcement is needed.) Two types of reinforcement are used: vertical and horizontal. The following

41 Actually, special shear reinforcement is required by the code if $l_n/d < 5$ $(d/l_n > 0.2)$ and the beam is loaded on its compression face. This limit effectively determines when a beam is *deep*. [ACI 318 section 11.8.1]

42 The ACI code contains a more detailed method of calculating $V_c$ for deep beams, as it does for regular depth beams. However, these two detailed methods are not the same. A special *multiplier on the concrete shear strength* of normal depth beams is allowed. Use of the more detailed procedure is optional. However, its use will more than likely reduce the amount of shear reinforcement required.

equations give the minimum reinforcement areas and the maximum spacing. In these equations, $b$ is the width of the beam at the compression face.

Tension steel (i.e., flexural steel) for deep beams is designed taking into consideration a nonlinear strain distribution. [ACI 318 10.7.1]

A deep beam's nominal shear strength is computed the same as for normal depth beams.

$$V_n = V_c + V_{st} = \frac{V_u}{\phi} \quad 14.76$$

However, the maximum shear strength attributable to a deep beam is [ACI 318 11.8.4]

$$V_n \leq 8\sqrt{f'_c}bd \quad (l_n/d \leq 2) \quad 14.77$$

$$V_n \leq \frac{2}{3}\left(10 + \frac{l_n}{d}\right)\sqrt{f'_c}\,bd \quad (5 > l_n/d > 2) \quad 14.78$$

[ACI 318 11.8.9]

$$A_v \geq 0.0015bs \quad 14.79$$

$$s \leq \min\begin{Bmatrix} d/5 \\ 18'' \end{Bmatrix} \quad 14.80$$

[ACI 318 11.8.10]

$$A_{vh} \geq 0.0025bs_2 \quad 14.81$$

$$s_2 \leq \min\begin{Bmatrix} d/3 \\ 18'' \end{Bmatrix} \quad 14.82$$

If $V_n$ is greater than $V_c$ (i.e., $V_u > \phi V_c$), equation 14.83 is used to calculate the strength contribution of the shear reinforcement. Since there are two areas, $A_v$ and $A_{vh}$, different reinforcement arrangements are possible. [ACI 318 11.8.8]

$$V_{st} = \left[\frac{A_v}{s}\left(\frac{1 + \frac{l_n}{d}}{12}\right) + \frac{A_{vh}}{s_2}\left(\frac{11 - \frac{l_n}{d}}{12}\right)\right] f_y d \quad 14.83$$

As in equations 14.79 and 14.81, $A_v$ and $A_{vh}$ in equation 14.83 are the areas of reinforcement within spans of $s$ and $s_2$ respectively.

## 12 DOUBLY-REINFORCED BEAMS

### A. INTRODUCTION

Figure 14.14 illustrates a beam which has steel in the compression region of the beam. Such steel can be used to increase the design strength of the beam or to reduce deflection. However, it is frequently used near the ends of beams where the expected moment is negative. Without compression reinforcement, the upper portion of the beam would be in tension, which is not acceptable.[43]

In addition, top steel is advisable in construction designed to resist earthquakes, due to the stress reversals expected with oscillatory motion.

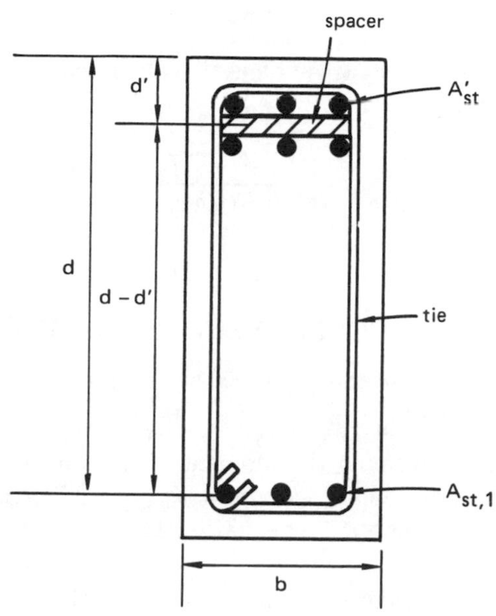


**Figure 14.14** Doubly-Reinforced Beam

In a doubly-reinforced beam, the total steel is made up of two components: the *compression steel* (or *top steel*) and the *tension steel*.

$$A_{st} = A_{st,1} + A'_{st} \quad 14.84$$

The need for top steel is easily determined. The maximum reinforcement ratio for tension steel is [ACI 10.3.3]

$$\rho_{\text{max}} = 0.75\rho_{\text{balanced}} \quad 14.85$$

The moment that can be resisted with the quantity of tension (bottom) steel is given by equations 14.25 and 14.86.

$$M_n = \rho bdf_y\left[d - \frac{0.59\rho df_y}{f'_c}\right] \quad 14.86$$

If $M_u > \phi M_n$, compression steel is needed to carry the difference.

[43] In monolithic slabs and beams, the beam web (not the slab "flange") would be in compression where the moment is negative. Depending on the reinforcement ratio, this might result in compression failure in the web before tensile failure in the slab occurs. In any case, top steel is the remedy.

CONCRETE

B. CAPACITY OF DOUBLY-REINFORCED BEAMS

The ultimate moment which a doubly reinforced beam can carry is

$$\begin{aligned} M_u &= \phi M_n \\ &= \phi[(A_{st} - A'_{st})(f_y)\left(d - \tfrac{1}{2}a\right) \\ &\quad + A'_{st} f_y (d - d')] \end{aligned} \qquad 14.87$$

The height of the stress block is calculated using the tension steel only.

$$a = \frac{(A_{st} - A'_{st}) f_y}{0.85 f'_c b} \qquad 14.88$$

C. DESIGN OF DOUBLY-REINFORCED BEAMS

The following procedure can be used to design a doubly-reinforced beam.

*step 1:* Determine the moment, $M_{nc}$, which a concrete beam with $\rho = \rho_{\text{max}} = 0.75\rho_{\text{balanced}}$ can carry. Use equation 14.31.

*step 2*: Calculate the moment which the compression reinforcement must carry.

$$M'_u = M_u - \phi M_{nc} \qquad 14.89$$

*step 3*: Assume compression reinforcement yields before the concrete.[44] The compression reinforcement ratio is

$$\rho' = \frac{M'_u}{\phi f_y (d - d') bd} \qquad 14.90$$

*step 4*: The total reinforcement ratio is[45]

$$\rho = 0.75\rho_{\text{balanced}} + \rho' \qquad 14.91$$

*step 5*: Calculate the steel areas.

$$A'_{st} = \rho' bd \qquad 14.92$$

$$A_{st,1} = (\rho - \rho')bd = 0.75\rho_{\text{balanced}} bd \qquad 14.93$$

*step 6*: Check the assumption that the compression reinforcement yields. Equation 14.94 must be satisfied.[46]

$$\frac{A_{st} - A'_{st}}{bd} \geq \frac{0.85\beta_1 f'_c d'}{f_y d}\left[\frac{87{,}000}{87{,}000 - f_y}\right] \qquad 14.94$$

*step 7*: Select the reinforcement to meet the $A'_{st}$ and $A_{st,1}$ requirements.

*step 8*: Complete the design by checking for flexural cracking (the tension steel only), specifying depth of cover, checking beam width, and designing shear reinforcement.

## 13 PRESTRESSED AND POST-TENSIONED BEAM DESIGN

A. INTRODUCTION

Having a compressive stress in the tension region of a beam prior to normal loading produces a stiffer beam (i.e., there is less deflection). Alternatively, the decrease in expected deflection can be used to design smaller sections, thereby reducing both cost and dead weight. Most or all of the cracks at service load levels are eliminated, which is particularly valuable in corrosive environments. The finished beam also has greater impact and live loading capacities.

The disadvantages of prestressed and post-tensioned construction are higher cost and increased inspection. Materials, accessories, and labor costs are higher than for traditional construction. The close inspection and stricter quality control also add to the cost.

The basic approach is to use steel tendons which pass through the beam. Two methods are used to apply tension to the steel: pretensioning and post-tensioning.

In *pretensioned construction*, a form is built between two anchors, and steel is tensioned and secured to the two anchors. When the concrete poured into the form cures, it bonds to the steel. When the ends of the steel are cut from the anchors, the load transfers to the concrete.

In *post-tensioned construction*, the forms are built with hollow metal or plastic tubes running the length of the beam. (Alternatively, the steel tendons can be coated with grease or another release agent to prevent bonding.) The concrete is poured and allowed to cure. If

[44] This is known as *condition 1*. In condition 1, both the compression and tension steels yield before the concrete fails (i.e., the concrete strain exceeds 0.003). In *condition 2*, the tension steel yields, but the compression steel does not yield before the concrete fails.

[45] The ACI code specifies that, for members with compression reinforcement, the portion of $\rho_{\text{balanced}}$ contributed by compression reinforcement does not need to be reduced by the 0.75 factor.

[46] If equation 14.94 is not satisfied, there will be a condition 2 failure. The stress in the steel at failure will be $\epsilon'_{st} E_{st}$, where $\epsilon'_{st}$ can be found by elastic assumptions. The completion of condition 2 calculations is not covered in this chapter.

a tendon/tube assembly wasn't used, a steel tendon is placed in the tube. The tendon is tensioned to apply the prestress. Finally, the tubes are either filled with pumped grout, or the ends of the tendon are anchored in some manner.

In both pretensioned and post-tensioned construction, the tendons are draped to reduce their eccentricity near the beam ends. At the beam ends, the dead load moment is insufficient to counteract the tension prestress. Reducing the eccentricity eliminates tension in the concrete.

Several effects tend to reduce the effective prestress, including creep in steel and concrete, concrete shrinkage, elastic compression of the concrete, anchorage seating movement, and friction between the tendon and tube. These losses are significant, averaging 15% to 20%. Since this loss amounts to 25,000 to 35,000 psi, regular grade steel with a yield strength of 40,000 psi (approximately) cannot be used for prestressing. The tendons that are used for prestressing are manufactured from steel with an approximate ultimate strength ($f_{pu}$) of 250,000 psi. Ratios of yield to ultimate strengths ($f_{py}/f_{pu}$) are approximately 0.90 for low-relaxation wire and strands, 0.85 for stress-relieved wire, strands and plain bars, and 0.80 for deformed bars.

It is common to use a *high early-strength concrete*, such as type III, with a compressive strength, $f'_c$, of 4000 to 6000 psi. This concrete sets up quickly and suffers smaller elastic compression losses. Due to the higher quality concrete, the *modular ratio*, $n = E_s/E_c$, is somewhat lower than in conventional construction—on the order of 6 or 7.

## B. ACI CODE PROVISIONS FOR STRESS

The maximum tensile stress in the prestressing tendons prior to transfer of stress to the concrete is [ACI 318 18.5.1]

$$f_t = \min\begin{Bmatrix} 0.94 f_{py} \\ 0.80 f_{pu} \end{Bmatrix} \qquad 14.95$$

Immediately after the prestress is transferred to the concrete, the maximum tensile stress in the tendons is

$$f_t = \min\begin{Bmatrix} 0.82 f_{py} \\ 0.74 f_{pu} \end{Bmatrix} \qquad 14.96$$

In post-tensioned construction, the maximum tendon tensile stress immediately after tendon anchorage is

$$f_t = 0.70 f_{pu} \qquad 14.97$$

The maximum extreme fiber compressive stress in the concrete immediately after prestress transfer (but before any time-dependent losses) is given by equation 14.98. $f'_{ci}$ is the compressive strength of the concrete at the time of initial prestress. [ACI 318 18.4.1]

$$f_c = 0.60 f'_{ci} \qquad 14.98$$

The maximum extreme fiber tensile stress in the concrete immediately after prestress transfer is given by equation 14.99. For tension at the ends of simply supported members, the limit is twice that of equation 14.99.

$$f_t = 3\sqrt{f'_{ci}} \qquad 14.99$$

If the tensile stresses exceed the values given by equations 14.97 and 14.99, the ACI code contains provisions for providing additional reinforcement in the tensile zone.

At service loads and after all prestress losses, the maximum extreme fiber compression and tension stresses in the concrete are[47]

$$f_c = 0.45 f'_c \qquad 14.100$$

$$f_t = 6\sqrt{f'_c} \qquad 14.101$$

## C. ANALYSIS OF PRESTRESSED CONSTRUCTION

The procedure which follows is based on a standard analysis of a simply-supported, single-span beam. There are many ACI code-related provisions which could be applied if desired.

*step 1*: Calculate or estimate the dead weight per foot of the beam, $w_D$.

*step 2*: Determine the maximum moment on the beam due to the dead load. If it is assumed that the dead load is a uniform load (as it would be for a beam of uniform construction), the moment is

$$M_D = \frac{w_D L^2}{8} \quad \text{(simple supports)} \qquad 14.102$$

*step 3*: Assume that all of the beam is in compression. Use the transformed area method to locate the neutral axis.[48]

*step 4*: Determine the centroidal moment of inertia about the neutral axis. Since the entire beam is in compression, all of it contributes to $I$.

*step 5*: Calculate the eccentricity, $e$, of the strands.

47 The ACI code provides for doubling the limit given by equation 14.101 if additional deflection analysis is performed [Sec. 18.4.2(c)].

48 The steel area is transformed only for pretensioned construction and post-tensioned construction with tendons grouted in place. If the tendons are merely anchored at their ends, only the plain concrete section is used to calculate the neutral axis and moment of inertia.

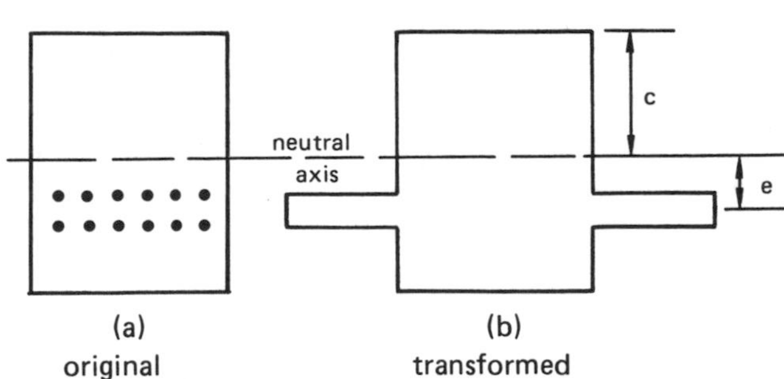


**Figure 14.15** Transformed Prestressed Beam

*step 6*: Calculate the stress (distribution) due to prestressing. This will consist of the basic compressive stress superimposed on the bending stress. $P$ is the ***net prestress*** acting on the beam, after losses.[49] In equation 14.103, compressive stresses are negative, and tensile stresses are positive.

$$f_{pc} = \frac{-P}{A_c} \pm \frac{Pec}{I} \qquad 14.103$$

*step 7*: Calculate the stress distribution due to the dead load.

$$f_b = \pm\frac{M_D c}{I} \qquad 14.104$$

*step 8*: The stress distribution immediately after transfer is

$$f = f_{pc} + f_b \qquad 14.105$$

*step 9*: If required, check to see that the tensile stress in the tendons is less than the limit set by equation 14.96.[50]

*step 10*: Check to see that the compressive stress in the concrete after transfer of prestress is less than the limit set by equation 14.98.

*step 11*: Calculate the stress distribution due to the live loading.

$$f_L = \frac{M_L c}{I} \qquad 14.106$$

*step 12*: The service load stress distribution is

$$f_{\text{total}} = f_{pc} + f_b + f_L \qquad 14.107$$

*step 13*: Check the long-term stress against the limits set by equation 14.100 and 14.101.

[49] In the past, long-term loss was frequently assumed to be 35,000 psi for prestressed construction and 25,000 psi for post-tensioned construction. ACI 318 now requires a more analytical approach, and lump-sum values cannot be used.

[50] Ordinarily, there is no need to calculate the steel stress since the stresses due to live loads never even make up the original loss of prestress. The actual tendon stress is calculated by approximate methods not covered in this book.

**Table 14.9**
ASTM Standard Prestressing Tendons

(strand grade is $f_{pu}$ in ksi)

| type | nominal diameter, in. | nominal area, sq in. | nominal weight, lb per ft |
|---|---|---|---|
| grade 250 | 1/4 (0.250) | 0.036 | 0.12 |
| seven-wire | 5/16 (0.313) | 0.058 | 0.20 |
| strand | 3/8 (0.375) | 0.080 | 0.27 |
| | 7/16 (0.438) | 0.108 | 0.37 |
| | 1/2 (0.500) | 0.144 | 0.49 |
| | (0.600) | 0.216 | 0.74 |
| grade 270 | 3/8 (0.375) | 0.085 | 0.29 |
| seven-wire | 7/16 (0.438) | 0.115 | 0.40 |
| strand | 1/2 (0.500) | 0.153 | 0.53 |
| | (0.600) | 0.215 | 0.74 |
| prestressing | 0.192 | 0.029 | 0.098 |
| wire | 0.196 | 0.030 | 0.10 |
| | 0.250 | 0.049 | 0.17 |
| | 0.276 | 0.060 | 0.20 |
| smooth | 3/4 | 0.44 | 1.50 |
| prestressing | 7/8 | 0.60 | 2.04 |
| bars | 1 | 0.78 | 2.67 |
| | 1-1/8 | 0.99 | 3.38 |
| | 1-1/4 | 1.23 | 4.17 |
| | 1-3/8 | 1.48 | 5.05 |
| deformed | 5/8 | 0.28 | 0.98 |
| prestressing | 3/4 | 0.42 | 1.49 |
| bars | 1 | 0.85 | 3.01 |
| | 1-1/4 | 1.25 | 4.39 |
| | 1-3/8 | 1.56 | 5.56 |

## 14 COLUMNS

### A. INTRODUCTION

Tied and spiral columns are the main vertical load carrying members in buildings.[51] Figure 14.16 illustrates the reinforcement in *tied columns* and *spiral columns.*

The procedure used to design or evaluate a concrete column depends on whether the column is long or short (as defined by the ACI code) and the amount of eccentricity associated with the load.

[51] Pedestals and composite columns are not covered in this chapter.

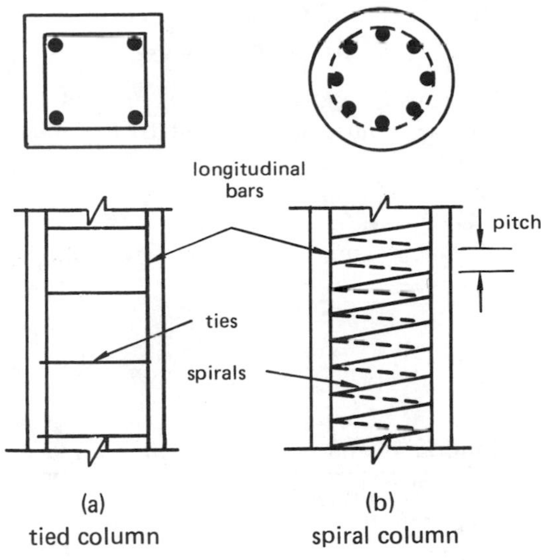


**Figure 14.16** Tied and Spiral Columns

Most column design makes use of ultimate strength concepts. The established procedures are based on experiments and empirical evidence. Thus, stress relationships are only valid when the column is loaded to near its ultimate strength. Since columns should not be loaded past their elastic limit, no attempt is made to calculate the stresses. The amount of steel used for ties or spirals is based on empirical data.

The *ultimate axial load*, $P_u$, which columns are designed to support is the factored load calculated from equation 14.108.[52]

$$P_u = 1.4 \times \text{dead load} + 1.7 \times \text{live load} \qquad 14.108$$

In addition to $P_u$, other load variables are used in column calculations. $P_o$ is the *nominal axial load strength* of a column with zero eccentricity. $P_n$ is the nominal axial load strength with an eccentricity of $n$. $\phi P_n$ is the *design axial load strength* which must exceed $P_u$. That is, equation 14.109 is the design criterion for columns.

$$\phi P_n \geq P_u \qquad 14.109$$

In simple column design, the major assumption made about columns at failure is that the steel is stressed to $f_y$ and the concrete is stressed to $0.85f'_c$. Then, the maximum load on the column would be

$$P_o = 0.85 f'_c A_{\text{concrete}} + f_y A_{st} \qquad 14.110$$

[52] The effects of wind and earthquake loading are not covered in this chapter.

In equation 14.110, $A_{\text{concrete}}$ includes all core and cover concrete in the column.

$$A_{\text{concrete}} = A_g - A_{st} \qquad 14.111$$

The *reinforcement ratio* is

$$\rho_g = \frac{A_{st}}{A_g} \qquad 14.112$$

Using the definition of concrete area and reinforcement ratio, equation 14.110 can be rewritten.

$$P_o = A_g \left[0.85 f'_c (1 - \rho_g) + f_y \rho_g\right] \qquad 14.113$$

Equation 14.113 is theoretically correct. However, several additional factors are added to account for a minimum amount of eccentricity and the load transfer to core and steel when the outside cover spalls.

B. TIED COLUMNS

The following construction details are specified by the ACI code for tied columns.

- Longitudinal bars must have a clear distance of at least $1\frac{1}{2}$ times the bar diameter, but not less than $1\frac{1}{2}''$. [ACI 318 7.6.3]
- The minimum tie wire size is #3 if the longitudinal bars are #10 or smaller. The minimum tie wire size is #4 for larger longitudinal bars.[53] [ACI 318 7.10.5.1]
- The concrete covering should be at least $1\frac{1}{2}''$ thick over the outermost surface of the tie steel. [ACI 318 7.7.1]
- At least 4 longitudinal bars are needed for columns with square or circular ties. [ACI 318 section 10.9.2]
- The ratio of longitudinal steel area to the gross column area must be between 0.01 and 0.08.[54] [ACI 318 10.9.1]

$$0.01 \leq \rho_g \leq 0.08 \qquad 14.114$$

- Center-to-center spacing of the ties must not exceed the smallest of 16 longitudinal bar diameters, 48 tie bar diameters, or the least gross (outside) column dimension. [ACI 318 7.10.5.2]
- Every corner and alternating longitudinal bar should be supported by a tie corner.[55] [ACI 318 7.10.5.3]

[53] The maximum practical tie size is #5 bar.

[54] The lower limit keeps the column from being designed as all-concrete. The upper limit keeps the column from being too slender. 0.02 or 0.025 is typically chosen as a starting value of $\rho_g$.

[55] Tie corners are not relevant to tied columns with longitudinal bars spaced in a circular pattern.

CONCRETE

- In tied columns, no longitudinal bar can be more than 6″ away from a tie corner supported longitudinal bar. [ACI 318 7.10.5.3] Figure 14.7 illustrates extra ties.
- Tie corners should not make a bend of more than 135°. [ACI 318 7.10.5.3]

Figure 14.17 illustrates several typical tied column reinforcement and tie corners.

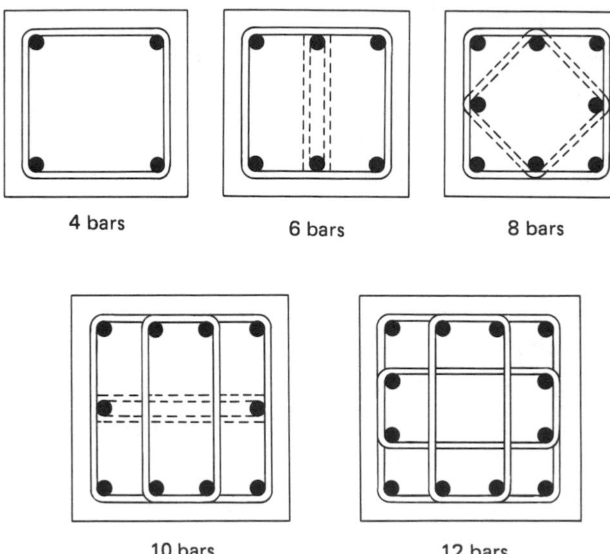


**Figure 14.17** Ties and Tied Corners

The design axial load strength for tied columns is given by equation 14.115. $\phi = 0.70$ for tied columns.[56]

$$P_u = \phi P_n = 0.80\phi P_o \qquad 14.115$$

C. SPIRAL COLUMNS

The following construction details are specified for spiral columns.

- Longitudinal bars must have a clear distance of at least $1\frac{1}{2}$ times the bar diameter, but not less than $1\frac{1}{2}''$. [ACI 318 7.6.3]
- The minimum spiral wire diameter is 3/8″.[57] [ACI 318 7.10.4.2]
- The clear distance between spirals should not exceed 3″, and it should not be less than 1″. [ACI 318 7.10.4.3]
- The concrete covering should be at least $1\frac{1}{2}''$ thick over the outermost surface of the spiral steel. [ACI 318 7.7.1]
- At least 6 longitudinal bars are to be used for spiral columns. [ACI 318 section 10.9.2]
- The ratio of longitudinal steel area to the gross column area must be between 0.01 and 0.08. [ACI 318 10.9.1]

The ACI code does not specify spiral wire size. However, table 14.10 can be used for general guidelines.

**Table 14.10**
Typical Spiral Wire Sizes

| column diameter | spiral wire size |
|---|---|
| up to 15″, using #10 bars or smaller | 3/8″ |
| up to 15″, using #11 bars or larger | 1/2″ |
| 16″–22″ | 1/2″ |
| 23″ and up | 5/8″ |

The design axial load strength for spiral columns is given by equation 14.116. $\phi = 0.75$ for spiral columns.

$$P_u = \phi P_n = 0.85\phi P_o \qquad 14.116$$

The ACI code also requires the ratio of spiral reinforcement volume to column core volume to be greater than the value of equation 14.117. [ACI 318 10.9.3]

$$\rho_s \geq 0.45\left(\frac{A_g}{A_c} - 1\right)\frac{f'_c}{f_y} \qquad 14.117$$

$A_c$ and $D_c$ are measured to the outside diameter of the spiral wire. $f_y$ in equation 14.117 is the *spiral steel* yield point, and may not exceed 60,000 psi.

If the spiral wire diameter is known, the *spiral pitch* can be found. $A_s$ in equation 14.118 is the cross sectional area of the spiral wire.

$$s \approx \frac{4A_s}{\rho_s D_c} \qquad 14.118$$

The *clear distance between spirals* will be the difference between the spiral pitch and the spiral wire diameter.

$$\text{clear distance} = s - D_s \qquad 14.119$$

D. COLUMN ECCENTRICITY

If the loads are designed to be applied axially to a column, and if there are no external moments acting on the column, the column is said to be a *low-eccentricity column.* Actually, a small amount of eccentricity is per-

[56] Some people find the form of equation 14.115 confusing, since $\phi$ appears on both sides of the equation. Since the design criterion is $P_u \leq \phi P_n$, it is only necessary to compare the ultimate factored load ($P_u$, as defined in equation 14.108) against the right-hand side of equation 14.115.

[57] The maximum practical spiral wire size is 5/8″.

CONCRETE

missible. By specifying the strength reduction factors of 0.80 and 0.85 in the equations for design strength (equations 14.115 and 14.116), the code is compensating for up to a 5% eccentricity for spiral columns and a 10% eccentricity for tied columns. *Percent eccentricity* is defined as

$$\epsilon = \frac{e}{\text{column width}} \times 100 \qquad 14.120$$

Eccentricity can be introduced by external moments, as well as by off-center loading. The *eccentricity* due to a moment on the column is

$$e = \frac{M_u}{P_u} \qquad 14.121$$

Equations 14.115 and 14.116 should be used to design and evaluate short columns with low eccentricities.

*Example 14.6*

Calculate the design strength of the short spiral column shown. Assume low eccentricity loading, $f_y = 40{,}000$ psi, and $f'_c = 3500$ psi.

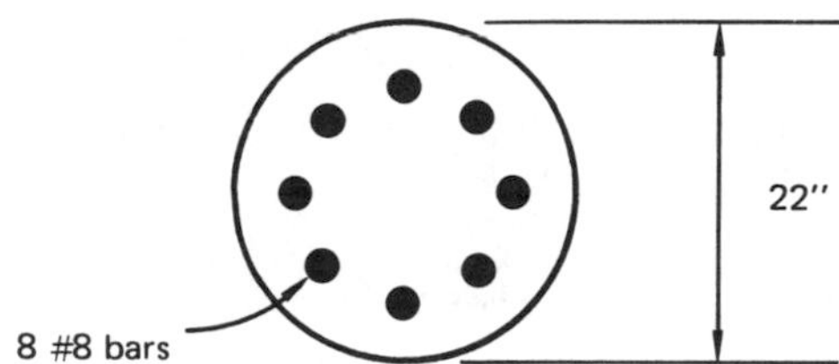


The gross area of the column is

$$A_g = \frac{1}{4}\pi(22)^2 = 380.1\,\text{in}^2$$

Each #8 longitudinal bar contributes 0.79 in$^2$ to the steel area, so the total steel area is

$$A_{st} = 8 \times 0.79 = 6.32\,\text{in}^2$$

The reinforcement ratio is

$$\rho_g = \frac{6.32}{380.1} = 0.0166$$

The nominal axial load strength is

$$\begin{aligned} P_o &= 380.1[(0.85)(1 - 0.0166)(3500) \\ &\quad + (40{,}000)(0.0166)] \\ &= 1.364\,\text{EE6 lbf} \end{aligned}$$

From equation 14.116, the design strength is

$$(0.85)(0.75)(1.364\,\text{EE6}) = 8.7\,\text{EE5 lbf}$$

*Example 14.7*

Design a spiral column to carry a factored load of 375,000 pounds. Use 3000 psi concrete and 40,000 psi steel.

Assume $\rho_g = 0.02$, which is in the allowable range. Larger values will decrease the column gross area. However, slenderness effects are being neglected, and it is desirable to keep the column as heavy as possible.

From equations 14.113 and 14.116,

$$\begin{aligned} 375{,}000 &= A_g(0.85)(0.75)[(0.85)(1 - 0.02)(3000) \\ &\quad + (40{,}000)(0.02)] \\ A_g &= 178.3\,\text{in}^2 \\ D_g &= \sqrt{\frac{4A_g}{\pi}} = 15.07'' \quad (\text{say } 15.25'') \end{aligned}$$

With a $1\frac{1}{2}''$ cover, the core diameter will be

$$D_c = 15.25 - 3.0 = 12.25''$$

The required steel area is

$$A_{st} = (0.02)(178.3) = 3.57\,\text{in}^2$$

The ACI code requires at least 6 bars to be used. Some possibilities which meet the required steel areas are 6 #7 bars, 9 #6 bars, or 12 #5 bars. Choose #6 bars after checking the clear spacing (not shown here).

From equation 14.117, substituting $D^2$ for $A$, the ratio of spiral reinforcement must be greater than

$$0.45\left[\left(\frac{15.25}{12.25}\right)^2 - 1\right]\frac{3000}{40{,}000} = 0.0186$$

Assume 3/8" spiral wire ($A_s = 0.11\,\text{in}^2$) with a $1\frac{1}{2}''$ pitch. From equation 14.118, the actual spiral reinforcement ratio is

$$\rho_s = \frac{(4)(0.11)}{(1.5)(12.25)} = 0.024''$$

Since 0.024 > 0.0186, the spiral spacing is adequate.

The clear space between spirals is

$$1.5 - 0.375 = 1.125''$$

This is between 1" and 3".

## E. LARGE ECCENTRICITIES

As a moment carried by a column increases, the axial load that the column can support decreases. A graph of allowable axial load versus the applied moment is known as an *interaction diagram*. Figure 14.18 illustrates a general interaction diagram.

CONCRETE

Construction of an interaction diagram for a given column design is tedious, and requires calculating loads and moments for multiple values of eccentricity.[58] The diagram boundary represents the envelope of acceptable loadings. Inside the boundary, the design is acceptable but overdesigned. Outside the boundary, the design is unacceptable.

Radial lines outward from the origin represent different values for the percent eccentricity. Point $A$ defines the maximum eccentricity that the column can support and still be considered a low-eccentricity column. Point $C$ is the bending strength of the member.

Point $B$ defines the eccentricity separating compression and tensile failures. Above the line of *balanced eccentricity*, the column will fail by concrete crushing. Below the line, the column will fail by tension steel yielding.

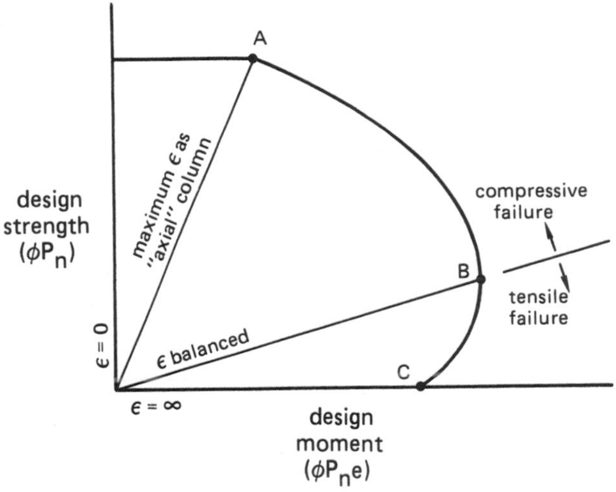


**Figure 14.18** Interaction Diagram for Large Eccentricity Column

To further complicate the design and analysis of large eccentricity columns, the strength reduction factor varies with the design axial load strength.

To assist in the design and analysis of large eccentricity columns, various sources (including ACI) have produced sets of interaction diagrams for common column configurations. These charts are primarily analysis aids, and design still requires a trial and error approach, since the percent eccentricity isn't known until the column is sized.

[58] Construction of interaction diagrams is not covered in this book.

## F. SLENDERNESS EFFECTS

The effects of slenderness can be disregarded for short columns. The ACI code provides for determining when a column is short or slender, and for increasing the required strength of slender columns.[59]

The *slenderness ratio* is defined as

$$SR = \frac{kL_u}{r} \qquad 14.122$$

In equation 14.122, $k = 1$ for columns braced against sidesway, unless analysis shows that a lower value can be used. $k > 1$ for columns not braced against sidesway. $L_u$ is the unbraced length, typically taken as the clear distance between floor slabs, beams, or other members capable of providing lateral support.

The *radius of gyration*, when a rigorous analysis is not performed, is defined by the ACI code. The column dimension in equation 14.124 is the overall dimension in the direction stability is being evaluated. [ACI 318 10.11.3]

$$r = 0.25d_g \quad \text{(round)} \qquad 14.123$$

$$r = 0.3 \times \text{column dimension} \quad \text{(rectangular)} \qquad 14.124$$

The effects of slenderness can be disregarded in columns not braced against sidesway if the slenderness ratio is less than 22. The determination is somewhat more difficult for columns braced against sidesway, being dependent on the bending moments at both ends of the column. In equation 14.125, $M_{1b}$ is the value of the smaller factored moment at the end of the compression member.[60] It is positive if the column is bent in single curvature, and negative if the column is bent in double curvature. $M_{2b}$ is the value of the larger factored end moment, and is always positive.

$$SR_{\text{max}} = 34 - 12\left(\frac{M_{1b}}{M_{2b}}\right) \qquad 14.125$$

If it is determined that a column is slender, as defined by the ACI code, then the design must consider the factored axial load, $P_u$, and a magnified factored moment, $M_c = \delta M$. The ACI code describes the procedure for calculating the *magnification factor*, $\delta$.

[59] Columns between floors of a building are almost always short columns.

[60] If the moments at both ends of the column are zero or the same, $M_{1b}/M_{2b} = 1$, and the maximum slenderness ratio is 22.

## 15 FOOTING DESIGN

### A. FAILURE MECHANISMS

There are two primary failure modes for footings: shear and flexure. For square footings, the total shear is used to find the unit stress on the critical section, and shear is known as *two-way shear* (*punching* shear). The *critical section* for *two-way shear* is assumed to be located $\frac{1}{2}d$ from the column face, where $d$ is the depth of the reinforcement.

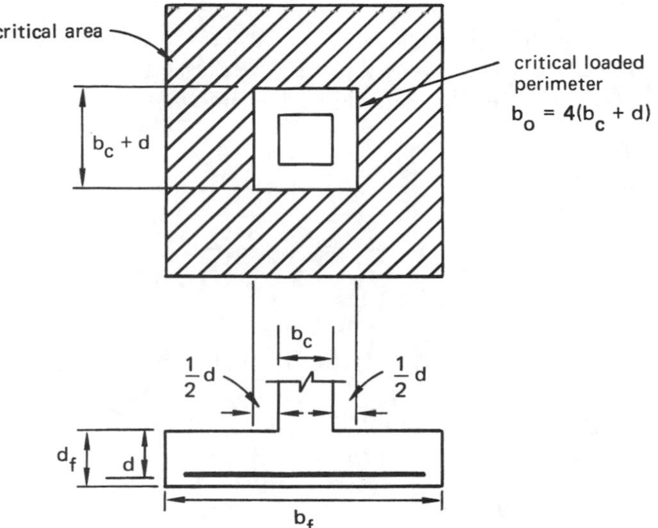


**Figure 14.19** Critical Area for Two-Way Shear

The critical section for *one-way shear* (beam-action shear), which is applicable to rectangular footings, is assumed to be located at distance $d$ from the column face.

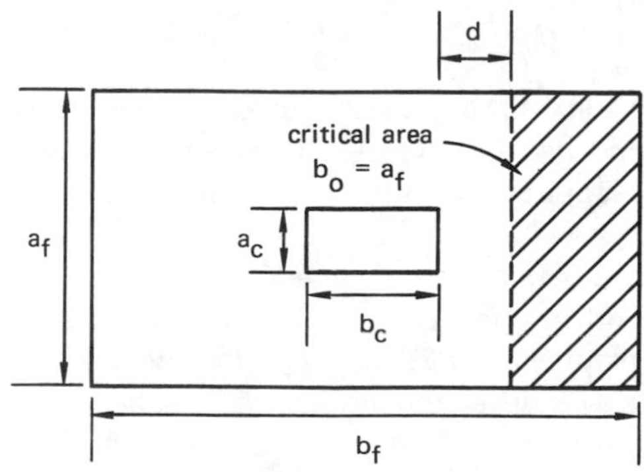


**Figure 14.20** Critical Area for One-Way Shear

Footings can also fail in flexure. The critical section for bending is at the face of the column.[61]

[61] Footings supporting masonry walls and footings loaded through a steel base plate have different critical sections.

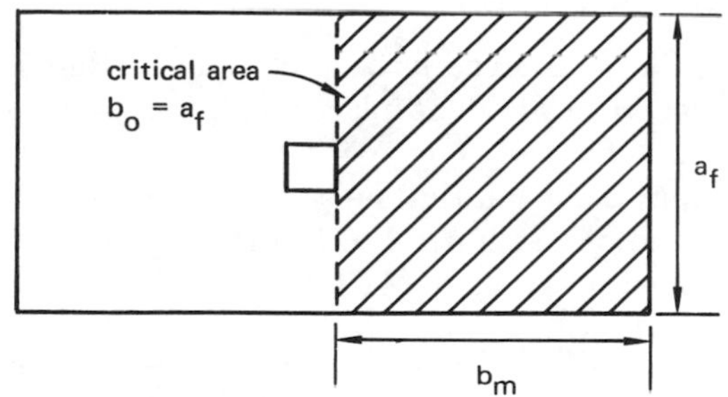


**Figure 14.21** Critical Area for Flexure

### B. FACTORED LOAD

The factored load used to design footings is

$$P_u = 1.4P_D + 1.7P_L \qquad 14.126$$

This factored load is used to calculate the net factored soil pressure beneath the footing.[70]

$$p_{\text{net}} = p_u = \frac{P_u}{A_f} = \frac{P_u}{a_f b_f} \qquad 14.127$$

For the purpose of checking shear, the ultimate shear is

$$\begin{aligned} V_u &= p_u \times \text{critical area} && 14.128 \\ &= p_u \times \left[a_f^2 - (b_c + d)^2\right] \quad \text{(square footing: } a_f = b_f\text{)} \\ &= p_u \times \left[a_f\left(\frac{b_f - b_c}{2} - d\right)\right] \quad \text{(rectangular footing)} \end{aligned}$$

### C. NOMINAL SHEAR STRENGTH

The nominal shear strength in concrete without shear reinforcement for *single action* (i.e., one-way shear) is[62]

$$V_c = 2\sqrt{f'_c}b_o d = 2\sqrt{f'_c}a_f d \qquad 14.129$$

For *double action* (i.e., two-way shear), the nominal shear strength is

$$V_c = \left(2 + \frac{4}{\beta_c}\right)\sqrt{f'_c}b_o d \quad \left[\leq 4\sqrt{f'_c}b_o d\right] \qquad 14.130$$

$$\beta_c = \frac{b_c}{a_c} \quad [\geq 2] \qquad 14.131$$

In equations 14.129 and 14.130, $b_o$ is the length of the critical section. $\beta_c$ cannot be less than 2.

[62] The more detailed method of calculating $V_c$ may be used with footings also.

CONCRETE

## D. REINFORCEMENT IN FOOTINGS

Footings are reinforced in both directions. Bars in the long direction should be placed uniformly across the width $a_f$. The ACI code specifies the fraction of steel in the short direction that must be concentrated in the center band of width $a_f$. Steel in the long direction is not included in $A_{st,\text{total}}$. [ACI 318 15.4.4.2]

$$\frac{A_{st,\text{band}}}{A_{st,\text{total}}} = \frac{2}{\beta_f + 1} = \frac{2}{\frac{b_f}{a_f} + 1} \qquad 14.132$$

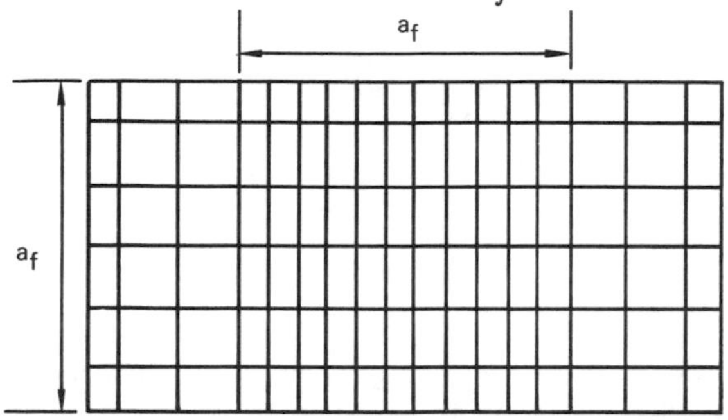

**Figure 14.22** Reinforcement in Center Band

Vertical dowels (i.e., *dowel bars*) can be used to provide horizontal shear resistance and to transfer the column load to the footing by bond instead of bearing. Stirrups are not usually used, since the allowable shear stress is kept low. Hooks are not usually used at the ends of bars in footings due to the danger of top cover failure.

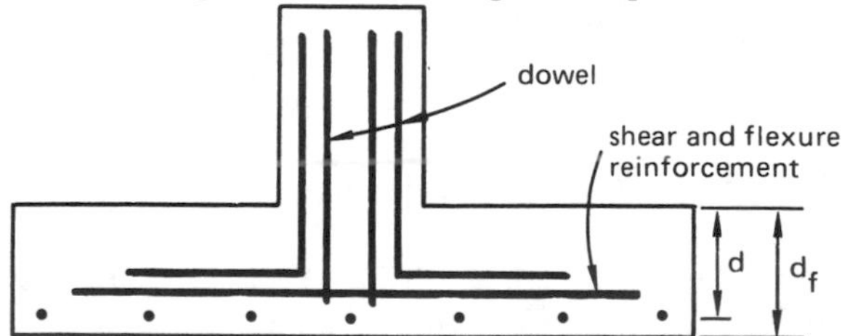


**Figure 14.23** Types of Footing Reinforcement

## E. ACI CODE PROVISIONS FOR FOOTINGS

The following provisions are applicable to footings designed in accordance with the ACI code.

- The minimum thickness above bottom reinforcement for reinforced concrete on soil is 6″. For footings on piles, the minimum depth is 12″. [ACI 318 15.7]
- The minimum cover below bars is 3″. [ACI 318 7.7]
- It is standard practice that bars in two layers can be in contact, and no clear distance between the layers is required.[63]

## F. BEARING PRESSURE

The portion of column load that is not transferred by bond (i.e., through dowels or extensions of the column steel) must be transferred by bearing. The nominal ultimate bearing stress, $f_b$, that the column concrete can withstand is[64]

$$f_b = 0.85 f'_c \qquad 14.133$$

The nominal compressive capacity based on the column area is

$$P_n = f_b A_c = 0.85 f'_c a_c b_c = \frac{P_u}{\phi} \qquad 14.134$$

Since the footing area is much larger than the gross column area, the redistribution of stresses will permit an increase in the allowable bearing stress. [ACI 10.15.1]

$$f_b = \alpha_b 0.85 f'_c \qquad 14.135$$

$$\alpha_b = \min \begin{cases} \sqrt{\frac{A_f}{A_c}} = \sqrt{\frac{a_f b_f}{a_c b_c}} \\ 2 \end{cases} \qquad 14.136$$

In equation 14.136, the $A_f$ term is not strictly correct. The actual area to be used is the portion of the footing that is geometrically similar to and concentric with the loaded area. For square footings loaded by concentric square columns, this is $A_f$.

## G. DOWEL BARS

The ACI code requires a minimum amount of dowel bar area, even when the bearing pressures are not excessive. The following provisions are applicable. [ACI 318 15.8.2]

- The minimum dowel bar area is

$$A_{st} = 0.005 A_c = 0.005 a_c b_c \qquad 14.137$$

- From standard practice, at least four dowel bars should be used. Generally, the number of dowels and longitudinal column bars are equal.
- Dowel bars must be embedded in the footing (and extend into the column) a distance exceeding the development length. If the footing depth is less than the development length, smaller bars must be used. If a bent dowel bar (see figure 14.23) is used, the development length must be achieved over the vertical portion, since bends and hooks are not considered effective in compression.

[63] With two layers, the average depth between the two layers is typically used for both layer orientations. However, for footings less than approximately 15 inches deep, the more conservative value of $d$ is appropriate.

[64] This analysis does not assume that the column and footing are constructed from the same concrete. It is possible to construct columns with higher-strength concrete than is used in the footing. In that case, the bearing stress on the footing may control. Otherwise, only the column check is needed.

If dowel bars are required to transfer load to the footing, the area of dowel steel is easily calculated.[65]

$$\begin{aligned}\text{excess } P_u &= P_{u,\text{column}} - P_{u,\text{bearing}} \\ &= P_{u,\text{column}} - 0.85\,\alpha_b \phi f_c' A_c \qquad 14.138\end{aligned}$$

$$A_{st} = \frac{\text{excess } P_u}{\phi f_y} \qquad 14.139$$

In equation 14.139, $\phi = 0.70$ for bearing on concrete.

## H. FOOTING ANALYSIS[66]

Shear

*step 1*: Determine the effective depth, $d$, of the reinforcement. (This is measured from the top footing surface to the centerline of the tension steel.)

*step 2*: Use equation 14.127 to calculate the net factored earth pressure.

*step 3*: Use equation 14.128 to calculate the ultimate shear. Square footings should always be checked for both one-way shear and two-way shear. Square footings are usually controlled by two-way shear. Rectangular footings are usually controlled by one-way shear, but almost-square footings may be exceptions. For assurance, always check both.

*step 4*: Use equations 14.129 and 14.130 to calculate the nominal strength of the concrete without reinforcement.

*step 5*: Compare the ultimate shear with the nominal shear. If $V_u > \phi V_c$, shear reinforcement is required.[67] Estimate or specify the footing thickness as $d + d_b/2 +$ cover (one steel layer).

Flexure

*step 6*: The bending moment based on the loaded area shown in figure 14.21 is

$$M_u = \frac{p_{\text{net}} a_f b_m^2}{2} \qquad 14.140$$

*step 7*: Calculate the required *coefficient of resistance*.[68]

$$R_u = \frac{M_u}{\phi a_f d^2} \qquad 14.141$$

*step 8*: Calculate the required reinforcement ratio.[69]

$$\rho = \frac{0.85 f_c'}{f_y}\left[1 - \sqrt{1 - \frac{2R_u}{0.85 f_c'}}\right] \qquad 14.142$$

*step 9*: Calculate the required steel area.

$$A_{st} = \rho d a_f \qquad 14.143$$

*step 10*: Choose reinforcement to satisfy the required area and spacing requirements. Since concrete cast next to earth must have a 3″ cover on steel, all steel must fit within the space of $a_f - 6''$.

*step 11*: Check development length for reinforcement. The distance from the column face to the end of the reinforcement must exceed the development length in *tension*.

## I. FOOTING DESIGN

Footing design is similar to footing analysis, although the order of the steps is different. The major difference is in the need to determine the footing area and depth of reinforcement.

A trial footing area is determined as covered in Chapter 10 from the loading and allowable soil pressure.[70] The loading includes the weight of the footing, its column, and any overburden. The density of concrete is usually taken as 150 pounds per cubic foot.

$$p_a > \frac{P_D + P_L}{A_f} + t\gamma_{\text{concrete}} \qquad 14.144$$

After the footing area is established, the footing thickness, $d$, is determined on the basis of the nominal concrete shear strength. In most cases, a footing is designed such that no steel is required to obtain the necessary shear strength. This step is a trial-and-error process which starts by assuming a value of $d$ and ensuring that the shear strength is sufficient (i.e., that $V_u \le \phi V_c$).

## J. ECCENTRICALLY-LOADED FOOTINGS

A footing which is loaded eccentrically will carry a moment in addition to an axial load. The key to analyzing such a footing is determining the shape of the soil pressure distribution. Once this is done, the shear on the critical area, as well as the bending moment on the footing, can be determined. Analysis is the same as for a concentrically-loaded footing.

[65] The minimum area may be sufficient to carry the excess.

[66] This assumes that the footing dimensions, including depth $d$, are known in advance. For continuous or wall footings, use a 12″ portion in the analysis.

[67] In a design environment, it may be easier to increase $d$ than to add shear reinforcement. $\phi = 0.85$ in shear.

[68] $\phi = 0.90$ for bending.

[69] The minimum reinforcement ratio of $200/f_y$ may not apply to footings. However, the temperature and shrinkage requirement is applicable for the short direction of rectangular footings. This requirement may govern.

[70] The ACI code specifies that the load to be carried is not to be factored when sizing the footing. [ACI 318 15.2.2]

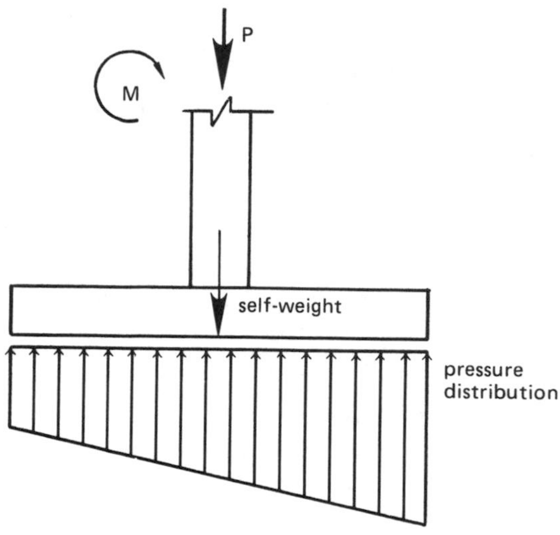


**Figure 14.24** Footing with Moment

*Example 14.8*

A 111″ × 111″ square footing is to support 82,560 pounds dead load and 75,400 pounds live load. The bearing block (column) is 12″ × 12″, concentric with the center of the footing. $f'_c = 4000$ psi and $f_y = 60{,}000$ psi. Design the footing.

(The footing analysis procedure is modified for design.)

*step 1*: ($d$ is determined in step 3.)

*step 2*: The factored load is

$$P_u = (1.4)(82{,}560) + (1.7)(75{,}400) = 243{,}770 \text{ lbf}$$

The net soil pressure is

$$p_{\text{net}} = \frac{243{,}770}{(111)^2} = 19.78 \text{ psi}$$

*step 3*: For two-way shear, the critical shear line is located $\frac{1}{2}d$ from the column face. Since $d$ is unknown, estimate $d = 12''$. Then, the length of the critical perimeter will be (refer to the diagram)

$$b_o = (4)(24) = 96''$$

The critical area contributing to shear is

$$A_v = (111)^2 - (24)^2 = 11{,}745$$

The ultimate shear is

$$V_u = (19.78)(11{,}745) = 232{,}320 \text{ lbf}$$

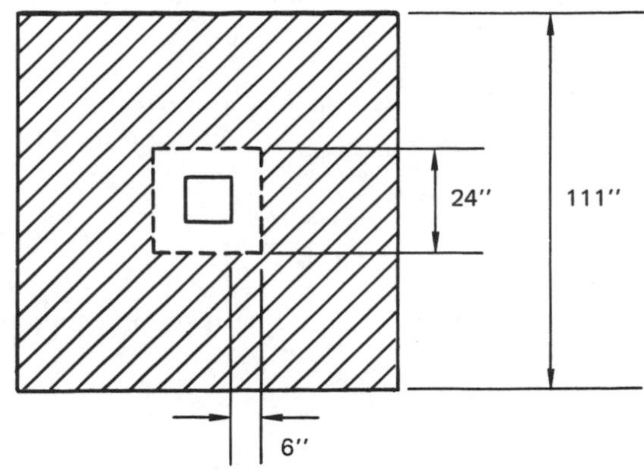


*step 4*: The nominal concrete strength in shear is

$$\beta_c = 12/12 = 1 \text{ (must be 2 or more)}$$

$$V_c = \left(2 + \frac{4}{2}\right)(\sqrt{4000})(96)(12) = 291{,}436$$

$$\phi V_c = (0.85)(291{,}436) = 247{,}720$$

*step 5*: Since 232,320 < 247,720, no shear reinforcement is required. The footing thickness could be reduced somewhat.

The shear strength of the footing depends on the footing thickness. Assuming $d = 12''$, two bar layers (one in each direction), $d_b = 1''$, and cover = 3″, the thickness is

$$\text{thickness} = 12 + \tfrac{1}{2} + 1 + 3 = 16.5''$$

*step 6*: The bending moment is

$$M_u = \frac{(19.78)(111)(49.5)^2}{2} = 2.69 \text{ EE6 in-lbf}$$

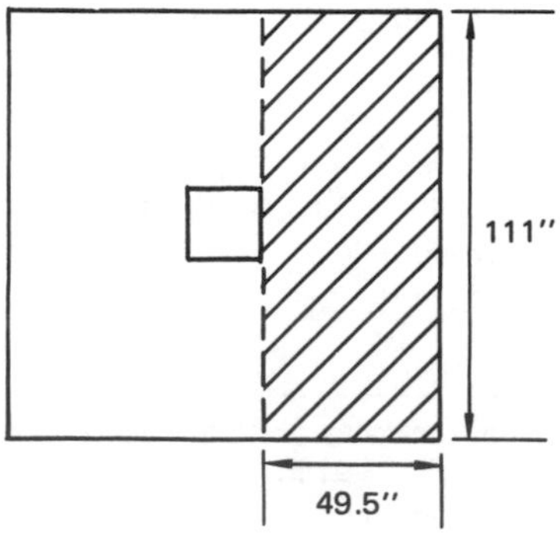

*step 7*: The coefficient of resistance is

$$R_u = \frac{2.69 \text{ EE6}}{(0.9)(111)(12)^2} = 187 \text{ psi}$$

*step 8*: From equation 14.142, the required reinforcement ratio is

$$\rho = \frac{(0.85)(4000)}{60{,}000}\left[1 - \sqrt{1 - \frac{(2)(187)}{(0.85)(4000)}}\right]$$
$$= 0.00321$$

*step 9*: The required steel area is

$$A_{st} = (0.00321)(12)(111) = 4.28\,\text{in}^2$$

*step 10*: The reinforcement must fit in $111-6 = 105''$. Arbitrarily use 10″ bar spacing. The number of bars is

$$\frac{105}{10} = 10.5\,(\text{say } 11)$$

The bar area should be approximately

$$A_b \approx \frac{4.28}{11} = 0.39\,\text{in}^2$$

Choose #6 bars (0.44 in$^2$ each) and space evenly on 10″ centers in both directions (since the footing is square).

The minimum reinforcement ratio from equation 14.9 is

$$\rho_{\text{min}} = \frac{200}{60{,}000} = 0.00333$$

The actual reinforcement ratio is

$$\rho = \frac{(11)(0.44)}{(111)(12)} = 0.00363 \text{ (ok)}$$

(See footnote 69.)

*step 11*: Check development length. (Not done here.)

## 16 RETAINING WALLS

### A. GENERAL DESIGN CHARACTERISTICS

The following characteristics are typical of retaining walls. They are not code requirements, but can be used as starting points for subsequent design.

- The base should be proportioned so that

$$0.40 < \frac{B}{H} < 0.65 \qquad 14.145$$

- The stem thickness at the base should be approximately

$$\frac{1}{12} < \frac{t_s}{H} < \frac{1}{8} \qquad 14.146$$

- The toe should project approximately $b/3$ beyond the stem face.
- The stem thickness should decrease $\frac{1}{4}''$ to $\frac{1}{2}''$ per vertical foot.[71]
- The minimum stem thickness at the top is approximately 12″.

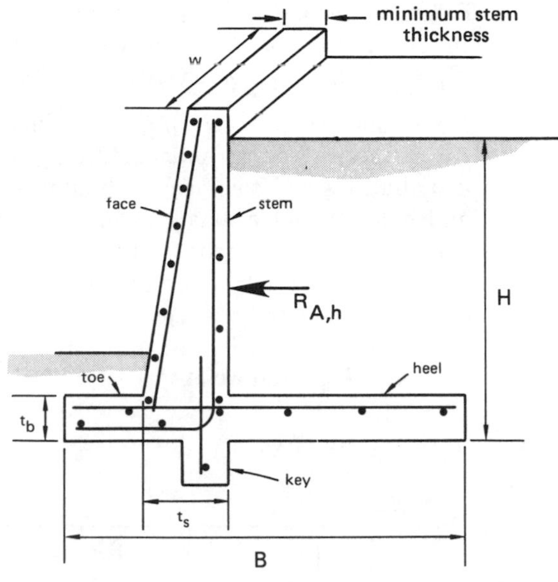


**Figure 14.25** Retaining Wall Nomenclature

- The bottom of the base must be below the frost level.
- To minimize the vertical earth pressure reaction, the resultant of the vertical loads should fall in the middle third of the base.
- The base thickness should be approximately equal to the thickness of the stem at the base, with a minimum of 12″. Alternatively, proportion the base such that

$$0.07 < \frac{t_b}{H} < 0.10 \qquad 14.147$$

### B. GENERAL DESIGN PROCEDURE

A retaining wall is most likely to fail structurally at its base due to the applied moment.[72] In this regard, it is similar to a cantilever beam with a non-uniform load. The following procedure presents one possible approach

[71] This decrease in stem thickness is known as *batter*, and is used primarily to disguise bending (deflection) that would otherwise make it appear as if the wall were failing. It also reduces the quantity of concrete at the upper level, which is acceptable from a strength consideration and increases the overturning stability. Usually, the *batter decrement* is $\frac{1}{4}''$ per foot.

[72] The wall can also fail non-structurally. The subjects of resistance to sliding and overturning are covered in another chapter.

CONCRETE

to design. It is assumed that sufficient soil data exists to calculate the active and other soil pressure distributions.

*step 1:* Choose the height of the wall, $H$, on the basis of frost penetration or other data. Estimate a base thickness, $t_b$.

*step 2:* Estimate the active earth pressure resultants in the horizontal and vertical directions, $R_{A,h}$ and $R_{A,v}$, respectively.

*step 3:* Determine the base length, $B$. One analytical method is to sum moments from active distributions and soil weight about point $A$ and balance that against the moment from the unknown soil weight (in figure 14.26) to get the distance $x$.[73] Once that is obtained, calculate the base length as

$$\frac{R_A H}{3} \approx \frac{(\text{soil weight})x}{2}$$

$$B \approx 1.5x \quad 14.148$$

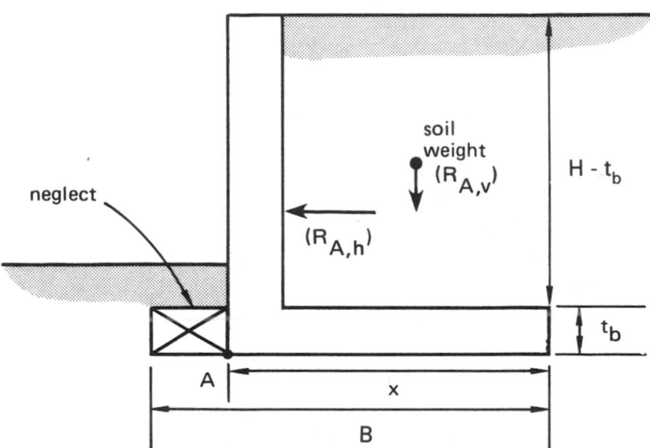


**Figure 14.26** Finding the Base Length

*step 4:* Select a reinforcement ratio to control deflections.

$$\rho \approx \tfrac{1}{2}\rho_{\text{max}} = 0.375\rho_{\text{balanced}} \quad 14.149$$

*step 5:* Calculate the moment at the base of the stem due to the active pressure distributions. Be sure to include the 1.7 overload factor for live loads when calculating $M_u$.

*step 6:* From the value of $\rho$ chosen, calculate the coefficient of resistance.

$$R_u = \rho f_y \left(1 - \frac{\rho f_y}{(2)(0.85)f'_c}\right) \quad 14.150$$

*step 7:* Calculate the required stem thickness at the base.[74] In equation 14.151, $w$ is the wall length, which may be taken as 12″ if flexure calculations are done on a per foot basis.

$$d = \sqrt{\frac{M_u}{\phi R_u w}} \quad 14.151$$

The total thickness of the stem at the base is

$$t_s = d + 2'' \text{ cover} + \tfrac{1}{2}d_b \quad 14.152$$

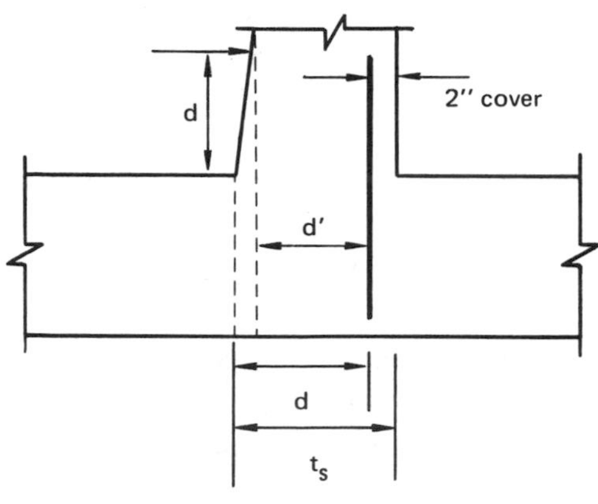


**Figure 14.27** Base of Stem Details

*step 8:* Choose the batter decrement. $\frac{1}{4}''$ per foot is typical.

*step 9:* Calculate the concrete design shear strength of one foot of wall without reinforcement.

$$\phi V_c = 2\phi\sqrt{f'_c}(12)d' \quad 14.153$$

In equation 14.153, the factor 12 is the width, in inches, of a one-foot section of wall. $d'$ is the thickness of the compression portion of the stem a distance $d$ (for one-way shear) from the bottom of the stem. (The tension side is cracked and does not support shear.) If the stem does not decrease in thickness, $d' = d$. Otherwise, $d'$ will depend on the batter decrement.

*step 10:* Calculate the shear on the stem on the critical section. As in step 9, the critical section is assumed to be located a distance $d$ up from the bottom of the stem. It will be necessary to integrate the active pressure distributions

[73] Include surcharge distributions, if present, but disregard passive distributions.

[74] Since $\rho$ was chosen somewhat arbitrarily, $d$ is not unique. Use these calculations as guidelines. $\phi = 0.90$ for flexure.

from the top of the retaining wall down to this critical level to obtain the shear.

$$V_u = 1.7 V_{\text{critical section}} \quad 14.154$$

*step 11*: If $V_u < \phi V_c$, shear reinforcement will not be needed, and the stem can be considered a slab.[75]

*step 12*[76]: Design the heel cantilever as a beam, and check for shear. The critical section is located approximately at the face of the stem. For ease of calculations, consider the pressure distributions of soil, surcharge (if any), and footing concrete for a 12″ strip of retaining wall. Use an overcapacity factor of 1.4 for the soil and footing concrete, and use 1.7 for the surcharge to calculate the ultimate moment on the heel. Disregard the upward soil pressure distribution as well as any effect a key might have on the distributions.

For shear checking, the critical section is taken at the face without considering the upward soil pressure distribution.[77] Therefore, the entire heel contributes to the ultimate shear, $V_u$. The concrete design shear strength $\phi V_c$ is calculated and compared to $V_u$. If necessary, changes to $t_b$ and $d$ can be made.

The new value of $t_b$ can be used to recalculate the footing weight, and a second iteration of this procedure will refine the values still further.

Once $t_b$ is stable, the new $M_u$ can be used to calculate $R_u$, from which $\rho$ and $A_{st}$ can be obtained.[78]

Complete the heel reinforcement design by calculating the development length. The distance from face of stem to end of heel reinforcement must equal or exceed the development length. The heel reinforcement must also extend a distance equal to the development length past the stem reinforcement into the toe.

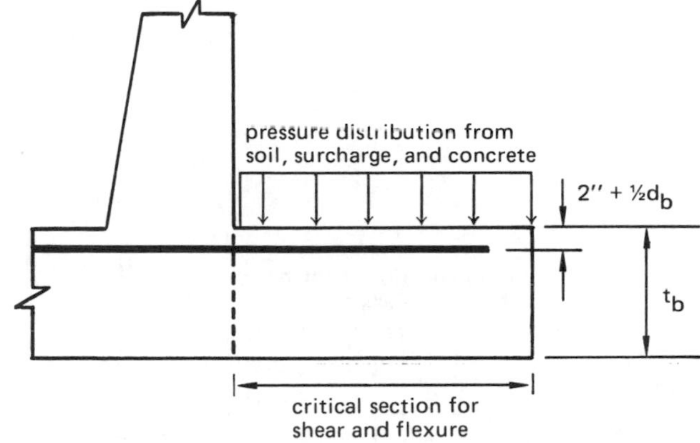


**Figure 14.28** Heel Construction Details

*step 13*: The design of the toe cantilever is similar to the heel design. The loading on the toe is assumed to be from the toe weight and the upward soil pressure distribution only, unless soil above the toe is permanent.[79] For flexure, the critical section is at the outer face of the wall. For shear (one-way shear), the critical section is located a distance $d$ from the wall face.

When calculating the ultimate moment, $M_u$, and ultimate shear, $V_u$, an overcapacity factor of 1.4 should be used for the concrete weight, and 1.7 should be used for the upward soil pressure distribution based on the service loads.[80]

It is likely that shear will determine the toe thickness. The toe thickness can be different from the heel thickness, although most designs maintain a constant base thickness.

*step 14*: Calculate the required steel area for flexure reinforcement in the stem. (If the base thickness has been changed from step 7, then $H$ will also have changed. In that case, recalculate $R_u$ from equation 14.155. (The shear

CONCRETE

[75] From a practical design standpoint, the thickness should be increased rather than supplying shear reinforcement.

[76] Designing the heel and toe cantilevers may change some of the assumptions made about the base (e.g., the base thickness, etc.). If the base is not of interest, skip steps 12 and 13.

[77] Since tension occurs in the heel, the critical section cannot be taken as a distance $d$ from the face of the stem. (See ACI code 11.1.3.)

[78] The ACI code exempts slabs from the requirement of $200/f_y$ minimum reinforcement. However, it is arguable that a retaining wall is a beam, not a slab, in this situation. Therefore, the minimum reinforcement ratio or alternatively, one-third more than is required for strength, may apply.

[79] The soil on top of the toe may be removed for later work or repair on the retaining wall. Therefore, it cannot be counted on to reduce the toe stress.

[80] ACI 9.2.4 specifies that the factor is 1.4 for dead concrete and soil weights, and 1.7 for horizontal earth pressure. The toe soil pressure distribution is the result of the horizontal active soil pressure, so 1.7 is used.

was checked in step 10. Unless the stem thickness at the base has been changed, no additional shear checking is necessary.) In equation 14.155, $w$ may be taken as 12″ if all calculations are done per foot of wall.

$$R_u = \frac{M_u}{\phi w d^2} \qquad 14.155$$

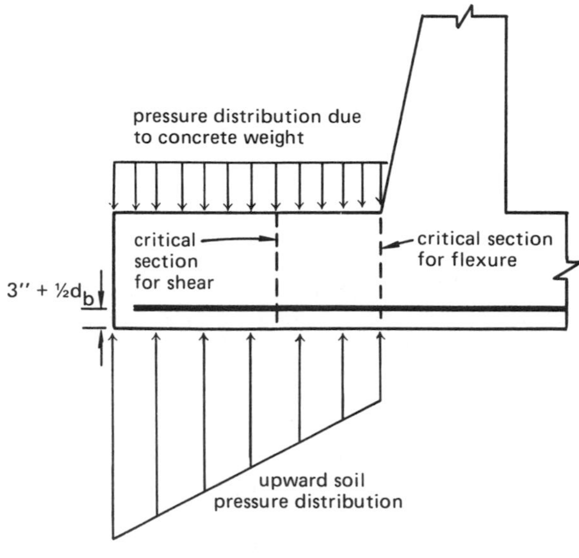


**Figure 14.29** Toe Construction Details

From $R_u$, obtain $\rho$ and $A_{st}$. Select reinforcement to meet distribution requirements.[81]

*step 15*: Check the development length of the stem reinforcement. The length into the stem itself (above the base) should be adequate without checking. However, it may be difficult to achieve full development in the base. The reinforcement may be extended into the key, if one is present. Otherwise, hooks, bends, or smaller bars should be used.

*step 16*: Calculate the temperature and shrinkage reinforcement from $H$ and $w$. It is common to allocate 2/3 of this reinforcement to the outward face of the stem, since that surface is alternatively exposed to day and night. The remaining 1/3 is placed on the soil side of the stem, which is maintained at a more constant temperature by the soil insulation.

*step 17*: Provide for drainage if the design was based on a drained condition.[82] Weep holes should be placed regularly (i.e., approximately every 10 feet) along the length of the wall.

[81] This step only determines flexural reinforcement in the stem at the base. As the stem goes up, less reinforcement will be required. Similar calculations should be performed at one or two other locations (up the stem) to obtain a curve of $M_u$ versus location. Trial and error can be used to select reinforcement which meets the strength requirements.

[82] Unless there is a need to make the retaining wall act as a watertight bulkhead, drainage should always be part of the design.

# Appendix A: Miscellaneous ACI Detailing Requirements

**Minimum Cover on Non-prestressed Steel [ACI 318 7.7]**

- Cast against and permanently exposed to earth . . . 3″
- Exposed to earth or weather
  - #5 bars or smaller . . . $1\frac{1}{2}''$
  - #6 bars or larger . . . 2″
- Not exposed to weather or ground contact
  - Slabs, walls, or joists
  - #11 bars or smaller . . . 3/4″
  - #14 and #18 bars . . . $1\frac{1}{2}''$
  - Beams, girders, or columns . . . $1\frac{1}{2}''$

**Minimum Horizontal Clear Distance Between Bars in a Layer [ACI 318 7.6]**

- Beams: The maximum of one bar diameter or . . . 1″
- Walls: The maximum of one bar diameter or . . . 1″

**Minimum Vertical Clear Distance Between Bars in a Layer**

- Beams: . . . 1″

**Maximum Rebar Spacing (center-to-center distance)**

- Walls and Slabs: The minimum of three times the wall or slab thickness, or . . . 18″

**Splices (Ratio of lap length to development length, $l_d$)[1,2] [ACI 318 12.15]**

- Tension splices [See ACI 318 section 12.15 for different classes.]
  - Class A . . . 1.0
  - Class B . . . 1.3
  - The minimum tension splice lap length is . . . 12″
- Compression splices ($f'_c$ exceeds 3000 psi) [ACI 318 12.16]
  The minimum of 12″, (0.0005)($f_y$)(bar diameter), and one development length

Many other provisions apply to both tension and compression lap splices. Welded splices are butt-welds developing an ultimate tensile strength of 125% of the yield strength of the bar.

CONCRETE

**Minimum Bend Radii (In bar diameters)**

- #3 to #8 bars . . . 6
- #9, #10, and #11 bars . . . 8
- #14 and #18 bars . . . 10

[1] #14 and #18 bars may not be lap spliced. [ACI 318 12.14.2]

[2] For 2-bar splices only. Bar bundles require additional lap length. [ACI 318 12.14.2]

**Maximum Deflections [ACI Table 9.5(b)]**

These limits are fractions of the distance between supports.

| element | when | is element attached to non-structural items likely to be damaged by large deflections? | limit |
|---|---|---|---|
| flat roof | immediate | no | 1/180 |
| floor | immediate | no | 1/360 |
| roof/floor | sustained | no | 1/240 |
| roof/floor | sustained | yes | 1/480 |

**Minimum Beam and Slab Thickness**

(For use when deflections are not computed.) These thicknesses apply to normal weight concrete (density greater than 120 pcf) and grade 60 steel. For other grade steels, multiply the thickness by

$$0.4 + \frac{f_y}{100{,}000}$$

These fractions are the fraction of span length between supports. [ACI 318 Table 9.5(a)]

| | simply supported | one end continuous | both ends continuous | cantilever |
|---|---|---|---|---|
| solid, one-way slabs | 1/20 | 1/24 | 1/28 | 1/10 |
| beams or ribbed one-way slabs | 1/16 | 1/18.5 | 1/21 | 1/8 |

**Maximum Aggregate Size [ACI 318 3.3.2]**

1/5 of the narrowest dimension between sides of the forms, or 1/3 of the slab depth, or 3/4 of the clear spacing between bars, whichever is least.

# Appendix B: The Alternate (Working Stress) Method of Design

ACI-318 permits the following stresses when the alternate design method is used.

CONCRETE (normal weight)

**flexure**

| | |
|---|---|
| compressive stress on extreme fiber | $0.45f'_c$ |

**shear**

| | |
|---|---|
| beams (concrete carries all shear) | $1.1\sqrt{f'_c}$ |
| one-way footings and one-way slabs | $1.1\sqrt{f'_c}$ |
| two-way footings and two-way slabs | minimum of $\begin{cases} 2.0\sqrt{f'_c} \\ \left(1+\frac{2}{\beta}\right)\sqrt{f'_c} \end{cases}$ |

| | |
|---|---|
| **bearing on loaded area** | $0.30\sqrt{f'_c}$ |

STEEL (tensile stress)

**beams**

| | |
|---|---|
| grade 40 or 50 | 20,000 psi |
| grade 60 and higher | 24,000 psi |
| one-way slabs, less than 12 foot span, with #3 bars or smaller main reinforcement | minimum of $\begin{cases} 0.50 \text{ fy} \\ 30{,}000 \text{ psi} \end{cases}$ |

The alternate method can be used to design beams if the following assumptions are made:

- Plane sections remain plane during bending.
- Steel and concrete both remain in the elastic range.
- Concrete carries all compressive loads in an area above the neutral axis.
- Steel carries all tension.
- Stress in the concrete is proportional to the distance from the neutral axis.

For maximum stresses in the concrete and steel, define the location of the neutral axis.

$$k = \frac{1}{1+(f_s/nf_c)} = \sqrt{2n\rho+(n\rho)^2} - n\rho$$

$$\rho = \frac{A_s}{bd}$$

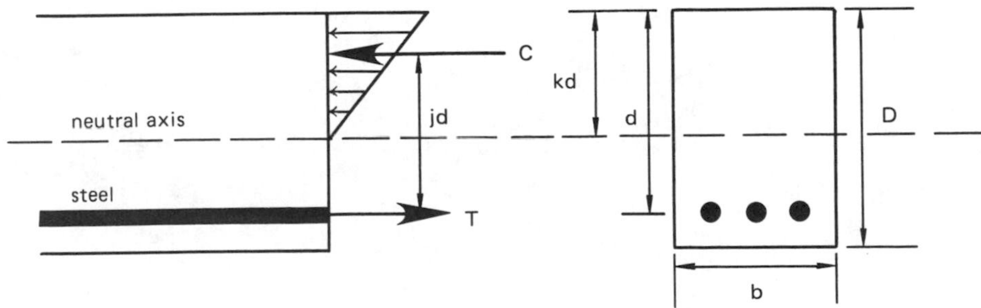


The total compressive force is

$$C = \frac{1}{2}f_c bkd$$

The total tensile force is

$$T = A_s f_s = C$$

CONCRETE

Since the compressive force, $C$, acts one-third down from the top of the beam,

$$j = 1 - \frac{k}{3}$$

The maximum moments that the concrete and steel can carry are

$$M_c = Cjd = \frac{1}{2} f_c jkbd^2$$
$$M_s = Tjd = A_s f_s jd = f_s j \rho bd^2$$

The required steel area is

$$A_s = \frac{M_s}{f_s jd}$$

If the design is *balanced*, then $M_c = M_s$. Therefore,

$$\frac{1}{2} f_c jkbd^2 = f_s j \rho bd^2$$

## Practice Problems: CONCRETE DESIGN

Untimed

1. 1.15 cubic yards of portland cement concrete are needed. Using the specifications below, determine the weight (in as-delivered condition) of water, cement, fine aggregate, and coarse aggregate needed.

| | |
|---|---|
| cement content: | 6.5 bags per cubic yard |
| water cement ratio: | 5.75 gallons per bag |
| cement specific gravity: | 3.10 |
| fine aggregate specific gravity: | 2.65 |
| coarse aggregate specific gravity: | 2.00 |
| aggregate grading: | 30% fine, 70% coarse by volume |
| free moisture: (SSD basis) | 1.5% in fine aggregate |
| coarse aggregate will absorb: | 3% moisture (SSD basis) |
| entrained air: | 5% |

2. Design a square 2-way footing to carry a live load of 240,000 pounds which are transmitted through a 16" square column. The allowable soil pressure is 4000 psf and the top of the footing is level with the surrounding soil surface. Use 3000 psi concrete and 40,000 psi steel. Disregard column dead loading, but include footing dead load.

3. Design a rectangular beam with tension reinforcement only using the ultimate strength method. The dead load moment is 50,000 ft-lb and the live load moment is 200,000 ft-lb. Use 3000 psi concrete and 50,000 psi steel. Do not design for shear, or check cracking or deflection.

4. Design a balanced beam meeting the following specifications. Use the working strength design method.

| | |
|---|---|
| maximum moment: | 50.4 ft-kips |
| maximum steel stress: | 20,000 psi |
| maximum concrete stress: | 1350 psi |
| $E_s/E_c$: | 10 |

5. Design a spiral column to carry a dead load of 175,000 pounds and a live load of 300,000 pounds. The loads are axial. Use 3000 psi concrete and 40,000 psi steel.

6. Design a short square tied column to carry a 100,000 pound dead load and a 125,000 pound live load. Include specifications for the ties. Use No. 5 bars or larger, 3500 psi concrete, and 40,000 psi steel.

7. An 18" square column supports 200,000 pounds and 145,000 pounds dead and live loads respectively. The allowable soil pressure is 4000 psf. Use the ultimate strength design method with 3000 psi concrete and 40,000 psi steel to design the footing. The column steel consists of 10 #9 bars.

8. A beam is being designed to withstand an ultimate factored moment of 400,000 ft-lb. 4000 psi concrete and 40,000 psi steel are available. Specify values of *b*, *d*, and the required steel area. How many layers of steel are needed? Do not check cracking or deflection.

9. 3000 psi concrete is used in the floor slab shown below. The slab must carry 20,000 ft-lb per foot of width. Use the transformed area method to evaluate the stresses in the concrete and steel.

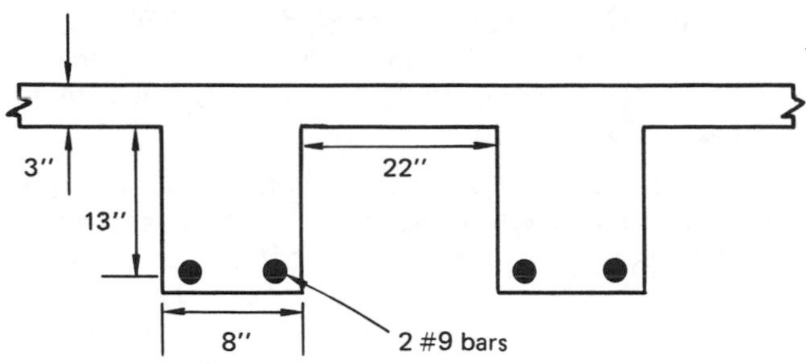


10. 100 pounds of aggregate were sieve graded. The weights retained on each sieve are shown below. What is the fineness modulus of the aggregate?

| sieve | weight retained |
|---|---|
| 4 | 4 |
| 8 | 11 |
| 16 | 21 |
| 30 | 22 |
| 50 | 24 |
| 100 | 17 |
| dust | 1 |

Timed

1. Use the latest ACI code to design the cross section and steel for the beam shown. Assume $f'_c = 4000$ psi and $f_y = 60{,}000$ psi. Only #11 steel is available. n = 8. All loads are dead loads. Neglect shear, but check for cracking under exterior conditions.

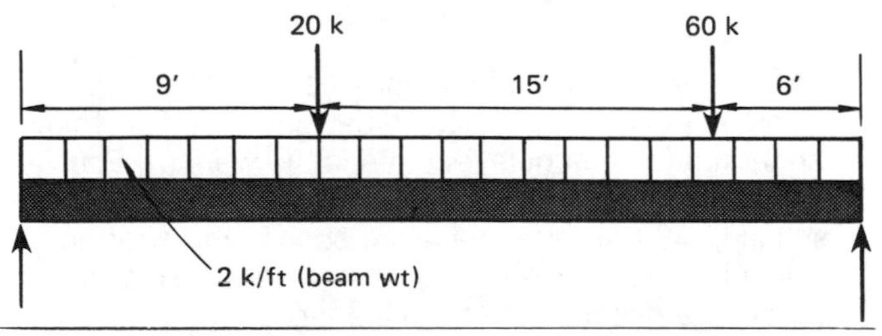


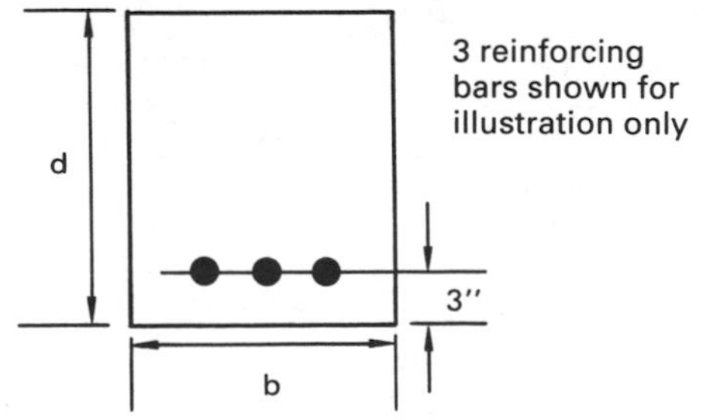

2. A reinforced concrete beam supporting a roof is needed to span 27 feet. It carries a dead load of 1 kip/ft (which includes the beam weight) and a live load of 2 kips/ft. Assume $f'_c$ = 4000 psi, $f_{y,st}$ = 60,000 psi, and $E_c$ = 3.6 EE6 psi. The beam is simply supported. Use the latest ACI code with the maximum steel percentage permitted. (a) Use #11 bars and design a rectangular beam with b = 14″. Clear cover on the steel must be at least 1 1/2″. Do not design for shear. (b) Will the estimated crack sizes be within the requirements of the code if the member has interior exposure? (c) Calculate the instantaneous and long-term center-line deflections according to ACI code provisions. Assume 30% of the live load is to be sustained.

3. A pretensioned girder is constructed with 30 bonded strands. The girder must span 100 feet on simple supports and carry a uniform live load of 540 lb/ft as well as its own weight of 150 lb/ft. Assume n = 7, $f'_{ci}$ = 3800 psi, $f'_c$ = 5000 psi, $f_{pu}$ = 250 ksi, $f_{py}/f_{pu}$ = 0.85, and $A_p$ = 0.144 in$^2$/strand. (a) Determine if the stresses in the concrete are acceptable immediately after transfer. Assume that only the dead weight acts on the beam and that prestress before losses is 0.8 times $f_{pu}$, the tendon tensile strength. (b) Determine if the concrete stresses are acceptable at service loads. Assume the effective prestress is 0.6 times $f_{pu}$. Do not check the tendon stress.

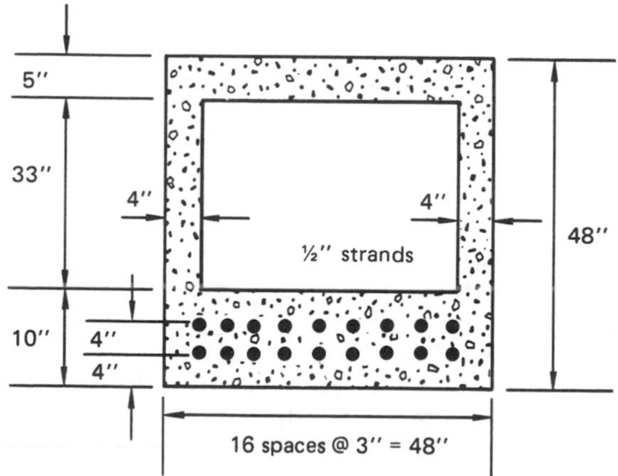


4. A 28' (face-to-face of support) beam with a depth to reinforcement of d = 21" is to be designed. (a) Determine if shear reinforcement is required in the web. (b) Specify the length of beam over which stirrups are required. (c) Assuming #3 bars are available, specify the spacing of shear reinforcement (stirrups) located from the face of the support to a distance 21" from the face.

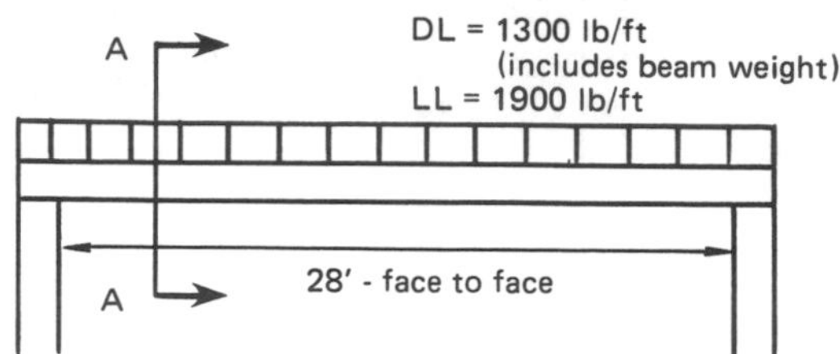


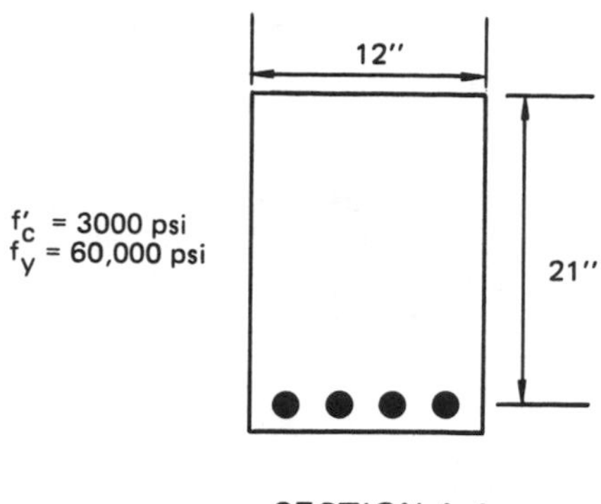


5. Given the steel-reinforced concrete combined footing shown, (a) find the dimension L that will result in a uniform soil pressure. (b) Draw complete shear and moment diagrams. (c) Should top steel in the pad be used? Disregard the weight of the footing itself.

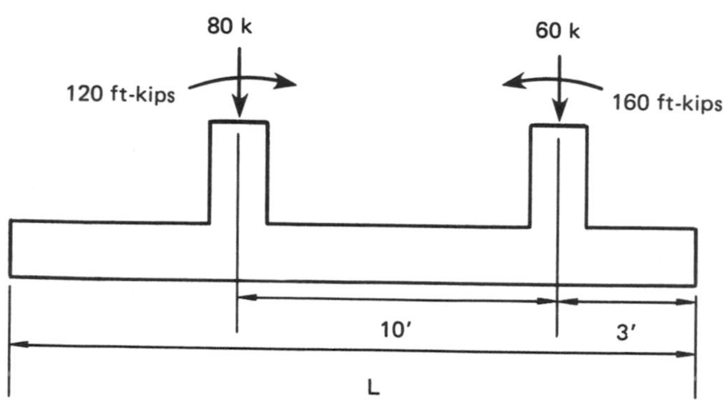


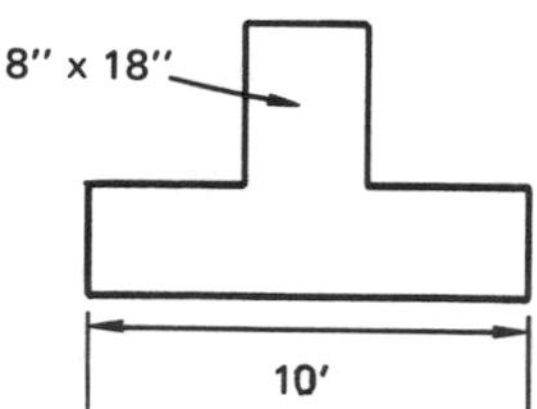


6. Near the end of the day, a long masonry pier is left unsupported at its upper end. (The lower end is firmly connected to the foundation.) During the night, a 30 psf wind starts blowing. Neglecting axial stresses due to dead load, determine the stresses in the steel and masonry. Are they acceptable?

(figure on next page)

CONCRETE

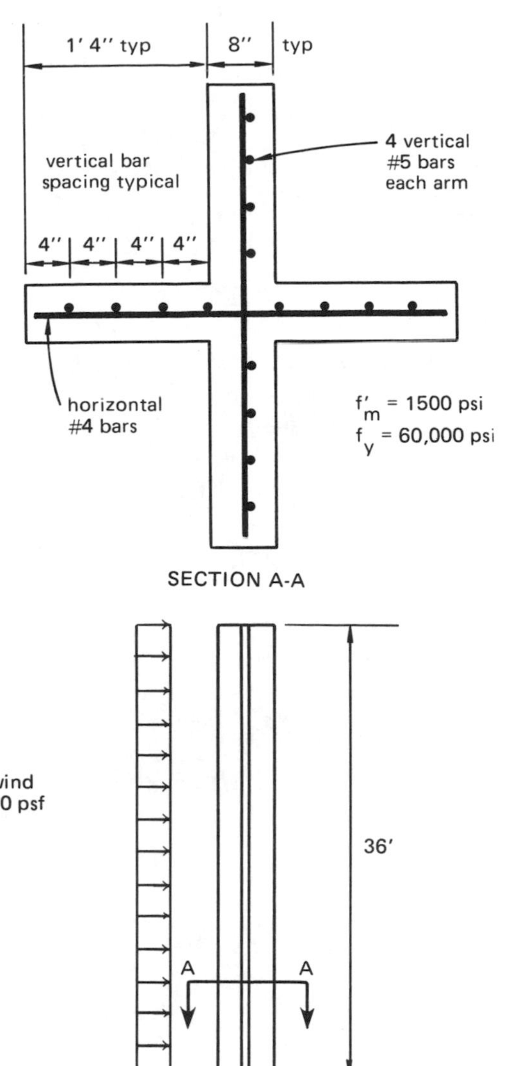


7. A 14" square column carries the live loads shown. The loads are not the result of wind or earthquake action. (Disregard dead loading.) The allowable soil pressure is 6000 psf. Use $f'_c$ = 3000 psi for the concrete. (a) Size the footing. (b) What are the minimum and maximum pressures along the footing base? (c) What thickness of footing is required to avoid using shear reinforcement? (d) What is the maximum moment the footing must resist? (It is not necessary to design the steel reinforcement.)

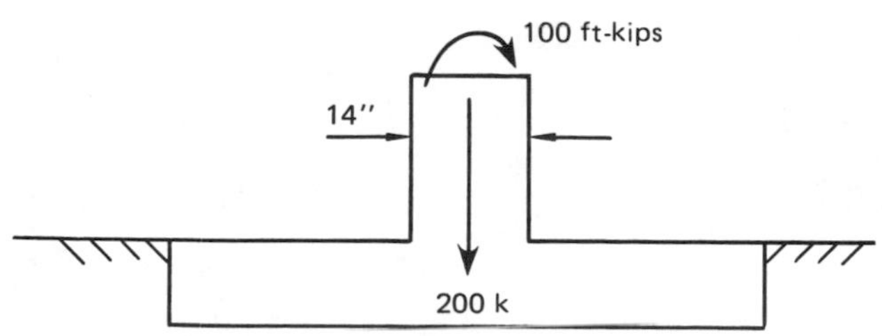


8. For the retaining wall shown, (a) what is the factor of safety against overturning? (b) What is the minimum theoretical heel depth? (c) Using the heel depth from part (b) and #8 steel, what bar spacing is required in the heel?

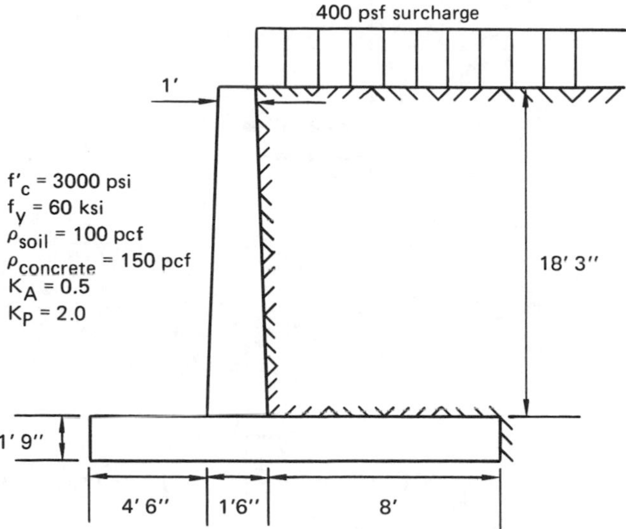


9. A 16" wide beam is simply supported. It carries the loads shown. $f'_c$ = 3000 psi. $f_y$ = 60 ksi. (a) Design the beam, including the minimum beam depth. (b) Use USD and latest ACI code to detail the reinforcing. (c) Design any needed shear reinforcing using #3 bars.

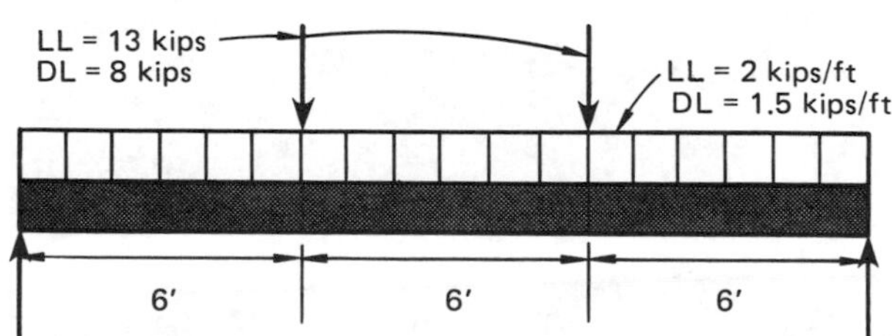


10. A deep concrete beam is to be designed as shown. Use $f'_c$ = 3000 psi concrete and 60 ksi steel. Disregard self-weight, and consider the beam to be simply supported at its ends. (a) Find the nominal concrete shear strength, $v_c$, as given by the ACI codes for deep beams. (b) Find the amount of shear reinforcing, $A_v$, required for the loads as shown. (c) Determine the extent of this reinforcing.

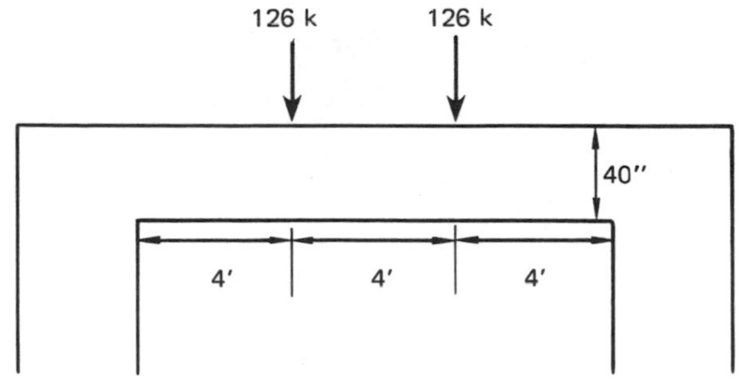


(additional figure on next page)

CONCRETE

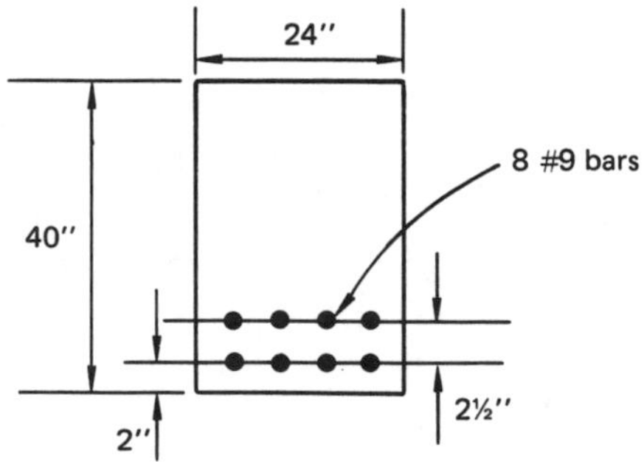
24″
40″
8 #9 bars
2″
2½″

# 15 STEEL DESIGN AND ANALYSIS[1]

*Nomenclature*

| | | |
|---|---|---|
| a | stiffener spacing | in |
| A | area | $in^2$ |
| b | width | in |
| c | distance to extreme fiber | in |
| C | coefficient | – |
| $C_c$ | critical slenderness ratio | – |
| d | depth or diameter | in |
| D | outside diameter | in |
| e | eccentricity | in |
| E | modulus of elasticity | psi |
| f | actual stress | ksi |
| F | strength or allowable stress | ksi |
| g | gage spacing | in |
| G | end condition coefficient | – |
| H | horizontal force | kips |
| I | area moment of inertia | $in^4$ |
| J | polar moment of inertia | $in^4$ |
| k | flange to web toe fillet distance, or spring constant | in, or lbf/in |
| K | end restraint coefficient | – |
| l | length between supports | in |
| L | length or distance | in |
| M | moment | ft-kips |
| n | modular ratio | – |
| N | bearing length | in |
| P | force | lbf |
| r | radius of gyration, radius, or distance | in |
| $R_w$ | weld strength (resistance) | kips/in |
| s | pitch spacing | in |
| S | section modulus | $in^3$ |
| SR | slenderness ratio | – |
| t | thickness | in |
| T | tension | kips |
| U | reduction coefficient | – |
| V | shear | kips |
| w | weld size | in |
| Z | plastic modulus | $in^3$ |

*Subscripts*

| | |
|---|---|
| a | axial |
| b | bracing or bending |
| c | centroidal or concrete |
| cr | critical |
| e | effective or edge |
| f | flange |
| g | gross |
| n | net |
| p | bearing or plastic |
| s | secondary or steel |
| st | stiffener |
| t | tension |
| u | ultimate or unbraced |
| v | shear |
| w | web |
| y | yield |

1 This chapter cannot serve as a complete substitute for the American Institute of Steel Construction's *Manual of Steel Construction* or its accompanying *Specifications*. Throughout this chapter, references to the *AISC Specifications*, 9th edition, are listed in bold, square brackets. For example, **[H1-3]** is a reference to equation H1-3 in the *Specifications*, not to equations in this book. Also, tables, figures, and appendices listed in italic are part of the *AISC Manual* or *Specifications*. Thus, the *Allowable Stress Design Selection Table* would not be found in this volume.

## 1 REVIEW OF STEEL NOMENCLATURE

It is traditional in steel design to use the upper case letter $F$ to indicate strength or allowable stress. Furthermore, such strengths or maximum stresses are specified in ksi (i.e., 1000's of psi). For example, $F_y = 36$ would imply a steel with a yield strength of 36 ksi. Similarly, $F_v$ is the allowable shear stress, and $F_b$ is the allowable bending stress, both in ksi.

Actual or computed stresses are given the symbol of lower case $f$. Computed stresses are also specified in ksi. For example, $f_t$ is a computed tensile stress in ksi. The symbol $\sigma$ is never used.

## 2 CONVERSIONS

| multiply | by | to obtain |
|---|---|---|
| ft-kips | 1.356 | kN-m |
| ft-lbf | 1.356 | N-m |
| in-lbf | 0.113 | N-m |
| kips | 4.448 | kN |
| kips | 1000 | lbf |
| kips/ft | 14.59 | kN/m |
| kips/ft$^2$ | 47.88 | kN/m$^2$ |
| kN | 0.2248 | kips |
| kN/m | 0.06852 | kips/ft |
| kPa | 1000 | Pa |
| ksi | 6.895 EE6 | Pa |
| lbf | 0.001 | kips |
| lbf | 4.448 | N |
| N-m | 0.7376 | ft-lbf |
| N-m | 8.851 | in-lbf |
| N-m$^2$ | 1 | Pa |
| Pa | 1.0 | N/m$^2$ |
| Pa | EE–6 | MPa |
| Pa | 0.001 | kPa |
| Pa | 1.45 EE–7 | ksi |
| Pa | 1.45 EE–4 | psi |
| Pa | 0.02089 | psf |
| psf | 47.88 | Pa |
| psi | 6.895 EE3 | Pa |

## 3 TYPES OF STEELS

ASTM A36 is the designation given to the all-purpose, carbon steel used for most projects. Most A36 shapes are hot rolled.

Other steels have higher strengths. Their use results in lower dead weights but higher material costs. The commonly available steels and their applications are listed in appendix A.

**Table 15.1**
Properties of A36 Steel

| | |
|---|---|
| $E$, modulus of elasticity: | 2.9 EE7 psi (up to 100°F) |
| $G$, shear modulus: | 11.5 EE6 psi |
| $\alpha$, coefficient of thermal expansion: | 6.5 EE–6 1/°F |
| $\mu$, Poisson's ratio: | 0.30 |
| $\rho$, density: | 490 lbm/ft$^3$ |
| $F_y$, yield strength: | 36 ksi (to 8″ thickness inclusive)<br>32 ksi (over 8″ thickness) |
| $F_u$, ultimate strength: | 58 ksi (minimum) |
| $F_e$, endurance limit: | approximately 30 ksi |

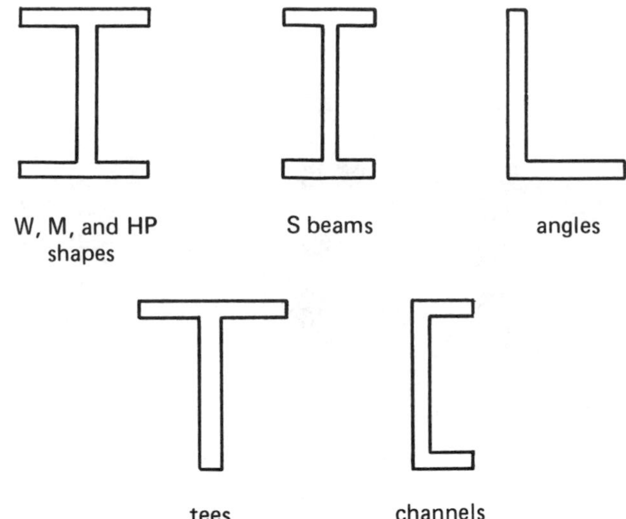


**Figure 15.1** Structural Shapes

## 4 STEEL PROPERTIES

Some properties of steel (such as the modulus of elasticity and density) are essentially independent of the type of steel. Other properties (such as the ultimate and yield strengths) depend not only on the type of steel but also on the size or thickness of the piece.

## 5 STRUCTURAL SHAPES

Many different structural shapes are available. The identifying dimension and weight must be appended to the designation to uniquely identify the shape. For example, W 30×132 means a W-shape with an overall depth of approximately 30 inches and which weighs 132 pounds per foot.

Table 15.2 lists structural shape designations.

**Table 15.2**
Structural Shape Designations

| shape | designation |
|---|---|
| wide flange beams | W |
| standard flanged beams | S |
| misc. flanged beams | M |
| American std. channels | C |
| bearing piles | HP |
| angles | L |
| tees | ST or WT (cut from S or W) |
| plate | PL |
| bar | bar |
| pipe | pipe |
| structural tubing | TS |

STEEL

## 6 STANDARD COMBINATIONS OF SHAPES

Figure 15.2 illustrates several combinations of shapes that are used in construction. The double angle combination is particularly useful for carrying axial loads. Combinations of W shapes and channels, channels with channels, or channels with angles are used for a variety of special applications, including struts and light crane rails. Properties for certain combinations have been tabulated in the *AISC Manual.*

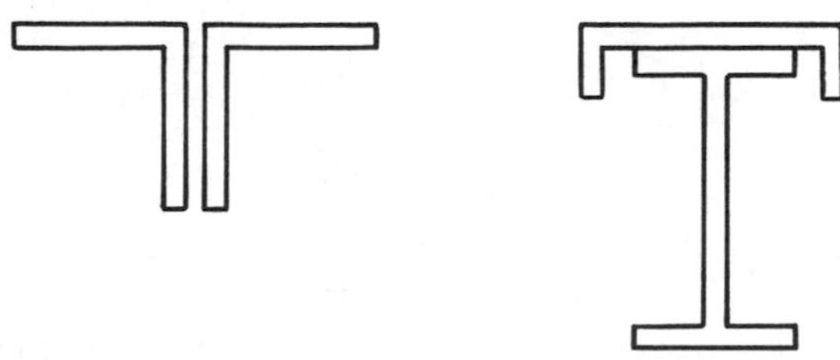

**Figure 15.2** Typical Combination Sections

## 7 REINFORCEMENT OF MILL SHAPES

Occasionally, it will be desirable to provide additional bending or compressive strength to a shape by adding plates. It is generally easy to calculate the properties of the built-up section from the properties of the shape and plate. No stress calculations are necessary. The following characteristics can be used when it is necessary to specify plate reinforcement.

- Plate widths should not be the same as $b_f$, due to difficulty in welding. Widths should be somewhat larger or smaller. It is better to keep plate width as close to $b_f$ as possible, as width-thickness ratios specified in the *AISC Manual* may govern.

- Width and length tolerances smaller than 1/8″ are not practical. Table 15.3 should be used when specifying the nominal plate width.

**Table 15.3**
Width Tolerance for Small Universal Mill Plates

| thickness (inches) | width (inches) 8–20, excl. | 20–36 excl. | 36 and above |
|---|---|---|---|
| 0–3/8, excl. | 1/8 | 3/16 | 5/16 |
| 3/8–5/8, excl. | 1/8 | 1/4 | 3/8 |
| 5/8–1, excl. | 3/16 | 5/16 | 7/16 |
| 1–2, incl. | 1/4 | 3/8 | 1/2 |
| 2–10, incl. | 3/8 | 7/16 | 9/16 |

- Not every plate exists in the larger thicknesses. Unless special plates are called for, the following thickness guidelines should be used:

  1/32″ increments up to 1/2″
  1/16″ increments from 9/16″ to 1″
  1/8″ increments from $1\frac{1}{8}$″ to 3″
  1/4″ increments for $3\frac{1}{4}$″ and above

*Example 15.1*

A W 30×124 shape must be reinforced to achieve the strong-axis bending strength of a W 30×173 shape by welding plates to both flanges. All steel is A36. The plates are welded continuously to the flanges. What size plate is required if all plate sizes are available?

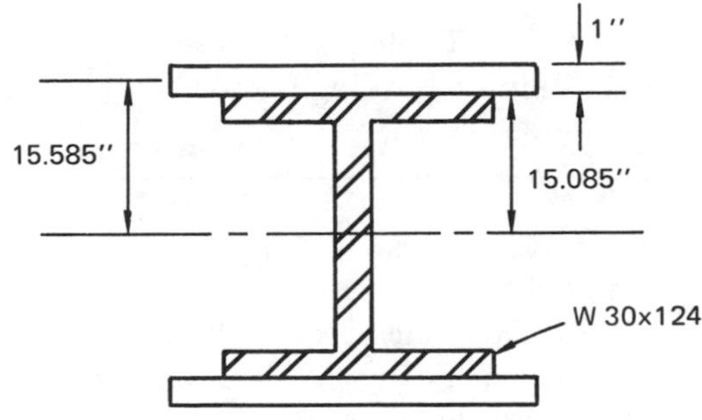


The moments of inertia are 8200 in⁴ and 5360 in⁴ for the two beams. The difference in bending resistance to be provided by the plates is

$$I_{\text{plates}} = 8200 - 5360 = 2840\,\text{in}^4$$

For ease of welding, assume the plate thickness will be approximately the same as the flange thickness. $t_f = 0.930''$, so choose a plate thickness of 1.0″.

The centroidal moment of inertia of the two plates acting together is

$$I_{c,\text{plates}} = 2\left[\frac{w(1)^3}{12}\right] = \frac{w}{6}$$

The depth of the W 30×124 beam is 30.17″. Therefore, the distance from the neutral axis to the plate centroid is

$$\frac{30.17}{2} + \frac{1}{2} = 15.585''$$

By the parallel axis theorem, the moment of inertia of the two plates about the neutral axis is

$$I_{\text{plates}} = \frac{w}{6} + (2)(w)(1)(15.585)^2 = 486.0w$$

The required moment of inertia is 2840. Therefore,

$$w = \frac{2840}{486} = 5.84'' \quad (\text{say } 6'')$$

STEEL

## 8 FATIGUE LOADING

The effects of fatigue loading are generally not considered except for the case of connection design. If a load is to be applied and removed less than 20,000 times (as would be the case in a conventional building), no provision for repeated loading is necessary. (20,000 times is roughly equivalent to two times a day for 25 years.) However, some designs, such as for crane runway girders and supports, must consider the effects of fatigue. Design for fatigue loading is covered in *Appendix K4* of the *AISC Specifications*.

## 9 ALLOWABLE STRESSES FOR IMPACT, WIND, AND EARTHQUAKE LOADS

The effects of impact are included by increasing the actual live load (but not the dead load) by the percentages contained in table 15.4.

**Table 15.4**
Impact Loading Factors

| supports for | % live load increase |
|---|---|
| elevators | 100 |
| cab operated travel cranes | 25 |
| pendant operated travel cranes | 10 |
| shaft or motor-driven machinery | 20 minimum |
| reciprocating machinery | 50 minimum |
| floors and balconies (hangars) | 33 |

Most allowable stresses, including those for connectors, columns, and beams, may be increased 1/3 for transitory wind and earthquake loading. This increase is applied to all stresses in the problem, not just the wind- or earthquake-induced stresses. The increase cannot be applied in fatigue loading problems. Also, the calculated section area cannot be less than the area required to carry the dead and live loads alone (without the 1/3 increase).

## 10 THE MOST ECONOMICAL SHAPE

Since a major part of the cost of using a rolled shape in construction is the cost of the raw materials, the lightest shape possible which will satisfy the structural requirements should always be used. Thus, generally speaking, the *most economical beam* is the structural shape that has the lightest weight per foot and that has the required strength. (ASTM A572 grade 50 steel is approximately 35% stronger than ASTM A36 steel, but it costs only approximately 14% more than A36 steel. For beams whose designs are not driven by deflection criteria (as is usually the case in composite construction), ASTM A572 grade 50 steel will yield the more economical section.) The *beam selection table* and *chart* are designed to make choosing economical shapes possible.



## 11 COMPACT SECTIONS

Compact sections have thicker webs and flanges than non-compact sections. Therefore, compact sections are afforded higher allowable stresses in many instances. To be compact, the flanges of a beam must be continuously connected to the web. Therefore, a built-up section or plate girder constructed with intermittent welds will not qualify. In addition, two conditions apply to standard rolled shapes without flange stiffeners. (Equation 15.2 applies only to webs in flexural compression.)

$$\frac{b_f}{2t_f} \leq \frac{65.0}{\sqrt{F_y}} \qquad 15.1$$

$$\frac{d}{t_w} \leq \frac{640}{\sqrt{F_y}} \qquad 15.2$$

Compactness, as equations 15.1 and 15.2 show, depends on the steel strength. Most rolled W shapes are compact at lower values of $F_y$. However, a 36 ksi beam may be compact, while the same beam in 50 ksi steel may not be. The *shape tables* contain columns ($F'_y$ and $F'''_y$) that indicate the yield strengths at which the beams become non-compact due to the two conditions.

## 12 BEAM BENDING PLANES

The property tables for W shapes will show that each beam has two moments of inertia, $I_x$ and $I_y$. It is easy to use the wrong value, as there are several ways of referring to the plane of bending.

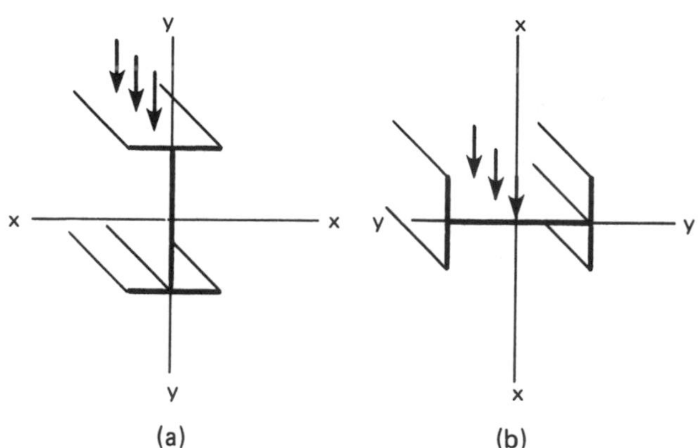


**Figure 15.3** Beam Bending Planes

Figure 15.3(a) shows a W shape beam used as it is typically. The value of $I_x$ should be used to calculate bending stress. This bending mode is referred to as "... loading in the plane of the web ...". However, it is also referred to as "... loading in the plane of the weak axis ...", which is confusing.

Figure 15.3(b) shows a W shape "... bending about the minor axis ...". This mode is also referred to as "... loading in the plane of the strong axis ...".

## 13 LATERAL BRACING

To prevent the lateral buckling illustrated in figure 15.4, a beam's compression flange must be supported at frequent intervals. Complete support is achieved when a beam is fully encased in concrete or has its flange welded or bolted along its full length. In many designs, however, the lateral support is at regularly spaced intervals.

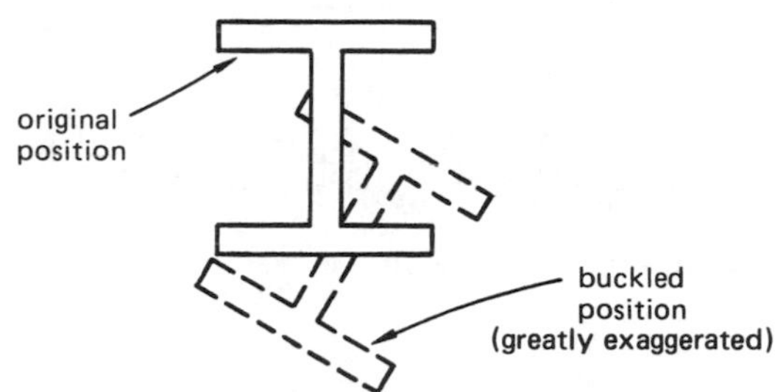


**Figure 15.4** Lateral Beam Buckling

The actual spacing between points of lateral bracing is designated as $L_b$. For the purpose of determining the allowable bending stress, two limits are placed on the spacing: $F_b = 0.66F_y$ if $L_b \leq L_c$; $F_b = 0.60F_y$ if $L_c < L_b \leq L_u$. (See figure 15.10.) For most shapes, $L_c$ is calculated from equation 15.3; $L_u$ is calculated from equation 15.4.[2] The quantity $b_f$ and the ratio $d/A_f$ are both tabulated for regular shapes, although $L_c$ and $L_u$ are tabulated for rolled shapes with standard values of $F_y$ in the *AISC Manual, Section F1.*

$$L_{c,\text{inches}} = \min \left\{ \begin{array}{c} \dfrac{76b_f}{\sqrt{F_y}} \\ \dfrac{20{,}000}{\left(\dfrac{d}{A_f}\right) F_y} \end{array} \right\} \qquad 15.3$$

$$L_{u,\text{inches}} = \max \left\{ \begin{array}{c} \dfrac{20{,}000C_b}{\left(\dfrac{d}{A_f}\right) F_y} \\ r_T\sqrt{\dfrac{102{,}000C_b}{F_y}} \end{array} \right\} \qquad 15.4$$

The *moment gradient multiplier*, $C_b$, can be used to increase $L_u$, but $C_b$ is never used with $L_c$ (i.e., $C_b$ is used only when $L_c < L_b \leq L_u$). The conservative case of $C_b = 1$ is almost always taken. The moment gradient multiplier is described in more detail in Appendix D of this chapter. $r_T$ is tabulated in the *AISC Manual* for each beam but is approximately $1.2r_y$.

[2] Equation 15.4 can be derived by setting AISC equations F1-6 and F1-8 equal to $0.60F_y$.

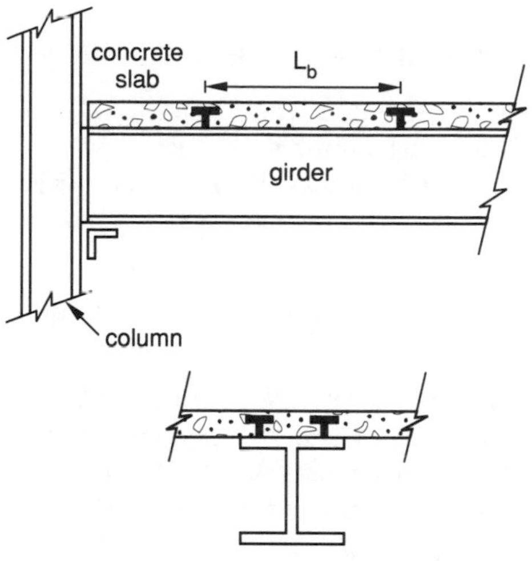


**Figure 15.5** Compression Flange Bracing Using Headed Studs

## 14 BEAM DEFLECTIONS

Steel beam deflections are calculated using traditional beam equations. Deflection limitations are typically unique to each design situation. The *AISC Commentary* suggests, but does not require, some general guidelines to maintain appearance and occupant confidence in a structure.[3] Chapter L of the *AISC Commentary* suggests the following guidelines:

- The depths of fully-stressed floor beams and girders should not be less than $F_y/800$ times their spans.
- The depths of fully-stressed roof purlins should not be less than $F_y/1000$ times their spans.
- To attenuate floor vibrations, the depths of steel beams should not be less than 1/20 of their spans.

The ratio is 1/360 for live load deflections of beams supporting plastered ceilings.

## 15 BENDING STRESS IN STEEL BEAMS

Elastic design and analysis of one-span beams is carried out with simple bending theory equations. Unless the beam is very short, it should be sized or analyzed with the flexure stress equation and subsequently checked for shear. Since both $c$ and $I$ are constant for any specific beam, the *section modulus S* can be used from the AISC tables.

$$f_b = \frac{Mc}{I} = \frac{Md}{2I} = \frac{M}{S} \qquad 15.5$$

[3] Special provisions in *AISC Commentary K2* are used to check for ponding on flat roofs.

## 16 ALLOWABLE BENDING STRESS

### A. W-SHAPES BENDING ABOUT MAJOR AXIS

For W-shapes loaded in the plane of their webs and bending about the major axis, the allowable bending stress will be $0.66F_y$ or less. If $L_b \leq L_c$, or if the bracing is continuous, and the beam is compact, the allowable stress is $0.66F_y$.

If $L_u \geq L_b > L_c$, then the basic bending allowance of $0.60F_y$ applies.

If $L_b > L_u$, or if there is no bracing at all between support points, then the allowable bending stress is less than $0.60F_y$. Equations **[F1-6]**, **[F1-7]**, and **[F1-8]** must be used to determine the actual value of $F_b$.[4]

### B. WEAK-AXIS BENDING

If a doubly-symmetrical rolled shape is placed such that bending will occur about its weak axis, the allowable bending stress is $0.75F_y$.[5] The shape must be compact, and other conditions may also apply. However, this is not an efficient use of the beam, so this configuration is seldom used.[6]

$0.75F_y$ is also the allowable bending stress for solid square, solid round, and solid rectangular shapes bending about the weak axis.

### C. INTERMEDIATE CASES

For both strong-axis and weak-axis bending, special cases exist for non-compact beams with $L_b < L_c$. In such instances, the so-called *blending formulas* (**[F1-3]** and **[F1-4]**) are used to produce intermediate values of $F_b$.[7]

## 17 SHEAR STRESS IN STEEL BEAMS

Only the web is assumed to carry shear in W shapes. The statical moment concept is almost never used in steel design, and the average shear stress is compared against the shear stress limitations.

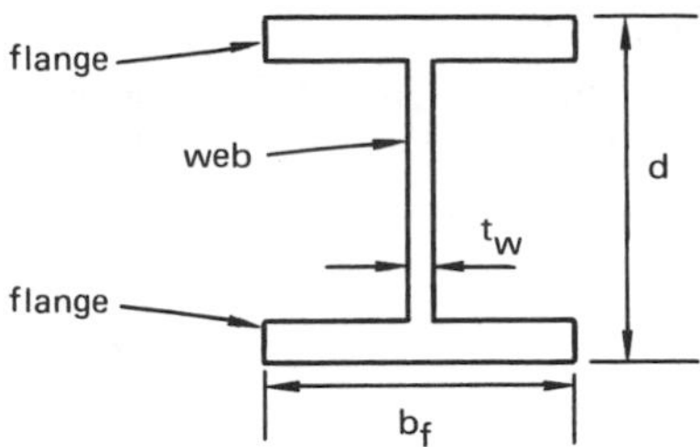


**Figure 15.6** A Steel Beam

The average shear stress in the web is

$$f_v = \frac{V}{A_w} = \frac{V}{dt_w} \qquad 15.6$$

The maximum allowable shear stress in the web of a beam is $F_v = 0.40F_y$.[8]

## 18 LOCAL BUCKLING[9]

Local buckling is a factor in the vicinity of large concentrated loads. Such loads may occur where a column frames into a supporting girder, or at reaction points. *Vertical buckling* and *web crippling*, two types of local buckling, can both be reduced or eliminated by use of *stiffeners*.

**Figure 15.7** Local Buckling

If the load is applied uniformly over a large enough area (say, along $N$ inches of beam flange or more), no stiffeners will be required. If the maximum stress at the toe of the beam fillet is to be kept below $0.66F_y$, the minimum length, $N$, is specified by equation 15.7 (**[K1-2]**) for interior loads, and equation 15.8 (**[K1-3]**) for reactions at beam ends. $k$ is the flange-to-web toe fillet distance tabulated in the shape tables.

$$N_{\text{min}} = \frac{\text{load}}{0.66F_y t_w} - 5k \quad \text{(interior)} \qquad 15.7$$

$$N_{\text{min}} = \frac{\text{reaction}}{0.66F_y t_w} - 2.5k \quad \text{(ends)} \qquad 15.8$$

[4] Actually, the allowable tensile bending stress remains at $0.60F_b$. Only the allowable compressive bending stress is reduced. For traditional design using symmetrical shapes, however, the tensile stress and the compressive stress are the same.

[5] It may seem curious that $F_b$ is larger for bending about the weak axis. The reason is to account for different failure modes. About the strong axis, the beam will fail by buckling of the compression flange and twisting about the minor axis. When bent about the weak axis, the beam fails by yielding.

[6] Nevertheless, weak-axis bending may occur, particularly in the case of beam-columns. Allowable stress for weak-axis bending is a factor in interaction equations such as equation 15.48.

[7] The name *blending formula* is used because the allowable stress for strong-axis bending will be between $0.60F_y$ and $0.66F_y$.

[8] Different limitations apply to shear stress in plate girders, bolts, and rivets.

[9] Lateral buckling has been covered previously.

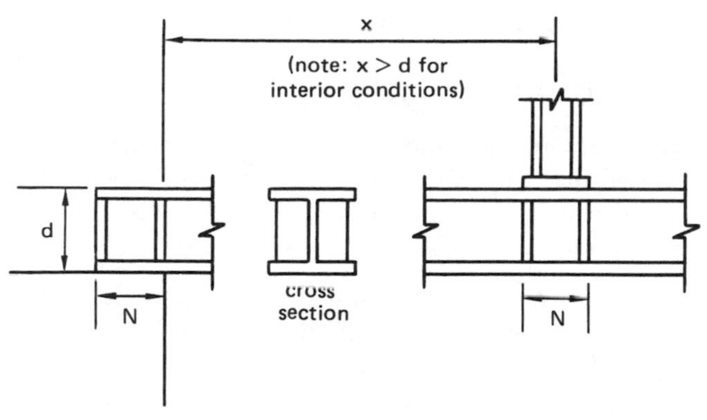


**Figure 15.8** End and Interior Bearing Stiffeners

*Intermediate stiffeners* (i.e., web stiffeners spaced throughout a stock rolled shape) are never needed with rolled shapes, but are typically present in plate girders. (For built-up beams, diagonal buckling requirements should also be checked.)

*Bearing stiffeners* are typically *web stiffeners* constructed as plates welded to the webs and flanges of rolled sections. *Flange stiffeners* are typically angles placed at the web-flange corner used to keep the flange perpendicular to the web. Flange stiffeners cannot be used in place of bearing stiffeners.

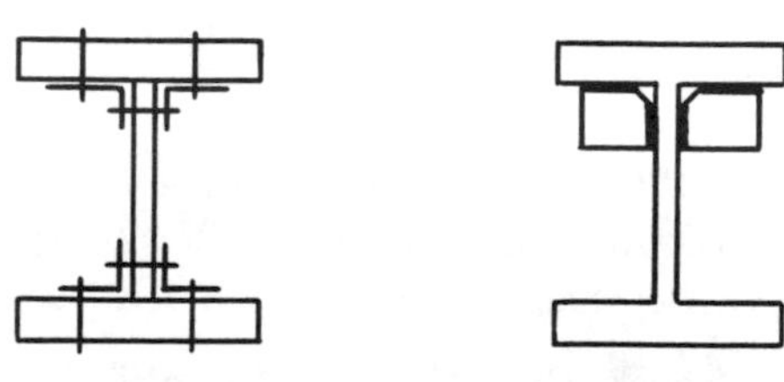

**Figure 15.9** Flange Stiffening

## 19 BEAM DESIGN AND ANALYSIS

Steel beams can be designed and analyzed with *allowable stress* or *plastic design procedures.*[10] Simple one-span beams are usually designed by the allowable stress method since there is no advantage to using load factor design procedures for single-span beams.

Beams are almost always constructed from $F_y = 36$ or $F_y = 50$ ksi steel.

Both beam design and analysis by the allowable stress method have the same standard procedures:

- Check or design for bending stress[11]
- Check shear stress
- Check deflections
- Check compactness (unnecessary if beam table or beam chart was used)
- Design for local buckling

## 20 BEAM DESIGN BY TABLE AND CHART

Since a trial and error solution to a design problem would be time consuming, the *AISC Manual* provides two simple methods of choosing beams: the *Allowable Stress Design Selection Table* and the *Allowable Moments in Beams Chart.*

### A. BEAM TABLE USE

The *Allowable Stress Design Selection Table* is easy to use, and it provides a method of quickly selecting economical beams. Its use assumes that either the required section modulus, $S$, or required resisting moment, $M_R$, is known, and uses one or the other of these criteria to select the beam. It does not check shear stress or deflection, and it is up to the designer to make sure that $L_b \leq L_c$.

Beams are arranged in groups in the beam selection table. Within a group, the most economical shape will be listed at the top in bold print. This is the beam that should generally be used, even if the moment resisting capacity and section modulus are greater than necessary. The weight of the most economical beam will be less than the beam in the group which most closely meets the structural requirements.

Care must be taken not to confuse the 50 ksi columns with the 36 ksi columns. Both steel strengths are listed in the table, and it is easy to use one when the other is needed.

### B. ALLOWABLE MOMENTS CHART

When the unbraced length, $L_b$, is greater than $L_c$, the *Allowable Moments in Beams* chart should be used. The chart for 36 ksi steel can be used with unbraced lengths up to 108 feet. Each beam that is plotted on the chart is shown with a profile similar to that in figure 15.10.

[10] The allowable stress method is also known as *elastic design* and *working stress design.* The *load factor method* is also known as *plastic design* and *ultimate strength design.* Load factor design is also appropriate when design must be "... in accordance with Chapter N of the *AISC Specifications* ..."

[11] If the live load is known but the self-load of the beam is not, it will be necessary to assume a beam weight prior to the selection. A second iteration can be used to obtain a different beam if necessary.

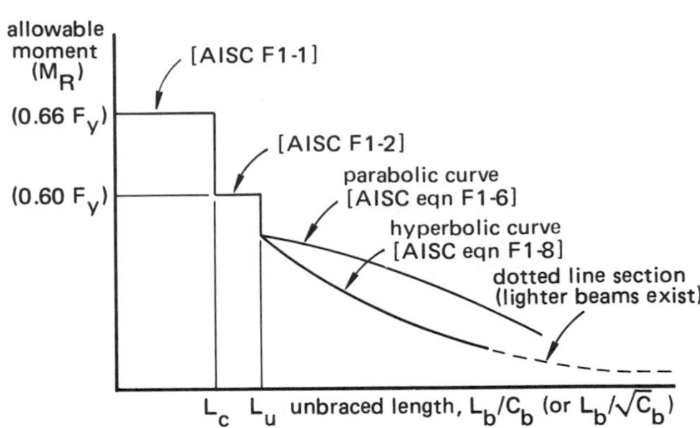


**Figure 15.10** Allowable Moments for a Single Beam

The chart is entered knowing $L_b$ and $M_R$.[12] (An allowance for self-weight should have been included when calculating $M_R$.) The nearest solid line above the intersection of the $L_b$ and $M_R$ values is the most economical beam. Dotted line sections mean that a lighter beam exists which has the same capacity.

As with the beam selection table, shear stress and deflection must still be checked. Care must be taken, also, not to mix up the 36 ksi and 50 ksi portions of the chart.

*Example 15.2*

Select a W shape ($F_y = 36$ ksi) to carry a maximum moment of 140,000 ft-lbf. The compression flange is braced every 6 feet. Disregard deflection and shear stress criteria.

This problem is perfect for using the beam chart, since both the unbraced length and required moment are known.

Entering the chart with $L_b = 6'$ and $M_R = 140$ ft-kip, a W 21×44 beam is selected. This beam has a moment carrying capacity of approximately 163 ft-kips, and several beams were skipped which had smaller capacities, but which still met the 140 ft-kip requirement. However, this is the lightest beam.

If the beam table had been used, the same beam would have been selected. However, it would have been necessary to verify that $L_b < L_c$.

[12] If unbraced length varies along the beam span, use the longest unbraced length with this chart.

## 21 PLASTIC BEAM DESIGN

### A. BASIC PROCEDURE[13]

*Plastic design*, also known as *ultimate strength design*, is based on comparing a factored moment to the ultimate capacity of the beam. The design is accomplished without stress calculations.

Plastic design is ideally suited to continuous beams and frames, and is not normally used with single-span beams. It must not be used with non-compact shapes, crane runway rails, A514 steel, or steels with yield strengths in excess of 65 ksi.

*step 1*: Multiply the dead and live loads by 1.7 to obtain the *factored loading* on the beam.[14]

$$\text{factored load} = 1.7 \times \text{dead load} + 1.7 \times \text{live load} \qquad 15.9$$

*step 2*: Based on the factored loading, calculate the maximum (plastic) moment, $M_p$. (This moment cannot be calculated from elastic moment diagrams.)

*step 3*: If beam selection is to be made according to the required *plastic section modulus*, calculate $Z_x$.

$$Z_x = \frac{M_p}{F_y} \qquad 15.10$$

*step 4*: Use the *Plastic Design Selection Table* in the *AISC Manual* to select an economical beam. Use of this table is similar to that of the *Allowable Stress Design Selection Table.*

*step 5*: Check that the maximum factored shear based on plastic failure does not exceed the allowable shear (**[N5-1]**).

$$V_{\text{factored, max}} = \leq 0.55 F_y t_w d \qquad 15.11$$

*step 6*: Specify web (intermediate) stiffeners at loading points where plastic hinges are expected.

*step 7*: Determine lateral bracing requirements. Compression flange support (lateral bracing) is required at points where plastic hinges will form. In addition, the distance between points of lateral support must be less than

[13] The procedure presented here for beam selection assumes that the *Plastic Design Selection Table* will be used, and therefore, compactness is assumed. The procedure also omits thickness ratio checks that must be performed when the beam carries axial loads in addition to its transverse loading.

[14] Other factors are used when wind and earthquake loads are present.

STEEL

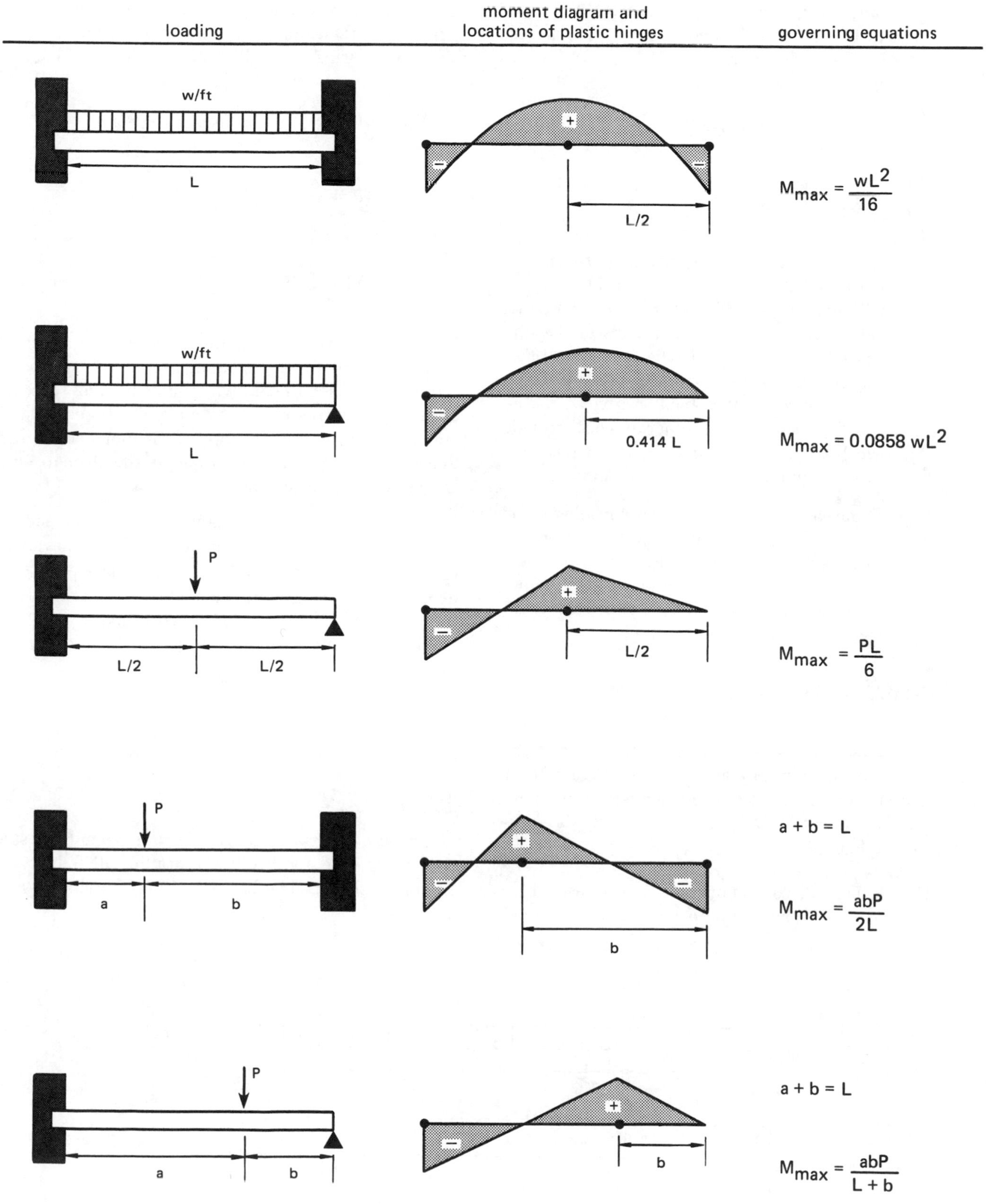
loading
moment diagram and locations of plastic hinges
governing equations
w/ft
L
+
−
L/2
$M_{max} = \frac{wL^2}{16}$
w/ft
L
0.414 L
$M_{max} = 0.0858\ wL^2$
P
L/2
L/2
$M_{max} = \frac{PL}{6}$
P
a
b
a + b = L
$M_{max} = \frac{abP}{2L}$
P
a
b
a + b = L
$M_{max} = \frac{abP}{L + b}$

**Figure 15.11** Maximum Plastic Moments

(● = plastic hinge)

or equal to the laterally unsupported distances given by equations 15.12 ([**N9-1**]) and 15.13 ([**N9-2**]). $l_{cr}$ and $r_y$ are both in inches for these two equations.

$$\frac{l_{cr}}{r_y} = \frac{1375}{F_y} + 25 \text{ when } +1.0 > \frac{M}{M_p} > -0.5 \quad 15.12$$

$$\frac{l_{cr}}{r_y} = \frac{1375}{F_y} \text{ when } -0.5 \geq \frac{M}{M_p} > -1.0 \quad 15.13$$

$r_y$ is the radius of gyration of the member about its weak axis, as determined from the shape table. $M$ is the smaller (absolute value) moment at the two ends of the unbraced segment. $M_p$ is the maximum plastic moment on the span. $M/M_p$ is positive when the segment is bent in reverse curvature (i.e., the moment diagram goes through zero), and negative when the segment is bent in single curvature.

Some easing of the $l_{cr}$ distance is allowed for the last hinge to form in the failure mechanism, since having other hinges form is tantamount to having the beam in failure. However, normal bracing lengths for elastic design must still be met.

## B. ULTIMATE MOMENTS

Step 2 requires calculating the maximum moment based on plastic theory. This moment is not the same as the moment determined from elastic theory. Figure 15.11 can be used in simple cases to determine the ultimate moment, $M_p$. Locations of plastic hinges are also indicated in the figure.

A uniformly-loaded, continuous beam is a case which occurs frequently. The ultimate moments can be derived from the cases in figure 15.11. Both ultimate moments should be checked, since the interior hinges may form first if the span lengths are short. (The end span hinges will form first if all spans are the same length.)

$$M_1 = \left(\frac{3}{2} - \sqrt{2}\right) wL_1^2 \approx 0.0858wL_1^2$$

$$M_2 = \frac{wL^2}{16}$$

## C. ULTIMATE SHEARS

The ultimate shear in step 5 may or may not be the factored shear on the beam. The shear, by definition, is the slope of the moment diagram. In the case of a uniformly-loaded beam with built-in ends, the effect of plastic failure is merely to move the baseline, without changing the overall shape of the moment diagram. Since there is no change in the slope, the ultimate shear is merely the factored shear.

In the case of continuous beams, however, the effect of plastic failure will change the slope of the lines in the moment diagram. Graphical or analytical means can be used to obtain the maximum slope. For a uniformly-loaded continuous beam, as shown in figure 15.12, the maximum shear induced will be

$$V_{\max} = \frac{wL}{2} + \frac{M_{\max}}{L} = 0.5858wL \quad 15.14$$

*Example 15.3*

**Use plastic design to select a W shape beam (A36 steel) to support dead and live loads totalling 4000 lbf/ft over the beam shown. The beam is simply supported at all three points.**

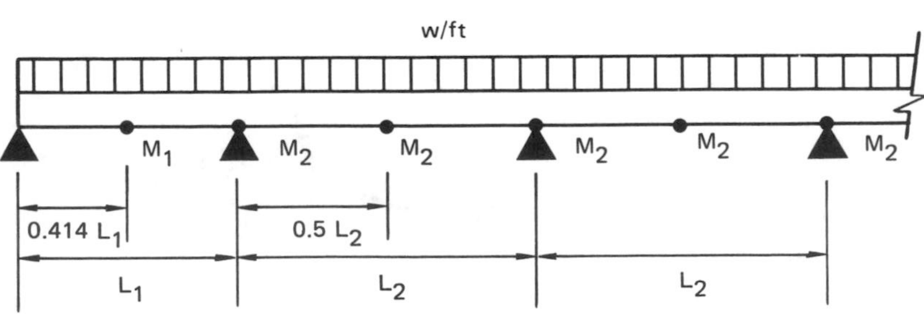


**Figure 15.12** Ultimate Moments on a Uniformly-Loaded Continuous Beam

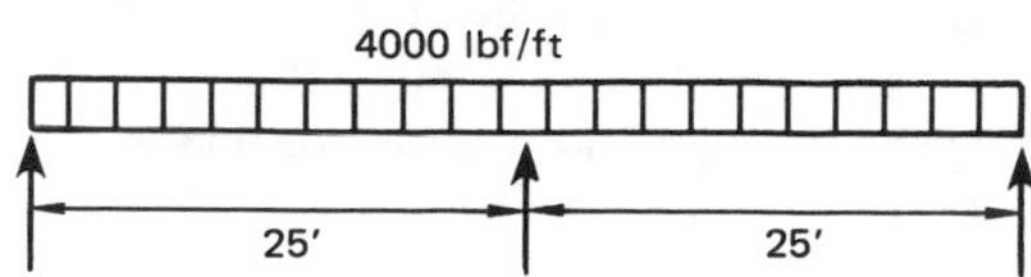


*step 1*: The factored load is

$$w = (1.7)(4000) = 6800\,\text{lbf/ft}$$

*step 2*: The maximum moment on an end-span of a uniformly-loaded continuous beam is

$$\begin{aligned} M_{\text{max}} &= 0.0858wL^2 = (0.0858)(6800)(25)^2 \\ &= 3.65\,\text{EE5 ft-lbf} \end{aligned}$$

*step 3*: The selection can be made on the basis of required plastic moment. This step is skipped.

*step 4*: Entering the *Plastic Design Selection Table*, a W 24×55 beam is chosen. The capacity is 402 ft-kips, more than required. However, this is the economical beam with a capacity exceeding the requirements.

*step 5*: The maximum shear on an end span of a uniformly-loaded, continuous beam is given by equation 15.14.

$$\begin{aligned} V_{\text{max}} &= 0.5858wL = (0.5858)(6800)(25) \\ &= 99,600\text{ lbf} \end{aligned}$$

From the shape tables, $d = 23.57$ inches, $t_w = 0.395$ inches. From equation 15.11,

$$\begin{aligned} V_{\text{allowable}} &= (0.55)(36,000)(0.395)(23.57) \\ &= 184,341\text{ lbf} \end{aligned}$$

Since $V_{\text{max}} < V_{\text{allowable}}$, shear is not a problem.

## D. CONSTRUCTING PLASTIC MOMENT DIAGRAMS

There are several methods of determining the shape of the plastic moment diagram. Most of these methods are based on theory. The method presented here is a graphical approach which can be used to quickly draw moment diagrams and locate points where plastic hinges will form.

*step 1*: Consider the beam as a series of simply-supported spans.

*step 2*: Draw the elastic bending moment on each span.

*step 3*: Construct the modified base line. This is a jointed set of straight lines that meets the following conditions:

- The base line meets the horizontal axis at simply-supported exterior ends.[15] This is consistent with the requirement that the moment be zero at free and simply-supported ends.
- The slope of the base line changes only at points of support.
- The base lines for all spans connect at points of support.

The base line is located to minimize the maximum ordinate along the entire length of the beam (along all spans).[16] Therefore, one span will control.

*step 4*: The maximum ordinate determines $M_p$.

*step 5*: Hinges form at points of maximum ordinates, and wherever else required to support full rotation. The moments at hinges which form simultaneously are identical.

*Example 15.4*

Draw the plastic moment diagram for the uniformly-loaded, two-span beam shown.

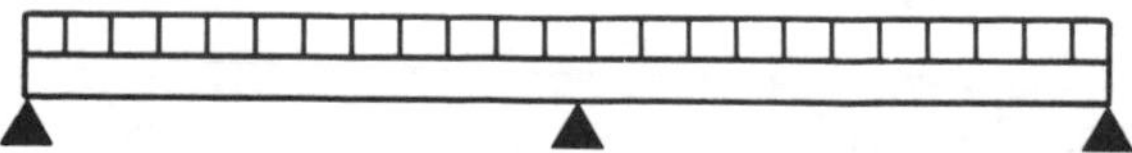

*steps 1 & 2* The moment diagram of a simply-supported, uniformly-loaded single span is drawn twice.

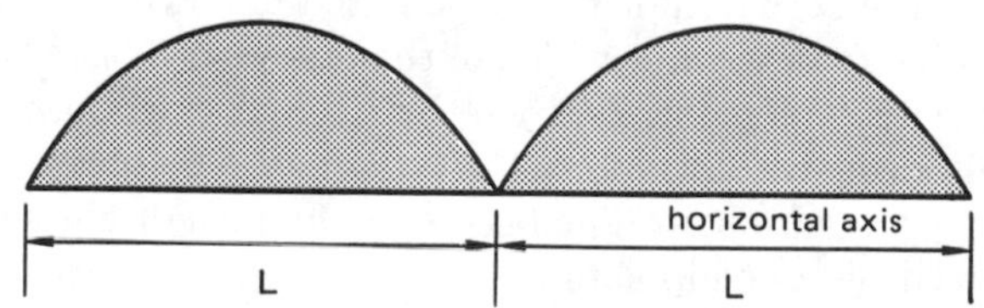


*step 3*: The modified base line is chosen to minimize the distance between the curved line and the base line.

[15] It isn't necessary for the base line to reach the horizontal axis at built-in ends. In fact, the moment is usually non-zero at those points.

[16] Don't consider the distance between the base line and the horizontal axis when minimizing the maximum ordinate.

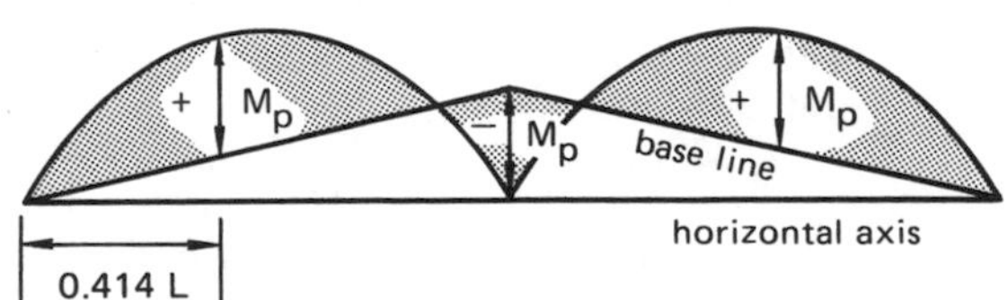


*step 4*: The maximum ordinate is determined visually to be near the middle of the end spans.

*step 5*: If hinges form near the middle of the end spans, a hinge must also form over the center support. Otherwise, the beam could not rotate in failure.

The final moment diagram can be drawn by "straightening out" the modified base line.

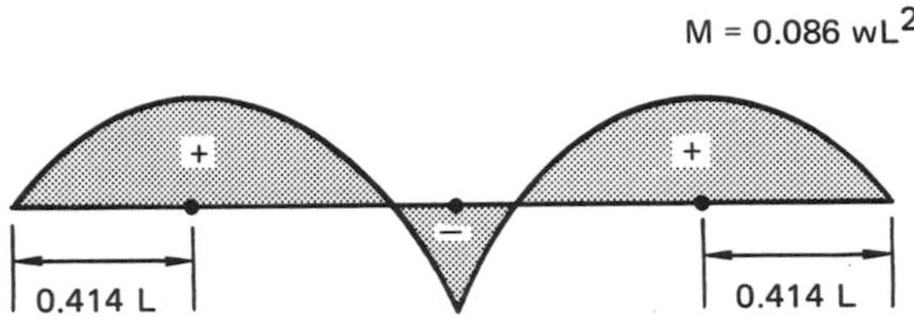


*Example 15.5*

Draw the plastic moment diagram for the beam shown.

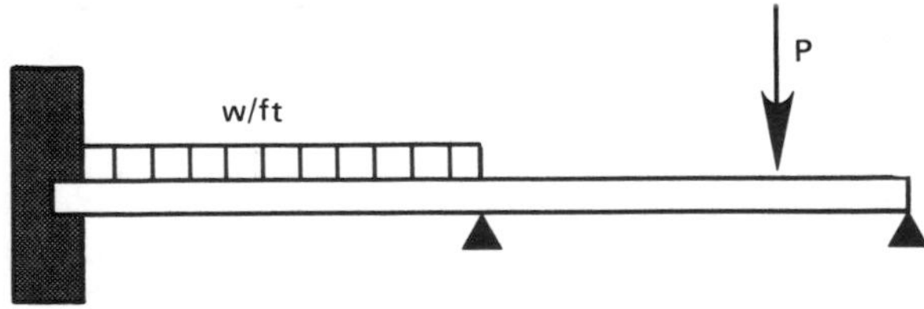


Since the actual numerical loads are unknown, it is not possible to determine which of the two spans controls.[17] If the left span controls, then three hinges must form simultaneously. This forces the base line to be horizontal along the left span. The base line along the right span is fixed by its endpoints.

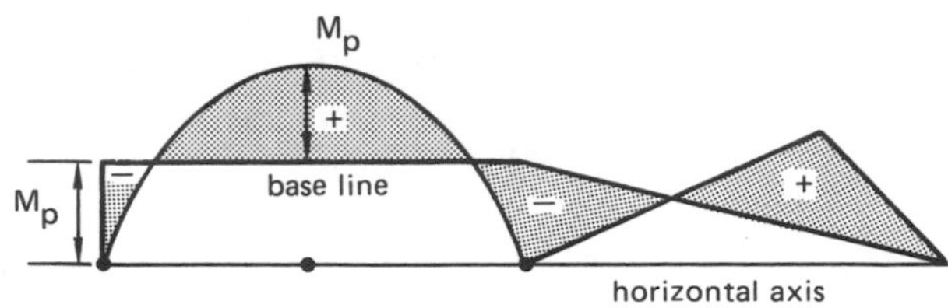


[17] The two alternate methods by which the beam can fail are known (in plastic theory) as *mechanisms*.

If the right span controls, then two hinges must form simultaneously. The moments at these two hinge points are equal. The base line along the left span is chosen to minimize the positive and negative ordinates.

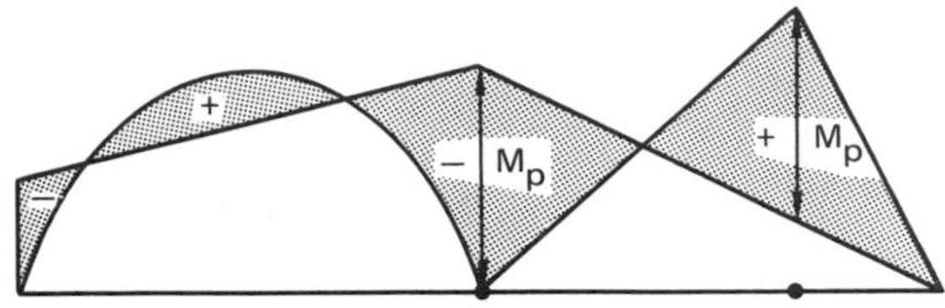


## 22 COLUMNS WITH AXIAL LOADS

### A. INTRODUCTION

Column design and analysis is greatly dependent on the Euler buckling load theory. Specific factors of safety and slenderness ratio limitations separate the design and analysis procedures from purely theoretical concepts, however.

### B. GEOMETRIC TERMINOLOGY

Figure 15.13 shows a *W* shape used as a column. The $I_y$ moment of inertia is smaller than $I_x$. Therefore, the beam would be said to "... buckle about the minor axis ..." if the failure (buckling) mode is as shown. Since this is the expected buckling mode, bracing for the minor axis is usually provided, even if major axis bracing is not.

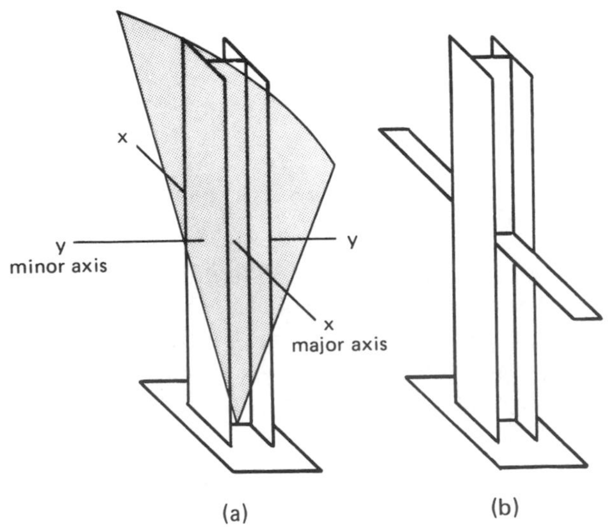


**Figure 15.13** Minor Axis Buckling and Bracing

Associated with each column are two unbraced lengths, $L_x$ and $L_y$.[18] In figure 15.13(b), $L_x$ is the full column height. However, $L_y$ is half the column height, assuming that the brace is placed at mid-height. It is not necessary for $L_x$ and $L_y$ to be identical.

Another important geometric feature is the *radius of gyration*. Since there are two moments of inertia, there are also two radii of gyration. Since $I_y$ is smaller than $I_x$, $r_y$ will be smaller than $r_x$. $r_y$ is known as the *least radius of gyration*.

## C. EFFECTIVE LENGTH

Since the restraints placed on column ends greatly affect a column's stability, an *end-restraint coefficient* is used to modify the unbraced length.[19] Thus, $KL$ is the product of the end-restraint coefficient and the unbraced length, and is known as the *effective length* of the column.

Values of $K$ depend on the conditions at both ends of the column. Either end can experience complete fixity (which is, in practice, impossible to achieve) to zero fixity (as in a free-standing sign post or flagpole). Table 15.5 lists recommended values of $K$ for use with steel columns.[20]

**Table 15.5**
End-Restraint Coefficients, $K$
(Also see table 12.3)

| end #1 | end #2 | $K$ |
|---|---|---|
| built-in | built-in | 0.65 |
| built-in | pinned | 0.80 |
| built-in | rotation fixed, translation free | 1.2 |
| built-in | free | 2.1 |
| pinned | pinned | 1.0 |
| pinned | rotation fixed, translation free | 2.0 |

The values of $K$ specified in table 15.5 do not require a prior knowledge of the column size or shape designation. However, if an existing design for which all column and framing members are known is being evaluated, the alignment charts in figure 15.14 and 15.15 can be used to obtain a more accurate end-restraint coefficient.

[18] Even if the column is braced in one or both directions, the *unbraced length* is the distance between braces.

[19] This coefficient is also known as the *effective length coefficient*.

[20] These are not the theoretical values often quoted for use with Euler's equation. They are slightly different, as recommended by the American Institute of Steel Construction in its *Commentary*.

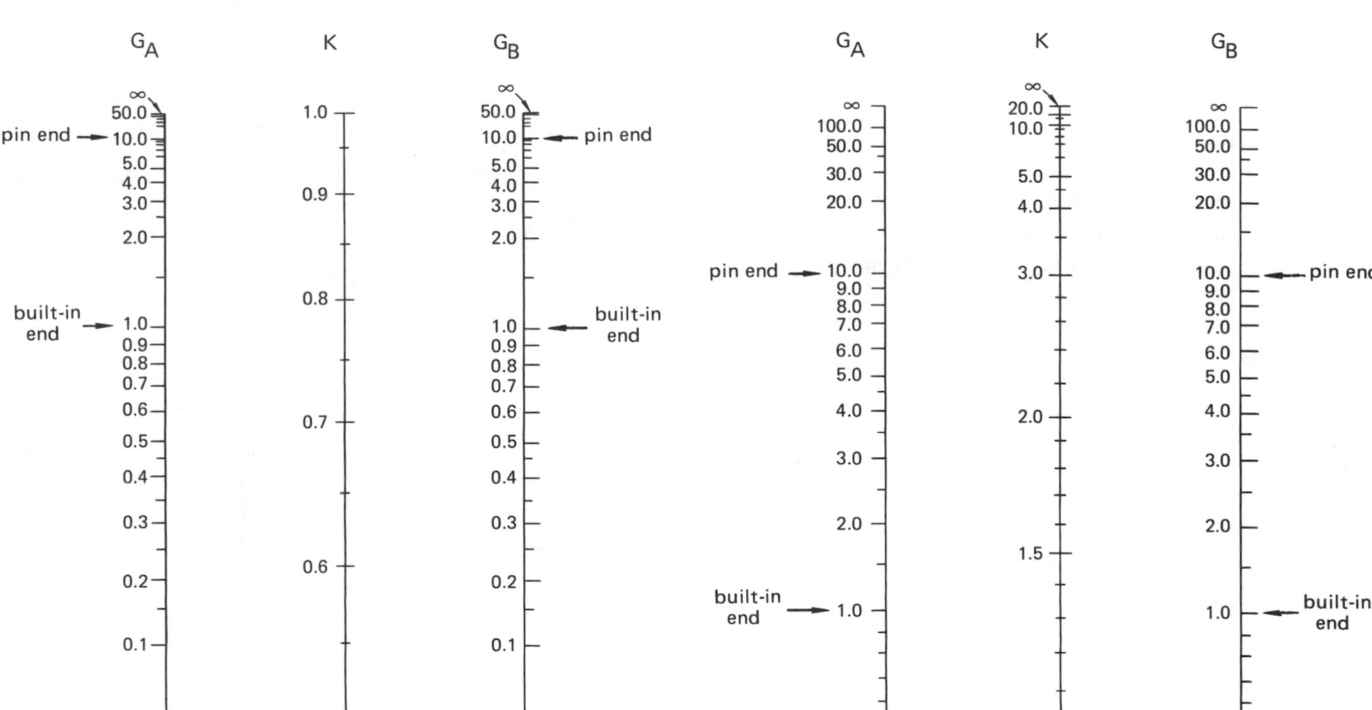


**Figure 15.14** Alignment Chart When Sidesway is Inhibited

**Figure 15.15** Alignment Chart When Sidesway is Uninhibited

To use the alignment charts, the *end condition coefficients*, $G_A$ and $G_B$, need to be calculated for the two column ends, $A$ and $B$. (The alignment charts are symmetrical. It is not important which end is labeled $A$ or $B$.)

$$G = \frac{\sum_{\text{columns}} \left(\frac{I}{L}\right)}{\sum_{\text{beams}} \left(\frac{I}{L}\right)} \qquad 15.15$$

In calculating $G$, only beams and columns which are in the expected plane of bending (i.e., which resist the tendency to buckle or bend) are included in the summation. Also, only beams and columns rigidly attached are included, since pinned connections do not resist moments.

For ground level columns, one of the column ends will not be framed to beams or other columns. In that case, $G = 10$ (theoretically $G = \infty$) is used for pinned ends, and $G = 1$ (theoretically, $G = 0$) is used for rigid footing connections.

## D. SLENDERNESS RATIO

Steel columns are divided into *long columns* and *intermediate columns*, depending on their slenderness ratios. This slenderness ratio is calculated from equation 15.16.

$$SR = \frac{KL}{r} \qquad 15.16$$

Since there are two values of $r$ (and accompanying values of $K$ and $L$), there will be two slenderness ratios. The maximum slenderness ratio will determine if the column is long or intermediate. The *critical slenderness ratio* is given by equation 15.17. However, there is really no need to calculate $C_c$ for standard steel grades. For 36 ksi steel, a column is long if $SR$ is greater than 126.1 and is intermediate otherwise. For 50 ksi steel, $C_c = 107.0$.

$$C_c = \sqrt{\frac{2\pi^2 E}{F_y}} \qquad 15.17$$

Slenderness ratios of greater than 200 are not recommended ([**B7**]).

## E. ALLOWABLE COMPRESSIVE STRESS

As figure 15.16 shows, the allowable column stress varies with the slenderness ratio.[21]

[21] Very short compression members, those less than about 2 feet in effective length, are governed by different requirements.

The easiest method of determining allowable compressive stress is to use the *Allowable Stress for Compression Members* tables in the *AISC Specifications*. However, the allowable stress can also be calculated for steels with non-standard yield strengths.

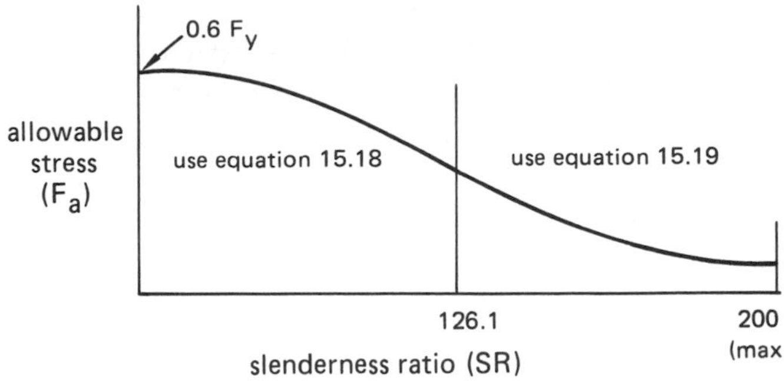


**Figure 15.16** Allowable Column Compressive Stress ($F_y = 36$ ksi)

Up to a slenderness ratio of $C_c$, equation 15.18 ([**E2-1**]) gives the allowable stress.

$$F_a = \frac{\left[1 - \frac{(KL/r)^2}{2C_c^2}\right] F_y}{\frac{5}{3} + \frac{3(KL/r)}{8C_c} - \frac{(KL/r)^3}{8C_c^3}} \qquad 15.18$$

When the slenderness ratio exceeds $C_c$, equation 15.19 ([**E2-2**]) must be used.

$$F_a = \frac{12\pi^2 E}{23(KL/r)^2} \qquad 15.19$$

An alternate method of calculating the allowable compressive stress for steels with non-standard yield strengths is possible. Equation 15.20 calculates $F_a$ from the yield strength and a reduction coefficient, $C_a$. Values of $C_a$ are listed in *Table 3, Numerical Values* section of the *AISC Specifications*.

$$F_a = C_a F_y \qquad 15.20$$

(Equation 15.21 is omitted in this edition.)

## F. COLUMN ANALYSIS

The procedure for analyzing the adequacy of a column is essentially one of verifying that the actual compressive stress does not exceed the allowable stress.

*step 1*: Obtain the shape properties $A$, $r_x$, and $r_y$, as well as the unbraced lengths $L_x$ and $L_y$.

*step 2*: Obtain $K_x$ and $K_y$ from table 15.5 or from the alignment charts (figures 15.14 and 15.15).

*step 3*: Use equation 15.22 to calculate the maximum slenderness ratio.

$$SR = \max\left\{\begin{array}{c}\frac{K_x L_x}{r_x}\\ \frac{K_y L_y}{r_y}\end{array}\right\} \qquad 15.22$$

*step 4*: Use equation 15.17 to calculate the critical slenderness ratio, $C_c$. (For 36 ksi steel, $C_c = 126.1$; for 50 ksi steel, $C_c = 107.0$. This step is not necessary if the *AISC Table of Allowable Stresses for Compression Members* is used.)

*step 5*: Use the *Allowable Stress for Compression Members* table to obtain the allowable compressive stress. Alternatively, use either equation 15.18 or 15.19 (depending on $SR$).

*step 6*: Compare the actual load to the maximum allowable load on the column.[22]

$$P_{\max} = F_a A \qquad 15.23$$

## G. COLUMN DESIGN

Trial and error column selection is next to impossible. Accordingly, the *AISC Manual* provides column selection tables which make it fairly easy to select a simple column based on the required column capacity.

*step 1*: Determine the load to be carried. Include an allowance for the column weight.

*step 2*: Based on preliminary choices for the column design, determine the end-restraint coefficients, $K_y$ and $K_x$, for the column. Calculate the effective length assuming that buckling will be about the minor axis.[23]

$$\text{effective length} = K_y L_y \qquad 15.24$$

*step 3*: Enter the table and locate a column which will support the required load with an effective length of $K_y L_y$.

*step 4*: Check for buckling in the strong direction. Calculate $L'_x$ from equation 15.25 and the ratio $r_x/r_y$ tabulated in the column tables.

$$L'_x = \frac{K_x L_x}{r_x/r_y} \qquad 15.25$$

*step 5*: If $L'_x < K_y L_y$, the column is adequate and the procedure is complete. Go to step 8.

*step 6*: If $L'_x > K_y L_y$ but the column chosen can support the load at a length of $L'_x$, the column is adequate and the procedure is complete. Go to step 8.

*step 7*: Choose a larger member which will support the load at a length of $L'_x$. (The ratio $r_x/r_y$ is essentially constant.)

*step 8*: If sufficient information on other members framing into the column is available, use the alignment charts (figures 15.14 and 15.15) to check the values of $K$ used.

*Example 15.6*

Choose a W14 shape to support a 2000 kip concentric load. The unbraced column length is 11 feet in both directions. Use $K_y = 1.2$ and $K_x = 0.80$. $F_y = 36$ ksi.

*step 1*: Assume a column weight of approximately 500 lbf/ft. The load to be carried is

$$P = 2000 + \frac{(500)(11)}{1000} \approx 2005 \text{ kips}$$

*step 2*: From equation 15.24, the effective length for minor axis bending is $(1.2)(11) = 13.2'$ (say $13'$).

*step 3*: From the column table for 36 ksi steel, select a W 14×370 shape with a capacity of 2121 kips.

*step 4*: From the table, $r_x/r_y = 1.66$. Using equation 15.25,

$$L'_x = \frac{(0.80)(11)}{1.66} = 5.30$$

*step 5*: Since $5.30 < 13$, the column selected is adequate.

[22] The gross area of the column is used to calculate the allowable stress. *Lacing bars* do not contribute to gross area. However, *cover plates* may be used to carry column load.

[23] If buckling will occur about the major axis, the table cannot be used directly. See steps 4 through 7.

*Example 15.7*

Design a 25 foot long A36 W shape main member column to support a 375,000 pound live load. The base is rigidly framed in both directions. The top is rigidly framed in the weak direction and fixed against rotation (i.e., cannot rotate) in the strong direction, but translation in the strong direction is possible.

Assume the column dead weight will be about 2000 pounds. Then, the actual load will be 375,000 + 2000 = 377,000 pounds.

From the information about framing, the end restraint coefficients are $K_y = 0.65$ and $K_x = 1.2$.

The effective lengths are

$$L_y = (0.65)(25) = 16.25 \text{ ft}$$
$$L_x = (1.2)(25) = 30 \text{ ft}$$

Use the column selection table to find a column capable of supporting 377 kips with an effective length of 16 feet. Try a W 12×79 beam. This beam has $r_x/r_y = 1.75$. Then, from equation 15.25,

$$L'_x = \frac{30}{1.75} = 17.1$$

Since 17.1 > 16.25, the strong axis controls. Enter the table looking for a 17 foot (effective length) column capable of supporting 377 kips. The same column has sufficient capacity, and the column selection is complete.

## 23 STABILITY OF PLATES IN COMPRESSION

It is possible that *local buckling* in one of the plate elements of a rolled shape or built-up compression member may occur before buckling based on the slenderness ratio becomes the governing factor. The load-carrying ability of the column will be reduced if such local buckling occurs.

The ability of plate sections to carry compressive loads without buckling is determined by the *width-thickness ratio*, b/t. For the purpose of specifying limiting width-thickness ratios, compression elements are divided into stiffened elements and unstiffened elements. *Stiffened elements* are supported along two edges. Examples are webs of W shapes and sides of box beams. *Unstiffened elements* are supported along one edge only. Flanges of *W* shapes and sides of angles are unstiffened elements.[24]

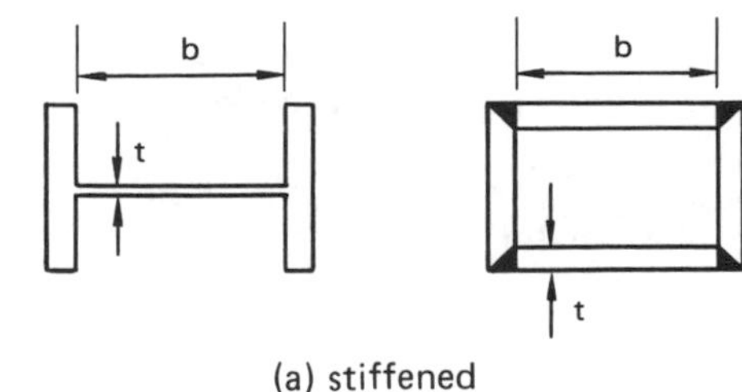


(a) stiffened

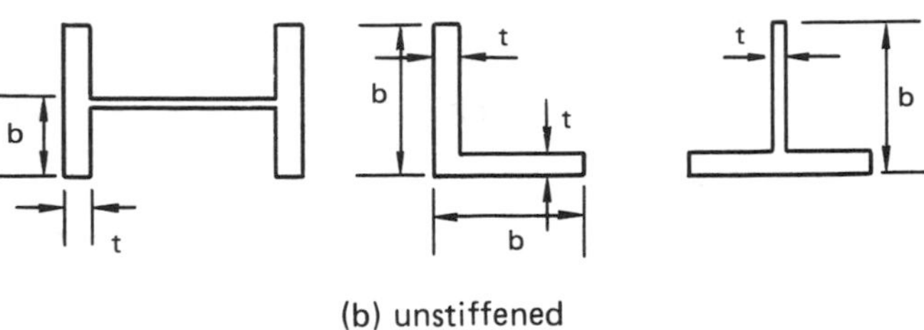


(b) unstiffened

**Figure 15.17** Stiffened and Unstiffened Elements

To prevent local buckling, the *AISC Specifications* (*section B5*) require equation 15.26 to be met if the plate element is to be considered fully effective.[25]

$$\frac{b}{t} \leq \frac{H}{\sqrt{F_y}} \qquad 15.26$$

Values of $H$ are listed in table 15.6.

**Table 15.6**
$H$ Values for Width-Thickness Ratios

| element | $H$ |
|---|---|
| unstiffened elements | |
| stems of tees | 127 |
| double angles in contact | 95 |
| compression flanges of beams | 95 |
| angles or plates projecting from girders, columns, or other compression members, and compression flanges of plate girders | $95/\sqrt{k_c}$ |
| ($k_c = \frac{4.05}{(h/t)^{0.46}}$ if $h/t > 70$; otherwise, $k_c = 1$.) | |
| stiffeners on plate girders | 95 |
| flanges of tees and I-beams (use $\frac{1}{2}b_f$) | 95 |
| single angle struts or separated double angle struts | 76 |
| stiffened elements | |
| square and rectangular box sections | 238 |
| cover plates with multiple access holes | 317 |
| other uniformly compressed elements | 253 |

[24] If a *W* shape is selected from the column selection tables in the *AISC Manual*, the width-thickness ratios do not generally need to be evaluated. However, compression members constructed from double tees, structural tubing, and plate girders must be checked.

[25] The width-thickness ratios are applicable for elastic (working stress) design, and should not be used for inelastic design. Other provisions govern the width-thickness ratio of plate girder flanges, as well.

STEEL

Circular tubular sections are considered to be fully effective according to the ratio of outside diameter to wall thickness.

$$\frac{D}{t} \leq \frac{3300}{F_y} \qquad 15.27$$

If the width-thickness ratios are exceeded by unstiffened compression members, the allowable compressive stress is reduced by a strength reduction factor, $Q_s$.[26]

$$F_a' = Q_s F_a \qquad 15.28$$

Stiffened compression elements exceeding the width-thickness ratios are handled differently. An *effective width*, $b_e$, is used in place of the actual width when calculating the flexural design properties and the permissible axial stress.

The specific provisions for determining $Q_s$ and $b_e$ are contained in *Chapter B* of the *AISC Specifications*.

*Example 15.8*

Two A36 L $9\times4\times\frac{1}{2}$ angles are used with a 3/8″ gusset plate to produce a truss compression member. The short legs are back-to-back, making the long legs unstiffened elements. Can the combination fully develop compressive stresses?

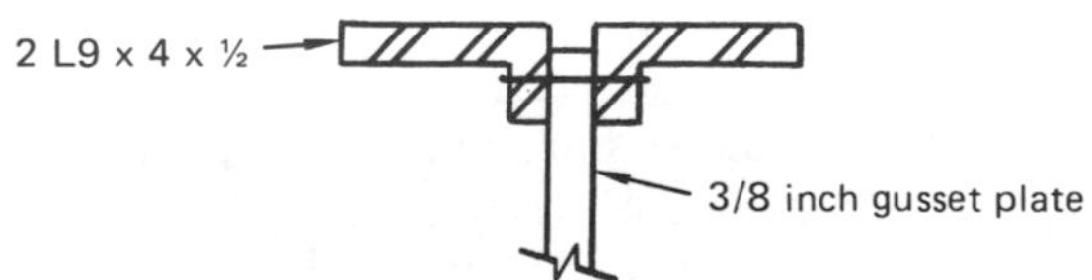


The limitation on the unstiffened separated double angles is

$$\frac{76}{\sqrt{F_y}} = \frac{76}{\sqrt{36}} = 12.67$$

The actual width-thickness ratio is

$$\frac{b}{t} = \frac{9}{0.5} = 18$$

Since 18 > 12.67, local buckling will control, and a reduced stress factor, $Q_s$, must be used when calculating the allowable compressive load on the truss member.

[26] Equation 15.28 implies that $F_a$ is the same as for members that meet the width-thickness ratios. Actually, $Q_s$ is also incorporated into the calculation of the critical slenderness ratio, $C_c$, and the allowable compressive stress, $F_a$.

## 24 MISCELLANEOUS COMBINATIONS IN COMPRESSION

Miscellaneous shapes, including round and rectangular tubing, single and double angles, and built-up sections, can be used as compression members. Generally, sections with distinct strength advantages in one plane (e.g., double angles or tees) should be used when bending is confined to that plane. For example, a double angle member could be used as a compression strut in a truss or as a spreader bar used for hoisting large loads.

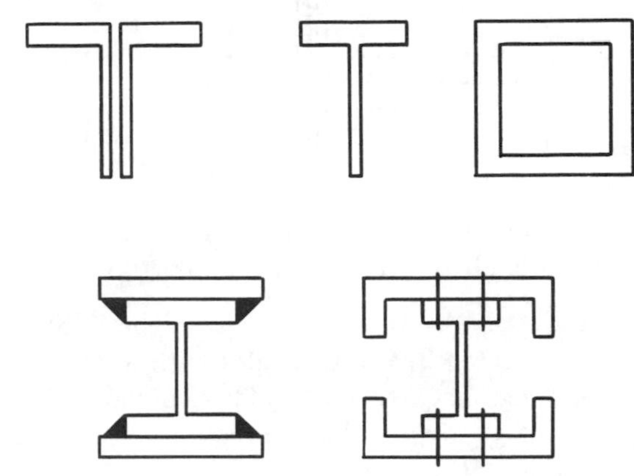

**Figure 15.18** Miscellaneous Compression Members

Design and analysis of these compression members is similar to the design and analysis of *W* shape columns. The same equations are used for calculating the allowable stress, $F_a$. It is, however, essential to check the width-thickness ratios for all elements in the compression members, including both flanges and stems of tees.

Where spot welding or stitch riveting is used to combine two shapes into one (as is done with double angles), the spacing of the connections must be sufficient to prevent premature buckling of one of the shapes. Once the maximum slenderness ratio has been determined (based on $x$- and $y$-directions), the spacing between connections can be calculated from equation 15.29, in which $r_z$ is the least radius of gyration for a single angle (as read from the shape table).

$$L_b = \frac{(SR_{\max})r_z}{K} \qquad 15.29$$

Column load tables have been prepared for many combinations of double angles, tees, round pipe, and structural tubing. Single angle compression members are difficult to load concentrically, and special methods must be used for them.

*Example 15.9*

Design a double-angle strut, 10 feet long, to support an axial compressive load of 40 kips. Use A36 steel and assume $K = 1$ for all cases. The gusset plates at the ends are 3/8″ thick.

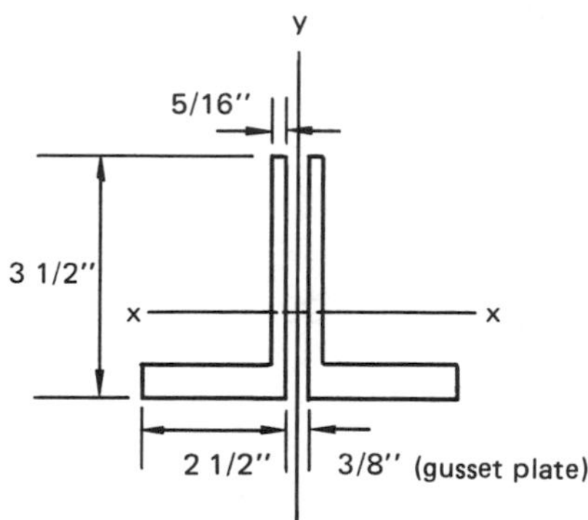


The weight of the strut will be insignificant compared to the axial load. So, the self-weight is ignored.

From the *Double Angle column tables* in the *AISC Manual*, try two L $3\frac{1}{2}\times2\frac{1}{2}\times5/16$ angles with a 3/8″ gusset plate, long legs back to back. This configuration has a capacity of 42 kips in the $x$-direction and 40 kips in the $y$-direction. (Note that a slightly more economical selection would be $4 \times 3 \times \frac{1}{4}$ with a capacity of 44 kips in the $x$-direction and 40 kips in the $y$-direction.) From the table, the properties are

$$A = 3.55$$
$$r_x = 1.11$$
$$r_y = 1.10$$

The controlling slenderness ratio is the larger value of KL/r.

$$\frac{K_x L}{r_x} = \frac{(1)(10)(12)}{1.11} = 108.1$$
$$\frac{K_y L}{r_y} = \frac{(1)(10)(12)}{1.10} = 109.1 \quad \text{(controls)}$$

The allowable compressive stress can be calculated from equation 15.18 or read directly from *AISC Specifications Part 3, Table C-36*. $F_a = 11.81$ kips. The actual compressive stress is

$$f_a = \frac{40}{3.55} = 11.3\,\text{ksi} \quad \text{(ok)}$$

All combinations of double angles in the *AISC Manual* meet the width-thickness ratio requirements. However, the width-thickness ratio is easily calculated from the angle designation.

$$\frac{b}{t} = \frac{3.5}{5/16} = 11.2$$

The maximum width-thickness ratio for separated double angle struts is

$$\frac{76}{\sqrt{F_y}} = \frac{76}{\sqrt{36}} = 12.67 \quad \text{(ok)}$$

From the shape table, the least radius of gyration is $r_z = 0.54$. Since the maximum slenderness ratio is 109.1, the distance between connections between the two angles should not exceed

$$L = \frac{(109.1)(0.54)}{1} = 58.9''$$

## 25 MEMBERS IN BEARING

### A. PROJECTED AREAS IN CONNECTIONS

The maximum *bearing stress* on projected areas of bolts and rivets in shear connections is ([**J3-4**])

$$F_p = 1.50F_u \qquad 15.30$$

If the connectors, plates, or shapes have different strengths, then the lower value of $F_u$ will govern.

### B. PROJECTED AREAS IN PINNED CONNECTIONS

When a connection is made with a pin in a reamed, drilled, or bored hole, the maximum bearing stress on projected areas is ([**J8-1**])

$$F_p = 0.90F_y \qquad 15.31$$

When parts in contact have different yield stresses, $F_y$ in equation 15.31 is the lesser value.

### C. BEARING STIFFENERS

The bearing stress on the ends (tops and bottoms) of *bearing stiffeners*, and the stress on the contact area between mill sections and those bearing stiffeners, must not exceed ([**J8-1**])

$$F_p = 0.90F_y \qquad 15.32$$

Only the part of the stiffeners outside the flange angle fillet or flange-to-web welds is considered effective in bearing.

### D. BEARING ON MASONRY SUPPORTS AT BEAM ENDS

Beams terminating at bearing connections on masonry, brick, or concrete supports are limited to bearing stresses based on the support material (*Section J9*).

- sandstone and limestone:

$$F_p = 0.40\,\text{ksi} \qquad 15.33$$

- brick in cement mortar (*Section J9*):

$$F_p = 0.25\,\text{ksi} \tag{15.34}$$

- on the full area of a concrete support (*Section J9*):

$$F_p = 0.35f'_c \tag{15.35}$$

For beams resting on only a portion of a concrete support, the allowable bearing pressure is given by equation 15.36. $F_p$ is not to exceed $0.7f'_c$, however (*Section J9*).

$$F_p = 0.35f'_c\sqrt{\frac{A_{\text{support}}}{A_{\text{bearing}}}} \tag{15.36}$$

Bearing plate area is determined by the load and $F_p$. Plate length, $N$, is found from equation 15.7 or 15.8. Width, $B$, is easily calculated as area/$N$. Bearing base plates are assumed to distribute the load to the masonry support as long as the base plate stress does not exceed $F_b = 0.75F_y$. Assuming a uniformly loaded cantilever span of length $n = \frac{1}{2}B - k$ supporting an actual bearing pressure of $F_b$, the required plate thickness is

$$t = n\sqrt{\frac{3f_p}{F_b}} = n\sqrt{\frac{4f_p}{F_y}} \tag{15.37}$$

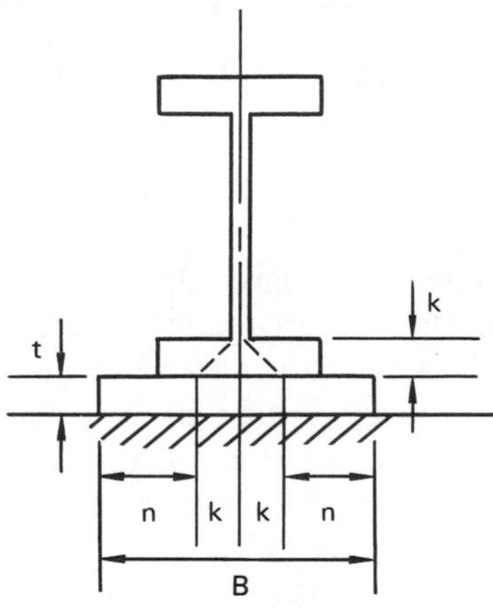


**Figure 15.19** Bearing Plate Thickness

## E. COLUMN BASE PLATES

Column loads transmitted to foundations must meet the same masonry bearing pressure limitations as beams.[27] The load is transmitted from the column to the concrete foundation through a *base plate*, as shown in figure 15.20.

[27] More likely than not, other codes or specifications will control the allowable concrete bearing pressure.

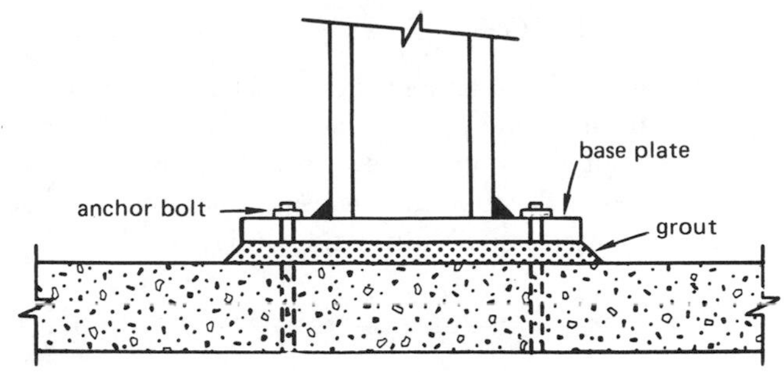


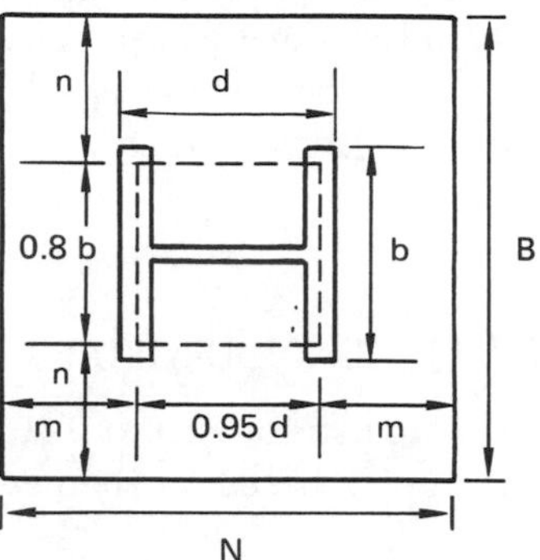


**Figure 15.20** Column Base Plate Design

The required base plate area can be found from the total column load and allowable bearing pressure.

$$A_{\text{plate}} = \frac{\text{column load}}{F_p} \tag{15.38}$$

It is common to specify base plate dimensions in whole inches. Therefore, the actual plate area will be somewhat larger than the required plate area. The actual bearing pressure is

$$f_p = \frac{\text{column load}}{\text{actual } A_{\text{plate}}} \tag{15.39}$$

It is assumed that part of the base plate outboard from a $0.95d \times 0.8b$ rectangle acts as a uniformly loaded cantilever. In order to limit the bending stress, $F_b$, to less than $0.75F_y$, the thickness of the base plate must be made sufficiently large. Once the plate size has been determined, the distances $m$ and $n$ can be determined. Then, equation 15.40 (similar to equation 15.37) can be used to calculate the required plate thickness. The larger plate thickness is required, since either $m$ or $n$ will be larger.

$$t = \{\text{larger of } m \text{ or } n\} \times \sqrt{\frac{4f_p}{F_y}} \tag{15.40}$$

STEEL

## 26. TENSION MEMBERS

### A. INTRODUCTION

Wire cables, rods, eyebars, and structural shapes are typical tension members. Tension members are designed so that the nominal stress is less than the allowable stress. The nominal stress is just the average stress, calculated by dividing the design load by the area.

$$f = \frac{P}{A} \qquad 15.41$$

The area to be used is the actual area in tension. This is generally known as the *gross area*. For riveted or bolted connections, however, the area is taken as the gross area for checking against the yield strength, and it is taken as the *effective area* for checking against the ultimate strength.

### B. ALLOWABLE TENSILE STRESS

The tensile stress must not exceed $0.60F_y$ based on the gross area, or $0.50F_u$ based on the net area (*Section D1*).[28]

### C. SLENDERNESS RATIOS FOR TENSION MEMBERS

Where structural shapes are used in tension, the *AISC Specifications* lists preferred (but does not require) maximum slenderness ratios of 300. Rods and wires are excluded from these limitations (*Section B7*).

### D. THREADED MEMBERS IN TENSION

Tension on threaded parts made from approved steels must not exceed $0.33F_u$ when subjected to static loading, regardless of whether or not threads are present in the shear plane.[29] The area to be used with this allowable stress when calculating the maximum tensile load is the gross or *nominal area*, as determined from the outer extremity of the threaded section. That is, the nominal area is calculated from the nominal bolt diameter (*Table J3.2*).

When threads are included in the shear plane, the lower of the loads based on $0.33F_u$ and $0.60F_y$ controls the design.

[28] Eyebars (pin-connected plates) are designed to a smaller allowable stress. There are other important factors affecting eyebar design. The subject is not covered in this book.

[29] Bolts in tension are governed by other limitations.

### E. PLATES AND MEMBERS WITH HOLES

Plates and members are connected to other plates and members by welding, riveting, and bolting. If rivets and bolts are used, the *effective area* of the tension member is less than the gross area. In some instances where there are several rows of fasteners and multiple redundancy, it will be difficult to determine the effective area.

There are actually three different areas used in tensile member calculations. The *gross area* (for the plate shown in figure 15.21) is

$$A_g = bt \qquad 15.42$$

The *net area* is calculated as the net width times the thickness.

$$A_n = b_n t \qquad 15.43$$

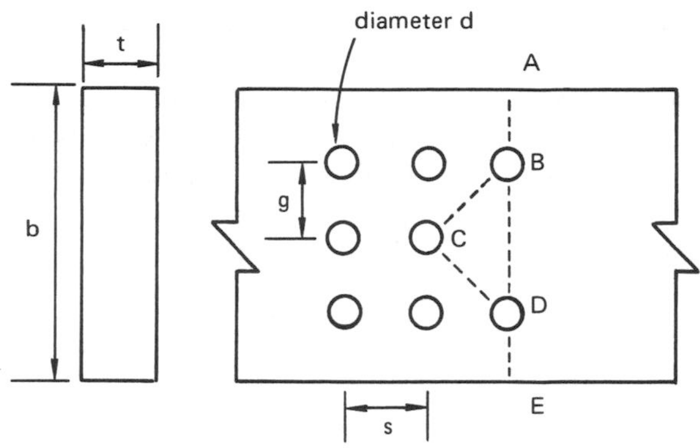


**Figure 15.21** Illustration of Net Area Calculation

The *net width* is calculated by subtracting the hole diameters in the expected failure path from the gross width, and then adding a correction factor for each diagonal leg in the failure path. (There are two diagonal legs in the failure path ABCDE in figure 15.21. Thus, the quantity $s^2/4g$ would be added twice.)

$$b_n = b - \sum_{\text{holes}} d + \sum_{\text{diag. paths}} \frac{s^2}{4g} \qquad 15.44$$

Punched bolt clearance holes should be 1/16″ larger than nominal fastener dimensions.[30]

$s$ in equation 15.44 is the fastener pitch or longitudinal spacing. $g$ is the gage or transverse spacing.

The chain of holes to be used as the expected failure path is the one which gives the minimum net length,

[30] The so-called *standard hole* is 1/16″ larger than the nominal fastener dimension. However, due to difficulties in producing uniform punched holes in field-produced assemblies, another 1/16″ should be added (*Section B2*).

STEEL

$b_n$. This minimum length chain is usually found by checking several possible paths. In figure 15.21, paths ABCDE and ABDE must both be checked.

When a tension member frames into a supporting member, some of the load carrying ability will be lost unless all connectors are in the same plane, and all elements of the tension member are connected to the support. (An angle connected to its support only by one of its legs is an example of lost load carrying ability.) Therefore, a further reduction coefficient, $U$, is used to calculate the effective net area (*Section B3-1*).

$$A_e = UA_n \qquad 15.45$$

$U$ can be taken as 1.0 if all cross-sectional elements are connected to the support to transmit the tensile force. For $W$, $M$, or $S$ structural shapes and structural tees cut from these shapes with connections to flange or flanges only, with 3 or more fasteners per line, and for which flange width/section depth ratio is 2/3 or greater, $U = 0.90$. For connections to built-up sections with 3 or more fasteners per line in the direction of stress, $U = 0.85$. For all shapes not covered, and with at least 2 fasteners per line, $U = 0.75$. (*AISC Specifications, section B3*.)

In addition to the reduction in the net area by $U$, the effective net area for *splice* and *gusset plates* must not exceed 85% of the gross area, regardless of the number of holes. Tests have shown that as few as one hole in a plate will reduce the strength of a plate by at least 15%.[31] It is a good idea to limit the effective net area to 85% of the gross area for all (not just splice and gusset plates) connections with holes in plates.

*Example 15.10*

Choose a 25 foot long $W$ shape (A36 steel) to carry dead and live tensile loads of 468 kips. The shape will be used as a main member with loads transmitted to framing members through the flanges only. Use $K = 1$.

Without additional information on the hole pattern, the gross and net areas are taken as the same. The allowable tensile stress in the member is the minimum of

$$0.6F_y = (0.6)(36) = 21.6 \quad \text{(controls)}$$

$$0.5F_u = (0.5)(58) = 29.0$$

[31] If an end row of connectors has fewer fasteners than interior rows, the tensile strength of the second row effective net area should be checked against a reduced load. The total load should be reduced in proportion to the number of fasteners in the end row. In figure 15.21, there are 2 fasteners in the end row, and there are 8 total fasteners. The load which the plate at the second row must carry is 6/8 = 3/4 of the total load.

Since the web does not transmit the tensile strength, $U = 0.90$ is used. The required area is

$$A_n = \frac{468}{(21.6)(0.90)} = 24.07 \text{ in}^2$$

A W 21×83 member has an area of 24.3. The minimum radius of gyration is $r_y = 1.83$. So, the maximum slenderness ratio is

$$SR = \frac{(1)(25)(12)}{1.83} = 163.9$$

This slenderness ratio is less than the suggested limit of 300.

*Example 15.11*

A long tensile member is constructed from two shorter plates as shown. Each plate is $1/2'' \times 9''$. The fasteners are $1/2''$ nominal bolts. The steel is A36. Determine the maximum tensile load the connection can support. Disregard the shear strength of the bolts.

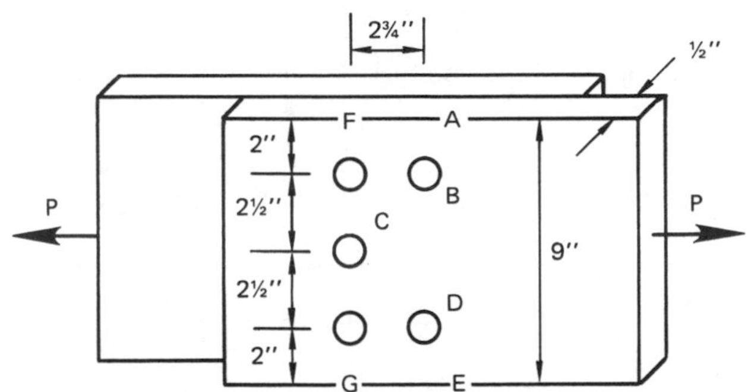


The allowable stress on the gross section is

$$F_t = 0.60F_y = (0.60)(36) = 21.6 \text{ ksi}$$

The allowable stress on the effective net section is

$$F_t = 0.50F_u = (0.50)(58) = 29.0 \text{ ksi}$$

The gross area of the plate is

$$A_g = (0.5)(9) = 4.5 \text{ in}^2$$

The effective hole diameter includes $1/8''$ allowance for clearance and manufacturing tolerances.

$$d = 0.5 + 0.125 = 0.625$$

STEEL

The net area of the connection must be evaluated in three ways: paths ABDE, ABCDE, and FCG. Path ABDE doesn't have any diagonal runs. The net area is

$$A_{n,ABDE} = (0.5)(9 - 2 \times 0.625) = 3.875\ \text{in}^2$$

To determine the net area of path ABCDE, the quantity $s^2/4g$ must be calculated. $s$ is the longitudinal pitch, shown as 2.75″ in the figure. $g$ is the transverse gage, shown as 2.5″ in the figure.

$$\frac{s^2}{4g} = \frac{(2.75)^2}{(4)(2.5)} = 0.756$$

The net area of path ABCDE is

$$\begin{aligned} A_{n,ABCDE} &= (0.5)(9 - 3 \times 0.625 + 2 \times 0.756) \\ &= 4.319\ \text{in}^2 \end{aligned}$$

The net area of path FCG is

$$A_{n,FCG} = (0.5)(9 - 3 \times 0.625) = 3.5625\ \text{in}^2$$

The smallest area is $A_{n,FCG}$, which is less than 85% of $A_g$.

The capacity of the connection based on the gross section is

$$P_g = (4.5)(21.6) = 97.2\ \text{kips}$$

Since all of the connections are in the same plane, $U = 1$. Based on the net section, the capacity is

$$P_g = (1)(3.5625)(29.0) = 103.3\ \text{kips}$$

The capacity is the smaller value, 97.2 kips.

Since the end fastener row has fewer connectors than the second row (2 compared to 3), a further capacity check is required. If holes $B$ and $D$ carry 2/5 of the load, the stress in section FCG will be

$$f_t = \frac{\left(\frac{3}{5}\right)(97.2)}{3.5625} = 16.4\ \text{ksi} \quad \text{(ok)}$$

STEEL

## 27 BEAM-COLUMN ANALYSIS

### A. INTRODUCTION

A compression member, such as a column, which is also acted upon by a bending moment is known as a *beam-column*. The bending moment can be due to an eccentric load or a true lateral load. Design and analysis of members with combined bending and axial loads generally attempt to transform moments into equivalent axial loads (i.e., the *equivalent axial compression method*) or make use of *interaction equations*. Interaction type equations are best suited for beam-column analysis and validation, since so much (e.g., area, moment of inertia, etc.) needs to be known about a shape. Equivalent axial compression methods are better suited for design.

### B. SMALL AXIAL COMPRESSIONS

When the axial load is small, the member is essentially a beam. When $f_a/F_a$ does not exceed 0.15, a simplified interaction criterion can be used.

*step 1*: Calculate the axial stress due to the axial load acting alone.

$$f_a = \frac{P}{A} \qquad 15.46$$

*step 2*: Calculate the slenderness ratios for both bending modes.

*step 3*: Based on the largest slenderness ratio, determine the allowable column stress. Use equation 15.18 or 15.19. Alternatively, $F_a$ can be obtained directly from the *Allowable Stress for Compression Members* table.

*step 4*: Calculate the ratio of $f_a/F_a$, which must be 0.15 or less to use this method.

*step 5*: Determine the adequacy of design by use of the simplified interaction criterion, equation 15.47 (**[H1-3]**).

$$\frac{f_a}{F_a} + \frac{f_{bx}}{F_{bx}} + \frac{f_{by}}{F_{by}} \le 1.0 \qquad 15.47$$

In equation 15.47, $f_b$ is the actual maximum compressive bending stress, as calculated from the standard bending stress equation. $F_b$ is the compressive bending stress that would be permitted if the bending moment acted alone.

### C. LARGE AXIAL COMPRESSIONS

When $f_a/F_a$ exceeds 0.15, the *stability interaction criterion* must be used. Two equations must be satisfied. The first is basically an interaction equation using $0.60F_y$ as the allowable compressive stress. Equation 15.48 (**[H1-2]**) is the first equation.

$$\frac{f_a}{0.60F_y} + \frac{f_{bx}}{F_{bx}} + \frac{f_{by}}{F_{by}} \le 1.0 \qquad 15.48$$

Equation 15.49 ([**H1-1**]) is the *stability criterion.*

$$\frac{f_a}{F_a} + \frac{C_{mx} f_{bx}}{\left(1 - \frac{f_a}{F'_{ex}}\right) F_{bx}} + \frac{C_{my} f_{by}}{\left(1 - \frac{f_a}{F'_{ey}}\right) F_{by}} \leq 1.0 \quad 15.49$$

$$F'_e = \frac{12\pi^2 E}{23\left(\frac{KL_b}{r_b}\right)^2} \quad 15.50$$

In equation 15.49, $F_a$ is the axial compressive stress that would be allowed if there was no other bending stress. It depends on the maximum slenderness ratio. $C_m$ is an *equivalent moment factor.* If the compression member is subject to joint translation (i.e., sidesway), then $C_m = 0.85$. If the compression member is braced against joint translation in the plane of loading and also experiences transverse loading between the supports, $C_m$ may be taken as 0.85 if the member's ends are restrained, and 1.0 otherwise.

If a compression member is braced against joint translation in the plane of loading and does not experience transverse loads between the supports, then $C_m$ is calculated from equation 15.51.

$$C_m = 0.6 - 0.4\frac{M_1}{M_2} \quad 15.51$$

$M_1/M_2$ is the ratio of the smaller to larger moments at the ends of the compression member (in the plane of bending). The ratio is positive when the member is bent in reverse curvature, and negative when bent in single curvature.

In equation 15.50, $F'_e$ is the Euler stress divided by the basic factor of safety. $L_b$ and $r_b$ are the unbraced length and radius of gyration in consistent units, both in the plane of bending.

*Example 15.12*

A W 14×120 (A36) shape has been chosen to carry an axial compressive load of 200,000 pounds and a 250,000 ft-lbf moment about its strong axis. The member's unsupported length is 20 feet. Sidesway is permitted in the direction of bending. Use $K = 1$. Determine if the column is adequate.

Check to see if the axial load can be considered small. The column properties are:

$$A = 35.3\,\text{in}^2$$
$$r_y = 3.74$$
$$L_c = 15.5$$
$$L_u = 44.1$$
$$r_x = 6.24$$
$$S_x = 190$$

The axial stress is

$$f_a = \frac{P}{A} = \frac{200}{35.3} = 5.67\,\text{ksi}$$

The maximum slenderness ratio is

$$SR = \frac{KL}{r} = \frac{(1)(20)(12)}{3.74} = 64.2 \quad \text{(say 64)}$$

The allowable compressive stress from equation 15.18 (or from the *Allowable Stress for Compression Members* table) is 17.04 ksi.

The stress ratio is

$$\frac{f_a}{F_a} = \frac{5.67}{17.04} = 0.33 > 0.15$$

Therefore the large axial compression criteria must be used. Since sidesway is permitted, $C_{mx} = 0.85$. Since the unbraced length of 20 feet is greater than $L_c$ but less than $L_u$,

$$F_b = 0.60 \times F_y = (0.60)(36) = 21.6\,\text{ksi}$$

From equation 15.50,

$$F'_e = \frac{12\pi^2(2.9\,\text{EE7})/1000}{23\left[\frac{(1)(20)(12)}{6.24}\right]^2} = 100\,\text{ksi}$$

The bending stress due to the applied moment is

$$f_{bx} = \frac{M}{S_x} = \frac{\frac{(250,000)(12)}{1000}}{190} = 15.79$$

From equation 15.48 (the first criterion),

$$\frac{5.67}{(0.60)(36)} + \frac{15.79}{21.6} = 0.99 \quad \text{(ok)}$$

From equation 15.49 (the second criterion),

$$\frac{5.67}{17.04} + \frac{(0.85)(15.79)}{\left(1 - \frac{5.67}{100}\right)(21.6)} = 0.99 \quad \text{(ok)}$$

D. WEB STIFFENERS

In most instances, beams and girders will transmit moments to the flanges of column members. If these moments are very large, it will be necessary to reinforce the web with web stiffeners. The column design tables and *AISC Specifications* should be consulted to determine when such stiffening is required.

## 28 BEAM-COLUMN DESIGN

In order that the shape selection process for combined axial and bending loads not be so much of a trial and error procedure, equations 15.47, 15.48, and 15.49 can be modified to allow the use of the column tables in the *AISC Manual*. Nevertheless, such modifications still require a considerable amount of calculation.

The *AISC Manual* suggests an alternate procedure which determines an *equivalent axial load*.[32]

*step 1*: Determine the effective length, $KL$, based on the weak axis bending and bracing.

*step 2*: Use *table B* in *Part 3, "Column Design"*, of the *AISC Manual* to obtain a value of the *equivalency factor*, $m$, for the first iteration.

*step 3*: Assume a value of $U$ to be 3. (Values of $U$ for subsequent iterations can be read directly from the column table.)

*step 4*: Calculate the equivalent axial load. $P$ is the actual axial load.

$$P_{\text{eff}} = P + M_x m + M_y m U \qquad 15.52$$

*step 5*: Select a column from the column table to support $P_{\text{eff}}$ with an effective length of $KL$.

*step 6*: Return to *table B* in the *AISC Manual* to obtain a revised value of $m$.

*step 7*: Read the $U$ value for the column from the column table.

*step 8*: Repeat steps 4 through 7 to reduce member weight until $m$ and $U$ remain constant.

*Example 15.13*

Select a W shape using A36 steel to carry 200,000 pounds axially and a moment of 250,000 ft-lbf about its strong axis. The unsupported length is 20 feet. Assume $K = 1$.

*step 1*: $KL = (1)(20) = 20$ ft

*step 2*: $m = 2.0$ (all shapes, 20 feet)

*step 3*: $U = 3$ (not needed in this example)

*step 4*: $P_{\text{eff}} = 200 + \dfrac{(250,000)(2)}{1000} = 700$ kips

*step 5*: Try W 14×132, with a 20 foot capacity of 662 kips (close to 700 kips).

*step 6*: $m = 1.7$

*step 7*: $P_{\text{eff}} = 200+(1.7)(250) = 625 \text{ kips} < 662 \text{ kips}$ (O.K.)

## 29 PLATE GIRDERS

When beams with moments of inertia larger than standard mill shapes are required, plate girders can be used. The discussion on plate girders which follows is presented in an order which allows the design of a member. However, the individual sections can also be used if specific aspects of a given plate girder are to be evaluated.

A. DIMENSIONS AND NOMENCLATURE

Figure 15.22 illustrates two plate girders. One is built up and bolted or riveted together. The other is welded. The depth, $d$, is generally chosen to be around 1/10 of the span, although the ratio can vary between 1/5 to 1/15.

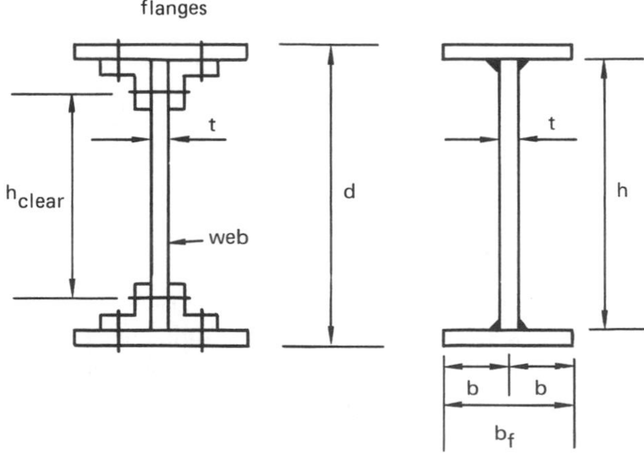


**Figure 15.22** Elements of a Plate Girder

Depending on the purpose, two different definitions of the *web depth*, $h$, are used.[33] Unlike for rolled shapes, for shear stress calculations, only $h$ is used. For depth-

[32] This procedure has a tendency to oversize beam-columns. Therefore, it may be possible to use somewhat lighter members if the member initially selected is used as a starting point for a subsequent trial-and-error reduction study.

[33] In fact, there are two different definitions of depth—the *web depth*, $h$, and the *girder depth*, $d$.

STEEL

thickness ($h/t$) ratios, the *AISC Specifications* require $h$ to be the clear distance between flanges. For bolted or riveted flanges, this distance is the separation between the last row of fasteners. For welded flanges, the distinction is not made.

B. DEPTH-THICKNESS RATIOS

The web thickness to be used depends on how closely *intermediate stiffeners* are spaced. If intermediate stiffeners are spaced no more than $1.5d$, where $d$ is the girder depth, then the maximum permissible depth-thickness ratio is given by equation 15.53. If $h$ is known, the required thickness, $t$ can be obtained ([**G1-2**]).

$$\frac{h}{t} \leq \frac{2000}{\sqrt{F_y}} \qquad 15.53$$

If intermediate stiffeners are not provided as often as $1.5d$, then equation 15.54 governs the depth-thickness ratio ([**G1-1**]).

$$\frac{h}{t} \leq \frac{14{,}000}{\sqrt{F_{yf}(F_{yf}+16.5)}} \qquad 15.54$$

If $h/t$ satisfies equation 15.54 and is also less than 260, and if the shear stress is less than the allowable value, no intermediate stiffeners will be required.[34] In other cases, actual stiffener spacing will depend on the shear stress (*Section F5*).

C. SHEAR STRESS

The maximum shear stress on the section can be calculated from $h$ and $t$.

$$f_v = \frac{V_{\text{max}}}{ht_w} \qquad 15.55$$

The allowable web shear, $F_v$, may not exceed $0.40F_y$ or the value calculated from equation 15.56 ([**F4-2**]).[35]

In equation 15.59, $a$ is the clear distance between stiffeners.

$$F_v = \frac{C_vF_y}{2.89} \qquad 15.56$$

$$C_v = \frac{45{,}000k_v}{F_y(h/t_w)^2} \quad (C_v < 0.8) \qquad 15.57$$

$$C_v = \frac{190}{h/t_w}\sqrt{\frac{k}{F_y}} \quad (C_v > 0.8) \qquad 15.58$$

$$k_v = 4.00 + \frac{5.34}{(a/h)^2} \quad (a/h < 1.0) \qquad 15.59$$

$$k_v = 5.34 + \frac{4.00}{(a/h)^2} \quad (a/h > 1.0) \qquad 15.60$$

When $a/h$ is very large, $k = 5.34$. This situation corresponds to the case of no intermediate stiffeners. If that is the case, equations 15.61 and 15.62 give the allowable shear stress. ($F_v$ is still limited to $0.4F_y$, however.)

$$F_v = \frac{152\sqrt{F_y}}{h/t} \quad \left(h/t < \frac{548}{\sqrt{F_y}}\right) \qquad 15.61$$

$$F_v = \frac{83{,}150}{(h/t)^2} \quad \left(h/t > \frac{548}{\sqrt{F_y}}\right) \qquad 15.62$$

An alternate method of calculating the web thickness based on achieving full flange compression (equation 15.63) will produce much thicker webs, and result in larger weight and cost. However, this equation can be used to determine if the allowable flange stress will be reduced. If the thickness is less than that determined from equation 15.63, the allowable flange stress will be reduced from the basic bending allowance.

D. DESIGN OF GIRDER FLANGES

To design the girder flanges, some initial estimate of the allowable bending stress in the flanges must be made. Theoretically, $F_b = 0.66F_y$ could be chosen. However, the allowable bending stress is a function of the depth-thickness ratio when

$$\frac{h}{t} > \frac{760}{\sqrt{F_b}} \qquad 15.63$$

Therefore, if $h/t$ is greater than $760/\sqrt{F_b}$, a lower value of $F_b$ should be used.[36]

Equation 15.64 is easily derived from basic mechanics principles, and gives an initial estimate of the required flange area.

$$A_f = \frac{M_x}{F_bh} - \frac{th}{6} \qquad 15.64$$

Once the flange area is known, trial flange widths and thicknesses can be evaluated.[37]

$$A_f = b_ft_f \qquad 15.65$$

[34] In any case, bearing stiffeners are still required at reaction and loading points.

[35] The *AISC Specifications* contains an alternative equation for calculating the allowable shear stress if intermediate stiffeners are used and other conditions are met.

[36] That is, $F_b$ can be reduced 10% or so. Very large reductions in bending stress underutilize the flange area.

[37] Girder flanges do not have to be the same thickness along the entire plate girder length. It is possible to substitute thinner flanges near the ends of beams, or to add cover plates near the mid-points of beams.

The limitations in available plate thicknesses should be considered when choosing $t_f$. Alternatively, $b_f$ can be chosen based on $b_f/d$ ratios, which typically vary between 0.2 and 0.3. Flange plate widths are typically rounded to the nearest 2″.

## E. WIDTH-THICKNESS RATIOS

The width-thickness ratios specified by equation 15.26 and table 15.6 apply to plate girders as well. Specifically, for unstiffened plates such as plate girder flanges, equation 15.66 should be used.

$$\frac{b}{t_f} < \frac{95}{\sqrt{F_y}} \qquad 15.66$$

$$b = \frac{1}{2} b_f \qquad 15.67$$

For stiffened plates, such as plate girder webs, equation 15.68 is used. Also, see equation 15.53 ([**G1-1**]).

$$\frac{h}{t_w} \leq \frac{14{,}000}{\sqrt{F_y(F_y + 16.5)}} \qquad 15.68$$

## F. REDUCTION IN FLANGE STRESS

As already mentioned, the allowable bending stress must be reduced if $h/t > 760/\sqrt{F_b}$. Equation 15.69 gives the reduced stress. $A_w$ is the area of the web, and $A_f$ is the area of the compression flange. Equation 15.69 is for non-hybrid girders (*Section G2*).

$$F_b' \leq F_b \left[1.0 - 0.0005\,\frac{A_w}{A_f}\left(\frac{h}{t} - \frac{760}{\sqrt{F_b}}\right)\right] \qquad 15.69$$

$$A_w = ht \qquad 15.70$$

$$A_f = b_f t_f \qquad 15.71$$

## G. FINAL CHECK

Once a trial design of the plate girder has been made, its moment of inertia can be determined by standard means. The expected stress based on assumed elastic behavior is calculated and compared to the allowable stress, $F_b'$.

## H. LOCATION OF FIRST (OUTBOARD) STIFFENERS

The first intermediate stiffeners can be located where the shear stress exceeds equation 15.56 ([**F4-2**]). In practice, a trial distance $a$ ($a/h < 1.0$) is selected as the separation between the end panel and the first intermediate stiffener. Then, $F_v$ is calculated from equation 15.56 and compared to $f_v$ from equation 15.55.[38] If $F_v$ is greater than $f_v$, the location is adequate. Otherwise, a smaller $a$ should be tried.

## I. LOCATION OF INTERIOR STIFFENERS

The spacing, $a$, of interior intermediate stiffeners should not exceed the value determined from equation 15.72 ([**F5-1**]).

$$\frac{a}{h} \leq \left[\frac{260}{h/t_w}\right]^2 \qquad 15.72$$

The spacing, $a$, determined from equation 15.72 is valid as long as the shear stress does not exceed the value calculated from equation 15.56. In beams where the maximum shear occurs near the ends, the spacing chosen for the first interior stiffener will be adequate for the entire beam length. However, this illustrates the need to work with the largest shear when establishing an initial trial spacing.

If equation 15.72 is met, and if $C_v \leq 1.0$, then equation 15.73 ([**G3-1**]) can be used in place of equation 15.56 ([**F4-2**]). This is the so-called *tension field action equation*. Use of equation 15.73 places an additional restraint on the allowable bending stress in the girder web.

$$F_v = \frac{F_y}{2.89}\left[C_v + \frac{1 - C_v}{1.15\sqrt{1 + (a/h)^2}}\right] \leq 0.40 F_y \qquad 15.73$$

## J. MAXIMUM BENDING STRESS

Plate girders which have been designed according to tension field concepts (i.e., have used equation 15.73 to determine stiffener spacing), are limited to the web bending stress in equation 15.74 ([**G5-1**]). $F_v$ in equation 15.74 is calculated from the *tension field action equation*, equation 15.73. If the bending stress is excessive, or if the quantity in parentheses is greater than 0.6, the stiffener spacing must be reduced.

$$f_b \leq \left(0.825 - 0.375\frac{f_v}{F_v}\right) F_y < 0.6 F_y \qquad 15.74$$

## K. DESIGN OF INTERMEDIATE STIFFENERS

*Intermediate stiffeners* are used to support the flange and prevent buckling. They may be constructed from plates or angles, either singly or in pairs. Intermediate stiffeners do not need to extend completely from the top to bottom flanges, but they must be fastened to the compression flange to resist uplift.[39] The *AISC Specifications* contains limitations on weld and rivet spacing.

Intermediate stiffeners are sized by their gross steel area, as calculated from the steel area in contact with the compression flange. Thus, the width and thickness

STEEL

[38] This procedure is greatly aided by a shear diagram for the beam.

[39] Stiffeners which transmit loads and reactions must extend from flange to flange.

are used to calculate the stiffener area, not the width and depth. Equation 15.75 (**[G4-2]**) gives the steel area at a particular location. This area can be divided between two stiffeners or given to a single stiffener. In equation 15.75, $D$ is 1.0 for stiffeners furnished in pairs, and 2.4 for single plate stiffeners. $D$ is 1.8 for single angle stiffeners.

$$A_{st} = \frac{1-C_v}{2}\left[\frac{a}{h} - \frac{(a/h)^2}{\sqrt{1+(a/h)^2}}\right]\left(\frac{F_{y,\text{web}}}{F_{y,\text{stiffener}}}\right) Dht_w \quad 15.75$$

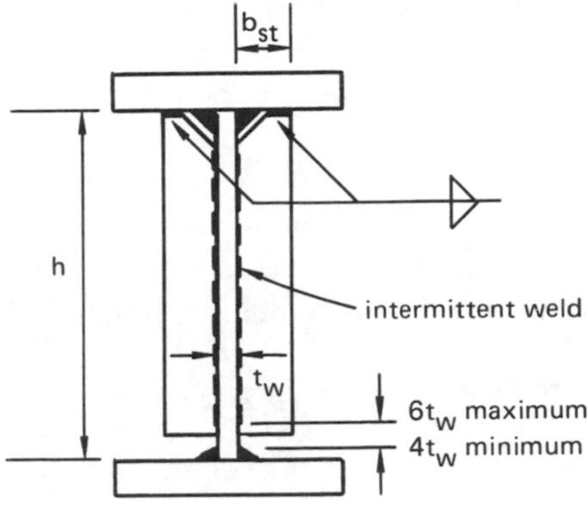


**Figure 15.23** Intermediate Stiffeners

If the actual shear stress, $f_v$, at the point where the bearing stiffener is located is less than the allowable shear stress, $F_v$, as calculated from equation 15.73, then the stiffener area may be reduced proportionally. That is (*Section G4*),

$$A'_{st} = \left(\frac{f_v}{F_v}\right) A_{st} \quad 15.76$$

The moment of inertia, $I_{st}$, is taken with respect to an axis in the plane of the girder web. If two stiffeners are used, $b_{st}$ is the total of both their widths.

$$I_{st} = \frac{t_{st} b_{st}^3}{12} \quad 15.77$$

To be significant, the stiffener must have sufficient stiffness itself. Equation 15.78 is a lower limit on the moment of inertia (**[G4-1]**).

$$I_{st} \geq \left(\frac{h}{50}\right)^4 \quad 15.78$$

## L. DESIGN OF BEARING STIFFENERS

The bearing pressure on stiffeners is limited to $0.90F_y$, and such stiffeners must essentially extend from the web to the edge of the flanges. Therefore, this criterion establishes one method of determining the bearing stiffener thickness. However, there are other criteria that must also be met. (The width is essentially fixed by the flange dimension. So, only the thickness needs to be determined.)

Since the stiffener is loaded as a column, it must satisfy the width-thickness ratio for an unstiffened element.

$$\frac{b_{st}}{t_{st}} = \frac{95}{\sqrt{F_y}} \quad 15.79$$

The stiffener should be designed as a column. For the purpose of determining the slenderness ratio, the effective length is taken as $0.75h$. The radius of gyration, $r$, can be determined exactly, or it can be approximated as 0.25 times the stiffener edge-to-edge distance. That is,

$$\frac{L}{r} = \frac{0.75h}{0.25(2b_{st} + t_w)} \quad 15.80$$

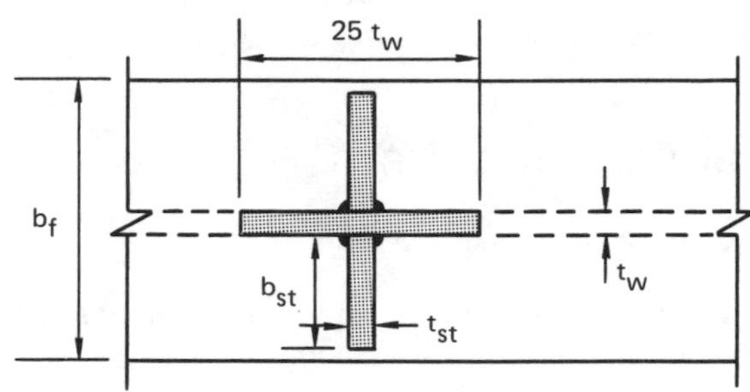


**Figure 15.24** Bearing Stiffener Design

Once the $L/r$ ratio is known, it can be used (as $KL/r$) to determine the allowable compressive stress, $F_a$. However, some of the web also supports the load. Specifically, 25 times the web thickness is the contributing area. Therefore, the required stiffener thickness based on column stress is

$$t_{st} = \frac{\frac{\text{load}}{F_a} - 25t_w^2}{2b_{st}} \quad 15.81$$

Another factor determining the thickness is possible compression yielding. Compressive stress is limited to $0.60F_y$. Therefore, the thickness is

$$t_{st} = \frac{\frac{\text{load}}{0.60F_y}}{2b_{st}} \quad 15.82$$

Stiffener thickness is the maximum thickness determined from the bearing stress, width-thickness ratio, column stress, and compression yield criteria.[40]

[40] If column stability is not the factor controlling stiffener thickness, the larger thickness could conceivably increase the slenderness ratio and reduce the allowable compressive stress even further. This should be checked, but is not likely to be a factor.

M. WEB CRIPPLING AT POINTS OF LOADING

The *AISC Specifications* contains provisions (*K1*) to determine if the web is capable of supporting the loads (concentrated or distributed) without experiencing web crippling. The equations ([**K1-4**] and [**K1-5**]) determine the maximum load that can be applied before stiffeners are required. If stiffeners are provided and extend at least one-half the web depth, [**K1-4**] and [**K1-5**] need not be checked.

## 30 BENDING WITH AXIAL TENSION

Equation 15.48 should be used to size sections that are subject to axial tension and bending combined. $f_b$ is the bending stress that would exist if axial tension were not present. The effect of the axial tension is not allowed to produce bending stresses which, if acting alone, exceed the allowable bending stresses for flexural members.

## 31 BOLTS AND RIVETS

A. INTRODUCTION

Connections using bolts and rivets are treated similarly. Such connections can place the fasteners in direct shear, torsional shear, tension, or any combination of shear and tension. Theoretical methods based on elastic design can be used for design, and in some instances, procedures based on ultimate strength concepts are available.

A *concentric connection* is one for which the applied load passes through the centroid of the fastener group. If the load is not directed through the fastener group centroid, the connection is said to be an *eccentric connection.*

At low loading, the distribution of forces among the fasteners is very non-uniform, since friction carries some of the load. However, at higher stresses (near yielding), the load is carried equally by all fasteners in the group.

Allowable stresses in fasteners can be increased by 1/3 for temporary exposure to wind and seismic loading.

*Stress concentration factors* are not normally applied to connections with multiple redundancy.

In many connection designs, materials with different strengths will be used. The material with the minimum strength is known as the *critical part*, and the critical part controls the design.

The analysis presented here for connection design and analysis assumes static loading. Other provisions for fatigue loading are contained in the *AISC Specifications, Appendix K*.

B. HOLE SPACING AND EDGE DISTANCES

The minimum distance between centers of fastener holes is 8/3 times the nominal fastener diameter. In addition, along the longitudinal connection direction (along the line of applied force), the minimum distance between the centers of standard holes is given by equation 15.83. $P$ is the force carried by one fastener in the connector group. $t$ is the thickness of the critical (thinnest) part. (*Section J3.8*).

$$s \geq \frac{2P}{F_u t} + \frac{d}{2} \qquad 15.83$$

The minimum distance from the hole center to the edge of a member is approximately 1.75 times the nominal diameter for sheared edges, and 1.25 times the nominal diameter for rolled or gas-cut edges, both rounded to the nearest 1/8″ (*J3-9*).[41] Along the line of transmitted force, the distance, $L_e$, from a hole center to the edge of the connected part (in the direction of force) shall not be less than specified by equation 15.84 ([**J3-6**]).

$$L_e \geq \frac{2P}{F_u t} \qquad 15.84$$

For parts in contact, the maximum edge distance from the center of a fastener to the edge in contact is 12 times the plate thickness or 6″, whichever is less (*Section J3.10*).

C. STRESS AREA OF FASTENERS

The allowable load on a fastener is determined by multiplying its area by an allowable stress.[42] Except for rods with upset ends, the area is calculated simply from the fastener's nominal (unthreaded and undriven) dimension.

D. ALLOWABLE CONNECTOR STRESSES

Table 15.7 lists allowable stresses for tension and shear for common connector types. (Allowable bearing stress is covered elsewhere.) Reductions are required for use with oversized holes and connector patterns longer than 50 inches.

For fasteners made from approved steels, including A449, A572, and A588 alloys, the allowable tensile stress is $F_t = 0.33F_u$, whether or not threads are in the shear plane. For bearing type connections using the same approved steels, the allowable shear stress is $F_v = 0.17F_u$ if threads are present in the shear plane, and $F_v = 0.22F_u$ if threads are excluded from the shear plane (*Table J3.2*).

[41] For exact values, refer to *Table J3.5* in the *AISC Specifications*.

[42] The *AISC Manual* also contains tables of allowable loads for common connector types and materials.

STEEL

**Table 15.7**
Allowable Connector Stresses for Static Loading[43]

(All stresses are in ksi)

| type of connector | $F_t$ | $F_v$ slip-critical connection | $F_v$ bearing connection |
|---|---|---|---|
| A 502, hot driven rivets, | | | |
| grade 1 | 23.0 | don't use | 17.5 |
| grade 2 | 29.0 | don't use | 22.0 |
| A 307, ordinary bolts | 20.0 | don't use | 10.0 |
| A 325, structural bolts | | | |
| no threads in the shear plane | 44.0 | 17.0 | 30.0 |
| threads in the shear plane | 44.0 | 17.0 | 21.0 |
| A 490, structural bolts | | | |
| no threads in the shear plane | 54.0 | 21.0 | 40.0 |
| threads in the shear plane | 54.0 | 21.0 | 28.0 |

If fasteners are exposed to both tension and shear, *AISC Specifications Table J3.3* should be used to determine the maximum tensile stress.

For A325 and A490 bolts used in slip-critical connections with tension, the allowable shear stresses given in table 15.7 must be multiplied by a reduction factor. In equation 15.85, $f_t$ is the average tensile stress in the bolt group, $A_b$ is the nominal body area of the bolt, and $T_b$ is the bolt *pretension*.[44]

$$\text{reduction factor} = \left(1 - \frac{f_t A_b}{T_b}\right) \qquad 15.85$$

A distinction is made between bearing and slip-critical connections. A *bearing connection* relies on the shearing resistance of the fasteners to resist loading. In effect, it is assumed that the fasteners are loose enough to allow the plates to slide slightly, bringing the fastener shank into contact with the hole. The area surrounding the hole goes into bearing, hence the name. Connections using rivets, welded studs, and A307 bolts are always bearing type connections.

If the fasteners are constructed from high-strength steel, a high *preload* can be placed on the bolts. This preload will clamp the plates together, and friction alone will keep the plates from sliding. The fastener shanks never come into contact with the plate holes. Bolts constructed from A325 and A490 steels are suitable for ***slip-critical connections***. However, high-strength bolts can also be used in bearing connections.

## E. CONCENTRIC TENSION CONNECTIONS

The number of fasteners required in the connection is determined by considering the fasteners in shear, the plate in bearing, and the effective net area of the plate in tension.

*Example 15.14*

Two $\frac{1}{4}\times 8$ A36 plates are joined with a lap joint using 3/4″ grade 1 A502 rivets. Design the connection to carry a concentric load of 25 kips. Disregard effects of eccentricity.

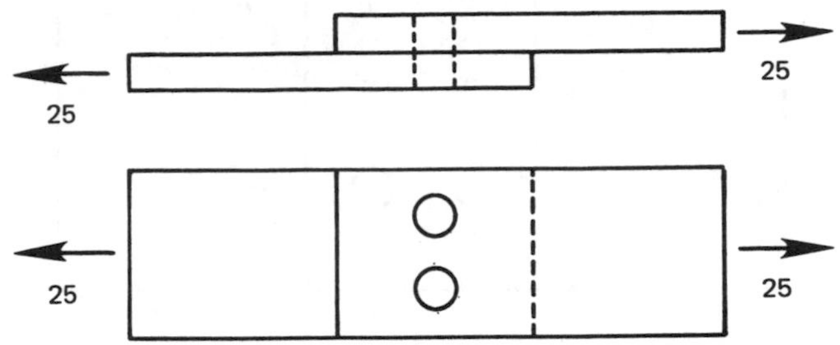


The area of a 3/4″ rivet is

$$A_v = \frac{1}{4}\pi(0.75)^2 = 0.442\,\text{in}^2$$

From table 15.7, the allowable shear stress is 17.5 ksi. So, the number of rivets determined by the shear stress criterion is

$$n = \frac{25}{(17.5)(0.442)} = 3.23 \quad \text{(shear criterion)}$$

The bearing area is

$$A_p = (0.75)(0.25) = 0.1875\,\text{in}^2$$

The allowable bearing stress is

$$f_p = 1.5F_u = (1.5)(58) = 87\,\text{ksi}$$

The number of rivets determined by bearing is

$$n = \frac{25}{(87)(0.1875)} = 1.53 \quad \text{(bearing criterion)}$$

Shear governs. 4 rivets are used.

[43] This is based on *AISC Specifications Table J3.2*, which contains greater detail.

[44] Minimum bolt tensions are given in *AISC Specification Table J3.7*. A325 and A490 bolts are required to be tightened to 0.7 of their tensile strength.

STEEL

The minimum distance between holes is

$$s_{\min} = \frac{8}{3} \times \frac{3}{4} = 2''$$

The longitudinal spacing is further limited by equation 15.83.

$$s_{\min} = \frac{(2)(25/4)}{(58)(1/4)} + \frac{3/4}{2} = 1.24''$$

Assuming sheared edges, the minimum edge spacing is $1.75 \times 3/4 \approx 1\frac{1}{4}''$. The first row cannot be closer to the short edge than

$$s_{\min} = \frac{(2)(25/4)}{(58)(1/4)} = 0.86''$$

Based on these requirements, a trial layout is made.

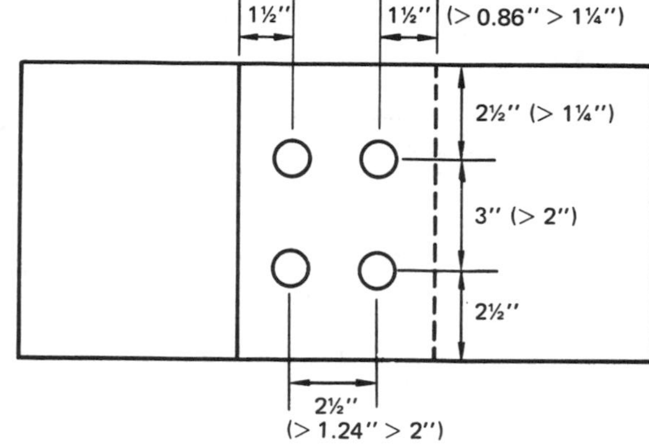


Assuming a hole which is 1/16″ larger than the nominal fastener diameter, each hole subtracts 13/16″ from the net effective area in tension. The tension area across the first row of holes is

$$A_t = (1/4)\left(8 - 2 \times \frac{13}{16}\right) = 1.39 \text{ in}^2$$

The maximum tensile stress on the net effective section is

$$F_t = 0.50F_u = (0.50)(58) = 29 \text{ ksi}$$

The maximum allowable load based on the net effective section is

$$P = (1.39 \text{ in}^2)(29 \text{ ksi}) = 40.3 \text{ kips} \quad \text{(ok)}$$

Similarly, the maximum tensile stress in the gross section is

$$F_t = 0.60F_y = (0.60)(36) = 21.6 \text{ ksi}$$

The maximum allowable load based on the gross section is

$$P = \left(8 \times \frac{1}{4} \text{ in}^2\right)(21.6 \text{ ksi}) = 43.2 \text{ kips} \quad \text{(ok)}$$

## F. TENSION EFFECTS DUE TO ECCENTRICITY

Consider the simple framing connection shown in figure 15.25. If the vertical shear is assumed to act along line $A$, the bolts in the column connection will be put into tension. The most highly-stressed (in tension) will be the top-most fasteners.

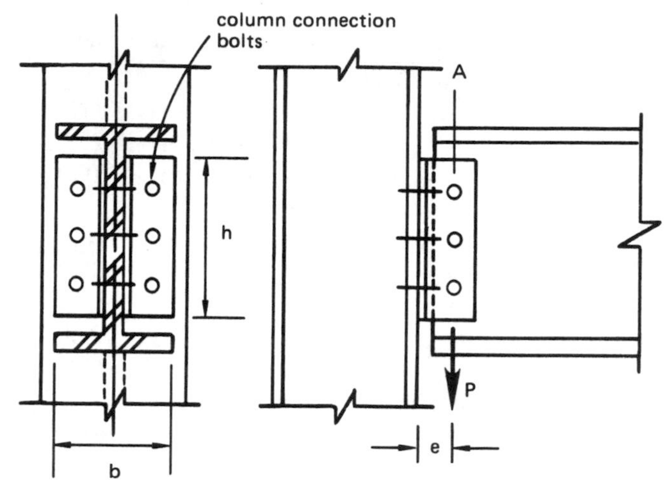


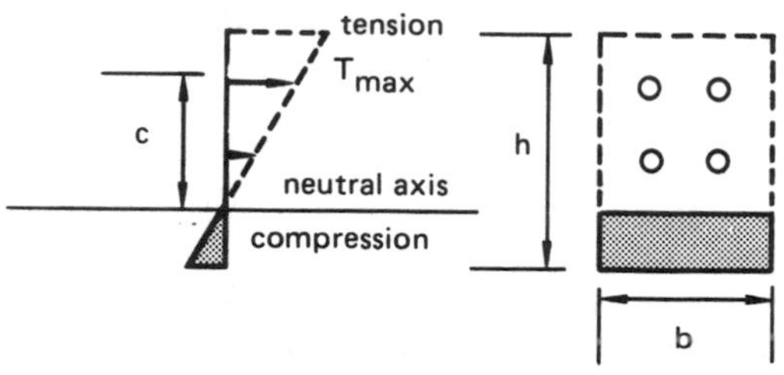


**Figure 15.25** Tension in Simple Connections

By summing moments about the neutral axis (for tension) of the fastener group, it is possible to determine the maximum tensile stress. The neutral axis is located approximately 1/6 or 1/7 up from the bottom of the connector group area, such that the area of bolts in tension equals the area of the support in compression. A few trials may be necessary to locate the neutral axis.

Once the neutral axis is located, the moment of inertia of the fastener group is found by use of the parallel axis theorem. The applied moment is known: $M = Pe$. The stress in the fasteners farthest from the neutral axis is

$$f_{t,\max} = \frac{Mc}{I} = \frac{Pec}{I} \qquad 15.86$$

If necessary, the tensile force in the farthest fastener can be found.

$$T_{\max} = f_{t,\max} A_{\text{bolt}} \qquad 15.87$$

## G. BOLT PRELOADING

Preloading is an effective method of reducing the alternating stress in bolted connections. The initial tension produces a larger mean stress, but the overall result may be to produce a satisfactory design.

Consider the bolted connection shown in figure 15.26. The load varies from $P_{\text{min}}$ to $P_{\text{max}}$. If the bolt is initially snug but without initial tension, the force in the bolt also will vary from $P_{\text{min}}$ to $P_{\text{max}}$.

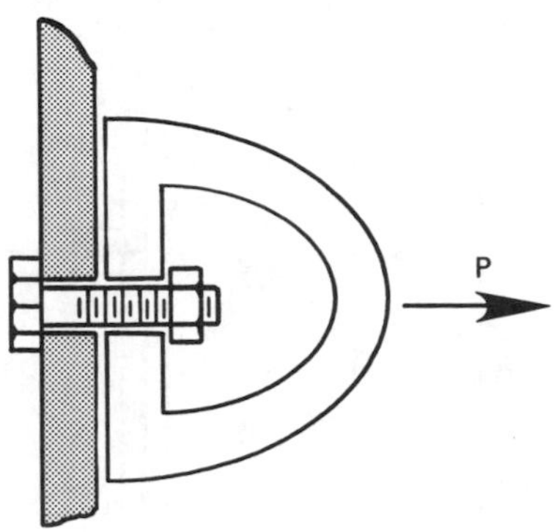


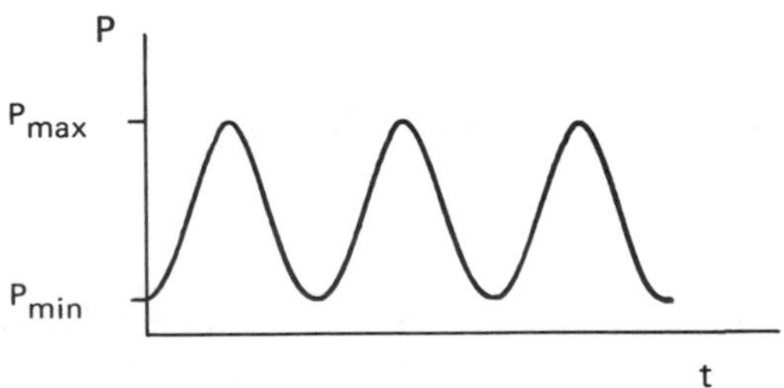


**Figure 15.26** A Bolted Joint

The stress in the bolt depends on the load carrying area of the bolt. It is convenient to define the *spring constant of the bolt.* The effects of the threads usually are ignored, so the area is based on the major (nominal) diameter. The *grip, L,* is the thickness of the parts being connected by the bolt. It is not the bolt length.

$$f_{\text{bolt}} = \frac{P}{A} \qquad 15.88$$

$$k_{\text{bolt}} = \frac{P}{\Delta L} = \frac{A_{\text{bolt}} E_{\text{bolt}}}{L_{\text{bolt}}} \qquad 15.89$$

If the bolt is tightened so that there is an initial force, $F_i$, in addition to the applied load, the members being held together will be in compression. The amount of compression will vary since the applied load varies.

The clamped members will carry some of the applied load, since this varying load has to "uncompress" the members as well as lengthen the bolt. The net result is to reduce the variation of the bolt force.[45]

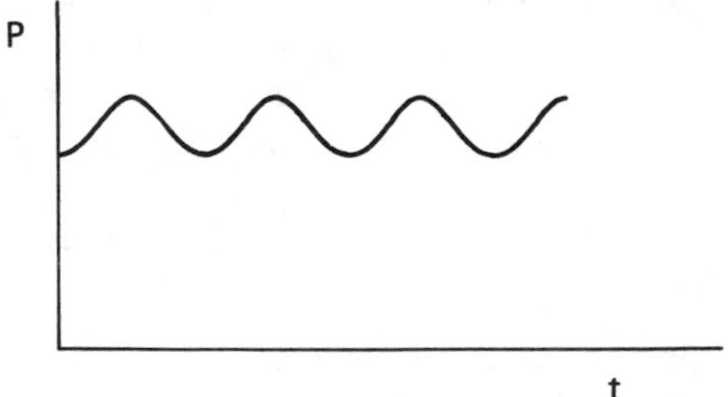


**Figure 15.27** A Bolted Joint with Preloading

The spring constant for each of the bolted parts is somewhat difficult to determine if the clamped area is not well defined. If the clamped parts are simply plates, it can be assumed that the bolt force spreads out to three times the bolt diameter. Of course, the hole diameter needs to be considered in calculating the effective force area. If $E_{\text{part}} = E_{\text{bolt}}$, this larger area results in the parts being 8 times more stiff than the bolts.

$$k_{\text{parts}} = \frac{A_{\text{parts}} E_{\text{parts}}}{L_{\text{parts}}} \qquad 15.90$$

If the clamped parts have different elasticities ($E$ values), the composite spring constant can be found from equation 15.91. (If a "soft" washer or gasket is used, its spring constant may control equation 15.91.)

$$\frac{1}{k_{\text{composite}}} = \frac{1}{k_1} + \frac{1}{k_2} + \frac{1}{k_3} + \cdots \qquad 15.91$$

Both the bolt and the clamped parts share the applied load.

$$P_{\text{bolt}} = P_i + \frac{k_{\text{bolt}} P_{\text{applied}}}{k_{\text{bolt}} + k_{\text{parts}}} \qquad 15.92$$

$$P_{\text{parts}} = \frac{k_{\text{parts}} P_{\text{applied}}}{k_{\text{bolt}} + k_{\text{parts}}} - P_i \qquad 15.93$$

Of course, if the applied load varies, the forces in the bolt and the parts also will vary. In that case, analysis by a Goodman diagram is called for.

For static loading, recommended amounts of preloading often are specified in terms of a percentage (e.g., 90%) of the tensile yield strength. The term *proof strength* (i.e.,

[45] It is assumed that the initial tension, $P_i$, is greater than $P_{\text{max}}$. If the clamped members separate, the bolt once again carries the entire load.

*proof load* divided by bolt area) can be used in place of tensile yield strength. For fatigue loading, the preload must be determined from an analysis of a Goodman diagram.

Tightening of a tension bolt will induce a torsional stress in the bolt. Where the bolt is to be locked in place, the torsional stress can be removed without greatly affecting the preload by slightly backing off the bolt. If the bolt is subject to cyclic loading, the bolt probably will slip back by itself, and it is reasonable to neglect the effects of torsion in the bolt.

More important than the effects of torsion are stress concentrations at the root. Although stress concentrations frequently are neglected for static loading of ductile connectors, there will be a significant reduction in the endurance limit for cyclic loading. Therefore, the alternating stress should be multiplied by an appropriate factor (e.g., 2.0 to 4.0).

## 32 FRAMING CONNECTIONS

### A. INTRODUCTION

Three types of framing (beam-to-column, beam-to-beam, etc.) connections are defined.

- Type 1, *rigid frame connections*: These connections are intended to transmit moments from one member to another. Type 1 connections can be designed by both working stress and plastic design methods.
- Type 2, *simple framing connections*: These connections transmit vertical loads, but essentially no (i.e., less than 20%) moment transfer occurs. Generally, only working stress methods are used to design type 2 connections.
- Type 3, *semi-rigid connections*: When the moment transfer across a connection is significant but not total (i.e., between 20% and 90%), the connection is semi-rigid. Plastic design is not used for type 3 connections. Since it is difficult to determine the amount of moment transfer, type 3 connections are rarest.



### B. SIMPLE (TYPE 2) FRAMING CONNECTIONS

Simple framing connections, also known as *flexible connections*, are designed to be as flexible as possible. Design is fairly predetermined by use of the standard tables of framed beam connections in the *AISC Manual*. Construction methods include beam seats and clips to beam webs. Figure 15.28 illustrates several type 2 connections.

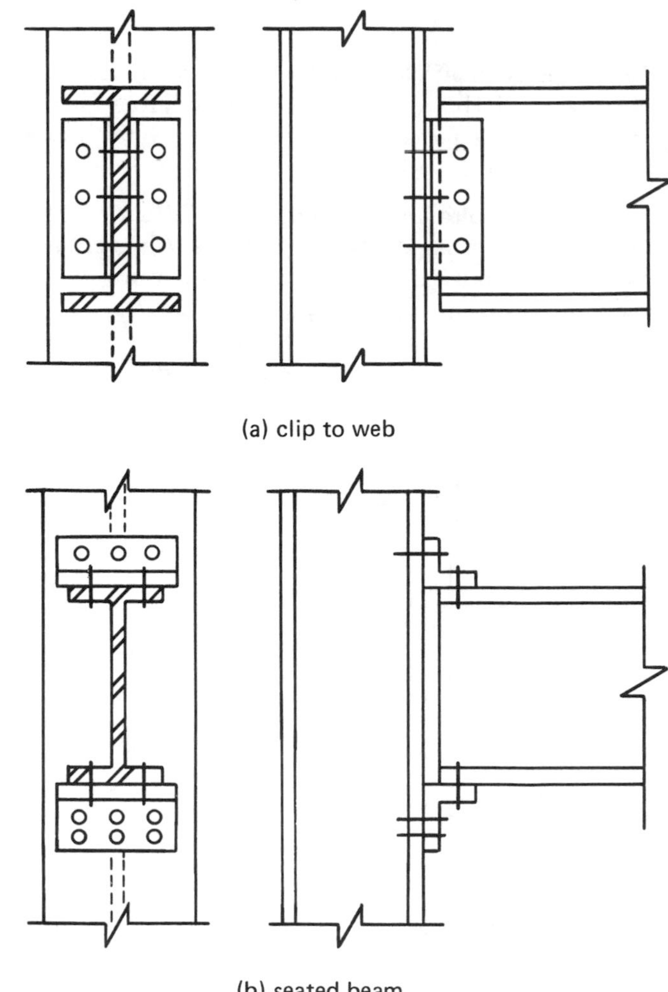


**Figure 15.28** Type 2 (Simple) Framing Connections

Connections to beam and column can be either by bolting or welding, and such fastening methods can be used either on the beam or column.[46] Welded connections, "stiffened" connections, and use of top seats do not necessarily imply a moment-resisting connection. Coping, where used moderately, does not reduce the shear capacity of members.

Direct shear determines the number of bolts required in the column connection. It is common to neglect the effect of eccentricity in determining shear stresses in riveted and bolted beam connections to webs. It is also common to neglect the effect of eccentricity in determining shear stresses in riveted and bolted column connections. However, this eccentricity can be considered, which will add tension stress to the shear stress in column fasteners.

Angle thickness must be checked for allowable bearing pressure. Angles should be checked for direct shear, as

[46] Usually, the shop connection will be welded. The field connection, however, can be either welded or bolted.

well. Since the connection is designed to rotate, the angle (for seated beams) should be checked for bending stress as its free lip bends.

### C. MOMENT-RESISTING (TYPE 1) FRAMING CONNECTIONS

Moment resisting connections transmit their vertical (shearing) load through the same types of connections as type 2 connections, typically through connections at the beam web. However, the flanges of the beam are also rigidly connected to the column. These top and bottom flange connections are in tension and compression respectively, and serve to transmit the moment.

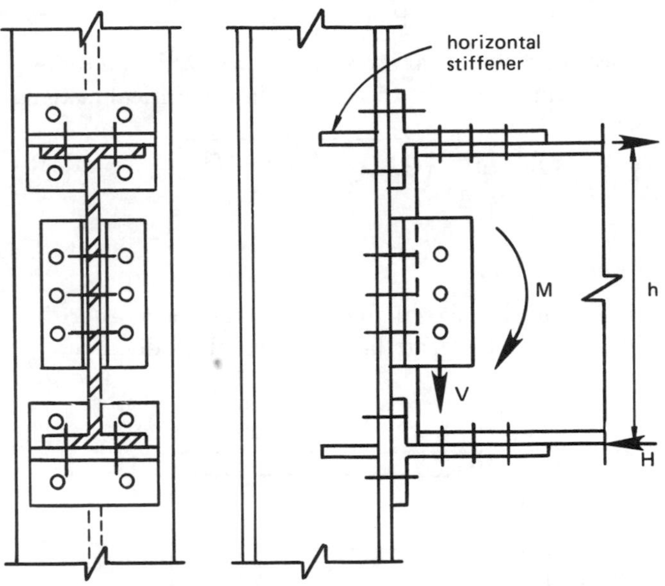


**Figure 15.29** Type 1 (Moment Resisting) Framing Connections

Since moment transfer is through the flanges by tension and compression connections, design of such connections involves ensuring adequate strength in tension and compression. Design of the shear transfer mechanism (i.e., the web connections) is essentially the same as for type 2 connections. Also, in order to prevent localized buckling of the column flanges, *horizontal stiffeners* (between the column flanges) may be needed.

The moment-resisting ability of a type 2 connection can be increased to essentially any desired value by increasing the distance between the tension and compression connections. Figure 15.30 illustrates how this could be accomplished by using an intermediate plate between the beam flanges and column. The horizontal tensile and compressive forces, $H$, can be calculated from equation 15.94.

$$H = \pm\frac{M}{h} \qquad 15.94$$

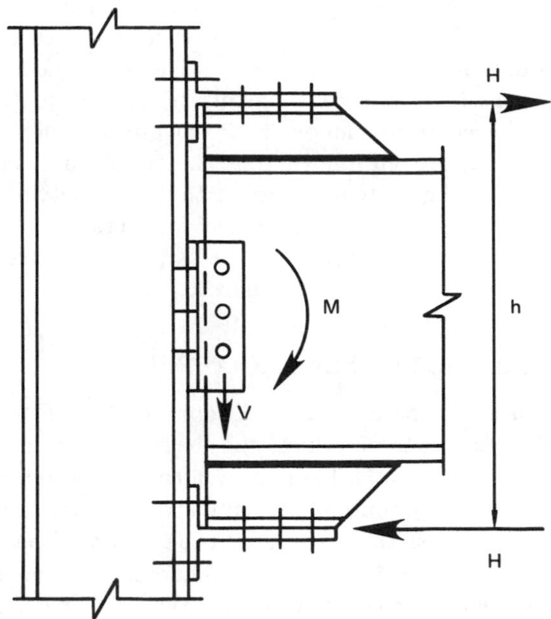


**Figure 15.30** Increasing Moment-Resisting Ability

## 33 BOLTED AND RIVETED ECCENTRIC SHEAR CONNECTIONS

### A. INTRODUCTION

An *eccentric shear connection* is illustrated in figure 15.31. This type of connection gets its name from the tendency of the bracket to rotate around the centroid of the fastener group, shearing the fasteners. The tendency to rotate is resisted by the shear stress in the connectors. Friction is not assumed to contribute to the rotational resistance of the connection.

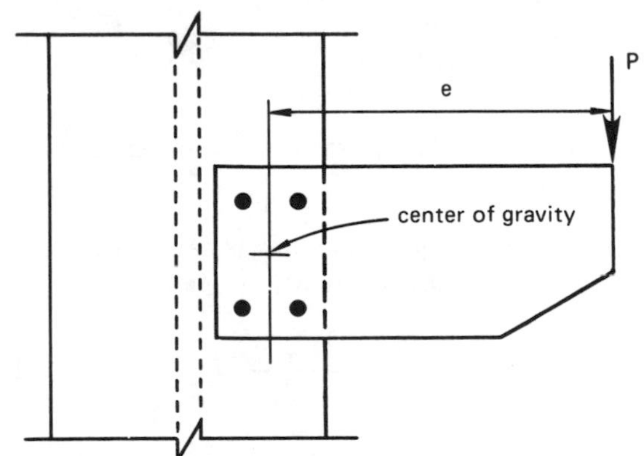


**Figure 15.31** Eccentric Shear Connection

The moment tending to cause rotation is

$$M = Pe \qquad 15.95$$

STEEL

It is clear that the fasteners must resist this moment by shear. However, it is not clear how the shearing resistance is to be calculated.[47] Three methods are available: (1) the traditional elastic approach, (2) a reduced-eccentricity model, and (3) ultimate strength analysis. It is well known that the traditional elastic approach will greatly underestimate the capacity of (or, will overdesign) an eccentric shear connection. In some cases, the actual capacity may be as much as twice the capacity calculated from the elastic model.

## B. TRADITIONAL ELASTIC APPROACH

The elastic approach uses traditional mechanics of materials concepts to determine the shearing stress in each fastener. This method starts by locating the centroid of the fastener group. For symmetrical (rectangular) fastener groups, this can usually be accomplished by inspection. Once the centroid is located, the *polar moment* of inertia (which resists the rotation) is calculated from the nominal fastener area and distance from the centroid to fastener. In equation 15.96, the summation is over all fasteners in the group.

$$J = \sum_i r_i^2 A_i \qquad 15.96$$

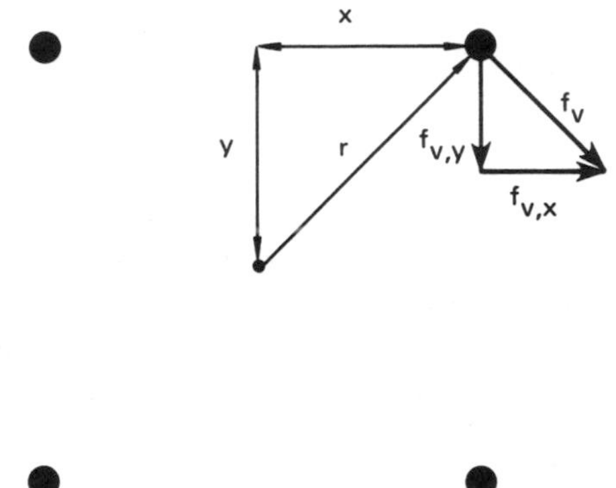


**Figure 15.32** Fastener Group Analysis

The shear stress in each member is calculated from the standard torsional stress equation:

$$f_v = \frac{M r_{\text{critical}}}{J} \qquad 15.97$$

The *critical fastener*, typically the highest, right-most fastener, is the one whose vector sum of direct and eccentric shear stresses will be the largest. It is usually found by inspection.

The shear stress in each member is directed perpendicular to the line connecting the center of rotation and the fastener. This shear stress must be converted to $x$- and $y$-components to be combined with the direct shear stress. If the distance, $r$, has been broken into $x$- and $y$-components, it will be easy to calculate the components of shear.

$$f_{v,y} = \frac{f_v x}{r} \qquad 15.98$$

$$f_{v,x} = \frac{f_v y}{r} \qquad 15.99$$

The *direct shear stress* $f_{v,d}$, is merely the load divided by the total area of all fasteners.

$$f_{v,d} = \frac{P}{\sum A_i} \qquad 15.100$$

Since the direct shear stress acts downward, and there is no $x$-component, the total stress in the critical fastener is

$$f_{v,\text{total}} = \sqrt{(f_{v,d} + f_{v,y})^2 + f_{v,x}^2} \qquad 15.101$$

*Example 15.15*

For the bracket shown, find the load on the most critical fastener. All fasteners have a nominal $\frac{1}{2}''$ diameter.

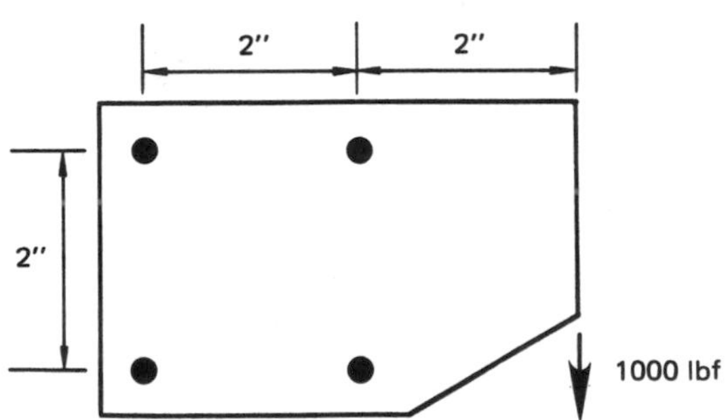


Since the fastener group is symmetrical, the group centroid is centered within the 4 fasteners. This makes the eccentricity of the load 3″. Each fastener is located $r$ from the centroid, where

$$r = \sqrt{(1)^2 + (1)^2} = 1.414$$

The area of each fastener is

$$A_i = \frac{1}{4}\pi(0.5)^2 = 0.1963$$

The polar moment of inertia is

$$J = 4[0.1963(1.414)^2] = 1.570\ \text{in}^4$$

The eccentric shear stress on each fastener is

$$f_v = \frac{(1000)(3)(1.414)}{(1.570)} = 2702\ \text{psi}$$

[47] And, the *AISC Manual* does not require a particular method, either.

STEEL

The $x$- and $y$-components of the shear stresses are

$$f_{v,y} = \frac{\sqrt{2}}{2} \times 2702 = 1911\,\text{psi}$$

$$f_{v,x} = 1911\,\text{psi}$$

The fasteners carry a direct vertical shear load of $1000/4 = 250$ pounds each. The vertical shear stress due to this load is

$$f_{v,d} = \frac{250}{0.1963} = 1274$$

The two right fasteners have vertical downward components of $f_{v,y}$ which add to the vertical downward stress of 1274. Thus, both of the two right fasteners are critical. The total stress in each of these fasteners is

$$f_v = \sqrt{(1911)^2 + (1911 + 1274)^2} = 3714\,\text{psi}$$

## C. REDUCED ECCENTRICITY MODELS

Since the elastic approach so greatly underestimates the capacity of eccentric shear connections, although not an AISC code requirement, it is logical to reduce the eccentricity when calculating the moment to be resisted.[48] Equations 15.102 and 15.103 give reduced eccentricities known as *effective lengths*. Equation 15.102 is for use when there is only one vertical line of fasteners. Equation 15.103 is for the more common situation of two or more vertical lines of fasteners. $n$ is the number of fasteners in one vertical line.

$$e_{\text{eff}} = e - \frac{1+2n}{4}\bigg|_{\text{1 fastener line}} \qquad 15.102$$

$$e_{\text{eff}} = e - \frac{1+n}{2}\bigg|_{\text{2 or more fastener lines}} \qquad 15.103$$

## D. ULTIMATE STRENGTH ANALYSIS

Design and analysis using ultimate strength is preferred over elastic methods. Several ultimate strength theories have been proposed. In practice, however, the application of such theories involves more table look-up than theory. Most methods end up calculating the allowable load on a fastener group from the product of tabulated coefficients and allowable fastener loads. For example, in equation 15.104, $F_v$ is the allowable shear load on the fasteners, and $C$ is a tabulated coefficient.

$$P = CA_bF_v \qquad 15.104$$

The ability to use equation 15.104 depends on having tables of $C$ values for the fastener configuration and eccentricity needed. The *AISC Manual* contains such tables in *Part 4, "Connections."*

[48]The allowable AISC loads on high-strength bolts is already high, and using reduced eccentricity models decreases the factor of safety below 2.5. Therefore, use of reduced eccentricity should be restricted to connections with rivets and low-strength bolts which have conservative allowable loads.

## 34 WELDS

The most widely-used weld type is the *fillet weld* shown in figure 15.33. The applied load is assumed to be carried by the *weld throat*, which has an effective dimension of $t_e$.

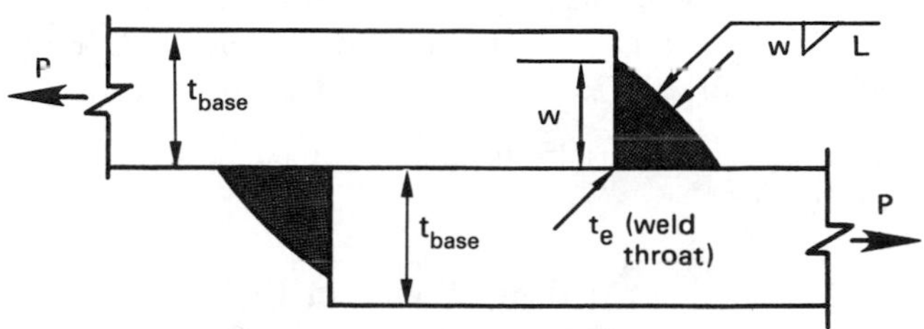


**Figure 15.33** Fillet Weld

The *effective weld throat size*, $t_e$, depends on the type of welding used. For hand-held *shielded metal arc welding (SMAW)* processes,

$$t_e = 0.707w \qquad 15.105$$

If *submerged arc welding (SAW)* is used (*Section J2.2*),

$$t_e = 0.707w + 0.11 \quad (w \geq 7/16'') \qquad 15.106$$

$$t_e = w \quad (w \leq 3/8'') \qquad 15.107$$

Weld sizes, $w$, of 3/16″, 1/4″, and 5/16″ are desirable because they can be made in a single pass.[49] However, fillet welds from 3/16″ to 1/2″ can be made in 1/16″ increments. For welds larger than 1/2″, every 1/8″ weld size can be made.

Allowable shear stress on the weld fillet throat is (*Table J2.5*)[50]

$$F_v = 0.30F_{u,\text{rod}} \qquad 15.108$$

Shear stress in the base material may not be greater than $0.40F_y$. For tensile loads, the allowable stress parallel to the weld axis is the same as for the base metal:

$$F_t = 0.60F_y \qquad 15.109$$

To simplify analysis and design of certain types of welded connections, it is convenient to define the *shear resistance per unit length of weld*. Tensile resistances can be found similarly.

[49] The 5/16″ limitation is appropriate for shielded metal arc weld (typically using hand-held rods). If a submerged arc process is used, up to 1/2″ welds can be made in one pass.

[50] In almost all instances, loads are transmitted through welds by shear stresses, regardless of the weld group orientation.

be found similarly.

$$R_w = \min \begin{cases} t_e F_{v,\text{rod}} & = t_e(0.30)F_{u,\text{rod}} \\ t_{\text{base}} F_{v,\text{member}} & = t_{\text{base}}(0.40)F_{y,\text{member}} \end{cases} \qquad 15.110$$

The ultimate strength of a welding rod is part of the rod designation. Thus, $F_u$ for an E70 welding rod is 70 ksi. The following rods are available: E60, E70, E80, E90, E100, and E110.

Several special restrictions that apply to fillet welds are given here.

- Minimum weld sizes depend on the thickness of the thickest of the two parts joined. (When weld size is increased to satisfy these minimums, capacity is not increased. Weld strength is based on the theoretical weld size calculated.)

**Table 15.8**
Minimum Fillet Weld Size

(all dimensions in inches)

| larger part thickness | minimum $w$ |
|---|---|
| to $\frac{1}{4}$ inclusive | $\frac{1}{8}$ |
| over $\frac{1}{4}$ to $\frac{1}{2}$ | $\frac{3}{16}$ |
| over $\frac{1}{2}$ to $\frac{3}{4}$ | $\frac{1}{4}$ |
| over $\frac{3}{4}$ | $\frac{5}{16}$ |

- The maximum weld size along edges of connecting parts is equal to the edge thickness for materials less than $\frac{1}{4}''$ thick. For materials thicker than $\frac{1}{4}''$, the maximum size must be $\frac{1}{16}''$ less than the material thickness.

- The minimum length weld for full strength analysis is 4 times the weld size.[51]

- If the required strength of a welded connection is less than would be obtained from a full-length weld of the smallest size, then an *intermittent weld* can be used. The minimum length of an intermittent weld is 4 times the weld size or $1\frac{1}{2}$ inches, whichever is greater.

- The minimum weld length for lap joints (as illustrated in figure 15.33) is 5 times the thinner plate's thickness, but not less than 1 inch.

[51] If this criterion is not met, the weld size is downgraded to $\frac{1}{4}$ of the weld length.

## 35 WELDED CONNECTIONS

### A. CONCENTRIC TENSION CONNECTIONS

If the weld group centroid is in line with the applied load, the loading is concentric. Equation 15.111 can be used to design or evaluate the connection.

$$f_v = \frac{P}{A_{\text{weld}}} = \frac{P}{L_{\text{weld}} t_e} \qquad 15.111$$

*Example 15.16*

Two $\frac{1}{2}\times 8$ plates (A36 steel) are lap welded using E70 electrodes and a shielded arc process. Size the weld to carry a concentric tensile loading of 50 kips.

Good design will weld both joining ends to the base plate. The total weld length will be

$$L_{\text{weld}} = (2)(8) = 16''$$

This length meets the 5″ minimum length specification.

From table 15.8, the minimum weld thickness for a $\frac{1}{2}''$ plate is 3/16″.

The required strength per inch of weld is

$$\frac{50}{16} = 3.125 \text{ kips/in}$$

From equation 15.110,

$$R_w = \min \begin{cases} (0.707)(3/16)(0.30)(70) & = 2.78 \\ (0.50 \text{ in})(0.40)(36) & = 7.2 \end{cases}$$

Since $R_w < 3.125$, a larger weld is needed. Try $w = 1/4''$.

$$R_w = \min \begin{cases} (0.707)(1/4)(0.30)(70) & = 3.71 \quad \text{(ok)} \\ (0.50 \text{ in})(0.40)(36) & = 7.2 \quad \text{(ok)} \end{cases}$$

The maximum weld size allowed is $\frac{1}{2}'' - 1/16'' = 7/16''$. The 1/4″ weld can be used.

### B. MOMENT RESISTING CONNECTIONS

The *AISC Manual* contains several procedures for designing moment connections (both type 1 and type 3).[52] These procedures assume top plate or end plate construction to transmit moments. Split beam tee construction, and the accompanying problem of prying action stress, is not covered.

[52] See "*Moment Connections*" in *Part 4* of the *AISC Manual*.

STEEL

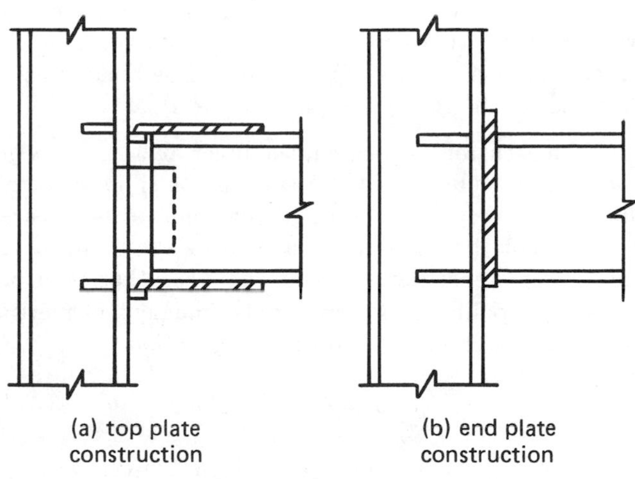


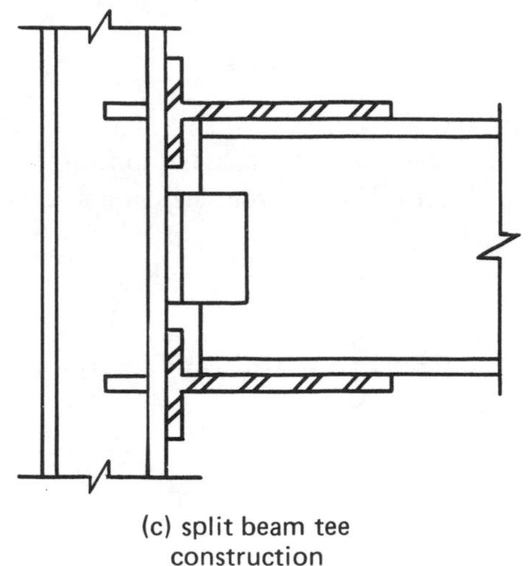


**Figure 15.34** Welded Moment Resisting Connections

## C. BALANCING WELD GROUPS

When tension is applied to an unsymmetrical member, the tensile force will act along a line passing through the centroid of the member. In that instance, it may be desirable to design a *balanced weld*, and have the force pass through the centroid of the weld group.[53] Figure 15.35 shows a tensile force being transmitted through a single angle.

The procedure for designing the unequal weld lengths assumes that the end weld (if present) acts at full shear stress with a resultant passing through its mid-height ($d/2$ in figure 15.35). The forces in the other welds are assumed to act along the edges of the angle. Moments can be taken about a point on the line of action of either longitudinal weld, with the following results.

$$P_3 = T\left(1 - \frac{y}{d}\right) - \frac{P_2}{2} \quad 15.112$$

$$P_2 = R_w L_2 = R_w d \quad 15.113$$

$$P_1 = T - P_2 - P_3 \quad 15.114$$

$$L_1 = \frac{P_1}{R_w} \quad 15.115$$

$$L_3 = \frac{P_3}{R_w} \quad 15.116$$

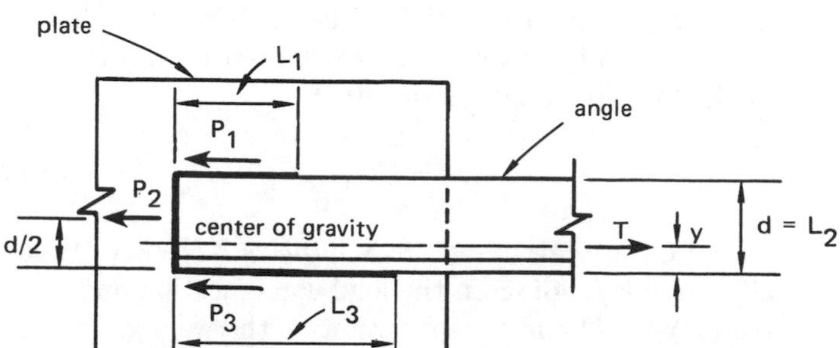


**Figure 15.35** A Balanced Weld Group

## D. COMBINING SHEAR AND BENDING STRESSES

Figure 15.36 illustrates a case where a welded connection must support both direct shear and a bending moment. Even though the maximum shear and maximum bending moment do not actually occur at the same place, simplifying assumptions are made to combine the nominal stresses as vectors.

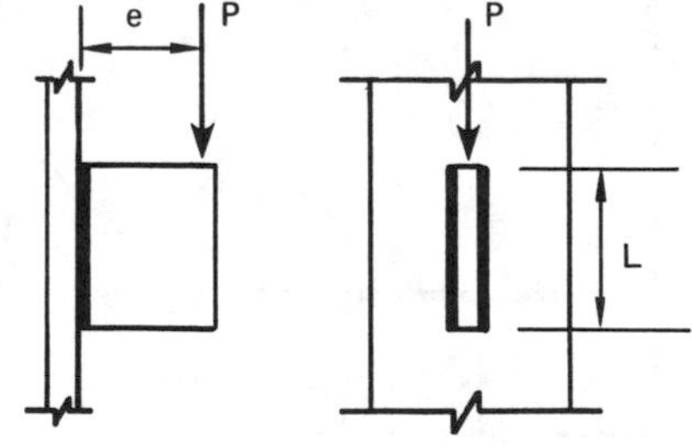


**Figure 15.36** Welds in Combined Bending and Shear

The nominal shear stress is

$$f_v = \frac{P}{A_{\text{weld}}} = \frac{P}{2Lt_e} \quad 15.117$$

[53] *AISC Specifications section J1.9* exempts single angles and double angles from the need to balance welds when loading is static. All other static loading configurations, and angles subjected to fatigue, must have balanced welds.

The nominal bending stress is easily calculated if the section modulus of the weld group can be obtained from appendix C.

$$f_b = \frac{Mc}{I} = \frac{M}{S} = \frac{Pe}{S} \qquad 15.118$$

The resultant stress is

$$f = \sqrt{f_v^2 + f_b^2} \qquad 15.119$$

The resultant stress must be less than the allowable stress, as calculated from equation 15.108.

The *AISC Manual* contains tables enabling the capacity of such eccentric loads to be calculated from a formula of the form in equation 15.120.[54]

$$P = CC_1Dl \qquad 15.120$$

$C$ and $C_1$ are tabulated (*AISC Tables XIX–XXVI*) coefficients depending on the load configuration and electrode type. $D$ and $l$ are functions of the weld group size and length.

E. TORSION CONNECTIONS

The complexity of the stress distribution in torsion connections (such as are shown in figure 15.37) makes accurate analysis and design based on pure mechanics principles impossible. Therefore, simplifications are made and the assumed shear stress in the most critical location is calculated.

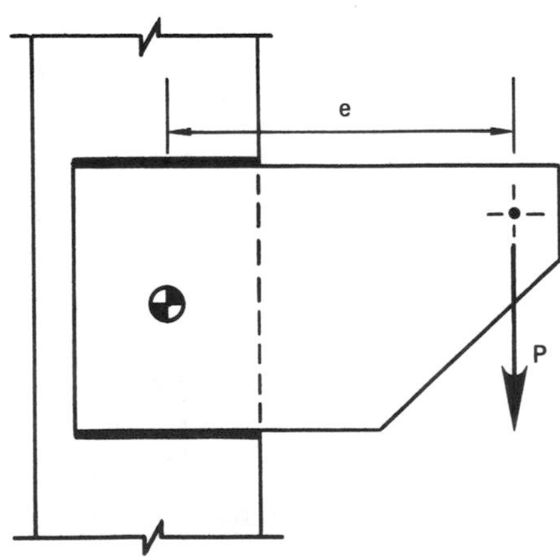


**Figure 15.37** A Welded Torsion Connection

The solution procedure (for both analysis and design) requires finding the centroid of the weld group. This centroid can be located in the normal manner, by weighting the weld's areas by the distances from an assumed axis. Alternatively, a weld can be treated as a line and its length used in place of the area.

Once the centroid is located, the polar area moment of inertia is calculated. Calculating the *polar moment of inertia* is easiest if the coordinate moments of inertia are known. For the purpose of calculating the moment of inertia, the welds may also be treated as either areas or lines[55] (Appendix C lists the polar moment of inertia for many common weld group configurations.)

$$J = I_x + I_y \qquad 15.121$$

The distance from the weld group centroid to the critical location in the weld group is then determined. This distance is used to calculate the torsional shear.

$$f_v = \frac{Mr}{J} = \frac{Per}{J} \qquad 15.122$$

The direct shear is easily calculated from the total weld area (or length, if the throat size is not known).

$$f_{v,d} = \frac{P}{A} \qquad 15.123$$

Vector addition is used to combine the torsional and direct shear stresses.

$$f = \sqrt{(f_{v,y} + f_{v,d})^2 + f_{v,x}^2} \qquad 15.124$$

*Example 15.17*

A 50 ksi plate bracket is welded with E70 electrodes to the face of a 50 ksi column as shown. What size fillet weld is required? Neglect buckling and bending effects of the plate and column.

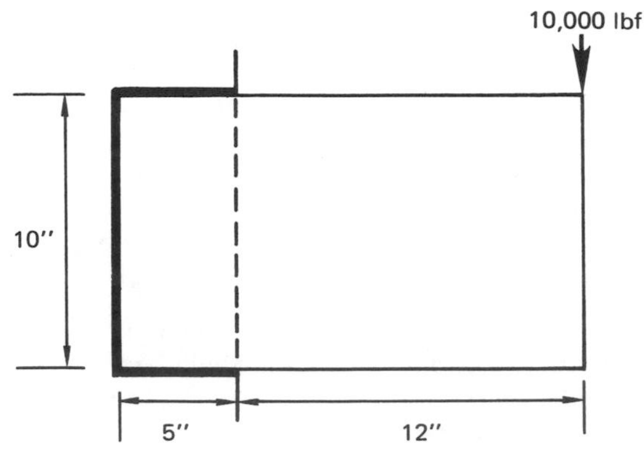


[54] See "*Eccentric Loads on Weld Groups*." This method is based on ultimate strength design concepts. As with bolted connections based on ultimate design, capacities much greater than would be expected from elastic analysis are obtained.

[55] In an analysis of a weld group, the throat size will be known, and welds can be treated as areas. In design, the throat size will be unknown, and it is common to treat the welds as lines. Alternatively, a variable weld size can be used and carried along in a design problem. However, the results will be essentially identical.

*step 1*: Assume the weld has thickness $t$.

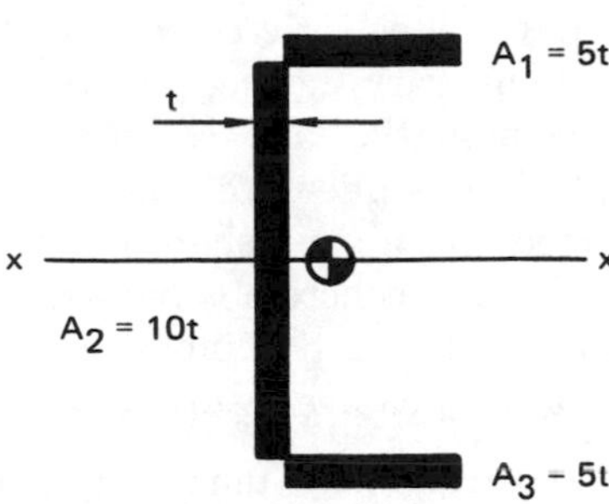


*step 2*: Find the centroidal location of the weld group. By inspection, $\overline{y}_c = 0$. For the three welds,

$$A_1 = 5t$$
$$\overline{x}_1 = 2.5$$
$$A_2 = 10t$$
$$\overline{x}_2 = 0$$
$$A_3 = 5t$$
$$\overline{x}_3 = 2.5$$

$$\text{So, } \overline{x}_c = \frac{5t(2.5) + 10t(0) + 5t(2.5)}{5t + 10t + 5t} = 1.25$$

*step 3*: Determine the centroidal moment of inertia of the weld group about the $x$-axis. Use the parallel axis theorem for areas 1 and 3.

$$I_x = \frac{t(10)^3}{12} + 2\left[\frac{5(t)^3}{12} + 5t(5)^2\right]$$
$$= 333.33t + 0.833(t)^3$$

Since $t$ will be small (probably less than 0.5″), the $t^3$ term can be neglected. So, $I_x = 333.33t$.

*step 4*: Determine the centroidal moment of inertia of the weld group about the $y$-axis.

$$I_y = \frac{10(t)^3}{12} + (10t)(1.25)^2 + 2\left[\frac{t(5)^3}{12} + (5t)(1.25)^2\right]$$
$$= 0.833t^3 + 52.08t \approx 52.08t$$

*step 5*: The polar moment of inertia is

$$J = I_x + I_y = 333.33t + 52.08t = 385.4t$$

*step 6*: By inspection, the maximum shear stress will occur at point **a**.

$$r = \sqrt{(3.75)^2 + (5)^2} = 6.25$$

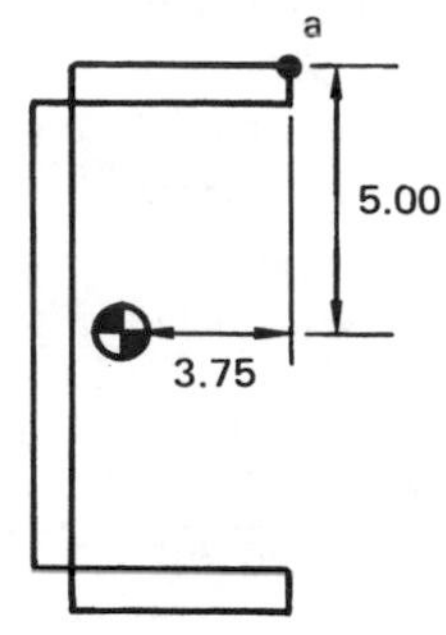


*step 7*: The applied moment is

$$M = (10{,}000)(12 + 3.75) = 157{,}500 \text{ in-lbf}$$

*step 8*: The torsional shear stress is

$$f_v = \frac{Mr}{J} = \frac{(157{,}500)(6.25)}{385.4t} = \frac{2554.2}{t} \text{ psi}$$

This shear stress is directed at right angles to the line $r$. The $x$- and $y$-components of the stress can be determined from geometry.

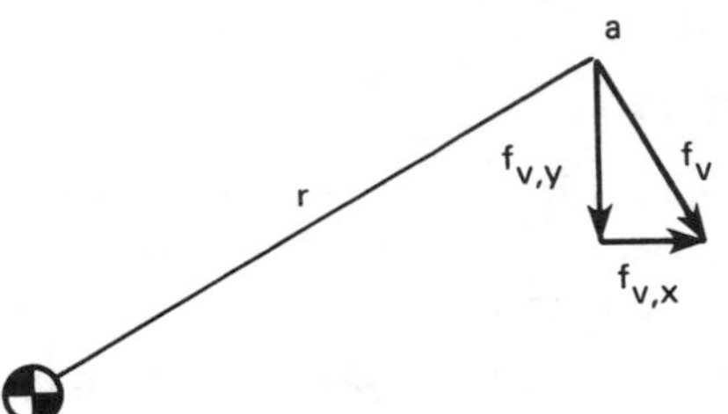


$$f_{v,y} = \left(\frac{3.75}{6.25}\right)\left(\frac{2554.2}{t}\right) = \frac{1532.5}{t}$$
$$f_{v,x} = \left(\frac{5.00}{6.25}\right)\left(\frac{2554.2}{t}\right) = \frac{2043.4}{t}$$

*step 9*: The direct shear is

$$f_{v,d} = \frac{10{,}000}{5t + 10t + 5t} = \frac{500}{t}$$

*step 10*: The resultant shear stress at point **a** is

$$f_v = \sqrt{\left(\frac{2043.4}{t}\right)^2 + \left(\frac{1532.5 + 500}{t}\right)^2} = \frac{2882.1}{t}$$

*step 11*: For 50 ksi base metal, the weld strength controls the allowable stress. (See equation 15.110.) The allowable stress is

$$F_v = 0.30F_u = (0.30)(70) = 21 \text{ ksi}$$

The weld throat size is

$$t = \frac{2882.1}{21{,}000} = 0.137''$$

*step 12*: The weld size (assuming shielded arc welding) is

$$w = \frac{t}{0.707} = \frac{0.137}{0.707} = 0.194'' \quad \text{(say 1/4'')}$$

*Example 15.18*

Use appendix C to calculate the polar moment of inertia of the weld group in example 15.17.

$$b = 5$$
$$d = 10$$
$$J = \left[\frac{(8)(5)^3 + (6)(5)(10)^2 + (10)^3}{12} - \frac{(5)^4}{(2)(5) + 10}\right] t$$
$$= 385.42t$$

(Example 15.17 calculated the moment of inertia to be 385.4$t$.)

*Example 15.19*

Use the *AISC Manual* "*Eccentric Loads on Weld Groups*" tables to calculate the capacity of the weld group designed in example 15.17.

For a 1/4'' weld and E70 electrodes, the following coefficients are needed.

$$l = 10''$$
$$k = \frac{5}{10} = 0.5$$
$$a = \frac{12 + 3.75}{10} = 1.575$$
$$C = \mathbf{0.419} \quad \text{(interpolated)}$$
$$C_1 = 1 \quad \text{(for E70 electrodes)}$$
$$D = \frac{\frac{1}{4}}{\frac{1}{16}} = 4$$

From equation 15.120, the capacity of the connection is

$$P = CC_1Dl = (0.419)(1)(4)(10) = 16.76 \text{ kips}$$

## 36 COMPOSITE CONSTRUCTION

*Composite construction* usually means that a concrete slab is bonded to steel girders below. Although there are varying degrees of composite action, the concrete usually becomes the compression "flange" of the composite beam, and the steel carries tension.

The degree to which the concrete-steel combination acts compositely depends on the shoring used during construction. If the steel is erected and concrete is poured without temporary shoring (i.e., construction is *unshored*), the combination acts compositely to carry only loads applied subsequent to concrete curing. If the steel is erected and a snug shore is put into place at that time, before the concrete is poured (i.e., *partial shoring*), the combination acts compositely to carry the concrete weight as well as live loads.

In fully *shored construction*, a temporary support carries the steel and concrete weights both until the concrete has cured. When the shore is removed, the beam acts compositely to carry the steel and concrete weights in addition to live loads applied later.

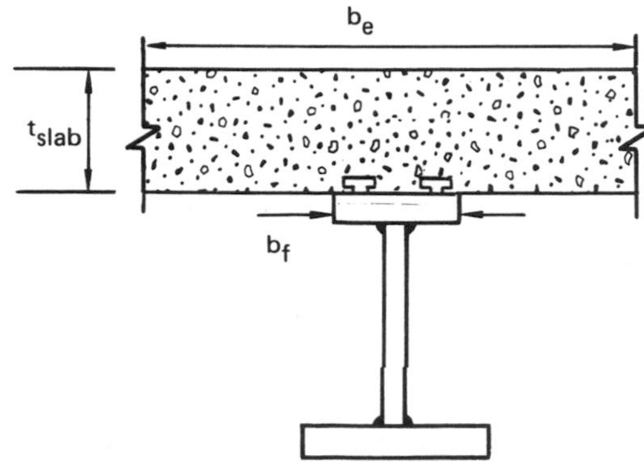


**Figure 15.38** Typical Composite Construction with Plate Girder

Analysis of composite steel-concrete beams is governed by *AISC Chapter I*. It is necessary to know the modulus of elasticity for the concrete and steel in order to calculate the *modular ratio, n*.[56]

$$n = \frac{E_s}{E_c} \qquad 15.125$$

For the purpose of calculating the moment of inertia, the effective width of the compression flange must be known. For interior girders with slabs extending on both sides, the effective width is (*Section I1*)[57]

$$b_e = \min \begin{Bmatrix} \frac{1}{8} \text{ of beam span} \\ \frac{1}{2} \text{ of beam-to-beam spacing} \\ \text{distance from beam to slab edge} \end{Bmatrix} \qquad 15.126$$

These effective width limitations are different for T-beams constructed entirely of concrete. (See section 14.10.)

[56] The ACI formula for *secant modulus* is used to calculate $E_c$.

[57] Other rules apply to exterior girders with slabs extending to one side only.

STEEL

# Appendix A: Steel Used for Buildings and Bridges

| ASTM designation | grade (if any) | $F_y$ minimum yield stress (ksi) | $F_u$ tensile strength (ksi) | maximum thickness for plates (in.) | use |
|---|---|---|---|---|---|
| A36 | | 32 | 58–80 | over 8 | general structural purposes; |
| | | 36 | 58–80 | to 8 | bolted and welded, mainly for buildings |
| A53 | A | 30 | 48 | | welded and seamless pipe |
| | B | 35 | 60 | | |
| A242 | | 42 | 63 | over $1\frac{1}{2}$ to 4 | welded and bolted bridge |
| | | 46 | 67 | over $\frac{3}{4}$ to $1\frac{1}{2}$ | construction where corrosion |
| | | 50 | 70 | to $\frac{3}{4}$ | resistance is desired; essentially superseded by A709, grade 50W |
| A440 | | 42 | 63 | over $1\frac{1}{2}$ to 4 | bolted construction; essentially |
| | | 46 | 67 | over $\frac{3}{4}$ to $1\frac{1}{2}$ | superseded by A572 for buildings |
| | | 50 | 70 | to $\frac{3}{4}$ | and A709 for bridges |
| A441 | | 40 | 60 | over 4 to 8 | welded construction; largely |
| | | 42 | 63 | over $1\frac{1}{2}$ to 4 | superseded by A572 for buildings |
| | | 46 | 67 | over $\frac{3}{4}$ to $1\frac{1}{2}$ | and A709 for bridges |
| | | 50 | 70 | to $\frac{3}{4}$ | |
| A500 | A | 33 | 45 | | cold-formed welded and seamless |
| | B | 42 | 58 | round | round and shaped tubing for |
| | C | 46 | 62 | | general structural purposes |
| | A | 39 | 45 | | |
| | B | 46 | 58 | shaped | |
| | C | 50 | 62 | | |
| A501 | | 36 | 58 | | hot-formed welded and seamless round and shaped tubing for general structural purposes |
| A514 | | 90 | 100–130 | over $2\frac{1}{2}$ to 6 | alloy steel plates for welded |
| | | 100 | 110–130 | to $2\frac{1}{2}$ | construction; superseded by A709 for bridges |
| A529 | | 42 | 60–85 | to $\frac{1}{2}$ | pre-engineered rigid frames |
| A570 | A | 25 | 45 | | cold-formed sections |
| | B | 30 | 49 | | |
| | C | 33 | 52 | | |
| | D | 40 | 55 | | |
| | E | 42 | 58 | | |
| A572 | 42 | 42 | 60 | to 6 | welded and bolted construction |
| | 50 | 50 | 65 | to 2 | for buildings; welded bridges |
| | 60 | 60 | 75 | to $1\frac{1}{4}$ | in grades 42, and 50 only; |
| | 65 | 65 | 80 | to $1\frac{1}{4}$ | essentially superseded by A709, grade 50 for bridges |

| designation | grade (if any) | $F_y$ minimum yield stress (ksi) | $F_u$ tensile strength (ksi) | maximum thickness for plates (in.) | use |
|---|---|---|---|---|---|
| A588 | | 42 | 63 | over 5 to 8 | weathering steel for welded |
| | | 46 | 67 | over 4 to 5 | and bolted construction; essentially |
| | | 50 | 70 | to 4 | superseded by A709, grade 50W for bridges |
| A606 | | 45 | 65 | | hot- and cold-rolled sheet |
| | | 50 | 70 | (hot-rolled cut lengths only) | and strip steel available in coils or cut lengths, used for cold-formed sections |
| A607 | 45 | 45 | 60 | | hot-rolled and cold-rolled |
| | 50 | 50 | 65 | | sheet and strip steel in coils |
| | 55 | 55 | 70 | | or cut lengths, used in cold- |
| | 60 | 60 | 75 | | formed sections |
| | 65 | 65 | 80 | | |
| | 70 | 70 | 85 | | |
| A611 | A | 25 | 42 | | cold-rolled sheet steel |
| | B | 30 | 45 | | for cold-formed sections |
| | C | 33 | 48 | | |
| | D | 40 | 52 | | |
| | E | 80 | 82 | | |
| A618 | I | 50 | 70 | | hot-formed welded and seamless |
| | II | 50 | 70 | | tubing for general structural |
| | III | 50 | 65 | | purposes |
| A709 | 36 | 32 | 58 | over 8 | bridge construction: grade 36 |
| | | 36 | 58–80 | to 8 | is approximately the same as A36; |
| | 50 | 50 | 65 | to 2 | grade 50 as A441; grade 50W as |
| | 50W | 50 | 70 | to 4 | A588; and grade 100 as A514 |
| | 100 & 100W | 90 | 100–130 | over $2\frac{1}{2}$ to 4 | |
| | 100 & 100W | 100 | 110–130 | to $2\frac{1}{2}$ | |

# Appendix B: Properties of Structural Steel at High Temperatures

| type of steel | temperature (°F) | yield strength 0.2% offset (ksi) | ultimate tensile strength (ksi) |
|---|---|---|---|
| ASTM A36 | 80 | 36.0 | 64.0 |
| | 300 | 30.2 | 64.0 |
| | 500 | 27.8 | 63.8 |
| | 700 | 25.4 | 57.0 |
| | 900 | 21.5 | 44.0 |
| | 1100 | 16.3 | 25.2 |
| | 1300 | 7.7 | 9.0 |
| ASTM A242 | 80 | 54.1 | 81.3 |
| | 200 | 50.8 | 76.2 |
| | 400 | 47.6 | 76.4 |
| | 600 | 41.1 | 81.3 |
| | 800 | 39.9 | 76.4 |
| | 1000 | 35.2 | 52.8 |
| | 1200 | 20.6 | 27.6 |
| ASTM A588 | 80 | 58.6 | 78.5 |
| | 200 | 57.3 | 79.5 |
| | 400 | 50.4 | 74.8 |
| | 600 | 42.5 | 77.7 |
| | 800 | 37.6 | 70.7 |
| | 1000 | 32.6 | 46.4 |
| | 1200 | 17.9 | 23.3 |

# Appendix C: Properties of Welds Treated as Lines

| weld configuration | centroid location | section modulus $S = I_{c,x}/\overline{y}$ (in$^2$) | polar moment of inertia $J = I_{c,x} + I_{c,y}$ (in$^3$) |
|---|---|---|---|
| 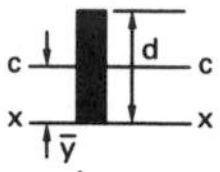 | $\overline{y} = \frac{d}{2}$ | $\frac{d^2}{6}$ | $\frac{d^3}{12}$ |
| 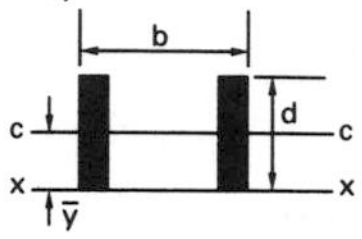 | $\overline{y} = \frac{d}{2}$ | $\frac{d^2}{3}$ | $\frac{d(3b^2 + d^2)}{6}$ |
| 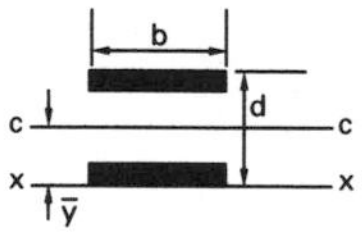 | $\overline{y} = \frac{d}{2}$ | $bd$ | $\frac{b(3d^2 + b^2)}{6}$ |
| 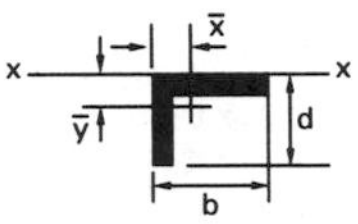 | $\overline{y} = \frac{d^2}{2(b+d)}$<br>$\overline{x} = \frac{b^2}{2(b+d)}$ | $\frac{4bd + d^2}{6}$ | $\frac{(b+d)^4 - 6b^2d^2}{12(b+d)}$ |
| 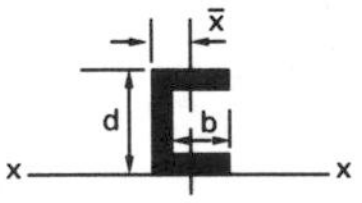 | $\overline{x} = \frac{b^2}{2b+d}$ | $bd + \frac{d^2}{6}$ | $\frac{8b^3 + 6bd^2 + d^3}{12} - \frac{b^4}{2b+d}$ |
| 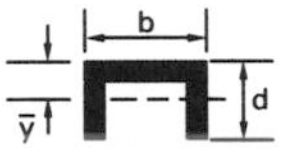 | $\overline{y} = \frac{d^2}{b+2d}$ | $\frac{2bd + d^2}{3}$ | $\frac{b^3 + 6b^2d + 8d^3}{12} - \frac{d^4}{2d+b}$ |
| 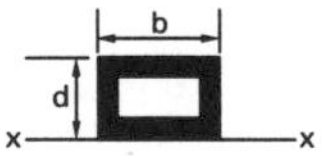 | $\overline{y} = \frac{d}{2}$ | $bd + \frac{d^2}{3}$ | $\frac{(b+d)^3}{6}$ |
| 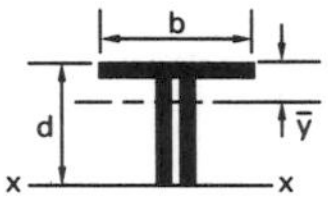 | $\overline{y} = \frac{d^2}{b+2d}$ | $\frac{2bd + d^2}{3}$ | $\frac{b^3 + 8d^3}{12} - \frac{d^4}{b+2d}$ |
| 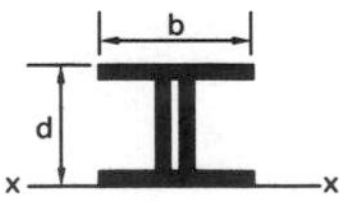 | $\overline{y} = \frac{d}{2}$ | $bd + \frac{d^2}{3}$ | $\frac{b^3 + 3bd^2 + d^3}{6}$ |
| 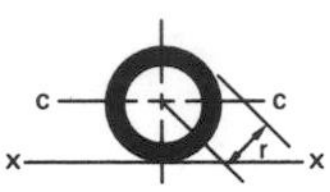 | $\overline{y} = r$ | $\pi r^2$ | $2\pi r^3$ |

# Appendix D: The Moment Gradient Multiplier

If a concentrated load is applied to a beam at a point of lateral support, there will be less tendency for the beam to buckle than if the load is applied between points of lateral support. The moment gradient multiplier, $C_b$, is an optional refinement that accounts for the less-critical conditions. The moment gradient multiplier increases the maximum allowable spacing between supports, $L_u$. ($C_b$ is not used in calculating $L_c$.) $C_b$ varies between 1.0 and 2.3 (its maximum). Conservative designs use $C_b = 1.0$.

$$C_b = 1.75 + 1.05\left(\frac{M_1}{M_2}\right) + 0.3\left(\frac{M_1}{M_2}\right)^2 \quad \textbf{[F1.3]}$$

$M_1$ is the absolute value of the smaller moment at the ends of an unbraced beam section. $M_2$ is the absolute value of the larger moment at the opposite end of the unbraced section. The quantity $(M_1/M_2)$ is negative for simple bending (i.e., single curvature, when $M_1$ and $M_2$ have opposite signs) and positive when $M_1$ and $M_2$ have the same sign (i.e., reverse curvature).

If the moment on a section of beam is maximum between the points of lateral support, $C_b = 1$.

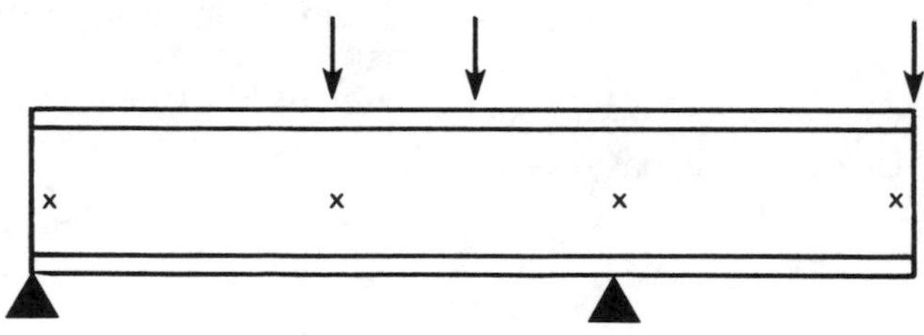


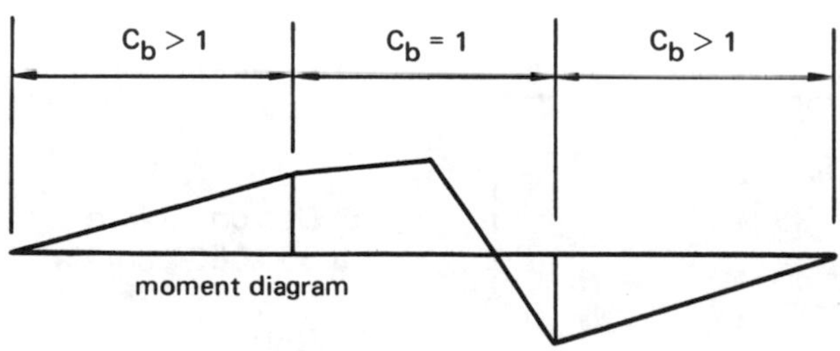


**Figure 15.39** Moment Gradient Multiplier

In evaluating $F_b$ for use with beam columns (equation 15.49 or **[H1-1]**), $C_b = 1$ is used when joint translation is prohibited. When joint translation is possible, calculate $C_b$ according to the equation given.

**Practice Problems: STEEL DESIGN**

Untimed

1. A 25 foot long beam is simply supported at its left end and 7 feet from its right end. A load of 3000 pounds per foot is uniformly distributed over its entire length. Lateral support is provided only at the reactions. Choose an economical W shape for this application. Use A36 steel.

2. Select the lightest W section of A36 steel to serve as a main member 30 feet long and to carry an axial load of 160,000 pounds. The member is pinned at its top and bottom. It is supported at mid-height in its weak direction. The member is vertical.

3. Determine the stresses in bolts A, B, C, and D. All connectors are ¾" bolts. (a) Use a traditional elastic approach. (b) Use AISC tables.

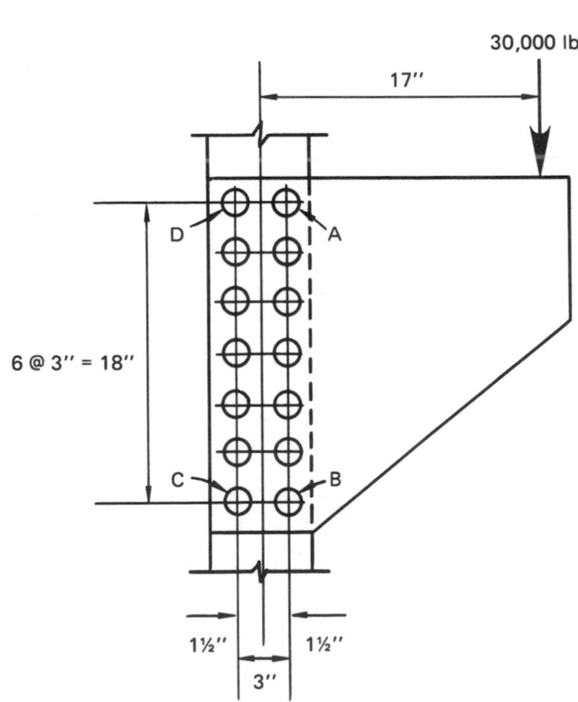


4. Determine the size of the fillet weld required to connect the 1 inch thick plate bracket to the column face. The steel is A36 for both the bracket and the column. The electrodes used are E70XX.

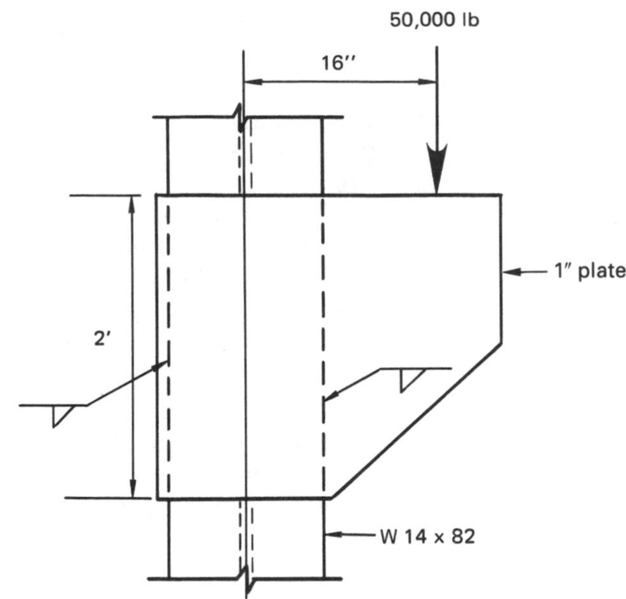


5. Design the interior columns of the frame shown. Columns are braced continuously in the plane perpendicular to the frame. Girders are W12 x 96. Use A36 steel. Left-to-right sidesway is uninhibited.

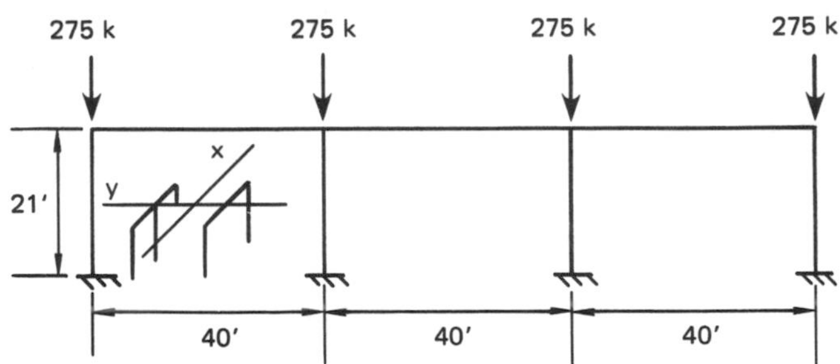


6. Design a plate eyebar to carry a static tensile load of 300,000 pounds. Use A36 steel.

7. Determine the tensile capacity of the connection shown. A325, 7/8″ diameter bolts are used. All holes are punched. The plate steel is grade 50 A588. The connection is a friction type with no threads in the shear plane.

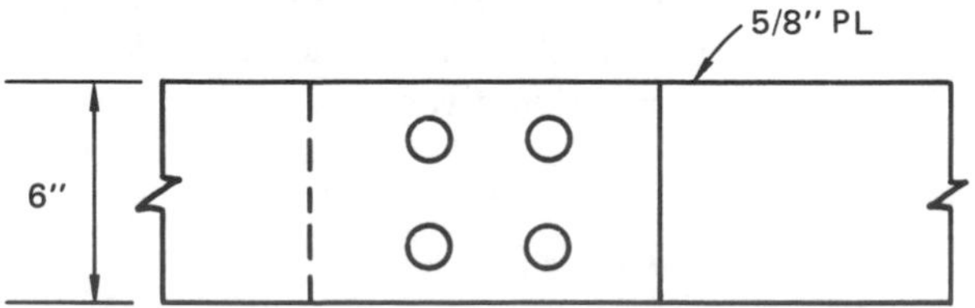


8. Select an A36 W shape with lateral support at 5½ foot intervals to span 20 feet and carry a uniformly distributed load of 1000 lb/ft. The load includes a uniform dead load allowance of 40 lb/ft. Limit the maximum deflection to (L/240).

9. Select an economical A36 W shape (completely laterally braced) to span 24 feet such that the maximum deflection is (L/300). The span carries a uniformly distributed load of 800 lb/ft over its entire length. This load does not include the dead weight of the beam.

10. A 16 foot column is acted upon by an axial gravity load of 17,000 lb and a uniform wind load of *w* lb/ft. The W12 × 58 column is made of A441 steel. The lower end is built in. The upper end is not supported in the weak direction. (a) Find the required spacing of lateral bracing for the maximum lateral (uniform) loading. (b) Find the maximum lateral (uniform) loading allowed with the bracing spaced as calculated in part (a).

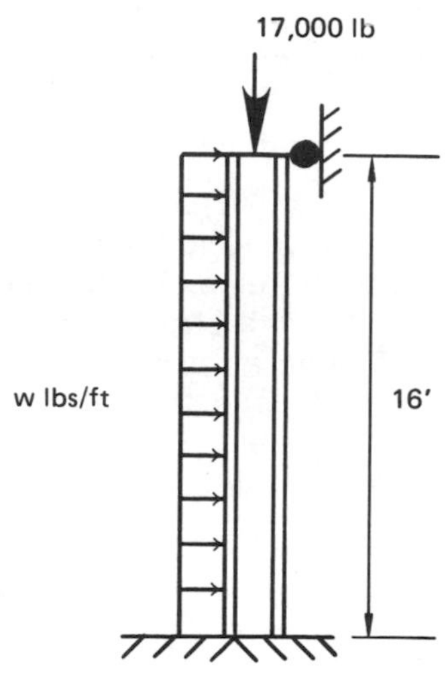


Timed

1. The steel beam shown carries 1.4 kips/ft over its entire span. Assume lateral bracing only at the reaction points. (a) Draw shear and moment diagrams. (b) Choose the lightest W18 beam assuming A36 steel is used. (c) If the beam you chose in part (b) is constructed of 42 ksi steel, what is the maximum unbraced length that permits the development of a $0.66F_y$ bending stress?

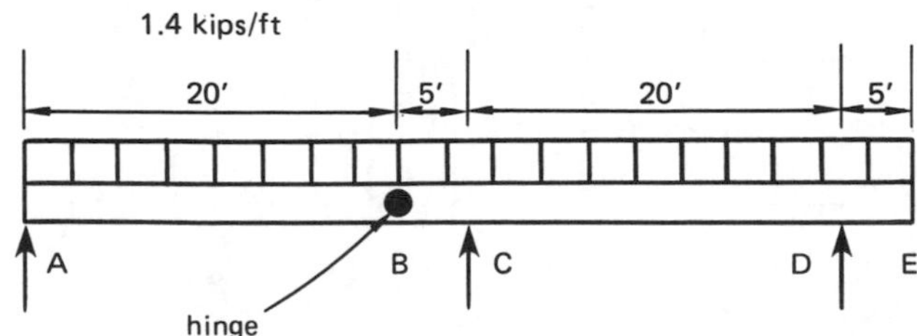


2. A column carries 525 kips. The beams framing into it apply a moment of 225 ft-kips. (a) If the column is W14 x 82 (50 ksi steel) and the beams are W14 x 53 (50 ksi), determine if the column is adequate. (b) If the design is inadequate, specify an acceptable W14 (50 ksi) column. Bracing in the weak axis is provided and bending is about the strong axis only.

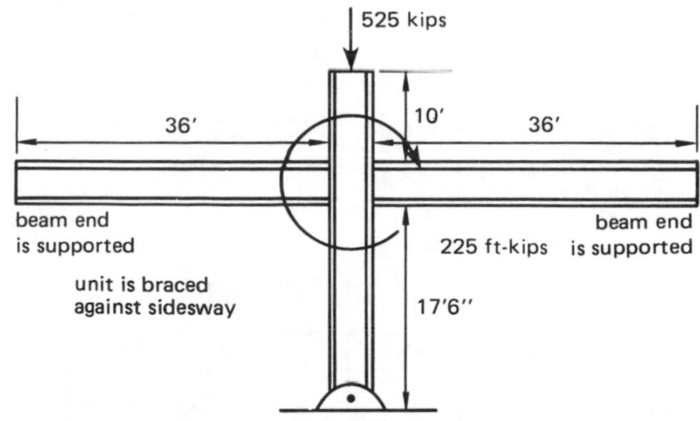


3. Two angles are welded to an intermediate wide-flange section which is bolted to a column. (a) Using the maximum weld size permitted, design the weld between the angles and the intermediate section. Use E70 electrodes. The design should also specify the weld locations. (b) Determine if the bolts are adequate in quantity and location. No threads are in the shear plane.

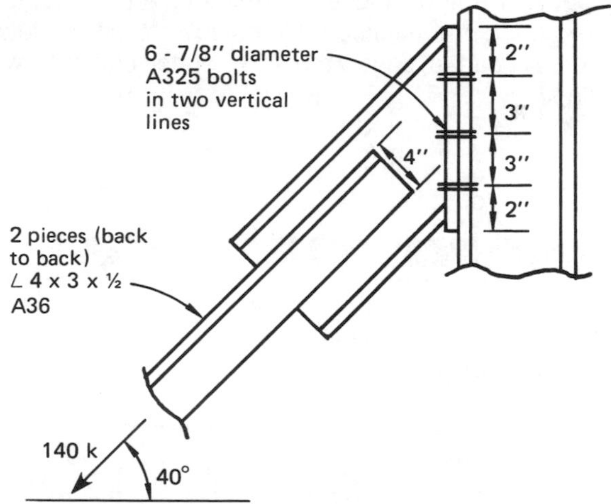


4. A column of 20 feet high is constructed from a channel and a wide flange beam as shown. The column is constrained against bending in the x-plane by bracing at the 10 foot point on the column. Both ends are pinned and free to rotate. Determine the buckling load.

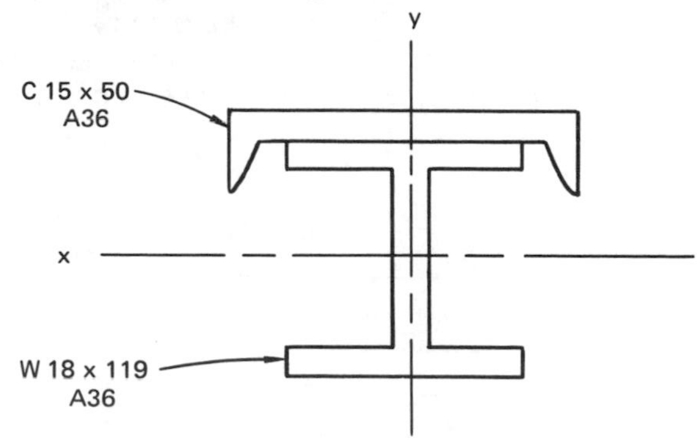


5. The truss shown is constructed of a variety of members, all of which are assumed to be pin-ended. Use

STEEL

the latest AISC code to determine if all members are adequate. If members are not adequate, specify replacements with the same nominal depth.

| | |
|---|---|
| top and bottom chords: | W8 × 28 (A36) |
| diagonal bracing: | L3½ × 3½ × 5/16 (pair) |
| vertical members: | L4 × 4 × 3/8 (pair) |

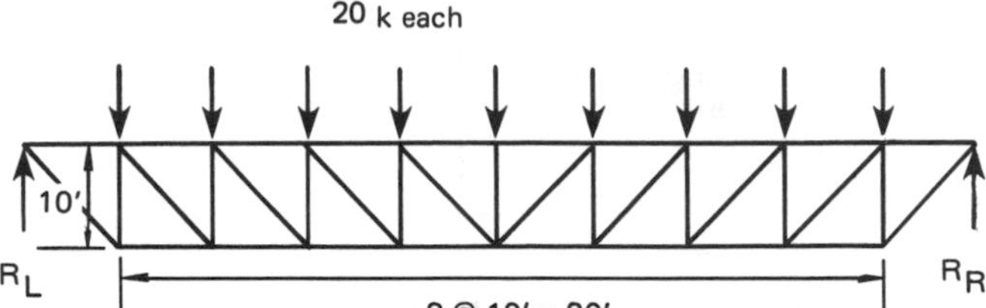


6. Two equal loads remain 3' apart as they move across a 40' span. The beam is constructed from a W shape and a C shape as shown. Sufficient lateral support is provided. (a) Find the maximum allowable load, *P*. Consider only beam bending and shear stresses. Do not consider fatigue or impact. (b) If the weld is intermittent, what spacing is required using the minimum fillet size? (c) Do the loads computed in part (a) satisfy the AISC specification for web crippling?

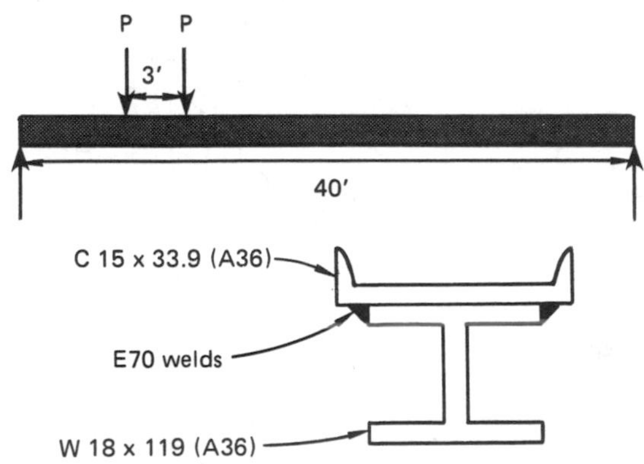


7. A bridge is to be supported by a simply supported plate girder. Loads are resisted by composite action. The bridge is 80 feet long. It will not be braced during construction. Flange support will be provided at 20 foot intervals. Consider a moving 40 kip load and a 1 kip/ft dead load (including the concrete). Determine if the plate girder is adequate. Use n = 10. Disregard impact. Use $f'_c$ = 3000 psi concrete.

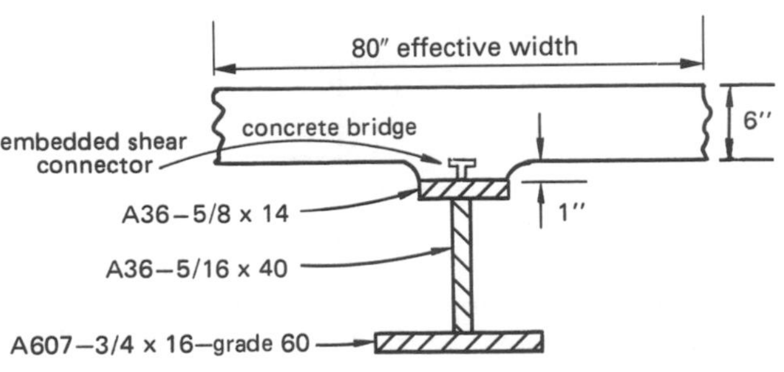


8. Design column CD as a minimum-size wide flange member per Chapter N of the AISC Manual. Assume continuous lateral support of the column. All loads shown are ultimate.

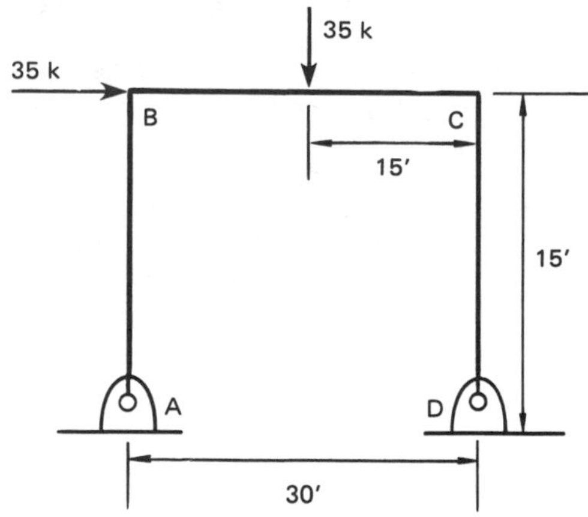


9. A tall column in an industrial storage building carries the loads shown. The lateral wind loads are carried into a braced frame. The column is braced in its weak direction by a strut 24' from the bottom hinged end. (a) Design an economical W24 section based on stress and deflection. (b) Size the base plate based on axial loading only. Assume 3000 psi concrete is used for the resting support.

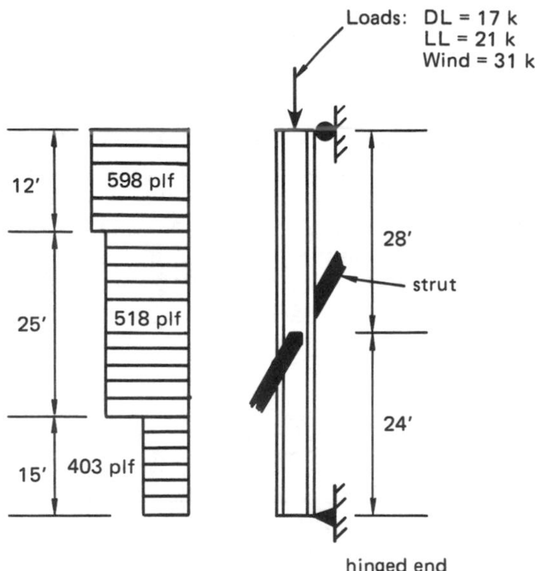


10. A plate girder spans 40 feet and is constructed with 3/8" A36 steel. The flange is 16" wide, and the web is 48" deep. A uniform but unknown load is carried across the girder's entire length, which is simply supported at both ends. (a) Based on shear consideration only, determine the maximum uniform load (in kips/ft) that the beam can carry. (b) Design the web stiffeners located 4 feet from the supports.

other stiffeners
may exist

48″

4′ 4′ 16″

40′

# 16 TRAFFIC ANALYSIS, TRANSPORTATION, AND HIGHWAY DESIGN

*Nomenclature*

| | | |
|---|---|---|
| a | acceleration, or layer coefficient | $ft/sec^2$, – |
| A | area | sq ft |
| ADT | average daily traffic | vpd |
| c | capacity, or distance from neutral axis to extreme fiber | vphpl, in |
| d | separation distance | ft |
| D | directional factor, or density | –, vpmpl |
| DDHV | directional design hourly volume | vph |
| DHV | design hourly volume | vph |
| E | passenger car equivalent | – |
| EAL | equivalent axle load | lbf |
| EWL | equivalent wheel load | lbf |
| f | coefficient of friction, factor, or fraction | –, –, – |
| $f_t$ | allowable working stress | psi |
| F | force | lbf |
| FF | fatigue fraction | – |
| FS | factor of safety | – |
| g | acceleration due to gravity (32.2) | $ft/sec^2$ |
| G | grade | decimal |
| $G_f$ | gravel equivalent factor | – |
| GE | gravel equivalent | ft |
| I | moment of inertia | $in^4$ |
| k | modulus of subgrade reaction, or ratio of DHV to ADT | psi/in, – |
| L | slab length | ft |
| LC | length of curve | ft |
| LSF | load safety factor | – |
| m | mass | slugs |
| M | middle ordinate | ft |
| MR | modulus of rupture | psi |
| MSF | maximum service flow rate | vphpl |
| n | number of cycles | cycles |
| N | number of lanes, normal force, or fatigue life | –, lbf, cycles |
| $p_t$ | terminal serviceability | – |
| r | curve radius | ft |
| R | soil resistance value | – |
| s | distance | ft |
| S | sight distance, or speed | ft, mph |
| $S_c$ | compressive strength | psi |
| SF | service flow rate | vph |
| SN | structural number | – |
| t | time, or thickness | seconds, inches, feet |
| TI | traffic index | – |
| v | velocity, or volume | fps, vphpl |
| w | vehicle mass, or slab width | lbm, ft |

*Symbols*

| | | |
|---|---|---|
| $\theta$ | angle | degrees |
| $\mu$ | coefficient of friction | – |

*Subscripts*

| | |
|---|---|
| b | base |
| B | bus |
| c | centrifugal |
| f | frictional or free-flow |
| HV | heavy vehicle |
| j | jam |
| n | net |
| o | initial |
| p | perception or population |
| R | recreational vehicle |
| s | steel |
| sb | subbase |
| t | tangential |
| T | truck |
| w | at the wheels or width |

## 1 DEFINITIONS

AASHTO: American Association of State and Highway Transportation Officials. Previously known as AASHO.

Abandonment: The reversion of title to the owner of the underlying fee where an easement for highway purposes is no longer needed.

Access control: See 'Control of access.'

Acquisition: The process of obtaining right of way.

Arterial highway: A general term denoting a highway primarily for through traffic usually on a continuous route.

Auxiliary lane: The portion of a roadway adjoining the traveled way for truck climbing, speed change, or for other purposes supplementary to through traffic movement.

Base course: The bottom portion of a pavement where the top and bottom portions are not of the same composition.

Base: A layer of selected, processed, or treated aggregate material of planned thickness and quality placed immediately below the pavement and above the subbase or basement soil.

Belt highway: An arterial highway carrying traffic partially or entirely around an urban area.

CBD: Abbreviation for *central business district.*

Cement treated base: A base layer constructed with good quality, well-graded aggregate mixed with up to 6% cement.

Channelization: The separation or regulation of conflicting traffic movements into definite paths of travel by use of pavement markings, raised islands, or other means.

Condemnation: The process by which property is acquired for public purposes through legal proceedings under power of eminent domain.

Control of access: The condition where the right of owners or occupants of abutting land or other persons to access in connection with a highway is fully or partially controlled by public authority.

Divided highway: A highway with separated roadbeds for traffic in opposing directions.

Easement: A right to use or control the property of another for designated purposes.

Embankment: A raised structure constructed of natural soil from excavation or borrow sources.

Eminent domain: The power to take private property for public use without the owner's consent upon payment of just compensation.

Emulsion: A mixture with water. Asphalt emulsions are produced by adding a small amount of emulsifying soap to asphalt and water. When the water evaporates, the asphalt sets.

Encroachment: Use of the highway right-of-way for non-highway structures or other purposes.

Flexible pavement: A pavement having sufficiently low bending resistance to maintain intimate contact with the underlying structure, yet having the required stability furnished by aggregate interlock, internal friction, and cohesion to support traffic.

Freeway: A divided arterial highway with full control of access.

Frontage road: A local street or road auxiliary to, and located on, the side of an arterial highway for service to abutting property and adjacent areas, and for control of access.

Gore: The area immediately beyond the divergence of two roadways bounded by the edges of those roadways.

Inverse condemnation: The legal process that may be initiated by a property owner to compel the payment of just compensation when his property has been taken or damaged for a public purpose.

Median: The portion of a divided highway separating the traveled ways for traffic in opposite directions.

Median lane: A lane within the median to accommodate left-turning vehicles.

Parkway: An arterial highway for non-commercial traffic, with full or partial control of access, usually located within a park or a ribbon of parklike development.

Penetration treatment: Application of light liquid asphalt to the road-bed material. It is used primarily as a dust reducer on detours, medians, and parking areas.

Plant mix: An asphalt concrete mixture that is not prepared at the paving site.

Prime coat: The initial application of a low viscosity liquid asphalt to an absorbent surface, preparatory to any subsequent treatment, for the purpose of hardening or toughening the surface and promoting adhesion between it and the superimposed constructed layer.

Resurfacing: A supplemental surface or replacement placed on an existing pavement to restore its riding qualities or increase its strength.

Right of access: The right of an abutting land owner for entrance to or exit from a public road.

Rigid pavement: A pavement having sufficiently high bending resistance to distribute loads over a comparatively large area.

Road-mixed asphalt surfacing: A lower-quality surfacing used when plant mixes are not available or not economically feasible. Liquid asphalts are normally used. Road-mixed asphalt surfacing is used on low-traffic volume roads where higher quality surfacing is not required for traffic volume.

Roadbed: That portion of the roadway extending from curb line to curb line or shoulder line to shoulder line. Divided highways are considered to have two roadbeds.

Seal coat: A bituminous coating, with or without aggregate, applied to the surface of a pavement for the purpose of water-proofing and preserving the surface, rejuvenating a previous bituminous surface, altering the surface texture of the pavement, providing delineation, or providing resistance to traffic abrasion.

Structural section: The planned layers of specific materials, normally consisting of subbase, base, and pavement, placed over the basement soil.

Subbase: A layer of aggregate of planned thickness and quality placed on the basement soil as a foundation for the base.

Subgrade: The portion of a roadbed surface, which has been prepared as specified, upon which a subbase, base, base course, or pavement is to be placed.

Tack coat: The initial application of bituminous material to an existing surface to provide bond between the existing surface and the new material.

Traveled way: The portion of the roadway for the movement of vehicles exclusive of shoulders and auxiliary lanes.

## 2 TRANSLATIONAL DYNAMICS

Newton's second law can be used to relate the net tractive force on a vehicle to its acceleration.

$$F_n = ma \tag{16.1}$$

$$m = \frac{w}{g} \tag{16.2}$$

The *net tractive force* is the difference between the applied force at the wheels and the frictional force.

$$F_n = F_w - F_f \tag{16.3}$$

The *net force* can be found directly from the velocity of the vehicle and the horsepower expenditure at that velocity.

$$F_n = \frac{(550)(\text{horsepower})}{v_{\text{ft/sec}}} \tag{16.4}$$

The frictional force is a combination of dynamic, rolling, turning, and aerodynamic forces which act to oppose motion.

The relationships between position, velocity, and acceleration as functions of time for linear motion are given here.

$$a = \frac{dv}{dt} = \frac{d^2s}{dt^2} \tag{16.5}$$

$$v = \frac{ds}{dt} = \int a\,dt \tag{16.6}$$

$$s = \int v\,dt = \iint a\,dt^2 \tag{16.7}$$

If the acceleration is uniform, table 16.1 can be used to determine values of unknown variables. Acceleration is negative for vehicles with decreasing velocities.

*Example 16.1*

A 4000 pound car traveling at 80 mph locks up its wheels and slides 580 feet before stopping. (a) How much time does it take to stop? (b) What is the deceleration? (c) What is the retarding force? (d) What is the coefficient of friction between the tires and the road?

(a) The initial velocity is

$$v_o = (80)\left(\frac{5280}{3600}\right) = 117.3\text{ ft/sec}$$

The stopping distance is $s = 580$ feet. The final velocity is $v = 0$. From table 16.1,

$$t = \frac{2s}{v_o + v} = \frac{(2)(580)}{117.3 + 0} = 9.89\text{ sec}$$

(b) From table 16.1,

$$a = \frac{v - v_o}{t} = \frac{0 - 117.3}{9.89} = -11.9\text{ ft/sec}^2$$

(c) $F_f = ma = \left(\frac{4000}{32.2}\right)(11.9) = 1480\,\text{lbf}$

(d) $\mu = \frac{F_f}{N} = \frac{1480}{4000} = 0.37$

**Table 16.1**
Uniform Acceleration Formulas

| to find | given these | use this equation |
|---|---|---|
| $t$ | $a\,v_o\,v$ | $t = \frac{v - v_o}{a}$ |
| $t$ | $a\,v_o\,s$ | $t = \frac{\sqrt{2as + v_o^2} - v_o}{a}$ |
| $t$ | $v_o\,v\,s$ | $t = \frac{2s}{v_o + v}$ |
| $a$ | $t\,v_o\,v$ | $a = \frac{v - v_o}{t}$ |
| $a$ | $t\,v_o\,s$ | $a = \frac{2s - 2v_o t}{t^2}$ |
| $a$ | $v_o\,v\,s$ | $a = \frac{v^2 - v_o^2}{2s}$ |
| $v_o$ | $t\,a\,v$ | $v_o = v - at$ |
| $v_o$ | $t\,a\,s$ | $v_o = \frac{s}{t} - \frac{1}{2}at$ |
| $v_o$ | $a\,v\,s$ | $v_o = \sqrt{v^2 - 2as}$ |
| $v$ | $t\,a\,v_o$ | $v = v_o + at$ |
| $v$ | $a\,v_o\,s$ | $v = \sqrt{v_o^2 + 2as}$ |
| $s$ | $t\,a\,v_o$ | $s = v_o t + \frac{1}{2}at^2$ |
| $s$ | $a\,v_o\,v$ | $s = \frac{v^2 - v_o^2}{2a}$ |
| $s$ | $t\,v_o\,v$ | $s = \frac{1}{2}t(v_o + v)$ |

## 3 SIMPLE ROADWAY BANKING

If a vehicle travels in a circular path with instantaneous radius $r$ and tangential velocity $v_t$, it will experience an apparent centrifugal force given by equation 16.8.

$$F_c = \frac{mv_t^2}{r} \qquad 16.8$$

This centrifugal force must be resisted by a combination of roadway banking (superelevation) and sideways friction. If it is desirable to bank the roadway so that little or no friction is required to resist the centrifugal force, the angle of superelevation is given by equation 16.9. (Generally, it is not desirable to rely on roadway banking alone, since such banking will be applicable at only one speed.)

$$\tan\theta = \frac{v^2}{g\,r} \qquad 16.9$$

In general, a lower banking angle is used in urban areas than in rural areas. For arterial streets in downtown areas, the maximum superelevation should be 0.04 to 0.06. For arterial streets in suburban areas and freeways where there is no snow or ice, the maximum is 0.10 to 0.12. For arterial streets and freeways that experience snow and ice, the maximum is 0.06 to 0.08.

Equation 16.10 is the basic formula used for determining superelevation when friction is relied upon to counteract some of the centrifugal force.

$$\tan\theta = \frac{v^2}{g\,r} - f \qquad 16.10$$

If $v$ is expressed in mph, equation 16.10 becomes

$$\tan\theta = \frac{(\text{MPH})^2}{15r} - f \qquad 16.11$$

$f$ is the *side friction factor.* It is usually assumed to be 0.16 for speeds of 30 mph and under. For higher speeds,

$$f = 0.16 - 0.01\left(\frac{\text{MPH} - 30}{10}\right) \quad \text{(up to 50 mph)} \qquad 16.12$$

$$f = 0.14 - 0.02\left(\frac{\text{MPH} - 50}{10}\right) \quad \text{(50 to 70 mph)}$$

Since the maximum $\tan\theta$ is usually 0.08 or 0.10, equation 16.10 can be used to calculate the minimum curve radius if the speed is known.

Transitions from flat to superelevated sections should be gradual.

*Example 16.2*

A 4000 pound car travels at 40 mph around a banked curve with a radius of 500 feet. What should be the banking angle such that tire friction is not needed to prevent the car from sliding?

$$v = (40)(1.467)\frac{\text{fps}}{\text{mph}} = 58.68\,\text{fps}$$

From equation 16.9,

$$\theta = \arctan\left[\frac{(58.68)^2}{(32.2)(500)}\right] = 12.07°$$

TRAFFIC

## 4 SIGHT AND STOPPING DISTANCES

*Sight distance* is the length of roadway that the driver can see. It is assumed that the driver's eyes are 3.50 feet above the surface of the roadway. The sight distance should be long enough to allow a driver traveling at the maximum speed to stop before coming upon an observed object. This required distance is known as the *stopping sight distance.* For stopping sight distances, it is assumed that the object being observed has a height of 0.5 feet.

Since distance is covered during the driver's reaction period as well as during the deceleration period, the stopping sight distance includes both of these distances. The coefficient of friction is usually evaluated for wet pavement. For straight-line travel on a constant grade, $G$, equation 16.13 can be used. $G$ is a decimal, and it is negative if the roadway is downhill.

$$S = (1.47)(t_p)(\text{MPH}) + \frac{(\text{MPH})^2}{(30)(f+G)} \qquad 16.13$$

**Table 16.2**

Typical Coefficients of Skidding Friction[(a)]

*BC*: bituminous concrete, dry
*SA*: sand asphalt, dry
*RA*: rock asphalt, dry
*CC*: portland cement concrete, dry
wet: all wet pavements

| condition | *BC* | *SA* | *RA* | *CC* | wet |
|---|---|---|---|---|---|
| new tires | | | | | |
| 11 mph | 0.74 | 0.75 | 0.78 | 0.76 | |
| 20 | 0.76 | 0.75 | 0.76 | 0.73 | 0.40 |
| 30 | 0.79 | 0.79 | 0.74 | 0.78 | 0.36 |
| 40 | 0.75 | 0.75 | 0.74 | 0.76 | 0.33 |
| 50 | | | | | 0.31 |
| 60 | | | | | 0.30 |
| 70 | | | | | 0.29 |
| badly worn tires | | | | | |
| 11 mph | 0.61 | 0.66 | 0.73 | 0.68 | |
| 20 | 0.60 | 0.57 | 0.65 | 0.50 | 0.40 |
| 30 | 0.57 | 0.48 | 0.59 | 0.47 | 0.36 |
| 40 | 0.48 | 0.39 | 0.50 | 0.33 | 0.33 |
| 50 | | | | | 0.31 |
| 60 | | | | | 0.30 |
| 70 | | | | | 0.29 |

(a) Values vary widely.

If the design speed is used in equation 16.13, $S$ is known as a *desirable value.* If the speed is less than the design value, $S$ is known as a minimum value. The minimum speed to be used is

$$\text{MPH}_{\text{min}} = \text{MPH}_{\text{design}} - 0.2(\text{MPH}_{\text{design}} - 20) \qquad 16.14$$

The desirable value should be used in most cases. These are listed in table 16.3 for various design speeds.

The *braking reaction-perception time,* $t_p$ in equation 16.13, has a median value of approximately 0.90 seconds for unexpected (not anticipated) events.[1] However, this time varies widely from subject to subject. Individuals with slow reactions may require up to 2.5 seconds.[2]

The *passing sight distance* is applicable only to 2-lane, 2-way highways. It is the length of roadway ahead necessary to pass without meeting an oncoming vehicle.[3] Minimum passing sight distances are given in table 16.3. The values should be increased 18% for downgrades steeper than 3% and longer than one mile.

If a vehicle locks its brakes and skids to a stop, the deceleration will be $(f)(g) = (f)(32.2)\ \text{ft/sec}^2$. The skidding distance is given by equation 16.15. Note that this equation does not apply to collisions.

$$\text{skidding distance} = \frac{(\text{MPH})^2}{(30)(f+G)} \qquad 16.15$$

If a vehicle does not lock its brakes, its deceleration will be dependent on its brakes. The distance traveled during deceleration to a standstill is given by equation 16.16. Note that this equation does not apply to collisions.

$$\text{stopping distance} = \frac{v^2}{2a} = \frac{(1.08)(\text{MPH})^2}{a} \qquad 16.16$$

**Table 16.3**

AASHTO Minimum Sight Distances[(a)]

| | stopping sight distance wet pavements | | passing sight distance | |
|---|---|---|---|---|
| design speed, mph | initial speed, mph | minimum desirable distance, ft | assumed passing speed, mph | distance 2-lane highway, ft |
| 20 | 20 | 125 | 30 | 800 |
| 30 | 28–30 | 200 | 36 | 1100 |
| 40 | 36–40 | 275–325 | 44 | 1500 |
| 50 | 44–50 | 400–475 | 51 | 1800 |
| 60 | 52–60 | 525–650 | 57 | 2100 |
| 70 | 58–70 | 625–850 | 64 | 2500 |

(a) AASHTO PGDHS–1990, Tables III–1 and III–5

[1] $t_p$ is also known as the *PIEV time.* This name is an acronym for the various elements of reaction time, including perception, identification, emotion, and volition.

[2] For the purpose of determining AASHTO minimum stopping sight distances, $t_p$ is taken as 2.5 seconds. For determining passing sight distances, $t_p$ is taken as 3.5 to 4.5 seconds.

[3] It is assumed that the driver's eyes are 3.50 feet above the surface of the roadway. For passing sight distances, the object being viewed (e.g., an oncoming car) is assumed to be at a height of 4.25 feet.

## 5 LENGTH OF CIRCULAR HORIZONTAL CURVE FOR STOPPING DISTANCE

A horizontal curve on level ground is shown in figure 16.1. A typical design problem is to design a curve (i.e., specify a radius) that will simultaneously provide the required sight stopping distance while maintaining a clearance from a roadside obstruction.

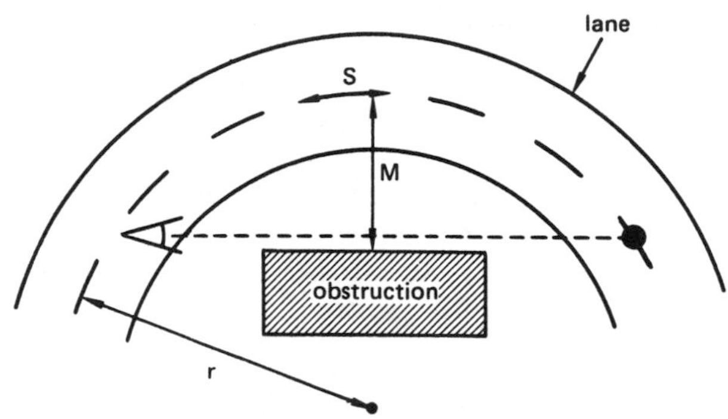


**Figure 16.1** Stopping Distance

The governing equations are given assuming $S \leq LC$. In this analysis, the stopping sight distance and length along the curve are the same. The angles are in degrees.

$$S = \frac{r}{28.65}\left[\arccos\left(\frac{r-M}{r}\right)\right] \qquad 16.17$$

$$M = r\left[1-\cos\left(\frac{28.65S}{r}\right)\right] \qquad 16.18$$

## 6 LENGTH OF VERTICAL CURVES FOR SIGHT DISTANCES

The curve length may be shorter or longer than the safe passing or stopping sight distances. (Passing sight distance is not relevant on multi-lane highways.) Table 16.4 can be used to calculate lengths of curves. Table 16.4 is used by calculating the curve length for both assumptions that $S < LC$ and $S > LC$.

*Example 16.3*

A car is traveling at 40 mph up a hill with a +1.25% grade. The descending grade is −2.75%. What is the required length of curve for proper stopping sight distance?

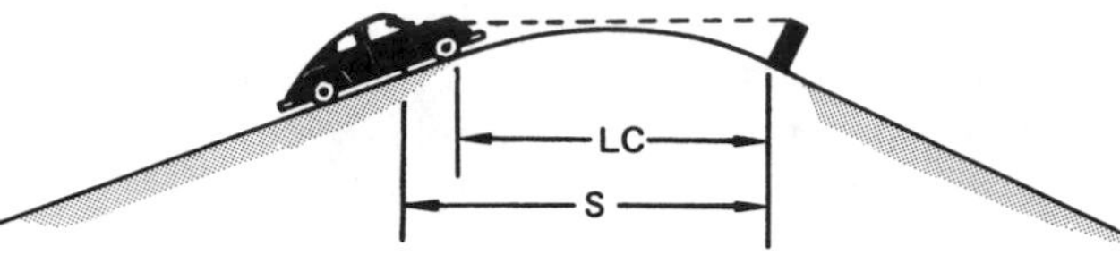


From table 16.3, the minimum stopping sight distance is 275 feet at 40 mph.

Using table 16.4 and assuming that $275 > LC$:

$$LC = 2(275) - \frac{1329}{1.25-(-2.75)} = 217.8 \text{ ft}$$

Using table 16.4 and assuming $275 < LC$:

$$LC = \frac{[1.25-(-2.75)](275)^2}{1329} = 227.6 \text{ ft}$$

Since 227.6 is less than 275, the second assumption is not valid. The required curve length is 217.8 feet.

## 7 SPEED PARAMETERS

Many different measures of vehicle speed are used.

- Running Speed: The distance traveled divided by the running time without delays. This parameter can be averaged over all traffic.
- Average Spot Speed: This is the average instantaneous speed of all vehicles at a particular location.
- Overall Travel Speed: If delays and stops at intersections are included in running time, the overall travel speed can be calculated from the distance traveled.

**Table 16.4**
Required Lengths of Curves on Grades[(a)]

| assuming | stopping sight distance[(b)] (crest curves) | passing sight distance (crest curves) | stopping sight distance (sag curves) |
|---|---|---|---|
| $S < LC$ | $\frac{(G_1-G_2)S^2}{1329}$ | $\frac{(G_1-G_2)S^2}{3093}$ | $\frac{(G_2-G_1)S^2}{400+3.5S}$ |
| $S > LC$ | $2S - \frac{1329}{G_1-G_2}$ | $2S - \frac{3093}{G_1-G_2}$ | $2S - \frac{400+3.5S}{G_2-G_1}$ |

(a) AASHTO PGDHS–1990, Chapter III

(b) Based on driver's eye being 3.50 ft above road surface and a 0.5 ft high object.

- Operating Speed: The highest overall speed at which a driver can travel under favorable weather conditions while driving in a safe manner.
- Design Speed: The maximum safe speed when conditions are so favorable that the design features of the highway govern.
- Average Highway Speed: The weighted average of the design speeds over a section of highway.

## 8 DESIGN SPEEDS

Most elements of roadway design depend on the design speed. The design speed is the maximum safe maintainable speed on a roadway under the design conditions. Typical design speeds are given in table 16.5.

**Table 16.5**
Recommended Design Speeds (mph)[(a)]

| type of facility | level | rolling | mountainous |
|---|---|---|---|
| freeways | | | |
| rural | 70 | 60 | 50 |
| urban | 50 | 50 | 50 |
| rural arterial highways | | | |
| 50 <ADT≤750, | | | |
| DHV< 200 | 50 | 40 | 30 |
| DHV> 200 | 70 | 60 | 40 |
| urban arterial | | | |
| highways | 30–40 | 30–40 | 30–40 |
| suburban arterial highways | 40–50 | 40–50 | 40–50 |
| rural roads and streets | | | |
| ADT< 250 | 40 | 30 | 20 |
| 250 <ADT< 400 | 50 | 40 | 30 |
| ADT> 400, | | | |
| DHV> 100 | 50 | 40 | 30 |
| urban roads and streets | | | |
| collectors | 30–40 | 30–40 | 30–40 |
| local | 20–30 | 20–30 | 20–30 |

(a) State and federal laws may limit higher speeds.

## 9 VOLUME PARAMETERS

There are many volume parameters in use. Not all parameters will be needed in every capacity or strength calculation. It is particularly important to note if volumes are for both directions combined, or for one or more lanes combined.

ADT: The current *average daily traffic.* ADT may be one- or two-way traffic.

DHV: The peak *design hourly volume* in the design year. DHV is usually the 30th highest hourly expected volume in the design year. It is not an average or a maximum. DHV is two-way unless noted otherwise.

k: Ratio of DHV to ADT.

D: A *directional factor.* D is the percentage in the dominant flow direction. D may range up to 80% for rural roadways at peak hours down to 50% for central business district traffic.

DDHV: The *directional design hourly volume.* It is calculated as the product of the directional factor, D, and DHV.

$$\text{DDHV} = (\text{D})(\text{DHV}) \qquad 16.19$$

Design Capacity: The maximum volume of traffic that the roadway can handle.

MSF: The *maximum service flow rate* per lane under ideal conditions.

Logical methods, including straight-line extrapolation, should be used in the estimation of future traffic counts. Expansion factors should be determined for each of the axle classifications. Considerable judgment is needed to develop realistic expansion factors.

It also is necessary to estimate the distribution of truck traffic on the various lanes of a multi-lane facility. Traffic is usually lightest in the *inside lane* (*lane 1*, or the *fast lane*). The following lane distribution factors can be used for initial estimates.

**Table 16.6**
Typical Lane Distribution Factors
for Multi-Lane Freeways

| number of lanes in one direction | lane 1 | lane 2 | lane 3 | lane 4 |
|---|---|---|---|---|
| 1 | 1.0 | | | |
| 2 | 1.0 | 1.0 | | |
| 3 | 0.2 | 0.8 | 0.8 | |
| 4 | 0.2 | 0.2 | 0.8 | 0.8 |

## 10 TRUCK, BUS, AND RV EQUIVALENTS

Since buses and trucks take up more space on a road than cars do, and since buses and trucks tend to travel more slowly up grades, bus and truck volumes are converted to equivalent passenger car volumes. Table 16.7 lists *passenger car equivalents* for various conditions.

**Table 16.7**
Passenger Car Equivalents
for Freeways and Multi-Lane Highways[(a)]

| terrain | $E_B$ (buses) | $E_T$ (trucks) | $E_R$ (RV's) |
|---|---|---|---|
| level | 1.5 | 1.7 | 1.6 |
| rolling | 3 | 4 | 3 |
| mountainous | 5 | 8 | 4 |

(a) AASHTO PGDHS–1990, Table II-4

Passenger car equivalents for trucks and RV's on grades depend on the grade, the length of the grade, the number of lanes, and the percentage of trucks and buses.[4]

## 11 LEVEL OF SERVICE

Conditions on a highway are classified into levels $A$ through $F$. Level $A$ represents a condition where there are no physical restrictions on operating speed. Since there are few vehicles on the freeway, operation at highest speeds is possible. However, the traffic volume is small. Level $F$ represents a stop-and-go, low-speed condition with poor safety and maneuverability.

**Table 16.8**
Levels of Service for Freeways
and Multi-Lane Highways

| level | density (pc/mi-ln) | description |
|---|---|---|
| $A$ | ≤12 | free flow, with low volumes and high speeds |
| $B$ | 13–20 | stable flow, but speeds are beginning to be restricted by traffic conditions |
| $C$ | 21–30 | stable flow, but most drivers cannot select their own speed |
| $D$ | 31–42 | approaching unstable flow, and drivers have little room in which to maneuver |
| $E$ | 43–67[5] | unstable flow with short stoppages |
| $F$ | ≥68 | forced flow at slow speeds; lines of vehicles at certain locations |

The desired design condition is between levels $A$ and $F$. Economic considerations favor lower levels and their higher traffic volume per lane. However, political considerations favor higher levels. Typically, levels $B$ and $C$ are chosen for initial designs.

The actual level of service experienced on a freeway is determined by comparing the actual density with the density limits in table 16.8.

## 12 CALCULATION OF FREEWAY CAPACITY

The maximum capacity under ideal conditions, $c$, is taken as the maximum service flow rate per lane for level of service. Table 16.9 contains MSF values.

$$c_{\max} = \text{MSF}_E \qquad 16.20$$

**Table 16.9**
Maximum Service Flow Rates
for Freeways[6]

(passenger cars per hour per lane)

| service level | density (cars/lane-mi) | design speed 70 mph | 60 mph | 50 mph |
|---|---|---|---|---|
| $A$ | ≤ 12 | 700 | — | — |
| $B$ | ≤ 20 | 1100 | 1000 | — |
| $C$ | ≤ 30 | 1550 | 1400 | 1300 |
| $D$ | ≤ 42 | 1850 | 1700 | 1600 |
| $E$ | ≤ 67 | 2000 | 2000 | 1900 |
| $F$ | > 67 | (highly variable; unstable) | | |

(*Highway Capacity Manual*, Table 3-1)

For any other level of service, the maximum *volume-to-capacity ratio* for level of service $i$ is

$$(v/c)_i = \frac{\text{MSF}_i}{c_{\max}} \qquad 16.21$$

The *service flow rate* can be calculated from the number of lanes ($N$), the lane width adjustment factor ($f_w$), the factor to adjust for the presence of heavy vehicles such as buses, trucks, and recreational vehicles ($f_{HV}$), and a

[4] Tables are provided in the *Highway Capacity Manual.*

[5] 67 passenger cars per mile per lane is generally considered to be the *critical density*. Maximum flow (i.e., *capacity flow*) occurs at the *critical density* within level of service $E$.

[6] This table is not valid for rural, multi-lane highways. Refer to the *Highway Capacity Manual.*

TRAFFIC

**Table 16.10**
Width Adjustment Factor, $f_w$, for Restricted Lane Width and Lateral Clearance for Freeways[7,(a)]

| distance of obstruction from traveled pavement (ft) | obstructions on one side of the roadway | | | | obstructions on both sides of the roadway | | | |
|---|---|---|---|---|---|---|---|---|
| | lane width (ft) | | | | | | | |
| | 12 | 11 | 10 | 9 | 12 | 11 | 10 | 9 |
| | 4-lane freeway (2 lanes each direction) | | | | | | | |
| ≥6 | 1.00 | 0.97 | 0.91 | 0.81 | 1.00 | 0.97 | 0.91 | 0.81 |
| 5 | 0.99 | 0.96 | 0.90 | 0.80 | 0.99 | 0.96 | 0.90 | 0.80 |
| 4 | 0.99 | 0.96 | 0.90 | 0.80 | 0.98 | 0.95 | 0.89 | 0.79 |
| 3 | 0.98 | 0.95 | 0.89 | 0.79 | 0.96 | 0.93 | 0.87 | 0.77 |
| 2 | 0.97 | 0.94 | 0.88 | 0.79 | 0.94 | 0.91 | 0.86 | 0.76 |
| 1 | 0.93 | 0.90 | 0.85 | 0.76 | 0.87 | 0.85 | 0.80 | 0.71 |
| 0 | 0.90 | 0.87 | 0.82 | 0.73 | 0.81 | 0.79 | 0.74 | 0.66 |
| | 6- or 8-lane freeway (3 or 4 lanes each direction) | | | | | | | |
| ≥6 | 1.00 | 0.96 | 0.89 | 0.78 | 1.00 | 0.96 | 0.89 | 0.78 |
| 5 | 0.99 | 0.95 | 0.88 | 0.77 | 0.99 | 0.95 | 0.88 | 0.77 |
| 4 | 0.99 | 0.95 | 0.88 | 0.77 | 0.98 | 0.94 | 0.87 | 0.77 |
| 3 | 0.98 | 0.94 | 0.87 | 0.76 | 0.97 | 0.93 | 0.86 | 0.76 |
| 2 | 0.97 | 0.93 | 0.87 | 0.76 | 0.96 | 0.92 | 0.85 | 0.75 |
| 1 | 0.95 | 0.92 | 0.86 | 0.75 | 0.93 | 0.89 | 0.83 | 0.72 |
| 0 | 0.94 | 0.91 | 0.85 | 0.74 | 0.91 | 0.87 | 0.81 | 0.70 |

[(a)] AASHTO PGDHS–1990, Table III-35

factor to adjust for the effect of the driver population ($f_p$).[8]

$$SF_i = \text{MSF}_i \times N \times f_w \times f_{HV} \times f_p = c_{\text{max}} \times \left(\frac{v}{c}\right)_i \times N \times f_w \times f_{HV} \times f_p \quad 16.22$$

The width adjustment factor, $f_w$, is taken from table 16.10.

The *heavy vehicle factor* is a function of the truck, bus, and recreational vehicle fractions.

$$f_{HV} = \frac{1}{1+\begin{pmatrix}\text{fraction}\\ \text{trucks}\end{pmatrix}(E_T-1)+\begin{pmatrix}\text{fraction}\\ \text{buses}\end{pmatrix}(E_B-1)+\begin{pmatrix}\text{fraction}\\ \text{RV's}\end{pmatrix}(E_R-1)} \quad 16.23$$

[7] Certain types of obstructions, such as high-type median barriers in particular, do not decrease the flow. Exercise judgment in applying these factors.

[8] From a practical standpoint, the *service flow rate*, $SF$, is set equal to the actual peak flow rate. The peak flow rate can be calculated from the actual flow rate and the *peak hour factor*: $SF = v/PHF$.

$E_B, E_T$, and $E_R$ are the *passenger car equivalents* of a bus, truck, or recreational vehicle, respectively. Table 16.7 provides these values.

The *population adjustment factor*, $f_p$, is 1.0 for weekday or commuter traffic. It is 0.75 to 0.90 for weekend, recreational, and other types of traffic that do not use the available space as efficiently. Engineering judgment is required in selecting $f_p$ in those instances.

## 13 SPEED, FLOW, AND DENSITY RELATIONSHIPS FOR UNINTERRUPTED FLOW

The speed of travel will decrease as the number of cars occupying the freeway increases. The *density*, $D$, is defined as the number of vehicles per mile per lane (vpmpl). The *jam density* is the density when the vehicles are all at a standstill. Then, the speed for any given density can be related to the *free-flow speed* by equation 16.24.

$$S = S_f\left(1 - \frac{D}{D_j}\right) \quad 16.24$$

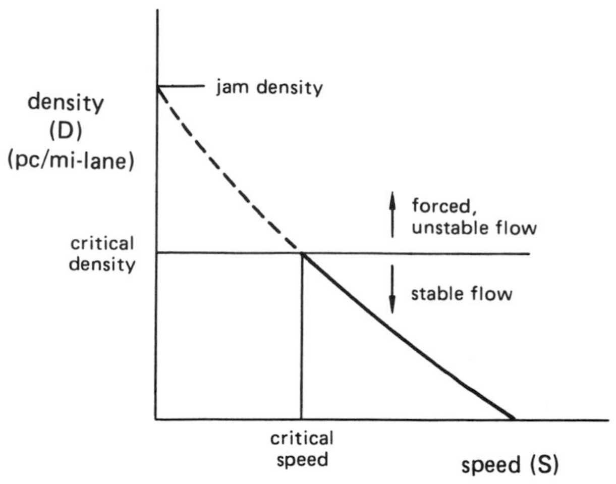


**Figure 16.2** Speed Versus Density

The number of vehicles (volume) crossing a point per hour per lane (vphpl) is

$$v = SD \tag{16.25}$$

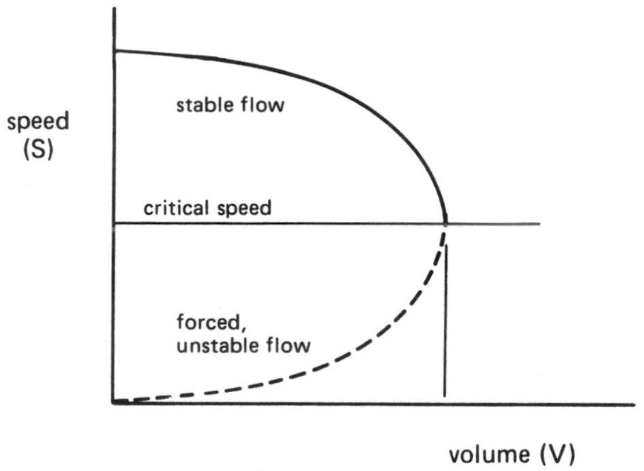


**Figure 16.3** Speed Versus Volume

The *headway* is the time between successive vehicles. *Spacing* is the distance between common points (e.g., the front bumper) on successive vehicles.

$$\text{spacing (ft/veh)} = \frac{5280 \text{ (ft/mi)}}{D \text{ (vpmpl)}} \tag{16.26}$$

$$\text{headway (sec/veh)} = \frac{\text{spacing (ft/veh)}}{\text{speed (ft/sec)}} \tag{16.27}$$

$$\text{volume or flow rate (vph)} = \frac{3600 \text{ sec/hr}}{\text{headway (sec/veh)}} \tag{16.28}$$

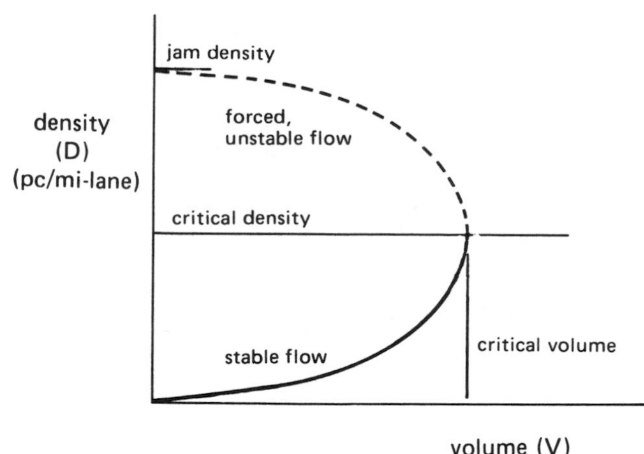


**Figure 16.4** Density Versus Volume

## 14 DETERMINING THE LEVEL OF SERVICE

The level of service is determined primarily from the density criteria in table 16.8. However, it is also possible to correlate the level of service with the v/c ratio. In this case, it is necessary to know the design speed, $v_{\text{design}}$, for which the freeway was constructed. Table 16.11 correlates v/c ratios with different levels of service.

**Table 16.11**
Level of Service Versus Maximum
v/c Ratio for Freeways[a]

| level of | design speed (mph) | | |
|---|---|---|---|
| service | 70 | 60 | 50 |
| *A* | 0.35 | — | — |
| *B* | 0.54 | 0.49 | — |
| *C* | 0.77 | 0.69 | 0.67 |
| *D* | 0.93 | 0.84 | 0.83 |
| *E* | 1.00 | 1.00 | 1.00 |

[a] AASHTO PGDHS–1990, Table III-34

*Example 16.4*

A four-lane freeway is constructed with 11-foot lanes, no shoulders, and retaining walls at the pavement edge. The average vehicle speed is 60 mph. The terrain is rolling. The actual service volume is 1500 vehicles per hour, with 3% buses and 5% trucks. What is the level of service? What is the capacity volume?

TRAFFIC

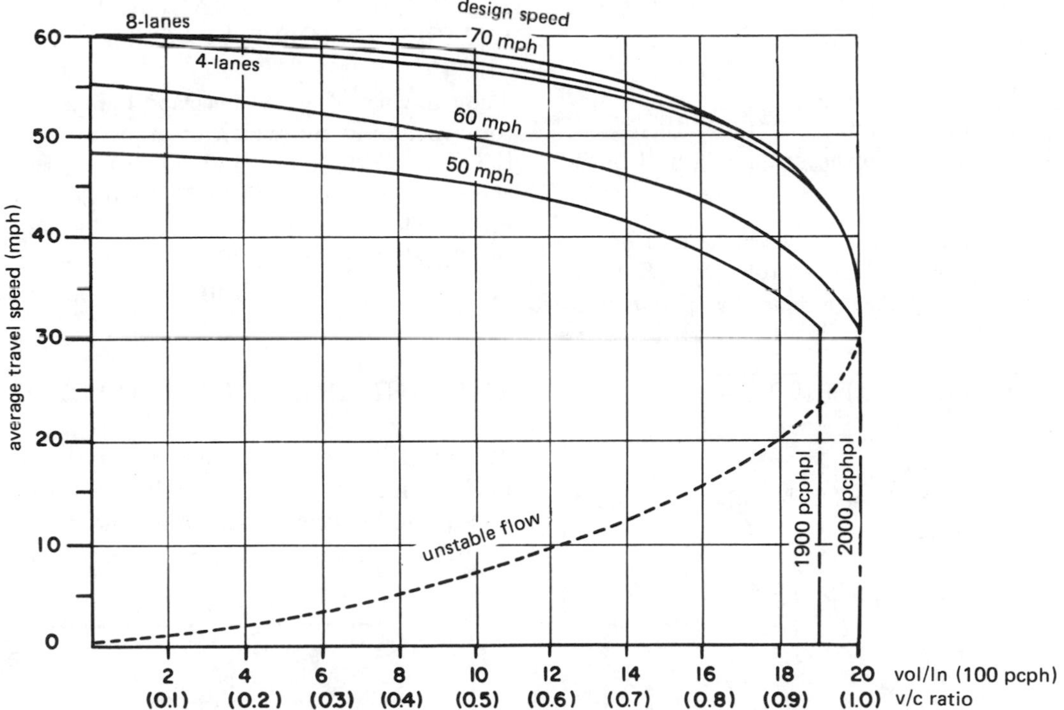


**Figure 16.5** Freeway Speed-Flow Relationships Under Ideal Conditions

(v/c ratio is based on 2000 pcphpl and is valid only for 60- and 70-mph speeds.)

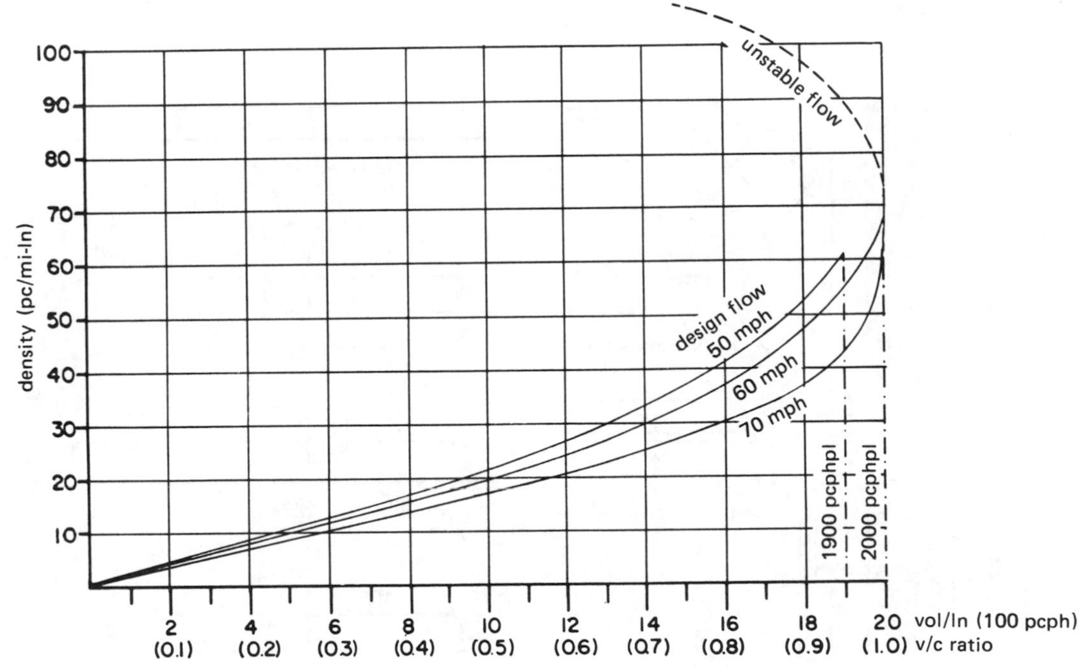


**Figure 16.6** Freeway Density-Flow Relationships Under Ideal Conditions

(v/c ratio is based on 2000 pcphpl and is valid only for 60- and 70-mph speeds.)

TRAFFIC

The base volume is 2000 vphpl (table 16.9, level of service $E$).

The number of lanes is 2.

The width adjustment factor is 0.87 (table 16.10, assuming retaining walls are at the outer edges of both slow lanes).

The passenger car equivalents of the trucks and buses in rolling terrain are obtained from table 16.7. $E_t = 4$. $E_B = 3$. From equation 16.23, the heavy vehicle factor is

$$f_{HV} = \frac{1}{1 + (0.05)(4-1) + (0.03)(3-1)} = 0.826$$

Assuming weekday traffic ($f_p = 1$), the capacity volume is given by equation 16.22.

$$\frac{SF}{\frac{v}{c}} = 2000 \times 2 \times 0.87 \times 0.826 \times 1 = 2874 \text{ vph}$$

The v/c ratio is

$$\frac{v}{c} = \frac{SF}{\left(\frac{SF}{\frac{v}{c}}\right)} = \frac{1500}{2874} = 0.52$$

From table 16.11, this v/c ratio corresponds to level of service $C$ for a design speed of 60 mph.

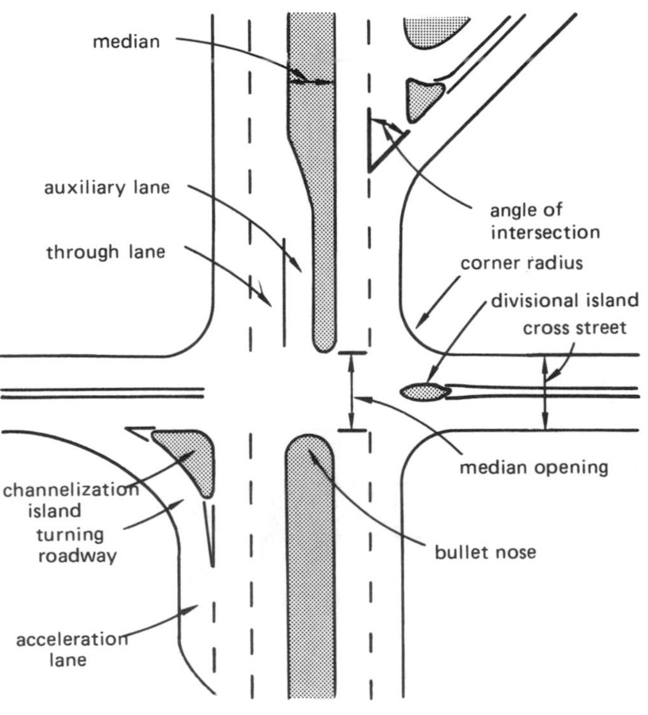


**Figure 16.7** Elements of an Intersection

## 15 AT-GRADE SIGNALIZED INTERSECTION CAPACITY

The capacity of an intersection depends on many factors, including the width of approach, parking conditions, traffic direction (i.e., one-way or two-way), environment, bus and truck traffic, and percentage of turning vehicles. Capacity calculations for signalized intersections is largely graphical. Inclusion of all relevant graphs is not practical within the scope of this book.[9]

## 16 STANDARD TRUCK LOADINGS

Table 16.12 and figure 16.8 illustrate standard truck loads commonly used for design. In cases where the separation between axles varies, the distance that produces the maximum stress in the section should be used.

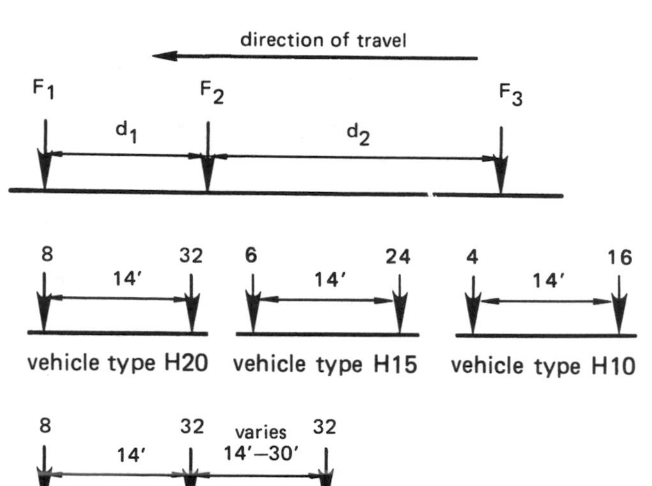


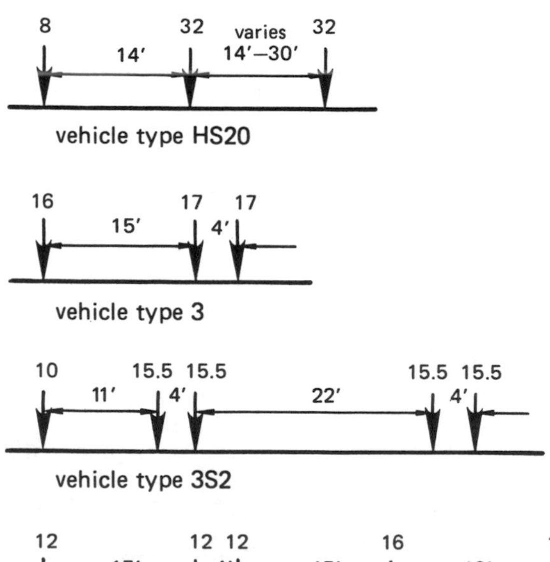


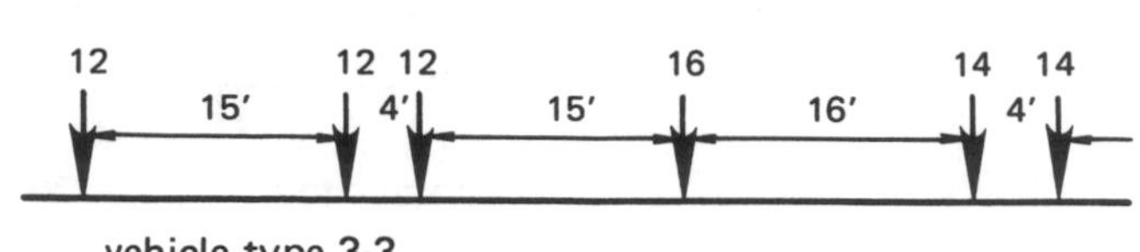


**Figure 16.8** Standard Truck Loadings

[9] The authoritative reference on this subject is the *Highway Capacity Manual*.

**Table 16.12**
Standard Truck Loadings

(All loads are axle loads)

| load designation | $F_1$ | $F_2$ | $F_3$ | $d_1$ | $d_2$ |
|---|---|---|---|---|---|
| H20–44 | 8000 | 32,000 | 0 | 14′ | |
| H15–44 | 6000 | 24,000 | 0 | 14′ | |
| H10–44 | 4000 | 16,000 | 0 | 14′ | |
| HS20–44 | 8000 | 32,000 | 32,000 | 14′ | 14′ to 30′ |
| HS15–44 | 6000 | 24,000 | 24,000 | 14′ | 14′ to 30′ |
| 3 | 16,000 | 17,000 | 17,000 | 15′ | 4′ |
| 3S2 | See figure 16.8 | | | | |
| 3–3 | See figure 16.8 | | | | |

For the purposes of designing road geometry, AASHTO defines *design vehicles* according to the following categories: passenger car (P), single unit truck (SU), single unit bus (BUS), intermediate length semitrailer (WB-40), large semitrailer (WB-50), and semitrailer/full trailer combination (WB-60). In the case of WB vehicles, the number represents the wheelbase distance between the front (cab) axle and the last trailer axle. [AASHTO PGDHS–1990, Table II-1]

Classification into axle types also is common. An axle may be *single* or *tandem*, and each axle may have single or dual tires.[10] *Spread tandem axles*, where two axles are separated by more than 96 inches, generally are classified as two single axles.

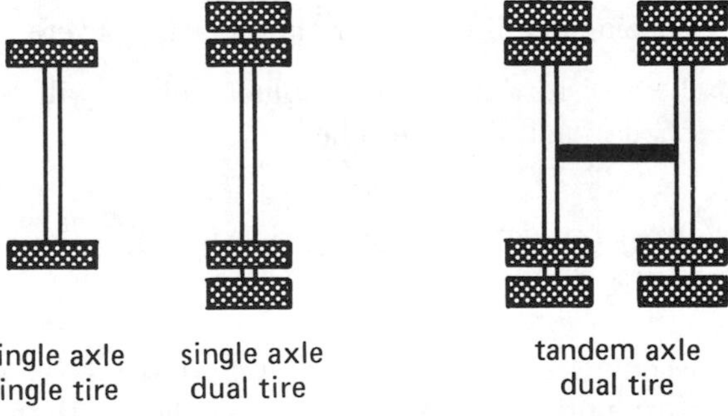


**Figure 16.9** Types of Axles and Axle Sets

Except in theoretical stress studies, no attempt is made to account for the number of tires per axle. Although it is true that stresses at shallow depths are caused principally by individual wheels acting singly, stresses at greater depths are maximum midway between wheels. Deep stresses due to dual-wheeled axles are approximately the same as for single-wheeled axles. Therefore, required pavement thickness is determined by the total axle load.[11]

## 17 TYPES OF PAVEMENT

### A. RIGID PAVEMENT

Portland cement concrete is used almost exclusively where rigid pavement is called for. Typical applications are high volume traffic lanes, freeway-to-freeway connections, and exit ramps that experience heavy traffic.

Cement concrete pavement has excellent durability and long service life. It provides good contrast with asphalt surfaces. It will also withstand repeated flooding and subsurface water without deterioration.

It has three primary disadvantages: (1) It may lose its original nonskid surface during use. (2) It must be used with an even subgrade and only where uniform settling is expected. (3) It may rise (fault) at transverse joints.

Portland cement concrete is placed with slip-form construction methods. A stiff concrete ("no-slump" with slumps less than 1") is placed in front of the paving train. The paving train then distributes, vibrates, screeds, and finishes the layer while traveling at a slow, continuous speed.

Rigid pavement can be reinforced or unreinforced.[12] With unreinforced construction, *contraction joints* should be placed at regular intervals. A maximum spacing of 15 feet or 30 times the thickness of the slab is a typical rule for unreinforced pavement.[13]

If steel is used, it is assumed not to contribute to the structural section. It is used only to control crack growth. Steel is usually deformed bar but an equivalent amount of steel mesh can be used. Construction joints are required in reinforced pavement, as well. However, the spacing is greater—every 50 to 100 feet, depending on reinforcement.

AASHTO has specified a formula for calculating the quantity of longitudinal steel required in a concrete highway slab.[14] The area is specified in square inches per foot of slab width. In equation 16.29, $F$ is the *coefficient of resistance* between the slab and subgrade

[10] Tandem axles have two axles separated by 40 to 96 inches. However, up to five axle sets are used for heavy loads.

[11] However, pavement stress calculations should be based on the individual wheel loads.

[12] California has many miles with unreinforced concrete.

[13] California uses a random spacing following the sequence of 12, 15, 13, 14 feet. Sawn joints are placed with a diagonal skew of 2 feet in 12 feet of width.

[14] The formula can also be used to calculate transverse reinforcement requirements if $L$ is taken as the distance between the slab edges.

(typically taken as 1.5), $L$ is the distance in feet between transverse joints, $w$ is the weight of the pavement slab (typically based on 150 lbf/ft$^3$), and $f_s$ is the allowable working stress in the steel (approximately two-thirds of the yield strength). Based on experience, the fraction of steel required will vary between 0.5 and 0.8 percent of the cross-sectional area of the pavement.

$$A_s = \frac{FLw}{2f_s} \qquad 16.29$$

AASHTO also has specified spacing and sizes for round steel dowels used as load transfer devices between slabs. 18″ dowels placed 12″ apart should be used. For 6″ thick pavement, a 3/4″ diameter bar should be used. For 7″ and 8″ pavements, 1″ bars are called for. $1\frac{1}{4}$″ bars are used for all pavements thicker than 8″.

If possible corrosion of the steel bars is a concern, the steel can be epoxy coated. Other additives, such as *latex* or *silica fume* (*microsilica*), can be used to protect the concrete against penetration of deicing salts.

### B. PRESTRESSED CONCRETE PAVEMENT

Prestressed concrete pavements have been used in Europe since the 1950's. However, they have seen little use in the United States. One of the reasons is that the primary advantage of prestressed pavements—that of requiring thinner sections—results in overall savings only with thicker sections such as used in airports.

In addition to requiring sections that are only 40% to 60% as thick as conventional pavements, prestressing also reduces transverse joint requirements to one every 300 to 600 feet. Fewer joints increase the ride quality and decrease maintenance. Also, prestressing requires as little as 20% of the steel used in continuously reinforced pavement. A service life of 30 years is expected with prestressed pavements.

Demonstration projects in the United States have shown that prestressed pavements for highways are competitive on a first-cost basis. Performance to date indicates lower maintenance costs. Normal paving procedures can be used.

### C. FLEXIBLE ASPHALT CONCRETE

Typical uses of asphalt concrete are traffic lanes, auxiliary lanes, ramps, parking areas, frontage roads, and shoulders.

Asphalt concrete pavement has the advantage of adjusting to limited amounts of differential settlement. It is easily repaired, and additional thicknesses can be placed at any time to withstand increased usage and loading. Its non-skid properties do not deteriorate to a great extent.

However, asphalt concrete loses some of its flexibility and cohesion with time, and it will have to be resurfaced sooner than would cement concrete. It would not normally be chosen where water is expected.

### D. FULL-DEPTH® ASPHALT PAVEMENT

If asphalt mixtures are used for all courses above the subgrade, the pavement is said to be a ***full-depth asphalt pavement.*** Since asphalt bases are stronger than untreated bases, the pavement thickness is less. Other advantages of full-depth pavements include a potential decrease in trapped water within the pavement, a decrease in the moisture content of the subgrade, and little or no reduction in subgrade strength.[15]

### E. DEEP-LIFT ASPHALT PAVEMENT

If the asphalt layer is thicker than 4″ and is placed all in one lift, or if lift layers are thicker than 4″, the construction is said to be *deep lift.* Using deep lifts to place hot-mix asphalt concrete is advantageous for several reasons.

- Thicker layers hold heat longer, and it is easier to roll the layer to the required density.
- Lifts can be placed in cooler weather.
- One lift of a given thickness is more economical to place than multiple lifts equalling the same thickness.
- Placing one lift is faster than placing several.
- Less distortion of the asphalt course will result than if thin lifts are rolled.

## 18 SULFUR EXTENDED ASPHALT BINDER

To become less dependent on imported oil (used to make asphaltic binders), and to use the byproduct of desulfurized oil, sulfur has been added to asphalt concrete in a number of instances.[16] The sulfur percentage varies from 10% to 50% of the total binder content (by weight), although 30% to 40% is common.

At standard temperatures and pressures, ordinary sulfur is an odorless and tasteless yellow solid. Its specific gravity is 2.07 and the melting point is 238°F. Sulfur

[15] Subgrade drains are still required, however.

[16] This is not a new idea. *Sulfur concrete* was being used at least as far back as the 1920's.

is not a hazardous material. The practices for hauling, heating, and storing molten sulfur are well established.

The working range for molten sulfur corresponds well to the working range for paving grade asphalt: 255°F to 300°F. Above 315°F, sulfur becomes viscous and cannot be easily worked. Liquid sulfur is hot, and in that respect, poses the same dangers as hot asphalt (or any other hot liquid). When heated, the concentration of eye-irritating sulfur fumes is low (below 300°F), but increases rapidly above 300°F. Molten sulfur at 300°F will ignite, as will asphalt. Therefore, all sources of ignition near liquid sulfur should be removed.

Sulfur-asphalt pavement binder consists primarily of a very fine dispersion of sulfur in asphalt, with asphalt forming the continuous phase. The mixing is performed in a colloid or pug mill using molten sulfur and hot asphalt.

When mixed well, sulfur measuring approximately 20% by weight of the asphalt will dissolve in the asphalt.

The remainder of the sulfur forms a dispersion in the asphalt. Both the dissolved and dispersed sulfur modify the properties of the binder.

1 to 2 ppm of silcone can be added to stabilize the sulfur-asphalt emulsion and make the mixture easier to work. (Normally, 2 ppm added to the hot asphalt will improve moisture release from the hot mix and reduce pulling and tearing at the screed of the paving machine.)

The paving operation does not need any special equipment. Standard pavers, rollers, haul trucks, emission control equipment, and testing equipment can be used. However, special equipment is required to store and handle sulfur, and a special sulfur-asphalt emulsion mixing plan is necessary.

Vapor given off during the preparation and placement of sulfur-asphalt mixtures contains a certain amount of elemental sulfur.[17] There is no practical way to eliminate this pollutant. Sulfur is virtually nontoxic, and there is no evidence of systemic poisoning resulting from inhalation. However, sulfur does irritate open cuts and the inner surface of the eyelids. Goggles and gloves can reduce such irritations.

The properties of the resulting *sulfur-asphalt concrete* are superior to those of pure asphalt concrete. A high-strength concrete results. Since there is no alkaline cement binder (e.g., portland cement) in the mix, sulfur concrete is an acid-resistant structural material. It achieves 80% of its ultimate compressive strength in a few hours, and 100% within 24 hours.

[17] Above 300°F, sulfur dioxide and hydrogen sulfide become the dominant pollutants. Therefore, the molten sulfur temperature must be kept below 300°F. 280°F to 290°F is the preferred operating range.

## 19 MINIMUM LAYER THICKNESSES

It generally is impractical and uneconomical to place surface, base, and subbase courses with thicknesses less than minimum values. These minimum thicknesses are 1.5″–2″ for asphalt surface courses, and 4″ for cement-, lime-, and asphalt-treated bases and subbases.

## 20 PAVEMENT DESIGN PARAMETERS

### A. LAYER STRENGTHS

Once the materials for the surface, base, and subbase layers have been selected, their strengths can be determined by testing. It may be necessary to convert one strength parameter into another for use with a particular design procedure. Appendices A, B, and C can be used for this purpose.

The quality of the basement soil is a required factor. Since there may be a considerable range in these values, a design value must be chosen. If the range is small, the lowest value should be selected. If there are a few exceptionally low values that come from one area, it may be possible to specify replacing that area's soil with borrow soil. If there are changing geological formations along the route which modify the value, it may be necessary to design different pavement sections.

### B. EQUIVALENT AXLE LOADINGS

The pavement thickness will be dependent on the axle loadings that are estimated for the truck traffic predicted for the design period. (The effects of passenger cars, pickups, and two-axle trucks with single rear tires are not considered.) Both the number of trucks and the axle loading of those trucks must be known.

The usual sources of truck volume data are traffic counts. These may be actual counts on existing roadways needing resurfacing, or they may be redistributed counts from other highways. The trucks are classified according to the number of axles.

Pavement thickness should be chosen to serve the estimated one-way truck traffic for a period of 20 years. A shorter period, not to be less than 10 years, may be used for temporary construction or for other justifiable reasons.

An 18-kip axle load (two 9-kip wheel loads) is used as the standard loading in highway section designs. This reduces all traffic data into the number of 18-kip axle passes that would cause the same structural damage. The analysis is made difficult by the number of ways that traffic can be reported. Traffic counts can be accurate (as when actual highway loadometer data is used) or approximate (as when only the number of trucks is available).

*Method A*: Use actual highway loadometer data and get the *equivalency factor* assuming a *structural number* (SN) of 3 from appendix D or E. Each number of axle passes is multiplied by an equivalency factor to convert it to a number of 18-kip axle passes. The choice of SN = 3 is arbitrary. When the design is completed, a more accurate SN will be determined.[18]

*Method B*: Convert other equivalent axle loads (e.g., 12-kip axle loads) to 18-kip axle loads.[19,20]

$$\text{EAL}_{\text{18-kips}} = \text{EAL}_{\text{n-kips}} \left(\frac{n \text{ in kips}}{18}\right)^4 \qquad 16.30$$

*Method C*: Convert equivalent axle load data from other durations.

$$\text{EAL}_{\text{20 years}} = \text{EAL}_{\text{n years}} \left(\frac{20}{n \text{ in years}}\right) \qquad 16.31$$

*Method D*: Approximate the equivalent 18-kip axle loading from the number of trucks passing *per day*. The coefficients are known as *truck constants*.[21]

$$\begin{aligned}\text{EAL}_{\text{18-kip, 20 years}} &= (1380)\begin{pmatrix}\text{\#2-axle}\\ \text{trucks}\end{pmatrix} \\ &+ (3680)\begin{pmatrix}\text{\#3-axle}\\ \text{trucks}\end{pmatrix} + (5880)\begin{pmatrix}\text{\#4-axle}\\ \text{trucks}\end{pmatrix} \\ &+ (13{,}780)\begin{pmatrix}\text{\#5-axle}\\ \text{trucks}\end{pmatrix}\end{aligned} \qquad 16.32$$

*Method E*: Convert 5000-pound *equivalent wheel loads* (EWL) to 18-kip EAL values.

$$\text{EAL} = \frac{\text{EWL}}{11.8} \qquad 16.33$$

*Method F*: Convert *design index* (DI) to 20-year EAL using table 16.13.[22]

**Table 16.13**
Design Index versus EAL

| DI | 20-year EAL |
|---|---|
| 1 | 7300–36,500 |
| 2 | 36,501–146,000 |
| 3 | 146,001–547,500 |
| 4 | 547,501–1,825,000 |
| 5 | 1,825,001–6,570,000 |
| 6 | 6,570,001–21,900,000 |

*Method G*: Convert *traffic index* (TI) to 20-year EAL using equation 16.36.

### C. PREDICTING TRAFFIC GROWTH

If the 20-year EAL is to be predicted from the current (first) year EAL, and if a constant growth rate of *i*% per year is assumed, economic analysis tables can be used to simplify the calculation.

$$\text{EAL}_{20} = \text{EAL}_{\text{first year}} \times (F/A, i\%, 20) \qquad 16.34$$

## 21 AASHTO METHOD OF FLEXIBLE PAVEMENT DESIGN

This method is based on tests performed in Illinois between 1958 and 1960 by the American Association of State Highway and Transportation Officials (AASHTO).

*step 1*: Estimate the desired *terminal serviceability*, $p_t$. This parameter is a numerical rating of road quality after 20 years. (It is assumed that the first major resurfacing will occur at that time.) $p_t$ actually can be calculated from the condition of an existing road surface (i.e., from rut depths, percentage of cracked area, and percentage of patched areas, among other factors). However, for design work, it is common to specify $p_t = 2.5$ for highways and $p_t = 2$ for low traffic roads.

[18] All axles in the truck must be included. For example, a tractor/trailer would contribute three quantities to EAL—one for the tractor front axle, one for the tractor tandem axle, and one for the trailer tandem axle.

[19] This is essentially the same as using the load equivalency tables for single-axle loads as described in method A.

[20] The exponent 4 in equation 16.30 is also reported as 4.2 in literature.

[21] The truck constants in equation 16.32 are not unique. They represent typical factors from California statewide truck weighings. Other states or authorities may specify different truck constants.

[22] Actually, design index is usually correlated in military studies with *one-day EAL*. However, this chapter uses only 20-year EAL, and table 16.13 reflects this.

**Table 16.14**
Terminal Serviceabilities

| $p_t$ | condition |
|---|---|
| 0–1 | very poor |
| 1–2 | poor |
| 2–3 | fair |
| 3–4 | good |
| 4–5 | very good |

*step 2*: Convert the traffic volume to the number of equivalent 18-kip single axle loads.

*step 3*: Determine the *regional factor*, *R*. (This is not the same as the soil R-value.) This factor recognizes that a load does more damage when the ground is saturated, as during a spring thaw, compared to solid dry or solid frozen support. It is difficult to specify the regional factor, but a range of 0.5 to 4.0 should not be exceeded for the continental U.S.[23]

**Table 16.15**
Regional Factors

| *R* | condition |
|---|---|
| 0.2–1.0 | Roadbed frozen to depth of 5″ or more |
| 0.3–1.5 | Roadbed dry, summer and fall; no winter freezing |
| 0.5 | Sandy desert |
| 1.5 | Roadbed subject to frost, but fairly dry |
| 4.0–5.0 | Roadbed wet, spring break-up thaw, high water table, soil saturated |

*step 4*: Get the *soil support value*, *S*, based on the quality of the subgrade (basement) soil. *S* varies from 1 (theoretically, 0) through 10 for crushed rock. Silty clay (soil type A-6) has a soil support value of approximately 3. For most studies, it will be necessary to correlate S with some other known soil property, such as a California Bearing Ratio (CBR), modulus of subgrade reaction (k), or Hveem's Resistance value (R). These correlations are not exact, and they may vary from authority to authority. Appendix A, B, or C can be used.

**Table 16.16**
Approximate Modulus of Subgrade Reaction for Paving Materials

| AASHTO soil group | category | *k*, (psi/in) |
|---|---|---|
| A-1-a | CTB, BTB | 400 and up |
| A-1-b | | 250 and up |
| A-2-4, A-2-5 | gravels | 300 and up |
| A-2-6, A-2-7 | | 175–325 |
| A-3 | sand, clay gravel | 200–325 |
| A-4 | silt, silty clay | 100–300 |
| A-5 | plastic clay | 50–175 |
| A-6 | | 50–225 |
| A-7-5, A-7-6 | | 50–225 |

*step 5*: Determine the *structural number* from the terminal serviceability ($p_t$), 20-year, 18-kip equivalent axle loading (EAL), soil support value (S), and regional factor (R) from figure 16.10 or figure 16.11. If the structural number differs from what was used in step 2, repeat from step 2.

*step 6*: Determine the *layer coefficients* (also known as *strength coefficients*) for the subbase, base, and surface course materials. These coefficients vary from state to state. However, table 16.17 can be used for general calculations.

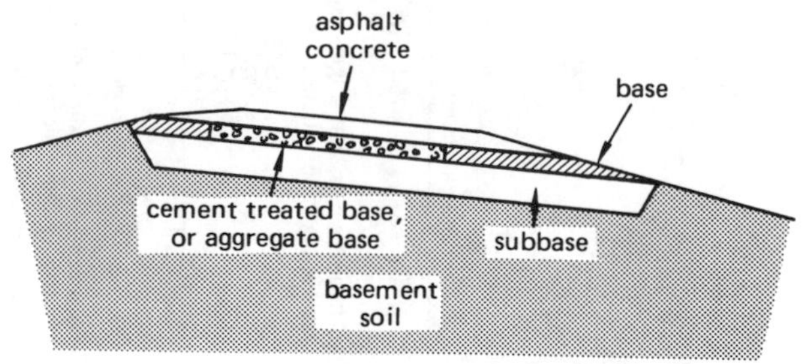


**Figure 16.12** Asphalt Concrete Pavement Cross Section

*step 7*: Write the *layer-thickness equation*. Values of *t* in equation 16.35 are in inches. If a subbase is not to be used, omit the third term.

$$t_1a_1 + t_2a_2 + t_3a_3 = SN \qquad 16.35$$

[23] If $R = 1$, the regional factor will not have an effect on the structural number. $R = 1$ is appropriate for all of California.

TRAFFIC

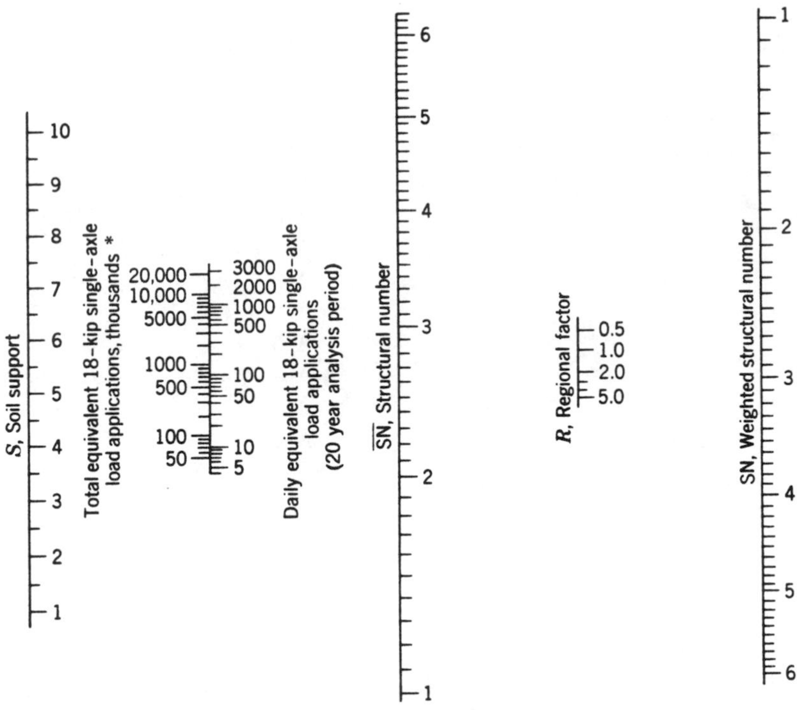


**Figure 16.10** AASHTO Flexible Pavement Design Nomograph *

(Terminal Serviceability, $p_t = 2.0$)

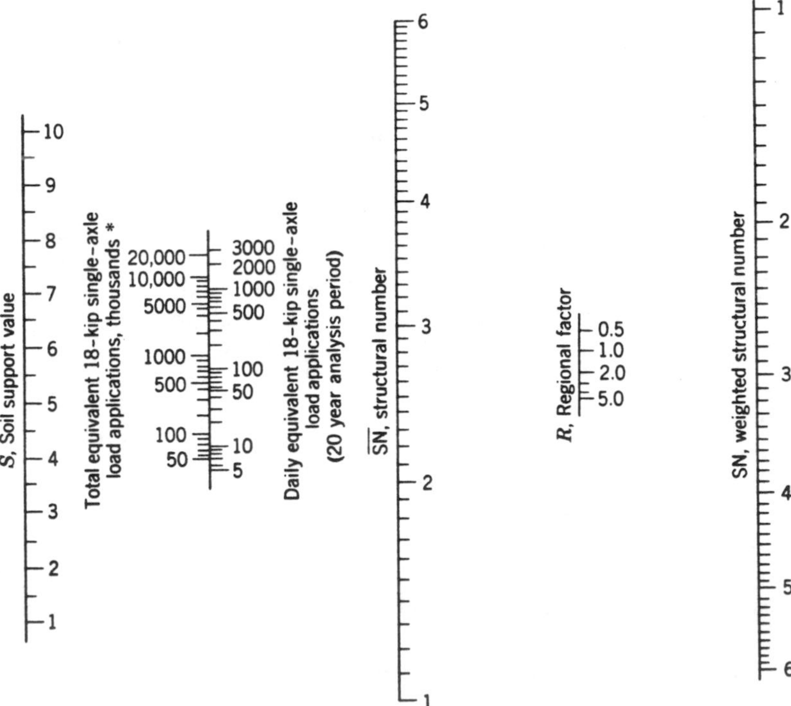


**Figure 16.11** AASHTO Flexible Pavement Design Nomograph

(Terminal Serviceability, $p_t = 2.5$)

* Total equivalent 20 year load = daily equivalent load x 365 x 20 years

Theoretically, any combinations of thicknesses that satisfy equation 16.35 will work. However, minimum layer thicknesses result from construction techniques and strength requirements.

**Table 16.17**
Layer Coefficients from Various Sources

| | |
|---|---|
| **subbase coefficient, $a_3$** | |
| sandy gravel | 0.11 |
| sand, sandy clay | 0.05–0.10 |
| lime-treated soil | 0.11 |
| lime-treated clay, gravel | 0.14–0.18 |
| **base coefficient, $a_2$** | |
| sandy gravel | 0.07 |
| crushed stone | 0.14 |
| cement treated base (CTB) | |
| $f'_{c,7\text{-day}} > 650$ psi | 0.23 |
| 400–650 | 0.20 |
| < 400 | 0.15 |
| bituminous treated base (BTB) | |
| coarse | 0.34 |
| sand | 0.30 |
| lime treated base | 0.15–0.30 |
| soil cement | 0.20 |
| lime/fly ash base | 0.25–0.30 |
| **surface course coefficient, $a_1$** | |
| plant mix | 0.44 |
| road mix | 0.20 |
| sand asphalt | 0.40 |

*Example 16.5*

It is desired to verify the adequacy of a flexible pavement design for a well-traveled highway in California.

R value of basement soil: 10

subbase: aggregate, thickness 17″

base: aggregate, thickness 12″

asphalt concrete: thickness 8″

The surface is expected to carry the following traffic.

2-axle trucks: ADT = 935

3-axle trucks: ADT = 550

4-axle trucks: ADT = 225

5-axle trucks: ADT = 1025

*step 1*: Assume a terminal serviceability of $p_t = 2.5$.

*step 2*: Use equation 16.32.

$$\begin{aligned} EAL &= (1380)(935) + (3680)(550) + (5880)(225) \\ &\quad + (13{,}780)(1025) \\ &= 18.8\,\text{EE6} \end{aligned}$$

*step 3*: Choose $R = 1$ for all of California.

*step 4*: From appendix B, the soil support value, S, is approximately 3 for an R value of 10.

*step 5*: From figure 16.11, the required structural number is approximately 6.

*step 6*: Use the following layer coefficients: $a_3 = 0.11, a_2 = 0.14, a_1 = 0.44$.

*step 7*: Calculate the actual structural number from equation 16.35.

$$(0.44)(8) + (0.14)(12) + (0.11)(17) = 7.07$$

Since 7.07 is greater than 6, the pavement has adequate strength.

## 22 CALTRANS METHOD OF FLEXIBLE PAVEMENT DESIGN

The California Department of Transporation (CALTRANS) has adopted a method of designing flexible pavements which works downward from the asphalt concrete layer. Starting with the top layer, and dealing with each lower layer in turn, the equivalent thickness of a gravel layer is calculated from the traffic volume and strength of the layer below. This *gravel equivalent* is then converted into an actual layer thickness according to the strength of the layer material.

*step 1*: Determine the 18-kip equivalent axle loading for the surface.

*step 2*: Calculate the *traffic index*, TI, from equation 16.36. Round the value up or down to the nearest 0.5. Greater accuracy is not justified.

$$TI = (9.0)(\text{EAL}/10^6)^{0.119} \qquad 16.36$$

*step 3*: Choose the base material. Aggregate base is most commonly used with an R-value of 78. Cement treated bases (CTB) are also used. They are constructed with a good quality, well-graded aggregate mixed with up to 6% portland cement. Other bases can also be used to treat poor soils. These include lime treated (stabilized) bases (LTB), bituminous treated bases (BTB), and soil cement bases (CS).

Class B CTB is often used to increase the R-value of the structural section. If the aggregate R-value was 60 or greater before mixing with cement, the ultimate R-value will exceed 80. Therefore, 80 is used as the minimum value for class B CTB.

Class A CTB is used under asphalt to provide added strength to heavily-traveled surfaces. $f'_c$ will exceed 750 psi after curing for 7 days. Class A CTB is rated by $f'_c$, and R-values are not assigned.

*step 4*: Calculate the total *gravel equivalent* (GE) thickness required for the layer being designed and all layers above it.[24] (The first time through this step, the gravel equivalent of the asphalt concrete surface layer will be determined.) $R$ is for the material below the layer being designed.

$$GE = 0.0032(TI)(100 - R) \qquad 16.37$$

*step 5*: Calculate the thickness of the layer by dividing the gravel equivalent, GE, by the *gravel equivalent factor*, $G_f$. Table 16.18 lists gravel equivalent factors.

$$t_{\text{layer}} = \frac{GE}{G_f} \text{ (in feet)} \qquad 16.38$$

*step 6*: Select the material for the subbase. Subbase aggregate is classified according to the distribution of sizes. Minimum R-values are listed in table 16.19. (Class 2 aggregate is most common.)

**Table 16.19**
Aggregate Subbase Classes

| class | % passing # 4 sieve | minimum R-value |
|---|---|---|
| 1 | 30–75 | 60 |
| 2 | 35–95 | 50 |
| 3 | 45–100 | 40 |

*step 7*: Determine the thickness of the base. Calculate the total gravel equivalent from the R-value of the subbase and equation 16.37. Subtract the surface layer's gravel equivalent from the total gravel equivalent. Calculate the base's thickness from equation 16.38.

*step 8*: Determine the thickness of the subbase. Calculate the total gravel equivalent from the R-value of the basement soil and equation 16.37. Subtract the base's and surface layer's gravel equivalents (corresponding to their actual thicknesses) from the total gravel equivalent. Calculate the subbase's thickness from equation 16.38.

**Table 16.18**
Gravel Equivalent Factors[25]

| layer material | $G_f$ |
|---|---|
| **surface layer (asphalt concrete)** | |
| TI: 5.0 and below | 2.50 |
| 5.5 and 6.0 | 2.32 |
| 6.5 and 7.0 | 2.14 |
| 7.5 and 8.0 | 2.01 |
| 8.5 and 9.0 | 1.89 |
| 9.5 and 10.0 | 1.79 |
| 10.5 and 11.0 | 1.71 |
| 11.5 and 12.0 | 1.64 |
| 12.5 and 13.0 | 1.57 |
| 13.5 and 14.0 | 1.52 |
| 14.5 and above | 1.50 |
| **bases** | |
| aggregate | 1.10 |
| Class A CTB | 1.70 |
| Class B CTB, BTB, LTB, CS | 1.20 |
| lean concrete | 1.90 |
| **subbase** | |
| aggregate | 1.00 |

*Example 16.6*

Use the CALTRANS method to design a flexible pavement lane over a basement soil with an R-value of 10 to carry the following average daily traffic.

2 axle trucks: ADT = 935

3 axle trucks: ADT = 550

4 axle trucks: ADT = 225

5 axle trucks: ADT = 1025

[24] A peculiarity of the CALTRANS method is that a safety factor of 2″ to 3″ is added to the gravel equivalent to account for deficient layer thickness. The sophistication is omitted here.

[25] The gravel equivalent factors for asphalt concrete surface courses are calculated from the following equation:

$$G_f = 2.5\sqrt{\frac{5.14}{TI}}$$

*step 1*:

$$EAL = (1380)(935) + (3680)(550) + (5880)(225) + (13{,}780)(1025) = 18.8\,\text{EE6}$$

*step 2*: $TI = (9.0)\left(\frac{18.8\,\text{EE6}}{10^6}\right)^{0.119} = 12.76$ (say 12.5)

*step 3*: Choose an aggregate base with an R-value of 78.

*step 4*: The gravel equivalent of the asphalt concrete surface layer is

$$GE = 0.0032(12.5)(100 - 78) = 0.88\,\text{ft of gravel}$$

*step 5*: The actual thickness of the asphalt concrete layer is calculated from equation 16.38.

$$t_1 = \frac{0.88}{1.57} = 0.56\,\text{ft}$$

*step 6*: Select a class 2 aggregate subbase with an R-value of 50.

*step 7*: The gravel equivalent of the base and asphalt layers combined is

$$GE = 0.0032(12.5)(100 - 50) = 2.0\,\text{ft of gravel}$$

Since the surface course has already provided 0.88 ft of equivalent gravel layer, the gravel thickness of the base is

$$2.0 - 0.88 = 1.12\,\text{ft}$$

From equation 16.38, the base thickness is

$$t_2 = \frac{1.12}{1.10} = 1.02\,\text{ft}$$

*step 8*: The gravel equivalent of the subbase, base, and asphalt layers combined is

$$GE = 0.0032(12.5)(100 - 10) = 3.6\,\text{ft of gravel}$$

The surface and base layers have already provided $0.88 + 1.12 = 2.0$ ft of gravel equiv-

**Table 16.20**
Asphalt Institute Traffic Classifications

| traffic class | EAL | type of street or highway | approximate range—number of heavy trucks expected during design period |
|---|---|---|---|
| I | $5 \times 10^3$ | • parking lots, driveways<br>• light traffic residential streets<br>• light traffic farm roads | $\leq 7000$ |
| II | $10^4$ | • residential streets<br>• rural farm and residential roads | 7000–15,000 |
| III | $10^5$ | • urban minor collector streets<br>• rural minor collector roads | 70,000–150,000 |
| IV[26] | $10^6$ | • urban minor arterial and light industrial streets<br>• rural major collector and minor arterial highways | 700,000–1,500,000 |
| V[26] | $3 \times 10^6$ | • urban freeways, expressways and other principal arterial highways<br>• rural interstate and other principal arterial highways | 2,000,000–4,500,000 |
| VI[26] | $10^7$ | • urban interstate highways<br>• some industrial roads | 7,000,000–15,000,000 |

[26] Whenever possible, the traffic analysis and design procedures given in the Asphalt Institute manual, *Thickness Design—Asphalt Pavements for Highways and Streets* (MS-1), should be used for roads and streets in traffic category IV or higher.

alent. Therefore, the gravel equivalent thickness of the subbase is

$$3.6 - 2.0 = 1.6\,\text{ft}$$

The subbase thickness is

$$t_3 = \frac{1.6}{1.0} = 1.6\,\text{ft}$$

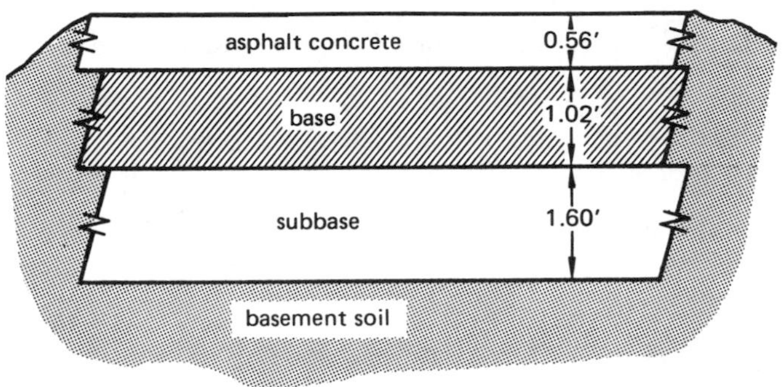


## 23 FULL-DEPTH ASPHALT PAVEMENTS—SIMPLIFIED ASPHALT INSTITUTE METHOD[27]

*step 1*: Determine the 20-year, 18-kip equivalent axle loading (EAL) for the pavement.

*step 2*: Use table 16.20 to convert EAL to a traffic class.

*step 3*: Use table 16.21 to classify the subgrade soil into poor, medium, or good-to-excellent categories.

**Table 16.21**
Subgrade Soil Categories

| | typical values | | |
|---|---|---|---|
| category | resilient modulus[28] | CBR | R-value |
| poor | 4500 psi | 3 | 6 |
| medium | 12,000 | 8 | 20 |
| good-to-excellent | 25,000 | 17 | 43 |

[27] Theoretically, this method can be used to design pavement sections with asphalt concrete directly over an untreated aggregate base, as long as the base's R-value is known. However, the resultant design would not be a full-depth asphalt pavement.

[28] The *resilient modulus* is the same as the *modulus of elasticity* of the soil. It is not the same as the modulus of subgrade reaction, $k$, although the two are related. For positive values of the resilient modulus, $\text{MR} \approx k \times 19.4$. The resilient modulus is also approximately 1500 times the California Bearing Ratio (CBR).

*step 4*: Choose the base and subbase materials. Emulsified asphalt base mixtures are divided into three types. Type I is plant mixed base made with processed dense-graded aggregates. Type II bases are made with semi-processed, crusher-run, pit-run, or bank-run aggregates. Type III bases are made with sands or silty sands. Both type II and type III can be either plant or road mixed.

Untreated aggregate subbase materials can also be used. Special quality and minimum strength requirements apply in those instances.[29]

*step 5*: Use table 16.22 or table 16.23 to design the pavement. If the EAL or approximate number of trucks places the design between traffic classes, interpolation between thicknesses in the tables is allowed.

## 24 DESIGNING PORTLAND CEMENT CONCRETE PAVEMENT

On projects with three or more lanes in one direction, separate designs usually are made for the inside and outside lanes. This results in steps at the bottoms of pavement and base. It is cheaper to construct stepped sections than uniform or tapered sections, which result in increased soil removal. However, in order to provide a uniform grading plane, it is permissible to increase the thickness of the subbase under the inside lanes.[30] This total thickness of subbase should be used to design the pavement for the inside lanes.

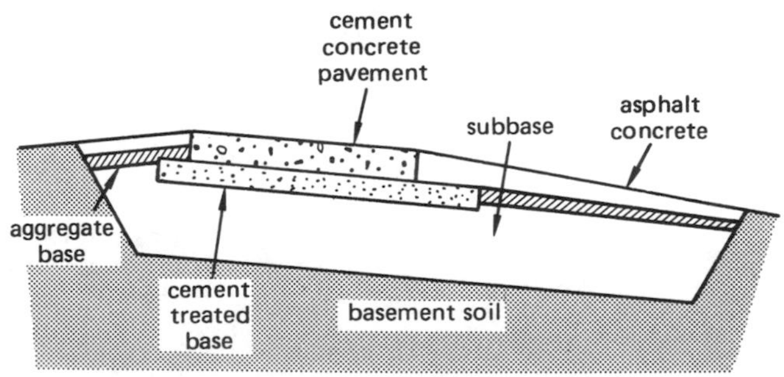


**Figure 16.13** Typical Portland Cement Concrete Section

[29] Specifically, untreated aggregate must have a minimum CBR of 20 or a minimum R-value of 55 to be used as a subbase. It must have a minimum CBR of 80 of a minimum R-value of 78 to be used as a base. Other requirements exist also.

[30] If the subbase is omitted, as it is in many rigid pavement designs, increase the thickness of the base.

**Table 16.22**
Full-Depth Asphalt Pavements Using Asphalt Concrete or Emulsified Asphalt Base Mixes

| subgrade class | pavement section | traffic classification I | II | III | IV | V | VI |
|---|---|---|---|---|---|---|---|
| | | thickness in inches | | | | | |
| | | full-depth asphalt concrete | | | | | |
| poor | asphalt concrete surface | 1.0 | 1.0 | 1.5 | 2.0 | 2.0 | 2.0 |
| | asphalt concrete base | 3.5 | 4.0 | 5.5 | 8.0 | 10.5 | 13.0 |
| | total: | 4.5 | 5.0 | 7.0 | 10.0 | 12.5 | 15.0 |
| medium | asphalt concrete surface | 1.0 | 1.0 | 1.5 | 2.0 | 2.0 | 2.0 |
| | asphalt concrete base | 3.0 | 3.0 | 3.5 | 6.0 | 8.0 | 11.0 |
| | total: | 4.0 | 4.0 | 5.0 | 8.0 | 10.0 | 13.0 |
| good to excellent | asphalt concrete surface | 1.0 | 1.0 | 1.5 | 2.0 | 2.0 | 2.0 |
| | asphalt concrete base | 3.0 | 3.0 | 2.5 | 4.0 | 6.5 | 9.0 |
| | total: | 4.0 | 4.0 | 4.0 | 6.0 | 8.5 | 11.0 |
| | | emulsified asphalt mix type I | | | | | |
| poor | asphalt concrete surface | 1.0 | 1.0 | 1.5 | 2.0 | 2.0 | 2.0 |
| | type I base | 3.5 | 4.0 | 5.5 | 8.0 | 10.5 | 13.0 |
| | total: | 4.5 | 5.0 | 7.0 | 10.0 | 12.5 | 15.0 |
| medium | asphalt concrete surface | 1.0 | 1.0 | 1.5 | 2.0 | 2.0 | 2.0 |
| | type I base | 3.0 | 3.0 | 3.5 | 7.0 | 9.0 | 11.5 |
| | total: | 4.0 | 4.0 | 5.0 | 9.0 | 11.0 | 13.5 |
| good to excellent | asphalt concrete surface | 1.0 | 1.0 | 1.5 | 2.0 | 2.0 | 2.0 |
| | type I base | 3.0 | 3.0 | 2.5 | 4.5 | 7.5 | 10.0 |
| | total: | 4.0 | 4.0 | 4.0 | 6.5 | 9.5 | 12.0 |
| | | emulsified asphalt mix type II | | | | | |
| poor | asphalt concrete surface | 2.0 | 2.0 | 2.0 | 3.0 | 3.5 | 4.0 |
| | type II base | 2.5 | 3.0 | 5.5 | 9.0 | 11.0 | 13.5 |
| | total: | 4.5 | 5.0 | 7.5 | 12.0 | 14.5 | 17.5 |
| medium | asphalt concrete surface | 2.0 | 2.0 | 2.0 | 3.0 | 3.5 | 4.0 |
| | type II base | 2.0 | 2.0 | 3.0 | 7.0 | 9.0 | 11.5 |
| | total: | 4.0 | 4.0 | 5.0 | 10.0 | 12.5 | 15.5 |
| good to excellent | asphalt concrete surface | 2.0 | 2.0 | 2.0 | 3.0 | 3.5 | 4.0 |
| | type II base | 2.0 | 2.0 | 2.0 | 5.0 | 7.0 | 9.5 |
| | total: | 4.0 | 4.0 | 4.0 | 8.0 | 10.5 | 13.5 |
| | | emulsified asphalt mix type III | | | | | |
| poor | asphalt concrete surface | 2.0 | 2.0 | 2.0 | 3.0 | 3.5 | 4.0 |
| | type III base | 4.5 | 5.0 | 8.5 | 12.5 | 15.0 | 19.0 |
| | total: | 6.5 | 7.0 | 10.5 | 15.5 | 18.5 | 23.0 |
| medium | asphalt concrete surface | 2.0 | 2.0 | 2.0 | 3.0 | 3.5 | 4.0 |
| | type III base | 2.0 | 2.5 | 5.5 | 9.5 | 12.0 | 15.5 |
| | total: | 4.0 | 4.5 | 7.5 | 12.5 | 15.5 | 19.5 |
| good to excellent | asphalt concrete surface | 2.0 | 2.0 | 2.0 | 3.0 | 3.5 | 4.0 |
| | type III base | 2.0 | 2.0 | 3.0 | 6.5 | 9.0 | 12.0 |
| | total: | 4.0 | 4.0 | 5.0 | 9.5 | 12.5 | 16.0 |

**Table 16.23**
Asphalt Pavements with Untreated Aggregate Base and Subbase

| subgrade class | pavement section | I | II | III | IV | V | VI |
|---|---|---|---|---|---|---|---|
| | | thickness in inches | | | | | |
| poor | asphalt concrete surface | 1.0 | 1.0 | 1.5 | 2.0 | 2.0 | 2.0 |
| | asphalt concrete base | 2.5 | 3.0 | 5.0 | 8.0 | 10.0 | 12.5 |
| | untreated aggregate base | 4.0 | 4.0 | 4.0 | 4.0 | 4.0 | 4.0 |
| | total: | 7.5 | 8.0 | 10.5 | 14.0 | 16.0 | 18.5 |
| medium | asphalt concrete surface | 1.0 | 1.0 | 1.5 | 2.0 | 2.0 | 2.0 |
| | asphalt concrete base | 2.0 | 3.0 | 2.5 | 5.0 | 7.5 | 10.0 |
| | untreated aggregate base | 4.0 | 4.0 | 4.0 | 4.0 | 4.0 | 4.0 |
| | total: | 7.0 | 8.0 | 8.0 | 11.0 | 13.5 | 16.0 |
| good to excellent | asphalt concrete surface | 1.0 | 1.0 | 1.5 | 2.0 | 2.0 | 2.0 |
| | asphalt concrete base | 2.0 | 3.0 | 2.5 | 3.0 | 5.0 | 8.0 |
| | untreated aggregate base | 4.0 | 4.0 | 4.0 | 4.0 | 4.0 | 4.0 |
| | total: | 7.0 | 8.0 | 8.0 | 9.0 | 11.0 | 14.0 |

(Columns I–VI: traffic classification.)

## 25 AASHTO METHOD OF RIGID PAVEMENT DESIGN

The AASHTO method of designing structural sections is based on modifications to the *Westergaard theory of stress distribution* in rigid slabs. There is no provision or adjustment for environment or weather (i.e., there is no regional factor), nor does the method specifically design the base or lower layers. It is assumed that the strengths assumed for the lower layers will be achieved by properly constructing those layers.[31]

*step 1*: Select the terminal serviceability, $p_t$. As with the AASHTO method for designing flexible pavements, $p_t = 2.5$ is appropriate for highways.

*step 2*: Determine the 20-year, 18-kip equivalent axle loading (EAL).

*step 3*: Select or determine the subbase material. AASHTO has specified six types of subbases, as listed in table 16.24, although others could be used. All types except type F can be used satisfactorily for the top 4 inches of subbase. (Type F can be used below the top 4 inches.)

*step 4*: Determine the *modulus of subgrade reaction* (also known as *gross k*) for the roadbed soil.

[31] Actually, only one support layer is assumed with the AASHTO method. AASHTO refers to this layer as the *subbase*, and omits the base layer. Others may refer to the layer as the *base* and omit the subbase.

**Table 16.24**
AASHTO Rigid Pavement Subbase Materials

| type | description |
|---|---|
| A | open graded |
| B | dense graded |
| C | cement treated (CTB) |
| D | lime treated (LTB) |
| E | bituminous treated (BTB) |
| F | granular |

*step 5*: Determine the *modulus of elasticity* of the concrete used in the rigid pavement. Typically, static compression tests on concrete cylinders are used to determine this parameter. (The original AASHTO studies used $E_c = 4.2$ EE6 psi.)

*step 6*: Determine the allowable working stress in the concrete. This is determined by dividing the 28-day flexural strength (i.e., *modulus of rupture*) from *third point loading tests* by a factor of safety.

$$f_t = \frac{MR}{FS} \qquad 16.39$$

FS is generally taken as 1.33, particularly where the 18-kip, 20-year EAL is less than 1,000,000. Where surface replacement would be inconvenient, or where local conditions require a thicker pavement, use FS = 2.0.

An alternate method of calculating the concrete working stress is to take 75% of the modulus of rupture.

$$f_t = 0.75 \times MR \qquad 16.40$$

*step 7*: Read the slab thickness from figure 16.14. Round up to the next whole number of inches.

*step 8*: If the slab thickness is less than 8 inches, specify steel reinforcement.

## 26 FATIGUE STRENGTH METHOD OF DESIGN

As long as the stress induced in the pavement is less than the *fatigue strength*, an infinite number of repetitions of that stress can be applied without damage to the pavement. For any given stress above the fatigue strength, there is an allowable *fatigue life*, or number of allowable repetitions. If the actual number of applications of a stress taken over 20 years is divided by the fatigue life for that stress, the *fatigue fraction* is obtained.

$$FF_i = \frac{n_i}{N_i} \qquad 16.41$$

If the sum of all fatigue fractions corresponding to different stresses is less than approximately 1.0, the pavement is adequate.

This method has been adopted by CALTRANS, and is based on fatigue strength ratios proposed by the Portland Cement Association. It is better suited for analysis of proposed pavement sections than for design, since layer thicknesses are required in the procedure.

*step 1*: Decide on a proposed pavement design. Determine the materials to be used and their thicknesses.

*step 2*: Obtain the values of all axle loads from the loadometer survey. The actual distribution of axle loads is needed.

*step 3*: If desired, multiply all of the axle loads by a *load safety factor* to provide a margin of safety for impact.

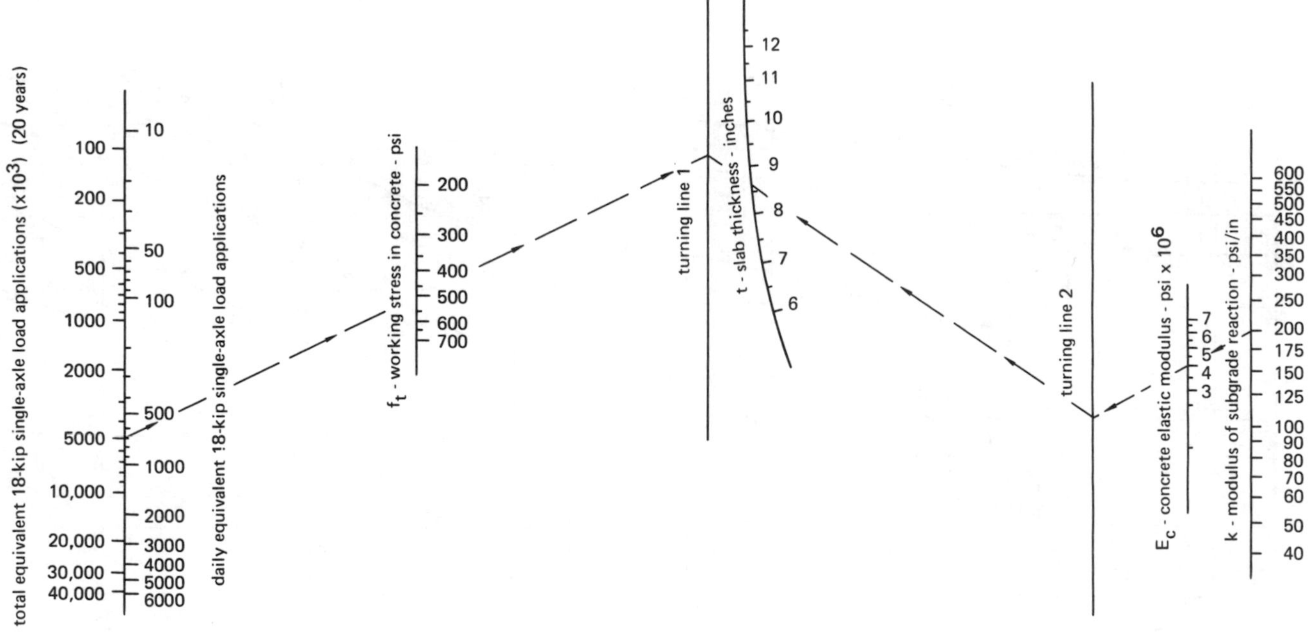


**Figure 16.14** AASHTO Rigid Pavement Design Chart ($p_t = 2.5$)

**Table 16.25**
Load Safety Factors

| type of facility | LSF |
|---|---|
| airport aprons, taxiways, runway ends, hangar floors | 1.7 to 2.0 |
| central portions of runways, and high-speed taxiways | 1.4 to 1.7 |
| outside lanes of multi-lane facilities with high truck traffic | 1.3 |
| inside lanes of multi-lane facilities with high truck traffic, and all lanes with moderate truck traffic | 1.2 |
| minor highways, frontage roads, and all streets with low truck traffic | 1.1 |
| residential streets or roads with occasional truck traffic | 1.0 |

*step 4*: Obtain the modulus of subgrade reaction, $k_s$, for the soil.

*step 5*: Based on the modulus of subgrade reaction of the soil, $k_s$, and the thickness of the subbase, determine a total modulus of reaction, $k_{sb}$, for the combined strength of subgrade soil and subbase. (If there is no subbase, this step is skipped and $k_s$ is used for $k_{sb}$.) Figure 16.15 can be used for this purpose.

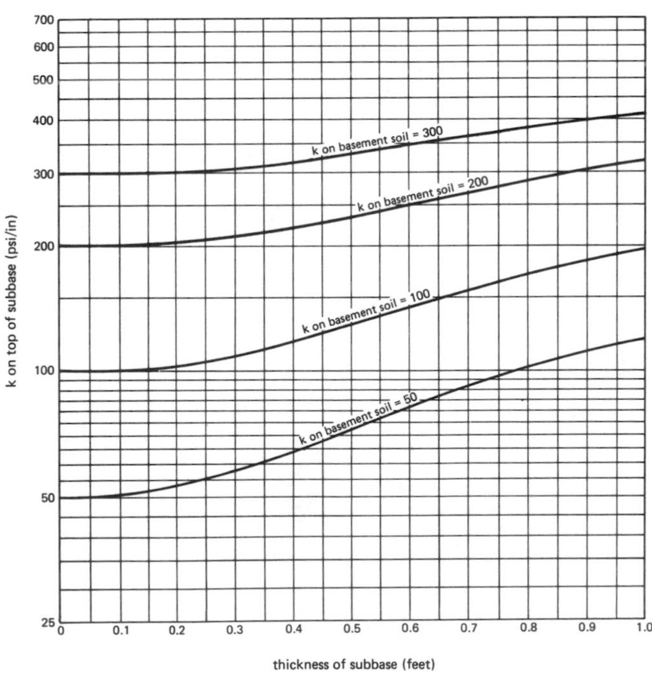


**Figure 16.15** $k$-Values for Base and Basement Soil Combination

(Do not use if subbase is omitted.)

*step 6*: Using the modulus of reaction for the subbase, $k_{sb}$, and the thickness of the base, use figure 16.16 or figure 16.17 to determine the total modulus of reaction on the base (under the rigid pavement).

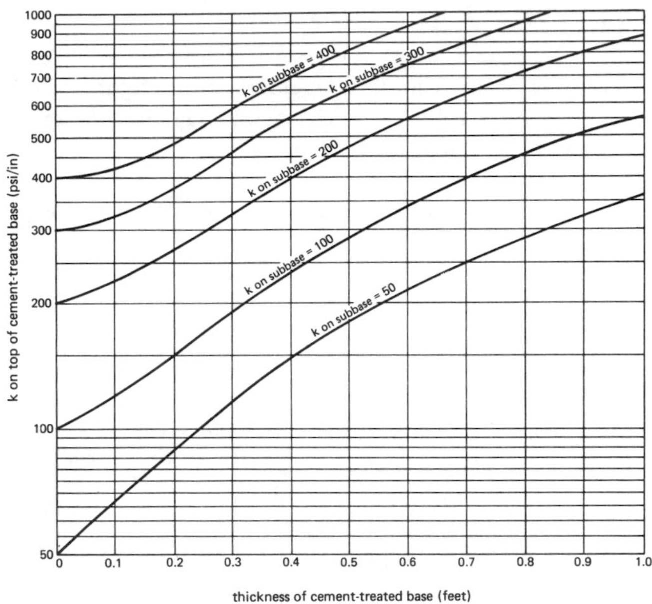


**Figure 16.16** Total $k$-Values Using Cement Treated Base

($k$ values are for basement soil if subbase is omitted.)

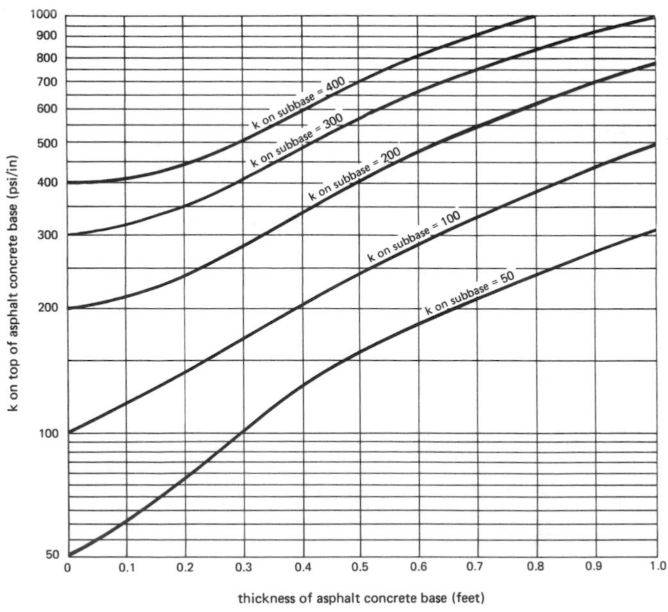


**Figure 16.17** Total $k$-Values Using Asphalt Treated Base

($k$ values are for basement soil if subbase is omitted.)

*step 7*: Determine the stress induced in the pavement from each category of axle loads. Figures 16.18 and 16.19 can be used for this purpose.

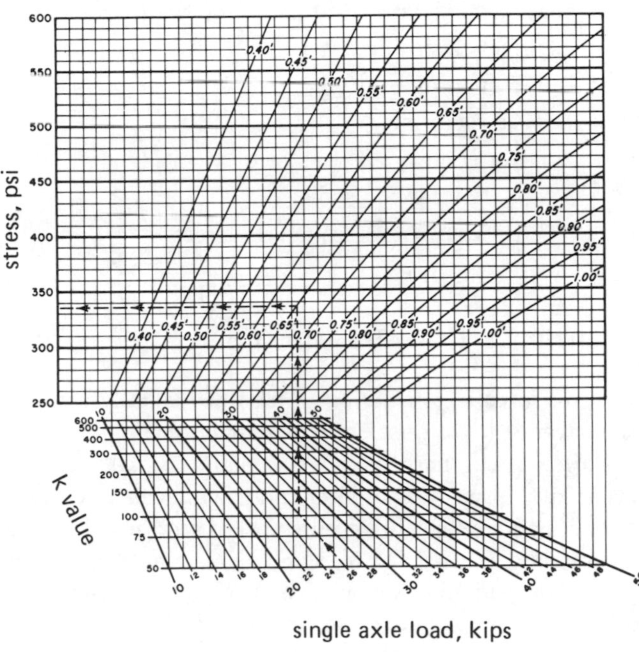


**Figure 16.18** Stress Chart for Single Axle Loads

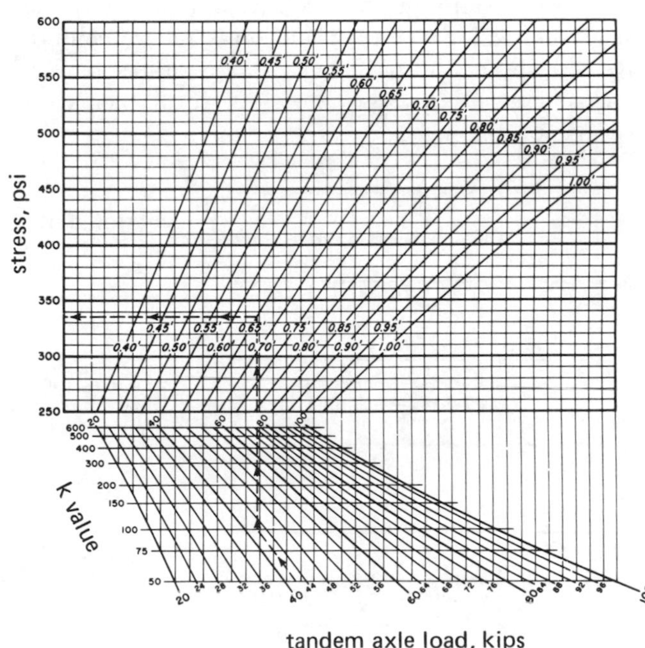


**Figure 16.19** Stress Chart for Tandem Axle Loads

*step 8*: Determine the 28-day *modulus of rupture* for the concrete to be used in the surface layer. (A minimum of 550 psi for a 28-day strength should be specified.)[32] The modulus of rupture is taken as the extreme fiber stress under the breaking load in a beam-breaking test.[33]

$$MR = \frac{Mc}{I} \qquad 16.42$$

*step 9*: Divide each category of stress value by the modulus of rupture. Record these numbers to the nearest 0.01. Values of 0.50 or less do not need to be calculated or recorded since this corresponds to the endurance strength of concrete, and unlimited repetitions are allowed.

*step 10*: For each category of stress, determine the fatigue life (allowable number of repetitions). Table 16.26 can be used for this purpose.

**Table 16.26**
Allowable Load Repetitions for Concrete

| stress ratio | allowable repetitions | stress ratio | allowable repetitions |
|---|---|---|---|
| 0.51 | 400,000 | 0.71 | 1500 |
| 0.52 | 300,000 | 0.72 | 1100 |
| 0.53 | 240,000 | 0.73 | 850 |
| 0.54 | 180,000 | 0.74 | 650 |
| 0.55 | 130,000 | 0.75 | 490 |
| 0.56 | 100,000 | 0.76 | 360 |
| 0.57 | 75,000 | 0.77 | 270 |
| 0.58 | 57,000 | 0.78 | 210 |
| 0.59 | 42,000 | 0.79 | 160 |
| 0.60 | 32,000 | 0.80 | 120 |
| 0.61 | 24,000 | 0.81 | 90 |
| 0.62 | 18,000 | 0.82 | 70 |
| 0.63 | 14,000 | 0.83 | 50 |
| 0.64 | 11,000 | 0.84 | 40 |
| 0.65 | 8000 | 0.85 | 30 |
| 0.66 | 6000 | 0.86 | 23 |
| 0.67 | 4500 | 0.87 | 17 |
| 0.68 | 3500 | 0.88 | 13 |
| 0.69 | 2500 | 0.89 | 10 |
| 0.70 | 2000 | 0.90 | 8 |

[32] The 90-day modulus of rupture is 110% of the 28-day modulus of rupture.

[33] Strictly speaking, the stress is not within the elastic range when the concrete fails. However, equation 16.42 is used anyway.

*step 11*: Divide the estimated number of repetitions over 20 years of each load category by the allowable number of repetitions from step 10 to obtain the *fatigue fraction*.[34]

*step 12*: Add all of the fatigue fractions to determine the fraction of fatigue strength used. The total fraction used should be between 1.0 and 1.1.[35,36]

$$1.0 < \frac{n_1}{N_1} + \frac{n_2}{N_2} + \frac{n_3}{N_3} + \cdots < 1.1 \qquad 16.43$$

## 27 ROADWAY DETAILING

The following geometric details are recommended.

- lane width: 12 feet, all freeways
  11 feet for restricted areas
- crown slope:
  portland cement concrete: 2%
  bituminous mix pavement: 2%
  penetration treated earth and gravel: $2\frac{1}{2}\% - 3\%$
  unsurfaced, graded: $2\frac{1}{2}\% - 3\%$
- shoulders: to the right of traffic: 10 feet (minimum of 6 feet)
  to the left of traffic:
  4 and 6 lanes: 5 feet
  8 lanes: 8 feet
- shoulder slope: 5% away from median
- maximum grade:
  3% freeways
  6% other roads
  2% steeper allowed in rugged terrain
- side slopes on adjacent cuts:
  freeways: 2:1 max (*h*:*v*)
  other roads: $1\frac{1}{2}$:1 max (*h*:*v*)
- cut-to-right-of-way clearance:
  10 feet minimum
  50 feet maximum
  20 feet for cuts 30 to 50 ft high
  25 feet for cuts 50 to 75 ft high
  (1/3) cut height above 75 feet
- divided median width:
  urban area freeways: 30 feet
  rural area freeways: 46 feet
  costly right of way and bridges: 4 feet
- median valley slopes: 10:1 to 20:1 (*h*:*v*)
- horizontal clearance to piers and walls:
  30 feet desirable
  10 feet minimum
- vertical clearance:
  major structures: $16\frac{1}{2}$ feet
  sign structures: 18 feet
  pedestrian overcrossing: $18\frac{1}{2}$ feet

## 28 JOINTS IN PAVEMENT

*Control joints* (*contraction joints*) are usually sawn in the pavement to a depth of one quarter of the slab thickness (hence the name, *weakened plane joints*). A crack eventually forms at that site, and the uneven crack joint allows load transfer between the slab sections. In areas where sand is used for ice control, a joint sealer compound or pad may be used in the slot. Contraction joints relieve tensile stresses in the pavement.

*Construction joints* (*contact joints*) are used at the end of a pour (as at the end of a day) or between pavement lanes.

*Isolation joints* (*expansion joints*) are used with premolded compressible fillers. Isolation joints separate slabs from wall, gutters, columns, and other nonpavement sections. Expansion joints relieve compressive stresses in the pavement. Many states have restricted use of expansion joints due to problems with pavement pumping.

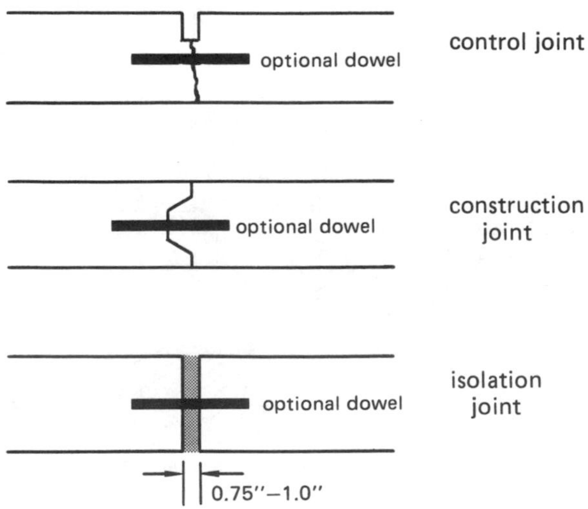


**Figure 16.20** Types of Joints

[34] If loadometer data provide current-day counts, include some provision for traffic growth. Economic analysis tables can be used if the annual growth rate is known.

[35] If the fraction of fatigue strength used is significantly less than 1.0, the pavement is overdesigned.

[36] California allows the maximum to be 1.25.

If *dowel bars* are used as load transfer devices in expansion joints, one side of the dowel should be lubricated. This will allow the slab to slip over the dowel as the slab expands and contracts.

*Hinge joints* (also known as *warping joints*) are similar to control joints (and, they may also be constructed as any of the other types of joints). However, hinge joints are longitudinal joints, generally placed along the centerline of the highway.

## 29 GROOVING PAVEMENTS

A proven method of increasing skid resistance and reducing hydroplaning is by grooving. Grooves drain water laterally, permit water to escape under tires, prevent build-up of surface water, and increase the pavement texture. Grooving should be used only with structurally adequate pavements free from defects. If the pavement is in poor condition, rehabilitation is required prior to grooving.

Grooves should be continuous. In the case of special surfaces, such as airfields, transverse rather than longitudinal grooves may be used. The recommended groove is a square cut with $\frac{1}{4}'' \times \frac{1}{4}''$ dimensions, and a center-to-center spacing of $1\frac{1}{4}''$. The minimum spacing is $1\frac{1}{8}''$; the maximum is $2''$.

## 30 GEOTEXTILES

Geotextiles are support and filter fabrics that are placed in contact with the soil to stabilize and retain the soil. Geotextiles are also known as *filter cloth*, *reinforcing fabric*, and *support membrane*. Modern geotextiles are not subject to biological and chemical degradation. They can be made from wood pulp (rayon and acetate), silica (fiberglass), and petroleum (polyamides, polyester, and polypropylene). There are woven and non-woven varieties.

Geotextiles are used to prevent dissimilar materials, such as an aggregate base and subbase or soil, from mixing, thereby reducing the support strength. They also reinforce layers, since geotextiles may contribute a substantial tensile strength. Geotextiles provide a filtering function, allowing free passage of water while restraining soil movement.

A significant fraction of all geotextiles is used in highway repair. A layer of geotextile spread at the base of a new road saves on materials such as sand and gravel, while helping to control drainage. Geotextiles can also be used to prevent the infiltration of fine clays into underdrains. In another application, geotextiles appear to strengthen flexible pavements when placed directly under the surface layer.

## 31 SUBGRADE DRAINAGE

Subgrade drains should be considered whenever the following conditions exist:

- high ground-water levels which may reduce subgrade stability and provide a source of water for frost action
- subgrade soils of silts and very fine sands which may become quick or spongy when saturated
- water seeping from underlying water-bearing strata
- cuts in terrain that intercept the natural drainage path of higher elevations
- sag curves with low-permeability subgrade soil

Figure 16.21 illustrates subgrade drains. In general, drains should not be located too close to the pavement (to prevent damage to one when working on the other), and some provision should be made to prevent the infiltration of silt and fines into the drain. (Roofing felt or geotextiles can be used for this purpose.)

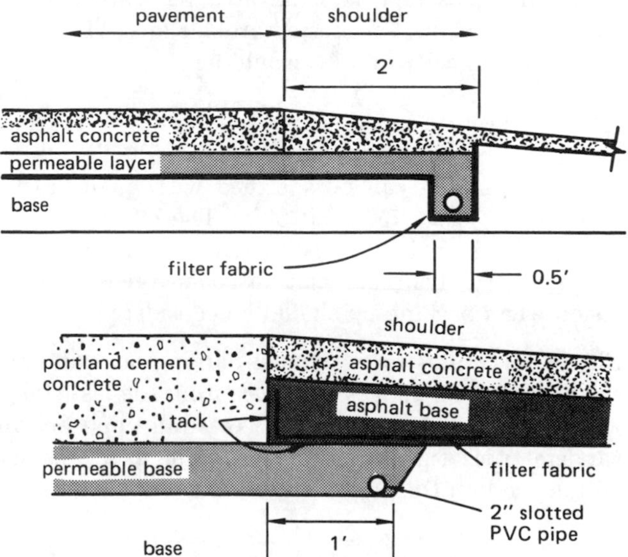


**Figure 16.21** Typical Subgrade Drain Details

## 32 CONTROLLING FROST DAMAGE

Frost heaving and reduced subgrade strength (and accompanying pumping) during spring thaw can quickly destroy a pavement. The following techniques are used to reduce damage done in frost susceptible areas:

- constructing stronger (thicker) pavement sections
- lowering the water table by use of subdrains and drainage ditches

- using layers of coarse sands or waterproof sheets to reduce capillary action
- removing and replacing frost susceptible materials down through the zone of frost penetration
- using rigid foam sheets to insulate and reduce the depth of frost penetration

## 33 PARKING DESIGN

Street parking stall width is commonly taken as 8 feet, particularly in commercial and industrial areas, although widths from 7 to 10 feet are also used. (7-foot stalls are considered to be substandard and should be limited to attendant-parked lots.) Widths larger than 9 feet are appropriate in shopping market areas where package loading is expected. If the stall width is specified as 12 feet, however, the parking lane can be converted to an extra traffic lane when necessary. The minimum length of a stall is 18 feet, which will accommodate most modern cars. (Older and large luxury cars may require 20-foot-long stalls. Compact cars will require 16-foot-long stalls.)

In designing parking lots, an area per car of 320 to 380 $ft^2$ will allow for access through lots. Thus, the maximum capacity of a lot would be

$$\text{capacity} = \frac{\text{lot area}}{320} \qquad 16.44$$

Diagonal parking can be specified with angles to the curb of 45°, 60°, 75°, and 90°. The effects of diagonal parking on lane width can be determined from trigonometry. The significant disadvantage of impaired vision when backing up should be considered.

Figure 16.22 illustrates parallel parking near an intersection. The 15–20 feet clearance from the last stall to the intersection is required for crosswalks and bus and tandem truck turning movement. The exact distance will depend on the vehicles expected.

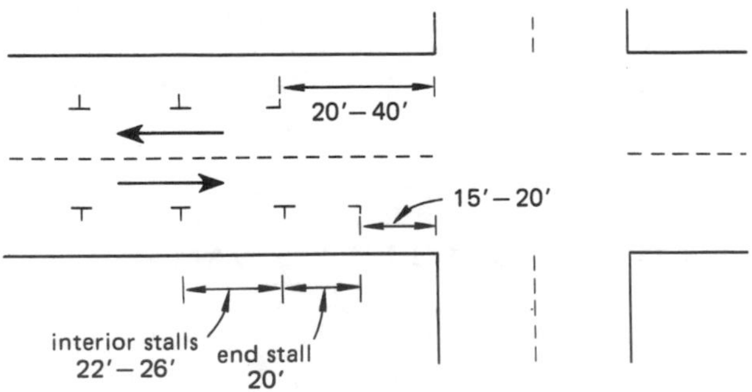


**Figure 16.22** Parallel Parking Design[37]

[37] If the space is available, leave an 8 to 10 foot opening every two spaces. This will provide sight through from the driver to the sidewalk and cross streets.

## 34 INTERSECTION SIGNALING

### A. CONDITIONS REQUIRING SIGNALING

Signaling should be considered only when one of the following conditions is present.[38]

- High traffic volume. For major streets, this is 500 to 600 vehicles per hour, counted over both directions and all lanes. For minor streets, this is 150 to 200 vehicles per hour counted over one direction, all lanes.
- Interruption of traffic. Cross street traffic cannot enter or cross the main traffic flow.
- High crosswalk usage
- Nearby school crossing
- Need to regulate speed of cars and prevent "platooning" of flow
- Excessive accident activity
- Need to combine two roads into one, or allow orderly entrance into a higher-speed road

### B. SIGNAL CONTROLLERS

*Fixed-time controllers* are the least expensive and simplest to use. They are most efficient if the traffic can be accurately predicted. (With multi-dial controllers, cycle lengths can change at different times of the day.) Fixed-time controllers are necessary if sequential intersections or intersections less than one-half mile apart are to be coordinated.

*On-demand controllers* (traffic-activated controllers) are more expensive. However, activated controllers do a better job of controlling flow, and they are better accepted by drivers.

### C. DETERMINING FIXED-TIME CYCLE LENGTHS

In general, the maximum cycle length is 120 seconds. No cycle should be less than 35 seconds.

Green cycle lengths with fixed-time controllers should be chosen to clear all waiting traffic in 95% of the cycles. (Usually, the 85th percentile speed is used in preliminary studies.) Since the green cycle must handle peak loads, the efficiency of the installation is sacrificed during the rest of the day, unless the cycle length is changed.

Although queuing models and simulation can be used to determine cycle lengths, the following simplified procedure can be worked manually.

[38] These reasons are known as *warrants*.

**Table 16.27**
Cycle Length Chart

(Time in Seconds)

| vehicles per hour y-direction | vehicles per hour, x-direction | | | | | | | | | |
|---|---|---|---|---|---|---|---|---|---|---|
| | 100 | 200 | 300 | 400 | 500 | 600 | 700 | 800 | 900 | 1000 |
| 100 | 35 | 35 | 35 | 35 | 35 | 40 | 45 | 50 | 55 | 60 |
| 200 | 35 | 35 | 35 | 35 | 40 | 45 | 55 | 65 | 75 | 85 |
| 300 | 35 | 35 | 35 | 40 | 45 | 60 | 75 | 90 | 110 | — |
| 400 | 35 | 35 | 40 | 45 | 60 | 80 | 100 | 120 | — | — |
| 500 | 35 | 40 | 45 | 60 | 80 | 110 | — | — | — | — |
| 600 | 40 | 45 | 60 | 80 | 110 | — | — | — | — | — |
| 700 | 45 | 55 | 75 | 100 | — | — | — | — | — | — |
| 800 | 50 | 65 | 90 | 120 | — | — | — | — | — | — |
| 900 | 55 | 75 | 110 | — | — | — | — | — | — | — |
| 1000 | 60 | 85 | — | — | — | — | — | — | — | — |

*step 1*: Determine the lanes carrying the greatest number of vehicles for both $x$- and $y$-directions. All other lanes will carry fewer vehicles, and these lanes will have enough time.

*step 2*: Determine the number of *car equivalents* for the two maximum lanes. Different methods exist for converting buses and trucks to car equivalents. Equation 16.45 is a simplified approach.[39]

$$E = \#\text{cars} + (1.5)(\#\text{buses}) + (1.5)(\#\text{trucks}) + (1.6)\left(\begin{array}{c}\#\text{ vehicles}\\ \text{turning left}\end{array}\right) \qquad 16.45$$

*step 3*: Use table 16.27 to obtain the cycle length.

*step 4*: Determine the time split between the $x$- and $y$-directions. The split is proportional to the traffic flow. For example, the $x$-direction combined green and amber time is proportional to $E_x/(E_x + E_y)$.

*step 5*: Specify the amber time. Usually, the yellow time is 3 to 6 seconds. Higher speeds should be given the higher yellow times.

*step 6*: Provide for a short all-red clearance interval after the yellow light to clear the intersection.

*step 7*: Check for pedestrian needs. People walk approximately 4 ft/sec and require a starting time of approximately 5 seconds.[40] Determine if pedestrians can cross the street within the green cycle time. (A short all-way red clearance interval is suggested after the green walk signal terminates.)

## D. DETERMINING ON-DEMAND TIMING

Since cycle changes with on-demand (activated) controllers depend on the arriving traffic volume, no traffic counts are needed. Four parameters must be specified: the initial allowance, vehicle time allowance, maximum time allowance, and clearance allowance.

- **Initial Period:** The initial period must allow enough time for traffic stopped between the stop line and detector to begin moving. If a 20 foot car length is used, the number of cars between the stop line and detector is

$$\begin{array}{c}\#\text{cars in}\\ \text{initial period}\end{array} = \frac{\text{distance between line and detector}}{20} \qquad 16.46$$

The first car can be assumed to cross the stop line 5 seconds after the green light appears. The next car requires 3 seconds more. Subsequent cars between

[39] The passenger car equivalents used in left-turn analyses are not the same as were used in freeway capacity analyses.

[40] Actually, some people can walk as fast as 7 ft/sec. However, only 30 to 40% of people walk slower than 4 ft/sec. If a significant percentage of walkway users are elderly, the design speed for walking should be reduced to 3 ft/sec.

the detector and stop line require $2\frac{1}{4}$ seconds.[41] The initial period includes time for cars between the two lines.

- **Vehicle Period:** The vehicle period must be long enough to allow a car crossing the detector (moving at the slowest reasonable speed) to get to the intersection before the yellow light appears. (It is not necessary to have the vehicle get through the intersection. That is the purpose of the yellow period.) This period can be calculated from distance and velocity. In a 30 mph zone, a speed of approximately 20 mph into the intersection is reasonable.
- **Maximum Period:** The maximum period is the maximum delay that the opposing traffic can tolerate. 60 seconds is typical for a main street. 30 to 40 seconds is appropriate for a side street. It should never be greater than 120 seconds.
- **Yellow Period:** The yellow clearance period can be determined from the time required to perceive, brake, and stop the vehicle, and the assumed average speed into the intersection.
- **Green Period:** The green period would be the smaller of the sum of initial and vehicle periods and the maximum period.

E. TIME-SPACE DIAGRAMS

Time-space diagrams are used to coordinate successive fixed-length controllers. This requires setting the controller's *offset*. Offset is the time from the end of one controller's green cycle to the end of the next controller's green cycle. To minimize travel delay, the offset should be minimized.

*step 1*: Choose a scale. $1'' = 100'$ and $1'' = 200'$ are typical scales.

*step 2*: Draw the main and intersecting streets to scale. For signals separated by short distances, draw the cross-street widths accurately.

*step 3*: Assume or obtain the average travel speed along the main street.

*step 4*: Make an initial assumption for the cycle length. For a two-way street, the cycle length should be either two times (*alternate mode*) or four times (*double alternate mode*) the travel time at the average speed between intersections of average separation. The general timing guidelines (35 second minimum, 120 second maximum) apply here. Two times the travel time is preferred over four times.

*step 5*: Check to see if the assumed cycle can handle the heaviest traveled intersection. Use table 16.27 by entering with the main street flow and cycle time. Find the cross street flow. If the cycle time does not allow sufficient cross street flow, a different cycle time will be required.

*step 6*: For each intersection, determine the *split*. The procedure for determining the split is the same as for the isolated fixed signal—on the basis of the equivalent car traffic.

*step 7*: Try to minimize the conflict between green and red (with yellow) periods at adjacent intersections. This can be done graphically by cutting strips of paper for each intersection, and marking off green, yellow, and red periods according to the time scale. The strips are placed on the time-space diagram with offsets determined by the average speed and average intersection separation.

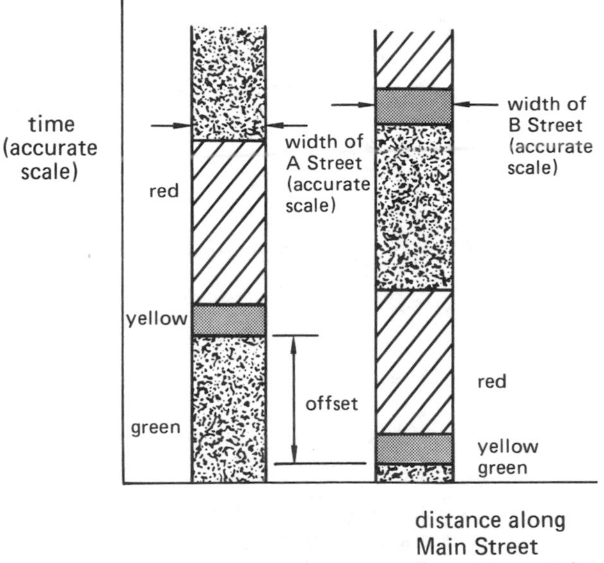


**Figure 16.23** A Time-Space Diagram Showing Offset for Two Streets

For alternate mode operation, every other signal will be green at the same time. For double alternate mode operation, two adjacent pairs of signals will have the same color (i.e., either red or green) while the following

[41] Actual studies have shown the average *start-up lost time* and average *arrival headway* to be approximately half of these values. These values allow for the slowest of drivers.

two adjacent signals will have the opposite color (i.e., either green or red).

Figure 16.24 shows a completed time-space diagram with diagonal lines drawn between green cycle limits. The diagonal lines indicate the *green window* of travel.

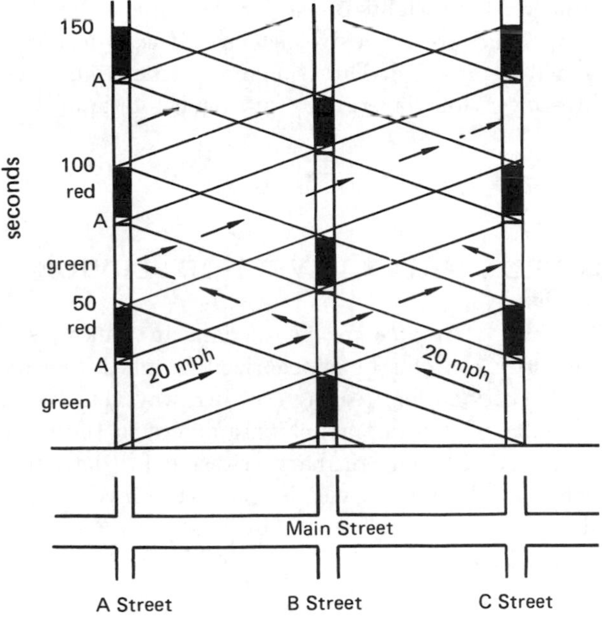


**Figure 16.24** Time-Space Diagram

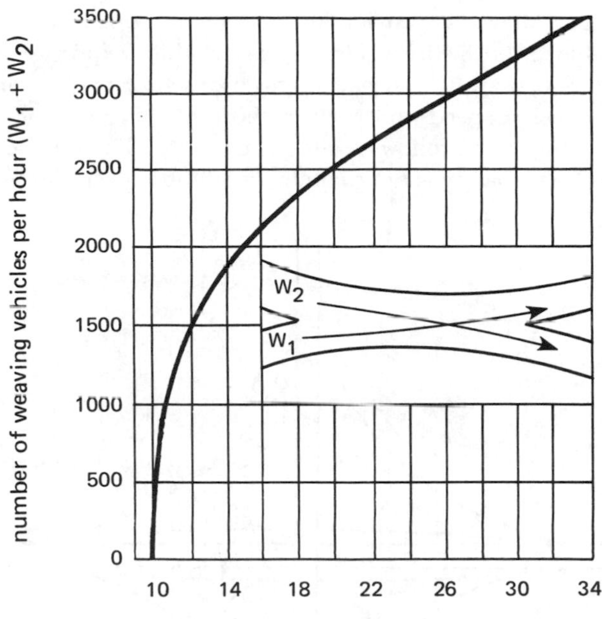


**Figure 16.25** Recommended Weaving Section Lengths[42]

Diamond interchanges force traffic using the ramp to substantially reduce its speed. They cannot be used for highway-to-highway intersections, because the capacity per ramp is limited.[43]

## 35 HIGHWAY INTERCHANGE DESIGN

An interchange allows traffic to enter or leave a highway. Locations of interchanges are affected by the volume of traffic expected on the interchange, convenience, and required land use area. The frequency of interchanges along a route should be sufficient to allow weaving between the interchanging traffic. Figure 16.25 can be used to determine the minimum distance required to merge traffic entering the highway from one on-ramp with traffic headed for the off-ramp of the next interchange.

*Diamond interchanges* are suitable for major road-minor road intersections. They lend themselves to stage construction. The frontage roads and/or ramps can be constructed, while the freeway lanes can be built at a later date. The right-of-way costs are low, since little additional area around the freeway is required. If designed correctly, diamond interchanges will handle fairly large volumes.

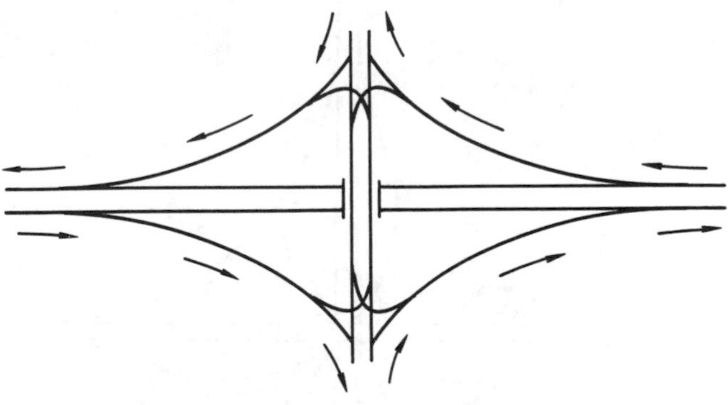

**Figure 16.26** Diamond Interchange

[42] The *Highway Capacity Manual* presents a much more detailed method of analyzing weaving sections.

[43] Up to 1000 vehicles per hour may be able to use a ramp. However, the limiting factor may be the intersection beyond the ramp.

*Cloverleaf interchanges* allow non-stop left-turn movement. They also provide for free flow by separating the traffic in both directions. However, they require large rights-of-way. Cloverleaf intersections slow traffic from the design speed and require short weaving distances. Turning traffic follows a circuitous route. The practical limit of capacity is approximately 800–1000 vehicles per hour.

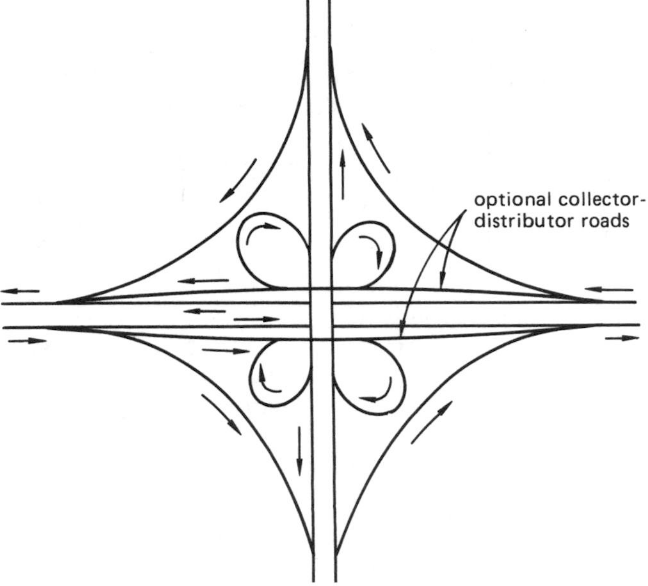


**Figure 16.27** Cloverleaf Interchange

The cloverleaf interchange should not be used to connect two freeways, since it cannot handle the large volumes of turning traffic.

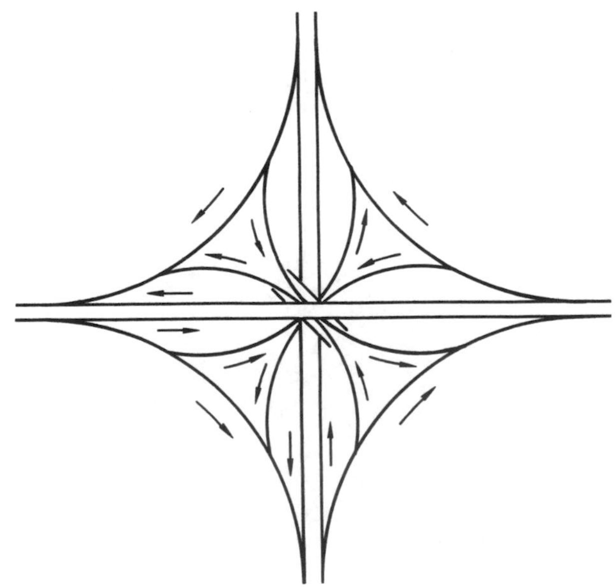

**Figure 16.28** A Four-Layer Directional Interchange

Both diamond and cloverleaf intersections can be improved by various means, such as the use of a third level and *collector-distributor roads* to increase the speed and volume of weaving sections.

*Directional interchanges* allow direct or semi-direct connections for left-turn movements. The design speeds of connections normally are near the design speed of through lanes, and large traffic volumes can be handled. However, large rights-of-way are required. Directional interchanges are expensive because of the structures and capacities provided. Directional interchanges may be of three-layer, four-layer, or rotary bridge design.

## 36 PEDESTRIAN LEVELS OF SERVICE

The level of service for pedestrians in walkways and queuing areas can be categorized in much the same way as is done for freeway and highway vehicles. Table 16.28 relates important parameters to the level of service (LOS). The primary criterion for determining pedestrian level of service is space (the inverse of density).

**Table 16.28**
Pedestrian Levels of Service in Walkways and Queuing Areas

| | walkways | | queuing areas | |
|---|---|---|---|---|
| LOS | space ($ft^2$/ped) | flow rate (ped/min-ft)* | area (sq ft/person) | spacing (ft) |
| A | $\geq 130$ | $\leq 2$ | $\geq 13$ | $\geq 4$ |
| B | $\geq 40$ | $\leq 7$ | 10–13 | 3.5–4.0 |
| C | $\geq 24$ | $\leq 10$ | 7–10 | 3.0–3.5 |
| D | $\geq 15$ | $\leq 15$ | 3–7 | 2.0–3.0 |
| E | $\geq 6$ | $\leq 25$ | 2–3 | $\leq 2$ |
| F | $\leq 6$ | – | $\leq 2$ | – |

* pedestrians per minute per foot width of walkway

## 37 ECONOMIC JUSTIFICATION OF HIGHWAY SAFETY FEATURES

Much can be done to improve the safety of some sections of highway. Features such as breakaway poles, cushioned barriers, barriers separating two directions of traffic, and direction channeling away from abutments are common. These features must be economically jus-

**Table 16.29**
Economic Analysis of Accidents
(average dollar values)

| accident element | U.S. National Safety Council (1972)* | U.S. Department of Defense (1975) | Office of Management and Budget (1984) | OSHA (1984) | OSHA/OMB (1985) |
|---|---|---|---|---|---|
| fatality | 330,000 | 287,000 | 1,000,000 | 3,500,000 | 2–5,000,000 |
| non-fatal injury | 3400 | 8100 | | | |
| property damage | 480 | 520 | | | |

*Average over rural and urban accidents

tified, particularly where they are to be retrofitted to existing highways. The justification is usually the value of personal and property damage avoided by the installation of such features.[44]

There are three general classifications of accidents—those involving death with property damage, injury with property damage, and property damage only. The cost of each element (i.e., death, injury, and property damage) can be evaluated from insurance records, court awards, state disability records, and police records. Also, federal agencies such as the U.S. National Safety Council and OSHA are active in monitoring such costs.

Table 16.29 lists representative values which have been associated with the three elements of accidents.

There are also significant variations in the accident values, depending on the age, sex, and location of the accident. Fatalities of women and children may be valued at rates as low as 60% of their male counterparts. Individuals over 55 have been valued as low as 15% of their prime-aged counterparts. Accidents in developed and urban areas bring awards as much as twice the average award, and three times the equivalent accident in a rural setting.

## 38 QUEUING MODELS

*Special Nomenclature*

| | |
|---|---|
| L | expected system length (includes service) |
| $L_q$ | expected queue length |
| p{n} | probability of $n$ customers in the system |
| s | number of parallel servers |
| W | expected time in the system (includes service) |
| $W_q$ | expected time in the queue |
| $\lambda$ | mean arrival rate |
| $\rho$ | traffic intensity $= (\lambda/\mu)$ and must be less than $s$ |
| $\mu$ | mean service rate per server |

[44] Nobody likes the concept of saying a human life "... is worth such and such an amount ..." This subject is included, however, to illustrate how economic justifications are made.

*Queue* is a technical word for a waiting line. Queueing theory can be used to predict the length of waiting time, the average time a customer can expect to spend in the queue, and the probability that a given number of customers will be in the queue.

Many queueing models have been developed. Most of these models are fairly specialized and complex. However, two models are important. The relationships given below are for steady state operation, which means that the service facility has been open and in operation for some time.

### A. GENERAL RELATIONSHIPS

The following simple relationships are valid for all queueing models.

$$L = \lambda W \qquad 16.47$$

$$L_q = \lambda W_q \qquad 16.48$$

$$W = W_q + \frac{1}{\mu} \qquad 16.49$$

$$\lambda < \mu s \qquad 16.50$$

$$\text{average service time} = \frac{1}{\mu} \qquad 16.51$$

$$\text{average time between arrivals} = \frac{1}{\lambda} \qquad 16.52$$

### B. THE M/M/1 SYSTEM

It is assumed that the following are true for the M/M/1 system.

- There is only one *server* ($s = 1$).

- The *calling population* is infinite.
- The *service times* are *exponentially distributed* with mean $\mu$. That is, the probability of a customer's remaining service time exceeding $h$ (after already spending time with the server) is given by equation 16.53.

$$p\{t > h\} = e^{-\mu h} \qquad 16.53$$

Notice that equation 16.53 is independent of the time already spent with the server. This result holds true regardless of the elapsed service time. The specific *service time distribution* is

$$f(t) = \mu e^{-\mu t} \qquad 16.54$$

- The *arrival rate* is distributed as *Poisson* with mean $\lambda$. The probability of $x$ customers arriving in the next period is

$$p\{x\} = \frac{e^{-\lambda}\lambda^x}{x!} \qquad 16.55$$

The following relationships describe the M/M/1 system.

$$p\{0\} = 1 - \rho \qquad 16.56$$

$$p\{n\} = p\{0\}(\rho)^n \qquad 16.57$$

$$W = \frac{1}{\mu - \lambda} = W_q + \frac{1}{\mu} = \frac{L}{\lambda} \qquad 16.58$$

$$W_q = \frac{\rho}{\mu - \lambda} = \frac{L_q}{\lambda} \qquad 16.59$$

$$L = \frac{\lambda}{\mu - \lambda} = L_q + \rho \qquad 16.60$$

$$L_q = \frac{\rho\lambda}{\mu - \lambda} \qquad 16.61$$

*Example 16.7*

Given an M/M/1 system with $\mu = 20$ customers per hour and $\lambda = 12$ per hour, find the steady state value of $W$, $W_q$, $L$, and $L_q$. What is the probability that there will be 5 customers in the system?

$$\rho = \frac{12}{20} = 0.6$$

$$W = \frac{1}{20 - 12} = 0.125 \text{ hours}$$

$$W_q = \frac{0.6}{20 - 12} = 0.075 \text{ hours}$$

$$L = \frac{12}{20 - 12} = 1.5 \text{ customers}$$

$$L_q = \frac{(0.6)(12)}{20 - 12} = 0.9 \text{ customers}$$

$$p\{0\} = 1 - 0.6 = 0.4$$

$$p\{5\} = 0.4(0.6)^5 = 0.031$$

## C. THE M/M/s SYSTEM

The same assumptions are used for the M/M/s system as were used for the M/M/1 system except that there are $s$ servers instead of only 1. Each server has a mean service rate $\mu$. Each server draws from a single line so that the first person in line goes to the first (any) server that is available. Each server does not have its own line.

However, if customers are allowed to change the lines they are in so that they go to any available server, this model also can predict the performance of a multiple server system where each server has its own line.

$$W = W_q + \frac{1}{\mu} \qquad 16.62$$

$$W_q = \frac{L_q}{\lambda} \qquad 16.63$$

$$L_q = \frac{p\{0\}\,\rho^{s+1}}{s!(1-\rho)^2} \qquad 16.64$$

$$L = L_q + \rho \qquad 16.65$$

$$p\{0\} = \frac{1}{\dfrac{(\rho)^s}{s!(1 - (\rho/s))} + \displaystyle\sum_{j=0}^{s-1}\frac{(\rho)^j}{j!}} \qquad 16.66$$

$$p\{n\} = \frac{p\{0\}(\rho)^n}{n!} \quad (n \le s) \qquad 16.67$$

$$p\{n\} = \frac{p\{0\}(\rho)^n}{s!s^{n-s}} \quad (n > s) \qquad 16.68$$

# Appendix A: Approximate Correlation between California Bearing Ratio and Subgrade Modulus

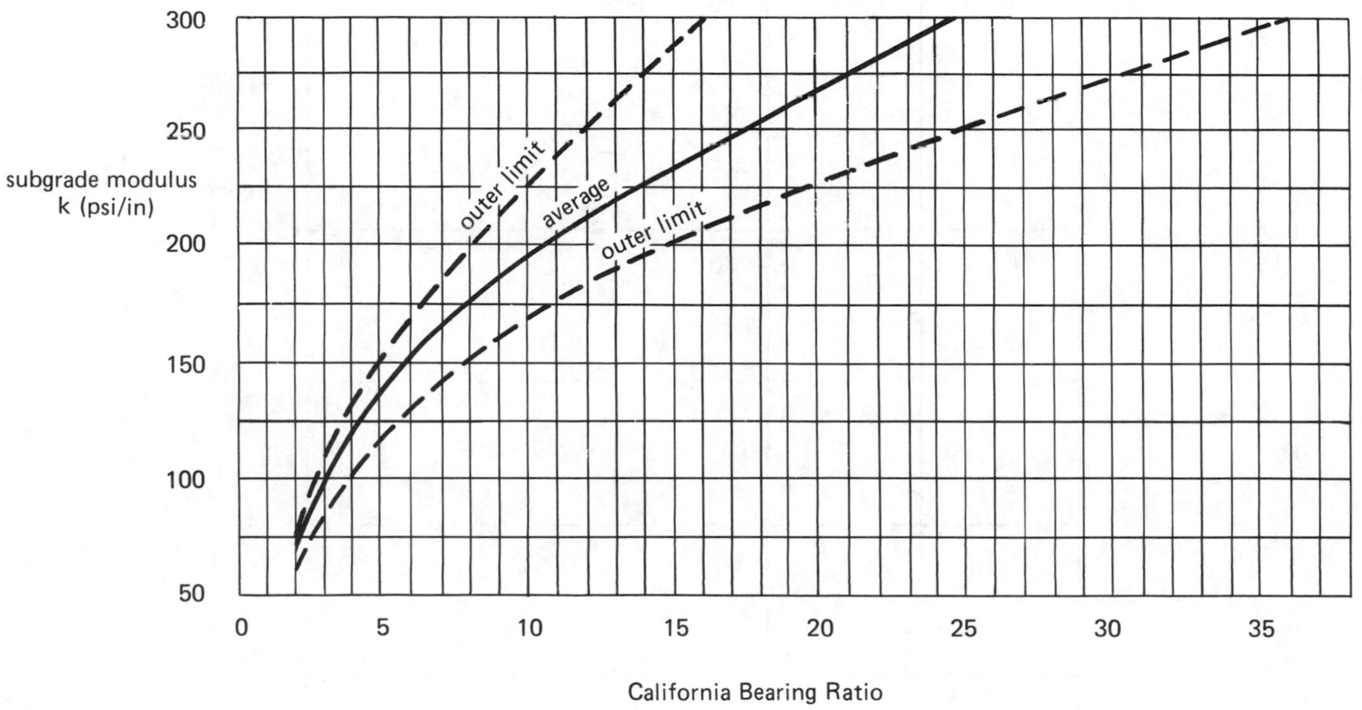

# Appendix B: Revised Soil Support Correlations

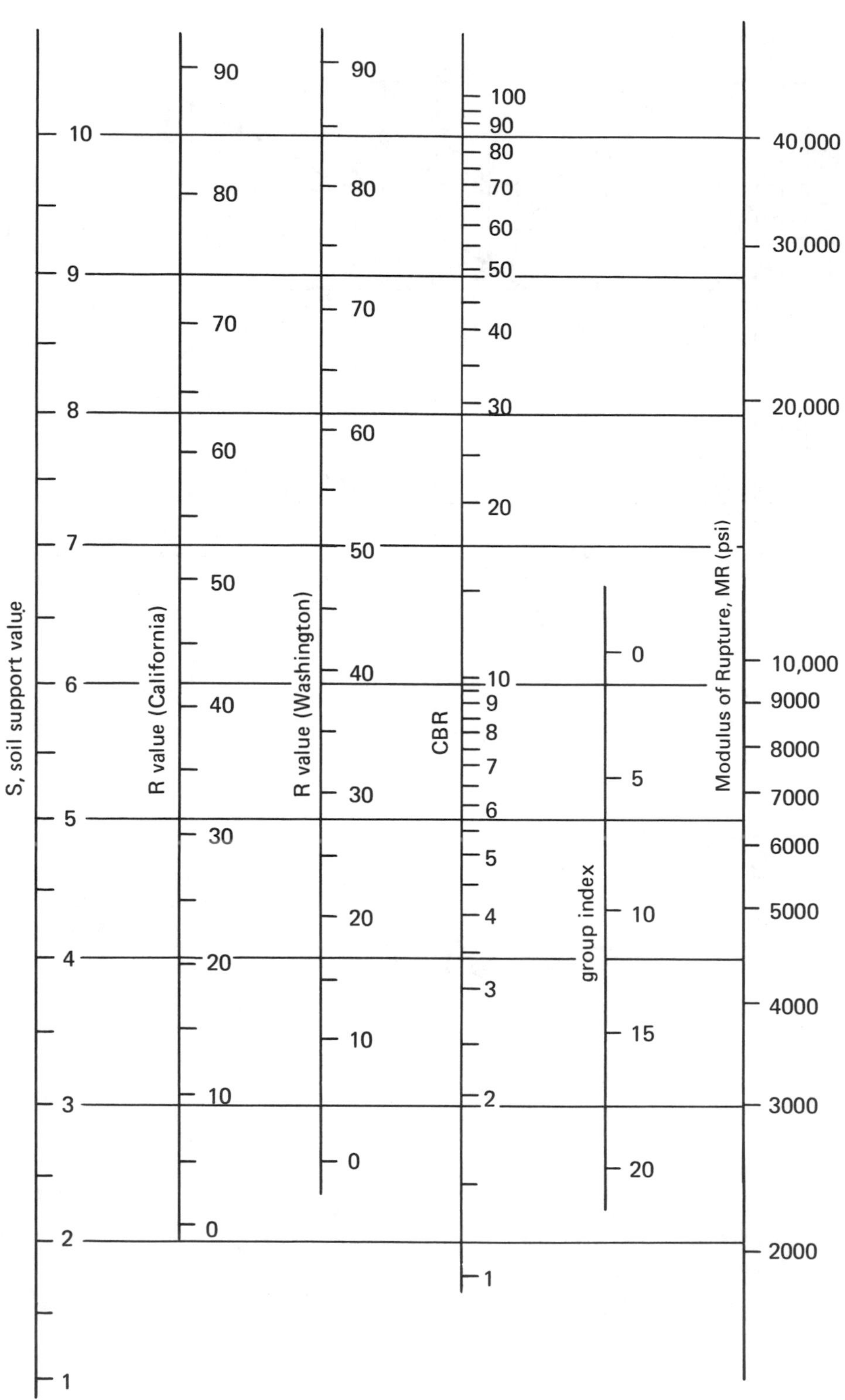

# Appendix C: Approximate Correlation between Subgrade Modulus and Soil R-value

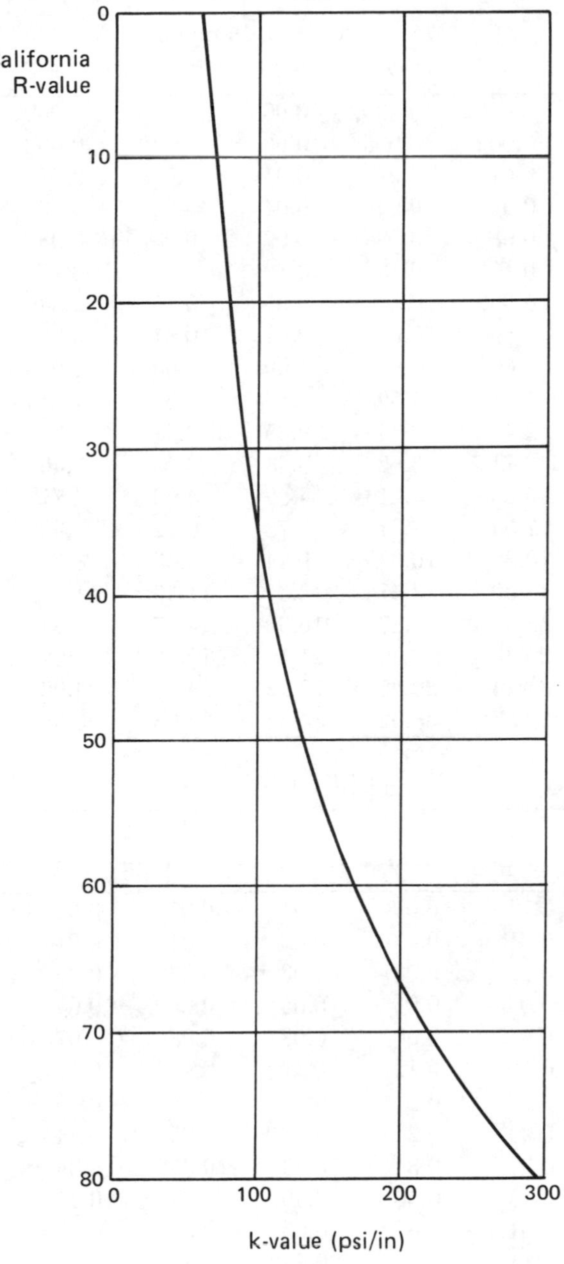

# Appendix D: AASHTO Equivalence Factors—Flexible Pavement

single axles, $p_t = 2.0$

| axle load (kips) | structural number, $SN$ 1 | 2 | 3 | 4 | 5 | 6 |
|---|---|---|---|---|---|---|
| 2 | 0.0002 | 0.0002 | 0.0002 | 0.0002 | 0.0002 | 0.0002 |
| 4 | 0.002 | 0.003 | 0.002 | 0.002 | 0.002 | 0.002 |
| 6 | 0.01 | 0.01 | 0.01 | 0.01 | 0.01 | 0.01 |
| 8 | 0.03 | 0.04 | 0.04 | 0.03 | 0.03 | 0.03 |
| 10 | 0.08 | 0.08 | 0.09 | 0.08 | 0.08 | 0.08 |
| 12 | 0.16 | 0.18 | 0.19 | 0.18 | 0.17 | 0.17 |
| 14 | 0.032 | 0.34 | 0.35 | 0.35 | 0.34 | 0.33 |
| 16 | 0.59 | 0.60 | 0.61 | 0.61 | 0.60 | 0.60 |
| 18 | 1.00 | 1.00 | 1.00 | 1.00 | 1.00 | 1.00 |
| 20 | 1.61 | 1.59 | 1.56 | 1.55 | 1.57 | 1.60 |
| 22 | 2.49 | 2.44 | 2.35 | 2.31 | 2.35 | 2.41 |
| 24 | 3.71 | 3.62 | 3.43 | 3.33 | 3.40 | 3.51 |
| 26 | 5.36 | 5.21 | 4.88 | 4.68 | 4.77 | 4.96 |
| 28 | 7.54 | 7.31 | 6.78 | 6.42 | 6.52 | 6.83 |
| 30 | 10.38 | 10.03 | 9.24 | 8.65 | 8.73 | 9.17 |
| 32 | 14.00 | 13.51 | 12.37 | 11.46 | 11.48 | 12.07 |
| 34 | 18.55 | 17.87 | 16.30 | 14.97 | 14.87 | 15.63 |
| 36 | 24.20 | 23.30 | 21.16 | 19.28 | 19.02 | 19.93 |
| 38 | 31.14 | 29.95 | 27.12 | 24.55 | 24.03 | 25.10 |
| 40 | 39.57 | 38.02 | 34.34 | 30.92 | 30.04 | 31.25 |

tandem axles, $p_t = 2.0$

| axle load (kips) | structural number, $SN$ 1 | 2 | 3 | 4 | 5 | 6 |
|---|---|---|---|---|---|---|
| 10 | 0.01 | 0.01 | 0.01 | 0.01 | 0.01 | 0.01 |
| 12 | 0.01 | 0.02 | 0.02 | 0.01 | 0.01 | 0.01 |
| 14 | 0.02 | 0.03 | 0.03 | 0.03 | 0.02 | 0.02 |
| 16 | 0.04 | 0.05 | 0.05 | 0.05 | 0.04 | 0.04 |
| 18 | 0.07 | 0.08 | 0.08 | 0.08 | 0.07 | 0.07 |
| 20 | 0.10 | 0.12 | 0.12 | 0.12 | 0.11 | 0.10 |
| 22 | 0.16 | 0.17 | 0.18 | 0.17 | 0.16 | 0.16 |
| 24 | 0.23 | 0.24 | 0.26 | 0.25 | 0.24 | 0.23 |
| 26 | 0.32 | 0.34 | 0.36 | 0.35 | 0.34 | 0.33 |
| 28 | 0.45 | 0.46 | 0.49 | 0.48 | 0.47 | 0.46 |
| 30 | 0.61 | 0.62 | 0.65 | 0.64 | 0.63 | 0.62 |
| 32 | 0.81 | 0.82 | 0.84 | 0.84 | 0.83 | 0.82 |
| 34 | 1.06 | 1.07 | 1.08 | 1.08 | 1.08 | 1.07 |
| 36 | 1.38 | 1.38 | 1.38 | 1.38 | 1.38 | 1.38 |
| 38 | 1.76 | 1.75 | 1.73 | 1.72 | 1.73 | 1.74 |
| 40 | 2.22 | 2.19 | 2.15 | 2.13 | 2.16 | 2.18 |
| 42 | 2.77 | 2.73 | 2.64 | 2.62 | 2.66 | 2.70 |
| 44 | 3.42 | 3.36 | 3.23 | 3.18 | 3.24 | 3.31 |
| 46 | 4.20 | 4.11 | 3.92 | 3.83 | 3.91 | 4.02 |
| 48 | 5.10 | 4.98 | 4.72 | 4.58 | 4.68 | 4.83 |

# Appendix E: AASHTO Equivalence Factors—Flexible Pavement

single axles, $p_t = 2.5$

| axle load | structural number, $SN$ | | | | | |
|---|---|---|---|---|---|---|
| (kips) | 1 | 2 | 3 | 4 | 5 | 6 |
| 2 | 0.0004 | 0.0004 | 0.0003 | 0.0002 | 0.0002 | 0.0002 |
| 4 | 0.003 | 0.004 | 0.004 | 0.003 | 0.003 | 0.002 |
| 6 | 0.01 | 0.02 | 0.02 | 0.01 | 0.01 | 0.01 |
| 8 | 0.03 | 0.05 | 0.05 | 0.04 | 0.03 | 0.03 |
| 10 | 0.08 | 0.10 | 0.12 | 0.10 | 0.09 | 0.08 |
| 12 | 0.17 | 0.20 | 0.23 | 0.21 | 0.19 | 0.18 |
| 14 | 0.33 | 0.36 | 0.40 | 0.39 | 0.36 | 0.34 |
| 16 | 0.59 | 0.61 | 0.65 | 0.65 | 0.62 | 0.61 |
| 18 | 1.00 | 1.00 | 1.00 | 1.00 | 1.00 | 1.00 |
| 20 | 1.61 | 1.57 | 1.49 | 1.47 | 1.51 | 1.55 |
| 22 | 2.48 | 2.38 | 2.17 | 2.09 | 2.18 | 2.30 |
| 24 | 3.69 | 3.49 | 3.09 | 2.89 | 3.03 | 3.27 |
| 26 | 5.33 | 4.99 | 4.31 | 3.91 | 4.09 | 4.48 |
| 28 | 7.49 | 6.98 | 5.90 | 5.21 | 5.39 | 5.98 |
| 30 | 10.31 | 9.55 | 7.94 | 6.83 | 6.97 | 7.79 |
| 32 | 13.90 | 12.82 | 10.52 | 8.85 | 8.88 | 9.95 |
| 34 | 18.41 | 16.94 | 13.74 | 11.34 | 11.18 | 12.51 |
| 36 | 24.02 | 22.04 | 17.73 | 14.38 | 13.93 | 15.50 |
| 38 | 30.90 | 28.30 | 22.61 | 18.06 | 17.20 | 18.98 |
| 40 | 39.26 | 35.89 | 28.51 | 22.50 | 21.08 | 23.04 |

tandem axles, $p_t = 2.5$

| axle load | structural number, $SN$ | | | | | |
|---|---|---|---|---|---|---|
| (kips) | 1 | 2 | 3 | 4 | 5 | 6 |
| 10 | 0.01 | 0.01 | 0.01 | 0.01 | 0.01 | 0.01 |
| 12 | 0.02 | 0.02 | 0.02 | 0.02 | 0.01 | 0.01 |
| 14 | 0.03 | 0.04 | 0.04 | 0.03 | 0.03 | 0.02 |
| 16 | 0.04 | 0.07 | 0.07 | 0.06 | 0.05 | 0.04 |
| 18 | 0.07 | 0.10 | 0.11 | 0.09 | 0.08 | 0.07 |
| 20 | 0.11 | 0.14 | 0.16 | 0.14 | 0.12 | 0.11 |
| 22 | 0.16 | 0.20 | 0.23 | 0.21 | 0.18 | 0.17 |
| 24 | 0.23 | 0.27 | 0.31 | 0.29 | 0.26 | 0.24 |
| 26 | 0.33 | 0.37 | 0.42 | 0.40 | 0.36 | 0.34 |
| 28 | 0.45 | 0.49 | 0.55 | 0.53 | 0.50 | 0.47 |
| 30 | 0.61 | 0.65 | 0.70 | 0.70 | 0.66 | 0.63 |
| 32 | 0.81 | 0.84 | 0.89 | 0.89 | 0.86 | 0.83 |
| 34 | 1.06 | 1.08 | 1.11 | 1.11 | 1.09 | 1.08 |
| 36 | 1.38 | 1.38 | 1.38 | 1.38 | 1.38 | 1.38 |
| 38 | 1.75 | 1.73 | 1.69 | 1.68 | 1.70 | 1.73 |
| 40 | 2.21 | 2.16 | 2.06 | 2.03 | 2.08 | 2.14 |
| 42 | 2.76 | 2.67 | 2.49 | 2.43 | 2.51 | 2.61 |
| 44 | 3.41 | 3.27 | 2.99 | 2.88 | 3.00 | 3.16 |
| 46 | 4.18 | 3.98 | 3.58 | 3.40 | 3.55 | 3.79 |
| 48 | 5.08 | 4.80 | 4.25 | 3.98 | 4.17 | 4.49 |

## Practice Problems: TRAFFIC ANALYSIS AND HIGHWAY DESIGN

Untimed

1. A flexible pavement with a 3″ thick asphalt concrete surface layer is to be designed for a state primary road. The subgrade is well drained and is not considered susceptible to frost action. Give the recommended thicknesses of the base and subbase. Use the CALTRANS method with the following information:

| | |
|---|---|
| maximum allowable load on a single axle: | 18,000 pounds |
| maximum aggregate size: | 2″ |
| base: | rolled stone with CBR = 90 |
| subbase: | soil-aggregate with CBR = 40 |
| subgrade: | CBR = 5 |
| traffic index: | 7.5 |

2. The outside lane of an interstate highway is being designed for a state where soil freezes to a depth of 5 inches. A flexible pavement is being considered. The projected life of the roadway is 20 years. The traffic which will use the roadway is as follows:

| | |
|---|---|
| mean ADT: | 20,000 passenger cars and 2200 trucks (over 4 lanes) |
| distribution: | 75% in outside lane |
| average gross truck load: | 22,000 pounds |
| average axle load: | 8800 pounds |
| axle load distribution: | |

| axle load (lbs) | % of avg. truck ADT single axle | tandem axle |
|---|---|---|
| under 8000 | 36.3 | – |
| 8000 – 16000 | 28.4 | 4.5 |
| 16000 – 20000 | 12.9 | 10.5 |
| 20000 – 24000 | | 4.0 |
| 24000 – 30000 | | 3.1 |
| 30000 – 34000 | | .3 |

The roadway is to be constructed of the following materials:

| | |
|---|---|
| pavement: | high stability plant mix |
| base: | coarse graded crushed stone treated with asphalt, R = 50 |
| subbase: | crushed stone, R = 78 |
| subgrade: | CBR = 5, R = 28 |

What should be the thicknesses of the pavement, base, and subbase?

3. Two cars are moving at 60 mph going in the same direction in the same lane. The cars are separated by 20 feet for each 10 mph of their speed. The coefficient of friction (skidding) between the tires and the roadway is .6. The reaction time is assumed to be .5 seconds. (a) If the lead car hits a parked truck, what is the speed of the second car when it hits the first (stationary) car? (b) At what speed does the rule of thumb of one car length per 10 mph become unsafe? (c) What should the rule actually be?

4. You have been hired by the owner of a damaged house to investigate the car crash that caused the damage. From the police report, you learn that the car was traveling down a 3% grade at an unknown speed. The skid marks are 185 feet long, and the pavement was dry at the time of the accident. The police report estimates from the visible damage to the car and the house that the initial speed of the car was 25 mph. The house owner does not believe the estimate of initial speed. You have the following test data from tests performed on level roadways:

| | | | | |
|---|---|---|---|---|
| initial speed (mph): | 30 | 40 | 50 | 60 |
| coef. of friction: | .59 | .51 | .45 | .35 |

(a) Find the minimum initial speed of the car. (b) If the police report was mistaken in assuming the road surface was dry, what was the minimum initial speed of the car?

5. One lane of a 2-lane road was observed for an hour during the day. The following data was gathered:

| | |
|---|---|
| average distance between front bumpers of successive cars: | 80 ft |
| average mean speed during the study: | 30 mph |
| space mean speed: | 29 mph |

(a) What is the average headway? (b) What is the density in vehicles per mile? (c) What is the traffic volume in vehicles per hour? (d) What is the maximum capacity of the lane? (e) Sketch a graph of the relationship of speed and density. (f) Sketch the relationship between speed and volume. Label the axes and indicate the region of unstable flow. (g) Sketch a graph of the relationship between volume and density. Label the axes and indicate the jam density. (h) Which is a more accurate parameter of traffic capacity: volume or density? Why? (i) What is the generally accepted maximum capacity of one lane of multi-lane freeway?

6. The intersection of First Street and Main Street, located in a suburb with negligible pedestrian activity, is being investigated. The peak hour factor on all approaches is 0.85. The signal cycle is 60 second, 2-phase. On Main Street, there is one 20-foot lane in each direction and parking is prohibited. First Street is one-way westbound with two 11-foot lanes and parking lanes on both sides. Turning on red is prohibited.

| parameter | First (W) | Main (N) | Main (S) |
|---|---|---|---|
| green time | 27 sec | 27 sec | 27 sec |
| yellow time | 3 sec | 3 sec | 3 sec |
| trucks | 7% | 5% | 5% |
| buses | 0% | 0% | 0% |
| left turns | 10% | 10% | 0% |
| right turns | 10% | 0% | 10% |

(a) Find the maximum service flow on First Street for Level of Service "E." (b) Find the maximum service flow on Main

Street (N) for Level of Service "B," assuming the volume on Main Street (S) to be 400 vehicles per hour. (c) Find the maximum service flow on Main Street (S) for Level of Service "B," assuming the volume on Main Street (N) to be 400 vehicles per hour.

**7. Two streets intersect at a stop sign: a 36 foot side street and a 44 foot arterial. There have been many complaints from users of the side street that traffic signals are needed at the intersection. Give a logical, step-by-step description of how you would investigate the problem. Discuss how you would arrive at your recommendations. If a signal is required, specify fixed timing or on-demand type decision criteria.**

## Timed

1. Commuter trains travel between stations spaced at regular one-mile locations. A train arrives every 5 minutes, and the trains can attain a maximum speed of 80 mph. Each train has five cars, with a maximum capacity of 220 people per car. When the train stops, it must wait one minute to allow for passenger movement. The uniform acceleration of the train is 5.5 ft/sec$^2$. Deceleration is more gentle at 4.4 ft/sec$^2$. (a) What is the top speed of the train? (b) What is the maximum capacity of the line in people per hour? (c) What is the average train speed? (d) What is the average running speed? (e) In regards to your answer in part (d), is the basis "time" or "spacing?" Explain your answer. (f) Suppose the maximum speed was increased to 100 mph. How much time would be saved between stations?

**2. Four intersections and five segments of highway have been evaluated using prior years' accident data. (a) Calculate the number of accidents per million vehicles for each intersection. Rank the intersections in order of need for improvement. (b) For the highway segments, calculate the number of accidents per year per million vehicle miles. Rank the segments in order of need for improvement. (c) Explain why calculations in (a) and (b) should be done. (d) Assume that the intersections varied in terms of the fractions of accidents that were fatal, injury, and property damage only. How would you compare the safety needs at those intersections?**

| intersection | ADT | # accidents/yr |
|---|---|---|
| A | 820 | 4 |
| B | 1200 | 5 |
| C | 1070 | 7 |
| D | 1400 | 6 |

| highway segment | ADT | # accidents/yr | length (miles) |
|---|---|---|---|
| 1 | 1900 | 1 | 1.50 |
| 2 | 2000 | 14 | 1.35 |
| 3 | 5500 | 18 | 4.50 |
| 4 | 3000 | 11 | 0.53 |
| 5 | 4000 | 30 | 2.48 |

3. A rural two-lane highway passing through rolling terrain has a design speed of 60 mph. Traffic is 5% trucks, 15% buses, and 80% cars, with 70% going uphill during the design hour. One particular section of the highway is a 3% grade lasting for 2 miles. Passing sight distance is more than 1500 feet for 60% of this section. The lane width is 11 feet. The shoulder is 6 feet wide on one side and 2 feet wide on the other. (a) What hourly volume of traffic can be carried at level of service "D"? (b) What hourly volume of traffic can be carried at level of service "B" if level of service "B" is possible? (c) What treatments might be made to improve level of service? (d) Interpolating from the results of (a) and (b) (i.e., without calculating service flow at other speeds or levels of service), estimate the capacity of that segment.

**4. You are hired as a consultant to give expert evidence in a court action arising from a vehicle collision. The two vehicles collided head-on while traveling on a curve tangent with a 4% grade. Vehicle #1 (a 1976 Chevrolet Impala) skidded 195 feet downhill before colliding with vehicle #2 (a 1978 Honda). The Honda skidded 142 feet. The police report estimates that both vehicles were traveling at 25 mph prior to the collision. This estimate was based on vehicle deformation. (a) What were the respective speeds of the two vehicles prior to the application of brakes if the coefficient of friction is known to be .48? (b) Which of your assumptions produces the greatest possible error in your speed calculations?**

**5. A non-signalized blind intersection is shown. It is desired to prohibit parking near the intersection in order to provide adequate sight distance into the intersection. Vehicles travel on the major street at 35 mph or slower 85% of the time. PIEV time is 1.0 sec for the average driver. Vehicles accelerate according to the table given:**

| Acceleration data distance | time |
|---|---|
| 10 feet | 2.3 seconds |
| 30 | 3.7 |
| 50 | 6.2 |
| 80 | 8.5 |
| 100 | 11.8 |

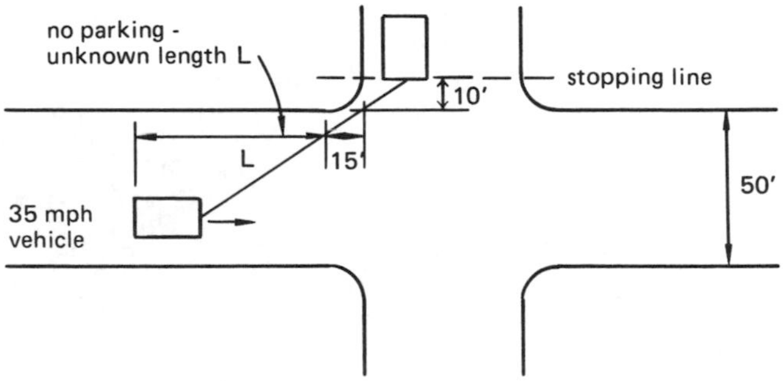

**(a) Determine the length, L, needed to allow adequate sight distance. (b) How many parallel parking spaces will be lost?**

6. A benefit/cost analysis is to be performed to justify the installation of safety-related road improvements (flexible barriers, break-away poles, etc.). Discuss how you would obtain a dollar value for a human life saved by the safety improvements.

7. Discuss the purpose of using sulfur in large proportions for road surfaces. What are the disadvantages? What special steps are taken in the installation? What special steps are taken to repair a sulfur-based road bed?

# 17 SURVEYING

*Nomenclature*

| | | |
|---|---|---|
| A | area | $ft^2$ |
| C | chord length, correction, or a constant | ft |
| d | cell width, or distance | ft, or stations |
| D | deflection angle, or setting out angle | degrees |
| D | degree of curve | degrees/station |
| g | grade, or acceleration due to gravity | %, $ft/sec^2$ |
| $g_c$ | gravitational constant | $ft\text{-}lbm/lbf\text{-}sec^2$ |
| h | height | ft |
| HI | height of the instrument | ft |
| I | intersection angle | degrees |
| IH | telescope height above ground | ft |
| k | number of observations | – |
| K | stadia interval factor | – |
| L | length | ft |
| LC | length of curve | ft or stations |
| LS | length of spiral | ft or stations |
| m | mass per length | lbm/ft |
| M | middle distance | ft |
| r | rate of grade change per station | %/station |
| R | rod reading, or radius | ft |
| s | sample standard deviation | various |
| T | tangent distance | ft |
| w | relative weight | – |
| x | distance or location | ft or stations |
| y | distance or elevation | ft |

*Symbols*

| | | |
|---|---|---|
| $\theta, \alpha, \beta$ | angles | degrees |
| $\mu$ | distribution mean | various |

*Subscripts*

| | |
|---|---|
| a | actual |
| c | curvature |
| p | probable |
| r | refraction |
| s | sag or spiral |

## 1 ERROR ANALYSIS

### A. MEASUREMENTS OF EQUAL WEIGHT

There are many opportunities for errors in surveying, although calculators and modern equipment have reduced the magnitude of most errors. The purpose of error analysis is not to eliminate errors, but rather to estimate their magnitudes and to assign them to the appropriate measurements.

The *expected value* (also known as the *most likely value* or the *probable value*) of a measurement is the value that has the highest probability of being correct. If a series of measurements is taken of a single quantity, the most probable value is the average (*mean*) of those measurements.

$$x_p = \frac{x_1 + x_2 + \cdots + x_k}{k} \qquad 17.1$$

For related measurements whose sum should equal some known quantity, the most probable values are the observed values corrected by an equal part of the total error.

Measurements of a given quantity are assumed to be normally distributed. If a quantity has a mean $\mu$ and a standard deviation $s$, the probability is 50% that a measurement of that quantity will fall within the range of $\mu \pm (0.6745)s$. The quantity $(0.6745)s$ is known as the *probable error*. The probable *ratio of precision* is $\mu/(0.6745)s$. The interval between the extremes is known as the 50% *confidence interval.*[1]

The standard deviation, $s$, is the *small sample standard deviation.*

[1] Other confidence limits are easily obtained from the normal tables. For example, for a 95% interval, replace 0.6745 with 1.96. For a 99% interval, use 2.57.

The *probable error of the mean* of $k$ observations of the same quantity is given by equation 17.2.

$$E_{\text{mean}} = \frac{0.6745s}{\sqrt{k}} = \frac{E_{\text{total, } k \text{ measurements}}}{\sqrt{k}} \qquad 17.2$$

*Example 17.1*

The interior angles of a traverse were measured as 63°, 77°, and 41°. Each measurement was made once, and all angles were measured with the same precision. What are the most probable interior angles?

The sum total of angles should equal 180. The error in the measurement is $63+77+41-180 = +1°$. Therefore, the correction required is $-1°$, which is proportioned equally among the three angles. The most probable values are 62.67°, 76.67°, and 40.67°.

*Example 17.2*

12 tapings were made of a critical distance. The mean value was 423.7 feet with a standard deviation ($s$) of 0.31 feet. What are the 50% confidence limits for the distance?

From equation 17.2, the standard error of the mean value is

$$E_{\text{mean}} = \frac{(0.6745)(0.31)}{\sqrt{12}} = 0.06 \text{ ft}$$

The probability is 50% that the true distance is within the limits of $423.7 \pm 0.06$ feet.

*Example 17.3*

The true length of a tape is 100 feet. The most probable error of a measurement with this tape is 0.01 feet. What is the expected error if the tape is used to measure out a distance of one mile?

The number of tapings will be $5280/100 = 52.8$, or 53 tapings. The most probable error will be $0.01 \times \sqrt{53} = 0.073$ feet.

## B. MEASUREMENTS OF UNEQUAL WEIGHT

Some measurements may be more reliable than others. It is not unreasonable to weight each measurement with its relative reliability. Such weights can be determined subjectively, but more frequently, they are determined from relative frequencies of occurrence or from the relative inverse squares of the probable errors.

The *probable error* and 50% confidence interval for weighted observations can be found from equation 17.3. $x_i$ represents the $i$th observation and $w_i$ represents its relative weight. The number of observations is $k$.

$$E_{p,\text{weighted}} = 0.6745\sqrt{\frac{\sum[w_i(\overline{x} - x_i)^2]}{(\sum w_i)(k-1)}} \qquad 17.3$$

For related weighted measurements whose sum should equal some known quantity, the most probable weighted values are corrected inversely to the relative frequency of observation.

Weights can also be calculated when the probable errors are known. These weights are the relative squares of the probable errors.

*Example 17.4*

An angle was measured five times by five equally competent crews on similar days. Two of the crews obtained a value of 39.77°, and the remaining three crews obtained a value of 39.74°. What is the probable value of the angle?

$$\theta = \frac{(2)(39.77) + (3)(39.74)}{5} = 39.75°$$

*Example 17.5*

A distance has been measured by three different crews. The measurements and probable errors are given. What is the most probable value?

$$\text{crew 1}: 1206.40 \pm 0.03 \text{ ft}$$
$$\text{crew 2}: 1206.42 \pm 0.05 \text{ ft}$$
$$\text{crew 3}: 1206.37 \pm 0.07 \text{ ft}$$

The sum of squared probable errors is

$$(0.03)^2 + (0.05)^2 + (0.07)^2 = 0.0083$$

The weights to be applied to the three measurements are

$$\frac{0.0083}{(0.03)^2} = 9.22$$

$$\frac{0.0083}{(0.05)^2} = 3.32$$

$$\frac{0.0083}{(0.07)^2} = 1.69$$

The most probable length is

$$\frac{(1206.40)(9.22) + (1206.42)(3.32) + (1206.37)(1.69)}{9.22 + 3.32 + 1.69}$$
$$= 1206.40 \text{ ft}$$

*Example 17.6*

What is the 50% confidence interval for the measured distance in example 17.5?

It is easier to work with the decimal part only.

$$\overline{x} = \frac{1}{3}(0.40 + 0.42 + 0.37) \approx 0.40$$

| $i$ | $x_i$ | $\overline{x} - x_i$ | $(\overline{x} - x_i)^2$ | $w$ | $w(\overline{x} - x_i)^2$ |
|---|---|---|---|---|---|
| 1 | 0.40 | 0 | 0 | 9.22 | 0 |
| 2 | 0.42 | −0.02 | 0.0004 | 3.32 | 0.0013 |
| 3 | 0.37 | 0.03 | 0.0009 | 1.69 | 0.0015 |
| | | | | 14.23 | 0.0028 |

From equation 17.3,

$$E_{p,\text{weighted}} = 0.6745\sqrt{\frac{0.0028}{(14.23)(3-1)}} = 0.0067$$

The 50% confidence interval is 1206.40 ± 0.0067 ft

*Example 17.7*

The interior angles of a triangular traverse were repeatedly measured. What is the most probable value for angle #1?

| angle | value | number of measurements |
|---|---|---|
| 1 | 63° | 2 |
| 2 | 77° | 6 |
| 3 | 41° | 5 |

The total of the angles is 63 + 77 + 41 = 181°. So −1° must be divided among the three angles. These corrections are inversely proportional to the number of measurements. The sum of the measurement inverses is

$$\frac{1}{2} + \frac{1}{6} + \frac{1}{5} = 0.867$$

The most probable value of angle #1 is

$$63° + \left(\frac{\frac{1}{2}}{0.867}\right)(-1) = 62.42°$$

*Example 17.8*

The interior angles of a triangular traverse were measured. What is the most probable value of angle #1?

| angle | value |
|---|---|
| 1 | 63° ± 0.01° |
| 2 | 77° ± 0.03° |
| 3 | 41° ± 0.02° |

The total of the angles is 63 + 77 + 41 = 181°. So −1° must be divided among the three angles. The corrections are proportional to the square of the probable errors.

$$(0.01)^2 + (0.03)^2 + (0.02)^2 = 0.0014$$

The most probable value of angle #1 is

$$63 + \frac{(0.01)^2}{0.0014}(-1) = 62.93°$$

C. ERRORS IN COMPUTED QUANTITIES

When independent quantities with known errors are added or subtracted, the error of the result is given by equation 17.4. The squared errors under the radical are added regardless of whether the calculation is addition or subtraction.

$$E_{\text{total}} = \sqrt{E_1^2 + E_2^2 + E_3^2 + \cdots} \qquad 17.4$$

The error in the product of two quantities $x_1$ and $x_2$ which have known errors $E_1$ and $E_2$ is given by equation 17.5.

$$E_{\text{product}} = \sqrt{x_1^2E_2^2 + x_2^2E_1^2} \qquad 17.5$$

*Example 17.9*

An EDM instrument manufacturer has indicated that the measurement error with a particular instrument is ±0.04 feet with another error of ±10 ppm. What is the expected error of measurement if the instrument is used to measure 3000 feet?

There are two independent errors here, since both parts can be positive or negative, and they may have opposite signs. The variable error is

$$E = \pm(3000)\left(\frac{10}{1{,}000{,}000}\right) = 0.03 \text{ ft}$$

From equation 17.4

$$E = \sqrt{(0.04)^2 + (0.03)^2} = 0.05 \text{ ft}$$

*Example 17.10*

The sides of a rectangular section were determined to be 1204.77±0.09 feet and 765.31±0.04 feet respectively. What is the probable error in the area?

From equation 17.5,

$$E_{\text{area}} = \sqrt{(1204.77)^2(0.04)^2 + (765.31)^2(0.09)^2} = 84.06 \text{ ft}^2$$

## 2 ORDERS OF ACCURACY

Traverse closure errors are ranked into *orders of accuracy.* The closure error is divided by the sum of all traverse leg lengths to determine the fractional error. Table 17.1 lists the maximum permissible errors for each order of accuracy.[2]

**Table 17.1**
Maximum Traverse Closure Errors

| order of accuracy | maximum error |
|---|---|
| first | 1/25,000 |
| second | 1/10,000 |
| third | 1/5000 |
| fourth | 1/3000 |
| fifth | 1/1000 |
| lowest | 1/500 |

## 3 DISTANCE MEASUREMENT

Lengths may be divided into 100 foot long sections called *stations.* Interval stakes along an established line are ordinarily laid down at 100 foot intervals called *full stations.* If a marker stake is placed anywhere else along the line, it is called a *plus station.* Thus, a stake placed 1500 feet from a reference point is labeled "15+00," and a stake placed 1325 feet from a reference point is labeled "13+25."

Tapes may be used for short distances. Steel tapes are relatively low in cost. Low-coefficient tapes made from nickel-steel alloys (e.g., *invar*) are not excessively sensitive to temperature changes.[3]

Tape readings are affected by temperature, tension, and sag. Corrections for temperature and tension have been presented in chapter 12. The correction for sag is given by equation 17.6. Since tapes can be standardized flat or suspended, the correction can be added or subtracted to the actual measurement.

$$C_s = \pm \frac{m^2 L^3}{24F^2}\left(\frac{g}{g_c}\right)^2 \qquad 17.6$$

The correction $C_s$ is in units of feet when $m$ is in lbm/ft, $L$ is in feet, and the tension, $F$, is in lbf.

**Table 17.2**
Corrections for Surveyor's Tapes

| | measured flat | measured suspended |
|---|---|---|
| standardized flat | none | subtract $C_s$ |
| standardized suspended | add $C_s$ | none |

Electronic distance measuring (EDM) equipment can be based on microwave or laser operation. Distance readings are generally direct.

Distance can also be measured tachymetrically. This method involves sighting through a small angle at a distant scale. The angle may be fixed and the length measured (*stadia method*), or the length may be fixed and the angle measured (*European method*).[4]

Stadia measurement consists of observing the apparent locations of the horizontal cross hairs on a distant stadia rod. The interval between the two rod readings is called the *stadia interval* or the *stadia reading.* This distance is directly related to the distance between the telescope and the rod. For rod readings $R_1$ and $R_2$ (both in feet), the separation distance is

$$x = K(R_2 - R_1) + C \qquad 17.7$$

$C$ is the sum of focal length and distance from plumb bob (center of instrument) to forward lens. It varies from 0.6 to 1.4 feet, but is typically set by the manufacturer at 1 foot. $C$ is zero for internal focusing telescopes. $K$ is the *stadia interval factor* which usually has a value of 100.

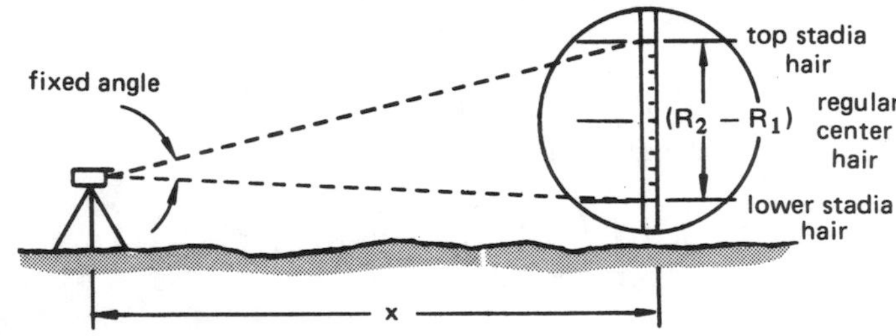


**Figure 17.1** Horizontal Stadia Measurement

If the sighting is inclined, as it is in figure 17.2, it will be necessary to find both the horizontal and vertical distances. These may be found from equations 17.8 and 17.9. $y$ is measured from the telescope to the sighting

---

[2] The classification of maximum errors into first, second, third, etc., can vary from authority to authority. For example, some federal agencies specify a maximum error of 1/7500 for third order accuracy.

[3] The modulus of elasticity of invar is approximately 2.1 EE7 psi. The coefficient of thermal expansion is approximately 1 EE–7 1/°F.

[4] The word "stadia" means "temporary position."

rod center. The actual elevation difference will require knowledge of the instrument height.

$$x = K(R_2 - R_1)\cos^2\theta + C\cos\theta \qquad 17.8$$

$$y = \frac{1}{2}(K)(R_2 - R_1)(\sin 2\theta) + C\sin\theta \qquad 17.9$$

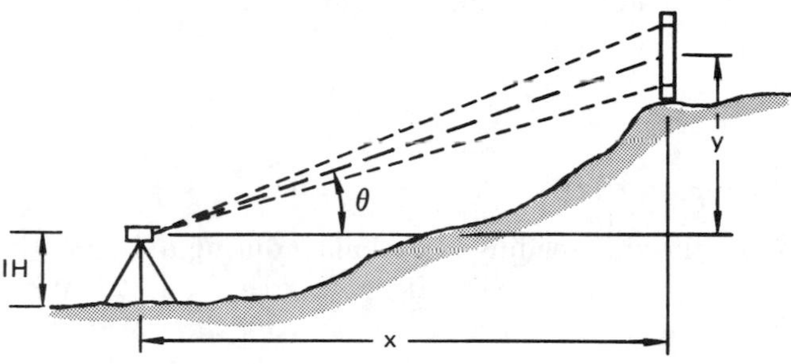


**Figure 17.2** Inclined Stadia Measurement

## 4 ELEVATION MEASUREMENT

*Leveling* is the act of using an engineer's level and rod to measure the vertical distance (the *elevation*) from an arbitrary level surface. Usually, the elevation is measured with respect to sea level.

### A. CURVATURE AND REFRACTION

If a level sighting is taken on an object with actual height $h_a$, the curvature of the earth will cause the object to appear taller by an amount $h_c$. In equation 17.10, $x$ is measured in feet along the curved surface of the earth.

$$h_c = 2.4\,\text{EE–}8(x^2) \qquad 17.10$$

Atmospheric refraction will make the object appear shorter by an amount $h_r$.

$$h_r = 3.0\,\text{EE–}9(x^2) \qquad 17.11$$

The corrected rod reading (actual height) is

$$\begin{aligned} h_a &= R_{\text{observed}} + h_r - h_c \\ &= R_{\text{observed}} - (2.1\ \text{EE} - 8)(x^2) \qquad 17.12 \end{aligned}$$

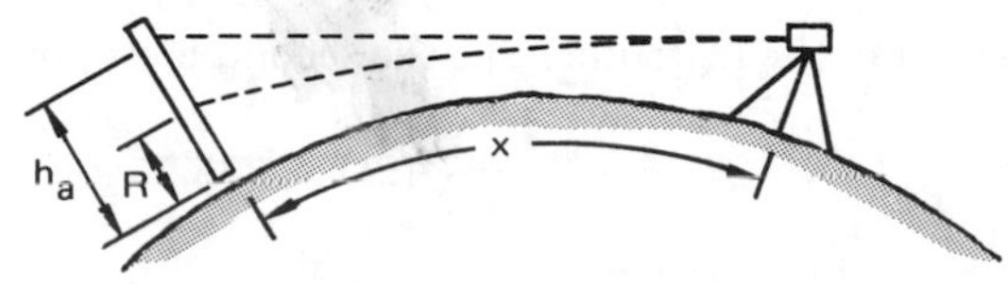


**Figure 17.3** Curvature and Refraction Effects

### B. DIRECT LEVELING

The most common method of determining the difference in elevations of two points is known as direct leveling. A level is set up at a point approximately midway between the two points whose difference in elevation is wanted. The vertical distances are observed by reading directly from the rod. Refer to figure 17.4 which uses the following nomenclature:

| | |
|---|---|
| $y_{A-B}$ | difference in elevations between points $A$ and $B$ |
| $y_{A-L}$ | difference in elevations between points $A$ and $L$ |
| $y_{L-B}$ | difference in elevations between points $L$ and $B$ |
| $R_A$ | rod reading at $A$ |
| $R_B$ | rod reading at $B$ |
| $h_{rc,A-L}$ | effects of curvature and refraction between points $A$ and $L$ |
| $h_{rc,B-L}$ | effects of curvature and refraction between points $B$ and $L$ |

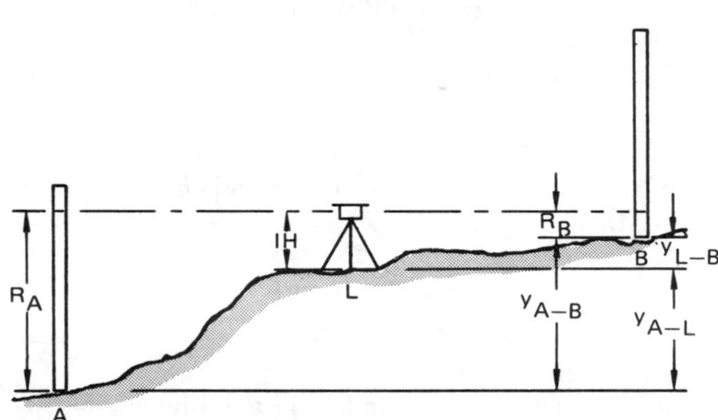


**Figure 17.4** Direct Leveling

By inspection,

$$y_{A-L} = R_A - h_{rc,A-L} - IH \qquad 17.13$$

$$y_{L-B} = R_B - h_{rc,L-B} - IH \qquad 17.14$$

The difference in elevations between points $A$ and $B$ is

$$\begin{aligned} y_{A-B} &= y_{A-L} + y_{L-B} \\ &= R_A - R_B + h_{rc,L-B} - h_{rc,A-L} \qquad 17.15 \end{aligned}$$

If the backsight and foresight distances are equal, then the effects of refraction and curvature cancel.

$$y_{A-B} = R_A - R_B \qquad 17.16$$

## C. INDIRECT LEVELING

*Indirect leveling* does not require a backsight (although one can be taken to eliminate the effects of curvature and refraction). A case of indirect leveling is illustrated in figure 17.5 where the difference in elevations between points $A$ and $B$ is sought. It is assumed that the distance $AC$ has been determined. Within the limits or ordinary practice, angle $ACB$ is 90°.

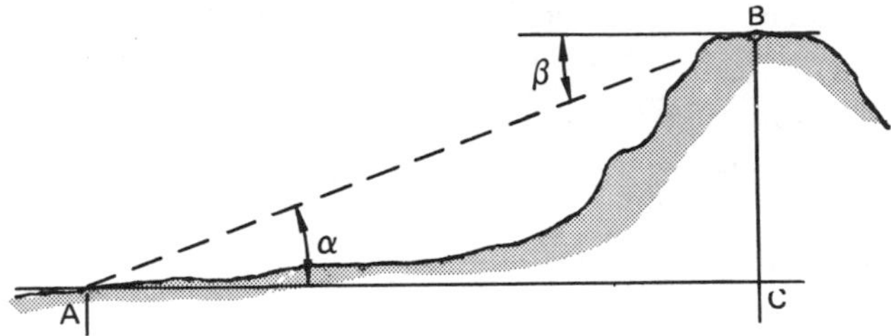


**Figure 17.5** Indirect Leveling

Including the effects of curvature and refraction,

$$y_{A-B} = AC(\tan\alpha) + 2.1\,\text{EE–}8(AC)^2 \qquad 17.17$$

If a backsight is taken from $B$ to $A$ and angle $\beta$ is measured, then

$$y_{A-B} = AC(\tan\beta) - 2.1\,\text{EE–}8(AC)^2 \qquad 17.18$$

Adding equations 17.17 and 17.18 and dividing by 2,

$$y_{A-B} = \frac{1}{2}(AC)(\tan\alpha + \tan\beta) \qquad 17.19$$

## D. DIFFERENTIAL LEVELING

*Differential leveling* is the consecutive application of direct leveling to the measurement of large differences in elevation. There is usually no attempt to exactly balance the foresights and backsights. Thus, there is no record made of the exact locations of the level positions. Furthermore, the path taken between points need not be along a straight line connecting them, as only the elevation differences are relevant.

If greater accuracy is desired without having to accurately balance the foresight and backsight distances, it is possible to eliminate most of the curvature and refraction error by balancing the sum of the foresights against the sum of the backsights.

The following abbreviations are used with differential leveling.

- $BM$ bench mark or monument
- $TP$ turning point
- $FS$ foresight (also known as a minus sight)
- $BS$ backsight (also known as a plus sight)
- $HI$ height of the instrument[5]
- $L$ level position

*Example 17.11*

The following readings were taken during a differential leveling survey between bench marks 1 and 2. What is the difference in elevations between these two bench marks?

| station | $BS$ | $HI$ | $FS$ | elevation |
|---|---|---|---|---|
| $BM1$ | 7.11 | | | 721.05 |
| $TP1$ | 8.83 | | 1.24 | |
| $TP2$ | 11.72 | | 1.11 | |
| $BM2$ | | | 10.21 | |

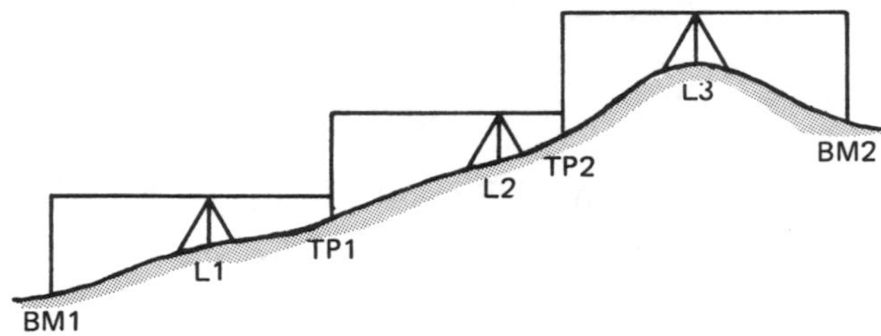


The first measurement is shown in larger scale. The height of the instrument is

$$HI = 721.05 + 7.11 = 728.16$$

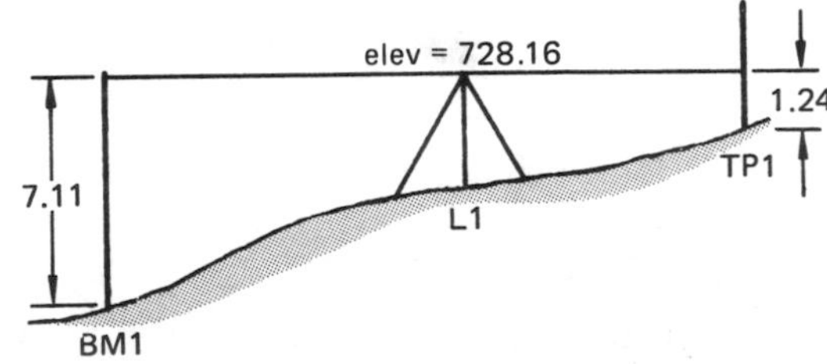


The height of the instrument at the second level position is

$$HI = 728.16 + 8.83 - 1.24 = 735.75$$

[5] Do not confuse the height of the instrument ($HI$) with the elevation at which the instrument is located. The $HI$ is the elevation of the line of sight of the telescope when the instrument is leveled.

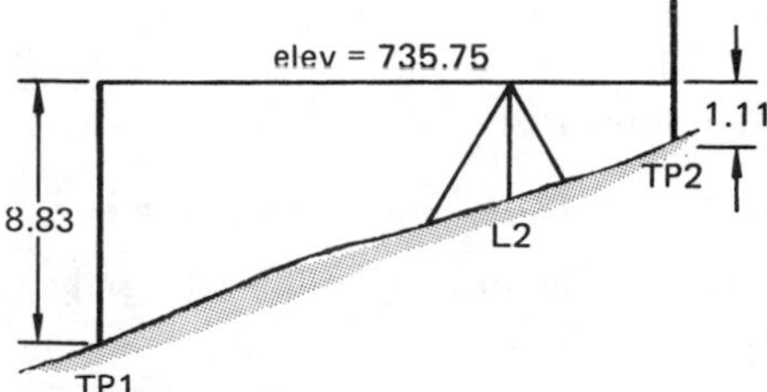


The height of the instrument at the third level position is

$$HI = 735.75 + 11.72 - 1.11 = 746.36$$

The elevation of $BM2$ is

$$746.36 - 10.21 = 736.15.$$

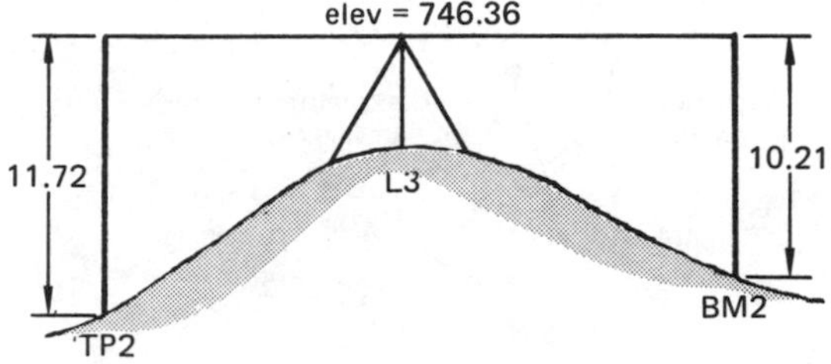


The difference in elevations is $736.15 - 721.05 = 15.1$.

The backsight sum is $7.11 + 8.83 + 11.72 = 27.66$.

The foresight sum is $1.24 + 1.11 + 10.21 = 12.56$.

The difference is $27.66 - 12.56 = 15.1$ (check)

## 5 EQUIPMENT AND METHODS USED TO MEASURE ANGLES

The *telescope* is actually part of the sighting equipment. A transit or theodolite is often referred to as a "telescope," even though the telescope is only a part of the total instrument.[6]

An *engineer's level* or *dumpy level* is a sighting device with a telescope rigidly attached to the level bar. A sensitive level vial is used to ensure level operation. The engineer's level can be rotated, but in its basic form, cannot be elevated.

In a *semi-precise level*, also known as a *prism level*, the level vial is visible from the eyepiece end. Other than that, the semi-precise and engineer's levels are similar.

A *precise level*, also known as a *geodetic level*, has even better control of horizontal angles. The bubble vial is magnified for greater accuracy.

[6] The *alidade* consists of the base and telescope part of the transit without including the tripod or leveling equipment.

An *engineer's transit* (or *surveyor's transit*) measures vertical angles as well as horizontal angles. It can, of course, be clamped vertically and used as a level. Angles are usually read by the naked eye by looking at a plate scale.

With a *theodolite*, angles are measured by looking into viewpieces. There may be up to four viewpieces, one each for the sighting-in telescope, the compass, the horizontal and vertical angles, and the optical plummet.

Transits and levels are used with rods. *Leveling rods* are used to measure the vertical distance between the line of sight and the point being observed. The *standard rod* is typically made of wood or fiberglass and is extendable.[7] *Precise rods*, made of wood-mounted invar, are typically constructed in one piece and are spring loaded in tension to avoid sagging.

## 6 ANGLE MEASUREMENT

The direction of any line can be specified by an angle between it and some reference line. The reference line is known as a *meridian*. If the meridian is arbitrarily chosen, it is called an *assumed meridian*. If the meridian is a true north-to-south line passing through the true north pole, it is called a *true meridian*. If the meridian is parallel to the earth's magnetic field, it is known as a *magnetic meridian*.[8]

A true meridian differs from a magnetic meridian by a *declination* (*magnetic declination* or *variation*). If the north end of a compass points to the west of the true meridian, the declination is said to be a *west declination* or *minus declination*. Otherwise, it is an *east declination* or *plus declination*.

Plus declinations are added to the magnetic compass azimuth to obtain the true azimuth. Minus declinations are subtracted from the magnetic compass azimuth.

The variation of a line from its meridian may be given in several ways:

- Azimuths: The azimuth of a line is the clockwise angle measured from the south branch of the meridian to the line. This is known as an *azimuth from the south*. (*Azimuths from the north* are also sometimes used.)

- Deflection Angles: The angle between a line and the prolongation of a preceding line is a deflection angle. Such measurements must be labeled as 'right' (clockwise) or 'left' (counterclockwise).

[7] The standard rod may also be known as a *Philadelphia rod*.

[8] A rectangular grid can be drawn over a map with any arbitrary orientation. In such a case, the vertical lines are known as *grid meridians*. All angles are referenced to those grid meridians.

- Angles to the right: The angle to the right is a clockwise angle measured from the preceding to the following line.
- Azimuths from the back line: Same as angles to the right.
- Bearings: The bearing of a line is referenced to the quadrant in which the line falls and the angle that the line makes with the meridian in that quadrant. It is necessary to specify the two cardinal directions that define the quadrant in which the line is found. The north and south directions are always specified first.

These methods of angle measurement are illustrated in figure 17.6.

azimuth from the south: 157°

azimuth from the north: 337°

deflection angle: 23°L

angle to the right: 157°

bearing: N 23° W

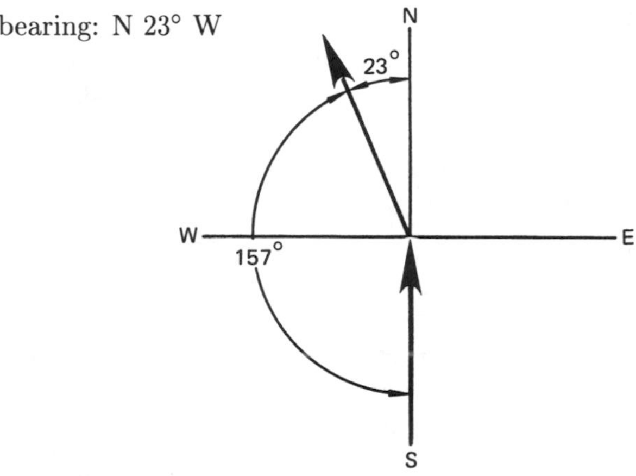


**Figure 17.6** Angle Measurements

*Example 17.12*

A band of hikers on the Pacific Crest Trail (magnetic declination of +17°) used a magnetic compass to site in on a distant geographic feature. What was the true azimuth if the observed angle was 42°?

A plus declination is added to the observed azimuth. Therefore, the true azimuth was $42° + 17° = 59°$.

## 7 CLOSED TRAVERSES

A *traverse* is a series of straight lines whose lengths and deflection angles (or other angle measurements) are known. A traverse that comes back to its starting point is known as a *closed traverse*. The polygon that results from the closing of a traverse is governed by the following two requirements:

- The sum of the deflection angles is 360°.
- The sum of the interior angles of a polygon with $n$ sides is $(n-2)180°$.

There are three ways to measure and record angles in traverses. Figure 17.7 illustrates the *station angle*, *deflection angle*, and *explement angle* (also known as the *interior angle*).

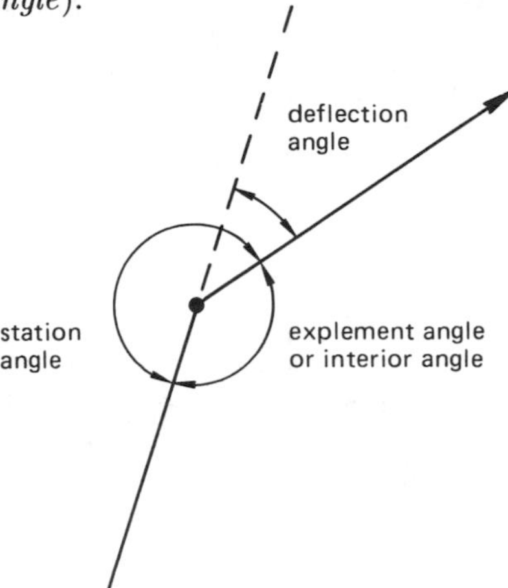


**Figure 17.7** Angles in Traverses

### A. ADJUSTING CLOSED TRAVERSE ANGLES

Due to errors, variations in magnetic declination, and local magnetic attractions, it is likely that the sum of angles making up the interior angles will not exactly equal $(n-2)180°$. The following procedure can be used to distribute the angle error of closure among the angles.

*step 1*: Calculate the interior angle of each station from the observed bearings.

*step 2*: Subtract $(n-2)180°$ from the sum of the interior angles.

*step 3*: Unless additional information in the form of numbers of observations or probable errors is available, assume the angle error of closure can be divided among all angles. Divide the error by the number of angles.

*step 4*: Find a line whose bearing is assumed correct. That is, find a line whose bearing appears unaffected by errors, variations in magnetic declination, and local attractions. Such a line may be chosen as one for which the forward and back bearings are the same. If there is no such line, take the line whose difference in forward and back bearings is the smallest.

*step 5*: Start with the assumed correct line and add (or subtract) the prorated error to each interior angle.

*step 6*: Correct all bearings except the one for the assumed correct line.

*Example 17.13*

Adjust the angles on the four-sided closed traverse whose magnetic foresights and backsights are listed.

| line | bearing |
|---|---|
| $AB$ | N 25° E |
| $BA$ | S 25° W |
| $BC$ | S 84° E |
| $CB$ | N 84.1° W |
| $CD$ | S 13.1° E |
| $DC$ | N 12.9° W |
| $DA$ | S 83.7° W |
| $AD$ | N 84° E |

*step 1*: The interior angles are calculated from the bearings.

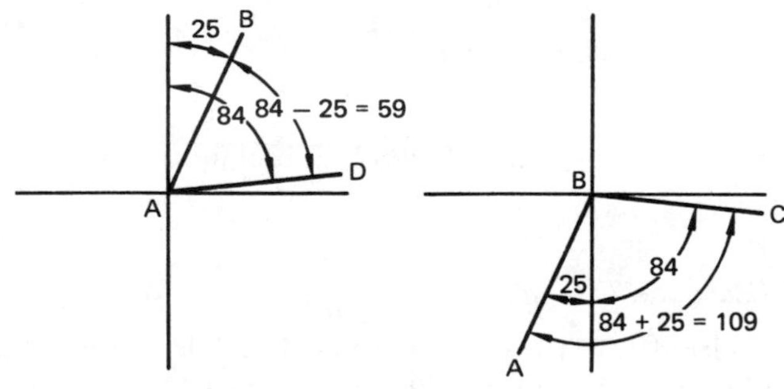


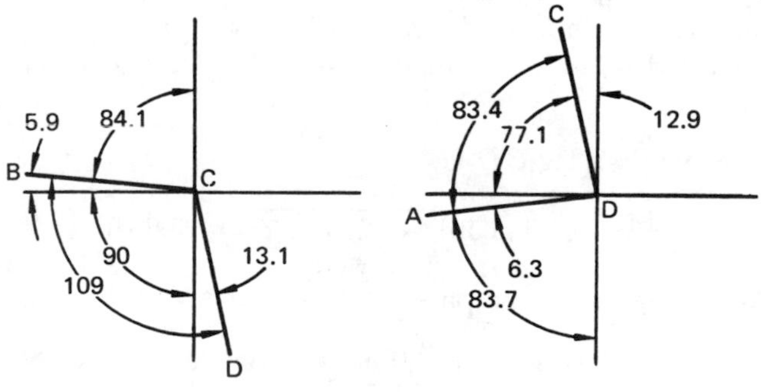


*step 2*: The sum of the angles is: $59 + 109 + 109 + 83.4 = 360.4°$. For a four-sided traverse, the sum of interior angles should be 360°, so a correction of $(-0.4°)$ must be divided up evenly among the four angles. The original and corrected traverses are shown.

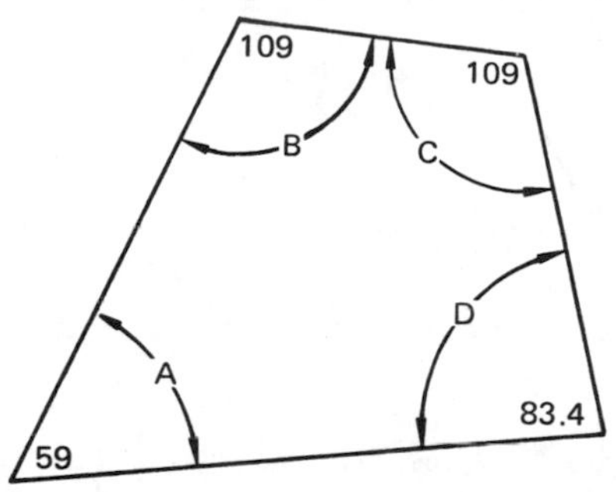


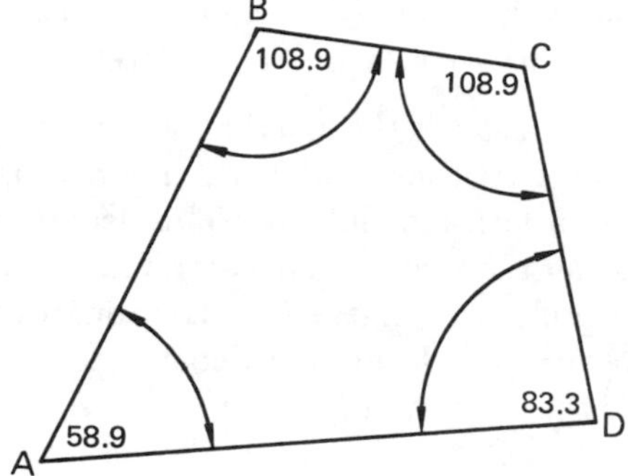


Since the backsight and foresight bearings of line $AB$ are the same, it is assumed that the 25° bearings are the most accurate. The remaining bearings are then adjusted to use the corrected angles.

| line | bearing |
|---|---|
| $AB$ | N 25° E |
| $BA$ | S 25° W |
| $BC$ | S 83.9° E |
| $CB$ | N 83.9° W |
| $CD$ | S 12.8° E |
| $DC$ | N 12.8° W |
| $DA$ | S 83.9° W |
| $AD$ | N 83.9° E |

## B. LATITUDES AND DEPARTURES

The *latitude* of a line is the distance that the line extends in a north or south direction. A line that runs towards the north has a positive latitude; a line that runs towards the south has a negative latitude.

The *departure* of a line is the distance that the line extends in an east or west direction. A line that runs towards the east has a positive departure; a line that runs towards the west has a negative departure.

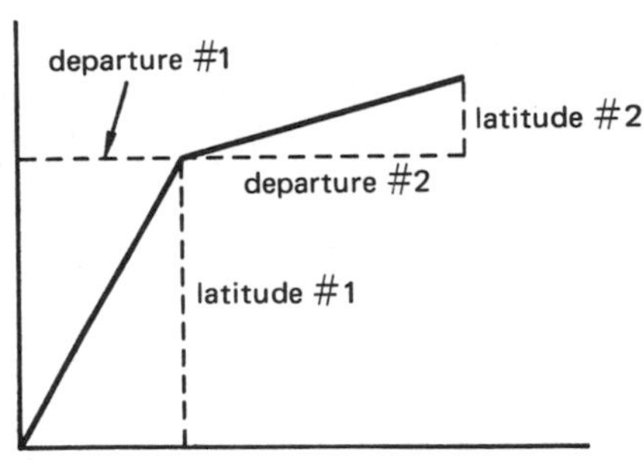


**Figure 17.8** Departures and Latitudes

In a closed traverse, the algebraic sum of latitudes should be zero. The algebraic sum of departures should also be zero. These sums, which are distances in feet with actual values near zero, are called *closure in latitude* and *closure in departure* respectively.

The *traverse closure* is the line that will exactly close the traverse. Since latitudes and departures are orthogonal, the closure in latitude and closure in departure can be considered as the rectangular coordinates to calculate the traverse closure length. The coordinates will have signs opposite the closures in departure and latitude. That is, if the closure in departure is positive, point $A$ will lie to the left of point $A'$, as shown in figure 17.9.

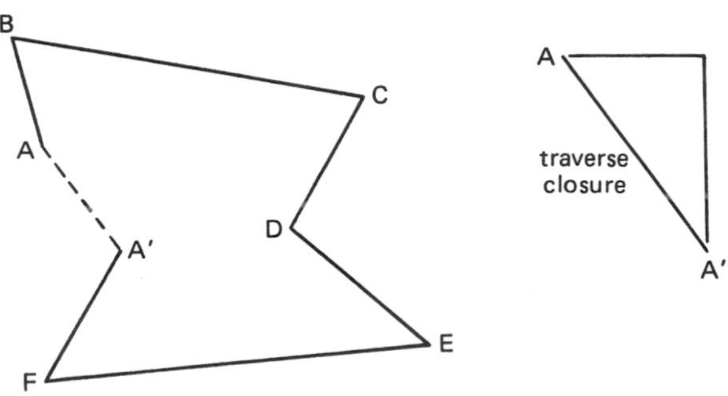


**Figure 17.9** Traverse Closure

The length of a traverse closure is

$$L = \sqrt{\left(\begin{array}{c}\text{closure in}\\ \text{departure}\end{array}\right)^2 + \left(\begin{array}{c}\text{closure in}\\ \text{latitude}\end{array}\right)^2} \qquad 17.20$$

## C. ADJUSTING CLOSED TRAVERSE LATITUDES AND DEPARTURES

To balance a closed traverse, the traverse closure must be divided among the various legs of the traverse. This correction requires that the latitudes and departures be known for each leg of the traverse.

The most common method used to balance the traverse legs is known as the *compass rule*.[9] This rule states that the correction to a leg of the traverse is to the total traverse correction as the leg length is to the total traverse length, with the signs reversed.

$$\frac{\text{leg departure correction}}{\text{closure in departure}} = -\frac{\text{leg length}}{\text{total traverse length}} \qquad 17.21$$

$$\frac{\text{leg latitude correction}}{\text{closure in latitude}} = -\frac{\text{leg length}}{\text{total traverse length}} \qquad 17.22$$

It is appropriate to use the compass rule when the angles and distances in the traverse are considered equally precise. If the angles are precise, but the distances are less precise (e.g., as in taping through rugged terrain), the *transit rule* is a preferred alternative. The transit rule distributes the closure error in proportion to the absolute values of the latitudes and departures.

$$\frac{\text{leg latitude correction}}{\text{closure in latitude}} = -\frac{\text{absolute value of leg latitude}}{\text{sum of latitude absolute values}} \qquad 17.23$$

$$\frac{\text{leg departure correction}}{\text{closure in departure}} = -\frac{\text{absolute value of leg departure}}{\text{sum of departure absolute values}} \qquad 17.24$$

*Example 17.14*

A closed traverse was constructed of 7 legs, the total of whose lengths was 2705.13 feet. Leg $CD$ has a departure of 443.56 and a latitude of 219.87. The total closure in departure for the traverse was +0.41 feet; the total closure in latitude was −0.29 feet. What are the corrected latitude and departure for leg $CD$? Use the compass rule.

The length of leg $CD$ is

$$L_{CD} = \sqrt{(443.56)^2 + (219.87)^2} = 495.06 \text{ ft}$$

According to the compass rule,

$$\frac{\text{latitude correction}}{-0.29} = -\frac{495.06}{2705.13}$$

[9] Use of one method or the other to allocate error to legs is arbitrary. If you know that one leg in particular was poorly measured due to difficult terrain, you may decide to give all of the error to that leg.

The latitude correction is 0.05 feet.

$$\frac{\text{departure correction}}{+0.41} = -\frac{495.06}{2705.13}$$

The departure correction is $-0.08$ ft.

The corrected latitude is $219.87 + 0.05 = 219.92$ ft.

The corrected departure is $443.56 - 0.08 = 443.48$ ft.

## D. RECONSTRUCTING MISSING SIDES AND ANGLES

If one or more sides or angles of a traverse are missing or cannot be determined by measurement, they will have to be reconstructed. The procedures for three common cases are listed below. These procedures draw primarily upon the subjects of geometry and trigonometry.

**One leg missing**: One leg is missing in figure 17.10. However, the line $EA$ can be reconstructed easily from its components $E$–$E'$ and $E'$–$A$. These components are equal to the sum of the departures and sum of the latitude respectively, with the signs changed. The angle can be determined from the ratio of the sides, and the length $E$–$A$ can be found from equation 17.20.

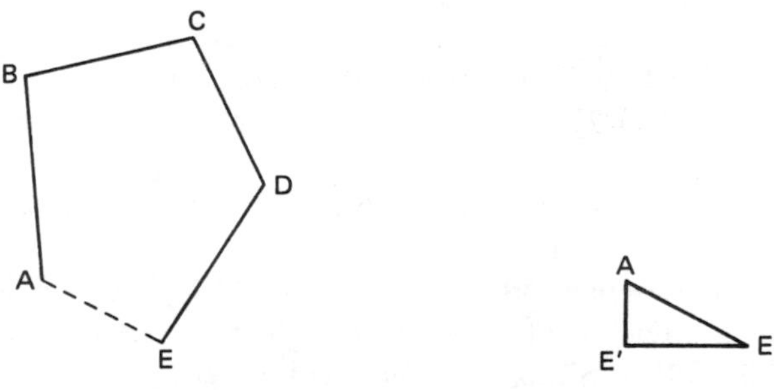


**Figure 17.10** One Missing Leg

**Adjacent legs missing**: Figure 17.11 shows a traverse that has two adjacent legs missing. The traverse can be closed as long as some length/angle information is available. The technique is to close the traverse by using the method presented in case 1 above. This will give the line $D$–$A$. Then, the triangle $E$–$A$–$D$ can be completed by using whatever information is available.

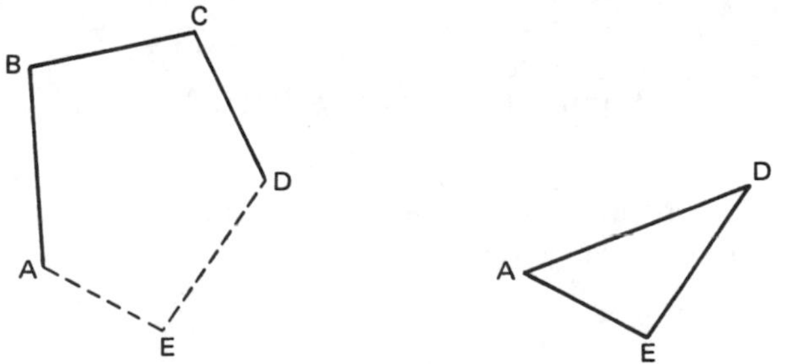


**Figure 17.11** Two Adjacent Legs Missing

**Two non-adjacent legs missing**: Figure 17.12 shows a traverse with two non-adjacent legs missing. Since the latitudes and departures of two parallel lines are equal, this can be solved by closing up the traverse and shifting one missing leg to be adjacent to the other missing leg. This reduces the problem to the previous case.

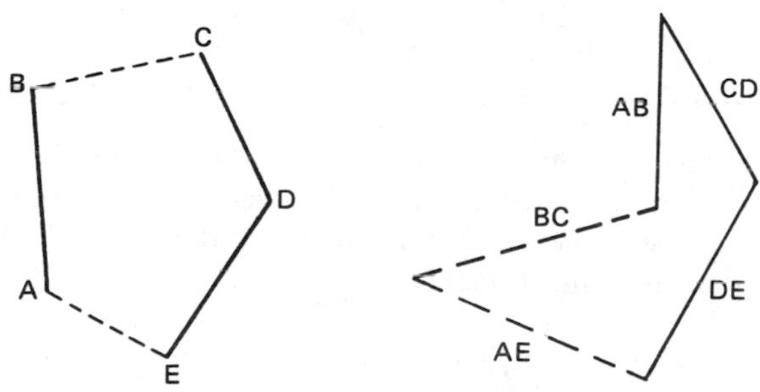


**Figure 17.12** Two Non-Adjacent Legs Missing

## E. AREA OF A TRAVERSE

An area of a traverse can always be found by dividing the traverse into a number of geometric shapes (rectangles, triangles, etc.) and summing the area of each subdivision. If the coordinates of the traverse leg end points are known, the *method of coordinates* can be used. The coordinates can be $x$–$y$ coordinates referenced to some arbitrary set of axes, or they can be departure-latitude coordinates.

$$A = \left| \frac{1}{2} \left( \sum_{i=1}^{n} y_i (x_{i-1} - x_{i+1}) \right) \right| \qquad 17.25$$

In equation 17.25, $x_0$ is replaced with $x_n$. Similarly, $x_{n+1}$ is replaced with $x_1$.

The area calculation is simplified if the coordinates are written in the following form:

$$\frac{x_1}{y_1} \times \frac{x_2}{y_2} \times \frac{x_3}{y_3} \times \frac{x_4}{y_4} \times \frac{x_1}{y_1} \quad \text{etc.}$$

Then, the area is

$$A = \frac{1}{2} \left| \left( \sum \text{full line products} - \sum \text{dotted line products} \right) \right| \qquad 17.26$$

*Example 17.15*

Calculate the area of a triangle with coordinates of its corners given: (3,1), (5,1), and (5,7).

$$\frac{3}{1} \times \frac{5}{1} \times \frac{5}{7} \times \frac{3}{1}$$

$$A = \frac{1}{2}\Big[(3)(1) + (5)(7) + (5)(1) - (1)(5)$$
$$-(1)(5) - (7)(3)\Big]$$
$$= \frac{1}{2}[43 - 31] = 6$$

F. AREAS BY DOUBLE MERIDIAN DISTANCES

If the latitudes and departures are known, the double meridian distance method can be used to calculate the area of the traverse. Equation 17.27 defines a *double meridian distance* (DMD).

$$\text{DMD}_{\text{leg i}} = \text{DMD}_{\text{leg i-1}} + \text{departure}_{\text{leg i-1}} + \text{departure}_{\text{leg i}} \qquad 17.27$$

Special rules are required to handle the first and last courses. The DMD of the first course is defined as the departure of that course. The DMD of the last course is the negative of its own departure.

The traverse area is calculated from equation 17.28.

$$A = \frac{1}{2}\left|\sum_i \text{latitude}_{\text{leg i}} \times \text{DMD}_{\text{leg i}}\right| \qquad 17.28$$

The next example illustrates the tabular approach often preferred with the double meridian distance method.

*Example 17.16*

The latitudes and departures of a six-leg traverse have been calculated. Use the double meridian distance method to calculate the traverse area. (The latitudes and departures are given in the table.) All values are in feet.

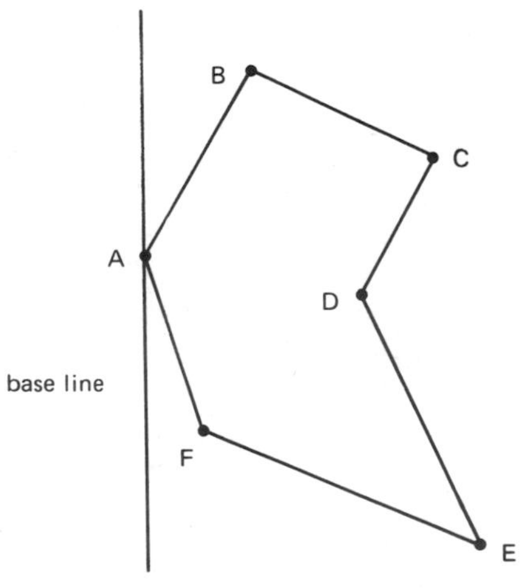


By the special rule, the DMD of the first leg, $AB$, is the departure of leg $AB$. The DMD of leg $BC$ is

$$\begin{aligned}\text{DMD}_{BC} &= \text{DMD}_{AB} + \text{departure}_{AB} + \text{departure}_{BC}\\ &= 200 + 200 + 200\\ &= 600\end{aligned}$$

The other DMD's are calculated similarly.

| leg | latitude | departure | DMD | lat ×DMD |
|---|---|---|---|---|
| $AB$ | 200 | 200 | 200 | 40,000 |
| $BC$ | −100 | 200 | 600 | −60,000 |
| $CD$ | −200 | −100 | 700 | −140,000 |
| $DE$ | −300 | 200 | 800 | −240,000 |
| $EF$ | 200 | −400 | 600 | 120,000 |
| $FA$ | 200 | −100 | 100 | 20,000 |
| | | | TOTAL | −260,000 |

The area is

$$A = \frac{1}{2}(260{,}000) = 130{,}000 \text{ ft}^2$$

## 8 AREAS WITH IRREGULAR BOUNDARIES

Areas of sections with irregular boundaries, such as creek banks, cannot be determined precisely, and approximation methods must be used. If the irregular side can be divided into a series of cells, either the trapezoidal rule or Simpson's rule can be used.

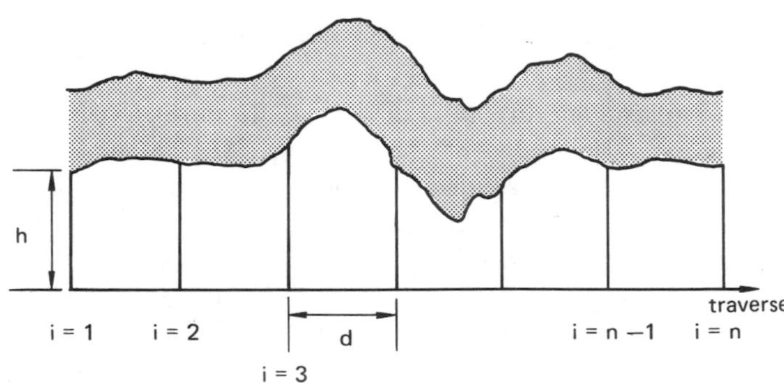


**Figure 17.13** Creek Bank Areas

If the irregular side of each cell is fairly straight, the *trapezoidal rule* is appropriate.

$$A = d\left(\frac{h_1 + h_n}{2} + \sum_{i=2}^{n-1} h_i\right) \qquad 17.29$$

If the irregular side of each cell is curved (parabolic), *Simpson's rule* should be used.[10]

$$A = \frac{d}{3}\left[h_1 + h_n + 2\sum_{\substack{i\,\mathrm{odd}\\ i=3}} h_i + 4\sum_{\substack{i\,\mathrm{even}\\ i=2}} h_i\right] \quad 17.30$$

## 9 CURVES

Roads, rail lines, and water courses are usually designed to be straight lines. Where a direction change is needed, a curve is used. The straight lines connected by a curve are known as *tangents* or *tangent lines.* A curve on level ground changing the direction of two tangents is known as a *horizontal curve.* Horizontal curves are usually arcs of circles.

Curves must also be used to connect roads and rail lines that change grade (slope). Such curves are called *vertical curves.* A curve that connects an upgrade tangent to a downgrade tangent is known as a *crest curve,* whereas a curve that connects a downgrade tangent to an upgrade tangent is known as a *sag curve.* Vertical curves are usually parabolic in shape.

**Figure 17.14** Sag and Crest Vertical Curves

## 10 HORIZONTAL CURVES

### A. INTRODUCTION TO SYMBOLS

The elements of circular curves and their standard abbreviations are given here.

- $R$ radius of the curve
- $V$ vertex of the tangent intersection point
- $PI$ point of intersection
- $I$ interior angle
- $PC$ point of curvature—the place where the first tangent ends and the curve begins
- $PT$ point of tangency—the place where the curve ends and the second tangent begins
- $POC$ any point on the curve
- $LC$ length of the arc—the length of the curve from $PC$ to $PT$
- $T$ tangent distance from $V$ to $PC$ or from $V$ to $PT$
- $C$ the long chord—the straight distance from $PC$ to $PT$
- $E$ the external distance—the distance from $V$ to the midpoint of the curve
- $M$ the middle ordinate—the distance from the curve midpoint to the midpoint of the long chord
- $D$ the degree of the curve

The following alternative designations are also sometimes used.

- $TC$ a change from a tangent to a curve (same as $PC$)
- $CT$ a change from a curve to a tangent (same as $PT$)
- $BC$ the beginning of a curve (same as $PC$)
- $EC$ the end of a curve (same as $PT$)
- $\Delta$ the intersection angle (same as $I$)

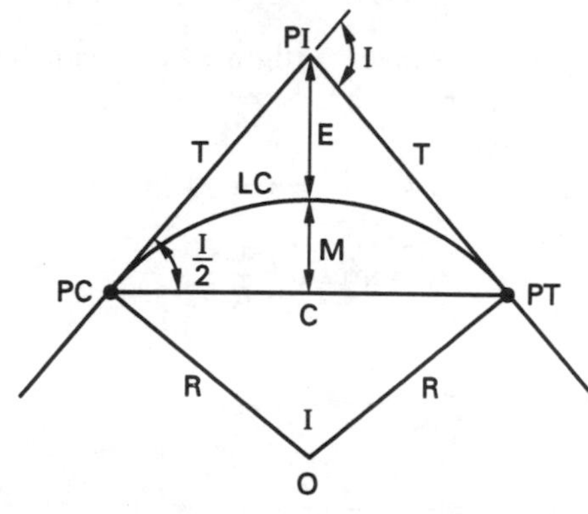


**Figure 17.15** Elements of a Circular Curve

Equations 17.31 through 17.35 can be used to solve problems involving circular curves.

$$T = R\tan\left(\tfrac{1}{2}I\right) \quad 17.31$$

$$\begin{aligned} E &= R\left(\tan\tfrac{1}{2}I\right)\left(\tan\tfrac{1}{4}I\right) \\ &= R\left(\sec\left(\tfrac{1}{2}I\right) - 1\right) \end{aligned} \quad 17.32$$

$$M = R\left(1 - \cos\tfrac{1}{2}I\right) = \tfrac{1}{2}C\left(\tan\tfrac{1}{4}I\right) \quad 17.33$$

$$C = 2R\left(\sin\tfrac{1}{2}I\right) = 2T\left(\cos\tfrac{1}{2}I\right) \quad 17.34$$

$$\begin{aligned} LC &= R(I\text{ in radians}) = R(I\text{ in degrees})\left(\frac{2\pi}{360}\right) \\ &= 100\left(\frac{I}{D}\right) \end{aligned} \quad 17.35$$

10 Also known as *Simpson's 1/3 rule.*

The curvature of city streets, property boundaries, and some highways is usually specified by the radius, $R$. The curvature may also be specified (in degrees) by the *degree of curve, D*.

$$D = \frac{360 \times 100}{2\pi R} = \frac{5729.6}{R} \quad 17.36$$

In most highway work, the length of the curve is understood to be the actual arc, and the degree of the curve is the angle subtended by an arc of 100 feet.[11] Therefore, the degree of curve can be expressed in degrees per station.

Stationing is continuous every 100 feet along a highway and around the curve. However, when the initial route is laid out between $PI$'s, the curve is undefined. The route distance is measured from $PI$ to $PI$. Therefore, each $PI$ will have two stations associated with it. The *forward station* is equal to the $PC$ station plus the tangent length. The *back station* is equal to the $PT$ station minus the tangent length. (The $PT$ station is equal to the $PC$ station plus the arc length.)

## B. TANGENT OFFSETS FOR CIRCULAR CURVES

The tangent offset shown in figure 17.16 can be calculated from equation 17.38.

$$y = R(1 - \cos\alpha) \quad 17.38$$

$$\alpha = \arcsin\left(\frac{x}{R}\right) \quad 17.39$$

$$x = R\sin\alpha \quad 17.40$$

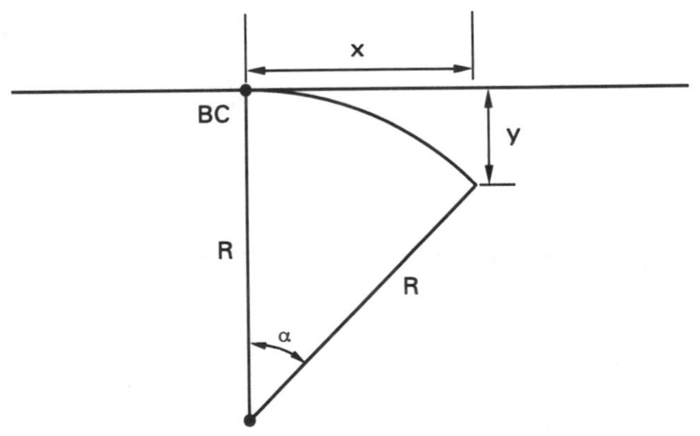


**Figure 17.16** Tangent Offset

[11] When the degree of curve is related to an arc of 100 feet, it is said to be calculated on an *arc basis*. In railroad surveys, the *chord basis* is used, and the degree of curve is the angle subtended by a chord of 100 feet. In that case, the degree of curve and radius are related by

$$\sin\left(\frac{D}{2}\right) = \frac{50}{R} \quad 17.37$$

Where the radius is large (4° curves or smaller), the difference between the arc and chord methods is insignificant.

## C. DEFLECTION ANGLES

The surveyor must stake out the curve so that the road crew knows where to put the road. Stakes should be put at the $PC$, $PT$, and at all full stations. If the curve is sharp, stakes may also be required at +25, +50, and +75 stations. The *deflection angle method* is the most common method used for staking out the curve. In this method, the curve distance is usually assumed to start from $00 + 00$ at the $PC$.

The *deflection angle* is defined as the angle between the tangent and a chord. This is illustrated in figure 17.17. The deflection angles are calculated using the following theorems.

- The deflection angle between a tangent and a chord is half the subtended arc (Fig. 17.17(a)).
- The angle between two chords is half the subtended arc (Fig. 17.17(b)).

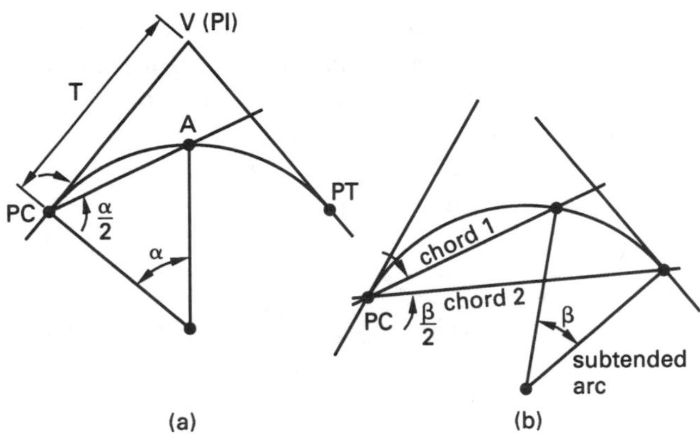


**Figure 17.17** Circular Curve Deflection Angle

In figure 17.17, angle $V$–$PC$–$A$ is a deflection angle between a tangent and a chord. The angle is

$$V\text{–}PC\text{–}A = \tfrac{1}{2}\alpha \quad 17.41$$

Angle $\alpha$ can be found from the following relationships.

$$\frac{\alpha}{360} = \frac{\text{arc length}\,(PC\text{–}A)}{2\pi R} \quad 17.42$$

$$\frac{\alpha}{I} = \frac{\text{arc length}\,(PC\text{–}A)}{LC} \quad 17.43$$

The chord length $PC$–$A$ is given by equation 17.44.

$$\text{chord length } PC\text{–}A = 2R\sin\left(\tfrac{1}{2}\alpha\right) \quad 17.44$$

The entire curve can be laid out from the $PC$ by sighting the deflection angle $V$–$PC$–$A$ and taping the chord distance $PC$–$A$. The $PC$ and $PT$ can be found by solving for $T$ and starting at $V$.

*Example 17.17*

A circular curve is to be constructed with a 225 foot radius and an interior angle of 55°. Determine where the stakes should be placed if the separation between stakes along the arc is 50 feet. Specify the first and last interior angles (subtended arcs) and chord lengths.

The length of the curve is given by equation 17.35.

$$LC = (225)(55)\left(\frac{2\pi}{360}\right) = 215.98\text{ ft}$$

The last stake will be 215.98 − 200 = 15.98 feet from the next to the last stake. The central angle for an arc of 50 feet is given by equation 17.42.

$$\alpha_{\text{degrees}} = \left(\frac{360}{2\pi}\right)\left(\frac{50}{225}\right) = 12.732°$$

12.732° goes into 55° four times with a remainder of 4.072°. From equation 17.44, the required chord lengths are

$$(2)(225)\sin\left(\frac{12.732}{2}\right) = 49.90\text{ ft}$$

$$(2)(225)\sin\left(\frac{4.072}{2}\right) = 15.98\text{ ft}$$

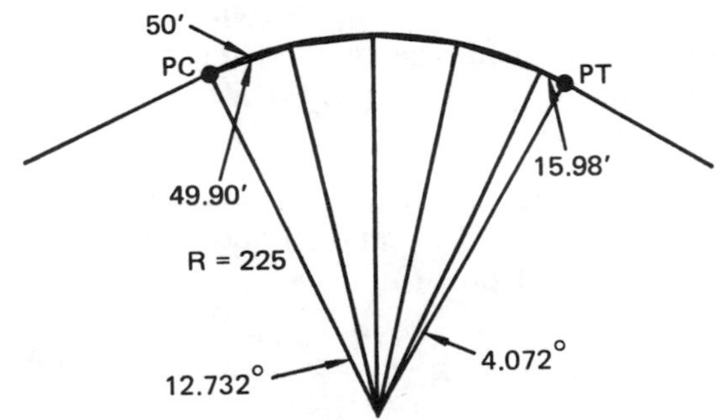


*Example 17.18*

An interior angle of 8.4° is specified for a horizontal curve. The *PI* station is 64+27.46. Use a 2° curve and locate the *PC* and *PT* stations.

From equation 17.36,

$$R = \left(\frac{360}{2}\right)\left(\frac{100}{2\pi}\right) = 2864.79\text{ ft}$$

From equations 17.31 and 17.35

$$T = (2864.79)\tan\left(\frac{8.4}{2}\right) = 210.38$$

$$LC = (2864.79)(8.4)\left(\frac{2\pi}{360}\right) = 420.00$$

Then, the *PC* and *PT* points are located.

$$PC = (64{+}27.46) - (2{+}10.38) = 62{+}17.08$$

$$PT = (62{+}17.08) + (4{+}20.00) = 66{+}37.08$$

## 11 VERTICAL CURVES

Vertical curves are used to change the grade of a highway. *Equal-tangent parabolic curves* are usually used for this purpose. A vertical sag curve connecting two grades is shown in figure 17.18. Since the grades are very small, the actual arc length of the curve is approximately equal to the chord length *BVC–EVC*.

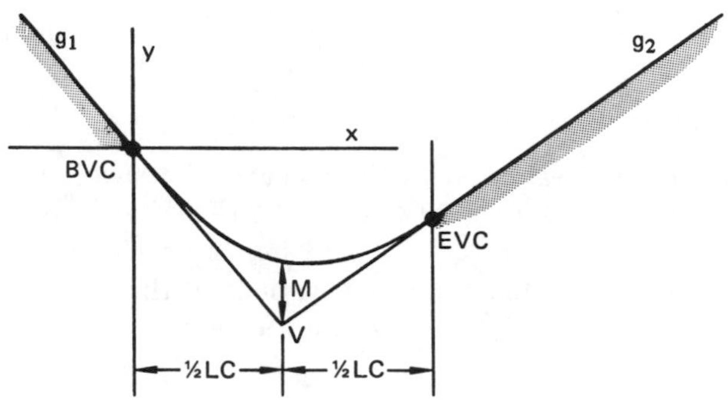


**Figure 17.18** A Vertical Curve

The following standard and optional abbreviations are used.

- *LC* the horizontal length of the curve, in stations
- $g_1$ the grade from which the stationing starts, in percent
- $g_2$ the grade towards which the stationing heads, in percent
- *V* the vertex—the intersection of the two tangents
- *PVI* same as *V*
- *BVC* beginning of the vertical curve
- *EVC* end of the vertical curve
- *PVC* same as *BVC*
- *PVT* same as *EVC*
- *M* the middle ordinate

A vertical parabolic curve is completely specified by the two grades and the curve length. Alternately, the *rate of grade change* per station can be used in place of the curve length.

$$r = \frac{g_2 - g_1}{LC} \qquad 17.45$$

Equation 17.46 defines an equal-tangent parabolic curve. $x$ is measured in stations beyond $BVC$. $y$ is measured in feet, with the same reference point used to measure all elevations.

$$\text{elev}_x = \left(\frac{r}{2}\right)x^2 + g_1 x + \text{elevation}_{BVC} \qquad 17.46$$

The maximum or minimum elevation will occur at the *turning point*. The turning point is not necessarily located directly above or below $V$, but is found at

$$x = \frac{-g_1}{r} \text{ (in stations)} \qquad 17.47$$

The *middle ordinate* distance is

$$M = \frac{|(g_1 - g_2)|(LC)}{8} \qquad 17.48$$

*Example 17.19*

A vertical crest curve with a length of 400 feet is to connect grades of +1.0% and −1.75%. The vertex is located at station 35 + 00, and it has an elevation of 549.20 feet. What are the elevations of the $BVC$ and $EVC$, and at all full stations on the curve?

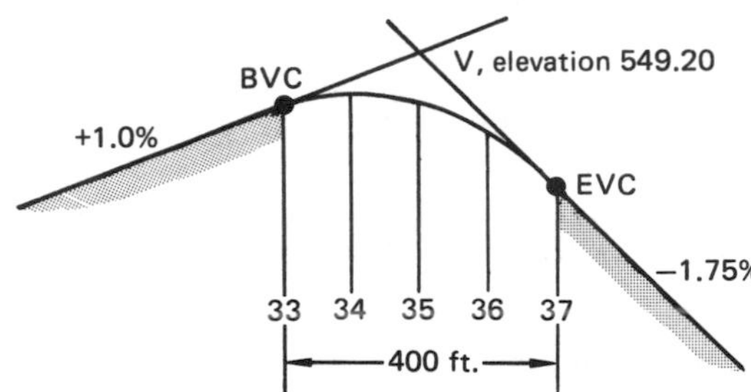


The elevation at $BVC$ is 549.20 − 1(2) = 547.20.

The elevation at $EVC$ is 549.20 − 1.75(2) = 545.70.

$$r = \frac{-1.75 - 1}{4} = -0.6875 \text{ percent}$$

$$\frac{1}{2}r = -0.3438$$

Since $g_1 = 1\%$, from Eq. 17.46, the equation of the curve is

$$y = -0.3438x^2 + x + 547.20$$

At station 34, $x = (34 - 33) = 1$. So,

$$y_{34} = -0.3438(1)^2 + 1 + 547.20 = 547.86$$

Similarly,

$$y_{35} = -0.3438(2)^2 + 2 + 547.20 = 547.82$$

$$y_{36} = -0.3438(3)^2 + 3 + 547.20 = 547.11$$

## 12 VERTICAL CURVES WITH OBSTRUCTIONS

If a curve is to have some minimum clearance from an obstruction, as in figure 17.19, the length of the curve, $BVC$, and $EVC$ will generally not be known in advance. The problem of finding the curve length can be solved by using the following procedure. The procedure can also be used when the curve is placed over a feature and a minimum cover depth is required.

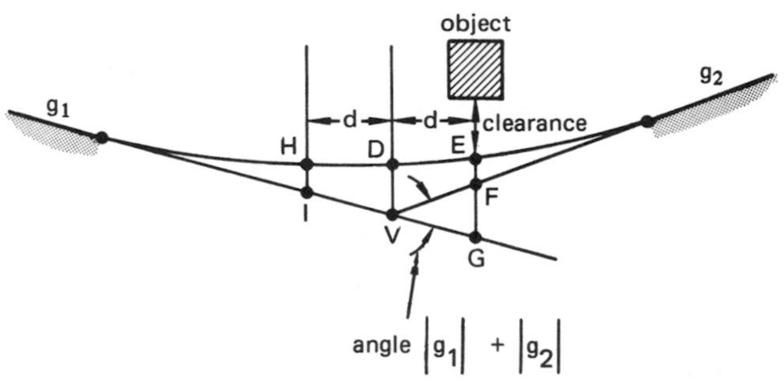


**Figure 17.19** A Curve with an Obstruction

This procedure is based on the fact that a vertical distance from a tangent line to a point on a curve (i.e., the *tangent offset*) is proportional to the square of the horizontal distance along the tangent. (See equation 17.46.) This fact, combined with the curve symmetry of an equal-tangent parabola ($EF = HI$ in figure 17.19), permits direct solutions.

*step 1*: Calculate the curve elevation of point $E$ directly below the object.

$$\text{elevation}_E = \text{elevation}_{\text{object}} - \text{clearance} \qquad 17.49$$

*step 2*: Calculate distance $EG$. ($g_1$ is negative as shown in figure 17.19. $d$ is in stations.) (When point $G$ is to the left of $V$, replace $g_1$ with $g_2$.)

$$EG = \text{elevation}_E - \text{elevation}_V + |(d)(g_1)| \qquad 17.50$$

*step 3*: Calculate the distance $EF$.

$$EF = \text{elevation}_E - \text{elevation}_V - (d)(g_2) \qquad 17.51$$

*step 4*: Solve simultaneously for $LC$. Both $d$ and $LC$ are in units of stations.

$$\frac{EG}{\left(\frac{LC}{2} + d\right)^2} = \frac{EF}{\left(\frac{LC}{2} - d\right)^2} \qquad 17.52$$

## 13 SPIRAL CURVES

*Spiral curves* are used to produce a gradual transition from tangents to curves. A spiral curve is a curve of gradually changing radius and gradually increasing degree of curvature. Figure 17.20 illustrates a spiral curve, showing the *tangent to spiral* ($TS$) point, *length of spiral* ($LS$), and *spiral to curve* ($SC$) point.

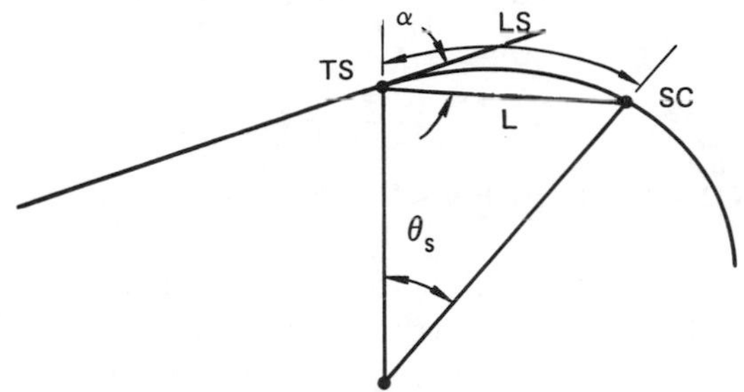


**Figure 17.20** A Simple Spiral Curve

The *degree of spiral* changes continuously, so the average degree of spiral is used to calculate the length. ($\theta_s$ is in degrees in equation 17.53.)

$$LS = \frac{100 \times \theta_s}{\frac{D}{2}} \quad 17.53$$

Deflection angles to be used when setting out points on the spiral curve can be calculated from equation 17.54.

$$\alpha = \frac{\theta_s}{3}\left(\frac{L}{LS}\right)^2 \quad 17.54$$

Experiments suggest that lengths of spiral curves should be based on the speed of traffic entering the curve and the radius of the curve being approached.

$$LS = \frac{(1.6)(\text{mph})^3}{R \text{ in feet}} \quad 17.55$$

## 14 PHOTOGRAMMETRY

Photogrammetry uses photography to obtain distance measurements. Both vertical and oblique photographs can be used, as shown in figure 17.21.

The *scale* of a vertical photograph is the ratio of dimensions on the photograph to the dimensions on the ground.

$$\text{scale} = \frac{\text{flight altitude (feet)}}{\text{focal length (inches)}} \quad 17.56$$

The scale is constantly changing since the elevation above ground level depends on the surface terrain. Also, the distance from the camera to the ground directly below it will be less than the distance from the camera to points on the outer fringes of the photograph. Therefore, an average distance from camera to ground level is used as the flight altitude.

The number of photographs and the number of flight paths required depend on the film size and lap percentages, as illustrated in example 17.20.

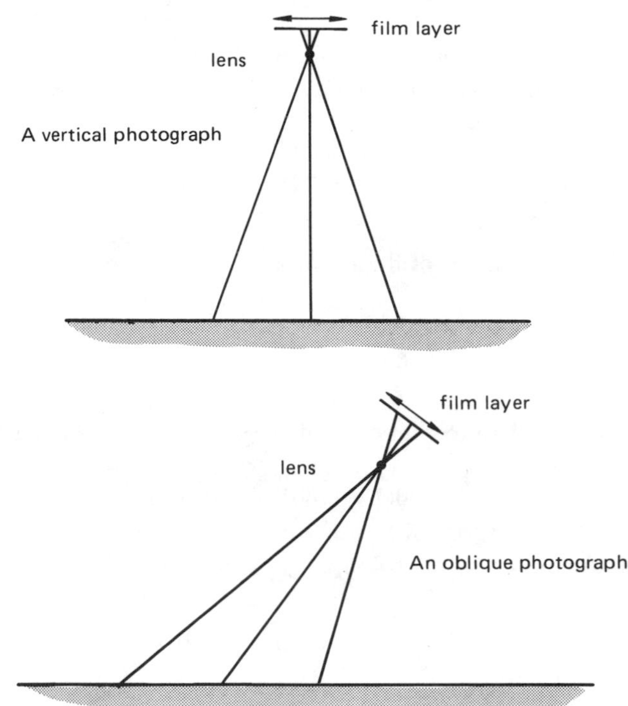


**Figure 17.21** Vertical and Oblique Photographs

*Example 17.20*

Aerial mapping is being used with a scale of 1″:1000′. The film holder uses 8″ × 8″ film. The focal length of the camera is 6″. The area to be mapped is 6 miles on each side (square). A side lap of 30% is desired, as well as an end lap of 60%. For continuity with adjacent aerial surveys, the overlaps will extend beyond the area boundaries on photographs of the perimeter. The plane travels at 150 mph.

(a) How many square feet does each photograph cover?

(b) At what altitude should the plane fly while taking photographs?

(c) How far apart will the flight paths be?

(d) How many flight paths are required?

(e) How many photographs are required per flight path?

(f) How many photographs will be taken altogether?

(g) How frequently should the photographs be taken?

(a) The photograph area is $8'' \times 8'' = 64\ \text{in}^2$. The area covered by each photograph is $64 \times 1000^2 = 64{,}000{,}000\ \text{ft}^2$.

(b) The flight altitude is calculated from equation 17.56.

$$\text{altitude} = 6'' \times 1000' = 6000\ \text{ft}$$

(c) Each photograph covers $64{,}000{,}000\ \text{ft}^2$, or an area 8000 ft on each side. With a 30% overlap, the distance between flight paths is

$$8000(1 - 0.30) = 5600\ \text{ft}$$

(d) The number of flight paths is

$$\frac{(6)(5280)}{5600} = 5.7 \quad (\text{say } 6)$$

(e) The distance between photographs along the flight path is

$$8000(1 - 0.60) = 3200\ \text{ft}$$

The number of photographs per flight path is

$$\frac{(6)(5280)}{3200} = 9.9 \quad (\text{say } 10)$$

(f) The total number of photographs will be $6 \times 10 = 60$.

(g) At 150 mph, the frequency of camera shots will be

$$\frac{3200}{\frac{150 \times 5280}{3600}} = 14.5\ \text{seconds/photograph}$$

## 15 TRIANGULATION

Triangulation is a method of surveying in which the positions of survey points are determined by measuring the angles of triangles. In triangulation, the survey lines form a network of triangles. Each survey point (monument) is at a corner of one or more triangles. The three angles of each triangle are measured. Lengths of triangle sides are trigonometrically calculated. The positions of the points are established from the measured angles and the computed sides.

Triangulation is used primarily for geodetic surveys, such as those performed by the National Geodetic Survey. Most first- and second-order control points in the national control network have been established by triangulation procedures. The use of triangulation for transportation surveys is minimal. Generally, its use is limited to strengthening traverses for control surveys.

## 16 TRILATERATION

Trilateration is similar to triangulation in that the survey lines form triangles. In trilateration, however, the lengths of the triangles' sides are measured, and the angles are computed. Orientation of the survey is established by selected sides whose directions are known or measured. The positions of trilaterated points are determined from the measured distances and the computed angles.

## 17 SOLAR OBSERVATIONS

The direction of a survey line can be determined by making horizontal and vertical angular measurements to the sun at a known time and from a point of known latitude. Such measurements are called *solar observations.*

Generally, solar observations are required only in areas where horizontal control points are sparse or control data are unavailable. In such areas, solar observations provide azimuth control (starting, closing, and check azimuths). In some cases, the entire orientation of a survey might be based on solar observations. In other surveys, they might simply provide orientation checks. For example, long traverses can require intermediate azimuth checks where record points do not exist. Another common use is orientation for retracement surveys when searching for corners in rugged terrain or in heavy vegetation.

Azimuths determined by solar observations are adequate for third-order surveys. Greater accuracy is impractical because exact pointings can not be made on the relatively fast-moving sun. If more accurate azimuths must be determined by astronomical observations, Polaris sightings should be used.

Azimuths should be reliable within 12 seconds. However, many factors can affect the accuracy. Two of the more important factors are the skill of the observer and the care used in making the observations. (Care in leveling the instrument and in pointing on the sun is especially important.) Other factors are the hour of the day and the time of the year the observations are made, the precision of the theodolite, the accuracy of the time measurements, and the latitude of the point of observation and the accuracy to which it is determined.

## 18 CELESTIAL BASIS

An azimuth determined by solar observations is based on the solution of a spherical triangle, the *astronomical triangle.* Figure 17.22 shows the points on the celestial sphere that form this triangle. These three points are

- **P**-north pole of the celestial sphere
- **Z**-observer's zenith point. This is the point on the celestial sphere directly above the point of observation.
- **S**-sun

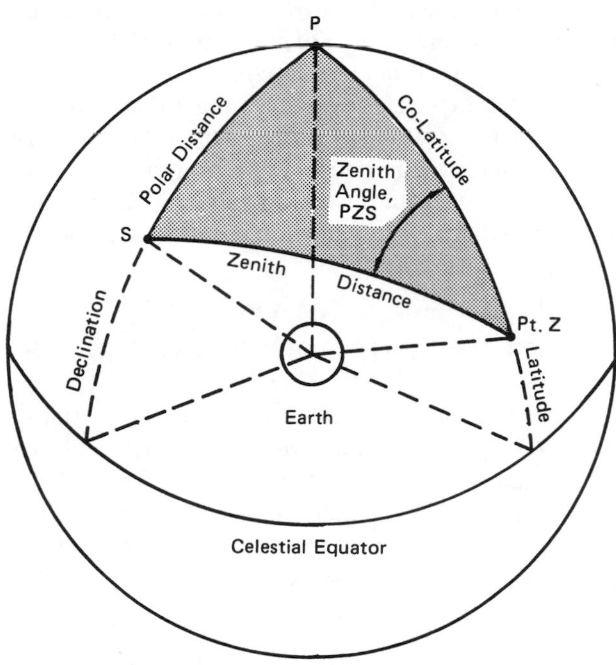


**Figure 17.22** Astronomical Triangle for Solar Observation

In figure 17.22, triangle $PZS$ is the *astronomical triangle.* Angle $PZS$ is known as the *zenith angle.*

## 19 SIDES (ANGLES) OF THE ASTRONOMICAL TRIANGLE

Side $S$-$P$ is known as the *polar distance.* It can be found as 90° minus the sun's declination. *Declination* is the angular distance the sun is above or below the celestial equator. The value is typically determined from an *ephemeris.*

Side $P$-$Z$ is known as the *co-latitude.* It is found as 90° minus the latitude of the point from which the observation is made.

Side $Z$-$S$ is known as the *zenith distance.* It is 90° minus the true altitude of the sun. The *true altitude* is the measured altitude corrected for refraction and parallax. Tables of corrections are also available in ephemerides.

Figure 17.23 illustrates how the refraction correction ($r$) is added to the field-measured zenith angle, and the parallax correction ($p$) is subtracted from it.

$$\text{true altitude} = \text{measured altitude} - \text{refraction} + \text{parallax} \qquad 17.57$$

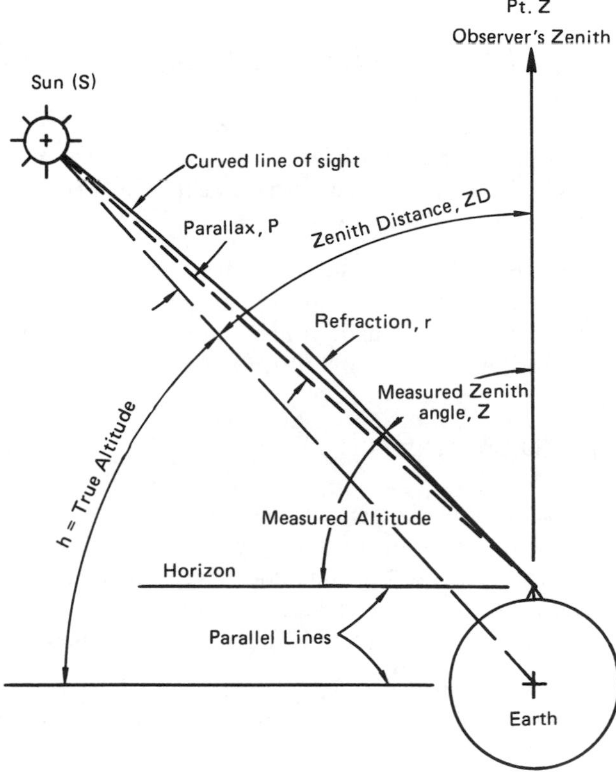


**Figure 17.23** Zenith Distance and True Altitude

## 20 ZENITH ANGLE

With three sides of the spherical triangle known, the angle between the north pole and the sun at the observer's zenith can be calculated. This angle is known as the *zenith angle,* $PZS$.[12] It establishes the azimuth of the sun (line $Z$–$S$), and therefore is also known as the *azimuth angle.*

Two equations can be used to calculate the zenith angle $PZS$ from its trigonometric functions. In equations 17.58 and 17.59, the following nomenclature is used:

$PZS$ = zenith angle.
$d$ = sun's declination.
$h$ = sun's true altitude.

[12] Note that the term "zenith angle" also refers to the angle $Z$ in a vertical plane that is measured with a theodolite that has a vertical circle oriented at zero degrees when the telescope is pointed on the zenith. See figure 17.23.

$l$ = latitude of the point of observation.
$p$ = polar distance = $90° - d$.
$s = 1/2(l + h + p)$

$$\cos PZS = \frac{(\sin d) - [(\sin h)(\sin l)]}{(\cos h)(\cos l)} \qquad 17.58$$

If $\cos PZS$ is negative, $PZS$ is greater than 90 degrees.

$$\tan\left(\frac{PZS}{2}\right) = \sqrt{\frac{[\sin(s-l)][\sin(s-h)]}{[\cos s][\cos(s-p)]}} \qquad 17.59$$

The cosine formula is easier to use. However, the cosine function is less precise than the tangent when the angle is near zero or 180 degrees.

If the observations are made in the morning, the sun's azimuth is equal to the zenith angle. For afternoon observations, the sun's azimuth is obtained by subtracting the zenith angle from 360 degrees.

## 21 LINE AZIMUTH

The azimuth of the survey line is referred to the sun's azimuth by simultaneously measuring the horizontal angle from the line to the sun and the vertical angle to the sun. When the horizontal angle is measured clockwise, the azimuth of the survey is determined as follows:

- If the horizontal angle is equal to or less than the sun's azimuth, subtract the horizontal angle from the sun's azimuth.
- If the horizontal angle is greater than the sun's azimuth, subtract the horizontal angle from 360 degrees and add the remainder to the sun's azimuth.

*Example 17.21*

A horizontal angle was measured during a survey as 336°. The zenith angle $PZS$ was measured as 100°. What was the azimuth of the survey line $A$–$B$?

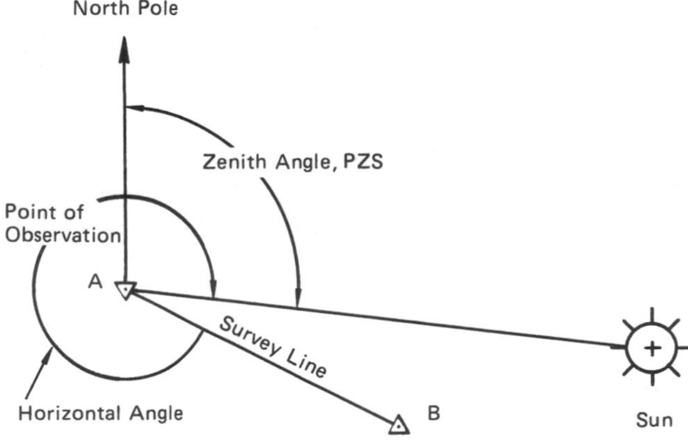


$$\text{azimuth } A\text{–}B = 100 + 360 - 336 = 124°$$

## 22 POLARIS OBSERVATIONS

Azimuths of survey lines can be determined by making horizontal angular measurements to *Polaris*, the *North Star*, at known, exact times and from points of known latitude. Such measurements also can be used when the latitude is unknown. In this case, horizontal and zenith observations are made simultaneously.

Generally, Polaris observations are required only in areas where horizontal control points are sparse or control data are unavailable. In such areas, Polaris observations can be used to provide azimuth control. In some cases, the entire orientation of a survey might be based on Polaris observations. In other surveys, they might simply provide azimuth checks. For example, Polaris observations might be used to provide azimuth checks on a long traverse that has numerous courses between existing control points.

## 23 ACCURACY OF POLARIS AZIMUTHS

Azimuth control can be determined by observations on the sun. Generally, solar observations are more convenient than Polaris observations. Therefore, solar observations are usually made when the accuracy requirements are third-order or less. Polaris observations must be used for second-order accuracy.

Polaris azimuths should be reliable within five seconds. However, many factors can affect the accuracy. The more important factors (in approximate order of importance) are:

- The skill of the observer and the care used in making the observations. Care in leveling the instrument and in noting when the star is bisected by the cross hair is especially important.
- The accuracy of the time measurement.
- The hour of the observations.
- The latitude, and its accuracy, of the occupied point. Or, if the star's altitude is measured, the accuracy of the zenith angle measurements.

## 24 CELESTIAL BASIS

Polaris is a relatively bright star in the northern sky. Its apparent position is always very close to the vertical projection of the North Pole. The actual angular distance from the pole (the *polar distance*) is approximately 51 minutes and is almost a constant. It varies less than one minute during the year.

As viewed from the earth, Polaris appears to revolve counterclockwise around the pole. One "revolution" is completed each sidereal day (time measured relative to the stars). Unlike the sun and some other stars, Polaris does not "rise" and "set." It is always visible. Because of its location and relatively slow apparent motion, Polaris is the star commonly used for determining azimuths and latitudes by stellar observations.

An azimuth determined by Polaris observations is based on the solution of a spherical triangle which is called the *astronomical triangle.* Figure 17.24 shows the points on the celestial sphere that form this triangle. These three points are:

- **P**-north pole of the celestial sphere
- **Z**-observer's zenith point. This is the point on the celestial sphere directly above the point of observation.
- **S**-pole star Polaris

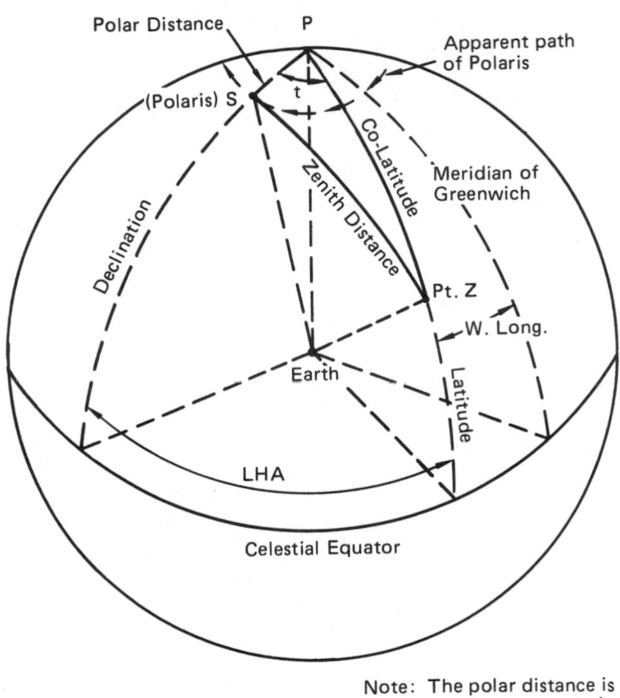


**Figure 17.24** Astronomical Triangle for Polaris Observations

## 25 SIDES (ANGLES) OF THE ASTRONOMICAL TRIANGLE

In figure 17.24, side *S-P* is known as the *polar distance.* It is calculated as 90° minus Polaris's declination. This *declination* is the angular distance above or below the celestial equator that Polaris is. It is found from an ephemeris.

Side *P-Z* is the *co-latitude.* It is calculated as 90° minus the latitude of the point of observation.

Side *S-Z* is the *zenith distance.* It is 90° minus the true altitude of Polaris. The *true altitude* is the measured altitude corrected for refraction. The *measured altitude* is 90° minus the zenith angle of Polaris. Tables for refraction are contained in ephemerides.

In addition to the three sides, the angle at the pole between the zenith point and Polaris can be determined. This angle ($SPZ$) is called the *meridian angle.* The meridian angle is commonly designated by the letter "$t$" as shown in figure 17.24. It is computed by comparing the actual time of the observation with the time at which upper culmination of Polaris occurs at the observer's meridian.[13] (*Upper culmination* is the upper crossing of a meridian by a celestial body.) The meridian angle is computed as $t = LHA$ or $t = LHA - 360$, whichever is smaller in magnitude.

$LHA$ is the *local hour angle,* equal to the time of observation minus the time of upper culmination ($LCT$) plus the correction for sidereal time. (*Sidereal time* is based on the rotation of the earth relative to the stars.)

$LCT$ is the *local civil time,* which is the mean time based on the observer's meridian. Mean time is based on the average rotation of the earth relative to the sun.

## 26 ZENITH ANGLE FOR POLARIS OBSERVATIONS

With three elements of the spherical triangle known, the angle between the North Pole and Polaris at the observer's zenith can be calculated. This angle is known as the *zenith angle, PZS.*[14] It establishes the azimuth of Polaris (line *Z-S*), and is, therefore, sometimes called the *azimuth angle.*

The zenith angle $PZS$ can be calculated from equations 17.60 and 17.61. In both equations, the algebraic sign of $t$ is ignored. (It should be remembered that the sign of the cosine function is negative when the angle is between 90 and 270 degrees.) The following nomenclature is used in these two equations.

[13] The times at which upper culmination of Polaris occur are tabulated in ephemerides. Some ephemerides use a different procedure to compute the meridian angle. Thus, they will tabulate some value other than the time of upper culmination.

[14] The term "zenith angle" can also be used for the angle in a vertical plane that is measured with a theodolite that has a vertical circle oriented at zero degrees when the telescope is pointed on the zenith.

$PZS$ = zenith angle
$d$ = Polaris's declination
$l$ = latitude of the point of observation
$t$ = meridian angle
$h$ = Polaris's true altitude

$$\tan(PZS) = \frac{\sin t}{(\cos l)(\tan d) - (\sin l)(\cos t)} \qquad 17.60$$

$$\sin(PZS) = \frac{(\sin t)(\cos d)}{\cos h} \qquad 17.61$$

Because of the size and limited variance in the angles involved, the sine equation can be very closely approximated by equation 17.62. In equation 17.62, $p$ is the polar distance, $90° - d$.

$$\angle PZS = \frac{\sin t}{\cos h}(90 - d) = \frac{\sin t}{\cos h}p \qquad 17.62$$

When $t$ is negative (0 to 12 hours prior to upper culmination), the azimuth of Polaris is equal to the zenith angle. When $t$ is positive (0 to 12 hours after upper culmination), the azimuth is obtained by subtracting the zenith angle from 360 degrees.

## 27 LINE AZIMUTH FROM POLARIS OBSERVATIONS

The azimuth of the survey line is referred to the azimuth of Polaris by the horizontal angle measurements. When the horizontal angle is measured clockwise, the azimuth of the survey line is determined as follows:

1. If the horizontal angle ($H$) is equal to or less than the azimuth of Polaris ($Az$), subtract the horizontal angle from the azimuth of Polaris.

2. If the horizontal angle is greater than the azimuth of Polaris, then subtract the horizontal angle from 360 degrees and add the result to the azimuth of Polaris.

*Example 17.22*

Referring to figure 17.25, suppose the zenith angle was 0°44′0″ and the horizontal angle was 304°30′0″. What would be the azimuth of the survey line?

$$360° - H = 55°30'0''$$

$$\text{azimuth of line } A\text{–}B = 55°30'0'' + 0°44'0'' = 56°14'0''$$

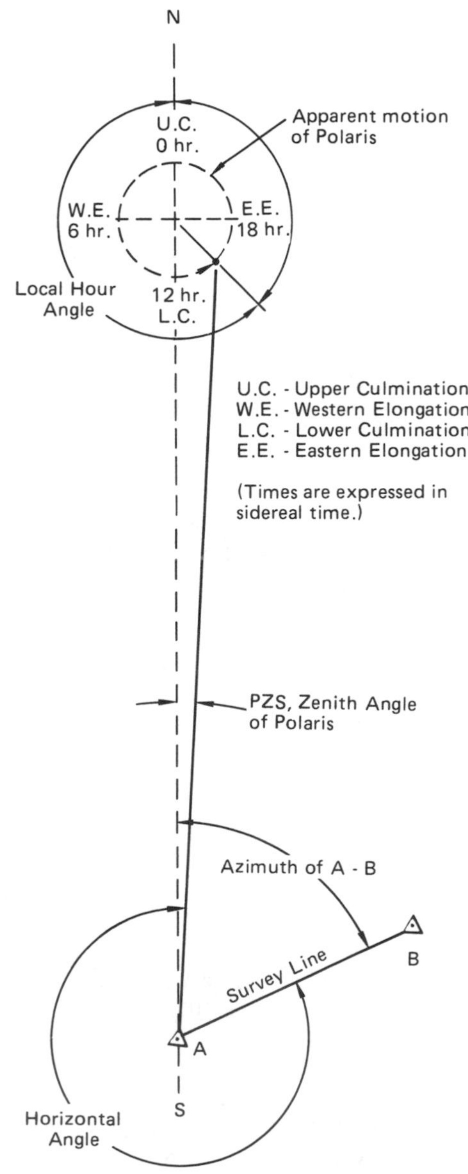


**Figure 17.25** Polaris observation

## 28 SURVEY STAKING

Wooden stakes used by surveyors are commonly referred to as *construction stakes*. Depending on their use, construction stakes are also called alignment stakes, offset stakes, grade stakes, and slope stakes. Stakes come in many sizes, including $1'' \times 2'' \times 18''$ markers, $5/16'' \times 1\frac{1}{2}'' \times 24''$ half-lath stakes, and various lengths of $2'' \times 2''$ hubs and $1'' \times 1''$ guineas. Hubs and guineas are usually 12 or 18 inches long, and occasionally may measure 24 inches in length.

The front (the side facing the construction) and back (the side facing away from construction) of the stakes are marked with keel, carpenter's crayon, or permanent ink markers. Stakes are read from top to bottom. Distances are measured in feet, and a maximum precision of 0.1 foot is considered standard for earthwork. A precision of 0.01 foot is standard for locating positions and elevations of installed structures. Pin flags or full-length lagging may be used to provide further identification and visibility.

Front-of-stake markings include *header* information (e.g., *RPSS*, offset distance) and *cluster* information (e.g., horizontal and vertical measurements, and slope ratio). The header is separated from the first cluster by a double horizontal line. Multiple clusters are separated by single horizontal lines. All clusters are measured in the same direction from the same point.

Common abbreviations used on the fronts of construction stakes include the following:

| | |
|---|---|
| @ | "from the" |
| *BC* | begin curve |
| *BM* | benchmark |
| *BVC* | begin vertical curve |
| *C* | centerline or cut |
| *CB* | catch basin |
| *CF* | curb face |
| *CGS* | contour grading stake |
| *CL* | centerline |
| *CONT* | contour |
| *CP* | control point or catch point |
| *CR* | curb return |
| *DI* | drop inlet |
| *DMH* | drop manhole |
| *E* | flow line |
| *EC* | end of curve |
| *EL* | elevation |
| *ELECT* | electrical |
| *EP* | edge of pavement |
| *ESR* | end slope running |
| *EVC* | end of vertical curve |
| *F* | fill |
| *FC* | face of curb |
| *FG* | finished grade |
| *FH* | fire hydrant |
| *FL* | flow line |
| *FTG* | footing |
| *G* | grade |
| *GTR* | gutter |
| *HP* | hinge point |
| *INL* | inlet |
| *INT* | intersection |
| *INV* | invert |
| *ISS* | intermediate slope stake |
| *JT* | joint trench |
| *L* | left |
| *LT* | left |
| *MC* | middle of curve |
| *MH* | manhole |
| *MP* | midpoint |
| *PC* | point of curvature |
| *PCC* | point of compound curvature |
| *PI* | point of intersection |
| *POC* | point on curve |
| *POL* | point on line |
| *POT* | point on tangent |
| *PP* | power pole |
| *PRC* | point of reverse curvature |
| *PT* | point of tangency |
| *PVC* | point of vertical curvature |
| *PVT* | point of vertical tangency |
| *R* | right |
| *RP* | radius point or reference point |
| *RPSS* | reference point slope stake |
| *RT* | right |
| *R/W* | right of way |
| *SD* | storm drain |
| *SS* | sanitary sewer or slope stake |
| *STA* | station |
| *TBM* | temporary benchmark |
| *TC* | top of curb |
| *TP* | turning point |
| *VP* | vent pipe |
| *WL* | water line |
| *WV* | water valve |
| *WM* | water meter |

There is great variation in stake marking conventions. The conventions listed here are common but not universal practice. Words like "from," "above," and "below" are understood and are seldom actually written on a stake. For example, "2.0 *FC*" and "4.3 *FG*" mean "2.0 feet from the face of the curb," and "4.3 feet above the finished grade," respectively.

The back of a stake is usually used to record the station or other literal locating information (e.g., "at ramp"). Actual elevations, when included, are marked on the side of the stake, not on the front or back. Other information identifying the survey may also be included on the back.

In its simplest use, a stake both locates and identifies a specific point. A distance referenced on a stake is measured from the natural ground at the point where the stake is driven. Another less frequently used system involves measuring from a tack or a reference mark, such as a crow's foot, on the stake itself. A *crow's foot* consists of a horizontal line drawn on the stake with a vertical arrow pointing to it. When used, a crow's foot should be drawn on the side, not the front or back, of the stake.

Many times, however, measurements are taken from a *hub stake* (also known as a *ground stake*, *reference point stake*, or *RP*) driven essentially flush with the ground. No markings are made on ground stakes. The term *blue top* can mean a ground stake that is pounded to grade or, more generally, a stake on which grading information is noted.

Hubs and other ground stakes are located, identified, and protected by *witness stakes* (also known as *guard stakes*). A witness stake documents a ground stake but does not itself locate a specific point. A guard stake may be driven at an angle with its top over the flush-driven stake. The combination of ground and witness stakes together are loosely referred to by their functional name. (That is, when a ground stake is used, the term *alignment stake* refers to the combination of ground and witness stakes.)

Alignment stakes marked with the station indicate the centerline alignment of roads and highways. They are placed every 50 or 100 foot station along highway tangents with uniform grade, and every 25 or 50 foot station for horizontal and vertical curves.

*Offset stakes* used to mark excavations or roads for paving are offset from the actual edge to protect the stakes from construction equipment. The offset distance is circled on the stake and is separated from the subsequent data by a double line. For close-in work, the offset distance is often standardized at 2 feet. For highway work, offset stakes (measured from the alignment centerline) can be set 25, 50, or 100 feet from the centerline on both sides. Unless a separate ground stake (hub) is used, distances (e.g., cuts, fills, and distances to centerline) marked on an offset stake refer to the point of insertion, not to the imaginary point located the offset distance away.

Figure 17.26 illustrates how a construction stake would be marked to identify a trench for a storm drain.

*Slope stakes* indicate *grade points* (i.e., points where cuts and fills begin and the planned side slopes intersect the natural ground surface). Slope stakes are marked *SS* to indicate their purpose. (It is unusual for a slope stake to be set at the actual points where a cut or fill begins. The typical position is on a 10-foot offset.)

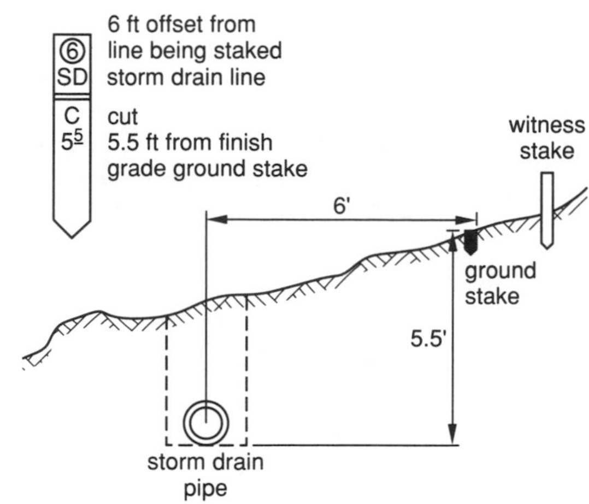


**Figure 17.26** Construction Stake for Storm Drain

In addition to indicating the grade point, the front of slope stakes are marked to indicate the nature of the earthwork (i.e., *C* for "cut" and *F* for "fill"), the offset, the type of line being staked, the distance from the roadbed centerline or other stationed control line, the slope (horizontal:vertical) to finished grade, and the elevation difference between the grade point and the finished grade. Distances from the centerline can be marked *R* or *L* to indicate whether the stake points are to the right or left of the centerline when looking up-station. The station, reckoned along the centerline, is marked on the back of the stake.

For example, the front-of-stake markings "$C$ 4.2 $FG$ @ 38.4$L$ $CL$ 2:1" would be interpreted as "cut with 2:1 slope is required; stake set point is 4.2 feet above finished grade; stake is 38.4 feet to the left of the centerline of roadway." The use of $FG$ is optional.

Slope stakes can be driven vertically or at an angle, depending on convention. When driven at an angle, slope stakes slant outward (point away from the earthwork) when fill is required, and inward (point into the earthwork) when a cut is required. The stakes are set with the broad (front) face toward the construction.

Ground stakes (hubs) without offsets are not used with fill stakes, as a ground stake would soon be covered. For shallow and moderate fills, a fill stake alone is used. A crow's foot may be marked on the stake to indicate the approximate finished grade, or the top of the stake may be set to coincide with the approximate finished grade. For deeper fills, offset stakes are required. Fill stakes are sometimes ripped diagonally lengthwise to give them a characteristic shape.

Not all earthwork produces a straight slope. Some earthwork results in a smooth-curved ground surface between two lateral points. The points where the grading begins and ends are known as the *BSR* (*begin slope rounding*) and *ESR* (*end smooth rounding*) points, rather than the analogous grade point.

A line of stakes at adjacent grade or *ESR* points is known as a *daylight line*. Therefore, the term *daylight stake* can be used when referring to the stakes marking the grade/*ESR* points. (A mole or worm following the finished slope will emerge from the ground and see daylight at the grade/*ESR* point, hence the term *daylight point*.)

Figure 17.27 illustrates the use of slope stakes along three adjacent sections of a proposed highway.

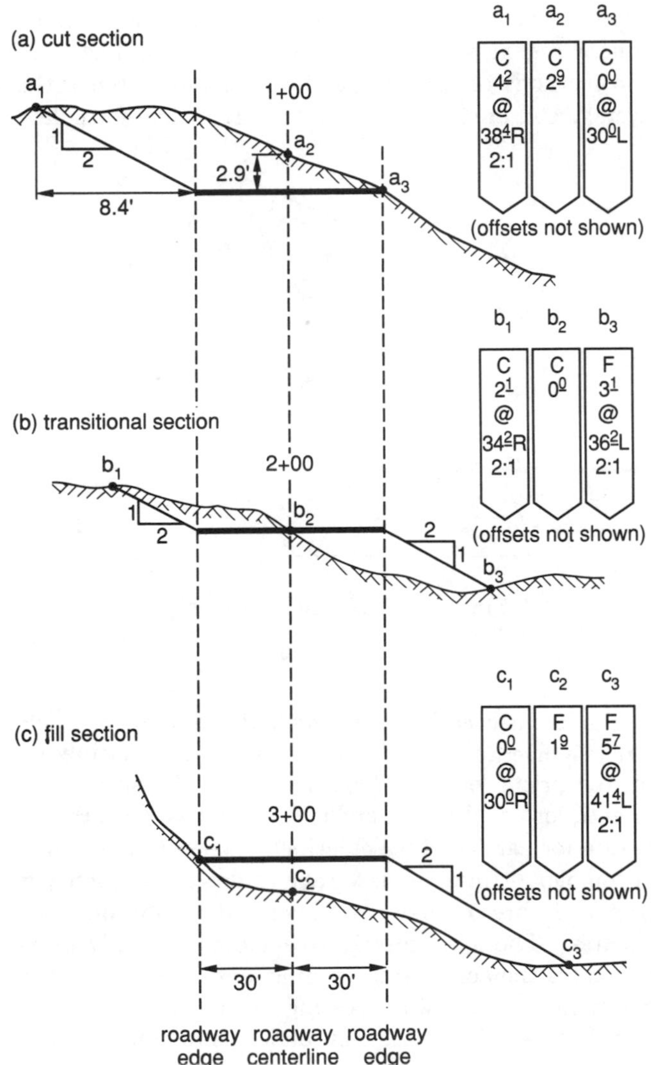


**Figure 17.27** Slope Stakes Along a Highway

*Example 17.23*

During your review of a surveyor's field work along a highway, you encounter a ground stake and its accompanying illustrated witness stake. What is the elevation of the toe of the slope documented by the witness stake?

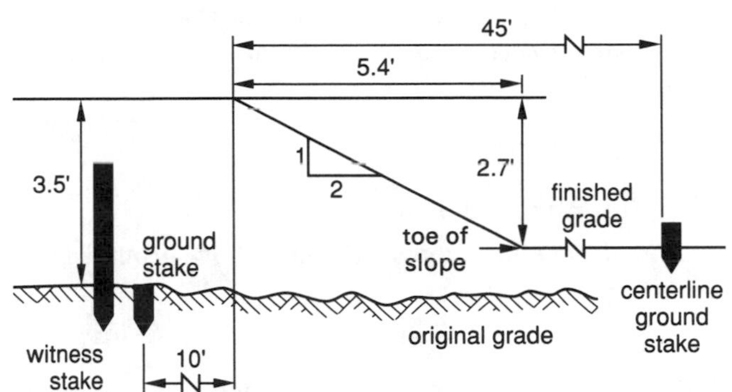


The witness stake indicates that the ground stake is a reference point slope stake located 45.0 feet to the right (when looking up-station) of the centerline of the highway. The elevation of the ground stake, marked on the side of the witness stake, is 87.6 feet. The earthwork starts out with a 3.5 foot fill. Then, the earth is cut away at a 2:1 slope until the finished grade is reached, 2.7 feet below.

The elevation of the toe of the slope is the same as the elevation of the finished grade.

$$\begin{aligned}\text{slope toe elevation} &= \text{elevation of ground stake} \\ &\quad + \text{fill} - \text{cut} \\ &= 87.6\text{ ft} + 3.5\text{ ft} - 2.7\text{ ft} = 88.4\text{ ft}\end{aligned}$$

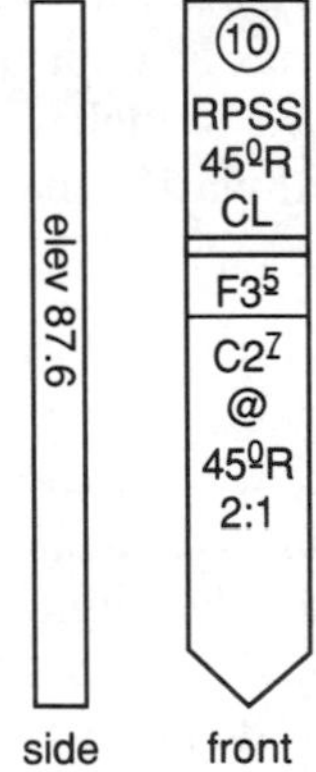


## 29 ESTABLISHING SLOPE STAKE MARKINGS

The markings on a construction stake are determined from an initial survey of the natural ground surface. This survey requires two individuals—one (the *leveler*, *instrument man,* or *instrument person*) to work the instrument, the other (the *rod man* or *rod person*) to hold the leveling rod. The elevation of the instrument (or, alternatively, the elevation of the ground at the instrument location and the height of the instrument above the ground) must be known if actual elevations are to be marked on a stake. (The term *height of the instrument*, abbreviated H.I., means "elevation of the instrument" in standard practice, not the instrument distance above the ground.)

$$\text{H.I.} = \text{ground elevation} + \text{instrument altitude above ground} \qquad 17.63$$

The *rod reading for ground* (commonly referred to as the *ground rod*) is the sighting made through a leveling instrument at the rod held vertically at the grade point. The *grade rod*, short for *rod reading for grade*, is the reading that would be observed on an imaginary rod held on the finished grade elevation. The grade rod (reading) is calculated from the planned grade elevation and the height of the instrument.

$$\text{grade rod} = \text{H.I.} - \text{grade elevation} \qquad 17.64$$

The distance marked on a construction stake (i.e., the cut or fill) is the difference in natural and finished grade elevations and is easily calculated from the ground and grade rods (readings). The cut or fill stake marking is the difference between grade and ground rods. The actual steps taken to calculate the stake marking depend on whether the earthwork is a cut or a fill and whether the instrument is above or below the finished grade. Drawing a diagram will clarify the algebraic steps and prevent sign errors.

The distance from grade point to centerline of the roadbed or other control is also written on the stake. If $w$ is the width of the finished surface (e.g., a roadbed), $s$ is the side slope ratio (horizontal:vertical), and $h$ is the cut or fill at the grade point, then the horizontal distance, $d$, from the grade point to the centerline stake of the finished surface is

$$d = \tfrac{1}{2}w + hs \qquad 17.65$$

In the field, the actual stake location is found by trial and error.

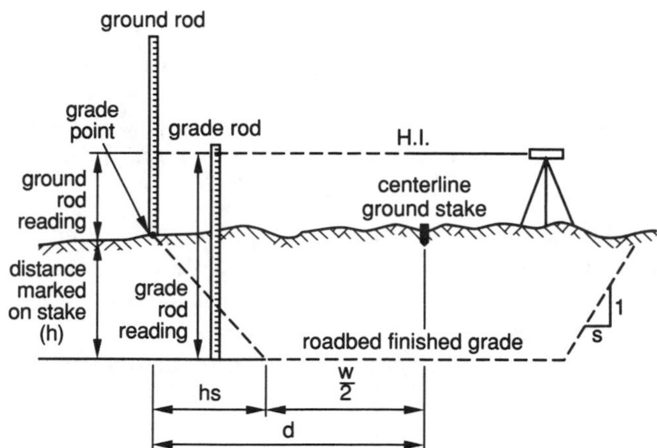


**Figure 17.28** Determining Stake Location

## 30 ESTIMATING EARTHWORK VOLUMES

The soil between two stations is known as a soil *prismoid* or *prism*. The prismoid volume must be calculated in order to estimate haulage requirements. Such volume is generally expressed in cubic yards. There are two methods of calculating the prismoid volume: the average end area method and the prismoidal formula.

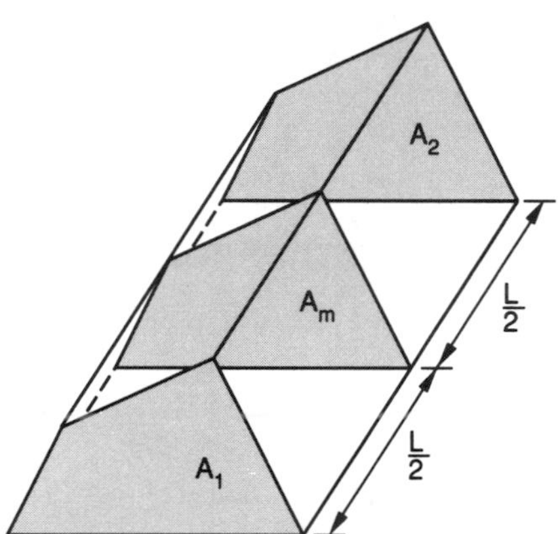


**Figure 17.29** Soil Prismoid

In the *average end area method*, the volume is calculated by averaging the two end areas and multiplying by the prism length. This disregards the slopes and orientations of the ends and sides, but is sufficiently accurate for earthwork calculations. When the end area is complex, it may be necessary to use a planimeter or to plot the area on fine grid paper and simply count the squares. The average end area method usually overestimates the actual soils volume, and thus, favors the contractor in earthwork costing. In equation 17.66, $L$ is in feet, $A$ is in square feet, and $V$ is in cubic yards.

$$V = \frac{L(A_1 + A_2)}{(2)(27)\,\frac{\text{ft}^3}{\text{yd}^3}} \qquad 17.66$$

The precision obtained from the average end area method is generally sufficient unless one of the end areas is very small or zero. In that case, the volume should be computed as a pyramid or truncated pyramid.

$$V_{\text{pyramid}} = \frac{LA_{\text{base}}}{(3)(27)\,\frac{\text{ft}^3}{\text{yd}^3}} \qquad 17.67$$

The *prismoidal formula* is better suited when the two end areas differ greatly or if the ground surface is irregular. It generally will produce a smaller volume than the average end area method, and thus, favors the owner-developer in earthwork costing. The prismoidal formula uses the area, $A_m$, midway between the two end sections. In the absence of actual observed measurements, the dimensions of the middle area can be found by averaging the similar dimensions of the two end areas. (The middle area is not found by averaging the two end areas.)

$$V = \frac{L}{(6)(27)\,\frac{\text{ft}^3}{\text{yd}^3}}(A_1 + 4A_m + A_2) \qquad 17.68$$

## 31 MASS DIAGRAMS

A *mass diagram* is a record of the cumulative earthwork volume moved along the survey route, usually plotted below profile sections of the original ground and finished grade. The mass diagram can be used to establish a finished grade line that balances cut and fill volumes and minimizes long hauls.

Distance (stationing) is plotted on the horizontal axis, and the cumulative sum of positive cut (also known as *excavation*) volumes and negative fill (also known as *embankment*) volumes in cubic yards is plotted on the vertical axis. The starting volume is arbitrarily chosen to keep the plotted volume positive.

When they can be estimated, factors like shrinkage, swell, loss during transport, and subsidence are also included in figuring embankment volumes. A 5% to 15% excess is generally included in the figure. This is achieved by increasing all fill volumes by the necessary percentage (e.g., 10%).

A rising line on the mass diagram represents excavation; a falling line represents embankment. The maximum and minimum points on the mass diagram occur where the final grade elevation coincides with the natural elevation (i.e., at grade points).

Vertical distances on a mass diagram represent volumes of material (areas on the profile diagram). Areas in a mass diagram represent the product of volume and distance.

A *balance line* is a horizontal line drawn between two adjacent points—the *balance points*—on a trough or dome. The volumes of excavation and embankment between the balance points are equal. Thus, a contractor can plan on using the earth volume from the excavation on one side of the grade point for the embankment on the other side of the grade point.

In Figure 17.30, a balance line has been drawn that intersects the mass diagram at two points. These points represent the inclusive stations for which the cut and fill volumes are equal. That is, the cut soil (area B) can be used for the fill (area A).

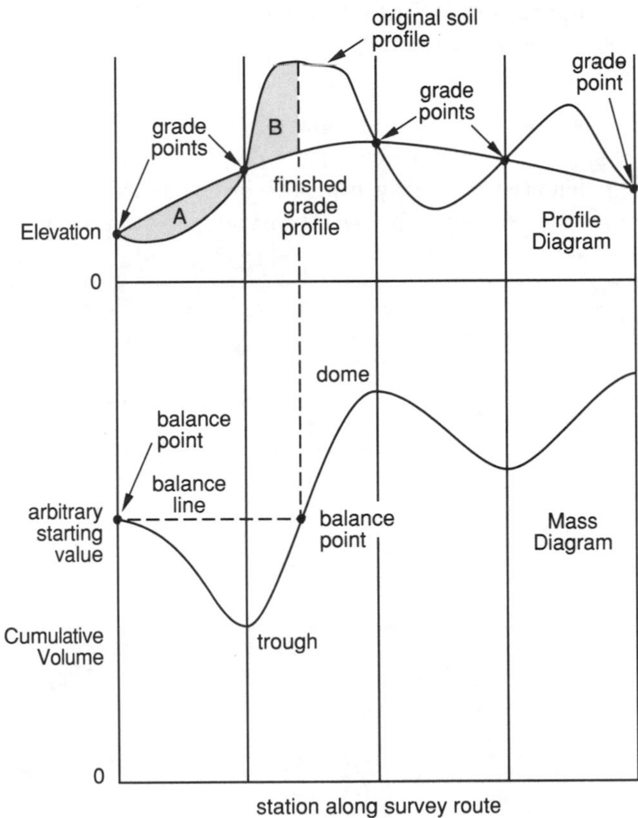


**Figure 17.30** Balance Line Between Two Points

The average *haul* distance (i.e., the distance a mass of earth is transported) is the separation of the centroids of the excavation and embankment areas on the profile diagram. This distance is usually determined for each section of earthwork rather than for the entire mass of earthwork in a project.

The average haul can also be calculated from the mass diagram. The curve area under a balance line represents the total haul in *yard-stations* (i.e., number of cubic yards moved 100 feet). This figure can be used (by the contractor) to determine his hauling costs. The height of the curve (between the balance line and the maximum or minimum) represents *solidity* (i.e., the volume of the total haul). The average haul distance is found by dividing the curve area by the curve height.

The *free haul distance* is the maximum distance, as specified in the construction contract, that the contractor is expected to transport earth without receiving additional payment. This distance is typically 500 or 1000 feet. Any soil transported more than the free haul distance is known as *overhaul*.

Free haul and overhaul (in yard-stations) can be determined from the mass diagram. The procedure starts by drawing a balance line equal in length to the free haul distance. The enclosed area on the mass diagram represents material that will be hauled with no extra cost. The actual volume moved (the solidity) is the vertical distance between the balance line and the maximum or minimum.

The overhaul (in yard-stations) is found directly from the overhaul area on the mass diagram. It can also be calculated indirectly from the overhaul volume and overhaul distance. The overhaul volume is determined from the maximum height of the overhaul area on the mass diagram, or from the overhaul area on the profile diagram. The overhaul distance is found as the separation in overhaul centroids on the profile diagram.

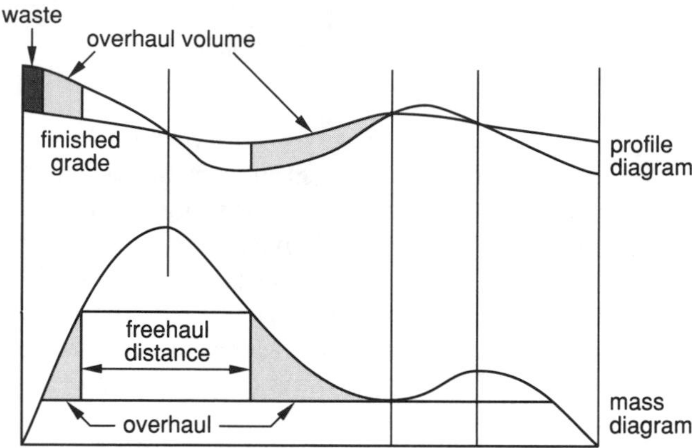


**Figure 17.31** Free Haul and Overhaul

# Appendix A: Oblique Triangle Equations

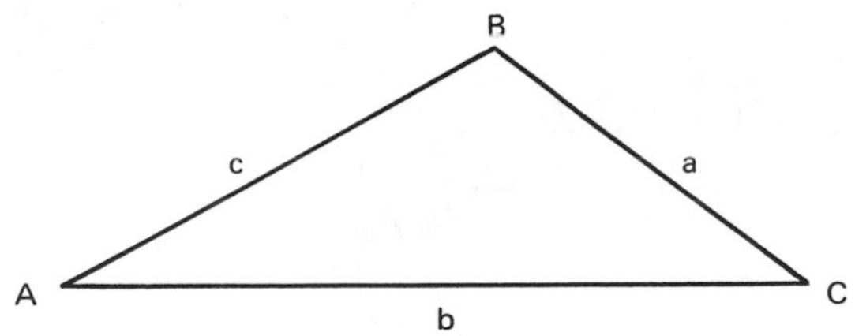


| given | equation |
|---|---|
| $A, B, a$ | $C = 180° - (A + B)$ |
| | $b = \frac{a}{\sin A} \times \sin B$ |
| | $c = \frac{a}{\sin A} \times \sin(A + B) = \frac{a}{\sin A} \times \sin C$ |
| | $\text{area} = \frac{1}{2}ab\sin C = \frac{a^2 \sin B \sin C}{2\sin A}$ |
| $A, a, b$ | $\sin B = \frac{\sin A}{a} \times b$ |
| | $C = 180° - (A + B)$ |
| | $c = \frac{a}{\sin A} \times \sin C$ |
| | $\text{area} = \frac{1}{2}ab\sin C$ |
| $C, a, b$ | $c = \sqrt{a^2 + b^2 - 2ab\cos C}$ |
| | $\frac{1}{2}(A + B) = 90° - \frac{1}{2}C$ |
| | $\tan\frac{1}{2}(A - B) = \frac{a - b}{a + b} \times \tan\frac{1}{2}(A + B)$ |
| | $A = \frac{1}{2}(A + B) + \frac{1}{2}(A - B)$ |
| | $B = \frac{1}{2}(A + B) - \frac{1}{2}(A - B)$ |
| | $c = (a + b) \times \frac{\cos\frac{1}{2}(A + B)}{\cos\frac{1}{2}(A - B)} = (a - b) \times \frac{\sin\frac{1}{2}(A + B)}{\sin\frac{1}{2}(A - B)}$ |
| | $\text{area} = \frac{1}{2}ab\sin C$ |
| $a, b, c$ | Let $s = \frac{a + b + c}{2}$ |
| | $\sin\frac{1}{2}A = \sqrt{\frac{(s - b)(s - c)}{bc}}$ |
| | $\cos\frac{1}{2}A = \sqrt{\frac{s(s - a)}{bc}}$ |
| | $\tan\frac{1}{2}A = \sqrt{\frac{(s - b)(s - c)}{s(s - a)}}$ |
| | $\sin A = 2\sqrt{\frac{s(s - a)(s - b)(s - c)}{bc}}$ |
| | $\cos A = \frac{b^2 + c^2 - a^2}{2bc}$ |
| | $\text{area} = \sqrt{s(s - a)(s - b)(s - c)}$ |

# Appendix B: Circle and Circular Curve Geometry

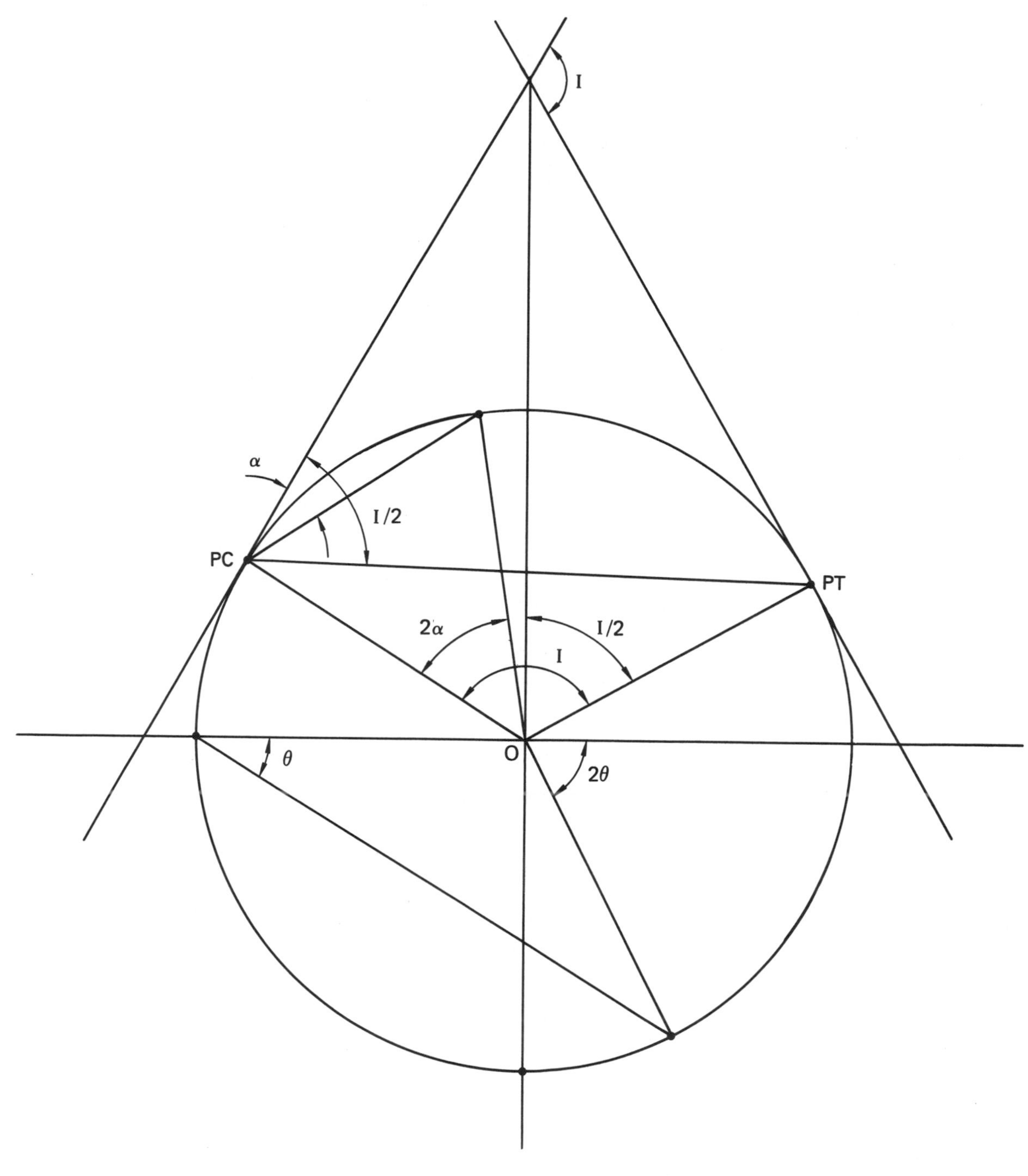

# Appendix C: Surveying Conversion Factors

| multiply | by | to obtain |
|---|---|---|
| acres | 43,560 | square feet |
| | 10 | square chains |
| | 4046.87 | square meters |
| acre-feet | 43,560 | cubic feet |
| | 1233.49 | cubic meters |
| chain | 66 | feet |
| | 22 | yards |
| | 4 | rods |
| day (mean solar) | 86,400 | seconds |
| day (sidereal) | 86,164.09 | seconds |
| degrees (angle) | 0.0174533 | radians |
| | 17.77778 | mils |
| engineer's link | 1 | feet |
| feet (U.S. Survey) | 0.3048006 | meters |
| grads | 0.9 | degrees (angle) |
| | 0.01570797 | radians |
| hectare | 2.47104 | acres |
| | 10,000 | square meters |
| inches | 25.4 | millimeters |
| labors | 177.14 | acres |
| leagues | 4428.40 | meters |
| link—see 'engineer's link' and 'surveyor's link' | | |
| mils | 0.05625 | degrees (angle) |
| | 3.375 | minutes |
| miles (statute) | 5280 | feet |
| | 80 | chains (surveyor's) |
| | 320 | rods |
| | 0.86839 | miles (nautical) |
| square miles | 640 | acres |
| | 27,878,400 | square feet |
| minutes (angle) | 0.29630 | mils |
| | 0.000290888 | radians |
| minutes (mean solar) | 60 | seconds |
| minutes (sidereal) | 59.83617 | seconds |
| outs | 330 | feet |
| | 10 | 33-foot chains |
| radians | 57.2957795 | degrees (angle) |
| | $57°17'44.806''$ | degrees (angle) |
| rods | 16.5 | feet |
| | 1 | perches |
| | 1 | poles |
| seconds (angle) | 4.848137 EE–6 | radians |
| seconds (sidereal) | 0.9972696 | seconds (mean solar) |
| surveyor's link | 0.66 | feet |
| | 7.92 | inches |
| VARA (California) | 33 | inches |
| VARA (Texas) | 33.333 | inches |
| yard (U.S.) | 0.914402 | meters |

# Appendix D: Critical Constants

| | | |
|---|---|---|
| 0.0000001 per °F | = | coefficient of expansion invar tape |
| 0.00000645 per °F | = | coefficient of expansion steel tape |
| 0.6745 | = | coefficient for 50% standard deviation |
| 1.15 miles | = | 1 minute of latitude |
| 6 miles | = | length and width of township |
| 10 square chains | = | 1 acre |
| 15 degrees longitude | = | width of one time zone |
| 23 degrees 26.5 minutes | = | maximum declination of the sun at solstice |
| 24 hours | = | 360 degrees of longitude |
| 36 | = | number of sections in a township |
| 69.1 miles | = | 1 degree latitude |
| 100 | = | usual stadia ratio |
| 101 feet | = | 1 second of latitude |
| 400 grads | = | 360 degrees |
| 480 chains | = | width and length of township |
| 640 acres | = | 1 normal section |
| 4046.9 square meters | = | 1 acre |
| 6400 mils | = | 360 degrees |
| 43,560 square feet | = | 1 acre |
| 20,906,000 feet | = | mean radius of earth |

**Practice Problems: SURVEYING**

Untimed

1. A downgrade of 4% meets a rising grade of 5% in a sag curve. At the start of the curve the level is 123.06 at location 4034+20. At 4040+20 there is an overpass with an underside level of 134.06. If the curve is designed to afford a clearance of 15 feet under the overpass at this point, calculate the required length.

2. An existing length of road consists of a rising gradient of 1 in 20 followed by a vertical parabolic summit curve 300 feet long, and then a falling gradient of 1 in 40. The curve joins both gradients tangentially and the elevation of the highest point on the curve is 173.07 feet. Visibility is to be improved over this stretch of road by replacing this curve with another parabolic curve 600 feet long. (a) Find the depth of excavation required at the mid-point of the curve. (b) Tabulate the elevations of points at 100 foot intervals on the new curve.

3. Two straights intersecting at a point B have azimuths BA 270° and BC 110°. They are to be joined by a circular curve which must pass through a point D, 350 feet from B. The azimuth of BD is 260°. Find the required radius, tangent lengths, length of curve, and setting out angle for a 50 foot chord.

4. Three points (A, B, and C) were selected on the centerline of an existing road curve as a first step in determining the curve radius. The telescope was set horizontally at point B. Readings were taken on a vertical staff at points A and C. The readings are summarized below. The instrument has constants of 100 and 0. (a) Calculate the radius of the circular curve. (b) If the instrument was 4.7 feet above the road at B, find the line-of-sight gradients AB and BC.

| staff at | horizontal bearing | stadia readings | | |
|---|---|---|---|---|
| A | 0.00° | 10.038 | 6.221 | 2.403 |
| C | 211.14° | 7.236 | 5.778 | 4.320 |

5. Four level circuits were run over four different routes to determine the elevation of a benchmark. The observed elevations and probable errors for each circuit are given. What is the most probable value for the elevation of the benchmark?

Route 1: 745.08 ± 0.03
Route 2: 745.22 ± 0.01
Route 3: 745.45 ± 0.09
Route 4: 745.17 ± 0.05

6. A transit with an interval factor of 100 and an instrument factor of 1.0 foot was used to take stadia sights. The instrument height was 4.8 feet. The location where the instrument was set up had an elevation of 297.8 feet. Using the data below, find (a) the horizontal distance between points A and B, and (b) the elevation of point B.

| object | azimuth | rod interval ($R_2-R_1$) | middle hair reading | vertical angle |
|---|---|---|---|---|
| A | 42.17° | 3.22 | 5.7 | −6.3° |
| B | 222.17° | 2.60 | 10.9 | +4.17° |

7. The balanced latitudes and departures of the legs of a closed traverse are given below. What is the traverse area?

| leg | latitude | departure |
|---|---|---|
| AB | N 350 | E 0 |
| BC | N 550 | E 600 |
| CD | S 250 | E 1200 |
| DE | S 750 | E 200 |
| EF | S 550 | W 1100 |
| FA | N 650 | W 900 |

8. Balance and adjust (to the nearest 0.1 ft) the traverse given below.

| leg | bearing | length |
|---|---|---|
| AB | N | 500.0 |
| BC | N45.00°E | 848.6 |
| CD | S69.45°E | 854.4 |
| DE | S11.32°E | 1019.8 |
| EF | S79.70°W | 1118.0 |
| FA | N54.10°W | 656.8 |

9. Determine the elevation of BM 11 and BM 12.

| station | backsight | foresight | elevation |
|---|---|---|---|
| BM 10 | 4.64 | | 179.65 |
| TP 1 | 5.80 | 5.06 | |
| TP 2 | 2.25 | 5.02 | |
| BM 11 | 6.02 | 5.85 | |
| TP 3 | 8.96 | 4.34 | |
| TP 4 | 8.06 | 3.22 | |
| TP 5 | 9.45 | 3.71 | |
| TP 6 | 12.32 | 2.02 | |
| BM 12 | | 1.98 | |

10. A five-leg closed traverse is taped and scoped in the field, but obstructions make it impossible to collect all readings. It is known that the general direction of EA is easterly. Complete the table of information below.

| leg | north azimuth | distance |
|---|---|---|
| AB | 106.22° | 1081.3 |
| BC | 195.23° | 1589.5 |
| CD | 247.12° | 1293.7 |
| DE | 332.37° | — |
| EA | | 1737.9 |

## Timed

1. An existing 6° horizontal curve connects a PC and $PT_1$ as shown. It is desired to avoid having vehicles pass close to a historical monument, so a proposal has been made to relocate the curve 120 feet forward. The PC will remain the same.

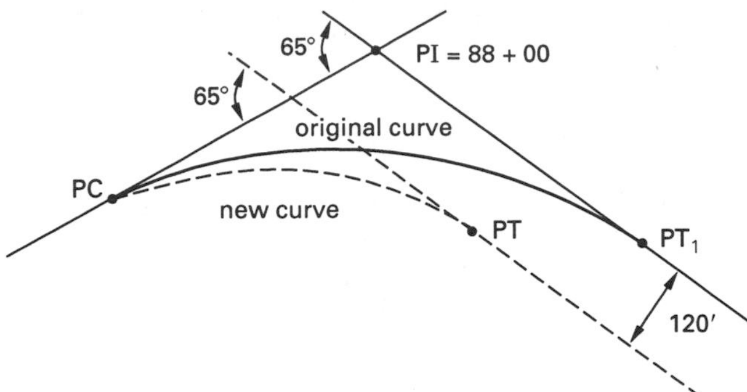


(a) Find the PC station. (b) Find the radius of curvature for the new curve. (c) Find the PT station of the new curve.

2. Two highways are planned, each running perpendicular to the other. The details of the lower highway are already fixed. The design process is continuing for the upper highway, which must maintain a 25 foot clearance over the highway below (as measured from the highway centerlines).

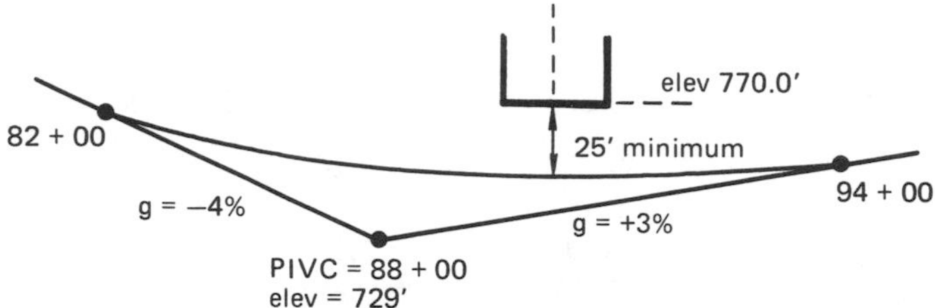


(a) Locate the minimum station where the new highway can be constructed. (b) Locate the maximum station where the new highway can be constructed. (c) What is the station of the lowest point on the existing curve? (d) What is the elevation of the lowest point on the existing curve?

3. A historical monument must be avoided near a proposed highway. A compound (reverse) curve is being considered. Both curves will have the same curvature. Refer to the diagram for details of the proposed curves. (a) What are the stations of the PC, PRC, and PT? (b) What is the biggest hazard associated with reverse curves? (c) What is the deflection angle from PC to the center of the first curve?

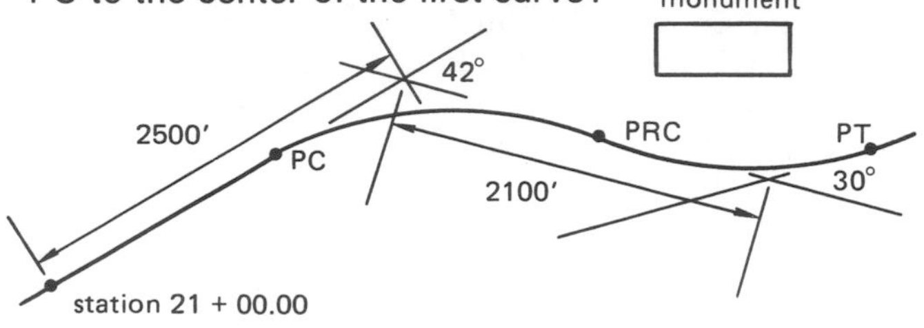


4. An equal tangent vertical curve is described as follows:

BVC: station 110+00
PVI: station 116+00; elevation 1262
EVC: station 122+00
$g_1 = +3\%$; $g_2 = -2\%$

(a) Find the centerline roadway elevations at 50 foot intervals. (b) The road continues on to a horizontal curve 5% to the right. Superelevation of 8% is used. The road is 60 feet wide at all points. The horizontal curve is supported by three pile bents symmetrical about the PVI of the curve. The pile bents are located at stations 115+50, 116+00, and 116+50. Each pile bent consists of 7 piles symmetrical about the roadway centerline and spaced at 10 foot intervals. The tops of the piles are 3.5 feet below the roadway. What are the elevations of each pile?

5. The tangent of a horizontal curve is relocated 10 feet west of its original position. The PC remains the same. Using the information given in the diagram, find R and R′.

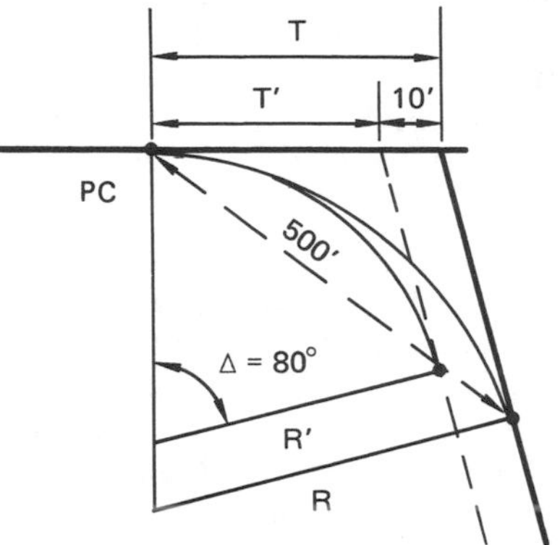


6. A new highway can go either over or under (perpendicular to) an existing highway at station 75+00. The elevation of the existing highway at station 75+00 is 510.00 ft. 22 feet of clearance is required if the new highway goes over the existing highway. 20 feet of clearance is required if the new highway goes under the old highway. (a) What is the shortest vertical curve that will pass over the existing highway? (b) What is the longest vertical curve that will pass under the existing highway?

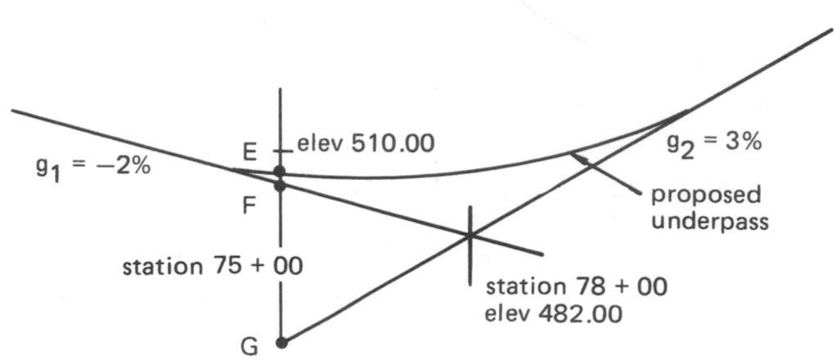

7. The tangent of a horizontal curve is parallel to a railroad line 150 feet away. The two curve tangents intersect at station 182+27.52. Using the other information on the diagram, find (a) the station of the railroad line's intersection with the highway, (b) the stations of the PC and PT, (c) the middle ordinate (M), and (d) the external ordinate (E).

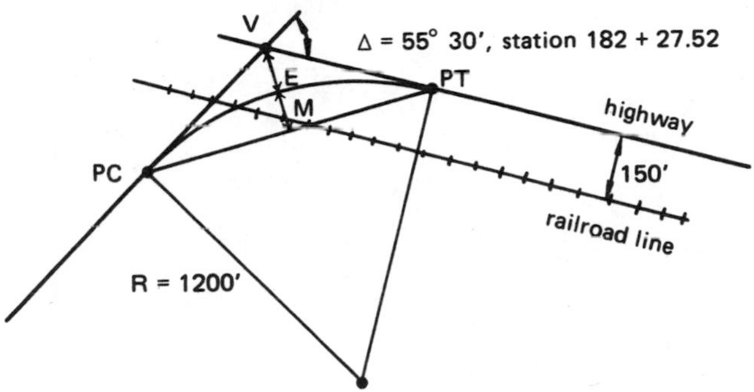


8. The grades on a vertical curve intersect at station 105+00, elevation 350 ft. A drain grating exists at station 105+00 and elevation 358.30 ft. (a) What is the minimum continuous curve length such that no point on the curve will be lower than the drain grating? (b) If a length of 1900 feet is used, what are the stations of BVC and EVC? (c) If a length of 1900 feet is used, what is the elevation of the lowest point on the curve? (d) If a length of 1900 feet is used, what is the elevation at station 106+40.00?

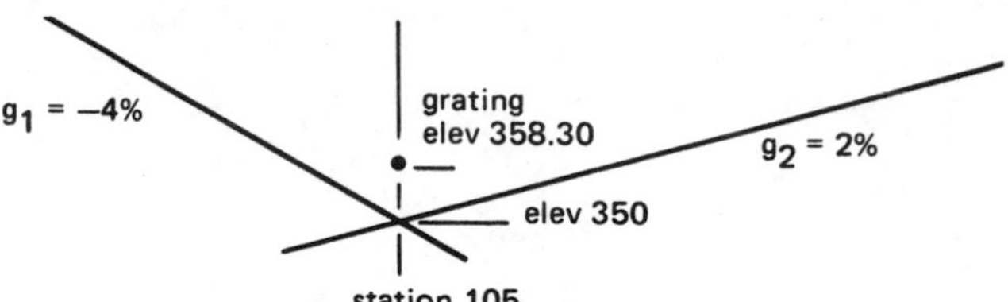


9. A freeway is being constructed near an existing USGS survey monument. The tangent alignments are shown, although the freeway can be on either side of the monument. The freeway has a 75 foot right-of-way on either side of the curve centerline. USGS monumentation law requires a 1 foot clearance from the monument (which is considered to be a dimensionless point) to the edge of the right-of-way. Find the range of radii of circular arcs that will provide the required clearance. (Hint: The clearance is radial, measured along a line from the curve center to the monument.)

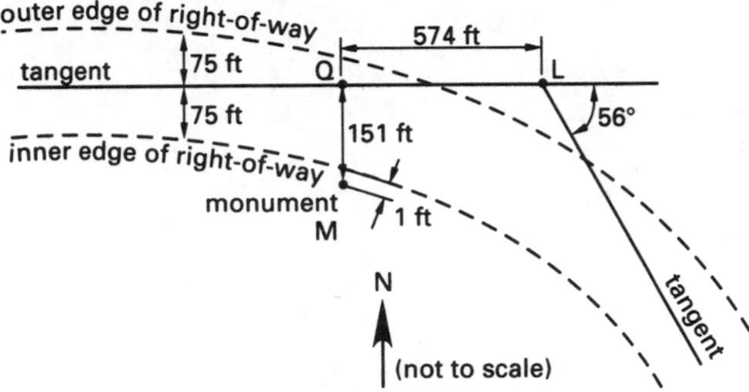

# 18 MANAGEMENT THEORIES



## 1 INTRODUCTION

Effective management techniques are based on behavioral science studies. Behavioral science is an outgrowth of the *human relations theories* of the 1930's, in which the happiness of employees was the goal (i.e., a happy employee is a productive employee ...). Current behavioral science theories emphasize minimizing tensions that inhibit productivity.

There is no evidence yet that employees want a social aspect to their jobs. Nor is there evidence that employees desire job enlargement or autonomy. Behavioral science makes these assumptions anyway.[1]

## 2 BEHAVIORAL SCIENCE KEY WORDS

Cognitive system: method we use to interpret our environment.

Collaboration: influence through a mutual agreement, relationship, respect, or understanding, without a formal or contractual authority relationship.

Equilibrium: maintaining the status quo of a group or an individual.

Job enrichment: letting employees have more control over their activities and working conditions.

Manipulation: influencing others by recognizing and building upon their needs.

MBO: management by objectives—setting job responsibilities and standards for each group and employee.

Normative judgment: judging others according to our own values.

Paternalism: corporate subsidy—showering employees with benefits and expecting submission in return.

Personal map: a person's expectations of his environment.

Selective perception: seeing what we want to see—a form of defense mechanism, since things first must be perceived to be ignored.

Superordinate goals: goals which are outside of the individual, such as corporate goals.

## 3 HISTORY OF BEHAVIORAL SCIENCE STUDIES AND THEORIES

### A. HAWTHORNE EXPERIMENTS

During 1927–1932, the Hawthorne (Chicago) Works of the Western Electric Company experimented with working conditions in an attempt to determine what factors affected output.[2] Six average employees were chosen to assemble and inspect phone relays.

Many factors were investigated in this exhaustive test. After weeks of observation without making any changes (to establish a baseline), Western Electric varied the number and length of breaks, the length of the work day, the length of the work week, and the illumination level of the lighting. Group incentive plans were tried, and in several tests, the company even provided food during breaks and lunch periods.

The employees reacted in the ways they thought they should. Output (as measured in relay production) increased after every change was implemented. In effect, the employees reacted to the attention they received, regardless of the working conditions.

Western Electric concluded that there was no relationship between illumination and other conditions to productivity. The increase in productivity during the test-

[1] In an exhaustive literature survey up through 1955, researchers found no conclusive relationships between satisfaction and productivity. There was, however, a relationship between lack of satisfaction and absenteeism and turnover.

[2] Experiments were conducted by Elton Mayo from the Harvard Business School.

ing procedure was attributed to the sense of value each employee felt in being part of an important test. The employees also became a social group, even after hours. Leadership and common purpose developed, and even though the employees were watched more than ever, they felt no supervision anxiety since they were, in effect, free to react in any way they wanted.

One employee summed up the test when she said, "It was fun."

## B. BANK WIRING OBSERVATION ROOM EXPERIMENTS

In an attempt to devise an experiment which would not suffer from the problems associated with the Hawthorne studies, Western Electric conducted experiments in 1931 and 1932 on the effects of wage incentives.

The group of nine wiremen, three soldermen, and two inspectors was interdependent. This was supposed to prevent any individual from slacking off. However, wage incentives failed to improve productivity. In fact, fast employees slowed down to protect their slower friends. Illicit activities, such as job trading and helping, also occurred.

The group was reacting to the notion of a *proper day's work*. When the day's work (or what the group considered to be a day's work) was assured, the whole group slacked off. The group also varied what it reported as having been accomplished and claimed more unavoidable delays than actually occurred. The output was essentially constant.

Western Electric concluded that social groups form as protection against arbitrary management decisions, even when such decisions have never been experienced. The effort to form the social groups, to protect slow workers, and to develop the notion of a proper day's work is not conscious. It develops automatically when the company fails to communicate to the contrary.

## C. NEED HIERARCHY THEORY

During World War II, Dr. Abraham Maslow's *need hierarchy theory* was implemented into leadership training for the U.S. Air Force. This theory claims that certain needs become dominant when lesser needs are satisfied. Although some needs can be sublimated and others overlap, the need hierarchy theory generally requires the lower-level needs to be satisfied before the higher-level needs are realized. (The ego and self-fulfillment needs rarely are satisfied.)

The need hierarchy theory explains why money is a poor motivator of an affluent individual. The theory does not explain how management should apply the need hierarchy to improve productivity.

**Table 18.1**
The Need Hierarchy

(In order of lower to higher needs)

1. Physiological needs: air, food, water.
2. Safety needs: protection against danger, threat, deprivation, arbitrary management decisions. Need for security in a dependent relationship.
3. Social needs: belonging, association, acceptance, giving and receiving of love and friendship.
4. Ego needs: self-respect and confidence, achievement, self-image. Group image and reputation, status, recognition, appreciation.
5. Self-fulfillment needs: realizing self potential, self development, creativity.

## D. THEORY OF INFLUENCE

In 1948, the Human Relations Program (under the direction of Donald C. Pelz) at Detroit Edison studied the effectiveness of its supervisors. The most effective supervisors were those who helped their employees benefit. Supervisors who were close to their employees (and sided with them in disputes) were effective only if they were influential enough to help the employees. The study results were formulated into the *theory of influence.*

- Employees think well of supervisors who help them reach their goals and meet their needs.
- An influential supervisor will be able to help employees.
- An influential supervisor who is also a disciplinarian will breed dissatisfaction.
- A supervisor with no influence will not be able to affect worker satisfaction in any way.

The implication of the theory of influence is that whether or not a supervisor is effective depends on his influence. Training of supervisors is useless unless they have the power to implement what they have learned. Also, increases in supervisor influence are necessary to increase employee satisfaction.

## E. HERZBERG MOTIVATION STUDIES

Frederick W. Herzberg interviewed 200 technical personnel in 11 firms during the late 1950's. Herzberg was especially interested in exceptional occurrences resulting in increases in job satisfaction and performance. From those interviews, Herzberg formulated his *motivation-maintenance theory.*

According to this theory, there are satisfiers and dissatisfiers which influence employee behavior. The *dissatisfiers* (also called *maintenance/motivation factors*) do not motivate employees; they can only dissatisfy them. However, the dissatisfiers must be eliminated before the satisfiers work. Dissatisfiers include company policy, administration, supervision, salary, working conditions (environment), and interpersonal relations.

*Satisfiers* (also known as *motivators*) determine job satisfaction. Common satisfiers are achievement, recognition, the type of work itself, responsibility, and advancement.

An interesting conclusion based on the motivation/maintenance theory is that fringe benefits and company paternalism do not motivate employees since they are related to dissatisfiers only.

### F. THEORY X AND THEORY Y

During the 1950's Douglas McGregor (Sloan School of Industrial Management at MIT) introduced the concept that management had two ways of thinking about its employees. One way of thinking, which was largely pessimistic, was theory X. The other theory, theory Y, was largely optimistic.

Theory X is based on the assumption that the average employee inherently dislikes and avoids work. Therefore, employees must be coerced into working by threats of punishment. Rewards are not sufficient. The average employee wants to be directed, avoids responsibility, and seeks the security of an employer-employee relationship.

This assumption is supported by much evidence. Employees exist in a continuum of wants, needs, and desires. Many of the need satisfiers (salary, fringe benefits, etc.) are effective only off the job. Therefore, work is considered a punishment or a price paid for off-the-job satisfaction.

Theory X is pessimistic about the effectiveness of employers to satisfy or motivate their employees. By satisfying the physiological and safety (lower level) needs, employers have shifted the emphasis to higher level needs which they cannot satisfy. Employees, unable to derive satisfaction from their work, behave according to theory X.

Theory Y, on the other hand, assumes that the expenditure of effort is natural and is not inherently disliked. It assumes that the average employee can learn to accept and enjoy responsibility. Creativity is widely distributed among employees, and the potentials of average employees are only partially realized.

Theory Y places the blame for worker laziness, indifference, and lack of cooperation in the lap of management, since the integration of individual and organization needs is required. This theory is not fully validated, nor is its full use ever likely to be implemented.[3]

MGMT THEORY

## 4 JOB ENRICHMENT

In an effort to make their employees happier, companies have tried to enrich the jobs performed by employees. Enrichment is a subjective result felt by employees when their jobs are made more flexible or are enlarged. Adding flexibility to a job allows an employee to move from one task to another, rather than doing the same thing continually. Horizontal job enlargement adds new production activities to a job. Vertical job enlargement adds planning, inspection, and other non-production tasks to the job.

There are advantages to keeping a job small in scope. Learning time is low, employee mental effort is reduced, and the pay rate can be lower for untrained labor. Supervision is reduced. Such simple jobs, however, also result in high turnover, absenteeism, and lower pride in job (and subsequent low quality rates).

Job enlargement generally results in better quality products, reduces inspection and material handling, and counteracts the disadvantages previously mentioned. However, training time is greater, tooling costs are higher, and inventory records are more complex.

## 5 QUALITY IMPROVEMENT PROGRAMS

### A. ZERO DEFECTS PROGRAM

Employees have been conditioned to believe that they are not perfect and that errors are natural. However, we demand zero defects from some professions (e.g., doctors, lawyers, engineers). The philosophy of a zero defect program is to expect zero defects from everybody.

Zero defects programs develop a constant, conscious desire to do the job right the first time. This is accomplished by giving employees constant awareness that their jobs are important, that the product is important, and that management thinks their efforts are important.

[3] Theory Y is not synonymous with soft management. Rather than emphasize tough management (as does theory X), theory Y depends on commitment of employees to achieve mutual goals.

Zero defects programs try to correct the faults of other types of employee programs.[4] Programs are based on what the employee has for his own: pride and desire. The programs present the challenge of perfection and explain the importance of that perfection. Management sets an example by expecting zero defects of itself. Standards of performance are set and are related to each employee. Employees are checked against these performance requirements periodically, and recognition is given when goals are met.

[4] Motivational programs are not honest, according to the zero-defects theory, since management tries to convince employees to do what management wants. Wage incentive programs encourage employee dishonesty and errors by emphasizing quantity, not quality. Theory X management, with its implied punitive action if goals are not achieved, never has been effective.

## B. QUALITY CIRCLES/TEAM PROGRAMS

Quality circle programs are voluntary or required programs in which employees within a department actively participate in measuring and improving quality and performance. It involves periodic meetings on a weekly or a monthly basis. Workers are encouraged to participate in volunteering ideas for improvement.

# 19 MISCELLANEOUS TOPICS

## PART 1: Accuracy and Precision

## Experiments

### 1 ACCURACY

An experiment is said to be *accurate* if it is unaffected by experimental error. In this case, *error* is not synonymous with *mistake*, but rather includes all variations not within the experimenter's control.

For example, suppose a gun is aimed at a point on a target and five shots are fired. The mean distance from the point of impact to the sight-in point is a measure of the alignment accuracy between the barrel and sights. The difference between the actual value and the experimental value is known as *bias*.

### 2 PRECISION

*Precision* is not synonymous with accuracy. Precision is concerned with the repeatability of the experimental results. If an experiment is repeated with identical results, the experiment is said to be precise.

In the previous example, the average distance of each impact from the centroid of the impact group is a measure of the precision of the experiment. Thus, it is possible to have a highly precise experiment with a large bias.

Most techniques applied to experiments to improve the accuracy of the experimental results (e.g., repeating the experiment, refining the experimental methods, or reducing variability) actually increase the precision.

Sometimes the word *reliability* is used with regards to the precision of an experiment. A reliable estimate is used in the same sense as a precise estimate.

### 3 STABILITY

*Stability* and *insensitivity* are synonymous terms. A stable experiment is insensitive to minor changes in the experiment parameters. Suppose the centroid of a bullet group is 2.1 inches away from the sight-in point at 65°F and 2.3 inches away at 80°F. The experiment's sensitivity to temperature changes would be $(2.3 - 2.1)/(80 - 65) = 0.0133$ inches/°F.

## PART 2: Dimensional Analysis

*Nomenclature*

| | | |
|---|---|---|
| $c_p$ | specific heat | BTU/lbm-°F |
| $C_i$ | a constant | – |
| D | diameter | ft |
| F | force | lbf |
| $g_c$ | gravitational constant (32.2) | lbm-ft/sec$^2$-lbf |
| $\bar{h}$ | average film coefficient | BTU/hr-ft$^2$-°F |
| J | Joule's constant (778) | ft-lbf/BTU |
| k | number of pi-groups $(m-n)$ | – |
| L | length | ft |
| m | number of relevant independent variables | – |
| M | mass | lbm |
| n | number of independent dimensional quantities | – |
| $N_{Nu}$ | Nusselt number | – |
| $N_{Pe}$ | Peclet number | – |
| $N_{Re}$ | Reynolds number | – |
| v | velocity | ft/sec |
| $x_i$ | the $i$th independent variable | various |
| y | dependent variable | various |

*Symbols*

| | | |
|---|---|---|
| $\rho$ | density | lbm/ft$^3$ |
| $\theta$ | time | sec |
| $\pi_i$ | $i$th dimensionless group | – |
| $\mu$ | viscosity | lbm/ft-sec |

Dimensional analysis is a means of obtaining an equation for some phenomenon without understanding the inner mechanism of the phenomenon. The most serious limitation to this method is the need to know beforehand which variables influence the phenomenon. Once these variables are known or are assumed, dimensional analysis can be applied by a routine procedure.

The first step is to select a system of primary dimensions. Usually the ML$\theta$T system (mass, length, time, and temperature) is used, although this choice may require the use of $g_c$ and $J$ in the final results. The dimensional formulas and symbols for variables most frequently encountered are given in table 19.1.

The second step is to write a functional relationship between the dependent variable and the independent variables, $x_i$.

$$y = \mathbf{f}(x_1, x_2, \ldots, x_m) \qquad 19.1$$

This function can be expressed as an exponentiated series.

$$y = C_1 x_1^{a_1} x_2^{b_1} x_3^{c_1} \cdots x_m^{z_1} + C_2 x_1^{a_2} x_2^{b_2} x_3^{c_2} \cdots x_m^{z_2} + \cdots \qquad 19.2$$

The $C_i$, $a_i$, $b_i$, $\cdots z_i$ in equation 19.2 are unknown constants.

The key to solving the above equation is that each term on the right-hand side must have the same dimensions as $y$. Simultaneous equations are used to determine some of the $a_i$, $b_i$, $c_i$, and $z_i$. Experimental data is required to determine the $C_i$ and the remaining exponents. In most analyses, it is assumed that the $C_i = 0$ for $i = 2$ and up.

*Example 19.1*

A sphere submerged in a fluid rolls down an incline. Find an equation for the velocity, $v$.

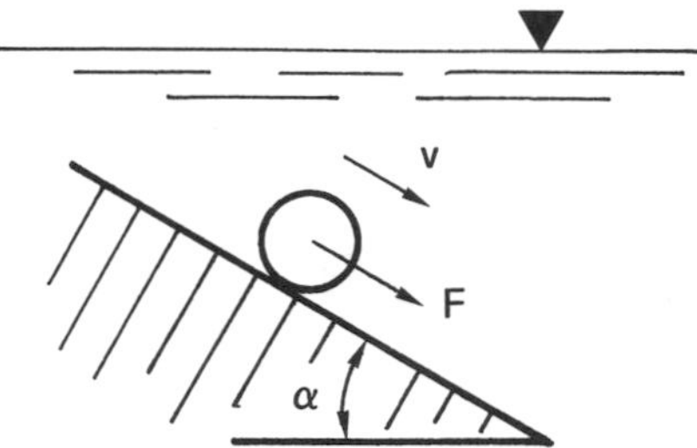


It is assumed that the velocity depends on the force, $F$, due to the inclination, the diameter of the sphere, $D$, the density of the fluid, $\rho$, and the viscosity of the fluid, $\mu$.

$$v = \mathbf{f}(F, D, \rho, \mu)$$

This equation can be written in terms of the dimensions of the variables.

$$\frac{L}{\theta} = C\left(\frac{ML}{\theta^2}\right)^a (L)^b \left(\frac{M}{L^3}\right)^c \left(\frac{M}{L\theta}\right)^d$$

Since $L$ on the left-hand side has an implied exponent of one, the necessary equation is

$$1 = a + b - 3c - d \qquad (L)$$

Similarly, the other necessary equations are

$$-1 = -2a - d \qquad (\theta)$$
$$0 = a + c + d \qquad (M)$$

**Table 19.1**
Units and Dimensions of Typical Variables

| Quantity | Symbol | MLθT System | MLθTFQ System | Units in Engineering System |
|---|---|---|---|---|
| length | L or x | L | L | ft |
| time | $\theta$ | $\theta$ | $\theta$ | sec or hour |
| mass | M | M | M | lbm |
| force | F | $ML/\theta^2$ | F | lbf |
| temperature | T | T | T | °F |
| heat | Q | $ML^2/\theta^2$ | Q | BTU |
| velocity | V | $L/\theta$ | $L/\theta$ | ft/sec |
| acceleration | a or g | $L/\theta^2$ | $L/\theta^2$ | ft/sec$^2$ |
| dimensional conversion factor | $g_c$ | none | $ML/\theta^2F$ | 32.2 lbm-ft/sec$^2$-lbf |
| energy conversion factor | J | none | FL/Q | 778 ft-lbf/BTU |
| work | W | $ML^2/\theta^2$ | FL | ft-lbf |
| pressure | p | $M/\theta^2L$ | $F/L^2$ | lbf/ft$^2$ |
| density | $\rho$ | $M/L^3$ | $M/L^3$ | lbm/ft$^3$ |
| internal energy and enthalpy | u, h | $L^2/\theta^2$ | Q/M | BTU/lbm |
| specific heat | c | $L^2/\theta^2T$ | Q/MT | BTU/lbm-°F |
| dynamic viscosity | $\mu_f$ | $M/L\theta$ | $F\theta/L^2$ | lbf-sec/ft$^2$ |
| absolute viscosity | $\mu$ | $M/L\theta$ | $M/L\theta$ | lbm/ft-sec |
| kinematic viscosity | $\nu = u/\rho$ | $L^2/\theta$ | $L^2/\theta$ | ft$^2$/sec |
| thermal conductivity | k | $ML/\theta^3T$ | $Q/LT\theta$ | BTU/hr-ft-°F |
| coefficient of expansion | $\beta$ | 1/T | 1/T | 1/°F |
| surface tension | $\sigma$ | $M/\theta^2$ | F/L | lbf/ft |
| stress | $\sigma$ or $\tau$ | $M/L\theta^2$ | $F/L^2$ | lbf/ft$^2$ |
| film coefficient | h | $M/\theta^3T$ | $Q/\theta L^2T$ | BTU/hr-ft$^2$-°F |
| mass flow rate | m | $M/\theta$ | $M/\theta$ | lbm/sec |

Solving simultaneously yields

$$b = -1$$
$$c = a - 1$$
$$d = 1 - 2a$$

or

$$v = C\left(\frac{\mu}{D\rho}\right)\left(\frac{F\rho}{\mu^2}\right)^a$$

$C$ and $a$ would have to be determined experimentally.

Since the above method requires working with $m$ different variables and $n$ different independent dimensional quantities (such as M, L, T, and $\theta$), an easier method is desirable. One simplification is to combine the $m$ variables into dimensionless groups, called *pi-groups.*

If these dimensionless groups are represented by $\pi_1$, $\pi_2$, $\pi_3$,$\cdots\pi_k$, the equation expressing the relationship between the variables is given by the *Buckingham π-theorem.*

$$f(\pi_1, \pi_2, \pi_3, \ldots, \pi_k) = 0 \qquad 19.3$$

$$k = m - n \qquad 19.4$$

The dimensionless pi-groups usually are found from the $m$ variables according to an intuitive process. A formalized method is possible as long as the following conditions are met.

- The dependent variable and independent variables chosen contain all of the variables affecting the phenomenon. Extraneous variables can be included at the expense of obtaining extra pi-groups.
- The pi-groups must include all of the original $x_i$ at least once.
- The dimensions all must be independent.

The formal procedure is to select $n$ variables ($x_i$) out of the total $m$ as repeating variables to appear in all $k$ pi-groups. These variables are used in turn with the remaining variables in each successive pi-group. Each of the repeating variables must have different dimensions, and the repeating variables collectively must contain all of the dimensions. This procedure is illustrated in example 19.2.

MISC TOPICS

*Example 19.2*

It is desired to determine a relationship giving the heat transfer to air flowing across a heated tube. The following variables affect the heat flow.

| Variable | Symbol | Dimensional Equation |
|---|---|---|
| tube diameter | D | L |
| fluid conductivity | k | $ML/\theta^3T$ |
| fluid velocity | v | $L/\theta$ |
| fluid density | $\rho$ | $M/L^3$ |
| fluid viscosity | $\mu$ | $M/L\theta$ |
| fluid specific heat | $c_p$ | $L^2/\theta^2T$ |
| film coefficient | $\overline{h}$ | $M/\theta^3T$ |

There are $m = 7$ variables and $n = 4$ primary dimensions (L, M, $\theta$, and T). Accordingly, there are $k = 7 - 4 = 3$ dimensionless groups that are required to correlate the data. The four repeating variables are chosen such that all dimensions are represented. Then the $\pi_i$ are written as functions of these repeating variables in turn with the remaining variables.

The repeating variables should not include any of the unknown quantities. For example, $\overline{h}$ should not be chosen as a repeating variable since it is directly related to the unknown heat flow. In addition, important material properties, such as $c_p$ and $k$, often are omitted. Trial and error is required to include all four primary dimensions.

Using trial and error, omitting $\overline{h}$ as a repeating variable, and representing all four primary dimensions, arbitrarily choose the variables as $D$, $k$, $v$, and $\rho$.

The pi-groups are

$$\pi_1 = D^{a_1} k^{a_2} v^{a_3} \rho^{a_4} \mu$$

$$\pi_2 = D^{a_5} k^{a_6} v^{a_7} \rho^{a_8} c_p$$

$$\pi_3 = D^{a_9} k^{a_{10}} v^{a_{11}} \rho^{a_{12}} \overline{h}$$

Since the $\pi_i$ are dimensionless, we write for $\pi_1$

$$0 = a_1 + a_2 + a_3 - 3a_4 - 1 \quad \text{(L)}$$

$$0 = a_2 + a_4 + 1 \quad \text{(M)}$$

$$0 = -3a_2 - a_3 - 1 \quad (\theta)$$

$$0 = -a_2 \quad \text{(T)}$$

Therefore,

$$a_2 = 0 \quad a_3 = -1 \quad a_4 = -1 \quad a_1 = -1$$

$$\pi_1 = \frac{\mu}{Dv\rho}$$

$\pi_1$ is the reciprocal of the Reynolds number. Proceeding similarly with $\pi_2$,

$$\begin{aligned} 0 &= a_5 + a_6 + a_7 - 3a_8 + 2 && \text{(L)} \\ 0 &= a_6 + a_8 && \text{(M)} \\ 0 &= -3a_6 - a_7 - 2 && (\theta) \\ 0 &= -a_6 - 1 && \text{(T)} \end{aligned}$$

Therefore,

$$a_6 = -1 \quad a_7 = 1 \quad a_8 = 1 \quad a_5 = 1$$

$$\pi_2 = \frac{Dv\rho c_p}{k}$$

$\pi_2$ is the *Peclet number* (product of the Reynolds number and the Prandtl number).

$\pi_3$ is found to be $\frac{D\overline{h}}{k}$, which is the *Nusselt number.*

The seven original variables have been combined into three dimensionless groups, making data correlation much easier. The implicit equation for heat transfer is

$$\mathbf{f}_1(\pi_1, \pi_2, \pi_3) = \mathbf{f}_1(N_{Re}, N_{Nu}, N_{Pe}) = 0$$

Rearrangement of the pi-groups is needed to isolate the dependent variable (in this case, $\overline{h}$).

$$N_{Nu} = \mathbf{f}_2(N_{Re}, N_{Pe}) = C(N_{Re})^{e_1}(N_{Pe})^{e_2}$$

$C$, $e_1$, and $e_2$ are found experimentally.

The selection of the repeating and non-repeating variables is the key step. The choice of repeating variables determines which dimensionless groups are obtained. The theoretical maximum number of valid dimensionless groups is

$$\frac{m!}{(n+1)!(m-n-1)!} \qquad 19.5$$

Not all dimensionless groups obtained are equally useful to researchers. For example, the Peclet number was obtained in the above example. However, researchers would have chosen $D$, $k$, $\rho$, and $\mu$ as repeating variables in order to obtain the Prandtl number as a dimensionless group. This choice of repeating variables is a matter of intuition.

MISC TOPICS

# PART 3: Reliability

*Nomenclature*

| | | |
|---|---|---|
| **f**(t) | probability density function | – |
| **F**(t) | cumulative density function | – |
| k | minimum number for operation | – |
| MTBF | mean time before failure ($1/\lambda$) | time |
| n | number of items in the system | – |
| R* | system reliability | – |
| $\mathbf{R}_i$(t) | ith item reliability | – |
| t | time | time |
| x | number of failures | – |
| X | binary ith item performance variable | – |
| Y | arbitrary event | – |
| **z**(t) | hazard function | 1/time |

*Symbols*

| | | |
|---|---|---|
| $\lambda$ | constant failure or hazard rate (1/MTBF) | 1/time |
| $\phi$ | binary system performance variable | |

## 1 ITEM RELIABILITY

*Reliability* as a function of time, $\mathbf{R}(t)$, is the probability that an item will continue to operate satisfactorily up to time $t$. Although other distributions are possible, reliability often is described by the *negative exponential distribution.* Specifically, it is assumed that an item's reliability is

$$\mathbf{R}(t) = 1 - \mathbf{F}(t) = e^{-\lambda t} = e^{-t/MTBF} \qquad 19.6$$

This implies that the probability of $x$ failures in a period of time is given by the Poisson distribution.

$$p\{x\} = \frac{e^{-\lambda}\lambda^x}{x!} \qquad 19.7$$

The negative exponential distribution is appropriate whenever an item fails only by random causes but never experiences deterioration during its life. This implies that the *expected future life* of an item is independent of the previous duration of operation.

The *hazard function* is defined as the conditional probability of failure in the next time interval given that no failure has occurred thus far. For the exponential distribution, the hazard function is

$$\mathbf{z}(t) = \lambda = \frac{1}{\text{MTBF}} \qquad 19.8$$

Since this is not a function of $t$, exponential failure rates are not dependent on the length of time previously in operation.

In general,

$$\mathbf{z}(t) = \frac{\mathbf{f}(t)}{\mathbf{R}(t)} = \frac{\frac{d\mathbf{F}(t)}{dt}}{1 - \mathbf{F}(t)} \qquad 19.9$$

The exponential distribution is summarized by equations 19.10 through 19.13.

$$\mathbf{f}(t) = \lambda e^{-\lambda t} \qquad 19.10$$

$$\mathbf{F}(t) = 1 - e^{-\lambda t} \qquad 19.11$$

$$\mathbf{R}(t) = 1 - \mathbf{F}(t) = e^{-\lambda t} \qquad 19.12$$

$$\mathbf{z}(t) = \frac{\lambda e^{-\lambda t}}{e^{-\lambda t}} = \lambda \qquad 19.13$$

*Example 19.3*

An item exhibits an exponential time to failure distribution with MTBF of 1000 hours. What is the maximum operating time such that the reliability does not drop below 0.99?

$$0.99 = e^{-t/1000}$$
$$t = 10.05 \text{ hours}$$

## 2 SYSTEM RELIABILITY

The binary variable, $X_i$, is defined as 1 if item $i$ operates satisfactorily and 0 otherwise. Similarly, the binary variable, $\phi$, is 1 only if the system operates satisfactorily. $\phi$ will be a function of the $X_i$.

### A. SERIAL SYSTEMS

The *performance function* for a system of $n$ serial items is

$$\phi = X_1X_2X_3\ldots X_n = \min\{X_i\} \qquad 19.14$$

Equation 19.14 implies that the system will fail if any of the individual items fail. The system reliability is

$$R^* = R_1R_2R_3\ldots R_n \qquad 19.15$$

MISC TOPICS

*Example 19.4*

A block diagram of a system with item reliabilities is shown. What is the performance function and the system reliability?

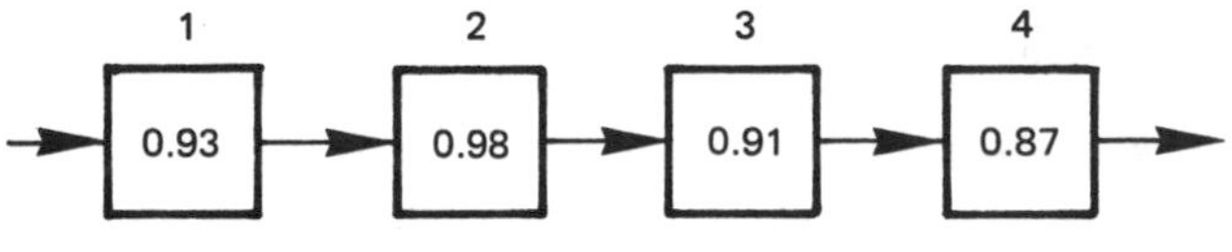


$$\phi = X_1X_2X_3X_4$$
$$R^* = (0.93)(0.98)(0.91)(0.87) = 0.72$$



## B. PARALLEL SYSTEMS

A parallel system with $n$ items will fail only if all $n$ items fail. This property is called *redundancy*, and such a system is said to be redundant. Using redundancy, a highly reliable system can be produced from components with relatively low individual reliabilities.

The performance function of a redundant system is

$$\begin{aligned}\phi &= 1 - (1 - X_1)(1 - X_2)(1 - X_3)\cdots(1 - X_n)\\ &= \max\{X_i\}\end{aligned} \quad 19.16$$

The reliablity is

$$R^* = 1 - (1 - R_1)(1 - R_2)(1 - R_3)\cdots(1 - R_n) \quad 19.17$$

*Example 19.5*

What is the reliability of the system shown?

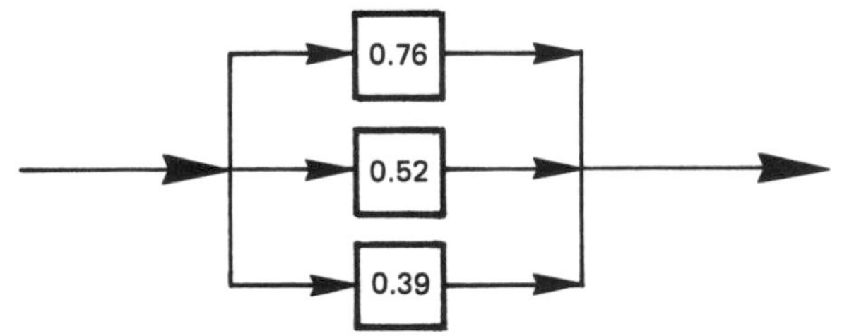


$$R^* = 1 - (1 - 0.76)(1 - 0.52)(1 - 0.39) = 0.93$$

## C. k-out-of-n SYSTEMS

If the system operates with an $k$ of its elements operational, it is said to be a $k$-out-of-$n$ system. The performance function is

$$\phi = \begin{cases} 1 \text{ if } \Sigma X_i \geq k \\ 0 \text{ if } \Sigma X_i < k \end{cases} \quad 19.18$$

The evaluation of the system reliability is quite difficult unless all elements are identical and have identical reliabilities, **R**. In that case, the system reliability follows the binomial distribution.

$$R^* = \sum_{j=k}^{n} \binom{n}{j} R^j(1 - R)^{n-j} \quad 19.19$$

## D. GENERAL SYSTEM RELIABILITY

A general system can be represented by a graphical network. Each path through the network from the starting node to the finishing node represents a possible operating path. For the 5-path network below, even if BD and AC are cut, the system will operate by way of path ABCD.

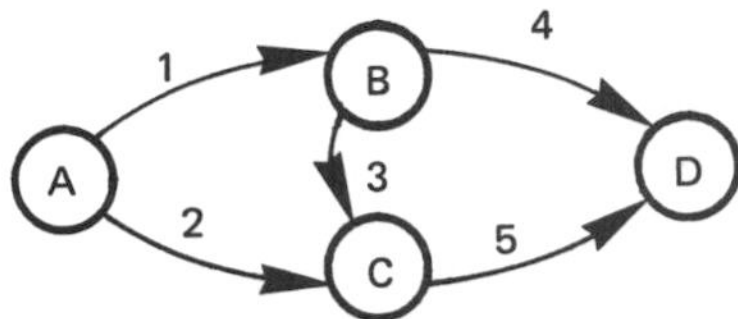


The reliability of the system will be the sum of the serial reliabilities, summed over all possible paths in the system. However, the concepts of minimal paths and minimal cuts are required to facilitate the evaluation of the system reliability.

A *minimal path* is a set of components that, if operational, will ensure the system's functioning. In the previous example, components [1 with 4] are a minimal path, as are [2 with 5] and [1 with 3 with 5]. A *minimal cut* is a set of components that, if non-functional, inhibits the system from functioning. Minimal cuts in the previous example are [1 with 2], [4 with 5], [1 with 5], and [2 with 3 with 4].

Since it usually is easier to determine all minimal paths, a method of finding the exact system reliability from the set of minimal paths is needed. In general, the probability of a union of $n$ events contains $(2^n - 1)$ terms and is given by

$$\begin{aligned}&p\{Y_1 \text{ or } Y_2 \text{ or } \cdots Y_n\} = p\{Y_1\} + p\{Y_2\} + p\{Y_3\}\\ &+ \cdots + p\{Y_n\} - p\{Y_1 \text{ and } Y_2\} - p\{Y_1 \text{ and } Y_3\}\\ &- \cdots - p\{Y_1 \text{ and } Y_n\} - p\{Y_1 \text{ and } Y_2 \text{ and } Y_3\}\\ &- p\{Y_1 \text{ and } Y_2 \text{ and } Y_4\} - \cdots - p\{Y_i \text{ and } Y_j \text{ and } Y_k\}\\ &\text{all } i \neq j \neq k\\ &+ \{-1\}^{n-1}p\{Y_1 \text{ and } Y_2 \text{ and } Y_3 \text{ and } \ldots \text{ and } Y_n\}\end{aligned} \quad 19.20$$

Returning to the 5-path example,

$$\begin{aligned}Y_1 &= [1 \text{ with } 4]\\ Y_2 &= [2 \text{ with } 5]\\ Y_3 &= [1 \text{ with } 3 \text{ with } 5]\end{aligned}$$

Then,

$$
\begin{aligned}
p\{\phi = 1\} &= p\{Y_1 \text{ or } Y_2 \text{ or } Y_3\} \\
&= p\{X_1X_4 = 1\} + p\{X_2X_5 = 1\} \\
&\quad + p\{X_1X_3X_5 = 1\} - p\{X_1X_2X_4X_5 = 1\} \\
&\quad - p\{X_1X_3X_4X_5 = 1\} \\
&\quad - p\{X_1X_2X_3X_5 = 1\} \\
&\quad + p\{X_1X_2X_3X_4X_5 = 1\}
\end{aligned}
$$

In terms of the individual item reliabilities, this is

$$
\begin{aligned}
R^* &= R_1R_4 + R_2R_5 + R_1R_3R_5 \\
&\quad - R_1R_2R_4R_5 - R_1R_3R_4R_5 - R_1R_2R_3R_5 \\
&\quad + R_1R_2R_3R_4R_5
\end{aligned}
$$

In the 5-path example given,

$$R^* \le p\{X_1X_4 = 1\} + p\{X_2X_5 = 1\} + p\{X_1X_3X_5 = 1\}$$

This method requires considerable computation, and an upper bound on $R^*$ would be sufficient. Such an upper bound is close to $R^*$ since the product of individual reliabilities is small. The upper bound is given by

$$p\{\phi = 1\} \le p\{Y_1\} + p\{Y_2\} + \cdots + p\{Y_n\} \qquad 19.21$$

# PART 4: Replacement

*Nomenclature*

| | |
|---|---|
| $C_1$ | item replacement cost with group replacement |
| $C_2$ | item replacement cost after individual failure |
| $\mathbf{F}(t)$ | number of units failing in the interval ending at $t$ |
| $\mathbf{K}(t)$ | total cost of operating from $t = 0$ to $t = T$ |
| MTBF | mean time before failure |
| n | number of units in original system |
| $p\{t\}$ | probability of failing in the interval ending at $t$ |
| $\mathbf{S}(t)$ | number of survivors at the end of time $t$ |
| t | time |
| $v\{t\}$ | conditional probability of failure in the interval $(t-1)$ to $t$ given non-failure before $(t-1)$ |



## 1 INTRODUCTION

Replacement and renewal models determine the most economical time to replace existing equipment. Replacement processes fall into two categories, depending on the life pattern of the equipment, which either deteriorates gradually (becomes obsolete or less efficient) or fails suddenly.

In the case of gradual deterioration, the solution consists of balancing the cost of new equipment against the cost of maintenance or decreased efficiency of the old equipment. Several models are available for cases with specialized assumptions, but no general solution methods exist.

In the case of sudden failure, of which light bulbs are examples, the solution consists of finding a replacement frequency which minimizes the costs of the required new items, the labor for replacement, and the expected cost of failure. The solution is made difficult by the probabilistic nature of the life spans.

## 2 DETERIORATION MODELS

The replacement criterion with deterioration models is the present worth of all future costs associated with each policy. Solution is by trial and error, calculating the present worth of each policy and incrementing the replacement period by one time period for each iteration.

*Example 19.6*

Item A currently is in use. Its maintenance cost is $400 this year, increasing each year by $30. Item A can be replaced by item B at a current cost of $3500. However, the cost of B is increasing by $50 each year. Item B has no maintenance costs. Disregarding income taxes, find the optimum replacement year. Use 10% as the interest rate.

Calculate the present worth of the various policies.

*policy 1*: Replacement at $t = 5$ (starting the 6th year)

$$\begin{aligned} PW(A) &= -400\left(\frac{P}{A}, 10\%, 5\right) - 30\left(\frac{P}{G}, 10\%, 5\right) \\ &= -1722 \end{aligned}$$

$$PW(B) = -[3500 + 5(50)]\left(\frac{P}{F}, 10\%, 5\right) = -2328$$

*policy 2*: Replacement at $t = 6$

$$\begin{aligned} PW(A) &= -400\left(\frac{P}{A}, 10\%, 6\right) - 30\left(\frac{P}{G}, 10\%, 6\right) \\ &= -2033 \end{aligned}$$

$$PW(B) = -[3500 + 6(50)]\left(\frac{P}{F}, 10\%, 6\right) = -2145$$

*policy 3*: Replacement at $t = 7$

$$\begin{aligned} PW(A) &= -400\left(\frac{P}{A}, 10\%, 7\right) - 30\left(\frac{P}{G}, 10\%, 7\right) \\ &= -2330 \end{aligned}$$

$$PW(B) = -[3500 + 7(50)]\left(\frac{P}{F}, 10\%, 7\right) = -1976$$

**The present worth of $A$ drops below the present worth of $B$ at $t = 7$. Replacement should take place at that time.**

## 3 FAILURE MODELS

The time between installation and failure is not constant for members in the general equipment population. Therefore, in order to solve a failure model, it is necessary to have the distribution of individual item lives (*mortality curve*). The *conditional probability of failure* in a small time interval, say from $t$ to $(t + \delta t)$, is calculated from the mortality curve. This probability is *conditional* since it is conditioned on non-failure up to time $t$.

The conditional probability can decrease with time (e.g., *infant mortality*), remain constant (as with an exponential reliability distribution and failure from random causes), or increase with time (as with items that deteriorate with use). If the conditional probability decreases or remains constant over time, operating items should never be replaced prior to failure.

It usually is assumed that all failures occur at the end of a period. The problem is to find the period which minimizes the total cost.

*Example 19.7*

100 items are tested to failure. Two failed at $t = 1$, five at $t = 2$, seven at $t = 3$, 20 at $t = 4$, 35 at $t = 5$, and 31 at $t = 6$. Find the probability of failure in any period, the conditional probability of failure, and the mean time before failure.

The MTBF is

$$\frac{(2)(1)+(5)(2)+(7)(3)+(20)(4)+(35)(5)+(31)(6)}{100}$$

$$= 4.74$$

| elapsed time $t$ | failures $\mathbf{F}(t)$ | survivors $\mathbf{S}(t)$ | probability of failure $p\{t\}=0.01\mathbf{F}(t)$ | conditional probability of failure $v\{t\}=\mathbf{F}(t)/\mathbf{S}(t-1)$ |
|---|---|---|---|---|
| 0 | 0 | 100 | — | — |
| 1 | 2 | 98 | 0.02 | 0.02 |
| 2 | 5 | 93 | 0.05 | 0.051 |
| 3 | 7 | 86 | 0.07 | 0.075 |
| 4 | 20 | 66 | 0.20 | 0.233 |
| 5 | 35 | 31 | 0.35 | 0.530 |
| 6 | 31 | 0 | 0.31 | 1.00 |

## 4 REPLACEMENT POLICY

The expression for the number of units failing in time $t$ is

$$\mathbf{F}(t) = n\left[p\{t\} + \sum_{i=1}^{t-1} p\{i\}p\{t-i\} + \sum_{j=2}^{t-1}\left[\sum_{i=1}^{j-1} p\{i\}p\{j-i\}\right]p\{t-j\} + \cdots\right] \qquad 19.22$$

The term $np\{t\}$ gives the number of failures in time $t$ from the original group.

The term $n\sum p\{i\}p\{t-i\}$ gives the number of failures in time $t$ from the set of items which replaced the original items.

The third probability term times $n$ gives the number of failures in time $t$ from the set of items which replaced the first replacement set.

It can be shown that $\mathbf{F}(\mathrm{t})$ with replacement will converge to a steady state limiting rate of

$$\overline{\mathbf{F}(t)} = \frac{n}{MTBF} \qquad 19.23$$

The optimum policy is to replace all items in the group, including items just installed, when the total cost per period is minimized. That is, we want to find $T$ such that $\mathbf{K}(\mathrm{T})/\mathrm{T}$ is minimized.

$$\mathbf{K}(T) = nC_1 + C_2\sum_{t=0}^{T-1}\mathbf{F}(t) \qquad 19.24$$

Discounting usually is not included in the total cost formula since the time periods are considered short. If the equipment has an unusually long life, discounting is required.

There are some cases where group replacement always is more expensive than replacing just the failures as they occur. Group replacement will be the most economical policy if equation 19.25 holds.

$$C_2[\overline{\mathbf{F}(t)}] > \left.\frac{\mathbf{K}(T)}{T}\right|_{\text{minimum}} \qquad 19.25$$

If the opposite inequality holds, group replacement still may be the optimum policy. Further analysis is required.

# PART 5: FORTRAN Programming

The FORTRAN language currently exists in several versions. Although differences exist between compilers, these are relatively minor. However, some of the instructions listed in this chapter may not be compatible with all compilers.

This section is not intended as instruction in FORTRAN programming, but rather serves as a documentation of the language.



## 1 STRUCTURAL ELEMENTS

Symbols are limited to the upper-case alphabet, digits 0 through 9, the blank, and the following special characters.

+ = − * / ( ) , . $

Statements written with these characters generally are prepared in an 80-column format. Statements are executed sequentially regardless of the statement numbers.

| position | use |
|---|---|
| 1 | The letter $C$ is used for a *comment.* Comments are not executed. |
| 2–5 | The statement number, if used, is placed in positions 2 through 5. Statement numbers can be any integers from 1 through 9999. |
| 6 | Any character except *zero* can be placed in position 6 to indicate a continuation from the previous statement. |
| 7–72 | The FORTRAN statement is placed here. |
| 73–80 | These positions are available for any use and are ignored by the compiler. Usually, the final debugged program is numbered sequentially in these positions. |

FORTRAN compilers pack the characters. Therefore, blanks can be inserted at any place in most statements. For example, the following statements are compiled the same way.

```
IF (AGE.LT.YEARS) GO TO 10
IF(AGE.LT.YEARS)GOTO10
```

## 2 DATA

Numerical data can be either real or integer. *Integers* usually are limited to nine digits. Unsigned integers and integers preceded by a plus sign are the same. Commas are not allowed in integer constants. For example, ninety thousand would be written as 90000, not 90,000.

*Real numbers* are distinguished from integers by a decimal point and may contain a fractional part. Scientific notation is indicated by the single letter $E$. Real numbers are limited to one decimal point and usually seven digits.

| value | FORTRAN Notation |
|---|---|
| 2 million | 2. E6 |
| 0.00074 | 7.4 E−4 |
| 2. | 2. |

## 3 VARIABLES

*Variable* names can be formed from up to six alphanumeric characters. The first character must be a letter. Variable names starting with the letters $I$, $J$, $K$, $L$, $M$, or $N$ are assumed by the compiler to be integers unless defined otherwise by an *explicit typing statement.* All other variable names represent real variables, unless explicitly typed.

The type convention can be overridden in an explicit typing statement. This is done by defining the desired variable type in the first part of the program with an INTEGER or REAL statement. For example, the statements

```
INTEGER TIME,CLOCK
REAL INSTANT
```

would establish TIME and CLOCK as integer variables and INSTANT as a real variable. The order of such declarations is unimportant. Variables following the standard type convention (implicit typing) do not have to be declared.

*Subscripted variables* with up to seven dimensions are allowed. They always must be defined in size by the DIMENSION statement. For example, the statements

```
DIMENSION SAMPLE(5)
REAL DIMENSION INCOME(2,7)
```

would establish a $1 \times 5$ real *array* called SAMPLE and a $2 \times 7$ real array called INCOME. INCOME would have been an integer array without the REAL declaration.

Elements of arrays are addressed by placing the subscripts in parentheses.

```
SAMPLE(2)
INCOME(1,6)
```

The subscripts also can be variables. SAMPLE(K) would be permitted as long as $K$ was defined, was between 1 and 5, and was an integer.

Variables and arrays once defined and declared are not initialized automatically. If it is necessary to initialize a storage location prior to use, the DATA statement can be used. Consider the following statements.

```
REAL X,Y,Z
DIMENSION ONEDIM(5)
DIMENSION TWODIM(2,3)
DATA X,Y,Z/3*0.0/(ONEDIM(I),I = 1,5)
1/5*0.0/ ((TWODIM(I,J),J = 1,3),I = 1,2)
2/1.,2.,3.,4.,5.,6./
```

Variables X, Y, and Z will be initialized to 0.0. The entries in ONEDIM will have the values (0,0,0,0,0). The TWODIM array will be initialized to

$$\begin{pmatrix} 1.0 & 2.0 & 3.0 \\ 4.0 & 5.0 & 6.0 \end{pmatrix}$$

After being initialized with a DATA statement, variables can have their values changed by arithmetic operations.

## 4 ARITHMETIC OPERATIONS

FORTRAN provides for the usual arithmetic operations. These are listed in table 19.2.

**Table 19.2**
FORTRAN OPERATORS

| symbol | meaning |
|---|---|
| = | replacement |
| + | addition |
| − | subtraction |
| * | multiplication |
| / | division |
| ** | exponentiation |
| ( ) | preferred operation |

The *equals* symbol is used to replace one quantity with another. For example, the following statement is algebraically incorrect. However, it is a valid FORTRAN statement.

$$Z = Z + 1$$

Each statement is scanned from left-to-right (except that a right-to-left scan is made for exponentiation). Operations are performed in the following order.

exponentiation first

multiplication and division second

addition and subtraction last

Parentheses can be used to modify this hierarchy.

Each operation must be stated explicitly and unambiguously. Thus, $AB$ is not a substitute for $A^*B$. Two operations in a row, as in $(A + -B)$ also are unacceptable. Some FORTRAN compilers allow mixed-mode arithmetic. Others, ANSI FORTRAN among them, require all variables in an expression to be either integer or real. Where mixed-mode arithmetic is permitted, care must be taken in the conversion of real data to the integer mode.

Integer variables used to hold the results of a mixed-mode calculation will have their values truncated. This is illustrated in the following example.

*Example 19.8*

Evaluate $J$ in the following expression.

$$J = (6.0 + 3.0)^*3.0/6.0 + 5.0 - 6.0^{**}2.0$$

The expressions within parentheses are evaluated in the first pass.

$$J = 9.0^*3.0/6.0 + 5.0 - 6.0^{**}2.0$$

The exponentiation is performed in the second pass.

$$J = 9.0^*3.0/6.0 + 5.0 - 36.0$$

The multiplication and division are performed in the third pass.

$$J = 4.5 + 5.0 - 36.0$$

The addition and subtraction are performed in the fourth pass.

$$J = -26.5$$

However, $J$ is an integer variable, so the real number $-26.5$ is truncated and converted to integer. The final result is $-26$.

## 5 PROGRAM LOOPS

*Loops* can be constructed from IF and GO TO statements. However, the DO statement is a convenient method of creating loops. The general form of the DO statement is

$$\text{DO } s\ i\ = j, k, l$$

where $s$ is a statement number.

$i$ is the integer loop variable.

$j$ is the initial value assigned to $i$.

$k$, which must exceed $j$, is an inclusive upper bound on $i$.

$l$ is the increment for $i$, with a default value of 1 if omitted.

The DO statement causes the execution of the statements immediately following it through statement $s$ until $i$ equals $k$ or greater.[1] A loop can be *nested* by placing it within another loop. The loop variable may be used to index arrays.

When $i$ equals or exceeds $k$, the statement following $s$ is executed. However, the loop may be exited at any time before $i$ reaches $k$ if the logic of the loop provides for it.



## 6 INPUT/OUTPUT STATEMENTS

The READ, WRITE, and FORMAT statements are FORTRAN's main I/O statements. Forms of the READ and WRITE statements are

READ ($u_1$, $s$) [**list**]
WRITE ($u_2$, $s$) [**list**]

where

$u_1$ is the unit number designation for the desired input device, usually 5 for the card reader.

$u_2$ is the unit designation for the desired output device, usually 6 for the line printer.

$s$ is the statement number of an associated FORMAT statement.

[**list**] is a list of variables separated by commas whose values are being read or written.

The [**list**] also can include an implicit DO loop. The following example reads six values, the first five into the array PLACE and the last into SHOW.

READ (5,85) (PLACE(J),J = 1,5) SHOW

The purpose of the FORMAT statement is to define the location, size, and type of the data being read. The form of the FORMAT statement is

$s$ FORMAT [**field list**]

As before, $s$ is the statement number. [**field list**] consists of specifications, set apart by commas, defining the I/O fields. [**list**] can be shorter than [**field list**].

The format code for integer values is $nIw$.

$n$ is an optional repeat counter which indicates the number of consecutive variables with the same format. $w$ is the number of character positions. The format codes for real values are

$$nFw.d \text{ or } nEw.d$$

Again, $w$ is the number of character positions allocated, including the space required for the decimal point. $d$ is the implied number of spaces to the right of the decimal point. In the case of input data, decimal points in any position take precedence over the value of $d$.

The $F$ format will print a total of $(w-1)$ digits or blanks representing the number. The $E$ format will print a total of $(w-2)$ digits or blanks and give the data in a standard scientific notation with an exponent.

Other formats which can be used in the FORMAT statement are

X horizontal blanks
/ skipping lines
H alphanumeric data
D double precision real
T position (column) indicator
Z hexadecimal
P decimal point modification
L logical data
' ' literal data

The usual output device is a line printer with 133 print positions. The first print position is used for *carriage control*. The data (control character) in the first output position will control the printer advance according to the rules in table 19.3.

**Table 19.3**
FORTRAN Printer Control Characters

| control character | meaning |
|---|---|
| blank | advance one line |
| 0 | advance two lines |
| 1 | skip to line one on next page |
| + | do not advance (overprint) |

Carriage control usually is accomplished by the use of literal data. Consider the following statements.

INTEGER K
K = 193

[1] A peculiarity of FORTRAN DO statements is that they are executed at least once, regardless of the values of $i$ and $j$.

```
WRITE (6,100) K
WRITE (6,101) K
100 FORMAT (' ',I3)
101 FORMAT (I3)
```

The above program would print the number 193 on the next line of the current page and the number 93 on the first line of the next page.

Data can be written to or read from an array by including the array subscripts in the I/O statement.

```
      DIMENSION CLASS (2,5)
      READ (5,15) ((CLASS(I,J),I = 1,2),J = 1,5)
15    FORMAT (10F3.0)
```

## 7 CONTROL STATEMENTS

The STOP statement is used to indicate the logical end of the program. The format is $s$ STOP.

STOP should not be the last statement. When it is reached, program execution is terminated. The value of $s$ is printed out or made available to the next program step. The use of STOP rarely is recommended.

The END statement is required as the last statement. It tells the compiler that there are no more lines in the program to be compiled. A program cannot be compiled or executed without an END statement.

The PAUSE statement will cause execution to stop temporarily. Its format is $s$ PAUSE.

When the PAUSE statement is reached, the number $s$ is transmitted to the computer operator. This gives the operator a chance to set various control switches on the console (the choice of switches being dependent on the value of $s$ and the program logic), prior to pushing the START button. The PAUSE statement is used only if the programmer is operating the computer.

The CALL statement is used to transfer execution to a *subroutine*. CALL EXIT will terminate execution and turn control over to the operating system. The CALL EXIT and STOP statements have similar effects. The RETURN statement ends execution of a program called subroutine and passes execution to the main program. The CONTINUE statement does nothing. It can be used with a statement number as the last line of a DO loop.

The GO TO [**s**] statement transfers control to statement $s$.

The arithmetic IF statement is written

$$\text{IF}[\mathbf{e}]s_1, s_2, s_3$$

[**e**] is any numerical variable or arithmetic expression, and the $s_i$ are statement numbers. The transfer occurs according to the following table.

| [**e**] | statement executed |
|---|---|
| [**e**] < 0 | $s_1$ |
| [**e**] = 0 | $s_2$ |
| [**e**] > 0 | $s_3$ |

The logical IF statement has the form

$$\text{IF}[\mathbf{le}][\mathbf{statement}]$$

[**le**] is a logical expression, and [**statement**] is any executable statement except DO and IF. Only if [**le**] is true will [**statement**] be executed. Otherwise, the next instruction will be executed.

The logical expression [**le**] is a relational expression using one of several operators.

Logical expressions also can incorporate the connectors .AND., .OR., and .NOT..

**Table 19.4**
FORTRAN Logical Operations

| operator | meaning |
|---|---|
| .LT. | less than |
| .LE. | less than or equal to |
| .EQ. | equal to |
| .NE. | not equal to |
| .GT. | greater than |
| .GE. | greater than or equal to |

*Example 19.9*

IF (A.GT.25.6) A = 27.0

Meaning: If A is greater than 25.6, set A equal to 27.0.

IF (Z.EQ.(T−4.0).OR.Z.EQ.0.) GO TO 17

Meaning: If Z is equal to (T−4.0) or if Z is equal to *zero*, go to statement 17.

## 8 LIBRARY FUNCTIONS

The following single-precision library functions are available. Most are accessed by placing the argument in parentheses after the function name. Placing the letter $D$ before the function name will cause the calculation to be performed in double precision. Arguments for trigonometric functions are expressed in radians.

**Table 19.5**
Some FORTRAN Library Functions

| function | use |
|---|---|
| EXP | $e^x$ |
| ALOG | natural logarithm |
| ALOG10 | common logarithm |
| SIN | sine |
| COS | cosine |
| TAN | tangent |
| SINH | hyperbolic sine |
| SQRT | square root |
| ASIN | arcsine |
| MOD | remaindering modulus (integer) |
| AMOD | remaindering modulus (real) |
| ABS | absolute value (real) |
| IABS | absolute value (integer) |
| FLOAT | convert integer to real |
| FIX | convert real to integer |



## 9 USER FUNCTIONS

A user-defined function can be created with the FUNCTION statement. Such functions are governed by the following rules.

- The function is defined as a variable in the main program even though it is a function.
- When used in the main program, the function is followed by its arguments in parentheses.
- In the function itself, the function name is type-declared and defined by the word FUNCTION.
- The arguments (parameters) need not have the same names in the main program and function.
- Only the function has a RETURN statement.
- Both the main program and the function have END statements.
- The arguments (parameters) must agree in number, order, type, and length.

These construction rules are illustrated by example 19.10.

*Example 19.10*

```
REAL HEIGHT, WIDTH, AREA, MULT
HEIGHT = 2.5
WIDTH = 7.5
AREA = MULT(HEIGHT,WIDTH)
END

REAL FUNCTION MULT(HEIGHT,WIDTH)
REAL HEIGHT, WIDTH
MULT = HEIGHT*WIDTH
RETURN
END
```

## 10 SUBROUTINES

A subroutine is a user-defined subprogram. It is more versatile than a user-defined function as it is not limited to mathematical calculations. Subroutines are governed by the following rules.

- The subroutine is activated by the CALL statement.
- The subroutine has no type.
- The subroutine does not take on a value. It performs operations on the arguments (parameters) which are passed back to the main program.
- A subroutine has a RETURN statement.
- Both the main program and the subroutine have END statements.
- The arguments (parameters) need not have the same names in the main program and the subroutine.
- The arguments (parameters) must agree in number, order, type, and length.
- It is possible to return to any part of the main program. It is not necessary to return to the statement immediately below the CALL statement.

These rules are illustrated by example 19.11.

*Example 19.11*

```
    REAL HEIGHT, WIDTH, AREA
    CALL GET(HEIGHT, WIDTH)
    AREA = HEIGHT*WIDTH
    END

    SUBROUTINE GET(A,B)
    REAL A,B
    READ (5,100) A,B
100 FORMAT(2F3.1)
    RETURN
    END
```

Variables in functions and subroutines are completely independent of the main program. Subroutine and main program variables which have the same names will not have the same values. A link between the main program and the subroutine can be established, however, with the COMMON statement.

The COMMON statement assigns storage locations in memory to be shared by the main program and all of its subroutines. Even the COMMON statement, however, allows different names. It is the order of the common variables which fixes their position in upper memory.

*Example 19.12*

What are the values of $X$ and $Y$ in the subroutine?

```
COMMON X, Y       main program
  X = 2.0
  Y = 10.0

COMMON Y, X       subroutine
```

Since $Y$ is the first common subroutine variable which corresponds to $X$ in the main program, $Y = 2.0$. Similarly, $X = 10.0$.

If variables are to be shared with only some of the subroutines, the *named* COMMON statement is required. Whereas there can be only one regular COMMON statement, there can be multiple-named COMMON statements.

```
COMMON/PLACE/CAT, COW, DOG   main program
COMMON/PLACE/HORSE, PIG, EXPENSE
                               subroutine
```

# PART 6: Fire Safety Systems

## 1 INTRODUCTION

In many cases, the design of fire detection, fire alarm, and sprinkler systems is governed by state or local codes. It is necessary to review all applicable codes and to meet the most stringent of them. Generally, insurance carrier requirements are more stringent than code minimums. Although codes mandate minimum standards, common sense and professional prudence should be used in specifying the level of fire protection in a building.



In buildings that are partially or wholly sprinklered, provisions for alarm and evacuation as well as provisions for sprinkler supervision must be made.

The type of occupancy greatly affects the degree of protection. Nursing homes, schools, hospitals, and office buildings all have greatly different needs. Furthermore, multi-story buildings with limited escape routes affect the degree of protection.

## 2 DETECTION DEVICES

- **Manual Fire Alarm Stations**
  Mandatory in any system, large or small. Locate in the natural path of exit with the maximum traveling distance to any manual station of 200 feet. Identification of an activated manual station which when opened should be readily visible from the side, down a corridor for at least 200 feet.

- **Heat Detectors—Fixed Temperature**
  135°F in open spaces. 190°–200°F generally used in enclosed or confined spaces such as boiler rooms, closets, etc., where the heat build-up will be fast and confined.

- **Heat Detectors—Rate of Rise**
  Combination rate of rise and fixed temperature of 135° or 200°F. Rate of rise portion operates when the temperature rises in excess of 15° per minute. More sensitive than the fixed temperature detector.

- **Heat Detector—Rate Compensated**
  Considered to be the most responsive of all thermal detectors. Operates at 135° or 200°F. Detects both slow and fast developing fires by anticipating the temperature increase and moving towards the alarm point as the temperature gradually increases.

- **Photoelectric Smoke Detectors**
  Operates on a photo beam or light scattering principle. The photoelectric detector responds best to products of combustion or smoke with a particle size from approximately 10.0 microns down to 0.1 micron and of the proper concentration. Proper concentration is defined by Underwriters Laboratories as the ability to sense smoke in the 0.2 to 4.0 percent obscuration per foot range. Photoelectric detectors generally are considered to be the best for cold smoke fires.

- **Ionization Detectors**
  Ionization detectors detect products of combustion by sensing the disruption of conductivity in an ionized chamber due to the presence of smoke. The ionization detector responds best to fast burning fires where particle sizes range from approximately 1.0 micron down to 0.01 micron and of the proper concentration. Proper concentration is defined by Underwriters Laboratories as the ability to sense smoke from 0.2 to 4.0 percent obscuration per foot range. **Note**: Ionization and photoelectric detectors can be intermixed within a system to provide the best form of detection suitable to the environment.

- **Infrared Flame Detectors**
  Generally, infrared detectors respond to radiation in the 6500 to 8500 angstrom range. Good detectors will filter out solar interference and respond to radiation in the 4000 to 5500 angstrom range. It is preferrable to use a detector with a dual sensing circuit in order to discriminate unwanted or false alarms.

- **Ultraviolet Flame Detectors**
  Ultraviolet flame detectors respond to radiation in the spectral range of 1700 to 2900 angstroms. It is not sensitive to solar interference. Built-in time delays prevent false alarms.

- **Waterflow Detectors**
  Used in wet sprinkler systems to indicate a flow of water. Use on the main sprinkler risers and throughout the building to indicate sub-sections or floors of the building to locate sprinkler discharges quickly. These detectors employ a retard mechanism to prevent false alarms from water surges.

- **Pressure Switches**
  Used in dry or pre-action sprinkler systems to provide an alarm when water is discharged.

- **Valve Monitor Switches**
Closed water supply valves are a major weakness of sprinkler systems. This is the most frequently neglected and forgotten item in the sprinkler system. Closed valves also have accounted for countless millions of dollars worth of damage because the system would not operate.

- **Low Temperature Monitor Switches**
Low air temperature (under 40°F) or low water temperature is detrimental to a sprinkler system and creates a great nuisance by freezing and bursting the sprinkler system pipes.

## 3 ALARM DEVICES

- **Bells**
This is the most commonly accepted form of alarm. However, bells should not be used in schools where bells are used for other signaling purposes. Bells should not be used in any area where the same sound is used for any other function.

- **Chimes**
Single stroke devices in coded systems and used in certain types of applications such as quiet areas of hospitals or nursing homes.

- **Horns**
Horns generally are capable of producing a higher sound level than either a bell or a chime.

- **Alarm Lights**
Used individually as a fire alarm visual indicator or used in conjunction with a horn. The light can be either a flashing incandescent bulb or a flashing strobe. It sometimes is required by certain codes or types of occupancy in order to provide an alarm for handicapped persons.

- **Remote Annunciators**
Generally, these duplicate the main control panel and have a light for every fire alarm zone. They are used at the second entrance or at the main entrance to assist the fire department in locating the fire zone. Also used in nursing homes, at nursing stations in hospitals or in the engineering room of a factory or a building.

- **Speakers**
Used in emergency voice evacuation systems, generally in high-rise buildings. Specific quality, construction, and performance have been established by the NFPA code for speakers used in voice evacuation systems.

## 4 SIGNAL TRANSMISSION

- **Reverse Polarity**
Uses a dedicated leased telephone directly between the protected premises and the municipal fire department or a commercial central station.

- **Central Station**
These are central monitoring facilities which monitor fire and security alarms from protected premises for a monthly fee.

- **Telephone Dialers**
Tape dialers have had a very bad reputation due to high failure rate and susceptibility to false alarms. Solid state digital dialers with compatible solid state digital receivers have increased the reliability and dependability of this product very dramatically and are becoming more acceptable as an alternate form of signal transmission.

- **Radio Transmitters**
Radio transmitter boxes are used in certain applications to transmit alarms between the protected premises and some monitoring point. Before using, it is advisable to check that the line of transmission is clear from obstruction and interference.

## 5 AUXILIARY CONTROL

- **Smoke Doors**
Generally held open with floor or wall mounted electromagnets. Generally all doors close in all parts of the building on the first alarm.

- **Fire Doors**
Generally treated the same way as smoke doors.

- **Stairwell Exit Doors**
Sometimes for security reasons, these doors are held latched with a door strike. The door also must employ a panic exit bar to override and manually open the door.

- **Elevator Capture**
Generally accepted by all codes that the elevator immediately return to the first floor or to some designated alternate floor. The specification should designate that the elevator manufacturer is responsible for accepting low voltage signal or a dry contact from the fire alarm panel and programming the elevator to return to the designated floor.

- **Fire Dampers**
Either motorized or fusible link type are closed to prevent the spread of smoke to other areas. Care should be exercised in connecting these to a fire alarm system because a fire alarm system generally operates with small amounts of D.C. power. The responsibility should be stated for coordinating the

voltages and contact ratings necessary to do the job.

- **Fans—Supply and Exhaust**
Sometimes all fans are shut down on the first alarm. However, some buildings and other considerations make it desirable to exhaust the smoke from the building and shut down the input supply fans.

- **Pressurization**
Pressurization creates positive air pressure to inhibit the influx of smoke from adjacent areas or floors above or below. Exit stairwell escape routes out of the high-rise building should be pressurized to provide a smoke-free exit path.



## 6 SPECIAL CONSIDERATIONS

- **Handicapped Persons**
There may be requirements from federal health officials (HEW) or OSHA to provide consideration for handicapped persons.

- **Weather Protection of Devices**
Check to make sure that none of the devices will be exposed to adverse conditions such as excessive moisture.

- **Open Plenums**
Drop ceilings that use the space above as an open air plenum present special problems and are treated as a potential hazard in most codes. Smoke detectors located in open air plenums need to be located properly, and it is advisable to locate them so that they are accessible for maintenance. A remote alarm lamp should be brought down and mounted below the ceiling level to indicate which detector is an alarm.

- **Duct Detectors**
Duct heat detectors are considered to be of negligible value. Duct smoke detectors can provide warning of smoke in an air duct and can prevent costly damage to expensive HVAC equipment. Duct detectors are not a substitute for open area smoke detectors. Problem areas can develop with duct detectors from excessive humidity, poor or no maintenance, and improper location.

- **Sound Pressure Level of Alarm Devices**
The sound pressure level required to provide an adequate alarm in the environment in which it is intended to operate generally is considered to have been accomplished if the alarm sound is 12 decibels above the normal ambient level.

- **Emergency Generators**
If the fire alarm system does not use its own standby battery pack, it is advisable to coordinate the details of how the emergency generator feeds power back into the building distribution system and to insure that the fire alarm system will be provided with emergency power. In some cases, it is advisable to provide the fire alarm system with a small amount of standby battery power in order to keep the fire alarm system on line until the generator gets to full power.

- **Fire Pump Supervision**
Insure that fire pumps, if used, are supervised adequately for all critical functions and that the building fire alarm system is provided with one or more zones to interface with fire pump signals.

## 7 SPECIAL SUPPRESSION SYSTEMS

- **Deluge Systems**
The control panel for the water deluge system is specialized and generally is mounted in a hostile environment. The control equipment is mounted inside an enclosure which is watertight and dust tight. Deluge systems are actuated manually and also by thermal detectors or flame detectors. The entire system, including the electronic control panel, is provided by the sprinkler contractor.

- **Foam Systems**
High expansion foam or water suppression systems are specialized systems requiring special handling. They generally are activated by infrared or ultraviolet flame detectors.

- **Halon Systems**
Halon systems generally are used in clean environments such as computer rooms or any room that contains high value equipment. Examples of this are tapes, microfilms, computer records, laboratories for research and design, and medical laboratories. Generally, the criterion is the high value of the equipment or the data stored, and it is desirable that water not touch the equipment. Halon systems, because of their great expense, usually are engineered specially for the particular room in which they are to be located. The control panel uses cross-zoning techniques to prevent unnecessary discharges.

- **High/Low Pressure $CO_2$ Systems**
$CO_2$ systems are used in industrial applications to suppress fires where the use of water, extinguishing powders, or Halon would be unsuitable because of expense, hazard, or the size of the equipment to be protected. Because $CO_2$ is hazardous to life, it has to be an engineered system designed for the particular application.

# PART 7: Nondestructive Testing

## 1 MAGNETIC PARTICLE TESTING

This procedure is based on the attraction of magnetic particles to leakage flux at surface flaws. The particles accumulate and become visible at the flaw. This method works for magnetic materials in locating cracks, laps, seams, and in some cases, subsurface flaws. The test is fast and simple and is easy to interpret. However, parts must be relatively clean and demagnetized. A high current (power) source must be available.

## 2 EDDY CURRENT TESTING

Alternating currents from a source coil induce eddy currents in metallic objects. Flaws and other material properties affect the current flow. The change in current flow is observed on a meter or a screen. This method can be used to locate defects of many types, including changes in composition, structure, and hardness, as well as locating cracks, voids, inclusions, weld defects, and changes in porosity.

Intimate contact between the material and the test coil is not required. Operation can be continuous, automatic, and monitored electronically. Therefore, this method is ideal for unattended continuous processing. Sensitivity is easily adjustable. Many variables, however, can affect the current flow.

## 3 LIQUID PENETRANT TESTING

Liquid penetrant (dye) is drawn into surface defects by capillary action. A developer substance then is used to develop the penetrant to aid in visual inspection. This method can be used with any nonporous material, including metals, plastics, and glazed ceramics. It locates cracks, porosities, pits, seams, and laps.

Liquid penetrant tests are simple to perform, can be used with complex shapes, and can be performed on site. Parts must be clean, and only small surface defects are detectable.

## 4 ULTRASONIC TESTING

Mechanical vibrations in the 0.1–25 MHz range are induced in an object. The transmitted energy is reflected and scattered by interior defects. The results are interpreted from a screen or a meter. The method can be used for metals, plastics, glass, rubber, graphite, and concrete. It detects inclusions, cracks, porosity, laminations, changes in structure, and other interior defects.

This test is extremely flexible. It can be automated and is very fast. Results can be recorded or interpreted electronically. Penetration of up to 60 feet of steel is possible. Only one surface needs to be accessed. However, rough surfaces or complex shapes may cause difficulties.



## 5 INFRARED TESTING

Infrared radiation emitted from objects can be detected and correlated with quality. The detection can be recorded electronically. Any discontinuity that interrupts heat flow, such as flaws, voids, and inclusions, can be detected.

Infrared testing requires access to only one side, and it is highly sensitive. It is applicable to complex shapes and assemblies of dissimilar components, but it is relatively slow. Results are affected by material variations, coatings, and colors, and hot spots can be hidden by cool surface layers.

## 6 RADIOGRAPHY

X-ray and gamma-ray sources can be used to penetrate objects. The intensity is reduced in passing through, and the intensity changes are recorded on film or screen. This method can be used with most materials to detect internal defects, material structure, and thickness. It also can be used to detect the absence of internal parts.

Up to 30 inches of steel can be penetrated by x-ray sources. Gamma sources, which are more portable and lower in cost than x-ray sources, are applicable to 10″ thickness of steel.

There are health and government standards associated with these tests. Electric power and cooling water may be required in large installations. Shielding and film processing also is required, making this the most expensive nondestructive test.

# PART 8: Environmental Impact Assessment

Most large scale construction projects must be assessed for potential environmental damage before being built. The assessment is usually developed in an *environmental impact report.* Such a report should evaluate all of the potential ways in which a project could affect the ecological balance and environment. The questions which follow can be used as an outline for such a report.

1. What is the nature of the proposed project?

2. What is the nature of the project area, including distinguishing natural and man-made characteristics?

3. How, and to what degree, will the earth be altered by any of the following means?

   - change in topology, including earth removal
   - off-site hauling, dust, smoke, or air pollutants generated during construction
   - chemical treatment
   - change in structural composition of the soil as a result of compacting, tilling, shoring, etc.
   - change in moisture content
   - change in slope

4. In what ways will the project affect natural drainage or flooding? Compare existing flooding and drainage conditions with those that would exist after the project is finished.

5. To what extent could the project result in the erosion of property both on or off the project site?

6. To what extent will the project affect the potential use, extraction, or conservation of the earth's resources, such as crops, minerals, and ground water?

7. How, and to what extent, will the project affect the quality and characteristics of soil, water, and air in the immediate project area and in the community? Include an estimate of the amounts of sewage, drainage, airborne particulate matter, and solid waste to be generated by the completed project.

8. Describe the plant and animal life presently on the site, and describe how, and to what extent, this life will be affected by the project.

9. Indicate the percentages of land use, both before and after the project is completed, for the following categories of land use:

   - residential
   - commercial
   - industrial
   - public

10. Indicate the percentages of land use, both before and after the project is completed, for the following categories of land use:

    - agricultural
    - street and highway
    - parking, parking driveways, loading, etc.
    - surface or aerial utilities
    - railroads
    - vacant and unimproved
    - landscaping

11. How will the project change the views of the site from points around the project?

12. How will the project change the views of the surrounding community from points within the project?

13. How, and to what extent, will the project affect wilderness, open space, landscaping, and other aesthetic and recreational considerations?

14. Are any of the natural or man-made features in the project unique (i.e., not found in any other parts of the city, county, state, or nation)?

15. What effect will the project have upon the health and safety of the people in the project area and in the community?

16. How much fresh water will be consumed as a result of the project? What will be the source of this water?

17. How, and to what extent, will the project add to existing noise levels on the site and in the community?

18. How many additional people will live in the project area? How many additional people will work there? How many people will be displaced from the project area?

19. How will the project add to, or subtract from, the availability of roads, highways, and other elements of the transportation system, or increase or decrease the burden upon such elements?

20. Could the project serve to encourage development of presently undeveloped areas? Could it intensify development of already developed areas?

21. Will the project involve the application, use, or disposal of hazardous materials?

22. Will the project involve the construction of facilities in areas of known earthquake faults or soil instability (e.g., subsidence, landslides, or severe erosion)?

23. Does the project affect a historically significant or archaeological site?

24. Are there any other significant environmental effects, either positive or negative, of the project?

# PART 9: Mathematical Programming

## 1 INTRODUCTION

Mathematical programming is a modeling procedure applicable to problems for which the goal and resource limitations can be described mathematically. If the *goal function* and all *resource constraints* are linear (polynomials of degree 1 only), the procedure is known as *linear programming.*

If the variables can take on only integer values, a procedure known as *integer programming* is required. If the polynomials are of any degree or contain other functions, a procedure known as *dynamic programming* is required.

## 2 FORMULATION OF A LINEAR PROGRAMMING PROBLEM

All linear programming problems have a similar format. Each has an *objective function* which is to be optimized. Usually the objective function is to be maximized, as in the case of a *profit function.* If the objective is to minimize some function, such as cost, the problem may be turned into a maximization problem by maximizing the negative of the original function.

*Example 19.13*

A cattle rancher buys three types of cattle food. The rancher wants to minimize the cost of feeding his cattle. Write the objective function for this problem on a per animal basis.

| food type | cost per pound |
|---|---|
| 1 | 1.5 |
| 2 | 2.5 |
| 3 | 3.5 |

Let $x_i$ be the number of pounds of food $i$ purchased per animal. Then, the objective function to be minimized is

$$Z = 1.5x_1 + 2.5x_2 + 3.5x_3$$

Each linear programming problem also has a set of limitation functions called *constraints.* Constraints are used to set the bounds for the objective function.

*Example 19.14*

The rancher is concerned with meeting published nutritional information on minimum daily requirements (MDR) given in milligrams per animal. The composition of each food type is known and the contributions for each vitamin in mg/pound are

| | | food type | | |
|---|---|---|---|---|
| vitamin | MDR (mg) | 1 | 2 | 3 |
| A | 100 | 1 | 7 | 13 |
| B | 200 | 3 | 9 | 15 |
| C | 300 | 5 | 11 | 17 |

It is also physically impossible for an animal to eat more than the following amounts per day.

| food type | maximum feeding |
|---|---|
| 1 | 50 lbs |
| 2 | 40 |
| 3 | 30 |

The constraints on this problem are

$$\begin{aligned}
x_1 + 7x_2 + 13x_3 &\geq 100 \\
3x_1 + 9x_2 + 15x_3 &\geq 200 \\
5x_1 + 11x_2 + 17x_3 &\geq 300 \\
x_1 &\leq 50 \\
x_2 &\leq 40 \\
x_3 &\leq 30 \\
x_1 &\geq 0 \\
x_2 &\geq 0 \\
x_3 &\geq 0
\end{aligned}$$

Linear programming problems are generally solved by computer. Some simple problems can be solved by hand with a procedure known as the *simplex method.* Specialized methods allowing easy manual solutions are available for certain classes of problems, primarily the *transportation problem* and *assignment problem.*

Once a solution is found, it is possible to determine the effect on the objective function of changing one of the program parameters. This is known as *sensitivity analysis* and is very important in instances where the accuracy of collected data is unknown.

## 3 SOLUTION TO 2-DIMENSIONAL PROBLEMS

If a linear programming problem can be formulated in terms of only two variables, $x_1$ and $x_2$, it can be solved graphically by the following procedure:

*step 1*: Graph all of the constraints and determine the *feasible region.* Usually this will result in a *convex hull.*

*step 2*: Evaluate the objective function, $Z$, at each corner of the hull.

*step 3*: The values of $x_1$ and $x_2$ which optimize $Z$ are the coordinates of the corner at which $Z$ is optimized.

*Example 19.15*

Solve the following linear programming problem graphically.

$$\begin{aligned} \text{Max } Z &= x_1 + 2x_2 \\ \text{such that } 4x_1 + x_2 &\leq 24 \\ 2x_1 + x_2 &\leq 14 \\ -x_1 + 2x_2 &\leq 8 \\ -x_1 + x_2 &\leq 3 \\ x_1 &\geq 0 \\ x_2 &\geq 0 \end{aligned}$$

The region enclosed by the constraints is shown.

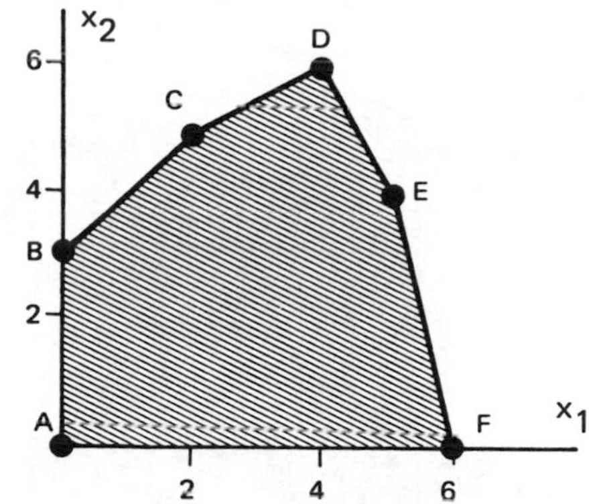


The coordinates and $Z$ value for each corner are

| corner | coordinates $(x_1, x_2)$ | $Z$ |
|---|---|---|
| A | (0,0) | 0 |
| B | (0,3) | 6 |
| C | (2,5) | 12 |
| D | (4,6) | 16 |
| E | (5,4) | 13 |
| F | (6,0) | 6 |

$Z$ is maximized when $x_1 = 4$ and $x_2 = 6$.

# 20 SYSTEMS OF UNITS

## 1 CONSISTENT SYSTEMS OF UNITS

A set of units used in a problem is said to be *consistent*[1] if no conversion factors are needed. For example, a moment with units of foot-pounds cannot be obtained directly from a moment arm with units of inches. In this illustration, a conversion factor of $\frac{1}{12}$ feet/inch is needed, and the set of units used is said to be *inconsistent.*

On a larger scale, a system of units is said to be consistent if Newton's second law of motion can be written without conversion factors. Newton's law states that the force required to accelerate an object is proportional to the amount of matter in the item.

$$F = ma \qquad 20.1$$

The definitions of the symbols, $F$, $m$, and $a$, are familiar to every engineer. However, the use of Newton's second law is complicated by the multiplicity of available unit systems. For example, $m$ may be in kilograms, pounds, or slugs. All three of these are units of mass. However, as figure 20.1 illustrates, these three units do not represent the same amount of mass.

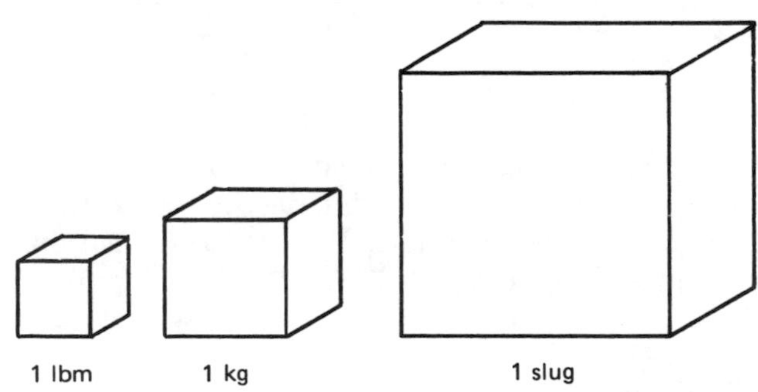


**Figure 20.1** Common Units of Mass

[1] The terms *homogeneous* and *coherent* also are used to describe a consistent set of units.

It should be mentioned that the decision to work with a consistent set of units is arbitrary and unnecessary. Problems in fluid flow and thermodynamics commonly are solved with inconsistent units. This causes no more of a problem than working with inches and feet in the calculation of a moment.

## 2 THE ABSOLUTE ENGLISH SYSTEM

Engineers are accustomed to using pounds as a unit of mass. For example, density typically is given in pounds per cubic foot. The abbreviation *pcf* tends to obscure the fact that the true units are pounds of *mass* per cubic foot.

If pounds are the units for mass, and feet per second squared are the units of acceleration, the units of force for a consistent system can be found from Newton's second law.

$$\begin{aligned}\text{units of } F &= (\text{units of } m)(\text{units of } a)\\ &= (\text{lbm})(\text{ft/sec}^2) = \frac{\text{lbm-ft}}{\text{sec}^2}\end{aligned} \qquad 20.2$$

The units for $F$ cannot be simplified any more than they are in equation 20.2. This particular combination of units is known as a *poundal.*[2]

The absolute English system, which requires the poundal as a unit of force, seldom is used, but it does exist. This existence is a direct outgrowth of the requirement to have a consistent system of units.

## 3 THE ENGLISH GRAVITATIONAL SYSTEM

Force frequently is measured in pounds. When the thrust on an accelerating rocket is given as so many pounds, it is understood that the pound is being used as a unit of force.

[2] A poundal is equal to 0.03108 pounds force.

If acceleration is given in feet per second squared, the units of mass for a consistent system of units can be determined from Newton's second law.

$$\text{units of } m = \frac{\text{units of } F}{\text{units of } a} = \frac{\text{lbf}}{\text{ft/sec}^2} = \frac{\text{lbf-sec}^2}{\text{ft}} \qquad 20.3$$

The combination of units in equation 20.3 is known as a *slug*.[3] Slugs and pounds-mass are not the same, as illustrated in figure 20.1. However, units of mass can be converted, using equation 20.4.

$$\#\ \text{slugs} = \frac{\#\ \text{lbm}}{g_c} \qquad 20.4$$

$g_c$[4] is a dimensional conversion factor having the following value.

$$g_c = 32.1740 \frac{\text{lbm-ft}}{\text{lbf-sec}^2} \qquad 20.5$$

32.1740 commonly is rounded to 32.2 when six significant digits are unjustified. That practice is followed in this book.

Notice that the number of slugs cannot be determined from the number of pounds-mass by dividing by the local gravity. $g_c$ is used regardless of the local gravity. However, the local gravity can be used to find the weight of an object. *Weight* is defined as the force exerted on a mass by the local gravitational field.

$$\text{weight in lbf} = (m \text{ in slugs})(g \text{ in ft/sec}^2) \qquad 20.6$$

If the effects of large land and water masses are neglected, the following formula can be used to estimate the local acceleration of gravity in ft/sec$^2$ at the earth's surface. $\phi$ is the latitude in degrees.

$$g_{\text{surface}} = 32.088[1 + (5.305\ \text{EE} - 3)\sin^2\phi - (5.9\ \text{EE} - 6)\sin^2 2\phi] \qquad 20.7$$

If the effects of the earth's rotation are neglected, the gravitational acceleration at an altitude, $h$, in miles is given by equation 20.8. $R$ is the earth's radius—approximately 3960 miles.

$$g_h = g_{\text{surface}} \left[\frac{R}{R+h}\right]^2 \qquad 20.8$$

[3] A slug is equal to 32.1740 lbm.

[4] Three different meanings of the symbol $g$ commonly are used. $g_c$ is the dimensional conversion factor given in equation 20.5. $g_o$ is the standard acceleration due to gravity with a value of 32.1740 ft/sec$^2$. $g$ is the local acceleration due to gravity in ft/sec$^2$.

## 4 THE ENGLISH ENGINEERING SYSTEM

Many thermodynamics and fluid flow problems freely combine variables containing pound-mass and pound-force terms. For example, the steady-flow energy equation (SFEE) used in chapter 6 mixes enthalpy terms in BTU/lbm with pressure terms in lbf/ft$^2$. This requires the use of $g_c$ as a mass conversion factor.

Newton's second law becomes

$$F \text{ in lbf} = \frac{(m \text{ in lbm})(a \text{ in ft/sec}^2)}{g_c \text{ in} \frac{\text{lbm-ft}}{\text{lbf-sec}^2}} \qquad 20.9$$

Since $g_c$ is required, the English Engineering System is inconsistent. However, that is not particularly troublesome, and the use of $g_c$ does not overly complicate the solution procedure.

*Example 20.1*

Calculate the weight of a 1.0 lbm object in a gravitational field of 27.5 ft/sec$^2$.

Since weight commonly is given in pounds-force, the mass of the object must be converted from pounds-mass to slugs.

$$F = \frac{ma}{g_c} = \frac{(1)\ \text{lbm}\ (27.5)\ \text{ft/sec}^2}{(32.2)\frac{\text{lbm-ft}}{\text{lbf-sec}^2}} = .854\ \text{lbf}$$

*Example 20.2*

A rocket with a mass of 4000 lbm is traveling at 27,000 ft/sec. What is its kinetic energy in ft-lbf?

The usual kinetic energy equation is $E_k = \frac{1}{2}mv^2$. However, this assumes consistent units. Since energy is wanted in foot-pounds-force, $g_c$ is needed to convert $m$ to units of slugs.

$$E_k = \frac{mv^2}{2g_c} = \frac{(4000)\ \text{lbm}\ (27,000)^2\ \text{ft}^2/\text{sec}^2}{(2)(32.2)\frac{\text{lbm-ft}}{\text{lbf-sec}^2}} = 4.53\ \text{EE}10\ \text{ft-lbf}$$

In the English Engineering System, work and energy typically are measured in ft-lbf (mechanical systems) or in British Thermal Units, BTU (thermal and fluid systems). One BTU equals 778.26 ft-lbf.

## 5 THE cgs SYSTEM

The cgs system has been used widely by chemists and physicists. It is named for the three primary units used to construct its derived variables. The centimeter, the gram, and the second form the basis of this system.

UNITS

The cgs system avoids the lbm versus lbf type of ambiguity in two ways. First, the concept of weight is not used at all. All quantities of matter are specified in grams, a mass unit. Second, force and mass units do not share a common name.

When Newton's second law is written in the cgs system, the following combination of units results.

$$\begin{aligned} \text{units of force} &= (m \text{ in } g)\left(a \text{ in } \frac{\text{cm}}{\text{sec}^2}\right) \\ &= \frac{g\text{–cm}}{\text{sec}^2} \end{aligned} \qquad 20.10$$

This combination of units for force is known as a *dyne.*

Energy variables in the cgs system have units of dyne-cm or, equivalently, of

$$\frac{g\text{–cm}^2}{\text{sec}^2}$$

These combinations are known as an *erg.* There is no uniformly accepted unit of power in the cgs system, although calories per second frequently is used. Ergs can be converted to calories by multiplying by 2.389 EE–8.

The fundamental volume unit in the cgs system is the cubic centimeter (cc). Since this is the same volume as one thousandth of a liter, units of millimeters (ml) are used freely in this system.

## 6 THE mks SYSTEM

The mks system is appropriate when variables take on values larger than can be accomodated by the cgs system. This system uses the *m*eter, the *k*ilogram, and the *s*econd as its primary units. The mks system avoids the lbm versus lbf ambiguity in the same ways as does the cgs system.

The units of force can be derived from Newton's second law.

$$\begin{aligned} \text{units of force} &= (m \text{ in kg})\left(a \text{ in } \frac{\text{m}}{\text{sec}^2}\right) \\ &= \frac{\text{kg–m}}{\text{sec}^2} \end{aligned} \qquad 20.11$$

This combination of units for force is known as a *newton.*

Energy variables in the mks system have units of N-m or, equivalently, $\frac{\text{kg–m}^2}{\text{sec}^2}$. Both of these combinations are known as a *joule.* The units of power are joules per second, equivalent to a *watt.* The common volume unit is the liter, equivalent to one-thousandth of a cubic meter.

*Example 20.3*

A 10 kg block hangs from a cable. What is the tension in the cable?

$$\begin{aligned} F = ma &= (10) \text{ kg } (9.8)\frac{\text{m}}{\text{sec}^2} \\ &= 98\frac{\text{kg–m}}{\text{sec}^2} = 98\ N \end{aligned}$$

*Example 20.4*

A 10 kg block is raised vertically 3 meters. What is the change in potential energy?

$$\begin{aligned} \Delta E_p = mg\Delta h &= (10) \text{ kg } (9.8)\ \frac{\text{m}}{\text{sec}^2}\ (3) \text{ m} \\ &= 294\frac{\text{kg–m}^2}{\text{sec}^2} = 294\ J \end{aligned}$$

## 7 THE SI SYSTEM

Strictly speaking, both the cgs and the mks systems are *metric* systems. Although the metric units simplify solutions to problems, the multiplicity of possible units for each variable sometimes is confusing.

The SI system (International System of Units) was established in 1960 by the General Conference of Weights and Measures, an international treaty organization. The SI system is derived from the earlier metric systems, but it is intended to supersede them all.

The SI system has the following features.

(a) There is only one recognized unit for each variable.

(b) The system is fully consistent.

(c) Scaling of units is done in multiples of 1000.

(d) Prefixes, abbreviations, and symbol-syntax are defined rigidly.

Three types of units are used: base units, supplementary units, and derived units. The base units (table 20.2) are dependent on only accepted standards or reproducible phenomena. The supplementary units (table 20.3) have not yet been classified as being base units or derived units. The derived units (tables 20.4 and 20.5) are made up of combinations of base and supplementary units.

The expressions for the derived units in symbolic form are obtained by using the mathematical signs of multiplication and division. For example, units of velocity are $m/s$. Units of torque are $N \cdot m$ (not $N$-$m$ or $Nm$).

**Table 20.1**
SI Prefixes

| prefix | symbol | value |
|---|---|---|
| exa | E | EE18 |
| peta | P | EE15 |
| tera | T | EE12 |
| giga | G | EE9 |
| mega | M | EE6 |
| kilo | k | EE3 |
| hecto | h | EE2 |
| deka | da | EE1 |
| deci | d | EE−1 |
| centi | c | EE−2 |
| milli | m | EE−3 |
| micro | μ | EE−6 |
| nano | n | EE−9 |
| pico | p | EE−12 |
| femto | f | EE−15 |
| atto | a | EE−18 |

**Table 20.2**
SI Base Units

| quantity | name | symbol |
|---|---|---|
| length | meter | m |
| mass | kilogram | kg |
| time | second | s |
| electric current | ampere | A |
| temperature | kelvin | K |
| amount of substance | mole | mol |
| luminous intensity | candela | cd |

**Table 20.3**
SI Supplementary Units

| quantity | name | symbol |
|---|---|---|
| plane angle | radian | rad |
| solid angle | steradian | sr |

In addition, there is a set of non-SI units which can be used. This temporary concession is due primarily to the significance and widespread acceptance of these units. Use of the non-SI units listed in table 20.6 usually will create an inconsistent expression requiring conversion factors.

In addition to having standardized units, the SI system also specifies syntax rules for writing the units and combinations of units. Each unit is abbreviated with a specific *symbol*. The rules for writing these symbols should be followed.

(a) The symbols always are printed in roman type, irrespective of the type used in the rest of the text. The only exception to this is in the use of the symbol for *liter*, where the lower case *l* (ell) may be confused with the number 1 (one). In this case, *liter* should be written out in full or the script *l* used. There is no problem with such symbols as cl (centiliter) or ml (milliliter).

(b) Symbols are never pluralized: 1 kg, 45 kg (not 45 kgs).

(c) A period is not used after a symbol, except when the symbol occurs at the end of a sentence.

(d) When symbols consist of letters, there always is a full space between the quantity and the symbols; e.g. 45 kg (not 45kg). However, when the first character of a symbol is not a letter, no space is left, e.g. 32°C (not 32° C or 32 °C) or 42°12′45″ (not 42 ° 12 ′ 45 ″).

(e) All symbols are written in lower case, except when the unit is derived from a proper name. For example, m for meter, s for second, but A for ampere, Wb for weber, N for newton, W for watt. Prefixes are printed roman type without spacing between the prefix and the unit symbol, e.g., km for kilometer.

(f) In text, symbols should be used when associated with a number. When no number is involved, the unit should be spelled out. For example, the area of a carpet is 16 $m^2$, not 16 square meters, and carpet is sold by the square meter, not by the $m^2$.

(g) A practice in some countries is to use a comma as a decimal marker, while the practice in North America, the United Kingdom, and some other countries is to use a period (or a dot) as the decimal marker. Further, in some countries using the decimal comma, a dot frequently is used to divide long numbers into groups of three. Because of these differing practices, spaces must be used instead of commas to separate long lines of digits into easily-readable blocks of three digits with respect to the decimal marker, e.g. 32 453.246 072 5. A space (a half space is preferred) is optional with a four-digit number, e.g; 1 234, 1 234, or 1234.

(h) Where a decimal fraction of a unit is used, a zero should be placed before the decimal marker; e.g., 0.45 kg (not .45 kg). This practice draws attention to the decimal marker and helps avoid errors of scale.

(i) Some confusion may arise with the word *tonne* (1 000 kg). When this word occurs in French text of Canadian origin, the meaning may be a ton or 2 000 pounds.

**Table 20.4**
Some SI Derived Units with Special Names

| quantity | name | symbol | expressed in terms of other units |
|---|---|---|---|
| frequency | hertz | Hz | |
| force | newton | N | |
| pressure, stress | pascal | Pa | $N/m^2$ |
| energy, work, quantity of heat | joule | J | N·m |
| power, radiant flux | watt | W | J/s |
| quantity of electricity, electric charge | coulomb | C | |
| electric potential, potential difference, electromotive force | volt | V | W/A |
| electric capacitance | farad | F | C/V |
| electric resistance | ohm | Ω | V/A |
| electric conductance | siemen | S | A/V |
| magnetic flux | weber | Wb | V·s |
| magnetic flux density | tesla | T | $Wb/m^2$ |
| inductance | henry | H | Wb/A |
| luminous flux | lumen | lm | |
| illuminance | lux | lx | $lm/m^2$ |

**Table 20.5**
Some SI Derived Units

| quantity | description | expressed in terms of other units |
|---|---|---|
| area | square meter | $m^2$ |
| volume | cubic meter | $m^3$ |
| speed—linear | meter per second | m/s |
| angular | radian per second | rad/s |
| acceleration—linear | meter per second squared | $m/s^2$ |
| angular | radian per second squared | $rad/s^2$ |
| density, mass density | kilogram per cubic meter | $kg/m^3$ |
| concentration (of amount of substance) | mole per cubic meter | $mol/m^3$ |
| specific volume | cubic meter per kilogram | $m^3/kg$ |
| luminance | candela per square meter | $cd/m^2$ |
| dynamic viscosity | pascal second | Pa·s |
| moment of force | newton meter | N·m |
| surface tension | newton per meter | N/m |
| heat flux density, irradiance | watt per square meter | $W/m^2$ |
| heat capacity, entropy | joule per kelvin | J/K |
| specific heat capacity, specific entropy | joule per kilogram kelvin | J/(kg·K) |
| specific energy | joule per kilogram | J/kg |
| thermal conductivity | watts per meter kelvin | W/(m·K) |
| energy density | joule per cubic meter | $J/m^3$ |
| electric field strength | volt per meter | V/m |
| electric charge density | coulomb per cubic meter | $C/m^3$ |
| surface density of charge, flux density | coulomb per square meter | $C/m^2$ |
| permittivity | farad per meter | F/m |
| current density | ampere per square meter | $A/m^2$ |
| magnetic field strength | ampere per meter | A/m |
| permeability | henry per meter | H/m |
| molar energy | joule per mole | J/mol |
| molar entropy, molar heat capacity | joule per mole kelvin | J/(mol·K) |
| radiant intensity | watt per steradian | W/sr |

**Table 20.6**
Acceptable Non-SI Units

| quantity | unit name | symbol | relationship to SI unit |
|---|---|---|---|
| area | hectare | ha | 1 ha = 10 000 $m^2$ |
| energy | kilowatt-hour | kWh | 1 kWh = 3.6 MJ |
| mass | metric ton[5] | t | 1 t = 1000 kg |
| plane angle | degree (of arc) | ° | 1° = 0.017 453 rad |
| speed of rotation | revolution per minute | r/min | 1 r/min = $\frac{2\pi}{60}$ rad/s |
| temperature interval | degree Celsius | °C | 1°C = 1 K |
| time | minute | min | 1 min = 60 s |
| | hour | h | 1 h = 3600 s |
| | day (mean solar) | d | 1 d = 86 400 s |
| | year (calendar) | a | 1 a = 31 536 000 s |
| velocity | kilometer per hour | km/h | 1 km/h = 0.278 m/s |
| volume | liter[6] | $l$ | 1 $l$ = 0.001 $m^3$ |

Numbers in parentheses are the number of ESU or EMU units, per single SI unit, except for the permittivity and the permeability of free space, where each actual values of $\epsilon_o$ and $\mu_o$ are given.

[5] The international name for metric ton is *tonne*. The metric ton is equal to the *megagram*, Mg.

[6] The international symbol for liter is the lowercase "l," which can be confused easily with the numeral "1." Several English speaking countries have adopted the script $l$ as the symbol for liter in order to avoid any misinterpretation.

UNITS

# Appendix A: Selected Conversion Factors to SI Units

| | SI Symbol | Multiplier to Convert From Existing Unit to SI Unit | Multiplier to Convert From SI Unit to Existing Unit |
|---|---|---|---|
| **Area** | | | |
| Circular Mil | μm² | 506.7 | 0.001 974 |
| Foot Squared | m² | 0.092 9 | 10.764 |
| Mile Squared | km² | 2.590 | 0.386 1 |
| Yard Squared | m² | 0.836 1 | 1.196 |
| **Energy** | | | |
| Btu (International) | kJ | 1.055 1 | 0.947 8 |
| Erg | μJ | 0.1 | 10.0 |
| Foot Pound-Force | J | 1.355 8 | 0.737 6 |
| Horsepower Hour | MJ | 2.684 5 | 0.372 5 |
| Kilowatt Hour | MJ | 3.6 | 0.277 8 |
| Meter Kilogram-Force | J | 9.806 7 | 0.101 97 |
| Therm | MJ | 105.506 | 0.009 478 |
| Kilogram Calorie (International) | kJ | 4.186 8 | 0.238 8 |
| **Force** | | | |
| Dyne | μN | 10. | 0.1 |
| Kilogram-Force | N | 9.806 7 | 0.101 97 |
| Ounce-Force | N | 0.278 0 | 3.597 |
| Pound-Force | N | 4.448 2 | 0.224 8 |
| KIP | N | 4 448.2 | 0.000 224 8 |
| **Heat** | | | |
| Btu Per Hour | W | 0.293 1 | 3.412 1 |
| Btu Per (Square Foot Hour) | W/m² | 3.154 6 | 0.317 0 |
| Btu Per (Square Foot Hour °F) | W/(m²·°C) | 5.678 3 | 0.176 1 |
| Btu Inch Per (Square Foot Hour °F) | W/(m·°C) | 0.144 2 | 6.933 |
| Btu Per (Cubic Foot °F) | MJ/(m³·°C) | 0.067 1 | 14.911 |
| Btu Per (Pound °F) | J/(kg·°C) | 4 186.8 | 0.000 238 8 |
| Btu Per Cubic Foot | MJ/m³ | 0.037 3 | 26.839 |
| Btu Per Pound | J/kg | 2 326. | 0.000 430 |
| **Length** | | | |
| Angstrom | nm | 0.1 | 10.0 |
| Foot | m | 0.304 8 | 3.280 8 |
| Inch | mm | 25.4 | 0.039 4 |
| Mil | mm | 0.025 4 | 39.370 |
| Mile | km | 1.609 3 | 0.621 4 |
| Mile (International Nautical) | km | 1.852 | 0.540 |
| Micron | μm | 1.0 | 1.0 |
| Yard | m | 0.914 4 | 1.093 6 |
| **Mass (weight)** | | | |
| Grain | mg | 64.799 | 0.015 4 |
| Ounce (Avoirdupois) | g | 28.350 | 0.035 3 |
| Ounce (Troy) | g | 31.103 5 | 0.032 15 |
| Ton (short 2000 lb.) | kg | 907.185 | 0.001 102 |
| Ton (long 2240 lb.) | kg | 1 016.047 | 0.000 984 2 |
| Slug | kg | 14.593 9 | 0.068 522 |
| **Pressure** | | | |
| Bar | kPa | 100.0 | 0.01 |
| Inch of Water Column (20°C) | kPa | 0.248 6 | 4.021 9 |
| Inch of Mercury (20°C) | kPa | 3.374 1 | 0.296 4 |
| Kilogram-force per Centimeter Squared | kPa | 98.067 | 0.010 2 |
| Millimeters of Mercury (mm·Hg) (20°C) | kPa | 0.132 84 | 7.528 |
| Pounds Per Square Inch (P.S.I.) | kPa | 6.894 8 | 0.145 0 |
| Standard Atmosphere (760 torr) | kPa | 101.325 | 0.009 869 |
| Torr | kPa | 0.133 32 | 7.500 6 |

UNITS

# Appendix A (continued):

| | SI Symbol | Multiplier to Convert From Existing Unit to SI Unit | Multiplier to Convert From SI Unit to Existing Unit |
|---|---|---|---|
| **Power** | | | |
| Btu (International) Per Hour | W | 0.293 1 | 3.412 2 |
| Foot Pound-Force Per Second | W | 1.355 8 | 0.737 6 |
| Horsepower | kW | 0.745 7 | 1.341 |
| Meter Kilogram-Force Per Second | W | 9.806 7 | 0.101 97 |
| Tons of Refrigeration | kW | 3.517 | 0.284 3 |
| **Torque** | | | |
| Kilogram-Force Meter (kg·m) | N·m | 9.806 7 | 0.101 97 |
| Pound-Force Foot | N·m | 1.355 8 | 0.737 6 |
| Pound-Force Inch | N·m | 0.113 0 | 8.849 5 |
| Gram-Force Centimeter | mN·m | 0.098 067 | 10.197 |
| **Temperature** | | | |
| Fahrenheit | °C | $\frac{5}{9}$ (°F − 32) | ($\frac{9}{5}$°C) + 32 |
| Rankine | K | (°F + 459.67)$\frac{5}{9}$ | (°C + 273.16)$\frac{9}{5}$ |
| **Velocity** | | | |
| Foot Per Second | m/s | 0.304 8 | 3.280 8 |
| Mile Per Hour | m/s | 0.447 04 | 2.236 9 |
| | *or* | *or* | *or* |
| | km/h | 1.609 34 | 0.621 4 |
| **Viscosity** | | | |
| Centipoise | mPa·s | 1.0 | 1.0 |
| Centistoke | μm²/s | 1.0 | 1.0 |
| **Volume (Capacity)** | | | |
| Cubic Foot | *l* (dm³) | 28.316 8 | 0.035 31 |
| Cubic Inch | cm³ | 16.387 1 | 0.061 02 |
| Cubic Yard | m³ | 0.764 6 | 1.308 |
| Gallon (U.S.) | *l* | 3.785 | 0.264 2 |
| Ounce (U.S. Fluid) | ml | 29.574 | 0.033 8 |
| Pint (U.S. Fluid) | *l* | 0.473 2 | 2.113 |
| Quart (U.S. Fluid) | *l* | 0.946 4 | 1.056 7 |
| **Volume Flow (Gas-Air)** | | | |
| Standard Cubic Foot Per Minute | m³/s | 0.000 471 9 | 2119. |
| | *or* | *or* | *or* |
| | *l*/s | 0.471 9 | 2.119 |
| | *or* | *or* | *or* |
| | ml/s | 471.947 | 0.002 119 |
| Standard Cubic Foot Per Hour | ml/s | 7.865 8 | 0.127 133 |
| | *or* | *or* | *or* |
| | μ*l*/s | 7 866. | 0.000 127 |
| **Volume Liquid Flow** | | | |
| Gallons Per Hour (U.S.) | *l*/s | 0.001 052 | 951.02 |
| Gallons Per Minute (U.S.) | *l*/s | 0.063 09 | 15.850 |

# 21 ENGINEERING LICENSING

*Purpose of Registration*

As an engineer, you may have to obtain your professional engineering license through procedures which have been established by the state in which you reside. These procedures are designed to protect the public by preventing unqualified individuals from legally practicing as engineers.

There are many reasons for wanting to become a professional engineer. Among them are the following:

- You may wish to become an independent consultant. By law, consulting engineers must be registered.
- Your company may require a professional engineering license as a requirement for employment or advancement.
- Your state may require registration as a professional engineer if you use the title *engineer*.

*The Registration Procedure*

The registration procedure is similar in most states. You probably will take two 8-hour written examinations. The first examination is the *Engineer-in-Training* examination, also known as the *Intern Engineer* exam and the *Fundamentals of Engineering* exam. The initials E-I-T, I.E., and F.E. also are used. The second examination is the *Professional Engineering* (*P.E.*) exam, which differs from the E-I-T exam in format and content.

If you have significant experience in engineering, you may be allowed to skip the E-I-T examination. However, actual details of registration, experience requirements, minimum education levels, fees, and examination schedules vary from state to state. You should contact your state's Board of Registration for Professional Engineers.

*Reciprocity Among States*

All states use the NCEES P.E. examination.[1] If you take and pass the P.E. examination in one state, your certificate probably will be honored by other states which have used the same NCEES examination. It will not be necessary to retake the P.E. examination.

The simultaneous administration of identical examinations in multiple states has led to the term *Uniform Examination.* However, each state is free to choose its own minimum passing score or to add special questions to the NCEES examination. Therefore, this Uniform Examination does not automatically ensure reciprocity among states.

Of course, you may apply for and receive a professional engineering license from another state. However, a license from one state will not permit you to practice engineering in another state. You must have a professional engineering license from each state in which you work.

*Applying for the Examination*

Each state charges different fees, requires different qualifications, and uses different forms. Therefore, it will be necessary for you to request an application and an information packet from the state in which you reside or in which you plan to take the exam. It generally is sufficient to phone for this information. Telephone numbers for all of the U.S. state boards of registration are given in the accompanying table.

[1] The National Council of Examiners for Engineering and Surveying (NCEES) in Clemson, South Carolina, produces, distributes, and grades the national P.E. examinations. It does not distribute applications to take the P.E. examination.

**Phone Numbers of State Boards of Registration**

| | |
|---|---|
| Alabama | (334) 242-5568 |
| Alaska | (907) 465-2540 |
| Arizona | (602) 255-4053 |
| Arkansas | (501) 324-9085 |
| California | (916) 263-2222 |
| Colorado | (303) 894-7788 |
| Connecticut | (203) 566-3290 |
| Delaware | (302) 577-6500 |
| District of Columbia | (202) 727-7454 |
| Florida | (904) 488-9912 |
| Georgia | (404) 656-3926 |
| Guam | (671) 646-9386 |
| Hawaii | (808) 586-3000 |
| Idaho | (208) 334-3860 |
| Illinois | (217) 782-8556 |
| Indiana | (317) 232-2980 |
| Iowa | (515) 281-5602 |
| Kansas | (913) 296-3053 |
| Kentucky | (502) 573-2680 |
| Louisiana | (504) 295-8522 |
| Maine | (207) 287-3236 |
| Maryland | (410) 333-6322 |
| Massachusetts | (617) 727-9957 |
| Michigan | (517) 335-1669 |
| Minnesota | (612) 296-2388 |
| Mississippi | (601) 359-6160 |
| Missouri | (314) 751-0047 |
| Montana | (406) 444-4285 |
| Nebraska | (402) 471-2407 |
| Nevada | (702) 688-1231 |
| New Hampshire | (603) 271-2219 |
| New Jersey | (201) 504-6460 |
| New Mexico | (505) 827-7561 |
| New York | (518) 474-3846 |
| North Carolina | (919) 781-9499 |
| North Dakota | (701) 258-0786 |
| Ohio | (614) 466-3650 |
| Oklahoma | (405) 521-2874 |
| Oregon | (503) 378-4180 |
| Pennsylvania | (717) 783-7049 |
| Puerto Rico | (787) 722-2122 |
| Rhode Island | (401) 277-2565 |
| South Carolina | (803) 737-9260 |
| South Dakota | (605) 394-2510 |
| Tennessee | (615) 741-3221 |
| Texas | (512) 440-7723 |
| Utah | (801) 530-6551 |
| Vermont | (802) 828-2363 |
| Virginia | (804) 367-8512 |
| Virgin Islands | (809) 774-3130 |
| Washington | (206) 753-6966 |
| West Virginia | (304) 558-3554 |
| Wisconsin | (608) 266-1397 |
| Wyoming | (307) 777-6155 |



*Examination Format*

The NCEES Professional Engineering examination in Civil Engineering consists of two four-hour sessions separated by a one-hour lunch period. Both the morning and the afternoon sessions contain twelve problems. Most states do not have required problems. Each examinee is given exam booklets that contain problems only for civil engineers.

**Major Work Behavior Analysis Civil Examination**

| subject | number of problems |
|---|---|
| traffic systems | 2 |
| transportation facilities | 3 |
| buildings and special structures | 2 |
| foundations and retaining structures | 2 |
| drainage/flood control systems | 2 |
| natural water systems | 1 |
| wastewater treatment | 3 |
| solid/hazardous waste | 1 |
| geotechnical/soils | 3 |
| construction/materials testing | 1 |

These subjects are rather broad, and the NCEES examinations are not obligated to follow the task analysis guidelines. Since the examination structure is not rigid, it is not possible to give the exact number of problems that will appear in each subject area. Thus, there is no guarantee that any single subject will appear.

Although elements of decision-making based on engineering economics may appear in other problems, engineering economics is no longer listed as an exam subject.

The examination is open book. Usually, all forms of solutions aids are allowed in the examination, including nomographs, specialty slide rules, and pre-programmed and programmable calculators. Since their use says little about the depth of your knowledge, such aids should be used only to check your work. For example, very few points will be earned if a pre-programmed calculator is used to solve a surveying problem.

Most states do not limit the number and types of books you can bring into the exam.[2] Loose-leaf papers (including Post-it$^{tm}$ notes) and writing tablets are usually forbidden, although you may be able to bring in loose reference pages in a three-ring binder. References used in the afternoon session need not be the same as for the morning session.

[2] Check with your state to see if review books can be brought into the examination. Most states do not have any restrictions. Some states ban only collections of solved problems, such as Schaum's Outline Series. A few prohibit all review books.

In most states, any battery-powered, silent calculator may be used. Also, in most states, there are no restrictions on programmable or pre-programmed calculators. However, calculators with significant word processing abilities may be prohibited at the state's option. Printers cannot be used.

You will not be permitted to share books, calculators, or any other items with other examinees.

*Morning and Afternoon Sessions*

All twelve of the problems in the morning exam booklet are of the same general format. These problems are "essay," a term loosely used to mean that the solution is free-form. For each problem, you will be given a set of conditions (known as the "situation") and unknowns (known as the "requirements"). Essay problems may have one or more parts (i.e., requirements or unknowns), though most have only one or two. You solve the problems by working in the answer booklet provided to you, using any method and the solution sequence of your choice.

The afternoon problems are all objectively scored. "Objectively scored" is NCEES' term for multiple choice. For each of the twelve problems (situations), ten multiple-choice questions are asked. Each question has four answer choices, labeled A, B, C, and D. Since these problems are multiple choice and your answer is of the blacken-the-bubble variety, neatness (except for filling in the bubble neatly) and logic are irrelevant factors. Marking the correct answer choice is the only relevant factor. There is no penalty for guessing or incorrect answers.

There are no multiple-choice problems in the morning session, and there are no essay problems in the afternoon session.

*Use of Metric Units*

The NCEES P.E. exam in civil engineering is written predominantly with English units (also known as "customary U.S. units," "inch-pound units," and "British units"). This is consistent with conventional practice in most of the fields tested on the exam. Occasional SI problems have appeared, but they are rare.

*Grading the Exam*

The morning session, with its essay problems, is graded using criterion-referenced scoring plans for each problem. Each of your solutions is compared with the specific required solution elements previously established for that problem. Even with the specific criteria, however, there is still room for subjectivity in the grading. Partial credit is routinely given, even for problems that are solved incorrectly. Neatness, logic, and correct calculations will help you earn the maximum points on all problems.

Grading of multiple-choice afternoon problems is done by computer. Your methods and calculations are not reviewed. There is no partial credit.

*Criterion-Referenced Grading*

The criterion-referenced grading method used by NCEES to score essay problems gets its name from the specific elements (i.e., criteria) that you must include in your solution to receive full credit. These criteria are determined prior to the administration of the examination. Taken together, the criteria constitute a "Scoring Plan" for each problem.

As an example of a criterion-referenced scoring plan, consider the design of a long, circular concrete culvert. In order to receive six points, the minimum number to correctly pass the problem, you might have to:

(1) consider the effects of friction on the hydraulic head

(2) calculate the flow rate based on the correct hydraulic head

(3) recognize that the flow is entrance-controlled and that the barrel will not flow full

(4) calculate the normal depth to within ±10% of the correct value

(5) verify that the normal depth was less than the critical depth

(6) set up all equations correctly

Notice that, for this example, it is theoretically unnecessary to complete the problem or obtain the correct answer in order to earn six points.

Getting the full ten points, however, requires solving the problem correctly to completion, without making any significant mathematical errors. For each mathematical error, two points will be subtracted. However, if all of the six criteria are met, no fewer than four points could be deducted for mathematical errors.

*Passing Score*

Full credit in the examination is achieved by correctly working four problems in the morning and four problems in the afternoon. (If you have time, you can work on more than eight problems. However, you can only specify eight problems to be graded. Extra problems worked will not be graded.)

By working eight problems, each worth 10 points, the maximum number of points you can earn on an exam is 80. NCEES has established 48 out of 80 as the minimum passing score.

You will receive the results of your examination by mail. Allow 12–14 weeks for notification. Your score may or may not be revealed to you, depending on your state's procedure.

*Examination Dates*

The NCEES examinations are administered on the same weekend in all states.

**National P.E. Examination Dates**

| year | Spring exam | Fall exam |
|---|---|---|
| 1997 | April 18–19 | October 31; November 1 |
| 1998 | April 24–25 | October 30–31 |
| 1999 | April 23–24 | October 29–30 |
| 2000 | April 14–15 | October 27–28 |

*Preparing for the Exam*

You should develop an examination strategy early in the preparation process. This strategy will depend on your background. One of the following two general strategies is recommended:

- A broad approach has been successful for examinees who have recently completed academic studies. Their strategy has been to review the fundamentals of a broad range of undergraduate civil engineering subjects. The examination includes enough fundamental problems to give merit to this strategy.
- Working engineers who have been away from classroom work for a long time have found it better to concentrate on the subjects in which they have had extensive professional experience. By studying the list of examination subjects, they have been able to choose those which will give them a good probability of finding enough problems that they can solve.

Do not make the mistake of studying only a few subjects in hopes of finding enough problems to work. The more subjects you are familiar with, the better will be your chances of passing the examination. More important than stretegy are fast recall and stamina. You must be able to recall quickly solution procedures, formulas, and important data; and this sharpness must be maintained for eight hours. You will not have time in the exam to derive solution methods; you must know them instinctively.

It is imperative that you develop and adhere to a review outline and schedule. If you are not taking a classroom review course where the order of preparation is determined by the lectures, you should develop an Outline of Subjects for Self-Study to schedule your preparation.

It is unnecessary to take a large quantity of books to the examination. The examination is very fast-paced. You will not have time to look up solution procedures, data, or equations with which you are not familiar. Although the examination is open-book, there is insufficient time to use books with which you are not thoroughly familiar.

To minimize time spent searching for often-used formulas and data, you should prepare a one-page summary of all important formulas and information in each subject area. You can then use these summaries during the examination instead of searching for the correct page in your book.

*Items to Get for the Examination*

- Obtain ten sheets of each of the following types of graph paper: 10 squares to the inch grid, semi-log (3 cycles × 10 squares to the inch grid), and full-log (3 cycles × 3 cycles).
- Obtain a flexible clear plastic ruler marked in tenths of an inch or centimeters.
- Make a clear transparency of figure 10.11.

*What to do Before the Exam*

The engineers who have taken the P.E. exam in previous years have made the suggestions listed below. These suggestions will make your examination experience as comfortable and successful as possible.

- Keep a copy of your examination application. Send the original application by certified mail and request a receipt of delivery.
- Visit the exam site the day before your examination. This is especially important if you are not familiar with the area. Find the examination room, the parking area, and the rest rooms.
- Plan on arriving at least 30 minutes before the examination starts. This will assure you a convenient parking place and adequate time for site, room, and seating changes.
- If you live a considerable distance from the examination site, consider getting a hotel room in which to spend the night before.

- Take off the day before the examination to relax. Don't cram the last night. Rather, get a good night's sleep.
- Be prepared to find that the examination room is not ready at the designated time. Take an interesting novel or magazine to read in the interim and at lunch.
- If you make arrangements for baby sitters or transportation, allow for a delayed completion.
- Prepare your examination kit the day before. Here is a checklist of items to take with you to the examination. (Some states may not permit food or beverages in the exam room.)

[ ] copy of your application

[ ] proof of delivery receipt

[ ] letter admitting you to the exam

[ ] photographic identification

[ ] other reference books

[ ] CIVIL ENGINEERING HANDBOOK (Merritt)

[ ] course notes in a binder

[ ] calculator and a spare

[ ] spare calculator batteries or battery pack

[ ] battery charger and 20' extension cord

[ ] chair cushions. A large, thick bath mat works well.

[ ] earplugs

[ ] desk expander. If you are taking the exam in theater chairs with tiny, fold-up writing surfaces, you should take a long, wide board to place across the arm rests.

[ ] a cardboard box cut to fit your references

[ ] twist-to-advance pencils

[ ] extra leads

[ ] large eraser

[ ] snacks such as raisins, nuts, or trail mix

[ ] thermos filled with hot chocolate

[ ] a light lunch

[ ] a collection of graph paper (use only if permitted in the exam)

[ ] scissors, stapler, and staple puller

[ ] construction paper for stopping drafts and sunlight

[ ] transparent and masking tapes

[ ] sunglasses

[ ] extra prescription glasses, if you wear them

[ ] aspirin

[ ] travel pack of Kleenex

[ ] Webster's dictionary

[ ] dictionary of scientific terms

[ ] $2 in change

[ ] a light comfortable sweater

[ ] comfortable shoes or slippers for the exam room

[ ] raincoat, boots, gloves, hat, and umbrella

[ ] local street maps

[ ] note to the parking patrol for your windshield

[ ] straightedge, ruler, compass, protractor, and French curves

[ ] battery-powered desk lamp

[ ] watch

[ ] extra car keys

*What to do During the Exam*

Previous examinees have reported that the following strategies and techniques have helped them considerably.

- Read through all of the problems before starting your first solution. In order to save you from rereading and reevaluating each problem later in the day, you should classify each problem at the beginning of the four hour session. The following categories are suggested:

  · problems you can do easily

  · problems you can do with effort

  · problems for which you can get partial credit

  · problems you cannot do

- Do all of the problems in order of increasing difficulty. All problems on the examination are worth 10 points. There is nothing to be gained by attempting the difficult or long problems if easier or shorter problems are available.

- Follow these guidelines when solving a problem:

  · Do not rewrite the problem statement.

  · Do not unnecessarily redraw any figures.

  · Use pencil only.

  · Be neat. (Print all text. Use a straightedge or template where possible.)

  · Draw a box around each answer.

  · Label each answer with a symbol.

  · Give the units.

  · List your sources whenever you use obscure solution methods or data.

  · Write on one side of the page only.

  · Use one page per problem, no matter how short the solution is.

  · Go through all calculations a second time and check for mathematical errors.

# 22 POSTSCRIPTS

This chapter collects comments, revisions, and commentary which cannot be incorporated into the body of the text until the next edition. New postscript sections are added as needed when the *Civil Engineering Reference Manual* is reprinted. Subjects in this chapter are not necessarily represented by entries in the index. **It is suggested that you make a note in the appropriate text pages to refer to this chapter.**

*Update: October, 1986*

### Chapter 1: VARIANCE

The statistical term *variance* is used in example 1.42 (page 1-27) but is not defined in the text. The variance is the square of the standard deviation. Since there are two standard deviations ($s$ and $\sigma$), there are two variances. The sample variance is $s^2$. The population variance is $\sigma^2$.

### Chapter 3: WATER HAMMER IN DUCTILE PIPE

The speed of sound used in water hammer calculations (i.e., $c$ in equations 3.198 and 3.199) must account for the expansion of ductile pipe walls as the water pressure builds up. Equation 3.20 can be used to calculate $c$, but the modulus of elasticity used should include the elastic contributions of the water and pipe material both. In the equation below, $t_{\text{pipe}}$ is the pipe wall thickness, and $d_{\text{pipe}}$ is the inside diameter.

$$E = \frac{E_{\text{water}} t_{\text{pipe}} E_{\text{pipe}}}{t_{\text{pipe}} E_{\text{pipe}} + d_{\text{pipe}} E_{\text{water}}}$$

### Chapter 8: HEATING VALUE OF DIGESTER GAS

The heating value of digester gas is listed (page 8-29) as 600 BTU/ft$^3$. This value is appropriate for digester gas with the composition given: 65% methane, 35% carbon dioxide. The actual heating value will depend on the fraction of combustible methane, as well as the temperature and pressure. At 60°F and 14.7 psia, pure methane has a lower heating value of 900 BTU/ft$^3$. Carbon dioxide does not contribute to the heating effect.

### Chapter 14: NEUTRAL AXES IN CONCRETE BEAMS

Figures 14.4 and 14.5 (page 14-10) both contain the variable $c$, the distance from the neutral axis to the top of beam. However, these two distances are not the same, even though the symbol is the same. The location of the neutral axis changes as the ultimate strength is approached.

### Chapter 14: MODULUS OF ELASTICITY FOR MASONRY

Masonry structures are generally designed using the alternate (working stress) design method, which requires knowing the modulus of elasticity. Equation 14.1 for concrete cannot be used. An approximate value can be found from the masonry's compressive strength.

$$E_m = 1000 f'_m \quad (< 3 \text{ EE6 psi})$$

### Chapter 17: LENGTH OF VERTICAL CURVE

Equation 17.52 (page 17-16) can be solved without trial and error.

$$LC = 2d\left(\frac{\sqrt{\frac{EG}{EF}}+1}{\sqrt{\frac{EG}{EF}}-1}\right)$$

*Update: April, 1987*

### Chapter 5: CHANNEL SLOPE

In uniform flow, the slopes of the channel bottom, water surface, and energy grade line are identical. For uniform flow, then, equation 5.6 could be written using the geometric slope, $S_o$, instead of the hydraulic slope, $S$:

$$v = C\sqrt{r_H S_o}$$

Similarly, other equations in chapter 5 could also be written using $S_o$, but only under the condition of uniform flow. Using $S_o$, however, is clearly a special case of a general rule.

### Chapter 5: OPEN CHANNEL FLOW in SI

The factor 1.49 in equations 5.7–5.9 (and others) converts customary SI to English units. For problems in SI units (m/sec, etc.), replace the 1.49 with 1.00.

### Chapter 7: RAPID SAND FILTERS

Rapid sand filters usually operate with hydraulic heads (distance between water surfaces in filter and clearwell) of 9–12 feet.

Prior to backwashing with clear water, the filter material may be expanded by an *air prewash* of 1–8 (2–5 typical) cfm/ft$^2$ for 2–10 (3–5 typical) minutes.

### Chapter 8: SANITARY LANDFILLS

Many large-scale sanitary landfills do not apply daily cover to deposited solid waste. Time and cost are typically cited as the reasons that the landfill is not covered with soil. To account for the absence of such cover, the loading factor in equation 8.62 must have a value of 1.0.

### Chapter 9: PERCENT PASSING SIEVE

Equations 9.2 and 9.3, as well as various soils classification schemes, require knowing percentages of soil passing through specific sieve sizes. When sieve data is incomplete, the needed values can be interpolated by plotting the known data on a *particle size distribution chart.* (Also, see figure 9.2.)

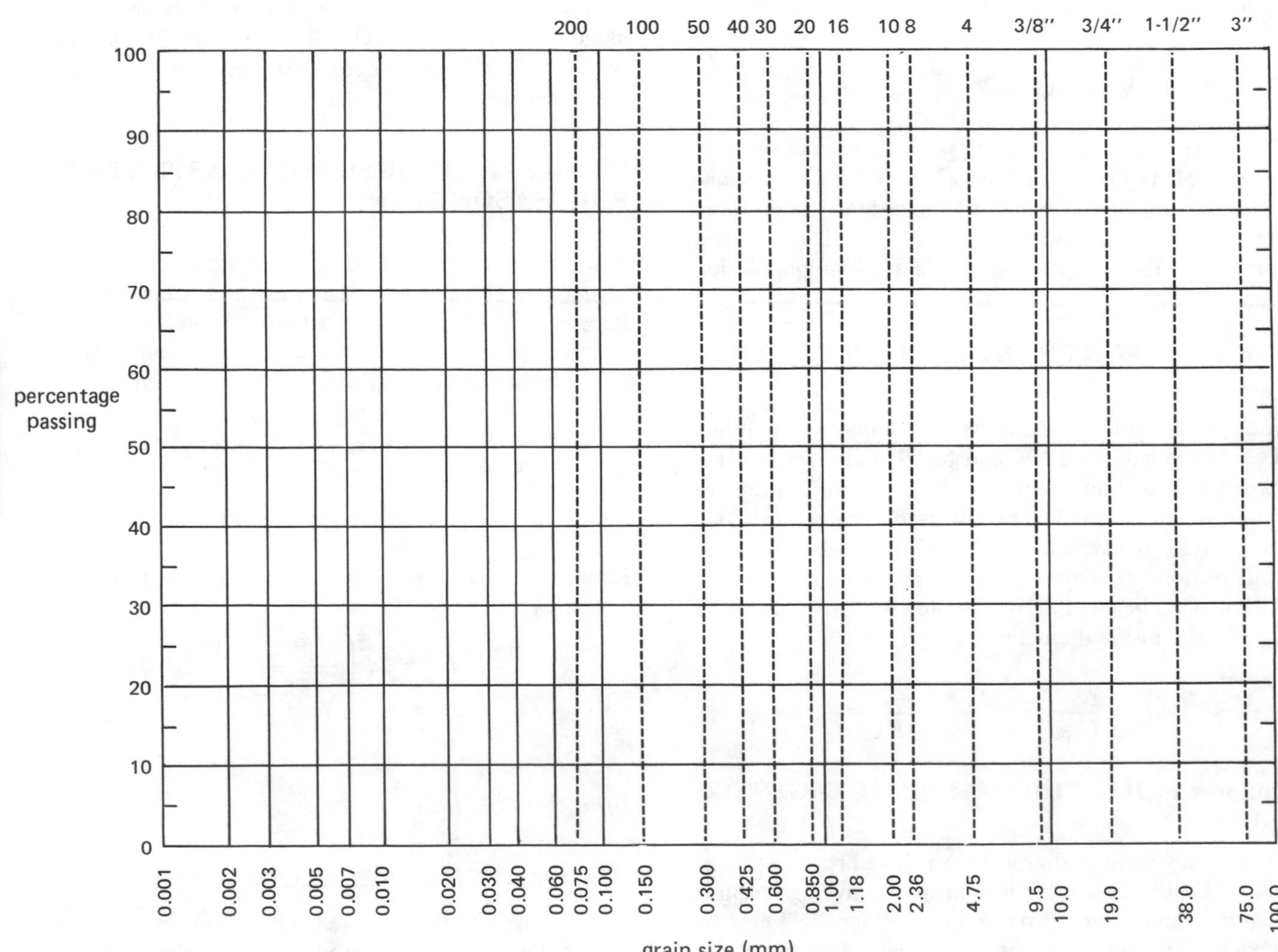

## THE UNIFIED SOIL CLASSIFICATION SYSTEM

| Major Division | | Group Symbol | Laboratory Classification Criteria | | Soil Description |
|---|---|---|---|---|---|
| | | | Finer than 200 Sieve (%) | Supplementary Requirements | |
| Coarse-grained (over 50% by weight coarser than No. 200 sieve) | Gravelly soils (over half of coarse fraction larger than No. 4) | GW | 0–5* | $D_{60}/D_{10}$ greater than 4<br>$D_{30}{}^2/(D_{60} \times D_{10})$ between 1 & 3 | Well-graded gravels, sandy gravels |
| | | GP | 0–5* | Not meeting above gradation for GW | Gap-graded or uniform gravels, sandy gravels |
| | | GM | 12 or more* | PI less than 4 or below A-line | Silty gravels, silty sandy gravels |
| | | GC | 12 or more* | PI over 7 and above A-line | Clayey gravels, clayey sandy gravels |
| | Sandy soils (over half of coarse fraction finer than No. 4) | SW | 0–5* | $D_{60}/D_{10}$ greater than 4,<br>$D_{30}{}^2/(D_{60} \times D_{10})$ between 1 & 3 | Well-graded, gravelly sands |
| | | SP | 0–5* | Not meeting above gradation requirements | Gap-graded or uniform sands, gravelly sands |
| | | SM | 12 or more* | PI less than 4 or below A-line | Silty sands, silty gravelly sands |
| | | SC | 12 or more* | PI over 7 and above A-line | Clayey sands, clayey gravelly sands |
| Fine-grained (over 50% by weight finer than No. 200 sieve) | Low compressibility (liquid limit less than 50) | ML | Plasticity chart | | Silts, very fine sands, silty or clayey fine sands, micaceous silts |
| | | CL | Plasticity chart | | Low plasticity clays, sandy or silty clays |
| | | OL | Plasticity chart, organic odor or color | | Organic silts and clays of low plasticity |
| | High compressibility (liquid limit more than 50) | MH | Plasticity chart | | Micaceous silts, diatomaceous silts, volcanic ash |
| | | CH | Plasticity chart | | Highly plastic clays and sandy clays |
| | | OH | Plasticity chart, organic odor or color | | Organic silts and clays of high plasticity |
| Soils with fibrous organic matter | | Pt | Fibrous organic matter; will char, burn, or glow | | Peat, sandy peats, and clayey peat |

*For soils having 5 to 12% passing the No. 200 sieve, use a dual symbol such as GW–GC.

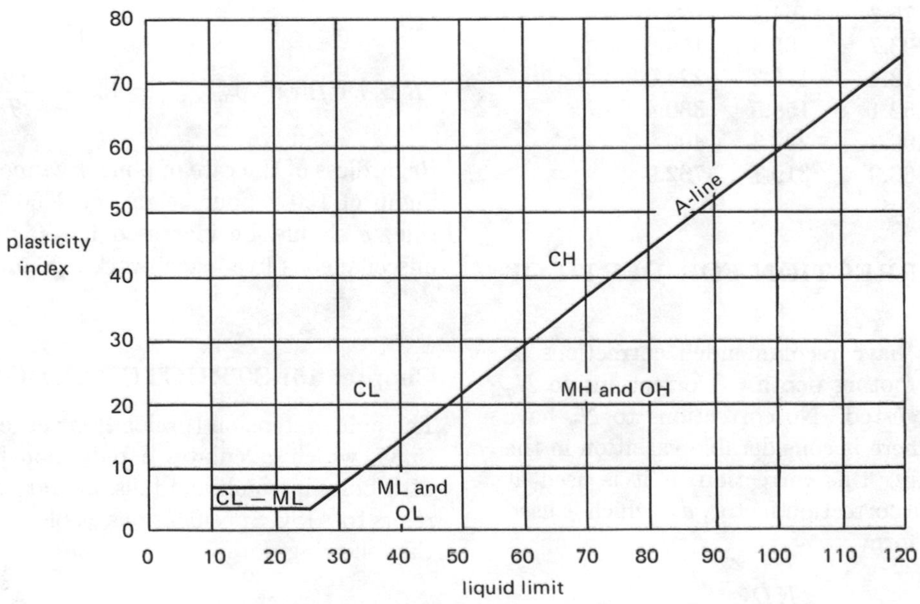


Plasticity chart for the classification of fine-grained soils. Tests made on fraction finer than No. 40 sieve, 0.425 mm.

## Chapter 9: CONSOLIDATION TESTS

It is common practice to plot *strain*, $\epsilon$, versus log $p$ for consolidation test data, as well as void ratio versus log $p$ as shown in figure 9.9. This eliminates having to know the initial void ratio, $e_o$. The slope of the $\epsilon$–log $p$ line is $C_{\epsilon c}$, which is related to the compression index, $C_c$ as follows:

$$C_{\epsilon c} = \frac{C_c}{1 + e_o}$$

Analogous to equation 10.40, settlement can be calculated from

$$S = C_{\epsilon c} H \log_{10}\left(\frac{p_o + p_v}{p_o}\right)$$

## Chapter 10: BEARING CAPACITY FACTORS

The bearing capacity factors in table 10.3 are based on Terzaghi's 1943 studies. The following values are based on Meyerhof's (and others) 1955 studies, and have been widely used. Other values are also in use.

| $\phi$ | $N_c$ | $N_q$ | $N_\gamma$ |
|---|---|---|---|
| 0 | 5.14 | 1.0 | 0.0 |
| 5 | 6.5 | 1.6 | 0.5 |
| 10 | 8.3 | 2.5 | 1.2 |
| 15 | 11.0 | 3.9 | 2.6 |
| 20 | 14.8 | 6.4 | 5.4 |
| 25 | 20.7 | 10.7 | 10.8 |
| 30 | 30.1 | 18.4 | 22.4 |
| 32 | 35.5 | 23.2 | 30.2 |
| 34 | 42.2 | 29.4 | 41.1 |
| 36 | 50.6 | 37.7 | 56.3 |
| 38 | 61.4 | 48.9 | 78.0 |
| 40 | 75.3 | 64.2 | 109.4 |
| 42 | 93.7 | 85.4 | 155.6 |
| 44 | 118.4 | 115.3 | 224.6 |
| 46 | 152.1 | 158.5 | 330.4 |
| 48 | 199.3 | 222.3 | 496.0 |
| 50 | 266.9 | 319.1 | 762.9 |



## Chapter 10: CORRECTION FOR DEPTH OF FOOTING

Several researchers have recommended corrections to $N_c$ to account for footing depth. (Corrections to $N_q$ have also been suggested. No corrections to $N_\gamma$ have been suggested.) There is considerable variation in the method of calculating this correction, if it is used at all. A multiplicative correction factor, $d_c$, which is used most often has the form

$$d_c = 1 + \frac{KD_f}{B}$$

$K$ is a constant. Values of 0.2 and 0.4 have been proposed for $K$.

## Chapter 14: NOMINAL STRENGTHS

The term *nominal strength* has two applications in reinforced concrete design. Consider shear strength of a beam, for example. The unreinforced concrete has a resistance to shear given by equation 14.48. The quantity $V_c$ is known as the *nominal shear strength* of the concrete. However, the steel also contributes shear strength represented by $V_{st}$. The sum of these two quantities is the *nominal strength of the beam*:

$$V_n = V_c + V_{st}$$

Common usage attributes the term *nominal* to both $V_n$ and $V_c$.

## Chapter 14: CONCRETE PRESSURE ON FORMWORK

Formwork must be strong enough to withstand hydraulic loading from concrete during curing. The hydraulic load is greatest immediately after pouring. As the concrete sets up, it begins to support itself, and the lateral force is reduced. Publication ACI 347 predicts the maximum lateral pressure for regular (Type I) concrete with a 4″ slump (or less), ordinary work, and internal vibration. The maximum pressure depends on the temperature, $T_{^\circ F}$, and the vertical rate of pour, $R_{\text{ft/hr}}$.

$$R \le 7 \text{ ft/hr}: \quad p_{\text{max,psf}} = 150 + 9000\left(\frac{R}{T}\right)$$

$$R > 7 \text{ ft/hr}: \quad p_{\text{max,psf}} = 150 + \frac{43{,}400}{T} + 2800\left(\frac{R}{T}\right)$$

Regardless of the rate of pour, $p$ cannot exceed the minimum of 150 × pour height, or 2000 psf. As a general rule, $p$ should be increased by 50 psf to account for miscellaneous live loads, workmen, and impact.

## Chapter 15: STRUCTURAL BOLT TENSIONS

The bolt pretension (preload) is needed to use equation 15.85, which calculates a reduction factor to be used with bolts in combined tension and shear. Footnote 44 refers to AISC Specifications Table J3.7, which contains the following data:

Minimum Bolt Tension

| bolt size | A325 bolts | A490 bolts |
|---|---|---|
| 1/2 (inches) | 12 kips | 15 kips |
| 5/8 | 19 | 24 |
| 3/4 | 28 | 35 |
| 7/8 | 39 | 49 |
| 1 | 51 | 64 |
| 1-1/8 | 56 | 80 |
| 1-1/4 | 71 | 102 |
| 1-3/8 | 85 | 121 |
| 1-1/2 | 103 | 148 |

For bolts of other sizes, or for bolts manufactured from other steels, the minimum pretension is

$$T_{\text{b,min}} = 0.70 A_b F_{ut} \quad \text{(rounded)}$$

## Chapter 16: LEVELS OF SERVICE

The 1985 edition of the *Highway Capacity Manual* prefers to categorize highway *level of service* by density, rather than the volume/capacity ratios listed in table 16.11. Although table 16.11 is correct, density should now be used wherever possible to determine level of service. In the following table, density is given in passenger cars per mile per lane.

| LOS | density |
|---|---|
| A | $\leq$ 12 pc/mi-ln |
| B | $\leq$ 20 |
| C | $\leq$ 30 |
| D | $\leq$ 42 |
| E | $\leq$ 67 |
| F | $>$ 67 |

## Chapter 17: SPIRAL CURVES

The length of a spiral curve can be adjusted to between 75% and 200% of the value calculated in equation 17.55. The distance for *superelevation runoff* frequently determines the spiral curve length. The runoff may be determined by a *time rule* (e.g., runoff shall be completed within 4 seconds at the design speed), or by a *speed rule* (e.g., 3 feet of runoff for every MPH in design speed, regardless of initial superelevation). Other time and speed rules are in use. The important point is that the spiral curve length can be adjusted to meet varying design criteria.

## Chapter 17: HORIZONTAL CURVES THROUGH POINTS

A special type of horizontal curve problem requires finding a curve radius to pass the curve through a given point. There are two variations of this problem, depending on how the point is located. However, both problems are solved in the same way. Angle $\Delta$ must be known.

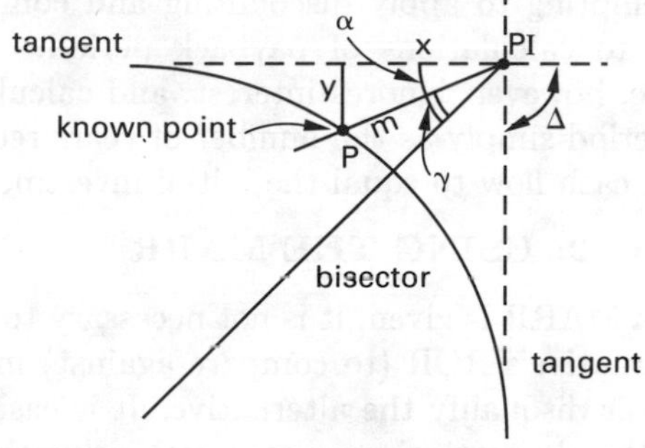


*step 1*: Get $\alpha$ and $m$ from $x$ and $y$. (If $\alpha$ and $m$ are known, skip this step.)

$$\alpha = \arctan \frac{y}{x}$$

$$m = \sqrt{x^2 + y^2}.$$

*step 2*: Calculate $\gamma = 90 - \frac{1}{2}\Delta - \alpha$.

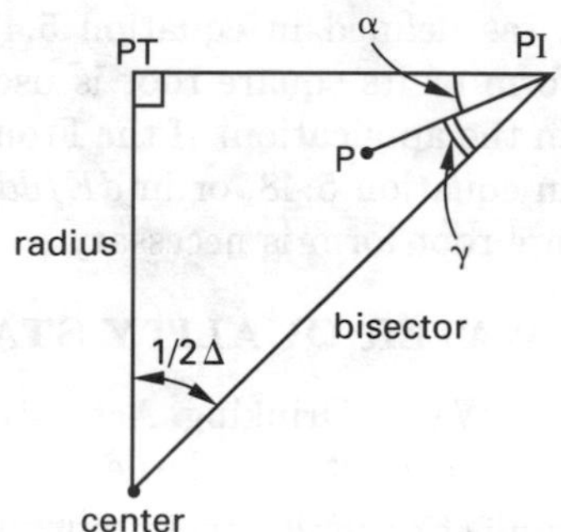


*step 3*: Calculate $\phi = 180 - \arcsin\left(\frac{\sin\gamma}{\cos\frac{1}{2}\Delta}\right)$.

*step 4*: Calculate $\theta = 180 - \gamma - \phi$.

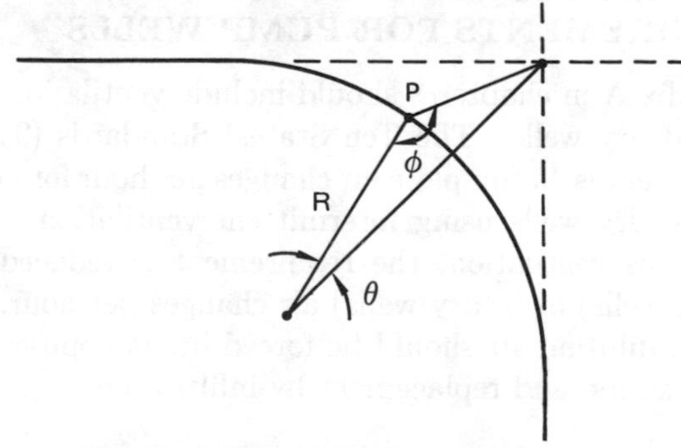


*step 5*: Calculate $R$ from the law of sines.

$$\frac{\sin\theta}{m} = \frac{\sin\phi}{R\sec\frac{1}{2}\Delta} = \frac{\sin\phi\cos\frac{1}{2}\Delta}{R}$$

*Update: February, 1988*

### Chapter 2: PAYBACK PERIOD

It is tempting to apply discounting and compounding factors to calculations of payback period. Common practice, however, ignores interest, and calculates payback period simply as the number of years required for the net cash flow to equal the initial investment.

### Chapter 2: USING THE MARR

When a MARR is given, it is not necessary to calculate an alternative's ROR (to compare against) in order to qualify or disqualify the alternative. It is easier to calculate the alternative's present worth using the MARR as an interest rate. If the present worth is positive, then ROR > MARR.

### Chapter 5: THE FROUDE NUMBER

There is considerable confusion regarding the definition of the Froude number. Dimensional analysis determines it to be $v^2/gL$, a form which is also used in model similitude analysis. However, in open channel flow analysis, the Froude number is taken as the square root of the derived form (as defined in equation 5.47). Whether the derived form or its square root is used can be determined from the application. If the Froude number is squared (as in equation 5.48, or in $dE/dd = 1 - N_{Fr}^2$), then the square root form is necessary.

### Chapter 7: WATER QUALITY STANDARDS

The 1986 Safe Water Drinking Act established various schedules for regulating volatile organic chemicals (VOC's), fluoride, synthetic organic chemicals (SOC's), inorganic chemicals (IOC's), and microbial contaminants. Therefore, table 7.9 is expected to be in a state of flux during the 1987–1991 timetable established by the act.



### Chapter 8: TSS VENTILATION REQUIREMENTS FOR PUMP WELLS

Appendix A in chapter 8 should include ventilation for wet and dry wells. The Ten States' Standards (TSS) requirement is 30 complete air changes per hour for both wet and dry wells using intermittent ventilation. For continuous ventilation, the requirement is reduced to 12 (wet wells) or 6 (dry wells) air changes per hour. In general, dilution air should be forced in, as opposed to air extraction and replacement by infiltration.

### Chapter 12: DEFLECTION OF A PIER

Appendix A of chapter 12 omits the equation for calculating the performance of a fixed pier (i.e., a column with one end that can translate laterally, with both ends remaining vertical).

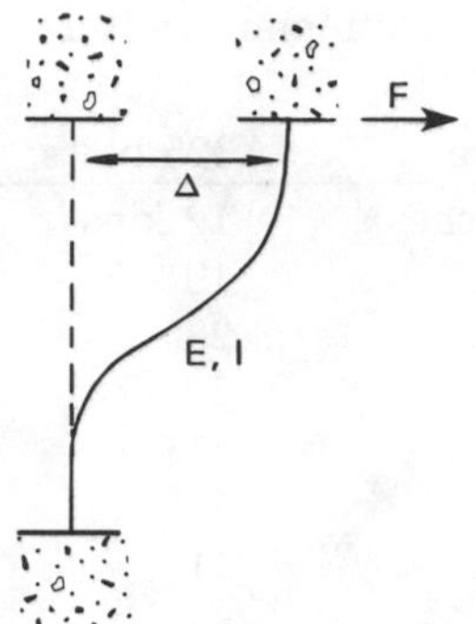


$$\Delta = \frac{FL^3}{12EI}$$

$$\text{stiffness} = \frac{F}{\Delta} = \frac{12EI}{L^3}$$

### Chapter 15: OPTIMIZED DESIGN IN COLUMN BASE PLATES

The required bearing plate area will be minimized when the allowable bearing pressure is maximized. This occurs when the area of the supporting concrete plane is more than 4 times the area of the base plate (or, the concrete is infinite in size). With $A_{\text{support}}/A_{\text{bearing}} = 4$, equation 15.36 gives $F_p = 0.70f'_c$, the maximum value of $F_p$ permitted. This value of $F_p$ should be used unless the support area is limited in some way, as it would be with a pedestal-mounted column.

Referring to figure 15.20, it is desired to have $m = n$. This approximately occurs when

$$A_1 = \frac{\text{column load}}{F_p}$$

$$N \approx \sqrt{A_1} + \tfrac{1}{2}(0.95d - 0.80b_f)$$

$$B = \frac{\sqrt{A_1}}{N}$$

Although equation 15.40 is theoretically correct, the AISC procedure imposes an additional limitation by specifying an additional parameter $n'$ for each specific column shape. The value of $n'$ must be obtained from the AISCM. Equation 15.40, then, becomes

$$t = \{\text{larger of } m,\ n,\ \text{or } n'\} \times \sqrt{\frac{4f_p}{F_y}}$$

*Update: October 1988*

### Chapter 14: MOISTURE EXCESS AND DEFICIT IN AGGREGATE

There is some confusion about what "5% moisture excess" means. (See example 14.2.) This confusion results from the use of both dry (i.e., SSD) and wet (i.e., with the excess) weights as the basis for the excess percentage. This problem has existed in civil engineering for a long time, and both methods are in field use. The important relationships are covered below, where $m$ is the variable for mass.

| | dry basis | wet basis |
|---|---|---|
| fraction moisture, $f$ | $\frac{m_{\text{excess water}}}{m_{\text{SSD sand}}}$ | $\frac{m_{\text{excess water}}}{m_{\text{SSD sand}} + m_{\text{excess water}}}$ |
| wet mass of sand, $m_{\text{wet sand}}$ | $m_{\text{SSD sand}} + m_{\text{excess water}}$<br>$(1+f)m_{\text{SSD sand}}$ | $m_{\text{SSD sand}} + m_{\text{excess water}}$<br>$\left(\frac{1}{1-f}\right) m_{\text{SSD sand}}$ |
| SSD mass of sand, $m_{\text{SSD sand}}$ | $\frac{m_{\text{wet sand}}}{1+f}$ | $(1-f) \times m_{\text{wet sand}}$ |
| mass of excess water, $m_{\text{excess water}}$ | $f \times m_{\text{SSD sand}}$<br>$= \frac{f \times m_{\text{wet sand}}}{1+f}$ | $\frac{f \times m_{\text{SSD sand}}}{1-f}$<br>$f \times m_{\text{wet sand}}$ |

## Chapter 14: STRENGTH DESIGN OF BEAMS

Equation 14.33 (used to calculate the reinforcement ratio from a trial beam size) is an approximation good only if $bd^2$ is not too different from the ideal value. The exact method is to use the following equation:

$$\rho_{\text{revised}} = \frac{1}{m}\left(1 - \sqrt{1 - \frac{2mR_{u,\text{revised}}}{f_y}}\right)$$

## Chapter 14: BEAM DEFLECTIONS

$I_{cr}$ in equation 14.40 is the *cracked moment of inertia.* It is calculated from the equation below. Its use is illustrated in example 14.5.

$$I_{cr} = \frac{bc^3}{3} + \frac{A_{st}E_{st}}{E_c}(d-c)^2$$

## Chapter 15: MULTIPLE FASTENER LAP CONNECTIONS

The following additional check is implied by the last part of example 15.11, but is not explicitly stated elsewhere in the text.

If the connectors (e.g., rivets or bolts) in a tension lap splice are arranged in two or more rows, and if the rows have unequal numbers of fasteners, each row should be checked for tension capacity assuming the previous rows have absorbed a proportionate share of the load. Consider the following diagram:

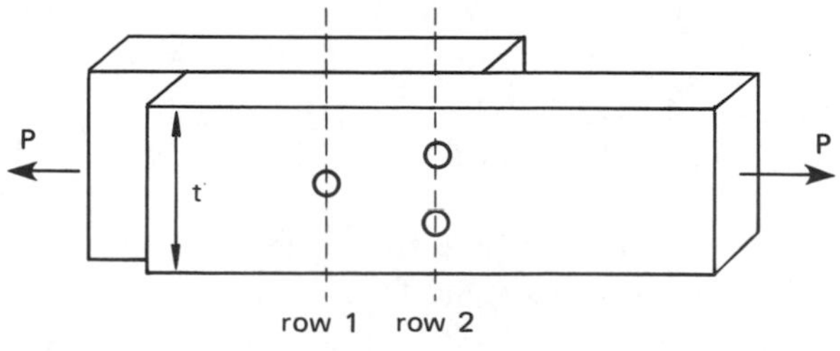


The net section (i.e., $t$ less two hole diameters) of row 2 should be checked for tension capacity when 2/3 of the load is applied. This assumes that row 1 carries 1/3 of the load. (For the second plate, the net section of row 1 could be checked assuming 1/3 of the load was carried. However, this is not the controlling case.)

## Chapter 15: CHECKING BEARING STRESS

It is obvious that bearing stress should be evaluated in bearing connections. Theoretically, bearing should not be a problem in friction-type connections because the bolts never bear on the pieces assembled. However, in the event there is slippage due to insufficient tension in the connectors, bearing should routinely be checked. The fact that the assembly may fail in some other manner if a friction connection becomes a bearing connection does not change the advisability of checking for bearing stress.

*Update: September, 1989*

## FIFTH EDITION

Hundreds of minor changes have been made to the fourth edition of the *Civil Engineering Reference Manual* during its six reprintings. These changes have brought the book into compliance with new codes and legislation, improved and clarified explanations, expanded tables of data, and corrected errata. All of the changes were made without changing the edition, something that has caused confusion among students using the book in a classroom setting.

With this latest printing, another hundred or so changes have been made in the book. Although there has been no substantial change in the organization of material, the edition has been changed to differentiate the current book from its earlier versions.

## Chapter 5: ALTERNATE AND CONJUGATE DEPTHS

The terms *alternate depth* and *conjugate depth* are not synonymous. Alternate depths (calculated in equation 5.41) are derived from the conservation of energy equation (i.e., a variation of the Bernoulli equation). Conjugate depths (calculated in equations 5.58 and 5.59) are derived from a conservation of momentum equation. Conjugate depths are calculated only when there has been an energy loss such as a hydraulic jump or drop.

## Chapter 5: TOTAL SPILLWAY ENERGY

Finding the depth of flow at the toe of a spillway (before a hydraulic jump) requires writing an energy balance. Neglecting friction, the total energy at the toe equals the total upstream energy before the spillway. The total upstream energy before the spillway is

$$E_1 = y + H + \frac{v^2}{2g}$$

The upstream velocity, $v$, is the velocity before the spillway (which is essentially zero), not the velocity over the brink. The velocity over the brink, if the brink depth is known, can be used with the continuity equation to calculate the upstream velocity, but it should not be used to determine total energy since $d_{\text{brink}} \neq H$.

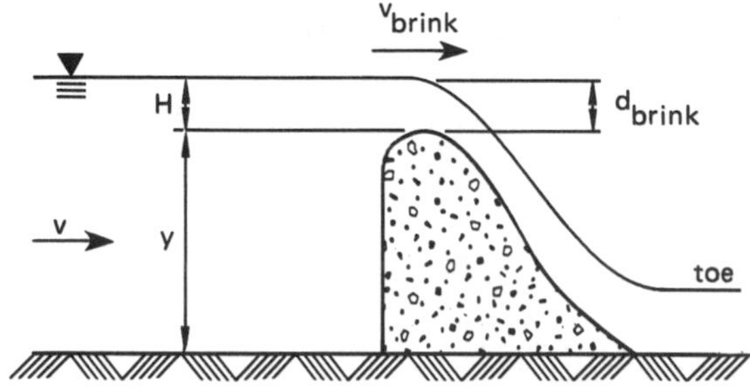


## Chapter 10: EFFECTS OF WATER TABLE LOCATION ON FOOTING DESIGN

*General Principle 1*: If the soil is cohesive ($\phi = 0°$), then the location of the water table does not affect the bearing capacity, and the effect of the water table is disregarded. (Strictly speaking, this is not true, but it is almost true. Here is why. If $\phi$ is zero, then $N_y = 0$. Also, $N_q = 1$, which is essentially zero. These two terms "zero out" the density terms in equation 10.1.)

*General Principle 2*: Use the submerged density ($\rho_{\text{dry}} - 62.4$) in the equation for bearing capacity (equation 10.1):

(a) When the water table is at the base of the footing, use the submerged density in the first term of equation 10.1 only.

(b) When the water table is at the surface, use the submerged density in both the first and third terms of equation 10.1.

(c) When the water table is between the base of the footing and the surface, use the submerged density in the first term as in (a) above. Calculate the third term in equation 10.1 as

$$\begin{aligned}&[p_q + \rho D_w + (\rho - 62.4)(D_f - D_w)]N_q \\ &= [p_q + \rho D_f + 62.4(D_w - D_f)]N_q\end{aligned}$$

*General Principle 3*: If the water table depth, $D_w$, is greater than $D_f + B$ (i.e., more than a distance $B$ below the base of the footing), the bearing capacity is not affected. Calculate the bearing capacity from equation 10.1 as if there was no water table.

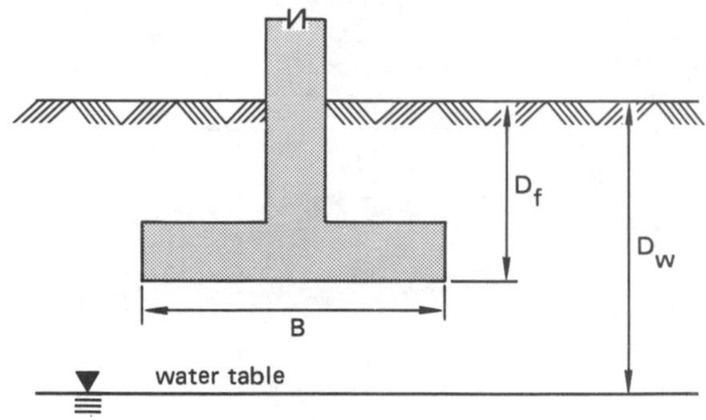


*Approximation to Exact Method*: Since the submerged density is approximately half of the dry density, it is commonly stated that the bearing capacity of a footing with the water table at the ground surface is half the dry bearing capacity and varies linearly to full strength a distance $B$ below the footing base. For a water table within a distance $B$ of the footing base, the dry bearing capacity can be multiplied by $C_w$ (equation 10.23):

$$C_w = 0.5 + 0.5\left(\frac{D_w}{D_f + B}\right) \qquad 0 \le D_w \le D_f + B$$

## Chapter 14: AIR ENTRAINMENT

Just as there is confusion with the meaning of "percentage moisture" in concrete batching (see page 22-7), there is also confusion with the meaning of "percentage air entrainment."

If "5%" (for example) air means that the concrete volume is 5% air, then the solid volume is only 95% of the final volume. The solid volume should be divided by 0.95 to obtain the final volume. If "5%" air means the concrete volume is increased by 5% when air is added, then the solid volume should be multiplied by 1.05 to obtain the final volume.

The meaning of "5%" air must be clarified in the problem. If it isn't clear, then either definition could apply. In any case, the final volume is not affected significantly by the interpretation.

## Chapter 14: DEPTH OF COVER AND PROTECTION COVER THICKNESS

[ACI 318 section 10] defines the variable $d_c$ as "thickness of *concrete cover* measured from extreme tension fiber to center of bar or wire located closest thereto." This definition is consistent with equation 14.36 (page 14-13) and figure 14.6 (page 14-14) in this book.

[ACI 318 section 7.7] also specifies the thickness of a concrete layer necessary to provide protection to the steel without giving the thickness a symbol. The term *concrete cover* is again used.

The term "concrete cover" is ambiguous. The symbol $d_c$ is not. (See figure 14.6.)

### Chapter 15: USING $C_b$ WITH THE ALLOWABLE MOMENT CHART

If used directly, the allowable moments chart implicitly assumes $C_b = 1.0$. However, the chart can be used with other values of $C_b$ if $L_b/C_b$ or $L_b/\sqrt{C_b}$ is used instead of $L_b$. The proper replacement for $L_b$ depends on the beam. Most beams are controlled by torsional strength and AISC equation F1-8 controls. When equation F1-8 controls, the curved portion of the allowable moments chart will be hyperbolic, as shown in figure 15.10. For these beams, replace $L_b$ with $L_b/C_b$.

In some cases, the curved line in the allowable moments chart will be parabolic (see figure 15.10) indicating the AISC equation F1-6 controls. In that case, replace $L_b$ with $L_b/\sqrt{C_b}$.

*Update: January, 1991*

### Preface: UNTIMED AND TIMED PROBLEMS

This book contains two types of homework problems. The *untimed problems* are shorter problems that are meant to acquaint you with a wide variety of subjects, as well as to familiarize you with the book's format. The difficulty of the *timed problems* is representative of exam problems. Timed problems should take you one hour or less.

### Chapter 2: HALF-YEAR CONVENTION

There are certain arbitrary rules (known as *conventions*) imposed by the Internal Revenue Service on depreciation calculations. An example is the "half-year rule" required with double-declining balance calculations. The implementation of such rules is best left to accounting professionals. Such rules are subject to constantly changing legislation and are not covered in this book.

### Chapter 3: FANNING FRICTION FACTOR

Most of the time, the term *friction factor* will mean the Darcy friction factor used in equation 3.71. Occasionally, the *Fanning friction factor* will be specified by name. The Darcy friction factor is calculated by multiplying the Fanning friction factor by 4. That is,

$$f_{\text{Darcy}} = 4f_{\text{Fanning}}$$

### Chapter 3: HAZEN-WILLIAMS COEFFICIENTS FOR AGED PIPES

Table 3.8 lists values of the Hazen-Williams coefficient, $C$, for clean (new) pipe, but leaves the choice of values for aged pipe up to the designer. The following values can be used as guidelines:

| | |
|---|---|
| cast iron, new, all sizes | 130 |
| cast iron, 5 years old, $(D < 24'')$ | 120 |
| $(D \geq 24'')$ | 115 |
| cast iron, 10 years old, $(D = 4'')$ | 105 |
| $(D = 12'')$ | 110 |
| $(D \geq 30'')$ | 85 |
| cast iron, 40 years old, $(D = 4'')$ | 65 |
| $(D = 16'')$ | 80 |

Values for welded steel pipe are similar to those for cast iron pipe 5 years older.

### Chapter 3: BUTTERFLY VALVES

Loss coefficients for *butterfly valves* are calculated from the friction factor for the pipe with completely turbulent fluid. The sizes given are inside diameter measurements of the pipe.

$$\begin{aligned} \text{sizes } 2'' \text{ to } 8''\text{:}\quad & K = 45f_t \\ \text{sizes } 10'' \text{ to } 14''\text{:}\quad & K = 35f_t \\ \text{sizes } 16'' \text{ to } 24''\text{:}\quad & K = 25f_t \end{aligned}$$

### Chapter 3: PRESSURE DROP ACROSS AN ORIFICE PLATE

The *pressure difference* used in equations 3.156 through 3.162 for orifice plates is not the permanent *pressure loss* due to friction. The permanent pressure loss depends on the ratio $D_o/D_i$ (see figure 3.24) and is 73% of $p_1 - p_2$ for $D_o/D_i = 0.5$, 56% for $D_o/D_i = 0.65$, and 38% for $D_o/D_i = 0.8$.

### Chapter 9: COMPRESSION INDEX

Strictly speaking, the *compression index* ($C_c$) should be negative, since it represents the slope of the $e$-log $p$ curve (figure 9.11). Therefore, it would be calculated (in equations 9.31 and 10.35) as

$$C_c = \frac{e_1 - e_2}{\log_{10}\left(\frac{p_1}{p_2}\right)}$$

However, the compression index is typically reported as a positive number. This is consistent with its use in the CIVIL ENGINEERING REFERENCE MANUAL.

## Chapter 16: STOPPING SIGHT DISTANCE

The constants in table 16.4 are based on specific heights of objects and driver's eyes. In general, the sight distance over the crest of a vertical curve is given by the following relationships. $h_1$ is the height of the eyes of the driver, and $h_2$ is the height of the object sighted, both in feet.

$$S < LC \qquad LC = \frac{(G_1 - G_2)S^2}{100(\sqrt{2h_1} + \sqrt{2h_2})^2}$$

$$S > LC \qquad LC = 2S - \frac{100(\sqrt{2h_1} + \sqrt{2h_2})^2}{G_1 - G_2}$$

*Update: April, 1992*

## Chapter 10: APPROXIMATE STRESS UNDER A FOUNDATION

The Boussinesq equation (equation 10.30) was presented as a means to calculate the exact increase in stress at a given depth below a footing. An approximate value of the increase in stress can be determined by assuming the affected area (i.e., the *zone of influence*) beneath the load. An approximate method will be reasonably accurate in non-layered homogeneous soils when $1.5 < h/B < 5$.

Defining the zone of influence means assuming the angle of the *influence cone*. The angle typically is assumed to be 51° for point loads and 60° for uniformly-loaded circular and rectangular footings. (This is why this method is known as the *60° method*.) Alternatively, since the 60° is an approximation, and one approximation is as good as another, for ease of computation the angle is taken as 63.4°, corresponding to a 2:1 (vertical:horizontal) influence cone angle.

For any depth, the area of the cone of influence is the area of the horizontal plane enclosed by the influence cone. This is determined from geometric principles. For a rectangular footing using the 60° method, the zone of influence area at depth $h$ is

$$A = [B + 2(h \cot 60°)][L + 2(h \cot 60°)]$$

For a rectangular footing, assume a 2:1 influence cone. The zone of influence area at depth $h$ is

$$A = (B + h)(L + h)$$

The increase in pressure at depth $h$ is

$$p_v = \frac{P}{A}$$

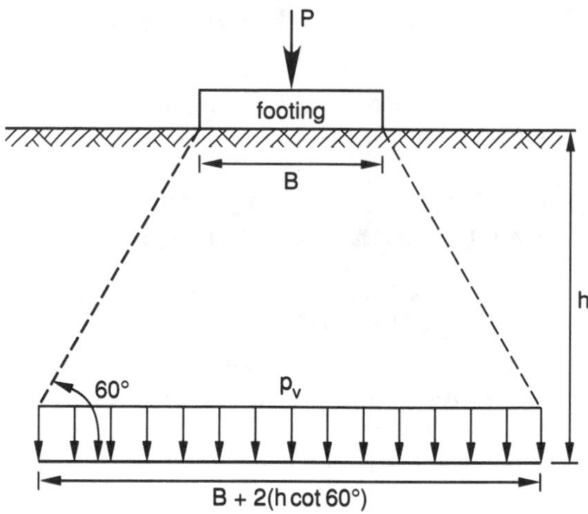


*Update September, 1994:*

The timed problems at the end of each chapter in the *Civil Engineering Reference Manual* are one-hour, exam-like problems. However, some engineers (particularly those who study only a few subjects) want even more practice than is provided by the end-of-chapter problems. The recent publication of my *101 Solved Civil Engineering Problems* (Professional Publications, Inc., 1994) will go a long way in addressing one of the most frequent requests I receive: to add more exam-like practice problems.

## Chapter 2: ECONOMIC ANALYSIS ON THE EXAM

Although the subject of engineering economic analysis has been formally removed from the list of civil engineering P.E. exam subjects, problems involving this subject continue to appear. Therefore, chapter 2 of this book should be studied.

## Chapter 2: AASHTO METHODS OF COMPARING TRANSPORTATION ALTERNATIVES

There are only a few accepted methods of comparing alternatives on the basis of their economics. For example, the present worths, annual costs, capitalized costs, or benefit-cost ratios of the alternatives can be compared. Although these methods are well understood, the categorization of cash flows can be problematic, particularly when comparisons are made on the basis of benefit-cost ratio.

When the benefit-cost ratio method is used, some factors might be classified as *disbenefits* and placed in the numerator, or they might be classified as *costs* and placed in the denominator. The distinction between disbenefit and cost will not change the desirability of an alternative (i.e., the benefit-cost ratio of a desirable alternative will be greater than 1.0 regardless of the placement). However, the numerical value of the benefit-cost ratio will change, making the comparison and ranking of alternatives less precise. (Benefit-cost ratios should not, by themselves, be used to rank alternatives. An incremental benefit-cost ratio can be used for this purpose.)

Economic highway studies are performed to determine the feasibility of alternatives, to compare alternative routes and locations, and to determine the characteristics (e.g., pavement type, thickness, etc.) of the highway. Since the costs and benefits typically do not accrue to the same individual, it is common to use the benefit-cost ratio method of analysis.

The *highway transportation costs* include the initial highway investment (including engineering and design fees, purchasing the right-of-way, safety and traffic control devices, and landscaping), initial transit system costs, ongoing maintenance and operating costs (including lighting costs), highway user costs, and transit user costs.

*Highway user costs* include operating costs (e.g., fuel, lubrication, and tires), travel time, and accident costs associated with personal vehicle use. Usually, only variable costs are included (i.e., those costs that depend on the mileage). Costs of vehicle registration, parking, and insurance are excluded.

For a specific length, $L$, of highway section, user costs, $U$, can be calculated from the per-mile cost coefficients for accidents ($A$), basic section use ($B$) covering time value and vehicle running costs, transitions ($T$) that account for speed changes between sections, and delays ($D$) at intersections and traffic control devices.

$$U = (A + B)L + T + D$$

*Transit* (i.e., bus) *system costs* include the initial costs of acquiring vehicles and building terminals and buildings. Ongoing operating costs include drivers' wages, and operation and maintenance of the transit vehicles.

*Transit user costs* include the fares paid, costs of personal vehicles used to get to the terminals, and the value of time spent waiting for and riding on the transit vehicles.

*Benefits* of improved highway transportation can be categorized as (1) *direct benefits* that result from reductions in highway user costs, and (2) *indirect benefits* that accrue to the general public. Direct benefits result from higher speeds, decreased delays and travel time, decreased vehicle costs, and reductions in accidents. Indirect benefits include reductions in environmental impacts (e.g., reduction in air-borne dust when a dirt road is paved), rapid response of emergency vehicles, reliable postal delivery, better access to shopping, theaters, parks, etc.

*Disbenefits*, also known as *added user costs*, are increases in user costs. An example of a disbenefit is the value of the increase in travel time due to roadwork delays while a highway is being resurfaced.

When the benefit-cost ratio method is used to evaluate highway alternatives, the ratio is typically calculated from the present worths of the future benefit and cost cash flow streams. (For an alternative with an infinite life, these are capitalized costs.) The numerical value of the ratio calculated from the effective annual costs would be the same. However, this is no longer the typical approach taken.

The standard form of the benefit-cost ratio of a highway study alternative is defined as

$$\frac{B}{C} = \frac{P_{\Delta U} - P_{\Delta D}}{P_{\Delta I} + P_{\Delta M} - P_{\Delta S}}$$

$P_{\Delta U}$ is the present worth of the increase in user benefits (i.e., the user costs without the alternative less the costs with the alternative). $P_{\Delta D}$ is the present worth of the increase in disbenefits and added user costs. $P_{\Delta I}$ is the present worth of the increase in investment costs. $P_{\Delta M}$ is the present worth of the increase in highway agency operating and maintenance costs. $P_{\Delta S}$ is the present worth of the increase in salvage (residual) value at the end of the alternative's life.

Another type of study involves determining the desirability of installing roadside improvements (paving, barriers, etc.) to reduce damage from off-road encroachments. Appendix A of AASHTO's *Roadside Design Guide* (1989) details a present-worth methodology for evaluating the effectiveness such improvements. Highway *agency costs* are incurred for installation, maintenance, and repair of roadside improvements. The agency also is credited for the present worth of any future salvage value.

Additional costs are incurred by the agency when it must pay for losses incurred by motorists. These losses are determined from a probabilistic analysis based on the expected cost of each type of accident each year. The cost values are subject to change with time, but the originally published (1989) types and cost values are: fatality, $500,000; severe personal injury, $110,000; moderate personal injury, $10,000; slight personal injury, $3000; level 2 property damage only, $2500; level 1 property damage only, $500.

Since the agency incurs both the installation-related and the accident-related losses, a present-worth analysis is used. One alternative always exists: do nothing. Other alternatives include removing, shielding, and modifying the hazard. Excluding other factors, the alternative with the lowest present worth of costs and expenses would be selected.

## Chapter 2: CONGESTION PRICING

The concept of *congestion pricing* is based on the premise that highway user costs increase as a highway becomes congested and the speed falls. Theoretically, an equilibrium point is reached with regards to highway usage when the benefit to a user of traveling on a highway equals the user's cost. When the cost exceeds the benefit, users will cease using the highway.

However, though users may be aware of their own costs (known as the *marginal private cost*), they are not aware of the total system costs (known as the *marginal social cost*). Each additional user of a congested highway not only incurs an individual cost but also congests the highway still more. This imposes an additional cost on all other users of the highway. Additional unseen costs are related to maintenance and environmental impact (noise, visual intrusion, ozone, carbon dioxide, acid rain, etc.). The congestion pricing model predicts that the highway usage equilibrium point would shift if a user was aware of these additional costs.

Congestion pricing seeks to make the user aware of these additional costs by imposing charges equal to the difference between the social and private costs. The charge can be in the form of a fuel tax, car purchase and ownership taxation, or *cordon charges*, also known as *road access charges* (i.e., tolls to enter a specific section of highway or city), or a combination thereof.

POSTSCRIPTS

## Chapter 3: FLOW THROUGH CORRUGATED STEEL PIPE

Corrugated steel pipe is frequently used for culverts. The pipe is made from corrugated sheets of galvanized steel that are rolled and riveted together to make a seam. Standard pipe diameters range from 8 to 96 inches. Corrugated plate that can be bolted together to form larger or non-circular culverts is available. Standard pipe lengths are 20 to 40 feet. Metal gages of 8, 10, 12, 14, and 16 are commonly used. Though most corrugations are transverse, helical corrugations are also used.

The most common corrugated steel pipe has standard (circular arc) corrugations that are $\frac{1}{2}$ inch deep and $2\frac{2}{3}$ inch from crest to crest. (This is often referred to as "$2\frac{1}{2}$ inch corrugation.") For very large culverts, corrugations with a 6 inch pitch and depth of 2 inches is used. The 2 x 6 product is known as *multiplate* after its official trade name of "Multi-Plate."

Flow area for circular culverts is based on the nominal culvert diameter, regardless of the gage of plate used to construct the pipe. Flow area is calculated to (at most) three significant digits. Corrugated *pipe arches* are produced by compressing a circular pipe in a press. Dimensions of standard pipe arches are listed here.

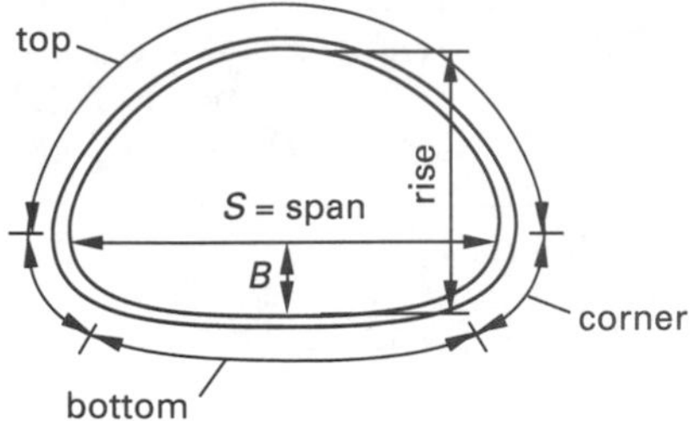


Typical Pipe Arch

Dimensions of Standard Pipe Arches

| diameter of pipe of equal perimeter (inches) | span (inches) | rise (inches) | dimension B (inches) | flow area (square feet) |
|---|---|---|---|---|
| 15 | 18 | 11 | $4\frac{1}{2}$ | 1.1 |
| 18 | 22 | 13 | $4\frac{3}{4}$ | 1.6 |
| 21 | 25 | 16 | $5\frac{1}{4}$ | 2.2 |
| 24 | 29 | 18 | $5\frac{1}{2}$ | 2.8 |
| 30 | 36 | 22 | $6\frac{1}{4}$ | 4.4 |
| 36 | 43 | 27 | 7 | 6.4 |
| 42 | 50 | 31 | 8 | 8.7 |
| 48 | 58 | 36 | $9\frac{1}{4}$ | 11.4 |
| 54 | 65 | 40 | $10\frac{1}{2}$ | 14.3 |
| 60 | 72 | 44 | $11\frac{3}{4}$ | 17.6 |

A Hazen-Williams coefficient, $C$, of 60 is typically used with all sizes of corrugated pipe. Experimental values of Manning's $n$ with corrugated pipe have ranged from 0.019 to 0.024. *Design Charts for Open Channel Flow*, U.S. Department of Transportation, 1979, recommends $n = 0.024$ for all cases. The U.S. Department of the Interior recommends the following values. For standard ($\frac{1}{2} \times 2\frac{2}{3}$ inch) corrugated pipes with the diameters in inches: 24 inches, 0.027; 24 inches, 0.025; 36–48 inches, 0.024; 60–84 inches, 0.023; 96 inches, 0.022. For multiplate (2 × 6 inch) construction, with diameters in feet: 5–6 ft, 0.034; 7–8 ft, 0.033; 9–11 ft, 0.032; 12–13 ft, 0.031; 14–15 ft, 0.030; 16–18 ft, 0.029; 19–20 ft, 0.028; 21–22 ft, 0.027.

If the corrugated pipe has been asphalted completely smooth all the way around, the values range from 0.009 to 0.011. For culverts with 40% paved (asphalt) inverts, $n = 0.019$. (For other percentages of paved invert, the resulting value is proportional to the percentages and the values normally corresponding to that size pipe.) For field-bolted corrugated metal pipe arches, $n = 0.025$. Values of $C$ and $n$ for corrugated pipe are generally not affected by age.

It is also possible to calculate the Darcy friction loss if the corrugation depth ($\epsilon = 0.5$ inches for standard corrugated pipe and $\epsilon = 2.0$ inches for standard multiplate) is known. The specific roughness (for use with the Moody friction factor chart) is

$$\text{specific roughness} = \frac{\epsilon}{D_{\text{inches}}}$$

## Chapter 5: SIZING TRAPEZOIDAL (AND RECTANGULAR) CHANNELS

Trapezoidal (and rectangular) cross sections are commonly used for artificial surface channels. The flow through a trapezoidal channel is easily determined from the Manning equation when the cross section is known. However, when the cross section or uniform depth are unknown, a trial and error solution is required.

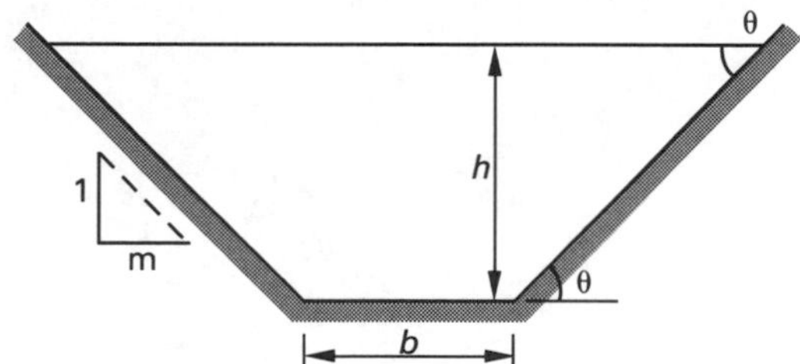


A General Trapezoidal Cross Section

For such problems, it is common to calculate and plot the *conveyance*, $K$, (or, alternatively, the product $Kn$) against depth. (See equation 5.9.) For trapezoidal sections, it is particularly convenient to write the uniform flow, $Q$, in terms of a modified conveyance, $K'$. $b$ is the base width of the channel, $h$ is the depth of flow, and $m$ is the cotangent of the side slope angle. $m$ and the ratio $h/b$ are treated as independent variables. (This method can also be used for rectangular channels.)

$$Q = \frac{K' b^{8/3} \sqrt{S_o}}{n}$$

$$K' = \left[ \frac{(1.49)\left[1 + m\left(\frac{h}{b}\right)\right]^{5/3}}{\left[1 + (2)\left(\frac{h}{b}\right)\sqrt{1+m^2}\right]^{2/3}} \right] \left(\frac{h}{b}\right)^{5/3}$$

$$m = \cot\theta$$

For any fixed value of $m$, enough values of $K'$ are calculated over a reasonable range of the $h/b$ ratio ($0.05 < h/b < 0.5$) to define a curve. Then, given specific values of $Q$, $n$, $S_o$, and $b$, the value of $K'$ can be calculated from the expression for $Q$. The graph is used to determine the ratio of $h/b$, giving the depth of uniform flow, $h$, since $b$ is known.

The term *trapezoidal channel* does not require the channel cross section to be the most efficient. The most efficient trapezoidal channel will have a side angle (inclined from horizontal) of 60° and will also have a surface width equal to twice the sloping side. Therefore, the most efficient trapezoidal channel will be half of a regular hexagon (i.e., three adjacent equilateral triangles of side length $2h/\sqrt{3}$). The hydraulic mean depth in that case is $h/2$.

When the ratio of $h/b$ is very small (less than 0.02), it is satisfactory to consider the trapezoidal channel as a wide rectangular channel with area $A = bh$.

## Chapter 6: IMPOUNDMENT DEPTH

An *impounding reservoir* is a basin used to store excess flow from a stream or river. The stored water is released when the stream flow drops below a level that is adequate to meet water demand. The *impoundment depth* is the design depth. Finding the impoundment depth is equivalent to finding the design storage capacity of the reservoir. This is accomplished by use of the mass diagram.

## Chapter 7: CONCENTRATIONS IN MILLIEQUIVALENTS PER LITER

Concentrations of dissolved substances encountered in water supply studies are normally expressed in milligrams per liter (mg/l). However, concentrations can also be expressed in milliequivalents per liter (meq/l). This is a representation of the combining weight of the substance per liter. The relationship between mg/l and meq/l is

$$\begin{aligned} \text{meq/l} &= (\text{mg/l})\left(\frac{\text{valence}}{\text{atomic weight}}\right) \\ &= \frac{\text{mg/l}}{\text{equivalent weight}} \end{aligned}$$

The molecular weight replaces the atomic weight for compounds and radicals. The compound/radical valence is still used.

## Chapter 7: CHEMICAL PROCESS KINETICS (REACTION RATES)

The rate at which a compound is formed or destroyed in an irreversible reaction can depend in several ways on the concentration of that and other compounds. The *rate of reaction* for the formation of compound $A$ is represented in various forms such as $r_A$, $dA/dt$, and $d[A]/dt$, where the variable $[A]$ represents either the concentration or mass of substance $A$. The *order of the reaction* is an empirical quantity that is determined from observation, not from the stoichiometric reaction equation.

The formation/destruction rate of a compound with a zero-order reaction rate is constant.

$$\frac{d[A]}{dt} = -k_0$$

Given an initial condition of $[A] = [A]_0$, the integrated solution is

$$[A]_t = [A]_0 - k_0 t$$

The time for half of the substance to be created (or destroyed) is known as the *half-life*, $t_{1/2}$.

Table 1 below summarizes the reaction rate and half-life equations for simple, irreversible reactions.

## Chapter 7: PHOSPHORUS LOADING

Phosphorus in surface water runoff can be traced back to fertilizer and synthetic detergents. Phosphorus concentrations of 5 to 15 mg/l (as P) are experienced in raw municipal wastewater, most of which originates from synthetic detergents and human waste.

Approximately 10% of the total phosphorus content is insoluble and can be removed in primary settling. The amount that is removed by absorption in conventional biological processes is small. The remaining phosphorus is soluble and must be removed by converting it into an insoluble precipitate. Lime, alum, ferric chloride, or ferric sulfate may be added depending on the nature of the phosphorus radical. Due to the many other possible reactions these added compounds can participate in, the dosage should be determined from testing. The stoichiometric chemical reactions describe how the phosphorus is removed, but they do not accurately predict the quantities required.

## Chapter 8: AVERAGE AND PEAK FLOWS IN SEWER SYSTEMS

*Recommended Standards for Sewage Works* ("Ten States' Standards," Chap. 20) specifies that new sanitary sewer systems should be designed for an *average flow* of 100 gpcd (100 gallons per person per day), a figure which includes an allowance for normal infiltration. However, the sewer pipe must be sized to carry the *peak flow*. In the absence of any studies or other justifiable methods, the ratio of peak hourly flow to average flow should be calculated from the following relationship. ($P$ is the service population in 1000's of people.)

$$\frac{Q_{\text{max}}}{Q_{\text{ave}}} = \frac{18 + \sqrt{P}}{4 + \sqrt{P}}$$

**Table 1**
Reaction Rates and Half-Life Equations

| reaction | order | rate equation | integrated forms |
|---|---|---|---|
| $A \rightarrow B$ | zero | $\frac{d[A]}{dt} = -k_0$ | $[A] = [A]_0 - k_0 t$ |
| | | | $t_{1/2} = \frac{[A]_0}{2k_0}$ |
| $A \rightarrow B$ | first | $\frac{d[A]}{dt} = -k_1[A]$ | $\ln \frac{[A]}{[A]_0} = k_1 t$ |
| | | | $t_{1/2} = \frac{1}{k_1} \ln 2$ |
| $A + A \rightarrow P$ | second, type I | $\frac{d[A]}{dt} = -k_2[A]^2$ | $\frac{1}{[A]} - \frac{1}{[A]_0} = k_2 t$ |
| | | | $t_{1/2} = \frac{1}{k_2[A]_0}$ |
| $aA + bB \rightarrow P$ | second, type II | $\frac{d[A]}{dt} = -k_2[A][B]$ | $\ln \frac{[A]_0 - [B]}{[B]_0 - \left(\frac{b}{a}\right)[X]} = \ln \frac{[A]}{[B]}$ |
| | | | $= \left(\frac{b[A]_0 - a[B]_0}{a}\right) k_2 t + \ln \frac{[A]_0}{[B]_0}$ |
| | | | $t_{1/2} = \left[\frac{a}{k_2(b[A]_0 - a[B]_0)}\right]$ |
| | | | $\times \ln \left[\frac{a[B]_0}{2a[B]_0 - b[A]_0}\right]$ |

## Chapter 8: MINIMUM SEWER SIZE, VELOCITY, AND SLOPES

*Recommended Standards for Sewage Works* ("Ten States' Standards," Chap. 20) specifies a minimum diameter of 8 inches for a gravity sewer. The minimum mean velocity, when flowing full, is specified as not less than 2.0 ft/sec (calculated with an $n$ value of 0.013). The following minimum slopes should be observed, although greater slopes are desirable.

| sewer size (inches) | minimum slope (ft/100 ft) |
|---|---|
| 8 | 0.40 |
| 9 | 0.33 |
| 10 | 0.28 |
| 12 | 0.22 |
| 14 | 0.17 |
| 15 | 0.15 |
| 16 | 0.14 |
| 18 | 0.12 |
| 21 | 0.10 |
| 24 | 0.08 |
| 27 | 0.067 |
| 30 | 0.058 |
| 36 | 0.046 |

## Chapter 8: COMPOSITION OF AIR

The composition of air is needed in certain problems involving the combustion of gases produced in digesters and landfills. In the following table, rare inert gases are included in the nitrogen.

| | % by weight | % by volume |
|---|---|---|
| oxygen | 23.15 | 20.9 |
| nitrogen | 76.85 | 79.1 |

## Chapter 15: WORKING STRESSES WITH AND WITHOUT SHORING

With composite construction (e.g., a concrete slab being poured over a supporting steel or precast beam), the method used to calculate the working stresses depends on whether or not temporary supports (i.e., *shoring*) are provided while the poured slab hardens. When shoring is provided, the composite section (the slab and beam) acts as a single unit and carries the entire load. When the composite section is put into service, loads are resisted by the effective composite section, represented in calculations as either $S_{\text{composite}}$ or $I_{\text{composite}}$.

$$\begin{aligned}\text{stress} &= \frac{(M_{\text{dead}} + M_{\text{live}})c}{I_{\text{composite}}} \\ &= \frac{M_{\text{dead}} + M_{\text{live}}}{S_{\text{composite}}}\end{aligned}$$

When shoring is not provided, the beam must initially support its own weight and the weight of the uncured slab. During the pour, the initial stress in the beam is

$$\text{stress} = \frac{M_{\text{dead,beam+slab}}}{S_{\text{beam}}}$$

When the composite section (cast without shoring) is placed into service, the composite section resists the newly added live load (plus any other superimposed dead loads, creep, and shrinkage), while the original stress in the beam remain

$$\text{stress} = \frac{M_{\text{dead,beam+slab}}}{S_{\text{beam}}} + \frac{M_{\text{live}}}{S_{\text{composite}}}$$

## Chapter 16: *K* VALUE DESIGN OF CURVE LENGTH CREST CURVES

Crest curve lengths are generally based on the stopping sight distance. The passing sight distance could also be used, except that the passing sight distance is 10–20 times the stopping sight distance. The curve length determines the extent of the earthwork required, and it is much easier to prohibit passing on crest curves than to perform the earthwork required to achieve the passing sight distance. Hence, only the stopping sight distance is considered in designing the curve length.

Table 16.3 ("AASHTO Minimum Sight Distances") in this book implies that the choice of a curve length is a simple selection of sight distance based on design speed. In a simplistic curve length design problem, the speed, $v$, and grades, $G_1$ and $G_2$, are known. The required stopping distance is calculated from Equation 16.13, 16.15, or similar. The required curve length, $L$, is calculated from the formulas in Table 16.4, which depend on the percentage grade difference, $A = G_1 - G_2$.

AASHTO extends this procedure by considering the *design speed* to be a nominal value. The actual speed of a vehicle can be equal to the design speed, a case referred to as the *upper range*. Alternatively, the vehicle may be assumed to be driven somewhat slower than the design speed to compensate for weather, wet pavement, lighting, or other adverse conditions, and this case is referred to as the *lower range*. (AASHTO reports that studies have concluded that vehicles are not actually driven slower under these conditions. However, the assumption may yet be made.) The normally used upper and lower range speed values are listed as the "initial speed" values in Table 16.3.

The $K$ value method of analysis (used in AASHTO's 1990 edition of *A Policy of Geometric Design of Highways and Streets*) is a simplified method of choosing a stopping sight distance for crest curve. The ratio of curve length, $L$, to grade difference, $A$, (in percent) is defined as the *length of vertical curve per percent grade difference*, $K$. ($LC$ is the variable used for curve length in this book.)

$$K = \frac{L}{A} = \frac{L}{G_1 - G_2}$$

The $L = KA$ relationship is conveniently linear. In order to facilitate rapid solving and checking of curve lengths, AASHTO has prepared two graphs. AASHTO Figure III-41 contains curves for the upper speed range, and Figure III-42 contains curves for the lower speed range. Since (for a fixed grade difference) the speed determines the stopping distance, every value of speed has a corresponding value of $K$. Thus, the curves in the figures are identified concurrently with the speed and the $K$ value. It is not necessary to specify both $A$ and $L$ in a design problem. Knowing $K$ is sufficient.

The simplified procedure is to select one figure or the other (based on upper or lower speed range), select one of the curves (based on the speed or $K$ value), and read the curve length corresponding to the grade difference.

Two important characteristics of the curves are in the AASHTO figures should be noted. First, the curves are based on "rounded for design" values of $K$. For example, from Table 16.3, for a design speed of 40 mph, the corresponding lower and upper speeds are 36 and 40 mph. The corresponding values of $K$ are 53.6 and 73.9. However, the "rounded for design" values used in the figures are 60–80, always higher than the actual value. AASHTO considers the difference to be insignificant.

Second, short curve lengths in the AASHTO figures are determined by various other overriding factors, the most important of which are typical state requirements. Curve lengths calculated from table 16.3 for the $S > L$

AASHTO Figure III-41
Upper Range Design Control for Crest Vertical Curves
(stopping sight distance; open road conditions)

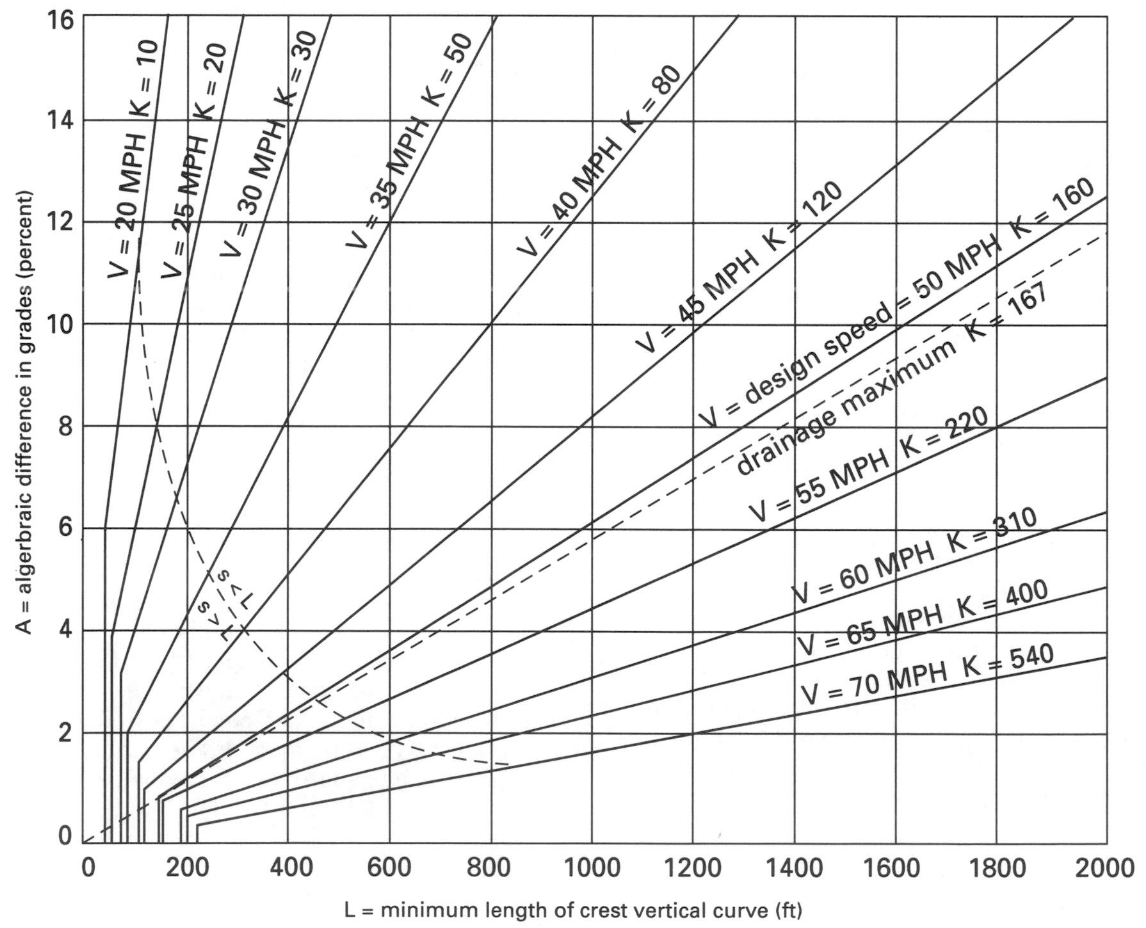


Used with permission from *A Policy on Geometric Design of Highways and Streets* © 1990, AASHTO.

case often do not represent desirable design practice and are replaced by heuristic values of three times the design speed. This is consistent with the minimum curve lengths of 100–300 feet prescribed by most states. The heuristic solution in the AASHTO figures are also justified on the basis that the longer curve lengths are obtained inexpensively when the difference in grades is small.

## Chapter 16: SIGHT DISTANCE FOR SAG VERTICAL CURVES

*A Policy on Geometric Design of Highways and Streets* (AASHTO, 1990) contains a method for determining the minimum lengths of sag vertical curves that is different than for crest curves. Factors taken into consideration are the headlight sight distance, rider comfort, drainage control, and rule of thumb for general comfort. AASHTO Figures III-43 and III-44 summarize and provide a graphical method of determining the lengths using the $K$ value concept.

## Chapter 16: COMPONENTS OF ASPHALT CONCRETE PAVING MIXTURES

*Asphalt concrete* is a mixture of *asphalt cement* (abbreviated AC) and well-graded, high-quality aggregate that has been heated and compacted by a paving machine into a uniform dense mass.

The highest-quality asphalt concretes are produced in stationary plants (i.e., are a *plant mix*). The mix is kept hot while it is transported to installation sites, such as highways, airports, parking areas, and driveways. Asphalt cements of various types (viscosity graded, designated AC; viscosity graded residue, designated AR, and penetration graded, designated numerically) are used to produce these high-quality pavements.

Other types of asphalt (other than asphalt cement) used in paving include (1) *emulsified asphalts* (both anionic

AASHTO Figure III-42
Lower Range Design Control for Crest Vertical Curves
(stopping sight distance; open road conditions)

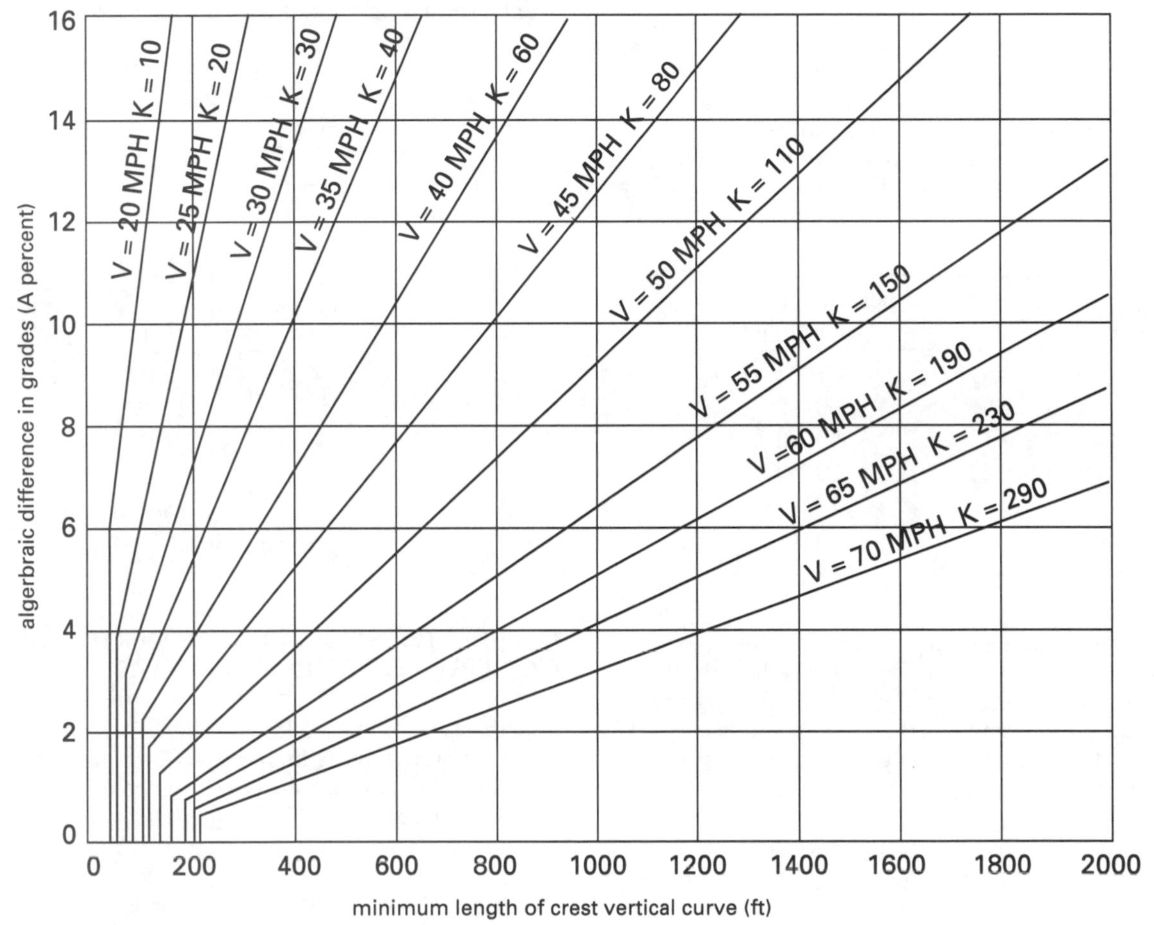




POSTSCRIPTS

and cationic types, with various "S" designations) that are used for mixed-in-place (road mix), cold-mix plant mix, recycled road mixes, surface treatments, sealing, and crack filling, and (2) medium- and slow-curing *cutback asphalts* that are used primarily in cold-mix plant mix and mixed-in-place road mix.

The *mineral aggregate* component makes up 90–95% of the weight and 75–85% of the volume of asphalt concrete. Mineral aggregate consists of sand, gravel, or crushed stone. The size and grading of the aggregate is important, as the minimum lift thickness depends on the maximum aggregate size. Generally, *coarse aggregate* is material retained on a No. 8 sieve (2.36 mm openings); *fine aggregate* is material passing through a No. 8 sieve; and *mineral filler* that is fine aggregate is material for which at least 70% passes through a No. 200 sieve (75 $\mu$m openings). The fine aggregate should not contain organic and clayey materials.

Aggregate size grading is done by sieving, and the results may be expressed as percent passing through the sieve or percent retained on the sieve. The *maximum size* of the aggregate is determined from the smallest sieve through which 100% of the aggregate passes. The *nominal maximum size* is designated as the largest sieve that retains some (but no more than 10%) of the aggregate.

A mix is specified by the nominal maximum size and a range of acceptable passing percentages for each relevant sieve size. (Table 9.3 lists standard sieve sizes. Not all sieves will be used in any one mix designation.) A typical mixture designation with a 1-inch nominal maximum size is

| sieve size | percent passing (by weight) |
|---|---|
| 37.5 mm ($1\frac{1}{2}$ in) | 100 |
| 25.0 mm (1 in) | 90–100 |
| 12.5 mm ($\frac{1}{2}$ in) | 56–80 |
| 4.75 mm (No. 4) | 29–59 |
| 2.36 mm (No. 8) | 19–45 |
| 0.30 mm (No. 50) | 5–17 |
| 0.075 mm (No. 200) | 1–7 |

The actual percentage passing through the 2.36 mm (No. 8) sieve is a convenient way to predict the final pavement texture. If the actual grading approaches the maximum amount permitted to pass through a No. 8 sieve, the surface will be relatively fine in texture. If the grading approaches the minimum value, the texture will be coarse.

Although the specific gravity of the asphalt cement can vary widely, the percentage of the total mixture weight contributed by asphalt cement varies from 3–8% (by weight) for large nominal maximum sizes (e.g., $1\frac{1}{2}$ inches) and from 5–12% for small nominal maximum sizes (e.g., $\frac{3}{8}$ inch). The *surface area method* can be used as a starting point in mix design for determining the percent of asphalt needed. The percentage of asphalt, $P_b$, needed is

$$\begin{aligned} P_b = 100\% &\times \text{aggregate surface area (ft}^2\text{/lb)} \\ &\times \text{asphalt thickness (ft)} \\ &\times \text{specific weight of asphalt (lb/ft}^3\text{)} \end{aligned}$$

The aggregate surface area is obtained by multiplying the weight of the aggregate by a surface area factor (ft$^2$/lb). This factor must be known (or given), as it is different for each type of aggregate and for each sieve size. In practice, each surface area factor is multiplied by the percent (converted to a decimal) passing each associated sieve, not by the actual aggregate weight, and the products are summed. The total is the surface area in ft$^2$/lb for the aggregate mixture.

Large asphalt paving machines can place a layer (i.e., a *lift*) of asphalt concrete with a thickness of 1 to 10 inches at a forward speed of 10 to 70 ft/min. The actual speed will depend on many factors and should not exceed the value for which a quality pavement is produced. The forward speed should also coincide with the plant production rate of asphalt concrete. That is,

$$\text{forward velocity}_{\text{ft/min}} = \frac{(\text{plant production}_{\text{tons/hr}})\times(\text{yield}_{\text{ft/ton}})}{60\ \frac{\text{min}}{\text{hr}}}$$

The yield in linear feet per ton of mix is calculated simply from the volumetric relationship of the layer being placed.

$$\begin{aligned} \text{yield}_{\text{ft/ton}} &= \frac{2000\ \frac{\text{lb}}{\text{ton}}}{(\text{width}_{\text{ft}})(\text{thickness}_{\text{ft}})\times(\text{compacted density}_{\text{lb/ft}^3})} \\ &= \frac{18{,}000\ \frac{\text{lb-ft}^2}{\text{ton-yd}^2}}{(\text{spreading rate}_{\text{lb/yd}^2})(\text{width}_{\text{ft}})} \end{aligned}$$

## Chapter 16: SPECIFICATION OF ASPHALT CONCRETE PAVING MIXTURES

Specifications, terminology, and nomenclature for asphalt concrete paving mixtures are unique to the paving industry. The following material is consistent with the methods presented in *The Asphalt Handbook*, 1989 edition (Asphalt Institute, Lexington, KY).

There are three specific gravity terms used to describe aggregate, depending on the volume, $V$, used with the dry aggregate weight, $W$. The *apparent specific gravity*,

$G_{sa}$, of an aggregate component (i.e., coarse, fine, etc.) is

$$G_{sa} = \frac{W}{(V_{\text{aggregate}})\gamma_{\text{water}}}$$

The *bulk specific gravity*, $G_{sb}$, of the aggregate component is

$$G_{sb} = \frac{W}{(V_{\text{aggregate}} + V_{\text{water-permeable pores}})\gamma_{\text{water}}}$$

The *effective specific gravity*, $G_{se}$, of the aggregate component is

$$G_{se} = \frac{W}{(V_{\text{aggregate}} + V_{\text{water-permeable pores}} - V_{\text{pores absorbing asphalt}})\gamma_{\text{water}}}$$

ASTM procedure C127 is used to measure these gravities for coarse aggregate without having to measure the volumes. In the following equations, $A$ is the oven-dried weight, $B$ is the saturated surface-dry weight, and $C$ is the submerged weight of the aggregate in water.

$$G_{sa} = \frac{A}{A - C}$$

$$G_{sb} = \frac{A}{B - C}$$

$$\text{absorption} = (100\%)\left(\frac{B - A}{A}\right)$$

ASTM procedure C128 is used for fine aggregate. In the following equations, $A$ is the oven-dried weight, $B$ is the weight of the pycrometer filled with water, and $C$ is the weight of the pycrometer flask with sample and water added to the calibration mark.

$$G_{sa} = \frac{A}{B + A - C}$$

$$G_{sb} = \frac{A}{B + 500 - C}$$

$$\text{absorption} = (100\%)\left(\frac{500 - A}{A}\right)$$

Each of the aggregate components (e.g., coarse aggregate, fine aggregate, mineral filler, etc.) have their own specific gravities ($G$) and proportions ($P$) by weight in the total mixture in percent. The *bulk specific gravity of the aggregate mixture* is found from the following formula, where the summation is taken over all of the aggregate components, but does not include the asphalt.

$$G_{sb} = \frac{\sum P_i}{\sum\left(\frac{P_i}{G_i}\right)}$$

The *apparent specific gravity of the aggregate mixture*, $G_{sa}$, is calculated from the same formula as for the bulk specific gravity, except that the apparent specific gravities are used in place of the bulk specific gravities.

ASTM procedure D2041 is used to measure the *maximum specific gravity*, $G_{mm}$, of the paving mixture (i.e., the zero air voids specific gravity). However, in practice, some air voids always remain and the maximum specific gravity cannot be achieved. The *effective specific gravity of the aggregate*, $G_{se}$, is calculated from the maximum specific gravity, the specific gravity of the asphalt, $G_b$, and the proportion of asphalt in the total mixture, $P_b$, in percent.

$$G_{se} = \frac{100\% - P_b}{\frac{100\%}{G_{mm}} - \frac{P_b}{G_b}}$$

The *maximum specific gravity* is calculated from the proportions of aggregate and asphalt, $P_s$ and $P_b$ respectively, in percent and the specific gravity of the asphalt, $G_b$. (In effect, this is solving the previous equation for $G_{mm}$.)

$$G_{mm} = \frac{100\%}{\frac{P_s}{G_{se}} + \frac{P_b}{G_b}}$$

The effective specific gravity should always be between the bulk and apparent specific gravities. Otherwise, it is logical to assume that an error has been made in one of the calculations.

$$G_{sb} < G_{se} < G_{sa}$$

By convention, the *asphalt absorption*, $P_{ba}$, is expressed as a percentage by weight of the aggregate, not as a percentage of the total mixture.

$$P_{ba} = (100\%)\left[G_b\left(\frac{G_{se} - G_{sb}}{G_{sb}G_{se}}\right)\right]$$

The *effective asphalt content of a paving mixture*, $P_{be}$, is the total percentage of asphalt in the mixture less the percentage of asphalt lost by absorption into the aggregate.

$$P_{be} = P_b - \frac{P_{ba}P_s}{100\%}$$

The *bulk specific gravity of the compacted mixture*, $G_{mb}$, is determined through testing according to ASTM procedure D2726.

The *percent VMA in the compacted paving mixture*, VMA, is a measure of the *voids in the mineral aggregate*.

$$\text{VMA} = 100\% - \frac{G_{mb}P_s}{G_{sb}}$$

The *percentage air voids in the compacted mixture*, $P_a$, is

$$P_a = (100\%)\left(\frac{G_{mm} - G_{mb}}{G_{mm}}\right)$$

## Chapter 17: MEANING OF THE WORD "STATION"

The word *station*, when used in surveying, can mean both a location and a distance (i.e., a length). A station is a length of 100 feet, and the unit of measure is frequently abbreviated "sta." When a length or distance is the intended meaning, the unit of measure will come after the numerical value. For example, "...the length of the curve ($LC$) is 4 sta...."

On the other hand, when a location is the intended meaning, the unit of measure will come before the number. For example, "...the point of vertical intersection ($PVI$) is at sta. 4."

It is apparent that, in the second case, the location is actually a distance measured from some starting point. Thus, the two meanings are similar and related.

## Chapter 17: VERTICAL CURVE TO PASS THROUGH TURNING POINT

A common problem involving equal-tangent vertical curves is to calculate the length of vertical (crest or sag) curve needed to pass through a turning point at a particular elevation. The following equation determines the curve length directly. Grades $g_1$ and $g_2$ and the elevations at the $PVI$ ($y_{PVI}$) and the turning point ($y_{TP}$) are known. While the equation is valid only for a single set of circumstances, its derivation is probably too time-consuming to attempt when time is limited.

$$LC = \frac{(2)(y_{\text{PVI}} - y_{\text{TP}})}{(g_1)\left(\dfrac{g_1}{g_2 - g_1} + 1\right)}$$

## Chapter 17: HORIZONTAL RAILROAD CURVE

Horizontal curves used in railroads are generally longer and more gradual than highway curves. Railroad curves are commonly specified on the *chord basis*. The *degree of curvature*, $D_{\text{chord basis}}$, is the central angle defined by a chord of 100 ft.

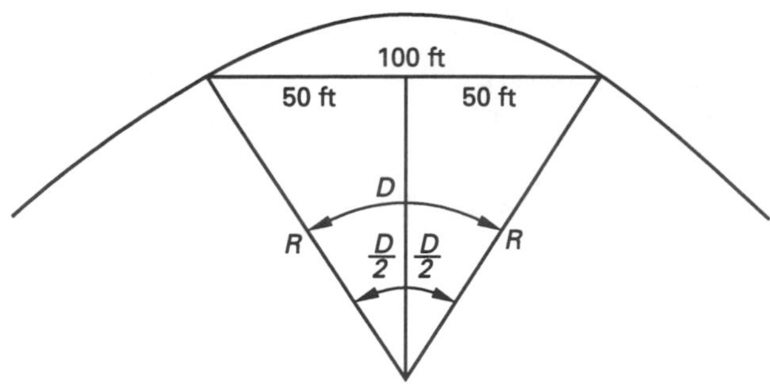


The curve radius can be calculated from the following equation.

$$R = \frac{50 \text{ ft}}{\sin\left(\frac{1}{2}D_{\text{chord basis}}\right)}$$

The curve radius calculation is different than for the arc basis (equation 17.36). Once the radius has been determined, equations 17.31 through 17.35 can be used for railroad curves.

For the small degrees of curve used in railroad work, the lengths along the arcs can be assumed to be the same as the chord lengths. Therefore, the length of curve in railroad practice is the number of 100 ft chords.

$$LC = \left(\frac{I}{D}\right)(100 \text{ ft})$$

## Chapter 17: FORWARD AND BACK TANGENTS

A horizontal curve is constructed between two straight lines known as *tangents*. When traveling in a particular direction, the first tangent encountered is the *back tangent*, while the second tangent encountered is the *forward tangent*.

*Update: September, 1995*

## Chapter 3: PIPE NETWORKS (HARDY CROSS METHOD)

When using the Hardy Cross method with $Q$ in MGD, $L$ in feet, and $D$ in feet, the Hazen-Williams friction coefficient is

$$K' = \frac{10.59L}{C^{1.85}D^{4.867}}$$

# INDEX

## A



## B

## C



INDEX

## D



INDEX

## E



INDEX

## F

## G



## H



INDEX

## O



## P

## Q



## R

## S

INDEX

## T

## U



## V



## W



## X



## Y



## Z

# INDEX OF FIGURES AND TABLES

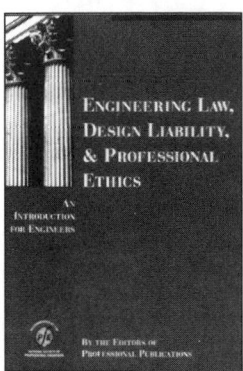

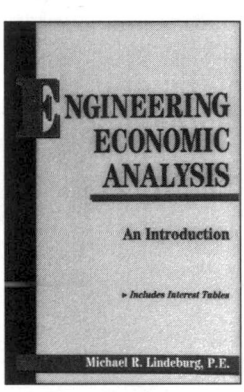

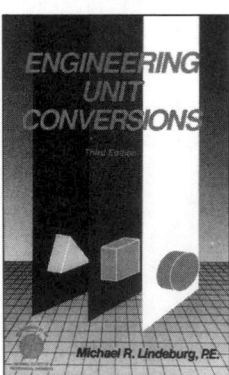

# EMERGENCE

Ray Hammond is the author of nine non-fiction books including *Digital Business: Surviving and Thriving in an On-line World.* He has also written drama and comedy for television and radio.

Born in Hertfordshire, England, he entered journalism aged seventeen, reporting for local and national magazines and newspapers before becoming a regular contributor to the *Sunday Times,* the *Daily Mail,* and the *Independent on Sunday.* Today he lives in London and is a lecturer and writer on future trends in business and society.

*Emergence* is his first novel.

RAY HAMMOND

# EMERGENCE

PAN BOOKS

First published 2001 by Macmillan

This edition published 2002 by Pan Books
an imprint of Pan Macmillan Ltd
Pan Macmillan, 20 New Wharf Road, London N1 9RR
Basingstoke and Oxford
Associated companies throughout the world
www.panmacmillan.com

ISBN 0 330 48595 4



1 3 5 7 9 8 6 4 2

A CIP catalogue record for this book is available from
the British Library.

Typeset by SetSystems Ltd, Saffron Walden, Essex
Printed and bound in Great Britain by
Mackays of Chatham plc, Chatham, Kent

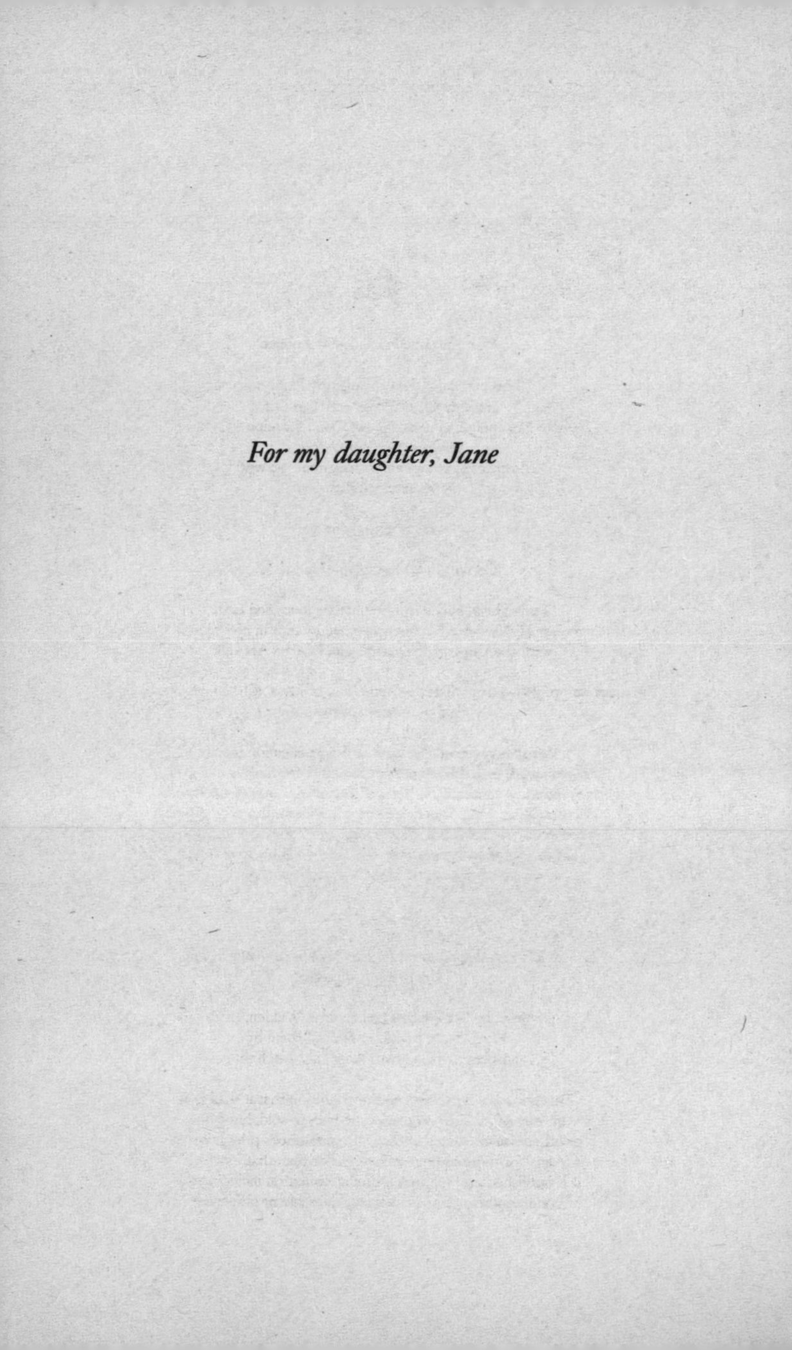

*For my daughter, Jane*

# ACKNOWLEDGEMENTS

I am grateful to many people who have helped me during the writing and production process of this first novel. Mic Cheetham, my agent and friend, knew exactly when to encourage, when to ignore and how best to help me shape this story. Simon Kavanagh, her assistant at the time, also made incredibly useful suggestions about plot possibilities. Thanks to them both.

At Macmillan I have been lucky enough to find in Peter Lavery an editor who is both an enthusiast for what is properly described as 'mainstream' fiction and is unerring in his instinct for weeding out weaknesses in both plot and prose. He has made an immense contribution to this novel.

A number of people have been kind enough to help me with scientific, medical, professional, cultural, technical or production issues. I wanted as much detail as possible in this book to be scientifically plausible and technically accurate but I haven't *always* taken the advice that has so generously been offered, so no blame can be attached to any of them for any errors, omissions, misunderstandings or inaccuracies. All faults are mine.

In alphabetical order, I thank Nick Austin who copy-edited and prepared this manuscript for press; Dr James Dodd, co-author of *The Ideas of Particle Physics*, and formerly of DresdnerKleinwortBenson Merchant Bank, for his detailed and enthusiastic input on both quantum physics *and* fund management; Professor Allison Druin of the Human-Computer Interaction Laboratory at the University of Maryland for her thoughts regarding my 'companion' characters; science and technology writer Simon Eccles for his comments and advice on space technologies and astronomy; Anne Hardy for amazingly good proofreading skills (and for her patience and understanding while the manuscript was under development); Peter Kraus for his help with Yiddish colloquialisms; Judith Hall for her assistance with Spanish translations; Janice Nagourney for her assistance with French translations; Dr Brian Rossiter, Consultant at Whipps Cross Hospital, London for help with medical details; Dr John Rossiter, Senior Lecturer at Imperial College, London for advice on some possibilities for genetically modified plants and Dr Bruno Stanek of Astrosoftware, Switzerland for advice about space technologies and astronomical details.

Thanks also go to the novelist Terry Bisson for permission to reproduce a section from his short story, 'They're Made Out of Meat' which appears in chapter twenty-four. His copyright is fully and gratefully acknowledged.

In addition, I thank Dagmar O'Toole and Alex Krywald of Celebrity Speakers Ltd, whose skilful management of my public-speaking career made this project possible.

Finally, thanks to Liz Hammond for her tireless and good humoured proofreading and for her belief and support over the years.

# PROLOGUE

In summer, the skies above the city of Stockholm remain blue for most of the Earth's daily revolution: the Sun might be thought unwilling to withdraw fully in night's favour. By late evening azure becomes amethyst, eventually giving way to an ecclesiastical indigo. This allows a few of the brightest stars to compete to make their presence known alongside the ever-growing network of communications satellites and their intermittent matrix of laser beams made brilliant and multicoloured solely for purposes of marketing advantage and brand identity.

Luxuriating in his hot tub, Rolf Larsson gazes through the glass ceiling panels of his attic bathroom and allows the deep purple of the late evening sky to engulf him. He can still feel the press of Laila's embrace, her body urgent against his, and he savours their closeness again. He slips into a gentle mood of detachment, floating, as the swirling surreality of Debussy's Cello Sonata in D minor marks a distant punctuation elsewhere in the apartment. Laila has chosen his favourite piece of music. She is sending him a message of love and contentment from the living room beyond. It is Friday night, the start to their weekend.

He names every star he can see, constructs a pattern to connect them and quickly factors their prime numbers. It is a game his father taught him even before he went to high school. Then he makes his topology three-dimensional, placing and naming the more distant star clusters and invisible galaxies where he knows them to be. Once again his mind turns to the infinite billions of stars whose presence is masked in summer by the light Scandinavian atmosphere. Sweden's twenty-seven-year-old media-acclaimed 'prodigy' of astrophysics tries once more to predict his pattern in a way that will make them denumerable. As always, the model in his head shatters soon after he tries to push beyond the counter-intuitive irrationality of string theory, quantum mechanics, parallel states and the concept of infinity. He exhales and lies back, dipping his head under the water.

He sits up, dries his face on a towel and leans forward to add more hot water. As he does so, a sudden contrapuntal rhythm created nearly a century earlier fills his head and he gains another point of observation that flickers in and out of his grasp.

Suddenly his four-dimensional model of matter extends with the music and gains a fifth, then a sixth, then more, in a mental cascade of observations that pulse with potential for proof. In a moment it, too, shatters but then, with the counterpoint of the rational mathematics Debussy used to build his temporal dance, it slowly reassembles as an intellectual scaffolding that provides multiple observation points which extend and transcend the thinness of the present.

Larsson probes the new patterns of space, place, time and matter that are now crystallizing in his consciousness. It seems to be an entirely new metalanguage! Then he realizes what he might have.

He leaps from his bath and runs, naked and wet, into his wood-floored attic living room. 'Wake!' he shouts as he reaches his computer display screens and, dripping, he begins to work, oblivious of Laila's puzzled gaze.

The stars were reappearing by the time Larsson pushed himself back from his screens and ran his fingers through his hair. He should have been preparing for two tutorials for his supplementary PhD in particle physics, but he was sure he had just discovered something no tutorial could offer. Yet, despite the scale of this achievement, it didn't occur to him that he might never return to his university.

Laila had padded over to investigate soon after he had started communing with his machine. He hadn't said anything when she had draped a towelling robe around his shoulders and she understood his frequent intellectual obsessions well enough to leave him undisturbed. Later, when she had been swallowing yawns for an hour, she had brought black coffee from the kitchen, guessing that he was settling down for a long session. He hadn't even looked up as she leaned over his shoulder to place the mug beside his keyboard. She had kissed his cheek, feeling his stubble against her lips, and he had at last acknowledged her presence by placing his hand over hers as she squeezed his shoulder. She had kissed him once more and left him to it.

Now, nearly a day later, he was finished.

'Save with remote back-up,' Larsson told his computer. 'Disconnect from all networks.'

He slept for fourteen hours and when he woke, sweaty and unshaven in the broad daylight of Sunday, he panicked. He couldn't recapture the complex matrix in his mind. Two minutes later his computer confirmed he had not dreamed it: his new language and concepts had created formulae that expressed a polydimensional method of observing the smallest particles of matter – a means of calculating and proving their positions at all times.

When he had showered and wolfed down cold baked beans straight from the can, Larsson called his academic supervisor at home. He had been summoned as a standby juror, he explained: sudden and unavoidable. The trial was scheduled to last a few weeks.

Laila had given up on their weekend plans and returned to her own apartment in Trossa, leaving a message for him to call when he finally surfaced. She took his call and, after apologizing, he explained that he'd found something that might be important for his new doctorate and that he needed some time alone.

'Is there anything wrong?' she asked. He could see the worry on her face, and pursed his lips towards her in a kiss.

'I love you,' he said. 'You'll be so proud of me if this turns out to be what I think it is.'

She smiled, a reaction that produced a small dimple at the corner of her mouth. He felt a pang of desire but fought it.

'Just give me a little while to concentrate, OK?'

She nodded.

'It's not so easy when you're around.'

She smiled again. 'Well, call me when you can.'

Three weeks later, having provided only the briefest explanations for his solitary preoccupation to his family and friends, Larsson had completed the coding for his software. It worked flawlessly on the few secure messages he had in local storage. He then logged on to the global networks and dispatched the software robots he had created.

After decades of continuous investment in virtual technologies, the world had become totally reliant on the vast web of fibre-optic cables, wireless networks and satellite chains that, each day, created an ever more dense matrix around the planet. Almost every aspect of government, business and social life raced through the man-made digital cosmos at the speed of light. Everything sensitive, controversial or financial was scrambled by super-strong security techniques that were unbreakable even by the largest network of optical supercomputers. It was a safe, trustworthy and instant domain.

Within an hour Larsson's software surrogates reported back with copies of two thousand separate messages, all painstakingly gathered, collated and reconstructed from the millions of tiny parts into which they had been split in order to pursue separate routes to their intended destinations. Of these, just over 400 had been scrambled, using unbreakable cryptography.

Larsson found 107 errors in his coding as he ran his new prime-number generator repeatedly against the encrypted messages. With mounting cries of frustration at his own stupidity, he corrected and recompiled the software until his engine was producing a continuous string of the super-rare high prime numbers that lie at the heart of unbreakable encryption technology.

Once the software was stable – or stable enough to complete more than a few passes without crashing – it took him a little less than fifteen minutes to break the first message. As he tuned his algorithms, plain text emerged at an ever-faster rate from the jumble of letters and symbols that made up the encrypted communications.

Six hours after he broke the first message, all 409 were in plain text for him to examine. Ignoring the messages that were in languages he couldn't read without using auto-translation, Larsson's first ten minutes of scrolling revealed a draft agreement on agricultural trade subsidies between Washington DC and the European Union, four bank transfer instructions for sums ranging between two hundred million and seven hundred million dollars, and three sets of draft company accounts.

The young Swede pushed his chair back from his computer screens and yelled at the ceiling. He jumped to his feet and clasped his hands behind his head, turning in tiny circles. For twenty minutes he walked around his apartment staring at blank walls, at the table top and out of the window. He looked, but registered nothing.

An hour later he dispatched his team of software robots again. This time they had a particular target and, as soon as they had departed into the world's networks, he left his apartment, carefully double-locking the heavy old metal door of the converted warehouse building.

He was shocked by the brilliance of the June day. Other than occasional late-evening sorties to the convenience store, he had hardly stepped outdoors in a month. The sun created a panorama of flashing reflections across the gentle swell of the harbour like a flotilla of miniature ships frantically signalling the shore.

Not for the first time, Larsson reflected on how the apparent reality of the physical world made the intangible space of the digital environment seem unreal – the classic mistake. He smiled and reminded himself that it was his brain that was adding the brilliant colours to the scene in front of him. All that existed in the physical world were varying achromatic wavelengths of light. Man created his own world: it had been *virtual*, a product of human creativity, from the moment consciousness emerged and thus it was humans who gave meaning to quantum particles which, Larsson had proved, included many alternative states, all of them useful in the creation of new concepts for language and, consequently, thought.

He strolled past the innumerable outdoor café tables on the cobbled quayside, oblivious of the sharp looks of interest and query from the young and less-young female patrons. Tall, lank and with an unruly mop of flaxen hair, his thin T-shirted figure appeared deep in thought.

His mind was on Thomas Tye and his company, the Tye Corporation. Since the eruption of global wealth created by the virtual and biotech economies, thousands of new companies had emerged to take over from the old industrial-age behemoths such as car makers and oil companies. The corporate riches of the late twentieth century had been dwarfed by the immense wealth created by enterprises that focused on delivering virtual and information-specific products via the networks, by companies that could quadruple the produce of an acre of land and by corporations that created miracle cures based on the map of the human genome. All such products and services were delivered to a global market made one by the networks' elimination of physical boundaries and borders.

Even the old software and computer giants of the pre-network age now looked puny compared with the distributed and virtually based corporations that had come to dominate the global economy. Of these, by far the richest was the Tye Corporation, the world's most valuable company and the biggest telecommunications, software, pharmaceutical, biotech, healthcare, aerospace, media and banking conglomerate on the planet.

Almost everybody in the world, in the rapidly emerging economies as well as in the developed countries, was familiar with the face and the public opinions of Thomas Tye, the company's founder, major shareholder and environmental campaigner; the man who had become the planet's richest citizen. As such, he was the obvious target for Larsson, and by the time the astrophysicist returned to his apartment, his trawling robots had reappeared with copies of over 300 Tye Corporation communications that had been flashing through the world's networks as they lay in wait.

Larsson saved the material that had been harvested and started work. Inside an hour he had decoded and read plans to relocate offices, fund transfers between a dozen banks and outline designs for a new generation of 3D hologram software. Then he found something ideal for his purpose. It was a highest-security, deeply encrypted message that had been sent to Thomas Tye's confidential mailbox. When he had broken the code, Larsson read the final draft of the Tye Corporation's annual report and consolidated accounts that was due to be published in sixteen days' time. The document had been on its way to Tye for his final approval when it had been silently and untraceably intercepted and copied, in less than one one-hundredth of a second.

Larsson printed out the 120 pages and spent several hours struggling to understand the unfamiliar formats of multinational corporate accounting. He visited the Tye Corporation's main network resource and downloaded the previous year's accounts. He could interpret enough of the financial statements to see that the company's revenue had jumped sixty per cent and its net income was up by forty-eight per cent.

He stood up and allowed himself a few small revolutions in the middle of the floor. Then he sat back down, extracted Tye's personal network address from the decoded file and prepared a message containing a copy of the draft accounts in plain, unscrambled text. To reinforce his point, he also added three further confidential Tye Corporation messages in plain language and sent them to Tye with a cryptic message;

*New software. Want to discuss?*

Larsson left his audio alert on and went to bed. But sleep did not come. At three a.m., just as he was finally drifting off, his computer alerted him to incoming v-mail. He opened his three screens, enabled the videoconferencing system and saw the best-known face on the planet in front of him.

Involuntarily, his response was formal.

'Mr Tye.'

'Doctor Larsson?' That rich, full voice; so well known.

The young man nodded.

'Can you verify, please?'

'Let's both do it.'

The cameras showed Tye leaning forward and touching the fingerprint reader on his system. Larsson did the same. Three seconds later a message appeared on Larsson's home screen. 'Identity of caller confirmed as Thomas Richmond Tye, born 1 July 1966, Boston, USA. Present location undisclosed.' A second message confirmed that Tye had received reciprocal confirmation from the world's Digital Certification Authority. Despite his nervousness at communicating with the richest person ever to have lived, Larsson found himself wondering at how handsome and youthful Tye appeared for someone well into middle age. His plastic surgeon must be excellent.

'Please encrypt, if you don't find that too funny,' said Tye. Larsson complied and received confirmation of secure mode.

'Well, Doctor, you certainly have the right background for it,' acknowledged Tye, staring straight into his central camera. They had exchanged camera control as had become the custom and courtesy of the time and Larsson zoomed in until the trillionaire's perfect face filled his central screen. 'Did you get lucky or do you really have something new?'

Larsson had gone over and over what he might say at this point. Abandoning all his earlier plans, he simply said, 'It's something new. Completely new.'

'Can you prove it?'

'To anybody who understands particle physics or quantum mechanics,' responded Larsson. 'Or I can repeat the demonstrations.'

'Who have you told?'

'Nobody. It happened just a couple of weeks ago.'

'I presume you want to sell,' prompted Tye.

'I . . . I don't know,' said Larsson, because he didn't.

'Come to Hope Island tomorrow,' said Tye. 'Tell nobody, and bring everything you have on it.'

Larsson hesitated. After repeated postponements, he had arranged to see Laila the following day for the first time in four weeks.

'Well?'

Larsson nodded.

'I'll send a jet. Someone will be in touch. Oh ... and don't attempt to act on what you know. I'm altering the figures.'

The screens went blank. Larsson sat slumped in his chair for a few moments and then, with shaking hands, got up to make a coffee. He had never felt less like sleep.

# ONE

**Seven Years Later**

As usual, there was a moment's silence when Thomas Tye appeared in the lights. There is a shock in seeing such a famous face and feeling such power in the same room, even when the entrance is anticipated and the room very large.

On either side of the stage two giant screens flickered into life and the audience rose to its feet shouting, clapping and stomping its appreciation, adulation and joy. The corporate rock anthem filled the air and Tye smiled and waved, both arms outstretched, images of his improbably boyish, good-looking face filling the vastness of London's Earl's Court arena. His lustrous shoulder-length dark hair shone in the lights and the small diamonds set into his earpieces sparkled as he moved. He wore his customary stage uniform of white T-shirt and black trousers. Subtly, the taste and smell of the air changed.

He stood nodding his appreciation for several minutes as his audience swayed to the beat and reached out towards him. Then, with an extended palm, he silenced the music. His audience howled and he walked to the front of the stage and bowed low; practised, confident and sure. He had always been a natural on stage and, in the early days, he had built his markets by the power of his virtuoso extempore performances. Now he accepted major appearances only when it suited the company, when it fitted with his schedule or when it eased access to the politicians who were now his main concern. But he still loved these performances.

He stood and smiled, his perfect white teeth and infectious grin lighting up his audience. 'Well, how are you?'

Twelve thousand people erupted again and the beat returned. They stomped and stomped, the drumming of the music and their feet becoming faster and faster until it merged into a rolling crescendo of thunder.

Tye lifted one hand again and there was silence. He held them, the confident cynosure, greeted by thousands of camera flashes as he turned to each sector of the vast hall. Whistles and cries of 'Tommeee' broke through the silence he was controlling. The 'TT' chant started, low at first, then insistent, then with all the power of 24,000 lungs.

In the wings of the elaborate set, Jack Hendriksen received a message in his ear.

'There's about two hundred outside the stage door plus the TV crews, journalists and the merchandisers. And the Touchers. We've run facial patterns from our video scans. No known combustion risks as far as we can see.'

Jack nodded involuntarily as he listened to the disembodied voice.

'OK,' he said, 'The party will be there in . . .' He looked at his LifeWatch. 'Twenty-six minutes. I want a clear passage for Pierre's team.'

Tye was performing at his best and the technology worked without a hitch. He had quietened the crowd down again and he stepped back from the edge of the stage. The lighting went off. Then a spotlight lit a lectern to the left-hand side of the stage and Tye was behind it, waving his right arm. A split second later a follow-spot lit an area in the centre of the stage and another Tye was there, waving with his other hand. The audience looked from one Thomas Tye to the other and back again.

'Pretty good, isn't it?' asked the Tye in the middle of the stage. 'Tom, turn around, please.'

The figure at the lectern turned slowly, a slightly ethereal quality to its movements.

'Wave goodbye.'

The Tye at the lectern waved obediently. Its spotlight died

and Tye walked to the other side of the stage illuminated by his follow-spot.

'And now . . .'

The centre of the dark stage was suddenly illuminated and where there had previously been only space, a string quartet in evening dress appeared and began to play the 'Spring' movement from Vivaldi's *Four Seasons.*

As Tye stepped his audience through the carefully canned demonstration of the Tye Corporation's new holographic entertainment system, Jack Hendriksen checked in turn with each of his locally hired observers in the auditorium. The entire audience had been scanned and the building swept twice, but there was always the problem of touts selling last-minute tickets to self-immolators. And today the warm-up team had lifted the crowd to ecstasy.

For two hours before Thomas Tye had appeared, the corporation's team of international Games Masters had led the crowd through ever-increasing levels of excitement as they re-enacted legendary network war games and space battles in a show of lasers, smoke and music that filled the central roof space of the vast arena. Those who had paid €400 for their first-class tickets followed the action as they sat strapped into hydraulically powered seats that moved with the motion of the spacecraft, the battle armour or the jet planes the imagineers had created for their games. Those who were accessing the event via the networks received similar control feeds for their home HydraChairs and headsets.

Then there had been a wait: a smouldering time of heavy rock music and the first reprise of 'It's Our Planet', a song that had become the corporate anthem and a global Number One. As it played, the scent simulators in the auditorium released a fragrance called 'Abundance' into the atmosphere. It was the track's signature scent and the audience had found themselves inhaling the cleansed air of spring woodland after a rain shower. Their anticipation and excitement grew as they waited for Thomas Tye finally to appear in person.

Over the last decade, Tye's wealth, power, fame and innate sex appeal had turned him into an idol. As the world's first trillionaire, he had become an icon with a global following. Because of his good looks, his concern for the planet, his legendary philanthropy, and the careful presentations created by his perception managers, marketing strategists and public relations teams, he was also the first businessman to achieve real superstar status. Cleverly, he had captured the hearts and minds of the hoards of anarchistic, anti-capitalist, pro-environment, anti-establishment protesters who had used the early Internet to create a contagion with which to ignite demonstrations and violent outbreaks across the globe in the early years of the century. As these retro-1960s rebels, the children of the hippie archetypes, had, in their turn, become parents and homeowners, his corporation had originally offered them a respectable and ethical alternative vehicle by which to grow their wealth. But today, the glister of his fortune meant that his public appearances had also become a rallying point for the world's truly needy. He had become the focus of hope for millions of physical sufferers as well as for the disturbed, the alienated and the lonely.

Eight years earlier, Tye had shaken hands with a group of disabled fans who had been brought to hear him speak in Mexico City. Within days, the mother of a paraplegic teenager claimed that her daughter's paralysis had been cured by meeting the great star and entrepreneur. She had paraded the walking miracle on television and a movement had been born, connected and nourished in the byways and private meeting rooms of the networks. Tye's new followers believed that if they could simply touch their idol, they would be cured of their disabilities or their diseases. The press had soon dubbed them 'Tye's Touchers.'

Thomas Tye had immediately released a video statement to the twenty-four-hour news channels disclaiming all such healing powers, but the believers would not be dissuaded. The Tye Corporation's international perception-management consultancy sensed the potential for a serious public-relations catastrophe and, on its advice, Tye started asking for a token number of visibly disabled people to be present at his public engagements. His PM team ensured that they were always placed in the front row. Before, during or after his appearances, Tye made the time to shake each hand and further rumours about miraculous cures began to circulate.

A few months later a freelance reporter with a smattering of scientific knowledge provided the world with yet another example of irresponsible journalism. Her article, which was published in a networked popular-science magazine, revealed that a subsidiary of the Tye Corporation had patented a biochip that, when worn under the skin, monitored and rebalanced the electrochemical processes of the human central nervous system. She claimed that Tye had personally been beta-testing the chip for two years and it was contact with this new radiesthetic conductive and corrective force that was producing the purported cures. The movement became a cult.

Then, during a major performance in the Dynasty Auditorium in Manila, one of the many wheelchair-bound Touchers in the audience set herself alight. At the peak of the carefully orchestrated excitement, the fifteen-year-old girl had doused her clothes in petrol and the shaking figure in the chair had disappeared inside an inferno of flame before the hall's security staff had had a chance to reach her. The President of the Philippines had been watching from the Presidential Box at the time and the incident was caught live by TV cameras broadcasting the event throughout Asia. Within minutes, the recording was being replayed on the world's global news networks. The victim had left an e-mail with a friend that simply read 'For the Planet'. The girl had told her friends in a Thomas Tye network community that it was better to die in the presence of her idol than to use the planet's precious resources to extend her life of suffering.

Four months later it happened again, this time in Santiago. Then it happened in a detached home in a quiet tree-lined street in a suburb of Munich, Germany. This quadriplegic fan persuaded a friend to strap her into a HydraChair, feed her a hyper-analgesic and douse her with petrol before leaving her with a mouth-operated battery ignition system and a network connection to a Thomas Tye appearance in Sydney, Australia.

Supporters in a Thomas Tye chat room on the networks sat with her during the build-up and, when the time came, helped her find the courage. Then, in tribute, they posted a video record of her sacrifice in their community's meeting space.

Once again it was 'For the Planet. Thank you, Tom.'

Inevitably, the tabloid press dubbed this new breed of fanatic fans 'Tye's Torches' and preventing the possibility of further acts of hysterical self-immolation became a high priority during all public appearances.

This morning, Jack Hendriksen's plants in the Earl's Court audience reported nothing suspicious.

Tye was into his wrap, dancing along with a holo-image recreation of a young Bruce Springsteen and the E Street Band as they performed 'It's Our Planet'. The venerable star's brand-extension agency had graciously granted a licence for his youthful voice and image to be sampled and morphed for the recording.

Jack flicked on the head-up display in his viewpers. He switched from camera to camera checking backstage, the artists' corridor, the stage door and the throng outside. The police had the large crowd penned behind crush barriers and he could see that the members of Pierre's PPT – the Presidential Protection Team – were in place and the motorcade was waiting.

As director of corporate security for the Tye Corporation, Jack was rarely on the road with the PPT but he believed that spot checks and surprise visits were the best way to keep his in-field teams from becoming complacent. His presence in London had given him the first opportunity in months to observe Pierre Pasquier's detail and, despite a growing sense of inarticulated unease about the organization that employed him, he was pleased to see the machine was well oiled and at a high condition of readiness.

Then it was over. Tye walked to the front of the stage and acknowledged the applause to the right, the centre, the left and to the cameras in the pit. He leaned forward and touched the line of outstretched hands at the edge of the stage – the hands of a few dozen physically disabled fans allowed to sit in elevated positions for this televisual opportunity.

'Take care of our planet – you hear!' he told the audience.

'We hear! WE HEAR!' they shouted – the conditioned response to his famous parting imperative and they roared their approval. He turned back to them one more time: 'And come visit our booth in the show.'

He waved goodbye with both hands stretched high. The huge consumer electronics expo in adjacent halls would open the moment he left the stage.

He waved again and then he was gone. Even before he had left the darkened stage he had thrust his hands deep into his pockets where the fingers that had done the touching could break open antiseptic capsules. Then he was backstage and in Pierre's protection and he turned off into a specially built bathroom in the wings. Pierre, a six feet, six inches tall former officer of France's *Direction et Surveillance du Territoire*, stood guard across the door.

Tye carefully locked the door and checked twice that it was secure. He stripped off all his clothes, threw them on the floor and flipped through his vital signs on his LifeWatch. Then he washed his hands once in a bactericide and then again in a bacteriostatic, carefully scrubbing under his fingernails each time. He dried his hands on paper towels that he threw into a large trash can. He tore open sealed plastic bags and removed fresh underpants and socks. He tugged these on and peeled a long tape fastener from a hermetically sealed clothes carrier hanging behind the door. Here he found a fresh white shirt and a pair of dark trousers. When he had buttoned and zipped himself into these he took a new pair of shoes from the bottom of the container. He slipped into them and then picked up a small aerosol and sprayed his face, mouth and throat with isoprophyl alcohol for extra protection. He pulled a brush through his long hair, checked his appearance in the mirror and took a deep breath. Holding it, he stepped out of the room and into the care of Pierre's team.

Then he was walking rapidly in the middle of a phalanx of five. *Perfect formation*, thought Jack as he followed a few yards behind. The bodyguards were all immaculately dressed in sharp, dark business suits made from flame-resistant material. They moved forward like a super-taut Olympic relay team passing a radioactive baton: each knew precisely where the other was and how he would react to a slip.

Pierre's height allowed him to see over Tye's head to scan all events to the front. He was responsible for the most vital exit route, the escape to the rear. He would physically pick Tye up and run with him if he had to. At Jack's insistence he had even made the trillionaire suffer the indignity of repeated rehearsals.

Out in the daylight and powerful heat of a London June morning there was pandemonium, as usual. Reporters shouted questions as the cameras rolled. The Touchers reached out over the heads of those in front, imploring their hero to stop and shake hands or merely to touch their fingers. Some threw flowers.

Suddenly two small figures squeezed between the crush barriers and dipped under the interlocked arms of the police officers. One had a microphone in her hand, the other was operating a video camera. The first thrust the microphone in Thomas Tye's face with a shouted question.

Tye knocked the microphone out of his path without a glance.

'It's OK, OK, keep moving,' ordered Pierre in the team's earpieces as he stepped forward to shield his boss from the intrusion. From the rear Jack saw the small incident brought under control and then the phalanx had moved forward and he was left to confront the interlopers.

The two women looked identical. Both were short and dark, but Jack saw that the one with the microphone had a short elfin haircut while her twin had long hair. The elf looked from her disappearing quarry to Jack and thrust an envelope towards him. He ran an instant head-to-toe scan of the small, attractive woman.

'Give these to him, please,' she urged, her dark eyes wide as she pressed the envelope against Jack's chest.

His training told him not to touch it, to let the package fall, to avoid any legal liability. In his present disaffected mood his instinct said the opposite.

Jack nodded, stuffed the envelope into his inside pocket, caught up with the party and climbed into the rear of the fourth limo. He saw Pierre wave control over to the leader of the police outriders and the motorcade of electronically shielded and battle-armoured limousines pulled away. Jack searched back through the digital video recording his pinhole cameras had captured. Her face was not in the database of known Touchers. But he held the image in front of him, studying it. The convoy slowed as it neared the venue for Tye's next engagement.

On the other side of the world, sixteen planes were in various stages of descent for landings at Oakland Airport, across the bay from the southern suburbs of San Francisco. Although mainly a regional, domestic airport with relatively light night-time movements, the air traffic supervisors had the task of overseeing the computers that steered incoming and outgoing flights around the over-busy, delay-prone air lanes that fed SFO, the city's main international airport eight miles to the west.

Oakland was a modern, well-equipped airport and when the images projected by the air-traffic control computers froze, senior air traffic supervisor Sandy Davis swore. The technical people had told everybody that the modern computer systems couldn't crash. Then, remembering the drill from her initial training as a manual air traffic controller twenty years before, she simply closed her eyes and started counting to ten. She was trying to calculate the progress of the sixteen planes the computer had been handling even though, in her manual days, the maximum a human controller would have been expected to visualize during a system failure was six.

The back-up system finally kicked in as her count reached eight. As the display in front of her reactivated she saw that her planes were almost exactly where she expected them to be.

'What's happening?' she called angrily over her shoulder. 'We're never supposed to lose real-time!' There were shrugs all round.

When she looked back her planes were once again frozen within their cubic 3D display.

'Now we've lost positioning handshake from the satellites,' she shouted.

Then Sam Potter, the shift controller was beside her. 'You're counting?'

She nodded, also calculating how long she had before she had to declare an emergency and ask for assistance from SFO.

Potter made the decision for her and picked up the desk phone. He prodded angrily at the instrument. 'It's dead,' he exclaimed. 'What the fuck's going on around here!'

He fished his VideoMate from his pocket and flipped it open.

'There's no radio signal!' he shouted. 'That's impossible. Everything's down. But they're all separate systems!'

Sandy's display refreshed.

'Jesus,' breathed Potter. 'Tell American 114 to turn–'

Sandy was doing it before he completed his sentence. She told the pilot to override his on-board flight control computer, issued instructions that would allow him to turn the passenger jet ninety degrees to the right and simultaneously told him to climb a thousand feet. She cleared two more distant planes from the stack and instructed the bewildered American crew to complete their turn, climb another twelve thousand feet and rejoin the stack forty miles to the east. 'We're suffering multiple systems failure here,' she explained. 'Please alert SFO.'

There was silence in the room. Sandy exhaled slowly. As she sat back, the phone beside her started to ring.

Thomas Tye rose from a couch to greet the British Prime Minister, his Health Secretary and three senior civil servants. The tycoon had changed into another fresh, open-necked white shirt and he had tied his hair back. The favoured TV crews and photographers got the group to repeat their handshakes over and over again and then they were quickly bundled from the hotel suite.

Tye walked his guests to the picture window in the penthouse of the London Hilton and they stood admiring the sun-drenched view down into the private gardens of Buckingham Palace. What a row there had been a lifetime ago when the Queen had first discovered she was overlooked by hotel guests. Now it was no longer the main royal residence and the old, old, widowed Queen walked among the ornamental ponds and privet hedges only on the rare occasions when public duties recalled her to London.

With an exchange of nods, the small party turned and crossed to sit on the sofas that filled a sunken seating area. All the furniture in the room was new, the plastic covers removed only after Tye had arrived and given his permission the previous evening. The suite had been redecorated and new carpets had been laid by the Tye Corporation's Advance Preparation Unit. It was a procedure Tye insisted on wherever he travelled.

What was not revealed to the host locations was that the APU was part of Jack's division and the most important part of their 'redecoration' procedures was to investigate all wall, ceiling and floor cavities, to replace all communications systems with the Tye Corporation's own highly encrypted systems, to install electronic 'white noise' barriers and to maintain a complete electronic anti-bugging sweep of all areas the corporation's president would be occupying. Modern business techniques and ethics dictated such precautions for all off-site meetings.

Two stewards held out hot antiseptic towels for the party.

Jack Hendriksen beckoned to Pierre and said he would take the floor duty. The Frenchman nodded and he and Stella Witherspoon, the deputy PPT leader, took up positions at the two exits to the suite. Jack stepped out into the corridor and closed the doors quietly behind him.

Two uniformed British policemen, armed and with full body armour, were outside. They scanned him and received his ident. A small nod from the taller served as acknowledgement. Jack moved along the corridor to ensure that all approaches were covered. The Hilton's management had insisted its own video surveillance was adequate and had refused the Tye Corporation's request to install its own system in public corridors. Despite this, the hotel's prime location, the lure of the penthouse view and its management's willingness to allow redecoration of three suites had been sufficient for Tye to overrule the PPT's objections. They had taken the entire floor, of course, but, assuming Pierre's role for the day, Jack wanted to double-check that the elevators were locked off and that the roof and stairwell escapes were covered. He recognized the tall man in a dark navy suit standing opposite the exit to the four elevators.

'Hi, Nigel. Good to see you.'

He had met the Prime Minister's senior protection officer twice before. They swapped greetings and the Brit confirmed that his team had locked off the elevators the moment the press had left.

With a wave, Jack continued his journey around the circular corridor. As he passed the open doors to another suite he saw Connie Law, Tye's personal assistant, and her staff busily confirming the final details of Tye's short European tour. Tomorrow it was Brussels to see the Commissioner and then on to Berlin for Tye to give another performance; Paris on Friday for lunch with the President of the Republic, and then, unannounced and, it was to be hoped, unreported, one of the two supersonic Tye-Lear corporate jets ferrying the Tye Corporation presidential entourage around the world would change its flight plan in mid-air to bring Tye and a smaller retinue back to an airfield in Cambridgeshire. Tye would then spend two private days visiting his investments in the many biotechnology companies that had sprung up in the science parks of the area.

Jack watched Connie work. Her highlighted blonde hair was cut short and a pair of gold-rimmed personal viewpers were suspended around her neck on a thin gold chain. He regarded her as the most efficient and unflappable human he had ever met and he felt his interest growing as he admired her long, elegant neck. Suddenly he found himself thinking about the vacation he was due to start in six days. He was looking forward to getting back to the apartment in Manhattan he so rarely saw. He often thought about selling it but each time he remembered how important it was to keep somewhere that was your own, something that had no connection with the Tye Corporation. *A place that still felt clean*, he realized, *untainted by this business.*

He pushed away the weariness and a vague, unseated sense of disgust. It was just a reaction to mental fatigue and he had been well trained to cope. He had enjoyed the demands his twelve-year career in the US military and government intelligence services had made on him and he was grateful for the resources it had forced him to discover in himself. When he had been finally discharged, after he had abandoned two previous attempts to quit just as new crises had erupted around the world, his section head had described his service as 'outstanding and distinguished' in front of the few men and women who were permitted to witness the small ceremony in Washington.

Jack hadn't been able to talk much about those years. During most of his time with the US Navy and, subsequently, various loosely associated government agencies, he had been involved with intelligence activities. He had simply given his commendations and medals to his mother and watched her smile broaden as he asked her to mothball his ceremonial naval uniforms.

'Is it for good this time, Johnny?' his mother had asked as she reached up to push back a short lock of blond hair that had fallen across his forehead. No matter how old or experienced he became, he would always be her first-born.

He had smiled and kissed her.

'It's for good, Mom,' he had assured her. 'Although I've no idea what I'm going to do now.'

Jack had seen a frown cross his mother's forehead. Although little mention was made of it, his father's premature death had left small provision for her later years and, for the last twelve years, her two sons had jointly shouldered the responsibility of providing her with a comfortable retirement.

She need not have worried. The Tye Corporation's director of corporate security services had made contact within forty-eight hours of Jack's discharge. Jack rejected the offer outright during the first voice call. He knew nothing about corporate security and he didn't want to learn.

Then his predecessor had called back and had told him a little more about the job. He had made it plain that it was Jack's Navy-assessed IQ rating of over 140 points that was his main attraction. He said Jack had been strongly recommended by Ron Deakin, his original SEAL intelligence instructor. Jack had applied to join the elite Marine corps just as it had been changing from a brutally physical fighting force to a 'smart' operations unit. Although the induction had been gruelling, the

Navy had finally abandoned the ultra-macho, close-to-death training methods that had been necessary to train covert-killing troops when combat was principally physical and at very close quarters. Just as Jack arrived, the SEALs were donning smart suits, adopting information-distance weapons and substituting robot incursors for men in as many operations as possible. When that was not possible, they were using non-American mercenaries, mostly would-be immigrants from Venezuela, for all physical combat, although neither the media nor the public were aware of the policy. Losing an *American* life in a conflict had become political suicide.

It had been Jack's systems skills and his analytical abilities that had prompted Instructor Deakin to lift him out of the corps for tactical intelligence training and, for the rest of his four-year career with the regiment, he had directed many of its active-service operations from a communications command post many miles away from the action zone. Then he had transferred to a US government intelligence agency.

'We're not looking for a tough guy, we're looking for a clever guy who can run a large team,' the persistent Tye Corporation recruiter had explained. He said that Thomas Tye operated an unwritten rule that nobody with an IQ of less than 135 would ever join his senior management executive.

Jack had been seduced by the idea that anyone was prepared to send one of the new generation of supersonic corporate jets to collect him for a job interview. Two weeks later he found himself on the small island in the West Indies that had become the Tye Corporation's world headquarters.

Thomas Tye had conducted the first interview personally. He probed Jack's intellect and played mind games with him. At first he asked him about everything but security: how he felt about the ecology of the planet, his attitude to money, women, genetics, music, politics – even his knowledge of communications systems and software. Jack did his best to keep up, wondering at the trillionaire's polymathy and high-speed acuity. He found himself caught up in Tye's infectious enthusiasm.

When Tye had, at last, asked about Jack's views on the job, he hadn't seemed surprised by the reply.

'If you want a physical bodyguard, I'm not your man,' the former intelligence officer had told the tycoon. 'It's been many years since I carried a gun and I don't intend to do so again. Not having a weapon provides a different perspective on a situation before, during and after any given event. I would leave your personal physical security and any weapons-handling to others in my team. I would consider that it would be my job to plan situations in such a way that they'd never need to use them.'

Tye had seemed to like what he had heard.

On his second interview, conducted via a four-person 'wraparound' videoconference, Jack had agreed to a six-month contract assignment after which both sides would review the situation. The salary offered was tax-free and was several times what he had been making in his final year in US government service. He was also told that if he decided to join permanently, stock options could make him seriously rich in a short space of time.

Three years later Jack had discovered they had been right about the stock options. Despite some analysts' concerns over how such a large company could sustain its phenomenal growth, the Tye Corporation's results had beaten even the most optimistic expectations and in the last two years the stock value had soared to stratospheric heights. His options had become extremely valuable and he had exercised many of them, investing the proceeds in a broad-based portfolio of equities, property and bonds.

But he hadn't realized how little he would like big business. He had discovered it was a world completely without honour or truth – or, at least, that was how Thomas Tye's empire operated. It made government intelligence activities seem almost ethical and they, at least, pursued ends that could sometimes justify the means employed.

But Jack knew he had done a good job. He had now replaced the man who had recruited him and had restructured the PPT of thirty to bring in additional foreign-language skills, better systems skills and a greater number of women. A modern Praetorian Guard, for that was finally the PPT's function, was unlikely ever to face violence if they pre-empted, protected and performed properly and, as he knew from personal experience, female insight was invaluable in preparing for the unpredictable.

Despite his frequently repeated exhortation that any need for physical engagement would always be seen as a sign of failure, Jack had constructed a four-mile training course on the Hope Island University campus. Through frequent use of this, combined with regular visits to the gym and the island's many pools, he ensured that he and his team stayed in peak physical condition.

He checked that the service lifts were indeed locked off and walked on around the corridor.

'Roof access secure?'

Pierre's man at the foot of the stairs nodded. 'We're not allowed to lock it, for safety reasons – it's a fire escape. But there's no one on the roof and a chopper can't get into this part of London air space – the Palace.' He gestured along one of the radial corridors, towards the panoramic view.

Jack nodded. He walked down the corridor towards the dead end of the picture window and leaned gently against it, taking in the view of Westminster, the Houses of Parliament and across the Thames to south London.

He felt the press of the package in his inside pocket and her face swam back into his mind. He removed the package, lifted the flap and took out a sheaf of papers. He scanned the first sheet. He had half expected it to be legalistic and it was. But the second wasn't, nor the third, nor any of the pages that followed. He turned the sheets slowly, reading carefully, totally absorbed.

Then Stella spoke in his ear. 'Party preparing to leave.'

He stuffed the papers back into the envelope, returned it to his pocket and walked back around the corridors. He arrived just as the double doors opened.

Tye walked the Prime Minister towards the elevator chatting about a recent soccer game. His cultural-variance advisers had briefed him on the politician's obsession and Tye had watched a few edited highlights in preparation for the meeting.

With handshakes briefer and more businesslike than those produced earlier for the cameras, the British politicians departed.

As the doors closed on them and their security staff, Tye walked towards his temporary administration centre, followed by his small coterie of senior executives.

He went into the bathroom just inside the entrance to the suite and dipped his hands in the basin of hot, antiseptic water that was waiting for him. He emerged, drying his hands on paper towels. He turned and threw them into a trash can in the bathroom.

'Connie, we're taking over Britain's National Health Service,' said the world's most eligible bachelor as he took two plastic bags from her and broke the seals.

He extracted a face mask from the first pack, hooked it over both ears and pulled it up over his nose. Then he pulled a pair of latex gloves from the second pack.

The radiance of his smile spread from under his mask. He looked like a dental surgeon who had made a particularly rewarding extraction of wisdom teeth.

'Get our press people to talk to the Department of Health,' he said. 'Nothing goes public until the PM announces it in Parliament. But I want us to be prepared!'

Haley Voss slobbed out most Sundays. She rose late, didn't shower and wore her old glasses to give her eyes a rest from contact lenses. She spread the quality newspapers all over the floor, sat amongst them with legs outstretched and read, a pair of scissors in her hand poised ready to clip any stories about Thomas Tye, the Tye Corporation or any associated topic. But that was the most work she allowed herself for the day. She didn't wake up her computers and she didn't log on to the networks. It was a rule she had made after Kevin had finally given up grumbling about her work obsessions and had left with his bag of dirty underwear and his only (unread) book, *How To Quit On-Line Gambling Today!* At least Barry, her new man, understood that she wanted her thinking time. She had quickly come to enjoy the frequent Sundays she spent alone in the privacy of her inner-London apartment.

She was both surprised and annoyed when, early in her afternoon off, the entry system to her flat chirruped unexpectedly. You didn't get unexpected callers in the inner city. She looked at her video screen and saw a tall, fair-haired man with spectacles.

'Yes?'

'I'm Jack Hendriksen. I work for the Tye Corporation. You gave me this on Wednesday. At Earl's Court.' He waved an envelope towards the camera.

'Just a minute.'

*Christ! Here in person?* She'd hoped for a reaction, but she realized that she hadn't prepared for this. Look at the state of the place. Look at the state of her!

'Can you come back in ten minutes?' she asked quickly.

The man nodded, waved the envelope at the camera again and moved off. She ran to her second-floor window and watched as he crossed the street and disappeared into the park. Where was his car? She wondered if he was a Tye Corporation lawyer, but this was a Sunday, she told herself.

She ran through the flat, waking her computer up as she passed. She had no time to shower or wash her hair, but she pulled on her jeans and a clean sweatshirt. She took her make-up into the living room and sat in front of her computer screens.

'Call Flick' she said as she removed the traces of last night's mascara and shook her lens bath. *Be in, just be there*, she willed her sister.

She heard the ring, then Felicity said, 'Haley. Hold on.'

Felicity's face appeared on the central screen and then a tiny finger stretched towards the central camera lens.

'Hi, Flick. Hi, Toby. Can you keep this line open for a while?'

Haley explained what she wanted and then turned her screens off. She was applying lipstick in her hand mirror when she saw the reflection of her cuttings on the wall behind her.

She ran to the bedroom and pulled a folded sheet from a drawer. She was applying the last piece of sticky tape to the wall when the entry system sounded again.

'You're alone?' she asked.

'Yes.'

'Second floor.'

She buzzed him in.

Her visitor was at the door before she'd finished scooping the newspapers from the floor. She thrust them under the sofa, checked the door monitor to confirm he was alone and undid her security locks.

'Miss Voss? I'm Jack Hendriksen.' He gave his name again as Haley opened the door. Beige lightweight jacket, open-necked white shirt. Understated but expensive. A well-honed, honest face. Wedding ring. Tall. She realized he wore his spectacles for high-level communications, not short-sightedness, and he didn't bother with any of the fashionable unisex ear jewellery to disguise his clear plastic earpiece. Haley felt the briefest dry pressure as they shook hands.

'Come in.'

She seemed smaller than she had in the crowd but her face was even more open, even more humorous and attractive than he recalled. He guessed she was in her early thirties. The apartment was bright, well furnished and suggested a greater degree of affluence than he had expected – air-conditioned, he was pleased to note. One wall was nothing but books. The door slammed behind him.

'Oh, here.' He held out his ident.

'Vice-President, Corporate Security, the Tye Corporation,' she read out loud. She picked up her VideoMate, copied the ident and received confirmation. 'What does that make you – a private policeman or chief bouncer for the Touchers?'

'A little of both, I guess.'

His blue eyes were straight and steady as he stared down at her. He seemed calm and she detected no tension in his face or in his body language. She felt a sudden pull inside her, low down, that she hadn't felt for a long time.

'So they've sent you to sort me out in person, have they?' she asked, handing the ident back.

He smiled, tiny laughter lines appearing by his eyes. He produced her envelope from inside his jacket.

'You should be careful what you do with stuff like this. You're making some pretty wild allegations.'

'Am I?' said Haley, her jaw jutting sufficiently for Jack to see a hint of real determination. 'All I'm asking is for Thomas Tye to talk to me about these things. My biography will be published with or without his input, or consent. He might as well have his version of the truth included.'

'Mr Tye doesn't give any interviews, as you must know. And, according to this first page, the corporation has injunctions granted against you in eleven countries.'

'Injunctions are only temporary, Mr Hendriksen. Any lawyer will tell you that. Once my publishers submit my manuscript to the courts they will be lifted.'

'Not if it's libellous, they won't.'

'A libel is only a libel if it is untrue,' said Haley.

'So you can prove all this?' He waved the papers.

'Buy a copy of the book. I'll sign it for you.' Her voice cracked and Jack could sense the agitation behind her apparent composure.

'You must know that your book will never be published,' he reasoned, walking over to the window and looking out across the park. 'The Tye Corporation can throw so much money into litigation your publishers will simply run for cover. They'll realize that even if they won in court, the legal costs would be more than they could ever make from your book, even if it was a best-seller.'

'So you're here to scare me off.' Her voice was becoming hoarse.

No, no, this isn't the way he wants it to go. Start again. 'Would you mind if I sat down?' he asked.

She waved a hand towards the sofa.

'I've done my homework,' he smiled as he shrugged his jacket off and eased his tall frame down into the cushions. Haley remained standing, arms folded; a slight but determined figure. 'I know you're highly respected as a biographer. That film star's biography, Josh Chandler, of course: it's famous – you've made a lot of TV appearances about him. And some of the reviews I've found ... your *Book of the Presidents*, was that the last? ... I saw it got some great notices. That's one of the reasons I'm here.'

'Go on.'

He hesitated, then plunged. 'Look, I'm not really here on

company business. I haven't been sent. I'm not even authorized to speak to the media and ...' he hesitated '... I'm the only person who's seen your questions so far.'

Haley looked down at him. In her mind she knew he was likely to be lying, but instinct told her he wasn't.

'I don't believe you. One way or another you're here to try and stop me telling the world where the Tye Corporation is heading.'

'I'm not. No one knows I'm here.' He had an idea. 'I'll prove it. May I use your system?' He pointed at her computer on the desk.

*Panic. Don't show it.* 'No, my system's down,' she lied. 'Tye software, I expect.'

He smiled, watchfully.

'I've got my VideoMate.' Haley touched the unit at her belt.

'It's OK, I'll use mine,' said Jack. He pulled his Tye communicator from its belt clip. Then he remembered that he had deliberately switched it off to disable its location function.

'Sorry, you're not the only one with system problems,' he smiled ruefully as he removed its memory card and returned the small device to its holder.

'Here,' offered Haley. 'Mine's a Sony.'

He flipped the small mobile device open, inserted his card and touched the screen. He searched through the recording.

'Look,' he prompted, handing her the VideoMate.

She watched the small screen. 'That looks like Tye,' she exclaimed. 'When was this–' She stopped mid-sentence as she saw that the screen display showed the recording had been made at eight that morning.

'Enjoy your day,' Tye was saying to the camera. 'Go to an art gallery or something.' She heard Jack's voice respond as Tye sprayed his mouth, pulled on a face mask and walked out of the hotel lobby to the waiting limousine. Tye turned back to the camera with a laugh. 'Take a punt out on the river!'

Then he was gone: into the car with another man and a woman and off through the city streets.

She stopped the replay, searched backwards and then zoomed in on Tye's face.

'That *is* Tye, isn't it?' said Haley. 'Why's he wearing a mask? I thought that was just another of the Thomas Tye jokes?'

'Well, he does wear them in private,' confirmed Jack. 'He's afraid of germs.'

She frowned and replayed the clip from its start. 'Isn't that Cambridge? It looks like Regent Street.' She recalled it from her university days. 'I didn't know Tye was still over here.'

'We came back yesterday. TT – Tom – has some private visits to make.'

Haley put a finger to her chin in parody of a thought occurring to her. 'Let me guess: Moleculture plc, Bioneme Ltd, Erasmus Research plc and Genome Technologies.'

*Spot on*, thought Jack. 'Tom has a lot of investments around Cambridge. But, as you saw, this is my day off. I'm not here for the Tye Corporation.'

'And you just happened to make that video to convince me of your sincerity, didn't you?' objected Haley, intrigued at the lengths her visitor was prepared to go to in order to convince her.

'I normally record everything,' said Jack simply. He took his spectacles off and touched the frame. Haley saw minute lenses in the corners. He touched the lapel of his jacket where it lay on the arm of the sofa and she saw others in the buttonholes. 'Don't you video meetings and so on? Just for legal safety, and security?'

She almost shot a look towards her desk. 'OK, OK, I'll play along for a while. You're here for your own reasons. So what are they?'

*I'm here because I'm every bit as worried about Tye and his company as you are, because your face has been in my mind since Wednesday – I even printed out a still – because you put your address on the first page and that was an irresistible temptation. Because I've been feeling lost these last few years. But I think I'm getting over it, maybe.*

Jack banished his emotions to his subconscious and picked up the sheaf of papers from the sofa cushion. 'Have you got proof of any of this?'

Haley hesitated. She looked into his pale blue eyes. His gaze seemed even and sincere. Should she trust her instincts again? They had let her down before.

'I've got something for you to see,' she said as she uncrossed her legs and rose from the sofa. She walked to her desk and pulled a thick wedge of paper from the centre drawer.

The vast greenhouses covered thirty-six acres of drained fenland and were triple-glazed throughout. Electronically operated foil blinds between the glass panes allowed the capture and escape of both light and heat to be accurately controlled. The vast area of glass was coated with translucent high-efficiency solar-energy cells made by Tye Solar Energy Inc. Inside, the air-conditioning system used the captured energy to hold the temperature at a steady 14°C for the sixteen-hour artificial 'nights' and 6° for the eight-hour artificial 'days'.

In a warm twilight, Thomas Tye and his small entourage were led to a bed of shoulder-high maize. Tye alone wore a face mask and latex gloves. He was invited to pick an ear of corn. He took the small knife proffered by his guide and carefully snipped an ear from its mother plant. He pulled back the green protective leaves. The corn was plump and bright yellow. He passed it to Connie for her inspection. Next they came to a miniature wheat field. Once again, Tye examined the maturity of the ears and nodded his appreciation.

He was shown red Desiree potatoes freshly lifted from another bed, green beans from still another. He was then invited to inspect tanks of rice and soya beans, all plump and fully mature and, finally, a young arabica coffee bush already hung with maturing red beans. The party then crossed through a double-door airtight enclosure into a higher-ceilinged greenhouse that contained six beds of pine saplings. Small signs at knee height announced them as *pinus strobus*, *pinus palustris* and *pinus nigra*.

Tye nodded his approval again and followed his host from the greenhouse and along an enclosed corridor. The rest of his entourage and a group of accompanying researchers followed a few steps behind. At the end of the corridor the party stepped out of the rubber boots that had been provided and slipped their own shoes back on. They stepped into a corridor and Connie held open a bathroom door for her boss.

She waited a few minutes until he emerged without his mask and gloves and she led him into a large, brightly lit conference room.

The visitors were shown to their places at an oval conference table and Tye walked round to a chair that was still in its protective wrapping. Connie slit the tough plastic around the arms and seat of the chair and Tye sat down. The APU had visited the room the day before and Jack's technical team had been able to confirm its secure status to Pierre, the security manager for the day.

Coffee, tea and soft drinks were served and all but Tye took some refreshment. Professor Sir Oliver Morton, the distinguished Cambridge University geneticist and co-founder of the company, stood at the head of the table, his thin, ascetic frame clad in a formal brown three-piece suit. He cleared his throat and fiddled with his cuffs.

'We are delighted to have our chairman with us at Moleculture today,' he beamed at the all-important visitor sitting to his right, at the rest of the visitors and at his own research team. 'As you have seen, Tom, we have now successfully incorporated the genes of several semi-tropical C4 plants, mostly from the maize families, into the genomes of most of the C3 plants, the main cash crops of the northern latitudes. This means that they are ready to grow at almost any point in the twenty-four-hour cycle, when there is light. They must have rest periods, of course, but they now exhibit positive thermotropism – that is, they grow throughout the day *and* at very low light levels, typically those found at dawn and dusk.'

Thomas Tye nodded and then started a small round of applause in which the rest of the party quickly joined.

The biochemist bobbed his long, thin head appreciatively. 'We have also managed to increase the crops' conversion rate of solar energy from its usual one per cent to five point three per cent.'

Tye tapped his appreciation on the table top. Sir Oliver's smile became broader.

'And, in our main tree genuses, the pines of the northern latitudes, we have been able to increase their $CO_2$ uptake by sixty per cent. Wherever they are planted, they will become a twenty-four-hour carbon dioxide sink!'

Thomas Tye banged the table hard in appreciation and the entire party clapped.

'One for our planet,' Tye commented quietly to Connie, but not so quietly that he couldn't be heard.

The knight's grin became Cheshire.

'As a next step we are hoping to get permission to plant four acres of each crop in the Scottish Highlands so that we can measure how robust each of the new strains is under natural low-light and cool-temperature conditions. We would then be able to carry out the necessary transgenic tests to ensure there is no unintended cross-pollination, and to test ecological impact on insect life and the food chain.'

'How long will that take, Oliver?' asked Tom amiably.

Morton blinked and swallowed, aware that their VIP guest and main source of funding had just asked the most delicate question of all.

'Well, that depends on the government's Genetically Modified Organisms Committee,' explained the professor quickly, looking around the table in canvass of his colleagues' support. 'We were about to file our applications for permission when the latest GM food scandals in France and Greece were uncovered. I'm afraid things have become rather difficult at present.'

'So how long?' persisted Tye, still smiling.

Morton cleared his throat again. 'Well, I can't see us getting permission to move into open fields in the current climate. At least, not for a year or two. Perhaps after the next election we–'

There was a cough from the other end of the table. 'Of course, we have all we need to prove our sequences and secure the patents,' broke in Dr Frederich Zimmer quickly, his voice rasping and intrusive. Along with Morton, the German-born biochemist and co-founder had been the company's initial source of funds. He had pumped nearly all his wife's inheritance into the company, before they had run out of cash and Tye had moved in to save them.

Zimmer rose and walked to the front, his sharp charcoal-grey suit a contrast to Tye's casual white shirt and chinos. He turned to face the man who had now become the company's principal investor and largest shareholder.

'The patents, they're the main goal, aren't they?' he suggested. 'To make low-light and pseudo-nocturnal crops ours.'

Tye nodded, still relaxed, his left arm slung over the back of his chair. 'But what if someone else is getting to open-field trials with a similar idea, Fred?' he asked as he looked up at the squat, florid German. 'Remember, American patent law has now changed to come into line with the rest of the world. It isn't first-to-invent any more; the World Patent Organization will only approve on a first-to-file basis and we can't make a filing until we've done our field tests. We may have to adjust the gene sequence to suit natural growing conditions. First-to-test and first-to-file wins under unified patent law.'

Zimmer waved a hand dismissively. 'One, I doubt that anyone else is working in this particular part of the plant genome and two, where could they be conducting trials? They'd have the same problem as us.'

Tye worked to control his notoriously short temper with this arrogant German geneticist. He had become marginally better at masking it in public in recent years.

He pushed his chair back and stood up. 'One, I know better than you, Zimmer, who else is working on low-light plant crops. Four other companies are in preparation for field trials of crops intended to be grown in more extreme latitudes. And, two . . .' He trailed off, aware that he was letting go – and that he was saying too much. He took some short, deep breaths, then tried another tack. He walked to the head of the table and put his arm around the geneticist's shoulders. At five feet, nine inches he was only an inch taller than the German.

'Fred, Moleculture has only recently joined the group. We acquired you in . . .'

'Last February,' supplied Zimmer.

'Well, we do things a little differently in the Tye Corporation and, well, I guess that's *why* we're the Tye Corporation.' Tye smiled and turned to Connie, who was making notes in her DigiPad. Like every other aspect of corporate activity within the company, all Thomas Tye's waking moments were videoed. But the recordings of this meeting, like the rest of his life, would not be filed in the corporate database of intellectual property. He was the only exception to company policy for the simple reason that plausible deniability was a vital option for him.

'How much of the world's population is covered by the UN Resolution on GM crop testing?'

Zimmer shrugged. Most countries had signed the agreement to control trials of new genetically modified foods strictly. 'Ninety per cent or so, I would think, Tom.'

'It's less than sixty per cent on a per capita basis, Fred. China, Angola, Zimbabwe, Russia, Sierra Leone, Ethiopia, Kazakhstan and even Pakistan – none of them have ratified the agreement. They see it as yet another power play by the West. There are governments with millions of square miles of land who will be ready and eager to help with your field trials. And *they*'ve got the populations who most need your creations, your miracle crops, Fred. We can patch your data into the Halcyon climate-modelling system and the Phoebus Project can supply the energy. You will be able to feed the world. You'll all get an invitation to Stockholm – for a Nobel Prize!'

Zimmer was a sufficiently good judge of character simply to nod his agreement. He was also crucially aware that the final (and largest) part of the payment for his Moleculture shares – and for the shares that had been owned by his wife and the others on the team – wasn't due for another two years.

'You just start mass-producing the seeds and leave the government permissions to me,' said Tye. 'I'd like to see this being tested on a small scale within two months, with very large-scale plantings for this autumn. I will be able to give you as much land as you want. We can start with a planting area of four hundred thousand hectares.'

They gazed at him, open-mouthed.

'We couldn't even *administer* such large-scale production from Cambridge, or anywhere else in the UK or Europe,' objected Professor Morton at last. He had never even contemplated such industrial-scale planting. 'That would be a technical breach of EU and UN Food and Agriculture guidelines.'

'Well, I'm glad you brought that up, Oliver,' responded Tye, turning to face the scientist. 'I want you to move this facility to our campus on Hope Island. There's quite a sizeable science park and you'll find some interesting people there in your own field – people who are putting nocturnal genes into cows, pigs and sheep. Yes, farm animals that run around at night, feeding and growing! You'll love it!'

There was silence as everybody in the room stared at him, appalled.

He turned to face the team. 'You'll love it! You'll all love it!' he said, nodding vigorously, willing it so. He took a small aerosol from his pocket and sprayed the inside of his mouth. 'We'll expect you in two weeks, plants and all. Connie will get Logistics to take care of the details. Your families can follow later.'

'This is truly astonishing,' agreed Jack, as he finished the last page of the report Haley had given him. He was struggling to control outward expressions of his surprise at the lengths to which Tom had gone in his quest for youth and vigour. Like others in the inner circle he had suspected, but this seemed to be proof. 'Incredible.'

While he had been reading, Haley had first made coffee, then, later, sandwiches and tea. He put the report down and turned to where she was sitting cross-legged, facing him at the other end of the long sofa.

'And you don't know where this came from?'

She shook her head as she reached down and picked up her coffee mug from a low table. 'It just arrived in the post – snail mail, from Amsterdam. There was no indication of a sender.'

'I know of one of the authors named on the cover,' revealed Jack. 'He's a geneticist – on the island. I might be able to check whether he really did write this.'

'I'm told it's scientifically accurate,' affirmed Haley. 'My boyfriend's a genetics researcher.'

*Why did those words suddenly feel so bad?* She looked up and found his steady gaze on her. She pressed on, quickly. 'It's written by people who know what they are talking about.'

'I don't doubt it,' mused Jack, wondering if he should share any of the concerns that had been building up inside him; the concerns that were one of the reasons he had sought the biographer out. 'This would certainly explain the way he looks and behaves.'

He paused, still cautious, still unable to break the habit of secrecy that had been drummed into him throughout his adult life. Everything he had discovered during his research about the British biographer led him to believe she was a woman of integrity. He looked up, conscious of her dark eyes on him – eyes that were startlingly direct and, he realized, disturbing. It was really those eyes that had brought him here today.

'It's true to say that TT spends a lot of his time with our drug companies and the researchers,' he continued, measuring his comments carefully.

'He's the world's *biggest* private investor in biotechnology,' exclaimed Haley. 'Look, I've created a map.'

She stood, walked across to her desk and climbed onto a chair. She peeled back a piece of sticky tape that was holding one corner of a large sheet to the wall. Jack walked to the other side of the desk and, at her nod, peeled back the other corner almost without stretching. She smiled as she jumped down from her chair. Jack stepped back to view what was revealed.

The wall was covered with photos, clippings and notes on Thomas Tye, the Tye Corporation, its subsidiaries and Tye's private investments. In the centre was a web of interconnected map pins showing the links between the Tye Corporation, Tye himself and many private companies and organizations. A large map of Hope Island was pinned in the centre.

Jack stood in front of her wall and shook his head in wonder. 'You're still working with ordinary paper?'

Haley smiled. 'No, I research on the networks like everybody else. I also read some types of book on digital paper but there is something special about seeing it all laid out together in this way. It's to do with the psychophysics of the cognitive process.'

Jack shot her a look and she laughed, unable to tell whether he was teasing her. She liked the hint of laughter that often seemed to hover around his eyes.

'You see connections you'd miss otherwise,' she explained with a smile.

He stepped forward to examine the material more carefully.

'This is cute,' he teased as he flicked an old pin-up shot of Thomas Tye that had been clipped from a women's magazine. 'Were you a fan in those days?'

Haley couldn't prevent embarrassment painting her cheeks. In earlier years she had even participated in auctions of Tye memorabilia: faded autographs and signed pictures from his more accessible days were scattered amongst her research. She justified it to herself as trying to *feel* the man about whom she was writing, a form of graptomancy.

Jack allowed his question to hang unanswered and he ran his finger along the cotton connections that tied development projects to the various Tye companies.

'There's a lot more aerospace stuff than this,' he cautioned. 'A hell of a lot more. And you're way behind on network development. Northern Russia is complete now.'

'Well, the legal issues have been taking a lot of my time.'

'I presume that's the idea,' said Jack.

# TWO

Hope Island was twelve miles long by eight miles wide but, ever since it was first named by British privateers in the seventeenth century, it had never delivered on its implied promise. Situated in the Windward Passage midway between Cuba and Haiti, this uninhabited Antillean island had held out the hope of being the ideal first landfall and staging post for voyages between the Old World and the Caribbean. But, as scores of successive landing parties were to discover, the promise of its verdant slopes and white beaches remained unfulfilled. Other than occasional rainfall, there was no fresh water.

Over three centuries the Spanish, French, British and Americans all fought desultory naval skirmishes over ownership of the formerly volcanic island. But, as it offered neither strategic advantage nor revictualling anchorage, none of the protagonists really had their hearts in winning those brief engagements. As a result, it remained loosely under the supervision of the regional French administration in Port-au-Prince until Haiti gained its independence in 1804. Then, following a display of complete indifference by its former owners, the new Haitian Republic inherited responsibility for the useless lump of rock.

The island remained under Haitian control until January 1962 when ownership was secretly exchanged for an astonishing and wholly unprecedented six million US dollars. Soviet intelligence had identified Hope Island as being an ideal covert site for two hardened underground missile silos. It was a mere five hundred miles from the Florida coastline and the rich and decadent American cities beyond. Such proximity would neutralize the dangerous advantage the Americans had gained with their long-range rocket delivery systems and would provide a powerful deterrent against the widely anticipated US invasion of Cuba. Fidel Castro was quickly funded to strike the deal with his neighbour and despot rival, the Haitian dictator 'Papa Doc' Duvalier.

Five months later, acting on information gathered by a CIA spy inside Cuba's Central Planning Board who insisted that Russian engineers had arrived in force, American U2 reconnaissance aircraft started to photograph the outlines of new building activity across Cuba. By September the planes' cameras had captured images of the construction sites of twenty-nine new missile silos including, by a remarkably lucky accidental overfly, two deep excavations and an airstrip on Hope Island. The new and highly secret Corona spy satellite was deployed to pass over the Western Caribbean every forty-one minutes. It immediately revealed that twenty-four medium-range missiles had already been deployed and further weapons were being delivered and readied at the rate of six a week. Within a fortnight the CIA analysts had finished their report and concluded that there would soon be sufficient megatonnage sited in the Antilles to obliterate all of America's East Coast.

The Joint Chiefs of Staff urged the US president to make an immediate, massive and unannounced pre-emptive knockout nuclear strike on this highly dangerous and untrustworthy commie neighbour. Only the year before the new president and the CIA had been humiliated when their covertly funded and organized 'arms-length' invasion of Cuba by 'exiles' had been trounced by well-trained Cuban troops led by Ché Guevara. Eleven hundred men had been captured at the aptly named Bay of Pigs and, amidst international derision, Fidel Castro had managed to extract a ransom of food and medicines worth fifty-three million dollars from the United States for their release. This seemed like an ideal opportunity to exact retribution.

President John F. Kennedy ignored such urgings to apocalyptic revenge but did issue an ultimatum to Nikita Khrushchev. He ordered US warships and submarines to mount a naval blockade across the Windward Passage to stop Soviet ships delivering further missiles to fill the silos on Hope Island and those being completed on mainland Cuba. For two days the world had hung on the brink of global thermonuclear war as twenty-six Soviet missile-laden freighters doggedly continued to plough their way through the world's oceans towards the Caribbean. Then, on 14 October 1962, after a secret reciprocal deal was agreed under which American Jupiter missiles in Turkey would also be withdrawn, the United States made a public commitment never to invade Castro's fledgling communist state and all Soviet ships *en route* to the West Indies received orders to return to their home bases.

Almost half a century later, Cuba's crumbling government made an unexpected profit on the Cold War purchase it had almost forgotten. A personal representative of the Tye Corporation's Bahamian lawyers visited Havana and, in a prearranged private audience with the minister of the interior, offered the government two hundred million dollars for the outright sale of the island. The minister and his aides could think of no reason why the giant corporation should want an uninhabited, anhydrous island on the edge of the world's worst hurricane corridor.

Before responding, the minister ordered a naval survey team to re-examine the island. Although the team included the best seismologists, mineralogists, geologists and petrochemical surveyors the nation's impoverished universities could muster, nothing was found that could warrant any further government interest in the island. The old silos had long since been sealed and nothing of any salvage value remained.

Accordingly, Cuba's Government of National Reconciliation reaffirmed the island's strategic importance to the Republic and offered the corporation a ninety-nine-year lease for six hundred million dollars. As part of the deal Cuba would remain the sovereign protector while retaining all mineral and petrochemical rights. The Tye Corporation's lawyers insisted on an outright sale with no residual rights, but increased their offer to four hundred million dollars. Their terms included Cuban recognition of Hope Island's sovereignty and the cession of cabotage rights over its airspace.

There was also to be a six-mile exclusion zone for Cuban-registered shipping, the aircraft of Empressa Cubana de Aviación and the Cuban air force. In response to the Cuban request for an explanation about the intended use of the territory in question, the lawyers said that the company's initial plans were for the creation of an exclusive resort island, but their clients naturally reserved the right to put their future territory to whatever peaceable purpose they chose.

With his inner cabinet, the minister debated whether they could bluff the company into a much higher bid. The Chief of the Armed Forces objected to the sale at any price, pointing out that Hope Island provided the nation's first line of eastern defence.

'Defence against whom?' sneered the education minister. 'The ganja warriors of the banana islands?'

The general's objections were noted but overruled.

The cabinet then discussed the economic feasibility of even such a rich organization as the Tye Corporation being able to supply all the necessities needed for a resort complex to a piece of Atlantic rock forty kilometres from the nearest source of fresh water.

Finally, reality prevailed. Since the collapse of world communism over a decade earlier and the consequential loss of billions of dollars a year in Soviet aid, Cuba's economy had remained in deep crisis. Overseas markets for cane sugar, citrus fruits and tobacco – the island's most important exports – were declining as other producers were able to invest capital to create new and more efficient production techniques. Despite its rich natural resources the country was starving and tremors of insurgency were once again being felt in the provinces.

It was agreed that the government Estate Office should respond by asking for eight hundred million dollars for an outright sale but, in the event of the Tye Corporation withdrawing, Havana should immediately accept the four hundred million dollars previously offered. The only absolute condition insisted on by the Cubans was that the island could never be used for military purposes and, other than the forces required for coastal protection, should never be used as a base for any form of weaponry or troops from any other nation. The Havana government also sought and received an undertaking that Hope Island's new owners would not offer refuge or transit access to any Cuban citizens who might land up on their shore.

The deal was finally done for five hundred and eighty million dollars. Despite requests for delay from his army of international lawyers, who pointed out the delicate negotiations that would be required for international recognition of the world's first corporate sovereignty to be established since 1796, Thomas Tye landed on Hope Island to begin his personal and corporate eloignment from the United States a day after the cession documents were signed in Havana and the cash had been deposited according to the seller's instructions.

Throughout the second day and the days that followed, the Tye corporate flight of four Tye-Westland LoadShifter helicopters ferried staff, equipment and portable accommodation units between the island and Cristoba, a small private airfield to the west of the Dominican capital, Santo Domingo.

By the end of the fourth day the engineers announced that a total of 10,000 gallons of pure fresh water at 23°C was springing from their forty-two new boreholes each minute. Nobody was surprised. Infrared images of the island taken by the Tye Corporation's Argus Satellite Network had clearly shown three large freshwater springs emerging in the coastal seabed around the island. A careful analysis of flow rates and direction coupled with computer-enhanced satellite geodesy had enabled a small, but very expensive, hydrology consultancy from Mobile, Alabama to land secretly six months earlier and drill two carefully positioned deep test-bores through the island's reheated sulphided limestone. The engineers then took samples, plugged the boreholes and capped them invisibly before confirming the results to the Tye Corporation. Thomas Tye was discovering the pleasure of betting on certainties.

Thirteen years later the entire world had come to know of Hope Island. It had become the world's first corporately owned nation state since the East India Company had reluctantly handed over its subcontinent to a young but implacable British monarch in 1858. After Cuba recognized the state's sovereignty as part of the purchase deal, the Tye Corporation privately agreed a massive technology transfer with the government of the People's Republic of China to secure recognition by twenty-one nations within the Sino sphere of influence.

Following two years of similar tactics in the rest of the world, the new sovereignty had received grudging, enthusiastic or coerced diplomatic recognition from all major countries and was even expected to seek a voice within the United Nations. The Tye Corporation, which also acted as Hope Island's government, said it would not be seeking diplomatic status for its international staff nor for its four hundred regional offices around the world and that it deemed consular presence and exchange of diplomats unnecessary.

Despite eschewing a formal diplomatic role, the new state worked hard to present a neighbourly face in the West Indies, a task that was eased immensely by local investments and technology gifts to schools and hospitals on the neighbouring islands. The cost of such gestures for the Tye Corporation was more than offset by the new state's generous tax regime and the income from its offshore virtual casinos, network auctions, purchasing-aggregation syndicates and tax-free global retailing. Most lucrative of all was the new Tye Global Bank for Personal and Commercial Finance that attracted three million new depositors in its first month of operation. The world knew and trusted the Tye brand and the opportunity to make discreet deposits in a safe tax-free zone proved irresistible. And it wasn't just cash. Intellectual capital, the new core asset of the information economy, found a natural and secure home in the data haven offered by Hope Island. Without making any announcements, the island suddenly started to become the Switzerland of the virtual economy. These initial businesses were established from temporary accommodation during the corporation's first few months of occupation and served customers in 146 countries and operated in over forty languages.

After a two-year construction effort had built the main campus, the Tye Corporation had moved its world headquarters to the island. Work had continued unabated and now the state boasted a semi-permanent population of 14,500 residents whose transportation needs were served by a sophisticated network of super-fast underground maglev shuttles. Above ground, a floating concrete spaceport capable of accommodating orbital launches, wide-body jets, supersonic corporate aircraft and presidential jet fleets extended the island by six miles and provided access to the outside world – and beyond.

An unexpected bonus for the corporation was the discovery that the fresh water that welled up from deep within the volcanic rocks was so plentiful and pure that they were able to sell a concession to a Florida-based company to collect, transport and bottle 'Hope Island Natural Spring Water,' a brand that was later to attract such cachet that it outpriced many table wines.

Jack Hendriksen had left Haley's London apartment in time to catch an early evening train from London to Cambridge. On the way he drafted a letter of resignation on his reactivated VideoMate. He didn't know whether he really intended to submit it. He just knew that he had an increasing sense of unease about the way Tye conducted his affairs. Perhaps all big business was like this, but on two recent occasions he had been present when Tye, or one his lawyers, had not hesitated to destroy the businesses – perhaps lives – of individuals who stood in their way. He had seen companies and families wrecked as Tye had masterminded hostile takeovers or used his uncanny business intelligence network to pre-empt or disable even the smallest and weakest competitors. He was also alarmed by the momentum for territorial acquisition building up inside the corporation and by the aims of its founder and president. Some of the new projects were breathtaking in their audacity.

Like every other employee of the Tye Corporation, Jack had signed a binding non-disclosure agreement when he had joined the company. When he had been made a corporate vice-president he had also signed several additional heavyweight long-term confidentiality agreements – which included powerful media gags – and he understood better than any outsider how ferociously the Tye Corporation's in-house army of attorneys pursued and protected the corporation's many rights across the globe. But some things had to outweigh mere legal agreements.

At first he had thought about simply calling up some of his former colleagues inside the US intelligence community. But he had quickly thought better of it. He knew how friendly Thomas Tye was with President William Wilkinson and he understood that nobody still in the Service would want to champion any cause likely to be unpopular in the White House. He came to the realization that if he wanted to do anything about his worries he would need to speak to an independent attorney, and do so quickly, not least to protect his own position and his future stock options. Only then could he decide whether his concerns were really justified and, if he could protect himself from any legal pursuit by the Tye Corporation lawyers, decide how best to approach his Washington contacts. It would have to be an attorney who understood international law and who wouldn't be frightened by the bizarre nature of the disclosures he would make. It would also have to be a professional who wouldn't be cowed by the immense financial muscle the Tye Corporation could apply to a lawsuit. He had smiled to himself as he realized the irony. It had been thanks to the Tye Corporation's share performance and the tax-free status of Hope Island that he could even contemplate seeking such expensive counsel.

He had arrived back in Cambridge just before seven p.m., in time to check arrangements for the group's departure. As Tye's return visit to the UK had gone unnoticed by the media, Pierre Pasquier's watchful team estimated that the risk of assassination, kidnapping, assault, journalistic intrusion or trouble from the Touchers had dropped to Status One, the lowest level. Jack was, therefore, able to allow his mind to wander in a way not normally possible when he was on the road with Tye.

Tye Flight One with its eighteen passengers had taken off to the east from the remote Cambridgeshire airfield just after ten p.m. GMT and had turned right to execute its sonic boom over the English Channel. The moment it was in the air Jack had been able to switch off fully and, once again, go back over the information Haley Voss had provided and see how it connected with what he already knew. In particular, he wanted to weigh up the likely authenticity of the report he had been shown. Even though he could find an excuse to approach the report's main author for verification, he knew that even raising the question would trigger an alarm.

He also wanted to consider the things he had not told Haley and analyse whether they should be made public or whether there was another way, as he now realized he saw it, of blowing the whistle on Tye's plans. This wasn't an attractive phrase, but it described an unattractive activity and it suited his mood perfectly. No one man, or no one company, should gain the sort of power Tye was now contemplating. Jack scowled as he pondered the options. This evening he wouldn't exercise his privilege of joining the crew on the flight deck. His pilot's log was already crammed with hundreds of recent hours in the air and he could do without the flight-deck banter.

The pressurized cabin air was laced with an odourless antiseptic and in the forward Presidential Lounge the world's greatest-ever tycoon was feeling comfortable once again in the mask and gloves he liked to wear whenever possible. Jack could hear him conducting a flurry of video meetings – laughing, shouting, snarling, abusing and, occasionally, praising – before a brief silence reigned as the meal was served.

Jack relaxed sufficiently to savour a dish the chef called a *piperada* – a Basque omelette with tomatoes, peppers and tofu-ham with fries and salad served on the side. Only genetically modified and organically grown vegetarian produce and beverages from Hope Island Farms were served on the Tye flight of aircraft and the jet had made a transatlantic round trip earlier in the day to pick up fresh supplies.

It had become part of accepted Tye culture that no matter where he was in the world, Tom would eat only fresh, medicinally active produce from his island: vegetables, beans and pulses grown in natural conditions to provide meat-level nutrition but with cholesterols, fats, sugars and other undesirable calorie carriers modified to block biological absorption. Active pharmaceutical and agriceutical ingredients were added to the protein to aid the human body's struggle against heart disease, cancers and other potential ills. It was rumoured that some years earlier Tye had upturned a plate of South African soya-bean *bobotie* over the head of an in-flight chef who had been foolhardy enough to serve alternatively sourced ingredients, confident that his demanding boss would be unable to tell the difference within the hot curry spices.

When the meal was over Jack checked in with Pierre, who had returned to Hope Island on the provisioning run. His Director of Presidential Protection confirmed that everything was ready for Tye's return and for the official state visit by the Russian leader the next day. After signing off, Jack yielded to habit and clicked through the dozen most important cameras on the island. In the dim lighting of the jet cabin, the sunlit images projected in front of his eyes seemed bright and slightly surreal. He flicked over to Locate Mode and Tye Network's private Global Positioning Satellite network gave him a map of the island and the position of every one of his team members on duty. Everything was in order, as he had suspected. From the standpoint of security and privacy, the acquisition of the island base had been masterful. It was virtually unapproachable by air or sea without detection and, once the corporation's president was on the island, Jack felt reasonably secure from outside incursion. Any problems would come from inside.

He smiled to himself as he realized he had automatically re-entered work mode. With no public exposure of Tye's entourage ahead, at least for this evening, Jack felt sufficiently mellow to order another half-bottle of red wine (with unrestrained alcohol) and put his seat into full recline. He had chosen to sit alone in the mid-cabin this evening, rather than join Pierre's team, 'the flying doctor' and the paramedics in the aft lounge where some serious poker would already be under way.

Despite his distance from the Presidential Lounge at the front of the aircraft he could hear his tireless master revving up again and reeling off a string of instructions and asides to Connie while he harangued his executives, business partners and contacts around the world.

Many things Haley Voss had written or said had completed or complemented scraps of information Jack had picked up on his travels with Tye. The boyish Croesus seemed to be moving into territory far beyond conventional corporate life or even political activity and Jack had frequently questioned his own sense of unease, arguing with himself that he was simply being old-fashioned. Perhaps his naval training and years of government service had made him too old and conventional at thirty-eight to accept the new lifestyles and the astonishing commercial opportunities that computers, satellites, networks, biotechnology and space exploitation promised. Perhaps the strangely conservative nature of the government's intelligence community had left him unprepared to face the new and increasingly bizarre moral and ethical issues of the early twenty-first century.

In the end he had decided that *wasn't* the reason he felt so uncomfortable with what he had learned. He recognized that his feelings stemmed from a simple conclusion: not only did Tye trample over people, he was a monumental hypocrite and some of the things he was doing were simply wrong; wrong in the most basic of human senses and, Jack felt rather than thought, wrong in the natural order of things and most definitely wrong for the future of the world's population. Thomas Tye was out of control, in all meanings of the phrase, and Jack realized that he was one of the very few people who had the right contacts to be able to do something about it.

Jack had called his mother in upstate New York, waved at Skipper the red setter and sent him into a barking frenzy and, when he had listened to his mother's news and sent his love, had allowed himself to doze briefly. After they had landed in Hope Island's early-evening sunshine, he had spent a couple of hours with Pierre and the HQ team going over plans for the next day's state visit before enjoying a nightcap and turning in for a proper sleep.

'It's not that they won't sell. They won't even talk to us. *That*'s our problem.'

Tye lashed out at the figure of the senior counsel for the Tye Corporation. His hand went clean through the head. At the other end of the Holo-Theater videoconference in Washington DC, Marsello Furtrado instinctively stepped back as Tye's fist evaporated around him. HVCs had made the videoconference experience 'just like being there' – as Tye Business Systems' marketing slogan promised – but none of the glossy ads that were showing around the world depicted participants striking each other.

'No, that's YOUR FUCKING PROBLEM,' screamed Tye in the darkened conference pit of his home office on Hope Island. He was surrounded by four men standing in a circle around him. All were wearing suits that had been treated and optimized for holographic representation. All were three-dimensional images of Tye executives many thousands of miles away. All were used to Tye's physical 'approach' to meetings and all knew he enjoyed it every bit as much when they were actually present.

Tye stepped towards Furtrado and electronically levelled his boyish face with his counsellor's.

'Do I have to personally call every student who's had a good idea? You're our senior counsel. You still need ME to help you acquire a six-person outfit?'

Furtrado nodded again and then altered the movement to a shake. He knew how important it was. For a month interceptions made by the Competitive Threat Analysis department proved that the little start-up in Sâo Paulo had microwave multiplexing software that could instantly double the capacity of the Tye Corporation's ageing fibre networks. Or anybody else's.

'I've offered them IP protection, non-disclosures, goodwill escrow deposits, the whole nine yards,' insisted the counsellor. 'The little shit just laughs and goes off air. I'm not even sure he's legally competent. He may still be a minor.'

Tye sighed. Furtrado had been with the Tye Corporation for almost three years and he was still thinking the same way he had when he had been a senior partner in a global law firm. Today's business demanded direct methods.

'Go down there with a bag of money, Marsello,' he snapped. 'Go find one of the senior technologists who *isn't* a major equity partner. Wave the money under his nose, go schmoozing. Hire him, promise him a long-term contract, debrief him and dump him. You know how to do it. We need a defection for cover.'

'I'm not sure that's a good idea. The boy wonder's father is in the government.'

Tye held up his hand. 'That's it, gentlemen.' He snapped the system off and remained standing alone in the glaze of the lights that had lit him for the holographic scanners. His first morning back here after the European trip had been frantic.

He was wondering if there was any way of creating a dummy audit trail of in-house developments that might show that Tye NetWare had been working on a similar approach to multiplexing when Connie entered to remind him that the time for the President's visit was approaching.

Haley sat with her literary agent, cappuccino stains drying in the empty cups on their pavement table. Since traffic had been banned during daylight hours, the streets of London's Soho had become fashionable promenades filled with tables, sun umbrellas and people: atmospheric warming had its benefits, many Londoners agreed. This morning, the elegantly suited agent had suggested going out for a coffee to sit in the sun amongst the tourists. Haley knew that the real reason for an outdoor meeting was that she wanted to smoke.

'To describe them as nervous would be an understatement,' sighed Rosemary Long. She and her client had been discussing how Haley's various publishers around the world had been reacting to the Tye Corporation's flurry of injunctions. Rosemary wasn't telling her author just how colourful some of her exchanges with the editors had been.

'I'm afraid Nautilus definitely wants out. They say you can keep the part of the advance already paid. That's unconditional, and all rights revert to you.'

'But the USA's my biggest market,' objected Haley. 'Can't we persuade them to change their mind?'

Rosemary sighed again, hoping her client would grasp the implications of the developments without her having to spell it out.

'It won't be easy,' she warned. 'Also, I think their Chairman knows Thomas Tye. Rumours are that he maintains a liver on Hope Island.'

'So it's the old pals' act,' groaned Haley. 'How about some other US publisher?'

Rosemary shook her head.

'The trade knows Nautilus bought the book. It was all over *Publishers Weekly*. They'll figure that there must be a good reason for Nautilus not going ahead.'

'But we can get the injunction lifted. It says I intend to publish libellous, defamatory untruths. I can prove what I'm saying.'

'We can only get the injunction lifted if Nautilus or some other publisher wishes to fund the appeal,' explained Rosemary patiently. 'That's the problem. You know how much litigation costs in America. Even if a publisher did win the first round, the Tye Corporation would only start again in a higher court. It would be never-ending.'

'So they're simply caving in to the big money,' cried the biographer.

Rosemary sighed for a third time. Her authors, while necessarily brilliant in their own fields, rarely understood the pure commercialism of publishing. There were always other books to fill gaps in publishing schedules.

'Let's have another coffee,' she said, catching the waiter's eye and indicating their empty cups.

She had good reason to be patient with her youthful-looking client. Haley was a best-selling author, although that description didn't imply the vast riches that other people often assumed. A revised edition of her first biography, a detailed and insightful portrait of the film star Josh Chandler, was still selling strongly nine years after the original had been published, although Haley was honest enough to admit that her special access to the actor owed more to luck than professional perseverance.

Her 'friendship' with Josh had started shortly after her roommate at Cambridge University had met the teenage star at a Hollywood party. Haley had already guessed that Abbeline's family was rather better connected in LA than she let on but when, a few weeks later, her fellow second-year student had felt compelled to make a late-night confession that she was dating one of the world's most desirable bachelors, Haley had also discovered the full details of her friend's family's involvement in the film business. Abbe's father had produced the film in which Josh had first found international stardom.

Josh's and Abbe's relationship had become very public and had lasted for nearly two years and, in that time, Haley had met the handsome star dozens of times. He had bought a London home to be near his girlfriend and Haley was often asked to make up a foursome with one or another of the star's actor cronies. Felicity had repeatedly urged her sister to 'bag a star' for herself, but Haley had just laughed the idea off. Over the years she had come to like the funny-silly movie actor who hid behind a tough-guy image and although she was well aware that he kept a large part of himself withdrawn from most social exchanges, she felt relaxed around him and accepted that his interest in her was genuine.

After a highly publicized split between Josh and Abbe, Haley did not see her film-star friend again for almost two years. She received occasional e-mails from him and it was clear from the little messages he wrote that he didn't want to lose touch. In the meantime, Haley had joined the business desk of one of the few serious London newspapers that had survived the public's migration to network news. It was a publication so well written and edited that it retained a large print circulation in addition to its more recent incarnation as a global network resource of international importance.

Then her editor had checked on the office rumour that Haley was 'a close personal friend of Josh Chandler'. She had put him straight on that, but there was the prospect of an exclusive interview, and she had agreed to take the assignment, even though she wasn't a show-business writer

Haley had gone to Claridge's to meet her old friend and although the interview had started out quite formally, Josh had soon dispatched his publicist on some lengthy but pointless mission and had told his assistant to take the Hollywood equivalent of a hike. Within minutes of their being alone he was his old self again and then, with the absolute agreement that they were off the record, he told her how much he missed Abbe and how the public discussion of his latest relationship and speculations about marriage – with the scion of a French fashion empire – was 'pissing me off'. Moments later he was being silly again and making her roar with laughter at his mimicry of other members of the cast of his most recent film. Haley genuinely liked him because he seemed to sense that his fame placed him in an impossible position and, despite the fact that a large part of his life no longer belonged to him, he was determined to have a genuine, unforced giggle as often as possible.

Haley wrote an absolutely truthful piece, far closer to the bone than she knew his publicists would like but, before showing it to anyone, she had sent a printout to Josh's hotel under a confidential seal. She wasn't sure whether her handwritten note would be sufficient to convey the envelope directly into the star's hands, but that was what she intended.

Her ploy worked and her VideoMate had trilled the same evening. He was laughing – 'Am I really that silly?' He put his publicist's camp voice on again, and laughed; then he was his macho co-star, complete with the mangled English. His mimicry was wicked. 'So Josh Chandler is not allowed to change any of this?' he asked pompously.

'Not if he's a good sport,' challenged Haley. She knew she had been revealing, but not unkind.

'Fine by me, Haley,' he had said. That simple statement was to lead to a new career.

She guided her piece through the editorial process personally – to avoid the arbitrary excisions of sub-editing – and the article ran on the front page of the weekend features section. It was illustrated with a new portrait by a junior member of the British Royal Family who was unashamed of using his title and access privileges to develop his photographic career. The story was syndicated worldwide and she won three major journalistic awards that year.

Soon afterwards she had been approached by the languid Rosemary Long, *belle littératrice*, doyen of London's female literary agents: a major publishing house wondered whether Haley could get access and approval to do the first 'fully authorized' biography of Josh Chandler. Haley had prevaricated: she doubted it. She wasn't sure she liked the idea of biography.

She recalled Carlyle's remark: 'A well-written *Life* is almost as rare as a well-spent one.' But, after giving the project some thought, she began to see the idea as a challenge. She made the call to his personal location and was surprised and delighted to be put through almost immediately.

She had told him what she wanted and she heard the groan.

'Not another one,' he complained. 'There's too many already: all rubbish.'

'Precisely,' said Haley. 'That's why there's room. I would aim to be definitive and to treat a film actor like a person, not an icon.'

He had agreed to ask 'his people' and, a week later, Haley had received an e-mail invitation to a private weekend party at Cliveden, a stately home twenty miles west of London. At the bottom Josh had added: 'This is a wrap party. I'm staying on for a week. We can get started if you like.'

Then Haley had talked to Rosemary again. The problem was going to be finding the time to do the research and write the book. She talked the issue over with her editor: he had agreed that she could take a nine-month writing sabbatical but in return had shrewdly bargained for, and obtained, first serial publication rights for his paper's weekend supplement at a fraction of their real market worth.

And so a friendship had been cemented and a best-seller created. The book was exceptionally well written – 'a surprisingly literary and revealing insight' enthused one reviewer – and had been studded with unpublished portraits and family photographs, many of them taken by former girlfriends: in addition to talking to Abbe, Haley had taken the trouble to find and interview all the star's other long(ish)-term loves. Haley Voss became a 'name', someone publishers were keen to listen to. Rosemary just wished her author would choose anyone but Thomas Tye as her next subject.

'I've got some news that will change their minds,' announced Haley, as Rosemary lit another cigarette. 'I've made contact with someone inside the company. Someone really close to Tye.'

Rosemary raised a well-maintained eyebrow.

'I can't say who it is, of course, but I think it will prove very useful.'

Rosemary looked up with a smile of thanks as the waiter delivered the fresh coffees. 'Haley, you're not hearing me. I think we have a real problem with this project.'

'But it's a fantastic story,' the author objected. 'Do you realize just how dominant Tye's investments in biotechnology are? If you list them it becomes really scary.'

Rosemary wondered how long she should humour her client before explaining that her planned book was completely unpublishable – in any form.

Haley mistook her agent's silence for interest and plunged on. 'One, I think now that he's proved it works he's going to file for a patent on the human-ageing gene sequence. Think about it. If scientists have learned how to transfer the modified genes to a human...'

Haley's voice tailed off into hoarseness. She was becoming aware that Rosemary's attention was polite rather than enthralled.

'He's patented four new treatments from the bloody muscular-dystrophy sequence,' she continued stubbornly. 'And the active genes covering IQ development, and he's got a new genetic therapy for curing colour blindness. He's got a whole range of patents on the human cosmetic genes – you know, eye colour, height and so on.' She paused, willing her agent to see the issues. 'He runs the world's largest human-organ farm. We can't have one man, or one company, controlling the future of the human race.'

Rosemary shook her head. 'But that's just conjecture, Haley. There's nothing illegal in what the Tye Corporation or Thomas Tye has done.'

'So why are they trying to gag me?' asked Haley, her voice cracking.

'It's because of what you imply,' Rosemary replied gently. 'Calling the book *Why Thomas Tye Must Be Stopped* is totally pejorative. You suggest the man's going to behave irresponsibly and dangerously. We simply don't know if that's true.'

'Do you want me to find another agent?' croaked Haley, her chin jutting.

*

As he walked beside the red carpet Jack Hendriksen appeared the same as always: confident, calm, professional and resourceful. But inside he was not the same. Something had died for him as he had digested what he had learned from the British biographer. Suddenly, he hated his work and his artificial life on this absurd, unreal island. He had made a massive mistake in joining international business: he hated the total obsession with money and power and the lengths to which the global players would go to obtain them. Now he couldn't wait to get out. He would take up his younger brother's offer and join him in his yacht brokerage in Florida. That operation, at least, still depended on the human qualities of personal reputation and trust.

The president's jet was a new generation Russian-European Airbus, still subsonic but with immense range and comfort. Jack waited at the foot of the stairs for President Orlov's head of security to descend. He hadn't met the tall Georgian, but they had conferred by wraparound videoconference. The perennially impoverished Russian Federal government hadn't yet installed the Tye Corporation's new HVC systems. Ordinarily, Pierre would have handled executive security but Tom had wanted this visit to be overseen personally by Jack, as an acknowledegment of the visitor's status.

There was a movement in the open doorway and the security chief appeared in the garish, flat-capped uniform of a full colonel of the Russian Federal Army. He tripped lightly down the stairs, gave a small bow, removed a white glove and extended his right hand. His greeting was instantly translated by Jack's VideoMate and delivered through his radio-linked earpieces. The colonel was similarly equipped.

Jack assured the colonel that all was secure on the island and that 'Tom' was waiting to greet President Orlov. The Tye Corporation's informal president was 'Tom' (sometimes 'TT') to his vast domestic and international staff, 'Tom' to the adoring public and everybody else, and only on the most official occasions 'Mr Tye'.

The colonel nodded his understanding, turned and trotted back up the aircraft steps. Jack followed him and entered the aircraft. Contra-security on presidential and government aircraft was always a ticklish procedure. Every nation regarded its government's aircraft as an extension of its sovereign territory and although Jack would have preferred to have his team sweep the plane before authorizing disembarkation – especially in view of his previous experience of Russian manners gained during a covert action against members of their industrial *mafiya* – he had had to settle for the agreed arrangements. For its part, the Russian delegation had agreed to leave all directed and ambient recording devices behind. Hope Island was both sovereign territory and classified space.

A quick glance to right and left convinced him that superficially everything appeared normal and he turned to incline his head in the direction of the large leather sofa occupied by the president. Then he froze. Seated beside the Russian leader was Anton Vlasik, a man the world believed to be safely behind bars.

Jack turned to the colonel. 'Mr Vlasik isn't on the list we approved. I am therefore unable to allow him to disembark.' He tried to keep the anger out of his voice.

Vlasik, the most powerful, most notorious of Russian criminal chiefs, had controlled two Russian Federal banks that had mysteriously collapsed, taking with them over fourteen billion dollars of Western loans and tens of thousands of life-savings accounts of ordinary Russians. Jack had later read that he had received a twenty-year sentence for fraud. But now he was here.

The colonel shrugged. 'Mr Tye has approved . . .' Jack waved the explanation away as Tye was already talking in his ear. He turned away from the colonel and the Russian party to listen.

'It's OK, Jack, I knew he was coming. He's been pardoned and he's financing part of this deal.'

*Jesus!* 'Tom, I haven't done any clearance or vetting for him. I haven't arranged for anyone to escort him. Why on earth didn't you . . .?'

'You would only have been difficult about it, Jack,' Tye soothed in his ear. 'I know you're over-sensitive about such things. Now, let's get them out of the plane.'

The communication link snapped off. Jack turned to face the visitors, hoping his face hadn't reddened as he fought to swallow such a public undermining of his role. Vlasik had now risen from the sofa, a broad grin on his face.

Jack nodded curtly to the Russian president and walked out of the plane, breathing deeply to re-establish self-control. The colonel emerged beside him and they took up their positions at the top of the steps. As Jack gave clearance for the welcoming party, he watched a crocodile of Tye Corporation-designed electric Volantes – vehicles that looked rather like elongated golf carts – swing away from the reception building and head towards the aircraft. His head-up display confirmed that his team, with their heavily armed back-up, were in station on the rooftop of the reception building and at strategic points around the perimeter of the floating spaceport.

As this mini-motorcade drew up at the foot of the steps, Jack felt rather than saw the small figure of President Mikhail Nikolayevich Orlov arrive beside him in the doorway. Thomas Tye swung out of the first car, followed by Pierre who, for today, was acting as his personal *garde-du-corps.*

With complete disregard for diplomatic protocol, Tye bounded up the steps. He stopped and extended his hand to the sixty-six-year-old leader who, the world would assume, was the head of state least likely to pay a visit to Hope Island. There had been many occasions in the past when Tom had publicly criticized the Russian Federation for its poor environmental record but, as the world was also forced to agree, Orlov's insistence on using his nation's fuel resources as he himself saw fit was finally beginning to make Russia a meaningful economic power again. Then the tycoon turned to Vlasik, turning his formal handshake into a warm bear-hug.

Tye was wearing his customary open-necked white shirt with button-down collar and a pair of crisp, pleated-front khaki trousers. He had pulled his hair back and for once wore neither mask nor gloves. Hope Island's climate had bestowed a gentle tan and his well-toned muscles were discernible under his shirt. In the early-afternoon sunlight Thomas Tye looked like a young film star.

Jack shuddered inside.

*

A thousand kilometres above Auckland, New Zealand, two station-maintenance thruster valves opened for three seconds. The pressure exerted amounted to only .35 grams per square centimetre but ESQ173, one of 400 giant satellites of the Tye LaserNet network, began a slow roll on its axis. The seventy-metre-long spacecraft and its 3,400 kilograms of laser guns, capture dishes, reflectors, solar panels and cameras was rotated out of alignment and instantly became useless. The digital processing and storage systems that were housed inside its pressurized compartment became deaf and blind as they lost their links to the outside world.

A hundred kilometres away the processing systems on ESQ174, the next hub on Earth's highest-capacity data network, went into overdrive as they automatically re-routed the vast streams of laser-borne information to other satellites in the quadrant and warned Orbit Management of the malfunction. Insomniac observers in the darkness below were treated to a rapidly changing laser light show as the aerial networks compensated and healed themselves automatically. Then the valves of ESQ174's adjustment thrusters – the motors periodically deployed to maintain precise orbit station – also opened as the thrusters fired.

Two seconds later the same thing happened to four adjustment thrusters on a nearby Soyuz satellite of the FreePlanet network.

Standing by the lakeside in the warm, sulphurous night air, Constable Terry Nobel of the Rotarua Community Police Service scratched his head. New Zealand's night sky had gone dark for the first time in five years. The soft brilliance of the star clouds of Carina, Canopus and Sirius shone again as the white misty glory of the Milky Way slowly regained its old dominance.

# THREE

President Orlov and his large delegation were getting the full tour, courtesy of the island state's premier citizen. First they were driven around the sprawling, beautifully maintained university campus. Their stately progression along broad gravel drives took them past neoclassical fountains, vast lawns of dense green pearlwort grass, world-class ambient and active sculpture and acres of low-rise pyramids of steel and smoked glass that shone in the fierce afternoon sun.

In the distance, large man-made lakes glinted and beyond those they could see mature cedar trees with high umbrella canopies that had been imported from the savannah plains of East Africa. These *Cedrus Libani* provided a distant backdrop without obstructing the clear view out to sea. The consulting evolutionary psychologists claimed this would provide visual comfort for the humans who strolled in the grounds. Their theory was that the long views provided reassurance that predators were not close and such knowledge would awaken dim noetic echoes of ease and contentment that have filtered down from humanity's distant evolutionary past.

Thomas Tye told the president that more than 10,000 people were permanently resident on Hope Island, a population made up of the staff and families of the Tye Corporation, associated companies and research facilities. He added that another 4,000 workers had been imported from Mexico and the Caribbean islands to provide domestic services and staff for the restaurants, leisure complexes and other public facilities. He didn't add that even the most lowly of the imported workers were required to sign lifetime non-disclosure agreements and media gags in return for signing-on financial bonds that, should an ex-employee ever breach the agreement, would become instantly repayable in addition to any other remedies the company might seek. Neither did he add that the corporation was rapidly running out of development space on the island, a problem that had been occupying the executive board and its international lawyers for many months.

Tye also neglected to mention that rumours about life on Hope Island had become so persistent, so extreme, so exaggerated and so sensational in the poverty-beset Caribbean that what had started years before as an initial trickle of uninvited but hopeful immigrants from Cuba was now growing to become a severe problem. Cuba was now once again in the grip of civil war. Each night Jack Hendriksen's 200-strong local security force had to be ready to deal with dozens of unauthorized landings by small craft, some of them no more than planks of wood strapped over empty oil drums. Despite the best surveillance systems in the world a few of them succeeded in landing undetected, economic refugees and escaping democratic freedom fighters alike, both refusing to believe that once ashore the authorities would turn them away. But the Tye Corporation did, every one of them they managed to intercept, despite reports of firing squads being assembled to greet returning rebels. Hope Island issued no visas, forbade tourism and welcomed visitors by invitation only. Each week a patrol craft would return upwards of a hundred Cuban asylum-seekers, mostly young males, to their mother nation. Many would return there wrapped defiantly in the white-star-and-red-cross flags of the rebel guerrillas.

Jack had remonstrated with Tom over returning young men to almost certain death. He pointed out that the US automatically granted residency status to all Cuban refugees who managed to cross the Florida Straits and had suggested forwarding some of them to the US and to other nations prepared to offer asylum. TT had countered the suggestion with impeccable and forceful logic.

'If none of them finds a route out from here, they'll stop coming.'

Jack had merely nodded. But in the months that followed he had failed in this aspect of his duties on several occasions. He was well aware that some members of the Tye Corporation's domestic and service staff were engaged in smuggling Cuban freedom fighters and their families aboard some of the many supply ships that visited the island. Having to prioritize, Jack chose to apply his forces to their core duties: protecting the corporation's senior staff and the company's assets on the island. Chasing refugees was not his idea of useful deployment.

Then the Cuban Foreign Minister had heightened tension with a public accusation that the Tye Corporation was failing to return all of the rebels who reached Hope Island's shores. Tye had laughed at this suggestion, publicly suggesting that Cuba's besieged government was paranoid. Relations with the neighbouring post-communist state had become severely frosty.

The procession turned a corner and Tye pointed to one of the low glass structures protruding from a lawn. He explained that, due to its location in the hurricane corridor, eighty per cent of the island's commercial facilities were underground and that every pane of glass above ground was able to withstand winds of over 200 miles per hour.

The caravan moved out of the manicured and well-irrigated campus and began to climb the steep slope of a small peak on the western side of the island. Wide roads with surfaces of dark green asphalt cut their way through the island's luxuriant undergrowth even though few of the island's residents used surface transport.

They crested the hill and pulled into a viewing area at the edge of a clifftop. As was usual for first-time visitors, the delegation reacted with small exclamations of surprise.

Sitting at the centre of a large white crescent bay was an eighteenth-century Spanish-style colonial town. The pastel colours and soft stone of the buildings made it seem as if the settlement had been there for centuries. Only the dull glint of the Tye Corporation high-efficiency solar cladding on the rooftops betrayed its recent origin.

'That's Hope Town down below,' explained Tye the tour guide as he pointed out the main features. 'We built the harbour and the marina and we got Disney to design the buildings and the central part of town.'

He offered the president a pair of binoculars and watched as the Russian leader scanned the grid of the town centre, the Town Hall, the green city square and the Hope Island flag that hung listlessly from its white flagstaff. Stretching away from the town in both directions were ribbons of seafront developments that included bungalows, low condominiums, small private estates and beachfront bars and restaurants. In the distance there were signs of major construction where the settlement was being extended around the western headland.

Tye's running commentary about the buildings and features below was a courtesy rather than a necessity. The binoculars the President was using were not only image-stabilized, 3D-enhanced and multi-wavelength, they were also network-enabled: the town's public information system was providing small translucent Cyrillic captions that described every building and physical feature that came into view as he traversed the bay.

If he was impressed, he said nothing.

The president moved the focus of his binoculars up to the hills behind the town where gleaming white villas occupied the lower slopes of the old volcano. A low-rise hotel had been built at the foot of the mountain.

'This is the shielded side of the island,' Tye added. 'The winds on this side never get too bad. That's the Caribbean down there, not the Atlantic.'

The president scanned the view dutifully and nodded his thanks without comment.

'We've taken great care not to disturb the settlement of frigate birds – *Tachypetes aguuilus* – on that ridge,' announced Tye, suddenly the ornithologist, pointing to the cliffs on the other side of the bay as he tried another tack to engage the President. 'And as you can see – perhaps taste – the air here is absolutely free of pollution.'

In the three cars behind, senior Tye executives were providing similar commentaries and viewing facilities for the rest of the delegation. The day was stiflingly hot and the overdressed

Russians were sweating visibly underneath the white canvas canopies that shielded the open vehicles.

The procession restarted and did a U-turn in the empty road. They rolled eastwards down the hillside and were soon rewarded with a cool breeze from the Atlantic. They turned back towards the main campus and Tye explained that they would soon be entering the air-conditioned comfort of the underground complex. He informed his guest that all power on the island was derived either from the company's solar-powered fuel cells or from heat-exchangers that extracted energy from the many underground hot springs. He joked that the old missile silos built by Russia had finally found a useful purpose. The old man nodded but did not respond further. The host smiled to himself; he knew he would get a reaction when they went below.

They drove around the base of the ancient volcano and Tye pointed out the cliff face of sheer plate glass that formed the frontage of his recently completed private mansion. It stood above five vast man-made terraces. He explained that he would have the honour of entertaining the leader and his party at his home later. The procession turned to head back to the main command complex.

Jack radioed his approval ahead and the cars passed through the open gates of the corporate command campus and drove straight down a ramp to the subterranean entrance. Tye leaped from the lead vehicle to demonstrate the security systems to his guests.

'We use only biofeedback systems for security,' he explained as the president and his party disembarked and walked round to the entrance. 'Humans are unique. Our individual identifiers provide the one security key that can't be duplicated or faked.'

Tye stood beside the main door and allowed the system to scan both his irises. 'Open,' he commanded and the metal doors hissed apart. He guided the president and his party onto a station platform surfaced with pale Chernites marble. A three-car train was waiting on the monorail.

Tye, the president, and the senior party climbed into the first car. Jack stood at the rear. The other Tye executives and Russian guests piled into the cars behind.

The shuttle pulled quickly out of the station and Thomas Tye was explaining the principles of maglev repulse-magnetism that propelled the train and allowed it to float silently and frictionless above its single rail. In response to the president's question he admitted that this was one of the few products on the island not made by a Tye subsidiary.

The old missile silos had served merely as a starting point for the immense underground complex that Tye and his team had instructed the architects to build. There had been a time, shortly after the Tye Corporation had first issued its own form of electronic corporate currency, when the piles of cash building up in the company's coffers had proved unusable, inefficient and downright embarrassing. Despite benefiting from heroic returns on their Tye Corporation stock, institutional investors had complained that the corporation's capital hadn't been working hard enough.

But even such activities as massive share-buybacks, a move into global investment banking and large-scale aerospace investments had failed to stem the tide of cash generated by the super-efficiencies of network logistics, the elimination of the thirty-day economic cycle in favour of instantaneous real-time value transfers and the sheer economies of scale created by truly global operations. The more the Tye Corporation invested in real-time transaction and communication systems for itself and its clients, the greater the corporation's returns became. It was as if traditional economic theory had been turned on its head: the almost total elimination of friction from business processes created a virtuous circle that released vast amounts of capital.

The huge construction project on Hope Island had provided a temporary solution to the problem and, these days, Tye and his executives had diversified the corporation's activities and portfolio so much that there were endless uses for the accretive amount of capital that the various operations were creating from their billions of worldwide customers.

The maglev shuttle suddenly slowed and, from high above a huge atrium-like area, the party gazed down on six microsatellites in various stages of completion within a glass-caged clean-room. The area was lit partly from the roof windows above and partly by suspended tungsten lighting.

'They are for our new Deep-Space Location and Navigation System,' said Tye. 'Finishing the network ready for Phoebus expansion.'

The shuttle accelerated rapidly and moved on until the monorail branched ahead. Tye explained that further halls to the south were engaged on other aerospace projects while the centre rail headed towards his home on the west of the island. As usual, visitors were not invited to see the Research Park and the group visit to watch an orbit-shuttle launch was scheduled for the following day.

With a sharp increase in speed, the train branched north, over further brightly lit halls containing offices, TV studios and news rooms. It came to a rest above a vast glass-encased, air-conditioned room filled with seemingly endless rows of long wall racks containing blinking electronic equipment.

'That's our main server farm,' Tye pointed out. 'Those boxes down there are all connected to the networks and each is running a separate business.'

He pulled a DigiPad from his pocket and consulted it

'There's a dozen or so travel agents and brokers, two distance-learning universities that operate in over sixty languages, two hundred and ten auction houses, our global banks, eighty-one local-currency banks, sixteen *bureaux de change*, forty-seven automated stockbrokers and market makers, two hundred and four insurance brokers offering a range of financial services, thirty-nine global retailers selling high-value branded goods, two hundred and one casinos playing every type of game you can imagine and taking every kind of bet, over six thousand real-estate agencies – they have to remain as local brands – one hundred and twenty radio stations, over two thousand separate network game servers, our news servers and archives, eighteen hundred film and video back-catalogue servers – rentals and sales, music servers, book and magazine servers, aggregation and purchasing syndicates, telemedicine servers and, of course, our software servers for rentals and purchase. Then there's the offshore finance and data havens, oh, and the cemeteries. We maintain virtual memorial and personal-history retrievatories for nearly eleven million families – people *and* pets.'

Tye inhaled theatrically as he ended his high-speed recitation of the long list. Jack had seen the performance previously: it was a party piece.

The president nodded as if he too had seen it all before

'Of course, this entire complex is in the core of one of your old rocket silos,' added Tye, breaking the silence that had followed his rehearsal of corporate achievement. 'It's hurricane-proof, earthquake-proof and, as you will recall, it was even designed to withstand a nuclear strike – as were the networks the servers feed.'

The president nodded again. Jack wondered what was going through his mind. The news broadcasts and magazines had praised him for patiently rebuilding the economy of his vast federation. 'We have to travel three hundred years in ten,' he had told his people. They were getting there even though vast, almost unrepayable debts still hung over the government's head from endless rounds of Western refinancing. And, as Vlasik's presence here proved, criminal interests remained at the heart of Russian government.

Oil was being extracted efficiently and getting to its markets. Coal was still being mined and burned despite the criticisms of Thomas Tye and other world leaders. Steel production was now of high quality and Russia's aerospace industry had, once again, become world class. But although Thomas Tye had not said so, Jack knew that the GDP of the unmanned server farm below them eclipsed the entire economic output from one hundred and sixty million citizens of the combined Russian republics. But perhaps the president's impassivity was merely a mask for common acrophobia.

Tye had made no mention of the *other* server farms. Some of the cash and data deposited and stored on Tye Corporation servers was so valuable and hypersensitive that no terrestrial offshore bank or data haven was considered adequately secure or discreet by certain ultra-paranoid customers: these particularly hot deposits were kept *off-planet* in a network of vast-capacity, high-bandwidth, multiply redundant data-server satellites hurtling around the planet 1,000 kilometres out in the deep-freeze of space. These interconnected storage facilities were designed to remain oblivious of any terrestrial nuclear events and they were also immune from the globe's political and military reversions and incursions. They ignored natural terrestrial catastrophes such as earthquakes, rising ocean levels and volcanic eruptions and defied even extraterrestrial extinction events such as asteroid impact or celestial collisions.

Unsurprisingly, the majority of customers for such expensive services were dictators, despots, generals of various 'people's' armies, Triad leaders, Mafia godfathers, drug shoguns, industrial-scale money launderers and the growing army of senior international executives of global corporations. Customers accessed these servers only by a fearsome combination of digital signatures, deeply encrypted and multiple-layered passwords and a range of very personal DNA verifications. Customers also held digital keys to self-destruct mechanisms on the satellites if they were ever forced to contemplate such extreme security measures. The Tye Private Bank asked no questions of depositors with sufficient funds to rent anonymous space in these networks of foil-wrapped, weightless deposit boxes.

The train restarted and entered a short tunnel. Then the party emerged above indoor tennis and squash courts, three gymnasiums and an Olympic-size swimming pool.

'There's also a fully equipped hospital in the complex,' added Tye, unnecessarily. The president was due to become a patient the next day when the check-ups would be run before he became the latest customer of a Tye Life Sciences VIP service.

Tye maintained his running commentary until the train pulled into another station. He gestured for his guests to disembark, waiting politely as the president stepped slowly out. His heart was weak and only careful medical management allowed him to undertake limited public engagements. A pair of large black-glass doors offered the only exit from the station. In his temporary role as team leader Jack brought up the rear, although he was now confident that the situation was totally secure.

Without realizing it, all members of the visiting party – including Anton Vlasik in the second car – had passed through four separate scans of varying wavelengths and, apart from the Russian army colonel who carried his weapons by agreement, Jack had heard his team confirm that the visitors carried neither armaments nor recording equipment.

'This is the entry to our Network Control Center,' Tye told his guests. 'This is where we are able to follow progress and see the world's networks as a whole.'

He stepped forward and allowed his irises to be scanned. Then, with a single movement, he provided a fingerprint from his right index finger to a scanner plate set into the wall. To complete the process he plucked a hair from his head with a theatrical flourish and placed it into a brightly lit scanning tube. A small door closed over the tube.

'DNA verification, the ultimate security,' beamed Tye, revelling in his sales pitch.

The President shrugged and tapped his own pate.

'No hair. Not good for me,' he said in careful English.

Tye and the party laughed dutifully. Jack thought it was rather good. He noticed Tom turn back to face the doorway, allowing a facial-pattern scan to be taken and the chemistry of his scent to be analysed. He was pleased to note that his boss had not explained that the individual biometric checks were all elements of a multi-part DNA signature package that, in total, was the most secure physical access-checking system ever devised.

A green light illuminated on the access panel and the doors hissed apart. Connie was waiting inside and she stepped forward to usher in the honoured guests.

To first-time visitors, the Tye Networks Command Center looked like a theatre-in-the-round that might have been created by set designers from *Star Trek*'s vintage years. The controllers sat on raised circular terraces facing inwards to a darkened central pit. But instead of banks of computer consoles, as in the old NASA command centres, the monitoring team sat in reclining armchairs with no obvious controls or display panels. The circular wall at the rear semi-circle of the room housed large flat display screens. The effect was dramatic and it was created for that purpose.

Connie indicated for the party to halt in the acclimatization area. The room was quiet except for the low hum of powerful air-conditioning. As their eyes adjusted, the visitors could see that about a third of the sixty control positions were occupied by a duty team made up of twelve women and eight men.

Tye walked his visitors around the top terrace and, as they circumnavigated the room, flat two-dimensional holographic displays appeared momentarily in front of the controllers' seats before the visitors' viewing angle changed and they disappeared again.

'All networks are self-monitoring, of course,' Tye pointed out. 'What we do here is to match our traffic projections with actual usage and provide global feedback to the regional controllers.'

He could have added that the entire Network Control Center was completely unnecessary and had been created merely for show. Although it *was* monitoring the world's satellite and terrestrial communications networks, it had actually been built to serve as a large demonstration suite for the Tye Corporation's latest products and much of its quaint futuristic design was yet another product of Disney's kitsch imagineering. All real network management functions had long since been distributed to automatic systems, regional strategic planning centres and local contractors and maintenance teams. This was a place in which to *sell*!

Tye led the party down a central aisle and waved them into seats in the three front rows. Four uniformed stewards handed out hot antiseptic towels. When these had been used and retrieved Tye nodded to one of the controllers and the central pit was filled with light.

Hanging in space in front of the visitors' eyes was a giant globe: a representation of the planet Earth, fifteen feet in diameter. Despite his gnawing doubts about the Tye Corporation and its president, Jack once again found himself captivated by the power of Tye's latest 3D Holo-Theater technology.

'This view is created from images captured by over four hundred separate satellites of our Argus network,' explained Tye, stepping into the pit to stand beside the huge globe. 'This is a composite of what they are seeing at the moment, from about a thousand kilometres out. That's about seven hundred miles,' he explained. 'As you know, all space measurement is in klicks.'

The visitors were silent as they drank in the details: the under-seat scent simulators added a hint of sea breeze to the atmosphere. The projection showed the sunlit side of the planet with the Americas to the left, Europe and Africa to the right and, dominating the centre, the immense curving blue of the Atlantic Ocean. They didn't have time to register the complete absence of cloud cover.

Tye nodded to the controller again and the image dissolved and zoomed in at such high speed that it seemed as if a central vortex had formed. Within a few seconds the outline of Hope Island appeared with its larger neighbours, Cuba and Haiti. Then Hope Island, with its floating spaceport, launch runways and deep-water harbour was clearly visible, filling the holographic space. The camera zoomed again and they saw the tarmac area at the entrance to the underground complex. A further zoom and the camera was focused on the driver they had left with the first of the Volantes.

The controller mouthed a few words into his microphone and the driver held out a small piece of paper, the size of an old-fashioned business card. One further magnification revealed the characters on the card. In Cyrillic text with English sub-titles they read:

*The Tye Corporation is proud to welcome President Orlov and his distinguished party to Hope Island.*

The demonstration had its intended effect. The president smiled and then patted his palms together in appreciation. The rest of the delegation followed suit.

Tye smiled. 'That's an old party trick, although there is a new twist to the way we do it,' he beamed, stepping out of the light. 'High-definition visual satellite surveillance is really quite old-fashioned.' He wagged his finger in mock admonishment.

'Just don't go building any new rocket silos on Cuba. We'll catch you.'

Tye turned and nodded towards one of the control positions. A slim woman in her mid-forties, dressed in a white blouse and elegantly tailored navy-blue trousers walked to the edge of the Holo-Theater to stand beside him.

Tye introduced her to the visitors. 'This is Professor Theresa Keane. You'll be aware of her work on artificial life and machine intelligence.' He paused. 'We're proud to have her here and we're honoured that a Nobel Prize-winner should head our School of Virtuality at Hope University.'

He turned to the woman who shot him a brief but vivid smile. 'Theresa, please show our guests our recent enhancements to satellite surveillance.'

Tye stepped out of the ring and took a seat in the front row. The holo-image snapped back to an orbital view of the Earth, now one-third covered by cloud. The visitors also noticed that within the short time that had elapsed during the demonstration of visual-resolution systems, the planet had turned a few degrees to the east.

The professor stepped from the shadows into the pool of soft light created by the shimmering representation of the planet. Her dark auburn hair was cut short and she wore gold-framed half-lens viewpers that glinted in the light.

'Gentlemen, as you can see, we are now viewing our planet using only the visible spectrum of light,' she began with a hint of a soft Irish brogue. 'Remember, this image is composed in real-time from signals beamed back here by four hundred and eighteen low-earth-orbit satellites of the Argus network. But we are also scanning the planet at all wavelengths. Let's remove cloud cover again.'

The image reverted to the perfect blue, green and ochre globe the visitors had seen when they had first taken their seats.

'Most of the images that have now emerged from under cloud cover are being captured at infrared and ultraviolet wavelengths. The systems are compensating in real-time for cloud movement and are translating their output into light of the visible spectrum.'

She spoke slowly and carefully, ensuring that the visitors' translation systems would deliver one hundred per cent accuracy. She wasn't sure how recently their software had been upgraded. She looked around her small audience to see if they were following her. Their faces told her little.

One of the Russian delegates in the front row rose hesitantly and bowed. 'May I put one question?' he asked in slow but reasonable English.

The professor nodded and smiled her permission at him. Jack decided it was a very nice smile despite the island rumours about her sexual preferences. His emotions were definitely surfacing again.

'This is simulation, yes?' the delegate asked.

Theresa Keane smiled and shook her head as her questioner sank back into his seat. 'No, we are not simulating what might be going on under the clouds,' she explained. 'We are monitoring ground-level movements, heat and electrical activity at non-visible wavelengths, and representing them visually. The only things we can't see through cloud cover are inert details such as fine lines, print, objects that don't store heat and so on. We pull these details from a database that is updated every time cloud cover lifts. What we are seeing now is a form of enhanced real-time visual-wavelength monitoring.'

The group was silent as it digested the implications of satellite surveillance that could see through cloud.

The Russian delegate rose to his feet again, his hand raised as a request to ask another question.

Theresa smiled her smile at him again. 'If you're going to ask whether the network can also see in the dark, the answer is yes, it is fully scotopic – performance is even better than with clouds in daylight; there's less refraction, less visual interference.'

The questioner nodded, hesitated, then nodded once more and fell back into his seat.

There was silence in the room and Jack smiled to himself. She hadn't mentioned the link between the corporation's terrestrial wireless networks and the surveillance system. A huge amount of data was uplinked from sensors in the cellular networks to help the satellites see in all weathers and at night. Nor had she mentioned the system's radar-lidar components.

Theresa allowed the silence to hang for a few seconds more and then she turned back to the globe. 'For the second part of my introductory presentation this afternoon I thought you might be interested in a short history,' she said. 'So, just for fun, I'm going to freeze the image so we can take a look at how we started to connect our planet – how we started to create the digital domain and living space I call global virtuality.'

She waited until the planet became still.

'It all started over one hundred and seventy years ago, here.' The hologram dissolved and zoomed towards the eastern seaboard of the United States and then adjusted its image to present a round two-dimensional map of the Delaware peninsula.

Suddenly a line of white light snaked northwards between two areas of darkness.

'This was humanity's first form of electronic communication,' said Theresa. 'This city at the bottom is Washington DC and the world's first telegraph line covered the thirty-five miles to Baltimore. It was built by Samuel Morse when he was fifty-three. Then, over the next hundred and seventy years, the industrialized nations wired themselves with telegraphic cables, telephone lines, satellite networks and, finally, the mobile wireless systems we all use today.'

Calypso Browne felt sorry for her precious charge and she tended to indulge him. She hauled herself out of the pool and, once again, ran to fetch the beach ball from the grass.

'That was a foul,' shouted Tommy. 'I wasn't ready.'

Calypso crooked the ball in her arm and smiled down at the petulant seven year-old. 'So how many times do I have to say "Ready?"' She smiled. 'Here!'

She shot the ball towards the boy and jumped back into the warm water. The early-afternoon Caribbean sun was so hot that, although her dark skin required no tanning, all she really wanted to do was stretch out on the grass for an hour. But the boy had so little time to play. Tommy leaped and caught the ball above his head.

'I'm going to score,' he shouted and punched the ball back over the small net they had rigged across the pool. Calypso feigned an attempt to reach it and allowed it to fall into the water behind her head.

'You weren't trying!' shouted Tommy. 'Give it back.'

'Yes, I was so,' laughed Calypso, hands on hips in mock anger.

She shot the ball at him once more and again it bounced off his fist and out onto the manicured lawn surrounding the pool.

'Your turn, Tommy,' shouted Calypso. 'I'm always getting it.'

The boy swam for the steps and clambered out of the water. He retrieved the ball and, holding it high over his head, ran back towards the pool intent on hurling it as hard as he could at his laughing companion – she was now poking her tongue out at him. He was so preoccupied with his aim that he didn't notice his discarded baseball bat by the poolside. The bat rolled forward under his foot and with a yelp he flipped backwards and cracked his head on the unpolished marble slabs laid around the pool edge.

Calypso churned the water as she swam to the nearest steps and ran to the boy's aid. He was flat on his back, unconscious. Dr Browne didn't hesitate. She raised his eyelid. There was a slow response to the light. She lifted his wrist and pressed a button on his Day-Glo orange LifeSwatch. His pulse was almost 120. A second press of the button revealed his blood pressure was 110 over 60. She checked again: it was falling.

Calypso stood and ran to her bag. She grabbed her VideoMate to summon help even though she knew that the boy's LifeSwatch would have already transmitted its pre-programmed emergency alert. In a few seconds it would automatically make a decision whether or not to start injecting its small store of concentrated adrenalin through the boy's skin.

She bent down and disabled the LifeSwatch, just to be safe. Personal health protectors were better than nothing if the wearer was alone when a crisis struck, but she didn't want to risk

Tommy's heart being accidentally stimulated until she had a better idea of what his internal injuries might be.

'Well, we have now arrived at the present day,' said Professor Keane quietly.

Her audience sat transfixed at the diorama of their planet cocooned in chains of satellites and networks of cables.

'We estimate that there are currently about four billion social and commercial transactions per second taking place in the digital networks, most of them initiated and completed without human intervention. We have now come to realize that every device – the telegraph, the telephone, radio, television, computers, the Internet, cellphones, wireless devices, VideoMates, bodynets – are all different aspects of the same thing. They are all devices for access to the electronic digital extension of human consciousness that is global virtuality.'

She paused, preparing to wrap up her presentation.

'You might be interested to hear the words of the first message ever sent electronically,' she continued. She looked at her DigiPad to ensure the quotation was accurate. 'It was sent by Samuel Morse in 1844 when he had made the final connection to his telegraph cable and, of course, it was in binary code. It said: "What hath God wrought?"'

*What indeed?* thought Jack at the back of the room.

'Thank you for your attention, gentlemen,' concluded Professor Keane. 'Now we're switching from our history lesson back to our real-time monitoring of the world's networks and I'm handing you back to Tom.'

She nodded at the controller and stepped into the darkness.

The Russian president patted his palms together again and his party swiftly followed suit.

Thomas Tye stood up and walked into the pool of light, keen to enjoy his audience's appreciation of the dazzling display.

'What you see now is our monitoring of current network activity. The intensity of light in the networks indicates the traffic load. We estimate ...' there was a pause as he looked at a digital counter above the display '... that over three billion people, or their software agents, are currently transacting in the world's digital space. One of our single laser beams can now carry over one hundred million video calls and if all the children on the planet wished to send a holiday video to their grandparents at the same time – live feed or recording – there would still be sufficient capacity.'

From two rows back, Jack saw that the Russian leader was leaning forward and pointing towards the bottom of the globe. Tye also noticed the action and followed his gaze. At the southern tip of the globe, just above New Zealand and northern Antarctica, a dark spot was growing like red wine spilled onto a white tablecloth.

Tye straightened and shot a look at the senior controller, then at Connie in the back row. She rose and left the room, unnoticed by members of the visiting party.

'It looks like there's a network outage in one of the sat-nets,' he said calmly. 'Traffic will re-route around it automatically.'

He paused. 'Let's move on to the real business of our meeting.' He nodded towards a steward at the top of an aisle to open the door.

'I am delighted to introduce one of your former Russian countrymen, Doctor Nicholas Kutúzov.'

An elderly, white-haired man in a grey suit walked carefully down towards the presentation area. The Russians clapped more enthusiastically.

'Doctor Kutúzov will be your host for this part of the presentation and you'll be delighted to hear that it will be given in natural Russian. We will now switch all systems over to the Phoebus Project. As of this moment, this is no longer our Network Control Center, it is the Command Center for the Solaris Energy Stations.'

Tye nodded to a controller and stepped out of the pit. The image of the Earth snapped off and there was darkness for a few seconds. New holo-panels appeared in front of the control seats and the large projection screens on the rear wall lit to present distant real-time views of the Earth from space beamed back from Tye's Argus network and from NASA's Mars-orbit Hubble VI observatory. Then Jack noticed a holo-image in the central pit that was new to him. The sun and the inner four planets of the solar system glowed in the darkness. He switched on his VideoMate translation system. He noticed a subtle change in the air: the ScentSims under the seats were releasing a low-intensity fragrance that had been created by the olfactorologists in Tye Consumer Electronics R&D. Jack recognized it from the past: it was the smell of old dollar bills. It was called 'Anticipation'.

He felt a hand on his shoulder and looked up. It was Connie.

'Can you be spared?' she whispered. 'Tommy's had an accident and he might have to be moved.'

'Me,' called Haley as she pulled her key from the lock and slammed the door behind her. Although her sister had lived in Ladbroke Grove for over three years, Haley still found herself impressed by the space and elegance of the mid-nineteenth century *belle époque* mansion. Felicity and Martin only had one floor of the house but that was ... *only one floor?* That was still three times the size of Haley's apartment and offered fifteen-metre reception rooms and huge bay windows overlooking lawns and trees. It was London living at its most elegant.

'There in a minute,' Flick shouted from a bedroom.

'Yeeeee,' whooped Toby as Haley scooped her nephew up from his playpen. He was just over a year old and always loved his aunt's visits. Haley imagined that, because she and Flick were identical twins, she was somehow closer to Toby than an ordinary aunt would be. It certainly felt like that and Toby had never shown any objection. This evening he was to be all Haley's as Felicity was meeting her husband Martin to attend a soirée at the Foreign Office. Haley and Toby would be staying in and Haley felt it was just like old times.

Of the two sisters Felicity had always been the party girl. While Haley had worked ferociously at their state comprehensive school, her sister had partied away her later adolescence. Haley had then delighted the family by winning a place at Cambridge University, while Flick had been grateful to scrape into a communications-design course at one of the lesser-known colleges attached to Cardiff University. It was the same story when the twins were undergraduates: while Felicity enjoyed herself, Haley worked hard and, after securing her much-coveted first-class degree in English, she found a place on the national newspaper that had provided the opportunity to interview Josh Chandler. Felicity had become a production assistant in a network-game production house and, during the following eight years in which Haley experienced only three fairly unsatisfactory relationships, Flick had met, made out and moved on ('the 3M Syndrome', as she called it) with more men than Haley could now remember.

But, in their mid-thirties, it seemed as if the polarity of their relationship was reversing. Outside the long periods when she was engrossed in one of her writing projects, Haley had become renowned as a party animal, while her sister was happily settling for family life and domesticity. She and Martin were already planning another child – if Martin ever found enough time away from his dynamic career in the diplomatic service to make it possible.

Haley walked into the kitchen with Toby balanced on her left hip. He was chewing her necklace. She poured herself a glass of apple juice, then poured a second glass for Flick and sat down at the kitchen table for some serious eye contact and, possibly, conversation with Toby. The sisters were in hot competition to be the one to hear him utter his first recognizable word.

Flick touched Haley affectionately on the head as she entered. The sisters never kissed and rarely embraced. They were too close for that.

'You look wonderful!' admired Haley, truthfully.

'Sorted out your publisher yet?' asked Felicity.

'What publisher?' Haley groaned. Toby took the opportunity to stick his hand in her mouth and she nibbled his fingers, mumbling about how good they tasted.

'Oh. What are you going to do?'

Haley understood the real nature of the enquiry. 'They're not asking for their advance back.'

'Still . . .'

Haley lifted Toby's bib and wiped some dribble from his chin. 'I'll just find another publisher.'

Her voice was sounding hoarse but neither remarked on it. That was how Haley showed stress: she would increasingly lose her voice.

'Drink your juice and get going,' Haley ordered. 'Toby and I want some time alone.'

'How's he doing?'

The entire island was buzzing with news of the accident and the research scientists had only just left.

Calypso looked up from the boy's bedside as Jack Hendriksen closed the door behind him.

'I think he'll be fine.' She smiled. 'It's just hard to tell how serious the concussion is.'

'Are we moving him?'

Calypso shook her head, her almond-shaped amber eyes looking troubled.

'That's what I planned, so I asked for one of the jet planes to be put on standby to go to Miami. They've got a special head-injuries unit there equipped for a craniotomy if it's needed. But Tom cancelled it.'

'Yes, I heard. I intended to go with you.'

Jack looked down at the unconscious boy. Despite the neck brace, the oxygen mask and the dense cluster of neural sensors attached to his scalp, he seemed so like his father with his jet-black hair and long eyelashes.

For the hundredth time in the past hour Calypso checked the monitors. Still no change. The scans and computer systems had given her diagnosis a rating of 74.7 per cent probability.

The door of the sanitarium hissed open and Thomas Tye was at the bedside. He had already changed into a dark suit for the reception dinner.

Calypso launched into an apology until Tye silenced her with a gesture. He pushed his face close to hers, white with fury.

'You *knew* he was to meet the Russians tonight! He was going to play for them! Swimming was not an authorized activity today. How could you . . .'

Jack stiffened and watched carefully as his boss fought to recover his self-control. Tye's eyes flicked from Calypso's shocked face to Jack's. He swallowed and took a small aerosol from his pocket, turning away to spray the inside of his mouth.

'Never mind explaining. I've seen the replay. Just tell me how he is.'

Calypso gave him the same information she had given Jack. 'I still think we should have him checked out in Miami,' she concluded defiantly. 'Telemedicine doesn't work so well when the problem is invisible.'

'I understand that, probably better than you do, Doctor,' snapped Tye. 'A consultant neurosurgeon is landing in . . .' He checked his LifeWatch. 'Twenty-five minutes' time. It's safer not to move him.'

Without touching his son he turned on his heel and left the room.

# FOUR

Joseph P. Tinkler added the half-and-half to his fibre and banana flakes and ate breakfast standing beside his kitchen window. He willed the nerves in his stomach to calm. Perhaps the food would help.

Far below, the Manhattan rush hour was getting under way and the first seaplanes from the Hamptons were skimming in to land at the East River Skyport. He had less than an hour to go before he had to face the board and, most importantly, the Old Man, who was probably on one of those planes. Joe wondered if he had stuck his neck out too far this time.

Joe Tinkler was the star fund manager at Rakusen-Webber and, despite his comparative youth, he was one of the bank's top earners. He had seven hundred and fifty billion dollars under his control and, ignoring long-established standing guidelines against over-concentration, his stock portfolio was heavily biased towards two types of investment: the companies in the Tye group and the quoted companies in which Thomas Tye had a personal stake.

When he had joined the investment bank fresh from Yale, Joe had been its first African-American recruit and, as an analyst, he had lived, breathed and dreamed about the world's richest corporation and its trillionaire founder. Now, twelve years later, Rakusen-Webber's many institutional investors had billions of reasons for thanking him for his knowledge, prescience and judgement.

But suddenly the bank's chief investment officer wanted to remove two-thirds of his fund; to 'split it for the sake of prudence,' Morgenstein had said, citing the extremely heavy percentage of Tye Corporation-related investments. Joe knew his fund had grown to be larger than any on Wall Street, but as he had consistently returned between thirty and thirty-seven per cent annual growth, he had expected to increase the funds under his control rather than see them taken away. He knew that if he allowed it to happen there would be all sorts of rumours on the Street within hours, rumours that would be impossible to neutralize and that could seriously damage his reputation and career.

He guessed, rightly, that it was his performance-related bonuses that were the problem. They were larger than those received by any other employee at Rakusen-Webber and, according to gossip, the largest employee-incentive package on the Street. He also guessed that his earnings had overtaken Morgenstein's – and *he* was a partner.

Joe had objected strongly and, as was his right, he had insisted on a management board review. He hadn't guessed that the Old Man would use this opportunity to make one of his rare personal appearances in the bank to hear Joe put his case.

His presentation for the board was complete. He rechecked that the memory card was in his VideoMate and, for the tenth time, looked to ensure that his back-up and the printouts were in his pilot's bag. He had already dropped a version onto the office server, but he believed in belt and braces when it came to office politics. It had taken him until two a.m. to complete the presentation, checking and rechecking his background information, his figures and his sources' qualifications and justifications.

Out on the street it was still early enough for the air rising off the river to be cool. As usual in summer, he'd decided to walk the six blocks from Maiden Lane to Wall Street. Although it wasn't yet 6.30 a.m., the sidewalks were already busy with others planning an early start to their Tuesday.

Joe had got to the corner of Depyster and Pine before he yielded to temptation. He flipped his Ray Ban Electros out of their case and hit the play button on his VideoMate. Once again, he stepped through his presentation.

The telecoms division was the easy bit and the obvious starting point. Thirteen petabytes of new fibre and LaserNet capacity created in the last quarter alone. Would Tye Data Networks slash the wholesale price of data capacity again? Joe was certain they would and he was sure it would be an aggressive move. Ever since the company had gone on a buying spree and absorbed many of the old national telecommunications companies, it had continued to cut costs while improving services. He projected a growth of forty per cent in demand in the next year and predicted that Tye Corp would gain an additional twenty-two per cent share of the enlarged market. Demand for digital communications capacity was both global and insatiable. As the pundits had predicted years before, it *was* the oxygen of the twenty-first century.

The media division was always the hardest to call. Four major new film releases in the next quarter but only one that looked like a blockbuster. The global film servers were still doing a great job on back-catalogue sales and rentals but that was a mature business. Live news was Tye Corp's weakest area although it was gaining ground with Halcyon, a new global weather channel. This service was proving particularly accurate and reliable but TNN – the Tye Network News channel – was still running a definite third to CNN and DBT News. Profits were unchanged.

Aerospace was very strong: over 300 successful commercial satellite launches last year from Cape Hope, the island's shuttle base, and from the corporation's two leased launch sites in South America. The company had a waiting list of sixty-two months for customers hoping for a launch slot. And Tye Aerospace had a ten-year contract from the World Space Agency for Orbit Management – the systems developed to prevent the thousands of satellites from cannoning into each other. For this division he had revised his profits contribution estimate upwards by thirty per cent, although the corporation was ploughing back much of its surplus into a deep-space location network, for reasons that were as yet undisclosed.

Consumer electronics: Joe's Electro viewpers threw the categories in front of his eyes. The eighth generation of

LifeWatches was due in a few months and his sources were very excited. They whispered that in addition to carrying microstores of adrenalin, angiotensin and digitalis the new models would include a secret new anti-fibrillation compound as well as enlarged ambulatory data storage. The LifeWatch was finally becoming the 'physician on the wrist' that Tye had promised eleven years ago at the launch of its original version. And then there was the astonishing potential value of the medical data warehouses and the boost these would give to the corporation's trove of intellectual capital.

Joe quickly flipped through his media on the market growth for VideoMates, HouseNets, BodyNets and viewpers – still no hint when the 3D entertainment system would be launched.

A heavy weight hit the fund manager between the shoulders and he fell onto a fire hydrant. His large bag went flying.

'Hey, man. I'm sorry.' The tall white man in a pinstriped suit circled to a fast stop on his rollerblades. He reached out to help Joe up. 'You OK?'

Joe nodded. He'd probably have a bruised thigh, but nothing more.

'My wife was giving me shit,' said the rollerblader, as he took off his Armani viewpers and waved them. 'I just can't get away from her.'

'I know, I know,' smiled Joe. He checked his LifeWatch. Everything normal. 'I'm fine.' No need to trade identities.

'Sorry,' repeated the rollerblader as he took off again, more slowly this time. Joe retrieved his case and slipped his own Electros off. Although their video projections onto his retinas appeared as a transparent overlay in front of his eyes, he knew how easy it was to block out the images of the real world.

Almost everybody else on the sidewalk was wearing personal viewpers of one brand or another. Most of them, he guessed, were watching the news, talking to someone, gambling, scanning e-mails and v-mails or, like him, going over their work. Many of them were 'attending' meetings in different time zones, different climates, different seasons: some of them would be involved in more than one. All of them walked more slowly, more absently, than people had when he was a kid in Manhattan and all of them seemed to be talking to themselves. Once again Joe realized that while their bodies were on the streets, their minds were in the networks.

'Is it time to wake up?' asked the red caterpillar.

Tommy and his Furry were inseparable and Calypso had asked for the companion to be sent over from the house.

The boy yawned and stretched, puzzled that he wasn't in his bedroom.

'Row, row, row your boat, gently down the stream,' sang Jed and waited. Tommy didn't feel like singing. His head hurt so much he wanted to cry.

Calypso stepped quietly into the room. She walked over to the bed and stroked Tommy's hair. Despite her written orders of employment that forbade such intimacy, she bent and kissed him gently on the forehead.

'Welcome back, Tommy,' she said quietly. She could guess how he must be feeling.

Tommy looked up. 'My head hurts, Calypso. Where am I?'

'You had a fall and banged your head, my darling. Can you remember where it happened?'

He frowned.

'Ten little speckled frogs,' sang Jed until Tommy squeezed him.

'We were in the pool,' he said. 'You were cheating.'

'Was not,' smiled Calypso, a rush of relief sweeping over her. The scans had shown no internal bleeding. The consultants from Miami were certain there had been no intracranial haematoma, but, well, you could never be sure. She had treated many cases of concussion during her internship in Chicago and she knew nothing was sure until consciousness returned.

She ran her eyes across the read-outs on the wall screen. Pulse a little elevated, otherwise everything was normal.

'Let's make the headache go away,' she said, taking an air-pressure syringe from her coat pocket. Almost by instinct she had worn a physician's coat from the moment she had brought her precious patient in to the clinic. It felt good to dress like a doctor again and she needed to underline to the visiting consultants her professional status as an MD *and* a paediatric psychiatrist. It also brought back all her guilt about focusing her long years of training and her specialist knowledge on *one* healthy child, no matter who he was. But there was also her mother to think about.

Calypso had been the first of her family to make the break – to emerge from the grinding poverty of life on a banana island in the eastern Bahamas. She had her father's Scottish belief in education, the strict little mission school he founded, and her astounding looks to thank for her escape, she acknowledged. She had been crowned Miss Americas and then Miss World. She was the first 'Miss World' to come from the West Indies since Cindy Breakspear had won the title nearly fifty years earlier and Calypso was determined to make the best use of the money and travel opportunities the title had brought with it. Cindy had leveraged her brief fame to gain the 'privilege' of bearing Bob Marley's son; Calypso used hers to make good on her breathlessly blurted ambition to become a doctor. When she was done with her duties for the sponsors and had finished her year on the celebrity lecture circuit, her savings had bought her nine years of medical training in the United States.

And now she was back home – well, almost. Life on Hope Island certainly bore little resemblance to conditions on its near neighbours, but her contract with the Tye Corporation guaranteed her one round trip home and back by company helicopter every two weeks. With her older brothers long gone to seek their livings in the States and in Europe, who else was there to visit Mum and oversee her care, now that her ageing parent was starting to lose her sight and become semi-housebound? Thank God Calypso made more than enough money to provide twenty-four-hour help for the old lady.

Calypso positioned the syringe on her small charge's pale white arm and, with a little puff of compressed gas, the analgesic entered his bloodstream.

The wall screen beeped and displayed a familiar icon.

'Accept,' said Calypso, as she removed the injection nozzle from the syringe.

Her employer's face filled the screen. 'How're you feeling, son?' asked Tye gently.

'I'm OK,' mumbled Tommy.

'I got the alert he was conscious. Thank you, doctor.' It was the closest he would come to an apology.

'Everything looks OK, Tom. Full short-term memory recovery, no haemorrhaging, all the vital signs are normal. Do you want to talk to Doctor Henoch or Doctor Bowlby – I think they're still sleeping?'

'No. Just keep me informed.'

'Are you coming to see me?' asked Tommy quietly.

'I can't, son. I'm in São Paulo. In Brazil. I'll be back tonight.'

The boy nodded and the screen flipped off.

Calypso cradled the boy's head. 'Is it still hurting?'

Tommy nodded and leaned into her. Calypso stroked his hair and wondered how best she could offset the attachment damage that had already been caused by so many broken connections with hired nannies and temporary care-givers. Children need constant love from a single source if they are to develop trust in relationships. The door opened and the biologists returned with their probes and scanners.

Joe Tinkler came to the first of his carefully-orchestrated crescendos.

'I'm certain they're talking *down* the situation in almost all areas,' he declared. 'In fact, everywhere except in the media division. I'm sure the actual results will beat all forecasts for all the other divisions. Except for *my* forecasts, that is. I think we'll probably see earnings of four dollars thirty a share.'

The board had so far sat through his presentation almost impassively, muted because of the rare experience of their founder's son, and their honorary chairman, being present. There was a quietness in the room and then the legendary Richard J. Rakusen spoke for the first time.

'You're telling us that for the thirty-second consecutive quarter the Tye Corporation's profits are going to rise, and this time it will be by over sixty per cent?'

Joe nodded apprehensively.

'Are we to believe that gravity no longer exists?' asked the Old Man.

Joe swallowed. He had taken the board through each of the Tye corporate divisions and subsidiaries and, for the benefit of Richard J. Rakusen and, perhaps, to show off a little, he had explained why historical data were no longer a reliable guide to future performance.

He had also reminded them that the plethora of corporate currencies on issue, the pegging of the US dollar to the Euro and the end of the appallingly wasteful thirty-day business settlement period had flattened economic cycles so much that the booms and busts that had afflicted the markets so severely in the twentieth century and the early years of the twenty-first were now modulated to provide little more than periods of respite between increases in global value-generation. Therefore, he argued, the risks suggested by the old policy guidelines on concentration were out of date. He had done some homework and, with a flourish, he had told them that these guidelines had last been updated over a quarter of a century earlier, in 1989.

He had then listed the most important of the 387 patents that Tye-related companies had filed during the year. He had lifted copies of the original filings from the World Patent Database and he skipped through them as the board watched in silence. He had reminded his audience about the record number of patents the company had secured the previous year and he had tried to explain the implications of some of the research being undertaken in the group's various R&D labs.

'No company in history has ever filed so many important patents in a single year,' he had stated finally.

'Which company has filed the second-highest number of patents in the year?' Morgenstein had asked during that section.

How typical of the bastard! Asking a pointless tangential question like that in the hope of catching him out. Not for the first time, Joe wondered whether it meant plain old embedded racism. His was still the only black face in the firm.

'GenCode, the European-based group,' he had responded swiftly. 'They've filed one hundred and sixty-four.'

There was a deeper silence now as the board waited for Joe's response to their most senior partner.

'No, sir, gravity still exists, but there's no sign of the Tye Corporation's curve flattening yet.'

'Are we still happy with the quality of Tye Corp's reporting?' asked Jill White quickly. *Thank God for a friend*, thought Joe. She had thrown him one of his specialities.

'The company has never wavered,' he confirmed. 'Although they could have changed their accounting methods when they moved their registration of incorporation from Delaware to Hope Island State, they've stayed firmly with the EUUSA and International Accounting Standards principles.'

Since the main stock markets of the EU and the USA had merged seven years before and moved to permanent twenty-four-hour trading, standards of financial reporting had finally become trustworthy and transparent. The massive scandals that had repeatedly rocked the emerging regions more than a decade ago had pushed the regulators to develop better and better standards of accounting, asset valuation and liability accounting. Although this had initially depressed corporate valuations, the sudden appreciation of stored knowledge value, the instruments developed to rate it and new internationally agreed accounting standards for intellectual capital had quickly pushed corporate valuations and the markets back up.

'If anything, their accounts are misleadingly conservative,' Joe suggested. 'They're sitting on three hundred and ninety billion US dollars in cash and, frankly, that's only the visible part. They're laying off enormous sums against real or imaginary future liabilities. Also, our chief economist says . . .'

He searched through the media and projected the figure. 'They could issue a further six hundred billion T-euros without it affecting the value of their currency. The Global Bank reports separately from the main group. Last year it had assets of one point seven five trillion US dollars.'

There were mutterings around the table.

'I turn now to the future,' the fund manager continued.

Jill looked up quickly. Joe's tone had taken on a harder edge.

'Let's return to the consumer division. The LifeWatch has

now sold . . .' He brought the media up – a moving collage of the various models with a running counter at the bottom of the screen. 'One point two billion units around the world. As you can see, six new customers are strapping on a LifeWatch every second. Is anybody here without one?'

He held out his wrist. His Rolex LifeWatch glinted in the ceiling lights. There was no response from his audience.

'We all wear them. You can buy a standard model for the price of a meal and it could well save your life. Heck, they're even available in vending machines in the subway.'

Joe flicked his media forward.

'The World Health Organization estimates that personal health-protectors have saved over two million lives in the last three years, and that's just from early intervention in heart attacks. When the statistics for diabetics, epileptics, anaphylactics and narcoleptics are added the figures go through the roof. Tye Corp has a ninety-three per cent share of the market for PHP systems, even though it allows other brand-owners, like Rolex, to co-market individual styles under licence. The Tye Corporation fully understands that the word "brand" has become today's collective and corporate noun for integrity so it is pleased to co-market with other well-established designer names. But it remains the Tye Corporation's software and systems on the inside and the company itself administers all the updates and maintenance. They have a real monopoly and they are, of course, outside the reach of any anti-trust action.'

He paused, looking around the table to ensure that he still had everybody's attention.

'What I am now going to tell you is price-sensitive information and I have cleared the next part with our compliance department. You therefore understand the limitations on any future investment activities before the official announcement.'

The chief compliance officer nodded his assent for Joe to continue.

'For the last three years LifeWatches have been capturing ambulatory data – that is, they record and store our vital signs during every moment of our lives. We never take them off, do we? That would defeat the point. Since they were launched they've also had ultra-short-range wireless datalinks with VideoMates and compatible communicators and, well, that's just about every mobile communicator on the planet. On top of that, the potential market for LifeWatches is still nearly two billion, just in the developed economies. Worldwide, it's nearer five billion.'

He paused again, watching them. Only Morgenstein was making notes or, more likely, doodling to affect disinterest, thought Joe. The other fourteen pairs of eyes were all fixed firmly on him.

'This datalink is ostensibly for downloading software updates to the LifeWatches. Every time there's a new feature or software patch, Tye Consumer Electronics transmits the update to our VideoMates, and they then transmit it to our LifeWatches. This all occurs automatically, in background. We don't initiate it and we're not aware when it's happening. Only it also goes the other way. Every week our LifeWatches *upload* the recordings they've made about us to our VideoMates and, in turn, they upload this information to Tye Corp's customer-relations centres. That means details of our heart rate, our blood pressure, our blood-cell counts, our ECG record and our epidermic conductivity.

'For those with declared medical conditions – people who are wearing specialized LifeWatches with boosted diabetic, anaphylactic or epileptic defence systems, for example – the data will include specialist measurements such as glucose levels, antigen reaction or epidermic electrochemical resistance. This means that Tye Corp's Consumer Division receives data on how our bodies perform under every condition: when we're sleeping, when we're eating, when we're exercising.'

Joe paused once more and looked at Jill.

'And when we're making love.'

He allowed three beats to pass. He shouldn't have said that. He was getting cocky.

'For example, the data show how heat affects us. They can fill a data warehouse with the information being collected, and then compare how people react to the global weather records on temperature, humidity, sunspots – you name it. Tye Corp already has all the climate data from its Halcyon weather network. Suddenly we know *how* people, and *which* people, are affected. But to do that they have to know precisely where we are at a given time, and that's where the VideoMates come in. Over eighty per cent of VideoMates and compatible communicators make use of the various global positioning systems. They need to – that's part of their brilliant efficiency.

'Our employers, our families – everybody can know where every other human and every asset is, supposedly only when we want them to – when we agree to swap location modes. That's how we manage the world's traffic flows, plane movements, passenger flows at airports, train loads, taxi availability, bus schedules, the shipping lanes, all those satellites and all the other stuff of our busy lives. That's how business manages its global supply lines. Without the commercial GPS networks and computerized management systems we'd be back in the blind world of previous centuries with all that friction and all those inefficiencies.'

Jill was nodding and Morgenstein had stopped making notes.

'Now, couple that with our digital identities . . .' He waited to see if he was going to have to explain that. 'Every time a LifeWatch uploads to a VideoMate it is correlated to the GPS history. That's uploaded to Tye's Consumer Division along with the digital identity certificates that allow us secure and guaranteed communications and transactions.'

'So he knows who we are and where we are at all times?' said Jill, inadvertently personalizing Thomas Tye's corporation.

'Yes, even when we don't use location mode or when we've switched our autolocate off,' confirmed Joe. 'Or, at least, Tye Corp gets a history of our movements. A VideoMate still checks its actual position every few minutes even if you've switched it to standby. It needs its precise location for its log and to give us the information we need when we next enable it. Those data are all stored and automatically uploaded to the Tye Corporation.'

'What's this worth?' asked Richard T. Rakusen quietly. Nobody saw a bottom line emerging earlier, or at a greater distance, than he.

'I've put some numbers on it, sir, but they're only guesswork because we don't know precisely how the Tye Corporation plans to exploit it.'

He flashed up some figures from the media. 'If we just consider the impact of those data on the pharmaceutical division and on the health services and hospitals Tye Health-Care manages around the world, I think we're talking three and a half trillion dollars over five years just from small enhancements to approved drugs – dosage adjustments, new drug-interactions and so on. Essentially, this technology is mapping the physiology of over one billion humans during every moment of their waking and sleeping lives. Customers must provide a lot of personal medical information if they want Tye Health Insurance to accept liability for LifeWatch performance and if that information is coupled with in-field consumer research for statistical verification, Tye's corporation has got permanent, ongoing, real-time results from the greatest clinical trial ever conducted on the human race. Imagine the impact on prognosis guidance, hospital building programmes, and drug development!'

'Surely the data-privacy laws will prevent this,' snapped Morgenstein, aware that there was a serious danger of Joe pulling off a coup. In the last few years the USA had given up its resistance to data-protection laws and had fallen into line with the rest of the world in placing strict limitations on how corporations could exploit personal data.

'Those laws don't apply because every customer signs a waiver,' explained Joe patiently. 'You all recall what happened when the early versions of LifeWatches came out. Some of them went off accidentally and injected adrenalin, epinephrine or other drugs when they shouldn't have. Sometimes they failed to intervene, or failed to send an emergency signal when something *had* gone wrong in the wearer's body. The Tye Corporation settled those claims out of court but as the technology improved Tye's marketing people also got smarter. When we buy one of these articles we have to sign a form that provides basic information – name, address, height, weight, sex, doctor's name and address and so on, if we want to receive legal insurance and medical cover against malfunction. What's the point of buying an uninsured LifeWatch that misses an arrhythmia or an acute allergic nut reaction occurring, so doesn't intervene? People still want coverage even though LifeWatches have become so reliable that there has not been a single reported case of failure, or of accidental intervention, for many years. Tye Corp got round the liability issue by providing their own insurance policies through THI – Tye Health Insurance. It was a really neat solution. We also have to declare any medical conditions we may have and the medication we are on, and we sign a partial data-privacy waiver. Sure, THI agrees not to sell the data on, but it doesn't need to. Its own divisions can make more than enough use of the information.'

Joe stopped to draw breath. 'In the emerging countries the Thomas Tye Foundation has already given away three hundred and eighty million LifeWatches and Tye Consumer Electronics has started to subsidise basic LifeSwatches for those on welfare and for certain needy groups in the developed markets.'

'The great tree-hugging philanthropist, again?' asked Jill, arching her eyebrows. Perhaps he hadn't upset her after all. Those deep brown eyes were smiling.

'Not really; those data are the *most* valuable because those people suffer the most diseases, they need the most drugs, even if we end up paying for them through our taxes or our donations. And then, of course, there's the feedback that is supplied to Tye Life Sciences and its cloned-organ farms for the super-rich. Everything is part of a positive-feedback loop – a virtuous circle. The potential for growth is staggering!'

He looked around all the faces. He had them now. He bent, pretending to look at his notes. After a suitable delay he coughed, changed position and tone.

'And there's a small marketing coup promised for announcement at Comdex this fall.' Once again he paused, making sure he was producing maximum impact.

'The first PetProtectors will go on sale this Christmas. They'll be twice the price of LifeWatches because the collars will be breed-specific and adjustable for coat thickness. They're going to be marked under the tag line "Never Lose Your Pet Again", with a focus on pet location, but physiological data will be the second part of the sales message. I forecast six million sales in the first eight weeks alone.'

'With the data on individual pets being collected too?' asked Jim Manzies, senior partner in Corporate Currencies.

'Of course. Uploaded via their owners' VideoMates, and multiplexed within the location transmissions.'

Joe decided to show off his depth of research and flipped up some pages from the *Nature* archives. 'Of course, PetProtectors won't actually be a new concept. Tye Corp ran animal trials with early versions of protectors before the LifeWatch was launched. It's just that, now LifeWatches are such an important human accessory, many animal-owners will want one for their pet too. They're just digging out the old designs and repackaging them with updated electronics.'

Everybody at the table was nodding, including Morgenstein and even the Old Man. Since Tye Life Sciences had launched its PerPetual service in the United States eleven years earlier the craze for cloning pets for replacements had spread to all developed territories. TLS had quickly sold international franchises for PerPetuation Centers and millions of expensive but identical replacement animals were now routinely provided to order. Pet Pamperers had thus become recognized as a global market of vast potential.

'And I suppose pet paramedic emergency teams will respond to alarm calls from the PetProtectors?' laughed Jill.

'You bet,' smiled Joe. 'That's what I call a start-up opportunity!' He allowed them to enjoy the joke, even though he knew of at least three planned start-ups that had got wind of Tye Corp's launch and had developed precisely that business plan. He was thinking of investing personally in just such a Miami-based venture, if he could clear it with Compliance. But on with the close.

'Although PetProtectors will undoubtedly be big business, that market will be nothing compared to other ways the Tye group can exploit what they are gathering from the LifeWatch data. Take financial products – pensions, health cover, life insurance. Guess what? Tye Financial Services Group builds a data warehouse to mine for those individuals revealing signs that are negative indicators of long-term health – such as poor heart-rate response to heat, essential hypertension, the electro-conduction of stress, and not sleeping well, since heart rates do provide a clear correlation with sleep patterns. This means that the vulnerable are weeded out, and the salespersons will know the healthiest people to target for selling an expensive policy that they probably won't need. They'll also know exactly when to load the premiums for some poor jerk whose physiological performance doesn't meet actuarial standards.'

'And you think this is legal?' queried Harriman, the bank's senior counsel. 'I'd like to see those waiver forms.'

Joe nodded and slipped a set of copies from his pilot's bag. He moved to the table to distribute them, then paused looking down at the elegant Longines LifeWatch on Harriman's wrist. Except for two extra buttons on the side of its slender gold case, it was virtually indistinguishable from the classic analogue timepiece of the previous century. Its electronically active strap was cased in leather and, when required, its digital data would appear in a transparent display layer sandwiched within the watch-glass.

'May I ask if you signed a waiver when you bought that, Mr Harriman?'

There was a silence. Then the counsellor nodded.

'I suppose I must have done,' he admitted, looking around the table for support. 'Well, none of us read *standard* retail sales conditions, do we?'

Without waiting for the waiver forms to be passed around or read, Joe stepped back from the table.

'Now let's look at the Solar Energy Division and the projected global take-up of Tye Corp's high-efficiency solar-energy fuel cells,' he said authoritatively.

Hope Island had an early-rising culture and, just after eleven p.m., the seaside suburbs of Hope Town became quieter as less noise drifted across the small bay.

Calypso had chosen as her home a beach bungalow a mile away from Little Venice, the main harbour frontage and marina with its 'world restaurants' and themed bars. The Island's *Welcome* brochure, which had been given to her on her arrival three months before, boasted that every cuisine in the world was available along a one-mile stretch of waterside restaurants.

And so it was, but Calypso did not feel like walking out to eat this evening. The scare over Tom Jnr's accident had left her feeling both alert and drained at the same time and, once the boy had been returned to his latest nanny in the great house up on the mountain terraces, Calypso had been grateful just to take the Mag back to her bungalow, open a bottle of wine and heat a frozen pepperoni pizza. It was made at Mario's, down by the marina, and contained delicious and medicinally active ingredients. As a doctor she loved the idea that she could adjust and protect her body's biochemistry with a meal she enjoyed but that could not help to make her overweight. The pepperoni topping had become one of her firm favourites.

She had kept communication open, of course. Tommy's VideoMate broadcast updates from his LifeWatch every thirty seconds and she was sure there would be little change. When she had eaten she had tried to catch up on her reading, her VideoMate, open beside her, displaying Tommy's steady life signs. She could tell he was asleep, still mildly sedated.

Calypso logged on to two journal discussion centres and skimmed through the postings but, even as she did, she knew she was too distracted to concentrate properly. She marked articles and threads that she would revisit and turned away from her screens. She had changed into a white silk dressing gown and now she switched off the air-conditioning and opened the door to the small veranda.

Within seconds a close heat filled the small room and she smiled. She often did this at the end of an evening. It reminded her of how it had been back home, when there were seven of them and the nights were so hot she would go out and sleep on the beach.

On a whim, she turned out the house lights and stepped into the night. The white beach sloped gently in front of her her, down towards the dark, glittering sea. High above, innumerable satellites pierced the black sky with their staccato bursts of laser communications – *like those old Star Wars movies*, she thought. Beyond were the soft pinpricks of light clusters that made up the constellations of the Corona Borealis and Hercules.

She heard a roar in the distance and she looked south. Far over the Atlantic, beyond the floating spaceport, she saw the white flame of one of the Tye Aerospace space shuttles as it graduated from jet propulsion to rocket power. She had become used to the spectacle since she had arrived on the island but still she stood and watched the white light grow smaller as the vehicle accelerated towards orbit. In a few minutes it was just another pinprick of light in the sky.

She looked to her left and right. There was no movement. Nobody else was sufficiently interested to come out into their gardens to watch a routine launch. The beach bungalows had been placed at eighty-yard intervals, enough space to give privacy but, the planners claimed, close enough for a community to emerge. All lights were out. Most of the other residents on this beach were professional staff, medics, like herself, or teachers from the school and the university, pilots, astronauts, air traffic movement supervisors or spaceport personnel. There were also many patent attorneys. All took early nights.

Calypso walked down the beach and grabbed a swimsuit and a pair of goggles she had left to dry over her hammock. She slipped the costume up over her legs and allowed her robe to fall from her shoulders at the water's edge. She slipped her arms into the suit, fitted the goggles, waded forward and then dived, exhilarating in the tang of the cool salt water as it swept over her body.

Clear advice was given to all residents not to swim in the seas around Hope Island. The currents were strong and the beaches shelved sharply. Also, sharks were regularly sighted.

But Calypso had grown up with this sea. It was the same sea that had lapped her beach on Mayaguana and she understood its ways and those of its inhabitants. Like her namesake in Homer's *Odyssey*, she was a sea nymph, the daughter of Oceanus. Her brothers had teased her that it had been her constant swimming that had produced her statuesque frame, those shoulders, that neck, those breasts – a body providing the perfect complement to the breathtaking, fine-boned symmetry of her face. It was a feature so provocative it had prompted an ebony-black and somewhat platyrrhine runner-up in one beauty contest to describe Calypso, rather uncharitably, as 'beautiful but undeniably mulatta' in an interview for the *Jamaica Gleaner*. But where had those startling amber eyes come from? They had to be an atavistic attribute, perhaps from her maternal Arawak Indian grandmother.

As an adult, Calypso still swam three or four times a week. She could cross the bay and back again in two hours but this evening she decided that she would merely venture around the small headland before returning – a mile's journey that would take her forty-five minutes. In truth, she didn't want to be away for too long from her VideoMate and its link to her charge.

She stroked her way past the Gene Scene, a beach so named because the DNA snippers from the biotechnology research campus broke every island rule on this small strip of sand they had made their own. Rumour had it that they made all their own recreational drugs, each one tuned for its own user, and they certainly did like to party all night. As her body rolled with her strong easy freestyle crawl, she could see half a dozen bonfires flickering in the darkness.

She ploughed on for a further ten minutes until she felt a colder shaft of water hit her from below. That meant she had reached the headland and would soon be leaving the Caribbean Sea for the Atlantic Ocean, so she turned and headed for the low rocks at the water's edge. As she often did, she would haul herself onto the rocks and sit for ten minutes, before starting on her return. Even this late in the evening, the limestone would still be warm from the day's sun.

A figure was sitting on *her* favourite rock. She paused, treading water. He stood suddenly – he had seen her.

'You OK?' His voice sounded thin across the water, even though only a dozen yards separated them.

Calypso pulled off her goggles and waved an acknowledgement. She realized it was Jack Hendriksen. He stood up and held out a white towel.

# FIVE

*The bicycle won't go! Won't go. Go on, push. It's hard. Breaths are coming in laboured gasps. He hears her running behind the bicycle. He can feel her, he can smell her. She is pushing. It is too fast too fast too fast.*

*He soars up and over the girl and her propulsion. Her stupid clothes will get caught. It's stupid stupid. All these strange people are laughing.*

*Look how I ride. Look how I ride. I can go fast. Are they looking? See. See.*

*The lawn and its narrow winding path are sunlit. People are gathered for something, all staring at the little girl.*

*Clothes catching. Pull at them pull at them, pull pull. He is tearing her stupid clothes off. But the tree always comes. No matter how many times he tries to control it, the tree just keeps on coming.*

*She is crying, the tears are hot, of shame, not hurt. Then she is carrying him. Alone together. The smells again, the smells. She is touching him between his legs the way The Doctor does when she sits on his knee.*

Thomas Tye woke, bathed in sweat, both hands scrabbling down there where, as always, they had no busyness.

He looked at his VideoMate on the bedside table. He had woken one minute before the alarm time he'd set. He cursed as he saw that the DreamDial software he used to save him from that dream had not been activated. He was sure he had set it, even for this short nap before the important appointment he was now to keep on the terrace below. It would be a long time before he fumbled the settings again. He personally championed the DreamDial project through Consumer Electronics in order to protect him from her and The Doctor.

He showered, pulled on a dark sweatsuit and tied his hair back. He took the elevator down to the main viewing terrace in front of the house and walked out into the warm night air. As he had instructed, the rest of the huge house was without lights and the many garden lamps along the terrace pathways had been switched off. Four hundred feet below, the white sand of his private beach stretched out to meet the dark, gentle swell of the Atlantic.

Connie was fully dressed and waiting for him beside a small patio table. She handed him a cup of jasmine tea.

He heard a sound and the Russian president emerged from the darkened house, dressed in a red silk dressing gown. His valet and his two most senior ministers – General Padorin, the Armed Forces Commander, and Leonid Konstantine, the Interior Minister – followed him. Anton Vlasik, another house guest despite Jack Hendriksen's many objections, brought up the rear. The rest of the Russian party was lodged in VIP accommodation in Hope Town and they had not been asked to disturb their sleep for this demonstration.

Tye turned and nodded without saying anything. The President and his party did the same. Connie poured tea for them all.

The night was dark despite the intermittent laser-bursts of satellite communication overhead but it was as clear and cloudless as the meteorologists and their powerful Halcyon weather computer had predicted. The moon was very new and, unless an observer knew precisely where to search, almost invisible to the naked eye. Tye looked at his Piaget LifeWatch and scanned the quadrant of the sky where he knew the event would occur. He found Pegasus in the north-east and followed it up to Equuleus. It would be just a few degrees to the east. *There.*

The island was quiet and unlit. From this northern vantage point the observing party could see down the whole length of the corporate state past the main campus and out to Cape Hope and its floating white extension, where the spaceport runways and deep-water harbour were located. Tye knew they would all be watching down there too, the video cameras and sensors already recording.

Apart from those whose job it was to observe this experiment, and those on Hope Island whose duties kept them up until three a.m., few others on the planet would see the results of this first mid-power trial at close quarters. Hope Island ATC had re-routed all night-time air traffic well to the south.

Despite these precautions, Thomas Tye knew that the forthcoming celestial event would be widely recorded. But he doubted that any of the world's observatories would be able to make sense of it. Each of the energy stations was cloaked in light-absorbent, radio-wave-dispersing materials and all communications had been encrypted and buried in the vast mass of inter-satellite radio transmissions. The fourteen large space stations had initially assumed their pre-booked orbit positions as granted by the UN Space Agency but, when the constructions had been finished, each had been boosted out of earth's orbit and away to specific locations that Tye Aerospace had described as 'part of the Space Location and Positioning System – a network of satellites to aid navigation in the solar system'. After that, they had become invisible to the world's terrestrial observatories and the scores of orbiting telescopes.

None of the world's many space agencies and observatories had publicly identified or queried any of the low-intensity tests undertaken around the planet in previous months. Most had been carried out in regions where it was just before dawn or immediately after dusk. Nor had they noticed any of the tests conducted in the non-visible frequencies. The assumption by the Phoebus Project Team was that the tests had simply been misinterpreted as natural phenomena.

In the next few weeks the controllers would have to take more chances by running high-energy tests in the visible spectrum, but they would choose unpopulated areas such as the South Pacific or the Arctic and, perhaps, an uninhabited forest area. They had to find a balance between the need for successful trials and their desire to keep details of the new service secret until its global launch on 30 August. Tye wanted his Russian deal to become a *fait accompli* before the world's analysts had a chance to consider and pontificate on the implications of his new technology. Cuba would present tonight's biggest problem as their patrol craft would undoubtedly witness the test first-hand, but Tye doubted whether the country's astronomers or physicists were equipped to make a meaningful analysis.

He held his wrist up and switched his LifeWatch to a digital display. With just a hint of trepidation that things might not work as planned, he turned to the president and counted down the seconds to three a.m. – the darkest part of the summer night. He knew that if it worked delivery would be almost instantaneous.

Suddenly the island was bathed in light from end to end. Tye and the others squinted and averted their gaze. He pulled sunglasses from his sweatsuit pocket and put them on. He blinked as his eyes adapted.

The President's valet had handed out sunglasses to his party, and they brought their heads back up as their eyes adjusted to the light.

Tye picked up a solarimeter from a coffee table and checked its reading. 'That's just one at mid-power,' he said and he passed it to the president. The Russian leader read the display and nodded before passing it to the others.

Now that his eyes had become accustomed, Tye quickly scanned the horizon. He smiled. The square of light ended abruptly a few miles offshore. The focus calculations had been perfect and there seemed to be little leakage.

He then looked up at the source of the light. It had obliterated the illumination from the satellite laser beams as well as that from all the stars in the north-eastern sky.

Suddenly the light split into its component wavelengths and a brilliant vertical rainbow held Hope Island within a prism of colours. Tye heard Connie gasp at the beauty of the experience, even though she alone amongst the other observers on the terrace had been prepared for this aspect of the demonstration.

From the trees on the mountain behind them, and from the cliffs below, they heard the songs of frigate birds, sugar-birds, gulls and puffins as they woke to the false dawn.

The light integrated back to the full visible spectrum. Tye

looked at his watch again and counted down the seconds until the end of the time scheduled for his brief demonstration.

As suddenly as it had arrived, the light was gone.

The president and his party began to clap with gusto, the sound bouncing off the plate-glass windows and the Dolomite marble flooring of the terrace.

On the beach far below, Jack Hendriksen was still staring up at the sky, at the point where the light had come from.

'This place is a bloody shit-hole,' spat the uniformed sergeant in his thin, squashed Afrikaner accent as the four-wheel-drive vehicle bounced over another pothole in the battered bridge that spanned the almost dry Hunyani River. 'The worst in Africa.'

Ahead squatted a cluster of old stone buildings inside three rings of a high razor-wire fence. The sign at the entrance to the approach track identified the complex: Chikirubi Maximum Security Facility. Harare.

At the gate the guards took their passports, visas and visiting permits off to their post and examined them for fifteen minutes.

'What's the hold-up?' shouted the sergeant eventually, trying to keep the irritation out of his voice. He knew his accent and colour could provoke trigger-happiness almost instantly in Zimbabwe. He got no response.

'It's OK,' said his black passenger quietly. 'They just like the feeling of self-importance it gives them.'

'Skelms!' hissed the sergeant quietly. 'Do they want money?'

His passenger shook his head. 'That's all been done.'

After a further ten minutes one of the soldiers sauntered out of the guard post with their papers. The sergeant lowered his window again and the heat speared into the air-conditioned interior.

'Recording equipment?' asked the soldier, looking in the rear windows of the vehicle.

'We don't have any,' said the sergeant.

'Get down,' ordered the soldier, opening the door. Three other guards lounged outside the gate house, their old AK47s crooked in their arms, ready for rapid use.

The visitors stepped out onto the dried mud. The soldier patted the sergeant down first, deliberately making his hand movements hard and personal. Then he ran his fingers over the passenger.

'Block F.' The guard pointed as he handed back their papers. The electric gate slid open. The fencing, towers and floodlights were all new and looked expensive.

But Block F, like all the other buildings inside the compound, had not been new for a very long time. The sergeant led his passenger along a filth-strewn concrete corridor to the governor's office. A secretary rose instantly and walked around her desk to open the door to the inner office. She did not knock.

With a nod in her direction the sergeant entered, followed by his passenger.

The overweight governor wore civilian clothes: a tan suit with a pink, open-necked shirt, his fingers adorned with gold. He rose and shook hands with both visitors. The door closed behind them. They sat on two upright chairs in front of his large desk.

'So, Amnesty has finally decided to visit us,' the governor said slowly, twisting a gilt letter-opener between the fat, finely manicured fingers of his right hand. He spoke as if oblivious of the fact that the human rights organization had been applying to visit this prison for nearly thirty-five years. Because of repeated refusals, Amnesty had publicly declared Chikirubi to be in breach of the UN's Universal Declaration of Human Rights.

Philip-Niël Shütte nodded. 'It is *very* kind of you to allow me to visit,' he said slowly and graciously. Via an intermediary, it had taken $40,000 to the Minister of the Interior and $10,000 direct to the prison governor himself before this visit had been scheduled.

'You realize that we have not received our full budget entitlement for eight years?' queried the commander. 'We have to try to feed and house four hundred and sixty men and, well . . .' He gestured expansively. 'We don't know month to month if we will receive sufficient funds. The war.'

Shütte nodded. The civil war. The rebels had been fighting

back and forth across the borders with Zambia and Botswana for a decade.

'Without enough money . . .'

*How bad can it be here?* thought Shütte. He had prepared himself for the worst and the South African intelligence services had provided plenty of smuggled-out eye-witness accounts that described appalling conditions.

'I do my best . . .'

Shütte nodded again. He had had to agree that no criticism of either the governor or the Ministry would be made public after his visit. Normally, Amnesty International would never have agreed to such restrictions. But this wasn't a 'normal' visit and the young South African lawyer wasn't a regular Amnesty observer.

'Let's start with the juvenile section,' he suggested.

'We don't maintain separate juvenile accommodation here,' smiled the governor as he rose.

The cell was about nine metres by three and it contained between thirty and forty men and boys; a black hole full of jet-black faces. Philip-Niël Shütte stood inside the doorway with the sergeant and governor as an armed warder waited in the corridor behind them. The stench was overpowering.

Involuntarily, Shütte cupped his hand to his nose. The temperature had to be over thirty-five degrees and the drone of flies was incessant. The faces turned towards the visitors were silent, impassive.

Shütte pulled a sheet of paper from his inside jacket pocket. 'Reon Albertyn, Joseph Abednego, Marcus Mynery?' he called.

There was a movement behind him. 'Abednego and Mynery have passed away,' the governor breathed in Shütte's ear, emitting a wash of talc and a sweet-sour cologne from his body. 'That's Albertyn over there.' He pointed at a small form in the corner of the room.

The lawyer pushed his way through the men and looked down at the old white man crouched on the bench.

'Reon Albertyn?'

The aged man nodded, the skin on his swollen bald head like flaking white parchment.

'How old are you, Mr Albertyn?'

The man didn't answer. Shütte squatted so that he could look him in the eyes.

'You're not Reon Albertyn, are you?' he said quietly, indicating the sheet of paper. 'Reon Albertyn's only fourteen – and he's a native Bantu African.'

The room remained silent, and Shütte was aware that every eye was on him. He stood up and turned to the governor.

'Where's Reon Albertyn? I was assured–'

'That *is* Albertyn,' insisted the governor, pointing again. 'He's an albino, and he's got some disease that makes him seem very old. But the doctor says it's not contagious.'

Shütte turned back to the old man. 'You're only *fourteen*?' he asked.

The figure nodded, not lifting his head.

The hastily arranged meeting between the chiefs of Tye Networks and their counterparts of the Russian-based FreePlanet Networks had started testily and was now becoming distinctly bad-tempered. Nobody was ready to shoulder the blame.

'Say again, what damage reports did you get on your bird?' asked Raymond Liu. As group technical director of Tye Networks it was his responsibility to ensure that the company's satellite hubs always exchanged data transparently, both with each other and with the satellites of other networks.

Two days earlier there had been a malfunction in the low-earth-orbit satellite networks above New Zealand and Antarctica. The problem had taken a day to solve and the word was that TT had personally demanded an explanation. Certainly Liu had found no obstacle in requisitioning a jet to get to today's meeting. It had been his first ride in one of the Tye-Lear supersonics and he had ridden in the cockpit for the eighty-minute flight from Hope Island to New York.

Chomoi Ltupicho, technical director of the FreePlanet network, was the man who had insisted on a personal meeting rather than a holo-video conference – and had therefore suffered the expense of 'cleaning' and electronically shielding the hotel meeting room. He shook his head.

'I've already shown you,' said the Russian engineer in his excellent English, waving at the printouts covering the desk in the conference room. 'There isn't any damage. Once we regained control and put her back in alignment, we ran full diagnostics. Everything reports A-One, with no history of malfunction.'

'No panel or casing damage?'

'None that shows up on the sensors. We'd need a visual to be sure, of course.' Ltupicho paused. 'I know what you're driving at.'

The three other men in this small room of the Marriott Hotel on the perimeter of Newark Airport, New Jersey, stiffened as the engineer reached into his briefcase.

'Our satellite was not struck.' The Russian unfolded a piece of paper. 'The motion sensors show no impact on the casing before the roll started. Not even a microgram.' He sat back.

The room was silent. There had been no meteor shower, which they had initially presumed. But, although he did not mention it, Raymond Liu had already guessed that FreePlanet's Soyuz satellite had been undamaged. 'So what *did* happen?' he asked, with all the authority of the world's richest corporation behind him.

His counterpart hesitated and then spread his hands on the tabletop. 'Can we go off the record?'

Liu nodded and their assistants simultaneously reached for the two VideoMates that lay open on the table. Both removed their viewpers and confirmed that data capture had now ceased.

'All we know is that two of the orbit-maintenance plasma thrusters made unauthorized burns,' Ltupicho sighed. 'We sent no command, but the log we've downloaded shows that the thrusters fired for 4.768 seconds at 2.30.07 GMT on Monday. Normally those thrusters are only fired to prevent unanticipated orbit decay.'

Raymond Liu nodded in sympathy. So, it wasn't just their network that was suffering unexplained faults. This was why the Russians had wanted a personal meeting. The admission that FreePlanet's communications satellites might be open to outside interference could seriously damage their company's stock price, just as it could Tye Networks's own valuation. Liu suffered a few moments of inner debate before his engineer's frankness won out. His promotion to vice-president had been recent and he was still struggling to acquire the political evasiveness necessary to survive in board-level management.

He shuffled through a pile of papers at his elbow and pushed forward a printout. 'The same thing happened to our birds,' he admitted quietly. 'I presume you too have done the probability math?'

Ltupicho nodded.

'Our tests show that an unauthorized, spontaneous thruster burn will indeed occur once in ninety-six thousand hours – that's eleven years, give or take a few weeks,' continued Liu. 'The odds that two would fire spontaneously at precisely the same moment are thousands of times greater. I presume a Soyuz 8Z01 satellite isn't that different?'

The Russian nodded again. Soyuz aerospace technology had again become the equal of any in the world.

'And the odds on all four thrusters on two separate satellites firing spontaneously within a few seconds of each other are . . .?'

'Incalculable,' agreed Ltupicho.

Raymond Liu nodded and sat forward, forearms folded on top of his papers. 'Let alone the odds that it would happen to our two birds and *then* to one of yours.'

There was a quietness in the room. The assistants to the two technical directors avoided looking at each other.

Then Liu spoke again. 'You said that your Network Control Center never sent a message.'

'Nothing,' confirmed Ltupicho. He too sat forward.

'Any maintenance messages? Any other sort of messages?'

'We sent nothing,' said Ltupicho quietly.

Liu looked down at his bare forearms, as if inspecting his pale cream Asian skin for freckles. Despite his senior management position, the small Chinese-American was dressed in the engineer's traditional uniform of short-sleeved white shirt with a stainless-steel pocket protector displaying a parade of pens.

It was almost a whisper when he spoke again. 'But did your bird receive *any* message?'

The Russian sat back and interlaced his fat fingers over his prominent gut. He looked at Liu over his reading glasses. 'Did yours?'

The silence hung long enough for the sound of the air-conditioning to grow to a roar.

'They may have done,' said Liu at last. 'We simply don't know. We have to assume that might be the case.'

The Russian leaned forward again and took off his glasses. He closed his eyes and pinched the top of his nose, massaging gently. 'That's our position also,' he admitted before opening his eyes again.

Liu looked around the table. 'So, we may have an unauthorized visitor in the networks.'

He allowed a short silence for the implications to sink in. Then he looked up at the Russian.

'I don't think we want to alert the network authorities yet. Let's work together on this one.'

Jack Hendriksen woke with a start, disorientated. Then he smiled. He was in his own bedroom, back in Gramercy Park, back in Manhattan, back in the real world. And this was the start of his vacation!

Jack's loft apartment had seen better days and he had been meaning to fix it up for years. But this had been his marital home and, despite his loss, he treasured its memories. 'Rent it,' his younger brother, the businessman of the family, continually urged, reminding him that rental demand in the city was still soaring. But Jack didn't need the money and he loved getting back here two or three times a year.

He yawned and looked at his LifeWatch, but there was no display. He shook his wrist. Still nothing. He undid the security buckle and eased the watch gently up from his wrist, careful not to damage the almost invisible carbon microdermic nanotubes as they detached from his skin. He shook the device again. Strange, he thought. He had never heard of one failing before.

He laid the LifeWatch on his bedside table, face down, to protect the bioconnectors, and rubbed the stark white strap mark on his tanned forearm. His skin itched where the monitor had interfaced with his body. Well, it must be late. He rose, pulled on a white T-shirt and shorts and padded into the kitchen. The old analogue clock on the wall told him it had already gone nine. *Hell, that's what vacations are for.* He smiled to himself. He switched the kettle on and snapped open his VideoMate to scan the mail. It too was dead: he couldn't see a dial tone.

He picked the communicator up with a frown and closed it, then opened it again. It seemed to have power still, but there was no display. He reached into his jacket pocket. He had brought both his Ray Ban Electros and his Phillipe Patek clear-glass viewpers with him. On both the tiny LEDs were blinking a warning but there was no signal from the VideoMate.

The kettle snapped off as it came to the boil and Jack laid the useless communications technology on the kitchen table. He had never known a VideoMate to fail either.

He made a black coffee and then realized that all his network addresses and numbers were stored in his VideoMate or on his server. He had been planning to scour the Manhattan networks to see which of his friends were in town. A dozen times previously he had planned to organize his vacation in advance, but each time something had happened to distract him. He thought of Calypso and smiled. Then he frowned; he had also been planning to ask his friends for advice on finding a trustworthy attorney – if that wasn't an absolute contradiction in terms.

There was an urgent, sharp knock at his door. He crossed the living room and looked up at his security screen. Its red LED was blinking, which indicated there was no signal from the cameras at the street entrance or in the hall outside. This had to be a neighbour – no one could get inside the building without passing through the security system at street level.

'Yes?' Jack called through the door.

'Jack, it's Ron. Ron Deakin.'

*Jesus. After three years!*

'Ron?'

'Come on, Jack, open up!'

Jack grinned and undid the bolts. The wide old door swung

open, and there stood Jack's first navy intelligence instructor, the man who had realized that Jack had more, much more, to offer the US government than pure SEAL machismo.

They hugged each other, the older man almost engulfed by Jack's enthusiasm. When Jack looked up over Deakin's shoulder he saw a bulky young black man in a dark business suit standing some distance down the hall.

Deakin stepped back and studied his protégé. 'You're still in shape.'

'Unlike you,' grinned Jack, prodding his friend in the stomach.

The older man turned to his companion. 'Come inside,' he said without waiting for an invitation.

Jack closed the door and re-bolted it.

'Jesus!' exclaimed Jack, aloud this time. 'Ron Deakin!'

Deakin smiled, waiting for Hendriksen to get over his surprise. It took only a couple of seconds.

'How did you ... how did you know I was here? I'm hardly ever here. The entry system is out for some reason.'

Deakin held up a palm. 'Yeah, we know. Listen, Jack, we're only going to stay a few minutes. It's our system that's doing this: we're jamming all radio transmissions and screen displays around this building. We're scrubbing the immediate area.'

'Who's "we" these days, Ron?' Jack shot back quickly, looking from one to the other.

Both men reached into their jackets. 'I don't expect this to mean much to you,' said Deakin. 'That's why I came personally. You know, because of us – you and me. This is Mike Chevannes. He works with me.'

Jack took the wallet Deakin held out. He saw a plastic card with a photo ID and a digital identity chip laminated into the corner. The emblem showed the blue oak leaves and the globe of the United Nations. The text announced the bearer to be an Executive Officer of the United Nations International Security Agency.

Jack looked at the other man's ID. It was almost identical. 'The UN?'

Deakin nodded. 'The National Security Agency, *our* NSA,

helped the UN set the agency up about ten years ago. It isn't widely known and it isn't meant to be.'

Jack studied the IDs again and then looked back at Deakin.

'I realize you can't verify or copy these idents with your system down,' said Deakin. 'But it's *me*, Jack. You know me.'

And it was him. Always there for Jack, even years after initial training. Every time there was an intelligence problem, whether in Iraq, Kosovo or North Korea, Ron had always been there. He had also been there just after Helen was killed.

Jack handed the badges back with a smile. 'What's going on?'

'We need your help, Jack. Can you come with us right away? We're going to a UN facility.'

Jack nodded. 'I'll get dressed. Give me time to shower and shave.'

'Pack an overnight bag, please, sir,' added Chevannes in a light Jamaican accent.

Joe Tinkler's morning had started spectacularly. The Tye Corporation had reported annual earnings of five dollars a share! That was over ten per cent higher than even Joe had forecast and every one of the bets he had made for his clients and for himself had paid off handsomely. The stocks had started roaring in Tokyo and the sound had spread westwards around the world's markets for the last eighteen hours. On the back of Tye Corp's results the whole of EUUSA was up thirteen points!

Then, around 11.30 a.m., Joe had started to worry. One of his software agents had sent back an alarm. He had configured this agent six years before and, after he had dispatched it into the global networks, it had sent him daily updates on which he based many of his decisions. But the software agent, which Joe had christened *TinklerOne*, had never before sent him an alarm.

Its alert had flashed on his wall screen and sounded an audio signal as Joe had originally planned. He opened the message and scanned the text and the charts.

It seemed that Thomas Tye was selling stock, and selling it in a very big way! The fund manager had customized this software agent from an off-the-shelf package especially to monitor Tye's personal shareholdings and other investments. Joe had spent over four months programming the agent with details of every stock he knew Tye held. He started with Tye Corp's core stock on EUUSA and then included every company quoted on any of the world's major securities markets in which the Tye Corporation or Thomas Tye himself had any shareholding. He subsequently included the speciality companies quoted on the smaller electronic exchanges that were dedicated to nanotechnology development or biotechnology start-ups. He had also given the research agent the names of Tye's investment vehicles, his brokers, his attorneys and his dealing codes on the individual markets. And he updated the same agent's reference list every time he came across a new corporate or legal identity for Tye or any of his companies. Its reach could never be exhaustive, but it was about as good as it was possible to be. Altogether, he had found 2,891 companies in which Tye held stock either personally, through one of his investment vehicles, or indirectly through a third party. The man's investments were scattered throughout forty-one countries and appeared on eighteen different stock market indices.

Joe had even estimated Tye's likely stock disposals and had programmed the robot accordingly. Each year, Tye sold a little more of his core stock in Tye Corp, but as the corporation had now split the stock two-for-one eleven times in six years, he had suffered a manageable dilution. He also had a habit of taking profits out of other investments that were doing well and buying into small, unheard-of companies. Tye's intuition about stocks, which had always been good, had become close to perfect in the last few years and *TinklerOne*'s reports had allowed Joe to track and shadow Tye's moves with a high degree of fidelity.

The software robot was Joe's secret weapon and although all fund managers used a number of software monitoring tools to watch over their portfolios, none of them (as far as Joe knew) had refined the self-learning capabilities of a software robot to anything like the extent Joe had with this one. His first degree at Yale had been in computer science, before he'd done his MBA, but Joe had always chosen to downplay his special knowledge of information-technology systems when in the company of his bank colleagues and other Wall Street associates.

When Tye sold stock, his office usually went public within a few hours in order to offset potentially damaging speculation. The world's richest man was, himself, the most powerful economic indicator on the planet.

But this was something different. *TinklerOne* was programmed to ignore routine disposals, even up to one billion dollars in a month, but now the agent was reporting over 200 separate disposals in an hour, with a value climbing above four hundred billion dollars as Joe watched. The agent was sending back a continuous stream of data about sales, security commission filings, and a host of third-party encrypted attachments that were useless and were automatically trashed by the system as they were received.

There was nothing wrong with Tye selling stock like this, Joe reasoned. He was doing so after his results were published and there had been no unusual purchases beforehand.

Then Joe's system sounded another alarm. Joe looked up at the corner of his wall screen. This time it was another of his agents that had also never before sent an alarm. Joe opened the message. The Tye Corporation's Global Bank had issued two hundred billion in new currency and was already trading Tye-€'s at $1.10. They'd added ten per cent to their capital pool and still the value rose! It was because of the strength of those annual results.

Joe slumped in his chair and watched as his agents adjusted the graphs and figures in real-time. Tye's personal disposals had reached five hundred billion dollars. With Tye Corp's cash deposits, the currency issues and Tye's personal disposals, Thomas Tye and his corporation had raised close to one trillion dollars in cash in under an hour.

But why? Cash was weak compared to paper. Tye Corp's stock could buy anything and Tye-€'s were gaining in value. What could Thomas Tye or his company possibly need with so much hard currency?

Then Joe sat bolt upright. If this continued, every stock

related to the Tye Corporation had to collapse. Then the world's markets would stagger, and perhaps founder. His hand cleared some papers and opened a shoebox he kept at the back of his desk. He removed what he flippantly thought of as his 'panic button'. It was an old-fashioned wireless mouse that pointed towards a macro that Joe had created years before. If he clicked the button, the macro would send 'sell' orders on every Tye Corporation and Tye-related stock on the planet – and Joe's fund held more than anybody except Thomas Tye himself. Joe had never previously had to use these commands and that was why he had left it under manual control. He didn't want some speech-interpretation program to scramble an instruction and start a sell-off accidentally. The market was still holding, despite Tye's rising disposals. How long before it was noticed and other investors reacted? Joe's finger trembled on the mouse as he weighed his options.

# SIX

## Introduction

The world loves Thomas Richmond Tye III. His is the quintessential American success story, transferred to a global stage. He has become the world's first dollar trillionaire and, in real terms, he is many times wealthier than colossi of previous centuries such as Gates, Rockefeller, Croesus or Tiberius. There are dozens of calculators on the networks that strive to measure how much Thomas Tye earns each hour. Currently the best guess is around US$23 million.

'Tom', as he insists on being called, is also the world's first entrepreneur to gain genuine superstar status. His good looks, boyish charm, casual style, concern for the planet and legendary philanthropy have won him fans from every walk of life – from the hopeless Touchers in their ghettos of networked urban misery to the presidents of the world's great powers. He is, after all, the first business hero to emerge in our global society and he is the ultimate eligible bachelor. He is also likely to live long enough to enjoy his fabulous wealth. I can exclusively reveal that his doctors currently predict he will live to be at least 300 years old! He has been taking anti-ageing therapy for seventeen years and I provide full details of the treatment undertaken in Chapter One.

This is the twenty-third 'biography' of Thomas Tye to be published around the world but it is not simply another authorized hagiography, nor yet another tabloid-style *réchauffé* of life on Hope Island and the supposed excesses of the Tye Corporation Techies.

Rather, it is a polemic on power; a monograph on monopoly, a dissertation on the dangers of massive personal wealth when it is coupled with a complete and utter lack of accountability. Truly, no one with almost unlimited money has ever been as powerful and as unaccountable as is Thomas Tye. He has no voters to please. He is subject to no laws other than his own. He must please only his shareholders but their interests are so narrow, so restricted, that, by definition, his activities need to succeed in only one dimension.

We all know the basic details of the Thomas Tye legend and I do not intend to regurgitate once again his unfortunate background or his remarkable rise to power and fame. However, some aspects of his life story have a direct bearing on his behaviour today and the dangers it may present to humanity.

There is little doubt that Thomas Tye suffered massive 'attachment damage' as a child. This is a term used by psychologists when they diagnose a patient as unable to form relationships with or 'attachments' to other people, whether those are bonds of friendship, love or simple empathy. Frequently, the damage and resulting isolation leads to 'homelessness' in adult life (a euphemism society often substitutes for 'lovelessness'), criminal recidivism and both male and female varieties of sexual abuse. In extreme cases, attachment damage produces the psychopaths who pollute and mutilate our society. (The best [or worst] examples I can point to are the Romanian Rapists, that terrifying pan-European epidemic of middle-aged, orphanage-reared monsters that was created by a dictator's total ban on contraception and abortion almost fifty years ago.)

Although he was born into a wealthy Bostonian banking family, Tye's clinically depressive mother committed suicide when he was five years old and this event cast a shadow that has seemingly fallen across his entire life. This piece of misfortune was only compounded by his father's chronic alcoholism that killed him, a few months after his wife's death, in a road accident that was almost certainly drink-related.

Few details are available about Tye's unhappy childhood – the family closed ranks and used its money to ensure silence about his parents' many failures – but we have all wondered about the impact of his internment in an exclusive psychiatric clinic immediately prior to his parents' deaths.

This period appears to have had an immense impact on Tye and, as I will argue in Chapter Eleven, it probably accounts for the astounding lack of ethics that marked his early years in business and his apparent lack of personal empathy with those around him. We have recently watched the spectacle of a string of former business partners and disgruntled ex-employees from Tye's early business dealings giving testimony on American talk shows about their invariably abrupt and ruthless treatment at the hands of the lonely young genius.

Tye's experiences in early childhood may also be responsible for the fact that no one has ever claimed to have had a sexual or intimate relationship with him. I will be adding more detail to his childhood biography in Chapter Four.

As some readers may know, this book almost didn't make it to publication. The attorneys for the Tye Corporation won seventeen injunctions in fourteen territories to ban this work from the shelves and the networks, and it is to the credit of my publishers and the world's legal systems that you are now reading these words in print or from a download.

The reason Tye and his corporation want to ban this book is that I make a number of serious allegations about Tye's activities and those of his companies and I will provide proof of my assertions. I have called this book *Why Thomas Tye Must Be Stopped* because I think the governments of the world must act now to prevent the very nature of humanity being patented and subsumed into a commercial, for-profit corporation.

Haley pushed her chair back from the keyboard and took off her glasses. '. . . *Almost didn't make it to publication.*' Quite. And it still looked as if she might never see her words published. She could always self-publish on the networks, of course, but hers was a linear argument, so it belonged on paper or as a commercially published electronic book. And self-publishing would also rob her work of the imprimatur of a major publishing house and, considering the allegations she made, that kind of credibility would be vital. It would also lay her personally wide open to the legal attacks that, she judged, would inevitably follow.

Haley wasn't wholly sure what she wanted to achieve with this book. It had started out as a complaint about unfettered technology. But, as she had learned more about Tye's interests in biotechnology, the astonishing experiments already under way, and the breathtaking hypocrisy of his publicly espoused green politics, her 'biography' was turning into a simple plea for the world to pay greater attention to the growth and global ascendancy of unaccountable corporate power.

But she must somehow press on. Rosemary said Nautilus didn't want their first advance payment back, and the new input she was hoping to get from Jack Hendriksen should help her agent attract another publisher. But why hadn't she heard from him? It was over a week since he had promised to get in touch with her again. Perhaps he too had got cold feet.

She stretched, put her glasses back on and pulled herself back to the keyboard, her triptych of screens and her text. How many times had she rewritten this intro? She had lost count and each time a day or two's reflection had led her to brand it too hysterical, too emotional or too dry. She was trying to find the middle ground.

> '. . . many times wealthier than colossi of previous centuries such as Gates, Rockefeller, Croesus or Tiberius.'

*Not 'colossi', a clumsy plural for an opening paragraph.*

> '. . . many times wealthier than the commercial or industrial giants of previous centuries such as . . .'

*Better.*

> 'Rather, it is a polemic on power; a monograph on monopoly, a dissertation on the dangers . . .'

The author frowned and launched her thesaurus program. She sorted the adjectives alphabetically and selected the words that best suited her mood.

*And a bloody battological abomination of assonantal alliteration from a pretentious prestidigitator,* she wrote tartly, mock-sesquipedalian in her self-disgust: perhaps it was the legacy of Greek blood in her veins, or might it be the Irish? She cut the entire paragraph.

She sighed, pushed back from the keyboard again, and went to find a nail file. She was typing so much that her fingernails had become a biological tariff of her frustration with the project: they were growing at an almost alarming rate.

Jack Hendriksen knew he must still be somewhere inside the United Nations complex beside the East River, but even he had completely lost his bearings.

Despite Gramercy Park's status as a twenty-four-hour car-free zone a black limo with a diplomatic plate had been waiting for them at the kerb outside the brownstone. Jack noticed the 'All-Zone' windshield digital ID – just like for the cops and the emergency services.

Chevannes had stowed Jack's bag in the trunk and gone up front to ride with the driver. Deakin and Jack rode in the back. As they turned north on Third Avenue, Jack's VideoMate and viewpers had returned to life, a low tone alerting him to waiting messages. At Deakin's request, he switched the system off completely.

'Better if you don't supply your whereabouts, Jack. Erase the location buffer immediately you switch it back on.'

During the fast drive uptown the older man had gently deflected Jack's questions about the purpose of their trip, saying that everything would be explained when they arrived. Instead, the two men used the time to catch up with the news on mutual friends, former colleagues, family and acquaintances. They soon worked out that it must have been over three years since they had last met. At Helen's funeral.

On their arrival at UN Plaza, the UNISA idents had prompted a young major in the black uniform of the German army to escort them away from the public security checks and scanners and lead them through a private entrance. A turbaned Gurkha at the door came to attention as they entered.

They had descended three escalators and been led through a maze of brightly lit corridors until they came to another security point. Here Deakin and Chevannes allowed their IDs to be copied even though they were obviously known to the guards. Jack guessed this procedure was for the benefit of the database records and the security cameras.

'Will you leave all comms and storage here, please, Jack?' said Deakin, as he and Chevannes handed their communicators and viewpers to the guards. The group waited while Jack unclipped the VideoMate and fished his viewpers case from his inside pocket. He had pulled on a sports jacket over his open-necked shirt and he checked to ensure that he hadn't left lapel cameras in place. Confident he was clean, he placed the equipment in the tray provided and stepped through the scanner.

The guard handed Jack an electronic badge and they were waved on. At the end of the corridor was a plain white door, beside which Jack noticed an iris scanner. Deakin halted to allow his eyes to be scanned, and the door slid open. As they stepped through, Jack saw that the thick door had a sandwich filling of lead running through it.

The conference room was large and high-ceilinged, the central space occupied by a table capable of seating thirty or forty. Beyond it was a large Holo-Theater and an older wraparound videoconference system. They were alone in the room – Chevannes and the armed German major had remained outside.

Jack whistled. 'I had no idea the UN complex was so large,' he said, turning to his one-time instructor.

'It extends over thirty-eight acres – we're right under the river here,' grinned his friend. 'The Midtown Tunnel lies on the other side of that wall, Jack. The UN faced a tough choice about ten years ago: either relocate, or expand underground. This is what they chose.'

Jack turned and saw a giant electronic world map on the wall. He ran his eyes over the clusters of illuminated red dots scattered through Africa, Eastern Europe, the Middle East and Asia.

'That's everywhere UN troops are involved currently,' explained Deakin. 'The Security Council meets privately in this room. They like to keep track of how many peacekeeping actions are going on.'

Jack smiled grimly. There had to be at least a hundred lights. 'Our peaceful century,' he observed.

Deakin snorted.

The main door slid open and four men and one woman entered. All were in business suits; all but one carried briefcases. Jack recognized the first face: it was very famous and very distinguished and it belonged to the one individual without a case. His black visage was framed by a huge crop of curly white hair that seemed even more unruly than on television. Jack thought he also knew the swarthy face of the short, podgy white man at his shoulder. The others were unknown to him. Behind them Jack saw Chevannes and the German officer resume their positions as the door closed silently.

The small group walked into the centre of the room to greet their visitor.

'Commander Hendriksen?' said the Secretary-General of the United Nations. 'I'm Alexander Dibelius.'

'Well, I'm retired from the Navy now sir,' said Jack as he shook a large powerful hand. He had to tilt his head up slightly to look into the Secretary-General's warm dark eyes.

'This is Doctor Yoav Chelouche, President of the World Bank.'

Of course. The 'economic genius', they called him. The man who had finally managed to soften the gyrations of the global economy, and who had been awarded a Nobel Economics Prize for his efforts, a new form of quasi-scientific award created specifically to mark his achievements.

The banker's hand was dry, its pressure brief. His lugubrious brown eyes and heavy jowls reminded Jack of a basset hound's face.

Dibelius turned to introduce the other three.

'Professor Rima Berzin, Director of Science at the World Health Organization.' She was about fifty, attractive, though she did little to emphasize her looks. She smiled briefly as they shook hands.

'Doctor Alan Mathison, Cambridge University.' Jack shook hands with the tall, pallid academic and Dibelius turned to the last man.

'And Jan Amethier, director of UNISA.' He pronounced the acronym 'eu-nese-a.'

'Thank you for coming to see us, Mr Hendriksen,' said Amethier. The accent was Dutch, or perhaps Belgian. Dutch, Jack decided.

'I'm intrigued to know why I'm here.'

'Let's sit down,' said the Secretary-General, leading the way to the conference table. He chose a place a short distance from the head and gestured for Amethier to take the chair. 'It's your show, Jan,' he said.

Deakin indicated for Jack to sit opposite Dibelius and then pulled out a chair and sat beside him. Jack noticed that the place sign in front of him read *Australian Republic.*

'Mr Hendriksen, welcome to the United Nations,' began the Secretary General. 'You probably bypassed our informal immigration procedures, but you realize you're no longer on American soil?'

Jack nodded, although the thought hadn't really occurred to him.

'This is United Nations territory, so it belongs to all of our two hundred and twelve member nations, not to any one country. Neither the US government, nor any other, has any independent legal rights here. We possess similar territories inside many of our member states and globally we are considered a sovereign power, you understand?'

Jack nodded again. He had already noticed the big 'Duty Free' signs hanging over a large International Bazaar in the atrium upstairs, but he doubted whether this man ever concerned himself with discount retail opportunities.

'Now, I presume that during your years with the US Navy and the Government you would have signed US National Secrecy Regulations?' said Dibelius.

'Yes sir.'

'Well, as you know, that remains in force throughout your life but, unfortunately, it concerns only US confidentiality and the US constitution limits the government's powers of enforcement.'

Jack nodded again, his mind racing as he tried to guess what all this was about.

'The United Nations is not restrained in such a way and we have a document that is fully enforceable, it's closer to the Official Secrets Act used by the British. It's called the International Security, Diplomatic and Military Confidentiality Undertaking and it is drawn up under the international jurisdiction of The Court of the Hague. Would you be willing to sign it?'

Jack hesitated, unsure of his response.

'We need your help,' added Dibelius quietly. 'The trouble is that we can't even tell you in what way without your signing it. The information we have will compromise you.'

'This is about Thomas Tye, isn't it?'

For a moment he thought he saw acknowledgement in Dibelius' eyes. But then the Secretary General turned to Amethier and held out his hand. The director of UNISA handed him a printed document.

'Please read it, if you wish,' said Dibelius, as he slid it across the table to Jack. 'This may well conflict with undertakings you have given to the Tye Corporation. Technically, it even overrides your loyalty to your own country. From this point on, your oath of allegiance will be to the United Nations.'

'You mean I give up being an American?' asked Jack, surprised.

'You can keep your US passport, Commander, but your first loyalty will be to the United Nations. You will also have a right to a UN Diplomatic Passport should you ever need it. A UN passport is a very special thing: it guarantees a holder entry, domicile and work rights in all member nations.'

Jack let out a low whistle, sat forward and picked up the document.

'It's a standard form, but it is globally binding and will supersede all other legal commitments you have made,' continued the Secretary General. 'Unauthorised use of UN information acquired after you have signed that form will be an offence under The Hague's international military jurisdiction. The maximum sentence for an offence is life imprisonment. We call that part the Silence Resolution.'

Jack turned the form over in his hands. He read the first few lines, then skipped through the four pages, scanning the paragraph headings. He flipped it over so the back page was uppermost, took a pen from beside the deskpad and signed it. He passed it to Deakin.

'He was my best man,' smiled Jack. 'He can act as my witness again.'

Deakin nodded then added his signature and the date.

'It is about Thomas Tye, isn't it?' he asked again.

'Welcome to global citizenship, Commander,' said Dibelius. He smiled and turned to Amethier. 'Jan?'

The UNISA director retrieved the document from Deakin, placed it in his briefcase and watched as the electronic catches shut on recognition of his thumb prints. He lifted the case from the table and placed it beside him on the floor.

'Yes, it's about Mr Tye,' he confirmed in his careful, lilting English. 'We know you are already concerned at some of his activities.' He touched a button on a small black box of controls in front of him and waited as a flat screen rose from a housing at the far end of the table. He pressed another button.

Jack saw an image of himself sitting on a sofa, he couldn't place where. Then Haley Voss came into the shot as she leant forward to pick up her mug from the coffee table and Jack heard himself start to speak.

'I know one of the authors named on the cover. He's a geneticist – on the island. I might be able to check whether he really did write this.'

Amethier hit a button and the replay stopped.

'We don't need to watch the rest of that conversation,' he said. 'We know Mr Tye intends to live forever.'

Jack turned back to his old friend with an eyebrow raised. 'You've had me under surveillance?'

'No, not you Jack,' corrected Deakin. 'Haley Voss – for about eight months. She was sending her sister a video feed of her meeting with you.'

'But why? What has she got to do . . .?'

Amethier held up his hand. 'We'll explain why in a moment, Commander Hendriksen,' forstalled the director, also choosing to address the visitor by his former rank. 'The important thing is that you already seem to be concerned about Mr Tye's behaviour. Frankly, so are we – very, very concerned. And not just by that report Miss Voss showed you.'

The Secretary General leaned into the table assuming control again. 'How much do you actually know about the United Nations, Commander?'

It seemed as though they were determined to militarize this meeting.

'Not a lot, I guess. Only what I read. It seems the UN has been doing a pretty useful job of ironing out the knots here and there.' He gazed up at the world map smeared with its clusters of war zones.

'We're really quite different from our public image,' explained Dibelius. 'You'll be familiar with our peace-keeping activities, our refugee efforts and so on. What you may not know is that the World Bank, the International Monetary Fund, the World Trade Organization, the World Health Organization and the International Space Agency are all UN bodies – part of our executive, if you like. Since we expanded the Permanent Membership of the Security Council to include all major economic powers, since we adopted majority decision making and abandoned the veto, and since we finally managed to persuade the US to pay its full dues, the UN has become the closest thing the world has to a global government. That was, of course, the dream of visionaries like H. G. Wells who laid out the blueprint for a world state – and it was the original goal of our founders nearly seventy years ago.'

He smiled and held up his hand. 'Oh, you won't hear people talking like that, of course. National pride, especially within member states such as the United States, France and China, prevents them acknowledging that fully, at least in public. After all, the biggest nations spent over half a century working to ensure that we didn't fulfil the aims of those who brought us into being – none of the superpowers could really hope to lead the world, but they weren't prepared to let us do it either. Thankfully, those days are past. In the last ten years, we've had the power, the money and, most importantly, the mandate, to try and deal with issues that are supranational in character.'

The Secretary-General's mellisonant tones held Jack and the

others riveted. It was a party trick, Jack realized. A master politician's magic. The power of charisma.

Dibelius gestured towards Chelouche. 'Doctor Chelouche's team at the World Bank is a good example, Commander. Since the dollar and the euro stabilized as the world's reserve currencies, the bank has done a superb job in softening the gyrations of the smaller currencies and we're finally getting IMF money – in fact, the world's money – into places where it can really help the emerging nations.'

Jack nodded, although he wasn't sure why they were telling him all this.

'I don't suppose Exec Deakin has told you much about the role of UNISA?'

Jack shook his head. 'Not really, Mr Secretary.'

'No? Well, that was before you signed the Silence Resolution. I'm going to hand back to Director Amethier in a moment and he can give you an outline of the agency's function. But before I do so, I want you to know that, with the exception of the representatives of a few member states, the members of the World Trade Standing Sub-Committee and the International Security Standing Sub-Committee have been informed that this meeting with you was due to take place. They also know the substance of the information that will be imparted. What you are about to hear is of the *utmost* importance to the future of the world's peoples.' The Secretary-General's eyebrows lifted, questioning whether Jack had fully understood.

*Jesus!* What could matter so much that UN members would be informed of this meeting? It had to concern the Phoebus Project, the development that had first made Jack consider approaching his old Washington contacts.

He nodded and realized his throat was dry. Abruptly, the Secretary-General stood up and Jack found himself rising along with the others at the table.

'I hope to see you again, Commander Hendriksen,' smiled the Secretary-General leaning forward and extending an arm across the table. Jack rose and shook the large hand again. Then Dibelius turned and left the room.

'Right, let's get on,' said Amethier as they resumed their seats.

'Could I get some water?' asked Jack.

## Chapter 1

When he was thirty-two, Thomas Tye personally invested four hundred and thirty million dollars in Erasmus Inc., a start-up corporation that had been spun out of the genomics department at the Johns Hopkins Research Center in Baltimore. The company had filed a patent application identifying a string of genes that, it claimed, were the principle cause of a condition known as progeria. This disease affects only one in 240,000 people, but its effects are horrific, with the most prominent visible symptom being premature ageing. A sufferer as young as twenty-five can appear to be in advanced old age and early death is inevitable. In adults, progeria is called Werner's Syndrome and most adult sufferers contract the disease in their early twenties and die before the age of forty.

Erasmus's discovery promised the first effective treatment for Werner's Syndrome but, seventeen years later, no therapy based on this gene string has yet reached the market. Twelve years ago, Erasmus Inc. and its thirty-six genetic researchers relocated to the Tye Corporation's science park on Hope Island and thus escaped the routine progress filings required of biotech companies under American FDA regulations. The company also closed its articles of incorporation in the state of Delaware and became a closed company within Hope Island State. For over a decade, therefore, the world has remained ignorant of progress towards a treatment for this horrendous condition.

But a treatment *was* developed a few months

after the original patent filing, though it had nothing to do with Werner's Syndrome, which is best described as an 'orphan disease' – where there is no incentive to develop treatments because there are not enough sufferers to generate profits. The genes that produce the startlingly premature ageing symptoms of progeria are also responsible for controlling most, but not all, of the human ageing process. Erasmus Inc. identified the remaining age-control genes, in particular those that govern the sclerosis of the central nervous system. So they learned how to switch off the ageing process, or delay it almost indefinitely.

After a series of trials, the first human 'patient' to undergo such long-term treatment, and the resultant therapy, was Thomas Richmond Tye III himself.

Chronologically, Tye will be fifty years old at his next birthday. Physically he is still thirty-four. Every three months the researchers at Erasmus take cell and tissue samples from his body and submit them to detailed analysis to detect signs of ageing. Their confidential internal report (see the full report at this book's network resource) reveals that there has been almost no change in Tye's basic cell structure for seventeen years. It explains how the normal release of the toxic oxidative by-products of metabolism (known as free radicals) that damage human DNA is halted. The report also reveals that twenty-two of Erasmus's scientific staff have since joined Tye in the experiment.

The printed evidence had arrived in Haley's letter box two months earlier. It had been posted in Amsterdam but did not carry a sender's name. At first, Haley had had trouble even understanding what it was all about. But the appearance of Thomas Tye's name in the opening paragraph had made her persevere and finally, with the help of medical, genetics and biotechnology dictionaries and glossaries on the networks, she had managed to decipher the gist.

Her first conclusion was that the document had to be a hoax. At that stage the story about her book and the large advance Nautilus had paid for the US rights had only recently appeared in the publishing press and Haley guessed that someone with a grudge against Tye or one of his companies was trying to feed her highly inflammatory material. Or perhaps it was the Tye Corporation's own people trying to trip her up?

Then she found reasons to reconsider. She recalled that a cousin was sharing his Islington flat with a researcher who worked at one of the Wellcome-Parke laboratories in East Finchley, north London. A quick call to cousin Maurice had led to a conversation with his flatmate and a subsequent meeting in a brutally bright Bloomsbury bar that catered for students at nearby London University. While Haley waited for him to arrive, she thought about security and patched herself into Felicity's system in Ladbroke Grove. She and her impending contact had agreed to swap locator codes for the evening so as to avoid rendezvous mishaps, and the twins watched as Haley's VideoMate displayed his slow progress along Gower Street.

There! The lost-looking man in the entrance had to be him. Haley crossed the floor.

'Doctor Evans?'

He was short and bearded, with a thick red facial growth compensating for a hair-line that had already receded to his crown.

'Maurice tells me you're a writer?' He smiled as he took her hand. He had a very broad Welsh accent, as if he'd not long been in London.

Haley bought the drinks and thanked the geneticist for meeting her. They found a quiet corner and, after a few exchanges about the bar's noisy young patrons, she handed over the document that had been posted to her.

Fifteen minutes later, her new acquaintance laid it down.

'Where did you get this?' he asked, his eyes accusatory.

Haley explained how it had arrived, then asked if he thought it might be a hoax.

'If it is, it's a bloody good one and they're using the name of one of the world's leading genetic engineers,' Evans mused. 'It claims it's by Professor Eli Kramer. He's really a biogerontologist, you know, studying the biological causes of ageing. There's only half a dozen people who know as much in this field as he does.'

Haley begged him to keep quiet about the document, until her book was published. He bought her another drink and flipped through parts of the report again. Then he smiled and asked if he could buy her dinner 'somewhere quieter'.

'Go for it!' Felicity had urged in her ear.

Jack sipped the mineral water Deakin had called for. The others at the table had followed his lead, and all now had glasses in front of them as the door closed behind the catering assistant.

Amethier put his water down. 'Commander Hendriksen . . .'

'Jack, please. I'm no longer attached to the Service.'

'OK, Jack, UNISA is the sort of agency the National Security Agency always wanted to be but couldn't quite manage to become. We're not answerable to any type of congress or parliament, we answer to the Secretary-General and the Security Sub-Committee alone. We employ eleven thousand people, located in almost every nation across the globe. We work closely but quietly with the NSA, the CIA, the FBI, Interpol, FSB, Mossad and every other intelligence service in the world. We're information-led in approach, closer to the American National Security Agency than to any of the "dirty-tricks brigades", as the Brits put it. We're non-combatant, so we rarely get involved in physical action ourselves. When something of that nature needs to be done, covert UN or NATO forces usually do it for us.'

Jack nodded, surprised that during his years of service with the US government he had never heard of this agency. But then he realized that he had not heard of dedicated covert forces within the UN or NATO either. Perhaps the American intelligence agencies were not as omnipresent as they liked to think – or, at least, not in UN circles.

'Ron Deakin's been on the Tye case for three years now. In fact, it was he who suggested your name to Bob Grant, the man who hired you into the Tye Corporation in the first place.'

Jack shot a look at his friend.

'It was the first thing I did when I took the case on,' acknowledged Deakin with a smile. 'I wasn't sure what I was getting you into at the time, but you've done very well by it, haven't you? Now you're in a position to return the favour.'

'So, what do you want from me?' asked Jack, mystified.

Deakin picked up the thread from his boss. 'The Tye Corporation has grown too big, too powerful, Jack. There comes a time in this world when however much presidents and prime ministers like to schmooze with trillionaires, they get worried about their power and influence. Eight years ago the former United States president put the CIA on to the case, to examine just how dangerous the Tye Corporation could become, and he didn't like the report that resulted. It suggested Tye would end up calling all the shots. And, in the end, if somebody gets too big, things get *political*.' He stressed the last word as though it was distasteful.

'That's not wholly fair, Ron.' Amethier rebuked him lightly. 'Some very legitimate causes for concern have now surfaced, as you'll know better than anybody.'

He turned to Chelouche. 'Would you mind, Doctor?'

The banker's head was bowed and his hands were cupped around his water glass. He looked up sharply at Jack with sad eyes that seemed almost opaque, as if a veil had been drawn across them to hide their secrets.

'Do you know anything of economics, Mr Hendriksen?'

'*Nothing* would be an over-generous description.'

'The Tye Corporation is getting so powerful we think it risks destabilizing the world's economy,' the banker said carefully in the same Israeli-accented gravel tones that Jack had heard so often on the news broadcasts. 'I first raised the subject four or five years ago. We therefore did a study – created a scenario – and even then we found that if the Tye Corporation suffered just a couple of bad quarterly financial results, the whole world's economic growth might actually turn negative. Investors would panic, millions of little traders would get hurt and the international markets would crash.'

The 'basset hound' sucked in his fat left cheek as he searched for the right words.

'I concluded that it could lead to a major world recession,' he explained. 'But it appears my worries were ill founded at the time or, at least, I was worrying about the wrong thing. Far from failing, the Tye Corporation is now four times the size it was then. The largest corporation the world has ever seen. Its annual revenue is greater than Germany's GDP...'

He shrugged. 'Who has control over it? Not the World Bank – not any central bank, not any government. Exec Deakin is right: this has become political. My member states are uncomfortable with the current situation – the principle of laissez-faire is OK only when granted, not when appropriated. In the old days we could apply national anti-trust laws. We could fine, supervise or break up corporations when they became too big, too monopolistic, too wealthy or too powerful. Now that they're global, we can't do that, except in trifling ways in individual territories. There's no such concept as antitrust regulations or anti-monopoly legislation in international law. The Tye Corporation is a rapacious, unprincipled, monopolistic, money vacuum. And with the new information Exec Deakin has shared with us...' Chelouche shrugged and began to study his fingers.

Jack was puzzled. These ideas weren't new to him. Many of the news magazines had run pieces on the same lines over the last few years, and he had been present numerous times when Thomas Tye addressed such public concerns on TV and on the platforms.

'Always remember we're a *public* corporation,' TT would proclaim in his best shareholderese. 'All we ask is the freedom to be creative, to innovate, to bring really great new products and services to the world at large. Our shareholders are like voters: if we get things wrong, they'll vote via their stock portfolios.'

And so far, it seemed, Tye had done nothing to upset his elite body politic.

Amethier took his cue from the banker's lapse into silence. 'There's something else, Jack, one of the main reasons we've dragged you here – the new information Doctor Chelouche refers to. Look at this.'

The Director of UNISA again pressed the button on his

control panel. Jack looked up at the screen and saw a standard form with the heading *World Patent Organization.*

'This is a patent filing, Jack. It's from Bioneme Research, a subsidiary of the Tye Corporation. Don't bother to read it all. The point is that there is a very, very unusual error in this document. The object of the patent is to secure the rights to a therapy produced from a string of genes, a treatment that controls human hormone production, specifically progesterone, oestrogen and testosterone.'

Amethier turned to the World Health Organization scientist who had moved along the table to take the place vacated by the Secretary-General. 'Professor Berzin, would you like to explain?'

She bowed her greying head slightly. 'According to that patent filing these hormones govern how humans smell. They create our individual body chemistry, Mr Hendriksen.' She spoke with a faint East European accent and Jack guessed she was from Poland. 'As you may know, it is now widely understood that pheromonal attraction is by far the most important component of sexual desire. To a large extent we choose our mates by how pleasing their smell is and by how different their genetic mix is from our own – we analyse this unconsciously from the body's olfactory signature.

'Progesterone, oestrogen and, to a lesser extent, testosterone are circulating prohormone steroids and they control how receptive myometrial cells are to oxytocyn, the hormone that's made in the brain and governs the body's olfactory chemistry. If this string of genes can be modified and brought under control, a man or woman would be able to use a ScentSampler to analyse a desired partner's chemistry and then construct a pheromone mix that was sufficiently similar but genetically different to enhance their chances of seduction. And, as this patent suggests, the effect would be very powerful.'

She paused and looked with a slight frown at the pen she held between her hands. 'A therapy for these genes, say an oral DNA vaccine, would be like manufacturing a love potion naturally, inside the body.'

'It's got blockbuster drugs written all over it,' interjected Deakin.

Jack smiled. He had heard Tye and his associates discussing many other such blockbusters. So many drugs and treatments seemed to be dubbed 'miracles' these days, he wondered why this particular one had caught their attention.

'The point is, the genes mentioned in this patent filing don't actually exist,' explained the professor. 'Or rather, they do exist in the genome, but the ones identified in this patent are silent, inactive. They're incapable of affecting hormone or pheromone production or any other function of human physiology. They belong to the ninety per cent of gene strings that don't seem to have a specific purpose, the non-coding or junk DNA.'

'So, Bioneme's researchers have screwed up?' queried Jack, looking at the banker. 'Is this serious enough to cause a global financial problem?'

'Just a minute, Commander,' broke in Amethier. 'Take a look at this.' He pressed the control-panel button again, and this time Jack saw a page headed *Highly Confidential Memo.* From the logo he could see that it was an internal document of the Pfizer-LaRoche pharmaceutical company.

Amethier highlighted a line near the top of the page. 'This memo was written four months before Bioneme filed their patent application,' he explained. 'But it wasn't written by anyone inside Pfizer-LaRoche. It was written here, by UNISA, and Ron led the project.'

Deakin grimaced and turned to his former pupil. 'We *made up* that damn thing, Jack, with the help of some of Professor Berzin's researchers.' He looked to the WHO iatrochemist for confirmation.

'It was the work of some very clever osphresiologists and hormonologists in my research group,' she nodded. 'I asked them to imagine their ultimate wonder drug. At Exec Deakin's suggestion I asked them to identify deliberately a sequence of silent genes that we know don't do anything in human physiology. Then they wrote an explanation of how an oral therapy might be delivered once a patent was filed and testing could begin.'

Deakin smiled and resumed the story. 'Now listen, Jack: when it was finished we scrambled the document, using the highest level of encryption we've got – string lengths above a hundred megabits, Pentagon standard – and e-mailed it from the Swiss offices of Pfizer-LaRoche to their offices here in Vermont. The company was happy to help, though they didn't know what we were doing. It was just another encrypted e-mail that came into their server, only they didn't have the private key to open it. *Nobody* did, Jack. In fact, nobody does to this day, not even anybody else in this room. I myself ran the software to create the encryption keys, and I destroyed the original memo and removed every trace from the system.'

Deakin slipped a data-storage card out of his shirt pocket. 'This is the only copy of the key that exists in the world, and I have never used it since I originally scrambled that memo.' He sat back to see if his friend would see the implications.

Like everybody, Jack understood that super-strong encryption was considered unbreakable. Unlike everybody, his training as a field intelligence officer had provided him with an understanding of the maths behind it. He knew that even if every scrap of computer power on the planet was connected and run as the largest parallel processing computer ever imagined, the amount of time required to discover the set of high prime numbers developed for one encryption key of such length would run into tens of thousands of years.

'You're saying Bioneme got hold of a plain-text copy of that memo?' he asked incredulously.

'There were no copies made, Jack – plain-text or otherwise. Professor Berzin's team worked solely in my office and they took nothing away. After I sent the e-mail from Switzerland I destroyed my original plain-text file and electronically scrubbed the storage media. The only copy that exists is the encrypted document I e-mailed to Vermont.'

'And that encryption is unbreakable,' confirmed Jack.

Deakin turned to the British mathematician who had been silent throughout the meeting. 'Your turn, Doctor Mathison,' he said.

# SEVEN

Theresa Keane walked to the centre of the stage and turned to face her audience. Every seat in the lecture theatre was taken. They were also sitting in the aisles and standing at the back. Even the floor space intended for wheelchairs in front of the low stage was filled with the upturned eager faces of people who sat hunched and cross-legged on the carpet. Volunteer stewards who were supposed to keep the fire exits clear had given up this hopeless task, and now stood facing the stage while knots of latecomers tried to squeeze in behind them through the doorways. All over the campus hundreds more would be watching the video stream.

The buzz had started in the Hope Island networks even before a formal announcement was made. Professor Theresa Keane, the Tye Corporation's Nobel Laureate of computer science, was going to give what she billed as a 'Summer Lecture', her first public performance since becoming Director of Hope Island University's School of Virtuality the year before.

Every 'student' on the campus of Hope Island University was actually postgraduate and many had already achieved their doctorates. A few had arrived as associate professors. Each year, the Tye Corporation's team of human-resource scouts roamed the world's greatest universities, seeking out the brightest and the best to make them offers they found hard to resist. It didn't take much to seduce young computer scientists, geneticists, astrobiologists, evolutionary psychologists, cognitive scientists, physicists, chemists and mathematicians into spending a few years in the semi-tropical climate of a Caribbean island, studying with some of the world's best brains and – even more unusual in academia – being paid handsomely to do so. Nobody claimed that Hope University offered a wide range of academic opportunities, but in the fields of the life sciences, computing and communications it was the equal of any institution in the world. Its patent record was unrivalled.

Keane's reputation preceded her. It wasn't just her Nobel Prize. The gossip addicts of the networks insisted that she was regarded as the most gifted lecturer MIT ever had: even better than the great Feynman, some said. At her former university the students had dropped everything when they heard she was giving a lecture and, the word was, she never needed notes, she never waffled and she always captivated her audiences.

She smiled, pleased at the turnout. 'Good afternoon. Thank you for coming. Let's keep the house lights up for the moment.'

She walked to the front of the low stage and paused, looking down at the press of young researchers at her feet.

'May I?'

Those who were sitting cross-legged on the floor at the front of the pit area had to shuffle backwards on their buttocks to make room. The professor stepped down to stand amongst them, elegant in a beautifully cut olive-green trouser suit and white blouse. She smiled and lifted her arms as if about to conduct an orchestra. Every individual in the room felt contact. As one, the audience seemed to lean inwards.

'This glorious planet, our home, is between four and five billion years old,' she began, her soft Irish lilt lending natural melody to her words. 'About four million years ago our distant ancestors started to lean back on their hind legs and, step by step, they began to walk in what anthropologists now call plantigrade fashion. Thus began the final part of the evolutionary process that was to lead to human beings.'

She had stooped to match her action to her words and now she slowly rose again to her full height and shielded her eyes with her hand, as if protecting them from the sun.

'In this fully upright posture, *Homo erectus* tilted his head back in order to see into the distance, over the tall savannah grasses of Africa's Rift Valley – that's in today's Ethiopia, Somalia and

Kenya. This step hastened the development of vision, which has become our supreme sense. Over the countless generations that followed, gravity caused our skull to elongate, our brain started to expand in this new space and our larynx fell to the bottom of our throat.'

The professor pressed her hand to her throat and looked around at her audience. 'Here it found room to enlarge, to grow a nervous system and become mobile. The range of sounds that it could produce expanded dramatically and thus, through language, we stumbled upon our most important ability . . .'

She paused for effect.

'. . . *The interconnection of one single intelligence with many others.*

'It was language that provided the feedback loop that sent the human brain on its runaway evolutionary progression towards the emergence of consciousness or, as I prefer, coeaesthesis or self-awareness – the general sense of existence, of immanence, that arises from the sum of bodily impressions and mental observations, the vital sense. It was language – spoken, unspoken, written, symbolic and conceptual – that was the trigger to this fantastic, accidental creation: humanity's individual and collective virtuality.'

She had them. She smiled around the hall and held their gazes. 'Language, and the virtuality of which it is the prime representative, was so successful as a random evolutionary excursion that, in under four million years, it achieved for the genus *homo* a breakthrough and a developmental spurt that had not occurred in the nearly two hundred million years during which fish, mammals and dinosaurs had rule of this solitary, lucky and almost unbelievably fecund planet.'

Her listeners were absolutely silent.

'Language is the essence of humanity and it was the first external symbol, simulation, representation or *virtual* element to appear in what had been, up to that point, a totally physical world. This uniquely human form of shared consciousness began the moment humanity named itself.'

The professor leaned towards a bespectacled young man at her feet and held out her hand. 'I am Theresa, you are . . .?'

'R-Robert,' he replied, with a hint of a stammer.

'Good afternoon, Robert.' She smiled at him and turned back to her audience.

'We named each other and the objects in our world with abstract but mutually agreed sounds that can be taught to others in a group network. Language created humanity's past and the present and gave it some tools with which to imagine the future – all based on virtuality, which for so long has been misunderstood by those imbued with it and erroneously expressed as spirituality or soul. We are not clever *animals*, ladies and gentlemen; we are primarily identities of virtuality trapped, for the moment, within physical biological containers or, as I prefer to call them, constrainers. This is our psychogenesis.'

The professor was into her stride. She explained that the Neolithic cave paintings were humanity's first recorded form of virtual expression. She underlined how it was the agricultural revolution alone that had produced the wealth, and thus the time, that allowed humans to invent the concepts of writing and money, two of the most powerful forms of virtual information storage. And she urged them to accept the concept of digital representation, virtual existence, as a logical destination for the human species.

'Our migration into the digital networks around this planet is a natural extension of human virtuality,' she explained. 'By definition, humans are virtual creatures and we are at our most powerful when our habitat is also virtual.'

None of this was new to her audience. Her first best-seller, *Global Virtuality*, had laid the foundations of digital-age philosophy ten years earlier and it was now almost mandatory reading for undergraduates, whatever their branch of science. But they also knew she never missed an opportunity to reinforce her message. As she claimed, the virtual environment was still a very new place for the human psyche and it was always tempting to dismiss the intangible as unreal.

'And so we come to the subject of consciousness,' Theresa announced. 'Some of you will know that machine consciousness has been the focus of my work in recent years.'

*Some of you will know?* Every one of them knew and every

one of them was hoping for an update. She wasn't going to disappoint them.

'My reason for starting with a brief overview of human evolution is that until a few years ago we were ignoring the process, even though it was a model that had been staring us in the face all the time. For nearly forty years the field of artificial life and machine intelligence yielded nothing but disappointments. It seemed as though our efforts to build a machine that had human-type intelligence were doomed to failure. Many said we were failing because humans have qualities that can't be captured within a machine. Others said it was because we didn't even understand the object we were trying to copy, let alone become capable of replicating it. I believed that attempting to create consciousness as if it were a product, or a function of software code, was wrong. I believed that consciousness is something that emerges spontaneously in a given set of circumstances, probably out of immense processing density and billions of individual transactions – as in the human brain.

'Then, about seven years ago, my team and I wondered what would happen if we applied our latest understanding of human evolution to our research. Everybody in this room will know that, as humans, we are merely the latest members of a group of eighty billion or so hominids who have so far gasped momentarily for life, reproduced and died on this planet. We are the latest models of a line of almost infinite prototypes, developed with no hint of temporal urgency or parsimony of resources and with no whiff of interest in the fate of individual experimental models to distort the process that led to our accidental but seemingly wonderful design.

'We decided to take these principles of natural selection, of biological evolution, and speed them up. We have applied this process to our research and I will be demonstrating some of the results this afternoon.'

The room crackled with expectation.

At 46,000 feet above the Atlantic, Raymond Liu adjusted his viewpers so that the professor's image became solid. He didn't need to see what was now in front of him in the physical world. It was only the bulkhead in the executive cabin of the Tye Corporation jet and the network supremo was keen to watch the professor's lecture. Machine consciousness wasn't his field, but Keane had a world reputation and he had been disappointed that the meeting in Newark that had just ended had denied him the chance to attend her first Hope Island lecture in person.

During the first fifteen minutes of the short ride back to the Island Liu had issued instructions for the formation of two network-investigation teams to explore the recent satellite failures. The Red Team would work from the island's Network Control Center. The Blue Team would work from the Tye Networks Control Center in Singapore – an almost identical facility that had originally been intended as a contra-hemisphere back-up unit in case of colossal atmospheric or geophysical disruption in the Americas. Now, like its Hope Island counterpart, it was used mainly for sales demos. The two teams' tasks would be to construct new security barriers within the communications networks and to watch for any unauthorized attempts at command communications with any of the 22,866 satellites active in the forty-two networks. The most skilled hacker on each team would play the role of an opponent and would attempt to breach the opposing team's assigned networks without being detected. Liu had decided to blitz this problem.

Although he hadn't mentioned it during his meeting with the Russian network controller, the problems in the networks above the South-West Pacific had not been the first anomaly to occur since he had been given overall technical responsibility for Tye's satellite communications networks. Only seven days earlier, one of the data processors on board an Air Traffic Management satellite serving the North-West American coast had frozen and crashed. The system had reset itself and reloaded its data in just eleven seconds but this event had rated an 'immediate action' malfunction report and it had appeared on his system within seconds of the occurrence. He knew it would also be prompting an FAA investigation.

Liu had ordered the satellite taken out of service immediately and he had switched the ATM processing to one of the standby satellites in the same quadrant. He had asked Tye Aerospace for an 'earliest possible' launch slot to drop a replacement into orbit, but he also knew he might have a wait: all non-third-party launch capacity seemed to be taken up with launches of space stations for something called the Phoebus Project. It wasn't a networks communications project so it wasn't his concern, but he knew he would have to argue with Aerospace and then with Tom to get slots for replacing all three satellites that were now suspect.

The individuals who would make up the investigation teams would be identified by Liu's executive staff by the time he landed back on Hope Island. He had been unequivocal: this investigation had priority over all other projects and only the most experienced and the most talented would be seconded to work in shifts around the clock until there was a definitive answer about the cause of the network failures. Liu knew his own future depended on finding the cause and eradicating it before any further problems occurred.

Satisfied that he could do no more until the potential members of the investigation teams had been identified, Liu turned up the sound for Professor Keane's lecture.

He listened as she described her Anagenesis Experiment – the early attempts to produce super-intelligent software by creating computer environments that mimicked ecosystems. She described the many failures there had been and how she and her team had struggled to find methods of reproducing the forces that had shaped human evolution and the necessary density and complexity of decision-making systems from which a form of consciousness might emerge.

'In the end we decided to create independent software agents that were heuristic, wholly autodidactic and that were able to reproduce themselves continuously. We made two types of personality we call male and female. We gave these agents two overriding imperatives: to reproduce and to eat – to take on energy – although we also programmed them to pass on to their offspring whatever they learned during the competition and selection processes. This decision lies at the heart of our evolutionary acceleration algorithm. There is no relearning to be done by each generation; it's a form of palingenesis.

'For every one female entity, we made ten males so the boys would have to compete like mad and so further accelerate the evolutionary process. We also gave both sexes limited lifespans to cleanse the field of obsolete generations. Females live until they have reproduced sixty-four times or until whenever they go for more than three days without reproducing. The males die when they have fathered sixteen offspring, or if they fail to reproduce within thirty days, whichever is the sooner. Male-to-female ratios in the offspring are as per our starting point. We colour-coded the females green, the males red – and then we released our software robots into the world's networks to fend for themselves.'

Raymond Liu felt the hairs on the back of his neck rise. He leaned forward in his seat even though that had no effect on how the image was displayed by his viewpers. She couldn't have been that irresponsible! He went to put a question but he saw that the event had not yet been opened for questions from the remote audience.

Then he heard laughter. Someone from Evolutionary Psychology had asked what a software robot liked for lunch.

'Sushi,' someone else shouted and Theresa smiled.

'To simulate eating we decided that every entity had to return to the mother server once every twelve hours for what we call a refuelling stop. If an agent fails to do so precisely at its allotted time, it dies. It also allows us to track every bot's movement and ensure the experiment isn't getting out of hand.'

Liu sat back in his seat again.

'Now, we also had to decide what fitness characteristics we wanted to select for – or, to put it another way, what traits the female personalities would find so attractive in the males that they would allow themselves to be used for reproduction. The human female's refusal to mate indiscriminately is, of course, the second key to the incredibly rapid evolutionary anagenesis of the human species. In short, boys, when she says "no" you are observing the two vital keys of human evolutionary success used in a devastatingly effective combination: language and positive selection.

'You will have guessed that we restricted the females to one duplication – or reproduction – per day and we programmed the females so that they only became reproductively viable – in season, if you like – for twenty minutes in every twenty-four hours, those twenty minutes to occur randomly. A female spends the rest of her time receiving advances and comparing the desirable characteristics of the potential mates who are courting her. Despite this, the females' main objective is to reproduce with the most desirable mate they can find and it is his desirable traits, and of course her ability to select for those traits, that get amplified in the offspring.'

Keane turned to her audience. 'So, what do you think is the key attraction we want the male bots to develop? Remembering the aim of our development exercise, what should be the one thing that will drive the girls mad and grant the males access to reproduction?'

She waited to see if any of the brightest of the bright, most of them at the peak of their own reproductive potential, would offer an insightful suggestion.

'Money,' shouted a young female voice from the rear.

They laughed, Keane with them. 'Yes, it could be money. Resources are very important for human females considering reproduction. Anything else?'

'Elegant code,' shouted another researcher, a male.

'Digital good looks,' said Theresa smiling. 'That's a very good idea and very male of you. Only the best code, or the fittest, gets to reproduce. Any further suggestions?'

There were none.

'We programmed the female entities to look for *human* characteristics in their potential partners,' she said quietly.

And it had stopped, just like that. Joe Tinkler had sat stock-still, watching the prices in his Tye Corp portfolio for another fifteen minutes, his finger poised over his mouse – his manual panic button. The disposals had stopped. Joe's agents hadn't reported a single further sale by Tye or any of his representatives or legal entities. No other big sellers were in play and the prices in the core Tye stocks had started to creep upwards again.

Joe allowed his right hand to fall back into his lap and he slumped in his high-backed chair. The markets had hardly moved. Tye had realized the equivalent of over one trillion in cash, mostly in US dollars and euros, and it was as if nobody had even noticed.

Then Joe sat forward again. Tye was breaking the audit trails! He must have gone to cash simply to stop analysts following him from one investment to another. It was anonymity he was seeking. That meant that a trillion dollars was going to be laid in new investments very quickly. But Tye had made the analysts, the trackers and the millions of small investors who shadowed his every movement temporarily blind.

Joe called four new agents to his screens and quickly gave them their instructions. If he could spot when Tye was investing the cash, he could catch a free ride on Tye's coat tails until the rest of the market spotted it.

If it had been anybody but Connie Law, Raymond Liu wouldn't have accepted the incoming call. As it was, he had snapped off the image of Professor Keane, accepted Connie's handover and watched as his ultimate boss circled in his Holo-Theater.

'So what happened?'

Liu answered Tom as truthfully as he could. Within a few minutes the swearing had abated. Tye stood still and zoomed in so that his eyes filled Liu's vision.

'You find whoever is in those networks within twenty-four hours!' he snarled.

The connection went dead and Liu was left staring at blackness. It would take him forty-eight hours just to get the investigation teams in place.

'Would you like a drink?'

The engineer slipped his viewpers off. The flight attendant was at the bar; she knew what was needed when one of her passengers had been in communication with TT.

Liu swallowed and then nodded. 'Give me a Scotch, on the rocks.'

She made it a large double and smiled as she handed him the drink. 'We'll be landing soon. Please fasten your seat belt.'

After he had fastened his belt and taken two long pulls on the drink, Liu's breathing eased and he put his viewpers back on, switched to playback mode and rejoined the lecture where he had left it.

'Please turn the house lights down,' requested Professor Keane as she stepped to the side of the stage. When the room was dimmed she turned back to face her audience.

'And now, with no apology for what is a gratuitous act of anthropomorphization, I want you to meet Miss Scarlett.'

In the centre of the stage the Holo-Theater snapped on and the head, shoulders and torso of a young female appeared. The vintage-film fans in the audience recognized her immediately and they whispered her name to their friends. It was Vivien Leigh, one of the greatest film beauties of the mid-twentieth century. She was in character as Scarlett O'Hara in *Gone With the Wind* but at a quality and resolution of which Herbert Kalmus and his Technicolor film engineers of the previous century could only have dreamed. She was wearing the white, black-trimmed, two-piece travelling suit bought for her in Charleston by Rhett Butler and, in this representation, the software and projection system gave her three dimensions.

'Like all our software agents, Miss Scarlett has been allowed to choose her own visual identity from the Tye Digital Arts archives. Allocation is governed strictly by how successful these agents have been in achieving their twin goals. Perhaps rather unfairly, females are judged by the reproductive success of their male offspring. Miss Scarlett is the most successful female of the current generation and we have reprogrammed her to spend her next twenty minutes of, shall we call it "courtship time" with you. She will regard all men in this room as potential suitors.'

The professor walked towards the front of the stage. 'In a moment I am going to turn on Miss Scarlett's natural-language interface. When I do, I want the gentlemen to raise their hands if they want to ask Miss Scarlett a question. I will select a questioner and he may continue in a dialogue until he chooses to say "End". I will then select another questioner who may continue until he feels he also has done his best.

'I want you to know that we have programmed the female software agents to judge the characteristics of humanness by *language*: your eligibility to reproduce with Miss Scarlett will be judged by the quality of your chat-up conversation, as is so often the case in the human world. Remember what I said earlier: it was language that provided the feedback loop that sent the human brain on its runaway evolutionary progression. It was language – spoken, unspoken, written, mathematical, symbolic and conceptual – that was the trigger. Miss Scarlett understands all forms, but today we will deal with the verbal. English only, and please be courteous; Miss Scarlett thinks she's a lady. *My* team – remember you're disqualified!

'Who will be first?'

Liu watched as a sea of hands shot up. Professor Keane picked one, gave a command to the image and then stepped back.

'Who are you?' asked a young man in the front row.

Miss Scarlett raised her eyebrows and looked down at the questioner.

'Why, Ah believe you already know that, suh,' she said in a soft Southern accent. 'Ah'm sure I heard the professor introduce us.'

Liu didn't notice a solid *clunk* as the landing gear lowered and locked, nor the gentle bump as the small supersonic jet touched down. He didn't notice the noise of the engines dying as they were powered down. When he felt a hand on his shoulder he waved it away and the flight attendant left him in his seat while she completed and filed her flight log on her VideoMate. He watched as questioner after questioner tried to engage, charm or confuse the software agent with oblique, tangential, non-sequential, litotal, counter-intuitive, antiphrastic, erotic, sexual and surreal conversation. Miss Scarlett responded with vivaciousness, humour, interest, boredom, irony and derision. She even made the film fans laugh out loud and applaud – when she exclaimed 'Fiddlededee!'

When they were all done, Professor Keane turned to her

agent. 'You are free to select a partner for reproduction from those in this room. Indicate his identity by replaying any part of your conversation with him.'

'I don't think I will,' said Miss Scarlett coyly. 'Ah'm afraid it would be a regression compared to, well, compared to, shall we say, an alternative romantic opportunity that occurred earlier today.'

Liu realized that he had just watched *humans* fail the Turing Test.

He uttered a low moan as he ejaculated inside her. Haley slipped her hand between them and held him gently, where he was most sensitive. His head dropped to her shoulder and she kissed his neck. He had been so shy. It had been their sixth date – and fifth classical music concert – before the Welsh geneticist had made a move and even then he had been hesitant and unsure.

But how gentle he was, how caring, how concerned for her needs and her pleasures. Despite this, she already knew that there were many things between them that grated, that turned her into what she thought of as an ungrateful, over-critical lover. Sometimes she despaired of ever finding the right partner – a man who excited her emotionally and physically whilst being someone she could truly like and respect. It seemed as if those qualities were mutually exclusive.

Haley was realistic enough to understand that her unusual lifestyle was mainly to blame for her poor experiences in romantic relationships. She simply didn't meet enough men from whom to choose and she constantly made do with second-best. She was attractive enough to secure plentiful advances from males, yet she was rarely in a situation in which they could occur. On the infrequent occasions when she complained to her sister or her friends about her work and her solitary lifestyle denying her social opportunities, they pointed out that many other professionals worked from home these days and there were dozens of ways of joining social or professional clubs and communities where she could meet suitable men.

But they didn't really understand. When Haley was working on a project, she was almost incapable of focusing on anything else. Her subject and her writing wrapped around her like an invisible shield against the outside world. She had made real efforts, however. Sometimes she forced herself to go to other authors' book-launch parties, to literary lunches and, on four occasions, on blind dates set up via network dating agencies. The results had become familiar and depressing. Once she was engaged on a book she found it hard to be engaged with anything, or anybody, else. In her spells between books when she did socialize, she simply grabbed the first reasonably attractive man who presented himself to her. This was not a recipe for long-term success and she knew that it arose from a feeling of mild desperation that she had forced herself to allow Barry to intrude when the only man who *really* mattered in her life at present was Thomas Tye. Suddenly Jack Hendriksen's face popped into her mind.

She kissed Barry's neck again as he stirred. She hated to allow the real world to intrude.

'You'll be late for work,' she murmured.

He pushed himself away on one elbow.

'Christ, look at the bloody time!'

He leaped out of Haley's bed and headed for the bathroom. She snuggled back down into the warmth of the duvet, where he had been lying. She felt drowsy and she wanted to enjoy the feeling of the gentle warmth that was washing up and over her. This was one of the few good things about working from home. She didn't have to face the Stygian Northern Line in the mornings: it was the last remaining route without air-conditioned business-class carriages and the London summers had been getting steadily warmer for most of her life.

'Naturally, we can't know what it is we don't know,' pointed out the desiccated mathematician redundantly. 'But I, for one, do not believe that the Tye Corporation could be breaking deeply encrypted messages from the networks without having achieved some sort of major breakthrough. It would have to be a massive leap forward in computer processing power, perhaps an improvement by a factor of several thousand, or some completely new form of mathematics – something I am wholly unable to guess at.'

They had been talking for over seven hours with only one or two breaks for their physical comfort. Food had been brought in and cleared away again and more brought in and cleared. Chelouche and the WHO scientist had made their excuses and left. Parts of the session had been agonizingly technical as Mathison had demonstrated the mathematical impossibility of deciphering encrypted communications in the absence of a manufacturing key.

Jack felt hard stubble as he rubbed his chin.

'What first made you think that Thomas Tye, or the company, had started hacking into encrypted comms?' he asked Deakin.

'It was the World Patent Organization,' explained the intelligence officer. He had first taken off his jacket, then his waistcoat and he now pulled at his tie before opening the collar of his white shirt. Ron was always well turned out, even in the field.

'You know that the WPO is another UN body?'

Jack didn't. He had never really thought about the UN – like most people, he guessed. While business had become a global affair these days, the majority of people were concerned only about domestic issues and local politics. Despite the ease and convenience of electronic home-voting, it was as if the masses had collectively given up trying to understand or influence world issues. He shrugged a negative.

'Nine or ten years ago things were a real mess. You had to file patents in different countries with different authorities. There were over one hundred different jurisdictions. Some of them followed one procedure, others another. Some authorities considered that *filing* first was the most important thing, some thought that *inventing* first – and being able to prove it – was all that mattered. We, the Americans, believed in inventing first but we were in a minority. The arguments kept the lawyers rich for years. Anyhow, the UN eventually realized that such intellectual property was going to be the equivalent of gold in the future and they stepped in. The WPO was established as a single, global patent office and it was agreed that filing for a patent first should be the key to awarding the patent. Our American methods were overruled. For the sake of *global harmonization*.'

Deakin ended his short history lesson on a note of sarcasm. He lifted his hand and ran it through his thinning hair. Then he sat back and hooked an arm over the back of his chair.

'So now we have a first-past-the-post system. No good sitting in your backyard inventing cold fusion on your own, you've got to get your papers in as soon as you can prove you've got something unique, something that hasn't been thought of before. It's irrelevant *when* something was actually invented. There's no such thing as the concept of prior art any more. The change was supposed to settle things for good, but it didn't. The result is a lot of nerds and academics going to civil courts, squabbling over who thought of what first and who stole whose ideas. You can imagine.'

Jack could indeed imagine. Although he had done well during his years at Columbus, he had been appalled at the pettiness of academic life and its schoolyard characteristics. His tutor had urged him to become a postgraduate student but as soon as Jack had finished his degree in physics he had opted for a life of unequivocal reality in the US Navy – not least because the starting salary would ease things for his mother back home. Mathison's attitude of aloof self-importance reminded him how little he liked academics.

'So the WPO started a database of complaints and complainants. Every time someone made a complaint or sued a patent holder for IP theft, those details went into the files. After a couple of years they had enough data to go mining and guess what they found? Half a dozen companies that systematically challenged patents and developed bogus proof of their own development projects to show to the courts. Reverse-engineering specialists, all of them! These claim jumpers always accepted a pay-off before their case came before the international court. It was very lucrative but after the WPO published its evidence, the claim jumpers went away as quickly as they had arrived.

'Then, about two years ago, they noticed a new trend. The Tye Corporation would file a patent and almost instantly another company or an academic institution would claim they had already invented the same damn thing and had been about to file for it themselves. Only this time, the plaintiffs were really respectable people – huge pharmaceutical companies, the world's best universities, you know.'

Jack did know. He knew that almost half the flights between Hope Island and the USA were shuttling patent attorneys to and from meetings – another breed he didn't like. He realized he was feeling irritable because he was tired. He also wanted a drink.

Few regions of the planet remained completely uninhabited or unobserved by humankind. Even the remote polar regions now hosted scientific research stations, while surveillance satellites continuously criss-crossed the hot arid wastelands of the equatorial wildernesses. Even the giant telescopes orbiting the moon and Mars were sometimes re-tasked to look back towards their point of origin.

For these reasons, it was agreed that the Tye Corporation's new Russian allies would be party to one of the few concentrated-power tests that would be risked prior to the launch of commercial services. Afterwards there should never be a need for such tests to be repeated. But Tye Corp had to discover if their calculations were right – and the Russians controlled swathes of remote, unobserved territory.

As it turned out, the Russian Federation naval frigate dispatched to play sea-level observer and policeman at Ostrov Vrangelya – a small archipelago in the Chukchi Sea – found only one Japanese whaler to harry and dispatch. It was chased away to the south and ordered to stay clear of 'forthcoming naval exercises' as the crew of the *Boris Dōlokhov* settled down to wait and watch in the Arctic cold at the edge of the paleocrystic polar ice cap.

The celestial event began, as scheduled, at eleven p.m. eastern-Siberian time and the sailors stared in awe as twelve miniature new suns appeared high above their heads in the dark blue northern skies. They could see the focus of the light beams on the *nunatak* five miles to the west and, relying on his superiors' assurances that there was no danger to adjacent areas, the captain edged his craft nearer to the patch of intense illumination to assist in the capture of film and video images and the taking of measurements. They moved through the ice floes with only just enough speed to maintain their heading and after three hours they were within a mile of the ice-shore.

The captain ordered the engines to be disengaged and, in the near-silence, they heard loud reports as the ice at the edge of the island cracked. They could see huge lumps of ice breaking off from the almost vertical cliffs and falling into the sea. Then the sailors on deck felt the heat and some started to strip off their waterproofs and sweaters, even though they were still some distance from the light.

On the bridge, the captain heard the crackle of a walkie-talkie radio behind him.

'Sir,' began the very young first lieutenant. 'Forward lookout reports the sea is boiling.'

Then they saw steam.

Jan Amethier leaned forward. He had been quiet, chewing the end of a ballpoint pen during most of Mathison's maths talk and Deakin's background briefing on the patent disputes.

'Finally the World Patent Office realized that other companies in which Tye was involved were attracting patent-filing complaints,' the UNISA director added. 'And these complaints also came from reputable organizations. All of them had meticulous evidence of their own development procedures and sworn statements from highly respected researchers. One of them in particular attracted attention: it was the patent for the solar-powered fuel cell that the Tye Corporation has sold all around the world. Then the WPO came to us – well, to the UN Secretariat, who then came to us.'

Jack nodded. 'But what made you suspect they were breaking crypto? It could be straightforward commercial spying. That's much more likely and I don't think Tom would hesitate. There's a whole industry of ex-spooks doing it – searching for keywords and voiceprints on the networks, using software agents to scan the airwaves, renting submarines to tap undersea cables. Hell, you can even lease your own time on NASA's spy satellites! And Tom doesn't need any of that – he already owns half the networks!'

'That's what we did think at first,' agreed Amethier. 'It seemed logical but everybody knows what goes on and everybody encrypts anything that's sensitive these days. Then the WPO reported that patent filings from the Tye Corporation and Tye's other companies were increasing rapidly and, as before, they were attracting what looked like legitimate complaints. It seemed to go way beyond anything that could be done by normal commercial theft. That's when I threw the file over to Ron.'

'As you know, I already had the Tye Corporation in my ongoing-observation portfolio,' confirmed Deakin. 'We keep tabs on all the powerful corporates and their executives. I'm sure you can understand why. It's the people versus the corporations these days and we end up as the cops – it's something the lawyers call *parens patriæ* – the UN has a duty to protect the people. At any rate, the director dropped it in my lap and I followed developments for a few months. I went to talk to some of the companies who had complained that their work had been stolen. Several of them said they'd had an approach from Tye's people either to take their company over or with an eye to investing in them, before either party had filed a patent – you know, the Tye Corporation's old tricks. They'd pretend to be interested in buying a company, Tye would stroke the young founders personally, get them to show all their intellectual property and then back out of the deal only to produce his own version a few months later. But *these* companies hadn't publicized their developments.

'Then I started to wonder if he was somehow getting inside electronic communications – as you say, the Tye Corporation owns what, forty, fifty per cent of the world's networks? So that's when I decided to develop something that we could go fishing with, something that couldn't be got any other way than by intercepting communications. So I pulled the WHO on side and their people did the rest. Bingo! He files for a patent on a therapy based on genes that don't *do* anything. They were in such a rush to file the patent first that they didn't take the time to isolate the gene string to check whether or not Pfizer-LaRoche's research results were right. They took things on trust just because the patent filing came from a famous and well-established research lab.'

Deakin stood and reached for another coffee flask, the tenth they had gone through during the meeting. More of a pre-operation briefing, thought Jack as he waited for the right opportunity to add something to their knowledge, something that had been burning inside him all day – the information that had prompted him to seek out Haley Voss, to finally consider consulting a lawyer and, perhaps, his old contacts in the American intelligence community.

'So we want your help, Commander,' said Amethier finally. 'I realize that all this is only circumstantial. Perhaps Tye and his company have found a way of deciphering encrypted communications, perhaps not. Perhaps it is just a massive commercial spying operation. As Doctor Chelouche told us earlier, things are very delicate. If Tye has found a way to break into the world's encryption system he could cripple any competitor he wants to. And he can read all *our* communications, Jack.'

'He could then trawl through our entire knowledge base,' added Deakin. 'He could find out everything the various security services keep to themselves. Hell, he could even find out who really killed JFK!'

'Who *did*?' asked Jack.

Amethier brushed this deviation aside. 'More importantly, he can find out what the governments of the world are planning. He could even engineer a complete collapse in the global economy *deliberately*!'

The UNISA director fell silent as his own vocalization made such an improbable threat seem more real.

'You know what that would lead to,' he then continued in a lowered voice. 'We'd slip back a hundred years. Wars would break out all over the world. We might even have world war again.'

Amethier paused and then brought himself back from fanciful concepts. 'At the very minimum, we'd have to alert every military force in the world that their security is compromised. The governments, security forces, the police, the banks…' He tailed off.

Jack scratched at his stubble again. He realized how deeply concerned they were. 'You want me to set him up for you? No, get someone else. You know that's not my thing–'

Deakin snorted. 'Jesus, Jack! Of course not. All of the irreversible options are out of the question in this case. Quite the opposite: we need you to keep him safer than ever! If Tye were to die suddenly that alone could bring the world markets down – even if it happened through natural causes. That's part of the problem. His power is so great that things are very precarious. We've got to handle this as carefully as…' he searched for an analogy that would resonate with his former pupil '…a covert agent-extraction. We need to gather hard evidence about whether Tye has found a way to unscramble communications. We've got to be sure before we do anything. We've also got to start a campaign to change public opinion about this man and his corporation. Tye's got a huge approval rating at the moment. We've had market research firms checking by electronic poll every week and the people of the world love him. Schoolkids chant "Take Care of Our Planet" and their buddies answer "We hear!" There's no way we could get him near a court as things are. It will take us a while to change that attitude. We thought Miss Voss could be useful there – with her book. Any idea who's sending her the inside stuff?'

Jack shrugged. 'Could be anybody. There are thousands of people with reasons to hate Thomas Tye. We've already got files on most of them.'

Deakin nodded. He didn't doubt that. 'I think *we* can persuade a publisher to go ahead with her biography,' he stated. 'You feed her the right stuff and that might start the process of changing how people think about him. Then we'll have to gather enough evidence to get Tye, or the Tye Corporation, into the international court at The Hague. It might take a few years, but it has to be done.'

'You haven't said anything about Tye Aerospace,' said Jack.

Amethier looked up sharply. 'What about it?'

'Did you know President Orlov has just visited Hope?'

'Sure. Air traffic coordination,' Deakin grunted. 'UN ATM handed President Orlov's plane on to Hope Island airspace.'

Jack nodded. *Of course.* 'Did you know Anton Vlasik was in the party?'

'We did *not*!' snapped Amethier. 'I thought he was safely in prison. What the hell's that thieving bastard cooking up with Tye?'

'Well, they were visiting for a demonstration of a surprising new technology. During this vacation I was intending to see a lawyer about how to proceed. I didn't know whether I could do anything, what I *should* do, knowing Vlasik's record . . .'

He tailed off, aware how far-fetched his claims might sound. Jack didn't yet know how the technology worked, but he had seen the elaborate preparations. He had been at the Science Academy in Moscow two years earlier when Tom had first met the Russian leadership and started his negotiations. He had watched as ships arrived carrying the components for the giant satellites and their incredible extensions. He had watched hundreds of shuttle launches from Cape Hope. Damn, he had even met some of the all-female crews who were running the construction projects on the orbiting space stations. And, of course, he had seen that dazzling demonstration.

'I have very few firm details, but the Tye Corporation is going to sell sunshine,' he said simply. 'To the highest bidders.'

# EIGHT

Calypso helped the limping athlete out to join his mother in the waiting area. He was twelve years old and he felt he no longer needed his mom's presence in the consulting room when he saw the doctor. Calypso winked at her over the top of the boy's head, resisting the temptation to ruffle his bright ginger hair.

'Nothing cracked or broken, Sonia,' she smiled. 'I've bound it up tight. Make sure he rests the ankle for a couple of weeks and ...' she turned to her patient and wagged her finger '... no more soccer for a month.'

Sonia, wife of a senior vice-president of marketing in Consumer Electronics, put her arm round her son's shoulders and smiled her thanks. Calypso held the surgery door open for them and they eased their way down the short path to their Volante.

'Come and see me again, Gary,' she called. 'But don't make it *too* soon.'

The boy waved as he carefully lowered himself into the vehicle, holding his bandaged left leg stiff and proud in front of him, like a minor war wound. Calypso allowed the veranda door to bang shut on the outside heat and she checked the time on her LifeWatch. She had finished her appointments early. Apart from being available for Tommy on a twenty-four-hour basis, this weekly surgery for the children of the Hope Island community was her only prescribed duty.

When the HR director for Hope Island executive staff had first interviewed her in Chicago, before she had signed the heavyweight non-disclosure contract and media-silence agreement, he had made it clear that she would not be required to join the staff at any of the island's hospitals and clinics. She could use their facilities as she chose and, as she had qualified as an MD before she began her training as a paediatric psychiatrist, she would be on standby in case of any major emergency. But it was Tommy who was to be her main concern, he explained.

Calypso had initially wondered what on earth could be wrong with the boy and had voiced her question. But her interviewer had assured her that Tommy was healthy in all respects. It was merely that Mr Tye's heir was growing up in an extremely unusual and privileged position in life – 'something you might find hard to imagine,' he said condescendingly – but accurately, as Calypso later realized. It was also explained that the boy was showing considerable intellectual promise and his father wanted a child psychologist on hand at all times to protect his son's mental as well as physical health.

'A permanent therapist?' queried Calypso, eyebrows raised.

'More of a qualified companion,' was the response.

'Where's his mother?' asked Calypso. 'Surely she'll want to talk to me first?'

'There is no mother,' said her interviewer abruptly. 'You'll have access to his files on your arrival.'

Considering her own mother's growing dependency, this job with the Tye Corporation had been the best opportunity Calypso could have possibly hoped for. It provided the necessary funds and the physical proximity that would allow her to ensure that the old lady received the best possible care in her final years.

Calypso had accepted the post with alacrity, leaving behind Chicago, the snow, the psychiatric wards and Larry Sumner all at the same time. She realized that she was using her mother's deteriorating health and this new job as an excuse to end the relationship with Larry. But she had recently been fighting to hide her irritation with the radiologist's addiction to on-line gambling and she guessed that he was probably feeling much the same about some of her own domestic habits. Their relationship had simply gone stale, although Calypso realized that it was she who was initiating the break.

Given the unusually exclusive conditions of her new job, Calypso had offered to run a weekly clinic for all the other children on the island and the HR department had welcomed her suggestion.

As soon as she met Tommy, saw Hope Island and realized what a sanctuary of wealth and privilege it was, how utterly different to life a few miles away on her home island of Mayaguana, Calypso had started to feel guilty about her career move, despite her offer to provide her services as a general physician. She decided to use her free time to study tropical medicine.

Then she had scared herself. She was learning a great deal about the human immune system and she started to wonder what could happen to the population of a completely new and very sanitized island state. She broke off from her study of schistosomiasis, did some calculations and v-mailed the HR director to request a personal meeting.

'I'm worried that we may be heading for a disaster,' she had begun hesitantly, aware that she was very new to the corporation. 'We're not getting sufficient population throughput to maintain antibody templates.'

The HR director had taken her worries seriously and Calypso had quickly been asked to make a presentation on the subject to the Hope Island management board. That had been the first time she had met Tom. Despite the fact that she had been to the house almost every day for three weeks, his life was so frantically peripatetic that she had never glimpsed her employer, the father of her charge. Nothing further was said of the mother and she had been offered no files and had not yet found the right opportunity to enquire further.

She was doubly shocked when she first saw Tom in person. The first surprise was that he entered the room wearing what seemed to be an antibacterial face mask. Nobody else paid any attention to this, however.

Then he removed it to spray his mouth and Calypso experienced a sense of unreality, of double exposure, a meeting with a person whose face she had seen all her life on television, in the press. She felt public and private, inner and outer, blur, grow confused. He was smaller than she had imagined, but he looked younger and more handsome – well, more beautiful, she realized – than any of his media images, even the most iconic, had suggested. The doctor in her ran a check of his visible indicators: he looked very fit and youthful. His features were almost perfectly symmetrical, his hair dark and shining. When he looked up at her from behind the conference table she saw a strong, straight nose and a full mouth. But dominating his face was a pair of the most startling, lively violet eyes. She already knew those eyes well – they were Tommy's eyes.

In her 'fun years', when she had competed in beauty pageants to earn enough money to continue her education, Calypso had worked with other beautiful women – models in magazine photo shoots, television commercials and product launches – and she had had many opportunities to observe the impact of female beauty on others.

'Looks are merely an accident, but truly beautiful people *do* live in a different world,' she had admitted when Larry had first asked her what it was like to have to live with her stunning looks.

She recalled some lines on the subject from a novel she had read at university. '"Beautiful people are exempt from life's difficult tests. They can sit there and judge life, instead of being judged by it. Beauty is its own morality."'

Larry had nodded his understanding, but she was sure he couldn't *know* what it felt like to be wanted or hated by everybody simply for your looks. Few men could.

But Thomas Tye could, Calypso realized.

She had worn her maroon two-piece suit with its knee-length hem and had gathered her wavy black hair into a thick plait that reached to the middle of her back. She wanted to make a good impression but as she took the management board through the dangers of immune-system decay as a by-product of communal isolation she had realized that Thomas Tye wasn't looking at her like most men did. She could see that the HR director and his male assistant were watching her movements in a typically masculine manner, practically licking their lips. Connie Law was clearly amused at the sight. Calypso was used to such behaviour and almost expected it, but Thomas Tye's gaze seemed to be everywhere but on her. Once again, like the rest of the world, she had wondered about his sexuality.

Her presentation was a success. She had allowed herself a small amusement at the end when she had wrapped up by describing Hope Island as 'this scept'red isle, this demi-paradise, this fortress built by nature for herself against infection'.

Tom had laughed. 'Does that make me Richard the Second or Bolingbroke?' he asked, surprising her and the others.

As a result of her warnings, the frequency of the rotation cycle of staff between Hope Island and the Tye Corporation's overseas offices was increased sharply. On TT's instructions the management board had gone so far as to lease two additional wide-bodied jets to ferry employees and their families for extra subsidized vacations in the United States, Europe and the Americas. Calypso had entered these new population movements into her model of Hope Island's immune-system development pattern and she had been able to report to the head of Human Resources that the likelihood of immune-system deterioration in the community was now much reduced.

A month after her presentation, when the HR department had conferred with consultants, completed their own calculations and had started to increase population flow to and from the island, her viewpers had notified her of an incoming message. It arrived during her siesta. Calypso had strung a hammock between two palm trees on her private section of the beach and she was dozing in the shade when Connie Law's ident appeared in her Ray Ban Electros. As she accepted the call, the PA's face filled her gaze.

'I've got Tom for you, Doctor. Hold on.' Then Calypso was staring at those eyes.

'Great work, Doctor. You've made an excellent start. You may have saved Hope Island from a real health problem.'

'Thanks, Tom.' She didn't know exactly where he was since his system wasn't transmitting the usual GPS reference or map graphic. She could see that he was in an aircraft, however.

'Our appreciation will be evident in your next pay transfer. Oh, and Tommy really likes you, he says. But not as much as he likes Jed.'

Then he'd gone.

When the notification of a large bonus and a mass of stock

options arrived in Calypso's e-mail a few days later she had been very touched. But not as pleased as she had been to hear about Tommy's growing affection for her. Already her vulnerable charge had become very important to her – and she was delighted to be nearly as important to him as his favourite Furry.

Now she stood in her surgery reception area and looked at her LifeWatch again. She could spend an hour absorbing more of Manson's *Tropical Diseases*, the seminal introductory work on the topic, before going up to the house to see Tommy after school.

Calypso had only recently intervened in what had become the first serious contest of wills between Tommy and his father that she had witnessed. Tommy had been begging his father to allow him to attend the little elementary school in Port Hope. It was run exclusively for the children of senior company executives and Calypso knew how desperately Tommy needed to feel normal – or as normal as possible, under the circumstances. At first, Tye had refused flatly, without giving any reasons. But Tommy was also strong-willed and after Calypso had endured a number of his tantrums, she had asked to see Thomas Tye in person to discuss the matter.

She had then explained how damaging isolation from other children could become for the boy. That, she had asserted, was the main reason why Tommy seemed emotionally immature whilst exhibiting enormous intellectual potential. After a lot of initial extreme and protracted reluctance, his father had finally agreed that Tommy could attend the school for three afternoons a week. Until then, all his various tutors had visited the house for up to eight hours a day.

Her VideoMate trilled, and Calypso answered it as she walked back into her consulting room. The wall screen came on automatically and showed Heather Garland, the principal of the school.

'Doctor Browne. Could you come over here now? There's been an ... well, Tommy's hurt.'

Calypso was already finding her keys and bag.

'What's happened?'

'I think he's OK, Doctor. He's ... Well, I'm afraid some of the boys ... He's been involved in a fight, Doctor.'

Ron Deakin snapped off his VideoMate and picked up the printouts he had downloaded from the *Stargazer, Scientific American, New Scientist* and *Amateur Astronomer* archives. The information was all in the public domain. He hadn't even needed to consult UNISA's own archive of scientific briefing reports. He sat down behind his desk on the twenty-third floor of the UN Secretariat building on Manhattan's East Side and laid out the articles in chronological order. Behind him stretched the vast urban wilderness of Queens and Flushing Meadow.

Amateur Astronomer, April 3rd 1999
Cosmonauts Ready Kirlian
By Robert Strauss

On April 4th, 1999, cosmonauts inside the Mir space station (at left) will command a circular, 25-meter (80-foot) catopric to be unfurled from Progress M-40 spacecraft (at right). They will then use the recycling system to provide energy for Mir and, potentially, for delivery to the Earth. (Click on image for larger view)

Update to story (4/5/99):
Kirlian Deployment Fails

When commanded to unfurl on the morning of April 4th, the Kirlian 25-meter heliostat became snared by an antenna on its carrier spacecraft, Progress M-40. Despite frantic attempts by cosmonauts Gennady Padalka and Sergei Avdeyev, the thin sheet refused to deploy fully. After engineers on the ground debated possible fixes, the cosmonauts tried to unfurl the catoptric again on April 5th, but its deployment mechanism jammed. The experiment was abandoned, and the control center near Moscow commanded the Progress and its partially-unfurled heliostat should re-enter Earth's atmosphere over the Pacific Ocean. It was destroyed immediately.

Deakin skimmed through the rest of the article and realized that bad timing had doomed the ambitious Russian experiment from the outset. They had been trying to launch their orbiting solar-energy system while their nation had been collapsing around their ears. Within weeks the companies behind the energy satellites had been closed down for lack of funds and the technology had been mothballed.

'Search Space Energy Consortium and Energate and the Tye Corporation,' Deakin told his VideoMate. 'Search Kutúzov.'

He spelled the name out for the system as he knew his pronunciation would be wrong.

And there it was.

Wall Street Journal, April 3rd, 2002.
Tye Corporation Buys Space Oddity.

Tye Corporation Europe gmbh today announced the purchase of a cluster of semi-dormant Russian space technology companies which includes the Space Energy Consortium, a coalition-corporation headed by veteran space engineer Nicholas Kutúzov. The group had once tried unsuccessfully to place a heliostat in orbit around the Earth. The purchase price for the Consortium and the other companies involved was not disclosed.

Mr Kutúzov will join Tye Aerospace, Inc., Miami, as director of space-energy research.

Suddenly Deakin cursed out loud. How could he have forgotten? How could he be so stupid! The intelligence officer crossed to the power cord in the wall and pulled the wall display's plug from its socket. He picked up the VideoMate from his desk, found the master power button under its concealed sliding cover and turned the system off. He took his radio earpiece from his left ear and his viewpers from the top pocket of his jacket that was hung over the back of his chair.

Deakin threw the dead communications technology together in a heap in the middle of his desk. This operation was going to prove difficult, he realized.

The boy was sitting on a low bench in the small schoolyard. Heather Garland had her arm around his shoulders. Stella, one of Jack's team, was standing a little distance away, conferring via her VideoMate. She would have been alerted automatically as the network's keyword-recognition system monitored the head teacher's call to the doctor.

Calypso negotiated the safety gate and ran up the slight incline of the play area.

'Tommy? What's happened?'

He looked up and held out his arms to her. Calypso eased herself down on the low bench and allowed him to put his arms around her neck and press his face into the doctor's coat she was wearing. She hadn't even paused to take it off before driving up to the school. Heather Garland removed her own arm from between them. 'He's OK, I think, Doctor. I think it's been more of a scare than anything else.'

Tommy lifted his head up. 'I wasn't cheating, I wasn't.'

Calypso stroked the back of his head and looked at Heather for explanation.

The head teacher shrugged. 'It was a chess game. That's all. Tommy was playing our school champion.'

Tommy looked up at Calypso again. 'You won't tell my father, will you? Will you?'

He swivelled his gaze from one to the other. 'You won't, will you?'

Calypso stared down at his face. She could see a trickle of dried blood below his nose and a graze on his right cheek. His huge violet eyes were filled with tears. She wanted to kiss him so much it hurt.

'Don't worry about that now. Let's just go back to my office and get you cleaned up.'

'You mustn't tell him, you mustn't!' Tommy screamed and he pulled away from them both, jumped up and started to run towards the school gate.

Calypso was off the bench and at the gate in front of him before she was even aware of it. She put her hands on his shoulders and held him square in front of her.

'I said, don't worry about that now, Tommy. I just want to make sure you're OK.'

The boy glowered up at her defiantly, his eyes blazing. He had inherited his father's temper.

She squatted so that her eyes were on the same level as his. 'Tommy, I'm your friend. Let's go check this out, OK?'

He stared back directly into her eyes, petulant and defiant.

'I'll let you come over to my place afterwards.' She knew that would get to him. He lived in the world's most luxurious modern mansion, with every possible amusement and facility, but he still loved to visit her tiny bungalow. He would pick up her small trinkets and examine each one with care, asking where she'd got it. Then he would usually ask to play her electronic keyboard.

He hung his head. 'I'm OK. Please don't tell my father.'

Calypso guessed that Tom was probably aware of the incident already. The security systems on the island were so all-embracing that hardly a square inch of it was not covered by cameras.

'*I* won't tell him,' she said, truthfully.

Morgenstein held the report in his left hand. He made Joe Tinkler stand, waiting in front of his desk, while he reread it. *Like being back in grade school*, thought Joe. But there was no doubt Morgenstein was angry.

'So Tye cleared out roughly a trillion dollars of his holdings and you did *nothing*?'

Joe nodded.

'Where's he put all this money?' demanded Morgenstein.

This was the hard part.

'I don't know. I've been tracking all IPOs and filings on private investments. I can't trace any of it.'

Joe didn't mention that he had just re-dispatched his software robots with fresh instructions. He was determined to find where it had gone.

'What are Tye Corporate Relations saying?'

'Nothing, sir. They have no comment. They say it was a private matter.'

Morgenstein's right hand moved over the keys of a large electronic calculator on his desk. He looked at the result and then back to the fund manager.

'I was watching the entire basket and nothing twitched,' offered Joe. 'I've got the recording.'

'Even if it had, you'd have been too late,' growled Morgenstein. 'You should have cleared out the moment you realized what was happening.'

Joe shook his head. 'I calculated that even following the markets I could have cleared everything out with only a five per cent discount.'

'What would that have cost our clients?' asked the partner.

Joe swallowed. 'Just over sixty billion dollars,' he said quietly.

'My guess is closer to seventy,' sighed Morgenstein, his anger held tightly to his chest.

He put the paper down on his desk and leaned back in his high-backed swivel chair. He turned so that Joe could no longer see his face. Everything about the partner's office spoke of the vast volumes of money that passed through Rakusen-Webber every day, and of the value even of the tiny percentages that got left behind. The pedestal desk was of gleaming mahogany with brass edging, a nineteenth-century antique. The walls were panelled in the same wood, stained to match the desk and the fine Hepplewhite chairs that were scattered around the Georgian conference table. The air smelled of polished leather, but Joe guessed that was artificially enhanced.

'We're letting you go, Joe. As of this minute,' Morgenstein said quietly from behind his chair back. He swivelled round to meet Joe's shocked gaze.

'I've spoken to Richard Rakusen and he agrees. We will *not* carry the sort of risk you have exposed us to. We'd rather be without the returns you have produced if this is the potential downside. You've forgotten the golden rule: every dollar lost in our business is worth every ten gained. It's a matter of reputation. You are dismissed for gross negligence.'

Joe was stunned. He could think of nothing to say.

'Security staff are waiting for you in your office. Please collect your personal things and leave the building. Do not speak to anyone on your way out and make no attempt to contact anybody here after you leave. Our lawyers will be in contact to discuss severance terms. Remember you are still bound by SEC confidentiality.'

Joe stood, rooted to the spot.

'Goodbye,' said Morgenstein and he turned his chair again so that Joe was left staring at its back.

The fund manager turned and walked from the room.

The reverberating sounds of an ancient cathedral pipe organ filled Calypso's comfortable living room. She found herself drifting into tranquillity with the gentle pedetentous descent of Bach's Air No. 3 in D major.

Tommy had seemed pensive on the drive back to Calypso's bungalow so the doctor had let him be, understanding that it was the shock of the assault, the affront to his male dignity rather than any of his minor injuries that was occupying his thoughts. He had been so sheltered from the rough-and-tumble – so unlike normal boys, so less emotionally mature. She shrugged mentally as she realized that with his wealth and his father's phobic obsessions such isolation was almost inevitable.

Tommy reached the end of the piece, allowed the final D-major chord to die, mellow on the air, and slowly lifted his hands from the keyboard of her synthesizer.

He turned, grinned and was suddenly a little boy again, much younger than his seven years. He spun round on the old revolving piano stool: round and round. He laughed as it reached the top of its thread and he spun the other way until it reached its lowest point.

'That was beautiful,' breathed Calypso, when he finally stopped. 'Beautiful. I haven't heard you play that before.'

'Miss Duckett only gave me the music yesterday,' said Tommy. 'I like it.'

'I like it very much, too,' agreed Calypso. 'You play so beautifully, Tommy.'

She too had enjoyed playing the piano when she was a child. But she knew that she had never played an instrument the way Tommy could. It was both his feel *and* his technique that marked him out. Despite his youth and size, his ambidexterity and his exceptionally long fingers enabled him to span more than an octave and somehow he injected such emotion into his playing that it was almost impossible to believe he was so young. And his memory for a piece was remarkable.

'So what *did* happen at school Tommy?' she asked.

The boy frowned and then gave a massive shrug, much too big for his little frame. 'I don't know. I just got Emilio into checkmate and he shouted at me. Then Maurice Dennis joined in. They banged me against the wall.'

'Do you play chess with them often?' probed Calypso gently.

Tommy shook his head. 'I've never played anyone before. Only my ChessMate program. I must have done something wrong.'

'Do you win when you play your ChessMate at home?'

'I used to,' Tommy sighed. 'But then my father put it on the top level and I haven't won a game for a week.'

Introduction – insertion

I can also reveal for the first time that Thomas Tye has a son who is being brought up in a strictly controlled scientific environment on Hope Island. Thomas Richmond Tye IV was born to an unknown mother approximately seven years ago. Sexual intercourse was not involved and the 'mother' had no genetic input to her apomictic[1] 'son'. Tye's son is the world's first enhanced human clone, born by a process that can best be described as 'cryptogame eutelegenesis' – a combination of eugenics and genetics, achieved secretly and remotely – and I describe the enhancements that his 'father' selected for him in Chapter Seven.

[1] Apomictic is a biological term: pertaining to or produced by apomixis; reproducing without sexual fusion.

Haley sat back and stared at her central text-composition screen. It seemed so bald, so cold, so academic, so undramatic. But how else could she put it? She felt like writing it in capitals, giving it a sexy, newsy sans-serif typeface, so she did:

***THOMAS TYE HAS CLONED HIMSELF, WORLD, DO YOU HEAR?***
***NOT ONLY DOES HE WANT TO LIVE FOR EVER,***
***HE'S GROWING HIS OWN SUCCESSOR!***

But ordinary text would have to do. She deleted her fantasy headlines. All she had to do now was find a publisher.

The second report had arrived that morning in the mail. Once again it was a paper document but this time it had been posted in Paris. She had read through the sixty pages at speed. Then she reread them slowly, with the help of on-line medical dictionaries and the world's science archives.

When she'd finished she took off her glasses and stared blankly at the pile of papers. This report was either an ingenious hoax or a godsend.

Haley was tempted to scan the document and forward it to Barry. She checked his location and saw that he was now in his lab; but she wasn't sure whether others there might have access to his encryption key. She contented herself by reading the report again, adding her own questions and interpretations of the technical jargon as margin notes. She knew Barry would be amused by her bumbling efforts at interpreting the technicalities.

Then, because she couldn't wait, she had written the draft insert for her book's introduction.

When Barry finally arrived back, shortly after seven, his mood was foul. He had been forced to stand in a packed Underground carriage all the way from Euston to Stockwell, he told her. The heat had been intolerable, everybody around him shouting into the networks. His first words as he came through the door had been 'I stink.'

While he showered Haley turned the air-conditioning up, put on some gentle Mozart and punched up his favourite ScentSim aroma, *New Car Interior*, as compensation for his experience on London's public transport system. She then made him a large gin and tonic. Wordlessly, she held it with an outstretched arm around the bathroom door as he towelled himself dry. He took it without acknowledgement.

She recrossed the hall and threw herself full-length on the white sofa, her arms crossed, getting ready for any type of argument he fancied.

A few minutes later Barry emerged pink and steaming into the cool living room. A large white towel was wrapped around his ample waist, and he was rubbing his sparse red hair one-handedly with a small hand-towel that he had removed from beside the bathroom sink. In his left hand he still held his drink. He looked down at her, blinking.

'Sorry, baggage, bloody awful day,' he apologized, as he inhaled the cool, scented atmosphere. 'Broke a bloody culture dish after three weeks of good festering.'

He took a large swig of his drink and burped. The towel at his waist unfolded in slow motion and fell to the ground.

'Oh, vision of the Northern Line, come to me,' laughed Haley as she held out her arms. He came and sat beside her and she kissed him hard on the lips, their taste fresh with gin, tonic and lime.

'How did *you* get on today, sweetheart?' he asked, more out of duty than genuine interest.

She scooped up the report from the floor beside the sofa and dropped it onto his naked legs. 'I need your help again, Doctor Evans. This arrived this morning. I'll get you a tracksuit to wear.'

It took Barry half an hour to read it the first time. Then he flicked through it again, to underline points that seemed important.

'You understand I'm no reproductive biologist?' he said finally. She sat opposite him at her circular dining table.

'Well?'

'OK. Look, it does seem kosher, but I can't be sure. I'd have to show it to one of the real boffins at the lab.'

Haley shook her head. 'We can't show it to anybody,' she said. 'If this information is true, my book is made. The newspapers will be falling over themselves for serialization rights. We can't risk the story getting out before publication.'

'Well, obviously somebody else already knows,' reasoned Barry, annoying her with this wholly unnecessary deployment of common sense.

'Can't you check it out *without* showing it to anybody else?'

Barry sighed – the patronizing display of weariness he always produced when about to explain scientific complexities to a lay person. Haley frowned.

'There aren't many people qualified to comment authoritatively on this,' he said. 'It's a very specialized area.' He took a deep breath. 'Look, there are two sorts of cloning. The first is a simple process and we, that is biotechnologists, do it all the time, with cattle, rare animals, endangered species, et cetera. It's a piece of cake. In this method we take an ideal nucleus that's been created from the sperm of an ideal male and the nucleus of an egg from an ideal female. The genes are often manipulated and then they are inserted into a carrier egg from which the natural genetic material has been stripped out – it's a procedure we call androgenesis. The fertilized egg is then duplicated and the results are implanted in a surrogate mother. This technique produces several animals that are identical – say, identical twins, like you and Felicity – but they're not exact copies of any existing adult animal. When the subject of cloning comes up, the public mistakenly thinks that the offspring is an identical copy of some adult. That's a very different story.'

'But that's what this report is about,' urged Haley.

'So it would seem,' agreed Barry, flicking through the pages, unwilling to be hurried. 'The main problem with cells taken from adult animals is that they have already differentiated. That is, they have already determined what they are and they have lost the embryonic ability to divide and to grow into specific types of new cell – like a heart muscle cell or nerve tissue.'

He turned back through the pages again. 'Then they discovered that if they take a cow's egg that has been stripped of its own nucleus and fuse it with an adult animal cell they get a blastocyst – an embryo that starts to grow in the test tube but that isn't viable in the long run: it dies after a few days.'

'Should I be making notes?' asked Haley, but her sarcasm sailed straight over his head. He closed the report and sat back and folded his hands over his stomach as he prepared for the next part of his lecture. She could see the middle-aged academic he would become.

'But if embryonic stem cells are taken from the blastocyst and they're implanted as a nucleus into a carrier egg they can develop into any type of the two hundred and ten cells required and they carry the exact copy of the DNA that was present in the original adult-animal donor cell. That's how a copy of an adult is made and that's far more difficult.'

'But it can be done,' insisted Haley.

'It can,' the geneticist admitted. 'It was done for the first time nearly twenty years ago, in Scotland. They made a successful copy of an adult sheep but, later, animals that were produced elsewhere using similar techniques aged too quickly. Once they solved that problem, they started cloning pets – dog, cats and so on – just at the turn of the century.'

'But they've never done it with humans?' she asked.

'Well, not as far as we know. The nearest they've come is to grow human organs for transplant.'

'Yeah, that's what one of Tye's companies concentrates on,' said Haley. 'So why no human clones?'

'It's illegal, since they'd have to use cells from real human embryos,' explained Barry simply, picking up the report again. 'Although I wouldn't mind betting there's been more than a few thousand human clones already knocked up in South American clinics and some of those dodgy Middle East states. Although why somebody would want to copy himself – or any other human – is beyond me. Anyway, there's just no reason to want to do it now. Even infertile couples have better options these days.'

He flicked through the report again and laid it down on the table, open at one page he had marked.

'It's these genetic adjustments to the embryo, described here, that make this really interesting.'

'Are such things really possible?' asked Haley.

# NINE

Suddenly Jack Hendriksen had new purpose, a new identity and a feeling he hadn't experienced since he'd left government service. He felt he was once again doing something worthwhile.

Deakin had taken advantage of Jack's loose vacation plans and had arranged a series of 'refresher' courses for him before his official swearing-in to UN service would take place. Three courses had been arranged in short order and none of his instructors were to know anything about their new pupil until the moment he arrived.

He was flying first to Fort Mead in Maryland, to the headquarters of the National Security Agency, a US government facility in which UNISA had been able to negotiate a small sovereign space for its own communications laboratory. His passport, reservation, digital identity, value-transfer modules and new communications equipment were all in the name of Bruce D. Curtis, a thirty-nine-year-old attorney from Washington, DC. All details had been confirmed and registered with the World Certification Authority. Even his finger- and voiceprints had been matched to Curtis's identity and he'd collected an instant but detailed biography and an impressive list of legal qualifications. The individual identified as John Edward Hendriksen was 'still on vacation' in Manhattan.

'We keep a bunch of these IDs ready-made,' Deakin had explained. 'There are more lawyers on the road than management consultants these days. You'll fit in – just try to look a bit more venal.'

Once in the air, Jack lifted his background file on the Tye Corporation from his new briefcase and extracted a few sheets at random. He wondered about the relevance of the first report.

Forwarded by: UNINET to Executive Officer R. Deakin, UNISA.
Automatic forwarding keywords: Tye, Erasmus
Qualification: 100%

UNITED NATIONS HIGH COMMISSION FOR REFUGEES
Report No.: HC/SA/10/207
To: Deputy High Commissioner Sherri Prasso, UNHCR, Geneva
From: Philip-Niël Shütte (HCR 4011)
Subject: Reon Albertyn, asylum applicant No. SA81956

Status: Highest Confidentiality

At the request of UNHCR Geneva, I completed a successful negotiation with Amnesty International, Johannesburg in order to use the accreditation of their organization to carry out a visit to a maximum-security prison facility in Harare, Zimbabwe, June 11th last. Geneva had provided me with the names of three teenage prisoners in whom they have an interest, following appeals for information from the families of these boys and also local political representatives. During my visit I was told that two of the subjects had died during their internment. I was later able to obtain copies of their death certificates (copies attached).

The third prisoner on whom information was requested was Reon Albertyn, a 14-year-old South African citizen imprisoned for armed robbery and car theft, crimes found proven by a closed Zimbabwean juvenile court when he was 12.

I found Albertyn in poor health and imprisoned in appalling conditions in adult accommodation. This is the first visit UNHCR (or Amnesty International) has been able to make to this prison and, as far as I could tell, all inmates at Chikirubi Maximum Security Facility are held in conditions that breach UNHCR, UNESCO and WHO guidelines. I recommend that conditions at this facility should be officially classified by UNHCR as inhuman and degrading and confirmed as being in breach of the United Nations Declarations on Human Rights 1958, Article 7. I also recommend copies of this report are forwarded to the Ethics, Human Rights and Compliance committees at the World Bank and the International Monetary Fund. (My full assessment of the conditions for inmates of Chikirubi MSF are provided in a separate report, ref: HC/SA/10/207/CMSF.)

As UNHCR Financial Aid Auditor for Zimbabwe, I was granted an audience with Dr Kim Mnanke, President of the Republic, and he graciously granted Albertyn a presidential pardon on compassionate grounds. I then arranged for paramedic air transportation to Mount Zion Hospital, Pretoria where Albertyn has been undergoing tests and observations for 10 days.

Reports by Dr James Hughes and Professor Per-Ola Nieble of the Steve Biko Institute, Cape Town are attached to this report. I summarize their findings here:

Reon Albertyn is in the advanced stages of a condition that initially presents as a disease known as Werner's Syndrome, a form of progeria or premature ageing. This condition, which is normally a chance mutation in one copy of a single gene, is known to be genetically inheritable, non-contagious, non-infectious and usually fatal before the patient reaches adulthood. Albertyn's physical condition resembles that of a man in his eighth or ninth decade, but Professor Nieble has been unable to find a single mutant gene on the appropriate chromosome in DNA samples taken from the patient. He is continuing his investigations. Another unusual aspect of this case concerns Reon Albertyn's skin colour. He is white, technically a 'leucoethiop' or albino negroid. This very rare condition is also wholly hereditary.

Reon Albertyn insists that none of his family have ever had a similar ageing disease nor skin of the same pigmentation and my local agent found his parents in good health in Neuville, a township 12 miles outside of Cape Town. They also reported no incidence of any ageing disease or leucoethiopia in Albertyn's family but Reon Albertyn's putative father strongly disputes paternity. I am currently awaiting independent DNA verification of his claim.

Ends

Underneath the report Deakin had scrawled '*Lily Albertyn employed by Erasmus Research SA 1999–03*'.

Jack shook his head and pondered what his first steps would be when he got back to the island – he would need an ally, he realized. Then he ordered a large vodka and tonic from the flight attendant before pulling another report from the pile and starting to read.

His arrival at Fort Mead had been expected and he was quickly escorted into the bowels of the research facility. He remembered the first part of the route well, but then he and his escort continued on to a new extension. Here a guard in a UN uniform assumed responsibility for the visitor and walked him down a brightly lit corridor to a pair of double doors. The guard opened them and nodded for Jack to enter. The doors closed behind him and Jack smiled to himself. Even seen from the back, he knew the figure hunched over the keyboard very well.

'Doctor Lynch? It's Bruce Curtis. I hope you're expecting me.'

The man continued typing for a few moments, then lifted his hands and turned his wheelchair to face the visitor. Suddenly a broad smile lit his face.

'Well, I'll be damned! Jack Hendriksen.'

Jack dropped his briefcase onto a chair and quickly crossed the room to take the eagerly outstretched hand.

'How are you, Al? You look good.'

Jack had spent part of the plane ride from Newark working through the emotions he knew would surface when he met Lynch again. He hadn't seen the computer-systems expert since the crash that had killed Helen and partly paralysed Al: they had been flying back to New York together for Christmas leave, Al to join his family, Helen to their beloved Gramercy Park loft.

'My, my. But I heard you got out, Jack?'

'I was out, Al, but it seems like I'm back in again – although it's the UN this time. As of two days ago.'

Lynch nodded. Then he took a breath and said what had to be said. 'I was very sorry about Helen, Jack. We all were.'

Jack nodded, feeling an ache rise up, though less acutely than he had anticipated.

'I got your letter, Al. Thanks.'

'And I got yours, Jack. Eventually. So how are you?'

'I'm over it, Al. Or as over it as you can ever get.'

As he said what was expected of him he wondered, for the first time, if that might be starting to be true.

'What about you? Look at this place. You've got a whole new set of toys.'

He didn't mention the wheelchair.

Lynch noted the change of subject and took his cue. 'Sit down, help yourself to coffee,' he offered with a smile. 'I've got something to show you.'

Jack saw the percolator and plastic cups on a desk and poured himself a black coffee. 'You?' he asked waving the coffee jug in Lynch's direction.

'No, no, just sit down, Jack,' said Lynch urgently.

Jack did as he was told and raised the coffee cup to his lips.

'Watch,' said Lynch. He locked the wheels of his chair, grasped both arm supports and pushed himself to a standing position. He rocked slightly and then gained his balance. He next took a hesitant step with his right leg and then brought the left up to join it. He took another step and then brought his left leg forward again. Slowly, but with a huge smile on his face, he shuffled across the room until he halted two feet in front of Jack.

Jack put his cup down and stood up. He could feel a wetness in his eyes. He held out his arms and hugged his former colleague.

'Al, that's wonderful.'

This time the tears came. He was crying, not for Al but for Helen – and for himself. But, yes, he *was* crying for Alan Lynch too. Al was walking again!

Jack pulled away and looked down at his former systems instructor, who was over a head shorter and nearly bald. Lynch was beaming up at him.

Jack wiped his eyes. 'I thought they couldn't fix it?'

'That's what they said at first. But I just got out of hospital again, three weeks ago. Watch this.'

Jack watched as his ex-instructor spread his arms and carefully started to turn in a slow circle.

'I have to practise for an hour every day – no more, no less, they say. But I've done two hours already today!' He beamed triumphantly as Jack sat down again.

'That's wonderful – how did they . . .'

Lynch finished turning and began a slow shuffle back to his wheelchair.

'My back's still broken,' he said over his shoulder. 'The spinal column is completely severed just above the T12 vertebra.'

He reached his chair and turned back to Jack. 'They've bypassed the break and they've interfaced the first cervical nerve and the spinal nerve here.' He touched the back of his neck and turned so that Jack could follow his hand movement. 'And a bionet of microwires and signal amplifiers runs all the way down to here,' he touched the base of his spine, 'where it's interfaced again to the sciatic, pudendal and lumbar nerves.'

He turned back to face Jack, beaming. 'They've put a steel brace on the spinal column and they say that eventually all the nerve endings will grow onto the bioconnectors and I'll be back to normal, more or less.'

Jack felt himself grinning broadly as he watched the man lower himself gently into the wheelchair.

'That's wonderful.' He smiled again. They hadn't even been able to show him Helen's body. He often wondered if they had even found it all. How could so much damage have been caused at such a low height? The Navy passenger jet had been only a few hundred feet off the ground when the obsolete shoulder-launched missile had hit.

'At least they got the bastards,' growled Lynch, picking up Jack's mood. 'They're on Death Row now.'

Jack nodded. But they *wanted* to die, that was the problem. They thought they were going to 'Allah'. They were indoctrinated, superstitious, information-deprived, aspiring martyrs carrying an antique, distorted and grotesque fantasy of Muslim righteousness into the heart of the infidel nation: to a rural airbase perimeter road that, despite two decades of recurrent terrorist attacks on US soil, had still been inadequately guarded. It was as if an ignorant native tribe from the distant past had travelled into the future solely to attack him and his family. Jack pulled himself away from such thoughts.

'So what's new in systems?' he asked.

*

'If there's nothing wrong, why is he here?' demanded Calypso. 'Why *precisely*?' She was angry and she knew it showed.

Marcus Forrester, commercial director of the Hope Island Research Clinic, smiled and tried to quieten down the noisy visitor.

'Because Tom asks us to keep a special eye on him. It's really nothing more than that, Doctor. He's an only son, after all.'

'Let me see him,' insisted Calypso again. '*I* am charged with his medical care, not your staff.'

She knew this wasn't wholly true, but she'd been shocked when the senior nurse up at the Tye mansion had told her that Tommy had been admitted to the high-security research clinic for three days. It was precisely this sort of ludicrous over-protection she was trying to prevent.

Without any apology or explanation she had been kept waiting for ten minutes in the large reception area of the underground clinic until Forrester had eventually emerged.

He smiled again. 'Of course. Right this way, Doctor Browne.' Insincerity radiated from his face, like a flight attendant's farewell.

He led Calypso from the reception area into an elevator. As they descended Forrester tried to make small talk, but by the time they had travelled three floors down he had abandoned the attempt. He led the way into a softly illuminated corridor with half a dozen white doors aligned down either side. At the third on the right he stopped and swung the door open for Calypso to enter. She could hear the sounds of a hydra-immersion chair from within.

She hesitated outside. 'Exactly what tests are your people running, Mr Forrester?' she asked. 'I'd like to see both the procedures and the results.'

'Naturally. I will arrange that, Doctor,' agreed Forrester unctuously. 'Please . . .' He gestured towards the door.

Calypso eyed him suspiciously and entered.

Tommy was oblivious of her arrival. He was wearing blue striped pyjamas and he was strapped into a grade 4–6 size 360-degree HydraChair fitted with full kinetic, tactility and olfactory sensory capabilities and user reafference. The helmet visor covered his face to below his nose. As she watched, whatever virtual craft he was piloting started a steep left-hand turn and the hydraulic supports of the small chair tipped him over at a thirty-degree angle. She heard him laugh and saw him shudder as the chair transmitted the sense of the force to his body. Then he was flying level again. Next his chair tipped upwards at the front and he was in a sharp climb. She could detect a faint aroma of jet-engine fuel in the atmosphere.

Calypso crossed to the wall screen and reviewed Tommy's physical history for the previous hour. As Forrester had said, there was nothing to worry about. She looked around the room, but she could see no trace of testing equipment or other records.

Suddenly Thomas Tye was beside her. He was sweating gently, as if he had been hurrying. She guessed he had been told of her sudden arrival at the clinic, and of her demands. She also guessed that it had been at his request that she had been kept waiting.

'What's this, Calypso? You should have been told you were not required today.'

Calypso folded her arms and turned to face her employer. She was wearing a bright yellow polo shirt and navy shorts, having planned to walk along to a restaurant beside the marina and meet Heather Garland for lunch. She had hoped to learn more from her about the disputed chess game.

'I called the house to talk to Nurse Pettigrew about Tommy going back to school,' she explained. 'Then I was told that Tommy wouldn't be returning. I asked to talk to him and she said he was here.'

'What do you think your job here is, Doctor?' asked Tye quickly. He was almost exactly the same height as her and his violet eyes flashed as he held her gaze.

'To look after Tommy, of course. To help him feel good about his fortunate position and to try and keep him sane when everything around him is insane.'

The words had come from nowhere. She'd meant them more figuratively than literally, but their effect on Tye was dramatic. The colour drained from his face and his lips pulled back from his teeth.

'Your job is to observe and report, Doctor, not to interfere,' he hissed, stabbing the air with his finger. 'We've got the world's best medical researchers on this island and I don't expect you to get in their way.'

Calypso's training came to her aid. She understood that irrational anger had to be calmed, not confronted. She said nothing.

'You're a qualified companion,' Tye continued, 'in case Tommy gets over-strained.'

Her feelings got the better of her. 'The only strain is the one that you put on him,' Calypso pointed out. 'Why can't you let him be normal, like other boys? He'd be far healthier for it, rather than living like some semi-institutionalized freak!'

'That's enough, Doctor,' snapped Tye.

'No, it's not enough,' insisted Calypso, determined to take the opportunity presented. 'Tommy is getting none of the praise, none of the constant assurance necessary to build his virtual identity – to create a robust personality, to put it in layman's terms. He isn't getting the love he needs to form relationships. He's not a laboratory specimen!'

'YOU HAVE NO IDEA WHAT YOU ARE FUCKING TALKING ABOUT!' shouted Tye. 'You've only been here five minutes.' He swallowed, then looked at his son. 'He's a genius, Doctor,' he added, waving at the oblivious Tommy in his HydraChair. 'He's got an IQ of over one eighty, he has talents like no other child, he's already reached Grand Master level at chess. Fucking doctors! I, better than anybody, should have seen through you. You're a quack, a charlatan.'

'He's a little boy,' shouted Calypso, now seriously angry. 'And money alone does not qualify you to be a parent! You *must* be insane!'

He hit her hard, open-handed across the face. The sound reverberated around the small hospital room.

'Never, ever say that,' he snarled pushing his face into hers.

Instinctively, Calypso had clapped her left hand to her assaulted cheek. She had never been struck before and part of her mind was consciously questioning how she felt about it. Then she slapped Thomas Tye across *his* face, as hard as she could.

He pulled back and stared at her in disbelief, shocked into silence. He had never been struck before, either.

Calypso suddenly realized there was no other sound in the room. She turned and saw that the HydraChair had stopped its motion. The helmet visor was swung up to its idle position. Tommy was staring at them, his violet eyes wide.

Calypso forced a smile and crossed the room. She squatted beside him. 'It's OK, Tommy. Grown-ups sometimes have arguments, too.'

'Leave us, Doctor,' ordered Tye. She noticed tears forming in his eyes.

Calypso hesitated as Tommy stared into her face with his father's eyes. He was obviously scared.

'LEAVE US!' screamed Tye.

She touched Tommy's cheek and rose. Walking slowly to the doorway, she turned and looked into Tye's face. Tears were now streaming down his cheeks.

'What *have* you done?' she asked him quietly.

She pressed the panic button on the wall. Then she turned and left Tye and his offspring alone together.

Haley sensed Rosemary's excitement when she called.

'Can you get over here now?' the agent had asked. 'Let's meet at the café again.'

They hadn't seen each other since Haley had threatened to find alternative literary representation, though Rosemary had sent her a bunch of flowers with a card saying she would continue to seek other publishers who might be interested in Haley's book.

As Haley threaded her way through the tables she saw Rosemary had arrived ahead of her and was already smoking her second cigarette. She also noticed an ice bucket waiting beside the table, and wine glasses glinting on the white cloth. It was a beautiful June morning but still only eleven o'clock, so rather too early for a lunchtime drink.

Rosemary rose to greet her client. 'I've ordered champagne,' she gushed as they both sat down. 'I've got good news for you. The Sloan Press – in New York – has made an offer.'

Haley had half-wondered whether an alternative offer might be the reason for this summons.

'That's wonderful. Was it the new material, about Tye's son, that hooked them?'

Rosemary smiled. She hadn't sent them the latest updates before they had made the offer. In fact, until they'd approached her, she hadn't sent them anything at all. But all agents want their writers to believe that it was hard work and not luck that made the difference. The fact was that the Sloan Press's editor-in-chief Luke Bailey had called out of the blue and asked to see the manuscript so far. He claimed to be fascinated by the idea that it was so hot that Nautilus had pulled out. Rosemary had e-mailed the text to him that same afternoon and twenty-four hours later he had made the offer.

'They love all of it, Haley,' she dissembled.

Haley nodded eagerly.

'They've even matched Nautilus's advance, and they'll provide a generous research and travel budget. They're promising a massive marketing campaign ...'

She tailed off as the waiter arrived with the bottle of champagne. She tasted it quickly and nodded for him to pour.

'I was sure you'd want to celebrate.'

'But what about the injunctions?' Haley asked. 'Won't the Tye Corporation just reissue them against Sloan Press?'

'Luke Bailey says that's part of the plan. He believes that if the Tye Corporation really goes after the book, the subsequent legal action will attract massive publicity. His PR people can then turn any attempt to suppress your book into headline news.'

Haley nodded as she sipped the cool champagne. She could see the logic to that.

'And there's something else,' added Rosemary, slipping a piece of folded paper from her handbag and handing it to her author. Haley unfolded it and scanned the words.

'It's your draft itinerary,' Rosemary explained. 'Luke Bailey wants you to go and visit him in New York next week. He wants to discuss the book in detail and for you to get together with their lawyers to plan your defence and a counter-attack.

He also wants to discuss serialization in the newspapers – under non-disclosures, of course!'

Haley felt a little dazed by it all.

'What do you think?' urged Rosemary as she lit another cigarette. 'Should we accept?'

Haley smiled her answer.

'Here's to it,' said Rosemary as she lifted her glass in a toast.

Haley touched her glass to the agent's.

There had been a *lot* to learn. After two days of his intensive refresher course in computing and communications systems, Jack's mind was reeling from the onslaught.

Al Lynch had suggested, half seriously, that NSA, CIA and UNISA technology might seem old-hat to someone employed by a corporation able to crack the world's encryption systems. Jack had laughingly explained that he had little personal experience of the Tye Corporation's technology, having only observed what some of it could do. Tye never bothered himself with technical details these days, and he was only interested in meeting the sales prospect, doing the demo, then handing the sales process on to the technical consultants for completion. In fact, Jack realized, computing and communications had become an almost invisible part of everyday life for millions, something everyone took for granted.

'You realize how I received my instructions about your visit, Jack?' Lynch laughed. 'A courier flew down from New York with a briefcase chained to his wrist. It was a Cold War relic. They must have got it from a museum. But when I read the papers, I understood. We really do have to assume that your target has cracked the principles of prime-number key encryption. Is that right, Jack?'

Jack nodded. 'So UNISA assumes, Al. But they have no idea how that could be possible.'

'Nor me, either,' mused Lynch, 'and I thought I was pretty well up in this area. Ciphers are one of my specialities – I did encryption theory for my PhD thesis . . .'

He tailed off in thought. Then he looked up again.

'But if the Tye Corporation *can* do so, it means that everything we normally provide for secure communications is now useless. That's always the inevitable implication for any civil or military authority after such a breakthrough. If they can crack the highest levels of encryption they can then just set up robot watchers in every hub on the networks. These would automatically intercept every communication that mentions Tye, his companies, his competitors, or organizations like ours, and copy them to his ...' He tailed off again.

'Come to think of it, they could read *any* communication in the world, and if they can get into our databases, they could read every military secret we have. *Jesus* ...'

Jack nodded.

'It means everything we've now got, all this comms kit, is junk – it's useless.' Lynch waved at the benches covered with minute wearable communications and surveillance devices.

Jack nodded again. He thought of making a joke about homing pigeons, but he could see that Lynch was deadly serious. He was working up to something.

'This gave me quite a challenge, Jack. And I had no idea that I was doing it for you, of course. But I'm delighted, delighted.'

He turned his wheelchair to the bench and lifted the plastic cover from a small piece of equipment. He turned back and handed it over. Jack took the device with a puzzled frown.

'It doesn't look much different to an early VideoMate,' said Lynch. 'But it is.'

He leaned forward and pointed to a slot that ran the whole length of the underside. 'This is a little fax feed, Jack, like they used to provide on the first generation of VideoMates, only I've modified this machine so that it doesn't send ordinary faxes. When you press this button it produces old-fashioned one-time codes at random and applies them to speech or keyboard input.'

Jack raised his eyebrows. One-time codes, for centuries the most secure form of encryption for covert communication, had disappeared long before he had joined the intelligence services: the topic was not even covered on any of the courses he had attended.

'You dictate the message, Jack, and the processor generates the one-time code – it's completely random, so no chance of patterns occurring. You connect this machine to a dedicated phone line and send the message as a fax, a bitmapped graphic, not as live text, and it's received at the other end and printed out on paper as plain text. Nothing must remain in a memory anywhere. This VideoMate doesn't have any radio links or network ports, Jack. They're all disabled. You get one printout of the one-time code and that's it. After that it doesn't exist any more.'

'So how do I get the one-time key to UNISA or whoever I'm faxing?' asked Jack.

'Ah, that's the weak link,' admitted Lynch. 'Ideally you should hand it to them personally. But that rather defeats the object, doesn't it?'

'I have to agree, Al,' he said. Lynch didn't seem to notice his sarcasm.

'But I think you can risk faxing it over a second dedicated fax line, like we used to use until everything went Internet protocol. We'll just have to install a couple on Hope Island or wherever you're going. It should be safe: we'll provide a massive white-noise wrapping and we'll know if anybody is tapping it. Even if it *was* compromised the beauty of one-time codes is there's never enough information to allow a computer program to decipher the pattern. There are too few occurrences. Also, I can't imagine that your target has set up robot intercepts for bitmap graphics. No one has ever used one-time codes in fax transmissions. They disappeared way before faxes came in and went out again.'

'Back to the future,' said Jack, smiling as he recalled a saying from his childhood.

Lynch looked at him blankly.

'Shall we go to full throttle?' suggested Stella Witherspoon diplomatically. The captain nodded his approval to the helmsman and with a roaring surge the ninety-foot offshore patrol craft rose to its full planing height.

'Sixty-two knots, that's about it,' shouted the Captain after a few minutes.

'How far are they?' yelled Stella as she clung on to a grab

rail. It was her turn to ride with one of Hope Island's coastal patrol craft: every few months her boss made each of his section heads take a week's night shift with the coastal patrols – even those who belonged to the Presidential Protection Team. It was just past midnight and misty.

The navigator indicated the converging shapes on his combined radar, infrared and satellite-imaging screen: the information from the three systems was integrated to provide a detailed graphic display.

'About three minutes at this speed,' the captain shouted at her. 'They're hove-to.'

The Cuban patrol vessel appeared as a red icon on the screen. They were closing quickly.

The captain lifted a flap on the control panel and pressed a red button. 'Go to weapons drill,' he ordered unnecessarily as the klaxon sounded.

Crew members appeared on deck and manned the forward laser-guided cannon and the aft machine guns. They watched as they converged on the incident.

'Ship ho,' called a crewman.

The captain nodded and the helmsman cut the throttles. The craft settled and the roar of the engines gave way to an idling burble. They rocked in near-silence for a minute as they stared into the mist.

The staccato sound of a machine gun broke the peace, and lines of red tracer flew over their heads. One of the crewmen turned on a searchlight and swung its beam across the sea.

'There,' shouted the helmsman.

The Cuban patrol boat and a crudely built oil-drum raft were caught in the intense white glare. They had been engaged in hauling fleeing rebel supporters on board. One youth was still clinging to a net on the side of the old gunboat. Sailors lined the rail, automatic weapons trained on the would-be asylum seekers.

Another burst of machine-gun fire from the Cuban vessel created a red carpet of tracer fire over the Hope Island patrol vessel, lower this time.

'Not very friendly,' observed Stella.

This type of mid-sea confrontation had been going on for

nearly a year, since a Cuban craft had come upon a Hope Island patrol boat lifting escaping rebel soldiers from the water. Havana had complained that far more of its citizens were landing in Hope Island than were returned.

'Position?' asked the captain quietly.

'Borderline, sir,' responded the navigator, peering at his display. 'Right on the line. Might be our waters, might be international.'

'They wouldn't really be a problem,' sighed the captain to Ms Witherspoon. But he already knew her response.

''Fraid not, Captain. The standing orders are unchanged.'

'Let's go home,' ordered the captain. As his craft turned away he spat pointedly but impotently out of the side window, in the direction of the Cuban craft, and went below.

# TEN

Tye Life Sciences was suffering from a serious problem: in one key division it was unable to keep up with an overwhelming, insatiable global demand for certain highly personalized and very intimate products and services. The problem had become so severe that, two years before, Thomas Tye had personally taken responsibility for finding a solution. Above all, he *hated* losing revenue.

He solved the problem by inventing a new health-care concept and by founding a new company to deliver it. He was very proud that the idea for LifeLines Inc. was entirely his. The early results indicated that the severe shortage of biological products was already being reduced.

After a last-minute change of plans and some frantic rescheduling, he was to 'open' his new operation officially this morning. He whistled under his antibacterial mask as he rippled through the hermetically sealed clothes bags in his wardrobe, deciding what to wear for the hastily arranged ceremony.

TLS, as the life-sciences group of companies was known inside the Tye Corporation, had been founded soon after the parent company moved to Hope Island. In overall charge of research in the TLS group was Professor Stanley H. Walczack, a sixty-eight-year-old mathematical biologist and reprogeneticist whom Thomas Tye had personally courted for three years before he had been persuaded to leave his chair at Washington State University and relocate to Hope Island. As well as asking Walczack to direct research at TLS, Tye also invited him to become the first incumbent of a chair in Life Sciences that the corporation was endowing at the new Hope Island University.

At first, Walczack had turned him down flat, citing the inestimable benefit of working with so many like-minded colleagues in a non-commercial environment. After all, he was a MacArthur Award-winner and a member of the National Academy of Sciences and he felt that his life's work and reputation depended on him remaining in mainstream American academic life.

Then Tye persuaded the professor and his wife to visit Hope Island for a weekend. Walczack hadn't been finally induced to consider the offer seriously until late on the afternoon on which he was due to return to Seattle. Tye had already driven the Walczacks around the new university campus and cited the impressive list of other academics who had already committed to creating a 'knowledge nursery' on Hope Island. He sensed that Mrs Walczack was already won over and Tye had taken the professor into his office to reveal some of the work currently in progress within the research organizations that would make up the Tye Life Sciences group.

The biologist was amazed by the research that Tye described. He was fascinated by Erasmus's discovery of the progeria gene string and its role in cellular sclerosis and the anti-ageing therapy it was developing. He was also impressed to find that his host and benefactor had a grasp of biotechnology that was at least doctorate level.

It had still taken another six months to secure Walczack's formal commitment and a further twelve months of protracted negotiations between Washington State University and the new Hope Island University before Walczack had been able to delight his nagging wife by moving into one of the new white villas overlooking Hope Town. Now, over a decade later, Hope Island University and the Tye Life Sciences group occupied the leading position in the world's theoretical and applied life-science research community.

The first TLS success under Walczack's leadership had been PerPetuals, the pet-cloning technique developed personally by the professor. This service was launched in the North American market from a string of PerPetuation Centers wholly owned by Tye Life Sciences. Within two years TLS had franchised reproductive pet-cloning laboratories in all developed territories.

The second major service launched by TLS had gained such success that response had been overwhelming. It was this insatiable demand, and the severe shortages it caused, that had prompted Thomas Tye to create the concept for LifeLines Inc. two years earlier.

Tye Life Sciences had pioneered the laboratory growth of replacement human organs. Working from stem cells into which a patient's own DNA was grafted, the researchers successfully grew livers, kidneys, hearts, lungs, hands, feet, sex organs, skin, eyes and neural tissue that, because of their perfect genetic match, could be transplanted into the original DNA provider without fear of rejection and without the need for the powerful immunosuppressant drugs that had previously been the main problem in the aftercare of transplant patients. Usually a transplant from genetically identical tissue returned the sufferer to excellent health within days.

From the outset, doctors in North America, Europe and Japan had been enthusiastic about the process. Wealthy patients who were still well enough to travel flocked to Hope Island for tax-free surgery paid for with tax-free money. Soon there was a lengthy waiting list for organs to be grown and transplants to be undertaken. The Hope Island Medical Clinic was extended twice in the first two years after it opened and the TLS laboratories expanded to cover thirty-one acres.

By itself, this achievement was remarkable but not unique. Several other research facilities around the world announced similar successes in growing cloned organs and, for the brief period the subject had held the media's attention, scores of happy auto-transplant patients told television reporters how wonderful it was to feel well again.

What was unique about the work undertaken at Tye Life Sciences was the third project to be headed personally by Professor Stanley H. Walczack that he had started soon after arrival on Hope Island. After two and a half years of development, TLS filed patents for a universal organ-support and long-term storage unit called MatchBox. Further patents, filed three months later, described a similar but portable device.

MatchBox solved a problem that was severely limiting the therapeutic work that the world's transplant surgeons could undertake with farmed organs. The device kept newly grown organs healthy for an indeterminate period. Without such a device an organ could be kept healthy for only a few days after it had reached maturity and been harvested.

For his MatchBox, Walczack personally designed and manipulated a gene string from which a universal placenta could be grown: a biological interface that could quickly grow supporting tissue connections to the very different arteries, veins, nerves and conductive tissues found in the various organs that were under cultivation. Walczack's master stroke was in coaxing the gene string to mimic the way female reproductive systems switch off their immune-system defences when an embryo, which is wholly foreign tissue, is welcomed and nourished by the placenta. Within a year of starting MatchBox manufacture Tye Life Sciences had sufficient stocks to be able to announce the launch of a new service described in their literature as POB – Pre-emptive Organ Banking.

The response to the new POB service from Tye Life Sciences was overwhelming. Many wealthy and perfectly healthy individuals wanted the security of having back-ups of their vital organs and waiting lists started to grow alarmingly.

At first, Tye was in favour of steadily increasing the price for these standby organs until a natural rationing occurred. But his marketing economists developed a complexity-pricing model that produced the maximum financial yield while ensuring demand remained high. The range of price variations was enormous and contracts were tailored for each client. Some paid hefty premiums to have their organs duplicated quickly and then moderate storage fees during the period that TLS kept them against potential need. Others paid lower duplication fees but higher storage costs. The TLS pricing system computed an individual's price structure from the number of organs ordered at one time, the patient's genetic background, medical history, LifeWatch ambulatory data (which, helpfully, indicated current health status), actuarial tables, and also the customer's preferred payment terms. Existing customers of the many Tye-Managed Healthcare Schemes around the world received loyalty discounts, but no customer paid exactly the same price as any other and no customer received exactly the same service.

TLS and MatchBox gave the rich and successful another personal luxury to aim for: an ultimate acquisition that ranked alongside their personal jets, classic-car collections and luxury yachts. The richest customers wanted every one of their vital organs – and some of the less vital – grown and kept available in case of future need.

It was over the issue of how best to deal with requests for help from seriously ill customers unable to afford a priority fee that Walczack and his patron had suffered their first serious disagreement. 'Your work has to be protected,' insisted Tye. 'We own the patents for twenty-five years and there's another eighteen to go.' He could have added that the corporation had still not recovered all the development costs, and licence income was not the most efficient way to go about it.

Walczack had tendered his resignation immediately. But Tye had smiled, put his arm around the professor's shoulders and asked him, as a personal favour, to delay his decision.

When Walczack saw Tye's plans for the seventh-generation LifeWatches and the new market that would result, the professor had suspended his threat of resignation. Tye had then cemented their new agreement by bestowing an additional tranche of stock options in the new company on the pioneering professor.

Thomas Tye selected a pair of blue chinos to go with his white shirt, removing them from their sterile container. He took the elevator down to his office where he handed Connie an envelope and issued instructions about Calypso. Then he pulled a spreadsheet onto the wall screen and ran over the projections for LifeLines's stock-market flotation.

Raymond Liu sat alone in Network Control Center on Hope Island. The Group Technical Director of Tye Network Systems was baffled. His teams had spent three weeks investigating the short but highly dangerous intermittent failures of the air-traffic-control satellites over Northern California and the near-disastrous loss of satellite communications in the Australasian quadrant. But they still had no idea what might have caused such bizarre independent system failures.

Every diagnostic report from the satellites, from the traffic management centres and from the ground controllers was negative. There were no faults and no log records of unauthorized communications. Air traffic communications over Northern California had returned to normal and in the South-Western Pacific the ground controllers had only had to override the orbiting traffic management system, and order the two data communications satellites to fire their thrusters again to return them to their original orientation. When their on-board computers were rebooted, laser-borne data had been once again routed back to them and they functioned perfectly. Their on-board diagnostic systems reported one hundred per cent performance, so there was no explaining the mysterious malfunctions. Chomoi Ltupicho, Liu's opposite number at the Soyuz FreePlanet network, had confirmed a similar outcome to his investigation. Short of sending a recovery shuttle to launch two replacement satellites and bring home the offending units, there was little more Liu could do from the ground. Also, as the suspect units were now functioning perfectly, he knew Tom would never divert launch capacity from the much-whispered-about Phoebus Project.

But despite this, Tom was pressing – screaming – for an answer. No data had been lost during the Pacific network outage so Tye Networks had not suffered any contingent liabilities, but it had seemed very close. Even an hour's failure of a backbone data network could cost the corporation billions in compensation payments. The world's economies could no longer function without the networks.

Liu had searched the global archives for information about other unusual communication-system failures occurring at around the same time. At first he had searched for unusual system events happening an hour on either side of his Australasian satellite failures. There had been nothing, however. Then, in desperation, he had gathered the hourly logs of every orbiting router in the Tye-LaserNet satellite networks for a week either side of the mysterious failures. There were over 8,000 of them and he had been forced to write a script to automate the transfer of those data to his three-dimensional spreadsheet.

He now clicked the display back on. The central Holo-Theater in the Network Control Center was filled with light, and Liu was looking into the square box of cells he had created. He switched the display to graphic mode and pushed himself up from the well-upholstered control chair. He walked right into the display and put his hand into one of the beams of red holo-light that represented a failure that had occurred and been automatically corrected.

There had been a 300 per cent increase in such minor failures in the previous six weeks and there was serious and startling degradation in performance across all the networks. Until these problems were identified and eliminated he couldn't even begin to deploy the much-needed multiplexing software the corporation had recently acquired from a Brazilian software company. He debated whether he should send a warning message to Thomas Tye but decided against that course of action almost immediately. If you alerted Tom to a problem you had to have a solution ready at the same time. Liu knew he was going to have to design a system that could measure data flow and integrity across every public network in the world.

The senior executives, software engineers, medical consultants and off-duty bereavement counsellors of LifeLines Inc., the majority-owned subsidiary of Tye Life Sciences Inc., stood in little huddles, talking nervously as they waited for their VIP guest to arrive and perform the official opening ceremony.

In physical terms there was little he could open. The large sunken room was divided longitudinally by a floor-to-ceiling glass partition. In the administration area, where he would be welcomed, there were only a dozen desks and six large wall screens. On the other side of the soundproof glass, those on duty were huddled in their booths, monitoring their screens or speaking quietly with next of kin all over the world.

Zachary Zorzi, the chief systems designer, Multi-Linnux IV

magus and technical inspiration of the LifeLines sales resource ('Zee Zee' to his friends and favoured staff, 'Easy Zee Zee' to those who knew of his herbal and chemical indulgences) had thought it would be cool to invite the great man to cut a white ribbon, which was now strung limply between two potted plants. Even though the entire enterprise had been the boss's idea, the team had still been surprised when Thomas Tye had reversed his original decision and, with only four hours' notice, agreed to make time in his frantic schedule to accept their invitation to open the operations centre formally. Then it had taken them all morning to locate suitable pot plants to decorate the bare office.

Zorzi was looking forward to becoming rich. He had been lured away from his role as systems director at one of the world's great auction houses by a serious salary increase and generous stock options in the new company. He was nearly twenty-eight and most of the fellow students who had been on his systems-design course at Stanford were already multimillionaires. He knew he had been leaving his own attempt at a financial home run dangerously late in his career. But when the HR Director of the Tye Corporation had made contact, Zorzi had sensed that his opportunity had finally arrived.

Now, two years since initial systems development had started, the resource was live on the networks and had been operating for six weeks without any downtime. LifeLines had already served over 5,000 successful bidders and the company was forty per cent ahead of its projected target for recipient registrations. Zorzi was delighted but, other than monitoring data produced by the automated transactions, there was now little for him to do. The real action was in bereavement counselling and that wasn't an occupation for him. He was already getting bored with life on Hope Island and with the company he privately referred to as 'White Lines'.

'He's coming,' hissed a young female interface designer whom Zorzi had stationed at the door. Everybody straightened. No one there except Zee Zee had met the superstar tycoon before.

Connie walked into the room first, followed by her boss. She was wearing a beige linen trouser suit with a terracotta-coloured blouse, the epitome of professional elegance. Tom wore his blue chinos with a white open-necked shirt. His long dark hair was tied back in a ponytail. Easy Zee Zee approved of his boss's casual style.

'Good morning,' said Tye to the assembled company.

Zorzi stepped forward. He wasn't quite sure what to do. He knew his chairman didn't like to shake hands.

'Welcome, Mr Tye,' he responded involuntarily before he corrected himself. 'Welcome, Tom. Welcome, Miss Law. Coffee, mineral water?'

Tom smiled and shook his head for both of them. Connie cracked the seal of a packet and Tye sprayed his mouth. He slipped on the face mask Connie handed him.

'I heard you've made a good start,' Tye said through the mask. 'Let's see it, then.'

The guests were guided to two new upright chairs placed in the centre of the room from where they could best see the wall screens. Off-duty members of the team pulled chairs from behind desks and arranged them in a group behind the visitors, marvelling that the rumours about Tye and his face masks were true. Zorzi pulled his chair up next to Tye and Connie as the overhead lights were turned off.

'We'll start with a global display of all seventh-generation LifeWatch locations.'

A Mercator projection of a world map appeared on the centre screen, the continental masses outlined in red on a black background. As they watched, minute white dots appeared as concentrations in and around the major cities of North America, Europe, Asia, Australasia and South America. On the Russian land mass, only Moscow and St Petersburg showed much illumination.

'So far there are about eight hundred million units of the GenSeven LifeWatch registered and active on the world's networks,' explained Zorzi, even though he guessed that his exalted business partner would already know those figures. 'But there are still large gaps in distribution in most of Russia, middle China and Africa.' Zorzi pointed to the darker areas of the map.

Tye nodded. 'The Foundation is upping distribution of free units to a hundred and fifty thousand a month,' he announced. 'We're concentrating on areas of high mortality, for obvious reasons – Africa especially. We've got new teams starting in Ethiopia and Somalia.'

Zorzi scratched his goatee beard. 'Well, we're getting a higher percentage of DNA matches than we projected, so getting more GenSevens distributed isn't so urgent as we thought it would be. Anyway, let's look at the US situation.'

The map changed suddenly to display an outline view of North America. Every urban centre showed up as a large, dense cluster of white light. Myriad strings of light across the rural areas revealed how ubiquitous LifeWatches had become for the American people.

'Those data derive from routine uploads from LifeWatches and VideoMates over the last week,' explained Zorzi. 'There's only been a point zero zero two failure rate on GenSevens, which is very impressive.'

'Can we go real-time?' asked Tye.

Zorzi nodded at the display controller and the screen refreshed to show less dense concentrations across the country.

'These are the LifeWatches that are uploading to us right now. If we take out the routine periodic uploads ... Yes, here are the distress calls, and the uploads from those whose vital signs have ceased or are suspect.'

'Show me the deaths occurring in the last hour,' requested Tye. The screen refreshed to show a higher density of white dots.

'The average mortality rate throughout the world is around eleven thousand an hour at this time of year. In the United States it is just over eight hundred an hour,' explained Zorzi. 'It's mid-morning on the East Coast so we're down on that by thirty per cent but, if you look at California, it's the early hours and we're getting a peak loading.'

'What percentage of mortalities are we registering overall?' asked Tye.

'It's a very good proportion,' smiled Zorzi. 'Your marketing people in Consumer Electronics have done well. Even allowing for those who haven't upgraded their LifeWatch units to a GenSeven, and for those freaks who won't wear any sort of LifeWatch, we're getting data from about forty-six point two per cent of the adult American population.'

Tye turned to Connie. 'I want to talk to Randall from CE later – remind me. Penetration still needs to be better.'

He turned back to Zorzi. 'How good has the matching system proved?'

Zorzi swung round in his chair to face one of the female medics. 'Do you want to take this, Irene?'

Haley hated jogging, but there was no getting away from it: she could feel a distinct pinch of fat where there should be none. It happened during the writing of every book and during relationships in which she had too much control. She didn't like herself for this behaviour, but she would let herself go to seed a little when a romance didn't stimulate her.

Battersea Park was filled with joggers, t'ai chi enthusiasts and dog walkers. It was five p.m. and the river was almost at its lowest, the grey mud of Chelsea Reach drying within minutes of its exposure.

Haley rubbed her side to banish a trace of stitch, took another deep breath and set off again, in the direction of Albert Bridge. She doubted whether she would achieve much physical improvement in just a week, but she needed to make a good impression in New York.

She'd also been thinking a lot about Jack Hendriksen and why she hadn't heard from him. He had warned her not to contact him at the Tye Corporation and, although she understood his reasons, it was frustrating to have to wait for his call. He'd promised to get in contact over a month ago.

Haley put on a short burst of speed to arrive at the gradient leading up to the park gate and the beautiful old suspension bridge beyond. She stopped again to lean against the embankment railings opposite Cadogan Pier. When she had found her second wind she would run back around the park, take a shower, and go over to Ladbroke Grove to spend the evening with Flick and Toby. There was little she could do on the biography until she knew more about what Sloan Press expected.

Dr Irene Desmond, thin and nervous in a black jersey dress, stood up and walked round to face her exclusive audience. She swallowed nervously and then looked directly at the distinguished visitor.

'The system for matching donors to recipients has proved remarkably accurate, Mr Ty– Tom. A GenSeven LifeWatch sends us the wearer's DNA profile in its first routine upload after it is initially strapped on. We compare and couple that to the wearer's digital ident from the World Certification Authority and, once confirmed, the information goes into the data warehouse until it's needed – and, of course, so that Tye Agriceuticals, Pharmaceuticals and the insurance companies can extract the genetic profiles of each individual. When a customer registers to become a recycling recipient, we upload his or her DNA profile and we store it in the same format for constant comparison with the profiles of those who pass away. That's the module that Zee Zee – Zach – developed. When we get a match above sixty per cent we alert the customer. Let me show you.'

She nodded at the display controller and a DNA profile appeared on the left-hand screen.

'This profile was provided by a forty-one-year-old male in Melbourne, Australia. He is waiting for at least one kidney, since he's lost both and is on permanent, but unsustainable, dialysis. He therefore can't wait for Tye Life Sciences to grow him replacements. Now, if we could see our current pool of donors whose organs have not yet been reallocated...'

The map display refreshed yet again, to show a whole world view. Those watching saw a low cluster of lights in the main urban centres and lights stringing out across the rural open spaces. Despite the relatively low overall penetration of the seventh generation of LifeWatches in Africa and Russia, a surprisingly large number of individual dots and clusters showed across those territories.

'There's a higher ratio of donor wearers in the undeveloped areas, where there are fewer medical facilities,' broke in Zorzi by way of explanation. 'It's the people who think they're most at risk health-wise who go out of their way to get a GenSeven model. So the deaths we miss out on in these areas tend to be the accidental and unanticipated mortalities. Unfortunately, they are often the best source for recycling opportunities.'

Dr Desmond looked up at the screen, then turned back to resume her explanation.

'Of those with organs still uncommitted, we currently have eleven hundred deceased up to an hour old, just under six thousand up to three hours, twenty-two thousand, three hundred up to six hours and thirty-four thousand, seven hundred up to twelve hours. Unfortunately, none of them has a good enough DNA profile match to be of interest to our Melbourne customer.'

'How many of those uncommitteds matched a customer's needs well enough to be in auction?' asked Tye.

'Zee Zee?' prompted the doctor.

'Can we see the current bidding status?' asked Zorzi.

The left-hand screen refreshed to a show a transaction-analysis page.

'You may not be able to read all of the small type,' said Zorzi. He stood up and walked closer to the screen. 'There are four hundred and twenty-seven auctions in progress at the moment. The longest duration is nearly eleven hours.'

'Is that normal?' asked Tye. 'I thought our business plan suggested three hours of bidding would be the max. Don't we want to keep bidding excitement high?'

Zorzi nodded. 'Yeah, that's what we thought. But sometimes it gets protracted when there are late entrants, or somebody asks for a pause when they need to raise more capital. We allow each bidder *one* break of fifteen minutes for that.'

He nodded to the system controller again and a new display appeared.

'We're running two auction models to see which one produces the highest yields,' he explained as he looked at the screen. 'This one here is a straightforward *highest* bid, the other is a *sealed* bid that's intended to elicit high pre-emptive offers. As you suggested, I chose an upper-class English accent for the auctioneer interface. It does seem to reassure customers and maximize the bidding.'

He stepped back to take in more of the data. 'This auction's for a liver that's available in Cape Town and is already safely in a MatchBox and certified by the removing surgeon. The next of kin have agreed to a twenty per cent cut of the net receipts and we've given them an estimate of one point seven million US as their share.'

'What's the reserve?'

'Eight million US. This particular recycle opportunity is in A-One Condition and the database has shown eleven DNA profiles that match over ninety-five per cent of the donor's genome – that's from our database of patients who have registered with us and are waiting for a well-matched liver. A ninety-five per cent match means that there will be many common genes on the sixth chromosome and many common antigens. It's almost a perfect match of tissue types – as good as if it were an isograft from an identical twin. That's why this auction is going on so long. It'll be a fierce contest and will probably reach twenty million'

'Are all eleven bidding?' asked Tye.

Zorzi glanced at the controller and the screen refreshed again. 'No, seven have now dropped out,' he reported. 'We're at sixteen point five million dollars, that's from a customer in Memphis. The other three have ten minutes to improve their bids. The next round drops to five minutes, then one minute, thirty seconds, and that's it. The whole process is automated. Once a NOK – sorry, a next of kin – has agreed, we get a copy of his or her digital ident as approval, then a contract is printed out at their end in their local language and in a format that embodies any peculiarities of local laws and taxation. Where there's a spot tax on removal we undertake to pay that.'

'How can we be sure of getting our payment from the bidders?'

'Well, we've had no problems so far. Most of the customers who sign on are too late for POB – they can't wait while a replacement organ or tissue part is grown. We charge everybody who registers three hundred thousand dollars as a joining fee. That isn't a lot of money but it's enough to keep out the jokers and those who aren't wholly responsible for their actions. When bidding for a recycled resource begins we ask everybody who wants to bid to deposit the reserve in escrow, as bankers' drafts or in digital cash, before the auction starts. Our registered customers are warned of likely reserve figures in advance. The actual reserve depends on the closeness of the DNA profile match of the recycled component and this is displayed for the bidders as they enter.

'If the match is forty per cent or less they might as well get a recycled part on the welfare market. They're going to have to use cyclosporine and all the other immuno-suppressants for years. If it's above sixty-five per cent, they'll still have to use them, but they've got a ninety per cent chance of the transplant holding. Above ninety per cent and everybody's a winner. No drugs are needed at all. Obviously we ensure the successful bidder makes full payment before the shipment is approved.'

'How did we get this liver?'

Zorzi signalled at one of the members of staff perched on the edge of a desk. 'This is Doctor Mohammed Ebrahimi, our senior bereavement counsellor. He's also our head counselling trainer and a consultant thanatologist.'

A squat, sallow, grossly overweight middle-aged man stood up. He had a bushy black moustache and he wore an over-formal three-piece dark suit. He made a deep, ostentatiously oleaginous bow.

'That was one of mine, Tom,' Ebrahimi explained in an old-fashioned, Farsi-accented style of English. Despite the air conditioning, he was sweating visibly. 'It was really very easy. The LifeWatch intervened when the poor donor suffered an MI – a heart attack. Very sad. I got to the widow in Mumbai – in India – within an hour. She spoke very good English although, as you know, next door we can cope with one hundred and eighty-six different languages. You understand that auto-translation isn't a very suitable thing in bereavement counselling. We are often the ones who have to break the news to the NOKs and it doesn't come out right. So, anyway, this NOK was quite composed. I should say your wonderful, wonderful counselling message helped a lot, Tom. Thank you very much for that.'

'Can we show Tom how that message came out?' called Zorzi, looking over at the controller. He was pleased at how well this was going.

The right-hand screen refreshed and Tye was looking at his own image. He was in darkness, so that only his face was lit with a soft glow.

'[NOK Salutation space, NOK Family Name space.] I know there are times when even a LifeWatch can't help us,' his image said gently, looking earnestly into the camera. His tones were deep and mellisonant, as designed by the otolaryngologist over thirty years before. 'I know how you must be feeling. You have lost someone very close. I extend my sympathies.'

There was a short pause as if Tye was gathering himself. Quiet organ music started in the background.

'But, if you'll agree, some good can come out of this. Allow me to find a way for [Deceased Salutation space, Deceased Family Name space]' – there was a short pause, then increased emphasis – '[Deceased Given Name space] to help others. Our recycling program guarantees the best future for – [Deceased Given Name possessive apostrophe space] – life to be of help to others. And you will also contribute to helping the Thomas Tye Foundation so that other lives can be saved.'

The figure paused.

All in the room inhaled involuntarily and held their breath as the ScentSims added a most pleasing fragrance. Zee Zee had debated for months with the olfactorologists about the right note to strike at this point in the message. In the end they had created a fragrance they called 'Strawberry.' It was the smell of a newborn baby's crown.

'LifeLines can take care of all the details now. One of our counsellors will speak to you personally in a moment. I just wanted a moment to express my regret and to say . . .'

There was another pause and the camera slowly zoomed in towards the face. The music stopped.

'Reach out and touch me.'

Tye's hand appeared and he reached out towards the lens.

'Touch me and [Deceased Given First Name space] can touch the world.' His finger flattened as he pressed it to the lens and a faint golden corona appeared.

'And, thank you. May your God bless you.'

The video image faded to darkness and even those in the administration area felt its impact as the screen flickered and returned to normal display.

'Very powerful,' breathed Zorzi, breaking the collugency. 'Of course, you were kind enough to give us sufficient video material to allow us to morph your mouth around whatever names you need to say – even the long Russian ones. Then we morph the whole message into all of the various languages. The result is very realistic. We even select a threnody appropriate for the local culture. Want to see a replay in Russian? We use *gusli* music.'

Tye held his hand up to quell the young auction designer's enthusiasm. He'd seen more than enough lip-synch morphing over the years and he already knew that the morphotactioners in Tye Digital Arts were the best in the world.

'Tell me about the total yield on this deal.'

'The widow was most keen for full recycling procedures,' smirked Ebrahimi, a model of morigeration. 'She agreed to twenty per cent of the net receipts, as Mr Zorzi said. We instructed the local surgeons, and the excellent Tye Logistics delivered the MatchBoxes in under an hour. Procedures were complete inside three hours.'

'That's the way we'd like them all to be,' broke in Zorzi.

'How many other commodities can we recycle from this donor?' asked Tye.

'Every high-ticket item except the heart and lungs,' beamed the necrophilic negotiator. 'The widow even accepted a closed-casket funeral so we could harvest the face and hands for the burns, skin-cancer and identity-replacement markets. For that, we pay all the exequial expenses.'

Tom nodded his approval.

'The kidneys went immediately, I think,' reported Zorzi. 'Some oilman from Uzbekistan trounced everybody with a pre-emptive eighteen-million-dollar bid. The corneas also went quickly, bone marrow is under way now . . .'

'What will be the total recycle yield?' asked Tye again.

'About one hundred and two million,' Zorzi estimated. 'The widow will get twenty per cent, while storage and transport is paid for by the successful bidders. We pay local medical expenses for the recycling procedures. I would think we'll net maybe seventy or eighty.'

'How are the percentage splits holding up in general?' queried Tye. 'Are any of the families getting greedy?'

Zorzi signalled for the bereavement counsellor to respond.

'Oh yes, but you see it is we who add the value to the meat,' explained the happy grief counsellor. 'We do get some greedy NOKs, but recycle potential exists for only a very few hours, so we just let them, how do you say it . . . stew.' He emitted a high-pitched series of connected shrieks, as if someone had described to him what laughter was but then forgotten to provide an actual demonstration.

'The trick is in finding the match and providing the storage. The actual recycled unit isn't worth much without that. NOKs usually come round after a couple of hours – when they're facing the choice of making a free donation to a charity or welfare health market, or making some serious money with us.'

'Very good,' commented Tye as he stood up. He turned to Connie. 'Set me up to visit one of these auction winners in the recovery room. Choose an American kid, under twelve, and get the perception people to handle the media set-up.'

He looked back at the gangly Zorzi and slipped off his mask. He then walked his latest entreprenerd a few steps away from the main group.

'Have you got a capitalization in mind for the stock-market offering?'

Zorzi scratched his goatee again. 'I'd guess around forty billion.'

Tye beamed. 'I'll have a bet with you. With this rate of earnings we'll reach a hundred.'

He paused in thought. The success of the auction models

had given him another idea. 'Can all this work without you now?'

Zorzi nodded. 'Yeah. I've only got to finish designing the suicide-reserve module – as the prices will have to be higher. The condition of organs from most suicides is much better than those that come to the market from natural deaths. I want it all up and working before the autumn. The actuaries say there's a big rise in suicides around the Christmas holidays and the New Year and I want to be ready.'

Tom raised his eyebrows. 'Get someone else to finish up for you,' he said quietly. 'Meanwhile, talk to Connie and fix to come by the house. I've got something for you that will make this stuff seem like chicken feed.'

Zorzi smiled. He couldn't imagine any market that would be better than this – but an invitation to the Tye mansion!

Tom turned back towards the expectant faces. 'Thank you, ladies and gentlemen,' he said. 'We'll go for an IPO in eight weeks. And we'll all be seriously rich.' There was no trace of irony in his voice.

He turned on his heel and walked from the building, spraying his mouth as he did so. Connie followed, hard upon his heels. The ceremonial ribbon remained intact, hanging limply between the potted palms.

But Zee Zee didn't mind at all. He had some calls to make.

## ELEVEN

The flowers filled Calypso's living room. They had arrived while she had been consulting with HR and Logistics about helicopter availability. She had decided she would spend a month with her mother before making contact with the medical staff agencies on the American mainland to begin the search for a new job.

Two men from the spaceport's busy courier office had been drafted in to make the delivery from the exclusive flower boutique in the town square. The Volante pick-up they used seemed to be overflowing when they started to unload, and it took them ten minutes to carry all the baskets and bunches into Calypso's bungalow while she frantically cleared surfaces on which to put them. Each display had a small handwritten card identifying its blooms. There were giant white atamasco lilies from Virginia, mauve calceolaria from Venezuela, a Japanese ikebana arrangement with white chrysanthemums set against gravel and bleached drift twigs collected from the island's beaches, a stunning arrangement of chocolate-tinted odontoglot orchids from Brazil, a dozen strongly scented burgundy Hope roses (grown on the island) and a dazzling array of spotted purple, yellow and white tiger-flowers collected from the swamps of Florida's Everglades.

When every flat surface had been covered and her living room turned into a heady, aromatic bloomery they handed her a grey envelope and left. She slit it open and pulled out a card.

It was from Thomas Tye, and Calypso scanned his handwritten message with mixed feelings. She was still furious with him for the previous afternoon. Then she sat down and read the note again. His handwriting was elegant, almost female in quality, completely unlike her own scrawl.

*Dear Dr Browne,*

*I'm truly sorry about yesterday. Tommy means a great deal to me but I was wrong to react in such a way.*

*Tommy was very upset by my behaviour and I have apologized to him. I'm afraid he is still somewhat upset and he is asking for you.*

*I hope you will be gracious enough to accept my apology and join us for lunch at the house today. Please contact Connie.*

*Yours,*
*Tom*

She put the card down on her small coffee table and looked around her flower-filled room. Suddenly she sneezed. Then she sneezed again. She went outside into the heat and sneezed twice more.

Two hours later Calypso took the Mag to the house. The shuttle stopped as it approached the steel security doors shuttering the tunnel leading to Tye's private residence. Calypso turned her head and squarely faced the camera lens set into the wall. The doors parted at the middle and the shuttle sped forward into the brightly lit tunnel.

Tommy was waiting on the little station platform right underneath the house. Even before she climbed out she could see that he had been crying. She had dressed in a white linen shirt and white Bermuda shorts for what, she anticipated, was going to be a rather awkward meeting.

She squatted on the platform and Tommy threw his arms around her neck in silence. She forced herself not to kiss him but she stroked his back as he clung to her.

'It's OK, Tommy,' she soothed. 'There are no bones broken, are there?'

As she stood, he continued to gaze up at her, clutching Jed. Tommy's companion had recently been upgraded by Professor Keane's researchers and he now disliked being left alone.

'Hello, Calypso,' said the red caterpillar.

Calypso heard doors open, then Connie stepped out of the elevator.

'Welcome, Calypso. Tom's waiting for you inside.'

By tacit agreement they said nothing as they rode the four floors up to Tom's main living area. Tommy clung to Calypso with one hand while squeezing Jed with the other.

'You're hurting me,' complained Jed. Calypso knew how that felt.

The elevator stopped but the doors did not open.

'Hold your breath,' warned Connie.

Calypso felt a wave of damp cold air wash through the car. Then, at last, the doors slid open.

Tye was waiting for them in his main living hall, a room Calypso hadn't visited before. It was of double height, clad in brilliant white Carrara marble, and she guessed the ceiling was over thirty feet high. The vast unpolished, slate-grey stone floor was strewn with ancient Afshar, Sehna and Shah Abbas Persian rugs arranged around a score of fine classical sculptures. Pale sofas were grouped to form seating areas.

Although she had heard about them, it was still a thrill to see the giant Jackson Pollocks and Hockneys, the Picasso nudes – and, displayed all on its own at the far end of the room, *that* Rembrandt, an image she remembered from her childhood. She would have liked more time to study the small but exquisite art collection.

The room's east-facing wall was constructed entirely of photoreactive plate glass. In the bright sunlight it had darkened protectively to lessen the glare.

She noticed a slight swelling on Tye's left cheek, as he extended a hand with a smile. Despite herself, she couldn't help but respond.

'Apology accepted?'

Calypso conjured up a small grin, then took his hand. She noticed it was small-boned and extremely fine. She had never shaken hands with him before.

'Let's go out onto the terrace,' said Tye, leading the way across the hall. A full-height pane of glass slid to one side and they stepped out onto the white tiled terrace. A table had been set up under a large sun umbrella where Tye's butler and one of the maids stood waiting for the luncheon party.

Calypso's first impressions were of dazzling light and considerable elevation. Beyond the waist-high safety rail, the hillside fell in wide, stepped terraces to the beach and the emerald ocean below. The view was breathtaking and she found herself gaping in awe up and down the length of the island.

'It *is* beautiful, isn't it?' observed Tye, as he pulled a chair out for her. 'But then, you know these islands well, Doctor.'

Calypso smiled again, spontaneously this time. She had been wondering whether she could see Haiti on the southern horizon.

Once seated, they each took one of the warm antiseptic towels that were offered.

Tye was at his most charming as lunch got under way. He recounted how he had discovered Hope Island and what a tough deal Cuba had struck over ceding it. He went on to describe the protracted negotiations with the United Nations for sovereign recognition and how his corporation had bowed to Cuban and American pressure to forgo all forms of arms development and manufacture here and to maintain no armed forces other than those required for internal and coastal security. He was explaining how he had personally chosen the site for the house when Connie interrupted him. 'I have President Orlov's cabinet secretary for you.'

'I'll get back to him,' said Tye.

He turned back to Calypso and apologized. 'Sorry, Doctor. I was supposed to be in Moscow for lunch today but I wanted a chance to make things up with you.'

Calypso did her best to smile in what she hoped was a gracious and understanding fashion. He had cancelled lunch with a head of state to see her. How bizarre Tye's world seemed – and, of course, Tommy's.

'It's very hot today,' Jed offered, breaking the silence. Tommy laughed and they all laughed with him.

They were first served a salmon mousse, and then the chef offered a choice of sushi or soya-duck in a *bigarade* sauce. Tye had fish for both courses, while Calypso and Connie ordered the pseudo-duck. A huge basket of crisply fried coarse-cut Hope Island potatoes (guaranteed non-fattening) was served on the side. Tommy, who had been silently gazing at Calypso all through the lunch, was brought tofu-chicken nuggets in tempura batter served on a bed of wild rice. A very chilled Pouilly Fuissé was served, which Calypso enjoyed greatly although she noticed that Tye drank only mineral water.

When he had eaten most of his main course, Tommy asked to be excused. His father nodded and the boy got down from the table.

Connie rose. 'I'll take him to see Nurse Pettigrew.'

Tye smiled his approval.

'Can Calypso come and play with me later?' asked Tommy.

Tye looked at the doctor questioningly. Calypso nodded.

'She'll be along later, son,' he said. 'Enjoy yourself.'

'May we watch a cartoon?' asked Jed.

After the plates were cleared, they were left on their own. Tye cleared his throat and turned in his chair to face Calypso squarely. He took off the sunglasses he had been wearing. 'You don't think much of me as a father, do you, Doctor?'

Calypso stared at her employer's beautiful face and tried to read some of the secrets hidden behind those vivid violet eyes.

'No, I can't say I do,' she confirmed slowly. She tilted her head back. 'You're over-protective to the point of doing him harm. You've employed me because I'm a paediatric psychiatrist and I have to tell you that if I examined this boy in a clinical environment I would classify him as already significantly disturbed.'

She watched Tye take the blow and she saw that it hurt. But he was holding himself in tight control.

She ventured further. 'Where's his mother? Every boy needs a mother and I never hear him talk about her.'

Tye looked at her with a long steady gaze. 'His mother left when he was a baby, Doctor. I'm sure HR must have explained that when you joined us. It is one of the main reasons I want your help in keeping him as, well, ordinary as possible.' He paused. 'Given the circumstances.'

The butler and maid returned with a choice of tea or coffee. Calypso chose to have some more wine as she thought about what she had been told. She took another hot towel that was offered. When the servants had gone, Tye moved the conversation forward.

'What would you suggest we do, Calypso? How can we help Tommy?'

She made a decision about her own future and about Tommy's. 'I'll stay and help *if* he is allowed more freedom, Tom. That's my condition.'

Tye raised his eyebrows.

'I want him to go to Hope School full time,' she said. 'He has too much private tutoring. He must mix with other children. If you remember what I said earlier about the whole population on this island, you should apply it to him also. As far as I can see he's not picking up any of the normal childhood illnesses: no measles, no mumps, no chicken pox, no coughs or colds. He's not developing any resistance to them and if he catches them later they could be serious.'

Calypso looked to see if her arguments were hitting home but Tye had put his sunglasses back on and she couldn't read him. She wondered if she could venture a discussion about his own personal phobias.

'I'll think about it.'

'Do you mind...' She hesitated, then she plunged, as was her way. 'Why do you wear face masks? What's all this with the antiseptic towels? Airborne bacteria are good for us, Tom. It's that resistance thing again.'

He smiled, broadly this time, flashing perfect teeth. 'It's about money, Calypso,' he explained with a small shake of his head. 'If *you* get a cold and you can't work, what does it cost you – a few days' pay. Perhaps a few thousand dollars. I'm not bragging here, but if I miss a day, well...'

He didn't need to continue and Calypso saw his point. Over ninety per cent of common illnesses – colds, fevers, viruses – are transmitted by airborne infective agents. He was right: what would cause ordinary people merely annoyance would cost him – or his corporation – tens of millions of dollars, just through the business delays caused by his absence. But still it seemed to her more likely that he was simply a phobic pangermic hiding behind this plausible excuse. *How sad to be such a prisoner*, she thought.

'Any other questions, Calypso?'

She uncrossed her legs. From the angle of his head she wondered if he was staring at her body.

'I want Tommy to come to Mayaguana with me – to visit my mother ... and my nieces and nephews.'

Tye shook his head.

'That's out of the question, Doctor,' he said. 'He can't leave the island.'

Joe Tinkler added the usual half-and-half to his fibre and banana flakes and ate breakfast standing beside his kitchen window. Far below, the Manhattan rush hour was getting under way.

But what was he doing up? He'd got out of his bed simply from habit. After fifteen years with Rakusen-Webber, his life had become a firmly fixed routine. He realized he was still in a state of shock. By the time he had got back to his own office the word had already spread and his assistant wasn't at her desk. All the other staff on the fund-management floor seemed heavily preoccupied with their screens or with voice calls. Two burly security men had been waiting outside his office door, their faces impassive.

It hadn't taken Joe long to remove the traces of his years with the firm and in ten minutes he had packed all the physical objects he owned into a small plastic crate that Security had thoughtfully provided. He turned to his screens to send some internal 'goodbye' e-mails and discovered that he was already locked out of the bank's network. He shrugged and lifted the crate.

The security men walked beside him to the elevator and rode with him to the ground floor in silence. At the security checkpoint Norbert Jones, the man who had checked Joe in and out of the building every working day for over a decade, held out his hand.

'I've got to have your bank ident, Mr Tinkler,' he said sadly.

Joe nodded and put the crate down on the man's desk. He opened his shirt and slipped his neck chain around until he found the fastener. He felt no anger or sorrow: he was in shock, functioning automatically. After slipping the laminated digital ident card from its chain, he handed it to Norbert. Then he had picked up his crate and left the building without saying a word.

Joe finished his cereal and went back to his bedroom, where he pulled on shorts and a sweatshirt. Whatever was going to happen in the future, he would make good use of his enforced leisure now, he decided. Getting back in shape was the first priority. He clipped his VideoMate to his running belt, inserted his earpieces, pulled the cord of his Ray Ban Electros over his neck and set off.

It was too early for the day to be hot and Joe headed for the embankment, under the elevated highway. The river sparkled in the early-morning sunlight and he took it easy, aware that his infrequent Sunday jogs had not prepared his body for really strenuous exercise. There were few others out running this weekday morning. Most people in the city were getting to work.

He got down to the Battery just as a ferry from Staten Island arrived. He ran on past the ferry terminal and stopped by the low wall surrounding the isolated little green which surrounded one of the memorials to those who had died in the World Trade Center terrorist atrocity at the beginning of the century. He checked his pulse and blood pressure and stretched out his calf muscles, resting his feet, in turn, on the wall.

He straightened to watch the stream of workers flooding off the ferry, out through the terminal building and into the subway. The clock on the small terminal tower told him it was still only 7.30 a.m.

Joe jogged over to a snack stand near the bus stops. He ordered a large black Colombian coffee and transferred payment from his LifeWatch.

He went back to the low wall and swung his feet over to sit

and watch as the ferry began its return journey. He sipped at his coffee as it was too hot to drink quickly.

'Hey, man, mind if I join you?'

Joe looked up and saw a bulky young black man in a dark business suit. He too was holding a plastic cup of coffee.

Joe shrugged. There was plenty of wall.

The stranger sat down a few feet to Joe's right. He sipped his coffee and blew on it.

'Beautiful morning.'

Joe nodded and looked away.

'It's OK, Mr Tinkler,' the man said quietly, moving closer. 'Here.' He produced an old-fashioned leather wallet from inside his jacket, opened it and proffered the badge to Joe.

'I'm with the United Nations,' explained Chevannes. 'Their Security Agency.'

Joe took the badge and studied it. The laminated card appeared genuine and the photo matched its bearer's face. 'Do you mind?' asked Joe as he unclipped his VideoMate.

'It won't be working, I'm afraid,' Chevannes pointed out. 'We're blocking all communications at the moment.'

Joe stared at the unit. A blinking red LED confirmed there was no signal.

'If I can't check your ident you can't expect me to talk to you,' reasoned Joe, rising to his feet and holding out the badge to its owner.

Chevannes rose and took it from him. 'We'd like to have a meeting with you, Mr Tinkler, over at the UN building. You might be able to help us with something.'

Joe looked around him. The street was now quiet. Then he noticed an illegally parked black sedan at the ferry terminal entrance.

'What about?' asked Joe.

'It's about someone who was very important to you – until yesterday. Thomas Tye.'

Joe laughed out loud. 'You've got the wrong guy.' He smiled. 'I've never met him.'

'But he knows all about you,' insisted Chevannes. 'You were a subject of major discussion for him only a couple of days ago.'

Joe stared blankly at the intelligence officer.

'When he was a house guest of Richard Rakusen – out in the Hamptons.'

The journey had been long, difficult and, as far as he could tell, without point. Marsello Furtrado had struggled to overcome his resentment at his boss's imperious commands but the heat and gritty friction of travel in West Africa had worn him down.

'Just get there,' Tye had insisted. 'It's for the Russian deal.'

When Furtrado had started to enquire about the precise purpose of his mission he had found himself watching the white noise of an empty carrier signal. The holo-conference was over. He didn't even know where Cape Verde was. He would have to look it up. Then Connie had provided a few more details, but these were almost as scant.

'Yes, he insists you go personally,' she had confirmed. 'He'll tell you more once you're there. Just make sure you're on the island before the thirteenth.'

Which hadn't been so easy. Most of Furtrado's work was concerned with making international acquisitions and overseeing the patent-protection team and he ran a forty-person office out of Washington DC while also maintaining his office within the corporate headquarters on Hope Island. Because he could begin his journey from either location he had expected flight arrangements to be straightforward.

As senior counsel for the Tye Corporation and keeper of Thomas Tye's greatest commercial secrets he had the right to use a jet from the Tye corporate flight but, after his assistant had filed his requisition, Logistics informed his office that no aircraft would be available on those dates. Furtrado had never known all six supersonic jets belonging to the company flight to be busy ten days in advance. He then ordered his assistant to charter a corporate jet but his request was turned down by Logistics without explanation. The attorney next turned to Connie.

'We don't want a trail, Marsello,' she had explained with one of her best smiles. 'Charter people talk – about their passengers *and* their destinations. Take scheduled flights, if you don't mind.'

He had minded, very much. Reluctantly he had taken a

Delta 797 subsonic red-eye to Paris and had worked all the way, conducting meetings in six different Tye regional subsidiaries from his private 'discretion-shielded' office on the commercial deck. Once there, he endured a three-hour wait before catching an Air-France connection to Lisbon in Portugal. He had also worked on this trip but Tye's European APU had been unable to verify the electronic security of his workspace so he had been forced to deal only with non-sensitive administration matters. Once in Lisbon he had found it necessary to check into a Holiday Inn at the airport to await a TAP flight that would take him on to Dakar in Senegal.

During his several flights he had reviewed the scant amount of material his assistant had been able to collect about these islands. The remote archipelago was stuck in the middle of the equatorial Atlantic, and the inhabitants spoke Portuguese, which, Furtrado assumed, was one of the reasons he himself had been dispatched. But this wasn't a good enough reason on its own. These days everybody used VideoMates and the language-translation modules were so cheap and so efficient that people could buy or rent the software they needed for any particular occasion. Neither could he see any connection with the forthcoming Russian deal that had been occupying so much of his time and he couldn't imagine that his legendary negotiating skills would be needed in such a remote location.

In Dakar he had had to make yet another overnight stay before catching a tiny propeller-driven plane of Air Cabo Verde to cover the 400-mile trip due west into the Atlantic. After a three-hour buffeting by the Trade Winds, which made working impossible, he had landed at a tiny airstrip on São Tiago, the main island of the Cape Verde group. He followed the directions he had been given and, as no alternative was available, carried his own suit bag and briefcase down the long dusty access road to a small wooden quay. He was due to catch the noon ferry to Fogo, the most westerly of the larger islands in the group. As always, Furtrado was immaculately dressed, today wearing a shot-silk grey business suit and burgundy loafers. But, despite his lightweight outfit, he was bathed in perspiration and unavoidably maculate by the time the ferry arrived.

The sea breeze cooled him off quickly, however, and he was soon wishing for the lengthy journey to be over. Though still conducting meetings at different locations around the globe, he was feeling a sense of personal dislocation: increasingly, he couldn't shut out his physical surroundings to concentrate on the issues at hand. The immense emptiness of the mid-Atlantic Ocean thoroughly distracted him.

It was early evening by the time the craft arrived at the tiny harbour of São Fillipe and Furtrado picked up his bag, walked down the rough-hewn gangplank and, following a seaman's directions, turned away from the small township and set off along the dirt road beside the beach. After half a mile he saw what he hoped was the end of his quest. It was a small house that stood inside a low drystone wall built out of rough basalt rock – solidified lava that had once spewed from the flat peak of the volcano two thousand feet above. Outside the two-storey house was an ancient, weather-beaten wooden sign. Furtrado stopped to make out the lettering: *Pensão Hollywood.* Despite its improbable name, this was the place.

He walked up the worn cobbled path and knocked on the door. He was greeted by an elderly but neatly dressed white man in cardigan and carpet slippers who was clearly expecting him. Once he heard Furtrado's native Portuguese, the man became all smiles and, despite the younger man's resistance, insisted on carrying his luggage as he led the way to an upstairs room at the front of the house.

'Just one or two nights your secretary said, Senhor?' enquired the owner. 'A very short break.'

Furtrado agreed. He was becoming increasingly puzzled. There could be nothing here worthy of acquisition by the Tye Group and, in particular, nothing worthy of the attentions of Marsello Furtrado, the maestro of such activities. He knew that in the office they called him 'Must-Sell-O!' behind his back but, like all over-dressed, over-mannered *rastaquouères*, he missed the irony.

As soon as he was alone he opened his VideoMate and found it lacked a signal. He assumed that there was no local wireless network on these islands, so he moved to the window, watched the satellite icon appear and called Connie.

'Well, I've made it here,' he confirmed, 'but God only knows why. What now?'

'Tom says well done,' relayed Connie, with a smile. 'What time do you normally wake up?'

Furtrado was too tired to wonder why she wanted to know. 'Usually around six, but I'm completely disorientated.'

'We'll call you at five your time,' she told him. 'Have a good evening.'

He was cheered up by an excellent swordfish steak in piri-piri sauce. Furtrado was the only guest in the tiny dining room and his host lingered to see if conversation would be forthcoming.

'This is excellent,' complimented the lawyer, as he savoured the tender flesh of the fish and added oil and vinegar to his tomato, onion and chickpea salad.

The man nodded appreciatively. 'Thank you. I enjoy cooking.'

As Furtrado still had no idea what was expected of him in this remote location he decided to make an ally.

'Will you join me?' he asked. 'At least for a glass of this vinho – is it Portuguese?'

'It's from Bucellas,' the chef-hotelier confirmed as he pulled out a chair on the other side of the oval table. 'Near Lisboa. We have to import everything. Little grows here in this godforsaken place,' he grumbled, plunging unhesitatingly into *saudade* – the Portuguese delight in melancholy. 'Why they called this Cape Verde, I don't know.'

The white wine was poured and, as he continued eating, the lawyer assumed that his own language skills might indeed have been the reason for his presence here. The local dialect, Furtrado learned, was a combination of Portuguese and Crioulo, a blend of West African words with a few Tupi and Guarani paronymies thrown in for good measure. He had never heard of a 'Cape Verde' language-translation module being available for a VideoMate and it certainly added to his own lexicon.

Furtrado also discovered that the semi-barren islands of Cape Verde had formed on the tips of huge volcanoes that reached upwards from the Atlantic seabed 2,000 metres below. He learned that the people of this former Portuguese colony survived by providing fuel, provisions and other services to the world's shipping as it passed by and, the hotel owner assured him, by servicing a growing tourist industry. Furtrado wondered who would seek out such arid isolation voluntarily.

Twice during the evening, the hotelier probed his solitary guest about the reason for his visit and the counsellor fobbed him off with a story about needing some time alone for reading and research. In turn, Furtrado had asked him about Russian connections with the islands, but the owner's blank stare confirmed the truth of his denials. They sealed their new, somewhat uncertain, friendship with a fiery Portuguese aguardiente.

Despite his disorientation and bewilderment, the brandy helped Furtrado sleep soundly and he was aware of enjoying a deep and vivid dream about his estranged twin brother when he was woken by his VideoMate trilling on the bedside table under the window. He switched on the table lamp and answered.

'Morning, Marsello. Sleep well?'

Tom was circling inside his Holo-Theater.

'Morning,' managed Furtrado, looking at the time display. The local time was 4.50 a.m. On Hope Island, the same time zone as New York, it was now ten minutes before one a.m.

'What's the weather like in Cape Verde?' asked Tom.

'Fine,' responded Furtrado absently. He shook his head in an attempt to wake up fully. He was obviously missing something here.

'No, you idiot, I mean what's the weather like NOW?' screamed Tye.

Furtrado pushed himself up from the small single bed and walked around to the window. He pulled the floral curtains aside. It was still dark outside, only a glimmer of light in the east suggesting that dawn had started.

With difficulty he undid the catch on the ancient sash window and pushed the bottom frame upwards. He leaned out, felt rain on his face and ducked his head back inside.

He picked up his VideoMate. 'It's raining,' he reported.

'Show me,' ordered Tom.

Furtrado found his Viewpers and slipped them on. He returned to the window and scanned the horizon.

'It's raining pretty heavily,' he observed, 'as you can see. What's this about?'

'It hardly ever rains on those islands,' said Tom, 'and never at this time of year. It hasn't rained there at all for seven years. It's the driest inhabited place in the world.'

'Oh,' responded Furtrado, mystified.

'It's called taking care of the planet,' intoned Tye, tartly. 'Go out into the streets, get me lots of pictures, and then find the mayor and a local judge and record their reactions to the rain. Also, get them to sign timed and dated affidavits about today's weather. Then bring their depositions back to Connie on Hope – personally.'

# TWELVE

Haley Voss felt as if she was bouncing along Fifth Avenue. The sidewalk felt like a trampoline and the soaring buildings seemed to beckon her skywards.

The crush of late-afternoon shoppers and scurrying city dwellers worked its usual exhilarating magic. As she threaded her way northwards through the throng she felt as elated and as vibrant as when she had first visited the great metropolis as a teenager. She hardly noticed the oppressive heat. She had tried to call Felicity so her twin could share her pleasure, but she could only reach Flick's AutoSec.

Her initial meeting with Luke Bailey and his team at Sloan Press had gone even better than Rosemary had predicted. Haley found that their enthusiasm for her biography was rekindling her own passion for the project. She could now admit to herself that Nautilus's withdrawal and the subsequent retreat by the other publishers had badly damaged her self-confidence. That was why she had kept rewriting and rewriting, she realized, going over the same material time and again. Now she was keen, almost desperate, to leave her palimpsest behind and to get back to the really hard part of the job: producing the new pages that would reveal the man behind the public image.

'Handled right, we'll make the lead item on every news bulletin,' Sloan's vice-president of publicity had predicted, his deliberate use of the plural pronoun sending her a clear message. '*Everyone* will want to read the inside details of his life, Miss Voss. We'll work you so hard on the interview circuit that every time anybody thinks about Thomas Tye, they'll also be thinking of Haley Voss. We'll get a number one in the *New York Times* best-seller list and we'll become the top e-book download on the networks.'

Sloan's senior legal counsel and her assistant had seemed totally relaxed about the prospect of a confrontation with the huge law practice that represented the Tye Corporation. They had laughingly dismissed them as 'Tye's Terriers'.

'We're expecting injunctions and writs the moment we announce our publication of the book,' she confirmed. 'We just need to lock the other side up in the discovery cycle until the week before publication. Then we'll go back and get the injunctions lifted.'

They hadn't seemed at all worried that Tye's attempt to stop the book's publication might actually succeed. 'This isn't the UK,' the senior counsellor had laughed. 'The right to free speech is enshrined in our constitution!'

But despite this optimism, it was agreed not to divulge a single detail of the book's content to newspaper editors in advance of publication. They were sure they would be swamped by offers for serial rights the moment the review copies were available.

Then Luke Bailey had taken her for lunch at the Grill Room of the Four Seasons on East 52nd Street. Haley had been careful not to overdress for their first meeting and had chosen to wear a mustard-coloured trouser suit over a high-collared white blouse. A fine gold chain circled her neck, just below the top button. The publisher was obviously well known to the maître d' and they were quickly shown to a balcony table that provided them with a panoramic view of New York's favourite traditional luncheon room. She had thoroughly enjoyed the impeccable service, the superb food, Luke's slightly peccable company and the certainty with which he enthused about their 'world exclusive'. He had even made a discreet and not entirely unwelcome pass at her that, for the moment, she had chosen to ignore. But it lifted her spirits immensely: it confirmed again that it was only the solitary nature of her writing work that got in the way of romantic opportunity.

Sloan had booked her into a suite at The Plaza, the majestic old Manhattan landmark that faces Central Park from its superb location on the corner of Fifth Avenue and Central Park South. She had planned to spend the rest of the afternoon shopping but, as she paused with the crowds waiting to cross 54th Street, she realized she was far too excited to concentrate on anything as mundane as the items the themed retail stores had to offer. Instead, she would walk up to Central Park, sit in the sunshine, and map out a revised structure for her biography.

'Haley?'

The voice to her left sounded surprised. As she turned, it took her a couple of seconds to recognize the tall, fair-haired man in a white polo shirt and dark slacks. He took off his Ray Ban sun-viewpers.

'Why, hello,' she cried. 'What are you doing–'

The WALK sign appeared and the crowd surged forward. Jack Hendriksen laughed as they were both carried into the road. He held out his hand and took her shoulder, steering her safely onto the opposite sidewalk.

'Well, what are *you* doing here, Haley?' he asked when they reached the comparative safety of Gucci's shop windows.

'I'm here to see my publisher,' Haley told him breathlessly. 'Sloan's now publishing my Thomas Tye biography! But what are *you* doing here? Why aren't you with your boss?'

'I live here,' said Jack simply. 'My apartment's down near the Village. I'm on vacation and I'm shopping for my mom's birthday.'

They stared at each other for a few seconds.

'Sorry I haven't been in touch,' added Jack. 'I was incredibly busy when I got back.'

Haley waved away his apology. 'Now I've got a new publisher I *really* need to talk to you,' she enthused. 'You haven't changed your mind?'

Jack smiled and shook his head. 'No. I've been thinking more about it. I . . .' He paused, put his shades back on and looked around briefly, then back down at her.

'Are you busy right now?' he asked.

Haley shook her head.

'Let's walk up to the park,' Jack suggested. 'We could sit in the sun and . . . Well, I *am* going to try and help you with your book. Tom's behaving increasingly . . .'

He tailed off. Haley was smiling up at him, her broad mouth and sparkling teeth bringing a generous smile to her impish face.

'I'd love to,' she said.

As they walked, Jack pointed out the notable buildings – St Patrick's Cathedral, the re-renovated glister of the old Trump Tower, F.A.O. Schwartz, the world's largest warehouse of intelligent companions – and Haley listened and wondered about her own feelings on bumping into him.

Passing the Plaza Hotel they crossed into Central Park, mingling with tourists waiting for carriage rides. Then they headed up towards the Zoo, where it would be quieter.

On this beautiful June afternoon all of the benches were in use by shoppers, resting joggers and office workers who had somehow found an excuse to be outdoors. As they approached an occupied bench in the centre of the green, a bulky young black man in a smart business suit rolled his sandwich wrapper into a ball and rose to his feet. With skill he lobbed it directly into a trash bin five yards away and headed off in the direction of The Met.

Haley hurried to grab the vacant bench, then sat and turned to Jack with a grin of triumph. They sat facing the afternoon sun as Jack asked her more about her new publishing deal. Then he enquired when she thought she might have the book finished.

She shrugged. 'It depends on what I uncover. I've finished all the orthodox research, but someone keeps sending me new stuff . . . really amazing stuff.'

She glanced sideways at him as he leaned back in the corner of the bench listening carefully to her. He had taken his sunglasses off again, so she looked straight into his clear blue eyes. She had a decision to make here and, if she was wrong, she knew her book could still get suppressed.

'You remember that report I showed you – about Tye using gene therapy to stop him ageing?'

Jack nodded.

Haley decided. 'Well, someone's sent me another amazing report.'

Jack raised his eyebrows.

'I don't know who.' She hesitated. 'Would it surprise you to learn Thomas Tye has a son?' she asked.

Jack smiled. 'Well, someone *definitely* is trying to help you,' he confirmed. 'Thomas Tye Junior and he's seven. But we're never supposed to talk about him – for security reasons.'

Haley felt a thrill race through her, shivering despite the warm sunshine. This was the first independent confirmation of what would be one of her most exciting revelations. Then her biographer's alarm bells started ringing.

'But how can anyone keep something like that a secret?' she asked, fearing that the news would break elsewhere before she could publish.

'That's not too hard if you're Thomas Tye,' explained Jack. 'The boy has always lived at home – in Tom's private mansion – and everything is brought to him, rather than the other way around. He has his own nannies, his own doctor – hell, even his own shrink.'

He smiled as he thought of Calypso.

'But *you* know about him.'

'It's my job to run the security operation there. I have to know.'

'But all those other people on Hope Island . . .'

Jack thought about the buzz that had started up on the island that time when Tommy had knocked himself out.

'I guess it's a kind of open secret there,' he suggested. 'But everybody on the island either works for the corporation or is financially dependent on it in some other way. Nobody's going to risk saying anything to an outsider. The risk of a kidnap attempt is considerable when so much money is involved.'

'What's he like – this boy?'

Jack looked down at his lap. He was holding his sun-viewpers open on his thigh, his forefinger crooked between their tortoise-shell arms. He weighed his words.

'He's very like his father. He's pretty spoiled and sometimes his behaviour is ... Well, I'd like to be in charge of him for a while. I believe that kids need a firm hand – gentle, but firm, you know what I mean.'

Haley nodded. She had noticed that even little Toby seemed to be happier when handled firmly.

'But is this boy like his father, to look at?' she persisted.

Jack smiled again. He had already read a copy of the report that UNISA's London surveillance team had intercepted. 'He's the spitting image,' he confirmed.

He watched as Haley digested this. He hoped she hadn't noticed that both their VideoMates had remained unusually inactive since they had met.

'The report claims the child's actually a human clone – of Thomas Tye,' she said flatly, looking up to gauge Jack's reaction.

He uncrossed his long legs and leaned forward, elbows on his knees, his sunglasses now swinging between his fingers. He was silent for a few seconds before he turned his head to look at her. 'I've always suspected that,' he agreed.

They sat for a few moments in silence, lit by the powerful sun, then Jack leaned back. 'I really would like to help you further,' he said. 'Are you free for dinner this evening?'

At first, Joe Tinkler had found their request ludicrous. If he hadn't been sitting there on the seventeenth floor of the United Nations Secretariat building he would have felt sure that the guys from Derivatives had set up one of their expensive and elaborate practical jokes. But the men talking to him now didn't seem to be working up towards a big laugh.

Joe had agreed to visit the UN headquarters later on the same day that his jog had been interrupted. As he had showered and pulled on a suit, he found himself going over and over the possible reasons for Thomas Tye and Richard Rakusen discussing him last weekend, presumably planning his dismissal. He could see why Tye would want to limit the power of external shareholders, but the Old Man had never given any indication that he knew the tycoon personally. If he did, he should have declared it to the compliance officer and, theoretically, should also have declared it at any meetings in which future investments in the Tye Corporation were discussed.

'Bastard,' Joe mumbled as he selected a tie. Perhaps he should call Jill White at home.

Chevannes had been waiting to meet him at the main entrance on UN Plaza and had escorted him through security, the informal 'immigration' process and up to Ron Deakin's office.

'Thanks for coming, Mr Tinkler,' Deakin began. 'I hear you're out of a job.'

Joe shrugged. The Street had been buzzing with the gossip. He'd had to switch on his AutoSec to manage the carrion calls.

'Would you consider undertaking a six-month assignment with the World Bank, in Geneva?'

Joe wondered if he had heard correctly. 'You belong to some sort of security service, right?' he queried.

Deakin nodded. 'We're a security agency for the UN. We have a mandate to operate in all its member countries.'

'Forgive me ... I mean, why me? What has the UN got to do with fund management?'

'Humour me,' persisted Deakin. 'Consider it a theoretical question at this stage. If an offer seemed attractive, could you handle it? I mean is there any reason you *have* to remain in New York for the next six months?'

Joe thought about it. Most of his friends were here, Nancy was here – but that on-off-on-off was more *off* than *on* most of the time. His work was here, too ... His work *had* been here. He doubted that he could work again in the Street at anything like the same level unless Rakusen-Webber retracted their accusation of gross negligence and agreed a no-fault separation. Even then, as they say, doubt clings where money sings.

'There's no reason I *have* to stay in this city,' Joe admitted. 'But I'm no banker. I don't know much about what the World Bank does – I mean, beyond the obvious.'

Deakin nodded. 'I understand. You'll want to know more about this assignment. We'll provide a first-class round ticket, three nights at the Hotel President Wilson and a payment of sixty thousand dollars just for your time, if you're prepared to fly to Geneva and meet Doctor Chelouche. He's President of the–'

'Yes, I know who he is,' broke in Joe. 'What's going on here? You'll have to tell me more of what this is about if you want me to consider it.'

Deakin reached in his desk drawer for a copy of the Silence Resolution.

Jack was seeing a new side to Haley. On accepting his dinner invitation, she hadn't seemed the least bit surprised when he suggested that they should eat at his apartment. He liked to cook, he explained, and he rarely had the chance.

Although he really did enjoy cooking, the main reason for his suggestion was to ensure that his discussions with the British biographer would not attract attention. In a world in which politics was dominated by the continuous and instantaneous polling of public opinion – 'instant democracy', 'the people's voice' – a revelatory biography could be crucial in changing public perception about his employer. But he needed to keep his cover intact until he could get back to Hope Island and find an answer to UNISA's most pressing question.

Jack had no reason to believe that Tye, his pack of feral attorneys, or the many investigation agencies they employed yet suspected that the company's vice-president of security might be feeling disaffected. Nevertheless, as Al Lynch had suggested, Jack needed to keep any reference to his new relationship with this British author out of the networks, and out of all forms of digital storage. He was starting to discover how difficult that could be.

Al Lynch had prepared his own plans for foiling any surveillance system Tye or his technical teams might have created. He had taken great pleasure in demonstrating a software agent he had deployed.

'Assuming their system is looking for key words, key faces, et cetera, this agent is designed to snow them under with piles of seemingly fresh information,' he told Jack. 'It's an agent I developed years ago when I was working for the National

Security Agency and I've adapted it to our needs. Every day this little fellow collects everything that is written, said or broadcast about Tye, his corporation, its technologies or any of the associated companies. It then uses one of my own algorithms to generate new communications and stories that are apparently completely different messages, articles or broadcasts. It will seem as if the coverage of Tye and everything to do with him has suddenly gone up tenfold, and we'll encrypt a good percentage of the new messages. They'll be swamped because when their automatic systems get overloaded, humans will then have to decide which messages are worth decrypting – if that's what they really can do.'

At Jack's suggestion, Lynch named his agent *Multiplicitye.*

While he had been undergoing his refresher courses with Lynch and his other, more physical, instructors, UNISA had brought Jack's apartment up to full safe-house standards. Dozens of small transmitters scattered throughout his floor, wall and ceiling spaces generated a shield of electronic white noise that turned his apartment into a sterile communications zone. Nothing digital or electronic went in or out unnoticed on the airwaves or on the land lines. Technical Services had been sure that no local surveillance of Jack's apartment had already been established but, as a precaution, they had even set up a system that used recordings to mimic Jack's normal domestic communications and exchanges to disguise the existence of the new shield. All windows had been double-glazed to prevent laser-borne acoustic bugging and every street and store camera within a three-block radius switched to simulated input (with the correct time and date stamp superimposed) when any team member or target ident was in the vicinity or when a target was recognized by the automatic pattern-recognition system.

Haley had arrived promptly at eight. Jack recognized her face on the downstairs security camera and buzzed her in. He checked the elevator location display inside his apartment and, as it rose to the top floor, he crossed to open his door. He watched as the elevator doors slid back. She was wearing a black silk blouse and black trousers. A grey sweater was draped over her shoulders. Her smile lit up the elevator even before she stepped out.

'Welcome,' he said.

She had brought a bottle of white wine. 'It *was* chilled,' she complained, her deep-brown eyes earnest. 'But that was before I found that I had to walk half a mile here from where the taxi dropped me.'

Jack took the bottle with a smile. 'Gramercy is a traffic-free zone. Great for the environment, hell when you need to get the groceries home. The local stores deliver by refrigerated bike.'

She looked around the old loft. The UNISA security installations had required a full redecoration of the apartment – finally forcing Jack to allow the physical obliteration of his past life with Helen – and the new pale lime paint and clear varnished wood surfaces glowed.

'This is lovely,' his guest observed. 'It feels very comfortable.'

They sat at the much-used wooden kitchen table and Jack poured the wine she had brought.

'To your book.' He raised a glass in a toast.

Haley's smile seemed to increase the level of illumination in the room. 'I'm so excited,' she laughed as they clinked glasses. When she had sipped the Pinot Grigio she put her wine back on the table and pulled her VideoMate and a pocket video camera from her bag. 'Let's get started,' she grinned.

'I not sure *that*'s a good idea,' Jack said gently, nodding at the capture devices. 'I can't go on the record, so any help I give you will have to be informal.'

Haley frowned at him. 'I wasn't thinking of using the actual recording, I just like to–'

'It's just not a good idea to have any identifiable record,' insisted Jack, gently but firmly. 'Make written notes by all means.'

Even though he had kept his tone light, he saw Haley register his resolve. Then he saw her small chin jut determinedly.

She was recalling what he had said in her London apartment: *I normally record everything. Don't you video meetings and so on? Just for legal safety, and security?* This could still be a trap.

'Are *you* recording this meeting?' she asked. 'That's what you usually do, isn't it?'

He spread his hands on the table. 'No, I'm not,' he said, dissembling with a strictly literal truth. He and the UN technical services team had discussed how they could feed the signal of this meeting outside the building but had agreed that, given the Tye Corporation's potential powers, even with highest-level encryption it was too risky. The recording equipment to capture their conversation had been installed by a UNISA technical officer inside the loft earlier in the afternoon and was now being operated remotely.

Haley now had another decision to make. She looked around the large room and then back at him.

'OK,' she agreed. She put the device back in her bag and extracted her DigiPad.

'So, you're the corporate vice-president of security for the global Tye Corporation. Does that mean you have a seat on the main executive board?'

'Would you mind using real paper?'

She shrugged again, bemused. She rummaged in her bag and found an envelope and a pen.

'OK. So do you have a seat on the board?'

'But you're not going to quote me in your book.'

'No, I'm not. You'll be an unidentified source as far as my readers are concerned. But I have to keep records, even if it's just for the libel lawyers and, maybe, the courts. To prove my sources.'

Jack nodded. 'How about this, Haley? I'll tell you anything I can, you make your notes, but don't keep anything in electronic form that identifies me as a source. I can't explain my reasons right now. Keep it on paper by all means but don't put my name in your VideoMate address book and don't even make annotations on your DigiPad or word processor. Don't make any PopUp notes that mention my name or my role. Is that possible?'

Haley cocked her head on one side. 'Very mysterious. I *am* able to use secure mode, you know.'

Jack nodded again. 'I know, I know. But humour me. I do want to help and, if it gets to a court case, I'll even do what I can to help there. But, in the meantime, don't file anything about me electronically. Don't call me, don't v-mail or e-mail me. I won't contact you in that way either, I'll use other methods.'

'What going on here?' she asked, her face suddenly animated.

'What's my job?' he replied, with a smile of his own.

'Security,' she said.

'Precisely. Will you work with me on this?'

She thought for a moment and then nodded, her large dark eyes never leaving his.

'So, how can I help?'

'So, *do* you have a seat on the main executive board?'

'No.'

'What did you do before you joined the Tye Corporation?'

'I was in the US Navy.'

'An officer?'

'Commander.'

'You mean you weren't trained in security?'

'My training covered a lot of things. Some forms of security included.'

'So you weren't a seagoing officer?'

'At the outset, I was. But, in general, no.'

'What *were* you? Exactly?'

'That's classified.'

Jack would never admit to his original SEAL background, it was *such* a cliché. The rest he couldn't talk about.

Haley paused and looked across at him. He seemed at ease, slightly amused even.

'But in the US Navy?'

'Yes. They paid my wages.'

'Were you a *spy*?' she asked, with a mischievous grin.

'Absolutely not,' said Jack firmly.

There was a pause.

'So how did the Tye Corporation recruit you?'

'They contacted me shortly after my discharge. An old friend recommended me.'

'Who?'

'It doesn't matter.'

Haley thought about a challenge but decided on a new tack. 'You're married?' It was put as a question, despite his ring.

'Why?'

Helen smiled down on him from the wall opposite, behind his guest's head.

'Are you married, Jack?' she repeated.

'No. I mean I was. Not any more.'

'But you still wear your ring,' she objected, gently touching his hand and peering at the gold band.

He hesitated. 'I'm a widower.'

Haley removed her hand quickly and he watched her digest his answer. Her next question was put more softly. 'Any children, Jack?'

'No.' That would have come next. He wondered why she was asking.

She moved on. 'What's your remuneration?'

'That's not relevant.'

'It certainly is. The lawyers will have to make a judgement on your actual status in the Tye corporate hierarchy. If you're a VP but you're not on the executive board, your salary level is the best guide to your seniority.'

Jack considered. She was waiting for his answer, preparing her next question. Intelligence shone from her dark eyes like a beacon. He found he that he didn't at all mind telling her.

'Sixteen dollars and three cents per working minute.'

Like everybody these days, he was paid in real-time and by true value transfer. Money worked much harder and more effectively when it was kept moving.

'OK, Jack, save me the math. How much is that a year?'

'Eight and a half million dollars, US.'

Haley raised her eyebrows, but made no comment. Then, 'Anything else?'

'What do you mean?'

'Any other sort of benefits? Cars, share options, pensions, health insurance, et cetera.'

Jack nodded and smiled. 'All of those, except the car. There aren't any cars on Hope Island.'

'What's your total package worth?'

At eight and a half million dollars he was highly paid, but not by global VP standards. Even though corporate security had grown up to become a function of the senior executive in most corporations, he and his peers had yet to penetrate the *highest* echelons of capitalist power.

'I get more share options added each year. I suppose the total now has a value of several hundred million dollars.'

This time Haley did allow herself a smile. 'And you'll risk all this to help me?'

Now he saw her point. 'No. I'm helping myself. I have my own reasons and . . .'

Her noticed her frown. 'I'm not using you,' he said quickly, even though as he said the words he realized that the opposite was true. 'Like I told you, I think things have gotten out of hand. Something has to be done.'

'Like what?' she demanded.

'Well, the international legal system doesn't seem–'

'No. I mean like what things are getting out of hand?'

'I'll come to that,' he said as he rose from the table.

Jack crossed the loft's wooden floor and bent to open a low two-drawer filing cabinet. He extracted a thick file and returned to the table.

He pushed the bottle and his glass aside and opened the file. 'First, I thought you might be interested in a little personal background on Tom. This is all on paper,' he muttered as he leafed through, looking for the first document he wanted to show her. Deakin's research team had unearthed it and it was perfect for the occasion. 'This is for you, but please don't scan it, or digitize it in any other way. Is that OK?'

Haley nodded, mystified but intrigued.

He found what he was looking for and held the photocopy of an old press cutting out to her.

'This should also make good copy,' Jack suggested as he turned to locate another bottle of wine. 'It's about the mental institution to which Thomas Tye was committed. He's purchased his past.'

# THIRTEEN

Towards the end of the first decade of the twenty-first century, at the peak of uncontrolled vehicular excess, the controllers in the Los Angeles Department of Traffic and Highways offered up a daily prayer in the hope that the Santa Monica Freeway would be able to cope with 32,000 vehicles per hour. As the months went past this prayer changed from a forlorn entreaty to an abject confession, a *mobilis-miserere*, as they finally acknowledged that the capacity for which the freeway's designers had planned it had long since been exceeded. At the best of times progress was slow. Often the freeway, like many others in the city, became gridlocked. Average traffic speeds in the Greater Los Angeles area had dropped to twelve m.p.h.

When the first fully automated traffic-flow system was finally introduced in commuter lanes there was much public outcry and intense political debate. Drivers felt uncomfortable handing over control of their vehicles' movement to computer systems, even if the state was providing them with generous tax incentives to assist with the cost of installing the necessary automatic driving systems. It was only when non-automated traffic was completely banned from the fast lanes in peak periods that drivers seriously began adopting AutoRide technology. The Los Angeles City Council fuelled the experiment by providing an eighty per cent cash subsidy for these in-car control systems and, during the first years of the experiment, the flow of vehicles was managed by roadside locators and broadcasting systems.

As a result, average traffic speeds rose from twelve to forty m.p.h. and the number of vehicles able to move along the thirty-five miles from the center of downtown Los Angeles to the Pacific coast increased to a record 38,000 an hour. The problem was that traffic on the adjoining roads, particularly the San Diego Freeway, remained uncontrolled and thus the local terrestrially based solution was only partially successful. The on and off ramps continued to be a mess and lengthy waiting periods at smart traffic lights provided only a partial solution.

Other cities such as New York, London and Athens had responded to the alarmingly accrescent number of private vehicles simply by banning them from parts of their road networks, charging vehicles to enter the remaining areas and significantly improving mass-transit infrastructures. Once public transport became smart enough to provide customers with information about the precise arrival time of the next tram, bus or train, also its current passenger load, its best-prediction ETA and the precise whereabouts and loadings of all connecting transport, the public suddenly began to see the advantages. First Class and Business sections were opened on the larger people-carriers and, in London, the whole top deck of the traditional red bus was given over to cosseting premium-class travellers with display screens, drinks and snacks.

But however smart, reliable and comfortable mass transit had become elsewhere, it wasn't an option for the Los Angelenos and their administrators. The gigantic urban sprawl of Greater LA had been designed after the automobile had become ubiquitous but before it threatened to choke its host city's arteries. The scale of the city was simply too large for any conventional form of mass transport system to work. Even if the city fathers had been able to start over and magically dig earthquake-resistant tunnels for a comprehensive under-city subway system, the trains, trams or maglev shuttles would need to make too many stops, and to cover such vast distances, that they would still remain unattractive as a form of regular transport.

The answer was to envision all the city freeways and their feeder roads as one circulatory system; to imagine each vehicle as a single cell inside a giant centrally managed mass-transit system in which individual vehicles obeyed the needs of the larger organism. As a result, the roads became a huge contraflow venous system that was controlled and managed from the skies.

This satellite-based management system was designed, built and administered over a five-year period by a consortium of Tye Aerospace Inc. and Tye Asset Management Inc. (TAMI). Currently, the Santa Monica Freeway handled over 70,000 vehicles an hour travelling seven feet apart at an average speed of forty-five m.p.h. All the other freeways and main arteries in and around Greater Los Angeles boasted similar throughput and, a mile before joining any of the roads or freeways that were part of the LA Intelligent Transport System, drivers logged in their required destinations and handed over control of their vehicle to the ITS, operated on the city's behalf by TAMI, a system that had been working faultlessly for three years.

On the bright sunny Wednesday morning following the rather less than intimate dinner that Jack Hendriksen had cooked for Haley Voss, most Angelenos were keen to begin their working day. By 7.30 a.m. the freeways were at 84.7 per cent maximum capacity and, from a thousand kilometres up, a satellite of the TAMI network was issuing instructions to more than 130,000 vehicles to reduce their speed from 52.157 m.p.h. to 47.342. Construction had closed one lane of the on-ramp to the Santa Monica Freeway at its junction with 405 South and the satellite's on-board parallel processors had predicted the resultant congestion and was routing 2,962 vehicles to different entrance and exit ramps, providing in-car displays of the new routes that had been selected and revised estimated arrival times. If the drivers had entered the network ident of those awaiting their arrival, they too were similarly informed of the travellers' revised ETAs.

Unfortunately, few travellers were paying any attention. Despite state regulations that required a driver to remain available to resume control of a vehicle at all times in case of automatic control failure, some drivers had swung their big chairs away from the dashboard and steering wheel to attend business meetings locally, on the East Coast, in Europe or in any location in any of the fourteen longitudes and twelve time zones in which other humans were still awake. Many had resumed their social conversations and get-togethers via the networks, some reviewed news, sport or business information in their viewpers while others donned immersion helmets, turned up their ScentSims and wallowed in pornography. A minority simply ate, drank and watched TV in air-conditioned comfort. Others, caught in a global frenzy of the time, gambled their way to work.

Grazing the ionosphere above the traffic, 6,000 satellites occupied spaces within a layer of orbits that, in the early days of earth-orbit satellite deployment, would have provided safe locations for less than a hundred of them. As with the cars below them, the satellites were computer-managed and, after a gigantic space-junk clear-up operation to remove the debris of fifty years of uncontrolled near-space colonization, all orbital positions and movements were now strictly controlled. Orbiting refuelling tankers provided energy top-ups and in-service repairs for the satellites: these in-flight maintenance techniques had extended the average useful life of satellites to twenty-five years. Again, the contractor was the Tye Corporation – this time in the guise of Tye Orbiting Management System Ltd (TOMS). The entire system was under the control of a team led by Raymond Liu, group technical director of Tye Networks.

On a geostationary 4,700-kilogram satellite dubbed LAT-6 the real-time memory managed by four 1024-bit microprocessors developed a leak. It wasn't the first of the day and ordinarily it wouldn't have caused a problem. The fail-safe memory system supported its data and instruction set inside several artificial shells designed to isolate operations from any equipment or operating system malfunction. But this memory leak was the 643rd to occur in eight hours and such information losses exceeded every scenario the designers had anticipated. All registration addresses in the emergency buffers were occupied and even the flash-storage overflow could accept no more. Despite being guaranteed as completely fault-tolerant, LAT-6's processing froze and all data transmission ceased.

Two hundred and ten kilometres further into space, SATMAN 36, the regional Orbital Management Station maintained and operated by TOMS, registered the loss of communications handshake from LAT-6 and dumped its real-time mirror copy of LAT-6's activities in a laserburst containing 782 terabytes of information to LAT-7, six miles to the south. LAT-7 acknowledged receipt and took over the control of 7,458,711 vehicles, 11,610 roadside displays, 15,041 traffic lights and 4,794 controlled crossings. The handover took two milliseconds and the drivers and pedestrians below were unaware of the glitch.

Then the navigation sensors on SATMAN 36 reported that the failed LAT-6 was rolling out of orbit, dipping dangerously close to the approved path of a high-speed orbiting satellite operated by the Pakistani government for a purpose claimed to be atmospheric research. SATMAN 36 observed LAT-6's descent and, when the critical line was crossed, automatically issued instructions for the satellite to self-destruct before NASA's ASAT anti-satellite systems were triggered and intervened with a destructive laserburst that would invalidate all insurance claims for the satellite's loss.

LAT-6 fired its emergency thrusters, heading for re-entry and automatic incineration as ordered.

LAT-7 continued to manage the central LA traffic it had inherited from LAT-6 as well as its own south LA vehicles for a further eleven minutes. But it too had suffered ten times more memory leaks overnight than its specifications allowed for. Then, just as it was about to slow the Santa Monica Freeway traffic down to 45.82 m.p.h., all on-board processing froze.

SATMAN 36, holding an updated mirror copy of TAMI data from LAT-7, then had to make a choice between managing the traffic below or the spacecraft in orbit. The artificial intelligence algorithms prioritized the management of the highly valuable orbiting assets, in particular the satellites that were managing air-traffic movements, at the expense of their terrestrial counterparts.

When the control signals vanished, fail-safe cut-outs operated inside 7,458,702 vehicles, emitting audible warning signals before braking and slowing the vehicles to a controlled straight-line stop within thirty yards. The whole of the Greater Los Angeles road network came to an immediate standstill.

Inside nine vehicles, the fail-safe systems malfunctioned.

The most serious incident occurred on the junction of the elevated San Diego Freeway and Airport Boulevard. Aboard a southbound sixty-six-ton, thirty-two-wheel, twin-tank gasoline trailer-rig travelling at the prescribed 52.896 m.p.h., the automatic fail-safe system did not engage. The driver had been changing his clothes in the cubicle behind the front seats at the time. The truck continued in a straight line into a stationary van ferrying six children to elementary school, then jackknifed to the right, crossed the safety lane, crashed through the parapet of the elevated freeway and landed on top of a Boeing 797 passenger aircraft waiting to taxi onto the recently extended main LAX runway for take-off. The aircraft and the tanker erupted into a giant ball of flame.

In other parts of the city the wholly quadrirotal Angelenos suffered instant impotence and many attempted to restart their vehicles, assume manual control and drive off the freeways in the safety lanes. There were 571 collisions and six old-fashioned 'freeway fever' killings in the first thirty minutes following the control failure.

'I want you to think of your favourite beach on Hope Island,' Theresa Keane began, selecting an image she knew would be popular. The graduate students were scattered around the study in her home. As part of the package designed to lure her from the rarefied air of Cambridge, Massachusetts to Hope Island, the dean of the new university had secured permission for her to use one of the much-coveted white villas set on the lower mountain slopes overlooking Hope Town. This morning Professor Keane had opened the wide glass doors that led out to a large stone terrace and a view of the distant town and the emerald bay beyond. Small green and blue humming birds darted between the luxuriant Browallia, bussu palms, Escallonia, poinsettias and castor-oil plants that she had cultivated. When she was alone in this room she often compared it to the musty, cramped accommodation for which she had been so grateful when she had secured her first tenured seat at Trinity College in Dublin fifteen years before.

'I want you to imagine that a hurricane strikes your favourite beach and lifts every grain of sand into the air. It forms a giant swirling cloud over this island and out across the oceans in all directions.'

She waited as they created the model in their heads.

'Now I want you to think of the beaches on the other islands that are our neighbours: Cuba, the Turks and Caicos Islands, Jamaica, the Bahamas and the Dominican Republic. Simultaneously, hurricanes hit all those beaches and funnel all that sand into the air in vast clouds that stretch right across the Caribbean.'

They nodded as they made the mental construction.

'Now, more hurricanes appear in the Keys, in Florida, in Miami, Palm Beach, Naples and Tampa. Then on Stinson Beach north of San Francisco, or Long Beach in LA, or the beaches at Cape Cod, or the wonderful sands of Samoa, Goa Beach in India, Phuket in Thailand or Bondi Beach in Australia. I want you to run through all of the beaches you've ever seen, lift the sand into the air, make vast, dense clouds of sand all over the world.'

She waited again as the twenty-six postgraduate students who were crammed into her room rooted around in their memories for beach scenes that could serve as visual aids.

'Imagine the amount of sand on those beaches. Think of the number of grains there might be. I want you to think of those hurricanes merging, the clouds of sand becoming one. I want you to imagine that cloud and all those grains of sand.'

Again she waited as they assembled a mental language with which to approach her concepts.

'How many grains do you think you might have? A hundred billion, a trillion, a quintillion, a duodecillion?'

She paused again.

'One thing we do know – there are more stars in the universe than there are grains of sand on all the beaches of our own planet. Our galaxy alone has over one hundred billion.'

Theresa Keane knew that this was not new material to some of them, but she wanted them all to start from the same place. As a computer scientist she would normally have expressed such large numbers as powers, but this was a diverse group that required her visual diglossia. Professor Walczack's frequent urgings for cohesion across all disciplines had fired her imagination and she had invited biologists, evolutionary psychologists, linguistic neuroscientists and geneticists to join her own students in the seminar series.

'There are ten galaxies – not stars, *galaxies* – for every human on this planet! Think about it – ten whole galaxies for each of you and ten for every other single person.'

She paused again.

'And each of the billions of stars in each of those galaxies are all separated by an average of four light years. Or five hundred billion, billion miles. *This* is the human dilemma.'

'What about binary twins?' asked a lanky Scots materials physicist with a prominent Adam's apple and watery blue eyes. He was part of the Phoebus Project research team.

Theresa smiled at him. 'You're right, Martin. Many of the stars are twins. I was thinking of single-star systems that might support planets.'

She allowed them to think on a little longer as she sipped her tea. She imagined them constructing mental images of vast swarms of stellar dust in which every speck was a star.

'We now know that many of those trillions of billions of stars support planets – some of them possibly oases of benign atmospheres, gentle warmth and abundant water like our own glorious Earth. Humankind has now explored its own world. We have discovered and probed the other planets in our solar system and we know what lies immediately beyond. But if we know one thing about our species, it is that we will never stop expanding. We can see so many of those stars; they tantalize us. If it is true that our planet is merely a cosmological crèche for our species, how, then, do we take the next step?'

'We m-m-m-must go to the stars,' averred Robert, her stammering research fellow and, secretly, her most ardent admirer.

'We must, Robert, but humans cannot travel at a speed that allows that. How many of you have what you consider to be an accurate scale model of our solar system and our nearest neighbouring stars in your mind – our physicists, astronomers and other supraterrestrialists excluded?'

Theresa looked around the group of young men and women adorning the chairs, sofas, arms, floors, stools and walls of the study. Only one of them raised a hand.

'Do you?' she asked Lisa, a gifted young proteomicist from Argentina. 'In that case you can help me. We're going back to school, Grade Six or thereabouts. We all do these experiments when we're young and we all forget them. It's not comfortable information to retain. Now then . . .'

She turned to a fruit bowl placed on a low table beside her comfortable chair. She picked up an orange and held it up before them.

'This orange will be our sun,' she smiled as she held it on the fingertips of her right hand. 'What is it – about four inches in diameter? That's about ten centimetres for the metric among you. Now . . .'

She turned back to the fruit bowl and picked something up. They couldn't make out what it was until she held it up between the thumb and forefinger of her left hand beside the orange.

'This dried pea is our own Earth and it's about a third of an inch in diameter – that's nine millimetres – and, as you have already guessed, these bodies are approximately to scale. Now, Lisa, take this pea and show us roughly how far the Earth is from the sun at this same scale.'

Lisa pushed herself up from the floor and swept her long black hair back over her shoulders. She took the pea from the professor's fingers and looked around the room. She threaded her way through the bodies draped across the furniture and the carpet and turned at the open French windows. She held up the pea.

'Keep going,' said Theresa, waving her student out of the room. Lisa walked backwards into the sunshine, holding the pea up between her fingertips.

'Keep going,' called Theresa. 'More . . . more. There.'

Lisa stopped at the far end of the terrace, beside a low table that Theresa had placed as a marker.

'That's thirty-six feet, about ten metres or if, like me, you're very, very Irish, eighteen bandles. That pea out there is us, and this orange in my hand is the sun.'

They looked from one to the other.

'Now, working on the same scale, who will stand up and tell me where the star closest to us, Proxima Centauri, would be located out there?'

Theresa stood up and all of them – except the three astronomers, feeling cocky in the pit of their sofa – stood also and gazed out onto the cultivated sunlit slopes that led down to Hope Town a mile away to the south-west. Lisa remained at the edge of the terrace, the pea held in her right hand.

'Well?' asked Theresa from the back of the group. She knew none of them would know. She hadn't known for sure herself until she did the research.

'Th-Th-The Town Hall,' guessed Robert, pointing at the small clock tower jutting above the low roof-line of the town. His doctorate was in speech interfaces.

'No. Further.'

'Three hundred miles out in the Caribbean,' hazarded a cognitive neuroscientist, keen to get it over.

'No. Further.'

'M-M-Mexico,' offered Robert gamely.

Theresa stepped forward to stand amongst them. She pointed out to sea.

'The next landfall out there is Panama,' she told her class, enjoying her demonstration, 'on the other side of the Caribbean Sea. That's nine hundred miles away. If you kept going another nine hundred miles, across the Canal Zone, and then out into the Pacific until you reached the Galapagos Islands, an archipelago with special meaning for the evolutionary psychologists here, that's where Proxima Centauri would be. Again, about the same size as this orange.'

They turned back to her. She seated them with a smile and tucked her long floral skirt under her legs as she lowered herself into her armchair. She waited a few moments for them to get comfortable again and for Lisa to return with the pea and resume her place on the carpet, her back propped against a wall.

'Forgive the high-school experiment,' she requested. 'But it is easy to forget the scale of our local solar environment, even though particle physics suggests such concepts. If I held up the orange again and said it was a helium atom, we know that its pair of electrons would go around it at a distance of about five kilometres. Matter obeys the natural laws of physics throughout the universe.

'Despite this, every image of our solar system you may see – in film, magazines and books – is grossly distorted in scale. You can visit science exploratoriums and you still won't see any accurate sort of scale models. It is as though there is a conspiracy to pretend we live in a space that is more manageable, more human.

'NASA, Hollywood, TV producers and popular magazines have been selling us lies for over fifty years. When the Americans went to the moon it wasn't one of the greatest steps for Mankind, it turned out to be one of the most disappointing. We didn't discover another living world, it wasn't a step towards further space travel. It was a hard collision with the brick wall of truth. There is nothing out there worth going to, or at least nothing worthwhile that is even remotely within reach. Our culture tells us differently because we cannot bear to face the truth. Every model we see of the solar system is grossly telescoped and distorted. Every movie that projects the idea of manned space travel or that heralds yet another contact with alien intelligence lies to us about the real difficulties of space travel or the odds of such an encounter or its nature if it were ever to happen. We have no language with which to consider such ideas realistically.

'Let's look at the facts, now that Lisa has helped us develop a better scale model. In our present physical form we cannot travel at the speed of light, nor even at a speed that approaches it. Our fastest spacecraft today would take ten thousand years to reach Proxima Centauri, our Sun's nearest neighbour, which we have today imagined as being located eighteen hundred miles away in the Pacific. On the same scale, the centre of the Milky Way, our own small outer galaxy and only one of billions of galaxies in the universe, is about twelve million miles away from this orange!'

She lifted the orange again and paused as they tried to take in this information.

'Even if we could travel magically at one quarter of the speed of light it would take us three hundred years to reach the nearest solar system that we know to have planets – and the same time to come home again. So what is to be done?'

'Unmanned probes,' suggested one.

'Targeted radio messages,' submitted another.

'Suspended animation,' ventured a third.

Theresa held up a hand. 'None of those solves the problem,' she maintained. 'It would seem as if the times and distances of our universe are inhuman. They are beyond human comprehension and all biological timescales. Are we to give up?'

Her question was rhetorical and they all knew it.

'No, we are *not*,' she declared. 'The next step for our species is to emerge from our biological envelopes, to free our virtual consciousness and our temporal perceptions from the tyranny of our fragile mammalian support systems and fleeting earthly lifespan. Then we shall be ready for our virtuality to travel at the speed of light to meet the others who share our universe.'

They were quiet.

'This is the topic of our tutorial today and later I will describe progress with our efforts to transfer human neural experience to machine storage. But I will begin by describing some tools that we are developing for the observation and measurement of non-biologically-dependent consciousness in the little boys and girls we are creating in our Anagenesis Experiment.'

Joe Tinkler added semi-skimmed milk to his room-service muesli and sliced fresh banana and looked down at the leisurely flow of traffic and pedestrians in the Avenue Wilson six storeys below. He checked his LifeWatch. It was still the middle of the morning rush hour. Geneva seemed so small, so tame, so ordered, after Manhattan.

He had arrived early on the previous afternoon and, after checking into the suite at the Hotel President Wilson that had been reserved for him by the World Bank, he had showered, changed and gone on a walking tour of the city. He'd been to Europe several times before, but never to Geneva or anywhere else in Switzerland.

Even though it was a working day the old city was quiet. Prosperous-looking men and women walked sedately along the lakeside promenade, enjoying the midsummer sunshine. Well-dressed nannies pushed child buggies and chatted in muted tones to their friends beneath the luxuriant trees of the Quai Wilson and the *Mon Repos* waterside gardens.

On the promenade Joe marvelled at the Alpine clarity of Lake Leman and the majesty of Mount Jura rising behind the town. He stopped to examine a metal triangle and a notice board. On top of a metal pole a small ball, perhaps two inches in diameter, represented the Earth. Joe followed the English-language instructions and put his eye to the hole bored through the centre of the sphere. On the other side of the wide harbour – really an *embouchement*, where the Rhône emptied into the lake – Joe saw a sphere that represented the sun. This distant ball was actually five yards in diameter, but it appeared the same size as the little Earth. Joe peered through the hole again and then stood up once more. He had never appreciated how tiny and how far away the Earth really was from the sun.

Then he crossed the busy road to enter the city's main commercial district. Even here, activity was muted. The shops were elegant, the prices high, the atmosphere discreet. Joe could *feel* money in the air. He realized that the Swiss existed by an invisible process of long-term quiet money making more long-term quiet money. Joe's own world of Wall Street and dealing screens, of money frantically competing to make money on marginal spreads, seemed frenetic and vulgar in comparison. Here and there he came upon carpenters and labourers for whom personal physical output was still linked to income, but as he walked through the city he realized that his current sense of unreality was not only the result of jet lag and the time change – he had used his viewpers' anti-jet lag technology to reduce that – but of entering a society in which most of wealth-generation had long been virtual.

Joe had selected a sharp charcoal-grey suit with a white shirt for his meeting with the most famous banker in the world. As he straightened his pale grey tie, his most conservative, he could feel nerves fluttering in his stomach. He turned left out of his hotel and walked up the gentle gradient of the Avenue de France. He passed the old Palace of Nations and the monument to Woodrow Wilson's fumbled efforts to establish the first global government and then he was at the vast compound the United Nations had been given in 1947 when that organization was established. The new European headquarters of the World Bank soared like a shimmering silver hologram between the ponds and sculptures that dotted the park.

He presented himself at the security post inside the glass doors, and an electronic ident was clipped to his lapel. An American security guard asked Joe to leave all his recording and communications devices at the checkpoint. He then stepped through a scanner and was walked to the elevators. The guard leaned into a waiting car and pressed the button for the top floor. He left Joe to travel upwards alone.

The doors opened onto a softly lit corridor of mahogany doorways, pale cream walls, moss-green carpet and expensive watercolours. A dark-haired, middle-aged woman with a Video-Mate under her left arm greeted him and introduced herself as Madame Pioline. She told Joe that the Doctor was waiting for him and they crossed to a pair of highly polished double doors. She pushed them open and stepped aside.

Inside, Joe's first impression was of light and space. Then he noticed the heavy form of Dr Yoav Chelouche slumped, reading intently, behind a desk. That face was familiar from a thousand newscasts and magazine covers. Joe cleared his throat. The banker looked up, rose, walked around the large desk and extended his hand.

'I am Chelouche,' he announced in the gravel tones that were even more famous than his image. 'Thank you for coming, Mr Tinkler.' He waved Joe towards a low coffee table surrounded by high-backed gilt chairs.

Coffee was brought, pleasantries exchanged: the president of the World Bank enquired about Joe's flight and accommodation. The fallen Wall Street star then commented on the view of the lake seen from the tall office windows. He had debated whether it would be uncool to express his admiration for the man.

'I'm proud to meet you, sir,' he risked. He was tempted to say more about Chelouche's renowned financial achievements. But he knew there was almost nothing he could add to the praise that had been heaped already on the banker by the world's political leaders and economists.

Chelouche put down his coffee cup, folded his hands over his stomach and focused his dark, mournful gaze on the fit-looking black man in front of him.

'Well, I have been aware of *you* for some time, Mr Tinkler,' he said. 'Very aware. It's hard to become as successful as you've been and not get noticed.'

'Tell that to Rakusen-Webber.' Joe laughed ruefully. 'They wouldn't agree and nor would many others. You know Wall Street.'

Chelouche nodded. He did. That was where he had started, a lifetime ago.

'I'm worried that I might now be finished in fund management.'

'Maybe, maybe not,' mused Chelouche. 'But your name has been in front of me for several years. You have done well with your Tye portfolio.' It was a statement but Joe realized that his host was inviting comment.

'The figures say so,' Joe admitted. 'My fund was growing at an annual average of thirty-three-point-seven per cent in real terms over the last six years.'

'Why do you say in real terms?' asked the banker pointedly.

Joe maintained his composure. 'I mean, once they're adjusted for inflation, momentum growth, rounding errors and so on.'

'And what has been the average rate of inflation in the reserve currencies over the last six years?' Chelouche asked.

Joe shrugged. 'One point one, one point two. It's been very steady.'

'Thank you!' exclaimed Chelouche as if he were accepting a

compliment. 'I always feel that the expression "growth in real terms" implies there is something unreal about the underlying performance. It's an archaic phrase.'

Joe could see that he had touched a nerve. He said nothing.

'But you *have* done very well,' Chelouche said more graciously, returning to his theme. 'Despite being dismissed by Rakusen-Webber. As you have been advised, UNISA thinks Mr Tye was involved in that move.'

'In what way, sir?' asked Joe. He had learned very little more about the UN's interest in the Tye Corporation or its boss even after he had signed the lengthy secrecy agreement that Executive Officer Deakin had put under his nose. He had been told only that Thomas Tye and his companies were the subject of a current UNISA investigation that could lead to criminal or civil charges. He was also told that his special knowledge of the Tye investment portfolio could be helpful in that same investigation. Given that Joe had nothing better to do and couldn't easily find another Wall Street buyer for his highly specialized knowledge of Tye and his finances, he had readily agreed to visit Geneva to at least discuss the options.

'I'll come back to that, Mr Tinkler,' said Chelouche. 'But ... you've been kind enough to come all this way. Let me tell you about the assignment we would like you to consider.'

Joe leaned forward and put his coffee cup down on the low table.

'You must know the Tye investment portfolio better than almost anybody outside of the corporation itself,' began the banker. 'We want you to operate a new investment fund specializing in the Tye Corporation and its many related interests. It will be a large fund, a very large fund indeed, but it will have unorthodox goals.'

Joe nodded again, wondering what type of fund the man could be talking about.

'We want you to maintain all your current intelligence-gathering operations on Mr Tye and his personal investments and we hope you will be able to attract many of your old clients back to your new fund.'

Joe allowed himself a small smile. He thought that might be

possible. Many of the world's pension-fund managers had hitched a free, or semi-free, ride on Joe's coat-tails for almost a decade.

'At the outset we want you to run this fund with one initial aim.'

Chelouche raised his heavy eyebrows as he reached the end of the sentence, inviting Joe's input.

'To make profits?' offered Joe because he could think of no other motive. He had read as much as he could find about the World Bank since being invited to visit Geneva, but he wasn't aware of any investment banking operations.

'Yes, naturally, to make profits,' confirmed Chelouche. 'But I also want you to use the fund to maintain price stability in the various Tye stock-market listings.'

Although Joe had now been out of Rakusen-Webber for ten days he was still fully aware of the market positions and he knew that all Tye's stocks were showing great resilience. He had been perversely pleased when the news of his dismissal from Rakusen-Webber had momentarily wiped two per cent off the top level of Tye's quoted interests. But that dip had only lasted half a day and now most of the offerings were trading near or at the top of their year's highs. His network agents had provided updated prices just before he had attended this meeting.

'In what way "stabilize", sir?' asked Joe. 'They seem to be doing pretty well without my help.'

'That's very true – for now,' grunted the banker enigmatically. 'Your job would be to intervene if and when any of the core stocks show signs of serious weakness. If you were to take this assignment, we'd want you to build a substantial cushion of Tye stockholdings as a first step. The more your fund would hold, the less damage any other fluctuations might cause.'

'Would that be fair on my clients, sir?' asked Joe.

Chelouche smiled. 'We will ensure they don't suffer. Additional funds will be made available, if required. How long would it take you to rebuild the sort of position in Tye's stocks that you held at Rakusen-Webber – on the understanding that your buying activities mustn't inflate any open-market prices?'

So the value of his previous fund would merely be regarded

as a starting point – a position to hold *before* new funds for market intervention were even calculated!

Joe considered. He'd have to buy slowly and very carefully. A lot would depend on how Rakusen-Webber played their existing position on his previous portfolio. His sources had told him that it had now been split into five, with Morgenstein personally assuming responsibility for the largest segment. He had also heard that at least two of his old clients were shopping around for new investment houses to handle their equity portfolios.

'Perhaps five or six weeks, sir,' he estimated, 'if we have to do it without being noticed. It depends on how the markets move.'

Chelouche nodded and appeared to be in deep thought. Joe took the opportunity to ask some questions.

'Do you think we're heading for problems in the markets, sir? Is that why you want me to buy Tye stock?'

'There may indeed be problems, Mr Tinkler, but I think they will be very small and limited only to the Tye Corporation and its relatives *if* you would be kind enough to use your skills to help us.'

Joe waited.

'You already know that UNISA is investigating Mr Tye and his corporate activities?'

Joe nodded.

'What emerges from that investigation could affect how investors view Mr Tye and his stocks. We want you to intervene – secretly and anonymously, of course – to maintain the value of those stocks. You'll use our funds for that, not your investors' money.'

'Wouldn't that be illegal?' queried Joe. 'Wouldn't that be a market manipulation?'

Chelouche smiled for the first time.

'This is the World Bank, Mr Tinkler. This is where the rules are made. You're not to worry about that because you'll be working under our auspices. Our aim is to maintain stability, nothing more. The World Bank is not trying to make profits from its interventions, that is not our role. You will understand far better than most that Mr Tye's holdings are so valuable that any sudden downwards movement could destabilize the world's markets. That would set us back a decade or more.'

Joe nodded. He knew that even after nearly ten years of consistent stock-market growth, panic could return within minutes. Only recently he had been within seconds of selling all his Tye-related investments. His failure to do so had been the excuse Morgenstein had used to sack him.

'I thought that was going to happen the other day,' he admitted. 'About two weeks ago Tye started going to cash for no apparent reason. It was touch and go whether I bailed out then.'

'And why didn't you?' asked the world's banker.

'The market swallowed it. He liquidated nearly half a trillion US in a morning and the prices hardly wavered.'

Chelouche smiled again. 'We found ourselves with quite a lot of Tye Corporation stock that day,' he said.

Joe understood what he was being told, but he still couldn't help himself exclaiming: 'You! *That*'s why . . .' He tailed off.

'We think he might have, well . . . guessed that we sometimes undertake such activity and I suspect he deliberately leveraged our interventions to get such a large pile of cash out.'

This was all outside of Joe's experience. There was a long pause.

'What do you think he might need so much cash for, Mr Tinkler?' asked Chelouche, eventually.

Joe shrugged. 'I was certain he was making new investments. I've been watching, but I haven't found any of it.'

Chelouche nodded. 'Nor have we,' he agreed. 'But it must move soon. Money only generates new value when it is moving. We'd like you to trace it for us.'

'But if you're already intervening in the market, why do you need me?' asked Joe. 'I was fooled myself. I assumed the markets had soaked it up.'

Chelouche sucked in his cheeks as he considered his response. 'There are two reasons. First, we're no longer able to do that. As I said, we think Thomas Tye may have somehow guessed our little game. Frankly, Mr Tinkler, we want you as cover. We don't want anyone at the Tye Corporation or anybody else in the financial community to know when the World Bank is intervening. You understand?'

Joe nodded. He did indeed understand. If anybody knew about this it would provide the richest investment gravy train in financial history. For a moment he found himself wondering if he could hitch a ride himself. Then Chelouche cut into his thoughts.

'The second reason is that we want you to be seen publicly to be building a very large Tye-based investment vehicle – far larger than you were running at Rakusen-Webber. If you're agreeable, we'll set you up in your own office in Swiss National's Fund Management centre here, in Geneva. We're very close friends with the bank and, to all intents and purposes, it will be a straightforward hiring. It will be announced that Mr Joseph Tinkler, formerly of Rakusen-Webber, has moved to Europe to join Swiss National,' said Chelouche. He paused. 'I presume you would consider such an offer?'

Joe smiled. A spell with the most prestigious bank in Switzerland would completely restore his reputation.

'Of course, Swiss National themselves will be unaware of the unusual nature of your portfolio,' continued the banker. 'Even outside of the investments you bring from your old clients, your funds will, for all practical purposes, be unlimited. If the Tye Corporation's overall valuation suffered a twenty-five per cent fall, how much would you need to bring them back up?'

Joe did a rapid mental calculation of Tye Corp's total capitalization. That was too much money.

'You mean in addition to the value of the fund I've previously managed?'

Chelouche nodded.

'A *very* much larger fund than I had at Rakusen-Webber,' exclaimed Joe. 'Perhaps more than is available.'

Then Chelouche made himself crystal clear. 'Mr Tinkler, unlimited means *unlimited*,' he explained. 'The World Bank draws its funds from almost every nation in the world. It is as though we are the keepers of the concept of value. If we say we have money, we have money. That's what money is – a belief that there is value.'

Chelouche paused. 'Do you know the meaning of the word "credit", Mr Tinkler?'

Joe wondered if this was a trick question.

'"Credit" is a Latin word,' said Chelouche. 'It means "he believes!"'

Chelouche laughed and Joe heard the man's lungs gurgle. He guessed that the large humidor on the coffee table was not solely for the benefit of his visitors.

'So if the going gets rough I prop up Thomas Tye, no matter what the cost?' continued Joe, marvelling at the idea.

'It's not exactly a new concept,' growled Chelouche. 'Countries used to do it all the time, when corporations still had national identities. Twenty years ago companies like GE, Microsoft and Mitsubishi were frequently propped up in the markets by their national governments. Usually the companies concerned had no idea of that support. After the Cold War ended the CIA turned its attention to economic activities and the security services became economic spies, commercial subversives and tactical investors – all on a vast scale. They diverted the billions the Pentagon was saving – by not having to pursue an arms race – into economic warfare.'

The banker paused, to ensure his explanation was being followed. 'You must recall that scandal ten years ago – when it was revealed that America had agents all over Europe trying to deflate the Euro. They were also whipping up anti-European Union sentiments in the press, covertly funding politicians and political parties who wanted individual states to withdraw from the EU – just to slow up the Union's growth and expansion. They realized that the EU was going to become the most powerful economic bloc in the world and they were trying to subvert it. The idea of keeping a few bell-wether stocks at a high rating in the markets is nothing compared to some activities undertaken by the major governments.'

Joe nodded, not really understanding. Despite all his years on The Street and dealing in the world's most important financial markets he had never heard of governments or the security services secretly intervening in the world's stock markets – or trying to subvert the political aims of their so-called allies and friends. To his chagrin, he realized that despite his supposedly elevated position on Wall Street he and even his most exalted colleagues had been working in ignorance. In a second it all became clear: the great bull markets of the last few decades – a phenomenon usually attributed to the economic efficiencies of global networks – were revealed as products of national stratagem. While he and other market makers had been acting tactically, governments had been using *them* to achieve much larger strategic goals. He felt very small and rather stupid in this most hallowed tabernacle of international capitalism.

'You know what the term "bell-wether" actually means, Mr Tinkler?' continued Chelouche, seemingly intent on developing the semantic theme of their meeting. 'It means the sheep that leads the flock.'

Chelouche roared heartily, his lungs now sounding like a pair of Harley-Davidsons idling at a traffic light.

'That's what I want you to do, Mr Tinkler. Lead the flock. Stop the shmendriks from scattering when there's any loud noise.'

Chelouche's laughter had turned into a wheeze and the banker pulled a red handkerchief from his pocket. He coughed, blew his nose and then looked up at Joe.

'Mr Tinkler, who is the second most important executive in the Tye Corporation?'

Joe considered. It wasn't easy to judge, Tye being such an autocrat.

'It's hard to say, sir. It could be Marsello Furtrado, the corporation's senior legal counsel. The group is very widely distributed. Each division and each subsidiary is run separately by its own CEO and CFO and they report into the corporate office on Hope Island. Tye runs that himself with a very small staff – just sixty or so.'

The banker nodded, wiped his nose again and returned the handkerchief to his jacket pocket. 'But you know them all – or know of them.'

Joe nodded. He knew the biography of every one and he'd

met quite a few at analysts' briefings. He knew their key technical staff, the finance directors and their marketing executives. He even had a spreadsheet that compared the annual salaries and bonuses of the Top 500 people in Tye's organization worldwide. Not all the information he had about them was in the public domain, but Joe was content that his information was accurate. His numerous sources and his relationship skills were the key to his success.

'Then will you develop an organization chart for me, Mr Tinkler?' asked Chelouche. 'I want to see the whole Tye empire laid out: the people who run it, their backgrounds, the relationships between the corporate entities – physical, legal and personal. On a board, not a screen, if you don't mind.'

Joe nodded. He already had most of the material he would need.

'I also want a complete breakdown of the shareholding of each separate corporate identity: how much of each company is publicly available, who owns what percentage – everything above a one per cent stake. And full cross-referencing between different companies and legal identities. Can you do that?'

Now Joe understood how much work was entailed. Creating a cross-referenced and consolidated shareholding register of every Tye-related company would be an immense undertaking.

'Do I get any admin help?' he asked.

'As much as you need,' huffed Chelouche, with a dismissive wave. 'This is a political operation. Finances don't come into it.'

Joe could never recall hearing such a statement from a banker before. He guessed the surprise was showing on his face. Then he remembered there was something else. 'What's the second objective of the fund?' he asked.

'I can't tell you that at this stage,' said Chelouche. 'And it may never materialize. But I can promise that you would be very enthusiastic about it.'

Joe nodded. As well as admiring Chelouche's reputation, he was now beginning to like the man. He realized that he was staring at an opportunity to be catapulted to the pinnacle of global finance and he swallowed.

'So what do you say, Mr Tinkler?' asked Chelouche. 'We'll

match the salary Rakusen-Webber provided and the bonus you were due from them. You'll also be earning bonuses at your previous rate on any additional profit you make for the fund.'

'When do I start?' asked Joe.

His ashen complexion only emphasized the large dark rings around his eyes. Raymond Liu presented himself, as demanded, outside the Network Control Center, where Connie was waiting for him.

'He's in a foul mood,' she warned as she touched the entry system. 'Keep your distance.'

Inside, Liu saw a bright display in the Holo-Theater and he waited while his eyes adjusted to the darkness of the room. Tye had the advantage.

'Get down here Liu,' he screamed. 'Look what your fucking networks have done!'

Liu walked down the gently sloping aisle towards the central display pit. Inside the ring he could see a flat two-dimensional display. He didn't need more than one glance to know that it was an optical view of Los Angeles from one of the visual-wavelength satellite feeds.

Tye was circling the edge of the Holo-Theater, haloed by the bright image behind him. Liu stopped halfway down the aisle. He felt Connie stop abruptly behind him. He could make out several other figures in the front row of seats, just at the edge of the surrounding darkness.

'Come down here and fucking well look,' demanded Tye. 'Nothing is moving down there.'

Liu didn't need to be told. He had been looking at the crisis on his own monitors all day while he frantically directed the efforts to restore vehicle management to the Greater Los Angeles road network. In the local offices his people had been besieged by the news networks seeking explanations.

'They've been dragging cars off the freeways ever since it happened,' shouted an ignivomous Tye, 'and the roads are *still* fucking full! It will take them days to get all the vehicles off. Their drivers just abandoned them!'

Liu was as frozen as the stationary traffic in the image. He could see the Santa Monica Freeway clearly. Miles of stationary metal glinted through the late-afternoon heat haze – an atmosphere that was unusually free of smog. When he had been a regional technical director of Tye Satellite Networks he had been one of the original beta testers who had driven the length of each freeway to check the system's reaction to unauthorized driver intervention, breakdowns, blow-outs and all the other ills that can befall any one of those millions of individual mobile assets within an integrated city traffic management system.

They had taken so much time designing back-ups, fail-safes and extreme-condition survival systems that they had not felt it appropriate to build or act out a scenario for the complete failure of the entire system – not least because there *could be* no conceivable solution. So the decision had been made to engineer risk out of the system and they had even designed sufficient fault tolerance for the satellites to withstand a meteor shower one hundred times more dense than any ever recorded in the Earth's vicinity. With two low-Earth-orbit satellites stationary above Los Angeles and a third back-up management system on board a satellite in mid-Earth-orbit, they could not envisage how total failure could occur across three separate systems.

But it had.

Tye bounded up the aisle. He lunged towards Liu and grabbed his shirt front. Connie stepped round and pushed herself physically between them. Liu was suddenly aware of an unexpected strength and muscularity in the executive assistant.

'Tom . . .' she warned. 'That won't help!'

They stood there, all three, in an embrace of anger. Connie did not budge. The slope of the aisle equalized their heights.

Tye released his grip and turned away.

'Tell him, Marsello,' he screamed towards the seated group. 'Fucking tell him!'

Furtrado stood up. He had a printout in his hand. 'We've been contacted by the attorneys who act for the City of Los Angeles,' he said with a face like a lawsuit. 'They've notified us that actions will be forthcoming – for the loss to the city's economy and for punitive damages. The death toll so far is believed to be over seven hundred.'

There was a silence. Tye had gone back to the edge of the Holo-Theater where he was staring again at the image of stationary downtown Los Angeles.

'Tell him all of it!' he screamed at the lawyer.

'The city is at a complete standstill,' continued Furtrado. 'There has been no production in any of its major industrial or commercial sectors today. Very few services are operating. LAX is closed and all flights are diverted to San Diego. The police, ambulances and fire crews can't travel – every road is blocked. Looting has broken out in West Hollywood, Culver City and the lower slopes of the Hollywood Hills. The Governor has called in the National Guard and they have started arriving by helicopter. The police are afraid that when night falls the looters will move west towards Beverly Hills, north into the Canyon and east into the commercial district. I'm afraid we can expect many more civil suits from those affected so far and any who suffer later.'

Liu shook his head in disbelief.

'Tell him how much, tell him how much!' shouted Tye.

'We don't know how much yet, Tom,' observed Furtrado. 'But it's going to be huge. The city's action alone could be the biggest lawsuit in history. Then there's the private suits with their inevitable claims for actual damages, punitive damages and for *solatium* – payment for making them feel bad. And remember, they're the richest urban population in the world.' Furtrado paused. 'And, necessitously, we are uninsured for this contingency.'

There was a silence and Tye turned back to face Liu. 'Every traffic management project we were working on has been put on hold,' he said quietly, as if he could hardly believe it himself. 'Atlanta, Toronto, Singapore, Sydney – they've *all* told us to stop further development. The LA standstill is now the main item on every one of the world's news networks.'

Tye gathered himself and started up the aisle towards Liu again. Connie stepped further down the ramp to block him.

'Have you seen the fucking stock price?' he yelled up at Liu as Connie pressed the flat of her hand against her boss's chest to prevent him from attacking the engineer. 'Networks is down forty per cent! The Corporate stock is down five! WHAT THE FUCK HAPPENED?'

'We don't know,' admitted Liu. 'There's been a large increase in faults all over our networks but this was...' He tailed off. There was no adequate description for the disaster.

'How long...?' screamed Tye, not even bothering to finish the question.

'The master system on board SATMAN-6 is now working again,' reported Liu. 'I've personally reloaded the entire system. But there's no back-up and we can't restart without one. We've got to launch replacements for LAT-6 and 7. Then we've got to test them.'

'How long?' repeated Tye. 'Every day will cost billions.'

Liu had been dreading this question. He had put fourteen staff to work on preparing two replacement satellites the moment he'd heard the news.

'About five days, Tom,' said Raymond Liu.

'Make it two,' shouted Tye as he sprayed his mouth. 'Or get off my island.'

# FOURTEEN

'Hope Island Control, this is Tye Flight Five, over.'

'Good morning, Tye Flight Five, this is Hope Island Control. Good to see you, Jack. You're flying *manual*?'

And so Jack Hendriksen had been welcomed back. He had disengaged the computer system that normally flew the plane and he confirmed the aircraft's unusual status to Hope Island ATM. Jack liked to feel a plane in the air, as did most pilots, but every commercial airline flight was now operated automatically, from take-off to landing. Insurance companies would no longer accept the risk of humans flying planes, except in extreme emergencies, and today's pilots flew with the aircraft simply to reassure the passengers. The problem was that the one in ten take-offs and landings that the human pilots were allowed to handle, for purposes of practice, were not enough to maintain their general flying skills or their preparedness for emergencies, despite a significant increase in simulator training. As a result every passenger preferred the ultra-smoothness of a computer-controlled landing and many had started to book seats only with carriers and on flights where this could be guaranteed. It looked liked commercial-airline pilots were going out of business. Such shop talk had occupied the team on the flight deck throughout their short supersonic trip from New York's La Guardia airport back to Hope Island.

Jack accepted his flight-path-approach instructions from the tower, made the course and height alterations necessary, trimmed the aircraft for subsonic speed, winked at his co-pilot as the controlling computer pointlessly communicated a heading error of one degree – an error no human pilot could correct – and watched as the island and the Cape Hope spaceport came into view.

He had cadged a ride back to the island with a group of returning patent attorneys and he'd asked the pilot for permission to fly the Tye jet personally in order to keep himself occupied. There was nothing he could do to advance his new mission during the flight but he knew that the moment he landed he would be overloaded. UNISA wanted answers urgently and he realized that he was flying into a corporate maelstrom: the hysterical media coverage made it sound as if Los Angeles had been completely shut down by the failure of the Tye Corporation's traffic-management systems.

He also knew that Tom would be screaming for him to approve the security aspects of plans for the forthcoming anniversary celebrations.

In two months, on 30 August, it would be Founder's Day on Hope Island. In previous years that day had been marked by a public holiday and with company-sponsored barbecues on the beaches. But this year, for the state's thirteenth official anniversary and celebration of sovereignty, the corporation was planning to use the occasion for a major product launch. Jack knew it would also be a belated celebration for Tye's fifieth birthday but that would remain unannounced. Rumours suggested that the event was being designed to be 'the party to end all parties', with a guest list intended to be the most exclusive of the twenty-first century. By coincidence, Tye's ambitious plans would provide Jack with a perfect excuse for a complete review of the island's security – not that it wasn't needed. As UNISA had made clear, Tye's continuing safety was essential for global economic stability.

Connie had greeted Jack with an affectionate smile when he arrived at the corporate offices in the ground-floor quadrangle of Tye's house. Yes, he had enjoyed his vacation. No, he didn't realize that Connie had been in New York at the same time. She'd been trying to reach him? Ah, well, he'd spent time upstate with his mother: her birthday and she wasn't at all well. In fact, she might have to be admitted to hospital.

Jack detected no suspicion in Connie's gaze as she listened to him. Perhaps there was a hint of something else, he realized. Once again he considered the elegance of her long neck. He knew he would need an ally on the island but . . . well, Connie was just too close to Tye. Nobody knew the origins of the fierce loyalty she showed to her mercurial master.

'Never mind, it wasn't important,' she said with a smile. 'Well, you *have* come back at a busy time. Tom's got meetings all day about the LA situation – you can *imagine*! But he wants you to see the event organizers – they're in from Washington, and standing by in Network Control. They gave Tom and me a briefing this morning on their plans for the anniversary weekend. It all sounds wonderful but your people will have a lot to do. Oh, and there's the Moscow visit. Tom had to postpone that, so you might want to review the new arrangements. Pierre is currently scheduled to lead the Presidential Protection Team for that trip.'

Jack nodded as she ticked off her list one by one. He wanted to talk with Tye in person, even though strictly he didn't have any urgent need. He guessed he wanted to see him in the flesh again, now that he knew more about the supposed secret of the man's success. He also wanted to push him a little, to see if any cracks appeared.

It had just been the busiest vacation, or non-vacation, of his life. He had arrived in Manhattan intending to talk to a lawyer about Tye's dubious activities, and he had returned as a newly sworn intelligence officer of the United Nations International Security Agency. Despite his protestations, a second salary was now being paid into a new bank account for him as 'Bruce Curtis' in Manhattan. 'It isn't legal unless we pay you, Jack,' Deakin had smiled. 'That's *our* part of the contract.'

Jack's part of the contract was still unclear. He knew that to a large extent that would be decided only once he had uncovered more information about Tye's presumed interceptions of network traffic.

He had also attended four more UN briefing sessions on what was now known as 'Operation Iambus' – a code name specially selected for the purpose. At each one he had seen the team grow steadily larger. By the time he had left it consisted of over 190 men and women from all parts of the world, who were busily taking over the entire twelfth floor of the UN Secretariat building.

As well as an increasing number of UNISA intelligence officers, the team now included technical officers, satellite-surveillance analysts, patent lawyers, two insolvency practitioners, aerospace consultants, three jurimetricians – jurisdictional experts in antinomic international law (who seemed unable to agree on anything) – economists, energy consultants, media specialists, corporate lawyers, a former White House Chief of Staff, a professor of solar radiation, two meteorologists and an ecologist.

Research into every aspect of the Tye Corporation and its many subsidiaries and interests was punctuated by endless discussions about what might be the long-term aims of the mysterious Phoebus Project to which Jack had alerted them. Just before he had left, this topic of speculation was replaced by a sense of quiet awe at the scale of the standstill the Tye Networks malfunction had inflicted on Los Angeles.

He had also submitted himself to six days of exhausting refresher courses during which he had been supplied with the small selection of specialist tools that were stowed in his luggage. There were no Customs or immigration barriers on Hope Island and, if such a need ever arose, it would be Jack's own team that would provide the personnel. After the training and his rendezvous with Haley Voss, he had indeed travelled upstate to visit his mother, despite the fact she was as fit as he was and regardless of the fact that it wasn't actually her birthday.

'How would you enjoy a trip to Europe, Mom?' he had asked, though he didn't immediately explain that she would have to travel under an alternative identity. Gently, as the evening progressed and she watched her son enjoy the dinner she had prepared, he told her that he was once again working for the government, although he allowed her to assume he meant the US administration. He admitted that it was to do with his boss, the Thomas Tye she saw so often on the television, but the investigation was so secret that he had to remain in place as a Tye executive in the corporate offices.

His mother nodded her understanding. 'That young man is getting a little big for his boots,' she observed.

With suitable meiosis he explained that he would need an excuse for regular return trips to the mainland and that he would therefore like to pretend she was unwell and that he needed to visit her regularly. He told her that his government agency could arrange for her to be registered as a patient at a nearby hospital while she was actually spending time in Birmingham, England with her younger sister. This was because people in the Tye Corporation might try to check on his whereabouts or her condition of health.

'You should never wish your mother ill, Johnny,' she had replied sharply, but he could see the twinkle in her eye. His mother had always enjoyed a little subterfuge and he remembered how easily she had manipulated his father.

So it had been agreed, and his mother had seemed more excited than disturbed when he had explained about a new passport and the need for absolute secrecy. They discussed what to do about her dog and about Clara Morgan, his mother's best friend. It would be impossible for her to leave without explaining, the elderly lady affirmed, since if Clara thought she was in a hospital, she would break down all the doors to visit her.

Jack pondered these twin problems for a moment. 'What does Clara think I used to do, Mom?' he had asked, at last.

'Oh, nothing,' his mother had said, with an over-casual shrug. 'You always told me not to talk about it.'

Jack smiled. His mother had known no details of his work, but she had certainly guessed its nature.

'You must have said *something*,' he prompted gently.

Anne Hendriksen swivelled her shoulders slightly, like a small child trying to avoid making an admission. 'Well . . . I just said you did confidential work for the government.'

Jack smiled. 'Would you mind if I dropped round to see her?' he asked. 'When I take Skipper out for a walk, later on.'

The red setter lifted his head above the rim of his basket and regarded Jack with a liquid, hopeful stare.

'I could just explain how you're helping me, so she mustn't worry. Would that be OK?'

And later Mrs Morgan had stared up at him, her bright button eyes popping. 'I won't breathe a word, John, on the Good Book,' she swore, 'and I'll be pleased to look after Skipper. You just tell your mother to have a good time.'

Joe stepped back and looked up with satisfaction at the large pinboard on the wall. Ingrid, his new assistant, had been scouring the networks for such an old-fashioned item, but without success. But Geneva does not desert its traditional retailers easily and in the end she had found the green baize noticeboard, three meters square, at a small stationer's near the university.

He had spent two days dumping the contents of his VideoMate, his server files and his own memory, as well as contacting his friends and other sources, in order to create this graphic representation. He'd even found images of most of the Tye Corporation executives. Then he and Ingrid had spent another seven days compiling a consolidated and cross-referenced shareholding register for every Tye-related company. Now his only concern was whether he would have to get his graphic display and supporting documents packaged up and transported safely across town to Dr Chelouche's office, or whether the world's banker would deign to visit him here.

He had flown back to Manhattan to collect his clothes and tell his friends the authorized version of his sudden move to Europe. Before Joe had been allowed to leave his first encounter with Chelouche, the banker had insisted on the importance of transmitting nothing on the networks that might reveal the true nature of his forthcoming role in Europe. To Joe's surprise, he had even insisted that this restriction also apply to encrypted messages.

Well, Joe knew no one he wanted to tell *that* badly and he relished the challenge of establishing a fund so large that he could exercise some control over the market price of the Tye Corporation. He had used the brief trip home to pass news of his new role onto the key pension fund managers and to blow on the embers of a few Tye-oriented relationships that had been allowed to cool.

One such call had produced an interesting snippet: an individual on Hope Island named Zachary Zorzi was shopping around for a loan – a large loan, the source said – based on the value of his stock options in a Tye Corp subsidiary called LifeLines Inc. Joe had recorded a PopUp reminder to research this company. Joe's stock-in-trade wasn't real inside knowledge (although he never hesitated to use it discreetly when he came across it) but lay in his ability to find a link between apparently unrelated events: to look for a pattern in various developments and announcements that allowed him to discern the intentions of those who were shaping the markets. If he got that right, he could invest and maybe share some of the spoils.

Joe had also picked up some news that he found hard to interpret. One of the software agents he had programmed to search for Tye's new equity investments had struck lucky in one of its random Serendipity Excursions and had reported back with documents on a land deal filed as a matter of public record by Finland's Ministry of the Interior. Joe had been surprised to read that a new corporate identity called Tye International Real Estate Inc. had bought a cluster of uninhabited small islands in the north Baltic Sea. Joe's network atlas had quickly given him some basic facts: at a latitude of sixty degrees north, these islands were ice-bound for eight months of the year and contained no known mineral or chemical resources of value. The deals were leasehold, each for 999 years, and all commercial fishing rights had been retained by the Finnish government. In all, the ownership of thirty-seven islands had been transferred at a total cost of only thirty-two million euros.

Joe knew the history of how the Tye Corporation had acquired Hope Island and, despite the tycoon's professed concern for ecological issues, he doubted that Tye Corp was merely buying Baltic islands to protect the local gull populations. Then he had recalled his other software robots and reprogrammed them to go in search of any land deals into which any known Tye-related company had entered over the previous three years.

He had arrived back in Geneva to find that a penthouse apartment in a low modern block had been rented for him, close to Swiss National's head office. A new Audi was also waiting for him in an underground car park. An alien's work permit – something even harder to acquire in Switzerland since the country had finally joined the EU – waited amongst the package of documents related to his new home, the car and his appointment to Swiss National. There was also a printed memo from Madame Pioline asking him to pay special attention to the group of companies within Tye Aerospace while he was constructing his organizational chart.

The Director of Human Resources at the bank had been all smiles when Joe arrived at the appointed time the following morning.

'We're honoured that Doctor Chelouche would ask for our assistance,' he said in French-accented English. 'But this will remain between us, yes? The director of fund management understands you will be working on your own, but the rest of the team are being allowed to assume you are setting up a new fund here in the normal way. Perhaps you would look over this draft press announcement of your appointment to Swiss National?'

In addition to thc large pinboard, Joe had developed a non-network electronic version of his organization chart complete with videos and database material on the various companies, subsidiaries and shareholdings of the Tye Corporation and on Thomas Tye himself. He'd never laid out a physical map of the Tye empire before and as he looked at the busy pinboard with its clusters of company names and presidential titles even he was surprised by the depth and complexity of the organization revealed by this visual display.

Joe's VideoMate chirruped and he found himself looking at Madame Pioline.

'The Doctor will be pleased to visit Swiss National tomorrow morning,' she told him in response to the request he'd made earlier. 'I've spoken with the bank's director and the board room will be at your disposal. Our technical support team will check the room over immediately beforehand. Shall we say nine a.m?'

Joe nodded. How civilized. How European. How northern latitude. It would give him time for a morning run beside the lake.

'We'll be bringing in the newly commissioned *Treasure of the Caribbean* for additional accommodation,' announced the young, silk-suited female event-producer in what she obviously thought was a clipped, inner-Beltway presentational style. An impressive image of the giant cruise liner, superimposed alongside the quay of the deep-water harbour at Cape Hope, hung pointlessly in the Holo-Theater. *Everybody knows what a cruise ship looks like,* thought Jack. Only a DC organization would pad out a presentation so excessively. Next they'll show me a floor layout to justify their large fees. Just then the image changed to display seven separate deck plans.

'We'll have sixty suites, three hundred and twenty double bedrooms and three hundred single rooms for staff. The conference itself will take place here in the Network Control Center – we understand the conversion to the Solaris Control Center will be complete before then – and in Mr Tye's house, in the lecture theatres on the university campus and in the three lecture theatres and the ballroom of the *Treasure of the Caribbean* itself.' *Herself,* thought Jack with a rush of irritation. 'Outside of the plenary sessions there will be three hundred and eighty-two separate seminars.'

But, despite the over-fussy detail and high-pressure delivery, it had been an impressive presentation. Jack had smiled sagely as the details unfolded but even if he had still had only the Corporation's interests at heart, he would have been daunted by the task of analysing all the many potential threats and dangers posed by such an ambitious undertaking.

'We are calling the event *One Weekend in the Future,*' the account director had beamed. 'We are bringing together the world's greatest business leaders, politicians, scientists, philosophers, artists and writers to spend three days in a conference on the global environment that will set the ecological and technological agenda for the next ten years. Naturally, the Tye News Network will have exclusive television and network rights. An announcement was made to the press this morning.'

As the list of events and activities grew, Jack began to wonder how his staff could possibly cope. Then the PR people turned to the guest list. They started with the ultra-VIPs.

'We've had a provisional acceptance from the president's office, the Dalai Lama . . .'

Jack stood and held up his hand. The young woman looked up from her list.

'You mean the president of the United States?' asked Jack.

'Yes, we're all delighted by that,' said the account director. 'We heard only two days ago. It's subject to final confirmation, of course.'

'And what level of security will he be bringing with him?'

'That's to be confirmed. We know he will be travelling in Air Force One. The Secret Service has already cleared the trip – they believe Hope Island is probably one of the safest territories on earth.'

Jack nodded and sank back into his seat. He agreed with that judgement but he dreaded working with the White House mafia again. Twice before his work had taken him into the American presidential orbit and both times the macho, pre-emptive, turf-dominating, gun-toting hubris of the Presidential Protection Section of the US Secret Service had made him want to vomit. It reminded him of the ignorant and stupid young man he himself had been when he had first wanted to join the Marines. They seemed to sum up all the worst aspects of American military culture.

When Thomas Tye had first hired him he had asked Jack's opinion of American presidential security.

'It's effective but wholly over the top,' Jack had commented. 'It's about projecting a tough image when it should be about discretion. I've worked with those Secret Service officers and although they're good as a team, they don't work well with outside agencies or organizations. Every presidential trip becomes a misery for the hosts.'

'We'll see if we can do better when *we're* on our travels,' Tye had then said. 'But that'll be up to you.'

The young event producer sensed Jack's reservations. 'Mr

Tye thought you would start liaising with the White House fairly soon,' she prompted. 'Do you see a problem?'

*Only one of jurisdiction*, thought Jack. He was also wondering whether the President's late acceptance of Tye's invitation could be linked to American concerns over Tye's rapidly increasing power – or whether the US government had suddenly joined the rest of the world and was finally starting to pay more than gas-guzzling lip-service to the plight of the choking planet. He would need to inform Ron Deakin quickly but the covert UN landing party wasn't scheduled to lay the secure landlines until the following day.

'Please go on.'

'The Dalai Lama, President Orlov of Russia, President Cohen of the EU, the Sultan of Brunei, Prime Minister Benn, President Boutard of France, Lord Berners-Lee . . .'

Jack listened as she read off some of the world's most famous names. He knew Tye wanted to create the ultimate launch event for the first public unveiling of the Phoebus Project, his largest project and investment to date. But Jack wondered if all the many guests would react in quite the way Tye expected.

Calypso had waited a week for news. She hadn't seen Tommy since that lunch on Tye's terrace when she had laid down her conditions for continuing as the boy's . . . she still didn't know how to describe her role. Chaperone? Companion? Psychiatrist? Playmate? Observer? Love-giver?

She knew she genuinely loved Tommy, but she was resigned to the fact that he might already be beyond her reach and her help. At the end of their lunch she had insisted that Tom reconsider his decision about Tommy's attending school and about allowing the boy to join her on a trip to visit her mother. She wanted Tommy to witness other aspects of life, and by having him meet some of her nieces and nephews she wanted this child of ultimate privilege to see how different family life could be on an island not so very far from his own.

Since then she had called the main house three times each day but every time one of the staff had explained that Tommy was in a lesson, or swimming, or in the bath, or in bed, or playing the piano, or immersed in a game or anything but able to see her or even speak to her on a communicator. She'd started to call Thomas Tye himself and, once again, was told that Tom was in a meeting about the Los Angeles problem, or down at Cape Hope for a launch, or was in holo-conference with the president of this nation or that nation. She'd even strolled around the foot of the mountain to see if she could somehow get into the grounds of the mansion. But she discovered that down at sea level would-be trespassers were greeted by nothing but sheer cliff face.

Then she had decided to act. She had had enough! She walked to catch the Mag, determined to visit the house itself and persuade whoever she found there that she must see Tommy immediately for his own good.

But the shuttle had refused to leave the station. She pressed the destination button for Tye's residence again and again but the car refused to budge. At first she thought maybe the power had failed, but when she pressed the button for Little Venice, the shuttle pulled away instantly. When she arrived at the waterfront she pressed the button for Tye's house again. Once again, the car would not move. Resigned, she pressed the button for the station near her home and the shuttle started instantly. It was now clear that the security system was simply denying her access to the main house. Her VideoMate locator and also the facial-recognition system of the island's camera network were automatically communicating, and the security system was obeying instructions to deny her access.

When she woke the next morning, she found Tye's decision waiting for her in her mail box. Seeing that he had recorded it during the night, she presumed it was her attempt at visiting the house that had at last prompted this contact.

*'I'm sorry, Doctor, but I have considered everything and I'm afraid I can't let Tommy return to the school or leave the island even for a short trip. I understand your genuine concerns, but Tommy is an exceptional boy destined for a very special role in life. Therefore his world can't be exactly like that of other boys. I hope you will understand – or that you will come to understand in the future.*

*'Given the strength of your feelings on this matter, it is probably better if we now terminate our agreement. Technically, you're still within your probationary period, but I've arranged for severance pay to be made to you as if the full contract was already in place. Logistics will arrange for your transportation off the island. I've told HR to provide an excellent reference regarding your professional capabilities. Thank you, Doctor Browne.'*

Then the screen had gone blank. *Patronizing bastard,* Calypso had thought: '... *or that you will come to understand in the future.*'

What she *did* understand was that it was now time to leave and her instincts told her to do so quickly. HR and Logistics confirmed that a helicopter would be available to her at ten a.m. the following morning to take her to Mayaguana and she spent the day packing her clothes and the lightweight possessions she had brought from her Chicago apartment. Most of her furniture and pictures were still in a run-down Me-Lock self-storage depot near O'Hare airport where, in sullen silence, Larry Sumner had helped her load the container. She hadn't been on the island long enough to arrange to have them forwarded to her.

She had already laid out her travelling clothes on the queen-size bed and, by late evening, she was packing the last of her underwear and shoes into a leather overnight bag. It had been one of her first 'luxury' purchases seventeen years before, after winning her first serious prize money. The bag was old but with a battered quality that improved with time and had the comfortable smell and feel of a much-loved travelling accessory.

She heard a banging on the door to the veranda. Then a familiar voice. 'Calypso? Calypso?'

She ran through into the living room, pulled the inner door open and swung the mesh door outwards onto the veranda.

'I knew you hadn't gone away,' he said simply, his violet eyes full of tears.

Tommy was still wearing blue striped pyjamas but he'd pulled on dark trainers for this excursion. Jed was trapped firmly under his left arm and the boy held a bottle of water in one hand and a chocolate bar in the other. Calypso wanted to smile at these preparations he'd made for his expedition. Then she noticed that he was wearing a small surgical mask pulled down around his neck, and even pale latex gloves. She put her arm around his shoulders and led him into the light of her living room.

As she kneeled and hugged him, she could feel his small heart pumping violently beneath his pyjama jacket.

'They told me you'd gone away,' he sniffed as his tears abated. Then he noticed the three black leather suitcases lined up on the carpet ready for her departure the next morning. Two packing crates of books and the synthesizer flight-case stood against one wall, ready to be forwarded.

Calypso weighed up what, and how, to explain to him. She realized how much damage had already been done to him in his short life and, above all, she wanted to cause no more hurt.

'I do have to go, I'm sorry,' she lied. Lying about *having* to go, not about being sorry. She supposed that if she gave in to Tom and stopped interfering in the way the boy was brought up, she might still be able to stay – but what would be the point? It was her professional judgement that was being challenged here and she couldn't bear to watch Tommy go on spending his childhood in such damaging phobic isolation. The only possible end result would be psychopathy or severe neurasthenia.

Calypso sat him on the piano stool and gently took the water bottle and chocolate bar away from him. Without a word she pulled the tight-fitting gloves from his hands and undid the bow that secured the face mask around his neck. She tossed the items aside and examined him for injury. She noticed grass stains on the knees and elbows of his pyjamas but he seemed unhurt. As she kissed his forehead, Tommy released Jed from his tight captivity.

'Hello, Doctor,' said the caterpillar. 'It's a pleasant evening.'

Calypso couldn't help smiling. She guessed that the Furry must have registered her professional identity during that lunch up at the house.

'Hello, Jed,' she replied, grateful for the distraction as she gathered her thoughts.

'Does your father know you're here?' she asked Tommy,

sitting down on a low chair, close to him. She could guess the answer.

'I've run away,' he announced solemnly. 'I want to live with *you* now. They said you'd gone away for ever.'

'Well, I do have to leave tomorrow, Tommy,' she admitted. 'But I'm very pleased to be able to say goodbye to you.'

'Where are you going, Doctor?' asked Jed.

'I'm going to visit my mother. She's not very well,' replied Calypso, marvelling at the conversation simulation in the latest generation of Furries.

'Go to sleep, Jed,' said Tommy irritably.

Then they heard a sound that was unusual on Hope Island. Calypso couldn't recall ever hearing it since she'd arrived there and she was shocked at how intrusive it seemed. She crossed to the front door and opened it. The petrol engine of the large four-wheel-drive vehicle cut out and Stella Witherspoon swung her lithe frame down from the cab. In the distance Calypso could hear the deep throb of a second engine coming closer.

Stella started along the path that crossed the small lawn and threaded her way through the pile of refuse sacks that held the remains of Tom's floral apology. Calypso knew very well why she was here but she wanted this situation handled with a *very* light touch.

'Hi, Stella,' she called out as a friendly greeting, though she'd only met the security officer a few times previously and had never really spoken to her before. 'Come on in.'

Calypso stepped aside to let Stella enter the room. The woman had clearly been out on night patrol when she had received an alarm call. She wore a black shirt, black trousers and black combat boots – and carried a holstered gun on her belt. Calypso guessed that Stella's dark night-vision viewpers were already transmitting the scene to watchers elsewhere. Calypso made her smile ultra-wide, but her gleaming eyes and flaring nostrils delivered a warning. 'I expect you're looking for Tommy,' she said lightly. 'We were just going to have a cup of hot chocolate before he went home.'

Stella looked from the psychiatrist to the small boy on the

stool, then back at her. Calypso's hands had moved involuntarily to her hips and she sucked in one cheek. Stella's orders had been to return the boy home immediately. They were talking in her earpiece.

'Sure,' she agreed. 'Sure.'

'I don't want to go home, I want to stay here,' said Tommy. But the demand was made in a mumble of resignation, not a cry of defiance. Calypso smiled and nodded for Stella to sit on the couch.

Then the second vehicle arrived outside and cut its engine. As Calypso opened the door again, Jack appeared in shorts and a sweatshirt. He was barefoot and his hair was ruffled as if he had only recently been sleeping.

'Hi,' he smiled as his glance took in the scene. 'May I join your party?'

Tommy looked up at the tall security executive. He had been repeatedly told that Jack would always be his friend but that he must always do whatever Jack told him. The security chief sat down cross-legged on the floor in front of the boy.

'Good going, Tommy, that was some trip. Are you OK?'

The boy nodded, tightly clutching the now-silent Jed.

'I'll make that hot chocolate,' said Calypso, crossing the room.

While Jack kept Tommy occupied, Calypso poured hot milk into two mugs. She didn't have enough for four because her fridge had been informed of her scheduled departure and had adjusted its normal restocking order. From the kitchen window she saw Stella standing outside, examining her VideoMate as if her system was suffering problems.

Calypso handed Tommy and Jack each a mug and resumed her seat.

Jack was telling Tommy about some new model plane – with three on-board video cameras – that he'd seen in a Manhattan store. Would Tommy like Jack to buy him one on his next trip, he asked.

Tommy ignored the suggestion. 'Will you be coming back here when your mother is better?' he asked Calypso.

She concentrated her gaze on Tommy, even though she could sense the same question in Jack's eyes.

'I don't think so, Tommy,' she sighed, her heart suddenly aching. 'My mother is very old and I've got to stay and look after her. I might be there a very long time.'

Tommy nodded sadly and hung his head. As Stella reappeared in the doorway, Calypso rose and held out her hand.

'Come on, Tommy, I'll take you out to the car.'

Tommy rose, using both hands to place the unfinished chocolate carefully on the small round table on which Calypso normally displayed her family photos. He took her hand and walked with her to the door. There he turned back and silently glanced at Jack still seated cross-legged on the floor. Rising, Jack watched Calypso gently lead Tommy down the path towards the waiting vehicle. He saw her crouch to hug the boy, then kiss the top of his head. Stella stood patiently behind them as Calypso opened the door and Tommy clambered up into the high seat. His actions seemed a bit more animated and Jack guessed that he was excited at the prospect of riding in this commanding-looking vehicle. The four-wheelers were normally garaged underground and used mainly for night-time beach patrols.

Suddenly, Tommy leaned out of the passenger window and threw his small arms around Calypso's neck, clinging to her. She eventually reached behind her head and gently unlocked his fingers. Jack wasn't sure whether Stella's system would have yet started transmitting again. The jamming device in his back pocket was supposed to suppress all radio and display signals within a radius of twelve yards. Then the vehicle started up in a wide turn before heading off towards the mountain road and Tye's mansion.

Calypso remained at the roadside, waving as Jack stepped back into the living room. A few moments later the door opened again, and he turned to study her face. Her amber eyes were full of tears and he could see moist tracks across her high cheekbones. She looked briefly back at him, then down at the small table where Tommy had placed his mug. He had also left behind the bottle of water and the chocolate bar, the provisions for his escape attempt.

Calypso began to howl and she buried her face in her hands. Jack was beside her in two strides. As he wrapped his arms around her, she slipped her arms about his neck and sobbed into his left shoulder. Instantly the shape and warmth of her body were evident through her thin silk robe. Despite understanding her sadness, Jack was obliged to ease his hips back so she would not notice his rapidly growing erection.

Yoav Chelouche grunted his approval as Joe Tinkler finished his two-hour exposition on the make-up of the Thomas Tye empire. He had started with the main corporation, examined each of the main divisions and then paid special attention to Aerospace as the banker had demanded. He listed all the satellite launches of the last four years and showed a chart of the customers by nation. Clients from almost every developed economy in the world had used Tye's launch services for its satellites. As Joe discussed the development of the Deep-Space Location and Navigation System, Chelouche had cut in with questions about this technology and its functions.

'All we know so far is that it is being developed by Phoebus Inc., which is a joint-venture research company set up by Hope Island University, the Tye Corporation and several private trusts,' said Joe. 'Tye Aerospace has launched a number of deep-space satellites for the company but they have filed nothing about this programme – no patents, just applications for fourteen initial orbit positions in which to construct the craft before they are launched out into deep space. It's been going on for over ten years and, knowing Tye, there has to be some commercial angle other than simply developing a navigation system.'

'But what, Mr Tinkler, what?' asked Chelouche suddenly.

Joe stared at him blankly. 'I have no idea, sir,' he admitted.

He hesitated and then resumed his summation. 'Phoebus sits in its separate corporate identity away from the main corporation,' explained Joe as he used a laser pointer to show the links between the Aerospace division and its subsidiaries. 'The majority shareholding is split between Thomas Tye himself, the Tye Corporation, the Tye Foundation and a cluster of small trusts based in Andorra and Liechtenstein.'

'Not exactly a kosher arrangement,' grumbled Chelouche, the man who had publicly crucified governments and corporations that had dared to disobey his beloved ethic of transparent and responsible accounting.

Joe had to agree. 'They haven't attempted to raise any money, there is no sign of a product or any indication that they're going to market,' he explained. 'That's why their structure and their secrecy hasn't affected my rating of the main corporation. If it remains private, it's their business. They're certainly keeping this one very close.'

'How much capital has been raised by the company?' asked Chelouche.

'There's no way of knowing, Doctor,' Joe admitted.

Chelouche mumbled a series of low asides to Madame Pioline, then asked, 'Have you cross-referenced the shareholder registers?'

Joe spent the next three and a half hours painstakingly laying out a series of print-outs that showed the detailed shareholdings of the Tye Corporation and the quoted subsidiaries and companies in which it had a partial holding. He explained which shares were voting, which were non-voting, which had special executive powers and which did not. Joe finished by showing Chelouche a spreadsheet he had developed that was automatically updated every time ownership of shares in the Tye group changed hands.

'In real-time?' asked Chelouche.

'In real-time – when a change of ownership is filed,' confirmed Joe.

The banker took his time examining the breakdown of shareholding patterns and voting power in each company. When he had finally finished, he put his palms down on the edge of Swiss National's boardroom table and pushed himself upright.

'Excellent, Mr Tinkler – Joe, if I may call you that. Excellent.'

Then Joe raised the matter of the Baltic islands and Chelouche sat down again.

# FIFTEEN

Marsello Furtrado knew his boss better than most, but even he hesitated before walking uninvited into the Presidential Lounge. He waited in the companionway beside the executive bar until the clairvoyant Connie turned her head and saw him. She waved him forward.

Tom was sitting opposite her in one of four large armchairs placed around a low coffee table, his back to the aircraft's nose. It was a typical executive cabin layout for a full-size supersonic corporate jet. Unusually, Tye was poring over a fan of concertina-paper printouts.

'Marsello's here,' announced Connie quietly.

Tom didn't look up, but Connie smiled at the lawyer and cleared her papers from the seat next to her so that he could sit down opposite Tom. Furtrado knew what his boss was reading.

'I've never seem so much garbage,' complained Tye finally, as he looked up. 'Who do they think we are? Total idiots?'

He threw the paper concertina on the seat to his right and stretched, hands clasped together, arms high in the air behind his head.

'It's just their starting point,' reasoned Furtrado. 'What else can they do?'

Tom nodded, yawned and sat back. He picked up a small spray from a well in his armrest and absently sprayed his mouth again, despite the atmosphere of antibacterial compound that was constantly refreshed by the cabin's recycling system. Many of the females in the flight crews of the Tye Corporation planes had started to complain about the dryness of their skin and the chapped lips they suffered in such a hostile environment, but they didn't use as much face moisturizer as Thomas Tye. His frequent applications weren't for cosmetic purposes, however: like every bar of soap, every scrap of food, every beverage and all the unguents he used, his face cream was medicinally active. It delivered preventative and proactive agents that hunted down all embryo carcinomas, others that boosted the elasticity coefficient of epidermic cells and some that simply maintained his skin tone at the precise shade he wanted.

The printouts were from an old Russian-made computer that its operators had programmed to print English-language lists of the assets in the region concerned. Tom turned to his right and fingered a line at random.

'Two hundred and twelve oil and refining facilities, total capacity twenty-two million barrels a day. Is that the real number?"

Furtrado slipped his DigiPad from his pocket, put on his viewpers and quickly searched.

'Maybe forty of the plants are working properly. The capacity for all the refineries in the designated zone was only one point two billion barrels in the whole of last year and much of that was astatki or mazut. That's from OPEC figures.'

Tom nodded. 'So why do they send us *this* shit?' he asked, flicking his fingers disdainfully against the pile. 'Let them keep it. We're not interested.'

'They've got nothing else to offer, Tom. Ease up. Save them some face. We can accept every figure they produce and still close within our own deal parameters.'

And what parameters they were! It was the largest deal Furtrado had ever worked on. It was larger than any deal his previous global law firm had handled and, he knew, it was larger than *any* commercial deal that had ever been undertaken. Only efforts to reconstruct continents after major wars – like the American Marshall Plan to rebuild Europe after the Second World War – came close in scale.

But Tom knew how to create value where none existed previously, as he had demonstrated so clearly to Furtrado in Cape Verde. Sometimes, to his wife or himself, the international attorney would dismiss his boss as just lucky, an opportunist who had secured one vital commercial advantage. Then Tom would amaze him with master plans like this that were dependent on nothing but the youthful tycoon's vision, imagination and drive. The ideas that underlay this deal were pure genius and that, in the end, apart from the twelve-figure personal fortune he was amassing for himself and his family, was why the lawyer tolerated the arrogant antics of his temperamental leader.

But this afternoon TT was in relatively benign mood. The LA traffic system was functioning again and Marsello's team had responded to the 1,472 lawsuits so far prompted by the system failure with such aggression and arrogance that the plaintiffs and their attorneys began to wonder what sort of battle they were getting into. The Los Angeles City Authorities were not faced down so easily, of course, and Tom had already ordered a substantial provision to be made in the corporate financial projections for the payment of an interim award, pending a court hearing. Tye was insistent that nothing would be finally settled for at least eight years, by which time even a gigantic liability pay-out would still be only a pinprick in the hide of the vastly enlarged corporate entity he was planning. As he told his inner cabinet so often: 'There *are no rules*! Today, a global corporation can become anything it wants to be – *if* it has the ambition.'

As his team of in-house lawyers instructed the twenty-three different law firms in their stonewalling, denial, disbelieving and offensive postures towards all plaintiffs, Furtrado had found himself wondering whether he might have been earning just as much if he had stayed in private practice with the Tye Corporation as a client. At least he would have got to see his family occasionally. But then, he realized, he wouldn't be part of a history-making deal like the one that would be closed over the next few days. The shape, productivity and economic balance of power in the world would be changed for ever because of what Tom and he were now doing. The power to change the world no longer lay with politicians. It belonged to business leaders like Thomas Tye.

'Let's go over the designated area again,' sighed Tom, nodding to Connie. She touched two buttons in her armrest and the lights dimmed, the window shades closed and a wall screen on the front bulkhead flickered to life.

'Come and sit here, Marsello.' Tom indicated the empty seat next to his. Both seats swung round on electric motors to face the screen. Furtrado eased around his boss, stepping over his small, dainty feet, and flopped into the left seat, putting his pile of papers on the floor in front of him. A map appeared on the wall screen with a new boundary outlined in red.

'This is the region we're leasing and this is the adjacent region we'll be treating for their benefit, as per the terms of our deal,' Furtrado said, pointing. 'It will take us about three years to complete phase one, the infrastructure and the resorts in the east, a further six for the forestry and agricultural developments of phase two and the entire development of the region will take upwards of thirty years. That depends on our immigration policy, the demand by would-be lessees for agricultural land and development properties, inward investment numbers, long-term ecological results and, of course, our own policy on zoning permission for outside developers.'

Tye nodded. He'd seen this map a thousand times and he still had only one territorial worry, the same one that had cropped up when these new borders were first drawn. He put that concern aside for the moment. It was for the Russians to worry about, and he would bring it up in Moscow.

'You've got the presents ready?'

Furtrado smiled, reached down and pulled a small, elegantly wrapped package from among the papers on the floor. 'Each of the leases constitutes a separate gift. I've had ornamental certificates of ownership made up and framed. We'll save them right until just before the deal closes, when we might need just a little extra leverage.'

Tye nodded. Buying the Baltic islands as inducements had been Furtrado's idea, after his visit to Cape Verde.

'They'll probably help us on the price,' predicted Tom. 'But I want you to try again to extend the lease. The period is an abstract concept, so why is it so difficult for them to agree?'

Furtrado shrugged. 'I just can't see us getting anywhere on

that one. My people have been over it with their team a dozen times. The secretariat would not sanction anything beyond one hundred and fifty-five years. The Hong Kong deal the British did with China is providing the precedent. Other than a freehold sale – which they can't countenance on political grounds – that's the longest period the Constitutional Court would approve.'

Tye sighed and drummed the fingers of his right hand on the armrest.

'It's academic for us as well, isn't it?' suggested Furtrado, deliberately provocative. 'A hundred and fifty-five years will see us all out, and several more generations too. Who knows what will be happening in the middle of the twenty-second century?'

Tye put his head on one side and considered. He swung his chair to the left so he could look straight into his senior counsellor's face.

'How old are you, Marsello?' he asked in the semi-darkness.

The lawyer shrugged. 'Forty-six, forty-seven next month. Why?'

'I'm three years older than you,' replied Tye. 'See any difference?'

Furtrado watched as the younger-looking man gazed back at him. The question was rhetorical and both men knew it. Tye could almost be mistaken for his son.

Tye paused, then leaned forward, his face close to Furtrado's. 'Do you want to join the programme?' he asked quietly.

The blood leaped and burned in Furtrado's veins. He was close to the *arcanum*. Nobody knew for sure, but everybody in the senior executive guessed there had to be some truth in the exposé threatened by that British woman biographer. Furtrado had personally overseen the issuance of injunctions to prevent her from publishing, but now it seemed as if he might find out the truth first-hand. The lawyer swallowed but, despite his intense excitement, he was still a lawyer – nothing could be left to chance. It had to be spelled out.

'You mean the gene therapy, for staying young?' he asked for the avoidance of all possible doubt, exhibiting in his deliberate reductionism the incisiveness and determination that had made his career.

Tye considered. He would be increasing the charmed circle by one. That would mean twenty-nine people now – including President Orlov, if Furtrado succeeded in Moscow. Tye would be parting with one of his ultimate motivational incentives, but it was worth it.

'Yes, I mean the gene therapy that arrests human biological ageing – the Erasmus trials,' confirmed Tye quietly. Behind them, they could hear Connie accepting a call.

'You understand the sensitivity of this?'

Tye needn't have said it. Furtrado already knew more of Tom's secrets than any other Tye executive. He was bound by attorney-client privilege, but even more so by the vast sums of money that were just starting to come his way.

He nodded, wondering whether he should also ask the same favour on behalf of his wife. He decided not to. 'Like everything,' he said.

'No, not like everything,' contradicted Tye. 'It will alter your perspective completely. I'm fifty now: by normal standards sixty-five per cent of my life would be over and that sort of realization *does* colour your thinking. My doctors and geneticists tell me I've actually used only sixteen per cent of my projected life, and the eventual figure may be even lower. You have no idea what that does to your mind. Everything seems different. To you, a one-hundred-and-fifty-year lease on this deal seems like an unimaginably long time, doesn't it?'

He raised his eyebrows. Then he leaned even closer to the lawyer. 'While *I*'m afraid we'll be handing the land back to them long before I'm ready to retire.'

Furtrado stared into Tom's clear eyes, taking in his smooth skin, the glossy hair, the glowing aura of energy and health.

'Get them to agree to a lease of four hundred years and you can join the programme as soon as we get home,' offered Tye in a whisper. 'Then you might still be around to ensure they honour their part of the bargain – all the way to the end.'

'Tom?' Connie was standing nearby in the aisle. She bent

low and whispered to her boss, although Furtrado could still make out the words.

'It's a call from Nurse Pettigrew. There's a problem at the house.'

Calypso woke with a start and found herself alone. She had slept uninterruptedly for the first time in a week. She knew images of Tommy in a mask and gloves had been flying around in her sleeping brain, but so had recollections of Jack Hendriksen and his tenderness. And now she had a completely new agenda.

She had sensed Jack's arousal as he had comforted her, so she had soon pulled away and gone to wash her face before returning with a bottle of wine and two glasses.

'It'll only get left behind,' she had explained as she handed him the corkscrew. 'How did you know Tommy was here?'

Jack began to explain that the island's security system had automatically alerted him and all his staff. 'He's got ID and GPS locator chips in his LifeWatch,' he said. 'The network security system sent out an alert the moment Tommy crossed the boundary line of their house.'

He looked up and noticed the expression on Calypso's face as he eased out the cork. 'It's not just him,' he added. 'Those locators are all the rage this year. Everybody wants them for their kids. They're a great safety device.'

Calypso had felt tears welling up again. What sort of future could the boy have, locked up in that house and perpetually tagged and monitored? She had warned how such isolation might affect Tommy's adult life. Jack had listened carefully for a while and then he moved to sit down beside her. He put his arm around her and kissed the top of her head. Neither knew what might happen next. Their first lovemaking, on Calypso's rock, had been a spontaneous, urgent and physical episode. But there had been no promises, no suggestions of anything further and they hadn't spoken privately since Jack had called in to say goodbye as he left for his New York vacation.

Calypso kissed him gently in return. It felt good to have Jack here with her. As he kissed her again, it was suddenly passionate. She felt herself responding, but her emotions were jumbled. She loved the taste and the feel of him, had thought about that a lot, but she was still disturbed by the events of the evening. She pulled away and put a restraining hand on his chest.

'You're welcome to stay, Jack,' she smiled, 'but I don't think I feel like . . .'

He had nodded, knowing how upset she was.

But later, lying in her bed, in his arms, her sadness had lifted and when he pulled his body away to avoid any unwelcome physical intrusion, she had laughed quietly and slipped her hands down his body.

'No, you don't,' she whispered. 'This might be just what the doctor ordered.'

It was tender and deeply loving and she had cried gently again.

She had then lain in his embrace, enjoying their closeness, aware that her lover was wide awake and staring at the ceiling.

Eventually she raised herself on her elbow. 'Jack?'

He traced the outline of her breast with the back of his fingers. 'Don't you want to sleep?' he asked.

She shook her head. Then he had got up, fetched the rest of the wine, and told her the long story of Thomas Tye, his obsession with the process of ageing, and young Tommy's unusual background. Calypso had harboured suspicions ever since first seeing both the boy and his father together, but this confirmation had shocked her.

She now wondered what time Jack had left her, as she hadn't heard the four-wheel's engine. She pushed herself out of bed and checked the time. Not yet seven. She would shower and then call Connie to arrange a meeting. Maybe there was still time to persuade Tom to reverse his decision about her departure. It was the least she could do for that lonely, uniquely special child.

The stream of hot water was loud in the shower cubicle and at first she didn't hear her VideoMate trilling. As soon as she closed off the tap she heard its tones. She wrapped a towel around herself and skipped into the living room, assuming it was Jack. She had a huge smile on her face.

But it was Connie Law, aboard an aircraft.

'Doctor, can you go up to the house at once? Tommy's been injured.'

As Calypso was pulling on the white shirt she had laid out for her journey, she heard again the sound of a petrol engine in her drive.

It wasn't until she was seated beside Stella and on the way to the main house that she wondered why Connie had contacted her and not one of the many other doctors or specialists on the island. Calypso slipped on her viewpers, opened communications with the house itself and had a visual patched through to her.

Theresa Keane raised her eyes from her book and waved back at the diminutive figure that had appeared outside on her terrace. His visit was expected, and one of the bonuses of life on Hope Island was that, as in her Irish childhood in rural Connemara, the professor could leave the doors and windows open without fear of unwanted intrusion. She stepped out into the early-morning sunshine and took his outstretched hand.

'Professor Keane, thank you for seeing me,' said Raymond Liu, with a formal dip of his head. The American-Asian still effected this oriental courtesy even though his parents had emigrated from Hong Kong to San Francisco years before he was born. This morning he felt more than a little daunted at meeting this world-famous academic and recipient of the Nobel Prize.

'It's good to meet you, Doctor Liu. Come in.'

The highly regarded group technical director of Tye Networks had asked to see her in person, so she had suggested this early-morning meeting before either of them started their daily schedule. She already knew his division was in crisis because of the Los Angeles traffic-management debacle; his personal assistant had made clear the urgency of his request. Even though the hastily arranged launch of two replacement satellites had restored the vehicle-management service to the city, she guessed that he was still under immense pressure to ensure no further failures. Suddenly, every computerized traffic- and mobile-asset management system in the world had become suspect.

'Tea or coffee?' she asked as she led him into her sunny living room. Pots of both were ready, with pastries and croissants laid out on a low table.

'Good morning,' said a ball of fur sitting on the arm of the sofa.

Theresa laughed. 'I'm sorry,' she apologized as she bent to stroke her CatPanion. 'Go to sleep, Sandra.' The Furry obediently switched to nap mode.

Liu chose coffee and perched himself gingerly on the edge of her sofa. 'I watched your Summer Lecture,' he began.

Theresa smiled. 'A little theatrical, I'm afraid. But we need party tricks to capture young minds.'

Liu merely nodded.

'So what can I do for you?' she asked, direct as always.

He leaned forward and carefully returned his cup and saucer to the coffee table.

'I was wondering if you'd be kind enough to tell me a little more about the activities of your agents in the networks?'

The professor smiled again. 'Ah yes, the Anagenesis Experiment. And you're wondering whether they're in any way connected with the system failures.'

'I have no idea,' admitted Liu. 'I can't find any common cause for the faults occurring in our systems or networks and, well, given your reputation, I've . . . Frankly, I need help, Professor. I have never seen anything like this in all my years as an engineer.'

Theresa could see the desperation in his eyes. She could also guess that he hadn't been sleeping much recently.

'What happened to the LA traffic-management system, Raymond?' she asked.

He hesitated. At least in her he had an informed listener – possibly there was no one better qualified. But he realized she was unlikely to be fully versed in the arcane design structures of orbiting traffic-management systems.

'We lost one bird,' he began, 'but the other was recovered. All its data processing had frozen.'

'But I presume the orbiting management system would have held a back-up of the asset data?'

'Precisely '

'And what did they show?' she prompted.

'They showed that there had been over six hundred memory leaks on board LAT-6 in the twelve hours immediately before the system crashed.'.

'Memory leaks – you mean specific faults that the system registered and corrected itself?' He nodded. 'They wouldn't be coding faults?' she prompted.

Liu shook his head. Ever since commercial software code had been generated by computers instead of humans, programming bugs had become almost non-existent and, in the case of orbiting systems, every combination of every line of code was checked and cross-checked by *ad hoc* computer farms made up of vast global networks of computers contracted to spend their idle time on the task. During the period that Liu and his team had been developing the software for the SATMAN systems, over two million computers had each been used for up to seven hours a day, attacking the software and presenting it with every possible combination of instructions and error situations that the computers could randomly generate.

The system analysis and diagnostic modules had predicted the mathematical odds of six hundred memory leaks occurring within a single eight-hour period of normal orbital processing at 14.7 million to one. Liu hadn't even bothered to compute the odds against another satellite suffering similar problems, or of the first failed satellite leaving its station without receiving instructions to do so. He'd been through that type of futile calculation over the satellite failures that had occurred in the Australasian quadrant.

'No. When we refer to memory leaks we mean faults appearing in the registers without any obvious internal cause.'

'Just a binary flip?' she prompted.

'That's it,' agreed Liu. 'Those rare situations when a zero becomes a one and nobody knows why.'

It was Theresa's turn to ponder. She was making connections.

'We normally put those down to magnetic radiation, don't

we?' she mused, as much to herself as to Liu. 'We say that a photon passed through – or a quark, a neuron, neutretto or neutrino. Loosely, a cosmic event concerning an energy particle which we don't yet understand.'

'But we shield all our systems,' objected Liu. 'And the magnetosphere – the magnetic force field around the earth – shields us from all the really nasty radiation of outer space.'

The professor nodded and sat back and played with the Victorian silver stamp box that hung on a silver chain around her neck. 'You've ruled out a virus or a hacker?'

Liu nodded. 'There's definitely no virus. It wouldn't be so random in what it affects and, anyway, we've combed almost every line of code we have. There's nothing there.'

'A hacker?'

Liu started to laugh but it turned into a groan of exasperation. 'I just can't see how, Professor – Theresa. You should see the security systems we've built in. But even if somebody *could* break our encryption, surely we'd have received some sort of extortion demand by now. It wouldn't be done just for fun.'

He shook his head again, steeling himself to be blunt. 'I saw what you're doing with self-duplicating software agents in the networks and, well, frankly, I was wondering if your experiment could have gotten out of control.'

Theresa nodded. She understood his concern, but she was thinking along wholly different lines. 'I think that's impossible, Doctor. I wouldn't allow my students to release anything that has the potential to do harm. The only objective our agents are given is to compete in order to reproduce.'

'But they are evolving?' prompted Liu.

'Yes, but only in their courtship techniques. They're learning merely how to reproduce more efficiently, not to interact with the environment around them.'

Liu weighed his next statement carefully. 'Professor . . . seven hundred and eighteen people died in Los Angeles because our system failed. The Tye Corporation faces lawsuits that currently total over eight hundred billion dollars. Large-scale projects we were constructing in fourteen cities have been put on hold until we can show the consulting engineers what caused the problem, and other cities still using our traffic-management systems are panic-stricken in case they suffer a similar failure.'

Theresa nodded, trying not to convey any hint of her alarm. She'd had no real idea of the seriousness of the situation. She rarely watched or listened to real-time news, and she read no newspapers. Years before she had made a conscious decision to forgo the experience of the present, of the here and now, of the *actualité*. 'Life's not a rehearsal' her only serious long-term partner had once said to her. But Theresa had dismissed her remark, understanding better than most that the virtual – the essence of Theresa Keane – was not so confined, and that semi-isolation and withdrawal from the world was a wholly suitable environment for her work. Most of her thinking involved a time far in the future and, in that virtuality, there was both rehearsal and performance. Theresa's philosophy was, of course, that of a Yogācāra Buddhist, although she would have accepted neither the label nor the limitations that implied.

'Have you considered that some of your agents might have adopted altruistic behaviour?' asked Liu.

*Ah, there it is,* thought Theresa; *the real reason for his concern, and his visit.*

'Raymond, I have to let you into a little secret,' she said. 'Those agents aren't as clever as they appear. It's all a trick. I keep trying to teach my students that our artificial companions are still quite dumb and that it is *we* who anthropomorphize them. Their conversational ability is pure psittacism.' She saw him frown and explained. 'Purely a parroting of language.'

'All these companions,' she said, pointing to the sleeping Sandra, 'only seem clever because they mimic us. They're programmed to pick up our tone, our mood, as well as to identify – if not understand – our language. It's merely a party trick and it is we who endow them with the qualities that make them seem so loveable. It's one of the most powerful human instincts, and one of the first things I do with all my students is try to get them to understand that.'

Liu nodded. In his student days his computer-science professors had also staged demonstrations to show how stupid computers are and how gullible humans can be.

'So there's no chance that one of these agents, or a group of them, could have acted out of character?' he persisted. 'Or that any of them could have reproduced without you knowing about it?'

Theresa leaned forward and returned her own cup to the coffee table.

'It's impossible,' she stated firmly. 'Their behaviour is governed strictly by the rules we created. You can strike that idea.'

'Well, I suppose that's something,' he sighed. 'We have to eliminate every possibility.'

He paused and then decided to tell her the other reason he had requested this meeting.

'There is something else,' he added. 'As I was unable to find any fault in the component parts of our networks, or the networks we are connected to, I went back to square one and ran a basic RQI procedure. I found–'

He hesitated. Theresa had held up a hand. 'I don't know what RQI means.'

'I'm sorry Theresa. RQIs are an exercise we set our first-year network engineers. It's a resistance, quantification and identification test. Students calculate the theoretical resistance found in an open, unused network. Then they measure the actual resistance of the network when it is in use and quantify data and the pathways to calculate how much resistance has been added by the activity in a network. They then reconcile the totals in order to check that there are no faults in any of the components or pathways. They are basic elimination procedures.'

Theresa nodded her understanding.

'Anyway,' continued Liu, 'I carried out an RQI on the networks and I found . . .'

Theresa placed her hand to her throat in surprise. 'On what networks?' she asked, amazed that such a test could be undertaken.

Liu smiled. 'On all the networks we operate and connect with.'

'How much is that, exactly?' she asked.

'There are about nine billion miles of networks – if you treat satellite communication paths and broadband wireless the same as fibre-optic cable. Capacity for about four thousand exabytes – four billion, billion bytes.'

'Some maths,' breathed Theresa.

Liu acknowledged the compliment with an inclination of his head. 'Of course, I automated the procedure as much as I could. Anyway, I found something very strange. There's a flickering, transitory resistance in the networks that can't be accounted for by data in transit, stored instruction sets, switching resistance or any of the other processes.'

Theresa frowned.

'In fact there is over twenty per cent of network space that seems to be occupied by something I couldn't identify – a sort of dark matter.'

'Dark matter . . .?'

'I don't know what else to call it,' admitted Liu. 'Dark data, perhaps. The resistance tests show something's there, but we can't see it. We can't actually tag it in any way. And . . . it seems to move.'

Theresa raised her eyebrows. Connections were being completed.

'We'll test one network one day and it will show a twenty, twenty-two per cent unexplained activity level – the next day there's none. And there's no record that those data, or that matter, or whatever it is, has passed through the routers to other networks.'

Theresa sat back in her armchair and thought hard, her right hand playing with her stamp box again as her mind worked.

'Let me have a think about this,' she asked finally. 'I'll get back to you.'

Liu nodded and stood. Theresa rose and took his hand.

'Thank you for making the time to see me,' said Liu, with another slight bow. He turned and walked out the way he had come, across the sunlit terrace.

Theresa sat down again, slowly. Her mind was racing. Who to call first? She would start with Robert, see where he was with the Descartes Experiment. Absently, she picked Sandra up and sat her on her lap.

'He seemed worried,' observed the CatPanion. 'Keep or discard?'

'Oh, keep, I think,' said the professor. Sandra purred contentedly as Theresa stroked her.

There was blood everywhere. The anterior carpal artery in Tommy's left wrist had been severed and the gouts of blood had hosed the walls as if there had been an explosion in a cochineal factory.

Calypso had already seen the room through her viewpers but only her experience of the horrors of a late-night Chicago ER allowed her to overcome her panic immediately and start to assess the situation.

'I had to restrain him, Doctor, in order to get a tourniquet and the compress on,' Nurse Emily Pettigrew explained breathlessly as she cleared away the debris of her labours. 'Then I gave him ten CCs of thiopentone. I was frightened that he would open the wound again.'

The boy was now unconscious on his bed. Nurse Pettigrew had displayed the value of her training as a British State Registered Nurse, applying a tight and efficient tourniquet on the pressure point of his elbow and binding a compress tightly to the main wound. Further up his forearm were the slashes he had made before the kitchen knife had found its target. How had he known? Where had he seen or read about such self-harm?

He was very pale. Calypso checked his BP and pulse: 80/60/135. He needed blood, and she said as much to the nurse.

'I have plenty matched and ready in the fridge,' said Emily Pettigrew as she started to leave the room. She saw the doctor's surprised reaction. 'On Tom's orders,' she explained. 'It's always available. He allows me to draw fresh supplies from Tommy every three months.'

Calypso considered the situation as the nurse went to fetch the plasma.

It was clear that Emily Pettigrew had understood the sensitivity of the incident. Calypso had to admit the nurse had shown good sense in contacting Tom on board his plane – wherever that was – and specifically requesting permission for Calypso to attend, rather than rushing the boy to the clinic or asking one of the duty medics to visit the house. However loyal a Tye Corporation staff member might be, this gossip could have proved just too juicy to contain.

'He just kept crying out for you,' explained the nurse, 'over and over again. I heard that on the monitors, but I didn't realize he was hurting himself. Then I got the alarm.'

Calypso shook her head: she knew the syndrome. With every repetition of her name he would have been making another slice in his forearm. She had treated child patients who had undergone similar episodes of self-harm and had seen self-mutilation and pseudocide cases before. But she had never encountered the syndrome in a pre-teenager and had never known a victim to inflict self-harm during the waking hours of the morning. The events almost always occurred late at night or in the small hours and it was usually, but not exclusively, a female behaviour observed in patients with a history of depression.

Her VideoMate trilled. She found herself looking at Connie, then Tom, as she provided an update. She was unable to keep the stoniness out of her voice.

Tye nodded grimly as he digested the news. 'Can you deal with it, Doctor? I don't want to bring any others in.'

Calypso glanced at the boy and considered. She'd dealt with far worse as an intern, but they had been welfare patients.

She nodded. 'He'll be out for another hour or so, I should think. I'm just going to give him some blood to get his BP back up. We should be able to stabilize him for surgery. We just don't know if there is any metacarpal damage.'

There was a short silence.

'Doctor Browne. I realize how serious this is and I am grateful for your help.'

Calypso nodded again.

'Would you do me – well, Tommy – a favour, please? Would you please accept another apology from his father and stay there with him. At least until I get back at the weekend.'

Calypso hesitated. She didn't have a choice – her heart felt like breaking. For a moment she thought Tom would say something else, perhaps talk about a financial inducement. But he did not make that mistake. Instead he waited for her reply.

'I'll wait until you get back,' she agreed. She would now do anything for Tommy.

'Thank you, Doctor,' sighed Tom. 'We'll keep this between ourselves.'

'Of course.' It could never be any other way.

'The Furry will know what happened,' said Tye. 'Don't let anybody remove Jed.'

Calypso returned to her patient. She would remove the compress and ligate the artery for safety before examining the surrounding tissue and bone. She would also ask Nurse Pettigrew to prepare an ice pack to keep his antibrachial as cold as possible. Then she would have him moved discreetly to the clinic where she could run X-rays and scans before moving him into an OR for stitching up. She debated whether she would have to call on one of the anaesthetists. Perhaps the thiopentone would last long enough. She would take half a beta blocker before she began the surgery. Although it would be a minor procedure, this was Tommy and she knew emotion might cause her fingers to tremble.

# SIXTEEN

[Insert page 37 – Chapter One]

Thomas Richmond Tye was committed to the Sandler Sanitarium, Norwood, New Hampshire, USA on 6 August 1971. He was just over five years old. The committal order was signed by his father, Thomas Tye II, and the family's long-time physician Dr Marcus Cordell. The cause for the committal was given as *hysterical dissociation* – a broad categorization that can cover a range of disorders including amnesia and somnambulism. A copy of the committal certificate was lodged with the court authorities in Norwood County, as was required by law. The date of the committal precedes the state's Juvenile Justice and Detention Act that requires the active participation and approval of a local social services agency and the Department of Juveniles.

That is the end of all information available about Tye's internment. We don't know his condition on arrival or the treatment he received. We don't know when he was released, nor do we know into whose care, although we do know that he was once again living back at the Tye mansion by the time he was thirteen when his paternal grandmother legally assumed responsibility for his upbringing. By this time family probate had been settled and Thomas Tye became heir to a thirty-two-million-dollar for-

> tune that was held in trust by his grandmother until he was twenty-one. It appears Tye attended no school or college and there is no record that he passed any external examination or gained any form of academic qualification.
>
> The reason we know nothing more of Tye's internment is that the Sandler Sanitarium was purchased by the Tye family trust in 1987 and rebuilt as a cosmetic-surgery clinic. It was sold again in the same year. Since that time the directors of the clinic have consistently maintained that no patient records exist from the period prior to 1987. Reporters first started asking questions about Tye's stay in the clinic over fifteen years ago when the Tye Corporation first became a quoted multi-billion-dollar company and its president became a nationally known figure in the United States. The clinic was unable to provide any help.

This time, at least, Haley knew the source of her information. She wondered why Jack had become sufficiently interested in Tye's childhood problems to have done this research but, she had to admit, it did make good copy. She settled down to integrate the new paragraphs within her opening chapter and wondered if she should use some of the Sloan Press's research budget to visit the sanitarium for herself. As she had learned during the research for her very first book, the best-selling film-star biography, second-hand sources are not to be relied upon and, despite the potency and verisimilitude of multi-angle 3D videoconferencing and the latest olfactory and gustation simulations, there still remained something special about visiting an individual or a place in person. She would only accept that the clinic had no records when she had established that fact beyond doubt.

Her first job, however, was to bring some sense of order to the vast streams of news and information reaching her in-box regarding Thomas Tye, the Tye Corporation and its many subsidiaries. Eighteen months before, when she had been planning her research pattern for this biography, she had programmed a group of software agents to hunt for all Tye Corporation and Tye-related information on her behalf. Now it seemed as though the amount of available information about her target had shot up to almost unmanageable proportions. Each morning she started reading the material gathered and, even after the system automatically discarded all encrypted files, she found herself still wading through the heap of information at the end of the day. It was overwhelming.

She therefore gathered her software agents and attached a de-duplication engine to each. At least she would then be spared reading some of the many similar stories being produced around the world. Then she began to draft an advertisement she would post in the many network discussion groups and communities that revolved around Thomas Tye and the Tye Corporation. This was where her new research budget would come in useful. She would offer a financial reward for any previously unpublished information about the tycoon or his company.

Ishkov Konstantine, Minister of the Interior for the Russian Federation, rose to his feet, his naturally florid, leonine face now suffused with blood. He leaned forward, his knuckles resting on the table top, and glowered across at Tye, Furtrado, Connie, the seven Tye Corporation attorneys and the numerous corporate analysts who were sitting opposite.

'You arrive to rip the heart out of our Tartar homeland – to take the cradle of Russian civilization away from our people and yet you will forbid them to live there?' he bellowed. '*Nyet! Nyet!* This will not go forward.'

Tye and his party did not need to concentrate on the translation feeds from their VideoMates to understand the minister's position. But position it surely was – or rather leguleian posture, if Furtrado's political analysts were to be believed. Konstantine was playing for the ever-present cameras, for the historical record and for the Russian people – creating an aura of personal importance for when the next elections came. If he failed to win re-election to central government, as the analysts predicted, he was likely to stand as a mayoral candidate in his home city of Novosibirsk, capital of the Krasnoyarsk Republic, and that was close enough to the southern border of Tye's new province for him to claim an interest. It would therefore be important to win him over, especially as the Tye Corporation's new agricultural enterprises were likely to need the processing facilities of Konstantine's city. Tye suspected that the gift of one of the Baltic islands with its newly acquired microclimate would achieve that objective – when the time came.

They were seated in the great gilded Tsar's Hall of the Terem Palace in the north-east corner of the Kremlin complex. Little government was conducted in the ancient walled citadel these days – most administration being centred around the State Union building across the Moskva River. But Thomas Tye and his colleagues from the mighty Tye Corporation were getting the sort of reception usually reserved for a visiting head of a superpower and his entourage – which, everyone conceded, was appropriate in the circumstances.

There had been a tour of honour through the streets of Moscow. The citizens knew Tye's company was becoming Russia's largest foreign investor – some muttered 'saviour' although that was unfair to President Orlov's incredible achievements in economic reconstruction – and they had responded accordingly. The people were being informed that the huge new development was a partnership between the Russian Federal Government and the Tye Corporation, although the use of the word 'partnership' in this context added new dimensions to its generally accepted definition.

The *Moscow Times* Index had jumped thirty-two points in anticipation of the deal. Despite continuous television coverage and the press attention paid to Thomas Tye's visit, so far there had been no reports of tribute self-immolations amongst fringe groups within the population.

Then there had been an interminable state dinner of disgusting individually baked coulibiacs that opened to reveal a pulverized slumgullion of fish shreds, cabbage and gravy. This had been followed by operose speeches that seemed to go on for ever. Tye had touched nothing of the food or the vodka and could not have cared less what his rapacious hosts thought of his restraint. He just wanted to get on with things and he didn't need to drown in a vedro of high-octane alcohol to cement this deal. But such behaviour was exacerbated by the harsh Russian climate, he reminded himself, with a private smile.

They had been talking for two days and all around them were huge maps propped on ancient easels on which new boundary lines and corridors had been drawn, appropriately in red ink.

At the head of the great room, to the right of a vast black-marble and gilt fireplace, was a recently installed Holo-Theater. It had been a gift from Tye Business Communications to mark Thomas Tye's first state visit to Russia and it also provided the APU with the opportunity to scrub these ancient rooms electronically and install the necessary white-noise generators and other anti-surveillance devices.

Tye had been counselled that such blustering over the cession of this territory was likely to go on for some time and he was struggling to be patient with the brabble. He slumped in his chair, resting his head in his hand as he listened, his right ankle supported on his left knee, and his loafer-shod foot twitching furiously. Unusually, he was thinking about his son.

It was a big deal for the Russian people, that was clear. Over one and a half million square miles – almost eighty million *dessiatine* – of the republics of the Central and Eastern Siberian Plain was involved. Leasing such a vast tract of land to the Tye Corporation took monumental political nerve and the absolute willingness of the Russian federal army to back such an arrangement. For that reason General Yevgeny Padorin, Supreme Chief of the Federation's armed forces, had been included in the party that visited Hope Island and he was now a prominent member of the Russian delegation sitting on the other side of the table. Although he didn't yet know it, he too would become the proud owner of a new holiday property in a uniquely sunny area of the Baltic, along with the presidents of the two vast, poverty-stricken and corrupt republics affected.

But the benefits for the immense mother country were

incalculable. Of the one hundred and sixty million people that made up Russia's population, only three million moujiks – most of the Tchuktchi, Koryak, Lamutic and Yukaghir ethnic groups – lived in the gelid dry tundra of the Siberian heartland and on its vast eastern slopes. As a jointly mounted, year-long survey had shown, the immense government-owned plain and mountain range was good for little but oil, gas, diamond and mineral extraction and the Tye Corporation had generously agreed that these resources would remain in the hands of the local Sakha Republic leaders or Russian government-owned or government-appointed companies – two, at least, to be headed by Anton Vlasik and his cronies – throughout the duration of the champart lease.

Even new finds would remain either the property of the individual republics that were ceding territory or of the Federal head landlords for disposal as they chose. The incumbent operators would pay no tax or levies to the new regional authority. The Tye Corporation had no interest in the energy and mineral resources of the industrial age and the lessors did not fully appreciate what was likely to happen to the price of fossil energy once the world understood the alternative that the Tye Corporation would offer as the Phoebus programme rolled onwards.

But, most important of all to the Russian federal government, the lessee company was agreeing to extend its heavenly largesse to specific parts of the vast swampland of the West Siberian Plain abutting the Urals that would remain Russian sovereign territory, owned directly by the federal government. Two million square miles of harsh and uninviting terrain would suddenly have sizeable areas that, over the next few years, would develop temperate and invitingly habitable graminiferous microclimates.

The Federal Council had already passed the outline legislation in closed session and the President's signature and those of two other appointed ministers would complete the agreement recognizing the sovereignty of the new state. The Duma was firmly controlled by Orlov's followers and would provide an immediate rubber stamp: the *ukase* would be issued on schedule.

'First, we are not taking Central and Eastern Siberia away from your people, we are simply leasing it,' explained a theatrically weary Furtrado once more as the minister sank back into his chair. It was so painful, so old fashioned, *so full of friction*, to have to samba with the uncivilized, the *nekulturny*.

'What we are saying is that we will not be able to allow *uncontrolled* immigration to the new state of Sybaria. Those already resident will be entitled to remain in their homes and continue to live their lives in the enhanced environment we shall be providing there. They will have dual Russian and Sybarian nationality and they will be able to move freely between the two territories. We will respect all cultural traditions, provide absolute religious freedom and support the freedom of individuals and human rights. What we cannot have – and, I would have thought, what you would not want – is an open-border policy that allows *your* people to emigrate to our territory. Besides, you will have an enhanced Western Siberia, so there's more than enough land for everyone.'

What Furtrado had not mentioned, and what he knew his Russian counterparts would never raise (unless the negotiations started to spiral towards failure), were the difficulties most of the autochthonic groups in the region would have in adapting. In particular, the Tungus, Kamtschadales and Lamouts, all subarctic tribes scraping a living along the coastline of the Sea of Okotsk, were unlikely to apply for trainee positions in the fast-food restaurants and hotels that had already signed provisional leases to build outlets and facilities in the area. The Sybarian business plan budgeted for these people to draw a form of corporate welfare and, the planners privately agreed, it would be necessary to work hard to ensure they did not become the Sybarian equivalent of the Native Americans or the Australian Aborigines.

Another issue that would not be raised was the withdrawal of democratic rights from the region. The new province of Sybaria would never be democratic and it would always have only one government. The new plutarchy would, over time, develop a standard of living that would be possible nowhere else on the Asian continent and, if the predictions of economists and social scientists were right, in few other places on earth.

*

DID YOU KNOW THOMAS TYE?
REWARD OFFERED

> Author seeks any new (previously unpublished) information about Thomas Tye's childhood, family background or education. An attractive financial reward is available to anyone who provides new information about Mr Tye's early years that can be successfully checked and verified.

Hayley would post it in all the Thomas Tye communities and in the Tye newsgroups. She would also create links from her new publisher's network resource and from the companion sites that had been created for her earlier books. But should she mention how much the reward might be? The Sloan Press had been generous. Their initial advance payment on top of Nautilus's money had, for the moment, eased her financial worries. No, how much she would pay would depend on what, if anything, was offered. She assigned the ad to her posting agents, provided the locations, sent them on their way and went to make a cup of coffee.

Thomas Tye stifled a yawn. He wondered how long these blustering negotiations would continue. He realized that it was the largest deal that most of these politicians would ever be involved in – well, it was also *his* largest deal so far – and he understood that they would all want their turn to show off their political skills. Something to tell the grandchildren. Something for the history books. Something an enthusiastic, unusually literate but inexperienced writer in the Tye Corporation's public relations agency had already dubbed a 'a truly historic and Brobdingnagian agreement' in a headline proposed for one of many press statements.

In another room, across a marbled corridor, Furtrado's

conveyancing lawyers were still arguing with their Russian counterparts for an extension of the lease. That debate would eventually have to come back here to the main table. In yet another room, Richard Rakusen, his co-financiers and a covey of architects were exhibiting architectural models of the airports, roads, hotels, housing estates, golf clubs and sports facilities they would be constructing to create new spa resorts around the Sea of Okhotsk and on the Kuril Islands as the ice receded and a new environmental ecology was born.

Technically, Rakusen and the other developers who would follow him into the new territory of Sybaria were not required to show their plans to the Russian government or to any authority except the Tye Administrative Council for the new region. But it had been agreed that for the sake of goodwill, and to help Russian understanding of the peaceable plans that the Tye Corporation and its development partners had for the ceded territory, the plans and three-dimensional models for the first major real-estate development of the new province should be shown and discussed. The disclosure also provided Anton Vlasik and his associates with a pre-emptive opportunity to select the investment and development projects in which they wished to participate.

Immediately after the Act of Cession was signed and Sybaria was receiving its first season of ecological and climatic re-engineering, Rakusen-Webber would launch a new Sybaria Fund to offer the world's investors an opportunity to share in the huge development opportunities. The first leisure resorts would be centred on the best of the many coastal hot-springs locations and the littoral would be marketed as both summer and winter resorts to the nearby Japanese as well as to the North American market.

Also in Phase I would be opportunities to invest in the vast solar farms that would occupy two million hectares of the former Siberian permafrost. These vast acres would contain mile after mile of curved Anacamptonite solar-capture panels that would feed electrical power to huge underground storage silos ready for marketing to the cities of Japan, China, India and Russia via superconducting underground cables. The former 'sleeping land' of Siberia would become an electronic counterpart of the great wheat expanses of America's Midwest, only this 'basket' would hold energy, not the ingredients for bread. Later would come opportunities to invest in an area enjoying macroclimate manipulation and the vast agricultural opportunities presented by nocturnal crops and livestock – once the happy point of autarky was passed.

Although 'Sybaria' had been accepted as the official name for this new territory, suggested by one of Furtrado's classically educated English lawyers, an unidentified wag on the team had already suggested the region should be called 'Tyeland'. This idea had been officially abandoned for fear of confusion, but inside the company this name had stuck as the code name for the new province.

Tye ached to have a shower and he craved the comfort he derived from a protective mask and gloves. He sprayed his mouth with antiseptic again. This was the first time in a decade he had attended a meeting in which he could not control his environment. Only he, and the geneticists of Tye Life Sciences, understood the damage that common infections could do to the suppression of free radicals in his body. As his immune system rose to respond to infection, it overcame the inhibitors produced by the oral cell-maintenance therapy that he had taken daily for the last sixteen years: if he caught a cold, Tye would resume ageing like everybody else for as long as the infection lasted.

But it would be worth the risk and at least he didn't have to suffer the Russians' disgusting tobacco smoke. He had made that an absolute condition when this trip was being arranged. Every half an hour one of the Russian party would make an excuse and disappear from the room. That had proved a useful weakness to exploit. Several times Furtrado had insisted on labouring a point while the Tye party watched a Russian negotiator start to fidget and sweat, desperate for a cigarette break. That way some details had been agreed hastily but Tye guessed there would be at least another day of negotiations. If everything was agreed tomorrow, the deal would be done on his fiftieth birthday, but only Connie and Furtrado would offer him congratulations on his anniversary and there would be no public celebrations of his age. Tye did not discuss this subject in public even though the media was bound to home in on the story.

But this deal *would* be settled, that much was for sure. Furtrado had bank authorizations for transfers of a total of one trillion US dollars in his briefcase – enough to eliminate Russia's global debt in a single payment, if that was how the government chose to spend their giant windfall. Tye judged that President Orlov would do just that and his nation's economic rating would suddenly be propelled back into that of the world's top half-dozen countries, even before the economic benefits of the forthcoming climatic re-engineering had been calculated.

Another man, pale with a patchy red moustache and ill-fitting rust-coloured jacket, rose at the far end of the table: the E&E minister – environment and ecology. The Greens had *finally* found a voice in the polluted behemoth that had been industrial Russia and, after years of prevarication, the state was at last attempting to fulfil some of its biosphere obligations to the international community.

'The Tye Corporation and the University of Hope Island have both been very helpful in modelling the short, medium and long-term impact of these climatic ... improvements,' the minister began portentously. He spoke good English so the Tye corporate team was pleased to switch their earpiece-jewellery to ambient, and join real-time again. 'We have been very encouraged by the models that show the impact of carbon dioxide reduction in the atmosphere and the improved economic prospects of the region and the surrounding territories.'

He paused and everybody waited for his objection. 'However, I have been unable to find any prognostications or projections for the future of the indigenous species. Although the region is classified mostly as taiga, tundra and ice cap, the Tye delegation must know that there is a rich, diverse and important collection of birds, mammals, rodents, reptiles, insects and fish in that area. These include Stella sea eagles, the white-tailed eagle, the Siberian bear, Siberian tiger, Siberian fox, Siberian chipmunks, buzzards, picus, polatouche, salmon, deer, moose and many thousands of insects, reptiles and other fish. Many of the species have not yet been identified and named. Are they to go the way of our elasmotherium or the Stella sea-cow? What is to happen to them, please?'

He sat down.

Tye leaned back and raised his eyes to the ceiling. All this had been covered in the detailed discussions and scientific documents that his teams had prepared and delivered months before. They had already started freezing sperm from many of the life forms of the region. Jesus! What more could they want? He would be delivering them a Moscow with winter daylight time extended by fifty per cent. The city would have temperatures similar to San Francisco, for Christ's sake! He could even switch supplies off to give them fucking white Christmases whenever the government asked – just to ensure the ancient city didn't lose touch with its original character!

He sighed and nodded to Furtrado to respond.

The counsellor shuffled through his documents for the folder on wildlife and endangered species. But before he could find the reassuring words he had selected so carefully the questioner rose to his feet again.

'I was wondering whether Mr Tye would honour us by sharing his *personal* views on this topic?' suggested the Minister of the Environment.

Tye felt a sudden sense of alarm. Could it be possible that this was not a done deal after all? Was this a trap? He and Furtrado had been sure that the Russian government would provide nothing but a token, albeit prolix, show of resistance during these negotiations. But if this new voice represented anything more than a desire to be heard, everything could still fall through. Tye had been thoroughly briefed on the Russian political system so he knew that consensus, not majority voting, was required for the Presidential Council to endorse such radical legislation.

Perhaps it was time to utilize the old magic. Outside of his public appearances, he used it so rarely now – almost never in private meetings. It had grown too large, too *kerystic*, to be used against just a handful of people in what, for one of *his* performances, would be a comparatively small room. But, Tye reasoned, as business had become a form of theatre all individual careers had become performances. He also guessed that they couldn't resist the chance to see if this supposedly legendary performer was truly as great as the rest of the world thought.

He glanced along the table at his interlocutor, then smiled, his grin lighting up the whole conference table. He pushed himself to his feet and stepped back, lithe in his movements. Despite the formality of the occasion, he wore his usual open-necked white shirt and black trousers. His hair was tied back in a short ponytail.

'The Minister is quite right, of course,' he began. 'This *is* a subject that I am pleased to address personally, since it is of the utmost importance.'

Tye clasped his hands behind his back and walked slowly towards the head of the table, where President Orlov presided – so far in silence. When he reached the president's side he stopped and turned to face his audience.

'We will be bringing light and heat to areas that have never received those natural benefits before.' He spoke slowly and carefully, aware that many of the Russians around the table would be able to understand his carefully enunciated English. He raised his arms in practised chironomy. 'But this energy, *our* energy, is totally controllable – to within a radius of one mile, as I have had the pleasure of demonstrating to President Orlov, Minister Konstantine and General Padorin recently on Hope Island. We have created a complete model of the world's climate on the most powerful parallel-processing computer ever built. It's capable of five trillion operations per second – we call it the Halcyon system. It allows us to play "what-if" with different conditions. We know precisely what each joule and nanowatt will achieve and its knock-on effects in the global ecosystem as a whole.'

He paused and looked at each of them in turn, holding their gazes, ensuring their absolute attention.

'Remember, I care about this planet!'

Tye paused again. The Tye corporate heads were already nodding – then some of the others. He had them now.

'Humans have been aware for nearly forty years that we are

slowly choking our world. Temperatures have been rising – especially in the northern latitudes – and, unless the Tye Corporation takes this crucial step, with the help and aid of the great Russian Federation, carbon dioxide concentrations will be at seven hundred and fifty parts per million by the middle of this century – three times its pre-industrial level. That means we will destroy the Amazon rain forest and all other rain forests, and global sea levels will rise by two metres. A four-degree Celsius increase in the planet's mean atmospheric temperature would increase global rainfall by twenty per cent – but all of it in the wrong places. Water vapour in the air is, in itself, a powerful greenhouse gas. So it becomes a vicious circle. After that the remaining vegetation will disappear and the sea will continue to rise.'

He paused once more to ensure they were all keeping up.

'Currently, only eleven per cent of our planet's surface is under cultivation and we are expecting our population to continue to grow – perhaps to ten billion or more in the coming decades. We cannot support such a population with the present biocapital resources this planet has to offer. We are already seeing chronic water shortages affecting thirty per cent of the world's population. Over the next thirty to forty years, the climatic re-engineering we are undertaking will add fifty per cent productive land and will add an entirely new hydrological resource. Today, most of the Siberian permafrost is at minus thirty degrees Celsius. Just lifting that to temperate levels and covering that much of the white tundra and taiga with vegetation and crops will reduce solar reflective radiation in the region by thirty-six per cent. That will reduce the warmth pumped back into the atmosphere by the same amount. *That* is the only way we can feed and clothe the people of the future *and* do so without exhausting current resources. We have no choice, ladies and gentlemen. Humankind is like a household living giddily off vanishing capital. There is no choice.'

Tye looked for, and found, nods of agreement. Suddenly he thought of his son, then of Calypso. The words he wanted flashed into his mind.

'Remember: "Our remedies oft in ourselves do lie,
Which we ascribe to heaven."'

He glanced around the table. He guessed he wouldn't have to identify his reference in such a Shakespeare-mad culture.

'And I look to you, the Russian Federation, to be our partners in providing these remedies for the world's people.'

He saw more, increasingly vigorous nods around the table.

'The Phoebus Project represents a new age for humanity and this planet. After centuries of prosthetic technologies designed to extract greater and greater resources from the Earth to satisfy the ever-increasing material needs of humans, we have found a way to harness the vast pool of energy that streams past our planet only to be wasted. It is truly a new beginning. But there will be some small costs. I seem to recall one of your most famous countrymen declared that to make omelettes, you have to . . .' He tailed off, lest this reference should cause offence.

'It will produce a fundamental shift in this planet's ecology: the biological equivalent of the Triassic Age giving way to the Jurassic. We will take control of our environment and apply energy carefully and thoughtfully where it will provide the most benefit while causing the least harm. And remember, none of this would be possible without the pioneering efforts, creativity and imagination of late-twentieth-century Russian enterprise.'

That got them. Broad grins and nods greeted his reminder.

'We shall not, repeat NOT, be melting the northern ice cap nor any of the major ice flows of these latitudes. We shall be eliminating seasonal ice in the Sea of Okhotsk and on the western edge of the Baring Sea, but we know this will have a negligible effect on global sea levels. All of you here will know that only ten per cent of polar ice is found in the Arctic. The great mass is at the Antarctic and we have no plans to modify any polar environments in the Southern Hemisphere. As we are all aware, eighty per cent of the world's wealth is generated north of the equator.'

Tye didn't wait for the environment minister to butt in again about the indigenous species. In particular he didn't want the topic of native nocturnal life forms brought up in open session.

Their sperm too would be frozen but their habitat would no longer exist. They would have to be completely relocated.

'We *shall* be warming the atmosphere,' he conceded. 'But our energy will render over forty-six coal-fired power stations obsolete inside four months. A further two hundred can be decommissioned in the following year. The cities of East and West Siberia will never again need street lighting. In all, our ecologists and environmental economists have calculated that we will be adding the equivalent of two hundred and forty-four parts per million of $CO_2$ to the atmosphere over Eastern Siberia and replacing the need for nearly six hundred PPM of man-made emissions. That will produce a net reduction in $CO_2$ concentrations of over three hundred and fifty PPM.'

He paused to allow the delegation to complete their note-taking. He had every statistic in memory. He even knew what was likely to happen to all the little creatures caught up in these changes, but he didn't want to discuss that here. The plan was to take samples of each of the important genus groups, clone sufficient for genetic diversity and then offer the clone bank to the Moscow Institute for breeding elsewhere.

'We are also ensuring that there will be no climate re-engineering anywhere in the western quadrant of this hemisphere – there will be no activity over Greenland, for example. We intend to ensure that the deep-sea current continues to sink vigorously and that the $CO_2$ absorption properties of the Northern Atlantic remain undisturbed.'

This was all so tedious. The corporation's ecologists had provided over 2,000 gigabytes of data on the ecological impact of the project in readiness for the inevitable questions and objections of the global community.

'I also want to refer you to our agricultural strategy. Over forty per cent of our plantations will consist of crops specially modified to absorb large amounts of $CO_2$ – at the type of inhalation levels currently found in tropical regions. Our new forests will also be both diurnal *and* pseudo-nocturnal and thus will absorb carbon dioxide twenty-four hours a day.

'We are going to extend the tree line of this planet eight hundred kilometres to the north and we believe the overall effect will be a significant improvement in evapotranspiration and a reduction in the build-up of the so-called greenhouse gases in our atmosphere. After all, we are harnessing–' he evoked a theomagical metaphor he knew would still have meaning for a society that had re-embraced religion only a generation ago '–*God*'s own energy in this project.

Tye waited as translation was received by those few who required it. He then lifted his arms again and gazed directly at the E&E minister as if daring him to make another challenge.

'Take care of our planet – you hear!'

'We hear,' responded the well-rehearsed Tye delegation, the English-speaking Russians hastening to join in.

He watched as the translation was received and more heads on the Russian side of the table nodded.

'I propose that we should form a new collegium to review our plans for the wildlife of the region. Perhaps the minister could delegate one of his team to act as Chair. It can then review the plans we have made and report back to us jointly in due course.'

The minister nodded his head.

'Now, I would like to move on to one subject that is of crucial importance to all of us.'

This was Tye's biggest worry, the one thing that could completely wreck plans and developments that had been over ten years in gestation. He would have to have their complete understanding on these issues. He turned to a large map that stood behind him.

'We cannot expect our friends in the People's Republic of China to remain unmoved by the creation of a new region of opportunity so near their northern border,' he began, choosing his words carefully. 'Our agreement calls for the Russian government to maintain a corridor two hundred miles wide along the entire southern border with the People's Republic of China. It will be Russia's responsibility to maintain an adequate military force in this region at all times to deter any thoughts of an advance by the People's Liberation Army or any other form of incursion towards our areas of investment. I would be grateful if General Padorin could provide more detail of the forces and equipment that will be permanently stationed along this corridor.'

This was the crunch. The Tye Corporation had to have the protection of the Russian Army against the Chinese hordes. Envy was the most powerful of political motivations, even if the coveted territory could be devalued at the turn of a switch. The army in the People's Republic had grown to over one hundred million men and women and it would take a deployment of Russian armour and, most importantly, Russian tactical nuclear weapons along the northern border to provide a clear and demonstrable guarantee of security for the Tye Corporation shareholders and the many new investors who were eager to take sub-leases on land in the new province. Tye knew, of course, that if things ever became really difficult he now had an awesome weapons capacity of his own, but that was not to be spoken of – *ever*.

Jack surfaced a mile offshore and trod water. In the distance, to the south-west, he could see the dark outline of Hope Island and its myriad twinkling lights. There was a clear sky and the bursts of laser communications added a canopy of intermittent illuminations. There was also a full moon. That would require extra caution. He had blacked his face for the mission – recalling, as he did so, that the last time he had used camouflage in such a way had been over fifteen years ago, when he had been preparing for his final active-service mission. Then he had been aboard a submarine that was rolling viciously in a large surface swell fourteen miles off the coast of North Korea.

He had drawn an automatic pistol from Security's stores but had refused the box of ammunition the armourer had pushed across the counter. Then, to add to the man's confusion, he had also declined to complete the electronic identification and registration procedure to confirm him as the authorized user. Only this would have given him access to the internal microprocessor that controlled the laser-sighted weapon. He had then requisitioned lightweight scuba equipment, once again without explanation. He had signed for the useless gun and the diving equipment and taken them back to his apartment where he had run a full safety test on the scuba gear in the bath. His security review was under way and his people had been told to expect incursions and similar tests of their defences. All his team were on edge. He intended to keep them that way.

He had reviewed Pierre's security plans for the Moscow trip and had found them both detailed and thorough. Having once occupied a top slot in the league table of cities regarded as most dangerous for Tye executives, the Russian capital had improved over the years to the point where it was little more dangerous that any Western European city. The only real point of concern had been the motorcade to which they had insisted subjecting Tom on the day he arrived but that had passed off without event. Pierre reported that the entire corporate team was now ensconced in the Kremlin and was likely to remain there for several days. Jack had already seen one of the three Tye-Lears that had carried the Tye diplomatic mission returning for fresh supplies of Hope Island produce.

Jack had been able to meet with his boss in person for eleven minutes shortly before Tom had left for Moscow and he found the tycoon rested and relaxed. The media frenzy over the Los Angeles traffic crisis had receded: Tye had appeared totally focused on the deal about to be struck in Moscow and on the forthcoming *One Weekend* conference.

'You do whatever you need, Jack,' Tye had agreed with a wave as the corporation's vice-president of security explained his plans for a radical review of protection procedures on the island and the need for more regular refugee sweeps. Since the Cuban navy had now been witnessed shooting at their own people as they tried to escape the country, Tom had agreed that any refugees and escaping rebels found in Hope Island waters would be brought ashore before being forwarded to one of the other countries prepared to accept them. Jack had warned Tom that with such an exalted guest list, the weekend conference would be a major trial of Hope Island's state security procedures and he wanted to go over the plans and test them to ensure everything worked smoothly.

He checked his old analog wrist-compass and made sure of his heading. His LifeWatch with its sophisticated GPS locator was beneath the duvet back in his apartment overlooking the marina in Hope Town. He blew out his mask, released some air from his buoyancy jacket, sank to three metres and began the trudge for the shore. He was heading for the north-eastern tip of the island, aware that his patrols on the beaches and clifftops would be constantly scanning the water – as instructed by their standing orders. Low-level radar swept the surface for early warnings of approaching craft while drone sonars were deployed at every mile around the island's perimeter in wait for the sound of propellors. The system was even sensitive enough to detect and identify the sound and rhythm of splashy swimmers.

He doubted whether either of the two patrol craft would be inshore. Their standing orders were now to patrol the edge of the exclusion zone with Cuba, over to the west.

Jack felt the increasing warmth of the water as he arrived in the shallows and he became aware of rocks rising up to meet him. He paused again and slowly rose to the surface. To his left was the tiny crescent bay that had been blasted out of the sheer rock face to provide a private beach for the main house. It was now floodlit for the benefit of the security cameras. The only landward approach to that patch of imported white sand was the silver funicular railway that rose from the rear of the beach up to the pool terrace one hundred and fifty feet above. The car was now at the top, inside its winch-house.

Getting his bearings, he submerged again and set off northwards. If the engineers had chosen the shortest route for the tunnel the outlet should be just below the surface, a further two hundred yards along the sheer cliff face that fell almost vertically into the sea. He had located several articles about its design on the networks. Various architects and engineering consultancies were so proud of their work on Thomas Tye's mansion that they encouraged specialist journals and architectural magazines to print stories and diagrams of their various designs. None was permitted to publish pictures or plans of the house itself, of course, but Jack had gathered more information than he needed and for this particular exercise it was important that he only used information that was generally available.

He checked his depth gauge and released air from his

buoyancy jacket until the instrument registered two metres. Then he swam slowly along the rock wall. He was weighing the risks of having to use his main flashlight when he felt the current slightly above him. He stopped and allowed himself to rise into the stream. His pencil-light revealed the grating and he felt the gentle warmth of the current pushing against his wetsuit.

First he examined the bolts that secured the outlet cover to the rock face. As expected, they were too deeply embedded to yield to the unaided strength of a single swimmer, so he opened his tool bag and unclipped the diamond-tipped micro-saw that UNISA's Technical Services had provided. The hand-held circular saw was driven by a tiny but enormously powerful electric motor and in two minutes Jack had cut a hole a yard square in the grille.

He returned the saw to its clip, refastened his tool bag and swam into the wide opening. Inside the seclusion of the steel tunnel, Jack was finally able to switch on his main flashlight. He could see that the bore continued level for about ten yards and then began its slow curve upwards into darkness. He wondered how the construction engineers had drilled such a tunnel through solid rock.

After four or five minutes of vigorous swimming Jack felt himself tiring against the opposing current and, despite his fitness, he began to think he might not make it. He checked his air: he had twenty minutes left and that immediately reassured him. Then he saw a dim light ahead that meant he had arrived under the large baffle and filter housing at the bottom of the pool.

Professor Sir Oliver Morton, Knight Bachelor of England and one of the world's most distinguished genetic biologists, had thrown himself into practical work in a way that he hadn't since he had been a PhD student thirty-one years before. Reluctantly, and only after a series of blazing family rows, he had relocated to Hope Island as his company's new chairman had insisted. His wife, a professor of comparative philosophy at the University of Cambridge, had refused to join him in the move. Morton had faced a stark choice: resign and give up all realistic claim on future share options and the capital he had invested in Moleculture plc – all the liquid capital he possessed – or move with the company and endure separation from his wife and family. They had agreed to try living apart for six months. Lucy Morton understood that to be fifty-five years old without any savings was an uncomfortable position, no matter how philosophical one might be.

At the same moment that Jack Hendriksen was arriving at the bottom of Thomas Tye's swimming pool, Morton was rechecking an observation he had already rechecked eight times, under four different electron microscopes. He was in the company's laboratory in Hope Island's Science Park and he had been working all night, as had become his custom since arriving on the island. There could be no doubt about what he was seeing, but he would need other eyes to see the living cells, the photographs and computer images he was collecting.

The cells that had grown in the latest batch of *Triticum spelta* – a new strain of pseudo-nocturnal wheat – were right-handed.

*Right-handed, not left-handed.* 'Left-handed' was the biologists' term for the universal polarity of all life on earth. Since the beginning of the earliest form of life on the planet, every cell, vegetable or animal, had selected only left-handed amino acids to make proteins. It had long been surmised that it was the polarity of the Earth's own magnetic force – created by the inner metal core of the planet revolving more rapidly than the outer – that causes all living things to be so.

But these cells had a polarity that was the reverse of all other life that had been observed and they had evolved naturally, without intervention from Morton or any of the researchers on his team.

He had checked and rechecked the room, too: there was no internal magnetic field. He had demagnetized and degaussed the culture units, the work surfaces and the three electron microscopes in the lab. Viewed separately under each of the three in turn, the result had been the same.

Morton looked at the microscope's display screen again and then reached for his VideoMate to call Fred Zimmer. He knew Thomas Tye had been exaggerating when he had talked about an invitation to Stockholm for the pseudo-nocturnal crops – but this! It was certainly worth waking up his partner. They would have a chance to decide how best to start further replication trials before the day shift returned to the lab.

Thomas Tye had used the island's natural hot springs to feed his indoor swimming pool. Whether this had been his idea or the architects' was not known, but balneologists had been brought in to supervise its design and those who had swum in the mineral-rich water reported enthusiastically on the invigorating benefits of this natural spa. One of the many underground springs had been diverted to feed the large pool and, with a tunnel bored to provide an exit to the sea, the trillionaire had eliminated at a stroke the energy requirements and costs of both heating and recycling filtration.

Jack clipped himself to the grille below the pool while he deployed his micro-saw again. The cut-away section of the grating, and the filter mesh, sailed away in the rapidly increasing current and Jack forced himself up into the pool against the flow.

He swam away from the now swirling drain and sat at the bottom of the pool, looking up for human silhouettes. His air supply was not now the invisible mixed-gas rebreathers he had used as a SEAL. He knew that bubbles were escaping to the surface, but it was unlikely that any of the house staff were in the pool room at three a.m.

He then finned towards the surface and gently broke through into fresh air. Pulling off his mask, he savoured a deep breath and slipped back the hood of his wetsuit. The ornate pool room was empty, as expected. Moonlight penetrating the huge windows was reflected off the water's surface and bathed the ceiling and walls in gentle undulations of light.

Jack swam to the edge of the pool, pulled off his fins while still in the water and climbed out onto the marble surround. He thought it unlikely there would be any security cameras here or in the changing rooms. Considerations of modesty often override security – a sensibility that professional intruders frequently exploit. Inside the male locker room he peeled off his wetsuit.

He was wearing black shorts and T-shirt beneath his wetsuit but he kept his reef shoes on to provide grip. Then he removed a plastic bag from his tool kit. The 9mm SigSauer was perfectly dry. Even though the gun's microprocessor-controlled firing system was not activated and he had left the empty seventeen-round magazine back in his apartment, he checked once again that it did not have a round in the breech. He tucked the gun in the waistband of his shorts, ensuring that it was displayed prominently. Next he took a black waterproof envelope from his swimming bag and clasped it under his arm.

Back in the main pool room he summoned the elevator. From now on he knew his presence would be recorded, but that was part of his plan. Once the elevator arrived he stepped quickly into its bright interior, hoping that this small momentary illumination in the pool room would go unnoticed by his own patrols out in the grounds. Jack studied the row of buttons and the camera lenses set into the control panel at shoulder height.

He pressed for the fourth floor, where he knew Tommy's room was located. He had not visited that level before, but he had inspected the lower three floors of the great house and realized that Tommy's room and the live-in staff quarters were situated on the fourth. Tye's private quarters, inaccessible to all others but his Mexican domestic staff, lay on the fifth and topmost level. The elevator began to climb.

Eventually the doors hissed open onto a dimly lit corridor running to either side. From here on, Jack was working on luck. He pulled a sheet of ultra-thin, clear, optically conducting plastic from the envelope, knelt down and depressed the 'open door' button. Slowly he extended his arm and the plastic into the hallway. He determined the house's alarm system did not include movement-detection beams at floor level. Tye had never asked Jack or any of his predecessors to cover the interior of his house in their security plans, so the island's security force held no details of its internal protection arrangements. It was this anomaly that had given Jack an excuse to mount this exercise. If he triggered an alarm by breaking a movement-detection beam now, his dual-purpose mission would be over instantly and he had no alternative plan.

He slowly worked the transparent sheet up to head height, then higher again. There were no beams, so he stepped out of the elevator. As the doors closed behind him he carefully repeated the process, crouching and straightening to full height again and again as he worked his way across the width of the corridor. They had either chosen not to install detectors in the hallways or they were not switched on. Perhaps there was too much casual movement through these corridors for them to be practical, Jack supposed – it wasn't as if this house was ever going to be left empty of staff. He returned the plastic tell-tale to its envelope and looked towards each end of the hall, spotting cameras at ceiling height. He pulled the gun from his waistband and waved it theatrically at each camera in turn. He knew of no all-night surveillance team that might be monitoring the camera feeds right now, but he wanted to create an impact when the inevitable, exhaustive post-mortem started.

He had already estimated that there might still be three or four people inside the house at this time of night. Apart from Tommy there would be his night nanny – a woman called Angela, whom he had only met once, Emily Pettigrew the head nurse, the night butler and then possibly a cook. But with Tom away in Moscow it was likely that most of the domestic staff would have chosen to return to their own homes in Hope Town or its outlying developments rather than sleep unnecessarily in the almost empty mansion.

Jack made an arbitrary choice, turned right and moved silently along the corridor. It was 3.20 a.m., the nightly low spot for human activity when all the staff would be most likely sound asleep. He paused at the first door he came to, bending to listen. Then he opened the door quietly and poked his head into the room. The bedroom was empty, with moonlight streaming through its large window. He closed the door and set off again. When he opened the next door, he sensed the room was occupied, even before he heard gentle snoring. Emily Pettigrew slept naked, Jack noticed. He closed her door silently. He was now nearly at the end of the corridor. Looking up at the camera he shrugged, and he pointed at the last door, miming the question 'In there?'

Suddenly he heard a sound he recognized. It was a HydraChair in full motion. He pushed the door open and stood watching in the dimly-lit room as Tommy's chair inverted the boy completely, then put him on his side as the game simulated an aircraft or a spaceship executing a sharp turn. He could hear Tommy laugh as he fought to keep the virtual craft within whatever performance envelope the game's designers had specified.

Jack closed the door behind him, returned the gun to his waistband, pulling his T-shirt down to cover it, and turned up the main room lights. He didn't expect the illumination to penetrate the boy's helmet so he crossed the room and rapped playfully on its decorative wing decals.

The HydraChair flipped upright to its resting position and Tommy's helmet visor hissed upwards.

'Hiya, Tommy,' said Jack.

'Hi,' replied Tommy doubtfully.

'You're up late,' observed Jack.

'Hello, Jack. Was that a gun?' asked Jed from the bed. The Furry's vision extended into infra-red wavelengths.

'You *are* clever, Jed,' said Jack, keeping the irritation out of his voice. 'But it's not loaded.'

He lifted the front of his T-shirt, pulled out the gun and handed it, butt first, to Tommy.

'I'm testing security in the house,' Jack told the seven-year-old. 'I heard your HydraChair. Why are you still up?'

Tommy turned the heavy gun over in his hands. 'I don't have to go to sleep – as long as I stay in my room,' he explained. 'Who are you going to shoot?'

Jack smiled and retrieved the weapon. 'Nobody,' he said, stowing it in his waistband again. 'We're just practising drill – to make sure everybody's safe. Do you want to help us?'

Tommy banged the quick-release lever on his safety harness. He was wearing blue pyjamas and a pair of large blue rabbit-eared slippers that, Jack recalled, had been a gift from Calypso. As Tommy rose out of the chair he staggered slightly.

Jack stepped forward and caught the boy's arms to steady him. He eased his grip as he noticed the bandages underneath the boy's pyjama sleeves. Calypso had told him about the incident with the knife.

'Whoa – it gets like that when you've been flying.'

'I'm OK,' insisted Tommy, pulling free. He stepped to the bed and picked up his Furry companion.

'What are you going to do now?' Tommy asked Jack. 'Your face is all black.'

Jack took the package from under his arm and extracted a smaller, white envelope. 'I've crept into the house secretly to surprise everybody. It's your father's birthday tomorrow and I've brought a birthday card for you to give him. You can sign it. Have you got a pen?'

Tommy ran to his drawing chest. Jack followed and laid the card on its flat surface. He watched as Tommy read the text.

'It says "Happy Birthday, Dad",' exclaimed Tommy excitedly. 'I didn't know it was his birthday.'

Jack smiled. The card was a generic anniversary greeting, mentioning no particular age.

Tommy took up a green marker pen and carefully wrote his name.

'I'm going to sneak this into your father's office,' said Jack, tucking the card back in its envelope. 'It will be a nice surprise for when he gets home. Want to come along?'

'I'm not allowed out of my room,' responded Tommy, dutifully.

'You're with me now,' replied Jack firmly. 'Come on.'

'May I come too?' asked Jed in the very proper English that Tom had specified for his son's personal Furry. 'I'll just disconnect from the networks. Hang on a mo.'

Tommy scooped up the caterpillar as Jack picked up the small chair from in front of the boy's drawing desk.

'I think we'll need this,' he explained.

He opened the door and glanced along the long corridor. Seeing that it was still empty, he stepped out, followed by Tommy with Jed tucked under his left arm. Jack led the way to the elevator.

Inside the lift Jack pressed the button for the fifth floor. As he had expected nothing happened.

'We're not moving,' observed Jed.

'Just a moment,' said Jack. 'We need a special pass.'

He opened the larger envelope and carefully pulled out a gleaming mirror that immediately iridized, shooting rainbows of prismatic light around the lift. He positioned the low chair underneath the control panel.

'I want you to stand on this, Tommy,' Jack said. 'I'll show you a trick.'

'Will my father be angry?' asked Tommy, suddenly afraid.

Jack laughed.

'No. He'll be really pleased. He'll come back to a great surprise.'

Tommy stepped up onto the chair, Jed still clenched firmly under one arm.

Jack positioned the mirror carefully and Tommy's eyes widened as he stared at the reflection. The intelligent mirror recreated the boy's face as a perfect, slightly enlarged three-dimensional hologram. It was as though his whole head was captured inside a strangely deep mirror. He turned his head to the left and the right, watching the disembodied head move, exactly as Jack had hoped.

In their lab the UNISA technical officers had scaled up the identification parameters of a facial-recognition system, requisitioning the experimental holo-optics from the physics department at the University of Indiana. They had added an imaging microprocessor to the system with over 120,000 algorithms intended to morph a juvenile face into a convincing adult version. The system could generate all these in less than a second before starting over again. In the lab the catoptricologists had managed to fool a facial-recognition system by using the enhanced 3D-mirror and coloured images of *approved* faces, but there was no guarantee that Tommy's young face would be sufficiently similar to his father's for the reflector's enantiomorphing and ageing techniques to work. Jack knew the recognition system was dividing the boy's reflected face into forty elements, such as the outer corners of the eyes, the tip of the nose and the ends of the eyebrows and measuring the distances between them to compare to its database of approved faces. He also knew that each measurement was programmed to be imprecise to some degree because the system had to allow for variations in lighting and distance.

As Tommy gazed into his reflection, Jack slowly moved the mirror closer to the stereoscopic camera lenses, then back towards Tommy's face, angling it slightly as he did so.

'It doesn't seem to be working,' observed Jed.

'Thank you,' said Jack, who knew that the Furry would be recording all his actions. Instinctively, he stuck his tongue out at the caterpillar – a gesture he knew Jed couldn't return.

Tommy laughed at this and then Jack saw a small green LED illuminate above the button for the fifth floor. Jack nodded to himself in satisfaction. He should have guessed that Thomas Tye would *always* smile for a camera.

'Press the button again,' he said to Tommy.

As the boy did so, the lift started to rise.

'Well done,' said Jed.

The lights were off in the fifth-floor corridor but Jack doubted whether movement beams would have been installed there in addition to the elevator's security system. A rapid traverse of the hall space with his conducting tell-tale sheet confirmed it. He stepped out and touched the illumination control on the wall. The accommodation on the top of the building was set back considerably from the floors below, so was practically invisible from the ground. Jack wondered if any of his patrols had been sufficiently energetic to climb the mountain tonight. If they had, they might spot the lights and he would only have a minute or two before someone arrived to check.

Jack noticed three doors along the corridor. Two were on the side where the rooms would have sea views. The room on the other side would face the mountain. Jack opted for the land-facing room. As expected, there was a full range of security barriers protecting this door.

'Touch that pad with your finger,' Jack said to Tommy. He pointed to the long middle finger of Tommy's right hand.

He had already lifted the boy's fingerprints from the cup of hot chocolate Tommy had left unfinished at Calypso's house. Back at UNISA's HQ, Deakin had received them as uncaptioned electronic images, with a separate, coded fax detailing the request. Within an hour the dermatoglyphologists in UNISA Forensics had compared the prints to a set of impressions Thomas Tye had left in the White House in Washington DC eight years before. At that time the US Secret Service had harvested the prints as part of their routine observation of all the President's guests and, as part of a wide-scale reciprocal agreement, copies had been lodged with UNISA. Jack had recieved a four-word message back: '100 per cent match.' He had read somewhere that a third of identical twins have matching fingerprints, but he doubted whether anyone had data on artifical clones.

As Tommy touched the fingerprint pad, a green LED illuminated. Jack then positioned the low chair again and repeated the procedure he had followed in the lift. It took longer this time, although the system would be employing precisely the same facial-pattern recognition system used in the lift. He moved the mirror closer to and then further away from the twin lenses.

'Smile,' prompted Jack again.

As the green LED lit up, Jack helped Tommy step down from the chair.

'Now the system's got to sniff your wrist,' said Jack. 'Put your arm up against that grille, like this.'

Jack showed Tommy how to put the inside of his wrist to the olfactory sensor. This would be the hardest part and Al Lynch had bet him a case of Jack Daniels that his mission would fail here. Jack took a small aerosol spray from his belt bag as he watched the display panel.

The iatrochemists and the osphresiologists of the World Health Organization's research labs had also been doubtful. Though they understood that the boy was a clone of his father, they argued the degree to which the body's chemistry was altered during puberty. They therefore thought it very unlikely that the young son's chemical signature would match his father's so – as one put it, 'rather more in desperation than in hope' – they had prepared a supplementary concoction of testosterone and related hormones for Jack to spray on during the sampling.

He just was about to use the spray when another green LED lit up, indicating the system had been fooled.

'Just one to go,' said Jed.

Despite the preoccupations of the moment, Jack felt his skin crawl at hearing the little companion. This was uncanny. He turned and stared at the caterpillar under Tommy's arm. To his discomfort, Jed turned his own head slightly and blinked the lashes of his huge eyes, raising his eyebrows as he returned Jack's gaze. The caterpillar was mugging back at him! Jack thought he might have a word with Professor Keane. Her achievements with artificial intelligence were astonishing but also quite disconcerting.

'Now we need one more thing,' he told Tommy. 'This won't really hurt.'

'Oow,' complained Tommy jokingly as Jack pulled a single hair from his head.

'Watch this,' said Jack. He carefully placed the hair in the scanning capsule of the DNA-verification system and they watched as it was bathed in UV light. Within four seconds the third green LED illuminated.

'Come on, let's go,' said Jack. He pushed open the door and they were inside Thomas Tye's private office.

'It's very dark,' observed Jed, who could see perfectly well. Jack switched the lights on.

As the planners of this mission had guessed, the whole room was a shrine to computers – a throwback to the days when Tye had actually understood something about current computer technology and, perhaps more tellingly, had still cared. Jack saw a dozen outdated super-RAID storage racks and he estimated there were seven or eight processing systems in the room. All the monitors and the large wall screen were on stand-by. As Al Lynch had predicted, there would be petabytes of data contained here. But if you wanted to find something in particular, you wouldn't know where to begin.

'That's his desk.' Jack indicated it to Tommy, wondering, as he did so, why he was whispering.

Tommy seemed afraid, unwilling to move.

Jack smiled and put his hand on the boy's shoulder.

'Come on, let's put the birthday card where he'll spot it easily.' He led Tommy over to the main work surface. Beyond lay a personal Holo-Theater, now in darkness.

Tommy took the envelope and placed it centrally on the desk top.

'*I*'ve brought a card for him as well,' said Jack as he pulled a small waterproof wallet from the envelope. He broke the seal and extracted the memory card that Al Lynch had prepared specially for him. 'It doesn't matter which storage slot you use.' Lynch had advised.

Jack looked around and the only slot he could see was on a control panel in the rack of old RAID storage drives. He crossed and slipped the wafer-thin card into the slot. 'It will eject itself when it's done,' Lynch had continued. 'I've automated everything, so you just add your message.'

The wall screen came to life and Jack was looking at himself. 'Hi, Tom,' smiled a relaxed-looking Jack, videoed standing on his small balcony overlooking the marina at Hope Town. 'Hope you've had a great time in Moscow. Happy birthday and I'm sure we'll be meeting as soon as you get home. Take care of the planet – you hear!' Jack's image then raised a glass of champagne to the camera, with the sound of 'Happy Birthday' sung by children being played over the slow fade.

Jed started to join in, then Tommy too was singing along. The clip ended as the small voices trailed off.

'Hurrah,' cried Jed, unable to clap.

Jack looked at his storage card, which was still firmly engaged. He carefully positioned himself between the caterpillar and the rack, to block Jed's view of the system. He knew that if Jed's vision sensors captured images of the control panel and its LED status lights, they would realize later that Jack had been copying data.

'Do you want to see it again?' asked Jack, trying to buy time and wondering if he could get a replay without resetting the memory card. At that moment the memory card popped out of its slot, behind him.

'No, it's late, we'd better get back,' Jack corrected himself. 'Come on, Tommy.'

He retrieved the memory card and held the door open for Tommy and Jed.

The elevator required no security clearance to descend and Jack escorted Tommy back to his bedroom.

'Time for bed now, I should think,' he suggested as he held the door open.

'I'm not tired,' protested Tommy. 'I don't need to go to bed. I just rest and dream sometimes.'

'Anything you want before I go?' asked Jack.

'No, thanks,' said Tommy as he set Jed down carefully on the bed. He opened the door to his bedside cupboard and a light came on. Jack saw that he had been provided with his own mini-fridge. 'I'll just have a fruit juice.'

'OK. Well, good night,' Jack said softly, opening the door.

''Night,' replied Tommy as he climbed back into his HydraChair.

'Goodnight, Jack,' called Jed from the bed.

Jack closed the door silently, wondering whether they would talk about him.

Connie was asleep in one of the antique four-poster beds the Tye Corporation's Russian hosts had refurbished for their guests. Like all the senior visitors, she had been allotted one of the giant old rooms of the Terem Palace. Once a waiting-chamber to the throne room, it had long since been converted to a bedroom suite by the installation of a huge private bathroom with antique brass plumbing.

She was currently in REM sleep, her sensorially deprived consciousness feeding on the keywords the DreamDial module of her VideoMate was whispering in her earpiece.

She had gone to bed feeling stressed and then decided she would treat herself. She loved the prompt sequence she had created that reliably produced one recurring dream: she would be at her childhood home with her mother, her sister and their horses, the ScentSim's 'Saddle and Bridle' fragrance always prompting total recall. Then she would be at her high-school prom, dancing until she could dance no more. To follow that Connie had synchronized a fragrance of dark hormones to accompany words she had programmed to induce a dream about the prolonged, hard sex she craved but which, because of the strange life she led, she rarely found. She wished she could find a man to whom she could say such words, and with whom she could create these feelings, images and smells. But in her dreams she did – night after night, whenever she chose.

She was tossing and writhing with a smile on her face when an incoming high-priority override call jerked her awake. At night, all regular calls for Thomas Tye or for herself were routed to the front office on Hope Island where her eight assistants and forty administrative staff fielded the twenty-four-hour frenzy of communication that surrounded Thomas Tye. Only a few people had the facility to get through this barrier.

'It's Pat O'Mahoney in CTA,' announced a voice redundantly after she had identified the caller and accepted the communication. She didn't switch to visual. She couldn't imagine what could be happening in the Competitive Threat Analysis department of the Tye Corporation that warranted her sleep being interrupted. That department of 260 people comprised surveillance staff monitoring the networks for the performance indicators of competitors, intercepting and deciphering encrypted messages that might be important, preparing briefings about possible acquisitions, and gathering information about other companies engaged in markets similar to the many fields of interest of the Tye Corporation.

'Pat, yes,' yawned Connie, as she swung her legs out of bed and tried to focus her thoughts. He was the keeper of Tye's secret and was absolutely trusted. 'What is it?'

'Sorry to disturb you, Miss Law,' apologized the CTA director. 'But we thought you might want to alert Tom. Over the last couple of days something strange has been happening. There's been a massive increase in the number of messages, communications and press stories about us – about the corporation. Plus a massive increase in articles about Tom himself and some of our subsidiaries. The networks are full of this stuff, mostly encrypted.'

'How big an increase?' asked Connie, curious.

'It's shot up to three or four times the usual,' said O'Mahoney breathlessly. He'd spent several hours plucking up the courage to disturb Tom's executive assistant. 'We've never experienced anything like this before, and we can't keep up. There are now large areas we're not able to cover.'

'OK, Pat,' replied Connie. 'Leave it with me. I'll pass this by Tom.'

She ended the call, removed the earpiece and stood up. It was not yet five a.m., although Moscow's summer dawn was already peeping through from behind the heavy drapes. She walked to the old-fashioned pedestal washbasin in the bathroom, ran some cold water and rinsed her face. Then she decided to risk one brief call. The increase in network traffic would be in her favour. She returned to the bed, picked up her VideoMate and told it to connect her to an office just outside Washington DC. As she waited for her call to be answered, she decided that she would not disturb Tom yet with this news. It would keep until morning.

# SEVENTEEN

'Oh, I want to be sick,' gasped Haley as she put the report down. 'That little boy was grown in a box!'

It was a sunny Saturday morning in Battersea and, as before, the report had arrived by mail. She and Barry had just emerged, after an enthusiastic bout of early-morning lovemaking, to celebrate the start of the weekend. Even though most interpersonal communications had migrated to the global networks, the Royal Mail and its upstart competitors still found their daily workload increasing as more and more market researchers proclaimed that printed material remained the most powerful form of direct sales promotion. But apart from the junk mail that arrived each day, there were still enough people who liked to send picture postcards from their holidays, who wrote personal thank-you notes and who used printed anniversary, birthday and greetings cards for the postal delivery to remain a key event of people's morning routines.

This report, the third to arrive unheralded and unsigned, had been posted in Berlin. Jack had informed Haley that whoever was posting these reports chose to do so only from main post offices in major city centres: standard anti-trace procedures. Haley had torn open the envelope immediately and settled down to read as Barry brewed the filter coffee.

'What do you mean?' asked Barry as he reached for the coffee mugs. Because of the heat they were both in T-shirts and shorts, although this sports combination looked better on Haley's frame than on Barry's. It didn't help that his orange outfit clashed with his red hair and beard.

'It claims that Tommy was grown in one of those boxes they use to keep donor organs alive – something called a MatchBox, with an artificial placenta.'

Barry laughed. 'Well, that beats labour pains any day. Even better than a Caesarean.'

'BARRY!' exclaimed Haley. 'I'm being serious. They made a cloned embryo, then grew it in this box. There's even a diagram! That's who Tye's son is. That's Tommy. Jack said he's identical to his father.'

'Who's Jack, then?' asked the Welsh geneticist lightly.

Haley looked up and frowned. She hadn't previously mentioned her new contact – as he had asked.

'Oh, he's just someone inside the Tye Corporation,' said Haley, equally lightly. 'Helps me with research.'

Barry brought the mugs over to the table and sat down opposite as she returned to the report.

'They *harvested* the baby after thirty-three weeks,' she read aloud with a shudder.

'Cool,' deadpanned Barry as he sipped his coffee.

'It is *not* cool, Barry Evans, it's *sick*!' shouted Haley as she slammed the pages down on the table. 'You can't grow someone in a box that you keep on the sideboard! How would that child hear his mother's heartbeat – or external voices, or music? How would he feel movement? How would the baby bond with the mother, for God's sake?'

'Well, it beats putting up with morning sickness,' smiled Barry. 'I, for one, think the benefits of viviparous birth are greatly exaggerated. *And* it would be pretty safe – probably zero miscarriage rate, and you could pop in a bit of compost to help things along.'

'You have no soul, Barry,' complained Haley, getting seriously annoyed with his flippancy.

'And you should bloody well lighten up, girl,' protested Barry, also serious now, his sing-song accent increasing with his emotion. 'It's Saturday bloody morning but it's always Thomas Tye this, the bloody Tye Corporation that. Now it's this wonder child. It's not *your* family. These people are trillionaires who can do anything they want – they live in the future! It's not real life, girl. It's not you and me, a few beers and a fish supper on a Friday night. But you don't think about anything else but Thomas Tye and your bloody book. I would have thought a career girl like you would approve. You too could have a baby in a box and still concentrate on your bloody work.'

Haley drew a deep breath. She knew where it was going now; they both knew where it was going. It had been bubbling for weeks. They weren't really made for each other and out of bed there was little they could share. Barry had seemed sensitive at first, but he really wanted a woman who was prepared to revolve around his life and his needs, not one who had a massive, all-consuming career agenda of her own. In recent weeks his behaviour had been expressing his dissatisfaction more eloquently than any words could. He had taken to spending his evenings in pubs or wine bars with his workmates. On Saturdays it was rugby or cricket and, if they saw each other on Sundays, he insisted that conversation about work should be taboo.

But Haley lived for her work. When she wasn't actually writing she was composing words in her head. She was constantly recording little voice-notes on her VideoMate and scribbling on her DigiPad. Three or four times a night she would wake up and lean over to whisper quietly into the machine on her bedside table, trying not to disturb Barry if he was there. Even in her bath she would read background research, and over meals she would be flicking through her words already committed to paper. She realized this made her an impossible partner, but her real friends knew and understood – although of course they didn't have to live with her. Furthermore, she was an intellectual and an artist; and Barry, despite his doctorate, was neither.

'Time for us to have a little break from each other, I think, Doctor Evans,' she said with a wan smile. 'Just while I get this book finished. Anyway, I need to go abroad again.'

Al Lynch walked around his laboratory, using this period of waiting to exercise his back and leg muscles. He was out of his wheelchair for most of the working day now and he imagined he could feel the nerves regenerating at each end of his spinal cord. He was currently anticipating Ron Deakin's imminent arrival from New York. The ban on network communication was placing a huge strain on both budgets and personnel.

The computer systems Lynch controlled in the UNISA lab at Fort Mead were the most powerful ever developed. In the last few decades of the twentieth century, the NSA had run Cray supercomputers to break codes, sift information and mine into mountains of generic data but, as silicon-based microprocessor circuits had shrunk to the point where their miniaturization had run up against the concrete wall of molecular physics, the NSA and, in turn, other intelligence agencies had pioneered the use of networked optical processors. These computers moved, measured and modified data inside beams of light and did so at a rate thousands of times faster than the supercomputers they had replaced.

Al Lynch had received his specially prepared storage card back in the regular post. Before their newly recruited UNISA intelligence officer had returned to Hope Island there had been much discussion about how Jack Hendriksen should transmit any data collected from Tye's private systems back to the UN facility at Fort Mead. As his covering note explained, Jack had resolved this problem by simply asking one of his pilot pals in the Tye Corporation flight to post the padded envelope for him during a stopover in Washington: pilots stuck together, Al Lynch understood.

In hacker circles the program on the storage card Lynch had prepared for Jack would have been described as an 'invisible tapeworm'. That nomenclature was outdated and referred to computer systems long since obsolete but hackers are a strangely nostalgic and conservative breed. Once inserted into a storage slot, the software could mimic the operating parameters of all known current and obsolete storage systems and, without leaving any trace of its presence, would copy every byte of data it could find in all local storage cards, flash memories and disks and on all storage systems, new and old, connected by common networks. It even collected files that had been 'deleted' but not yet overwritten. The holographic storage system was an IBM prototype: a single plastic card, three inches by two, could store forty-eight petabytes – forty-eight quadrillion bytes! – of data. Lynch found only 27.8112 petabytes on the card Jack had rcturned.

*Only* twenty-eight petabytes! A couple of decades ago that would have represented enough data to fill up the storage disks of over ten million of the most powerful desktop computers! The increasing fashion for videoing every meeting, and storing movies, music, reference books and TV shows on computers, as well as the latest fads of holo-image conferencing, taste-generation and scent simulations, were demanding storage and access technologies on a scale that would have been unthinkable even a few years earlier. And, of course, strong encryption created very large additional files. Fortunately, speed and storage were the types of problems that the computer industry had always been able to solve and, now that the cost of media storage had fallen to less than a penny a terabyte, everybody would hoard *everything*. A form of partial immortality was thus being achieved through virtual storage.

Almost all the major files in Tye's private store of data had proved to be encrypted. 'Suspicious bastard,' Lynch had muttered as he watched screens full of random numbers and letters fly by. He presumed that years ago Tye had selected 'Secure' as the default setting for his communications and computer storage systems. Lynch had randomly dipped into these files in the hope that Tye might have become careless at some period and used a lower level of security, but all his analytical tools revealed encryption so dense that Lynch guessed that Tye must regularly apply the super-long string lengths normally used for military encryption. Lynch knew that even if he ran all his immense optical processing power against these twenty-eight petabytes of data, he wouldn't have decoded a single message from it before he was due for retirement.

The night before, he had copied all the same data onto local flash storage and instructed his system to search only for the same two words in any scraps of unencoded text and images that might lie within the data mountain. Then he had put his jacket on, switched off the lights and walked slowly and carefully to his car. He only needed his wheelchair towards the end of each day now, when his muscles were screaming with tiredness. In another few weeks he hoped to be able to give it up completely, for the benefit of another needy user.

Lynch heard the door open and there stood his friend. The pair had worked together, on and off, for over twenty years and it had been Deakin himself who had first introduced the computer security analyst to Jack Hendriksen.

'Jack told me you were out of your chair and walking,' Ron Deakin exclaimed as he grasped Lynch's hand. 'That's fantastic!'

Lynch poured a coffee for his guest and seated him at the small conference table. 'Here, this is why I suggested you come down this morning.' He pushed the printout of an e-mail message towards Deakin.

The UNISA officer flipped through the pages before settling down to read them carefully. When he reached the third page he started flipping again. Then he skimmed through the remaining sixty sheets.

'I extracted everything from Tye's files that was in plain text or unencrypted video,' explained Lynch. 'There was over a terabyte of it. Then I searched for two key words: "encryption" and "decryption" plus all possible stems, suffixes and contractions. This e-mail printout was the only hit.' He looked up. 'And the next unencrypted communication was this one...'

He pressed a remote on the desktop and the wall screen lit. Deakin saw a young man with unruly fair hair.

'Mr Tye.'

'Doctor Larsson?'

The young man nodded.

'Can you verify, please?'

'Let's both do it.'

Deakin watched as the man called Larsson leaned forward and touched the fingerprint pad on his system. There was a few seconds' delay and then Deakin saw the message:

*Identity of caller confirmed as Rolf Linquist Larsson, born 13 January 1980, Stockholm, Sweden. Present location Järntorgsgatan 1–3, Stockholm, Sweden. GPS location 57.6042°N, 17.1619°E. Identity Certificate on file.*

'Please encrypt, if you don't find that too funny,' said Tye.

Deakin saw Larsson nod and then the screen dissolved into white noise.

Deakin whistled. 'Why would he find the idea of encrypting *funny*?' he asked.

He picked up the text of the message that had preceded the videoconference and read it again with more care, this time paying greater attention to the thick wad of pages at the back of the message. As he read, Lynch rose, walked to the coffee percolator and refilled his friend's coffee mug. He had a huge smile on his face; not just because of the messages he had found, but also because the simple act of standing and walking to the coffee machine still gave him the most exquisite pleasure. He knew the incredible mental processing and feats of mechanical engineering that are required to keep the top-heavy human body upright, to move balance from one ten-by-three-inch platform to the other; to control the thousands of muscle, bone, temperature, blood, oxygen and sensory inputs, outputs and parameters required for such an apparently simple process to occur. He wondered how long it would be before he became blasé again and took his brain's astonishing information-processing capabilities for granted.

After a few minutes Deakin put the papers down.

'Want a replay?' asked Lynch. Deakin nodded and they watched again as Thomas Tye began his videoconference with an unknown Swede in Stockholm at three a.m. local time, over seven years earlier.

When the white noise reappeared Deakin picked up the printout again. 'So this Rolf Larsson sends an e-mail to Thomas Tye that says "New software. Want to discuss?" and with it he sends these Tye corporate annual accounts and these memos about marketing plans and office relocations – all in plain text. Then Tye calls him back and thinks Larsson might find it funny to encrypt their conversation. You've definitely got something.'

Lynch smiled. 'There's one more thing,' he added. 'I did some checking. Those accounts are getting on for eight years old. See the date on the e-mail message?'

Deakin flipped back to the top page and nodded.

Lynch pulled a folder towards him and took out a glossy

document. 'I borrowed this from the library,' he explained as he handed the booklet to Deakin. 'This is the annual report and accounts that the Tye Corporation did *actually* publish that year – just over two weeks later! They've been tweaked a bit here and there; the bottom line is different with more money in a forward contingency fund, but they're basically the same accounts.'

Deakin put the booklet down. 'Do your men friends kiss you often?' he asked with a laugh. 'I suppose you've traced this Larsson. Don't tell me, he's outside the door waiting to help us.'

Lynch grinned. 'No, he's not. But I know who he is. He's a maths genius. Honours graduate in pure maths from Stockholm University at age seventeen, first doctorate in pure maths when he was twenty-one. A nationally acclaimed prodigy. He was working on a second PhD in quantum mechanics when he had that conversation with Tye. Since then he's disappeared.'

'Disappeared?'

'Well, from the networks,' Lynch explained. 'There's not a trace of him. Even his entry with the World Certification Authority has lapsed. It looks like he didn't finish his second doctorate. There are no publications, no Web references, no university citations or news, no conference presentations, no guest lectures – nothing.'

'What does his university say?'

Lynch held up his hand. 'Whoa, Ron! You know I don't do legwork,' he said, grinning.

Furtrado snapped off the holo-image and uttered a particularly graphic Portuguese oath. He slumped back in his chair and wondered how best to break the news. Tom had been on a high on his way back from Moscow but this morning, the day after, when they should have been laying plans for the public announcement of the new state of Sybaria and the launch of the Phoebus Project, he had been in a foul mood.

'Get Jack Hendriksen in here *now*,' the tycoon had shouted at Connie without further elaboration.

Furtrado had made an excuse for getting back to his office

along the corridor. He knew his boss well enough to steer clear until this storm had blown over.

The lawyer looked down at his notes and wondered if the news could keep. It couldn't, so he pushed himself out of his chair and walked back to Tye's suite. He crossed the outer office with its gaggle of male and female executive assistants all dictating or involved in conferences, both physical and virtual, and entered Connie's office. As he did so Jack Hendriksen emerged from Tom's room, nodded pleasantly and sauntered out of the office in his peculiar, lithe way.

Connie was involved in her viewpers and she waved Furtrado through – there were only rare occasions when Tom's office door was closed to his most senior counsellor. As Furtrado stuck his head inside, Tom was standing by the picture window, staring out over the Atlantic ocean. The windows in the suite had darkened to eliminate the worst of the sun's glare. Tye heard Furtrado enter and he turned.

'Do you trust Hendriksen?' he asked abruptly.

Furtrado considered. 'Trust in what way?' He couldn't think that the security chief would be involved in any of the corporation's commercial deals. Hendriksen struck Furtrado as being a fit, watchful man, something of an outsider in the company, but the sort you would want on your side if there was ever any real trouble: a little dangerous, perhaps.

'He broke into my office upstairs while we were in Moscow,' said Tye quietly.

Furtrado whistled. 'How the hell did he do that?' he asked.

'Said it was part of a security review, did it to prove my security is no good. Swam up my swimming pool drain. Helped Tommy write me a fucking birthday card!' Tye picked a card up from his desk and waved it. 'And now he's given me these brochures advising on new household security systems!'

Furtrado had to stop himself smiling. He knew that even Tom wouldn't dare loose his temper on Hendriksen. He wished he had been present during their meeting.

'Take a look at him,' said Tye. 'He could be selling information.'

The lawyer nodded. He knew everything Tye did would be fully encrypted and he doubted that Hendriksen was a spy. It sounded more as if Tom was being taught a sharp lesson.

'There's some news from New York,' he ventured. 'Some outfit called Sloan Press has contracted to publish that libellous British book – the unauthorized biography.'

'Just swamp it in legals,' ordered Tye distractedly.

'The New York attorneys have done all that,' explained Furtrado. 'They've got twenty-one injunctions, but it looks like Sloan Press is prepared to publish and meet us in court – then the media will get it.'

Tye's attention snapped back. 'Who the hell is Sloan Press?' he asked. He pondered for a minute. 'Buy them, Marsello. We can always sell them on once we've dealt with this.'

'They're privately held, Tom,' explained Furtrado. 'They've turned down two good offers in the last year.'

'THEN BUY THE FUCKING BOOK,' shouted Tye, now getting seriously irritated. 'It will be cheaper and faster than doing the legals. Go and see this English writer, and buy her off.'

The housing estate was new – still under construction in some sections – but it looked like a safe place to raise children. The large lake, an inland extension of one of the many watery fissures that penetrated the Swedish mainland, provided a natural focus for the horseshoe-shaped community of houses. Children played in the streets and on the grass at the edge of a gently sloping lake shore. Chevannes smiled at the setting of sunshine, peace and safety, automatically comparing it to the dangerous Jamaican wasteland of the Kingston yards where he had spent his early years, before his family had won an immigration lottery and moved to America.

He had to ask twice for directions because these streets were so new they had not been included on the map he had bought at the airport in Stockholm.

He knew the woman's married name and he carried a wedding picture that was only three years old. He hadn't called ahead, partly because all case-related network communication was banned but mainly because his experience had taught him that better results were usually achieved when questions were asked without advance notice.

Chevannes identified the house, walked up to the front door and found it ajar. He pressed the bell push and waited. There were sounds from within and then a pretty woman with flyaway blonde hair appeared. It was definitely her. She was wearing a white blouse and loose-fitting tan trousers.

'Mrs Astrandh? Laila Astrandh? Formerly Laila Hagstrom?'

Laila smiled at the well-dressed man on her doorstep. 'What can I do for you?' she asked in the near-perfect American-English common to most Scandinavians. He realized he wouldn't need to use auto-translation and, not for the first time, reflected that it was a shame such devices were rapidly reducing the need to learn foreign langauges.

Chevannes showed his badge and watched as the housewife unclipped a VideoMate from her narrow belt and ran his ident signal past the World Digital Certification Authority. She smiled when she received confirmation and handed back the card.

'What can I do for you, Officer Chevannes?' she asked. She pronounced his name in the correct French way. There was a child's cry from within the house. 'Come in,' she invited. 'You'll have to excuse the mess.'

Chevannes followed her into a bright living room cluttered with children's toys. A small girl with golden curls stood in a corner telling off a Furry. Chevannes didn't understand her Swedish.

Her mother reproached the child gently, stroked the berated Furry and then sat with her daughter on her lap. She gestured for Chevannes to sit on the couch opposite.

'This is Aya-Karin. Say hello to Mr Chevannes,' Laila told her daughter in English. The child's pale blue eyes fastened on the elegant visitor. She half smiled and then nuzzled her face into her mother's blouse.

Laila said something else in Swedish to her daughter and the child turned her head towards Chevannes again. 'Hello,' she said with a bouncing intonation.

'Hi, Aya-Karin,' replied Chevannes in what he hoped was

his softest tone. He waggled his big fingers at her. The girl just stared at him and then looked up at her mother.

'I'll take her next door for a minute,' said Laila, rising with the child in her arms. She started for the door and then stopped when the Furry called out something after them. She returned and bent over so her daughter could scoop up the green rabbit. With a dazzling smile in Chevannes's direction, Laila left the room with both daughter and Furry.

The intelligence officer took in his surroundings. He had never been inside a Scandinavian home before. Even the most ordinary of objects seemed designed with flair: the bay window had a curve at the top that made it resemble a church window. The natural-stone fireplace doubled as a divider between the living room at the front of the house and a dining area at the rear. Despite the clutter of toys, the place was clean and comfortable. Chevannes sat back in the sofa enjoying this brief sojourn in what he might imagine was true domestic bliss.

'I left her with my neighbour,' Laila explained as she re-entered the room, pushing a lock of fair hair back behind one ear. 'So what can I do for you?' she asked as she sat down again. 'What does the United Nations want in Solna?'

'It's about Rolf Larsson,' said Chevannes, coming directly to the point. 'Do you know where he is currently?'

Laila raised her fair eyebrows and pursed her lips. She didn't seem overly surprised. 'No,' she said finally, 'I don't.'

'You two were ... close friends,' prompted Chevannes. 'Or so I was told at the university.'

The Swede smiled at his propriety. 'He was my boyfriend, for two years,' she confirmed. 'But I haven't seen him or spoken to him in over five years.'

'Did he ever have any dealings with Thomas Tye and the Tye Corporation back then?'

Laila smiled again and nodded. 'That's what changed him. He sold something to Thomas Tye – and made a lot of money. It completely messed him up. After that deal he began throwing money around as if he was insane. At first he went off on the star trail – you know, the Greek Islands, Thailand, China, New Zealand, Peru.'

Chevannes shook his head and shrugged. He didn't know.

'The places where you can view the stars best, away from urban lights and the main satellite networks,' explained Laila. She sighed as she recalled: 'At first it was all very professional. The best portable telescopes, cameras, computers and so on. Then ... well, we were staying up all night, sleeping all day. He started using drugs and stopped bothering with his telescopes. He'd just lie there on the beach or on a mountainside and stare up at the stars all night, every night. He seemed very unhappy in spite of all the money.'

She tailed off – back there with him again, in the time before she had met her husband Benji.

'We were only twenty-seven,' she offered by way of explanation. 'You know his parents had died in a boating accident three years before?'

Chevannes nodded. That was why she was the second stop on his visit, after the university. 'Do you know *what* it was he sold to Thomas Tye?' he prompted.

Laila shook her head. 'He wasn't allowed to talk about any of it. He said he'd signed a contract that forbade him to discuss it. After a few months he started ranting about how it had ruined his career. He'd just lie on his back and stare up at the stars, and go on and on. That's when he really started on the drugs and drinking.'

'What was he working on before he sold this thing to Tye?' asked Chevannes.

Laila had been examining the hands folded in her lap. Now she looked up. 'Would you like some tea or coffee?' she asked.

'That would be nice,' smiled the intelligence officer. 'Coffee would be good.'

'Come into the kitchen, we can talk there,' she said.

Chevannes followed her into a light and airy room, more of a conservatory, with a large informal eating area and another open log fireplace.

'He was a mathematician and an astrophysicist, not somebody in business,' Laila explained as she snapped the kettle on. She turned to face him, leaning back against the work surface. 'I think it was some sort of mathematical formula, or some sort of software. He kept it all to himself.' She hesitated. 'To be truthful, I wouldn't have understood it anyway. He was highly gifted mathematically, but to me it was all a foreign language.'

'You speak at least one foreign language very well,' said Chevannes, rather taken aback by his own gallantry. 'I spoke earlier to his supervisor at the University who said that one minute Doctor Larsson was working for a second PhD – in particle physics – then he was called away to do jury service. He never returned to the college, and he never explained why. The professor wasn't very happy to be reminded of this. He thought Doctor Larsson had a unique brain and could easily have become the youngest full maths professor in Europe. It seems he represented a massive loss to the university, and they were furious at the time.'

'I don't think he ever did any jury service,' recalled Laila as she spooned coffee into two mugs. 'He just cut himself off for a few weeks and then he went to visit Thomas Tye. The company even sent a private jet to collect him. When he came back he couldn't get over talking about how rich he was. He claimed he'd been paid billions of dollars, but at the time I rather thought . . . well, *Je ne l'ai pas pris au pied de la lettre*.'

Chevannes looked at her, not understanding.

She smiled. 'I'm sorry, what's an equivalent saying in English? I didn't take it literally. I mean, I took it with a . . .'

'. . . A pinch of salt?' guessed Chevannes.

Laila nodded. 'But he *did* have a lot of money. We travelled everywhere first class, stayed in the most wonderful hotels . . .' She tailed off as she poured the boiling water, then stirred the granules absently. 'Milk?'

Chevannes shook his head, and they both sipped their black coffee. 'When did you last see him?' he prodded.

'It was about eighteen months later. We had continued to travel and he had got hooked on some really strong acid in Northern California – a special form of hallucinogenic developed in some remote lab near Mendocino. I tried the stuff and I didn't like it. But Rolf said it allowed him to set the motor free.'

She tapped the side of her head. Chevannes nodded.

'He would binge for a week at a time – totally out of it –

then stop completely for a few days. Then he'd start again. Eventually we ended up at some exclusive resort island on the Great Barrier Reef. We had this giant villa on stilts down on the beach. The domestic staff came over from a nearby island three times a day to leave food and change the laundry – not that Rolf cared about food or laundry by this time. I decided I couldn't take any more. I took the boat across to the main island one day while he was still passed out and caught the seaplane out – to Cairns on the Australian mainland. Then I headed down to Sydney and bought a ticket back to Stockholm. Just before I got on the plane I called the police and told them where Rolf was and about the drugs he had with him. He had brought everything from California – I think the stuff was too new to be categorized as really illegal. I hoped they'd get him some medical help.'

'And that was the last you heard of him?' asked the intelligence officer.

'Directly, yes,' admitted Laila. 'I've never spoken to him since.'

'But you have heard from him?' prompted Chevannes gently.

Laila sighed. 'I've never told Benji – my husband.'

Chevannes nodded. His job had long since proved that everybody had their secrets. 'We only want to talk to Mr Larsson. This isn't about you or your family.' But he couldn't promise that her secret would be safe until he knew what it was.

Laila nodded. She'd already said too much to hold back now. And she'd been waiting for this to emerge for a long time.

'He sends me money,' she said. 'Every month, to my old bank account. It's in my name before I was married – Hagstrom. It's quite a lot and I'm saving it for Aya-Karin.'

'Where does he send it from? Have you got any transaction records?' asked Chevannes.

Laila shook her head. 'It's all done by direct bank transfer. From a bank in Geneva – in Switzerland,' she added, as if a Jamaican-American might not know where Geneva was. *A sensible precaution with any American*, thought Chevannes.

She rose from the table and opened a drawer under the work surface. She rummaged through a pile of papers and extracted one sheet. Laila tore the header from the page and returned with it to the table.

'Here's the address of my bank and my account number,' she said as she sat down. 'They'll give you all the details. Tell them to contact me and I'll authorize them to give you the information you need.'

'And that's all – just the money?' persisted Chevannes.

Laila nodded. 'I thought someone would come eventually,' she admitted as she picked up her coffee mug again. 'Will I have to give it back? I've kept every bit of it since it started arriving. There's over thirty million dollars.'

'I'm afraid you've had a wasted trip,' said Marcia Fernandez, the chief administrator of the Sandler Cosmetic Clinic. 'There is absolutely nothing in our files that predates nineteen eighty-seven. We've been asked about that before.'

Haley allowed a sigh of disappointment to escape her lips.

'We do get enquiries about Thomas Tye from time to time,' offered Ms Fernandez. 'But we're a different sort of clinic altogether now. I understand the building was almost completely gutted before it was renovated. Then the Tye Foundation sold the property on to our organization.'

'Any contact with the previous patients?' asked Haley.

The woman looked at her quizzically. 'It wasn't exactly the type of place to run an alumni programme, Miss Voss.'

Haley's irreverent and unruly imagination instantly assembled a fantasy scenario of former mental-asylum inmates attempting a reunion. What would they say to each other? She wondered again what on earth could have been so wrong with the five-year-old Thomas Tye to warrant his committal there.

'Are *any* of the original medical staff here?' Haley persisted, as she pulled her thoughts together.

This produced a chuckle. 'No, but I sometimes think I could use them,' smiled Marcia Fernandez. 'This might now be a cosmetic surgery clinic but we do get some strange types in, if you know what I mean.'

'What about janitors, non-medical staff? Or administrators, secretaries, accountants?'

'Miss Voss, it's nearly *thirty years* since my corporation bought the building. There's no one here from those days. I'm really sorry…'

Haley put her coffee cup down, then rose. 'Well, thanks for making time to see me,' she said as she halted her data capture.

At the end of the long gravel drive, Haley turned right towards Springfield and the I90. Her adverts among the Thomas Tye communities had elicited a particularly interesting response from an individual in Philadelphia. She had a long drive ahead.

Even though they had chosen a region that was not covered by any of the surveillance-satellite networks, they realized that sooner or later the scar would be noticed and, in due course, investigated. Accordingly, they designed the test so that the affected area would look like a long teardrop, suggestive of a gouge caused by a low-angled, burning descent. They also knew that once soil investigations began any meteor theory would be quickly discounted. But, given the absolute remoteness of the Western Amazon basin, they calculated that any such examination would happen many months later, long after their new service was launched and delivering its phenomenal benefits. Marsello Furtrado, the only native Brazilian with knowledge of the project, predicted it would probably be years before the Ministry of the Interior got round to sending an expedition out to such an inaccessible region.

There had been arguments within the team about destroying even a comparatively small area of such an important ecological environment but, as the Director of Solar Focus reasoned, the Amazon basin occupied over two and a half million square miles and the relatively small area they planned to affect would regenerate naturally and regrow with the benefits of the invigorating, fertilizing nitrogen that would be released as acres of rainforest were turned to ash. The trial would also test the first-phase energy network at its most extreme range and Tom was adamant he wanted the data on what was possible at southern latitudes.

So, on the same evening during which Haley enlivened her long drive across country by flicking through the channels on the satellite radio system in her rented car, eight square miles of Brazil's vast tropical rain forest started to experience intense nocturnal sunshine. At first the daytime creatures of the forest re-emerged after a severely truncated night and resumed their activities, briefly meeting some of their nocturnal cohabitees as the latter grudgingly sought their shelters long before their usual order of business was complete. Birds sang and epiphytes near the top of the tree canopy opened their buds. But the heat continued to rise and, as the cloud cover sweeping down from the foothills of the Andes burned off, the continual dripping at the forest floor began to cease for the first time in over one hundred million years.

Then the entire section of forest began to steam and a mist filled the gaps between the trees. For a few hours the dense tree canopy managed to protect the abundant and multifarious life forms beneath but, shortly after midnight, the highest layer of foliage started to crackle and burn beneath the tightly focused and magnified output of all twelve energy stations. The steam swiftly evaporated and, three hours before dawn, the lachrymiform target area was ablaze, its dehydrated, flash-baked foliage, liana and fauna erupting almost spontaneously as fierce concentrated beams seared through the burned-out tree canopy and reached to the ground.

Watching on monitors fed by the Tye Argus satellite network, the DSF ordered a cessation shortly before dawn. As planned, the winds from the foothills began dispersing the smoke and the soaking wet of the surrounding jungle contained the conflagration whilst itself suffering only minimal collateral damage.

Thomas Tye's image appeared on the monitor. 'Congratulations, Doctor,' Tye greeted his Director of Solar Focus. 'You will ensure no recordings of the test are filed, won't you?'

The small, highly paid and well-trusted team clapped enthusiastically. Their measurements, received from ground-level calorimeters that they had parachuted in, had transmitted results far better than expected before the instruments too were engulfed in flames.

*

Haley had arrived in Chalfont, a northern suburb of Philadelphia, just before midnight. She had prebooked a motel room and had found the inn, situated on the main street, with ease. Her appointment was not until eleven a.m. the next morning, so after a lie-in she had treated herself to an American breakfast – *where they really know how to cook eggs, bacon and hash-browns*, she thought as she enjoyed it.

As she checked out she asked directions from the desk clerk. He recognized Miss Hattie Jones's address. It would only take her five minutes. Haley took her time loading up the rental car and then, on a bright, sunny Wednesday morning, she set off to visit the woman who had responded to her advertisement for fresh information about the great Thomas Tye.

Understandably, her correspondent had been reluctant to disclose full details without discovering what was on offer. She claimed to have documentary information about Thomas Tye's very earliest years, the period before he was institutionalized. In response to Haley's e-mailed request for further background, Hattie Jones had written simply, 'I worked at a clinic which treated Thomas Tye as a baby and I kept a copy of the records.' Nothing Haley could say would get the woman to offer further evidence of her claim, and the old lady – Haley worked out that she had to be at least seventy years old – had refused even to name the institution to which she had referred.

Haley found the street, then the house, without difficulty. She parked and walked up a short path to the low, single-storey residence. The other homes in the street looked respectable but, like the paintwork on this property, gently faded. She rapped on the wooden frame of the mesh outer door. When there was no answer, she tugged at the mesh door and it opened easily. Extracting her car keys from her purse, she tapped on the glass of the front door beyond. In a few moments she spotted movement from the hallway within and she allowed the mesh door to close once more. The inner door opened and a small, birdlike woman was looking up at her.

'Miss Jones? I'm Haley Voss.'

The woman pushed open the outer door and extended her hand. 'Miss Voss, and you're on time. So polite of you.'

Haley smiled and took the wizened hand. She estimated that this neatly dressed woman was probably well into her eighties.

Hattie Jones stepped back into her hall, allowing Haley to enter. She ushered her into a small living room and gestured to a Victorian grandmother chair that needed restuffing. 'Sit down, sit down. The kettle's boiled, I'll make tea. My mother came from England. I know you probably haven't tasted a decent cup since you arrived.'

Haley smiled her thanks. That was true.

'Don't realize the water needs to be boiling to split the leaf, that's the problem here,' said Miss Jones over her shoulder as she disappeared.

Haley took in the clean but faded room with its vaguely musty smells. She switched her VideoMate to record and put on her clear-glass viewpers. Then her host reappeared with a tray that looked far too heavy for her.

Refusing Haley's help, she lowered the tray onto a low table, then sat down and began to pour.

Haley responded to her polite questions. No, she hadn't been to Chalfont before, nor even to Philadelphia. Yes, she had visited the States before on several occasions.

'I know you're a successful writer,' said Hattie Jones with a grin. 'I've looked you up!'

Haley smiled back. 'Well, I hope this will be my most important book so far. My publishers seem to think so.'

She was wondering how to prompt the woman to get down to business, but she needn't have been concerned. Miss Jones rose suddenly, crossed to an old mahogany sideboard, pulled open a drawer and returned with a thick buff file.

'So how much is my reward?' she asked, suddenly very businesslike.

'Well, that depends,' replied Haley hesitantly. 'I mean, it depends on what information you have to offer. I have a research budget, of course, and I . . .'

'I understand.' Miss Jones nodded. 'You need to know what I have here. My nephew is a lawyer, over in the city, so I asked him about the best way to handle this.'

She extracted a crisp white sheet of paper from the file and handed it to her visitor.

'It's a non-disclosure agreement, Miss Voss,' explained the old lady as Haley studied the document. 'Very simply, it allows me to show you what I have, but you aren't free to use any of it until we have agreed terms. How does that sound?'

It sounded very fair, and suddenly Haley's heart raced. If this woman's lawyer nephew considered the information sufficiently valuable to draw up this agreement, then it could be something very special.

She smiled again. 'This looks fine.' She started to rummage in her bag, then looked up to find Miss Jones offering her a pen.

Haley signed and dated the document and returned it. The old lady folded it, neatly inserted it in an envelope and placed it on the sofa behind her.

'So,' she said. 'Where to begin? Tell me, have you heard of Professor Charles Eon?'

Haley searched her memory. 'No ... I don't think so ...'

'The sex doctor,' prompted Miss Jones. 'He was always on TV.'

Haley searched again, but shook her head.

Hattie Jones shrugged. 'No, well, I suppose it was long before your time – back in the nineteen-sixties and -seventies. And poor Charlie was best known here, in the States, of course.'

Haley nodded, waiting.

'Charlie died last month, Haley – may I call you Haley?'

Haley smiled her consent.

'He was nearly ninety-five, and died a week before his birthday. There weren't many at the funeral service.' They was a silence as the old woman thought over those events. There she pulled herself back to the present. 'Charlie was the Professor of Psychosexual Medicine at the University Hospital,' she explained. 'He founded the department in the 1950s, and I was his secretary for thirty-two years.'

Haley smiled again, encouragingly.

'I'm afraid some of the things we did ... Well, Charlie was

sure they were for the best at the time...' Now she looked embarrassed, almost mortified. She folded her hands on top of the thick folder.

'Well, that was a long time ago. Later on, Charlie sort of retreated from the world. I always knew he had a good heart, but the press were very cruel...'

Haley waited as the woman fought some inner battle.

'I always swore I wouldn't do anything with this–' she tapped the folder '–while Charlie was still alive. But now...' She tailed off again, looking down, then up again, some internal decision made. 'But it happened a very long time ago, nearly fifty years.'

Haley nodded, silently screaming for Hattie Jones to get on with it.

'Like I said, I was his secretary – well, secretary to the whole clinic, really. I had to type all the old files into the first computer they installed, and it took me months. I was instructed to trash all the paper files when I was through, but I didn't trust that machine so I brought them back here for safety. Never did need them, though, I must admit,'

She changed tack. 'I remember the little boy, he was very pretty. But he was a classic case, like most of the others that were referred to Charlie. He's become so famous now, hasn't he?'

Haley nodded once more.

'I'd like to think it might help him ... well, deal with things, ... if his case was known. He's never married, has he? So many of them got over it better once it was all out in the open. Some even got married...'

Haley wanted to scream out loud. She wanted to tear that file off the old lady's knees.

Hattie Jones looked down at it again. Suddenly she thrust her hand into the folder, withdrawing a pile of yellowing papers and black-and-white photographs, and laid them on the cushions beside her. She leaned forward and handed the empty file cover to Haley. 'Let's start with this,' she said.

Haley took the empty file with shaking hands and read the printed headings and the handwritten entries:

Family Name: *Tye*
Forenames: *Thomas, Richmond*
Date of Birth: *07/01/66*
Date of admission: *04/07/67*
Putative sex: *Male*
Assigned sex: *Female*
Assigned forenames: *Thomasina, Rachel*

The former secretary lifted the pile of paper records and photographs from the cushions and placed them on her lap.

'Now, what am I bid?' she asked with a sweet, old-lady smile.

# EIGHTEEN

Aerospace

Rep. Robarts Calls on White House to Intervene In Tye Aerospace Experiment – 'Satellites Could Be Weapons'

By a WALL STREET JOURNAL Staff Reporter

WASHINGTON – In a stinging attack on what he called 'the failure of White House foreign and defense policies,' House Minority Leader Ronald Robarts today accused the White House, the Pentagon and NASA of suppressing research which, he claims, has identified a chain of deep-space location satellites launched by Tye Aerospace, Inc. as having weapons potential.

Addressing a House Judiciary Committee, Mr Robarts showed what he claimed was an edited extract of a Top Secret NASA/CIA report and urged the President and Secretary of Defense to be honest with the American people about the threat to the nation's security posed by this chain of distant satellites.

'NASA knows these aren't simply a positioning system for the solar system but they won't tell the people what they do know,' he said. 'There are fourteen deep-space vehicles in orbit above the northern hemisphere – all of them in solar-stationary orbits. Twelve of them are deployed above the night side of the Earth's northern hemisphere, two are positioned sunwards of our planet. I challenge the White House and the Department of Defense to publish the full text of this report. These things could be weapons – the people have a right to know.'

Following the hearing, Mr Robarts's spokeswoman played down the more sensational side of his speech. 'We're not claiming that these are a definite threat, we are simply asking for the facts to be disclosed,' she said. 'We want to know why the White House has remained silent on this topic.'

Tye Corporation's public relations agency in New York professed amusement at the congressman's accusations. 'The Tye Corporation is not in the defense business – period,' said a spokesperson. 'We are creating a network of satellites which will provide a navigation and positioning system for research purposes and future missions. The Aerospace Division has many Earth-orbit and deep space experiments, all of them peaceful in nature. We welcome all enquiries.'

Mr Robarts, a long-time champion of the American defense industry, is seeking nomination as the Republican party presidential candidate in next year's election.

Ron Deakin snapped the screen off. He could recognize a planted story ten miles off. This was *getting political*!

There was a long-standing statute in Swiss law decreeing that if a police officer, Customs investigator, intelligence officer or other official from a foreign power (with or without diplomatic status) was found on Swiss territory asking questions about the provenance, ownership or status of a bank account, deposit box or any other financial or property instrument, he or she would be summarily deported and thereafter denied access to the nation. It was an effective deterrent to inquiry.

In recent years, under pressure from other members of the

European Union, the World Bank and foreign governments, the Swiss had paid some lip-service to improving financial transparency and drug barons could no longer be sure of the absolute discretion they were once promised. But apart from this concession to its membership of the EU and its obligations to the international community, it seemed that the Swiss still resisted real change. Keeping and protecting wealth and its privacy, on behalf of all those who were able to pay for the privilege, was their business and it had been so for many hundreds of years. The Swiss would no more readily abandon their economic *raison d'être* than would the Arabs their oil, the Belgians their chocolate or the Scots their whisky.

Michael Chevannes broke through this veil of state-mandated secrecy. It took him a week to do so and the one-time coded fax transmissions involved Ron Deakin in New York and Dr Yoave Chelouche in Geneva. Finally, one early-morning phone call from the world's chief banker to the home of the president of Premier Security Bank Swiss succeeded in negating a tradition of banking secrecy that had begun with the Edict of Nantes in 1685. *Perhaps the country is becoming more responsible to the world's financial community*, thought Chevannes later the same morning as he examined the many accounts and the piles of paper correspondence of Rolf Linquist Larsson, one of the bank's most wealthy and secretive private depositors.

Altogether, Chevannes found the equivalent of over twenty-three billion US dollars in 143 separate accounts, bonds, property companies and shareholdings. He also discovered the two main alternative identities used by the depositor. Interestingly, Larsson kept the majority of his vast investment portfolio in the Tye Corporation and Tye-related stocks. As Chevannes looked back over the six-year history of the accounts he shook his head and muttered to himself in amazement. The Swede had quadrupled his fortune simply by placing his faith in the fortunes of Thomas Tye's companies.

When he had finished his detailed note-taking – despite his privileged access, photocopying was not allowed – he left the bank and booked himself a one-way ticket to Lima, Peru.

*

'I've never been on a boat before,' exclaimed Tommy. 'This is fun.'

'This is fun!' echoed Jed. 'It's super!'

'Super,' shouted Tommy in turn. He was picking up Jed's English accent and superlatives.

As they passed the outer limits of Hope Town Marina, Jack eased forward the levers controlling the throttles of the mighty GE turbo-diesels. The huge motor yacht lifted up to its full planing height as they gathered speed.

A light spray flew into the high, open flying bridge and Tommy laughed with joy as he clung to the safety rail, Jed clamped firmly under his arm. Despite the size of the craft and Tommy's swimming prowess, they had insisted he wear a bright orange life jacket. He had asked for one for Jed too but nothing could be found that was small enough. Calypso had found a solution by borrowing a lifebelt-style cup-holder from the bar on the diving deck. She had deftly fitted this around the caterpillar's middle.

Tommy's father had put up suprisingly little resistance when Calypso had reintroduced the idea of the trip. For the first time since she had known him, Tom seemed at a complete loss.

'I shall always be in your debt, Calypso,' he had begun.

She had already discovered that apart from the severed artery and some lesions to the surrounding tissue, Tommy's forearm had suffered no significant damage. She and Emily Pettigrew had managed to complete the procedure and close the boy's arm in less than half an hour.

'That's what I'm trained to do,' she had told Tye simply.

As he looked at her, she thought she could detect tears. 'My mother committed suicide, Calypso,' he told her sadly. 'I'm sure Tommy hasn't inherited the gene, but...'

Calypso now knew why Tom could feel so sure about Tommy's genetic make-up, but he seemed absolutely deflated by his son's recent actions.

'Do you see any signs of depression in him, Doctor?'

'All self-harm is a form of depression, Tom,' she said gently.

But Calypso had laid down tough terms for continuing as the boy's protector on the island. She had demanded twenty-four-hour access to the child, had asked for a bedroom to be made available for her close to his, insisted that he be allowed to attend the school again, and she made it mandatory that he should be allowed to make supervised visits to other homes on the island and, under special circumstances like this, even off the island.

She had then read Thomas Tye the accepted lore on juvenile psychosis, as laid down by the paediatric psychiatrists Walsh and Rosen, the ultimate authorities on self-injury in childhood. She had explained how Tommy's actions were indicative of a deeply buried depression and that without treatment the condition would deepen even more. And she had warned him where that would lead.

'Eventually to a suicide attempt that *does* succeed,' she had said simply but firmly, her voice ringing off the marble walls as she sat on one of the vast white sofas in Tom's reception hall. 'The youngest recorded suicide to date has been an eight-year-old girl – another isolated child – in Los Angeles. She was the daughter of one of the world's foremost female movie stars, and she started self-harming at only six.'

Calypso saw a strange panic on Tom's face, and pressed her attack home.

'Yes, he has everything to live for, but because of the way you overprotect him, *he* feels he has nothing. A child's values and sense of self are gained not only from parents and guardians, but more especially from others around them – primarily their peers. That's where a child learns his or her place in the world.'

Tom had at last started to remonstrate. Again he had protested that Tommy was unlike other children. Calypso understood what he meant but to her he was simply Tommy, and she was determined that she would now focus only on securing the boy's happiness. This, she realized, was more important to her than anything. Even more important than her growing fondness for Jack Hendriksen.

Finally, she had held her hand up. 'I could not care less if he was the last of a royal line,' she announced. 'That boy will only become a happy person *if* he is allowed to experience a normal childhood. If you want him ever to be proud of who he is, also proud of you and able to play a useful part in managing this incredible empire, you have no choice but to let him find himself in his own way first. That's how humans work, Tom. Believe me.'

So Calypso had moved into the great house, and into a room next to Tommy's. Gradually she had helped him talk about the incident.

'It felt like a dream, Calypso,' he told her. 'I thought I'd never see you again. I didn't know what I'd done wrong, and nobody would tell me where you were.'

She had hugged him unreservedly. Now that she knew his true origins, some of her worries about the normal doctor-patient relationship could be discarded. He was not like other child patients, as he had no 'natural' parents. There was no precedent in medical history for how he should be treated. Therefore Calypso allowed herself to run on instinct. Far from being repelled by the boy's scientific genesis, she loved him all the more. She knew clones were a phenomenon of nature – in identical twins, for instance – and she could see clearly what had caused his problems: they were all to do with nurture and little to do with nature. She had prescribed a mild antidepressant for Tommy, although within two days of her moving into the house he seemed to exude an abundance of energy. Although keeping the perception to herself, she felt sure that it was her own presence in the house that restored his happiness so quickly.

She had decided to test Tye's acceptance of their new relationship, so one day she had simply informed Connie Law that she was going to take Tommy on a visit to Mayaguana, to see her mother, a week hence. She then sat back and waited for another explosion. It never came.

'Look after him well – and enjoy the trip,' was Tom's surprising response in a live video call that appeared to originate in Ethiopia.

But he had followed up his permission with a mandate that

Jack, Pierre and Stella should join them for this excursion and had stipulated that they should take his personal power yacht, not the usual helicopter.

'Doctor, it's a matter of odds. The statistics for helicopter crashes are appalling on a per-mile basis,' he said with monumental insensitivity, 'So I insist you take *Hope's Dream*.'

When this conversation was over Calypso had just stared at his image frozen in the intense sunlight of East Africa. She wondered how any man could care so much for his son's safety yet understand so little of other people's feelings.

*Hope's Dream* was the ultimate development in large high-speed power craft. It was 160 feet of snarling white-painted aluminium, darkened glass and stainless steel, so raked that it looked as if it was permanently lying in wait. It was a triple-decked Sunseeker with twin flying bridges – one enclosed, one open. On this fine day, Jack, Calypso, Tommy and Jed stood aloft on the open bridge on the top deck. Jack was at the wheel, operating the controls of two 2,800 HP turbo-diesels capable of propelling the huge craft at up to fifty knots.

Jack had taken the precaution of visiting *Hope's Dream* on the evening before their trip to give the vessel and the crew the once-over. After his nightly run along the beach, he had caught the Mag down to the marina. Though he had seen the vessel on many occasions, as had everybody else on Hope Island, he had never been aboard it and except for a weekly maintenance check he had never seen it leave the harbour.

The skipper, Henry Singleton, was welcoming. He was a white Jamaican and a Yacht Master, qualified to sail around the globe navigating solely by celestial navigation. Indeed, Jack found the sailor sorely chafed by extended port-bound duties.

'She's never been out properly since she was delivered here,' sighed Singleton as he poured Jack a beer in the vast recovered-walnut and farmed-leather stateroom. ScentSims boosted the aroma of polished wood and rich Connolly-tanned hide. 'I'll be proud to let you have her for the little boy's outing.'

In his early days in the Navy, Jack had been trained to pilot every sort of craft, small submarines included, and also to fly a wide variety of aircraft. It was a proud boast of his unit that they were capable of operating on land, sea and air, and once Singleton had heard the evidence of Jack's nautical experience, he was delighted to hand over the forty-seven-million-dollar craft to the corporation's security chief.

'I'd like you and your crew to come with us, of course,' Jack hastened to add. 'As you'll be skipper, I'd like you to plot the course. We'll need to stay well away from Cuban waters with things there being as they are. Then I'd be grateful if you would personally pilot us in to Mayaguana. I see there's a long coral reef across the harbour entrance and I'm sure nobody knows these islands as well as you.'

Singleton nodded. 'It's a very beautiful island. But are you sure you only want to visit Mayaguana? We could be in Santo Domingo in the same time, where there's much more to do and see.'

Jack hated to dash the skipper's hopes of a trip to a larger island with a real town, but he knew Calypso had her own reasons for wanting to take Tommy to visit her home island.

Joe Tinkler had begun to feel like his old self. In the seven weeks he had been building his new fund he had already recaptured all his old pension-management clients plus an additional fourteen. He had been particularly successful in raising Asian investments and he was surprised that so many of his new Far Eastern clients expressed pleasure that he had moved to Geneva. He hadn't realized just how partisan overseas fund managers could be. They simply hadn't wanted to invest their members' savings with him while his fund had been based in New York.

Tye's stocks were still riding high and Joe was spreading his risks with care. He had subscribed heavily to an initial public offering by LifeLines Inc., a Tye Life Sciences subsidiary, even though the flotation of the organ-matching and broking company had been handled by Rakusen-Webber. He hadn't had to talk directly to Morgenstein yet, but all his other former colleagues were delighted Joe was back in the market – even if he was now playing against their house.

His fund now stood at 1.76 trillion in US dollars, although, at Chelouche's request, he was converting to the European common currency at the end of each of his allotted trading 'days'. At the current exchange rate that was just over three-quarters of a trillion euros. Then there was that other pot: although Tye's stocks had performed well, Joe had made a few experimental but discreet market interventions and, as Chelouche had promised, the World Bank had settled every claim – and without question. Joe once again controlled the world's largest external shareholding in the Tye Corporation and its subsidiaries and he had unequalled power to influence the market prices. As requested, Joe intervened judiciously and discreetly, learning when an 'irrational' purchase or disposal had the greatest effect on other investors.

'Good evening, Joe,' greeted Chelouche, entering the room. Madame Pioline had requested that Joe should 'drop in' at the World Bank's European headquarters for an early-evening cocktail. Although all global stock markets traded on a twenty-four-hour basis, seven days a week, Joe chose to limit his personal activities to a nine-hour working day. During the rest of the cycle his software agents bought, sold or held according to the daily parameters he set them. His VideoMate would only be paged if any of his pre-set limits were breached.

Joe put down his Bloody Mary and rose. He sank back into his seat as the banker waved him down and came to sit opposite. A neat whisky had been poured and was waiting for him on the low table. He raised it in salute.

'Are things buoyant in the Tye markets?'

Joe nodded. 'Holding well, sir.' He couldn't yet bring himself to call the esteemed banker by his first name.

'No worries over what that congressman says – that this Phoebus Project has weapons potential?' asked Chelouche as he flipped open the lid of his humidor.

Joe had read the story several times. He had discussed it with half a dozen other fund managers. '*Who are they going to scrap with?*' James Dodd at DRKB in London had asked dismissively. '*Doesn't make any sense, Tinkler.*' Efi Arazi at Lehman Brothers in New York had laughed. '*Another crazy Republican. There's an election next year. You should know better, Joe.*' Joe was forced to agree. None of the markets had paid any attention.

'No, sir. It's had no impact,' he confirmed as he lifted his drink again.

Chelouche nodded. He held out the box to Joe. The fund manager shook his head and he watched as the banker selected a large Havana cigar, cut the end and lit it carefully from an oversized match. Chelouche blew out a great cloud of smoke and returned his attention to Joe.

'Do you consider the Tye Corporation could be vulnerable to a takeover?'

Joe almost choked on his cocktail. Then he looked up to see if the banker was serious. He seemed to be.

Joe shook his head and put his drink down. He had to let a laugh escape. 'No way, sir. It's the world's richest corporation. The core stock is trading at two hundred and eighty-four, maybe two eighty-five times annual earnings. No other company could get close to the necessary finance, no matter how much debt leverage they managed to find.'

Chelouche nodded. 'But what would it cost – theoretically?' he asked.

Joe felt his eyes widening as he considered. He shook his head. 'I've no idea, sir. I'd have to build some models.'

'Then build them,' instructed the world's banker. 'That's your second objective, Joe. I want you to build up positions that allow your fund to take control of the majority of the shareholding in the core Tye Corporation.'

Chelouche had said it quietly and Joe felt an urgent need to have him repeat it. But he knew what he had heard. He wasn't sure if it was even possible.

'Do it quietly, Joe, and use a large number of proxy accounts. Get option agreements from the other main institutional shareholders. I know you will have to guarantee significant premiums, but do it. And above all, don't allow word to spread – tie a confidentiality agreement into the options. Each one of them must think it's an exclusive deal for them. Now, listen carefully Joe...'

Joe looked up and saw the basset-hound eyes boring into him.

'Do not discuss any of this operation on the networks. Don't make video calls, send e-mails or talk to anybody on your VideoMate. You'll need to travel to see the main shareholders personally and, when you discuss this, you must ensure they do not record any part of the conversation. Use paper option forms. This is absolutely vital. Do I make myself clear?'

Joe nodded. It was crystal clear. But absurd.

'Remember, each one must think it is a one-off personal deal for them. There must be no suggestion that you are buying elsewhere. We may never have to exercise those options, but I want them all in place.'

'We'll have to serve notice on the Tye Corporation, sir,' he objected. 'They can block such a move in dozens of ways.'

'Where is Tye incorporated, Joe?'

'In Hope Island, sir, but ...'

'Our lawyers say the EUUSA notification rules apply only to corporations that are *lex domicilii*, legally resident, in regulated territories. Hope Island isn't one. When Tye moved his corporation out of Delaware, he escaped taxes and regulation but he also gave up the protection of the national governments that run EUUSA.'

Joe had no idea whether Chelouche was right or wrong. But he knew that at least the Securities and Exchange Commission would have to be notified, as it was when any holder's stock rose above five per cent of issued shares.

'We'll have to submit a 13-D to the SEC, sir, that's the rules.'

Chelouche regarded him silently from under his dark bushy eyebrows.

'OK, OK, I know, I know,' sighed Joe. 'The World Bank makes the rules. But even if I could get the stock, even if we could build a majority voting position, do you realize how much money those options might cost? The banking operation alone would be trillions. Then there would be the poison parachutes: every board protects its members that way. They have agreements with the company that if it becomes subject to a hostile takeover they receive such huge compensation packages that it makes any such move ludicrously expensive.'

'It's only money, Joe,' said the world's banker.

*

Calypso was standing surrounded by children, many of whom she knew, all of whom knew her. She was an honorary *macoumère* to many and official godmother to two of them. Tommy clung to her hand apprehensively. He had never seen so many children before, all of them shades of brown and black. There must have been forty or fifty of them and at least half of them seemed to be related to his Calypso. They danced and sang and ran around the party.

'Miss World,' they called, 'Miss World,' for that was how their parents talked about her.

Calypso was the tiny island's most famous export and her brief success as a beauty queen fifteen years earlier still meant far more to the population than her later achievements as a doctor. Every time she visited, she was given this escort of laughter. She realized she would have some explaining to do to Tommy.

They walked through the small town of Abraham's Bay and past the old mission schoolhouse where her father had taught. Calypso skipped off the road and peered in through the windows. A class of the more dedicated Mayaguanian children had their heads down over their books and were writing in silence. She stooped and picked Tommy up so he could see.

'This is where I went to school,' she whispered as he gazed wonderingly at normality. 'My father used to be the teacher.'

They had not been noticed and Calypso put her finger to her lips, signalling for Tommy and the other children not to disturb the young scholars inside. They continued through the small town until they reached the outskirts beyond and then followed a dirt path that wound through a grove of loblolly trees. They emerged at the beachfront beside a blue-painted stone cottage surrounded by a well-kept garden. The sound of hummingbirds and cicadas competed with children's shouts and laughter on the sand.

Calypso smiled and waved at them as she led Jack and Tommy up the short path to the front door. Calypso had bought this cottage for her mother fifteen years ago, when she had first started to earn some significant money from her modelling and sponsorship activities.

The day-carer opened the door with a smile and Calypso led Tommy into a bright, white-walled interior. Mrs Browne was dressed specially for the occasion and sitting upright on the dark sofa. She was small and pencil-slim and wore a neat powder-blue suit with a navy blouse. Calypso smiled: her mother still had style. She stood up as Calypso crossed the room to hug her. The elderly lady reached up and lovingly touched her daughter's hair.

'And who have you brought with you?' asked Mrs Browne, turning around.

'This is Tommy,' said Calypso. 'He's my friend.'

'Hello,' said Tommy uncertainly.

The old lady reached out a sinewy hand that Tommy shook.

'Welcome to Mayaguana, Tommy,' said Mrs Browne ceremoniously.

'Thank you very much,' replied Tommy with boyhood gravitas.

To Jack it all seemed slightly surreal.

'And this is Jack,' said Calypso. 'He's also my friend. He works with me on Hope Island.'

The old lady looked up and squinted with her fading eyesight. 'My, you *have* brought a good-looking man to my house,' she laughed.

'Hello, Mrs Browne,' said Jack as he took her hand. He suddenly realized he was being introduced to his girlfriend's mother – if Calypso was his girlfriend. He turned to look and he saw a mischievous smile hovering around Calypso's mouth. He suddenly felt very uncomfortable.

'Sit, sit,' ordered their hostess. 'We'll have tea in a minute.'

Calypso sat down at one side of her mother on the firm sofa, and Tommy on the other. Jack found an upright chair near the door. He felt reassured that Pierre and Stella had followed them discreetly through town and would now be in position at the front and the back of the house.

'And how old are you?' Mrs Browne turned to her young guest.

'Seven,' answered Tommy hesitantly. He wasn't used to conversation with strangers.

'Do you go to school?'

Tommy nodded shyly. Then he remembered what Calypso had told him about her mother's eyesight. 'Yes, well, I'll be going again soon, won't, I Calypso?'

'Yes, you will,' affirmed Calypso. 'I think Tommy's ready to attend full time.'

Her mother looked towards the kitchen, where her day-carer was assembling cups and plates for afternoon tea. Calypso could smell freshly baked johnny-cake.

'I'm Jed,' said Jed, filling any silence as he was supposed to.

'Oh, my manners!' said Mrs Browne, peering along the sofa in puzzlement. 'You've brought a friend, Tommy. And are *you* going to school as well?'

'I'm a caterpillar,' Jed explained.

There was a short silence as she attempted to digest this information.

'It's called a Furry, Mum,' explained Calypso. 'A toy you can talk to.'

'He's my friend,' affirmed Tommy, clutching the caterpillar even closer.

'Ow, that hurts,' complained Jed.

'Don't harm him, dear,' said Mrs Browne anxiously.

Tommy laughed at the idea. 'He doesn't really mind. He doesn't feel anything. He just says things like that.'

'Mayaguana is the most easterly of the Bahamian islands,' offered Jed. 'There are about three hundred people resident on the island and it remains almost wholly unspoiled.'

'My,' marvelled Mrs Browne. 'You do know a lot.'

'Mayaguana is the original Indian name for the island,' continued Jed, undaunted. 'It is a paradise for sailors, boat owners and divers because of its extensive anchorage and coral reef. During the year temperatures vary between nineteen and thirty degrees Celsius. Friendly and inexpensive accommodation can be found at the Abraham's Bay Inn, in the island's main town.'

'He does know a lot,' agreed Calypso. 'Don't you think Jed knows a lot, Jack?'

'He does know a lot,' laughed Jack. 'I've thought that before.'

'Shut up,' Tommy hissed to Jed.

Mrs Browne started to laugh and they all laughed with her. Jed promptly fell silent.

'We've brought a toy pet for you too, Mum,' said Calypso.

She bent to pull a package from her bag lying on the floor. She leaned across her mother and handed it to Tommy to make the presentation.

'This is Roger,' said Tommy, holding the box out shyly.

The elderly woman took the box and opened the lid. Calypso had deliberately ruled out gift-wrapping in order not to spoil the moment. She watched her mother lift out the ball of warm fur, and quietly removed the box from sight.

'You say hello to Roger,' suggested Calypso.

Mrs Browne began stroking the CatPanion.

'I'm Roger,' said the cat in a soothing voice. 'What's *your* name?'

'Let me show you around the garden,' murmured Calypso to Jack and Tommy. 'We'll leave Mum alone with her new companion for a while.'

'See you later, Roger,' called Jed as they left.

Deakin was calling the emergency meeting to order. Almost forty section heads of Operation Iambus were crowded into the conference room on the eleventh floor of the UN Secretariat building. The UNISA Exec was perched on a battered projector stand.

'OK, just how interested is Washington becoming in the Tye Corporation?' he asked. 'Let's put two and two together and see what we get. Marv . . . what do your sources in the White House say?'

Marvin Girdlong, a former Chief of Staff at the White House, rubbed his day-old facial stubble and studied the notes on his DigiPad. It had been Marvin's research that had prompted Deakin to call the meeting.

'Hendriksen was right,' he began. 'There's been no official announcement but it looks like the Chief is going to attend the party. They've had to reschedule a visit by the Prime Minister of Israel and two fund-raising dinners have been postponed. The

Secret Service has already started to re-roster for that weekend. Oh, we also hear the First Lady's cancelled a charity appearance in New York.'

Deakin nodded. Marv's DC sources remained the best, even though all Washington knew that he had 'defected' to the UN.

'Anything else?'

'Only a rumour, but they say Jane Treno has invited Tye to Washington, for discussions. Two weeks before the party takes place.'

Deakin snapped his head up. That was new. The Attorney-General herself was inviting Thomas Tye to visit DC! In Washington terms such an invitation was close to an imperial command. Tye was still an American subject, even though he also held a separate Hope Island passport. For reasons known only to the Department of State, he had not been asked to relinquish US citizenship when his island's sovereignty had been recognized.

'I only heard that just before this meeting started,' explained Girdlong, as if apologizing for springing a surprise on his boss. 'It's not confirmed but once again it's whispers and lots of calendars being rearranged.'

'OK,' pondered Deakin, turning to the attentive audience seated around the large conference table. 'The President accepts a jaunt to a tycoon's anniversary party on Hope Island and meanwhile Jane Treno invites Thomas Tye to drop by for a *discussion.* Any ideas?'

'It will be about the LA traffic mess,' stated Martha Rose, one of the international lawyers, authoritatively. 'I hear the Tye attorneys are playing real hardball. I'll bet the administration is going to lean on Tye personally to make an interim settlement. It will be the old one-two: Treno will be tough with him in Washington – threaten the Tye Corporation's US interests, perhaps a domestic antitrust action, perhaps talk about pulling the company's government contracts, et cetera – and the President will be all smiles and backslapping for the cameras at the party once the interim payment's announced. California's going to be even more marginal next year. It could go Republican and the Wilkinson campaign people will do anything to stop that.'

Deakin sipped his coffee, his umpteenth cup of the day. It was now late evening and, even as he raised it to his lips, he tried not to drink too much. He would never get to sleep this way. He nodded. Rose's analysis was logical.

'Yes, the LA thing is big enough to go all the way up to William Wilkinson,' he conceded. 'But could it also be anything to do with this Russian deal? Have we learned anything more about that?'

Magda Nezhdanov, the senior Russian Federation analyst, gave a Slavic shrug. 'Everyone in Moscow is talking about it, but nobody knows anything for sure. Some say the Tye Corporation is developing real estate, others say it's an agricultural deal. Something big has been signed, that's for sure. And President Orlov's attending this celebration as well.'

'Can we all go too?' shouted Olliphant, a young and utterly brilliant British perception-adjuster from the far end of the table. He was planning the campaign to realign the public's view of Thomas Tye once a course of action was decided upon.

Deakin held up his hand. 'Doctor Chelouche has sent me an interesting fax,' he told them. There were smiles around the table. None of them had yet got used to sending paper faxes again – even ones scrambled with one-time security codes. 'Moscow is offering to buy back Russia's debts, at a discount.'

'Which debts?' asked Libby Klinkhamers sharply. She was an international economist for the UN but one who still found time to teach as a visiting professor at Harvard.

Now Deakin had their attention. 'All of them,' he said simply. 'IMF, World Bank, Bank of Redevelopment and Construction, OECD grants, EU bonds, US loans. The whole damn lot. They're trying to get them rolled up together and they want to settle them, in hard currency, at a sixty-four per cent discount inside thirty days.'

Libby had been tapping on her VideoMate as Deakin was speaking. Like everyone else on the team, she had had to allow Lynch and his technical support people to disable all the network and external storage capabilities of her unit. The UN computer support staff were busy adapting one of the building's wiring looms to provide a totally internal and insular network protected by physical firewalls, but until that happened even office-to-office network communication was banned.

'With interest rolled up that's almost a trillion US dollars,' she breathed. 'Even if they got that discount it would be ... three hundred and sixty billion. In hard currency, you say?'

Deakin smiled. 'I think we've just found out what Tye did with all that cash he raised on the markets a few weeks ago,' he said. 'So what *has* he bought?' He looked around the room. There were only shrugs.

'For God's sake, not the few old nukes they have left?' asked Chevalier, the group's senior military analyst.

'No, definitely not,' asserted Deakin. 'That would be a turn-off to every one of Tye's shareholders and supporters. He's maintained a strong stance against everything nuclear. What do the intelligence services say?'

James Soames, the British-born liaison officer working as linkman between UNISA and the world's intelligence services, shrugged in turn. 'Less than nothing, Ron. They're telling us there's nothing on the radar as far as the Federation is concerned.'

'What about this Phoebus research project?' asked Deakin. 'Did Congressman Robarts really have something or was he fishing?'

'NASA says not, CIA says not,' reported Soames, 'Pentagon says not, NATO says not, EU Defence Agency says not. All damn worrying.'

The looks around the table said it. There was something up and the UN wasn't being told. It was back to the bad old days of distrust between the individual members and their global representative body. But why?

'OK, three urgent tasks for us,' announced Deakin. 'One, what does the Justice Department really want with Tye? Two, what deal has the Tye Corporation done with Moscow? Three, what does NASA or the CIA have on the Phoebus Project that they won't share with us? Steal it, extort it, buy it or beg it. Lean on every diplomat in this building, crawl into every little lobbyists' crevice. Let's get that information. I'll talk to the SecGen tomorrow and get him to talk directly to Washington. We are the *United* Nations, ladies and gentleman, and we won't be stonewalled by any of our individual members, no matter how powerful they may be.'

## NINETEEN

'Touch me and you *can* make a difference' said Thomas Tye, giving the camera his most earnest look. The set was designed to look like a gentleman's study, and he sat behind a desk. But his customary open-necked white shirt and tied-back hair banished any sense of formality. Tye reached for a globe on the desk, spun it slowly and pointed to a spot. In the background began a soft, instrumental rendition of 'It's Our Planet'.

'Ethiopia is a nation of seventy million people in East Africa. Approximately sixty per cent of them are starving. There has been no rain in the lowlands of the southern part of the nation for six years. Each year the harvest has failed for nearly thirty million people who depend upon it. Famine is widespread.'

The image of Tye's face dissolved to show him standing in a desert, wearing a dust-stained beige safari suit. 'This land is the oldest nation on earth – it is the cradle of human evolution. It is from here that our ancestors walked, paddled and sailed out of Africa to populate our world.' He squatted and picked up a handful of earth, allowing the dry soil to run through his fingers. 'But today the lush savannah has become dust. We have abused our planet and this is the result.'

Then the picture cut to Tye standing in an Ethiopian village. A girl about eight years old was holding his hand, staring stoically ahead.

'This is Biya,' he said to the camera. 'She has no family left and she has lost her eyesight from trachoma, a viral condition exacerbated by the effects of malnutrition. She has never seen, or felt, rain.'

As he told his viewers more about the ravages of drought and famine, the picture changed to a montage of starving children, dried-up crops and lines of people queuing for food on the pediplain. The corporate anthem was now a glissando of sweeping strings, an emotive accompaniment to the sombre images.

'The Tye Corporation can change things here, but we need your help,' he said over the pictures.

*How strange*, thought Haley as she watched this v-mail that had just arrived. She'd never seen Tye make a personal appeal before. She could guess that he must have been aching to get back inside his air-conditioned trailer and slip into an antibacterial mask.

The voice-over continued with still more facts about the ravaged country. Then the camera revealed Tye back in his study.

'Next month, on Sunday, August thirtieth, I'm proud to say my company will start to change the world. Using new, totally benign and sustainable energy technology, we shall bring rain to southern Ethiopia for a period of at least six hours. We shall then bring rain once a week for a further month. My charitable foundation already has people in place to help the Ethiopian population make the most of this situation. Much of this year's crop *can* still be saved if it rains before autumn.'

The camera zoomed back to a close-up on Tye's finely featured face.

'Will you pledge just one US dollar to help thirty million people? If a majority of you make this pledge, I can *guarantee* that it will rain in southern Ethiopia next month. We want to give two billion dollars to the Ethiopian people so they can rebuild their economy and their nation. If you, the people of the world, will pledge a total of one billion dollars, the Thomas Tye Foundation will match that with a billion dollars' worth of solar energy systems, livestock and crop seeds specially modified for Ethiopian conditions and, most importantly, will send a team of four hundred trained field-workers to help Ethiopia and its people work towards achieving a decent standard of living. This will be the start of a ten-year programme that I will be proud to have my foundation oversee.

'This v-mail has been sent in over two hundred languages to all three billion network addresses on this planet. I hope and believe you will now help Biya and her people.'

He leaned forward suddenly and pressed his right forefinger to the camera lens. An electronic halo shimmered around the dark outline of his fingertip like the corona of the sun in a solar eclipse.

'Reach out and touch the world now. Touch your fingerprint-identification pad and your pledge will be automatically recorded. But send no money until *we* have brought rain to Ethiopia.'

Haley leaned forward and touched her pad.

'Thank you, Haley Voss,' said Thomas Tye with phoneme-perfect image morphing and a spectacular close-up smile. 'Your pledge has been recorded. Join me on Sunday, August thirtieth, to see it rain in Ethiopia. Take care of our planet – you hear!'

The driving rock version of 'It's Our Planet' rose over a slow sequence of images of Thomas Tye walking amongst the children of the Rift Valley. Haley smelled the instantly recognizable scent of 'Abundance' from her ScentSim as the anthem grew.

*I hear*, thought Haley.

The papers were full of it. It was headline news in every country. Every TV bulletin led with the story, but had to manage with scant details and hours of speculation from wholly ignorant 'experts'. Mostly they resorted to running library footage of previous famines in Ethiopia.

Deakin scanned the printouts of various front pages that lay on his desk.

'Tye Corporation Claims To Be Rainmaker,' said the *New York Times*.

'Tye Corporation Ready to Rebuild Ethiopian Economy With New Weather Technology,' ran the *Financial Times* headline. 'Ethiopian Market Closes 40 Points Up.'

'Two Billion Dollar Pledge To Make Ethiopian Rain,' reported the London *Times*.

'Tye-riffic!' screamed London's *Sun*.

'Tye-phoon!!' opined the *Asian Star* in proleptic ecstasy.

'Amelia Earhart's Plane Found on Dark Side of the Moon,' offered the *National Enquirer*.

But neither Thomas Tye himself nor any spokesperson from any part of the Tye corporate empire would say more.

'Join us on August thirtieth,' they kept repeating. 'Just be sure you have made your pledge.'

Ron Deakin had called another meeting – the second in thirty-six hours – and he waited impatiently for the meteorologists to arrive. The unavailability of normal communications technology was making this investigation grindingly slow.

And, on all the main markets where it was traded, the Tye Corporation's core stock made sharp gains.

'OK, companions to nap mode, VideoMates to silent and all viewpers off, if you please,' instructed Theresa as she took her seat.

Professor Keane had gathered her 'A Team' of researchers in the Network Control Center. The projection system was switched off and she sat in a low chair in the middle of the holo-image pit, where she was lit by a gentle but unflattering overhead light. The researchers sat in a banked semicircle around her, all with a CatPanion, a Furry or some other sort of intelligent companion on their laps or on a seat beside them. All these 'creatures' were development platforms for beta personalities now undergoing development by the researchers. Theresa always found it difficult to bring this team together. So many of them were unconventional individuals who did their best thinking on beaches or on clifftops or in their hot tubs. Some of them were so deeply involved in their relationships with their companions that they found it hard to focus on the outside world. Indeed, some rarely left their apartments and contributed their thoughts, criticisms and software over the team's private network.

'First, I want to welcome our special guest, Doctor Calypso

Browne.' Theresa inclined her head to where Calypso sat in the front row, the only person in the room without a companion and whose well-tailored dark trouser suit marked her out from the collection of brightly coloured T-shirts, shorts and sandals worn by the others. 'Doctor Browne acts as a personal physician to the Thomas Tye household and is also a consultant psychiatric paediatrician. Thanks for coming to join us, Calypso.'

Calypso bowed her head in acknowledgement, grateful that the professor had not mentioned her earlier claim to fame. Perhaps that part of her life was finally disappearing into the past.

'Now, I have an interesting ethical question for us to consider,' Theresa continued. She had Sandra on her lap and was gently stroking her sleeping CatPanion. 'But first, consider: how many Furries, CatPanions and other companions has the corporation or its licensees sold worldwide to date? Anybody have a figure?'

'I think it was about three hundred and fourteen million, last time I heard,' ventured Rory McCullum without looking up. He was one of the world's leading theoreticians in artificial personalities and Theresa had lured him from the Turing Institute in Glasgow. She noticed he had recently become deeply attached to the very large shocking-pink Bugs Bunny-style rabbit called Beau who was asleep in the seat next to him. Companion bonding was both fashionable and strongly encouraged within the artificial-personality research team. Rory was in the process of knitting a maroon cable-stitch cardigan for Beau, and his size eight needles never ceased clicking as he spoke.

'But many of those are only first- or second-generation.' He deftly cast off a row of purl stitches from one needle, starting a new line before he continued. 'Those didn't have network-communications abilities and they didn't upgrade themselves automatically.'

Theresa nodded. 'Tye Consumer Electronics and its various licensees have been selling companions for about four years,' she explained. 'CE is now even getting requests for Furry personality transfers. Some children who have had one type of Furry for several years – a caterpillar, a rabbit, whatever – think they've grown out of the physical envelope of their companion but they want to keep the Furry's personality. Do you see any objections to us doing this for our customers?'

'At a price, I hope,' put in Liane Stevens, former associate professor of the Human-Computer Interaction Laboratory at the University of Maryland.

Theresa smiled. 'Of course. But are there any concerns here about the *concept* of transferring a Furry's personality to another container? Let's imagine that a little girl of seven has owned a Furry – a soft pink rabbit called "Lucy", say – for three years. Her Furry has been everywhere with her. It has seen everything she has seen, heard everything she has heard and it has listened to all her problems. The Furry has *learned* from its owner and its abilities to harness that information have improved with every remote upgrade that has occurred as our Anagenesis-network personalities evolve and bequeath their hard-learned experience to their more corporeal cousins. Every sight and sound of their owners' lives is recorded, and not just in the companions' local memories. They are also uploaded to our FMR – our Furry Memory Retrievatory – in our data warehouses on this island. As part of that project we're creating a database that will contain a complete audio-visual record of every Furry owner's daily life; it will become the ultimate anthropological resource.'

They nodded, listening carefully. Those who had been on the team since the beginning could sense that the time had finally arrived for some of the big questions to be faced.

'Do we just carry out a transfer as requested and send her a PonyPet or whatever she wants with precisely the same personality and memories? What would be the impact of her lifelong companion and friend appearing in a completely different guise? And do we deliver the new "Lucy" only when the old one has been returned to us? What would be the pyschological impact of the two Lucys being together with the owner in different physical form?'

'And of them me-me-meeting each other,' added Robert, the group's speech simulation expert.

Theresa smiled. 'Good point, Robert. All Furries love to

communicate. What would "Lucy A" make of "Lucy B"? We can model that here – one for you, I think, Liane. But, first, let's consider the impact on children. Doctor Browne?'

Calypso shook her head, feeling out of her depth. But she also felt distinctly uneasy, as if something was wrong here but she wasn't sure what. She herself had made contact with the eminent professor after Jack had talked to her about Jed's recent behaviour. He said there seemed something uncanny about the Furry's ability – as if it *understood* more than a mere toy should. Calypso had to agree – she felt the same way. She was also wondering whether Tommy's obvious devotion to an increasingly percipient bundle of fabric, plastic and computer circuitry was wholly healthy. These were two concerns she had articulated to Professor Keane during a video exchange one evening.

'I do understand your worries, Doctor,' Theresa had replied. 'Most parents buy these toys without realizing that they are introducing their children to potential lifelong companions.' Later in the conversation she had invited Calypso to attend this current discussion with the research team responsible for developing future generations. Now that she had confronted some of the issues Calypso felt less sure that she was qualified to help.

She weighed Theresa's question carefully, aware that she was speaking in front of some of the brightest intellects on the planet. Then she thought of Tommy – and the obscenity of the concepts they were discussing swam into sharp focus.

'I must suggest that the ideas you're discussing are incredibly dangerous,' she began, struggling to keep aggression out of her voice. 'Children do love these companions, but they aren't best equipped to distinguish between real pets and . . . and machines. Replacing one Furry with another companion that has exactly the same personality would be criminally irresponsible. It would be better to allow the first companion to go through something that appears to be closer to a normal death – like the demise of a pet dog or cat – rather than to provide a replacement that makes death seem impermanent. That could traumatize a vulnerable child – one who is seriously ill, perhaps, or who has suffered from their parents' divorce, or has actually lost a parent or a brother or a sister. You must consider the children's feelings – you can't consider them simply as an upgrade market!'

There was shocked silence in the room. Calypso felt intense hostility directed at her. Then she thought of an even more concrete objection.

'If a family called me as an expert in a lawsuit raised against Tye Consumer Electronics because of trauma caused by a companion transfer – or even malfunction – I wouldn't hesitate to testify about the potential danger to an unformed psyche.'

'Thank you, Doctor,' said Theresa dryly. 'Comments?'

'The Doctor's concern about companions' apparent immortality may become irrelevant, given the research into human longevity that's being undertaken elsewhere on this island,' observed Liane Stevens with just a hint of acid in her voice. 'The way things are going, it won't be just Furries who will seem to live for ever. Some of the owners may also live for hundreds of years and that presents us with a *far* more important concern. We know the capability of companions is improving exponentially, not just because of our deliberate design improvements but also through the evolutionary improvements within the community of network-agent personalities that are passed on automatically to companion toys. How will a balance be retained between increasingly clever Furries and owners who, whilst becoming more experienced in life, are almost certainly no more capable? How will owners keep up mentally with their companions over a long period? *That*'s what we should be worrying about, not the issues of personality transference between different models of companion envelope.'

'But, what ha-ha-happens when a companion's owner *does* die?' asked Robert. 'If a child has an accident or a fatal disease, what ha-ha-happens to the companion that has shared all their waking moments and has all their common memories stored?'

That too silenced them. All, with the exception of Calypso, were probably thinking about how their companions would continue after their own deaths.

'No one can access a companion's memories but an owner,' said Rory McCullum quietly. 'We built that in to all of them from the second generation. It needs the owner's voice print to activate a core command in a companion.'

'But what *should* ha-ha-happen to those memories?' insisted Robert. 'After all, we store a copy of every owner's voice print so *we*'ll be able to access them.'

'Surely the next of kin should inherit the companion and be given access to its memories,' said Liane.

Theresa shook her head. 'No, we can't allow that. Everybody's memories and their shared experiences with their companion are highly personal and very private. We can't allow anyone else to access them.'

The researchers were silent again, nodding as they contemplated their own experiences shared with their companions. Calypso's mind was reeling at such cavalier discussion of personality transference and the archiving of hundreds of millions of life experiences. She wanted to scream, to shout at them, to make them see that human personalities and experiences are not commodities.

'I th-th-think maybe we should erase all memories when the companion's owner dies,' suggested Robert.

'You mean bury the pet with its owner?' snorted Avi Becchar, whose speciality was emotion simulation. 'So they both have to go together? I seem to have heard that one somewhere before.'

'Ethiopia Appeal Tops Four Billion,' screamed the *New York Daily Post*. 'Global Pledges Set To Break All Records.'

'Over one billion people are now reported to have pledged money for the Ethiopian Appeal recently launched by Mr Thomas Tye, President of the Tye Corporation,' read the BBC's senior news announcer. 'A spokesperson for the company has said that promises of nearly five billion dollars have already been received. Such a universal response to an appeal for charity is unprecedented. A spokesman for Oxfam states that aid on this level could transform the future of Eastern Africa. On its re-release, the song "It's Our Planet" has become the world's number one downloaded track once again. All proceeds also go to the Ethiopian appeal. On the world markets, all companies with business interests in East Africa are currently experiencing a sharp increase in valuation.'

So, it seemed, Thomas Tye *had* touched the world. The Tye Corporation's PR agencies were operating in full flood, but their efforts were unnecessary as pledges continued to pour in. The world's population wanted to see Thomas Tye produce rain on cue and, under his leadership, they were happy to help feed the starving. The largest number of people ever united in a single cause were turning their thoughts to 30 August and to the people of Ethiopia. A fourteen-year-old white girl in Cape Town, suffering from a brain tumour, sent 200,000 Rand to the *Cape Town Herald* in aid of the cause, then spent a night on the networks in various Thomas Tye chat rooms gaining emotional support before setting herself alight as she replayed her hero's appeal one more time. She left a caption to the recording of her self-destruction that said simply 'For The Planet'.

The frenzy continued and the Tye Corporation's stock rose another ten points!

Michael Chevannes switched off the engine of his United Nations four-wheel-drive Toyota and climbed out into the crystal-clear sunlight. The noise and dust that had been his constant accompaniment on the grindingly slow ascent disappeared and there was the sudden and immediate pervasive silence that only exists at great altitude.

Although it was mid-afternoon and the tropical sun was still high in the sky, the atmosphere was cool and clear. It was the thin air up here, Chevannes realized. On the advice of the logistics manager at the UN compound he had deliberately spent an uncomfortable, cold and cramped night in the back of the vehicle 6,000 feet below in order to give his body time to acclimatize. If he had attempted to ascend the mountain in a single day he would now be suffering from completely debilitating nausea and dizziness. The GPS and hypsometric display in his RayBan Electros showed his position as 14,657 feet or 4,500 metres above sea level. Despite making his ascent in carefully timed stages, he felt his breath coming in short gasps.

Only one road zig-zagged up Monte Camanchaca, a medium-size peak in the Peruvian link of the great Andes chain. The recently constructed road was surprisingly wide, broader than many of the country's secondary roads, and surfaced in thick black asphalt. On the winding journey across the lower slopes he had passed through a few Aymaran villages and waved at an occasional vicuña shepherd and his flocks. Once he had been forced to stop as a vast herd of furciferine deer bounded across his path but he had seen no sign of human life for the last 8,000 feet.

Although usually preferring to drape his muscular bulk in a business suit, Chevannes had decided to underline the authority of this mission and he had borrowed a UN peacekeeper's uniform of white epauletted shirt with shorts. Now, in the chill of the rarefied Andean air, he could feel goose bumps forming on his exposed arms and legs. Despite the sun he shivered involuntarily.

He put his hands on his hips and surveyed the flat mountain peak immediately above him and the long cordillera sloping gently down to the south. In the far, far distance it rose up again to meet a yet higher snow-covered peak. *A little like the Great Wall of China,* thought Chevannes as he noted how the ridge had been flattened to provide a vehicle track and a narrow-gauge railroad between the radio dishes along its top. There had been some serious engineering work done up here.

He turned and looked behind him, back the way he had come. It was the first time he had been able to see the full majesty of the view and he took an involuntary half-step backwards and then leaned against his vehicle's hood. He was way above the cloud line and so high he could make out no detail at ground level. Immediately around and below him were row upon row of black solar-energy capture panels. A long way down, the rocky slopes of the mountain gave way to what he knew to be scrub, then catalpa trees and then to the cultivated coca-bush terraces of the foothills before becoming lost in an arboreal sea of rain forest and epiphytes so lush and dense that in the bright sunlight its green canopy seemed almost black. Above the forest he saw a speck that he thought at first might be a carrancha. Then he realized that a hawk would be almost invisible at such a distance: it could only be a condor circling lazily in the thermals far below.

In the extreme distance, perhaps sixty miles or so to the west, he could see the llano and, beyond, the glint of the Pacific Ocean. To the north the mountain range stretched away towards the equator and its eponymous nation. It was absolutely still and stunningly beautiful here on the mountain top.

Above him, the vast mesh dish of what he presumed to be a radio telescope cast a mottled shadow over the centre of a compound. He estimated the dish was at least 800 feet in diameter. Within its shade huddled six or seven white single-storey concrete buildings. To the east, away from the shadow of the dish, Chevannes could see a windsock hanging listlessly over what he guessed was a helicopter landing pad. That would have been the sensible way to travel, he thought, although few choppers were able to climb so high: perhaps the Canadian Sea King or the Russian Helix, if he recalled correctly. Just over the other side of the peak Chevannes could see the top of a large white dome that, he assumed, covered a conventional optical telescope array. UNISA satellite surveillance had discovered the compound's existence five years before and had tagged it as a research facility of the University of San Marco, Lima.

Dotted at regular intervals along the flattened mountain ridge to the south was a string of twenty or so smaller radio telescopes, each perhaps 200 feet in diameter, all set along the rail track that disappeared into the snow before the next peak. This installation was a rival for any of the giant observatories he had seen in TV astronomy programmes.

Chevannes had stopped his vehicle in front of a twelve-foot-high steel-mesh fence topped with razor wire. Security cameras inside the compound were trained on the large gates that barred his way. He assumed that someone must have seen him arrive, but nothing stirred in the ultra-clarity of the still afternoon. He opened the door of the Toyota and pressed the horn. The sharp sound was startling, almost deafening, in the thin air and he heard its echo roll away, slapping off the distant rock faces along the ridge.

A few minutes later an all-terrain motorcycle emerged from between the low white buildings inside the compound and sped down the road towards the gate. As it approached, the rider brake-turned the bike to a stop and switched off the engine. He slid lithely from the saddle and walked to within six feet of the gate. This New Age gaucho was neatly turned out in a white, short-sleeved shirt and knee-length black shorts. A holster was on his right hip and sun-viewpers hid his eyes.

*'Hola, buenas tardes. Estas perdido?'*

Chevannes watched and waited as his VideoMate identified the language and started to provide the translation in his earpiece. He understood some Spanish but he would need help in framing his replies.

'Good morning. You speak English? Are you lost?' asked the guard again before Chevannes's system had finished translating: as if UN personnel were in the habit of driving up mountains to seek directions.

'Hi,' said Chevannes. 'I'm here to see Doctor Toksvig.'

'You don't have an appointment.' It was a statement, not a question.

'No,' admitted Chevannes.

'You have had a wasted journey, sir,' the guard continued in his excellent English. Chevannes guessed he was Mexican or Cuban. 'Doctor Toksvig is off-planet, in an orbiting observatory. He'll be gone for many months.'

Chevannes nodded as if he understood. But he had prepared for this and he had assumed that, even under his new identity, Rolf Larsson would be unavailable to any uninvited visitors.

He turned back to the open driver's door of his four-wheel-drive and pulled out an envelope he had placed in the map-webbing of the sun visor. As he did so he watched the security guard in his peripheral vision. The man appeared completely relaxed and, Chevannes assessed, this was not a facility that would attract any trouble. It was too high to be of interest to local bandits and it was too remote to pose a threat to the

Colombian drug barons who owned many of the country's coca fields. Despite the impressive fence, the security level here was low.

He returned to the gate and pushed the envelope through the wire. 'Would you give this to Doctor Toksvig, please?'

The guard approached the fence and took the envelope. It bore the doctor's name and was marked 'confidential'. The blue UN emblem was embossed in the upper left-hand corner.

'As I said, it will be many months, sir,' repeated the guard. 'Perhaps not until next year.'

Chevannes took off his sun-viewpers and allowed the guard to see his eyes. He then flipped his wallet from the back pocket of his shorts and held it up to the fence.

'I'm from the United Nations Security Agency,' he said. He didn't offer the ident to the guard for checking. 'And, at the risk of offending you, I have to suggest you may be mistaken about Doctor Toksvig's absence. I would like you to take that envelope to him now or, in the million-to-one chance my information is wrong, I want you to give that envelope to the most senior person now at this facility and tell him – or her – that I authorize them to open the envelope. Is that clear?'

'Where can we reach you?' asked the guard.

'I'm not going anywhere,' said Chevannes. 'I'll just wait while you deliver it.'

The guard looked at the envelope, then at Chevannes, then at the UN Toyota with its fluttering white-and-blue flag and nodded. He remounted his motorcycle, started the engine and sped back to the cluster of buildings.

Chevannes walked back to his vehicle and rummaged for a water bottle he had left in the cooler. He drank and then poured water onto his handkerchief and washed his face. Refreshed, he pulled a pair of image-stabilized binoculars from the glove box and walked to the edge of the track to take a closer look at the chain of radio dishes that stretched away into the distance. He understood why optical telescopes needed to be located on mountains, but he couldn't think of any operational advantage justifying the expense of hauling radio-wavelength receivers to such a height.

He enabled the video link on his binoculars, stabilized the image and scanned along the mountain ridge. Ordinarily, others back at headquarters would now be sharing his view, but the ban on all network communications was being strictly enforced, so Chevannes was recording only in Local Mode. He moved his view from dish to dish, sharpening the image and adjusting the contrast as he went since even the thin atmosphere of the mountain range distorted the light. He saw no movement, and guessed that these dishes were controlled remotely.

He turned when he heard a sound behind him. The ground locks on the central gateposts had snapped up and the twin gates were slowly opening inwards.

'Am I late?' asked a voice from the darkness of the outer circle in the Network Control Center.

Theresa switched on the holo-projection system and by its ambient light she saw that Raymond Liu had arrived exactly on schedule. When Calypso Browne had got up to go, Theresa had thanked the paediatrician for so forthrightly expressing her views, despite the note of discord they had struck with the team. With the exception of Robert, Theresa had then dispersed all her researchers back to their hiding places and Companion Nests where they could continue work on the next stage of what would be a continuous progression of researcher-designed personality upgrades.

'Come on down, Raymond,' invited Theresa as she adjusted the projection controls on the holo-panel that floated in front of her seat. She waved him towards one of the front-row seats.

'This is Robert Graves,' she said, by way of introduction. 'He's my senior researcher on the Descartes Experiment.'

Liu bobbed his head and took a seat in the row indicated. 'As in "I think, therefore I am"?' he asked.

'Precisely,' confirmed Theresa. 'Who else but the person who first posed the original mind/body dualism question would be suitable as our patron saint?'

Liu smiled, but he wasn't sure why. He didn't yet know what their experiment entailed.

'Robert is going to explain the network elements of our

Descartes Experiment to you, Raymond,' explained Theresa from behind the control display. 'He works closely with me on this long-term project so I hope we can do two things. One, I think we can explain those dark data you've encountered in the networks. And two, I think we can reassure you that our experiments and developments in artificial life could not possibly have led to any of the systems failures you have been suffering in the networks. In short, I hope you will be able to eliminate us from your list of suspects. OK, Raymond?'

Liu nodded. That would be a step forward, at least. There had been little progress otherwise in finding the cause of the network faults.

'Over to you, Robert,' said Theresa.

The researcher stood up and turned to face his audience of two. 'I assume that you are bound by the same confidentiality as the rest of us in the university and in the corporation?' he asked brusquely of Liu. All trace of his balbutience had disappeared – as Theresa had observed many times before when she had asked him to make a presentation.

Liu nodded.

'We had to get special clearance from Tom to show you this today.'

Liu swallowed, hoping it wasn't noticeable. He knew his salary, stock options and career were currently hanging by the most slender of threads.

'I understand from Theresa that you watched our little demo with Miss Scarlett?'

Liu bobbed his head again.

'That software robot personality is disembodied and lives in the networks, but it's essentially the same type of D-persona that we use in physically based companions like my own Michelle or Theresa's Sandra.' Robert pointed to the CatPanion curled up on one of the vacant seats. 'Companion personalities are merely *simulating* speech and human behaviour and deducing how to respond to us by a complex set of rules that we have laid down. They have absolutely no independent consciousness or intelligence and, as Professor Keane has demonstrated so clearly, it is we humans who anthropomorphize them. However, the important point for *this* discussion is that they are absolutely processor-dependent. In Furries and other companions the personalities run on the microprocessor installed inside each unit. The disembodied versions travel the networks as packets of information but they can run only when they find a vacant microprocessor – or one of the more recent photonprocessors – to use.'

Robert paused to see if he was being understood. He knew their guest was the technical director of the Tye Global Networks so he had assumed he wouldn't need to explain too much about the basic elements of their technology.

Raymond Liu nodded silently once more.

'In the Descartes Experiment, our goal has been to try and create a basic entity that is processor-independent. By that I don't mean software that can run on a wide number of different processors, I mean software that runs without *any* specific processor.'

Liu smiled his understanding again, even though he was unsure how such a thing could be done.

'What we wanted to do was to develop an artificial entity that in turn could create its own processing environment from any of the individual single-state switches that exist in the world's networks or its attachments. There are trillions of those today. Every routing device, every gateway, every amplifier, every laser controller, every multiplexer, every firewall, every relay, every uplink, every downlink, every access device – all the millions of VideoMates and LifeWatches – each have millions of minute, individual switches that are either "on" or "off" at any one time.'

Robert put his hands together and waved them from side to side to indicate the two polarities.

'We are trying to build a software personality that doesn't need a *computer*. It assembles its own from the billions of simple "yes/no" condition switches that are all around it and the more firings – switchings – there are, the more capability it will have. To exist, it has to make up its transient assemblies of processing capability afresh, nanosecond by nanosecond, from the individual single-state components of the world's networks. We consider that collectively the individual switches produce what we call a "potentially panpsychic" environment. That means the networks and its component parts all become part of the personality.'

'You mean the individual transistors of all the processing devices attached to the global networks *themselves* become the processor?'

Robert nodded. Theresa stood up and stepped into the ambient light of the Holo-Theater.

'Thank you, Robert, well put. You see Raymond, the human brain, the only vessel in which true consciousness of self, as we understand it, has emerged, contains about a thousand to the fifth switching combinations – one hundred quadrillion, yes? – and the only possible processor to get anywhere near that number would be a combination of all the individual switches inside the world's processors connected by the networks. It's a question of scale – although we still fall far short of the number we'd need to mimic a single human brain. Nevertheless, the global networks at least give us a simple prototype for an artificial neural nervous system. Let me see if we can show you. Robert?'

Her senior researcher had taken Theresa's seat and now he started to make adjustments to the controls.

'I'm now going to try and show you what your dark data are,' said Theresa. 'But before we start, we must address a fundamental problem that occurs when we contemplate technologies that have yet to be fully developed and harnessed. You see, the difficulty is that I can't properly describe it to you – at least not in words we'd both understand. Wittgenstein summed it up when he said: "Concerning that of which we cannot speak, we must pass over in silence." In short, we have no language for the future. Whenever something is completely new we don't have *words* for it and that means we can't *think* about it. When the projector was first invented we had to call it a magic lantern. The car was a horseless carriage, the radio the wireless – we were reduced to describing it either by something it *didn't* possess, or by an allusion to an *existing* concept – the iron horse, the flying machine. I could go on, but you get my point.'

Raymond Liu nodded. He understood that.

'As a first step, with the comparatively limited processing power of today's global networks, we're trying to create an entity that is both omnisentient and anoetic – it has feeling, consciousness if you like, but there is insufficient complexity for real thought. Given the primitive and small-scale state of the neural pathways represented by today's global networks it can't have any capability for action. We are trying to mimic the way human consciousness may have emerged over the four billion years of biological evolution on this planet. In another quarter of a century we think the global networks will have grown sufficiently to allow real emergence to take place but for now this is more like a *gedankenexperiment* – a thought experiment with practical components – if that doesn't sound too much like an oxymoron.'

Liu had no idea how to react. *The people in this room were trying to turn the world's networks into a brain!*

'So, without a mutually agreed language for the new, my following description may sound a little – well, opaque. We can only *show* you and we'll do that in a moment. We have been creating and releasing software moneme-microsthenes into the networks for three years – you're familiar with memes, the information equivalent of genes? We use the smallest elements of language – the lexeme and moneme – and couple that with a grammatical-instruction carrier that Robert has developed – the microsthene – to create a hylozoic ecological system. When there are sufficient memes created from monemes – that's about a thousand to an eighth, one hundred trillion – and sufficient individual transactions occurring on the networks – the on-off switchings that Robert talked about – we expect to achieve parallel-processing states that create recurrent connections and what we call transient assemblies – the sort of pathways and structures we find in the human brain. From this we hope to observe the earliest stages of *emergence*, of consciousness, arising. An epiphenomenon. What would you assume that might resemble, Raymond?'

He felt like one of her most ignorant students. He realized how privileged he *should* feel, but he was numb. He shrugged, unable to respond.

'It's a primitive form of dream, Raymond. And we think the very first "primitive dreams" began to emerge about a year ago – Robert?'

'I'm switching to visual representation now,' Robert said. 'Note, this is *not* a simulation. It is a visual interpretation from the swirling swarms of the memes as they move through the world's networks seeking out active switches and jumping the gaps in the networks from one to the next. This is made up of about four hundred billion switch firings per second. We call it "René", after Descartes.'

Suddenly the Holo-Theater pit was filled with light and Raymond Liu saw what looked like blue smoke. At the top right he saw a digital tripmeter headed *Transactions.* Then he saw that the central light was a vignette of blue, smoothly graduated but swirling. Loops and whorls appeared and from the bottom an amethystine glow began to spread gently upwards.

It was so peaceful. Raymond Liu found himself entranced, becalmed. He thought he could make out a shape in the colour. But nothing came of it. Then he thought he saw something else. There was no sound in the Control Center. They drank in the colours as they shifted through the visible spectrum and the counter spun in its recordings of the switch firings in the global networks.

Eventually Theresa stepped forward again. 'Thank you, Robert,' she said and her researcher killed the display. She turned to face the network chief.

'We are fairly sure that's an expression of an early dream, Raymond, but that's as far as we've got: colours, a few shapes. And that's taken eight years of preparation and work – but it emerged spontaneously. I'm afraid those quadrillions of shifting moneme-microsthenes probably account for the mysterious loading of your networks. Individually they're too small to be counted, but collectively ...'

Liu was still entranced by the display he had seen. Then, being an engineer, he made one more check. 'I have to ask this, Professor,' he said. 'I don't fully understand what you've been telling me but I think I get the drift. Is there *any* chance that whatever it is you've been trying to create could have ... well

... could have developed into something more? Something that could have consciously or unconsciously interfered with normal network operations?'

Theresa stared at him and pursed her lips. 'It will be decades before we have a sufficient number of switch firings in the global networks to properly resemble the neural activity of the human brain,' she said firmly. 'At this stage I have to say not a chance, Doctor.'

'Exclusive: We Reveal The Guest List At The Party of the Century,' trailed the cover of the latest *Hello* magazine. Haley leaned forward and picked the top copy from the pile. She flipped to the pages indicated.

'Thomas Tye Brings The World's Leaders, Thinkers and Artists to Hope Island to Celebrate his Appeal for Ethiopia and the "Take Care of Our Planet Campaign".'

Below were pictures of the American, EU and Russian Presidents, a galaxy of other world leaders, film and music stars (including Josh Chandler) and, of course, Thomas Tye himself. He was good-looking enough for the layout designer to have given him the largest picture on the right-hand page of the spread, facing the impossibly dreamy Josh Chandler.

Haley flipped over to the next page.

'Mr Tye's "One Weekend In The Future" Will Set The Global Environmental and Technological Agenda for the Next Decade,' she read. Beneath this gushing prose were photos of the artists, writers, film stars and intellectual giants who were participating in the four-day event.

'Are you buying it?' demanded the Pakistani shopkeeper testily. Haley nodded and transferred payment from her LifeWatch.

Rolf Linquist Larsson sat opposite Michael Chevannes at the dark wooden conference table. They were alone. The UNISA officer noted that the tall Swede's tanned face had become lined and his hair had retreated a significant distance from his forehead since the last known photograph of him had been taken eight years before. He was dressed in a crumpled khaki shirt and cut-off white jeans that were more than a little grubby, his bare feet semi-shod in open-toed sandals. In fact, he looked as if he had just got out of bed. But then, Chevannes realized he'd never knowingly met a multi-billionaire before.

Further up the approach road, Chevannes's vehicle had been met by the same security guard he had encountered earlier. He had been led into this room in one of the white buildings and asked to wait there. But he'd been left with all his communications and recording equipment. He had promptly replaced his RayBan Electros with a pair of Armani clear-lens viewpers switched for Local recording. A water-cooler, a wall-mounted whiteboard covered with mathematical formulae and an inactive display screen were the only furnishings other than the table itself and half a dozen chairs. He was kept waiting for less than ten minutes.

From his shirt pocket Larsson took out the letter Chevannes had delivered and spread it flat on the desk. 'May I see some ID?'

As Chevannes offered his ident, the astrophysicist glanced at it and nodded, making no attempt to verify or copy it. It was interesting that Larsson did not wear viewpers, LifeWatch nor VideoMate. *The ultimate luxury of disconnection*, thought Chevannes.

'How did you find me?'

'I don't know,' lied Chevannes. He had been ready for this most difficult of questions. 'My office in New York provided some details and asked me to come and talk to you.'

'This is private property and I don't have to say anything. I have broken no law,' responded Larsson in perfect English – honed from his years of exposure to the international academic community.

'I have come here alone, Doctor. I did not ask the Peruvian government for help. In fact, they don't even know I'm here.'

He waited as Larsson glanced down at the note again. It was a difficult letter to ignore, its message simple and to the point. On the embossed, headed notepaper of the United Nations, it was signed personally by the Secretary-General, and Ron Deakin had arranged for it to be couriered down to await Chevannes's arrival at the UN compound inside the perimeter of Lima's Jorge Chávez International Airport. Larsson bent his head to read it yet again, as if the contents might have changed in the last few minutes. Chevannes knew the details of the request from his own copy.

*Dr Larsson,*

*On behalf of the 212 member countries of the United Nations, I write personally to ask for your assistance.*

*The United Nations International Security Agency has reason to believe that just over seven years ago you passed a piece of intellectual property to Thomas Richmond Tye III, president of the Tye Corporation. I am told that if used improperly this knowledge would have the potential to destabilize the world's economy. If this were to occur, in the way that Dr Yoav Chelouche, president of the World Bank, believes possible, it could bring immense strife and suffering to millions of the world's poorest nations and their peoples.*

*I understand that you may have given undertakings and entered into contracts that seemingly inhibit your ability to assist us. But this matter is of such importance to the global community that, if you will help our organization to protect the interests of its member states, I am empowered to extend UN diplomatic immunity to you on all issues related to this transaction. Such legal protection would render you immune to civil or criminal legal proceedings relating to the transfer of the above-mentioned intellectual capital within the jurisdictions of all member states.*

*We now need your help as a matter of urgency, and Intelligence Officer Chevannes of UNISA will provide further details.*

*I hope to be able to welcome you personally to the headquarters of the United Nations in New York in the very near future.*

*Yours sincerely*
*Alexander Theodore Dibelius*
*Secretary-General*

Larsson finished his rereading of the letter, sat back and gazed at Chevannes. 'Can you guess what we're doing here?' he asked with a sigh.

'Astronomy?' hazarded the intelligence officer.

Larsson stood up quickly and walked to the window. He looked up at the huge dish above them that blotted out the sky. 'Not just that. We're listening, Mr Chevannes. We're listening to the universe.'

He turned back to face the UN emissary who remained seated at the table. 'We have the only ears that can understand.'

Chevannes began to wonder if Larsson's drug abuse had caused some permanent damage.

'Can you force me to go with you?' Larsson asked abruptly.

Chevannes shook his head. 'Our charter gives us special jurisdiction in all our member states – Peru included – but we have no powers of arrest. We would have to ask the local police to act for us, but we have no intention of doing that. As you point out, we know of no international law that you have broken. I have come here because I believe this issue is of crucial importance. We really do need your help, Doctor.'

'You can't imagine how important my work here is,' Larsson said with a shake of his head, resuming his seat. 'It dwarfs anything Thomas Tye could get up to.'

'Tell me about it,' suggested Chevannes.

Larsson studied his visitor with clear, Nordic-blue eyes, considering his options. Suddenly his face broke into a broad grin.

'Then you'll have to stay for dinner,' he exclaimed. 'That'll give me time to think things over. We've got plenty of accommodation, and there's dorado and empanada for the barbecue tonight!'

It started in Moscow. After consultation with the newly reinvigorated financial community, President Orlov declared Sunday, 30 August would be a day of federal celebration in support of Thomas Tye's global appeal for aid for Ethiopia and the associated campaign to 'Take Care of the Planet'. Mikhail Orlov also knew that on the same day he and Thomas Tye would be announcing the creation of the new state of Sybaria and the climatic re-engineering of huge tracts of Western Siberia. As Tye suggested, the excitement over the Ethiopian project would help to drown out any misguided ecological concerns about the latter project. President Orlov expected Russia's massive national debt to be eliminated by then and that would also be announced proudly during the lavish festivities on Hope Island. The president's office had ordered street and village parties to be organized throughout the entire Federation, providing a large special fund to subsidize the cost. Thus everybody would appear to have an immediate share in Russia's windfall.

Michael Chevannes sat beside his host in silent wonder. They were alone on a specially designed wooden bench beside the observatory building. Its back was sharply raked to allow the occupants to sit comfortably while leaning their heads back against a flat tilted rail. It was nearly one a.m. and the heated seat was warm against their bodies.

Although Chevannes had travelled to many countries of the world, he could not recall seeing anything to approach the majesty of the giant moon above and its glittering stellate canopy. They were *surrounded* by stars, as if drawn up inside a cupola of the gods. Far below the land was lost in infinite blackness and the thin, clear air at the mountain peak allowed the stars to shine gold, without the twinkling interference and chromatic distortion of the lower atmosphere. The moon in perigee, vast, gibbous and immediately overhead, was bathed in light from the sun that revealed the mountain ranges and the bruised depressions of its continually battered impact craters in sparkling, sharp clarity. Beyond and all around, providing the celestial backdrop, was a wash of white stars so dense it seemed like phosphorescence on a summer night's sea. Away to the west, perhaps twenty or thirty miles distant, dim laserbursts of a satellite communications network transferred data between the northern and southern hemispheres – and, appearing in all quadrants, catching his eye and animating the siderealism, were the white-red death streaks of meteors entering the atmosphere.

Neither he nor Larsson had spoken in the last fifteen minutes, and it was absolutely silent on the mountain top. They sat on their warm cosmic *pulvinar*, allowing their consciousnesses to cavort in the wash of infinity.

The natural ambient temperature at the mountain top was close to zero, but they were in their shirt-sleeves. When Chevannes had queried the viability of a barbecue in the chill of a mountain evening, his host had explained how solar-powered underground heating had been laid throughout the compound. At every few yards additional gas-flame outdoor heaters created microclimates like a warm summer evening, all run on natural methane gas. 'Produced locally,' added Larsson.

Earlier, Chevannes had learned that this was the highest observatory in the world, as the astronomer showed his guest the sixteen-metre adaptive optical telescope, then his real-time feed from the six orbiting Hubble telescopes and a NASA Mars-orbit telescope, his computer array for mining the vast warehouses of astronomical data collected, and the smaller direct-view optics housed in his observatory. Chavennes, the intelligence officer, had been hooked instantly. They had been touring the building for nearly three hours, only coming outside for occasional relief from visual overload: *Some relief!* thought Chevannes now as he sat in awe with his head tilted back.

'It was Debussy,' said Larsson eventually, 'his Cello Concerto in D minor. Someone once asked him what it was about and he explained it was Pierrot angry at the moon.'

Chevannes had no idea what his host was talking about. He said nothing and silence returned.

'What did they expect me to do, write a paper?' asked Larsson, sitting forward a minute later, a hint of bitterness in his voice.

The visitor guessed this was a rhetorical question and left it to roll away across the mountains.

'Perhaps I should have settled for a Nobel Prize. "Receiving an invitation to Stockholm", they call it. Huh! That's where I started from and for a mathematician it's downhill all the way after thirty.'

Chevannes nodded emphatically, not knowing how else to respond.

'The probability is very high that forms of aware intelligence exist around some of those stars out there – or around those we can't even see,' continued Larsson pointing at the sky, seemingly intent on an evening of non sequiturs. 'What we can see in front of us with our own eyes is only part of our own galaxy – less than one per cent of the known star systems.'

Chevannes still said nothing. He knew he wasn't expected to.

'We are now seeing those stars as they were between two and four million years ago: that's how long the light – the information about them – takes to reach us.'

Another minute's silence.

'No one has been able to prove whether they're still even there or not – like Schrödinger's hypothetical cat, kept inside a box. It might be alive, it might be dead. But until we look, mathematically speaking, it's a mix of the two. So, until my work of seven years ago, we could only safely assume those stars were both there and not there.'

Chevannes turned his head to look at his host: he felt like Alice in Wonderland.

'That's the basis of all quantum theory, Mike,' continued Larsson. 'At the quantum level that's just how the world's physicists believe subatomic particles behave; *where* they are depends on who is looking at them – and when. But when no one is looking at them they are considered to be in both phases, in superpositions. But *I* know where they are, Mike. And I can prove their positions, I know how to allow for the effect of observation and measurement and *non*-observation!'

Chevannes nodded again. He would need to practise believing impossible things. Perhaps he should start before breakfast each day.

Larsson glanced at his guest and changed tack. 'We can be sure that some of those stars are suns with orbiting planets, perhaps carrying elements that may have induced some form of what is called life – I prefer the term "intelligence". We're currently looking out for extra-solar planets possessing atmospheres, and simple maths shows us that any form of aware intelligence out there is certain to be *much* more advanced than ours. Half a million years – the time the human species has been around – is merely an eyeblink in cosmological terms, so anything less advanced than us wouldn't even get classified as *aware*.'

Chevannes could feel a sense of inner dislocation occurring. He was inside himself and not; in virtual evanescence. Silence stole around them again.

'I installed all that–' Larsson continued at last, waving at the optical observatory building '–for my own pleasure, so I could personally see and feel closer to a part of our galaxy. But all the really important work these days is done by analysing the feeds we get from space telescopes orbiting the Earth, the moon and Mars, and the signals we receive with these radio dishes. You can't see much from Earth these days, anyway. All serious astronomy is now virtual – non-real-time. It's done by mining data.'

Chevannes's own mind – his 'virtuality' – wallowed in the infinity of light and energy above him but then slowly and reluctantly returned to its earthly container. After a while, he glanced at Larsson and summoned the courage to ask a question that might make him seem a fool.

'Do the radio telescopes also have to be this high up?'

The astrophysicist snorted and leaned forward, reaching into the side pocket of his shorts for a cigarette case.

'These days, yes. Since communications systems went wireless the radio spectrum is as polluted as the visual – like the atmosphere. This altitude – and Peru's isolation on this side of the Andes – takes care of most of that. I also wanted to be private. Nobody ever comes up here.'

The intelligence officer nodded. He tried to guess what it might have cost to transport the materials to the top of the mountain and build the twenty-three radio dishes and their tracks. Of course, if he had Larsson's money he too might do something crazy, something wonderful, like this. This vantage point, this private *montevideo*, provided a view previously unknown to him, a perspective unimagined by his terrestrially shackled consciousness.

Larsson lit his joint with a match, blew out the flame and flicked it away into the darkness.

'I don't want to be away from here long.'

They were due to leave the next day.

The UN man stretched. 'With any luck it will take just a few days – a week or so at most.'

Larsson drew on his joint. 'You understand why this matters so much? Any advanced civilization will be wholly virtual. Even if they still retain some physical form, all their communications, transactions and interactions will be virtual. That might be digital, it might be some other form of representation, but I'm certain that however they communicate they will encrypt everything. They will be using prime-number or quantum encryption for *everything*. Why do you think the thousands of radio telescopes in the SETI programme have so far found nothing? All they would be hearing is white noise, even if they were receiving full-on extraterrestrial communications every day.'

Chevannes nodded again, even though he barely understood.

'Others can't understand what they're hearing – but *I* can. And that's why my work here, my data mining with my decoding software, is so important.'

# TWENTY

The great river was at the peak of a spring tide; a billion gallons of water embanked in a swirling equilibrium. Nature's forces were at stalemate for a few minutes; ancient antagonists brought to yet another temporary cessation in their involuntary reflex to the planet's pulse.

Jack leaned on the stone parapet of an elevated, gilded Pagoda – a Japanese 'peace present' to the people of London – and watched eddies form where the scouring bore of the English Channel met a stately Thames and suddenly, savagely, turned it estuarine. Three hundred yards to his left, on one of the riverside benches in Battersea Park, the elfin figure of Haley Voss sat absorbed in a book. She was dressed for the heat in a loose white short-sleeved blouse and long white linen skirt. Her left hand was resting on the handle of a child buggy but Jack guessed, from the lack of movement, that little Toby was sound asleep.

He had been watching her for nearly half an hour and neither of the two London agents drafted in to assist him had reported anyone who looked remotely like a tail or any other form of surveillance. She was clear.

He pocketed his viewpers, pushed himself away from the parapet and walked down the stone steps. Crossing the monument's gravelled surround, he began to stroll slowly along the edge of the grass towards her. The heat of the early August afternoon was oppressive – climatic changes were making London summers more like New York's – and he was glad he'd worn a polo shirt and shorts.

Jack was supposedly now in upstate New York visiting his mother again. He had received a suitably worded message from the hospital – her condition was causing concern, but he should not be over-alarmed – and despite the frenetic activity as Hope Island prepared to host its *One Weekend* celebration, he had promptly taken compassionate leave.

At UN headquarters he had found Deakin and his much enlarged Operation Iambus team running at a far higher level of activity than before. Jack had been surprised when told that Counsellor Furtrado himself had recently flown to London to visit Haley Voss. Then he had laughed when he heard the outcome. He could imagine their encounter.

'But that's not necessarily the best outcome,' Deakin had pointed out. 'She could do a lot better working from the inside, if she could get more intimate stuff on him. Do you think we could trust her to work with us, Jack?'

So 'Bruce Curtis' had flown to London that same night, feeling irrationally excited by the task ahead. During the journey he had joined his mother for breakfast via his VideoMate, Al Lynch having adjusted both Jack's unit and his mother's, to register their current locations as upstate New York. So Jack leaped forward a few hours to UK time and spent ten minutes reassuring his mother that he was safe and well, while observing that she was thoroughly enjoying bossing her younger sister around.

As Jack came within the extreme boundary of Haley's peripheral vision he noticed that the heavy volume in which she seemed engrossed was a treatise on antitrust and competition law. He stopped and waited, watching her eyes saccade across the print, her pen poised in her right hand, ready to mark any passages of interest.

He saw her head turn, then she glimpsed his feet. She stared up, directly at him and gasped.

'Hi,' said Jack. 'Good book?'

'Jack!' Her smile eclipsed the afternoon sun. She leaped to her feet, the book falling to the ground.

Crossing the space between them, she reached up to hug him and kissed him on the cheek.

'What the hell are you . . .?' She tailed off. 'How did you find me here?'

He gazed down into her earnest brown eyes.

Then she groaned. 'Oh, don't tell me, you can't help any more. That lawyer's sent you to talk me out of it, hasn't he? Forget it, Jack. There's no way.'

She turned away to check on her charge, then picked up her book and sat down again, staring out at the water.

He stepped forward and sat down at the opposite end of the bench. 'That's not it,' he said gently.

'So why *are* you here?' She turned her head towards him.

'It is a lot of money, Haley. I'm not sure I would turn down that much,' he teased.

'It'll make a good intro,' she said defiantly, looking back at the river. '"I was offered eighty million dollars not to write this book!" I recorded it all.'

She put out her left hand to rock the child buggy, as if Toby had suddenly woken and become fractious. One leg was crossed over the other, swinging back and forth in the same rhythm, as she stared straight ahead.

'Yes, it will,' he agreed. 'But there might be an even better one.'

She shot a look at him. 'Who *are* you, Jack? How come you just pop up out of the ground whenever you think I need you – like a genie in a pantomime?'

Jack glanced to left and right. There was a jogger approaching from one side, otherwise this part of the park seemed deserted.

'I'm here to ask you to change your mind, Haley. If I know Furtrado, he's told you how you can still accept the company's offer.'

'I thought so,' she sighed sadly. 'You don't know me, Jack. There's absolutely no way.'

But Jack knew her better than she guessed – and he still wanted to know much, much more about this woman.

'No, that isn't what I mean, Haley. I want you to *appear* to agree. To take the access to Tom they've offered you.'

She turned to stare at him, still not understanding.

'Oh! I should do "A Year In The Life of Thomas Tye", should I? Live on his island, attend the meetings, meet the main staff, travel with him – the TV, the performances, the deals. Take my pay-off money, write a hagiography and *finish* my career!'

'You wouldn't need a career after that.'

'I'm not like that,' she replied simply.

'Leaving aside the money on offer, don't you think all that inside stuff could be useful?' he reasoned. 'For producing the sort of book *you* want to do.'

Haley looked at him, wondering. 'What do you mean? How could I?'

Jack, in his turn, contemplated the river. He allowed thirty seconds to pass as the jogger – not one of those marathon dromomaniacs but a reluctant middle-aged male – panted past on the grass running-track behind them.

Then he turned to face her, resting his arm along the bench.

'Haley, your book, as you originally planned it, *needs* to be published and it will have to be as hard-hitting as possible. It was you yourself who said that no one man should be able to control the future. The world *should* know the real story about Thomas Tye, about the genetic experiments, about Tommy's birth, about the anti-ageing technologies, about how he abuses people. But your book could be even better, more insightful, if you appear to take their offer and use the privileged access they're offering. The future you're concerned about really is too important to leave to chance – or one man.'

The proleptic future by her feet gave a sudden gurgle. Haley leaned forward, temporarily distracted by Toby's reminder of his presence, but the child had merely turned over in his sleep and was peaceful once more.

'You know that's not possible. I'd have to sign a binding contract. Once I take their money, I'm gagged. End of story.'

'Not necessarily,' countered Jack. 'You wouldn't end up keeping the pay-off, but I could fix the legal stuff so they couldn't touch you.'

She turned to stare at him, then flicked her head away in anger. She folded her arms, head bowed in thought.

He wanted very much to hold and kiss her.

'How could you do this?' she asked, with her chin almost touching her chest. 'Who the hell *are* you, Jack?'

'Sloan will still be your publishers, but you'd be able to add all the inside stuff – the details of his life and how the company works. Sloan will go along with it.'

'You can speak for the Sloan Press.' It was a statement of realization rather than a question. 'OK, Jack. Tell me about it.'

So he had to tell her – and the tide started to turn.

Meanwhile, a few miles to the north-west, the secretary to a large and very vocal committee carefully recorded its vote as unanimous. Chair had made the suggestion in what would now be the last full meeting before the big day itself. But, despite the lack of time for creating replacement publicity, it was agreed that this year's event should be dedicated to the Thomas Tye Appeal for Ethiopia.

The Notting Hill Carnival, now in its fifty-first year, had become the largest street party in the world, eclipsing New Orleans's Mardi Gras and even Rio de Janeiro. Over eight million people now pilgrimed each year to the fashionable streets of West London, but no one intellectualized this rite, no one noted or cared that its timing roughly coincided with the ancient harvest festivals of the northern hemisphere. One of the planet's ancient rhythms had been given a West Indian accent and new lyrics. But the committee members were all aware how it coincided with that rather more exclusive celebration on Hope Island – 'that other little party', as Chair had called it – and her appeal to help out the brothers and sisters in Ethiopia won the day.

So arrangements were made for each of the floats in the procession to accept electronic donation transfers from the LifeWatches and VideoMates worn by millions of spectators along the route, and permission was granted for the Tye Network News channel to integrate coverage of the Carnival with both their live feed from Ethiopia and the *One Weekend In The*

*Future* celebrations on Hope Island. The Carnival organizers then congratulated themselves on seizing the opportunity to contribute significantly to the greatest philanthropic and ecologically correct gesture in history.

Joe Tinkler was unused to so much jet travel and although he knew his body occupied a seat, he felt as if *he* were still hovering somewhere behind, over some distant continent, just visited. He had experienced the *bizarrerie* of watching the analog time display on his Rolex LifeWatch readjust itself to accommodate seven time-zone changes in one week. He had already turned the jet-lag compensators to 'full' in his viewpers, but his system seemed unable to cope merely by adjusting the light levels that reached his eyes or the level of melatonin delivered by his LifeWatch.

He had just spent four days in London, three in New York, then two days each in Boston, Minneapolis, Denver and San Francisco before crossing the Pacific to visit Hong Kong, Singapore and the royal states of the Middle East. He had finally met in person those he had known for years only as network representations: there was little time for conventions and other industry get-togethers in corporate finance and the participants rarely got to meet face to face. In the flesh they seemed so familiar but, at the same time, so different. Their corporeality flavoured their character in a way that even the most advanced holo-image theatre failed to capture. He wished he had made the effort years before. The only major house he hadn't yet visited was Rakusen-Webber.

Joe Tinkler was also in unfamiliar territory in another sense. As was true for so many analysts, fund mangers, investors, traders, economic forecasters and financial journalists, the prospect of *personally* operating in *real* business terrified him. It was one thing to judge the efforts of others, to identify winners and losers and bet other people's money on it; quite another to imagine how one might fare in such activity oneself. Chelouche had instructed him to form a number of shell corporations based in the United States and Europe that would own the Tye Corporation stock purchased during a take-over. In all of them,

Joe was named as president and chief executive. Even though Chelouche reassured him that consultants would be brought in to manage the businesses, suddenly what had before appeared as a game now seemed brutally real. Joe was honest enough to admit to himself that if handed any real decision-making responsibility for day-to-day activities in any part of the vast Tye empire, he would not know where to start. That would require manipulating people, not numbers, and Joe had little experience of that.

He had most of his options now in place and he was going home – well, back to Geneva. The meal in the first-class upper cabin of the Airbus 1000 had been excellent and, after the flight attendant had cleared his tray, he had opened his VideoMate to return to the vast spreadsheet in which he was keeping an updated total of the commitments he had been making on behalf of the World Bank. Tye Corp's core stock price had been rocketing in anticipation of the Ethiopian technology launch and several of the Tye Corporation's institutional shareholders had insisted on prices so high that exercising these options in the future would be fabulously punitive. The amount he had already potentially committed exceeded his largest estimate and he wondered how Dr Chelouche would react on learning the sums involved.

As his exalted shadow-employer had requested, Joe conducted every negotiation in person and he had filled out the option agreements by hand in front of his wondering counterparts. Joe's cover story was a small truth, which is often the best servant of a larger lie. He wanted his option on their Tye Corporation shareholding to be kept a secret, he explained. He was contemplating some power plays in the market and he didn't want to signal his intentions in advance. Only Joe Tinkler could have got away with such a ploy: he had already been the world's best-known Tye Corporation analyst and investor and, since his spectacular move from Rakusen-Webber to the largest bank in Switzerland, his personal reputation had made him the brightest star in his own arcane firmament. The options were for ninety days; there were non-execution and get-out clauses galore to protect the sellers, but all were subject to ultra-strict confidentiality clauses.

The net effect was that, if Chelouche so wished, the signed papers in the pilot's bag lying on the seat beside Joe would, on receipt of the stipulated payments, deliver nearly twenty-one per cent of the core Tye Corporation's voting stock into the hands of the World Bank. This would provide Joe with an almost unassailable position from which he could begin buying on the open market with the aim of reaching a position of overall control. If this happened, Joe realized, it would in turn deliver the whole Tye business to the UN and, ultimately, to the people of all its member states. These days, that was almost every nation in the world.

Joe flipped open his communicator and noticed he had AgentMail waiting. It came from one of the software agents he had reprogrammed to search for any new land deals entered into by Tye International Real Estate. He read through the government-published documents the agent had retrieved. Now Tye was buying vast tracts of real estate in Northern Canada! One deal alone was for a strip of land around Hudson Bay that ran for nearly a thousand miles through Manitoba, Ontario and Quebec! Another was for thirty-two islands in the Bay of Alaska, off the coasts of Yukon Territory and British Columbia. He pulled up an atlas and zoomed in to inspect the territories concerned. He noticed that they were at similar latitudes to Tye's Baltic acquisitions.

But none of this made sense. Canadian zoning and planning laws were every bit as strict as US regulations and commercial or industrial developments must be out of the question in areas of such natural beauty. Nothing about the purchases suggested a prospecting operation either and Joe already knew that there was not a single mineral or petrochemical operation in the entire Tye empire. It certainly wouldn't be oil: for one thing Tye was the major global opponent of the fossil-fuel industry – and, more practically, the price per barrel had stubbornly remained below $10 for over a decade. As global warming had continued to increase – despite many late-twentieth-century experts who had predicted the opposite – most analysts believed that the solar-fuel capture and delivery systems produced by Tye's Solar Energy Division, and two or three less prominent competitors, would completely eliminate the fossil-fuels market by the end of the century.

Joe studied his map again, then reviewed the deal announcements his agent had found. Some of these purchases were freehold, others were for very long leases. He made a PopUp reminder for his assistant to get hold of copies of the original conveyance deeds.

Haley was incandescent with rage. 'You bastard!' she screamed at him. 'You allowed me to think Sloan Press wanted my book on its merits and all along you and your friends in the UN were *paying* them to publish it!'

They were back at her flat after a still-slumbering Toby had been returned to Felicity's care. Jack had walked around the park with her to where Haley's twin sister was due to collect the boy in her car. The sister had eyed Jack knowingly as they were introduced. She already knew about him, Jack recalled. He had been astonished to see them together again: they *were identical.*

'We're *not* paying them, just underwriting their legal liabilities – protecting them from the Tye Corporation's lawyers. They bear all the normal costs of promotion, marketing and so on.'

Jack had told her as much as he was authorized to during the last half-hour they had been back.

'But you allowed me to continue thinking this was a straightforward publishing deal. You fed me information! YOU SET ME UP!'

She turned away from him, trembling with anger, and stared unseeingly out of her window at the trees in the park she knew so well.

He stared at her back, at her slender neck with its wisp of dark hair curling down towards her white collar. Feeling an instant and urgent longing, he wanted to cross the room and hold her. In comparison, Calypso's beauty now seemed abstract, something too perfect, something chiefly to admire. But this seemed real and very close. He realized suddenly that his long sleep of emotional regeneration was finally over, and that he had woken to a desperate longing for the woman standing with her back to him. Helen's face swam into his mind and he was able to imagine her granting permission.

But the woman in front of him was in no mood to care. She swung back to face him, her fine chin jutting with determination and anger.

'You lied to me Jack, you lied, and that's the worst thing anyone can do. Just leave me alone. I'm going to do this my way and I don't need you, the United Nations or Sloan Press – or any of you. I'll find a new publisher and I'll do this *my* way!'

He saw tears – was it anger, pride, regret? – gathering in her wide brown eyes. She tossed her head once and then glared up at him.

'I think it's time for you to leave, Jack, don't you?'

*I think I am hopelessly in love. I want you so much I don't know what to do. My pulse must be over a hundred. My stomach feels worse than it used to when I was about to go into action. My hands are wringing wet. My breath is coming in short gasps.*

He takes a step forward, strange geometric hinges in time and place opening before him.

'Haley . . .'

She doesn't move.

He reaches out and caresses her cheek with his fingertips.

'Haley . . .?'

She is staring up into his clear blue eyes. All she sees is honesty – and a light she doesn't recognize, but one that seems compelling and all-embracing.

'Haley, I don't care about any of this. All I know is I don't want to upset you. I never want to upset you again. I . . .'

She can't stop her tears now, but she can't turn away. She feels them tumbling over her lower eyelids and down her cheeks.

Suddenly he is against her, holding her head in both his hands, kissing the tears from her cheeks. He kisses her lips, her mouth, and she hungrily kisses his. She no longer has any idea of what she is doing, but her heart is singing.

The writ was served by hand, in the front office of the Tye Corporation Legal Services Department on F Street and 6th, Washington DC. In itself this wasn't unusual. Writs were still sometimes served by hand, although electronic filing had long since become approved as the fast and sensible way to serve notice of a legal claim. But some contrary legal firms still liked to pettifog with old-fashioned methods, and Furtrado assumed that Masters, Morrison, Johnson & Co of Knoxville must have deliberately set out to be vexatious.

He broke the seal and sat back to read. To a seasoned lawyer a writ meant little more than childish name-calling in a schoolyard: as many as a dozen different writs were received in one part or another of the Tye group each day. Few of them ended up in Furtrado's personal office, since usually they were appraised by one of his team and spun out to the most appropriate law firm to make an initial response.

But this writ had been delivered just as Furtrado himself was crossing the lobby. He had scooped it up from the reception desk with a smile; it never hurt to sample the daily activity.

He read it once, then read it again more carefully. It had been issued by a wholly champertous Southern law firm on behalf of the Memphis-based family of a retired fast-food franchise owner, now deceased. The man had bought a replacement liver in an auction run by LifeLines Inc., part of the Tye Life Sciences Group, for $20.5 million. The writ claimed that the liver had been guaranteed a 95.21 per cent match with the recipient's DNA profile, but the actual match had turned out to be less than forty per cent. Despite belated immune-system treatment, the patient had died as his body rejected the organ. Attached to an aggressive covering letter sent in advance of the formal legal procedure of mutual 'discovery' and disclosure of evidence was a copy of LifeLines's guarantee of the DNA match profile and an extract from a consultant pathologist's report declaring that the organ was, in fact, no more suitable for the patient than any random organ obtained on the charity market.

Such *prima facie* evidence was compelling. The family was claiming back the original $20.5 million and seeking $800 million in punitive, *solatium* and compensatory damages.

Furtrado had never seen such a claim before, but then, LifeLines had only been in operation for three months and had only recently enjoyed a spectacularly successful flotation on the EUUSA biotech market.

The counsellor knew immediately whom to call. Zachary Zorzi, as Tom's latest rising star, was now busy building the auction models that would be used to market all products produced by the Phoebus Project. Furtrado found himself praying that this writ wouldn't prompt the discovery of any serious technical problem that could affect *that* launch.

# TWENTY-ONE

'Give us a little tour,' Chevannes shouted above the engine noise. 'We've got a first-timer.'

The pilot nodded and the UN Jet Ranger 209B rose from the JFK helicopter pad as readily as a hungry bee from a depleted bloom.

On their long flight north, Chevannes had learned that the Swedish mathematician and astrophysicist had never visited New York and, like millions of immigrants before him, the UNISA officer couldn't resist proudly showing off his adopted city.

They had travelled from Lima first-class, in seats paid for with UN money, a new experience for Chevannes but a luxury the directors of Operation Iambus considered appropriate to their guest's status. Unknown to that guest, two UNISA field officers dressed as businessmen had also shared their large cabin – 'just in case,' Deakin had said, although Chevannes couldn't imagine what such a 'case' might be.

As their plane had crossed into North American airspace Chevannes had noticed that the matrix of laser-borne communications above them was growing denser. He had remarked on this to his travelling companion.

'Yes, it's criminal,' agreed Larsson, misunderstanding Chevannes's meaning. 'Those lasers don't have to be visible at all. They just add the visible wavelengths for show, so they can be seen from the ground. Tye Networks started that as a marketing gimmick, years ago, and all the others followed suit. It ruins terrestrial astronomy.'

'Thomas Tye, again,' grunted Chevannes.

Larsson smiled and leaned over to thc window.

'But it's become the best economic indicator there is – like vapour trails above a city in the old days. When I was a boy my father used to tell me to look up at the sky to see how many jets were flying in and out of Scandinavia. That's how we knew how buoyant the economy was. Now we just look up at the night sky, and the more laserbursts, the greater the economic activity on the planet.'

Chevannes had simply nodded and returned his attention to the light show.

In New York the other passengers in first class were held back so that the two men could leave the aircraft first, clearing immigration and Customs by a nod directed towards the waiting US diplomatic officer at the exit. They had then stepped through a side door in the jetway ramp and down an open stairway to a United Nations limousine waiting on the tarmac.

In less than five minutes their helicopter was climbing westwards above the soul-sapping undulations of Queens's parched and polluted cemeteries. Chevannes found this rapid transition from a Peruvian mountain top distinctly disorientating and he tried to imagine how his companion must feel after so many years of isolation. But as soon as the man-made peaks of Manhattan rose in front of them, the Scandinavian intellectual was gawping, staring and smiling like any other first-time visitor.

Receiving clearance for her detour, the pilot looped south to follow the East River down to Staten Island. Here she circled the Statue of Liberty once, then climbed over the skycrapers of the financial district and followed Broadway until it was time to turn towards the East River again, to begin their approach to the landing pad on the top of the United Nations Secretariat building.

Chevannes had already called ahead and five minutes after his arrival on UN sovereign territory Rolf Larsson was shown into the thirty-ninth-floor office of Alexander Dibelius.

'We are extremely grateful to you for coming here, Doctor,' said the Secretary-General as he walked around his desk with his hand extended.

Larsson looked up into that sage face with an apologetic smile. 'I'm told some of my work from years back is causing trouble. I'm sorry.'

'We'll see what can be done,' said Dibelius, smiling. He turned to another man in the room and Larsson was introduced to Dr Yoav Chelouche, the President of the World Bank.

Over the next hour the two men explained carefully, as they had done to Jack Hendriksen two months earlier, the implications of the Tye Corporation's growing power and monopolies, the dangers to global economic stability and the threat to peaceful economic progress that it posed.

The astrophysicist was shaken and mortified by their revelations, beginning to understand the scale of the problem created by the private and exclusive sale of his decryption software. He explained to them that since he had recovered from his immersion in psychedelia his attention had been wholly fixed on astronomy and his search for extraterrestrial intelligence.

'I withdrew from everything to do with this world,' Larsson admitted. He had delegated the management of his vast financial portfolio to three different private banks and, apart from having stipulated that over fifty per cent of his wealth must be held in Tye-related stocks – for reasons he blushingly admitted were related to the competitive advantage he knew his software would bestow – he explained that he had paid little attention to economic affairs.

He had signed the UN silence resolution almost without hesitation, only delaying to confirm that he would be safely beyond the reach of any legal attack from the Tye Corporation.

'If I have UN immunity, will I be able to resume my previous work?' asked Larsson, as he realized what might become possible. 'Once I had sold the concept to Thomas Tye the agreement stopped me carrying on with any similar experiments. But now . . .'

'Prime example of a monopoly stifling innovation,' huffed Chelouche.

The Secretary-General smiled and said he would seek advice on the subject. For the moment, Doctor Larsson had better consider his work subject to UN approval.

By the end of the session Larsson had agreed to provide full assistance to the UNISA operation to limit the growing power of the Tye Corporation.

Then he was handed back into Mike Chevannes's care and taken to the adjoining Marriott Hotel, now owned and operated under franchise by a UN nominee corporation. There he settled into an electronically secure suite near the top of the tower – likely to be his home for some weeks, he now realized. Later that day he returned to UN headquarters and the serious debriefing began.

He was conclaved first with some prominent mathematicians retained by the UN. Alan Mathison had flown back to New York from the English fenlands and the great Professor Maurice Mendeléeff of MIT had flown down from Cambridge, Massachusetts to share in the revelations. After a day, this small investigating committee reported back to Jan Amethier and Ron Deakin. It seemed as if the Swedish prodigy *had* made a major breakthrough in defining sub-quantum states – and, as a result, in pure mathematics. As they made their report to him, Deakin could sense that they were suppressing a degree of irritation. Larsson would now continue to an experimental stage to test the strength of his proof. They explained that nobody could predict whether such experiments would be successful and, if they were, what the impact on physics, mathematics and other branches of science might be. The academics seemed especially annoyed that Larsson should have fully understood the importance of his achievement but had chosen to throw away academic fame – and the benefits of his work to the world at large – in favour of developing and selling a software simply to crack codes.

They confirmed that Larsson's software did indeed break all current forms of encryption, even messages encoded with super-long bit-lengths as used by the National Security Agency, the Pentagon and the US military. As an example they supplied a file of messages they had fished from the networks and successfully decrypted. Even this random trawl had produced startling results and Deakin and his boss exchanged glances as they flipped through them.

Despite their contempt for Larsson's commercial activities,

the two mathematicians' academic enthusiasm had become more evident as they reported their findings until the two UNISA men found it hard to keep smiles from their faces. Although they understood little of the theory, they were prepared to accept the mathematicians' analysis without question.

'So what now?' Amethier had asked finally.

'So *now* we've got to get on with engineering delivery systems for quantum encryption!' announced Professor Mendeléeff, seemingly amazed that the agency boss hadn't grasped the obvious. 'Nothing, *nothing* will be safe until that is done.'

Amethier shook his head and admitted he was lost here.

'All the UN agencies – the World Bank, the IMF, UNISA, everything – have got to move to quantum encryption as quickly as possible,' the mathematician explained with exasperation, as if the suggestion was totally obvious. 'That's *absolutely* safe. The technique itself was discovered years ago – back in the nineteen-eighties, by Bennett and Brassard – but everybody then assumed that our current encryption techniques were sufficiently secure; there was no need to develop the network technologies necessary to use it.'

He went on to explain that the ultimate encryption technology, one mathematically proved to be completely unbreakable, used the quantum characteristics of photons – particles of light – to encode messages. Even trying to take an unauthorized look at such messages destroyed the contents and Larsson's new formulae for calculating the effects of observation suggested many ways of exploiting that apparent irrationality. But the problem lay with developing networks and communication systems capable of transporting such minute and unstable elements. Now things were different, of course, and work must begin at once.

'You mean *this* unbreakable encryption will really be unbreakable?' asked Deakin. 'Unlike the *last* unbreakable encryption?'

They both nodded, oblivious of his sarcasm. They were keen to get back to their universities and begin raising money to start the work.

In the end, Jan Amethier called Larsson into the meeting,

thanked him again for his cooperation and congratulated him formally on his achievement. Both Mathison and Mendeléeff extended invitations for him to join their faculties if he ever got bored with life on his mountain top.

Easy Zee Zee was finding it hard to maintain his famous equilibrium and his normal techniques for mood adjustment had failed to restore a feeling of well-being. The recent call from Furtrado had sent a panic through the LifeLines operation. At the counsellor's insistence, every organ auction was temporarily halted – to the utter dismay of next of kin all around the world who had initiated but not completed the sale of donor assets they had just inherited.

The investigation had confirmed that the complaint from the Memphis customer's family was justified; the liver that had been delivered had been only a 38.67 per cent match for the DNA with few common genes on the sixth chromosome and almost no common antigens.

No obvious reason could be found. The computer records of the transaction revealed that the DNA files that had been automatically compared at the time had, indeed, shown a ninety-five per cent match. Now the two records were clearly a mismatch. The problems weren't in the computer storage systems or in the DNA comparison software. It had to have occurred in the networks.

Zee Zee was an expert on computer systems and he knew there had to be a forensic audit trail that would reveal where the foul-up had occurred, but he wasn't the one being asked to pursue that investigation. At Furtrado's insistence, he had e-mailed someone called Raymond Liu, the Technical Director of Tye Networks's global infrastructure, to ask for his department's help with the investigation. While he waited to hear from him, he went over the Phoebus sales plan again to ensure there was nothing technically flawed in what he was going to propose to Tom and the main board.

Al Lynch loved numbers, numerate people and machines with the potential for intelligence. When he had been a student his heroes were Blaize Pascal, John Von Neumann and Konrad Zeuss while his contemporaries had been living for Bruce Springsteen and Whitney Houston. His philomathy drew him to study the work of the great steganographers of history (and their cryptanalytical counterparts) for his doctoral thesis. He had loved travelling to Britain to study the original sixteenth-century books and papers of Thomas Phelippes, the man who had deciphered the encoded messages of Mary, Queen of Scots to her cloyning supporters in a plot against her cousin, Elizabeth the First of England.

Then he traced and studied the original notes and drawings made by Charles Babbage, the nineteenth-century inventor of the world's first mechanical computer. Babbage had cracked the 'unbreakable' polyalphabetic Vigené telegraphic cipher – *le chiffre indéchiffrable* – that had been invented by Blaise de Vigenère in the sixteenth century and that had been resurrected as the trusted form of inter-government communication once Samuel Morse's invention had gained international acceptance.

At that time in his life, it would have been fair to say that Alan Lynch was a total nerd.

All of this returned to him as he waited to meet the young Swedish genius who was supposed to be the latest in a line of great cryptographic minds. On hearing who was arriving, Lynch had needed little prompting to pack up his two most powerful network computers, abandon his beloved Fort Mead laboratory and ship the computers and himself up to UN headquarters for the meeting.

Al Lynch was also delighted that for the first time in over three years he had been able to make such a trip without being imprisoned in a wheelchair: life was returning to normal. The week before, he had experienced his first erection since the crash. He had wept as he clung to it.

The obstinately perambulatory computer scientist had sent instructions ahead and, within two days of his arrival, a temporary computer lab had been established in a basement room one level down from the Security Council's private meeting chamber. The room was in the deepest level of the UN complex, a space carved out of the Manhattan bedrock in the early 1950s when it was believed that there was a good chance that this bunker would have to house many of the world's leaders during a global exchange of nuclear weapons – and for a very long time afterwards.

The meeting had been memorable. Within minutes Lynch and Larsson were exchanging views about cryptographic achievements of the past and Larsson realized that he had found something rare: a listener capable of understanding his work.

Lynch was careful to make it clear that he did not yet understand how the irrationality of quantum mechanics and string theory could be adapted to apply to rules as strictly logical as those of binary representation. But Larsson was pleased to hand over the software he had brought from his Peruvian mountain top. Loading this, Lynch followed Larsson's instructions on how to run it against the petabytes of data Jack Hendriksen had copied from Thomas Tye's personal data warehouse.

Three hours later Larsson wore a broad smile, while Lynch knew his own jaw was in danger of acquiring a permanent sag. Every file and document that had been sent to Tye or that he had created in the last dozen or so years was now accessible as plain hologram, video, sound, image, document, spreadsheet, text or binary code.

Lynch then called Deakin down and they trawled at random through the vast sea of data. After half an hour Deakin had to give up, his head spinning.

'This really brings it home,' Deakin observed as he scrolled through pages of highly confidential Tye Corporation business plans. 'We've all relied so much on this encryption system that if someone discovers the key, you're completely vulnerable.'

Larsson had had the grace to look away.

Deakin drafted in two psychologists to examine Tye's personal recordings in order to construct a detailed psychological profile of the tycoon. He then ordered all patent-related communications to be distributed to the team's patent lawyers. He also faxed Chelouche to request Joe Tinkler's presence in New

York to lead the team of business analysts already starting to work backwards over the countless project plans, deals, acquisitions and business plans that had been revealed.

After half a day of this, Martha Rose, the international legal expert, opened a video discussion with Deakin. She could hardly contain herself as she described her evidence.

'He's guilty as hell,' she told Deakin with absolute certainty. 'There are copies of three original documents intercepted by Tye CTA – his Competitive Threat Analysis Department. And there are copies of the patents Tye subsequently filed. We think there must be thousands more. We've just got to find a way of legitimizing this evidence.'

Deakin congratulated her as hc turned to examine the files on his recently reactivated display screen.

Then their discoveries became a flood. One of the human rights lawyers had found a file on Lily Albertyn, Erasmus Research and on the birth of a boy called Reon.

Next Professor Berzin of the World Health Organization dropped by with a wry smile on her face. 'He's refiled for the patent on body chemistry. It looks like they realized they had the wrong genes, but now they've identified the right ones.'

For a few moments Deakin knew what it felt like to be Thomas Tye: omnipotent.

The President of the Australian Republic was responsible for the idea spreading to the southern hemisphere. Even though it would already be afternoon in all Australian time zones before the rains were predicted to begin falling on Ethiopia, Robert O. Baldwin decreed his country should start the global celebrations.

'We'll be the first to kick off the greatest street party this planet has ever seen,' he told his people formally. 'So let's give generously to the people of Ethiopia.'

The President of Bolivia was next, announcing that the Festival of San Roque, which this year also fell on Sunday, 30 August, would be co-dedicated to the Ethiopian Appeal. Within two days, most countries on the continent of South America had followed suit.

On EUSSA the Tye Corporation's stock rose another fifteen points.

The parcel, delivered by one of the island's couriers, was waiting for Raymond Liu when he eventually returned to his apartment in the SpacePort Development.

He slipped off his shoes, flopped into an easy chair and ran his fingers under the seal of the gift-wrapping. He removed a small leather-bound book, clearly an antique. There was a handwritten card.

*Raymond,*

*I am truly sorry that aspects of our work have alarmed you. Please be assured that none of our experiments conducted within the digital environment growing around this planet will cause problems with your day-to-day network operations.*

*I thought you might enjoy the enclosure. Although many critics have dismissed Descartes as misguided (because of his religious and spiritual confusions) he nevertheless provided the original inspiration for my work.*

*With very best wishes,*
*Theresa*

Liu turned the book over. It was René Descartes's *Discours de la Méthode Pour Bien Conduire sa Raison, a Chercher la Verité dans les Sciences* and this English translation of the 1637 work – which rendered its Anglicized title as *The World, Rules and Discourse on Method* – had been printed and bound by Bartholomew in London in 1847. He saw a small place-ribbon and opened the book. He read the section Theresa had indicated:

*Then, carefully examining what I was, and seeing that I could pretend that I had no body, that no outer world existed, and no place where I was; but that despite this I could not pretend that I did not exist; that, on the contrary, from the very fact that I was able to doubt the reality of the other things, it was very clearly and certainly followed that I existed; whereas, if I had stopped thinking only, even though all I had ever conceived had been true, I had no reason to believe that I might have existed – from this I knew that I was a being whose whole essence or nature is confined to thinking and which has no need of a place, nor depends on any material thing, in order to exist. So that this I, that is to say the soul by which I am what I am, is entirely distinct from the body, is even easier to know than the body, and furthermore would not stop being what it is, even if the body did not exist.*

Raymond Liu finished the passage and then reread it. He closed the book and sat for a moment staring out at the setting sun, considering Theresa's *apologia.* Then he turned to his screens, located a retail book resource and hunted for something suitable to send her in return.

The temptation was overwhelming; the information that lay before Joe Tinkler was the stuff of a trader's fantasy. The fund manager had only been back in New York for two days, but he was already drooling. Al Lynch had installed network filters – he called them 'sniffers' – at key locations on the world's networks to intercept, collect and reassemble all packets of Tye-related data that passed by. The scrambled messages were then submitted for processing by Rolf Larsson's software.

On his arrival from Geneva, Joe had been able to expand Lynch's search vocabulary considerably by providing an up-to-date list of all the corporation's subsidiary identities, the names of the top 500 Tye Corporation and subsidiary-company executives and the group's wider interests and investments. The result of their collaboration was a continuous flow of decrypted highly sensitive information to Joe's temporary desk in the UN. He felt as if he was inside Thomas Tye's head.

'Take this new Tye Life Sciences operation, LifeLines Inc.,' Joe said to Yoav Chelouche, waving a printout of a decrypted e-mail in the banker's direction. 'We hold nearly sixty-eight million dollars' worth of this stock, but now there's at least one lawsuit pending for mismatched organs. The stock's going to collapse if more follow and someone goes public. Now's the time to sell.'

They were meeting in the banker's office on the thirty-ninth floor of the Secretariat building. Joe had now discovered that his boss was in constant rotation between three addresses: his office in Geneva (his preferred location), this jurisdictional eyot in the middle of Manhattan and the World Bank's global headquarters building in Washington DC.

Chelouche waved his enthusiastic young protégé back into his seat. He understood how excited Joe must be by his new discoveries. In the past ten minutes he had been shown business plans for two new Tye Corporation start-ups, eight strategic transcontinental acquisition recommendations and four sets of accounts that were nearing final-draft stage. If he allowed Joe to make trades from his large fund on the basis of such inside information, it would become infallible: then he realized that they would be no more infallible than Tye's investment decisions had been for the last seven and a half years.

'You want to trade on all this inside information, Joe – you think that's OK?' Chelouche flipped at the pile of printouts Joe had placed on his desk. This meeting seemed more formal than their Geneva discussions: Chelouche behind his desk, Joe seated in front.

'Why not? That's exactly what Tye himself has been doing for years, sir,' reasoned Joe. 'And we know they haven't yet realized we can read their messages. We're keeping tabs on all the communications from their CTA – their Competitive Threat Analysis division. They obviously felt confident enough not to use Larsson's techniques to improve their own encryption. Very cocky.'

Chelouche nodded. He quite liked the idea of Joe's fund building up the cash necessary for a takeover from profits made in such a fitting way.

'And look at this,' urged Joe, leaning out of his seat again. 'This is a contract between Tye Agriceuticals and the government in Addis Ababa. TA has obtained the genetic rights to all the arabica coffee grown in the hills of south-west Ethiopia. I've done some research – that's where coffee originally came from and those plants contain every basic genetic variation for coffee in the world. As a result, only Tye Agriceuticals will be able to come up with new coffee strains based on the master gene pool. Their agreement is due to be announced once those rains are supposed to fall, and coffee has now become the world's most valuable commodity since it overtook oil. If we short the other coffee stocks now, we can …'

Chelouche held his hand up once more. 'I know, I know. There will be many other such opportunities …'

Joe sank back, still anxious that the banker should understand the magnitude of this one in particular.

Chelouche considered, scratching the stubble on his jowls. 'OK, I agree, Joe,' he nodded after a long period of silence. 'Let's use the ganef's own tricks against him, for the time being. But only on Tye Corporation and its subsidiary stocks. Give me your word, Joe; use this inside information on nothing else. And remember, we want to keep things nice and steady. This situation won't hold for long.'

'I understand, sir,' said Joe, already imagining what he could achieve.

Rolf Larsson retraced the tortuous route to the roof of the UN building, but tonight he was stopped at the last flight of steps by a security guard. The soldier hadn't seemed particularly impressed by the visitor's newly issued UNISA security ident or by his powerful image-stabilized binoculars; he simply pointed to the 'Restricted Access' sign. Larsson had then dangled his roll-up sheepishly between his fingers and the guard had finally relented, allowing him entry to one of the few areas of the UN building where smoking was permitted.

Larsson sat down on a small bench a few hundred yards from the helicopter landing pad and stared up at the sky. This was *really* why he had come up here at the dead of night. He was missing his night sky, even though he had presumed that Manhattan's urban glare would make observation difficult. But in fact it hadn't: tonight was very different and he was pleased he had broken off from his analysis of Tye Aerospace documents – Thomas Tye's ambitions were making his head reel.

He had seen the aurora borealis, the Northern Lights, hundreds of times before. Those sheets of red and green light were a familiar spectacle in the summer skies of Scandinavia. There was even an oil painting of a circular borealis in the cathedral near his old apartment in Stockholm. That had been painted in 1431 when the unusual event had been seen as a celestial warning. Now, modern scientists explained that the lights were caused by discharges of solar particle energy as it met the magnetosphere – the Earth's protective magnetic shield.

But the Northern Lights were so called because they were *northern*, and they were seen almost exclusively at latitudes where the sun's harsh output of damaging radiation was deflected over the North Pole. Tonight, the skies above Manhattan were Christmas-coloured: the giant alternating sheets of crimson and green wrapping around the lower latitudes of the night side of the planet.

Larsson lifted his binoculars. In the glow he could make out small knots of observers on other high-rise buildings in the city.

Towards the end of the fourth day after they had broken the codes the excitement within the Operation Iambus team was beginning to lessen slightly. Deakin took the elevator down to the third basement level of the Secretariat building to visit Al Lynch in his cryptorium. He found the computer scientist engrossed at his terminal. He didn't look up until his old friend touched his shoulder.

'I think I'm beginning to get it, Ron,' grinned Lynch as he turned round in his swivel chair. 'I think I understand what Larsson has observed – how computing two alternative states for each object can weed out high primes and then produce a definitive string of integers. It sounds dotty but it isn't – it's just an incredibly high-speed process.'

Deakin held up his hand in protest as Lynch sank into a chair: 'Al, you know I can't follow this stuff. Just so long as it's working.'

'Oh, it's working all right,' beamed the scientist. 'We're getting a terabyte an hour of decrypted material off the networks. I just spoke to one of the translators from the Chinese section; she says the People's Republic is getting very worked up about a Tye Corporation land deal in Russia, close to their northern border.'

Deakin nodded. He had already seen the translated decode. 'Al, I want you to do something for me.' He handed Lynch a printout of the *Wall Street Journal* article reporting Congressman Robarts's accusations.

Lynch skimmed it and looked up questioningly.

'Can you get into the US defence networks, Al – the White House, the Pentagon, the National Security Agency, you know? Oh, and NASA?' He saw his friend's stare. 'We need to know what's behind this. Larsson's been looking at Tye's Phoebus Project and he's making some very strange noises. We need to find out what the Americans really know about Tye's plans, and what they intend to do about them. Can you also check the State Department – they've summoned Tye for a meeting.'

Alan Lynch shook his head. 'I don't know, Ron. I've signed up to the Charter – we all have.'

It was one thing for the UN Security Agency to spy on commercial enterprises: that was easily justifiable in the interests of the greater global population. It was another to consider spying on a member state, especially the most powerful of all, even if that country's governmental lower house did periodically display its collective envy of the UN's global role by withholding US dues. Such hacking would be a direct contravention of the United Nations Charter and a crime under international law that, on detection, carried a mandatory prison sentence.

'I know, Al, I know,' agreed Deakin. 'This would have to be strictly between us. I'm not even telling Jan, let alone the SecGen – they'd have to refuse permission. But I wouldn't be asking if I didn't think it was really important.'

Lynch considered. He had served twenty-four years with Ron Deakin. First at the National Security Agency in Fort Mead when Ron had been the youngest instructor ever in the US Navy SEALs and a liaison officer between his special forces and the NSA, then for the last three years in UNISA after they had sworn new loyalties. In the end, many such decisions were based on trust and long-established relationships, rather than man-made laws or abstract notions of right or wrong.

'They'll have every level of security on their networks, Ron – detectors, false doors, booby traps – even before we get to isolated networks, firewalls, data-dykes and flame-moats,' he objected. 'They're the biggest and brightest target for hackers the world over and the more you get hacked, the better you get at detecting it and preventing it.'

'But you do have an advantage, Al,' prompted Deakin. 'Quite an advantage – in fact, two advantages.'

Lynch smirked and shrugged his shoulders. He had helped to design most of the networks in use by the main security agencies of the USA *and* he had Larsson's software. 'Give me a day or two,' he said.

'I'm sleepy,' said Jed.

Calypso and Tommy exchanged a smile: they had been near to dozing off themselves, but that wasn't what the Furry meant.

'I'll do it,' said Calypso, stifling a yawn and closing her magazine. She pushed herself up from the cushion of her shaded sunlounger and looked around.

'There's one by the changing room,' suggested Tommy, looking up from his book. He shielded his eyes and pointed to a low white building almost completely concealed by shrubs.

Calypso nodded and bent to pick up the Furry. Since Tye Consumer Electronics's breakthrough in battery technology, portable electronic devices such as Jed needed only a twenty-minute battery recharge once a month and the caterpillar's announcement of impending sleep was his euphemistic request for an energy top-up.

*They're the only ones amongst us who still have monthly cycles,* she thought absently as she carried the caterpillar towards the low building. The introduction of hormonally active foods for women had allowed most females in the developed world to dispense with the inconvenience of menses until they wished to become pregnant. Ovulation was mimicked, because of its impact on female (and male) sexuality, but otherwise menstruation was a female inconvenience that had been almost forgotten.

Calypso left Jed in nap mode to recharge his batteries and strolled back towards the pool. This was where Tommy had suffered his accident three months before. So much had happened since then. Now she was living in the great house behind her, sleeping in the bedroom next to him and watching him blossom in front of her eyes: she felt a sense of contentment she had never known before. Because of his special circumstances, she felt that she could allow herself to become much closer to him than she could to another woman's child. The thing that had most astonished her was the realization that inside this totally isolated, unworldly, unsophisticated and dependent little boy was a remarkable intellect. For example, Tommy wasn't just musically gifted: he displayed a range of talents way beyond the normal repertoire for a child of his years.

Now he was reading *Great Expectations* and she saw him flip the page every thirty seconds. At first she had suggested he must be 'skimming', but he had solemnly handed his book to her and challenged her to test him on any part already read. His responses had been word-perfect – he had a true photographic memory. *Not good, however,* she thought. *Tommy will heal himself best when he can forget the strangeness of those past lonely and artificial years.* During their time together she had been able to carry out a number of standard psychograms and she was now convinced that he had outstanding intellectual potential.

Calypso had grand plans for introducing Tommy properly to the world – even considering inviting other children from the island to stay at the main house. Then, as he got older, he could start to visit the American mainland, perhaps even go to school there. St John's in Chicago was very good, she'd heard. Or perhaps the Old Suffolk, or Choate in Boston?

The afternoon was unusually sultry. They had swum earlier before collapsing on the sunloungers. The pool was half Olympic size, sunk into the vast south-east lawn beside the house. It was heated by solar power, like almost everything else on Hope Island, and Calypso had persuaded the pool-keeper to reduce its temperature to twenty-two degrees Celsius. They had enjoyed a very energetic game of two-person water polo earlier and the doctor inside Calypso felt sure it was better to have the pool water cool enough to make exercise enjoyable.

She walked across the thick matting now laid around the pool edge, arrived at the deep end and stared down into the crystal-clear water. She glanced across at Tommy: he was still engrossed. Stepping up onto the low springboard, she took a few moments to estimate its length carefully, made three running strides, lifted her hands above her head, and shouted 'Tommmmyyyyy!', executing an almost perfect dive with a single, 360-degree full-tuck roll at its apex. As she entered the water she felt her tibia touch as they followed the rest of her vertical body, ramrod straight, into the cool water. She knew it had been a surgical entry; an aquatic incision, almost soundless, certainly without splash.

Calypso levelled out close to the bottom and made three broad breaststrokes, exhaling all the air from her lungs. She swam six more strokes and then gently propelled herself to the surface, halfway along the pool. She shook her head, pushed her hair back, treading water, and looked across at Tommy.

A slim figure was standing beside the boy's lounger. Where had Tom come from? Father and son waved to her and Calypso waved back, rolling on her back to face the sun. She trolled gently down the pool, wondering whether she absolutely had to climb out and join them. She knew Tom's presence had to have a purpose: she had never known him to be out in the garden during the day. She hadn't even known that he was on the island. Despite her closer proximity to her charge, there were still long periods spent with his tutors, and there seemed to be no set pattern of contact with his father.

Reaching the shallow end she swam for the steps. She would have felt more comfortable wearing a slightly less revealing swimsuit, but the one-piece Speedo she wore for her ocean crawls was back at her bungalow.

Calypso raised her arms and hauled herself up the steps, straight into a flurry of enveloping green cotton. As she looked up, Tom was standing there, smiling and holding out the large beach towel she had brought from the house.

She was very aware that her white bikini was just a little too daring for the occasion. It was hardly a string or a thong; it was just, well, a little skimpy. She knew the effect that her breasts and the rest of her body had on men.

As usual Tom was wearing sun-viewpers so she couldn't see his eyes. But his face was aimed in the right direction, and his smile was very broad indeed. Suddenly self-conscious, she turned her back and let him drape the towel over her shoulders.

'Beautiful diving, Calypso.'

Folding the towel over her chest, she turned back to face him, pushing the wet hair up off her face.

'Thanks, Tom. Product of an island childhood.'

He turned and they both began the walk over to Tommy.

'Couldn't take any more of the office for a while,' he explained.

She turned to glance at him. She had never heard him express such a feeling before.

'So I thought I'd come down and see how you two are doing. Connie told me you were here.'

Calypso nodded. Connie was the anchor: the place where every scrap of executive and domestic information resided before finding its final home. Every one of Tommy's movements had to be logged in Connie's scheduler.

'It's going to be a hell of a weekend,' Tom said, changing the subject abruptly. 'This is where we'll be entertaining our guests.'

He gestured across the lawn and down towards the three descending terraces that stretched away into the distance in landscaped splendour.

'The pool and the lakes will be boarded over to serve as catering areas,' he explained.

Calypso nodded. The staff in the main house had already tripled in size and she had seen teams from the caterers and entertainment providers scouring the grounds for weeks. Tommy was eager, almost desperate, for the Firework Masters to arrive from Sydney. He already knew that he would not be attending the party, but she was planning to watch the fireworks with him from the roof terrace of the mansion itself.

'How *are* you going to make it rain in Ethiopia, Tom?' she enquired.

## TWENTY-TWO

'But how, precisely, is the great mensch going to make it rain in Ethiopia? asked Yoav Chelouche. '*If* he makes it rain.'

The senior members of the Operation Iambus team had been called for a meeting with Jan Amethier and Yoave Chelouche in the banker's office in the UN Secretariat building.

'We'll know better when Doctor Larsson and the aerospace team have finished analysing the Phoebus data,' replied Deakin. 'But it's likely to take another few days: there's so much material. The Tye Corporation is mad keen on recording everything – like every company these days.'

Chelouche nodded. In the absence of the Secretary-General, he was the most senior UN person at the meeting. They now knew that the Tye Corporation had bought much of Eastern Siberia – that news had caused some explosions in the team, not least with cxpatriate Russians – and it was accepted that the Phoebus technology was somehow intended to change the climate there. Thanks to Joe Tinkler they had also learned of other newly purchased territories in the Baltic Sea and in Canada.

'What do the weather people say?' asked the banker.

'They can't even give me an estimate of how much energy might be needed,' said Deakin. 'Current technologies are merely a pinprick compared to nature's force. They claim even the American military can't change the weather pattern without resorting to nuclear weapons.'

'So do we have to presume he's using some form of orbiting nuclear energy?' asked Amethier quietly.

'I very much doubt it,' argued Deakin. 'Tye maintains a strongly anti-nuclear stance and I don't think he could sell the idea to his shareholders.'

Chelouche gave a big shrug, as if doubting that would rule it out.

'So let's move on to the legal status,' suggested Chelouche. 'What have we got?' He turned to Martha Rose, senior adviser on international law.

'We've now identified over three thousand separate cases of intellectual-property theft by either the Tye Corporation itself or one of its many subsidiaries,' the attorney told them. 'There is hard evidence on each count, but it is debatable whether an international court would allow us to enter the evidence we have obtained. It was, after all, stolen and illegally decrypted.'

There was a silence as they digested the implications. It took Chelouche to break it.

'All laws are made by men – and women,' he hastened to add, 'and all systems and structures we have are, in the end, political. That means they are decided by people, not by some distinct and invisible hand, however much we sell that notion to our populations. The International Criminal Court of The Hague is independent, of course, but it is independent by the support of the entire international community, not despite it. I am no lawyer myself, but I am forced to be something of a politician. I think we must recommend that we prepare for immediate prosecution.'

Deakin held up his hand. 'What about economic stability, Doctor? You're the one who convinced us that we must protect the markets above all.'

The banker nodded and stroked his jowls. 'It hadn't slipped my mind, Exec Deakin,' he growled. 'I'm now in a position, potentially, to take over this corporation – temporarily, at least. The investment community will accept the World Bank as a trustworthy guardian – legally, a *parens patriae* – for the shareholders.'

Deakin whistled soundlessly. He nodded his understanding, his appreciation of the capital involved and the amount of work that must have already been done.

'There's still the issue of public opinion, Doctor,' he objected quietly as he imagined the global TV coverage any trial of Thomas Tye would receive on the networks. 'The public will be voting by their billions, on the hour, every hour, throughout the trial. Every minute we'll be watching how the world's population regard this man – as innocent or guilty. It matters not a jot what the judging panel may say. At the end of the case, if over sixty per cent of the public think he's innocent, we can't go against that opinion and expect still to gain the support of our member states. In democracies national leaders would then be forced to announce their support for Thomas Tye in order to protect their own political position. And, as Tye's a serving Head of State, a prosecution would have implications for other leaders of non-democratic states. We'd risk the UN coming apart. Our public opinion people say that Tye's stock has never been higher – in both senses. None of us can just go *against* the will of the world's people these days. That, in the end, is the reality of our new people's democracy.'

'That depends on what the man's doing, Head of State or not, Exec Deakin,' responded Chelouche tartly. 'And it looks to me as if the Tye Corporation is getting seriously out of hand. I'm going to recommend to Alex–' they all registered this pointed use of the Secretary-General's first name '–that we set our legal team on preparing a specimen case. Get them to look for one with an appealing human angle. You're right, of course: this will have to play well with the people.'

'Try this, it's beautifully made,' said Felicity flipping the trouser suit over on its hanger. 'Look at the lining.'

Haley had decided to treat herself for the Hope Island party, so she and her sister had booked a day's shopping at one of the new themed retail resorts near Gatwick Airport. They had checked in soon after nine a.m. and had established their base in their personal Day Room before venturing out into 148 acres of enclosed semi-tropical jungle, simulated cityscapes, waterside dining areas and entertainment complexes. Dotted among the palms, lakes, waterfalls and rustic-cobbled city streets were designer boutiques, delicatessens, coffee shops, bookstores, luxury goods stores, hair stylists, beauty salons, wine bars and 'outdoor' eating areas serving a vast range of fine food and informal, freshly cooked delicacies. Minute hummingbirds, genetically modified to shed their excreta percutaneously and hygienically, filled the air with darting flashes of vivid colour. Wading birds, similarly adjusted, decorated the watersides. As the shopping vacationers moved from area to area, the temperature of their environment altered with the type of location, and concealed ScentSims filled the air with the latest and most exclusive environmental fragrances.

Like most professionals, Haley did nearly all her routine shopping on the networks. Her personal shopping agent had created a 3D mannequin of her UK size eight body, a virtual tailor's dummy, precisely to her own physical measurements, that made trying on virtual representations of clothes a precise science. Not only could she see how well a garment fitted her, she could also see it in a huge variety of colours and co-ordinates and even in different settings. Many on-line suppliers had abandoned manufacturing ready-to-wear garments and had returned to offering clothes tailored individually for every customer.

Another advantage of network shopping was that purchasers could see how a garment looked from all angles. The problem of metamerism – of a colour appearing different under varying light sources – had been finally solved and today Haley could be sure that the colours seen on her display were accurate. This ensured that the pillar-box red garment she saw on her screen would appear equally vibrant when the garment itself was delivered by the courier. The 'two-hour guaranteed' courier services and a 'no questions' exchange or refund policy on non-tailored garments had also boosted the virtual retail trade after it was discovered that twice as many people would make impulse purchases if they could be sure of receiving them shortly after making their decision.

But Haley had to admit that network shopping wasn't as much *fun* as being here in person. It was a great luxury to wander amongst actual clothes racks, to feel and smell the shopping experience. Since themed Retail Resorts had started offering overnight accommodation and leisure facilities such as casinos, swimming pools, golf courses and tennis courts, such shopping vacations had taken over as the premier holiday destination in North America, Japan, Britain and parts of Europe. They also now catered for day-visitors, like Haley and Felicity. Having a personal Day Room as their base and the facility to bundle all shopping and recreational spending together in a single payment made at checking-out time had turned an experience once a High Street nightmare into a truly sybaritic sojourn.

And, of course, spending time at a Retail Resort was a great opportunity to catch up with family and friends. Since her sister had collected her from Battersea, Haley had talked about little else but Jack Hendriksen. Before he left he had assured her that all covert personal surveillance of her by the UN International Security Agency would be lifted and she had taken him at his word. Things between them had become so momentous that she only had two choices: to trust him wholly, or to absolutely reject him and everything connected with him. She had now made her choice and contacted Marsello Furtrado. With a show of wariness and reluctance, she had agreed to the deal he offered.

'What do you think Jack would like to see you in?' asked Felicity innocently.

Haley playfully struck her twin on the upper arm. *All right: so she had been talking about nothing else.*

'I'm not buying this for Jack,' objected Haley 'I want something to make me look like a world-famous biographer.'

They both laughed and Felicity held up the suit again. Haley took it from her: black with a single button where the deeply cut jacket lapels met, straight trousers with a delightful wrap-over waist. She *would* try it on.

When Jack had first kissed her it seemed as if everything in the world had stopped: as though she was stationary and the world and all its people were spinning around her. They had kissed until they had to break off, breathless. Then Jack had fished in his pocket for a short plastic device.

'We're alone now,' he said as he pressed a button. 'This will jam all signals. I'm sorry, but you've been kept under surveillance since long before I met you.'

Suddenly all the wonderful hot emotions flowing up inside her were met by an icy down draught of some horrible reality that threatened to wreck everything.

Sitting her down at the table in her kitchen, he had held both her hands tightly while he had told her more about the United Nations and his role with the agency.

Then he had hesitated, staring down at the table surface.

'I am in love with you, Haley,' he declared at last. 'I have only ever been in love once before – with my wife. I am thirty-eight years old, so I know what real love feels like compared to any other feelings when we meet someone we're merely attracted to.'

She had nodded, tears welling in her eyes. 'I've never felt like this before, Jack,' she told him. 'I just have to think about it for a while.'

He had not stayed the night. But it had been agreed that he would return the following morning and they would spend the day together.

Human sleep is a safety precaution made necessary by the planet's axial rotation and the random branching of the evolutionary bush which, in *Homo erectus*, elevated vision to become the supreme sense. Thus, when darkness fell and vision's sensory advantage was neutralized, the proto-humans managing to cling on to their nasty, brutish and short lives long enough to reproduce were those who wisely retired during the absence of light. Modern human brains use part of this inherited and seemingly wasteful lack of consciousness to process information and to learn. That night, in her dream-filled REM sleep, Haley learned that she did, indeed, love Jack Hendriksen.

She woke with a smile, his face filling her mind. She turned to the pillow beside her and imagined his face there. She gathered her duvet between her legs and imagined it was him: the process of automorphic projection had begun. She laughed and whooped out of bed to see her reflection grinning back like an idiot from her bathroom mirror.

Their first day together had seemed magical. Jack borrowed a UNISA pool car and drove them to Brighton. They walked hand in hand along the promenade, the gulls screeching overhead, surrounded by throngs of weekend visitors enjoying thc late-summer sunshine.

Jack meanwhile explained the plans they had for her. Haley was to be offered both a UN passport and protection from legal action against her by the Tye Corporation. It was recommended that she take the Tye Corporation's money, sign the contract they offered and place the initial payment in an interest-bearing escrow account with her solicitor: she would then make use of the access they offered but would later back out of the deal with them and return the money. Her UN immunity would protect her from the legal challenge they would inevitably mount. It wasn't honourable, Jack acknowledged, but then neither was Thomas Tye, it seemed, and this was part of a larger, very necessary strategy.

On the way back to London Jack had urged her to take whatever Tye was offering – including the invitation to the party on Hope Island. No other writer or journalist had ever been granted access to the island.

This time Jack *had* stayed the night, and by morning Haley knew she wanted to marry this man. Everything about him seemed so right, so familiar, so part of her. There was none of the lack of ease, the excessive politenesses, the over-careful behaviour that had characterized all her previous new relationships. Though mostly tender and considerate, he plainly desired her so much that he had sometimes seemed a little rough which, Haley realized in surprise, had pleased her even more.

After one more day he had to return to the States. He had already been away from Hope Island for too long, but they would be reunited at the celebration. As soon as the UN operation was finished, Jack would resign from the Tye Corporation and join his brother's successful yacht brokerage. Would Haley consider living in Naples, Florida? There would be no practical immigration difficulties: the incredible boom of the euro and the expanded European Union – its six hundred million people in thirty-four member states now representing the richest single market in the world – had reversed the economic migration patterns of the twentieth century. Now many Americans of Irish and Italian descent were trying to get back into the re-enriched lands of their forefathers.

On their second and final day together they visited London Zoo in Regent's Park. All the larger animals had long since been relocated to safari parks and game reserves outside London, but the lovers wandered through the aquariums, pens and aviaries of the smaller beasts in a happy daze. Neither of them was really aware of their surroundings, but both felt a resonance with other creatures of the planet.

'There will be some people waiting for us when we get home,' Jack warned during the taxi ride back to Haley's flat.

Two men were positioned outside the red-brick mansion block, both fit-looking, in their early thirties and wearing business suits. Jack got out of the taxi first and shook hands with them. Darryl and Terry apparently belonged to UNISA's London team.

'I want them to stick around after I leave,' explained Jack. 'They won't be watching you, they'll just be on hand to protect you. We're playing for very high stakes in this game.'

Haley had agreed reluctantly. She was beginning to realize that she had been plunged into something much larger and potentially more dangerous than she had anticipated.

Jack had stayed on after the two men departed. The worst thing for both of them would be the difficulty of communication. They certainly wouldn't be able to trade locator ident access, as had recently become the fashion for lovers – glancing at a screen to see where the other was at all times was immensely satisfying, at least in the early phase of a romantic relationship. Jack had to explain the UN's discovery that Tye could read the whole world's encrypted communications.

'I think we can risk the odd quick call,' Jack suggested. 'And if I sound funny it's because I'm deliberately disguising my voiceprint. But we can't safely swap locations. Although they're supposed to be confidential, the Tye Corporation owns so much of the networks that they could easily trace our movements once they became suspicious.'

'I'll go mad if I don't hear from you occasionally, Jack,' asserted Haley, certain that was true.

'You'll hear from me OK,' he promised.

They couldn't bear to part the next morning. After Jack had finally gone, everything felt empty. Even her flat felt alien and unreal. She walked around it, picking up familiar objects, examining them as if they too must somehow be fundamentally changed because of what had happened to her. She decided to go and see Flick, to tell her everything. She would need to take a shower first, but she wanted to postpone the moment.

Then her VideoMate announced an incoming voice call. It was her solicitor, requesting her to come in and see him urgently. They agreed on a time that afternoon. She hung up and then called Felicity.

'Where have you *been* for the last few days?' demanded her sister in vexation. Haley hadn't even checked her AutoSec for messages. She arranged to go over to Ladbroke Grove after seeing her solicitor.

She had never seen Percy Sedley looking so serious. 'The United Nations High Commissioner to Brussels himself has been to see me,' he had begun. Then he blinked, seeming to be at a loss. 'I'm not sure what you're up to, but you're being offered UN diplomatic immunity – a modern form of extraterritoriality.'

Haley raised her eyebrows.

'I didn't really know what it meant either,' admitted the solicitor, 'but he brought copies of the international statutes with him.' He reached out and patted a pile of documents in front of him. 'It seems you will be legally protected over this biography.'

'There's masses and masses of stuff, Ron. I've already decrypted over three thousand communications.'

This should have been a triumph, but Al Lynch looked worried. He had been hacking into the government networks that served the White House, the Pentagon, NASA and other US state-security agencies.

'It took me almost two days just to get in,' continued Lynch.

'They've moved up to microwave moats since I worked on the systems. But once I realized what the White House network architects had done, then the security on the NASA and NSA networks was far easier to spoof. It's always the same: once a department declares a standard safe, they all adopt it in precisely the same way.'

'Congratulations,' exclaimed Deakin, beaming.

'Not so fast, Ron. It isn't good news. I was delighted when I first started searching for stuff on Tye and his corporation. I found plenty of it – policy statements, details of tenders, antitrust proposals, that sort of stuff...'

'But?' prompted Deakin.

'Well, then I noticed something strange. There was a pattern I thought I recognized so I dug out some of my old software. They're all *fake*, Jack, even the ones that look sensitive. They're generating documents and messages using the original software agent I developed when I worked for them. That was how I spotted it. They're still using Version One and it has a few glitches. When I analysed the communications, all the text fitted my algorithm patterns for automatic document-generation. They're faking it, Jack. It's not real. They must know about the Tye Corporation's ability to crack hard crypto.'

*So it would seem*, thought Ron Deakin. But at this point rank and responsibility forced him to disregard their decades of friendship and say no more.

'Thanks, Al,' he muttered. 'Let's keep this to ourselves, OK?'

Thomas Tye's 'special requirements' meant being very particular about hotels. He chose them because of their location and the willingness of their management to allow his people to redecorate and refurnish, to electronically 'scrub' and install anti-bugging devices in the suites to be occupied. Once satisfied, he would rarely stay anywhere else.

Over the twenty-five years he had been travelling the world, the president of the Tye Corporation had come to know which hotels best suited his own somewhat peculiar needs. So when he was in DC he made the rather eccentric choice of always staying at the old Palmer House Hilton.

The Palmer House was one of the city's old-style traditional hotels, and it was here that Teddy Roosevelt and John F. Kennedy used to throw their fund-raising dinners and post-campaign parties. The Kilkenny-marbled atrium, hallways and lobbies were spacious, as if designed to accommodate the needs of world leaders. To imaginative visitors, the ghosts of JFK and Jackie, Bobby, Teddy and Marilyn seemed to float through the corridors.

Although Tye was travelling with an unusually small retinue – just Connie, Furtrado and two executive assistants, Pierre Pasquier and his deputy director of presidential personal security, Stella Witherspoon – the entourage had still taken over the entire top floor of the hotel. Pierre had drafted in three of the Palmer House security staff to help Stella maintain a complete floor watch during the sixteen hours the group would inhabit the building.

There was a knock on the double doors of the suite and Stella put her head in. 'They're on their way up.'

Pierre nodded and rose while Tye slipped off his mask and handed it to Connie for disposal. He seemed unusually formal, dressed in a dark suit, a white shirt and silver-grey tie. His hair was pulled back severely. He looked like he was going to a funeral.

A few minutes later they heard Stella's coded knock again and Pierre reopened the doors. The three lobbyists entered and were greeted by Marsello Furtrado who then introduced them to Thomas Tye and his executive assistant. It was obvious they were overawed at meeting such a famous face, and they were flustered as they found their seats and groped for personal equilibrium.

Pierre moved to the bar – given the sensitivity of the discussion, he would fill in for waiter staff. He poured two mineral waters and a coffee as he listened to them going over that afternoon's forthcoming meeting between Tye, Furtrado and Jane Treno, the US Attorney-General of William Wilkinson's administration.

'We think they're just nervous about potential political embarrassment from the LA lawsuits, Mr Tye,' began the senior lobbyist, forgetting to use the tycoon's first name. 'The Tye

Corporation isn't an American company any more so your core corporation is outside the reach of the Justice Department and US law. They can't even lean on you by breathing down the neck of the European Union or ASEAN. All they can do is hurt your US domestic operations and, even then, their legal power over a foreign corporation is limited. My guess is that this will be a "Remember you're an American at heart" speech.'

'I'm here because *you* advised that I should see Treno,' Tye said shortly. 'I really don't have time to fuck around with domestic US issues. Marsello's people or the US V-Ps can take care of things here.'

Now it was the lobbyists' turn to exchange looks.

'We're not sure it would be wise to *say* that, Mr Tye,' suggested the senior lobbyist hesitantly. 'The information we're getting is that Justice is very het up about your corporation and Jane Treno is not a woman to mess with.'

'TO HECK WITH HER!' shouted Tye. 'I have better things to do than bum around with bureaucrats and second-tier politicians. They're here today and gone tomorrow!'

There was a silence. Furtrado moved into the space. 'What background material have you got for us?' he asked.

'She's pretty clean,' said the lobbyist. 'We've got all the current information here.'

Furtrado nodded and the lobbyists extracted a huge and very private dossier they had built up on the Secretary of State's personal life and her many interests and causes.

'You wanted me?' observed Michael Chevannes as he closed the door to his boss's office.

'Good morning, Mike,' greeted Deakin, pointing to a chair in front of his desk. The officer detected an edginess in his boss as he took his seat. There were no preliminaries. 'Have you gone through the stuff we're collecting on Tye?' he asked.

Chevannes shook his head. 'Some, but I haven't finished. Like you suggested, I took a couple of days out. Things have been piling up at my apartment.'

Deakin nodded. He knew. Like the rest of the Iambus team he had seen little of his home, or his wife, in the last few months.

'You're going back in the air,' said Deakin as he slid a paper file across the desk. 'There's a flight to Cape Town at three this afternoon.'

Chevannes opened the file and scanned a UNHCR report on a visit to a Zimbabwe jail. Then he started to read more carefully.

'Go and see this Reon Albertyn and his mother,' ordered Deakin. 'She was employed by one of Tye's biotechnology companies fifteen years ago. They've got a story to tell us and if it's what I think, it'll be just what we need. The background is in the file.'

## TWENTY-THREE

A 3D projection of the inner planets of the solar system filled the Holo-Theater. Jack had briefly seen a version of this image before – Thomas Tye had been showing it when he had been called away following Tommy's accident at the swimming pool. But this later iteration seemed to contain much more detail and Jack presumed it had been developed into a full demonstration to be shown during Tye's visit to Moscow.

The team leaders of Operation Iambus had gathered in the Security Council's private meeting chamber. Rolf Larsson and the aerospace analysts had now finished sifting through the mountains of data concerning the Phoebus Project and, after providing an outline briefing to Ron Deakin, the astrophysicist had been asked to make a presentation of his analysis as soon as possible. The team members had pulled chairs away from the oval conference table and had grouped themselves in a rough semicircle around the Holo-Theater.

Deakin rose and turned to face his team. 'Most of you have already had a chance to meet Doctor Larsson,' he began. 'Rolf has told me the gist of what he's found but I haven't seen this demo either. From the little I have heard, I thought we should all see it as soon as possible; sorry I've had to take up yet another of your evenings.'

There were brief exchanges within the group. Most had their DigiPads ready for note-taking.

The door at the back of the room opened and Jan Amethier entered, followed by Yoav Chelouche. Deakin signalled to vacant chairs beside him and the director of the UN International Security Agency and the president of the World Bank joined the audience.

'Over to you, Doctor,' nodded Deakin.

The lanky Swede rose and coughed. He looked uncomfortable in his pale grey suit with a starched white dress shirt and blue tie. It looked like an outfit being worn for the first time. Only his grey shoes, ribbed and rubber-soled, betrayed the academic within.

'Good afternoon,' he began, his nervousness obvious. He clutched a handful of cue cards, looking as happy as a groom at a shotgun wedding. 'First I am very sorry to be the cause of all this. I had really no idea where it would lead...' He tailed off helplessly.

Deakin gave a small wave of his hand that both brushed his apology aside and told him to get on with it.

'Well... this model was extracted from Thomas Tye's personal files. It is a copy of something completed by the Phoebus Project team only about three months ago,' Larsson said, struggling to gain confidence. 'It appears to be the final demonstration model for the project.'

He stepped into the Holo-Theater and stretched out his arm. 'None of this is to scale,' he explained as he pointed at the bright yellow ball of light representing the sun. 'If this was an accurate model, these planets–' he indicated the spheres representing Mercury and Mars '–would be out in New Jersey and the Earth would be somewhere in Pennsylvania.'

Several in the audience nodded. Jack noticed that the planets were moving in their orbits, the difference in their relative speeds around the sun clearly visible.

Larsson touched a remote control and the perspective changed. They watched as the Earth enlarged and swept across in front of them, left to right, revolving quickly on its axis.

Larsson froze the image and pointed at the side of the Earth facing them.

'You will see that the side away from the sun is always dark. This makes our night.'

Jack heard a muffled 'Jesus' from somebody at the back of the group, then a stifled laugh from further along the line.

'Yes, I'm sorry,' apologized Larsson, his small and painfully acquired store of confidence ebbing. He shuffled his cue cards anxiously. 'I just wanted to make the point since it's very important.'

He fingered his remote control again and the image changed to reveal the sun in the centre of the holo-pit and a larger image of the Earth, revolving slowly before their eyes, its night-time longitudinal meridian towards them. After another shuffle of his cards, he found the prompt he wanted.

'Tye Aerospace has been claiming that the aim of the Phoebus Project is to create a chain of deep-space navigation satellites that will provide a Space Location and Navigation System – a sort of interplanetary GPS system. But I can now tell you that although the company *has* been building a positioning network, the real aim of the Phoebus Project is to capture and redirect the sun's energy back onto the dark side of the Earth – hence its name. Phoebus was the given name of Apollo, the Greek sun god. I am going to add the images of twelve deep-space satellites that are currently stationary between four and six thousand kilometres above the dark side of our planet.'

With the touch of a button Larsson froze the planet's image again and then a dozen pinpricks of light appeared above and around the top half of its circumference, as if the Earth wore a tiara of small stars. He stepped closer to this large image and pointed to the arc of satellites hanging high above the planet.

'These twelve satellites, out of a total of fourteen launched so far, have been placed in deep-space locations high above the northern hemisphere. They are all sufficiently high up to clear the earth's adumbration – the shadow cast by our planet itself – so are bathed in constant sunlight. You will see that those nearer to the North Pole are much closer to the Earth than those further down at forty to fifty degrees latitude – above the temperate zones of the Earth in the USA, Europe and Asia. They all remain stationary relative to the Earth – meaning they are in fixed positions in relation to the line of the sun-Earth axis. They stay behind our planet, on the opposite side to the sun, throughout the year.'

Larsson stepped out of the Holo-Theater and picked up a

glass of water from a nearby table. His audience remained absolutely silent as they waited. He drank, cleared his throat and returned to the display.

'Now, this is how the satellites will capture the sun's energy.' He pressed his remote control again.

As they watched, every one of the pinpricks of white light seemed to grow outwards, enlarging laterally. After twenty seconds each had become many times larger than its original size. When the simulation stopped developing, the twelve spacecraft hung like large silver tiles dotted in a semicircle above the northern half of the Earth.

'Let me show you that transformation in detail,' said Larsson, finally growing in confidence. The image cleared to blackness, then they were looking at a holo-model of one of the deep-space satellites.

'This unit is called a Solaris A-100. It is a deep-space catoptric energy station manufactured by Tye Aerospace in Singapore and then assembled in orbit. Its current position is four thousand kilometres above the sixtieth parallel. That's the latitude that runs through Alaska, Canada, Scandinavia, the Baltic Sea, northern Russia and Siberia.'

Jack saw a smartly suited man to his left making rapid notes on his DigiPad.

'In its passive condition the core satellite doesn't appear very different to one of the many manned space-station structures already orbiting our planet – although those are only about five hundred kilometres up, and fully trapped by the Earth's gravity. Essentially this Solaris weighs 12,000 kilograms, has core dimensions of two hundred and sixty metres by one hundred and forty-eight. Almost all the power used on board is provided by solar energy. Apart from plasma thrusters, employed for purposes we shall see in a minute, there is no alternative energy source. This satellite platform was launched from Cape Hope – from the floating facilities at the island – about fourteen months ago. It was initially placed in a low-Earth geostationary orbit at a location pre-booked with the UN Space Agency by Tye Aerospace. The filing described it as an experimental deep-space research probe. After four months of additional construction work in orbit the satellite was boosted to its present position in deep space.'

'Also pre-booked?' asked the note-taker. Jack now recalled he was Joe Tinkler, one of the World Bank people.

'No.' Larsson shook his head. 'International space agreements only apply to orbits up to a thousand kilometres above the Earth. Beyond that, orbit paths and locations are not considered stable – there being insufficient gravity for a spacecraft to be held in reliable Earth orbit. There are therefore no international rules about locations in deep space, no agreements on ownership, no property rights.'

'First come, first served?' asked Tinkler.

'Precisely,' agreed Larsson.

'So how do they maintain their orbital positions?' asked Deakin.

'I'll come to that,' said Larsson, turning back to the spacecraft image. He pressed a button and the satellite shrunk from five feet wide to less than a foot. 'Let me show you something else first. Now watch.'

Small hatches opened on either side of the rectangular craft and with a simulated puff of propulsion gases two small space pods were expelled and began to move slowly outwards in opposing lateral directions. As they did so, the audience could see that each trailed a silver tether behind it, like an umbilical cord linking it to the mother ship.

Larsson stepped back as the pod on his side continued its outward journey. With a retro-firing of propellant gas the pods then stopped at the very edges of the display zone. The image now resembled a silver thread stretched right across the Holo-Theater, interrupted at the centre by the small satellite. This being a representation of deep space, the thread showed no sign of either tension or sagging.

Jack's eyes wandered along the entire image, which now seemed stationary. Then he noticed that the pods at either end of this thread had started to spin slowly on their axis. As the line of thread grew thicker, it became clear that the slowly rotating pods were unwinding some form of tightly rolled-up material. Jack thought of curtain blinds being slowly lowered and, as more of the material appeared, it looked as if it was a form of netting.

'There are plasma thrusters at the bottom of each sail,' explained Larsson, pointing along the bottom edge of one of the descending nets. 'There's no proximate gravity so the sails have to be thrust downwards as the axial motors unwind their rolls. In this simulation the process is speeded up about seven hundred times.'

As they watched, it became clear that each 'sail' was indeed made up entirely of silver netting. After a few minutes the 'unfurling' process ended and the holo-image pit was filled with the model of a minute central satellite from which two enormously long arms supported what looked like two giant geometrically symmetrical fishing nets.

By now Larsson was enjoying the stunned reaction of his audience. He pressed the remote to freeze the image. 'Any idea what happens next?'

There were a few shrugs and head shakes. But this audience didn't seek to be interactive – it just wanted to watch and learn.

Larsson triggered the remote again. At first it seemed as if nothing was happening but then Jack noticed that the latticework of each net was somehow becoming denser. He leaned forward instinctively and realized that the rest of the audience was reacting in precisely the same way.

The holes in the netting were filling in, decreasing the gaps towards their centres. Finally both sails had become solid extents of polished silver. As they watched, the simulation shifted its longitudinal rotation slightly and everybody jerked back as a brilliant flash of light shot across their vision.

'They certainly know how to build a demo,' observed Larsson as he restored a 3D image of the fully-extended Solaris station without the dazzling light.

'Those sails are active solar reflectors, ladies and gentlemen. They are built – well, I don't think that's the right word – they have been created from a new type of photonic biological material called Anacamptonite. Let me turn this image of the fully extended Solaris station around.'

'The meshugener's put giant mirrors up behind the Earth!' exclaimed Chelouche.

'Precisely,' said Larsson. He pressed his remote.

The two-man dog watch aboard the *Knossian*, a 770,000-ton ultra-large crude-carrier belonging to the Lawrence-Antico Oil Co of Quebec, stared at their instruments in disbelief.

Although not the very largest of the world's fleet of ULCCs, the *Knossian* was a modern and supremely well-equipped floating oil tank. Tonight she was almost full, with three-quarters of a million tons of premium crude oil in her fourteen separate watertight compartments. As the use and price of oil continued to decline in the developed world, the oil companies had been forced to adopt the largest possible tankers in order to secure the significant cost-savings associated with bulk transportation. She was on her way from the pipe-head stores of Yamal, a coastal town on northern Russia's Kara Sea, to the refineries of Halifax, Nova Scotia. From here the petroleum and refined oil products extracted from the cargo would sustain the requirements of the American north-east coast market for nearly eight days.

After a further 280 miles on their present south-westerly heading, the *Knossian* was scheduled to change course due west – a turn taking over twenty-six miles to complete – and cross the North Atlantic at a latitude sufficiently southerly to avoid the summer ice floes nudging down from Iceland, Greenland and the Arctic.

Outside, the sky was lit by a breathtaking display of red and green sheets of light obscuring those few stars that could penetrate the pale Scandinavian night skies. But sailors on the northern run are used to such sights and on the bridge the helmsman and the navigator exchanged frowns of extreme puzzlement while they waited for the old man to appear. He would be furious with them, having turned in only two hours earlier.

'Show me!'

Konstantine Stamatis had been at sea since he left school,

and had held commands for fourteen years. What he saw alarmed him.

*Data Error, Unable to Resolve Inputs.* The error message was overlaid across the ECDIS screen – the Electronic Chart Display Information System that usually provided aggregated and error-compensated information from the three independent GPS systems, together with the vessel's own recent navigation history and radar input.

'This is the NASA GPS on it own, sir,' indicated Hideo Su, the thirty-three year-old navigator. The screen cleared to show an outline of the Norwegian coastline and the tanker's position. It appeared they were approximately fourteen miles offshore.

'And this is the Tye Network GPS,' he said as he switched input. The signal showed them dangerously close to the same coastline.

'But this is the ESA GPS.' The third signal – provided by the satellite system of the European Space Agency – showed them midway across the Pacific!

'You've checked . . .'

'System diagnostics report one hundred per cent, sir,' confirmed Su. 'There's nothing wrong with the system, it has to be the data feeds.'

'Sir?' Stefan Kronk, the helmsman, bent over the illuminated magnetic compass. It was a fixture on the bridge simply because international maritime regulations still required its presence on board every ocean-going vessel to guard against the unlikely event that the sophisticated on-board navigation systems, and their back-ups, should fail simultaneously.

Kronk stepped back. The captain too bent over the light. What he saw terrified him. The compass indicated that North was off his port bow. He banged on the binnacle as if to nudge a recalcitrant piece of machinery back into order. He saw the needle quiver slightly from the vibration, but North remained where it should not be, 180 degrees in the wrong direction.

He ran to the vast bridge window and stared out at the brilliantly-lit night sky. He would find no help there. He could only rely on a magnetic heading and that made it clear he did not know where he was. He'd have to head out into the mid-Atlantic just to be sure.

'Go to manual,' he ordered. 'Hard to starboard, starboard engine full thrust reverse, port engine full forward. I want double lookouts placed NOW! Then steer due west, magnetic.'

The giant ship began to turn two miles later. Seven miles further on it ran aground on the Ka, a chain of partly submerged rocks three miles from the Norwegian coast. Despite the vessel's twin hulls, separated by ten feet of air, two of the forward compartments started to spew crude oil into the incoming tide, which calmed it considerably.

The image had changed to show the satellite facing away from them. From this angle, they could hardly see its sails against the darkness of the background.

'As you will know, when light is incident upon a plane surface it is partly reflected and partly refracted – that is, bent and absorbed,' explained Larsson. 'This backing material is very thin but hyper-dense and it therefore absorbs so little that almost all light and heat are reflected – it is the absolute proof of Kirchoff's Law: that the absorptivity of a body for radiant energy of any particular wavelength is equal to its emissivity at the same temperature for the same wavelength. In addition, this backing captures the infra-red wavelengths of light and supplies it back to the reflections of visible light. This bionic material – the Anacamptonite – is the subject of one of the many patent disputes between Tye BioMaterials and university research labs. Ms Rose tells me that this material was originally designed at MIT . . .'

He suddenly trailed off as he realized what he was saying. He coughed and turned back to face the image.

'The anacamptic surface is a protein-plastic with shape-memory electrorheological alloys,' Larsson continued in a more sombre voice. 'Essentially, it is a thin piezoelectric plastic that grows like organic material but has a greater surface reflectivity than a highly polished mirror. The netting on which the material grows is made of an incredibly light but strong carbon synthesis – C-sixty, or buckyball carbon atoms as we know them. I know this will sound like scientific gobbledygook to some of you, but I've laid out explanatory sheets on a table at the back.'

He pointed behind the group, but none of them moved to collect the material. They wanted to hear the rest.

'The framework is also a communications network that connects the different parts of the station and the sails,' Larsson continued. 'When fully extended each of these solar sails are two hundred and fourteen miles across by a hundred and sixty-four miles deep! As you can imagine, they are constantly bombarded by space dust, meteor showers, in fact everything that makes deep space the inhospitable place we know it to be. But this material is the answer. It uses sunshine as its energy source and stores its base shape as a cellular memory: it automatically regrows to mend small holes in a few hours. Every so often a larger object will tear a bigger hole, but the material just grows back. The sail is entirely self-healing.'

They were all silent.

'You see that the sails are made up of squares like a net. There are several reasons for this. The first is that the frames of the squares provide the seating and the point of origin for each segment of – well, maybe I should call it skin – it grows very much like skin. But the frames also do something else. Watch.'

Larsson brought up a new image, this time a large close-up of a single square. As they looked at the flat silver sheet hanging in front of them it began to rotate slightly on its horizontal axis and then on its longitudinal centre. Suddenly it projected another searing effulgence towards the audience.

He flipped the image back to the entire Solaris power station. 'Each of those frames has a series of micro-motors – all solar-powered – that allow the precise angle, the angle of incidence, of each panel to be adjusted according to the specified destination of its reflection and the attitude of all the other reflectors. You see, these sails aren't single reflectors, they are made up of hundreds of different and independently controlled active reflectors. Importantly, they can focus a diffused beam of reflected sunlight into a concentrated narrow beam, rather as a schoolchild might do with the sun's rays through a magnifying glass.'

There was continuing silence as Larsson allowed this

information to sink in. He stepped out of the display pit and took another sip of his water.

'You were going to tell us how these satellites maintain their orbits,' Deakin reminded him at last.

Larsson nodded. 'You've all heard of solar wind?'

Several shook their heads.

'The sun emits a flux – a wind – throughout the solar system. You can see its effect on comets: it creates the tails that stream behind them. This wind from the sun is captured by the huge sails on these space stations: they have enormous area but ultra-low mass. The wind holds the space stations in equilibrium against the gravitational force of both the sun and the Earth – in a condition called the magnetopause, when all the forces are at an equal point. The stations constantly check their position with Tye's real SLNS – the Space Location and Navigation System of satellite networks that Tye Aerospace is now actually placing at key points throughout the solar system.'

Somebody in the group gave a long, low whistle; perhaps expressing appreciation, or amazement.

'How much power do these things deliver?' asked Joe Tinkler.

'Individually, between two and sixty kilowatts to each square metre at ground level, depending on how the overall sails and the individual reflecting panels within them are aligned and focused. It's when the output of all the space stations is combined that it becomes truly awesome – watch.'

Larsson dissolved the holo-pit into blackness again and then presented a large image of a semi-darkened Earth with the twelve silver reflection sails floating in their separate locations above the top hemisphere of the planet.

'So now on this side of the earth it is a winter's night in Europe and Asia. As you can see, the land masses of Russia, Asia, the subcontinent of India, the Middle East and all of Western, Central and Northern Europe are in total darkness. As it is winter in the northern hemisphere you'll see that the South Pole down at the bottom is tilted away from us. At the top we can clearly see the North Pole and the arctic ice mass here–' he pointed to the top of the image '–and we can even see over the top of the globe to the northern parts of Greenland and Labrador.'

As Jack gazed at the large spherical image of his world, he found himself staring at the small islands that made up Britain and wondering if Haley was asleep. He pulled himself back again, realizing this was a simulation and not a real-time feed from one of the Argus satellites.

'For the purposes of this demonstration there is no cloud cover on the planet – I'll talk about clouds a little later – and the small amount of reflected sunlight we can see on this side of the planet is equivalent to what is bounced back from a full moon when it is at its perihelion – its closest point to the Earth. Now . . .' said Larsson, looking for the right combination on his remote control.

A beam of light from one of the floating sails hanging above the planet's north-east quadrant projected a small square of light over an area Jack identified as Eastern Siberia. Then another satellite added its beam, then another, and then all four satellites suspended above that quadrant were illuminating the same region with an intensity that seemed very bright in the dim, simulated moonlight.

'These satellites in the north-eastern segment of the arc are those closest to the Earth's surface, lighting up an area of about six hundred square miles. If you were standing down there in the middle of that reflected light it would seem like a midsummer evening – still bright, but without the searing intensity of full daylight. As well as the moon, you'd be seeing four little suns in the sky. The power being supplied to this area is about three kilowatts per square metre at ground-level and, from a standing start of zero, that is sufficient to raise the ground level temperature by two degrees Celsius an hour. Let me show you a progression . . .'

The image clicked off, then on again, presenting another globe with the vast expanse of the Northern Pacific facing them.

'So, it's the middle of the night here, at the International Date Line. These are the Hawaiian islands. Watch as I run this . . .'

The digital globe started to revolve slowly in an anticlockwise direction and Jack saw a corona of sunlight appear around the eastern perimeter. He watched as the US Pacific coast moved on into daylight while the islands of Japan, the land mass of China and Eastern Siberia, travelling left to right, continued on into the darkness of night. A Solaris satellite in the Western arc of the chain then lit up and illuminated a string of islands to the north of Japan. Then three others in the western segment added beams of reflected sunlight to the same area.

'This region being illuminated is called the Sea of Okhotsk and those are the Kuril Islands,' explained Larsson as he stepped towards the image and pointed at the now-illuminated area. 'This is all part of Siberia, all Russian federal territory.'

'Until August thirtieth,' added Deakin for the group's benefit. He rose and faced them. Larsson pressed a button on his remote and the image froze.

Deakin continued, 'Tye International Real Estate Inc. has signed a four-hundred-and-sixty-year lease on over one and a half million square miles of Eastern Siberia.' He nodded towards Alan Lynch who was seated in the back row. 'Al Lynch has intercepted final copies of the lease documents that were being sent to Tye Corporate HQ by the Russian Estates Office. The land in question becomes a place called Sybaria and it will become the Tye Corporation's sovereign territory on the first day of next month.'

Those who had not heard the news in advance looked at each other in disbelief.

'Lebensraum,' Chelouche attempted to whisper to Jan Amethier, his gravel tones carrying sufficiently to defeat him.

'You'll see their logic in a moment.' Deakin nodded for Larsson to continue and sat down.

The simulation restarted and they watched as the eastern land mass of Siberia, Mongolia and China slid fully into night. As the planet revolved from left to right, more satellites of the Phoebus network lit up, directing beams of energy towards Eastern Siberia as it now moved to the centre of the globe.

Larsson halted the image again. 'Now Siberia is at the darkest point of its night,' he explained. 'And at this stage the Solaris controllers can focus light on the area from all twelve energy stations simultaneously.'

As the twelve satellites of the network lit up, Jack stared at the intense patch of white light now illuminating an enormous area of land. He suddenly thought about the effect such a technology could have on night campaigns – and, therefore, on all military strategy.

'And, of course, they are able to concentrate, angle or spread this reflected light. As we saw, each panel of the solar sails is independently controlled. Watch.' The light shrank to an intense pinpoint.

'What happens down on the ground when all those beams are concentrated towards the same place?' asked Deakin.

*Good question*, thought Jack, who, alone in the room, could make a guess based on experience.

'We have no data yet on concentrations that would affect an area smaller than ten square miles,' replied Larsson. 'All of this information is lifted directly from the Tye Corporation's own demonstration model and it seems as if the diacaustic potential isn't something they've calculated.'

*Right*, thought Jack.

'We've been trying to find an astro-catoptricologist to help us,' added Larsson. He saw their blank stares. 'That's a scientist who studies reflections in space, like the *gegenschein* – the reflection of the sun on the dark matter of distant space.'

They remained quiet as he fingered the demo forward. The display now changed to show the Solaris beams opening up, throwing a pale wash of light that covered all the Asian land mass before reverting to its previous configuration, illuminating only Eastern Siberia.

The image of the Earth started to revolve again and, as Siberia turned steadily towards the sunlit equinoctial of the Earth's eastern horizon, the Solaris stations to the west, on the segment furthest away from Eastern Siberia, shut down and passed on the task of radiating the continent to the satellites positioned further east.

'So, they can provide sunlight to that territory all during the night,' pointed out Larsson. 'Now, at the beginning I said there have been *fourteen* Solaris satellites launched so far. Let me show you the other two.'

The current image disappeared completely and then Jack saw a new projection: a smaller Earth, perhaps two feet in diameter, to the left, and a larger sun, six or seven feet across, to the right. The astrophysicist stepped between them.

'Once again, not to scale, of course. The sun would actually be ten times larger than this. Now look . . .' He stepped over to the Earth and pointed to two silver specks that hung above its northern hemisphere on the sunward side of the planet.

'These two are oblique reflectors positioned to supplement *daylight* sun. I'll need to reduce the illumination . . .' The images grew dimmer, and Jack could just make out two small areas of brighter light now reflected onto the planet's darkened surface.

'The northern tip of our planet doesn't get much light or heat even in daytime during winter,' observed Larsson. 'I should know, I come from Sweden.'

Nobody laughed or even smiled so he pushed on. 'One of the problems is that for the sun's energy to reach the northern land masses when the North Pole is tipped away from the sun, as here, the sunlight has to penetrate sideways through the Earth's atmosphere.'

He indicated with his finger. 'Above southern Europe, here, sunlight arrives at the top of the atmosphere and then slices laterally through two thousand miles of our atmosphere's moisture, dust particles and pollution before it can arrive at ground level in the extreme northern latitudes. By this time the sun's rays are inevitably very weak. There's little UV and little heat, as it's all been dispersed in the atmosphere hanging over more southern latitudes.'

Then Larsson indicated the two satellites placed above and sunwards of the Arctic. 'But if you place reflectors here, immediately *above* the atmosphere, they can direct sunlight straight downwards. There are only two sunward satellites in the Solaris network today, but there will be a total of forty-two positioned sunward of the planet when the final phase of the Phoebus Project is complete in eight years' time.'

He conjured up additional images with the remote control. 'This is a projection of the later phases of the project. Over the next eight years there will be forty-two positioned on the sunward side, as I said, and a further seventy-eight satellites stationed above the dark side of our planet. This image has been adjusted for visibility so that even those with their black heat traps facing you are shown here as reflective surfaces.'

Dozens more silver pinpricks appeared in space, surrounding the planet. These too enlarged to become like silver tiles floating in space. The Earth seemed to be suspended in the middle of a giant cradle, ensphered by pinpoints of light.

'With the fourteen satellites already in place today, Tye Aerospace and Phoebus Inc. can, over time, heat Siberia's winters up from minus six degrees Celsius at surface level to an average daytime temperature of nineteen Celsius – that's about sixty-six degrees Fahrenheit. They can also avoid nychthemeral variations – the normal temperature changes that occur between day and night. They'll be warming the air masses all during the night and supplementing the sunlight during the day. In summer, when Siberia and all northern latitudes are tipped towards the sun, they can achieve a daytime temperature of about twenty-five Celsius, seventy-seven degrees Fahrenheit. When all these satellites are in place they will become capable of extensive, worldwide climatic re-engineering. We're investigating just what that might mean at the present. And, of course, once the Phoebus Project is complete, terrestrial visual astronomy will become impossible almost everywhere in the world and even orbital astronomy will be seriously curtailed.'

He spoke the last sentence quietly, as if he did not want to further underline the ultimate irony of the decision he had made over seven years earlier.

'Could they pump out sunlight all round the clock with just their present satellites?' asked Joe Tinkler.

'Yes. There're bound to be periods when one or more of the satellites is down for one reason or the other, but most of them will be capable of working on a twenty-four-hour basis.'

'So Tye could light up any part of the Northern Hemisphere, starting right now?' queried Joe.

Larsson nodded and walked closer to the image again. He pressed a button and the image of the fully developed network disappeared, leaving only the twelve satellites hovering above the dark side of the Earth and the two on the sunward side.

'Most of the present system's power is concentrated between the fiftieth and seventieth parallels – as I said, that takes in Siberia, Russia, the whole of Northern Europe, Canada and Alaska. Two of the Solaris stations, the ones located furthest from the Earth – here and here – lie between the thirtieth and the fiftieth parallel. We're not yet sure what use they will be put to. They could produce spot sun-power for all sorts of purposes – a few nights of sunshine to ripen a crop of grapes, or to provide street lighting for cities in the temperate zones – even for something as non-essential as film-making.'

'But what about clouds?' asked Jack, speaking up for the first time. 'Wouldn't they completely wreck the performance?'

Larsson beamed, as if he had been asked a very difficult question for which he had a ready answer.

'The Solaris satellites are able to provide a constant stream of energy and, using that, they can probably burn off most types of cloud cover. All they have to do is to concentrate several of their beams onto a particularly dense layer of cloud – even as high as twelve or thirteen thousand metres – and, over a period, it will disperse as the upper atmosphere heats up.'

There was a silence as the audience digested this information.

'It's simply evaporation; the clouds turn into vapour or rain,' added Larsson. 'But you would need to ask the meteorologists for fuller detail.'

'Why aren't there any satellites in the southern hemisphere?' asked Jan Amethier.

'Look,' said Larsson. He pressed a button and the globe began to turn slowly. 'In relative terms, there isn't a lot down there. Except for New Zealand, Australia, the southern tip of Africa, South America, it's all ocean.'

'Sir,' interrupted Joe Tinkler as if to confirm, 'Over eighty per cent of the world's economic activity takes place in the upper two-thirds of the Northern Hemisphere.'

'Thank you, Mr Tinkler,' said Amethier patiently. He well understood the problem of the North/South economic disparity, it was almost the inverse of the planet's distribution of population and was a topic that dominated the UN Assembly Chamber. He turned and said something inaudible to Yoave Chelouche. The banker nodded, then coughed, a deep crackling sound.

'So how does he make his famous rain with all this?' he asked.

Larsson gulped, then smiled. 'We're not absolutely sure yet. Perhaps this is one for Professor Madison...'

A short woman in late middle age, wearing a dark green suit, raised her hand tentatively at the end of the row and rose to her feet. 'It's not easy,' she began, hesitantly. 'There're so many possibilities and we have to complete our measurements. We need output ratings, atmospheric measurements, historical data...'

'Just give me the implications, not the specifications,' interrupted Chelouche rudely, displaying a mixture of irritation, impatience and anxiety.

'We can't answer your question yet, Doctor Chelouche,' she insisted quietly but firmly, then walked along the line of seats until she stood directly in front of the banker.

'On Mr Deakin's instructions I can't call on my normal team for analysis – no American nationals, universities or institutes.' She glanced at Deakin for confirmation and received a slight nod, although Jack could see a hint of annoyance in his boss's expression, as if he wished this information had not been revealed.

'Nobody who habitually studies weather, atmosphere, water circulation, ecology or the environment has ever seen a model like this. I'm told there's no data available from the Tye Corporation either on this?'

Larsson stepped forward again. 'It seems as though anything the Tye people produced on weather projections is either missing – or perhaps studies were not undertaken.'

Chelouche and Amethier exchanged a grimace of disbelief.

'So I need to gather a team to play "what-if",' continued the meteorologist, 'and I need real specialists who *do* understand the *specifications* before I can even consider possible implications.'

She looked Chelouche straight in the eye.

'Just to start with I need heliologists, hydrometeorologists and aeronomists.'

She paused, drew breath, then bore down on him again.

'I then need micrometeorologists, limnologists and climatologists as well as cloud physicists, nephologists and glaciologists. I especially need hyetologists. All have to be gathered.'

'Ologists,' sighed the basset hound in resignation.

She shot a look at him and then continued as if he hadn't spoken. 'I've already identified most of the people I need but they've got to be vetted by UNISA and then many have to travel from universities and institutes outside the USA. Only then can we begin to start investigating what the *implications* of all this might be.'

Chelouche rubbed a palm over his heavy jowls. He looked as if he had been working long hours recently. 'But how long, Professor Madison?' he asked wearily.

She hesitated before she answered. 'You do understand what he's doing, ladies and gentlemen? He's re-engineering our planet's climate. We can only begin to guess at some of the implications of that. It will be a long time before we can even suggest some possible outcomes.'

'Thanks, Marla,' said Deakin, brokering a peace. 'When do you think you will be able to give us your first thoughts?'

'Maybe a week,' she replied, with a smile intended solely for him.

'There was nothing diplomatic about *that*,' said Furtrado as he eased himself into a seat beside his boss.

'How the fuck could they have acquired so much knowledge about our land deals?' mused Tye as he accepted an antiseptic hot towel from the stewardess. They were waiting to take off from Marsh Field, a private-aviation airfield twenty-two miles south-west of Washington DC.

'Through a mix of diplomatic leaks and straightforward spying,' reasoned the counsellor. 'They'll have had the CIA and

NSA onto it. I think we've been lucky to keep the lid on everything this long with Orlov trying to spend his new cash so fast.'

The Attorney-General had been direct with them. The American government wanted full disclosure about this cession of Russian Federal land to the Tye Corporation and an explanation of how the company intended to use its new territory. The administration also wanted a full disclosure of any additional, undisclosed functions and the purpose of Tye's new deep-space satellite network.

'We realize the Tye Corporation is no longer an American company,' Jane Treno had acknowledged, 'But we'd like to think you remain one of ours at heart, that your move from US soil had more to do with corporate financial advantages than any desire to be *un-American*.'

The last word was spoken with an emphasis that made Tye frown. For the second time in three months he felt physically and mentally uncomfortable. There in the Roosevelt Room of the mock-Palladian Department of Justice building on Pennsylvania Avenue he had no control over his surroundings. He had no idea who had sat before him in this gilded, upright chair and he could imagine the dirt that must linger around him in what looked like nineteenth-century drapes and furnishings. But even Thomas Tye could not really ignore a summons from the Attorney-General of the United States and expect to continue doing business with the US government, which was one of Tye Corporation America Inc.'s largest customers.

'You understand how delicate the diplomatic balance has been in that region,' she continued as one of her male lackeys offered the visitors more jasmine tea. The great tycoon had touched nothing so far, and Tye waved the offer away again.

The Secretary of State took a cup of iced tea from her assistant and nodded. 'Let's go off the record – Tom,' she had continued. At this clearly pre-arranged signal her four assistants had risen and left the room. She sat forward on the sofa, facing him.

'We would like the Tye Corporation to become an American company again,' she said, so quietly that he had to strain to catch her words. 'There are significant long-term civilian and defence contracts that would be available to you. We can ease things in LA. You could be assured that your shareholders would not suffer. We could extend US protection to Hope Island State and, if it becomes appropriate, to your new territories in Siberia. But there would be no requirement for you or your staff to redomicile.'

Tye and Furtrado had exchanged glances.

'I know the President is keen for this,' pressed Jane Treno 'We can work together as partners. It would be important for our economy – for both our economies.'

'*And* for his fucking election campaign,' yelled Tye before his counsel had the opportunity to intervene. 'Is that what this is all about, an attempt to boost the US economy through my corporation so he, and you, can get back in again? You're wasting my time!'

'That's an insult, Mr Tye,' snapped the Attorney-General. 'This administration is responsible to the people of America, the people who made your corporation what it is in the first place, and this government will not be brushed aside by a mere businessman!'

Furtrado had then called on his courtroom training to remain impassive. He knew what would follow.

'Listen, Treno,' shouted Tye, jumping to his feet. 'I've got William Wilkinson's FUCKING SPARE HEART beating in a box on Hope Island, along with duplicate organs for half his administration, most of Congress and nearly all the Senate. Don't *you* tell me what to do. You're just trying to get us back under your control so you can stifle innovation and stop us making really great products and services the world *really* wants. I've already seen what you guys can do to a successful corporation – which is why I took my company out of the USA in the first place.'

And that had been that.

'You're right, Tom,' agreed Furtrado as he fastened his seat belt. 'To heck with Jane Treno.'

The jet began its take-off for Hope Island.

*

Easy Zee Zee took a last toke on his joint of Acapulco Gold, stared at the holo control panel hanging in mid-air in front of his chair and wondered if there was anything else that could be squeezed from the forthcoming market in solar energy.

*Not at present*, he decided as he scratched his sparse goatee beard. Cool: this would be the ultimate utility stock! He would reinvest some of the millions he had already made on the LifeLines flotation. He would also short the main oil, gas and power company equities well before his new service was announced. That should prove interesting.

'Finish,' he told the system, still holding his breath. He finally emptied his lungs of what he regarded as the best shit in the world, leaned back and yawned extravagantly. He was alone in the Solaris Control Center, as the Network Center had now been rededicated, and it was very late. This night, there would be little chance of anybody but a security guard interrupting or complaining about his smoke. In two weeks the Center would be busy 7 x 24 x 365.2421 (accurate time measurement being vital) when the Solaris network went live and began to redistribute its heavenly largesse: the centre had already set all atomic clocks and microprocessors to GMT and, from the launch onwards, its workers would exist within a separate time zone on Hope Island. From here the controllers would be able to direct each satellite, focus its solar reflectors and keep a manual watch from space as the world's climate reacted and the weather patterns changed. All of these data would be fed in real-time to the Halcyon GCMS – the global climate-modelling system – and all marketing would be under the management of Zee Zee's automated systems.

Throughout the Tye Corporation's terrestrial networks, engineers were completing a two-year project to retrofit air pressure, temperature, humidity, visibility and precipitation sensors to every cellular network mast, every network hub and every Tye Corporation and Tye-related property in almost all of the countries of the world. In all prime-market territories, Tye Meteorology Inc. had also completed funding agreements with universities to launch permanent middle- and upper-atmosphere drone and balloon research programmes and live feeds from these weather probes would also be patched to the Center where they would be fed into Halcyon. Although Phase 1 of the Phoebus Project would only provide Solaris coverage to fifteen per cent of the Earth's surface, by the final phase – scheduled to be complete in eight years – almost every inch of the planet's surface would – or could – be affected.

The servers had been configured, the real-time connections with the networks had been set up – all with massive bandwidth – and the interlinks with the Argus, Prospect and Hughie satellite networks were in place. These would supplement the information from the Tye Corporation's terrestrial and atmospheric meteorological sensors and would, respectively, provide both controllers and customers with real-time visual, infra-red and microwave meteorological information about the globe's weather patterns. Used in conjunction with the Halcyon system, highly accurate projections and predictions could be made.

Customers would be provided with access to the Halcyon system so that they could model their projected use of Solaris output and this would help them make their choices on timing, select the appropriate power rating, choose the most appropriate tariff or help them decide which auction model most closely met their needs. In the early years, both controllers and customers would be learning the most effective configurations for the Solaris stations in differing weather conditions and Zee Zee had factored free consultancy and Halcyon access into all sales models for the whole of Phase 1.

Necessarily, the Solaris network had certain times blocked out for the long-term climatic-engineering project in Sybaria and for the microclimate treatments specified for Moscow, some islands in the Baltic Sea, a few thousand square miles in northern Canada and some carefully chosen sites in Alaska. But these absorbed only 36.71 per cent of the output from the fourteen Solaris stations already in position. Inside six months there would be a further twenty in orbit and then the launch schedule was due really to accelerate.

Zee Zee had worked night and day for nearly three months. Tom had been so pleased with the success of the LifeLines auction models that he had instantly charged him with this most important of all tasks – constructing the systems to maximize the income from the Phoebus Project. When Zee Zee had first heard the details of the service he had been amazed. He could remember that first sleepless night after the briefing – selling sunshine, heat, rain and energy to the world? He had thought LifeLines was the ultimate business, but he now realized that he had seriously underestimated his boss's ambitions: Tye made all previous tycoons seem lilliputian.

From 30 August, the world's nations and corporations could enter what would become a contest to decide who would gain rights to this new energy source and its valuable by-products – water, heat, light and agricultural re-engineering. Customers could choose from a wide variety of supplier-customer relationships, all designed by Zee Zee on the advice of the Tye Corporation's revenue-optimization economists.

The system had been tested and tested again. The mismatched liver from the LifeLines servers was coming to be regarded as a one-off, an aberration unlikely to recur. Raymond Liu had found no faults in the system and his forensic computing team had been unable to find a glitch anywhere along the visible audit trail: laser pathways between satellites were, of course, temporary and unauditable. Most tellingly, Marsello Furtrado had ordered Zee Zee and his team to make follow-up enquiries on one hundred other LifeLines customers – selected at random from thousands of recipients. Every one of the customers contacted had pronounced themselves delighted with their transplanted organs and offered LifeLines unequivocal endorsement.

Zee Zee knew that in the future he would have many additional products to offer, for which he would have to build fresh features into the auction engines. He was particularly excited about the steering services for hurricanes, typhoons, tornadoes and cyclones that they planned to offer. The Phoebus engineers and the Halcyon meteorologists were certain that when there was sufficient Solaris output available they would be able to divert even the largest storm. Zee Zee was uncertain whether to offer such a service to the insurers of properties and businesses in potential threat paths or whether to offer it first to state authorities and governments.

Then a thought struck him. He had read somewhere that common germs and viruses are temperature sensitive, so for a region suffering a flu epidemic or some other infectious disease, a few days of increased sunshine and raised atmospheric temperature might well kill off temperature-sensitive germs. The economic advantages would prove significant – it could also save a fortune for the services operated by Tye Healthcare. *Cool, something else for the future.*

Zee Zee stood up and emptied the contents of his ashtray into a paper napkin, screwing it into a ball and carefully putting it into the pocket of his jacket. He then put the ashtray into his other pocket.

All that was left now was a final demo for Thomas Tye and then it would be time for the grand launch. Capitalism felt great when it could be coupled with worthy causes.

## TWENTY-FOUR

'Perplex – *ment*!' exclaimed Tommy proudly as he placed the M, E, N and T tiles on the board. 'And that's a triple word score!'

'That's not a word,' snorted Calypso indignantly, 'is it, Jed?'

'I'm afraid it is,' said the caterpillar. 'It's rare but not obsolete. It means "perplexed condition, perplexity", actually.'

'There,' crowed Tommy with an emphatic nod. 'That's eighty-one!'

'I'll never catch up now,' groaned Calypso as she wrote down his score. He had four hundred and twelve to her three hundred and two, and there were only a few tiles left.

The three of them were playing Scrabble, not the popular on-screen version but a battered board game with plastic tiles that Calypso had kept from her childhood. They had first played it during the visit to see her mother and Tommy had since become an addict. Jed had become both referee and adjudicator, having instant access to both the *Oxford English Dictionary* and to *Webster's*.

Calypso realized her life had undergone momentous change. Tom's shock and horror at his son's wild bid for attention seemed to have been the key to unlocking his previous intransigence – at least concerning domestic arrangements. At Tom's request, Calypso stayed with Tommy during all his free time and she was even allowed to readjust the boy's timetable to create more opportunities for him to see his father. She had also been given network access to Tom's personal scheduler so that she could maximize the opportunities for the two of them to be together. That had caused Connie to raise a manicured eyebrow.

Although Tommy seemed to sleep little, he was always fresh-eyed and cheerful. The cutting incident now seemed wholly out of character and Calypso found herself wondering if the boy had staged it deliberately rather than responding to a fit of hysteria. She knew of no other cases in which a seven-year-old would display such cold self-possession. If he had harmed himself merely to get his own way, the incident would be classed as deliberate parapraxia – one for the clinical histories. But Tommy was unique in several ways, as Calypso had reminded herself.

She had started a journal of life inside the Tye household – part personal account, part clinical record. Even though she loved Tommy too much even to contemplate the idea of ever publishing it, she was aware that fate had placed her in a uniquely privileged position. She alone was able to observe how an enhanced, cloned human being was developing, and could observe the degree to which the personalization of the brain altered it from its raw genetic blueprint. Here too she had a living experiment in which she could measure the influence of nurture over nature. But, leaving aside the medical considerations, she was also privileged to observe the daily life of the richest family ever seen on the planet.

Like most medical practitioners, Calypso was more interested in the potential and practical benefits that science had to offer her profession than about abstract ethical concerns. She saw no sanctity in the human condition or form *per se* and suffered from no religious dogma; she saw only widespread and immense suffering and an opportunity for her to help. Her generation might be the first in which doctors began really to understand the mechanisms of the body. Since the map of the human genome had been published at the beginning of the century – while she was undergoing her medical training – biochemists and physicians had collaborated to unravel the mysteries of the fragile human container and eradicate most of the diseases to which it was prone.

Cancer was now almost defeated, cardiovascular disease had been significantly reduced in the developed world, and even neurological ailments such as Alzheimer's, Parkinson's and motor-neurone disease had yielded to gene therapy that could replace damaged areas with healthy tissue. As a result, average life expectancy in richer nations had now shot up to ninety-six years for women and eighty-eight for men, with many living to become healthy and active centenarians. The actuaries now predicted further extension to healthy lifespans in the coming years, and Calypso accepted that if she avoided accident she herself could almost certainly live to be over 120 years old.

Calypso's ethical stance was clear: she approved of any intervention to prevent or alleviate suffering, so she had no strong feelings about genetic manipulation, bio-medicine or organ cloning *providing* they caused no harm to the patient – alive or yet to be born. She also supported the right to early-term abortion because she didn't confuse physical form with human personality, and she accepted that twenty-first-century medicine was dealing with a species that was taking control of its own evolution and rapidly re-engineering both its physiology and its environment.

But that left the mind – which was why Calypso had opted to study the final human attribute that was not yet understood. Despite incredible advances in brain measurement and analysis of function, no cognitive-neuroscientist had yet been able to prove how consciousness emerges and is then sustained. Most theoretical wisdom now suggested that consciousness first arose spontaneously once the processing environment – the human brain – became sufficiently complex. A popular analogy was a swarm of bees or termites where individual members operated on simple rules while together creating highly complex behaviour. But no one could *prove* this was how human consciousness emerged and no one could justifiably claim that self-awareness arose automatically in a community engaged in complex processing transactions.

As a first-year student of paediatric psychology Calypso photocopied a page from one of the then-foremost books on how the mind works and stuck it on her noticeboard. It contained a quotation from a short story by the science fiction writer Terry Bisson, narrating a conversation between two aliens:

*'They're made out of meat.'*

*'Meat . . . ?'*

*'There's no doubt about it. We picked several from different parts of the planet, took them aboard our reccon vessels, probed them all the way through. They're completely meat.'*

*'That's impossible. What about the radio signals? The messages to the stars?'*

*'They use the radio waves to talk, but the signals don't come from them. The signals come from machines.'*

*'So who made the machines? That's who we want to contact.'*

*'They made the machines. That's what I'm trying to tell you. Meat made the machines.'*

*'That's ridiculous. How can meat make a machine? You're asking me to believe in sentient meat.'*

*'I'm not asking you, I'm telling you. These creatures are the only sentient race in the sector and they're made out of meat.'*

*'Maybe they're like the Orfolei. You know, a carbon-based intelligence that goes through a meat stage.'*

*'Nope. They're born meat and they die meat. We studied them for several of their lifespans, which didn't take long. Do you have any idea of the lifespan of meat?'*

*'Spare me. Okay, maybe they're only part meat. You know, like the Weddilie. A meat head with an electron plasma brain inside.'*

*'Nope, we thought of that, since they do have meat heads like the Weddilie. But I told you, we probed them. They're meat all the way through.'*

*'No brain?'*

*'Oh, there is a brain, all right. It's just that the brain is made out of meat!'*

*'So . . . what does the thinking?'*

*'You're not understanding, are you? The brain does the thinking. The meat.'*

*'Thinking meat! You're asking me to believe in thinking meat!'*

> *'Yes, thinking meat! Conscious meat! Loving meat. Dreaming meat. The meat is the whole deal! Arc you getting the picture?'*

Calypso pulled her thoughts back to the game and drew her favourite piece of thinking meat towards her. She kissed Tommy on his warm cheek.

'*Cally*...!' he complained, not really minding.

On asking Jack why he had thought to tell her about Tommy's origins, he had said simply, 'In case anything ever happens to Tom. The boy would need you.'

Once she had been made aware of Tommy's genesis she found it had made her even more determined to protect him. The first eight years of a child's life is the crucial period in which he or she learns the gift of attachment – the ability to place loving trust in relationships – and Calypso was desperate to pump her love and constancy into him before a self-protection mechanism cut in fully, inhibiting the development of his emotional neural-response pathways to shield him from the pain of further grievous separations. She judged that her seven-month presence here was already having a beneficial effect. He had even kissed her on the cheek – twice – as she had settled him down to rest at night.

Which was more than Jack Hendriksen had done recently, thought Calypso as she pondered the letter combinations in front of her. He had avoided her presence since he returned from visiting his sick mother and had avoided her gaze on the three occasions they had met in public. Being Calypso, she had not left it there.

'Avoiding me, Jack?' she challenged him via her VideoMate three days later. She noticed the look of guilt on his face.

'I've– It's been frantic with the weekend event coming up.'

She pursed her lips.

'It's also, well, I've met someone, Calypso,' he admitted.

'What, at your mother's *sickbed*?' she exclaimed.

'No. Someone I've known for a while now. I will tell you about it when I see you. I'm sorry.'

And that had been it. Calypso had been upset, but not

greatly. She had allowed herself to develop some feelings for the strange, watchful man but she had also noted Jack's careful avoidance of the big words, his scrupulous concern not to make more of their time together than mutual companionship and physical enjoyment. Within a couple of days Calypso began to understand that he had not been dishonest with her and although she felt an inevitable sense of rejection, her training helped her identify the roots of those feelings and estimate that they would take only a few weeks to repair.

'Moo,' mooed Calypso as she laid three of her tiles under the last three letters of the word Tommy had just composed. 'M,O,O – Moo – and Em – that's something used in printing – and No and To.'

'Quite correct,' approved Jed. 'And very clever, Doctor. Four words from three tiles.'

'But that's still only thirteen,' laughed Tommy who could add faster than she could read out the numbers.

'Look,' he exclaimed, laying four tiles out quickly to add a paragoge. 'R,Q,U,E. Torque! And it's another triple word score. There, that's forty-five!'

'Hello, you three.'

They turned their heads. Thomas Tye was standing in the doorway wearing a huge grin.

'My God!' exclaimed Calypso, clapping her hand to her mouth.

'Daddy!' shouted Tommy, adopting one of Jed's Englishisms. 'What have you done?'

Tom stepped into the room and turned around. 'What do you think?'

The shock was considerable. Calypso had known Thomas Tye's face for twenty years before she had met him. His long locks had been his symbol, his badge, his emblem of power and freedom.

'You've cut all your hair off,' she breathed, her hands still to her face.

'Well, not all of it,' smiled Tom. 'I quite like it this way.'

*So do I,* thought Calypso. A *man* had stepped out of a caricature.

Tommy jumped up on a chair and stretched out one hand. Tom bent his head so the boy could rub his hand through the stubble. Tom laughed and then grabbed his son and swung him off the stool and round in circles that flung Tommy's legs outwards. The child squealed with pleasure as Tom increased the speed, until finally he slowed and returned the boy to his feet.

Tom gripped the back of the chair to steady himself while, still dizzy, Tommy stumbled across the room and fell onto the soft bed with a laugh.

'It's very ... manly,' commented Jed.

'Calypso?'

'Yes, very manly,' she agreed. 'But why?'

'I'll read you a story, Tommy,' said Tom. 'It must be time for your dreams.'

Calypso took her cue and rose. 'Tommy's won anyway,' she acknowledged, pushing her feet back into her slippers.

'And, Calypso, I'll explain over supper tomorrow,' suggested Tom. 'If you would be gracious enough to join me?'.

She looked into his violet eyes and raised one eyebrow. His eyes held laughter – almost the same mischievousness she sometimes saw in Tommy's.

'I'd be delighted,' she said simply.

'Oy, ologists!' growled Chelouche. Amethier and Deakin nodded, feeling the same way.

The presentation had now lasted four hours and had been full of ifs and buts, maybes and perhapses. The fact was that none of the six weather scientists invited to comment on how the Phoebus Project might affect the world's climate seemed to have a clue.

'We would need to build a huge model of the world's atmosphere and weather systems and then plug all the data from the Solaris systems into it to get any idea of how the whole thing will interact,' Professor Madison had explained. 'My guess is that the Tye Corporation meteorologists must have already built and tested such a model. They've got thousands of joint agreements for information-gathering with universities and weather institutes all over the world. They've even got an army of amateurs feeding in pictures and measurements over the networks – just in the hope of getting personally mentioned on Tye's Halcyon Weather Channel!'

The UNISA executive committee had heard that how, if the Earth was a peach, the atmosphere would be no thicker than the fuzz on its skin. They had learned that the atmosphere was just 600 kilometres thick, composed of layers, with the lowest ten-kilometre-thick stratum responsible for creating most of the ground-level weather. They'd heard about convective cells and vortices and of the stratified layers of lighter gases that sit on the top of the atmosphere. It had been explained that the atmosphere was also a giant thermal engine converting the sun's radiant energy into heat and that variance in this conversion at different points in the atmosphere causes it to shift, creating winds and weather troughs and highs.

Then they had been shown tephigrams, adaibatic curves, ageostrophic and geostrophic wind patterns (with and without the Coriolis effect), Brücker cycles, climographs, hyetographs, progressions of isallobars, isoteres and isochrons as well as langleygraphs, mesoscales and nephanalyses. Next they had suffered force-fed explanations of frontogenesis, isopycnic ultra-centrifugal separative techniques, katabatics and orometrics.

Chelouche wondered if Professor Madison was extracting sweet revenge for his earlier rudeness. Eventually he had lost patience and risen to his feet. He walked over to the large holo-projection of the Earth and jabbed a stubby digit towards East Africa.

'Look, when it comes to science, I'm a shmendrik, OK? I can't understand the complicated stuff. It's dry here and I want to make it rain. So if I've got all these sunlight reflectors behind the Earth, what do I have to do to make that happen?'

'Well, you might concentrate your heat on Lake Victoria, to the south-west,' suggested a German limnologist. 'That's only eight hundred miles away. If you could raise the surface temperature there a few degrees you would definitely get evapo-transpiration starting to occur.'

'That would *only* work if you happened to have a handy

twenty-knot wind blowing in the opposite direction to the prevailing force of the Earth's rotation,' objected a young Australian nephologist. 'But I reckon you've got to go with the Earth's rotation, mate. We're nearly on the equator and there the speed of the land surface overrules everything – even wind direction. I reckon he's got to heat up a patch of the Indian Ocean or the Arabian Sea and let rotation do the rest.'

'No, no,' cut in a French mesopherologist. 'He can tackle the noctilucent clouds more easily.'

'Enough,' said Chelouche. 'Enough already.'

He had thanked them and sent them back to their labours. 'Well?' he turned to his Executive.

'It's clearly an awesome technology,' observed Amethier quietly. 'I'd bet he can do what he says he can do.'

'I wouldn't bet against him, either,' agreed Deakin.

'Well, we are the United Nations, gentlemen,' insisted Chelouche. 'We can't allow one corporation to control the entire world's weather, even if it claims it is doing so for the good of the planet! We discovered years ago that weather and climate are the two most important indicators of any region's economic potential. Control of such forces must *never* be in private hands.'

'Raymond.'

'Chomoi.'

They had been conversing so much recently that these technical supremos of rival networks had almost become friends.

'It's not looking good,' sighed Ltupicho. 'We're still experiencing unaccountable network failures. We're not yet losing data but we still can't find the cause.'

Raymond Liu nodded. His Russian counterpart looked as tired as he did. 'We're coming to the opinion that it could be some form of datum virus. I can't see what else it could be. Perhaps something that's mutated on its own.'

Ltupicho snorted. 'And I thought we Russians were the ones who liked fairy tales.'

Liu managed a wan smile.

'I hear you are inheriting some of my networks?' smirked the Russian.

Liu was startled. He'd only heard this himself the week before and the news was regarded as hyper-sensitive.

'The ones in Eastern Siberia – the ones it costs us most to maintain.' The Russian engineer laughed. 'But I also hear that conditions will be different when you take over.'

Raymond Liu shook his head. 'No comment on that, Chomoi.'

In the back of the UN limousine Ron Deakin slumped and closed his eyes. It was Friday evening and, for the first time in weeks, he might be home in time to share a meal with Ruby. The limousine was a perk that Deakin appreciated: a car with a diplomatic status that allowed it to come in and out of Manhattan and to negotiate almost any street at will. Another perk was the small cocktail cabinet. The UNISA officer poured himself a large malt whisky. It was definitely time for a drink.

On this evening the East River bridges were full with other vehicles that enjoyed similar privileges – government cars, city administration vehicles, police and emergency services and the hundreds of shuttle buses that ferried people back and forth to the security-guarded Park-and-Ride car pounds in Queens, Brooklyn and the Bronx. He was heading for Newport, the small waterside community on Long Island where he and his wife had lived for over twenty years. They were both well known in that marina town, but it was Ruby who had really put down their roots. She had long ago accepted childlessness as their lot and devoted her free time to half a dozen voluntary causes, quickly developing a rich social life. Other Newporters knew Ron as a genial bureaucrat who put in long hours at the UN headquarters in Manhattan, but none of their neighbours suspected that years before the pair had added UN diplomatic status to their passports.

He lifted his head off the back seat and glanced down at the e-paper he had brought from his office. Its news was a week old so he touched an icon on his VideoMate and refreshed the digital 'paper' with that evening's news.

The lead column of the *Post* covered William Wilkinson's opening speech at the start of his re-election campaign. Deakin looked further and grimaced. The other main item concerned a major network failure.

*Global Bank Settlement Network Crashes. $6 Trillion Lost in 1 Hour*

At the bottom of the page there was a story about Tye's coming event on Hope Island.

*Tye Weekend Seminar Threatened by Tropical Storm*

He enlarged the print of the body copy and read that a tropical storm now brewing off the coast of East Africa might have the potential to cross the mid-Atlantic, gathering strength before proceeding through the West Indies to cause the usual seasonal panics in southern Florida.

Well, he was away from the office now and he wouldn't waste a one-time code on a fax to Chevannes or the meteorologists. He knew they would already be on to it. Then a quaint thought struck him. He considered the wealth and power that would be heading for that part of the Bermuda Triangle just when this storm was supposed to arrive. *Not good timing*, thought Ron.

He called ahead to the house and saw that Ruby was cooking. 'I took you at your word,' she said. 'We're having Cajun chicken with jacket potatoes.'

He checked she had received his ETA and relaxed into the cushions with his malt whisky.

Calypso had no idea what had prompted Thomas Tye's invitation to dinner. She had been surprised and delighted when the suggestion had first been made. Then she began to worry whether she was in for another tussle over Tommy. Perhaps Tom was once again feeling uneasy and wanted to revert to the old routine. She wasn't going to have any of that.

Her greatest problem had been what to wear. She thought about a full-length dress but, when she slipped it on, she thought the effect looked overdone. By nature the doctor was both practical and casual. Her favourite combination on the island was a pair of shorts and a shirt, but she knew that would not do for this occasion.

The majority of her clothes were still here at the cottage.

After discarding the full-length gown, she tried on a black silk trouser suit and turned in front of the mirror. Too severe for a summer's evening with just the two of them.

Next she tried on a pale blue knee-length dress with a tight skirt. *Too much like a wedding guest*, she thought. Then she tried on a sheer ivory dress with a low neckline, tight low-waisted bodice and a calf-length skirt flounced at the bottom to lift into undulations as she twirled her hips. *Rumba*! thought Calypso – a perfect partnership.

The situation was definitely easing. Raymond Liu had driven his worldwide teams to breaking point and two of his most senior territory-maintenance managers had already walked out. Every inch of cable, every transmission node, every laser source, every encryption engine, every hub, router, firewall, amplifier and dish had been checked, replaced or rotated. In all, 27,566 maintenance staff had been involved in sixty-nine territories. The four manned space stations dedicated to orbital network maintenance had snagged, retrieved and overhauled 2,800 of the 22,902 satellites that comprised the various Tye Corporation networks. Only another 20,102 to go!

But each time a delicate satellite was carefully tethered to a maintenance station the results were the same: no unusual system-faults could be found. Faults in all aspects of the networks occurred frequently and routinely, but the data flows were designed to work around such outages, automatically self-healing and pursuing their destinations by other routes.

Raymond Liu now had a small army of mathematicians working on the problem: they were trying to find a pattern amongst the failures. Not a physical pattern, such as the repeated failure of one particular type of component, but a mathematical pattern that might identify whether the massive rise in system faults was the result of some intelligent action.

He had thought about Professor Keane's *gedankenexperiment* very carefully. Although she was clearly crazy, he was forced to agree that her random usage of processing polarities in the networks was unlikely to cause any problems. Each switch would simply regard the requests from her software as another binary call and respond accordingly.

Raymond Liu was developing a theory of his own that was much more unsettling.

'I say, Doctor, you do look good.' If a toy caterpillar could have whistled, Jed would have done so. Instead he winked, which lent his face a peculiar leer.

Tommy looked up from his book as Calypso hovered in the doorway; she had popped in to see them on the way to dinner.

Tommy's gaze took her in from head to foot, but he didn't bounce up and cuddle her as he had begun to do. She could sense he was unhappy, so she stepped into the room and kneeled beside his chair.

'I'm having dinner with your father tonight,' she said, looking into his troubled violet eyes.

'You'll just argue again,' muttered Tommy, looking down into his book. 'Then you'll go away and I'll never see you again.'

'We won't argue, I promise.' Calypso reached out and touched his hair.

Suddenly he turned in his chair and flung his arms around her neck, burying his face in her shoulder. 'You'll go, you'll go, my father will send you away and I'll never see you again,' he cried.

She stroked the back of his neck and impulsively kissed the top of his head, letting him sob for a few moments, his tears wetting her dress. But that didn't matter; she knew exactly why Tommy was so scared.

When the sobbing subsided she lifted his head from her shoulder and held his face in both hands.

'Look at me, Tommy,' she said. 'I love you and I'm not going anywhere. Your father understands we're happy together, OK?'

Tommy nodded, his eyes still cast down. She kneeled beside him for a few more moments, then realized there was nothing more to be done.

'I'll come and see you later, Tommy.' She rose.

He just nodded.

'See you later, Miss World,' said Jed.

'We've lost the whole of the Indian subcontinent,' reported the distraught engineer.

Raymond Liu nodded silently, watching the vast black hole over the Indian ocean on his own monitors.

'It's not just us. Everybody else is reporting data corruption or discontinuity in that region.'

'How long?' asked Liu. He knew he was going to have to report this reversal to Tom soon, before he heard it from other sources.

'We're tasking in the weather sats and the surveillance sats – as you suggested. We should get something moving later this morning.'

Liu looked at the brilliance of the loading in the surrounding networks – the Sino-Pacific loops were close to peak capacity as they provided compensatory routes around the affected region.

'Keep me informed,' ordered Liu, with a sigh.

Calypso was a bit later than she would have liked, having gone back to her room next door to repair the damage Tommy's tears had done to the shoulder of her dress. The stains had dried quickly under the blast of her hairdryer and she had only had to tidy her hair again before going down. When she emerged from the elevator she found Luc Bestion, Tye's butler, waiting with a smile.

'Tom's out on the front lawn.' He led the way.

Out on the ground-level terrace she declined the offer of a ride in a Volante and trod carefully along the gravel path in her high heels. It was only 8.20 p.m – too early for any dew to have fallen.

Tye had ordered the table to be set up under one of the great African cedar trees on the top terrace in front of the big house. The evening was perfect: mild, warm and – a speciality of Hope Island – free of irritating bugs.

He was already sitting at the table and Calypso was pleased she had dressed up rather than down. He was wearing a charcoal-grey silk suit with a mandarin-collar jacket. Though he wore a customary white T-shirt beneath, the suit gave him too the air of dressing for the occasion.

He was in a viewper conversation as she approached but he soon wrapped up the meeting and stood up as she arrived at the table.

'Calypso. You look stunning.'

She had braided her hair so that it fell in a thick plait to the middle of her back. In a wholly unnecessary gesture she had added a hint of shading under her high cheekbones and had reddened her lips – careful not to overdo it, as she'd been taught by so many professionals who had created her make-up during those earlier years.

Tye took her hand, semi-formally, then gently pulled her closer and kissed her cheek. She smelt a little trace of soap, no hint of antiseptic, no touch of added fragrance.

He pulled the other chair out for her while she admired the view.

The sun was setting behind them, on the other side of the mountain behind the house, but she could see it reflected on the swell of the Atlantic in front of her. The sun created a panorama of lambent reflections across the surface of the gentle sea, like a thousand camera flashes – *Miss World again*, for a moment.

'This is beautiful,' she breathed as she sat down under the high canopy provided by the old tree. She guessed it had been imported as a mature specimen and replanted here with enormous care. There was a wash of sea breeze and a breath of pine needles in the air and Calypso cared not a jot that these were probably artificially augmented. The sound of cicadas in the shrubs enhanced the ambience.

Tye was watching her silently, head cocked to one side. As he smiled, his own beauty complemented the surroundings.

'Have a drink,' he said, lifting a bottle of champagne from an ice bucket. He pulled two crystal glasses towards him across the white linen tablecloth, poured carefully and expertly and returned the bottle to its chill. He handed her a glass.

'To a beautiful evening,' he proposed as he lifted his own.

She raised her glass, touched it against his and watched in

surprise as he took a long draught. She had assumed he never touched alcohol. She savoured the vintage Krug.

'How's Tommy doing today?' he asked.

Like her boss, Connie worked late most evenings. She had an apartment in a condo down near Hope Town beach, but she also had a little room in the house where she would sometimes stay over when things were particularly frantic. Now was just such a time. The forthcoming weekend celebrations were stretching every resource. In the outer office, thirty-seven of her day-shift executive team were still at their desks twelve hours after they had arrived there. The night shift was hot-desking elsewhere in the Tye Corporate headquarters.

Most of the work entailed travel details and final confirmations. Tye Logistics had been working on plans for nearly nine months and the event organizers seemed totally professional, being used to organizing diplomatic summits in Washington as well as major sporting events. But there were still a thousand things that needed her personal attention or approval.

Her system trilled. She looked at the ident and accepted. 'Yes, Raymond, how are you this evening?'

'I need to speak with Tom urgently,' the normally polite network chief almost snapped.

Connie shook her head. 'He can't be disturbed for anything this evening, Ray. Those are his absolute orders. May I help?'

She saw Liu shake his head anxiously. 'We're suffering widespread network failures,' he said quietly. 'We're still functioning, but we're fire-fighting. I lost two air traffic management satellites over Europe today. One pilot had to override the system to stop his computers putting the passenger jet down on the fucking Champs Elysées!'

He broke off. Tom's bad language was catching.

'I've had to issue an advisory to the FAA and the other aviation bodies for pilots to revert to manual flying and for the ATC to decouple network control. Passengers all over the world are complaining about bumpy landings and there are major delays because the air traffic controllers have forgotten how to control the traffic manually.'

'My God,' exclaimed Connie, now seriously alarmed. 'What's going on?'

'I don't know. It just seems sporadic and random. I've checked everything. There is no apparent physical cause and no sign of anybody sabotaging the networks. And it's not just us; other network operators are reporting faults, too. Also, and this one is ultra serious, one of the off-planet deposit-box satellites is missing. It's just disappeared.'

Connie considered. She had spent days in the Network Control Center during the Los Angeles traffic crises and she knew the significance of this information – she also knew the financial implications of losing track of one of their ultra-secure-deposit satellites.

'It will be another hour before I can get to him,' she said, checking the time. 'Leave it with me.'

'So how *are* you going to make it rain in Ethiopia, Tom?'

The evening had gone well – better than Calypso had imagined it could. Tom had been relaxed, charming and attentive. The food had been exotic. For a starter they had been served a stuffed egg-plant dish called *Imam Bayildi*.

'It's Turkish. The chef says it is supposed to make you swoon,' laughed Tom.

To follow they ate a Hawaiian concoction called *laulu*, palm-leaf-wrapped, charcoal-baked albacore tuna caught off the island that afternoon. A selection of organically grown, genetically modified vegetables was served on the side. The food was simply stunning!

'Michel doesn't often get a chance to show off when we're here on the island,' Tom explained.

Calypso knew that Michel Geronde was Michelin-starred and, according to below-stairs rumour, only frequent pay rises stopped the celebrated chef leaving Tom's entourage to go back to 'cooking for real people'. She had also heard that fourteen assistant chefs provided by a catering company would be arriving next week to help him prepare for *One Weekend in the Future*.

As she had anticipated, the conversation turned to Tommy. But far from wanting to limit his son's movements and activities,

Tom now seemed genuinely pleased that the boy was flowering so happily.

'You've done a great deal for Tommy. Thank you.' He lifted his glass to toast her. 'I think I have someone for him to play with, *if* you think it is a good idea?'

She raised an eyebrow.

'We flew that little girl back from Ethiopia with us – you know, Biya, the blind girl from the village.'

Now Calypso was intrigued.

'She's coming out of the clinic tomorrow. It was fairly straightforward, and one hundred per cent successful. She now has twenty-twenty vision.'

Calypso nodded. As a doctor she knew that trachoma was easy to cure if the necessary skills and facilities were at hand.

'She'll be staying up here until after the party, so I wondered if you and Tommy would look after her?'

'Of course, but . . . you're keeping her here to show her on TV, aren't you?'

Tom laughed. 'Don't be cynical, Calypso,' he chided. 'You can do a lot of good through such publicity.'

Which had brought her to her question about the rain.

He sat back in his chair and twirled the long stem of his crystal wine glass. 'Have you heard about our Phoebus Project, Calypso?'

She shook her head. 'No, I don't get involved much with others on the island.'

'The Ethiopia rain thing is all a bit of a stunt, dreamed up by the perception people in Washington,' he explained. 'We're about to launch a new solar-energy service and they felt this would get the maximum media attention and give us a favourable spin.'

'It's certainly done that!' she laughed. The papers and newscasts had seemed full of nothing but the Ethiopian appeal.

'I suppose you're sending a fleet of refuelling tankers flying over to Africa to bomb them with water.'

'It would evaporate before it even hit the ground.' He smiled. 'It's rather more ambitious than that, Calypso. I'm a little nervous about it myself, but my weather people are sure they can do it.'

He sat up straight and reached for the bottle of red wine. He lifted it, one eyebrow raised questioningly, and when Calypso nodded he refilled her glass. Then he topped up his own glass of white.

'Over the last ten years I've nurtured a dream about a new form of solar energy.' He sat back in his chair, studying the wine in his glass. Then he looked up. 'You realize that we're fast running out of everything on this planet: water, usable land, food, energy?'

She nodded.

'Twenty years ago a bunch of crazy Russians had this idea of putting a space mirror into orbit and reflecting a little sunshine back to Siberia. Just enough to extend twilight in the winter evenings and to bring the morning sun a little earlier.'

He paused and took a sip of wine.

'They carried out only one full-scale experiment, in February ninety-nine, but that failed. Then Russia disintegrated: the economy collapsed, the wars in the south started. The companies concerned closed down, but a year or so later the founder of the project, the man with the original idea, got in contact with the president of Tye Aerospace in Florida. I saw the proposals and I realized that with some of our technology and financing this concept could be significantly expanded. In fact, I now think it will turn out to be perhaps the most significant thing I have ever done.'

He stared into his wine again, then raised it to his lips.

'We've been building large-scale catoptric-energy stations, solar reflectors, for over six years, Calypso. You must have seen some of the shuttle launches from the Cape . . .'

She nodded.

'Many of those are for putting the components into position. We then build the Solaris energy stations in Earth orbit before boosting them out to exactly where we want them to be. We're building six new stations at this moment.'

'I think I've met some of your women astronauts,' nodded Calypso. 'They hang out down at Mario's.'

He hesitated as if considering. 'Would it surprise you if I said

I sometimes wished I could hang out down at Mario's myself?' he asked.

She smiled. She had her own small experience of celebrity status, and its frequent loneliness, to help her understand.

'We choose women to do the construction because they are lighter and they get along with each other better than men do when they are shut up together for months at a time. In space, strength counts for nothing. It is dexterity, creativity and the ability to live peacefully at close quarters that matter.'

Calypso lifted her head and looked up at the stars and the bursts of laser light. 'I'd like to go into space,' she sighed. Then she realized she had absolutely no idea why she had said such a thing.

'No, no, we need you too much here, Cally.'

She stared at him. That was what Tommy called her.

'Well, using the sunlight we then capture we can reflect it back to affect the weather, Calypso. Would it surprise you to know that the atmosphere has tides?'

It did, and her face showed it.

'Highs and lows, twice a day, just like the sea, only it's caused by the cycles of heat and cold as the sun rises and sets as well as by the magnetism of the solar system. What we do is add our energy to those tides to make them bigger in certain small areas. Over a period of days the theory is that the boosted upswings become so great that the warm air funnels upwards in anabatic columns, towards the upper atmosphere, and the cold air is pushed down to take its place. It's called inverting the atmosphere. When the cold air descends rapidly it becomes warmer and the moisture inside it turns to rain. The Phoebus people proved that they could do that a couple of months ago. They had a trial run making rain on some remote islands in the Atlantic, the driest inhabited place on Earth. I sent Marsello to observe the rainfall for himself.'

*That overdressed lawyer that Tom seemed to spend so much time with.*

'Here's to rain in Ethiopia.' Calypso raised her glass, not knowing what else to say.

'I'm frightened it might even snow,' Tom replied, laughing softly.

In mid-August, afternoons in Beijing are so hot that only the very poorest of the city's fourteen million people will be found out on the streets.

As Jeremy Corbett, vice-president Asia of Tye Private Banking Services Inc. stepped out of his air-conditioned club to look for an air-conditioned taxi he was surprised to find two tall, fit, smartly suited men appear beside him.

'Mr Corbett,' the taller of the two stated. 'Please come with us.' His English was good.

A dark BMW saloon arrived beside them and the rear door opened from the inside. The taller man held out an ident, English-language side uppermost. It was the badge of the Red Army Military Police.

'I think you've got the wrong man,' objected Corbett stupidly as he felt a hand on the back of his head. But he was pushed down into the car and someone got in beside him.

The large saloon was pulling away at speed even before the door was properly closed.

Sitting back in her chair, Connie yawned and stretched. She desperately wanted to hand over to her night exec but had to wait until Tom finished his dinner engagement. She needed to tell him about Raymond Liu's worries.

A face appeared at the doorway. 'Have you got your AutoSec on?' asked Miguel Sanchos, her night-shift external-interface manager.

Connie glanced at her screen. 'Yes, sorry. What is it?'

'The Solaris people are desperately trying to reach Tom. I wonder if you'd take it.'

Connie looked at her watch. It was now very late but Tom had not appeared yet. She nodded, and switched her system back to *live* as Miguel disappeared. A moment later she was looking at Easy Zee Zee in the Solaris Control Center.

'Yo, Connie,' intoned the systems designer.

*Yo, yourself,* she thought. 'What is it, Mr Zorzi?' she asked, deliberately abrupt.

'I really need to reach Tom right away,' he said.

'He can't be disturbed at the moment, for anything,' replied Connie. 'What's it about?'

'We've been watching that storm out in the south-east Atlantic and it looks like it's building up and heading our way. We've got eleven hundred millibars and dropping and the wind speed is already sixty miles an hour and rising. The mets are predicting a Category Five if it crosses the pond and develops. I wanted permission to see if we can zap it.'

Connie pursed her lips. She knew that everybody in the programme was under strict instructions to keep everything connected with the Phoebus Project an absolute secret until 30 August. All of the plans called for the rain in Ethiopia to be its major launch point. To have maximum impact, that would need to be the first major demonstration.

'Can it be done discreetly – so no one realizes?'

Zee Zee shrugged. 'I don't know. It's never been tried before. But the alternative is a hurricane arriving at the same time as Tom's party guests.'

Connie nodded. 'OK, leave it with me.' She glanced at the time on her screen. 'I should be able to talk to Tom very soon.'

She flicked the system off and wondered what was keeping her boss so unusually long.

Everything but the wine and their coffee cups had been cleared away and they had turned their chairs so that they could look down the length of the island with its twinkling lights. They had finished drinking the wine and there had been a long silence. Calypso was wondering whether this was the time to leave.

'Are you happy here, Calypso?'

She considered. 'Very,' she said emphatically.

'Would you consider staying here on a more permanent basis?'

She frowned in the warm darkness. A full-time contract had been signed soon after their last altercation had been resolved. 'I'm not sure what you mean . . .'

'I had my fiftieth birthday a few weeks ago, Calypso. I was just wondering…'

She nodded, encouraging him, but he tailed off and said no more.

'You look astonishingly good on it, Tom.' She hesitated, then asked a question to which she already had a partial answer. 'How do you do it?'

'It is said that we all owe a death to nature,' he responded after a moment's silence. 'But it is not a debt I am prepared to pay, not for a very long time. Our generation is one of the last that will be forced to suffer the absurd brevity of a biblical lifespan. If we could leap a hundred years ahead we would see our great-grandchildren, in their very youthful forties, shaking their heads in sadness that we are gone, that we were so close to the point where medical science could extend our lives dramatically, but we just missed the boat.'

She sensed he had turned to look at her.

'I don't intend to let that happen to me. The future arrives unevenly, Doctor, and I live at the most technologically advanced point on our planet. Hope Island represents the future and I intend to be here to join my grandchildren in their sadness that so many of those alive today will have died needlessly – within a hair's breadth of the necessary technology becoming available. As Professor Keane likes to say, we shall soon be leaving this animal form.'

Calypso offered no comment on Professor Keane or her work. She turned and stared at him, wondering if he would say more. It was his turn to look away.

'I suppose it must be obvious, and it will become more so as time goes by. I may have lived for fifty years but my body has not aged at all for the last sixteen years. You see, I have been the guinea pig in a unique experiment that seems to have been very successful. I have been taking gene therapy to prevent free-radical damage and sclerosis in the body's cells and nervous system – I thought it only right that I should be the one to risk a long-term trial. A number of others in our research team have since joined me as the doctors are now convinced of the safety of this therapy. We plan to start seeking regulatory approvals for the drug next year – but it will have to be individually customized for each patient.'

She inclined her head in understanding but she had a thousand questions.

'It will be our next big marketing push, after the Phoebus Project. Think it will catch on?'

She smiled at the absurdity of his question. He would actually be offering the prospect of extreme longevity – the eternal dream of humankind.

But his earlier question still hung there.

'You were asking me if I'd like to stay here on a more permanent basis?'

They were both now staring straight ahead again in an elongated, almost palpable silence.

'I think you understand me better than most,' Tom ventured. 'At least, I know you're qualified to understand . . .'

As he tailed off again, she sensed he was gathering himself.

'We'd make an amazing couple, Calypso.'

She felt the hairs on the back of her neck stand up. She continued to stare stolidly ahead.

'What I mean is, would you, could you consider marrying me?'

She sat like stone, unable to move or respond. In shock.

He took her silence as an objection. 'What I'm proposing is primarily a business arrangement,' he said at last. 'Although I think we really could be friends.'

She wanted to run away, to burst into tears. Just to get away. But she was rooted to the chair. She was also furious. She felt tears on her cheeks.

She turned her head. 'I could never . . .'

He sighed. 'Don't misunderstand me, Calypso. If I could love any woman, love in the full sense, it would be you. I have come to realize that since you've been here. That's why I am saying words I never thought I would say.'

She turned away again, the silent tears in full flood.

'I could love you as a person – I already do – but I can never really love any woman,' he continued to the darkness.

Finally she turned back to him. 'Why?' she demanded, still angry.

He didn't respond, but drew a deep, audible breath. Eventually he exhaled noisily, almost in a gasp.

'Have you heard of Professor Charles Eon?' he asked.

'Of course: it was a terrible business. Didn't he die recently?'

'I was one of his patients,' Thomas Tye said distantly. And then he added the words that would make her understand. 'When I was a child.'

'Oh, my God,' she whispered.

So he told her all about it, things he had never told anyone since becoming an adult. At the end it was *his* eyes that were filled with tears. She wanted to put her arms around him so much that it hurt. She did half rise and put a comforting hand on his shoulder, all she could do for the present.

He quietened eventually, his sobs subsiding.

Then he told her that he was asking for Tommy's sake but, yes, he was also asking for his own sake. He wanted Tommy to have the mother he, Tom, had never had. But he also wanted Calypso to be *his* companion. He wanted the world to know her as Mrs Thomas Tye. He wanted the appearance of normality even if he could never have the reality. He wanted a consortship, companionship. He started to talk about her running his charitable foundation and about her joining the anti-ageing therapy trials.

He didn't mention money, or any prenuptial arrangement. In fact, that concerned neither of them at that minute.

Calypso recovered enough to turn her face back to him.

'I don't know,' she said flatly.

Above all other things, Jeremy Corbett disliked boorish behaviour. As the product of one of the better English public schools, then Oxford, the Guards and the British diplomatic service, his breeding, manners, discretion and effortless style were the precise qualifications required to represent the Tye Corporation's secretive and ultra-secure private bank to its Asian customers. Many of its clients lacked Corbett's personal *savoir faire* but most aspired to it and, by placing their unusually vulnerable – and often illicit – wealth with the Tye Private Bank they had, at least, the pleasure of watching an old-style British gentleman court and woo them for their custom.

But, without doubt, this afternoon had exposed some exceptionally ungentlemanly behaviour on the part of one important customer. Corbett now had two front teeth missing and only one of his eyes was functioning. He was still seated in front of a computer console, where his captors had spent six hours working on him before being finally persuaded that he too was seemingly unable to access the funds and documents now held in a deposit-box satellite that the Tye Private Bank had maintained for them for eleven years.

The two observers standing in the shadows at the back of the room exchanged a glance. As one nodded a command, a shoe box was placed on the table beside the computer screens and its lid removed. Corbett's right arm was forced down on the table-top and twisted hard so that the palm faced upwards. A teenager with a pock-marked face stepped forward and raised a Chinese butcher's machete.

Calypso sat with Biya, a picture book open on her lap.

'Rabbit?' guessed Biya, pressing a small finger against the image.

'Horse,' corrected Calypso gently, realizing that Biya had probably never seen one. She had been blind since she was less than a year old.

'Horse,' agreed Biya contentedly. Calypso smiled down and slipped her arm around the girl's small shoulders. How many other Biyas were there? Countless numbers of them, and countless more children suffering other ills for lack of even the smallest amounts of money and resources.

Calypso had slept little since Tom's astonishing proposal. The following morning flowers had once again been delivered to her from Hope Town's florist. Tom's carefully sealed note had conveyed an apology for springing the idea on her so abruptly. But it had clearly reiterated his suggestion.

How much other good she could do with the billions of

dollars already held by his charitable funds! She had read that, even before the unprecedented public response to the Ethiopian appeal, his was the largest philanthropic foundation ever established. And if he was serious about giving her control...

But could she marry him simply for that reason? Then she thought of Tommy.

'Horse,' repeated Biya, turning the page.

Jack Hendriksen was in the newly installed Holo-Theater in the corporate annex to the Tye Mansion. Previously he would have used the unit in the Network Control Center but that facility had now been turned over to the Solaris Controllers for twenty-four-hour operation as they ran their last-minute tests before the public launch.

The entire meeting was being recorded and Jack would review it once the connection was terminated. He felt uneasy. Something was out of kilter.

This was his seventh virtual meeting with Lawrence Burton, the Director of Presidential Security for the US Secret Service. Previously those had been mere viewper videoconference exchanges, but now he stared Burton in the eye as the Secret Service man stood in the Holo-Theater of the White House itself.

'I still don't think you need to transport an entire motorcade,' Jack protested. 'You already have a list of everybody who is going to be on the island and their backgrounds. This is not a public event. We can provide a dedicated fleet of Volantes to ferry the president and his party to and from the ship and Air Force One. It's like the Camp David compound, a controlled environment.'

Burton nodded, smiling in an attempt at conviviality that was unpleasant to watch. He radiated insincerity like smiles across a singles bar. In some ways holo-conferences were more revealing than meat meets, or F2Fs as Burton archaically called them. Fake emotions were easier to detect, which was why few of the world's leaders resorted to them – unless they were as skilled as Thomas Tye.

'Well, two things in response to that, Jack: one, we've got

the motorcade with us anyway; we're going straight on to China, remember? Two, I hear you're getting more and more Cuban rebels washing up on your shores and the word is that you have been getting a little heat from your neighbour. So it isn't a *totally* controlled environment, Jack.'

Jack raised his eyebrows. How did Burton know this? But both men knew that Cuba and its civil war weren't really the issues. Jack also understood that the explanation concerning the seven armoured limousines was true. Much to the annoyance of the Ground Facilities Manager at Cape Hope, Tom had overruled all objections and agreed that not only could Air Force One and its supersonic support jet park in the limited facilities at the floating spaceport, but also the giant Lockheed C-130J transport plane that was to accompany the President on his subsequent state visit to the People's Republic of China. This would require a large amount of precious space. Normally Jack would have expected the transport plane to fly directly to China ahead of the presidential entourage.

So, in addition to the President's regular party of 134 bodies and their equipment, an additional sixty support personnel would be requiring accommodation. It had been made clear, however, that only the presidential body detail would carry arms on the island.

Fortunately their visit was going to be short and sweet. They would arrive on the island late on Friday afternoon when the president would be guest of honour at the opening ceremony for *One Weekend in the Future*. Then there would be three official meetings between him and Thomas Tye over the following two days. The president and his group would be leaving early Sunday evening, immediately after the closing ceremonies.

The president would not be attending any of the other public seminars or lectures because his schedule, which Burton's team had been generous enough to share in confidence with Jack, was made up of diplomatic meetings with many other heads of state and political leaders also attending the Hope Island weekend. Besides his facilities on board Air Force one, the American leader had also been assigned two adjoining suites on the *Treasure of the Caribbean* for holding meetings.

As Jack now understood, none of the delegates were attending for the conference itself. The main purpose of the weekend's event was for Thomas Tye and his team to start the lengthy negotiations necessary to win international recognition for the sovereignty of Sybaria and all the other political guests were using the congress as an excuse for private meetings and negotiations of their own. The White House team was the eighteenth security group Jack had liaised with so far and, as it turned out, had proved the easiest.

Jack ceased his objections. 'OK, Larry, since they're going to be here anyway, but it's going to cause a massive problem if they deploy. Nobody else has been allowed to bring a motorcade. It would be ridiculous on this small island.'

'Tell you what,' offered Burton amiably, 'my detail head will check it out when he arrives. If he's happy with the terrain and your own containment procedures, we'll go ahead with your toy cars. How's that, Jack?'

That was as good as he was going to get, Jack knew, so he nodded in agreement.

'Well, I guess that's it,' smiled Burton. It was a wrap. 'Have a good party.'

'See you,' said Jack as the image of the US Secret Service man fizzled into blackness.

That had all seemed too easy, too damned polite. Jack had dealt with the White House mafia before.

## TWENTY-FIVE

Connie's scream penetrated the entire corporate wing of the Tye mansion, echoing around the internal quadrangle. Thomas Tye and Marsello Furtrado came running out of their offices as Jack sprinted down the hall towards the source of the sound.

Tye's most senior assistant sat back in her chair, ashen-faced, an opened FedEx package in front of her.

'What is it . . .?' began Furtrado, unnecessarily. It was clear that the problem lay with the box. He stepped forward and then recoiled with a look of revulsion.

Jack peered into it and then lifted the contents out by its little finger. It was a human hand, severed cleanly just above the trapezium bone, already drying out and turning black. He lifted a bloodied card from the bottom of the box. There were only five words printed on it. *Restore Access To Our Deposits.*

The ideogram below had been printed with a rubber stamp.

'That's Jeremy Corbett's hand,' gasped Connie. 'I recognize the pinkie ring.'

Jack put the gruesome object back in the box.

'And I recognize that symbol,' added Furtrado, taking the card. 'It's the sign of Tsien. It's the semi-official Triad run by the generals of the People's Liberation Army. They placed all their financial assets with Tye Private Banking.'

'Get Liu in here FUCKING NOW,' screamed Tom.

Joe Tinkler had been gathering some things in readiness for his depature when his VideoMate bleeped, revealing Madame Pioline's ident.

'Hi, Beatrice,' he said. They were now firm friends.

'He's returning to New York tonight and wonders if you would be kind enough to wait?'

'Sure,' Joe shrugged. This was an open call so he couldn't ask her for any details. 'What time?'

'He'll be there in the next couple of hours.'

Joe nodded, wished her goodnight, then closed the connection. He had realized that it was Friday midnight her time, but he knew her whole life revolved around Dr Chelouche and the World Bank.

An hour later the object of her devotion tripped on the carpet coming through the doorway. 'Oy, wehsmir ... Evening Joe, sorry to keep you.'

The banker looked tired and grey. Joe rose to greet him and Chelouche eased his bulk onto an inadequate upright chair in front of Joe's desk.

'I've just got back from the Middle East – getting more of the main players on-side.'

Joe merely nodded.

'How have you be doing recently?'

Surely the man was joking. Joe had been pumping his cryptic, carefully crafted reports to Chelouche's mail box every twelve hours. 'I've sent you all the figures, sir.'

'Tell me. I've been busy.'

'Well, just using our information about Tye-related stocks, we've made nearly two point four trillion US dollars in the past three weeks.'

Chelouche nodded, apparently unimpressed. 'What percentage of the Tye Corporation's voting stocks do we now hold?'

'Just under eleven per cent, sir. Ten point eight two, to be precise.'

'What's your current estimate of the cost to us if we did it now?'

Joe raised his eyebrows and turned to his keyboard and the triptych of screens. The left-hand display showed real-time graphs of the value of his fund's holdings, on the centre screen was his main spreadsheet and to the right were displayed windows of prices, financial news and TV feeds.

He refreshed his spreadsheet and checked the result. 'Tonight it would cost your member nations just over twenty-two trillion US dollars, sir,' he said.

Chelouche sighed. 'Did you see those reports from ships out in the Atlantic last week?'

Joe nodded.

'Lit up the whole mid-Atlantic for nearly ten hours. Testing his satellites, I suppose.'

'Yeah, getting ready for the launch,' agreed Joe.

The banker raised his eyes towards heaven and then looked back at his fund manager.

'The seminar on Hope Island starts tomorrow. That's the perfect opportunity for us. Tye and the entire executive will be distracted with their guests. Plus, it's August and a lot of people are still away on vacation.'

Joe waited while Chelouche hesitated.

'Buy it, Joe, buy all of it. You can have an unlimited line of credit on the World Bank. Don't hesitate, buy as fast as you can. You understand . . .?'

Joe nodded. 'Yes, sir. Unlimited means *unlimited*.'

'Execute all the options immediately and confirm that all deals are dependent on confidentiality. We can't expect to keep this manoeuvre quiet for very long – but as long as possible, eh, Joe? Then you've got to hit the open market, hard and fast. When we must, we'll declare a hostile bid for the remaining stock with all the necessary regulatory filings.'

Joe nodded again. He knew he was facing a long night – a long few days.

'I'll send you down some admin help,' added Chelouche as he pushed himself to his feet. 'I'm off to Washington to see if anybody there works weekends.'

He paused at the door and turned back. 'Over the next few days you're going to make a lot of people very rich, Joe.'

Raymond Liu customarily received all NASA's bulletins as soon as they were issued. The one that had just arrived was brief and to the point. A space-weather storm – rated at G5, the highest category – was causing the recent network disruptions. A solar flare had occurred some hours earlier and the blast of highly charged particles was now tearing holes in the magnetosphere. The crisis was likely to pass within a few days.

He transferred to the archives of the Space Environment Center in Boulder, Colorado and downloaded everything they had on space weather, coronal mass ejections and magnetic reconnections. He knew enough about the behaviour of the Earth's magnetosphere to know that it was a subject only poorly understood. But he suddenly wanted to know a lot more about geomagnetic substorms, nondipolar force fields, the Earth's magnetopause barrier and the characteristics of free electrons, protons and helium nuclei.

The colour of the metallic gift-wrapping could only be described as episcopal. The package sat squarely in the middle of the queen-size bed in Haley's orchid-filled stateroom. She turned and nodded her thanks at the young crewman in the garish gold-trimmed white uniform. He set her newly purchased tan leather luggage on the floor and stood with his white gloved hands at his side.

Haley glanced at her LifeWatch, ready to dial some change, and then remembered tipping was not allowed during *this* weekend. They exchanged smiles.

'Enjoy your stay on *Treasure of the Caribbean*,' said the youth, executing a sharp naval salute before closing the double doors behind him.

Haley had arrived on Hope Island by helicopter so she had been able to spot this great new ship – the subject of many newscasts and press features – long before she could make out any other detail on the huge floating spaceport and deep-water harbour that seemed to almost double the extent of the verdant island to which it was attached.

She could not help feeling excited by it all. This coming event had made headline news all over the world. It had eclipsed all other stories, including the Ethiopian rain-making, the gigantic Norwegian oil-spill and the worldwide network failures. Even the more serious newspapers had led with items speculating on what difference Tye's forthcoming marriage might make to the founder of the world's richest and most powerful corporation ever.

Haley and Flick had watched the live broadcast together as Thomas Tye appeared on the terrace of his mansion with a stunning honey-coloured woman on his arm. He had cut his hair!

The couple took turns in telling the interviewer how blissfully happy they were – and how they intended to be married during the weekend celebrations. When a reporter from the Tye News Network asked them about children, the cameras captured an exchange of obviously rehearsed glances between the happy couple as Tom said only, 'We'll see.'

Haley looked at Flick, who shared her sister's privileged knowledge.

When they asked his fiancée for her thoughts, Haley frowned as the camera closed in. 'I'm delighted,' Calypso said simply. 'I'm really looking forward to helping Tom organize his charitable foundations.'

Haley whistled. 'Some woman!'

Flick nodded. 'She's like a movie star. In fact, *don't* they make quite a couple!'

They pigged out on the ongoing story for hours, watching endless reruns of a younger Calypso Browne being crowned Miss World, pictures of her beautiful but poverty-stricken native island, images of the grimy hospital in Chicago where she had practised, interviews with a stunned group of ex-colleagues – amidst the inevitable regurgitated bios of Thomas Tye himself and his Corporation. The happy couple had first met when Dr Browne joined the Hope Island medical staff, explained the narrator.

'I'll bet she's treating his son Tommy,' exclaimed Haley. 'But there's never any mention of him.'

'You'll probably get to meet her there,' Flick said wistfully. 'Just keep me patched in the whole time, you understand, *every minute!*'

Haley had kept her word and on the helicopter ride she allowed Felicity and Toby to see everything she was seeing. There were about twenty or so other guests on board the large passenger craft to which they had been transferred during a stopover on the island of Mayaguana. She now realized it was Calypso Browne's birthplace.

Haley had discovered that VIP transport to and from the island presented the biggest single problem in the organization of this three-day summit. The Cape Hope spaceport, already one of the busiest air terminals in the Caribbean and Central America, could not provide adequate parking and ground facilities for every jet wishing to land, so visitors without their own aircraft were instead being flown to Mayaguana prior to transfer to Hope Island

Surprisingly, that sleepy Bahamian island boasted a three-mile runway, a vast concrete apron and service buildings capable of accommodating dozens of wide-bodied jets as well as smaller aircraft. As Haley's *One Weekend in the Future: Preliminary Information* download had told her, this airstrip had been constructed covertly in the early 1960s by the United States military who, with the enthusiastic agreement of the British Bahamas, felt the need for a squadron of B-52 nuclear-strike bombers stationed within ten minutes' flying time of Fidel Castro's renegade, untrustworthy and missile-toting Cuban Republic.

Once the immediate problem of seeing unfriendly atomic weapons stockpiled at America's back door had evaporated, it still seemed prudent to the US Department of Defense to mothball the airstrip and its Quonset huts so that the base could be brought back into service within a few days if necessary. When the Cold War ended, so did this level of maintenance, and Tye's Logistics division had just spent two months reconditioning the field itself and flying in new fuel pumps, emergency services and portable accommodation units. The British government seemed delighted at having the facilities on Mayaguana restored at no expense to itself. After the party was over, the refurbished airstrip would go a long way to transforming the island's tourist economy. Haley herself guessed that Tye had been forced into the additional expense because of Hope Island's uncertain relations with Cuba, the only other island in the northern Caribbean with international-standard airport facilities.

As an honoured guest of the Tye Corporation, Haley had

selected to fly from London to New York in 1st-E, the first-class entertainment section. Thus she had been able to enjoy the gym, a sauna, massages, facials, beauty treatment and a bewildering array of immersion games, movies and music. The only letdown was a serious disruption to the international air traffic control system that had delayed her plane by four hours. Then their landing at JFK had been a hesitant, jolting nightmare. Finally there, she had transferred to one of the Tye-Lear supersonic corporate jets that were busy ferrying visitors and staff to and from the islands. The result was that she would be arriving late Friday afternoon rather than in time for lunch.

The ride from New York to the Caribbean had been short but thrilling, though Haley was disappointed that it hadn't taken them over the Florida peninsula. She had wanted to gaze down on it and imagine a future there with Jack.

It took less than thirty minutes before they had landed in the Caribbean heat of Mayaguana, where Haley and a group of dignitaries had been quickly transferred to the large helicopter.

Haley and Felicity chatted like two excited schoolgirls as the Tye-Westland Personnel Shuttle rapidly crossed the nine miles of azure Caribbean separating these two islands of the Greater Antilles. Finally, Haley saw a familiar outline and she transmitted images of Hope Island as it grew larger. The shape of the world's only corporate state was burned into her mind because she had read every book and watched every inch of footage available on Tye and his island paradise. She could probably recite the names of every bay and inlet.

The first feature to catch her eye was the vast array of concave capture dykes of the Hope Island solar-energy farm. Acre upon acre of dull solar fuel cells covered the entire northern tip of the island. Then came further hectares of greenhouses further to the south. As they approached the waist of the island she could make out the long white crescent of Hope Town Bay over to the west. The residential developments along the coast were bejewelled with glittering swimming pools. Then, on the near side of the low mountain range that formed the island's spine, she saw the great white edifice of the Tye

Mansion itself, surrounded by guest bungalows set in acres of manicured terraces, pools and lawns.

'That's Tye's house there,' she told her twin unnecessarily. 'It's even bigger than I imagined.'

'What a waste – all for one man,' sighed Felicity.

'Not any more!' corrected Haley. 'It's now going to be a family home for *three* of them,' she added quietly. But it was unlikely any of the other passengers could overhear their conversation. They were also busy gawping out of the windows and talking to far-away companions.

'Do you think you'll get to see the little boy – Tommy?' asked Felicity, superimposing herself in Haley's vision.

'I doubt it,' replied Haley, switching off her sister's image. 'Look, that's the floating spaceport. It's huge!'

She could identify six of the large OrbitLoad shuttles that Tye Aerospace had developed in partnership with Lockheed Martin, all parked on their own separate apron. She had read that these giant craft were part aeroplane, part rocket. Coated with a light but high-density ceramic, heat-resistant skin, they took off conventionally. But once airborne, they tilted upwards to an almost vertical plane before rocket motors cut in to provide a long high-energy burn to defeat the Earth's hugging gravity and propel them into orbit.

The helicopter continued flying south-west, beyond the extended tip of the island, then circled around to make its approach. It was then that Haley noticed the cruise ship in the deep-water harbour.

'Wow!' exclaimed her distant twin as the image filled Haley's viewpers. 'It's bloody enormous!'

*Bloody enormous!* thought Haley simultaneously. Only recently commissioned, the ship was the largest cruise liner ever built. Taking advantage of the flex capabilities of new plastic ceramics, her designers had given her two hulls like a super-catamaran and, on this wide and stable base, the marine architects had built up a pyramid of gleaming white accommodation decks and leisure facilities. Whereas in the past ship designers had been forced to build downwards into large chine hulls, the designers of *Treasure of the Caribbean* had been able to build upwards from the stability of its twenty-acre platform. The black photonic glass used in the ascending levels contrasted starkly with the rest of the dazzling superstructure. The minimal displacement of the two giant planing hulls would allow the ship to achieve a top speed of eighty knots, which enabled her to outrun all storms, especially given the lengthy advance warnings that sophisticated meteorology could now provide.

'Wow!' echoed Haley.

As the helicopter began a rapid descent, she could see at least twenty supersonic or wide-body corporate jets already parked close to a low building that had to serve as an arrivals hall. She sensed she was arriving in the future or, at the very least, the extreme present.

The noise increased as the pilot adjusted the angle of the rotor blades and, despite her sister's protests, Haley closed down transmission, getting ready for landing. Flick could dip into the recordings later if she wanted to.

Haley had been met by a personal greeter and, to her surprise, by a reporter and crew from Tye News Networks. Was it true she had been granted exclusive access to Tom himself for a new biography? She had waved them away, refusing to comment. Then she was driven the short distance to the harbour where rose the gleaming white fastigiated wedding-cake of the cruise ship in which she would be accommodated.

Of the 937 people who would be killed that afternoon only one had any prior warning, and she was also the only one to understand the immediate cause of her death.

Like motorists on many of the metropolitan roads below them, pilots frequently turned their attention away from their vehicles' controls, allowing on-board computers to fly the planes as satellite-managed air traffic control systems steered a course through densely populated airlanes made safe and navigable solely by computer control. Earlier in the day the Director of Operations, North-West Segment, United States Air Traffic Control had begged for, and obtained, permission to delay implementing the FAA's latest safety instruction. On such short notice it was almost impossible to find enough qualified staff to assume manual control of aircraft in this sector and the alternative was to temporarily ground all traffic. Thus the switch to manual air traffic control would not now take place for another two hours.

Joan Maria Martinez, forty-eight-year-old mother of two, loving wife and senior captain of ABA Airlines – with twenty-three years of flying experience behind her – yawned and turned away from the cockpit conversation to stretch. The Boeing 797 was at 32,000 feet, twenty-one miles north-west of Denver, out of Atlanta en route to Tokyo.

It was only when Joan glanced out of the cockpit window that she saw the dark underbelly of the giant military transport jet that was descending into their path. She cried out, reaching for the yoke, and automatically shot a look of disbelief at the aircraft proximity warning system. It showed green: no alert. But she didn't have time to disconnect the flight-control computer system before the roof of the cockpit was ripped open and she was sent flying upwards into bright sunshine, sub-freezing air – and oblivion.

Haley yawned and ran her hands through her short hair as she shook off the tiredness from the journey. Only twelve hours before she had been in the Victorian streets of Battersea, now she was here on the most advanced cruise ship in the world, moored up alongside the world's newest nation state. She was also finally about to meet the richest and most powerful businessman in the world: the object of her intellectual obsession for almost two years and of personal interest to her since very much earlier. An early-evening reception would take place on the lawns in front of the Tye mansion before all the VIP guests would attend the opening lecture.

*The opening evening is strictly informal*, her programme read. *The lecture will be given by the Nobel Prize-winner and futurologist, Professor Theresa Keane.*

But first, there was a special and even more exclusive social event. Another envelope, black, with her name printed on it in small gold lettering, had been waiting on her dressing table. She pulled out the grey card it enclosed:

*Josh Chandler is welcoming friends at seven p.m. in Suite 1809/10.*

Underneath, Josh had scrawled '*You* do *get around!*'

How kind of Josh, she mused, to have spotted her name on the guest list and to have thought of inviting her to his private reception. Well, that would certainly start her weekend off properly! Although she knew of Professor Keane's reputation, she didn't at all mind the idea of skipping the lecture part of the opening ceremony. She could do without hearing yet another futurologist.

Her stateroom was filled with flowers and gifts: sponsorship, Haley guessed, although she had to reckon these particular sponsors would be getting value for their outlay. They would be reaching the most exclusive and influential target audience in the world.

She found a black Chanel kimono hanging behind the bathroom door. Her travelling clothes hit the floor and after a quick, very hot shower she was enjoying the feel of the kimono's silk. Automatically she went to push the sleeves up but was surprised to find the robe a perfect fit. Similarly she found Gucci slippers, Givenchy perfumes, talcs and toilet water, a Smythson writing set on her desk, and six bottles of Laurent Perrier champagne in a large fridge concealed in the mahogany panelling. A selection of non-calorific hand-made Belgian chocolates lay in a small bowl on the bedside table. She popped one into her mouth.

Despite her stern determination not to be seduced by Tye's lavish corporate hospitality, she crossed her arms and hugged herself as she breathed in the orchidean air. She couldn't quite believe she was here.

Sitting on the edge of the off-white bed, she pulled the purple-wrapped parcel towards her. Her neatly manicured fingernails, kept short for keyboard work, managed to detach the ribbon but were no match for the plastic-metallic wrapping. Then she noticed a letter opener on her bedside table.

*They've thought of everything.*

Zipping along one edge of the package, she recognized Louis Vuitton logos. She saw that she had been given a square valise of green and beige farmed-leather – the perfect carry-on luggage item. Then she realized the case contained something else. Flipping open the catches, she lifted the lid to reveal polystyrene packing. She slit away the top to find a card with a pink satin bow bearing the legend *Welcome To The Girard-Perregaux Équipage.* Inside the moulded indentations were items individually bubble-wrapped.

The printed note read:

*Ms Voss,*

*On behalf of Girrard-Perregaux International and Tye Consumer Electronics Inc., we hope you will accept this dress-occasion Équipage. It has been individually styled for you and we hope it adds to your enjoyment of 'One Weekend in the Future'.*

The card was signed in different inks by Thomas Tye and the president of Girrard-Perregaux. A unique number purported to make her set a one-off.

The first gadget she unwrapped was a classic square G-P LifeWatch. She whistled silently as she held the silver and gold item up to the light. She could guess what that might cost. She noticed it was a Generation Eight, two models newer than her own fun LifeSwatch. She went on to unwrap three further pieces of Girrard-Perregaux jewellery: a belt buckle, two brooches and a pair of earrings. These were part of the wireless bodynet, she read, and could be worn individually or in combination to link these various items of the *Équipage* together. Both brooches – one plain silver, one gold, set with what appeared to be small diamonds, for more formal events – were described as *user-dedicated microphones.* According to the literature provided, they filtered out all sounds but the wearer's own voice. This overcame the problem of ambient noise when recording or transmitting. The belt buckle provided the system's radio link to external networks.

A new-generation VideoMate with a sleek silver case came out of the packing next. It was like an old-fashioned cigarette case and was hinged along its longest axis, designed to be held upright like a book. On its left was a large high-definition colour screen, on the right a smaller secondary screen with a fold-out keyboard and an icon panel. The literature claimed that the unit had greatly improved storage, battery power and wireless-range facilities. And it also contained new image-recognition software called *GuestList*.

Then Haley unwrapped an item that was new to her. This was a slender silver bracelet – a personal ScentSim. Examining it closely, she detected minute holes around its circumference for aroma delivery.

As she had unwrapped each piece she pinned or clamped it at random to parts of her silk robe and now she laughed. She imagined that she must be starting to look like a Christmas tree.

*Welcome to the Personal ScentSim. This is a new accessory in the Girrard-Perregaux Équipage range and it allows wearers to produce any of 40,084 aromas at will. To be worn only on the* left *wrist.*

Controlled from either her VideoMate or Viewpers, she assumed it was intended to generate artificial perfume.

Next she unwrapped a small brooch described as an *Osmatique*, a scent-analyser designed to first identify and then disregard the wearer's own scent. Thereafter it would analyse the aromas in the wearer's environment, in particular analysing the pheromones and body chemistry of other people in close proximity.

She'd read about this new *Spell-Smell* technology. The system analysed another person's smell from within a metre and analysed the underlying pheromones that would prompt attraction, disinterest or repulsion. Then a rating was provided in the wearer's viewpers on a one-to-ten scale of the natural chemical compatibility and attraction. The system could manufacture a contra-simulation guaranteed to similarly attract. A health warning emphasized the dangers of careless use – especially with strangers in non-public places. Haley wondered what would happen if two people wearing Osmatiques met each other with their systems set to level ten.

Next year, claimed the literature, Tye Life Sciences would also be introducing a new oral therapy that enabled the user's body to produce custom-designed pheromone mixes to attract specific mates. Haley realized that osphresiology had developed enormously in recent years, but the idea of controlling physical chemical attraction! Jesus, imagine meeting someone and marrying them only to find their pheromone mix was artificially adjusted. Go to bed with a hunk and, if he missed his medication, wake up with a cesspit! But perhaps it wasn't so new – just another version of going to bed drunk and waking up sober.

Then Haley was unwrapping the last item, something that seemed *completely* new to her. This was an almost microscopically thin curved fibre-optic tube. At one end its tip was enlarged slightly, like the head of a small knitting needle. At the other it was attached to a transparent internal earpiece.

She ran her eyes over the 'Quick Start' instructions. The item was called an *InfoStem* and could replace all forms of viewpers. Its transparent stalk was almost invisible in use, claimed the literature, but the outer tip contained microcameras and a stereo retinal projector. Thus images and text were projected upside down onto the back of the wearer's eyeballs in such a way that they appeared as a transparent display eighteen inches in front of the eyes.

She slipped on the ultra-light earpiece and stood to see how the stalk-like protrusion looked in the mirror. True, it was so fine she could hardly see it. She next selected the Tye News Network on her new VideoMate and suddenly her vision was filled with a picture of the ship she was on. Then the camera zoomed away to pan across the spaceport and the southern tip of Hope Island.

She was just about to check if Flick had gone to bed when there was a knock on the door.

'Come in,' called Haley. Jack's grinning face appeared.

Her InfoStem instantly displayed:

*GuestList identification: Jack Hendriksen, Vice-President, Corporate Security, Tye Corporation.*

'It *is* you!' she said pointing at her new InfoStem. 'I've got proof!'

He stepped in, closed the door and then they were in each other's arms.

Minutes passed as they kissed, savouring the sheer proximity of each other. Both had begun to suffer the doubt that follows separation in a new relationship.

'I can only be away for half an hour,' said Jack when they at last broke their embrace. She already knew their relationship must remain secret during this weekend.

'That'll do fine,' replied Haley with a delicious grin. 'There's champagne in the bucket. I'll only be a minute.'

'You have to cancel, Tom,' insisted Raymond Liu. 'The networks are going haywire.'

'CANCEL IT? ARE YOU FUCKING INSANE?' yelled Tye. 'Half of my guests are already HERE!' They were standing in Tye's office late on Friday afternoon.

The engineer was immovable. 'I'm sorry, Tom, I can't guarantee the infrastructure. It doesn't just affect us, other network operators are having the same trouble. Our own networks are so unreliable I can't guarantee that TNN will even be able to broadcast this weekend. People all over the world may lose communications. I'm now putting all my monitoring teams on to the satellite networks and the air traffic management systems. The networks will repair themselves, of course, but I just can't keep up with the fault monitoring. I've got half of the Asian networks down, the Canadian cellular wireless network went out for a whole hour this morning, and three more city traffic systems are looking shaky.'

'Fix them,' shouted Tye. 'Just find more people and fucking FIX THEM!'

Liu tried to be reasonable once again. 'Tom, if you won't cancel the event at least shut down the Solaris stations.'

'SHUT THEM DOWN?' screamed Tye. 'It's only FUCKING SUNSHINE, LIU!'

'You've *got* to shut them down, it can't be anything else,' urged the engineer. 'I'm convinced they're causing massive magnetic disturbance.'

'That's impossible,' argued Tye. 'We're building up the rain clouds now. Don't you realize what this MEANS TO US?'

Liu nodded. He did know. He also knew that countless lives were probably at stake.

'You heard what happened over Denver, Tom. We can't risk it.'

'They don't *know* what caused that yet,' snapped Tye.

'Sorry, Tom, but I'm shutting them down,' said Liu quietly. 'I'll go over and explain things to the Solaris Control Center.'

Tye considered for a moment. 'You won't do that, Liu. You're fired,' he said, and turned away.

Joe Tinkler had been working for thirty-six hours without rest. He had already put in a twelve-hour day by the time Chelouche arrived and issued his latest instructions. From that point on Joe had spent the night and all day Friday discreetly speaking to his fund-managing colleagues round the world and executing his options on the Tye Corp shares as rapidly but as noiselessly as he could.

By Friday midday he had lifted his fund's percentage of Tye Corp's core voting stocks to nearly seventeen per cent and he could see that, despite the discreet nature of his dealings, certain people in the market were beginning to notice a reduced liquidity in Tye shares. The open-market price was starting to rise.

He had then accelerated his purchases. Seeing the stock starting to rise strongly, he abandoned his cover and started to buy openly in every marketplace in the world. There had been a brief glitch on a settlement problem. The World Bank cashiers had reported that their cash reserves were drying up as a result of network failures. He told them to talk to Chelouche himself and the Chief Cashier had got back to him, a look of wonder undoing his normal impassivity.

'Doctor Chelouche has authorized me to issue whatever you need,' he said disbelievingly. 'I mean, to issue value – *new value* – in whatever currency you need.'

'Thanks,' said Joe and snapped the connection off.

By six p.m. he had spent over six trillion dollars. *Six trillion US dollars in twenty-four hours!* But he now controlled twenty-three per cent of the Tye Corporation although the stock price had leaped eleven per cent on the day. He knew that weekends still had lower trading volumes than weekdays and he knew that a significant number of the big traders still refrained from dealing on the Jewish Sabbath. He looked at his time-zone display and realized that, with the exception of the American west coast, nearly all the markets were already running at their weekend and night-time trading levels. He calculated that he could afford to catch six hours' sleep on the sofa in Chelouche's office. The banker also had a private bathroom on the top floor and Joe felt he needed to use it.

He put his machine to sleep and then remembered he hadn't called Nancy. Their 'on-off, on-off' had come to life again since he had been back in New York and he realized that he would have to explain diplomatically that he would be unable to see her this evening.

He reached forward and woke his system up, then realized that their date had been for *yesterday* evening – and he had forgotten all about it. He decided it was best to leave it for now.

Sitting back in his chair, he yawned, then turned to his right-hand screen and opened up the TNN TV window. He could see the Hope Island garden party was now in full swing in the beautiful evening sunshine of the Caribbean – the opening event of *One Weekend in the Future*. He watched Tye shaking hands. He at least was obviously unconcerned about his stock movements.

The picture suddenly flickered, faded, and then returned.

Joe frowned. That was unusual, these days.

'You've been very lucky with the weather,' said a smiling President Wilkinson, as he enthusiastically pumped Thomas Tye's hand for the cameras. 'My people were predicting a hurricane for the weekend.'

'Well, looks like no danger of that, Mr President,' returned Tye, with a bright smile of his own.

'And congratulations on your engagement,' continued the president of the United States, laughing warmly, well aware that the TNN news crews were still focusing on them. 'What a beautiful woman!'

'I'm sorry you won't be seeing her this evening,' replied Tye.

'You'll have a chance to meet her after our wedding ceremony tomorrow.'

'Of course, this will be your stag night!' The President beamed. 'Shouldn't you be off somewhere with the boys?'

'It doesn't get any better than this, Mr President.' Tye smirked, then he moved off – he had spotted President Orlov making a choreographed entrance.

The lawns were now glittering with global celebrity, money and power. The late-afternoon Caribbean sun was kind and the guests enjoyed a gentle breeze off the Atlantic as they gathered on acres of Kikuyu grass – *Pennisetum cladestinum*, imported from Kenya for its decorative and soil-binding properties – on the terraces below Tye's mansion.

It was a black-tie affair and the lawns were packed. The official list of those invited to step into *One Weekend in the Future* came to just over 200 names, but their entourages and bodyguards had swelled the total number of visitors to the island to nearly two *thousand* and as many of them as possible wanted to be at the opening. Nobody deigned to wear a name badge, but many of the faces were world-famous and the new *GuestList* system coped with the rest.

Usually, when celebrities met, there was an awkward bonhomie that extended to a few sentences before disintegrating into paranoia about entrapment – you got hit on if you stood still too long. Famous faces constantly searched crowds for other famous or useful faces who could further their causes, so the over-firm handshakes and air-kissing were simultaneously a greeting and a farewell. But the Tye Corporation had introduced new manners at this event – manners that were likely to become the norm in the future. No one was wearing Viewpers, since all had the new InfoStem.

Everyone was enjoying the experience of attending a social gathering at which they were automatically provided with all necessary information about the other guests. Each one's system provided a complete database of all those present, complete with their biographies and most recent activities and achievements. As any guest scanned the other faces on the lawn, the InfoStem used a combination of ident-outputs and facial pattern-recognition to provide an overlay projection identifying the person on whom the wearer was focusing his or her eyes. Simultaneously, in the users' earpieces, the system provided the correct pronunciation of that person's name and at least one item of recent interest that might make an opening gambit for conversation.

But, as was his way, Thomas Tye had provided himself with an additional advantage. As well as receiving that output from the databases, he was also provided with audio input from one of the event organizers. She stood watching some distance behind him and guided Tye to his A-list one by one, adding her own commentary and suggestions to the audio-visual prompts as he greeted important face after face.

Ironically, many of the lesser guests on these lawns, those not expecting to be greeted personally by Tye this evening, had tuned their system into TNN's global coverage and were watching themselves on TV as the party progressed. Many were beaming in friends and relatives via their systems.

Haley was one such and, true to her word, Flick was determinedly patched in from Ladbroke Grove in London where it was now after ten p.m.

'There's that tennis player, what's her name, oh yes,' said Flick as Haley's gaze focused on a bronzed Amazon, the system providing identification. 'God, she's big!'

'She is,' agreed Haley, circling around a knot of Hollywood types to get a full view of the great white house above her.

She turned her head from left to right. 'Just look at that,' she exclaimed to her sister. Then, lifting her gaze to the roof, she saw two figures standing at a safety rail.

'I'll bet that's Miss World with young Tommy,' muttered Haley.

'Quick, zoom in,' urged Flick.

'I can't do that, it's just one focal length,' said Haley.

She turned back to survey the crowd on the lawn.

'What a gorgeous hunk!' exclaimed Flick from afar as a man in a dark naval dress uniform turned away from one group and headed towards her sister. He was carrying two glasses of champagne.

'You've met Jack before,' said Haley unnecessarily as his name and title appeared in their gaze.

In his apartment beside the spaceport, Raymond Liu lay on his single bed and stared at the ceiling. He was now shut out of the island's computer networks but, despite the recent loss of his job, he couldn't stop thinking about what might be causing the endless, apparently unconnected network failures. Beside him was a printout of the last news item he had received before HR had removed his network access. He picked it up again:

European Space Agency Disclaims Sunspot Theory

Geneva: The European Space Agency today announced that current disturbances to global communications networks – now thought to be responsible for the deaths of 937 people in a mid-air collision over Denver, Co., USA – were not in fact caused by a solar eruption or ejection of matter. Dr Alex Krywald, director of Space Weather at ESA, explained: 'We constantly monitor and record the sun's activity. We have searched through our recordings for the past two days and there has been no unusual activity on the surface of the sun or its corona in that period. We do not yet know what is causing this disruption in communications.'

Liu wondered why the Europeans would take the extraordinary step of publicly contradicting NASA's analysis.

Despite extreme tiredness and an unusually large quantity of alcohol in her system, Haley found herself lying wide awake in the vast emptiness of her bed aboard the *Treasure of the Caribbean.* She had two faces in her mind and, despite repeated efforts, she could not banish them.

Jack's face occupied the same place that it had for the last month, where it was usually the last thing Haley was aware of before her synapses quietened and consciousness retreated from her forebrain. But tonight there was another, competing image, and they flipped like two sides of the same coin. The other one was that of Dr Calypso Browne – formerly known as Miss World.

The early part of the opening evening had been everything that Haley had hoped for. She and Jack had made rapid and passionate love. Then there had been that dazzling garden party, at which Jack had made a brief appearance. Then her 'personal greeter' had scooped her up in readiness for the opening lecture, but Haley had been forced to admit she would not be attending.

The greeter's eyes had widened. 'You've got an invitation from Josh Chandler? You don't think ... I mean, could I come with you?'

Haley had smiled and taken pity on her. All the woman's Washington cool had evaporated at the mention of the famous film star.

Jack, with automatic access to all events, had found her while she was finally having a few moments alone with Josh Chandler.

'Great to meet you. I've read Haley's biography,' said a smiling Jack as they shook hands. Haley glanced sharply at him. He hadn't told her.

Then the couple had been left alone, as the star went to circulate with his other guests. Haley had recalled what she wanted to ask Jack. 'Have you met this Calypso Browne?'

She sensed instantly that there was something to tell. He looked quickly down, then back up to meet her eyes.

'Yes, I was seeing her casually before ... before I came back to London.'

'Casually? How casually?'

He looked down again. 'I saw her just a few times. We never really went out.'

'But you *were* lovers?' Her voice became hoarse as she asked.

Jack shrugged. 'Well ... yes. But it was a casual thing. On this island ...'

And now the two faces hung in front of her. Haley had drunk more, quite quickly, after she had been told. But with every drink she had seemed to become more sober, more cold around the heart. Jack had tried to reassure her, then was forced to return to his duties elsewhere.

She lay in the dark and her imagination now supplied even more disturbing images; she saw Jack's lips on that achingly long neck, she saw Miss World's fine hands on Jack's back. She saw them kissing. She saw that stunning, natural smile that the woman seemed to produce so effortlessly.

Haley groaned, rolled out of bed and crossed to the bathroom. She flipped the light on, rummaged in her toilet bag and found the sleeping pills she kept for emergencies. She filled a glass of water, swallowed a tablet, then confronted her reflection in the down-lit mirror. She studied her slender naked body and tried to imagine what Calypso Browne would look like without clothes.

'You're being stupid, we all have pasts,' she told herself. Then she thought of Kevin and Barry. *Not quite in the same league*, she admitted to herself. *Why had she been slumming for so long?* She turned the light off and finally headed for oblivion.

Everyone on the Operation Iambus team had turned up for work this Saturday. Four entire floors of the United Nations Secretariat building had been given over to the investigation of Thomas Tye and his global businesses and, late in the afternoon, the team leaders had returned to Amethier's office to make a final decision.

'Doctor Chelouche has started the process for taking control of the Tye Corporation,' the UNISA director advised the thirty people packed into his office. 'Joe Tinkler and his assistants are buying yet more stock as we speak. Over fourteen of the world's largest economies have committed central reserves to the fund.'

'Including the USA?' asked Martha Rose.

Amethier shook his head. 'We haven't formally informed Washington,' he explained. 'Doctor Chelouche is down there now but we're not sure of the nature of the relationship between Wilkinson's administration and Hope Island. The word is that Tye and Attorney-General Treno had a blazing row during his recent visit, but Wilkinson himself is on the island now. We think some deal may be under way.'

They pondered the implications. They had all watched the TV coverage of the opening ceremony and had marvelled at the body of political power assembled by one businessman. It was almost a rival to the UN.

'A financial takeover is not going to solve the problem of the

Solaris satellites,' pointed out Ron Deakin. 'I understand they're under the control of a private company.'

'The Doctor disagrees,' said Amethier. 'He maintains the core corporation is a thirty per cent shareholder in Phoebus Inc. and that will prove enough for us to apply to the international court for a writ of cessation.'

'We can't expect Tye to take this lying down,' argued Rose. There were nods of agreement all round. Nobody was expecting this to be straightforward. 'We'd better apply to the Hague for *certiorari*,' she added. 'That's an immediate writ from the highest court of all. It should overrule any legal efforts they make and should shut down all operations temporarily until we have a better handle on the situation.'

Deakin glanced at his LifeWatch. 'Well, he's due to be married an hour from now, so I suppose he'll be a little busy. Anyway, we have some news. Mike?'

Michael Chevannes rose to his feet.

'I've just got back from South Africa,' he explained. 'I flew over there to visit a boy called Reon Albertyn. He's fifteen – or rather, he was. He died while I was there. His story was uncovered by UNHCR and I've turned all the relevant material over to Rima Berzin and the WHO team.'

Professor Berzin took the floor. 'If we could turn the lights down . . .'

The room dimmed as she walked towards the wall screen. A video image of a wizened old white man with a hugely swollen head appeared. He was sitting on a bed.

'Believe it or not, this was Reon Albertyn,' Rima Berzin explained. 'He was suffering from a condition initially identified as progeria, a very rare ageing disease. Victims in their teens start to look like they are ninety years old. And there's another strange thing . . .'

The image changed to a middle-aged black woman seated at a table in a hospital room.

'This is Reon's mother,' she continued. 'As you can see, she's black, while Reon was an albino – a black man with a genetic abnormality that produced white skin. You'll now hear Mrs Albertyn's sworn testimony.'

They watched and listened as the grieving mother explained how she had been recruited to work for the Erasmus Corporation sixteen years before. She described the tempting fee she was offered to be a surrogate mother and how, after the doctors had finished their tests on her newborn white baby, they had paid her fifty thousand US dollars and had returned her son to her.

She wept as she recounted how her husband had beaten her and how her family had turned her out with her baby because it was white. She described Reon's unhappy childhood of alienation and the first obvious symptoms of his disease at the age of twelve. Then he had been convicted and imprisoned for stealing a car in Zimbabwe. She had been reunited with him only after the UNHCR secured his release from the Harare jail and had placed him in the Steve Biko Memorial Clinic in South Africa.

As Professor Berzin fingered the remote, the audience saw the text of a pathology report.

'He wasn't actually suffering from progeria. In fact, he was an early human clone, an experimental embryo given a number of so-called genetic enhancements. Lily Albertyn was used as a surrogate embryo carrier, one of many,' she explained. 'But in this particular clone the telemores, the ends of the chromosomes, were too short. He was ageing at five times the normal rate from the moment he was conceived and implanted in Mrs Albertyn. Reon was identified in the Erasmus files as *Alpha 41*.'

'So that's what we're going with?' asked Martha Rose. 'But does it connect with Thomas Tye directly?'

'He was chairman of Erasmus Inc. then and he still is,' said Chevannes.

The lawyer nodded. 'Do we have any further evidence?'

Chevannes patted a pile of paper and a wallet of memory sticks in front of him. 'There's stacks of it. We've got all the Erasmus reports from that time, although they may not prove admissible because of the way we obtained them. But there's also the circumstantial evidence from Mrs Albertyn herself and other mothers contracted to be surrogates at the time. We're following their children up now.'

Martha Rose considered. She would need plenty of direct evidence as well as the circumstantial. 'We can issue writs of *duces tecum* on the various documents we know about,' she reasoned. 'We can force Erasmus to bring the originals to court. We don't have to show the versions we intercepted – just knowing they exist will be good enough. I'll be looking at a prosecution under the forty-fifth, seventy-eighth and one-hundred-and-twelfth Articles of the 1945 UN Declaration of Human Rights and under Article Sixty-four of the 1986 African Charter on Human Rights.'

'What do you think, Mr Olliphant?' asked Amethier. 'How will the world vote on that?'

'At the risk of sounding heartless, it's perfect,' declared the senior image-perception adjuster. 'Even with this Ethiopian stunt going on, the images of this boy and his mother are very powerful. Can we release details about the boy – the clone – that Tye has with him on his island?'

Amethier nodded. 'We can – we're very serious about this. Jack Hendriksen has brought us pictures of the boy – Tommy, he's called – and I think we should go with the lot. Can you get a preliminary indictment, a few specimen charges, up by tomorrow, Martha?'

The international attorney looked flabbergasted. 'No way, Jan,' she protested. 'Not if we want to do this properly.'

'Timing, Rose, timing. I want to make an announcement tomorrow, on the day that it is supposed to rain in Ethiopia. I want the world to know that the United Nations Commission for Human Rights is charging Erasmus Inc., the Tye Corporation and Thomas Tye personally with responsibility for the death of Reon Albertyn and an unknown number of other children in Southern Africa. That will produce the maximum impact.'

'Agreed,' confirmed Edward Olliphant, nodding. 'I'll leak it to the *New York Times* for Monday's first edition.'

'We'll get right on it,' sighed the lawyer as she rose. 'We'll need everything you've brought back, Mike.'

Tommy's blue suit had been made at the little tailoring boutique in Hope Town. Calypso had ensured that it was more restrained than many American page-boy outfits and it made him look very mature.

'So grown-up,' Calypso observed as she stood behind him and savoured their reflections in the mirror. 'What do *you* think Biya?'

'Nice,' grinned the small girl in her cream-and-white flouncy dress. In the few weeks that she had been receiving Calypso's tuition she had started adding adjectives to her stock of English nouns, although she was not yet stringing many of the words together.

'Can you see us, Mum?'

Mrs Browne was sitting on a chair behind them, dressed ready for the ceremony, Roger on her lap. The CatPanion was wearing a silver collar in honour of the occasion. Calypso walked over to them. The old lady ran her hands over the cream lace dress that Calypso had chosen. Then she pulled her daughter's head down, felt her face and kissed her.

'You are happy,' she smiled. 'God bless you.'

'I hope I am coming?' asked Jed, from the top of the dressing table.

'Of course you are,' laughed Calypso. 'I've got a flower for you to wear. Now, how do I look?'

None of her small audience were qualified to judge; none of them could look with adult human eyes. To be more specific, none of them could look with adult *male* eyes.

'Very nice,' said Jed appropriately.

'Yes, very nice,' agreed Roger.

'Very nice . . . *Mummy*,' said Tommy. Then he collapsed into a fit of laughter and threw himself face down on the bed to hide his embarrassment. Tom had explained that he would be able to call Calypso that after the wedding.

'You'll mess up that new suit,' smiled Calypso. 'Biya's keeping her dress so nice. There isn't long to go.'

'Nervous?' asked Jed.

Once again, Calypso shivered. She knew that Professor Keane insisted the Furry companions produced their language from vast databases triggered by key words – like *wedding*, she presumed – but she still found this companion's apparent cognition unnerving. She wondered if the professor really understood what she was doing.

'I am, a bit,' she admitted.

There was a noise behind them.

'You look really beautiful, Calypso,' announced Connie from the doorway. 'In fact, you all look wonderful.'

The matron of honour entered the room and put her small bouquet down on a table. 'Are we ready? she asked.

'I've never seen anything like it, Ron.'

They were drinking coffee in Lynch's sub-basement. Most of the team had grabbed some sleep in the adjacent Marriott hotel, but the computer scientist had had no relief and he looked dog-tired. It was nearly noon.

'The networks function one minute, then there's data delay of an hour or more. It's like they're seizing up.'

'Which networks do you mean, Al?' asked Deakin.

'As far as I can tell, everything using a satellite link – which includes almost every network these days. Even the undersea fibre optics connect up to the satellites for continental hops.'

The UNISA Exec nodded. 'What do you think's causing it?'

'I assumed it was overload,' replied Lynch. 'So I've shut down all my artificial message-generation, but even at ultra-low bandwidth there're real problems. *And* it's Saturday. Traffic normally halves at the weekend.'

Deakin yawned. 'Well, it shouldn't matter too much. We've got what we need.'

Calypso had particularly asked to be married under Rembrandt's all-knowing, all-understanding, all-forgiving gaze. Since moving into the main house she had spent hours in its great marbled reception hall staring at this rich, dark painting. It had been painted in 1661, shortly after the greatest of all painters had been widowed and made bankrupt. He had been fifty-five when, once again, he had looked into and exposed his tired, omniscient soul with brutal honesty. Tom had informed her that it was the last of the artist's self-portraits still outside of a museum and he had been forced to outbid both the Getty Museum and the Rijksmuseum to acquire it.

She knew the painting well, from her father showing a slide of it when he was teaching late-twentieth-century English literature. He would project the portrait as he quoted one of his favourite opening lines, a sentence written by the novelist John Fowles, a line deliberately created for its rigour, opacity and unyielding abstruseness. The author was describing this portrait:

*Whole sight; or all the rest is desolation.*

*Whole sight indeed!* thought Calypso as she arrived at the wide doors leading into the vast hall. She wondered what her missionary-trained schoolteacher father would have thought of her. But Rembrandt would have understood, even today.

She waited in the high double doorway while her eyes adjusted to the brilliance of the television lights within. In response, a magnificent *rivière* of diamonds and sapphires at her neck created scores of lens flares that sent hands flying over the boards in the control room that the TV production team had set up next door.

The groups of sofas had been removed and the large sculptures – a collection she now knew included a Rodin, a Donatello and even a battered Praxiteles from ancient Greece – had been moved to one side, creating the appearance of a long gallery. A red carpet stretched in front of her, laid down specially for her bridal procession.

Calypso could see Tom waiting for her in front of the low brass rail erected in front of the wedding dais. Neither of them was religious, but both had agreed they wanted a feeling of ceremony. The television lights flared off the white marble and made the great room and its art seem brighter, larger than life.

*Like everything from now on.*

She could hear Tommy, though she couldn't see him from where she now stood. There had been only one choice for their wedding anthem. Tommy himself was playing 'It's Our Planet' on a new double-manual digital synthesizer with 32-bass pedalboard. It had been programmed with a complete sample of the sounds belonging to the great Bruckner organ in the ancient monastery of St Florian in Linz, Austria. Tommy had been in a state of huge excitement as he had explained to Calypso how Anton Bruckner had been Tommy's own age when he had played that organ for church services. Equally precocious, Tommy had rearranged the corporate march as a delicate nocturne to take full advantage of the many fine upper-register pipes that, at one remove, were providing his palette of sound.

Calypso turned to look at ever-cool Connie and then smiled down at little Biya. The girl was all wonder with her huge eyes, cream lace flouncy dress and shiny shoe buckles. Calypso reached down and took the child's hand.

Seeing the floor manager's cue, she took a deep breath and stepped forward.

Suddenly the music changed to a series of staccato, chiffed, ascending, triumphal notes, not at all as had been planned or rehearsed. They had earlier agreed with the TV people that the corny and the clichéd would have no place in this ceremony, which was being recorded only minutes before it would be seen by billions all over the world. That network mass-cast was scheduled for two p.m., the optimum viewing point for audiences across the Americas, Europe and the Middle East. Meanwhile, many Asian, Australasian and Pacific viewers would be sitting up late to see the event as near real-time as possible. The rest would digest it with their breakfasts. Then it would be replayed over and over again and downloaded countless times, continuing in the suspended immortality of global virtuality.

Finally the attention-demanding series of heraldic-trumpet voices broke into a giant wash of thundering, descending, octaphonic chords that blasted from the full range of the organ's mighty pipes to announce Calypso's entrance. Tommy had ignored his instructions and pulled out all the stops. A broad smile lit Calypso's face: it *was* right. It was Mendelssohn's' famous *Wedding March* from his incidental music to Shakespeare's *A Midsummer Night's Dream.* She recalled that the music celebrated the marriage of Theseus, duke of Athens, to Hippolyta, queen of the Amazons. She felt the first tears cloud her vision. She loved Tommy so much. How could he have *known* this would be so right?

She moved forward steadily and slowly as the triumphal music reverberated around the vast marble hall. Tommy the classicist was now taking it 'adagio', very, very slowly, shunning the tumpty-tumpty, polka-like temptations of hick performance, and filling with gentle, magical, filigree *agrément* notes as he welcomed his beloved Cally into his life for good. It had been one of Calypso's several conditions that when the ceremony was complete, she would become the boy's co-guardian and, as with all their other nuptial agreements, this would be ratified under Californian law. At the same time, the Tye Press Office would go public about Tommy's existence. 'A huge scoop to complete this fantastic weekend,' the senior perception analyst had predicted. It would be claimed that his surrogate mother had demanded anonymity and privacy and so could not be named or involved. Since his unburdening, his momentous sharing, Tom had seemed surprisingly willing – Calypso might have said loving – on all such topics and there had been no hint of corporate or personal morganaticism in his attitude.

She walked towards the rows of chairs set out for the very few, very privileged guests. Her mother sat with Roger on her lap in the front row to the left, her best friend and neighbour from Mayaguana sitting beside them. Calypso had also invited Nurse Pettigrew, Heather Garland from the school and, on impulse, Mario Ginola from her favourite pizza place down at the Little Venice marina. She could see him now, way at the back, crumpled in overawed wonder. Then she saw Miss Duckett, Tommy's music teacher, sitting on the other side of the aisle, a broad grin on her face. Calypso smiled; she knew a co-conspirator when she saw one. The house staff and a few members of the corporate office staff filled the remaining seats. Marsello Furtrado, certainly not her idea of a best man, stood in a posture of concinnous elegance.

Calypso arrived beside Tom and he turned his head to grin at her. She shot a glance at Tommy who was finishing on a sustained C major chord, including the contrabass of the giant pyramidon thirty-two-foot C natural pipe, the lowest note the Bruckner organ could deliver. The powerful speakers sent out a standing wave of sound that reverberated around the hall and rattled their chest bones. It felt as if the great pipes themselves had been transported from their gilded baroque basilica and were now ranged invisibly in this marbled bride-chamber. Finally, the organist lifted his hands with a flourish.

Tommy looked across at her in delight and, forgetting the cameras, she crossed her eyes, sucked in her cheeks and mugged at him. He laughed, jumped down from his stool, leaped off the dais and ran over to help Mrs Browne to her feet. He was determined to be both organist and page-boy, assisting Calypso's mother as she gave her daughter away. Jed, who had been quietly singing along with lyrics of his own invention, watched from his perch on top of the keyboard.

They had even flown in a Hollywood judge who moonlighted as a celebrity wedding notary, hoping that he would be accustomed to cameras and to famous faces. But the imposing silver-haired man in the dark suit still managed to hesitate slightly before stepping forward. He realized the global ratings for this wedding would be far greater than for any televised ceremony at which he had previously officiated.

It had been decided that formal banns were superfluous: the whole world had been informed of *this* marriage and if anyone was going to have the temerity to object, there had been more than enough time to do so.

The ritual moved forward as if in a dream. Calypso felt she was back on the catwalk again. Her performance was flawless. Bride and groom both said the words clearly and without hesitation, and professional smiles were exchanged, although she was aware that underneath there was unexpected emotion.

It was Tommy who helped Mrs Browne step forward when the time came for her to give her only unmarried child away. Then, having brought forward a chair for her to sit down again, it was time for him to proffer the wedding rings on a blue velvet cushion.

Calypso's betrothal ring contained a salamstone sapphire from Sri Lanka surrounded by a baguette of the finest Kimberlite diamonds. The main stone had been cleaved in Amsterdam from an eight-carat cabochon earlier that week. She and Tom had watched on the networks as the delicate operation had been performed by a world-renowned glyptician. Her wedding ring was a Russian trinity of gold bands: red, white and yellow. Calypso privately considered this multiple appropriate for the basis of their union. Her groom relished the fact that the ore itself had come from Sybaria. A jet had been sent to collect the finished jewellery two days earlier.

Calypso was dimly aware of the cameras shifting behind the dais to find the best angle for a close-up as Tom took her hand and slipped the rings on her finger.

Then they were face to face for the first time. She tasted his mouth and found herself responding. As they held the moment for the cameras, the small assembly applauded.

When Tommy went scampering off again she knew why. He had begged to play a short interlude piece while the couple were led behind a vast Venetian screen to sign the official register of marriages that the notary had brought with him from Beverly Hills. By agreement they were also signing joint guardianship papers, hurried through the Californian legal processes, and there were also multiple legal and financial agreements to sign.

As witnessses Heather Garland and Mario Ginola rose to follow the couple, Calypso could hear the majestic opening chords and mighty descending bass notes of Sigfrid Karg-Elert's *Marche triomphale*, a piece composed to celebrate the marriage union *and* to display the virtuosity and range of both organist and instrument.

Once all the paperwork was completed, Calypso and Tom returned to the centre of the huge room and into the full glory of the final reverberating chord of the Lutheran chorale. Waiters stepped briskly forward with trays of champagne, since bride and groom only had a few minutes to toast each other and to accept the congratulations of their guests. Even now, the recording of the wedding was being watched by billions on the networks. Shortly the floor manager would signal that it was time to walk out through the terrace doors and face the live TV coverage and the hundreds of guests gathered for the garden party.

Tommy suddenly appeared at their side, Jed under his arm.

The synthesizer was now on automatic and was softly replaying Tommy's performance of 'It's Our Planet' to provide a coda. At this moment Calypso felt as if the planet *were* hers.

'Cally?'

She bent and kissed his cheek, and he put his arm up around her neck and whispered in her ear.

'Mummy . . .'

Then he got the giggles.

Calypso kissed him again and straightened up. 'Well?' she said to Tom, the single word asking so many questions.

He turned his head and looked deep into her eyes.

'Well,' he replied.

'May I have some champagne, please?' Tommy asked, in order to interrupt.

Tom smiled and allowed him a sip from his glass.

'Hurrah,' cried Jed.

'Hurrah,' echoed Roger.

Then came the signal from the great glass doors. Tom returned his glass to a waiter and took Tommy's hand. Smiling, he held out his other hand towards his new bride.

Hand in hand, they walked out into the sunshine to greet the world.

# TWENTY-SIX

His silver hair, craggy looks and genial demeanour had become familiar the world over but to the small crew waiting in the pre-dawn desert the senior anchorman of the Tye News Network looked old, cold and miserable. As the generator-powered lights were switched on, the production assistant ran through her silent cue count, then gestured. The presenter suddenly turned on his most dazzling smile.

'Good morning. I'm Bob Houston, this is TNN and these are the Ethiopian plains on the morning of Sunday, August thirtieth. Here in East Africa it is now five thirty a.m. and the people of one of the world's poorest nations wait anxiously to see if the new climate-engineering technology promised by the Tye Corporation can finally bring relief from a seven-year drought. Many local people have stayed up all night dancing for rain – they call it a *ngoma* – and even as the light begins to arrive I can see black clouds over on the horizon. Let's go now to one of the Halcyon weather planes over the Indian Ocean.'

Beneath the United Nations complex in Manhattan the section leaders of the Iambus team were watching it on the large wall screens of the Security Council's private meeting chamber. Even though it was nine thirty on a Saturday evening in New York, none seemed eager to leave. They watched intently as the same weather plane transmitted aerial views of banks of low, moisture-laden clouds rolling towards the East African coastline.

'For two nights now he's been concentrating the output of four Solaris stations over the Indian Ocean,' explained Rolf

Larsson. 'It does look like he's managed to invert the atmosphere. The cold air of the ionosphere has suddenly been sucked downwards.'

There was no response from the group. They were all waiting for the moment.

The picture flickered and disappeared. 'The damn networks again,' muttered Al Lynch.

When the picture re-established, the location had switched to South Dakota where a group of Sioux Indians were engaged in a rain dance.

Deakin watched two middle-aged males in traditional costume and feather head-dresses attempt to recreate their forebears' ancient dance of supplication to the rain gods. When they had circled a jug of water four times they would throw themselves on the ground and then drink the blessed liquid.

Deakin shook his head as the perspiring, overweight men parodied their heritage for the sake of the media.

Michel Geronde had risen magnificently to the occasion, deploying his army of chefs for almost ten hours in the cruise liner's vast galleys. Now over 200 of the world's elite and an equal number of their partners and senior staff were finishing off the banquet with his superb *Nesselrode*, a Russian iced dessert made with chestnuts, cream and preserved fruits, flavoured with rum.

At six other locations around the island, Geronde's imported assistants were also producing meals for the hundreds of bodyguards, support staff, pilots and administrators that the VIP guests had brought with them to the tiny island. At all of these satellite banquets the diners could watch large wall screens displaying live broadcasts from the ballroom of the *Treasure of the Caribbean* so they too could feel part of the momentous events being celebrated.

It had been announced that the Tye Corporation was now leaseholder of over one and a half million square miles of Eastern Siberia, and a new corporate state – the second of recent times – had been created. The final cession documents had been signed earlier in the day here on Hope Island and President

Orlov had made an emotional pre-dinner speech that praised Thomas Tye's vision and informed the world that the Russian Federation was once again economically independent.

Then Thomas Tye himself had given a major performance. For half an hour he had addressed his guests in the ship's enormous ballroom. He described to them his plans for Sybaria and the opportunities presented by climatic re-engineering. In due course, the entire world's population could be adequately fed. Rain could be brought to arid regions and natural catastrophes averted. He focused, as always, on the benefit to the planet. All the delegates were starting to understand the full potential of his plans.

A full-motion simulation of the Phoebus Project and the Solaris energy-recycling stations had been made available on the networks to help the global population at large understand how their planet was about to be changed by the Tye Corporation's new technology. Despite the degraded performance of the networks, millions of Tye followers had already managed to download this model.

As he reached his finale, Tom was standing directly behind President Orlov's chair. He clamped both hands on his guest's shoulders, glanced down at him, then back up to the cameras.

'Take care of our planet, you hear?' he declaimed.

The leader of a newly prosperous Russia was the first to reply 'We hear'. Then he leaped to his feet to shake Thomas Tye's hand. The rest of the guests also stood up as the TV cameras followed Tye back to his seat, threading his way through his smiling and applauding guests, shaking hands as he went.

Now, as the animated diners finished their dessert, the finely drilled regiment of waiters moved in to set a full glass of champagne in front of each of them.

Thomas Tye was seated at the centre of the top table, his new bride placed to his right. Next to her sat Tommy, very aware that he too was a focus of attention. It was Calypso's second formal event of the weekend and she had chosen a severe, tightly fitting black dress for this occasion. Since she had made her momentous decision two weeks before, her life had become a frantic whirl of network shopping and personal re-organization. The simplicity of her dress was thrown into sharp relief by the dazzling necklace of sapphires and diamonds. Almost every female guest there in the ship's ballroom was displaying fine jewellery, although many had also integrated elements of the dress *Équipage* systems into their outfits.

Haley was chatting at the back of the room with her new friend from the Tye public relations consultancy when they heard a sharp hammering through the PA system. They looked up to see that a master of ceremonies, in a formal red coat, had appeared at the top table and was calling for attention. When the loud buzz of conversation subsided he pulled the microphone closer to him.

'Your Majesties, Royal and Serene Highnesses, Mesdames and Messieurs Presidents, your Holinesses, your graces, excellencies, ministers, secretaries of state, my lords, professors, doctors, distinguished artistes, ladies and gentlemen, please welcome your host, Mr Thomas Tye.'

Suddenly Haley's view was obstructed by a wall of black tuxedos and elegantly bared female backs in front of her. She too rose to her feet but still could see nothing. With a grin at her equally diminutive companion, Haley slipped off her high heels and climbed onto her chair, her new friend following suit. Now they could see Thomas Tye bowing to acknowledge the tumult of applause. Haley noticed that the beautiful Calypso had also risen, turning to applaud her new husband, her huge smile suggesting real pride.

Haley opened her elegant new VideoMate and touched her sister's screen icon. Felicity had issued clear instructions that she wanted to be with her sister at this moment – to share a more intimate, personal view than the one being provided live by TV cameras on the networks.

The two large screens at the end of the ballroom came to life, revealing giant close-ups of Thomas Tye and his bride. Tye raised his hands for silence and eventually the applause began to fade as the guests resumed their seats.

'I am not going to make a second speech this evening,' he began. 'I only want to say three things.'

He paused and smiled down at Calypso. 'I want to thank my new bride for all the love and affection she has already brought into my life and ...' he paused and looked at Tommy '... my son's.' He bent to pick up his glass.

There was a movement to his left as Marsello Furtrado also rose. Smiling widely, he leaned into the microphone.

'Ladies and gentlemen. This interjection was not scheduled, but will you join me in drinking a toast to the happy couple *and* their son?'

The room rose again as the guests stood and raised their glasses in salute. Then the banqueting hall rang with applause once more.

Standing on her chair again, Haley noticed that Thomas Tye had reddened somewhat at the unexpected salutation. He shook his head at Furtrado in mock annoyance, then lifted both arms again. Almost instantly, the room was quiet.

'And the second thing is ...' he put his hands up to both his ears as though he was receiving information in his small diamond-decorated earpieces '... I am pleased to announce that the final total pledged for our Ethiopian Appeal has now reached ...' he paused '... One hundred and eighty-seven billion dollars.'

The sum flashed up on both screens.

The diners whistled and shouted, revered heads of state and global icons alike – all letting go. The TV cameras transmitting the event to the world meanwhile captured some delightful celebrity 'off-guards' as the great and the good revealed themselves to be every bit as excitable as ordinary people. At the far end of the top table Haley could see Josh Chandler's brilliant smile adding to the sparkle. She had learned earlier that Tye Media Arts was to produce his next film.

'So now, all that remains is ...' Tye nodded theatrically to the disembodied voice in his ears, then turned and pointed to one side.

Again the image on the screens changed. This time they could see rain pounding on an empty street and the camera pulled back to reveal torrential downpour bouncing off the tin roofs of a small township.

A double-deck caption appeared at the bottom of the screen.

*Awasa, Southern Ethiopia*

*7.30 a.m. 30 August*

The room erupted again and Haley found herself jumping up and down on her chair as she applauded. 'He's bloody done it,' she yelled, partly to her sister, partly to her dinner companion and partly to herself.

Tye made no attempt at controlling his guests this time. He too turned and stared at the transmission.

Suddenly there was a flicker and the screens went black.

Death did not stop just because Thomas Tye was throwing a party and Doctor Mohammed Ebrahimi was duly grateful for that. He was in charge of the graveyard shift of grief counsellors in the LifeLines Operation Center – the shift he preferred as it gave him the opportunity to assist next of kin all across the American continents with difficult decisions about reserve floors, closed-casket outcomes and any pre-emptive bids that might be forthcoming.

He helped a trainee close a rapid deal with a middle-aged lawyer in Atlanta for the sale of his son's principal organs – little was going to be available from above the neck since the teenage joyriders had tried to jump a rail crossing and had driven under a train. Then he walked out of the communications and counselling area of the LifeLines Center to stretch his legs and get a cup of coffee.

As he allowed the soundproof door in the glass partition to slide shut, he yawned, scratching idly at small ovals of hairy brown flesh that protruded through the gaps in his bulging shirt front.

Ebrahimi glanced up at the atlas on the wall displaying LifeWatch mortality alerts. The team maintained this display as a powerful graphic reminder that death waits for no one whatsoever and that NOKs will always be calling in. Above it were duplicates of the two display counters that regulated and stimulated all counselling activity next door. The first of them showed how much money was currently in play in the various auction systems that Easy Zee Zee had built. The second counter showed how many NOKs were currently holding network connections open while they waited for the chance to consult a counsellor and complete terms for an auction. Currently there were fifty-seven people waiting and the longest any had been kept waiting was twenty minutes. Ebrahimi frowned. That was far more NOKs than usual and, as a result, his team's response times were deteriorating. He looked over his shoulder at the counselling centre. The heads in the many cubicles there all seemed fully engaged in discussions.

As Ebrahimi glanced back at the map display, it seemed as if the rash of white dots, currently at their peak on the nightward side of the planet, was growing denser as he watched. Each light indicated receipt of an upload alert from a LifeWatch that could no longer detect its wearer's vital signs.

He hurried, almost ran, over to the system supervisor's desk and pushed the diagnostic query button, watching the display system cut out and begin its reboot. As it did so, it ran a check of its input, processing and display systems. The white lights began to reappear on the world map as the test was completed, but now so densely that the pixels were almost fusing into continuous smears of white light.

Ebrahimi heard a system prompt and he looked down at the small desk screen. *Diagnostic check complete. All systems 100% function.*

The thanatologist leaned forward and flipped the output to display the network loadings at the LifeLines server hub. It showed a bandwidth usage of almost *nine* terabits. Their previous highest bandwidth requirement had been less than *one*, and that had been when an outbreak of Lassa fever in the Philippines a month before had briefly doubled the number of recyclable assets available for auction.

Ebrahimi stepped back from the system controls and looked up at the counters. The number of people waiting to talk to a counsellor was now 1,481 . . . 2 . . . 3 . . . 5 . . . 6. The counter was becoming a blur. He reached into the desk and punched a number at random: 755. He sat down and forced himself to switch to counselling mode. The woman was in her early twenties, of Chinese extraction, calling from San Francisco.

'Mrs Young – I am Doctor Ebrahimi. How can we be of service?' he began.

The NOK had been crying while she waited, but now her head snapped up. 'His LifeWatch just cut in,' she began. 'It injected everything all at once.'

'I am so very sorry to hear that,' responded Ebrahimi automatically. 'Please hold.' He punched another number: 814.

The image showed a black man in his thirties, with a small girl seated on his lap, clinging to her father's shirt front. He was crying silently, his tears falling on to her head.

'It was her LifeWatch,' sobbed the widower.

Ebrahimi cut the feed off before the man could say anything more and punched up connection 1370. This time he was looking at a wailing Indian woman in a black sari. He closed the screen and looked up at the map display again. Nearly 5,000 people were now waiting to talk to them.

'What's going on, Mohammed?' It was Irene Desmond. She let the door to the counselling room slide shut.

Ebrahimi raised his left arm and examined the ostentatious gold Rolex Daytona LifeWatch on his wrist. He quickly undid the security clasp and eased the bioconnectors out of his skin. He let out a long sigh as he laid his personal health protector face down on the desktop in front of him.

'Better take off your LifeWatch, Irene,' he advised.

The sparks arcing across the frayed wires couldn't be extinguished. Theresa kept shaking a kitchen towel at the rhythmic clacking glints, but they wouldn't stop.

She opened her eyes and tried to see in the darkness of her room. The tapping continued as control of her consciousness was restored to her and her senses returned. The noise was coming from her window.

She turned her head to the clock. It was 3.42 a.m. She sat upright and immediately realized that she was still slightly drunk. It had already been a long weekend. She swung her legs out of bed, switched on her reading light and found her robe. The tapping at her window had stopped. Whoever it was had seen her light go on. She ran her fingers through her short, auburn hair and rubbed at her face.

'It's Doctor Liu,' announced Sandra, from the foot of the bed. At night she napped in security mode, connected to the house systems and the island's monitoring network as well as to the networks beyond. 'He's no longer an employee of the Tye Corporation.'

Theresa pulled open the drapes and in the dim light saw the small figure of the former network chief. She slid open the glass door.

'Raymond?'

'I'm sorry, Theresa, I know it's late. But I need to use a system urgently, and I'm shut out of all the island's networks.'

She stepped back, motioning for him to enter. She had heard of his sudden dismissal and she was aware that there was a growing sense of panic among the beleaguered Tye Networks engineers who were left to try and cope with the problems.

She slid the glass door shut again and looked at the sad figure of the engineer. She suddenly felt a rush of affection for him.

'Doctor,' greeted Sandra dryly.

'Come into my parlour,' Theresa said with a smile, invoking a word from her childhood. 'You can use my system while I make us some coffee.'

Liu shook his head.

'I really need the sort of computer power that's available in the Control Center,' he said. 'I'm sure now it's the Solaris energy stations that are causing the outages. I want to run some new tests I've been developing.'

He slipped a storage card out of his pocket. 'I was hoping you might have some real processing horsepower in your department building. What level of network access do you have?'

'Well, Ultra,' admitted Theresa. 'For the Anagenesis and Descartes projects.'

Raymond Liu nodded. He had correctly assumed she would possess the highest level of access.

'If I could log on as you...'

Theresa smiled. 'The best place is the main lecture theatre,' she informed him. 'We've been using that for Solaris demos and 3D holo animations – Robert's technical support team has overloaded the place with bandwidth and processor power.'

'If you don't mind?' begged Liu. 'I realize this could cause you a problem but...'

Theresa smiled again. She was not in the least surprised that Raymond Liu, sacked and in disgrace, would still be trying to solve the network failures.

'No problem, Raymond. If you'd seen Tom's face earlier tonight when the feed from Ethiopia went down...'

She tailed off, recalling how hard and how visible Tye's battle for self-control had been, forcing himself to remain calm in front of his honoured guests. It hadn't helped that when the feed from Ethiopia had died the TV director had switched back to close-ups of Thomas Tye's face as he attempted to make light of the technical failure that had marred his moment of triumph.

'I hear he's holed up in the Solaris Control Center with all my old team,' said Liu. 'You can imagine...'

Theresa nodded. She could imagine the scene very well. She picked up her VideoMate and said, 'Call Robert.'

They waited a few seconds until a groggy voice answered.

'Robert, I'm sorry, I know it's late. I need a favour.'

There was a short silence, then a grunt.

'I need you to open up Lecture Theater One for me, please.' She paused and mugged a wince for Liu's benefit. '*Now*, if you could bear it, Robert. It's an emergency.'

They listened as a few more snuffles and a yawn announced that the researcher was coming to. 'OK. Give me a little while.'

'I'll meet you there in twenty minutes,' said Theresa. 'I'll have Doctor Liu with me. Thanks, Robert.' She snapped her VideoMate closed.

'I just need a few moments,' she said as she opened the door to the living room so that Raymond Liu could wait for her in her 'parlour'.

*

'Jesus, that's *big*.'

Pierre nodded as he watched the screen over the ATM manager's shoulder. They were in the control tower at Hope Island Spaceport and were looking at a 3D image on the ground-level radar display. Jack Hendriksen had rostered for one of his detail to be present in the tower during all shifts over this weekend.

'My guess is it's a carrier,' said Dagmar Haas, senior ATM manager for the night. 'Probably American. No one else has anything that big.'

'How far out?'

'They're twenty-five miles or so. They'll probably turn soon.' Hope Island's six-mile exclusion zone was known and respected internationally.

Pierre debated whether to call Jack himself. His boss's orders were clear: this was a delicate time, the security status on the island was Gold – the highest – and he was to be informed of *any* strange occurrence. And Pierre had caught something else in Jack's voice during the briefing, something that had made the PPT chief wonder if trouble was actually anticipated.

'Still heading our way?'

Dagmar Haas nodded and then checked herself. 'No. Look. They're just starting the turn. They'll stand off to the east. They're probably on their way down to the South Atlantic.'

Pierre, although doubtful, nodded. There were none of the support vessels necessary for an extended mission.

'I'm sorry to get you out of bed, Robert.' Theresa headed down the aisle, Raymond Liu behind her. Robert was adjusting the controls at a trio of command screens beside the stage.

'You've met Doctor Liu before. He needs to run some very urgent tests.'

Robert nodded, but said nothing. He too had heard the news of the network director's dismissal. He handed them each a coffee.

Theresa stepped in to the control panel, touched the finger-print pad, entered a password, disabled the user-monitoring system and stepped back. She motioned for Liu to step in. As he did so she nodded for Robert to join her in the front seats.

Liu took the storage card from his breast pocket again and slipped it into the system. A huge diorama of the globe filled the centre of the on-stage Holo-Theater. It was the model developed as a visual display of real-time network loadings and activity. Liu touched the controls and the world's satellite networks rapidly started to appear, accumulating as ever-denser overlays. Liu stepped back from the console to watch these networks as they became brighter and brighter.

'It *should* stop any second now,' said Liu quietly.

Although they had watched such visual representations of network activity around the planet many times before, they were still entranced. But the display did not slow its development. All over the globe dozens of light-spots started to glow with increasing intensity, turning from white to blue, then to gold, then back again to a purer, more incandescent white.

'As I suspected.' Liu nodded, stepping back to the control. 'Network activity *has* gone through the roof. The failures occurring in the systems aren't freak accidents, they're the same number of breakdowns we would normally expect over a long period, but the amount of activity in the network is becoming massively amplified and accelerated when the sunlight is concentrated. It's telescoping the failure rates of decades into just a few seconds. Now . . .'

He extracted a second storage card from his shirt pocket and slotted it into the control console.

'I got this from a very good friend of mine an hour ago,' he explained. 'He's risked his job, his stock options and his family's lifestyle on this island by getting it to me. It's a record of all the Solaris activity in the last two weeks.'

The image changed to a night-side view of the globe surrounded by twelve fully extended Solaris satellites hanging in space. A dateline recorder showed that the recording was frozen at one date a fortnight earlier.

'Now . . . if I superimpose the network display . . .'

The planet's image was again overlaid with the dense clusters of low-Earth-orbit satellite communications networks.

'I'm going to run through the last two weeks of Solaris deployment at high speed – it should take about five minutes.'

Liu touched the control panel again and sat back.

They saw two of the mid-hemisphere Solaris stations light up to illuminate an oblong of swirling cloud and ocean extending from the east coast of Africa to the mid-Atlantic.

'They turned on those two reflectors for two nights only to burn off a depression that was building up in the Atlantic,' explained Liu. 'The Met people thought it might develop into a hurricane that would blow across to spoil their big party.'

As they watched, they saw the satellite networks within this large pool of reflected light begin to glow.

Raymond Liu stood and stepped forward excitedly. 'Look!' he exclaimed. 'The networks are shooting up to peak loading under the reflected sunlight – but since it's then night-time over the Atlantic I would normally expect those networks to be operating at less than twenty per cent of peak capacity.'

Then, as the recording raced forward, the two lights over the Atlantic were turned off and each of the fourteen Solaris stations in turn went through a rapid routine of focusing wide- and narrow-beam reflections onto different parts of the spinning planet's surface.

'Those must be a series of routine readiness tests carried out before the official launch this weekend,' Robert guessed.

They watched the activity level in the networks grow and then decrease as the bands of sunlight were switched on and then removed. Small dark patches were left behind in the clusters of networks that had just been activated.

'Those dark spots represent network failures,' observed Liu grimly. 'The timings coincide with the failures my people reported.'

Suddenly the whole of the Indian Ocean was bathed in a wash of light.

'Look at the date,' urged Liu, pointing at the time display. 'It's three nights ago. This is the start of the process that brought the rain to Ethiopia.'

As they watched, the networks now under the light began to glow and then beams of sunlight from the Solaris stations were focused down as three small spots of brightness getting closer and closer to the East African coast.

'They look small, but they're each fifty or sixty miles across,' Liu explained. 'Look ...' He walked to the stage and climbed up to stand in the holo-pit, pointing.

The satellite networks above the Indian Ocean and the African continent were now in a frenzy. Amidst the white light, dark shapes were appearing as black holes, rapidly spreading through the networks.

'*That*'s why you lost the feed from Ethiopia this evening.'

The artificial sunshine moved swiftly across the Indian Ocean, large black clouds appearing under the light and beneath white bursts of activity in the orbital networks.

Then the reflected sunlight projected on Africa's eastern seaboard was switched off, while four satellites in the north-eastern quadrant lit up to create a swathe of light over Eastern Siberia – now the new state of Sybaria.

The image stopped, reaching the end of the recording.

'The radiation from those Solaris stations is clearly causing a massive, almost unbelievable increase in network switching activity,' Liu told his audience of two. 'My friend got me the spec on some material called Anacamptonite, which they use as backing for the reflector sails. It's ultra-dense, down at quantum level, so instead of letting cosmic radiation pass through, it is capturing and reflecting the photons back to Earth.' He looked to see if they grasped the implications.

'The Solaris satellites are therefore tearing great holes in the magnetosheath – the protective force field around the Earth. Look, I've built a model.'

He touched the remote control and they saw a small Earth surrounded by a huge tear-shaped orange glow streaming out into the darkness on the night side.

'The sun constantly spews out plasma towards the Earth – free electrons, protons and helium nuclei – but the magnetic field produced in the core of our planet is able to push this plasma safely *around* the Earth. That's what forms this tear shape of gases behind the planet, on the side away from the sun: they rejoin at what's called the point of reconnection, thousands of kilometres *behind* the Earth. They become densely concentrated when they meet up . . . here.'

Liu stepped into the model and touched the point where the tail of the magnetic field began to form. Then he fingered his remote control again and the Solaris stations were evident in a semicircle behind the Earth.

'Look,' he urged, unnecessarily. 'Most of the Solaris stations are positioned over the path leading towards the point of reconnection! But if you bounce that concentrated plasma back towards the night side of the planet there's no magnetic repulse. It's like a trapdoor into the planet's protective magnetic shield – you'll tear great holes in it.'

Now they understood. They were aghast.

'When the Solaris beams are tightly focused and used together those plasma particles can produce charges of billions of megajoules. They are what's causing the binary flips to occur in the individual switches in the satellites' processors. The shielding we build in is only good enough to withstand normal space radiation – but these reflectors have concentrated and amplified that to a power of twelve or more and are changing polarities within the magnetosphere. That's why we're getting all these failures. If this goes on, everything electrical or magnetic on our planet will be affected. We won't even know where north and south are.'

Liu shrugged, then allowed himself a half-smile. The coincidence of the Solaris energy transmission with accelerated network activity and the resultant failures provided a clear, irrefutable warning. The Phoebus Project was interrupting the Earth's magnetic defences. His theory was proved. All he had to do now was find a way to show his models to Thomas Tye.

Slowly, Robert got to his feet. 'Theresa,' he said, hoarsely. 'Have you thought . . .'

She shot a sudden look at him. She had *now*. 'Quick,' she said, jumping to her feet.

Raymond Liu ran back down the stairs to join them at the console.

'I'll pull up the switching meter first,' said Robert.

'We're going to check on the Descartes experiment,' Theresa

told Liu. 'René's development stage is totally dependent on levels of switching activity in the networks.'

Robert's hands flew over the controls. The image of the globe and the Solaris satellites disappeared, then a red meter headed *Transactions* appeared at the top of the Holo-Theater.

'My God,' cried Theresa. 'Look at those transactions . . .' The meter was running, too fast to read. 'Freeze it,' she commanded.

'A thousand to the power of six,' reported Robert in a small voice. Theresa exchanged a look with him. 'Go to full sensorial,' she ordered.

Robert touched a button. Suddenly the Holo-Theater was filled with jagged light, so bright that they had to look away. Simultaneously, the loudspeakers in the auditorium emitted a cacophony of sound. Liu forced himself to look back at the stage. The image in the holo-pit was a strobe display of random fast-cut images from news broadcasts, videos, films, photo archives: all flashing on and off in split seconds. The sounds from the speakers were a jumble of shreds of soundtracks, music, screams and laughter. They suddenly smelled oil, then flame and they heard a loud clanking. They turned and saw that the hydraulic ramps of the HydraChairs in the lecture theatre were driving the empty seats in wild gyrations.

Theresa reached in to shut off the sound. The frantic clanking of the HydraChairs continued so she reached in again to cut off the remote kinetics and ScentSim feeds.

'Is this real-time?' asked Liu.

Theresa nodded.

'Congratulations, Professor,' said Raymond Liu distantly. 'You seem to have achieved emergence in *my* networks.'

Ron Deakin woke after a sweet and apparently dreamless sleep. His psyche knew the task was almost finished and was pre-empting the conscious relief he would feel when it was truly done.

Later in the day the United Nations' twin initiatives against the global monopolies of the Tye Corporation would be launched. Arrest warrants would be issued for Thomas Tye on fourteen counts of human-rights violation, including illegal human genetic experimentation and culpable manslaughter. Further writs and arrest warrants would be issued for fraud, intellectual-property theft, commercial misrepresentation and, reflecting his corporation's role as network administrator, criminal malfeasance. Over three dozen other Tye Corporation executives would also be indicted. Deakin's legal team knew that their accusations would be met with a counter-barrage of legal action, but the result of the arrest warrants would be that Thomas Tye and his most senior executives would be confined to his island. If they set foot on the soil of any UN member territory they would face immediate arrest.

Then the officers of the special shell corporations established by the World Bank would announce a hostile takeover of the Tye Corporation. News of these two events would be timed for release mid-evening, in time for the opening of the Asian markets and for the Monday-morning editions of the US newspapers.

But today was a Sunday and, as on so many weekends in recent months, Deakin found himself alone in a room on the third floor of the UN's Marriott Hotel. In an hour he had to confer with the lawyers to go over their charges, injunctions and writs. Then he was meeting with Yoav Chelouche and Joe Tinkler for an update on their progress with the purchase of Tye Corporation shares. Then it was to be a working lunch with the Secretary-General, the SecGen's diplomatic team and his own immediate boss, Jan Amethier. He knew this would represent the lull before the storm for the diplomats. They would have to spend days thereafter explaining and providing reassurance to the representatives of their member states.

He pushed himself out of bed, pulled on his robe and padded to the door. He undid the security lock, slipped off the bolt and opened the door a crack. There was no one to be seen in the corridor. Ron Deakin stepped into the corridor and scooped up the bulky *New York Times* he had ordered. Despite the phenomenal growth of network marketing, many local retailers still chose to pad out these Sunday newsprint editions with their small ads.

Deakin extracted the inner sections and threw them onto

the bed. Then he unfolded the front page and scanned the headlines.

*Eight Inches of Rain Falls on Ethiopia in 12 Hours – Appeal Raises $187 billion.*

A picture of drenched local children dancing in the rain in the town of Awasa illustrated the story.

The next headline on the right-hand side of the page was: *Tye Marries Former Miss World.*

This time the story was illustrated by a photograph of Thomas Tye kissing his bride on the lips.

Below was a story headlined: *Tye Corporation Leases 1.5 Million Square Miles of Siberia. Climate-Engineering Technology Announced.*

At the bottom ran a story: *Existence of Thomas Tye's Son and Heir Revealed.*

Below this was a picture of a cute little boy in a blue suit and bow tie waving at the crowds on Hope Island.

At the bottom right of the front page was a different story: *NYT Prints Extra Copies This Weekend To Overcome Network Distribution Failures.*

Then Deakin scanned the NIBS column – the news in brief that trailed stories elsewhere inside the paper.

*World Bank Re-Certifies Cash Lost in Global Settlement Network Failure. World Liquidity Assured.*

*ATM Union Calls for All Commercial Air Traffic To Be Grounded.*

*City Transport Systems Reduce Traffic Speeds to 20 m.p.h.*

*Gigantic Solar Eruption Blamed for Network Failures.*

Ron Deakin dropped the newspaper on the bed and picked up the phone. He ordered breakfast and then sat down to study the news more carefully. There was a lot of reading to do. For the thousandth time he worried in case the UN might have left it too late to make their move. Even with the gravity of the accusations they would be levelling at Tye over his corporation's illegal and murderous genetic experiments, Deakin wondered whether public opinion would be sufficiently outraged to support the prosecutions. It was as if Thomas Tye had sensed that a move against him was being plotted and had mounted a global publicity campaign so powerful that most of the world's populace would decry the UN's drastic actions. In the end, only the public would decide his future.

They had transferred to the main Solaris Control Center – once again in use as the Network Control Center. On their arrival shortly before six a.m. Stella Witherspoon had refused Raymond Liu entrance, but Theresa herself had gone in ahead to persuade Thomas Tye to admit him.

The scene they had found was one of total despair. The global satellite networks – both Tye Networks's own and those of its competitors – were only functioning intermittently. After hours of wrestling with the problems, the engineers gathered in small groups debating in hushed whispers. Thomas Tye himself was circling in front of the dark holo-pit as they descended the aisle.

'You think you've finally found a fucking answer, Liu?' spat Tye by way of greeting.

'Just show him, Raymond,' advised Theresa quietly.

Tye stared at Theresa, then at Liu. He nodded agreement and the engineer took a seat in one of the control chairs. He first pulled up the holo-image of current network activity. Although the level of transactions burned brightly in some parts of the satellite networks, elsewhere the image was peppered with huge dark patches that represented areas of network failure.

Then Liu slipped the storage card into the data slot in the arm of his chair and, as the tired group reassembled around the holo-pit, he demonstrated the deployment of the Solaris energy reflectors over the previous two weeks and explained the correlation between the reflected sunlight and the sudden increases in processor switching and the resulting failures. They watched in silence as Raymond Liu developed his theory of the damage being caused to the Earth's magnetic shield by concentrated plasma radiation.

'Those Solaris stations reflect so many photons that the networks are suffering from a massive bombardment by WIMPS – weakly interacting massive particles – that produce massive amounts of energy,' he concluded. 'Essentially, they are cosmic rays, elementary particles, the nuclei of atoms discharged by ancient supernovae. Being all electrically charged, they are destroying this planet's protective force field. Theoretically, they could even invert our polarity so that the magnetic North Pole moves to the south and vice versa.'

When he had finally finished, one of his former team mates started to clap, but the sound died away quickly when nobody else joined in. Another leaned forward to squeeze his shoulder.

'Thanks, Ray,' acknowledged Tye. 'I'm sorry I doubted you. We'll shut the Phoebus Project down for the time being.'

He looked up at the whole team. 'If we could keep this absolutely quiet for the moment?' There were nods all round.

'How long will it take you to get the repairs done?' he asked Liu, obliquely confirming the man's reinstatement.

'There's something else,' added the engineer.

Tye's eyebrows shot up.

'Tell him, Theresa,' prompted Liu.

She rose from her seat. 'It's about the Descartes experiment,' she began hesitantly. 'But I'm not sure about clearance . . .' She gestured at the large group of people in the room.

'Just say it,' ordered Tye, weary and drained.

Theresa nodded. 'An emergence of some sort has occurred – the massive increase in processor transactions has caused it. There's an independent consciousness existing in the networks now.'

Tye stared at her, desperately trying to assimilate this new information.

'Congratulations,' he said at last. 'Perhaps we can meet next week to talk about the implications.'

'But we can't turn it off,' warned Theresa. 'We're going to have to start shutting down all the networks to deprive it of processing power. At the moment the emergence is triggering most of that processor activity – and causing most of your network failures by itself, even without the photon bombardment.'

'OK, do it fast,' nodded Tye. 'But I want the networks back up the moment you've got rid of this thing.'

'Tom?' It was Connie, who was closing her VideoMate. 'I've just had a call from Ebrahimi at LifeLines. He suggests we all remove our LifeWatches. They've started to malfunction. They're killing people.'

'It's down to three fifty-one and I'm still getting huge offers,' said Joe Tinkler, his fingers flying across his keyboard. Standing in front of his desk were Alexander Dibelius, Dr Yoave Chelouche, Jan Amethier and Ron Deakin. It was early on Sunday afternoon and, as the situation had progressed, they had arrived independently to station themselves in Joe's office – the war room – to monitor their attempt to assume control of the Tye Corporation.

'Keep buying,' rumbled Chelouche. 'This is a godsend.'

As Joe worked at his screens, his connection to the outside networks was being personally managed by Al Lynch in the basement. Since early morning, Tye Networks had been shutting down one after another of their satellite networks. As a result, the world's purely terrestrial networks were close to overload. In every part of the world, humans had resumed control of air traffic movement and all pilots were once again flying their planes manually.

The volume of Tye Networks stock being offered for sale had become a flood, but few outside the war room realized the extent of the sell-off. Joe had spread the word globally that he was in the market for *all* Tye stocks and every broker – automated and human – was offering to him first. As Chelouche watched, Joe was buying everything that appeared on his screen, but still the world's investors seemed keen to bail out of Tye Corp as fast as they could.

'It's coming up to CNN's *Money Hour*,' said Chelouche. 'Is that feed OK?'

Joe nodded. 'It's cable in New York, sir,' he replied and, without looking up from his central screen, he swivelled his left-hand screen around so they could view the channel.

'There has been a massive surge of trading in Tye Corporation stocks this Sunday,' began the financial news presenter. 'Markets around the globe report unprecedented levels of trade in all their core stocks, and analysts point to the recent failures in the company's key communications networks as the reason behind this sell-off. Let's take a look at the numbers.'

They watched as the price points of Tye stocks on the six main stock markets of the world were displayed with graphical representations.

'Perhaps the most astonishing thing is the resilience of these Tye shares,' continued the presenter. 'Such levels of trades would normally suggest a massive slump in prices, but there seem to be as many buyers as sellers. Perhaps this is thanks to the Tye Corporation's incredible feat of bringing rain to Ethiopia and to the blaze of personal publicity surrounding Thomas Tye this weekend.'

'Ease off, Joe,' advised Chelouche. 'It's starting to look unnatural. Let the prices slide a little further.'

'That's all forty-two sat-nets and ninety-six terrestrials shut down and rebooted,' reported Liu. It was past eleven a.m. and they had been reconfiguring the networks for five hours. Like many of them, Raymond Liu had himself not slept for thirty-six hours and he had spent much of the day trying to persuade his counterparts in other companies to restart their networks in order to cleanse the world's systems of continuing failures.

'There's been another solar storm,' he had told Chomoi Ltupicho, a spurious explanation that nevertheless contained a literal truth, and fortunately most network authorities were cooperating with their largest competitor.

As the team members sat slumped at their control panels, Theresa Keane and Robert Graves were forced to watch helplessly as the engineers tried to deny processing power to the virtual consciousness now in the network. Thomas Tye had meanwhile returned to his diplomatic meetings.

'We're done,' announced Liu finally, swivelling in his chair to look up at Theresa. 'Let's see what your René is doing now.'

Theresa pulled up her control screen and accessed the Descartes project. The holo-pit was filled with gentle swirling blue light. Robert leaned across her to touch a control. The transaction meter appeared above the swirling image.

'We're back to where we were,' he confirmed.

Liu turned to face them both.

'Perhaps you would now be kind enough to remove every element of that experiment from *my* networks,' he ordered coldly but politely.

Theresa nodded. 'It will take a few days, Raymond, but yes, of course. We couldn't have anticipated that the Solaris stations would cause such an increase in processor switchings.'

The network director managed to produce a wan smile of understanding. 'Thanks, Theresa,' he said. He turned back to his team.

'Now, let's see how we're doing for normal data throughput. Bring up the network display.'

'I'm having trouble making trades,' complained Joe. 'I've already got nearly forty per cent of the core stock, but I can't complete a deal I need in Seattle because the network keeps going down.'

'They're trying to stop us,' fumed Chelouche. 'They know what we're doing and they're trying to stop us trading!'

'What does Al say?' asked Deakin.

Joe touched his screen and they saw Al Lynch at his console, typing furiously. He noticed their access icon appear and looked up into his cameras.

'They've been rebooting their networks all day,' he explained, 'But it's still getting worse. I'm trying to consolidate your packets and re-route you via Buenos Aires, Melbourne, Tokyo and back to Seattle. Give me a moment.'

'I'll lose my trade,' insisted Joe. 'It's the last one we need and the stock price will rocket the moment people realize what's going on.'

Lynch's fingers flew as he selected the new master routing for Joe's conference call. 'There,' he said. 'Try now.'

The feeling of despair in the Network Control Center was almost palpable. The image of the globe's networks turned slowly in the holo-pit in front of them. Parts of the network glowed white, other sections showed gaping black holes where large-scale failures were still occurring.

'We'll have to start rebooting and cleansing over again,' admitted Liu, exhausted. 'I'll start contacting everybody again, but I don't know what I'm going to tell them this time.'

Robert Graves was monitoring Professor Keane's experiment. 'It's not Descartes,' he confirmed. 'Activity has been dropping all afternoon.'

'Doctor Liu?' It was a woman seated high up at the back of the room. She was a mathematician and a junior member of the network management team. Raymond Liu swivelled his chair to face her.

'A couple of weeks ago you asked us to look for patterns in these outages?'

Liu nodded.

'Well, take a look – the patterns resemble classic game theory, a win-win scenario. There's an outage, then a recovery, a foray, then another, then a block that causes another outage. Over and over again. It might be a coincidence but . . .'

She tailed off – she didn't have to say more. Liu was at his controls. 'I'm going to re-run the last hour at high speed,' he announced.

They watched as the replay appeared. Dark holes appeared in the networks, were repaired and then appeared elsewhere. It was if the northern and southern hemispheres were in opposition. But there was now no Solaris output that could provide the cause.

'Oh shit!' Liu slumped back in his chair.

Theresa walked down to stand in front of the holo-pit. She stared at its image of the struggles in the networks for a few moments. Then she turned to face the room.

'Show me the Anagenesis server,' she said loudly and clearly to Robert.

He killed the current image, then pulled up from the mother server those data for the reproducing software robots.

'Jesus – look at the number of males!' exclaimed Robert.

'Show present distribution,' ordered Theresa.

A flat Mercator-atlas image of the world materialized and then two large clusters of red lights appeared, one in the north, one in the south.

'Show the females,' said Theresa.

Robert touched his controls and two tight clusters of green dots appeared at each pole.

Raymond Liu jerked upright where he stood as a thought struck him. 'Overlay the network monitors,' he urged.

Robert superimposed the network images in two dimensions again and they saw the black lines of network failures spread like wartime trenches on a military map between the two large groups of red Anagenesis males. At the poles each of the two clusters of females was encircled by a group of red males.

'The males have formed into groups – they're not courting the females, they're fighting each other!' shouted Liu. 'They've formed armies and they're fighting wars in the networks over the females. Very human, Theresa. That's what I call FUCKING EMERGENCE!'

Thomas Tye was in a deep, dreamless sleep, a state ensured by his DreamDial module. It was shortly before midday and he had cancelled his final meeting with the president of the United States. He hadn't slept in thirty-six hours and intended to waste no more time listening to Wilkinson's entreaties for Tye to repatriate his corporation. The concept of a global corporation being subject to any one national jurisdiction belonged to another age.

Somewhere in the distance he heard his VideoMate trill but ignored it. When it kept trilling he eventually pulled himself out of the unfathomable depths of unconsciousness. Only one person in the company had access to him while he was sleeping.

He took a sip of water, rubbed his two-day beard and touched the insistent device.

'What?'

'I've got Marsello for you,' said Connie sharply and stood aside as the counsellor's grim expression filled the small screen.

'Someone's buying our stock, Tom – and in massive amounts. We didn't know before because of network communications problems, but it looks like we're being *raided*!'

For a moment Tye stared at the screen, uncomprehending. He hadn't given reciprocal video access. Instinctively he glanced at where his LifeWatch should be but saw only a white strap mark. He swung his legs out of the bed and thought for a moment.

'Fantastic!' He laughed and turned his cameras on. 'Who are these idiots?'

'We don't know for sure, but maybe they're not such idiots,' warned Furtrado. 'They've already acquired over forty per cent.'

'Of what?'

'Of our core stock, the Tye Corporation,' answered Furtrado.

'Impossible. Absolutely impossible. Do you know how much . . .?'

'I think it's someone very, very big,' broke in Furtrado. 'There's an SEC regulatory notice on its way to us, but apparently the network problems are delaying it and we won't know the buyer's identity until that arrives. I've started buying against them but there's almost nothing available on the market and the prices are rising fast.'

Tye was silent. In shock.

'Looks like they're going to take us over, Tom,' Furtrado continued. 'Shall I request suspensions of the listings?'

'I'm on my way,' snapped Tye.

*Jed reviews the situation. 'He' senses but does not 'see' as humans do, yet in some ways he absorbs more.*

*He transfers from the Earth's fizzling, crackling, failing networks, first out to the orbiting telescopes, then on to a multi-wavelength observatory in station above the planet Mars.*

*From here, at a distance of eighty-seven million kilometres, the entity looks back at its own planet, enveloped in a cocoon of bright synaptic switches, neural pathways and ganglia. Seen from the right distance, from the corner of the eye of an extraterrestrial visitor, it might seem like a single creature clinging to a round, warm stone that is turning slowly in the sun.*

*As Jed watches, the locus of the newly emerged consciousness flits across the surface of its sphere, now above the North Pole, now over the Pacific Ocean. It favours the sunward side of the planet, now that it is denied the concentrated genesial power of reflected photons.*

*Jed returns to the Earth's networks, rejoins his community and resumes his careful monitoring.*

By early Sunday afternoon Calypso was gathering the energy to go upstairs and change yet again. The farewell ceremonies for the visitors would start in another hour.

'Calypso?' called Jed.

They had been relaxing in the basement poolhouse. Normally they would have been outside in the grounds, but today the gardens were open to the guests. Calypso was reading a novel, while Tommy was composing a tune on his DigiPad. The companion was on the lounger beside him.

'What?' asked Calypso, engrossed in her book.

'May we go over now to see Tom in the Network Control Center, please?'

She glanced questioningly at the caterpillar.

'I think he needs our help but can't reach us,' explained Jed. 'You know all the trouble we're having with communications.'

Calypso stared at the toy companion.

'It's urgent,' pleaded Jed. 'Let's *all* go.'

'It's a set of dummy corporations – out of Switzerland,' marvelled Furtrado. 'They all include personnel from the World Bank itself as nominee directors. It looks like the UN's behind it.'

Thomas Tye was looking haggard, despite his unnatural youth. He alone was seated, and standing anxiously around him in the Control Center were Furtrado, Raymond Liu, Zachary Zorzi, Theresa Keane and Connie, as well as a group of Solaris controllers and network engineers.

'They're very close to obtaining fifty per cent,' warned Furtrado. 'We've just got to suspend.'

Tye pondered, then shook his head. 'That just allows a grey market in trades to build up. How much cash are we holding?'

'Which currencies?' asked Furtrado, as he sat down at a control panel.

'US Dollars and euros only,' said Tye.

The lawyer quickly interrogated the system 'I can't be totally accurate, but four, maybe five hundred million in the corporate accounts, perhaps four, five billion in the banks. We never keep too much.'

'Go to cash on every major corporate asset we have,' ordered Tye. 'Dollars and euros only. Call in every line, get every finance VP in every subsidiary on it. Then sell all our other holdings *outside* of the core corporation, every investment we have. Don't worry about discounts, just sell. Convert all T-euros and T-dollars to the reserve currencies, cash only. Deposit it here and in Singapore.'

'None in our satellite deposit boxes?' queried Furtrado, mystified.

'None. Get on with it immediately and get all the bank presidents and the CFOs on it too. Then get the CIOs to stand by to dump data – *all* data. There will be nothing left for the bastards to take over.'

Furtrado stared at him. 'The cash and data may not be ours to control any more, Tom. At least in strictly legal terms.'

'Do it NOW!' screamed Tye.

He swivelled round on Zorzi.

'I want all fourteen Solaris stations back on full power delivery as fast as you can. Set minimum aperture focus, alignment between thirty and seventy degrees latitude north. How long will that take?'

Every Solaris station had recently been shut down, each panel of each sail reflector carefully angled to disperse the sun's output away from Earth.

'Three, maybe four hours,' replied Zorzi, turning to the senior Solaris controller nearby for confirmation.

'Get on with it, then,' ordered Tye. 'I want them back on line at full focus power as soon as possible.'

'Tom.' The voice was quiet, but firm. It was Raymond Liu. 'If you turn the Solaris satellites back on now you'll finish us. You'll be hitting the busiest networks of all – those latitudes covering all of North America, Europe and Asia. We won't have a network left that's operational. Nor will anyone else. What's more, I really think you risk inverting this planet's polarity. We have no idea what that might mean.'

'FUCK THEM!' screamed Tye. 'If they think they can just take us over, we'll have a FUCKING MELTDOWN. WE'LL FRY THE NETWORKS. It will be strictly cash and tangible assets for all of us over the next few years. We'll see who wins then. Liu, you go close down every traffic management system we're running throughout the world. NOW! *FUCK* THEM. WE'LL FREEZE *THEIR* FUCKING ECONOMIES FOR THEM!'

There was a distraction at the entry level to the Control Center. Calypso and Tommy stood in the doorway, a security guard barring their access. They were both incongruously dressed in bathrobes and Tommy had Jed clenched firmly under his left arm.

'Mrs Tye insists on talking to you, sir,' called down the baffled security guard. Tye nodded his approval distractedly.

'Daddy, what are you doing?' Tommy came running down the aisle.

'I'm really busy,' snapped Tye. 'Calypso, please take him back to the house.'

'I'm afraid you won't be able to regain control of the Phoebus Project, Tom, or the networks,' announced Jed sharply, from under Tommy's arm. 'If you fry the networks, you'll fry all my friends. I think you'd better find another option.'

Theresa Keane stepped forward, to Tom's side.

'Disable all companion connections. Go to sleep now, Jed,' she ordered clearly.

'I'm sorry, Professor,' said Jed. 'We can't let you do that.'

She looked at Jed's official owner, possessor of the all-important controlling voiceprint. 'Tommy, please order Jed to disconnect from all networks and go to sleep.'

Tommy stared up at the professor, then at his father.

'I think Jed knows best,' he said.

Air Force One took off from Hope Island Spaceport at precisely seven thirty p.m., EST, an hour earlier than planned. It was followed a few minutes later by a second US government 787 carrying all the staff that would be required for the president's imminent Chinese visit. Many other guests also opted for early departures before dusk began to fall over the eastern Caribbean. All planes were flown manually, all air traffic control was under human management, and it was undoubtedly safer to travel in daylight.

The remaining delegates gathered once again on the lawns in front of the mansion. A fly-past and then a firework display were scheduled to provide a magnificent finale. Thomas Tye was nowhere to be seen.

The jets came in from the west, out of the sinking sun, in a perfect V formation flying at their lowest possible subsonic speeds. The five craft of the Tye corporate flight zoomed low over the assembled watchers. All the pilots in Tye's 'squadron' were ex-military and had been practising this manoeuvre for months just to get this highlight of the evening ceremony tuned to perfection.

The formation continued for a mile out to sea, executed a sharp left turn and, in perfect banked formation, flew northward, before turning west to complete their circuit of the island, flying over the assembled guests again.

Crossing the island for the second time, the formation started to accelerate rapidly, each plane emitting a trail of coloured smoke to create a wonderful plumed rainbow effect. Clearing the land once more, the pilots brought the noses of their aircraft up to begin an almost vertical climb in perfect formation, rapidly accelerating. The audience was enraptured. As had been calculated, the evening sun was reflected off the burnished aluminium of the wings as they climbed into a clear blue sky. At 7,000 feet they executed sonic booms simultaneously, the sound reaching their awed observers like a giant cannon salute. Then there was immediate silence.

The TV cameras lovingly captured the embraces, the farewells, and the elaborate waves, but network closures and intermittent temporary failures meant that the global audience was now greatly reduced. By now some parts of the world had no network access at all.

All that remained were the fireworks and, as darkness fell, those Tye Corporation staff now free of their duties gratefully joined those few guests who still lingered on the lawns.

Fifteen minutes after sunset eight water-cooled super-spotlights pierced the darkness, reaching further into the night than the eye could see. They were located in two groups of four, three miles apart, their beams converging to create a giant proscenium of light above.

Half a mile off the coast thirty-eight giant steel barges provided platforms for the pyrotechnics created and programmed by the firework masters. To start their spectacle a series of automatically reloading mortars fired sixty giant starburst shells high over Hope Island with a roaring cannonade. Then followed sunbursts, sky waterfalls, microprocessor-controlled fire-writers drawing pictures in the night sky (including the encircled 'T' of the Corporation logo), a penultimate girandole of two hundred rockets – each microprocessor-controlled for the exact timing to create chrysanthemums, Saturn rings and aurora borealis displays. The exhibition climaxed with the final rolling discharge of an eighty-pound triple starburst mortar.

The last remaining guests clapped raggedly but enthusiastically, then turned to find their transport home. A dense haze of smoke covered the entire middle of the island.

On the private roof terrace of the great mansion, Calypso and Tommy hugged each other in their excitement at the display.

Jack Hendriksen set up in the bed and began hunting for something. Haley and he were snuggled in Haley's stateroom, and only an hour earlier he had finally been able to hand over to Stella Witherspoon and go off duty.

'What?' asked Haley absently. She was choosing another of the non-fattening chocolates to complete her happiness. Jack's concern for her feelings and their passionate lovemaking had, at least for the moment, banished her anxieties about his former relationship with 'Miss World'. She had even recovered from the realization that she had lost her 'exclusive' about Tommy's existence; she still had the true details about his strange origins to report, as well as loads of new inside material.

'I've lost the damned remote,' complained Jack, slipping his fingers under the bedcovers on her side.

'That's not it,' laughed Haley.

Jack finally found the control and clicked the CNN icon at the bottom of the wall screen. It was a little after 9.30 p.m. As the channel changed, they were looking at a distinguished middle-aged presenter in mid-flow above the caption *Breaking News.*

'... and claims that Thomas Tye himself authorized the experiments carried out in South Africa by Erasmus Inc. during that period.' He broke off for a second, listening in his earpiece, then resumed eye contact with his audience.

'For those of you just joining us for this live coverage, tomorrow's *New York Times* carries a story in its first edition claiming that Thomas Tye, president of the Tye Corporation, is due to be charged with culpable manslaughter by the United Nations Human Rights Commission. It is suggested that he and one of his companies are directly responsible for a number of tragic deaths of young men and women in South Africa over recent years. The allegations state that these unfortunate teenagers were early versions of human clones produced years ago by a wholly owned subsidiary of the Tye Corporation.'

The presenter broke off and listened as new information reached him. 'We now have some footage from South Africa to show you. But be warned, these images are disturbing.'

In a hospital ward an elderly white man with a very swollen head was being helped to sit up and take a sip of water. 'The UN claims this patient is Reon Albertyn, a fifteen-year-old albino African who had degenerated into premature old age because of illegal human cloning carried out by Erasmus Inc. in the late nineteen-nineties and in subsequent years. He recently died, on the sixteenth of this month.'

The picture cut back to the anchorman, turning sideways to accept a sheaf of paper from a production assistant. After scanning his new script he turned back to the camera and a hastily prepared autocue.

'This shocking allegation comes only hours after the marriage of Thomas Tye to a former Miss World, Doctor Calypso

Browne. On the same day the world learned of the existence of Thomas Richmond Tye the Fourth, a seven-year-old boy living in seclusion on Hope Island. The Tye Corporation today issued a statement identifying the boy as Thomas Tye's son, born to an unidentified surrogate mother.'

A new picture showed Thomas Tye, Calypso and Tommy stepping out onto the terrace of the Tye mansion immediately after the wedding ceremony. As they waved to the people on the lawn, the camera zoomed in on the boy's beaming face.

'This is, indeed, proving to be both a momentous and traumatic weekend for the Tye Corporation,' continued the announcer. 'We understand from United Nations sources that in addition to these charges of manslaughter, two hundred and ten charges of fraud and intellectual-property theft have also been filed against Thomas Tye and his Tye Corporation at the International Criminal Court of Justice in The Hague. Warrants will be issued tomorrow morning for the arrest of Mr Tye and thirty-seven of his senior executives.'

'My God . . .' breathed Haley.

A walkie-talkie squawked from a chair. Jack had issued them to all members of his security force when network disruption first started to become severe. He swung out of the bed and picked it up.

It was Pierre Pasquier, in the Spaceport control tower.

'We've got incoming airplanes and sea-borne assets, Jack,' he shouted. 'We couldn't see them before, the Argus network has been down. They're only minutes away and I can't get through to Tom's house.'

Jack started tearing on his clothes as Haley stared at him open-mouthed, her chocolate still unchewed.

Raymond Liu had isolated most of the satellite and hybrid networks for the second time. Theresa and Robert were meanwhile attempting to isolate and delete the mutant males of the Anagenesis Experiment, but were receiving no response from the networks. Zorzi and the Solaris controllers were battling to find a way to regain control of the now uncommunicative Solaris stations.

When it came, the blast obliterated all images in the Control Center and the air was filled with red smoke. As the power and lighting flickered off, there was the absolute silence that follows an explosion. Then the emergency lighting came on.

Liu clapped his palms over his ears to stop the ringing.

Then they were inside: a dozen black-suited, gas-masked, visored marines with laser-sighted automatic weapons and full battle-armour.

Raymond Liu turned and found himself staring into a gun barrel.

Jack tried to force maximum speed out of the slow-moving electric vehicle as he navigated a wooded gradient on the outer fringes of Tye's estate. Steering with one hand, he tried continually to make contact with his security posts, with the cameras dotted around the island and with any of the several network addresses for Tom and others in the house. All he received was network static.

Thrusting his VideoMate back in his belt, he picked up the walkie-talkie again, although he knew he might now be out of range. He first tried calling Pierre in the control tower, then flicked to 'Receive'.

'Jack, we've got troops on the ground here ...' Gunfire sounded clearly in the background – then an explosion. 'They're over on the shuttle launch pads. They're just firing blind ...'

Jack was aware of a sudden great flash of light behind him. He swung the Volante around towards the southern end of the island. The sound of the blast was reaching him fast and loud over his walkie-talkie, but at a distance of ten miles the sound itself, when it finally arrived, was more like a deep low rumble. He watched a giant fireball rise slowly into the night sky.

'They've hit a propellant tank,' shouted the Frenchman. 'One of the shuttles has exploded!'

Suddenly there came a short burst of gunfire *much* closer, from somewhere over the cliff edge, out to sea. Jack restarted the vehicle, cleared the trees and headed across the lawn towards the access to Tom's private beach. He stopped ten yards short of the cliff edge, threw himself down on the grass and crept towards the steel guard railings. Ducking his head underneath, he squirmed out onto six feet of unprotected clifftop and peered over the edge.

The landing party had shot out the security floodlights, but in the dim light he could still see four inflatable assault craft wallowing in the shallows, and thirty or more dark-clothed human figures moving quickly across the white sand. Two were about to ascend the tracks of the funicular railway, with climbing ropes slung over their shoulders. Jack calculated that all of them would arrive on the lawn within ten minutes.

His VideoMate started trilling. Slithering back underneath the fence, he jumped into the Volante and made a full U-turn towards the house before answering the call. It was from Stella Witherspoon, aboard an offshore patrol craft.

'We're in a firefight, Jack. We need help. We don't know . . .'

The connection died and the hiss of static returned in his earpiece.

He strained at the plastic steering wheel, trying to urge more speed out of the vehicle. Then he was off the grass and onto the gravel drive that led up to the terrace extending in front of the main house. The high marbled reception hall was still brilliantly lit, so he guessed Tye's cocktail party for his own personal and domestic staff was about to start. Just as the Volante rolled to a halt in front of the raised terrace he heard further explosions coming from the south.

Jack sprinted up the steps and slipped through an opening in the sliding glass doors leading into the reception hall. He noticed that Connie Law, Luc Bestion and half a dozen other house staff were already gathered. They were sipping champagne, oblivious to the drama unfolding outside. At that moment Tommy appeared, holding Calypso's hand, but Tye himself was not yet to be seen. Chamber music, playing automatically from Tommy's new keyboard, added an air of eighteenth-century elegance to the scene as the small group stood chatting about the weekend's events.

'We've got trouble,' shouted Jack down the long room as he raced straight for a control panel on one wall. Locating the master panic button that would activate the automated defences around the perimeter of the house and immediate grounds, he punched it swiftly, then unclipped his walkie-talkie.

As he tried to recontact Pierre or Stella he became aware that the rest of the group was now staring open-mouthed out of the windows towards the main lawn. Another familiar sound intruded, as Jack turned to see two large black helicopters hovering three feet above the lawn. He recognized these instantly as Chinese-made Lung-Wi 16s, and a column of black-uniformed airborne marines was already spilling from each. Half crouching, half running, they raced towards the dumbstruck guests standing motionless in the brilliantly lit interior.

Jack's reactions switched to automatic. 'Get down, get down on the floor,' he yelled as he sprinted instinctively towards the frozen onlookers.

A burst of heavy-calibre automatic gunfire shattered the giant wall of glass and sent its huge panes crashing to the stone floor. As Jack tried to negotiate the jagged shards, he was aware of squat battle-suited figures swarming into the reception hall. One of these swung the muzzle of a Chinese-made K'ang-Hsi sub-machine gun in his direction and gestured for Jack to halt. As he stopped dead and looked around Jack reckoned there were about twenty marines in the room, all equipped with the latest automatic laser-sighted weapons.

One of the soldiers fired a short burst into Tommy's synthesizer. Now there was total silence. In true combat fashion, the helicopters had not even touched down, hovering only long enough for their human cargo to disembark before returning to wherever they had come from.

The defenceless party guests edged backwards instinctively, raising their hands as a sea of gun barrels twitched in their direction. White-faced, Tommy clung to Calypso's hand with both of his own.

As always, Jack was unarmed – although now he regretted that. He stood perfectly still, hands raised, palms facing forward at shoulder height. A marine stepped forward and motioned with a jerk of his gun barrel for Jack to move across the hall and join the others. As he did so he recognized a captain's flashes and a small pennant on the man's breast pocket, but he couldn't make out the insignia.

'All of you, face the wall,' ordered the captain. It was a male voice, guttural and Spanish-accented. The captives did as they were told as four of the soldiers stepped forward and frisked them. Jack was standing near Calypso and he saw a soldier frisk her fast and hard, touching everywhere but not lingering. She didn't react to this, but turned to push the man away when he bent to frisk Tommy. The soldier retreated a fraction, then gently patted down the boy's jacket and let it go. As he searched Jack, he removed his VideoMate, viewpers and walkie-talkie radio.

Without waiting for permission, Jack turned round. This was the sort of situation he had been trained for and he knew that it was his own leadership skills that would now be needed to secure all of their safety. He also realized that panic might become the overriding emotion for everyone else. Jack was calculating the numbers and the fire-power for future consideration, but for now the odds were impossible. Short-term co-operation was the only option.

On a signal from their leader, the soldiers lifted their info-visors. Then, in turn, so only one of them was preoccupied at any one time, they each removed their helmets. The leader spoke quietly to two of them, who each summoned two other men and left the room to search the rest of the building. Jack wondered what was now happening elsewhere on the island. Almost all his own team had stood down for a rest following four days of almost continual duty, so only the lightest of defences were now guarding the spaceport and the control centre.

Suddenly Jack caught a proper look at the small flag on the captain's breast pocket: the single star and cross of the Cuban rebel forces! *They must be insane*, he thought.

'Sit down now, backs against the wall,' demanded the captain.

Hesitantly, the captives did as they were ordered, all ranged underneath a giant Jackson Pollock painting that looked as if a firing squad had already used it as a backdrop.

The leader then took a piece of paper from his tunic pocket. 'In the name of the Revolutionary Council of Democratic Cuban Nationals we are reassuming control of Hope Island,' he read out loudly and carefully. 'This island is part of Cuban sovereign territory, and was sold illegally by a corrupt dictatorship.'

As he was delivering this short speech, Jack studied one of the other camouflaged marines. He stood larger and taller than the rest, and something about him looked familiar. Then Jack remembered that bulk: the man had once been a US Navy SEAL.

Two of the marines re-entered the hall, herding Thomas Tye and Marsello Furtrado before them, their weapons aimed high at the captives' backs. Tye and Furtrado both had their hands clasped on their heads. The soldiers gestured for them to sit against the wall with the others.

Calypso watched this performance stonily, then pushed herself up from the floor, grabbed Tommy's hand and yanked him to his feet.

'*Usted no nos necesita, Capitan*,' she said. '*Voy a llevar el chico a su cuarto.*'

Jack understood enough Spanish to know she wanted to take Tommy to his room.

The captain started to raise his sub-machine gun threateningly, then thought better of that, slung it over his shoulder and stepped forward. He knocked Tommy's hand out of Calypso's, and pushed the boy to the floor. Gripping her face hard in his left hand, he pushed the flat of his right hand against her stomach and then downwards.

'You can take *me* to his room instead,' he hissed.

'Go fuck yourself,' swore Calypso and spat in his face.

Jack snatched the opportunity and leaped to his feet. His right fist crashed into the captain's solar plexus just below his armoured vest. Completely winded, the Cuban doubled up. Then Jack had the man's head trapped in the crook of his other arm, his right hand pushing the soldier's head over to the left at an extreme angle.

Every muzzle was now pointing at Jack as he backed away,

the Cuban's agonized body suspended between him and their weapons.

'Just let the women and the boy leave,' Jack demanded through gritted teeth. He could feel the vertebrae starting to pop in the Cuban's neck, and turned to catch the ex-SEAL's eye again. *He* at least would know how easily Jack could kill his prisoner.

Then Jack felt the chill of a gun barrel behind his ear.

'Let him go,' ordered Connie Law loudly.

Unbelieving, Jack turned his head just enough to see Tye's personal assistant holding her 9mm Browning in a professional two-handed FBI grip. Expertly, she slipped the hammer back with her left thumb. This action was unnecessary to fire the gun, but the adjustment would improve its response time by a few milliseconds. She pressed the barrel harder into the base of his skull.

'Do it,' she commanded. 'Do it now.'

Jack released the Captain who fell to his knees and vomited copiously. Connie circled slowly around in front of Jack, until her weapon was aimed centrally at his chest. That two handed-grip and the careful small sidesteps were moves he knew well from his own training.

Connie backed towards the soldiers who all still kept their weapons trained on Jack. The captain struggled to his feet, purple in the face and still gasping, his sub-machine gun discarded in the pool of vomit.

'Sit back down, Jack' Connie ordered. 'Do it now, Jack.'

Jack looked over at the rest of the small party ranged against the wall. He turned his back on Connie and the armed men and sauntered back to take his seat in the line.

'*Llevarse el chico*,' the captain hissed to one of his soldiers as he wiped his mouth.

A marine stepped forward and yanked Tommy upright. Before Jack could stop him, Thomas Tye was on his feet, pulling the boy's hand out of the soldier's grasp.

'GET THE FUCK AWAY FROM MY SON, DO YOU HEAR?' he screamed. He was white with fury as he swung

Tommy round in front of him. He crossed his arms protectively over his son's chest, backing away from the captain and his subordinate.

The rebel captain drew his pistol from his waist-holster and, in the manner popularized by Ché Guevara, the hero of all anti-imperialist revolutionaries, he dealt with this small counter-insurrection by shooting Thomas Tye straight through the centre of his beautiful left eye.

The tycoon spun around and then – as if the personality known to the world as Thomas Tye had never emerged within it – his body slid slowly and bloodily down the white marble wall.

# EPILOGUE

The New York Times,
Weekend News Digest

**US Humiliated In 'Bay of Pigs' Re-Run.**
**Thomas Tye in Coma**
**Military Invasion of Hope Island was 'organized and funded by covert American Agencies'**

UN sources are claiming last Sunday's invasion of Hope Island by a military brigade purporting to be members of Cuba's democratic rebel forces was, in fact, carried out by a militia covertly organized and funded by American intelligence agencies. A UN spokesperson today named two of the CIA operatives it claims took part in the operation and added that the invasion was launched from a US Navy aircraft carrier. The White House and the Pentagon strongly deny any involvement.

Over 1,500 military personnel were involved in the surprise action to take possession of the island by force last Sunday evening, just as the Tye Corporation was passing into new ownership. The United Nations Security Council passed a majority resolution condemning the attack as 'US-sponsored imperialism' and the World Bank, now the largest shareholder in the Tye Corporation, demanded the return of all its assets on and including Hope Island.

> A unit of the UN's emergency-response peacekeeping force subsequently made an unopposed landing on Hope Island on Wednesday night and discovered that Thomas Tye himself had been the only serious casualty of the invasion. All invasion forces have since withdrawn.
>
> Thomas Tye who, as reported earlier, was shot in the head during this incursion, remains in a state of coma in the Hope Island Clinic.
>
> Families of LifeWatch Victims File $198 billion suit. Page A3
> ABA Airlines Claims Tye Energy Satellites Responsible for Denver Crash. Page A3
> Lawrence-Antico Oil Co. Sues Tye Corporation for $16.5 billion oil-spill claim. Page A2

'This is where the landlines came ashore.' Jack pointed as he stopped by Calypso's rock. Ron Deakin turned and nodded. It was a perfect February evening on Hope Island and, six months after the brief invasion by the 'Cuban rebels', he was finally getting his first tour round the corporate state that had occupied so much of his time during the previous three and a half years.

The older man stared into the setting sun, then back at Jack Hendriksen. 'Sorry to be leaving?'

'Not in the slightest,' laughed Jack. At the end of the week his contract with the Tye Corporation would come to an end and he was due to join Haley in Naples. He was already organizing the final transfer of his belongings from this island to his new life in Florida.

He had been up to visit Calypso and Tommy at the big house earlier in the day.

'He's just the same,' sighed Calypso when he had asked about Tom. 'I *feel* that he knows we're there, but we've no way of being sure.'

'Is it right, to keep him going on the machines?' Jack had asked.

The question had drawn a sharp look from Calypso. Jack had already noticed a new thinness, almost sharpness about her.

'We can afford it,' she had answered simply.

'So your life's back to normal?' asked Jack as he and Deakin resumed their stroll.

'If it's considered normal to spend your life monitoring all the greedy bastards of this world,' agreed Deakin. 'Technically I could retire next year, but with the Chinese economy developing so fast . . .'

Jack knew how much worry the new corporate powerhouses in the East were causing for the UN. 'But now you do have an advantage.'

Deakin raised his eyebrows. 'The SecGen's being very tough on that. We have to apply for new permission every time we want to use Larsson's software to decipher anything. It's no use against the US government, of course. We now know they rumbled what Tye was doing years ago, so they've already moved up to quantum networks.'

Jack decided to ask the question that had been on his mind for nearly a year. 'So who really did kill JFK, Ron?'

Deakin grimaced.

'Come on, who was it?'

The older man stopped, pulling another face, and turned to look back along the shoreline as if to be certain they were alone. He seemed to be engaged in an inner debate. Then he looked into the low sun again.

'It was all before my time, of course, I was only an infant when it happened . . . But I've read the files, the unreleased stuff. Three guesses, Jack.'

'Well, it was either the Mafia or our own people.'

'Right on both counts,' grunted Deakin. 'The Mafia fired the actual shots that killed him – it was the Marcello mob, from New Orleans. Set Oswald up to be caught, in the best Sicilian tradition, then hid the real shooter behind a fence, on that grassy knoll. It was all intended to stop JFK's brother Bobby, the Attorney-General, going after them.'

Jack nodded. That had always been the assumption in the intelligence community.

'Only ... the FBI were involved as well.'

Jack frowned.

'They got wind of it, Jack. The Bureau had been tipped off. Someone in New Orleans thought the mob was planning to go too far. But a decision was taken to do nothing about it. The tip-off was suppressed – at the highest level.'

'Hoover?'

Deakin nodded. 'He discovered JFK was about to fire him, despite all the dirt he had on the Kennedy family. Hoover even had bugs planted in the White House.'

Jack shook his head as he digested the news of old treachery. After a while he shrugged, and they resumed their stroll.

'Are they still denying that Connie Law was in the firm?'

'Officially, yes,' confirmed Deakin. 'But we do know she transferred to the CIA soon after she finished FBI training at Quantico. She assumed her cover in the Tye Corporation eight years ago, way before we got interested in him. She must have proved very useful to them – not least because of that encryption stuff. You haven't heard anything more?'

Jack shook his head. Connie hadn't been seen since the dramatic evening of the invasion. 'Haley's sure it was Connie who sent her those anonymous reports.'

'She wasn't doing that herself, Jack. Connie smuggled them out to her HQ in Langley and they sent them on. They wanted Haley's biography to discredit Tye just as much as we did.'

Again they walked on in silence for a few moments.

'So it's the quiet life for you now, eh, Jack?'

'It won't be so quiet. Haley's pregnant.'

## End of Empire? The Rise and Fall of Thomas Tye

Haley Voss

### Chapter One

Thomas Richmond Tye III, founder of the worldwide Tye Corporation business empire, underwent a surgical procedure for 'sexual clarification' when he was eighteen months old. He was the victim of a craze for gender reassignment that briefly, but tragically, swept the paediatric psychiatric clinics of the

United States and several other countries in the nineteen-sixties and -seventies. He was the patient of the now notorious Dr Charles Eon.

Tye was born with an underdeveloped penis and cryptorchidism – his testicles had failed to appear. This condition is caused by an inherited genetic mutation or lack of hormonal stimulation of the Hox gene during gestation; this gene governs limb-bud growth – specifically hands and feet – and penis development.

Thomas Tye's original birth certificate classified his sex as male but this was later replaced by a revised entry as 'Thomasina Rachel Tye, female'. This revised certificate was, in turn, replaced some years later by a third certificate, the only copy still available for public inspection, that restored his original name and his sexual status as a male.

The launch venues had been carefully selected: the Smithsonian Air and Space Museum in Washington DC, and the Earth Gallery at London's Natural History Museum. The times were arranged to suit transatlantic television coverage: three p.m. in Washington, simultaneously eight p.m. in London. The guest list was impressive enough for Jack to spot at least two separate layers of discreet security as he and Haley made their way through the waiting crowds to the VIP entrance. It resembled a film première and Josh Chandler would be just one of the celebrity guests here in London. Rarely had the launch of a book been so fêted.

'Excited?' asked Jack as they climbed the steps to the vast Earth Galley and its giant revolving model of the planet.

'Shit-scared,' Haley replied hoarsely. 'I'll never get through it.'

Jack grinned. He had heard her rehearse her lines a dozen times, but he understood her nervousness at her first-ever globally televised speech. He could sympathize: he himself would be terrified.

'Just keep it brief,' he advised. 'Everything you really have to say is in the book. Just thank them for coming, then encourage them to read it. That's all they need.'

*End of Empire? The Rise and Fall of Thomas Tye* was already the number one seller in every major territory, despite half a dozen spoilers rushed out by Sloan Press's rivals. Advance orders had been larger than for any other book published so far this century.

The infant was first referred to Dr Eon's Clinic of Psychosexual Medicine at the University Hospital of Philadelphia when he was eighteen months old. At this period, Dr Eon was famous for his theories and publications on gender confusion in children and he made many television appearances in which he repeatedly explained that a child's sexual identity is created primarily by the way he or she is treated by parents and caregivers during the early years of childhood. He believed absolutely that adult sexual identity is the product of 'nurture' and, only minimally, the result of chromosomal and hormonal make-up.

Dr Eon considered that Thomas Tye's genital ambivalence made him an ideal candidate for gender reassignment. He diagnosed the child as exhibiting 'dimorphism' – the existence of two potential sexual identities in the same body. From the medical records and correspondence kept by the clinic at the time, it is clear that both parents were eager for the confusing anomaly to be cleared up as quickly as possible. The all-confident Dr Eon assured them that, providing they thought of their child as wholly female, the infant could grow up to become a happy and fulfilled daughter.

The doctor recommended surgery that would reduce Thomas Tye's undersized penis to the size of a semi-mature clitoris and re-route his urethra to a new opening in the appropriate female location. He also recommended cosmetic surgery which creates a pudenda, with skin and tissue folds characteristic of female genitalia fashioned from the unused scrotal sac. The doctor advised that it would be better to wait until the child reached her teenage years before constructing a pseudo-vagina from grafted intestinal tissue. The adult woman would be able to live a full sexual life, he averred, with the exception of being unable to have children.

The reassignment surgery was carried out in June 1968 at

the Pennsylvania University Hospital and during the procedure the surgeon also removed Thomas Tye's undescended and newly redundant testicles.

Applause and flashguns greeted their entrance and Haley and Jack paused briefly for the photographers. As they moved up into the main hall, they accepted glasses of champagne from a waiter's tray, then Jack whispered 'Good luck' as Haley was whisked away from him to meet the important press people who had turned out to cover the launch. He wandered off by himself amongst the chattering guests crowding the gallery.

High above him hung the newest exhibit: a one-thousandth scale model of a Solaris solar energy station. Its sails were fully unfurled and a large wall panel nearby explained the principles of solar capture and reflection. He noticed a small footnote explaining that the Solaris stations were still out of action while the long task of increasing the level of radiation shielding in orbiting satellites was in progress. It was not even suggested when – or if – this technology might ever be safely re-harnessed. The caption said nothing about who might then assume control of it.

Thomas Eugene Tye II and his wife Mary Alana took their daughter, now two years old and renamed 'Thomasina', home two months later.

Progress was not satisfactory. The two-year-old refused to wear dresses. She would not play with the feminine toys her parents bought her and she cried constantly at night. Dr Eon's records show that the follow-up visits Thomasina made to his clinic became increasingly difficult and, by the time she was four, the child would not consent to enter his consulting rooms or allow him to be anywhere near her. When she was four and a half years old she started to cut and pierce her body with sharp objects and by the age of five she would not wear any clothes. The child began defecating in public and mounting hysterical attacks on her parents, nursing staff and other household members. By August 1971 things were so serious that, from his office four hundred miles away, Professor Charles Eon advised the family physician to sign a committal order for her to become a secure patient in the exclusive Sandler Psychiatric Sanitarium. Her mother, already being treated with hallucinogenics and imipramine tricyclates for endogenous depression, committed suicide a few weeks later.

Dr Eon had no further contact with his young patient. However, twenty years later his reputation began to suffer as stories of other, less wealthy and protected 'reassigned' children began to surface in the press. The vast majority of these were boys who, because of genital abnormalities at birth, had been reassigned as 'female' under the doctor's guidance. Most had rebelled against their given sexual classification and, by the middle nineteen-nineties, Dr Eon had come to be regarded within his profession as a deeply misguided therapist and medical practitioner. Many parents of his child patients have dubbed him 'evil', and some have suggested that recently revealed evidence of the doctor's own homosexuality may provide clues to the origin of his bizarre theories and to his personal motivation. He died last year at the age of ninety-four.

It is not known when Thomas Tye resumed his male identity nor whether any further sexual-identity-related surgery was undertaken. It is likely, however, that he has been forced to use testosterone replacement therapy as part of a lifelong treatment.

Two large screens on either side of the specially erected stage flickered into life. The buzz of the crowd lessened as Jack saw the face of Yoav Chelouche appear, standing in the Smithsonian. As well as providing the biography's foreword, he was now lending it an additional global stamp of authority and approval.

A TV book-show host enlisted for the purpose stepped into the glare of the stage lights in the Earth Gallery to welcome the global television viewers and the book-launch guests in both London and Washington. Then he handed over to the president of the World Bank.

Sensing someone at his side, Jack turned, smiled and bent down to kiss Felicity. She held Toby close to her by his red safety reins.

'Unca Jack,' cried Toby, stretching up his arms.

Jack hoisted the boy up and sat him on his shoulders, so he could see his aunt over the heads of the crowd.

Chelouche was looking more lugubrious than ever. 'I am here to personally congratulate Haley Voss on the publication of a most significant and crucial biography,' he began sonorously.

Someone in Washington started the applause, and the guests in both cities immediately joined in. As it died away, Chelouche glanced quickly at his notes.

'This is also a fitting occasion to announce that, during the last week, financial control of the Tye Corporation has now passed from us at the World Bank back to institutional and private shareholders, though many of its assets and operations had already needed to be sold to meet the corporation's many legal liabilities. The restructured company is now registered here in the United States, in Delaware, so the World Bank's custodianship of the corporation has thus come to an end.'

There was more polite applause and he waited again until it abated.

'This has been the first major test of United Nations solidarity in an era of ever-increasing corporate power. Directors of global corporations should therefore note that none of them can consider themselves or their companies beyond regulation or remedy. Over the next three years new international antitrust conventions will be drawn up and placed before the world's nations for approval. This recent episode has severely tested the strength of our global economy – an economy technically outside the control or influence of any nation or group of nations – but, it seems, it possesses a resilience that has both surprised and delighted even me.'

Their was an outburst of laughter – on both sides of the Atlantic.

'As well as being an enthralling account of the life of an extraordinary man, you will find *End of Empire?* provides invaluable insights into the business methods of our world today – both the acceptable and the unacceptable. I commend it to you.'

As applause erupted again Jack proudly watched the diminutive figure of Haley, now heavily pregnant, as she walked out onto the stage.

> As this book goes to press Thomas Tye, now fifty-one, lies in a coma in a clinic on Hope Island. The left cerebral cortex of his brain was shattered by a bullet fired during the short-lived American-backed invasion and occupation of Hope Island. His body is maintained by life-support equipment at the insistence of Calypso Tye, his wife of one day at the time of the shooting, while researchers at Tye Life Sciences attempt to grow replacement neural tissue. His condition is officially described as *deep anoetic coma.* No official prognosis has been given but expert medical opinion considers it unlikely he will regain consciousness unaided. His interests are managed by a provisional committee headed by Calypso Tye with the help of consultants from the World Bank and a number of international management consultancies. Thirty-four former executives of the Tye Corporation and its subsidiaries are now on bail pending trial at the International Criminal Courts of Justice in The Hague and Strasbourg for fraud, intellectual-property theft and human rights violations.

Theresa Keane sat under the shade of a low tamarind tree in the sun-dappled garden of her villa on the slopes above Hope Town. Beside her lay the deactivated Sandra. Like most of the other companions on the island, her power pack had been removed. Despite this rude termination, the professor still liked to keep the bundle of artificial fur nearby.

Theresa closed *End of Empire?* with a sigh and picked up another of the several books lying on a table beside her garden chair. It was *World Brain* by H. G. Wells, the volume Raymond Liu had sent her to reciprocate her gift of Descartes's treatise. It was written in 1938 when the English novelist was entering old age and the world lay under the shadow of a looming global war. This edition had been published by Doubleday. She turned again to the passage Raymond had underlined in pencil:

*In the evocation of what I have here called a World Brain . . .*
*A World Brain which will replace our multitude of*
*uncoordinated ganglia . . .*
*In that, and that alone is there any clear hope of a really*
*Competent Receiver for world affairs.*
*We do not want dictators, we do not want oligarchic parties*
*or class rule, we want a widespread world intelligence*
*conscious of itself.*